Exploring QSAR
Hydrophobic, Electronic, and Steric Constants

Corwin Hansch
Pomona College

Albert Leo
Pomona College

David Hoekman
Pomona College

Stephen R. Heller, Consulting Editor
(Computer Applications in Chemistry Books)

ACS Professional Reference Book
American Chemical Society, Washington, DC 1995

Library of Congress Cataloging-in-Publication Data

Exploring QSAR.

p. cm. — (ACS professional reference book)

Includes bibliographical references and index.

Contents: [1]. Fundamentals and applications in chemistry and biology/Corwin Hansch, Albert Leo—[2]. Hydrophobic, electronic, and steric contstants/Corwin Hansch, Albert Leo, David Hoekman.

ISBN 0–8412–2993–7 (set).—ISBN 0–8412–3060–9 (set: pbk.).—ISBN 0–8412–2987–2 (v. 1).—ISBN 0–8412–2988–0 (v. 1: pbk.).—ISBN 0–8412–2991–0 (v. 2).—ISBN 0–8412–2992–9 (v. 2: pbk.)

1. QSAR (Biochemistry) I. Hansch, Corwin. II. Leo, Albert. III. Hoekman, D. H. IV. Series.

QP517.S85S97 1995
574.19′24—dc20

94–23811
CIP

The paper used in this publication meets the minimum requirements of American National Standard for Information Sciences—Permanence of Paper for Printed Library Materials, ANSI Z39.48–1984. ∞

About the Authors

CORWIN HANSCH received his Ph.D. in 1944 from New York University in the field of synthetic organic chemistry, studying under H. G. Lindwall. After a brief postdoctoral period with H. R. Snyder at the University of Illinois, he joined DuPont and worked first on the Manhattan Project at the University of Chicago and Richland, Washington, and then at the experimental station in Wilmington, Delaware. In 1946 he joined the chemistry department at Pomona College, where he has remained except for two sabbatical leaves, one in Vladimir Prelog's laboratory in Zurich and the other in Rolf Huisgen's laboratory in Munich. His main interests in research have been the high-temperature dehydrocyclization reaction and the correlation of chemical structure with biological activity.

ALBERT LEO was born in 1925 in Winfield, Illinois, and educated in southern California. He spent two years in the U.S. Army Infantry, serving in the ETO from 1944 to 1945. He received his B.S. in chemistry from Pomona College (1948; Phi Beta Kappa, Magna Cum Laude) and M.S. and Ph.D. in physical organic chemistry from the University of Chicago, studying reaction kinetics under Frank Westheimer. After 15 years in industrial research and development in food chemistry, he returned to Pomona College to initiate and direct the MedChem Project under the leadership of his former professor Corwin Hansch. The project provides software and databases useful in the design of bioactive chemicals and is distributed worldwide. Leo was given an Excellence in Science award by Sigma Xi in 1980 and was chairman of the Gordon Conference on QSAR in Biology in 1981.

DAVID HOEKMAN was born in 1961 in Holland, Michigan. He attended Pomona College for five years starting in 1980 and initially majored in physics. After three years of study he switched to biology and received his B.S. in biology in 1985. He spent a year working on ecological wood anatomy with Sherwin Carlquist at Rancho Santa Ana Botanic Garden. After a year of study in the botany department at the University of California at Berkeley, he turned his attentions toward computer programming. Since 1987 he has been the head of computer operations at the MedChem Project, where he designed and implemented a QSAR database and analysis package.

Contents

Preface

The motivations for this book are to provide an introduction to what is commonly referred to as "quantitative structure–activity relationships" (QSAR) and to provide substituent constants for their construction. As straightforward as this might sound, it is now virtually an impossible task. There are so many ways to approach the problem, so many computer programs, so many different types of parameters, that no small group can hope to attain a mastery of all of the various methodologies. QSAR is a highly active area of research in which many companies already offer software of all kinds. Thus, this field is often referred to as "computer-assisted drug design" or CADD.

For good reasons, science has been divided into major disciplines, and these disciplines have been divided into innumerable compartments and subcompartments. The subject we are attempting to present resides in a number of major divisions: chemistry, biology, medicine, statistics, computer science, and environmental science. These divisions break down into many large, highly complex, and fast-changing compartments. For example, some of the essential chemical compartments are physical organic, biological, medicinal, quantum, and computational chemistry. Some examples under the heading of biology are molecular, cell, whole animal biology, neurobiology, and psychobiology. Under medicine, there are the diverse areas of chemotherapy, pharmacology, metabolism, and epidemiology. In the computational area, there are the compartments of molecular mechanics and dynamics, pattern recognition, statistics, and database management. Environmental science is concerned with bioconcentration, the distribution of chemicals in various ecosystems as well as toxicology. All of these various specialties play a role in an as yet undefined science: that of how chemicals react with the vast number of forms of life ranging from viruses to bacteria, plants, insects, fish, reptiles, mammals, and finally, humans. One must keep in mind the reactivity to subunits such as enzymes, organelles, cells, and membranes. Toxicology has progressed in a few decades from a very necessary, but dreary subject, to a fascinating area where all of the advanced ideas from chemistry, biology, and medicine come together.

A person or an animal might be regarded as an immensely complex swarm of chemical reactions evolving from zygote to embryo, to fetus, to child, and finally to adult. During this evolution, various aspects of the swarm of reactions wax and wane. The swarm changes character constantly as uncounted numbers of chemicals enter the system as food, drugs, drink, and air and influence it. We have by no means characterized all of the elements in the many spices, herbs,

ix

fruits, vegetables, cigarette smoke, engine exhaust, etc. The rather recent discovery of the "food mutagens" that are carcinogenic and that we consume daily came as a complete surprise. They are much more potent than the polycyclic aromatic hydrocarbons long known to occur in burnt food and coal smoke.

Clearly, there are many points of departure in the study of chemical–biological interactions, and one can feel as though one is working at the center of a problem in many different compartments. What we are attempting in this relatively small book is obviously superficial, but this seems to be the only way to make a start.

Our approach will be largely limited to that stemming from physical organic chemistry, and initiated by L. P. Hammett about 1935. That is, the use of experimentally determined parameters from model systems. The basic philosophy is that the structural changes that affect the biological activities of a set of congeners are of three major types: electronic, steric, and hydrophobic. Other factors, such as hydrogen bonding, polarizability, and dipole moment, appear to play less important roles, at least with the tools we now have available. To date, electronic variations have been largely treated using Hammett σ constants or pK_a values, although molecular-orbital parameters are gaining in importance. Hydrophobic changes are modeled with partition coefficients (log P or π) from the octanol–water system. The most formidable problem of all, that of accounting for the role of geometry or three-dimensional (3-D) shape of the chemicals, has not been addressed in a general way. We cannot do so without a good 3-D model of the receptor's active site. In principle, by studying the perturbations in a cell or mouse from "enough" molecular probes, some salient features of the receptors should be deducible. In a rough analogy, it would be like analyzing the very large number of diffractions from X-rays interacting with a crystal of protein to deduce the protein structure. In the bioreceptor, our problems encounter countless side reactions with other receptors, metabolism, elimination, etc. If specific electronic effects, hydrophobicity, and hydrogen bonding can be more or less accounted for, then the remaining variance can be distributed among steric interactions. Ideally, the problem should be approached on three levels: isolated enzyme, intact cell, and whole animal. As yet, the resources have not been available for anyone to study a well-designed set of, for example, 100 congeners in all three systems.

At present, small steric effects of ordinary substituents can be accounted for rather well, but little information can be obtained about the overall shape of the active site.

We make no attempt to deal with the problem of predicting inactivity as such. Because we cannot find the log of O, such data cannot be included in a QSAR; however, when the QSAR is obtained, if the inactive compounds are "congeners", the predicted activity should obviously be very low. The problem of predicting inactive compounds in a global sense is not soluble, at least at present. Among the universe of chemicals, the vast majority will be inactive in almost any specific assay, and thus it will be a long time before the multitude of reasons behind inactivity can be sorted out. In the meantime, if one simply guesses that all chemicals will be inactive for a specific type of activity, one will find a good correlation if a large enough set of compounds is considered.

Linear free-energy relationships (LFER) in general and the Hammett equation in particular have been criticized because of their empirical nature and because of their inaccurate results, that is, for lack of perfection. Applying the same high standards for acceptance would undoubtedly cause much, if not all, of science to be rejected. Some medicinal chemists have lost faith in QSAR when it failed to accurately predict the biological activity of a "near congener". One must bear in mind that all LFERs are bound to fail sooner or later as incremental changes in structure accumulate to differentiate an analog from those upon which the relationship was based. This failure does not mean that "all is lost". The point of failure is a new point of departure to increase one's understanding of the SAR.

Probably no scientific discipline has provided more intellectual satisfaction combined with practical benefit to humankind than has organic chemistry. Yet the successful organic chemist must often rely on a bit of "art" to cover the areas of his science that have yet to be fully explored. The most distinguished of organic chemists have related disastrous experiences when attempting to extend in the laboratory the predictions made in equations contained in textbooks. It is not unusual to take weeks, months, or even years to "work out" the optimum methods for carrying our a multistep synthesis. What teacher of organic chemistry has not carried out a "one-line" reaction in a textbook, found only a 10% yield of the designated product, and only wondered what the other 90% consisted of?

In attacking the problem of inhibiting a receptor inside a pathogenic cell inside an animal, there are no basic laws of chemistry to start with, and one is often in doubt about the numerical values of the biological end point. How pure a number is it? To what extent is it a composite of an interaction with an important receptor and many less important side reactions of all kinds including metabolism? From the beginning, the Pomona school based its approach on time-honored principle of organic chemistry that substituent changes in one series of compounds often parallel those in another series. This approach had been Hammett's guiding principle in formulating his famous equation. He stuck to his position despite objections by the theoretically minded chemists that in correlating free-energy-related processes, he seemed to be ignoring the fact that Gibbs free energy, ΔG, depends on both change in enthalpy ΔH, and organization energy, $T\Delta S$. We still do not completely understand just why the Hammett equation is so generally successful. From the very beginning, the Pomona school had to face the fact that although nonlinear relationships with the Hammett equation are relatively rare, they are very common in biological QSAR; in fact, they are generally expected.

We began with the concept of additivity of the contributions from various physicochemical properties as the first approximation, but progress in QSAR has depended on the use of nonlinear terms, especially hydrophobic and steric. The interaction of the independent variables is an area that has still not been adequately explored. Even as early as 1965, it was all too clear that, in the calculation of log P from π, the effect of the electronic interactions of substituents on each other had to be taken into account. This need for study of the independent variables became even more apparent as the CLOGP program for the calculation of partition coefficients from molecular fragments began to evolve. The success of the concept of additivity of fragments to explain the properties of organic molecules, limited as it is, has started many research programs in the area of structure–activity relationships.

The normal test of a QSAR's ability to account for reality is to make and test new congeners with properties both inside and outside of explored data space to assess its predictability. This in-depth validation is important, but it is not the ultimate test of whether or not we are on the right track in a general sense. There are so many published examples for which two or more apparently different approaches can be used to produce QSAR with more or less the same explanatory power. Lateral validation is, we believe, the only ultimate test. By this statement we mean the establishment of large self-consistent matrices of QSAR not only from widely different biological systems but also including appropriate QSAR from basic organic reactions in simple solvents. To accomplish a paradigm change in the way we do structure–activity analysis is going to be a huge task. Wherever possible in our writing, we have drawn attention to the adage that similar reactions will have similar characteristics (similar terms in the QSAR).

Of course, to do this one needs a common set of parameters. In developing the book, we have tried to limit our examples and discussion to the most generally accepted parameters, realizing that changes are occurring and will continue to occur in just how this should be done.

To make significant progress in the development of a cohesive approach to QSAR requires the ready availability of large numbers of QSAR. One cannot search the literature every time a new idea or area of QSAR is to be analyzed. Only with a computerized database with a smoothly interacting model-building program can we begin seriously to work on the problem of lateral validation. Currently, our program is based on only about 6000 QSAR, but even so, it was invaluable in writing this book.

A crucial factor in advancing QSAR is the best possible database of parameters. In constructing our Tables of Physicochemical Parameters, we tried to be comprehensive rather than critical; that is, we have not attempted to pass judgment on the quality of all of the values, although we do indicate the ones we normally prefer. When there is an obvious discrepancy, we leave it up to the user to consult the original literature to decide whether the model system is appropriate, or whether experimental conditions cast doubt on its validity. Finally, even after considerable cross-checking, we know there must be a number of outright errors in these tables. We would greatly appreciate our readers taking the time to point them out to us.

In conclusion, one cannot expect more success in the rationalization of chemical reactions in the highly heterogeneous milieu of an animal or even a cell than can be achieved with ordinary organic reactions in solution.

Acknowledgments

We are indebted to Teri Klein of the Molecular Graphics Laboratory at the University of California in San Francisco for composing the stereo pictures in Chapter 7. We would not have been able to make many of the interesting comparisons of QSAR were it not for the help of an excellent computer program written for this purpose by David Hoekman. We are especially thankful to Toshio Fujita for helpful suggestions and for correcting many errors. In addition Mathew Ames, William Denny, Teri Klein, Hugo Kubinyi, C. A. Ramsden, Cynthia Selassie, Peter and Jacqueline Sinclair, Carlo Silipo, Robert Taft, Chiyozo Takayama, and Richard Weinshilboum read and commented on various parts of the manuscript.

We thank Patricia Arms for the many hours of word and formula processing necessary to complete the manuscript.

Dedication

We dedicate this book to our wives, Gloria and Georganna, without whose support this book would not have been possible.

How To Use the Tables in this Volume

Octanol LogP

The values in this table are arranged by molformula within the standard categories: carbon, hydrogen, and any other elements in alphabetical order. Reading left to right, the table is arranged as follows:

- The first column gives arbitrary index numbers, which are separated in groups of five to facilitate reading across the row.
- The second column may contain the symbol *, which designates the preferred value measured as or converted to the neutral form, or the symbol √, which designates a "good" value measured as the ion or a tautomeric mixture.
- The third column contains the log P value. If there is more than one value for the same solute, they are in ascending order.
- The fourth contains the pH of measurement.
- The fifth contains the reference number.
- The sixth contains the footnote number. Because the footnote may contain very important information, such as methodology (for example, HPLC) or buffer type, it should always be consulted.
- The seventh column contains the molformula. If the initial molformula ends with a period, it is followed by the formula for the counter ion.
- The eighth column is the solute name. An NSC# (National Cancer Society number) is considered a valid name, as are some commercial laboratory numbers, such as SKF#525. When available, the CAS Registry number follows in boldface type. Because it was not always possible to list the IUPAC name, several shortcuts have been used:
  1. Sometimes common groups are abbreviated, such as NO_2, and CF_3, DIAM for diamino, or DIME for dimethyl.
  2. Some peptides are listed by three-letter abbreviations, such as ALA-LEU-PHE, and blocked peptides are listed by AC-ALA-ALA-N.
  3. In a few complex solutes the typewritten structure is given, such as #4977 4-ME-1,2,5-

THIADIAZOLE-3-C(=NOH)SCH$_2$CH$_2$N(ET)$_2$. Often these are depicted at the end of the table.

4. Lacking Greek symbols, the following abbreviations are used: A- for alpha; B- for beta; G- for gamma; W- for omega; D- for delta (unsaturation).
5. Saturation positions are indicated but the level is abbreviated; for example, 2,3,4,5-H$_4$, is a tetrahydro analog.

If the name is appended with a heavy cross (+), the structure is depicted in the section at the end of the volume, and the structures are arranged by access number. If depiction is not indicated, and the name leaves some doubt as to the structure, one should look in the immediate vicinity for a similar name that has the structure depicted.

Hammett Sigmas

This table of electronic, steric, and hydrophobic parameters is based, to a large degree, on the concept of *substituents* on a benzene ring that act upon a particular "probe" moiety (initally Hammett used the ionization of a carboxyl group). This probe is attached to the hydrocarbon ring and is sensitive to differences in the polar character of the substituents under study. Data for a given substituent is accessed via its empirical formula, which is given in the standard order: C$_x$H$_y$ and other elements in alphabetical order. A structural formula is also provided to distinguish substituents with identical empirical formulas. This is given in ordinary typewriter symbols beginning at the attachment bond.

There are two exceptions to the simple *substituent* concept that must be carefully noted and understood:

1a. A number of sigma para+ values have been determined for various positions on polycyclic aromatic rings by pyrolysis or hydrogen exchange of tritiated derivatives (*see* references 699–710). Thus there is neither a "probe" as such nor a "substituent". The position to which the parameter applies is noted by providing a *name* for the entire ring system that takes the place of the structural designation and is followed by a colon. The carbon and hydrogen count for the ring containing the probe is subtracted from the total to get the substituent empirical formula for access. For example, one may want the electronic parameters for the 3-position of perylene

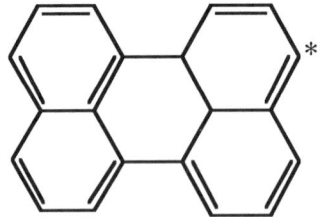

The 3-position establishes which benzene ring is "parent", and the balance of the atom count comes to $C_{14}H_9$, which represents the "substituent". The appropriate data is easily found, because the only structural entry besides 3-perylene under this empirical formula is 6-(1,2-benzopyrenyl):

1b. When the "active" substituent is on a ring fused to the one with the probe, such as in the bromonaphthoic acid illustrated below, the empirical formula entry is only the benzo-portion, $C_4H_3BR_1$, and the structural designation reflects the molecular name [i.e., 2-(7-bromonaphthyl]:

2. In some heteroaromatic solutes the probe may be on the same ring as the heteroatom that is providing the electronic influence. In this case the electronic parameters may be considered to apply to the heteroatom as a "fused-in" substituent, but the empirical formula used for access will be for the entire ring (minus the probe) and will be followed by the symbol @. A name for the intact ring will be used as the structural formula. It is followed by a colon. Note the difference where the probe is (for example, carboxylic acid ionization):

MOLFORMULA = $C_5H_4N_1$@ $C_5H_4N_1$
STRUCT.FORM = 3-pyridyl   3-pyridyl

In the "fused-in" case, only sigma meta values appear, while in the "normal" case, all the positional values are possible. When the probe is on a different ring in a fused hetero, such as 7-quinolinyl (shown below), the data is accessed by the substituent empirical formula, e.g. $C_3H_3N_1$@

If the probe were at the 2-, 3-, or 4-positions, access would be through the empirical formula of the entire molecule: $C_9H_6N_1@$.

Parameter Labels

The left-hand side of each page of the Hammett sigmas table contains those values most frequently used. For more details, see *Exploring QSAR: Fundamentals and Applications in Chemistry and Biology* in the sections noted below:

Pi (p) as defined by Fujita and Hansch (4-4-1)
Sigma meta (S. Meta) as defined by Hammett (1-1)
Sigma para (S. Para) as defined by Hammett (1-1)
Molar refractivity (M. Refr) as defined by Hansch et al. (3-5)
$\mathbf{E_s}$ **(ES)** as defined by Taft (3-2)
$\mathbf{E_s^c}$ Taft steric parameter with Hancock's hyperconjugation correction (3-3)

The labels and description of the other parameters listed on the right side of each page are given below. A full discussion for many may be found in *Exploring QSAR: Fundamentals and Applications in Chemistry and Biology* Chapters 1 and 3 or in the appropriate reference in the list at the end of this table.

σ_I	sigma inductive (51)
σ'	sigma prime: field effect through bicyclooctane ring (112)
σ^Q	sigma quinuclidine: field effect on protonated nitrogen (384)
σ^*	sigma star: aliphatic inductive effect (100; p 587)
σ_m^o	sigma zero *meta*: phenylacetic acid ionization or ester hydrolysis (28)
σ_p^+	sigma plus *para*: delocalizing +charge; solvolysis of *t*-cumyl chlorides (15)
σ_m^+	sigma plus *meta* (as sigma plus *para*) (15)
σ_p^-	sigma minus *para*: delocalizing conjugated negative charge; ionization of phenols (^{P}) or anilines (^{A})
σ_m^-	sigma minus *meta* (as sigma minus *para*)
σ_p^o	sigma zero *para*: phenylacetic acid ionization or ester hydrolysis (28)
σ_p^N	sigma normalized (22)
σ_o	sigma ortho: various methods (11)
σ_o^+	sigma plus *ortho*: delocalizing +charge (355)
σ_o^-	sigma minus ortho (92)

σ_R^o	sigma zero resonance: pi-electron delocalization from $C^{13}NMR$ (296)
σ_R^B	sigma benzoic acid resonance: pi-electron delocalization, benzoic acid model (143)
σ_R^A	sigma aniline resonance: pi-electron delocalization, aniline model (143)
σ_R^+	sigma plus resonance: pi-electron delocalization from t-cumyl chlorides.
ES-A	steric constant: calculated from atomic distances, Austel (667)
ES-CH	steric constant: calculated from atomic distances, Charton (382)
E_s^H	steric constant: hydroboration of substituted ethylenes (252)
L_{stm}	length of substituent; sterimol calculation (383)
B_{1stm}	width of substituent; sterimol calculation (383)
B_{5stm}	width of substituent; sterimol calculation (383)
F	field effect: of Swain & Lupton (modified)
R	resonance effect: of Swain & Lupton (modified)
R^+	+resonance effect: resonance delocalization of +charge
R^-	−resonance effect: resonance delocalization of −charge
σ_L	sigma localized: Charton (382)
σ^F	sigma phosporus: ionization of phosphorous acids (132)
σ_{Nu}	sigma nucleophilic: solvolysis, nucleophilic assistance (491)
σ_I^F	sigma inductive phosphorus: attached to phosphorus (842)
σ_R^F	sigma resonance phosphorus: attached to phosphorus (842)
σ_T^o	sigma zero twist: resonance effect diminished by twisting 90° out of ring plane (469)
$\sigma_m^{\cdot}$	sigma radical *meta*: various models (88, 672, 783, 822, 841)
$\sigma_p^{\cdot}$	sigma radical *para*: various models (88, 672, 783, 822, 841)
σ_m^c	sigma carbocation *meta* (678, 754)
σ_p^c	sigma carbocation *para* (678, 754)
σ_{Cg}	sigma contiguous: delocalization when attached directly to oxygen or nitrogen (645)
K_{ap}	charge transfer: contribution to electron-donor/acceptor interaction (653)
DP	group dipole: from dipole moment (698)
ELN	group electronegativity (760)
$\sigma^{\Phi+}$	sigma phosphonium: C acidity adjacent to phosphonium
E_s^P	steric pyridinium: quaternerized with MeI

Octanol LogP

	logP	pH	Ref.	Note	MF	Name / CAS / activity
1 ★	0.74		552	400	Ar1	ARGON 7440-37-1
2	-2.95		2358	200	(Cl1)2.Cd1	CADMIUMCHLORIDE
3 ✓	-2.59		2358	200	"	CADMIUMCHLORIDE
4	-2.47		2019		Cl2Sn1	STANNOUS CHLORIDE 62171-37-3
5	-3.26		2019		Cl4Sn1	STANNIC CHLORIDE 7446-78-8
6 ★	1.68		547	400	F6S1	SULFUR HEXAFLUORIDE 2551-62-4
7 ★	0.28		552	400	He1	HELIUM 7440-59-7
8 ★	-0.22	7.4	2071		Hg1.(Cl1)2	MERCURIC CHLORIDE 7487-94-7
9 ✓	2.49		536		I2	IODINE 7553-56-2 anti-infective (topical)
10 ★	0.89		552	400	Kr1	KRYPTON 7439-90-9
1 ★	0.67		552	400	N2	NITROGEN 7727-37-9
2	0.36		552	500	N2O1	NITROUS OXIDE 10024-97-2
3 ✓	0.43		552	400	N2O1	NITROUSOXIDE
4 ★	0.28		552	400	Ne1	NEON 7440-01-9
15 ★	0.65		552	400	O2	OXYGEN 7782-44-7
6	-3.88	7.4	903	1	Rh1.(Cl1)3	RHODIUM CHLORIDE 10049-07-7
7 ★	1.51		2581		Rn1	RADON 10043-92-2
8	1.15	7.0	1232	1	Xe1	XENON 7440-63-3
9 ★	1.28		552	400	"	XENON
20 ★	0.45		552	400	H2	HYDROGEN 1333-74-0
1	-1.06		1237		H2O1	TRITIUM OXIDE 13670-17-2
2 ★	-1.38		500		H2O1	WATER 7732-18-5
3	-1.15	7.4	2306	1	"	WATER
4	-3.40	7.4	1237	1	H2O4P1.Na1	SODIUM ION (PHOSPHATE BUFFER)
25	-1.86		1670		H3O4P1	PHOSPHORIC ACID 7664-38-2
6 ✓	-2.07		579		H4N2	HYDRAZINE 302-01-2
7 ✓	-2.19		2426	537	H6Cl2N2Pt1	CISPLATIN 15663-27-1 antineoplastic
8	-1.43		564		"	CISPLATIN
9 ✓	-1.43		564	452	H6Cl2N2Pt1	DIAMINE-DICHLOROPLATINUM-CIS
30 ✓	-3.36		2426		H10N2O2Pt1.(N1O3)2	PLATINUM,DIAMMONIO-DIAQUA-DINITRATE
1 ★	1.86		547	400	C1Br1F3	BROMOTRIFLUOROMETHANE 75-63-8
2 ★	2.67		594		C1Br3	BROMOFORM 75-25-2 sedative, antitussive
3 ★	3.42		594	400	C1Br4	CARBON TETRABROMIDE 558-13-4
4 ★	1.08		547	400	C1Cl1F2	CHLORODIFLUOROMETHANE 75-45-6
35 ★	1.65		547	400	C1Cl1F3	CHLOROTRIFLUOROMETHANE 75-72-9
6 ★	1.55		547	400	C1Cl2F1	DICHLOROFLUOROMETHANE 75-43-4
7 ★	2.16		547	400	C1Cl2F2	DICHLORODIFLUOROMETHANE 75-71-8
8 ★	1.97		503		C1Cl3	CHLOROFORM 67-66-3 solvent, anesthetic (vet)
9	1.95		547	160	C1Cl3	DEUTEROCHLOROFORM 1111-89-3
40 ★	2.53		547	400	C1Cl3F1	TRICHLOROFLUOROMETHANE 75-69-4
1 ★	2.09		579		C1Cl3N1O2	CHLOROPICRIN 76-06-2
2 ★	2.83		547	400	C1Cl4	CARBON TETRACHLORIDE 56-23-5 solvent, anthelmintic
3 ★	0.64		547	400	C1F3	FLUOROFORM 75-46-7
4 ★	1.18		552	400	C1F4	TETRAFLUOROMETHANE 75-73-0
45 ★	-0.25		547	400	C1N1	HYDROCYANIC ACID 74-90-8
6	0.32		1262		C1N1S1	HYPOTHIOCYANOUS ACID
7 ★	0.83		2466	716	C1O2	CARBONDIOXIDE 124-38-9
8	-2.81		536		C1S1.Na1	SODIUMTHIOCYANATE 540-72-7
9 ★	1.94		579		C1S2	CARBON DISULFIDE 75-15-0
50 ★	1.41		1560	100	C1H2Br1Cl1	BROMOCHLOROMETHANE 74-97-5
1 ★	1.88		594		C1H2Br2	DIBROMOMETHANE 74-95-3
2 ★	1.25		547	400	C1H2Cl2	METHYLENECHLORIDE 75-09-2
3 ★	0.20		547	400	C1H2F2	DIFLUOROMETHANE 75-10-5
4 ★	2.30		592		C1H2I2	DIIODOMETHANE 75-11-6
55 ★	-0.60		924		C1H2N4	TETRAZOLE 288-94-8
6 ★	0.35		1322		C1H2O1	FORMALDEHYDE 50-00-0
7 ★	-0.54		510		C1H2O2	FORMIC ACID 64-18-6
8 ★	1.19		547	400	C1H3Br1	METHYLBROMIDE 74-83-9
9 ★	0.91		547	400	C1H3Cl1	METHYLCHLORIDE 74-87-3 anesthetic (local), refrigerant
60 ★	0.41	7.4	2071		C1H3Cl1Hg1	METHYL MERCURIC CHLORIDE 115-09-3 fungicide
1 ★	0.51		547	400	C1H3F1	METHYLFLUORIDE 593-53-3
2 ★	0.55	1.0	590		C1H3F1O2S1	METHANESULFONYL FLOURIDE 558-25-8
3 ★	1.51		547	400	C1H3I1	METHYL IODIDE 74-88-4
4 ★	-1.51		572		C1H3N1O1	FORMAMIDE 75-12-7
65 ★	-0.35		547	400	C1H3N1O2	NITROMETHANE 75-52-5
6 ★	1.09		547	400	C1H4	METHANE 74-82-8
7 ★	-2.05		1137		C1H4N2O1	FORMIC ACID HYDRAZIDE 624-84-0
8 ★	-2.11		547		C1H4N2O1	UREA 57-13-6 diuretic
9	-1.66	7.4	1237	1	"	UREA
70 ★	-1.80		401		C1H4N2O2	HYDROXYUREA 127-07-1 antineoplastic
1	-1.27	7.4	1184	1	"	HYDROXYUREA
2 ★	-1.02		1087		C1H4N2S1	THIOUREA 62-56-6
3	-0.89		1173	537	C1H4N4O2	2-NITROGUANIDINE 556-88-7
4 ★	-0.77		547	60	C1H4O1	METHANOL 67-56-1 pharmaceutic aid (solvent)
75	-0.50	7.4	2306	1	"	METHANOL
6	-0.32	7.4	1758	1	"	METHANOL
7	-3.10		1189		C1H5As1O3	MONOSODIUM METHANEARSONATE 2163-80-6
8 ★	-0.57		510		C1H5N1	METHYLAMINE 74-89-5
9 ★	-2.75		1394		C1H5N3O1	SEMICARBAZIDE 563-41-7
80 ★	-1.05	13.0	579	224	C1H6N2	METHYLHYDRAZINE 60-34-4
1 ★	2.82		547	400	C2Cl2F4	1,2-DICHLOROTETRAFLUOROETHANE 76-14-2
2 ★	3.16		1369	459	C2Cl3F3	1,1,2-TRICHLOROTRIFLUOROETHANE 76-13-1
3 ★	2.09		579		C2Cl3N1	TRICHLOROACETONITRILE 545-06-2 insecticide
4 ★	3.40		579		C2Cl4	TETRACHLOROETHYLENE 127-18-4 anthelmintic
85 ★	4.14		2000		C2Cl6	HEXACHLOROETHANE 67-72-1
6 ★	2.00		552	400	C2F6	HEXAFLUOROETHANE 76-16-4
7 ★	0.07		579		C2N2	CYANOGEN 460-19-5
8	1.70	7.0	1232	1	C2H1Br1Cl1F3	HALOTHANE 151-67-7 anesthetic (inhalation)
9 ★	2.30		547	60	"	HALOTHANE

	logP	pH	Ref.	Note	MF	Name / CAS / activity
90 ★	2.24		509		C2H1Br2N3	1,2,4-TRIAZOLE,3,5-DIBROMO 7411-23-6
1 ★	3.20		1683		C2H1Br3	TRIBROMOETHENE 598-16-3
2 ★	2.61		2605		C2H1Cl3	TRICHLOROETHYLENE 79-01-6 *analgesic (inhalation), anesthetic (inhalant,vet)*
3 ★	1.33		572		C2H1Cl3O2	TRICHLOROACETIC ACID 76-03-9 *caustic*
4 ★	3.22		579		C2H1Cl5	PENTACHLOROETHANE 76-01-7
95 ★	0.37		547	400	C2H2	ACETYLENE 74-86-2
6 ★	0.45		579		C2H2Cl1N1	CHLOROACETONITRILE 107-14-2
7 ★	2.13		572		C2H2Cl2	1,1-DICHLOROETHYLENE 75-35-4
8 ★	1.86		572		C2H2Cl2	CIS-1,2-DICHLOROETHYLENE 156-59-2
9 ★	2.09		572		C2H2Cl2	TRANS-1,2-DICHLOROETHYLENE 156-60-5
100 ★	0.92		572		C2H2Cl2O2	DICHLOROACETIC ACID 79-43-6
1 ★	1.04		508		C2H2Cl3N1O1	TRICHLOROACETAMIDE 594-65-0
2 ★	2.39		1179		C2H2Cl4	1,1,2,2-TETRACHLOROETHANE 79-34-5
3 ★	1.24		547	400	C2H2F2	1,1-DIFLUOROETHYLENE 75-38-7
4 ★	0.12		528		C2H2F3N1O1	TRIFLUOROACETAMIDE 354-38-1
5 ★	0.16		2166		C2H2N2O1	FURAZAN 288-37-9
6 ★	-0.33		2592		C2H2N4O2	1H-1,2,4-TRIAZOLE,3-NITRO
7 ★	1.57		572		C2H3Br1	VINYLBROMIDE 593-60-2
8 ★	0.41		510		C2H3Br1O2	BROMOACETIC ACID 79-08-3
9 ★	2.10		591		C2H3Br3O1	TRIBROMOETHANOL 75-80-9 *anesthetic (inhalation)*
110	2.79		579		C2H3Cl1	CHLOROETHYLENE 75-01-4
1 ★	0.22		572		C2H3Cl1O2	CHLOROACETIC ACID 79-11-8
2 ★	0.19		594		C2H3Cl2N1O1	ACETAMIDE,2,2-DICHLORO 683-72-7
3 ★	2.49		547	400	C2H3Cl3	1,1,1-TRICHLOROETHANE 71-55-6
4 ★	1.89		590		C2H3Cl3	1,1,2-TRICHLOROETHANE 79-00-5
15 ★	1.42	7.4	404	1	C2H3Cl3O1	2,2,2-TRICHLOROETHANOL 115-20-8
6 ★	0.99		579		C2H3Cl3O2	2,2,2-TRICHLORO-1,1-ETHANEDIOL 302-17-0 *hypnotic, sedative*
7	1.61	7.4	404	1	"	2,2,2-TRICHLORO-1,1-ETHANEDIOL
8 ★	0.41		506		C2H3F3O1	2,2,2-TRIFLUOROETHANOL 75-89-8
9 ★	-0.34		503		C2H3N1	ACETONITRILE 75-05-8
120 ★	0.94		1947		C2H3N1S1	METHYLISOTHIOCYANATE 556-61-6 *pesticide*
1 ★	-0.58		590		C2H3N3	1H-1,2,4-TRIAZOLE 288-88-0
2 ★	-0.29		924		C2H3N3	2H-1,2,3-TRIAZOLE 288-36-8
3 ★	-1.44		1137		C2H3N3O1	2-AMINO-1,3,4-OXADIAZOLE 3775-60-8
4 ★	-0.57	7.4	401	1	C2H3N3S1	2-AMINO-1,3,4-THIADIAZOLE 4005-51-0 *antineoplastic*
25 ✓	-1.18	7.4	903	1	(C2H3O2)4.(Rh1)2	RHODIUM(II)ACETATE
6 ★	1.13		547		C2H4	ETHYLENE 74-85-1
7 ★	-0.52		528		C2H4Br1N1O1	BROMOACETAMIDE 683-57-8
8 ★	1.96		579		C2H4Br2	1,2-DIBROMOETHANE 106-93-4
9 ★	-0.53		508		C2H4Cl1N1O1	CHLOROACETAMIDE 79-07-2
130 ★	1.10		594		C2H4Cl1N1O2	ETHANE,1-CHLORO-1-NITRO 598-92-5
1 ★	1.79		547	400	C2H4Cl2	1,1-DICHLOROETHANE 75-34-3
2 ★	1.48		547	400	C2H4Cl2	1,2-DICHLOROETHANE 107-06-2
3 ★	0.37		425		C2H4Cl2O1	2,2-DICHLOROETHANOL 598-38-9
4 ★	-1.05		528		C2H4F1O1	FLUOROACETAMIDE 640-19-7
35 ★	0.75		547	400	C2H4F2	1,1-DIFLUOROETHANE 75-37-6
6 ★	0.24		1701		C2H4F3N1	ETHYLAMINE,2,2,2-TRIFLUORO 753-90-2
7 ★	-0.19		528		C2H4I1O1	IODOACETAMIDE 144-48-9
8 ★	2.71	7.4	579	1	C2H4I2	1,2-DIIODOETHANE 624-73-7
9 ★	-1.37	13.0	579		C2H4N2	ACETONITRILE,AMINO 540-61-4
140	1.16		1358		C2H4N2O6	ETHYLENEGLYCOLDINTRATE 628-96-6
1 ★	-1.15		1173	537	C2H4N4	2-CYANOGUANIDINE 461-58-5
2 ★	-0.87		2660		C2H4N4	AMITROLE 61-82-5 *herbicide*
3	-0.84	7.0	2667	226	"	AMITROLE
4 ★	-0.49		2166		C2H4N4O1	1,2,5-OXADIAZOLE,3,4-DIAMINO
45 ★	-0.90	5.4	264		C2H4N4O2S2	2-AMINO-1,3,4-THIADIAZOLE-5-SULFONAMIDE
6 ★	-0.30		547	400	C2H4O1	ETHYLENEOXIDE 75-21-8
7 ★	-0.17		505		C2H4O2	ACETIC ACID 64-19-7
8 ★	0.03		599		C2H4O2	METHYLFORMATE 107-31-3
9 ★	0.09		547		C2H4O2S1	MERCAPTOACETIC ACID 68-11-1
150 ★	-1.11		510		C2H4O3	HYDROXYACETIC ACID 79-14-1
1 ★	1.61		547	400	C2H5Br1	ETHYLBROMIDE 74-96-4
2 ★	0.18		594		C2H5Br1O1	2-BROMOETHANOL 540-51-2
3 ★	1.43		547	400	C2H5Cl1	ETHYLCHLORIDE 75-00-3 *anesthetic (topical)*
4 ★	-0.06		594		C2H5Cl1O1	2-CHLOROETHANOL 107-07-3
55	-0.20		2019		C2H5Cl3Sn1	TRICHLOROSTANNANE,ETHYL 1066-57-5
6 ★	-0.76		594	400	C2H5F1O1	2-FLUOROETHANOL 371-62-0
7	2.00		503		C2H5I1	ETHYL IODIDE 75-03-6
8 ★	-0.13		508		C2H5N1O1	ACETALDOXIME 107-29-9
9 ★	-1.26		1087		C2H5N1O1	ACETAMIDE 60-35-5
160 ★	-0.97		594	400	C2H5N1O1	METHYLFORMAMIDE 123-39-7 *antineoplastic*
1 ★	-1.59		579		C2H5N1O2	ACETOHYDROXAMIC ACID 546-88-3 *enzyme inhibitor (urease)*
2 ★	-3.21		547	463	C2H5N1O2	GLYCINE 56-40-6 *nutrient*
3 ★	0.18		547	400	C2H5N1O2	NITROETHANE 79-24-3
4 ★	-0.66	7.4	172	1	C2H5N1O2	O-METHYLCARBAMATE 598-55-0
65 ★	-0.42		547		C2H5N1O3	2-NITROETHANOL 625-48-9
6 ★	-0.26		579		C2H5N1S1	THIOACETAMIDE 62-55-5
7 ★	-0.03		401		C2H5N3O2	1-METHYL-1-NITROSOUREA 684-93-5
8 ★	-1.61	7.4	401	1	C2H5N5	GUANAZOLE 1455-77-2 *antineoplastic*
9	-1.61	7.4	401	1	"	GUANAZOLE
170 ★	1.81		547	400	C2H6	ETHANE 74-84-0
1	-0.96	7.0	2625	1	C2H6Cl1O3P1	ETHEPHON 11672-87-0 *growth regulator*
2	-3.03		2019		C2H6Cl2Sn1	DICHLOROSTANNANE,DIMETHYL 753-73-1
3 ★	2.59		547		C2H6Hg1	DIMETHYLMERCURY 593-74-8
4 ★	-1.58		1137		C2H6N2O1	ACETIC ACID HYDRAZIDE 1068-57-1

	logP	pH	Ref.	Note	MF	Name / CAS / activity
75 ★	-1.40		2198		C2H6N2O1	METHYLUREA 598-50-5
6 ★	-0.57		871		C2H6N2O1	N-NITROSODIMETHYLAMINE 62-75-9
7 ★	-0.46	7.4	1184	1	C2H6N2O2	1-METHYL-1-HYDROXYUREA 7433-43-4
8 ★	-0.68		1173	537	C2H6N2S1	METHYLTHIOUREA 598-52-7
9 ★	0.10		547	400	C2H6O1	DIMETHYLETHER 115-10-6
180	-0.37	7.0	1232	1	C2H6O1	ETHANOL 64-17-5 *solvent,anti-infective*
1 ★	-0.31		547	400	"	ETHANOL
2	-0.22	7.2	2016		"	ETHANOL
3 ★	-1.35		540		C2H6O1S1	DIMETHYLSULFOXIDE 67-68-5 *anti-inflammatory (topical)*
4 ★	-1.36		572		C2H6O2	ETHANE-1,2-DIOL 107-21-1
85 ★	-1.34		594	400	C2H6O2S1	METHYLSULFONE 67-71-0
6 ★	1.77		536		C2H6S2	DIMETHYL DISULFIDE 624-92-0
7 ★	-0.38	13.0	1013	821	C2H7N1	DIMETHYLAMINE 124-40-3
8 ★	-0.13	13.0	588	224	C2H7N1	ETHYLAMINE 75-04-7
9 ★	-1.31		510		C2H7N1O1	ETHANOLAMINE 141-43-5 *sclerosing agent*
190 ★	-1.15	13.0	594		C2H7N1O1	METHANAMINE,N-HYDROXY,N-METHYL 5725-96-2
1 ★	-2.20		1394		C2H7N3O1	4-METHYLSEMICARBAZIDE 17696-95-6
2	-0.66		600		C2H8N1O2P1S1	METHAMIDOPHOS 10265-92-6 *insecticide*
3 ★	-2.04	13.0	579		C2H8N2	ETHYLENE DIAMINE 107-15-3 *urinary acidifier (vet)*
4 ✓	-1.68		2426	537	C2H10Cl2N2Pt1	PLATINUM,BIS-METHYLAMMONIO-DICHLORO
95 ✓	-3.28		2426		C2H14N2O2Pt1.(N1O3)2-	PLATINUM,BIS-METHYLAMONIO-DIAQUA-DINITRATE
6 ★	1.46		579		C3F6O1	HEXAFLUOROACETONE 10057-01-1
7 ★	2.82		2137		C3H1Br2N3O2	IMIDAZOLE,2,4-DIBROMO-5-NITRO
8 ★	1.96		509		C3H1Br3N2	IMIDAZOLE,2,4,5-TRIBROMO 2034-22-2
9 ★	1.85		509		C3H1Cl3N2	IMIDAZOLE,2,4,5-TRICHLORO 7682-38-4
200 ★	2.78		509		C3H1I3N2	IMIDAZOLE,2,4,5-TRIIODO 1746-25-4
1 ★	0.71		2387		C3H2Br1N3O2	IMIDAZOLE,2-BROMO-4-NITRO 65902-59-2
2 ★	0.85		2387		C3H2Br1N3O2	IMIDAZOLE,4-NITRO-5-BROMO 6963-65-1
3 ★	2.10		547	400	C3H2Cl1F5O1	2-CL-1,1,2-TRIFL-ET-DIF-ME-ETHER 13838-16-9 *anesthetic (inhalation)*
4 ★	2.06		591		C3H2Cl1F5O1	ISOFLURANE 26675-46-7 *anesthetic (inhalation)*
5 ★	0.66		2387		C3H2Cl1N3O2	IMIDAZOLE,2-CHLORO-4-NITRO 57531-37-0
6 ★	0.78		2387		C3H2Cl1N3O2	IMIDAZOLE,4-NITRO-5-CHLORO 57531-38-1
7 ★	1.66		590		C3H2F6O1	HEXAFLUORO-2-PROPANOL 920-66-1
8	0.28	7.4	1975	1	C3H2I1N3O2	IMIDAZOLE,4-NITRO-5-IODO 76529-48-1
9 ★	-0.50		579		C3H2N2	MALONONITRILE 109-77-3
210 ★	0.20		2387		C3H2N4O4	IMIDAZOLE,2,4-DINITRO 5213-49-0
1 ★	0.65		2387		C3H2N4O4	IMIDAZOLE,4,5-DINITRO 19183-14-3
2 ★	1.02		2166		C3H3Cl1N2O2	FURAZAN-2-OXIDE,3-METHYL-4-CHLORO
3 ★	0.90		2166		C3H3Cl1N2O2	FURAZAN-2-OXIDE,4-METHYL-3-CHLORO
4 ★	2.03		594		C3H3Cl3O2	TRICHLOROACETIC ACID,METHYL ESTER 598-99-2
15	0.20		579		C3H3F3O1	1,1,1-TRIFLUOROACETONE 421-50-1
6 ★	0.60		594	400	C3H3F3O2	TRIFLUOUROACETIC ACID,METHYL ESTER 431-47-0
7 ★	1.23		506		C3H3F5O1	PROPANOL,2,2,3,3,3-PENTAFLUORO 422-05-9
8 ★	1.70		547		C3H3I1N2	4-IODOPYRAZOLE 3469-69-0
9 ★	0.25		570	400	C3H3N1	ACRYLONITRILE 107-13-1
220 ★	0.08		508		C3H3N1O1	ISOXAZOLE 288-14-2
1 ★	0.12		1627		C3H3N1O1	OXAZOLE 288-42-6
2 ★	-0.76	1.0	594	342	C3H3N1O2	CYANOACETIC ACID 372-09-8
3 ★	0.44		505		C3H3N1S1	THIAZOLE 288-47-1
4	-0.73	7.4	1163	1	C3H3N3	SYM-TRIAZINE 290-87-9
25	0.66	7.4	1094	1	C3H3N3O1S1	1,2,3-THIADIAZOLE-5-CARBOXALDOXIME 61444-94-8
6 ★	0.86	7.4	1094	2	"	1,2,3-THIADIAZOLE-5-CARBOXALDOXIME
7	-0.34	7.4	2206	306	C3H3N3O2	2-NITROIMIDAZOLE 527-73-1
8 ★	0.15		1008	354	"	2-NITROIMIDAZOLE
9 ★	-0.11		1008	354	C3H3N3O2	4-NITROIMIDAZOLE 3034-38-6
230 ★	0.59		547		C3H3N3O2	4-NITROPYRAZOLE 2075-46-9
1 ★	-1.87		924	400	C3H3N3O2	5-AZAURACIL 71-33-0
2	-0.17	7.4	2206	306	C3H3N3O2	5-NITROIMIDAZOLE 3034-38-6
3 ★	-0.01	9.0	1409	2	"	5-NITROIMIDAZOLE
4	-1.37	7.4	579	1	C3H3N3O2	6-AZAURACIL 461-89-2
35 ★	-0.59	7.4	579	2	"	6-AZAURACIL
6 ★	0.83		594		C3H3N3O2S1	THIAZOLE,2-AMINO-5-NITRO 121-66-4 *antihistomonad in fowl*
7 ★	0.97		2166		C3H3N3O3	1,2,5-OXADIAZOLE,3-NITRO-4-METHYL
8 ★	0.78		2166		C3H3N3O4	FURAZAN-2-OXIDE,3-METHYL-4-NITRO
9 ★	1.45		547	400	C3H4	ALLENE 463-49-0
240 ★	0.94		547	400	C3H4	METHYLACETYLENE 74-99-7
1 ★	0.18		1896		C3H4Cl1N1	3-CHLOROPROPIONITRILE 1617-17-0
2 ★	2.21		530		C3H4Cl2F2O1	METHOXYFLURANE 76-38-0 *anesthetic (inhalation)*
3	0.78		1178		C3H4Cl2O2	DALAPON 75-99-0 *herbicide*
4 ★	-0.08		259		C3H4N2	IMIDAZOLE 288-32-4
45 ★	0.26		547		C3H4N2	PYRAZOLE 288-13-1
6 ★	-0.18		1625		C3H4N2O1	OXAZOLE-2-AMINE 4570-45-0
7 ★	-1.69		508		C3H4N2O2	HYDANTOIN 461-72-3
8 ★	0.38		572		C3H4N2S1	2-AMINOTHIAZOLE 96-50-4 *thyroid inhibitor*
9 ★	-0.68	7.4	2273	1	C3H4N4O2	1,2,4-TRIAZOLE,1-METHYL-3-NITRO 26621-45-4
250 ★	-0.38		579		C3H4O1	PROPARGYL ALCOHOL 107-19-7
1 ★	-0.01		572		C3H4O1	PROPENAL 107-02-8 *herbicide*
2 ★	0.35	1.0	594		C3H4O2	ACRYLIC ACID 79-10-7
3 ★	-0.81	1.0	594	342	C3H4O4	MALONIC ACID 141-82-2
4 ★	1.79		1560	459	C3H5Br1	ALLYLBROMIDE 106-95-6
55 ★	0.85		2219		C3H5Br1O1	EPIBROMOHYDRIN 3132-64-7
6 ★	0.92		510		C3H5Br1O2	A-BROMOPROPIONIC ACID 598-72-1
7 ★	2.00		579	400	C3H5Cl1	PROPENE,2-CHLORO 557-98-2
8 ★	0.45		2219		C3H5Cl1O1	EPICHLOROHYDRIN 106-89-8
9 ★	0.41	1.0	594	400	C3H5Cl1O2	B-CHLOROPROPIONIC ACID 107-94-8

	logP	pH	Ref.	Note	MF	Name / CAS / activity
260 ★	-0.16		594	400	C3H5F1O1	EPIPFLUOROHYDRIN 503-09-3
1	-0.39		579		C3H5F1O1	FLUOROACETONE 430-51-3
2 ★	0.71		425		C3H5F3O1	1,1,1-TRIFLUORO-2-PROPANOL 374-01-6
3 ★	0.39		1942		C3H5F3O1	3,3,3-TRIFLUOROPROPANOL
4 ★	-0.43	13.0	594		C3H5N1	2-PROPYN-1-AMINE 2450-71-7
65 ★	0.16		503		C3H5N1	PROPIONITRILE 107-12-0
6 ★	-0.94		1896		C3H5N1O1	3-HYDROXYPROPIONITRILE 78-97-7
7 ★	-0.67		572		C3H5N1O1	ACRYLAMIDE 79-06-1
8	0.34	7.4	1094	1	C3H5N1O2	HYDROXYIMINOACETONE 306-44-5
9	0.34	7.8	2334		"	HYDROXYIMINOACETONE
270 ★	0.39	7.4	1094	2	"	HYDROXYIMINOACETONE
1 ★	-1.42		594	400	C3H5N1O2S1	METHYLSULFONYLACETONITRILE 2274-42-2
2 ★	1.47		850		C3H5N1S1	ETHYLISOTHIOCYANATE 542-85-8
3 ★	-0.55		2619		C3H5N3	1,2,4-TRIAZOLE,3-METHYL
4 ★	-1.09	13.0	547	224	C3H5N3	4-AMINOPYRAZOLE 28466-26-4
75	-1.64		594	299	C3H5N3	IMIDAZOLE,2-AMINO 1450-93-7
6 ★	0.09		2166		C3H5N3O1	1,2,5-OXADIAZOLE,3-AMINO-4-METHYL
7 ★	-0.92		1137		C3H5N3O1	2-AMINO-5-METHYL-1,3,4-OXADIAZOLE 52838-39-8
8 ★	-0.17		2166		C3H5N3O2	FURAZAN-2-OXIDE,3-METHYL-4-AMINO
9 ★	-0.32		2166		C3H5N3O2	FURAZAN-2-OXIDE,4-METHYL-3-AMINO
280	1.62		1358		C3H5N3O9	GLYCERYLTRINITRATE 55-63-0
1 ★	1.72		547	400	C3H6	CYCLOPROPANE 75-19-4 anesthetic (inhalation)
2 ★	1.77		547	400	C3H6	PROPYLENE 115-07-1
3	2.18		1560	100	C3H6Br1Cl1	1-BROMO-3-CHLOROPROPANE 109-70-6
4 ★	2.37		594		C3H6Br2	1,3-DIBROMOPROPANE 109-64-8
85 ★	0.57		401		C3H6Cl1N3O2	1-(2-CHLOROETHYL)-1-NITROSOUREA(NCS47547) 2365-30-2
6 ★	2.00		547		C3H6Cl2	1,3-DICHLOROPROPANE 142-28-9
7 ★	-0.83		594		C3H6F1O1	A-FLUORO-N-METHYLACETAMIDE 367-49-7
8 ★	-0.36		579		C3H6F2O1	1,3-DIFLUORO-2-PROPANOL 453-13-4
9 ★	0.55		1701		C3H6F3N1	N-METHYL-2,2,2-TRIFLUOROETHYLAMINE
290 ★	3.02		579		C3H6I2	1,3-DIIODOPROPANE 627-31-6
1 ★	-0.15		579		C3H6N2	DIMETHYLCYANAMIDE 1467-79-4
2 ★	-0.66		507		C3H6N2S1	IMIDAZOLIDONE,2-THIO 96-45-7
3 ★	-0.77		1173	537	C3H6N4	1-METHYL-2-CYANOGUANIDINE 1609-07-0
4 ★	-1.37		599		C3H6N6	1,3,5-TRIAZINE,2,4,6-TRIAMINO 108-78-1
95	0.87		1179		C3H6N6O6	1,3,5-TRIAZA-1,3,5-TRINITROCYCLOHEXANE 121-82-4 rodenticide, explosive
6 ★	-0.24		510		C3H6O1	ACETONE 67-64-1
7 ★	0.17		508		C3H6O1	ALLYL ALCOHOL 107-18-6
8 ★	0.59		572		C3H6O1	PROPIONALDEHYDE 123-38-6
9 ★	0.03		572		C3H6O1	PROPYLENEOXIDE 75-56-9
300 ★	-0.14		594	400	C3H6O1	TRIMETHYLENE OXIDE 503-30-0
1 ★	-0.37		572		C3H6O2	1,3-DIOXOLANE 646-06-0
2 ★	0.18		510		C3H6O2	ACETIC ACID,METHYL ESTER 79-20-9
3 ★	-0.95		2219		C3H6O2	GLYCIDOL 556-52-5
4 ★	0.33		505		C3H6O2	PROPIONIC ACID 79-09-4 antimicrobial
5 ★	0.43		547		C3H6O2S1	3-MERCATPTOPROPIONIC ACID 107-96-0
6 ★	-0.72		986		C3H6O3	A-HYDROXYPROPIONIC ACID 598-82-3
7 ★	-0.43		579		C3H6O3	TRIOXANE 110-88-3
8 ★	2.10		503		C3H7Br1	1-BROMOPROPANE 106-94-5
9 ★	2.14		594	400	C3H7Br1	2-PROMOPROPANE 75-26-3
310 ★	0.02	7.4	2071		C3H7Br1Hg1O1	1-BROMOMERCURY-2-HYDROXYPROPANE 18832-83-2
1 ★	2.04		547	400	C3H7Cl1	1-CHLOROPROPANE 540-54-5
2 ★	1.90		547	400	C3H7Cl1	2-CHLOROPROPANE 75-29-6
3 ★	0.50		594	400	C3H7Cl1O1	3-CHLORO-1-PROPANOL 627-30-5
4 ★	-0.28		594	400	C3H7F1O1	PROPANOL,3-FLUORO 462-43-1
15 ★	2.89		594	400	C3H7I1	2-IODOPROPANE 75-30-9
6 ★	0.07	13.0	588	224	C3H7N1	ALLYLAMINE 107-11-9
7 ★	0.07		579		C3H7N1	CYCLOPROPYLAMINE 765-30-0
8 ★	0.12		579		C3H7N1O1	ACETONEOXIME 127-06-0
9 ★	-1.01		541		C3H7N1O1	DIMETHYLFORMAMIDE 68-12-2
320 ★	-1.05		536		C3H7N1O1	N-METHYLACETAMIDE 79-16-3
1 ★	-0.66	1.0	2226		C3H7N1O1	PROPIONAMIDE 79-05-0
2 ★	0.87		547	60	C3H7N1O2	1-NITROPROPANE 108-03-2
3 ★	0.80		2571		C3H7N1O2	2-NITROPROPANE 79-46-9
4 ★	-3.05		2654	448	C3H7N1O2	3-AMINOPROPIONIC ACID 28854-76-4
25 ★	-2.96		547	463	C3H7N1O2	ALANINE 302-72-7
6	-2.74	7.0	1207	1	"	ALANINE
7 ★	-0.06		702		C3H7N1O2	O,N-DIMETHYLCARBAMATE 6642-30-4
8	-0.15	7.4	172	1	C3H7N1O2	O-ETHYLCARBAMATE 51-79-6 antineoplastic
9 ★	-0.15		505			O-ETHYLCARBAMATE
330	0.04	7.0	1232	1	"	O-ETHYLCARBAMATE
1 ★	-2.78		2654	448	C3H7N1O2	SARCOSINE 107-97-1
2 ★	-2.49		1590	459	C3H7N1O2S1	CYSTEINE 52-90-4 amino acid
3	4.32		1323	820	C3H7N1O3	SERINE,TETRAHEXYL AMMONIUM SALT
4 ★	-3.07	7.0	1207	1	C3H7N1O3	SERINE 56-45-1 amino acid
35 ★	0.23	8.9	1498		C3H7N3O2	1-ETHYL-1-NITROSOUREA 759-73-9
6 ★	2.36		547	400	C3H8	PROPANE 74-98-6
7	-1.70	5.3	1365		C3H8N1O5P1	N-(PHOSPHONOMETHYL)GLYCINE 1071-83-6 herbicide
8 ★	-0.49		547		C3H8N2O1	DIMETHYLUREA,SYM. 96-31-1
9 ★	-0.74		547		C3H8N2O1	ETHYLUREA 625-52-5
340	-0.02		2570		C3H8N2O1	METHYLETHYLNITROSAMINE 10595-95-6
1 ★	-1.00		1137		C3H8N2O1	PROPIONIC ACID HYDRAZIDE 24535-11-3
2 ★	-0.10	7.4	1184	1	C3H8N2O2	1-ETHYL-1-HYDROXYUREA 7433-42-3
3 ★	-0.76	7.4	1184	1	C3H8N2O2	3-ETHYLHYDROXYUREA 5710-11-2
4 ★	-0.24		1173	537	C3H8N2S1	1,3-DIMETHYLTHIOUREA 534-13-4

	logP	pH	Ref.	Note	MF	Name / **CAS** / *activity*
45 ★	-0.21	6.5	2013		C3H8N2S1	ETHYLTHIOUREA **625-53-6**
6 ★	-0.70		1173	537	C3H8N4O2	1,3-DIMETHYL-2-NITROGUANIDINE
7 ★	0.05		425		C3H8O1	ISOPROPYL ALCOHOL **67-63-0** *anti-infective (topical)*
8 ★	0.25		547	60	C3H8O1	PROPANOL **71-23-8**
9 ★	-0.92		1397	122	C3H8O2	1,2-PROPANEDIOL **57-55-6** *phramaceutic aid (humectant)*
350 ★	-1.04		579		C3H8O2	1,3-PROPANEDIOL **504-63-2**
1 ★	0.18		579	400	C3H8O2	DIMETHOXYMETHANE **109-87-5**
2 ★	-0.77		768		C3H8O2	METHOXYETHANOL **109-86-4**
3	-1.79	7.4	1758	1	C3H8O3	GLYCEROL **56-81-5** *antiglaucoma*
4 ★	-1.76	7.4	1010	1	"	GLYCEROL
55 ★	1.54		594	400	C3H8S1	ETHYLMETHYL SULFIDE **624-89-5**
6 ★	1.81		579		C3H8S1	PROPANETHIOL **107-03-9**
7	-0.29		1959		C3H9Cl1Sn1	CHLOROSTANNANE,TRIMETHYL **1066-45-1**
8 ★	0.26	13.0	588	224	C3H9N1	I-PROPYLAMINE **75-31-0**
9 ★	0.15	13.0	588	224	C3H9N1	METHYLETHYLAMINE **624-78-2**
360 ★	0.48	13.0	588	224	C3H9N1	PROPYLAMINE **107-10-8**
1 ★	0.16	2.7	547	340	C3H9N1	TRIMETHYLAMINE **75-50-3**
2 ★	-0.96		510		C3H9N1O1	2-PROPANOL,1-AMINO **78-96-6**
3 ★	-0.94	13.0	579		C3H9N1O1	ETHANOLAMINE,N-METHYL **109-83-1**
4 ★	-1.12	13.0	594		C3H9N1O1	PROPANOL,3-AMINO **156-87-6**
65 ★	-1.57		1394		C3H9N3O1	4,4-DIMETHYLSEMICARBAZIDE **22718-51-0**
6 ★	-1.62		1394		C3H9N3O1	4-ETHYLSEMICARBAZIDE **13050-41-4**
7 ★	-0.30		594	400	C3H9N3S1	4,4-DIMETHYLTHIOSEMICARBAZIDE **6926-58-5**
8	-0.60		2182	459	C3H9O2P1	METHYLPHOSPHINIC ACID,ETHYL ESTER **16391-07-4**
9 ★	-0.66		579		C3H9O3P1	METHYLPHOSPHONIC ACID,DIMETHYL ESTER **756-79-6**
370	-0.52		508		C3H9O4P1	PHOSPHORIC ACID,TRIMETHYL ESTER **512-56-1**
1	-0.78		547		C3H9O4P1	TRIMETHYLPHOSPHATE
2 ★	-0.07		600		C3H10N1O2P1S1	O,S-DIME-N-ME-PHOSPHORAMIDOTHIOATE **28167-49-9**
3 ★	-1.43	13.0	579		C3H10N2	1,3-PROPANEDIAMINE **109-76-2**
4 ★	4.78		1179		C4Cl6	HEXACHLORO-1,3-BUTADIENE **87-68-3**
75 ★	3.53		1814	459	C4H2Br2S1	2,3-DIBROMOTHIOPHENE **3140-93-0**
6 ★	1.15		2499		C4H2Cl1F1N2	PYRAZINE,2-CHLORO-6-FLUORO
7 ★	1.58		2499		C4H2Cl2N2	PYRAZINE,2,5-DICHLORO
8 ★	1.53		2499		C4H2Cl2N2	PYRAZINE,2,6-DICHLORO
9 ★	0.74		2499		C4H2F2N2	PYRAZINE,2,6-DIFLUORO
380 ★	-0.25		594		C4H2N2	FUMARONITRILE **764-42-1**
1 ✓	-2.70		1966		C4H3Au1O4S1.(Na1)2	GOLD SODIUM THIOMALONATE
2 ★	0.66		2412		C4H3Br1N2	5-BROMOPYRIMIDINE **4595-59-9**
3 ★	0.93		2412		C4H3Br1N2	PYRAZINE,2-BROMO
4 ★	0.50		2613		C4H3Br1N2	PYRIMIDINE,2-BROMO
85 ★	-0.21		536		C4H3Br1N2O2	5-BROMOURACIL **51-20-7**
6 ★	2.18		1055		C4H3Br1O1	3-BROMOFURAN **22037-28-1**
7 ★	2.75		1478		C4H3Br1S1	2-BROMOTHIOPHENE **1003-09-4**
8 ★	2.62		1478		C4H3Br1S1	3-BROMOTHIOPHENE **872-31-1**
9 ★	0.29		579		C4H3Cl1N2	2-CHLOROPYRIMIDINE **1722-12-9**
390 ★	0.70		2412		C4H3Cl1N2	PYRAZINE,2-CHLORO
1 ★	0.10		2412		C4H3Cl1N2	PYRIDAZINE,3-CHLORO **1120-95-2**
2 ★	0.47		2412		C4H3Cl1N2	PYRIMIDINE,4-CHLORO
3 ★	0.47		2412		C4H3Cl1N2	PYRIMIDINE,5-CHLORO **17180-94-8**
4 ★	-0.35		536		C4H3Cl1N2O2	5-CHLOROURACIL **1820-81-1**
95 ★	2.54		1478		C4H3Cl1S1	2-CHLOROTHIOPHENE **96-43-5**
6 ★	0.29		2412		C4H3F1N2	PYRAZINE,2-FLUORO
7 ★	0.02		2412		C4H3F1N2	PYRIMIDINE,2-FLUORO
8 ★	-0.03		2412		C4H3F1N2	PYRIMIDINE,5-FLUORO **475-21-8**
9	-0.99	7.4	401	1	C4H3F1N2O2	5-FLUOROURACIL **51-21-8** *antineoplastic*
400	-0.96	7.4	2270	1	"	5-FLUOROURACIL
1 ★	-0.89	7.4	401	2	"	5-FLUOROURACIL
2	-0.82	7.2	1797	314	"	5-FLUOROURACIL
3 ★	1.94		579	400	C4H3F7O1	BUTANOL,2,2,3,3,4,4,4-HEPTAFLUORO **375-01-9**
4 ★	0.04		536		C4H3I1N2O2	5-IODOURACIL **696-07-1**
5 ★	-0.29		572		C4H3N1O2	MALEIMIDE **541-59-3**
6 ★	1.55		1478		C4H3N1O2S1	2-NITROTHIOPHENE **609-40-5**
7 ★	1.55		1478		C4H3N1O2S1	3-NITROTHIOPHENE **822-84-4**
8 ★	0.66		594		C4H3N1O3	FURAN,2-NITRO **609-39-2**
9 ★	0.24		547		C4H3N3	4-CYANOPYRAZOLE **31108-57-3**
410 ★	-1.97		1200	460	C4H3N5O1	8-AZAHYPOXANTHINE
1 ★	0.33		2387		C4H4Br1N3O2	IMIDAZOLE,1-METHYL-2-BROMO-4-NITRO **16681-63-3**
2 ★	0.77		2387		C4H4Br1N3O2	IMIDAZOLE,1-METHYL-4-BROMO-5-NITRO **59177-47-8**
3 ★	0.36		2387		C4H4Br1N3O2	IMIDAZOLE,1-METHYL-4-NITRO-5-BROMO **933-87-9**
4 ★	0.86		2387		C4H4Br1N3O2	IMIDAZOLE,2-BROMO-4-NITRO-5-METHYL
15 ★	1.14		2387		C4H4Br1N3O2	IMIDAZOLE,2-METHYL-4-NITRO-5-BROMO **933-87-9**
6 ★	0.35		1519	400	C4H4Cl1N3	3-CL-6-PYRIDAZINEAMINE
7 ★	0.67		2499		C4H4Cl1N3	PYRAZINE,2-AMINO-5-CHLORO
8 ★	0.95		2499		C4H4Cl1N3	PYRAZINE,2-AMINO-6-CHLORO
9 ★	0.30		1008	354	C4H4Cl1N3O2	1-METHYL-4-NITRO-5-CHLOROIMIDAZOLE **4897-25-0**
420 ★	0.19		2387		C4H4Cl1N3O2	IMIDAZOLE,1-METHYL-2-CHLORO-4-NITRO **63634-21-9**
1 ★	0.62		2387		C4H4Cl1N3O2	IMIDAZOLE,1-METHYL-2-CHLORO-5-NITRO **86072-07-3**
2 ★	0.69		2387		C4H4Cl1N3O2	IMIDAZOLE,1-METHYL-4-CHLORO-5-NITRO **4897-31-8**
3 ★	1.06		2387		C4H4Cl1N3O2	IMIDAZOLE,2-METHYL-4-NITRO-5-CHLORO **4897-25-0**
4 ★	2.01		594	400	C4H4Cl2	2-BUTYNE,1,4-DICHLORO **821-10-3**
25 ★	2.15		871		C4H4F6N2O1	N-NITROSO-BIS(2,2,2-TRIFLUOROETHYL)AMINE
6 ★	1.00	7.4	1975	1	C4H4I1N3O2	IMIDAZOLE,1-METHYL-4-NITRO-5-IODO **35681-63-1**
7 ★	1.29	7.4	1975	1	C4H4I1N3O2	IMIDAZOLE,1-METHYL-5-NITRO-4-IODO **76529-47-0**
8 ★	-0.23		924	400	C4H4N2	PYRAZINE **290-37-9**
9 ★	-0.72		547		C4H4N2	PYRIDAZINE **289-80-5**

	logP	pH	Ref.	Note	MF	Name / **CAS** / *activity*
430 ★	-0.40		507		C4H4N2	PYRIMIDINE **289-95-2**
1 ★	-0.99		2069		C4H4N2	SUCCINODINITRILE **110-61-2**
2 ✓	-1.76		590	815	C4H4N2O1	2-PYRIMIDONE **557-01-7**
3 ★	-1.62		924	400	"	2-PYRIMIDONE
4 ★	-1.38		924		C4H4N2O1	4-PYRIMIDONE **4562-27-0**
35 ★	-1.49		924		C4H4N2O1	PYRAZINE-2-ONE **6270-63-9**
6 ★	0.54		1625		C4H4N2O1S1	FORMAMIDE,N-2-THIAZOLYL **25602-39-5**
7 ★	-0.28		536		C4H4N2O1S1	THIOURACIL **141-90-2** *anti-anginal, thyroid inhibitor*
8 ★	-0.84		570	815	C4H4N2O2	MALEIC ACID HYDRAZIDE **123-33-1** *herbicide~growth regulator*
9 ★	-1.07		536		C4H4N2O2	URACIL **66-22-8**
440	-1.00	7.2	1797	314	"	URACIL
1 ★	-1.47		505		C4H4N2O3	BARBITURIC ACID **67-52-7**
2 ★	1.02		2166		C4H4N2O3	FURAZAN-2-OXIDE,2-METHYL-3-CARBOXALDEHYDE
3 ★	1.29		2166		C4H4N2O3	FURAZAN-2-OXIDE,3-CARBOXALDEHYDE-4-METHYL
4 ★	-1.66		527		C4H4N2O4	3-CARBOXYMETHYLSYDNONE **26537-53-1**
45 ★	-1.11		924	400	C4H4N2S1	2-(1H)-PYRIMIDINTHIONE **1450-85-7**
6 ✓	-0.97		590	815	"	2-(1H)-PYRIMIDINTHIONE
7 ✓	-0.97		590		C4H4N2S1	PYRIMIDINE,2-THIOL
8 ★	0.04		871		C4H4N4O1	N-NITROSO-BIS(CYANOMETHYL)AMINE
9 ★	-0.17		2387		C4H4N4O4	IMIDAZOLE,1-METHYL-2,4-DINITRO **5213-50-3**
450 ★	0.41		2387		C4H4N4O4	IMIDAZOLE,1-METHYL-4,5-DINITRO **19183-15-4**
1 ★	0.77		2387		C4H4N4O4	IMIDAZOLE,2-METHYL-4,5-DINITRO **19183-16-5**
2 ★	-0.96		1200		C4H4N6	8-AZAADENINE **1123-54-2**
3 ★	-0.71	4.2	401		C4H4N6O1	8-AZAGUANINE **134-58-7** *antineoplastic*
4 ★	1.34		547	400	C4H4O1	FURAN **110-00-9**
55 ★	-0.60		579		C4H4O2	2(5H)-FURANONE **497-23-4**
6 ★	0.46	1.0	594		C4H4O4	FUMARIC ACID **110-17-8**
7 ★	-0.34	1.0	594		C4H4O4	MALEIC ACID **110-16-7**
8 ★	1.81		502		C4H4S1	THIOPHENE **110-02-1**
9 ★	2.39		599		C4H5Cl3O2	ETHYLTRICHLOROACETATE
460 ★	1.18		505		C4H5F3O2	ACETIC ACID,TRIFLUORO-ETHYL ESTER **383-63-1**
1 ★	0.40		1896		C4H5N1	ALLYLNITRILE **109-75-1**
2 ★	0.29		579		C4H5N1	CYCLOPROPYLCYANIDE **5500-21-0**
3 ★	0.68		1896		C4H5N1	METHACRYLONITRILE **126-98-7**
4 ★	0.75		503		C4H5N1	PYRROLE **109-97-7**
65 ★	0.45		594		C4H5N1O1	ISOXAZOLE,5-METHYL **5765-44-6**
6 ★	0.46	1.0	574	342	C4H5N1O2	3-HYDROXY-5-METHYLISOXAZOLE **10004-44-1** *fungicide*
7 ★	-0.47		594		C4H5N1O2	CYANOACETIC ACID,METHYL ESTER **105-34-0**
8 ★	0.97		594		C4H5N1S1	THIAZOLE,4-METHYL **693-95-8**
9 ★	-0.07		163	459	C4H5N3	2-AMINOPYRAZINE **5049-61-6**
470 ★	-0.53		2412		C4H5N3	PYRIDAZINE,4-AMINO **20744-39-2**
1 ★	-0.22		505		C4H5N3	PYRIMIDINE,2-AMINO **5469-70-5**
2 ★	-0.25		2412		C4H5N3	PYRIMIDINE,4-AMINO **591-54-8**
3 ★	-0.99		924	400	C4H5N3O1	2-AMINO-4-PYRIMIDONE **108-53-2**
4	-2.40	7.2	1797	314	C4H5N3O1	CYTOSINE **71-30-7**
75 ★	-1.73		536		"	CYTOSINE
6	0.74	7.4	1094	1	C4H5N3O1S1	2-METHYL-1,3,4-THIADIAZOLE-5-CARBOXALDOXIME
7 ★	0.78	7.4	1094	2	"	2-METHYL-1,3,4-THIADIAZOLE-5-CARBOXALDOXIME
8	0.72	7.4	1094	1	C4H5N3O1S1	3-METHYL-1,3,4-THIADIAZOLE-5-CARBOXALDOXIME **61444-96-0**
9 ★	1.29	7.4	1094	2	"	3-METHYL-1,3,4-THIADIAZOLE-5-CARBOXALDOXIME
480	0.47	7.6	1769	212	C4H5N3O2	1,2,4-OXADIAZOLE-3-METHYL-5-HYDROXIMIC ACID **90507-21-4**
1 ★	0.55	7.4	2298	2	"	1,2,4-OXADIAZOLE-3-METHYL-5-HYDROXIMIC ACID
2 ★	0.30		2166		C4H5N3O2	1,2,5-OXADIAZOLE,3-METHYL-4-CARBOXAMIDO
3 ★	1.02	1.0	2196		C4H5N3O2	1-(H)-PYRAZOLE,3-METHYL-4-NITRO **5334-39-4**
4 ★	0.33		1731		C4H5N3O2	2-METHYL-5-NITROIMIDAZOLE
85 ★	0.06		1008	354	"	2-METHYL-5-NITROIMIDAZOLE **696-23-1**
6	0.10	7.4	2206	306	"	2-METHYL-5-NITROIMIDAZOLE
7 ★	0.18	7.4	2206	306	C4H5N3O2	4-METHYL-5-NITROIMIDAZOLE **14003-66-8**
8 ★	-0.17		2387		C4H5N3O2	IMIDAZOLE,1-METHYL-2-NITRO **1671-82-5**
9 ★	-0.44		2387		C4H5N3O2	IMIDAZOLE,1-METHYL-4-NITRO **3034-41-1**
490 ★	0.16	7.4	2206	306	C4H5N3O2	IMIDAZOLE,1-METHYL-5-NITRO **3034-42-2**
1 ★	0.13		2387		C4H5N3O2	IMIDAZOLE,2-METHYL-4-NITRO
2 ★	0.23		2387		C4H5N3O2	IMIDAZOLE,4-NITRO-5-METHYL
3 ★	0.03		2166		C4H5N3O3	FURAZAN-2-OXIDE,3-METHYL-4-CARBOXAMIDO
4 ★	0.36		2166		C4H5N3O3	FURAZAN-2-OXIDE,4-METHYL-3-CARBOXAMIDO
95 ★	0.13		2387		C4H5N3O3	IMIDAZOLE,4-NITRO-5-METHOXY **68019-78-3**
6 ★	1.99		547	400	C4H6	1,3-BUTADIENE **106-99-0**
7 ★	1.46		547	400	C4H6	2-BUTYNE **503-17-3**
8 ★	0.56		1896		C4H6Cl1N1	4-CHLOROBUTYRONITRILE **628-20-6**
9 ★	0.83		871		C4H6Cl2N2O1	3,4-DICHLORO-N-NITROSOPYRROLIDINE **59863-59-1**
500 ★	0.87		594		C4H6F2O2	DIFLUOROACETIC ACID,ETHYL ESTER **454-31-9**
1 ★	0.23	9.0	547	1	C4H6N2	4-METHYLIMIDAZOLE **822-36-6**
2 ★	-0.06	8.4	579	1	C4H6N2	IMIDAZOLE,1-METHYL **616-47-7**
3 ★	0.24		594		C4H6N2	IMIDAZOLE,2-METHYL **693-98-1**
4 ★	0.23		594	400	C4H6N2	PYRAZOLE,1-METHYL **930-36-9**
5 ★	0.37		547		C4H6N2O1	4-METHOXYPYRAZOLE
6 ★	0.26		871		C4H6N2O1	DELTA-3-N-NITROSOPYRROLIDINE **10552-94-0**
7 ★	0.71		2166		C4H6N2O1	FURAZAN,DIMETHYL **4975-21-7**
8 ★	0.23		2166		C4H6N2O2	FURAZAN,DIMETHYL,2-OXIDE **2518-42-5**
9	-2.39	7.0	1207	1	C4H6N2O2	MUSCIMOL **2763-96-4** *agonist-gabba-a*
510 ★	1.28		2166		C4H6N2S1	1,2,5-THIADIAZOLE,3,4-DIMETHYL
1 ★	0.03		547		C4H6N2S1	2,5-DIMETHYL-1,3,4-THIADIAZOLE **27464-82-0**
2 ★	-0.34		599		C4H6N2S1	METHIMAZOLE **60-56-0** *thyroid inhibitor*
3 ★	0.89		2166		C4H6N2Se1	1,2,5-SELENODIAZOLE,3,4-DIMETHYL
4 ★	-0.45		2499		C4H6N4	PYRAZINE,2,6-DIAMINO

	logP	pH	Ref.	Note	MF	Name / CAS / activity
15 ★	-1.10	9.0	594	1	C4H6N4O1	4-IMIDAZOLECARBOXAMIDE,5-AMINO
6 ★	0.38		57		C4H6N4O1S1	3-METHYLTHIO-4-AMINO-1,2,4-TRIAZINE-5-ONE
7 ★	-0.19		2592		C4H6N4O2	1,2,4-TRIAZOLE,1-ETHYL-3-NITRO
8 ★	-0.23		2166		C4H6N4O2	1,2,5-OXADIAZOLE,3-AMINO-4-ACETYLAMINO
9	-1.00		2387		C4H6N4O2	IMIDAZOLE,1-METHYL-4-NITRO-5-AMINO 4531-54-8
520 ★	-1.26	7.4	830	1	C4H6N4O2	N-DIAZOACETYLGLYCINEAMIDE 817-99-2
1	-0.45	7.2	2036	1	C4H6N4O3S2	ACETAZOLAMIDE 59-66-5 carbonic anhydrase inhibitor
2 ★	-0.26		505		"	ACETAZOLAMIDE
3 ★	-0.74		2279		C4H6N4O4S1	IMIDAZOLE,1-METHYL-4-NITRO-5-SULFONAMIDO 6339-55-5
4 ★	0.46		547	400	C4H6O1	2,5-DIHYDROFURAN 1708-29-8
25 ★	0.60		589		C4H6O1S1	G-THIOBUTYROLACTONE 39700-44-2
6 ★	-1.34		541		C4H6O2	2,3-BUTANEDIONE 431-03-8
7 ★	0.80		1402		C4H6O2	ACRYLIC ACID,METHYL ESTER 96-33-3
8 ★	0.72		300		C4H6O2	CROTONIC ACID 3724-65-0
9 ★	0.63	1.0	579	342	C4H6O2	CYCLOPROPANECARBOXYLIC ACID 1759-53-1
530 ★	-0.52		594		C4H6O2	ERYTHRITOLANHYDRIDE 564-00-1
1 ★	-0.64		547		C4H6O2	G-BUTYROLACTONE 96-48-0
2 ★	0.93		997		C4H6O2	METHACRYLIC ACID 79-41-4
3 ★	0.73		572		C4H6O2	VINYLACETATE 108-05-4
4 ★	-0.41		599		C4H6O3	PROPYLENECARBONATE 108-32-7 gelling agent
35 ★	-0.17		579		C4H6O4	OXALIC ACID,DIMETHYL ESTER 553-90-2
6 ★	-0.59		510		C4H6O4	SUCCINIC ACID 110-15-6
7	-0.57	7.4	2288	1	"	SUCCINIC ACID
8 ★	-1.26		510		C4H6O5	MALIC ACID 6915-15-7 pharmaceutic acid
9	1.11		1966	459	C4H7Au1Cl1N1	I-PROPYLISOCYANIDE GOLD CHLORIDE
540	2.53		1560	459	C4H7Br1	4-BROMO-1-BUTENE 5162-44-7
1	1.11		1560	459	C4H7Br1O2	2-BROMOETHYLACETATE 927-68-4
2 ★	1.42		510		C4H7Br1O2	A-BROMOBUTYRIC ACID 80-58-0
3	1.38		600		C4H7Br2Cl2O4P1	NALED 300-76-5 insecticide
4 ★	0.94	5.0	590		C4H7Cl1O2	CHLOROACETIC ACID,ETHYL ESTER 105-39-5
45 ★	1.40		51		C4H7Cl2O3P1	DICHLOROVINYLPHOSPHONATE,O,O-DIMETHYL 1185-97-3
6 ★	1.43		2049		C4H7Cl2O4P1	DICHLOROVINYLPHOSPHATE,O,O-DIMETHYL 62-73-7 anthelmintic~insecticide
7 ★	2.03		505		C4H7Cl3O1	CHLOROBUTANOL 57-15-8 analgesic (dental), antimicrobial
8 ★	1.04		1942		C4H7F3O1	3,3,3-TRIFLUORO-2-METHYLPROPANOL-2
9 ★	0.71		1942		C4H7F3O1	4,4,4-TRIFLUOROBUTANOL-2
550 ★	0.90		1942		C4H7F3O1	4,4,4-TRIFLUOROBUTANOL
1 ★	0.53		1896		C4H7N1	BUTYRONITRILE 109-74-0
2 ★	0.46		1896		C4H7N1	I-BUTYRONITRILE 78-82-0
3	-0.85	5.0	2425	1	C4H7N1O1	2-PYRROLIDINONE 616-45-5
4 ★	-2.84		2654	448	C4H7N1O2	AZACYCLOBUTANE,A-CARBOXYLIC ACID
55 ★	-2.78		2654	448	C4H7N1O2	CYCLOPROPANECARBOXYLIC ACID,A-AMINO
6 ★	1.21	7.8	2334		C4H7N1O2S1	2-OXOPROPANAL OXIME,1-METHYLTHIO 112740-61-1
7	-1.02	7.8	2334	1	C4H7N1O3S1	2-OXOPROPANAL OXIME,1-METHYLSULFOXIDE 112740-56-4
8 ★	-0.72	7.8	2334	2	"	2-OXOPROPANAL OXIME,1-METHYLSULFOXIDE
9	-4.25		1590	459	C4H7N1O4	ASPARTATE,MONOSODIUM SALT
560	-0.84	7.8	2334	1	C4H7N1O4S1	2-OXOPROPANAL OXIME,1-METHYLSULFONE 112740-58-6
1 ★	-0.69	7.8	2334	2	"	2-OXOPROPANAL OXIME,1-METHYLSULFONE
2 ★	-0.05		561		C4H7N1S1	2-AZACYCLOPENTANTHIONE 2295-35-4
3 ★	-0.30		1137		C4H7N3O1	2-AMINO-5-ETHYL-1,3,4-OXADIAZOLE
4 ★	-1.76	7.4	1010	1	C4H7N3O1	CREATININE 60-27-5
65	-0.76	7.4	830	1	C4H7N5O2	N-DIAZOACETYLGLYCINE HYDRAZIDE 820-75-7
6 ★	2.40		547	400	C4H8	1-BUTENE 106-98-9
7 ★	2.33		547	400	C4H8	2-BUTENE-CIS 590-18-1
8 ★	2.31		547	400	C4H8	2-BUTENE-TR 624-64-6
9 ★	2.34		547	400	C4H8	I-BUTYLENE 115-11-7
570 ★	0.34		507		C4H8Br1N1O1	BROMOACETAMIDE,N-ETHYL
1 ★	1.29		579		C4H8Cl2O1	2,2'-DICHLOROETHYLETHER 111-44-4
2 ★	0.51		51		C4H8Cl3O4P1	DIME-1-OH-2,2,2-TRICLETHYLPHOSPHONATE 52-68-6 anthelmintic (vet)
3 ★	1.06		1701		C4H8F3N1	N,N-DIMETHYL-2,2,2-TRIFLUOROETHYLAMINE 819-06-7
4 ★	0.74		1701		C4H8F3N1	N-ETHYL-2,2,2-TRIFLUOROETHYLAMINE
75 ★	0.52		507		C4H8N2	2-IMIDAZOLINE,2-METHYL 534-26-9
6 ★	-0.19		871		C4H8N2O1	N-NITROSOPYRROLIDINE 930-55-2
7 ★	0.40		871		C4H8N2O1S1	N-NITROSOTHIOMORPHOLINE 26541-51-5
8 ★	-0.80		599		C4H8N2O2	1-DIMETHYLAMINO-2-NITROETHYLENE
9 ★	-1.83	7.1	1591		C4H8N2O2	GLYCIN-AMIDE,N-ACETYL
580	-1.68	7.0	2309	1	"	GLYCIN-AMIDE,N-ACETYL
1 ★	-0.70		2198		C4H8N2O2	METHYLUREA,N'-ACETYL 623-59-6
2 ★	-0.44		871		C4H8N2O2	N-NITROSOMORPHOLINE 59-89-2
3 ★	-3.41		1591		C4H8N2O3	ASPARAGINE 70-47-3
4 ★	-2.92		547		C4H8N2O3	GLYCINE,GLYCYL 556-50-3
85 ★	-0.68		1157		C4H8N2S1	N,N'-TRIMETHYLENETHIOUREA 2055-46-1
6 ★	-0.40		1173	537	C4H8N4	1,3-DIMETHYL-2-CYANOGUANIDINE
7 ★	0.51		2166		C4H8N4O1	1,2,5-OXADIAZOLE,3-AMINO-4-ETHYLAMINO
8 ★	-0.85		871		C4H8N4O2	N,N'-DINITROSOPIPERAZINE 140-79-4
9 ★	0.29		503		C4H8O1	2-BUTANONE 78-93-3
590 ★	0.88	7.4	547	1	C4H8O1	BUTYRALDEHYDE 123-72-8
1 ★	1.04		530		C4H8O1	ETHYLVINYLETHER 109-92-2
2 ★	0.46		547	400	C4H8O1	TETRAHYDROFURAN 109-99-9
3 ★	-0.81		594		C4H8O2	2-BUTENE-1,4-DIOL 110-64-5
4 ★	0.73		503		C4H8O2	ACETIC ACID,ETHYL ESTER 141-78-6
95 ★	0.79		510		C4H8O2	BUTYRIC ACID 107-92-6
6 ★	-0.27		547	400	C4H8O2	DIOXANE 123-91-1
7 ★	0.83		508		C4H8O2	FORMIC ACID,PROPYL ESTER 110-74-7
8	1.10	3.5	2391	1	C4H8O2	I-BUTYRIC ACID 79-31-2
9	0.11		2596		C4H8O2	METHYL GLYCIDYL ETHER 930-37-0

	logP	pH	Ref.	Note	MF	Name / CAS / activity
600 ★	0.82		579	400	C4H8O2	PROPIONIC ACID,METHYL ESTER 554-12-1
1 ✓	-3.20		508		C4H8O2	SODIUM BUTRYATE
2 ★	-0.77		572		C4H8O2S1	TETRAHYDROTHIOPHENE-,1,1-DIOXIDE 126-33-0
3 ★	-0.36		510		C4H8O3	A-HYDROXY-I-BUTYRIC ACID 594-61-6
4 ★	2.75		1560	100	C4H9Br1	1-BROMOBUTANE 109-65-9
5 ★	1.15		594	400	C4H9Br1	PROPANE,2-BROMO-2-METHYL 507-19-7
6 ★	2.64		547	60	C4H9Cl1	1-CHLOROBUTANE 109-69-3
7 ★	2.33		594	400	C4H9Cl1	BUTANE,2-CHLORO 78-86-4
8 ★	1.82		579	400	C4H9Cl1	T-BUTYL CHLORIDE 507-20-0
9 ★	0.98		579		C4H9Cl1O1	2-CHLOROETHYLETHER 628-34-2
610 ★	0.85		590		C4H9Cl1O1	BUTANOL,4-CHLORO 928-51-8
1	0.41		2019		C4H9Cl3Sn1	TRICHLOROSTANNANE,BUTYL 1118-46-3
2 ★	2.58		579	60	C4H9F1	1-FLUOROBUTANE 2366-52-1
3	-1.60		2084		C4H9Hg1N1O2S1	CYSTEINE,S-METHYL MERCURY
4 ★	0.46	13.0	588	224	C4H9N1	PYRROLIDINE 123-75-1
15 ★	-0.21		503		C4H9N1O1	BUTYRAMIDE 541-35-5
6 ★	-0.86		547		C4H9N1O1	MORPHOLINE 110-91-8
7 ★	-0.77		536		C4H9N1O1	N,N-DIMETHYLACETAMIDE 127-19-5
8 ★	1.47		547	400	C4H9N1O2	1-NITROBUTANE 627-05-4
9 ★	1.17		535		C4H9N1O2	2-METHYL-2-NITROPROPANE 594-70-7
620 ★	1.20		2571		C4H9N1O2	2-NITROBUTANE 600-24-8
1 ★	-2.53		2654	448	C4H9N1O2	A-AMINOBUTYRIC ACID 80-60-4
2	-2.48	7.2	2323	468	"	A-AMINOBUTYRIC ACID
3 ★	-3.17	7.0	1207	1	C4H9N1O2	G-AMINOBUTYRIC ACID 56-12-2
4 ★	-2.91		2654	448	C4H9N1O2	GLYCINE,N,N-DIMETHYL 1118-68-9
25 ★	1.40		2571		C4H9N1O2	ISO-NITROBUTANE 625-74-1
6 ★	0.34		572		C4H9N1O2	N-METHYLCARBAMIC ACID,ETHYL ESTER 105-40-8
7 ★	0.36	7.4	172	1	C4H9N1O2	O-PROPYLCARBAMATE 627-12-3
8 ★	2.15		536		C4H9N1O3	BUTYLNITRATE 928-45-0
9 ★	-2.94	7.0	1207	1	C4H9N1O3	THREONINE 72-19-5 *amino acid*
630 ★	-1.55		1394		C4H9N3O1	4-ALLYLSEMICARBAZIDE 57421-73-5
1 ★	0.18		871		C4H9N3O1	N-NITROSOPIPERAZINE 5632-47-3
2 ★	0.36		871		C4H9N3O2	1-NITROSO-TRIMETHYLUREA 3475-63-6
3	0.30	4.0	979	364	C4H9N3O2	1-PROPYL-1-NITROSOUREA 816-57-9
4 ★	2.89		547	400	C4H10	BUTANE 106-97-8
35 ★	2.76		547	400	C4H10	I-BUTANE 75-28-5
6	-3.15		2019		C4H10Cl2Sn1	DICHLOROSTANNANE,DIETHYL 866-55-7
7 ★	-0.85		1027		C4H10N1O3P1S1	ORTHENE 30560-19-1 *insecticide*
8 ★	-1.50	13.0	594		C4H10N2	PIPERAZINE 110-85-0 *anthelmintic*
9 ★	-0.62		1137		C4H10N2O1	BUTYRIC ACID HYDRAZIDE 3538-65-6
640 ★	-0.63		1137		C4H10N2O1	I-BUTYRIC ACID HYDRAZIDE 3619-17-8
1 ★	0.48		871		C4H10N2O1	N-NITROSODIETHYLAMINE 55-18-5
2	0.51		2570		C4H10N2O1	N-NITROSOMETHYLPROPYL AMINE 924-46-9
3 ★	0.20	7.4	1184	1	C4H10N2O2	3-I-PROPYLHYDROXYUREA 60165-07-3
4 ★	-0.22	7.4	1184	1	C4H10N2O2	3-PROPYLHYDROXYUREA 5710-12-3
45 ✓	-4.64	7.2	2323	468	C4H10N2O2	BUTYRIC ACID,2,4-DIAMINO 305-62-4
6 ★	0.10		579		C4H10N2S1	THIOUREA,TRIMETHYL 2489-77-2
7	0.88	7.4	2306	1	C4H10O1	BUTANOL 71-36-3
8 ★	0.88		532		"	BUTANOL
9 ★	0.89		547	400	C4H10O1	ETHYLETHER 60-29-7 *anesthetic*
650 ★	0.76		425		C4H10O1	I-BUTANOL 78-83-1
1 ★	1.21		594	400	C4H10O1	METHYLPROPYLETHER 557-17-5
2 ★	0.61		503		C4H10O1	S-BUTANOL 78-92-2
3 ★	0.35		547	400	C4H10O1	T-BUTANOL 75-65-0
4 ★	-0.83		579	400	C4H10O2	1,4-BUTANEDIOL 110-63-4
55 ★	-0.92		510		C4H10O2	2,3-BUTANEDIOL 513-85-9
6 ★	-0.21		1819		C4H10O2	ETHANE,1,2-DIMETHOXY 110-71-4
7 ★	-0.32		2562		C4H10O2	ETHOXYETHANOL 110-80-5
8 ★	-0.59		594		C4H10O2S1	DIETHYLSULFONE 597-35-3
9 ★	-0.63		594	400	C4H10O2S1	THIODIGLYCOL 111-48-8
660 ★	0.25	13.0	579	224	C4H10O3	TRIMETHYLORTHOFORMATE 149-73-5
1 ★	-2.29		1087		C4H10O4	ERYTHRITOL 149-32-6
2 ★	1.14		579		C4H10O4S1	DIETHYLSULFATE 64-67-5
3 ★	2.28		536		C4H10S1	BUTANETHIOL 109-79-5
4 ★	1.95		503		C4H10S1	DIETHYL SULFIDE 352-93-2
65	-1.22	7.0	834	1	C4H11N1	BUTYLAMINE 109-73-9
6 ★	0.97	13.0	588	224	"	BUTYLAMINE
7 ★	0.58	13.0	588	224	C4H11N1	DIETHYLAMINE 109-89-7
8 ★	0.70		567		C4H11N1	ETHANAMINE,N,N-DIMETHYL 598-56-1
9 ★	0.73	13.0	588	224	C4H11N1	I-BUTYLAMINE 78-81-9
670 ★	0.84	13.0	579		C4H11N1	METHYLPROPYLAMINE 627-35-0
1	0.74	13.0	588	224	C4H11N1	S-BUTYLAMINE 513-49-5
2 ★	0.40		505		C4H11N1	T-BUTYLAMINE 75-64-9
3 ★	-1.43		510		C4H11N1O2	DIETHANOLAMINE 111-42-2
4 ★	-1.30		1394		C4H11N3O1	4-ISOPROPYLSEMICARBAZIDE 57930-20-8
75 ★	-1.10		1394		C4H11N3O1	4-PROPYLSEMICARBAZIDE 57421-72-4
6	-1.43	7.3	761	314	C4H11N5	N,N-DIMETHYLBIGUANIDE 657-24-9 *antidiabetic*
7 ★	3.31		547		C4H12Ge1	TETRAMETHYLGERMANE 865-52-1
8 ★	0.15		600		C4H12N1O2P1S1	O,S-DIETHYLPHOSPHORAMIDOTHIOATE 16271-10-6
9 ★	0.07		600		C4H12N1O2P1S1	O,S-DIME-N-ET-PHOSPHORAMIDOTHIOATE 52067-48-8
680	-3.92		2173		C4H12N1.I1	TETRAMETHYL AMMONIUM IODIDE
1	-2.52		1901	820	"	TETRAMETHYL AMMONIUM IODIDE
2 ★	-0.62	13.0	594		C4H12N2	1,2-ETHANDIAMINE,N,N'-DIMETHYL 110-70-3
3 ★	-2.66	7.4	1010	1	C4H12N2	TETRAMETHYLENEDIAMINE 110-60-1
4 ★	2.97		547		C4H12Pb1	TETRAMETHYLLEAD 75-74-1

	logP	pH	Ref.	Note	MF	Name / **CAS** / *activity*
85 ★	3.24		594	400	C4H12Si1	TETRAMETHYLSILANE **75-76-3**
6 ★	3.48		547	400	C4H12Sn1	TETRAMETHYLTIN **594-27-4**
7	3.53		529	831	C5Cl5N1	2,3,4,5,6PENTACHLOROPYRIDINE **2176-62-7**
8 ★	5.04		1736		C5Cl6	HEXACHLORO-1,3-CYCLOPENTADIENE
9	5.04		1369	459	C5Cl6	HEXACHLOROCYCLOPENTADIENE **77-47-4**
690	2.34		579		C5F12	PERFLUOROPENTANE
1	4.40		579	400	"	PERFLUOROPENTANE
2	3.08	7.3	509	2	C5H1Br3N4	PURINE,2,6,8-TRIBROMO
3	3.90	7.3	509	2	C5H1Cl3N4	PURINE,2,6,8,-TRICHLORO **2562-52-9**
4 ★	3.32		529	831	C5H1Cl4N1	2,3,5,6-TETRACHLOROPYRIDINE **2402-79-1**
95 ★	0.92		2499		C5H2Cl1N3	PYRAZINE,2-CHLORO-5-CYANO
6 ★	0.79		2499		C5H2Cl1N3	PYRAZINE,2-CHLORO-6-CYANO
7 ★	3.11		529	831	C5H2Cl3N1	2,3,5-TRICHLOROPYRIDINE **16063-70-0**
8 ★	2.77		529	831	C5H2Cl3N1	2,3,6-TRICHLOROPYRIDINE **6515-09-9**
9 ★	2.68		529	831	C5H2Cl3N1	2,4,6-TRICHLOROPYRIDINE **16063-69-7**
700 ★	3.21		1190		C5H2Cl3N1O1	3,5,6-TRICHLORO-2-PYRIDINOL **6515-38-4**
1	3.21		1178		C5H2Cl3N1O1	3,5,6-TRICHLOROPYRIDINOL
2 ★	2.11		529	831	C5H3Cl2N1	2,3-DICHLORPYRIDINE **2402-77-9**
3 ★	2.40		529	831	C5H3Cl2N1	2,5-DICHLORPYRIDINE **16110-09-1**
4 ★	2.15		529	831	C5H3Cl2N1	2,6-DICHLORPYRIDINE **2402-78-0**
5 ★	2.56		529	831	C5H3Cl2N1	3,5-DICHLORPYRIDINE **2457-47-8**
6 ★	1.17		594	400	C5H3F2N1	PYRIDINE,2,6-DIFLUORO **1513-65-1**
7 ★	0.04		536		C5H3F3N2O2	5-TRIFLUOROMETHYLURACIL **54-20-6**
8 ★	0.96		1478		C5H3N1O1	2-CYANOFURAN **617-90-3**
9 ★	1.01	7.4	594	1	C5H3N1O4	2-FURALDEHYDE,5-NITRO **698-63-5**
710	-1.72		1008	354	C5H3N1O5	5-NITRO-2-FUROIC ACID
1 ★	1.27		1478		C5H3N1S1	2-CYANOTHIOPHENE **1003-31-2**
2 ★	-0.01		2412		C5H3N3	PYRAZINE,2-CYANO
3 ★	-0.63		2412		C5H3N3	PYRIDAZINE,4-CYANO
4 ★	-0.31		2412		C5H3N3	PYRIMIDINE,2-CYANO
15 ★	-0.08		2412		C5H3N3	PYRIMIDINE,4-CYANO **42839-04-3**
6 ★	-0.93	1.0	547	342	C5H3N3O2	5-CYANOURACIL **4425-56-3**
7 ★	1.42		260		C5H4Br1N1	2-BROMOPYRIDINE **109-04-6**
8 ★	1.60		260		C5H4Br1N1	3-BROMOPYRIDINE **626-55-1**
9 ★	1.54		260		C5H4Br1N1	4-BROMOPYRIDINE **1120-87-2**
720 ★	1.22		594	400	C5H4Cl1N1	2-CHLOROPYRIDINE **109-09-1**
1	1.34	7.0	566	1	"	2-CHLOROPYRIDINE
2 ★	1.33		163	459	C5H4Cl1N1	3-CHLOROPYRIDINE **626-60-8**
3 ★	1.28		2412		C5H4Cl1N1	4-CHLOROPYRIDINE **626-61-9**
4 ★	0.93		547		C5H4Cl1N1O1	PYRIDINE,2-CHLORO-6-HYDROXY
25 ★	0.00		1519	400	C5H4Cl1N3O1	3-CHLORO-6-PYRIDAZINECARBOXAMIDE
6 ★	0.28		2499		C5H4Cl1N3O1	PYRAZINE,2-CHLORO-6-CARBOXAMIDO
7 ★	0.84		2412		C5H4F1N1	PYRIDINE,2-FLUORO **372-48-5**
8 ★	0.77		2412		C5H4F1N1	PYRIDINE,3-FLUORO **372-47-4**
9 ★	1.80		1519	459	C5H4I1N1	3-IODOPYRIDINE **1120-90-7**
730 ★	1.13		2221	400	C5H4N2	PYRROLE,2-CYANO **4513-94-4**
1 ★	0.48		163	459	C5H4N2O2	2-NITROPYRIDINE **15009-91-3**
2 ★	0.60		163	459	C5H4N2O2	3-NITROPYRIDINE **2530-26-9**
3 ★	0.33		570		C5H4N2O2	4-NITROPYRIDINE **1122-61-8**
4	-0.74	7.4	547	1	C5H4N2O3	4-NITROPYRIDINE-1-OXIDE **1124-33-0**
35	-0.57	7.0	452	1	"	4-NITROPYRIDINE-1-OXIDE
6 ★	-0.55		547		"	4-NITROPYRIDINE-1-OXIDE
7 ★	-1.03		547		C5H4N2O3	5-FORMYLURACIL **1195-08-0**
8 ★	0.30		1731		C5H4N2O4	NIFUROXIME **6236-05-1** *antibacterial (topical)*
9	-1.70		2447	300	C5H4N2O4	OROTIC ACID **65-86-1** *uricosuric*
740 ★	-0.37		547		C5H4N4	PURINE **120-73-0**
1	-0.21	7.0	1793	1	"	PURINE
2 ★	-0.09		1627	400	C5H4N4	TETRAZOLO(1,5A)PYRIDINE **274-87-3**
3 ★	-0.55	8.0	2070	306	C5H4N4O1	ALLOPURINOL **315-30-0** *xanthine oxidase inhibitor*
4 ★	-1.11	7.4	293	1	C5H4N4O1	HYPOXANTHINE **68-94-0**
45 ★	-0.73		572		C5H4N4O2	XANTHINE **69-89-6**
6	-2.92	7.4	293	1	C5H4N4O3	URIC ACID **69-93-2**
7	-2.90	7.4	1200	1	"	URIC ACID
8	-2.17		1200	460	"	URIC ACID
9	0.18	2.0	2312	337	"	URIC ACID
750 ★	0.01	7.4	401	2	C5H4N4S1	6-PURINETHIOL **50-44-2** *antineoplastic*
1 ★	1.01		1478		C5H4O1S1	3-THIOPHENECARBOXALDEHYDE **498-62-4**
2 ★	1.02		1478		C5H4O1S1	THIOPHENE-2-CARBOXALDEHYDE **98-03-3**
3 ★	0.51		1478		C5H4O2	FURAN-3-CARBOXALDEHYDE **498-60-2**
4 ★	0.41		572		C5H4O2	FURFURAL **98-01-1**
55 ★	1.50		1478		C5H4O2S1	THIOPHEN-3-CARBOXYLIC ACID **88-13-1**
6 ★	1.57		1478		C5H4O2S1	THIOPHENE-2-CARBOXYLIC ACID **527-72-0**
7	0.50		1937	459	C5H4O3	4-HYDROXY-2-PYRONE **675-10-5**
8 ★	1.03		1478		C5H4O3	FURAN-3-CARBOXYLIC ACID **488-93-7**
9	0.64		1055		C5H4O3	FURANE-2-CARBOXYLIC ACID **88-14-2**
760 ★	-1.12		1564		C5H5Br1N2O3	PYRIMIDINE-2,4-DIONE,3-HYDROXY-5-BROMO-6-METHYL **77317-64-7**
1 ★	1.08		2499		C5H5Cl1N2	PYRAZINE,2-CHLORO-5-METHYL
2 ★	1.03		2499		C5H5Cl1N2	PYRAZINE,2-CHLORO-6-METHYL
3 ★	0.89		1519	459	C5H5Cl1N2O1	3-CHLORO-6-METHOXYPYRIDAZINE
4 ★	1.52		2499		C5H5Cl1N2O1	PYRAZINE,2-CHLORO-5-METHOXY
65 ★	1.65		2499		C5H5Cl1N2O1	PYRAZINE,2-CHLORO-6-METHOXY
6	-0.82		1375		C5H5Cl1N2O1	PYRIDAZINE,3-CHLORO-4-HYDROXY-6-METHYL
7 ★	2.55		2506		C5H5Cl3N2O1S1	ETHAZOL **2593-15-9** *fungicide*
8	0.63	7.4	2362	1	C5H5F1N2	PYRIDINE,3-FLUORO-4-AMINO
9 ★	2.12		505		C5H5F5O2	PENTAFLUOROPROPIONIC ACID,ETHYL ESTER **426-65-3**

	logP	pH	Ref.	Note	MF	Name / CAS / activity
770	0.59		1734		C5H5N1	PYRIDINE,PENTADEUTERO **7291-22-7**
1	0.60	7.2	2317	1	C5H5N1	PYRIDINE **110-86-1**
2 ★	0.65		502		"	PYRIDINE
3 ★	0.48		572		C5H5N1O1	3-HYDROXYPYRIDINE **109-00-2**
4 ✓	-1.31		590	815	C5H5N1O1	4-PYRIDONE **626-64-2**
75 ★	-1.30		924	400	"	4-PYRIDONE
6 ★	-0.58		548	815	C5H5N1O1	A-PYRIDONE **109-10-4**
7	-1.27	7.4	1940	1	C5H5N1O1	PYRIDINE,1-OXIDE **694-59-7**
8 ★	-1.20		2615		"	PYRIDINE,1-OXIDE
9 ★	0.64		1625		C5H5N1O1	PYRROLE,2-CARBOXALDEHYDE **1003-29-8**
780	-0.11	7.0	332	1	C5H5N1O2	FURAN-2-CARBOXAMIDE **609-38-1**
1 ★	-0.11		1055		"	FURAN-2-CARBOXAMIDE
2 ★	0.09		1478		C5H5N1O2	FURANE-3-CARBOXAMIDE **609-35-8**
3 ★	-0.61		594		C5H5N1O2	PYRIDINE-N-OXIDE,3-HYDROXY **6602-28-4**
4 ★	0.85	2.0	1625		C5H5N1O2	PYRROLE,2-CARBOXYLIC ACID **634-97-9**
85 ★	0.68		1346		C5H5N1O3	5-METHYLISOXAZOLE-3-CARBOXYLIC ACID **4857-42-5**
6 ✓	-0.13		590	815	C5H5N1S1	2-PYRIDINETHIONE **2637-34-5**
7 ✓	0.20		590	815	C5H5N1S1	4-PYRIDINETHIONE **4556-23-4**
8 ★	0.52		1478		C5H5N1S1	THIOPHEN-2-CARBOXAMIDE **5813-89-8**
9 ★	0.50		1478		C5H5N1S1	THIOPHEN-3-CARBOXAMIDE **51460-47-0**
790 ★	1.34		2221	457	C5H5N1S2	THIOPHENE,2-THIOCARBOXAMIDE **20300-02-1**
1 ★	-0.60		1519	400	C5H5N3O1	2-PYRAZINECARBOXAMIDE **98-96-4** antibacterial (tuberculostatic)
2	-0.59	7.2	2317	1	"	2-PYRAZINECARBOXAMIDE
3 ★	-0.73		2412		C5H5N3O1	PYRIDAZINE,3-CARBAMOYL
4 ★	-0.96		2412		C5H5N3O1	PYRIDAZINE,4-CARBAMOYL **88511-47-1**
95 ★	-1.20		2412		C5H5N3O1	PYRIMIDINE,2-CARBOXAMIDO
6 ★	-0.68		2412		C5H5N3O1	PYRIMIDINE,4-CARBOXAMIDE
7 ★	-0.92		2412		C5H5N3O1	PYRIMIDINE,5-CARBOXAMIDE **40929-49-5**
8 ★	-1.02		594		C5H5N3O2	2-PYRIDINEAMINE,N-NITRO **26482-54-2**
9	0.65		2138	459	C5H5N3O2	PYRIDINE,2-AMINO-5-NITRO **4214-76-0**
800	-0.69	9.0	1409		C5H5N3O3	5-NITROIMIDAZOLE,1-ME-2-FORMYL **4750-57-6**
1 ★	-0.10		2474		C5H5N3O3	IMIDAZOLE-5-CARBOXALDEHYDE,1-METHYL-2-NITRO **39928-74-0**
2 ★	1.30		594		C5H5N3O3S1	NITHIAMIDE **140-40-9** antibacterial (vet)
3	-1.52	7.4	694	1	C5H5N3O4	1-B-CARBOXYMETHYL-5-NITROIMIDAZOLE, SODIUM SALT
4	-0.34		2474		C5H5N3O4	IMIDAZOLE-1-ACETIC ACID,2-NITRO
5 ★	-0.34		1519	400	C5H5N3S1	PYRIMIDINE-2-THIOCARBOXAMIDE
6	-0.16	7.4	293	1	C5H5N5	ADENINE **73-24-5**
7 ★	-0.09		536		"	ADENINE
8	-0.05	7.4	2348	1	"	ADENINE
9 ★	-0.96	7.4	594	1	C5H5N5O1	GUANINE **73-40-5**
810 ★	-0.07	8.2	401	2	C5H5N5S1	THIOGUANINE **154-42-7** antineoplastic
1 ★	2.66		541		(C5H5)2.Fe1	FERROCENE **102-54-5**
2 ★	0.87		2387		C5H6Br1N3O2	IMIDAZOLE,1,2-DIMETHYL-4-BROMO-5-NITRO
3 ★	0.44		2387		C5H6Br1N3O2	IMIDAZOLE,1,2-DIMETHYL-4-NITRO-5-BROMO **21117-52-2**
4 ★	0.82		2387		C5H6Br1N3O2	IMIDAZOLE,1,4-DIMETHYL-2-BROMO-5-NITRO
15 ★	0.39		2387		C5H6Br1N3O2	IMIDAZOLE,1,5-DIMETHYL-2-BROMO-4-NITRO
6 ★	0.87		2387		C5H6Cl1N3O2	IMIDAZOLE,1,2-DIMETHYL-4-CHLORO-5-NITRO **91027-94-0**
7 ★	0.34		2387		C5H6Cl1N3O2	IMIDAZOLE,1,2-DIMETHYL-5-NITRO-4-CHLORO **91027-93-9**
8 ★	-0.21	7.4	2361	1	C5H6F2N4O3	TRIAZINE,3-NITRO-1-(3-HYDROXY-2,2-DIFLUORO)PROPYL
9	0.49		548	815	C5H6N2	2-AMINOPYRIDINE
820 ★	0.49		548	815	"	2-AMINOPYRIDINE **504-29-0**
1 ★	0.23		163	459	C5H6N2	2-METHYLPYRAZINE **109-08-0**
2 ★	0.11		260		C5H6N2	3-AMINOPYRIDINE **462-08-8**
3	-0.76	7.4	2362	1	C5H6N2	4-AMINOPYRIDINE **504-24-5**
4 ★	0.32		590		"	4-AMINOPYRIDINE
25 ★	0.16		260		C5H6N2	4-METHYLPYRIMIDINE **3438-46-8**
6 ★	-0.72		2069		C5H6N2	PENTANEDINITRILE **544-13-8**
7 ★	-0.35		2412		C5H6N2	PYRIDAZINE,3-METHYL **1632-76-4**
8 ★	-0.32		2412		C5H6N2	PYRIDAZINE,4-METHYL **1120-88-3**
9 ★	-0.05		2412		C5H6N2	PYRIMIDINE,2-METHYL
830 ★	0.01		2412		C5H6N2	PYRIMIDINE,5-METHYL **2036-41-1**
1 ★	-0.34		579		C5H6N2O1	1-ACETYLIMIDAZOLE **2466-76-4**
2	-0.87	2.0	447		C5H6N2O1	2-AMINOPYRIDINE-N-OXIDE **14150-95-9**
3	-0.87	5.0	447	1	"	2-AMINOPYRIDINE-N-OXIDE
4	-0.87	7.4	447	1	"	2-AMINOPYRIDINE-N-OXIDE
35 ★	-1.45		594		C5H6N2O1	2-PYRIMIDONE,4-METHYL **15231-48-8**
6 ★	-1.45		594		C5H6N2O1	2-PYRIMIDONE,6-METHYL
7 ★	0.73		2412		C5H6N2O1	PYRAZINE,2-METHOXY **3149-28-8**
8 ★	0.08		2412		C5H6N2O1	PYRIDAZINE,3-METHOXY **19064-65-4**
9 ★	-0.31		2412		C5H6N2O1	PYRIDAZINE,4-METHOXY **20733-11-3**
840	-1.32	7.4	2362	1	C5H6N2O1	PYRIDINE-3-HYDROXY-4-AMINO
1 ★	0.23		2412		C5H6N2O1	PYRIMIDINE,2-METHOXY
2 ★	0.54		2412		C5H6N2O1	PYRIMIDINE,4-METHOXY **6104-41-2**
3 ★	0.07		2412		C5H6N2O1	PYRIMIDINE,5-METHOXY
4	0.22		536	815	C5H6N2O1S1	4-HYDROXY-2-METHYLTHIO-PYRIMIDINE **20651-23-4**
45 ★	0.76		594		C5H6N2O1S1	THIAZOLE,2-ACETAMIDO **2719-23-5**
6 ★	-1.20		534		C5H6N2O2	1-METHYLURACIL **615-77-0**
7	-0.77	7.4	594	1	C5H6N2O2	2,4-PYRIMIDINEDIONE,6-METHYL **626-48-2**
8 ★	-0.30		1375		C5H6N2O2	PYRIDAZINE,3,4-DIHYDROXY-6-METHYL
9 ★	-0.62		536		C5H6N2O2	THYMINE **65-71-4**
850 ★	-0.49		1519	400	C5H6N2O2S1	3-SULFONAMIDOPYRIDINE **2438-76-8**
1 ★	0.94		2166		C5H6N2O3	1,2,5-OXADIAZOLE,3-METHYL-4-METHOXYCARBONYL
2 ★	-1.94		547		C5H6N2O3	5-HYDROXYMETHYL-D2-URACIL **52565-39-6**
3 ★	-0.85	0.7	962		C5H6N2O3	BARBITURIC ACID,N-METHYL **2565-47-1**
4 ★	0.80		2166		C5H6N2O3	FURAZAN-2-OXIDE,3-METHYL-4-ACETYL

	logP	pH	Ref.	Note	MF	Name / CAS / activity
55 ★	0.98		2166		C5H6N2O3	FURAZAN-2-OXIDE,4-METHYL-3-ACETYL
6 ★	0.56		2166		C5H6N2O4	FURAZAN-2-OXIDE,3-METHYL-4-METHOXYCARBONYL
7 ★	1.17		2412		C5H6N2S1	PYRAZINE,2-METHYLTHIO 21948-70-9
8 ★	1.01		2412		C5H6N2S1	PYRIMIDINE,2-METHYLTHIO 823-09-6
9	-1.22		1404		C5H6N4O3	2(2-NO2-1-IMIDAZOLYL)ACETAMIDE
860 ★	0.29	7.4	2206	306	C5H6N4O3	2-IMIDAZOLECARBOXAMIDE,1-METHYL-5-NITRO
1	-0.43	7.4	2561	1	C5H6N4O3	2-NITROIMIDAZOLE-1-ACETALDOXIME
2 ★	-0.38	1.0	2196		C5H6N4O3	PYRAZOLE,3-METHYL-4-NITRO-5-CARBOXAMIDE
3 ★	-0.38	7.4	2273	1	C5H6N4O4	1-TRIAZOLEACETIC ACID-3-NITRO,METHYL ESTER 70965-23-0
4 ★	0.49		2387		C5H6N4O4	IMIDAZOLE,1,2-DIMETHYL-4,5-DINITRO 19183-17-6
65 ★	1.85		1478		C5H6O1	2-METHYLFURAN 534-22-5
6 ★	0.87		1478		C5H6O1S1	2-HYDROXYMETHYLTHIOPHENE 636-72-6
7 ★	1.44		1478		C5H6O2	2-METHOXYFURAN 25414-22-6
8 ★	0.30		1478		C5H6O2	3-HYDROXYMETHYLFURAN 4412-91-3
9	0.22	7.0	2505	1	C5H6O2	FURFURYL ALCOHOL 98-00-0
870 ★	0.28		1478		"	FURFURYL ALCOHOL
1 ★	0.38		1625		C5H6O2S2	THIOPHENE,2-MESYL 38695-58-8
2 ★	2.33		1478		C5H6S1	2-METHYLTHIOPHENE 554-14-3
3 ★	2.34		1478		C5H6S1	3-METHYLTHIOPHENE 616-44-4
4 ★	0.05	7.4	2273	1	C5H7Cl1N4O3	1,2,4-TRIAZOLE,3-NITRO-1-(3-CHLORO-2-HYDROXY)PROPYL 104958-94-3
75 ★	-0.59	7.4	2361	1	C5H7F1N4O3	TRIAZOLE,3-NITRO-1-(3-FLUORO-2-HYDROXY)PROPYL
6 ★	-0.80	7.4	2361	1	C5H7F1N4O3	TRIAZOLE,3-NITRO-1-(3-HYDROXY-2-FLUORO)PROPYL
7 ★	1.21		570		C5H7N1	N-METHYLPYRROLE 96-54-8
8 ★	0.37		1478		C5H7N1O1	2-AMINOMETHYLFURAN 617-89-0
9 ★	-0.95		594		C5H7N1O2	N-METHYLSUCCINIMDE 1121-07-9
880 ★	-1.80	7.2	2323	468	C5H7N1O2	PROPARGYLGLYCINE
1	-0.25	7.4	401	1	C5H7N3	2,3-DIHYDRO-1H-IMIDAZO-PYRAZOLE 6714-29-0 antineoplastic
2 ★	0.19		1519		C5H7N3	6-METHYL-4-PYRIMIDINAMINE 3435-28-7
3 ★	0.35		2499		C5H7N3	PYRAZINE,2-AMINO-6-METHYL
4 ★	0.56		2444		C5H7N3	PYRAZINE,2-METHYLAMINO 32111-28-7
85	-1.44	7.4	2362	1	C5H7N3	PYRIDINE,3,4-DIAMINO 54-96-6
6	-0.85		1375		C5H7N3O1	3-AMINO-4-HYDROXY-6-METHYLPYRIDAZINE
7 ★	-0.38		547		C5H7N3O1	4-ACETYLAMINOPYRAZOLE
8 ★	0.63		2499		C5H7N3O1	PYRAZINE,2-AMINO-5-METHOXY
9 ★	0.73		2499		C5H7N3O1	PYRAZINE,2-AMINO-6-METHOXY
890 ★	0.56		2166		C5H7N3O2	1,2,5-OXADIAZOLE,3-METHYL-4-(N-METHYLCARBOXAMIDO)
1 ★	0.17		2166		C5H7N3O2	1,2,5-OXADIAZOLE,3-METHYL-4-ACETYLAMINO
2 ★	0.22		505		C5H7N3O2	5-ETHYL-6-AZAURACIL 19213-65-1
3 ★	0.14		2387		C5H7N3O2	DIMETRIDAZOLE 551-92-8 antiprotozoal (vet)
4	0.30	7.4	2206	306	"	DIMETRIDAZOLE
95	0.64	7.4	2380	1	"	DIMETRIDAZOLE
6 ★	-0.24		2387		C5H7N3O2	IMIDAZOLE,1,2-DIMETHYL-4-NITRO 13230-04-1
7	-0.29	7.4	2206	306	C5H7N3O2	IMIDAZOLE,1,5-DIMETHYL 4-NITRO 7464-68-8
8 ★	-0.11		2387		"	IMIDAZOLE,1,5-DIMETHYL 4-NITRO
9 ★	0.43	7.4	2206	306	C5H7N3O2	IMIDAZOLE,2,4-DIMETHYL-5-NITRO
900 ★	0.94	7.4	2206	306	C5H7N3O2S1	IMIDAZOL,1-METHYL-2-METHYLTHIO-5-NITRO 1615-41-4
1 ★	0.42		2166		C5H7N3O3	1,2,5-OXADIAZOLE,3-METHYL-4-METHOXYCARBOXAMINO
2 ★	-0.47		1003		C5H7N3O3	1-(2-HYDROXYETHYL)-2-NITROIMIDAZOLE 5006-67-7
3 ★	-0.35	7.4	1022	1	C5H7N3O3	1-METHYL-2-NITRO-5-HYDROXYMETHYLIMIDAZOLE
4	-0.03	9.0	1409		C5H7N3O3	5-NITROIMIDAZOLE,1-ME-2-HYDROXYMETHYL 936-05-0
5	0.19		2387		C5H7N3O3	IMIDAZOLE,1-METHYL-4-NITRO-5-METHOXY 85012-71-1
6 ★	0.21		2387		C5H7N3O3	IMIDAZOLE,2-METHYL-4-NITRO-5-METHOXY 35687-42-4
7	0.13		1375		C5H7N3S1	3-AMINO-4-SULFHYDRYL-6-METHYLPYRIDAZINE
8 ★	2.40	7.4	2648	448	C5H8	1,3-PENTADIENE,CIS
9 ★	2.44	7.4	2648	448	C5H8	1,3-PENTADIENE,TRANS
910 ★	2.48		547	60	C5H8	1,4-PENTADIENE 591-93-5
1 ★	1.98		503		C5H8	1-PENTYNE 627-19-0
2 ★	1.23		871		C5H8Br2N2O1	3,4-DIBROMO-N-NITROSOPIPERIDINE 57541-73-8
3	1.15		2550	459	C5H8Cl1N5	DESISOPROPYLATRAZINE
4 ★	1.04		871		C5H8Cl2N2O1	3,4-DICHLORO-N-NITROSOPIPERIDINE 57541-72-7
15 ★	1.84		591		C5H8Cl3N1O3	CARBOCLORAL 541-79-7 hypnotic
6 ✓	0.75		2366	122	C5H8Cu1N6S2	COPPER(II)PYRUVALDEHYDE BIS(THIOSEMICARBAZONE)
7 ★	0.47		594	400	C5H8N2	ACRYLONITRILE,3-DIMETHYLAMINO
8 ★	1.01		594		C5H8N2	PYRAZOLE,3,5-DIMETHYL 67-51-6
9 ★	0.80		547		C5H8N2O1	4-ETHOXYPYRAZOLE
920 ★	1.18		871		C5H8N2O1	DELTA-3-N-NITROSOPIPERIDINE
1 ★	1.18		2166		C5H8N2O1	FURAZAN,METHYLETHYL 17647-69-7
2	-0.48	7.0	1405	1	C5H8N2O2	HYDANTOIN,5,5-DIMETHYL 77-71-4
3 ★	-0.47		871		C5H8N2O2	N-NITROSO-4-PIPERIDONE 55556-91-7
4	-1.05		1728		C5H8N2O2	N-NITROSOPIPERIDONE
25 ★	0.51		2188		C5H8N2O2S2	METHYLCARBAMIC ACID,4-OXIMINO-1,3-DITHIOLANE ESTER
6 ★	-0.68		505		C5H8N2O3	UREA,1,3-DIACETYL 638-20-0
7 ★	0.92		2166		C5H8N2O3S1	1,2,5-OXADIAZOLE,3-METHYL-4-ETHYLSULFONYL
8 ★	0.06		2188		C5H8N2O3S1	METHYLCARBAMIC ACID,4-OXIMINO-1,3-OXATHIOLANE ESTER
9 ★	0.65		2166		C5H8N2O4S1	FURAZAN-2-OXIDE,3-METHYL-4-ETHYLSULFONYL
930 ★	0.62		1519		C5H8N4	4-HYDRAZINO-6-METHYLPYRIMIDINE
1 ★	-0.16		57		C5H8N4O1S1	3-METHIO-4-AMINO-6-ME-1,2,4-TRIAZINE-5-ONE 18826-96-5
2 ★	0.30		2592		C5H8N4O2	1,2,4-TRIAZOLE,1-PROPYL-3-NITRO
3 ★	0.30		2166		C5H8N4O2	1,2,5-OXADIAZOLE,3-(N'-METHYLUREIDO)-4-METHYL
4	-1.10		2387		C5H8N4O2	IMIDAZOLE,1,2-DIMETHYL-4-NITRO-5-AMINO
35 ★	-0.87	7.4	830	1	C5H8N4O2	N-DIAZOACETYLGLYCINE-N'-METHYLAMIDE
6	0.32	7.2	2036		C5H8N4O3S2	ACETAZOLAMIDE,N-METHYL 68300-47-0
7 ★	0.49	7.2	2036	2	"	ACETAZOLAMIDE,N-METHYL
8 ★	0.13		264		C5H8N4O3S2	METHAZOLAMIDE 554-57-4 diuretic, carbonic anhydrase inhibitor
9 ★	1.24		1705		C5H8N4S1	GUANIDINE,N2-(5-METHYLTHIAZOL-2-YL)

	logP	pH	Ref.	Note	MF	Name / **CAS** / *activity*
940 ★	1.27		595		C5H8N4S1	THIAZOLE,2-(3,3-DIMETHYL)TRIAZENYL
1 ★	1.24		1120		C5H8O1	2-ETHYLPROPENAL **922-63-4**
2 ★	0.28		579		C5H8O1	2-METHYL-3-BUTYN-2-OL **115-19-5**
3 ★	0.52		599		C5H8O1	3-PENTEN-2-ONE **625-33-2**
4 ★	0.49		579		C5H8O1	CYCLOPROPYLMETHYL KETONE **765-43-5**
45 ★	0.69		594	298	C5H8O1	PYRAN,2,3-DIHYDRO **110-87-2**
6 ★	0.40		579		C5H8O2	ACETYLACETONE **123-54-6**
7 ★	1.32		579		C5H8O2	ACRYLIC ACID,ETHYL ESTER **140-88-5**
8 ★	0.97		594	400	C5H8O2	ALLYLACETATE **591-87-7**
9 ★	-0.35		547		C5H8O2	D-VALEROLACTONE **542-28-9**
950 ★	1.38		1402		C5H8O2	METHACRYLIC ACID,METHYL ESTER **80-62-6**
1 ★	-0.21		1402		C5H8O3	ACRYLIC ACID,2-HYDROXYETHYL ESTER **818-61-1**
2	-2.18	7.4	594	1	C5H8O3	LEVULINIC ACID **123-76-2**
3 ★	-0.49		570		"	LEVULINIC ACID
4 ★	0.39	1.0	594	342	C5H8O4	DIMETHYLMALONIC ACID **595-46-0**
55 ★	-0.29		570		C5H8O4	GLUTARIC ACID **110-94-1**
6 ★	-0.05		579		C5H8O4	MALONIC ACID,DIMETHYL ESTER **108-59-8**
7	1.24		1966	459	C5H9Au1Cl1N1	T-BUTYLISOCYANIDE GOLD CHLORIDE ✚
8 ★	0.96		401		C5H9Cl1F1N3O2	NITROSOUREA,3-(2-CHLOROETHYL)-1-(2-FLUOROETHYL) **13908-92-4**
9 ★	1.53		401		C5H9Cl2N3O2	1,3-BIS(2-CHLOROETHYL)-1-NITROSOUREA **154-93-8** *antineoplastic*
960 ★	1.39		1942		C5H9F3O1	4,4,4-TRIFLUORO-2-METHYLBUTANOL
1 ★	1.37		1942		C5H9F3O1	4,4,4-TRIFLUORO-3-METHYLBUTANOL
2 ★	1.15		1942		C5H9F3O1	5,5,5-TRIFLUOROPENTANOL
3 ★	1.87		599		C5H9N1	1-PYRROLINE,2-METHYL
4 ★	1.10	1.0	579	400	C5H9N1	2-METHYLBUTYRONITRILE **18936-17-9**
65 ★	1.07	1.0	579	400	C5H9N1	I-VALERONITRILE **625-28-5**
6	0.98	7.0	579		C5H9N1	TRIMETHYLACETONITRILE **630-18-2**
7 ★	1.08	1.0	579	400	"	TRIMETHYLACETONITRILE
8 ★	1.12	7.0	579	400	C5H9N1	VALERONITRILE **110-59-8**
9 ★	-0.46		579		C5H9N1O1	D-VALEROLACTAM **675-20-7**
970 ★	-0.54		599		C5H9N1O1	N-METHYL-2-PYRROLIDINONE **872-50-4**
1 ★	-0.32		594		C5H9N1O1	PYRROLIDINE,N-FORMYL **3760-54-1**
2	-2.54	7.0	1207	1	C5H9N1O2	PROLINE **147-85-3** *amino acid*
3 ★	-2.50	7.0	2197		"	PROLINE
4 ★	-3.17		1590	459	C5H9N1O3	HYDROXYPROLINE **51-35-4**
75	-4.19		1590	459	C5H9N1O4	GLUTAMIC ACID, MONOSODIUM SALT
6 ✓	-3.69	7.0	1207	1	C5H9N1O4	GLUTAMIC ACID **56-86-0** *acidifier (gastric)*
7	-0.22		2114		"	GLUTAMIC ACID
8 ★	0.13		561		C5H9N1S1	2-AZACYCLOHEXANTHIONE **13070-01-4**
9 ★	2.82		1947		C5H9N1S1	2-METHYLPROPANE,ISOTHIOCYANATO **591-82-2**
980 ★	2.81		1947		C5H9N1S1	BUTANE,2-ISOTHIOCYANATO
1 ★	2.92		1947		C5H9N1S1	BUTANE,ISOTHIOCYANATO **592-82-5**
2 ★	2.03		502		C5H9N1S1	THIOCYANIC ACID,BUTYL ESTER **628-83-1**
3 ★	-0.70		1853		C5H9N3	HISTAMINE **51-45-6** *stimulant (gastric)*
4 ★	0.20		1137		C5H9N3O1	2-AMINO-5-I-PROPYL-1,3,4-OXADIAZOLE
85 ★	0.23		1137		C5H9N3O1	2-AMINO-5-PROPYL-1,3,4-OXADIAZOLE
6 ★	1.40		2097		C5H9N3O1	DIAZINECARBONITRILE,T-BUTYL,OXIDE
7	-0.49		723		C5H9N5O1	2,4-DIAMINO-6-METHYLAMINO-PYRIMIDINE-3-OXIDE
8 ★	3.00		547	400	C5H10	CYCLOPENTANE **287-92-3**
9 ★	0.30		977		C5H10Cl1N3O3	HYDROXYETHYL-CNU **60784-46-5** *antineoplastic*
990 ★	0.36		1728		C5H10N2O1	N-NITROSOPIPERIDINE **100-75-4**
1 ★	-0.03		1728		C5H10N2O2	2-METHYL-N-NITROSOMORPHOLINE
2 ★	0.04		673		C5H10N2O2	2-METHYLPROPANOYLUREA
3 ★	-0.47		871		C5H10N2O2	3-HYDROXY-N-NITROSOPIPERIDINE
4 ★	-0.89		871		C5H10N2O2	4-HYDROXY-N-NITROSOPIPERIDINE
95	-0.13	7.1	1624		C5H10N2O2	ACETONE,O-((MEAMINO)CARBONYL)OXIME **10520-34-0**
6 ★	-1.52	7.1	1591		C5H10N2O2	ALANIN-AMIDE,N-ACETYL
7	-1.41	7.0	2309	1	"	ALANIN-AMIDE,N-ACETYL
8 ★	-0.01		2198		C5H10N2O2	ETHYLUREA,N'-ACETYL
9 ★	-1.56		1558		C5H10N2O2	GLYCINE,N-ACETYL-N'-METHYLAMINO-AMIDE
1000 ★	-0.29	7.1	1591		C5H10N2O2S1	CYSTEIN-AMIDE,N-ACETYL
1 ★	0.60		1629		C5H10N2O2S1	METHOMYL **16752-77-5** *insecticide~nematocide*
2 ★	-3.15		1590	459	C5H10N2O3	GLUTAMINE **56-85-9**
3 ★	-1.87	7.1	1591		C5H10N2O3	SERIN-AMIDE,N-ACETYL
4 ★	-0.28		871		C5H10N4O2	2-METHYL-N,N'-DINITROSOPIPERAZINE **55556-94-0**
5 ★	-0.25		1728		C5H10N4O2	2-METHYL-N,N'-DINITROSOPIPERAZINE
6 ★	-0.51		871		C5H10N4O2	N,N'-DINITROSOHOMOPIPERAZINE **55557-00-1**
7 ★	-0.30	4.0	979	364	C5H10N6O4	1,1'-PROPYLEN-BIS(1-NITROSOUREA) **27640-19-3**
8 ★	0.84		594	400	C5H10O1	2-BUTANONE,3-METHYL **563-80-4**
9 ★	0.91	7.0	566	1	C5H10O1	2-PENTANONE **107-87-9**
1010	0.99		1560	100	C5H10O1	3-PENTANONE **96-22-0**
1 ★	0.95		579		C5H10O1	TETRAHYDROPYRAN **142-68-7**
2 ★	0.46		579	400	C5H10O2	2,2-DIMETHYL-1,3-DIOXOLANE **2916-31-6**
3 ★	1.18	1.0	579	400	C5H10O2	2-METHYLBUTYRIC ACID **116-53-0**
4 ★	1.24		1560	100	C5H10O2	ACETIC ACID,PROPYL ESTER **109-60-4**
15 ★	1.47	1.0	579		C5H10O2	ACETIC ACID,TRIMETHYL **75-98-9**
6 ★	-0.10		2562		C5H10O2	ENTYLENE GLYCOL,MONOALLYL ETHER **111-45-5**
7	0.36		2596		C5H10O2	ETHYL GLYCIDYL ETHER **4016-11-9**
8 ★	1.16	1.0	579	400	C5H10O2	I-VALERIC ACID **503-74-2**
9 ★	1.29		594	400	C5H10O2	METHYLBUTYRATE **623-42-7** *flavoring (artificial rum & fruit essences)*
1020 ★	1.21		503		C5H10O2	PROPIONIC ACID,ETHYL ESTER **105-37-3**
1 ★	1.39		1135	537	C5H10O2	VALERIC ACID **109-52-4**
2 ★	1.21		579		C5H10O3	DIETHYLCARBONATE **105-58-8**
3	-0.06		1266		C5H10O3	METHYL-4-HYDROXYBUTYRATE **925-57-5**
4 ★	-3.02		1087		C5H10O5	ARABINOSE **147-81-9**

	logP	pH	Ref.	Note	MF	Name / CAS / activity
25 ★	-2.32	7.4	293	1	C5H10O5	RIBOSE 50-69-1
6 ★	3.37		1560	100	C5H11Br1	1-BROMOPENTANE 110-53-2
7 ★	2.52		594	400	C5H11Cl1	BUTANE,2-CHLORO-2-METHYL 594-36-5
8 ★	2.33		503		C5H11F1	1-FLUOROPENTANE 592-50-7
9 ★	0.81	13.0	588	224	C5H11N1	ALLYLETHYLAMINE 2424-02-4
1030 ★	0.92	12.0	594	400	C5H11N1	N-METHYLPYRROLIDINE 120-94-5
1 ★	0.84	13.0	588	224	C5H11N1	PIPERIDINE 110-89-4
2 ★	-0.11		594		C5H11N1O1	DIMETHYLPROPIONAMIDE 758-96-3
3 ★	-0.33		505		C5H11N1O1	MORPHOLINE,N-METHYL 109-02-4
4 ★	2.01		547	400	C5H11N1O2	1-NITROPENTANE 628-05-7
35 ★	-2.63		2018		C5H11N1O2	5-AMINOPENTANOIC ACID 660-88-8
6 ★	-2.11		2654	448	C5H11N1O2	A-AMINOVALERIC ACID 6600-40-4
7	-1.93	7.2	2323	468	"	A-AMINOVALERIC ACID
8 ★	0.89		594	400	C5H11N1O2	N,N-DIMETHYLETHYLCARBAMATE 687-48-9
9 ★	0.85	7.4	172	1	C5H11N1O2	O-BUTYLCARBAMATE 592-35-8
1040 ★	0.65	7.4	172	1	C5H11N1O2	O-I-BUTYLCARBAMATE 543-28-2
1 ★	0.95		702		C5H11N1O2	O-PROPYL-N-METHYLCARBAMATE 17671-76-0
2 ★	0.48	7.4	172	1	C5H11N1O2	O-T-BUTYLCARBAMATE 4248-19-5
3 ★	-2.26	7.0	1207	1	C5H11N1O2	VALINE 72-18-4 amino acid
4	-2.08	7.0	2197		"	VALINE
45 ★	-1.87		405		C5H11N1O2S1	METHIONINE 63-68-3 amino acid
6	-1.78		1590	459	C5H11N1O2S1	PENICILLAMINE 52-67-5 chelating agent
7	-2.51	7.2	2323	468	C5H11N1O3	ETHYLSERINE
8 ★	2.84		579		C5H11N1O3	ISOAMYLNITRATE 543-87-3
9 ★	-3.10	7.2	2323	468	C5H11N1O4S1	METHIONINESULFONE
1050 ★	0.20		871		C5H11N3O1	4-METHYL-N-NITROSOPIPERAZINE 16339-07-4
1	1.04	4.0	979	364	C5H11N3O2	1-(I-BUTYL)-1-NITROSOUREA 760-60-1
2	1.04	4.0	979	364	C5H11N3O2	1-BUTYL-1-NITROSOUREA 869-01-2
3 ★	0.71		871		C5H11N3O2	1-NITROSO-1-ETHYL-3,3-DIMETHYLUREA 50285-71-7
4 ★	0.86		579		C5H11O4P1	PHOSPHONIC ACID,2-OXOPROPYL,DIMETHYL ESTER 4202-14-6
55 ★	3.11		547	400	C5H12	NEOPENTANE 463-82-1
6 ★	3.39		547	822	C5H12	PENTANE 109-66-0
7 ★	0.78		1693		C5H12N1O3P1S2	DIMETHOATE 60-51-5 insecticide
8 ★	-0.28	9.0	547	400	C5H12N2	N-AMINOPIPERIDINE 2213-43-6
9 ★	-0.35		1137		C5H12N2O1	2,2-DIMETHYLPROPIONIC ACID HYDRAZIDE 17883-59-9
1060 ★	-0.33		1137		C5H12N2O1	2-METHYLBUTYRIC ACID HYDRAZIDE
1 ★	-0.31		1137		C5H12N2O1	I-VALERIC ACID HYDRAZIDE 24310-18-7
2	1.02		2570		C5H12N2O1	METHYLBUTYLNITROSAMINE 7068-83-9
3 ★	0.41		547		C5H12N2O1	N-BUTYLUREA 592-31-4
4 ★	0.90		2570		C5H12N2O1	N-NITROSOETHYL-I-PROPYL AMINE 16339-04-1
65 ★	0.76		2570		C5H12N2O1	N-NITROSOMETHYL-T-BUTYL AMINE 2504-18-9
6 ★	0.19		579		C5H12N2O1	TETRAMETHYLUREA 632-22-4
7 ★	-0.11		1137		C5H12N2O1	VALERIC ACID HYDRAZIDE 38291-82-6
8 ★	0.32	7.4	1184	1	C5H12N2O2	3-BUTYLHYDROXYUREA 5681-57-2
9 ★	-4.41	7.2	2323	468	C5H12N2O2	ORNITHINE 70-26-8 anticholestermemic
1070 ★	0.57	6.5	2013		C5H12N2S1	THIOUREA,N,N'-DIETHYL 105-55-5
1 ★	0.49	6.5	2013		C5H12N2S1	THIOUREA,TETRAMETHYL 2782-91-4
2	-1.28	10.1	401	1	C5H12N8	MITOGUAZONE 7059-23-6 antineoplastic ✚
3 ★	1.31		2126		C5H12O1	1-PROPANOL,2,2-DIMETHYL 75-84-3
4 ★	1.19		2126		C5H12O1	2-PENTANOL 6032-29-7
75 ★	0.89		504		C5H12O1	2-PROPANOL,2-ETHYL 75-85-4 solvent
6 ★	1.28		425		C5H12O1	3-METHYL-2-BUTANOL 598-75-4
7 ★	1.21		547	400	C5H12O1	3-PENTANOL 584-02-1
8 ★	1.66		594	400	C5H12O1	BUTYL METHYLETHER 628-28-4
9 ★	1.16		2126		C5H12O1	I-PENTANOL 123-51-3
1080 ★	0.94		1952		C5H12O1	METHYL-T-BUTYLETHER 1634-04-4 cholelitholytic agent
1 ★	1.56		547	400	C5H12O1	PENTANOL 71-41-0
2 ★	0.84		505		C5H12O2	DIETHOXYMETHANE 462-95-3
3 ★	0.05		768		C5H12O2	I-PROPOXYETHANOL 109-59-1
4 ★	-0.09		594	400	C5H12O2	PROPANE,1,2-DIMETHOXY 7778-85-0
85	-0.86	7.0	834	1	C5H13N1	AMYLAMINE 110-58-7
6 ★	1.49	13.0	588	224	"	AMYLAMINE
7 ★	0.93	13.0	588	224	C5H13N1	ETHYL-I-PROPYLAMINE 19961-27-4
8 ★	1.33		505		C5H13N1	METHYLBUTYLAMINE 110-68-9
9 ★	0.41	13.0	594	2	C5H13N3	TETRAMETHYLGUANIDINE 80-70-6
1090 ★	-0.62		1394		C5H13N3O1	4,4-DIETHYLSEMICARBAZIDE
1 ★	-0.60		1394		C5H13N3O1	4-BUTYLSEMICARBAZIDE 20605-19-0
2 ★	-0.84		1394		C5H13N3O1	4-ISOBUTYLSEMICARBAZIDE
3 ★	-0.81		1394		C5H13N3O1	4-T-BUTYLSEMICARBAZIDE
4	0.71		2182	459	C5H13O2P1S1	METHYLPHOSPHONIC ACID,O-ETHYL,S-ETHYL ESTER 2511-10-6
95	2.08		2182	459	C5H13O2P1S1	METHYLTHIOPHOSPHONIC ACID,DIETHYL ESTER 6996-81-2
6 ★	3.22		512		C5H13Si1	SILANE,DIMETHYL-PROPYL 18143-31-2
7	-3.71		549	466	C5H14N1O1.Br1	CHOLINEBROMIDE 1927-06-6
8 ★	0.65		600		C5H14N1O2P1S1	O,S-DIMETHYL-N-PROPYL-PHOSPHORAMIDOTHIOATE 16271-16-2
9	-3.00		524	820	C5H14N1.I1	TRIMETHYL-ETHYL-AMMONIUM IODIDE 51-93-4
1100 ★	-0.20	13.0	594	400	C5H14N2	N,N,N'-TRIMETHYLETHYLENEDIAMINE 142-25-6
1 ★	-0.67		1908		C5H15N2O1P1	PHOSPHONICDIAMIDE,PENTAMETHYL 2511-17-3
2 ★	6.07		1814	459	C6Br6	HEXABROMOBENZENE 87-82-1
3 ★	4.22		1364		C6Cl5N1O2	PENTACHLORONITROBENZENE 82-68-8 fungicide (soil)
4 ★	5.73		2293	400	C6Cl6	HEXACHLOROBENZENE 118-74-1 fungicide
5 ★	2.55		594	400	C6F6	HEXAFLUOROBENZENE 392-56-3
6 ★	2.91		579		C6F12	CYCLOHEXANE,PERFLUORO 355-68-0
7 ★	0.95		2499		C6H1Cl1N4	PYRAZINE,2,3-DICYANO-5-CHLORO
8 ★	3.89		1710		C6H1Cl4N1O2	2,3,5,6-TETRACHLORONITROBENZENE 117-18-0
9 ★	4.17	7.3	509	2	C6H1Cl4N3	4,5,6,7-TETRACHLOROBENZOTRIAZOLE 2338-10-5

	logP	pH	Ref.	Note	MF	Name / CAS / activity
1110 ★	5.18		2293	400	C6H1Cl5	PENTACHLOROBENZENE **608-93-5**
1 ★	5.12	1.4	536	342	C6H1Cl5O1	PENTACHLOROPHENOL **87-86-5** *herbicide~preservative(wood)*
2 ★	2.53		594		C6H1F5	PENTAFLUOROBENZENE **363-72-4**
3 ★	3.23		359	824	C6H1F5O1	PENTAFLUOROPHENOL **771-61-9**
4 ★	5.13		1600		C6H2Br4	1,2,4,5-TETRABROMOBENZENE **636-28-2**
15 ★	3.61		1710		C6H2Cl3N1O2	2,3,4-TRICHLORONITROBENZENE **17700-09-3**
6 ★	3.48		1710		C6H2Cl3N1O2	2,4,5-TRICHLORONITROBENZENE **89-69-0**
7 ★	3.93		2134		C6H2Cl3N1O3	PHENOL,2,3,6-TRICHLORO-4-NITRO **20404-02-8**
8 ★	4.64		2293	400	C6H2Cl4	1,2,3,4-TETRACHLOROBENZENE **634-66-2**
9 ★	4.66		2293	400	C6H2Cl4	1,2,3,5-TETRACHLOROBENZENE **634-90-2**
1120 ★	4.60		2293	400	C6H2Cl4	1,2,4,5-TETRACHLOROBENZENE **95-94-3**
1 ★	4.21		1709		C6H2Cl4O1	2,3,4,5-TETRACHLOROPHENOL **4901-51-3**
2 ★	4.45		1554		C6H2Cl4O1	2,3,4,6-TETRACHLOROPHENOL **58-90-2**
3 ★	3.88		1709		C6H2Cl4O1	2,3,5,6-TETRACHLOROPHENOL **935-95-5**
4 ★	4.29		1554		C6H2Cl4O2	TETRACHLORO-1,2-BENZENEDIOL **1198-55-6**
25 ★	5.08		2293	400	C6H2Cl5N1	PENTACHLOROANILINE **527-20-8**
6 ✓	0.13		567	820	C6H2N3O7.C4H12N1	TETRAMETHYL AMMONIUM PICRATE
7 ✓	1.01		1701	820	C6H2N3O7.C5H11F3N1	N,N,N-TRIME-2,2,2-TRIFLUOROETAMININE PICRATE
8	0.48		567	820	C6H2N3O7.C5H14N1	ETHYLTRIMETHYL AMMONIUM PICRATE
9 ✓	0.60		1701	820	"	ETHYLTRIMETHYL AMMONIUM PICRATE
1130 ★	0.38		2499		C6H2N4	PYRAZINE,2,3-DICYANO
1	3.25		1263		C6H3Br1Cl1N1O2	3-CHLORO-4-BROMONITROBENZENE **29682-39-1**
2 ★	3.57		2134		C6H3Br2N1O3	2,6-DIBROMO-4-NITROPHENOL **99-28-5**
3 ★	4.51		1600		C6H3Br3	1,3,5-TRIBROMOBENZENE **626-39-1**
4	3.96		359	824	C6H3Br3O1	2,4,6-TRIBROMOPHENOL **118-79-6**
35	4.13		1400	459	"	2,4,6-TRIBROMOPHENOL
6	4.23		757		"	2,4,6-TRIBROMOPHENOL
7 ★	4.37	1.4	536	342	C6H3Br3O2	2,4,6-TRIBROMORESORCINOL **2437-49-2**
8 ★	2.18		2410		C6H3Cl1N2O4	1,2-DINITROBENZENE,4-CHLORO
9	2.17		2163	459	C6H3Cl1N2O4	BENZENE,1-CHLORO-2,4-DINITRO **97-00-7**
1140	-1.30		1008	837	C6H3Cl1N2O6S1	2,4-DINITROBENZENESULFONYLCHLORIDE
1 ★	1.79		332		C6H3Cl1N2S1	THIENO(2,3-D)-PYRIMIDINE,4-CHLORO **16269-66-2**
2 ★	0.84		599		C6H3Cl1O2	2-CHLOROBENZOQUINONE
3 ★	3.05		1710		C6H3Cl2N1O2	2,3-DICHLORONITROBENZENE **3209-22-1**
4 ★	3.09		1710		C6H3Cl2N1O2	2,4-DICHLORONITROBENZENE **89-61-2**
45 ★	3.12		1710		C6H3Cl2N1O2	3,4-DICHLORONITROBENZENE **99-54-7**
6 ★	2.90		2305	459	C6H3Cl2N1O2	NITROBENZENE,2,5-DICHLORO
7 ★	3.13		2305	459	C6H3Cl2N1O2	NITROBENZENE,3,5-DICHLORO
8 ★	2.94		359	824	C6H3Cl2N1O3	2,6-DICHLORO-4-NITROPHENOL **618-80-4**
9 ★	4.14		2293	400	C6H3Cl3	1,2,3-TRICHLOROBENZENE **87-61-6**
1150 ★	4.02		2000		C6H3Cl3	1,2,4-TRICHLOROBENZENE **120-82-1**
1 ★	4.19		2293	400	C6H3Cl3	1,3,5-TRICHLOROBENZENE **108-70-3**
2	0.30		1178		C6H3Cl3N2O2	PICLORAM **1918-02-1** *herbicide*
3	3.51		1400	459	C6H3Cl3O1	2,3,4-TRICHLOROPHENOL **15950-66-0**
4	3.54		1505	459	"	2,3,4-TRICHLOROPHENOL
55	3.80		1709		"	2,3,4-TRICHLOROPHENOL
6	3.84		1709		C6H3Cl3O1	2,3,5-TRICHLOROPHENOL **933-78-8**
7	4.56		1400	459	"	2,3,5-TRICHLOROPHENOL
8 ★	3.77		1709		C6H3Cl3O1	2,3,6-TRICHLOROPHENOL **933-75-5**
9 ★	3.72		508		C6H3Cl3O1	2,4,5-TRICHLOROPHENOL **95-95-4** *fungicide*
1160 ★	3.69		508		C6H3Cl3O1	2,4,6-TRICHLOROPHENOL **88-06-2**
1 ★	4.01		579		C6H3Cl3O1	3,4,5-TRICHLOROPHENOL **609-19-8**
2 ★	4.28		1822	459	"	3,4,5-TRICHLOROPHENOL
3	3.71		1738	459	C6H3Cl3O2	3,4,5-TRICHLOROCATECHOL **56961-20-7**
4	3.60		1738	459	C6H3Cl3O2	3,4,6-TRICHLOROCATECHOL **32139-72-3**
65 ★	4.04		1505	459	C6H3Cl4N1	2,3,4,5-TETRACHLOROANILINE **634-83-3**
6 ★	4.57		2293	400	"	2,3,4,5-TETRACHLOROANILINE
7 ★	4.46		2293	400	C6H3Cl4N1	2,3,5,6-TETRACHLOROANILINE **3481-20-7**
8 ★	3.41		1190		C6H3Cl4N1	NITRAPYRIN
9 ★	2.41		594	400	C6H3F3	1,2,4-TRIFLUOROBENZENE **367-23-7**
1170	0.62		1412		C6H3F6N5O2	1,2,4-TRIAZOLE,3,5-DI(TRIFLUOROACETAMIDO)
1	1.68		848		C6H3I3O2	2,4,6-TRI-IODO-RESORCINOL
2	-2.46		1463		C6H3N2O5	2,4-DINITROPHENOLATEANION
3 ★	1.18		536		C6H3N3O6	1,3,5-TRINITROBENZENE **99-35-4**
4 ✓	1.39		1701	820	C6H3N3O7	2,2,2-TRIFLUOROETHYLAMINE PICRATE
75 ★	0.89	1.0	579	342	C6H3N3O7	2,4,6-TRINITROPHENOL **88-89-1**
6 ✓	2.05		567	820	C6H3N3O7	BUTYLAMINE PICRATE
7 ✓	1.29		567	820	C6H3N3O7	DIETHYLAMINE PICRATE
8 ✓	0.76		567	820	C6H3N3O7	DIMETHYLAMINE PICRATE
9 ✓	1.03		567	820	C6H3N3O7	ETHYLAMINE PICRATE
1180 ✓	1.89		567	820	C6H3N3O7	I-BUTYLAMINE PICRATE
1 ✓	1.37		567	820	C6H3N3O7	I-PROPYLAMINE PICRATE
2 ✓	0.70		576	820	C6H3N3O7	METHYLAMINE PICRATE
3 ✓	1.29		1701	820	C6H3N3O7	N,N-DIMETHYL-2,2,2-TRIFLUOROETHYLAMINE PICRATE
4 ✓	1.51		1701	820	C6H3N3O7	N-ETHYL-2,2,2-TRIFLUOROETHYLAMINE PICRATE
85 ✓	1.37		1701	820	C6H3N3O7	N-METHYL-2,2,2-TRIFLUOROETHYLAMINE PICRATE
6 ✓	1.50		567	820	C6H3N3O7	PROPYLAMINE PICRATE
7 ✓	1.80		567	820	C6H3N3O7	S-BUTYLAMINE PICRATE
8 ✓	1.68		567	820	C6H3N3O7	T-BUTYLAMINE PICRATE
9 ✓	0.60		567	820	C6H3N3O7	TRIMETHYLAMINE PICRATE
1190	-1.31	7.4	1966	1	C6H4Au1Cl2N1O2	1,1-DICHLORO-1,3,2-OXAZAURO(3,4)PYRIDIN-3-ONE ✚
1 ★	1.12		538		C6H4Br1N1O1	N-BROMOBENZOQUINONEMONIMINE
2 ★	2.64		501		C6H4Br1N1O2	BENZENE,3-BROMO-1-NITRO **585-79-5**
3	2.38	7.0	925	1	C6H4Br1N1O2	BENZENE,4-BROMO-1-NITRO **586-78-7**
4 ★	2.55		897		"	BENZENE,4-BROMO-1-NITRO

	logP	pH	Ref.	Note	MF	Name / **CAS** / *activity*
95 ★	2.52		547		C6H4Br1N1O2	O-BROMONITROBENZENE **577-19-5**
6 ★	2.45		1946		C6H4Br1N1O3	FURAN,2-(2-BROMOETHENYL)-5-NITRO **67363-72-8**
7 ★	3.75		536		C6H4Br2	M-DIBROMOBENZENE **108-36-1**
8 ★	3.64		536		C6H4Br2	O-DIBROMOBENZENE **583-53-9**
9 ★	3.79		1600		C6H4Br2	P-DIBROMOBENZENE **106-37-6**
1200 ★	3.22		359	824	C6H4Br2O1	2,4-DIBROMOPHENOL **615-58-7**
1 ★	3.36		2645	225	C6H4Br2O1	2,6-DIBROMOPHENOL **608-33-3**
2 ★	1.26		538		C6H4Cl1N1O1	N-CHLOROBENZOQUINONEMONIMINE **637-61-6**
3	-1.70		1190		C6H4Cl1N1O2	6-CHLOROPICOLINIC ACID **4684-94-0**
4 ★	2.24		531		C6H4Cl1N1O2	BENZENE,2-CHLORO-1-NITRO **88-73-3**
5 ★	2.46		501		C6H4Cl1N1O2	BENZENE,3-CHLORO-1-NITRO **121-73-3**
6 ★	2.39		501		C6H4Cl1N1O2	BENZENE,4-CHLORO-1-NITRO **100-00-5**
7	0.73	7.4	1075	324	C6H4Cl1N1O3	2-CHLORO-4-NITROPHENOL **619-08-9**
8 ★	2.46	1.5	2329	314	C6H4Cl1N1O3	PHENOL,2-NITRO-4-CHLORO **89-64-5**
9 ★	3.53		2293	400	C6H4Cl2	1,3-DICHLOROBENZENE **541-73-1**
1210 ★	3.43		2293	400	C6H4Cl2	O-DICHLOROBENZENE **95-50-1** *herbicide~insecticide*
1 ★	3.44		2293	400	C6H4Cl2	P-DICHLOROBENZENE **106-46-7** *fumigant*
2	1.80		2506		C6H4Cl2N2O2	DICHLORAN **99-30-9** *fungicide*
3 ★	2.84		1505	459	C6H4Cl2O1	2,3-DICHLOROPHENOL **576-24-9**
4 ★	3.06		359	824	C6H4Cl2O1	2,4-DICHLOROPHENOL **120-83-2**
15 ★	3.06		1400	459	C6H4Cl2O1	2,5-DICHLOROPHENOL **583-78-8**
6 ★	2.75		2215		C6H4Cl2O1	2,6-DICHLOROPHENOL **87-65-0**
7 ★	3.33		1690	459	C6H4Cl2O1	3,4-DICHLOROPHENOL **95-77-2**
8 ★	3.62		572		C6H4Cl2O1	3,5-DICHLOROPHENOL **591-35-5**
9 ★	3.68		2293	400	C6H4Cl3N1	2,3,4-TRICHLOROANILINE **634-67-3**
1220 ★	3.69		2293	400	C6H4Cl3N1	2,4,5-TRICHLOROANILINE **636-30-6**
1 ★	3.69		2293	400	C6H4Cl3N1	2,4,6-TRICHLOROANILINE **634-93-5**
2 ★	3.32		1505	459	C6H4Cl3N1	3,4,5-TRICHLOROANILINE **634-91-3**
3	4.27		1190		C6H4Cl3N1O1	2-METHOXY-3,5,6-TRICHLOROPYRIDINE **31557-34-3**
4 ★	4.31		1140		C6H4Cl6	HEXACHLOROCYCLOHEXENE **57722-15-3**
25 ★	4.34		1140		"	HEXACHLOROCYCLOHEXENE **57722-16-4**
6 ★	4.35		1140		"	HEXACHLOROCYCLOHEXENE **57722-17-5**
7 ★	4.12		1140		"	HEXACHLOROCYCLOHEXENE **59229-56-0**
8 ★	1.90	7.0	925	1	C6H4F1O2	M-FLUORONITROBENZENE **402-67-5**
9 ★	1.69		547		C6H4F1O2	O-FLUORONITROBENZENE **1493-27-2**
1230 ★	1.80	7.0	925	1	C6H4F1O2	P-FLUORONITROBENZENE **350-46-9**
1 ★	1.91	1.5	2329	314	C6H4F1O3	PHENOL,2-NITRO-5-FLUORO **446-36-6**
2 ★	2.37		594	400	C6H4F2	BENZENE,1,2-DIFLUORO **367-11-3**
3	1.46	7.4	2646	448	C6H4F2O1	PHENOL,2,6-DIFLUORO **28177-48-2**
4 ★	2.05		594	400	"	PHENOL,2,6-DIFLUORO
35	1.54		1412		C6H4F7N5O1	1,2,4-TRIAZOLE,3-AMINO,5-HEPTAFLUOROBUTYRAMIDO
6 ★	2.94		562		C6H4I1O2	M-IODONITROBENZENE **645-00-1**
7 ★	4.11		594	400	C6H4I2	BENZENE,1,4-DIIODO **624-38-4**
8 ★	0.40		594	400	C6H4N2	2-CYANOPYRIDINE **100-70-9**
9	-0.44	7.2	2317	1	C6H4N2	3-CYANOPYRIDINE **100-54-9**
1240 ★	0.23		2412		"	3-CYANOPYRIDINE
1 ★	0.46		163	459	C6H4N2	4-CYANOPYRIDINE **100-48-1**
2 ★	-0.94		1716		C6H4N2O1	4-CYANOPYRIDINEOXIDE **14906-59-3**
3 ★	1.69		2166		C6H4N2O1	BENZOFURAZAN **273-09-6**
4 ★	1.43		2166		C6H4N2O2	BENZOFURAZAN-2-OXIDE
45 ★	1.49		531		C6H4N2O4	M-DINITROBENZENE **99-65-0**
6 ★	1.69		547		C6H4N2O4	O-DINITROBENZENE **528-29-0**
7 ★	1.46		501		C6H4N2O4	P-DINITROBENZENE **100-25-4**
8	-0.39		1463		C6H4N2O5	2,4-DINITROPHENOLATE,POTASSIUM SALT
9 ★	1.67	1.5	2329	314	C6H4N2O5	2,4-DINITROPHENOL **51-28-5**
1250 ★	1.75		503		C6H4N2O5	2,5-DINITROPHENOL **329-71-5**
1 ★	1.37	1.0	572	342	C6H4N2O5	2,6-DINITROPHENOL **573-56-8**
2 ★	2.36		505		C6H4N2O5	3,5-DINITROPHENOL **586-11-8**
3 ★	1.30		1946		C6H4N2O5	FURAN,2-(2-NITROETHENYL)-5-NITRO
4 ★	2.10		995		C6H4N2O6	2,4-DINITRORESORCINOL **519-44-8**
55 ★	2.01		590	60	C6H4N2S1	2,1,3-BENZOTHIADIAZOLE **273-13-2**
6 ★	1.64		2166		C6H4N2Se1	2,1,3-BENZOSELENADIAZOLE **273-15-4**
7 ✓	-0.13		505		C6H4N4	ISOPROPENYLAMINE,1,1,3-TRICYANO **868-54-2**
8 ★	-0.58		1627		C6H4N4	PTERIDINE **91-18-9**
9 ★	1.95	7.3	509	2	C6H4N4O2	5-NITROBENZTRIAZOLE **2338-12-7**
1260 ★	0.20		2649		C6H4O2	QUINONE **106-51-4**
1 ★	2.99		501		C6H5Br1	BROMOBENZENE **108-86-1**
2	-2.50	7.4	2071		C6H5Br1Hg1O3S1	P-BROMOMERCURIPHENYLSULFONIC ACID ✚
3	-0.68		1375		C6H5Br1N4O1	1,2,4-TRIAZOLO(4,3-B)PYRIDAZINE-6-ME-7-BR-8-OH
4 ★	2.63		501		C6H5Br1O1	M-BROMOPHENOL **591-20-8**
65 ★	2.35		501		C6H5Br1O1	O-BROMOPHENOL **95-56-7**
6 ★	2.59		501		C6H5Br1O1	P-BROMOPHENOL **106-41-2**
7 ★	1.84		2523		C6H5Br1O3	FURAN,2-BROMO-5-METHOXYCARBONYL
8 ★	2.89		594	400	C6H5Cl1	CHLOROBENZENE **108-90-7**
9 ★	2.06		1772		C6H5Cl1F1N1	3-CHLORO-4-FLUOROANILINE **367-21-5**
1270 ★	1.78		594		C6H5Cl1Hg1	PHENYLMERCURIC CHLORIDE **100-56-1** *antimicrobial~fungicide*
1 ★	1.51		1135	537	C6H5Cl1Hg1O1	CHLOROMERCURIPHENOL **90-03-9** *disinfectant*
2 ★	2.72		2159		C6H5Cl1N2O2	2-NITRO-4-CHLOROANILINE **89-63-4**
3 ★	2.06		594		C6H5Cl1N2O2	ANILINE,4-CHLORO-3-NITRO **635-22-3**
4 ★	0.47		2642		C6H5Cl1N2O2	PYRAZINE,2-CHLORO-5-METHOXYCARBONYL
75 ★	0.47		2499		C6H5Cl1N2O2	PYRAZINE,2-CHLORO-6-METHOXYCARBONYL
6 ★	1.53		1549		C6H5Cl1N2O3	2-AMINO-6-CHLORO-4-NITROPHENOL **6358-09-4**
7 ★	2.50		501		C6H5Cl1O1	M-CHLOROPHENOL **108-43-0**
8 ★	2.15		501		C6H5Cl1O1	O-CHLOROPHENOL **95-57-8**
9 ★	2.39		501		C6H5Cl1O1	PARACHLORPHENOL **106-48-9** *antibacterial (topical)*

	logP	pH	Ref.	Note	MF	Name / **CAS** / *activity*
1280	1.80		1568		C6H5Cl1O2	4-CHLORORESORCINOL **95-88-5**
1 ★	1.40		1153		C6H5Cl1O2	CHLOROHYDROQUINONE **615-67-8**
2 ★	2.86		2293	400	C6H5Cl2N1	2,3-DICHLOROANILINE **608-27-5**
3 ★	2.91		2293	400	C6H5Cl2N1	2,4-DICHLOROANILINE **554-00-7**
4 ★	2.92		2293	400	C6H5Cl2N1	2,5-DICHLOROANILINE **95-82-9**
85 ★	2.82		2293	400	C6H5Cl2N1	2,6-DICHLOROANILINE **608-31-1**
6 ★	2.69		530		C6H5Cl2N1	3,4-DICHLOROANILINE **95-76-1**
7 ★	2.90		1505	459	C6H5Cl2N1	3,5-DICHLOROANILINE **626-43-7**
8	1.44	2.0	264		C6H5Cl2N1O2S1	3,4-DICHLOROBENZENESULFONAMIDE **23815-28-3**
9 ★	3.85		1140		C6H5Cl5	PENTACHLOROCYCLOHEXENE-3,4,6/5 **54083-25-9**
1290 ★	3.60		1140		C6H5Cl5	PENTACHLOROCYCLOHEXENE-3,4/5,6 **54083-24-8**
1 ★	3.80		1140		C6H5Cl5	PENTACHLOROCYCLOHEXENE-3,5,6/4 **51795-30-3**
2 ★	3.61		1140		C6H5Cl5	PENTACHLOROCYCLOHEXENE-3,5/4,6 **643-15-2**
3 ★	3.95		1140		C6H5Cl5	PENTACHLOROCYCLOHEXENE-3,6/4,5 **319-94-8**
4 ★	2.27		501		C6H5F1	FLUOROBENZENE **462-06-6**
95	-0.68	7.4	2270	1	C6H5F1N2O3	5-FLUOROURACIL-3-ACETYL **75410-15-0**
6 ★	-0.34	4.0	2270	302	"	5-FLUOROURACIL-3-ACETYL
7	-1.38	7.4	2266	1	C6H5F1N2O4	URACIL,1-METHOXYCARBONYL-5-FLUORO **71759-43-8**
8 ★	-0.68	4.0	2010		"	URACIL,1-METHOXYCARBONYL-5-FLUORO
9 ★	1.93		501		C6H5F1O1	M-FLUOROPHENOL **372-20-3**
1300 ★	1.71		501		C6H5F1O1	O-FLUOROPHENOL **367-12-4**
1 ★	1.77		501		C6H5F1O1	P-FLUOROPHENOL **371-41-5**
2 ★	1.54		1772		C6H5F2N1	2,4-DIFLUOROANILINE **367-25-9**
3 ★	3.36		536		C6H5F5S1	PHENYLSULFURPENTAFLUORIDE **2557-81-5**
4 ★	3.25		508		C6H5I1	IODOBENZENE **591-50-4**
5 ★	-1.61		508		C6H5I1O1	IODOSOBENZENE **536-80-1**
6 ★	2.93		501		C6H5I1O1	M-IODOPHENOL **626-02-8**
7 ★	2.65		501		C6H5I1O1	O-IODOPHENOL **533-58-4**
8 ★	2.91		501		C6H5I1O1	P-IODOPHENOL **540-38-5**
9 ★	-1.33		541		C6H5I1O2	IODOXYBENZENE
1310 ★	0.85		1478		C6H5N1O1	2-CYANOMETHYLFURAN **2745-25-7**
1 ★	2.01		508		C6H5N1O1	NITROSOBENZENE **586-96-9**
2 ★	0.44		594	400	C6H5N1O1	PYRIDINE,2-CARBOXALDEHYDE **1121-60-4**
3	0.14		594		C6H5N1O1	PYRIDINE,3-CARBOXALDEHYDE **500-22-1**
4	0.29		2247	459	"	PYRIDINE,3-CARBOXALDEHYDE
15 ★	0.43		547		C6H5N1O1	PYRIDINE-4-CARBOXALDEHYDE **872-85-5**
6	-1.15	3.8	547		C6H5N1O2	ISONICOTINIC ACID **55-22-1**
7	-0.57		547	827	"	ISONICOTINIC ACID
8 ★	1.08		538		C6H5N1O2	N-HYDROXYBENZOQUINONEMONIMINE **637-62-7**
9	-0.82	4.5	2107	537	C6H5N1O2	NICOTINIC ACID **59-67-6** *vitamin*
1320	-0.66	3.0	2107	537	"	NICOTINIC ACID
1	-0.46		1135	537	"	NICOTINIC ACID
2	0.66	2.0	2312	337	"	NICOTINIC ACID
3 ★	1.85		501		C6H5N1O2	NITROBENZENE **98-95-3**
4 ★	1.29		547		C6H5N1O2	P-NITROSOPHENOL **104-91-6**
25	-1.98	2.0	447		C6H5N1O2	PICOLINIC ACID **98-98-6**
6	-1.50	7.4	447	1	"	PICOLINIC ACID
7 ★	1.96		1478		C6H5N1O2S1	2-(B-NITROVINYL)THIOPHENE **874-84-0**
8 ★	1.94		1478		C6H5N1O2S1	3-(B-NITROVINYL)THIOPHENE **28783-31-5**
9 ★	1.56		1478		C6H5N1O3	2-(B-NITROVINYL)FURAN **699-18-3**
1330 ★	1.41		1478		C6H5N1O3	3-(B-NITROVINYL)FURAN **53916-74-8**
1	-2.53	7.4	447	1	C6H5N1O3	3-HYDROXYPICOLINIC ACID **874-24-8**
2	-1.27	2.0	447		"	3-HYDROXYPICOLINIC ACID
3	1.74	7.4	2646	448	C6H5N1O3	M-NITROPHENOL **554-84-7**
4 ★	2.00		501		"	M-NITROPHENOL
35	1.68	7.4	2646	448	C6H5N1O3	O-NITROPHENOL **88-75-5**
6 ★	1.79		501		"	O-NITROPHENOL
7	1.38	7.4	772	1	C6H5N1O3	P-NITROPHENOL **100-02-7**
8	1.77	7.4	2646	448	"	P-NITROPHENOL
9 ★	1.91		501		"	P-NITROPHENOL
1340 ✓	-1.31	12.8	534	1	C6H5N1O3	SODIUM P-NITROPHENOXIDE **14609-74-6**
1	-0.84	10.2	534	310	"	SODIUM P-NITROPHENOXIDE
2 ★	1.66	1.0	594		C6H5N1O4	1,2-DIHYDROXYBENZENE,4-NITRO **3316-09-4**
3 ★	1.56	1.4	536	342	C6H5N1O4	2-NITRORESORCINOL **601-89-8**
4	1.05	6.8	448	2	C6H5N1O4	4-NITRORESORCINOL **3163-07-3**
45 ★	-0.14		731	604	C6H5N1O4	N-MALEOYLGLYCINE **25021-08-3**
6 ★	1.26		1478		C6H5N1S1	2-CYANOMETHYLTHIOPHENE **20893-30-5**
7 ★	1.26		1478		C6H5N1S1	3-CYANOMETHYLTHIOPHENE **13781-53-8**
8 ★	0.30		594		C6H5N3	1H-IMIDAZO-[4,5-B]-PYRIDINE **273-21-2**
9 ★	0.18		2031		C6H5N3	5-PYRIDOIMIDAZOLE **272-97-9**
1350 ★	1.44		559		C6H5N3	BENZOTRIAZOLE **95-14-7**
1 ★	2.59		541		C6H5N3	PHENYLAZIDE **622-37-7**
2 ★	0.26		2499		C6H5N3	PYRAZINE,2-CYANO-5-METHYL
3 ★	0.44		2499		C6H5N3	PYRAZINE,2-CYANO-6-METHYL
4	0.86		897	459	C6H5N3O1	2-PYRIMIDINENITRILE,4-METHOXY
55 ★	0.69		2592		C6H5N3O1	BENZOTRIAZOLE,2-HYDROXY
6 ★	0.95		2499		C6H5N3O1	PYRAZINE,2-CYANO-6-METHOXY
7 ★	0.52	7.4	2206	306	C6H5N3O2	IMIDAZOLE,5-NITRO-1-PROPARGYL
8 ★	1.79		547		C6H5N3O4	ANILINE,2,6-DINITRO **606-22-4**
9 ★	1.89		547		C6H5N3O4	ANILINE,3,5-DINITRO **618-87-1**
1360 ★	0.93		1002		C6H5N3O5	PICRAMIC ACID **96-91-3**
1	0.96	7.0	332	1	C6H5N3S1	THIENO(2,33-D)-PYRIMIDINE,4-AMINO **14080-56-9** ✦
2	-1.10		1375		C6H5N5O2	1,2,4-TRIAZOLO(4,3-B)PYRIDAZIN-8-ONE-6-ME-7-OH-IMINYL ✦
3 ★	2.24		579	400	C6H6	2,4-HEXADIYNE **2809-69-0**
4 ★	2.13		501		C6H6	BENZENE **71-43-2**

	logP	pH	Ref.	Note	MF	Name / **CAS** / *activity*
65	2.03		1734		C6H6	HEXADEUTEROBENZENE **1076-43-3**
6 ★	3.28		421	456	C6H6Br1Cl4F1	2-F-3-BR-TETRACL-CYCLOHEXANE(2,3,5,6/1,4)
7 ★	3.81		421	456	C6H6Br1Cl5	1-BR-PENTACL-CYCLOHEXANE(2,3,5,6/1,4)
8 ★	3.74		421	456	C6H6Br1Cl5	2-BR-PENTACL-CYCLOHEXANE(2,3,5,6/1,4)
9 ★	2.10		272		C6H6Br1N1	M-BROMOANILINE **591-19-5**
1370 ★	2.11		558		C6H6Br1N1	O-BROMOANILINE **615-36-1**
1 ★	2.26		272		C6H6Br1N1	P-BROMOANILINE **106-40-1**
2 ★	1.39	2.1	2265		C6H6Br1N1O2S1	BENZENESULFONAMIDE,3-BR
3 ★	1.36	2.0	264		C6H6Br1N1O2S1	P-BROMOBENZENESULFONAMIDE **701-34-8**
4 ★	0.25	7.4	725	1	C6H6Br1N3O1	2-BROMOISONIAZID **29849-15-8**
75 ★	3.88		421	456	C6H6Br2Cl4	1,2-DIBR-TETRACL-CYCLOHEXANE(2,3,5,6/1,4)
6 ★	3.99		421	456	C6H6Br2Cl4	2,3-DIBR-TETRACL-CYCLOHEXANE(2,3,5,6/1,4)
7 ★	1.88		501		C6H6Cl1N1	M-CHLOROANILINE **108-42-9**
8 ★	1.90		530		C6H6Cl1N1	O-CHLOROANILINE **95-51-2**
9	1.83	7.4	1737	1	C6H6Cl1N1	P-CHLOROANILINE **106-47-8**
1380 ★	1.88		2293	400	"	P-CHLOROANILINE
1 ★	1.81		2163	459	C6H6Cl1N1O1	PHENOL,2-AMINO-4-CHLORO **95-85-2**
2 ★	1.29	2.0	264		C6H6Cl1N1O2S1	M-CHLOROBENZENESULFONAMIDE **17260-71-8**
3 ★	0.74	2.0	264		C6H6Cl1N1O2S1	O-CHLOROBENZENESULFONAMIDE **6961-82-6**
4 ★	1.24	7.2	2036	1	C6H6Cl1N1O2S1	P-CHLOROBENZENESULFONAMIDE **98-64-6**
85 ★	0.11	7.4	725	1	C6H6Cl1N3O1	2-CHLOROISONIAZID **58481-04-2**
6 ★	0.56		2499		C6H6Cl1N3O1	PYRAZINE,2-ACETYLAMINO-5-CHLORO
7 ★	1.10		2499		C6H6Cl1N3O1	PYRAZINE,2-CHLORO-6-ACETYLAMINO
8 ★	2.00		594		C6H6Cl2N2	O-PHENYLENEDIAMINE,4,5-DICHLORO **5348-42-5**
9 ★	3.65		1553		C6H6Cl4	TETRACHLOROCYCLOHEXENE(3/45)
1390 ★	3.52		1553		C6H6Cl4	TETRACHLOROCYCLOHEXENE(34/5)
1 ★	3.72		1553		C6H6Cl4	TETRACHLOROCYCLOHEXENE(35/4)
2 ★	3.74		1140		C6H6Cl4	TETRACHLOROCYCLOHEXENE
3 ★	3.08		1140		"	TETRACHLOROCYCLOHEXENE **1782-00-9**
4 ★	3.08		1140		"	TETRACHLOROCYCLOHEXENE **319-81-3**
95 ★	3.15		1140		"	TETRACHLOROCYCLOHEXENE **33875-95-5**
6 ★	3.40		1140		"	TETRACHLOROCYCLOHEXENE **41992-55-6**
7 ★	0.56		564		C6H6Cl4N2Pt1	DICHLORO-4,5-DICHLORO-O-PHENYLENEDIAMINOPLATINUM
8 ★	3.19		421	456	C6H6Cl5F1	2-FLUOROPENTACHLORO-CYCLOHEXANE(2,3,5,6/1,4)
9 ★	3.96		421	456	C6H6Cl5I1	1-IODO-PENTA-CL-CYCLOHEXANE(2,3,5,6/1,4)
1400 ★	4.05		421	456	C6H6Cl5I1	3-IODO-PENTACL-CYCLOHEXANE(2,3,5,6/1,4)
1 ★	3.80		516		C6H6Cl6	HEXACHLOROCYCLOHEXANE, ALPHA ISOMER **319-84-6**
2 ★	3.78		516		C6H6Cl6	HEXACHLOROCYCLOHEXANE, BETA ISOMER **319-85-7**
3 ★	4.14		516		C6H6Cl6	HEXACHLOROCYCLOHEXANE, DELTA ISOMER **319-86-8**
4 ★	3.72		421		C6H6Cl6	LINDANE **58-89-9** *antiparasitic (topical)*
5 ★	1.30		501		C6H6F1N1	M-FLUOROANILINE **372-19-0**
6 ★	1.26		272		C6H6F1N1	O-FLUOROANILINE **348-54-9**
7 ★	1.15		501		C6H6F1N1	P-FLUOROANILINE **371-40-4**
8	-0.11	7.4	725	1	C6H6F1N3O1	2-FLUOROISONIAZID **369-24-4**
9 ★	-0.20	4.0	2009		C6H6F1N3O3	URACIL,1-METHYLAMINOCARBONYL-5-FLUORO
1410 ★	2.32		558		C6H6I1N1	O-IODOANILINE **615-43-0**
1 ★	2.34		572		C6H6I1N1	P-IODOANILINE **540-37-4**
2	0.52	7.4	725	1	C6H6I1N3O1	2-IODOISONIAZID **29247-87-8**
3 ★	1.62	2.1	2265		C6H6I1O2S1	BENZENESULFONAMIDE,3-IODO
4 ★	1.59	2.1	2265		C6H6I1O2S1	BENZENESULFONAMIDE,4-IODO **825-86-5**
15 ★	0.15		2412		C6H6N2O1	2-PYRIDINECARBOXAMIDE **1452-77-3**
6 ★	-0.28		1519	400	C6H6N2O1	I-NICOTINAMIDE **1453-82-3**
7	-0.44	7.2	2317	1	C6H6N2O1	NICOTINAMIDE **98-92-0** *vitamin*
8 ★	-0.37		547		"	NICOTINAMIDE
9 ★	-0.34		1519	400	"	NICOTINAMIDE
1420 ★	0.20		2412		C6H6N2O1	PYRAZINE,2-ACETYL
1	0.68	7.2	2317	1	C6H6N2O1	PYRIDINE-3-ALDOXIME **1193-92-6**
2 ★	0.77		579		C6H6N2O1	PYRIDINE-4-ALDOXIME **696-54-8**
3	0.03		1456		C6H6N2O2	2-AMINO-4-NITROSOPHENOL
4 ★	0.99		163	459	C6H6N2O2	2-METHYL-5-NITROPYRIDINE **21203-68-9**
25 ★	-0.17		1519	400	C6H6N2O2	2-PYRAZINECARBOXYLIC ACID,METHYL ESTER **6164-79-0**
6 ★	0.65	7.4	447	1	C6H6N2O2	3-HYDROXYPICOLINAMIDE **933-90-4**
7 ★	1.49		594	400	C6H6N2O2	BENZOQUINONE,1,4-DIOXIME **105-11-3**
8 ★	1.37		501		C6H6N2O2	M-NITROANILINE **99-09-2**
9 ★	1.85		547		C6H6N2O2	O-NITROANILINE **88-74-4**
1430	1.30	7.4	2646	448	C6H6N2O2	P-NITROANILINE **100-01-6**
1 ★	1.39		501		"	P-NITROANILINE
2 ★	-0.43		2412		C6H6N2O2	PYRIDAZINE-3-CARBOXYLIC ACID,METHYL ESTER
3 ★	-0.27		2412		C6H6N2O2	PYRIDAZINE-4-CARBOXYLIC ACID,METHYL ESTER
4 ★	-0.71		2412		C6H6N2O2	PYRIMIDINE,2-METHOXYCARBONYL **34253-03-7**
35 ★	-0.27		2412		C6H6N2O2	PYRIMIDINE-4-CARBOXYLIC ACID,METHYL ESTER
6 ★	0.03		2412		C6H6N2O2	PYRIMIDINE-5-CARBOXYLIC ACID,METHYL ESTER
7 ★	1.12		547		C6H6N2O2S1	1,3-DIHYDRO-BENZOTHIADIAZOLE-2,2-DIOXIDE **1615-06-1**
8 ★	1.53		579		C6H6N2O3	2-AMINO-4-NITROPHENOL **99-57-0**
9 ★	0.96		1909		C6H6N2O3	4-AMINO-2-NITROPHENOL **119-34-6**
1440 ★	1.55		594		C6H6N2O3	PYRIDINE,2-METHOXY-5-NITRO **5446-92-4**
1	0.63		2447		C6H6N2O4	4-PYRIMIDINECARBOXYLIC ACID,TETRAHYDRO-2,6-DIOXO,METHYL ESTER **6153-44-2**
2 ★	0.55	2.0	264		C6H6N2O4S1	M-NITROBENZENESULFONAMIDE **121-52-8**
3 ★	0.34	2.0	264		C6H6N2O4S1	O-NITROBENZENESULFONAMIDE **5455-59-4**
4 ★	0.64	2.0	264		C6H6N2O4S1	P-NITROBENZENESULFONAMIDE **6325-93-5**
45 ★	1.24		1519		C6H6N2S1	2-PYRIDINETHIOCARBOXAMIDE **5346-38-3**
6 ★	0.67		1519	400	C6H6N2S1	3-PYRIDINETHIOCARBOXAMIDE **4621-66-3**
7	-0.21		1375		C6H6N4O1	1,2,4-TRIAZOLO(4,3-B)PYRIDAZINE-6-ME-8-OH ✚
8 ★	-0.07	7.4	2206	306	C6H6N4O2	ACETONITRILE,2-METHYL-5-NITROIMIDAZOL-2YL
9 ★	-0.27	3.0	2191		C6H6N4O2	XANTHINE,1-METHYL **6136-37-4**

	logP	pH	Ref.	Note	MF	Name / CAS / activity
1450 ★	-0.72		2537		C6H6N4O2	XANTHINE,3-METHYL **1076-22-8**
1 ★	-0.89	3.0	2191		C6H6N4O2	XANTHINE,7-METHYL **552-62-5**
2 ★	1.70		611		C6H6N4O2S2	1(5-NO2-2-THIAZOLYL)-2-IMIDAZOLIDINETHIONE
3 ★	-0.76	7.4	725	1	C6H6N4O3	2-NITROISONIAZID **58481-05-3**
4 ★	-0.57	3.0	2191		C6H6N4O3	URIC ACID,1-METHYL **708-79-2**
55 ★	-1.08	3.0	2191		C6H6N4O3	URIC ACID,3-METHYL **605-99-2**
6 ★	-1.18	3.0	2191		C6H6N4O3	URIC ACID,7-METHYL **612-37-3**
7	0.93	7.4	2206	306	C6H6N4O3S1	1-(5-NO2-2-THIAZOLYL)-2-IMIDAZOLIDINONE **61-57-4** *antischistosomal*
8 ★	0.95		555		"	1-(5-NO2-2-THIAZOLYL)-2-IMIDAZOLIDINONE
9 ★	0.57	7.4	553	1	C6H6N4O3S1	2(1,3-OXAZOLIDINYLIDEN-2N)-5-NO2-THIAZOLE
1460 ★	1.64		1731		C6H6N4O3S1	5-NITRO-2-FURALDEHYDETHIOSEMICARBAZONE
1 ★	1.46	7.4	595	1	C6H6N4O4	2,4-DINITROPHENYLHYDRAZINE **119-26-6**
2	1.48		547		"	2,4-DINITROPHENYLHYDRAZINE
3	0.23	7.4	1497	1	C6H6N4O4	NITROFURAZONE **59-87-0** *antiinfective (topical)*
4 ★	0.23	7.4	1022	1	"	NITROFURAZONE
65	0.62	7.4	772	1	C6H6O1	PHENOL **108-95-2**
6 ★	1.46		501		"	PHENOL
7 ★	1.25		1478		C6H6O1S1	2-ACETYLTHIOPHENE **88-15-3**
8 ★	1.24		1478		C6H6O1S1	3-ACETYLTHIOPHENE **1468-83-3**
9 ★	0.52		1478		C6H6O2	2-ACETYLFURAN **1192-62-7**
1470 ★	0.67		1006		C6H6O2	5-METHYLFURFURAL **620-02-0**
1 ★	0.80		501		C6H6O2	M-DIHYDROXYBENZENE **108-46-3** *keratolytic*
2 ★	0.88		508		C6H6O2	O-DIHYDROXYBENZENE **120-80-9**
3 ★	0.59		516		C6H6O2	P-DIHYDROXYBENZENE **123-31-9** *deipegmentor*
4 ✓	-3.54		547	820	C6H6O2S1	BENZENESULFINIC ACID,SODIUM SALT **873-55-2**
75 ★	1.76		1478		C6H6O2S1	THIOPHEN-3-CARBOXYLIC ACID,METHYL ESTER **22913-26-4**
6	0.68		2163	459	C6H6O3	1,2,3-TRIHYDROXYBENZENE **87-66-1**
7 ★	0.16	1.4	536	342	C6H6O3	1,3,5-TRIHYDROXYBENZENE **108-73-6** *antispasmodic*
8 ★	1.28		2523		C6H6O3	FURAN,3-METHOXYCARBONYL
9 ★	1.00		1055		C6H6O3	FUROIC ACID,METHYL ESTER **611-13-2**
1480 ★	2.52		536		C6H6S1	THIOPHENOL **108-98-5**
1 ★	0.06		541	606	C6H7As1O3	PHENYLARSONIC ACID **98-05-5**
2 ★	1.58		502		C6H7B1O2	PHENYLBORONIC ACID **98-80-6**
3	-2.91	7.4	921	22	C6H7Br1N1.I1	3-BROMO-N-METHYLPYRIDINIUM IODIDE **32222-42-7**
4	0.46		579		C6H7Br1N2	P-BROMOPHENYLHYDRAZINEHCL **589-21-9**
85	1.39	13.0	579	224	"	P-BROMOPHENYLHYDRAZINEHCL
6 ★	-0.30		1564		C6H7Br1N2O3	PYRIMIDINE-2,4-DIONE,3-METHOXY-5-BROMO-6-METHYL **77317-65-8**
7 ★	1.02		2200		C6H7Br2N3O4	MISONIDAZOLE,4,5-DIBROMO
8 ★	0.85		1909		C6H7Cl1N2	4-CHLORO-M-PHENYLENEDIAMINE **5131-60-2**
9 ★	1.28	7.4	594	314	C6H7Cl1N2	O-PHENYLENEDIAMINE,4-CHLORO **95-83-0**
1490 ★	1.50		2499		C6H7Cl1N2	PYRAZINE,3-CHLORO-2,5-DIMETHYL
1 ★	1.99		2499		C6H7Cl1N2O1	PYRAZINE,2-CHLORO-5-ETHOXY
2 ★	2.22		2499		C6H7Cl1N2O1	PYRAZINE,2-CHLORO-6-ETHOXY
3 ★	0.50		564		C6H7Cl2N3O2Pt1	DICHLORO-4-NITRO-O-PHENYLENEDIAMINOPLATINUM
4 ★	2.84		1553		C6H7Cl3	TRICHLOROCYCLOHEXENE(34/5) **56994-25-3**
95 ★	0.45		564		C6H7Cl3N2Pt1	DICHLORO-4-CHLORO-O-PHENYLENEDIAMINOPLATINUM
6 ★	3.53		421		C6H7Cl5	1-H-PENTACHLOROCYCLOHEXANE **22138-39-2**
7 ★	3.37		421		C6H7Cl5	3-H-PENTACHLOROCYCLOHEXANE
8 ★	2.54		421		C6H7Cl5O1	1-HYDROXYPENTACHLOROCYCLOHEXANE **53861-64-6**
9 ★	0.04	7.4	2361	1	C6H7F2N3O3	IMIDAZOLE,2-NITRO-1-(3-HYDROXY-2,2-DIFLUORO)PROPYL
1500	1.46		1501	100	C6H7F3N4O1S1	THIAZAFLURON **25366-23-8**
1	1.85		1211		"	THIAZAFLURON
2 ★	1.99		1895		C6H7F7N2S1	THIOUREA,N-2,3,4-HEPTAFLUOROBUTYL-N'-METHYL
3 ★	1.11		536		C6H7N1	2-PICOLINE **109-06-8**
4 ★	1.20		260		C6H7N1	3-METHYLPYRIDINE **108-99-6**
5 ★	1.22		260		C6H7N1	4-METHYLPYRIDINE **108-89-4**
6	0.84		1734		C6H7N1	PENTADEUTEROBENZEAMINE **4165-61-1**
7	0.84		1734		"	PENTADEUTEROBENZEAMINE **62-53-3**
8 ★	0.90		501		"	PENTADEUTEROBENZEAMINE
9	0.95	7.4	1737	1	"	PENTADEUTEROBENZEAMINE
1510	0.98	7.5	2201		"	PENTADEUTEROBENZEAMINE
1 ★	1.36		579		C6H7N1O1	2-METHOXYPYRIDINE **1628-89-3**
2 ★	1.00		163	459	C6H7N1O1	4-METHOXYPYRIDINE **620-08-6**
3 ★	0.06		1474		C6H7N1O1	4-PYRIDINEMETHANOL **586-95-8**
4 ★	0.21	5.6	726	41	C6H7N1O1	M-AMINOPHENOL **591-27-5**
15 ★	-0.23		924	400	C6H7N1O1	N-METHYL-A-PYRIDONE **694-85-9**
6 ★	-1.22		924	400	C6H7N1O1	N-METHYL-G-PYRIDONE **695-19-2**
7 ★	-0.02		1474		C6H7N1O1	NICOTINYL ALCOHOL **100-55-0** *vasodilator (peripheral)*
8 ★	0.62		508		C6H7N1O1	O-AMINOPHENOL **95-55-6**
9	0.72	7.4	772	1	"	O-AMINOPHENOL
1520	0.04	7.4	772	1	C6H7N1O1	P-AMINOPHENOL **123-30-8**
1 ★	0.04		531		"	P-AMINOPHENOL
2 ★	0.79		541		C6H7N1O1	PHENYLHYDROXYLAMINE **100-65-2**
3 ★	0.06		1474		C6H7N1O1	PICONOL **586-98-1**
4 ★	0.99		2412		C6H7N1O1	PYRIDINE,3-METHOXY **7295-76-3**
25 ★	0.93		1625		C6H7N1O1	PYRROLE,2-ACETYL **1072-83-9**
6	0.70	5.0	2153	2	C6H7N1O2	2-(1-PYRROLYL)ACETIC ACID
7	-0.49	7.4	2685	1	C6H7N1O2	4-PYRIDONE,3-HYDROXY-2-METHYL *iron chelator*
8 ★	0.23		2523		C6H7N1O2	FURAN-2-CARBOXAMIDE,N-METHYL
9 ★	0.34		2523		C6H7N1O2	FURAN-3-CARBOXAMIDE,N-METHYL
1530 ★	1.17		1625	457	C6H7N1O2	PYRROLE-2-CARBOXYLIC ACID,METHYL ESTER **1193-62-0**
1 ★	0.31		501		C6H7N1O2S1	BENZENESULFONAMIDE **98-10-2**
2 ★	-2.05	7.2	2323	468	C6H7N1O2S1	THIENYLGLYCINE
3	-2.16		2457	300	C6H7N1O3S1	ANILINE-4-SULFONIC ACID **121-57-3** *antibacterial*
4 ★	0.48		590		C6H7N1O3S1	N-HYDROXYBENZENESULFONAMIDE **599-71-3**

	logP	pH	Ref.	Note	MF	Name / CAS / activity
35 ★	0.06	5.6	726	41	C6H7N1O3S1	P-HYDROXYBENZENESULFONAMIDE **1576-43-8**
6 ★	0.89	7.4	1022	1	C6H7N1O4	FURASPOR *antifungal*
7 ★	1.81		1478		C6H7N1S1	1-(THIOPHEN-2-YL)ACETALDEHYDEOXIME **59445-83-9**
8 ★	1.10		1478		C6H7N1S1	2-ACETAMIDOTHIOPHENE **13053-81-1**
9 ★	1.71		1519	459	C6H7N1S1	2-METHYLTHIOPYRIDINE **18438-38-5**
1540 ★	0.31		1478		C6H7N1S1	3-ACETAMIDOTHIOPHENE **13781-66-3**
1	1.43	7.4	547	1	C6H7N1S1	O-AMINOTHIOPHENOL **137-07-5**
2 ★	0.33		1478		C6H7N1S1	THIOPHEN-2-YL-ACETAMIDE **4461-29-4**
3 ★	1.31		594		C6H7N1S1	THIOPHENOL,4-AMINO **1193-02-8**
4 ★	-0.75		1519	400	C6H7N3O1	2-ACETAMIDOPYRIMIDINE **13053-88-8**
45 ✓	0.88	7.0	538	701	C6H7N3O1	2-AMINONICOTINAMIDE **13438-65-8**
6 ✓	0.70	7.0	538	701	C6H7N3O1	6-AMINONICOTINAMIDE **329-89-5**
7	-1.14	7.4	725	1	C6H7N3O1	ISONIAZID **54-85-3** *antibacterial (tuberculostatic)*
8 ★	-0.70		1519	400	"	ISONIAZID
9 ★	-0.03		2412		C6H7N3O1	PYRAZINE,2-ACETYLAMINO
1550 ★	-0.25		2499		C6H7N3O1	PYRAZINE,2-CARBOXAMIDO-5-METHYL
1 ★	-0.13		2499		C6H7N3O1	PYRAZINE,2-CARBOXAMIDO-6-METHYL
2 ★	-0.41		2412		C6H7N3O1	PYRIDAZINE,4-ACETYLAMINO
3 ★	0.03		2412		C6H7N3O1	PYRIMIDINE,4-ACETYLAMINO
4 ★	-0.22		2412		C6H7N3O1	PYRIMIDINE,5-ACETYLAMINO
55 ★	0.64	7.4	1022	1	C6H7N3O2	1-METHYL-2-NITRO-5-VINYLIMIDAZOLE
6 ★	0.53		1909		C6H7N3O2	2-NITRO-P-PHENYLENEDIAMINE **5307-14-2**
7 ★	1.27		547		C6H7N3O2	3-NITRO-O-PHENYLENEDIAMINE **3694-52-8**
8 ★	0.88		579		C6H7N3O2	4-NITRO-O-PHENYLENEDIAMINE **99-56-9**
9 ★	0.62	7.4	2206	306	C6H7N3O2	IMIDAZOLE,1-METHYL-5-NITRO-2-VINYL
1560 ★	1.41	13.0	594		C6H7N3O2	P-NITROPHENYLHYDRAZINE **100-16-3**
1 ★	0.13		2499		C6H7N3O2	PYRAZINE,2-CARBOXAMIDO-6-METHOXY
2 ★	-0.34	7.4	1022	1	C6H7N3O3	1-(ACETYLMETHYL)-2-NITROIMIDAZOLE
3 ★	-0.65		2352	2	C6H7N3O3S1	3-PYRIDINESULFONYLUREA
4 ★	-0.09	7.4	1022	1	C6H7N3O4	1-(CH2CO2CH3)-2-NITROIMIDAZOLE
65 ★	0.51	7.4	1022	1	C6H7N3O4	1-METHYL-2-NITRO-5-METHOXYCARBONYLIMIDAZOLE
6	-1.70	7.4	2206	1	C6H7N3O4	ACETIC ACID,2-METHYL-5-NITROIMIDAZOL-2YL **1010-93-1**
7 ★	-0.03	7.4	2348	1	C6H7N5	ADENINE,9-METHYL
8	-0.52		1375		C6H7N5O1	1,2,4-TRIAZOLO(4,3-B)PYRIDAZINE-6-ME-7-NH2-8-OH
9 ★	1.38	7.4	553	1	C6H7N5O2S1	2(1,3-IMIDAZOLINYLIDEN-2-AMINO)-5-NO2-THIAZOLE **24240-69-5**
1570 ✓	-0.65	3.0	541		C6H7O3P1	PHENYLPHOSPHONIC ACID **1571-33-1**
1 ★	0.54	3.0	541		"	PHENYLPHOSPHONIC ACID
2 ★	2.47		547	400	C6H8	1,3-CYCLOHEXADIENE **592-57-4**
3 ★	2.30		547	400	C6H8	1,4-CYCLOHEXADIENE **628-41-1**
4 ★	0.16		2200		C6H8Br1N3O4	MISONIDAZOLE,5-BROMO
75 ★	2.12	7.4	1401	1	C6H8Cl1N1S1	CLOMETHIAZOLE **533-45-9** *sedative, tranquilizer*
6 ★	0.98		1519	400	C6H8Cl1N3	3-CHLORO-6-PYRIDAZINAMINE,N,N-DIMETHYL
7 ★	1.70		2499		C6H8Cl1N3	PYRAZINE,2-CHLORO-5-DIMETHYLAMINO
8 ★	1.95		2499		C6H8Cl1N3	PYRAZINE,2-CHLORO-6-DIMETHYLAMINO
9 ★	0.88	6.0	2417	1	C6H8Cl1N3O2	IMIDAZOLE,1-CHLOROETHYL-2-METHYL-5-NITRO
1580	0.15		1008	354	C6H8Cl1N3O3	1(3-CL-2-HYDROXYPROPYL)-2-NITROIMIDAZOLE
1	0.18		1467		"	1(3-CL-2-HYDROXYPROPYL)-2-NITROIMIDAZOLE
2 ★	0.12	7.0	1783	1	C6H8Cl1N5O3	2-((6-AM-3-CL-5-NO2-PYRAZIN-2-YL)AMINO)ETHANOL **86845-62-7**
3 ★	-0.85		564		C6H8Cl2N2Pt1	DICHLORO-O-PHENYLENEDIAMINOPLATINUM
4 ★	2.82		421		C6H8Cl4	TETRACHLOROCYCLOHEXANE **60067-92-7**
85 ★	-0.52		2354		C6H8F1N3O3	1-(3-FLUORO-2-HYDROXYPROPYL)-2-NITROIMIDAZOLE
6	-0.44	7.4	2361	1	"	1-(3-FLUORO-2-HYDROXYPROPYL)-2-NITROIMIDAZOLE
7	-0.39	7.4	1022	1	"	1-(3-FLUORO-2-HYDROXYPROPYL)-2-NITROIMIDAZOLE
8 ★	-0.38	7.4	2361	1	C6H8F1N3O3	IMIDAZOLE,2-NITRO-1-(3-HYDROXY-2-FLUORO)PROPYL
9 ✓	-3.30	7.4	921	22	C6H8N1.I1	N-METHYLPYRIDINIUM IODIDE **930-73-4**
1590 ★	0.63		163	459	C6H8N2	2,5-DIMETHYLPYRAZINE **123-32-0**
1 ★	1.07		547		C6H8N2	2-(METHYLAMINO)PYRIDINE **4597-87-9**
2 ★	0.56	7.0	566	1	C6H8N2	2-AMINO-4-PICOLINE **695-34-1**
3 ★	1.02		260		C6H8N2	2-AMINO-5-METHYLPYRIDINE **1603-41-4**
4 ★	0.69		2444		C6H8N2	2-ETHYLPYRAZINE
95 ★	-0.21		1474		C6H8N2	2-PYRIDINEMETHANEAMINE
6	-1.67	2.7	1473	340	C6H8N2	3-PYRIDINEMETHANEAMINE HYDROCHLORIDE
7 ★	0.62		260		C6H8N2	4,6-DIMETHYLPYRIMIDINE **1558-17-4**
8	-1.51	2.7	1473		C6H8N2	4-PYRIDINEMETHANEAMINE HYDROCHLORIDE **3731-53-1**
9 ★	-0.38		1474		C6H8N2	4-PYRIDINEMETHANEAMINE
1600 ★	-0.32		2069		C6H8N2	HEXANEDINITRILE **111-69-3**
1 ★	-0.33	7.4	594	314	C6H8N2	M-PHENYLENEDIAMINE **108-45-2**
2 ★	0.15		531		C6H8N2	O-PHENYLENEDIAMINE **95-54-5**
3	-0.75	7.4	579	1	C6H8N2	P-PHENYLENEDIAMINE **624-18-0**
4 ★	-0.30		1909		"	P-PHENYLENEDIAMINE
5 ★	1.25	8.9	534	2	C6H8N2	PHENYLHYDRAZINE **100-63-0**
6 ★	-0.32		1474		C6H8N2	PICOLAMINE **3731-52-0**
7 ★	0.54		2499		C6H8N2	PYRAZINE,2,3-DIMETHYL
8 ★	0.54		2499		C6H8N2	PYRAZINE,2,6-DIMETHYL **108-50-9**
9 ★	0.35		594	400	C6H8N2	PYRAZOLE,1-ALLYL
1610	-0.51	7.4	2362	1	C6H8N2	PYRIDINE,3-METHYL-4-AMINO
1	-1.56	2.7	1473	340	C6H8N2	PYRIDINEMETHANEAMINEHYDROCHLORIDE **3731-51-9**
2 ★	1.28		2412		C6H8N2O1	PYRAZINE,2-ETHOXY
3 ★	1.24		2499		C6H8N2O1	PYRAZINE,2-METHOXY-3-METHYL
4 ★	1.29		2499		C6H8N2O1	PYRAZINE,2-METHYL-6-METHOXY
15 ★	0.63		2412		C6H8N2O1	PYRIDAZINE,3-ETHOXY
6 ★	0.20		2412		C6H8N2O1	PYRIDAZINE,4-ETHOXY
7	-0.06	7.4	2362	1	C6H8N2O1	PYRIDINE,3-METHOXY-4-AMINO
8 ★	0.74		2412		C6H8N2O1	PYRIMIDINE,2-ETHOXY
9 ★	0.97		2412		C6H8N2O1	PYRIMIDINE,4-ETHOXY

	logP	pH	Ref.	Note	MF	Name / CAS / activity
1620 ★	0.56		2412		C6H8N2O1	PYRIMIDINE,5-ETHOXY
1 ★	1.14		2499		C6H8N2O2	PYRAZINE,2,5-DIMETHOXY
2 ★	1.58		2499		C6H8N2O2	PYRAZINE,2,6-DIMETHOXY
3 ★	-0.54		594	400	C6H8N2O2	PYRIMIDINE,1,3-DIMETHYL-2,4-DIOXO **874-14-6**
4 ★	0.22	5.9	590		C6H8N2O2S1	BENZENESULFONAMIDE,2-AMINO **3306-62-5**
25 ★	-0.14	3.5	547	1	C6H8N2O2S1	BENZENESULFONYL HYDRAZIDE **80-17-1**
6 ★	-0.38		2467	448	C6H8N2O2S1	M-AMINOBENZENESULFONAMIDE **98-18-0**
7 ★	0.40		541		C6H8N2O2S1	PHENYLSULFAMIDE **15959-53-2**
8	-0.89	7.5	652	1	C6H8N2O2S1	SULFANILAMIDE **63-74-1** antibacterial
9	-0.76	7.4	1737	1	"	SULFANILAMIDE
1630 ★	-0.62	5.5	2265		"	SULFANILAMIDE
1 ★	0.50	1.0	2202	342	C6H8N2O2S1	THIOBARBITURIC ACID,5,5-DIMETHYL
2 ★	-0.35	0.7	962		C6H8N2O3	BARBITURIC ACID,5-ETHYL **2518-72-1**
3 ★	-0.55		547		C6H8N2O4S2	1,3-BENZENEDISULFONAMIDE **3701-01-7**
4 ★	-0.96	2.1	2265		C6H8N2O4S2	1,4-BENZENEDISULFONAMIDE **16993-45-6**
35 ★	1.31		1358		C6H8N2O8	ISOSORBIDEDINITRATE **87-33-2** vasodilator (coronary)
6 ★	1.81		2499		C6H8N2S1	PYRAZINE,2-METHIO-3-METHYL
7 ★	1.85		1519	459	C6H8N2S2	3,6-DIMETHYLTHIOPYRIDAZINE
8 ★	-0.98	7.4	725	1	C6H8N4O1	2-AMINOISONIAZID **58481-01-9**
9 ★	0.08		871		C6H8N4O1	N-NITROSO-BIS(2-CYANOETHYL)AMINE
1640 ★	0.04	7.4	1022	1	C6H8N4O3	1-METHYL-2-NITRO-5-(CH=N(O)CH3)IMIDAZOLE
1	0.29	7.4	2561	1	C6H8N4O3	2-NITROIMIDAZOLE-1-(ACETALDOXIME-METHYL ETHER)
2 ★	-0.23	1.0	2196		C6H8N4O3	PYRAZOLE,3-METHYL-4-NITRO-5-(N-METHYLCARBOXAMIDE)
3	-0.74	7.4	2561	1	C6H8N4O4	2-NITROIMIDAZOLE-1-(N-HYDROXY-N-METHYLACETAMIDO)
4	-1.15	7.4	2561	1	C6H8N4O4	2-NITROIMIDAZOLE-1-(N-HYDROXYPROPIONAMIDO)
45	-1.08	7.4	2561	1	C6H8N4O4	2-NITROIMIDAZOLE-1-(N-METHOXYACETAMIDO)
6 ★	-0.37	7.4	2206	306	C6H8N4O4	5-NITROIMIDAZOLE,1-ME-2-(ME-CARBAMATE) **7681-76-7** antiprotozoal
7	0.81		1731		C6H8N4O5	1-(B-O2NO-ET)2-METHYL-5-NITROIMIDAZOLE
8 ★	0.98		1967		C6H8N4S1	THIOPHENE-2-CARBOXALDEHYDE,GUANYLHYDRAZONE **97183-52-3**
9 ★	0.76	6.0	2417	1	C6H8N6O2	IMIDAZOLE,1-AZIDOETHYL-2-METHYL-5-NITRO
1650 ★	0.65	7.4	2206	306	C6H8N6O2S1	CARBOXALDEHYDETHIOSEMICARBAZONE,1-METHYL-5-NITROIMIDAZOL-2YL **4994-21-2**
1 ★	0.58		513		C6H8O1	1-HEXYN-5-ONE
2 ★	0.61		1625		C6H8O1	2-CYCLOHEXENE-1-ONE **930-68-7**
3 ★	2.40		1478		C6H8O1	2-ETHYLFURAN **3208-16-0**
4 ★	2.24		2619		C6H8O1	FURANE,2,5-DIMETHYL **625-86-5**
55 ★	1.33		1135	537	C6H8O2	SORBIC ACID **110-44-1** antimicrobial agent
6 ★	0.04		1059		C6H8O3S1	5,6-DIHYDRO-2-METHYL-1,4-OXATHIIN-3-CARBOXYLIC ACID ✚
7 ★	0.74		579		C6H8O4	FUMARIC ACID,METHYL ESTER **624-49-7**
8 ★	0.22		579		C6H8O4	MALEIC ACID,METHYL ESTER **624-48-6**
9 ★	-1.64	3.0	594		C6H8O6	ASCORBIC ACID **50-81-7** vitamin (antiscorbutic)
1660 ★	-1.72		510		C6H8O7	CITRIC ACID **77-92-9** acidifier, flavoring agent
1 ★	2.87		1478		C6H8S1	2-ETHYLTHIOPHENE **872-55-9**
2 ★	2.82		1478		C6H8S1	3-ETHYLTHIOPHENE **52006-63-0**
3 ★	-0.04	7.4	2361	1	C6H9F1N4O3	TRIAZINE,3-NITRO-1-(3-METHOXY-2-FLUORO)PROPYL
4 ★	2.31		2488		C6H9F2N5S1	1,3,5-TRIAZINE,2-DIFLUOROMETHIO-4,6-BIS-METHYLAMINO
65 ★	1.47		2619		C6H9N1	PYRROLE,2,5-DIMETHYL **625-84-3**
6 ★	1.59		2619		C6H9N1	PYRROLE,2-ETHYL
7 ★	0.37		572		C6H9N1O1	N-VINYL-2-PYRROLIDINONE **88-12-0**
8 ★	-0.05	5.0	2425	1	C6H9N1O2	2-PYRROLIDONE,1-ACETYL **932-17-2**
9 ★	-0.70		213		C6H9N1O2	N-FORMYLCYCLOBUTANECARBOXAMIDE **23046-86-8**
1670 ★	-0.22		1059		C6H9N1O2S1	5,6-DIHYDRO-2-METHYL-1,4-OXATHIIN-3-CARBOXAMIDE ✚
1 ★	-0.35		579		C6H9N1O2S1	CITIOLONE **1195-16-0** treatment of hepatic disorder
2 ★	0.38	7.4	2490	1	C6H9N1O3S3	THIOPHENE-2-SULFONAMIDE,5-HYDROXYETHYL **104437-96-9**
3 ★	-0.54	7.4	2490	1	C6H9N1O5S3	THIOPHENE-2-SULFONAMIDE,5-(HYDROXYETHYL)SULFONYL **104438-00-8**
4 ★	-0.40		2073		C6H9N1O6	ISOSORBIDE-2-NITRATE **16106-20-0** vasodilator (coronary)
75 ★	-0.15		2073		C6H9N1O6	ISOSORBIDE-5-NITRATE **16051-77-7**
6 ★	0.39		1519		C6H9N3	2,6-DIMETHYL-4-PYRIMIDINAMINE **461-98-3**
7 ★	0.29		1519	400	C6H9N3	3-PYRIDAZINAMINE,N,N-DIMETHYL **17258-31-0**
8	-0.32		1375		C6H9N3	3-PYRIDAZINEAMINE,4,6-DIMETHYL
9 ★	-0.09		1519	459	C6H9N3	4-PYRIDAZINAMINE,N,N-DIMETHYL **17258-38-7**
1680 ★	0.93		2412		"	PYRAZINEAMINE,N,N-DIMETHYL **5214-29-9** stimulant (central)
1	0.96	7.4	579	1	"	PYRAZINEAMINE,N,N-DIMETHYL
2	-0.85	7.4	2362	1	C6H9N3	PYRIDINE,3-METHYLAMINO-4-AMINO
3 ★	1.07		2412		C6H9N3	PYRIMIDINE,2-DIMETHYLAMINO **5621-02-3**
4 ★	0.58		2412		C6H9N3	PYRIMIDINE,4-DIMETHYLAMINO **31401-45-3**
85 ★	0.46		2412		C6H9N3	PYRIMIDINE,5-DIMETHYLAMINO **31401-46-4**
6	-0.85		1375		C6H9N3O1	3-AMINO-4-METHOXY-6-METHYLPYRIDAZINE
7 ★	-3.56		1590	459	C6H9N3O2	HISTIDINE **71-00-1** amino acid
8	-1.95	7.0	1207	1	"	HISTIDINE
9 ★	0.62	7.4	2206	306	C6H9N3O2	IMIDAZOLE,2-ETHYL-1-METHYL-5-NITRO
1690 ★	0.76	7.4	2206	306	C6H9N3O2	IMIDAZOLE,2-ETHYL-4-METHYL-5-NITRO
1	-0.60	7.4	2206	306	C6H9N3O3	1(2-OH-ETHYL)-2-METHYL-4-NITROIMIDAZOLE
2 ★	-0.59		1008	354	"	1(2-OH-ETHYL)-2-METHYL-4-NITROIMIDAZOLE
3 ★	-0.16		1003		C6H9N3O3	1-(2-HYDROXYPROPYL)-2-NITROIMIDAZOLE
4 ★	0.12	7.4	1022	1	C6H9N3O3	1-(2-METHOXYMETHYL)-2-NITROIMIDAZOLE
95 ★	-0.09		2387		C6H9N3O3	IMIDAZOLE,1,2-DIMETHYL-4-NITRO-5-METHOXY **35687-44-6**
6 ★	0.18		2474		C6H9N3O3	IMIDAZOLE-1-ETHANOL,5-METHYL-2-NITRO **23571-38-2**
7	-0.16	7.4	2380	1	C6H9N3O3	METRONIDAZOLE **443-48-1** antiprotozoal (trichomonas)
8	-0.11	7.4	694	1	"	METRONIDAZOLE
9	-0.09	7.4	2206	306	"	METRONIDAZOLE
1700 ★	-0.02		1080		"	METRONIDAZOLE
1 ★	-0.08	7.4	830	1	C6H9N3O3	N-DIAZOACETYLGLYCINE,ETHYL ESTER **999-29-1**
2	-1.22		1467		C6H9N3O3S1	1-(2-METHYLSULFINYLETHYL)-2-NITROIMIDAZOLE
3 ★	-0.96		1080		C6H9N3O4	1-(2,3-DIHYDROXYPROPYL)-2-NITROIMIDAZOLE **13551-92-3** antineoplastic
4 ★	-0.89	7.4	1022	1	C6H9N3O4	1-METHYL-2-NITRO-5-(1,2-DIHYDROXYETHYL)IMIDAZOLE

	logP	pH	Ref.	Note	MF	Name / CAS / activity
5	-0.37		2279		C6H9N3O4S1	IMIDAZOLE,1-METHYL-4-NITRO-5-ETHYLSULFONATO **13755-80-1**
6 ★	-1.30	7.4	1022	1	C6H9N3O4S1	METHYL(2-(2-NITRO-1-IMIDAZOLYL)ETHYL)SULFONE
7 ★	0.02	1.4	536	342	C6H9N3O4S2	ANILINE,N-(SO2NHSO2NH2)
8	-2.05	6.0	2417	1	C6H9N3O5S1	IMIDAZOLE,1-ETHANESULFONIC ACID-2-METHYL-5-NITRO
9	-1.57	7.4	694	1	C6H9N3O5S2	IMIDAZOLE,2-METHYL-5-NITRO-1-(2-ETHYLTHIO)SULFONIC ACID, SODIUM SALT
1710 ✓	-1.01		1412		C6H9N5O2	1,2,4-TRIAZOLE,3,5-DIACETAMIDO
1 ★	-1.59	7.4	2273	1	C6H9N5O4	(3-NITRO-1,2,4-TRIAZOL-1-YL)ACETAMIDE,N-(2-HYDROXY)ETHYL **104958-85-2**
2 ★	2.87		594	400	C6H10	1,5-HEXADIENE **592-42-7**
3	2.73		1560	100	C6H10	1-HEXYNE **693-02-7**
4 ★	2.80		594	400	C6H10	CIS-2-TRANS-4-HEXADIENE **5194-50-3**
15 ★	2.86		547	400	C6H10	CYCLOHEXENE **110-83-8**
6 ★	3.16		594	400	C6H10	TRANS-1,3-HEXADIENE **592-48-3**
7 ★	3.01		594	400	C6H10	TRANS-2-TRANS-4-HEXADIENE **5194-51-4**
8	-1.82	7.4	401	1	C6H10Cl1N3O4	UREA,1-(2CLET)-1-NO-3-(2-PROPIONIC ACID)
9	1.84	7.4	401	2	"	UREA,1-(2CLET)-1-NO-3-(2-PROPIONIC ACID)
1720	1.51		2550	459	C6H10Cl1N5	DESETHYLATRAZINE
1 ★	3.18		1553		C6H10Cl2	1,2-DICHLOROCYCLOHEXANE-C
2 ★	3.21		1553		C6H10Cl2	1,2-DICHLOROCYCLOHEXANE-TRANS
3 ★	1.13		547		C6H10N2O1	4-I-PROPOXYPYRAZOLE
4 ★	1.42		547		C6H10N2O1	4-PROPOXYPYRAZOLE
25 ★	1.71		2166		C6H10N2O1	FURAZAN,METHYLPROPYL **77580-78-0**
6 ★	-0.66		1728		C6H10N2O2	N-NITROSOPIPERIDONE,3-METHYL
7 ★	0.49		2188		C6H10N2O2S2	METHYLCARBAMIC ACID,2-OXIMINO-1,4-DITHIANE ESTER
8 ★	0.91		2188		C6H10N2O2S2	METHYLCARBAMIC ACID,4-OXIMINO-5-METHYL-1,3-DITHIANE ESTER
9 ★	0.00		2188		C6H10N2O3S1	METHYLCARBAMIC ACID,3-OXIMINO-1,4-OXATHIANE ESTER
1730 ★	0.46		2188		C6H10N2O3S1	METHYLCARBAMIC ACID,4-OXIMINO-5-METHYL-1,3-OXATHIOLANE ESTER
1 ★	-2.60	7.1	1591		C6H10N2O4	ASPARTIC ACID-MONOAMIDE,N-ACETYL
2	3.25		509		C6H10N2O6S3	IMIDAZOLE,2,4,5-TRIMETHYLSULFONYL **59800-93-0**
3	-3.00	7.6	2003	314	C6H10N3O2.Cl1	IMIDAZOLIUM CHLORIDE,1-METHOXYMETHYL-2-HYDROXYIMINOMETHYL-3-METHYL **91900-09-3**
4 ★	0.14		953		C6H10N4	METRAZOLE **54-95-5** *cns stimulant, narcotic antagonist*
35 ★	1.22		57		C6H10N4O1S1	3-MERCAPTO-4-AMINO-6-I-PROPYL-1,2,4-TRIAZIN-5-ONE **22278-77-9**
6 ★	0.46		57		C6H10N4O1S1	3-METHIO-4-AMINO-6-ETHYL-1,2,4-TRIAZINE-5-ONE **21087-59-2**
7 ★	-0.62	7.4	830	1	C6H10N4O2	N-DIAZOACETYLGLYCINE-N'-ETHYLAMIDE
8 ★	-1.35	7.4	830	1	C6H10N4O3	N-DIAZOACETYLGLYCINE-N'-HYDROXYETHYLAMIDE
9 ★	-0.49	7.4	2273	1	C6H10N4O4	1,2,4-TRIAZOLE,3-NITRO-1-(2-HYDROXY-3-METHOXY)PROPYL **104958-86-3**
1740 ★	-0.36		2567		C6H10N4O4	4-NITRO-1,2,3-TRIAZOLE,1-(N-3-METHOXY-2-HYDROXY)PROPYL
1 ★	-0.07		2567		C6H10N4O4	4-NITRO-1,2,3-TRIAZOLE,2-N-(3-METHOXY-2-HYDROXY)PROPYL
2 ★	-0.12		563		C6H10N6O1	3-(3,3-DIMETHYL-1-TRIAZENO)PYRAZOLE-4-CARBOXAMIDE **21466-00-2**
3 ★	-0.24		525		C6H10N6O1	IMIDAZOLE,4-CARBOXAMIDE-5-(3,3-DIMETHYLTRIAZENYL) **4342-03-4** *antineoplastic*
4	1.26	7.4	1821	1	C6H10O1	1,2-EPOXYCYCLOHEXANE **286-16-8**
45 ★	1.02		502		C6H10O1	1-HEXEN-5-ONE **109-49-9**
6 ★	0.81		527		C6H10O1	CYCLOHEXANONE **108-94-1**
7 ★	-0.27		579		C6H10O2	2,5-HEXANEDIONE **110-13-4**
8	0.46		2596		C6H10O2	ALLYL GLYCIDYL ETHER **106-92-3**
9 ★	1.94		1402		C6H10O2	METHACRYLIC ACID,ETHYL ESTER **97-63-2**
1750 ★	-0.13		502		C6H10O3	4-KETOVALERIC ACID,METHYL ESTER **624-45-3**
1 ★	0.35		1402		C6H10O3	ACRYLIC ACID,2-HYDROXYPROPYL ESTER **999-61-1**
2 ★	0.24		579		C6H10O3	ETHYLACETOACETATE **141-97-9**
3 ★	0.47		1402		C6H10O3	METHACRYLIC ACID,2-HYDROXYETHYL ESTER **868-77-9**
4 ★	-1.29		401		C6H10O4	1,2-5,6-DIANHYDROGALACTITOL **23261-20-3**
55 ★	0.08		510		C6H10O4	ADIPIC ACID **124-04-9**
6 ★	0.35		594		C6H10O4	BUTANEDIOIC ACID,METHYL ESTER **106-65-0**
7 ★	0.56		592		C6H10O4	OXALIC ACID,DIETHYL ESTER **95-92-1**
8 ★	-2.57		1144		C6H10O7	GLUCURONIC ACID **6556-12-3**
9 ✓	-2.50		1966		C6H11Au1O5S1	GOLD THIOGLUCOSE **12192-57-3** *antirheumatic*
1760 ★	3.20		1683		C6H11Br1	BROMOCYCLOHEXANE **108-85-0**
1 ★	1.14		843		C6H11Br1N2O2	BROMISOVALUM **496-67-3** *sedative, hypnotic*
2 ★	1.00		843		C6H11Cl1N2O2	A-CHLORO-I-VALERYLUREA
3 ★	1.70		1942		C6H11F3O1	1,1,1-TRIFLUOROHEXANOL-3
4 ★	1.64		1942		C6H11F3O1	6,6,6-TRIFLUOROHEXANOL
65 ✓	1.87		567	820	C6H11N1	DIALLYLAMINE PICRATE
6 ★	1.11	13.0	588	224	C6H11N1	DIALLYLAMINE **124-02-7**
7 ★	1.66		1896		C6H11N1	HEXANENITRILE **628-73-9**
8 ★	1.54		1896		C6H11N1	I-HEXANENITRILE
9 ★	-2.58		401		C6H11N1O2	1-AMINOCYCLOPENTANECARBOXYLIC ACID **52-52-8** *antineoplastic*
1770 ★	-2.89		2654	448	C6H11N1O2	3-CARBOXYPIPERIDINE **498-95-3**
1	-2.66	7.0	1207	1	"	3-CARBOXYPIPERIDINE
2 ★	-3.05		2654	448	C6H11N1O2	4-CARBOXYPIPERIDINE
3 ★	-2.31		2654	448	C6H11N1O2	HOMOPROLINE
4 ★	1.09		504		C6H11N1O2	O-(1-ETHYL-ALLYL)CARBAMATE
75 ★	-0.57	2.2	2228	314	C6H11N1O3	ALANINE-N-ACETYL,METHYL ESTER **3619-02-1**
6 ★	0.75		561		C6H11N1S1	2-AZACYCLOHEPTANTHIONE **7203-96-5**
7 ★	2.42		1693		C6H11N2O4P1S3	METHIDATHION **950-37-8** *insecticide*
8 ★	0.65		1137		C6H11N3O1	2-AMINO-5-BUTYL-1,3,4-OXADIAZOLE **52838-38-7**
9 ★	0.40		1137		C6H11N3O1	2-AMINO-5-I-BUTYL-1,3,4-OXADIAZOLE
1780 ★	0.62		1137		C6H11N3O1	2-AMINO-5-S-BUTYL-1,3,4-OXADIAZOLE
1 ★	0.48		1137		C6H11N3O1	2-AMINO-5-T-BUTYL-1,3,4-OXADIAZOLE
2 ★	-2.41	7.1	1591		C6H11N3O3	ASPARAGIN-AMIDE,N-ACETYL
3 ★	0.36		536		C6H11N3O4	2-ME-2-NITROPROPIONALDEHYDE-N-ME-CARBAMOYLOXIME **6129-11-9**
4 ★	-2.68		547	463	C6H11N3O4	GLY-GLY-GLY **556-33-2**
85	0.16		723		C6H11N5O1	2,4-DIAMINO-6-DIMETHYLAMINO-PYRIMIDINE-3-OXIDE
6	-0.04		723		C6H11N5O1	2,4-DIAMINO-6-ETHYLAMINO-PYRIMIDINE-3-OXIDE
7 ★	3.39		1560	100	C6H12	1-HEXENE **592-41-6**
8 ★	3.44		547	60	C6H12	CYCLOHEXANE **110-82-7**
9 ★	3.37		579		C6H12	METHYLCYCLOPENTANE **96-37-7**

	logP	pH	Ref.	Note	MF	Name / CAS / activity
1790 ★	-0.29		401		C6H12Br2O4	1,6-DIBROMO-1,6-DIDEOXYGALACTITOL **10318-26-0** *antineoplastic*
1 ★	-0.24		401		C6H12Br2O4	NSC94100 **488-41-5**
2 ✓	1.53		1701	820	C6H12F3N1	N,N-DIETHYL-2,2,2-TRIFLUOROETHYLAMINE PICRATE
3 ★	1.88		1701		C6H12F3N1	N-ETHYL-2,2,2-TRIFLUOROETHYLAMINE **37174-09-7**
4 ✓	1.78		1701	820	C6H12F3N1	N-PROPYL-N-METHYL-2,2,2-TRIFLUOROETHYLAMINE PICRATE
95 ★	2.18		1701		C6H12F3N1	N-PROPYL-N-METHYL-2,2,2-TRIFLUOROETHYLAMINE
6 ★	1.48		2506		C6H12N1O4P1S2	FORMOTHION **2540-82-1** *insecticide*
7	-1.93		532	820	C6H12N2O1	1-(2-HYDROXYETHYL)-2-METHYLIMIDAZOLINEHCL
8 ★	0.04		532		C6H12N2O1	1-(2-HYDROXYETHYL)-2-METHYLIMIDAZOLINE **695-94-3**
9 ★	0.86		871		C6H12N2O1	2,5-DIMETHYL-N-NITROSOPYRROLIDINE **55556-86-0**
1800 ★	0.71		871		C6H12N2O1	2-METHYL-N-NITROSOPIPERIDINE **7247-89-4**
1 ★	0.99		871		C6H12N2O1	3-METHYL-N-NITROSOPIPERIDINE **13603-07-1**
2 ★	1.05		871		C6H12N2O1	4-METHYL-N-NITROSOPIPERIDINE **15104-03-7**
3 ★	0.92		871		C6H12N2O1	N-NITROSO-HEXAMETHYLENEIMINE **932-83-2**
4 ★	0.32		871		C6H12N2O2	2,6-DIMETHYL-N-NITROSOMORPHOLINE **1456-28-6**
5 ★	-1.21		1558		C6H12N2O2	ALANINE,N-ACETYL,N'-METHYLAMINO-AMIDE
6 ★	-0.87	7.0	2309	1	C6H12N2O2	GLYCINAMIDE,2,2-DIMETHYL-N-ACETYL
7 ★	0.45		673		C6H12N2O2	I-VALERYLUREA **2274-08-0**
8 ★	-1.57	7.1	1591		C6H12N2O3	THREONIN-AMIDE, N-ACETYL
9 ★	0.73		594	400	C6H12N2S1	1-PIPERIDINETHIOCARBOXAMIDE **14294-09-8**
1810 ★	-0.62		2083		C6H12N3O1P1	PHOSPHINEOXIDE,TRIS-(1-AZIRIDINYL) **545-55-1** *antineoplastic*
1 ★	0.53		401		C6H12N3P1S1	THIOTEPA **52-24-4** *antineoplastic*
2 ★	0.42		1705		C6H12N4	GUANIDINE,N1-PROPYL-N2-CYANO-N3-METHYL
3	1.29	3.5	2391	1	C6H12N4	TETRAZOLE,5-I-AMYL
4 ★	0.15		871		C6H12N4O2	2,5-DIMETHYL-N,N'-DINITROSOPIPERAZINE **55556-88-2**
15 ★	0.08		871		C6H12N4O2	2,6-DIMETHYL-N,N'-DINITROSOPIPERAZINE **55380-34-2**
6 ★	1.93		2219		C6H12O1	1,2-EPOXYHEXANE
7 ★	1.38		502		C6H12O1	2-HEXANONE **591-78-6**
8 ★	1.31		2047		C6H12O1	4-METHYL-2-PENTANONE **108-10-1** *alcohol denaturant*
9 ★	1.23		503		C6H12O1	CYCLOHEXANOL **108-93-0**
1820 ★	1.78	7.0	566	1	C6H12O1	HEXALDEHYDE **66-25-1**
1 ★	1.20		579	400	C6H12O1	PINACOLONE **75-97-8**
2 ★	1.22		1952		C6H12O1	TETRAHYDROFURAN,2,5-DIMETHYL (CIS) **2144-41-4**
3 ★	1.34		1952		C6H12O1	TETRAHYDROFURAN,2,5-DIMETHYL (TRANS) **2390-94-5**
4	1.90	3.5	2391	1	C6H12O2	2,2-DIMETHYLBUTYRIC ACID **595-37-9**
25 ★	1.68		1815		C6H12O2	2-ETHYL-BUTANOIC ACID **88-09-5**
6 ★	1.80		1815		C6H12O2	2-METHYL-PENTANOIC ACID **97-61-0**
7 ★	1.78		594	400	C6H12O2	ACETIC ACID, BUTYL ESTER **123-86-4**
8 ★	1.72		594	400	C6H12O2	ACETIC ACID, S-BUTYL ESTER **105-46-4**
9 ★	1.76		594	400	C6H12O2	ACETIC ACID,T-BUTYL ESTER **540-88-5**
1830 ★	0.23		508		C6H12O2	CYCLOHEXANEDIOL-CIS **1792-81-0**
1 ★	0.08		508		C6H12O2	CYCLOHEXANEDIOL-TRANS **1460-57-7**
2	1.98		1533	820	C6H12O2	HEXANOIC ACID, HEXYLAMINE SALT
3 ★	1.92		505		C6H12O2	HEXANOIC ACID **142-62-1**
4 ★	1.78		594	400	C6H12O2	I-BUTYLACETATE **106-63-8**
35	0.80		2596		C6H12O2	I-PROPYL GLYCIDYL ETHER **4016-14-2**
6 ★	1.82		594	400	C6H12O2	ISOVALERIC ACID,METHYL ESTER
7	0.59		2596		C6H12O2	PROPYL GLYCIDYL ETHER **3126-95-2**
8 ✓	-2.17		508		C6H12O2	SODIUM HEXANOATE
9 ★	1.83		594	400	C6H12O2	TRIMETHYLACETIC ACID,METHYL ESTER **598-98-1**
1840 ★	1.96		594	400	C6H12O2	VALERIC ACID,METHYL ESTER **624-24-8**
1	0.43		1266		C6H12O3	ETHYL-4-HYDROXYBUTYRATE **10191-10-0**
2 ★	0.67	7.5	1510	1	C6H12O3	PARALDEHYDE **123-63-7** *hypnotic, sedative*
3 ★	3.80		1560	100	C6H13Br1	1-BROMOHEXANE **111-25-1**
4 ★	1.33	13.0	588	224	C6H13N1	ALLYLPROPYLAMINE **5666-21-7**
45 ✓	2.47		567	820	C6H13N1	CYCLOHEXYLAMINE PICRATE
6 ★	1.49	13.0	588	224	C6H13N1	CYCLOHEXYLAMINE **108-91-8**
7 ★	1.30	11.0	1443	322	C6H13N1	N-METHYLPIPERIDINE **626-67-5**
8 ★	0.34		579		C6H13N1O1	DIETHYLACETAMIDE **685-91-6**
9 ★	2.70		2571		C6H13N1O2	1-NITROHEXANE **646-14-0**
1850 ★	-2.95		2018		C6H13N1O2	6-AMINOHEXANOIC ACID **60-32-2** *hemostatic*
1	-1.60	7.2	2323	468	C6H13N1O2	A-AMINOCAPROIC ACID **327-57-1**
2 ★	-1.54		2654	448	"	A-AMINOCAPROIC ACID
3 ★	1.38		1969		C6H13N1O2	DIMETHYLCARBAMATE,I-PROPYL ESTER **38580-89-1**
4 ★	-1.72	7.0	2197		C6H13N1O2	I-LEUCINE **73-32-5** *amino acid*
55	-1.61	7.0	2197		C6H13N1O2	LEUCINE **61-90-5** *amino acid*
6 ★	-1.52		405		"	LEUCINE
7 ★	1.35	7.4	172	1	C6H13N1O2	O-PENTYLCARBAMATE **638-42-6**
8 ★	0.94	7.4	172	1	C6H13N1O2	O-T-PENTYLCARBAMATE
9 ★	-1.77		1233		C6H13N1O2	T-BUTYLGLYCINE **33105-81-6**
1860 ★	0.73		871		C6H13N3O1	N-NITROSO-3,5-DIMETHYLPIPERAZINE
1 ★	1.11		871		C6H13N3O2	1-NITROSO-3,3-DIETHYL-1-METHYLUREA **50285-72-8**
2	-3.19		1590	459	C6H13N3O3	CITRULLINE **372-75-8** *anti-asthenia*
3 ★	3.42		579	400	C6H14	2,3-DIMETHYLBUTANE **79-29-8**
4 ★	3.90		1389		C6H14	HEXANE **110-54-3**
65	3.82		547	400	C6H14	NEOHEXANE **75-83-2**
6	-3.83		1257	820	C6H14N1O2.I1	FORMYLCHOLINE IODIDE
7 ✓	-3.61		1704	820	C6H14N1O2.C2H3.Br1	(DEUTEROACETYLOXYME)TETRAMETHYLAMMONIUMBROMIDE
8 ✓	0.63		567	820	C6H14N1.C6H2N3O7	N,N-DIMETHYLPYRROLIDINIUM PICRATE
9 ★	0.08	13.0	594		C6H14N2	1-PYRROLIDINETHANEAMINE **7154-73-6**
1870 ★	-0.40	13.0	594		C6H14N2	N-METHYLHOMOPIPERAZINE **4318-37-0**
1 ★	-0.40	13.0	594	400	C6H14N2	PIPERAZINE,1,4-DIMETHYL **106-58-1**
2 ★	1.38		2570		C6H14N2O1	DI-I-PROPYLNITROSAMINE **601-77-4**
3 ★	1.36		871		C6H14N2O1	DIPROPYLNITROSAMINE **621-64-7**
4	1.57		2570		C6H14N2O1	ETHYLBUTYLNITROSAMINE **4549-44-4**

	logP	pH	Ref.	Note	MF	Name / CAS / activity
75 ★	0.48		1137		C6H14N2O1	HEXANOIC ACID HYDRAZIDE 2443-62-1
6 ★	1.34		871		C6H14N2O1	N-NITROSO-METHYL-NEOPENTYLAMINE
7 ★	-3.05	7.0	1207	1	C6H14N2O2	LYSINE 56-87-1 amino acid
8 ★	-0.38		871		C6H14N2O3	N-NITROSO-BIS(2-METHOXYETHYL)AMINE
9 ★	-4.20		1590	459	C6H14N4O2	ARGININE 74-79-3 ammonia detoxicant
1880	-4.08	7.0	1207	1	"	ARGININE
1 ★	1.76		2126		C6H14O1	2-HEXANOL 626-93-7
2 ★	1.48		425		C6H14O1	3,3-DIMETHYL-2-BUTANOL 464-07-3
3 ★	1.65		2126		C6H14O1	3-HEXANOL 623-37-0
4 ★	2.03		503		C6H14O1	BUTYL-ETHYLETHER 628-81-9
85 ★	1.52		1952		C6H14O1	DI-I-PROPYLETHER 108-20-3
6 ★	2.03		505		C6H14O1	DIPROPYLETHER 111-43-3
7 ★	2.03		508		C6H14O1	HEXANOL 111-27-3
8 ★	0.66		1819		C6H14O2	1,2-DIETHOXYETHANE 629-14-1
9 ★	0.83		768		C6H14O2	BUTOXYETHANOL 111-76-2
1890 ★	0.84		505		C6H14O2	DIETHYLACETAL 105-57-7
1 ★	0.75		2562		C6H14O2	ETHYLENE GLYCOL,MONOISOBUTYL ETHER 4439-24-1
2 ★	0.19		2562		"	ETHYLENE GLYCOL,MONOISOBUTYL ETHER 7580-85-0
3	0.39		1464		C6H14O2S1	PROPYLSULFONE 598-03-8
4 ★	-0.54		1819		C6H14O3	DIETHYLENEGLYCOLMONOETHYLETHER 111-90-0
95 ★	-0.36		1819		C6H14O3	DIETHYLETHER,1,5-DIMETHOXY 111-96-6
6 ★	-0.67		579		C6H14O3	DIPROPYLENEGLYCOL 110-98-5
7 ★	-3.10		1010		C6H14O6	GALACTITOL 608-66-2
8 ★	-3.10	7.4	2306	1	C6H14O6	MANNITOL 69-65-8 diagnostic aid
9 ★	-0.52		401		C6H14O6S2	MYLERAN 55-98-1 antineoplastic
1900 ★	2.84		594	400	C6H14S1	DI-I-PROPYL SULFIDE 625-80-9
1 ★	0.75		1966		C6H15Au1Cl1P1	CHLORO(TRIETHYLPHOSPHINE)GOLD
2	-1.74		2019		C6H15Cl1Sn1	STANNANE,CHLOROTRIETHYL 994-31-0
3	0.57		1959		"	STANNANE,CHLOROTRIETHYL
4 ✓	1.74		567	820	C6H15N1	DI-I-PROPYLAMINE PICRATE
5 ★	1.40	13.0	599		C6H15N1	DI-I-PROPYLAMINE 108-18-9
6 ★	1.70		505		C6H15N1	DIMETHYLBUTYLAMINE 927-62-8
7 ✓	2.18		567	820	C6H15N1	DIPROPYLAMINE, PICRATE
8 ★	1.67	13.0	588	224	C6H15N1	DIPROPYLAMINE 142-84-7
9	-0.42	7.0	834	1	C6H15N1	HEXYLAMINE 111-26-2
1910 ★	2.06	13.0	588	224	"	HEXYLAMINE
1 ✓	1.44		567	820	C6H15N1	S-BUTYLDIMETHYLAMINE PICRATE
2 ✓	1.25		567	820	C6H15N1	TRIETHYLAMINE PICRATE
3 ★	1.45	13.0	588	224	C6H15N1	TRIETHYLAMINE 121-44-8
4 ★	0.41	13.0	594	400	C6H15N1O1	ETHANOL,2-(T-BUTYLAMINO)
15 ★	-0.82		510		C6H15N1O2	DI-I-PROPANOLAMINE 110-97-4 alkalizing agent
6 ★	-1.00	9.0	594	307	C6H15N1O3	TROLAMINE 102-71-6 analgesic
7 ✓	2.18		567	820	C6H15N1.C6H3N3O7	DIPROPYLAMINE PICRATE
8 ✓	1.54		567	820	C6H15N1.C6H3N3O7	I-BUTYLDIMETHYLAMINE PICRATE
9 ✓	1.54		567	820	"	I-BUTYLDIMETHYLAMINE PICRATE
1920 ✓	-3.90		1704	820	C6H15N2O2.Br1	CARBAMIC ACID,O-ETHYLENE(TRIME AMMONIUM)BR ✚
1 ★	-0.04		1394		C6H15N3O1	4-PENTYLSEMICARBAZIDE 50405-18-0
2 ★	-1.20	7.3	761	314	C6H15N5	BUTYLBIGUANIDE 692-13-7 antidiabetic
3 ★	0.45		579		C6H15O3P1	DI-I-PROPYLPHOSPHITE 1809-20-7
4	0.66		2182	459	C6H15O3P1	ETHYLPHOSPHONIC ACID,DIETHYL ESTER 78-38-6
25 ★	1.02		2344		C6H15O3P1S2	DEMETON-S-METHYL 919-86-8
6 ★	0.80		547		C6H15O4P1	TRIETHYLPHOSPHATE 78-40-0
7 ★	3.57		512		C6H15Si1	SILANE,BUTYL-DIMETHYL 1001-52-1
8	-2.62		1257	820	C6H16N1O1.I1	METHOXYETHYLTRIMETHYL AMMONIUM IODIDE
9 ★	0.95		600		C6H16N1O2P1S1	O,S-DI-ME-N-S-BU-PHOSPHORAMIDOTHIOATE 52067-50-2
1930 ★	0.94		600		C6H16N1O2P1S1	O,S-DIME-N-BU-PHOSPHORAMIDOTHIOATE 52067-49-9
1 ★	0.97		600		C6H16N1O2P1S1	O,S-DIME-N-I-BU-PHOSPHORAMIDOTHIOATE 52067-51-3
2 ★	0.96		600		C6H16N1O2P1S1	O,S-DIME-N-T-BU-PHOSPHORAMIDOTHIOATE 52067-44-4
3 ★	1.23		600		C6H16N1O2P1S1	O,S-DIPROPYLPHOSPHORAMIDOTHIOATE 28167-45-5
4 ✓	0.60		567	820	C6H16N1.C6H3N3O7	I-PROPYLTRIMETHYL AMMONIUM PICRATE
35 ✓	0.82		567	820	C6H16N1.C6H3N3O7	PROPYLTRIMETHYL AMMONIUM PICRATE
6 ★	0.21	13.0	594		C6H16N2	1,2-ETHANDIAMINE,N,N'-DIETHYL 111-74-0
7 ★	0.21	13.0	594		C6H16N2	1,2-ETHANDIAMINE,N,N-DIETHYL 100-36-7
8 ★	0.30		508		C6H16N2	ETHYLENEDIAMINE,N,N,N',N'-TETRAMETHYL 110-18-9
9 ★	0.02		1908		C6H16N3O1P1	1,3,2-DIAZAPHOSPHOLIDEN-2-AM,N,N,1,3-TETRAME-2-OXIDE 7778-06-5 ✚
1940 ★	0.61		594	400	C6H16O2Si1	DIMETHYLDIETHOXYSILANE 78-62-6
1 ★	3.84		512		C6H16Si1	SILANE,PROPYL-TRIMETHYL 3510-70-1
2 ★	-0.92	13.0	594		C6H17N3	1,1-DIMETHYLDIETHYLENETRIAMINE 34066-96-1
3 ✓	-0.32		2426	537	C6H18Cl2N2Pt1	PLATINUM,BIS-ISOPROPYLAMMONIO-DICHLORO
4 ★	0.28		508		C6H18N3O1P1	HEXAMETHYLPHOSPHORICTRIAMIDE 680-31-9
45	4.20		1740	459	C6H18O1Si2	HEXAMETHYLDISILOXANE 107-46-0
6 ★	4.53		509		C7H1Cl5N2	BENZIMIDAZOLE,2,4,5,6,7-PENTACHLORO 7682-34-0
7 ★	2.50		2410		C7H2Cl1F3N2O4	1,3-DINITROBENZENE,2-CHLORO-5-TRIFLUOROMETHYL
8	3.21	7.3	509	2	C7H2Cl2F3N3	4-PYRIDINEIMIDAZOLE,2-TRIFLOROME-6,7-DICL 19918-41-3
9	2.46	7.3	509	2	C7H2Cl1F3N3	4-PYRIDINEIMIDAZOLE,2-TRIFLUOROME-6-CL 13577-71-4
1950 ★	2.50		1586		C7H3Cl1N2O6	BENZOIC ACID,4-CHLORO-3,5-DINITRO 118-97-8
1 ★	1.29		2499		C7H3Cl1N4	PYRAZINE,2,3-DICYANO-5-METHYL-6-CHLORO
2 ★	2.74		2159		C7H3Cl2N1	BENZONITRILE,2,6-DICHLORO 1194-65-6 herbicide
3	1.11		1263		C7H3F5O2S1	PENTAFLUOROPHENYLMETHYLSULFONE 651-85-4
4 ★	3.35		591		C7H3F12N3	MIDAFLUR 23757-42-8 sedative
55	0.90	6.5	2220	1	C7H3I2N1O1	4-HYDROXY-3,5-DIIODOBENZONITRILE 1689-83-4 herbicide
6	2.00	5.5	2220	1	"	4-HYDROXY-3,5-DIIODOBENZONITRILE
7 ★	0.24		594		C7H3N3	3,4-PYRIDINEDICARBONITRILE 1633-40-7
8	-2.00	7.4	1966	1	C7H4Au1Cl2N1O4	4-CARBOXY-1,1-DICHLORO-1,3,2-OXAZAURO(3,4)PYRIDIN-3-ONE
9	-1.51	7.4	1966	1	C7H4Au1Cl2N1O4	6-CARBOXY-1,1-DICHLORO-1,3,2-OXAZAURO(3,4)PYRIDIN-3-ONE

	logP	pH	Ref.	Note	MF	Name / CAS / activity
1960 ★	4.12		1947		C7H4Br1N1S1	BROMOBENZENE,3-ISOTHIOCYANATO 2131-59-1
1 ★	4.03		850		C7H4Br1N1S1	P-BROMOPHENYLISOTHIOCYANATE 1985-12-2
2 ★	2.56		2097		C7H4Br1N3O1	DIAZINECARBONITRILE,(3-BROMOPHENYL),OXIDE 62825-09-6
3 ★	2.50		2097		C7H4Br1N3O1	DIAZINECARBONITRILE,(4-BROMOPHENYL),OXIDE 62825-10-9
4 ★	3.58		2645	225	C7H4Br2O2	BENZOIC ACID,2,6-DIBROMO
65 ★	2.11		2645	225	C7H4Cl1F1O2	BENZOIC ACID,2-CHLORO-6-FLUORO
6 ★	3.53		594		C7H4Cl1F3	2-CHLOROBENZOTRIFLUORIDE 88-16-4
7 ★	2.03		1586		C7H4Cl1N1O4	BENZOIC ACID,2-CHLORO-4-NITRO 99-60-5
8 ★	2.03		1586		C7H4Cl1N1O4	BENZOIC ACID,2-CHLORO-5-NITRO 2516-96-3
9 ★	1.56		2645	225	C7H4Cl1N1O4	BENZOIC ACID,2-CHLORO-6-NITRO
1970 ★	2.08		1586		C7H4Cl1N1O4	BENZOIC ACID,2-NITRO-4-CHLORO 6280-88-2
1 ★	2.13		1586		C7H4Cl1N1O4	BENZOIC ACID,2-NITRO-5-CHLORO 2516-95-2
2	3.36		1947		C7H4Cl1N1S1	CHLOROBENZENE,3-ISOTHIOCYANATO 2392-68-9
3 ★	3.91		1947		C7H4Cl1N1S1	CHLOROBENZENE,4-ISOTHIOCYANATO
4 ★	2.40		2097		C7H4Cl1N3O1	DIAZINECARBONITRILE,(3-CHLOROPHENYL),OXIDE 62825-07-4
75 ★	2.33		2097		C7H4Cl1N3O1	DIAZINECARBONITRILE,(4-CHLOROPHENYL),OXIDE 54797-22-7
6 ✓	-1.30	7.4	2360	1	C7H4Cl1N3O3	BENZOTRIAZINE,1,4-DI-N-OXIDE-3-HYDROXY-7-CHLORO
7 ★	3.88		607	606	C7H4Cl2F3N1O2S1	TRIFLUOROMETHANESULFONANILIDE,2,4-DI-CHLORO 23383-96-2
8	2.30	7.7	1986	1	C7H4Cl2N2O2S2	BENZOTHIAZOLE,2-SULFONAMIDE-4,6-DICHLORO
9 ★	2.60	7.7	1986	2	"	BENZOTHIAZOLE,2-SULFONAMIDE-4,6-DICHLORO
1980 ★	2.82	2.0	2522	342	C7H4Cl2O2	BENZOIC ACID,2,5-DICHLORO 50-79-3
1 ★	2.23	1.0	547		C7H4Cl2O2	BENZOIC ACID,2,6-DICHLORO 50-30-6
2 ★	3.46		1920		C7H4Cl2O2	BENZOIC ACID,3,4-DICHLORO 51-44-5
3	3.00		1586		C7H4Cl2O2	BENZOIC ACID,3,5-DICHLORO 51-36-5
4	4.42		1822	459	C7H4Cl4O2	TETRACHLOROGUAIACOL 2539-17-5
85	4.76		1738	459	"	TETRACHLOROGUAIACOL
6 ★	1.47		594		C7H4F1N1	BENZONITRILE,M-FLUORO 403-54-3
7 ★	1.85		2097		C7H4F1N3O1	DIAZINECARBONITRILE,(3-FLUOROPHENYL),OXIDE +
8 ★	1.80		2097		C7H4F1N3O1	DIAZINECARBONITRILE,(4-FLUOROPHENYL,OXIDE
9 ★	1.59		2645	225	C7H4F2O2	BENZOIC ACID,2,6-DIFLUORO
1990 ★	2.62	7.0	925	1	C7H4F3N1O2	M-TRIFLUOROMETHYLNITROBENZENE 98-46-4
1 ★	2.55		2097		C7H4F3N1O2	NITROBENZENE,4-TRIFLUOROMETHYL 402-54-0
2 ★	2.58		547		C7H4F3N1O2	O-NITRO-A,A,A-TRIFLUOROTOLUENE 384-22-5
3 ★	2.34	1.5	2329	314	C7H4F3N1O3	PHENOL,2-NITRO-4-TRIFLUOROMETHYL 400-99-7
4 ★	1.38	7.3	509	2	C7H4F3N3	4-PYRIDINEIMIDAZOLE,2-TRIFLUOROMETHYL 13797-63-2
95 ★	0.94	7.3	509	2	C7H4F3N3	5-PYRIDINEIMIDAZOLE,2-TRIFLUOROMETHYL 19918-36-6
6	2.29		1283		C7H4F3N3O4	2,6-DINITRO-A,A,A-TRIFLUORO-P-TOLUIDINE 445-66-9
7 ★	3.06		607	606	C7H4F5N1O2S1	TRIFLUOROMETHANESULFONANILIDE,2,4-DI-FLUORO 23384-22-7
8 ★	4.22		1947		C7H4I1S1	IODOBENZENE,4-ISOTHIOCYANATO 2059-76-9
9 ★	4.56		543		C7H4I2O3	3,5-DIIODOSALICYLIC ACID 133-91-5 growth promotant (animal), iodine source
2000 ★	1.17		501		C7H4N2O2	BENZENE,3-CYANO-1-NITRO 619-24-9
1 ★	1.19		501		C7H4N2O2	BENZENE,4-CYANO-1-NITRO 619-72-7
2 ★	1.02		2410		C7H4N2O2	NITROBENZENE,2-CYANO
3 ★	3.62		850		C7H4N2O2S1	4-NITROPHENYLISOTHIOCYANATE 2131-61-5
4 ★	1.46	5.0	1625	1	C7H4N2O3	BENZISOXAZOLE,5-NITRO 39835-28-4
5	1.55		1586		C7H4N2O6	BENZOIC ACID,3,5-DINITRO 99-34-3
6	1.71		1586		C7H4N2O7	SALICYLIC ACID,3,5-DINITRO 609-99-4
7 ★	1.45		2097		C7H4N4O3	DIAZINECARBONITRILE,(3-NITROPHENYL),OXIDE 62825-12-1
8 ★	1.51		2097		C7H4N4O3	DIAZINECARBONITRILE,(4-NITROPHENYL),OXIDE 60142-50-9
9 ★	1.60		260		C7H4N4O4	5,7-DINITROBENZPYRAZOLE 31208-76-1
2010 ★	1.00		2410		C7H5Br1Cl1N1O1	BENZAMIDE,2-CHLORO-6-BROMO
1 ★	0.63		2410		C7H5Br1F1O1	BENZAMIDE,2-FLUORO-6-BROMO
2 ★	3.82		607	606	C7H5Br1F3N1O2S1	TRIFLUOROMETHANESULFONANILIDE,M-BR 23384-08-9
3 ★	0.28		2410		C7H5Br1N2O3	BENZAMIDE,2-BROMO-6-NITRO
4 ★	1.21		2452		C7H5Br1O2	BENZOQUINONE,2-METHYL-6-BROMO 6293-55-6
15 ★	2.87		501		C7H5Br1O2	M-BROMOBENZOIC ACID 585-76-2
6 ★	2.20		1185		C7H5Br1O2	O-BROMOBENZOIC ACID 88-65-3
7 ★	2.86		501		C7H5Br1O2	P-BROMOBENZOIC ACID 586-76-5
8 ★	3.23		1185		C7H5Br1O3	5-BROMOSALICYLIC ACID 89-55-4
9 ★	1.00		2410		C7H5Br2N1O1	BENZAMIDE,2,6-DIBROMO
2020 ★	0.49		2410		C7H5Cl1F1O1	BENZAMIDE,2-FLUORO-6-CHLORO
1 ★	3.06		594		C7H5Cl1F3N1	BENZOTRIFLUORIDE,3-AMINO-4-CHLORO 121-50-6
2 ★	4.00		607	606	C7H5Cl1F3N1O2S1	TRIFLUOROMETHANESULFONANILIDE,M-CL 23384-03-4
3 ★	3.96		607	606	C7H5Cl1F3N1O2S1	TRIFLUOROMETHANESULFONANILIDE,P-CL 23384-04-5
4	-1.54	7.4	2071		C7H5Cl1Hg1O2	P-CHLOROMERCURIBENZOIC ACID
25 ★	2.46		505		C7H5Cl1N2O1	BENZOXAZOLE,2-AMINO-5-CHLORO 61-80-3 muscle relaxant (skeletal)
6 ★	1.01		522		C7H5Cl1N2O2S1	1,2,4-BENZOTHIADIAZINE-1,1-DIOXIDE-6-CHLORO 19477-31-7
7	1.02	7.4	601	1	"	1,2,4-BENZOTHIADIAZINE-1,1-DIOXIDE-6-CHLORO
8	1.16	7.4	601	2	"	1,2,4-BENZOTHIADIAZINE-1,1-DIOXIDE-6-CHLORO
9 ★	1.07		522		C7H5Cl1N2O2S1	1,2,4-BENZOTHIADIAZINE-1,1-DIOXIDE-7-CHLORO
2030	1.11	7.4	601	1	"	1,2,4-BENZOTHIADIAZINE-1,1-DIOXIDE-7-CHLORO
1	1.22	7.4	601	2	"	1,2,4-BENZOTHIADIAZINE-1,1-DIOXIDE-7-CHLORO
2 ★	2.10	7.2	1685	314	C7H5Cl1N2O2S2	6-CHLOROBENZOTHIAZOLE-2-SULFONAMIDE 88946-20-7
3 ★	0.12		2410		C7H5Cl1N2O3	BENZAMIDE,2-CHLORO-6-NITRO
4 ★	1.48		541		C7H5Cl1N4	1-PHENYL-5-CHLOROTETRAZOLE 14210-25-4
35 ★	0.41	7.4	2360	1	C7H5Cl1N4O2	BENZOTRIAZINE-1,4-DI-N-OXIDE-3-AMINO-6-CHLORO
6 ★	2.33		547		C7H5Cl1O1	BENZALDEHYDE,O-CHLORO 89-98-5
7 ★	2.10	7.0	925	1	C7H5Cl1O1	BENZALDEHYDE,P-CHLORO 104-88-1
8 ★	2.68		501		C7H5Cl1O2	M-CHLOROBENZOIC ACID 535-80-8
9 ★	2.05		1185		C7H5Cl1O2	O-CHLOROBENZOIC ACID 118-91-2
2040 ★	2.65		501		C7H5Cl1O2	P-CHLOROBENZOIC ACID 74-11-3
1 ★	3.09		1185		C7H5Cl1O3	5-CHLOROSALICYLIC ACID 321-14-2
2	2.41		1586		C7H5Cl1O3	BENZOIC ACID,4-HYDROXY-3-CHLORO 3964-58-7
3 ★	2.54		541		C7H5Cl2N1	PHENYLISOCYANIDE,DICHLORIDE 622-44-6
4 ★	0.77		2410		C7H5Cl2N1O1	2,6-DICHLOROBENZAMIDE 2008-58-4

	logP	pH	Ref.	Note	MF	Name / CAS / activity
45	1.82		1586		C7H5Cl2N1O2	BENZOIC ACID,2-AMINO-3,5-DICHLORO **2789-92-6**
6 ★	3.82		1143	459	C7H5Cl3	2,4-DICHLOROBENZYLCHLORIDE **94-99-5**
7	2.92		536		C7H5Cl3	A,A,A-TRICHLOROTOLUENE **98-07-7**
8	2.30		2661		C7H5Cl3N2O2	PICLORAM,METHYL ESTER
9 ★	4.04		2159		C7H5Cl3O1	ANISOLE,2,4,6-TRICHLORO **87-40-1**
2050 ★	3.77		1554		C7H5Cl3O2	3,4,5-TRICHLORO-2-METHOXYPHENOL **57057-83-7**
1	3.72		1822	459	C7H5Cl3O2	4,5,6-TRICHLOROGUAIACOL **2668-24-8**
2	3.92		1738	459	"	4,5,6-TRICHLOROGUAIACOL
3 ★	3.78		536		C7H5Cl3S1	TRICHLOROMETHYLTHIOBENZENE **701-65-5**
4 ★	2.15		501		C7H5F1O2	M-FLUOROBENZOIC ACID **455-38-9**
55 ★	1.77		1185		C7H5F1O2	O-FLUOROBENZOIC ACID **445-29-4**
6 ★	2.07		501		C7H5F1O2	P-FLUOROBENZOIC ACID **456-22-4**
7 ★	0.25		2198		C7H5F2N1O1	BENZAMIDE,2,6-DIFLUORO **18063-03-1**
8 ★	3.01		536		C7H5F3	BENZENE,TRIFLUOROMETHYL **98-08-8**
9 ★	4.39		607	606	C7H5F3I1O2S1	TRIFLUOROMETHANESULFONANILIDE,P-IODO **23384-10-3**
2060 ★	1.73	2.0	264		C7H5F3N2O4S1	3-TRIFLUOROMETHYL-4-NITROBENZENESULFONAMIDE **21988-05-6**
1 ★	3.17		508		C7H5F3O1	BENZENE,TRIFLUOROMETHOXY **456-55-3**
2 ★	2.95		501		C7H5F3O1	M-TRIFLUOROMETHYLPHENOL **98-17-9**
3 ★	2.80		300		C7H5F3O1	O-TRIFLUOROMETHYLPHENOL **444-30-4**
4 ★	2.82		1941		C7H5F3O1	P-TRIFLUOROMETHYLPHENOL **402-45-9**
65 ★	2.68		536		C7H5F3O2S1	SULFONE,PHENYL-TRIFLUOROMETHYL **426-58-4**
6 ★	3.57		536		C7H5F3S1	TRIFLUROMETHYLTHIOBENZENE **456-56-4**
7 ★	3.42		607	606	C7H5F4N1O2S1	TRIFLUOROMETHANESULFONANILIDE,M-F **23384-01-2**
8 ★	3.25		607	606	C7H5F4N1O2S1	TRIFLUOROMETHANESULFONANILIDE,P-F **23384-00-1**
9 ★	3.13		501		C7H5I1O2	M-IODOBENZOIC ACID **618-51-9**
2070 ★	2.40		527		C7H5I1O2	O-IODOBENZOIC ACID **88-67-5**
1 ★	3.02		501		C7H5I1O2	P-IODOBENZOIC ACID **619-58-9**
2 ★	1.56		501		C7H5N1	BENZONITRILE **100-47-0**
3 ★	1.63		1627		C7H5N1O1	1,2-BENZISOXAZOLE **271-95-4**
4 ★	1.52		547		C7H5N1O1	BENZISOXAZOLE **271-58-9**
75 ★	1.59		519		C7H5N1O1	BENZOXAZOLE **273-53-0**
6 ★	1.70		501		C7H5N1O1	M-CYANOPHENOL **873-62-1**
7 ★	1.61		825		C7H5N1O1	O-CYANOPHENOL **611-20-1**
8 ★	1.60		501		C7H5N1O1	P-CYANOPHENOL **767-00-0**
9 ★	1.16		924	400	C7H5N1O2	O-PHENYLENECARBAMATE **59-49-4**
2080 ★	0.83	6.8	448	2	C7H5N1O3	5-NITROSOTROPALONE **2297-94-1**
1 ★	1.47	7.0	925	1	C7H5N1O3	M-NITROBENZALDEHYDE **99-61-6**
2 ★	1.74		547		C7H5N1O3	O-NITROBENZALDEHYDE **552-89-6**
3 ★	1.56	7.0	925	1	C7H5N1O3	P-NITROBENZALDEHYDE **555-16-8**
4 ★	0.91		527		C7H5N1O3S1	SACCHARIN **81-07-2** *non-nutritive sweetener*
85 ★	0.97	7.4	594	1	C7H5N1O4	2-FURAN-ACROLIEN,5-NITRO **1874-22-2**
6 ★	1.83		501		C7H5N1O4	M-NITROBENZOIC ACID **121-92-6**
7 ★	1.46		1185		C7H5N1O4	O-NITROBENZOIC ACID **552-16-9**
8 ★	1.89		501		C7H5N1O4	P-NITROBENZOIC ACID **62-23-7**
9 ★	1.48	1.5	2329	314	C7H5N1O4	PHENOL,2-NITRO-4-FORMYL **3011-34-5**
2090 ★	0.68		1946		C7H5N1O5	2-PROPENOIC ACID,3-(5-NITRO-2-FURANYL) **6281-23-8**
1 ★	2.34		892		C7H5N1O5	5-NITROSALICYLIC ACID **96-97-9**
2 ★	1.98		1586		C7H5N1O5	BENZOIC ACID,3-HYDROXY-4-NITRO **619-14-7**
3	1.85		1586		C7H5N1O5	BENZOIC ACID,4-HYDROXY-3-NITRO **616-82-0**
4 ★	2.74		850		C7H5N1S1	4-HYDROXYPHENYLISOTHIOCYANATE **2131-60-4**
95 ★	1.74		572		C7H5N1S1	6-AZATHIANAPHTHENE **271-06-7**
6 ★	2.01		503		C7H5N1S1	BENZOTHIAZOLE **95-16-9**
7 ★	3.23		1947		C7H5N1S1	PHENOL,3-ISOTHIOCYANATO **3125-63-1**
8 ★	3.28		505		C7H5N1S1	PHENYLISOTHIOCYANATE **103-72-0**
9 ★	2.54		541		C7H5N1S1	PHENYLTHIOCYANATE **5285-87-0**
2100 ★	1.76		924	400	C7H5N1S1	S-ORTHOPHENYLENETHIOCARBAMATE **934-34-9** ✚
1	1.61	7.4	772	1	C7H5N1S2	MERCAPTOBENZOTHIAZOLE **149-30-4**
2 ★	2.41		2585		"	MERCAPTOBENZOTHIAZOLE
3 ★	2.30		2097		C7H5N3	DIAZINECARBONITRILE,PHENYL **622-83-3**
4 ★	-0.04		547		C7H5N3	PYRIDO-[2,3-B]-PYRAZINE **322-46-3**
5 ★	0.83		924	400	C7H5N3O1	BENZOTRIAZIN-4-ONE **90-16-4**
6 ★	1.87		2097		C7H5N3O1	DIAZINECARBONITRILE,PHENYL,OXIDE **54797-20-5**
7 ★	-0.80	7.4	2360	1	C7H5N3O2	1.2.4-BENZOTRIAZINE,1,4-DIOXIDE
8 ★	1.64		509		C7H5N3O2	BENZIMIDAZOLE,5-NITRO **94-52-0**
9	1.50		2139	459	C7H5N3O2	BENZIMIDAZOLE,6-NITRO
2110 ★	2.03	7.4	594	1	C7H5N3O2	INDAZOLE,5-NITRO **5401-94-5**
1 ★	2.06	7.4	594	1	C7H5N3O2	INDAZOLE,6-NITRO **7597-18-4**
2 ★	2.54	2.0	2522	342	C7H5N3O2	P-TRIAZOBENZOIC ACID
3	2.38	7.4	2298	1	C7H5N3O2S1	1,2,4-OXADIAZOL-5-ALDOXIME,3-(2-THIENYL) **103499-05-4**
4 ★	2.47	7.4	2298	2	"	1,2,4-OXADIAZOL-5-ALDOXIME,3-(2-THIENYL)
15 ✓	-2.70	7.4	2360	1	C7H5N3O3	BENZOTRIAZINE,1,4-DI-N-OXIDE-3-HYDROXY
6	0.90	7.7	1986	1	C7H5N3O4S2	BENZOTHIAZOLE,2-SULFONAMIDE-6-NITRO
7 ★	1.36	7.7	1986	2	"	BENZOTHIAZOLE,2-SULFONAMIDE-6-NITRO
8 ★	0.83		547		C7H5N3O5	3,5-DINITROBENZAMIDE **121-81-3** *coccidiostat (poultry)*
9 ★	1.60		579		C7H5N3O6	2,4,6-TRINITROTOLUENE **118-96-7**
2120 ★	1.65		541		C7H6Br1N1O1	M-BROMOBENZAMIDE **22726-00-7**
1 ★	0.77		2410		C7H6Br1N1O1	O-BROMOBENZAMIDE **4001-73-4**
2 ★	1.76		541		C7H6Br1N1O1	P-BROMOBENZAMIDE **698-67-9**
3	1.49		866		C7H6Br1N1O2	2-BROMO-4-AMINOBENZOIC ACID **2486-52-4**
4	2.23		866		C7H6Br1N1O2	3-BROMO-4-AMINOBENZOIC ACID **6311-37-1**
25 ★	-0.30	6.5	321	1	C7H6Br1N3O4S2	BROMOTHIAZIDE **19367-61-4**
6 ★	2.77		1560	459	C7H6Cl1F1	M-FLUOROBENZYLCHLORIDE **456-42-8**
7	2.67		1560	459	C7H6Cl1F1	O-FLUOROBENZYLCHLORIDE **345-35-7**
8 ★	0.02	4.0	2010		C7H6Cl1F1N2O4	URACIL,1-(2-CHLOROETHOXY)CARBONYL-5-FLUORO
9 ★	2.84		607	606	C7H6Cl1F2N1O2S1	DIFLUOROMETHANESULFONANILIDE,P-CL **1513-31-1**

	logP	pH	Ref.	Note	MF	Name / CAS / activity
2130 ★	2.67		594	400	C7H6Cl1N1O1	BENZALDOXIME,P-CHLORO 3848-36-0
1 ★	0.64		2410		C7H6Cl1N1O1	BENZAMIDE,2-CHLORO
2 ★	1.51		541		C7H6Cl1N1O1	M-CHLOROBENZAMIDE 618-48-4
3 ★	1.55		541		C7H6Cl1N1O1	P-CHLOROBENZAMIDE 619-56-7
4 ★	2.04		453		C7H6Cl1N1O1	P-CHLOROFORMANILIDE 2617-79-0
35	1.33		866		C7H6Cl1N1O2	2-CHLORO-4-AMINOBENZOIC ACID 2457-76-3
6	2.35		866		C7H6Cl1N1O2	3-CHLORO-4-AMINOBENZOIC ACID 2486-71-7
7 ★	1.57		1586		C7H6Cl1N1O2	BENZOIC ACID,2-AMINO-5-CHLORO 635-21-2
8 ★	3.09		2305	459	C7H6Cl1N1O2	TOLUENE,2-CHLORO-6-NITRO
9 ★	3.05		2305	459	C7H6Cl1N1O2	TOLUENE,4-CHLORO-2-NITRO
2140 ★	2.93	1.5	2329	314	C7H6Cl1N1O3	PHENOL,2-NITRO-4-CHLORO-5-METHYL 7147-89-9
1 ★	2.40		2390		C7H6Cl1N1S1	THIOBENZAMIDE,4-CHLORO 2521-24-6
2 ★	1.88		595		C7H6Cl1N3	BENZIMIDAZOLE,2-AMINO-5-CHLORO 5418-93-9
3 ★	0.91		522		C7H6Cl1N3O2S1	1,2,4-BENZOTHIADIAZINE-1,1-DIOXIDE-3-AMINO-6-CHLORO 37157-99-6
4 ★	-0.24		522		C7H6Cl1N3O4S2	CHLOROTHIAZIDE 58-94-6 diuretic, antihypertensive
45 ★	2.39		509		C7H6Cl1N3S1	4-PYRIDINEIMIDAZOLE,2-METHYLTHIO-6-CHLORO 40852-07-1
6 ★	4.24		1143	459	C7H6Cl2	2,4-DICHLOROTOLUENE 95-73-8
7 ★	4.29		1143	459	C7H6Cl2	2,6-DICHLOROTOLUENE 118-69-4
8	3.18		2363		C7H6Cl2	BENZYLCHLORIDE,4-CHLORO
9 ★	2.64		1263		C7H6Cl2N2O1	3,4-DICHLOROPHENYLUREA 2327-02-8
2150 ★	3.80		1568		C7H6Cl2O1	3,5-DICHLOROANISOLE 33719-74-3
1 ★	3.26		1554		C7H6Cl2O2	4,5-DICHLORO-2-METHOXYPHENOL 2460-49-3
2	3.19		1822	459	C7H6Cl2O2	4,5-DICHLOROGUAIACOL
3 ★	2.30		1816		C7H6Cl2O2	BENZYL ALCOHOL,2-HYDROXY-3,5-DICHLORO 6641-02-7
4 ★	-0.70	6.5	321	1	C7H6F1N3O4S2	FLUOROTHIAZIDE 1535-61-1
55 ★	2.03		594	400	C7H6F1O1	BENZALDOXIME,P-FLUORO 459-23-4
6 ★	0.91		541		C7H6F1O1	M-FLUOROBENZAMIDE 455-37-8
7 ★	0.59		2410		C7H6F1O1	O-FLUOROBENZAMIDE 445-28-3
8 ★	0.91		541		C7H6F1O1	P-FLUOROBENZAMIDE 824-75-9
9 ★	1.27	7.0	925	1	C7H6F1O1	P-FLUOROFORMANILIDE 459-25-6
2160	1.29		866		C7H6F1O2	2-FLUORO-4-AMINOBENZOIC ACID 446-31-1
1	1.41		866		C7H6F1O2	3-FLUORO-4-AMINOBENZOIC ACID
2 ★	1.12		2646	448	C7H6F2O1	BENZYL ALCOHOL,2,6-DIFLUORO
3 ★	2.44		579		C7H6F2O1	DIFLUOROMETHOXYBENZENE 458-92-4
4 ★	2.41		594	400	C7H6F3N1	2-AMINOBENZOTRIFLUOURIDE 88-17-5
65 ★	2.29	7.4	594	1	C7H6F3N1	M-TRIFLUOROMETHYLANILINE 98-16-8
6 ★	2.39	7.4	590		C7H6F3N1	P-TRIFLUOROMETHYLANILINE 455-14-1
7 ★	3.05		607	606	C7H6F3N1O2S1	TRIFLUOROMETHANESULFONANILIDE 456-64-4
8 ★	2.51		607	606	C7H6F3N1O3S1	TRIFLUOROMETHANESULFONANILIDE,M-OH 23375-12-4
9 ★	0.93		2410		C7H6I1O1	BENZAMIDE,2-IODO
2170 ★	1.99		547		C7H6I1O1	P-IODOBENZAMIDE 3956-07-8
1	1.65		866		C7H6I1O2	2-IODO-4-AMINOBENZOIC ACID
2	2.36		866		C7H6I1O2	3-IODO-4-AMINOBENZOIC ACID 2122-63-6
3 ★	2.06		1586		C7H6I1O2	BENZOIC ACID,2-AMINO-5-IODO 5326-47-6
4 ★	0.84		163	459	C7H6N2	2-CYANO-6-METHYLPYRIDINE 1620-75-3
75 ★	0.81		163	459	C7H6N2	4-CYANO-2-METHYLPYRIDINE 2214-53-1
6 ★	1.82		532		C7H6N2	7-AZAINDOLE 271-63-6
7 ★	1.32		594		C7H6N2	BENZIMIDAZOLE 51-17-2
8 ★	1.40		594	400	C7H6N2	BENZONITRILE,2-AMINO 1885-29-6
9 ★	1.77		1627		C7H6N2	INDAZOLE 271-44-3
2180	1.07		547		C7H6N2	M-AMINOBENZONITRILE 2237-30-1
1 ★	1.87		547		C7H6N2	PHENYLCYANAMIDEHEMIHYDRATE 622-34-4
2	1.31		1625	457	C7H6N2O1	BENZISOXAZOLE,3-AMINO
3 ★	1.53	2.0	1625		C7H6N2O1	BENZOXAZOLE,2-AMINO 4570-41-6
4 ★	1.12		305		C7H6N2O1	O-PHENYLENEUREA 615-16-7
85 ★	0.16		522		C7H6N2O2S1	1,2,4-BENZOTHIADIAZINE,1,1-DIOXIDE 359-85-3
6	0.18	7.4	601	1	"	1,2,4-BENZOTHIADIAZINE,1,1-DIOXIDE
7 ★	0.28	2.1	2265		C7H6N2O2S1	BENZENESULFONAMIDE,3-CYANO 3118-68-1
8 ★	0.23	2.0	264		C7H6N2O2S1	P-CYANOBENZENESULFONAMIDE 3119-02-6
9 ★	1.75		2390		C7H6N2O2S1	THIOBENZAMIDE,4-NITRO 26060-30-0
2190 ★	1.33	7.2	1685	314	C7H6N2O2S2	BENZOTHIAZOLE-2-SULFONAMIDE 433-17-0
1 ★	1.95		594		C7H6N2O3	BENZALDOXIME,P-NITRO 1129-37-9
2 ★	-0.15		594		C7H6N2O3	BENZAMIDE,2-NITRO 610-15-1
3 ★	0.77		541		C7H6N2O3	M-NITROBENZAMIDE 645-09-0
4 ★	1.40	7.0	925	1	C7H6N2O3	M-NITROFORMANILIDE 102-38-5
95 ★	0.82		536		C7H6N2O3	P-NITROBENZAMIDE 619-80-7
6 ★	1.43		453		C7H6N2O3	P-NITROFORMANILIDE 16135-31-2
7 ★	1.13	7.2	1685	314	C7H6N2O3S2	6-HYDROXYBENZOTHIAZOLE-2-SULFONAMIDE 29927-14-8
8 ★	2.08		2410		C7H6N2O4	1,2-DINITROBENZENE,4-METHYL 610-39-9
9 ★	1.98		536		C7H6N2O4	2,4-DINITROTOLUENE 121-14-2
2200	1.21		866		C7H6N2O4	2-NITRO-4-AMINOBENZOIC ACID 610-36-6
1 ★	0.65		1946		C7H6N2O4	2-PROPENOIC AMIDE,3-(5-NITRO-2-FURANYL) 710-25-8
2 ★	1.91		1586		C7H6N2O4	BENZOIC ACID,2-AMINO-4-NITRO 619-17-0
3 ★	2.10		2410		C7H6N2O4	TOLUENE,2,6-DINITRO
4 ★	2.13		2215		C7H6N2O5	2-METHYL-4,6-DINITROPHENOL 534-52-1 insecticide~fungicide
5 ★	1.66		305		C7H6N2S1	O-PHENYLENETHIOUREA 583-39-1
6 ★	1.09		541		C7H6N4	1-PHENYLTETRAZOLE 5378-52-9
7 ★	1.65	1.0	579	342	C7H6N4	2H-TETRAZOLE,5-PHENYL 18039-42-4
8 ★	-0.30	7.4	2360	1	C7H6N4O2	BENZOTRIAZINE,1,4-DI-N-OXIDE-3-AMINO radiosensitizer
9 ★	2.33		2128		C7H6N4O4	FORMALDEHYDE,2,4-DINITROPHENYLHYDRAZONE 1081-15-8
2210 ★	-0.70	6.5	321	1	C7H6N4O6S2	1,2,4-BENZOTHIDIAZINE-1,1-DIOXIDE-6-NITRO-7-SULFONAMIDO 23141-81-3
1 ★	1.48		536		C7H6O1	BENZALDEHYDE 100-52-7 flavoring agent
2 ★	2.08		536		C7H6O2	1,2-METHYLENEDIOXYBENZENE 274-09-9
3 ★	0.72		2649		C7H6O2	2-METHYL-1,4-BENZOQUINONE 553-97-9
4	0.81	6.8	448	2	C7H6O2	B-TROPALONE 3324-76-3

	logP	pH	Ref.	Note	MF	Name / **CAS** / *activity*
15	0.24	7.4	2288	1	C7H6O2	BENZOIC ACID **65-85-0** *antifungal*
6 ★	1.87		501		"	BENZOIC ACID
7 ★	1.15		1478		C7H6O2	FURAN-2YL-ACROLIEN **623-30-3**
8 ★	1.38		853		C7H6O2	M-HYDROXYBENZALDEHYDE **100-83-4**
9 ★	1.81	5.4	530		C7H6O2	O-HYDROXYBENZALDEHYDE **90-02-8**
2220 ★	1.35		825		C7H6O2	P-HYDROXYBENZALDEHYDE **123-08-0**
1 ★	1.26		163		C7H6O2	PHENYLFORMATE **1864-94-4**
2 ★	0.53		506		C7H6O2	TROPOLONE **533-75-5**
3 ★	2.39	1.0	577	342	C7H6O2S1	THIOSALICYLIC ACID **147-93-3**
4 ★	-0.06		1963		C7H6O3	BENZOQUINONE,2-METHOXY **2880-58-2**
25 ★	1.50		501		C7H6O3	M-HYDROXYBENZOIC ACID **99-06-9**
6	1.28	7.4	2312	337	C7H6O3	P-HYDROXYBENZOIC ACID **99-96-7**
7	1.54	7.4	2539	2	"	P-HYDROXYBENZOIC ACID
8 ★	1.58		501		"	P-HYDROXYBENZOIC ACID
9	-2.11	7.4	2288	1	C7H6O3	SALICYLIC ACID **69-72-7** *keratolytic*
2230	-1.70	7.0	2230	1	"	SALICYLIC ACID
1	-0.90	7.4	1838	1	"	SALICYLIC ACID
2 ★	2.26		503		"	SALICYLIC ACID
3 ✓	-1.43	8.1	541	314	C7H6O3	SODIUM SALICYLATE **63-36-5**
4	-0.85	7.2	532	314	"	SODIUM SALICYLATE
35 ★	1.20		1480		C7H6O4	2,3-DIHYDROXYBENZOIC ACID **303-38-8**
6 ★	1.63		1145		C7H6O4	2,4-DIHYDROXYBENZOIC ACID **89-86-1**
7	2.06	7.4	340	2	"	2,4-DIHYDROXYBENZOIC ACID
8	1.40		2626	537	C7H6O4	2,5-DIHYDROXYBENZOIC ACID **490-79-9** *analgesic, anti-inflammatory*
9	1.61		1586		"	2,5-DIHYDROXYBENZOIC ACID
2240	1.74	7.4	340	2	"	2,5-DIHYDROXYBENZOIC ACID
1 ★	2.20		508		C7H6O4	2,6-DIHYDROXYBENZOIC ACID **303-07-1**
2 ★	1.15	7.4	2539	2	C7H6O4	3,4-DIHYDROXYBENZOIC ACID **99-50-3**
3 ★	0.86		1480		C7H6O4	3,5-DIHYDROXYBENZOIC ACID **99-10-5**
4 ★	1.05		1145		C7H6O5	2,3,4-TRIHYDROXYBENZOIC ACID **610-02-6**
45 ★	0.70		1335		C7H6O5	3,4,5-TRIHYDROXYBENZOIC ACID **149-91-7**
6 ★	2.92		505	20	C7H7Br1	A-BROMOTOLUENE **100-39-0**
7 ★	1.26		1057		C7H7Br1N2O1	BENZOYLHYDRAZINE,M-BROMO
8 ★	0.26		1057		C7H7Br1N2O1	BENZOYLHYDRAZINE,O-BROMO
9 ★	1.28		1057		C7H7Br1N2O1	BENZOYLHYDRAZINE,P-BROMO
2250 ★	2.08		1263		C7H7Br1N2O1	M-BROMOPHENYLUREA **2989-98-2**
1 ★	1.98		1263		C7H7Br1N2O1	P-BROMOPHENYLUREA **1967-25-5**
2 ★	2.86		2410		C7H7Br1O1	ANISOLE,2-BROMO
3 ★	1.72		1816		C7H7Br1O2	BENZYL ALCOHOL,2-HYDROXY-5-BROMO **2316-64-5** *anti-inflammatory*
4 ★	1.38		2452		C7H7Br1O2	HYDROQUINONE,2-METHYL-6-BROMO
55 ★	2.30		516		C7H7Cl1	A-CHLOROTOLUENE **100-44-7**
6 ★	3.28		531		C7H7Cl1	M-CHLOROTOLUENE **108-41-8**
7 ★	3.42		531		C7H7Cl1	O-CHLOROTOLUENE **95-49-8**
8 ★	3.33		531		C7H7Cl1	P-CHLOROTOLUENE **106-43-4**
9 ★	2.25		607	606	C7H7Cl1F1O2S1	FLUOROMETHANESULFONANILIDE,P-CL **50585-76-7**
2260 ★	1.18		1057		C7H7Cl1N2O1	BENZOYLHYDRAZINE,M-CHLORO
1 ★	0.14		1057		C7H7Cl1N2O1	BENZOYLHYDRAZINE,O-CHLORO
2 ★	1.12		1057		C7H7Cl1N2O1	BENZOYLHYDRAZINE,P-CHLORO
3 ★	1.82		1263		C7H7Cl1N2O1	M-CHLOROPHENYLUREA **1967-27-7**
4 ★	1.27		1263		C7H7Cl1N2O1	O-CHLOROPHENYLUREA **114-38-5**
65	1.60	7.0	925	1	C7H7Cl1N2O1	P-CHLOROPHENYLUREA **140-38-5**
6 ★	1.80		1624		"	P-CHLOROPHENYLUREA
7 ★	1.39	7.4	1184	1	C7H7Cl1N2O2	3-(P-CHLOROPHENYL)HYDROXYUREA **30085-34-8**
8 ★	0.12		1519	400	C7H7Cl1N2O2	3-CL-6-PYRIDAZINECARBOXYLIC ACID,ETHYL ESTER
9	-1.22		1375		C7H7Cl1N2O3	4-PYRIDAZINYLOXYACETIC ACID,3-CL-6-METHYL
2270	-0.85	7.4	971	1	C7H7Cl1N4O2	8-CHLOROTHEOPHYLLINE **85-18-7**
1	3.70	7.3	509	2	C7H7Cl1N4O4S2	PURINE,2,6-DI-(METHYLSULFONYL)-8-CHLORO
2	0.63		509		C7H7Cl1N4S2	PURINE,2,8-DIMETHYLTHIO,6-CHLORO
3 ★	-0.47		1472		C7H7Cl1N6O2	AZOLASTONE **85622-95-3** *antineoplastic*
4 ★	2.78		757		C7H7Cl1O1	2-METHYL-4-CHLOROPHENOL **1570-64-5**
75 ★	2.78		1568		C7H7Cl1O1	4-CHLOROANISOLE **623-12-1**
6 ★	2.68		2410		C7H7Cl1O1	ANISOLE,2-CHLORO
7 ★	2.98		2163	459	C7H7Cl1O1	ANISOLE,3-CHLORO
8 ★	3.10		300		C7H7Cl1O1	CHLOROCRESOL **59-50-7** *antiseptic*
9 ★	1.94		501		C7H7Cl1O1	M-CHLOROBENZYL ALCOHOL **873-63-2**
2280 ★	1.96		501		C7H7Cl1O1	P-CHLOROBENZYL ALCOHOL **873-76-7**
1 ★	2.80		2645	225	C7H7Cl1O1	PHENOL,2-CHLORO-6-METHYL
2 ★	2.63		2163	459	C7H7Cl1O1	PHENOL,3-CHLORO-6-METHYL **1570-64-5**
3 ★	1.44		1816		C7H7Cl1O2	BENZYL ALCOHOL,2-HYDROXY-5-CHLORO **5330-38-1**
4 ★	1.24		594		C7H7Cl1O2S1	CHLOROMETHYLPHENYLSULFONE **7205-98-3**
85 ★	1.06	7.0	925	1	C7H7Cl1O2S1	P-METHYLSULFONYLCHLOROBENZENE **98-57-7**
6 ★	2.91		547		C7H7Cl2N1	ANILINE,N-METHYL-2,6-DICHLORO **56462-00-1**
7 ★	2.53		354		C7H7Cl3N1O3P1	O-METHYL-O-(2,4,5-TRICHLOROPHENYL)PHOSPHORAMIDATE **2214-34-8**
8 ★	4.31		928		C7H7Cl3N1O3P1S1	METHYLCHLORPYRIFOS **5598-13-0** *insecticide*
9 ★	1.29		1263		C7H7F1N2O1	M-FLUOROPHENYLUREA **770-19-4**
2290 ★	0.88		1263		C7H7F1N2O1	O-FLUOROPHENYLUREA **656-31-5**
1 ★	1.04		1263		C7H7F1N2O1	P-FLUOROPHENYLUREA **659-30-3**
2	-0.21	7.4	2266	1	C7H7F1N2O3	5-FLUOROURACIL-3-PROPIONYL **75410-16-1**
3 ★	0.19	4.0	2266	302	"	5-FLUOROURACIL-3-PROPIONYL
4 ★	0.11	4.0	2271	302	C7H7F1N2O4	5-FLUOROURACIL-3-ETHOXYCARBONYL
95	-0.79	7.4	2266	1	C7H7F1N2O4	URACIL,1-ETHOXYCARBONYL-5-FLUORO **21839-33-8**
6 ★	-0.17	4.0	2010		"	URACIL,1-ETHOXYCARBONYL-5-FLUORO
7 ★	2.14		2410		C7H7F1O1	ANISOLE,2-FLUORO
8 ★	1.31		2646	448	C7H7F1O1	BENZYL ALCOHOL,2-FLUORO
9 ★	1.36		2646	448	C7H7F1O1	BENZYL ALCOHOL,4-FLUORO

	logP	pH	Ref.	Note	MF	Name / CAS / activity
2300 ★	2.74		527		C7H7F1O2S1	P-FLUOROSULFONYLTOLUENE **455-16-3**
1 ★	1.95		607	606	C7H7F2N1O2S1	DIFLUOROMETHANESULFONANILIDE **658-43-5**
2 ★	1.53		1057		C7H7I1N2O1	BENZOYLHYDRAZINE,M-IODO
3 ★	0.47		1057		C7H7I1N2O1	BENZOYLHYDRAZINE,O-IODO
4 ★	1.55		1057		C7H7I1N2O1	BENZOYLHYDRAZINE,P-IODO
5 ★	0.85	7.0	566	1	C7H7N1O1	2-ACETYLPYRIDINE **1122-62-9**
6	0.34	7.2	2317	1	C7H7N1O1	3-ACETYLPYRIDINE **350-03-8**
7 ★	0.43		163	459	"	3-ACETYLPYRIDINE
8	0.40	7.2	2317	1	C7H7N1O1	4-ACETYLPYRIDINE **1122-54-9**
9 ★	0.48	7.0	566	1	"	4-ACETYLPYRIDINE
2310	1.79		594		C7H7N1O1	BENZALDOXIME,ANTI **932-90-1**
1 ★	1.85		594		C7H7N1O1	BENZALDOXIME,SYN **622-31-1**
2 ★	1.75		536		C7H7N1O1	BENZALDOXIME
3 ★	0.64		501		C7H7N1O1	BENZAMIDE **55-21-0**
4 ★	1.15		508		C7H7N1O1	FORMANILIDE **103-70-8**
15	1.71		1174		C7H7N1O2	2-METHYL-4-NITROSOPHENOL **6971-38-6**
6	1.71		1198			2-METHYL-4-NITROSOPHENOL
7 ★	0.36		2412		C7H7N1O2	2-PYRIDINECARBOXYLIC ACID,METHYL ESTER **2459-07-6**
8 ★	1.75		562		C7H7N1O2	A-NITROTOLUENE **622-42-4**
9	0.26		536	815	C7H7N1O2	BENZOHYDROXAMIC ACID **495-18-1**
2320 ★	0.26		536	815	"	BENZOHYDROXAMIC ACID
1 ★	1.08		508		C7H7N1O2	CARBAMIC ACID,PHENYL ESTER **622-46-8**
2	0.40	7.2	2317	1	C7H7N1O2	I-NICOTINIC ACID,METHYL ESTER **2459-09-8**
3 ★	0.87		163	459	"	I-NICOTINIC ACID,METHYL ESTER
4	-0.79	2.0	2522	342	C7H7N1O2	M-AMINOBENZOIC ACID **99-05-8**
25	0.07		547		"	M-AMINOBENZOIC ACID
6	0.28		1136		"	M-AMINOBENZOIC ACID
7 ★	0.39		541		C7H7N1O2	M-HYDROXYBENZAMIDE **618-49-5**
8 ★	2.42		2410		C7H7N1O2	M-NITROTOLUENE **99-08-1**
9	0.72	7.2	2317	1	C7H7N1O2	NICOTINIC ACID,METHYL ESTER **93-60-7** rubifacient
2330 ★	0.83	7.0	332	1	"	NICOTINIC ACID,METHYL ESTER
1	0.71	7.4	772	1	C7H7N1O2	O-AMINOBENZOIC ACID **118-92-3**
2 ★	1.21		508		"	O-AMINOBENZOIC ACID
3	0.89	7.0	353	1	C7H7N1O2	O-HYDROXYBENZAMIDE **65-45-2** analgesic
4	1.15	7.4	2145	1	"	O-HYDROXYBENZAMIDE
35 ★	1.28		503		"	O-HYDROXYBENZAMIDE
6 ★	2.30				"	O-NITROTOLUENE **88-72-2**
7	-1.62	7.4	1051	1	C7H7N1O2	P-AMINOBENZOIC ACID **150-13-0** ultraviolet screen
8 ★	0.83		866		"	P-AMINOBENZOIC ACID
9	0.85	7.4	772	1	"	P-AMINOBENZOIC ACID
2340 ★	0.33		541		C7H7N1O2	P-HYDROXYBENZAMIDE **619-57-8**
1 ★	2.37		501		C7H7N1O2	P-NITROTOLUENE **99-99-0**
2	-3.17	12.2	2600		C7H7N1O2	PYRIDYLACETIC ACID vasodilator
3	-2.69	7.4	2600	1	"	PYRIDYLACETIC ACID
4	-2.42	2.0	2600	342	"	PYRIDYLACETIC ACID
45	0.50		866		C7H7N1O3	3-HYDROXY-4-AMINOBENZOIC ACID **2374-03-0**
6	1.84	7.4	1075	324	C7H7N1O3	3-METHYL-4-NITROPHENOL **2581-34-2**
7 ★	2.48	1.5	2329	314	"	3-METHYL-4-NITROPHENOL
8	-0.15	7.0	353	1	C7H7N1O3	B-RESORCYLAMIDE **3147-45-3**
9	1.24		594		C7H7N1O3	BENZYL ALCOHOL,2-NITRO **612-25-9**
2350	0.60	7.0	353	1	C7H7N1O3	G-RESORCYLAMIDE **3147-50-0**
1 ★	2.16		501		C7H7N1O3	M-NITROANISOLE **555-03-3**
2 ★	1.21		501		C7H7N1O3	M-NITROBENZYL ALCOHOL **619-25-0**
3	-0.16		1586		C7H7N1O3	MESALAMINE **89-57-6** anti-inflammatory
4 ★	1.73		547		C7H7N1O3	O-NITROANISOLE **91-23-6**
55 ★	1.32		866		C7H7N1O3	P-AMINOSALICYLIC ACID **65-49-6** antibacterial (tuberculostatic)
6	2.00	7.4	820	1	C7H7N1O3	P-NITROANISOLE **100-17-4**
7 ★	2.03		501		"	P-NITROANISOLE
8	2.05	7.0	925	1	"	P-NITROANISOLE
9 ★	1.26		501		C7H7N1O3	P-NITROBENZYL ALCOHOL **619-73-8**
2360	1.31	7.0	2505	1	"	P-NITROBENZYL ALCOHOL
1 ★	2.29	1.5	2329	314	C7H7N1O3	PHENOL,2-NITRO-3-METHYL **4920-77-8**
2	2.15	7.4	1075	2	C7H7N1O3	PHENOL,2-NITRO-4-METHYL **119-33-5**
3 ★	2.37	1.5	2329	314	"	PHENOL,2-NITRO-4-METHYL
4 ★	2.31	1.5	2329	314	C7H7N1O3	PHENOL,2-NITRO-5-METHYL **700-38-9**
65	0.74		2088	1	C7H7N1O3S3	THIENO-4-OXO-7-THIOPYRAN,2-SULFONAMIDE **105951-31-3**
6	0.91		2088	1	C7H7N1O3S3	THIENO-4-THIOPYRAN-7-ONE,2-SULFONAMIDE **105951-32-4**
7 ★	0.04		731	604	C7H7N1O4	N-MALEOYL-3-AMINOPROPIONIC ACID **7423-55-4**
8 ★	2.02	1.5	2329	314	C7H7N1O4	PHENOL,2-NITRO-4-METHOXY **1568-70-3**
9	0.50	1.0	547	342	C7H7N1O4S1	CARZENIDE **138-41-0** carbonic anhydrase inhibitor
2370 ★	0.50	7.0	925	1	C7H7N1O4S1	M-METHYLSULFONYLNITROBENZENE **2976-32-1**
1	-0.04		2088	1	C7H7N1O5S3	THIENO-4-OXO-7-THIOPYRAN,2-SULFONAMIDE-7-DIOXIDE **105951-35-7**
2	-0.20		2088	1	C7H7N1O5S3	THIENO-4-THIOPYRAN-7-ONE,2-SULFONAMIDE-4-DIOXIDE **105951-39-1**
3 ★	1.49		547		C7H7N1S1	THIOBENZAMIDE **2227-79-4**
4 ★	1.13		559		C7H7N3	1-METHYLBENZOTRIAZOLE **13351-73-0**
75 ★	0.91		572		C7H7N3	2-AMINOBENZIMIDAZOLE **934-32-7**
6 ★	1.64		559		C7H7N3	2-METHYLBENZOTRIAZOLE **16584-00-2**
7	0.65		2139	459	C7H7N3	INDAZOLE,5-AMINO **19335-11-6**
8	-0.38	7.4	595	1	C7H7N3O1	BENZIMIDAZOLE,2-AMINO-5-HYDROXY
9 ★	0.72	13.0	594		C7H7N3O2	BENZAMIDINE,3-NITRO **56406-50-9**
2380 ★	0.58		2097		C7H7N3O2	DIAZINECARBOXAMIDE,2-OXO-2-PHENYL **60142-49-6**
1	0.38	7.7	1986	1	C7H7N3O2S2	BENZOTHIAZOLE-2-SULFONAMIDE-6-AMINO
2 ★	0.52	7.7	1986	2	"	BENZOTHIAZOLE-2-SULFONAMIDE-6-AMINO
3	-1.07		1008	354	C7H7N3O2.H1Cl1	M-NITROBENZAMIDINEHYDROCHLORIDE
4 ★	0.60		1008	354	C7H7N3O3	1(2-HYDROXYETHYL)-5-CYANO-2-NITROPYRROLE
85 ★	0.23		1057		C7H7N3O3	BENZOYLHYDRAZINE,M-NITRO
6 ★	-0.54		1057		C7H7N3O3	BENZOYLHYDRAZINE,O-NITRO
7 ★	0.35		1057		C7H7N3O3	BENZOYLHYDRAZINE,P-NITRO
8 ★	0.98		611		C7H7N3O3S1	1-(5-NITRO-2-FURYL)-2-IMIDAZOLIDINETHIONE **53207-62-8** +
9 ★	0.25		611		C7H7N3O4	1(5-NO2-2-FURYL)-2-IMIDAZOLIDINONE **53207-61-7**

	logP	pH	Ref.	Note	MF	Name / **CAS** / *activity*
2390 ★	-0.70	6.5	321	1	C7H7N3O4S2	1,2,4-BENZOTHIADIAZINE-1,1-DIOXIDE-7-SULFONAMIDO **23141-75-5**
1 ★	1.43	7.0	332	1	C7H7N3S1	THIENO(2,3-D)-PYRIMIDINE,4-(METHYLAMINO) **56844-20-3**
2 ★	2.63		1772		C7H8	CYCLOHEPTATRIENE **544-25-2**
3	2.57		1734		C7H8	TOLUENE,PERDEUTERO **2037-26-5**
4 ★	2.73		508		C7H8	TOLUENE **108-88-3**
95 ★	0.30		1564		C7H8Br1Cl1N2O3	PYRIMIDINE-2,4-DIONE,3-CHLORETHOXY-5-BROMO-6-ME **77317-67-0** ✚
6	2.53		1263		C7H8Br1N1	3-METHYL-4-BROMOANILINE **6933-10-4**
7 ★	0.08	6.5	321	2	C7H8Br1N3O4S2	HYDROBROMOTHIAZIDE **23141-83-5**
8	1.85		1263		C7H8Cl1N1O1	3-CHLORO-4-METHOXYANILINE **5345-54-0**
9 ★	2.55		594		C7H8Cl1N1O1	PYRIDINE,2-CHLORO-6-ETHOXY
2400 ★	1.92	7.2	2036	1	C7H8Cl1N1O2S1	BENZENESULFONAMIDE,N-METHYL-4-CHLORO **6333-79-5**
1 ★	1.85		607	606	C7H8Cl1N1O2S1	METHANESULFONANILIDE,P-CL **4284-51-9**
2 ★	-0.07		505		C7H8Cl1N3O4S2	HYDROCHLOROTHIAZIDE **58-93-5** *diuretic*
3 ★	-0.37	4.0	2009		C7H8F1N3O3	URACIL,1-DIMETHYLAMINOCARBONYL-5-FLUORO
4 ★	0.35	4.0	2009		C7H8F1N3O3	URACIL,1-ETHYLAMINOCARBONYL-5-FLUORO
5 ★	1.35		607	606	C7H8F1O2S1	FLUOROMETHANESULFONANILIDE **2070-61-3**
6 ★	1.86		579		C7H8Hg1	PHENYLMETHYLMERCURY **21392-61-0**
7 ★	0.57	7.4	1975	1	C7H8I1N3O4	4-NITRO-5-IODOIMIDAZOLE-1-ACETIC ACID, ETHYL ESTER **96258-80-9**
8 ★	0.93	7.4	1975	1	C7H8I1N3O4	5-NITRO-4-IODOIMIDAZOLE-1-ACETIC ACID, ETHYL ESTER **96258-81-0**
9	-1.59		548	370	C7H8N2	BENZAMIDINE HYDROCHLORIDE **1670-14-0**
2410	-1.29	7.4	594	1	C7H8N2	BENZAMIDINE
1 ★	0.65	13.0	594		"	BENZAMIDINE
2 ★	0.61		163	459	C7H8N2O1	2-ACETAMINOPYRIDINE **5231-96-9**
3 ★	-0.65		1474		C7H8N2O1	2-PYRIDINEACETAMIDE **5451-39-8**
4 ★	0.41		163	459	C7H8N2O1	3-ACETAMINOPYRIDINE **5867-45-8**
15 ★	-0.71		1474		C7H8N2O1	3-PYRIDINEACETAMIDE **3724-16-1**
6 ★	0.50		1519	459	C7H8N2O1	4-ACETAMINOPYRIDINE **5221-42-1**
7 ★	-0.65		1474		C7H8N2O1	4-PYRIDINEACETAMIDE **39640-62-5**
8 ★	0.19		547		C7H8N2O1	BENZOYLHYDRAZINE **613-94-5**
9	0.23	7.5	547	1	"	BENZOYLHYDRAZINE
2420	-0.33		541		C7H8N2O1	M-AMINOBENZAMIDE **3544-24-9** *radiosensitizer*
1	0.00	7.2	2317	1	C7H8N2O1	N-METHYLNICOTINAMIDE
2 ★	0.00		1035		"	N-METHYLNICOTINAMIDE
3 ★	0.35		547		C7H8N2O1	O-AMINOBENZAMIDE **88-68-6**
4	-0.44	7.4	1737	1	C7H8N2O1	P-AMINOBENZAMIDE **2835-68-9**
25 ★	-0.41		547		"	P-AMINOBENZAMIDE
6 ★	0.83		508		C7H8N2O1	PHENYLUREA **64-10-8**
7	-0.31		866		C7H8N2O2	2,4-DIAMINOBENZOIC ACID **611-03-0**
8	0.13		866		C7H8N2O2	3,4-DIAMINOBENZOIC ACID **619-05-6**
9 ★	0.41	7.4	1184	1	C7H8N2O2	3-PHENYLHYDROXYUREA **7335-35-5**
2430 ★	1.87		2163	459	C7H8N2O2	ANILINE,2-METHYL-5-NITRO **99-55-8**
1	1.82		2139	459	C7H8N2O2	ANILINE,4-METHYL-2-NITRO **89-62-3**
2	-1.76		1586		C7H8N2O2	BENZOIC ACID,3,5-DIAMINO **535-87-5**
3 ★	-0.08		1057		C7H8N2O2	BENZOYLHYDRAZINE,M-HYDROXY
4 ★	0.60		1057		C7H8N2O2	BENZOYLHYDRAZINE,O-HYDROXY
35 ★	-0.33		1057		C7H8N2O2	BENZOYLHYDRAZINE,P-HYDROXY
6 ★	2.18		547		C7H8N2O2	N-METHYL-O-NITROANILINE **612-28-2**
7 ★	2.04		562		C7H8N2O2	P-NITRO-N-METHYLANILINE **100-15-2**
8 ★	1.06	13.0	547	224	C7H8N2O2	P-NITROBENZYLAMINE **7409-30-5**
9 ★	0.10		2499		C7H8N2O2	PYRAZINE,2-METHOXYCARBONYL-6-METHYL
2440 ★	0.17		2499		C7H8N2O2	PYRAZINE,2-METHYL-5-METHOXYCARBONYL
1 ★	0.28		2412		C7H8N2O2	PYRAZINE-2-CARBOXYLIC ACID, ETHYL ESTER
2 ★	0.23		2412		C7H8N2O2	PYRIDAZINE-4-CARBOXYLIC ACID,ETHYL ESTER **39123-39-2**
3 ★	-0.31		2613		C7H8N2O2	PYRIMIDINE-2-CARBOXYLIC ACID,ETHYL ESTER
4 ★	0.52		2412		C7H8N2O2	PYRIMIDINE-5-CARBOXYLIC ACID,ETHYL ESTER
45 ★	1.47		594		C7H8N2O3	ANILINE,2-METHOXY-5-NITRO **99-59-2**
6 ★	1.86		2163	459	C7H8N2O3	ANILINE,4-METHOXY-2-NITRO **96-96-8**
7 ★	0.69		2499		C7H8N2O3	PYRAZINE,2-METHOXY-6-METHOXYCARBONYL
8 ★	-0.70	2.1	2265		C7H8N2O3S1	BENZENESULFONAMIDE,3-CARBOXAMIDO
9 ★	-0.79	2.1	2265		C7H8N2O3S1	BENZENSULFONAMIDE,4-CARBOXAMIDO
2450 ★	0.28		2352	1	C7H8N2O3S1	PHENYLSULFONYLUREA **35207-08-0**
1	0.68		2447		C7H8N2O4	4-PYRIDINECARBOXYLIC ACID,TETRAHYDRO-2,6-DIOXO,ETHYL ESTER **1747-53-1**
2 ★	0.99		547		C7H8N2S1	O-AMINOTHIOBENZAMIDE **2454-39-9**
3 ★	0.73		547		C7H8N2S1	PHENYLTHIOUREA **103-85-5**
4 ★	1.12		2499		C7H8N4	PYRAZINE,2-CYANO-6-DIMETHYLAMINO
55	0.69	7.4	401	1	C7H8N4O1S1	5-HYDROXYPICOLINALDEHYDETHIOSEMICARBAZONE **19494-89-4** *antineoplastic*
6	0.93	7.4	401	2	"	5-HYDROXYPICOLINALDEHYDETHIOSEMICARBAZONE
7 ★	1.04		527		"	5-HYDROXYPICOLINALDEHYDETHIOSEMICARBAZONE
8	0.02		1375		C7H8N4O2	1,2,4-TRIAZOLO(4,3-B)PYRIDAZINE-6-ETHOXY-8-OH
9 ★	-0.78		505		C7H8N4O2	THEOBROMINE **83-67-0** *diuretic, vasodilator, cardiac depressant*
2460	-0.02	7.4	971	1	C7H8N4O2	THEOPHYLLINE **58-55-9** *bronchodilator*
1 ★	-0.02		505		"	THEOPHYLLINE
2 ★	-0.22	3.0	2191		C7H8N4O2	XANTHINE,1,7-DIMETHYL **611-59-6**
3 ★	-0.11		2537		C7H8N4O2	XANTHINE,3-ETHYL **41078-01-7**
4 ★	1.78		611		C7H8N4O2S2	1-(5-NITRO-2-THIAZOLYL)-2-METHYLTHIO-2-IMIDAZOLINE **37422-15-4** ✚
65 ★	-0.52	3.0	2191		C7H8N4O3	URIC ACID,1,3-DIMETHYL **944-73-0**
6 ★	-0.22	3.0	2191		C7H8N4O3	URIC ACID,1,7-DIMETHYL **33868-03-0**
7 ★	-1.33	3.0	2191		C7H8N4O3	URIC ACID,3,7-DIMETHYL **13087-49-5**
8 ★	1.34	7.4	553	1	C7H8N4O3S1	2(4-ME-1,3-OXAZOLIDINYLIDEN-2N)-5-NO2-THIAZOLE **24240-83-3** ✚
9	-0.47	7.4	2561	1	C7H8N4O4	2-NITROIMIDAZOLE-1-(ACETALDOXIME-ACETATE)
2470 ★	0.20	6.5	321	2	C7H8N4O6S2	6-NITROHYDROTHIAZIDE **23141-88-0**
1 ★	0.11	7.4	2028	1	C7H8N6	1,2,4-TRIAZOLE,3-AMINO-5-(2-AMINOPYRIDIN-4-YL) **77314-75-1**
2 ★	2.11		501		C7H8O1	ANISOLE **100-66-3**
3	1.01	7.0	2505	1	C7H8O1	BENZYL ALCOHOL **100-51-6** *antibacterial*
4 ★	1.10		501		"	BENZYL ALCOHOL

	logP	pH	Ref.	Note	MF	Name / CAS / activity
75 ★	1.96		501		C7H8O1	M-METHYLPHENOL 108-39-4 *disinfectant*
6 ★	1.95		532		C7H8O1	O-METHYLPHENOL 95-48-7
7 ★	1.94		501		C7H8O1	P-METHYLPHENOL 106-44-5
8	2.00	7.4	2582	1	"	P-METHYLPHENOL
9 ★	0.55		536		C7H8O1S1	METHYLPHENYLSULFOXIDE 1193-82-4
2480	1.78		853		C7H8O1S1	P-METHIOPHENOL 1073-72-9
1 ★	1.37	7.4	2018		C7H8O2	BENZENE,1,2-DIHYDROXY-4-METHYL 452-86-8
2 ★	0.10		594	400	C7H8O2	GAMMA-PYRONE,2,6-DIMETHYL 1004-36-0
3 ★	0.98		2452		C7H8O2	HYDROQUINONE-2-METHYL 95-71-6
4 ★	0.49		501		C7H8O2	M-HYDROXYBENZYL ALCOHOL 620-24-6
85 ★	1.58		501		C7H8O2	M-METHOXYPHENOL 150-19-6
6 ★	1.34		501		C7H8O2	NEQUINOL 150-76-5 *tmt of hyperpigmentation*
7 ★	1.32		558		C7H8O2	O-METHOXYPHENOL 90-05-1 *expectorant*
8 ★	0.25		501		C7H8O2	P-HYDROXYBENZYL ALCOHOL 623-05-2
9 ★	0.73		260		C7H8O2	SALICYL ALCOHOL 90-01-7 *anesthetic (local)*
2490 ★	0.50		508		C7H8O2S1	SULFONE,METHYLPHENYL 3112-85-4
1 ★	1.51		1478		C7H8O2S1	THIOPHEN-2-ACETIC ACID,METHYL ESTER 19432-68-9
2 ★	1.46		1478		C7H8O2S1	THIOPHEN-3-ACETIC ACID,METHYL ESTER 58414-52-1
3 ★	2.33		525		C7H8O2S1	THIOPHENE,2-CARBOXYLIC ACID,ETHYL ESTER 2810-04-0
4 ★	1.51		1478		C7H8O3	3-METHYLFUROIC ACID,METHYL ESTER 6141-57-7
95	0.74		1937	459	C7H8O3	4-HYDROXY-2-PYRONE,3,6-DIMETHYL 5192-62-1
6 ★	1.78		1055		C7H8O3	FURAN-3-CARBOXYLIC ACID,ETHYL ESTER 614-98-2
7 ★	1.52		525		C7H8O3	FUROIC ACID,ETHYL ESTER 614-99-3
8 ★	0.47		2452		C7H8O3	HYDROQUINONE,METHOXY 824-46-4
9 ★	1.25		536		C7H8O3S1	PHENYLMETHANESULFONATE 16156-59-5
2500 ★	2.74		508		C7H8S1	METHYLTHIOBENZENE 100-68-5
1 ★	2.87		536		C7H8Se1	METHYLPHENYLSELENIDE 4346-64-9
2 ★	0.11		1564		C7H9Br1N2O3	PYRIMIDINE-2,4-DIONE,3-ETHOXY-5-BROMO-6-METHYL 77317-66-9
3 ★	2.71		2499		C7H9Cl1N2O1	PYRAZINE,2-CHLORO-6-PROPOXY
4 ★	4.04		421		C7H9Cl5	1-METHYLPENTACHLOROCYCLOHEXANE 56421-44-4
5 ★	2.94		421		C7H9Cl5O1	1-HYDROXYMETHYLPENTACHLOROCYCLOHEXANE 56400-43-2
6 ★	3.14		421	456	C7H9Cl5O1	2-MEO-PENTACL-CYCLOHEXANE(2,3,5,6/1,4) 56421-31-9
7 ★	3.51		421		C7H9Cl5O1	3-METHOXYPENTACHLOROCYCLOHEXANE 56046-08-3
8 ★	3.75		421		C7H9Cl5S1	1-METHIOPENTACHLOROCYCLOHEXANE 60132-42-5
9 ★	3.85		421		C7H9Cl5S1	3-METHIOPENTACHLOROCYCLOHEXANE 56046-09-4
2510 ★	-0.77		2567		C7H9F2N5O4	3-NITRO-1,2,4-TRIAZOLE,1-(N-HYDROXYETHYL)-B,B-DIFLUOROPROPIONAMIDO ✚
1 ★	1.68	7.0	566	1	C7H9N1	2,6-LUTIDINE 108-48-5
2 ★	1.69	7.0	566	1	C7H9N1	2-ETHYLPYRIDINE 100-71-0
3 ★	1.78	6.0	590		C7H9N1	3,5-DIMETHYLPYRIDINE 591-22-0
4 ★	1.66		2412		C7H9N1	3-ETHYLPYRIDINE 536-78-7
15 ★	1.65		2412		C7H9N1	4-ETHYLPYRIDINE 536-75-4
6 ★	1.66		505		C7H9N1	ANILINE,N-METHYL 100-61-8
7	-2.13	2.0	1473		C7H9N1	BENZYLAMINE HYDROCHLORIDE 3287-99-8
8	-1.96	1.0	541	342	"	BENZYLAMINE HYDROCHLORIDE
9	-1.37	7.0	1473		C7H9N1	BENZYLAMINE 100-46-9
2520 ★	1.09		502		"	BENZYLAMINE
1 ★	1.40		501		C7H9N1	M-TOLUIDINE 108-44-1
2 ★	1.32		531		C7H9N1	O-TOLUIDINE 95-53-4
3	1.42	7.5	2201			O-TOLUIDINE
4	1.39	7.5	2201		C7H9N1	P-TOLUIDINE 106-49-0
25 ★	1.39		501		"	P-TOLUIDINE
6	1.44	7.4	1737	1	"	P-TOLUIDINE
7 ★	0.12		1474		C7H9N1O1	2-PYRIDINEETHANOL 103-74-2
8 ★	0.12		1474		C7H9N1O1	3-PYRIDINEETHANOL 6293-56-7
9 ★	0.10		1474		C7H9N1O1	4-PYRIDINEETHANOL 5344-27-4
2530 ★	-0.05		501		C7H9N1O1	M-AMINOBENZYL ALCOHOL 1877-77-6
1 ★	0.93		501		C7H9N1O1	M-METHOXYANILINE 536-90-3
2	1.00		2163	459	"	M-METHOXYANILINE
3 ★	1.18		547		C7H9N1O1	O-METHOXYANILINE 90-04-0
4	1.23	7.0	566	1	"	O-METHOXYANILINE
35 ★	-0.22	7.4	1737	1	C7H9N1O1	P-AMINOBENZYL ALCOHOL 61224-32-6
6 ★	1.16	7.4	594	314	C7H9N1O1	P-CRESOL,2-AMINO 95-84-1
7	0.81	7.4	1737	1	C7H9N1O1	P-METHOXYANILINE 104-94-9
8 ★	0.95		272		"	P-METHOXYANILINE
9 ★	1.81		2412		C7H9N1O1	PYRIDINE,2-ETHOXY
2540 ★	1.44		2412		C7H9N1O1	PYRIDINE,3-ETHOXY
1	0.05	7.4	2685	1	C7H9N1O2	4-PYRIDONE,3-HYDROXY-2-ETHYL *iron chelator*
2 ★	0.61		2523		C7H9N1O2	FURAN-2-CARBOXAMIDE,N-ETHYL
3 ★	0.72		2523		C7H9N1O2	FURAN-3-CARBOXAMIDE,N-ETHYL
4 ★	0.24		1055		C7H9N1O2	N,N-DIMETHYL-3-FURANCARBOXAMIDE
45 ★	0.41		1055		C7H9N1O2	N,N-DIMETHYL-FURAN-2-CARBOXAMIDE 13156-75-7
6	-0.77	7.4	2685	1	C7H9N1O2	PYRIDIN-4-ONE,1,2-DIMETHYL-3-HYDROXY
7	-0.68	7.4	2472	324	"	PYRIDIN-4-ONE,1,2-DIMETHYL-3-HYDROXY
8 ★	2.30		590		C7H9N1O2	PYRIDINE,2,6-DIMETHOXY 6231-18-1
9 ★	1.96		2221	459	C7H9N1O2	PYRROLE-3-CARBOXYLIC ACID-5-METHYL,METHYL ESTER 40611-76-5
2550 ★	0.13		594		C7H9N1O2S1	A-TOLUENESULFONAMIDE 4563-33-1
1 ★	0.92		570		C7H9N1O2S1	BENZENESULFONAMIDE,N-METHYL 5183-78-8
2 ★	0.85	2.0	264		C7H9N1O2S1	M-METHYLBENZENESULFONAMIDE 1899-94-1
3 ★	0.95		536		C7H9N1O2S1	METHANESULFONANILIDE 1197-22-4
4 ★	0.84	2.0	264		C7H9N1O2S1	O-METHYLBENZENESULFONAMIDE 88-19-7
55 ★	0.82	2.0	264		C7H9N1O2S1	P-METHYLBENZENESULFONAMIDE 70-55-3
6 ★	-0.12		853		C7H9N1O2S1	P-METHYLSULFONYLANILINE 5470-49-5
7 ★	-0.11		1478		C7H9N1O3	FURAN,2-CH2OCH2CONH2
8 ★	0.57	2.1	2265		C7H9N1O3S1	BENZENESULFONAMIDE,3-METHOXY- 58734-57-9
9 ★	0.47	2.0	264		C7H9N1O3S1	P-METHOXYBENZENESULFONAMIDE 1129-26-6

	logP	pH	Ref.	Note	MF	Name / CAS / activity
2560	0.48		2088	1	C7H9N1O3S3	THIENO-4-THIOPYRAN,2-SULFONAMIDE-7-HYDROXY 106319-44-2
1	0.68		2088	1	C7H9N1O3S3	THIENO-7-THIOPYRAN,2-SULFONAMIDE-4-HYDROXY 106319-38-4
2	-1.01		2088	1	C7H9N1O4S3	THIENO-4-THIOPYRAN,2-SULFONAMIDE-4-OXIDE-7-HYDROXY 105951-36-8
3	-0.92		2088	1	C7H9N1O4S3	THIENO-7-THIOPYRAN,2-SULFONAMIDE-4-HYDROXY-7-OXIDE 106335-79-9
4	-0.16		2088	1	C7H9N1O4S3	THIENO-7-THIOPYRAN,2-SULFONAMIDE-7-DIOXIDE 105951-71-1
65	-0.70		2088	1	C7H9N1O5S3	THIENO-4-THIOPYRAN,2-SULFONAMIDE-4-DIOXIDE-6-HYDROXY 106319-46-4
6	-0.59		2088	1	C7H9N1O5S3	THIENO-4-THIOPYRAN,2-SULFONAMIDE-4-DIOXIDE-7-HYDROXY 106319-45-3
7	-0.38		2088	1	C7H9N1O5S3	THIENO-7-THIOPYRAN,2-SULFONAMIDE-4-HYDROXIDE-7-DIOXIDE (4R) 105951-84-6
8	-0.38		2088	1	C7H9N1O5S3	THIENO-7-THIOPYRAN,2-SULFONAMIDE-4-HYDROXIDE (4S)-7-DIOXIDE 105951-48-2
9	-0.35		2088	1	C7H9N1O5S3	THIENO-7-THIOPYRAN,2-SULFONAMIDE-4-HYDROXY-7-DIOXIDE 106400-04-8
2570	-0.44		2088	1	C7H9N1O5S3	THIENO-7-THIOPYRAN,2-SULFONAMIDE-5-HYDROXY-7-DIOXIDE 106319-42-0
1 ★	1.45		853		C7H9N1S1	M-METHIOANILINE 1783-81-9
2	1.20		853		C7H9N1S1	O-METHIOANILINE 2987-53-3
3 ★	0.75		1478		C7H9N1S1	THIOPHEN-2-CARBOXAMIDE,N,N-DIMETHYL 30717-57-8
4 ★	0.55		1478		C7H9N1S1	THIOPHENE-3-CARBOXAMIDE,N,N-DIMETHYL 59906-37-5
75	-2.02		301		C7H9N2O1.Cl1	N1-METHYLNICOTINAMIDECHLORIDE
6	-3.24	7.4	1094	1	C7H9N2O1.C1H3O3S1	2((HYDROXYIMINO)ME)-1-ME-PYRIDINIUMMESULFONATE 154-97-2
7 ★	0.53	13.0	1625	224	C7H9N3	2-PHENYLGUANIDINE 930-68-7
8 ★	-0.49	13.0	594		C7H9N3	BENZAMIDINE,4-AMINO 2498-50-2
9 ★	-0.37	7.4	725	1	C7H9N3O1	2-METHYLISONIAZID 3758-59-6
2580 ★	-0.86		1057		C7H9N3O1	3-AMINOBENZOIC ACID HYDRAZIDE 14062-34-1
1 ★	-0.75		1057		C7H9N3O1	4-AMINOBENZOIC ACID HYDRAZIDE 5351-17-7
2 ★	-0.18		1057		C7H9N3O1	ANTHRANILIC ACID HYDRAZIDE 1904-58-1
3 ★	0.13		594		C7H9N3O1	PHENICARBAZIDE 103-03-7 antipyretic
4 ★	0.38		2499		C7H9N3O1	PYRAZINE,2-ACETYLAMINO-6-METHYL
85	-0.80		2444		C7H9N3O1	PYRAZINE,N,N-DIMETHYLCARBOXAMIDO
6 ★	-0.10	7.4	725	1	C7H9N3O2	2-METHOXYISONIAZID
7	-0.89		1375		C7H9N3O2	3-ACETAMIDOPYRIDAZINE,4-OH-6-METHYL
8	1.00	9.0	1409		C7H9N3O2	5-NITROIMIDZOLE,1-METHYL-2-CYCLOPROPYL 74550-87-1
9	-2.88	7.2	2323	468	C7H9N3O2	B-PYRAZINYL-L-ALANINE
2590 ★	-2.79		1924		"	B-PYRAZINYL-L-ALANINE
1 ★	0.82		2499		C7H9N3O2	PYRAZINE,2-ACETYLAMINO-6-METHOXY
2 ★	0.61		2499		C7H9N3O2	PYRAZINE,2-CARBOXAMIDO-6-ETHOXY
3 ★	-0.45		2352	2	C7H9N3O3S1	3-PYRIDINESULFONYLUREA,N'-METHYL
4 ★	-0.26		1003		C7H9N3O4	1-(3-METHOXY-2-OXOPROPANYL)2-NITROIMIDAZOLE
95 ★	0.17	7.4	2206	306	C7H9N3O4	ACETIC ACID,2-METHYL-5-NITROIMIDAZOL-2YL,METHYL ESTER 1013-51-0
6 ★	-0.52	6.5	321	2	C7H9N3O4S2	HYDROTHIAZIDE 23141-82-4
7	-2.27	7.4	2037	1	C7H9N5O1	ISOGUANINE,1,7-DIMETHYL
8 ★	1.80	7.4	553	1	C7H9N5O2S1	2(1-ME-1,3-IMIDAZOLINYLIDEN-2N)-5-NO2-THIAZOLE 31052-76-3 ✚
9 ★	1.82	7.4	553	1	C7H9N5O2S1	2(4-ME-1,3-IMIDAZOLINYLIDEN-2N)-5-NO2-THIAZOLE 37385-10-7
2600 ★	0.74	7.4	553	1	C7H9N5O3S1	2(H6-5-OH-PYRIMIDINYLIDEN-2-N)-5-NO2-THIAZOLE 37394-79-9 ✚
1 ★	0.46		1500		C7H10Br1N3O4	4-BROMOMISONIDAZOLE 83200-90-2
2	0.58	7.4	2206	306	C7H10Cl1N3O3	ORNIDAZOLE 16773-42-5 anti-infective
3 ★	0.60		1008	354	"	ORNIDAZOLE
4	0.62	7.4	1022			ORNIDAZOLE
5 ★	-0.22	7.0	1783	1	C7H10Cl1N5O4	3((6-AM-3-CL-5-NO2-PYRAZIN-2YL)AM)-1,2-PROPANEDIOL 88793-46-8 ✚
6 ★	-0.57		564		C7H10Cl2N2O1Pt1	DICHLORO-4-METHOXY-O-PHENYLENEDIAMINOPLATINUM
7 ★	-0.06		564		C7H10Cl2N2Pt1	DICHLORO-4-METHYL-O-PHENYLENEDIAMINOPLATINUM
8 ★	2.99		421		C7H10Cl4O1	1-H-2-METHOXYTETRACHLOROCYCLOHEXANE 56933-31-4
9 ★	0.18	7.4	2361	1	C7H10F1N3O3	IMIDAZOLE,2-NITRO-1-(3-METHOXY-2-FLUORO)PROPYL
2610 ★	0.41	7.4	1975	1	C7H10I1N3O4	IMIDAZOLE,1-(3-METHOXY-2-HYDROXYPROPYL)-4-NITRO-5-IODO 96258-78-5
1 ★	0.89	7.4	1975	1	C7H10I1N3O4	IMIDAZOLE,1-(3-METHOXY-2-HYDROXYPROPYL)-5-NITRO-4-IODO 96258-79-6
2 ★	0.63	7.4	594	314	C7H10N2	1,2-BENZENEDIAMINE,4-METHYL 496-72-0
3 ★	0.14	7.4	594	314	C7H10N2	1,3-BENZENEDIAMINE,4-METHYL 95-80-7
4 ★	1.65	9.2	2412		C7H10N2	2-PYRIDINEAMINE,N,N-DIMETHYL 5683-33-0
15	-1.92	2.7	1473	340	C7H10N2	2-PYRIDINEETHANEAMINEHYDROCHLORIDE
6 ★	0.08		1474		C7H10N2	2-PYRIDINEETHANEAMINE 2706-56-1
7	-1.94	2.7	1473	340	C7H10N2	3-PYRIDINEETHANEAMINE HYDROCHLORIDE 20173-24-4
8 ★	-0.11	11.0	533	322	C7H10N2	3-PYRIDINEETHANEAMINE
9 ★	1.34		163	459	C7H10N2	4-DIMETHYLAMINOPYRIDINE 1122-58-3
2620	-0.16	7.4	2362	1	C7H10N2	4-PYRIDINEAMINE,3-ETHYL
1	-1.88	2.7	1473	340	C7H10N2	4-PYRIDINEETHANEAMINE HYDROCHLORIDE 13258-63-4
2 ★	-0.01		1474		C7H10N2	4-PYRIDINEETHANEAMINE
3 ★	0.05		2069		C7H10N2	HEPTANEDINITRILE 646-20-8
4 ★	0.39	11.0	533	322	C7H10N2	N-METHYL-3-PYRIDYLMETHYLAMINE 20173-04-0
25 ★	0.95		2499		C7H10N2	PYRAZINE,2,3,5-TRIMETHYL
6 ★	1.07		2499		C7H10N2	PYRAZINE,2-ETHYL-3-METHYL
7 ★	1.41		2412		C7H10N2	PYRIDINE,3-DIMETHYLAMINO
8 ★	1.80		2499		C7H10N2O1	2-ETHYL-3-METHOXYPYRAZINE 25680-58-4
9 ★	1.82		2499		C7H10N2O1	PYRAZINE,2-ETHOXY-3-METHYL
2630 ★	1.84		2412		C7H10N2O1	PYRAZINE,2-PROPOXY
1 ★	1.98		2499		C7H10N2O2	PYRAZINE,2-ETHOXY-6-METHOXY
2	-0.34	7.4	2307	1	C7H10N2O2	THPO,O-METHYL
3 ★	-0.04		2307	2	"	THPO,O-METHYL
4 ★	-0.15		2544		C7H10N2O2	THYMINE,1-ETHYL
35 ★	-0.01	6.0	479		C7H10N2O2S1	N1-METHYLSULFANILAMIDE 1709-52-0
6	0.08	5.4	264		C7H10N2O2S1	P-METHYLAMINOBENZENESULFONAMIDE 16891-79-5
7 ★	-0.02	1.0	2202	342	C7H10N2O3	BARBITURIC ACID,5-I-PROPYL 7391-69-7
8 ★	0.02		1829		C7H10N2O3	BARBITURIC ACID,5-METHYL-5-ETHYL 27653-63-0
9 ★	0.15	0.7	962		C7H10N2O3	BARBITURIC ACID,N-METHYL,5-ETHYL
2640 ★	-0.73		2467	448	C7H10N2O4S2	BENZENESULFONAMIDE,4-METHYLSULFONAMIDO
1 ★	1.86		1519	459	C7H10N2S1	2-METHYLTHIO-4,6-DIMETHYLPYRIMIDINE 14001-64-0
2 ✓	-3.10		2397	101	C7H10N3O1	IMIDAZOLIUM CHLORIDE,1-METHYL-2-HYDROXYIMINOMETHYL-3-VINYL ✚
3 ★	0.38		2499		C7H10N4O1	PYRAZINE,2-CARBOXAMIDO-6-DIMETHYLAMINO
4 ★	-1.22		508		C7H10N4O2S1	SULFAGUANIDINE 57-67-0 antibacterial
45 ★	0.13	1.0	2196		C7H10N4O3	PYRAZOLE,3-METHYL-4-NITRO-5-(N-ETHYLCARBOXAMIDO)
6 ★	-0.64	3.0	2191		C7H10N4O3	URACIL,6-AMINO-1,3-DIMETHYL-5-FORMYLAMINO 7597-60-6
7	-1.30	3.0	2191	459	C7H10N4O3	URACIL,6-AMINO-1-METHYL-5-(N-METHYLFORMYLAMINO) 33130-54-0
8 ✓	-1.30	3.0	2191		"	URACIL,6-AMINO-1-METHYL-5-(N-METHYLFORMYLAMINO)
9 ★	-2.03	3.0	2191		C7H10N4O3	URACIL,6-AMINO-3-METHYL-5-(N-METHYLFORMYLAMINO) 55782-76-8

	logP	pH	Ref.	Note	MF	Name / CAS / activity
2650 ★	-0.49	7.0	1783	1	C7H10N4O4	2-(2,3-DIHYDROXYPROPYL)AMINO-3-NITROPYRAZINE 87885-48-1
1	-1.68	7.4	2361	1	C7H10N4O4	2-(2-NITROIMIDAZOL-1-YL)-N-ETHAN-2-OL-ACETAMIDE 22668-01-5 *antineoplastic*
2 ★	-1.34	7.4	1428	1	"	2-(2-NITROIMIDAZOL-1-YL)-N-ETHAN-2-OL-ACETAMIDE
3	-1.32	7.4	2561	1	"	2-(2-NITROIMIDAZOL-1-YL)-N-ETHAN-2-OL-ACETAMIDE
4	-1.11	7.4	2561	1	C7H10N4O4	2-NITROIMIDAZOLE-1-(N-HYDROXYBUTYRAMIDO)
55 ★	-0.29	7.4	2206	306	C7H10N4O4	BAMNIDAZOLE 31478-45-2 *antiprotozoal*
6 ★	-1.59	7.4	1428	1	C7H10N4O5	2(2-NITROIMIDAZOL-1-YL)-N,N-DIMETHANOLACETAMIDE
7 ★	-1.54	7.4	2273	1	C7H10N4O6	1,2,4-TRIAZOLE,3-NITRO-1-D-RIBOFURANOSE 105086-85-9
8 ★	1.05		1967		C7H10N4S1	THIOPHENE-2-ACETYL,GUANYLHYDRAZONE 72189-64-1
9 ✓	-2.44		2397	101	C7H10Br1N3O1	IMIDAZOLIUM CHLORIDE,1-METHYL-2-HYDROXYIMINOMETHYL-3-BROMOETHYL 132540-13-7
2660	-2.58		2397	101	C7H11F1N3O1	IMIDAZOLIUM CHLORIDE,1-METHYL-2-HYDROXYIMINOMETHYL-3-FLUOROETHYL 132540-09-1
1 ★	2.68		2488		C7H11F2N5S1	1,3,5-TRIAZINE,2-DIFLUOROMETHIO-4-ETHYLAMINO-6-METHYLAMINO
2 ★	1.19		1055		C7H11N1O1	FURANMETHAMINE,N,N-DIMETHYL 14496-34-5
3 ★	-0.40		213		C7H11N1O2	N-ACETYLCYCLOBUTANECARBOXAMIDE 6932-05-4
4 ★	0.85	7.4	2490	1	C7H11N1O3S3	THIOPHENE-2-SULFONAMIDE,5-(3-HYDROXYPROPYL)THIO 104437-99-2
65 ★	-0.36	7.4	2490	1	C7H11N1O5S3	THIOPHENE-2-SULFONAMAIDE,5-(3-HYDROXYPROPYL)SULFONYL 104438-02-0
6 ★	1.67		1478		C7H11N1S1	THIOPHEN,2-(N,N-DIMETHYLAMINOMETHYL) 26019-17-0
7 ★	1.57		2499		C7H11N3	PYRAZINE,2-DIMETHYLAMINO-6-METHYL
8	0.24		1375		C7H11N3O1	3-DIMETHYLAMINOPYRIDAZINE,4-OH-6-METHYL
9 ★	0.90		897		C7H11N3O1	4-PYRIDAZINEAMINE,3-METHOXY-N,N-DIMETHYL 38717-24-7
2670 ★	1.65		2499		C7H11N3O1	PYRAZINE,2-METHOXY-5-DIMETHLAMINO
1 ★	1.99		2499		C7H11N3O1	PYRAZINE,2-METHOXY-6-DIMETHYLAMINO
2	2.03	7.4	2298	1	C7H11N3O2	1,2,4-OXADIAZOL-5-OXIME,3-T-BUTYL 99764-47-3
3 ★	2.15	7.4	2298	2	"	1,2,4-OXADIAZOL-5-OXIME,3-T-BUTYL
4 ★	1.09	7.4	1022	1	C7H11N3O2	1-METHYL-2-NITRO-5-I-PROPYLIMIDAZOLE
75 ★	1.08	7.4	2206	306	C7H11N3O2	5-NITROIMIDAZOLE,1-ME-2-I-PR 14885-29-1 *antiprotozoal (histomonas)*
6 ★	0.53	7.4	1022	1	C7H11N3O3	1-(2-ETHOXYETHYL)-2-NITROIMIDAZOLE
7 ★	0.43	7.4	1022	1	C7H11N3O3	1-METHYL-2-NITRO-5-(1-HYDROXY-1-METHYLETHYL)IMIDAZOLE
8	0.15	7.4	2530	1	C7H11N3O3	SECNIDAZOLE 3366-95-8 *anti-amebic, antiprotozoal (trichomonas)*
9 ★	0.22	7.4	2206	306	"	SECNIDAZOLE
2680	0.44	7.4	2380	1	"	SECNIDAZOLE
1 ★	-0.59	7.4	1022	1	C7H11N3O4	1-(2,3-DIHYDROXYPROPYL)-2-METHYL-5-NITROIMIDAZOLE 62580-80-7
2	-0.46	7.4	2361	1	C7H11N3O4	MISONIDAZOLE 13551-87-6 *antiprotozoal, antineoplastic*
3	-0.37	7.4	1470	1	"	MISONIDAZOLE
4	-0.37	7.4	2380	1	"	MISONIDAZOLE
85	-0.37	7.4	2561	1	"	MISONIDAZOLE
6 ★	-0.37		1003		"	MISONIDAZOLE
7	-0.32	7.0	1783	1	"	MISONIDAZOLE
8	-0.80	7.4	1022	1	C7H11N3O4S1	1-(2-ETHYLSULFONYLETHYL)-2-NITROIMIDAZOLE
9 ★	-0.58		1003		"	1-(2-ETHYLSULFONYLETHYL)-2-NITROIMIDAZOLE
2690 ★	-0.82		835		C7H11N3O4S1	ME(2-(2-ME-5-NITRO-1-IMIDAZOLYL)ETHYL)SULFONE 19387-94-1
1 ★	-1.17		2563		C7H11N3O5	2-NITROIMIDAZOLE,1-(1,3-DIHYDROXYPROP-2YL)OXYMETHYL
2	-0.11	7.4	1948	1	C7H11N3S1	1-(2-THIAZOLYL)PIPERAZINE 42270-37-1
3	0.61	7.4	1948	2	"	1-(2-THIAZOLYL)PIPERAZINE
4 ★	-0.72	7.4	2273	1	C7H11N5O3	1,2,4-TRIAZOLE,3-NITRO-1-(3-AZIRIDINYL-2-HYDROXY)PROPYL 104958-92-1
95	-0.10		2547		C7H11N5O3S1	1,2,4-TRIAZOLE,3-NITRO-1-(N-METHIOETHYLCARBAMYL)METHYL
6	0.00	7.4	2547	1	"	1,2,4-TRIAZOLE,3-NITRO-1-(N-METHIOETHYLCARBAMYL)METHYL
7 ★	0.25		2547		C7H11N5O3S1	3-NITROTRIAZOLE-1-(N-METHOXYETHYL)THIOACETAMIDO
8	0.49	7.4	2547	1	"	3-NITROTRIAZOLE-1-(N-METHOXYETHYL)THIOACETAMIDO
9	-0.85	7.4	2273	1	C7H11N5O4	1,2,4-TRIAZOLE,3-NITRO-1-(N-METHOXYETHYL)ACETAMIDO 104958-90-9
2700	-0.78		2547		"	1,2,4-TRIAZOLE,3-NITRO-1-(N-METHOXYETHYL)ACETAMIDO
1	-1.27	7.4	2273	1	C7H11N5O4	1,2,4-TRIAZOLE,3-NITRO-1-N-(3-HYDROXYPROPYL)ACETAMIDO 104958-91-0
2	-1.08		2547		"	1,2,4-TRIAZOLE,3-NITRO-1-N-(3-HYDROXYPROPYL)ACETAMIDO
3 ★	-1.32		2567		C7H11N5O4	3-NITRO-1,2,4-TRIAZOLE,1-(N-HYDROXYETHYL)PROPIONAMIDO
4 ★	2.08		401		C7H12Cl1N3O2S2	3-CYCLOHEXYL-3,5-DITHIAN)-1-(2CLET)-1-NITROSOUREA 33022-04-7
5	-1.75	7.4	401	1	C7H12Cl1N3O4	UREA,1-(2CLET)-1-NO-3-(2-I-BUTYRIC ACID) 52320-87-3
6	1.74	7.4	401	2	"	UREA,1-(2CLET)-1-NO-3-(2-I-BUTYRIC ACID)
7 ★	-0.94		401		C7H12Cl1N3O6S2	3-(M-DI-SO2-PYRAN)-1-(2-CLETHYL)-1-NITROSOUREA 33022-05-8 ✚
8 ★	2.31		964		C7H12Cl1N5	2-CL-N4,N4,N6,N6-TETRAMETHYLMELAMINE 3140-74-7 ✚
9 ★	2.18		1084		C7H12Cl1N5	SIMAZINE 122-34-9 *herbicide*
2710 ✓	1.95		2366	122	C7H12Cu1N6S2	COPPER(II)PYRUVALDEHYDE-BIS(N4-METHYLTHIOSEMICARBAZONE)
1 ★	1.54		401		C7H12F1N3O2S2	3-(CYCLOHEXYL-3,5-DITHIAN)-1-(2FLET)-1-NITROSOUREA 33024-47-4
2 ★	-0.17	13.0	594		C7H12N2	DBN 3001-72-7 ✚
3	0.52		1407		C7H12N2O1	1,2,3,6-H4-PYRIDINE,1-METHYLCARBONYLAMINO
4 ★	0.64		401		C7H12N2O1	1,3-DIALLYLUREA 1801-72-5
15 ★	-0.86	7.0	2309	1	C7H12N2O2	GLYCINAMIDE,N-ACETYL-2-ALLYL
6 ★	-1.34	7.1	1591		C7H12N2O2	PROLIN-AMIDE,N-ACETYL
7 ★	1.33		2188		C7H12N2O2S2	METHYLCARBAMIC ACID,4-OXIMINO-2,2-DIMETHYL-1,3-DITHIOLANE ESTER ✚
8 ★	1.31		2188		C7H12N2O2S2	METHYLCARBAMIC ACID,4-OXIMINO-5,5-DIMETHYL-1,3-DITHIOLANE ESTER
9 ★	0.96		2188		C7H12N2O3S1	METHYLCARBAMIC ACID,4-OXIMINO-5,5-DIMETHYL-1,3-OXATHIOLANE ESTER
2720 ★	0.82		2188		C7H12N2O3S1	METHYLCARBAMIC ACID,4-OXIMINO-5,5-DIMETHYL-1,3-OXATHIOLANE ESTER
1 ★	-2.47	7.1	1591		C7H12N2O4	GLUTAMIC ACID-MONOAMIDE,N-ACETYL
2 ✓	-3.00		2397	101	C7H12N4O2	IMIDAZOLIUM CHLORIDE,1-METHYL-2-HYDROXYIMINOMETHYL-3-METHOXYMETHYL 91900-09-3
3 ★	1.01		57		C7H12N4O1S1	3-METHIO-4-AMINO-6-I-PR-1,2,4-TRIAZINE-5-ONE 21087-61-6
4 ★	0.93		57		C7H12N4O1S1	3-METHIO-4-AMINO-6-N-PR-1,2,4-TRIAZINE-5-ONE 23141-81-3
25 ★	1.34		2592		C7H12N4O2	1,2,4-TRIAZOLE,1-I-AMYL-3-NITRO
6 ★	-0.06		57		C7H12N4O2	3-METHOXY-4-AMINO-6-I-PR-1,2,4-TRIAZINE-5-ONE 18826-97-6
7 ★	-0.41	7.4	830	1	C7H12N4O2	N-DIAZOACETYLGLYCINE-N'-I-PROPYLAMIDE
8 ★	-0.24	7.4	830	1	C7H12N4O2	N-DIAZOACETYLGLYCINE-N'-PROPYLAMIDE ✚
9 ★	-0.07	7.4	2273	1	C7H12N4O4	1,2,4-TRIAZOLE,3-NITRO-1-(3-ETHOXY-2-HYDROXY)PROPYL 104958-88-5
2730 ★	1.50		502		C7H12O1	2-BUTANONE,4-CYCLOPROPYL 2046-23-3
1 ★	1.38		579		C7H12O1	CYCLOHEXANONE,4-METHYL 589-92-4
2	1.12		1231		C7H12O2	3,5-HEPTANEDIONE 7424-54-6
3 ★	2.36		1402		C7H12O2	ACRYLIC ACID,BUTYL ESTER 141-32-2
4 ★	2.22		1402		C7H12O2	ACRYLIC ACID,I-BUTYL ESTER 106-63-8
35	1.96	2.0	2312	337	C7H12O2	CYCLOHEXANECARBOXYLIC ACID
6 ★	2.25		1402		C7H12O2	METHACRYLIC ACID,I-PROPYL ESTER 4655-34-9
7 ★	0.97		1402		C7H12O3	METHACRYLIC ACID,2-HYDROXYPROPYL ESTER 923-26-2
8 ★	0.67	1.0	594		C7H12O4	GLUTARIC ACID,2,2-DIMETHYL 681-57-2
9 ★	0.62		579		C7H12O4	GLUTARIC ACID,DIMETHYL ESTER 1119-40-0

	logP	pH	Ref.	Note	MF	Name / CAS / activity
2740 ★	0.96		592		C7H12O4	MALONIC ACID,DIETHYL ESTER **105-53-3**
1 ★	1.54		843		C7H13Br1N2O2	A-BROMO-A-ETHYLBUTYRLUREA **77-65-6** *sedative, hypnotic*
2 ★	1.40		843		C7H13Cl1N2O2	A-CHLORO-A-ETHYLBUTANOYLUREA
3 ✓	1.33		567	820	C7H13N1	QUINUCLIDINE PICRATE
4 ★	1.38	13.0	588	224	C7H13N1	QUINUCLIDINE **100-76-5**
45 ★	0.24		589		C7H13N1O1	2-AZACYCLOOCTANONE **673-66-5**
6 ★	-1.87		2654	448	C7H13N1O2	CYCLOHEXANECARBOXYLIC ACID-1-AMINO **2756-85-6**
7 ★	-1.31	7.2	2323	468	C7H13N1O2	CYCLOPENTYLGLYCINE
8	-1.05	7.4	1443	1	C7H13N1O2	PIPERIDINE-1-ACETIC ACID **55049-18-8**
9 ★	-0.17	2.2	2228	314	C7H13N1O3	BUTANOIC ACID-2-ACETYLAMINO,METHYL ESTER **3619-01-0**
2750 ★	1.00		561		C7H13N1S1	2-AZACYCLOOCTANTHIONE **22928-63-8**
1 ★	1.10		1137		C7H13N3O1	2-AMINO-5-AMYL-1,3,4-OXADIAZOLE
2 ★	-2.05	7.1	1591		C7H13N3O3	GLUTAMIN-AMIDE,N-ACETYL
3 ★	-0.47		1263		C7H13N3O3S1	OXAMYL **23135-22-0** *nematocide~insecticide*
4 ★	0.66		536		C7H13N3O4	3-NITRO-3-ME-2-BUTANONE-(N-ME-CARBAMOYL)OXIME **20417-83-8**
55 ★	0.30		57		C7H13N5O1	6-I-PROPYL-4-AMINO-3-MEAMINO-1,2,4-TRIAZIN-5-ONE **21087-57-0**
6 ★	1.20		1903		C7H13O6P1	MEVINPHOS **7786-34-7** *insecticide*
7 ★	3.99		1560	100	C7H14	1-HEPTENE **592-76-7**
8 ★	4.00		594	400	C7H14	CYCLOHEPTANE **291-64-5**
9 ★	3.61		594		C7H14	METHYLCYCLOHEXANE **108-87-2**
2760	1.94		1717		C7H14Cl3N2O2P1	3-CHLORO-CYTOXAN **87154-30-1**
1 ✓	1.92		1701	820	C7H14F3N1	N-PROPYL-N-ETHYL-2,2,2-TRIFLUOROETHYLAMINE PICRATE
2 ★	2.38		1701		C7H14F3N1	N-PROPYL-N-ETHYL-2,2,2-TRIFLUOROETHYLAMINE
3 ★	0.48	7.0	2483	1	C7H14N1O2	DOXO ✚
4 ★	-0.20		2049		C7H14N1O5P1	MONOCROTOPHOS **6923-22-4** *insecticide*
65 ★	1.36		871		C7H14N2O1	2,6-DIMETHYL-N-NITROSOPIPERIDINE **17721-95-8**
6 ★	1.53		871		C7H14N2O1	3,5-DIMETHYL-N-NITROSOPIPERIDINE
7 ★	1.48		871		C7H14N2O1	N-NITROSO-HEPTAMETHYLENEIMINE **20917-49-1**
8 ★	1.43		871		C7H14N2O1	N-NITROSO-METHYLCYCLOHEXYLAMINE
9 ★	0.91		673		C7H14N2O2	2-ETHYLBUTANOYLUREA **2274-01-3**
2770 ★	-0.45	7.0	2309	1	C7H14N2O2	GLYCINAMIDE,N-ACETYL-2-PROPYL
1 ★	-0.61	7.1	1591		C7H14N2O2	VALIN-AMIDE,N-ACETYL
2 ★	1.13		1693		C7H14N2O2S1	ALDICARB **116-06-3** *insecticide*
3 ★	-0.60	7.1	1591		C7H14N2O2S1	METHIONIN-AMIDE,N-ACETYL
4	-0.34	7.0	2309	1	"	METHIONIN-AMIDE,N-ACETYL
75	-2.84		1590	459	C7H14N2O3	FORMYLLYSINE **1190-48-3**
6 ★	-0.57		1624		C7H14N2O4S1	ALDICARBSULFONE **1646-88-4** *insecticide*
7 ★	1.13		1157		C7H14N2S1	N,N'-DIETHYL-ETHYLENETHIOUREA
8 ✓	-3.52		1704	820	C7H14N3O2.Br1	AZOACETYLOXYMETHYL-TETRAMETHYLAMMONIUMBROMIDE
9 ★	1.20		964		C7H14N6	N2,N2,N4,N4-TETRAMETHYLMELAMINE **2827-47-6**
2780 ★	1.33		579		C7H14O1	1-METHYLCYCLOHEXANOL **590-67-0**
1 ★	1.86		594	400	C7H14O1	2,4-DIMETHYL-3-PENTANONE **565-80-0**
2 ★	1.98		1560	459	C7H14O1	2-HEPTANONE **110-43-0**
3 ★	1.84		2126		C7H14O1	2-METHYLCYCLOHEXANOL-CIS **7443-70-1**
4 ★	1.82		2126		C7H14O1	2-METHYLCYCLOHEXANOL-TRANS **7443-52-9**
85 ★	1.88		2047		C7H14O1	5-METHYL-2-HEXANONE
6 ★	1.79		579		C7H14O1	CYCLOHEXANOL,4-METHYL **589-91-3**
7 ★	2.23		1815		C7H14O2	2-ETHYL-PENTANOIC ACID **20225-24-5**
8	0.63		2596		C7H14O2	BUTYL GLYCIDYL ETHER **2426-08-6**
9	2.19	3.5	2391	1	C7H14O2	PENTANOIC ACID,3-ETHYL
2790 ★	-0.17		530		C7H14O4	GLYCERYLMONOBUTYRATE **557-25-5**
1 ★	4.36		1560	100	C7H15Br1	1-BROMOHEPTANE **629-04-9**
2 ✓	-2.81		1704	820	C7H15Br1N1O2.Br1	2-(BROMOACETYLOXY)ETHYL,TRIMETHYL AMMONIUM BROMIDE
3 ★	4.15		1560	100	C7H15Cl1	1-CHLOROHEPTANE **629-06-1**
4 ✓	-3.12		1704	820	C7H15Cl1N1O2.Br1	2-(CHLOROACETYLOXY)ETHYL,TRIMETHYL AMMONIUM BROMIDE
95 ★	0.63		401		C7H15Cl2N2O2P1	CYTOXAN **50-18-0** *antineoplastic, immunosuppressive*
6 ★	0.86		401		C7H15Cl2N2O2P1	IFOSFAMIDE **3778-73-2** *antineoplastic*
7 ✓	-3.41		1704	820	C7H15F1O2.Br1	2-(FLUOROACETYLOXY)ETHYL,TRIMETHYL AMMONIUM BROMIDE
8 ★	4.70		1560	100	C7H15I1	1-IODOHEPTANE **4282-40-0**
9 ✓	-2.63		1704	820	C7H15I1O2.Br1	2-(IODOACETYLOXY)ETHYL,TRIMETHYL AMMONIUM BROMIDE
2800	-1.52	7.4	1443	1	C7H15N1	N-ETHYLPIPERIDINE **766-09-6**
1 ★	1.53	13.0	594	400	C7H15N1	PIPERIDINE,2,6-DIMETHYL-CIS **766-17-6**
2	-1.30	7.4	1443	1	C7H15N1O1	2-PIPERIDINOETHANOL **3040-44-6**
3 ★	0.96	11.0	1443	322	"	2-PIPERIDINOETHANOL
4 ★	1.98		1969		C7H15N1O2	2-ETHYLBUTYL CARBAMATE (ESTER) **24847-58-3**
5 ★	-3.03		2654	448	C7H15N1O2	7-AMINOHEPTANOIC ACID
6 ★	2.00		1969		C7H15N1O2	DIMETHYLCARBAMATE,I-BUTYL ESTER **52113-78-7**
7 ★	-1.42		573		C7H15N1O2	NEOPENTYLGLYCINE
8 ★	1.85	7.4	172	1	C7H15N1O2	O-HEXYLCARBAMATE **2114-20-7**
9 ★	1.96		702		C7H15N1O2	O-PENTYL-N-METHYLCARBAMATE **2594-17-4**
2810 ★	1.45	7.4	172	1	C7H15N1O2	O-T-HEXYLCARBAMATE **52642-50-9**
1 ★	0.15		1394		C7H15N3O1	4-CYCLOHEXYLSEMICARBAZIDE **52662-76-7**
2 ★	1.54		871		C7H15N3O2	1-NITROSOTRIETHYLUREA **50285-70-6**
3	4.66		1560		C7H16	HEPTANE **142-82-5**
4 ★	-4.12		549	466	C7H16N1O2.Br1	ACETYLCHOLINEBROMIDE **66-23-9**
15 ✓	-3.03		1704	820	C7H16N1O2.Br1	DIOXOLAN-3YL-TETRAMETHYLAMMONIUMBROMIDE
6 ✓	-2.52	7.2	2016		C7H16N1O2.Cl1	ACETYLCHOLINECHLORIDE **60-31-1** *cardiac depressant, vasodilator*
7 ✓	-3.03		2033		C7H16N1O2.I1	OXAPROPANIUM IODIDE **541-66-2** *cholinergic* ✚
8 ✓	-3.58		1704	820	C7H16N1O3.Br1	(METHOXYACETYLOXYETHYL)TRIMETHYL AMMONIUM BROMIDE
9 ✓	-2.70		1704	820	C7H16N1S1.Br1	OXATHIOLAN-5YL-TETRAMETHYLAMMONIUMBROMIDE ✚
2820 ✓	-2.70		2033		C7H16N1S1.I1	1,3-THIOXOLANE-4-TRIMETHYLAMMONIOMETHYL IODIDE
1 ✓	0.86		567	820	C7H16N1.C6H2N3O7	N,N-DIMETHYLPIPERIDINIUM PICRATE
2 ★	0.97		1137		C7H16N2O1	HEPTANOIC ACID HYDRAZIDE **22371-32-0**
3	2.03		2570		C7H16N2O1	N-NITROSOPROPYLBUTYL AMINE **25413-64-3**
4	-4.21		1590	459	C7H16N2O2	N'-METHYLLYSINEION **1188-07-4**

	logP	pH	Ref.	Note	MF	Name / CAS / activity
25	-3.90		1590	459	C7H16N4O2	HOMOARGININEION 156-86-5
6 ★	2.31		579		C7H16O1	HEPTAN-2-OL 543-49-7
7 ★	2.24		579		C7H16O1	HEPTAN-3-OL 589-82-2
8 ★	2.22		579		C7H16O1	HEPTAN-4-OL 589-55-9
9 ★	2.72		579		C7H16O1	HEPTANOL 111-70-6
2830 ★	1.20	13.0	594	400	C7H16O3	TRIETHYLORTHOFORMATE 122-51-0
1	0.15	7.0	834	1	C7H17N1	HEPTYLAMINE 111-68-2
2 ★	2.57	9.3	588	2	"	HEPTYLAMINE
3 ★	2.07	13.0	588	224	C7H17N1	PROPYL-I-BUTYLAMINE
4 ★	1.91	13.0	588	224	C7H17N1	PROPYL-S-BUTYLAMINE 39190-67-5
35 ✓	2.16		567		C7H17N1	PROPYL-T-BUTYLAMINE PICRATE
6 ★	2.12	13.0	588	224	C7H17N1	PROPYLBUTYLAMINE 20193-21-9
7 ★	1.56	13.0	594		C7H17N1	T-BUTYL-I-PROPYLAMINE 7515-80-2
8 ★	2.40	13.0	1806	224	C7H17N1	TUAMINOHEPTANE 123-82-0 adrenergic (vascoconstrictor)
9 ✓	-3.46		1704	820	C7H17N2O1.Br1	CH3CH=NOCH2CH2N(CH3)3-BROMIDE
2840 ★	0.46		1394		C7H17N3O1	4-HEXYLSEMICARBAZIDE
1	-1.02	7.3	761	314	C7H17N5	PENTYLBIGUANIDE 21306-55-8
2	3.26		2182	459	C7H17O2P1S1	METHYTHIOPHOSPHORIC ACID,DIPROPYL ESTER
3 ★	1.48		1108		C7H17O2P1S2	O-ETHYL-S-ETHYLTHIOETHYL-METHYLPHOSPHONATE
4 ★	3.56		579		C7H17O2P1S3	PHORATE 298-02-2 insecticide
45	1.03		2182	459	C7H17O3P1	METHYLPHOSPHONIC ACID,BIS-I-PROPYL ESTER 1445-75-6
6 ★	2.07		579		C7H17O3P1S2	O,O-DIETHYL,S-ETHYLTHIOMETHYLPHOSPHATE 2600-69-3
7 ★	1.78		1693		C7H17O4P1S2	PHORATESULFOXIDE 2588-05-8
8 ★	1.98		1693		C7H17O5P1S2	PHORATESULFONE 2588-06-9
9	-2.22		1257	820	C7H18N1O1.I1	ETHOXYETHYLTRIMETHYL AMMONIUM IODIDE
2850	-1.61		1901	820	C7H18N1.I1	S-BUTYLTRIMETHYL AMMONIUM IODIDE 40916-78-7
1	-2.08		1901	820	C7H18N1.I1	T-BUTYLTRIMETHYL AMMONIUM IODIDE 25728-37-4
2	-2.13		1901	820	C7H18N1.I1	TRIETHYLMETHYL AMMONIUM IODIDE 302-57-8
3	-2.60		524	820	C7H18N1.I1	TRIMETHYLBUTYL AMMONIUM IODIDE 7722-19-2
4	-1.49		1901	820	"	TRIMETHYLBUTYL AMMONIUM IODIDE
55 ✓	1.41		567	820	C7H18N1.C6H2N3O7	BUTYLTRIMETHYL AMMONIUM PICRATE
6 ✓	1.21		567	820	C7H18N1.C6H2N3O7	I-BUTYLTRIMETHYL AMMONIUM PICRATE
7 ✓	0.87		567	820	C7H18N1.C6H2N3O7	METHYLTRIETHYL AMMONIUM PICRATE
8 ✓	1.03		567	820	C7H18N1.C6H2N3O7	S-BUTYLTRIMETHYL AMMONIUM PICRATE
9 ✓	0.79		567	820	C7H18N1.C6H2N3O7	T-BUTYLTRIMETHYL AMMONIUM PICRATE
2860	0.91		1463		C7H18N1.C6H3N2O5	BUTYLTRIMETHYL AMMONIUM-2,4-DINITROPHENOLATE
1 ★	4.20		512		C7H18Si1	SILANE,BUTYL-TRIMETHYL 1000-49-3
2 ★	2.90		2300	20	C8Cl4N2	CHLOROTHALONIL 1897-45-6 fungicide, nematocide
3 ★	4.81		509		C8H1Br4F3N2	BENZIMIDAZOLE,4,5,6,7-TETRABROMO-2-TRIFLUOROMETHYL 2338-30-9
4	2.62	8.0	786	1	C8H1Cl4F3N2	BENZIMIDAZOLE-4,5,6,7-TETRACHLORO-2-TRIFLUOROMETHYL 2338-29-6
65	5.58	8.0	786	2	"	BENZIMIDAZOLE-4,5,6,7-TETRACHLORO-2-TRIFLUOROMETHYL
6 ★	4.08		509		C8H2Br3F3N2	BENZIMIDAZOLE,4,5,6-TRIBROMO-2-TRIFLUOROMETHYL 7682-32-8
7 ★	3.78		509		C8H2Cl3F3N2	BENZIMIDAZOLE,4,5,7-TRICHLORO-2-TRIFLUOROMETHYL 3393-59-7
8 ★	3.20		2300	20	C8H2Cl4O2	FTHALIDE 27355-22-2 pesticide
9 ★	2.16		1946		C8H3Br1N2O1	PROPANEDINITRILE,(5-BROMO-2-FURANYL)METHYLENE 56656-96-3
2870 ★	4.15		509		C8H3Br2F3N2	BENZIMIDAZOLE,2-TRIFLUOROMETHYL-5,6-DIBROMO 6587-21-9
1 ★	3.21		509		C8H3Cl1F3N3O2	BENZIMIDAZOLE,5-CHLORO-6-NITRO-2-TRIFLUOROME 6609-40-1
2 ★	2.17		2141		C8H3Cl1N2O1	PROPANEDINITRILE,(5-CHLORO-2-FURANYL)METHYLENE
3 ★	3.49		509		C8H3Cl2F3N2	BENZIMIDAZOLE,2-TRIFLUOROME-4,5-DICHLORO 3615-21-2
4 ★	3.49		509		C8H3Cl2F3N2	BENZIMIDAZOLE,2-TRIFLUOROME-4,6-DICHLORO 4228-88-0
75 ★	2.87		509		C8H3Cl2F3N2	BENZIMIDAZOLE,2-TRIFLUOROME-4,7-DICHLORO 4228-89-1
6 ★	3.99		509		C8H3Cl2F3N2	BENZIMIDAZOLE,2-TRIFLUOROME-5,6-DICHLORO 2338-25-2
7	3.89		509		C8H3F3N4O4	BENZIMIDAZOLE,2-TRIFLUOROME-5,6-DINITRO 2338-24-1
8 ★	2.42		1946		C8H3I1N2O1	PROPANEDINITRILE,(5-IODO-2-FURANYL)METHYLENE 69527-40-8 ✚
9 ★	3.57		509		C8H4Br1F3N2	BENZIMIDAZOLE,2-TRIFLUOROME-5-BROMO 3671-60-1
2880 ★	3.39		509		C8H4Cl1F3N2	BENZIMIDAZOLE,2-TRIFLUOROME-5-CHLORO 656-49-5
1 ★	2.93		509		C8H4Cl1F3N2	BENZIMIDAZOLE,2-TRIFLUROME-4-CHLORO 2338-31-0
2 ★	1.65		522		C8H4Cl1F3N2O2S1	1,2,4-BENZOTHIADIAZINE-1,1-DIOXIDE-3-TRIFLUOROMETHYL-6-CHLORO 2251-64-1
3 ★	3.90		1792		C8H4Cl2F3N1O1	TRIFLUOROACETAMIDE,N-(3,4-DICHLOROPHENYL)
4	2.83	8.0	786	1	C8H4Cl4N2	4,5,6,7-TETRACHLORO-2-ME-BENZIMIDAZOLE 18237-94-0
85 ✓	4.62		536		C8H4Cl6	P-DI(TRICHLOROMETHYL)BENZENE 68-36-0
6 ★	2.52		2097		C8H4F3N3O1	DIAZINECARBONITRILE,(3-TRIFLUOROMETHYLPHENYL),OXIDE
7 ★	2.47		2097		C8H4F3N3O1	DIAZINECARBONITRILE,(4-TRIFLUOROMETHYLPHENYL),OXIDE
8 ★	2.68		509		C8H4F3N3O2	BENZIMIDAZOLE,2-TRIFLUOROME-5-NITRO 327-19-5
9 ✓	-0.90	7.4	2360	1	C8H4F3N3O3	BENZOTRIAZINE-1,4-DI-N-OXIDE-3-HYDROXY-7-TRIFLUOROMETHYL
2890 ★	1.70		2410		C8H4F5N1O1	ACETANILIDE,2,3,4,5,6-PENTAFLUORO
1 ★	3.83		594	400	C8H4F6	1,3-BIS-(TRIFLUOROMETHYL)BENZENE 402-31-3
2 ★	3.83		594		C8H4F6	1,4-BIS(TRIFLUOROMETHYL)BENZENE 433-19-2
3 ★	0.80		594		C8H4N2	BENZENE,1,3-DICYANO 626-17-5
4 ★	0.93		594		C8H4N2	BENZENE,1,4-DICYANO 623-26-7
95 ★	0.99		2410		C8H4N2	PHTHALONITRILE 91-15-6
6 ★	1.40		1946		C8H4N2O1	PROPANEDINITRILE,(2-FURANYLMETHYLENE) 3237-22-7
7 ★	0.47	7.4	594	1	C8H4N2O4	ISATIN,5-NITRO 611-09-6
8 ★	0.74		494		C8H4N2O4S2	5(5-NO2-2-FURFURYLIDINE)-2-S-THIAZOLIDINE-4-ONE 13410-84-9 ✚
9 ★	0.63		494		C8H4N2O5S1	5-(5-NO2-2-FURFURILIDINE)THIAZOLIDINE-2,4-DIONE 27564-47-2
2900 ★	0.55		494		C8H4N2O5S1	6-(5-NO2-2-FURYL)-1,3-THIAZIN-2,4-DIONE 52661-48-0 ✚
1 ★	3.13		1947		C8H4N2S1	BENZONITRILE,3-ISOTHIOCYANATO 3125-78-8
2	2.65		1947		C8H4N2S1	BENZONITRILE,4-ISOTHIOCYANATO 2719-32-6
3 ★	1.15		2097		C8H4N4O1	DIAZINECARBONITRILE,(3-CYANOPHENYL),OXIDE
4 ★	1.18		2097		C8H4N4O1	DIAZINECARBONITRILE,(4-CYANOPHENYL),OXIDE
5 ★	3.34		505		C8H5Br1F3N1O1	BROMOBENZENE,P-TRIFLUOROACETAMIDO 24568-11-4
6 ★	1.72		1519	459	C8H5Br1N2	3-BROMOCINNOLINE
7 ★	2.23		163	459	C8H5Cl1N2	2-CHLOROQUINOXALINE 1448-87-9
8 ★	1.57		1519	459	C8H5Cl1N2	3-CHLOROCINNOLINE
9 ★	1.75		163	459	C8H5Cl1N2	5-CHLOROQUINOXALINE

	logP	pH	Ref.	Note	MF	Name / **CAS** / *activity*
2910 ★	2.10		163	459	C8H5Cl1N2	6-CHLOROQUINOXALINE **5448-43-1**
1 ★	1.74		1519	459	C8H5Cl1N2	7-CHLOROQUINAZOLINE **7556-99-2**
2 ★	3.20		2166		C8H5Cl1N2O1	1,2,5-OXADIAZOLE,3-CHLORO-4-PHENYL
3 ★	2.70		2166		C8H5Cl1N2O2	FURAZAN-2-OXIDE,3-CHLORO-4-PHENYL
4 ★	2.82		2166		C8H5Cl1N2O2	FURAZAN-2-OXIDE,3-PHENYL-4-CHLORO
15 ★	2.23		520		C8H5Cl1N2O4	STYRENE,2-CHLORO-5-NITRO,B-NITRO **15851-93-1**
6 ★	1.79		2499		C8H5Cl1N4	PYRAZINE,2,3-DICYANO-5-ETHYL-6-CHLORO
7 ★	2.90		2410		C8H5Cl2F2N1O1	ACETANILIDE,2,4-DIFLUORO-3,5-DICHLORO
8 ★	3.18		1792		C8H5Cl2F2N1O1	DIFLUOROACETAMIDE,N-(3,4-DICHLOROPHENYL)
9 ★	3.26		306		C8H5Cl2N1O2	STYRENE,2,4-DICHLORO-B-NITRO **18984-21-9**
2920 ★	1.81		522		C8H5Cl3N2O2S1	1,2,4-BENZOTHIADIAZINE-1,1-DIOXIDE-3-DIFLUOROMETHYL-6-CHLORO **37157-97-4**
1 ★	3.20		2456	346	C8H5Cl3O2	CHLORFENAC **85-34-7** *herbicide*
2 ★	3.31		2456	821	C8H5Cl3O3	PHENOXYACETIC ACID,2,4,5-TRICHLORO **93-76-5** *herbicide*
3 ★	2.67		509		C8H5F3N2	BENZIMIDAZOLE,2-TRIFLUOROMETHYL **312-73-2**
4 ★	2.47		547		C8H5F3O1	M-(TRIFLUOROMETHYL)BENZALDEHYDE **454-89-7**
25 ★	2.15		536		C8H5F3O1	TRIFLUOROACETOPHENONE **434-45-7**
6 ★	3.10	1.0	590	342	C8H5F3O2	BENZOIC ACID,4-TRIFLUOROMETHYL **455-24-3**
7 ★	2.95		501		C8H5F3O2	M-TRIFLUOROMETHYLBENZOIC ACID **454-92-2**
8	1.46		1273		C8H5F3O2S1	THENOYLTRIFLUOROACETONE **326-91-0**
9 ★	3.36		547		C8H5F5	PENTAFLUOROETHYLBENZENE **309-11-5**
2930 ★	3.57		547		C8H5F6N1	3,5-BIS(TRIFLUOROMETHYL)ANILINE **328-74-5**
1 ★	4.50		607	606	C8H5F6N1O2S1	TRIFLUOROMETHANESULFONANILIDE,M-CF3 **23384-11-4**
2 ★	4.47		607	606	C8H5F6N1O2S1	TRIFLUOROMETHANESULFONANILIDE,P-CF3 **23384-12-5**
3 ★	1.18		594		C8H5N1O1	BENZALDEHYDE,M-CYANO **24964-64-5**
4 ★	0.83	7.4	594	1	C8H5N1O2	INDOLE,2,3-DIONE **91-56-5**
35 ★	1.48		501		C8H5N1O2	M-CYANOBENZOIC ACID **1877-72-1**
6 ★	1.56		501		C8H5N1O2	P-CYANOBENZOIC ACID **619-65-8**
7 ★	1.15		506		C8H5N1O2	PHTHALIMIDE **85-41-6**
8 ★	3.52		850		C8H5N1O2S1	4-CARBOXYPHENYLISOTHIOCYANATE **2131-62-6**
9	2.40		1947		C8H5N1O2S1	BENZOIC ACID,3-ISOTHIOCYANATO **2131-63-7**
2940 ★	2.92		2221	459	C8H5N1O2S1	BENZOTHIOPHENE,5-NITRO **4965-26-8**
1 ★	0.75	1.0	547	342	C8H5N1O6	3-NITROPHTHALIC ACID **603-11-2**
2 ★	0.73		1519	459	C8H5N3O2	3-NITROCINNOLINE **21905-82-8**
3 ★	2.63		2166		C8H5N3O3	1,2,5-OXADIAZOLE,3-NITRO-4-PHENYL
4 ★	1.09		494		C8H5N3O4S1	2-AMINO-5-(5-NITRO-2-FURFURYL)THIAZOLONE **52661-38-8**
45 ★	2.53		502		C8H6	BENZENE,ETHYNYL **536-74-3**
6 ★	2.86		2410		C8H6Br1Cl1F1O1	ACETANILIDE,2-FLUORO-3-CHLORO-4-BROMO
7 ★	1.65		522		C8H6Br1Cl1N2O2S1	1,2,4-BENZOTHIADIAZINE,1,1-DIOXIDE-3-METHYL-5-BROMO-7-CHLORO **5108-54-3**
8 ★	2.75		588		C8H6Br1Cl1O3	PHENOXYACETIC ACID,3-BROMO-4-CHLORO
9 ★	3.04		2410		C8H6Br1Cl2N1O1	ACETANILIDE,2-BROMO-3,5-DICHLORO
2950 ★	2.11		2410		C8H6Br1F1N2O3	ACETANILIDE,2-FLUORO-4-BROMO-5-NITRO
1 ★	3.00		519		C8H6Br1N1	5-BROMOINDOLE **10075-50-0**
2 ★	2.42		2645		C8H6Cl1F1O2	BENZOIC ACID,2-CHLORO-6-FLUORO,METHYL ESTER
3 ★	2.20		588		C8H6Cl1F1O3	PHENOXYACETIC ACID,3-CHLORO-5-FLUORO
4 ★	1.81		522		C8H6Cl1I1N2O2S1	1,2,4-BENZOTHIADIAZINE,1,1-DIOXIDE-3-METHYL-5-IODO-7-CHLORO
55 ★	3.10		588		C8H6Cl1I1O3	PHENOXYACETIC ACID,4-CHLORO-3-IODO
6 ★	3.25		2221	459	C8H6Cl1N1	INDOLE,6-CHLORO **53294-05-6**
7 ★	2.47		2363		C8H6Cl1N1	PHENYLACETONITRILE,4-CHLORO
8 ★	2.85		520		C8H6Cl1N1O2	STYRENE,2-CHLORO-B-NITRO **3156-34-1**
9 ★	2.93		520		C8H6Cl1N1O2	STYRENE,3-CHLORO-B-NITRO **3156-35-2**
2960 ★	2.44		520		C8H6Cl1N1O2	STYRENE,4-CHLORO,B-NITRO **706-07-0**
1 ★	2.21		2645		C8H6Cl1N1O4	BENZOIC ACID,2-CHLORO-6-NITRO,METHYL ESTER
2 ★	1.85		588		C8H6Cl1N1O5	PHENOXYACETIC ACID,4-CHLORO-3-NITRO
3 ★	0.85		522		C8H6Cl1N3O4S1	1,2,4-BENZOTHIADIAZINE,1,1-DIOXIDE-3-METHYL-5-NITRO-7-CHLORO **37157-79-2**
4 ★	1.42		522		C8H6Cl1N3O4S1	1,2,4-BENZOTHIADIAZINE-1,1-DIOXIDE-3-METHYL-6-NITRO-7-CHLORO **31365-75-0**
65 ★	1.68		522		C8H6Cl2N2O2S1	1,2,4-BENZOTHIADIAZINE-1,1-DIOXIDE-3-CHLOROMETHYL-6-CHLORO **37157-57-6**
6 ★	2.79		2645		C8H6Cl2O2	BENZOIC ACID,2,6-DICHLORO,METHYL ESTER
7 ★	2.94		2645		C8H6Cl2O2	PHENYLACETATE,2,6-DICHLORO
8	0.65		2328		C8H6Cl2O3	2,4-D (DIMETHYLAMINE SALT)
9 ★	2.21		547		C8H6Cl2O3	DICAMBA **1918-00-9** *herbicide*
2970 ★	2.81		501		C8H6Cl2O3	PHENOXYACETIC ACID,2,4-DICHLORO **94-75-7** *herbicide*
1 ★	2.81		505		C8H6Cl2O3	PHENOXYACETIC ACID,3,4-DICHLORO **588-22-7**
2 ★	2.84		2393	345	C8H6Cl2O3	PHENOXYACETIC ACID,3,5-DICHLORO **587-64-4**
3 ★	2.88		2410		C8H6Cl3N1O1	ACETANILIDE,2,3,4-TRICHLORO
4 ★	3.76		2410		C8H6Cl3N1O1	ACETANILIDE,3,4,5-TRICHLORO
75 ★	3.29		1792		C8H6Cl3N1O1	CHLOROACETAMIDE,N-(3,4-DICHLOROPHENYL) **20149-84-2**
6 ★	2.42		588		C8H6F1O3	PHENOXYACETIC ACID,5-FLUORO-3-IODO
7 ★	1.96		2645		C8H6F2O2	BENZOIC ACID,2,6-DIFLUORO,METHYL ESTER
8 ★	0.68		2410		C8H6F3N1O1	BENZAMIDE,2-TRIFLUOROMETHYL
9 ★	1.71		541		C8H6F3N1O1	P-TRIFLUOROMETHYLBENZAMIDE **1891-90-3**
2980 ★	2.21		541		C8H6F3N1O1	TRIFLUOROACETANILIDE **404-24-0**
1 ★	-0.15	6.5	321	1	C8H6F3N3O4S2	FLUMETHIAZIDE **148-56-1**
2 ★	0.93		924	400	C8H6N2	CINNOLINE **253-66-7**
3	1.00	7.4	594	1	"	CINNOLINE
4 ★	0.57		547		C8H6N2	PHTHALAZINE **253-52-1**
85 ★	1.00		547		C8H6N2	QUINAZOLINE **253-82-7**
6 ★	1.32		536		C8H6N2	QUINOXALINE **91-19-0**
7 ★	0.75		1519	400	C8H6N2O1	1,2(H)-PHTHALAZINONE **119-39-1**
8	1.31		1519	459	C8H6N2O1	1,3,4-OXADIAZOLE,2-PHENYL **825-56-9**
9 ★	0.68		1627	400	C8H6N2O1	3-CINNOLONE **31777-46-5**
2990 ★	0.20		1439		C8H6N2O1	4H-PYRIDO(1,2-A)PYRIMIDIN-4-ONE **23443-10-9**
1 ★	0.82		924		C8H6N2O1	CINNOLINE-4-ONE **18514-84-6**
2 ★	0.52		547		C8H6N2O1	M-CYANOBENZAMIDE **3441-01-8**
3 ★	0.48		547		C8H6N2O1	P-CYANOBENZAMIDE **3034-34-2**
4 ★	1.08		453		C8H6N2O1	P-CYANOFORMANILIDE **6321-94-4**

	logP	pH	Ref.	Note	MF	Name / CAS / activity
95 ✓	0.94		590		C8H6N2O1	QUINAZOLINE,4-HYDROXY 112-30-1
6 ★	0.77		924	400	C8H6N2O1	QUINAZOLINE-4-ONE 491-36-1
7 ★	0.80		924	400	C8H6N2O1	QUINOXALINE-2-ONE 1196-57-2
8 ★	0.27		924	400	C8H6N2O2	2-AMINO-1,3-BENZOXAZIN-4-ONE 771-39-1
9 ★	-0.45		924	400	C8H6N2O2	4-(3H)-QUINAZOLINONE,3-HYDROXY 5319-71-1
3000 ★	2.53		1625	457	C8H6N2O2	INDOLE,5-NITRO 6146-52-7
1 ★	2.63		2221	459	C8H6N2O2	INDOLE-7-NITRO 6960-42-5
2 ★	1.37		2256		C8H6N2O2	PHENYLACETONITRILE,4-NITRO
3 ★	0.77		2222		C8H6N2O2	QUINAZOLINE-2-4-DIONE, 86-96-4
4 ★	0.20		924		C8H6N2O2	QUINOXALINE-1,4-DIHYDRO-2,3-DIONE 15804-19-0
5 ★	1.74	3.5	2458	2	C8H6N2O3	OXIMIDAZOLIDINE-3,5-DIONE-2-PHENYL
6 ★	0.16		594		C8H6N2O3	PHTHALIMIDE,5-AMINO-N-HYDROXY
7 ★	1.80		520		C8H6N2O4	STYRENE,2-NITRO,B-NITRO 3156-39-6
8 ★	1.82		520		C8H6N2O4	STYRENE,3-NITRO,B-NITRO 882-26-8
9 ★	1.89		520		C8H6N2O4	STYRENE,4-NITRO,B-NITRO 3156-41-0
3010 ★	1.84		2128		C8H6N2O6	3,5-DINITROBENZOIC ACID,METHYL ESTER 2702-58-1
1 ★	2.42		594		C8H6N2S1	1,2,3-THIADIAZOLE,4-PHENYL 25445-77-6
2 ★	2.13	5.0	594	312	C8H6N2S2	1,2,4-THIADIAZOLE,3-PHENYL-5-THIOL 20069-34-5
3 ★	1.95		547		C8H6N2S2	3-PHENYL-1,2,4-THIADIAZOLE-5(4H)-THIONE
4 ★	1.13		595		C8H6N4	BENZIMIDAZOLE,2-AMINO-5-CYANO 63655-40-3
15 ★	1.10		555		C8H6N4O3	3-(5-NITRO-2-FURFURILIDINE)1,2,4-TRIAZOLE 50832-71-8
6 ★	1.63		260		C8H6N4O4	1-METHYL-5,7-DINITROBENZPYRAZOLE
7 ★	1.30		611		C8H6N4O4	2-(5-NITROFUR-2YLVINYL)-5-AMINO-3-OXOIMIDAZOLE 21959-57-9
8 ★	-0.47		1008	354	C8H6N4O5	NITROFURANTOIN 67-20-9 antibacterial (urinary)
9	-0.19	7.4	1497	1	"	NITROFURANTOIN
3020 ★	2.67		505		C8H6O1	BENZOFURAN 271-89-6
1 ★	0.80		1524		C8H6O2	1(3H)-ISOBENZOFURANONE 87-41-2
2 ★	0.88		1625		C8H6O2S1	BENZOTHIOPHENE-1,1-DIOXIDE 825-44-5
3 ★	0.57		1524		C8H6O3	1(3H)-ISOBENZOFURANONE-3-HYDROXYL 16859-59-9 ✦
4 ★	1.76	2.0	2522	342	C8H6O3	P-CARBOXYBENZALDEHYDE 619-66-9
25 ★	1.05		505		C8H6O3	PIPERONAL 120-57-0 flavoring, pediculicide
6 ★	1.66		501		C8H6O4	M-PHTHALIC ACID 121-91-5
7 ★	0.73	1.0	547	342	C8H6O4	O-PHTHALIC ACID 88-99-3
8 ★	2.00		572		C8H6O4	TEREPHTHALIC ACID 100-21-0
9	-0.50	5.0	447	1	C8H6O5	STIPITATIC ACID 4440-39-5
3030 ★	3.12		505		C8H6S1	BENZOTHIOPHENE,(B) 95-15-8
1 ★	3.19		536		C8H6Se1	SELANONAPHTHENE 272-30-0
2 ★	1.37		522		C8H7Br1N2O2S1	1,2,4-BENZOTHIADIAZINE-1,1-DIOXIDE-3,METHYL-7-BROMO 13460-15-6
3 ★	1.37		522		C8H7Br1N2O2S1	1,2,4-BENZOTHIADIAZINE-1,1-DIOXIDE-3,METHYL-6-BROMO 37148-00-8
4 ★	-0.10		1564		C8H7Br1N2O3	PYRIMIDINE-2,4-DIONE,3-PROPARGYLOXY-5-BROMO-6-ME 77317-71-6
35	2.43	7.0	925	1	C8H7Br1O1	P-BROMOACETOPHENONE 99-90-1
6 ★	2.43		495		"	P-BROMOACETOPHENONE
7 ★	2.20		558		C8H7Br1O2	2-BROMOPHENYLACETATE 1829-37-4
8 ★	2.95		2645		C8H7Br1O2	BENZOIC ACID,P-BROMO,METHYL ESTER
9 ★	2.37		501		C8H7Br1O2	M-BROMOPHENYLACETIC ACID 1878-67-7
3040 ★	2.31		501		C8H7Br1O2	P-BROMOPHENYLACETIC ACID 1878-68-8
1 ★	2.22		501		C8H7Br1O3	PHENOXYACETIC ACID,M-BROMO 1798-90-9
2 ★	2.04	1.0	547	342	C8H7Br1O3	PHENOXYACETIC ACID,O-BROMO 1879-56-7
3 ★	2.45		501		C8H7Br1O3	PHENOXYACETIC ACID,P-BROMO 1878-91-7
4 ★	1.53		2410		C8H7Br2N1O1	ACETANILIDE,2,6-DIBROMO
45 ★	3.42		354		C8H7Br3O2	2-(2,4,6-TRIBROMOPHENOXY)ETHANOL 23976-66-1
6 ★	1.83		2410		C8H7Cl1F1O1	ACETANILIDE,2-FLUORO-3-CHLORPO
7 ★	1.91		2410		C8H7Cl1F1O1	ACETANILIDE,2-FLUORO-4-CHLORO
8 ★	0.08	6.5	321	2	C8H7Cl1F3N3O4S2	3-TRIFLUMETHYLHYDROCHLOROTHIAZIDE 1547-10-0
9 ★	0.72		522		C8H7Cl1N2O2S1	1,2,4-BENZOTHIADIAZINE,1,1-DIOXIDE,3-METHYL-5-CHLORO 31363-85-6
3050 ★	1.21		522		C8H7Cl1N2O2S1	1,2,4-BENZOTHIADIAZINE-1,1-DIOXIDE-3-METHYL-6-CHLORO 14559-54-7
1	1.08	7.4	601	1	C8H7Cl1N2O2S1	1,2,4-BENZOTHIADIAZINE-1,1-DIOXIDE-3-METHYL-7-CHLORO 364-98-7 antihypertensive
2 ★	1.20		522		"	1,2,4-BENZOTHIADIAZINE-1,1-DIOXIDE-3-METHYL-7-CHLORO
3 ★	0.62		522		C8H7Cl1N2O2S1	1,2,4-BENZOTHIADIAZINE-1,1-DIOXIDE-3-METHYL-8-CHLORO 22680-31-5
4 ★	1.78		2410		C8H7Cl1N2O3	ACETANILIDE,2-NITRO-4-CHLORO
55 ★	3.22		509		C8H7Cl1N2S1	BENZIMIDAZOLE,5-CHLORO-2-(METHYLTHIO) 7692-57-1
6 ★	2.51	7.0	925	1	C8H7Cl1O1	M-CHLOROACETOPHENONE 99-02-5
7 ★	2.09		547		C8H7Cl1O1	O-CHLOROACETOPHENONE 2142-68-9
8 ★	2.32		564		C8H7Cl1O1	P-CHLOROACETOPHENONE 99-91-2
9 ★	2.18		558		C8H7Cl1O2	2-CHLOROPHENYLACETATE 4525-75-1
3060 ★	2.32		558		C8H7Cl1O2	3-CHLOROPHENYLACETATE 13031-39-5
1 ★	2.31		2645	225	C8H7Cl1O2	BENZOIC ACID,2-CHLORO-6-METHYL
2 ★	2.85		2645		C8H7Cl1O2	BENZOIC ACID,M-CHLORO,METHYL ESTER
3 ★	2.76		2645		C8H7Cl1O2	BENZOIC ACID,P-CHLORO,METHYL ESTER
4 ★	2.09		501		C8H7Cl1O2	M-CHLOROPHENYLACETIC ACID 1878-65-5
65 ★	2.38		547		C8H7Cl1O2	O-CHLOROBENZOIC ACID,METHYL ESTER 610-96-8
6 ★	2.12		501		C8H7Cl1O2	P-CHLOROPHENYLACETIC ACID 1878-66-6
7 ★	2.03		501		C8H7Cl1O3	PHENOXYACETIC ACID,M-CHLORO 588-32-9
8 ★	1.86	1.0	547	342	C8H7Cl1O3	PHENOXYACETIC ACID,O-CHLORO 614-61-9
9 ★	2.25		2393	345	C8H7Cl1O3	PHENOXYACETIC ACID,P-CHLORO 122-88-3
3070	2.54		1263		C8H7Cl2N1O1	3,4-DICHLOROACETANILIDE 2150-93-8
1	2.96		1792		"	3,4-DICHLOROACETANILIDE
2 ★	2.09		2410		C8H7Cl2N1O1	ACETANILIDE,2,3-DICHLORO
3 ★	2.18		2410		C8H7Cl2N1O1	ACETANILIDE,2,4-DICHLORO
4 ★	2.19		2410		C8H7Cl2N1O1	ACETANILIDE,2,5-DICHLORO
75 ★	1.32		2410		C8H7Cl2N1O1	ACETANILIDE,2,6-DICHLORO
6 ★	3.17		2410		C8H7Cl2N1O1	ACETANILIDE,3,5-DICHLORO
7 ★	2.48		536		C8H7Cl2N1O2	2,3-DICHLORO-N-METHYLPHENYLCARBAMATE
8 ★	2.44		536		C8H7Cl2N1O2	2,5-DICHLORO-N-METHYLPHENYLCARBAMATE 18315-62-3
9 ★	3.54		2094		C8H7Cl2N1O2	METHYL-N-(3,4-DICHLOROPHENYL)CARBAMATE 1918-18-9

	logP	pH	Ref.	Note	MF	Name / **CAS** / *activity*
3080 ★	2.80		273		C8H7Cl2N1O2	N-METHYL-3,4-DICHLOROPHENYLCARBAMATE **18315-50-9**
1 ★	3.03		273		C8H7Cl2N1O2	N-METHYL-3,5-DICHLOROPHENYLCARBAMATE **13538-26-6**
2 ★	4.23		547		C8H7Cl2N1S1	3,4-DICHLOROPHENYLTHIOCARBAMATE,S-METHYL ESTER
3 ★	2.98		594	1	C8H7Cl2N3	BENZIMIDAZOLE,1-METHYL-2-AMINO-5,6-DICHLORO
4 ★	3.74		1554		C8H7Cl3O3	3,4,5-TRICHLORO-2,6-DIMETHOXYPHENOL
85 ★	0.52	6.5	321	2	C8H7Cl4N3O4S2	TECLOTHIAZIDE **4267-05-4** *diuretic*
6 ★	0.62		522		C8H7F1N2O2S1	1,2,4-BENZOTHIADIAZINE-1,1-DIOXIDE-3-METHYL-7-FLUORO **31365-74-9**
7 ★	1.77	7.0	925	1	C8H7F1O1	M-FLUOROACETOPHENONE **455-36-7**
8 ★	1.72	7.0	925	1	C8H7F1O1	P-FLUOROACETOPHENONE **403-42-9**
9 ★	1.76		558		C8H7F1O2	2-FLUOROPHENYLACETATE **29650-44-0**
3090 ★	1.74		558		C8H7F1O2	3-FLUOROPHENYLACETATE **701-83-7**
1 ★	1.82		2645		C8H7F1O2	BENZOIC ACID,O-FLUORO,METHYL ESTER
2 ★	2.28		2645		C8H7F1O2	BENZOIC ACID,P-FLUORO,METHYL ESTER
3 ★	1.65		501		C8H7F1O2	M-FLUOROPHENYLACETIC ACID **331-25-9**
4 ★	1.50		501		C8H7F1O2	O-FLUOROPHENYLACETIC ACID **451-82-1**
95 ★	1.55		501		C8H7F1O2	P-FLUOROPHENYLACETIC ACID **405-50-5**
6 ★	1.48		501		C8H7F1O3	PHENOXYACETIC ACID,M-FLUORO **404-98-8**
7 ★	1.39	1.0	547	342	C8H7F1O3	PHENOXYACETIC ACID,O-FLUORO **348-10-7**
8 ★	1.64		2393	345	C8H7F1O3	PHENOXYACETIC ACID,P-FLUORO **405-79-8**
9 ★	1.86		527		C8H7F1O4S1	P-FLUOROSULFONYLPHENYLACETIC ACID
3100 ★	1.82		527		C8H7F1O5S1	P-FLUOROSULFONYLPHENOXYACETIC ACID
1 ★	0.69		2410		C8H7F2N1O1	ACETANILIDE,2,6-DIFLUORO
2 ★	2.31		1263		C8H7F3N2O1	M-TRIFLUOROMETHYLPHENYLUREA **13114-87-9**
3 ★	2.24		547		C8H7F3O1	M-(TRIFLUOROMETHYL)BENZYL ALCOHOL **349-75-7**
4 ★	2.78		501		C8H7F5O3S1	PHENOXYACETIC ACID,O-PENTAFLUOROTHIO
5 ★	1.61		522		C8H7I1N2O2S1	1,2,4-BENZOTHIADIAZINE-1,1-DIOXIDE-3-METHYL-7-IODO
6 ★	2.55		558		C8H7I1O2	2-IODOPHENYLACETATE **32865-61-5**
7 ★	2.74		2645		C8H7I1O2	BENZOIC ACID,O-IODO,METHYL ESTER
8 ★	2.62		501		C8H7I1O2	M-IODOPHENYLACETIC ACID **1878-69-9**
9 ★	2.64		501		C8H7I1O2	P-IODOPHENYLACETIC ACID **1798-06-7**
3110 ★	2.33	0.3	2061		C8H7I1O2	PHENYLACETIC ACID,2-IODO **18698-96-9**
1 ★	2.44		501		C8H7I1O3	PHENOXYACETIC ACID,M-IODO **1878-93-9**
2 ★	2.33	1.0	547	342	C8H7I1O3	PHENOXYACETIC ACID,O-IODO **1878-92-8**
3 ★	2.69		501		C8H7I1O3	PHENOXYACETIC ACID,P-IODO **1878-94-0**
4 ★	2.14		503		C8H7N1	INDOLE **120-72-9**
15 ★	2.49		1453	459	C8H7N1	INDOLIZINE **274-40-8**
6 ★	2.21		547		C8H7N1	O-TOLUNITRILE **529-19-1**
7 ★	1.56		502		C8H7N1	PHENYLACETONITRILE **140-29-4**
8 ★	1.16		924	400	C8H7N1O1	OXINDOLE **59-48-3**
9 ★	2.11		520		C8H7N1O2	STYRENE,B-NITRO **102-96-5**
3120 ★	1.42		501		C8H7N1O3	M-NITROACETOPHENONE **121-89-1**
1 ★	1.28		495		C8H7N1O3	O-NITROACETOPHENONE **577-59-3**
2 ★	1.53		495		C8H7N1O3	P-NITROACETOPHENONE **100-19-6**
3 ★	2.07		520		C8H7N1O3	STYRENE,3-HYDROXY,B-NITRO **3156-44-3**
4 ★	2.12		520		C8H7N1O3	STYRENE,4-HYDROXY,B-NITRO **3179-08-6**
25 ★	1.89		2645		C8H7N1O4	BENZOIC ACID,M-NITRO,METHYL ESTER
6 ★	1.66		2645		C8H7N1O4	BENZOIC ACID,O-NITRO,METHYL ESTER
7 ★	1.82		558		C8H7N1O4	M-NITROPHENYLACETATE **1523-06-4**
8 ★	1.45		501		C8H7N1O4	M-NITROPHENYLACETIC ACID **1877-73-2**
9 ★	1.94		2645		C8H7N1O4	METHYL-P-NITRO BENZOATE **619-50-1**
3130 ★	1.55		558		C8H7N1O4	O-NITROPHENYLACETATE **610-69-5**
1 ★	1.50		541		C8H7N1O4	P-NITROPHENYLACETATE **830-03-5**
2 ★	1.39		501		C8H7N1O4	P-NITROPHENYLACETIC ACID **104-03-0**
3 ★	1.65		1946		C8H7N1O5	2-PROPENOIC ACID,3-(5-NITRO-2-FURANYL),METHYL ESTER **1874-24-4**
4 ★	1.37		501		C8H7N1O5	PHENOXYACETIC ACID,M-NITRO **1878-88-2**
35 ★	0.97	1.0	547	342	C8H7N1O5	PHENOXYACETIC ACID,O-NITRO **1878-87-1**
6 ★	1.48		501		C8H7N1O5	PHENOXYACETIC ACID,P-NITRO **1798-11-4**
7 ★	3.92		850		C8H7N1S1	4-METHYLPHENYLISOTHIOCYANATE **622-59-3**
8 ★	3.68		1947		C8H7N1S1	ANISOLE,3-ISOTHIOCYANATO **3125-64-2**
9 ★	3.58		1947		C8H7N1S1	ANISOLE,4-ISOTHIOCYANATO **2284-20-0**
3140 ★	1.98		594		C8H7N1S1	BENZOTHIAZOL-2-ONE,3-METHYL **2786-62-1**
1 ★	3.16		850		C8H7N1S1	BENZYLISOTHIOCYANATE **622-78-6**
2 ★	1.99		541		C8H7N1S1	BENZYLTHIOCYANATE
3 ★	3.10		2585		C8H7N1S2	METHYLTHIOBENZOTHIAZOLE **615-22-5**
4 ★	1.71		2619		C8H7N3	1,2,4-TRIAZOLE,3-PHENYL
45 ★	0.82		1519	459	C8H7N3	1-PHTHALAZINAMINE **19064-69-8**
6 ★	1.50		163	459	C8H7N3	2-AMINOQUINOXALINE **5424-05-5**
7 ★	1.00		1519	459	C8H7N3	3-CINNOLINAMINE **17372-79-1**
8 ★	1.04		1519	459	C8H7N3	4-CINNOLINAMINE **5152-83-0**
9 ★	1.28		1519	459	C8H7N3	4-QUINAZOLINAMINE **15018-66-3**
3150 ★	1.29		163	459	C8H7N3	5-AMINOQUINOXALINE **16566-20-4**
1 ★	0.90		163	459	C8H7N3	6-AMINOQUINOXALINE **6298-37-9**
2 ★	1.81		2166		C8H7N3O1	1,2,5-OXADIAZOLE,3-AMINO-4-PHENYL
3 ★	0.46		931		C8H7N3O1	2-(2-PYRIDYL)-5-METHYL-1,3,4-OXADIAZOLE
4 ★	0.60		924	400	C8H7N3O1	2-AMINOQUINAZOLINE-4-ONE **20198-19-0**
55 ★	2.09		931		C8H7N3O1	3-(2-PYRIDYL)-4-METHYL-1,2,5-OXADIAZOLE
6 ★	0.79		931		C8H7N3O1	3-(2-PYRIDYL)-5-METHYL-1,2,4-OXADIAZOLE
7 ★	0.84		924	400	C8H7N3O1	3-METHYLBENZOTRIAZIN-4-ONE **22305-44-8**
8 ★	0.89		931		C8H7N3O1	5-(2-PYRIDYL)-3-METHYL-1,2,4-OXADIAZOLE
9 ★	2.49		2097		C8H7N3O1	DIAZINECARBONITRILE,(3-METHYLPHENYL),OXIDE **64462-06-2**
3160 ★	2.39		2097		C8H7N3O1	DIAZINECARBONITRILE,(4-METHYLPHENYL,OXIDE
1	-1.88		559		C8H7N3O2	1-CARBOXYMETHYLBENZOTRIAZOLE **4144-64-3**
2 ✓	-1.64		559		C8H7N3O2	2-CARBOXYMETHYLBENZOTRIAZOLE **4144-68-7**
3 ★	1.94	7.4	594	1	C8H7N3O2	BENZIMIDAZOLE,2-METHYL-5-NITRO **1792-40-1**
4 ★	2.24		2097		C8H7N3O2	DIAZINECARBONITRILE,(3-METHOXYPHENYL),OXIDE **62825-14-3**

	logP	pH	Ref.	Note	MF	Name / **CAS** / *activity*
65 ★	2.15		2097		C8H7N3O2	DIAZINECARBONITRILE,(4-METHOXYPHENYL)OXIDE **62825-15-4**
6 ★	1.42		2166		C8H7N3O2	FURAZAN-2-OXIDE,3-AMINO-4-PHENYL
7 ★	1.47		2166		C8H7N3O2	FURAZAN-2-OXIDE,4-AMINO-3-PHENYL
8 ★	-0.95	7.4	594	1	C8H7N3O2	HYDANTOIN,5-(3-PYRIDYL)
9 ★	1.95	7.4	594	1	C8H7N3O2	INDAZOLE,1-METHYL-6-NITRO **6850-23-3**
3170 ★	1.76	7.4	594	1	C8H7N3O2	INDAZOLE,2-METHYL-6-NITRO **6850-22-2**
1 ★	-0.30	7.4	2360	1	C8H7N3O3	BENZOTRIAZINE,1,4-DI-N-OXIDE-3-METHOXY
2 ✓	-2.00	7.4	2360	1	C8H7N3O4	BENZOTRIAZINE,1,4-DI-N-OXIDE-3-HYDROXY-7-METHOXY
3 ★	0.65		522		C8H7N3O4S1	1,2,4-BENZOTHIADIAZINE-1,1-DIOXIDE-3-METHYL-7-NITRO **37148-07-5**
4 ★	0.68		354		C8H7N3O5	3,5-DINITRO-4-METHYLBENZAMIDE **4551-76-2**
75 ★	0.56		2410		C8H7N3O5	ACETANILIDE,2,6-DINITRO
6 ★	-0.05	7.4	594	1	C8H7N3O5	FURAZOLIDONE **67-45-8** *anti-infective (topical), antiprotozoal (trichomonas, topical)*
7	-3.68	7.4	1377	1	C8H7N3O6	N-(2,4-DINITROPHENYL)GLYCINE **1084-76-0**
8 ★	1.84	7.0	547		C8H7N3S1	1,2,4-TRIAZOLE-3-THIONE,1,2-DIHYDRO-5-PHENYL **3414-94-6**
9	1.28		555		C8H7N5O3	3-AMINO-5-(5-NO2-2-FURFURILIDENE)-1,2,4-TRIAZOLE
3180 ★	3.08		594		C8H8	CYCLOOCTATETRAENE **629-20-9**
1 ★	2.95		536		C8H8	STYRENE **100-42-5**
2 ★	5.21		2293	400	C8H8Br1Cl2O3P1S1	BROMOPHOS **2104-96-3** *insecticide (vet)*
3 ★	2.32		1894		C8H8Br1N1O1	3-BROMOACETANILIDE **621-38-5**
4 ★	1.40		1894		C8H8Br1N1O1	ACETANILIDE,2-BROMO **614-76-6**
85 ★	2.29	7.2	415	314	C8H8Br1N1O1	P-BROMOACETANILIDE **103-88-8** *analgesic, antipyretic*
6 ★	2.25		2142		C8H8Br1N1O2	2-BROMOPHENYLCARBAMATE,O-METHYL
7 ★	2.78		2142		C8H8Br1N1O2	3-BROMOPHENYLCARBAMATE,O-METHYL
8 ★	2.76		1818		C8H8Br1N1O2	BENZAMIDE,2-HYDROXY-3-METHYL-5-BROMO **40912-73-0**
9 ★	1.77		273		C8H8Br1N1O2	N-METHYL-2-BROMOPHENYLCARBAMATE **13538-27-7**
3190 ★	2.25		273		C8H8Br1N1O2	N-METHYL-3-BROMOPHENYLCARBAMATE **13538-60-8**
1 ★	2.17		273		C8H8Br1N1O2	N-METHYL-4-BROMOPHENYLCARBAMATE **13538-50-6**
2 ★	2.90		2159		C8H8Br1N1O2	PHENETHYLBROMIDE,4-NITRO **5339-26-4**
3 ★	2.15		2410		C8H8Cl1N1O1	3-CHLOROACETANILIDE **588-07-8**
4 ★	1.63		547		C8H8Cl1N1O1	A-CHLOROACETANILIDE **587-65-5**
95 ★	0.89		2410		C8H8Cl1N1O1	BENZAMIDE,2-CHLORO-6-METHYL
6 ★	1.28		579		C8H8Cl1N1O1	O-CHLOROACETANILIDE **587-65-5**
7	1.87	7.2	415	314	C8H8Cl1N1O1	P-CHLOROACETANILIDE **539-03-7**
8	2.05	7.0	925	1	"	P-CHLOROACETANILIDE
9 ★	2.12		1894		"	P-CHLOROACETANILIDE
3200 ★	2.13		2142		C8H8Cl1N1O2	2-CHLOROPHENYLCARBAMATE,O-METHYL
1 ★	2.57		2142		C8H8Cl1N1O2	4-CHLOROPHENYLCARBAMATE,O-METHYL
2 ★	0.59		2410		C8H8Cl1N1O2	BENZAMIDE,2-CHLORO-6-METHOXY
3 ★	3.30		594		C8H8Cl1N1O2	METHYL BENZOATE,2-AMINO-5-CHLORO **5202-89-1**
4 ★	2.58		1263		C8H8Cl1N1O2	METHYL-N-(M-CHLOROPHENYL)CARBAMATE **2150-88-1**
5 ★	1.64		273		C8H8Cl1N1O2	N-METHYL-2-CHLOROPHENYLCARBAMATE **3942-54-9** *insecticide*
6 ★	2.03		273		C8H8Cl1N1O2	N-METHYL-3-CHLOROPHENYLCARBAMATE **4090-00-0**
7 ★	2.01		273		C8H8Cl1N1O2	N-METHYL-4-CHLOROPHENYLCARBAMATE **2620-53-3**
8 ★	1.97		1969		C8H8Cl1N1O2	P-CHLOROBENZYLCARBAMATE (ESTER) **2621-80-9**
9 ★	1.16		588		C8H8Cl1N1O3	PHENOXYACETIC ACID,3-AMINO-4-CHLORO
3210 ★	2.09		595		C8H8Cl1N3	BENZIMIDAZOLE,1-METHYL-2-AMINO-5-CHLORO **103748-25-0**
1 ★	2.01		594	1	C8H8Cl1N3	BENZIMIDAZOLE,1-METHYL-2-AMINO-6-CHLORO
2	0.77	4.0	1732		C8H8Cl1N3O2	1-ME-1-NITROSO-3-(P-CHLOROPHENYL)UREA **25355-61-7**
3 ★	0.63		522		C8H8Cl1N3O2S1	1,2,4-BENZOTHIADIAZINE-1,1-DIOXIDE-3-METHYL-6-AMINO-7-CHLORO **37148-08-6**
4 ★	-0.05	6.5	321	1	C8H8Cl1N3O4S2	CHLOROTHIAZIDE,3-METHYL **1025-75-8**
15	0.53		2123	22	C8H8Cl1N5O2	BENZALDEHYDE,4-CHLORO-3-NITRO,GUANYLHYDRAZONE **102632-30-4**
6 ★	1.79	9.9	704	821	C8H8Cl1N5O3	4-NITRO-3-CHLOROPHENYLAMIDINOUREA **46833-92-5**
7 ★	2.72		1143	459	C8H8Cl2	1,3-(BIS-CHLOROMETHYL)BENZENE **626-16-4**
8 ★	5.51		2293	400	C8H8Cl2I1O3P1S1	IODOFENPHOS **18181-70-9** *insecticide*
9 ★	2.94		1263		C8H8Cl2N2O1	1-METHYL-3-(3,4-DICHLOROPHENYL)UREA **3567-62-2**
3220 ★	2.39	9.9	704	821	C8H8Cl2N4O1	3,4-DICHLOROPHENYLAMIDINOUREA **21724-58-3**
1 ★	0.56	6.5	321	2	C8H8Cl3N3O4S2	TRICHLORMETHIAZIDE **133-67-5** *diuretic, antihypertensive*
2 ★	5.07		2293	400	C8H8Cl3O3P1S1	RONNEL **299-84-3** *insecticide (systemic)*
3 ★	4.06		421		C8H8Cl5F3O1	1-(3,3,3-TRIFLUOROETHOXY)PENTACHLOROCYCLOHEXANE
4 ★	4.15		421		C8H8Cl5F3O1	3-(3,3,3-TRIFLUOROETHOXY)PENTACHLOROCYCLOHEXANE **56400-11-4**
25 ★	1.88	8.9	1498		C8H8F1N3O2	1-METHYL-1-NITROSO-3-(P-FLUROROPHENYL)UREA **777-59-3**
6 ★	1.61		1894		C8H8F1O1	3-FLUOROACETANILIDE **351-28-0**
7 ★	0.96		1894		C8H8F1O1	ACETANILIDE,2-FLUORO **399-31-5**
8 ★	1.47	7.2	415	314	C8H8F1O1	P-FLUOROACETANILIDE **351-83-7**
9 ★	1.66		2142		C8H8F1O2	2-FLUOROPHENYLCARBAMATE,O-METHYL
3230 ★	2.25		2142		C8H8F1O2	3-FLUOROPHENYLCARBAMATE,O-METHYL
1 ★	1.95		2142		C8H8F1O2	4-FLUOROPHENYLCARBAMATE,O-METHYL
2 ★	1.25		273		C8H8F1O2	N-METHYL-2-FLUOROPHENYLCARBAMATE **704-73-4**
3 ★	1.48		273		C8H8F1O2	N-METHYL-3-FLUOROPHENYLCARBAMATE **705-48-6**
4 ★	1.28		273		C8H8F1O2	N-METHYL-4-FLUOROPHENYLCARBAMATE **705-70-4**
35 ★	2.17		527		C8H8F1O3S1	P-ACETAMIDO-BENZENESULFONYLFLUORIDE **329-20-4**
6 ★	2.91		594		C8H8F3N1O2S1	TRIFLUOROMETHANESULFONAMIDE,N-BENZYL
7 ★	3.60		607	606	C8H8F3N1O2S1	TRIFLUOROMETHANESULFONANILIDE,P-ME **37595-73-6**
8 ★	3.74		607	606	C8H8F3N1O2S2	TRIFLUOROMETHANESULFONANILIDE,P-METHIO **23375-06-6**
9 ★	3.13		607	606	C8H8F3N1O3S1	TRIFLUOROMETHANESULFONANILIDE,M-METHOXY **23384-33-0**
3240 ★	3.00		607	606	C8H8F3N1O3S1	TRIFLUOROMETHANESULFONANILIDE,P-METHOXY **23384-34-1**
1 ★	1.85		607	606	C8H8F3N1O4S2	TRIFLUOROMETHANESULFONANILIDE,M-MESULFONYL **23375-08-8**
2 ★	1.99		607	606	C8H8F3N1O4S2	TRIFLUOROMETHANESULFONANILIDE,P-MESULFONYL **23375-10-2**
3 ★	0.36		1780		C8H8F3N3O4S2	HYDROFLUMETHIAZIDE **135-09-1** *antihypertensive, diuretic*
4 ★	0.71		547		C8H8Hg1O2	PHENYL MERCURIC ACETATE **62-38-4** *antimicrobial~fungicide*
45 ★	1.53		1894		C8H8I1O1	ACETANILIDE,2-IODO **19591-17-4**
6 ★	2.64		1894		C8H8I1O1	ACETANILIDE,3-IODO **19230-45-6**
7	2.46	7.2	415	314	C8H8I1O1	P-IODOACETANILIDE **622-50-4**
8 ★	2.71		1894		"	P-IODOACETANILIDE
9 ★	2.44		2142		C8H8I1O2	2-IODOPHENYLCARBAMATE,O-METHYL

	logP	pH	Ref.	Note	MF	Name / CAS / activity
3250 ★	3.19		2142		C8H8I1O2	3-IODOPHENYLCARBAMATE,O-METHYL
1 ★	1.94		536		C8H8I1O2	N-METHYL-2-IODOPHENYLCARBAMATE **13538-28-8**
2 ★	2.52		273		C8H8I1O2	N-METHYL-3-IODOPHENYLCARBAMATE **13941-09-8**
3 ★	2.46		273		C8H8I1O2	N-METHYL-4-IODOPHENYLCARBAMATE **13538-51-7**
4	1.43	8.0	786	1	C8H8N2	2-METHYLBENZIMIDAZOLE **615-15-6**
55 ★	0.70		1625		C8H8N2	INDOLE,5-AMINO **5192-03-0**
6 ★	1.82		1625	457	C8H8N2O1	BENZOXAZOLE,2-METHYLAMINO
7 ★	0.61		594		C8H8N2O1S1	AZETIDIN-2-ONE,4-(4-PYRIDYLTHIO)
8	1.81		2161	468	C8H8N2O1S1	AZETIDINE-2-ONE,4-(2-PYRIMIDINYL)
9 ★	2.07	7.4	594	1	C8H8N2O2	INDOLINE,5-NITRO **32692-19-6**
3260 ★	1.92	7.4	594	1	C8H8N2O2	INDOLINE,6-NITRO **19727-83-4**
1 ★	-0.21		547		C8H8N2O2	ISO-PHTHALAMIDE **1740-57-4**
2 ★	-1.73		547		C8H8N2O2	O-PHTHALDIAMIDE **88-96-0**
3 ★	0.29		522		C8H8N2O2S1	1,2,4-BENZOTHIADIAZINE,1,1-DIOXIDE-3-METHYL **360-81-6**
4	-1.43	7.4	476	1	C8H8N2O2S1	8-ETHYL-THIAZOLO(3,2A)PYRIMIDIN-5,7-DIONE
65 ★	0.10		2410		C8H8N2O3	BENZAMIDE,2-METHYL-6-NITRO
6 ★	1.47	7.0	925	1	C8H8N2O3	M-NITROACETANILIDE **122-28-1**
7	1.59	7.4	2327	1	"	M-NITROACETANILIDE
8 ★	1.00		547		C8H8N2O3	O-NITROACETANILIDE **552-32-9**
9 ★	1.66	7.2	415	314	C8H8N2O3	P-NITROACETANILIDE **104-04-1**
3270 ★	2.07		2142		C8H8N2O4	2-NITROPHENYLCARBAMATE,O-METHYL
1 ★	0.98		1946		C8H8N2O4	2-PROPENOIC N-METHYLAMIDE,3-(5-NITRO-2-FURANYL)
2 ★	1.95		2142		C8H8N2O4	3-NITROPHENYLCARBAMATE,O-METHYL
3 ★	2.25		2142		C8H8N2O4	4-NITROPHENYLCARBAMATE,O-METHYL
4 ★	-0.26		2410		C8H8N2O4	BENZAMIDE,2-NITRO-6-METHOXY
75 ★	1.02		273		C8H8N2O4	N-METHYL-2-NITROPHENYLCARBAMATE **7374-06-3**
6 ★	1.39		273		C8H8N2O4	N-METHYL-3-NITROPHENYLCARBAMATE **6132-21-4**
7 ★	1.47		273		C8H8N2O4	N-METHYL-4-NITROPHENYLCARBAMATE **5819-21-6**
8 ★	-0.02		2499		C8H8N2O4	PYRAZINE,2,3-BIS-METHOXYCARBONYL
9 ★	2.67		2215		C8H8N2O5	PHENOL,2-ETHYL-4,6-DINITRO
3280 ★	1.05		547		C8H8N4	1-PHENYL-3-CYANOGUANIDINE **41410-39-3**
1 ★	1.00	13.0	547	224	C8H8N4	2,4-DIAMINOQUINAZOLINE **1899-48-5**
2 ★	1.67		1519	459	C8H8N4	4-HYDRAZINOQUINAZOLINE
3	1.00		1401		C8H8N4	HYDRALAZINE **86-54-4** *antihypertensive*
4 ★	1.36		2606	424	C8H8N4	TETRAZOLE,1-BENZYL
85 ★	1.03		2606	424	C8H8N4	TETRAZOLE,2-BENZYL
6 ★	1.36	2.2	2018		C8H8N4	TETRAZOLE,5-BENZYL **18489-25-3**
7 ★	-0.10		559		C8H8N4O1	1-CARBAMYLMETHYLBENZOTRIAZOLE
8 ★	0.28		559		C8H8N4O1	2-CARBAMYLMETHYLBENZOTRIAZOLE
9 ★	1.40	7.0	1621	1	C8H8N4O1	PYRAZOLE-3-ONE,1-(4-PYRIMIDINYL)-5-METHYL **18697-64-8**
3290 ★	0.20	7.4	2360	1	C8H8N4O2	BENZOTRIAZINE,1,4-DI-N-OXIDE-3-AMINO-7-METHYL
1 ★	-1.65	7.4	1793	1	C8H8N4O2S1	6-PURINETHIONE,4-CARBOXYMETHYL
2 ★	-0.35		2070		C8H8N4O3	ALLOPURINOL,1-ACETYLOXYMETHYL
3 ★	0.00	7.4	2360	1	C8H8N4O3	BENZOTRIAZINE-1,4-DI-N-OXIDE-3-AMINO-7-METHOXY
4 ★	2.14		570		C8H8O1	2,3-DIHYDROBENZOFURAN **496-16-2**
95 ★	1.58		501		C8H8O1	ACETOPHENONE **98-86-2**
6 ★	2.09		594	400	C8H8O1	BENZALDEHYDE,O-METHYL **529-20-4**
7 ★	1.78		547		C8H8O1	PHENYLACETALDEHYDE **122-78-1**
8 ★	1.76		547		C8H8O1	PHTHALAN **496-14-0**
9	1.51	7.4	1821	1	C8H8O1	STYRENEOXIDE **96-09-3**
3300 ★	1.61		547		"	STYRENEOXIDE
1 ★	1.91		514		C8H8O1S1	PHENYLTHIO-ACETIC ACID
2 ★	2.23		530		C8H8O1S1	THIOACETIC ACID,S-PHENYL ESTER **934-87-2**
3 ★	2.05		514		C8H8O1Se1	PHENYLSELENO-ACETIC ACID
4 ★	2.01		547		C8H8O2	1,4-BENZODIOXAN **493-09-4**
5 ★	1.28		2649		C8H8O2	2,5-DIMETHYL-1,4-BENZOQUINONE **137-18-8**
6 ★	1.22		2649		C8H8O2	2,6-DIMETHYL-1,4-BENZOQUINONE **527-61-7**
7 ★	1.35		1478		C8H8O2	2-(B-ACETYLVINYL)FURAN **623-15-4**
8 ★	1.92		825		C8H8O2	2-HYDROXYACETOPHENONE **118-93-4**
9 ★	1.49		501		C8H8O2	ACETIC ACID,PHENYL ESTER **122-79-2**
3310 ★	1.70		599		C8H8O2	BENZALDEHYDE,O-METHOXY **135-02-4**
1	1.61	7.0	925	1	C8H8O2	BENZALDEHYDE,P-METHOXY **123-11-5**
2 ★	1.76	5.6	726	41	"	BENZALDEHYDE,P-METHOXY
3 ★	2.12		501		C8H8O2	BENZOIC ACID,METHYL ESTER **93-58-3**
4 ★	1.24		2452		C8H8O2	BENZOQUINONE,2,3-DIMETHYL **526-86-3**
15 ★	1.39		501		C8H8O2	M-HYDROXYACETOPHENONE **121-71-1**
6 ★	2.37		501		C8H8O2	M-TOLUIC ACID **99-04-7**
7 ★	2.46	1.0	590	342	C8H8O2	O-TOLUIC ACID **118-90-1**
8 ★	1.35		501		C8H8O2	P-HYDROXYACETOPHENONE **99-93-4**
9 ★	2.27		501		C8H8O2	P-TOLUIC ACID **99-94-5**
3320 ★	1.41		501		C8H8O2	PHENYLACETIC ACID **103-82-2**
1 ★	2.74		853		C8H8O2S1	P-METHIOBENZOIC ACID **13205-48-6**
2 ★	1.88	1.0	547	342	C8H8O2S1	S-PHENYLMERCAPTOACETIC ACID **103-04-8**
3 ★	0.97		763		C8H8O3	3-HYDROXY-4-METHOXYBENZALDEHYDE **621-59-0**
4 ★	2.99		1185		C8H8O3	4-METHYLSALICYLIC ACID **50-85-1**
25 ★	2.78		1185		C8H8O3	5-METHYLSALICYLIC ACID
6 ★	1.21		825		C8H8O3	BENZALDEHYDE,3-METHOXY-4-HYDROXY **121-33-5** *pharmaceutic aid (flavor)*
7 ★	0.93		2555		C8H8O3	BENZOIC ACID,P-HYDROXYMETHYL
8 ★	1.05		503		C8H8O3	BENZYL ALCOHOL,3,4-METHYLENEDIOXY **495-76-1**
9 ★	1.89		300		C8H8O3	M-HYDROXYBENZOIC ACID,METHYL ESTER **19438-10-9**
3330	1.23		853		C8H8O3	M-HYDROXYPHENYLACETATE **102-29-4** *antiseborrheic, keratolytic*
1 ★	0.85		501		C8H8O3	M-HYDROXYPHENYLACETIC ACID
2 ★	2.02		501		C8H8O3	M-METHOXYBENZOIC ACID **586-38-9**
3 ★	2.34		594	400	C8H8O3	METHYL SALICYLATE **119-36-8** *pharmaceutic aid (flavor)*
4	2.36	7.4	2268	1	"	METHYL SALICYLATE

	logP	pH	Ref.	Note	MF	Name / CAS / activity
35 ★	0.85		501		C8H8O3	O-HYDROXYPHENYLACETIC ACID 614-75-5
6 ★	1.59	7.4	340	2	C8H8O3	O-METHOXYBENZOIC ACID 579-75-9
7 ★	1.37		763		C8H8O3	O-VANILLIN 148-53-8
8 ★	1.96		536		C8H8O3	P-HYDROXYBENZOIC ACID,METHYL ESTER 99-76-3 *antifungal*
9	1.98	7.5	1259	1	"	P-HYDROXYBENZOIC ACID,METHYL ESTER
3340 ★	0.75	1.2	1077		C8H8O3	P-HYDROXYPHENYLACETIC ACID 156-38-7
1	0.14	7.5	1259	1	C8H8O3	P-METHOXYBENZOIC ACID 100-09-4
2	1.92	7.4	2539	2	"	P-METHOXYBENZOIC ACID
3 ★	1.96		501		"	P-METHOXYBENZOIC ACID
4 ★	1.34	1.0	547	342	C8H8O3	PHENOXYACETIC ACID 122-59-8
45 ★	0.62	1.0	603	342	C8H8O3	PHENYLACETIC ACID,A-HYDROXY 90-64-2
6 ★	2.86		1586		C8H8O3	SALICYLIC ACID,3-METHYL 83-40-9
7 ★	1.77		1478		C8H8O3S1	THIOPHEN-2-COCO2ET 4075-58-5
8	-3.20	7.4	1168	1	C8H8O4	3,4-DIHYDROXYPHENYLACETIC ACID 102-32-9
9	0.98	2.0	2312	337	"	3,4-DIHYDROXYPHENYLACETIC ACID
3350 ★	1.43		1586		C8H8O4	BENZOIC ACID,4-HYDROXY-3-METHOXY 121-34-6
1	1.69	7.4	2539	2	"	BENZOIC ACID,4-HYDROXY-3-METHOXY
2 ★	0.34		1963		C8H8O4	BENZOQUINONE,2,3-DIMETHOXY 3117-02-0
3 ★	-0.06		1963		C8H8O4	BENZOQUINONE,2,6-DIMETHOXY 530-55-2
4 ★	0.76		501		C8H8O4	PHENOXYACETIC ACID,M-HYDROXY 1878-83-7
55 ★	0.85		588		C8H8O4	PHENOXYACETIC ACID,O-HYDROXY 6324-11-4
6 ★	0.65		501		C8H8O4	PHENOXYACETIC ACID,P-HYDROXY 1878-84-8
7	0.86	2.0	2312	337	C8H8O4	PHENYLACETIC ACID,2,5-DIHYDROXY 451-13-8
8 ★	1.94		1586		C8H8O4	SALICYLIC ACID,3-METHOXY 877-22-5
9 ★	0.67		853		C8H8O4S1	P-METHYLSULFONYLBENZOIC ACID 4052-30-6
3360	-4.63	7.4	1168	1	C8H8O5	3,4-DIHYDROXYMANDELIC ACID 775-01-9
1 ★	0.86		1335		C8H8O5	METHYLGALLATE 99-24-1
2 ★	3.09		502		C8H9Br1	ETHYLBROMIDE,B-PHENYL 103-63-9
3 ★	2.49		1263		C8H9Br1N2O1	3-METHYL-4-BROMOPHENYLUREA
4	1.65		2030		C8H9Br1N2O1	NITROSOAMINE,N-METHYL-N-(3-BROMOBENZYL)
65	1.57		2030		C8H9Br1N2O1	NITROSOAMINE,N-METHYL-N-(4-BROMOBENZYL)
6 ★	1.14	7.4	775	1	C8H9Br1N2O2	3-(N,N-DIMETHYLCARBAMYLOXY)-2-BROMOPYRIDINE 51581-35-2
7 ★	0.32		1564		C8H9Br1N2O3	PYRIMIDINE-2,4-DIONE,3-ALLYLOXY-5-BROMO-6-METHYL 77317-70-5
8	1.30		2123	22	C8H9Br1N4	BENZALDEHYDE,4-BROMO,GUANYLHYDRAZONE 37873-43-1
9 ★	3.82		547		C8H9Cl1	3,4-DIMETHYLCHLOROBENZENE 615-60-1
3370 ★	2.95		502		C8H9Cl1	ETHYLCHLORIDE,B-PHENYL 622-24-2
1 ★	3.45		945		C8H9Cl1N1O5P1S1	CHLORTHION 500-28-7
2 ★	3.58		928		C8H9Cl1N1O5P1S1	DICAPTHON 2463-84-5
3 ★	1.83		945		C8H9Cl1N1O6P1	2-CHLORO-DIMETHYLPARA-OXON 17650-76-9
4 ★	1.83		945		C8H9Cl1N1O6P1	3-CHLORO-DIMETHYLPARA-OXON 2255-15-4
75 ★	2.18		1263		C8H9Cl1N2O1	1-METHYL-3-(3-CHLOROPHENYL)UREA 20940-42-5
6	1.46		2030		C8H9Cl1N2O1	NITROSOAMINE,N-METHYL-N-(3-CHLOROBENZYL)
7 ★	2.19		2030		C8H9Cl1N2O1	NITROSOAMINE,N-METHYL-N-(4-CHLOROPHENYL)
8	1.04	7.4	775	1	C8H9Cl1N2O2	3-(N,N-DIMETHYLCARBAMYLOXY)-2-CHLOROPYRIDINE 51581-34-1
9 ★	1.37		1263		C8H9Cl1N2O2	3-CHLORO-4-METHOXYPHENYLUREA 25277-05-8
3380 ★	1.16	7.4	2004	1	C8H9Cl1N2O3S1	N2-METHYL-N1-P-CHLOROBENZENESULFONYLUREA 52102-43-9
1	0.96		2123	22	C8H9Cl1N4	BENZALDEHYDE,4-CHLORO,GUANYLHYDRAZONE 13308-88-8
2 ★	1.78	9.9	704	821	C8H9Cl1N4O1	4-CHLOROPHENYLAMIDINOUREA 58247-24-8
3 ★	2.69		2410		C8H9Cl1O2	1,4-DIMETHOXYBENZENE,2-CHLORO
4 ★	0.11	6.5	321	2	C8H9Cl2N3O4S2	3-CHLOROMETHYLHYDROCHLOROTHIAZIDE 1824-47-1
85 ★	0.46	6.5	321	2	C8H9Cl2N3O4S2	3-DICHLOROMETHYLHYDROTHIAZIDE 23141-87-6
6 ★	3.06		354		C8H9Cl3N1O3P1	METHYLPHOSPHORAMIDATE,O-METHYL-O-(2,4,5-TRICHLOROPHENYL) 2213-70-9
7 ★	3.40		421		C8H9Cl5O2	1-ACETOXYPENTACHLOROCYCLOHEXANE 60423-87-2
8 ★	1.59		1263		C8H9F1N2O1	3-METHYL-4-FLUOROPHENYLUREA
9 ★	1.67		2030		C8H9F1N2O1	NITROSOAMINE,N-METHYL,N-(4-FLUOROBENZYL)
3390 ★	0.67	4.0	2270	302	C8H9F1N2O3	5-FLUOROURACIL-3-BUTYRYL 94452-21-8
1	-0.27	7.4	401	2	C8H9F1N2O3	FTORAFUR 17902-23-7 *antineoplastic*
2	-0.50	7.4	2266	1	C8H9F1N2O4	5-FLUOROURACIL-1-CARBOXYLIC ACID,I-PROPYL ESTER
3 ★	0.20	4.0	2266	302	"	5-FLUOROURACIL-1-CARBOXYLIC ACID,I-PROPYL ESTER
4	-0.04	4.0	2269	302	C8H9F1N2O5	5-FLUOROURACIL-1-ETHOXYCARBONYLOXYMETHYL
95 ★	0.15	4.0	2269	302	C8H9F1N2O5	5-FLUOROURACIL-3-ETHOXYCARBONYLOXYMETHYL
6 ★	1.26	7.4	775	1	C8H9I1N2O2	3-(N,N-DIMETHYLCARBAMYLOXY)-2-IODOPYRIDINE 51581-36-3
7 ★	1.64		564		C8H9N1	N-BENZILIDENEMETHYLAMINE 622-29-7
8 ★	1.16		501		C8H9N1O1	ACETANILIDE 103-84-4 *analgesic, antipyretic*
9 ★	2.40		594	400	C8H9N1O1	BENZALDOXIME,P-METHYL 3235-02-7
3400 ★	0.76		2410		C8H9N1O1	BENZAMIDE,2-METHYL
1 ★	0.77		538		C8H9N1O1	ETHYL-4-PYRIDYL KETONE 1701-69-5
2 ★	1.18		541		C8H9N1O1	M-METHYLBENZAMIDE 618-47-3
3 ★	0.86		541		C8H9N1O1	N-METHYLBENZAMIDE 613-93-4
4 ★	1.09		603		C8H9N1O1	N-METHYLFORMANILIDE 93-61-8
5 ★	1.63		547		C8H9N1O1	O-AMINOACETOPHENONE 551-93-9
6 ★	2.53		547		C8H9N1O1	O-METHYLBENZALDOXIME 3376-32-7
7 ★	0.83	7.4	1737	1	C8H9N1O1	P-AMINOACETOPHENONE 99-92-3
8 ★	1.18		536		C8H9N1O1	P-METHYLBENZAMIDE 619-55-6
9 ★	1.61		453		C8H9N1O1	P-METHYLFORMANILIDE 3085-54-9
3410 ★	0.45		502		C8H9N1O1	PHENYLACETAMIDE 103-81-1
1 ★	2.95		527		C8H9N1O2	1,3-DIMETHYL-2-NITROBENZENE 81-20-9
2	1.14		866		C8H9N1O2	2-METHYL-4-AMINOBENZOIC ACID 2486-75-1
3	1.37		866		C8H9N1O2	3-METHYL-4-AMINOBENZOIC ACID 2486-70-6
4	-2.07	7.4	547	1	C8H9N1O2	A-PHENYLGLYCINE 69-91-0
15 ★	-1.70	7.2	2323	468	"	A-PHENYLGLYCINE
6	0.56	1.0	547	342	C8H9N1O2	ANTHRANILIC ACID,N-METHYL 119-68-6
7	2.12		1818		C8H9N1O2	BENZAMIDE,2-HYDROXY-3-METHYL 14008-60-7
8 ★	2.08		306		C8H9N1O2	BENZENE,B-NITROETHYL 6125-24-2
9 ★	1.88		594	298	C8H9N1O2	BENZOIC ACID,2-AMINO,METHYL ESTER 134-20-3

	logP	pH	Ref.	Note	MF	Name / **CAS** / *activity*
3420	-1.57	7.4	594	1	C8H9N1O2	BENZYLAMINE-4-CARBOXYLIC ACID **56-91-7**
1 ★	-1.55	6.0	594	1	"	BENZYLAMINE-4-CARBOXYLIC ACID
2 ★	0.62		514		C8H9N1O2	GLYCINE,N-PHENYL **103-01-5**
3 ★	0.71		564		C8H9N1O2	GLYCOLANILIDE **4746-61-6**
4	0.78	7.2	2317	1	C8H9N1O2	I-NICOTINIC ACID,ETHYL ESTER **1570-45-2**
25 ★	1.43		525		"	I-NICOTINIC ACID,ETHYL ESTER
6 ★	0.73		853		C8H9N1O2	M-HYDROXYACETANILIDE **621-42-1**
7 ★	0.85		547		C8H9N1O2	M-METHOXYBENZAMIDE **5813-86-5**
8 ★	1.25		453		C8H9N1O2	M-METHOXYFORMANILIDE **27153-17-9**
9 ★	1.76		1263		C8H9N1O2	METHYL-N-PHENYLCARBAMATE **2603-10-3**
3430	0.39	3.0	1268		C8H9N1O2	N-(HYDROXYMETHYL)BENZAMIDE **6282-02-6**
1	0.41	9.2	1268	1	"	N-(HYDROXYMETHYL)BENZAMIDE
2 ★	1.24		503		C8H9N1O2	N-METHYLPHENYLCARBAMATE **1943-79-9**
3 ★	1.32		525		C8H9N1O2	NICOTINIC ACID,ETHYL ESTER **614-18-6**
4	1.36	7.0	332	1	"	NICOTINIC ACID,ETHYL ESTER
35 ★	2.83		2305	459	C8H9N1O2	NITROBENZENE,2,3-DIMETHYL
6 ★	2.91		2305	459	C8H9N1O2	NITROBENZENE,3,4-DIMETHYL
7 ★	3.03	7.4	594	314	C8H9N1O2	NITROBENZENE,4-ETHYL **100-12-9**
8 ★	1.20		547		C8H9N1O2	O-BENZYLCARBAMATE **621-84-1**
9	1.23	7.4	172	1	"	O-BENZYLCARBAMATE
3440 ★	0.72		853		C8H9N1O2	O-HYDROXYACETANILIDE **614-80-2**
1 ★	0.84		547		C8H9N1O2	O-METHOXYBENZAMIDE **2439-77-2**
2 ★	1.35		2184		C8H9N1O2	P-AMINOBENZOIC ACID,METHYL ESTER **619-45-4**
3 ★	0.45		594	400	C8H9N1O2	P-AMINOPHENYLACETATE **13871-68-6**
4	0.31	7.2	803	314	C8H9N1O2	P-HYDROXYACETANILIDE **103-90-2** *analgesic, antipyretic*
45 ★	0.51	2.0	1838		"	P-HYDROXYACETANILIDE
6	0.36		726	837	"	P-HYDROXYACETANILIDE
7 ★	0.86		260		C8H9N1O2	P-METHOXYBENZAMIDE **3424-93-9**
8 ★	1.03		453		C8H9N1O2	P-METHOXYFORMANILIDE **5470-34-8**
9 ★	0.76		546		C8H9N1O2	PHENOXYACETAMIDE **621-88-5**
3450 ★	0.87		525		C8H9N1O2	PICOLINIC ACID,ETHYL ESTER **2524-52-9**
1	0.89	7.0	590		"	PICOLINIC ACID,ETHYL ESTER
2 ★	1.12		1213		C8H9N1O3	2-(P-NITROPHENYL)ETHANOL **100-27-6**
3	0.45		866		C8H9N1O3	2-METHOXY-4-AMINOBENZOIC ACID **2486-80-8**
4 ★	2.68		2410		C8H9N1O3	ANISOLE,2-METHYL-3-NITRO
55 ★	2.53	7.4	820	1	C8H9N1O3	P-NITROPHENETOLE **100-29-8**
6	-0.82	7.4	1074	324	C8H9N1O3	PYRIDOXAL **66-72-8**
7 ★	0.25	2.1	2265		C8H9N1O3S1	BENZENESULFONAMIDE,3-ACETYL
8 ★	0.20	2.0	264		C8H9N1O3S1	P-ACETYLBENZENESULFONAMIDE **1565-17-9**
9 ★	0.45		2209		C8H9N1O4	MALEIMIDE,2-METHYL,N-ACETYLOXYMETHYL
3460	0.40		731	604	C8H9N1O4	N-MALEOYL-4-AMINOBUTYRIC ACID
1 ★	0.49		731		C8H9N1O4	N-MALEOYLGLYCINE,ETHYL ESTER **1585-79-1**
2	-1.30	7.4	1074	324	C8H9N1O4	PYRIDOXIC ACID **82-82-6**
3 ★	0.64		773	601	C8H9N1O4S1	4-SULFAMYLBENZOIC ACID,METHYL ESTER **22808-73-7**
4 ★	2.40		547		C8H9N1S1	PHENYLTHIOCARBAMATE,S-METHYL ESTER **13509-38-1**
65 ★	1.71		547		C8H9N1S1	THIOACETANILIDE **637-53-6**
6 ★	1.63		2390		C8H9N1S1	THIOBENZAMIDE,4-METHOXY **2362-64-3**
7 ★	2.01		2390		C8H9N1S1	THIOBENZAMIDE,4-METHYL **2362-62-1**
8 ★	1.58		559		C8H9N3	1-ETHYLBENZOTRIAZOLE **16584-05-7**
9 ★	2.10		559		C8H9N3	2-ETHYLBENZOTRIAZOLE **16584-04-6**
3470 ★	0.95		595		C8H9N3	BENZIMIDAZOLE,1-METHYL-2-AMINO **1622-57-7**
1 ★	1.11		595		C8H9N3	BENZIMIDAZOLE,2-AMINO-5-METHYL **6285-68-3**
2 ★	0.50		559		C8H9N3O1	1-(2-HYDROXYETHYL)BENZOTRIAZOLE **938-56-7**
3 ★	0.92		559		C8H9N3O1	2-(2-HYDROXYETHYL)BENZOTRIAZOLE **939-72-0**
4 ★	1.27		547		C8H9N3O1	BENZALDEHYDESEMICARBAZONE **1574-10-3**
75	-0.21	7.4	595	1	C8H9N3O1	BENZIMIDAZOLE,1-METHYL-2-AMINO-5-HYDROXY
6 ★	0.57		595		C8H9N3O1	BENZIMIDAZOLE,2-AMINO-5-METHOXY **6232-91-3**
7 ★	1.61		260		C8H9N3O1S1	M-HYDROXYBENZALDEHYDETHIOSEMICARBIZONE **7420-37-3**
8 ★	-0.57		400		C8H9N3O2	1-ACETYL-2-PICOLINOYLHYDRAZINE **17433-31-7** *antineoplastic*
9 ★	-0.50		547		C8H9N3O2	1-BENZOYL-2-CARBAMYLHYDRAZINE **2845-79-6**
3480 ★	1.72	8.9	1498		C8H9N3O2	1-METHYL-1-NITROSO-3-PHENYLUREA **21561-99-9**
1	0.82		2030		C8H9N3O3	NITROSOAMINE,N-METHYL,N-(4-NITROBENZYL)
2 ★	1.17		2030		C8H9N3O3	NITROSOAMINE,N-METHYL-N-(3-NITROBENZYL)
3 ★	1.19		611		C8H9N3O3S1	1(5-NO2-2-FURYL)-2-METHYLTHIO-2-IMIDAZOLINE **53207-66-2**
4 ★	0.54	7.4	775	1	C8H9N3O4	3-(N,N-DIMETHYLCARBAMYLOXY)-2-NITROPYRIDINE **51581-33-0**
85 ★	-0.40	6.5	321	1	C8H9N3O4S2	1,2,4-BENZOTHIAZIDE-1,1-DIOXIDE-6-METHYL-7-SULFONAMIDO **3256-99-3**
6 ★	1.86		547		C8H9N3S1	BENZALDEHYDETHIOSEMICARBAZONE **1627-73-2**
7 ★	1.82	7.0	332	1	C8H9N3S1	THIENO(2,3-D)-PYRIMIDINE,4-(DIMETHYLAMINO)
8 ★	-0.41		2544		C8H9N5O1	9-METHYL-N-ACETYLADENINE
9	0.53		2123	22	C8H9N5O2	BENZALDEHYDE,2-NITRO,GUANYLHYDRAZONE **102632-31-5**
3490	0.01		2123	22	C8H9N5O2	BENZALDEHYDE,3-NITRO,GUANYLHYDRAZONE **90792-54-4**
1	0.87		2123	22	C8H9N5O2	BENZALDEHYDE,4-NITRO,GUANYLHYDRAZONE **30068-29-2**
2 ★	1.28	9.9	704	821	C8H9N5O3	3-NITROPHENYLAMIDINOUREA **16018-79-4**
3 ★	1.42		594	400	C8H9O3P1	1,3,2-DIOXAPHOSPHOLANE,2-PHENOXY **1077-05-0**
4 ★	2.67		572		C8H9O3P1S1	SALITHION **3811-49-2** *insecticide*
95	5.60		2129	471	C8H10	CYCLOOCTATRIENE
6 ★	3.15		502		C8H10	ETHYLBENZENE **100-41-4**
7 ★	3.20		531		C8H10	M-XYLENE **108-38-3**
8 ★	3.12		536		C8H10	O-XYLENE **95-47-6**
9 ★	3.15		531		C8H10	P-XYLENE **106-42-3**
3500 ★	1.71		504		C8H10Cl1N1O2	O-(1-ETHYL-1-ETHYNYL-3-CHLOROALLYL)CARBAMATE
1 ★	3.03	7.4	541	1	C8H10Cl1N3	1-(O-CHLOROPHENYL)-3,3-DIMETHYLTRIAZINE **20241-00-3**
2 ★	3.44		538		C8H10Cl1N3	3,3-DIMETHYL-1-(3-CHLOROPHENYL)TRIAZINE **20241-05-8**
3 ★	0.08	6.5	321	2	C8H10Cl1N3O4S2	3-METHYLHYDROCHLOROTHIAZIDE **890-67-5**
4	-0.17	7.4	1350	1	C8H10Cl1N3S1	TIAMENIDINE *antihypertensive*

	logP	pH	Ref.	Note	MF	Name / CAS / activity
5 ★	2.41		354		C8H10Cl2N1O3P1	METHYLPHOSPHORAMIDATE,O-ME-O-(2,4-DICL-PHENYL) 10363-40-3
6 ★	0.72	7.4	2490	1	C8H10F1O3S2	BENZENESULFONAMIDE,3-FLUORO-4-HYDROXYETHYLTHIO 108966-70-7
7 ★	-0.60	7.4	2490	1	C8H10F1O5S2	BENZENESULFONAMIDE,3-FLUORO-4-(2-HYDROXYETHYL)SULFONYL 108966-76-3
8 ★	-0.52	7.4	2361	1	C8H10F2N4O4	IMIDAZOLE,2-NITRO-1-(3-HYDROXYETHYLAMINO-3-OXO-2,2-DIFLUORO)PROPYL ✚
9 ★	0.52		1008	354	C8H10F3N3O4	1(2-OH-6-CF3-ETOPROPYL)-2-NITROIMIDAZOLE
3510 ★	3.50	2.0	541		C8H10I1N3	1-(O-IODOPHENYL)-3,3-DIMETHYLTRIAZINE
1 ★	0.32		2354		C8H10I1N3O5	AZOMYCIN,6-IODORIBOSIDE
2 ★	2.86		572		C8H10N1O5P1S1	DIMETHYLPARATHION 298-00-0 insecticide
3 ★	1.33		251		C8H10N1O6P1	DIMETHYLPARA-OXON 950-35-6
4 ★	1.12		1213		C8H10N2O1	1-PHENYL-3-METHYLUREA 1007-36-9
15 ★	-0.27		1474		C8H10N2O1	2-PYRIDINEPROPANEAMIDE
6 ★	-0.26		1474		C8H10N2O1	3-PYRIDINEPROPANEAMIDE
7 ★	-0.25		1474		C8H10N2O1	4-PYRIDINEPROPANEAMIDE
8 ★	-0.14		1439		C8H10N2O1	4H-PYRIDO(1,2-A)PYRIMIDIN-4-ONE,6,7,8,9-H4 ✚
9 ★	0.74		1057		C8H10N2O1	BENZOYLHYDRAZINE,M-METHYL
3520 ★	0.22		1057		C8H10N2O1	BENZOYLHYDRAZINE,O-METHYL
1 ★	0.73		1057		C8H10N2O1	BENZOYLHYDRAZINE,P-METHYL
2 ★	0.73		547		C8H10N2O1	BENZYLUREA 538-32-9
3 ★	1.29		1263		C8H10N2O1	M-TOLYLUREA 63-99-0
4 ★	1.55		2030		C8H10N2O1	METHYLBENZYLNITROSAMINE 937-40-6
25 ★	-0.48		1519	400	C8H10N2O1	N,N-DIMETHYLNICOTINAMIDE 6972-69-6
6 ★	0.31		1035		C8H10N2O1	N-ETHYLNICOTINAMIDE
7	-0.31	7.4	1737	1	C8H10N2O1	P-AMINOACETANILIDE 122-80-5
8 ★	0.08	7.2	415	314	"	P-AMINOACETANILIDE
9 ★	-0.45		2444		C8H10N2O1	PYRIDINE,2-N,N-DIMETHYLCARBOXAMIDO
3530 ★	0.42		503		C8H10N2O1	UREA,1-METHYL-1-PHENYL 4559-87-9
1 ★	1.57		1624		C8H10N2O1S1	3-(METHYLTHIO)PHENYLUREA
2	2.07		2139	459	C8H10N2O2	ANILINE,2-NITRO-4,5-DIMETHYL 6972-71-0
3 ★	2.16		2441	525	C8H10N2O2	ANILINE,N,N-DIMETHYL-3-NITRO 619-31-8
4 ★	0.34		2649		C8H10N2O2	BENZOQUINONE,2,5-BIS-METHYLAMINO
35 ★	0.40		1057		C8H10N2O2	BENZOYLHYDRAZINE,M-METHOXY
6 ★	0.25		1057		C8H10N2O2	BENZOYLHYDRAZINE,O-METHOXY
7 ★	0.25		1057		C8H10N2O2	BENZOYLHYDRAZINE,P-METHOXY
8 ★	-0.11		1035		C8H10N2O2	N-HYDROXYETHYLNICOTINAMIDE
9 ★	2.27	7.0	925	1	C8H10N2O2	P-NITRO-N,N-DIMETHYLANILINE 100-23-2
3540 ★	0.51		2499		C8H10N2O2	PYRAZINE,2-ETHOXYCARBONYL-6-METHYL
1 ★	0.61		2499		C8H10N2O2	PYRAZINE,2-METHYL-5-ETHOXYCARBONYL
2 ★	1.20		547		C8H10N2O2S1	BENZALHYDRAZONE,N-METHYLSUFONYL
3	-0.54	7.4	2461	1	C8H10N2O2S3	THIENO[2,3-B]THIOPHENE-2-SULFONAMIDE,5-METHAMINOMETHYL 122266-89-1
4	-0.49	7.4	2461	1	C8H10N2O2S3	THIENO[3,2-B]THIOPHENE-2-SULFONAMIDE,5-METHYLAMINOMETHYL 127025-27-8
45 ★	1.20		2499		C8H10N2O3	PYRAZINE,2-METHOXY-6-ETHOXYCARBONYL
6 ★	-0.12		1624		C8H10N2O3S1	3-MESYLPHENYLUREA
7 ★	-0.31		773	601	C8H10N2O3S1	4-SULFAMYLBENZAMIDE,N-METHYL
8 ★	0.31	3.5	2458	302	C8H10N2O3S1	BENZYLAMIDE,N-SULFAMIDO
9	-0.26	7.4	2004	1	C8H10N2O3S1	N2-METHYL-N1-BENZENESULFONYLUREA 52102-38-2
3550 ★	0.44		2352	2	"	N2-METHYL-N1-BENZENESULFONYLUREA
1 ★	0.00	2.1	2265		C8H10N2O3S1	N4-ACETYLSULFANILAMIDE 121-61-9
2 ★	0.58		594		C8H10N2O3S1	P-TOLUENESULFONAMIDE,N-METHYL,N-NITROSO 80-11-5
3	-2.21	7.5	542	1	C8H10N2O3S1	SULFANILACETAMIDE 144-80-9 antibacterial
4 ★	-0.96		480	601	"	SULFANILACETAMIDE
55 ★	0.24	3.5	2458	2	C8H10N2O3S1	SULFONUREA,S-BENZYL
6	-0.89	7.4	2525	1	C8H10N2O3S2	5-(METHYLAMINOMETHYL)THIENO[2,3-B]FURAN-2-SULFONAMIDE
7	0.81		2447		C8H10N2O4	4-PYRIMIDINECARBOXYLIC ACID,TETRAHYDRO-2,6-DIOXO,I-PROPYL ESTER
8	1.10		2447		C8H10N2O4	4-PYRIMIDINECARBOXYLIC ACID,TETRAHYDRO-2,6-DIOXO,PROPYL ESTER
9 ★	-0.27		572		C8H10N2O4S1	ASULAM 3337-71-1 herbicide
3560	-1.59	7.5	652	1	C8H10N2O4S1	SULFANILYLGLYCINE 5616-30-8
1 ★	0.38	7.4	2490	1	C8H10N2O5S2	BENZENESULFONAMIDE,3-NITRO-4-HYDROXYETHYLTHIO 108966-58-1
2 ★	0.81		594		C8H10N2O6S2	NITROBENZENE,2-(N,N-BIS-METHYLSULFONYL)
3 ★	0.85		503		C8H10N2S1	UREA,1-METHYL-1-PHENYL-2-THIO 4104-75-0
4 ★	1.18		1967		C8H10N4	BENZALDEHYDE,GUANYLHYDRAZONE 3357-37-7
65 ★	0.81		547		C8H10N4O1	BENZALDEHYDE,(4-AMINOSEMICARBAZONE)
6 ★	0.48	6.6	2590	459	C8H10N4O1	BENZAMIDE,4-(3-METHYLTRIAZENYL) antineoplastic
7 ★	-0.22	6.6	2590	459	C8H10N4O2	1-(4-CARBAMOYLPHENYL)-3-METHYL-3-HYDROXYTRIAZENE 42548-73-2 antineoplastic
8 ★	-1.25	7.4	725	1	C8H10N4O2	2-ACETAMIDOISONIAZID 58481-02-0
9 ★	2.75		564		C8H10N4O2	3,3-DIMETHYL-1-(3-NITROPHENYL)TRIAZENE 20241-06-9
3570	-0.16	7.4	1940	1	C8H10N4O2	CAFFEINE 58-08-2 stimulant
1	-0.10	7.4	971	1	"	CAFFEINE
2 ★	-0.07		505		"	CAFFEINE
3	-0.03	7.4	2357	1	"	CAFFEINE
4	0.01	7.4	919	1	"	CAFFEINE
75 ★	0.33		2537		C8H10N4O2	ENPROFYLLINE 41078-02-8 bronchodilator
6	0.58		547		C8H10N4O2	M-DIURIEDOBENZENE 1135-58-6
7 ★	3.25	6.6	2590	459	C8H10N4O2	NITROBENZENE,4-(3,3-DIMETHYLTRIAZENYL) antineoplastic
8 ★	0.25	1.0	2196		C8H10N4O3	PYRAZOLE,3-METHYL-4-NITRO-5-(N-CYCLOPROPYLCARBOXAMIDO)
9 ★	-0.37	3.0	2191		C8H10N4O3	URIC ACID,1,3,7-TRIMETHYL 5415-44-1
3580 ★	1.28	7.4	553	1	C8H10N4O3S2	3-(1-HYDROXYETHYL-1,3-THIAZOLIDINYLIDEN-2N)-5-NITROTHIAZOLE 24240-67-3 ✚
1 ★	0.43	7.4	553	1	C8H10N4O4S1	(3-HYDROXYETHYL-1,3-OXAZOLIDINYLIDEN-2N)-5-NITROTHIAZOLE 24240-65-1 ✚
2	3.58	7.3	509	2	C8H10N4O6S3	PURINE,2,6,8-TRI-METHYLSULFONYL
3	0.69	7.4	2028	459	C8H10N6	1,2,4-TRIAZOLE,3-AMINO-5-(2-METHYLAMINO)PYRIDIN-4-YL 77314-76-2
4	0.14	7.4	401	1	C8H10N6O1	IMIDAZOLE,5-CARBOXAMIDE-4-(3-METHYL-3-PROPYNYLTRIAZENYL)
85 ★	0.33	7.4	401	2	"	IMIDAZOLE,5-CARBOXAMIDE-4-(3-METHYL-3-PROPYNYLTRIAZENYL)
6 ★	1.42		579		C8H10O1	1-PHENYLETHANOL 1517-69-7
7 ★	2.30		886		C8H10O1	2,4-DIMETHYLPHENOL 105-67-9
8 ★	2.33		886		C8H10O1	2,5-DIMETHYLPHENOL 95-87-4
9 ★	2.36		300		C8H10O1	2,6-DIMETHYLPHENOL 576-26-1

	logP	pH	Ref.	Note	MF	Name / CAS / activity
3590 ★	2.23		886		C8H10O1	3,4-DIMETHYLPHENOL **95-65-8**
1 ★	2.35		300		C8H10O1	3,5-DIMETHYLPHENOL **108-68-9**
2 ★	1.92		579		C8H10O1	BENZYLMETHYLETHER **538-86-3**
3 ★	2.40		501		C8H10O1	M-ETHYLPHENOL **620-17-7**
4 ★	2.66		578		C8H10O1	M-METHYLANISOLE **100-84-5**
95 ★	1.60		501		C8H10O1	M-METHYLBENZYL ALCOHOL **587-03-1**
6 ★	2.47		558		C8H10O1	O-ETHYLPHENOL **90-00-6**
7 ★	2.74		547		C8H10O1	O-METHYLANISOLE **578-58-5**
8	2.40	7.4	2582	1	C8H10O1	P-ETHYLPHENOL **123-07-9**
9 ★	2.58	5.6	726	41	"	P-ETHYLPHENOL
3600 ★	2.66		579		C8H10O1	P-METHYLANISOLE **104-93-8**
1	1.40	7.0	2505	1	C8H10O1	P-METHYLBENZYL ALCOHOL **589-18-4**
2 ★	1.58		501		"	P-METHYLBENZYL ALCOHOL
3 ★	2.51		541		C8H10O1	PHENETOLE **103-73-1**
4 ★	1.36		502		C8H10O1	PHENYLETHYL ALCOHOL **60-12-8** *antimicrobial*
5 ★	1.16		508		C8H10O2	2-PHENOXYETHANOL **122-99-6** *antiseptic (topical)*
6 ★	2.21		536		C8H10O2	BENZENE,1,3-DIMETHOXY **151-10-0**
7 ★	2.03		2163	459	C8H10O2	BENZENE,1,4-DIMETHOXY **150-78-7**
8 ★	0.97		1816		C8H10O2	BENZYL ALCOHOL,2-HYDROXY-5-METHYL **4383-07-7**
9 ★	1.24		2452		C8H10O2	HYDROQUINONE,2,3-DIMETHYL **608-43-5**
3610 ★	0.98		2452		C8H10O2	HYDROQUINONE,2,6-DIMETHYL **654-42-2**
1 ★	1.98	5.6	726	41	C8H10O2	M-ETHOXYPHENOL **621-34-1**
2 ★	1.60		579		C8H10O2	O-DIMETHOXYBENZENE **91-16-7**
3 ★	1.68		558		C8H10O2	O-ETHOXYPHENOL **94-71-3**
4 ★	1.13	7.0	538	701	C8H10O2	O-METHOXYBENZYL ALCOHOL **612-16-8**
15 ★	1.81		300		C8H10O2	P-ETHOXYPHENOL **622-62-8**
6 ★	1.10		501		C8H10O2	P-METHOXYBENZYL ALCOHOL **105-13-5**
7 ★	1.15		2645	225	C8H10O3	2,6-DIMETHOXYPHENOL
8 ★	1.64		572		C8H10O3	3,5-DIMETHOXYPHENOL **500-99-2**
9	0.36		1937	459	C8H10O3	4-HYDROXY-2-PYRONE,3,5,6-TRIMETHYL **50405-44-2**
3620 ★	0.86	7.0	925	1	C8H10O3S1	METHYLSULFONYLANISOLE **43032-67-3**
1 ★	0.76		594		C8H10O3S1	PHENYLMETHYLSULFONE,4-METHOXY
2 ★	-1.01	7.4	1168	1	C8H10O4	3,4-DIHYDROXYPHENYLGLYCOL **3343-19-9**
3 ★	0.59		2452		C8H10O4	HYDROQUINONE,2,3-DIMETHOXY **52643-52-4**
4 ★	0.17		2452		C8H10O4	HYDROQUINONE,2,6-DIMETHOXY **15233-65-5**
25 ★	1.51		2011		C8H10O4S2	PROPANEDIOIC ACID,1,3-DITHIOLAN-2-YLIDINE,DIMETHYL ESTER **19723-86-5** ✚
6 ★	3.20		541		C8H10S1	PHENYLETHYL SULFIDE **622-38-8**
7 ★	0.45		1564		C8H11Br1N2O3	PYRIMIDINE-2,4-DIONE,3-I-PROPOXY-5-BROMO-6-ME **77317-69-2**
8 ★	0.76		1564		C8H11Br1N2O3	PYRIMIDINE-2,4-DIONE,3-PROPOXY-5-BROMO-6-ME **77317-68-1**
9 ★	1.61		354		C8H11Cl1N1O3P1	METHYLPHOSPHORAMIDATE,O-ME-O-(2-CL-PHENYL) **19608-64-1**
3630 ★	1.85		354		C8H11Cl1N1O3P1	METHYLPHOSPHORAMIDATE,O-ME-O-(4-CL-PHENYL) **19670-19-0**
1 ★	0.82	1.0	2202	342	C8H11Cl1N2O3	BARBITURIC ACID, 4-CHLORO-5-T-BUTYL
2 ★	0.37	7.4	401	2	C8H11Cl1N4O4	PCNU **13909-02-9** *antineoplastic*
3 ★	1.02	7.4	593	1	C8H11Cl3O6	CHLORALOSE-ALPHA **15879-93-3** *sedative*
4 ★	1.12	7.4	593	1	C8H11Cl3O6	CHLORALOSE-BETA **16376-36-6**
35 ★	3.69		421	456	C8H11Cl5O1	2-ETO-PENTACL-CYCLOHEXANE(2,3,5,6/1,4) **56421-32-0**
6 ★	3.97		421		C8H11Cl5O1	3-ETHOXYPENTACHLOROCYCLOHEXANE **56421-35-3**
7	-0.31	7.4	1948	1	C8H11F1N4	1-(2-(4-FLUOROPYRIMIDINYL))PIPERAZINE
8	0.97	7.4	1948	2	"	1-(2-(4-FLUOROPYRIMIDINYL))PIPERAZINE
9 ★	0.66		2547		C8H11F2N5O3S1	3-NITROTRIAZOLE,N-METHOXYETHYL-DIFLUOROPROPIOTHIOAMIDE ANALOG ✚
3640 ★	0.41		2547		C8H11F2N5O3S1	3-NITROTRIAZOLE,TRIFLUOROPROPIONAMIDE ANALOG ✚
1 ★	-0.63		2547		C8H11F2N5O4	3-NITROTRIAZOLE,N-HYDROXYPROPYL-DIFLUOROPROPIONAMIDE ANALOG
2 ★	-0.48		2567		C8H11F2N5O4	FNT-2 ✚
3 ★	-0.15	7.4	2361	1	C8H11F2N5O4	TRIAZINE,3-NITRO-1-(3-METHOXYETHYLAMINO-3-OXO-2,2-DIFLUORO)PROPYL
4 ★	1.88		572		C8H11N1	2,4,6-COLLIDINE **108-75-8**
45 ★	1.68	7.5	2201		C8H11N1	2,4-DIMETHYLANILINE **95-68-1**
6 ★	1.83	7.4	594	314	C8H11N1	2,5-DIMETHYLANILINE **95-78-3**
7 ★	1.84		594	400	C8H11N1	2,6-DIMETHYLANILINE **87-62-7**
8 ★	2.10		506		C8H11N1	4-PROPYLPYRIDINE **1122-81-2**
9	-1.39	7.0	2046	1	C8H11N1	ETHYLAMINE,2-PHENYL **64-04-0**
3650	-1.27	7.0	1473	1	"	ETHYLAMINE,2-PHENYL
1	-1.18	7.4	1258	1	"	ETHYLAMINE,2-PHENYL
2	-1.14	7.4	1168	1	"	ETHYLAMINE,2-PHENYL
3	1.16	7.4	1258	2	"	ETHYLAMINE,2-PHENYL
4 ★	1.41		502		"	ETHYLAMINE,2-PHENYL
55 ★	2.31		501		C8H11N1	N,N-DIMETHYLANILINE **121-69-7**
6 ★	2.16		324		C8H11N1	N-ETHYLANILINE **103-69-5**
7 ★	1.52	13.0	547	224	C8H11N1	N-METHYLBENZYLAMINE **103-67-3**
8 ★	1.74		558		C8H11N1	O-ETHYLANILINE **578-54-1**
9 ★	2.16		324		C8H11N1	O-TOLUIDINE,N-METHYL **611-21-2**
3660 ★	1.96	7.4	1737	1	C8H11N1	P-ETHYLANILINE **589-16-2**
1	1.97	7.5	2201		"	P-ETHYLANILINE
2 ★	2.15		324		C8H11N1	P-TOLUIDINE,N-METHYL **623-08-5**
3 ★	0.82		570		C8H11N1O1	2-ANILINOETHANOL **122-98-5**
4 ★	0.58		1474		C8H11N1O1	2-PYRIDINEPROPANOL **2859-68-9**
65 ★	0.60		1474		C8H11N1O1	3-PYRIDINEPROPANOL **2859-67-8**
6 ★	0.58		1474		C8H11N1O1	4-PYRIDINEPROPANOL **2629-72-3**
7 ★	1.74	7.4	594	314	C8H11N1O1	ANILINE,2-METHOXY-5-METHYL **120-71-8**
8 ★	1.23	7.4	594	314	C8H11N1O1	ANILINE,2-METHYL-4-METHOXY **102-50-1**
9 ★	1.56		501		C8H11N1O1	M-DIMETHYLAMINOPHENOL **99-07-0**
3670 ★	0.10	7.4	1737	1	C8H11N1O1	P-AMINOPHENYLETHANOL **122-98-5**
1 ★	1.24	7.4	1737	1	C8H11N1O1	P-ETHOXYANILINE **156-43-4**
2 ★	2.38		2412		C8H11N1O1	PYRIDINE,2-PROPOXY
3	-2.12	7.4	1168	1	C8H11N1O1	TYRAMINE **51-67-2** *adrenergic*
4	-2.07	7.4	1258	1	"	TYRAMINE

	logP	pH	Ref.	Note	MF	Name / CAS / activity
75	-2.05	7.0	2046	1	"	TYRAMINE
6	-0.42	10.0	2046	1	"	TYRAMINE
7 ★	1.20	7.4	1737	1	C8H11N1O2	2,5-DIMETHOXYANILINE 102-56-7
8 ★	0.01		1478		C8H11N1O2	2-(N,N-DIMETHYLACETAMIDO)FURAN
9	-0.20	7.4	2685	1	C8H11N1O2	4-PYRIDONE,3-HYDROXY-2-ETHYL-1-METHYL iron chelator
3680	-3.40		2475	100	C8H11N1O2	DOPAMINE 51-61-6 adrenergic
1	-2.48	7.4	1168	1	"	DOPAMINE
2	-2.38	7.4	1258	1	"	DOPAMINE
3	-2.36	7.4	2261	1	"	DOPAMINE
4	-0.98	7.4	551	1	"	DOPAMINE
85	-0.97	7.4	1258	2	"	DOPAMINE
6	0.52	7.5	2254	2	"	DOPAMINE
7	2.80	7.4	2282	1	"	DOPAMINE
8 ★	1.09		504		C8H11N1O2	O-(1-ETHYL-1-VINYL-2-PROPYNYL)CARBAMATE
9	-2.39	7.4	1258	1	C8H11N1O2	OCTOPAMINE 104-14-3 adrenergic
3690	-2.24	7.4	1168	1	"	OCTOPAMINE
1	-1.06	7.4	1258	2	"	OCTOPAMINE
2	-0.40	7.4	2472	324	C8H11N1O2	PYRIDIN-4-ONE,1-ETHYL-2-METHYL-3-HYDROXY
3	-0.31	7.4	2685	1	"	PYRIDIN-4-ONE,1-ETHYL-2-METHYL-3-HYDROXY
4 ★	2.14		2221	459	C8H11N1O2	PYRROLE-3-CARBOXYLIC ACID-4-METHYL,ETHYL ESTER 2199-49-7
95 ★	1.35		546		C8H11N1O2S1	N,N-DIMETHYLBENZENESULFONAMIDE 14417-01-7
6 ★	1.02		579		C8H11N1O2S1	N-METHYL-N-PHENYLMETHANESULFONAMIDE 13229-35-1
7 ★	1.31		773	601	C8H11N1O2S1	P-ETHYLBENZENESULFONAMIDE 138-38-5
8 ★	1.28	7.4	2272	1	C8H11N1O2S1	P-TOLUENESULFONAMIDE-N-METHYL 640-61-9
9	-3.01	7.4	1168	1	C8H11N1O3	NORADRENALINE 51-41-2 adrenergic
3700	-2.97	7.4	1258	1	"	NORADRENALINE
1	-1.24	7.4	551	1	"	NORADRENALINE
2	-0.64		1135	537	"	NORADRENALINE
3	-1.10	7.4	2685	1	C8H11N1O3	PYRIDIN-4-ONE,1-HYDROXYETHYL-2-METHYL-3-HYDROXY iron chelator
4	-0.77	7.4	1074	324	C8H11N1O3	PYRIDOXINE 65-23-6 vitamin (enzyme co-factor)
5 ★	0.36	7.4	2490	1	C8H11N1O3S2	BENZENESULFONAMIDE,4-HYDROXYETHIO 108966-48-9
6 ★	0.62		536		C8H11N1O4S2	N,N-DI-(METHYLSULFONYL)-ANILINE
7 ★	-0.85	7.4	2490	1	C8H11N1O5S2	BENZENESULFONAMIDE,4-HYDROXYETHYLSULFONYL 108966-49-0
8 ★	0.11	7.4	2490	1	C8H11N1O6S3	THIOPHENE-2-SULFONAMIDE,5-(2-ACETYLOXY)ETHYLSULFONYL 135832-36-9
9 ★	0.65		1478		C8H11N1S1	3-ACETAMIDOTHIOPHENE,N,N-DIMETHYL
3710 ★	0.65		1478		C8H11N1S1	THIOPHEN,2-(N,N,-DIMETHYLACETAMIDO)
1	-2.02		301		C8H11N2O1.Cl1	N1-ETHYLNICOTINAMIDECHLORIDE 14596-52-2
2 ★	1.27		354		C8H11N2O5P1	METHYLPHOSPHORAMIDATE,O-METHYL-O-(4-NITROPHENYL) 54267-24-2
3 ★	2.59	7.4	536	1	C8H11N3	N,N-DIMETHYLTRIAZENOBENZENE 7227-91-0
4 ★	1.80		541	1	C8H11N3O1	1-PHENYL-3,3-DIMETHYLTRIAZINE OXIDE 59477-92-8
15 ★	0.12	7.4	725	1	C8H11N3O1	2-ETHYLISONIAZID 4608-25-7
6 ★	-0.22		1394		C8H11N3O1	4-BENZYLSEMICARBAZIDE 16956-42-6
7 ★	0.48	7.4	725	1	C8H11N3O2	2-ETHOXYISONIAZID 58481-00-8
8 ★	0.90		2499		C8H11N3O2	PYRAZINE,2-DIMETHYLAMINO-6-METHOXYCARBONYL
9 ★	-0.93	7.0	2440	1	C8H11N3O3S1	3'-THIO-DIDEOXYCYTIDINE antiviral (aids treatment)
3720 ★	0.30	7.4	694	1	C8H11N3O4	METRONIDAZOLEACETATE 13182-82-6
1 ★	-0.22	6.5	321	2	C8H11N3O4S2	6-METHYLHYDROTHIAZIDE 1824-46-0
2 ★	-2.14		401		C8H11N3O6	6-AZAURIDINE(NCS32074) 54-25-1 antineoplastic
3 ★	-1.15		1551		C8H11N3O6	AZOMYCIN RIBOSIDE 17306-43-3
4 ★	-2.43		594		C8H11N3O6	URICYTIN 40919-33-3
25 ★	1.17		594		C8H11N3S1	3-THIOSEMICARBAZIDE,4-METHYL-1-PHENYL 13207-50-6
6 ✓	-3.30		2397	101	C8H11N4O1	IMIDAZOLIUM CHLORIDE,1-METHYL-2-HYDROXYIMINOMETHYL-3-CYANOETHYL 132540-04-6
7 ★	0.74		526		C8H11N5	ADENINE,9-PROPYL 707-98-2
8 ★	-1.08		2669		C8H11N5O2	6-DEOXYACYCLOVIR 84408-37-7 antiviral (hsv)
9	-2.41	7.4	2037	1	C8H11N5O2	ISOGUANINE,1-METHYL-7-METHOXYMETHYL
3730	-1.22	7.4	2037	1	C8H11N5O2	ISOGUANINE,1-METHYL-9-METHOXYMETHYL
1 ★	0.92	7.4	553	1	C8H11N5O2S1	2(1,3-DIME-1,3IMIDAZOLINYLIDEN-2N)-5NO2-THIAZOLE 31052-79-6
2 ★	1.95	7.4	553	1	C8H11N5O2S1	2(1,4-DIME-1,3IMIDAZOLINYLIDEN-2N)-5NO2-THIAZOLE 31052-78-5
3 ★	2.02	7.4	553	1	C8H11N5O2S1	2(1,5-DIME-1,3IMIDAZOLINYLIDEN-2N)-5NO2-THIAZOLE 31052-84-3
4 ★	2.28	7.4	553	1	C8H11N5O2S1	2(1-ET-1,3-IMIDAZOLINYLIDEN-2N)-5-NO2-THIAZOLE 31052-77-4
35 ★	-1.56		2669		C8H11N5O3	ACYCLOVIR 59277-89-3 antiviral (hsv)
6 ★	1.23	7.4	553	1	C8H11N5O3S1	2(1(2-HOET)-1,3-IMIDAZOLINYLIDEN-2N)-5NO2THIAZOL 24240-70-8
7 ★	-0.96	7.4	2273	1	C8H11N5O4	MORPHOLINE,-1-(3-NITROTRIAZIN-1-YL)METHYLCARBONYL 104958-84-1
8	0.97	7.3	2170	2	C8H11N7O1	AMBAZONE,OXY ANALOG
9	1.56	6.7	2170	2	"	AMBAZONE,OXY ANALOG
3740	0.72	7.3	2170	1	C8H11N7S1	AMBAZONE 6011-12-7 antimicrobial
1	1.23	6.7	2170	1	"	AMBAZONE
2 ★	0.95		547		C8H11O3P1	O,O-DIMETHYLPHENYLPHOSPHONATE 2240-41-7
3 ★	2.97		2303		C8H11O3P1S1	PHOSPHOROTHIOTIC ACID,O,O-DIMETHYL-O-PHENYL ESTER
4 ★	1.22		558		C8H11O4P1	O,O-DIMETHYL-O-PHENYLPHOSPHATE 10113-28-7
45 ★	2.57		541		C8H11P1	(DIMETHYL)-PHENYLPHOSPHINE 672-66-2
6 ✓	0.17	5.9	590		C8H11P1S1.C1H3I1	DIMETHYLPHENYLPHOSPHINE SULFIDE METH IODIDE
7 ★	3.99		512		C8H11Si1	SILANE,DIMETHYL-PHENYL 766-77-8
8 ★	3.16		1772		C8H12	CYCLOOCTA-1,5-DIENE 111-78-4
9	-2.21	7.4	401	1	C8H12Cl1N3O6	N-(CHLOROETHYLAMINOCARBONYL)GLUTAMIC ACID 52320-85-1
3750 ★	0.60	4.2	401	2	"	N-(CHLOROETHYLAMINOCARBONYL)GLUTAMIC ACID
1 ★	2.13		2094		C8H12Cl1N5	SIMAZINE,2-CYCLOPROPYLAMINE ANALOG 22936-85-2
2 ★	-0.47		564		C8H12Cl2N2O2Pt1	DICHLORO-4,5-DIMEO-O-PHENYLENEDIAMINOPLATINUM
3 ★	-0.23		564		C8H12Cl2N2Pt1	DICHLORO-4,5-DIMETHYL-O-PHENYLENEDIAMINOPLATINUM
4 ★	4.40		421		C8H12Cl4	1,4-DIMETHYLTETRACHLOROCYCLOHEXANE 56421-45-5
55 ★	3.30		421		C8H12Cl4O1S1	2-METHOXY-3-METHIOTETRACHLOROCYCLOHEXANE 60067-85-8
6 ★	2.55		421		C8H12Cl4O2	1,2-DIMETHOXYTETRACHLOROCYCLOHEXANE 60132-40-3
7 ★	2.82		421		C8H12Cl4O2	2,3-DIMETHOXYTETRACHLOROCYCLOHEXANE
8 ★	3.15		421		C8H12Cl4O2	3,6-DIMETHOXY-1,2,4,5-TETRACHLOROCYCLOHEXANE 56400-36-3
9 ✓	-1.20		2397	101	C8H12F3N4O3S1	IMIDAZOLIUM CHLORIDE,1-METHYL-2-HYDROXYIMINOMETHYL-3-TRIFLUOROMETHYLSULFONAMIDOETHYL
3760	-0.64	7.4	1074	324	C8H12N1O6P1	PYRIDOXOLPHOSPHATE 1883-15-4
1	-1.84	2.7	1473	340	C8H12N2	2-PYRIDINEPROPANEAMINEHYDROCHLORIDE 15583-16-1
2 ★	0.49		1474		C8H12N2	2-PYRIDINEPROPANEAMINE
3	-1.90	2.7	1473	340	C8H12N2	3-PYRIDINEPROPANEAMINE HYDROCHLORIDE 41038-69-1
4 ★	0.44		1474		C8H12N2	3-PYRIDINEPROPANEAMINE

	logP	pH	Ref.	Note	MF	Name / **CAS** / *activity*
65	-1.92	2.7	1473		C8H12N2	4-PYRIDINEPROPANEAMINE HYDROCHLORIDE
6 ★	0.40		1474		C8H12N2	4-PYRIDINEPROPANEAMINE
7 ★	0.49	11.0	533	322	C8H12N2	N,N-DIMETHYL-3-PYRIDYLMETHYLAMINE **2055-21-2**
8 ★	0.76	11.0	533	322	C8H12N2	N-ETHYL-3-PYRIDYLMETHYLAMINE **3000-75-7**
9 ★	0.59		2069		C8H12N2	OCTANEDINITRILE **629-40-3**
3770 ★	1.51		2499		C8H12N2	PYRAZINE,2,3-DIETHYL
1 ★	1.57		2499		C8H12N2	PYRAZINE,2-PROPYL-3-METHYL
2 ★	1.28		2499		C8H12N2	PYRAZINE,TETRAMETHYL **1124-11-4**
3 ★	2.29		1519	459	C8H12N2O1	2-METHYL-6-PROPOXYPYRAZINE
4 ★	1.80		1519	459	C8H12N2O1	3-BUTOXYPYRIDAZINE
75 ★	2.24		2499		C8H12N2O1	PYRAZINE,2-I-PROPOXY-3-METHYL
6	-1.55	7.4	2685	1	C8H12N2O2	4-PYRIDONE,3-HYDROXY-2-METHYL-1-(2-AMINO)ETHYL *iron chelator*
7 ★	2.55		2499		C8H12N2O2	PYRAZINE,2,6-DIETHOXY
8	-1.30	7.4	1074	324	C8H12N2O2	PYRIDOXAMINE **524-36-7**
9 ★	0.40	7.4	2307	1	C8H12N2O2	THPO-O,N-DIMETHYL *agonist-muscarinic* ✦
3780 ★	1.50	1.0	2202	342	C8H12N2O2S1	2-THIO-5,5-DIETHYLBARBITURIC ACID **77-32-7** *thyroid inhibitor*
1	1.57	7.4	1269	1	"	2-THIO-5,5-DIETHYLBARBITURIC ACID
2 ★	0.76	7.4	925	1	C8H12N2O2S1	N',N'-DIMETHYLSULFANILAMIDE **6162-21-6**
3	0.67	6.0	479		C8H12N2O2S1	N1-DIMETHYLSULFANILAMIDE **1709-59-7**
4 ★	0.81		508		C8H12N2O2S1	PHENETHYLSULFAMIDE **710-15-6**
85	0.61	7.4	1873	1	C8H12N2O3	5,5-DIETHYLBARBITURIC ACID **57-44-3** *sedative, hypnotic*
6 ★	0.65		504		"	5,5-DIETHYLBARBITURIC ACID
7 ★	0.77	2.0	300		C8H12N2O3	BARBITURIC ACID,5-BUTYL
8 ★	0.54	1.0	2202	342	C8H12N2O3	BARBITURIC ACID,5-ISOPROPYL-5-METHYL **53943-59-2**
9 ★	0.28	1.0	2202	342	C8H12N2O3	BARBITURIC ACID,5-T-BUTYL **90197-63-0**
3790 ★	0.30		1564		C8H12N2O3	PYRIMIDINE-2,4-DIONE,3-PROPOXY-6-METHYL
1 ★	0.07		547		C8H12N2O4S2	M-N-BIS(METHYLSULFONAMIDO)BENZENE
2 ★	-0.60	7.4	2490	1	C8H12N2O5S2	BENZENESULFONAMIDE,3-AMINO-4-HYDROXYETHYLSULFONYL **135832-45-0**
3 ✓	-2.70	7.4	2397	101	C8H12N3O2	IMIDAZOLIUM CHLORIDE,1-METHYL-2-HYDROXYIMINOMETHYL-3-ACETYLMETHYL
4	-0.59	7.4	1948	1	C8H12N4	1-(2-PYRIMIDINYL)PIPERAZINE **20980-22-7**
95	0.49	7.4	1948	2	"	1-(2-PYRIMIDINYL)PIPERAZINE
6 ✓	-1.15		1008	354	C8H12N4O2	4(1-PYRROLIDYLMETHYL)-2-NITROIMIDAZOLE
7	-1.23		1559		C8H12N4O2	5-(N-PIPERIDINO)-2-NITRO-IMIDAZOLE
8 ★	-1.70	7.1	1591		C8H12N4O2	HISTIDIN-AMIDE,N-ACETYL
9 ★	1.06	7.4	536	1	C8H12N4O2S1	4-(3,3-DIMETHYLTRIAZENO)BENZENESULFONAMIDE **55469-64-2**
3800	2.15		1355		C8H12N4O2S2	6-T-BUIMIDAZO(2,1-B)1,3,4-THIADIAZOLE-2-SULFONAMIDE ✦
1 ★	-0.47	7.4	694	1	C8H12N4O3	1-B-AMINOACETYLETHYL-2-METHYL-5-NITROIMIDAZOLE
2 ★	-0.55		541		C8H12N4O3	5-DIAZO-6-BUTOXY-6-H-URACIL
3	0.22	7.4	2273	1	C8H12N4O3	IMIDAZOLE,2-NITRO-1-(3-AZIRIDINYL-2-HYDROXY)PROPYL **88876-88-4**
4	0.33	7.4	2477	1	"	IMIDAZOLE,2-NITRO-1-(3-AZIRIDINYL-2-HYDROXY)PROPYL
5 ★	-0.48	7.4	830	1	C8H12N4O3	N-(DIAZOACETYLAMINOACETYL)MORPHOLINE
6 ★	0.65		2196		C8H12N4O3	PYRAZOLE,3-METHYL-4-NITRO-5-(N-PROPYLCARBOXAMIDO)
7 ★	-1.32	3.0	2191		C8H12N4O3	URACIL,6-AMINO-1,3-DIMETHYL-5-(N-METHYLFORMYLAMINO) **33130-55-1**
8	1.09	7.4	2206	306	C8H12N4O3S1	CARNIDAZOLE **42116-76-7** *antiprotozoal*
9 ★	1.15	7.4	1022	1	"	CARNIDAZOLE
3810 ★	1.12	7.4	2206	306	C8H12N4O3S1	METHYLTHIOCARBAMATE,(2-METHYL-5-NITROIMIDAZOL-2-YL)ETHYL
1 ★	0.02	7.4	2206	306	C8H12N4O4	(2-METHYL-5-NITROIMIDAZOL-2YL)ETHYLCARBAMIC ACID, METHYL ESTER
2 ★	-1.08		2567		C8H12N4O4	2-NITROIMIDAZOLE,1-(N-HYDROXYETHYL)PROPIONAMIDO
3	-0.74	7.4	2561	1	C8H12N4O4	2-NITROIMIDAZOLE-1-(N-METHOXYBUTYRAMIDO)
4	-1.16		1404		C8H12N4O4	2-NO2-IMIDAZOLE,1-CH(ME)CONHCH2CH2OH
15	-1.85		1404		C8H12N4O5	2-NO2-IMIDAZOLE,1-CH2CONHCH2CHOHCH2OH **74141-75-6**
6 ★	-0.89	7.0	1783	1	C8H12N4O5	5-(1,3,4-TRIHYDROXYBUT-2YLAMINO)-2-NITROPYRAZINE **89690-76-6**
7 ★	-2.17	7.4	594	1	C8H12N4O5	AZACITIDINE **65886-71-7** *antineoplastic*
8	-1.06	13.0	1943	224	C8H12N6O2	ADENINE,2-AMINO-9-(2-HYDROXYETHOXY)METHYL
9	2.08	7.4	1821	1	C8H12O1	1-VINYL-3,4-EPOXYCYCLOHEXANE
3820 ★	1.73		504		C8H12O1	CYCLOHEXANOL,1-ETHYNYL **78-27-3**
1	0.44	7.4	1821	1	C8H12O2	1-EPOXYETHYL-3,4-EPOXYCYCLOHEXANE
2	2.27	3.5	2391	1	C8H12O2	2-(1'-PROPENYL)-2-PENTENOIC ACID **93238-15-4**
3	-1.80		1466	42	C8H13Br1N1O1.Br1	2-FURANYLME-(5-BR-N,N,N-TRIMETHYL)AMMONIUMBROMIDE **75523-39-6**
4 ✓	-1.80		2033		C8H13Br1N1O1.I1	FURANE,2-BROMO-5-TRIMETHYLAMMONIOMETHYL IODIDE
25 ✓	-1.28		2397	101	C8H13Br1N3O1	IMIDAZOLIUM CHLORIDE,1-METHYL-2-HYDROXYIMINOMETHYL-3-(3-BROMOPROPYL)
6 ✓	-2.30		2397	101	C8H13Br1N3O2	IMIDAZOLIUM CHLORIDE,1-METHYL-2-HYDROXYIMINOMETHYL-3-(2-BROMOETHOXY)METHYL
7 ✓	-1.72		1466	42	C8H13Cl1N1O1.Br1	2-FURANYLME-(5-CL-N,N,N-TRIME)AMMONIUMBROMIDE **75523-40-9**
8 ✓	-1.72		2033		C8H13Cl1N1O1.I1	FURANE,2-CHLORO-5-TRIMETHYLAMMONIOMETHYL IODIDE
9 ✓	-2.84		2397	101	C8H13F1N3O2	IMIDAZOLIUM CHLORIDE,1-METHYL-2-HYDROXYIMINOMETHYL-3-(2-FLUOROETHOXY)METHYL
3830 ★	3.28		2488		C8H13F2N5S1	1,3,5-TRIAZINE,2-DIFLUOROMETHIO-4,6-BIS-ETHYLAMINO
1 ★	3.27		2488		C8H13F2N5S1	1,3,5-TRIAZINE,2-DIFLUOROMETHIO-4-I-PROPYLAMINO-6-METHYLAMINO
2 ★	3.28		2488		C8H13F2N5S1	1,3,5-TRIAZINE,2-DIFLUOROMETHIO-4-PROPYLAMINO-6-METHYLAMINO
3 ★	1.63		1478		C8H13N1O1	3-ME,2-(N,N-DIMETHYLAMINOMETHYL)FURAN **20863-54-1**
4 ★	1.18	5.0	2425	1	C8H13N1O2	2-PYRROLIDONE,1-BUTYRYL **22707-38-6**
35	-0.20	7.4	2307	1	C8H13N1O2	ARECOLIN **63-75-2** *anthelmintic*
6 ★	0.35		2307	2	"	ARECOLIN
7 ★	0.23		505	20	C8H13N1O2	GLUTARIMIDE,2-METHYL,2-ETHYL
8 ★	-0.04		213		C8H13N1O2	N-PROPIONYLCYCLOBUTANECARBOXAMIDE **23046-87-9**
9 ★	1.52		505		C8H13N1O2S1	THIAMORPHOLINE-3,5-DIONE,2,2-DIETHYL
3840 ★	0.95	7.4	2490	1	C8H13N1O3S3	THIOPHENE-2-SULFONAMIDE,5-(4-HYDROXYBUTYL)THIO **135832-38-1**
1 ★	-0.34	7.4	2490	1	C8H13N1O5S3	THIOPHENE-2-SULFONAMIDE,5-(4-HYDROXYBUTYL)SULFONYL **135832-39-2**
2 ★	1.99		1478		C8H13N1S1	THIOPHENE,2-CH2CH2NME2 **26019-18-1**
3 ★	1.94		1478		C8H13N1S1	THIOPHENE,3-CH2CH2NME2
4 ✓	-2.25		2033		C8H13N1O3.I1	FURANE,2-NITRO-5-TRIMETHYLAMMONIOMETHYL IODIDE
45 ★	0.85		2397		C8H13N3O1	IMIDAZOLE,1-T-BUTYL-2-HYDROXYIMINOMETHYL
6 ★	0.04		1008	354	C8H13N3O4	1(2-OH-3-MEO-PROPYL)-2-ME-5-NITROIMIDAZOLE
7 ★	0.04		1008	354	C8H13N3O4	1-(3-ETHOXY-2-HYDROXYPROPYL)-2-NITROIMIDAZOLE
8	0.10	7.4	1428	1	"	1-(3-ETHOXY-2-HYDROXYPROPYL)-2-NITROIMIDAZOLE
9 ★	-0.36	7.4	1022	1	C8H13N3O4	1-(3-MEO-2-HYDROXYPROPYL)-2-ME-4-NITROIMIDAZOLE

	logP	pH	Ref.	Note	MF	Name / CAS / activity
3850	-0.02	7.4	1497	1	"	1-(3-MEO-2-HYDROXYPROPYL)-2-ME-4-NITROIMIDAZOLE
1 ★	-0.84	7.4	2397		C8H13N3O4S1	IMIDAZOLE,1-(2-METHYLSULFONYLETHOXY)METHYL-2-HYDROXYIMINOMETHYL
2 ★	-1.00		835		C8H13N3O4S1	METHYL-(3-(2-METHYL-5-NITROIMIDAZOL-1YL)PROPYL)-SULFONE 28795-24-6
3 ★	-0.36		1008	354	C8H13N3O4S1	TINIDAZOL 19387-91-8 antiprotozoal
4	-0.33	7.4	2206	306	"	TINIDAZOL
55	0.25	7.4	2380	1	"	TINIDAZOL
6	0.84		723		C8H13N5O1	2,4-DIAMINO-6-PYRROLIDINO-PYRIMIDINE-3-OXIDE 55921-65-8
7 ✓	-0.07		1412		C8H13N5O2	1,2,4-TRIAZOLE,3,5-DIPROPIONAMIDO
8	0.36		723		C8H13N5O2	2,4-DIAMINO-6-MORPHOLINO-PYRIMIDINE-3-OXIDE 55921-64-7
9	-2.41	7.4	2037	1	C8H13N5O2	2H-PURIN-2-ONE,1,3-DIHYDRO-1-METHYL-6-AMINO-7-METHOXYMETHYL ✚
3860 ★	0.44		563		C8H13N5O2	IMIDAZOLE,4-ETHOXYCARBONYL-5-(3,3-DIMETHYLTRIAZENYL) 36137-88-9
1 ★	0.89	7.4	2273	1	C8H13N5O3	1,2,4-TRIAZOLE,3-NITRO-1-(N-BUTYLACETAMIDO) 104958-89-6
2	-0.07		2547		C8H13N5O3S1	1,2,4-TRIAZOLE,3-NITRO-1-(N-METHIOETHYLCARBAMYL)ETHYL
3	0.26		2547		C8H13N5O3S1	1,2,4-TRIAZOLE,3-NITRO-1-(N-METHIOPROPYLCARBAMYL)METHYL
4 ★	0.63		2547		C8H13N5O3S1	3-NITROTRIAZOLE,1-(N-METHOXYPROPYL)THIOACETAMIDO
65	-0.51	7.4	2547	1	C8H13N5O4	1,2,4-TRIAZOLE,3-NITRO-1-(N-(3-METHOXYPROPYL)ACETAMIDO) 104958-93-2
6	-0.49	7.4	2273	1	"	1,2,4-TRIAZOLE,3-NITRO-1-(N-(3-METHOXYPROPYL)ACETAMIDO)
7	-0.39		2547		"	1,2,4-TRIAZOLE,3-NITRO-1-(N-(3-METHOXYPROPYL)ACETAMIDO)
8	-0.65		2547		C8H13N5O4	1,2,4-TRIAZOLE,3-NITRO-1-(N-METHOXYETHYLCARBAMYL)ETHYL
9 ★	2.20		401		C8H14Cl1N3O2	3-CYCLOPENTYL-1-(2-CHLOROETHYL)-1-NITROSOUREA 13909-03-0
3870 ★	2.07	7.4	401	2	C8H14Cl1N3O2S1	1-(2-CHLOROETHYL)-3-(4-SULFAPYRANYL)-1-NITROSOUREA 33022-01-4
1 ★	0.90		401		C8H14Cl1N3O4S1	3-(DIOXOTETRAHYDROTHIOPYRAN-4YL)-1-(2-CHLOROETHYL)-1-NITROSOUREA 33022-02-5 ✚
2 ★	2.61		2094		C8H14Cl1N5	ATRAZINE 1912-24-9 herbicide
3 ★	1.43	7.4	401	2	C8H14F1N3O2S1	1-(2-FLUOROETHYL)-3-(4-SULFAPYRANYL)-1-NITROSOUREA 32319-90-7
4	1.31	7.0	2484	1	C8H14N1O1	2,5-DIHYDROPYRROLE,1-OXY-2,2,5,5-TETRAMETHYL
75	-2.40		1466	42	C8H14N1O1.Br1	2-FURANYLME-(N,N,N-TRIMETHYL)AMMONIUMBROMIDE 541-64-0
6 ✓	-2.40		1704	820	C8H14N1O1.Br1	3-FURANYL-TETRAMETHYLAMMONIUMBROMIDE
7 ✓	-2.40		2033		C8H14N1O1.I1	FURANE,2-TRIMETHYLAMMONIOMETHYL IODIDE 541-64-0 cholinergic
8 ✓	-2.40		2033		C8H14N1O1.I1	FURANE,3-TRIMETHYLAMMONIOMETHYL IODIDE
9 ★	2.96		547		C8H14N2	4-PENTYLPYRAZOLE 52222-71-6
3880	1.09		1407		C8H14N2O2	1,2,3,6-H4-PYRIDINE,N-ETHOXYCARBONYLAMINO
1	-0.84		401		C8H14N2O2	PIPERAZINE,1,4-BIS(2,3-EPOXYPROPYL) 2917-98-8
2 ★	-1.19		1558		C8H14N2O2	PROLINE,N-ACETYL-N'-METHYLAMINO-AMIDE
3 ★	1.51		2188		C8H14N2O2S2	METHYLCARBAMIC ACID,2-OXIMINO-3,3-DIMETHYL-1,4-DITHIANE ESTER
4 ★	-2.71	6.4	2017		C8H14N2O3	PROLINE,D-ALANYL 13485-59-1
85 ★	1.48		2188		C8H14N2O3S1	METHYLCARBAMIC ACID,4-OXIMINO-5-I-PROPYL-1,3-OXATHIOLANE ESTER ✚
6 ★	1.52		2188		C8H14N2O3S1	METHYLCARBAMIC ACID,4-OXIMINO-5-PROPYL-1,3-OXATHIOLANE ESTER
7 ★	0.51		2188		C8H14N2O4S1	METHYLCARBAMIC ACID,4-OXIMINO-5-METHOXYETHYL-1,3-OXATHIOLANE ESTER
8 ✓	-2.98		2397	101	C8H14N3O2	IMIDAZOLIUM CHLORIDE,1-METHYL-2-HYDROXYIMINOMETHYL-3-(2-METHOXYETHYL)
9 ✓	-2.70		2397	101	C8H14N3O2	IMIDAZOLIUM CHLORIDE,1-METHYL-2-HYDROXYIMINOMETHYL-3-ETHOXYMETHYL 117941-45-4
3890 ✓	-3.50		2397	101	C8H14N3O3S1	IMIDAZOLIUM CHLORIDE,1-METHYL-2-HYDROXYIMINOMETHYL-3-(2-METHYLSULFONYLETHYL) 132540-19-3
1 ✓	0.00	7.4	2397	101	C8H14N3O4S1	IMIDAZOLIUM CHLORIDE,1-METHYL-2-HYDROXYIMINOMETHYL-3-PROPANESULFONIC ACID
2 ★	0.99		1519	459	C8H14N4	3,6-PYRIDAZINEDIAMINE,N,N-TETRAMETHYL
3 ★	0.84		897		C8H14N4	PYRIDAZINE-3,5-DIAMINE,N,N'-TETRAMETHYL 38717-44-1
4	1.59	3.5	2391	1	C8H14N4	TETRAZOLE,5-CYCLOHEXYLMETHYL
95 ★	1.52		57		C8H14N4O1S1	3-ETHYLTHIO-4-AMINO-6-I-PR-1,2,4-TRIAZINE-5-ONE
6 ★	1.39		57		C8H14N4O1S1	3-METHIO-4-AMINO-6-I-BU-1,2,4-TRIAZINE-5-ONE 21087-62-7
7 ★	1.70		57		C8H14N4O1S1	METRIBUZIN 21087-64-9 herbicide
8 ★	0.37	7.4	830	1	C8H14N4O2	N-DIAZOACETYLGLYCINE-N'-I-BUTYLAMIDE
9 ★	0.15	9.2	1853	1	C8H14N4S2	THIABURIMAMIDE 38603-23-5
3900 ★	0.78	4.6	401	2	C8H14N6O1	IMIDAZOLE-5-CONH2,4-(3,3-DIET-1-NNN)- 4574-37-2
1 ★	-1.09	4.2	401	2	C8H14N6O3	IMIDAZOLE,5-CARBOXAMIDE-4-(3,3-BIS(2-HYDROXYETHYL)TRIAZENYL) 21244-66-6
2	2.43	7.4	1821	1	C8H14O1	EPOXYETHYLCYCLOHEXANE
3	2.44	3.5	2391	1	C8H14O2	2-PROPYL-2-PENTENOIC ACID 60218-41-9
4	1.67		1231		C8H14O2	5,5-DIMETHYL-2,4-HEXANEDIONE 7307-04-2
5	2.31	3.5	2391	1	C8H14O2	CYCLOHEXANEACETIC ACID
6	2.38	3.5	2391	1	C8H14O2	CYCLOHEXANECARBOXYLIC ACID-1-METHYL 1123-25-7
7 ★	2.88		1402		C8H14O2	METHACRYLIC ACID,BUTYL ESTER 97-88-1
8 ★	2.66		1402		C8H14O2	METHACRYLIC ACID,I-BUTYL ESTER 97-86-9
9 ★	2.54		1402		C8H14O2	METHACRYLIC ACID,T-BUTYL ESTER 585-07-9
3910	1.31	3.5	2391	1	C8H14O3	4-OXOPENTANOIC ACID,2-PROPYL 688-04-0
1	0.55		502		C8H14O3	HEPTANOIC ACID,6-KETOMETHYL ESTER 2046-21-1
2 ★	1.03		579		C8H14O4	ADIPIC ACID,DIMETHYL ESTER 627-93-0
3 ★	1.19		1819		C8H14O4	BUTANEDIOIC ACID,ETHYL ESTER 123-25-1
4 ★	-0.29		508		C8H14O6	TARTARIC ACID,DIETHYL ESTER 87-91-2
15 ★	2.75		1896		C8H15N1	OCTANENITRILE 124-12-9
6 ★	0.67		589		C8H15N1O1	2-AZACYCLONONANONE
7 ★	1.14		814		C8H15N1O1	ALLYL-I-PROPYL-ACETAMIDE 3829-78-5
8 ★	0.65		589		C8H15N1O1	N-ETHYL-EPSILON-CAPROLACTAM
9 ★	0.75		590		C8H15N1O2	3-CYANOPROPIONALDEHYDE-DIETHYLACETAL 18381-45-8
3920 ★	-0.90	7.2	2323	468	C8H15N1O2	CYCLOHEXYLGLYCINE
1 ★	-2.45		2654	448	C8H15N1O2	PIPERIDINE-1-ACETIC ACID
2 ★	0.33	2.2	2228	314	C8H15N1O3	PENTANOIC ACID-2-ACETYLAMINO,METHYL ESTER
3	2.27	3.5	2391	1	C8H15N1O3	SUCCINAMIC ACID,N,N-DIETHYL
4 ★	1.44		561		C8H15N1S1	2-AZACYCLONONANTHIONE
25 ★	1.43		1137		C8H15N3O1	2-AMINO-5-HEXYL-1,3,4-OXADIAZOLE
6 ★	-2.00		2377		C8H15N3O3	AC-ALA-ALA-N
7 ★	1.30		1629		C8H15N3O3S2	N-(N"-METHYL-MECARBAMYLTHIO)METHOMYL
8 ★	1.40		1629		C8H15N3O4S2	N-(N'-O-DIMETHYLCARBAMYLTHIO)METHOMYL
9 ★	-1.08	7.4	401	1	C8H15N3O6	XYLOPYRANOSIDE,ME-3-DEOXY-3-(3-ME-3-NITROSOUREIDO) 52019-05-3
3930 ★	-1.45		401		C8H15N3O7	STREPTOZOCIN 18883-66-4 antineoplastic
1	1.02		723		C8H15N5O1	2,4-DIAMINO-6-BUTYLAMINO-PYRIMIDINE-3-OXIDE
2 ★	1.16		723		C8H15N5O1	2,4-DIAMINO-6-DIETHYLAMINO-PYRIMIDINE-3-OXIDE
3 ★	1.82		2094		C8H15N5O1	2-METHOXY-4,6-BIS(ETHYLAMINO)-S-TRIAZINE 673-04-1
4 ★	2.54		2094		C8H15N5S1	SIMETRYN 1014-70-6 herbicide

	logP	pH	Ref.	Note	MF	Name / CAS / activity
35	-0.57	9.2	1940	1	C8H15N7O2S3	FAMOTIDINE 76824-35-6 antagonist (histamine-2 receptors)
6 ★	4.57		1560	100	C8H16	1-OCTENE 111-66-0
7 ★	1.00	7.0	2483	1	C8H16N1O1	PROXO
8 ✓	-3.63		1704	820	C8H16N1O2.Br1	TRIMETHYL-PROPENOYLOXYETHYL AMMONIUM BROMIDE
9 ★	1.04		2049		C8H16N1O3P1S2	MEPHOSFOLAN 950-10-7 insecticide
3940 ★	0.00		1903		C8H16N1O5P1	DICROTOPHOS 141-66-2 insecticide
1 ✓	0.97		567	820	C8H16N1.C6H2N3O7	N-METHYLQUINUCLIDINIUM PICRATE
2 ★	2.04		871		C8H16N2O1	N-NITROSO-OCTAMETHYLENEIMINE
3 ★	-0.03	7.1	1591		C8H16N2O2	ISOLEUCIN-AMIDE,N-ACETYL
4 ★	-0.13	7.1	1591		C8H16N2O2	LEUCIN-AMIDE,N-ACETYL
45	0.06	7.0	2309	1	"	LEUCIN-AMIDE,N-ACETYL
6 ★	-0.35		1558		C8H16N2O2	VALINE,N-ACETYL,N'-METHYLAMINO-AMIDE
7 ★	0.96		843		C8H16N2O3	1-(2-METHOXY-2-ETHYLBUTANOYL)UREA
8	2.01	3.5	2391	1	C8H16N4	TETRAZOLE,5-HEPTYL 92712-47-5
9 ★	0.91		871		C8H16N4O2	2,3,5,6-TETRAMETHYL-N,N'-DINITROSOPIPERAZINE 23264-57-5
3950 ★	1.83		964		C8H16N6	PENTAMETHYLMELAMINE 16268-62-5 antineoplastic
1 ★	1.61		964		C8H16N6O1	N2-HYDROXY-N2,N4,N6,N6-PENTAMETHYLMELAMINE
2 ★	0.58	4.0	979	364	C8H16N6O4	1,1'-HEXAMETHYLENE-BIS(1-NITROSOUREA) 27640-22-8
3 ★	2.37		2047		C8H16O1	2-OCTANONE 111-13-7
4 ★	2.10		2126		C8H16O1	CYCLOHEXANOL,2,6-DIMETHYL-,A,A,A 39170-83-7
55 ★	2.38		2126		C8H16O1	CYCLOHEXANOL,2,6-DIMETHYL-,A,B,B 42846-29-7
6 ★	2.37		2126		C8H16O1	CYCLOHEXANOL,2,6-DIMETHYL-A,A,A 39170-84-8
7 ★	2.06		594	400	C8H16O1	TETRAHYDROFURAN,2,2,5,5-TETRAMETHYL 15045-43-9
8 ★	2.64		1815		C8H16O2	2-ETHYL-HEXANOIC ACID 149-57-5
9 ★	3.05		1135	537	C8H16O2	OCTANOIC ACID 124-07-2 antifungal
3960	0.13	7.5	1510	1	C8H16O2	VALPROIC ACID 99-66-1 anticonvulsant
1 ★	2.75		1815		"	VALPROIC ACID
2 ★	-0.04		2308		C8H16O4	12-CROWN-4-ETHER 294-93-9
3	-1.51		1294		C8H16O5S1	1-ETHYLTHIO-B-GALACTOPYRANOSIDE
4	-2.28		1294		C8H16O6	B-ETHYLGALACTOPYRANOSIDE 18997-88-1
65 ★	4.89		1560	100	C8H17Br1	1-BROMOOCTANE 111-83-1
6 ✓	2.10		567	820	C8H17N1	CYCLOHEXYLDIMETHYLAMINE PICRATE
7 ★	1.48		814		C8H17N1O1	PROPYL-I-PROPYLACETAMIDE
8 ★	2.36	7.4	172	1	C8H17N1O2	O-HEPTYLCARBAMATE 4248-20-8
9 ★	-2.65		2018		C8H17N1O2	OCTANOIC ACID,8-AMINO
3970 ★	1.81		505		C8H17N1S1	PROPIONAMIDE,2-BUTYLTHIO-2-METHYL
1	-0.70	7.0	2484	1	C8H17N2O1	1-OXYPYRROLINE,2,2,5,5-TETRAMETHYL-3-AMINO
2	1.03	12.0	2484	1	"	1-OXYPYRROLINE,2,2,5,5-TETRAMETHYL-3-AMINO
3 ★	-2.82	7.1	1591		C8H17N3O2	LYSIN-AMIDE,N-ACETYL
4 ★	-2.84	7.1	1591		C8H17N5O2	ARGININ-AMIDE,N-ACETYL
75	5.18		1560	100	C8H18	OCTANE 111-65-9
6 ✓	0.36		2426	537	C8H18Cl2N2Pt1	PLATINUM,BIS-CYCLOBUTYLAMMONIO-DICHLORO
7	1.56		2019		C8H18Cl2Sn1	DICHLOROSTANNANE,DIBUTYL 683-18-1
8 ✓	-2.30		1704	820	C8H18N1O1.Br1	(2-TETRAHYDROFURANYL)TETRAMETHYLAMMONIUMBROMIDE
9 ✓	-2.65		1704	820	C8H18N1O1.Br1	(TETRAHYDROFURAN-3YL)TETRAMETHYLAMMONIUMBROMIDE
3980	-1.62		1257	820	C8H18N1O1.I1	CYCLOPROPOXYETHYLTRIMETHYL AMMONIUM IODIDE
1 ✓	-2.30		2033		C8H18N1O1.I1	TETRAHYDROFURANE,2-TRIMETHYLAMMONIOMETHYL IODIDE
2 ✓	-2.65		2033		C8H18N1O1.I1	TETRAHYDROFURANE,3-TRIMETHYLAMMONIOMETHYL IODIDE
3 ✓	-2.33		1704	820	C8H18N1O2.Br1	(2-ME-DIOXOLAN-4YL)-TETRAMETHYLAMMONIUMBROMIDE
4 ✓	-2.33		2033		C8H18N1O2.I1	1,3-DIOXOLANE-2-METHYL-4-TRIMETHYLAMMONIOMETHYL IODIDE
85 ✓	-2.96		1257	820	C8H18N1O2.I1	PROPIONYLCHOLINE IODIDE
6 ✓	-2.10		1704	820	C8H18N1S1.Br1	(2-ME-OXATHIOLAN-5YL)-TETRAMETHYLAMMONIUMBROMIDE ✚
7 ✓	-2.10		2033		C8H18N1S1.I1	1,3-THIOXOLANE-2-METHYL-4-TRIMETHYLAMMONIOMETHYL IODIDE ✚
8 ★	2.29		2570		C8H18N2O1	N-NITROSO-DI-I-BUTYL AMINE 997-95-5
9 ★	2.63		2570		C8H18N2O1	N-NITROSODIBUTYLAMINE 924-16-3
3990 ★	1.00		1137	828	C8H18N2O1	OCTANOIC ACID HYDRAZIDE 6304-39-8
1 ★	0.79		871		C8H18N2O3	N-NITROSO-BIS(2-ETHOXYETHYL)AMINE
2 ★	2.31		1895		C8H18N2S1	THIOUREA,N-HEXYL-N'-METHYL 53393-06-9
3 ★	1.73		1967		C8H18N4	4-OCTANONE,GUANYLHYDRAZONE 97183-51-2
4 ★	2.90		2126		C8H18O1	2-OCTANOL 123-96-6
95 ★	2.68		2126		C8H18O1	4-OCTANOL 589-62-8
6 ★	2.78		1952		C8H18O1	DI-I-BUTYL ETHER 628-55-7
7 ★	3.21		1819		C8H18O1	DIBUTYLETHER 142-96-1
8 ★	3.00		594	400	C8H18O1	OCTANOL 111-87-5
9	3.22		1560	100	C8H18O2	2-ETHYL-1,3-HEXANEDIOL 94-96-2 repellant,insect
4000 ★	1.86		1819		C8H18O2	HEXYLOXYETHANOL 112-25-4
1 ★	0.39		1819		C8H18O3	BIS(2-ETHOXYETHYL)ETHER 112-36-7
2 ★	0.56		1819		C8H18O3	DIETHYLENEGLYCOL,MONOBUTYLETHER 112-34-5
3 ★	2.82	9.3	588	2	C8H19N1	2-ETHYLHEXYLAMINE 104-75-6
4 ✓	2.77		567	820	C8H19N1	DI-I-BUTYLAMINE PICRATE
5 ★	2.83	9.3	588	2	C8H19N1	DIBUTYLAMINE 111-92-2
6 ★	2.68		567		C8H19N1	ETHYL-DIPROPYLAMINE 20634-92-8
7	0.76	7.0	834	1	C8H19N1	OCTYLAMINE 111-86-4
8	-3.07	7.4	401	1	C8H19N1O6S2	1-PROPANOL,3,3'-IMINODI-DIME-SO4 13425-98-4 antineoplastic
9	-1.71	7.4	401	2	C8H19N1O6S2	1-PROPANOL,3,3'-IMINODI-DIME-SO4
4010 ✓	3.19		567	820	C8H19N1.C6H3N3O7	DIBUTYLAMINE PICRATE
1	-0.03		1394		C8H19N3O1	4-HEPTYLSEMICARBAZIDE
2 ★	3.59		2506		C8H19O2P1S2	ETHOPROP 13194-48-4 nematocide
3 ★	2.03		508		C8H19O2P1S2	ETHYLPHOSPHONATE,O-ET-S-(2-ET-THIOETHYL) 50728-06-8
4 ★	4.02		1693		C8H19O2P1S3	DEMETONTHIOL 298-04-4 insecticide
15 ★	1.73		1693		C8H19O3P1S3	DISULFOTONSULFOXIDE 2497-07-6
6 ★	1.87		1693		C8H19O4P1S3	DISULFOTONSULFONE 2497-06-5
7	-1.72		1257	820	C8H20N1O1.I1	I-PROPOXYETHYLTRIMETHYL AMMONIUM IODIDE
8	-1.97		1257	820	C8H20N1O1.I1	PROPOXYETHYLTRIMETHYL AMMONIUM IODIDE
9 ★	1.11		600		C8H20N1O2P1S1	O,S-DIME-N-HEXYL-PHOSPHORAMIDOTHIOATE

	logP	pH	Ref.	Note	MF	Name / CAS / activity
4020 ✓	-2.82	7.4	921	22	C8H20N1.Br1	TETRAETHYL AMMONIUM BROMIDE 71-91-0
1	-2.59		1087		C8H20N1.Cl1	TETRAETHYLAMMONIUMCHLORIDE 56-34-8 *ganglionic blocker*
2 ★	-2.82		524	820	C8H20N1.I1	TETRAETHYL AMMONIUM IODIDE 68-05-3
3 ✓	-2.22		1257	820	C8H20N1.I1	TRIMETHYL-PENTYL AMMONIUM IODIDE
4 ✓	0.98		567	820	C8H20N1.C6H2N3O7	DI-I-PROPYL-DIMETHYL AMMONIUM PICRATE
25	0.94		567	820	C8H20N1.C6H2N3O7	TETRAETHYL AMMONIUM PICRATE
6 ✓	1.07		1701	820	"	TETRAETHYL AMMONIUM PICRATE
7	-1.00	7.5	1230	1	C8H20N2Se1	BIS(B-(N,N-DIMETHYLAMINO)ETHYLSELENIDE
8 ★	0.82		1908		C8H20N3O1P1	PHOSPHORIC TRIAMIDE,N,N'-TETRAMETHYL,N''-TETRAMETHYLENE 40725-71-1
9	-3.82		1108		C8H20O2P1S2.C1H3O4S1	O-ET-S-(B-ME-ET-S+-ET)METHYLTHIOPHOSPHONATE,MESO-
4030 ✓	-1.71		2426	537	C8H22N2O2Pt1.(N1O3)2	PLATINUM,BIS-CYCLOBUTYLAMMONIO-DIAQUA-DINITRATE ✚
1 ★	0.66		1908		C8H22N3O1P1	PHOSPHORICTRIAMIDE,N-DIETHYL,N',N'-TETRAMETHYL
2	4.80		1740	459	C8H24O2Si3	OCTAMETHYLTRISILOXANE 107-51-7
3	2.00	7.2	2212	1	C9H4Cl1N5O2	CARBONYLCYANIDE,2-CHLORO-4-NITROPHENYLHYDRAZONE 3780-92-5
4	5.05	7.2	2212	2	"	CARBONYLCYANIDE,2-CHLORO-4-NITROPHENYLHYDRAZONE
35	5.10	4.0	2264	1	C9H4Cl2F6N2	HEXAFLUOROACETONE,3,4-DICHLOROPHENYLHYDRAZONE
6	1.80	7.2	2212	1	C9H4Cl2N4	CARBONYLCYANIDE,2,3-DICHLOROPHENYLHYDRAZONE 3720-50-1
7 ★	3.85	7.2	2212	2	"	CARBONYLCYANIDE,2,3-DICHLOROPHENYLHYDRAZONE
8	2.01	7.2	2212	1	C9H4Cl2N4	CARBONYLCYANIDE,2,6-DICHLOROPHENYLHYDRAZONE 3780-85-6
9 ★	4.51	7.2	2212	2	"	CARBONYLCYANIDE,2,6-DICHLOROPHENYLHYDRAZONE
4040 ★	2.85		536		C9H4Cl3N1O2S1	N-(TRICHLOROMETHYLTHIO)PHTHALIMIDE 133-07-3 *fungicide*
1	3.80	4.0	2264	1	C9H4F6N4O4	HEXAFLUOROACETONE,2,4-DINITROPHENYLHYDRAZONE
2	2.80	4.0	2264	1	C9H4F6N4O4	TRIFLUOROMETHYLACETALDEHYDE,(2,6-DINITRO-4-TRIFLUOROMETHYL)PHENYLHYDRAZONE
3	2.20	7.2	2212	1	C9H5Br1N4	CARBONYLCYANIDE,2-BROMOPHENYLHYDRAZONE 101398-30-5
4 ★	3.11	7.2	2212	2	"	CARBONYLCYANIDE,2-BROMOPHENYLHYDRAZONE
45 ★	1.33	7.0	452	1	C9H5Cl1N2O3	3-CHLORO-4-NITROQUINOLINE-1-OXIDE 14100-52-8
6 ★	1.41	7.0	452	1	C9H5Cl1N2O3	6-CHLORO-4-NITROQUINOLINE-1-OXIDE 3741-12-6
7	1.75	7.2	2212	1	C9H5Cl1N4	CARBONYLCYANIDE,2-CHLOROPHENYLHYDRAZONE 55653-03-7
8 ★	3.12	7.2	2212	2	"	CARBONYLCYANIDE,2-CHLOROPHENYLHYDRAZONE
9	2.19	7.2	1985	314	C9H5Cl1N4	CARBONYLCYANIDE,M-CHLORO-PHENYLHYDRAZONE 555-60-2
4050 ★	3.39	7.2	2212	2	"	CARBONYLCYANIDE,M-CHLORO-PHENYLHYDRAZONE
1	2.23	7.2	1985	314	C9H5Cl1N4	PROPANEDINITRILE,4-CHLOROPHENYLHYDRAZONE 946-76-9
2 ★	3.39	3.0	1985		"	PROPANEDINITRILE,4-CHLOROPHENYLHYDRAZONE
3 ★	3.57	7.4	2211	1	C9H5Cl2N1	QUINOLINE,4,7-DICHLORO 86-98-6
4 ★	1.85	7.4	2211	1	C9H5Cl2N1O1	QUINOLINE-N-OXIDE,4,7-DICHLORO
55 ★	3.00		2300	20	C9H5Cl3N4	DYRENE 101-05-3 *fungicide*
6 ★	1.00	7.0	452	1	C9H5F1N2O3	8-FLUORO-4-NITROQUINOLINE-1-OXIDE 19789-69-6
7 ★	1.01		163	459	C9H5N3	6-CYANOQUINOXALINE 23088-24-6
8 ★	0.95	7.0	452	1	C9H5N3O5	4,5-DINITROQUINOLINE-1-OXIDE 16238-73-6
9 ★	0.90	7.0	452	1	C9H5N3O5	4,6-DINITROQUINOLINE-1-OXIDE 1596-52-7
4060 ★	0.76	7.0	452	1	C9H5N3O5	4,8-DINITROQUINOLINE-1-OXIDE 14753-19-6
1	1.27	7.2	1985	314	C9H5N5O2	PROPANEDINITRILE,2-NITROPHENYLHYDRAZONE 3722-12-1
2 ★	2.90	3.0	1985		"	PROPANEDINITRILE,2-NITROPHENYLHYDRAZONE
3	1.17	7.2	1985	314	C9H5N5O2	PROPANEDINITRILE,4-NITROPHENYLHYDRAZONE 55653-13-9
4 ★	2.37	3.0	1985		"	PROPANEDINITRILE,4-NITROPHENYLHYDRAZONE
65 ★	3.26		2410		C9H6Br1Cl1F3N1O1	ACETANILIDE,2-BROMO-3-TRIFLUOROMETHYL-5-CHLORO
6 ★	3.41		2410		C9H6Br1Cl1F3N1O1	ACETANILIDE,2-CHLORO-3-TRIFLUOROMETHYL-5-BROMO
7 ★	2.22		2410		C9H6Br1F4N1O1	ACETANILIDE,2-BROMO-3-TRIFLUOROMETHYL-6-FLUORO
8 ★	3.17		2410		C9H6Br1F4N1O1	ACETANILIDE,2-FLUORO-4-BROMO-5-TRIFLUOROMETHYL
9 ★	3.24	7.4	593	1	C9H6Br1N1	3-BROMOQUINOLINE 5332-24-1
4070 ★	2.83	7.0	538	701	C9H6Br1N1	6-BROMOQUINOLINE 5332-25-2
1 ★	2.92		163	459	C9H6Br1N1	7-BROMOQUINOLINE 4965-36-0
2 ★	2.41		1946		C9H6Br1N1O3	2-PROPENOIC ACID,2-CYANO-3-(5-BROMO-2-FURYL),METHYL ESTER
3 ★	2.86		2410		C9H6Cl1F4N1O1	ACETANILIDE,2-CHLORO-3-TRIFLUOROMETHYL-5-FLUORO
4 ★	3.15		2410		C9H6Cl1F4N1O1	ACETANILIDE,2-FLUORO-3-TRIFLUOROMETHYL-5-CHLORO
75 ★	3.02		2410		C9H6Cl1F4N1O1	ACETANILIDE,2-FLUORO-4-CHLORO-5-TRIFLUOROMETHYL
6 ★	2.71		163	459	C9H6Cl1N1	2-CHLOROQUINOLINE 612-62-4
7 ★	2.73		532		C9H6Cl1N1	6-CHLOROQUINOLINE 612-57-7
8 ★	2.44	7.4	595	314	C9H6Cl1N1	8-CHLOROQUINOLINE 611-33-6
9 ★	2.67		826		C9H6Cl1N1O1	4-CHLORO-8-QUINOLINOL 57334-36-8
4080 ★	1.08	7.0	452	1	C9H6Cl1N1O1	4-CHLOROQUINOLINE-1-OXIDE 4637-59-6
1 ★	2.88		1315		C9H6Cl1N1O1	8-QUINOLINOL,5-CHLORO 130-16-5 *antibacterial*
2 ★	2.29		1946		C9H6Cl1N1O3	2-PROPENOIC ACID,2-CYANO-3-(5-CHLORO-2-FURANYL),METHYL ESTER
3 ★	1.56		588		C9H6Cl1N1O3	PHENOXYACETIC ACID,3-CYANO-4-CHLORO
4 ★	3.27		2410		C9H6Cl2F3N1O1	ACETANILIDE,2,3-DICHLORO-5-TRIFLUOROMETHYL
85 ★	3.24		2410		C9H6Cl2F3N1O1	ACETANILIDE,2,4-DICHLORO-5-TRIFLUOROMETHYL
6 ★	3.27		2410		C9H6Cl2F3N1O1	ACETANILIDE,2,5-DICHLORO-3-TRIFLUOROMETHYL
7 ★	3.71		2410		C9H6Cl2F3N1O2	ACETANILIDE,2,4-DICHLORO-5-TRIFLUOROMETHOXY
8 ★	2.72		1792		C9H6Cl2N2O1	CYANOACETAMIDE,N-(3,4-DICHLOROPHENYL) 15386-80-8
9	3.62		2340		C9H6Cl6O3S1	ENDOSULFAN(ALPHA) 33213-65-9
4090 ★	3.83		1555		"	ENDOSULFAN(ALPHA) 959-98-8
1 ★	3.66		2340		C9H6Cl6O4S1	ENDOSULFAN SULFATE 1031-07-8
2 ★	2.19	7.4	595	314	C9H6F1N1	QUINOLINE,8-FLUORO 394-68-3
3 ★	1.52	7.4	2206	306	C9H6F1N3O2	IMIDAZOLE,5-NITRO-2-(P-FLUOROPHENYL)
4 ★	2.34		2410		C9H6F5N1O1	ACETANILIDE,2,4-DIFLUORO-5-TRIFLUOROMETHYL
95 ★	0.43	6.5	321	1	C9H6F5N3O4S2	FLUETHIAZIDE 23141-80-2
6 ★	3.41		547		C9H6F6O1	1,1,1,3,3,3-HEXAFLUORO-2-PHENYL-2-PROPANOL
7 ★	3.04		1519	459	C9H6I1N1	2-IODOQUINOLINE 6560-83-4
8 ★	2.75		1946		C9H6I1O3	2-PROPENOIC ACID,2-CYANO-3-(5-IODO-2-FURANYL),METHYL ESTER
9 ★	2.32		1625	457	C9H6N2	INDOLE,3-CYANO 5457-28-3
4100 ★	2.46		2221	459	C9H6N2	INDOLE-4-CYANO 16136-52-0
1 ★	2.36		2221	459	C9H6N2	INDOLE-5-CYANO 15861-24-2
2 ★	2.38		2221	459	C9H6N2	INDOLE-6-CYANO 15861-36-6
3 ★	2.03		2141		C9H6N2O1	PROPANEDINITRILE,(2-(5-METHYLFURANYL)METHYLENE)
4 ★	2.45		1946		C9H6N2O1S1	PROPANEDINITRILE,(5-METHIO-2-FURANYL)METHYLENE 76542-53-5

	logP	pH	Ref.	Note	MF	Name / **CAS** / *activity*
5 ★	1.97		163	459	C9H6N2O2	3-NITROQUINOLINE **17576-53-3**
6 ★	2.06		163	459	C9H6N2O2	4-NITROQUINOLINE **3741-15-9**
7	2.14	7.0	452	1	"	4-NITROQUINOLINE
8 ★	1.86		532		C9H6N2O2	5-NITROQUINOLINE **607-34-1**
9 ★	1.84		532		C9H6N2O2	6-NITROQUINOLINE **613-50-3**
4110 ★	1.82		532		C9H6N2O2	7-NITROQUINOLINE **613-51-4**
1 ★	1.40		532		C9H6N2O2	8-NITROQUINOLINE **607-35-2**
2 ★	0.56	7.0	452	1	C9H6N2O3	3-NITROQUINOLINE-1-OXIDE **7433-86-5**
3	1.02	7.0	452	1	C9H6N2O3	4-NITROQUINOLINEOXIDE **56-57-5**
4 ★	1.09		1716		"	4-NITROQUINOLINEOXIDE
15 ★	0.49	7.0	452	1	C9H6N2O3	5-NITROQUINOLINE-1-OXIDE **7613-19-6**
6 ★	0.39	7.0	452	1	C9H6N2O3	6-NITROQUINOLINE-1-OXIDE **13675-92-8**
7 ★	0.36	7.0	452	1	C9H6N2O3	7-NITROQUINOLINE-1-OXIDE **14753-17-4**
8 ★	0.04	7.0	452	1	C9H6N2O3	8-NITROQUINOLINE-1-OXIDE **14753-18-5**
9 ★	0.60		2458	302	C9H6N2O3	IMIDAZOLIDINETRIONE,1-PHENYL
4120	1.99		1315		C9H6N2O3	NITROXOLINE **4008-48-4** *antibacterial*
1 ★	1.97		555		C9H6N2O3S1	2-(5-NO2-2-FURFURILIDENE)THIAZOLE **49561-47-9**
2 ★	1.41		494		C9H6N2O4S2	2-METHIO-5(5-NO2-2-FURFURYLIDINE)THIAZOLINE-4-ONE **52661-45-7**
3 ★	1.06		494		C9H6N2O4S2	3-ME-5(5-NO2-2-FURFURYLIDINE)-2-S-THIAZOLIDINE-4-ONE **15913-35-6** +
4 ★	1.78		494		C9H6N2O5S1	3-ME-5-(5-NO2-2-FURFURILIDINE)THIAZOLIDINE-2,4-DIONE
25 ★	1.71		594		C9H6N2O7S1	PHTHALIMIDO METHYLSULFONE,5-NITRO +
6 ★	2.17		1625	457	C9H6N2S1	2-BENZOTHIOPHENAMINE,3-CYANO **18774-47-5**
7 ★	3.12		2585		C9H6N2S3	2-(THIOCYANOMETHYLTHIO)BENZOTHIAZOLE
8	1.85	7.2	1985	314	C9H6N4	CARBONYLCYANIDE,PHENYLHYDRAZONE **6017-21-6**
9 ★	2.67		505		"	CARBONYLCYANIDE,PHENYLHYDRAZONE
4130	1.48	7.2	1985	314	C9H6N4O1	PROPANEDINITRILE,2-(3-HYDROXYPHENYLHYDRAZONE) **96382-70-6**
1 ★	2.28	3.0	1985		"	PROPANEDINITRILE,2-(3-HYDROXYPHENYLHYDRAZONE)
2 ★	0.61		505		C9H6O2	1,3-INDANDIONE **606-23-5**
3 ★	1.38		1941		C9H6O2	4H-1-BENZOPYRAN-4-ONE **491-38-3**
4 ★	1.39		505		C9H6O2	COUMARIN **91-64-5**
35 ★	2.41	1.0	594		C9H6O3	2-BENZOFUROIC ACID **496-41-3**
6	-0.82	7.4	2438	1	C9H6O3	4-HYDROXYCOUMARIN **1076-38-6**
7 ★	1.44		1941		C9H6O3	4H-1-BENZOPYRAN-4-ONE,7-HYDROXY **59887-89-7**
8 ★	1.58		1941		C9H6O3	COUMARIN,7-HYDROXY
9	0.65		536		C9H6O4	1,2,3-INDANTRIONE **485-47-2**
4140	0.70		547		"	1,2,3-INDANTRIONE
1 ★	2.09		576		C9H7Br1Cl1N5	2'-BROMO-4'-CHLOROBENZOGUANAMINE **57381-50-7**
2 ★	1.85		576		C9H7Br1Cl1N5	2'-BROMO-5'-CHLOROBENZOGUANAMINE **57381-38-1**
3 ★	2.04		576		C9H7Br1Cl1N5	2'-CHLORO-5'-BROMOBENZOGUANAMINE **57381-45-0**
4 ★	1.45		576		C9H7Br1F1N5	2'-BROMO-5'-FLUOROBENZOGUANAMINE **57381-40-5**
45 ★	1.94		576		C9H7Br1F1N5	2'-FLUORO-5'-BROMOBENZOGUANAMINE **57381-60-9**
6 ★	2.09		576		C9H7Br2N5	2',5'-DIBROMOBENZOGUANAMINE **57381-42-7**
7 ★	1.38		576		C9H7Cl1F1N5	2'-CHLORO-5'-FLUOROBENZOGUANAMINE **57381-57-4**
8 ★	1.81		576		C9H7Cl1F1N5	2'-FLUORO-5'-CHLOROBENZOGUANAMINE **57381-35-8**
9 ★	1.98		2410		C9H7Cl1F3N1O1	ACETANILIDE,2-TRIFLUOROMETHYL-4-CHLORO
4150 ★	3.33		2410		C9H7Cl1F3N1O1	ACETANILIDE,3-TRIFLUOROMETHYL-4-CHLORO
1 ★	1.96		1519	459	C9H7Cl1N2	4-METHYL-6-CHLOROCINNOLINE
2 ★	1.98		1519	459	C9H7Cl1N2	4-METHYL-7-CHLOROCINNOLINE
3 ★	2.11		489	220	C9H7Cl1O3	5-CL-2,3-DIHYDRO-BENZOFURAN-2-CO2H **34385-94-9**
4 ★	1.07		588		C9H7Cl1O5	PHENOXYACETIC ACID,3-CARBOXY-4-CHLORO
55 ★	2.54		2665		C9H7Cl2N3	1-BENZYLTRIAZOLE,3',4'-DICHLORO *fungicide*
6 ★	2.07		576		C9H7Cl2N5	2',4'-DICHLOROBENZOGUANAMINE **57381-46-1**
7 ★	1.97		576		C9H7Cl2N5	2',5'-DICHLOROBENZOGUANAMINE **57381-26-7** *anti-ulcer*
8 ★	1.38		576		C9H7Cl2N5	2',6'-DICHLOROBENZOGUANAMINE **57381-54-1**
9 ★	3.80		2456	821	C9H7Cl3O3	SILVEX **93-72-1** *herbicide*
4160 ★	1.59		522		C9H7F3N2O2S1	1,2,4-BENZOTHIADIAZINE-1,1-DIOXIDE-3-METHYL-6-TRIFLUOROMETHYL **723-57-9**
1 ★	1.51		522		C9H7F3N2O2S1	1,2,4-BENZOTHIADIAZINE-1,1-DIOXIDE-34-METHYL-7-TRIFLUOROMETHYL **20246-63-3**
2	3.30	4.0	2264	1	C9H7F3N4O4	1,1,1-TRIFLUOROACETONE,2,4-DINITROPHENYLHYDRAZONE
3 ★	2.63		558		C9H7F3O2	ACETIC ACID,M-TRIFLUOROMETHYLPHENYL ESTER
4 ★	2.59		558		C9H7F3O2	ACETIC ACID,O-TRIFLUOROMETHYLPHENYL ESTER
65 ★	2.62		501		C9H7F3O2	M-TRIFLUOROMETHYLPHENYLACETIC ACID **351-35-9**
6 ★	2.45		853		C9H7F3O2	P-TRIFLUOROMETHYLPHENYLACETIC ACID **32857-62-8**
7 ★	2.36		501		C9H7F3O3	M-TRIFLUOROMETHYLPHENOXYACETIC ACID **349-82-6**
8 ★	2.86		501		C9H7F3O3S1	M-TRIFLUOROMETHYLTHIOPHENOXYACETIC ACID
9 ★	2.48		501		C9H7F3O4	M-TRIFLUOROMETHOXYPHENOXYACETIC ACID
4170 ★	2.19		501		C9H7F3O5S1	M-TRIFLUOROMETHYLSULFONYLPHENOXYACETIC ACID
1 ★	2.24		2410		C9H7F4N1O1	ACETANILIDE,2-FLUORO-5-TRIFLUOROMETHYL
2 ★	1.96		306		C9H7N1	CINNAMONITRILE **1885-38-7**
3 ★	2.08		503		C9H7N1	I-QUINOLINE **119-65-3**
4 ★	2.03		502		C9H7N1	QUINOLINE **91-22-5**
75	2.09	7.4	2211	1	"	QUINOLINE
6 ★	2.08		1519		C9H7N1O1	3-QUINOLINOL **580-18-7**
7 ✓	0.58		924	815	C9H7N1O1	4-HYDROXYQUINOLINE **611-36-9**
8 ✓	0.75	7.4	590	1	"	4-HYDROXYQUINOLINE
9 ★	0.58		924	400	C9H7N1O1	4-QUINOLONE **529-37-3**
4180 ★	1.80		1519	459	C9H7N1O1	6-HYDROXYQUINOLINE **580-16-5**
1 ★	1.98	7.4	579	1	C9H7N1O1	7-QUINOLINOL **580-20-1**
2 ★	2.02		525		C9H7N1O1	8-QUINOLINOL **148-24-3** *disinfectant*
3 ★	1.26		527	815	C9H7N1O1	A-QUINOLONE **59-31-4**
4 ★	0.97		594		C9H7N1O1	BENZOYLACETONITRILE
85 ★	0.25		547		C9H7N1O1	I-QUINOLINE-N-OXIDE **1532-72-5**
6 ★	1.68		2221	459	C9H7N1O1	INDOLE-3-CARBOXALDEHYDE **487-89-8**
7 ★	1.16	7.0	925	1	C9H7N1O1	M-CYANOACETOPHENONE **6136-68-1**
8 ★	1.22	7.0	925	1	C9H7N1O1	P-CYANOACETOPHENONE **1443-80-7**
9 ★	2.27		590		C9H7N1O1	QUINOLINE,5-HYDROXY **578-67-6**

	logP	pH	Ref.	Note	MF	Name / CAS / activity
4190	0.36	7.4	2211	1	C9H7N1O1	QUINOLINE-1-OXIDE **1613-37-2**
1 ★	0.36	7.0	538	701	"	QUINOLINE-1-OXIDE
2	0.39	7.0	452	1	"	QUINOLINE-1-OXIDE
3 ★	1.33		558		C9H7N1O2	2-CYANOPHENYLACETATE **5715-02-6**
4 ★	2.55	1.0	574	342	C9H7N1O2	3-HYDROXY-5-PHENYLISOXAZOLE
95 ✓	0.62		826		C9H7N1O2	4,8-DIHYDROXYQUINOLINE **14959-84-3**
6 ★	-0.47	7.0	452	1	C9H7N1O2	4-HYDROXYQUINOLINE-1-OXIDE **3039-74-5**
7 ★	2.31	2.0	1625		C9H7N1O2	INDOLE,2-CARBOXYLIC ACID **1477-50-5**
8 ★	1.99		2221	457	C9H7N1O2	INDOLE-3-CARBOXYLIC ACID **771-50-6**
9 ★	1.18		501		C9H7N1O2	M-CYANOPHENYLPHENYLACETIC ACID **1878-71-3**
4200 ★	0.58		924	400	C9H7N1O2	N-METHYLINDOL-2,3-DIONE **2058-74-4**
1 ★	1.31		547		C9H7N1O2	PHTHALIMIDE,N-METHYL **550-44-7**
2 ★	1.65		1946		C9H7N1O3	2-PROPENOIC ACID,2-CYANO-3-(2-FURANYL),METHYL ESTER **3695-86-1**
3	0.93		1464		C9H7N1O3	3-PHENYL-2,4-OXAZOLIDINEDIONE **3759-90-8**
4	0.26		2177		C9H7N1O3	ACRYLIC ACID-3-(NICOTINOYL)
5 ★	0.54	3.5	2458	2	C9H7N1O3	ISOXAZOLIDINE-3,5-DIONE,2-PHENYL
6 ★	1.09	3.5	2458	302	C9H7N1O3	OXAZOLIDINE-2,4-DIONE,5-PHENYL
7 ★	0.95		501		C9H7N1O3	PHENOXYACETIC ACID,M-CYANO **1879-58-9**
8 ★	0.93		501		C9H7N1O3	PHENOXYACETIC ACID,P-CYANO **1878-82-6**
9	-0.20	7.4	594	1	C9H7N1O4	P-NITROCINNAMIC ACID-TRANS **882-06-4**
4210 ★	2.12	1.0	594	342	"	P-NITROCINNAMIC ACID-TRANS
1	-1.66		508		C9H7N1.C7H8O3S1	QUINOLINIUM-P-TOLUENESULFONATE
2 ★	1.93		1519	459	C9H7N3	2-PHENYL-S-TRIAZINE **1722-18-5**
3 ★	0.98		1519	459	C9H7N3O1	2-QUINOXALINECARBOXAMIDE **5182-90-1**
4	2.33	7.4	2298	1	C9H7N3O2	1,2,4-OXADIAZOL-3-ALDOXIME,5-PHENYL **103499-08-7**
15 ★	2.40	7.4	2298	2	"	1,2,4-OXADIAZOL-3-ALDOXIME,5-PHENYL
6	2.61	7.6	1769	212	C9H7N3O2	1,2,4-OXADIAZOLE-3-PHENYL-5-HYDROXIMIC ACID **90507-20-3** ✚
7 ★	2.72	7.4	2298	2	"	1,2,4-OXADIAZOLE-3-PHENYL-5-HYDROXIMIC ACID
8 ★	1.31		1519	459	C9H7N3O2	4-QUINOLINAMINE,8-NITRO
9	1.96	7.6	1769	212	C9H7N3O2	5-PHENYL-1,3,4-OXADIAZOLE-3-HYDROXIMIC ACID **90507-19-0**
4220	1.61		2138	459	C9H7N3O2	QUINOLINE,5-AMINO-6-NITRO **35975-00-9**
1 ★	1.32		555		C9H7N3O3	2-(5-NO2-2-FURFURILIDENE)IMIDAZOLE **6756-33-8**
2 ★	1.16		494		C9H7N3O4S1	2-METHYLAMINO-5-(5-NITRO-2-FURFURYL)THIAZOLONE **25603-06-9** ✚
3 ★	0.86		494		C9H7N3O4S1	2-METHYLAMINO-6-(5-NITROFUR-2YL)-1,3-THIAZIN-4-ONE **52661-53-7** ✚
4 ★	1.44		494		C9H7N3O4S1	3-METHYL-2-IMINO-5-(5-NITRO-2-FURFURILIDINO)THIAZOLIN-4-ONE **25580-69-2** ✚
25 ✓	-0.43	9.0	2027	1	C9H7N3O4S2	THIAZOLE,2-AMINO-5-(P-NITROPHENYL)SULFONE **39565-05-4**
6 ★	1.70		2300	20	C9H7N3S1	TRICYCLAZOLE **41814-78-2** pesticide
7	0.02	7.4	1528	1	C9H7N7O2S1	AZATHIOPRINE **446-86-6** immunosuppressant
8	0.08	7.4	1497	1	"	AZATHIOPRINE
9 ★	0.10		1420		"	AZATHIOPRINE
4230 ★	2.92		534		C9H8	INDENE **95-13-6**
1	1.97	7.4	976	1	C9H8Br1Cl2N3	2-(4-BROMO-2,6-DICHLOROPHENYLIMINO)IMIDAZOLINE **26827-71-4**
2	2.31	7.4	1350	1	"	2-(4-BROMO-2,6-DICHLOROPHENYLIMINO)IMIDAZOLINE
3 ★	2.66	7.4	976	2	"	2-(4-BROMO-2,6-DICHLOROPHENYLIMINO)IMIDAZOLINE
4 ★	1.26		576		C9H8Br1N5	2'-BROMOBENZOGUANAMINE **30530-48-4**
35 ★	2.44		576		C9H8Br1N5	3'-BROMOBENZOGUANAMINE **30101-52-1**
6	2.24	7.4	976	1	C9H8Br3N3	2-(2,4,6-TRIBROMOPHENYLIMINO)IMIDAZOLIDINE **22683-09-6**
7	2.51	7.4	1350	1	"	2-(2,4,6-TRIBROMOPHENYLIMINO)IMIDAZOLIDINE
8 ★	2.74	7.4	976	2	"	2-(2,4,6-TRIBROMOPHENYLIMINO)IMIDAZOLIDINE
9	2.31		2161	468	C9H8Cl1N1S1	AZETIDINE-2-ONE,4-(4-CHLORO)PHENYLTHIO
4240	1.84		2665		C9H8Cl1N3	1-BENZYLTRIAZOLE,4'-CHLORO fungicide
1 ★	1.19		576		C9H8Cl1N5	2'-CHLOROBENZOGUANAMINE **29366-77-6**
2 ★	2.33		576		C9H8Cl1N5	3'-CHLOROBENZOGUANAMINE **4514-54-9**
3 ★	2.33		576		C9H8Cl1N5	4'-CHLOROBENZOGUANAMINE **4514-53-8**
4 ★	1.90	9.9	704	821	C9H8Cl1N5O1	CLOGUANAMIL **21702-93-2** ✚
45 ★	2.89		1624		C9H8Cl2N2O2	3,4-DICHLOROBENZALDEHYDE,O-((MEAMINO)CARBONYL)OXIME ✚
6 ★	1.68	7.4	1616	1	C9H8Cl2N2O2	BENZAMIDE,O-DICHLOROACETYLAMINO
7	1.36	7.4	1350	1	C9H8Cl2N2S1	IMIDAZOLINE,2-(2,6-DICHLOROPHENYL)THIO
8	1.92	7.4	976	1	C9H8Cl2N4O2	2-(2,6-CL2-4-NO2-PHENYLIMINO)IMIDAZOLINE
9 ★	2.10	7.4	976	2	"	2-(2,6-CL2-4-NO2-PHENYLIMINO)IMIDAZOLINE
4250 ★	3.43		2625	448	C9H8Cl2O3	DICHLORPROP **120-36-5** herbicide
1 ★	2.35		536		C9H8Cl3N1O2S1	CAPTAN **133-06-2** fungicide
2	1.71	7.4	1547	1	C9H8Cl3N3	2-(2,3,6-TRICHLOROPHENYLIMINO)IMIDAZOLIDINE **82780-90-3**
3 ★	2.46		2400	2	"	2-(2,3,6-TRICHLOROPHENYLIMINO)IMIDAZOLIDINE
4	1.63	7.4	821	1	C9H8Cl3N3	2-(2,4,5-TRICHLOROPHENYLIMINO)IMIDAZOLIDINE **59465-40-6**
55 ★	2.11	7.4	821	2	"	2-(2,4,5-TRICHLOROPHENYLIMINO)IMIDAZOLIDINE
6	1.38	7.4	821	1	C9H8Cl3N3	2-(2,4,6-TRICHLOROPHENYLIMINO)IMIDAZOLIDINE **59465-51-9**
7	1.44	7.4	1398	1	"	2-(2,4,6-TRICHLOROPHENYLIMINO)IMIDAZOLIDINE
8	1.47	7.4	976	1	"	2-(2,4,6-TRICHLOROPHENYLIMINO)IMIDAZOLIDINE
9	1.79	7.4	821	2	"	2-(2,4,6-TRICHLOROPHENYLIMINO)IMIDAZOLIDINE
4260 ★	2.18	7.4	976	2	"	2-(2,4,6-TRICHLOROPHENYLIMINO)IMIDAZOLIDINE
1	1.10		2665		C9H8F1N3	1-BENZYLTRIAZOLE,4'-FLUORO fungicide
2 ★	1.73		576		C9H8F1N5	3'-FLUOROBENZOGUANAMINE **30530-43-9**
3 ★	1.61		576		C9H8F1N5	4'-FLUOROBENZOGUANAMINE **30530-44-0**
4 ★	1.11		1894		C9H8F3N1O1	ACETANILIDE,2-TRIFLUOROMETHYL **344-62-7**
65 ★	2.52		1894		C9H8F3N1O1	ACETANILIDE,4-TRIFLUOROMETHYL **349-97-3**
6	2.73	7.4	2327	1	"	ACETANILIDE,4-TRIFLUOROMETHYL
7 ★	2.56		547		C9H8F3N1O1	M-(TRIFLUOROMETHYL)ACETANILIDE **351-36-0**
8 ★	2.13		2142		C9H8F3N1O2	2-TRIFLUOROMETHYLPHENYLCARBAMATE,O-METHYL
9 ★	2.74		2142		C9H8F3N1O2	3-TRIFLUOROMETHYLPHENYLCARBAMATE,O-METHYL
4270 ★	2.62		2410		C9H8F3N1O2	ACETANILIDE,4-TRIFLUOROMETHOXY
1 ★	2.37		273		C9H8F3N1O2	N-METHYL-3-TRIFLUOROMETHYLPHENYLCARBAMATE **14061-27-9**
2 ★	2.78		607	606	C9H8F3N1O3S1	TRIFLUOROMETHANESULFONANILIDE,M-ACETYL **23375-11-3**
3 ★	2.83		607	606	C9H8F3N1O3S1	TRIFLUOROMETHANESULFONANILIDE,P-ACETYL **23383-94-0**
4 ★	1.41		576		C9H8I1N5	2'-IODOBENZOGUANAMINE

	logP	pH	Ref.	Note	MF	Name / CAS / activity
75 ★	0.97	0.3	2061		C9H8I1O3	HIPPURIC ACID,2-IODO 147-58-0
6 ★	1.88	0.3	2061		C9H8I1O3	HIPPURIC ACID,3-IODO 52386-94-4
7 ★	1.94	0.3	2061		C9H8I1O3	HIPPURIC ACID,4-IODO
8 ★	1.90		1519	459	C9H8N2	1-ISOQUINOLINAMINE 1532-84-9
9 ★	2.20		579		C9H8N2	1-PHENYLPYRAZOLE 1126-00-7
4280 ★	1.87		163	459	C9H8N2	2-AMINOQUINOLINE 580-22-3
1 ★	1.61		163	459	C9H8N2	2-METHYLQUINOXALINE 7251-61-8
2 ★	1.88		547		C9H8N2	2-PHENYLIMIDAZOLE 670-96-2
3	0.40	7.0	452	1	C9H8N2	4-AMINOQUINOLINE 578-68-7
4 ★	1.63		163	459	"	4-AMINOQUINOLINE
85 ★	1.16		532		C9H8N2	5-AMINOQUINOLINE 611-34-7
6 ★	2.04		163	459	C9H8N2	5-METHYLQUINOXALINE 13708-12-8
7 ★	1.28		163	459	C9H8N2	6-AMINOQUINOLINE 580-15-4
8 ★	1.79		532		C9H8N2	8-AMINOQUINOLINE 578-66-5
9 ★	1.63		163	459	C9H8N2	QUINOLINE,3-AMINO 580-17-6
4290 ★	0.61		924	400	C9H8N2O1	1(2H)-PHTHALAZINONE,2-METHYL 6091-81-2
1 ★	2.31		163	459	C9H8N2O1	2-METHOXYQUINOXALINE 39209-88-6
2 ★	0.69		924	400	C9H8N2O1	3-METHYL-QUINAZOLINE-4-ONE 2436-66-0
3 ★	0.05	7.0	452	1	C9H8N2O1	4-AMINOQUINOLINE-1-OXIDE 2508-86-3
4 ★	1.41		1519	459	C9H8N2O1	4-METHOXYCINNOLINE
95 ★	1.97		1519	459	C9H8N2O1	4-METHOXYQUINAZOLINE 16347-95-8
6 ★	0.57		1439		C9H8N2O1	4H-PYRIDO(1,2-A)PYRIMIDIN-4-ONE,2-METHYL 1693-94-3
7 ★	0.77		1439		C9H8N2O1	4H-PYRIDO(1,2-A)PYRIMIDIN-4-ONE,3-METHYL 39080-57-4
8 ★	0.74		1439		C9H8N2O1	4H-PYRIDO(1,2-A)PYRIMIDIN-4-ONE,6-METHYL 23443-11-0
9	0.61		1725	459	C9H8N2O1	4H-PYRIDO(1,2-A)PYRIMIDIN-4-ONE,7-METHYL
4300	0.56		1725	459	C9H8N2O1	4H-PYRIDO(1,2-A)PYRIMIDIN-4-ONE,8-METHYL
1	0.79		1725	459	C9H8N2O1	4H-PYRIDO(1,2-A)PYRIMIDIN-4-ONE,9-METHYL
2	-0.11		826		C9H8N2O1	5-AMINO-8-HYDROXYQUINOLINE 13207-66-4
3 ★	1.20		163	459	C9H8N2O1	5-METHYLQUINOXALINE-1-OXIDE
4 ★	0.57		1894		C9H8N2O1	ACETANILIDE,2-CYANO 25116-00-1
5 ★	1.20		1894		C9H8N2O1	ACETANILIDE,3-CYANO 58202-84-9
6	1.36	7.4	2327	1	C9H8N2O1	ACETANILIDE,4-CYANO 35704-19-9
7 ★	1.37		1894		"	ACETANILIDE,4-CYANO
8 ★	2.59		2166		C9H8N2O1	FURAZAN,3-METHYL-4-PHENYL 10349-09-4
9 ★	0.66		2221	459	C9H8N2O1	INDOLE-4-CARBOXAMIDE 1670-86-6
4310 ★	0.87		2221	459	C9H8N2O1	INDOLE-5-CARBOXAMIDE 1670-87-7
1 ★	1.07		2221	459	C9H8N2O1	INDOLE-5-FORMAMIDO
2 ★	0.79		924		C9H8N2O1	N-METHYLQUINOXALINE-2-ONE
3 ★	2.92		2166		C9H8N2O1S1	1,2,5-OXADIAZOLE,3-METHYL-4-PHENYLTHIO
4 ★	1.77		2142		C9H8N2O2	3-CYANOPHENYLCARBAMATE,O-METHYL
15 ★	1.82		2142		C9H8N2O2	4-CYANOPHENYLCARBAMATE,O-METHYL
6	0.00	7.0	452	1	C9H8N2O2	4-HYDROXYAMINOQUINOLINE-1-OXIDE 4637-56-3
7	-0.35		163	459	C9H8N2O2	6-METHYLQUINOXALINE-1,4-DIOXIDE 33368-89-7
8	-1.85	7.4	1616	1	C9H8N2O2	BENZOIC ACID,O-CYANOMETHYLAMINO
9 ★	1.80	7.4	1616	2	"	BENZOIC ACID,O-CYANOMETHYLAMINO
4320 ★	2.16		2166		C9H8N2O2	FURAZAN,3-METHYL-4-PHENYL-2-OXIDE 6898-86-8
1 ★	2.17		2166		C9H8N2O2	FURAZAN,3-PHENYL-4-METHYL-2-OXIDE 6898-87-9
2 ★	0.46		2458	302	C9H8N2O2	IMIDAZOLIDINEDIONE,5-PHENYL
3 ★	0.83		2458	302	C9H8N2O2	IMIDAZOLINEDIONE,3-PHENYL
4 ★	0.86		273	60	C9H8N2O2	N-METHYL-2-CYANOPHENYLCARBAMATE 942-79-0
25 ★	0.97		273		C9H8N2O2	N-METHYL-3-CYANOPHENYLCARBAMATE 943-49-7
6 ★	0.95		273		C9H8N2O2	N-METHYL-4-CYANOPHENYLCARBAMATE 18315-52-1
7 ★	0.55		924		C9H8N2O2	QUINAZOLINE-2,4-DIONE,7-METHYL
8 ★	0.55		1705		C9H8N2O2	QUINAZOLINE-2-4-DIONE,6-METHYL
9 ★	1.76		2166		C9H8N2O2S1	1,2,5-OXADIAZOLE,3-METHYL-4-PHENYLSULFOXIDE
4330	0.79		163	459	C9H8N2O2S1	2-METHYLSULFONYLQUINOXALINE 16310-37-5
1 ★	0.56		1519	400	C9H8N2O2S1	6-SULFONAMIDOQUINOLINE
2 ★	0.42		595		C9H8N2O2S1	8-SULFONAMIDOQUINOLINE 35203-91-9
3 ★	2.63		2166		C9H8N2O2S1	FURAZAN-2-OXIDE,3-METHYL-4-PHENYLTHIO
4 ★	2.74		2166		C9H8N2O2S1	FURAZAN-2-OXIDE,4-METHYL-3-PHENYLTHIO
35 ★	0.49		1524		C9H8N2O3	1(3H)-ISOBENZOFURANONE-3-UREYL
6 ★	2.36		2166		C9H8N2O3S1	1,2,5-OXADIAZOLE,3-METHYL-4-PHENYLSULFONE
7 ★	1.50		2166		C9H8N2O3S1	FURAZAN-2-OXIDE,3-METHYL-4-PHENYLSULFOXIDE
8 ★	1.75		2166		C9H8N2O3S1	FURAZAN-2-OXIDE,4-METHYL-3-SULFOXIDE
9	1.85		2161	468	C9H8N2O4	AZETIDINE-2-ONE,4-(4-NITRO)PHENOXY
4340 ★	2.03		2166		C9H8N2O4S1	FURAZAN-2-OXIDE,3-METHYL-4-PHENYLSULFONYL
1 ★	2.26		2166		C9H8N2O4S1	FURAZAN-2-OXIDE,4-METHYL-3-PHENYLSULFONYL
2 ★	0.80		594		C9H8N2O5S1	PHTHALIMIDO METHYLSULFONATE,5-AMINO
3	2.79		163	459	C9H8N2S1	2-METHIOQUINOXALINE 21948-73-2
4 ★	2.50		1519	459	C9H8N2S1	2-METHYLTHIOQUINAZOLINE 6141-18-0
45 ★	1.29		594		C9H8N4	BENZIMIDAZOLE,1-METHYL-2-AMINO-6-CYANO
6 ★	1.11		595		C9H8N4	BENZIMIDAZOLE,2-AMINO-5-CYANO-1-METHYL
7 ★	1.30		1625	457	C9H8N4O1	QUINOXALINE,2-UREYL
8 ★	1.76		594		C9H8N4O2	1,2,4-OXADIAZOLE,3-PHENYL-5-UREIDO
9 ★	1.41		2166		C9H8N4O2	1,2,5-OXADIAZOLE,3-AMINO-4-BENZOYLAMINO
4350 ★	0.86	7.4	2206	306	C9H8N4O2	IMIDAZOLE,5-NITRO-2-(P-AMINOPHENYL)
1 ★	-0.60	7.4	2360	1	C9H8N4O3	BENZOTRIAZINE,1,4-DI-N-OXIDE-3-ACETYLAMINO
2 ★	-0.02		536		C9H8N4O5	5-(1-AZIRIDINYL)-2,4-DINITROBENZAMIDE 21919-05-1
3	0.20	7.4	1497	1	"	5-(1-AZIRIDINYL)-2,4-DINITROBENZAMIDE
4 ★	0.49		576		C9H8N6O2	2'-NITROBENZOGUANAMINE 29366-71-0
55 ★	1.57		576		C9H8N6O2	3'-NITROBENZOGUANAMINE 29366-72-1
6 ★	1.60		576		C9H8N6O2	4'-NITROBENZOGUANAMINE 29366-73-2
7 ★	1.88		507		C9H8O1	ACRYLOPHENONE 768-03-6
8 ★	1.90		547	406	C9H8O1	CINNAMALDEHYDE 104-55-2 flavoring agent
9 ★	1.41		1524		C9H8O2	1(3H)-ISOBENZOFURANONE-3-METHYL 3453-64-3

	logP	pH	Ref.	Note	MF	Name / CAS / activity
4360 ★	2.13		505		C9H8O2	CINNAMIC ACID, TRANS **140-10-3**
1 ★	3.25		530		C9H8O2S1	5,7-DIMETHYL-2-OXO-1,3-BENZOXATHIOL **15062-75-6**
2 ★	1.00		1524		C9H8O3	1(3H)-ISOBENZOFURANONE-3-METHOXYL **4122-57-0 ✚**
3	0.81		547	815	C9H8O3	3-HYDROXY-3-METHYLPHTHALIDE **577-56-0**
4 ★	1.56		2527		C9H8O3	BENZOIC ACID,M-ACETYL
65 ★	1.61		2527		C9H8O3	BENZOIC ACID,P-ACETYL **586-89-0**
6 ★	1.46	2.0	2312	337	C9H8O3	CINNAMIC ACID,P-HYDROXY **7400-08-0**
7 ★	0.81	1.0	547	342	C9H8O3	O-ACETYLBENZOIC ACID **577-56-0**
8 ★	1.40		489	220	C9H8O4	1,4-BENZODIOXANE-2-CARBOXYLIC ACID
9	-1.20	7.4	1838	1	C9H8O4	ACETYLSALICYLIC ACID **50-78-2** *analgesic, antipyretic*
4370	-1.15	7.4	693	1	"	ACETYLSALICYLIC ACID
1	1.07	7.4	2217	1	"	ACETYLSALICYLIC ACID
2 ★	1.19		505		"	ACETYLSALICYLIC ACID
3 ★	1.15	2.0	2312	337	C9H8O4	CINNAMIC ACID,3,4-DIHYDROXY
4 ★	1.83		501		C9H8O4	ISOPHTHALIC ACID,METHYL ESTER **1877-71-0**
75 ★	1.14		501		C9H8O4	M-CARBOXYPHENYLACETIC ACID **2084-13-1**
6 ★	1.13	1.0	547	342	C9H8O4	MONOMETHYLPHTHALATE
7 ★	0.79		853		C9H8O4	P-FORMYLPHENOXYACETIC ACID **22042-71-3**
8 ★	1.11		501		C9H8O5	PHENOXYACETIC ACID,M-CARBOXY
9 ★	2.73		2410		C9H9Br1Cl1N1O1	ACETANILIDE,2-CHLORO-3-METHYL-4-BROMO
4380 ★	2.69		2410		C9H9Br1Cl1N1O1	ACETANILIDE,2-METHYL-3-CHLORO-4-BROMO
1	0.87	7.4	976	1	C9H9Br1Cl1N3	2(2-BR-6-CL-PHENYLIMINO)IMIDAZOLINE
2	1.73	7.4	976	2	"	2(2-BR-6-CL-PHENYLIMINO)IMIDAZOLINE
3	0.47	7.4	1992	1	C9H9Br1Cl1N3	2-(2-BROMO-3-CHLOROPHENYLIMINO)IMIDAZOLINE
4	0.67	7.4	1992	1	C9H9Br1Cl1N3	2-(2-CHLORO-3-BROMOPHENYLIMINO)IMIDAZOLINE **67563-96-6**
85	1.04	7.4	2400	1	"	2-(2-CHLORO-3-BROMOPHENYLIMINO)IMIDAZOLINE
6 ★	2.05		2400	2	"	2-(2-CHLORO-3-BROMOPHENYLIMINO)IMIDAZOLINE
7	1.21	7.4	2400	1	C9H9Br1Cl1N3	2-(2-CHLORO-5-BROMOPHENYLIMINO)IMIDAZOLIDINE **15327-50-1**
8 ★	2.05		2400	2	"	2-(2-CHLORO-5-BROMOPHENYLIMINO)IMIDAZOLIDINE
9	-0.02	7.4	1992	1	C9H9Br1F1N3	2-(2-BROMO-3-FLUOROPHENYLIMINO)IMIDAZOLINE
4390	0.89	7.4	1992	1	C9H9Br2N3	2-(2,3-DIBROMOPHENYLIMINO)IMIDAZOLINE **78834-88-5**
1	1.14	7.4	2400	1	"	2-(2,3-DIBROMOPHENYLIMINO)IMIDAZOLINE
2 ★	2.16		2400	2	"	2-(2,3-DIBROMOPHENYLIMINO)IMIDAZOLINE
3	1.15	7.4	821	1	C9H9Br2N3	2-(2,6-DIBROMOPHENYLIMINO)IMIDAZOLIDINE **4205-93-0**
4	1.17	7.4	1398	1	"	2-(2,6-DIBROMOPHENYLIMINO)IMIDAZOLIDINE
95	1.21	7.4	976	1	"	2-(2,6-DIBROMOPHENYLIMINO)IMIDAZOLIDINE
6	1.73	7.4	821	2	"	2-(2,6-DIBROMOPHENYLIMINO)IMIDAZOLIDINE
7 ★	1.96	7.4	976	2	"	2-(2,6-DIBROMOPHENYLIMINO)IMIDAZOLIDINE
8	0.52	7.4	976	1	C9H9Cl1F1N3	2(2-CL-6-F-PHENYLIMINO)IMIDAZOLINE
9 ★	1.45	7.4	976	2	"	2(2-CL-6-F-PHENYLIMINO)IMIDAZOLINE
4400	-0.49	7.4	1992	1	C9H9Cl1F1N3	2-(2-CHLORO-3-FLUOROPHENYLIMINO)IMIDAZOLINE **81060-10-8**
1	0.00	7.4	2400	1	"	2-(2-CHLORO-3-FLUOROPHENYLIMINO)IMIDAZOLINE
2 ★	1.52		2400	2	"	2-(2-CHLORO-3-FLUOROPHENYLIMINO)IMIDAZOLINE
3	0.00	7.4	2400	1	C9H9Cl1F1N3	2-(2-CHLORO-5-FLUOROPHENYLIMINO)IMIDAZOLIDINE **81060-11-9**
4 ★	1.52		2400	2	"	2-(2-CHLORO-5-FLUOROPHENYLIMINO)IMIDAZOLIDINE
5 ★	3.41		2303		C9H9Cl1N1O3P1S1	METHYLISOCYANOTHION ✚
6 ★	2.27		1624		C9H9Cl1N2O2	P-CHLOROBENZALDEHYDE,O-((MEAMINO)CARBONYL)OXIME
7 ★	1.61		522		C9H9Cl1N2O2S1	1,2,4-BENZOTHIADIAZINE-1,1-DIOXIDE-3-ETHYL-6-CHLORO **14559-55-8**
8 ★	1.62		522		C9H9Cl1N2O2S1	1,2,4-BENZOTHIADIAZINE-1,1-DIOXIDE-3-ETHYL-7-CHLORO **1207-63-2**
9 ★	1.12		522		C9H9Cl1N2O3S1	1,2,4-BENZOTHIADIAZINE-1,1-DIOXIDE-3-METHOXYMETHYL-6-CHLORO **37157-71-4**
4410 ★	1.22		2463		C9H9Cl1N4O1	BENZOTRIAZOLE,2-METHYL-5-CHLOROACETAMIDO
1 ★	2.59		1521		C9H9Cl1O2	ACETIC ACID,P-CHLOROBENZYL ESTER **5406-33-7**
2 ★	2.71		2645		C9H9Cl1O2	PHENYLACETATE,2-CHLORO-6-METHYL
3 ★	2.31		489	220	C9H9Cl1O3	2-(P-CHLOROPHENOXY)-PROPIONIC ACID **3307-39-9**
4 ★	3.07		547		C9H9Cl2N1O1	PROPANIL **709-98-8** *herbicide*
15 ★	3.00		536		C9H9Cl2N1O2	2,4-DICHLORO-5-METHYL-N-METHYLPHENYLCARBAMATE
6 ★	2.41	7.4	1616	1	C9H9Cl2N1O2	ANISOLE,O-DICHLOROACETYLAMINO
7 ★	2.77		1792		C9H9Cl2N1O2	METHOXYACETAMIDE,N-(3,4-DICHLOROPHENYL)
8	0.53	7.4	1398	1	C9H9Cl2N3	2(2,3-DICHLOROPHENYLIMINO)IMIDAZOLINE **15327-44-3**
9	0.53	7.4	1992	1	"	2(2,3-DICHLOROPHENYLIMINO)IMIDAZOLINE
4420	0.57	7.4	976	1	"	2(2,3-DICHLOROPHENYLIMINO)IMIDAZOLINE
1	0.85	7.4	2400	1	"	2(2,3-DICHLOROPHENYLIMINO)IMIDAZOLINE
2 ★	1.94		2400	2	"	2(2,3-DICHLOROPHENYLIMINO)IMIDAZOLINE
3	2.01	7.4	976	2	"	2(2,3-DICHLOROPHENYLIMINO)IMIDAZOLINE
4	0.45	7.4	1097	1	C9H9Cl2N3	2-(2,4-DICHLOROPHENYLIMINO)IMIDAZOLIDINE **23830-88-8**
25	0.45	7.4	1398	1	"	2-(2,4-DICHLOROPHENYLIMINO)IMIDAZOLIDINE
6	0.69	7.4	821	1	"	2-(2,4-DICHLOROPHENYLIMINO)IMIDAZOLIDINE
7	1.90	7.4	976	2	"	2-(2,4-DICHLOROPHENYLIMINO)IMIDAZOLIDINE
8 ★	1.96	7.4	821	2	"	2-(2,4-DICHLOROPHENYLIMINO)IMIDAZOLIDINE
9	0.95	7.4	821	1	C9H9Cl2N3	2-(2,5-DICHLOROANILINO)-2-IMIDAZOLINE **56514-55-7**
4430	0.65	7.4	976	1	C9H9Cl2N3	2-(2,5-DICHLOROPHENYLIMINO)IMIDAZOLIDINE **56514-55-7**
1	0.79	7.4	1398	1	"	2-(2,5-DICHLOROPHENYLIMINO)IMIDAZOLIDINE
2	0.95	7.4	821	1	"	2-(2,5-DICHLOROPHENYLIMINO)IMIDAZOLIDINE
3	1.00	7.4	2400	1	"	2-(2,5-DICHLOROPHENYLIMINO)IMIDAZOLIDINE
4	1.65	7.4	821	2	"	2-(2,5-DICHLOROPHENYLIMINO)IMIDAZOLIDINE
35 ★	1.94		2400	2	"	2-(2,5-DICHLOROPHENYLIMINO)IMIDAZOLIDINE
6	2.04	7.4	976	2	"	2-(2,5-DICHLOROPHENYLIMINO)IMIDAZOLIDINE
7	1.43	7.1	2155	2	C9H9Cl2N3	2-(2,6-DICHLOROANILINO)-2-IMIDAZOLE **57066-25-8**
8	0.48	7.0	2230	1	C9H9Cl2N3	CLONIDINE **4205-90-7** *antihypertensive*
9	0.48	7.4	1398	1	"	CLONIDINE
4440	0.51	7.4	1097	1	"	CLONIDINE
1	0.62	7.4	976	1	"	CLONIDINE
2	0.83	7.4	821	1	"	CLONIDINE
3	0.85	7.4	1350	1	"	CLONIDINE
4	0.91	7.4	2400	1	"	CLONIDINE
45	1.43	7.1	2155	2	"	CLONIDINE
6 ★	1.57	7.4	821	2	"	CLONIDINE
7	1.59	7.4	976	2	"	CLONIDINE
8	-0.40	7.4	821	1	C9H9Cl2N3O1	2-(2,6-DICL-4-OH-PHENYLIMINO)IMIDAZOLIDINE
9 ★	1.52	7.4	821	2	"	2-(2,6-DICL-4-OH-PHENYLIMINO)IMIDAZOLIDINE

	logP	pH	Ref.	Note	MF	Name / **CAS** / *activity*
4450	3.16		401	335	C9H9Cl2N3O2	1-(2-CLET)-3-(P-CLPHENYL)-1-NITROSUUREA **13907-78-3**
1	-0.16	7.4	976	1	C9H9F2N3	2-(2,6-DIFLUOROPHENYLIMINO)IMIDAZOLIDINE **59772-33-7**
2	0.02	7.4	821	1	"	2-(2,6-DIFLUOROPHENYLIMINO)IMIDAZOLIDINE
3 ★	0.93	7.4	976	2	"	2-(2,6-DIFLUOROPHENYLIMINO)IMIDAZOLIDINE
4	1.06	7.4	821	2	"	2-(2,6-DIFLUOROPHENYLIMINO)IMIDAZOLIDINE
55 ★	3.31		1701		C9H9F3	PROPYLBENZENE,3,3,3-TRIFLUORO
6	1.99		2030		C9H9F3N2O1	NITROSOAMINE,N-METHYL-N-(3-TRIFLUOROMETHYLBENZYL)
7	2.68		2061		C9H9I1O2	BENZENEPROPIONIC ACID,3-IODO
8	2.32		2061		C9H9I1O2	PROPIONIC ACID,3-(2-IODOPHENYL)
9	2.81		2061		C9H9I1O2	PROPIONIC ACID,3-(4-IODOPHENYL) **1643-29-4**
4460 ★	2.72		572		C9H9N1	1-METHYLINDOLE **603-76-9**
1 ★	2.53	7.4	339	1	C9H9N1	2-METHYLINDOLE **95-20-5**
2 ★	1.27		579	400	C9H9N1	2-PROPYN-1-AMINE,N,N-DI-2-PROPYNYL **6921-29-5** ✚
3 ★	2.60		519		C9H9N1	3-METHYLINDOLE **83-34-1**
4 ★	2.68		519		C9H9N1	5-METHYLINDOLE **614-96-0**
65 ★	1.72		502		C9H9N1	BENZYLACETONITRILE **645-59-0**
6 ★	2.56		2221	459	C9H9N1	INDOLE-7-METHYL **933-67-5**
7	1.62		2363		C9H9N1	PHENYLACETONITRILE,4-METHYL
8 ★	2.74		579		C9H9N1O1	2,6-DIMETHYLBENZONITRILE-N-OXIDE **19111-74-1**
9 ★	1.21		1817		C9H9N1O1	3-PYRIDINEMETHANOL,A-2-PROPYNYL **89242-75-1**
4470 ★	2.06	7.4	339	1	C9H9N1O1	5-METHOXYINDOLE **1006-94-6**
1 ★	1.61		570		C9H9N1O1	B-PHENOXYPROPIONITRILE **3055-86-5**
2 ★	2.74		2077		C9H9N1O1	BENZONITRILE,N-OXIDE,2,6-DIMETHYL **19111-74-1**
3	1.70		2363		C9H9N1O1	BENZYLNITRILE,4-METHOXY
4 ★	1.43		572		C9H9N1O1	CINNAMAMIDE **621-79-4**
75 ★	1.95	7.0	1016	1	C9H9N1O1	N-FORMYL-STYRYLAMINE
6	1.94		2161	468	C9H9N1O2	AZETIDINE-2-ONE,4-PHENOXY
7 ★	0.94		453		C9H9N1O2	P-ACETYLFORMANILIDE **41656-75-1**
8	1.25		415	837	C9H9N1O2	P-FORMYLACETANILIDE **122-85-0**
9 ★	2.63		157		C9H9N1O2	STYRENE,2-METHYL,B-NITRO **34222-71-4**
4480 ★	2.66		306		C9H9N1O2	STYRENE,4-METHYL-B-NITRO **7559-36-6**
1 ★	2.52		520		C9H9N1O2	STYRENE,B-METHYL-B-NITRO **705-60-2**
2 ★	0.69		2157		C9H9N1O3	2-(BENZOYLOXY)ACETAMIDE
3 ★	1.32	2.0	2522	342	C9H9N1O3	BENZOIC ACID,3-ACETAMIDO
4	-3.93	12.2	2600		C9H9N1O3	G-(3-PYRIDYL)-G-OXOBUTYRIC ACID
85	-2.70	7.4	2600	1	"	G-(3-PYRIDYL)-G-OXOBUTYRIC ACID
6	-1.06	2.0	2600	342	"	G-(3-PYRIDYL)-G-OXOBUTYRIC ACID
7 ★	0.31	1.0	547	342	C9H9N1O3	GLYCINE,N-BENZOYL **495-69-2**
8 ★	0.92		273		C9H9N1O3	N-METHYL-3-FORMYLPHENYLCARBAMATE **54335-82-9**
9 ★	0.99		273		C9H9N1O3	N-METHYL-4-FORMYLPHENYLCARBAMATE **54335-83-0**
4490	-2.20	7.4	1616	1	C9H9N1O3	O-AMINOBENZOIC ACID,N-ACETYL **89-52-1**
1 ★	1.88		508		"	O-AMINOBENZOIC ACID,N-ACETYL
2 ★	0.27	7.0	925	1	C9H9N1O3	P-ACETOXYBENZAMIDE **51074-95-4**
3	0.56	7.2	415	314	C9H9N1O3	P-CARBOXYACETANILIDE **556-08-1** *immunostimulant*
4 ★	1.31		853		"	P-CARBOXYACETANILIDE
95 ★	2.37		306		C9H9N1O3	STYRENE,3-METHOXY,B-NITRO **3179-09-7**
6 ★	2.20		157		C9H9N1O3	STYRENE,4-METHOXY,B-NITRO **3179-10-0**
7 ★	0.80	3.5	2458	2	C9H9N1O3S1	THIAZOLIDINE-1,1,4-TRIONE-2-PHENYL
8 ★	1.73		1521		C9H9N1O4	ACETIC ACID,P-NITROBENZYL ESTER **619-90-9**
9 ★	-0.27		572		C9H9N1O4	BENZADOX **5251-93-4** *herbicide*
4500 ★	1.62		2336	346	C9H9N1O4	BENZOIC ACID,2-HYDROXY-4-ACETAMINO
1 ★	2.35	7.0	925	1	C9H9N1O4	ETHYL-M-NITRO BENZOATE **618-98-4**
2 ★	2.33	7.0	925	1	C9H9N1O4	ETHYL-P-NITRO BENZOATE **99-77-4**
3 ★	0.95		2626	537	C9H9N1O4	HIPPURIC ACID,2-HYDROXY
4 ★	1.88		520		C9H9N1O4	STYRENE,4-HYDROXY-3-METHOXY,B-NITRO **6178-42-3**
5 ★	2.10		1946		C9H9N1O5	2-PROPENOIC ACID,3-(5-NITRO-2-FURANYL),ETHYL ESTER **1874-12-0**
6	0.28		2626	537	C9H9N1O5	GENTISURIC ACID
7 ★	1.06	8.0	603	1	C9H9N1O7	2-NITRO-5-DI-(ACETOXYMETHYL)FURANE **92-55-7**
8 ★	3.46		850		C9H9N1S1	1-PHENETHYLISOTHIOCYANATE **4478-92-6**
9 ★	3.47		850		C9H9N1S1	2-PHENETHYLISOTHIOCYANATE **2257-09-2**
4510 ★	1.58		594		C9H9N1S1	AZETIDINE-2-ONE,4-PHENYLTHIO
1 ★	1.27		2606	424	C9H9N3	1,2,3-TRIAZOLE,1-BENZYL **4368-68-7**
2 ★	1.56		2606	424	C9H9N3	1,2,3-TRIAZOLE,2-BENZYL
3 ★	0.92		2665		C9H9N3	1,2,4-TRIAZOLE,1-BENZYL **6085-94-5** *fungicide*
4 ★	0.74		2606	424	C9H9N3	1,2,4-TRIAZOLE,4-BENZYL **16227-13-7**
15 ★	1.57		1519	459	C9H9N3	8-METHYL-4-CINNOLINAMINE
6 ★	0.95		1137		C9H9N3O1	2-AMINO-5-BENZYL-1,3,4-OXADIAZOLE
7 ★	0.50	7.4	1616	1	C9H9N3O1	BENZAMIDE,O-CYANOMETHYLAMINO
8 ★	0.95		580		C9H9N3O1	N-NITROSO-N-METHYL-(4-CYANOBENZYL)AMINE
9 ★	1.46		574		C9H9N3O1S1	1,3,4-TRIAZIN-2-ONE,1,3-DIHYDRO-5-BENZYLTHIO
4520	-0.63		559		C9H9N3O2	1-(2-CARBOXYETHYL)BENZOTRIAZOLE
1 ★	-0.03		559		C9H9N3O2	2-(2-CARBOXYETHYL)BENZOTRIAZOLE
2	-1.50	5.6	2156	1	C9H9N3O2	BENZOTRIAZOLE,1-(2-PROPANOIC ACID)
3	-1.16	5.6	2156	1	C9H9N3O2	BENZOTRIAZOLE,2-(2-PROPANOIC ACID)
4 ★	1.52		1641		C9H9N3O2	CARBENDAZIM **10605-21-0** *fungicide*
25 ★	0.87		547		C9H9N3O2S1	4-BENZENESULFONAMIDOPYRAZOLE
6	-0.43	7.5	652	1	C9H9N3O2S2	SULFATHIAZOLE **72-14-0** *antibacterial*
7 ★	0.05		508		"	SULFATHIAZOLE
8	0.68	7.7	1986	1	C9H9N3O3S2	BENZOTHIAZOLE,2-SULFONAMIDE-6-ACETAMIDO
9 ★	0.86	7.7	1986	2	"	BENZOTHIAZOLE,2-SULFONAMIDE-6-ACETAMIDO
4530	-2.70	7.4	1377	1	C9H9N3O6	N-(2,4-DINITROPHENYL)ALANINE **1655-52-3** ✚
1 ★	1.35	7.1	2155	2	C9H9N3S1	1,2,3,4-H4-ISOQUINOLINE,7,8-THIADIAZOLYL ✚
2 ★	2.65		574		C9H9N3S1	1H-1,2,3-TRIAZOLE,5-BENZYLTHIO
3 ★	1.45		574		C9H9N3S1	1H-1,3,4-TRIAZOLE,2-BENZYLTHIO
4 ★	1.36		576		C9H9N5	BENZOGUANAMINE **91-76-9**

	logP	pH	Ref.	Note	MF	Name / **CAS** / *activity*
35	3.35		506		C9H10	1-PROPENE,1-PHENYL **637-50-3**
6 ★	3.23		536		C9H10	ALLYLBENZENE **300-57-2**
7 ★	3.27		557		C9H10	CYCLOPROPYLBENZENE **873-49-4**
8 ★	3.18		547		C9H10	INDANE **496-11-7**
9 ★	3.48		579		C9H10	STYRENE,ALPHA-METHYL **98-83-9**
4540 ★	3.09		1263		C9H10Br1Cl1N2O2	MALORAN **13360-45-7** *herbicide*
1 ★	2.03		2410		C9H10Br1N1O1	ACETANILIDE,2-METHYL-4-BROMO
2 ★	2.80		2410		C9H10Br1N1O1	ACETANILIDE,3-METHYL-4-BROMO
3 ★	2.17		558		C9H10Br1N1O2	2-BROMOPHENYLDIMETHYLCARBAMATE **7305-04-6**
4 ★	2.02		2410		C9H10Br1N1O2	ACETANILIDE,2-METHOXY-4-BROMO
45	3.51		1818		C9H10Br1N1O2	N-METHYLBENZAMIDE,2-HYDROXY-3-METHYL-5-BROMO **40912-87-6**
6	-0.50	7.4	821	1	C9H10Br1N3	2-(4-BROMOPHENYLIMINO)IMIDAZOLIDINE
7 ★	1.75	7.4	821	2	"	2-(4-BROMOPHENYLIMINO)IMIDAZOLIDINE
8 ★	1.46		2105		C9H10Br1N3O2	HYDRAZINECARBOXAMIDE,1-ACETYL-N-(3-BROMOPHENYL)
9 ★	1.41		2105		C9H10Br1N3O2	HYDRAZINECARBOXAMIDE,1-ACETYL-N-(4-BROMOPHENYL) ✚
4550 ★	-0.28		2521		C9H10Cl1F1N2O4	5-CHLOROURACIL,1-(2,3-DIDEOXY-2-FLUOROPENTOFURANOSYL)
1 ★	1.22		2410		C9H10Cl1N1O1	ACETANILIDE,2-CHLORO-6-METHYL
2 ★	1.82		2410		C9H10Cl1N1O1	ACETANILIDE,2-METHYL-3-CHLORO
3 ★	1.85		2410		C9H10Cl1N1O1	ACETANILIDE,2-METHYL-5-CHLORO
4 ★	2.69		2410		C9H10Cl1N1O1	ACETANILIDE,3-CHLORO-4-METHYL
55 ★	1.81		2410		C9H10Cl1N1O2	3-CHLORO-4-METHOXYACETANILIDE **7073-42-9**
6 ★	2.27		558		C9H10Cl1N1O2	4-CHLOROPHENYLDIMETHYLCARBAMATE **7305-03-5**
7 ★	1.68		2410		C9H10Cl1N1O2	ACETANILIDE,2-METHOXY-3-CHLORO
8 ★	2.42		1969		C9H10Cl1N1O2	METHYLCARBAMATE,P-CHLOROBENZYL ESTER
9 ★	2.57		273		C9H10Cl1N1O2	N-METHYLCARBAMATE,3-METHYL-4-CHLOROPHENYL **2589-65-3**
4560 ★	-0.48		547		C9H10Cl1N1O2	P-CHLOROPHENYLALANINE,DL **7424-00-2** *serotonin inhibitor*
1 ★	2.04	7.4	1426	1	C9H10Cl1N1S1	THIAZOLIDINE,2-(M-CHLOROPHENYL) **60980-82-7**
2 ★	2.10	7.4	1426	1	C9H10Cl1N1S1	THIAZOLIDINE,2-(P-CHLOROPHENYL) **7738-99-0**
3	-0.67	7.4	976	1	C9H10Cl1N3	2-(2-CHLOROPHENYLIMINO)IMIDAZOLINE **4749-68-2**
4 ★	1.36	7.4	976	2	"	2-(2-CHLOROPHENYLIMINO)IMIDAZOLINE
65 ★	2.49		401	335	C9H10Cl1N3O2	1-(2-CHLOROETHYL)-1-NITROSO-3-PHENYLUREA **13206-67-2**
6 ★	1.28		2105		C9H10Cl1N3O2	HYDRAZINECARBOXAMIDE,1-ACETYL-N-(3-CHLOROPHENYL)
7 ★	1.21		2105		C9H10Cl1N3O2	HYDRAZINECARBOXAMIDE,1-ACETYL-N-(4-CHLOROPHENYL)
8 ★	-3.10		1979		C9H10Cl1N3O3	5'-CHLOROCYCLOCYTOSINE
9 ★	3.07		429		C9H10Cl2N2O1	1,1-DIMETHYL-3-(3,5-DICHLOROPHENYL)UREA **10290-38-7**
4570 ★	2.68		1263		C9H10Cl2N2O1	DIURON **330-54-1** *herbicide*
1 ★	3.20		1070		C9H10Cl2N2O2	LINURON **330-55-2** *herbicide*
2 ★	2.77		1213		C9H10F1N3O1	1-ME-1-ACETYL-3-(P-FLUOROPHENYL)TRIAZENE
3 ★	0.42		2105		C9H10F1N3O2	HYDRAZINECARBOXAMIDE,1-ACETYL-N-(4-FLUOROPHENYL)
4 ★	0.71		2105		C9H10F1N3O2	HYDRAZINECARBOXAMIDE,1-ACETYL-N-(3-FLUOROPHENYL)
75 ★	-1.89		547		C9H10F1O2	P-FLUOROPHENYLALANINE,DL **60-17-3**
6 ★	-0.72		2521		C9H10F2N2O4	5-FLUOROURACIL,1-(2,3-DIDEOXY-2-FLUOROPENTOFURANOSYL)
7 ★	3.78		538		C9H10F3N3	3,3-DIMETHYL-1-(3-TRIFLUOROMETHYLPHENYL)TRIAZENE **402-38-0**
8 ★	4.00	6.6	2590	459	C9H10F3N3	TRIFLUOROMETHYLBENZENE,4-(3,3-DIMETHYLTRIAZENYL) *antineoplastic*
9 ★	1.70		2105		C9H10I1N3O2	HYDRAZINECARBOXAMIDE,1-ACETYL-N-(3-IODOPHENYL)
4580 ★	1.75		2105		C9H10I1N3O2	HYDRAZINECARBOXAMIDE,1-ACETYL-N-(4-IODOPHENYL)
1 ★	2.75		2303		C9H10N1O3P1S1	CYAP **2636-26-2** *insecticide* ✚
2 ★	2.35		509		C9H10N2	BENZIMIDAZOLE,5,6-DIMETHYL **582-60-5**
3 ★	1.72	8.0	1625		C9H10N2	INDOLE-2-AMINE,N-METHYL **36092-88-3**
4 ★	2.20		594		C9H10N2	P-CYANO-N,N-DIMETHYLANILINE **1197-19-9**
85 ★	0.89		572		C9H10N2O1	1-PHENYL-3-PYRAZOLIDINONE **92-43-3**
6 ★	1.30	7.4	1616	1	C9H10N2O1	ANISOLE,O-CYANOMETHYLAMINO
7 ★	1.46		924	400	C9H10N2O1	BENZIMIDAZOL-2-ONE,1,3-DIMETHYL **3097-21-0**
8 ★	2.10		1625	457	C9H10N2O1	BENZOXAZOLE-2-AMINE,N,N-DIMETHYL **13858-89-4**
9	-2.69	2.0	2600	342	C9H10N2O1	DEMETHYLCOTININE
4590	-0.40	12.2	2600		"	DEMETHYLCOTININE
1	-0.30	7.4	2600	1	"	DEMETHYLCOTININE
2 ★	0.87		603		C9H10N2O2	1-PHENYLACETYLUREA **63-98-9** *anticonvulsant*
3 ★	1.49		1624		C9H10N2O2	BENZALDEHYDE,O-(METHYLAMINO)CARBONYL OXIME **2426-12-2**
4 ★	0.32	7.4	1616	1	C9H10N2O2	BENZAMIDE,O-ACETYLAMINO **33809-77-7**
95	-1.01	12.2	2600		"	DEMETHYLCOTININE-N-OXIDE
6	-1.00	7.4	2600	1	"	DEMETHYLCOTININE-N-OXIDE
7	-0.99	2.0	2600	342	"	DEMETHYLCOTININE-N-OXIDE
8 ★	1.27		2198		C9H10N2O2	METHYLUREA,N'-BENZOYL
9 ★	0.01		547		C9H10N2O2	P-N-ACETYLAMINOBENZAMIDE **58202-83-8**
4600 ★	1.60		2198		C9H10N2O2	PHENYLUREA,N'-ACETYL **102-03-4**
1 ★	0.52		522		C9H10N2O2S1	1,2,4-BENZOTHIADIAZINE,1,1-DIOXIDE-3,5-DIMETHYL **31365-88-5**
2 ★	0.74		522		C9H10N2O2S1	1,2,4-BENZOTHIADIAZINE,1,1-DIOXIDE-3,6-DIMETHYL **31363-88-9**
3 ★	0.81		522		C9H10N2O2S1	1,2,4-BENZOTHIADIAZINE-1,1-DIOXIDE-3,7-DIMETHYL **31363-89-0**
4 ★	1.59	7.4	1426	1	C9H10N2O2S1	THIAZOLIDINE,2-(M-NITROPHENYL)
5 ★	1.68	7.4	1426	1	C9H10N2O2S1	THIAZOLIDINE,2-(P-NITROPHENYL) **831-25-4**
6 ★	0.61		2410		C9H10N2O3	ACETANILIDE,2-METHYL-6-NITRO
7 ★	1.49		2410		C9H10N2O3	ACETANILIDE,2-NITRO-4-METHYL
8	-2.25	7.4	1616	1	C9H10N2O3	BENZOIC ACID,O-CARBOXAMIDOMETHYLAMINO
9	0.95	7.4	1616	2	"	BENZOIC ACID,O-CARBOXAMIDOMETHYLAMINO
4610 ★	2.53		2097		C9H10N2O3	DIAZINE OXIDE,1-PHENYL-2-ETHOXYCARBONYL **56751-20-3**
1 ★	0.91	7.4	2327	1	C9H10N2O3	P-NITROACETANILIDE,N-METHYL
2 ★	0.74		522		C9H10N2O3S1	1,2,4-BENZOTHIADIAZINE-1,1-DIOXIDE-3-METHYL-6-METHOXY **6451-55-4**
3 ★	2.01	7.2	1685	314	C9H10N2O3S2	ETHOXYZOLAMIDE **452-35-7** *diuretic*
4 ★	0.82		1946		C9H10N2O4	2-PROPENOIC N,N-DIMETHYLAMIDE,3-(5-NITRO-2-FURANYL)
15 ★	1.39		1946		C9H10N2O4	2-PROPENOIC N-ETHYLAMIDE,3-(5-NITRO-2-FURANYL)
6 ★	1.50		558		C9H10N2O4	4-NITROPHENYLDIMETHYLCARBAMATE **7244-70-4**
7 ★	1.09		2410		C9H10N2O4	ACETANILIDE,2-NITRO-4-METHOXY
8 ★	1.35		558		C9H10N2O4	O-NITROPHENYLDIMETHYLCARBAMATE **3373-86-2**
9	-1.34	7.2	2323	468	C9H10N2O4	P-NITROPHENYLALANINE **2922-40-9**

	logP	pH	Ref.	Note	MF	Name / CAS / activity
4620 ★	-1.25		547		"	P-NITROPHENYLALANINE
1 ★	0.26		501		C9H10N2O4	PHENOXYACETIC ACID,3-UREIDO
2 ★	-1.07		2252		C9H10N2O4	URIDINE,2',3'-DEHYDRO-2'-3'-DEOXY 5974-93-6
3	0.38	7.7	1986	1	C9H10N2O4S2	BENZOTHIAZOLE,2-SULFONAMIDE-6-HYDROXYETHOXY
4 ★	0.59	7.7	1986	2	"	BENZOTHIAZOLE,2-SULFONAMIDE-6-HYDROXYETHOXY
25	-2.85		1979		C9H10N2O5	2,2'-CYCLOURIDINE ✚
6 ★	3.10		2215		C9H10N2O5	PHENOL,2-I-PROPYL-4,6-DINITRO
7 ★	3.96		1947		C9H10N2S1	DIMETHYLANILINE,4-ISOTHIOCYANATO 2131-64-8
8 ★	2.39		563		C9H10N4	3,3-DIMETHYL-1-(4-CYANOPHENYL)TRIAZENE 23456-95-3
9 ★	0.27		559		C9H10N4O1	1-(2-CARBAMYLETHYL)BENZOTRIAZOLE
4630 ★	0.66		559		C9H10N4O1	2-(2-CARBAMYLETHYL)BENZOTRIAZOLE
1 ★	0.35	5.6	2156	1	C9H10N4O1	BENZOTRIAZOLE,1-(2-PROPIONAMIDE)
2 ★	0.86	5.6	2165	1	C9H10N4O1	BENZOTRIAZOLE,2-(2-PROPIONAMIDE)
3 ★	2.30		2097			DIAZINECARBONITRILE,(4-DIMETHYLAMINOPHENYL),OXIDE
4 ★	2.21	7.0	1621	1	C9H10N4O1	PYRIMIDINE,4-(3-MEO-5-ME-PYRAZOL-1-YL) 23903-40-4
35 ★	0.50	7.0	1621	1	C9H10N4O1	PYRIMIDINE,4-(5-MEO-3-ME-1H-PYRAZOL-1-YL) 18694-41-2
6 ★	0.56	7.4	2360	1	C9H10N4O2	BENZOTRIAZINE,1,4-DI-N-OXIDE-3-AMINO-6,7-DIMETHYL
7 ★	0.99	7.0	1621	1	C9H10N4O2	PYRAZOL-3-ONE,1,2-H2-2-(6-MEO-4-PYRIMIDINYL)-5-ME 23917-23-9 ✚
8	-1.11	7.5	652	1	C9H10N4O2S2	SULFAMETHIZOLE 144-82-1 antibacterial
9 ★	0.54		508		"	SULFAMETHIZOLE
4640 ★	3.05		2128		C9H10N4O4	ACETONE,2,4-DINITROPHENYLHYDRAZONE 1567-89-1
1 ★	0.21		2070		C9H10N4O4	ALLOPURINOL,1-ETHOXYCARBONYLOXYMETHYL ✚
2 ★	0.62		2105		C9H10N4O4	HYDRAZINECARBOXAMIDE,1-ACETYL-N-(3-NITROPHENYL)
3 ★	0.77		2105		C9H10N4O4	HYDRAZINECARBOXAMIDE,1-ACETYL-N-(4-NITROPHENYL)
4 ★	1.63		574		C9H10N4S1	1-METHYL-5-THIOBENZYLTETRAZOLE
45 ★	1.44		502		C9H10O1	2-PROPANONE,1-PHENYL 103-79-7
6 ★	2.94		505		C9H10O1	ALLYLPHENYLETHER 1746-13-0
7 ★	1.95		508		C9H10O1	CINNAMYL ALCOHOL 104-54-1
8 ★	2.10		1093		C9H10O1	P-METHYLACETOPHENONE 122-00-9
9 ★	2.19		562		C9H10O1	PROPIOPHENONE 93-55-0
4650 ★	1.80		2452		C9H10O2	2,3,5-TRIMETHYL-1,4-BENZOQUINONE 935-92-2
1 ★	1.04		594		C9H10O2	2-PROPANONE-1-PHENOXY 621-87-4
2 ★	2.23	1.0	782	342	C9H10O2	3-METHYL-4-METHOXYBENZALDEHYDE 32723-67-4
3	1.80	3.5	1542	1	C9H10O2	A-PHENYLPROPIONIC ACID 492-37-5
4 ★	1.96		502		C9H10O2	ACETIC ACID,BENZYL ESTER 140-11-4
55 ★	1.82		594	400	C9H10O2	ACETOPHENONE,2-METHOXY 579-74-8
6 ★	1.84		502		C9H10O2	B-PHENYLPROPIONIC ACID 501-52-0
7 ★	2.21		2645	225	C9H10O2	BENZOIC ACID,2,6-DIMETHYL 632-46-2
8 ★	2.89	2.0	2236		C9H10O2	BENZOIC ACID-4-ETHYL 619-64-7
9 ★	2.64		541		C9H10O2	ETHYL BENZOATE 93-89-0
4660 ★	2.09		853		C9H10O2	M-CRESYL ACETATE 122-46-3 antiseptic (topical), antifungal
1 ★	1.84	7.0	925	1	C9H10O2	M-METHOXYACETOPHENONE 586-37-8
2 ★	1.95		501		C9H10O2	M-METHYLPHENYLACETIC ACID 621-36-3
3 ★	2.77		2645			M-TOLUIC ACID,METHYL ESTER
4 ★	2.70		2645			METHYL-P-METHYL BENZOATE 99-75-2
65 ★	2.54		825		C9H10O2	O-HYDROXYPROPIOPHENONE 610-99-1
6 ★	2.75		547		C9H10O2	O-TOLUIC ACID,METHYL ESTER 89-71-4
7 ★	1.93		558		C9H10O2	O-TOLYLACETATE 533-18-6
8 ★	2.03		825		C9H10O2	P-HYDROXYPROPIOPHENONE 70-70-2 inhibitor (pituitary)
9 ★	1.74		547		C9H10O2	P-METHOXYACETOPHENONE 100-06-1
4670 ★	1.86		501		C9H10O2	P-METHYLPHENYLACETIC ACID 622-47-9
1 ★	2.11		853		C9H10O2	P-TOLYLACETATE 140-39-6
2 ★	1.83		502		C9H10O2	PHENYLACETIC ACID,METHYL ESTER 101-41-7
3 ★	2.10		547		C9H10O2S1	ACETIC ACID,ALPHA-BENZYLTHIO 103-46-8
4 ★	1.12		1895		C9H10O3	1,2-ETHANEDIOL, MONOBENZOATE 94-33-7
75 ★	1.61		2645		C9H10O3	3-METHOXYPHENYLACETATE 5451-83-2
6 ★	1.54		558		C9H10O3	4-METHOXYPHENYLACETATE 1200-06-2
7 ★	1.98	2.2	2018		C9H10O3	ACETOPHENONE,2-HYDROXY-4-METHOXY 352-41-0
8 ★	2.04		2645		C9H10O3	BENZOIC ACID,O-METHOXY,METHYL ESTER
9 ★	2.95	1.2	2586		C9H10O3	ETHYL SALICYLATE
4680 ★	1.50		501		C9H10O3	M-METHOXYPHENYLACETIC ACID 1798-09-0
1 ★	1.78		501		C9H10O3	M-METHYLPHENOXYACETIC ACID 1643-15-8
2 ★	1.38		558		C9H10O3	O-METHOXYPHENYLACETATE 613-70-7
3 ★	1.98	1.0	547	342	C9H10O3	O-METHYLPHENOXYACETIC ACID 1878-49-5
4 ★	2.27	7.0	925	1	C9H10O3	P-ANISIC ACID,METHYL ESTER 121-98-2
85 ★	2.39	1.2	1077		C9H10O3	P-ETHOXYBENZOIC ACID 619-86-3
6 ★	1.16	1.2	1077		C9H10O3	P-HYDROXY-B-PHENYLPROPIONIC ACID 501-97-3
7	2.44	7.5	1259	1	C9H10O3	P-HYDROXYBENZOIC ACID,ETHYL ESTER 120-47-8 pharmaceutic aid
8 ★	2.47		508		"	P-HYDROXYBENZOIC ACID,ETHYL ESTER
9 ★	1.42		501		C9H10O3	P-METHOXYPHENYLACETIC ACID 104-01-8
4690 ★	1.86	1.0	530	342	C9H10O3	P-METHYLPHENOXYACETIC ACID 940-64-7
1 ★	1.41	1.0	782	342	C9H10O3	PHENOXYACETIC ACID,METHYL ESTER 2065-23-8
2 ★	0.77	1.0	547	342	C9H10O3	TROPIC ACID 529-64-6
3 ★	1.90		501		C9H10O3S1	PHENOXYACETIC ACID,M-METHYLTHIO
4 ★	0.66		1586		C9H10O4	2,6-DIMETHOXYBENZOIC ACID 1466-76-8
95 ★	1.61		1586		C9H10O4	3,4-DIMETHOXYBENZOIC ACID 93-07-2
6	1.97	7.4	2539	2	"	3,4-DIMETHOXYBENZOIC ACID
7 ★	2.19		1586		C9H10O4	3,5-DIMETHOXYBENZOIC ACID 1132-21-4
8	-2.75	7.4	1168	1	C9H10O4	3-METHOXY-4-HYDROXYPHENYLACETIC ACID 306-08-1
9	1.18	2.0	2312	337	"	3-METHOXY-4-HYDROXYPHENYLACETIC ACID
4700 ★	1.38		501		C9H10O4	PHENOXYACETIC ACID,M-METHOXY 2088-24-6
1 ★	0.98	1.0	547	342	C9H10O4	PHENOXYACETIC ACID,O-METHOXY 1878-85-9
2 ★	1.23		501		C9H10O4	PHENOXYACETIC ACID,P-METHOXY 1877-75-4
3 ★	0.78		2452		C9H10O4	UBIQUINONE-0 605-94-7
4 ★	0.06		501		C9H10O4S1	M-METHYLSULFONYLPHENYLACETIC ACID

	logP	pH	Ref.	Note	MF	Name / CAS / activity
5	-4.61	7.4	1168	1	C9H10O5	3-METHOXY-4-HYDROXYMANDELIC ACID 55-10-7
6 ★	1.04		1586		C9H10O5	BENZOIC ACID,4-HYDROXY-3,5-DIMETHOXY 530-57-4
7 ★	0.22		1963		C9H10O5	BENZOQUINONE,2,3,5-TRIMETHOXY 3117-05-3
8 ★	1.30		1335		C9H10O5	ETHYLGALLATE 831-61-8
9 ★	0.06		2393	345	C9H10O5S1	PHENOXYACETIC ACID,4-MESYL
4710 ★	0.01		501		C9H10O5S1	PHENOXYACETIC ACID,M-METHYLSULFONYL
1 ★	3.51		579		C9H10S1	ALLYLPHENYL SULFIDE 5296-64-0
2 ★	3.72		502		C9H11Br1	PROPYLBROMIDE,G-PHENYL 637-59-2
3 ★	2.19		429		C9H11Br1N2O1	1,1-DIMETHYL-3-P-BROMOPHENYLUREA 20940-43-6
4 ★	2.38		1263		C9H11Br1N2O2	METOBROMURON 3060-89-7 herbicide
15 ★	-0.29		401		C9H11Br1N2O5	2'-DEOXY-5-BROMOURIDINE 59-14-3 antineoplastic
6 ★	3.55		502		C9H11Cl1	3-PHENYL-1-CHLOROPROPANE 104-52-9
7 ★	-0.46		2521		C9H11Cl1F1N3O3	5-CHLOROCYTOSINE,1-(2,3-DIDEOXY-2-FLUOROPENTOFURANOSYL)
8 ★	2.00		429		C9H11Cl1N2O1	1,1-DIMETHYL-3-M-CHLOROPHENYLUREA 587-34-8
9 ★	1.94		429		C9H11Cl1N2O1	1,1-DIMETHYL-3-P-CHLOROPHENYLUREA 150-68-5 herbicide
4720 ★	2.61		1263		C9H11Cl1N2O1	1-METHYL-3-(3-CL-4-METHYLPHENYL)UREA 22175-22-0
1 ★	1.76		1263		C9H11Cl1N2O2	1-METHYL-3-(3-CL-4-METHOXYPHENYL)UREA 20782-57-4
2 ★	1.26	7.4	2026	1	C9H11Cl1N2O2	BENZAMIDE,2-(2-HYDROXYETHYL)AMINO-5-CHLORO
3 ★	2.30		1263		C9H11Cl1N2O2	MONOLINURON 1746-81-2 herbicide +
4 ★	1.79	7.4	2004	1	C9H11Cl1N2O3S1	N2-ETHYL-N1-P-CHLOROBENZENESULFONYLUREA 24535-70-4
25 ★	2.19	9.9	704	821	C9H11Cl1N4O1	4-CHLORO-3-METHYLPHENYLAMIDINOUREA 56189-70-9
6	1.42	6.5	321	2	C9H11Cl2N3O4S2	2-METHYL-3-CHLOROMETHYLHYDROCHLOROTHIAZIDE 135-07-9 diuretic, antihypertensive
7 ★	3.18		354		C9H11Cl3N1O3P1	ETHYLPHOSPHORAMIDATE,O-ME-O-(2,4,5-TRICL-PHENYL) 2213-84-5
8 ★	5.27		2293	400	C9H11Cl3N1O3P1S1	CHLORPYRIFOS 2921-88-2 insecticide
9 ★	2.95		502		C9H11F1	PROPYLFLUORIDE,G-PHENYL
4730 ★	1.37		1263		C9H11F1N2O1	1,1-DIMETHYL-3-M-FLUOROPHENYLUREA 330-39-2
1 ★	1.13		1263		C9H11F1N2O1	1,1-DIMETHYL-3-P-FLUOROPHENYLUREA 332-33-2
2 ★	1.68		2612		C9H11F1N2O2	1-(3'-FLUOROPHENYL)-3-METHOXY-3-METHYLUREA 28170-26-5
3 ★	1.46		2612		C9H11F1N2O2	1-(4-FLUOROPHENYL)-3-METHOXY-3-METHYLUREA 88132-24-5
4	0.70	7.4	2461	1	C9H11F1N2O2S3	THIENO[2,3-B]THIOPHENE-2-CARBOXAMIDE,5-(2-FLUOROETHYL)AMINOMETHYL 122267-00-9
35	0.34	7.4	2525		C9H11F1N2O3S2	5-(2-FLUOROETHYLAMINO)METHYLTHIENO[2,3-B]FURAN-2-SULFONAMIDE
6	0.06	7.4	2266	1	C9H11F1N2O4	5-FLUOROURACIL-1-BUTYRYLOXYMETHYL 66542-37-8
7 ★	0.47	4.0	2266	302	"	5-FLUOROURACIL-1-BUTYRYLOXYMETHYL
8 ★	-0.87		2521		C9H11F1N2O4	URACIL,1-(2,3-DIDEOXY-2-FLUOROFURANOSYL)
9	0.25	7.4	2266	1	C9H11F1N2O4	URACIL,1-(I-BUTOXY)CARBONYL-5-FLUORO 71759-45-0
4740 ★	0.87	4.0	2010		"	URACIL,1-(I-BUTOXY)CARBONYL-5-FLUORO
1	0.19	7.4	2266	1	C9H11F1N2O4	URACIL,1-BUTOXYCARBONYL-5-FLUORO 85326-32-5
2 ★	0.89	4.0	2010		"	URACIL,1-BUTOXYCARBONYL-5-FLUORO
3 ★	-0.49		2252		C9H11F1N2O4	URIDINE,2',3'-DIDEOXY-3'-FLUORO
4	-1.47	7.4	401	1	C9H11F1N2O5	2'-DEOXY-5-FLUOROURIDINE 50-91-9 antiviral, antineoplastic
45 ★	-1.16	7.4	401	2	C9H11F1N2O5	2'-DEOXY-5-FLUOROURIDINE
6 ★	-0.92		2521		C9H11F2N3O3	5-FLUOROCYTOSINE,1-(2,3-DIDEOXY-2-FLUOROPENTOFURANOSYL)
7 ★	-0.95		2567		C9H11F2N5O4	FNI-1 +
8 ★	3.90		508		C9H11I1	PROPYL IODIDE,G-PHENYL 4119-41-9
9 ★	2.29		324		C9H11N1	1,2,3,4-TETRAHYDROQUINOLINE 635-46-1
4750 ★	1.49	13.0	1806	224	C9H11N1	CYCLOPROPANEAMINE-2-PHENYL(CIS) 69684-88-4
1 ★	1.58	13.0	1806	224	C9H11N1	CYCLOPROPANEAMINE-2-PHENYL(TRANS) 155-09-9 antidepressant (inhibitor-mao)
2 ★	1.68		1894		C9H11N1O1	3-METHYLACETANILIDE 537-92-8
3 ★	0.90		2410		C9H11N1O1	BENZAMIDE,2,6-DIMETHYL
4 ★	1.17		2410		C9H11N1O1	BENZAMIDE,2-ETHYL
55 ★	0.62		541		C9H11N1O1	N,N-DIMETHYLBENZAMIDE 611-74-5
6 ★	0.95		579		C9H11N1O1	N-BENZYLACETAMIDE 588-46-5
7	0.97	7.2	415	314	C9H11N1O1	N-METHYLACETANILIDE 579-10-2
8 ★	1.12	7.0	538	701	"	N-METHYLACETANILIDE
9 ★	0.86		2410		C9H11N1O1	O-ACETAMIDOTOLUENE 120-66-1
4760 ★	1.81	7.0	925	1	C9H11N1O1	P-FORMYL-N,N-DIMETHYLANILINE 100-10-7
1	1.40	7.2	415	314	C9H11N1O1	P-METHYLACETANILIDE 103-89-9
2 ★	1.70		1894		"	P-METHYLACETANILIDE
3 ★	0.91		502		C9H11N1O1	PROPIONAMIDE,G-PHENYL 102-93-2
4 ★	1.61		541		C9H11N1O1	PROPIONANILIDE 620-71-3
65 ★	1.56	7.4	447	1	C9H11N1O1	W-PHENYLPROPIONALDEHYDEOXIME 1197-50-8
6 ★	0.17		803		C9H11N1O2	2-METHYL-4-HYDROXYACETANILIDE 39495-15-3
7 ★	1.69		2142		C9H11N1O2	2-METHYLPHENYLCARBAMATE,O-METHYL
8	0.22	2.0	2522	342	C9H11N1O2	3-AMINOBENZOIC ACID,N,N-DIMETHYL 99-64-9
9 ★	0.79		803		C9H11N1O2	3-METHYL-4-HYDROXYACETANILIDE 16375-90-9
4770 ★	2.22		2142		C9H11N1O2	3-METHYLPHENYLCARBAMATE,O-METHYL
1 ★	2.27		2142		C9H11N1O2	4-METHYLPHENYLCARBAMATE,O-METHYL
2 ★	1.28		1894		C9H11N1O2	ACETANILIDE,3-METHOXY 588-16-9
3	1.03	7.2	415	314	C9H11N1O2	ACETANILIDE,P-METHOXY 51-66-1
4 ★	1.05		1894		"	ACETANILIDE,P-METHOXY
75 ★	1.76	8.2	2018		C9H11N1O2	ACETOPHENONE,2-AMINO-4-METHOXY 42465-53-2
6 ★	1.50		1969		C9H11N1O2	CARBAMIC ACID,PHENETHYL ESTER 6326-19-8
7 ★	2.30		503		C9H11N1O2	ETHYLCARBAMATE,N-PHENYL 101-99-5
8 ★	1.56		558		C9H11N1O2	N,N-DIMETHYLPHENYLCARBAMATE 6969-90-0
9 ★	1.69		702		"	N,N-DIMETHYLPHENYLCARBAMATE
4780 ★	0.55		564		C9H11N1O2	N-BENZYLGLYCOLAMIDE 19340-77-3
1 ★	0.61	7.2	415	314	C9H11N1O2	N-ME-P-HYDROXYACETANILIDE 579-58-8
2 ★	1.70		273		C9H11N1O2	N-METHYL-M-TOLYLCARBAMATE 1129-41-5 pesticide +
3 ★	1.46		273		C9H11N1O2	N-METHYL-O-TOLYLCARBAMATE 1128-78-5
4 ★	1.66		273		C9H11N1O2	N-METHYL-P-TOLYLCARBAMATE 1129-48-2
85 ★	1.68		1969		C9H11N1O2	N-METHYLCARBAMATE,BENZYL ESTER
6 ★	1.02		547		C9H11N1O2	N-METHYLPHENOXYACETAMIDE 15422-25-0
7 ★	1.94	7.0	332	1	C9H11N1O2	NICOTINIC ACID,PROPYL ESTER 7681-15-4
8 ★	3.34		2410		C9H11N1O2	NITROBENZENE,2-I-PROPYL
9 ★	3.45	7.4	594	314	C9H11N1O2	NITROBENZENE,4-I-PROPYL 1817-47-6

	logP	pH	Ref.	Note	MF	Name / CAS / activity
4790 ★	3.45	7.4	594	314	C9H11N1O2	NITROBENZENE,4-PROPYL 10342-59-3
1 ★	2.57		525		C9H11N1O2	O-AMINOBENZOIC ACID,ETHYL ESTER 87-25-2
2	0.67		2410		C9H11N1O2	O-METHOXYACETANILIDE 93-26-5
3	0.94		1894		"	O-METHOXYACETANILIDE
4	0.98		260		"	O-METHOXYACETANILIDE
95	0.98	7.4	1616	1	"	O-METHOXYACETANILIDE
6 ★	1.86		577		C9H11N1O2	P-AMINOBENZOIC ACID,ETHYL ESTER 94-09-7 anesthetic (topical)
7	1.97	7.5	1259	1	"	P-AMINOBENZOIC ACID,ETHYL ESTER
8 ★	1.30		541		C9H11N1O2	P-ETHOXYBENZAMIDE
9 ★	1.76		1969		C9H11N1O2	P-METHYLABENZYLCARBAMATE (ESTER) 54850-42-9
4800 ★	-1.52	7.0	547	463	C9H11N1O2	PHENYLALANINE 63-91-2 amino acid
1	-1.46	7.0	2197		"	PHENYLALANINE
2 ★	-1.40	5.0	1953	1	C9H11N1O2	PROPANOIC ACID,3-AMINO-3-PHENYL 614-19-7
3 ★	0.80		594		C9H11N1O2	PROPIONAMIDE,3-PHENOXY
4 ★	0.44		599		C9H11N1O2	SALICYLAMIDE,N,N-DIMETHYL
5 ★	1.51		536		C9H11N1O2S1	2-METHYLTHIOPHENYLCARBAMATE,N-METHYL
6 ★	1.92		273		C9H11N1O2S1	N-METHYL-4-METHYLTHIOPHENYLCARBAMATE 3938-34-9
7 ★	0.99		866		C9H11N1O3	2-ETHOXY-4-AMINOBENZOIC ACID
8 ★	1.84		2142		C9H11N1O3	2-METHOXYPHENYLCARBAMATE,O-METHYL
9 ★	1.92		2142		C9H11N1O3	3-METHOXYPHENYLCARBAMATE,O-METHYL
4810 ★	1.69		2142		C9H11N1O3	4-METHOXYPHENYLCARBAMATE,O-METHYL
1 ★	-0.22		2410		C9H11N1O3	BENZAMIDE,2,6-DIMETHOXY
2	-1.05	7.4	1616	1	C9H11N1O3	BENZOIC ACID,O-(2-HYDROXYETHYL)AMINO 25784-00-3
3 ★	1.35	7.4	1616	2	"	BENZOIC ACID,O-(2-HYDROXYETHYL)AMINO
4	-3.65	12.2	2600		C9H11N1O3	G-(3-PYRIDYL)-G-HYDROXYBUTYRIC ACID
15	-2.66	7.4	2600	1	"	G-(3-PYRIDYL)-G-HYDROXYBUTYRIC ACID
6	-2.13	2.0	2600	342	"	G-(3-PYRIDYL)-G-HYDROXYBUTYRIC ACID
7 ★	0.81		273		C9H11N1O3	N-METHYL-2-METHOXYPHENYLCARBAMATE 3938-24-7
8 ★	1.30		273		C9H11N1O3	N-METHYL-3-METHOXYPHENYLCARBAMATE 3938-28-1
9 ★	1.20		273		C9H11N1O3	N-METHYL-4-METHOXYPHENYLCARBAMATE 3938-29-2
4820 ★	2.80	7.4	820	1	C9H11N1O3	P-NITROPHENYL-I-PROPYLETHER 26455-31-2
1 ★	-2.26		508		C9H11N1O3	TYROSINE 60-18-4 amino acid
2	-2.05	7.4	1168	1	"	TYROSINE
3 ★	0.58		1894		C9H11N1O3S1	ACETANILIDE,4-METHYLSULFONYL 22821-80-3
4 ★	0.29		547		C9H11N1O3S1	N-ACETYL-METHANESULFONANILIDE
25 ★	-2.74	7.2	2323	468	C9H11N1O4	DOPA 63-84-3
6	-2.39	7.4	1168	1	"	DOPA
7 ★	0.65		2209		C9H11N1O4	MALEIMIDE,2,3-DIMETHYL,N-ACETYLOXYMETHYL
8 ★	0.41		731		C9H11N1O4	N-MALEOYL-3-AMINOPROPIONIC ACID,ETHYL ESTER 57079-05-1
9 ★	0.60		731	604	C9H11N1O4	N-MALEOYL-5-AMINOPENTANOIC ACID 57078-99-6
4830	1.17		773	601	C9H11N1O4S1	4-SULFAMYLBENZOIC ACID,ETHYL ESTER 5446-77-5
1 ★	0.34		273	60	C9H11N1O4S1	N-METHYL-4-METHYLSULFONYLPHENYLCARBAMATE
2	-2.00	7.4	2490	1	C9H11N1O5S2	BENZENESULFONAMIDE,3-CARBOXY-4-HYDROXYETHYLTHIO 108966-65-0
3 ★	1.53		1894		C9H11N1S1	ACETANILIDE,2-METHIO 6310-41-4
4 ★	2.14		1894		C9H11N1S1	ACETANILIDE,3-METHIO 2524-78-9
35 ★	2.03		1894		C9H11N1S1	ACETANILIDE,4-METHIO 10352-44-0
6 ★	1.98		559		C9H11N3	1-I-PROPYLBENZOTRIAZOLE
7 ★	2.13		559		C9H11N3	1-PROPYLBENZOTRIAZOLE 16584-02-4
8 ★	2.53		559		C9H11N3	2-I-PROPYLBENZOTRIAZOLE
9	-1.92	7.4	976	1	C9H11N3	2-PHENYLIMIDAZOLIDINE
4840	-1.80	7.4	821	1	"	2-PHENYLIMIDAZOLIDINE
1 ★	1.01	7.4	976	2	"	2-PHENYLIMIDAZOLIDINE
2	1.20	7.4	821	2	"	2-PHENYLIMIDAZOLIDINE
3 ★	2.53		559		C9H11N3	2-PROPYLBENZOTRIAZOLE
4 ★	1.47		594		C9H11N3	BENZIMIDAZOLE-2-AMINE,1,5-DIMETHYL
45 ★	0.91		559		C9H11N3O1	1-(3-HYDROXYPROPYL)BENZOTRIAZOLE
6 ★	1.40		559		C9H11N3O1	2-(3-HYDROXYPROPYL)BENZOTRIAZOLE
7 ★	0.87		595		C9H11N3O1	BENZIMIDAZOLE,1-METHYL-2-AMINO-5-METHOXY
8 ★	1.58	8.9	1498		C9H11N3O2	1,3-DIMETHYL-1-NITROSO-3-PHENYLUREA 72586-68-6
9 ★	2.25	8.9	1498		C9H11N3O2	1-ETHYL-1-NITROSO-3-PHENYLUREA 54680-35-2
4850	0.72	4.0	1732		C9H11N3O2	1-METHYL-1-NITROSO-3-(P-METHYLPHENYL)UREA 23139-00-6
1	-3.00	7.4	1398	1	C9H11N3O2	2(2,6-DIHYDROXYPHENYLIMINO)IMIDAZOLINE
2	-1.60	7.4	1398	1	C9H11N3O2	2-(3,4-DIHYDROXYPHENYLIMINO)IMIDAZOLIDINE
3 ★	2.03		538		C9H11N3O2	3,3-DIMETHYL-1-(3-CARBOXYPHENYL)TRIAZENE
4	-0.44		1214		C9H11N3O2	3,3-DIMETHYL-1-(4-CARBOXYPHENYL)TRIAZENE antineoplastic
55	-0.43	3.3	2590	331	"	3,3-DIMETHYL-1-(4-CARBOXYPHENYL)TRIAZENE
6	-0.28	3.3	2590	331	C9H11N3O2	BENZOIC ACID,2-(3,3-DIMETHYLTRIAZENYL) antineoplastic
7 ★	0.25		2105		C9H11N3O2	HYDRAZINECARBOXAMIDE,1-ACETYL-N-PHENYL
8 ★	0.67		2397		C9H11N3O2	IMIDAZOLE,1-BUTYN-3YL-2-HYDROXYIMINOMETHYL
9 ★	-0.38		1439		C9H11N3O2	PYRIDO(1,2A)PYRIMIDIN-4-ONE,3-CONH2,4,6,7,8,9-H5 ✚
4860 ★	1.46		2612		C9H11N3O3	1,1-DIMETHYL-3-(M-NITROPHENYL)UREA 7159-98-0
1 ★	1.51		429		C9H11N3O3	1,1-DIMETHYL-3-P-NITROPHENYLUREA 7159-97-9
2 ★	-1.57	7.0	2440	1	C9H11N3O3	CYTIDINE,2',3'-DIDEHYDRO-2',3'-DIDEOXY antiviral (hiv)
3 ★	-0.88		2105		C9H11N3O3	HYDRAZINECARBOXAMIDE,1-ACETYL-N-(4-HYDROXYPHENYL)
4	0.64	4.0	1732		C9H11N3O3	N-ME-N-NITROSO-N'-(P-METHOXYPHENYL)-UREA 25355-59-3
65 ★	1.76		611		C9H11N3O3S1	N-I-PROPYL-3-(5-NO2-2-THIAZOLYL)ACRYLAMIDE 53207-64-0
6 ★	-0.22		1564		C9H11N3O3S1	PYRIMIDINE-2,4-DIONE,3-PROPOXY-5-SCN-6-METHYL 77317-86-3
7 ★	1.64		2612		C9H11N3O4	1-(3'-NITROPHENYL)-3-MEO-3-METHYLUREA 88132-19-8
8 ★	1.74		2612		C9H11N3O4	1-METHOXY-1-METHYL-3-(P-NITROPHENYL)UREA 88132-15-4
9 ★	-1.64		1979		C9H11N3O4	2',5'-ANHYDRO-ARA-C
4870 ★	-2.37	11.0	401	322	C9H11N3O4	ANCITABINE 31698-14-3 antineoplastic
1 ★	2.36		595		C9H11N5	INDAZOLE,5-(3,3-DIMETHYL)TRIAZENYL
2 ★	-0.41		2544		C9H11N5O1	9-ETHYL-N-ACETYLADENINE
3	-0.57	7.4	2037	1	C9H11N5O1	ISOGUANINE,1-METHYL-9-ALLYL
4	-0.37	7.0	2440	1	C9H11N5O4	URIDINE,2',3'-DIDEOXY-3'-AZIDO 84472-85-5
75 ★	-0.32		2252		C9H11N5O4	URIDINE,2',3'-DIDEOXY-3'-AZIDO
6	3.55		1560	459	C9H12	1,2,3-TRIMETHYLBENZENE 526-73-8
7	3.59		594	400	"	1,2,3-TRIMETHYLBENZENE
8	3.66		1505	459	"	1,2,3-TRIMETHYLBENZENE
9	3.63		2146	458	C9H12	1,2,4-TRIMETHYLBENZENE 95-63-6

	logP	pH	Ref.	Note	MF	Name / **CAS** / *activity*
4880	3.78		1505	459	"	1,2,4-TRIMETHYLBENZENE
1 ★	3.42		536		C9H12	1,3,5-TRIMETHYLBENZENE **108-67-8**
2 ★	3.53		1560	459	C9H12	1-ETHYL-2-METHYLBENZENE **611-14-3**
3	3.88		2253	459	C9H12	BENZENE,1-ETHYL-3-METHYL **620-14-4**
4	3.90		2253	459	C9H12	BENZENE,1-ETHYL-4-METHYL **622-96-8**
85 ★	3.66		508		C9H12	ISOPROPYLBENZENE **98-82-8**
6 ★	3.72		1265		C9H12	PROPYLBENZENE **103-65-1**
7 ★	-0.48	7.0	2440	1	C9H12Br1N3O3	DIDEOXYCYTIDINE,5-BROMO
8 ★	1.59	7.4	2490	1	C9H12Cl1N1O3S2	BENZENESULFONAMIDE,3-CHLORO-4-HYDROXYPROPYLTHIO **108966-73-0**
9 ★	-0.71		1979		C9H12Cl1N3O4	5'-CHLORO-ARA-C
4890	-1.05	7.0	2440	1	C9H12Cl1N3O4	5'-CHLOROCYTIDINE
1 ★	-1.05		1979		C9H12Cl1N3O4	5'-CHLOROCYTIDINE
2 ★	0.30	6.5	321	2	C9H12Cl1N3O4S2	3-ETHYLHYDROCHLOROTHIAZIDE **1824-58-4** *diuretic*
3	0.15	6.5	321	1	C9H12Cl1N3O4S2	HYDROCHLOROTHIAZIDE,3,3-DIMETHYL
4 ★	1.58	11.5	831	314	C9H12Cl1N5	N1-P-CHLOROPHENYL-N5-METHYLBIGUANIDE **60221-92-3**
95 ★	2.75		2094		C9H12Cl1N5	SIMAZINE,2,6-BIS(ALLYLAMINE) ANALOG **15468-86-7** ✚
6 ★	0.00	7.0	1783	1	C9H12Cl1N5O5	N(5-CL-3-NO2-6(2,3-DIOHPR)AMINO)PYRAZIN)ACETAMIDE **89083-18-1** ✚
7 ★	4.82		1693		C9H12Cl1O2P1S3	CARBOPHENOTHION-METHYL **953-17-3**
8 ★	3.63		1521		C9H12Cl2O2	(1R)T-CYPROPANE-CO2ME-2,2-DIME-3(2,2-DICLETHENYL) **59897-94-8** ✚
9 ★	-0.92		2252		C9H12F1N3O3	CYTIDINE,DIDEOXY-3'-FLUORO
4900 ★	-1.18	7.0	2440	1	C9H12F1N3O3	DIDEOXY-ARA-C,2'-BETA-FLUORO
1 ★	-1.14	7.0	2440	1	C9H12F1N3O3	DIDEOXYCYTIDINE,2'-ALPHA-FLUORO
2 ★	-1.09	7.0	2440	1	C9H12F1N3O3	DIDEOXYCYTIDINE,5-FLUORO
3 ★	1.44	4.0	2009		C9H12F1N3O3	URACIL,1-BUTYLAMINOCARBONYL-5-FLUORO
4 ★	0.91	7.4	2490	1	C9H12F1O3S2	BENZENESULFONAMIDE,3-FLUORO-4-(3-HYDROXYPROPYL)THIO **108966-74-1**
5 ★	-0.41	7.4	2490	1	C9H12F1O5S2	BENZENESULFONAMIDE,3-FLUORO-4-(3-HYDROXYPROPYL)SULFONYL **108966-77-4**
6 ★	-0.64		2567		C9H12F2N4O4	2-NITROIMIDAZOLE,1-(N-HYDROXYPROPYL-B,B-DIFLUORO)PROPIONAMIDO
7 ★	-0.31	7.0	2440	1	C9H12I1N3O3	DIDEOXYCYTIDINE,5-IODO
8 ★	3.30		945		C9H12N1O5P1S1	FENITROTHION **122-14-5** *insecticide*
9 ★	1.69		945		C9H12N1O6P1	3-METHYL-DIMETHYLPARA-OXON **2255-17-6**
4910 ★	0.17	11.0	533	322	C9H12N2	NORNICOTINE **494-97-3** *insecticide*
1 ★	1.64		547		C9H12N2O1	1-PHENYL-3-ETHYLUREA **621-04-5**
2 ★	-0.01		1474		C9H12N2O1	2-PYRIDINEBUTANEAMIDE
3 ★	-0.08		1474		C9H12N2O1	3-PYRIDINEBUTANEAMIDE
4 ★	-0.09		1474		C9H12N2O1	4-PYRIDINEBUTANEAMIDE
15 ★	0.12		1724		C9H12N2O1	4H-PYRIDO(1,2-A)PYRIMIDIN-4-ONE,6,7,8,9-H4,3-ME ✚
6	0.30		1439		C9H12N2O1	4H-PYRIDO(1,2-A)PYRIMIDIN-4-ONE,6,7,8,9-H4,6-ME
7 ★	1.31		594		C9H12N2O1	ACETIC ACID HYDRAZIDE,1-METHYL-2-PHENYL **38604-70-5**
8 ★	0.06		1622		C9H12N2O1	BENZAMIDE,N-(2-AMINOETHYL) **1009-17-2**
9 ★	0.95		541		C9H12N2O1	M-DIMETHYLAMINOBENZAMIDE **33225-17-1**
4920 ★	0.59		1035		C9H12N2O1	N-I-PROPYLNICOTINAMIDE
1 ★	0.66		1035		C9H12N2O1	N-PROPYLNICOTINAMIDE
2 ★	1.96		2030		C9H12N2O1	NITROSOAMINE,N-METHYL-N-(3-METHYLBENZYL) **62783-49-7**
3 ★	2.14		2030		C9H12N2O1	NITROSOAMINE,N-METHYL-N-(4-METHYLBENZYL)
4 ★	1.14		541		C9H12N2O1	P-DIMETHYLAMINOBENZAMIDE **6083-47-2**
25 ★	0.98		503		C9H12N2O1	UREA,1,1-DIMETHYL-3-PHENYL **101-42-8**
6 ★	1.02		505		C9H12N2O1	UREA,1,3-DIMETHYLPHENYL **938-91-0**
7 ★	0.51		580		C9H12N2O2	1,1-DIMETHYL-3-(3-HYDROXYPHENYL)UREA **4849-46-1**
8 ★	1.29		2612		C9H12N2O2	1-PHENYL-3-METHOXY-3-METHYLUREA **1576-17-6**
9 ★	0.46		1969		C9H12N2O2	3-(3-PYRIDYL)PROPANOL, CARBAMATE (ESTER) **98154-96-2**
4930 ★	0.78	7.4	1616	1	C9H12N2O2	ANISOLE,O-CARBOXAMIDOMETHYLAMINO
1 ★	-0.05	7.4	2026	1	C9H12N2O2	BENZAMIDE,2-(2-HYDROXYETHYL)AMINO
2	0.66		844	837	C9H12N2O2	BICYCLO(2,2,1)HEPT-5-EN,2,3-DICARBOXAMID
3 ★	1.22		594		C9H12N2O2	HYDRAZINECARBOXYLIC ACID,2-PHENYL,ETHYL ESTER **6233-02-9**
4 ★	1.62		2030		C9H12N2O2	N-NITROSO-N-METHYL-(4-METHOXYBENZYL)AMINE
35 ★	1.60		2030		C9H12N2O2	NITROSOAMINE,N-METHYL-N-(3-METHOXYBENZYL)
6 ★	-2.20		547		C9H12N2O2	P-AMINOPHENYLALANINE **2410-24-4**
7	0.26	7.4	2307	1	C9H12N2O2	THPO,O-ALLYL ✚
8 ★	0.56		2307	2	"	THPO,O-ALLYL
9 ★	1.26		2094		C9H12N2O2	UREA,1-METHYL-3-(3-METHOXYPHENYL **23138-98-9**
4940 ★	0.83		2612		C9H12N2O3	1-(3-HYDROXYPHENYL)-3-METHOXY-3-METHYLUREA **30087-17-3**
1 ★	0.95		96		C9H12N2O3	BARBITURIC ACID,5-ALLYL,5-ETHYL **2373-84-4**
2 ★	-2.44	7.2	2323	468	C9H12N2O3	PYRIDINYLMETHYLSERINE
3 ★	0.03		773	601	C9H12N2O3S1	4-SULFAMYLBENZAMIDE,N-ETHYL **59777-62-7**
4 ★	0.95		2004		C9H12N2O3S1	BENZENESULFONAMIDE,4-METHYL-N-METHYLAMINOCARBONYL **13909-69-8**
45 ★	0.90		2352	2	C9H12N2O3S1	N2-ETHYL-N1-BENZENESULFONYLUREA **32324-41-7**
6 ★	0.28		2352	2	C9H12N2O3S1	PHENYLSULFONYLUREA,N,N-DIMETHYL
7	-0.59	7.4	2525	1	C9H12N2O3S2	5-(ETHYLAMINOMETHYL)THIENO[2,3-B]FURAN-2-SULFONAMIDE
8	-0.72	7.4	2461	1	C9H12N2O3S3	THIENO[2,3-B]THIOPHENE-2-SULFONAMIDE,5-(2-HYDROXYETHYL)AMINOMETHYL **122266-93-7**
9	1.33		2447		C9H12N2O4	4-PYRIDINECARBOXYLIC ACID,TETRAHYDRO-2,6-DIOXO,BUTYL ESTER
4950	1.01		2447		C9H12N2O4	4-PYRIMIDINECARBOXYLIC ACID,TETRAHYDRO-2,6-DIOXO,I-BUTYL ESTER
1	-1.00	7.0	2440	1	C9H12N2O4	URIDINE,2',3'-DIDEOXY **69400-58-4**
2 ★	-0.89		2252		C9H12N2O4	URIDINE,2',3'-DIDEOXY
3 ★	-1.62		1979		C9H12N2O5	DEOXYURIDINE **951-78-0**
4	-1.54	7.0	2440	1	"	DEOXYURIDINE
55 ★	0.97		1705		C9H12N2O5S1	P-NITRO-(3-SULFONAMIDO)-PROPOXYBENZENE
6 ★	-1.73	7.4	594	1	C9H12N2O5S1	TIAZOFURIN **60084-10-8** *antineoplastic* ✚
7 ★	0.96	7.4	2490	1	C9H12N2O5S2	BENZENESULFONAMIDE,3-NITRO-4-(3-HYDROXYPROPYL)THIO **108966-59-2**
8	-2.52	7.2	1797	314	C9H12N2O6	URIDINE **58-96-8** *antineoplastic*
9 ★	-1.98		1739		"	URIDINE
4960	-1.94	7.0	2440	1	"	URIDINE
1 ★	-0.05	7.4	2490	1	C9H12N2O7S2	BENZENESULFONAMIDE,3-NITRO-4-(3-HYDROXYPROPYL)SULFONYL **108966-60-5**
2 ★	1.35		1173	537	C9H12N2S1	1-METHYL-3-BENZYLTHIOUREA **2740-94-5**
3 ★	1.42		547		C9H12N2S1	1-PHENYL-3-ETHYLTHIOUREA **2741-06-2**
4 ★	1.04	5.6	590		C9H12N2S1	THIOUREA,N,N-DIMETHYL-N'-PHENYL **705-62-4**

	logP	pH	Ref.	Note	MF	Name / CAS / activity
65 ✓	-2.90		2397	101	C9H12N3O2	IMIDAZOLIUM CHLORIDE,1-METHYL-2-HYDROXYIMINOMETHYL-3-(2-PROPYNYLOXY)METHYL **117982-94-2**
6 ★	1.73	7.4	536	2	C9H12N4O1	2-(3,3-DIMETHYL-1-TRIAZINO)-BENZAMIDE **33330-89-1**
7 ★	1.20		563		C9H12N4O1	4-(3,3-DIMETHYL-1-TRIAZENO)BENZAMIDE **33330-91-5** antineoplastic
8	0.29		2123	22	C9H12N4O1	BENZALDEHYDE,4-METHOXY,GUANYLHYDRAZONE **13308-82-2**
9 ★	1.30	6.6	2590	459	C9H12N4O1	BENZAMIDE,3-(3,3-DIMETHYLTRIAZENYL) antineoplastic
4970 ★	1.19	6.6	2590	459	C9H12N4O1	BENZAMIDE,4-(3-METHYLTRIAZENYL) antineoplastic
1	0.32	1.0	591	342	C9H12N4O1S1	GUANIDINE,1-HYDROXY-2-(P-METHYLTHIOPHENYL)METHYLIMINO
2 ★	1.19	6.6	2590	459	C9H12N4O2	1-(4-CARBAMOYLPHENYL)-3-METHYL-3-METHOXYTRIAZENE **66974-76-3** antineoplastic
3 ★	-1.58	7.4	725	1	C9H12N4O2	2-(ACETAMIDOMETHYL)ISONIAZID **58481-03-1**
4 ★	0.56	9.9	704	821	C9H12N4O2	4-METHOXYPHENYLAMIDINOUREA **58247-23-7**
75 ★	1.02		2537		C9H12N4O2	XANTHINE,1-METHYL-3-PROPYL
6 ★	0.84		2537		C9H12N4O2	XANTHINE,3-BUTYL **41078-03-9**
7 ★	0.25	1.0	2196		C9H12N4O3	PYRAZOLE,3-METHYL-4-NITRO-5-(N,N-TETRAMETHYLENECARBOXAMIDO)
8 ★	1.88	7.4	553	1	C9H12N4O3S1	2(4,5,5-ME3-1,3-OXAZOLIDINYLIDEN-2N)5-NO2THIAZOL **30122-47-5** ✚
9 ★	-0.10	7.0	1783	1	C9H12N4O6	6((2,3-DIOHPR)AMINO)-3-NITROPYRAZINOICAC.,ME.ESTER **87885-53-8** ✚
4980	0.97	9.2	2638		C9H12N6	1,2,4-TRIAZOLE,3-AMINO-5-(2-ETHYLAMINO)PYRIDIN-4-YL **77314-77-3**
1	1.13	7.4	2028	1	"	1,2,4-TRIAZOLE,3-AMINO-5-(2-ETHYLAMINO)PYRIDIN-4-YL
2	-0.54	7.4	594	1	C9H12N6	TRIETHYLENEMELAMINE **51-18-3** antineoplastic
3	0.24	7.4	2028	459	C9H12N6O1	1,2,4-TRIAZOLE,3-AMINO-5-(2-HYDROXYETHYLAMINO)PYRIDIN-4-YL **77314-62-6**
4	-0.70	7.0	2440	1	C9H12N6O3	CYTIDINE,DIDEOXY-3'-AZIDO **84472-89-9**
85 ★	-0.64		2252		C9H12N6O3	CYTIDINE,DIDEOXY-3'-AZIDO
6 ★	1.97		594		C9H12O1	1-PHENYL-2-PROPANOL **698-87-3**
7 ★	3.02		2410		C9H12O1	2,6-DIMETHYLANISOLE **1004-66-6**
8 ★	3.18		2410		C9H12O1	ANISOLE,2-ETHYL
9	2.16		2363		C9H12O1	BENZENE,ETHOXYMETHYL
4990 ★	2.88		558		C9H12O1	O-I-PROPYLPHENOL **88-69-7**
1 ★	2.93		558		C9H12O1	O-PROPYLPHENOL **644-35-9**
2 ★	2.90		1941		C9H12O1	P-I-PROPYLPHENOL **99-89-8**
3	2.88	7.4	2582	1	C9H12O1	P-PROPYLPHENOL **645-56-7**
4 ★	3.20		1875		"	P-PROPYLPHENOL
95 ★	3.18		541		C9H12O1	PHENYLPROPYLETHER **622-85-5**
6 ★	1.88		502		C9H12O1	PROPANOL,3-PHENYL **122-97-4**
7 ★	2.76		2410		C9H12O2	1,3-DIHYDROXYBENZENE,4-METHYL
8 ★	2.87		2410		C9H12O2	1,3-DIMETHOXYBENZENE,2-METHYL
9 ★	1.73		547		C9H12O2	2-METHOXYETHOXYBENZENE **41532-81-4**
5000 ★	1.69		2452		C9H12O2	HYDROQUINONE,TRIMETHYL **700-13-0**
1 ★	2.59		1477		C9H12O2	M-PROPOXYPHENOL **16533-50-9**
2 ★	2.09		558		C9H12O2	O-I-PROPOXYPHENOL **4812-20-8**
3 ★	2.33		1940	459	C9H12O2	P-PROPOXYPHENOL **18979-50-5**
4 ★	0.47	7.4	1168	1	C9H12O2	3-METHOXY-4-HYDROXYPHENYLETHANOL **2380-78-1** ✚
5 ★	0.84		2098		C9H12O3	4-IPOMEANOL **32954-58-8** antifungal, antineoplastic
6 ★	1.53		507		C9H12O3	BENZENE,1,2,3-TRIMETHOXY **634-36-6**
7 ★	0.70		507		C9H12O3	PHENYLGLYCEROL **538-43-2**
8 ★	1.81		547		C9H12O3S1	ETHYL-P-TOLUENESULFONATE **80-40-0**
9	-0.58	7.4	1168	1	C9H12O4	3-METHOXY-4-HYDROXYPHENYLGLYCOL **534-82-7**
5010 ★	1.02		2452		C9H12O4	HYDROQUINONE,2,3-DIMETHOXY-5-METHYL **3066-90-8**
1 ★	2.11		1564		C9H13Br1N2O2	BROMACIL **314-40-9** herbicide
2 ★	1.26		1564		C9H13Br1N2O3	PYRIMIDINE-2,4-DIONE,3-BUTOXY-5-BROMO-6-ME **77317-72-7**
3 ★	1.23		1564		C9H13Br1N2O3	PYRIMIDINE-2,4-DIONE,3-I-BUTOXY-5-BROMO-6-ME **77317-74-9**
4 ★	0.72		1564		C9H13Br1N2O3	PYRIMIDINE-2,4-DIONE,3-PROPOXY-5-BR-1,6-DIMETHYL **77317-88-5**
15 ★	1.11		1564		C9H13Br1N2O3	PYRIMIDINE-2,4-DIONE,3-S-BUTOXY-5-BROMO-6-ME **77317-73-8**
6	0.85	7.4	401	1	C9H13Cl1F1N3O2	FLURODOPAN **5767-43-1** antineoplastic
7	-2.30		2173		C9H13Cl1N1.I1	M-CHLORO-TRIMETHYLANILINIUM IODIDE **2373-41-3**
8 ★	1.89		1188		C9H13Cl1N2O2	TERBACIL **5902-51-2** herbicide ✚
9 ★	2.22		1084		C9H13Cl1N6	CYANAZINE **21725-46-2** herbicide
5020	-1.66	3.0	1797		C9H13Cl1N6O2	ACNU **55661-38-6**
1 ★	0.39		1634		C9H13Cl1N6O2	NIMUSTINE **42471-28-3** antineoplastic
2	1.04	7.2	1797	314	"	NIMUSTINE
3	3.51		1263		C9H13F1O3P1S1	PHOSPHOROTHIONICAC,O,O-DIET-O-(6-F-2-PYRIDINYL) **39624-86-7**
4 ★	0.76		2547		C9H13F2N5O3S1	3-NITROTRIAZOLE,N-METHIOPROPYL-DIFLUOROPROPIONAMIDE ANALOG ✚
25 ★	1.10		2547		C9H13F2N5O3S1	3-NITROTRIAZOLE,N-METHOXYPROPYL-DIFLUOROPROPIOTHIOAMIDE ANALOG ✚
6 ★	0.01		2547		C9H13F2N5O4	3-NITROTRIAZOLE,N-METHOXYPROPYL-DIFLUOROPROPIONAMIDE ANALOG
7 ★	1.77	7.4	2397		C9H13F3N4O3S1	IMIDAZOLE,1-(3-(N-METHYL)TRIFLUOROMETHYLSULFONAMIDO)PROPYL-2-HYDROXYIMINOMETHYL
8	-1.13	2.0	1473		C9H13N1	3-PHENYLPROPYLAMINEHYDROCHLORIDE **30684-05-0**
9	-1.09		530	820	"	3-PHENYLPROPYLAMINEHYDROCHLORIDE
5030	-0.88	7.0	1473	1	"	3-PHENYLPROPYLAMINEHYDROCHLORIDE
1	2.42		2247	459	C9H13N1	4-T-BUTYLPYRIDINE **3978-81-2**
2	-1.18	1.0	547	342	C9H13N1	AMPHETAMINE HYDROCHLORIDE
3	-0.94	7.4	1258	1	C9H13N1	AMPHETAMINE **300-62-9** stimulant (central)
4	-0.84	7.4	1168	1	"	AMPHETAMINE
35	-0.68	7.2	969	314	"	AMPHETAMINE
6	0.31	7.4	1015	1	"	AMPHETAMINE
7	1.54	7.4	1258	2	"	AMPHETAMINE
8 ★	1.76		551		"	AMPHETAMINE
9 ★	2.12		594		C9H13N1	ANILINE,2-I-PROPYL **643-28-7**
5040 ★	2.40	7.5	2201		C9H13N1	ANILINE,4-PROPYL **99-88-7**
1 ★	2.80		853		C9H13N1	M-TOLUIDINE-N,N-DIMETHYL **121-72-2**
2 ★	1.98	13.0	588	224	C9H13N1	N,N-DIMETHYLBENZYLAMINE **103-83-3**
3 ★	2.45		272		C9H13N1	N-PROPYLANILINE **622-80-0**
4 ★	2.85		853		C9H13N1	O-TOLUIDINE,N,N-DIMETHYL **609-72-3**
45	3.13		442	2	C9H13N1	P-BUTYLPYRIDINE **5335-75-1**
6 ★	2.23	7.4	1737	1	C9H13N1	P-I-PROPYLANILINE **99-88-7**
7	2.31	7.5	2201		"	P-I-PROPYLANILINE
8 ★	2.49	7.4	1737	1	"	P-I-PROPYLANILINE
9 ★	2.81		324		C9H13N1	P-TOLUIDINE,N,N-DIMETHYL **99-97-8**

	logP	pH	Ref.	Note	MF	Name / CAS / activity
5050 ★	1.83		502		C9H13N1	PROPYLAMINE,3-PHENYL **2038-57-5**
1 ★	2.40		1519	459	C9H13N1	PYRIDINE,2-T-BUTYL **5944-41-2**
2 ★	0.86		1474		C9H13N1O1	2-PYRIDINEBUTANOL **17945-79-8**
3 ★	0.92		1474		C9H13N1O1	3-PYRIDINEBUTANOL **60753-14-2**
4 ★	0.90		1474		C9H13N1O1	4-PYRIDINEBUTANOL
55 ★	0.79	13.0	594		C9H13N1O1	ETHANOLAMINE,N-BENZYL **104-63-2**
6 ★	0.83	11.0	2648	448	C9H13N1O1	NORPSEUDOEPHEDRINE
7 ★	0.67	11.0	2648	448	"	NORPSEUDOEPHEDRINE **14838-15-4**
8	1.63		853		C9H13N1O1	O-ANISIDINE,N,N-DIMETHYL **700-75-4**
9 ★	2.42		1625	457	C9H13N1O1	PENTANONE,PYRROL-2-YL
5060	-2.46	7.4	2261	1	C9H13N1O2	1,2-BENZENEDIOL,4-(2-METHYLAMINO)ETHYL **501-15-5**
1	-1.16	7.4	2556	1	"	1,2-BENZENEDIOL,4-(2-METHYLAMINO)ETHYL
2	0.05	7.4	2685	1	C9H13N1O2	4-PYRIDONE,3-HYDROXY-2-METHYL-1-I-PROPYL *iron chelator*
3	1.15	7.4	1616	1	C9H13N1O2	ANISOLE,O-(2-HYDROXYETHYL)AMINO
4 ★	1.50	7.4	1616	2	"	ANISOLE,O-(2-HYDROXYETHYL)AMINO
65 ★	0.62		1478		C9H13N1O2	FURAN,2-CH2CH2CONME2
6	-1.68	7.4	1168	1	C9H13N1O2	METARAMINOL **54-49-9** *adrenergic*
7	-2.22	7.4	1168	1	C9H13N1O2	O-METHYLDOPAMINE **554-52-9**
8	-2.10	7.4	2261	1	"	O-METHYLDOPAMINE
9	-1.89	7.4	1168	1	C9H13N1O2	PHENYLEPHRINE **59-42-7** *adrenergic (ophthalmic)*
5070	-1.52	7.5	1563	1	"	PHENYLEPHRINE
1	0.28	7.1	2155	2	"	PHENYLEPHRINE
2	-0.08	7.4	2472	324	C9H13N1O2	PYRIDIN-4-ONE,1,2-DIETHYL-3-HYDROXY *iron chelator*
3	0.25	7.4	2685	1	"	PYRIDIN-4-ONE,1,2-DIETHYL-3-HYDROXY
4	0.13	7.4	2472	324	C9H13N1O2	PYRIDIN-4-ONE,1-PROPYL-2-METHYL-3-HYDROXY *iron chelator*
75	0.18	7.4	2685	1	"	PYRIDIN-4-ONE,1-PROPYL-2-METHYL-3-HYDROXY
6	-2.17	7.4	1168	1	C9H13N1O2	SYNEPHRINE **94-07-5** *adrenergic, vasopressor*
7 ✓	-0.62	9.6	594	310	"	SYNEPHRINE
8 ★	1.70	2.1	2265		C9H13N1O2S1	BENZENESULFONAMIDE,3-I-PROPYL
9 ★	1.75	2.1	2265			BENZENESULFONAMIDE,4-I-PROPYL
5080 ★	1.64		773	601	C9H13N1O2S1	P-PROPYLBENZENESULFONAMIDE
1 ★	-0.10	7.4	1074	324	C9H13N1O3	3-O-METHYLPYRIDOXOL
2	-0.66	7.4	2685	1	C9H13N1O3	4-PYRIDONE,3-HYDROXY-2-ETHYL-1-(2-HYDROXY)ETHYL *iron chelator*
3	-0.88	7.4	2685	1	C9H13N1O3	4-PYRIDONE,3-HYDROXY-2-METHYL-1-(3-HYDROXYPROPYL) *iron chelator*
4	-2.61	7.4	1168	1	C9H13N1O3	A-METHYLNORADRENALINE **6539-57-7**
85	-1.75		1768		C9H13N1O3	EPINEPHRINE BITARTRATE **51-42-3**
6	-1.52		465		"	EPINEPHRINE BITARTRATE
7	-1.52		465		C9H13N1O3	EPINEPHRINE MALEATE
8	-2.59	7.4	1168	1	C9H13N1O3	EPINEPHRINE **51-43-4** *adrenergic (vascoconstrictor)*
9	-2.49	7.4	1168	1	C9H13N1O3	O-METHYLNORADRENALINE **97-31-4**
5090	-0.52	7.4	2472	324	C9H13N1O3	PYRIDIN-4-ONE,1-METHOXYETHYL-2-METHYL-3-HYDROXY *iron chelator*
1	-0.41	7.4	2685	1	"	PYRIDIN-4-ONE,1-METHOXYETHYL-2-METHYL-3-HYDROXY
2 ★	0.17	7.4	2490	1	C9H13N1O3S1	BENZENESULFONAMIDE,4-(3-HYDROXYPROPYL) **135832-46-1**
3 ★	0.80	7.4	2490	1	C9H13N1O3S2	BENZENESULFONAMIDE,4-(3-HYDROXYPROPYL)THIO **108966-51-4**
4	-1.52		465		C9H13N1O3.C4H6O5	EPINEPHRINEMALATE
95 ★	-0.64	7.4	2490	1	C9H13N1O5S2	BENZENESULFONAMIDE,4-(3-HYDROXYPROPYL)SULFONYL **108966-48-9**
6 ★	0.17	7.4	2490	1	C9H13N1O6S3	THIOPHENE-2-SULFONAMIDE,5-(3-ACETYLOXY)PROPYLSULFONYL **104438-04-2**
7	-1.43		301		C9H13N2O1.Cl1	N1-PROPYLNICOTINAMIDECHLORIDE **2254-97-9**
8	-3.01		2173		C9H13N2O2.I1	M-NITRO-TRIMETHYLANILINIUM IODIDE **27389-55-5**
9	-0.39	7.4	1948	1	C9H13N3	1-(2-PYRIDYL)PIPERAZINE **34803-66-2**
5100	0.74	7.4	1948	2	"	1-(2-PYRIDYL)PIPERAZINE
1 ★	2.85		538		C9H13N3	3,3-DIMETHYL-1-(3-TOLYL)TRIAZENE **20241-03-6**
2 ★	-0.10		580		C9H13N3O1	1,1-DIMETHYL-3-(3-AMINOPHENYL)UREA **39938-79-9**
3 ★	2.34	7.4	541	1	C9H13N3O1	1-(O-ANISYL)-3,3-DIMETHYLTRIAZENE **20240-99-7**
4 ★	1.73	12.0	541		C9H13N3O1	1-(O-HYDROXYMEPHENYL)-3,3-DIMETRIAZENE
5 ★	1.71	12.0	541		C9H13N3O1	1-PHENYL-3-ME-3-(2-HYDROXYETHYL)TRIAZENE
6 ★	0.63	7.4	725	1	C9H13N3O1	2-PROPYLISONIAZID **14339-54-9**
7 ★	2.99	6.6	2590	459	C9H13N3O1	ANISOLE,4-(3,3-DIMETHYLTRIAZENYL) *antineoplastic*
8 ★	0.37		534		C9H13N3O1	IPRONIAZID **54-92-2** *antidepressant, mao inhibitor*
9	0.37	7.4	534	1	C9H13N3O1.H3O4P1	IPRONIAZIDPHOSPHATE **305-33-9**
5110 ★	-0.16		2612		C9H13N3O2	1-METHOXY-1-METHYL-3-(P-AMINOPHENYL)UREA **88132-16-5**
1	-0.06		1375		C9H13N3O2	3-(N-MORPHOLINO)-4-OH-6-METHYLPYRIDAZINE
2 ★	1.24		2499		C9H13N3O2	PYRAZINE,2-DIMETHYLAMINO-6-ETHOXYCARBONYL
3 ★	1.59	6.6	2590	459	C9H13N3O2S1	METHYLPHENYLSULFONE,4-(3,3-DIMETHYLTRIAZENYL) *antineoplastic*
4	-1.33	7.0	2440	1	C9H13N3O3	DIDEOXYCYTIDINE **7481-89-2** *antiviral, antineoplastic*
15 ★	-1.30		2252		"	DIDEOXYCYTIDINE
6 ★	0.40	7.4	1022	1	C9H13N3O4	1-(2-HYDROXY-3-ALLYLOXYPROPYL)-2-NITROIMIDAZOLE
7	0.37	7.4	1470	1	C9H13N3O4	1-(2-OH-PR)-4-ACETYL-5-ME-2-NITROIMIDAZOLE
8	-1.89	7.0	2440	1	C9H13N3O4	DEOXYCYTIDINE **951-77-9**
9 ★	-1.77		2252		"	DEOXYCYTIDINE
5120	-2.29	7.4	2440	1	C9H13N3O5	1-B-D-ARABINOFURANOSYLCYTOSINE **147-94-4** *antineoplastic, antiviral*
1 ★	-2.13	7.4	401	1	"	1-B-D-ARABINOFURANOSYLCYTOSINE
2	-2.10	7.0	801	1	"	1-B-D-ARABINOFURANOSYLCYTOSINE
3	-1.92	7.2	1797	314	"	1-B-D-ARABINOFURANOSYLCYTOSINE
4 ★	-2.51		1739		C9H13N3O5	1-B-D-RIBOFURANOSYLCYTOSINE **65-46-3**
25 ★	-0.66		1551		C9H13N3O6S1	1-(B-D-GLUCOTHIOPYRANOSYL)2-NITROIMIDAZOLE **83107-47-5**
6 ★	-1.40		1551		C9H13N3O7	1-(B-D-GLUCOPYRANOSYL)2-NITROIMIDAZOLE **83107-46-4**
7 ★	3.13	7.4	541	1	C9H13N3S1	1-(O-THIOANISYL)-3,3-DIMETHYLTRIAZINE **52416-13-4**
8 ★	2.98		538		C9H13N3S1	3,3-DIMETHYL-1-(3-METHIOPHENYL)TRIAZENE **52416-14-5**
9 ✓	0.00		2397	101	C9H13N4O1	IMIDAZOLIUM CHLORIDE,1-METHYL-2-HYDROXYIMINOMETHYL-3-(3-CYANOPROPYL) **132540-05-7**
5130	1.25		526		C9H13N5	ADENINE,9-BUTYL **2715-70-0**
1 ★	1.25		563		C9H13N5O1	4-(3,3-DIMETHYL-1-TRIAZINO)PHENYLUREA
2 ★	0.14		526		C9H13N5O1	ADENINE,9-(1-HYDROXYMETHYL-PROPYL)
3 ★	1.28		563		C9H13N5O1	BENZOIC ACID HYDRAZIDE,O-(3,3-DIME-TRIAZINO)
4	-1.49	7.4	2037	1	C9H13N5O2	ISOGUANINE,1-METHYL-7-METHOXYETHYL
35	-0.96	7.4	2037	1	C9H13N5O2	ISOGUANINE,1-METHYL-9-METHOXYETHYL
6 ★	2.35	7.4	553	1	C9H13N5O2S1	2(1,5,5-ME3-1,3-IMIDAZOLINYLIDEN-2N)5-NO2THIAZOL **37385-14-1**
7 ★	0.16	7.4	553	1	C9H13N5O3S1	2(H6-5-OH-1,3DIMEPYRIMIDINYLIDEN-2N)-5NO2THIAZOL ✚
8 ★	-0.74	7.4	2273	1	C9H13N5O4	4-(3-NITRO-1,2,4-TRIAZOL-1-YL)PROPIONYLMORPHOLINE **104987-39-5**
9 ★	1.35		591		C9H13N5S1	GUANIDINE,1-AMINO-2-(P-METHYLTHIOPHENYL)METHYLIMINO

	logP	pH	Ref.	Note	MF	Name / CAS / activity
5140 ★	0.89		2218		C9H13N5S2	GUANIDINE,N-METHYL-N'-CYANO-N''-THIAZOL-2YLMETHYLTHIOETHYL
1	1.24	7.3	2170	1	C9H13N7S1	TOLUQUINONE,QUANYLHYDRAZONE-THIOSEMICARBAZONE
2	1.66	6.7	2170	1	"	TOLUQUINONE,QUANYLHYDRAZONE-THIOSEMICARBAZONE
3 ★	3.64		2303		C9H13O3P1S2	FENTHION-3-DEMETHYL
4	-2.49		1466		C9H14Br1N2.Br1	2-PYRIDYLME-(6-BR-N,N-TRIMETHYL)AMMONIUMBRIDE 75523-37-4
45	-2.60		1466		C9H14Cl1N2.Br1	2-PYRIDYLME-(6-CL-N,N-TRIMETHYL)AMMONIUMBRIDE 75523-36-3
6 ★	2.45		401		C9H14Cl1N3O2	3-(2-CYCLOHEXEN-1YL)-1-(2-CLOROETHYL)-1-NITROSOUREA 33021-94-2
7 ★	2.77		2094		C9H14Cl1N5	SIMAZINE,2-CYCLOBUTYLAMINE ANALOG 102587-50-8
8 ★	2.85		2094		C9H14Cl1N5	SIMAZINE,2-CYCLOPROPYLMETHYLAMINE ANALOG 40533-52-6
9	-3.20		1466		C9H14F1N2.Br1	6-FL-2-PYRIDYLME-TRIMETHYLAMMONIUMBRIDE 75523-35-2
5150 ✓	-0.82		2397	101	C9H14F3N4O3S1	IMIDAZOLIUM CHLORIDE,1-METHYL-2-HYDROXYIMINOMETHYL-3-(2-TRIFLUOROMETHYLSULFONAMIDO)PROPYL
1 ✓	-1.26		2397	101	C9H14F3N4O3S1	IMIDAZOLIUM CHLORIDE,1-METHYL-2-HYDROXYIMINOMETHYL-3-(3-TRIFLUOROSULFONAMIDOPROPYL)
2 ✓	-1.50		2397	101	C9H14F3N4O3S1	IMIDAZOLIUM CHLORIDE,1-METHYL-2-HYDROXYIMINOMETHYL-3-(N-METHYL-TRIFLUOROMETHYLSULFONAMIDOETHYL
3	-2.68		2173		C9H14N1O1.I1	TRIMETHYLANILINIUM IODIDE,M-HYDROXY 2498-27-3
4	-2.51	7.4	921	22	"	TRIMETHYLANILINIUM IODIDE,M-HYDROXY
55	1.12	2.5	2484		C9H14N1O3	2,5-DIHYDROPYRROLE,1-OXY-2,2,5,5-TETRAMETHYL-3-CARBOXY
6 ★	1.81		354		C9H14N1O3P1S1	METHYLPHOSPHORAMIDATE,O-ME-O-(4-METHIOPHENYL)
7	-1.52	7.4	1074	324	C9H14N1O3.Cl1	N-METHYLPYRIDOXINIUMCHLORIDE
8 ★	1.06		354		C9H14N1O4P1	METHYLPHOSPHORAMIDATE,O-METHYL-O-(4-MEO-PHENYL) 17795-32-3
9 ✓	-2.69		527	404	C9H14N1.Br1	BUTYLPYRIDINIUMBRIDE 874-80-6
5160	-2.74		2173		C9H14N1.I1	PHENYLTRIMETHYL AMMONIUM IODIDE 98-04-4
1	-2.52	7.4	921	22	"	PHENYLTRIMETHYL AMMONIUM IODIDE
2	-1.60	2.7	1473	340	C9H14N2	2-PYRIDINEBUTANEAMINEHYDROCHLORIDE 34974-00-0
3 ★	0.86		1474		C9H14N2	2-PYRIDINEBUTANEAMINE
4	-1.63	2.7	1473	340	C9H14N2	3-PYRIDINEBUTANEAMINE HYDROCHLORIDE 6021-23-4
65 ★	0.88		1474		C9H14N2	3-PYRIDINEBUTANEAMINE
6	-1.67	2.7	1473	340	C9H14N2	4-PYRIDINEBUTANEAMINEHYDROCHLORIDE
7 ★	0.86		1474		C9H14N2	4-PYRIDINEBUTANEAMINE
8 ★	0.82	11.0	533	322	C9H14N2	N,N-DIMETHYL-2-(3-PYRIDYL)ETHYLAMINE 20173-26-6
9 ★	0.54	11.0	533	322	C9H14N2	N-ETHYL-2-(3-PYRIDYL)ETHYLAMINE 19730-15-5
5170	0.82	11.0	533	322	C9H14N2	N-I-PROPYL-3-PYRIDYLMETHYLAMINE 19730-12-2
1	-1.49	7.4	785	1	C9H14N2	N-PHENYLPROPYLENEDIAMINE 4742-01-2
2 ★	0.90	11.0	533	322	C9H14N2	N-PROPYL-3-(3-PYRIDYL)METHYLAMINE 19730-13-3
3 ★	2.10		2499		C9H14N2	PYRAZINE,2-BUTYL-3-METHYL
4 ★	1.96		2613		C9H14N2	PYRAZINE,2-METHYL-3-I-BUTYL
75 ★	1.95		2499		C9H14N2	PYRAZINE-2,3-DIETHYL-5-METHYL
6	1.12		1407		C9H14N2O1	1,2,3,6-H4-PYRIDINE,1-CYCLOPROPYLCARBONYLAMINO
7	-2.10	7.4	2685	1	C9H14N2O2	4-PYRIDONE,3-HYDROXY-2-METHYL-1-(3-AMINO)PROPYL *iron chelator*
8 ★	0.91		2544		C9H14N2O2	THYMINE,1-BUTYL
9 ★	1.43		1205		C9H14N2O2S1	N,N-DIMETHYL-4-METHYLAMINOBENZENESULFONAMIDE 71398-48-6
5180 ★	1.15	0.7	962	1	C9H14N2O3	BARBITURIC ACID,5,5-DIETHYL,1-METHYL 50-11-3 *anticonvulsant*
1 ★	0.97		96		C9H14N2O3	BARBITURIC ACID,5-ETHYL-5-I-PROPYL 76-76-6 *sedative, hypnotic*
2 ★	0.80	1.0	2202	342	C9H14N2O3	BARBITURIC ACID,5-METHYL-5-T-BUTYL
3	0.66	7.4	899	1	C9H14N2O3	BARBITURIC ACID,5-PROPYL-5-ETHYL 33376-25-9
4 ★	1.10		308		C9H14N2O3	N-METHYL-5-BUTYLBARBITURIC ACID 34569-18-1
85 ★	-0.48	7.4	2490	1	C9H14N2O5S2	BENZENESULFONAMIDE,3-AMINO-4-(3-HYDROXYPROPYL)SULFONYL 108966-61-6
6 ✓	-2.32		2397	101	C9H14N3O2	IMIDAZOLIUM CHLORIDE,1-(1-PROPENYL)-2-HYDROXYIMINOMETHYL-3-METHOXYMETHYL
7 ✓	-2.50		2397	101	C9H14N3O2	IMIDAZOLIUM CHLORIDE,1-METHYL-2-HYDROXYIMINOMETHYL-3-ALLYLOXYMETHYL 117982-93-1
8 ★	0.38		57		C9H14N4O2S1	6-(2-PERHYDROPYRANYL)-4-AMINO-3-METHIO-1,2,4-TRIAZINONE 21087-58-1
9 ✓	-0.15		2218		C9H14N4O2S2	1,1-ETHENEDIAMINE,N-METHYL,N'-THIAZOL-2YLMETHYLTHIOETHYL-2-NITRO ✚
5190 ★	1.28	6.0	2417	1	C9H14N4O2S2	IMIDAZOLE,1-(DIMETHYLDITHIOCARBAMOYL)ETHYL-2-METHYL-5-NITRO
1	0.07	9.0	1409		C9H14N4O3	1-(2(N-MORPHOLINO)ET)-5-NO2-IMIDAZOLE 6506-37-2 *antitrichomonal*
2	0.10	7.4	2206	306	"	1-(2(N-MORPHOLINO)ET)-5-NO2-IMIDAZOLE
3	0.15	7.4	1022	1	"	1-(2(N-MORPHOLINO)ET)-5-NO2-IMIDAZOLE
4	1.30	7.4	2380	1	"	1-(2(N-MORPHOLINO)ET)-5-NO2-IMIDAZOLE
95	-0.38	7.4	1022	1	C9H14N4O3	1-(2-N-MORPHOLINO)ETHYL)-4-NITROIMIDAZOLE
6	-0.10	7.4	1022	1	C9H14N4O3	2-NITROIMIDAZOLE,1-(2-(N-MORPHOLINO)ETHYL)
7	-0.09		1467		"	2-NITROIMIDAZOLE,1-(2-(N-MORPHOLINO)ETHYL)
8	-1.51	7.4	2477	1	C9H14N4O3	IMIDAZOLE,2-NITRO-1-(3-AZETIDINYL-2-HYDROXYPROPYL) 134419-50-4
9 ★	1.27	1.0	2196		C9H14N4O3	PYRAZOLE,3-METHYL-4-NITRO-5-(N-BUTYLCARBOXAMIDO)
5200 ★	1.50	7.4	2206	306	C9H14N4O3S1	(2-METHYL-5-NITROIMIDAZOL-2YL)ETHYLTHIOCARBAMIC ACID,ETHYL ESTER
1 ★	1.46	7.4	2206	306	C9H14N4O3S1	SULNIDAZOLE 51022-76-5 *antiprotozoal (trichomonas)*
2 ★	0.46	6.0	2417	1	C9H14N4O4	IMIDAZOLE,1-(N,N-DIMETHYLCARBAMOYLETHYL)-2-METHYL-5-NITRO
3	0.74	7.0	2233	459	C9H14N4O4S1	CARMETHIZOLE 117120-88-4
4	-1.64		1404		C9H14N4O5	2-NO2-IMIDAZOLE,1-CH(ME)CONHCH2CHOHCH2OH
5 ★	-1.59		1404		C9H14N4O5	2-NO2-IMIDAZOLE,1-CH2CON(CH2CH2OH)2 74141-74-5
6	-1.77		1404		C9H14N4O6	2-NO2-IMIDAZOLE,1-CH2CONHC(CH2OH)3
7 ★	3.84		503		C9H14O1Si1	PHENOL,P-(TRIMETHYLSILYL) 13132-25-7
8 ★	0.25		579		C9H14O6	GLYCERYLTRIACETATE 102-76-1 *antifungal*
9 ★	4.72		512		C9H14Si1	SILANE,PHENYL-TRIMETHYL 768-32-1
5210 ✓	-2.20		2397	101	C9H15Br1N3O2	IMIDAZOLIUM CHLORIDE,1-METHYL-2-HYDROXYIMINOMETHYL-3-(2-BROMOETHOXY)ETHYL
1 ✓	-1.76		2397	101	C9H15Br1N3O2	IMIDAZOLIUM CHLORIDE,1-METHYL-2-HYDROXYIMINOMETHYL-3-(3-BROMOPROPOXY)METHYL 132540-14-8
2	3.71		1930		C9H15Br6O4P1	TRIS-(2,3-DIBROMOPROPYL)-PHOSPHATE 126-72-7
3 ✓	-2.00		2033		C9H15Cl1N1O1.I1	FURANE,2-CHLOROMETHYL-5-TRIMETHYLAMMONIOMETHYL IODIDE
4 ✓	-2.32		2397	101	C9H15Cl1N3O2	IMIDAZOLIUM CHLORIDE,1-METHYL-2-HYDROXYIMINOMETHYL-3-(2-CHLOROETHOXY)ETHYL 132540-12-6
15 ✓	-1.93		2397	101	C9H15Cl1N3O2	IMIDAZOLIUM CHLORIDE,1-METHYL-2-HYDROXYIMINOMETHYL-3-(3-CHLOROPROPYOXY)METHYL
6 ★	2.73	7.4	401	2	C9H15Cl1N3O2	1-(2-CHLOROETHYL)-3-(2-CHLOROCYCLOHEXYL)-1-NITROSOUREA 13909-11-0
7 ★	1.87		421		C9H15Cl3O3	1,2,3-TRIMETHOXY-4,5,6-TRICHLOROCYCLOHEXANE 60067-82-5
8 ✓	-2.62		2397	101	C9H15F1N3O2	IMIDAZOLIUM CHLORIDE,1-METHYL-2-HYDROXYIMINOMETHYL-3-(2-FLUOROETHOXY)ETHYL 132540-10-4
9 ★	3.70		2488		C9H15F2N5S1	1,3,5-TRIAZINE,2-DIFLUOROMETHIO-4-BUTYLAMINO-6-METHYLAMINO
5220 ★	3.70		2488		C9H15F2N5S1	1,3,5-TRIAZINE,2-DIFLUOROMETHIO-4-I-PROPYLAMINO-6-ETHYLAMINO
1 ★	3.80		2488		C9H15F2N5S1	1,3,5-TRIAZINE,2-DIFLUOROMETHIO-4-PROPYLAMINO-6-ETHYLAMINO
2 ★	3.91		2488		C9H15F2N5S1	1,3,5-TRIAZINE,2-DIFLUOROMETHIO-4-T-BUTYLAMINO-6-METHYLAMINO
3 ✓	-3.01		2397	101	C9H15F3N3O2S1	IMIDAZOLIUM CHLORIDE,1,2-DIMETHYL-3-(2-N-METHYL-TRIFLUOROMETHYLSULFONAMIDO)ETHYL
4 ★	2.59	13.0	579	224	C9H15N1	TRIALLYLAMINE 102-70-5
25 ★	2.08		1478		C9H15N1O1	2-(3-N,N-DIMETHYLPROPYL)FURAN 49547-83-3
6	0.41		213		C9H15N1O2	N-BUTYROYLCYCLOBUTANECARBOXAMIDE 6815-52-7
7 ★	0.26		213		C9H15N1O2	N-I-BUTYROYLCYCLOBUTANECARBOXAMIDE 23046-88-0
8	0.25	7.0	2484	1	C9H15N2O2	2,5-DIHYDROPYRROLE,1-OXY-2,2,5,5-TETRAMETHYL-3-CARBOXAMIDO
9	-3.40		1466		C9H15N2.Br1	2-PYRIDYLMETHYLENE-TRIMETHYLAMMONIUMBRIDE 19004-42-3

	logP	pH	Ref.	Note	MF	Name / CAS / activity
5230	-3.50		1466		C9H15N2.Br1	3-PYRIDYLME-(N,N,N-TRIMETHYL)AMMONIUMBROMIDE 16593-50-3
1	-3.95		1466		C9H15N2.Br1	4-PYRIDYLME-(N,N,N-TRIMETHYL)AMMONIUMBROMIDE 19067-63-1
2	0.05	7.4	1350	1	C9H15N3O1	AZEPEXOLE 36067-73-9
3	0.99	7.4	416	1	C9H15N3O2	1,3-DIME-5-DIMEAMINO-6-METHYLURACIL 38507-32-3
4 ★	0.51	7.4	1022	1	C9H15N3O4	1-(3-I-PROPOXY-2-HYDROXYPROPYL)-2-NITROIMIDAZOLE
35 ★	0.13		835		C9H15N3O4S1	I-PROPYL(2-(2-ME-5-NITRO-1-IMIDAZOLYL)ETHYL)SULFONE 19390-40-0
6 ★	0.23		835		C9H15N3O4S1	PROPYL(2-(2-ME-5-NITRO-1-IMIDAZOLYL)ETHYL)SULFONE 19387-93-0
7 ✓	-2.65		2397	101	C9H15N4O4	IMIDAZOLIUM CHLORIDE,1-METHYL-2-HYDROXYIMINOMETHYL-3-(3-NITROPROPOXY)METHYL 132566-51-9
8 ★	1.24		723		C9H15N5O1	MINOXIDIL 38304-91-5 antihypertensive, antialopecia
9	-1.49	7.4	2037	1	C9H15N5O2	2H-PURIN-2-ONE,1,3-DIHYDRO-1-METHYL-6-AMINO-7-METHOXYETHYL
5240	-0.68	7.4	579	324	C9H15N5O2	TETRAHYDROPTERIDINE,2-AMINO-4-HYDROXY-6-ETHOXYMETHYL
1 ★	2.83		401		C9H16Cl1N3O2	LOMUSTINE 13010-47-4 antineoplastic
2 ★	1.11		401		C9H16Cl1N3O3	1-(2-CLET)-3-(4-OH-CYHEX)-1-NITROSOUREA-C 52049-26-0
3 ★	1.00		401		C9H16Cl1N3O3	1-(2-CLET)-3-(4-OH-CYHEX)-1-NITROSOUREA-T 56239-24-8
4 ★	1.75		547		C9H16Cl1N3O3	1-NO-1-(2-CLET)-3-(2-HYDROXYCYCLOHEXYL)UREA-CIS 56323-43-4
45 ★	1.34		547		C9H16Cl1N3O3	1-NO-1-(2-CLET)-3-(2-HYDROXYCYCLOHEXYL)UREA-TR 58494-43-2
6 ★	0.16		547		C9H16Cl1N3O4	1-NO-1-(2-CLET)-3-(2,6-DIHYDROXYCYCLOHEXYL)UREA 58484-09-6
7	-1.02	7.4	401	1	C9H16Cl1N3O7	1-(2-CHLOROETHYL)-1-NITROSO-3-(2-GLUCOPYRANOSYL) UREA 58484-07-4
8 ★	-0.66		547		"	1-(2-CHLOROETHYL)-1-NITROSO-3-(2-GLUCOPYRANOSYL) UREA
9 ★	-1.02		1634		C9H16Cl1N3O7	CHLOROZOTOCIN 54749-90-5 antineoplastic
5250 ★	2.93		1084		C9H16Cl1N5	PROPAZINE 139-40-2 herbicide
1 ★	3.06		2550	459	C9H16Cl1N5	TERBUTHYLAZINE 5915-41-3 herbicide
2 ★	3.34		1190		C9H16Cl1N5	TRIETAZINE 1912-26-1 herbicide
3 ✓	2.70		2366	122	C9H16Cu1N6S2	COPPER(II)PYRUVALDEHYDE-BIS(N4-DIMETHYLTHIOSEMICARBAZONE)
4	1.78	7.0	2484		C9H16N1O1	1,2,3,6-TETRAHYDROPYRIDINE,1-OXY-2,2,6,6-TETRAMETHYL
55	-1.85		1466	42	C9H16N1O1.Br1	2-FURANYLME-(5-N,N,N-TETRAMETHYL)AMMONIUMBROMIDE 1197-60-0
6	-2.00		1466	42	C9H16N1O1.Br1	3-FURANYLME-(N,N,N-TRIME)AMMONIUMBROMIDE 68724-15-2
7 ✓	-2.33		2033		C9H16N1O1.I1	FURANE,2-METHYL-4-TRIMETHYLAMMONIOMETHYL IODIDE
8 ✓	-1.85		2033		C9H16N1O1.I1	FURANE,2-METHYL-5-TRIMETHYLAMMONIOMETHYL IODIDE
9	0.12	7.0	2484		C9H16N1O2	TEMPONE
5260 ✓	-2.30		2033		C9H16N1O2.I1	FURANE,2-HYDROXYMETHYL-5-TRIMETHYLAMMONIOMETHYL IODIDE
1	0.78	2.5	2484		C9H16N1O3	1-OXYPYRROLINE,2,2,5,5-TETRAMETHYL-3-CARBOXY
2	0.48	2.5	2484		C9H16N1O3	1-OXYPYRROLINE,2,2,5-TRIMETHYL-5-ACETIC ACID
3 ★	1.38	13.0	594	2	C9H16N2	DBU 6674-22-2
4 ★	1.40		504		C9H16N2O3	UREA,1,3-DIBUTYRYL
65	1.82	7.4	1269	1	C9H16N2O3S1	METHYLDIETHYLTHIOMALONURATE
6	0.95	3.5	1267	1	C9H16N2O4	METHYL-2,2-DIETHYLMALONURATE
7 ✓	-2.72		2397	101	C9H16N3O1	IMIDAZOLIUM CHLORIDE,1-(T-BUTYL)-2-HYDROXYIMINOMETHYL-3-METHYL
8 ✓	-2.17		2397	101	C9H16N3O2	IMIDAZOLIUM CHLORIDE,1-METHYL-2-HYDROXYIMINOMETHYL-3-I-PROPOXYMETHYL 91900-10-6
9 ✓	-2.56		2397	101	C9H16N3O2	IMIDAZOLIUM CHLORIDE,1-PROPYL-2-HYDROXYIMINOMETHYL-3-METHOXYMETHYL
5270	-2.17	7.6	2003	314	C9H16N3O2.Cl1	IMIDAZOLIUM CHLORIDE,1-I-PROPOXYMETHYL-2-HYDROXYIMINOMETHYL-3-METHYL 91900-10-6
1 ✓	-2.81		2397	101	C9H16N3O3	IMIDAZOLIUM CHLORIDE,1-METHYL-2-HYDROXYIMINOMETHYL-3-(2-METHOXYETHOXY)METHYL 132540-00-2
2 ✓	-2.96		2397	101	C9H16N3O3S1	IMIDAZOLIUM CHLORIDE,1-METHYL-2-HYDROXYIMINOMETHYL-3-(2-ETHYLSULFONYL)ETHYL 132566-53-1
3 ★	-0.06		1173	537	C9H16N4O1S1	1-ME-3-(2-(5-ME-IMIDAZOL-4-YL)METHIOETHYLUREA
4 ★	2.06		57		C9H16N4O1S1	3-I-PRTHIO-4-AMINO-6-I-PR-1,2,4-TRIAZINE-5-ONE 50917-24-3
75 ★	1.85		57		C9H16N4O1S1	3-METHIO-4-AMINO-6-I-PENT-1,2,4-TRIAZINE-5-ONE 21085-18-7
6 ★	2.12		57		C9H16N4O1S1	3-PROPYLTHIO-4-AMINO-6-I-PR-1,2,4-TRIAZINE-5-ONE 50917-23-2
7	0.74	7.6	1769	212	C9H16N4O1S2	1,2,5-THIADIAZOLE-3-C(=NOH)SCH2CH2N(ET)2 90507-24-7 ✚
8 ★	0.40	9.2	1853	1	C9H16N4S1	BURIMAMIDE 34970-69-9
9 ★	0.50		1173	537	C9H16N4S2	METIAMIDE 34839-70-8 antagonist (histamine h-2 receptors)
5280	0.22		1764		C9H16N4S2	N-METHIOUREA,N'-2(1-MEIMIDAZOL-4YL)METHYLTHIOETHYL 34839-70-8
1	0.41		1764		C9H16N4S2	N-METHIOUREA,N'-2(1-MEIMIDAZOL-5YL)METHYLTHIOETHYL
2	0.65	7.4	401	1	C9H16N6O1	IMIDAZOLE-4-(3-T-BUTYL-3-METHYLTRIAZENYL)-5-CARBOXAMIDO
3 ★	0.94	7.4	401	2	"	IMIDAZOLE-4-(3-T-BUTYL-3-METHYLTRIAZENYL)-5-CARBOXAMIDO
4 ★	1.08	4.6	401	2	C9H16N6O1	IMIDAZOLE-5-CONH2,4(3-S-BU-3-ME-NNN) 39980-81-9
85 ★	0.13		2302		C9H16N6O2S1	1-ME-2-NO2-3-(2-(5-ME-IMIDAZOL-4YL)ME-S-ETGUANIDINE ✚
6	2.22		1231		C9H16O2	2,6-DIMETHYL-3,5-HEPTANEDIONE 18362-64-6
7	2.23		1231		C9H16O2	4,6-NONANEDIONE 14090-88-1
8 ★	1.57		510		C9H16O4	AZELAIC ACID 123-99-9 anti-acne
9 ★	1.38		594		C9H16O4	PIMELIC ACID,DIMETHYL ESTER 1732-08-7
5290	3.82		1501	100	C9H17Cl1N3O4P1	ISAZOPHOS
1 ★	2.29	9.2	536	2	C9H17N1	METHYL-I-PROPYL-(1,1-DIMETHYLPROPYN-3-YL)AMINE
2 ★	3.12		1896		C9H17N1	NONANENITRILE 2243-27-8
3 ★	-0.58	7.2	2323	468	C9H17N1O2	CYCLOHEXYLALANINE
4 ★	1.68		1555		C9H17N1O2S1	LETHANE384 112-56-1 insecticide ✚
95	1.14	0.5	1997	346	C9H17N1O3	GLYCINE,N-HEXANOYL
6 ★	0.79	2.2	2228	314	C9H17N1O3	HEXANOIC ACID-2-ACETYLAMINO,METHYL ESTER
7 ★	0.84	2.2	2228	314	C9H17N1O3	ISOHEXANOIC ACID-2-ACETYLAMINO,METHYL ESTER
8 ★	3.21		1364		C9H17N1S1	MOLINATE 2212-67-1 herbicide
9	-0.21	7.0	2484	1	C9H17N2O2	1-OXYPYRROLINE,2,2,5,5-TETRAMETHYL-3-CARBOXAMIDO
5300	-1.33		2377		C9H17N3O3	AC-GLY-VAL-N
1 ★	1.60		1629		C9H17N3O3S2	N-(N"-METHYL-ETCARBAMYLTHIO)METHOMYL ✚
2 ★	1.90		1629		C9H17N3O4S2	N-(N"-METHYL-ETHOXYCARBAMYLTHIO)METHOMYL
3	-1.45	7.4	401	1	C9H17N3O7	GLUCOPYRANOSIDE,ME-3-DEOXY-3-(3-ME-3-NO-UREIDO) 52019-12-2
4	-0.82	7.4	401	1	"	GLUCOPYRANOSIDE,ME-3-DEOXY-3-(3-ME-3)-NO-UREIDO
5	-1.57	7.4	401	1	C9H17N3O7	GLUCOPYRANOSIDE,ME-6-DEOXY-6-(3-ME-3-NO-UREIDO) 41110-59-2
6	-3.06	7.4	2397	101	C9H17N4O1	IMIDAZOLIUM CHLORIDE,1-METHYL-2-HYDROXYIMINOMETHYL-3-DIMETHYLAIMINOETHYL 132540-21-7
7 ✓	-3.31		2397	101	C9H17N4O3S1	IMIDAZOLIUM CHLORIDE,1-METHYL-2-HYDROXYIMINOMETHYL-3-(N-METHYL-METHYLSULFONYLETHYL
8 ★	2.69		2550	459	C9H17N5O1	ATRATON 1610-17-9 herbicide
9 ★	2.98		1084		C9H17N5S1	AMETRYN 834-12-8 herbicide
5310 ★	5.15		1560	100	C9H18	1-NONENE 124-11-8
1 ★	3.32		579		C9H18Ge1O1	4,4-DIETHYL-4-GERMA-CYCLOHEXANONE
2 ★	1.85	7.0	2483	1	C9H18N1O1	2,2,6,6-TETRAMETHYLPIPERIDINYL-N-OXY 2564-83-2
3	-2.19		1901	820	C9H18N1O1.I1	METHYLTROPINE IODIDE
4	0.62	7.0	2484		C9H18N1O1	1-OXYPIPERIDINE,2,2,6,6-TETRAMETHYL-4-HYDROXY
15	0.34	7.0	2484	1	C9H18N1O2	1-OXYPYRROLINE,2,2,5,5-TETRAMETHYL-3-HYDROXYMETHYL
6	-2.70		1257	820	C9H18N1O2.I1	CHOLINECYCLOPROPANECARBOXYLATE IODIDE
7 ✓	1.17		567	820	C9H18N1.C6H2N3O7	N-ETHYLQUINUCLIDINIUM PICRATE
8 ✓	2.49		871		C9H18N2O1	2,2,6,6-TETRAMETHYL-N-NITROSOPIPERIDINE 6130-93-4
9 ★	2.54		871		C9H18N2O1	4-T-BUTYL-N-NITROSOPIPERIDINE 46061-25-0

	logP	pH	Ref.	Note	MF	Name / CAS / activity
5320 ★	0.14		1558		C9H18N2O2	ISOLEUCINE,N-ACETYL,N'-METHYLAMINO-AMIDE
1 ★	0.14		1558		C9H18N2O2	LEUCINE,N-ACETYL,N'-METHYLAMINO-AMIDE
2 ★	-2.60	7.0	2374		C9H18N2O3	ALANYLISOLEUCINE
3	2.75		2616	459	C9H18N2O3S1	THIOFANOX 39196-18-4
4 ★	-2.49	7.0	2374		C9H18N2O4	LEUCINE,SERYL 6665-16-3
25 ★	0.70		505		C9H18N2O4	MEPROBAMATE 57-53-4 sedative
6 ★	1.20		1629		C9H18N4O3S2	N-(N"DIME-N'-ME)-URYLTHIO)METHOMYL
7	2.52	7.4	401	1	C9H18N6	HEXAMETHYLMELAMINE 645-05-6 antineoplastic
8 ★	2.73		430		"	HEXAMETHYLMELAMINE
9 ★	0.40		964		C9H18N6O3	N2,N4,N6-TRIMETHYL-N,N,N-HYDROXYMETHYLMELAMINE
5330 ★	3.14		2047		C9H18O1	2-NONANONE 821-55-6
1 ★	2.92		2047		C9H18O1	2-OCTANONE,5-METHYL 58654-67-4
2 ★	3.00		594		C9H18O1	PENTANONE,2,2,4,4-TETRAMETHYL 815-24-7
3 ★	3.53		599	60	"	PENTANONE,2,2,4,4-TETRAMETHYL
4 ★	3.01		1815		C9H18O2	2-PROPYL-HEXANOIC ACID 3274-28-0
35	-1.26		1294		C9H18O5S1	1-I-PROPYLTHIO-B-GALACTOPYRANOSIDE
6	-1.18		1294		C9H18O5S1	1-PROPYLTHIO-B-GALACTOPYRANOSIDE
7	-1.66		1294		C9H18O6	B-PROPYLGALACTOPYRANOSIDE 62178-32-9
8 ★	2.96	13.0	1806	224	C9H19N1	CYCLOHEXANE,2-AMINOPROPYL 54704-34-6
9 ★	2.15	13.0	594	400	C9H19N1	PIPERIDINE,2,2,6,6-TETRAMETHYL 768-66-1
5340 ★	2.82		1969		C9H19N1O2	DIETHYLCARBAMATE,I-BUTYL ESTER 93677-64-6
1 ★	2.87		1969		C9H19N1O2	DIMETHYLCARBAMATE,2-ETHYLBUTYL ESTER 98154-99-5
2 ★	2.84	7.4	172	1	C9H19N1O2	O-OCTYLCARBAMATE
3 ★	3.21		594		C9H19N1S1	EPTAM 759-94-4 herbicide
4	-1.40	7.0	2484	1	C9H19N2O1	1-OXYPIPERIDINE,2,2,6,6-TETRAMETHYL-4-AMINO
45	0.54	12.0	2484	1	"	1-OXYPIPERIDINE,2,2,6,6-TETRAMETHYL-4-AMINO
6	-1.70	7.0	2484	1	C9H19N2O1	1-OXYPYRROLINE,2,2,5,5-TETRAMETHYL-3-AMINOMETHYL
7	0.36	12.0	2484	1	"	1-OXYPYRROLINE,2,2,5,5-TETRAMETHYL-3-AMINOMETHYL
8 ✓	-1.80		1704	820	C9H20N1O1.Br1	(5-ME-TETRAHYDROFURAN-2YL)TETRAMEAMMONIUMBROMIDE
9 ✓	-2.15		1704	820	C9H20N1O1.Br1	(5-ME-TETRAHYDROFURAN-3YL)TETRAMEAMMONIUMBROMIDE
5350 ✓	-2.15		2033		C9H20N1O1.I1	TETRAHYDROFURANE,2-METHYL-4-TRIMETHYLAMMONIOMETHYL IODIDE
1 ✓	-1.80		2033		C9H20N1O1.I1	TETRAHYDROFURANE,2-METHYL-5-TRIMETHYLAMMONIOMETHYL IODIDE
2 ✓	-2.95		2033		C9H20N1O2.I1	TETRAHYDROFURAN-3-OL,2-METHYL-5-TRIMETHYLAMMONIOMETHYL IODIDE
3 ✓	-2.00		1704	820	C9H20N1.Br1	CYCLOPENTYL-TETRAMETHYLAMMONIUMBROMIDE
4 ✓	-2.00		2033		C9H20N1.I1	CYCLOPENTANE,TRIMETHYLAMMONIOMETHYL IODIDE
55 ✓	1.64		567	820	C9H20N1.C6H2N3O7	CYCLOHEXYLTRIMETHYL AMMONIUM PICRATE
6 ★	2.75	6.5	2013		C9H20N2S1	THIOUREA,N,N'-DIBUTYL 109-46-6
7 ★	2.08		1967		C9H20N4	2-OCTANONE,GUANYLHYDRAZONE 93484-23-2
8 ★	4.26		579		C9H20O1	NONANOL 143-08-8
9 ★	1.97	13.0	594	400	C9H20O3	TRIETHYL ORTHOPROPIONATE 115-80-0
5360	0.93		2019		C9H21Cl1Sn1	STANNANE,CHLOROTRIPROPYL 2279-76-7
1	1.72		1959		"	STANNANE,CHLOROTRIPROPYL
2 ✓	2.34		567	820	C9H21N1	METHYL-DI-I-BUTYLAMINE
3	1.33	7.0	834	1	C9H21N1	NONYLAMINE 112-20-9
4 ✓	1.94		567	820	C9H21N1	PROPYL-DI-I-PROPYLAMINE PICRATE
65 ✓	2.45		567	820	C9H21N1	TRIPROPYLAMINE PICRATE
6 ★	2.79	9.3	588	2	C9H21N1	TRIPROPYLAMINE 102-69-2
7 ✓	1.94		567	820	C9H21N1.C6H3N3O7	PROPYL-DI-I-PROPYLAMINE PICRATE
8	-0.43		1394		C9H21N3O1	4-OCTYLSEMICARBAZIDE
9	-0.88		530		C9H21N3.H2O4S1	OCTYLGUANIDIUMSULFATE
5370	-0.61	7.3	761	314	C9H21N5	HEPTYLBIGUANIDE 3374-71-8
1 ★	4.48		1639		C9H21O2P1S3	TERBUFOS 13071-79-9 insecticide
2	3.26		2182	459	C9H21O3P1	METHYLPHOSPHONIC ACID,DIBUTYL ESTER 2404-73-1
3 ★	2.26		594	400	C9H21O3P1	TRIPROPYLPHOSPHITE 923-99-9
4 ★	2.21		1639		C9H21O3P1S3	TERBUFOS-SULFOXIDE 10548-10-4
75 ★	1.87		547		C9H21O4P1	TRIPROPYLPHOSPHATE 513-08-6
6	2.66		1649		C9H21O4P1S2	TERBUFOSSULFOXIDE
7 ★	2.48		1639		C9H21O4P1S3	TERBUFOS-SULFONE 56070-16-7
8	2.76		1649		C9H21O5P1S2	TERBUFOSSULFONE
9 ★	-1.84		524	820	C9H22N1.I1	HEXYL AMMONIUM IODIDE, N,N,N-TRIMETHYL 15066-77-0
5380 ✓	1.33		567	820	C9H22N1.C6H2N3O7	PROPYLTRIETHYL AMMONIUM PICRATE
1 ★	5.07		1693		C9H22O4P2S4	ETHION 563-12-2 insecticide~miticide
2 ★	6.42		2169		C10Cl8	NAPHTHALENE,OCTACHLORO 2234-13-1
3 ★	5.41		579		C10Cl10O1	KEPONE 143-50-0 insecticide
4 ★	5.28		579		C10Cl12	MIREX 2385-85-5 insecticide
85	6.00		2343	20	C10F18	PERFLUNAFENE 306-94-5
6	5.18	7.3	509	2	C10H3Cl2F3N4	QUINOXALINEIMIDAZOLE,2-TRIFLUOROME-5,7-DICL 30466-63-8
7	3.60		2264	1	C10H3F9N4O4	HEXAFLUOROACETONE,2,6-DINITRO-4-TRIFLUOROMETHYLPHENYLHYDRAZONE
8	3.91	7.3	509	2	C10H4Cl1F3N4	QUINOXALINEIMIDAZOLE,2-TRIFLUORO-6-CL 30466-62-7
9 ★	5.75		2169		C10H4Cl4	NAPHTHALENE,1,2,3,4-TETRACHLORO 20020-02-4
5390 ★	5.77		2169		C10H4Cl4	NAPHTHALENE,1,2,3,5-TETRACHLORO 53555-63-8
1 ★	6.19		2169		C10H4Cl4	NAPHTHALENE,1,3,5,7-TETRACHLORO 53555-64-9
2 ★	5.76		2169		C10H4Cl4	NAPHTHALENE,1,3,5,8-TETRACHLORO 31604-28-1
3 ★	5.81		2169		C10H4Cl4	NAPHTHALENE,1,4,6,7-TETRACHLORO 55720-43-9
4	3.40	4.0	2264	1	C10H4F6N4O2	HEXAFLUOROACETONE,4-CYANO-2-NITROPHENYLHYDRAZONE
95	4.80	4.0	2264	1	C10H4F9N3O2	HEXAFLUOROACETONE,2-NITRO-4-TRIFLUOROMETHYLPHENYLHYDRAZONE
6	4.60	4.0	2264	1	C10H4F9N3O2	HEXAFLUOROACETONE,4-NITRO-2-TRIFLUOROMETHYLPHENYLHYDRAZONE
7 ★	2.44		2649		C10H5Br1O2	1,4-NAPHTHOQUINONE,2-BROMO 2065-37-4
8 ★	2.29		2649		C10H5Cl1O2	1,4-NAPHTHOQUINONE,2-CHLORO 1010-60-2
9	0.48		1556	820	C10H5Cl2O1.K1	2,4-DICHLORO-1-NAPHTHOLATE,POTASSIUM SALT
5400 ★	5.35		2169		C10H5Cl3	NAPHTHALENE,1,3,7-TRICHLORO 55720-37-1
1 ★	5.12		2169		C10H5Cl3	NAPHTHALENE,2,3,6-TRICHLORO 55720-40-6
2	5.27		1369	459	C10H5Cl7	HEPTACHLOR 76-44-8
3	5.58		2185	459	"	HEPTACHLOR
4	2.00	7.2	2212	1	C10H5F3N4	CARBONYLCYANIDE,2-TRIFLUOROMETHYLPHENYLHYDRAZONE 101398-31-6
5 ★	4.20	7.2	2212	2	"	CARBONYLCYANIDE,2-TRIFLUOROMETHYLPHENYLHYDRAZONE
6	3.08	7.3	509	2	C10H5F3N4	QUINOXALINEIMIDAZOLE,2-TRIFLUOROMETHYL 30504-88-2
7	2.42	7.2	1985	314	C10H5F3N4O1	CARBONYLCYANIDE,P-TRIFLUOROMETHOXYPHENYLHYDRAZONE 370-86-5
8 ★	3.68	3.0	1985		"	CARBONYLCYANIDE,P-TRIFLUOROMETHOXYPHENYLHYDRAZONE
9 ★	2.29		593		C10H5N3O6	NAPHTHALENE,1,3,8-TRINITRO

	logP	pH	Ref.	Note	MF	Name / CAS / activity
5410 ★	3.51		2410		C10H6Br1F6N1O1	ACETANILIDE,2-BROMO-3,5-DI(TRIFLUOROMETHYL)
1 ★	2.93		1176		C10H6Br1N1O2	3-BROMO-2-NITROSO-1-NAPHTHOL 30922-52-2
2 ★	2.08		2649		C10H6Cl1N1O2	1,4-NAPHTHOQUINONE,2-CHLORO,3-AMINO 2797-51-5
3	0.78	7.4	1255	1	C10H6Cl1N1O3	7-CHLORO-4-HYDROXYQUINOLINE-3-CARBOXYLIC ACID
4 ★	1.31		555		C10H6Cl1N3O3	4-CHLORO-2-(5-NO2-2-FURFURILIDENE)PYRIMIDINE
15 ★	4.42		2169		C10H6Cl2	NAPHTHALENE,1,2-DICHLORO 2050-69-3
6 ★	4.66		2169		C10H6Cl2	NAPHTHALENE,1,4-DICHLORO 1825-31-6
7 ★	4.67		2169		C10H6Cl2	NAPHTHALENE,1,5-DICHLORO 1825-30-5
8 ★	4.56		2169		C10H6Cl2	NAPHTHALENE,1,7-DICHLORO 2050-73-9
9 ★	4.19		2169		C10H6Cl2	NAPHTHALENE,1,8-DICHLORO 2050-74-0
5420 ★	4.51		2169		C10H6Cl2	NAPHTHALENE,2,3-DICHLORO 2050-75-1
1 ★	4.17		2506		C10H6Cl2N2	FENCLORIM herbicide safener
2 ★	4.76	6.5	1556	1	C10H6Cl2O1	2,4-DICHLORO-1-NAPHTHOL 2050-76-2
3 ★	5.15		331		C10H6Cl3N3O2	2,4,5-TRICL-C6H2NHN=C(CN)COOME-CIS +
4 ★	5.22		331		C10H6Cl3N3O2	3,4,5-TRI-CL-C6H2NHN=C(CN)COOME 36865-56-2
25 ★	4.40		579		C10H6Cl4O4	DACTHAL 1861-32-1
6	-1.10	7.4	2266		C10H6F1N3O3	5-FLUOROURACIL-3-NICOTINYL 88740-57-2
7 ★	-0.06	4.0	2266	302	"	5-FLUOROURACIL-3-NICOTINYL
8	0.29	7.4	1255	1	C10H6F1O3	7-FLUORO-4-HYDROXYQUINOLINE-3-CARBOXYLIC ACID
9 ★	3.02	7.0	538	701	C10H6F3N1	7-TRIFLUOROMETHYLQUINOLINE
5430 ★	2.50		532		C10H6F3N1	8-TRIFLUOROMETHYLQUINOLINE
1 ✓	2.05		532		C10H6F3N1O1	4-HYDROXY-7-TRIFLUOROMETHYLQUINOLINE 322-97-4
2 ★	3.44		2221	459	C10H6F3N1O1	INDOLE-3-TRIFLUOROACETYL
3	3.00	4.0	2264	1	C10H6F6N4O4	1,1,1-TRIFLUOROACETONE,2,6-DINITRO-4-TRIFLUOROMETHYLPHENYLHYDRAZONE
4 ★	1.60		1519	459	C10H6N2	2-CYANOQUINOLINE 1436-43-7
35 ★	1.57		1519	459	C10H6N2	3-CYANOQUINOLINE 34846-64-5
6 ★	1.98		163	459	C10H6N2	4-CYANOQUINOLINE 2973-27-5
7 ★	2.18		547	500	C10H6N2	MALONONITRILE,BENZAL 2700-22-3
8 ★	1.39	7.4	595	314	C10H6N2	QUINOLINE,8-CYANO
9 ★	2.83		593		C10H6N2O4	NAPHTHALENE,1,3-DINITRO 606-37-1
5440 ★	2.58		595		C10H6N2O4	NAPHTHALENE,1,5-DINITRO 605-71-0
1 ★	2.52		579		C10H6N2O4	NAPHTHALENE,1,8-DINITRO 602-38-0
2 ★	0.93		494		C10H6N2O4S2	5-(5-NO2-2-FURYL-PROPENILIDINE)-2-S-THIAZOLIDEN-4ONE 27464-57-9 +
3	-0.17	7.4	1255	1	C10H6N2O5	7-NITRO-4-HYDROXYQUINOLINE-3-CARBOXYLIC ACID
4	-0.03		555		C10H6N2O5S1	4-CARBOXY-2-(5-NO2-2-FURFURILIDENE)THIAZOLE
45 ★	1.71		2649		C10H6O2	1,4-NAPHTHOQUINONE 130-15-4
6 ★	1.38		520		C10H6O3	1,4-NAPHTHOQUINONE,2-HYDROXY 83-72-7
7 ★	1.92		547		C10H6O3	5-HYDROXY-1,4-NAPHTHOQUINONE 481-39-0
8 ★	1.24		1478		C10H6O4	BIS-FURANE-2-CARBONYL 492-94-4
9 ★	1.63	0.8	2017		C10H6O4	CHROMONE-2-CARBOXYLIC ACID 4940-39-0 tmt of capillary fragility
5450	-1.08		520		C10H6O5S1	1,4-NAPHTHOQUINONE-2-SULFONATE,POTASSIUM SALT 34169-62-5
1 ★	4.24		547	401	C10H7Cl1	NAPHTHALENE,1-CHLORO 90-13-1
2 ★	4.14		594		C10H7Cl1	NAPHTHALENE,2-CHLORO 91-58-7
3 ★	2.27		576		C10H7Cl1F3N5	2'-CHLORO-5'-TRIFLUOROMETHYLBENZOGUANAMINE
4 ★	3.94		2641		C10H7Cl1O1	4-CHLORO-1-NAPHTHOL 604-44-4
55 ★	1.77		489	220	C10H7Cl1O4	6-CL-4-KETOCHROMAN-2-CARBOXYLIC ACID 33607-91-9
6	3.19		509		C10H7Cl2F3N2	BENZIMIDAZOLE,2-TRIFLUORMETHYL-4,7-DICL-5,6-DIME 7692-54-8
7	1.40		2060		C10H7Cl2N1O2	SUCCINIMIDE,N-(3,5-DICHLOROPHENYL) 24096-53-5
8 ★	4.56		331		C10H7Cl2N3O1	BUTANENITRILE,2[(3,4-DICHLOROPHENYL)HYDRAZONO]-3-OXO 28317-61-5
9 ★	4.68		331		C10H7Cl2N3O1	BUTANENITRILE,2[(3,5-DICHLOROPHENYL)HYDRAZONO]-3-OXO 28317-62-6
5460 ★	4.50		331		C10H7Cl2N3O2	3,5-DICL-C6H3NHN=C(CN)COOME 28313-58-8
1	4.34		2661		C10H7Cl5O1	TRIDIPHANE 58138-08-2 herbicide
2	4.92		2616	459	C10H7Cl5O2	ACETOFENATE
3 ★	3.72		2410		C10H7F6N1O1	ACETANILIDE,3,5-DI(TRIFLUOROMETHYL)
4 ★	2.14		1519	459	C10H7N1O1	2-QUINOLINECARBOXALDEHYDE 5470-96-2
65 ★	1.77		2649		C10H7N1O2	1,4-NAPHTHOQUINONE,2-AMINO 2348-81-4
6 ★	2.59		1142		C10H7N1O2	1-NITROAZULENE 7206-56-6
7 ★	3.19		562		C10H7N1O2	1-NITRONAPHTHALENE 86-57-7
8 ★	2.28	2.0	890		C10H7N1O2	1-NITROSO-2-NAPHTHOL 131-91-9
9 ★	2.46	3.5	594		C10H7N1O2	2-NITROSO-1-NAPHTHOL 132-53-6
5470	-2.03	7.4	447	1	C10H7N1O2	A-CARBOXY-I-QUINOLINE 486-73-7
1	-0.53	2.0	447		"	A-CARBOXY-I-QUINOLINE
2 ★	1.09		594		C10H7N1O2	MALEIMIDE,N-PHENYL 941-69-5
3 ★	2.27		538		C10H7N1O2	N-HYDROXY-1,4-NAPHTHOQUINONEMONIMINE 4965-30-4
4 ★	3.28	7.4	593	1	C10H7N1O2	NAPHTHALENE,2-NITRO 581-89-5
75	0.22	7.4	545	1	C10H7N1O3	3-CARBOXY-4-HYDROXYQUINOLINE 34785-11-0
6	0.23	7.4	1255	1	C10H7N1O3	4-HYDROXYQUINOLINE-3-CARBOXYLIC ACID
7 ★	3.50	6.5	1556	1	C10H7N1O3	6-NITRO-1-NAPHTHOL 38397-06-7
8 ★	2.26	3.5	2458	2	C10H7N1O3	IMIDAZOLINE-2,3,5-TRIONE-4-PHENYL
9	0.29	7.4	1255	1	C10H7N1O4	4,7-DIHYDROXYQUINOLINE-3-CARBOXYLIC ACID
5480 ★	1.34		555		C10H7N3O3	2-(5-NO2-2-FURFURILIDENE)PYRIMIDINE 1083-59-6 +
1	1.03		494		C10H7N3O5S1	2-ACETAMIDO-5-(5-NITRO-2-FURFURYL)THIAZOLONE
2	1.01		494		C10H7N3O5S1	2-ACETAMIDO-6(5-NO2-2-FURYL)-1,3-THIAZIN-4-ONE
3 ★	2.47	6.0	2560	1	C10H7N3S1	THIABENDAZOLE 148-79-8 anthelmintic~fungicide
4	1.27		1723		C10H7N5O1	2-PHENYL-8-AZAPURIN-6-ONE
85	1.85	7.2	2212	1	C10H7N5O2	CARBONYLCYANIDE,2-METHYL4-NITROPHENYLHYDRAZONE 64691-86-7
6 ★	3.55	7.2	2212	2	"	CARBONYLCYANIDE,2-METHYL4-NITROPHENYLHYDRAZONE
7	1.30	7.2	2212	1	C10H7N5O2	CARBONYLCYANIDE,2-NITRO-4-METHYLPHENYLHYDRAZONE 64691-87-8
8 ★	2.50	7.2	2212	2	"	CARBONYLCYANIDE,2-NITRO-4-METHYLPHENYLHYDRAZONE
9	3.41		331		C10H7N5O5	2,4-DINO2-C6H3NHN=C(CN)CO-ME 28317-71-7
5490 ★	3.20		525		C10H8	AZULENE 275-51-4
1 ★	3.30		541		C10H8	NAPHTHALENE 91-20-3
2 ★	4.70		1810	459	C10H8Br1F6N1	N,N-BIS(2,2,2-TRIFLUOROETHYL)ANILINE,P-BROMO
3 ★	1.09		1156		C10H8Br1N1O2	M-BROMO-N-PHENYLSUCCINIMIDE 58714-54-8
4 ★	0.65		1156		C10H8Br1N1O2	O-BROMO-N-PHENYLSUCCINIMIDE

	logP	pH	Ref.	Note	MF	Name / CAS / activity
95 ★	1.18		1156		C10H8Br1N1O2	P-BROMO-N-PHENYLSUCCINIMIDE 41167-74-2
6	2.80	7.4	1350	1	C10H8Br2F3N3	2(2,4-BR2-6-CF3-PHENYLIMINO)IMIDAZOLINE
7 ★	2.03		2649		C10H8Br2N2O2	BENZOQUINONE,2,5-BIS-AZIRIDINYL-3,6-DIBROMO
8 ★	0.98		1156		C10H8Cl1N1O2	M-CHLORO-N-PHENYLSUCCINIMIDE 15386-99-9
9 ★	0.56		1156		C10H8Cl1N1O2	O-CHLORO-N-PHENYLSUCCINIMIDE 7402-22-4
5500 ★	0.96		1156		C10H8Cl1N1O2	P-CHLORO-N-PHENYLSUCCINIMIDE 6943-00-6
1 ★	4.01		331		C10H8Cl1N3O1	BUTANENITRILE,2[(2-CHLOROPHENYL)HYDRAZONO-3-OXO 28317-59-1
2 ★	3.91		331		C10H8Cl1N3O1	BUTANENITRILE,2[(3-CHLOROPHENYL)HYDRAZONO]-3-OXO 28317-60-4
3 ★	4.13		331		C10H8Cl1N3O1	BUTANENITRILE,2[(4-CHLOROPHENYL)HYDRAZONO]-3-OXO 28317-58-0
4	1.14		1353		C10H8Cl1N3O1	PYRAZON 1698-60-8 herbicide
5	3.56		331		C10H8Cl1N3O2	3-CL-C6H4NH=C(CN)COOME 36874-69-8
6 ★	2.60		2506		C10H8Cl1N3O2	DRAZOXOLON 51450-97-6 fungicide ✚
7 ★	1.83	7.4	2206	306	C10H8F1N3O2	IMIDAZOLE,1-METHYL-5-NITRO-2-(P-FLUOROPHENYL)
8 ★	0.38		1156		C10H8F1O2	M-FLUORO-N-PHENYLSUCCINIMIDE 60693-35-8
9 ★	0.22		1156		C10H8F1O2	O-FLUORO-N-PHENYLSUCCINIMIDE
5510 ★	0.29		1156		C10H8F1O2	P-FLUORO-N-PHENYLSUCCINIMIDE 60693-37-0
1 ★	1.08	7.4	2504	2	C10H8F2N2O1S1	IMIDAZOLE-2-THIONE,3-(3,5-DIFLUORO-4-HYDROXYBENZYL) 95333-60-1
2 ★	1.98	7.4	2504	2	C10H8F2N2S1	IMIDAZOLE-2-THIONE,3-(3,5-DIFLUOROBENZYL) 95333-81-6
3 ★	1.74		2665		C10H8F3N3	1-BENZYLTRIAZOLE,3'-TRIFLUOROMETHYL fungicide
4 ★	2.67		576		C10H8F3N5	3'-(TRIFLUOROMETHYL)BENZOGUANAMINE 30508-78-2
15 ★	3.20		576		C10H8F3N5S1	3'-(TRIFLUOROMETHIO)BENZOGUANAMINE
6 ★	1.39		1156		C10H8I1O2	M-IODO-N-PHENYLSUCCINIMIDE
7 ★	0.80		1156		C10H8I1O2	O-IODO-N-PHENYLSUCCINIMIDE
8 ★	1.73		594		C10H8N2	2,2'-BIPYRIDINE 366-18-7
9 ★	1.92		579		C10H8N2	4-PHENYLPYRIMIDINE 3438-48-0
5520 ★	1.46	7.2	2317	1	C10H8N2	PYRIDINE,2-(3-PYRIDYL) 581-50-0
1 ★	1.58	7.2	2317	1	C10H8N2	PYRIDINE,2-(4-PYRIDYL) 581-47-5
2 ★	1.44		1519	459	C10H8N2O1	3-ACETYLCINNOLINE
3 ★	1.06		1519	459	C10H8N2O1	3-QUINOLINECARBOXAMIDE 6480-67-7
4 ★	1.80		1519	459	C10H8N2O1	8-FOMAMIDOQUINOLINE
25 ★	1.56		547		C10H8N2O1	BENZALCYANOACETAMIDE 709-79-5
6 ★	1.62	7.0	1016	1	C10H8N2O1	N-FORMYL-P-CYANOSTYRYLAMINE
7 ★	2.30		594		C10H8N2O1S2	BENZOTHIAZOLE,2-(4-AZETIDINONYL)THIO
8	1.77		2138	459	C10H8N2O2	NAPHTHALENE,1-AMINO-4-NITRO 776-34-1
9 ★	1.99	7.4	594	1	C10H8N2O2	QUINALDINE,8-NITRO 881-07-2
5530 ★	1.25	7.0	452	1	C10H8N2O3	2-METHYL-4-NITROQUINOLINE-1-OXIDE 4831-62-3
1 ★	1.06	7.0	452	1	C10H8N2O3	3-METHYL-4-NITROQUINOLINE-1-OXIDE 14073-00-8
2 ★	1.92		1519	459	C10H8N2O3	4-METHOXY-8-NITRO-QUINOLINE
3 ★	1.36	7.0	452	1	C10H8N2O3	5-METHYL-4-NITROQUINOLINE-1-OXIDE 14094-43-0
4 ★	1.43	7.0	452	1	C10H8N2O3	6-METHYL-4-NITROQUINOLINE-1-OXIDE 715-48-0
35 ★	1.42	7.0	452	1	C10H8N2O3	7-METHYL-4-NITROQUINOLINE-1-OXIDE 14753-13-0
6 ★	1.59	7.0	452	1	C10H8N2O3	8-METHYL-4-NITROQUINOLINE-1-OXIDE 14094-45-2
7 ★	1.87	7.4	594	1	C10H8N2O3	QUINOLINE,6-METHOXY-8-NITRO 85-81-4
8 ★	1.53		555		C10H8N2O3S1	4-ME-(5-NO2-2-FURFURILIDENE)THIAZOLE 6448-55-1
9	1.66		866		C10H8N2O3S2	3-RHODANO-4-AMINOBENZOIC ACID ✚
5540 ★	0.38		1156		C10H8N2O4	M-NITRO-N-PHENYLSUCCINIMIDE 31036-66-5
1 ★	0.36		1156		C10H8N2O4	O-NITRO-N-PHENYLSUCCINIMIDE 18377-52-1
2 ★	0.46		1156		C10H8N2O4	P-NITRO-N-PHENYLSUCCINIMIDE 35488-92-7
3 ★	2.29		494		C10H8N2O5S1	3-ET-5(5-NO2-2-FURFURILIDINE)THIAZOLIDINE-2,4-DIONE 25603-13-8
4	-1.13	7.4	1255	1	C10H8N2O5S1	7-SULFAMYL-4-HYDROXYQUINOLINE-3-CARBOXYLIC ACID
45 ★	1.38	7.4	593	1	C10H8N4	2-AMINODIPYRIDO[1,2-A:3',2'-D]IMIDAZOLE ✚
6	1.90	7.2	2212	1	C10H8N4	CARBONYLCYANIDE,2-METHYLPHENYLHYDRAZONE 55653-08-2
7 ★	2.38	7.2	2212	2	"	CARBONYLCYANIDE,2-METHYLPHENYLHYDRAZONE
8	2.10	7.2	1985	314	C10H8N4	PROPANEDINITRILE,4-METHYLPHENYLHYDRAZONE 40257-94-1
9 ★	3.20	3.0	2212		"	PROPANEDINITRILE,4-METHYLPHENYLHYDRAZONE
5550	1.20		2279		C10H8N4O6S1	IMIDAZOLE,1-METHYL-4-NITRO-5-O-NITROPHENYLSULFONATO
1 ★	2.84	7.4	772	1	C10H8O1	1-NAPHTHOL 90-15-3
2	2.95	7.4	2582	1	"	1-NAPHTHOL
3 ★	2.70		359	824	C10H8O1	2-NAPHTHOL 135-19-3
4 ★	2.78		1625	457	C10H8O1S1	BENZOTHIOPHENE,3-ACETYL 1128-05-8
55	2.09		2040	459	C10H8O2	BUTENELACTONE,5-(HEX-4-ENE-2-YNILIDINE)
6 ★	1.97		1113		C10H8O2	NAPHTHALENE-1,3-DIOL 132-86-5
7 ★	1.82		1113		C10H8O2	NAPHTHALENE-1,5-DIOL 83-56-7
8 ★	1.94		1113		C10H8O2	NAPHTHALENE-1,7-DIOL 575-38-2
9 ★	1.53		1105		C10H8O3	3-BENZOYLACRYLIC ACID 17812-07-6
5560	1.58	7.4	772	1	C10H8O3	4-METHYLUMBELLIFERONE 90-33-5 choleretic, antispasmodic
1 ★	1.90	7.4	2582	1	"	4-METHYLUMBELLIFERONE
2 ★	2.53		2642		C10H8O3	BENZOFURAN,2-METHOXYCARBONYL
3 ★	1.47	3.5	2458	2	C10H8O3	BUTYROLACTONE,3-OXO-4-PHENYL
4 ★	0.63	0.5	579	346	C10H8O3S1	NAPHTHALENE-2-SULFONIC ACID 120-18-3
65 ★	1.26	0.8	2017		C10H8O4	CHROMANONE-2-CARBOXYLIC ACID
6	2.59		1167	459	C10H8O4	CINNAMIC ACID,3,4-METHYLENEDIOXY 2373-80-0
7 ★	0.54		1478		C10H8O4	FURAN,2-CH(OH)CO-FURYL 552-86-3
8 ★	-0.18		1604		C10H8O4S1	1-NAPHTHOL-3-SULFONIC ACID 3771-14-0
9	-0.44		1054		C10H8O4S1	1-NAPHTHOL-4-SULFONIC ACID 84-87-7
5570	-0.17		1054		C10H8O4S1	1-NAPHTHOL-5-SULFONIC ACID 117-59-9
1 ★	3.74		541		C10H8S1	2-PHENYLTHIOPHENE 825-55-8
2 ★	1.96		522		C10H9Cl1N2O2S1	1,2,4-BENZOTHIADIAZINE-1,1-DIOXIDE-3-(PROPEN-1-YL)-6-CHLORO 37157-59-8
3 ★	1.98		522		C10H9Cl1N2O2S1	1,2,4-BENZOTHIADIAZINE-1,1-DIOXIDE-3-CYCLOPROPYL-6-CHLORO 13460-17-8
4 ★	2.89		2499		C10H9Cl1N4	PYRAZINE,2,3-DICYANO-5-CHLORO-6-BUTYL
75 ★	1.72	7.4	2341	1	C10H9Cl1N4O1	3[1-(2-BENZOXAZOLYL)HYDRAZINO]PROPANENITRILE,5-CHLORO
6	2.40		489	220	C10H9Cl1O3	6-CHLOROCHROMAN-2-CARBOXYLIC ACID 40026-24-2
7 ★	3.63		1792		C10H9Cl2N1O1	2-BUTENEAMIDE,N-(3,4-DICHLOROPHENYL)
8 ★	3.83		1792		C10H9Cl2N1O1	CYLOPROPANAMIDE,N-(3,4-DICHLOROPHENYL) 2759-71-9
9 ★	2.65	7.4	1616	1	C10H9Cl2N1O2	ACETOPHENONE,O-DICHLOROACETYLAMINO

	logP	pH	Ref.	Note	MF	Name / CAS / activity
5580 ★	4.56		354		C10H9Cl3	A-(2,2,2-TRICHLOROETHYL)STYRENE 20057-31-2
1 ★	3.83		1263		C10H9Cl4N1O2S1	DIFOLITAN 2425-06-1 fungicide
2 ★	3.53		1772		C10H9Cl4O4P1	TETRACHLORVINPHOS 961-11-5 antineoplastic~insecticide
3 ★	1.12	7.4	2504	2	C10H9F1N2O1S1	IMIDAZOLE-2-THIONE,3-(3-FLUORO-4-HYDROXYBENZYL) 95333-49-6
4 ★	0.10		306		C10H9F1N2O2	MALONAMIDE,3-FLUOROBENZAL 15948-56-8
85 ★	1.77	7.4	2504	2	C10H9F1N2S1	IMIDAZOLE-2-THIONE,3-(3-FLUOROBENZYL) 95333-80-5
6 ★	1.46		541		C10H9F3N2O4	1-(P-NITROPHENYL)-2-(A,A,A-TRIFLACETAMIDO)-ETHANOL
7	3.20	4.0	2264	1	C10H9F3N4O4	ACETONE,2,6-DINITRO-4-TRIFLUOROMETHYLPHENYLHYDRAZONE
8	3.54		579		C10H9F6N1	N,N-BIS(2,2,2-TRIFLUOROETHYL)ANILINE
9	3.69		1810	459	"	N,N-BIS(2,2,2-TRIFLUOROETHYL)ANILINE
5590 ★	1.64		2222		C10H9N1	1-H-TETRAZOLE,5-PHENYL
1 ★	2.59		324		C10H9N1	2-METHYLQUINOLINE 91-63-4 anesthetic (transport of fish)
2 ★	2.61	7.4	593	1	C10H9N1	4-METHYLQUINOLINE 491-35-0
3 ★	2.57		532		C10H9N1	6-METHYLQUINOLINE 91-62-3
4 ★	2.47		532		C10H9N1	7-METHYLQUINOLINE 612-60-2
95 ★	2.60		532		C10H9N1	8-METHYLQUINOLINE 611-32-5
6 ★	2.25		324		C10H9N1	A-NAPHTHYLAMINE 134-32-7
7	2.27	7.5	2201		"	A-NAPHTHYLAMINE
8 ★	2.28		562		C10H9N1	B-NAPHTHYLAMINE 91-59-8
9	2.40	7.5	2201		"	B-NAPHTHYLAMINE
5600 ★	2.68		594		C10H9N1	ISOQUINOLINE,3-METHYL 1125-80-0
1 ★	3.08		547		C10H9N1	N-PHENYLPYRROLE 635-90-5
2 ★	2.53		593		C10H9N1	QUINOLINE,3-METHYL 612-58-8
3 ★	2.37		826		C10H9N1O1	5-METHYL-8-QUINOLINOL 5541-67-3 anti-amebic, antifungal
4 ★	2.20		532		C10H9N1O1	6-METHOXYQUINOLINE 5263-87-6
5 ★	0.95	7.0	538	701	C10H9N1O1	6-METHYLQUINOLINE,N-OXIDE
6 ★	2.37		163	459	C10H9N1O1	7-METHOXYQUINOLINE 4964-76-5
7 ★	1.84		532		C10H9N1O1	8-METHOXYQUINOLINE 938-33-0
8 ★	2.33		137		C10H9N1O1	8-QUINOLINOL,2-METHYL 826-81-3
9 ★	2.41		137		C10H9N1O1	8-QUINOLINOL,4-METHYL 3846-73-9
5610 ★	2.06		1625	457	C10H9N1O1	INDOLE,3-ACETYL 25314-91-4
1 ★	2.22		1625	457	C10H9N1O1	ISOQUINOLINE,6-METHOXY 52986-70-6
2 ★	1.45		924	400	C10H9N1O1	N-METHYL-A-QUINOLONE 606-43-9
3 ★	0.44		924	400	C10H9N1O1	N-METHYL4-QUINOLONE 83-54-5
4 ★	1.29	7.4	595	314	C10H9N1O1	QUINOLLINE,7-METHANOL
15	-0.64	7.4	447	1	C10H9N1O2	2-CARBOXY-N-METHYLINDOLE 32387-21-6
6 ★	2.00		2221	459	"	2-CARBOXY-N-METHYLINDOLE
7 ★	1.59		826		C10H9N1O2	4-METHYL-5,8-DIHYDROXYQUINOLINE 57334-34-6
8 ★	2.06		826		C10H9N1O2	5-METHOXY-8-QUINOLINOL 57334-35-7
9 ★	1.59		536		C10H9N1O2	CARFIMATE 3567-38-2 ✚
5620	2.45		2642		C10H9N1O2	INDOLE,2-METHOXYCARBONYL
1 ★	2.57		2642		C10H9N1O2	INDOLE,3-METHOXYCARBONYL
2	-0.52	7.0	1073	1	C10H9N1O2	INDOLE-3-ACETIC ACID 87-51-4
3	1.41	7.0	1073	2	"	INDOLE-3-ACETIC ACID
4 ★	1.41		505		"	INDOLE-3-ACETIC ACID
25 ★	2.01		2221	459	C10H9N1O2	INDOLE-3-OL,ACETATE ESTER 608-08-2
6 ★	2.58		2221	449	C10H9N1O2	INDOLE-5-CARBOXYLIC ACID,METHYL ESTER 942-24-5
7 ★	0.06		1156		C10H9N1O2	N-PHENYLSUCCINIMIDE 83-25-0
8	0.48	3.5	2458	302	C10H9N1O2	VALEROLACTAM-3-OXO-4-PHENYL
9 ★	0.70		1519	400	C10H9N1O2S1	6-METHYLSULFONYLQUINOLINE
5630	-0.77	7.0	1073	1	C10H9N1O2S1	INDOL-3-YL-THIOACETIC ACID 54466-88-5
1 ★	1.78	7.0	1073	2	"	INDOL-3-YL-THIOACETIC ACID
2 ★	1.98	3.5	2458	2	C10H9N1O2S1	THIAZOLIDINE,2,4-DIONE-5-PHENYL
3 ★	2.09		1946		C10H9N1O3	2-PROPENOIC ACID,2-CYANO-3-(5-METHYL-2-FURANYL),METHYL ESTER 73403-31-3
4 ★	1.40		572		C10H9N1O3	3,4-DIOXYMETHYLENECINNAMAMIDE
35 ★	1.06		870		C10H9N1O3	3-ME-5-PHENYLOXAZOLIDINE-2,4-DIONE 5841-66-7
6 ★	1.37		594		C10H9N1O3	AZETIDIN-2-ONE,4-BENZOYLOXY
7	1.17	2.0	2312	337	C10H9N1O3	INDOLE-3-ACETIC ACID,5-HYDROXY
8	-1.18	5.8	547	312	C10H9N1O3S1	1-AMINO-2-NAPHTHALENSULFONIC ACID 81-06-1
9	-1.08	6.3	547	362	"	1-AMINO-2-NAPHTHALENSULFONIC ACID
5640 ✓	-0.97		547		"	1-AMINO-2-NAPHTHALENSULFONIC ACID
1	0.17	6.0	547	314	"	1-AMINO-2-NAPHTHALENSULFONIC ACID
2 ★	-1.16		579		C10H9N1O3S1	2-AMINO-1-NAPHTHALENESULFONIC ACID 81-16-3
3 ★	2.61		1946		C10H9N1O3S1	2-PROPENOIC ACID,2-CYANO-3-(5-METHYLTHIO-2-FURANYL),METHYL ESTER 76542-54-6
4	-2.34		579		C10H9N1O3S1	4-AMINO-NAPHTHALENE-1-SULFONIC ACID(AS SODIUM SALT)
45 ★	0.12	7.0	452	1	C10H9N1O3S1	4-METHYLSULFONYLQUINOLINE-1-OXIDE 20872-53-1
6 ★	1.40		2300	20	C10H9N1O3S1	PROBENAZOLE 27605-76-1 pesticide
7	1.11	7.4	2542	1	C10H9N1O5S3	THIOPHENE-2-SULFONAMIDE,4-(3-HYDROXYPHENYLSULFONYL)
8 ★	1.11	7.4	2542	1	C10H9N1O5S3	THIOPHENE-2-SULFONAMIDE,4-(4-HYDROXYPHENYL)SULFONYL
9 ★	-0.10	7.4	2542	1	C10H9N1O6S2	FURAN-2-SULFONAMIDE,4-(4-HYDROXYPHENYL)SULFONYL
5650 ★	3.04		1519		C10H9N1S1	2-METHYLTHIOQUINOLINE 40279-26-3
1	1.53		163	459	C10H9N3O1	2-ACETAMINOQUINOXALINE
2	1.06		163	459	C10H9N3O1	5-ACETAMINOQUINOXALINE
3	1.00		163	459	C10H9N3O1	6-ACETAMINOQUINOXALINE
4 ★	0.35		1519	459	C10H9N3O1	6-QUINOLINECARBOXYLIC ACID HYDRAZIDE
55 ★	1.62		595		C10H9N3O1	QUINOLINE,8-UREIDO 32451-61-9
6	2.18	7.4	2298	1	C10H9N3O2	1,2,4-OXADIAZOL-5-ALDOXIME,3-BENZYL 103499-06-5
7 ★	2.29	7.4	2298	2	"	1,2,4-OXADIAZOL-5-ALDOXIME,3-BENZYL
8 ★	0.19		547		C10H9N3O2	4-BENZYL-1,2,4-TRIAZIN-3-ONE-1-OXIDE
9 ★	2.59		331		C10H9N3O2	C6H5NHN=C(CN)COOME 36874-74-5
5660 ★	2.59		331		"	C6H5NHN=C(CN)COOME
1 ★	1.36	7.4	2206	306	C10H9N3O2	IMIDAZOLE,1-BENZYL-4-NITRO 13230-13-2
2 ★	1.58	7.4	2206	306	C10H9N3O2	IMIDAZOLE,1-METHYL-5-NITRO-4-PHENYL
3 ★	2.21		2166		C10H9N3O3	1,2,5-OXADIAZOLE,3-METHYL-4-(P-NITROBENZYL)
4 ★	1.14		494		C10H9N3O4S1	2-(N,N-DIMETHYL)-5-(5-NO2-2-FURFURYL)THIAZOLONE 25694-31-9

	logP	pH	Ref.	Note	MF	Name / CAS / activity
65 ★	1.16		494		C10H9N3O4S1	2-ETHYLAMINO-5-(5-NITRO-2-FURFURYL)THIAZOLONE 27472-85-1
6 ★	0.55		494		C10H9N3O4S1	2-N,N-DIME-6(5-NO2-2-FURYL)-1,3-THIAZINE-4-ONE
7 ★	1.44		494		C10H9N3O4S1	3-ET-2=NH-5-(5-NO2-2-FURFURILIDINE)THIAZOLINE-4-ONE 25603-08-1 ✢
8 ★	1.42		494		C10H9N3O4S1	3-ME-2-MEIMINO-5-(5-NO2-2-FURFURILIDINE)THIAZOL-4ONE 25603-09-2
9	1.20		2279		C10H9N3O4S1	IMIDAZOLE,1-METHYL-4-NITRO-5-PHENYLSULFONATO 80348-54-5
5670	1.08	7.4	1497	1	C10H9N3O5S1	4-NITROIMIDAZOLE,1-ME-5-SO3PH
1 ★	2.18		547		C10H9N5	1-(3,5-DICYANOPHENYL)-3,3-DIMETHYLTRIAZENE
2 ★	1.08		1570		C10H10Br1N1O3	BROMOACETIC ACID, P-ACETYLAMINOPHENYL ESTER
3 ★	1.53		576		C10H10Br1N5O1	2'-METHOXY-5'-BROMOBENZOGUANAMINE
4 ★	0.32	7.0	2440	1	C10H10Cl1F1N4O2	DIDEOXYPURINE,6-CHLORO-2'-FLUORO
75 ★	0.75		588		C10H10Cl1N1O4	PHENOXYACETIC ACID,3-ACETAMIDO-4-CHLORO
6 ★	0.53		522		C10H10Cl1N3O3S1	1,2,4-BENZOTHIADIAZINE-1,1-DIOXIDE-3-METHYL-6-ACETYLAMINO-7-CHLORO
7	-0.53	7.4	1350	1	C10H10Cl2N2	2-(2,6-DICHLOROBENZYL)IMIDAZOLINE 52115-81-8
8	2.12	7.4	1350	1	C10H10Cl2N4	KUM-32, CLONIDINE ANALOG ✢
9 ★	3.53		2456	346	C10H10Cl2O3	2,4-DICHLOROPHENOXYBUTYRIC ACID 94-82-6 *herbicide*
5680	0.00	7.4	821	1	C10H10F3N3	2-(2-TRIFLUOROMETHYLPHENYLIMINO)IMIDAZOLIDINE 40065-00-7
1 ★	1.42	7.4	821	2	"	2-(2-TRIFLUOROMETHYLPHENYLIMINO)IMIDAZOLIDINE
2 ★	1.63		2105		C10H10F3N3O2	HYDRAZINECARBOXAMIDE,1-ACETYL-N-(3-TRIFLUOROMETHYLPHENYL)
3	3.65		1283		C10H10F3N3O4	2,6-DINITRO-N-PR-A,A,A-TRIFLUORO-P-TOLUIDINE 2077-99-8
4 ✓	-2.64		527	820	C10H10N1.Br1	N-METHYLQUINOLINIUMBROMIDE 2516-72-5
85 ★	2.29		163	459	C10H10N2	6,7-DIMETHYLQUINOXALINE 7153-23-3
6 ★	1.60		2606	424	C10H10N2	IMIDAZOLE,1-BENZYL 4238-71-5
7	0.89		2138	459	C10H10N2	NAPHTHALENE,1,5-DIAMINO 2243-62-1
8	1.78		2139	459	C10H10N2	NAPHTHALENE,1,8-DIAMINO 479-27-6
9	1.54		2139	459	C10H10N2	NAPHTHALENE,2,3-DIAMINO 771-97-1
5690 ★	1.78		2606	424	C10H10N2	PYRAZOLE,1-BENZYL 10199-67-4
1 ★	2.74		594		C10H10N2O1	1,2,4-OXADIAZOLE,5-METHYL-3-(P-TOLYL) 55752-22-2
2 ★	1.05		1439		C10H10N2O1	4H-PYRIDO(1,2-A)PYRIMIDIN-4-ONE,2,6-DIMETHYL 16867-28-0
3 ★	1.26		1439		C10H10N2O1	4H-PYRIDO(1,2-A)PYRIMIDIN-4-ONE,3,6-DIMETHYL 39080-46-1
4 ★	2.22		163	459	C10H10N2O1	6-ETHOXYQUINOXALINE
95 ★	1.58	7.4	1616	1	C10H10N2O1	ACETOPHENONE,O-CYANOMETHYLAMINO
6 ★	1.74		574		C10H10N2O1S1	2-THIOBENZYL-5-ME-1,3,4-OXADIAZOLE
7 ★	1.02	7.4	2504	2	C10H10N2O1S1	IMIDAZOLE-2-THIONE,3-(4-HYDROXYBENZYL) 95333-64-5
8 ★	2.22		1519	459	C10H10N2O2	1,4-DIMETHOXYPHTHALAZINE 57315-37-4
9 ★	1.34		594		C10H10N2O2	2,3-QUINOXALINEDIONE,1,4-DIHYDRO-6,7-DIMETHYL 2474-50-2
5700 ★	0.18		2649		C10H10N2O2	BENZOQUINONE,2,5-BIS-(1-AZIRIDINYL)
1 ★	2.08		2166		C10H10N2O2	FURAZAN-2-OXIDE,3-METHYL-4-BENZYL
2 ★	2.22		2166		C10H10N2O2	FURAZAN-2-OXIDE,4-METHYL-3-BENZYL
3 ★	1.02		2434		C10H10N2O2	HYDANTOIN,5-METHYL-5-PHENYL
4 ★	0.60		2458	302	C10H10N2O2	IMIDAZOLIDINEDIONE,5-BENZYL
5 ★	0.76		2458	302	C10H10N2O2	IMIDAZOLINEDIONE,3-BENZYL
6 ★	-0.13		306		C10H10N2O2	MALONAMIDE,BENZAL 19411-83-7
7 ★	0.08		924		C10H10N2O2	N,N-DIMETHYL-QUINOXALINE-2,3-DIONE 58175-07-8
8 ★	1.27		594		C10H10N2O3	PHTHALIMIDE,5-(N,N-DIMETHYLAMINO)-N-HYDROXY
9 ★	0.98		547		C10H10N2O5	P-NITRO-METHYLHIPPURATE
5710 ★	2.35		574		C10H10N2S2	2-THIOBENZYL-5-METHYL-1,3,4-THIADIAZOLE
1 ★	0.90	7.4	2341	1	C10H10N4O1	3[1-(2-BENZOXAZOLYL)HYDRAZINO]PROPANENITRILE
2 ★	1.66		57		C10H10N4O1S1	3-METHIO-4-AMINO-6-PHENYL-1,2,4-TRIAZINE-5-ONE 21087-63-8
3 ★	0.37		2105		C10H10N4O2	HYDRAZINECARBOXAMIDE,1-ACETYL-N-(4-CYANOPHENYL)
4 ★	1.06	7.4	2206	306	C10H10N4O2	IMIDAZOLE,1-METHYL-5-NITRO-2-(P-AMINOPHENYL)
15 ★	-0.24	4.0	360	2	C10H10N4O2S1	4-SULFANILAMIDOPYRIMIDINE 599-82-6
6	-1.26	7.5	652	1	C10H10N4O2S1	SULFADIAZINE 68-35-9 *antibacterial*
7	-1.00	7.4	1157	324	"	SULFADIAZINE
8 ★	-0.09		480	601	"	SULFADIAZINE
9	1.19		1895		C10H10N4O3	1,2,3-TRIAZOLE,1-(P-NITROPHENOXY)ETHYL
5720 ★	-0.40	7.4	2360	1	C10H10N4O3	BENZOTRIAZINE,1,4-DI-N-OXIDE-3-ACETYLAMINO-7-METHYL
1 ★	0.76	7.4	2360	1	C10H10N4O3	BENZOTRIAZINE,1,4-DI-N-OXIDE-3-AMINO-7-ALLYLOXY
2	0.91		2279		C10H10N4O4S1	IMIDAZOLE,1-METHYL-4-NITRO-5-P-AMINOPHENYLSULFONATO
3 ★	0.44		1499		C10H10N4O5	5-(1-AZIRIDINYL)-2,4-DINITROBENZAMIDE,N-METHYL 24570-14-7
4	-2.11		401		C10H10N4O5	A-(OHME)-A'-(6-OH-9H-PURIN-9YL)-DIGLYCOLALDEHYDE
25	0.85	9.0	1409		C10H10N6O2	AZANIDAZOLE 62973-76-6 *antiprotozoal*
6	-0.30	9.0	1409		C10H10N6O3	5-NITROIMIDAZOLE,1-(N-2-PYRIMIDIN)ACETAMID-2-ME
7 ★	2.07		306		C10H10O1	METHYLSTYRYL KETONE 122-57-6
8 ★	1.73		1524		C10H10O2	1(3H)-ISOBENZOFURANONE-3,3-DIMETHYL
9	2.89		1167	459	C10H10O2	2-PHENYL-METHACRYLIC ACID, 1199-77-5
5730	2.52		896		C10H10O2	BENZOYLACETONE 93-91-4
1 ★	2.62		306		C10H10O2	CINNAMIC ACID,METHYL ESTER 103-26-4
2 ★	1.43	7.0	925	1	C10H10O2	M-ACETYLACETOPHENONE 6781-42-6
3 ★	1.34	7.0	925	1	C10H10O2	P-ACETYLACETOPHENONE 1009-61-6
4 ★	1.87		572		C10H10O3	3-KETOCARBOFURANPHENOL 17781-16-7 ✢
35 ★	1.30	3.5	2179	302	C10H10O3	BUTANOIC ACID,4-OXO-4-PHENYL
6 ★	1.90		489	220	C10H10O3	CHROMAN-2-CARBOXYLIC ACID 51939-71-0
7	2.37	2.0	2312	337	C10H10O3	M-METHOXYCINNAMIC ACID-TRANS 6099-04-3
8 ★	1.81	1.8	2648	448	C10H10O3	O-METHOXYCINNAMIC ACID,CIS
9 ★	2.54	1.8	2648	448	C10H10O3	O-METHOXYCINNAMIC ACID-TRANS 6099-03-2
5740 ★	1.29	7.0	925	1	C10H10O3	P-ACETOXYACETOPHENONE
1	2.68		1167	459	C10H10O3	P-METHOXYCINNAMIC ACID-TRANS 830-09-1
2 ★	1.46		2234		C10H10O4	ACETYLSALICYLIC ACID,METHYL ESTER 580-02-9 *perfume fixative*
3 ★	1.51	2.0	2312	337	C10H10O4	CINNAMIC ACID,3-METHOXY-4-HYDROXY 1135-24-6
4 ★	1.56		547		C10H10O4	DIMETHYLPHTHALATE 131-11-3 *repellant,insect*
45 ★	2.25	7.0	925	1	C10H10O4	DIMETHYLTEREPHTHALATE 120-61-6
6	2.22		1167	459	C10H10O4	METHYLCINNAMATE,3,4-DIHYDROXY 3843-74-1
7 ★	0.98		501		C10H10O4	PHENOXYACETIC ACID,M-ACETYL 1878-80-4
8 ★	1.25		501		C10H10O4	PHENOXYACETIC ACID,O-ACETYL 1878-62-2
9 ★	0.87		501		C10H10O4	PHENOXYACETIC ACID,P-ACETYL 1878-81-5

	logP	pH	Ref.	Note	MF	Name / **CAS** / *activity*
5750	1.37	7.4	899	1	C10H11Br1N2O3	BARBITURIC ACID,5-ALLYL-5(2-BROMALLYL) **561-86-4** *hypnotic*
1 ★	0.35	7.0	2411	1	C10H11Br1N4O2	2',3'-DIDEOXYPURINE,6-BROMO
2	-2.44	7.4	2357	200	C10H11Br1N5O6P1	CYCLIC AMP,8-BROMO
3	-2.64	7.4	2357	200	C10H11Br1N5O7P1	CYCLIC GMP,8-BROMO
4	1.87		458		C10H11Br1O2Se1	P-BROMOBENZYLSELENOPROPIONIC ACID **43169-99-9**
55	-1.71	7.4	1204	1	C10H11Br2N3	N-ME-N(2,6-BR2-PH)AMINO)-2-IMIDAZOLINE
6 ★	0.28	7.0	2440	1	C10H11Cl1F1N5O2	DIDEOXY-ARA-G,6-CHLORO-2'-FLUORO
7	-0.05	7.0	239	1	C10H11Cl1N2	5-CHLOROTRYPTAMINE **3764-94-1**
8 ★	2.00		522		C10H11Cl1N2O2S1	1,2,4-BENZOTHIADIAZINE-1,1-DIOXIDE-3-I-PROPYL-6-CHLORO **37148-20-2**
9 ★	2.04		522		C10H11Cl1N2O2S1	1,2,4-BENZOTHIADIAZINE-1,1-DIOXIDE-3-PROPYL-6-CHLORO **37148-19-9**
5760 ★	2.26		522		C10H11Cl1N2O2S2	1,2,4-BENZOTHIAZIDE-1,1-DIOXIDE-3-ETHYLTHIOMETHYL-6-CHLORO **37158-00-2**
1 ★	0.24	7.0	2411	1	C10H11Cl1N4O2	2',3'-DIDEOXYPURINE,6-CHLORO
2	1.90		458		C10H11Cl1O2Se1	M-CHLOROBENZYLSELENOPROPIONIC ACID **57239-19-7**
3	1.86		458		C10H11Cl1O2Se1	P-CHLOROBENZYLSELENOPROPIONIC ACID **57239-20-0**
4 ★	2.57		489	220	C10H11Cl1O3	CLOFIBRIC ACID **882-09-7** *antihyperlipoproteinemic*
65 ★	3.13		2625	448	C10H11Cl1O3	MCPP **7085-19-0** *herbicide*
6 ★	3.72		1792		C10H11Cl2N1O1	1-METHYL-PROPANEAMIDE,N-(3,4-DICHLOROPHENYL) **882-14-4**
7 ★	3.77		1792		C10H11Cl2N1O1	BUTANEAMIDE,N-(3,4-DICHLOROPHENYL) **2150-95-0**
8	1.06	7.4	1547	1	C10H11Cl2N3	2-(2,3-DICL-6-METHYLPHENYLAMINO)IMIDAZOLIDINE **82801-84-1**
9 ★	2.35		2400	2	"	2-(2,3-DICL-6-METHYLPHENYLAMINO)IMIDAZOLIDINE
5770	0.47	7.4	976	1	C10H11Cl2N3	2-(2,4-CL2-6-ME-PHENYLIMINO)IMIDAZOLINE
1 ★	2.38	7.4	976	2	"	2-(2,4-CL2-6-ME-PHENYLIMINO)IMIDAZOLINE
2	0.73	7.4	976	1	C10H11Cl2N3	2-(2,6-CL2-4-ME-PHENYLIMINO)IMIDAZOLINE
3 ★	1.92	7.4	976	2	"	2-(2,6-CL2-4-ME-PHENYLIMINO)IMIDAZOLINE
4	-0.34	7.4	1350		C10H11Cl2N3	ST-404 ✚
75	0.15	7.4	976	1	C10H11Cl2N3O1	2-(2,6-CL2-4-MEO-PHENYLIMINO)IMIDAZOLINE
6 ★	1.60	7.4	976	2	"	2-(2,6-CL2-4-MEO-PHENYLIMINO)IMIDAZOLINE
7 ★	3.58		2479		C10H11F1N3O2	PHENYLBIURET,2-FLUORO-5-IODO
8	-0.05	7.0	2440	1	C10H11F1N4O2	2',3'-DIDEOXYPURINE,6-FLUORO
9 ★	0.00	7.0	2411	1	C10H11F1N4O2	2',3'-DIDEOXYPURINE,6-FLUORO
5780 ★	-0.40	7.0	2440	1	C10H11F1N4O2	DIDEOXY-ARA-PURINE,2'-FLUORO
1 ★	-1.21	7.0	2440	1	C10H11F1N4O3	DIDEOXY-ARA-I,2'-FLUORO
2 ★	2.30		2479		C10H11F1N4O4	PHENYLBIURET,2-FLUORO-5-NITRO
3	1.63		458		C10H11F1O2Se1	M-FLUOROBENZYLSELENOPROPIONIC ACID **57239-16-4**
4	1.63		458		C10H11F1O2Se1	P-FLUOROBENZYLSELENOPROPIONIC ACID **57239-17-5**
85 ★	2.40		2479	459	C10H11F2N3O2	PHENYLBIURET,2,5-DIFLUORO
6 ★	2.42		1263		C10H11F3N2O1	1,1-DIME-3-(M-CF3-PHENYL)UREA **2164-17-2** *herbicide* ✚
7 ★	2.64		2612		C10H11F3N2O2	1-METHOXY-1-METHYL-3-(3-CF3-PHENYL)UREA **838-89-1**
8 ★	-0.46	4.2	401		C10H11F3N2O5	TRIFLURIDINE **70-00-8** *antiviral (ophthalmic)*
9 ★	1.86	6.6	2590	459	C10H11F3N4O1	BENZAMIDE,4-(3-METHYL-3-TRIFLUOROETHYLTRIAZENYL) *antineoplastic*
5790 ★	0.53	7.0	2411	1	C10H11I1N4O2	2',3'-DIDEOXYPURINE,6-IODO
1	2.03		458		C10H11I1O2Se1	P-IODO-BENZYLSELENOPROPIONIC ACID
2 ★	2.21		502		C10H11N1	G-PHENYLPROPYLCYANIDE **2046-18-6**
3 ★	2.82		519		C10H11N1	INDOLE,1,2-DIMETHYL **875-79-6**
4	-0.36	7.4	2648	1	C10H11N1O1	2-EXO-AMINO-9-OXABENZOBICYCLO(2,2,1)HEPTENE **73159-84-9**
95	0.66	7.4	2648	2	"	2-EXO-AMINO-9-OXABENZOBICYCLO(2,2,1)HEPTENE
6 ★	0.75	13.0	1623	224	"	2-EXO-AMINO-9-OXABENZOBICYCLO(2,2,1)HEPTENE
7	-0.21	7.4	2648	1	C10H11N1O1	9-OXABENZOBICYCLO(2,2,1)HEPTENE,2-ENDO-AMINO **73208-84-1**
8 ★	0.40	13.0	1623	224	"	9-OXABENZOBICYCLO(2,2,1)HEPTENE,2-ENDO-AMINO
9	0.51	7.4	2648	2	"	9-OXABENZOBICYCLO(2,2,1)HEPTENE,2-ENDO-AMINO
5800 ★	2.53		2221	459	C10H11N1O1	INDOLE-6-ETHOXY
1	1.81		572		C10H11N1O1	N-METHYLCINNAMAMIDE **2757-10-0**
2	1.23		2363		C10H11N1O1	PHENYLACETONITRILE,4-METHOXYMETHYL
3 ★	1.24		1894		C10H11N1O2	ACETANILIDE,4-ACETYL **2719-21-3**
4 ★	1.50	7.4	1616	1	C10H11N1O2	ACETOPHENONE,O-ACETYLAMINO **5234-26-4**
5 ★	1.15		594		C10H11N1O2	BENZYLCYANIDE,3,4-DIMETHOXY
6 ★	1.06	7.0	925	1	C10H11N1O2	M-ACETYLAMINOACETOPHENONE **7463-31-2**
7 ★	1.10		1894		"	M-ACETYLAMINOACETOPHENONE
8 ★	1.94	7.0	1016	1	C10H11N1O2	P-METHOXYSTYRYLAMINE,N-FORMYL (T) **53643-53-1**
9 ★	2.86		306		C10H11N1O2	STYRENE,B-ETHYL-B-NITRO **1202-32-0**
5810 ★	0.99		2157		C10H11N1O3	2-(BENZOYLOXY)-(N-METHYL)ACETAMIDE
1 ★	1.63		2142		C10H11N1O3	3-ACETYLPHENYLCARBAMATE,O-METHYL
2	1.25		866		C10H11N1O3	3-ALLYLOXY-4-AMINOBENZOIC ACID **2486-77-3**
3 ★	1.65		1894		C10H11N1O3	BENZOIC ACID,2-ACETYLAMINO,METHYL ESTER **2719-08-6**
4 ★	1.46		1894		C10H11N1O3	BENZOIC ACID,3-ACETYLAMINO,METHYL ESTER **52189-36-3**
15	-1.15	7.4	1616	1	C10H11N1O3	BENZOIC ACID,O-ACETYLMETHYLAMINO
6 ★	1.85	7.4	1616	2	"	BENZOIC ACID,O-ACETYLMETHYLAMINO
7 ★	1.03	7.0	925	1	C10H11N1O3	M-ACETOXYACETANILIDE **6317-89-1**
8 ★	0.82		547		C10H11N1O3	METHYLHIPPURATE **1205-08-9**
9	0.90		273		C10H11N1O3	N-METHYL-3-ACETYLPHENYLCARBAMATE
5820 ★	1.01		273		C10H11N1O3	N-METHYL-4-ACETYLPHENYLCARBAMATE **1135-43-9**
1 ★	1.48		594		C10H11N1O3	OXANILIC ACID,ETHYL ESTER **1457-85-8**
2 ★	0.80		1894		C10H11N1O3	P-ACETOXYACETANILIDE **2623-33-8**
3	1.05	7.0	925	1	"	P-ACETOXYACETANILIDE
4	1.64	7.0	925	1	C10H11N1O3	P-METHOXYCARBONYLACETANILIDE **17012-22-5**
25 ★	1.79		1894		"	P-METHOXYCARBONYLACETANILIDE
6 ★	1.42		273		C10H11N1O4	N-METHYL-3-CARBOMETHOXYPHENYLCARBAMATE
7 ★	1.50		273		C10H11N1O4	N-METHYL-4-CARBOMETHOXYPHENYLCARBAMATE
8 ★	0.48		501		C10H11N1O4	PHENOXYACETIC ACID,M-ACETAMIDO
9	0.52	7.4	2145	1	C10H11N1O4	SALICYLAMIDE,ACETYLOXYMETHYL ESTER
5830 ★	1.63		458		C10H11N1O4Se1	P-NITROBENZYLSELENOPROPIONIC ACID **34835-05-7**
1 ★	2.64		1946		C10H11N1O5	2-PROPENOIC ACID,3-(5-NITRO-2-FURANYL),I-PROPYL ESTER
2 ★	2.67		1946		C10H11N1O5	2-PROPENOIC ACID,3-(5-NITRO-2-FURANYL),PROPYL ESTER
3 ★	1.10		2665		C10H11N3O1	1-BENZYLTRIAZOLE,4'-METHOXY *fungcide*
4	1.65	7.0	332	1	C10H11N3O1S1	THIENO(2,3-D)-PYRIMIDINE,4-(N-MORPHOLINO) **18740-23-3**
35 ★	0.01		559		C10H11N3O2	1-(3-CARBOXYPROPYL)BENZOTRIAZOLE **654-19-3**
6 ★	1.33		2612		C10H11N3O2	1-(3-CYANOPHENYL)-3-METHOXY-3-METHYLUREA **88132-17-6**
7	1.30		559	818	C10H11N3O2	1-CARBETHOXYMETHYLBENZOTRIAZOLE
8 ★	0.60		559		C10H11N3O2	2-(3-CARBOXYPROPYL)BENZOTRIAZOLE **4144-70-1**
9	1.85		559	818	C10H11N3O2	2-CARBETHOXYMETHYLBENZOTRIAZOLE

	logP	pH	Ref.	Note	MF	Name / CAS / activity
5840	0.37	3.3	2590	331	C10H11N3O2	BENZOIC ACID,2-(3-AZETIDINYL TRIAZINE) *antineoplastic*
1	-1.20	5.6	2156	1	C10H11N3O2	BENZOTRIAZOLE,1-(2-BUTANOIC ACID)
2	0.20	5.6	2156	1	C10H11N3O2	BENZOTRIAZOLE,1-(2-METHYL-2-PROPIONIC ACID)
3	-0.20	5.6	2165	1	C10H11N3O2	BENZOTRIAZOLE,1-(3-BUTANOIC ACID)
4	-0.79	5.6	2156	1	C10H11N3O2	BENZOTRIAZOLE,2-(2-BUTANOIC ACID)
45	0.64	5.6	2156	1	C10H11N3O2	BENZOTRIAZOLE,2-(2-METHYL-2-PROPIONIC ACID)
6	0.34	5.6	2165	1	C10H11N3O2	BENZOTRIAZOLE,2-(3-BUTANOIC ACID)
7	1.63		479		C10H11N3O2S2	N1-METHYL-N1-(2-THIAZOLYL)SULFANILAMIDE 51203-19-1
8	0.26	4.0	1732		C10H11N3O3	1-ME-1-NITROSO-3-(P-ACETYLPHENYL)-UREA 72586-67-5
9 ★	0.89		480	601	C10H11N3O3S1	SULFAMETHOXAZOLE 723-46-6 *antibacterial*
5850	-0.97	7.4	1377	1	C10H11N3O6	N-(2,4-DINITROPHENYL)NORVALINE 31356-37-3
1 ★	1.68		574		C10H11N3S1	1-H-2-THIOBENZYL-5-ME-1,3,4-TRIAZOLE
2 ★	1.26		576		C10H11N5	2'-METHYLBENZOGUANAMINE 30508-25-9
3 ★	1.92		576		C10H11N5	4'-METHYLBENZOGUANAMINE 19388-12-6
4 ★	0.58		576		C10H11N5O1	2'-METHOXYBENZOGUANAMINE
55 ★	0.80		563		C10H11N5O1	N-CYANO-2(3,3-DIMETHYL-1-TRIAZENO)BENZAMIDE
6 ★	1.28	9.9	704	821	C10H11N5O2	4-CYANO-3-METHOXYPHENYLAMIDINOUREA 58247-25-9
7	-0.53	7.0	2440	1	C10H11N5O2	ADENOSINE,2',3'-DIDEOXYDIDEHYDRO 7057-48-9
8 ★	-0.36	7.5	2252	1	C10H11N5O2	ADENOSINE,2',3'-DIDEOXYDIDEHYDRO
9 ★	0.29		576		C10H11N5O2S1	3'-(METHYLSULFONYL)BENZOGUANAMINE
5860 ★	-1.21	7.4	2252	1	C10H11N5O3	2',3'-DIDEOXYDIDEHYDROGUANOSINE
1 ★	2.16		576		C10H11N5S1	3'-METHIOBENZOGUANAMINE
2 ★	1.08	5.7	1996	2	C10H11N5S1	GUANIDINE,N2-(4-(P-AMINOPHENYL)THIAZOL-2-YL)
3 ★	0.80	7.5	2252	1	C10H11N11O2	2',3'-DIAZIDO-DIDEOXYADENOSINE
4 ★	3.49		579		C10H12	TETRAHYDRONAPHTHALENE 119-64-2
65	0.24	7.4	1966	1	C10H12Au1Cl2N1O2	6-BUTYL-1,1,-DICHLORO-1,3,2-OXAZAURO(3,4)PYRININ-3-ONE
6 ★	6.15		2293	400	C10H12Br1Cl2O3P1S1	BROMOPHOS-ETHYL 4824-78-6 *insecticide & acaricide*
7 ★	2.12		2410		C10H12Br1N1O1	ACETANILIDE,2,6-DIMETHYL-4-BROMO
8 ★	2.27		2410		C10H12Br1N1O1	ACETANILIDE,2-ETHYL-4-BROMO
9 ★	3.33		2410		"	ACETANILIDE,2-ETHYL-4-BROMO
5870 ★	3.28		2410		C10H12Br1N1O1	ACETANILIDE,3,5-DIMETHYL-4-BROMO
1 ★	3.01		1681		C10H12Br1N1O2	N,N-DIMETHYLCARBAMATE,P-BROMOBENZYL ESTER 84640-33-5
2	3.61		1818		C10H12Br1N1O2	N-ETHYLBENZAMIDE,2-HYDROXY-3-METHYL-5-BROMO 40912-88-7
3	-0.19	7.4	1992	1	C10H12Br1N3	2-(2-METHYL-3-BROMOPHENYLIMINO)IMIDAZOLINE 16822-94-9
4	-0.02	7.4	2400	1	"	2-(2-METHYL-3-BROMOPHENYLIMINO)IMIDAZOLINE
75 ★	1.94		2400	2	"	2-(2-METHYL-3-BROMOPHENYLIMINO)IMIDAZOLINE
6	-0.25	7.4	2400	1	C10H12Br1N3	2-(2-METHYL-5-BROMOPHENYLIMINO)IMIDAZOLIDINE
7 ★	1.94		2400	2	"	2-(2-METHYL-5-BROMOPHENYLIMINO)IMIDAZOLIDINE
8 ★	0.34	7.0	2411	1	C10H12Br1N5O2	2',3'-DIDEOXYPURINE,2-AMINO-6-BROMO
9 ★	2.40	7.4	2392	2	C10H12Cl1N1	BENZAZEPINE,6-CHLORO 26232-35-9
5880 ★	2.95		536		C10H12Cl1N1O2	2,5-DIMETHYL-4-CHLORO-N-METHYLPHENYLCARBAMATE
1	-0.96		1592		C10H12Cl1N1O2	BACLOFEN 1134-47-0 *relaxant (muscle), thermogenic (bat)*
2 ★	3.51		2094		C10H12Cl1N1O2	CHLOROPHAM 101-21-3 *herbicide*
3 ★	2.82		1681		C10H12Cl1N1O2	N,N-DIMETHYLCARBAMATE,M-CHLOROBENZYL ESTER 84640-26-6
4 ★	2.93		1681		C10H12Cl1N1O2	N,N-DIMETHYLCARBAMATE,P-CHLOROBENZYL ESTER 84640-27-7
85	3.55		1818		C10H12Cl1N1O2	N-ETHYLBENZAMIDE,2-HYDROXY-3-METHYL-5-CHLORO
6	2.30		2661		C10H12Cl1N1O2	N-METHYLCARBAMATE,3,4-DIMETHYL,6-CHLOROPHENYL 671-04-5
7	-0.57	7.4	1992	1	C10H12Cl1N3	2-(2-CHLORO-3-METHYLPHENYLIMINO)IMIDAZOLINE 16822-97-2
8	-0.13	7.4	2400	1	"	2-(2-CHLORO-3-METHYLPHENYLIMINO)IMIDAZOLINE
9 ★	1.83		2400	2	"	2-(2-CHLORO-3-METHYLPHENYLIMINO)IMIDAZOLINE
5890	-0.11	7.4	2400	1	C10H12Cl1N3	2-(2-CHLORO-5-METHYLPHENYLIMINO)IMIDAZOLIDINE 16822-82-5
1 ★	1.83		2400	2	"	2-(2-CHLORO-5-METHYLPHENYLIMINO)IMIDAZOLIDINE
2	-0.96	7.4	1398	1	C10H12Cl1N3	2-(2-CL-4-ME-PHENYLIMINO)IMIDAZOLIDINE 4201-22-3
3	-0.48	7.4	976	1	"	2-(2-CL-4-ME-PHENYLIMINO)IMIDAZOLIDINE
4	-0.28	7.4	821	1	"	2-(2-CL-4-ME-PHENYLIMINO)IMIDAZOLIDINE
95 ★	1.80	7.4	976	2	"	2-(2-CL-4-ME-PHENYLIMINO)IMIDAZOLIDINE
6	1.87	7.4	821	2	"	2-(2-CL-4-ME-PHENYLIMINO)IMIDAZOLIDINE
7	-0.57	7.4	976	1	C10H12Cl1N3	2-(2-CL-6-ME-PHENYLIMINO)IMIDAZOLIDINE 4201-24-5
8	-0.40	7.4	821	1	"	2-(2-CL-6-ME-PHENYLIMINO)IMIDAZOLIDINE
9	1.41	7.4	821	2	"	2-(2-CL-6-ME-PHENYLIMINO)IMIDAZOLIDINE
5900 ★	1.71	7.4	976	2	"	2-(2-CL-6-ME-PHENYLIMINO)IMIDAZOLIDINE
1	-0.79	7.4	1992	1	C10H12Cl1N3	2-(2-METHYL-3-CHLOROPHENYLIMINO)IMIDAZOLINE
2	-1.06	7.4	976	1	C10H12Cl1N3	2-(4-CL-2-ME-PHENYLIMINO)IMIDAZOLIDINE 4201-26-7
3	-0.58	7.4	821	1	"	2-(4-CL-2-ME-PHENYLIMINO)IMIDAZOLIDINE
4	1.77	7.4	821	2	"	2-(4-CL-2-ME-PHENYLIMINO)IMIDAZOLIDINE
5 ★	1.80	7.4	976	2	"	2-(4-CL-2-ME-PHENYLIMINO)IMIDAZOLIDINE
6	-0.30	7.4	1398	1	C10H12Cl1N3	2-(5-CL-2-ME-PHENYLIMINO)IMIDAZOLIDINE 16822-85-8
7	-0.30	7.4	821	1	"	2-(5-CL-2-ME-PHENYLIMINO)IMIDAZOLIDINE
8	-0.27	7.4	2400	1	"	2-(5-CL-2-ME-PHENYLIMINO)IMIDAZOLIDINE
9	1.80	7.4	821	2	"	2-(5-CL-2-ME-PHENYLIMINO)IMIDAZOLIDINE
5910 ★	1.83		2400	2	"	2-(5-CL-2-ME-PHENYLIMINO)IMIDAZOLIDINE
1 ★	2.66		401	335	C10H12Cl1N3O3	1-(2-CHLOROETHYL)-3-(3-METHOXYPHENYL)-1-NITROSOUREA 13909-21-2
2 ★	1.15	6.5	321	1	C10H12Cl1N3O4S2	CHLOROTHIAZIDE-3-PROPYL 2854-99-1
3 ★	0.21	7.0	2411	1	C10H12Cl1N5O2	2',3'-DIDEOXYPURINE,2-AMINO-6-CHLORO
4 ✓	-1.06		2426		C10H12Cl2N2O2Pt1.(N1O3)2	PLATINUM,BIS-(P-CHLOROPYRIDONIO)-DIAQUA-DINITRATE
15 ★	5.23		1693		C10H12Cl3O2P1S1	TRICHLORONATE 327-98-0 *insecticide*
6	-0.80	7.4	1398	1	C10H12F1N3	2-(2-METHYL-5-FLUOROPHENYLIMINO)IMIDAZOLIDINE 28125-87-3
7	-0.77	7.4	2400	1	"	2-(2-METHYL-5-FLUOROPHENYLIMINO)IMIDAZOLIDINE
8 ★	1.41		2400	2	"	2-(2-METHYL-5-FLUOROPHENYLIMINO)IMIDAZOLIDINE
9 ★	2.12		2479		C10H12F1N3O2	PHENYLBIURET,2-FLUORO
5920 ★	-0.05	7.0	2411	1	C10H12F1N5O2	2',3'-DIDEOXYPURINE,2-AMINO-6-FLUORO
1 ★	0.08	7.5	2252	1	C10H12F1N5O2	3'-FLUORO-2',3'-DIDEOXYADENOSINE 87418-35-7 *antiviral*
2 ★	-0.18	7.0	2440	1	C10H12F1N5O2	DIDEOXY-ARA-A,2'-FLUORO
3 ★	-0.12	7.0	2440	1	C10H12F1N5O2	DIDEOXYADENOSINE,2'-ALPHA-FLUORO
4 ★	-0.03	7.0	2440	1	C10H12F1N5O2	DIDEOXYPURINE,2-AMINO-8-FLUORO
25 ★	-0.66	7.5	2252	1	C10H12F1N5O3	3'-FLUORO-2',3'-DIDEOXYGUANOSINE 92562-88-4
6 ★	-1.38	7.0	2440	1	C10H12F1N5O3	DIDEOXY-ARA-A,N1-OXIDE-2'-FLUORO
7 ★	-1.18	7.0	2440	1	C10H12F1N5O3	DIDEOXY-ARA-G,2'-FLUORO
8 ★	1.42		1182		C10H12F1O1	2-(M-FLUOROPHENYL)MORPHOLINE
9	-1.60		1592		C10H12F1O2	BAFLOFEN ✦

		logP	pH	Ref.	Note	MF	Name / CAS / activity
5930	★	2.30		1681		C10H12F1O2	N,N-DIMETHYLCARBAMATE,P-FLUOROBENZYL ESTER **84640-22-2**
1	★	0.52	7.0	2411	1	C10H12I1N5O2	2',3'-DIDEOXYPURINE,2-AMINO-6-IODO
2	★	3.32		1681		C10H12I1O2	N,N-DIMETHYLCARBAMATE,P-IODOBENZYL ESTER **84640-34-6**
3	★	1.13	11.0	533	322	C10H12N2	4,5-DIHYDRONICOTYRINE **525-74-6**
4	★	1.48	12.2	2600	322	C10H12N2	DIHYDRONICOTYRINE
35	★	2.65		2136		C10H12N2	TOLAZOLINE **59-98-3** *vasodilator (peripheral)*
6		-1.02	7.0	239	1	C10H12N2	TRYPTAMINE **61-54-1**
7	★	1.55	13.0	594		"	TRYPTAMINE
8	★	1.25		260		C10H12N2O1	2-(G-HYDROXYPROPYL)-BENZIMIDAZOLE **2403-66-9**
9	★	1.65		1263		C10H12N2O1	3-PHENYL-1-CYCLOPROPYLUREA
5940	★	-0.32		599	60	C10H12N2O1	COTININE **486-56-6** *antidepressant*
1		0.07	7.4	2600	1	"	COTININE
2	★	0.22		1725	459	C10H12N2O1	PYRIDO-PYRIMIDINONE ANALOG(RINGS=565) ✚
3		-1.75	7.0	2046	1	C10H12N2O1	SEROTONIN **50-67-9** *neurotransmitter*
4	★	0.21		505		"	SEROTONIN
45		1.88		1407		C10H12N2O1S1	1,2,3,6-H4-PYRIDINE,N-(2-THIENYL)CONH-
6		0.81		1407		C10H12N2O2	1,2,3,6-H4-PYRIDINE,N-(2-FURANYL)CONH-
7		-2.06	7.0	477	1	C10H12N2O2	1-(O-CYANOPHENOXY)-3-AMINOPROPANOL-2
8		-2.93	2.0	2600	342	C10H12N2O2	3'-HYDROXYCOTININE,CIS
9		-1.48	12.2	2600		C10H12N2O2	3'-HYDROXYCOTININE,CIS
5950		-1.22	7.4	2600	1	C10H12N2O2	3'-HYDROXYCOTININE,CIS
1		-2.97	2.0	2600	342	C10H12N2O2	3'-HYDROXYCOTININE,TRANS
2		-1.45	12.2	2600		C10H12N2O2	3'-HYDROXYCOTININE,TRANS
3		-1.20	7.4	2600	1	C10H12N2O2	3'-HYDROXYCOTININE,TRANS
4		-2.02	2.0	2600	342	C10H12N2O2	5'-HYDROXYCOTININE
55		-0.57	12.2	2600		C10H12N2O2	5'-HYDROXYCOTININE
6		-0.16	7.4	2600	1	C10H12N2O2	5'-HYDROXYCOTININE
7	★	-0.03		1894		C10H12N2O2	ACETANILIDE,2-ACETYLAMINO **2050-85-3**
8	★	0.61		1616	1	C10H12N2O2	BENZAMIDE,O-ACETYLMETHYLAMINO
9		-2.87	12.2	2600		C10H12N2O2	COTININE-N-OXIDE
5960		-2.61	2.0	2600	342	"	COTININE-N-OXIDE
1		-2.54	7.4	2600	1	"	COTININE-N-OXIDE
2	★	1.92		2198		C10H12N2O2	ETHYLUREA,N'-BENZOYL
3		-2.20	2.0	2600	342	C10H12N2O2	G-(3-PYRIDYL)-G-OXO-N-METHYLBUTYRAMIDE
4		-0.69	12.2	2600		"	G-(3-PYRIDYL)-G-OXO-N-METHYLBUTYRAMIDE
65		-0.13	7.4	2600	1	"	G-(3-PYRIDYL)-G-OXO-N-METHYLBUTYRAMIDE
6	★	0.50		547		C10H12N2O2	M-DIACETAMIDOBENZENE **10268-78-7**
7	★	2.67		520		C10H12N2O2	STYRENE,4-DIMETHYLAMINO,B-NITRO **2604-08-2**
8	★	1.25		522		C10H12N2O2S1	1,2,4-BENZOTHIADIAZINE-1,1-DIOXIDE-3-METHYL-6-ETHYL **37148-03-1**
9	★	-0.25		1439		C10H12N2O3	3-MEOCO-PYRIDO(1,2A)PYRIMIDIN-4-ON,4,6,7,8,9-H5 ✚
5970	★	1.18		2410		C10H12N2O3	ACETANILIDE,2,4-DIMETHYL-6-NITRO
1		0.94	7.4	899	1	C10H12N2O3	BARBITURIC ACID,DIALLYL **52-43-7** *hypnotic*
2		0.99	7.4	1873	1	"	BARBITURIC ACID,DIALLYL
3	★	1.05	0.7	962		"	BARBITURIC ACID,DIALLYL
4	★	1.21		2030		C10H12N2O3	NITROSOAMINE,N-METHYL,N-(4-METHOXYCARBONYLBENZYL)
75	★	-0.23		547		C10H12N2O3	P-AMINOHIPPURIC ACID,METHYL ESTER **5259-86-9**
6	★	-1.90	7.2	2323	468	C10H12N2O3	PHENYLASPARAGINE
7	★	2.80		579		C10H12N2O3S1	BENTAZON **25057-89-0** *herbicide*
8	★	2.59	7.2	1685	314	C10H12N2O3S2	BENZOTHIAZOLE,2-N-METHYLSULFONAMIDO-6-ETHOXY **88946-19-4**
9	★	1.86		1946		C10H12N2O4	2-PROPENOIC N-PROPYLAMIDE,3-(5-NITRO-2-FURANYL)
5980	★	1.98	7.4	2659	459	C10H12N2O4	M-NITROBENZYLCARBAMATE,N,N-DIMETHYL
1	★	1.95		1681		C10H12N2O4	N,N-DIMETHYLCARBAMATE,P-NITROBENZYL ESTER **84640-31-3**
2	★	1.34		611		C10H12N2O4	N-I-PROPYL-3-(5-NO2-2-FURYL)ACRYLAMIDE **1951-56-0**
3	★	0.65		2499		C10H12N2O4	PYRAZINE,2,3-BIS-ETHOXYCARBONYL
4	★	-0.81		2252		C10H12N2O4	THYMIDINE,2',3'-DIDEHYDRO **3056-17-5** *antiviral (hiv)*
85		-0.15	7.4	2461	1	C10H12N2O4S4	THIENO[2,3-B]THIOPHENE-2-SULFONAMIDE,5-N-(METHYLSULFOXYETHYL)CARBOXAMIDO **129949-91-3**
6	★	-2.19	7.0	2440	1	C10H12N2O5	CPE-U
7	★	3.56		2215		C10H12N2O5	DINOSEB **88-85-7** *herbicide*
8		0.30	6.5	2220	1	C10H12N2O5	I-DINOSEB **530-17-6**
9		1.66	5.5	2220	1	"	I-DINOSEB
5990	★	1.72		2072	2	C10H12N2O5S1	BENZENESULFONAMIDE,2-NITRO-N-(I-BUTANOYL)
1		-0.35	7.4	2461	1	C10H12N2O5S3	THIENO[2,3-B]THIOPHENE-2-SULFONAMIDE,5-N-(2,3-DIHYDROXYPROPYL)CARBOXAMIDO **129949-90-2**
2	★	0.72		1049		C10H12N3O3P1	ETHYLPHOSPHORAMIDATE,A-CYANOBENZALDOXIME ESTER
3	★	2.75		572		C10H12N3O3P1S2	METHYLAZINPHOS **86-50-0** *insecticide*
4	★	0.78		1693		C10H12N3O4P1S1	METHYLAZINPHOS,O-ANALOG **961-22-8**
95	★	0.37		559		C10H12N4O1	1-(3-CARBAMYLPROPYL)BENZOTRIAZOLE
6	★	0.78		559		C10H12N4O1	2-(3-CARBAMYLPROPYL)BENZOTRIAZOLE
7	★	2.91	6.6	2590	459	C10H12N4O1	BENZAMIDE,2-(3-AZETIDINYL TRIAZENE) *antineoplastic*
8	★	0.88	5.6	2156	1	C10H12N4O1	BENZOTRIAZOLE,1-(2-BUTYRAMIDE)
9	★	0.76	5.6	2156	1	C10H12N4O1	BENZOTRIAZOLE,1-(2-METHYL-2-PROPIONAMIDE)
6000	★	0.75	5.6	2165	1	C10H12N4O1	BENZOTRIAZOLE,1-(3-BUTANAMIDE)
1	★	1.25	5.6	2156	1	C10H12N4O1	BENZOTRIAZOLE,2-(2-BUTYRAMIDE)
2	★	1.16	5.6	2165	1	C10H12N4O1	BENZOTRIAZOLE,2-(2-METHYL-2-PROPIONAMIDE)
3	★	1.00	5.6	2165	1	C10H12N4O1	BENZOTRIAZOLE,2-(3-BUTANAMIDE)
4	★	0.72	7.0	1621	1	C10H12N4O1	PYRAZOL-3-ONE,1,2-H2-1-(4,6-DIME-2-PYRIMIDINYL)-5-ME **18597-53-0**
5	★	0.89	7.0	1621	1	C10H12N4O1	PYRAZOL-3-ONE,1,2-H2-2-(2,6-DIME-4-PYRIMIDINYL)-5-ME **18597-57-4**
6	★	1.29	7.0	1621	1	C10H12N4O1	PYRAZOLE-3-ONE,1(4,6-DIME-2-PYRIMIDINYL)-5-ME **18697-50-2**
7	★	1.37	6.6	2590	459	C10H12N4O2	BENZAMIDE,4-(3-METHYL-3-ACETYLTRIAZENYL) *antineoplastic*
8	★	1.83		2463		C10H12N4O2	BENZOTRIAZOLE,2-METHYL-5-ETHOXYCARBAMIDO
9	★	-0.46	7.0	2440	1	C10H12N4O2	DIDEOXYPURINE
6010	★	0.72	7.0	1621	1	C10H12N4O2	PYRAZOL-3-ONE,H2-2-(2-ME-6-MEO-4-PYRIMIDINYL)-5-ME **18597-55-2**
1	★	1.03	7.0	1621	1	C10H12N4O2	PYRAZOL-3-ONE,H2-2-(2-MEO-6-ME-4-PYRIMIDINYL)-5-ME **23898-90-0**
2	★	0.30	7.0	1621	1	C10H12N4O2	PYRAZOL-3-ONE,H2-2-(6-MEO-4-PYRIMIDINYL)-1,5-DIME **23906-03-8**
3	★	1.53	7.0	1621	1	C10H12N4O2	PYRIMIDINE,4(5-MEO-3-ME-1H-PYRAZOL-1-YL)6-MEO **18694-42-3**
4	★	1.01	7.0	1621	1	C10H12N4O2	PYRIMIDINE,4-MEO-2-(5-MEO-3-ME-1H-PYRAZOL-1YL) **23917-24-0**

	logP	pH	Ref.	Note	MF	Name / **CAS** / *activity*
15 ★	-0.66	7.0	2440	1	C10H12N4O2S1	DIDEOXYPURINE,6-THIOL
6 ★	1.01	3.0	883		C10H12N4O2S2	SULFAETHIDOLE **94-19-9** *antibacterial*
7 ★	-1.24	7.0	2440	1	C10H12N4O3	2',3'-DIDEOXY-INOSINE **69655-05-6** *antiviral, antineoplastic*
8 ★	0.60		2070		C10H12N4O3	ALLOPURINOL,1-BUTYRYLOXYMETHYL
9 ★	0.33		2070		C10H12N4O3	ALLOPURINOL,2-BUTYRYLOXYMETHYL
6020 ★	1.16	7.0	1621	1	C10H12N4O3	PYRAZOL-3-ONE,H2-1-(4,6-DIMEO-2-PYRIMIDINYL)-5-ME **23905-85-3**
1 ★	-1.71	7.0	2440	1	C10H12N4O4	2'-DEOXYINOSINE
2 ★	-0.60	7.4	401	2	C10H12N4O4S1	6-MERCAPTOPURINERIBOSIDE **4988-64-1**
3 ★	1.88		553		C10H12N4O4S2	2(1-ACETOXET-1,3-THIAZOLIDINYLIDEN-2N)5-NO2THIAZOLE **37427-69-3**
4 ★	-1.10	7.4	2360	1	C10H12N4O5	BENZOTRIAZINE-1,4-DI-N-OXIDE-3-AMINO-7-(2,3-DIHYDROXY)PROPOXY
25 ★	-2.10	7.0	2440	1	C10H12N4O5	INOSINE **58-63-9** *cardiotonic*
6	-2.08	7.4	293	1	"	INOSINE
7	-2.96	7.4	2357	200	C10H12N5O6P1	CYCLIC AMP **60-92-4**
8	-3.52	7.4	2357	200	C10H12N5O7P1	CYCLIC GMP
9	1.26	7.4	2028	459	C10H12N6	1,2,4-TRIAZOLE,3-AMINO-5-(2-ALLYLAMINO)PYRIDIN-4-YL **77314-61-5**
6030 ★	0.33	7.4	2349	1	C10H12N6O2	2,6-DIAMINOPURINE,2',3'-DIDEOXYDIDEHYDRORIBOSIDE
1	-0.76	3.9	401	2	C10H12N8O1	IMIDAZOLE,5-CARBOXAMIDE-4-(3,3-BIS(CYANOETHYL)TRIAZENYL)
2 ★	0.35	7.5	2252	1	C10H12N8O2	3'-AZIDO-2',3'-DIDEOXYADENOSINE **66323-44-2**
3 ★	-0.33	7.5	2252	1	C10H12N8O3	3'-AZIDO-2',3'-DIDEOXYGUANOSINE **66323-46-4**
4	2.77	2.0	2312	337	C10H12O1	BUTYROPHENONE **495-40-9**
35 ★	2.73		579		C10H12O1	ISOBUTYROPHENONE **611-70-1**
6 ★	1.95		508		C10H12O1	TR-2-PHENYLCYCLOPROPYLCARBINOL **936-98-1**
7 ★	2.23		2649		C10H12O2	2,3,5,6-TETRAMETHYL-1,4-BENZOQUINONE **527-17-3**
8 ★	2.42		558		C10H12O2	2-ETHYLPHENYLACETATE **3056-59-5**
9 ★	2.18	3.5	1542	1	C10H12O2	3-PHENYL-N-BUTYRIC ACID
6040 ★	2.56		558		C10H12O2	4-ETHYLPHENYLACETATE **3245-23-6**
1	0.96	3.9	448	2	C10H12O2	4-I-PROPYLTROPALONE **499-44-5**
2 ★	2.42		502		C10H12O2	4-PHENYLBUTYRIC ACID **1821-12-1**
3 ★	1.89	5.0	447	1	C10H12O2	5-I-PROPYLTROPALONE **672-76-4**
4	1.82	5.0	447	1	C10H12O2	6-I-PROPYLTROPALONE
45 ★	2.30		502		C10H12O2	ACETIC ACID,B-PHENYLETHYL ESTER **103-45-7**
6 ★	2.44		1521		C10H12O2	ACETIC ACID,P-METHYLBENZYL ESTER **2216-45-7**
7 ★	2.32		502		C10H12O2	B-PHENYLPROPIONIC ACID,METHYL ESTER **103-25-3**
8 ★	2.55		2645		C10H12O2	BENZOIC ACID,2,6-DIMETHYL,METHYL ESTER
9 ★	3.18		1895		C10H12O2	BENZOIC ACID,I-PROPYL ESTER **939-48-0**
6050 ★	3.42	2.0	2522	342	C10H12O2	BENZOIC ACID,P-PROPYL
1 ★	3.01		594		C10H12O2	BENZOIC ACID,PROPYL ESTER **2315-68-6**
2 ★	3.40	2.0	2236		C10H12O2	BENZOIC ACID-4-ISOPROPYL **536-66-3**
3	2.20		1927		C10H12O2	BENZOQUINONE,2-I-PROPYL-5-METHYL
4 ★	2.08		572		C10H12O2	CARBOFURANPHENOL
55 ★	2.32		2645		C10H12O2	PHENYLACETATE,2,6-DIMETHYL
6 ★	2.28		2222		C10H12O2	PHENYLACETIC ACID,ETHYL ESTER **101-97-3**
7	1.67		458		C10H12O2Se1	BENZYLSELENOPROPIONIC ACID **6926-05-2**
8 ★	1.95		558		C10H12O3	4-ETHOXYPHENYLACETATE
9 ★	2.02		1521		C10H12O3	ACETIC ACID,M-METHOXYBENZYL ESTER **35480-26-3**
6060 ★	1.78		1705		C10H12O3	BENZOIC ACID,2-METHOXYETHYL ESTER **57453-98-2**
1 ★	3.09	2.0	2522	342	C10H12O3	BENZOIC ACID,4-PROPOXY **5438-19-7**
2 ★	1.53		572		C10H12O3	MANDELIC ACID,ETHYL ESTER **774-40-3**
3 ★	2.08	1.0	782	342	C10H12O3	O-METHYLPHENOXYACETIC ACID,METHYL ESTER
4 ★	1.95	1.2	1077		C10H12O3	P-ETHOXYPHENYLACETIC ACID **4919-33-9**
65	3.01	7.5	1259	1	C10H12O3	P-HYDROXYBENZOIC ACID,PROPYL ESTER **94-13-3** *pharmaceutic aid (antifungal agent)*
6 ★	3.04		508		"	P-HYDROXYBENZOIC ACID,PROPYL ESTER
7 ★	1.90	1.2	1077		C10H12O3	P-METHOXY-B-PHENYLPROPIONIC ACID **1929-29-9**
8 ★	2.25		501		C10H12O3	PHENOXYACETIC ACID,M-ETHYL **1878-51-9**
9 ★	2.53	1.0	547	342	C10H12O3	PHENOXYACETIC ACID,O-ETHYL **1798-03-4**
6070 ★	2.08		2645		C10H12O3	PHENYLACETATE,M-ETHOXY
1 ★	3.80	1.2	2586		C10H12O3	PROPYL SALICYLATE
2 ★	0.58		1898		C10H12O4	(4-HYDROXYETHYL)VANILLIN
3 ★	1.26		2452		C10H12O4	BENZOQUINONE,2,3-DIMETHOXY-5,6-DIMETHYL **483-54-5**
4	1.33	2.0	2312	337	C10H12O4	MANDELIC ACID,3-METHOXY
75 ★	2.90		2186		C10H12O4S2	DITHIOLANYLIDENEMALONIC ACID,DIETHYL ESTER
6 ★	2.72		2186		C10H12O4S2	DITHIOLANYLIDINEMALONIC ACID,METHYLPROPYL ESTER
7 ★	1.54		1335		C10H12O5	I-PROPYLGALLATE **1138-60-9**
8	0.84	2.0	2312	337	C10H12O5	MANDELIC ACID,4-HYDROXY-3-METHOXY
9 ★	1.80		1335		C10H12O5	PROPYLGALLATE **121-79-9** *antioxidant*
6080 ★	0.54		2393	345	C10H12O5S1	PHENOXYACETIC ACID,3-METHYL-4-MESYL **15267-77-3**
1 ★	0.41		1963		C10H12O6	BENZOQUINONE,2,3,5,6-TETRAMETHOXY **3117-06-4**
2	1.47	7.4	899	1	C10H13Br1N2O3	BARBITURIC ACID,5-(2-BROMALLYL)-5-I-PROPYL **545-93-7** *sedative, hypnotic*
3 ✓	-1.53		2397	101	C10H13Br1N3O2	IMIDAZOLIUM CHLORIDE,1-METHYL-2-HYDROXYIMINOMETHYL-3-(4-BROMOBUTYN-3YLOXY)METHYL
4	1.05	7.4	1948	1	C10H13Cl1N2	1-(M-CHLOROPHENYL)PIPERAZINE **6640-24-0** *serotonin-agonist (5-ht1b)*
85	2.11	7.4	1948	2	"	1-(M-CHLOROPHENYL)PIPERAZINE
6	0.75	7.4	1948	1	C10H13Cl1N2	1-(O-CHLOROPHENYL)PIPERAZINE **39512-50-0**
7	2.13	7.4	1948	2	"	1-(O-CHLOROPHENYL)PIPERAZINE
8	0.91	7.4	1948	1	C10H13Cl1N2	1-(P-CHLOROPHENYL)PIPERAZINE
9	2.01	7.4	1948	2	"	1-(P-CHLOROPHENYL)PIPERAZINE
6090 ★	2.89		1693		C10H13Cl1N2	CHLORDIMEFORM **6164-98-3** *acaricide*
1 ★	2.41		2094		C10H13Cl1N2O1	CHLORTOLURON **15545-48-9** *herbicide*
2 ★	1.64		1263		C10H13Cl1N2O2	METOXURON **19937-59-8** *herbicide*
3 ★	1.73		1019		C10H13Cl1N2O2	N'-ETHYL-(4-CHLOROPHENOXY)ACETIC ACID HYDRAZIDE
4 ★	1.87		2072	2	C10H13Cl1N2O3S1	BENZENESULFONAMIDE,2-(I-BUTYROYLAMINO)-4-CHLORO
95 ★	2.27	7.4	2004	2	C10H13Cl1N2O3S1	CHLORPROPAMIDE **94-20-2** *antidiabetic*
6 ★	3.92		1135	537	C10H13Cl1O1	CHLOROTHYMOL **89-68-9**
7 ★	3.95		2506		C10H13Cl2F1N2O2S2	TOLYLFLUANID **731-27-1** *fungicide*
8 ★	2.90	7.4	401	1	C10H13Cl2N1	N,N-DI-B-CHLOROETHYLANILINE **553-27-5** *antineoplastic*
9 ★	5.14		928		C10H13Cl2O3P1S1	DICHLOFENTHION **97-17-6** *nematocide~insecticide*

	logP	pH	Ref.	Note	MF	Name / CAS / activity
6100 ★	3.70		354		C10H13Cl3N1O3P1	PROPYLPHOSPHORAMIDATE,O-ME-O-(2,4,5-TRICLPHENYL) 2213-85-6
1	0.22	7.4	1948	1	C10H13F1N2	1-(P-FLUOROPHENYL)PIPERAZINE 2252-63-3
2 ★	1.28	7.4	1948	2	"	1-(P-FLUOROPHENYL)PIPERAZINE
3 ★	-0.28		2252		C10H13F1N2O4	THYMIDINE,3'-DEOXY-3'-FLUORO 25526-93-6 antiviral
4 ★	-0.44		2521		C10H13F1N2O4	THYMINE,1-(2,3-DIDEOXY-2-FLUOROPENTOFURANOSYL)
5 ★	0.05	7.5	2252	1	C10H13F1N6O2	2,6-DIAMINOPURINE,3'-FLUORO-2',3'-DIDEOXYRIBOSIDE
6 ★	2.08	13.0	1806	224	C10H13N1	2-NAPHTHYLENEAMINE-1,2,3,4-TETRAHYDRO 2954-50-9
7 ★	1.93	13.0	1806	224	C10H13N1	3,4-BENZOPIPERIDINE-6-METHYL
8 ★	3.56	13.0	594	400	C10H13N1	PYRROLIDINE,1-PHENYL 4096-21-3
9 ★	1.11		1182		C10H13N1O1	2-PHENYLMORPHOLINE 23972-41-0
6110 ★	1.35		2410		C10H13N1O1	ACETANILIDE,2,3-DIMETHYL
1 ★	1.46		2410		C10H13N1O1	ACETANILIDE,2,4-DIMETHYL
2 ★	1.47		2410		C10H13N1O1	ACETANILIDE,2,5-DIMETHYL
3 ★	1.07		2410		C10H13N1O1	ACETANILIDE,2,6-DIMETHYL
4 ★	1.33		1894		C10H13N1O1	ACETANILIDE,2-ETHYL 33098-65-6
15 ★	2.10		2410		C10H13N1O1	ACETANILIDE,3,4-DIMETHYL
6 ★	2.17		2410		C10H13N1O1	ACETANILIDE,3,5-DIMETHYL
7 ★	2.18		2222		C10H13N1O1	ACETANILIDE,4-ETHYL 3663-34-1
8 ★	1.39		2410		C10H13N1O1	BENZAMIDE,2-I-PROPYL
9 ★	1.41		502		C10H13N1O1	BUTYRAMIDE,4-PHENYL
6120 ★	2.05		603		C10H13N1O1	BUTYRANILIDE 1129-50-6
1 ★	1.95		547		C10H13N1O1	I-BUTYRANILIDE 4406-41-1
2 ★	1.36		547		C10H13N1O1	N-PHENYLMORPHOLINE 92-53-5
3 ★	2.10	7.0	925	1	C10H13N1O1	P-ACETYL-N,N-DIEMTHYLANILINE 2124-31-4
4 ★	2.14		541		C10H13N1O1	P-I-PROPYLBENZAMIDE 619-76-1
25	-0.80	7.4	2648	1	C10H13N1O1	TETRALIN,1-HYDROXY-2-AMINO
6	-0.47	7.4	2648	1	"	TETRALIN,1-HYDROXY-2-AMINO
7 ★	0.81	7.4	2648	2	"	TETRALIN,1-HYDROXY-2-AMINO
8 ★	0.86	7.4	2648	2	"	TETRALIN,1-HYDROXY-2-AMINO
9 ★	2.14	7.4	447	1	C10H13N1O1	W-PHENYLPROPIONALDEHYDEOXIME,METHYLETHER
6130 ★	0.57		803		C10H13N1O2	2,3-DIMETHYL-4-HYDROXYACETANILIDE
1 ★	0.60		803		C10H13N1O2	2,5-DIMETHYL-4-HYDROXYACETANILIDE
2 ★	0.31		803		C10H13N1O2	2,6-DIMETHYL-4-HYDROXYACETANILIDE 6337-56-0
3 ★	1.11		803		C10H13N1O2	3,5-DIMETHYL-4-HYDROXYACETANILIDE 22900-79-4
4 ★	1.31		803		C10H13N1O2	3-ETHYL-4-HYDROXYACETANILIDE 28026-77-9
35 ★	2.72		2142		C10H13N1O2	3-ETHYLPHENYLCARBAMATE,O-METHYL
6 ★	2.72		2142		C10H13N1O2	4-ETHYLPHENYLCARBAMATE,O-METHYL
7 ★	1.49		1894		C10H13N1O2	ACETANILIDE,2-ETHOXY 581-08-8
8 ★	1.28		2410		C10H13N1O2	ACETANILIDE,2-METHOXY-3-METHYL
9 ★	1.13		2410		C10H13N1O2	ACETANILIDE,2-METHYL-3-METHOXY
6140	1.06	7.2	415	314	C10H13N1O2	ACETANILIDE,P-ETHOXY 62-44-2 analgesic, antipyretic
1	1.57	7.4	919	1	"	ACETANILIDE,P-ETHOXY
2 ★	1.58		503		"	ACETANILIDE,P-ETHOXY
3	1.65	7.4	1838	1	"	ACETANILIDE,P-ETHOXY
4 ★	1.59	7.4	1616	1	C10H13N1O2	ACETOPHENONE,O-(2-HYDROXYLETHYL)AMINO
45 ★	1.49	7.4	1616	1	C10H13N1O2	ANISOLE,O-ACETYLMETHYLAMINO
6	-1.29	7.4	447	1	C10H13N1O2	FUSCARIC ACID 536-69-6
7	0.68	2.0	447		"	FUSCARIC ACID
8 ★	-1.20	7.2	2323	468	C10H13N1O2	HOMOPHENYLALANINE
9 ★	2.60		1263		C10H13N1O2	I-PROPYL-N-PHENYLCARBAMATE 122-42-9 herbicide
6150 ★	1.82	7.0	925	1	C10H13N1O2	M-ETHOXYACETANILIDE 591-33-3
1 ★	1.00		260		C10H13N1O2	M-METHOXY-N,N-DIMETHYLBENZAMIDE 7290-99-5
2 ★	2.05		558		C10H13N1O2	M-TOLYLDIMETHYLCARBAMATE 7305-07-9
3 ★	2.09		273		C10H13N1O2	MEOBAL 2655-12-1 insecticide
4 ★	1.69		594	400	C10H13N1O2	METHYLCARBAMATE,N-(2,6-XYLYL) 20642-93-7
55 ★	2.16		1681		C10H13N1O2	N,N-DIMETHYLCARBAMATE,BENZYL ESTER 10507-52-5
6 ★	0.80		547		C10H13N1O2	N,N-DIMETHYLPHENOXYACETAMIDE 10397-59-8
7 ★	0.72	7.2	415	314	C10H13N1O2	N-ME-P-METHOXYACETANILIDE 35813-38-8
8 ★	1.95		273		C10H13N1O2	N-METHYL-2,3-DIMETHYLPHENYLCARBAMATE 2655-12-1
9 ★	2.03		273		C10H13N1O2	N-METHYL-2,5-DIMETHYLPHENYLCARBAMATE 3971-99-1
6160	1.93		273		C10H13N1O2	N-METHYL-2-ETHYLPHENYLCARBAMATE 2631-42-7
1 ★	2.23		273		C10H13N1O2	N-METHYL-4-ETHYLPHENYLCARBAMATE 2631-30-3
2 ★	2.23		273		C10H13N1O2	N-METHYLCARBAMATE,3,5-DIMETHYLPHENYL 2655-14-3
3 ★	2.20		273		C10H13N1O2	N-METHYLCARBAMATE,M-ETHYLPHENYL 4043-23-6
4	-2.23	7.4	2261		C10H13N1O2	NAPHTHALENE,TETRAHYDRO-2-AMINO-6,7-DIHYDROXY
65	1.37	7.5	2254	2	"	NAPHTHALENE,TETRAHYDRO-2-AMINO-6,7-DIHYDROXY
6 ★	2.27	7.0	332	1	C10H13N1O2	NICOTINIC ACID,BUTYL ESTER 6938-06-3
7 ★	3.89	7.4	594	312	C10H13N1O2	NITROBENZENE,4-T-BUTYL 3382-56-7
8 ★	0.71		260		C10H13N1O2	O-METHOXY-N,N-DIMETHYLBENZAMIDE
9 ★	1.86		558		C10H13N1O2	O-TOLYLDIMETHYLCARBAMATE 7305-06-8
6170 ★	0.96		260		C10H13N1O2	P-METHOXY-N,N-DIMETHYLBENZAMIDE 7291-00-1
1 ★	2.03		558		C10H13N1O2	P-TOLYLDIMETHYLCARBAMATE 7305-08-0
2 ★	1.96		1969		C10H13N1O2	PHENPROBAMATE 673-31-4 tranquilizer, muscle relaxant
3 ★	-1.53		2018		C10H13N1O2	PHENYLALANINE,N-METHYL 4415-69-4
4 ★	2.80		1263		C10H13N1O2	PROPYL-N-PHENYLCARBAMATE 5532-90-1
75 ★	2.43		2630		C10H13N1O2	RISOCAINE 94-12-2 anesthetic (local)
6 ★	1.64	8.0	1011	2	C10H13N1O2	TENAMFETAMINE 51497-09-7 psychotomimetic
7	-2.02	7.4	2261	1	C10H13N1O2	TETRALIN,2-AMINO-5,6-DIHYDROXY
8 ★	2.47		273		C10H13N1O2S1	N-METHYLCARBAMATE,3-METHYL-4-METHYLTHIOPHENYL
9 ★	2.40		2142		C10H13N1O3	2-ETHOXYPHENYLCARBAMATE,O-METHYL
6180 ★	1.53		866		C10H13N1O3	2-PROPOXY-4-AMINOBENZOIC ACID 2486-79-5
1 ★	2.34		2142		C10H13N1O3	3-ETHOXYPHENYLCARBAMATE,O-METHYL
2 ★	1.60		558		C10H13N1O3	3-METHOXYPHENYLDIMETHYLCARBAMATE 7305-09-1
3	3.65		579		C10H13N1O3	3-T-BUTYL-4-NITROPHENOL 5722-68-9
4 ★	2.11		2142		C10H13N1O3	4-ETHOXYPHENYLCARBAMATE,O-METHYL

	logP	pH	Ref.	Note	MF	Name / **CAS** / *activity*
85 ★	1.53		558		C10H13N1O3	4-METHOXYPHENYLDIMETHYLCARBAMATE **7305-10-4**
6 ★	-0.96	7.2	2323	468	C10H13N1O3	BENZYLSERINE
7 ★	1.24		273		C10H13N1O3	N-METHYL-2-ETHOXYPHENYLCARBAMATE **23409-17-8**
8 ★	1.75		273		C10H13N1O3	N-METHYL-3-ETHOXYPHENYLCARBAMATE **7225-96-9**
9 ★	1.63		273		C10H13N1O3	N-METHYL-4-ETHOXYPHENYLCARBAMATE **13538-54-0**
6190 ★	3.50	7.4	820	1	C10H13N1O3	P-NITROPHENYLBUTYLETHER **7244-78-2**
1 ★	3.84	1.5	2329	314	C10H13N1O3	PHENOL,2-NITRO-4-S-BUTYL **3555-18-8**
2	-1.89		1590	459	C10H13N1O3	TYROSINE,A-METHYL **658-48-0**
3 ★	3.24		1895		C10H13N1O3S1	NITROBENZENE,4-(3-METHIO)PROPOXY **75032-34-7**
4 ★	1.73	7.4	2272	1	C10H13N1O3S1	P-TOLUENESULFONAMIDE,N-METHYL-N-ACETYL **16697-83-9**
95	0.82		1898		C10H13N1O4	(4-HYDROXYETHYL)VANILLINOXIME
6	-0.40	7.4	2062	1	C10H13N1O4	L-DOPA,METHYL ESTER **7101-51-1**
7 ★	-0.21	7.4	2062	2	"	L-DOPA,METHYL ESTER
8 ★	0.67		731		C10H13N1O4	N-MALEOYL-4-AMINOBUYRIC ACID,ETHYL ESTER
9 ★	1.15		731	604	C10H13N1O4	N-MALEOYL-6-AMINOHEXANOIC ACID
6200	-2.54		1590	459	C10H13N1O4	TYROSINE,3-METHOXY
1 ★	1.75		773	601	C10H13N1O4S1	4-SULFAMYLBENZOIC ACID,PROPYL ESTER **59777-58-1**
2 ★	0.93		1705		C10H13N1O4S1	METHYL-3-(P-NITROPHENOXY)PROPYL-SULFONE
3 ★	1.10		1705		C10H13N1O5S1	P-NITRO-(3-METHYLSULFONYL)PROPOXYBENZENE
4	-2.00	7.4	2490	1	C10H13N1O5S2	BENZENESULFONAMIDE,3-CARBOXY-4-(3-HYDROXYPROPYL)THIO **108966-62-7**
5	-3.00	7.4	2490	1	C10H13N1O7S2	BENZENESULFONAMIDE,3-CARBOXY-4-(3-HYDROXYPROPYL)SULFONYL **108966-68-3**
6 ★	-0.47	7.4	2490	1	C10H13N1O7S2	BENZENESULFONAMIDE,3-METHOXYCARBONYL-4-HYDROXYETHYLSULFONYL **135832-43-8**
7	1.05	7.4	2535	1	C10H13N1S1	G-(2-THIOPHENOYL)PROPYL-ETHANOLAMINE (DEHYDRO-FUSED) +
8 ★	2.51	7.5	2545	2	C10H13N1S1	MPTP-2-THIENYL ANALOG
9 ✓	-2.19	2.0	2600	342	C10H13N2	NICOTINE IMINIUM ION
6210	0.19	7.4	2600	1	"	NICOTINE IMINIUM ION
1	1.48	12.2	2600		"	NICOTINE IMINIUM ION
2 ★	2.57		559		C10H13N3	1-BUTYLBENZOTRIAZOLE
3 ★	2.31		559		C10H13N3	1-S-BUTYLBENZOTRIAZOLE **63936-04-9**
4	-1.82	7.4	976	1	C10H13N3	2-(2-TOLYLIMINO)IMIDAZOLINE
15 ★	1.28	7.4	976	2	"	2-(2-TOLYLIMINO)IMIDAZOLINE
6 ★	3.12		559		C10H13N3	2-BUTYLBENZOTRIAZOLE
7 ★	2.95		559		C10H13N3	2-S-BUTYLBENZOTRIAZOLE **66382-00-1**
8 ★	0.75	13.2	2018		C10H13N3	DEBRISOQUIN **1131-64-2** *antihypertensive*
9 ★	1.03		559		C10H13N3O1	1-(4-HYDROXYBUTYL)BENZOTRIAZOLE
6220 ★	1.57		559		C10H13N3O1	2-(4-HYDROXYBUTYL)BENZOTRIAZOLE
1 ★	3.23		547		C10H13N3O1	P-(3-METHYL-3-ACETYL-1-TRIAZENO)-TOLUENE
2 ★	2.32	7.4	447	1	C10H13N3O1	W-PHENYPROPIONALDEHYDESEMICARBIZONE
3 ★	2.50	8.9	1498		C10H13N3O2	1-I-PROPYL-1-NITROSO-3-PHENYLUREA **78326-57-5**
4 ★	2.61		1213		C10H13N3O2	1-ME-1-ACETYL-3-(P-METHOXYPHENYL)TRIAZENE
25	0.03	3.3	2590	331	C10H13N3O2	3-ETHYL-1-(4-CARBOXYPHENYL)-3-METHYLTRIAZENE **74109-20-9**
6	0.04	7.4	1214	1	"	3-ETHYL-1-(4-CARBOXYPHENYL)-3-METHYLTRIAZENE
7	0.11	3.3	2590	331	C10H13N3O2	BENZOIC ACID,2-(3-METHYL-3-ETHYLTRIAZENYL) *antineoplastic*
8 ★	0.77		2105		C10H13N3O2	HYDRAZINECARBOXAMIDE,1-ACETYL-N-(3-METHYLPHENYL)
9 ★	0.76		2105		C10H13N3O2	HYDRAZINECARBOXAMIDE,1-ACETYL-N-(4-METHYLPHENYL)
6230 ★	2.51	6.6	2590	459	C10H13N3O2	METHYL BENZOATE,2-(3,3-DIMETHYLTRIAZENYL) *antineoplastic*
1 ★	3.12	6.6	2590	459	C10H13N3O2	METHYL BENZOATE,3-(3,3-DIMETHYLTRIAZENYL) *antineoplastic*
2 ★	3.22	6.6	2590	459	C10H13N3O2	METHYL BENZOATE,4-(3,3-DIMETHYLTRIAZENYL) *antineoplastic*
3	-1.49	7.4	2607	1	C10H13N3O2	PHENYLACETIC ACID,P-(3,3-DIMETHYL-1-TRIAZENO)
4	-1.10	7.0	2607	1	"	PHENYLACETIC ACID,P-(3,3-DIMETHYL-1-TRIAZENO)
35 ★	1.38	7.0	2607	2	"	PHENYLACETIC ACID,P-(3,3-DIMETHYL-1-TRIAZENO)
6 ★	0.04		1439		C10H13N3O2	PYRIDO(1,2A)PYRIMIDIN-4-ONE,3-CONH2,4,6,7,8,9-H5,6-ME
7	-0.07	6.0	479		C10H13N3O2S2	3-METHYL-2-SULFONILAMIDE-2,3-DIHYDROTHIAZOLE **51203-20-4**
8 ★	0.39		2105		C10H13N3O3	HYDRAZINECARBOXAMIDE,1-ACETYL-N-(3-METHOXYPHENYL)
9 ★	0.06		2105		C10H13N3O3	HYDRAZINECARBOXAMIDE,1-ACETYL-N-(4-METHOXYPHENYL)
6240 ★	-2.36	7.0	2440	1	C10H13N3O4	ARA-CPE-C
1 ★	-2.45	7.0	2440	1	C10H13N3O4	CPE-C **90597-22-1**
2 ★	0.32		522		C10H13N3O4S2	1,2,4-BENZOTHIADIAZINE-1,1-DIOXIDE-3-METHYL-7-DIMETHYLSULFONAMIDO **37157-82-7**
3 ★	-0.05		1003		C10H13N3O6	1-(2,3-BIS-ACETYLOXYPROPYL)-2-NITROIMIDAZOLE
4 ★	0.09		2544		C10H13N5O1	9-PROPYL-N-ACETYLADENINE
45	-0.29	7.0	2411	1	C10H13N5O2	2',3'-DIDEOXYADENOSINE **4097-22-7** *antiviral, antineoplastic*
6 ★	-0.22	7.5	2252	1	C10H13N5O2	2',3'-DIDEOXYADENOSINE
7 ★	0.22		563		C10H13N5O2	5-(3,3-DIMETHYL-1-TRIAZENO)ISOPHTHALAMIDE
8 ★	-0.20	9.0	2027	1	C10H13N5O2	IMIDAZOLE-1,2,METHYL-PYRAZIN-1-ETHYL-5-NITRO **74571-56-5**
9	-0.78	7.4	2037	1	C10H13N5O2	ISOGUANINE,1-METHYL-7-VINYLOXYMETHYL
6250	-0.24	7.4	2037	1	C10H13N5O2	ISOGUANINE,1-METHYL-9-VINYLOXYETHYL
1 ★	-0.59	7.0	2440	1	C10H13N5O2S1	DIDEOXYPURINE,2-AMINO-6-THIOL
2	-1.09	7.0	2411	1	C10H13N5O3	2',3'-DIDEOXYGUANOSINE
3	-1.07		2440	5	C10H13N5O3	2',3'-DIDEOXYGUANOSINE
4 ★	-1.01	7.4	2252	1	C10H13N5O3	2',3'-DIDEOXYGUANOSINE **85326-06-3**
55	-0.65	7.0	2440	1	C10H13N5O3	2'-DEOXYADENOSINE **958-09-8**
6 ★	-0.55	7.5	2252	1	C10H13N5O3	2'-DEOXYADENOSINE
7 ★	-0.49	8.0	2267	306	C10H13N5O3	ALLOPURINOL,1-N,N-DIMETHYLGYCYLOXYMETHYL **98204-07-0**
8	-1.20	7.4	2037	1	C10H13N5O3	ISOGUANINE,1-METHYL-7-ACETYLOXYETHYL
9 ★	-0.79	7.4	401	2	C10H13N5O3S1	A-2'-DEOXYTHIOGUANOSINE **2133-81-5**
6260 ★	-0.56	7.4	401	2	C10H13N5O3S1	B-2'-DEOXYTHIOGUANOSINE **1688-22-6**
1	4.69	6.6	2590	459	C10H13N5O4	1,3-DINITIROBENZENE,5-(3,3-DIETHYLTRIAZENYL) *antineoplastic*
2 ★	-1.30	7.4	2252	1	C10H13N5O4	2'DEOXYGUANOSINE **961-07-9**
3 ★	-1.30		2669		C10H13N5O4	ACYCLOVIR,N2-ACETYL
4	-1.10		293	1	C10H13N5O4	ADENOSINE **58-61-7**
65 ★	-1.05	7.0	2440	1	"	ADENOSINE
6 ★	-1.11		401		C10H13N5O4	VIDARABINE **24356-66-9** *antiviral*
7	0.05	7.0	2440	1	C10H13N5O4	ZIDOVUDINE **30516-87-1** *antiviral*
8 ★	0.05	7.0	2411	1	"	ZIDOVUDINE
9	0.08	7.4	2591	1	"	ZIDOVUDINE
6270 ★	-1.89	7.0	2440	1	C10H13N5O5	GUANOSINE **118-00-3**
1 ★	-1.80	7.4	2037	1	C10H13N5O5	ISOGUANINE,1-METHYL-9-(2-FURANOSYL)
2 ★	0.00	7.4	2349	1	C10H13N9O2	2,6-DIAMINOPURINE,3'-AZIDO-2',3'-DIDEOXYRIBOSIDE
3	0.24	7.4	2349	1	C10H13N9O2	2,6-DIAMINOPURINE,3'-AZIDO-2',3'-DIDEOXYRIBOSIDE
4 ★	0.24	7.4	2349	1	C10H13N9O2	2,6-DIAMINOPURINE,3'-AZIDO-2',3'-DIDEOXYRIBOSIDE **114753-52-5**

	logP	pH	Ref.	Note	MF	Name / CAS / activity
75 ★	0.00	7.4	2349	1	C10H13N9O2	2,6-DIAMINOPURINE,3'-AZIDO-2',3'-DIDEOXYRIBOSIDE 121231-92-3
6 ★	3.98		2146	458	C10H14	1,2,3,4-TETRAMETHYLBENZENE 488-23-3
7 ★	4.04		2146	458	C10H14	1,2,3,5-TETRAMETHYLBENZENE 527-53-7
8 ★	4.00		536		C10H14	1,2,4,5-TETRAMETHYLBENZENE 95-93-2
9 ★	3.72		594	401	C10H14	1,2-DIETHYLBENZENE 135-01-3
6280	4.45		2253	459	C10H14	1,4-DIETHYLBENZENE 105-05-5
1	4.44		2253	459	C10H14	2-BUTYLBENZENE 135-98-8
2	4.44		2253	459	C10H14	BENZENE,1,3-DIETHYL 141-93-5
3	4.22		2253	459	C10H14	BENZENE,1-ETHYL-2,3-DIMETHYL 933-98-2
4	4.35		2253	459	C10H14	BENZENE,1-ETHYL-2,4-DIMETHYL 874-41-9
85	4.43		2253	459	C10H14	BENZENE,1-ETHYL-3,5-DIMETHYL 934-74-7
6	4.26		2253	459	C10H14	BENZENE,1-METHYL-2-I-PROPYL 527-84-4
7	4.38		2253	459	C10H14	BENZENE,1-METHYL-2-PROPYL 1074-17-5
8	4.38		2253	459	C10H14	BENZENE,1-METHYL-3-I-PROPYL 535-77-3
9	4.53		2253	459	C10H14	BENZENE,1-METHYL-3-PROPYL 1074-43-7
6290	4.56		2253	459	C10H14	BENZENE,1-METHYL-4-PROPYL 1074-55-1
1	4.17		2253	459	C10H14	BENZENE,2-ETHYL-1,3-DIMETHYL 2870-04-4
2	4.31		2253	459	C10H14	BENZENE,2-ETHYL-1,4-DIMETHYL 1758-88-9
3	4.38		2253	459	C10H14	BENZENE,4-ETHYL-1,2-DIMETHYL 934-80-5
4	4.54		2253	459	C10H14	BENZENE,I-BUTYL 538-93-2
95 ★	4.11		508		C10H14	BENZENE,T-BUTYL 98-06-6
6 ★	4.38		2293	400	C10H14	BUTYLBENZENE 104-51-8
7 ★	4.10		1179		C10H14	P-I-PROPYLTOLUENE 99-87-6
8 ★	2.03	8.0	868	2	C10H14Br1N1O2	3,5-DIMETHOXY-4-BROMOPHENETHYLAMINE 61367-72-4
9	0.40	7.4	1155	1	C10H14Cl1N1	CHLORPHENTERMINE 461-78-9 *anorexic, stimulant*
6300 ★	2.60		1155		"	CHLORPHENTERMINE
1 ★	2.14	11.5	831	314	C10H14Cl1N5	N1-P-CHLOROPHENYL-N5-ETHYLBIGUANIDE 60221-93-4
2 ★	-0.66	7.0	1783	1	C10H14Cl1N5O6	N(5-CL-3-NO2-6(1,3,4-TRIOHBU)AM)PYRAZIN)ACETAMIDE 89690-74-4 ✚
3 ★	4.30		354		C10H14Cl2N1O2P1S1	ZYTRON 299-85-4
4	0.85		1231		C10H14Cu1O4	COPPERDIACETYLACETONECHELATE 13395-16-9
5 ★	1.28	7.4	2490	1	C10H14F1O3S2	BENZENESULFONAMIDE,3-FLUORO-4-(4-HYDROXYBUTYL)THIO 108966-75-2
6 ★	-0.26	7.4	2490	1	C10H14F1O5S2	BENZENESULFONAMIDE,3-FLUORO-4-(4-HYDROXYBUTYL)SULFONYL 108966-78-5
7 ★	3.89		945		C10H14N1O5P1S1	3,5-DIMETHYL-DIMETHYLPARATHION 50590-05-1
8 ★	3.74		945		C10H14N1O5P1S1	3-ETHYL-DIMETHYLPARATHION 13074-09-4
9 ★	3.83		572		C10H14N1O5P1S1	PARATHION 56-38-2 *insecticide*
6310 ★	2.57		945		C10H14N1O6P1	3,5-DIMETHYL-DIMETHYLPARA-OXON 50590-06-2
1 ★	2.19		945		C10H14N1O6P1	3-ETHYL-DIMETHYLPARA-OXON
2 ★	1.98		1693		C10H14N1O6P1	PARAOXON 311-45-5
3	1.10	7.4	1948	2	C10H14N2	1-PHENYLPIPERAZINE 92-54-6
4 ★	1.11	13.0	594		"	1-PHENYLPIPERAZINE
15 ★	1.10	11.0	533	322	C10H14N2	3-PYRIDYLMETHYL-N-PYRROLIDINE 370-09-2
6 ★	1.13	11.0	533	322	C10H14N2	METANICOTINE 538-79-4
7	-0.20	7.4	594	1	"	NICOTINE,BROMTHYMOLBLUE SALT
8	-0.13	7.4	594	1	"	NICOTINE,BROMTHYMOLBLUE SALT
9	-0.02	7.4	594	1	"	NICOTINE,BROMTHYMOLBLUE SALT
6320	-0.01	7.4	594	1	"	NICOTINE,BROMTHYMOLBLUE SALT
1	0.28	7.4	2600	1	C10H14N2	NICOTINE 54-11-5
2	0.45	7.4	593	1	"	NICOTINE
3	0.55	7.4	594	1	"	NICOTINE
4 ★	1.17	11.0	533	322	"	NICOTINE
25	0.93	7.2	2317	1	C10H14N2	PIPERIDINE,2-(3-PYRIDYL) 494-52-0 *insecticide*
6 ★	0.97	11.0	533	322	"	PIPERIDINE,2-(3-PYRIDYL)
7 ★	1.00	7.4	447	1	C10H14N2	W-PHENYLPROPIONALDEHYDE,N-METHYLHYDRAZONE
8 ★	1.33		1263		C10H14N2O1	1,1-DIMETHYL-3-P-TOLYLUREA 7160-01-2
9 ★	0.38		1474		C10H14N2O1	2-PYRIDINEPENTANEAMIDE
6330 ★	0.40		1474		C10H14N2O1	3-PYRIDINEPENTANEAMIDE
1 ★	0.04	11.0	533	322	C10H14N2O1	3-PYRIDYLMETHYL-N-MORPHOLINE 17751-47-2
2 ★	0.42		1474		C10H14N2O1	4-PYRIDINEPENTANEAMIDE
3	0.60		1439		C10H14N2O1	4H-PYRIDO(1,2-A)PYRIMIDIN-4-ONE,6,7,8,9-H4,2,3-DIME ✚
4	0.67		1439		C10H14N2O1	4H-PYRIDO(1,2-A)PYRIMIDIN-4-ONE,6,7,8,9-H4,2,6-DIME
35 ★	0.82		1439		C10H14N2O1	4H-PYRIDO(1,2-A)PYRIMIDIN-4-ONE,6,7,8,9-H4,3,6-DIME
6 ★	0.69		1724		C10H14N2O1	4H-PYRIDO(1,2-A)PYRIMIDIN-4-ONE,6,7,8,9-H4,3,9-DIME
7 ★	1.29		1894		C10H14N2O1	ACETANILIDE,3-DIMETHYLAMINO 7474-95-5
8	0.30		1622		C10H14N2O1	BENZAMIDE,N-(3-AMINOPROPYL) 6108-74-3
9 ★	2.07	7.4	1616	1	C10H14N2O1	BENZAMIDE,O-ISOPROPYLAMINO
6340 ★	0.41		1983		C10H14N2O1	N-(AMINOACETYL)-2,6-XYLIDIDE 18865-38-8
1	-2.30	3.0	1268		C10H14N2O1	N-(DIMETHYLAMINOMETHYL)BENZAMIDE 59917-58-7
2	0.42	7.4	1268	1	"	N-(DIMETHYLAMINOMETHYL)BENZAMIDE
3	0.85	9.2	1268	1	"	N-(DIMETHYLAMINOMETHYL)BENZAMIDE
4	-1.24	3.0	1268		C10H14N2O1	N-(ETHYLAMINOMETHYL)BENZAMIDE 52387-57-2
45	0.15	7.4	1268	1	"	N-(ETHYLAMINOMETHYL)BENZAMIDE
6	0.97	9.2	1268	1	"	N-(ETHYLAMINOMETHYL)BENZAMIDE
7 ★	0.90		1035		C10H14N2O1	N-BUTYLNICOTINAMIDE
8	-3.20	2.0	2600	342	C10H14N2O1	NICOTINE OXIDE 491-26-9
9	-1.80	12.2	2600		"	NICOTINE OXIDE
6350	-1.80	7.4	2600	1	"	NICOTINE OXIDE
1 ★	0.33		505		C10H14N2O1	NIKETHAMIDE 59-26-7 *stimulant (respiratory)*
2 ★	1.00		1894		C10H14N2O1	P-ACETAMIDO-N,N-DIEMTHYLANILINE 7463-28-7
3	1.13	7.0	925	1	"	P-ACETAMIDO-N,N-DIEMTHYLANILINE
4 ★	2.10		2094		C10H14N2O1	UREA,1-METHYL-3-(3-ETHYLPHENYL) 23138-95-6
55 ★	0.83		1263		C10H14N2O2	1,1-DIMETHYL-3-P-METHOXYPHENYLUREA 7160-02-3
6 ★	0.62	7.4	2026	1	C10H14N2O2	BENZAMIDE,2-(2-HYDROXYPROPYL)AMINO
7 ★	0.70	7.4	2026	1	C10H14N2O2	BENZAMIDE,2-(3-HYDROXYPROPYL)AMINO
8 ★	1.03		2649		C10H14N2O2	BENZOQUINONE,2,5-BIS-DIMETHYLAMINO
9 ★	1.00		1019		C10H14N2O2	N'-ETHYL-PHENOXYACETIC ACID HYDRAZIDE
6360 ★	1.06		1681		C10H14N2O2	N,N-DIMETHYLCARBAMATE,M-AMINOBENZYL ESTER 84640-25-5
1 ★	1.43		273		C10H14N2O2	N-METHYL-3-DIMETHYLAMINOPHENYLCARBAMATE 2631-39-2
2 ★	0.00		1622		C10H14N2O2	O-METHOXYBENZAMIDE,N-(2-AMINOETHYL)
3 ✓	-1.59		2426		C10H14N2O2Pt1.(N1O3)2	PLATINUM,BIS-PYRIDONIO-DIAQUA-DINITRATE
4 ★	2.19		505		C10H14N2O2S1	BARBITURIC ACID,5-ETHYL-5-METHYLALLYL-2-THIO 115-56-0 *antiemetic*

	logP	pH	Ref.	Note	MF	Name / **CAS** / *activity*
65	1.18	7.4	2461	1	C10H14N2O2S4	THIENO[2,3-B]-THIOPHENE-2-SULFONAMIDE,5-METHIOETHYLAMINOMETHYL **122266-99-3**
6 ★	0.62		2612		C10H14N2O3	1-(3'-HYDROXYMETHYLPHENYL)-3-MEO-3-METHYLUREA **88132-18-7**
7 ★	0.28		547		C10H14N2O3	1-PHENYL-3-(2,3-DIHYDROXYPROPYL)UREA
8	-1.10	7.4	2685	1	C10H14N2O3	4-PYRIDONE,3-HYDROXY-2-METHYL-1-(2-METHYLAMINOCARBONYL)ETHYL *iron chelator*
9	1.35	0.7	962		C10H14N2O3	BARBITURIC ACID,5-ALLYL-5-ETHYL,1-METHYL
6370 ★	1.15		1935		C10H14N2O3	BARBITURIC ACID,5-ALLYL-5-I-PROPYL **77-02-1** *sedative, hypnotic*
1	1.22	7.4	899	1	"	BARBITURIC ACID,5-ALLYL-5-I-PROPYL
2 ★	-0.80	7.4	2026	1	C10H14N2O3	BENZAMIDE,2-(2,3-DIHYDROXYPROPYL)AMINO
3 ★	0.51		773	601	C10H14N2O3S1	4-SULFAMYLBENZAMIDE,N-PROPYL
4 ★	0.97		2072	2	C10H14N2O3S1	BENZENESULFONAMIDE,2-(I-BUTYROYLAMINO)
75 ★	1.17		2072	2	C10H14N2O3S1	BENZENESULFONAMIDE,2-AMINO-N-(I-BUTANOYL)
6 ★	1.14	7.4	2004	1	C10H14N2O3S1	BENZENESULFONAMIDE,4-METHYL-N-ETHYLAMINOCARBONYL **1467-23-8**
7 ★	0.40		1205		C10H14N2O3S1	N,N-DIMETHYL-4-(N-ME-N-FORMYL)BENZENESULFONAMIDE
8 ★	1.34		2352	2	C10H14N2O3S1	N2-I-PROPYL-N1-BENZENESULFONYLUREA **3149-01-7**
9 ★	1.44		2352	2	C10H14N2O3S1	N2-PROPYL-N1-BENZENESULFONYLUREA **4932-53-0**
6380	0.03	7.4	2525	1	C10H14N2O3S2	5-(PROPYLAMINOMETHYL)THIENO[2,3-B]FURAN-2-SULFONAMIDE ✚
1	0.32	7.4	2461	1	C10H14N2O3S3	THIENO[2,3-B]THIOPHENE-2-SULFONAMIDE,5-METHOXYETHOXYAMINOMETHYL **122266-90-4**
2	0.51	7.4	2461	1	C10H14N2O3S3	THIENO[3,2-B]THIOPHENE-2-SULFONAMIDE,5-METHOXYETHYLAMINOMETHYL **127025-29-0**
3	-0.44	7.4	2461	1	C10H14N2O3S4	THIENO[2,3-B]THIOPHENE-2-SULFONAMIDE,5-(2-METHYLSULFOXO)ETHYLAMINOMETHYL **122267-01-0**
4	1.48		2447		C10H14N2O4	4-PYRIDINECARBOXYLIC ACID,TETRAHYDRO-2,6-DIOXO,I-PENTYL ESTER
85	1.62		2447		C10H14N2O4	4-PYRIDINECARBOXYLIC ACID,TETRAHYDRO-2,6-DIOXO,PENTYL ESTER
6 ★	-0.63		2252		C10H14N2O4	DIDEOXYTHYMIDINE **3416-05-5**
7	-0.61	7.0	2440	1	"	DIDEOXYTHYMIDINE
8	0.02	7.4	2525	1	C10H14N2O4S2	5-(METHOXYETHYLAMINOMETHYL)THIENO[2,3-B]FURAN-2-SULFONAMIDE
9	-0.26	7.4	2461	1	C10H14N2O4S4	THIENO[2,3-B]THIOPHENE-2-SULFONAMIDE,5-METHYLSULFONYLETHYLAMINOMETHYL **133445-74-6**
6390 ★	-1.17		2252		C10H14N2O5	THYMIDINE **50-89-5** *antineoplastic*
1	-1.15	7.0	2440	1	"	THYMIDINE
2	1.28		2535	1	C10H14N2S1	PYRROLINO[1.2-A]IMIDAZOLINE,7A-THION-2YL ✚
3 ★	1.59	5.6	590		C10H14N2S1	THIOUREA,N,N-DIMETHYL-N'-BENZYL **2741-14-2**
4 ✓	-1.94		2397	101	C10H14N3O1S1	IMIDAZOLIUM CHLORIDE,1-METHYL-2-HYDROXYIMINOMETHYL-3-BUTYN-3YLTHIOMETHYL
95 ✓	-2.14		2397	101	C10H14N3O2	IMIDAZOLIUM CHLORIDE,1-METHYL-2-HYDROXYIMINOMETHYL-3-(1-METHYLPROPYN-2-YLOXY)METHYL
6 ✓	-2.75		2397	101	C10H14N3O2	IMIDAZOLIUM CHLORIDE,1-METHYL-2-HYDROXYIMINOMETHYL-3-(2-PROPYN-2YLOXY)ETHYL
7 ✓	-2.36		2397	101	C10H14N3O2	IMIDAZOLIUM CHLORIDE,1-METHYL-2-HYDROXYIMINOMETHYL-3-BUTYN-3-YLOXYMETHYL **117983-25-2**
8 ✓	-1.76		2397	101	C10H14N3O3	IMIDAZOLIUM CHLORIDE,1-METHYL-2-HYDROXYIMINOMETHYL-3-BUTEN-2OYLOXYMETHYL
9 ★	1.69	6.6	2590	459	C10H14N4O1	1-(4-CARBAMOYLPHENYL)-3-METHYL-3-ETHYLTRIAZENE **59708-19-9** *antineoplastic*
6400 ★	1.61		538		C10H14N4O1	3,3-DIMETHYL-1-(3-ACETYLAMINOPHENYL)TRIAZENE
1	0.24		2123	22	C10H14N4O1	BENZALDEHYDE,4-ETHOXY,GUANYLHYDRAZONE **82530-96-9**
2 ★	2.59	6.6	2590	459	C10H14N4O1	BENZAMIDE,2-(3-METHYL-3-ETHYLTRIAZENYL) *antineoplastic*
3 ★	1.83		563		C10H14N4O1	N-METHYL-2-(3,3-DIMETHYL-1-TRIAZINO)BENZAMIDE
4 ★	0.26	6.3	2589	459	C10H14N4O2	1-(4-CARBAMOYLPHENYL)-3-METHYL-3-HYDROXYETHYLTRIAZENE **59708-20-2** *antineoplastic*
5	1.15		1161		C10H14N4O2	1-METHYL-3-I-BUTYLXANTHINE **28822-58-4**
6	1.36	7.4	971	1	"	1-METHYL-3-I-BUTYLXANTHINE
7 ★	0.44		547		C10H14N4O2	4-METHOXY-3-(3,3-DIMETHYLTRIAZENO)BENZAMIDE
8 ★	4.17	6.6	2590	459	C10H14N4O2	NITROBENZENE,4-(3,3-DIETHYLTRIAZENYL) *antineoplastic*
9 ★	1.29		2537		C10H14N4O2	XANTHINE,1-METHYL-3-BUTYL **31542-48-0**
6410 ★	-0.77	7.4	2111	2	C10H14N4O3	PROXYPHYLLINE **603-00-9** *bronchodilator, vasodilator, smooth muscle relaxant*
1 ★	0.79	1.0	2196		C10H14N4O3	PYRAZOLE,3-METHYL-4-NITRO-5-(N,N-PENTAMETHYLENECARBOXAMIDO)
2 ★	1.24	1.0	2196		C10H14N4O3	PYRAZOLE,3-METHYL-4-NITRO-5-(N-CYCLOPENTYLCARBOXAMIDO)
3	2.40	7.4	553	1	C10H14N4O3S1	2(4,5-ME4-1,3-OXAZOLIDINYLIDEN-2N)-5-NO2THIAZOL **24229-59-2**
4	-2.02	7.4	2477	1	C10H14N4O5	1-[2-HYDROXY-3-(2-NITROIMIDAZOL-1-YL)PROPYL]-2-AZETIDINECARBOXYLIC ACID **134419-51-5**
15	-3.52	1.0	594		C10H14N5O7P1	3'-ADENYLIC ACID **84-21-9**
6 ★	-3.52	1.0	594	400	C10H14N5O7P1	AMP **61-19-8** *nutrient*
7	1.61	7.4	2028	459	C10H14N6	1,2,4-TRIAZOLE,3-AMINO-5-(2-PROPYLAMINO)PYRIDIN-4-YL **77314-58-0**
8	0.94	7.4	2028	459	C10H14N6O1	1,2,4-TRIAZOLE,3-AMINO-5-(2-METHOXYETHYLAMINO)PYRIDIN-4-YL **77314-64-8**
9	-0.49	7.4	2440	1	C10H14N6O2	2,6-DIAMINOPURINE,2',3'-DIDEOXYRIBOSIDE
6420 ★	-0.46	7.5	2252	1	C10H14N6O2	2,6-DIAMINOPURINE,2',3'-DIDEOXYRIBOSIDE
1 ★	-0.52	7.5	2252	1	C10H14N6O3	2,6-DIAMINOPURINE,2'-DEOXYRIBOSIDE **4546-70-7**
2 ★	-1.42		401		C10H14N6O4	NSC#128,668 **13263-99-5** ✚
3 ★	2.35		1474		C10H14O1	BUTANOL,4-PHENYL **3360-41-6**
4	3.27		558		C10H14O1	O-S-BUTYLPHENOL **89-72-5**
25 ★	3.31		558	60	C10H14O1	O-T-BUTYLPHENOL **88-18-6**
6	3.64		1875		C10H14O1	P-BUTYLPHENOL **1638-22-8**
7 ★	3.08	7.4	2582	1	C10H14O1	P-S-BUTYLPHENOL **99-71-8**
8 ★	3.31		508		C10H14O1	P-T-BUTYLPHENOL **98-54-4**
9 ★	3.03		2215		C10H14O1	PHENOL,2,6-DIETHYL
6430 ★	2.70		502		C10H14O1	PROPANE,1-METHOXY-3-PHENYL **2046-33-5**
1 ★	3.30		503		C10H14O1	THYMOL **89-83-8** *stabilizer*
2 ★	2.96		2487		C10H14O2	1,2-DIHYDROXYBENZENE,4-BUTYL
3 ★	2.76		2098		C10H14O2	1-(3-FURYL)-4-METHYLPENTAN-1-ONE **553-84-4**
4	1.28		2363		C10H14O2	BENZENE,1,4-BIS(METHOXYMETHYL)
35 ★	1.52		503		C10H14O2	CAMPHORQUINONE **465-29-2**
6 ★	1.80		2452		C10H14O2	DUROHYDROQUINONE **527-18-4**
7 ★	3.03		1477		C10H14O2	M-BUTOXYPHENOL **18979-72-1**
8 ★	2.42		599		C10H14O2	O-DIETHOXYBENZENE
9 ★	2.90		1940	459	C10H14O2	P-BUTOXYPHENOL **122-94-1**
6440 ★	1.41		505		C10H14O3	1,2-PROPANEDIOL,3-(2-TOLYLOXY) **59-47-2** *muscle relaxant (skeletal)*
1 ★	2.27	13.0	579	224	C10H14O3	TRIMETHYLORTHO BENZOATE **707-07-3**
2 ★	1.53	1.0	594		C10H14O4	3,5-DIMETHYL-1-CYCLOHEXENE,1,2-DICARBOXYLIC ACID
3 ★	1.60		2452		C10H14O4	HYDROQUINONE,2,3-DIMETHOXY-5,6-DIMETHYL
4 ★	2.45		2011		C10H14O4S2	PROPANEDIOIC ACID,1,3-DITHIOLAN-2-YLIDINE,DIETHYL ESTER **59937-29-0**
45 ★	0.39		2452		C10H14O6	HYDROQUINONE,TETRAMETHOXY **52092-59-8**
6	-1.81		571		C10H15Br1N1.Br1	M-BROMOBENZYLTRIMETHYLAMMONIUMBROMIDE
7 ★	1.69		1564		C10H15Br1N2O3	PYRIMIDINE-2,4-DIONE,3-PENTOXY-5-BROMO-6-ME **77317-75-0**
8 ★	1.73		1564		C10H15Br1N2O3	PYRIMIDINE-2,4-DIONE,3-PROPOXY-5-BR-6-PROPYL
9	-1.81		571		C10H15Cl1N1.Br1	M-CHLOROBENZYLTRIMETHYLAMMONIUMBROMIDE **25251-56-3**
6450	-2.20		571		C10H15F1N1.Br1	M-FLUOROBENZYLTRIMETHYLAMMONIUMBROMIDE
1	-1.51		571		C10H15I1N1.Br1	M-IODOBENZYLTRIMETHYLAMMONIUMBROMIDE
2 ★	2.28		1508		C10H15N1	1-BENZYLPROPYLAMINE **53309-89-0**
3	-0.70	2.0	1473		C10H15N1	4-PHENYLBUTYLAMINE HYDROCHLORIDE **30684-06-1**
4	-0.67	7.4	785	1	C10H15N1	4-PHENYLBUTYLAMINE **13214-66-9**

	logP	pH	Ref.	Note	MF	Name / CAS / activity
55 ★	2.40		1474		"	4-PHENYLBUTYLAMINE
6 ★	3.05	7.5	2201		C10H15N1	ANILINE,4-BUTYL 104-13-2
7	2.47	7.5	2201		C10H15N1	ANILINE,4-T-BUTYL 769-92-6
8 ★	2.70	7.4	597	314	"	ANILINE,4-T-BUTYL
9 ★	2.07	13.0	547	224	C10H15N1	METHAMPHETAMINE 537-46-2
6460 ★	3.31		324		C10H15N1	N,N-DIETHYLANILINE 91-66-7
1	3.58		272	837	C10H15N1	N-BUTYLANILINE 1126-78-9
2	3.75		442	2	C10H15N1	P-AMYLPYRIDINE 2961-50-4
3	-0.71	7.5	1588	1	C10H15N1	PHENETHYLAMINE,N,A-DIMETHYL *stimulant (central)*
4	-0.80	7.4	1155	1	C10H15N1	PHENTERMINE 122-09-8 *anorexic*
65 ★	1.90		1155		"	PHENTERMINE
6 ★	1.28	13.0	594		C10H15N1O1	2-AMINOPROPANOL,N-BENZYL 3217-09-2
7	1.04	1.0	547	342	C10H15N1O1	2-N,N-DIMETHYLAMINO-1-PHENYLETHANOL HYDROCHLORIDE 1797-76-8
8 ★	1.54	13.0	547	224	C10H15N1O1	2-N,N-DIMETHYLAMINO-1-PHENYLETHANOL 6853-14-1
9 ★	1.28		1474		C10H15N1O1	2-PYRIDINEPENTANOL
6470 ★	1.41		1474		C10H15N1O1	3-PYRIDINEPENTANOL
1 ★	1.77	7.0	538	701	C10H15N1O1	4-METHOXYAMPHETAMINE 64-13-1
2 ★	1.39		1474		C10H15N1O1	4-PYRIDINEPENTANOL
3 ★	1.43	11.0	2648	448	C10H15N1O1	EPHEDRINE -D
4	-2.45		1768	820	C10H15N1O1	EPHEDRINE HYDROCHLORIDE 50-98-6
75	-1.48	7.4	1157	324	C10H15N1O1	EPHEDRINE 299-42-3 *adrenergic, bronchodilator*
6 ★	0.93		505		"	EPHEDRINE
7 ★	2.10	7.4	594	1	C10H15N1O1	ETHANOLAMINE,N-ETHYL-N-PHENYL 92-50-2
8	1.40		2371		C10H15N1O1	PSEUDOEPHEDRINE 90-82-4
9 ★	0.77	8.0	1011	2	C10H15N1O2	3,4-DIMETHOXYPHENETHYLAMINE
6480	0.73	7.4	2685	1	C10H15N1O2	4-PYRIDONE,3-HYDROXY-2-ETHYL-1-I-PROPYL *iron chelator*
1	0.70	7.4	2685	1	C10H15N1O2	4-PYRIDONE,3-HYDROXY-2-ETHYL-1-PROPYL *iron chelator*
2	-1.15	7.4	2261	1	C10H15N1O2	DOPAMINE,N,N-DIMETHYL 21581-37-3
3	0.30	7.4	2472	324	C10H15N1O2	PYRIDIN-4-ONE,1-BUTYL-2-METHYL-3-HYDROXY *iron chelator*
4	0.70	7.4	2685	1	"	PYRIDIN-4-ONE,1-BUTYL-2-METHYL-3-HYDROXY
85 ★	2.45		773	601	C10H15N1O2S1	P-BUTYLBENZENESULFONAMIDE 1135-00-8
6	-0.26	7.4	2685	1	C10H15N1O3	4-PYRIDONE,3-HYDROXY-2-METHYL-1-(3-METHOXYPROP-2-YL) *iron chelator*
7	-0.74	7.4	2685	1	C10H15N1O3	4-PYRIDONE,3-HYDROXY-2-METHYL-1-(4-HYDROXYBUTYL) *iron chelator*
8	-2.27	7.4	1168	1	C10H15N1O3	METANEPHRINE 5001-33-2
9	-0.08	7.4	2472	324	C10H15N1O3	PYRIDIN-4-ONE,1-METHOXYETHYL-2-ETHYL-3-HYDROXY *iron chelator*
6490	0.04	7.4	2685	1	"	PYRIDIN-4-ONE,1-METHOXYETHYL-2-ETHYL-3-HYDROXY
1	2.09		2467	448	C10H15N1O3S1	BENZENESULFONAMIDE,3-BUTYLOXY
2	2.10	2.1	2265			BENZENESULFONAMIDE,3-BUTYLOXY
3 ★	0.57	7.4	2490	1	C10H15N1O3S1	BENZENESULFONAMIDE,4-(4-HYDROXYBUTYL) 135832-48-3
4 ★	2.09	2.1	2265		C10H15N1O3S1	BENZENESULFONAMIDE,4-BUTYLOXY
95 ★	0.93	7.4	2490	1	C10H15N1O3S2	BENZENESULFONAMIDE,4-(4-HYDROXYBUTYL)THIO 135832-41-6
6 ★	-0.60	7.4	2490	1	C10H15N1O5S2	BENZENESULFONAMIDE,4-(4-HYDROXYBUTYL)SULFONYL 135832-42-7
7	-2.28	7.4	401	1	C10H15N1O6	3-DEAZAURIDINE 39935-49-4
8 ★	-0.86	7.4	401	2	"	3-DEAZAURIDINE
9 ★	-0.14	7.4	2490	1	C10H15N1O7S3	THIOPHENE-2-SULFONAMIDE,5-(3-METHOXYACETYLOXY)PROPYLSULFONYL 104438-05-3
6500	-1.39		301		C10H15N2O1.Cl1	N1-BUTYLNICOTINAMIDECHLORIDE 2255-05-2
1	-2.66		571		C10H15N2O2.Br1	M-NITROBENZYLTRIMETHYLAMMONIUMBROMIDE
2	1.15	7.4	2331	1	C10H15N3	2H-PYRAZOLO[3,4-A]-QUINOLIZINE,1,2,3,6,7,10B-HEXAHYDRO 112114-04-2
3 ★	1.02	7.4	725	1	C10H15N3O1	2-I-BUTYLISONIAZID 58480-99-2
4	0.51		1439		C10H15N3O2	PYRIDO(1,2A)PYRIMIDIN-4-ONE,3-CONH2,-H7,6-ME
5 ★	1.15	6.6	2590	459	C10H15N3O3S1	HYDROXYETHYLPHENYLSULFONE,4-(3,3-DIMETHYLTRIAZENYL) *antineoplastic*
6	-2.05	7.0	1812	1	C10H15N3O4	5'-AMINO-5'-DEOXYTHYMIDINE
7 ★	-2.45	7.0	2440	1	C10H15N3O4	CARBODINE
8 ★	-2.41	7.0	2440	1	C10H15N3O4	ISOCARBODINE
9 ★	0.14		2612		C10H15N3O4S1	1-(4'-MESULFONYLAMINOPHENYL)-3-MEO-3-METHYLUREA 88132-27-8
6510	1.38		2279		C10H15N3O4S1	IMIDAZOLE,1-METHYL-4-NITRO-5-CYCLOHEXYLSULFONATO
1	0.03	7.4	1470	1	C10H15N3O5	1-(3-MEO-2-OH-PR)-4-ACETYL-5-ME-2-NITROIMIDAZOLE
2 ★	-2.01	7.0	2440	1	C10H15N3O5	5-METHYLCYTIDINE 2140-61-6
3	0.39	7.4	2331	1	C10H15N3S1	5H-THIAZOLO[4,5-A]QUINOLIZIN-2-AMINE,4,7,8,9,10,10A-HEXAHYDRO 112114-07-5
4	1.09	7.4	1350	1	C10H15N3S1	TALIPEXOLE 101626-70-4 *hypotensive*
15 ★	1.48	10.6	594		"	TALIPEXOLE
6 ✓	-1.63		2397	101	C10H15N4O2	1,2,4-TRIAZOLIUM CHLORIDE,1-METHYL-3-HYDROXYIMINOMETHYL-4-(3-(PROPYN-2YLOXY)PROPYL
7 ★	1.79		526		C10H15N5	ADENINE,9-PENTYL 2002-36-0
8	0.28		2123	22	C10H15N5	BENZALDEHYDE,4-DIMETHYLAMINO,GUANYLHYDRAZONE 38407-85-1
9 ★	-1.48	6.0	1327		C10H15N5	PHENETHYLBIGUANIDE 114-86-3 *antidiabetic*
6520	-0.83	7.3	761	314	"	PHENETHYLBIGUANIDE
1	1.60		723	459	C10H15N5O1	2,4-DIAMINO-6-DIALLYLAMINO-PYRIMIDINE-3-OXIDE 55921-61-4
2 ★	0.66		526		C10H15N5O1	ADENINE,9-(1-HYDROXYMETHYL-BUTYL)
3 ★	-0.50		2302		C10H15N5O1S1	CIMETIDINE,IMIDAZOLIDONE ANALOG
4	-0.13	7.4	2037	1	C10H15N5O1S1	ISOGUANINE,1-METHYL-7-ETHYLTHIOETHYL
25	0.00	7.4	2037	1	C10H15N5O1S1	ISOGUANINE,1-METHYL-9-ETHYLTHIOETHYL
6	-0.78	7.4	2037	1	C10H15N5O2	2H-PURIN-2-ONE,1,3-DIHYDRO-1-METHYL-6-AMINO-7-VINYLOXYETHYL
7	-1.10	7.4	2037	1	C10H15N5O2	ISOGUANINE,1-METHYL-7-ETHOXYETHYL
8	-0.40	7.4	2037	1	C10H15N5O2	ISOGUANINE,1-METHYL-9-ETHOXYETHYL
9	-1.20	7.4	2037	1	C10H15N5O3	2H-PURIN-2-ONE,1,3-DIHYDRO-1-METHYL-6-AMINO-7-ACETYLOXYETHYL
6530	-1.52	7.4	2037	1	C10H15N5O3	ISOGUANINE,1-METHYL-9-METHOXYETHOXYMETHYL
1 ★	3.94		572		C10H15O1S2	O-ET-S-PH-ETHYLPHOSPHONOTHIOATE 944-22-9 *insecticide*
2 ★	2.11		1693		C10H15O2P1S1	FONOFOS-O-ANALOG 944-21-8
3 ★	3.46		336		C10H15O3P1S1	O,O-DIETHYL-O-PHENYLPHOSPHOROTHIOATE 32345-29-2
4 ★	4.09		1693		C10H15O3P1S2	FENTHION 55-38-9 *insecticide*
35 ★	1.64		336		C10H15O4P1	O,O-DIETHYL-O-PHENYLPHOSPHATE 2510-86-3
6 ★	0.42		599		C10H16Br2N2O2	PIPOBROMAN 54-91-1 *antineoplastic*
7 ★	2.98		401		C10H16Cl1N3O2	3-(2-NORBORNYL)-1-(2-CLOROETHYL)-1-NITROSOUREA 13909-13-2
8	-0.95	7.4	401	1	C10H16Cl1N3O4	1-(2-CLET)-3-(4-COOH-CYHEXYL)-1-NITROSOUREA-C 42558-93-0
9 ✓	1.68	4.6	401	2	"	1-(2-CLET)-3-(4-COOH-CYHEXYL)-1-NITROSOUREA-C
6540 ★	1.53	4.6	401	2	C10H16Cl1N3O4	1-(2-CLET)-3-(4-COOH-CYHEXYL)-1-NITROSOUREA-T 42558-94-1
1	-1.37	7.4	401	1	C10H16Cl1N3O4	3(1-CO2H-1-CYPEN)-1-NO-1-(2-CLET)UREA
2	-1.37	7.4	401	1	C10H16Cl1N3O4	UREA,1-(2-CHLOROET)-1-NITROSO-3-(1-CARBOXYCYCLOPENT-1-YL) 52320-88-4
3 ★	2.01	7.4	401	2	"	UREA,1-(2-CHLOROET)-1-NITROSO-3-(1-CARBOXYCYCLOPENT-1-YL)
4 ✓	-1.19		2397	101	C10H16F3N4O3S1	IMIDAZOLIUM CHLORIDE,1-ETHYL-2-HYDROXYIMINOMETHYL-3-(2-(N-METHYL)TRIFLUOROMETHYLSULFONAMIDO)ETHYL

	logP	pH	Ref.	Note	MF	Name / CAS / activity
45 ✓	-0.42		2397	101	C10H16F3N4O3S1	IMIDAZOLIUM CHLORIDE,1-METHYL-2-HYDROXYIMINOMETHYL-3-(2-TRIFLUOROMETHYLSULFONAMIDO)BUTYL
6 ✓	-0.82		2397	101	C10H16F3N4O3S1	IMIDAZOLIUM CHLORIDE,1-METHYL-2-HYDROXYIMINOMETHYL-3-(2-TRIFLUOROMETHYLSULFONAMIDO)PROPYL
7 ✓	-1.03		2397	101	C10H16F3N4O3S1	IMIDAZOLIUM CHLORIDE,1-METHYL-2-HYDROXYIMINOMETHYL-3-(N-ETHYL-TRIFLUOROMETHYLSULFONYL)ETHYL
8 ✓	-1.37		2397	101	C10H16F3N4O3S1	IMIDAZOLIUM CHLORIDE,1-METHYL-2-HYDROXYIMINOMETHYL-3-(N-METHYL-TRIFLUOROMETHYLSULFONAMIDOPROPYL)
9 ✓	-1.47		2397	101	C10H16F3N4O4S1	IMIDAZOLIUM CHLORIDE,1-METHYL-2-HYDROXYIMINOMETHYL-3-(N-METHYL-TRIFLUOROMETHYLSULFONAMIDOETHOXYMETHYL
6550	-2.40		571		C10H16N1O1.Br1	M-HYDROXYBENZYLTRIMETHYLAMMONIUMBROMIDE
1	-2.54		2173		C10H16N1O1.I1	M-METHOXY-TRIMETHYLANILINIUM IODIDE 53290-32-7
2 ★	2.60		1693		C10H16N1O3P1S1	PARATHION-AMINO 3735-01-1
3	-0.49	2.5	2484		C10H16N1O5	1-OXYPRROLINE,2,2,5,5-TETRAMETHYL-3,4-DICARBOXY
4	0.46	2.5	2484		"	1-OXYPRROLINE,2,2,5,5-TETRAMETHYL-3,4-DICARBOXY
55	-2.38		571		C10H16N1.Br1	BENZYLTRIMETHYLAMMONIUMBROMIDE 5350-41-4
6	-3.38		1635	820	C10H16N1.Cl1	BENZYLTRIMETHYLAMMONIUMCHLORIDE 56-93-9
7 ✓	-2.33	7.4	594	314	"	BENZYLTRIMETHYLAMMONIUMCHLORIDE
8	-2.17		570			BENZYLTRIMETHYLAMMONIUMCHLORIDE
9	-1.04		1901	820	C10H16N1.I1	BENZYLTRIMETHYL AMMONIUM IODIDE 5400-94-2
6560	-2.31		2173		C10H16N1.I1	M-METHYL-TRIMETHYLANILINIUM IODIDE 33046-97-8
1	-1.85		1506		C10H16N1.C7H7O3S1	BENZYL-TRIMETHYLAMMONIUMTOSYLATE 19362-89-1
2	-1.23	2.7	1473	340	C10H16N2	2-PYRIDINEPENTANEAMINEHYDROCHLORIDE 59082-57-4
3 ★	1.32		1474		C10H16N2	2-PYRIDINEPENTANEAMINE
4	-1.26	2.7	1473	340	C10H16N2	3-PYRIDINEPENTANEAMINE HYDROCHLORIDE
65 ★	1.41		1474		C10H16N2	3-PYRIDINEPENTANEAMINE
6 ★	0.91	11.0	533	322	C10H16N2	4-(N-METHYL)-3-PYRIDYLBUTYLAMINE 3000-74-6
7	-1.32	2.7	1473	340	C10H16N2	4-PYRIDINEPENTANEAMINE HYDROCHLORIDE 59082-52-9
8 ★	1.40		1474		C10H16N2	4-PYRIDINEPENTANEAMINE
9 ★	1.75		2069		C10H16N2	DECANEDINITRILE 1871-96-1
6570	-1.39	7.4	785	1	C10H16N2	N'-PHENYL-N-METHYLPROPYLENEDIAMINE 2095-73-0
1 ★	1.01	11.0	533	322	C10H16N2	N,N-DIETHYL-3-PYRIDYLMETHYLAMINE 2055-14-3
2 ★	1.34	11.0	533	322	C10H16N2	N-BUTYL-3-PYRIDYLMETHYLAMINE 20173-12-0
3 ★	0.05	7.0	2309	1	C10H16N2O2	GLYCINAMIDE,2,2-DIALLYL-N-ACETYL
4 ★	2.66	1.0	2202	342	C10H16N2O2S1	THIOBARBITURIC ACID,5,5-DIPROPYL
75 ★	1.56	1.0	2202	342	C10H16N2O3	BARBITURIC ACID, 5,5-DI-ISOPROPYL
6 ★	1.75	1.0	2202	342	C10H16N2O3	BARBITURIC ACID,5,5-DIPROPYL 2217-08-5
7	1.26	7.4	899	1	C10H16N2O3	BARBITURIC ACID,5-BUTYL-5-ETHYL 77-28-1 sedative, hypnotic
8 ★	1.73	1.0	2202	342	"	BARBITURIC ACID,5-BUTYL-5-ETHYL
9 ★	1.38	1.0	2202	342	C10H16N2O3	BARBITURIC ACID,5-ETHYL-5-T-BUTYL
6580 ★	1.65	0.7	962		C10H16N2O3	BARBITURIC ACID,5-PROPYL-5-ETHYL,1-METHYL 56344-90-2
1 ★	1.65	1.0	2202	342	C10H16N2O3	BARBITURIC ACID,5-S-BUTYL-5-ETHYL 125-40-6 hypnotic, sedative
2 ✓	-1.90		2397	101	C10H16N3O2	IMIDAZOLIUM CHLORIDE,1-METHYL-2-HYDROXYIMINOMETHYL-3-(1-METHYLPROPEN-2YLOXY)METHYL
3 ✓	-1.83		2397	101	C10H16N3O2	IMIDAZOLIUM CHLORIDE,1-METHYL-2-HYDROXYIMINOMETHYL-3-(2-METHYLPROPEN-2-YLOXY)METHYL
4 ✓	-1.54		2397	101	C10H16N3O2	IMIDAZOLIUM CHLORIDE,1-METHYL-2-HYDROXYIMINOMETHYL-3-BUTEN-2YLOXYMETHYL 117983-07-0
85 ✓	-1.93		2397	101	"	IMIDAZOLIUM CHLORIDE,1-METHYL-2-HYDROXYIMINOMETHYL-3-BUTEN-2YLOXYMETHYL 117983-08-1
6 ✓	-1.84		2397	101	C10H16N3O2	IMIDAZOLIUM CHLORIDE,1-METHYL-2-HYDROXYIMINOMETHYL-3-BUTEN-3YLOXYMETHYL 117983-15-0
7 ✓	-2.10		2397	101	C10H16N3O2	IMIDAZOLIUM CHLORIDE,1-METHYL-2-HYDROXYIMINOMETHYL-3-CYCLOBUYLOXYMETHYL 117941-54-5
8 ✓	-2.22		2397	101	C10H16N3O2	IMIDAZOLIUM CHLORIDE,1-PROPYL-2-HYDROXYIMINOMETHYL-3-ACETYLMETHYL
9 ✓	-1.53		2397	101	C10H16N3O3	IMIDAZOLIUM CHLORIDE,1-METHYL-2-HYDROXYIMINOMETHYL-3-BUTYROYLOXYMETHYL
6590 ★	1.12	7.4	725	1	C10H16N4O1	2-DIETHYLAMINOISONIAZID 19353-98-1
1 ★	2.14		57		C10H16N4O1S1	3-METHIO-4-AMINO-6-CYCLOHEXYL-1,2,4-TRIAZINE-5-ONE 21085-19-8
2 ★	1.94	7.4	2206	306	C10H16N4O3S1	(2-PROPYL-5-NITROIMIDAZOL-2YL)ETHYLTHIOCARBAMIC ACID,METHYL ESTER
3 ★	0.88	7.4	2206	306	C10H16N4O4	(2-METHYL-5-NITROIMIDAZOL-2YL)ETHYLCARBAMIC ACID,ISOPROPYL ESTER
4	-0.66	7.4	2037	1	C10H16N6O1	ISOGUANINE,1-METHYL-7-DIMETHYLAMINOETHYL
95	-1.80	7.4	2037	1	C10H16N6O1	ISOGUANINE,1-METHYL-9-DIMETHYLAMINOETHYL
6 ★	-1.54		1045		C10H16N6O2	6(BIS(HOCH2)ME-AMINO-2,4-DIAZIRIDINYLTRIAZINE
7	0.00	7.2	2416	1	C10H16N6S1	CIMETIDINE 51481-61-9 antagonist (histamine h2 receptors)
8	0.33	7.4	2481	1	"	CIMETIDINE
9 ★	0.40		1173		"	CIMETIDINE
6600 ★	0.67	9.2	1940	1	C10H16N8S2	TIOTIDINE 69014-14-8 antagonist (histamine h-2)
1 ★	2.14		505		C10H16O1	ADAMANTANE,1-HYDROXY 768-95-6
2	1.97		527		C10H16O1	DECAHYDRO-2-NAPHTHALENONE 4832-17-1
3 ★	4.13		547		C10H16Si1	BENZYLTRIMETHYLSILANE 770-09-2
4 ✓	-1.23		2397	101	C10H17Br1N3O2	IMIDAZOLIUM CHLORIDE,1-METHYL-2-HYDROXYIMINOMETHYL-3-(2-METHYL-3-BROMOPROPOXY)METHYL
5 ✓	-1.80		2397	101	C10H17Br1N3O2	IMIDAZOLIUM CHLORIDE,1-METHYL-2-HYDROXYIMINOMETHYL-3-(3-BROMOPROPOXY)METHYL
6 ✓	-1.56		2397	101	C10H17Cl1N3O2	IMIDAZOLIUM CHLORIDE,1-ETHYL-2-HYDROXYIMINOMETHYL-3-(3-CHLOROPROPOXY)METHYL
7 ✓	-1.88		2397	101	C10H17Cl1N3O2	IMIDAZOLIUM CHLORIDE,1-METHYL-2-HYDROXYIMINOMETHYL-3-(3-CHLOROPROPOXY)METHYL
8 ✓	-1.47		2397	101	C10H17Cl1N3O2	IMIDAZOLIUM CHLORIDE,1-METHYL-2-HYDROXYIMINOMETHYL-3-(4-CHLOROBUTOXY)METHYL 132540-11-5
9 ★	4.25		2488		C10H17F2N5S1	1,3,5-TRIAZINE,2-DIFLUOROMETHIO-4,6-BIS-I-PROPYLAMINO
6610 ★	4.51		2488		C10H17F2N5S1	1,3,5-TRIAZINE,2-DIFLUOROMETHIO-4-BUTYLAMINO-6-ETHYLAMINO
1 ★	4.30		2488		C10H17F2N5S1	1,3,5-TRIAZINE,2-DIFLUOROMETHIO-4-I-BUTYL-6-ETHYLAMINO
2 ★	4.23		2488		C10H17F2N5S1	1,3,5-TRIAZINE,2-DIFLUOROMETHIO-4-I-PROPYLAMINO-6-PROPYLAMINO
3 ★	4.57		2488		C10H17F2N5S1	1,3,5-TRIAZINE,2-DIFLUOROMETHIO-4-T-BUTYL-6-ETHYLAMINO
4 ★	4.42		2488		C10H17F2N5S1	1,3,5-TRIAZINE,2-DIFLUOROMETHLTHIO-4-AMYLAMINO-6-METHYLAMINO
15 ✓	-2.79		2397	101	C10H17F3N3O2S1	IMIDAZOLIUM CHLORIDE,1,2-DIMETHYL-3-(2-N-TRIFLUOROMETHYLSULFONAMIDO)ETHYL
6 ✓	-2.67		2397	101	C10H17F3N3O2S1	IMIDAZOLIUM CHLORIDE,1-METHYL-2-ETHYL-3-(2N-METHYL-TRIFLUOROMETHYLSULFONAMIDO)ETHYL
7 ✓	-2.68		2397	101	C10H17F3N3O3S1	IMIDAZOLIUM CHLORIDE,1-METHYL-2-(1-HYDROXYETHYL)-3-(2-N-METHYL-TRIFLUOROMETHYLSULFONAMIDO)ETHYL
8 ★	2.44		1422	462	C10H17N1	AMANTADINE 768-94-5 antiviral
9 ✓	0.78		2213		C10H17N1O2	METHYPRYLON 125-64-4 sedative
6620	0.48		213		C10H17N1O2	N-I-PENTANOYLCYCLOBUTANECARBOXAMIDE 25031-86-1
1	0.75		213		C10H17N1O2	N-PENTANOYLCYCLOBUTANECARBOXAMIDE 25031-80-5
2	0.53		213		C10H17N1O2	N-T-PENTANOYLCYCLOBUTANECARBOXAMIDE 25031-87-2
3	-2.90		1466		C10H17N2.Br1	2-PYRIDYLME-(6,N,N-TETRAMETHYL)AMMONIUMBROMIDE 22337-35-5
4	-3.20		1466		C10H17N2.Br1	4-PYRIDYLME-(6-N,N,N-TETRAMETHYL)AMMONIUMBROMIDE 75523-38-5
25	-2.50		571		C10H17N2.Br1	M-AMINOBENZYLTRIMETHYLAMMONIUMBROMIDE
6 ★	1.70		1629		C10H17N3O3S2	N-(N'-METHYL-CY-PROPYLCARBAMYLTHIO)METHOMYL ✚
7	1.32	7.4	1497	1	C10H17N3O4	2-NITROIMIDAZOLE, 1-(3-BUTOXY-2-OH)PROPYL
8	-1.40		1890		C10H17N3O6S1	GLUTATHIONE,CADMIUM(II) SALT
9 ✓	-1.96		2397	101	C10H17N4O4	IMIDAZOLIUM CHLORIDE,1-METHYL-2-HYDROXYIMINOMETHYL-3-(2-METHYL-2-NITROPROPOXY)METHYL
6630	1.62		723	459	C10H17N5O1	2,4-DIAMINO-6-CYCLOHEXYLAMINOPYRIMIDINE-3-OXIDE 55921-57-8
1	-0.13	7.4	2037	1	C10H17N5O1S1	2H-PURIN-2-ONE,1,3-DIHYDRO-1-METHYL-6-AMINO-7-ETHYLTHIOETHYL
2 ✓	0.79		1412		C10H17N5O2	1,2,4-TRIAZOLE,3,5-BUTYRAMIDO
3	-1.10	7.4	2037	1	C10H17N5O2	2H-PURIN-2-ONE,1,3-DIHYDRO-1-METHYL-6-AMINO-7-ETHOXYETHYL
4	0.49		57		C10H17N5O2	3-MORPHOLINO-4-AMINO-6-I-PR-1,2,4-TRIAZINE-5-ONE 50917-19-6

	logP	pH	Ref.	Note	MF	Name / CAS / activity
35	-0.32	9.0	1409		C10H17N5O2	5-NITROIMIDAZOLE,1-ME-2-(4-ME-1-PIPERIDNYL)METHYL 54387-29-0
6 ★	-0.40	9.0	1949		C10H17N5O2S1	CIMETIDINE,NITRODIAMINOETHENE ANALOG 55884-23-6 ✚
7 ★	-0.74	9.0	1949		C10H17N5O2S2	CIMETIDINE,4-THIOPYRIMIDIN-2YL-DIOXIDE ANALOG 54855-72-0 ✚
8	-0.89	7.4	2273	1	C10H17N5O3	1-((3-NITRO-1,2,4-TRIAZOL-1-YL)-2-HYDROXYPROPYL)PIPERIDINE 104958-87-4
9	-3.31		547	463	C10H17N5O6	L-PENTAGLYCINE
6640 ★	3.30		401		C10H18Cl1N3O2	SEMUSTINE 13909-09-6 antineoplastic
1	3.94		1283		C10H18Cl1N5	2-CL-4-I-PR-AMINO-6-DIETAMINO-S-TRIAZINE 1912-25-0
2 ★	0.90		977		C10H18Cl2N6O4	1,1'-TETRAMETHYLENE-BIS-CNU 60784-43-2
3	1.46		171	251	C10H18Cu1N6S3	B-2844COPPERBISTHIOSEMICARBAZONE
4	1.72		171	250	"	B-2844COPPERBISTHIOSEMICARBAZONE
45	-1.38		1466	42	C10H18N1O1.Br1	2-FURANYLME-(5-ET-N,N,N-TRIME)AMMONIUMBROMIDE 75523-41-0
6 ✓	-1.38		2033		C10H18N1O1.I1	FURANE,2-ETHYL-5-TRIMETHYLAMMONIOMETHYL IODIDE
7	-2.35		1144		C10H18N1O2.I1	ARECAIDINE,ETHYL ESTER, METH IODIDE ✚
8	1.12	2.5	2484		C10H18N1O3	1-OXYPIPERIDINE,2,2,6,6-TETRAMETHYL-4-CARBOXY
9	1.03		1407		C10H18N2O1	1,2,3,6-H4-PYRIDINE,1-T-BUTYLCARBONYLAMINO
6650 ★	-2.52	5.4	2017	1	C10H18N2O3S1	PROLINE,METHIONYL 59227-86-0
1	1.45	3.5	1267	1	C10H18N2O4	ETHYL-2,2-DIETHYLMALONURATE
2	1.00	3.5	1267	1	C10H18N2O5	METHOXYMETHYL-2,2-DIETHYLMALONURATE
3 ✓	-2.58		2397	101	C10H18N3O2	IMIDAZOLIUM CHLORIDE,1-(T-BUTYL)-2-HYDROXYIMINOMETHYL-3-METHOXYMETHYL
4 ✓	-1.39		2397	101	C10H18N3O2	IMIDAZOLIUM CHLORIDE,1-METHYL-2-HYDROXYIMINOMETHYL-3-BUTOXYMETHYL 91900-13-9
55	-1.39	7.6	2003	314	C10H18N3O2.Cl1	IMIDAZOLIUM CHLORIDE,1-BUTOXYMETHYL-2-HYDROXYIMINOMETHYL-3-METHYL 91900-13-9
6 ✓	-2.60		2397	101	C10H18N3O3S1	IMIDAZOLIUM CHLORIDE,1-ETHYL-2-HYDROXYIMINOMETHYL-3-(3-METHYLSULFONYLPROPYL 132566-55-3
7 ✓	0.00		2397	101	C10H18N3O4S1	IMIDAZOLIUM CHLORIDE,1-ETHYL-2-HYDROXYIMINOMETHYL-3-(2-METHYLSULFONYLETHOXY)METHYL
8 ✓	0.00		2397	101	C10H18N3O4S1	IMIDAZOLIUM CHLORIDE,1-METHYL-2-HYDROXYIMINOMETHYL-3-(2-ETHYLSULFONYLETHOXY)METHYL
9 ✓	-2.50		2397	101	C10H18N3O4S1	IMIDAZOLIUM CHLORIDE,1-METHYL-2-HYDROXYIMINOMETHYL-3-(3-METHYLSULFONYLPROPOXY)METHYL
6660 ✓	-2.50	7.4	2397	101	C10H18N3O4S1	IMIDAZOLIUM CHLORIDE,1-PROPYL-2-HYDROXYIMINOMETHYL-3-PROPANESULFONIC ACID
1 ★	3.21		57		C10H18N4O1S1	3-BUTYLTHIO-4-AMINO-6-I-PROPYL-1,2,4-TRIAZINE-5-ONE 50917-25-4
2 ★	2.68		57		C10H18N4O1S1	3-METHIO-4-AMINO-6-N-HEXYL-1,2,4-TRIAZINE-5-ONE 21085-20-1 ✚
3 ★	1.02	7.6	1769	212	C10H18N4O1S2	4-ME-1,2,5-THIADIAZOLE-3-C(=NOH)SCH2CH2N(ET)2 90507-25-8 ✚
4 ★	1.51	7.4	830	1	C10H18N4O2	N-DIAZOACETYLGLYCINE-N'-HEXYLAMIDE 60141-99-3
65	-0.41	7.4	2298	1	C10H18N4O2S1	1,2,4-OXADIAZOLE-5-(DIETHYLAMINOETHYL)THIOHYDROXAMATE,3-METHYL 90507-31-6 ✚
6 ★	0.71	7.4	2298	2	"	1,2,4-OXADIAZOLE-5-(DIETHYLAMINOETHYL)THIOHYDROXAMATE,3-METHYL
7	1.10	7.6	1769	212	C10H18N4O2S1	4-ME-1,2,5-OXADIAZOLE-3-C(=NOH)SCH2CH2N(ET)2 90507-23-6
8 ★	-1.70	7.1	1591		C10H18N4O4S2	CYSTINE-AMIDE,N-ACETYL
9 ★	1.70		1629		C10H18N4O4S3	THIODICARB 59669-26-0 insecticide
6670 ★	-0.07		1018		C10H18N4O5S2	N-(N'-O-DIMETHYLCARBAMYLTHIO)OXAMYL
1 ★	0.85	9.2	1853	1	C10H18N4S1	METHYLBURIMAMIDE 51264-00-7
2	1.36		171	252	C10H18N6Ni1S3	B-2844NICKELBISTHIOSEMICARBAZONE
3	1.43		171	250	"	B-2844NICKELBISTHIOSEMICARBAZONE
4	1.45		171	251	"	B-2844NICKELBISTHIOSEMICARBAZONE
75	-0.66	7.4	2037	1	C10H18N6O1	2H-PURIN-2-ONE,1,3-DIHYDRO-1-METHYL-6-AMINO-7-DIMETHYLAMINOETHYL
6	-1.19		964		C10H18N6O2	N2-CARBOXYMETHYL-N2,N4,N4,N6,N6-PENTAMETHYLMELAMINE
7	-0.35		964		C10H18N6O2S1	2-(S-CYSTEINYL)-TETRAMETHYLMELAMINE
8 ★	0.03	4.2	401	2	C10H18N6O3	IMIDAZOLE,5-CARBOXAMIDE-4-(3,3-BIS-(2-METHOXYETHYL)TRIAZENYL)
9	1.46		171	250	C10H18N6S3Zn1	B-2844-ZINC-BIS-THIOSEMICARBAZONE ✚
6680	1.62		171	251	"	B-2844-ZINC-BIS-THIOSEMICARBAZONE
1	1.74		171	252	"	B-2844-ZINC-BIS-THIOSEMICARBAZONE
2 ★	2.72		594	400	C10H18O1	BORNEOL 507-70-0
3 ★	2.50		594	400	C10H18O1	EUCALYPTOL 470-82-6 pharmaceutic flavor, cockroach repellant
4 ★	2.32		594	400	C10H18O1	ISOBORNEOL
85 ★	2.72		579		C10H18O2	DECALACTONE-GAMMA 706-14-9
6 ★	2.16		1819		C10H18O4	BUTANEDIOIC ACID,PROPYL ESTER 925-15-5
7	-1.70	2.0	1443		C10H19N1O1	3-METHYL-4-PIPERIDINOBUTAN-2-ONE 42327-99-1
8	1.28	7.4	1443	1	"	3-METHYL-4-PIPERIDINOBUTAN-2-ONE
9 ★	1.28	2.2	2228	314	C10H19N1O3	HEPTANOIC ACID-2-ACETYLAMINO,METHYL ESTER
6690 ★	-1.13		2377		C10H19N3O3	AC-ALA-VAL-N
1 ★	-0.78		2377		C10H19N3O3	AC-GLY-LEU-N
2 ★	-1.14		2377		C10H19N3O3	AC-VAL-ALA-N
3 ★	-3.09		594		C10H19N3O3	METHANETRICARBOXAMIDE,HEXAMETHYL
4 ★	2.20		1629		C10H19N3O3S2	N-(N"-METHYL-I-PRCARBAMYLTHIO)METHOMYL
95 ★	2.20		1629		C10H19N3O3S2	N-(N"-METHYL-PRCARBAMYLTHIO)METHOMYL
6 ★	-1.53		2377		C10H19N3O4	AC-SER-VAL-N
7	-2.81	7.4	2397	101	C10H19N4O1	IMIDAZOLIUM CHLORIDE,1-METHYL-2-HYDROXYIMINOMETHYL-3-(2-DIMETHYLAMINO)PROPYL 132566-74-6
8	-3.14	7.4	2397	101	C10H19N4O1	IMIDAZOLIUM CHLORIDE,1-METHYL-2-HYDROXYIMINOMETHYL-3-(3-DIMETHYLAMINOPROPYL) 131206-89-8
9	-3.01		2397	101	C10H19N4O2	IMIDAZOLE,1-TRIMETHYLAMMONIOETHOXYMETHYL-2-HYDROXYIMINOMETHYL
6700	2.16		723	459	C10H19N5O1	2,4-DIAMINO-6-DIPROPYLAMINO-PYRIMIDINE-3-OXIDE 55921-60-3
1 ★	2.99		2550	459	C10H19N5O1	2-MEO-4,6-BIS(I-PROPYLAMINO)S-TRIAZINE 1610-18-0 herbicide
2 ★	1.78		57		C10H19N5O1	3-N-BUTYLAMINO-4-AMINO-6-I-PR-1,2,4-TRIAZINE-5-ONE 33665-71-3
3 ★	3.10		2550	459	C10H19N5O1	TERBUMETON 33693-04-8
4 ★	3.51		2094		C10H19N5S1	PROMETRIN 7287-19-6 herbicide
5 ★	3.74		1211		C10H19N5S1	TERBUTRYN 886-50-0 herbicide
6 ★	2.36		1189		C10H19O6P1S2	MALATHION 121-75-5 pediculicide~insecticide
7	-1.02		2358	200	(C10H20N1O5S2)2.Cd1	N-PROPYL-N-DITHIOCARBOXY-D-GLUCAMINE
8 ✓	-0.91		2358	200	"	N-PROPYL-N-DITHIOCARBOXY-D-GLUCAMINE
9 ✓	-1.85		2033		C10H20N1.I1	CYCLOPENTENE,1-METHYL-3-TRIMETHYLAMMONIOMETHYL IODIDE
6710 ✓	-1.95		2033		C10H20N1.I1	CYCLOPENTENE,3-METHYL-1-TRIMETHYLAMMONIOMETHYL IODIDE
1 ✓	1.60		567	820	C10H20N1.C6H2N3O7	N-PROPYLQUINUCLIDINIUM PICRATE
2 ★	0.75	7.0	2309	1	C10H20N2O2	GLYCINAMIDE,2,2-DIPROPYL-N-ACETYL
3 ★	-2.82	7.0	2197		C10H20N2O3	VALINYLVALINE 3918-94-3
4 ★	-2.53	7.0	2374		C10H20N2O3S1	METHIONYLVALINE
15 ★	2.83		305		C10H20N2S1	TRIMETHYLENETHIOUREA,N,N-DIPROPYL 30826-88-1
6 ★	3.88		1555		C10H20N2S4	DISULFIRAM 97-77-8 alcohol deterrent
7	-2.61	7.4	2397	101	C10H20N4O1	IMIDAZOLIUM CHLORIDE,1-METHYL-2-HYDROXYIMINOMETHYL-3-TRIMETHYLAMMONIOETHYL
8 ★	1.96		964		C10H20N6O1	N2-2-HYDROXYETHYL-N2,N4,N4,N6,N6-PENTAMETHYLMELAMINE
9 ★	1.38		171		C10H20N6S3	DIPROPYL SULFIDE,2,2'-BIS-1-(4-METHYL-THIOSEMICARBIZID)
6720 ★	3.73		2047		C10H20O1	2-DECANONE 693-54-9
1 ★	3.02		2126		C10H20O1	CYCLOHEXANOL,4-T-BUTYL-CIS 937-05-3
2 ★	3.09		2126		C10H20O1	CYCLOHEXANOL,4-T-BUTYL-TRANS 21862-63-5
3 ★	3.20		1815		C10H20O2	2-BUTYL-HEXANOICAICD 3115-28-4
4 ★	3.20		1815		C10H20O2	2-PROPYL-HEPTANOIC ACID 31080-39-4

	logP	pH	Ref.	Note	MF	Name / CAS / activity
25 ★	4.09		505		C10H20O2	DECANOIC ACID 334-48-5
6	4.32		1560	100	C10H20O2	METHYLNONANOATE 98-01-1
7 ★	-0.48		2308		C10H20O5	15-CROWN-5-ETHER 33100-27-5
8	-0.54		1294		C10H20O5S1	1-BUTYLTHIO-B-GALACTOPYRANOSIDE
9	-1.04		1294		C10H20O6	B-BUTYLGALACTOPYRANOSIDE
6730	6.43		1369	459	C10H21Br1	1-BROMODECANE 112-29-8
1 ★	3.84		2506		C10H21N1S1	PEBULATE 591-62-8 herbicide
2 ★	3.84		2506		C10H21N1S1	VERNOLATE 1929-77-7 herbicide
3	-1.30	7.0	2484	1	C10H21N2O1	1-OXYPIPERIDINE,2,2,6,6-TETRAMETHYL-4-METHYLAMINO
4	0.68	12.0	2484	1	"	1-OXYPIPERIDINE,2,2,6,6-TETRAMETHYL-4-METHYLAMINO
35 ★	1.02		401		C10H21N2O2P1	PHOSPHINIC ACID,BIS(2,2-DIMEAZIRIDINYL),ETHYL ESTER 14984-65-7
6	0.38		2083		C10H21N2O3P1S1	DIAZIRIDINTHIOPHOSPHORAMIDATE,TRIOXYETHYLENE ESTER 101347-42-6
7	0.10		2083		C10H21N2O4P1	DIAZIRIDINYLPHOSPHORAMIDATE,TRIOXYETHYLENE ESTER 101347-40-4
8 ✓	0.81		2426	537	C10H22Cl2N2Pt1	PLATINUM,BIS-CYCLOPENTYLAMMONIO-DICHLORO
9 ✓	-2.60		2033		C10H22N1O1.I1	CYCLOPENTANOL,2-METHYL-4-TRIMETHYLAMMONIOMETHYL IODIDE
6740 ✓	-1.50		1704	820	C10H22N1.Br1	(3-METHYLCYCLOPENTYL)-TETRAMETHYLAMMONIUMBROMIDE
1 ✓	-1.50		2033		C10H22N1.I1	METHYLCYCLOPENTANE-3-TRIMETHYLAMMONIOMETHYL IODIDE
2 ✓	1.69		567	820	C10H22N1.C6H2N3O7	N-ET-N-I-BUTYLPYRROLIDINIUM PICRATE
3 ✓	1.60		567	820	C10H22N1.C6H2N3O7	N-ME-N-S-BUTYLPIPERIDINIUM PICRATE
4 ★	1.88		1908		C10H22N3O1P1	PHOSPHONICAMIDE,N,N-DIME-P,P-DI-1-PYRROLIDINYL 53439-65-9
45	2.17		1967		C10H22N4	4-NONANONE,GUANYLHYDRAZONE 97183-50-1
6 ★	4.57		579		C10H22O1	DECANOL 112-30-1
7 ★	2.48		1819		C10H22O2	1,2-DIBUTOXYETHANE
8 ★	1.70		1819		C10H22O3	ETHANOL,2-(2-HEXOXYETHOXY) 112-59-4
9	1.92	7.0	834	1	C10H23N1	DECYLAMINE 2016-57-1
6750	-0.90		1394		C10H23N3O1	4-NONYLSEMICARBAZIDE
1	-0.29	7.3	761	314	C10H23N5	OCTYLBIGUANIDE 20709-45-9
2 ★	0.69		579		C10H23O5P1	DIETHYLPHOSPHONOACETALDEHYDE,DIETHYLACETAL 7598-61-0 ✚
3	4.02		512		C10H23Si1	SILANE,OCTYL-DIMETHYL 40934-68-7
4	-1.90		1966	459	C10H24Au2Cl2P2	1,2-BIS(DIETHYLPHOSPHINO)ETHANE,BIS-CHLOROGOLD ✚
55	-1.33		1966		"	1,2-BIS(DIETHYLPHOSPHINO)ETHANE,BIS-CHLOROGOLD
6	1.51		600		C10H24N1O2P1S1	O,S-DIME-N-OCTYL-PHOSPHORAMIDOTHIOATE 52067-53-5
7	-0.98		1901	820	C10H24N1.I1	TRIPROPYLMETHYL AMMONIUM IODIDE 3531-14-4
8 ✓	1.80		567	820	C10H24N1.C6H2N3O7	BUTYLTRIETHYL AMMONIUM PICRATE
9 ✓	2.05		567	820	C10H24N1.C6H2N3O7	DI-I-BUTYLDIMETHYL AMMONIUM PICRATE
6760 ✓	1.96		567	820	C10H24N1.C6H2N3O7	METHYLTRIPROPYL AMMONIUM PICRATE
1 ★	-1.43	13.0	594	400	C10H24N4	PIPERAZINE,1,4-BIS-(3-AMINOPROPYL) 7209-38-3
2 ✓	-4.10		2085		C10H24N4.O2Tc1	CYCLAM,TECHNICIUM OXIDE COMPLEX
3	-1.15	3.0	2661		C10H24N6	GUAZATINE 13516-27-3
4 ✓	-1.14		2426	537	C10H26N2O2Pt1.(N1O3)2	PLATINUM,BIS-CYCLOPENTYLAMMONIO-DIAQUA-DINITRATE
65	5.40	4.0	1740	459	C10H30O3Si4	DECAMETHYLTETRASILOXANE 141-62-8
6	5.60	4.0	2264	1	C11H4F12N2	HEXAFLUOROACETONE,(3,5-DI-TRIFLUOROMETHYL)PHENYLHYDRAZONE
7 ★	4.36		331		C11H6Cl2F3N3O2	2,6-DICL-4-CF3-C6H2NHN=C(CN)COOME 28313-69-1
8 ★	4.30		2506		C11H6Cl2N2	FENPICLONIL ✚
9	2.08		1566		C11H6O3	ANGELICIN ✚
6770	2.41		1353		C11H7Cl1F3N3O1	5-AMINO-4-CL-2(3-CF3-PHENYL)-3-PYRIDAZINONE
1 ★	5.08		331		C11H7Cl1F3N3O1	BUTANENITRILE,2-[(2-CL-5-CF3PHENYL)HYDRAZONO]-3-OXO 28317-56-8
2 ★	4.66		331		C11H7Cl1F3N3O2	2-CF3-4-CL-C6H3NHN=C(CN)COOME 36865-72-2
3 ★	4.42		331		C11H7Cl1F3N3O2	2-CL-5-CF3-C6H3NHN=C(CN)COOME 28313-77-1 ✚
4	1.29	7.2	1985	314	C11H7Cl1N4O1	PROPANEDINITRILE,4-CHLOROACETYLPHENYLHYDRAZONE
75 ★	2.15	3.0	1985		"	PROPANEDINITRILE,4-CHLOROACETYLPHENYLHYDRAZONE
6	0.16	7.4	2266		C11H7F1N2O3	5-FLUOROURACIL-3-BENZOYL 61251-77-2
7 ★	0.80		2266	302	"	5-FLUOROURACIL-3-BENZOYL
8	-0.06	7.4	2266	1	C11H7F1N2O4	5-FLUOROURACIL-1-CARBOXYLIC ACID,PHENYL ESTER 66999-97-1
9 ★	0.64		2266	302	"	5-FLUOROURACIL-1-CARBOXYLIC ACID,PHENYL ESTER
6780 ★	1.73	4.0	2271	302	C11H7F1N2O4	5-FLUOROURACIL-3-PHENOXYCARBONYL
1 ★	3.24	7.4	595	1	C11H7F3N2O1	QUINOLINE,8-TRIFLUOROACETAMIDO
2 ★	3.44		331		C11H7F3N4O4	2-NO2-4-CF3-C6H3NHN=C(CN)COOME 28313-79-3
3	-2.57	7.4	1255	1	C11H7N1O5	4-HYDROXYQUINOLINE,3,7-DICARBOXYLIC ACID
4 ★	4.34		850		C11H7N1S1	1-ISOTHIOCYANONAPHTHALENE 551-06-4
85 ★	4.34		850		C11H7N1S1	2-ISOTHIOCYANONAPHTHALENE 1636-33-5
6	-1.07	2.7	541	340	C11H7O2.Na1	2-NAPHTHOIC ACID,SODIUM SALT 17273-79-9
7	0.17	7.2	541	314	C11H7O3.Na1	3-HYDROXY-2-NAPHTHOIC ACID,SODIUM SALT 14206-62-3
8	3.83		331		C11H8Cl1F2N3O2	3-CHF2-4-CL-C6H3NHN=C(CN)COOME 36874-60-9
9	2.62		163	459	C11H8Cl1N1O1	4-ACETYL-7-CHLOROQUINOLINE
6790	3.20		2300		C11H8Cl2N2O1	MONGARD ✚
1	2.02	7.4	1350	1	C11H8Cl2N2S1	SANDOZ-WANDER-44-549 ✚
2	5.21		331		C11H8Cl3N3O2	2,4,5-TRI-CL-C6H2NHN=C(CN)COOET 28322-78-3
3	5.03		331		C11H8Cl3N3O2	3,4,5-TRI-CL-C6H2NHN=C(CN)COOET 36865-51-7
4	1.50	4.0	2009	459	C11H8F1N3O3	URACIL,1-PHENYLAMINOCARBONYL-5-FLUORO
95	1.26		1156		C11H8F3N1O2	M-TRIFLUOROMETHYL-N-PHENYLSUCCINIMIDE 60050-38-6
6	0.76		1156		C11H8F3N1O2	O-TRIFLUOROMETHYL-N-PHENYLSUCCINIMIDE
7	1.45		1156		C11H8F3N1O2	P-TRIFLUOROMETHYL-N-PHENYLSUCCINIMIDE
8 ★	5.04		331		C11H8F3N3O1S1	BUTANENITRILE,2[(2-SCF3-PHENYL)HYDRAZONO]-3-OXO 28317-78-4
9 ★	3.72		331		C11H8F3N3O2	2-CF3-C6H4NHN=C(CN)COOME 28384-50-1
6800 ★	3.78		331		C11H8F3N3O2	3-CF3-C6H4NHN=C(CN)COOME 28313-74-8
1 ★	3.79		331		C11H8F3N3O2	4-CF3-C6H4NHN=C(CN)COOME 28313-76-0
2 ★	4.27		331		C11H8F3N3O2S1	4-SCF3-C6H4NHN=C(CN)COOME 28313-92-0
3 ★	4.22		331		C11H8F3N3O4S1	4-SO2CF3-C6H4NHN=C(CN)COOME 28313-60-2
4	3.06	7.4	1941	1	C11H8N2	1-NORHARMAN
5 ★	3.17	13.0	2326		"	1-NORHARMAN
6 ★	2.40		306		C11H8N2	MALONONITRILE,4-METHYLBENZAL 2826-25-7
7 ★	2.10		306		C11H8N2	MALONONITRILE,A-METHYLBENZAL 5447-87-0
8	0.02		1156		C11H8N2O2	M-CYANO-N-PHENYLSUCCINIMIDE
9 ★	1.68		522		C11H8N2O2S2	1,2,4-BENZOTHIADIAZINE-1,1-DIOXIDE-3-(2-THIENYL) 37157-95-2
6810 ★	1.06		522		C11H8N2O3S1	1,2,4-BENZOTHIADIAZINE-1,1-DIOXIDE-3-(2-FURYL) 37157-96-3
1	-0.95	7.4	1255	1	C11H8N2O4	7-CARBAMYL-4-HYDROXYQUINOLINE-3-CARBOXYLIC ACID
2 ★	0.15		1008	354	C11H8N2O5	2(2-FURYL)-3-(5-NO2-2-FURYL)ACRYLAMIDE
3 ★	2.22		494		C11H8N2O5S1	3-ALLYL-5(5-NO2-FURFURILIDINE)THIAZOLIDIN-2,4-DIONE 25603-14-9
4 ★	1.40		494		C11H8N2O5S1	3-ME-5(5-NO2-2-FURYLPROPENILIDINE)THIAZOLID-2,4-DION 25580-68-1

	logP	pH	Ref.	Note	MF	Name / **CAS** / *activity*
15	-2.22		1008	354	C11H8N3O4.Cl1	N-(2,4-DINITROPHENYL)PYRIDINIUMCHLORIDE
6	1.15	7.2	1985	314	C11H8N4O1	PROPANEDINITRILE,4-ACETYLPHENYLHYDRAZONE **55653-16-2**
7 ★	2.10	3.0	1985		"	PROPANEDINITRILE,4-ACETYLPHENYLHYDRAZONE
8 ★	1.93	7.3	538	2	C11H8N6O2	8-(3-NITROPHENYL)-ADENINE **17659-57-3**
9	2.64		1142		C11H8O1	1-AZULENECARBOXALDEHYDE **7206-61-3**
6820	2.20	7.4	1497	1	C11H8O2	1,4-NAPHTHOQUINONE,2-METHYL **58-27-5** *vitamin (prothrombogenic)*
1 ★	2.20		520		"	1,4-NAPHTHOQUINONE,2-METHYL
2 ★	2.10		520		C11H8O2	1,4-NAPHTHOQUINONE,6-METHYL **605-93-6**
3	2.88	3.4	1142		C11H8O2	1-AZULENECARBOXYLIC ACID **1201-25-8**
4 ★	3.10	1.0	547	342	C11H8O2	1-NAPHTHOIC ACID **86-55-5**
25 ★	3.28	1.0	579	342	C11H8O2	2-NAPHTHOIC ACID **93-09-4**
6	3.06		1533	820	C11H8O2.C6H15N1	1-NAPHTHOIC ACID:HEXYLAMINEIONPAIR
7 ★	1.35		520		C11H8O3	1,4-NAPHTHOQUINONE,2-METHOXY **2348-82-5**
8 ★	1.20		520		C11H8O3	1,4-NAPHTHOQUINONE,2-METHYL,3-HYDROXY **483-55-6**
9	3.05	2.0	2312	337	C11H8O3	2-NAPHTHOIC ACID,3-HYDROXY
6830	2.34		1478		C11H8O3	FURAN,2-COCH=CH-FURYL **3988-76-9**
1	1.42		1353		C11H9Br1N2O2	4-BR-5-METHOXY-2-PHENYL-3-PYRIDAZINONE
2	-1.15	7.4	401	1	C11H9Br1O4	3-P-ANISOYL-3-BROMOACRYLIC ACID **16170-76-6** *antineoplastic*
3 ★	2.74		328		C11H9Br2N3O2S1	N1-(3,5-DIBR-2-PYRIDYL)SULFANILAMIDE **30961-41-2**
4	2.44	7.2	1985	314	C11H9Cl1N4	PROPANEDINITRILE,4-(2-CHLOROETHYL)PHENYLHYDRAZONE **81865-11-4**
35 ★	3.58	3.0	1985		"	PROPANEDINITRILE,4-(2-CHLOROETHYL)PHENYLHYDRAZONE
6	0.42	7.4	1687	1	C11H9Cl1O2S1	TIANAFAC **51527-19-6**
7 ★	3.92	7.4	1687	2	"	TIANAFAC
8 ★	4.66		331		C11H9Cl2N3O2	3,4-DICL-C6H3NHN=C(CN)COOET **36865-77-7**
9 ★	3.82		331		C11H9Cl2N3O2	3,5-DICL-C6H3NHN=C(CN)COOET **28313-59-9**
6840 ★	1.97		328		C11H9Cl2N3O2S1	N1-(3,5-DICL-2-PYRIDYL)SULFANILAMIDE **30961-38-7**
1	3.60	7.4	1485	1	C11H9Cl3N4	2,4-DIAMINOPYRIMIDINE,5-(2,4,5-TRICHLORO)BENZYL
2 ★	3.79	7.4	1485	2	"	2,4-DIAMINOPYRIMIDINE,5-(2,4,5-TRICHLORO)BENZYL
3	3.45	7.4	1485	1	C11H9Cl3N4	2,4-DIAMINOPYRIMIDINE,5-(3,4,5-TRICHLORO)BENZYL
4 ★	3.64	7.4	1485	2	"	2,4-DIAMINOPYRIMIDINE,5-(3,4,5-TRICHLORO)BENZYL
45	0.73	7.4	2635	1	C11H9F1N2O3	SORBINIL **68367-52-2** *enzyme inhibitor (aldose reductase)*
6	2.35		2635	459	"	SORBINIL
7 ★	1.72	7.4	2341	1	C11H9F3N4O1	3[1-(2-BENZOXAZOLYL)HYDRAZINO]PROPANENITRILE,5-TRIFLUOROMETHYL
8	-1.05		1082		C11H9I3N2O4	DIATRIZOIC ACID **117-96-4** *radiopaque medium*
9	-1.15		1082		C11H9I3N2O4	IOTHALAMIC ACID **2276-90-6** *diagnostic aid (radiopaque)*
6850 ★	2.63		562	459	C11H9N1	2-PHENYLPYRIDINE **1008-89-5**
1 ★	2.59		547		C11H9N1	4-PHENYLPYRIDINE **939-23-1**
2	2.54		163	459	C11H9N1O1	2-ACETYLQUINOLINE **1011-47-8**
3 ★	2.39		570		C11H9N1O1	2-PHENOXYPYRIDINE **4783-68-0**
4 ★	1.76		1519	459	C11H9N1O1	3-ACETYLQUINOLINE **33021-53-3**
55 ★	1.58		532		C11H9N1O1	6-ACETYLQUINOLINE
6 ★	2.25		1519	459	C11H9N1O2	3-QUINOLINECARBOXYLIC ACID,METHYL ESTER **53951-84-1**
7	0.54	5.0	2153	1	C11H9N1O2	BENZOIC ACID,2-(1-PYRROLYL)
8	2.11	5.0	2153	1	C11H9N1O2	BENZOIC ACID,4-(1-PYRROLYL)
9 ★	1.74		2649		C11H9N1O2	NAPHTHOQUINONE,2-METHYLAMINO
6860 ★	2.03	7.4	595	314	C11H9N1O2	QUINOLINE,8-ACETYLOXY **2598-29-0**
1	0.72	7.4	1255	1	C11H9N1O4	7-METHOXY-4-HYDROXYQUINOLINE-3-CARBOXYLIC ACID
2 ★	1.55	1.0	594		C11H9N1O4	BENZOIC ACID,2-SUCCINIMIDO
3 ★	1.15		2209		C11H9N1O4	PHTHALIMIDE,N-ACETYLOXYMETHYL
4 ★	1.41	7.4	2542	1	C11H9N1O4S2	THIOPHENE-2-SULFONAMIDE,4-(P-HYDROXYBENZOYL)
65	-1.16	7.4	1255	1	C11H9N1O5S1	7-MESULFONYL-4-OH-QUINOLINE-3-CARBOXYLIC ACID
6 ★	3.15		1519	459	C11H9N1S1	3-PHENYLTHIOPYRIDINE **28856-77-1**
7 ★	2.60	7.4	593	1	C11H9N3	2-AMINO-9H-PYRIDO[2,3-B]INDOLE ✚
8 ★	0.14	7.4	594	1	C11H9N3O3S1	MERBARONE **97534-21-9** *antineoplastic (topoisomerase ii inhib.)*
9 ★	1.35		494		C11H9N3O4S1	2-METHYLAMINO-5-(5-NO2-FURYLPROPENILIDENE)THIAZOLONE **27472-83-9**
6870 ★	1.05		494		C11H9N3O5S1	2-(N-ME-N-ACETAMIDO)-5-(5-NO2-2-FURFURYL)THIAZOLONE **52661-43-5**
1 ★	1.21		494		C11H9N3O5S1	3-ME-2-ACETIMINO-5(5-NO2-FURFURILIDINE)THIAZOL-4-ONE **52661-66-2**
2 ★	2.31	7.3	538	2	C11H9N5	8-PHENYLADENINE **17720-22-8**
3	1.07	7.2	1985	314	C11H9N5O1	PROPANEDINITRILE,4-ACETAMIDOPHENYLHYDRAZONE **55121-29-4**
4 ★	1.83	3.0	1985		"	PROPANEDINITRILE,4-ACETAMIDOPHENYLHYDRAZONE
75 ★	4.14		331		C11H9N5O6	2,4-DINO2-C6H3NHN=C(CN)COOET **28313-64-6**
6	8.00	8.0	1927	1	C11H9O5S1.Na1	MENADIONE,SODIUMBISULFITE **130-37-0**
7 ★	3.87		560		C11H10	1-METHYLNAPHTHALENE **90-12-0**
8 ★	3.86		560		C11H10	2-METHYLNAPHTHALENE **91-57-6**
9 ★	1.24		328		C11H10Br1N3O2S1	N1-(3-BROMO-2-PYRIDYL)SULFANILAMIDE **30961-39-8**
6880 ★	2.30		328		C11H10Br1N3O2S1	N1-(5-BROMO-2-PYRIDYL)SULFANILAMIDE **16805-99-5**
1	1.19		1353		C11H10Cl1N3O1	4-CL-5-(MEAMINO)-2-PHENYL-3-PYRIDAZINONE
2 ★	4.28		331		C11H10Cl1N3O1	BUTANENITRILE,2[(2-ME-4-CHLOROPHENYL)HYDRAZONO]-3-OXO **28317-64-8**
3 ★	3.38		331		C11H10Cl1N3O2	2-CL-C6H4NHN=C(CN)COOET **3994-20-8**
4 ★	3.94		331		C11H10Cl1N3O2	3-CL-C6H4NHN=C(CN)COOET **36874-67-6**
85 ★	1.58		328		C11H10Cl1N3O2S1	N1-(2-CL-3-PYRIDYL)SULFANILAMIDE **34392-79-5**
6 ★	0.82		328		C11H10Cl1N3O2S1	N1-(3-CHLORO-2-PYRIDYL)SULFANILAMIDE **26807-64-7**
7 ★	1.82		328		C11H10Cl1N3O2S1	N1-(5-CHLORO-2-PYRIDYL)SULFANILAMIDE **30961-36-5**
8 ★	2.24		328		C11H10Cl1N3O2S1	N1-(6-CL-3-PYRIDYL)SULFANILAMIDE **34392-82-0**
9 ★	1.80	7.4	2206	306	C11H10Cl1N3O3	IMIDAZOLE,1-METHYL-5-NITRO-2-(P-CHLOROPHENOXY)METHYL
6890	2.56	7.2	1162	314	C11H10Cl2N4	METOPRINE **7761-45-7** *antineoplastic*
1 ★	2.82		884		"	METOPRINE
2 ★	2.81	13.0	547		C11H10Cl2N4	PYRIMIDINE,2,4-DIAMINO-5-(2,3-DICHLORO)BENZYL
3 ★	1.33	7.4	2206	306	C11H10F1N3O3	FLUNIDAZOLE **4548-15-6** *antiprotozoal*
4 ★	2.25		1895		C11H10F3N1O3	TRIFLUORACETAMIDE,N-BENZOYLETHYL **24568-14-7**
95 ★	0.46	7.4	1975	1	C11H10I1N5O3	IMIDAZOLE,1-(3-PICOLYLACETAMIDE)-4-NITRO-5-IODO **96258-82-1**
6 ★	0.88	7.4	1975	1	C11H10I1N5O3	IMIDAZOLE,1-(3-PICOLYLACETAMIDE)-5-NITRO-4-IODO **96258-83-2**
7 ★	-1.40		1975		C11H10I1N5O4	IMIDAZOLE,1-(3-N-OXO-PICOLYLACETAMIDE)-5-NITRO-4-IODO **96258-84-3**
8 ★	2.75		1519	459	C11H10N2	2-PYRIDINEAMINE,N-PHENYL **6631-37-4**
9 ★	1.92		163	459	C11H10N2O1	3-ACETAMINOQUINOLINE
6900	0.71		163	459	C11H10N2O1	5-ACETAMINOQUINOLINE **42464-80-2**
1 ★	1.55		163	459	C11H10N2O1	6-ACETAMINOQUINOLINE **22433-76-7**
2 ★	1.55		1519	459	C11H10N2O1	7-ACETYLAMINOQUINOLINE
3	0.97		1725	459	C11H10N2O1	PYIDO-PYRIMIDINE ANALOG(RINGS:566)
4	1.19		1725	459	C11H10N2O1	PYRIDO-PYRIMIDINE ANALOG(RINGS:566) ✚

	logP	pH	Ref.	Note	MF	Name / CAS / activity
5 ★	1.91	7.4	595	314	C11H10N2O1	QUINOLINE,8-ACETYLAMINO 33757-42-5
6 ★	1.87	7.4	595	314	C11H10N2O1	QUINOLINE,8-CARBOXAMIDOMETHYL
7 ★	-0.11	7.0	452	1	C11H10N2O2	4-ACETAMIDOQUINOLINE-1-OXIDE 23484-11-9
8 ★	0.97	7.0	452	1	C11H10N2O2	4-ACETYLOXYAMINOQUINOLINE 32654-59-4
9 ★	0.25		1439		C11H10N2O3	4H-PYRIDO(1,2-A)PYRIMIDIN-4-ONE,3-ETHOXYCARBONYL ✚
6910	2.42		494		C11H10N2O5S1	3-IPR-5(5-NO2-FURFURILIDINE)THIAZOLIDINE-2,4-DIONE 27550-11-4
1	2.40		494		C11H10N2O5S1	3-PR-5(5-NO2-2-FURFURILIDINE)THIAZOLIDINE-2,4-DIONE 52661-71-9
2 ★	1.66		572		C11H10N2S1	A-NAPHTHYLTHIOUREA 86-88-4 rodenticide
3 ★	1.46	7.4	593	1	C11H10N4	2-AMINO-3-METHYLIMIDAZ[4,5-F]QUINOLINE ✚
4 ★	1.75	7.4	593	1	C11H10N4	2-AMINO-6-METHYLDIPRYIDO[1,2-A,3',2'-D]IMIDAZOLE ✚
15	2.60	7.2	2212	1	C11H10N4	CARBONYLCYANIDE,2,5-DIMETHYLPHENYLHYDRAZONE 101398-33-8
6 ★	3.08	7.2	2212	2	"	CARBONYLCYANIDE,2,5-DIMETHYLPHENYLHYDRAZONE
7	2.10	7.2	2212	1	C11H10N4	CARBONYLCYANIDE,2,6-DIMETHYLPHENYLHYDRAZONE 104901-18-0
8 ★	2.80	7.2	2212	2	"	CARBONYLCYANIDE,2,6-DIMETHYLPHENYLHYDRAZONE
9 ★	1.34	7.4	595	314	C11H10N4	IMIDAZO[4,5,F]QUINOLINE,2-AMINO-1-METHYL
6920 ★	0.98	7.4	2206	306	C11H10N4O3	IMIDAZOLE,1-METHYL-5-NITRO-2-(P-CARBOXAMIDOPHENYL)
1 ★	1.81	1.0	2196		C11H10N4O3	PYRAZOLE,3-METHYL-4-NITRO-5-CARBOXANILIDE
2 ★	-1.40	7.4	2360	1	C11H10N4O4	BENZOTRIAZINE,1,4-DI-N-OXIDE-3-(1,3-DIOXOBUTYL)AMINO ✚
3	3.21		331		C11H10N4O6S1	2-SO2ME-4NO2-C6H3NHN=C(CN)COOME 36865-82-4
4 ★	2.39		594		C11H10O1	NAPHTHALENE,1-HYDROXYMETHYL 4780-79-4
25	2.53	7.0	2505	1	"	NAPHTHALENE,1-HYDROXYMETHYL
6 ★	3.63		594		C11H10O1	NAPHTHALENE,1-METHOXY 2216-69-5
7 ★	2.21	7.0	2505	1	C11H10O1	NAPHTHALENE,2-METHANOL
8 ★	3.47		594		C11H10O1	NAPHTHALENE,2-METHOXY 93-04-9
9	2.45		2040	459	C11H10O2	2,8-DECADIENE-4,6-DIYNOIC ACID,METHYL ESTER
6930 ★	2.31		1941		C11H10O2	4H-1-BENZOPYRAN-4-ONE,2,3-DIMETHYL 17584-90-6
1	3.15		1167	459	C11H10O2	5-PHENYLPENT-2,4-DIENOIC ACID 1552-94-9
2 ★	1.36		2649		C11H11Br1N2O2	BENZOQUINONE,2,5-BIS-AZIRIDINYL-3-BROMO-6-METHYL
3	-1.05	1.0	594		C11H11Br1N4O2	PYRIMIDINE,2,4-DIAMINO-5-(3-BROMO-4,5-DIHYDROXY)BENZYL
4 ★	2.17		537		C11H11Br3N2O5	TRIBROMAMPHENICOL 49648-42-2
35 ★	2.76		2410		C11H11Cl1F1O2	ACETANILIDE,2-FLUORO-4-CHLORO-5-ALLYLOXY
6 ★	1.20		2649		C11H11Cl1N2O2	BENZOQUINONE,2,5-BIS-AZIRIDINYL-3-CHLORO-6-METHYL
7 ★	2.24		522		C11H11Cl1N2O2S1	1,2,4-BENZOTHIADIAZINE-1,1-DIOXIDE-3-CYCLOBUTYL-7-CHLORO 37148-24-6
8	-0.36	1.0	547	342	C11H11Cl1N4	2,4-DIAMINO-5-(3-CHLOROBENZYL)PYRIMIDINE
9 ★	2.24	13.0	547		C11H11Cl1N4	PYRIMIDINE,2,4-DIAMINO-5-(2-CHLORO)BENZYL
6940	4.55		2404	459	C11H11Cl1O1	BENZOPYRAN,2,2-DIMETHYL-7-CHLORO 80055-55-6
1	3.87		547	20	C11H11Cl1O1	BENZOPYRAN,2,2-DIMETHYL-8-CHLORO 80055-56-7
2	4.06		2404	459	"	BENZOPYRAN,2,2-DIMETHYL-8-CHLORO
3	-0.21	7.4	693	1	C11H11Cl1O3	ALCLOFENAC 22131-79-9 anti-inflammatory
4 ★	2.48	7.4	539	2	"	ALCLOFENAC
45 ★	3.49		2094		C11H11Cl2N1O1	1-BUTENE,3-(3,4-DICHLOROCARBOXANILIDE) 74054-79-8
6 ★	3.63		1792		C11H11Cl2N1O1	1-ME-CYLOPROPANAMIDE,N-(3,4-DICHLOROPHENYL)
7	3.49		2094		C11H11Cl2N1O1	1-METHYLBUTENE-3YL ANILIDE, 3,4 DICHLORO-
8	3.94		2572	459	C11H11Cl2N1O2	PHENOPYLATE 40575-34-6
9	2.98		2572	459	C11H11Cl2N1O3	O-PHENYLPYRROLIDINOCARBAMATE,2,4-DICHLORO-5-HYDROXY 143121-08-8
6950 ★	1.97		537	270	C11H11Cl3N2O5	TRICHLORAMPHENICOL 19934-51-1
1	3.72		2616	459	C11H11Cl3O3	METHOXYFENATE
2 ★	-1.07		2347		C11H11F1N2O2	TRYPTOPHAN,4-FLUORO
3 ★	-0.71		2347		C11H11F1N2O2	TRYPTOPHAN,5-FLUORO
4 ★	-0.78		2347		C11H11F1N2O2	TRYPTOPHAN,6-FLUORO
55	-0.80	1.0	547	342	C11H11F1N4	2,4-DIAMINO-5-(3-FLUOROBENZYL)PYRIMIDINE
6 ★	1.07		537	270	C11H11F3N2O5	TRIFLUORAMPHENICOL 42583-67-5
7 ★	3.10		2072	2	C11H11F3N2O5S1	BENZENESULFONAMIDE,2-NITRO-4-TRIFLUOROMETHYL-N-(I-BUTANOYL)
8 ★	1.27		2366	122	C11H11I1N2O1	4-IODOANTIPYRINE 129-81-7 radioactive agent
9 ★	2.88		579		C11H11N1	N-METHYL-1-NAPHTHYLAMINE 2216-68-4
6960 ★	2.71		2606	424	C11H11N1	PYRROLE,1-BENZYL 2051-97-0
1 ★	2.93	7.4	595	314	C11H11N1	QUINOLINE,8-ETHYL
2 ★	2.81		1519	459	C11H11N1O1	2-METHYL-4-METHOXY-QUINOLINE 31835-53-7
3 ★	2.71		826		C11H11N1O1	4,5-DIMETHYL-8-QUINOLINOL 15011-28-6
4 ★	1.60		2300		C11H11N1O1	CORATOP 57369-32-1
65 ★	1.95		2221	459	C11H11N1O1	INDOLE-1-METHYL-3-ACETYL
6 ★	2.25	7.4	595	314	C11H11N1O1	QUINOLINE,8-ETHOXY
7	0.00	7.0	1073	1	C11H11N1O2	3-INDOLYLPROPIONIC ACID 830-96-6
8 ★	1.75	7.0	1073	2	"	3-INDOLYLPROPIONIC ACID
9	2.75		826		C11H11N1O2	4-METHYL-5-METHOXY-8-QUINOLINOL 57334-38-0
6970 ★	3.18		2221	459	C11H11N1O2	INDOLE-2-CARBOXYLIC ACID,ETHYL ESTER 3770-50-1
1 ★	3.04		2221	459	C11H11N1O2	INDOLE-3-CARBOXYLIC ACID,ETHYL ESTER 776-41-0
2 ★	0.57		1156		C11H11N1O2	N-(M-TOLYL)SUCCINIMIDE
3 ★	0.29		1156		C11H11N1O2	N-(O-TOLYL)SUCCINIMIDE
4 ★	0.58		1156		C11H11N1O2	N-(P-TOLYL)SUCCINIMIDE 2314-79-6
75	0.76	3.5	2458	302	C11H11N1O2	PYRROLIDINE-2,4-DIONE,1-BENZYL
6	0.16	7.0	1073	1	C11H11N1O2S1	INDOL-3-YL-2-THIOPROPIONIC ACID
7 ★	2.15	7.0	1073	2	"	INDOL-3-YL-2-THIOPROPIONIC ACID
8	1.37		458		C11H11N1O2Se1	P-CYANOBENZYLSELENOPROPIONIC ACID
9 ★	0.99		2209		C11H11N1O3	DIHYDROISOINDOLE,2-OXO-N-ACETYLOXYMETHYL
6980	0.18		1156		C11H11N1O3	M-METHOXY-N-PHENYLSUCCINIMIDE 16141-40-5
1	1.32	7.0	1016	1	C11H11N1O3	N-FORMYL-P-ACETOXYSTYRLAMINE
2	-0.02		1156		C11H11N1O3	O-METHOXY-N-PHENYLSUCCINIMIDE
3	0.58	7.4	594	1	C11H11N1O3	OXAZOLIDINEDIONE,5-ETHYL-5-PHENYL 92288-54-5
4 ★	0.15		1156		C11H11N1O3	P-METHOXY-N-PHENYLSUCCINIMIDE 2314-80-9
85 ★	1.63	7.4	2542	1	C11H11N1O4S3	THIOPHENE-2-SULFONAMIDE,4-(4-METHYLPHENYL)SULFONYL
6 ★	0.22		2234		C11H11N1O5	ACETYLSALICYLIC ACID,GLYCOLAMIDE ESTER 50785-22-3
7	-0.62	7.4	353	1	C11H11N1O5	O,O'-DIACETYLRESORCYLAMIDE
8 ★	1.49	7.4	2542	1	C11H11N1O5S2	FURAN-2-SULFONAMIDE,4-(4-METHYL)PHENYLSULFONYL
9 ★	1.36	7.4	2542	1	C11H11N1O5S3	THIOPHENE-2-SULFONAMIDE,4-(4-METHOXYPHENYL)SULFONYL
6990 ★	1.17	7.4	2542	1	C11H11N1O6S2	FURAN-2-SULFONAMIDE,4-(4-METHOXYPHENYLSULFONYL)
1	2.62		759		C11H11N1S1	2-(P-HYDROXYBENZYL)-5-METHYLTHIAZOLE ✚
2 ★	2.28		536		C11H11N3O1	1-PHENYL-3,5-DIMETHYL-4-NITROSOPYRAZOLE 715-99-1
3	0.51		1059		C11H11N3O1S1	2-AMINO-5-THIAZOLECARBOXANILIDE
4	-0.21	3.3	2590	331	C11H11N3O2	3(2-PROPYNYL)-1-(4-CARBOXYPHENYL)-3-METHYLTRIAZEN 74109-31-2 antineoplastic

	logP	pH	Ref.	Note	MF	Name / CAS / activity
95	0.00	7.4	1214	1	"	3(2-PROPYNYL)-1-(4-CARBOXYPHENYL)-3-METHYLTRIAZEN
6	0.20	7.2	556		C11H11N3O2S1	5-METHYL-2-BENZENESULFONAMIDOPYRIMIDINE 13428-53-0
7 ★	0.00		508			SULFAPYRIDINE 144-83-2 suppressant (dermititis herpetiformis)
8	0.90	7.5	652	1	"	SULFAPYRIDINE
9 ★	1.89	7.4	1022	1	C11H11N3O3	1-(2-PHENOXYETHYL)-2-NITROIMIDAZOLE
7000 ★	1.69		1705		C11H11N3O3	N-(2-(P-NITROPHENOXY)ETHYL)IMIDAZOLE
1	1.29		494		C11H11N3O4S1	2-I-PROPYL-5-(5-NITRO-2-FURFURYL)THIAZOLONE 27472-92-0
2	1.24		494		C11H11N3O4S1	2-PROPYLAMINO-5-(5-NITRO-2-FURFURYL)THIAZOLONE 27472-90-8
3	1.32		494		C11H11N3O4S1	3-PR-2=NH-5-(5-NO2-2-FURFURILIDINE)THIAZOLINE-4-ONE 52661-59-3
4	1.23		2279		C11H11N3O4S1	IMIDAZOLE,1-METHYL-4-NITRO-5-BENZYLSULFONATO 3024-15-5
5 ★	0.52	7.4	1022	1	C11H11N3O4S1	PHENYL(2-(2-NITRO-1-IMIDAZOLYL)ETHYL)SULFONE
6 ★	1.01	7.4	593	1	C11H11N5	2-AMINO-3,8-DIMETHYLIMIDAZO[4,5-F]QUINOXALINE ✚
7	-1.03	1.0	547	342	C11H11N5O2	2,4-DIAMINO-5-(4-NITROBENZYL)PYRIMIDINE
8 ★	1.87		537	800	C11H12Br1Cl2N1O3	CHLORAMPHENICOL,P-BROMO ANALOG 23885-59-8
9 ★	3.38	7.6	2007	1	C11H12Br1N1O2S1	BENZOYLTHIOFORMOHYDROXIMATE,4-BROMO-S-PROPYL 99481-58-0
7010	0.09	7.4	594	1	C11H12Br1N1O3	GLUTARAMIC ACID,P-BROMOPHENYL
1 ★	1.12	1.0	594			GLUTARAMIC ACID,P-BROMOPHENYL
2 ★	1.36		537	270	C11H12Br2N2O5	DIBROMAMPHENICOL 17371-30-1
3	1.15	7.4	2495	1	C11H12Br2N3O2P1	PHOSPHORIC TRIAMIDE,N-(3,5-DIBROMOBENZOYL)-N',N"-DIETHYLENE
4	3.13		2572	459	C11H12Cl1N1O2	O-PHENYLPYRROLIDINOCARBAMATE,2-CHLORO 143121-06-6
15	3.38		2572	459	C11H12Cl1N1O2	O-PHENYLPYRROLIDINOCARBAMATE,4-CHLORO 1759-02-0
6 ★	2.22		537	800	C11H12Cl2I1O3	P-IODO-AMPHENICOL 49648-53-5
7	0.73	7.4	1350	1	C11H12Cl2N2O1	LOFEXIDINE 31036-80-3 antihypertensive
8	1.00	7.4	2582	1	C11H12Cl2N2O5	CHLORAMPHENICOL 56-75-7 antibacterial, antirickettsial
9	1.09	7.0	1295	1	"	CHLORAMPHENICOL
7020 ★	1.14		506		"	CHLORAMPHENICOL
1 ★	1.64		537		C11H12Cl3N1O3	CHLORAMPHENICOL,P-CHLORO ANALOG
2 ★	2.37	7.5	2545	2	C11H12F1N1	4'-FLUORO-DESMETHYL-MPTP ✚
3 ★	0.42		537		C11H12F2N2O5	DIFLUORAMPHENICOL 49648-37-5
4	2.18		1182		C11H12F3N1O1	2-(M-TRIFLUOROMETHYLPHENYL)MORPHOLINE 31599-68-5
25 ★	3.08		1681		C11H12F3N1O2	N,N-DIMETHYLCARBAMATE,P-(CF3)BENZYL ESTER 84640-35-7
6	1.35		547		C11H12I2N3O2P1	BIS(1-AZIRIDINYL)2,5-I2-BENZAMIDOPHOSPHINEOXIDE
7	1.52	7.4	2495	1	"	BIS(1-AZIRIDINYL)2,5-I2-BENZAMIDOPHOSPHINEOXIDE
8	2.52	7.4	2495	1	C11H12I2N3O2P1	PHOSPHORIC-TRIAMIDE,N-(3,5-DI-IODOBENZOYL)-N',N"-DIETHYLENE
9 ★	2.78		1693		C11H12N1O4P1S2	PHOSMET 732-11-6 insecticide
7030	-2.76		527	820	C11H12N1.Br1	ETHYLQUINOLINIUMBROMIDE 26670-42-8
1 ★	2.81		1519	459	C11H12N2	2-QUINOLINAMINE,N,N-DIMETHYL 21154-18-7
2	2.71		163	459	C11H12N2	7-DIMETHYLAMINOQUINOLINE
3	2.73		163	459	C11H12N2	8-DIMETHYLAMINOQUINOLINE 29526-42-9
4	0.26	7.4	579	1	C11H12N2	TETRAHYDRO-B-CARBOLINE
35 ★	1.59		1439		C11H12N2O1	4H-PYRIDO(1,2-A)PYRIMIDIN-4-ONE,2-ET,6-ME ✚
6 ★	1.63		1439		C11H12N2O1	4H-PYRIDO(1,2-A)PYRIMIDIN-4-ONE,2-ME,6-ET 38326-28-2
7 ★	1.79		1439		C11H12N2O1	4H-PYRIDO(1,2-A)PYRIMIDIN-4-ONE,3-ET,6-ME 57773-19-0
8 ★	0.38	7.4	651	1	C11H12N2O1	ANTIPYRINE 60-80-0 analgesic
9 ★	1.21		2221	459	C11H12N2O1	INDOLE-2-METHYL-5-ACETYLAMINO
7040 ★	0.07	6.9	2175	1	C11H12N2O2	3-METHYL-3-(4-PYRIDYL)PIPERIDINE-2,6-DIONE
1	2.80		2363		C11H12N2O2	BUTYRONITRILE,3-METHYL-2-(4-NITROPHENYL)
2	1.33	7.0	1405	1	C11H12N2O2	HYDANTOIN,5-ETHYL-5-PHENYL 5696-06-0 anticonvulsant
3 ★	1.53		505		"	HYDANTOIN,5-ETHYL-5-PHENYL
4	1.72		2031		C11H12N2O2	IMIDAZOLE-2-(2,4-DIMETHOXYPHENYL)
45 ★	1.69		1969		C11H12N2O2	INDOLE-3-ETHANOL,CARBAMATE (ESTER) 91350-72-0
6 ★	0.40		306		C11H12N2O2	MALONAMIDE,3-METHYLBENZAL 15888-02-5
7 ★	1.67		1681		C11H12N2O2	N,N-DIMETHYLCARBAMATE,P-CYANOBENZYL ESTER 84640-32-4
8	-1.75	7.0	1590	1	C11H12N2O2	TRYPTOPHAN 73-22-3 amino acid
9	-1.11	7.0	1207	1	"	TRYPTOPHAN
7050	-1.06	7.0	2197		"	TRYPTOPHAN
1 ★	-1.06		547	463		TRYPTOPHAN
2 ★	0.01		306		C11H12N2O3	MALONAMIDE,3-METHOXYBENZAL 15804-68-9
3	0.62	2.0	2312	337	C11H12N2O3	OXITRIPTAN 4350-09-8 antidepressant
4 ★	0.09		2157		C11H12N2O4	2-(BENZOYLOXY)-(N-ACETAMIDO)ACETAMIDE
55	-1.02	7.2	1685	314	C11H12N2O4S2	BENZOTHIAZOLE,2-N-ACETYLSULFONAMIDO-6-ETHOXY 88946-18-3
6 ★	1.65		594		C11H12N2O5S1	PHTHALIMIDO METHYLSULFONE,5-(N,N-DIMETHYLAMINO)
7	1.67	7.4	1940	1	C11H12N2S1	LEVAMISOLE 14769-73-4 anthelmintic
8 ★	1.84	9.8	2639	306	"	LEVAMISOLE
9 ★	1.58	13.0	547	224	C11H12N4	2,4-DIAMINO-5-BENZYLPYRIMIDINE 7319-45-1
7060 ★	2.30		2499		C11H12N4	PYRAZINE,2,3-DICYANO-5-BUTYL-6-METHYL
1	-1.34	1.0	579		C11H12N4O1	2,4-DIAMINOPYRIMIDINE,5-(3-HYDROXYBENZYL)
2	-1.47	1.0	594		C11H12N4O1	2,4-DIAMINOPYRIMIDINE,5-(4-HYDROXYBENZYL)
3 ★	1.48	7.4	2341	1	C11H12N4O1	3[1-(2-BENOXAZOLYL)HYDRAZINO]PROPANENITRILE,5-METHYL
4	-0.98	1.0	594		C11H12N4O1	PYRIMIDINE,2,4-DIAMINO-5-(2-HYDROXY)BENZYL
65	0.82	7.4	594	1	"	PYRIMIDINE,2,4-DIAMINO-5-(2-HYDROXY)BENZYL
6 ★	1.50	7.4	2341	1	C11H12N4O1S1	3[1-(2-BENZOXAZOLYL)HYDRAZINO]PROAPANENITRILE,5-METHIO
7	-2.08	1.0	594		C11H12N4O2	2,4-DIAMINOPYRIMIDINE,5-(3,4-DIHYDROXYBENZYL)
8 ★	0.98	7.4	2341	1	C11H12N4O2	3[1-(2-BENZOXAZOLYL)HYDRAZINO]PROPANENITRILE,5-METHOXY
9 ★	0.06	7.4	2206	306	C11H12N4O2	PANIDAZOLE 13752-33-5
7070	-0.12	7.5	652	1	C11H12N4O2S1	SULFAMERAZINE 127-79-7 antibacterial
1 ★	0.14		508		"	SULFAMERAZINE
2 ★	0.34	4.0	360	2	C11H12N4O2S1	SULFAPERIN 599-88-2 antibacterial
3	0.70	4.0	360	2	C11H12N4O3S1	SULFALENE 152-47-6 antibacterial
4	1.01		328		"	SULFALENE
75 ★	0.41	4.0	360	2	C11H12N4O3S1	SULFAMETER 651-06-9 antibacterial
6 ★	0.32		480	601	C11H12N4O3S1	SULFAMETHOXYPYRIDAZINE 80-35-3 antibacterial
7 ★	0.76		480	220	C11H12N4O3S1	SULFAMONOMETHOXINE 1220-83-3
8 ★	0.63		1499		C11H12N4O5	5-(1-AZIRIDINYL)-2,4-DINITROBENZAMIDE,N,N-DIMETHYL 27221-03-0
9 ★	3.31		2404		C11H12O1	BENZOPYRAN,2,2-DIMETHYL 2513-25-9
7080 ★	2.10		1524		C11H12O2	1(3H)-ISOBENZOFURANONE-3-I-PROPYL 64002-57-9
1 ★	2.99		306		C11H12O2	CINNAMIC ACID,ETHYL ESTER 103-36-6
2 ★	2.33		501		C11H12O3	5-INDANOXYACETIC ACID 1878-58-6
3 ★	1.87		594	400	C11H12O3	BENZOYLACETIC ACID,ETHYL ESTER 94-02-0
4 ★	1.62	3.5	2179	302	C11H12O3	BUTANOIC ACID,2-METHYL-4-OXO-4-PHENYL

	logP	pH	Ref.	Note	MF	Name / **CAS** / *activity*
85 ★	1.85		1705		C11H12O4	BENZOIC ACID,2-ACETYLOXYETHYL ESTER
6 ★	1.38	3.5	2179	302	C11H12O4	BUTANOIC ACID-4-OXO-4-(4-METHOXYPHENYL)
7	2.34		1167	459	C11H12O4	CINNAMIC ACID,3,4-DIMETHOXY **2316-26-9**
8 ★	1.62		757		C11H12O4	DIACETOXYTOLUOL **581-55-5**
9	2.56		1167	459	C11H12O4	ETHYLCINNAMATE,3,4-DIHYDROXY **102-37-4**
7090 ★	1.94		2234		C11H12O4S1	ACETYLSALICYLIC ACID,METHYLTHIOMETHYL ESTER **76432-30-9**
1 ★	0.11		2234		C11H12O5S1	ACETYLSALICYLIC ACID,METHYLSUFOXYMETHYL ESTER **76432-33-2**
2 ★	0.57		2234		C11H12O6S1	ACTYLSALICYLIC ACID, METHYLSULFONYLMETHYL ESTER **76432-35-4**
3	1.78	7.4	899	1	C11H13Br1N2O3	BARBITURIC ACID,ALLYL,BROMALLYL,N-METHYL **64889-77-6**
4 ★	0.66		537	270	C11H13Br1N2O5	MONOBROMAMPHENICOL **40027-72-3**
95	0.88	7.4	2495	1	C11H13Br1N3O2P1	PHOSPHORIC-TRIAMIDE,N-M-BROMOBENZOYL-N',N"-DIAZIRIDINYL ANALOG
6 ★	1.43	7.4	2495	2	C11H13Br1N3O2P1	PHOSPHORIC-TRIAMIDE,N-M-BROMOBENZOYL-N',N"-DIAZIRIDINYL ANALOG
7	0.45	7.4	2495	1	C11H13Br1N3O2P1	PHOSPHORIC-TRIAMIDE,N-O-BROMOBENZOYL-N',N"-DIAZIRIDINYL ANALOG
8 ★	1.08	7.4	2495	2	C11H13Br1N3O2P1	PHOSPHORIC-TRIAMIDE,N-O-BROMOBENZOYL-N',N"-DIAZIRIDINYL ANALOG
9	1.01	7.4	2495	1	C11H13Br1N3O2P1	PHOSPHORIC-TRIAMIDE,N-P-BROMOBENZOYL-N',N"-DIAZIRIDINYL ANALOG
7100 ★	1.63	7.4	2495	2	C11H13Br1N3O2P1	PHOSPHORIC-TRIAMIDE,N-P-BROMOBENZOYL-N',N"-DIAZIRIDINYL ANALOG
1	-0.63	7.4	1204	1	C11H13Br2N3	N-CYCLOPROPYLMETHYL,N-(2,6-DIBROMOPHENYL)-GUANIDINE
2	-1.01	7.4	1204	1	C11H13Br2N3	N-ET-N(2,6-BR2-PH)AMINO)-2-IMIDAZOLINE
3	1.13	6.5	321	2	C11H13Cl1F3N3O4S3	POLYTHIAZIDE **346-18-9** *diurectic, antihypertensive*
4	1.90		1780	"		POLYTHIAZIDE
5 ★	2.35		522		C11H13Cl1N2O2S1	1,2,4-BENZOTHIADIAZINE-1,1-DIOXIDE-3-I-BUTYL-6-CHLORO **37148-21-3**
6 ★	2.47		522		C11H13Cl1N2O2S1	1,2,4-BENZOTHIADIAZINE-1,1-DIOXIDE-3-S-BUTYL-7-CHLORO **37148-22-4**
7 ★	2.40		522		C11H13Cl1N2O2S1	1,2,4-BENZOTHIADIAZINE-1,1-DIOXIDE-3-T-BUTYL-7-CHLORO **13460-16-7**
8 ★	2.52		522		C11H13Cl1N2O2S1	1,2,4-BENZOTHIADIZINE-1,1-DIOXIDE-3-BUTYL-7-CHLORO **37157-85-0**
9 ★	0.59		537	270	C11H13Cl1N2O5	MONOCHLORO-AMPHENICOL **17278-57-8**
7110	0.46	7.4	2495	1	C11H13Cl1N3O2P1	PHOSPHORIC-TRIAMIDE,N-P-CHLOROBENZOYL-N',N"-DIAZIRIDINYL ANALOG
1 ★	0.97	7.4	2495	2	C11H13Cl1N3O2P1	PHOSPHORIC-TRIAMIDE,N-P-CHLOROBENZOYL-N',N"-DIAZIRIDINYL ANALOG
2 ★	2.50		1049		C11H13Cl1N3O3P1	N-ME-PHOSPHORAMIDATE,O-ET,O-A-CN-4CLBENZALDOXIME +
3 ★	3.47		2625	448	C11H13Cl1O3	2-(2-METHYL-4-CHLOROPHENOXY)BUTYRIC ACID
4 ★	2.90		489	220	C11H13Cl1O3	2-(P-CHLOROPHENOXY)-2-ET-PROPIONIC ACID **17413-90-0**
15 ★	3.88		1792		C11H13Cl2N1O1	1,1-DIMETHYLPROPANEAMIDE,N-(3,4-DICHLOROPHENYL)
6 ★	4.11		1792		C11H13Cl2N1O1	2-METHYLBUTANEAMIDE,N-(3,4-DICHLOROPHENYL) **7160-25-0**
7 ★	4.17		1792		C11H13Cl2N1O1	PENTANEAMIDE,N-(3,4-DICHLOROPHENYL) **2150-96-1**
8 ★	0.94		537		C11H13Cl2N1O3	1-PHENYL-1,3-PROPANEDIOL,2-DICHLOROACETAMIDO **25126-19-6**
9	0.72	7.4	2266	1	C11H13F1N2O4	5-FLUOROURACIL-1-CARBOXYLIC ACID,CYCLOHEXYL ESTER
7120 ★	1.42	4.0	2266	302	"	5-FLUOROURACIL-1-CARBOXYLIC ACID,CYCLOHEXYL ESTER
1 ★	0.15		537	270	C11H13F1N2O5	MONOFLUORO-AMPHENICOL **49648-38-6**
2	0.13	7.4	2495	1	C11H13F1N3O2P1	PHOSPHORIC-TRIAMIDE,N-P-FLUOROBENZOYL-N',N"-DIAZIRIDYL ANALOG
3 ★	0.86	7.4	2495	2	C11H13F1N3O2P1	PHOSPHORIC-TRIAMIDE,N-P-FLUOROBENZOYL-N',N"-DIAZIRIDYL ANALOG
4	1.38	7.4	1948	1	C11H13F3N2	1-(M-TRIFLUOROMETHYLPHENYL)PIPERAZINE **15532-75-9** *serotonin-agonist (5-ht1b)*
25	2.43	7.4	1948	2	"	1-(M-TRIFLUOROMETHYLPHENYL)PIPERAZINE
6 ★	2.45		2072	2	C11H13F3N2O3S1	BENZENESULFONAMIDE,2-(I-BUTYROYLAMINO)-4-TRIFLUOROMETHYL
7 ★	2.56		2072	2	C11H13F3N2O3S1	BENZENESULFONAMIDE,2-AMINO-4-TRIFLUOROMETHYL-N-(I-BUTANOYL)
8	4.30		2506	20	C11H13F3N4O4	DINITRAMINE **29091-05-2** *herbicide*
9 ★	1.03		537	270	C11H13I1N2O5	MONOIODOAMPHENICOL **40027-73-4**
7130	1.37	7.4	2495	1	C11H13I1N3O2P1	PHOSPHORIC-TRIAMIDE,N-M-IODOBENZOYL-N',N"-DIAZIRIDINYL ANALOG
1 ★	1.91	7.4	2495	2	C11H13I1N3O2P1	PHOSPHORIC-TRIAMIDE,N-M-IODOBENZOYL-N',N"-DIAZIRIDINYL ANALOG
2	0.44	7.4	2495	1	C11H13I1N3O2P1	PHOSPHORIC-TRIAMIDE,N-O-IODOBENZOYL-N',N"-DIAZIRIDINYL ANALOG
3 ★	1.05	7.4	2495	2	C11H13I1N3O2P1	PHOSPHORIC-TRIAMIDE,N-O-IODOBENZOYL-N',N"-DIAZIRIDINYL ANALOG
4	1.63	7.4	2495	1	C11H13I1N3O2P1	PHOSPHORIC-TRIAMIDE,N-P-IODOBENZOYL-N',N"-DIAZIRIDINYL ANALOG
35 ★	2.25	7.4	2495	2	C11H13I1N3O2P1	PHOSPHORIC-TRIAMIDE,N-P-IODOBENZOYL-N',N"-DIAZIRIDINYL ANALOG
6	-0.73	1.0	547	342	C11H13N1	2-EXO-AMINOBENZOBICYCLO(2,2,1)HEPTENE
7	-0.64	1.0	547	342	C11H13N1	ANTI-9-AMINOBENZOBICYCLO(2,2,1)HEPTENE
8 ★	-0.23	7.4	2648	2	C11H13N1	BENZOBICYCLO(2,2,1)HEPTENE,2-EN-AMINO **58742-04-4**
9 ★	2.00	13.0	1623	224	"	BENZOBICYCLO(2,2,1)HEPTENE,2-EN-AMINO
7140	2.01	7.4	2648	2	"	BENZOBICYCLO(2,2,1)HEPTENE,2-EN-AMINO
1	0.02	7.4	2648	1	C11H13N1	BENZOBICYCLO(2,2,1)HEPTENE,2-EX-AMINO **62624-26-4**
2 ★	2.09	13.0	1623	224	"	BENZOBICYCLO(2,2,1)HEPTENE,2-EX-AMINO
3	2.10	7.4	2648	2	"	BENZOBICYCLO(2,2,1)HEPTENE,2-EX-AMINO
4 ★	2.08	13.0	1623	224	C11H13N1	BENZOBICYCLO(2,2,1)HEPTENE,9-EN-AMINO **72597-35-4**
45 ★	2.13	13.0	1623	224	C11H13N1	BENZOBICYCLO(2,2,1)HEPTENE,9-EX-AMINO **14098-20-5**
6 ★	2.33		2256		C11H13N1	BUTYRONITRILE,3-METHYL-2-PHENYL
7 ★	1.99	7.5	2545	2	C11H13N1	DESMETHYL-MPTP
8 ★	1.43		757		C11H13N1O1	1-PHENYL-2-ACETAMIDOCYCLOPROPANE **38954-41-5**
9 ★	0.91	13.0	1623	224	C11H13N1O1	9-OXA-BENZOBICYCLO(2,2,1)HEPTENE,2-EXO-METHYLAMINO **86943-80-8**
7150 ★	0.59	13.0	1623	224	C11H13N1O1	9-OXABENZOBICYCLO(2,2,1)HEPTENE,2-ENDO-METHYLAMINO **86992-70-3**
1 ★	1.73		572		C11H13N1O1	N,N-DIMETHYLCINNAMAMIDE **13156-74-6**
2 ★	1.25		1213		C11H13N1O1	N-BENZYL-2-PYRROLIDINONE **5291-77-0**
3 ★	2.53		2256		C11H13N1O1	PHENYLACETONITRILE,4-METHOXY
4	2.73		2572	459	C11H13N1O2	O-PHENYLPYRROLIDINOCARBAMATE **55379-71-0**
55 ★	1.07		2157		C11H13N1O3	2-(BENZOYLOXY)-(N,N-DIMETHYL)ACETAMIDE
6 ★	1.28		2157		C11H13N1O3	2-(BENZOYLOXY)-(N-ETHYL)ACETAMIDE
7 ★	0.93		558		C11H13N1O3	2-ACETYLPHENYLDIMETHYLCARBAMATE **29230-99-7**
8 ★	1.18		558		C11H13N1O3	3-ACETYLPHENYLDIMETHYLCARBAMATE **2689-47-6**
9 ★	1.49		273		C11H13N1O3	N-METHYL-3-PROPIONYLPHENYLCARBAMATE **10051-63-5**
7160 ★	1.55		273		C11H13N1O3	N-METHYL-4-PROPIONYLPHENYLCARBAMATE **54266-28-3**
1 ★	1.70		2506		C11H13N1O4	BENDIOCARB **22781-23-3** *insecticide*
2 ★	1.15		1521		C11H13N1O4	CYCLOHEXENE-1,2-DICARBOXIMIDE,N-ACETYLOXYMETHYL
3 ★	0.97		2209		C11H13N1O4	CYCLOHEXENE-4,5-DICARBOXIMIDE,N-ACETYLOXYMETHYL
4	0.67		572		C11H13N1O4	DIOXACARB **6988-21-2** *insecticide*
65 ★	1.21	7.4	2268	1	C11H13N1O4	SALICYLIC ACID,N,N-DIMETHYLGLYCOLAMIDE ESTER
6 ★	2.83		1946		C11H13N1O5	2-PROPENOIC ACID,3-(5-NITRO-2-FURANYL),BUTYL ESTER
7 ★	3.14		1946		C11H13N1O5	2-PROPENOIC ACID,3-(5-NITRO-2-FURANYL),I-BUTYL ESTER
8 ★	3.09		1946		C11H13N1O5	2-PROPENOIC ACID,3-(5-NITRO-2-FURANYL),SEC-BUTYL ESTER
9 ★	3.06		1946		C11H13N1O5	2-PROPENOIC ACID,3-(5-NITRO-2-FURANYL),T-BUTYL ESTER
7170 ★	2.61	5.6	2165	1	C11H13N3	BENZOTRIAZOLE,1-CYCLOPENTYL
1 ★	3.04	5.6	2156	1	C11H13N3	BENZOTRIAZOLE,2-CYCLOPENTANYL
2	-0.32		1407		C11H13N3O2	1,2,3,6-H4-PYRIDINE,1-OXIDO-2-PYRIDINYL-CONH-
3 ★	1.60		559		C11H13N3O2	1-(2-CARBETHOXYETHYL)BENZOTRIAZOLE
4	0.49		559		C11H13N3O2	1-(4-CARBOXYBUTYL)BENZOTRIAZOLE

	logP	pH	Ref.	Note	MF	Name / CAS / activity
75	2.12		559		C11H13N3O2	2-(2-CARBETHOXYETHYL)BENZOTRIAZOLE
6	0.94		559		C11H13N3O2	2-(4-CARBOXYBUTYL)BENZOTRIAZOLE
7	0.31	3.3	2590	331	C11H13N3O2	3-ALLYL-1-(4-CARBOXYPHENYL)-3-METHYLTRIAZENE **74109-30-1** *antineoplastic*
8	0.32	7.4	1214	1	"	3-ALLYL-1-(4-CARBOXYPHENYL)-3-METHYLTRIAZENE
9	0.37	2.6	2590	331	C11H13N3O2	BENZOIC ACID,2-(3-METHYL-3-ALLYLTRIAZENYL) *antineoplastic*
7180	-0.70	5.6	2156	1	C11H13N3O2	BENZOTRIAZOLE,1-(2-PENTANOIC ACID)
1 ★	1.87	5.6	2156	1	C11H13N3O2	BENZOTRIAZOLE,1-(2-PROPIONIC ACID,ETHYL ESTER)
2	0.15	5.6	2165	1	C11H13N3O2	BENZOTRIAZOLE,1-(4-PENTANOIC ACID)
3	-0.41	5.6	2156	1	C11H13N3O2	BENZOTRIAZOLE,2-(2-PENTANOIC ACID)
4 ★	2.38	5.6	2156	1	C11H13N3O2	BENZOTRIAZOLE,2-(2-PROPIONIC ACID,ETHYL ESTER)
85	0.69	5.6	2165	1	C11H13N3O2	BENZOTRIAZOLE,2-(4-PENTANOIC ACID)
6	0.11	7.4	2607	1	C11H13N3O2	CINNAMIC ACID,P-(3,3-DIMETHYL-1-TRIAZENO)
7	0.29	7.0	2607	1	"	CINNAMIC ACID,P-(3,3-DIMETHYL-1-TRIAZENO)
8 ★	2.66	7.0	2607	2	"	CINNAMIC ACID,P-(3,3-DIMETHYL-1-TRIAZENO)
9	1.75	6.0	479		C11H13N3O3S1	N1-METHYL-N1-(5-ME-3-ISOXAZOLYL)SULFANILAMIDE **51543-31-8**
7190	-0.87	7.5	652	1	C11H13N3O3S1	SULFISOXAZOLE **127-69-5** *antibacterial*
1 ★	1.01		508		"	SULFISOXAZOLE
2	0.95	7.4	1948	1	C11H13N3S1	1-(1,2-BENZOTHIAZOL-3-YL)PIPERAZINE
3	2.14	7.4	1948	2	"	1-(1,2-BENZOTHIAZOL-3-YL)PIPERAZINE
4 ★	1.65		576		C11H13N5	2'-ETHYLBENZOGUANAMINE
95	0.04	7.4	1485	1	C11H13N5	2,4-DIAMINOPYRIMIDINE,5-(P-AMINOBENZYL)
6 ★	0.22	7.4	1485	2	"	2,4-DIAMINOPYRIMIDINE,5-(P-AMINOBENZYL)
7	1.79		547		C11H13N5O1	(CH3)2NNN-C6H4-2(CONHCH2CN)
8 ★	1.03		576		C11H13N5O1	2'-ETHOXYBENZOGUANAMINE
9	-1.37	3.3	2590	331	C11H13N5O2S2	N-THIAZO-2-YL-BENZENESULFONAMIDE,4-(3,3-DIMETHYLTRIAZENYL)
7200	1.39	6.6	2590	459	"	N-THIAZO-2-YL-BENZENESULFONAMIDE,4-(3,3-DIMETHYLTRIAZENYL)
1	4.19		1818		C11H14Br1N1O2	N-I-PROPYLBENZAMIDE,2-HYDROXY-3-METHYL-5-BROMO
2	-0.28	7.4	976	1	C11H14Br1N3	2-(2,6-DIME-4-BR-PHENYLIMINO)IMIDAZOLINE
3	2.80	7.4	976	2	"	2-(2,6-DIME-4-BR-PHENYLIMINO)IMIDAZOLINE
4	-2.69	7.4	594		C11H14Br1N5	1-(3-BROMOPHENYL)-2,2-DIMETHYL-4,6-DIAMINO-S-TRIAZINE
5 ★	1.09	13.0	547	224	C11H14Br1N5	4,6-DIAMINO-2,2-DIMETHYL-1-(4-BROMOPHENYL)-SYM-TRIAZINE
6	2.29	7.1	2155	1	C11H14Cl1N1	BENZAZEPINE,6-CHLORO-3-METHYL *antagonist-adrenoceptor-a2*
7 ★	2.92	7.4	2392	2	"	BENZAZEPINE,6-CHLORO-3-METHYL
8 ★	2.18		579		C11H14Cl1N1O1	PROPACHLOR **1918-16-7** *herbicide*
9	1.63	7.4	2462	1	C11H14Cl1N1O2	P-CHLOROBENZOIC ACID,N,N-DIMETHYLAMINOETHYL ESTER
7210 ★	2.63	7.4	2462	2	"	P-CHLOROBENZOIC ACID,N,N-DIMETHYLAMINOETHYL ESTER
1	-0.36	7.4	976	1	C11H14Cl1N3	2-(2,4-DIME-6-CL-PHENYLIMINO)IMIDAZOLINE
2	2.10	7.4	976	2	"	2-(2,4-DIME-6-CL-PHENYLIMINO)IMIDAZOLINE
3	-0.62	7.4	976	1	C11H14Cl1N3	2-(2,6-DIME-4-CL-PHENYLIMINO)IMIDAZOLINE
4	2.50	7.4	976	2	"	2-(2,6-DIME-4-CL-PHENYLIMINO)IMIDAZOLINE
15	-1.37	7.4	821	1	C11H14Cl1N3O2	2-(5-CL-2,4-DIMETHOXYPHENYLIMINO)IMIDAZOLIDINE **63346-71-4**
6	1.38	7.4	821	2	"	2-(5-CL-2,4-DIMETHOXYPHENYLIMINO)IMIDAZOLIDINE
7 ★	4.48		2094		C11H14Cl2N2O1	UREA,1-ISOBUTYL-3-(3,4-DICHLOROPHENYL) **5006-90-6**
8 ★	-0.29		537		C11H14Cl2N2O3	CHLORAMPHENICOL,P-AMINO ANALOG
9 ✓	0.95		2406	102	C11H14Cl3N2S1	TETRAMETHYLTHIOURONIUM IODIDE,S-(2,3,4-TRICHLOROPHENYL) ✚
7220 ✓	0.68		2406	102	C11H14Cl3N2S1	TETRAMETHYLTHIOURONIUM IODIDE,S-(2,4,5-TRICHLOROPHENYL)
1 ✓	0.05		2406	102	C11H14Cl3N2S1	TETRAMETHYLTHIOURONIUM IODIDE,S-(2,4,6-TRICHLOROPHENYL)
2 ✓	0.74		2406	102	C11H14Cl3N2S1	TETRAMETHYLTHIOURONIUM IODIDE,S-(3,4,5-TRICHLOROPHENYL)
3 ★	2.14		2479	459	C11H14F1N3O3	PHENYLBIURET,2-FLUORO-5-METHOXY
4	0.37	13.0	547	224	C11H14F1N5	4,6-DIAM-2,2-DIME-1(4-FLUOROPH)-S-TRIAZINE
25 ★	0.12	7.0	2440	1	C11H14F1N5O2	DIDEOXY-ARA-A,2-METHYL-2'-FLUORO
6 ★	0.10	7.0	2440	1	C11H14F1N5O2	DIDEOXY-ARA-A,8-METHYL-2'-FLUORO
7 ★	0.27	7.0	2440	1	C11H14F1N5O2	DIDEOXY-ARA-A,N6-METHYL-2'-FLUORO
8 ★	1.42		1718		C11H14F1O1	P-FLUOROBENZYLIDENE-T-BUTYLAMINE,N-OXIDE
9 ★	1.39	13.0	547	224	C11H14I1N5	4,6-DIAM-2,2-DIME-1(4-IODOPH)-S-TRIAZINE
7230	-0.69	7.0	239	1	C11H14N2	5-METHYLTRYPTAMINE **1821-47-2**
1	0.30	7.4	2545	1	C11H14N2	MPTP-4-(PYRID-4YL) ANALOG ✚
2 ★	1.25	7.5	2545	2	"	MPTP-4-(PYRID-4YL) ANALOG
3 ★	2.59		871		C11H14N2O1	4-PHENYL-N-NITROSOPIPERIDINE **6652-04-6**
4	-1.57	7.0	239	1	C11H14N2O1	5-METHOXYTRYPTAMINE **608-07-1**
35	1.28	7.4	1350	1	C11H14N2O1	BAY-C-6014, CLONIDINE ANALOG ✚
6	0.71		1725	459	C11H14N2O1	PYRIDO-PYRIMIDINONE ANALOG(RINGS=566) ✚
7	0.81		1725	459	"	PYRIDO-PYRIMIDINONE ANALOG(RINGS=566)
8 ★	1.80	7.4	447	1	C11H14N2O1	W-PHENYLPROPIONALDEHYDE,N-ACETYLHYDRAZONE
9 ★	0.95		2612		C11H14N2O2	1,1-DIMETHYL-3-(M-ACETYLPHENYL)UREA **42865-65-6**
7240	-1.90	7.0	477	1	C11H14N2O2	1-(O-CYANOPHENOXY)-3-MEAMINOPROPANOL-2
1	0.58		2541		C11H14N2O2	GLY-ETHYLAMIDE,N1-BENZOYL
2 ★	0.13		594		C11H14N2O2	MALONAMIDE,2-ETHYL-2-PHENYL **80866-90-6**
3 ★	3.26		547		C11H14N2O2	N-(P-NITROPHENYL)PIPERIDINE **6574-15-8**
4 ★	2.09		871		C11H14N2O2	N-NITROSO-PHENMETRAZINE
45 ★	-0.04	7.1	1591		C11H14N2O2	PHENYLALANIN-AMIDE,N-ACETYL
6	0.12	7.0	2309	1	"	PHENYLALANIN-AMIDE,N-ACETYL
7	0.40	7.4	2461	1	C11H14N2O2S3	THIENO[2,3-B]THIOPHENE-2-SULFONAMIDE,5-(CYCLOPROPYLMETHYL)AMINOMETHYL **122266-95-9**
8	1.14	7.4	2461	1	C11H14N2O2S4	THIENO[2,3-B]THIOPHENE-2-SULFONAMIDE,5-(1-THIOMORPHOLINO)METHYLAMINOMETHYL **133445-75-7**
9 ★	1.22		2612		C11H14N2O2	1-(4-ACETYLPHENYL)-3-METHOXY-3-METHYLUREA **88132-23-4**
7250 ★	0.15		1439		C11H14N2O3	3-ETOCO-PYRIDO(1,2A)PYRIMIDIN-4-ONE,4,6,7,8,9-H5
1 ★	0.16		1439		C11H14N2O3	3-MEOCO-PYRIDO(1,2A)PYRIMIDIN-4-ON,4,6,7,8,9-H5,6-ME ✚
2 ★	1.55	0.7	962		C11H14N2O3	ALLOBARBITAL,N-METHYL **780-59-6**
3	1.02	7.4	899	1	C11H14N2O3	NORHEXOBARBITAL **718-67-2**
4 ★	0.77	7.4	2659	459	C11H14N2O3	P-CARBOXAMIDOBENZYLCARBAMATE,N,N,-DIMETHYL
55 ★	-0.87	7.1	1591		C11H14N2O3	TYROSIN-AMIDE,N-ACETYL
6	1.17	7.4	2461	1	C11H14N2O3S3	THIENO[2,3-B]THIOPHENE-2-SULFONAMIDE,5-(1-MORPHOLINO)METHYLAMINOMETHYL **122266-97-1**
7	0.11	7.4	2461	1	C11H14N2O3S4	THIENO[2,3-B]THIOPHENE-2-SULFONAMIDE,5-(1-THIOMORPHOLIN-4-OXO)METHYLAMINOMETHYL **133445-65-5**
8 ★	2.36		1946		C11H14N2O4	2-PROPENOIC N-BUTYLAMIDE,3-(5-NITRO-2-FURANYL)
9 ★	2.23		1946		C11H14N2O4	2-PROPENOIC-N-I-BUTYLAMIDE,3-(5-NITRO-2-FURANYL)
7260 ★	2.28		1946		C11H14N2O4	2-PROPENOIC-N-SEC-BUTYLAMIDE,3-(5-NITRO-2-FURANYL)
1 ★	2.33		1946		C11H14N2O4	2-PROPENOIC-N-T-BUTYLAMIDE,3-(5-NITRO-2-FURANYL)
2 ★	0.70	7.4	2462	1	C11H14N2O4	P-NITROBENZOIC ACID,N,N-DIMETHYLAMINOETHYL ESTER
3 ★	1.70	7.4	2462	2	"	P-NITROBENZOIC ACID,N,N-DIMETHYLAMINOETHYL ESTER
4	0.88	7.4	2525	1	C11H14N2O4S2	5-(1-MORPHOLINO)THIENO[2,3-B]FURAN-2-SULFONAMIDE

	logP	pH	Ref.	Note	MF	Name / CAS / activity
65 ★	-0.03		537		C11H14N2O5	N-ACETYL-A-HYDROXYMETHYL-B-OH-4-NITROPHENETHYLAMINE 4423-58-9
6 ★	2.17		2072	2	C11H14N2O5S1	BENZENESULFONAMIDE,2-NITRO-4-METHYL-N-(I-BUTANOYL)
7	-0.15	7.4	2495	1	C11H14N3O2P1	PHOSPHORIC TRIAMIDE,N-BENZOYL-N'-N''-DIAZIRIDINYL ANALOG
8 ★	0.81	7.4	2495	2	C11H14N3O2P1	PHOSPHORIC TRIAMIDE,N-BENZOYL-N',N''-DIAZIRIDINYL ANALOG
9	0.58		559	818	C11H14N4O1	1-(4-CARBAMOYLBUTYL)BENZOTRIAZOLE
7270	1.01		559		C11H14N4O1	2-(4-CARBAMYLBUTYLBENZOTRIAZOLE
1	2.09		547		C11H14N4O1	4-(3-ME-3-(2-ALLYL)TRIAZENO)BENZAMIDE
2 ★	2.91	6.3	2590	459	C11H14N4O1	BENZAMIDE,2-(3-METHYL-3-ALLYLTRIAZENYL) antineoplastic
3 ★	1.38	5.6	2165	1	C11H14N4O1	BENZOTRIAZOLE,1-(2-PENTANAMIDE)
4 ★	1.15	5.6	2165	1	C11H14N4O1	BENZOTRIAZOLE,1-(4-PENTANAMIDE)
75 ★	1.75	5.6	2165	1	C11H14N4O1	BENZOTRIAZOLE,2-(2-PENTANAMIDE)
6 ★	1.35	5.6	2165	1	C11H14N4O1	BENZOTRIAZOLE,2-(4-PENTANAMIDE)
7 ★	0.24	7.0	1621	1	C11H14N4O1	PYRAZOL-3-ONE,H2-2(2,6-DIME-4-PYRIMIDINYL)1,5-DIME 23898-86-4
8 ★	0.51	7.0	1621	1	C11H14N4O1	PYRAZOLE-3-ONE,1-(4,6-DIME-2-PYRIMIDINYL)-2,5-DIME 23898-92-2
9 ★	1.82	7.0	1621	1	C11H14N4O1	PYRIMIDINE,2-(3-MEO-5-ME-PYRAZOL-1-YL)-4,6-DIME 23905-77-3
7280 ★	0.73	7.0	1621	1	C11H14N4O1	PYRIMIDINE,2-(5-MEO-3-ME-1H-PYRAZOL-1-YL)4,6-DIME 18694-43-4
1 ★	2.60	7.0	1621	1	C11H14N4O1	PYRIMIDINE,4-(5-MEO-5-ME-PYRAZOL-1-YL)-2,6-DIME 23903-41-5
2 ★	-0.12	7.0	2440	1	C11H14N4O2	DIDEOXYPURINE,6-METHYL
3 ★	0.70	7.0	1621	1	C11H14N4O2	PYRAZOL-3-ONE,2(2-ME-6-MEO-4-PYRIMIDINYL)1,5-DIME 23898-89-7
4 ★	0.38	7.0	1621	1	C11H14N4O2	PYRAZOLE-3-ONE,1-(6-MEO-2-ME-4-PYRIMIDINYL)-2,5-DIME 23898-95-5
85 ★	1.26	7.0	1621	1	C11H14N4O2	PYRIMIDINE,2(5-MEO-3-ME-PYRAZOL-1-YL)4-MEO-6-ME 18694-40-1 analgesic, anti-inflammatory
6 ★	1.54	7.0	1621	1	C11H14N4O2	PYRIMIDINE,6(5-MEO-3-ME-1H-PYRAZOL-1-YL)2-ME-4-MEO 23903-42-6
7 ★	1.07		2070		C11H14N4O3	ALLOPURINOL,1-PIVALOYLOXYMETHYL
8 ★	0.79		2070		C11H14N4O3	ALLOPURINOL,2-PIVALOYLOXYMETHYL
9 ★	2.22	7.4	2206	306	C11H14N4O3	HEXAHYDROISOXAZOLE,3-(1-METHYL-5-NITROIMIDAZOLYL)
7290	-1.68	3.3	2590	331	C11H14N4O3	N-CARBOXYMETHYLBENZAMIDE,4-(3,3-DIMETHYLTRIAZENYL) antineoplastic
1 ★	1.86	7.0	1621	1	C11H14N4O3	PYRIMIDINE,6(5-MEO-3-ME-1H-PYRAZOL-1-YL)2,4-DIMEO 23905-98-8
2	-0.80		401		C11H14N4O4	TUBERCIDIN 69-33-0 antineoplastic, antifungal
3	0.09		400		C11H14N4O4S1	6-METHIOPURINERIBOSIDE 342-69-8 antineoplastic
4	-1.66		1404		C11H14N4O7	2-NO2-IMIDAZOLE,1-CH2CON(CH2CH2O2CH)2
95	-1.97	7.4	594	1	C11H14N6O2	4,6-DIAMINO-2,2-DIMETHYL-1-(3-NITROPHENYL)-SYM-TRIAZINE
6 ★	0.06	13.0	547	224	"	4,6-DIAMINO-2,2-DIMETHYL-1-(3-NITROPHENYL)-SYM-TRIAZINE
7 ★	0.63		2289		C11H14N6O2	TETRAZOLE,2-METHYL-5-(4-DIMETHYLUREYL)PHENOXY 117121-41-2
8	0.31	9.0	1409		C11H14N6O5	2-PROPANOL,1,3-BIS-(2-METHYL-5-NITRO-IMIDAZOL-1-YL) 74550-92-8
9 ★	3.00		579		C11H14O1	2,2-DIMETHYLPROPIOPHENONE 938-16-9
7300 ★	2.42		502		C11H14O1	2-PENTANONE,5-PHENYL 2235-83-8
1 ★	2.98		1093		C11H14O1	P-I-PROPYLACETOPHENONE
2 ★	2.78		558		C11H14O2	2-I-PROPYLPHENYLACETATE 1608-68-0
3 ★	2.77		502		C11H14O2	4-PHENYLBUTYRIC ACID,METHYL ESTER 2046-17-5
4 ★	2.70		2430		C11H14O2	5-PHENYLVALERIC ACID 2270-20-4
5 ★	2.73		306		C11H14O2	B-PHENYLPROPIONIC ACID,ETHYL ESTER 2021-28-5
6 ★	3.97		2431	459	C11H14O2	BENZOIC ACID,4-BUTYL
7 ★	3.85	2.0	2236		C11H14O2	BENZOIC ACID-4-T-BUTYL 98-73-7
8 ★	3.84		594		C11H14O2	BUTYL BENZOATE 136-60-7
9 ★	3.18		2168		C11H14O2	ISOVALERIC ACID,PHENYL ESTER
7310	1.91		458		C11H14O2Se1	M-METHYLBENZYLSELENOPROPIONIC ACID
1	1.79		458		C11H14O2Se1	P-METHYLBENZYLSELENOPROPIONIC ACID 43169-97-7
2 ★	4.63	1.2	2586		C11H14O3	BUTYL SALICYLATE
3 ★	2.84		558		C11H14O3	O-I-PROPYLPHENOXYACETIC ACID
4 ★	3.57	1.2	1077		C11H14O3	P-BUTOXYBENZOIC ACID 1498-96-0
15	3.32	7.5	1259	1	C11H14O3	P-HYDROXYBENZOIC ACID,BUTYL ESTER 94-26-8 antifungal
6 ★	3.57		508		"	P-HYDROXYBENZOIC ACID,BUTYL ESTER
7 ★	2.33	1.2	1077		C11H14O3	P-METHOXY-G-PHENYLBUTYRIC ACID
8 ★	2.59		501		C11H14O3	PHENOXYACETIC ACID,M-ISOPROPYL
9 ★	2.71		501		C11H14O3	PHENOXYACETIC ACID,M-PROPYL 2084-11-9
7320 ★	2.69		501		C11H14O3	PHENOXYACETIC ACID,P-ISOPROPYL
1 ★	1.63		594	400	C11H14O4	2,3,4-TRIMETHOXYACETOPHENONE 13909-73-4
2 ★	2.37	1.0	599	220	C11H14O4	BENZOIC ACID,3,4-DIETHOXY
3 ★	1.16		1895		C11H14O4	DIETHYLENEGLYCOL, MONOBENZOATE
4 ★	2.41		1335		C11H14O5	BUTYLGALLATE 1083-41-6
25 ★	2.46		1335		C11H14O5	I-BUTYLGALLATE 3856-05-1
6 ★	4.68		1211		C11H15Br1Cl1O3P1S1	PROFENOFOS 41198-08-7 insecticide
7 ★	2.92		1263		C11H15Br1N2O1	1,1-DIMETHYL-3-(3,5-DIME-4-BROMOPHENYL)UREA
8	2.11	7.4	899	1	C11H15Br1N2O3	BARBITURIC ACID,I-PROPYL,BROMALLYL,N-METHYL 125-55-3
9	1.72	7.4	899	1	C11H15Br1N2O3	BARBITURIC ACID,S-BUTYL,B-BROMALLYL 1142-70-7 sedative, hypnotic
7330 ★	2.01		1019		C11H15Cl1N2O2	IPROCLOZIDE 3544-35-2 monoamineoxidase inhibitor
1 ★	1.82		1019		C11H15Cl1N2O2	N'-I-PROPYL-(2-CL-PHENOXY)ACETIC ACID HYDRAZIDE
2 ★	1.91		1019		C11H15Cl1N2O2	N'-I-PROPYL-(3-CL-PHENOXY)ACETIC ACID HYDRAZIDE
3 ★	2.81	7.4	2004	1	C11H15Cl1N2O3S1	N2-BUTYL-N1-P-CHLOROBENZENESULFONYLUREA 13909-64-3
4 ✓	-0.09		2406	102	C11H15Cl1N3O2S1	TETRAMETHYLTHIOURONIUM IODIDE,S-(2-NITRO-4-CHLOROPHENYL)
35 ★	0.50	7.0	1783	1	C11H15Cl1N6O3	N(5-CL-6(4-ME-PIPERIZIN)3-NO2-PHENYL)ACETAMIDE 89083-22-7 ✚
6	1.26	7.4	2092	1	C11H15Cl2N1O1	1-(3,4-DICLPHENYL)-2-I-PROPYLAMINOETHANOL 59-61-0
7 ★	3.32		1984		"	1-(3,4-DICLPHENYL)-2-I-PROPYLAMINOETHANOL
8 ✓	0.29		2406	102	C11H15Cl2N2S1	TETRAMETHYLTHIOURONIUM IDODE,S-(2,3-DICHLOROPHENYL)
9 ✓	0.47		2406	102	C11H15Cl2N2S1	TETRAMETHYLTHIOURONIUM IODIDE,S-(2,4-DIOCHLOROPHENYL)
7340 ✓	0.00		2406	102	C11H15Cl2N2S1	TETRAMETHYLTHIOURONIUM IODIDE,S-(2,6-DICHLOROPHENYL)
1 ✓	0.49		2406	102	C11H15Cl2N2S1	TETRAMETHYLTHIOURONIUM IODIDE,S-(3,4-DICHLOROPHENYL)
2 ✓	0.52		2406	102	C11H15Cl2N2S1	TETRAMETHYLTHIOURONIUM IODIDE,S-(3,5-DICHLOROPHENYL)
3 ✓	0.05		2406	102	C11H15Cl2N2S1	TETRAMETHYLTHLIOURONIUM IODIDE,S-(2,5-DICHLOROPHENYL)
4 ★	4.92		2519		C11H15Cl2O3P1S1	PROTHIOPHOS 34643-46-4
45 ★	3.72		336		C11H15Cl2O4P1S1	O,O-DIET-O-(2,6-DICL-4-METHIOPHENYL)PHOSPHATE 26707-54-0
6 ★	4.18		354		C11H15Cl3N1O3P1	BUTYLPHOSPHORAMIDATE,O-ME-O-(2,4,5-TRICL-PHENYL)
7 ★	3.86		354		C11H15Cl3N1O3P1	I-BUTYLPHOSPHORAMIDATE,O-ME-O-(2,4,5-TRICLPHENYL
8 ★	3.99		354		C11H15Cl3N1O3P1	S-BUTYLPHOSPHORAMIDATE,O-ME-O-(2,4,5-TRICLPHENYL
9 ★	3.83		354		C11H15Cl3N1O3P1	T-BUTYLPHOSPHORAMIDATE,O-ME-O-(2,4,5-TRICLPHENYL

	logP	pH	Ref.	Note	MF	Name / **CAS** / *activity*
7350	1.34	7.4	2266	1	C11H15F1N2O4	5-FLUOROURACIL-1-CARBOXYLIC ACID,HEXYL ESTER
1 ★	2.04	4.0	2266	302	"	5-FLUOROURACIL-1-CARBOXYLIC ACID,HEXYL ESTER
2	-1.35		571		C11H15F3N1.Br1	M-TRIFLUOROMETHYLBENZYLTRIMETHYLAMMONIUMBROMIDE
3 ★	2.38	13.0	1806	224	C11H15N1	2-NAPHTHALENEAMINE-1,2,3,4-TETRAHYDRO-N-METHYL **19485-85-9**
4 ★	2.61	7.5	2545	2	C11H15N1	CINNAMYLAMINE,N,N-DIMETHYL
55 ★	2.98	13.0	547	224	C11H15N1	N-PHENYLPIPERIDINE **4096-20-2**
6 ★	1.52		1182		C11H15N1O1	2-(M-TOLYL)MORPHOLINE
7 ★	1.58		1182		C11H15N1O1	4-BENZYLMORPHOLINE **10316-00-4**
8 ★	1.76		2410		C11H15N1O1	ACETANILIDE,2,3,4-TRIMETHYL
9 ★	1.41		2410		C11H15N1O1	ACETANILIDE,2-ETHYL-6-METHYL
7360 ★	1.71		1894		C11H15N1O1	ACETANILIDE,2-I-PROPYL **19246-04-9**
1 ★	2.50		2410		C11H15N1O1	ACETANILIDE,3,4,5-TRIMETHYL
2 ★	2.75		1894		C11H15N1O1	ACETANILIDE,4-PROPYL **20330-99-8**
3 ★	1.26		579		C11H15N1O1	BENZYLIDENE-T-BUTYLAMINE,N-OXIDE
4 ★	2.51		603		C11H15N1O1	I-VALERANILIDE **2364-50-3**
65 ★	1.25		603		C11H15N1O1	N,N-DIETHYLBENZAMIDE **1696-17-9**
6 ★	2.51		547		C11H15N1O1	P-T-BUTYLBENZAMIDE **56108-12-4**
7 ★	2.56		603		C11H15N1O1	SEC-VALERANILIDE **54394-78-4**
8 ★	1.99		603		C11H15N1O1	TERT-VALERANILIDE **6625-74-7**
9 ★	2.61		603		C11H15N1O1	VALERANILIDE **10264-18-3**
7370	1.03	7.1	2155	2	C11H15N1O2	1,2,3,4-(H4)-ISOQUINOLINE,5,8-DIMETHOXY
1 ★	0.82		803		C11H15N1O2	2,3,5-TRIMETHYL-4-HYDROXYACETANILIDE
2	0.75		803		C11H15N1O2	2,3,6-TRIMETHYL-4-HYDROXYACETANILIDE **36592-59-3**
3 ★	2.26	6.5	1556	1	C11H15N1O2	2-BUTYRAMIDO-5-METHYLPHENOL
4 ★	1.71		803		C11H15N1O2	3-I-PROPYL-4-HYDROXYACETANILIDE **13780-91-1**
75	-1.46	7.4	2261	1	C11H15N1O2	6,7-ADTN,7-METHOXY ✦
6	0.86	7.4	2462	1	C11H15N1O2	BENZOIC ACID,DIMETHYLAMINOETHYL ESTER
7 ★	2.06	7.4	2462	2	"	BENZOIC ACID,DIMETHYLAMINOETHYL ESTER
8	0.26	7.4	1616	1	C11H15N1O2	BENZOIC ACID,O-BUTYLAMINO
9	2.25	2.5	1616	1	"	BENZOIC ACID,O-BUTYLAMINO
7380 ★	3.30		1263		C11H15N1O2	BUTYL-N-PHENYLCARBAMATE **1538-74-5**
1 ★	2.40		1969		C11H15N1O2	DIMETHYLCARBAMATE,PHENETHYL ESTER **98183-16-5**
2 ★	1.99		594	400	C11H15N1O2	METHYLCARBAMATE,N-(2-ETHYL-6-METHYLPHENYL)
3	1.54	7.4	2353	2	C11H15N1O2	MORPHOLINE,2-HYDROXY-2-PHENYL-4-METHYL
4 ★	2.63		1681		C11H15N1O2	N,N-DIMETHYLCARBAMATE,P-METHYLBENZYL ESTER **84640-21-1**
85	2.78	7.4	2659	459	"	N,N-DIMETHYLCARBAMATE,P-METHYLBENZYL ESTER
6 ★	2.31		273		C11H15N1O2	N-METHYL-2-I-PROPYLPHENYLCARBAMATE **2631-40-5** *pesticide*
7 ★	2.40		273		C11H15N1O2	N-METHYL-2-PROPYLPHENYLCARBAMATE **15482-11-8**
8 ★	2.64		273		C11H15N1O2	N-METHYL-3-PROPYLPHENYLCARBAMATE
9 ★	2.57		536		C11H15N1O2	N-METHYL-4-I-PROPYLPHENYLCARBAMATE **4089-99-0**
7390 ★	2.80		273		"	N-METHYL-4-I-PROPYLPHENYLCARBAMATE
1 ★	2.72		273		C11H15N1O2	N-METHYL-4-PROPYLPHENYLCARBAMATE
2 ★	2.52		536		C11H15N1O2	N-METHYLCARBAMATE,2,4,5-TRIMETHYLPHENYL **671-03-4**
3 ★	2.66		702		C11H15N1O2	N-METHYLCARBAMATE,3,4,5-TRIMETHYLPHENYL **2686-99-9**
4 ★	2.63		273		C11H15N1O2	N-METHYLCARBAMATE,M-I-PROPYLPHENYL **64-00-6**
95 ★	2.87		2630		C11H15N1O2	P-AMINOBENZOIC ACID,BUTYL ESTER **94-25-7** *anesthetic (topical)*
6	3.05	7.5	1259	1	"	P-AMINOBENZOIC ACID,BUTYL ESTER
7 ★	2.48		547		C11H15N1O2	P-BUTOXYBENZAMIDE
8 ★	-0.36		502		C11H15N1O2	VALERIC ACID,2-AMINO-5-PHENYL **2046-19-7**
9 ★	2.86	7.4	2659	459	C11H15N1O2S1	N,N-DIMETHYLCARBAMATE,P-METHIOBENZYL ESTER **84640-39-1**
7400 ★	2.92		1263		C11H15N1O2S1	N-METHYLCARBAMATE,3,5-DIMETHYL-4-METHYLTHIOPHENYL **2032-65-7** *acaricide~insecticide*
1	1.24		866		C11H15N1O3	2-BUTOXY-4-AMINOBENZOIC ACID **61566-62-9**
2 ★	1.52		273		C11H15N1O3	BAYGON **114-26-1** *insecticide*
3 ★	2.09		1681		C11H15N1O3	N,N-DIMETHYLCARBAMATE,M-METHOXYBENZYL ESTER **84640-23-3**
4	2.31	7.4	2659	459	"	N,N-DIMETHYLCARBAMATE,M-METHOXYBENZYL ESTER
5 ★	2.20		1681		C11H15N1O3	N,N-DIMETHYLCARBAMATE,P-METHOXYBENZYL ESTER **84640-24-4**
6 ★	1.96		273		C11H15N1O3	N-METHYL-3-I-PROPOXYPHENYLCARBAMATE **3938-20-3**
7	-0.35		844	837	C11H15N1O4	6-HYDROXYNORBORNANE-2-CARBOXAMIDE-3,5-LACTONE
8	0.01	7.4	2062	1	C11H15N1O4	L-DOPA,ETHYL ESTER **74985-73-2**
9 ★	0.28	7.4	2062	2	"	L-DOPA,ETHYL ESTER
7410	-0.62	7.4	2472	324	C11H15N1O4	PYRIDIN-4-ONE,2-METHYL-3-HYDROXY-1-PROPIONIC ACID ETHYL ESTER
1 ★	2.34		773	601	C11H15N1O4S1	4-SULFAMYLBENZOIC ACID,BUTYL ESTER **59777-59-2**
2 ★	0.98	7.4	2490	1	C11H15N1O4S1	BENZENESULFONAMIDE,4-(3-ACETYLOXY)PROPYL **135832-47-2**
3 ★	0.78	7.4	2490	1	C11H15N1O5S2	BENZENESULFONAMIDE,3-METHOXYCARBONYL-4-(3-HYDROXYPROPYL)THIO **108966-66-1**
4 ★	-0.30	7.4	2490	1	C11H15N1O7S2	BENZENESULFONAMIDE,3-METHOXYCARBONYL-4-(3-HYDROXYPROPYL)SULFONYL **108966-67-2**
15	1.51	7.4	2331	1	C11H15N1S1	5H-THIENO[2,3-A]QUINOLIZINE,4,7,8,9,10,10A-HEXAHYDRO **112114-08-6**
6	1.18	7.4	2535	1	C11H15N1S1	G-(2-THIOPHENOYL)PROPYL-(2-PROPANOL)AMINE (DEHYDRO-FUSED)
7	1.20	7.4	2535	1	C11H15N1S1	G-(3-METHYLTHION-2OYL)PROPYL-ETHANOLAMINE (DEHYDRO-FUSED)
8	-3.05		571		C11H15N2.Br1	M-CYANOBENZYLTRIMETHYLAMMONIUMBROMIDE
9 ★	3.22		559		C11H15N3	1-PENTYLBENZOTRIAZOLE
7420 ★	2.95		559		C11H15N3	1-S-PENTYLBENZOTRIAZOLE
1	-1.66	7.4	976	1	C11H15N3	2-(2,4-DIMETHYLPHENYLIMINO)IMIDAZOLIDINE **4794-83-6**
2	-1.46	7.4	821	1	"	2-(2,4-DIMETHYLPHENYLIMINO)IMIDAZOLIDINE
3	1.78	7.4	976	2	"	2-(2,4-DIMETHYLPHENYLIMINO)IMIDAZOLIDINE
4 ★	1.79	7.4	821	2	"	2-(2,4-DIMETHYLPHENYLIMINO)IMIDAZOLIDINE
25	-1.65	7.4	821	1	C11H15N3	2-(2,6-DIMETHYLPHENYLIMINO)IMIDAZOLIDINE **4859-06-7**
6	-1.54	7.4	976	1	"	2-(2,6-DIMETHYLPHENYLIMINO)IMIDAZOLIDINE
7	-1.40	7.4	1398	1	"	2-(2,6-DIMETHYLPHENYLIMINO)IMIDAZOLIDINE
8 ★	1.45	7.4	821	2	"	2-(2,6-DIMETHYLPHENYLIMINO)IMIDAZOLIDINE
9	1.86	7.4	976	2	"	2-(2,6-DIMETHYLPHENYLIMINO)IMIDAZOLIDINE
7430 ★	3.63		559		C11H15N3	2-PENTYLBENZOTRIAZOLE
1 ★	3.42		559		C11H15N3	2-S-PENTYLBENZOTRIAZOLE
2	1.30		1407		C11H15N3O1	1,2,3,6-H4-PYRIDINE,N-(1-ME-2-PYRROLYL)CONH-
3 ★	1.51		559		C11H15N3O1	1-(5-HYDROXYPENTYL)BENZOTRIAZOLE
4	3.36		1213		C11H15N3O1	1-ME-1-ACETYL-3-(P-ETHYLPHENYL)TRIAZENE
35	-1.56	7.4	821	1	C11H15N3O1	2-(2-MEO-4-ME-PHENYLIMINO)IMIDAZOLIDINE
6 ★	1.64	7.4	821	2	"	2-(2-MEO-4-ME-PHENYLIMINO)IMIDAZOLIDINE
7 ★	1.93		559		C11H15N3O1	2-(5-HYDROXYPENTYL)BENZOTRIAZOLE
8 ★	0.48	7.4	365	1	C11H15N3O1	PIPERAZINE-2-CARBOXANILIDE **36385-57-6**
9	-1.78	7.4	821	1	C11H15N3O2	2-(2,5-DIMETHOXYPHENYLIMINO)IMIDAZOLIDINE **50531-51-6**

	logP	pH	Ref.	Note	MF	Name / **CAS** / *activity*
7440 ★	0.97	7.4	821	2	"	2-(2,5-DIMETHOXYPHENYLIMINO)IMIDAZOLIDINE
1	0.34	7.4	1214	1	C11H15N3O2	3-I-PROPYL-1-(4-CARBOXYPHENYL)-3-METHYLTRIAZENE **74109-28-7** *antineoplastic*
2	0.38	3.3	2590	331	"	3-I-PROPYL-1-(4-CARBOXYPHENYL)-3-METHYLTRIAZENE
3	0.46	3.3	2590	331	C11H15N3O2	3-PROPYL-1-(4-CARBOXYPHENYL)-3-METHYLTRIAZENE **74109-21-0** *antineoplastic*
4	0.46	7.4	1214	1	"	3-PROPYL-1-(4-CARBOXYPHENYL)-3-METHYLTRIAZENE
45	1.49	3.3	2590	331	C11H15N3O2	BENZOIC ACID,2-(3,3-DIETHYLTRIAZENYL) *antineoplastic*
6 ★	3.66	6.6	2590	459	C11H15N3O2	ETHYL BENZOATE,4-(3,3-DIMETHYLTRIAZENYL) *antineoplastic*
7	0.12		1439		C11H15N3O2	PYRIDO(1,2A)PYRIMIDIN-4-ON,3-CONH2,-H7,1,6-DIME
8	-1.66	12.0	541		C11H15N3O4	SODIUM BENZOATE, ORTHO-METHYLPROPYLTRIAZINYL
9 ★	0.55		2105		C11H15N3O3	HYDRAZINECARBOXAMIDE,1-ACETYL-N-(4-ETHOXYPHENYL)
7450	-0.48	6.0	479		C11H15N3O3S1	2,5-DIME-3-SULFANILAMIDE-2,3-DIHYDROISOXAZOLE **51543-32-9**
1	-0.40	7.4	2461	1	C11H15N3O3S3	THIENO[2,3-B]THIOPHENE-2-SULFONAMIDE,5-N-(DIMETHYLAMINOETHYL)CARBOXAMIDO **129949-85-5**
2 ★	0.24		931		C11H15N3O4	PYRICARBATE **1882-26-4** *antiarteriosclerotic*
3	-1.40	7.0	801	1	C11H15N3O6	1-(5-ACETYLARABINOSYL)CYTOSINE **31088-09-2**
4	-1.57	7.0	2440	1	C11H15N3O6	N-ACETYL-ARA-C
55 ★	-1.35		1739		"	N-ACETYL-ARA-C
6	-2.75	7.4	547	1	C11H15N5	4,6-DIAM-1,2-H2-2,2-DIME-1-PHENYL-S-TRIAZINE **4022-58-6**
7 ★	0.28	13.0	547	224	"	4,6-DIAM-1,2-H2-2,2-DIME-1-PHENYL-S-TRIAZINE
8	-1.21	7.4	547	1	C11H15N5O1	1,2-DIHYDROTRIAZINE,2,2-DIMETHYL-4.6-DIAMINO-1-(3'HYDROXY)PHENYL
9 ★	0.55		2544		C11H15N5O1	9-BUTYL-N-ACETYLADENINE
7460 ★	0.02	9.0	1949		C11H15N5O1S1	CIMETIDINE,PYRIMID-4ONE-2YL ANALOG **54855-84-4**
1 ★	-0.40		2302		C11H15N5O1S1	CIMETIDINE,PYRIMIDIN-2-ONE ANALOG
2 ★	1.18		563		C11H15N5O2	BENZAMIDE,N(2-ACETAMIDO)-O-(3,3-DIME)-1-TRIAZINO
3 ★	1.49		547		C11H15N5O2	N-ACETYLBENZOYLHYDRAZINE,2-(3,3-DIMETHYLTRIAZENE)
4 ★	1.04	9.0	1949		C11H15N5O2S1	CIMETIDINE,3-NITROPYRROLE ANALOG **74188-69-5** ✦
65 ★	-0.19	7.0	2440	1	C11H15N5O3	ADENOSINE,2'-DEOXY-N6-METHYL
6	-0.45	7.4	2037	1	C11H15N5O3	ISOGUANINE,1-METHYL-9-PROPIONIC ACID,ETHYL ESTER
7 ★	4.56		1505	459	C11H16	PENTAMETHYLBENZENE **700-12-9**
8	4.90		1560	459	C11H16	PENTYLBENZENE **538-68-1**
9 ★	2.44	8.0	1011	2	C11H16Br1N1O2	3,5-DIMETHOXY-4-BROMOAMPHETAMINE
7470 ★	2.58	7.0	538	701	C11H16Br1N1O2	BROLAMFETAMINE **64638-07-9** *psychodysleptic*
1 ✓	0.06		2406	102	C11H16Br1N2S1	TETRAMETHYLTHIOURONIUM IODIDE,S-(2-BROMOPHENYL)
2 ✓	0.01		2406	102	C11H16Br1N2S1	TETRAMETHYLTHIOURONIUM IODIDE,S-(3-BROMOPHENYL)
3 ✓	0.06		2406	102	C11H16Br1N2S1	TETRAMETHYLTHIOURONIUM IODIDE,S-(4-BROMOPPHENYL)
4 ✓	-0.20		2406	102	C11H16Cl1N2S1	TETRAMETHYLTHIOURONIUM CHLORIDE,S-(CHLOROPHENYL)
75 ✓	-0.19		2406	102	C11H16Cl1N2S1	TETRAMETHYLTHIOURONIUM IODIDE,S-(2-CHLOROPHENYL)
6 ✓	-0.17		2406	102	C11H16Cl1N2S1	TETRAMETHYLTHIOURONIUM IODIDE,S-(3-CHLOROPHENYL)
7 ★	0.36	6.5	321	2	C11H16Cl1N3O4S2	3-T-BUTYLHYDROCHLOROTHIAZIDE **23141-84-6**
8 ★	2.53	13.0	594	1	C11H16Cl1N5	CHLOROGUANIDE **500-92-5** *antimalarial*
9	0.21	5.0	831	1	C11H16Cl1N5	N1-P-CHLOROPHENYL-N5-PROPYLBIGUANIDE **49871-96-7**
7480	2.63	11.5	831	314	"	N1-P-CHLOROPHENYL-N5-PROPYLBIGUANIDE
1 ★	5.33		1639		C11H16Cl1O2P1S3	CARBOFENOTION **786-19-6** *insecticide*
2	4.40		354		C11H16Cl1O3P1S2	O,O-DIET-O-(2-CL-4-METHIOPHENYL)PHOSPHOROTHIOATE **26512-71-0**
3	5.06		354		C11H16Cl1O3P1S2	O,O-DIET-O-(3-CL-4-METHIOPHENYL)PHOSPHOROTHIOATE **21840-66-4**
4	3.05		354		C11H16Cl1O4P1S1	O,O-DIET-O-(2-CL-4-METHIOPHENYL)PHOSPHATE **24493-78-5**
85	2.94		336		C11H16Cl1O4P1S1	O,O-DIET-O-(3-CL-4-METHYLTHIOPHENYL)PHOSPHATE **26798-03-8**
6	3.48		354		"	O,O-DIET-O-(3-CL-4-METHYLTHIOPHENYL)PHOSPHATE
7 ✓	-0.55		2406	102	C11H16F1N2S1	TETRAMETHYLTHIOURONIUM IODIDE,S-(2-FLUOROPHENYL)
8 ✓	-0.53		2406	102	C11H16F1N2S1	TETRAMETHYLTHIOURONIUM IODIDE,S-(3-FLUOROPHENYL)
9 ✓	-0.58		2406	102	C11H16F1N2S1	TETRAMETHYLTHIOURONIUM IODIDE,S-(4-FLUOROPHENYL)
7490 ★	2.63	4.0	2436	1	C11H16F1N3O3	CARMOFUR **61422-45-5** *antineoplastic*
1 ✓	-1.33		2397	101	C11H16F3N4O3S1	IMIDAZOLIUM CHLORIDE,1-METHYL-2-HYDROXYIMINOMETHYL-3-(1-TRIFLUOROMETHYLSULFONYLPYRROLIDIN-2-YL)METHYL
2 ✓	0.30		2406	102	C11H16I1N2S1	TETRAMETHYLTHIOURONIUM IODIDE,S-(2-IODOPHENYL)
3 ✓	0.29		2406	102	C11H16I1N2S1	TETRAMETHYLTHIOURONIUM IODIDE,S-(3-IODOPHENYL)
4 ✓	0.28		2406	102	C11H16I1N2S1	TETRAMETHYLTHIOURONIUM IODIDE,S-(4-IODOPHENYL)
95	2.70		2616	459	C11H16N1O4P1S1	OPTUNAL **747-45-5**
6 ★	2.20		525		C11H16N1O5P1	PHOSPHONATE,O-(P-NITROPHENYL)-O-PROPYL,ETHYL
7 ★	4.05		945		C11H16N1O5P1S1	3-I-PROPYL-DIMETHYLPARATHION **1592-82-1**
8 ★	2.57		945		C11H16N1O6P1	3-I-PROPYL-DIMETHYLPARA-OXON **13074-11-8**
9	0.73	7.4	1948	1	C11H16N2	1-(O-METHYLPHENYL)PIPERAZINE **39512-51-1**
7500	2.14	7.4	1948	2	"	1-(O-METHYLPHENYL)PIPERAZINE
1 ★	2.22	13.0	594		"	1-(O-METHYLPHENYL)PIPERAZINE
2 ★	1.34	11.0	533	322	C11H16N2	3-PYRIDYLETHYL-2-(N-PYRROLIDINE) **20173-28-8**
3 ★	1.49	11.0	533	322	C11H16N2	3-PYRIDYLMETHYL-N-PIPERIDINE **13552-35-7**
4 ★	0.96	11.0	533	322	C11H16N2	METHYLANABASINE **24380-92-5**
5 ★	1.68	11.0	533	322	C11H16N2	N,N-DIMETHYL-(3-PYRIDYLBUTENE-1-YL)-AMINE
6	2.05	7.4	447	1	C11H16N2	W-PHENYLPROPIONALDEHYDE,N,N-DIMETHYLHYDRAZONE
7 ★	1.90		1263		C11H16N2O1	1,1-DIMETHYL-3-(3,5-DIMETHYLPHENYL)UREA **36627-56-2**
8	-0.30	7.4	1948	1	C11H16N2O1	1-(O-METHOXYPHENYL)PIPERAZINE **35386-24-4**
9	1.09	7.4	1948	2	"	1-(O-METHOXYPHENYL)PIPERAZINE
7510	0.62		1352		C11H16N2O1	2-AMINOPROPIONYL-2',6'-XYLIDIDE
1 ★	1.10		1622		C11H16N2O1	BENZAMIDE,N-(2-DIMETHYLAMINOETHYL) **63224-18-0**
2 ★	0.58		1622		C11H16N2O1	BENZAMIDE,N-(4-AMINOBUTYL) **5692-23-9**
3 ★	2.33	7.4	1616	1	C11H16N2O1	BENZAMIDE,O-BUTYLAMINO
4 ★	1.11		1147		C11H16N2O1	N-(N',N'-DIMETHYLAMINO)ACETYL)BENZYLAMINE
15	0.76		1352		C11H16N2O1	TOCAINIDE **41708-72-9** *cardiac depressant (anti-arrhythmic)*
6 ★	1.44		1969		C11H16N2O2	DIMETHYLCARBAMATE,3-(3-PYRIDYL)PROPYL ESTER **98154-97-3**
7 ★	1.24		1019		C11H16N2O2	N'-I-PROPYL-PHENOXYACETIC ACID HYDRAZIDE
8 ★	1.90	9.0	1693		C11H16N2O2	N-ME-CARBAMATE,3-ME,4-DIMEAMINOPHENYL **2032-59-9** *insecticide*
9 ★	0.20		1622		C11H16N2O2	O-METHOXYBENZAMIDE,N-(3-AMINOPROPYL)
7520	0.00	7.4	1956	2	C11H16N2O2	PILOCARPINE **92-13-7** *antiglaucoma agent, cholinergic (ophthalmic)*
1	0.22	7.4	1758	1	"	PILOCARPINE
2	0.92	7.4	2461	1	C11H16N2O2S3	THIENO[2,3-B]THIOPHENE-2-SULFONAMIDE,5-I-BUTYLAMINOMETHYL **122266-88-0**
3	-0.08	7.4	2461	1	C11H16N2O2S3	THIENO[2,3-B]THIOPHENE-2-SULFONAMIDE,5-T-BUTYLAMINOMETHYL **122266-91-5**
4	1.24	7.4	2461	1	C11H16N2O2S3	THIENO[3,2-B]THIOPHENE-2-SULFONAMIDE,5-I-BUTYLAMINOMETHYL **127025-26-5**
25	0.57		1439		C11H16N2O3	3-ETOCO-PYRIDO(1,2A)PYRIMIDIN-4-ON,1,4,6,7,8,9,10-H7
6	-0.82	7.4	2685	1	C11H16N2O3	4-PYRIDONE,3-HYDROXY-2-METHYL-1-(2-ETHYLAMINOCARBONYL)ETHYL *iron chelator*
7	-1.16	7.4	2685	1	C11H16N2O3	4-PYRIDONE,3-HYDROXY-2-METHYL-1-(N,N-DIMETHYLPROPIONAMIDO) *iron chelator*
8 ★	1.64	1.0	2202	342	C11H16N2O3	BARBITURIC ACID,5-T-BUTYL-5-ALLYL
9	1.35	7.4	899	1	C11H16N2O3	BARBITURIC ACID,ALLYL,S-BUTYL **115-44-6** *sedative*

	logP	pH	Ref.	Note	MF	Name / **CAS** / *activity*
7530 ★	1.47	7.4	899	2	"	BARBITURIC ACID,ALLYL,S-BUTYL
1	-0.57	7.0	474	1	C11H16N2O3	NIFENANOL **7413-36-7** *adrenergic (beta-receptor)*
2	0.03	7.4	1628	1	"	NIFENANOL
3 ★	1.28		474		"	NIFENANOL
4 ★	1.05		773	601	C11H16N2O3S1	4-SULFAMYLBENZAMIDE,N-BUTYL **59777-63-8**
35 ★	1.45		2072	2	C11H16N2O3S1	BENZENESULFONAMIDE,2-(I-BUTYROYLAMINO)-4-METHYL
6 ★	1.65		2072	2	C11H16N2O3S1	BENZENESULFONAMIDE,2-AMINO-4-METHYL-N-(I-BUTANOYL)
7 ★	0.23		1205		C11H16N2O3S1	N,N-DIMETHYL-4-(N-METHYL-N-ACETYL)BENZENESULFONAMIDE
8 ★	1.95		2352		C11H16N2O3S1	N2-T-BUTYL-N1-BENZENESULFONYLUREA **52102-41-7**
9 ★	0.99		2352		C11H16N2O3S1	PHENYLSULFONYLUREA,N,N-DIETHYL
7540	0.84	7.4	2525	1	C11H16N2O3S2	5-(I-BUTYLAMINOMETHYL)THIENO[2,3-B]FURAN-2-SULFONAMIDE
1	0.20	7.4	2461	1	C11H16N2O3S3	THIENO[2,3-B]THIOPHENE-2-CARBOXAMIDE,5-METHOXYPROPYLAMINOMETHYL **122289-58-1**
2	-0.68	7.4	2461	1	C11H16N2O3S3	THIENO[2,3-B]THIOPHENE-2-SULFONAMIDE,5-OMEGAHYDROXYBUTYLAMINOMETHYL **122266-92-6**
3 ★	1.74	8.0	1011	2	C11H16N2O4	2,5-DIMETHOXY-4-NITROAMPHETAMINE
4	1.76		2447		C11H16N2O4	4-PYRIDINECARBOXYLIC ACID,TETRAHYDRO-2,6-DIOXO,HEXYL ESTER
45	1.37	3.5	1267	1	C11H16N2O4	METHYL-2,2-DIALLYLMALONURATE
6 ★	0.43		547		C11H16N2O4	N-(TRIHYDROXYMETHYLMETHANE)-N'-PHENYLUREA
7	-2.21	7.2	846	314	C11H16N2O4S1	6-(N-PROPIONYLAMINO)PENICILLANIC ACID **33947-21-6**
8 ★	1.37		2072	2	C11H16N2O4S1	BENZENESULFONAMIDE,2-(I-BUTYROYLAMINO)-4-METHOXY
9 ★	1.58		2072	2	C11H16N2O4S1	BENZENESULFONAMIDE,2-AMINO-4-METHOXY-N-(I-BUTANOYL)
7550	-0.85	7.4	2525	1	C11H16N2O4S2	5-(4-HYDROXYBUTYLAMINO)METHYLTHIENO[2,3-B]FURAN-2-SULFONAMIDE
1	0.53	7.4	2525	1	C11H16N2O4S2	5-(ETHOXYETHYLAMINOMETHYL)THIENO[2,3-B]FURAN-2-SULFONAMIDE
2	-0.38	7.4	2525	1	C11H16N2O4S2	5-(METHOXYPROPYLAMINO)METHYLTHIENO[2,3-B]FURAN-2-SULFONAMIDE
3	0.23	7.4	2461	1	C11H16N2O4S3	THIENO[2,3-B]THIOPHENE-2-SULFONAMIDE,5-(N,N-2-HYDROXYETHYL)AMINOMETHYL **133445-76-8**
4 ✓	-2.02		2397	101	C11H16N3O1	IMIDAZOLIUM CHLORIDE,1-METHYL-2-HYDROXYIMINOMETHYL-3-HEXYN-5-YL
55 ✓	-2.27		2397	101	C11H16N3O2	IMIDAZOLIUM CHLORIDE,1-METHYL-2-HYDROXYIM,INOMETHYL-3-(3-PROPYN-2YLOXY)PROPYL
6 ✓	-1.62		2397	101	C11H16N3O2	IMIDAZOLIUM CHLORIDE,1-METHYL-2-HYDROXYIMINOMETHYL-3-(1,1-DIMETHYLPROPYN-2-YLOXY)METHYL
7 ✓	-2.10		2397	101	C11H16N3O2	IMIDAZOLIUM CHLORIDE,1-METHYL-2-HYDROXYIMINOMETHYL-3-(1-METHYLBUTYN-3-YLOXY)METHYL
8 ✓	-1.74		2397	101	C11H16N3O2	IMIDAZOLIUM CHLORIDE,1-METHYL-2-HYDROXYIMINOMETHYL-3-(1-METHYLBUTYN-3YLOXY)METHYL **117983-22-9**
9 ✓	-1.99		2397	101	C11H16N3O2	IMIDAZOLIUM CHLORIDE,1-METHYL-2-HYDROXYIMINOMETHYL-3-PENTYN-3-YLOXY)METHYL **117983-26-3**
7560 ✓	-2.10		2397	101	C11H16N3O2	IMIDAZOLIUM CHLORIDE,1-METHYL-2-HYDROXYIMINOMETHYL-3-PENTYN-4-YLOXYMETHYL **117983-30-9**
1 ✓	-0.92		2406	102	C11H16N3O2S1	TETRAMETHYLTHIOURONIUM IODIDE,S-(2-NITROPHENYL)
2 ✓	-0.88		2406	102	C11H16N3O2S1	TETRAMETHYLTHIOURONIUM IODIDE,S-(3-NITROPHENYL)
3 ★	2.95		2150	459	C11H16N4	ADAMANTANE,1-(2-TETRAZOLYL)
4 ★	2.95		2606	459	C11H16N4	ADAMANTANE,1-(TETRAZOL-1YL)
65 ★	2.26	6.6	2590	459	C11H16N4O1	1-(4-CARBAMOYLPHENYL)-3,3-DIETHYLTRIAZENE *antineoplastic*
6 ★	2.02	6.3	2589	459	C11H16N4O1	1-(4-CARBAMOYLPHENYL)-3-METHYL-3-I-PROPYLTRIAZENE **59708-23-5** *antineoplastic*
7 ★	2.17	6.6	2590	459	C11H16N4O1	1-(4-CARBAMOYLPHENYL)-3-METHYL-3-PROPYLTRIAZENE **89529-99-7** *antineoplastic*
8 ★	2.96	6.6	2590	459	C11H16N4O1	BENZAMIDE,2-(3,3-DIETHYLTRIAZENYL) *antineoplastic*
9 ★	0.11		2105		C11H16N4O2	HYDRAZINECARBOXAMIDE,1-ACETYL-N-(4-DIMETHYLAMINOPHENYL)
7570	0.37	7.4	2477	1	C11H16N4O3	A[(2-NITROIMIDAZOL-1-YL)METHYL]-6-AZABICYCLO[3.1.0]-6-ETHANOL **120277-93-2**
1 ★	1.15	1.0	2196		C11H16N4O3	PYRAZOLE,3-METHYL-4-NITRO-5-(N-HEXAMETHYLENECARBOXAMIDO)
2 ★	1.71	1.0	2196		C11H16N4O3	PYRAZOLE,3-METHYL-4-NITRO-5-(N-CYCLOHEXYLCARBOXAMIDO)
3	0.07	7.4	2477	1	C11H16N4O4	A[(2-NITROIMIDAZOL-1-YL)METHYL]-3-OXA-7-AZABICYCLO[4.1.0]HEPTANE-7-ETHANOL **120277-97-6**
4 ★	-2.09	7.4	594	1	C11H16N4O4	PENTOSTATIN **63677-95-2** *antineoplastic*
75	-1.74	7.4	401	1	C11H16N4O4	RAZOXANE **21416-87-5** *antineoplastic*
6	-0.29	7.4	2477	1	C11H16N4O5	METHYL 1-[2-HYDROXY-3-(2-NITROIMIDAZOL-1-YL)-PROPPYL]-2-AZETIDINECARBOXYLATE **134419-52-6**
7	2.15	7.4	2028	459	C11H16N6	1,2,4-TRIAZOLE,3-AMINO-5-(2-BUTYLAMINO)PYRIDIN-4YL **77314-79-5**
8 ★	-0.59	13.0	547	224	C11H16N6	4,6-DIAMINO-2,2-DIMETHYL-1-(4-AMINOPHENYL)-SYM-TRIAZINE
9 ★	1.25	5.2	2566	302	C11H16N6S2	2-ACETYLPYRIDINE-THIOCARBONOHYDRAZONE,N,N-DIMETHYL **96860-23-0** *inhib.ribonucleotide reductase*
7580	4.06		579		C11H16O1	P-PENTYLPHENOL **14938-35-3**
1	3.13		2316	337	C11H16O1	PHENOL,3-PENTYL
2	3.13		2316	337	C11H16O2	1,2-DIHYDROXYBENZENE,4-PENTYL
3 ★	3.63		1477		C11H16O2	M-PENTOXYPHENOL **18979-73-2**
4 ★	3.50		1940	459	C11H16O2	P-PENTOXYPHENOL
85 ★	2.90		2011		C11H16O4S2	PROPANEDIOIC ACID,1,3-DITHIOLAN-2-YLIDINE,ETHYL-I-PROPYL ESTER **50780-76-2** ✚
6 ★	1.05		1589		C11H16O5S1	ETHYL-(3,4,5-TRIMETHOXY)PHENYL-SULFONE
7 ★	0.18		1589		C11H16O6S1	(2-HYDROXYETHYL)-(3,4,5-TRIMETHOXY)PHENYL-SULFONE
8 ★	2.48		1564		C11H17Br1N2O3	PYRIMIDINE-2,4-DIONE,3-HEXYLOXY-5-BROMO-6-ME **77317-76-1**
9	0.43	7.4	2477	1	C11H17Br1N4O3	TR-A-[[(2-BROMOCYCLOPENTYL)AMINO]METHYL]-2-NITROIMIDAZOLE-1-ETHANOL **134419-54-8**
7590	-0.17	7.4	2477	1	C11H17Br1N4O4	TR-A-[[[(4-BROMOTETRAHYDRO-2H-PYRAN-3-YL)AMINO]-METHYL]2-NITROIMIDAZOLE-1-ETHANOL **134419-55-9**
1	-0.60	7.4	785	1	C11H17N1	4-PHENYL-N-METHYLBUTYLAMINE **4265-99-0**
2	-0.18	2.0	1473		"	5-PHENYLPENTANEAMINE HYDROCHLORIDE **17734-21-3**
3	-0.04	7.0	1473		"	5-PHENYLPENTANEAMINE HYDROCHLORIDE
4 ★	3.39	7.4	597	314	C11H17N1	ANILINE,4-PENTYL **33228-44-3**
95 ★	4.41	13.0	594	400	C11H17N1	ANILINE,N,N,2,4,6-PENTAMETHYLANILINE **13021-15-3**
6 ★	2.59		1508		C11H17N1	BENZENEETHANEAMINE,A-I-PROPYL **46114-16-3**
7 ★	2.79		1508		C11H17N1	BENZENEETHANEAMINE,A-PROPYL **63951-01-9**
8 ★	2.73		502		C11H17N1	N,N-DIMETHYL-N-G-PHENYLPROPYLAMINE **1199-99-1**
9	4.35		442	2	C11H17N1	P-HEXYLPYRIDINE **27876-24-0**
7600	-0.07	7.5	1588	1	C11H17N1	PHENETHYLAMINE,N-METHYL-A-ETHYL
1	2.82	7.4	1588	2	"	PHENETHYLAMINE,N-METHYL-A-ETHYL
2 ★	1.87	13.0	1806	224	C11H17N1O1	2-(METHYLAMINO)-1-METHOXY-1-PHENYLPROPANE(ERYTHRO)
3 ★	2.05	13.0	1806	224	C11H17N1O1	2-(METHYLAMINO)-1-METHOXY-1-PHENYLPROPANE(THREO)
4 ★	2.85	7.4	1616		C11H17N1O1	ANISOLE,O-BUTYLAMINO
5 ★	1.70	11.0	2648	448	C11H17N1O1	METHYLEPHEDRINE **552-79-4** *sympathomimetic*
6	2.68	11.0	2648	448	"	METHYLEPHEDRINE
7	0.35	7.4	2375	1	C11H17N1O1	MEXILETINE **31828-71-4** *cardiac depressant (anti-arrhythmic)*
8 ★	2.15	7.4	2375	2	"	MEXILETINE
9 ★	2.29		588		C11H17N1O1	P-DIETHYLAMINOBENZYL ALCOHOL
7610	1.49	7.0	538	701	C11H17N1O2	2,3-DIMETHOXYAMPHETAMINE **15402-81-0**
1 ★	1.75	7.0	538	701	C11H17N1O2	2,4-DIMETHOXYAMPHETAMINE **23690-13-3**
2 ★	1.72	8.0	1011	2	C11H17N1O2	2,5-DIMETHOXYAMPHETAMINE **2801-68-5**
3	1.88	7.0	538	701	"	2,5-DIMETHOXYAMPHETAMINE
4	1.00	7.0	538	701	C11H17N1O2	3,4-DIMETHOXYAMPHETAMINE **120-26-3**
15	1.20	8.0	1011	2	"	3,4-DIMETHOXYAMPHETAMINE
6	1.24	7.4	2685	1	C11H17N1O2	4-PYRIDONE,3-HYDROXY-2-METHYL-1-PENTYL *iron chelator*
7	1.22	7.4	2685	1	C11H17N1O2	4-PYRIDONE,3-HYDROXY-2-ETHYL-1-BUTYL *iron chelator*
8	1.81	8.0	1011	2	C11H17N1O2S1	2,5-DIMETHOXY-4-METHIOPHENETHYLAMINE
9	1.74	7.4	2535	1	C11H17N1O2S1	N,N,N-(G-THIONOYL)PROPYL-METHYL-ETHANOLAMINE

	logP	pH	Ref.	Note	MF	Name / CAS / activity
7620	0.12	7.4	2685	1	C11H17N1O3	4-PYRIDONE,3-HYDROXY-2-METHYL-1-(3-ETHOXYPROPYL) *iron chelator*
1	-2.69		2475	100	C11H17N1O3	ISOPROTERENOL 7683-59-2 *adrenergic, bronchodilator*
2	-1.88	7.4	1168	1		ISOPROTERENOL
3	0.78	8.0	868	2	C11H17N1O3	MESCALINE 54-04-6
4 ★	1.04	7.4	2490	1	C11H17N1O3S1	BENZENESULFONAMIDE,4-(5-HYDROXY)PENTYL 135832-50-7
25 ★	1.40	7.4	2490	1	C11H17N1O3S2	BENZENESULFONAMIDE,4-(3-HYDROXY-3-METHYLBUTYL)THIO 108966-53-6
6 ★	1.04	7.4	2490	1	C11H17N1O3S2	BENZENESULFONAMIDE,4-(5-HYDROXYPENTYL)THIO 108966-54-7
7 ★	-0.09	7.4	2490	1	C11H17N1O5S2	BENZENESULFONAMIDE,4-(3-HYDROXY-3-METHYLBUTYL)SULFONYL 108966-56-9
8 ✓	-0.88		2406	102	C11H17N2O1S1	TETRAMETHYLTHIOURONIUM IODIDE,S-(4-HYDROXYPHENYL)
9	-2.80		571		C11H17N2O1.Br1	M-CARBAMYLBENZYLTRIMETHYLAMMONIUMBROMIDE
7630 ✓	-0.68		2406	102	C11H17N2S1	TETRAMETHYLTHIURONIUM IODIDE,S-PHENYL
1 ★	1.59	12.5	2377	2	C11H17N3	GUANIDINE,4-PHENYLBUTYL
2 ★	0.83		2612		C11H17N3O2	1-(4-ETHYLAMINOPHENYL)-3-METHOXY-3-METHYLUREA 88132-21-2
3	-0.37	7.4	775	1	C11H17N3O2	3-(N,N-DIMECARBAMYLOXY)-2-N,N-DIMEAMINOMEPYRIDINE 51581-37-4
4	0.22		1439		C11H17N3O2	PYRIDO(1,2A)PYRIMIDIN-4-ONE,3-CONH2,-H7,1,6-DIME(AX)
35	0.43		1439		"	PYRIDO(1,2A)PYRIMIDIN-4-ONE,3-CONH2,-H7,1,6-DIME(AX)
6 ★	2.60	6.6	2590	459	C11H17N3O2S1	METHYLPHENYLSULFONE,4-(3,3-DIETHYLTRIAZENYL) *antineoplastic*
7 ★	1.01	7.4	2004	1	C11H17N3O3S1	CARBUTAMIDE 339-43-5 *hypoglycemic*
8	1.16		526		C11H17N5O1	ADENINE,9-(1-HYDROXYMETHYL-PENTYL)
9 ★	-0.87	9.0	1949		C11H17N5O2S1	CIMETIDINE,3-NITRO-4,5-DIHYDROPYRROL-YL ANALOG 74188-64-0 ✚
7640	-1.82		536		C11H17N5O6	5-(3,3-DIMETHYL-1-TRIAZENYL)URIDINE 38099-11-5
1 ★	4.16		1693		C11H17O3P1S2	FENSULFOTHION-SULFIDE 3070-15-3
2 ★	2.79		354		C11H17O4P1S1	O,O-DIETHYL-O-(4-METHYLTHIOPHENYL)PHOSPHATE 3070-13-1
3 ★	2.23		1693		C11H17O4P1S2	FENSULFOTHION 115-90-2 *insecticide*
4 ★	2.56		1693		C11H17O5P1S2	O,O-DIET-O-(4-METHYLSULFONYLPHENYL)PHOSPHOROTHIOATE 14255-72-2
45	0.00		336		C11H17O6P1S1	O,O-DIETHYL-O-(4-METHYLSULFONYLPHENYL)PHOSPHATE
6	1.89		547		C11H18Cl1N3O4	1(2'CLET)1-NO-3(3'CARBOMETHOXYCYCLOHEXYL)UREA
7 ★	1.98		547		C11H18Cl1N3O4	1-NO-1-(2-CLET)-3-(2-ACETYLOXYCYCLOHEXYL)UREA-CIS
8 ★	1.93		547		C11H18Cl1N3O4	1-NO-1-(2-CLET)-3-(2-ACETYLOXYCYCLOHEXYL)UREA-TR 59414-38-9
9 ★	1.89		547		C11H18Cl1N3O4	1-NO-1-(2-CLET)-3-(4-ACETYLOXYCYCLOHEXYL)UREA-CIS
7650	-0.63	7.4	401	1	C11H18Cl1N3O4	A-(3-2CLET,3-NO-UREIDO)CY-HEX-AC.ACID
1 ★	3.63		2094		C11H18Cl1N5	SIMAZINE,2-CYCLOHEXYLAMINE ANALOG 84712-77-6
2	2.99		2162	459	C11H18Cl1N5S1	MEZILAMINE 50335-55-2
3 ★	1.33		547		C11H18F1N3O4	1-NO-1-(2-FLET)-3-(4-ACETYLOXYCYCLOHEXYL)UREA-TR
4 ✓	-1.10		2397	101	C11H18F3N4O3S1	IMIDAZOLIUM CHLORIDE,1-ETHYL-2-HYDROXYIMINOMETHYL-3-(3-N-METHYL-TRIFLUOROMETHYLSULFONAMIDO)PROPYL
55 ✓	-0.77		2397	101	C11H18F3N4O3S1	IMIDAZOLIUM CHLORIDE,1-ETHYL-2-HYDROXYIMINOMETHYL-3-(N-ETHYL-TRIFLUOROMETHYLSULFONAMIDO)ETHYL
6 ✓	-0.69		2397	101	C11H18F3N4O3S1	IMIDAZOLIUM CHLORIDE,1-METHYL-2-HYDROXYIMINOMETHYL-3-(2-N-PROPYL)TRIFLUOROMETHYLSULFONAMIDO)ETHYL
7 ✓	-1.02		2397	101	C11H18F3N4O3S1	IMIDAZOLIUM CHLORIDE,1-METHYL-2-HYDROXYIMINOMETHYL-3-(3-(N-ETHYL)TRIFLUOROMETHYLSULFONAMIDO)PROPYL
8 ✓	-1.13		2397	101	C11H18F3N4O3S1	IMIDAZOLIUM CHLORIDE,1-METHYL-2-HYDROXYIMINOMETHYL-3-(5-TRIFLUOROMETHYLSULFONAMIDOPENTYL)
9 ✓	-1.63		2397	101	C11H18F3N4O4S1	IMIDAZOLIUM CHLORIDE,1-METHYL-2-HYDROXYIMINOMETHYL-L3-(N-METHYL-TRIFLUOROMETHYLSULFONAMIDOETHOXYETHYL
7660	-1.90		571		C11H18N1O1.Br1	M-METHOXYBENZYLTRIMETHYLAMMONIUMBROMIDE
1	-1.74		571		C11H18N1S1.Br1	M-METHYLTHIOBENZYLTRIMETHYLAMMONIUMBROMIDE
2	-3.38		527	820	C11H18N1.Br1	BENZYLDIMETHYLETHYLAMMONIUMBROMIDE 50328-48-8
3	-2.03		527	404	C11H18N1.Br1	HEXYLPYRIDINIUMBROMIDE
4	-1.99		571		C11H18N1.Br1	M-METHYLBENZYLTRIMETHYLAMMONIUMBROMIDE 21949-11-1
65	-2.99	7.4	1635	820	C11H18N1.Cl1	BENZYLDIMETHYLETHYLAMMONIUMCHLORIDE 5197-80-8
6 ✓	-2.22	7.4	921	22	C11H18N1.I1	DIETHYLMETHYLPHENYL AMMONIUM IODIDE 1007-67-6
7 ✓	-1.79		1729	820	C11H18N1.I1	N-HEXYLPYRIDINIUM IODIDE 46122-12-7
8 ★	1.49	11.0	533	322	C11H18N2	4-(N,N-DIMETHYL)-3-PYRIDYLBUTYLAMINE 1441-44-7
9	0.45	7.4	785	1	C11H18N2	N'-PHENYL-N,N-DIMETHYLPROPYLENEDIAMINE 13658-95-2
7670 ★	1.23	11.0	533	322	C11H18N2	N,N-DIETHYL-3-PYRIDYLETHYLAMINE 20173-34-6
1	1.92		1407		C11H18N2O1	1,2,3,6-H4-PYRIDINE,1-CYCLOPENTYLCARBONYLAMINO
2 ★	1.93		2544		C11H18N2O2	THYMINE,1-HEXYL
3 ★	2.98		505		C11H18N2O2S1	BARBITURIC ACID,5-ETHYL-5-I-AMYL-2-THIO 4388-79-8
4	2.36	7.0	1232	1	C11H18N2O2S1	THIOPENTAL 76-75-5 *anesthetic (intravenous), anticonvulsant*
75 ★	2.85		1135	537	"	THIOPENTAL
6 ★	2.36		1727	459	C11H18N2O3	5,5-DIETHYLBARBITURIC ACID,N1-I-PROPYL
7 ★	1.44		1727	459	C11H18N2O3	5,5-DIETHYLBARBITURIC ACID,O2-I-PROPYL
8 ★	1.86		1727	459	C11H18N2O3	5,5-DIETHYLBARBITURIC ACID,O4-I-PROPYL
9	1.78	7.4	899	1	C11H18N2O3	BARBITURIC ACID,5-AMYL-5-ETHYL 115-58-2
7680 ★	2.24		505		"	BARBITURIC ACID,5-AMYL-5-ETHYL
1	1.58	7.4	899	1	"	BARBITURIC ACID,5-ETHYL-5-I-AMYL 57-43-2 *sedative*
2	1.76	7.4	1873	1	"	BARBITURIC ACID,5-ETHYL-5-I-AMYL
3 ★	2.07		96		"	BARBITURIC ACID,5-ETHYL-5-I-AMYL
4	1.73	7.4	899	1	C11H18N2O3	PENTOBARBITAL 76-74-4 *sedative*
85	1.75	7.0	1232	1	"	PENTOBARBITAL
6 ★	2.07	1.0	2202	342	"	PENTOBARBITAL
7 ★	2.10	2.0	300		"	PENTOBARBITAL
8	2.18	7.4	551	1	"	PENTOBARBITAL
9 ★	1.70		594	400	C11H18N2O3S1	BENZENESULFONAMIDE,N,N-DIETHYL-3-AMINO-4-METHOXY
7690	0.35		505		C11H18N2O4	BARBITURIC ACID,5-ET-5-(3-HYDROXY-1-MEBUTYL) 4241-40-1
1	0.99	3.5	1267	1	C11H18N2O6	METHOXYCARBONYLMETHYL-2,2-DIETHYLMALONURATE
2 ✓	-1.54		2397	101	C11H18N3O2	IMIDAZOLIUM CHLORIDE,1-METHYL-2-HYDROXYIMINOMETHYL-3-(1,1-DIMETHYLPROPEN-2YLOXY)METHYL
3 ✓	-1.40		2397	101	C11H18N3O2	IMIDAZOLIUM CHLORIDE,1-METHYL-2-HYDROXYIMINOMETHYL-3-(3-METHYLBUTEN-2YLOXY)METHYL 117983-12-7
4 ✓	-1.41		2397	101	C11H18N3O2	IMIDAZOLIUM CHLORIDE,1-METHYL-2-HYDROXYIMINOMETHYL-3-(3-METHYLBUTEN-3YLOXY)METHYL 117983-17-2
95 ✓	-1.46		2397	101	C11H18N3O2	IMIDAZOLIUM CHLORIDE,1-METHYL-2-HYDROXYIMINOMETHYL-3-CYCLOPENTYLOXYMETHYL 117941-55-6
6 ✓	-1.30		2397	101	C11H18N3O2	IMIDAZOLIUM CHLORIDE,1-METHYL-2-HYDROXYIMINOMETHYL-3-PENTEN-2YL-OXYMETHYL 117983-09-2
7 ✓	-1.40		2397	101	"	IMIDAZOLIUM CHLORIDE,1-METHYL-2-HYDROXYIMINOMETHYL-3-PENTEN-2YL-OXYMETHYL 117983-10-5
8 ✓	-2.59		2397	101	C11H18N3O2	IMIDAZOLIUM CHLORIDE,1-T-BUTYL-2-HYDROXYIMINOMETHYL-3-ACETYLMETHYL
9 ✓	-1.21		2397	101	C11H18N3O3	IMIDAZOLIUM CHLORIDE,1-METHYL-2-HYDROXYIMINOMETHYL-3-I-BUTYRYLOXYMETHYL
7700 ✓	-1.36		2397	101	C11H18N3O3	IMIDAZOLIUM CHLORIDE,1-METHYL-2-HYDROXYIMINOMETHYL-3-PIVALOYLOXYMETHYL
1 ★	1.70		2506		C11H18N4O2	PIRIMACARB 23103-98-2 *aphicide*
2	-0.40	7.4	2491	1	C11H18N4O3	PIMONIDAZOLE 70132-50-2 *radiosensitizer*
3	0.93	7.4	1497	1	"	PIMONIDAZOLE
4 ★	1.00	7.4	2491	2	"	PIMONIDAZOLE
5 ★	2.46	1.0	2196		C11H18N4O3	PYRAZOLE,3-METHYL-4-NITRO-5-(N-HEXYLCARBOXAMIDO)
6 ★	1.22	7.4	2206	306	C11H18N4O4	(2-METHYL-5-NITROIMIDAZOL-2YL)ETHYLCARBAMIC ACID,BUTYL ESTER
7	0.47	7.4	1497	1	C11H18N4O4	2-NITROIMIDAZOLE,1-(2-OH-3-N-MORPHOLINO)PROPYL
8	-1.37		1559		C11H18N4O4	5(2-OH-3-(N-OXOPIPERIDINO)PROPYL)-2-NITRO-IMIDAZOLE
9	-0.96		1404		C11H18N4O5	2-NO2-IMIDAZOLE,1-CH2CON(CH2CHOHCH3)2

	logP	pH	Ref.	Note	MF	Name / CAS / activity
7710	-2.47	7.4	2037	1	C11H18N6O1	ISOGUANINE,1-METHYL-7-(3-DIMETHYLAMINO)PROPYL
1	-2.00	7.4	2037	1	C11H18N6O1	ISOGUANINE,1-METHYL-9-(3-DIMETHYLAMINO)PROPYL
2	-0.75		1045		C11H18N6O2	6(1,1-BIS(HOCH2)ETAMINO-2,4-DIAZIRIDINYLTRIAZINE
3	-1.76		1045		C11H18N6O3	6(TRI(HOCH2)ME-AMINO-2,4-DIAZIRIDINYLTRIAZINE
4 ★	3.76		1521		C11H18O2	(1R)T-CYPROPANE-CO2H-2,2-DIME-3-(2-MEPROPENYL),ME ESTER 24141-52-4
15 ✓	-0.94		2397	101	C11H19Br1N3O2	IMIDAZOLIUM CHLORIDE,1-ETHYL-2-HYDROXYIMINOMETHYL-3-(1-METHYL-3-BROMOPROPOXY)METHYL
6 ✓	-0.81		2397	101	C11H19Br1N3O2	IMIDAZOLIUM CHLORIDE,1-METHYL-2-HYDROXYIMINOMETHYL-3-(2,2-DIMETHYL-3-BROMOPROPOXYMETHYL
7 ✓	-1.09		2397	101	C11H19Cl1N3O1	IMIDAZOLIUM CHLORIDE,1-METHYL-2-HYDROXYIMINOMETHYL-3-(6-CHLOROHEXYL)
8 ✓	-1.00		2397	101	C11H19Cl1N3O2	IMIDAZOLIUM CHLORIDE,1-METHYL-2-HYDROXYIMINOMETHYL-3-((2,2-DIMETHYL-3-CHLOROPROPOXY)METHYL
9 ✓	-1.12		2397	101	C11H19Cl1N3O2	IMIDAZOLIUM CHLORIDE,1-METHYL-2-HYDROXYIMINOMETHYL-3-(1-ETHYL3-CHLOROPROPOXY)METHYL
7720 ✓	-1.23		2397	101	C11H19Cl1N3O2	IMIDAZOLIUM CHLORIDE,1-METHYL-2-HYDROXYIMINOMETHYL-3-(1-METHYL-4-CHLOROBUTOXYL)METHYL
1 ★	4.71		2488		C11H19F2N5S1	1,3,5-TRIAZINE,2-DIFLUOROMETHIO-4-AMYLAMINO-6-ETHYLAMINO
2 ★	4.75		2488		C11H19F2N5S1	1,3,5-TRIAZINE,2-DIFLUOROMETHIO-4-I-PROPYLAMINO-6-BUTYLAMINO
3 ★	4.70		2488		C11H19F2N5S1	1,3,5-TRIAZINE,2-DIFLUOROMETHIO-4-I-PROPYLAMINO-6-T-BUTYLAMINO
4 ★	3.01		1422	462	C11H19N1	1-AMINOADAMANTANE,3-METHYL 33103-93-4
25 ★	3.04	5.0	2425	1	C11H19N1O2	2-PYRIDONE,1-HEPTANOYL 100400-78-0
6	0.89		213		C11H19N1O2	N-HEXANOYLCYCLOBUTANECARBOXAMIDE 25031-81-6
7 ★	1.90		2506		C11H19N3O1	DIMETHIRIMOL 5221-53-4 fungicide
8	-2.97	7.4	2397	101	C11H19N4O1	IMIDAZOLIUM CHLORIDE,1-METHYL-2-HYDROXYIMINOMETHYL-3-(1-METHYLPYRROLIDIN-2YL)METHYL
9 ✓	-2.73	7.4	2397	101	C11H19N4O1	IMIDAZOLIUM CHLORIDE,1-METHYL-2-HYDROXYIMINOMETHYL-3-(1-PYRROLIDINYL)ETHYL 132540-22-8
7730 ✓	-2.91		2397	101	C11H19N4O2	IMIDAZOLIUM CHLORIDE,1-METHYL-2-HYDROXYIMINOMETHYL-3-(1-MORPHOLINYL)ETHYL 132566-80-4
1 ✓	-2.18		2397	101	C11H19N4O3	IMIDAZOLIUM CHLORIDE,1-METHYL-2-HYDROXYIMINOMETHYL-3-(4-METHYL-4-NITROPENTYL) 132566-39-3
2 ✓	-1.77		2397	101	C11H19N4O4	IMIDAZOLIUM CHLORIDE,1-ETHYL-2-HYDROXYIMINOMETHYL-3-(2-NITROPROPOXY)METHYL 132566-48-4
3	-2.50		2397	101	C11H19N4O4	IMIDAZOLIUM CHLORIDE,1-METHYL-2-CARBOXAMIDO-3-(1-METHYL-2-NITROPROPOXY)METHYL
4 ✓	-1.75		2397	101	C11H19N4O4	IMIDAZOLIUM CHLORIDE,1-METHYL-2-HYDROXYIMINOMETHYL-3-(1,2-DIMETHYL-2-NITROPROPOXY)METHYL
35 ✓	-2.11		2397	101	C11H19N4O4	IMIDAZOLIUM CHLORIDE,1-METHYL-2-HYDROXYIMINOMETHYL-3-(1-METHYL-4-NITROBUTOXY)METHYL
6 ✓	-0.11		2397	101	C11H19N4O4	IMIDAZOLIUM CHLORIDE,1-METHYL-2-HYDROXYIMINOMETHYL-3-(A-ETHYL-2-NITROPROPOXY)METHYL
7 ✓	-1.71		2397	101	C11H19N4O4	IMIDAZOLIUM CHLORIDE,1-METHYL-2-HYDROXYIMINOMONMETHYL-3-(1-METHYL-2-NITROBUTOXY)METHYL
8	0.45	7.4	579	324	C11H19N5O2	TETRAHYDROPTERIDINE,2-AMINO-4-HYDROXY-6-I-BUTOXYMETHYL
9 ★	2.57		547		C11H20Cl1N3O2	1-NITROSO-1-(2-CHLOROETHYL)-3-(1,6-DIMETHYLCYCLOHEXYL)UREA
7740 ★	1.67		2238		C11H20Cl1N4O3	NITROSOUREA,1-(2-CHLOROETHYL)-3-(1-OXO-2,2,5,5-TETRAMETHYLPYRROLIDIN-3-YL)
1 ★	-1.79		2377		C11H20N2O3	ILE-PRO
2 ★	-1.76		2377		C11H20N2O3	LEU-PRO
3 ★	-2.56		2377		C11H20N2O3	PRO-ILE
4 ★	-2.41		2377		C11H20N2O3	PRO-LEU
45	1.99	3.5	1267	1	C11H20N2O4	I-PROPYL-2,2-DIETHYLMALONURATE
6 ★	-0.35	7.4	2490	1	C11H20N2O4S3	THIOPHENE-2-SULFONAMIDE,5-(3-I-BUTYLAMINOPROPYL)SULFONYL 135832-40-5
7 ✓	-0.72	7.4	2397	101	C11H20N3O1	IMIDAZOLIUM CHLORIDE,1-METHYL-2-HYDROXYIMINOMETHYL-3-HEXYL
8 ✓	-1.12		2397	101	C11H20N3O2	IMIDAZOLIUM CHLORIDE,1-METHYL-2-HYDROXYIMINOMETHYL-3-(1-METHYLBUTOXY)METHYL 117941-50-1
9 ✓	-0.91		2397	101	C11H20N3O2	IMIDAZOLIUM CHLORIDE,1-METHYL-2-HYDROXYIMINOMETHYL-3-(2,2-DIMETHYLPROPOXY)METHYL 91900-11-7
7750 ✓	-1.08		2397	101	C11H20N3O2	IMIDAZOLIUM CHLORIDE,1-METHYL-2-HYDROXYIMINOMETHYL-3-I-PENTOXYMETHYL 117941-49-8
1	-0.91	7.6	2003	314	C11H20N3O2.Cl1	IMIDAZOLIUM CHLORIDE,1-(2,2-DIMETHYLPROPOXY)METHYL-2-HYDROXYIMINOMETHYL-3-METHYL 91900-11-7
2 ✓	-2.22		2397	101	C11H20N3O3	IMIDAZOLIUM CHLORIDE,1-METHYL-2-HYDROXYIMINOMETHYL-3-(3-METHOXYBUTOXY)METHYL 132540-01-3
3 ★	4.20		1693		C11H20N3O3P1S1	PIRIMIPHOS-METHYL 29232-93-7
4 ✓	-3.11		2397	101	C11H20N3O4S1	IMIDAZOLIUM CHLORIDE,1-METHYL-2-HYDROXYIMINOMETHYL-3-(4-METHYLSULFONYLBUTYOXY)METHYL
55 ★	-0.08	9.0	2027	1	C11H20N4O3	IMIDAZOLE,1-(3-DIETHYLAMINO-2-HYDROXYPROPYL)-2-METHYL-5-NITRO 74550-94-0
6 ★	2.00		1629		C11H20N4O3S2	N-(N"DIET-N'-CYCLOPR)-URYLTHIO)METHOMYL ✚
7 ★	-1.85		2377		C11H20N4O4	AC-ASN-VAL-N
8	-2.47	7.4	2037		C11H20N6O1	2H-PURIN-2-ONE,1,3-DIHYDRO-1-METHYL-6-AMINO-7-DIMETHYLAMINOPROPYL
9	1.14	4.2	401	2	C11H20N6O2	IMIDAZOLE,5-CARBOXAMIDE-4-(3-T-BUTYL-3-(2-METHOXY)ETHYLTRIAZENYL)
7760	3.56		1231		C11H20O2	5,7-UNDECANEDIONE 1942-48-9
1 ★	3.86		1135	537	C11H20O2	UNDECYLENIC ACID 112-38-9 antifungal
2 ★	1.70	2.2	2228	314	C11H21N1O3	OCTANOIC ACID-2-ACETYLAMINO,METHYL ESTER
3 ★	4.11		2506		C11H21N1S1	CYCLOATE 1134-23-2 herbicide
4	0.30	7.0	2484	1	C11H21N2O1	2,5-DIHYDROPYRROLE,1-OXY-2,2,5,5-TETRAHYDRO-3-DIMETHYLAMINOMETHYL
65	0.70	12.0	2484	1	"	2,5-DIHYDROPYRROLE,1-OXY-2,2,5,5-TETRAHYDRO-3-DIMETHYLAMINOMETHYL
6 ★	0.52	7.4	1342	324	C11H21N2O2	4-ACETAMIDO-2,2,6,6-TETRAMETHYLPIPERIDINE-N-OXIDE 14691-89-5
7 ★	-0.54		2377		C11H21N3O3	AC-ALA-LEU-N
8	1.05		2541		C11H21N3O3	GLY-VAL-ETHYLAMIDE,N1-ACETYL
9 ★	2.40		1629		C11H21N3O3S2	N-(N"-BUTYL-MECARBAMYLTHIO)METHOMYL
7770 ★	-1.25		2377		C11H21N3O4	AC-THR-VAL-N
1 ★	2.50		1629		C11H21N3O4S2	N-(N"-BUTYL-METHOXYCARBAMYLTHIO)METHOMYL
2 ★	2.80		1629		C11H21N3O4S2	N-(N"-METHYL-T-BUTOXYCARBAMYLTHIO)METHOMYL
3	-2.57	7.4	2397	101	C11H21N4O1	IMIDAZOLIUM CHLORIDE,1-METHYL-2-HYDROXYIMINOMETHYL-3-(2-DIMETHYLAMINO)BUTYL 132566-75-7
4	-2.55	7.4	2397	101	C11H21N4O1	IMIDAZOLIUM CHLORIDE,1-METHYL-2-HYDROXYIMINOMETHYL-3-DIETHYLAMINOETHYL
75 ✓	-2.25		2397	101	C11H21N4O3S1	IMIDAZOLIUM CHLORIDE,1-METHYLSULFONYLETHYL-2-HYDROXYIMINOMETHYL-3-DIMTHYLAMINOETHYL
6 ★	3.90		2506		C11H21N5S1	DIMETHAMETRYN 22936-75-0 herbicide
7	-0.01		2358	22	(C11H22N1O5S2)2.Cd1	N-BUTYL-N-DITHIOCARBOXY-D-GLUCAMINE
8 ✓	0.21		2358	200	"	N-BUTYL-N-DITHIOCARBOXY-D-GLUCAMINE
9 ★	-2.05	7.0	2374		C11H22N2O3	LEUCINYLVALINE
7780 ★	-2.07	7.0	2197		C11H22N2O3	VALINYLLEUCINE 3989-97-7
1 ★	-1.87	7.0	2374		C11H22N2O3S1	LEUCINYLMETHIONINE
2 ★	-1.84	7.0	2374		C11H22N2O3S1	METHIONYLLEUCINE
3 ✓	-0.34		2397	101	C11H22N3O2Si1	IMIDAZOLIUM CHLORIDE,1-METHYL-2-HYDROXYIMINOMETHYL-3-(2-TRIMETHYLSILYLETHOXY)METHYL
4	-3.18	7.4	2397	101	C11H22N4O2	IMIDAZOLIUM CHLORIDE,1-METHYL-2-HYDROXYIMINOMETHYL-3-(TRIMETHYLAMMONIOETHOXYMETHYL
85 ★	3.20		1629		C11H22N4O3S2	N-(N"DIET-N'-ME)-URYLTHIO)METHOMYL
6	3.45		964		C11H22N6	N6,N6-DIETHYL-N2,N2,N4,N4-TETRAMETHYLMELAMINE 16268-75-0
7	1.60		964		C11H22N6O2	N2,N2-(2-HOET)-N4,N4,N6,N6-TETRAMETHYLMELAMINE
8 ★	4.09		2047		C11H22O1	2-UNDECANONE 112-12-9
9	4.41		1560	100	C11H22O2	METHYLDECANOATE 110-42-9
7790	0.39		1294		C11H22O5S1	1-PENTYLTHIO-B-GALACTOPYRANOSIDE
1	-0.18		1294		C11H22O5S1	1-THIO-3-PENTYLGALACTOPYRANOSIDE
2	-0.80		1294		C11H22O6	B-3-PENTYLGALACTOPYRANOSIDE
3	-0.42		1294		C11H22O6	B-PENTYLGALACTOPYRANOSIDE
4 ★	3.70	13.0	1806	224	C11H23N1	CYCLOOCTYLETHANEAMINE,A-METHYL
95 ★	3.60		1969		C11H23N1O2	DIETHYLCARBAMATE,2-ETHYLBUTYL ESTER 98155-00-1
6 ★	4.15		2506		C11H23N1S1	BUTYLATE 2008-41-5 herbicide
7	-1.30	7.0	2484	1	C11H23N2O1	1-OXYPIPERIDINE,2,2,6,6-TETRAMETHYL-4-DIMETHYLAMINO
8	0.72	12.0	2484	1	"	1-OXYPIPERIDINE,2,2,6,6-TETRAMETHYL-4-DIMETHYLAMINO
9 ✓	2.12		567	820	C11H24N1.C6H2N3O7	CYCLOHEXYLPROPYLDIMETHYL AMMONIUM PICRATE

	logP	pH	Ref.	Note	MF	Name / **CAS** / *activity*
7800 ★	1.68		536		C11H24O2Sn1	TRIPROPYLTINACETATE **3267-78-5**
1	-1.40		1394		C11H25N3O1	4-DECYLSEMICARBAZIDE
2	3.74		2182	459	C11H25O3P1	METHYLPHOSPHONIC ACID,DIPENTYL ESTER **1000-36-8**
3	-1.21		2173		C11H26N1.I1	TRIMETHYL-OCTYL-AMMONIUM IODIDE **14251-76-4**
4	-1.07		524	820	"	TRIMETHYL-OCTYL-AMMONIUM IODIDE
5	-2.19		524	820	C11H26N1.I1	TRIPROPYL-ETHYL-AMMONIUM IODIDE **15066-80-5**
6 ✓	2.05		567	820	C11H26N1.C6H2N3O7	ETHYLTRIPROPYL AMMONIUM PICRATE
7 ✓	2.22		567	820	C11H26N1.C6H2N3O7	METHYL-I-BUTYLDIPROPYL AMMONIUM PICRATE
8	1.81	7.0	1970	1	C11H26N2O1S2.Te1	1-(HYDROXYMETHYL)-N,N'-BIS(2-METHYL-2-MERCAPTOPROPYL)ETHYLENEDIAMINE,TECHNICIUM COMPLEX
9	4.24		512		C11H26Si1	SILANE,OCTYL-TRIMETHYL **3429-76-3**
7810	1.67	7.0	1970	1	C11H27N3S2.Tc1	N,N'-BIS(2-METHYL-2-MERCAPTOPROPYL)-3-AMINOPROPANE,1,2-DIAMINE,TECHNICIUM COMPLEX **96929-44-1**
1	8.58		2146	458	C12Br10	DECABROMOBIPHENYL **13654-09-6**
2	7.97		2146	458	C12Cl8O1	DIBENZOFURAN-OCTACHLORO **39001-02-0**
3	7.59		2146	458	C12Cl8O2	DIBENZODIOXIN-OCTACHLORO **3268-87-9**
4	8.60		2008	459	"	DIBENZODIOXIN-OCTACHLORO
15	11.82		1808	459	"	DIBENZODIOXIN-OCTACHLORO
6 ★	8.27		2293	400	C12Cl10	2,2',3,3',4,4',5,5',6,6'-PCB **2051-24-3**
7 ★	7.10		2343	20	C12F23N1	PERFLUORO-P-METHYLCYCLOHEXYLPIPERIDINE
8 ★	7.92		2395	401	C12H1Cl7O1	DIBENZOFURAN,1,2,3,4,6,7,8-HEPTACHLORO
9	11.03		1808	459	C12H1Cl7O2	DIBENZODIOXIN-1,2,3,4,6,7,8-HEPTACHLORO **58200-70-7**
7820	8.09		2149	467	C12H1Cl9	2,2',3,3',4,4',5,5',6-NONACHLOROBIPHENYL **40186-72-9**
1	7.52		2149	458	C12H1Cl9	2,2',3,3',4,4',5,6,6'-NONACHLOROBIPHENYL **52663-79-3**
2	7.74		2149	467	C12H1Cl9	2,2',3,3',4,4',5,6,6'-NONACHLOROBIPHENYL
3 ★	8.16		1697	458	C12H1Cl9	2,2',3,3',4,5,5',6,6'-PCB **52663-77-1**
4	7.79		2008	459	C12H2Cl6O2	DIBENZO-P-DIOXIN,1,2,3,4,7,8-HEXACHLORO **39227-28-6**
25	10.22		1808	459	"	DIBENZO-P-DIOXIN,1,2,3,4,7,8-HEXACHLORO
6	7.67		2149	458	C12H2Cl8	2,2',3,3',4,4',5,5'-OCTACHLORO-BIPHENYL **35694-08-7**
7	7.80		2149	467	C12H2Cl8	2,2',3,3',4,4',5,5'-OCTACHLORO-BIPHENYL
8	9.35		1529	459	C12H2Cl8	2,2',3,3',4,4',5,5'-OCTACHLORO-BIPHENYL
9	7.65		2149	467	C12H2Cl8	2,2',3,3',4,4',5,6'-OCTACHLOROBIPHENYL **42740-50-1**
7830	7.56		2149	467	C12H2Cl8	2,2',3,3',4,4',5,6-OCTACHLORO-BIPHENYL **52663-78-2**
1	9.25		1529	459	C12H2Cl8	2,2',3,3',4,4',5,6-OCTACHLORO-BIPHENYL
2	7.30		2149	467	C12H2Cl8	2,2',3,3',4,4',6,6'-OCTACHLOROBIPHENYL **33091-17-7**
3	7.21		2149	458	C12H2Cl8	2,2',3,3',4,5',6,6'-OCTACHLOROBIPHENYL **40186-71-8**
4	7.27		2149	467	C12H2Cl8	2,2',3,3',4,5',6,6'-OCTACHLOROBIPHENYL
35	7.62		2149	467	C12H2Cl8	2,2',3,3',4,5,5',6-OCTACHLOROBIPHENYL **52663-75-9**
6	7.62		2149	467	C12H2Cl8	2,2',3,3',4,5,5',6-OCTACHLOROBIPHENYL **68194-17-2**
7	7.20		2149	467	C12H2Cl8	2,2',3,3',4,5,6,6'-OCTACHLOROBIPHENYL **52663-73-7**
8 ★	7.73		2293	400	C12H2Cl8	2,2',3,5',6,6'-PCB **2136-99-4**
9	7.65		2149	467	C12H2Cl8	2,2',3,4,4',5,5',6-OCTACHLOROBIPHENYL **52663-76-0**
7840	7.30		2149	467	C12H2Cl8	2,2',3,4,4',5,6,6'-OCTACHLOROBIPHENYL **74472-52-9**
1	8.00		2149	467	C12H2Cl8	2,3,3',4,4',5,5',6-OCTACHLOROBIPHENYL **74472-53-0**
2 ★	6.26		2395	401	C12H3Cl5O1	DIBENZOFURAN,1,2,3,6,7-PENTACHLORO **57117-42-7**
3 ★	6.27		2395	401	C12H3Cl5O1	DIBENZOFURAN,1,2,4,6,7-PENTACHLORO **83704-50-1**
4 ★	6.34		2395	401	C12H3Cl5O1	DIBENZOFURAN,1,2,4,6,8-PENTACHLORO **69698-57-3**
45 ★	6.26		2395	401	C12H3Cl5O1	DIBENZOFURAN,1,2,4,7,8-PENTACHLORO **58802-15-6**
6 ★	6.24		2395	401	C12H3Cl5O1	DIBENZOFURAN,1,3,4,6,8-PENTACHLORO **83704-55-6**
7 ★	6.34		2395	401	C12H3Cl5O1	DIBENZOFURAN,1,3,4,6,9-PENTACHLORO **70648-15-6**
8 ★	6.51		2395	401	C12H3Cl5O1	DIBENZOFURAN,1,3,4,8,9-PENTACHLORO
9 ★	6.47		2395	401	C12H3Cl5O1	DIBENZOFURAN,2,3,4,6,7-PENTACHLORO **57117-43-8**
7850 ★	6.92		2395	401	C12H3Cl5O1	DIBENZOFURAN,2,3,4,7,8-PENTACHLORO **57117-31-4**
1 ★	6.42		2395	401	C12H3Cl5O1	DIBENZOFURAN,2,3,4,8,9-PENTACHLORO
2 ★	6.30		2395	401	C12H3Cl5O2	DIBENZODIOXIN,1,2,3,4,6-PENTACHLORO **67028-19-7**
3 ★	6.74		2395	401	C12H3Cl5O2	DIBENZODIOXIN,1,2,3,6,7-PENTACHLORO **71925-15-0**
4 ★	6.53		2395	401	C12H3Cl5O2	DIBENZODIOXIN,1,2,3,6,8-PENTACHLORO **71925-16-1**
55 ★	6.24		2395	401	C12H3Cl5O2	DIBENZODIOXIN,1,2,3,6,9-PENTACHLORO **82291-34-7**
6 ★	6.64		2395	401	C12H3Cl5O2	DIBENZODIOXIN,1,2,3,7,8-PENTACHLORO **40321-76-4**
7 ★	6.40		2395	401	C12H3Cl5O2	DIBENZODIOXIN,1,2,3,7,9-PENTACHLORO **71925-17-2**
8 ★	6.20		2395	1	C12H3Cl5O2	DIBENZODIOXIN,1,2,4,7,8-PENTACHLORO **58802-08-7**
9 ★	6.20		2395	401	C12H3Cl5O2	DIBENZODIOXIN,1,2,4,7,8-PENTACHLO
7860	7.27		2149	467	C12H3Cl7	2,2',3,3',4,4',5-HEPTACHLOROBIPHENYL **35065-30-6**
1	6.68		1697	458	C12H3Cl7	2,2',3,3',4,4',6-PCB **52663-71-5**
2	7.11		2149	467	C12H3Cl7	2,2',3,3',4,4',6-PCB
3	7.08		2149	467	C12H3Cl7	2,2',3,3',4,5',6'-HEPTACHLOROBIPHENYL **52663-70-4**
4	7.17		2149	467	C12H3Cl7	2,2',3,3',4,5',6-HEPTACHLOROBIPHENYL **40186-70-7**
65	7.33		2149	467	C12H3Cl7	2,2',3,4,5,5'-HEPTACHLOROBIPHENYL **52663-74-8**
6	7.11		2149	467	C12H3Cl7	2,2',3,3',4,5,6-HEPTACHLOROBIPHENYL **38411-25-5**
7	7.02		2149	467	C12H3Cl7	2,2',3,3',4,5,6-HEPTACHLOROBIPHENYL **68194-16-1**
8	6.76		2149	467	C12H3Cl7	2,2',3,3',4,6,6'-HEPTACHLOROBIPHENYL **52663-65-7**
9	7.14		2149	467	C12H3Cl7	2,2',3,3',5,5'-HEPTACHLOROBIPHENYL **52663-67-9**
7870	6.73		2149	467	C12H3Cl7	2,2',3,3',5,6,6'-PCB **52663-64-6**
1	7.17		2149	467	C12H3Cl7	2,2',3,4',5,5',6-HEPTACHLOROBIPHENYL **52663-68-0**
2	6.82		2149	467	C12H3Cl7	2,2',3,4,5,6,6'-HEPTACHLOROBIPHENYL **74487-85-7**
3	7.20		2149	467	C12H3Cl7	2,2',3,4,4',5,6-HEPTACHLOROBIPHENYL **52663-69-1**
4	7.36		2149	467	C12H3Cl7	2,2',3,4,4',5,5'-HEPTACHLOROBIPHENYL **35065-29-3**
75	7.20		2149	467	C12H3Cl7	2,2',3,4,4',5,6'-HEPTACHLOROBIPHENYL **60145-23-5**
6	7.11		2149	467	C12H3Cl7	2,2',3,4,4',5,6-HEPTACHLOROBIPHENYL **74472-47-2**
7	6.85		2149	467	C12H3Cl7	2,2',3,4,4',6,6'-HEPTACHLOROBIPHENYL **74472-48-3**
8	7.11		2149	467	C12H3Cl7	2,2',3,4,5,5',6-HEPTACHLOROBIPHENYL **52712-05-7**
9	6.69		2149	467	C12H3Cl7	2,2',3,4,5,6,6'-HEPTACHLOROBIPHENYL **74472-49-4**
7880	7.52		2149	467	C12H3Cl7	2,3,3',4,4',5',6-HEPTACHLOROBIPHENYL **69782-91-8**
1	7.55		2149	467	C12H3Cl7	2,3,3',4,4',5',6-HEPTACHLOROBIPHENYL **74472-50-7**
2	7.71		2149	467	C12H3Cl7	2,3,3',4,4',5,5'-HEPTACHLOROBIPHENYL **39635-31-9**
3	7.46		2149	467	C12H3Cl7	2,3,3',4,4',5,6-HEPTACHLOROBIPHENYL **41411-64-7**
4	7.52		2149	467	C12H3Cl7	2,3,3',4,5,5',6-HEPTACHLOROBIPHENYL **74472-51-8**
85	6.14		2008	459	C12H4Cl4	BIPHENYLENE,2,3,6,7-TETRACHLORO **7090-41-7**
6 ★	6.23		2395	401	C12H4Cl4O1	DIBENZOFURAN,1,2,7,8-TETRACHLORO **58802-20-3**
7 ★	6.23		2395	401	C12H4Cl4O1	DIBENZOFURAN,1,3,4,7-TETRACHLORO **70648-16-7**
8 ★	6.13		2395	401	C12H4Cl4O1	DIBENZOFURAN,1,3,4,8-TETRACHLORO **92341-04-3**
9 ★	5.89		2395	401	C12H4Cl4O1	DIBENZOFURAN,1,3,4,9-TETRACHLORO **83704-28-3**

	logP	pH	Ref.	Note	MF	Name / **CAS** / *activity*
7890 ★	6.37		2395	401	C12H4Cl4O1	DIBENZOFURAN,1,3,6,8-TETRACHLORO **71998-72-6**
1 ★	5.60		2395	401	C12H4Cl4O1	DIBENZOFURAN,1,4,6,9-TETRACHLORO **70648-19-0**
2 ★	6.11		2395	401	C12H4Cl4O1	DIBENZOFURAN,2,3,4,6-TETRACHLORO **83704-30-7**
3 ★	6.31		2395	401	C12H4Cl4O1	DIBENZOFURAN,2,3,6,7-TETRACHLORO **57117-39-2**
4 ★	6.73		2395	401	C12H4Cl4O1	DIBENZOFURAN,2,3,6,8-TETRACHLORO **57117-37-0**
95 ★	6.53		2395	401	C12H4Cl4O1	DIBENZOFURAN,2,3,7,8-TETRACHLORO **51207-31-9**
6 ★	6.25		2395	401	C12H4Cl4O1	DIBENZOFURAN,2,4,6,7-TETRACHLORO **57117-38-1**
7 ★	6.39		2395	401	C12H4Cl4O2	DIBENZODIOXIN,1,2,3,9-TETRACHLORO **71669-26-6**
8 ★	6.43		2395	401	C12H4Cl4O2	DIBENZODIOXIN,1,2,6,8-TETRACHLORO **67323-56-2**
9 ★	6.30		2395	401	C12H4Cl4O2	DIBENZODIOXIN,1,3,7,8-TETRACHLORO **50585-46-1**
7900 ★	6.39		2395	401	C12H4Cl4O2	DIBENZODIOXIN,1,4,7,8-TETRACHLORO **40581-94-0**
1 ★	7.18		2008	459	C12H4Cl4O2	DIBENZODIOXIN,1,2,3,4-TETRACHLORO **30746-58-8**
2 ★	8.22		1808	459	C12H4Cl4O2	DIBENZODIOXIN,1,2,3,7-TETRACHLORO **67028-18-6**
3 ★	6.29		2395	401	C12H4Cl4O2	DIBENZODIOXIN,1,3,6,8-TETRACHLORO **33423-92-6**
4 ★	6.39		2395	401	C12H4Cl4O2	DIBENZODIOXIN,1,3,7,9-TETRACHLORO **62470-53-5**
5 ★	6.42		2395	401	C12H4Cl4O2	DIBENZODIOXIN,2,3,7,8-TETRACHLORO **1746-01-6**
6 ★	7.32		2293	400	C12H4Cl6	2,2',3,3',4,4'-PCB **38380-07-3**
7 ★	7.39		1661	459	C12H4Cl6	2,2',3,3',4,5'-PCB **52663-66-8**
8 ★	7.32		1661	459	C12H4Cl6	2,2',3,3',4,5HEXACHLOROBIPHENYL **55215-18-4**
9 ★	7.04		2536	472	C12H4Cl6	2,2',3,3',4,6'-HEXACHLORO-BIPHENYL **38380-05-1**
7910 ★	6.82		2536	472	C12H4Cl6	2,2',3,3',4,6-HEXACHLOROBIPHENYL **61798-70-7**
1 ★	7.07		2536	472	C12H4Cl6	2,2',3,3',5,5'-HEXACHLOROBIPHENYL **35694-04-3**
2 ★	7.15		1661	459	C12H4Cl6	2,2',3,3',5,6'-HEXACHLORO-BIPHENYL **52744-13-5**
3 ★	6.81		2536	472	C12H4Cl6	2,2',3,3',5,6-HEXACHLOROBIPHENYL **52704-70-8**
4 ★	7.12		2293	400	C12H4Cl6	2,2',3,3',6,6'-PCB **38411-22-2**
15 ★	6.86		2536	472	C12H4Cl6	2,2',3,4',5',6-HEXACHLORO-BIPHENYL **38380-04-0**
6 ★	7.12		2536	472	C12H4Cl6	2,2',3,4',5,5'-HEXACHLOROBIPHENYL **51908-16-8**
7 ★	6.87		2536	472	C12H4Cl6	2,2',3,4',5,6'-HEXACHLOROBIPHENYL **74472-41-6**
8 ★	6.93		2536	472	C12H4Cl6	2,2',3,4',5,6-HEXACHLOROBIPHENYL **68194-13-8**
9 ★	6.75		2536	472	C12H4Cl6	2,2',3,4',6,6'-HEXACHLOROBIPHENYL **68194-08-1**
7920 ★	7.25		2536	472	C12H4Cl6	2,2',3,4,4',5'-HEXACHLORO-BIPHENYL **35065-28-2**
1 ★	7.43		2536	472	C12H4Cl6	2,2',3,4,4',5-PCB **35694-06-5**
2 ★	6.97		2536	472	C12H4Cl6	2,2',3,4,4',6'-HEXACHLOROBIPHENYL **59291-64-4**
3 ★	6.95		2536	472	C12H4Cl6	2,2',3,4,4',6-HEXACHLOROBIPHENYL **56030-56-9**
4 ★	6.79		2536	472	C12H4Cl6	2,2',3,4,5',6-HEXACHLOROBIPHENYL **68194-14-9**
25 ★	7.19		2536	472	C12H4Cl6	2,2',3,4,5,5'-HEXACHLOROBIPHENYL **52712-04-6**
6 ★	6.92		2536	472	C12H4Cl6	2,2',3,4,5,6'-HEXACHLOROBIPHENYL **68194-15-0**
7 ★	6.97		2536	472	C12H4Cl6	2,2',3,4,5,6-HEXACHLOROBIPHENYL **41411-61-4**
8 ★	6.71		2536	472	C12H4Cl6	2,2',3,4,6,6'-HEXACHLOROBIPHENYL **74472-40-5**
9 ★	6.85		2536	472	C12H4Cl6	2,2',3,5,5',6-HEXACHLOROBIPHENYL **52663-63-5**
7930 ★	6.51		2536	472	C12H4Cl6	2,2',3,5,6,6'-HEXACHLOROBIPHENYL **68194-09-2**
1 ★	6.89		2536	472	C12H4Cl6	2,2',4,4',5,6'-HEXACHLOROBIPHENYL **60145-22-4**
2 ★	7.29		2293	400	C12H4Cl6	2,2',4,4',6,6'-PCB **33979-03-2**
3 ★	7.17		2536	472	C12H4Cl6	2,3',4,4',5',6-HEXACHLOROBIPHENYL **59291-65-5**
4 ★	7.50		2536	472	C12H4Cl6	2,3',4,4',5,5'-HEXACHLOROBIPHENYL **52663-72-6**
35 ★	7.20		2536	472	C12H4Cl6	2,3,3',4',5',6-HEXACHLOROBIPHENYL **74472-45-0**
6 ★	7.47		2536	472	C12H4Cl6	2,3,3',4',5,5'-HEXACHLOROBIPHENYL **39635-34-2**
7 ★	7.25		2536	472	C12H4Cl6	2,3,3',4',5,6-HEXACHLOROBIPHENYL **74472-44-9**
8 ★	7.60		2536	472	C12H4Cl6	2,3,3',4,4',5'-HEXACHLOROBIPHENYL **69782-90-7**
9 ★	7.57		2536	472	C12H4Cl6	2,3,3',4,4',5-HEXACHLOROBIPHENYL **38380-08-4**
7940 ★	7.25		2536	472	C12H4Cl6	2,3,3',4,4',6-HEXACHLOROBIPHENYL **74472-42-7**
1 ★	7.10		2536	472	C12H4Cl6	2,3,3',4,5',6-HEXACHLOROBIPHENYL **74472-43-8**
2 ★	7.43		2536	472	C12H4Cl6	2,3,3',4,5,5'-HEXACHLOROBIPHENYL **39635-35-3**
3 ★	7.30		2536	472	C12H4Cl6	2,3,3',4,5,6-HEXACHLOROBIPHENYL **41411-62-5**
4 ★	7.37		1661	459	C12H4Cl6	2,3,3',5,5',6-PCB **74472-46-1**
45 ★	7.31		2536	472	C12H4Cl6	2,3,4,4',5,6-HEXACHLOROBIPHENYL **41411-63-6**
6 ★	7.16		2536	472	C12H4Cl6	2,4,5,2',4',5'-PCB **35065-27-1**
7 ★	7.41		2293	400	C12H4Cl6	3,3',4,4',5,5'-HEXACHLOROBIPHENYL **32774-16-6**
8 ★	7.10		2146	458	C12H5Br5	2,4,5,2',5'-PENTABROMOBIPHENYL **67888-96-4**
9 ★	7.47		1808	459	C12H5Cl3O2	DIBENZODIOXIN-1,2,4-TRICHLORO **39227-58-2**
7950	5.75		2008	459	C12H5Cl4N1	9H-CARBAZOLE,1,3,6,8-TETRACHLORO **58910-96-6**
1 ★	7.12		1661	459	C12H5Cl5	2',3,4,4',5'-PENTACHLORO-BIPHENYL **31508-00-6**
2	6.74		2149	467	C12H5Cl5	2',3,4,4',5-PENTACHLOROBIPHENYL **65510-44-3**
3 ★	6.67		1661	459	C12H5Cl5	2,2',3',4,5-PCB **41464-51-1**
4	6.20		2149	467	C12H5Cl5	2,2',3,3',4-PENTACHLROPBIPHENYL **52663-62-4**
55	6.26		2149	467	C12H5Cl5	2,2',3,3',5-PENTACHLOROBIPHENYL **60145-20-2**
6 ★	6.04		1661	459	C12H5Cl5	2,2',3,3'6-PCB **52663-60-2**
7	6.36		2149	467	C12H5Cl5	2,2',3,4',5-PENTACHLOROBIPHENYL **68194-07-0**
8	6.13		2149	467	C12H5Cl5	2,2',3,4',6-PENTACHLOROBIPHENYL **60233-25-2**
9 ★	6.31		1661	459	C12H5Cl5	2,2',3,4',6-PCB **68194-05-8**
7960 ★	6.61		1661	459	C12H5Cl5	2,2',3,4,4'-PCB **65510-45-4**
1 ★	6.37		1661	459	C12H5Cl5	2,2',3,4,5'-PENTACHLOROBIPHENYL **38380-02-8**
2	6.23		2149	467	C12H5Cl5	2,2',3,4,5-PENTACHLOROBIPHENYL **55312-69-1**
3	6.07		2149	467	C12H5Cl5	2,2',3,4,6'-PENTACHLOROBIPHENYL **73575-57-2**
4	6.07		2149	467	C12H5Cl5	2,2',3,4,6-PENTACHLOROBIPHENYL **55215-17-3**
65 ★	6.55		1661	459	C12H5Cl5	2,2',3,5',6-PENTACHLORO-BIPHENYL **38379-99-6**
6 ★	6.97		1661	459	C12H5Cl5	2,2',3,5,5'-PCB **52663-61-3**
7	6.13		2149	467	C12H5Cl5	2,2',3,5,6' PENTACHLOROBIPHENYL **73575-55-0**
8	6.04		2149	467	C12H5Cl5	2,2',3,5,6-PENTACHLOROBIPHENYL **73575-56-1**
9	5.71		2149	467	C12H5Cl5	2,2',3,6,6'-PENTACHLOROBIPHENYL **73575-54-9**
7970 ★	7.21		1661	459	C12H5Cl5	2,2',4,4',5-PCB **38380-01-7**
1	6.23		2149	467	C12H5Cl5	2,2',4,4',6-PENTACHLOROBIPHENYL **39485-83-1**
2	6.22		2149	467	C12H5Cl5	2,2',4,5',6-PENTACHLOROBIPHENYL **60145-21-3**
3	6.16		2149	467	C12H5Cl5	2,2',4,5,6'-PENTACHLOROBIPHENYL **68194-06-9**
4 ★	5.81		2149	467	C12H5Cl5	2,2',4,6,6'-PENTACHLOROBIPHENYL **56558-16-8**

	logP	pH	Ref.	Note	MF	Name / CAS / activity
75	6.51		2149	467	C12H5Cl5	2,3',4',5',6-PENTACHLOROBIPHENYL 74472-39-2
6	6.73		2149	467	C12H5Cl5	2,3',4,5,5'-PENTACHLOROBIPHENYL 70424-70-3
7	6.58		2149	467	C12H5Cl5	2,3',4,4',6-PENTACHLOROBIPHENYL 56558-17-9
8	6.64		2149	467	C12H5Cl5	2,3',4,5',6-PENTACHLOROBIPHENYL 56558-18-0
9	6.79		2149	467	C12H5Cl5	2,3',4,5,5'-PCB 68194-12-7
7980	6.64		2149	467	C12H5Cl5	2,3,3',4',5'-PENTACHLOROBIPHENYL 76842-07-4
1	6.71		2149	467	C12H5Cl5	2,3,3',4',5-PENTACHLOROBIPHENYL 70424-68-9
2	6.48		2149	467	C12H5Cl5	2,3,3',4',6-PENTACHLOROBIPHENYL 38380-03-9
3 ★	6.65		2149	467	C12H5Cl5	2,3,3',4,4'-PENTACHLOROBIPHENYL 32598-14-4
4	6.71		2149	467	C12H5Cl5	2,3,3',4,5'-PENTACHLOROBIPHENYL 70362-41-3
85	6.64		2149	467	C12H5Cl5	2,3,3',4,5-PENTACHLOROBIPHENYL 70424-69-0
6	6.48		2149	467	C12H5Cl5	2,3,3',4,6-PENTACHLOROBIPHENYL 74472-35-8
7	6.54		2149	467	C12H5Cl5	2,3,3',5',6-PENTACHLOROBIPHENYL 68194-10-5
8	6.76		2149	467	C12H5Cl5	2,3,3',5,5'-PENTACHLOROBIPHENYL 39635-32-0
9	6.45		2149	467	C12H5Cl5	2,3,3',5,6-PENTACHLOROBIPHENYL 74472-36-9
7990	6.46		2149	467	C12H5Cl5	2,3,4',5,6-PENTACHLOROBIPHENYL 68194-11-6
1	6.65		2149	467	C12H5Cl5	2,3,4,4',5-PENTACHLOROBIPHENYL 74472-37-0
2	6.49		2149	467	C12H5Cl5	2,3,4,4',6-PENTACHLOROBIPHENYL 74472-38-1
3 ★	6.74		2293	400	C12H5Cl5	2,3,4,5,6PENTACHLOROBIPHENYL 18259-05-7
4 ★	6.50		2146	458	C12H5Cl5	2,4,5,2',5'-PCB 37680-73-2
95	6.89		2149	467	C12H5Cl5	3,3',4,4',5-PENTACHLOROBIPHENYL 57465-28-8
6	6.95		2149	467	C12H5Cl5	3,3',4,5,5'-PENTACHLOROBIPHENYL 39635-33-1
7	7.41		1068	459	C12H6Br4	3,5,3',5'TETRABROMOBIPHENYL 16400-50-3
8 ★	5.65		1808	459	C12H6Cl2O1	DIBENZOFURAN-2,8-DICHLORO 5409-83-6
9 ★	6.38		1808	459	C12H6Cl2O2	DIBENZODIOXIN-2,7-DICHLORO 33857-26-0
8000 ★	4.70		2300		C12H6Cl3N1O3	CHLORNITROFEN 1836-77-7
1	5.76		2149	467	C12H6Cl4	2,2',3,4'-TETRACHLOROBIPHENYL 36559-22-5
2 ★	6.11		1661	459	C12H6Cl4	2,2',3,4TETRACHLOROBIPHENYL 52663-59-9
3 ★	5.81		1661	459	C12H6Cl4	2,2',3,5'-TETRACHLORO-BIPHENYL 41464-39-5
4	5.75		2149	467	C12H6Cl4	2,2',3,5-TETRACHLOROBIPHENYL 70362-46-8
5	5.53		2149	467	C12H6Cl4	2,2',3,6'-TETRACHLOROBIPHENYL 41464-47-5
6	5.53		2149	467	C12H6Cl4	2,2',3,6-TETRACHLOROBIPHENYL 70362-45-7
7 ★	6.36		2293	400	C12H6Cl4	2,2',4,5'-PCB 41464-40-8
8	5.78		2149	467	C12H6Cl4	2,2',4,5-TETRACHLOROBIPHENYL 70362-47-9
9	5.63		2149	467	C12H6Cl4	2,2',4,6'-TETRACHLOROBIPHENYL 68194-04-7
8010	5.63		2149	467	C12H6Cl4	2,2',4,6-TETRACHLOROBIPHENYL 62796-65-0
1 ★	6.09		1661	459	C12H6Cl4	2,2',5,5'-TETRACHLORO-BIPHENYL 35693-99-3
2 ★	5.62		2149	467	C12H6Cl4	2,2',5,6'-TETRACHLOROBIPHENYL
3	6.13		2149	467	C12H6Cl4	2,3',4',5'-TETRACHLOROBIPHENYL 70362-48-0
4 ★	6.23		1661	459	C12H6Cl4	2,3',4',5-TETRACHLORO-BIPHENYL 32598-11-1
15	5.98		2149	467	C12H6Cl4	2,3',4',6-TETRACHLOROBIPHENYL 41464-46-4
6 ★	6.31		1661	459	C12H6Cl4	2,3',4,4'-TETRACHLORO-BIPHENYL 32598-10-0
7	6.26		2149	467	C12H6Cl4	2,3',4,5'-TETRACHLOROBIPHENYL 73575-52-7
8	6.20		2149	467	C12H6Cl4	2,3',4,5-TETRACHLOROBIPHENYL 73575-53-8
9	6.04		2149	467	C12H6Cl4	2,3',4,6-TETRACHLOROBIPHENYL 60233-24-1
8020	6.04		2149	467	C12H6Cl4	2,3',5',6-TETRACHLOROBIPHENYL 74338-23-1
1	6.26		2149	467	C12H6Cl4	2,3',5,5'-TETRACHLOROBIPHENYL 41464-42-0
2 ★	6.18		2293	400	C12H6Cl4	2,3,2',3'-PCB 38444-93-8
3	6.11		2149	467	C12H6Cl4	2,3,3',4'-TETRACHLOROBIPHENYL 41464-43-1
4	6.11		2149	467	C12H6Cl4	2,3,3',4-TETRACHLOROBIPHENYL 74338-24-2
25 ★	6.17		2149	467	C12H6Cl4	2,3,3',5'-TETRACHLOROBIPHENYL 41464-49-7
6	6.17		2149	467	C12H6Cl4	2,3,3',5-TETRACHLOROBIPHENYL 70424-67-8
7	5.95		2149	467	C12H6Cl4	2,3,3',6-TETRACHLOROBIPHENYL 74472-33-6
8	6.17		2149	467	C12H6Cl4	2,3,4',5-TETRACHLOROBIPHENYL 74472-34-7
9	5.95		2149	467	C12H6Cl4	2,3,4',6-TETRACHLOROBIPHENYL 52663-58-8
8030	5.84		1661	459	C12H6Cl4	2,3,4,4'-PCB 33025-41-1
1	6.11		2149	467	C12H6Cl4	2,3,4,4'-PCB
2 ★	6.41		2293	400	C12H6Cl4	2,3,4,5-TETRACHLOROBIPHENYL 33284-53-6
3	5.89		2149	467	C12H6Cl4	2,3,4,6-TETRACHLOROBIPHENYL 54230-22-7
4 ★	5.86		2149	467	C12H6Cl4	2,3,5,6PCB 33284-54-7
35 ★	6.29		1661	459	C12H6Cl4	2,4,2',4'-PCB 2437-79-8
6 ★	6.67		1661	459	C12H6Cl4	2,4,4',5-PCB 32690-93-0
7	6.05		2149	467	C12H6Cl4	2,4,4',6-TETRACHLOROBIPHENYL 32598-12-2
8 ★	5.94		2293	400	C12H6Cl4	2,6,2',6'-PCB 15968-05-5
9	6.42		2149	467	C12H6Cl4	3,3',4,5'-TETRACHLOROBIPHENYL 41464-48-6
8040	6.35		2149	467	C12H6Cl4	3,3',4,5-TETRACHLOROBIPHENYL 70362-49-1
1 ★	6.63		2293	400	C12H6Cl4	3,4,3',4'-PCB 32598-13-3
2	6.36		2149	467	C12H6Cl4	3,4,4',5-TETRACHLOROBIPHENYL 70362-50-4
3	6.48		2149	467	C12H6Cl4	3,5,3',5'PCB 33284-52-5
4	6.85		1068	459	C12H6Cl4	3,5,3',5'PCB
45	4.13		1962		C12H6Cl4O3S1	BITHIONOLOXIDE 844-26-8 anti-infective, topical
6 ★	1.95		590		C12H6O2	ACENAPHTHENEQUINONE 82-86-0
7	6.41		1068	459	C12H7Br3	3,5,4'TRIBROMOBIPHENYL
8 ★	5.05		1808	459	C12H7Cl1O2	DIBENZODIOXIN-1-CHLORO 39227-53-7
9 ★	5.45		1808	459	C12H7Cl1O2	DIBENZODIOXIN-2-CHLORO 39227-54-8
8050	4.64		2675	459	C12H7Cl2N1O3	NITROFEN 1836-75-5 herbicide
1 ★	5.31		1661	459	C12H7Cl3	2,2',3'-TRICHLORO-BIPHENYL 38444-78-9
2 ★	5.76		1661	459	C12H7Cl3	2,2',4-PCB 37680-66-3
3 ★	5.60		1662	458	C12H7Cl3	2,2',5-PCB 37680-65-2
4 ★	5.48		1661	459	C12H7Cl3	2,2',6-PCB 38444-73-4
55 ★	5.87		2293	400	C12H7Cl3	2,3',4'-TRICHLOROBIPHENYL 38444-86-9
6	5.67		2149	467	C12H7Cl3	2,3',4-TRICHLOROBIPHENYL 55712-37-3
7	5.66		2149	467	C12H7Cl3	2,3',5'-TRICHLOROBIPHENYL 37680-68-5
8 ★	5.76		1661	459	C12H7Cl3	2,3',5-PCB 38444-81-4
9	5.44		2149	467	C12H7Cl3	2,3',6-TRICHLOROBIPHENYL 38444-76-7
8060 ★	5.57		1661	459	C12H7Cl3	2,3,3'TRICHLOROBIPHENYL 38444-84-7
1 ★	5.42		1661	459	C12H7Cl3	2,3,4'PCB 38444-85-8
2 ★	5.86		2293	400	C12H7Cl3	2,3,4-TRICHLOROBIPHENYL 55702-46-0
3	5.57		2149	467	C12H7Cl3	2,3,5-TRICHLOROBIPHENYL 55720-44-0
4 ★	5.67		1661	459	C12H7Cl3	2,3,6-PCB 55702-45-9

	logP	pH	Ref.	Note	MF	Name / CAS / activity
65 ★	5.79		1662	458	C12H7Cl3	2,4',5-PCB 16606-02-3
6 ★	5.75		1661	459	C12H7Cl3	2,4',6-PCB 38444-77-8
7 ★	5.62		1578		C12H7Cl3	2,4,4'PCB 7012-37-5
8 ★	5.90		2293	400	C12H7Cl3	2,4,5-PCB 15862-07-4
9 ★	5.71		2293	400	C12H7Cl3	2,4,6-PCB 35693-92-6
8070	5.82		2149	467	C12H7Cl3	3,3',4-TRICHLOROBIPHENYL 37680-69-6
1	5.88		2149	467	C12H7Cl3	3,3'5-PCB 38444-87-0
2	5.89		2149	467	C12H7Cl3	3,4',5-TRICHLOROBIPHENYL 38444-88-1
3	5.83		2149	467	C12H7Cl3	3,4,4'-PCB 38444-90-5
4	5.76		2149	467	C12H7Cl3	3,4,5-TRICHLOROBIPHENYL 53555-66-1
75	5.31		331		C12H7F4N3O3	4,5-(-OCF2CF2O-)C6H3NHN=C(CN)CO-ME 36865-60-8
6	4.86		331		C12H7F4N3O4	4,5-(-OCF2CF2O-)C6H3NHN=C(CN)COOME 36865-53-9
7 ★	5.02		331		C12H7F6N3O2	3,5-DI-CF3-C6H3NHN=C(CN)COOME-CIS 36865-54-0
8 ★	5.16		331		C12H7F6N3O2	3,5-DI-CF3-C6H3NHN=C(CN)COOME-TRANS 36905-00-7
9	2.81		331		C12H7N5O2	2,5-CN-C6H3NHN=C(CN)COOME 36874-72-3
8080	5.78		1068	459	C12H8Br2	3,5DIBROMOBIPHENYL
1 ★	5.72		2146	458	C12H8Br2	4,4'-DIBROMOBIPHENYL 92-86-4
2	2.39		1353		C12H8Cl1F3N2O2	4-CL-5-METHOXY-2-(3-CF3-PHENYL)3-PYRIDAZINONE
3	2.61		544		C12H8Cl1N1O1	4-(4-CHLOROBENZOYL)PYRIDINE 6318-51-0
4	0.71		1571	702	C12H8Cl1N3O3	2'-HYDROXY-5'-CHLORO-4-NITROAZOBENZENE 24624-05-3
85 ★	4.97		2293	400	C12H8Cl2	2,2'-PCB 13029-08-8
6 ★	5.02		1661	459	C12H8Cl2	2,3'PCB 25569-80-6
7	4.97		2149	467	C12H8Cl2	2,3-DICHLOROBIPHENYL 16605-91-7
8 ★	5.10		1578		C12H8Cl2	2,4'-DICHLORO-BIPHENYL 34883-43-7
9 ★	5.30		1661	459	C12H8Cl2	2,4-PCB 33284-50-3
8090 ★	5.16		1697	458	C12H8Cl2	2,5-PCB 34883-39-1
1 ★	4.98		2293	400	C12H8Cl2	2,6-PCB 33146-45-1
2	5.28		2149	467	C12H8Cl2	3,3'-DICHLOROBIPHENYL 2050-67-1
3	5.29		2149	467	C12H8Cl2	3,4'-DICHLOROBIPHENYL 2974-90-5
4 ★	5.29		2146	458	C12H8Cl2	3,4-DICHLOROBIPHENYL 2974-92-7
95	5.28		2149	467	C12H8Cl2	3,5-DICHLOROBIPHENYL 34883-41-5
6 ★	5.58		928		C12H8Cl2	4,4'-PCB 2050-68-2
7 ★	6.50		2293	400	C12H8Cl6	ALDRIN 309-00-2 insecticide
8 ★	5.20		2293	400	C12H8Cl6O1	ENDRIN 72-20-8 insecticide
9 ★	1.78		567		C12H8N2	1,10-PHENANTHROLINE 66-71-7
8100 ★	2.51		567		C12H8N2	1,7-PHENANTHROLINE 230-46-6
1 ★	2.05		567		C12H8N2	4,7-PHENANTHROLINE 230-07-9
2 ★	2.40		599	60	C12H8N2	5,6-DIAZAPHENANTHRENE 230-17-1
3 ★	2.84		505		C12H8N2	PHENAZINE 92-82-0
4 ★	2.24		594		C12H8N2O1	BNEZO(C)CINNOLINE,N-OXIDE
5	2.00		1725	459	C12H8N2O1	PYRIDO-PYRIMIDINE ANALOG (RINGS:666)
6 ★	1.76		538		C12H8N2O3	4-(4-NITROBENZOYL)-PYRIDINE 27693-38-5
7 ★	3.37		594		C12H8N2O4S2	2-NITROPHENYL DISULFIDE 1155-00-6
8 ★	4.06		594		C12H8N2O4S2	3-NITROPHENYL DISULFIDE 537-91-7 coccidiostat
9	2.22	7.4	594	1	C12H8N4O7	5-NITRO-2-FURANACRYLIC-N-(5-NTIRO-2-FURFURYLIDINE) HYDRAZIDE 37962-27-9 ✦
8110 ★	2.79	7.4	594	1	"	5-NITRO-2-FURANACRYLIC-N-(5-NTIRO-2-FURFURYLIDINE) HYDRAZIDE
1 ★	4.12		505		C12H8O1	DIBENZOFURAN 132-64-9
2 ★	4.54		594		C12H8O1S1	PHENOXATHIIN 262-20-4
3 ★	4.38		1808	459	C12H8O2	DIBENZO-1,4-DIOXIN 262-12-4
4	2.38		1566		C12H8O3	ANGELICIN,4'-METHYL
15 ★	2.32		1991		C12H8O3	ANGELICIN,4-METHYL 6457-92-7
6 ★	2.55		1991		C12H8O3	ANGELICIN,5'-METHYL
7 ★	2.41		1991		C12H8O3	ANGELICIN,5-METHYL 73459-03-7
8 ★	1.98	1.0	594		C12H8O4	1,8-NAPHTHALENEDICARBOXYLIC ACID 518-05-8
9 ★	4.38		1742		C12H8S1	DIBENZOTHIOPHENE 132-65-0 keratolytic
8120	5.15		2040	459	C12H8S2	2,2'-BITHIOPHINE,5-(BUT-1-YN-3-ENYL)
1 ★	4.57		594		C12H8S2	THIANTHRENE 92-85-3
2	5.57		2040	459	C12H8S3	2,2',5',2"-TERTHIOPHENE 1081-34-1
3 ★	4.59		2146	458	C12H9Br1	2-BROMOBIPHENYL 2052-07-5
4	4.85		2146	458	C12H9Br1	3-BROMOBIPHENYL 2113-57-7
25	4.95		2146	458	C12H9Br1	4-BROMOBIPHENYL 92-66-0
6 ★	4.53		2293	400	C12H9Cl1	2-CHLOROBIPHENYL 2051-60-7
7 ★	4.58		1662	458	C12H9Cl1	3-CHLOROBIPHENYL 2051-61-8
8 ★	4.61		1663		C12H9Cl1	P-CHLOROBIPHENYL 2051-62-9
9 ★	2.30		1353		C12H9Cl1F3N3O1	NORFLURAZON 27314-13-2 herbicide
8130 ★	4.45		331		C12H9Cl1F3N3O2	2-CL-5-CF3-C6H3NHN=C(CN)COOET 28313-52-2
1 ★	4.86		331		C12H9Cl1F3N3O2	3-CF3,4-CLC6H3NHN=C(CN)COOET 28313-57-7
2 ★	2.52		2489		C12H9Cl1N2O1	NICOTINANILIDE,4'-CHLORO
3	4.04		2675	459	C12H9Cl1N2O3	ACLONIFEN
4	4.08		1060		C12H9Cl1O1	P-MONOCHORODIPHENYLOXIDE
35 ★	3.12		2467		C12H9Cl2N1O2S1	ANILINE,P-(2,4-DICHLOROPHENYL)SULFONYL
6 ★	3.10		2300	20	C12H9Cl2N1O3	VINCLOZOLIN 50471-44-8 fungicide
7 ★	2.64		1881	606	C12H9Cl2O4P1	PHOSPHORIC ACID,BIS(P-CHLOROPHENYL) ESTER 4795-31-7
8 ★	1.42	4.0	2271	302	C12H9F1N2O4	5-FLUOROURACIL-3-BENZYLOXYCARBONYL
9	0.48	7.4	2266	1	C12H9F1N2O4	URACIL,1-BENZYLOXYCARBONYL-5-FLUORO 66999-98-2
8140 ★	1.18	4.0	2010		"	URACIL,1-BENZYLOXYCARBONYL-5-FLUORO
1 ★	1.11	4.0	2269	302	C12H9F1N2O5	5-FLUOROURACIL,3-PHENOXYCARBONYLOXYMETHYL
2 ★	3.72		536		C12H9N1	CARBAZOLE 86-74-8
3	1.90	7.2	2317	1	C12H9N1O1	4-BENZOYLPYRIDINE 14548-46-0
4 ★	1.98		538		"	4-BENZOYLPYRIDINE
45 ★	3.85		547		C12H9N1O1	PHENOXAZINE 135-67-1
6 ★	1.88	8.0	603	1	C12H9N1O1	PHENYL-A-PYRIDYL KETONE 91-02-1
7 ★	1.88	8.0	603	1	C12H9N1O1	PHENYL-B-PYRIDYL KETONE 5424-19-1
8 ★	2.13		538		C12H9N1O2	4-(2-HYDROXYBENZOYL)-PYRIDINE 22526-29-0
9 ★	1.37		544		C12H9N1O2	4-(4-HYDROXYBENZOYL)PYRIDINE 51246-77-6

	logP	pH	Ref.	Note	MF	Name / CAS / activity
8150 ★	3.85	7.4	593	1	C12H9N1O2	ACENAPHTHENE,5-NITRO
1 ★	3.87		2159		C12H9N1O2	BIPHENYL,3-NITRO 2113-58-8
2 ★	3.82		579		C12H9N1O2	BIPHENYL,4-NITRO 92-93-3
3 ★	1.71		2649		C12H9N1O2	NAPHTHOQUINONE,2-(N-AZIRIDINYL)
4 ★	1.29		520		C12H9N1O3	1,4-NAPHTHOQUINONE,2-ACETAMIDO
55 ★	3.83		594		C12H9N1O3	DIPHENYL ETHER,4-NITRO 620-88-2
6 ★	3.71	1.5	2329	314	C12H9N1O3	PHENOL,2-NITRO-4-PHENYL 885-82-5
7	-0.16	7.4	1255	1	C12H9N1O4	7-ACETYL-4-HYDROXYQUINOLINE-3-CARBOXYLIC ACID
8	0.05	7.2	2446	1	C12H9N1O6	MILOXACIN 37065-29-5 antibacterial
9 ★	4.42		850		C12H9N1S1	1-NAPHTHYLMETHYLISOTHIOCYANATE 17112-82-2
8160 ★	4.42		716		C12H9N1S1	B-NAPHTHYLMETHYLISOTHIOCYANATE 19495-05-7
1 ★	4.15		508		C12H9N1S1	PHENOTHIAZINE 92-84-2 anthelmintic~insecticde
2 ★	2.05	6.0	2560	1	C12H9N3O2S1	THIABENDAZOLE,1-METHOXYCARBONYL
3	1.06		1571	702	C12H9N3O3	4'-HYDROXY-4-NITROAZOBENZENE 1435-60-5
4 ★	1.82		2489		C12H9N3O3	NICOTINANILIDE,4'-NITRO
65	0.95		494		C12H9N3O5S1	2-ACETAMIDO-5-(5-NO2-FURYLPROPENILIDENE)THIAZOLONE 52717-67-6
6 ★	1.84		2467	448	C12H9N3O6S1	ANILINE,P-(2,4-DINITROPHENYL)SULFONYL 75333-79-8
7 ★	0.27		2070		C12H9N5O3	ALLOPURINOL-1-(3-NICOTINOYL)OXYMETHYL
8	-0.61	2.7	562	340	C12H9O2.Na1	2-PHENYLETHYNYLCYCLOPROPANECARBOXYLATE-T 56892-91-2
9 ✓	-0.67	2.7	541	340	C12H9O3S1.Na1	P-BIPHENYLSULFONIC ACID,SODIUM SALT 2217-82-5
8170 ★	3.92		1179		C12H10	ACENAPHTHENE 83-32-9
1 ★	4.01		2293	400	C12H10	BIPHENYL 92-52-4 fungicide
2 ★	2.85		2467		C12H10Br1N1O2S1	ANILINE,P-(4-BROMOPHENYL)SULFONYL 6626-22-8
3 ★	2.30	5.0	2153	2	C12H10Cl1N1O2	2-(4-CHLOROPHENYL)-2-(1-PYRROLYL)ACETIC ACID
4 ★	2.57		2467		C12H10Cl1N1O2S1	ANILINE,P-(4-CHLOROPHENYL)SULFONYL 7146-68-1
75	3.91		2040	459	C12H10Cl1O1	3,11-TRIDECADIENE-5,7,9-TRIYNOL,2-CHLORO
6 ★	3.36		2506		C12H10Cl2F3N1O1	FLUROCHLORIDONE 61213-25-0 herbicide (carotenoid synth.inhib.)
7 ★	3.51		1179		C12H10Cl2N2	3,3'-DICHLOROBENZIDINE 91-94-1
8 ★	2.49	5.0	2153	2	C12H10F1O2	2-(2-FLUOROPHENYL)-2-(1-PYRROLYL)ACETIC ACID
9 ★	2.17		2467	448	C12H10F1O2S1	ANILINE,P-(4-FLUOROPHENYL)SULFONYL 312-35-6
8180 ★	2.11		571		C12H10N2	1,2-DI-(A-PYRIDYL)ETHYLENE 1437-15-6
1 ★	3.82		1179		C12H10N2	AZOBENZENE 103-33-3 acaricide (ticks & mites)
2	3.10	7.4	1941	1	C12H10N2	HARMANE 486-84-0
3	3.50	13.0	2326		"	HARMANE
4 ★	3.13		1179		C12H10N2O1	DIPHENYLNITROSAMINE 86-30-6
85 ★	2.18	7.4	1941	1	C12H10N2O1	HARMOL 525-57-5
6	2.19	7.4	772	1	"	HARMOL
7 ★	1.73		548		C12H10N2O1	NICOTINANILIDE 1752-96-1
8 ★	2.73	7.2	2317	1	C12H10N2O2	4-(4-NITROBENZYL)PYRIDINE
9 ★	3.74		594		C12H10N2O2	ANILINE,4-NITRO-N-PHENYL 836-30-6
8190 ★	3.66		547		C12H10N2O2	DIPHENYLAMINE,2-NITRO
1 ★	2.85	8.9	1498		C12H10N2O3	N-ME-N-NITROSOCARBAMIC ACID,1-NAPHTHYL ESTER 7090-25-7
2 ★	0.56		538		C12H10N2O3S1	4-(BENZOYLSULFONAMIDE)-PYRIDINE 51246-76-5
3 ★	2.87		2097		C12H10N2O3S1	DIAZINESULFONE,1,3-DIPHENYL,OXIDE
4	1.37		555		C12H10N2O4	NIFURPIRINOL 13411-16-0 antibacterial
95 ★	2.03		2467	448	C12H10N2O4S1	ANILINE,P-(4-NITROPHENYL)SULFONYL 1948-92-1
6	-1.52	7.2	2446	1	C12H10N2O5	CINOXACIN 28657-80-9 antibacterial
7	0.17	4.7	2314	1	"	CINOXACIN
8	0.50	4.7	2314	2	"	CINOXACIN
9	1.30		555		C12H10N2O5S1	4-CARBETHOXY-2-(5-NO2-2-FURFURILIDENE)THIAZOLE
8200	0.15		328		C12H10N4O2S1	N1-(3-CYANO-2-PYRIDYL)SULFANILAMIDE 30961-35-4
1	1.67		555		C12H10N4O4	3-ACETYLAMINO-2-(5-NO2-2-FURFURILIDENE)PYRIMIDINE
2 ★	2.87		1142		C12H10O1	1-ACETYLAZULENE 7206-57-7
3 ★	3.23		853		C12H10O1	3-PHENYLPHENOL 580-51-8
4 ★	4.21		508		C12H10O1	DIPHENYLETHER 101-84-8
5 ★	3.09		558		C12H10O1	O-PHENYLPHENOL 90-43-7 acaricide~fungicide
6 ★	3.20		853		C12H10O1	P-PHENYLPHENOL 92-69-3
7 ★	2.06		541		C12H10O1S1	PHENYLSULFOXIDE 945-51-7
8	2.49		520	20	C12H10O2	1,4-NAPHTHOQUINONE,6,7-DIMETHYL 2202-79-1
9 ★	3.35		596		C12H10O2	4-PHENOXYPHENOL 831-82-3
8210 ★	2.78		757		C12H10O2	NAPHTHALENE-1-ACETOXY 830-81-9
1	-0.36	7.4	594	1	C12H10O2	NAPHTHYL-2-ACETIC ACID 581-96-4
2 ★	2.81	1.0	594		"	NAPHTHYL-2-ACETIC ACID
3 ★	3.34		594		C12H10O2S1	4,4'-THIODIPHENOL 2664-63-3
4 ★	2.40		503		C12H10O2S1	SULFONE,DIPHENYL 127-63-9 ovicide (mites)
15 ★	2.14		520		C12H10O3	1,4-NAPHTHOQUINONE,2-METHYL,3-METHOXY 5416-18-2
6	-1.50	10.0	2012	459	C12H10O3	1-NAPHTHOXYACETIC ACID,SODIUM SALT
7 ★	2.80	1.0	594		C12H10O3	1-NAPHTHOXYACETIC ACID 2976-75-2
8 ✓	-1.10	13.0	590		C12H10O3	2-NAPHTHYLOXYACETIC ACID, SODIUM SALT
9 ★	2.53		501		C12H10O3	ACETIC ACID,2-NAPHTHYLOXY 120-23-0
8220 ★	2.32		2487		C12H10O3	NAPHTHALENE,2,3-DIHYDROXY-6-ACETYL
1 ★	3.06		536		C12H10O3S1	PHENYLBENZENESULFONATE 4358-63-8
2 ★	4.45		536		C12H10S1	DIPHENYL SULFIDE 139-66-2
3 ★	4.41		547		C12H10S2	DIPHENYL DISULFIDE 882-33-7
4	2.30	7.4	1485	1	C12H11Br1N4O2	2,4-DIAMINOPYRIMIDINE,5(2-BR-4,5-OCH2O-)BENZYL ✦
25 ★	2.48	7.4	1485	2	"	2,4-DIAMINOPYRIMIDINE,5(2-BR-4,5-OCH2O-)BENZYL
6	2.75		306		C12H11Br1O2	ACETYLACETONE,4-BROMOBENZAL 15795-19-4
7	-0.18	7.4	1204	1	C12H11Br2N3	N(2-PROPYN-1YL)-N(2,6-DIBRPH)AMINO)2-IMIDAZOLINE
8 ★	3.90		2506		C12H11Cl1F3N1O3	FLUXOFENIM herbicide safener
9	2.73		163	459	C12H11Cl1N2O1	4-PROPIONYLAMINO-7-CHLOROQUINOLINE
8230 ★	2.00		328		C12H11Cl1N2O2S1	N1-(2-CHLOROPHENYL)SULFANILAMIDE 19837-85-5
1 ★	2.71		328		C12H11Cl1N2O2S1	N1-(4-CHLOROPHENYL)SULFANILAMIDE 16803-92-2
2	0.75		536		C12H11Cl1N2O3	N-(4-CHLOROPHENYL)-3-ACETYLAMINOSUCCINIMIDE 37107-20-3
3	-1.12	7.0	1526	1	C12H11Cl1N2O5S1	FUROSEMIDE 54-31-9 diuretic
4	-0.92	7.4	2637	1	"	FUROSEMIDE

	logP	pH	Ref.	Note	MF	Name / **CAS** / *activity*
35	-0.83		1343		"	FUROSEMIDE
6 ★	2.19		306		C12H11Cl1O2	ACETYLACETONE,2-CHLOROBENZAL **15725-14-1**
7 ★	2.26		306		C12H11Cl1O2	ACETYLACETONE,3-CHLOROBENZAL **15725-15-2**
8 ★	2.32		306		C12H11Cl1O2	ACETYLACETONE,4-CHLOROBENZAL **19411-75-7**
9 ★	2.32	7.0	2292	1	C12H11Cl2N3O2	AZACONAZOLE **60207-31-0** *antifungal*
8240 ★	1.65		306		C12H11F1O2	ACETYLACETONE,4-FLUOROBENZAL **15851-94-2**
1	-1.44		547	370	C12H11F1O2S1.Br1	N-(4-SO2F-BENZYL)--PYRIDINIUMBROMIDE
2	0.06	1.0	594		C12H11F3N4O1	2,4-DIAMINOPYRIMIDINE,5-(4-TRIFLUOROMETHOXYBENZYL)
3	2.83	7.4	1940	1	C12H11N1	2-AMINOBIPHENYL **90-41-5**
4 ★	2.84		558		"	2-AMINOBIPHENYL
45 ★	2.86	7.5	2201		C12H11N1	4-AMINOBIPHENYL **92-67-1**
6 ★	3.50		536		C12H11N1	DIPHENYLAMINE **122-39-4**
7 ★	2.62	7.2	2317	1	C12H11N1	PYRIDINE,4-BENZYL
8 ★	2.48		1142		C12H11N1O1	1-ACETAMIDOAZULENE **23702-21-8**
9 ★	1.29		1817		C12H11N1O1	3-PYRIDINEMETHANOL,A,A-DI-2-PROPYNYL **89242-77-3**
8250 ★	2.46		853		C12H11N1O1	O-PHENOXYANILINE **2688-84-8**
1 ★	2.93	7.4	594	314	C12H11N1O1	P-PHENOXYANILINE **139-59-3**
2 ★	2.82	6.4	2018		C12H11N1O1	PHENOL,4-PHENYLAMINO **122-37-2**
3 ★	1.52	8.0	603	1	C12H11N1O1	PHENYL-A-PYRIDYLCARBINOL **14159-57-0**
4	0.72	5.0	2153	1	C12H11N1O2	2-(4-(1-PYRROLYL)PHENYL)ACETIC ACID
55 ★	1.81		1478		C12H11N1O2	2-(N-BENZYLCARBOXAMIDE) **10354-48-0**
6 ★	1.57	5.0	2153	2	C12H11N1O2	2-PHENYL-2-(1-PYRROLYL)ACETIC ACID
7 ★	2.64		1519	459	C12H11N1O2	6-QUINOLINECARBOXYLIC ACID,ETHYL ESTER
8	1.83		163	459	C12H11N1O2	8-ETHOXYCARBONYLQUINOLINE **25635-22-7**
9 ★	2.43		306		C12H11N1O2	BENZALCYANOACETIC ACID,ETHYL ESTER **2025-40-3**
8260 ★	2.36		273		C12H11N1O2	N-METHYLCARBAMATE,1-NAPHTHYL **63-25-2** *insecticide~parasiticide (vet)*
1 ★	2.56		273		C12H11N1O2	N-METHYLCARBAMATE,2-NAPHTHYL **4089-04-7**
2 ★	1.90		2649		C12H11N1O2	NAPHTHOQUINONE,2-DIMETHYLAMINO
3 ★	1.84		2467	448	C12H11N1O2S1	ANILINE,P-PHENYLSULFONYL **7019-01-4**
4 ★	2.58		536		C12H11N1O2S1	BENZENESULFANILIDE **1678-25-7**
65	2.74		163	459	C12H11N1O2S1	QUINOLINE,8-ETHYLTHIOCARBONATE
6	2.28	2.1	2265		C12H11N1O2S1	[1,1'-BIPHENYL]-4-SULFONAMIDE **4371-23-7**
7	2.60		2467	448	C12H11N1O2S1	[1,1'-BIPHENYL]-4-SULFONAMIDE
8 ★	2.84	6.5	1556	1	C12H11N1O3	2-(2-FURAMIDO)-5-METHYLPHENOL
9 ★	1.23		2649		C12H11N1O3	NAPHTHOQUINONE,2-HYDROXYETHYLAMINO
8270	1.73		163	459	C12H11N1O3	QUINOLINE,8-ETHYLCARBONATE **42322-29-2**
1	1.37		2060		C12H11N1O3	SUCCINIMIDE,N-(P-ACETYLPHENYL)
2 ★	1.57		2467		C12H11N1O3S1	ANILINE,P-(4-HYDROXYPHENYL)SULFONYL **25963-47-7**
3 ★	2.16	7.4	2542	1	C12H11N1O3S2	THIOPHENE-2-SULFONAMIDE,4-(P-METHYLBENZOYL)
4	0.36		1156		C12H11N1O4	M-METHOXYCARBONYL-N-PHENYLSUCCINIMIDE
75 ★	1.29		2467		C12H11N1O4S1	ANILINE,P-(2,4-DIHYDROXYPHENYL)SULFONYL
6 ★	0.52	1.0	590		C12H11N1O4S2	BENZENESULFONAMIDE,N-PHENYLSULFONYL **2618-96-4**
7 ★	1.75	7.4	2542	1	C12H11N1O4S2	THIOPHENE-2-SULFONAMIDE,4-(P-METHOXYBENZOYL)
8 ★	1.53		2467		C12H11N1O5S1	ANILINE,P-(2,4,6-TRIHYDROXYPHENYL)SULFONYL
9 ★	1.52	7.4	2542	1	C12H11N1O5S1	FURAN-2-SULFONAMIDE,4-(4-METHOXYBENZOYL)
8280 ★	1.21		2620		C12H11N1O6	NITECAPONE *antioxidant~comt inhibitor*
1 ★	1.24		574		C12H11N1S1	2-THIOBENZYLPYRIDINEOXIDE
2 ★	2.90	7.4	593	1	C12H11N3	2-AMINO-3-METHYL-9H-PYRIDO[2,3-B]INDOLE
3 ★	1.97	7.4	593	1	C12H11N3	3-AMINO-1-METHYL-5H-PYRIDO[4,3-B]INDOLE ✦
4 ★	3.41		596		C12H11N3	P-AMINOAZOBENZENE **60-09-3**
85 ★	1.36	7.4	725	1	C12H11N3O1	2-PHENYLISONIAZID **58481-06-4**
6 ★	3.13		2379		C12H11N3O1S1	FURFURALTHIOSEMICARBAZONE,N-PHENYL
7 ★	2.52		2379		C12H11N3O2	FURFURAL-SEMICARBAZONE,N-PHENYL
8 ★	1.55		555		C12H11N3O3	5-ET-2-(5-NO2-2-FURFURILIDENE)PYRIMIDINE
9 ★	1.63	7.4	2206	306	C12H11N3O4	IMIDAZOLE,1-METHYL-5-NITRO-2-(P-METHOXYCARBONYLPHENYL)
8290 ★	2.11		1136		C12H11N3O4S1	N1-(2-NITROPHENYL)SULFANILAMIDE **19837-88-8**
1 ★	2.14		985		C12H11N3O4S1	N1-(4-NITROPHENYL)SULFANILAMIDE **6829-82-9**
2	1.21		494		C12H11N3O5S1	3-ET-2-ACETIMINO-5(5-NO2-FURFURILIDINE)THIAZOL-4ONE **52661-67-3**
3 ★	1.70	7.4	2348	1	C12H11N5	ADENINE,9-BENZYL
4 ★	0.98		505		C12H11N7	PTERIDINE,2,4,7-TRIAMINO-6-PHENYL **396-01-0** *diuretic*
95	1.11	7.4	2035	324	"	PTERIDINE,2,4,7-TRIAMINO-6-PHENYL
6	0.92	7.5	2035	324	C12H11N7O1	TRIAMTERENE,P-HYDROXY
7	-1.52	7.4	2035	324	C12H11N7O4S1	TRIAMTERENE,P-SULFATE
8 ★	4.31		560		C12H12	1,2-DIMETHYLNAPHTHALENE **573-98-8**
9 ★	4.42		560		C12H12	1,3-DIMETHYLNAPHTHALENE **575-41-7**
8300 ★	4.37		560		C12H12	1,4-DIMETHYLNAPHTHALENE **571-58-4**
1 ★	4.38		560		C12H12	1,5-DIMETHYLNAPHTHALENE **571-61-9**
2 ★	4.44		560		C12H12	1,7-DIMETHYLNAPHTHALENE **575-37-1**
3 ★	4.26		560		C12H12	1,8-DIMETHYLNAPHTHALENE **569-41-5**
4 ★	4.39		2000		C12H12	1-ETHYLNAPHTHALENE **1127-76-0**
5 ★	4.40		560		C12H12	2,3-DIMETHYLNAPHTHALENE **581-40-8**
6 ★	4.31		560		C12H12	2,6-DIMETHYLNAPHTHALENE **581-42-0**
7 ★	4.38		560		C12H12	2-ETHYLNAPHTHALENE
8 ★	2.96		579		C12H12Br1N1O2	N-(4-BROMOBUTYL)PHTHALIMIDE **3236-48-4**
9	1.79		328		C12H12Br1N3O2S1	N1-(3-BR-5-ME-2-PYRIDYL)SULFANILAMIDE **30961-43-4**
8310	2.05		328		C12H12Br1N3O2S1	N1-(3-ME-5-BR-2-PYRIDYL)SULFANILAMIDE **30961-42-3**
1	3.17		1479	459	C12H12Br2N4O1	2,4-DIAMINO-5-(2,4-DIBRMO-5-MEO-BENZYL)PYRIMIDINE
2	1.55		1353		C12H12Cl1N3O1	4-CL-5-(DIMEAMINO)-2-PHENYL-3-PYRIDAZINONE
3	-0.64		328		C12H12Cl1N3O2S1	N1-ME-N1-(3-CL-2-PYRIDYL)SULFANILAMIDE
4	0.59		328		C12H12Cl1N3O2S1	N1-ME-N1-(5-CL-2-PYRIDYL)SULFANILAMIDE
15	-1.34	7.0	2418	1	C12H12Cl1N5O4S1	CHLORSULFURON **64902-72-3** *herbicide*
6	0.74	5.0	2418	1	"	CHLORSULFURON
7 ★	0.78		537		C12H12Cl2N2O3	N-DICLACETYL-A-OHME-B-OH-P-CYANOPHENETHYLAMINE **23885-61-2**
8 ★	3.09	7.4	594	1	C12H12Cl2N4	ETOPRINE **18588-57-3** *antineoplastic*
9	2.59	7.4	1485	1	C12H12Cl2N4O1	2,4-DIAMINOPYRIMIDINE,5(3,5-DICL-4-MEO-)BENZYL

	logP	pH	Ref.	Note	MF	Name / **CAS** / *activity*
8320 ★	2.78	7.4	1485	2	"	2,4-DIAMINOPYRIMIDINE,5(3,5-DICL-4-MEO-)BENZYL
1 ★	0.14		2649		C12H12F1N3O2	BENZOQUINONE,2,3,5-TRIS-AZIRIDINYL-6-FLUORO
2 ★	1.61		2649		C12H12F2N2O2	BENZOQUINONE,3,5-BIS-(2-METHYLAZIRIDIN-1YL)-3,6-DIFLUORO
3 ★	2.91	13.0	1623	224	C12H12F3N1	BENZOBICYCLO(2,2,1)HEPTENE,6-CF3-2-ENDO-AMINO **83118-48-3**
4 ★	3.21	13.0	1623	224	C12H12F3N1	BENZOBICYCLO(2,2,1)HEPTENE,6-CF3-2-EXO-AMINO **83118-50-7**
25 ★	2.85	13.0	1623	224	C12H12F3N1	BENZOBICYCLO(2,2,1)HEPTENE,7-CF3-2-ENDO-AMINO **86022-72-2**
6 ★	3.19	13.0	1623	224	C12H12F3N1	BENZOBICYCLO(2,2,1)HEPTENE,7-CF3-2-EXO-AMINO **83118-51-8**
7	-2.62		527	820	C12H12N1.Br1	BENZYLPYRIDINIUMBROMIDE **2589-31-3**
8	-2.28		2646	448	C12H12N1.I1	PYRIDINIUM IODIDE,1-METHYL-4-PHENYL
9	-1.04		2646		"	PYRIDINIUM IODIDE,1-METHYL-4-PHENYL
8330 ★	2.80	13.0	547	224	C12H12N2	1,1-DIPHENYLHYDRAZINE **530-47-2** ✚
1 ★	1.34		879		C12H12N2	BENZIDINE **92-87-5**
2 ★	2.94		536		C12H12N2	HYDRAZOBENZENE **122-66-7**
3 ★	0.26		594		C12H12N2O1	1,2,3,4-TETRAHYDROCARBOLINE,9-FORMYL
4 ★	0.99		1519	459	C12H12N2O1	3-QUINOLINECARBOXAMIDE,N,N-DIMETHYL
35	1.49		2032	459	C12H12N2O1	4-PYRIMIDONE,(2,3)-CYCLOHEXYL-(5,6)-BENZO-
6	1.59		2032	459	C12H12N2O1	4-PYRIMIDONE,(2,3)-CYLCOPENTYL-(5,6)2-METHYLBENZO-
7	1.37	7.5	2492	1	C12H12N2O1	9-AMINO-4-OXO-1,2,3,4-TETRAHYDROACRIDINE
8	1.86		2492	2	"	9-AMINO-4-OXO-1,2,3,4-TETRAHYDROACRIDINE
9 ★	1.36	7.4	594	314	C12H12N2O1	DIPHENYLETHER,4,4'-DIAMINO **101-80-4**
8340	1.63		1725	459	C12H12N2O1	PYIDO-PYRIMIDINE ANALOG(RINGS:666) ✚
1	1.77		1725	459	C12H12N2O1	PYRIDO-PYRIMIDINE ANALOG,6-ME(RINGS:566) ✚
2	1.61		1725	459	C12H12N2O1	PYRIDO-PYRIMIDINE ANALOG,7-ME(RINGS:566) ✚
3	1.65		1725	459	C12H12N2O1	PYRIDO-PYRIMIDINE ANALOG,8-ME(RINGS:566)
4	1.71		1725	459	C12H12N2O1	PYRIDO-PYRIMIDINE ANALOG,9-ME(RINGS:566)
45	1.08		1059		C12H12N2O1S1	2,4-DIMETHYL-5-THIAZOLECARBOXANILIDE
6	1.17		2089		C12H12N2O2	N-(3-DELTA-2-OXOPIPERIDIN-1-YL)BENZAMIDE **104642-89-9**
7 ★	1.69		522		C12H12N2O2S1	1,2,4-BENZOTHIADIAZINE-1,1-DIOXIDE-3-CYCLOPENTEN-1-YL **37157-89-4**
8 ★	1.53		522		C12H12N2O2S1	1,2,4-BENZOTHIADIAZINE-1,1-DIOXIDE-3-CYCLOPENTEN-3-YL **37157-90-7**
9 ★	1.56		2467		C12H12N2O2S1	ANILINE,P-(2-AMINOPHENYL)SULFONYL **27147-69-9**
8350 ★	1.03		2467		C12H12N2O2S1	ANILINE,P-(3-AMINOPHENYL)SULFONYL **34262-32-3**
1 ★	0.97		547		C12H12N2O2S1	DI(P-AMINOPHENYL)SULFONE
2 ★	1.55		985		C12H12N2O2S1	SULFABENZ **127-77-5** *antibacterial, coccidiostat*
3 ★	0.96		1439		C12H12N2O3	4H-PYRIDO(1,2-A)PYRIMIDIN-4-ONE,3-ETO-CO,6-ME
4 ★	0.90		1439		C12H12N2O3	4H-PYRIDO(1,2-A)PYRIMIDIN-4-ONE,3-ETO-CO,7-ME
55 ★	0.87		1439		C12H12N2O3	4H-PYRIDO(1,2-A)PYRIMIDIN-4-ONE,3-ETO-CO,8-ME
6 ★	1.16		1439		C12H12N2O3	4H-PYRIDO(1,2-A)PYRIMIDIN-4-ONE,3-ETO-CO,9-ME
7	1.33	7.4	1255	1	C12H12N2O3	7(DIETAMINO)-4-HYDROXYQUINOLINE-3-CARBOXYLIC ACID
8 ★	-0.33		536	1	C12H12N2O3	N-PHENYL-3-ACETYLAMINOSUCCINIMIDE **30820-34-9**
9	0.59	7.2	2446	1	C12H12N2O3	NALIDIXIC ACID **389-08-2** *antibacterial*
8360 ★	1.41		1135	537	"	NALIDIXIC ACID
1	1.14	7.4	1873	1	C12H12N2O3	PHENOBARBITAL **50-06-6** *anticonvulsant, hypnotic, sedative*
2	1.15	7.0	919	1	"	PHENOBARBITAL
3	1.32	7.0	2230	1	"	PHENOBARBITAL
4 ★	1.47	2.0	300		"	PHENOBARBITAL
65	1.53	7.4	551	1	"	PHENOBARBITAL
6	1.71	7.0	1232	1	"	PHENOBARBITAL
7 ★	0.88		2467		C12H12N2O3S1	ANILINE,-P-(4-HYDROXYLAMINOPHENYL)SULFONYL **32695-27-5**
8 ★	2.18	7.4	594	1	C12H12N2S1	4,4'-THIODIANILINE **139-65-1**
9 ★	3.05		594	400	C12H12N2S2	DIPHENYL DISULFIDE,2,2'-DIAMINO **1141-88-4**
8370	-3.55	7.4	921	22	C12H12N2.(Cl1)2	DIQUATDICHLORIDE **85-00-7**
1	-0.68	7.4	1004	1	C6H5OP(S)(OH)(NHN=CH(2-PYRIDYL-N-OXIDE)	
2 ★	1.98	7.4	593	1	C12H12N4	2-AMINO-3,4-DIMETHYLIMIDAZO[4,5-F]QUINOLINE ✚
3 ★	1.82	7.4	595	1	C12H12N4	IMIDAZO[4,5-F]QUINOLINE,3-METHYL-2-METHYLAMINO **102408-26-4**
4 ★	2.96		2379		C12H12N4O1	PYRRAL-SEMICARBAZONE,N'-PHENYL
75	1.30	7.4	1485	1	C12H12N4O2	2,4-DIAMINOPYRIMIDINE,5-(3,4-OCH2O-)BENZYL
6 ★	1.51	7.4	1485	2	C12H12N4O2	2,4-DIAMINOPYRIMIDINE,5-(3,4-OCH2O-)BENZYL
7	0.85	7.4	2206	306	C12H12N4O3	BENZNIDAZOLE **22994-85-0** *antiprotozoal*
8 ★	0.91	7.4	1022	1	"	BENZNIDAZOLE
9 ★	0.55	7.4	2206	306	C12H12N4O3	IMIDAZOLE,1-METHYL-5-NITRO-2-(P-CARBOXAMIDOBENZYL)
8380	1.49	1.0	2196		C12H12N4O3	PYRAZOLE,3-METHYL-4-NITRO-5-(N-BENZYLCARBOXAMIDO)
1 ★	2.34	1.0	2196		C12H12N4O3	PYRAZOLE,3-METHYL-4-NITRO-5-(N-P-TOLYLCARBOXAMIDO)
2 ★	0.74		538		C12H12N4O5	5-(1-AZIRIDINYL)-2,4-DINITROBENZAMIDE,N-CYCLOPROPYL **24570-17-0**
3	0.20		2279		C12H12N4O5S1	IMIDAZOLE,1-METHYL-4-NITRO-5-M-ACETAMIDOPHENYLSULFONATO
4	-0.06		2279		C12H12N4O5S1	IMIDAZOLE,1-METHYL-4-NITRO-5-O-ACETAMIDOPHENYLSULFONATO
85	1.60		2279		C12H12N4O6S1	IMIDAZOLE,1-METHYL-4-NITRO-5-O-CARBOMETHOXYPHENYLSULFONAMIDO
6 ★	3.73		2379		C12H12N4S1	PYRRAL-THIOSEMICARBAZONE,N'-PHENYL
7 ★	1.96		1967		C12H12N4S1	THIENYL-2-PHENONE,GUANYLHYDRAZONE **97183-49-8**
8 ★	2.56		1524		C12H12O2	1(3H)-ISOBENZOFURANONE-3-SPIROPENTYL ✚
9	2.50		2012	459	C12H12O2	2-(1-NAPHTHOXY)ETHANOL
8390 ★	1.59		306		C12H12O2	ACETYLACETONE,BENZAL **4335-90-4**
1 ★	3.25		2487		C12H12O2	NAPHTHALENE,2,3-DIHYDROXY-6-ETHYL
2 ★	2.70		2487		C12H12O3	NAPHTHALENE,2,3-DIHYDROXY-6-ETHOXY
3 ★	2.24		306		C12H12O4	DIMETHYLMALONATE,BENZAL **6626-84-2**
4 ★	1.42		2234		C12H12O6	ACETYLSALICYLIC ACID,ACETYLOXYMETHYL ESTER **118247-06-6**
95 ★	-0.70	3.0	1638		C12H13Br1N2O6S1	CEPHALOSPORANIC ACID,7(BROMOACETYLAMINO) **26973-80-8**
6	2.00		1479	459	C12H13Br1N4O1	2,4-DIAMINO-5-(2-BR-5-MEO-BENZYL)PYRIMIDINE
7	2.44		1479	459	C12H13Br1N4O1	2,4-DIAMINO-5-(2-MEO-5-BR-BENZYL)PYRIMIDINE
8	1.88		1479	459	C12H13Br1N4O1	2,4-DIAMINO-5-(3-BR-4-MEO-BENZYL)PYRIMIDINE **56183-32-5**
9	-0.75	7.4	1204	1	C12H13Br2N3	N-ALLYL-N-(2,6-BR2-PH)AMINO)-2-IMIDAZOLINE
8400 ★	3.00		1070		C12H13Cl1N2O1	BUTURON **3766-60-7**
1	1.61	7.4	2111	1	C12H13Cl1N4	PYRIMETHAMINE **58-14-0** *antimalarial*
2	2.44	7.2	1162	314	"	PYRIMETHAMINE
3 ★	2.69		884		"	PYRIMETHAMINE
4	1.65		1479	459	C12H13Cl1N4O1	2,4-DIAMINO-5-(3-CL-4-MEO-BENZYL)PYRIMIDINE **73275-70-4**
5	1.87		1479	459	C12H13Cl1N4O1	2,4-DIAMINO-5-(4-CL-3-MEO-BENZYL)PYRIMIDINE **73264-25-2**
6 ★	3.53		1524		C12H13Cl1O2	1(3H)-ISOBENZOFURANONE-3-BUTYL-6-CHLORO
7 ★	2.17	3.5	2179	302	C12H13Cl1O4	PENTANOIC ACID,5-OXO-5-(3-CHLORO-4-METHOXY)PHENYL
8 ★	4.47		1792		C12H13Cl2N1O1	CYCLOPENTANAMIDE,N-(3,4-DICHLOROPHENYL)
9 ★	2.30		565		C12H13Cl2N1O3	ET-N-CHLOROACETYL-N-(2-CHLOROPHENYL)GLYCINATE **51114-25-1**

	logP	pH	Ref.	Note	MF	Name / CAS / activity
8410	-1.66	7.4	1204	1	C12H13Cl2N3	ALINIDINE **33178-86-8** *anti-arrhythmic*
1	1.27		2089		C12H13F1N2O1	N-(3-DELTA-3-FLUOROPIPERIDIN-1-YL)BENZAMIDE **104642-69-5**
2	1.08		1479	459	C12H13F1N4O1	2,4-DIAMINO-5-(3-FLUORO-4-MEO-BENZYL)PYRIMIDINE **73264-24-1**
3 ★	4.27		509		C12H13F3N2	BENZIMIDAZOLE,5-BUTYL-2-(TRIFLUOROMETHYL)
4	2.70		1479	459	C12H13I1N4O1	2,4-DIAMINO-5-(2-MEO-5-IODO-BENZYL)PYRIMIDINE
15	2.09		1479	459	C12H13I1N4O1	2,4-DIAMINO-5-(3-IODO-4-MEO-BENZYL)PYRIMIDINE
6 ★	3.46		579		C12H13N1	1-NAPHTHYLAMINE,N,N-DIMETHYL **86-56-6**
7	1.17	7.4	1990	1	C12H13N1O1	ETHYLAMINE,2-(1-NAPHTHOXY) **50882-68-3**
8 ★	2.03		2221	459	C12H13N1O1	INDOLE-1,2-DIMETHYL-3-ACETYL
9	0.39	7.0	1073	1	C12H13N1O2	3-INDOLEBUTYRIC ACID **133-32-4**
8420	1.92	7.0	1073	2	"	3-INDOLEBUTYRIC ACID
1 ★	2.30		579		"	3-INDOLEBUTYRIC ACID
2 ★	2.04		1705		C12H13N1O2	BENZOIC ACID,4-CYANOBUTYL ESTER
3 ★	1.04		1156		C12H13N1O2	M-ETHYL-N-PHENYLSUCCINIMIDE
4	1.63	7.4	1163	1	C12H13N1O2	N-PHENYL-3,4-DIME-2-OH-5-OXO-2,5-H2-PYRROLONE
25 ★	0.64		1156		C12H13N1O2	O-ETHYL-N-PHENYLSUCCINIMIDE
6 ★	1.10		1156		C12H13N1O2	P-ETHYL-N-PHENYLSUCCINIMIDE
7 ★	2.14		536		C12H13N1O2S1	VITAVAX **5234-68-4** *fungicide*
8 ★	1.20		2157		C12H13N1O3	2-(BENZOYLOXY)-(N-TRIMETHYLENE)ACETAMIDE
9 ★	0.64		1059		C12H13N1O3S1	CARBOXIN S-OXIDE *fungicide* ✚
8430 ★	2.92		1524		C12H13N1O4	1(3H)-ISOBENZOFURANONE-3-BUTYL-6-NITRO
1 ★	0.74		1059		C12H13N1O4S1	OXYCARBOXIN *fungicide*
2 ★	1.24		547		C12H13N1O4S2	ETHYL-4-SULFONAMIDONAPHTHYLSULFONE **842-00-2**
3 ★	0.67	2.0	2157		C12H13N1O5	2-(BENZOYLOXY)-(N-METHYL-N-ACETIC ACID)ACETAMIDE
4 ★	2.50		1946		C12H13N1O7	PROPANDIOIC ACID,(5-NITRO-2-FURANYL)METHLENE, DIETHYL ESTER **69513-13-9**
35 ★	1.51		594		C12H13N2O3P1	PHOSPHOROHYDRAZIDIC ACID,DIPHENYL ESTER **33862-44-1**
6	0.85		2653	459	C12H13N3O1	O-PHENYLENEUREA,N-(D3-PIPERAZIN-4YL)
7 ★	1.49	7.4	534	1	C12H13N3O2	ISOCARBOXAZID **59-63-2** *antidepressant*
8 ★	-0.13		2649		C12H13N3O2	TRIAZIQUONE **68-76-8** *antineoplastic*
9	0.88	7.2	556		C12H13N3O2S1	5-ETHYL-2-BENZENESULFONAMIDOPYRIMIDINE **13428-54-1**
8440 ★	0.38		2467		C12H13N3O2S1	ANILINE,P-(2,4-DIAMINOPHENYL)SULFONYL
1	0.26		328		C12H13N3O2S1	N1-(3-METHYL-2-PYRIDYL)SULFANILAMIDE
2	0.49		328		C12H13N3O2S1	N1-(5-METHYL-2-PYRIDYL)SULFANILAMIDE **3731-45-1**
3 ★	1.30	6.0	479		C12H13N3O2S1	N1-METHYL-N1-2-PRYIDYLSULFANILAMIDE **51543-29-4**
4	2.38	7.4	1022	1	C12H13N3O3	1-(3-PHENOXYPROPYL)-2-NITROIMIDAZOLE
45	0.53	7.2	556		C12H13N3O3S1	5-ETHOXY-2-BENZENESULFONAMIDOPYRIMIDINE **13419-02-8**
6 ★	2.50	7.4	2206	306	C12H13N3O3S1	FEXINIDAZOLE **59729-37-2** *antiprotozoal*
7	1.43	7.4	1497	1	C12H13N3O4	1(3-PHO-2-HYDROXYPROPYL)-2-NITROIMIDAZOLE
8	1.49		1080		"	1(3-PHO-2-HYDROXYPROPYL)-2-NITROIMIDAZOLE
9	1.22		494		C12H13N3O4S1	2-(N,N-DIETHYL)-5-(5-NO2-2-FURFURYL)THIAZOLONE **27473-73-0**
8450	1.34		494		C12H13N3O4S1	3ET-2-ETIMINO-5-(5-NO2-2-FURFURILIDENE)THIAZOL-4ONE **52661-61-7**
1	-0.70		835		C12H13N3O4S1	PHENYL(2-(2-ME-5-NITRO-1-IMIDAZOLYL)ETHYL)SULFONE **28795-25-7**
2	0.35		985		C12H13N3O4S2	N1-(4-SO2NH2-PHENYL)SULFANILAMIDE **547-52-4** *antibacterial (topical)*
3 ★	-0.22		537		C12H13N3O5	CYANAMPHENICOL **23885-71-4**
4 ✓	-1.62		2397	101	C12H13N4O3	IMIDAZOLIUM CHLORIDE,1-METHYL-2-HYDROXYIMINOMETHYL-3-(M-NITROBENZYL)
55 ✓	-1.42		2397	101	C12H13N4O3	IMIDAZOLIUM CHLORIDE,1-METHYL-2-HYDROXYIMINOMETHYL-3-(P-NITROBENZYL)
6	-0.72	7.4	2683	1	C12H13N5O1	TRYPTAMINE,5-(3-AMINOOXADIAZOLYL) *5ht-agonist*
7	-1.57		2418	1	C12H13N5O6S2	HARMONY **79277-27-3**
8	0.52	5.0	2418	1	"	HARMONY
9 ★	4.53		594		C12H14	CYCLOHEXENE,1-PHENYL **31017-40-0**
8460 ★	3.32	7.5	2545	2	C12H14Cl1N1	4'-CHLORO-MPTP
1	3.41		2363		C12H14Cl1N1	BUTYRONITRILE,3-METHYL-2-(4-CHLOROBENZYL)
2 ★	1.60		565		C12H14Cl1N1O3	ETHYL-N-CHLOROACETYL-N-PHENYLGLYCINATE **51114-26-2**
3 ★	1.20		757		C12H14Cl1N1O3	P(N-2-CHLOROETHYL-N-ACETYL)AMINOPHENYLACETIC ACID **32562-52-0**
4 ★	-0.48		536	800	C12H14Cl2N2O4	P-AMIDO-AMPHENICOL **39960-99-1**
65 ★	3.10		1263		C12H14Cl3O4P1	CLOFENVINFOS **470-90-6** *insecticide*
6 ★	1.82		2479	459	C12H14F1N3O3	PHENYLBIURET,2-FLUORO-5-ACETYL
7 ★	2.28		2479		C12H14F1N3O4	PHENYLBIURET,2-FLUORO-5-METHOXYCARBONYL
8 ★	1.22	13.0	547	224	C12H14F3N5	4,6-DIAMINO-2,2-DIMETHYL-1-(4-CF3-PHENYL)SYM-TRIAZINE
9 ★	3.48		354		C12H14N1O4P1S1	N-(O,O-DIETHYLTHIOPHOSPHORYL)PHTHALIMIDE **5131-24-8** *fungicide*
8470	-2.52		527	820	C12H14N1.Br1	N-PROPYLQUINOLINIUMBROMIDE **6294-92-4**
1	3.95		303	820	C12H14N1.C8H17O4S1	1,2,6-TRIMETHYLQUINOLINIUMOCTYLSULFATE
2	4.50		303	820	C12H14N1.C9H19O4S1	1,2,6-TRIMETHYLQUINOLINIUMNONYLSULFATE
3	5.05		303	820	C12H14N1.C10H21O4S1	1,2,6-TRIMETHYLQUINOLINIUMDECYLSULFATE
4	5.45		303	820	C12H14N1.C11H23O4S1	1,2,6-TRIMETHYLQUINOLINIUMUNDECYLSULFATE
75 ★	1.86		1439		C12H14N2O1	4H-PYRIDO(1,2-A)PYRIMIDIN-4-ONE,2,6-DIME,3-ET
6 ★	1.87		1439		C12H14N2O1	4H-PYRIDO(1,2-A)PYRIMIDIN-4-ONE,2-ET-3,6-DIME
7 ★	1.75		1439		C12H14N2O1	4H-PYRIDO(1,2-A)PYRIMIDIN-4-ONE,2-ME,-3-PR
8 ★	0.58	6.8	2175	1	C12H14N2O2	3-ETHYL-3-(4-PYRIDYL)PIPERIDINE-2,6-DIONE
9 ★	1.07		2649		C12H14N2O2	BENZOQUINONE,2,5-BIS-(2-METHYLAZIRIDIN-1YL)
8480	0.51	7.4	1163	1	C12H14N2O2	N(3-AMINOPH)3,4-DIME-2-OH-5-OXO-2,5-H2-PYRROLONE
1	0.43	7.4	2462	1	C12H14N2O2	P-CYANOBENZOIC ACID,N,N-DIMETHYLAMINOETHYL ESTER
2 ★	1.43	7.4	2462	2	"	P-CYANOBENZOIC ACID,N,N-DIMETHYLAMINOETHYL ESTER
3 ★	0.91		591		C12H14N2O2	PRIMIDONE **125-33-7** *anticonvulsant*
4 ★	1.83		522		C12H14N2O2S1	1,2,4-BENZOTHIADIAZINE-1,1-DIOXIDE-3-CYCLOPENTYL **20434-64-4**
85 ★	2.01	7.0	538	701	C12H14N2O2S1	1-DIMETHYLAMINO-5-SULFONAMIDONAPHTHALENE **1431-39-6**
6	1.51	7.4	899	1	C12H14N2O3	BARBITURIC ACID,5-ALLYL,5-(2-CYCLOPENTEN-1-YL) **76-68-6** *sedative, hypnotic*
7 ★	0.42		2157		C12H14N2O4	2-(BENZOYLOXY)-(N-2-PROPIONAMIDO)ACETAMIDE
8 ★	0.08		2157		C12H14N2O4	2-(BENZOYLOXY)-(N-METHYL,N-(2-ACETAMIDO))ACETAMIDE
9 ★	0.17		2649		C12H14N2O4	BENZOQUINONE,2,5-BIS-AZIRIDINYL-3,6-DIMETHOXY
8490 ★	1.33		1705		C12H14N2O4	N-(2-(P-NITROPHENOXY)ETHYL)-2-PYRROLIDINONE
1	-5.00	7.4	921	22	C12H14N2.(I1)2	PARAQUAT DIIODIDE **1983-60-4**
2	-1.30	7.6	2003	314	C12H14N3O1.Cl1	IMIDAZOLIUM CHLORIDE,1-BENZYLOXYMETHYL-2-HYDROXYIMINOMETHYL-3-METHYL **91900-15-1**
3	-0.51	1.0	547	342	C12H14N4	2,4-DIAMINO-5-(3-METHYLBENZYL)PYRIMIDINE
4 ★	0.30	5.0	2402	40	C12H14N4	NAPHTHO[1,2C5,6C']DIPYRAZOLE,TR-2,3B,4,5,7,8B,9,10-OCTAHYDRO **66818-22-2** ✚
95	1.75		1479	459	C12H14N4O1	2,4-DIAMINO-5(2-MEO-BENZYL)PYRIMIDINE
6	1.34		1479	459	C12H14N4O1	2,4-DIAMINO-5(4-MEO-BENZYL)PYRIMIDINE **20285-70-5**
7	1.35	7.4	1485		C12H14N4O1	2,4-DIAMINO-5(3-METHOXYBENZYL)PYRIMIDINE **59481-28-6**
8 ★	1.54	7.4	1485	2	"	2,4-DIAMINO-5-(3-METHOXYBENZYL)PYRIMIDINE
9	0.55	7.4	1485	1	C12H14N4O2	2,4-DIAMPYRIMIDINE,5-(3-HYDROXY-4-MEO-)BENZYL

	logP	pH	Ref.	Note	MF	Name / **CAS** / *activity*
8500 ★	0.74	7.4	1485	2	"	2,4-DIAMPYRIMIDINE,5-(3-HYDROXY-4-MEO-)BENZYL
1	0.45	7.4	1485	1	C12H14N4O2	2,4-DIAMPYRIMIDINE,5-(3-MEO-4-HYDROXY)BENZYL
2 ★	0.64	7.4	1485	2	"	2,4-DIAMPYRIMIDINE,5-(3-MEO-4-HYDROXY)BENZYL
3	-0.92	7.4	651	1	C12H14N4O2	ANTIPYRYLUREA **2620-71-5**
4 ★	0.90	4.0	360	2	C12H14N4O2S1	2-SULFANILAMIDO-5-ETHYLPYRIMIDINE **3271-01-0**
5	-0.72	4.0	360	2	C12H14N4O2S1	5,6-DIMETHYL-4-SULFANILAMIDOPYRIMIDINE **1740-04-1**
6	0.59	4.0	360	2	C12H14N4O2S1	5-SULFANILAMIDO-2,4-DIMETHYLPYRIMIDINE **7411-79-2**
7	-0.43	7.4	1051	1	C12H14N4O2S1	SULFAMETHAZINE **57-68-1** *antibacterial*
8 ★	0.28	4.0	360	2	"	SULFAMETHAZINE
9 ★	-0.33	4.0	360	2	C12H14N4O2S1	SULFISOMIDINE **515-64-0** *antibacterial*
8510	-0.54	7.4	1485	1	C12H14N4O3	2,4-DIAMINOPYRIMIDINE,5(3,4-DIHYDROXY-5-MEO)BENZYL
1 ★	-0.31	7.4	1485	2	"	2,4-DIAMINOPYRIMIDINE,5(3,4-DIHYDROXY-5-MEO)BENZYL
2	-0.23	7.4	1485	1	C12H14N4O3	2,4-DIAMINOPYRIMIDINE,5-(3,5-DIHYDROXY-4-MEO)BENZYL
3 ★	-0.04	7.4	1485	2	"	2,4-DIAMINOPYRIMIDINE,5-(3,5-DIHYDROXY-4-MEO)BENZYL
4 ★	0.71	13.0	547	224	C12H14N4O3S1	4,6-DIAM-1-(2-MESULFONYLOXY)BENZYLPYRIMIDINE
15 ★	0.61	4.0	360	2	C12H14N4O3S1	SULFAMETOMIDINE **3772-76-7** *antibacterial*
6 ★	-0.90		2529		C12H14N4O4	DDI,5'-ACETATE ✚
7 ★	1.63	4.0	360	2	C12H14N4O4S1	SULFADIMETHOXINE **122-11-2** *antibacterial*
8	0.70	4.0	360	2	C12H14N4O4S1	SULFADOXINE **2447-57-6** *antibacterial*
9	1.06		328		"	SULFADOXINE
8520 ★	1.40		2506		C12H14N4O4S2	THIOPHANATEMETHYL **23564-25-2** *fungicide*
1 ★	-0.24	13.0	547	224	C12H14N6	4,6-DIAM-2,2-DIME-1-(3-CYANOPH)-S-TRIAZINE
2	1.54	4.2	401	2	C12H14N6O1	IMIDAZOLE,5-CARBOXAMIDE-4-(3-METHYL-3-BENZYLTRIAZENYL) **39942-91-1**
3	1.44		547		C12H14N6O2	(CH3)2NNN-C6H4-2-(CONHNHCOCH2CN)
4	-0.06	7.5	652	1	C12H14N6O2S1	SULFADIAZINE,N,N-DIMETHYLTRIAZENE **49638-47-3**
25	1.23	3.3	2590	331	"	SULFADIAZINE,N,N-DIMETHYLTRIAZENE
6	1.32	6.6	2590	459	"	SULFADIAZINE,N,N-DIMETHYLTRIAZENE
7	4.03		2404	459	C12H14O1	BENZOPYRAN,2,2,7-TRIMETHYL **82305-02-0**
8	4.32		2404	459	C12H14O1S1	BENZOTHIOPYRAN,2,2-DIMETHYL-7-METHOXY **86778-10-1**
9 ★	2.69		1524		C12H14O2	1(3H)-ISOBENZOFURANONE,3,3-DIETHYL
8530 ★	1.89		306		C12H14O2	3-BENZYL-2,4-PENTANEDIONE **1134-87-8**
1 ★	3.35		2404		C12H14O2	BENZOPYRAN,2,2-DIMETHYL-7-METHOXY **17598-02-6**
2 ★	2.80		1524		C12H14O2	RS-1(3H)-ISOBENZOFURANONE,3-BUTYL **6066-49-5**
3	2.67		501		C12H14O3	2-(5,6,7,8-TETRAHYDRONAPHTHYLOXY)-ACETIC ACID **1878-59-7**
4 ★	2.61		2487		C12H14O3	PROPYLSTYRYL KETONE,3,4-DIHYDROXY
35 ★	2.47		572		C12H14O4	DIETHYLPHTHALATE **84-66-2**
6	1.70	7.6	2007	1	C12H15Br1N2O2S1	BENZOYLTHIOFORMOHYDROXIMATE,4-BROMO-S-(2-DIMETHYLAMINO)ETHYL **102976-77-2**
7	-0.82	7.4	1204	1	C12H15Br2N3	N-I-PR-N(2,6-BR2-PH)AMINO)-2-IMIDAZOLINE
8	-0.55	7.4	1204	1	C12H15Br2N3	N-PR-N(2,6-BR2-PH)AMINO)-2-IMIDAZOLINE
9 ★	4.38		1639		C12H15Cl1N1O4P1S2	PHOSALONE **2310-17-0** *insecticide*
8540 ★	2.71		522		C12H15Cl1N2O2S1	1,2,4-BENZOTHIADIAZINE-1,1-DIOXIDE-3-NEOPENTYL-7-CHLORO **37157-54-3**
1	1.74	7.6	2007	1	C12H15Cl1N2O2S1	BENZOYLTHIOFOMOHYDROXIMATE,4-CHLORO-S-(2-DIETHYLAMINO)ETHYL **102941-48-0**
2 ★	0.98		537		C12H15Cl1N6O2	MONOCHLORMETHYLAMPHENICOL **49648-35-3**
3 ★	1.50		2289		C12H15Cl1N6O2	TETRAZOLE,1-ETHYL-5-(2-CHLORO-4-DIMETHYLUREYL)PHENOXY **117121-39-8**
4 ★	1.70		2289	459	C12H15Cl1N6O2	TETRAZOLE,2-ETHYL-5-(2-CHLORO-4-DIMETHYLUREYL)PHENOXY **117121-51-4**
45 ★	0.26		511		C12H15Cl1O6	GLUCOPYRANOSIDE,4-CHLOROPHENYL(BETA) **4756-30-3**
6	3.38		1050		C12H15Cl2N1O2	3-N,N-BIS(2-CHLOROET)AMINO-4-METHYL-BENZOIC ACID
7 ★	-0.27		537	800	C12H15Cl2N1O5S1	RACEPHENICOL **847-25-6** *antibacterial*
8 ★	-0.60		536	800	C12H15Cl2N3O4	UREIDOAMPHENICOL **39961-07-4**
9	1.15	7.6	2007	1	C12H15F1N2O2S1	BENZOYLTHIOFORMOHYDROXIMATE,4-FLUORO-S-(2-DIMETHYLAMINO)ETHYL **102941-37-7**
8550	0.27		511		C12H15I1O6	GLUCOPYRANOSIDE,2-IODOPHENYL(BETA) **7234-29-9**
1 ★	0.75		511		C12H15I1O6	GLUCOPYRANOSIDE,4-IODOPHENYL(BETA) **20838-40-8**
2 ★	2.29	13.0	1623	224	C12H15N1	2-ENDO-AMINOBENZOBICYCLO(2,2,2)OCTENE **14342-36-0**
3 ★	2.32	13.0	1623	224	C12H15N1	2-ENDO-METHYLAMINOBENZOBICYCLO(2,2,1)HEPTENE **58742-05-5**
4 ★	2.32	13.0	1623	224	C12H15N1	2-EXO-AMINOBENZOBICYCLO(2,2,2)OCTENE **15537-20-9**
55	-0.81	1.0	541	342	C12H15N1	2-EXO-METHYLAMINOBENZOBICYCLO(2,2,1)HEPTENE
6	-0.83	1.0	547	342	C12H15N1	ANTI-9-METHYLAMINOBENZOBICYCLO(2,2,1)HEPTENE
7 ★	2.41	13.0	1623	224	C12H15N1	BENZOBICYCLO(2,2,1)HEPTTENE,2-EXO-METHYLAMINO **62624-27-5**
8 ★	2.37	13.0	1623	224	C12H15N1	BENZOBICYCLO(2,2,1)HEPTTENE,9-ENDO-METHYLAMINO **86992-69-0**
9 ★	2.47	13.0	1623	224	C12H15N1	BENZOBICYCLO(2,2,1)HEPTTENE,9-EXO-METHYLAMINO **86943-79-5**
8560	0.38	7.4	2648	1	C12H15N1	BENZOCYCLOHEPTANE,5,8-METHENO-10-AMINO
1	0.42	7.4	2648	1	"	BENZOCYCLOHEPTANE,5,8-METHENO-10-AMINO
2 ★	1.63	7.4	2648	2	"	BENZOCYCLOHEPTANE,5,8-METHENO-10-AMINO
3 ★	2.04	7.4	2648	2	"	BENZOCYCLOHEPTANE,5,8-METHENO-10-AMINO
4 ★	2.71	7.5	2545	2	C12H15N1	MPTP **28289-54-5** *parkinson-inducer (as mpp+)* ✚
65	-0.28	7.4	2617	1	C12H15N1	TETRAHYDROCYCLOPENTENINDANE-3A-AMINE
6	2.58		2617	2	"	TETRAHYDROCYCLOPENTENINDANE-3A-AMINE
7	2.72	7.0	538	701	C12H15N1O1	1,3,3-TRIMETHYL-2,3-DIHYDRO-4-QUINOLONE **53207-52-6**
8 ★	2.68		2256		C12H15N1O1	BUTYRONITRILE,2-(4-METHOXYPHENYL)-3-METHYL
9	1.04	7.4	2535	1	C12H15N1O1	G-BENZOYLPROPYL-ETHANOL AMINE (DEHYDRO-FUSED) ✚
8570 ★	2.42		1524		C12H15N1O2	1(3H)-ISOBENZOFURANONE,3,3-DIMETHYL-6-DIMETHYLAMINO
1 ★	2.86		1524		C12H15N1O2	1(3H)-ISOBENZOFURANONE,3-BUTYL-6-AMINO
2 ★	1.73		2157		C12H15N1O3	2-(BENZOYLOXY)-(N-ISOPROPYL)ACETAMIDE
3 ★	1.88		2157		C12H15N1O3	2-(BENZOYLOXY)-(N-PROPYL)ACETAMIDE
4 ★	1.27		2157		C12H15N1O3	2-(BENZYLOXY)-(N-METHYL-N-ETHYL)ACETAMIDE
75 ★	1.39		1705		C12H15N1O3	BENZOIC ACID,4-CARBOXAMIDOBUTYL ESTER
6 ★	2.32		1189		C12H15N1O3	CARBOFURAN **1563-66-2** *insecticide*
7 ★	0.74		546		C12H15N1O3	N-PHENOXYACETYLMORPHOLINE **18495-00-6**
8 ★	0.92	2.2	2228	314	C12H15N1O3	PHENYLALANINE-N-ACETYL,METHYL ESTER **21156-62-7**
9 ★	0.58		2157		C12H15N1O4	2-(BENZOYLOXY)-(N-METHYL,N-(2-ETHANOL)ACETAMIDE
8580	-2.00	7.4	2282	1	C12H15N1O4	DOPAMINE,DIACETYL
1	1.66	7.4	2145	1	C12H15N1O4	SALICYLAMIDE,BUTYRYLOXYMETHYL ESTER
2 ★	0.33	2.2	2228	314	C12H15N1O4	TYROSINE-N-ACETYL,METHYL ESTER **2440-79-1**
3 ★	3.40		1946		C12H15N1O5	2-PROPENOIC ACID,3-(5-NITRO-2-FURANYL),PENTYL ESTER
4 ★	0.86	7.4	2268	1	C12H15N1O5	SALICYLIC ACID,(N-HYDROXYETHYL-N-METHYL)GLYCOLAMIDE ESTER
85 ★	-0.44		511		C12H15N1O8	GALACTOPYRANOSIDE,P-NITROPHENYL(BETA)
6	-0.78		511		C12H15N1O8	GLUCOPYRANOSIDE,2-NITROPHENYL(BETA) **2816-24-2**
7 ★	-0.51		511		C12H15N1O8	GLUCOPYRANOSIDE,3-NITROPHENYL(BETA) **20838-44-2**
8	-0.46		2629		C12H15N1O8	GLUCOPYRANOSIDE,P-NITROPHENYL(BETA) **3767-28-0**
9	-0.39		2648	448	"	GLUCOPYRANOSIDE,P-NITROPHENYL(BETA)

	logP	pH	Ref.	Note	MF	Name / CAS / activity
8590	-0.39		511		"	GLUCOPYRANOSIDE,P-NITROPHENYL(BETA)
1	-0.18		511		"	GLUCOPYRANOSIDE,P-NITROPHENYL(BETA)
2 ★	4.39		1693		C12H15N2O3P1S1	PHOXIM 14816-18-3 insecticide
3 ★	4.44		2506		C12H15N2O3P1S1	QUINALPHOS 13593-03-8 insecticide
4 ★	3.11	5.6	2165	1	C12H15N3	BENZOTRIAZOLE,1-CYCLOHEXYL
95 ★	3.63	5.6	2165	1	C12H15N3	BENZOTRIAZOLE,2-CYCLOHEXYL
6	0.05		1407		C12H15N3O1	1,2,3,6-H4-PYRIDINE,3-PYRIDINYLMETHYL-CONH- ✛
7	1.14		2653	459	C12H15N3O1	O-PHENYLENEUREA,N-PIPERIDIN-4YL
8	1.86		559		C12H15N3O2	1-(3-CARBETHOXYPROPYL)BENZOTRIAZOLE
9	0.86		559		C12H15N3O2	1-(5-CARBOXYPENTYL)BENZOTRIAZOLE
8600	2.41		559		C12H15N3O2	2-(3-CARBETHOXYPROPYL)BENZOTRIAZOLE
1	1.51		559		C12H15N3O2	2-(5-CARBOXYPENTYL)BENZOTRIAZOLE
2 ★	2.34	5.6	2156	1	C12H15N3O2	BENZOTRIAZOLE,1-(2-BUTANOIC ACID,ETHYL ESTER)
3	-0.16	5.6	2156	1	C12H15N3O2	BENZOTRIAZOLE,1-(2-HEXANOIC ACID)
4 ★	2.98	5.6	2165	1	C12H15N3O2	BENZOTRIAZOLE,1-(2-METHYL-2-PROPANOIC ACID,ETHYL ESTER)
5 ★	2.60	5.6	2156	1	C12H15N3O2	BENZOTRIAZOLE,1-(2-METHYL-2-PROPIONIC ACID,ETHYL ESTER)
6 ★	2.27	5.6	2165	1	C12H15N3O2	BENZOTRIAZOLE,1-(3-BUTANOIC ACID,ETHYL ESTER)
7 ★	2.88	5.6	2156	1	C12H15N3O2	BENZOTRIAZOLE,2-(2-BUTANOIC ACID,ETHYL ESTER)
8	0.14	5.6	2156	1	C12H15N3O2	BENZOTRIAZOLE,2-(2-HEXANOIC ACID)
9 ★	2.70	5.6	2165	1	C12H15N3O2	BENZOTRIAZOLE,2-(3-BUTANOIC ACID,ETHYL ESTER)
8610	-0.58	6.0	479		C12H15N3O2S1	1-METHYL-2-SULFANILAMIDE-1,2-DIHYDROPYRIDINE 51543-30-7
1	0.30	7.4	1377	1	C12H15N3O6	N-(2,4-DINITROPHENYL)NORLEUCINE 29854-66-8
2	1.76		1049		C12H15N4O5P1	N-ET-PHOSPHORAMIDATE,O-ET,O-A-CN-4-NO2-BENZALDOXIME
3	0.76		1479	459	C12H15N5O1	2,4-DIAMINO-5-(3-AMINO-4-MEO-BENZYL)PYRIMIDINE
4	0.49		1479	459	C12H15N5O1	2,4-DIAMINO-5-(4-AMINO-3-MEO-BENZYL)PYRIMIDINE
15	1.73		57		C12H15N5O1	3-PHENYLAMINO-4-AMINO-6-I-PR-1,2,4-TRIAZINE-5-ONE 33665-75-7
6	0.10	7.4	2037	1	C12H15N5O1	ISOGUANINE,1-METHYL-7-CYCLOHEX-2-ENYL
7	1.83	7.4	1485		C12H15N5O2	2,4-DIAMINOPYRIMIDINE,5(2,6-DIMEO-PYRIDIN-3-YL)CH2
8 ★	2.01	7.4	1485	2	"	2,4-DIAMINOPYRIMIDINE,5(2,6-DIMEO-PYRIDIN-3-YL)CH2
9	1.48	7.4	1485		C12H15N5O2	2,4-DIAMINOPYRIMIDINE,5(2,6-DIMEO-PYRIDIN-4YL)CH2-
8620 ★	1.64	7.4	1485	2	"	2,4-DIAMINOPYRIMIDINE,5(2,6-DIMEO-PYRIDIN-4YL)CH2-
1	-0.20		2302		C12H15N5O3S1	CIMETIDINE,2-(3-NITROPYRIDIN-4-ONE) ANALOG
2	-0.67	3.3	2590	331	C12H15N5O3S1	SULFAMETHOXAZOLE,N,N-DIMETHYLTRIAZENE 58489-34-2
3	0.35	7.5	652	1	"	SULFAMETHOXAZOLE,N,N-DIMETHYLTRIAZENE
4	1.51	6.6	2590	459	"	SULFAMETHOXAZOLE,N,N-DIMETHYLTRIAZENE
25 ★	-0.19	7.0	2440	1	C12H15N5O4	2'-DEOXYADENOSINE,N6-BENZOYL
6 ★	3.39		2410		C12H16Br1N1O1	ACETANILIDE,2-BROMO-4-BUTYL
7	4.21		1818		C12H16Br1N1O2	N-BUTYLBENZAMIDE,2-HYDROXY-3-METHYL-5-BROMO 40912-89-8
8 ★	3.43		2506		C12H16Cl1N1S1	ORBENCARB herbicide (protein synth.inhib.)
9 ★	3.40		1679		C12H16Cl1N1S1	THIOBENCARB 28249-77-6 herbicide
8630	0.84	7.4	2004	1	C12H16Cl1N3O3S1	CHLOROPENTAZIDE 1224-71-1
1 ★	2.01	6.5	321	1	C12H16Cl1N3O4S2	CHLOROTHIAZIDE-3-AMYL 2854-98-0
2	0.97	7.4	321	1	C12H16Cl1N3O4S2	HYDROCHLOROTHIAZIDE,3-SPIROCYCLOHEXYL
3 ★	3.80		1070		C12H16Cl2N2O1	NEBURON 555-37-3 herbicide
4 ★	3.23		2479		C12H16F1N3O2	PHENYLBIURET,2-FLUORO-5-ETHYL
35 ★	0.64	7.0	2440	1	C12H16F1N5O2	DIDEOXY-ARA-A,2,N6-DIMETHYL-2'-FLUORO
6 ★	3.36	13.0	1806	224	C12H16F3N1	FENFLURAMINE 458-24-2 anorexic
7 ★	1.28	2.7	547	340	C12H16N1O1	A-(2-PIPERIDYL)PHENYLCARBINOL
8 ★	1.47	7.5	2545	2	C12H16N2	3'-AMINO-MPTP
9 ★	1.48	7.5	2545	2	C12H16N2	4'-AMINO-MPTP
8640	0.39	7.4	2331	1	C12H16N2	6H-PYRIDO[1,2-H]1,7-NAPHTHYRIDINE,5,8,9,10,11,11A-HEXAHYDRO 112114-03-1
1	0.42		258		C12H16N2	N,N-DIMETHYLTRYPTAMINE 61-50-7 halucinogen
2 ★	3.04		547		C12H16N2	N-BENZALAMINOPIPERIDINE 1885-87-6
3	1.08	7.4	2535	1	C12H16N2	PYRROLINO[1,2-A]IMIDAZOLINE,7A-PHENYL ✛
4 ★	2.65		1263		C12H16N2O1	3-PHENYL-1-CYCLOPENTYLUREA 13140-89-1
45	1.22		2032	459	C12H16N2O1	4-PYRIMIDONE,(5,6)CYCLOHEXYL-(2,3)METHYCYLCYLOPENTYL
6 ★	0.35		1622		C12H16N2O1	PIPERAZINE,1-METHYL,4-BENZOYL 75349-23-4
7	1.21		1725	459	C12H16N2O1	PYRIDO-PYRIMIDINE ANALOG,6-ME(RINGS:566)
8	1.19		1725	459	C12H16N2O1	PYRIDO-PYRIMIDINE ANALOG,7-ME(RINGS:566) ✛
9	1.20		1725	459	C12H16N2O1	PYRIDO-PYRIMIDINE ANALOG,8-ME(RINGS:566)
8650	1.21		1725	459	C12H16N2O1	PYRIDO-PYRIMIDINE ANALOG,9-ME(RINGS:566)
1	1.24		1725	459	C12H16N2O1	PYRIDO-PYRIMIDINONE ANALOG(RINGS=567) ✛
2	1.25		1725	459	"	PYRIDO-PYRIMIDINONE ANALOG(RINGS=567)
3	1.17		1725	459	C12H16N2O1	PYRIDOPYRIMIDINE ANALOG (RINGS:666) ✛
4	-1.94	7.0	477	1	C12H16N2O2	1-(O-CYANOPHENOXY)-3-ETAMINOPROPANOL-2 41397-54-0
55 ★	0.40		1558		C12H16N2O2	PHENYLALANINE,N-ACETYL,N'-METHYLAMINO-AMIDE
6 ★	1.88		522		C12H16N2O2S1	1,2,4-BENZOTHIADIAZINE-1,1-DIOXIDE-3-(1-ETHYLPROPYL) 7752-09-2
7 ★	0.62		1439		C12H16N2O3	3-ETOCO-PYRIDO(1,2A)PYRIMIDIN-4-ON,4,6,7,8,9-H5,6-ME
8 ★	0.71		1439		C12H16N2O3	3-ETOCO-PYRIDO(1,2A)PYRIMIDIN-4-ON,4,6,7,8,9-H5-7-ME
9 ★	0.66		1439		C12H16N2O3	3-ETOCO-PYRIDO(1,2A)PYRIMIDIN-4-ON,4,6,7,8,9-H5-8-ME
8660 ★	3.10		2410		C12H16N2O3	ACETANILIDE,2-NITRO-4-BUTYL
1 ★	1.77	7.4	899	1	C12H16N2O3	CYCLOBARBITAL 52-31-3 hypnotic, sedative
2 ★	1.49		505		C12H16N2O3	HEXOBARBITAL 56-29-1 sedative
3	1.98	7.4	899	1	"	HEXOBARBITAL
4 ★	-0.32		1558		C12H16N2O3	TYROSINE,N-ACETYL,N'-METHYLAMINO-AMIDE
65 ★	2.24		2612		C12H16N2O4	1-(4-ETHOXYCARBONYLPHENYL)-3-METHOXY-3-METHYLUREA 88132-25-6
6 ★	2.61		1946		C12H16N2O4	2-PROPENOIC-N-T-AMYLAMIDE,3-(5-NITRO-2-FURANYL)
7	1.54		1705		C12H16N2O4	N-(2-(P-NITROPHENOXY)ETHYL)MORPHOLINE 65300-53-0
8 ★	2.17		2072	2	C12H16N2O5S1	BENZENESULFONAMIDE,2-NITRO-N-(2-ETHYLBUTANOYL)
9 ★	1.00	7.4	1097	1	C12H16N2S1	XYLAZINE 7361-61-7 analgesic, muscle relaxant (vet)
8670	1.34	7.4	1350	1	"	XYLAZINE
1 ★	3.29		1049		C12H16N3O3P1	N,N-DIME-PHOSPHORAMIDATE,O-ET,O-A-CN-BENZALDOXIME
2	-0.02	7.4	2495		C12H16N3O3P1	PHOSPHORIC TRIAMIDE,N-P-METHOXYBENZOYL-N',N*-DIETHYLENE
3 ★	1.08	7.4	2495	2	C12H16N3O3P1	PHOSPHORIC TRIAMIDE,N-P-METHOXYBENZOYL-N',N*-DIETHYLENE
4	0.04	7.4	2495	1	C12H16N3O3P1	PHOSPHORIC-TRIAMIDE,N-M-METHOXYBENZOYL-N',N*-DIETHYLENE
75 ★	0.86	7.4	2495	2	C12H16N3O3P1	PHOSPHORIC-TRIAMIDE,N-M-METHOXYBENZOYL-N',N*-DIETHYLENE
6 ★	3.55		1639		C12H16N3O3P1S1	TRIAZOPHOS 24017-47-8 insecticide~miticide~nematocide
7 ★	3.40		1693		C12H16N3O3P1S2	AZINPHOS-ETHYL 2642-71-9 insecticide
8 ★	1.63		1693		C12H16N3O4P1S1	AZINPHOS-ETHYL,O-ANALOG 39923-25-6
9 ✓	-1.10		2406	102	C12H16N3S1	TETRAMETHYLTHIOURONIUM IODIDE,S-(3-CYANOPHENYL)

	logP	pH	Ref.	Note	MF	Name / CAS / activity
8680 ✓	-1.13		2406	102	C12H16N3S1	TETRAMETHYLTHIOURONIUM IODIDE,S-(4-CYANOPHENYL)
1 ★	0.79		559		C12H16N4O1	1-(5-CARBAMYLPENTYL)BENZOTRIAZOLE
2 ★	1.41		559		C12H16N4O1	2-(5-CARBAMYLPENTYL)BENZOTRIAZOLE
3 ★	1.88	5.6	2156	1	C12H16N4O1	BENZOTRIAZOLE,1-(2-HEXANAMIDE)
4 ★	2.28	5.6	2165	1	C12H16N4O1	BENZOTRIAZOLE,2-(2-HEXANAMIDE)
85 ★	0.80		2302		C12H16N4O1S1	CIMETIDINE,2(PYRIDIN-4-ONE) ANALOG ✚
6 ★	0.79	9.0	1949		C12H16N4O1S1	CIMETIDINE,2-PYRIDON-6YL ANALOG 54855-69-5
7 ★	0.13		2649		C12H16N4O2	BENZOQUINONE,2,5-BIS-AZIRIDINYL-3,6-BIS-METHYLAMINO
8 ★	2.44		2424		C12H16N4O2S1	BENZOTRIAZOLE,2-CYCLOPENTYL-5-METHYLSULFONAMIDO
9	-0.08	9.0	1949		C12H16N4O2S1	CIMETIDINE,(4-METHYLAMINOCYCLOBUTENEDIONE)-ANALOG 99035-21-9
8690 ★	-0.08		2302		C12H16N4O2S1	CIMETIDINE,(METHYLAMINOCLYCLOBUTANEDIONE) ANALOG ✚
1	-0.77		1404		C12H16N4O7	2-NO2-IMIDAZOLE,1-CH2CONHCH2CH(OAC)CH2OAC
2 ★	-2.48	7.4	594	1	C12H16N6O1	1,2-DIHYDROTRIAZINE,2,2-DIMETHYL-4,6-DIAMINO-1-(3'-AMIDO)PHENYL
3 ★	1.10		2289		C12H16N6O2	TETRAZOLE,2-ETHYL-5-(4-DIMETHYLUREYL)PHENOXY 117121-42-3
4 ★	4.22		570		C12H16O1	P-CYCLOHEXYLPHENOL 1131-60-8
95 ★	3.82		2487		C12H16O2	1,2-DIHYDROXYBENZENE-4-HEXEN-1YL
6 ★	3.27		2430		C12H16O2	6-PHENYLCAPROIC ACID 5581-75-9
7	4.10	2.0	2312	337	C12H16O2	BENZOIC ACID,I-PENTYL ESTER
8 ★	3.26		2168		C12H16O2	ISOVALERIC ACID,BENZYL ESTER
9 ★	3.33		2168		C12H16O2	ISOVALERIC ACID-2-PHENYL,METHYL ESTER
8700 ★	3.40		2645		C12H16O2	PHENYLACETATE,2,6-DIETHYL
1 ★	3.32		558		C12H16O3	O-SEC-BUTYLPHENOXYACETIC ACID
2 ★	3.33		558		C12H16O3	O-T-BUTYLPHENOXYACETIC ACID 19271-90-0
3 ★	3.18		501		C12H16O3	PHENOXYACETIC ACID,M-BUTYL 1878-53-1
4 ★	2.96		501		C12H16O3	PHENOXYACETIC ACID,M-T-BUTYL 1878-55-3
5 ★	3.12		501		C12H16O3	PHENOXYACETIC ACID,P-S-BUTYL 4917-89-9
6 ★	2.33		1521		C12H16O3S1	(1R)CIS-CYPR-CO2ME-2,2-DIME-3(2-OXOTHIEN-3-YLIDEN-ME) ✚
7 ★	1.75	3.4	1708		C12H16O4	PHENYLACETIC ACID,3-METHOXY-4-I-PROPOXY
8 ★	3.93		2186		C12H16O4S2	DITHIOLANYLIDENEMALONIC ACID,DIPROPYL ESTER ✚
9 ★	3.75		2186		C12H16O4S2	MALOTILATE 59937-28-9 liver disorder treatment
8710 ★	2.67		1335		C12H16O5	AMYLGALLATE 4568-93-8
1 ★	2.80		1335		C12H16O5	I-AMYLGALLATE 2486-02-4
2	-0.56		1294		C12H16O5S1	1-PHENYLTHIO-B-GALACTOPYRANOSIDE
3 ★	1.34		1589		C12H16O5S1	ALLYL-(3,4,5-TRIMETHOXY)PHENYL-SULFONE
4 ★	-0.89		2648	448	C12H16O6	BETAPHENYL-GALACTOSIDE
15 ★	-0.75		547		C12H16O6	GLUCOPYRANOSIDE,PHENYL(BETA) 1464-44-4
6	-1.64		2629		C12H16O7	B-D-GLUCOPYRANOSIDE,P-HYDROXYPHENYL 497-76-7 diuretic, urinary anti-infective
7	-1.35		511		"	B-D-GLUCOPYRANOSIDE,P-HYDROXYPHENYL
8	2.09	7.4	899	1	C12H17Br1N2O3	BARBITURIC ACID,5-(1-MEBUTYL)-5-BROMALLYL 1216-40-6
9	2.43	7.4	899	1	C12H17Br1N2O3	BARBITURIC ACID,S-BUTYL,BROMALLYL,N-METHYL
8720 ★	1.13		1205		C12H17Br1N2O3S1	N,N-DIMETHYLBENZENESULFONAMIDE,4-(N-METHYL-2-BROMOPROPIONAMIDO)
1 ★	2.00		1019		C12H17Cl1N2O2	N-I-PR-1(2-CL-PHENOXY)PROPIONIC ACID HYDRAZIDE
2 ★	2.18		1019		C12H17Cl1N2O2	N-I-PR-1(3-CL-PHENOXY)PROPIONIC ACID HYDRAZIDE
3 ★	2.14		1019		C12H17Cl1N2O2	N-I-PR-1(4-CL-PHENOXY)PROPIONIC ACID HYDRAZIDE
4 ★	2.55		2072	2	C12H17Cl1N2O3S1	BENZENESULFONAMIDE,2-(2-ETHYLBUTANOYLAMINO)-4-CHLORO
25 ★	2.53		2072	2	C12H17Cl1N2O3S1	BENZENESULFONAMIDE,2-AMINO-4-CHLORO-N-(2-ETHYLBUTANOYL)
6 ★	0.98		1205		C12H17Cl1N2O3S1	N,N-DIMETHYLBENZENESULFONAMIDE,4-(N-METHYL-2-CHLOROPROPIONAMIDO)
7 ★	-0.64		2518	122	C12H17F1N4O3	2'-DEOXY-3'-FLUORO-CYTIDINE,N4-(DIMETHYLAMINO)METHYLENE
8 ★	1.39		1205		C12H17I1N2O3S1	N,N-DIMETHYLBENZENESULFONAMIDE,4-(N-METHYL-2-IODOPROPIONAMIDO)
9	3.31	7.5	2201		C12H17N1	4-CYCLOHEXYLANILINE 6373-50-8
8730 ★	3.65		579		"	4-CYCLOHEXYLANILINE
1	-0.69	7.4	560	1	C12H17N1O1	A-(2-PIPERIDYL)BENZYL ALCOHOL.HCL 5583-32-4
2 ★	1.58	7.4	560	2	"	A-(2-PIPERIDYL)BENZYL ALCOHOL.HCL
3 ★	1.85		2410		C12H17N1O1	ACETANILIDE,2,6-DIETHYL
4 ★	3.30		2410		C12H17N1O1	ACETANILIDE,4-BUTYL 3663-20-5
35 ★	3.05		1894		C12H17N1O1	ACETANILIDE,4-T-BUTYL 20330-45-4
6 ★	1.38		473		C12H17N1O1	B-PHENYL-B-HYDROXY-N-ETHYLPYRROLIDINE 5407-61-4
7 ★	3.28		603		C12H17N1O1	CAPRYLANILIDE 621-15-8
8 ★	2.02		2328		C12H17N1O1	DIETHYLTOLUAMIDE 134-62-3 repellant-arthropod
9 ★	1.74		1718		C12H17N1O1	P-METHYLBENZYLIDENE-T-BUTYLAMINE,N-OXIDE
8740 ★	0.95		803		C12H17N1O2	2,3,5,6-TETRAMETHYL-4-HYDROXYACETANILIDE
1 ★	2.65		558		C12H17N1O2	2-I-PROPYLPHENYLDIMETHYLCARBAMATE 25007-30-1
2 ★	1.87		803		C12H17N1O2	3,5-DIETHYL-4-HYDROXYACETANILIDE 55205-89-5
3 ★	2.36		803		C12H17N1O2	3-T-BUTYL-4-HYDROXYACETANILIDE 4151-47-7
4 ★	3.68		2142		C12H17N1O2	4-BUTYLPHENYLCARBAMATE,O-METHYL
45 ★	2.72		2410		C12H17N1O2	ACETANILIDE,3-PROPYL-4-METHOXY
6 ★	2.93		1969		C12H17N1O2	DIETHYLCARBAMATE,BENZYL ESTER 51170-56-0
7 ★	2.60		1969		C12H17N1O2	DIMETHYLCARBAMATE,(1-METHYL-2-PHENYL)ETHYL ESTER 98154-93-9
8 ★	2.93		1969		C12H17N1O2	DIMETHYLCARBAMATE,3-PHENYLPROPYL ESTER 91564-01-1
9	1.92	7.4	2353	2	C12H17N1O2	MORPHOLINE,2-HYDROXY-2-PHENYL-3,4-DIMETHYL
8750	1.79	7.4	2353	2	C12H17N1O2	MORPHOLINE,2-HYDROXY-2-PHENYL-4,6-DIMETHYL
1 ★	2.04		757		C12H17N1O2	N-ETHYL-N-(2-ACETOXYETHYL)ANILINE
2 ★	2.65		273		C12H17N1O2	N-METHYL-2-T-BUTYLPHENYLCARBAMATE 2626-81-5
3 ★	3.11		273		C12H17N1O2	N-METHYL-3-METHYL-4-I-PROPYLPHENYLCARBAMATE 3766-82-3
4 ★	2.84		273		C12H17N1O2	N-METHYL-3-METHYL-6-I-PROPYLPHENYLCARBAMATE 18659-24-0
55 ★	3.20		273		C12H17N1O2	N-METHYL-4-S-BUTYLPHENYLCARBAMATE 3942-51-6
6 ★	3.06		273		C12H17N1O2	N-METHYL-4-T-BUTYLPHENYLCARBAMATE 2626-83-7
7 ★	3.10		273		C12H17N1O2	N-METHYLCARBAMATE,3-I-PROPYL-5-METHYLPHENYL 2631-37-0 insecticide (?)
8 ★	2.93		273		C12H17N1O2	N-METHYLCARBAMATE,M-T-BUTYLPHENYL 780-11-0
9 ★	2.78		273		C12H17N1O2	N-METHYLCARBAMATE,O-S-BUTYLPHENYL 3766-81-2 insecticide
8760 ★	3.59		2107	537	C12H17N1O2	NICOTINIC ACID,HEXYL ESTER 23597-82-2
1 ★	3.48		2630		C12H17N1O2	P-AMINOBENZOIC ACID,PENTYL ESTER 13110-37-7
2 ★	1.21		1718		C12H17N1O2	P-METHOXYBENZYLIDENE-T-BUTYLAMINE,N-OXIDE
3	1.33	7.4	2462	1	C12H17N1O2	P-TOLUIC ACID,N,N-DIMETHYLAMINOETHYL ESTER
4 ★	2.53	7.4	2462	2	"	P-TOLUIC ACID,N,N-DIMETHYLAMINOETHYL ESTER

	logP	pH	Ref.	Note	MF	Name / CAS / activity
65 ★	3.80		1263		C12H17N1O2	PENTYL-N-PHENYLCARBAMATE
6	-0.93	7.4	2261	1	C12H17N1O2	TETRALIN,6,7-DIHYDROXY-2-DIMETHYLAMINO
7 ★	2.98		536		C12H17N1O2S1	2-BUTYLTHIOPHENYLCARBAMATE,N-METHYL
8 ★	2.30		866	838	C12H17N1O3	2-I-PENTOXY-4-AMINOBENZOIC ACID 61566-64-1
9 ★	2.06		558		C12H17N1O3	2-I-PROPOXYPHENYLDIMETHYLCARBAMATE
8770 ★	2.38		866		C12H17N1O3	2-PENTOXY-4-AMINOBENZOIC ACID 61566-63-0
1	1.47	1.4	693	2	C12H17N1O3	BUFEXAMIC ACID 2438-72-4 *anti-inflammatory, analgesic*
2	2.08	7.4	693	1	"	BUFEXAMIC ACID
3 ★	2.96		273		C12H17N1O3	N-METHYL-3-BUTOXYPHENYLCARBAMATE 3978-68-5
4 ★	2.86		273		C12H17N1O3	N-METHYL-4-BUTOXYPHENYLCARBAMATE 3978-69-6
75	1.01	7.4	2462	1	C12H17N1O3	P-METHOXYBENZOIC ACID,N,N-DIMETHYLAMINOETHYL ESTER
6 ★	2.21	7.4	2462	2	"	P-METHOXYBENZOIC ACID,N,N-DIMETHYLAMINOETHYL ESTER
7	1.67		2222		C12H17N1O3S1	ETHYLSULFONAMIDE,N-(4-PHENYL)BUTYRYL
8 ★	2.09		2142		C12H17N1O4	3,4-DIETHOXYPHENYLCARBAMATE,O-METHYL
9	0.55	7.4	2062	1	C12H17N1O4	L-DOPA,PROPYL ESTER 39638-51-2
8780 ★	0.76	7.4	2062	2	"	L-DOPA,PROPYL ESTER
1 ★	1.50		731		C12H17N1O4	N-MALEOYL-6-AMINOHEXANOIC ACID,ETHYL ESTER
2	-0.81		731	605	C12H17N1O4	N-MALEOYL-8-AMINOOCTANOIC ACID
3	1.99		731	604	"	N-MALEOYL-8-AMINOOCTANOIC ACID
4 ★	2.71		773	601	C12H17N1O4S1	4-SULFAMYLBENZOIC ACID,PENTYL ESTER 59777-60-5
85	-0.71	7.4	2062	1	C12H17N1O5	L-DOPA,2-HYDROXYPROPYL ESTER
6 ★	-0.49	7.4	2062	2	"	L-DOPA,2-HYDROXYPROPYL ESTER
7	-1.23		511		C12H17N1O6	GLUCOPYRANOSIDE,2-AMINOPHENYL(BETA) 7265-01-2
8	-2.67		511		C12H17N1O6	GLUCOPYRANOSIDE,4-AMINOPHENYL(BETA) 20818-25-1
9	-1.40		2406	102	C12H17N2O2S1	TETRAMETHYLTHIOURONIUM IODIDE,S-(2-CARBOXYPHENYL)
8790 ✓	-1.28		2406	102	C12H17N2O2S1	TETRAMETHYLTHIOURONIUM IODIDE,S-(4-CARBOXYPHENYL)
1 ★	3.70		559		C12H17N3	1-HEXYLBENZOTRIAZOLE
2 ★	3.44		559		C12H17N3	1-S-HEXYLBENZOTRIAZOLE
3	-2.30	7.4	1398	1	C12H17N3	2-(2,4,6-TRIMETHYLPHENYLIMINO)IMIDAZOLIDINE
4	-1.28		976	1	"	2-(2,4,6-TRIMETHYLPHENYLIMINO)IMIDAZOLIDINE
95 ★	2.32	7.4	976	2	"	2-(2,4,6-TRIMETHYLPHENYLIMINO)IMIDAZOLIDINE
6 ★	4.24		559		C12H17N3	2-HEXYLBENZOTRIAZOLE
7 ★	3.78		559		C12H17N3	2-S-HEXYLBENZOTRIAZOLE
8 ★	1.71		559		C12H17N3O1	1-(6-HYDROXYHEXYL)BENZOTRIAZOLE
9 ★	2.26		559		C12H17N3O1	2-(6-HYDROXYHEXYL)BENZOTRIAZOLE ✦
8800	0.26	7.4	365	2	C12H17N3O1	PIPERAZINE-2-CARBOXANILIDE,2'-ME 36385-59-8
1 ★	2.26		592		C12H17N3O1S1	2'-HYDROXY-5'-METHYLACETOPHENONE,(N,N-DIMETHYLTHIO)SEMICARBIZONE
2	0.96	3.3	2590	331	C12H17N3O2	3-BUTYL-1-(4-CARBOXYPHENYL)-3-METHYLTRIAZENE 74109-22-1 *antineoplastic*
3	0.97	7.4	1214	1	"	3-BUTYL-1-(4-CARBOXYPHENYL)-3-METHYLTRIAZENE
4	2.34		781		C12H17N3O2	6-T-BU-3-I-PR-ISOXAZOLO(3,4D)PYRIMIDINONE 38897-15-3
5	-0.08	7.4	2607	1	C12H17N3O2	PHENYLBUTYRIC ACID,P-(3,3-DIMETHYL-1-TRIAZENO)
6	0.32	7.0	2607	1	"	PHENYLBUTYRIC ACID,P-(3,3-DIMETHYL-1-TRIAZENO)
7 ★	2.62	7.0	2607	2	"	PHENYLBUTYRIC ACID,P-(3,3-DIMETHYL-1-TRIAZENO)
8	0.60		1439		C12H17N3O2	PYRIDO(1,2A)PYRIMIDIN-4-ON,3-CONH2,-H7,1-ET,6-ME ✦
9 ★	2.66	6.3	2590	459	C12H17N3O3	ETHYL BENZOATE,4-(3-METHYL-3-HYDROXYETHYLTRIAZENYL) *antineoplastic*
8810 ★	0.65	7.4	2004	1	C12H17N3O3S1	BENZENESULFONAMIDE,4-METHYL-N(1-PYRROLIDINE)AMINOCARBONYL 1220-55-9
1	-1.49	7.0	1812	1	C12H17N3O5	5'-AMINO-3'O-ACETYL-5'-DEOXYTHYMIDINE
2 ★	0.30		1205		C12H17N3O5S1	N,N-DIME-4(2-NITROPROPIONYL)BENZENESULFONAMIDE
3	-1.83	7.4	594	1	C12H17N5	1,2-DIHYDROTRIAZINE,2,2-DIMETHYL-4,6-DIAMINO-1-(3'-METHYL)PHENYL
4 ★	0.85	13.0	547	224	C12H17N5	4,6-DIAM-2,2-DIME-1(4-METHYLPH)-S-TRIAZINE
15 ★	2.18	7.4	1776	1	C12H17N5	N(2,5DIME-PYRROL-1YL)-6(DIMEAM)-3-PYRIDAZINEAMINE 75842-06-7
6	-2.57		594	314	C12H17N5O1	1,2-DIHYDROTRIAZINE,4,6-DIAMINO-2,2-DIMETHYL-1-(3-METHOXY)PHENYL
7 ★	0.40	13.0	547	224	C12H17N5O1	4,6-DIAM-2,2-DIME-1(4-MEOPH)-S-TRIAZINE
8 ★	0.98		2544		C12H17N5O1	9-PENTYL-N-ACETYLADENINE
9 ★	0.97		2218		C12H17N5O1S1	GUANIDINE,N-METHYL-N'-CYANO-N"-(3-METHOXYPYRID-2YL)METHYLTHIOETHYL
8820	0.19	8.0	2267	306	C12H17N5O3	ALLOPURINOL,1-LEUCINYLOXYMETHYL 98204-06-9
1 ★	0.20	8.0	2267	306	C12H17N5O3	ALLOPURINOL,1-N,N-DIETHYLGYCYLOXYMETHYL 98204-08-1
2	-0.90	7.4	2037		C12H17N5O3	ISOGUANINE,1-METHYL-7-BUTYRIC ACID,ETHYL ESTER
3 ★	-0.10		2474		C12H17N5O4	NIFURPIPONE 24632-47-1 *antibacterial*
4 ★	3.69		2506		C12H17O4P1S2	PHENTHOATE 2597-03-7 *insecticide (cholinesterase inhib.)*
25	5.65		1369	459	C12H18	1,5,9-CYCLODODECATRIENE
6	4.12		1772	459	C12H18	CYCLODODECATRIENE 4904-61-4
7 ★	4.61		2146	458	C12H18	HEXAMETHYLBENZENE 87-85-4
8 ★	5.52		1560	459	C12H18	HEXYLBENZENE 1077-16-3
9 ✓	0.45		2406	102	C12H18Cl1N2S1	TETRAMETHYLTHIOURONIUM IODIDE,S-(3-CHLORO-4-METHYLPHENYL) ✦
8830	0.78	5.0	831	1	C12H18Cl1N5	N1-P-CHLOROPHENYL-N5-BUTYLBIGUANIDE 60221-94-5
1	4.27	6.6	2590	459	C12H18Cl2N4	ANILINE,N,N-BIS(2-CHLOROETHYL),4-(3,3-DIMETHYLTRIAZENYL) *antineoplastic*
2 ✓	-1.02		2397	101	C12H18F3N4O3S1	IMIDAZOLIUM CHLORIDE,1-METHYL-2-HYDROXYIMINOMETHYL-3-(1-TRIFLUOROMETHYLSULFONYLPIPERIDIN-3YL)METHYL
3 ✓	-0.69		2397	101	C12H18F3N4O3S1	IMIDAZOLIUM CHLORIDE,1-METHYL-2-HYDROXYIMINOMETHYL-3-(1-TRIFLUOROMETHYLSULFONYLPIPERIDN-2YL)METHYL
4 ✓	-2.48		2180		C12H18N1O2.Br1	BENZYLTRIMETHYL AMMONIUM BROMIDE,3-METHOXYCARBONYL
35 ★	2.40		579		C12H18N1O5P1	T-BUTYLPHOSPHONIC ACID,ETHYL,P-NITROPHENYL ESTER 13538-14-2
6 ★	1.66	11.0	533	322	C12H18N2	3-PYRIDYLETHYL-2-(N-PIPERIDINE) 13450-66-3
7 ★	2.87		580		C12H18N2O1	1,1-DIMETHYL-3-(4-I-PROPYLPHENYL)UREA 34123-59-6 *herbicide*
8	0.94	7.4	1414	1	C12H18N2O1	2(DIMEAMINO)-N(2,6-DIMEPHENYL)ACETAMIDE 21236-54-4
9	0.84		1352		C12H18N2O1	2-AMINOBUTYRYL-2',6'-XYLIDIDE
8840	0.71		1352		C12H18N2O1	3-AMINOBUTYRYL-2',6'-XYLIDIDE
1	1.43		1439		C12H18N2O1	4H-PYRID(1,2A)PYRIMIDIN-4-ON,6,7,8,9-H4,2,6-DIME,3-ET ✦
2 ★	1.27		1622		C12H18N2O1	BENZAMIDE,N-(3-DIMETHYLAMINOPROPYL) 40948-30-9
3 ★	1.32		1983		C12H18N2O1	DESETHYL LIDOCAINE 7728-40-7
4	-0.50	3.0	1268		C12H18N2O1	N-(I-BUTYLAMINOMETHYL)BENZAMIDE 52387-58-3
45	1.17	7.4	1268	1	"	N-(I-BUTYLAMINOMETHYL)BENZAMIDE
6	1.73	9.2	1268	1	"	N-(I-BUTYLAMINOMETHYL)BENZAMIDE
7 ★	2.55		2094		C12H18N2O1	UREA,1,1-DIMETHYL-3-(4-PROPYLPHENYL) 102587-47-3
8 ★	2.95		2612		C12H18N2O2	1-(4-I-PROPYLPHENYL)-3-METHOXY-3-METHYLUREA 34861-40-0
9	1.95		29		C12H18N2O2	BENZOIC ACID,P-ME-AMINO,N,N-DIMETHYLAMINOETHYL ESTER
8850	1.83		1019		C12H18N2O2	N'-I-PROPYL-(2-TOLYLOXY)ACETIC ACID HYDRAZIDE
1 ★	1.77		1019		C12H18N2O2	N'-I-PROPYL-(3-TOLYLOXY)ACETIC ACID HYDRAZIDE
2 ★	1.81		1019		C12H18N2O2	N'-I-PROPYL-(4-TOLYLOXY)ACETIC ACID HYDRAZIDE
3 ★	1.39		1019		C12H18N2O2	N'-I-PROPYL-1-PHENOXYPROPIONIC ACID HYDRAZIDE
4 ★	2.28		1681		C12H18N2O2	N,N-DIMETHYLCARBAMATE,M-(DIMEAMINO)BENZYL ESTER 84640-28-8

	logP	pH	Ref.	Note	MF	Name / CAS / activity
55	2.61	7.4	2659	459	"	N,N-DIMETHYLCARBAMATE,M-(DIMEAMINO)BENZYL ESTER
6 ★	1.14		1622		C12H18N2O2	O-METHOXYBENZAMIDE,N-(2-DIMETHYLAMINOETHYL)
7 ★	0.58		1622		C12H18N2O2	O-METHOXYBENZAMIDE,N-(4-AMINOBUTYL)
8 ★	3.23		505		C12H18N2O2S1	THIAMYLAL 77-27-0 anesthetic (intravenous)
9	0.79		1439		C12H18N2O3	3-ETOCO-PYRIDO(1,2A)PYRIMIDIN-4-ON,-H7,6-ME
8860	-0.47	7.4	2685	1	C12H18N2O3	4-PYRIDONE,3-HYDROXY-2-METHYL-1-(2-I-PROPYLAMINOCARBONYL)ETHYL iron chelator
1	-0.40	7.4	2685	1	C12H18N2O3	4-PYRIDONE,3-HYDROXY-2-METHYL-1-(2-PROPYLAMINOCARBONYL)ETHYL iron chelator
2	-0.69	7.4	2685	1	C12H18N2O3	4-PYRIDONE,3-HYDROXY-2-METHYL-1-(N-ETHYL,N-METHYLPROPIONAMIDO) iron chelator
3 ★	1.97	7.4	899	1	C12H18N2O3	BARBITURIC AICD, 5-ALLYL-5-(1-METHYLBUTYL) hypnotic, sedative
4 ★	1.11		1019		C12H18N2O3	N'-I-PROPYL-(2-METHOXYPHENOXY)ACETIC ACID HYDRAZIDE
65 ★	1.33		1019		C12H18N2O3	N'-I-PROPYL-(3-METHOXYPHENOXY)ACETIC ACID HYDRAZIDE
6 ★	1.21		1019		C12H18N2O3	N'-I-PROPYL-(4-METHOXYPHENOXY)ACETIC ACID HYDRAZIDE
7	1.71	7.4	899	1	C12H18N2O3	NEALBARBITAL 561-83-1 sedative
8 ★	2.06	1.0	2202	342	"	NEALBARBITAL
9 ★	1.51		773	601	C12H18N2O3S1	4-SULFAMYLBENZAMIDE,N-PENTYL 59777-64-9
8870 ★	1.62		2072	2	C12H18N2O3S1	BENZENESULFONAMIDE,2-(2-ETHYLBUTANOYLAMINO)
1 ★	1.64		2072	2	C12H18N2O3S1	BENZENESULFONAMIDE,2-AMINO-N-(2-ETHYLBUTANOYL)
2 ★	0.71		1205		C12H18N2O3S1	N,N-DIME-4(N-ME-N-PROPIONYL)BENZENESULFONAMIDE
3	2.30	5.9	1351		C12H18N2O3S1	N-(5(1-PIPERIDINO)THIAZOLIDON-2-YLIDENE)AC.AC.ET.ES
4	2.34	7.4	1351	1	"	N-(5(1-PIPERIDINO)THIAZOLIDON-2-YLIDENE)AC.AC.ET.ES
75	0.70	5.9	1351		C12H18N2O3S1	N-ME(5-PYRROLIDIN-THIAZOLIDON-2-YLIDENE)AC.AC.ET.ES
6	2.00	7.4	1351	1	"	N-ME(5-PYRROLIDIN-THIAZOLIDON-2-YLIDENE)AC.AC.ET.ES
7 ★	2.34	7.3	538	2	C12H18N2O3S1	TOLBUTAMIDE 64-77-7 antidiabetic
8	2.52	7.4	2004		"	TOLBUTAMIDE
9	1.87		2447		C12H18N2O4	4-PYRIDINECARBOXYLIC ACID,TETRAHYDRO-2,6-DIOXO, HEPTYL ESTER
8880 ✓	-2.13		2426		C12H18N2O4Pt1.(N1O3)2	PLATINUM,BIS-(P-HYDROXYMETHYLPYRIDONIO)-DIAQUA-DINITRATE
1	-2.12	7.2	846	314	C12H18N2O4S1	6-(N-BUTYRYLAMINO)PENICILLANIC ACID 54661-85-7
2 ★	-0.26		1205		C12H18N2O4S1	N,N-DIME-4(N-ME-N-(2-OH-PROPIONYL)BENZENESULFONAMIDE
3 ★	3.03	2.1	2265		C12H18N2O5S1	BENZENESULFONAMIDE,3-NITRO
4 ✓	-1.71		2397	101	C12H18N3O2	IMIDAZOLIUM CHLORIDE,1-METHOXYMETHYL-2-HYDROXYIMINOMETHYL-3-HEXYN-5YL
85 ✓	-1.36		2397	101	C12H18N3O2	IMIDAZOLIUM CHLORIDE,1-METHYL-2-HYDROXYIMINOMETHYL-3-(1,1-DIMETHYLBUTYN-2YLOXY)METHYL
6 ✓	-1.67		2397	101	C12H18N3O2	IMIDAZOLIUM CHLORIDE,1-METHYL-2-HYDROXYIMINOMETHYL-3-(1,2-DIMETHYLBUTYN-3YLOXY)METHYL
7 ✓	-1.55		2397	101	"	IMIDAZOLIUM CHLORIDE,1-METHYL-2-HYDROXYIMINOMETHYL-3-(1,2-DIMETHYLBUTYN-3YLOXY)METHYL
8 ✓	-1.10		2397	101	C12H18N3O2	IMIDAZOLIUM CHLORIDE,1-METHYL-2-HYDROXYIMINOMETHYL-3-(1-PROPYLPROPYN-2-YLOXY)METHYL
9 ✓	-1.25		2397	101	C12H18N3O2	IMIDAZOLIUM CHLORIDE,1-METHYL-2-HYDROXYIMINOMETHYL-3-(2,2-DIMETHYLBUTYN-3YLOXY)METHYL
8890 ✓	-1.42		2397	101	C12H18N3O2	IMIDAZOLIUM CHLORIDE,1-METHYL-2-HYDROXYIMINOMETHYL-3-(CYCLOHEXEN-3YLOXY)METHYL
1 ✓	0.23		2406	102	C12H18N3O3S1	TETRAMETHYLTHIOURONIUM IODIDE,S-(2-NITRO-4-METHYLPHENYL)
2 ✓	-0.22		2406	102	C12H18N3O3S1	TETRAMETHYLTHIOURONIUM IODIDE,S-(2-NITRO-4-METHOXYPHENYL)
3 ★	2.46		563		C12H18N4O1	4-(3-ME-3-BUTYL-1-TRIAZENO)BENZAMIDE 59708-21-5
4	1.38		2123	22	C12H18N4O1	BENZALDEHYDE,4-T-BUTOXY,GUANYLHYDRAZONE 102632-28-0
95 ★	3.52	6.6	2590	459	C12H18N4O1	BENZAMIDE,2-(3-METHYL-3-BUTYLTRIAZENYL) antineoplastic
6 ★	2.61		547		C12H18N4O1	P-(3-METHYL-3-T-BUTYLTRIAZENO)BENZAMIDE 59708-25-7
7 ★	-0.95		2518	122	C12H18N4O3	2',3'-DIDEOXYCYTIDINE,N4-(DIMETHYLAMINO)METHYLENE
8	1.07	7.4	2477	1	C12H18N4O3	A-[(2-NITROIMIDAZOL-1YL)-METHYL]-7-AZABICYCLO[4.1.0]HEPTANE-7-ETHANOL 120277-90-9
9 ★	1.68	1.0	2196		C12H18N4O3	PYRAZOLE,3-METHYL-4-NITRO-5-(N-METHYL-N-CYCLOHEXYLCARBOXAMIDO)
8900	-0.43	7.4	2491	1	C12H18N4O6S1	N-(2-AMINOETHYL)-N-ME-3-NITRO-4(1-ME-1-NITROETHYL)BENZENESULFONAMIDE 135685-04-0
1 ★	0.62	7.4	2491	2		N-(2-AMINOETHYL)-N-ME-3-NITRO-4(1-ME-1-NITROETHYL)BENZENESULFONAMIDE
2	3.73		2506		C12H18N4O6S1	ORYZALIN 19044-88-3 herbicide
3 ★	3.79		594		C12H18O1	PROPOFOL 2078-54-8 anesthetic (intravenous)
4 ★	3.45		1135	537	C12H18O2	RESORCINOL,4-HEXYL 136-77-6 anthelmintic
5 ★	2.32		2487		C12H18O3	PENTANOL,1-(3,4-DIHYDROXY)PHENYL
6 ★	1.12		2308		C12H18O4	1,2-BIS(2-METHOXYETHOXY)BENZENE
7 ★	2.67		2168		C12H18O4	CYCLOPROPANECARBOXYLIC ACID-2,2-DIMETHYL-3-(2-METHYLPROPENOIC ACID),DIMETHYL ESTER
8 ★	3.35		2011		C12H18O4S2	PROPANEDIOIC ACID,1,3-DITHIOLAN-2-YLIDINE,DIPROPYL ESTER 50780-68-2
9 ★	3.32		2011		C12H18O4S2	PROPANEDIOIC ACID,1,3-DITHIOLAN-2-YLIDINE-DI-I-PROPYL ESTER 50512-35-1
8910 ★	1.56		1589		C12H18O5S1	PROPYL-(3,4,5-TRIMETHOXY)PHENYL-SULFONE
1 ★	1.60		2011		C12H18O5S2	PROPANEDIOIC ACID,1,3-DITHIOLAN-2-YLIDINE-1-OXO,DI-I-PROPYL ESTER 52303-69-2
2	1.21	7.4	2477	1	C12H18Br1N4O3	TR-A-[[(2-BROMOCYCLOHEXYL)AMINO]METHYL]-2-NITROIMIDAZOLE-1-ETHANOL 134419-53-7
3 ★	3.42		354		C12H19Cl1N1O3P1	RUELENE 299-86-5 anthelmintic~insecticide
4 ✓	-1.26		2397	101	C12H19F3N5O3S1	IMIDAZOLIUM CHLORIDE,1-METHYL-2-HYDROXYIMINOMETHYL-3-(4-TRIFLUOROMETHYLSULFONYLPIPERRAZIN-1YL)ETHYL
15 ★	3.18		579		C12H19N1	2,6-DI-I-PROPYLANILINE 24544-04-5
6	0.58	7.4	785	1	C12H19N1	4-PHENYL-N,N-DIMETHYLBUTYLAMINE 1202-55-7
7 ★	3.36		1508		C12H19N1	BENZENEETHANEAMINE,A-BUTYL 67309-36-8
8 ★	3.29		1508		C12H19N1	BENZENEETHANEAMINE,A-I-BUTYL 67309-38-0
9 ★	2.91		1508		C12H19N1	BENZENEETHANEAMINE,A-T-BUTYL 67309-37-9
8920	5.00		442	2	C12H19N1	P-HEPTYLPYRIDINE 40089-90-5
1	0.33	7.5	1588	1	C12H19N1	PHENETHYLAMINE,N-METHYL-A-ISOPROPYL
2	3.12	7.4	1588	2	"	PHENETHYLAMINE,N-METHYL-A-ISOPROPYL
3	0.75	1.0	547	342	C12H19N1O1	2-N,N-DIETHYLAMINO-1-PHENYLETHANOL HYDROCHLORIDE
4 ★	2.02	13.0	547	224	C12H19N1O1	2-N,N-DIETHYLAMINO-1-PHENYLETHANOL 4249-64-3
25	-0.77	1.0	547	342	C12H19N1O1	2-N-BUTYLAMINO-1-PHENYLETHANOL
6	2.08	7.0	538	701	C12H19N1O2	2,5-DIMETHOXY-4-METHYLAMPHETAMINE 15588-95-1
7 ★	2.24	8.0	1011	2	"	2,5-DIMETHOXY-4-METHYLAMPHETAMINE
8	-0.28	7.4	487	1	C12H19N1O2	2-OH-I-PROPYLAMINOPROPYL-PHENYLETHER 7695-63-8
9 ★	1.72	7.4	487	2	"	2-OH-I-PROPYLAMINOPROPYL-PHENYLETHER
8930	1.90	7.4	2685	1	C12H19N1O2	4-PYRIDONE,3-HYDROXY-2-METHYL-1-HEXYL iron chelator
1	-1.08	7.4	2261	1	C12H19N1O2	DOPAMINE,N,N-DIETHYL
2 ★	2.17	8.0	1011	2	C12H19N1O2S1	2,5-DIMETHOXY-4-METHIOAMPHETAMINE
3	1.36	7.0	538	701	C12H19N1O3	2,3,4-TRIMETHOXYAMPHETAMINE 1082-23-1
4 ★	1.51	8.0	1011	2	"	2,3,4-TRIMETHOXYAMPHETAMINE
35 ★	1.74		538		C12H19N1O3	2,4,5-TRIMETHOXYAMPHETAMINE 1083-09-6
6 ★	1.57	7.0	538	701	C12H19N1O3	2,4,6-TRIMETHOXYAMPHETAMINE 15402-79-6
7 ★	1.21	8.0	1011	2	C12H19N1O3	3,4,5-TRIMETHOXYAMPHETAMINE 1082-88-8
8	1.48	7.0	538	701	"	3,4,5-TRIMETHOXYAMPHETAMINE
9	1.11	8.0	868	2	C12H19N1O3	3,5-DIMETHOXY-4-ETHOXYPHENETHYLAMINE 39201-82-6
8940	1.34	7.4	1074	324	C12H19N1O3	4-BUTOXYMETHYLPYRIDOXOL
1	0.72	7.4	2685	1	C12H19N1O3	4-PYRIDONE,3-HYDROXY-2-ETHYL-1-(3-ETHOXYPROPYL) iron chelator
2	-1.44	7.4	1168	1	C12H19N1O3	METIPRENALINE 1212-03-9 bronchodilator
3	0.27		1168		"	METIPRENALINE
4 ★	2.93	2.1	2467		C12H19N1O3S1	BENZENESULFONAMIDE,4-HEXYLOXY

	logP	pH	Ref.	Note	MF	Name / CAS / activity
45 ✓	-0.52		2406	102	C12H19N2O1S1	TETRAMETHYLTHIOURONIUM IODIDE,S-(2-METHOXYPHENYL)
6 ✓	-0.50		2406	102	C12H19N2O1S1	TETRAMETHYLTHIOURONIUM IODIDE,S-(3-METHOXYPHENYL)
7 ✓	-0.51		2406	102	C12H19N2O1S1	TETRAMETHYLTHIOURONIUM IODIDE,S-(4-METHOXYPHENYL)
8 ✓	-1.26		1729	820	C12H19N2O1.I1	3-CARBAMOYLPYRIDINIUM IODIDE,N-HEXYL 65481-90-5
9 ✓	-0.14		2406	102	C12H19N2S1	TETRAMETHYLTHIOURONIUM IODIDE,S-(3-METHYLPHENYL)
8950 ✓	-0.10		2406	102	C12H19N2S1	TETRAMETHYLTHIOURONIUM IODIDE,S-(4-METHYLPHENYL)
1 ✓	-0.10		2406	102	C12H19N2S1	TETRAMETHYLTHIOURONIUM IODIDE,S-2-METHYLPHENYL
2 ★	0.06	10.1	401	1	C12H19N3O1	PROCARBAZINE 671-16-9 antineoplastic
3 ★	0.48	7.4	2004	1	C12H19N3O3S1	BENZENESULFONAMIDE,4-METHYL-N-(2,2-DIMETHYLHYDRAZINOCARBONLYL)
4	0.56		2123	22	C12H19N5	BENZALDEHYDE,4-DIETHYLAMINO,GUANYLHYDRAZONE 102632-29-1
55	-0.90	7.4	2037	1	C12H19N5O3	2H-PURIN-2-ONE,1,3-DIHYDRO-1-METHYL-6-AMINO-7-CARBETHOXYPROPYL
6	-0.55	7.4	2491	1	C12H19N5O6S1	N-ME-N-(2-METHYLAMINO)ETHYL)3-NITRO-4-(UREIDO-OXY)ME)BENZENESULFONAMIDE 135685-12-0
7 ★	0.45	7.4	2491	2	"	N-ME-N-(2-METHYLAMINO)ETHYL)3-NITRO-4-(UREIDO-OXY)ME)BENZENESULFONAMIDE
8	1.78	7.3	2170	2	C12H19N7S1	AMBAZONE,THYMOL ANALOG ✚
9	2.29	6.7	2170	2	"	AMBAZONE,THYMOL ANALOG
8960 ★	4.90		2516		C12H19O2P1S3	SULPROFOS 35400-43-2 insecticide
1 ★	4.16		354		C12H19O3P1S2	O,O-DIET-O-(2-ME-4-METHIOPHENYL)PHOSPHOROTHIOATE 2670-77-1
2 ★	3.21		354		C12H19O4P1S1	O,O-DIET-O-(2-METHYL-4-METHIOPHENYL)PHOSPHATE 26512-63-0
3 ★	3.28		354		C12H19O4P1S1	O,O-DIET-O-(3-METHYL-4-METHIOPHENYL)PHOSPHATE 4799-59-1
4	2.24		547		C12H20Cl1N3O4	1-NO-1-(2-CLET)-3-(4-ACETYLOXYME)-CYCLOHEXYLUREA-T
65	2.74		401	335	C12H20Cl2N6O4	1,1'-(T-1,4-CYHEX)BIS(3-(2CLET)-3-NITROSO)UREA 13907-57-8
6 ★	1.34		547		C12H20F2N6O4	CYCLOHEXANE,1,4-BIS(1-NO-1-(2-FLET)-3-UREYL) 13908-98-0
7 ✓	-0.69		2397	101	C12H20F3N4O3S1	IMIDAZOLIUM CHLORIDE,1-ETHYL-2-HYDROXYIMINOMETHYL-3-(3-N-ETHYL-TRIFLUOROMETHYLSULFONAMIDO)PROPYL
8 ✓	-0.52		2397	101	C12H20F3N4O3S1	IMIDAZOLIUM CHLORIDE,1-METHYL-2-HYDROXYIMINOMETHYL-3-(3-N-PROPYL-TRIFLUOROMETHYLSULFONAMIDO)PROPYL
9 ✓	-0.68		2397	101	C12H20F3N4O3S1	IMIDAZOLIUM CHLORIDE,1-PROPYL-2-HYDROXYIMINOMETHYL-3-(3-N-METHYL-TRIFLUOROMETHYLSULFONAMIDO)PROPYL
8970 ✓	-0.40		2397	101	C12H20F3N4O3S1	IMIDAZOLIUM CHLORIDE,1-PROPYL-2-HYDROXYIMINOMETHYL-3-(N-ETHYL-TRIFLUOROMETHYLSULFONAMIDO)ETHYL
1 ✓	-1.52		2397	101	C12H20F3N4O5S2	IMIDAZOLIUM CHLORIDE ANALOG ✚
2 ✓	-1.83		2180		C12H20N1O1.Br1	BENZYLTRIMETHYL AMMONIUM BROMIDE,3-ETHOXY
3 ★	2.65		354		C12H20N1O3P1	METHYLPHOSPHORAMIDATE,O-ME-O-(3-T-BUTYLPHENYL) 19590-05-7
4 ★	2.71		354		C12H20N1O3P1	METHYLPHOSPHORAMIDATE,O-ME-O-(4-T-BUTYLPHENYL) 19590-04-6
75 ✓	-1.33		2180		C12H20N1S1.Br1	BENZYLTRIMETHYL AMMONIUM BROMIDE,3-ETHYLTHIO
6	-1.62		571		C12H20N1.Br1	M-ETHYLBENZYLTRIMETHYLAMMONIUMBROMIDE
7	-2.60		1635	820	C12H20N1.Cl1	BENZYLDIMETHYLPROPYLAMMONIUMCHLORIDE
8 ✓	-1.71		1729	820	C12H20N1.I1	2-METHYLPYRIDINIUM IODIDE,N-HEXYL 14402-24-5
9	-2.02		502		C12H20N1.I1	G-PHENYLPROPYL-TRIMETHYL-AMMONIUM IODIDE 2125-48-6
8980 ★	2.27	11.0	533	322	C12H20N2	N,N-DI-I-PROPYL-3-PYRIDYLMETHYLAMINE 20173-18-6
1	1.46	11.0	533	322	C12H20N2	N,N-DIPROPYL-3-PYRIDYLMETHYLAMINE 20173-16-4
2 ★	1.81		447	815	C12H20N2O2	ASPERGILLIC ACID 1873-43-4
3	1.12	7.0	2309	1	C12H20N2O2	N-ACETYL-2,2-BIS(METHALLYL)GLYCINAMIDE
4 ★	2.70	1.0	2202	342	C12H20N2O3	BARBITURIC ACID,5,5-DIBUTYL 17013-41-1
85 ★	2.45	0.7	962		C12H20N2O3	BARBITURIC ACID,5-ET-5-(1-MEBU),1-METHYL 57562-99-9
6 ★	2.45	0.7	962		C12H20N2O3	BARBITURIC ACID,5-ET-5-I-AMYL,1-METHYL 6270-45-7
7 ★	2.02		899		C12H20N2O3	BARBITURIC ACID,5-ETHYL-5-(2,3-DIMETHYLBUTYL) 36380-48-0
8 ★	2.46	7.4	899	1	C12H20N2O3	BARBITURIC ACID,5-HEXYL-5-ETHYL 77-30-5
9	1.48	5.9	1351		C12H20N2O3S1	N-ME-5(N,N-DIET)THIAZOLIDON-2-YLACETICAC.ET.EST.
8990	2.20	7.4	1351	1	"	N-ME-5(N,N-DIET)THIAZOLIDON-2-YLACETICAC.ET.EST.
1	-2.00	7.4	2399	1	C12H20N2O3S1	SOTALOL 3930-20-9 anti-adrenergic (beta-receptor)
2	-1.96	7.0	1362	1	"	SOTALOL
3	-1.96	7.0	474	1	"	SOTALOL
4	-1.82	7.4	2473	1	"	SOTALOL
95	-1.80	7.4	1799	1	"	SOTALOL
6	-1.52	7.4	1799	1	"	SOTALOL
7	-1.51	8.0	474	1	"	SOTALOL
8	-1.50	7.4	2465	41	"	SOTALOL
9	-1.41	7.4	1362	1	"	SOTALOL
9000	-1.38	7.4	2399	1	"	SOTALOL
1	-0.95	10.2	2465	41	"	SOTALOL
2	-0.79	7.4	1799	2	"	SOTALOL
3	-0.65	9.2	2465	41	"	SOTALOL
4	0.23	7.4	2490	1	C12H20N2O4S2	BENZENESULFONAMIDE,4-(I-BUTYLAMINO)ETHYLSULFONYL 135832-44-9
5 ★	2.67		1895		C12H20N2S1	THIOUREA,N-1-ADAMANYTYL-N'-METHYL 25444-83-1
6 ✓	-0.87	7.4	2397	101	C12H20N3O2	IMIDAZOLIUM CHLORIDE,1-METHYL-2-HYDROXYIMINOMETHYL-3-(1-PROPYLPROPEN-2YLOXY)METHYL
7 ✓	-1.02		2397	101	C12H20N3O2	IMIDAZOLIUM CHLORIDE,1-METHYL-2-HYDROXYIMINOMETHYL-3-CYCLOHEXYLOXYMETHYL 117941-56-7
8 ✓	-0.82		2397	101	C12H20N3O2	IMIDAZOLIUM CHLORIDE,1-METHYL-2-HYDROXYIMINOMETHYL-3-CYCLOPENTYLMETHOXYMETHYL
9 ✓	-0.71		2397	101	C12H20N3O2	IMIDAZOLIUM CHLORIDE,1-METHYL-2-HYDROXYIMINOMETHYL-3-HEXEN-2-YLOXYMETHYL 117983-11-6
9010 ✓	-0.76		2397	101	C12H20N3O2	IMIDAZOLIUM CHLORIDE,1-METHYL-2-HYDROXYIMINOMETHYL-3-HEXEN-3YLOXYMETHYL 117983-16-1
1 ✓	-1.40		2406	102	C12H20N3O2S2	TETRAMETHYLTHIOURONIUM IODIDE,S-(4-N-METHYLSULFONAMIDO)PHENYL
2 ✓	-2.21		2397	101	C12H20N3O3	IMIDAZOLIUM CHLORIDE,1-METHYL-2-HYDROXYIMINOMETHYL-3-HEXANOIC ACID,METHYL ESTER 132540-07-9
3	-0.81		547		C12H20N5O1P1	1(2,2-DIMEAZIRIDINYL)2N(2-PYRIMIDINYL)PHOSPHORAMIDE
4	-0.32		1045		C12H20N6O2	6(1,1-BIS(HOCH2)-AM-2,4-DIAZIRIDINYLTRIAZINE
15 ★	2.99		594		C12H20O3Si1	TRIETHOXYPHENYLSILANE 780-69-8
6 ★	3.38		2011		C12H20O4S2	PROPANEDIOIC ACID,BIS(METHYLTHIO)METHYLENE,DI-I-PROPYL ESTER 55084-16-7
7 ✓	-0.57		2397	101	C12H21Br1N3O2	IMIDAZOLIUM CHLORIDE,1-ETHYL-2-HYDROXYIMINOMETHYL-3-(2,2-DIMETHYL-3-BROMOPROPOXY)METHYL
8 ★	5.20		2488		C12H21F2N5S1	1,3,5-TRIAZINE,2-DIFLUOROMETHIO-4-I-PROPYLAMINO-6-AMYLAMINO
9 ★	3.28		1422	462	C12H21N1	1-AMINOADAMANTANE,3,5-DIMETHYL 19982-08-2 antiparkinsonian
9020 ★	3.40		1422	462	C12H21N1	1-AMINOADAMANTANE,3-ETHYL 41100-45-2
1	1.44	7.4	2535	1	C12H21N1O1	G-CYCLOHEXANOYLPROPYL-ETHANOLAMINE (DEHYDRO-FUSED)
2 ★	3.81		1693		C12H21N2O3P1S1	DIAZINON 333-41-5 insecticide
3 ★	2.07		1693		C12H21N2O4P1	DIAZINON-O-ANALOG 962-58-3
4 ★	2.53		2397	101	C12H21N3O2	IMIDAZOLE,1-METHYL-2-HYDROXYIMINOMETHYL-4-(1,2-DIMETHYLPROPOXY)METHYL
25 ✓	-0.40	3.0	594		C12H21N3O6	N-T-BOC-ALA-GLY-GLY
6 ✓	-2.32		2397	101	C12H21N4O1	IMIDAZOLIUM CHLORIDE,1-METHYL-2-HYDROXYIMINOMETHYL-3-(1-PIPERIDINYL)ETHYL 132566-79-1
7 ✓	-2.77		2397	101	C12H21N4O1	IMIDAZOLIUM CHLORIDE,1-METHYL-2-HYDROXYIMINOMETHYL-3-(N-ETHYLPYRROLIDIN-2YL)METHYL
8 ✓	-1.80		2397	101	C12H21N4O3	IMIDAZOLIUM CHLORIDE,1-ETHYL-2-HYDROXYIMINOMETHYL-3-(4-METHYL-4-NITROPENTYL) 132566-40-6
9 ✓	-1.61		2397	101	C12H21N4O3	IMIDAZOLIUM CHLORIDE,1-METHYL-2-HYDROXYIMINOMETHYL-3,4-DIMETHYL-4-NITROPENTYL) 132566-41-7
9030 ✓	-1.54		2397	101	C12H21N4O4	IMIDAZOLIUM CHLORIDE,1-ETHYL-2-HYDROXYIMINOMETHYL-3-(1,2-DIMETHYL-2-NITROPROPYLOXY)METHYL
1 ✓	-1.46		2397	101	C12H21N4O4	IMIDAZOLIUM CHLORIDE,1-METHYL-2-HYDROXYIMINOMETHYL-3-(1-ETHYL-2-METHYL-2-NITROPROPOXY)METHYL
2	0.80		547		C12H22Cl1N3O7	N1-NO-N1-CLET-N3-2(3,4,6-TRIMEO-GLUCOSYL)UREA
3 ★	1.58		2238		C12H22Cl1N4O3	NITROSOUREA,1-(2-CHLOROETHYL)-3-(1-OXO-2,2,6,6-TETRAMETHYLPIPERIDIN-4-YL)
4 ★	4.74		2094		C12H22Cl1N5	SIMAZINE,2-(HEPTAN-2-AMINE) ANALOG 102587-54-2

	logP	pH	Ref.	Note	MF	Name / CAS / activity
35	1.39	7.4	1350	1	C12H22N2O1	BAY-A-6781, CLONIDINE ANALOG ✚
6 ✓	-0.56		2397	101	C12H22N3O2	IMIDAZOLIUM CHLORIDE,1-HEXYL-2-HYDROXYIMINOMETHYL-3-METHOXYMETHYL
7 ✓	-0.68		2397	101	C12H22N3O2	IMIDAZOLIUM CHLORIDE,1-METHYL-2-HYDROXYIMINOMETHYL-3-(1,2,2-TRIMETHYLPROPOXY)METHYL
8 ✓	-0.76		2397	101	C12H22N3O2	IMIDAZOLIUM CHLORIDE,1-METHYL-2-HYDROXYIMINOMETHYL-3-(1,2-DIMETHYLBUTOXY)METHYL 117941-52-3
9 ✓	-0.86		2397	101	C12H22N3O2	IMIDAZOLIUM CHLORIDE,1-METHYL-2-HYDROXYIMINOMETHYL-3-(3-HEXYLOXY)METHYL 117941-51-2
9040 ✓	-0.11		2397	101	C12H22N3O2	IMIDAZOLIUM CHLORIDE,1-METHYL-2-HYDROXYIMINOMETHYL-3-HEXYLOXYMETHYL 117941-46-5
1	-0.68	7.6	2003	314	C12H22N3O2.Cl1	IMIDAZOLIUM CHLORIDE,1-(1,2,2-TRIMETHYLPROPOXY)METHYL-2-HYDROXYIMINOMETHYL-3-METHYL 91900-12-8
2 ✓	-2.40		2397	101	C12H22N3O3	IMIDAZOLIUM CHLORIDE,1-METHYL-2-ETHYL-3-(1,2-DIMETHYL-2-NITROPROPOXY)METHYL
3	-0.38	9.0	1409		C12H22N4O3	5-NITROIMIDZOLE,1-(DIET-AMETHOXYET)-2-ME
4 ★	-1.43		2377		C12H22N4O4	AC-ASN-ILE-N
45 ★	-1.85		2377		C12H22N4O4	AC-GLN-VAL-N
6 ★	-1.56		2377		C12H22N4O4	AC-GLY-ALA-VAL-N
7 ★	-1.23		2377		C12H22N4O4	AC-GLY-LEU-GLY-N
8 ★	-1.41		2377		C12H22N4O4	AC-ILE-ASN-N
9 ★	-1.30		2377		C12H22N4O4	AC-LEU-ASN-N
9050 ★	-1.82		2377		C12H22N4O4	AC-VAL-GLN-N
1	-1.00	7.2	170	314	C12H22N6O2	DIAZINEDICARBOXYLICACIDBIS(N'-METHYLPIPERAZIDE) 53202-52-1
2	4.10		1369	459	C12H22O1	CYCLODODECANONE 830-13-7
3	4.70		1369	459	C12H22O1	CYCLODODECENEEPOXIDE
4	-0.63		511		C12H22O6	GLUCOPYRANOSIDE,CYCLOHEXYL(BETA) 5284-99-1
55 ★	-3.70		1144		C12H22O11	SUCROSE 57-50-1
6	-3.01	7.4	1237	1	"	SUCROSE
7	2.30	3.5	2391	1	C12H23N1O3	SUCCINAMIC ACID, N,N-DIBUTYL
8 ✓	-1.71		2397	101	C12H23N2O1	IMIDAZOLIUM CHLORIDE,1,2-DIMETHYL-3-(1,2,2-TRIMETHYLPROPOXY)METHYL
9 ★	-0.32		2377		C12H23N3O3	AC-VAL-VAL-N
9060 ★	-0.86		2377		C12H23N3O4	AC-THR-ILE-N
1	-1.67	7.4	2397	101	C12H23N4O1	IMIDAZOLIUM CHLORIDE ANALOG ✚
2 ✓	-2.21		2397	101	C12H23N4O1	IMIDAZOLIUM CHLORIDE,1-METHYL-2-HYDROXYIMINOMETHYL-3-(2-DIETHYLAMINO)PROPYL
3	-2.78	7.4	2397	101	C12H23N4O1	IMIDAZOLIUM CHLORIDE,1-METHYL-2-HYDROXYIMINOMETHYL-3-(3-DIETHYLAMINOPROPYL) 132566-76-8
4 ✓	-2.13		2397	101	C12H23N4O4S1	IMIDAZOLIUM CHLORIDE,1-(2-METHYLSULFONYLETHOXY)METHYL-2-HYDROXYIMINOMETHYL-3-DIMETHYLAMINOETHYL
65	3.19		723		C12H23N5O1	2,4-DIAMINO-6-DIBUTYLAMINO-PYRIMIDINE-3-OXIDE 55921-62-5
6	0.63		2358	200	(C12H24N1O5S2)2.Cd1	N-PENTYL-N-DITHIOCARBOXY-D-GLUCAMINE
7 ✓	0.98		2358	200	"	N-PENTYL-N-DITHIOCARBOXY-D-GLUCAMINE
8 ★	1.87	7.0	2309	1	C12H24N2O2	GLYCINAMIDE,2,2-DI-I-BUTYL-N-ACETYL
9	0.08	7.0	1230	1	C12H24N2O2Se1	BIS(B-MORPHOLINOETHYL)SELENIDE
9070	0.32	7.5	1230	1	"	BIS(B-MORPHOLINOETHYL)SELENIDE
1 ★	-1.82	7.0	2374		C12H24N2O3	ISOLEUCINYLISOLEUCINE
2 ★	-1.64	7.0	2374		C12H24N2O3	LEUCINYLISOLEUCINE
3 ★	-1.46	7.0	2197		C12H24N2O3	LEUCINYLLEUCINE 3303-31-9
4	1.67		1908		C12H24N3O1P1	PYRROLIDINE,1,1',1"-PHOSPHINYLIDINE,TRIS 6415-07-2
75 ✓	0.43		2397	101	C12H24N3O2Si1	IMIDAZOLIUM CHLORIDE,1-METHYL-2-HYDROXYIMINOMETHYL-3-(3-TRIMETHYLSILYLPROPOXY)METHYL
6	-3.08	7.4	2397	101	C12H24N4O1	IMIDAZOLIUM CHLORIDE,1-METHYL-2-HYDROXYIMINOMETHYL-3-(N,N,N-DIETHYL,METHYLAMMONIOETHYL
7	3.90		964		C12H24N6	N2,N4,N6-TRIMETHYL-N,N,N,N-TRIETHYLMELAMINE
8 ★	3.75		594		C12H24O1	CYCLODODECANOL 1724-39-6
9 ★	4.20		505	20	C12H24O2	DODECANOIC ACID 143-07-7
9080	0.86		1294		C12H24O5S1	1-HEXYLTHIO-B-GALACTOPYRANOSIDE
1 ★	-0.68		2308		C12H24O6	18-CROWN-6-ETHER 17455-13-9
2	0.16		1294		C12H24O6	B-HEXYLGALACTOPYRANOSIDE
3 ✓	1.60	5.4	530		C12H25O4S1.Na1	DODECYLSULFATE, SODIUM SALT 151-21-3
4	5.90		303	820	C12H25O4S1.C12H14N1	1,2,6-TRIMETHYLQUINOLINIUMDODECYLSULFATE
85 ★	6.10		1906		C12H26	DODECANE 112-40-3
6	0.61		1966	459	C12H26Au1O5P1S1	GLUCOSE,2-TRIETHYLPHOSPHINAUROTHIO
7 ★	0.63		715		C12H26Cl2N2Pt1	DICHLORODI-(CYCLOHEXYLAMINO)PLATINUM-C 38780-35-7
8 ★	5.13		505		C12H26O1	1-DODECANOL 112-53-8
9 ★	1.92		1819		C12H26O3	DIETHYLETHER,1,5-DIBUTOXY
9090 ✓	0.96		541		C12H26O3S1	DODECYLSULFONATE, SODIUM SALT 2386-53-0
1 ★	4.25	1.0	590		C12H27Cl1Sn1	STANNANE, CHLOROTRIBUTYL 1461-22-9
2 ★	3.15		1957		C12H27F1Sn1	TRIBUTYLTIN FLUORIDE
3	1.85		530	820	C12H27N1	DODECYLAMINEHYDROCHLORIDE 929-73-7
4	-1.95		1394		C12H27N3O1	4-UNDECYLSEMICARBAZIDE
95 ★	3.23		2506		C12H27O1S3	DEF 78-48-8 defoliant
6	2.50		547		C12H27O4P1	TRIBUTYLPHOSPHATE 126-73-8
7	4.00		1101	421	"	TRIBUTYLPHOSPHATE
8	3.20	6.0	1958		C12H27Sn1	TRIBUTYLTIN 36643-28-4
9	-2.13		2173		C12H28N1.I1	TETRAPROPYL AMMONIUM IODIDE 631-40-3
9100 ✓	2.48		567	820	C12H28N1.C6H2N3O7	METHYLPROPYL-DI-I-BUTYL AMMONIUM PICRATE
1 ✓	2.35		567	820	C12H28N1.C6H2N3O7	TETRAPROPYL AMMONIUM PICRATE
2 ★	2.39		1908		C12H28N3O1P1	PHOSPHONICDIAMIDE,N,N-TETRAETHYL-P-1-PYRROLIDINYL 69981-38-0
3	-2.63		983		C12H28O2P1S2.C1H3O4S1	(C6H13)OP(O)(CH3)SCH2CH2S+(ME)(ET),MESO4- 21044-32-6
4	-2.63		983		C12H28O2P1S2.C1H3O4S1	(CH3)3CCH(CH3)OP(O)(CH3)SCH2CH2S+(ME)(ET),MESO4-
5 ✓	-1.46		1966		C12H30Au1P2.Cl1	CHLORO-BIS(TRIETHYLPHOSPHINE)GOLD
6 ✓	-1.46		1966		"	CHLORO-BIS(TRIETHYLPHOSPHINE)GOLD
7 ✓	-0.91		2426	537	C12H30N2O2Pt1.(N1O3)2	PLATINUM,BIS-CYCLOHEXYLAMMONIO-DIAQUA-DINITRATE
8 ★	2.53		1908		C12H30N3O1P1	PHOSPHORICTRIAMIDE,HEXAETHYL 2622-07-3
9	6.00		1740	459	C12H36O4Si5	DODECAMETHYLPENTASILOXANE 141-63-9
9110 ★	2.40		595		C13H4N4O9	9-FLUORENONE,2,4,5,7-TETRANITRO 746-53-2
1 ★	2.42		593		C13H5N3O7	9-FLUORENONE,2,4,7-TRINITRO 129-79-3
2 ★	7.54	12.5	430	808	C13H6Cl6O2	HEXACHLOROPHENE 70-30-4 antibacterial~fungicide~acaricide
3 ★	2.84		595		C13H6N2O5	9-FLUORENONE,2,7-DINITRO 31551-45-8
4 ★	3.30		2506		C13H7Br2N3O6	BROMOFENOXIME 13181-17-4 herbicide
15	4.54		2675	459	C13H7Cl1F3N1O3	NITROFLUORFEN herbicide
6 ★	3.65		1501		C13H7F3N2O5	FLUORODIFEN 15457-05-3
7 ★	3.06		595		C13H7N1O3	9-FLUORENONE,3-NITRO 3096-52-4
8	4.91		2040	459	C13H8	BENZENE,1,3,5-HEPTATRIYNYL 4300-27-0
9 ★	4.04	2.5	2207		C13H8Cl1F1O2	BENZOPHENONE,5-FLUORO-2-HYDROXY-4'-CHLORO
9120 ★	3.10	7.4	401	2	C13H8Cl2O3	2-HYDROXYNAPHTHOQUINONE,3-(3,3-DICLALLYL) 36417-16-0 antineoplastic
1	0.76	7.4	1838	1	C13H8F2O3	DIFLUNISAL 22494-42-4 anti-inflammatory, analgesic
2 ★	4.44	2.0	1838		"	DIFLUNISAL
3 ★	3.35		595		C13H8N2O4	FLUORENE,2,7-DINITRO
4	3.09		2040	459	C13H8O1	2-HEPTENE-4,6-DIYNAL,7-PHENYL 20252-42-0

	logP	pH	Ref.	Note	MF	Name / CAS / activity
25 ★	3.58		547		C13H8O1	9-FLUORENONE 486-25-9
6 ★	3.99		547		C13H8O1S1	THIOXANTHONE 492-22-8
7 ★	3.39		596		C13H8O2	XANTHONE 90-47-1
8	5.16		2601	300	C13H9Br1Cl1N1O2	4-CHLOROANTHRANILIC ACID,N-(3-BROMOPHENYL)
9	5.37		2601	300	C13H9Cl1I1O2	4-CHLOROANTHRANILIC ACID,N-(4-IODOPHENYL)
9130	4.05		2601	300	C13H9Cl1N2O4	4-CHLOROANTHRANILIC ACID,N-(3-NITROPHENYL)
1	4.04		2601	300	C13H9Cl1N2O4	4-CHLOROANTHRANILIC ACID,N-(4-NITROPHENYL)
2 ★	2.16		579		C13H9Cl2N1O1	2,6-DICHLOROBENZILIDINE-ANILINE-N-OXIDE
3 ★	4.42		1792		C13H9Cl2N1O1	BENZAMIDE,N-(3,4-DICHLOROPHENYL) 10286-75-6
4	4.67		2601	300	C13H9Cl2N1O2	4-CHLOROANTHRANILIC ACID,N-(2-CHLOROPHENYL)
35	5.00		2601	300	C13H9Cl2N1O2	4-CHLOROANTHRANILIC ACID,N-(3-CHLOROPHENYL)
6	4.97		2601	300	C13H9Cl2N1O2	4-CHLOROANTHRANILIC ACID,N-(4-CHLOROPHENYL)
7	0.54		536		C13H9Cl2N1O2	N-(2,5-DICHLOROPHENYL)ANTHRANILIC ACID, SODIUM SALT
8 ★	4.27		822		C13H9Cl2N1O3	4-AMINOSALICYLIC ACID,2,4-DICL-PHENYL ESTER
9	2.91		885		C13H9Cl3N2O1	TRICLOCARBAN 101-20-2 disinfectant
9140	1.59	7.4	693	1	C13H9F3N2O2	NIFLUMIC ACID 4394-00-7 anti-inflammatory
1 ★	3.54		594		C13H9N1	1-AZAPHENANTHRENE 230-27-3
2 ★	3.48		594		C13H9N1	5-AZAPHENANTHRENE 229-87-8
3	-0.50		532	820	C13H9N1	ACRIDINE HYDROCHLORIDE 17784-47-3
4	3.29	7.4	1209	1	C13H9N1	ACRIDINE 260-94-6
45 ★	3.40		517		"	ACRIDINE
6 ★	2.95	7.4	597	314	C13H9N1O1	9-FLUORENONE,2-AMINO
7 ★	4.26		541		C13H9N1S1	2-PHENYLBENZTHIAZOLE 883-93-2
8	4.66		850		C13H9N1S1	4-BIPHENYLISOTHIOCYANATE 1510-24-3
9	4.75		850		C13H9N1S1	4-ISOTHIOCYANODIPHENYLETHER 3529-87-1
9150	4.40	8.0	850		C13H9N1S2	4-ISOTHIOCYANODIPHENYLSULFOXIDE 19988-99-9
1	-0.79		824	1	C13H9N2O4S1	PYOCYANINE-3-SULFONATE
2	1.40		555		C13H9N3O3	2-(5-NO2-2-FURANVINYL)BENZIMIDAZOLE 6450-22-2
3	5.55		850		C13H9N3S1	4-ISOTHIOCYANOAZOBENZENE 7612-96-6
4	4.76		2040	459	C13H10	BENZENE,HEPT-1,3-DIYN-5-ENYL 13678-98-3
55 ★	4.18		536		C13H10	FLUORENE 86-73-7
6 ★	6.31		928		C13H10Br1Cl2O2P1S1	LEPTOPHOS 21609-90-5 insecticide
7 ★	4.58		1693		C13H10Br1Cl2O3P1	LEPTOPHOS-O-ANALOG 25006-32-0
8 ★	3.92	7.4	594	1	C13H10Br1N1	FLUORENE,2-AMINO-7-BROMO 6638-60-4
9 ★	3.74		822		C13H10Br1N1O3	4-AMINOSALICYLIC ACID,2-BROMOPHENYL ESTER 56356-31-1
9160 ★	3.84		822		C13H10Br1N1O3	4-AMINOSALICYLIC ACID,3-BROMOPHENYL ESTER 56356-32-2
1 ★	3.46		822		C13H10Br1N1O3	4-AMINOSALICYLIC ACID,4-BROMOPHENYL ESTER 56356-33-3
2	0.83	8.0	545	1	C13H10Br1N1O4S1	N-(P-BROMOPHENYLSULFONYL)-ANTHRANILATE,NA 51012-29-4
3	4.25		2601	300	C13H10Cl1N1O2	4-CHLOROANTHRANILIC ACID,N-PHENYL
4	3.94		2601	300	C13H10Cl1N1O2	N-(2-CHLOROPHENYL)ANTHRANILIC ACID
65	4.24		2601	300	C13H10Cl1N1O2	N-(4-CHLOROPHENYL)ANTHRANILIC ACID
6	2.43	7.0	426	1	C13H10Cl1N1O2	N-PHENYLANTHRANILIC ACID,3'-CHLORO 13278-36-9
7 ★	5.57	2.0	2236		C13H10Cl1N1O2	N-PHENYLANTHRANILIC ACID,3'-CHLORO
8 ★	3.72		822		C13H10Cl1N1O3	4-AMINOSALICYLIC ACID,2-CHLOROPHENYL ESTER 56356-28-6
9 ★	3.90		822		C13H10Cl1N1O3	4-AMINOSALICYLIC ACID,3-CHLOROPHENYL ESTER 56356-29-7
9170 ★	3.60		822		C13H10Cl1N1O3	4-AMINOSALICYLIC ACID,4-CHLOROPHENYL ESTER 56356-30-0
1	0.54	8.0	545	1	C13H10Cl1N1O4S1	N-(P-CHLOROPHENYLSULFONYL)-ANTHRANILATE,NA 51012-31-8
2	0.23	7.5	1344	1	C13H10Cl1N1O6	3-DESETHYL-9-HYDROXYMETHYLQUINCARBATE
3	2.59		401		C13H10Cl1N3O1	DEMTHYL-DEPHENYL-5-PYRROL-2-YL-DIAZAPAM 30195-30-3 antiviral
4 ★	1.10	6.5	321	1	C13H10Cl1N3O4S2	CHLOROTHIAZIDE,3-PHENYL 1163-51-5
75	0.08	12.0	2640		C13H10Cl1N3O4S2	TENOXICAM,1-CHLORO anti-arthritic
6	0.61	7.4	2640	1	"	TENOXICAM,1-CHLORO
7	1.71	2.1	2640	1	"	TENOXICAM,1-CHLORO
8	2.62	7.4	2640	2	"	TENOXICAM,1-CHLORO
9 ★	4.70		505		C13H10Cl2N2O1	UREA,1-(3,4-DICHLOROPHENYL)-3-PHENYL 2008-73-3
9180 ★	4.26		1135	537	C13H10Cl2O2	DICHLOROPHENE 97-23-4 anthelmintic
1 ★	3.29		822		C13H10F1O3	4-AMINOSALICYLIC ACID,2-FLUOROPHENYL ESTER 56356-25-3
2 ★	3.42		822		C13H10F1O3	4-AMINOSALICYLIC ACID,3-FLUOROPHENYL ESTER 56356-26-4
3 ★	3.27		822		C13H10F1O3	4-AMINOSALICYLIC ACID,4-FLUOROPHENYL ESTER 56356-27-5
4	0.03	8.0	545	1	C13H10F1O4S1	N-(P-FLUOROPHENYLSULFONYL)-ANTHRANILATE,NA 51012-30-7
85 ★	4.97		607	606	C13H10F3N1O2S1	TRIFLUOROMETHANESULFONANILIDE,P-PHENYL 50585-77-8
6 ★	2.47		513		C13H10N2	1-AMINOACRIDINE 578-06-3
7 ★	2.62		513		C13H10N2	2-AMINOACRIDINE 581-28-2
8 ★	3.24		579		C13H10N2	2-PHENYLBENZIMIDAZOLE 716-79-0 anthelmintic (obsolete)
9 ★	2.77	12.0	599	220	C13H10N2	3-AMINOACRIDINE 581-29-3
9190 ★	3.26		513		C13H10N2	4-AMINOACRIDINE 578-07-4
1 ★	3.00		594		C13H10N2	9-AMINOACRIDINE 90-45-9 anti-infective (topical)
2 ★	3.33		594	400	C13H10N2	INDOLE,2-(2-PYRIDYL)
3	0.15	8.0	824	1	C13H10N2O1	PYOCYANINE 85-66-5
4 ★	3.06	7.4	595	1	C13H10N2O2	FLUORENE,2-AMINO-7-NITRO 1214-32-0
95 ★	1.78		522		C13H10N2O2S1	1,2,4-BENZOTHIADIAZINE-1,1-DIOXIDE-3-PHENYL 18818-44-5
6 ★	1.63		2467		C13H10N2O2S1	ANILINE,P-(4-CYANOPHENYL)SULFONYL
7	-0.97	8.0	824	1	C13H10N2O3S1	N-METHYLPHENAZONIUM-3-SULFONATE 40735-77-1
8	-0.57	8.0	978	340	C13H10N2O4	3-NITRO-N-PHENYLANTHRANILIC ACID,SODIUM SALT 54420-95-0
9	1.30	7.0	426	1	C13H10N2O4	N-PHENYLANTHRANILIC ACID,3'-NITRO 27693-70-5
9200 ★	4.57		426		C13H10N2O4	N-PHENYLANTHRANILIC ACID,3'-NITRO
1 ★	0.33		505		C13H10N2O4	THALIDOMIDE 50-35-1 sedative, hypnotic
2	-0.79	8.0	824	1	C13H10N2O4S1	PYOCYANINE-3-SULFONATE
3	-0.04	8.0	545	1	C13H10N2O6S1	N-(P-NITROPHENYLSULFONYL)-ANTHRANILATE,NA 38957-45-8
4	0.32	7.4	545	1	"	N-(P-NITROPHENYLSULFONYL)-ANTHRANILATE,NA
5	4.94		850		C13H10N2S1	4-ISOTHIOCYANODIPHENYLAMINE 23246-36-8
6 ★	1.50		2070		C13H10N4O3	ALLOPURINOL,1-BENZOYLOXYMETHYL
7	2.48		2040	459	C13H10O1	2-HEPTENE,4,6-DIYN-1-OL,7-PHENYL 13641-62-8
8 ★	3.18		536		C13H10O1	BENZOPHENONE 119-61-9
9 ★	2.89	1.0	579	342	C13H10O2	2-BIPHENYLCARBOXYLIC ACID 947-84-2
9210 ★	3.38		579		C13H10O2	BENZALDEHYDE,3-PHENOXY 39515-51-0
1 ★	3.75		2248		C13H10O2	BENZOIC ACID,4-PHENYL
2 ★	3.52		825		C13H10O2	O-HYDROXYBENZOPHENONE 117-99-7
3 ★	3.07		825		C13H10O2	P-HYDROXYBENZOPHENONE 1137-42-4
4	-0.38	10.2	534	310	C13H10O2	P-PHENYLBENZOIC ACID,SODIUM SALT 17264-58-3

	logP	pH	Ref.	Note	MF	Name / CAS / activity
15 ✓	-0.27	12.8	534	1	"	P-PHENYLBENZOIC ACID,SODIUM SALT
6	0.77	7.3	534	1	"	P-PHENYLBENZOIC ACID,SODIUM SALT
7 ★	3.59		508		C13H10O2	PHENYL BENZOATE 93-99-2
8	3.02		1566		C13H10O3	ANGELICIN,4'5-DIMETHYL ✚
9	3.08		1566		C13H10O3	ANGELICIN,4,4'-DIMETHYL
9220	2.54		1991		C13H10O3	ANGELICIN,4,5'-DIMETHYL
1	2.56		1991		C13H10O3	ANGELICIN,5,5'-DIMETHYL
2 ★	3.28		547		C13H10O3	DIPHENYLCARBONATE 102-09-0
3 ★	3.91	1.0	579	342	C13H10O3	M-PHENOXYBENZOIC ACID 3739-38-6
4 ★	3.11	1.0	579	342	C13H10O3	O-PHENOXYBENZOIC ACID 2243-42-7
25 ★	3.91	1.0	547	342	C13H10O3	P-PHENOXYBENZOIC ACID 2215-77-2
6	0.43		1937	459	C13H10O4	4-HYDROXY-2-PYRONE,6-BENZOYLMETHYL 20851-38-1
7 ★	3.86		594		C13H10S1	THIOXANTHENE 261-31-4
8 ★	3.70		1624		C13H11Br1N2O2	4-(4-BROMOPHENOXY)PHENYLUREA
9	3.53	7.4	1163	1	C13H11Cl1F3N1O2	N(3CL-4CF3PH)3,4-DIME-2OH-2,5-H2-PYRROL-5ONE
9230	2.67	7.4	1353		C13H11Cl1F3N3O1	4-CL-5(DIMEAMINO)-2(3-CF3-PH)-3-PYRIDAZINONE
1 ★	3.61	7.4	2004	1	C13H11Cl1N2O3S1	N2-PHENYL-N1-P-CHLOROBENZENESULFONYLUREA 25270-44-4
2	3.10		2482		C13H11Cl2F4N3O1	TETRACONAZOLE ✚
3 ★	3.00		2300	20	C13H11Cl2N1O2	PROCYMIDONE 32809-16-8 fungicide (systemic)
4 ★	3.14	7.4	594	314	C13H11N1	2-AMINOFLUORENE 153-78-6
35 ★	2.82		572	60	C13H11N1	BENZALANILINE 538-51-2
6 ★	3.18	7.4	594	1	C13H11N1	FLUORENE,1-AMINO 6344-63-4
7	2.89		2012	459	C13H11N1O1	3-(1-NAPHTHOXY)PROPIONITRILE
8 ★	2.51		538		C13H11N1O1	4-(4-METHYLBENZOYL)-PYRIDINE 14548-30-2
9 ★	2.62		508		C13H11N1O1	BENZANILIDE 93-98-1
9240 ★	3.20		579		C13H11N1O1	BENZOPHENONEOXIME 574-66-3
1 ★	1.35		538		C13H11N1O1	BENZYL-4-PYRIDYL KETONE 1017-24-9
2 ★	1.88		579		C13H11N1O1	BENZYLIDINEANILINE-N-OXIDE
3 ★	3.09		506	20	C13H11N1O1	SALICYLALDEHYDE-ANIL 779-84-0
4 ★	1.86		538		C13H11N1O1	4-(2-METHOXYBENZOYL)-PYRIDINE 51246-73-2
45 ★	1.94		544		C13H11N1O2	4-(4-METHOXYBENZOYL)PYRIDINE 14548-47-1
6	2.29		2161	468	C13H11N1O2	AZETIDINE-2-ONE,4-NAPHTHALEN-2-YLOXY
7	2.75		594		"	AZETIDINE-2-ONE,4-NAPHTHALEN-2-YLOXY
8	2.66		2161	468	C13H11N1O2	AZETIDINE-2-ONE,4-NAPHTHYLEN-1-YLOXY
9	-0.72		536		C13H11N1O2	N-PHENYLANTHRANILIC ACID,SODIUM SALT 6232-32-2
9250	0.02	2.7	978	340	"	N-PHENYLANTHRANILIC ACID,SODIUM SALT
1	1.54	7.0	426	1	C13H11N1O2	N-PHENYLANTHRANILIC ACID 91-40-7
2 ★	4.36		426		"	N-PHENYLANTHRANILIC ACID
3 ★	2.17		2649		C13H11N1O2	NAPHTHOQUINONE,2-(2-METHYLAZIRIDIN-1YL)
4 ★	2.40		2107	537	C13H11N1O2	NICOTINIC ACID, BENZYL ESTER 94-44-0
55 ★	3.30		571		C13H11N1O2	O,N-DIPHENYLCARBAMATE 4930-03-4
6 ★	3.27		507		C13H11N1O2	SALICYLANILIDE 87-17-2 antifungal (topical)
7	1.43	7.0	353	1	C13H11N1O3	2,6-DIHYDROXYBENZANILIDE 437-69-4
8 ★	1.96		1941		C13H11N1O3	4H-1-BENZOPYRAN-4-ONE,2,3-DIMETHYL-7-CYANOMETHOXY
9	0.81	7.0	426	1	C13H11N1O3	N-PHENYLANTHRANILIC ACID,3'-HYDROXY 21003-78-1
9260 ★	3.49		426		C13H11N1O3	N-PHENYLANTHRANILIC ACID,3'-HYDROXY
1 ★	3.15		822		C13H11N1O3	PHENYL-4-AMINO SALICYLATE 133-11-9 antibacterial (tuberculostatic)
2	-0.46	8.0	545	1	C13H11N1O4S1	N-PHENYLSULFONYLANTHRANILATE, SODIUM SALT 34837-67-7
3	-0.12	7.4	545	1	"	N-PHENYLSULFONYLANTHRANILATE, SODIUM SALT
4	0.35	7.2	2446	1	C13H11N1O5	OXOLINIC ACID 14698-29-4 antibacterial
65	2.64		2161	468	C13H11N1S1	AZETIDINE-2-ONE,4-NAPHTHLEN-2-YLTHIO
6 ★	3.03		579		C13H11N1S1	THIOBENZANILIDE 636-04-4
7	-0.81		824	1	C13H11N2.C1H3O4S1	N-METHYLPHENAZONIUMMETHYLSULFATE
8 ★	1.83	12.0	599		C13H11N3	PROFLAVINE 92-62-6 antiseptic (topical)
9 ★	4.31		594		C13H11N3O1	DROMETRIZOLE 2440-22-4 ultraviolet screen
9270 ★	1.33	4.9	2018		C13H11N3O1S1	TIMOPRAZOLE 57237-97-5
1 ★	1.83		985		C13H11N3O2S1	N1-(4-CYANOPHENYL)SULFANILAMIDE 25612-07-1
2 ★	2.51	6.0	2560	1	C13H11N3O2S1	THIABENDAZOLE,1-ETHOXYCARBONYL
3	0.95		1571	702	C13H11N3O3	2'-HYDROXY-5'-METHYL-4-NITROAZOBENZENE 1435-68-3
4	-0.28	7.4	693	1	C13H11N3O4S2	SUDOXICAM 34042-85-8 anti-inflammatory
75 ★	1.64	1.4	693	2	"	SUDOXICAM
6	-0.64	12.0	2640		C13H11N3O4S2	TENOXICAM 59804-37-4 anti-inflammatory
7	-0.32	7.4	2640	1	"	TENOXICAM
8	0.81	2.1	2640	1	"	TENOXICAM
9	-1.00	4.5	1409		C13H11N3O5	5-NITROIMIDAZOLE,1-ME-2-(2-PHENOXY)PROPENOICAC. 74550-88-2
9280	1.21		494		C13H11N3O5S1	2-(N(ME)COCH3)-5-5(NO2-FURYLPROPENILIDENE)THIAZOLON
1 ★	4.63		2146	458	C13H12	4-METHYLBIPHENYL 644-08-6
2 ★	4.14		536		C13H12	DIPHENYLMETHANE 101-81-5
3	1.34	5.0	2153	1	C13H12Cl1N1O2	3-(4-CHLOROPHENYL)-2-(1-PYRROLYL)PROPANOIC ACID
4	1.38		1051	837	C13H12Cl2N2	METHYLENE-N,N'-BIS-2-CHLOROANILINE
85 ★	3.67		306		C13H12Cl2O3	ETHYLACETOACETATE,2,4-DICHLOROBENZAL 15725-29-8
6 ★	2.86		306		C13H12Cl2O3	ETHYLACETOACETATE,2,6-DICHLOROBENZAL 15725-30-1
7 ★	3.92		306		C13H12Cl2O3	ETHYLACETOACETATE,3,4-DICHLOROBENZAL 15725-31-2
8	5.86		331		C13H12Cl3N3O1	2,4,5-TRICL-C6H2NHN=C(CN)CO-T-BU 28343-28-4
9 ★	2.04	5.0	2153	2	C13H12F1O2	3-(4-FLUOROPHENYL)-2-(1-PYRROLYL)PROPANOIC ACID
9290	3.13	7.4	1163	1	C13H12F3N1O2	N(4-CF3-PH)3,4-DIME-2-OH-2,5-H2-PYRROL-5ONE
1	4.22		331		C13H12F3N3O3S1	2-SO2ET-5-CF3-C6H3NHN=C(CN)CO-ME 28317-77-3
2 ★	2.79		579		C13H12N2	BENZOPHENONEHYDRAZONE 5350-57-2
3 ★	1.53		603		C13H12N2O1	1,1-DIPHENYLUREA 603-54-3
4 ★	3.00		599		C13H12N2O1	1,3-DIPHENYLUREA 102-07-8
95	2.87	7.4	1941	1	C13H12N2O1	HARMINE 442-51-3
6 ★	3.56	13.0	2326		"	HARMINE
7	3.38		547		C13H12N2O1	N-BENZYL-N-NITROSOANILINE 612-98-6
8 ★	2.08		2489		C13H12N2O1	NICOTINANILIDE,4'-METHYL
9 ★	0.83	7.4	2224	1	C13H12N2O1	PHENYLACETAMIDE,ALPHA-(2-PYRIDYL)
9300 ★	1.57		2410		C13H12N2O2	ACETANILIDE,4-(2-PYRIDYLOXY)
1	0.34	7.0	426	1	C13H12N2O2	N-PHENYLANTHRANILIC ACID,3'-AMINO 25293-29-2
2 ★	2.80		1263		C13H12N2O2	P-PHENOXYPHENYLUREA
3 ★	2.08		822		C13H12N2O3	4-AMINOSALICYLIC ACID,3-AMINOPHENYL ESTER 56356-23-1
4 ★	2.04		822		C13H12N2O3	4-AMINOSALICYLIC ACID,4-AMINOPHENYL ESTER 56356-24-2

	logP	pH	Ref.	Note	MF	Name / CAS / activity
5 ★	0.90		2467		C13H12N2O3S1	ANILINE,P-(4-CARBOXAMIDOPHENYL)SULFONYL **24454-46-4**
6	-1.61	7.5	542	1	C13H12N2O3S1	SULFABENZAMIDE **127-71-9** antibacterial
7	1.04	7.0	452	1	C13H12N2O4	O,O'-DIACETYL-4-HYDROXYAMINOQUINOLINE-1-OXIDE
8 ★	1.61		1941		C13H12N2O5	4H-1-BENZOPYRAN-4-ONE-2,3-DIMETHYL-7-ACETYLAMINO-8-NITRO
9 ★	2.60		1920		C13H12N2O5S1	NIMESULIDE **51803-78-2** anti-inflammatory
9310 ★	2.17		547		C13H12N2S1	1,3-DIPHENYL-2-THIOUREA **102-08-9**
1 ★	2.23	7.4	594	1	C13H12N4	PHIP ✚
2 ★	2.19		2612		C13H12N4O2	1-(4-(B,B-DICYANOVINYL)PH)-3-METHOXY-3-METHYLUREA **88132-20-1**
3 ★	0.39	7.4	2636	1	C13H12N4O2S1	ISOMAZOLE,(5,6)PYRIDAZINE ANALOG cardiotonic
4 ★	0.66	7.4	2636	1	C13H12N4O2S1	ISOMAZOLE,PYRAZINE ANALOG
15 ★	0.23	7.4	2636	1	C13H12N4O2S1	ISOMAZOLE,PYRIDAZINE ANALOG cardiotonic
6 ★	0.53	7.4	2636	1	C13H12N4O2S1	ISOMAZOLE,PYRIMIDYL ANALOG cardiotonic
7 ★	-0.67	1.0	2196		C13H12N8O4S3	CEFTEZOLE **26973-24-0** antibacterial
8 ★	3.79		546		C13H12O1	BENZYLPHENYLETHER **946-80-5**
9 ★	2.67		537		C13H12O1	DIPHENYLCARBINOL **91-01-0**
9320 ★	3.83		1142		C13H12O2	1-AZULENECARBOXYLIC ACID,ETHYL ESTER **1206-88-8**
1 ★	2.91		759		C13H12O2	4,4'-DIHYDROXYDIPHENYLMETHANE **620-92-8**
2 ★	2.92		853		C13H12O2	O-PHENOXYANISOLE **1695-04-1**
3 ★	2.05		547		C13H12O2S1	BENZYLPHENYLSULFONE **3112-88-7**
4	-0.90	10.0	2012	459	C13H12O3	1-NAPHTHOXYPROPIONIC ACID, SODIUM SALT
25	3.00		2012	459	C13H12O3	1-NAPHTHOXYPROPIONIC ACID
6 ★	1.32	0.6	2017		C13H12O6	CHROMONE-2-CARBOXYLIC ACID,5-(2-HYDROXYPROPOXY)
7 ★	2.84		306		C13H13Cl1O3	ETHYLACETOACETATE,2-CHLOROBENZAL **15725-22-1**
8 ★	2.97		306		C13H13Cl1O3	ETHYLACETOACETATE,3-CHLOROBENZAL **15725-23-2**
9 ★	3.04		306		C13H13Cl1O3	ETHYLACETOACETATE,4-CHLOROBENZAL **19411-80-4**
9330	3.95		2572	459	C13H13Cl2N1O4	O-PHENYLPYRROLIDINOCARBAMATE,2,4-DICHLORO-5-METHOXYCARBONYL **133636-94-9**
1 ★	1.76	7.0	2292	1	C13H13Cl2N3O3	DIOXOLANE,2-(2,4-DICHLOROPHENYL)-2-(1,2,4-TRIAZOL-1-YL)-4-HYDROXYMETHYL
2 ★	3.10		2506		C13H13Cl2N3O3	IPRODIONE **36734-19-7** fungicide
3 ★	4.12		2424		C13H13Cl3N4O1	BENZOTRIAZOLE,2-CYCLOPENTYL-5-TRICHLOROACETAMIDO
4	-1.09		547	370	C13H13F1O2S1.Br1	4-METHYL-N-(4-SO2F-BENZYL)-PYRIDINIUMBR **33803-13-3**
35 ★	2.46		306		C13H13F1O3	ETHYLACETOACETATE,3-FLUOROBENZAL **15725-21-0**
6	-0.18	1.0	594		C13H13F3N4O1	2,4-DIAMINOPYRIMIDINE,5-(3-TRIFLUOROMETHYL-4-METHOXYBENZYL)
7	-3.98	2.9	2192	537	C13H13F3N6O4S3	CEFAZAFLUR **58665-96-6** antibacterial
8	4.93		2404	459	C13H13F3O2	BENZOPYRAN,2,2-DIMETHYL-7-(2,2,2-TRIFLUOROETHOXY) **75423-04-0**
9	-1.69		594	314	C13H13F6N5	1,2-DIHYDROTRIAZINE,4,6-DIAMINO-2,2-DIMETHYL-1-(3,5-DI-TRIFLUOROMETHYL)PHENYL
9340 ★	3.90	13.0	547	224	C13H13N1	DIPHENYLMETHYLAMINE **552-82-9**
1 ★	3.13		548		C13H13N1	N-BENZYLANILINE **103-32-2**
2 ★	3.43		594	400	C13H13N1O1	PYRIDINE,2-BENZYLOXY-6-METHYL
3 ★	2.34		2153	2	C13H13N1O2	3-PHENYL-2-(1-PYRROLYL)PROPANOIC ACID
4 ★	-0.22	7.2	2323	468	C13H13N1O2	ALPHA-NAPHTHYLALANINE
45 ★	-0.15	7.2	2323	468	C13H13N1O2	BETA-NAPHTHYLALANINE
6 ★	2.40		2467		C13H13N1O2S1	ANILINE,P-(4-METHYLPHENYL)SULFONYL **4094-38-6**
7 ★	1.56	5.0	2153	2	C13H13N1O3	3-(4-HYDROXYPHENYL)-2-(1-PYRROLYL)PROPANOIC ACID
8 ★	1.79		1941		C13H13N1O3	4H-1-BENZOPYRAN-4-ONE,2,3-DIMETHYL-6-ACETYLAMINO
9 ★	1.96		2467		C13H13N1O3S1	ANILINE,P-(4-METHOXYPHENYL)SULFONYL **17078-72-7**
9350 ★	0.79	2.0	2157		C13H13N1O5	AZETIDINE,N-(2-BENZOYLOXYACETYL)-2-CARBOXY
1	-0.09	7.4	2621	1	C13H13N1O5	MITOSENE(5,5,6),DIHYDROXY ANALOG
2 ★	1.97	7.4	593	1	C13H13N3	3-AMINO-1,4-DIMETHYL-5H-PYRIDO[4,3-B]INDOLE
3	-0.05		2623		C13H13N3	DIPHENYLGUANIDINE
4	1.01	7.4	725	1	C13H13N3O1	2-BENZYLISONIAZID **58481-07-5**
55 ★	1.64	7.4	2206	306	C13H13N3O2	IMIDAZOLE,1-METHYL-2-(P-METHYLSTYRYL)-5-NITRO
6 ★	2.25	7.4	2206	306	C13H13N3O3	IMIDAZOLE,1-HYDROXYETHYL-2-STYRYL-5-NITRO
7	0.44	7.4	2525	1	C13H13N3O3S2	5-(2-PYRIDYLMETHYLAMINO)METHYLTHIENO[2,3-B]FURAN-2-SULFONAMIDE
8 ★	1.27		555		C13H13N3O5	1-IMIDAZOLEACETIC ACID-2-(5-NITROFURFURILIDINO),ETHYL ESTER
9 ★	1.17		494		C13H13N3O5S1	3-PR-2-ACETIMINO-5(5-NO2-FURFURILIDINE)THIAZOL-4ONE ✚
9360 ★	-0.45		1980		C13H13N3O6S1	CEPHACETRILE **10206-21-0** antibacterial
1	0.07	7.4	2037	1	C13H13N5O1	ISOGUANINE,1-METHYL-7-BENZYL
2	0.65	7.4	2037	1	C13H13N5O1	ISOGUANINE,1-METHYL-9-BENZYL
3	0.75	7.4	1722	1	C13H13N5O4S1	SULFACHRYSOIDINE **485-41-6** antibacterial (topical)
4	1.19	7.4	2035	324	C13H13N7O1	TRIAMTERENE,P-METHOXY ✚
65 ★	4.73		560		C13H14	2,3,6-TRIMETHYLNAPHTHALENE **829-26-5**
6 ★	4.90		2001		C13H14	NAPHTHALENE,1,4,5-TRIMETHYL
7 ★	0.68		2649		C13H14Br1N3O4	BENZOQUINONE,2,5-BIS-AZIRIDINYL-3-BROMO-6-CARAMYLOXYETHYL
8 ★	4.62		331		C13H14Cl1N3O1	3-CL-C6H4NHN=C(CN)CO-T-BU **28317-91-1**
9	0.06	7.4	2018	1	C13H14Cl1N3O2S1	TIARAMIDE,DESETHANOL ✚
9370 ★	1.35	10.8	2018		"	TIARAMIDE,DESETHANOL
1 ★	3.40		2424	459	C13H14Cl2N4O1	BENZOTRIAZOLE,2-CYCLOPENTYL-5-(DICHLOROACETAMIDO)
2	2.32	7.4	1485	1	C13H14Cl2N4O2	2,4-DIAMINOPYRIMIDINE,5(2,6-DICL-3,5-DIMEO)BENZYL
3 ★	2.55	7.4	1485	2	"	2,4-DIAMINOPYRIMIDINE,5(2,6-DICL-3,5-DIMEO)BENZYL
4	1.84	7.4	2648	2	C13H14F3N1	BENZOCYCLOHEPTANE,5,8-METHENO-10-AMINO-1-TRIFLUOROMETHYL
75	2.02	7.4	2648	1	"	BENZOCYCLOHEPTANE,5,8-METHENO-10-AMINO-1-TRIFLUOROMETHYL
6 ★	3.41	7.4	2648	2	"	BENZOCYCLOHEPTANE,5,8-METHENO-10-AMINO-1-TRIFLUOROMETHYL
7 ★	3.65	7.4	2648	2	"	BENZOCYCLOHEPTANE,5,8-METHENO-10-AMINO-1-TRIFLUOROMETHYL
8	1.96	7.4	2648	1	C13H14F3N1	BENZOCYCLOHEPTANE,5,8-METHENO-10-AMINO-2-TRIFLUOROMETHYL
9	2.01	7.4	2648	2	"	BENZOCYCLOHEPTANE,5,8-METHENO-10-AMINO-2-TRIFLUOROMETHYL
9380 ★	3.54	7.4	2648	2	"	BENZOCYCLOHEPTANE,5,8-METHENO-10-AMINO-2-TRIFLUOROMETHYL
1 ★	3.64	7.4	2648	2	"	BENZOCYCLOHEPTANE,5,8-METHENO-10-AMINO-2-TRIFLUOROMETHYL
2	1.80	7.4	2648	1	C13H14F3N1	BENZOCYCLOHEPTANE,5,8-METHENO-10-AMINO-3-TRIFLUOROMETHYL
3	1.87	7.4	2648	1	"	BENZOCYCLOHEPTANE,5,8-METHENO-10-AMINO-3-TRIFLUOROMETHYL
4 ★	3.45	7.4	2648	2	"	BENZOCYCLOHEPTANE,5,8-METHENO-10-AMINO-3-TRIFLUOROMETHYL
85 ★	3.59	7.4	2648	2	"	BENZOCYCLOHEPTANE,5,8-METHENO-10-AMINO-3-TRIFLUOROMETHYL
6 ★	5.11		2506		C13H14F3N3O4	ETHALFLURALIN **55283-68-6** herbicide (cell division inhib.)
7	-2.35		527	820	C13H14N1.Br1	B-PHENYLETHYLPYRIDINIUMBROMIDE **6324-18-1**
8	-1.90		2646	448	C13H14N1.I1	PYRIDINIUM IODIDE,1-METHYL-4-(4-METHYLPHENYL)
9 ★	1.59		879		C13H14N2	DI-(P-AMINOPHENYL)METHANE **101-77-9**
9390	1.62	7.4	1737	1	"	DI-(P-AMINOPHENYL)METHANE
1	0.46	7.5	2492	1	C13H14N2	TACRINE **321-64-2** stimulant-respiratory~acche-inhibitor (cognition enhancer
2	2.71		2492	2	"	TACRINE
3	1.63		2032	459	C13H14N2O1	4-PYRIMIDONE,(1,2,B)PYRIDYL-(5,6)CYCLOHEPTYL
4	1.63	7.5	2492	1	C13H14N2O1	9-AMINO-4-OXO-8-METHYL-1,2,3,4-TETRAHYDROACRIDINE

	logP	pH	Ref.	Note	MF	Name / CAS / activity
95	2.28		2492	2	"	9-AMINO-4-OXO-8-METHYL-1,2,3,4-TETRAHYDROACRIDINE
6	1.37		2089		C13H14N2O1	N-(3-DELTA-5-METHYLENEPIPERIDIN-1-YL)BENZAMIDE 104642-71-9
7	1.98		1725	459	C13H14N2O1	PYRIDO-PYRIMIDINE ANALOG (RINGS:667) ✚
8	1.91		1725	459	C13H14N2O1	PYRIDO-PYRIMIDINE ANALOG,6-ME(RINGS:666)
9	2.07		1725	459	C13H14N2O1	PYRIDO-PYRIMIDINE ANALOG,7-ME(RINGS:666)
9400	1.86		1725	459	C13H14N2O1	PYRIDO-PYRIMIDINE ANALOG,9-ME(RINGS:666)
1	2.44		1725	459	C13H14N2O1	PYRIDO-PYRIMIDINE ANALOG,9-ME(RINGS:666)
2 ★	2.31		1519	459	C13H14N2O2	7-ETHYL-3-CINNOLINECARBOXYLIC ACID,ETHYL ESTER
3 ★	2.64		2222		C13H14N2O2	BENZOIC ACID,3-(5-IMIDAZOLYL)PROPYL ESTER
4 ★	2.05		522		C13H14N2O2S1	1,2,4-BENZOTHIADIAZINE-1,1-DIOXIDE-3-CYCLOHEXEN-3-YL 37157-91-8
5 ★	2.03		328		C13H14N2O2S1	N1-(2-METHYLPHENYL)SULFANILAMIDE 16803-96-6
6 ★	2.11		1136		C13H14N2O2S1	N1-(4-METHYLPHENYL)SULFANILAMIDE 16803-95-5
7	2.08	7.0	452	1	C13H14N2O3	6-BUTYL-4-NITROQUINOLINE-1-OXIDE 21070-32-6
8	1.51	7.0	452	1	C13H14N2O3	6-T-BUTYL-4-NITROQUINOLINE-1-OXIDE 23484-01-7
9 ★	1.84	2.0	919		C13H14N2O3	BARBITURIC ACID,1-ME-5-ET,5-PHENYL 115-38-8 anticonvulsant, sedative
9410	1.86	7.4	899	1	"	BARBITURIC ACID,1-ME-5-ET,5-PHENYL
1	2.01		897	459	C13H14N2O3	CINNOLINE-4-ONE,3-ETHOXYCARBONYL-6-ETHYL
2	-2.01	7.0	2197		C13H14N2O3	TRYPTOPHAN,N-ACETYL 1218-34-4
3 ★	1.56		328		C13H14N2O3S1	N1-(2-METHOXYPHENYL)SULFANILAMIDE 19837-84-4
4 ★	1.51		328		C13H14N2O3S1	N1-(4-METHOXYPHENYL)SULFANILAMIDE 19837-74-2
15	-0.85	7.4	2542	1	C13H14N2O3S2	THIOPHENE-2-SULFONAMIDE,4-(P-METHYLAMINOMETHYLBENZOYL)
6 ★	-0.43		2234		C13H14N2O6	ACETYLSALICYLIC ACID,(N-(2-ACETAMIDO)GLYCOLAMIDE) ESTER 118247-02-2
7	0.93		1375		C13H14N4	3-(2-BENZALHYDRAZINO)PYRIDAZINE,4,6-DIME
8 ★	2.28	7.4	595	1	C13H14N4	IMIDAZO[4,5-F]QUINOLINE,3-METHYL-2-DIMETHYLAMINO 102408-27-5 ✚
9 ★	2.71		2379		C13H14N4O1	N-METHYLPYRRAL,SEMICARBAZONE,N'-PHENYL ✚
9420	-0.37	7.4	2683	1	C13H14N4O1	TRYPTAMINE,5-(3-METHYLOXADIAZOLYL) 5ht-agonist
1	1.70	7.4	1485	1	C13H14N4O2	2,4-DIAMINOPYRIMIDINE-5(2-ME-4,5-OCH2O-)BENZYL
2 ★	1.89	7.4	1485	2	"	2,4-DIAMINOPYRIMIDINE-5(2-ME-4,5-OCH2O-)BENZYL
3	0.52		1713		C13H14N4O2	6-PH-2-(2-UREIDOETHYL)-3(2H)-PYRIDAZINONE
4 ★	2.67		594	400	C13H14N4O2	PYRRAL SEMICARBAZONE,N'-(4-METHOXY)PHENYL
25	1.19	7.4	1485	1	C13H14N4O3	2,4-DIAMINOPYRIMIDINE,5-(3-MEO-4,5-OCH2O-)BENZYL
6 ★	1.38	7.4	1485	2	"	2,4-DIAMINOPYRIMIDINE,5-(3-MEO-4,5-OCH2O-)BENZYL
7 ★	1.65	1.0	2196		C13H14N4O3	PYRAZOLE,3-METHYL-4-NITRO-5-(N-METHYL-N-BENZYLCARBOXAMIDO)
8 ★	1.82	1.0	2196		C13H14N4O3	PYRAZOLE,3-METHYL-4-NITRO-5-(N-PHENETHYLCARBOXAMIDO)
9 ★	3.64		2379		C13H14N4S1	N-METHYLPYRRAL,THIOSEMICARBAZONE,N'-PHENYL
9430 ★	1.00		2289	459	C13H14N6O2	TETRAZOLE,1-PROPYNYL-5-(4-DIMETHYLUREYL)PHENOXY 117121-38-7
1 ★	1.65		2289	459	C13H14N6O2	TETRAZOLE,2-PROPYNYL-5-(4-DIMETHYLUREYL)PHENOXY 117144-73-7
2 ★	1.60	9.0	2195		C13H14N8S1	GUANIDINE,N-METHYL-N'-CYANO-N"-3-(2-N'-GUANIDYL)THIAZOL-4-YL)PHENYL
3 ★	1.98		306		C13H14O2	ACETYLACETONE,2-METHYLBENZAL 19411-71-3
4 ★	2.14		306		C13H14O2	ACETYLACETONE,4-METHYLBENZAL 15818-09-4
35 ★	1.73		306		C13H14O3	ACETYLACETONE,2-METHOXYBENZAL 15725-16-3
6 ★	1.79		306		C13H14O3	ACETYLACETONE,3-METHOXYBENZAL 15818-10-7
7 ★	1.73		306		C13H14O3	ACETYLACETONE,4-METHOXYBENZAL 15725-17-4
8 ★	2.21		306		C13H14O3	ETHYLACETOACETATE,BENZAL 620-80-4
9 ★	1.67		2234		C13H14O6	ACETYLSALICYLIC ACID,ETHOXYCARBONYLMETHYL ESTER 50785-24-5
9440 ★	3.90		1481		C13H15Br1Cl1N1O1	P-CHLOROCINNAMAMIDE,B-BROMO,N-S-BUTYL
1	-0.79	7.4	1204	1	C13H15Br1Cl1N3	N-CYPROPYLME-N(2BR-6CL-PH)AMINO)2-IMIDAZOLINE
2	-1.03	7.4	1204	1	C13H15Br1F1N3	N-CYPROPYLME-N(2-BR-6-F-PH)AMINO)2-IMIDAZOLINE
3	1.64		1479	459	C13H15Br1N4O2	2,4-DIAMINO-5-(2-BR-4,5-DIMEO-BENZYL)PYRIMIDINE
4	1.82	7.4	1485	1	C13H15Br1N4O2	BRODIMOPRIM 56518-41-3 antibacterial
45 ★	2.01	7.4	1485	2	"	BRODIMOPRIM
6 ★	2.23	13.0	594		C13H15Br1N4O2	PYRIMIDINE,2,4-DIAMINO-5-(3-BROMO-4,5-DIMETHOXY)BENZYL
7	5.15		2404	459	C13H15Br1O2	BENZOPYRAN,2,2-DIMETHYL-6-BROMO-7-ETHOXY 67015-38-7
8	-0.41	7.4	1204	1	C13H15Br2N3	N(2,3-BR2-PH)-N(CYPROPYLME)AMINO)-2-IMIDAZOLINE
9	-0.31	7.4	1204	1	C13H15Br2N3	N(2,4-BR2-PH)-N(CYPROPYLME)AMINO)-2-IMIDAZOLINE
9450	-0.52	7.4	1204	1	C13H15Br2N3	N(2,5-BR2-PH)-N(CYPROPYLME)AMINO)-2-IMIDAZOLINE
1	-1.12	7.4	1204	1	C13H15Br2N3	N(2,6-BR2-PH)-N(CYPROPYLME)AMINO)-2-IMIDAZOLINE
2	-0.49	7.4	1204	1	C13H15Br2N3	N(2-BUTEN-1-YL)N(2,6DIBRPH)AMINO)-2-IMIDAZOLINE
3	-0.66	7.4	1204	1	C13H15Br2N3	N(2-ME-ALLYL)-N(2,6-DIBRPH)AMINO)-2-IMIDAZOLINE
4	-0.63	7.4	1204	1	C13H15Br2N3	N(3-BUTEN-1-YL)N(2,6DIBRPH)AMINO)-2-IMIDAZOLINE
55	-0.98	7.4	1204	1	C13H15Cl1F1N3	N-CYPROPYLME-N(2CL-6-F-PH)AMINO)2-IMIDAZOLINE
6 ★	2.78		2424		C13H15Cl1N4O1	BENZIMIDAZOLE,2-CYCLOPENTYL-5-(CHLOROACETAMIDO)
7	1.68	7.4	1485	1	C13H15Cl1N4O2	2,4-DIAMINOPYRIMIDINE,5(3,5-DIMEO-4-CHLORO)BENZYL
8 ★	1.87	7.4	1485	2	"	2,4-DIAMINOPYRIMIDINE,5(3,5-DIMEO-4-CHLORO)BENZYL
9	1.73	7.4	1485	1	C13H15Cl1N4O2	2,4-DIAMINOPYRIMIDINE,5(4,5-DIMEO-3-CHLORO)BENZYL
9460 ★	1.92	7.4	1485	2	"	2,4-DIAMINOPYRIMIDINE,5(4,5-DIMEO-3-CHLORO)BENZYL
1	1.94	7.4	1485	1	"	2,4-DIAMINOPYRIMIDINE,5-(2-CL-3,5-DIMEO-)BENZYL
2 ★	2.13	7.4	1485	2	"	2,4-DIAMINOPYRIMIDINE,5-(2-CL-3,5-DIMEO-)BENZYL
3	4.89		2404	459	C13H15Cl1O2	BENZOPYRAN,2,2-DIMETHYL-6-CHLORO-7-ETHOXY 101185-73-3
4 ★	2.60	3.5	2179	302	C13H15Cl1O4	PENTANOIC ACID,3-METHYL-5-OXO-5-(3-CHLORO-4-METHOXY)PHENYL
65 ★	4.72		1792		C13H15Cl2N1O1	CYCLOHEXANAMIDE,N-(3,4-DICHLOROPHENYL)
6 ★	3.61		1481		C13H15Cl2N1O1	P-CHLOROCINNAMAMIDE,B-CHLORO,N-S-BUTYL
7 ★	0.57		536	800	C13H15Cl2N1O4	CETOPHENICOL 735-52-4 antibacterial
8	-0.92	7.4	1204	1	C13H15Cl2N3	N-CYPROPYLME-N-(2,6-DICLPH)AMINO)-2-IMIDAZOLINE
9	3.40		2420		C13H15Cl2N3	PENCONAZOLE 66246-88-6
9470	3.50		2482		"	PENCONAZOLE
1 ★	3.20		1629		C13H15Cl2N3O3S2	N-(N"-ME-(2,4-DICL-PHENYL)CARBAMYLTHIO)METHOMYL
2 ★	2.60		1629		C13H15Cl2N3O3S2	N-(N"-ME-(2,6-DICL-PHENYL)CARBAMYLTHIO)METHOMYL
3 ✓	-1.13		2397	101	C13H15F1N3O2	IMIDAZOLIUM CHLORIDE,1-METHYL-2-HYDROXYIMINOMETHYL-3-(P-FLUOROBENZYLOXYMETHYL 117983-34-3
4	-1.41	7.4	1204	1	C13H15F2N3	N-CYPROPYLME-N-(2,6-DIFLPH)AMINO)-2-IMIDAZOLINE
75 ★	3.57		2072	2	C13H15F3N2O5S1	BENZENESULFONAMIDE,2-NITRO-4-TRIFLUOROMETHYL-N-(2-ETHYLBUTANOYL)
6 ★	0.49		511		C13H15F3O6	GLUCOPYRANOSIDE,3-TRIFLUOROMETHYLPHENYL(BETA) 20772-25-2
7	2.14	7.4	1485	1	C13H15I1N4O2	2,4-DIAMINOPYRIMIDINE,5(3,4-DIMEO-5-IODO)BENZYL
8 ★	2.33	7.4	1485	2	"	2,4-DIAMINOPYRIMIDINE,5(3,4-DIMEO-5-IODO)BENZYL
9	2.21	7.4	1485	1	C13H15I1N4O2	2,4-DIAMINOPYRIMIDINE,5(4-IODO-3,5-DIMEO)BENZYL
9480 ★	2.40	7.4	1485	2	"	2,4-DIAMINOPYRIMIDINE,5(4-IODO-3,5-DIMEO-)BENZYL
1	1.01	7.4	1990	1	C13H15N1O1	PROPYLAMINE,3-(1-NAPHTHOXY)
2 ★	1.90		505		C13H15N1O2	GLUTETHIMIDE 77-21-4 sedative
3 ★	1.44		2157		C13H15N1O3	2-(BENZOYLOXY)-(N-TETRAMETHYLENE)ACETAMIDE
4	2.41		1973	459	C13H15N1O3	5-(CYCLOBUTYLCARBONYL)-H2-PYRROLO-PYRROLE-1-CARBOXYLIC ACID 96327-63-8

	logP	pH	Ref.	Note	MF	Name / CAS / activity
85 ★	0.90		2157		C13H15N1O4	MORPHOLINE,N-(2-BENZOYLOXYACETYL)
6 ★	0.90		2234		C13H15N1O5	ACETYLSALICYLIC ACID,(N-ETHYLGLYCOLAMIDE) ESTER 118247-01-1
7 ★	0.38		2234		C13H15N1O5	ACETYLSALICYLIC ACID,N,N-DIMETHYLGLYCOLAMIDE ESTER 118247-04-4
8 ★	2.61		1519	459	C13H15N3	2-N-PIPERIDINYL-QUINAZOLINE
9 ★	2.92		1519	459	C13H15N3	4-N-PIPERIDINYLQUINAZOLINE 41229-10-1
9490	1.03	7.4	1948	1	C13H15N3	QUIPAZINE 4774-24-7 antidepressant, oxytocic
1	2.04	7.4	1948	2	"	QUIPAZINE
2	1.40	7.4	416	1	C13H15N3O2	1-PHENYL-5-DIMEAMINO-6-METHYLURACIL 31992-01-5
3	0.62	3.3	2590	331	C13H15N3O2	BENZOIC ACID,2-(3,3-DIALLYLTRIAZENYL) antineoplastic
4 ★	1.55	7.0	2292	1	C13H15N3O2	DIOXOLANE,2-P-TOLYL-2-(1,2,4-TRIAZOL-1-YL)METHYL
95 ★	0.42	7.1	1591		C13H15N3O2	TRYPTOPHAN-AMIDE,N-ACETYL
6	1.26	7.2	556		C13H15N3O2S1	5-PROPYL-2-BENZENESULFONAMIDOPYRIMIDINE 55078-60-9
7	-2.40	7.0	2014		C13H15N3O3	ARSENAL ✦
8	-0.93	4.0	2014		"	ARSENAL
9	0.86	7.2	556		C13H15N3O3S1	2-I-PROPOXY-2-BENZENESULFONAMIDOPYRIMIDINE 4462-47-9
9500	1.14	7.2	556		C13H15N3O3S1	5-PROPOXY-2-BENZENESULFONAMIDOPYRIMIDINE 13418-64-9
1 ✓	-1.46		2397	101	C13H15N4O3	IMIDAZOLIUM CHLORIDE,1-METHYL-2-HYDROXYIMINOMETHYL-3-(P-NITROPHENETHYL)
2	-1.94	1.0	547	342	C13H15N5O1	2,4-DIAMINO-5-(4-N-ACETYLBENZYL)PYRMIDINE
3	0.07	7.4	2037	1	C13H15N5O1	2H-PURIN-2-ONE-1,3-DIHYDRO-1-METHYL-6-AMINO-7-BENZYL
4 ★	-1.90	7.4	2683	1	C13H15N5O1	TRYPTAMINE,5-(3-AMINOOXADIAZOLYLMETHYL) 5ht-agonist
5 ★	-0.34	13.0	547	224	C13H15N5O2	4,6-DIAM-1(2-CARBOXAMIDOMETHOXY)BENZYLPYRIMIDINE
6 ★	0.28	13.0	594		C13H15N5O2	PYRIMIDINE,2,4-DIAMINO-5-(3-CARBONAMIDOMETHOXY)BENZYL
7	-1.11	3.3	2590	331	C13H15N5O2S1	SULFAPYRIDINE,N,N-DIMETHYLTRIAZENE 58489-35-3
8	1.49	7.5	652	1	"	SULFAPYRIDINE,N,N-DIMETHYLTRIAZENE
9	1.54	6.6	2590	459	"	SULFAPYRIDINE,N,N-DIMETHYLTRIAZENE
9510 ★	0.28		2289		C13H15N7O2	TETRAZOLE,2-PROPIONITRILE-5-(4-DIMETHYLUREYL)PHENOXY 117121-50-3
1 ★	2.66		1481		C13H16Br1N1O1	A-BROMO-N-S-BUTYLCINNAMAMIDE,CIS
2 ★	3.56		1481		C13H16Br1N1O1	A-BROMO-N-S-BUTYLCINNAMAMIDE,TRANS
3	-1.30	7.4	1204	1	C13H16Br1N3	N-CYPROPYLME-N(2-BRPH)AMINO)-2-IMIDAZOLINE
4 ★	3.30	7.4	2392	2	C13H16Cl1N1	BENZAZEPINE,6-CHLORO-3-ALLYL 90047-59-9
15 ★	2.40		1481		C13H16Cl1N1O1	A-CHLORO-N-S-BUTYLCINNAMAMIDE,CIS
6 ★	3.52		1481		C13H16Cl1N1O1	A-CHLORO-N-S-BUTYLCINNAMAMIDE,TRANS
7 ★	2.18	13.0	547	224	C13H16Cl1N1O1	KETAMINE 6740-88-1 anesthetic
8 ★	3.37		1481		C13H16Cl1N1O1	P-CHLOROCINNAMAMIDE,N-S-BUTYL
9 ★	1.78		565		C13H16Cl1N1O4	ETHYL-N-CHLOROACETYL-N-(2-MEOPHENYL)GLYCINATE 60145-78-0
9520 ★	2.70		1629		C13H16Cl1N3O3S2	N-(N"-ME-(3-CL-PH)CARBAMYLTHIO)METHOMYL
1 ★	2.50		1629		C13H16Cl1N3O3S2	N-(N"-METHYL-(2-CL-PHENYL)CARBAMYLTHIO)METHOMYL
2 ★	2.60		1629		C13H16Cl1N3O3S2	N-(N"-METHYL-(4-CL-PHENYL)CARBAMYLTHIO)METHOMYL
3 ★	3.65	7.4	1851	1	C13H16Cl2N2O1	1-PIPERIDINEACETAMIDE,N-(3,4-DICHLOROPHENYL) 22010-25-9
4 ★	4.45		2512		C13H16Cl3N1O3	TRICLAMIDE ✦
25	2.51		2662	459	C13H16Cl3N1O4	N-(3,4-DIETHOXYPHENYL)CARBAMATE,2,2,2-TRICHLOROETHYL ESTER fungicide
6 ★	3.10		1481		C13H16F1O1	A-FLUORO-N-S-BUTYLCINNAMAMIDE,TRANS
7 ★	5.29	7.0	2506		C13H16F3N3O4	BENEFIN 1861-40-1 herbicide
8 ★	5.34		1190		C13H16F3N3O4	TRIFLURALIN 1582-09-8 herbicide
9	-0.90	7.4	1350	1	C13H16N2	TETRAHYDROZOLINE 84-22-0 adrenergic (vasoconstrictor)
9530	1.73		1725	459	C13H16N2O1	4H-PYRIDO(1,2-A)PYRIMIDIN-4-ONE,2,6,8-TRIME-3-ET
1 ★	2.22		1439		C13H16N2O1	4H-PYRIDO(1,2-A)PYRIMIDIN-4-ONE,2,6-DIME,3-PR ✦
2	1.37		2089		C13H16N2O1	N-(3-DELTA-3-METHYLPIPERIDIN-1-YL)BENZAMIDE 104642-66-2
3	1.20		2089		C13H16N2O1	N-(3-DELTA-4-METHYLPIPERIDIN-1-YL)BENZAMIDE 104642-67-3
4	0.70		2089		C13H16N2O1	N-(3-DELTA-6-METHYLPIPERIDIN-1-YL)BENZAMIDE 104642-65-1
35 ★	1.13	6.8	2175	1	C13H16N2O2	3-PROPYL-3-(4-PYRIDYL)PIPERIDINE-2,6-DIONE
6 ★	1.35		2649		C13H16N2O2	BENZOQUINONE,2,5-BIS-AZIRIDINYL-3-METHYL-6-ETHYL
7 ★	2.54		1969		C13H16N2O2	DIMETHYLCARBAMATE,2-(INDOLE-3YL)ETHYL ESTER 98154-94-0
8	1.62		2089		C13H16N2O2	N-(3-DELTA-3-METHOXYPIPERIDIN-1-YL)BENZAMIDE 104642-75-8
9	0.64		2089		C13H16N2O2	N-(3-DELTA-4-HYDROXYMETHYLPIPERIDIN-1-YL)BENZAMIDE 104642-72-0
9540	1.16		2089		C13H16N2O2	N-(3-DELTA-4-METHOXYPIPERIDIN-1-YL)BENZAMIDE 104642-76-4
1	1.48		2288	2	C13H16N2O2	PIRMAGREL 85691-74-3 inhibitor (thromboxane synthetase)
2 ★	2.20		522		C13H16N2O2S1	1,2,4-BENZOTHIADIAZINE-1,1-DIOXIDE-3-CYCLOHEXYL 37157-88-3
3	1.89	7.4	899	1	C13H16N2O3	BARBITURIC ACID,5-ALLYL,5-(2-CYPENTEN-1-YL)N-ME
4 ★	0.49		2649		C13H16N2O3	BENZOQUINONE,2,5-BIS-AZIRIDINYL-3-METHYL-6-HYDROXYETHYL
45	1.58	3.5	1267	1	C13H16N2O4	METHYL-2-ETHYL-2-PHENYLMALONURATE
6	-0.68	7.4	2542	1	C13H16N2O5S3	THIOPHENE-2-SULFONAMIDE,4-(4-HYDROXY-3-DIMETHYLAMINOMETHYL)PHENYLSULFONYL
7 ✓	-1.41		2426		C13H16N2O7Pt1.(N1O3)2	PLATINUM,BIS-(P-ACETYLOXYPYRIDONIO)-DIAQUA-DINITRATE
8 ✓	-1.19		2397	101	C13H16N3O2	IMIDAZOLIUM CHLORIDE,1-BENZYL-2-HYDROXYIMINOMETHYL-3-METHOXYMETHYL
9 ✓	-1.30		2397	101	C13H16N3O2	IMIDAZOLIUM CHLORIDE,1-METHYL-2-HYDROXYIMINOMETHYL-3-BENZYLOXYMETHYL 91900-15-1
9550	0.49	7.4	651	1	C13H16N3O4S1.Na1	SULPYRINE 68-89-3 analgesic, antipyretic
1	1.55	13.0	547	224	C13H16N4O1	2,4-DIAMINO-5-(2-METHOXYMETHYLBENZYL)PYRIMIDINE
2	1.91		1479	459	C13H16N4O1	2,4-DIAMINO-5-(3-METHYL-4-MEO-BENZYL)PYRIMIDINE 73264-18-3
3	1.95	7.4	1485	1	C13H16N4O1	2,4-DIAMINOPYRIMIDINE,5-(3-MEO-4-METHYL)BENZYL 73264-19-4
4 ★	2.14	7.4	1485	2	"	2,4-DIAMINOPYRIMIDINE,5-(3-MEO-4-METHYL)BENZYL
55 ★	2.45		2424		C13H16N4O1	BENZOTRIAZOLE,2-CYCLOPENTYL-5-ACETAMIDO
6	1.73	7.4	1485	1	C13H16N4O1S1	2,4-DIAMINOPYRIMIDINE,5(3-MEO-4-MES)-BENZYL
7 ★	1.92	7.4	1485	2	"	2,4-DIAMINOPYRIMIDINE,5(3-MEO-4-MES)-BENZYL
8	0.78	7.4	1485	1	C13H16N4O2	2,4-DIAMINO-5-(3,4-DIMETHOXYBENZYL)PYRIMIDINE 5355-16-8 antibacterial
9 ★	0.97	7.4	1485	2	"	2,4-DIAMINO-5-(3,4-DIMETHOXYBENZYL)PYRIMIDINE
9560 ★	1.62	13.0	579	224	C13H16N4O2	2,4-DIAMINOPYRIMIDINE,5-(2,3-DIMETHOXYBENZYL)
1	1.38	7.4	1485	1	C13H16N4O2	2,4-DIAMINOPYRIMIDINE,5-(3,5-DIMETHOXY)BENZYL
2 ★	1.57	7.4	1485	2	"	2,4-DIAMINOPYRIMIDINE,5-(3,5-DIMETHOXY)BENZYL
3 ★	1.70	13.0	547		C13H16N4O2	PYRIMIDINE,2,4-DIAMINO-5-(2,5-DIMETHOXY)BENZYL
4 ★	1.23		328		C13H16N4O2S1	N1-(5-I-PR-2-PYRIMIDYL)SULFANILAMIDE
65	0.18	7.4	1485	1	C13H16N4O3	2,4-DIAMINOPYRIMIDINE,5-(3,5-DIMEO-4-HYDROXY)BENZYL
6 ★	0.37	7.4	1485	2	"	2,4-DIAMINOPYRIMIDINE,5-(3,5-DIMEO-4-HYDROXY)BENZYL
7 ★	1.00	4.0	360	2	C13H16N4O3S1	5-ET-6-MEO-4-SULFANILAMIDOPYRIMIDINE 7756-44-7
8 ★	0.90	13.0	594		C13H16N4O4	2,4-PYRIMIDINEDIAMINE,5-(3',4',5'-TRIMETHOXYPHENOXY)
9	-1.92	1.0	594		C13H16N4O4S1	2,4-DIAMINOPYRIMIDINE,5-(3-METHYLSULFONYL-4-METHOXYBENZYL)
9570 ★	1.03	4.0	360	2	C13H16N4O4S1	5-ETO-6-MEO-4-SULFANILAMIDOPYRIMIDINE 5532-46-7
1 ★	1.18	4.0	360	2	C13H16N4O4S1	5-MEO-6-ETO-4-SULFANILAMIDOPYRIMIDINE 5018-56-4
2 ★	0.99	4.2	401	2	C13H16N6O2	IMIDAZOLE,5-CARBOXAMIDE-4-(3-BENZYL-3-(2-HYDROXYETHYL)TRIAZENYL)
3 ★	1.00		2289	459	C13H16N6O2	TETRAZOLE,1-ALLYL-5-(4-DIMETHYLUREYL)PHENOXY 117121-35-4
4 ★	1.65		2289	459	C13H16N6O2	TETRAZOLE,2-ALLYL-5-(4-DIMETHYLUREYL)PHENOXY 117121-47-8

	logP	pH	Ref.	Note	MF	Name / CAS / activity
75	-0.93	3.3	2590	331	C13H16N6O2S1	N-(4-METHYLPYRIMIDIN-2YL)BENZENESULFONAMIDE,4-(3,3-DIMETHYLTRIAZENYL) *antineoplastic*
6	1.75	6.6	2590	459	"	N-(4-METHYLPYRIMIDIN-2YL)BENZENESULFONAMIDE,4-(3,3-DIMETHYLTRIAZENYL)
7	0.78	7.5	652	1	C13H16N6O2S1	SULFAMERAZINE,N,N-DIMETHYLTRIAZENE 58489-36-4
8	-0.07		401		C13H16N6O4	NSC154020 35943-35-2 ✚
9	3.52		2404	459	C13H16O2S1	BENZOTHIOPYRAN,2,2-DIMETHYL-6,7-DIMETHOXY 70441-95-1
9580	2.77		2404	459	C13H16O3	BENZOPYRAN,2,2-DIMETHYL-6,7-DIMETHOXY 644-06-4
1 ★	2.52		306		C13H16O3	BENZYLACETOACETIC ACID,ETHYL ESTER 620-79-1
2 ★	3.41		501		C13H16O3	PHENOXYACETIC ACID,4-CYCLOPENTYL
3	2.35		1636	40	C13H16O4	2-HYDROXY-3-PHENOXYPROPYLMETH ACRYLATE 16926-87-7
4 ★	3.53		2487		C13H16O4	CINNAMIC ACID,3,4-DIHYDROXY BUTYL ESTER
85 ★	1.33		1898		C13H16O5	(4-HYDROXYETHYL)VANILLIN,PROPIONYL ESTER
6	2.93	7.4	934	1	C13H17Br1N2O1	N-(M-BROMOPHENYL)-3-N'-PIPERIDINOACETAMIDE 53316-92-0
7 ★	3.62	7.4	934	2	C13H17Br1N2O1	N-(M-BROMOPHENYL)-3-N'-PIPERIDINOACETAMIDE
8	2.94	7.4	934	1	C13H17Br1N2O1	N-(O-BROMOPHENYL)-3-N'-PIPERIDINOACETAMIDE 58479-89-3
9	2.88	7.4	934	1	C13H17Br1N2O1	N-(P-BROMOPHENYL)-3-N'-PIPERIDINOACETAMIDE 58479-86-0
9590 ★	3.57	7.4	934	2	C13H17Br1N2O1	N-(P-BROMOPHENYL)-3-N'-PIPERIDINOACETAMIDE
1	-0.64	7.4	1204	1	C13H17Br2N3	N(S-BU)-N-(2,6-BR2-PH)AMINO)-2-IMIDAZOLINE
2	-0.25	7.4	1204	1	C13H17Br2N3	N-BU-N-(2,6-BR2-PH)AMINO)-2-IMIDAZOLINE
3 ★	1.64	7.4	2392	1	C13H17Cl1N2O1	BENZAZEPINE,6-CHLORO-3-AMIDOETHYL
4	2.76	7.4	934	1	C13H17Cl1N2O1	N-(M-CHLOROPHENYL)-3-N'-PIPERIDINOACETAMIDE 38367-19-0
95 ★	3.45	7.4	934	2	C13H17Cl1N2O1	N-(M-CHLOROPHENYL)-3-N'-PIPERIDINOACETAMIDE
6	2.76	7.4	934	1	C13H17Cl1N2O1	N-(O-CHLOROPHENYL)-3-N'-PIPERIDINOACETAMIDE 38367-23-6
7	2.65	7.4	934	1	C13H17Cl1N2O1	N-(P-CHLOROPHENYL)-3-N'-PIPERIDINOACETAMIDE 27471-82-5
8	3.26	7.4	1851	1	C13H17Cl1N2O1	N-(P-CHLOROPHENYL)-3-N'-PIPERIDINOACETAMIDE
9 ★	3.34	7.4	934	1	C13H17Cl1N2O1	N-(P-CHLOROPHENYL)-3-N'-PIPERIDINOACETAMIDE
9600 ★	2.94		1481		C13H17Cl1N2O1	P-CHLOROCINNAMAMIDE,B-AMINO,N-S-BUTYL
1 ★	3.38	7.4	2004	1	C13H17Cl1N2O3S1	N2-CYCLOHEXYL-N1-P-CHLOROBENZENESULFONYLUREA 963-03-1
2 ★	4.26		2506		C13H17Cl1O3	MCPB-ETHYL 10443-70-6 ✚
3	2.16	7.4	934	1	C13H17F1N2O1	N-(M-FLUOROPHENYL)-3-N'-PIPERIDINOACETAMIDE 58479-90-6
4 ★	2.85	7.4	934	2	C13H17F1N2O1	N-(M-FLUOROPHENYL)-3-N'-PIPERIDINOACETAMIDE
5	2.16	7.4	934	1	C13H17F1N2O1	N-(O-FLUOROPHENYL)-3-N'-PIPERIDINOACETAMIDE 58479-88-2
6	2.01	7.4	934	1	C13H17F1N2O1	N-(P-FLUOROPHENYL)-3-N'-PIPERIDINOACETAMIDE 37163-41-0
7 ★	2.70	7.4	934	2	C13H17F1N2O1	N-(P-FLUOROPHENYL)-3-N'-PIPERIDINOACETAMIDE
8 ★	3.06		2072	2	C13H17F3N2O3S1	BENZENESULFONAMIDE,2-(2-ETHYLBUTANOYLAMINO)-4-TRIFLUOROMETHYL
9 ★	3.04		2072	2	C13H17F3N2O3S1	BENZENESULFONAMIDE,2-AMINO-4-TRIFLUOROMETHYL-N-(2-ETHYLBUTANOYL)
9610	3.23	7.4	934	1	C13H17I1N2O1	N-(M-IODOPHENYL)-3-N'-PIPERIDINOACETAMIDE 58479-91-7
1 ★	3.92	7.4	934	2	C13H17I1N2O1	N-(M-IODOPHENYL)-3-N'-PIPERIDINOACETAMIDE
2	3.21	7.4	934	1	C13H17I1N2O1	N-(P-IODOPHENYL)-3-N'-PIPERIDINOACETAMIDE 58479-87-1
3 ★	3.90	7.4	934	2	C13H17I1N2O1	N-(P-IODOPHENYL)-3-N'-PIPERIDINOACETAMIDE
4 ★	3.10	7.5	2545	2	C13H17N1	2'-METHYL-MPTP
15	0.53	7.4	1015	1	C13H17N1	2-AMINOBENZOBICYCLO(2,2,2)OCTENE(ENDO)
6	-0.26	7.4	1015	1	C13H17N1	2-AMINOBENZOBICYCLO(2,2,2)OCTENE(EXO)
7	-1.65		307	1	C13H17N1	2-METHYL-6,7-BENZOMORPHANHYDROCHLORIDE
8 ★	3.18	7.5	2545	2	C13H17N1	3-METHYL-MPTP
9 ★	2.62	13.0	1623	224	C13H17N1	BENZOBICYCLO(2,2,1)HEPTENE,2-ENDO-ETHYLAMINO 86992-68-9
9620 ★	2.72	13.0	1623	224	C13H17N1	BENZOBICYCLO(2,2,1)HEPTENE,2-EXO-ETHYLAMINO 86943-78-4
1	0.30	7.4	2617	1	C13H17N1	HEXAHYDROFLUOREN-9A-AMINE
2	2.94		2617	2	"	HEXAHYDROFLUOREN-9A-AMINE
3 ★	3.18	7.5	2545	2	C13H17N1	N-ETHYL-PTP ✚
4 ★	2.54	13.0	594		C13H17N1	QUINUCLIDINE,4-PHENYL
25 ★	3.13	13.0	594		C13H17N1O1	4'-PIPERIDINOACETOPHENONE 10342-85-5
6	1.09	7.4	2535	1	C13H17N1O1	G-(O-METHYLBENZOYL)PROPYL-ETHANOL AMINE (DEHYDRO-FUSED)
7	1.46	7.4	2535	1	C13H17N1O1	G-(PHENACETYL)PROPYL-ETHANOLAMINE (DEHYDRO-FUSED)
8	1.91	7.4	2535	1	C13H17N1O1	G-BENZOYLPROPYL-(2-PROPANOL)AMINE (DEHYDRO-FUSED)
9 ★	2.00	7.5	2545	2	C13H17N1O1	N-HYDROXYETHYL-PTP
9630 ★	2.85		1481		C13H17N1O1	N-S-BUTYLCINNAMAMIDE
1 ★	1.81		546		C13H17N1O2	N-PHENOXYACETYLPIPERIDINE 36405-75-1
2 ★	2.06		2157		C13H17N1O3	2-(BENZOYLOXY)-(N,N-DIETHYL)ACETAMIDE
3 ★	2.36		2157		C13H17N1O3	2-(BENZOYLOXY)-(N-BUTYL)ACETAMIDE
4 ★	2.26		2157		C13H17N1O3	2-(BENZOYLOXY)-(N-T-BUTYL)ACETAMIDE
35	2.60		1973	459	C13H17N1O3	5-(T-BUTYLCARBONYL)-H2-PYRROLO-PYRROLE-1-CARBOXYLIC ACID 96327-64-9
6 ★	1.59		656		C13H17N1O3	A,A-DIETHYLPHENYLACETAMIDE,3,4-DIOXYMETHYLENE
7	0.57	7.4	2462	1	C13H17N1O3	P-ACETYLBENZOIC ACID,N,N-DIMETHYLAMINOETHYL ESTER
8 ★	1.57	7.4	2462	2	"	P-ACETYLBENZOIC ACID,N,N-DIMETHYLAMINOETHYL ESTER
9	1.11	0.5	1997	346	C13H17N1O3	PHENYLALANINIE,N-BUTYRYL, HYDROCHLORIDE
9640 ★	0.93		2157		C13H17N1O4	2-(BENZOYLOXY)-(N-ETHYL,N-(2-ETHANOL))ACETAMIDE
1	-0.48	7.4	2556	1	C13H17N1O4	DA-EPN ✚
2	2.00	7.4	2145	1	C13H17N1O4	SALICYLAMIDE,PIVALOYLOXYMETHYL ESTER
3 ★	1.99	7.4	2268	1	C13H17N1O4	SALICYLIC ACID,N,N-DIETHYLGLYCOLAMIDE ESTER
4 ★	1.52		1898		C13H17N1O5	(4-HYDROXYETHYL)VANILLINOXIME,PROPIONYL ESTER
45 ★	0.17		2157		C13H17N1O5	2-(BENZOYLOXY)-(N,N-DI-(2-ETHANOL))ACETAMIDE
6 ★	0.25		2157		C13H17N1O6	2-(BENZOYLOXY)-(N-TRICARBINOLYLMETHYL)ACETAMIDE
7 ★	2.81		2665		C13H17N3	1-BENZYLTRIAZOLE,4'-BUTYL *fungicide*
8	-0.62	7.4	1350	1	C13H17N3	TRAMAZOLINE 1082-57-1 *adrenergic*
9	0.97	7.4	651	1	C13H17N3O1	AMINOPYRINE 58-15-1 *antipyretic, analgesic*
9650 ★	1.00	7.4	919	1	"	AMINOPYRINE
1 ★	2.20		559		C13H17N3O2	1-(4-CARBETHOXYBUTYL)BENZOTRIAZOLE
2 ★	2.85		559		C13H17N3O2	2-(4-CARBETHOXYBUTYL)BENZOTRIAZOLE
3 ★	2.82	5.6	2165	1	C13H17N3O2	BENZOTRIAZOLE,1-(2-PENTANOIC ACID,ETHYL ESTER)
4 ★	2.64	5.6	2165	1	C13H17N3O2	BENZOTRIAZOLE,1-(4-PENTANOIC ACID,ETHYL ESTER)
55 ★	3.38	5.6	2156	1	C13H17N3O2	BENZOTRIAZOLE,2-(2-PENTANOIC ACID,ETHYL ESTER)
6 ★	3.05	5.6	2165	1	C13H17N3O2	BENZOTRIAZOLE,2-(4-PENTANOIC ACID,ETHYL ESTER)
7 ★	1.04		871		C13H17N3O2	N-NITROSO-4-BENZOYL-2,6-DIMETHYLPIPERAZINE
8	1.41		871		C13H17N3O2	N-NITROSO-4-BENZOYL-3,5-DIMETHYLPIPERAZINE
9 ★	-0.56		2377		C13H17N3O3	AC-GLY-PHE-N
9660 ★	-0.50		2377		C13H17N3O3	AC-PHE-GLY-N
1	0.11		2661		C13H17N3O3	IMAZAPYR 81334-34-1
2	2.21	7.4	934	1	C13H17N3O3	N-(M-NITROPHENYL)-3-N'-PIPERIDINOACETAMIDE 35763-43-0
3 ★	2.59	7.4	934	2	C13H17N3O3	N-(M-NITROPHENYL)-3-N'-PIPERIDINOACETAMIDE
4	2.75	7.4	934	1	C13H17N3O3	N-(O-NITROPHENYL)-3-N'-PIPERIDINOACETAMIDE 35204-11-6

	logP	pH	Ref.	Note	MF	Name / CAS / activity
65	2.25	7.4	934	1	C13H17N3O3	N-(P-NITROPHENYL)-3-N'-PIPERIDINOACETAMIDE 38367-22-5
6	2.70	7.4	1851	1	C13H17N3O3	N-(P-NITROPHENYL)-3-N'-PIPERIDINOACETAMIDE
7 ★	2.73	7.4	934	2	C13H17N3O3	N-(P-NITROPHENYL)-3-N'-PIPERIDINOACETAMIDE
8	-0.17		1205		C13H17N3O3S1	N,N-DIME-4(N-ME-N(2-CYANOPROPIONYL)C6H4SULFONAMIDE
9	0.84		1205		C13H17N3O3S2	N,N-DIME-4(N-ME-N(2-NCS-PROPIONYL)BZSULFONAMIDE
9670 ★	2.20		1629		C13H17N3O3S2	N-(N"-METHYL-PHENYLCARBAMYLTHIO)METHOMYL
1	1.50	7.4	2461	1	C13H17N3O3S4	THIENO[2,3-B]THIOPHENE-2-SULFONAMIDE,5-N-(THIOMORPHOLIN-1-YL)ETHYLCARBOXAMIDO 129949-95-7
2	0.59	7.4	2461	1	C13H17N3O4S3	THIENO[2,3-B]THIOPHENE-2-SULFONAMIDO,5-N-(MORPHOLIN-1-YL)ETHYLCARBOXAMIDO 129949-93-5
3 ★	-3.51	7.0	594	314	C13H17N3O5	TYR-GLY-GLY
4 ✓	-1.92		2397	101	C13H17N4O3S1	IMIDAZOLIUM CHLORIDE,1-METHYL-2-HYDROXYIMINOMETHYL-3-PHENYLSULFONAMIDOETHYL
75 ✓	1.12	13.0	579		C13H17N5	2,4-DIAMINO-5-(3,5-DIMETHYL-4-AMINOBENZYL)PYRIMIDINE
6 ★	1.82	13.0	547	224	C13H17N5	2,4-DIAMINO-5-(4-DIMETHYLAMINOBENZYL)PYRIMIDINE
7 ★	-0.22	13.0	547	224	C13H17N5O1	4,6-DIAM-2,2-DIME-1(3-ACETYLPH)-S-TRIAZINE
8 ★	-0.04	13.0	547	224	C13H17N5O1	4,6-DIAM-2,2-DIME-1(4-ACETYLPH)-S-TRIAZINE
9	0.60	7.4	1485	1	C13H17N5O2	ADITEREN 56066-19-4
9680 ★	0.83	7.4	1485	2	"	ADITEREN
1	-3.07	7.4	2530	1	C13H17N5O8S2	AZTREONAM 78110-38-0 antimicrobial
2	2.78		2662	459	C13H18Br1N1O4	N-(3,4-DIETHOXYPHENYL)CARBAMATE,2-BROMOETHYL ESTER fungicide
3	-0.55	7.6	2007	1	C13H18Br1N2O2S1.I1	BENZOYLTHIOFORMOHYDROXIMATE,4-BROMO-S-(2-TRIMETHYLAMMONIO)ETHYL IODIDE 99481-59-1
4 ★	3.50	7.4	2392	2	C13H18Cl1N1	BENZAZEPINE,6-CHLORO-3-PROPYL
85 ★	3.27		1182		C13H18Cl1N1O1	2-(M-CHLOROPHENYL)-4-I-PROPYLMORPHOLINE
6 ★	3.00	7.4	2392	2	C13H18Cl1N1O1	BENZAZEPINE,6-CHLORO-3-METHOXYETHYL 73943-10-9
7 ★	3.83		2094		C13H18Cl1N1O1	MONALIDE 7287-36-7 herbicide
8 ★	4.31		2094		C13H18Cl1N1O1	SOLAN 2307-68-8 herbicide
9	2.62		2662	459	C13H18Cl1N1O4	N-(3,4-DIETHOXYPHENYL)CARBAMATE,2-CHLOROETHYL ESTER fungicide
9690	-0.80	7.6	2007	1	C13H18Cl1N2O2S1.Cl1	BENZOYLTHIOFORMOHYDROXIMATE,4-CHLORO-S-(2-TRIMETHYLAMMONIO)ETHYL CHLORIDE 102941-34-4
1 ★	2.07	7.4	2392	2	C13H18Cl1N3O1	BENZAZEPINE,6-CHLORO-3-UREIDOETHYL
2 ★	3.73		547		C13H18Cl1N3O2	1-NO-1-(2-CLET)-3-(2-(2-BENZYL)PROPYL)UREA 33021-93-1
3	1.64	7.4	2004	2	C13H18Cl1N3O3S1	AZEPINAMIDE 1228-19-9 antidiabetic
4 ★	1.27	6.5	321	2	C13H18Cl1N3O4S2	3-CYCLOPENTYLMETHYLHYDROCHLOROTHIAZIDE 742-20-1 antihypertensive
95	0.85	6.5	321	1	C13H18Cl1N3O4S2	HYDROCHLOROTHIAZIDE,3-(4-METHYLSPIROCYCLOHEXANE)
6 ★	3.76	7.4	1804	2	C13H18Cl2N2O1	BENZAMIDE,N(DIETAMET),3,5-DICHLORO
7	3.77	7.4	1413	2	"	BENZAMIDE,N(DIETAMET),3,5-DICHLORO
8	-1.55	7.6	2007	1	C13H18F1N2O2S1.Cl1	BENZOYLTHIOFORMOHYDROXIMATE,4-FLUORO-S-(2-TRIMETHYLAMMONIO(ETHYL CHLORIDE 102941-39-9
9 ★	3.53		2479		C13H18F1N3O2	PHENYLBIURET,2-FLUORO-5-I-PROPYL
9700 ★	3.08		2479		C13H18F1N3O3	PHENYLBIURET,2-FLUORO-5-I-PROPOXY
1 ★	1.87		2479	459	C13H18F1N3O4S1	PHENYLBIURET,2-FLUORO-5-I-PROPYLSULFONYL
2	2.27		2662	459	C13H18F1O4	N-(3,4-DIETHOXYPHENYL)CARBAMATE,2-FLUOROETHYL ESTER fungicide
3 ★	2.66		1780		C13H18F3N3O4S2	PENFLUZIDE 1766-91-2 diuretic
4 ★	2.90		2150	459	C13H18N2	ADAMANTANE,1-(1-IMIDAZOLYL) 69380-11-6
5 ★	3.05		2150	459	C13H18N2	ADAMANTANE,1-(1-PYRAZOLYL)
6 ★	2.33		572		C13H18N2O1	2,5-DIME-1-PYRROLIDINECARBOXANILIDE-CIS 35631-98-2
7	2.77		1263		C13H18N2O1	3-PHENYL-1-CYCLOHEXYLUREA 886-59-9
8	1.60		2032	459	C13H18N2O1	4-PYRIMIDONE,(5,6)-CYCLOHEPTYL-(2,3)-CYCLOHEXYL ✦
9	1.61		2032	459	C13H18N2O1	4-PYRIMIDONE,(5,6)-CYCLOPENTYL-(2,3)-CYCLOOCTYL
9710 ★	1.69		1147		C13H18N2O1	N-(PYRROLIDINOACETYL)BENZYLAMINE
1	1.76	7.4	934	1	C13H18N2O1	N-PHENYL-3-N'-PIPERIDINOACETAMIDE 4671-97-0
2	2.41	7.4	934	2	C13H18N2O1	N-PHENYL-3-N'-PIPERIDINOACETAMIDE
3 ★	2.71	7.4	1851	1	C13H18N2O1	N-PHENYL-3-N'-PIPERIDINOACETAMIDE
4	1.67		1725	459	C13H18N2O1	PYRIDO-PYRIMIDINE ANALOG,6-ME(RINGS:666)
15	1.70		1725	459	C13H18N2O1	PYRIDO-PYRIMIDINE ANALOG,7-ME(RINGS:666)
6	1.67		1725	459	C13H18N2O1	PYRIDO-PYRIMIDINE ANALOG,8-ME(RINGS:666)
7	1.76		1725	459	C13H18N2O1	PYRIDO-PYRIMIDINE ANALOG,9-ME(RINGS:666)
8	1.28		1725	459	C13H18N2O1	PYRIDO-PYRIMIDINONE ANALOG(RINGS=568)
9	1.73		1725	459	C13H18N2O1	PYRIDOPYRIMIDINE ANALOG (RING:667) ✦
9720 ★	0.67		1147		C13H18N2O2	N-(MORPHOLINOACETYL)BENZYLAMINE
1	-1.05	7.0	477	1	C13H18N2O2	N-I-PR-N-(3-(O-CYANOPHO)-2-PROPANOL)AMINE 38555-07-6
2	-0.48	7.4	1354	1	"	N-I-PR-N-(3-(O-CYANOPHO)-2-PROPANOL)AMINE
3 ★	0.29		1622		C13H18N2O2	PIPERAZINE,1-METHYL,4-(O-METHOXY)BENZOYL
4	1.52	7.6	2007	1	C13H18N2O2S1	BENZOYLTHIOFORMOHYDROXIMATE,4-METHYL-S-(2-DIMETHYLAMINO)ETHYL 102941-51-5
25 ★	0.99		1439		C13H18N2O3	3-I-PROPOXYCARBONYL-PYRIDO(1,2A)PYRIMIDIN-4-ONE,4,6,7,8,9-H5,6-ME
6 ★	1.14		1439		C13H18N2O3	3-PROPOXYCARBONYL-PYRIDO(1,2A)PYRIMIDIN-4-ONE,4,6,7,8,9-H5,ME6
7 ★	2.25	7.4	899	1	C13H18N2O3	CYCLOBARBITAL,N-METHYL 726-78-3
8 ★	2.03	2.0	300	1	C13H18N2O3	HEPTABARBITAL 509-86-4 hypnotic, sedative
9	2.17	7.4	899	1	"	HEPTABARBITAL
9730	1.07	7.6	2007	1	C13H18N2O3S1	BENZOYLTHIOFORMOHYDROXIMATE,4-METHOXY-S-(2-DIMETHYLAMINO)ETHYL 102941-29-7
1 ★	1.59		1681		C13H18N2O4	N,N-DIMETHYLCARBAMATE,P-(OCONME2)BENZYL ESTER 84640-38-0
2 ★	0.79		537	270	C13H18N2O5	DIMETHYLAMPHENICOL 18048-95-8
3 ★	0.87		537		C13H18N2O5	ETHYLAMPHENICOL 23885-69-0
4	-0.61		547		C13H18N2O5	N1-PHENYL-N3-2-(3-DEOXYGLUCOSYL)UREA
35 ★	2.67		2072	2	C13H18N2O5S1	BENZENESULFONAMIDE,2-NITRO-4-METHYL-N-(2-ETHYLBUTANOYL)
6 ✓	-0.94		2397	101	C13H18N3O2	IMIDAZOLIUM CHLORIDE,1-(1-PROPENYL)-2-HYDROXYIMINOMETHYL-3-(1-ETHYLPROPYN-2YLOXY)METHYL
7 ✓	-1.37		2397	101	C13H18N3O2	IMIDAZOLIUM CHLORIDE,1-(1-PROPENYL)-2-HYDROXYIMINOMETHYL-3-(PENTYN-4YLOXY)METHYL
8 ✓	-0.99		2397	101	C13H18N3O2	IMIDAZOLIUM CHLORIDE,1-METHYL-2-HYDROXYIMINOMETHYL-3-(1-ETHYNYLCYCLOPENTYLOXY)METHYL
9 ✓	-1.28		2397	101	C13H18N3O2	IMIDAZOLIUM CHLORIDE,1-METHYL-2-HYDROXYIMINOMETHYL-3-(2-ETHYNYLCYCLOPENTOXY)METHYL
9740	0.42	7.4	2495	1	C13H18N3O3P1	PHOSPHORIC-TRIAMIDE,N-M-ETHOXYBENZOYL-N',N"-DIETHYLENE
1 ★	1.24	7.4	2495	2	C13H18N3O3P1	PHOSPHORIC-TRIAMIDE,N-M-ETHOXYBENZOYL-N',N"-DIETHYLENE
2	2.39		1049		C13H18N3O4P1	N-ET-PHOSPHORAMIDATE,O-ET,O-A-CN-4-MEO-BENZALDOXIME
3	0.29		657		C13H18N4O3	5-ACETYLBUTYL-3,7-DIMETHYLXANTHINE 6493-05-6 vasodilator
4	1.60	7.4	2477	1	C13H18N4O3	A[(2-NITROIMIDAZOL-1-YL)METHYL]-3-AZATRICYCLO[3.2.1.0]OCTANE-3-ETHANOL 120278-00-4
45	-0.64		1404		C13H18N4O7	2(2-NO2-1-IMIDAZOLYL)ACETAMIDE,N,N(ACETYL)
6	0.25	7.4	1485	1	C13H18N6	2,4-DIAMINOPYRIMIDINE,5-(3,5-BIS-METHYLAMINO)BENZYL
7 ★	0.44	7.4	1485	2	"	2,4-DIAMINOPYRIMIDINE,5-(3,5-BIS-METHYLAMINO)BENZYL
8 ★	1.45		2289	459	C13H18N6O2	TETRAZOLE,2-I-PROPYL-5-(4-DIMETHYLUREYL)PHENOXY 117121-43-4
9	1.07	7.4	1838	1	C13H18O2	IBUPROFEN 15687-27-1 anti-inflammatory
9750	1.66	7.4	1687	1	"	IBUPROFEN
1 ★	3.50	2.0	1838		"	IBUPROFEN
2	4.50	7.4	1687	2	"	IBUPROFEN
3 ★	4.20		2168		C13H18O2	ISOVALERIC ACID-2-ETHYL,PHENYL ESTER
4 ★	3.63		2430		C13H18O2	PHENYLHEPTANOIC ACID

	logP	pH	Ref.	Note	MF	Name / CAS / activity
55	2.74		2404	459	C13H18O3	DIHYDROBENZOPYRAN,2,2-DIMETHYL-6,7-DIMETHOXY 69888-42-2
6 ★	4.35		508		C13H18O3	P-HYDROXYBENZOIC ACID,HEXYL ESTER 1083-27-8
7 ★	2.75	3.5	2250	302	C13H18O3	PROPANOIC ACID,2-HYDROXY-2-(4-I-BUTYL)PHENYL
8 ★	3.17		1335		C13H18O5	HEXYLGALLATE 1087-26-9
9 ★	1.09		1895		C13H18O5	TRIETHYLENEGLYCOL, MONOBENZOATE
9760	-0.47		1294		C13H18O5S1	1-BENZYLTHIO-B-GALACTOPYRANOSIDE
1	-0.16		511		C13H18O6	GLUCOPYRANOSIDE,2-METHYLPHENYL(BETA) 7234-31-3
2	-0.20		511		C13H18O6	GLUCOPYRANOSIDE,3-METHYLPHENYL(BETA) 6092-25-7
3	-0.16		511		C13H18O6	GLUCOPYRANOSIDE,4-METHYLPHENYL(BETA) 20274-94-6
4 ★	-0.70		511		C13H18O6	GLUCOPYRANOSIDE,BENZYL(BETA) 4304-12-5
65 ★	-1.04		511		C13H18O7	GLUCOPYRANOSIDE,2-METHOXYPHENYL(BETA) 6092-24-6
6 ★	-0.52		511		C13H18O7	GLUCOPYRANOSIDE,3-METHOXYPHENYL(BETA) 14062-61-4
7 ★	-0.73		511		C13H18O7	GLUCOPYRANOSIDE,4-METHOXYPHENYL(BETA) 6032-32-2
8 ★	-1.22		511		C13H18O7	SALICIN 138-52-3 analgesic
9	1.36	7.4	2477	1	C13H19Br1N4O3	TR-A-[[(2-BROMOBICYCLO[2.2.1]HEPTAN-1-YL)AMINO]METHYL]-2-NITROIMIDAZOLE-1-ETHANOL 134419-56-0
9770 ★	3.81		2506		C13H19Cl1N1O3P1S2	ANILOFOS herbicide ✚
1	0.41	5.7	2241		C13H19Cl1N2O1	BUTANILICAINE 3785-21-5 anesthetic (local)
2 ★	2.99	7.4	2026	1	C13H19Cl1N2O2	BENZAMIDE,2-(2-BUTOXYETHYL)AMINO-5-CHLORO
3	0.95	7.4	2409	1	C13H19Cl1N2O2	CHLOROPROCAINE 133-16-4 anesthetic (local)
4	1.24	7.4	2409	1	"	CHLOROPROCAINE
75 ★	2.86		2409	2	"	CHLOROPROCAINE
6	2.56		1019		C13H19Cl1N2O2	N'-I-PR-2-ME-2(4-CL-PHENOXY)PROPIONICAC.HYDRAZIDE
7 ★	2.51	7.4	2392	2	C13H19Cl1N2O2S1	BENZAZEPINE,6-CHLORO-3-METHYLSULFONAMIDOETHYL
8 ★	2.30		1182		C13H19N1O1	2-PHENYL-4-I-PROPYLMORPHOLINE 23222-62-0
9 ★	2.10		2410		C13H19N1O1	ACETANILIDE,2-ETHYL-6-I-PROPYL
9780 ★	3.10		2410		C13H19N1O1	ACETANILIDE,2-METHYL-4-BUTYL
1 ★	3.64		2410		C13H19N1O1	ACETANILIDE,3-METHYL-4-BUTYL
2	1.88		473		C13H19N1O1	B-PHENYL-B-METHOXY-N-ETHYLPYRROLIDINE 6577-49-7
3 ★	2.19		1718		C13H19N1O1	P-ETHYLBENZYLIDENE-T-BUTYLAMINE,N-OXIDE
4	2.18		870		C13H19N1O2	2-METHYL-5-T-BUTYL-4-HYDROXYACETANILIDE
85 ★	3.31		558		C13H19N1O2	2-S-BUTYLPHENYLDIMETHYLCARBAMATE 55379-70-9
6 ★	3.20		1969		C13H19N1O2	DIETHYLCARBAMATE,PHENETHYL ESTER 98183-17-6
7	1.38	9.0	1347		C13H19N1O2	IBUPROXAM 53648-05-8 anti-inflammatory
8 ★	-0.27		2018		C13H19N1O2	ISOLEUCINE,N-BENZYL 69410-51-1
9	2.52		594	400	C13H19N1O2	ISOPROPYLCARBAMATE,N-(2-ETHYL-6-METHYLPHENYL)
9790	1.04	7.4	2535	1	C13H19N1O2	N,N,N-(G-BENZOYL)PROPYL-METHYL-ETHANOLAMINE
1 ★	3.38		273		C13H19N1O2	N-METHYL-3-METHYL-4-T-BUTYLPHENYLCARBAMATE
2 ★	3.35		273		C13H19N1O2	N-METHYL-3-METHYL-5-T-BUTYLPHENYLCARBAMATE
3 ★	3.14		273		C13H19N1O2	N-METHYL-3-METHYL-6-T-BUTYLPHENYLCARBAMATE
4 ★	3.95		2630		C13H19N1O2	P-AMINOBENZOIC ACID,HEXYL ESTER 13476-55-6
95	1.95	7.4	2462	1	C13H19N1O2	P-ETHYLBENZOIC ACID,N,N-DIMETHYLAMINOETHYL ESTER
6 ★	3.15	7.4	2462	2	"	P-ETHYLBENZOIC ACID,N,N-DIMETHYLAMINOETHYL ESTER
7 ★	3.28		2506		C13H19N1O2S1	FENOTHIOCARB acaricide ✚
8 ★	3.22		1136		C13H19N1O3	2-HEXYLOXY-4-AMINOBENZOIC ACID 24397-14-6
9 ★	2.71		2487		C13H19N1O3	BENZAMIDE,N-HEXYL-3,4-DIHYDROXY
9800 ★	1.64		2045		C13H19N1O3	MOXISYLYTE,DES-I-PROPYL ✚
1 ★	3.03	7.4	2659	459	C13H19N1O3	N,N-DIMETHYLCARBAMATE,M-I-PROPOXYBENZYL ESTER 84640-29-9
2 ★	4.19		547	20	C13H19N1O3	PHENOL,2-I-BUTYL-6-I-PROPYL-4-NITRO
3 ★	2.50		2142		C13H19N1O4	3,4-DIETHOXYPHENYLCARBAMATE,O-ETHYL
4 ★	0.15		844		C13H19N1O4	N-ET-6-HYDROXYNORBORNANE-2-CARBOXAMIDE-3,5-LACTONE
5 ★	1.86	7.4	2490	1	C13H19N1O4S1	BENZENESULFONAMIDE,4-(5-ACETYLOXY)PENTYL 135832-51-8
6 ★	3.21	1.0	538	342	C13H19N1O4S1	PROBENECID 57-66-9 uricosuric
7	-0.09	7.4	2062	1	C13H19N1O5	L-DOPA,(3-METHOXY-2-PROPYL) ESTER
8 ★	0.08	7.4	2062	2	"	L-DOPA,(3-METHOXY-2-PROPYL) ESTER
9 ✓	-0.81		2406	102	C13H19N2O1S1	TETRAMETHYLTHIOURONIUM IODIDE,S-(2-ACETYL)
9810 ✓	-0.88		2406	102	C13H19N2O1S1	TETRAMETHYLTHIOURONIUM IODIDE,S-(3-ACETYLPHENYL)
1 ✓	-0.85		2406	102	C13H19N2O1S1	TETRAMETHYLTHIOURONIUM IODIDE,S-(4-ACETYLPHENYL)
2 ✓	-0.75		2406	102	C13H19N2O2S1	TETRAMETHYLTHIOURONIUM IODIDE,S-(3-METHOXYCARBONYLPHENYL)
3 ✓	-0.74		2406	102	C13H19N2O2S1	TETRAMETHYLTHIOURONIUM IODIDE,S-(4-METHOXYCARBONYLPHENYL)
4 ✓	-0.70		2406	102	C13H19N2O2S1	TETRAMETHYLTHIOURONIUM IODIDE,S-(METHOXYCARBONYL)
15	-1.30	7.4	1097	1	C13H19N3	2-(2,6-DIETHYLPHENYLIMINO)IMIDAZOLIDINE
6	-1.20	7.4	1398	1	"	2-(2,6-DIETHYLPHENYLIMINO)IMIDAZOLIDINE
7	-0.88	7.4	976	1	"	2-(2,6-DIETHYLPHENYLIMINO)IMIDAZOLIDINE
8	-0.84	7.4	821	1	"	2-(2,6-DIETHYLPHENYLIMINO)IMIDAZOLIDINE
9	2.36	7.4	821	2	"	2-(2,6-DIETHYLPHENYLIMINO)IMIDAZOLIDINE
9820	2.60	7.4	976	2	"	2-(2,6-DIETHYLPHENYLIMINO)IMIDAZOLIDINE
1	0.38	7.4	365	2	C13H19N3O1	PIPERAZINE-2-CARBOXANILIDE,2'6'-DIME 36371-18-3
2	-0.68	7.4	579	1	C13H19N3O1S1	ROCASTINE 91833-77-1
3 ★	1.08		2050		"	ROCASTINE
4	1.33		1214		C13H19N3O2	3(2-MEBU)-1-(4-CARBOXYPHENYL)-3-METHYLTRIAZENE 85514-39-2 antineoplastic
25	1.35	3.3	2590	331		3(2-MEBU)-1-(4-CARBOXYPHENYL)-3-METHYLTRIAZENE
6	1.46	3.3	2590	331	C13H19N3O2	3-PENTYL-1-(4-CARBOXYPHENYL)-3-METHYLTRIAZENE 74109-23-2 antineoplastic
7	1.49	7.4	1214	1	"	3-PENTYL-1-(4-CARBOXYPHENYL)-3-METHYLTRIAZENE
8 ★	2.14		2105		C13H19N3O2	HYDRAZINECARBOXAMIDE,1-ACETYL-N-(4-T-BUTYLPHENYL)
9 ★	1.34	7.4	2004	1	C13H19N3O3S1	BENZENESULFONAMIDE,4-METHYL-N(1-PIPERIDINE)AMINOCARBONYL 1443-94-3
9830 ★	5.18		2506		C13H19N3O4	PENDIMETHALIN 40318-45-4 herbicide
1	-0.56		1205		C13H19N3O4S1	N,N-DIME-4(N-ME(2-AMIDOPROPIONYL)BENZSULFONAMIDE
2 ★	1.24	6.5	321	2	C13H19N3O4S2	3-CYCLOPENTYLMETHYLHYDROTHIAZIDE 23141-86-8
3 ★	-0.70	7.4	2350	1	C13H19N3O4S2	DEOXYSPARSOMYCIN ✚
4	1.56	6.5	321	1	"	HYDROTHIAZIDE,3-(4'-METHYLSPIROCYCLOHEXANE)
35	-1.49	7.0	1812	1	C13H19N3O5	5'-AMINO-3'O-PROPIONYL-5'-DEOXYTHYMIDINE
6 ★	-1.71	7.4	2350	1	C13H19N3O5S2	SPARSOMYCIN 1404-64-4 antineoplastic
7	0.30	7.0	801	1	C13H19N3O6	1-(5-I-BUTYRYLARABINOSYL)CYTOSINE 31088-20-7
8	-1.47	7.4	594	1	C13H19N5	S-TRIAZINE,4,6-DIAMINO-1,2-DIHYDRO-2,2-DIMETHYL-1-(3'ETHYLPHENYL)
9 ★	1.50		2218		C13H19N5O1	GUANIDINE,N-METHYL-N'-CYANO-N"-(3-METHOXYPYRID-2YL)BUTYL
9840	-1.76	7.4	594	1	C13H19N5O1	S-TRIAZINE,4,6-DIAMINO-1,2-DIHYDRO-2,2-DIMETHYL-1-(3'-ETHOXYPHENYL)
1 ★	0.72	8.0	2267	306	C13H19N5O3	ALLOPURINOL,1-N,N-DIETHYLALANYLOXYMETHYL 98204-10-5
2 ★	1.70	7.4	1804	2	C13H20Cl1N3O1	PROCAINAMIDE,3-CHLORO 891-60-1
3 ★	3.56		401		C13H20Cl1N3O2	3-(1-ADAMANTYL)-1-(2-CLOROETHYL)-1-NITROSOUREA 14039-10-2
4 ★	3.90	4.0	2436	1	C13H20F1N3O3	URACIL,1-OCTYLAMINOCARBONYL-5-FLUORO

	logP	pH	Ref.	Note	MF	Name / CAS / activity
45 ✓	-0.61		2397	101	C13H20F3N4O3S1	IMIDAZOLIUM CHLORIDE,1-METHYL-2-HYDROXYIMINOMETHYL-3-(1-TRIFLUOROMETHYLSULFONYLPIPERIDIN-2YL)ETHYL
6 ✓	-1.98		2180		C13H20N1O2.Br1	BENZYLTRIMETHYL AMMONIUM BROMIDE,3-ETHOXYCARBONYL
7 ★	3.23		2612		C13H20N2O1	1,1-DIMETHYL-3-(M-BUTYLPHENYL)UREA **88132-40-5**
8 ★	3.10		2612		C13H20N2O1	1,1-DIMETHYL-3-(P-T-BUTYLPHENYL)UREA **32745-69-0**
9 ★	2.26		594		C13H20N2O1	1-(2,6-DI-I-PROPYLPHENYL)UREA
9850	1.53		1352		C13H20N2O1	1-(DIMETHYLAMINO)PROPIONYL-2',6'-XYLIDIDE
1	1.23	7.4	1414	1	C13H20N2O1	2(DIMEAMINO)-N-(2,6-DIMEPHENYL)PROPIONAMIDE
2	1.29	7.4	1414	1	C13H20N2O1	2(ET,ME-AMINO)N(2,6-DIMEPHENYL)ACETAMIDE
3	1.08		1352		C13H20N2O1	3-AMINOPENTANO-2',6'-XYLIDIDE
4	1.02		1352		C13H20N2O1	4-AMINOPENTANO-2',6'-XYLIDIDE
55 ★	1.99	7.4	833	2	C13H20N2O1	N-(2-DIETHYLAMINOETHYL)-BENZAMIDE **3690-53-7**
6	2.11	7.4	1804	2	"	N-(2-DIETHYLAMINOETHYL)-BENZAMIDE
7 ★	2.14		1147		C13H20N2O1	N-(N',N'-DIETHYLAMINO)ACETYL)BENZYLAMINE
8	1.40	7.4	2409	1	C13H20N2O1	PRILOCAINE **721-50-6** anesthetic (local)
9 ★	2.11		2409	2	"	PRILOCAINE
9860 ★	3.21		2612		C13H20N2O2	1-(3-T-BUTYLPHENYL)-3-METHOXY-3-METHYLUREA **70477-14-4**
1 ★	2.27	7.4	2026	1	C13H20N2O2	BENZAMIDE,2-(2-BUTOXYETHYL)AMINO
2 ★	0.99		844	837	C13H20N2O2	BICYCLO(2,2,1)HEPT-2-EN-5,6-DICARBOXAMIDE,N,N'-DIETHYL
3 ★	2.36		1969		C13H20N2O2	DIETHYLCARBAMATE,3-(3-PYRIDYL)PROPYL ESTER **98154-98-4**
4 ★	1.88		1019		C13H20N2O2	N'-I-PR-2-ME-(3-TOLYLOXY)PROPIONIC.HYDRAZIDE ✚
65 ★	1.88		1019		C13H20N2O2	N'-I-PR-2-ME-2-PHENOXYPROPIONIC ACID HYDRAZIDE
6 ★	1.27		1622		C13H20N2O2	O-METHOXYBENZAMIDE,N-(3-DIMETHYLAMINOPROPYL)
7 ★	2.62		29		C13H20N2O2	P-ETHYLAMINOBENZOIC ACID,DIMETHYLAMINOETHYL ESTER
8	-0.32	7.0	345	1	C13H20N2O2	PROCAINE **59-46-1** anesthetic (local)
9	0.00	7.4	2217	1	"	PROCAINE
9870	0.23	7.4	2409	1	"	PROCAINE
1	0.30	7.5	1259	1	"	PROCAINE
2	0.49	7.4	2409	1	"	PROCAINE
3	0.60	7.4	2284	1	"	PROCAINE
4 ★	1.92		505		"	PROCAINE
75	0.45		1439		C13H20N2O3	3-ETOCO-PYRIDO(1,2A)PYRIMIDIN-4-ON,-H7,1,6-DIME
6	0.08	7.4	2685	1	C13H20N2O3	4-PYRIDONE,3-HYDROXY-2-METHYL-1-(2-BUTYLAMINOCARBONYL)ETHYL iron chelator
7	-0.34	7.4	2685	1	C13H20N2O3	4-PYRIDONE,3-HYDROXY-2-METHYL-1-(N,N-DIETHYLPROPIONAMIDO) iron chelator
8	2.00	0.7	962		C13H20N2O3	BARBITURIC ACID,5-ET-5-CYCLOHEXYL,1-METHYL **4309-13-1**
9 ★	2.05		773		C13H20N2O3S1	4-SULFAMYLBENZAMIDE,N-HEXYL
9880 ★	2.13		2072	2	C13H20N2O3S1	BENZENESULFONAMIDE,2-(2-ETHYLBUTANOYLAMINO)-4-METHYL
1 ★	2.18		2072	2	C13H20N2O3S1	BENZENESULFONAMIDE,2-AMINO-N-(2-ETHYLBUTANOYL)-4-METHYL
2	2.23	5.9	1351		C13H20N2O3S1	DEXETOZOLINE **77519-25-6** diuretic
3	2.30	7.4	1351	1	"	DEXETOZOLINE
4	1.19		1205		C13H20N2O3S1	N,N-DIME-4(N-ME-N-BUTYROYL)BENZENESULFONAMIDE
85	0.23		1205		C13H20N2O3S1	N,N-DIME-4-(N-ME-N-BUT-3-ONYL)BENZENESULFONAMIDE
6 ★	0.99		1205		C13H20N2O3S2	N,N-DIME-4(N-ME-N(2-METHIOPROPIONYL)BZSULFONAMIDE
7	1.89		2447		C13H20N2O4	4-PYRIDINECARBOXYLIC ACID,TETRAHYDRO-2,6-DIOXO,OCTYL ESTER
8	1.75	3.5	1267	1	C13H20N2O4	METHYL-2-ME-2-CYHEXENYL-6-METHYLMALONURATE
9	-1.80	7.2	846	314	C13H20N2O4S1	6-(N-PENTANOYLAMINO)PENICILLANIC ACID **4704-53-4**
9890 ★	2.01		2072	2	C13H20N2O4S1	BENZENESULFONAMIDE,2-(2-ETHYLBUTANOYLAMINO)-4-METHOXY
1 ★	2.03		2072	2	C13H20N2O4S1	BENZENESULFONAMIDE,2-AMINO-4-METHOXY-N-(2-ETHYLBUTANOYL)
2 ★	0.33		1205		C13H20N2O4S1	N,N-DIME-4(N-ME-(2-MEO-PROPIONYL)BENZENESULFONAMIDE
3 ★	-0.67		1205		C13H20N2O4S2	N,N-DIME-BENZENESULFONAMIDE,P-N-ME-N-(2-ME-SULFONYLPROPIONYL)AMINO
4	0.71	7.4	2461	1	C13H20N2O4S3	THIENO[2,3-B]THIOPHENE-2-SULFONAMIDE,5-(2-(METHOXYETHOXYETHYL)AMINOETHYL **133445-80-4**
95	-0.14	7.4	2461	1	C13H20N2O4S3	THIENO[2,3-B]THIOPHENE-2-SULFONAMIDE,5-(METHOXYETHOXYPROPYL)AMINOMETHYL **122266-96-0**
6	1.17	7.4	2461	1	C13H20N2O4S3	THIENO[2,3-B]THIOPHENE-2-SULFONAMIDE,5-(N,N-BIS-2-HYDROXYPROPYL)AMINOMETHYL **133445-77-9**
7	1.58	7.4	2461	1	C13H20N2O4S3	THIENO[2,3-B]THIOPHENE-2-SULFONAMIDE,5-(N,N-BIS-METHOXYETHYL)AMINOMETHYL **122267-02-1**
8	1.56	7.4	2461	1	C13H20N2O4S3	THIENO[3,2-B]THIOPHENE-2-SULFONAMIDE,5-(N,N-BIS-METHOXYETHYL)AMINOMETHYL **127025-31-4**
9	-0.52	7.4	2525	1	C13H20N2O5S2	5-(METHOXYETHOXYPROPYLAMINO)METHYLTHIENO[2,3-B]FURAN-2-SULFONAMIDE
9900	1.29	7.4	2525	1	C13H20N2O5S2	5-(N,N-DIMETHOXYETHYLAMINO)METHYL-THIENO[2,3-B]FURAN,2-SULFONAMIDE
1	-0.68		1205		C13H20N2O6S2	N,N-DIME-4(N-ME-N-(2-MESO2O-PROPIONYL)C6H4SULFONAMID
2 ✓	-1.58		2406	102	C13H20N3O1S1	TETRAMETHYLTHIOURONIUM IODIDE,S-(4-ACETYLAMINOPHENYL)
3 ✓	-1.17		2397	101	C13H20N3O2	IMIDAZOLIUM CHLORIDE,1-METHYL-2-HYDROXYIMINOMETHYL-3-(1-PROPYLBUTYN-3YLOXY)METHYL **117983-29-6**
4 ✓	-1.84		2397	101	C13H20N3O2	IMIDAZOLIUM CHLORIDE,1-T-BUTYL-2-HYDROXYIMINOMETHYL-3-BUTYN-3YLOXYMETHYL
5 ★	2.05		1967		C13H20N4	2-HEXANONE,N-PHENYLGUANYLHYDRAZONE **97183-47-6**
6 ★	3.18	6.6	2590	459	C13H20N4O1	1-(4-CARBAMOYLPHENYL)-3-METHYL-3-PENTYLTRIAZENE **59708-22-4** antineoplastic
7 ★	3.00	6.6	2590	459	C13H20N4O1	BENZAMIDE,4-(3,3-DI-I-PROPYLTRIAZENYL) antineoplastic
8	1.20		657		C13H20N4O2	PENTIFYLLINE **1028-33-7** vasodilator
9	1.55	7.4	2477	1	C13H20N4O3	2-METHYL-A[(2-NITROIM,IDAZOL-1-YL)METHYL]-7-AZABICYCLO[4.1.0]HEPTANE-7-ETHANOL **120277-94-3**
9910	1.44	7.4	2477	1	C13H20N4O3	A[(2-NITROIMIDAZOL-1-YL)METHYL]-8-AZABICYCLO[5.1.0]OCTANE-8-ETHANOL **120277-95-4**
1 ★	2.01	7.4	2206	306	C13H20N4O4	(2-METHYL-5-NITROIMIDAZOL-2YL)ETHYLCARBAMIC ACID,CYCLOHEXYL ESTER
2	3.30	7.4	2028	459	C13H20N6	1,2,4-TRIAZOLE,3-AMINO-5-(2-HEXYLAMINO)PYRIDIN-4-YL **77314-59-1**
3	0.11		1045		C13H20N6O2	6(2,2-DIME-1,3-DIOXAN-5YL)AM-2,4-DIAZIRIDINYLTRIAZINE
4	3.85		1225	459	C13H20O1	2-S-BUTYL-6-I-PROPYLPHENOL
15	4.45		2316	337	C13H20O1	PHENOL,2-HEPTYL
6	4.15		2316	337	C13H20O1	PHENOL,4-HEPTYL
7	3.59		2316	337	C13H20O2	1,2-DIHYDROXYBENZENE,3-HEPTYL
8	2.90		2316	337	C13H20O2	1,3-DIHYDROXYBENZENE,4-HEPTYL
9 ★	2.70		2300	20	C13H20O4S2	ISOPROTHIOLANE **50512-35-1** pesticide
9920 ★	3.38		1564		C13H21Br1N2O3	PYRIMIDINE-2,4-DIONE,3-OCTYLOXY-5-BR-6-METHYL **77317-78-3**
1 ★	4.14		260		C13H21Cl1N2	N-DIETHYLAMINOETHYLANILINE,3-CL-4-METHYL **2519-75-7**
2	5.42		442	2	C13H21N1	P-OCTYLPYRIDINE **40089-91-6**
3	0.84	7.5	1588	1	C13H21N1	PHENETHYLAMINE,N-METHYL-A-ISOBUTYL
4	3.76	7.4	1588	1	"	PHENETHYLAMINE,N-METHYL-A-ISOBUTYL
25 ★	4.14		594	400	C13H21N1	PYRIDINE,2,6-DI-T-BUTYL **585-48-8**
6 ★	0.83	13.0	579	224	C13H21N1O1	5,5-DIMETHYL-3-(PIPERIDIN-1-YL)-2-CYCLOHEXEN-1-ONE **13358-76-4**
7 ★	2.55		594		C13H21N1O1	ADAMANTANE,1-(N,N-DIMETHYLCARBOXAMIDE)
8 ★	2.81	7.0	538	701	C13H21N1O2	2,5-DIMETHOXY-4-ETHYLAMPHETAMINE **22004-32-6**
9	2.28	7.4	2685	1	C13H21N1O2	4-PYRIDONE,3-HYDROXY-2-ETHYL-1-HEXYL iron chelator
9930 ★	0.43		573		C13H21N1O2	ADAMANTYLALANINE
1 ★	2.35		2045		C13H21N1O2	MOXISYLYTE,N-DESMETHYL,DESACETYL ✚
2	-0.28	7.0	474	1	C13H21N1O2	TOLIPROLOL **2933-94-0** adrenergic blocker (beta receptors)
3	0.22	7.4	1354	1	"	TOLIPROLOL
4	0.58	7.4	2245	1	"	TOLIPROLOL

	logP	pH	Ref.	Note	MF	Name / CAS / activity
35 ★	1.93		474		"	TOLIPROLOL
6	2.78	7.4	2245	2	"	TOLIPROLOL
7 ★	1.52	8.0	868	2	C13H21N1O3	3,5-DIMETHOXY-4-I-PROPOXYPHENETHYLAMINE 61367-70-2
8 ★	1.70	8.0	868	2	C13H21N1O3	3,5-DIMETHOXY-4-PROPOXYPHENETHYLAMINE 39201-78-0
9 ★	0.83	7.4	1074	324	C13H21N1O3	4-(I-PENTOXYMETHYL)PYRIDOXOL
9940	-0.64	7.4	2501	1	C13H21N1O3	MOPROLOL 5741-22-0 adrenergic (beta-receptor)
1 ★	1.69	13.0	2501			MOPROLOL
2 ★	2.56		2413	459	C13H21N1O4	TETRAHYDROPYRAN-2,4-DIONE,3[1-(ETHOXYIMINO)BUTYL]-6,6-DIMETHYL
3 ✓	-0.06		2406	102	C13H21N2O1S1	TETRAMETHYLTHIOURONIUM IODIDE,S-(2-ETHOXYPHENYL)
4 ✓	-0.10		2406	102	C13H21N2O1S1	TETRAMETHYLTHIOURONIUM IODIDE,S-(3-ETHOXYPHENYL)
45 ✓	-0.06		2406	102	C13H21N2O1S1	TETRAMETHYLTHIOURONIUM IODIDE,S-(4-ETHOXYPHENYL)
6	-0.67		301		C13H21N2O1.Cl1	N1-HEPTYLNICOTINAMIDECHLORIDE 2255-02-9
7 ✓	0.34		2406	102	C13H21N2S1	TETRAMETHYLTHIOURONIUM IODIDE,S-(3-ETHYLPHENYL)
8 ✓	0.29		2406	102	C13H21N2S1	TETRAMETHYLTHIOURONIUM IODIDE,S-(4-ETHYLPHENYL)
9 ✓	0.32		2406	102	C13H21N2S1	TETRAMETHYLTHIOURONIUM IODIDE,S-2-ETHYLPHENYL
9950	-1.16	7.4	2375	1	C13H21N3O1	PROCAINAMIDE 51-06-9 cardiac depressant (anti-arrhythmic)
1	-1.15	7.4	2409	1	"	PROCAINAMIDE
2	-1.09	7.4	1157	324	"	PROCAINAMIDE
3	0.79	7.4	1413	2	"	PROCAINAMIDE
4	0.79	7.4	1804	2	"	PROCAINAMIDE
55 ★	0.88	7.4	833	2	"	PROCAINAMIDE
6 ★	2.33		2612		C13H21N3O2	1-(4-DIETHYLAMINOPHENYL)-3-METHOXY-3-METHYLUREA 88132-22-3
7	2.53	7.4	459	1	C13H21N3O2	1-CYCLOHEXYL-3-H-5-DIMEAMINO-6-METHYLURACIL 32150-40-6
8	2.65	7.4	459	1	C13H21N3O2	1-ME-3-CYCLOPENTYL-5-DIMEAMINO-6-METHYLURACIL 32150-39-3
9 ★	3.21		1364		C13H21O3P1S1	PHOSPHOROTHIOTIC ACID,O,O-BIS(I-PROPYL),-S-BENZYL ESTER 26087-47-8 fungicide
9960 ★	3.67		2506		C13H21O4P1S1	PROPAPHOS 7292-16-2
1 ✓	0.01		2397	101	C13H22F3N4O3S1	IMIDAZOLIUM CHLORIDE ANALOG ✚
2 ✓	-1.89		2397	101	"	IMIDAZOLIUM CHLORIDE ANALOG
3 ✓	-1.17		2397	101	"	IMIDAZOLIUM CHLORIDE ANALOG
4 ✓	-1.88		2397	101	"	IMIDAZOLIUM CHLORIDE ANALOG
65 ★	3.23		1693		C13H22N1O3P1S1	FENAMIPHOS 22224-92-6 nematicide~insecticide
6 ✓	-0.88		2180		C13H22N1S1.Br1	BENZYLTRIMETHYL AMMONIUM BROMIDE,3-PROPYLTHIO
7	-1.85		527	820	C13H22N1.Br1	BENZYLDIMETHYLBUTYLAMMONIUMBROMIDE 58965-48-3
8 ✓	-1.13		2180		C13H22N1.Br1	BENZYLTRIMETHYL AMMONIUM BROMIDE,3-ISOPROPYL
9 ✓	-1.03		2180		C13H22N1.Br1	BENZYLTRIMETHYL AMMONIUM BROMIDE,3-PROPYL
9970	-0.95		527	404	C13H22N1.Br1	OCTYLPYRIDINIUMBROMIDE 2534-66-9
1 ✓	-1.51		1729	820	C13H22N1.I1	3,4-DIMETHYLPYRIDINIUM IODIDE,N-HEXYL
2 ✓	-1.19		1729	820	C13H22N1.I1	3,5-DIMETHYLPYRIDINIUM IODIDE,N-HEXYL
3	-1.80	7.4	921	22	C13H22N1.I1	METHYLPHENYLDIPROPYL AMMONIUM IODIDE 52111-77-0
4 ✓	-0.84		1729	820	C13H22N1.I1	N-OCTYLPYRIDINIUM IODIDE 10291-06-2
75	1.50		1407		C13H22N2O1	1,2,3,6-H4-PYRIDINE,1-CYCLOHEPTYLCARBONYLAMINO
6 ★	3.04		2544		C13H22N2O2	THYMINE,1-OCTYL
7	2.91	7.4	899	1	C13H22N2O3	BARBITURIC ACID,5-HEPTYL-5-ETHYL
8 ★	0.73	7.4	1956	2	C13H22N2O3	PILOCARPIC ACID,ETHYL ESTER 96914-10-2
9 ✓	-0.28		2397	101	C13H22N3O2	IMIDAZOLIUM CHLORIDE,1-(1-PROPENYL)-2-HYDROXYIMINOMETHYL-3-(2,2-DIMETHYLPROPOXY)METHYL
9980 ✓	-0.67		2397	101	C13H22N3O2	IMIDAZOLIUM CHLORIDE,1-HEXYL-2-HYDROXYIMINOMETHYL-3-ACETYLMETHYL
1 ✓	-0.47		2397	101	C13H22N3O2	IMIDAZOLIUM CHLORIDE,1-METHYL-2-HYDROXYIMINOMETHYL-3-(4,4-DIMETHYLPENTYN-2-YLOXY)METHYL
2 ✓	-0.12		2397	101	C13H22N3O2	IMIDAZOLIUM CHLORIDE,1-METHYL-2-HYDROXYIMINOMETHYL-3-CYCLOHEXYLMETHOXYMETHYL 117941-61-4
3 ✓	-0.56		2397	101	C13H22N3O2	IMIDAZOLIUM CHLORIDE,1-METHYL-2-HYDROXYIMINOMETHYL-3-CYLCOHEPTYLOXYMETHYL 117941-57-8
4 ✓	-0.81		2406	102	C13H22N3O2S2	TETRAMETHYLYTHIOURONIUM IODIDE,S-(4-N,N-DIMTHYLSUFONAMIDO)PHENYL
85 ✓	-2.59		2397	101	C13H22N3O4S1	IMIDAZOLIUM CHLORIDE,1-METHYL-2-HYDROXYIMINOMETHYL-3-(1-METHYLSULFONYLCYCLOHEXYLOXY)METHYL
6 ★	3.48	1.0	2196		C13H22N4O3	PYRAZOLE,3-METHYL-4-NITRO-5-(N-OCTYLCARBOXAMIDO)
7 ★	0.27	10.5	2101		C13H22N4O3S1	RANITIDINE 66357-35-5 antagonist (histamine h2-receptors)
8 ✓	-0.11		2397	101	C13H23Br1N3O2	IMIDAZOLIUM CHLORIDE,1-PROPYL-2-HYDROXYIMINOMETHYL-3-(2,2-DIMETHYL-3-BROMOPROPOXY)METHYL
9 ✓	-1.99		2397	101	C13H23Br1N3O4S1	IMIDAZOLIUM CHLORIDE ANALOG ✚
9990 ✓	-1.64		2397	101	"	IMIDAZOLIUM CHLORIDE ANALOG
1 ★	4.01		1422	462	C13H23N1	1-AMINOADAMANTANE,3,5,7-TRIMETHYL 15210-60-3
2 ★	3.76		1422	462	C13H23N1	1-AMINOADAMANTANE,3-PROPYL
3	0.85		213		C13H23N1O2	N-OCTANOYLCYCLOBUTANECARBOXAMIDE 25173-87-9
4 ★	3.60		1629		C13H23N3O4S2	N-(N"-METHYL-CYCLOHEXOXYCARBAMYLTHIO)METHOMYL
95 ✓	-1.09		2397	101	C13H23N4O4	IMIDAZOLIUM CHLORIDE,1-ETHYL-2-HYDROXYIMINOMETHYL-3-(1-ETHYL-2-NITROPROPOXY)METHYL
6 ✓	-1.06		2397	101	C13H23N4O4	IMIDAZOLIUM CHLORIDE,1-PROPYL-2-HYDROXYIMINOMETHYL-3-(1,2-DIMETHYL-2-NITROPROPOXY)METHYL
7 ✓	-1.72		2397	101	C13H23N4O6S1	IMIDAZOLIUM CHLORIDE,1-METHYLSULFONYLETHYL-2-HYDROXYIMINOMETHYL-3-(1,2-DIMETHYL-2-NITROPROPOXY)METHYL
8 ✓	-2.54		2397	101	C13H23N4O7S1	IMIDAZOLIUM CHLORIDE ANALOG ✚
9 ★	1.70		2238		C13H24Cl1N4O3	NITROSOUREA,1-(2-CHLOROETHYL)-3-METHYL-3-(1-OXO-2,2,6,6-TETRAMETHYLPIPERIDIN-4-YL)
10000 ✓	-0.66		2397	101	C13H24N3O2	IMIDAZOLIUM CHLORIDE,1,3-DIMETHYL-2-HYDROXYIMINOMETHYL-4-(1,2-DIMETHYLPROPOXY)METHYL
1 ✓	-0.25		2397	101	C13H24N3O2	IMIDAZOLIUM CHLORIDE,1-METHYL-2-HYDROXYIMINOMETHYL-3-(1,1,2,2-TETRAMETHYLPROPOXY)METHYL
2 ✓	-0.36		2397	101	C13H24N3O2	IMIDAZOLIUM CHLORIDE,1-METHYL-2-HYDROXYIMINOMETHYL-3-(1-ETHYL-2,2-DIMETHYLPROPOXY)METHYL
3 ✓	-0.28		2397	101	C13H24N3O2	IMIDAZOLIUM CHLORIDE,1-METHYL-2-HYDROXYIMONMETHYL-3-(1,3,3-TRIMETHYLBUTOXY)METHYL 117941-48-7
4 ✓	-0.28		2397	101	C13H24N3O2	IMIDAZOLIUM CHLORIDE,1-PROPYL-2-HYDROXYIMINOMETHYL-3-(2,2-DIMETHYLPROPOXY)METHYL
5 ★	4.85		1693		C13H24N3O3P1S1	PIRIMIPHOS ETHYL 23505-41-1 insecticide
6 ✓	-2.50		2397	101	C13H24N3O4S1	IMIDAZOLIUM CHLORIDE,1-PROPYL-2-HYDROXYIMINOMETHYL-3-(1,1-DIMETHYL-2-METHYLSULFONYLETHOXY)METHYL ✚
7 ✓	-2.47		2397	101	C13H24N3O6S2	IMIDAZOLIUM CHLORIDE ANALOG ✚
8	1.16	7.4	2298	1	C13H24N4O2S1	1,2,4-OXADIAZOLE-5-(DIETHYLAMINOETHYLTHIO)HYDROXIMATE,3-T-BUTYL 103499-09-8
9 ★	2.02	7.4	2298	2	"	1,2,4-OXADIAZOLE-5-(DIETHYLAMINOETHYLTHIO)HYDROXIMATE,3-T-BUTYL
10010 ★	2.70		2510		C13H24N4O3S1	BUPIRIMATE 41483-43-6 fungicide
1	-0.55	7.0	1362	1	C13H24N4O3S1	TIMOLOL 26839-75-8 anti-adrenergic (beta-receptor)
2	-0.52	7.4	2465	41	"	TIMOLOL
3	-0.23	7.4	2399	1	"	TIMOLOL
4	-0.15	7.2	2016		"	TIMOLOL
15	-0.04	7.4	2244	1	"	TIMOLOL
6	0.06	7.4	1362	1	"	TIMOLOL
7	0.09	7.4	2399	1	"	TIMOLOL
8	0.11	7.4	2178	1	"	TIMOLOL
9	0.20	7.4	1746	1	"	TIMOLOL
10020 ★	1.83	10.2	2465	41	"	TIMOLOL
1	1.91	7.4	1700	2	"	TIMOLOL
2 ★	-1.32		2377		C13H24N4O4	AC-GLN-LEU-N
3 ★	-1.40		2377		C13H24N4O4	AC-VAL-ALA-ALA-N
4	3.22		537		C13H24N6O1	IMIDAZOLE,4-CARBOXAMIDE-5-(3-METHYL-3-OCTYLTRIAZENYL) 56010-86-7

	logP	pH	Ref.	Note	MF	Name / CAS / activity
25	4.88		1231		C13H24O2	6,8-TRIDECANEDIONE 32743-88-7
6 ★	3.71	13.0	594		C13H25N1	N-METHYLDICYCLOHEXYL AMINE 7560-83-0
7	3.34		567	820	C13H25N1.C6H3N3O7	METHYL-DICYCLOHEXYLAMINE PICRATE
8 ★	1.35	7.4	1342	324	C13H25N2O2	4-BUTYRAMIDO-2,2,6,6-TETRAMETHYLPIPERIDINE-N-OXIDE 21270-93-9
9 ★	1.77		2093		C13H25N3O2P1S1	THIOTEPA,4-(2,2,6,6-TETRAMETHYL-1-OXYPIPERIDINYL) ESTER 51526-59-1
10030 ★	0.16		2377		C13H25N3O3	AC-ILE-VAL-N
1 ★	0.26		2377		C13H25N3O3	AC-LEU-VAL-N
2 ★	4.20		1629		C13H25N3O4S2	N-(N"-ME-HEXYLOXYCARBAMYLTHIO)METHOMYL
3 ✓	-1.79		2397	101	C13H25N4O1	IMIDAZOLIUM CHLORIDE,1-METHYL-2-HYDROXYIMINOMETHYL-3-DI--I-PROPYLAMINOETHYL 132566-73-5
4 ★	-0.48		2093		C13H25N5O3P1	TEPA,3-(2,2,6,6-TETRAMETHYL-1-OXYPYRROLIDINYL)URYL 103981-99-3
35 ★	1.57		2093		C13H26N4O1S1	THIOTEPA,4-(2,2,6,6-TETRAMETHYL-1-OXYPIPERIDINYL)AMIDE 33683-34-0 ✚
6 ★	2.60		1629		C13H26N4O3S2	N-(N"-DI-I-PR-N'-ME)-URYLTHIO)METHOMYL
7 ★	-0.60		2093		C13H26N5O3P1	TEPA,3-(2,2,6,6-TETRAMETHYL-1-HYDROXYPYRROLIDINYL)URYL 103981-95-9
8	1.50		1294		C13H26O5S1	1-HEPTYLTHIO-B-GALACTOPYRANOSIDE
9	0.73		1294		C13H26O6	B-HEPTYLGALACTOPYRANOSIDE
10040	2.18		871		C13H28N2O1	N-NITROSO-METHYLDODECYLAMINE
1 ★	3.68		579		C13H28N2S1	THIOUREA,TRIBUTYL 2422-88-0
2 ✓	1.00		530	450	C13H29N3	DODECYLGUANIDINEHYDROBROMIDE 112-65-2
3	-2.17		1394		C13H29N3O1	4-DODECYLSEMICARBAZIDE
4	1.15		530	450	C13H29N3.C2H4O2	DODECYLGUANIDIUMACETATE 2439-10-3
45	4.29		2182	459	C13H29O3P1	METHYLPHOSPHONIC ACID,BIS-(1,2,2-TRIMETHYL)PROPYL ESTER 7040-58-6
6	-0.74		549	60	C13H30N1.Br1	DECYLTRIMETHYLAMMONIUMBROMIDE 2082-84-0
7	0.26		1901	820	C13H30N1.I1	TRIBUTYLMETHYL AMMONIUM IODIDE 3085-79-8
8	-0.16		524	820	C13H30N1.I1	TRIMETHYL-DECYL-AMMONIUM IODIDE 7447-24-7
9	2.76		567	820	C13H30N1.C6H2N3O7	BUTYLTRIPROPYL AMMONIUM PICRATE
10050 ✓	2.64		567	820	C13H30N1.C6H2N3O7	I-BUTYLTRIPROPYL AMMONIUM PICRATE
1 ★	2.84		2506		C14H4N2O2S2	DITHIANON 3347-22-6 fungicide
2	2.17		2515		C14H5Cl6N1O3	TECLOFTALAM ✚
3	3.12	1.0	2471	459	C14H7N1O6	XANTHONE,6-CARBOXY-2-NITRO
4	2.83	1.0	2471	459	C14H7N5O4	XANTHONE,6-(5-TETRAZOLYL)-2-NITRO
55	2.05		2470	306	C14H8Br2N2O1S1	TETRAHYDROQUINAZOLINE-4-OXO-2-THIO,3-(4-BROMOPHENYL),6-BROMO
6 ★	3.20		2661		C14H8Cl2N4	CLOFENTEZINE 74115-24-5 acaricide
7 ★	6.96		2293	400	C14H8Cl4	DDE 72-55-9
8	1.55	2.7	978	340	C14H8F3N2O4.Na1	N-(3-CF3-PHENYL)-5-NITROANTHRANILIC ACID,NA+ 39053-08-2
9 ★	5.50		716		C14H8N2S2	4,4'-DIISOTHIOCYANATEBIPHENYL 18705-45-8
10060	2.86	1.0	2471	459	C14H8N4O2	XANTHONE,6-(5-TETRAZOLYL)
1 ★	3.39		570		C14H8O2	ANTHRAQUINONE 84-65-1 bird repellant
2 ★	2.52		594		C14H8O2	PHENANTHRENEQUINONE 84-11-7
3 ★	3.53		2203		C14H8O3	ANTHRAQUINONE,1-HYDROXY 129-43-1
4	3.16	1.0	2471	459	C14H8O4	XANTHONE,6-CARBOXY
65 ★	3.12		575		C14H8O4	XANTHONE-2-CARBOXYLIC ACID 40274-67-7
6 ★	2.71	7.4	874	1	C14H9Br1N2O1	4-PHENYL-6-BROMOQUINAZOLIN-2-ONE 33443-53-7
7	1.43		2470	306	C14H9Br1N2O1S1	TETRAHYDROQUINAZOLINE-4-OXO-2-THIO,3-(2-BROMOPHENYL)
8	2.05		2470	306	C14H9Br1N2O1S1	TETRAHYDROQUINAZOLINEJ-4-OXO-2-THIO,3-PHENYL,6-BROMO
9 ★	3.88		2198		C14H9Cl1F2N2O2	DIFLUBENZURON 35367-38-5 insecticide
10070	1.19	2.7	978	340	C14H9Cl1F3N1O2	N(3-CF3-PH)3-CHLOROANTHRANILIC ACID, SODIUM SALT 59425-32-0
1	2.29	2.7	978	340	C14H9Cl1F3N1O2	N(3-CF3-PH)4-CHLOROANTHRANILIC ACID, SODIUM SALT 579-87-3
2 ★	4.47		594	400	C14H9Cl1N2	AM-EX-OL 6484-25-9
3	2.53	7.4	874	1	C14H9Cl1N2O1	4-PHENYL-6-CHLOROQUINAZONIN-2-ONE 4797-43-7
4	1.62	7.4	2325	1	C14H9Cl2F2N1O2	PHENYLACETIC ACID,5-FLUORO-2-(2',6'-DICHLORO-4'-FLUORO)ANILINO 127792-20-5
75	5.00	7.4	2325	2	C14H9Cl2F2N1O2	PHENYLACETIC ACID,5-FLUORO-2-(2',6'-DICHLORO-4'-FLUORO)ANILINO
6	2.93		885		C14H9Cl2F3N2O1	CLOFLUCARBAN 369-77-7 disinfectant
7 ★	4.48		2038		C14H9Cl2N1O5	BIFENOX 42576-02-3 herbicide
8	4.62		460		C14H9Cl3F2	DFDT 475-26-3
9 ★	6.91		2293	400	C14H9Cl5	DDT 50-29-3 insecticide, pediculicide
10080 ★	4.28		2506		C14H9Cl5O1	KELTHANE 115-32-2 miticide
1 ★	3.20		2198		C14H9F3N2O2	PHENYLUREA,4-FLUORO,N'-(2,6-DIFLUOROBENZOYL)
2	0.58	2.7	978	340	C14H9F3N2O4	N-(3-CF3-PHENYL)-3-NITROANTHRANILIC ACID,SODIUM SALT 39053-10-6
3	1.88	2.7	978	340	C14H9F3N2O4	N-(3-CF3-PHENYL)-4-NITROANTHRANILIC ACID,NA+
4	2.08		2470	306	C14H9I1N2O1S1	TETRAHYDROQUINAZOLINE-4-OXO-2-THIO,3-(3-IODOPHENYL)
85	2.09		2470	306	C14H9I1N2O1S1	TETRAHYDROQUINAZOLINE-4-OXO-2-THIO,3-(4-IODOPHENYL)
6	2.84		2470	306	C14H9I1N2O1S1	TETRAHYDROQUINAZOLINE-4-OXO-2-THIO,3-PHENYL-6-IODO
7	2.87		579		C14H9N1O2	1-AMINOANTHRAQUINONE 82-45-1
8 ★	4.78		595		C14H9N1O2	ANTHRACENE,9-NITRO
9	2.03	7.4	712	1	C14H9N1O2	G-PYROPHTHALONE ✚
10090 ★	2.40		594		C14H9N1O2	N-PHENYLPHTHALIMIDE 520-03-6
1	4.90		850		C14H9N1O2S1	4-ISOTHIOCYANOPHENYL BENZOATE 49540-85-4
2	2.30	1.0	2471	459	C14H9N1O4	XANTHONE,6-CARBOXY-2-AMINO
3	4.88		850		C14H9N1S1	4-ISOTHIOCYANOBENZOPHENONE 26328-59-6
4	1.99		2470	306	C14H9N3O3S1	TETRAHYDROQUINAZOLINE-4-OXO-2-THIO,3-(3-NITROPHENYL)
95	1.88		2470	306	C14H9N3O3S1	TETRAHYDROQUINAZOLINE-4-OXO-2-THIO,3-(4-NITROPHENYL)
6	2.07		2470	306	C14H9N3O3S1	TETRAHYDROQUINAZOLINE-4-OXO-2-THIO,,3-(2-NITROPHENYL)
7	1.41		494		C14H9N3O4S1	2-PHENYLIMINO-5-(5-NO2-FURFURILIDINE)THIAZOL-4-ONE 27472-71-5
8	2.18	1.0	2471	459	C14H9N5O2	XANTHONE,6-(5-TETRAZOLYL)-2-AMINO
9 ★	4.45		517		C14H10	ANTHRACENE 120-12-7
10100 ★	4.78		547		C14H10	DIPHENYLACETYLENE 501-65-5
1 ★	4.46		517		C14H10	PHENANTHRENE 85-01-8
2	1.87	7.4	2325	1	C14H10Br1Cl2N1O2	PHENYLACETIC ACID,6-FLUORO-2-(2',6'-DICHLORO)ANILINO 122794-97-2
3 ★	5.31		1453	459	C14H10Br1N1	2-(3-BROMOPHENYL)INDOLIZINE
4 ★	5.43		1453	459	C14H10Br1N1	2-(4-BROMOPHENYL)INDOLIZINE
5	1.48	7.4	1688	1	C14H10Br1N3O1	BROMAZEPAM 1812-30-2 tranquilizer (minor)
6	1.60	7.4	461	1	"	BROMAZEPAM
7 ★	1.69	8.0	1874		"	BROMAZEPAM
8 ★	5.14		1453	459	C14H10Cl1N1	2-(3-CHLOROPHENYL)INDOLIZINE
9 ★	5.20		1453	459	C14H10Cl1N1	2-(4-CHLOROPHENYL)INDOLIZINE
10110	1.27	7.4	2325	1	C14H10Cl2F1O2	PHENYLACETIC ACID,2-(2',6'-DICHLORO-4-FLUORO)ANILINO 127792-22-7
1	1.48	7.4	2325	1	C14H10Cl2F1O2	PHENYLACETIC ACID,5-FLUORO-2-(2',6'-DICHLORO)ANILINO 127792-21-6
2	4.86	7.4	2325	1	C14H10Cl2F1O2	PHENYLACETIC ACID,5-FLUORO-2-(2',6'-DICHLORO)ANILINO
3 ★	4.48		2456	346	C14H10Cl2O2	ACETIC ACID,BIS-(4-CHLOROPHENYL)
4	1.73	7.4	1687	1	C14H10Cl2O3	FENCLOFENAC 34645-84-6 anti-inflammatory

	logP	pH	Ref.	Note	MF	Name / **CAS** / *activity*
15 ★	4.80	7.4	1687	2	"	FENCLOFENAC
6	1.10		2393	1	C14H10Cl2O3	PHENOXYACETIC ACID,4-(2,4-DICHLOROPHENOXY)
7	1.97	7.4	2325	1	C14H10Cl3N1O2	PHENYLACETIC ACID,5-CHLORO-2-(2',6'-DICHLORO)ANILINO **15307-85-4**
8 ★	6.22		2293	400	C14H10Cl4	DDD **72-54-8**
9	2.02	7.4	1838	1	C14H10F3N1O2	FLUFENAMIC ACID **530-78-9** *anti-inflammatory*
10120	2.08	7.4	693	1	"	FLUFENAMIC ACID
1 ★	5.25	2.0	1838			FLUFENAMIC ACID
2	1.29	2.7	978	340	C14H10F3N1O2	N-(M-TRIFLMEPHENYL)ANTHRANILICAICD,SODIUM SALT
3	-0.12		536		C14H10F3N1O2	N-(O-TRIFLMEPHENYL)ANTHRANILIC ACID,SODIUM SALT **32621-47-9**
4 ★	3.61		822		C14H10F3N1O3	4-AMINOSALICYLIC ACID,3-TRIFLUOROMETHYLPHENYL ESTER **56356-34-4**
25 ★	4.28		607	606	C14H10F3N1O3S1	TRIFLUOROMETHANESULFONANILIDE,P-BENZOYL **22731-28-8**
6	5.00	7.4	1735	1	C14H10F4	1,2-DIPHENYLTETRAFLUOROETHANE **425-32-1**
7 ★	2.71		1519	459	C14H10N2	4-PHENYLQUINAZOLINE **17629-01-5**
8	1.91	7.4	874	1	C14H10N2O1	4-PHENYLQUINAZOLIN-2-ONE **23441-75-0**
9 ★	3.83		2166		C14H10N2O1	FURAZAN,DIPHENYL **19768-02-6**
10130	1.72		2470	306	C14H10N2O1S1	TETRAHYDROQUINAZOLINE-4-OXO-2-THIO,3-PHENYL
1 ★	3.64		2166		C14H10N2O2	FURAZAN,DIPHENYL,2-OXIDE **5585-14-8**
2	1.85		2470	306	C14H10N2O2S1	TETRAHYDROQUINAZOLINE-4-OXO-2-THIO,3-(3-HYDROXYPHENYL)
3	1.32		2470	306	C14H10N2O2S1	TETRAHYDROQUINAZOLINE-4-OXO-2-THIO,3-(4-HYDROXYPHENYL)
4 ★	3.66		594		C14H10O1	9-ANTHRONE **90-44-8**
35 ★	2.64		1524		C14H10O2	1(3H)-ISOBENZOFURANONE-3-PHENYL **5398-11-8**
6	2.80		631		C14H10O2	9-CARBOXYFLUORENE **1989-33-9**
7 ★	3.38		547		C14H10O2	BENZIL **134-81-6**
8 ★	2.57		631		C14H10O2S1	9-CARBOXYTHIOXANTHENE **17394-14-8**
9	1.00		631		C14H10O3	9-CARBOXY-9-HYDROXYFLUORENE **467-69-6**
10140 ★	2.23		631		C14H10O3	9-CARBOXYXANTHENE **82-07-5**
1	2.26		579		C14H10O3	BENZOICANHYDRIDE **93-97-0**
2	-2.83	7.4	2288	1	C14H10O4	DIPHENIC ACID **482-05-3**
3 ★	2.07	1.0	579	342	"	DIPHENIC ACID
4	1.44	7.4	2325	1	C14H11Br1Cl1N1O2	PHENYLACETIC ACID,2-(2'-CHLORO-6'-BROMO)ANILINO **127792-23-8**
45 ★	4.64	7.4	2325	2	C14H11Br1Cl1N1O2	PHENYLACETIC ACID,2-(2'-CHLORO-6'-BROMO)ANILINO
6	0.56	7.4	2325	1	C14H11Cl1F1O2	PHENYLACETIC ACID,2-(2'-CHLORO-6'-FLUORO)ANILINO **100754-93-6**
7 ★	3.80	7.4	2325	2	C14H11Cl1F1O2	PHENYLACETIC ACID,2-(2'-CHLORO-6'-FLUORO)ANILINO
8	1.64	7.4	2325	1	C14H11Cl1I1O2	PHENYLACETIC ACID,2-(2'-CHLORO-6'-IODO)ANILINO **127792-24-9**
9 ★	4.80	7.4	2325	2	C14H11Cl1I1O2	PHENYLACETIC ACID,2-(2'-CHLORO-6'-IODO)ANILINO
10150 ★	3.13		2463		C14H11Cl1N4O1	BENZOTRIAZOLE,2-METHYL-5-P-CHLOROBENZAMIDO
1	3.42		2572	459	C14H11Cl2F2N1O3	RH-0978 **143121-10-2** ✚
2 ★	4.47		1792		C14H11Cl2N1O1	BENZENEACETAMIDE,N-(3,4-DICHLOROPHENYL)
3	1.13	7.4	1687	1	C14H11Cl2N1O2	DICLOFENAC **15307-86-5** *anti-inflammatory*
4	1.13	7.4	2325	1	"	DICLOFENAC
55	1.22	7.4	1838	1	"	DICLOFENAC
6 ★	4.40	2.0	1838		"	DICLOFENAC
7	4.49	7.4	2325	2	"	DICLOFENAC
8	4.75	7.4	1687	2	"	DICLOFENAC
9	1.66	7.4	2325	1	C14H11Cl2N1O2	PHENYLACETIC ACID,2-(2',3'-DICHLORO)ANILINO **70172-32-6**
10160 ★	4.86	7.4	2325	2	C14H11Cl2N1O2	PHENYLACETIC ACID,2-(2',3'-DICHLORO)ANILINO
1	1.33	7.4	2325	1	C14H11Cl2N1O2	PHENYLACETIC ACID,2-(2',4'-DICHLORO)ANILINO **70172-31-5**
2	1.64	7.4	2325	1	C14H11Cl2N1O2	PHENYLACETIC ACID,2-(2'.5'-DICHLORO)ANILINO **127792-31-8**
3 ★	4.94	7.4	2325	2	C14H11Cl2N1O2	PHENYLACETIC ACID,2-(2'.5'-DICHLORO)ANILINO
4	1.80	7.4	2325	1	C14H11Cl2N1O2	PHENYLACETIC ACID,2-(3',4'-DICHLORO)ANILINO **127792-33-0**
65 ★	4.90	7.4	2325	2	C14H11Cl2N1O2	PHENYLACETIC ACID,2-(3',4'-DICHLORO)ANILINO
6	1.98	7.4	2325	1	C14H11Cl2N1O2	PHENYLACETIC ACID,2-(3',5'-DICHLORO)ANILINO **127792-30-7**
7 ★	5.07	7.4	2325	2	C14H11Cl2N1O2	PHENYLACETIC ACID,2-(3',5'-DICHLORO)ANILINO
8	0.72	7.4	2325	1	C14H11Cl2N1O3	PHENYLACETIC ACID,2-(2',6'-DICHLORO-3'-HYDROXY)ANILINO **69002-85-3**
9	0.70	7.4	2325	1	C14H11Cl2N1O3	PHENYLACETIC ACID,2-(2',6'-DICHLORO-4'-HYDROXYANILINO) **64118-84-9**
10170 ★	3.70	7.4	2325	2	C14H11Cl2N1O3	PHENYLACETIC ACID,2-(2',6'-DICHLORO-4'-HYDROXYANILINO)
1	0.76	7.4	2325	1	C14H11Cl2N1O3	PHENYLACETIC ACID,5-HYDROXY-2-(2',6'-DICHLORO)ANILINO **69002-84-2**
2 ★	3.94	7.4	2325	2	C14H11Cl2N1O3	PHENYLACETIC ACID,5-HYDROXY-2-(2',6'-DICHLORO)ANILINO
3	1.73	7.4	2325	1	C14H11Cl2N1O3	PHENYLACETIC ACID,6-HYDROXY-2-(2',6'-DICHLOROANILINO) **127792-35-2**
4 ★	4.65	7.4	2325	2	C14H11Cl2N1O3	PHENYLACETIC ACID,6-HYDROXY-2-(2',6'-DICHLOROANILINO)
75	0.27	7.4	2325	1	C14H11Cl2N1O4	PHENYLACETIC ACID,5-HYDROXY-2-(2',6'-DICHLORO-4'-HYDROXY)ANILINO **69002-86-4**
6 ★	4.87		1523		C14H11Cl3	2,2-DIPHENYL-1,1,1-TRICHLOROETHANE **2971-22-4**
7	0.55	7.4	2325	1	C14H11F2N1O2	PHENYLACETIC ACID,2-(2',3'-DIFLUORO)ANILINO
8	0.26	7.4	2325	1	C14H11F2N1O2	PHENYLACETIC ACID,2-(2',6'-DIFLUORO)ANILINO **90233-40-2**
9 ★	3.57	7.4	2325	2	C14H11F2N1O2	PHENYLACETIC ACID,2-(2',6'-DIFLUORO)ANILINO
10180	0.55	7.4	2325	1	C14H11F2N1O2	PHENYLACETIC ACID,2-(2,4-DIFLUORO)ANILINO **127792-32-9**
1 ★	2.86	5.4	538		C14H11F2N1O3S1	DIFLUMIDONE **22736-85-2** *anti-inflammatory*
2 ★	3.69	7.4	594	314	C14H11N1	1-ANTHRACENEAMINE **610-49-1**
3 ★	4.29		1453	459	C14H11N1	2-PHENYLINDOLIZINE ✚
4 ★	3.03		2256		C14H11N1O1	PHENYLACETONITRILE,M-PHENOXY
85 ★	2.85		2649		C14H11N1O2	NAPHTHOQUINONE,2-(3-PYRROLIN-1YL)
6	-0.13	7.0	547	1	C14H11N1O3	BENZOPHENONE,2-AMINO-2'-CARBOXY
7 ✓	0.95	5.0	547	312	C14H11N1O3	BENZOPHENONE,2-AMINO-2'-CARBOXY
8 ★	1.43		538		C14H11N1O3	N-(P-ACETOXYPHENYL)BENZOQUINONEMONIMINE
9	0.34	7.0	353	1	C14H11N1O3	N-BENZOYLANTHRANILIC ACID **579-93-1**
10190	1.38	7.0	353	1	C14H11N1O3	N-BENZOYLSALICYLAMIDE **5663-74-1**
1	0.74	7.0	353	1	C14H11N1O4	N-SALICYLANTHRANILIC ACID **13316-98-8**
2	1.92	7.0	353	1	C14H11N1O4	SALICYLIMIDE **1972-71-0**
3	0.32	7.0	353	1	C14H11N1O5	N-(G-RESORCYL)-ANTHRANILIC ACID **13317-01-6**
4	0.40	7.0	353	1	C14H11N1O5	N-(G-RESORCYL)-P-AMINOBENZOIC ACID
95	0.86	7.0	353	1	C14H11N1O6	B-RESORCYLIMIDE
6	1.15	7.0	353	1	C14H11N1O6	G-RESORCYLIMIDE
7	4.40		850		C14H11N1S1	4-ISOTHIOCYANODIPHYLMETHANE
8	5.09		850		C14H11N1S1	DIPHENYLMETHYLISOTHIOCYANATE **3350-21-8**
9 ★	2.95		1519	459	C14H11N3	2-AMINO-4-PHENYLQUINAZOLE
10200	-0.52	7.4	2206	1	C14H11N3O4	STIRIMAZOLE **30529-16-9**
1 ★	2.53		2463		C14H11N5O3	BENZOTRIAZOLE,2-METHYL-5-P-NITROBENZAMIDO
2 ★	4.25		536		C14H12	9,10-DIHYDROANTHRACENE **613-31-0**
3 ★	4.52		594	400	C14H12	9,10-DIHYDROPHENANTHRENE **776-35-2**
4 ★	4.97		2001		C14H12	FLUORENE,1-METHYL **1730-37-6**

	logP	pH	Ref.	Note	MF	Name / **CAS** / *activity*
5 ★	4.81		547		C14H12	STILBENE-T **103-30-0**
6 ★	3.77		1988		C14H12Br1N1O1	ACETANILIDE,N-(3-BROMOPHENYL) **13140-73-3**
7 ★	3.70		1980		C14H12Br1N1O1	ACETANILIDE,N-(4-BROMOPHENYL) **7495-11-6**
8 ★	3.54		1988		C14H12Cl1N1O1	ACETANILIDE,N-(4-CHLOROPHENYL) **2990-06-9**
9	4.73		2601	300	C14H12Cl1N1O2	4-CHLOROANTHRANILIC ACID,N-(2-METHYLPHENYL)
10210	4.81		2601	300	C14H12Cl1N1O2	4-CHLOROANTHRANILIC ACID,N-(4-METHYLPHENYL)
1	0.91	7.4	2325	1	C14H12Cl1N1O2	PHENYLACETIC ACID,2-(2'-CHLORO)ANILINO **127792-34-1**
2 ★	5.17	2.0	594	400	C14H12Cl1N1O2	TOLFENAMIC ACID **13710-19-5** *analgesic, anti-inflammatory*
3	4.32		2601	300	C14H12Cl1N1O3	4-CHLOROANTHRANILIC ACID,N-(2-METHOXYPHENYL)
4	4.41		2601	300	C14H12Cl1N1O3	4-CHLOROANTHRANILIC ACID,N-(4-METHOXYLPHENYL)
15 ★	1.56	6.5	321	1	C14H12Cl1N3O4S2	CHLOROTHIAZIDE-3-BENZYL **3211-40-3**
6 ★	4.51		1523		C14H12Cl2	1,1-DICHLORO-2,2-DIPHENYLETHANE **2387-16-8**
7 ★	2.51		2506		C14H12Cl2N2O1	PYRIFENOX *fungicide*
8 ★	2.62		1987		C14H12F1O1	ACETANILIDE,N-(2-FLUOROPHENYL)
9 ★	3.10		1988		C14H12F1O1	ACETANILIDE,N-(3-FLUOROPHENYL) **5215-27-0**
10220 ★	2.86		1988		C14H12F1O1	ACETANILIDE,N-(4-FLUOROPHENYL)
1	1.11	7.2	2446	1	C14H12F1O3	FLUMEQUINE **42835-25-6** *antibacterial*
2 ★	3.91		1988		C14H12I1O1	ACETANILIDE,N-(3-IODOPHENYL)
3 ★	3.02		547		C14H12N2O1	2-ACETAMIDOCARBIZOLE
4 ★	2.56		547		C14H12N2O1	BENZALDEHYDEBENZOYLHYDRAZONE
25 ★	3.39		2198		C14H12N2O2	PHENYLUREA,N'-BENZOYL **1821-33-6**
6 ★	1.89		522		C14H12N2O2S1	1,2,4-BENZOTHIADIAZINE-1,1-DIOXIDE-3-BENZYL **20434-66-6**
7 ★	2.21	7.4	2636	1	C14H12N2O2S2	ISOMAZOLE,PYRIDYLTHIAZOLE ANALOG
8 ★	2.42	7.4	2636	1	C14H12N2O3	2(2,4-DIMETHOXYPHENYL)OXAZOLO[4,5-C]PYRIDINE
9 ★	2.73		1987		C14H12N2O3	ACETANILIDE,N-(2-NITROPHENYL)
10230 ★	2.93		1988		C14H12N2O3	ACETANILIDE,N-(3-NITROPHENYL) **13140-76-6**
1 ★	3.18		1988		C14H12N2O3	ACETANILIDE,N-(4-NITROPHENYL) **13140-77-7**
2 ★	2.99		501		C14H12N2O3	PHENOXYACETIC ACID,P-PHENYLAZO
3	-0.40	7.4	2636	1	C14H12N2O3S1	2(2-METHOXY-4-SULPHINYLPHENYL)OXAZOLO[4,5-B]PYRIDINE *cardiotonic*
4	3.06	7.7	1986	1	C14H12N2O3S2	BENZOTHIAZOLE,2-SULFONAMIDE-6-BENZYLOXY
35 ★	3.18	7.7	1986	2	"	BENZOTHIAZOLE,2-SULFONAMIDE-6-BENZYLOXY
6	1.73		1158		C14H12N2O4	O-(P-NITROBENZYL)BENZAMIDE **24367-68-8**
7 ★	2.44		2463		C14H12N4O1	BENZOTRIAZOLE,2-METHYL-5-BENZAMIDO
8 ★	1.68		2434		C14H12N4O2S1	SULFAQUINOXALINE **59-40-5** *antimicrobial (vet)*
9 ★	3.18		547		C14H12O1	2-PHENYLACETOPHENONE **451-40-1**
10240 ★	2.13		594		C14H12O2	BENZOIN **8050-35-9** *protectant (topical)*
1 ★	3.97		536		C14H12O2	BENZYL BENZOATE **120-51-4** *scabicide, pediculicide*
2 ★	3.03		1895		C14H12O2	BICYCLO(2.2.1)HEPTA-2,5-DIEN-7-OL BENZOATE **4796-68-3**
3 ★	3.09	1.0	547	342	C14H12O2	DIPHENYLACETIC ACID **117-34-0**
4 ★	2.30	1.0	570	342	C14H12O3	BENZILIC ACID **76-93-7**
45 ★	2.83		558		C14H12O3	O-PHENYLPHENOXYACETIC ACID **5348-75-4**
6 ★	3.18		501		C14H12O3	PHENOXYACETIC ACID,M-PHENYL **1878-57-5**
7 ★	4.28		260		C14H12O3S1	2-OH-3-CARBOXYBENZTHIOPHENYLETHER
8	0.78		1937	459	C14H12O4	4-HYDROXY-2-PYRONE,3-METHYL-6-BENZOYLMETHYL **24607-33-8**
9 ★	3.40	1.4	2018	2	C14H12O4	CYCLOHEXANO(6,7)CHROMONE-2-CARBOXYLIC ACID
10250	-0.52	7.4	1204	1	C14H13Br2N3O1	N(2-C4H3O-ME)N-(2,6-DIBRPH)AMINO)2-IMIDAZOLINE
1	-0.10	7.4	1204	1	C14H13Br2N3S1	N(2-C4H3S-ME)N(2,6-DIBRPH)AMINO)-2-IMIDAZOLINE
2 ★	5.67		331		C14H13Cl1F3N3O1	2-CF3-4-CL-C6H3NHN=C(CN)CO-T-BU **28317-94-4**
3 ★	5.31		331		C14H13Cl1F3N3O1	2-CL-5-CF3-C6H3NHN=C(CN)CO-T-BU **28343-25-1**
4 ★	2.63		2419		C14H13Cl1N2O4	PYRROLE,2-METHYL-3,4-DI(METHOXYCARBONYL)-5-(4-CHLOROPYRID-2YL)
55 ★	2.70		2419		C14H13Cl1N2O4	PYRROLE,2-METHYL-3,4-DI(METHOXYCARBONYL)-5-(5-CHLOROPYRID-2YL)
6 ★	2.65		2419		C14H13Cl1N2O4	PYRROLE,2-METHYL-3,4-DI(METHOXYCARBONYL)-5-(6-CHLOROPYRID-2YL)
7 ★	1.51		2419		C14H13Cl1N2O4	PYRROLE,2-METHYL-3,4-DI-(METHOXYCARONYL)-5-(3-CHLOROPYRID-2YL)
8	2.85	7.4	2028	459	C14H13Cl1N6	1,2,4-TRIAZOLE,3-AMINO-5-(2-P-CHLOROBENZYLAMINO)PYRIDIN-4-YL **77314-72-8**
9	3.82		2572	459	C14H13Cl2N1O3	O-PHENYLPYRROLIDINOCARBAMATE,2,4-DICHLORO-5-(PROPYN-2-YL)OXY **87365-63-7**
10260 ★	2.90		260		C14H13N1	DIHYDROMORPHANTHRIDINE ✚
1 ★	3.21		870		C14H13N1	O-METHYLBENZOPHENONEIMINE **22627-00-5**
2 ★	2.02		2410		C14H13N1O1	ACETANILIDE,2-PHENYL
3 ★	2.70		1988		C14H13N1O1	ACETANILIDE,N-PHENYL **621-06-7**
4 ★	2.02		570		C14H13N1O1	N,N-DIPHENYLACETAMIDE **519-87-9**
65	2.75		1213		C14H13N1O1	N-(M-TOLYL)BENZAMIDE **582-77-4**
6 ★	2.62		547		C14H13N1O1	N-BENZYL-N-FORMYLANILINE
7 ★	2.36		603		C14H13N1O1	N-METHYLBENZANILIDE **1934-92-5**
8 ★	2.41		2077		C14H13N1O1	P-METHYLBENZYLIDINEANILINE-N-OXIDE
9 ★	3.08	6.5	1556	1	C14H13N1O2	2-BENZAMIDO-5-METHYLPHENOL
10270	-0.12	2.7	978	340	C14H13N1O2	3-METHYL-N-PHENYLANTHRANILIC ACID, SODIUM SALT **40961-11-3**
1 ★	2.37		2410		C14H13N1O2	ACETANILIDE,2-PHENOXY
2 ★	3.13		1894		C14H13N1O2	ACETANILIDE,4-PHENOXY **6312-87-4**
3 ★	2.40		1987		C14H13N1O2	ACETANILIDE,N-(2-HYDROXYPHENYL)
4 ★	2.22		1988		C14H13N1O2	ACETANILIDE,N-(3-HYDROXYPHENYL)
75 ★	1.82		1988		C14H13N1O2	ACETANILIDE,N-(4-HYDROXYPHENYL)
6 ★	1.94		579		C14H13N1O2	BENZYLIDINEANILINE-N-OXIDE,4-METHOXY
7	-0.32		536		C14H13N1O2	N-(2-TOLYL)-ANTHRANILIC ACID,SODIUM SALT **16610-44-9**
8	4.01		2601	300	C14H13N1O2	N-(2-TOLYL)ANTHRANILIC ACID
9	4.09		2601	300	C14H13N1O2	N-(4-TOLYL)ANTHRANILIC ACID
10280 ★	2.55		1705		C14H13N1O2	N-(O-HYDROXYPHENYL)METHYLBENZAMIDE
1 ★	1.87		1705		C14H13N1O2	N-(P-HYDROXYPHENYL)METHYLBENZAMIDE
2	1.95	7.0	426	1	C14H13N1O2	N-PHENYLANTHRANILIC ACID,3'-METHYL **16524-22-4**
3	4.88		426		C14H13N1O2	N-PHENYLANTHRANILIC ACID,3'-METHYL
4 ★	2.89		2649		C14H13N1O2	NAPHTHOQUINONE,2-(N-PYRROLIDINYL)
85	1.61		1158		C14H13N1O2	O-BENZYLBENZAMIDE **29579-11-1**
6	2.44		1213		C14H13N1O2	P-AMINOBENZOIC ACID,N-BENZYL
7 ★	2.73		546		C14H13N1O2	PHENOXYACETANILIDE **18705-01-6**
8	0.19	7.4	2325	1	C14H13N1O2	PHENYLACETIC ACID,2-ANILINO **70172-33-7**
9 ★	3.49	7.4	2325	2	"	PHENYLACETIC ACID,2-ANILINO

	logP	pH	Ref.	Note	MF	Name / CAS / activity
10290	-1.04	2.7	978	340	C14H13N1O3	3-METHOXY-N-PHENYLANTHRANILIC ACID, SODIUM SALT 59425-27-3
1 ★	3.14		822		C14H13N1O3	4-AMINOSALICYLIC ACID,2-TOLYL ESTER 56356-13-9
2 ★	3.64		822		C14H13N1O3	4-AMINOSALICYLIC ACID,3-TOLYL ESTER 56356-14-0
3 ★	3.38		822		C14H13N1O3	4-AMINOSALICYLIC ACID,4-TOLYL ESTER 56356-15-1
4	3.60		2601	300	C14H13N1O3	N-(2-METHOXYPHENYL)ANTHRANILIC ACID
95	3.68		2601	300	C14H13N1O3	N-(4-METHOXYPHENYL)ANTHRANILIC ACID
6	1.41	7.0	426	1	C14H13N1O3	N-PHENYLANTHRANILIC ACID,3'-METHOXY 27693-73-8
7 ★	4.56	2.0	2236		C14H13N1O3	N-PHENYLANTHRANILIC ACID,3'-METHOXY
8 ★	1.67		2467		C14H13N1O3S1	ANILINE,P-(4-ACETYLPHENYL)SULFONYL
9 ★	2.88		822		C14H13N1O4	4-AMINOSALICYLIC ACID,2-METHOXYPHENYL ESTER 56356-17-3
10300 ★	3.25		822		C14H13N1O4	4-AMINOSALICYLIC ACID,3-METHOXYPHENYL ESTER 56356-18-4
1 ★	3.07		822		C14H13N1O4	4-AMINOSALICYLIC ACID,4-METHOXYPHENYL ESTER 4465-61-6
2 ★	2.25		2467		C14H13N1O4S1	ANILINE,P-(4-METHOXYCARBONYLPHENYL)SULFONYL 78297-66-2
3	0.95		634		C14H13N1O5	6,7-DIMETHYLENEDIOXY-4-OH-3-QUINOLINE-CO2ET
4 ★	0.50		2234		C14H13N1O6	ACETYLSALICYLIC ACID,(N-SUCCINIMIDYL)METHYL ESTER 32620-72-7
5 ★	2.83		594	400	C14H13N1S1	N-PHENYLTHIOACETANILIDE
6 ★	3.12		594		C14H13N1S1	THIOBENZANILIDE,N-METHYL 2628-58-2
7 ★	4.55		595		C14H13N3O1	DIBENZOFURANE,5-(3,3-DIMETHYL)TRIAZENYL
8 ★	2.57		2031		C14H13N3O2	1H-IMIDAZO(4,5,B)PYRIDINE,2-(2,4-DIMETHOXYPHENYL) ✚
9 ★	2.56		2031		C14H13N3O2	1H-IMIDAZO(4,5,C)PYRIDINE,2-(2,4-DIMETHOXYPHENYL) cardiotonic
10310 ★	1.66	7.4	2636	1	C14H13N3O2	ISOMAZOLE,2',6'-DIMETHOXY ANALOG
1 ★	1.11		2031	457	C14H13N3O2S1	IMIDAZO(1,2,A)PYRAZINE,2-(2-METHOXY-4-METHYLSULFOXYPHENYL) ✚
2 ★	0.06		2031		C14H13N3O2S1	IMIDAZO(1,2A)PYRIMIDINE,2-(2-METHOXY-4-METHYLSULFOXYPHENYL)
3	1.23	7.4	2636	1	C14H13N3O2S1	ISOMAZOLE 86315-52-8 cardiotonic
4 ★	1.25		2031		"	ISOMAZOLE
15 ★	1.17	8.0	2005		C14H13N3O2S1	SULMAZOLE 73384-60-8 cardiotonic
6	1.18	7.4	1940	1	"	SULMAZOLE
7	1.22	7.4	2636		"	SULMAZOLE
8 ★	2.98	6.0	2560	1	C14H13N3O2S1	THIABENDAZOLE,1-PROPOXYCARBONYL
9	0.79		1571	702	C14H13N3O3	2'-HYDROXY-5'-ETHYL-4-NITROAZOBENZENE
10320	1.46		494		C14H13N3O4S1	3-ALL-2-ALL-NH-5(5-NO2-FURFURILIDINE)THIAZOL-4-ONE 25603-12-7
1	0.09	7.4	2640	1	C14H13N3O4S2	MELOXICAM 71125-38-7 anti-arthritic
2 ★	3.01	2.1	2640	1	"	MELOXICAM
3	2.03	9.0	1409		C14H13N3O5	5-NITROIMIDAZOLE,1-ETOH-2-(3,4-DIOXYME-PH)CH=CH- 74550-86-0
4	-0.64	12.0	2640		C14H13N3O5S1	ISOXICAM 34552-84-6 anti-inflammatory (withdrawn)
25	-0.32	7.4	2640	1	"	ISOXICAM
6	2.72	3.1	2640	1	"	ISOXICAM
7	-0.57	7.4	2573	1	C14H13N3O7S1	3-METHYLCEPHALOSPORIN,5-NITROFURYL ANALOG ✚
8 ★	-0.15	8.0	2267	306	C14H13N5O3	ALLOPURINOL,PHENYLGLYCYLOXYMETHYL 98204-04-7
9 ★	5.09		2146	458	C14H14	4,4'-DIMETHYLBIPHENYL 613-33-2
10330 ★	4.79		505		C14H14	BIBENZYL 103-29-7
1	1.25	7.4	2037	1	C14H14Br1N5O2	ISOGUANINE,1-METHYL-7-(P-BROMOPHENOXY)ETHYL
2	1.73	7.4	2037	1	C14H14Br1N5O2	ISOGUANINE,1-METHYL-9-(P-BROMOPHENOXY)ETHYL
3	4.26		520		C14H14Cl1N1O2	1,4-NAPHTHOQUINONE,2-CHLORO,3-BUTYLAMINO 22272-30-6
4 ★	0.93	6.5	321	2	C14H14Cl1N3O4S2	3-BENZYLHYDROCHLOROTHIAZIDE 1824-50-6 diuretic, antihypertensive
35 ★	3.82		2506		C14H14Cl2N2O1	ENILCONAZOLE 35554-44-0 antifungal
6 ✓	3.49		306		C14H14Cl2O4	DIETHYLMALONATE,2,4-DICHLOROBENZAL 15725-40-3
7	2.69		306		C14H14Cl2O4	DIETHYLMALONATE,2,6-DICHLOROBENZAL 15725-41-4
8	3.85		1364		C14H14N1O4P1S1	O-ET-O-(P-NITROPHENYL)PHENYLPHOSPHOTHIONATE 2104-64-5 acaricide~insecticide
9	5.34		1814	459	"	O-ET-O-(P-NITROPHENYL)PHENYLPHOSPHOTHIONATE
10340	-1.56		527	820	C14H14N1.Br1	CINNAMYLPYRIDINIUMBROMIDE
1	-0.52	7.4	1350	1	C14H14N2	NAPHAZOLINE 835-31-4 adrenergic (vasoconstrictor)
2	-0.64	7.4	1097	1	"	NAPHAZOLINE
3 ★	2.79		2294		C14H14N2O1	1-VINYLIMIDAZOLE,ALPHA-(2-ALLYLOXY)PHENYL
4 ★	1.45		1987		C14H14N2O1	ACETANILIDE,N-(2-AMINOPHENYL)
45 ★	1.59		1988		C14H14N2O1	ACETANILIDE,N-(3-AMINOPHENYL) 85856-32-2
6 ★	2.02		538		C14H14N2O1	N-(P-DIMETHYLAMINOPHENYL)BENZOQUINONEMONIMINE
7 ★	2.06		522		C14H14N2O2S1	1,2,4-BENZOTHIADIAZINE-1,1-DIOXIDE-3-(5-NORBORNEN-2-YL) 37157-92-9
8 ★	3.11	7.4	2004	1	C14H14N2O3S1	BENZENESULFONAMIDE,4-METHYL-N-PHENYLAMINOCARBONYL 13909-63-2
9	0.18	7.4	2573	1	C14H14N2O4S2	CEPHALOTHIN ANALOG(3-METHYL) 34691-02-6
10350	0.61	3.0	1638		"	CEPHALOTHIN ANALOG(3-METHYL)
1	-0.30	7.4	2573	1	C14H14N2O5S1	3-METHYLCEPHALOSPORIN,FURYL ANALOG
2 ✓	0.78	12.0	2367		C14H14N3O3S1.Cs1	METHYL ORANGE,CESIUM SALT
3 ✓	0.73	11.9	2367		C14H14N3O3S1.K1	METHYL ORANGE,POTASSIUM SALT
4 ✓	0.68	12.0	2367		C14H14N3O3S1.Na1	METHYL ORANGE,SODIUM SALT
55 ✓	-0.88		2397	101	C14H14N3O5	IMIDAZOLIUM CHLORIDE, 1-METHYL-2-HYDROXYIMINOMETHYL-3-(3,4-METHYLENEDIOXYBENZOYLOXY)METHYL
6	-1.78		536	823	C14H14N3.Cl1	ACRIFLAVINE 86-40-8 anti-infective
7	1.49		2123	22	C14H14N4	BENZALDEHYDE,4-PHENYL,GUANYLHYDRAZONE 61072-53-5
8	1.87		2123	22	C14H14N4O1	BENZALDEHYDE,4-PHENOXY,GUANYLHYDRAZONE 100871-36-1
9	0.23	7.4	2024	1	C14H14N4O4	1,2-AZIRIDINOMITOSINE,7-AMINO 103422-25-9
10360	-0.77	7.2	846	314	C14H14N4O7S1	6-(N-2,4-DINITROPHENYLAMINO)PENICILLANIC ACID 1976-41-6
1	2.15	7.4	2028	459	C14H14N6	1,2,4-TRIAZOLE,3-AMINO-5-(2-BENZYLAMINO)PYRIDIN-4-YL 77314-71-7
2	-3.14	7.4	2464	41	C14H14N8O4S3	CEFAZOLIN 26970-89-8 antibacterial (systemic)
3	-3.09	7.4	2530	1	"	CEFAZOLIN
4	-0.90	7.4	2373	1	"	CEFAZOLIN
65 ★	-0.58	2.0	2464	41	"	CEFAZOLIN
6 ★	3.31		599		C14H14O1	BENZYLETHER 103-50-4
7 ★	3.06		603		C14H14O1	PHENYL-O-TOLYLCARBINOL 5472-13-9
8 ★	3.13		603		C14H14O1	PHENYL-P-TOLYLCARBINOL 1517-63-1
9 ★	1.99		594	400	C14H14O1S1	DIBENZYLSULFOXIDE
10370 ★	3.81		547		C14H14O2	1,2-DIPHENOXYETHANE 104-66-5
1 ★	1.56		2648		C14H14O2	ETHYLENE GLYCOL,1,2-DIPHENYL
2 ★	1.91		2648		"	ETHYLENE GLYCOL,1,2-DIPHENYL
3 ★	3.29		520	20	C14H14O2S1	1,4-NAPHTHOQUINONE,2-BUTYLTHIO
4	2.90		2012	459	C14H14O3	1-NAPHTHOXYACETIC ACID,ETHYL ESTER 41643-81-6
75	0.28	7.4	2277	1	C14H14O3	2-(6-MEO-2-NAPHTHYL)PROPIONIC ACID 22204-53-1 anti-inflammatory, analgesic, antipyretic
6	0.28	7.4	693	1	"	2-(6-MEO-2-NAPHTHYL)PROPIONIC ACID
7	0.33	7.4	1838	1	"	2-(6-MEO-2-NAPHTHYL)PROPIONIC ACID
8 ★	3.34	2.0	1838		"	2-(6-MEO-2-NAPHTHYL)PROPIONIC ACID
9 ★	3.24		1521		C14H14O3	3-FURANEMETHANOL-5-BENZYL,ACETATE 39856-64-9

	logP	pH	Ref.	Note	MF	Name / **CAS** / *activity*
10380 ★	3.23		1587		C14H14O4	DIALLYLPHTHALATE **131-17-9**
1	0.90	7.4	2559	1	C14H14O5	CHROMONE-2-CARBOXYLIC ACID,5-HYDROXY-6,8-DIETHYL
2 ★	2.35		306		C14H14O5	ETHYLACETOACETATE,3,4-METHYLENEDIOXYBENZAL
3 ★	4.02		306		C14H15Br1O4	DIETHYLMALONATE,4-BROMOBENZAL
4 ★	3.15		306		C14H15Cl1O4	DIETHYLMALONATE,2-CHLOROBENZAL
85 ★	3.49		306		C14H15Cl1O4	DIETHYLMALONATE,3-CHLOROBENZAL
6 ★	3.96		306		C14H15Cl1O4	DIETHYLMALONATE,4-CHLOROBENZAL
7	3.95		2572	459	C14H15Cl2N1O3	O-PHENYLDIETHYLCARBAMATE,2,4-DICHLORO-5-(PROPYN-2YL)OXY **143121-07-7**
8 ★	3.24		2421		C14H15Cl2N3O1	1-(2,5-DICHLOROPHENYL)-2-(1,2,4-TRIAZOL-1-YL)CYCLOHEXANOL
9 ★	3.03	7.0	2292	1	C14H15Cl2N3O2	DIOXOLANE,2-(2,4-DICHLOROPHENYL)-2-(1,2,4-TRIAZOL-1-YL)METHYL-4-ETHYL
10390	2.21	7.0	2292	1	C14H15Cl2N3O3	DIOXOLANE,2-(2,4-DICHLOROPHENYL)-2-(1,2,4-TRIAZOL-1-YL)METHYL-4-METHOXYMETHYL
1 ★	3.04		306		C14H15F1O4	DIETHYLMALONATE,2-FLUOROBENZAL
2 ★	3.19		306		C14H15F1O4	DIETHYLMALONATE,3-FLUOROBENZAL
3 ★	3.10		306		C14H15F1O4	DIETHYLMALONATE,4-FLUOROBENZAL
4 ✓	-0.18		2397	101	C14H15F3N3O2	IMIDAZOLIUM CHLORIDE,1-METHYL-2-HYDROXYIMINOMETHYL-3-(P-TRIFLUOROMETHYLBENZYLOXYMETHYL
95	4.49		2404	459	C14H15F3O3	BENZOPYRAN,2,2-DIMETHYL-6-METHOXY-7-TRIFLUOROETHOXY **75413-17-1**
6	4.79		2404	459	C14H15F3O3	BENZOPYRAN,2,2-DIMETHYL-6-TRIFLUOROETHOXY-7-METHOXY **75413-18-2**
7	-0.95	7.4	1204	1	C14H15F4N3	N-CYPROPYLME-N(2F-4-CF3-PH)AMINO)2-IMIDAZOLINE
8 ★	2.67	12.0	599		C14H15N1	DIBENZYLAMINE **103-49-1**
9 ★	4.22	13.0	547	224	C14H15N1	N-METHYL-N-BENZYLANILINE **614-30-2**
10400	3.11		520		C14H15N1O2	1,4-NAPHTHOQUINONE,2-BUTYLAMINO
1	1.98	7.4	1074	324	C14H15N1O2	3,4(TRIHYDRO)BENZOPYRANOPYRIDINE,2-ME,5-CH2OH
2 ★	2.73		2467		C14H15N1O2S1	ANILINE,P-(2,4-DIMETHYLPHENYL)SULFONYL
3 ★	1.63		2467		C14H15N1O4S1	ANILINE,P-(2,4-DIMETHOXYPHENYL)SULFONYL
4	1.14	7.4	2621	1	C14H15N1O5	MITOSENE(5,5,6),METHYL-DIHYDROXY ANALOG
5	0.37	7.4	2621	1	C14H15N1O5	MITOSENE,DIHYDROXY ANALOG
6 ★	1.02		2157		C14H15N1O5	PYRROLIDINE,N-(2-BENZOYLOXYACETYL)-2-CARBOXY
7 ★	2.54		306		C14H15N1O6	DIETHYLMALONATE,2-NITROBENZAL
8 ★	2.71		306		C14H15N1O6	DIETHYLMALONATE,3-NITROBENZAL
9 ★	3.03		306		C14H15N1O6	DIETHYLMALONATE,4-NITROBENZAL
10410 ★	4.58		588		C14H15N3	AZOBENZENE,4-DIMETHYLAMINO **60-11-7**
1	2.04		555		C14H15N3O4	4-BUTOXY-2-(5-NO2-2-FURFURILIDENE)PYRIMIDINE
2 ★	1.03	6.5	321	2	C14H15N3O4S2	3-BENZYLHYDROTHIAZIDE
3	0.64	7.4	606	1	C14H15N3O5	1-HYDROXY-7-AMINOMITOSENE
4	1.20	7.4	1485	1	C14H15N5O2	2,4-DIAMINOPYRIMIDINE,5-(3,5-DIMEO-4-CYANO)BENZYL
15 ★	1.39	7.4	1485	2	"	2,4-DIAMINOPYRIMIDINE,5-(3,5-DIMEO-4-CYANO)BENZYL
6	-1.85	7.0	2418	1	C14H15N5O6S1	METSULFURON-METHYL **74223-64-6** *antihistaminic*
7	0.00	5.0	2418	1	"	METSULFURON-METHYL
8	-1.95		1525		C14H15N5O7S3	7B-(THIADIAZ-3YL)THIOACETYLAMINO-CEPHAMYCIN **57792-84-4**
9 ★	-0.62		1980		C14H15N7O4S3	CEFMETAZOLE-7-DESMETHOXY
10420	1.79		1069		C14H15O2P1S1	O-ETHYL-O-PHENYL-PHENYLPHOSPHONOTHIOATE **57856-13-0**
1	1.25	7.4	2037	1	C14H16Br1N5O2	2H-PURIN-2-ONE,1,3-DIHYDRO-1-METHYL-6-AMINO-7-(P-BROMOPHENOXY)ETHYL
2	1.73		1479	459	C14H16Br2N4O3	2,4-DIAM-5(2,6-DIBR-3,4,5-TRIMEO-BENZYL)PYRIMIDINE
3 ★	2.68		2421		C14H16Cl1N3O1	1-(2-CHLOROPHENYL)-2-(1,2,4-TRIAZOL-1-YL)CYCLOHEXANOL
4	4.31		331		C14H16Cl1N3O1	2-ME-4-CLC6H3NHN=C(CN)CO-T-BU
25	2.32	7.4	416	1	C14H16Cl1N3O2	1-(P-CL-PHENYL)-3-ME-5-DIMEAMINO-6-ME-URACIL
6 ★	2.77		2665		C14H16Cl1N3O2	TRIADIMEFON **43121-43-3** *fungicide (systemic)*
7	1.80	6.5	321	2	C14H16Cl1N3O4S2	3(5-NORBORNEN-2-YL)HYDROCHLOROTHIAZIDE
8 ★	1.95		1780		C14H16Cl1N3O4S2	CYCLOTHIAZIDE **2259-96-3** *diuretic, antihypertensive*
9 ★	3.87		2463		C14H16Cl2N4O1	BENZOTRIAZOLE,2-CYCLOHEXYL-5-DICHLOROACETAMIDO
10430 ★	2.50		1629		C14H16F3N3O3S2	N-(N''-ME-(3-CF3-PHENYL)CARBAMYLTHIO)METHOMYL
1	6.34		1501	100	C14H16F3N3O4	PROFLURALIN **26399-36-0** *herbicide* ✦
2 ✓	-0.53		2397	101	C14H16F3N4O3S1	IMIDAZOLIUM CHLORIDE,1-METHYL-2-HYDROXYIMINOMETHYL-3-(2-N-PHENYL)TRIFLUOROMETHYLSULFONAMIDO)ETHYL
3	-1.86		527	820	C14H16N1.Br1	G-PHENYLPROPYLPYRIDINIUMBROMIDE
4 ★	2.34		1706		C14H16N2	BIPHENYL,4,4'-DIAMINO-3,3'-DIMETHYL **119-93-7**
35	-0.85	2.0	1039	210	C14H16N2	DIDESMETHYLPHENIRAMINE
6	1.86		2032	459	C14H16N2O1	4-PYRIMIDONE,(1,2,B)-4-METHYLPYRIDYL-(5,6)CYCLOHEPTYL
7	2.07		2032	459	C14H16N2O1	4-PYRIMIDONE,(1,2,B)-5-METHYLPYRIDYL-(5,6)-CYLCOHEPTYL
8	1.91		2032	459	C14H16N2O1	4-PYRIMIDONE,(1,2,B)-6-METHYLPYRIDYL-(5,6)-CYCLOHEPTYL
9	2.29		2032	459	C14H16N2O1	4-PYRIMIDONE,(1,2,B)-PYRIDYL-(5,6)-CYCLOOCTYL
10440	2.44		2032	459	C14H16N2O1	4-PYRIMIDONE,(2,3)CYCLOHEPTYL-(5,6)-2-METHYLBENZO-
1 ★	3.05		591		C14H16N2O2	ETOMIDATE **33125-97-2** *hypnotic, anticonvulsant*
2	-0.10		2089		C14H16N2O2	N-(3-DELTA-3-ACETYLPIPERIDIN-1-YL)BENZAMIDE **104642-77-5**
3	0.38		2089		C14H16N2O2	N-(3-DELTA-4-ACETYLPIPERIDIN-1-YL)BENZAMIDE **104642-78-6**
4 ★	2.04		2467		C14H16N2O2S1	ANILINE,P-(4-DIMETHYLAMINOPHENYL)SULFONYL **86552-09-2**
45 ★	1.98		2467		C14H16N2O2S1	ANILINE,P-(4-ETHYLAMINOPHENYL)SULFONYL **3572-34-7**
6 ★	2.52	6.4	252	817	C14H16N2O2S1	N-SULFANILYL-3,4-XYLAMIDE
7	0.68		1407		C14H16N2O3	1,2,3,6-H4-PYRIDINE,3,4-(METHYLENEDIOXY)PH-ME-CONH-
8 ★	0.20		2157		C14H16N2O4	PYRROLIDINE,N-(2-BENZOYLOXYACETYL)-2-CARBOXAMIDE
9	-0.44	7.4	2542	1	C14H16N2O4S2	THIOPHENE-2-SULFONAMIDE,4-(4-HYDROXY-3-DIMETHYLAMINOMETHYL)BENZOYL
10450	-0.46		2234		C14H16N2O6	ACETYLSALICYLIC ACID,N-(2-ACETAMIDO)GLYCOLAMIDE ESTER **116482-78-1**
1 ✓	-1.26		2397	101	C14H16N3O2	IMIDAZOLIUM CHLORIDE,1-BENZYL-2-HYDROXYIMINOMETHYL-3-ACETYLMETHYL
2	0.61		1375		C14H16N4O1	3-(2-FORMYL-2-BENZYL)HYDRAZINO)PYRIDAZINE,4,6-DIME
3 ★	2.31	13.0	547		C14H16N4O1	PYRIMIDINE,2,4-DIAMINO-5-(2-ALLYLOXY)BENZYL
4	0.13	7.4	2683	1	C14H16N4O1	TRYPTAMINE,5-(3-ETHYLOXADIAZOLYL) *5ht-agonist*
55	-1.70	7.4	2683	1	C14H16N4O1	TRYPTAMINE,5-(3-METHYLOXADIAZOLYLMETHYL) *5ht-agonist*
6	0.39		1713		C14H16N4O2	6-PH-2-(2-N3-ME-UREIDOET)-3(2H)-PYRIDAZINONE
7	0.69		1713		C14H16N4O2	6-PH-2-(3-UREIDOPROPYL)-3(2H)-PYRIDAZINONE
8 ★	2.07		57		C14H16N4O2S1	3-BENZOYLMETHIO-4-AMINO-6-I-PR-1,2,4-TRIAZINE-5-ONE
9	1.07	7.2	2446	1	C14H16N4O3	PIROMIDIC ACID **19562-30-2** *antibacterial*
10460	1.52	7.0	2292	1	C14H16N4O4	DIOXOLANE,2-(4-NITROPHENYL)-2-(1,2,4-TRIAZOL-1-YL)METHYL-4-ETHYL
1 ★	1.69	7.4	2206	306	C14H16N4O4	O-BENZYLCARBAMATE,N-(2-METHYL-5-NITRO-IMIDAZOL-1-YL)ETHYL
2	-0.20	7.4	606	1	C14H16N4O5	1-HYDROXY-2,7-DIAMINOMITOSENE
3 ★	3.34		1524		C14H16O2	1(3H)-ISOBENZOFURANONE-3-CYCLOHEXYL
4 ★	3.39		1941		C14H16O3	4H-1-BENZOPYRAN-4-ONE,2-I-PROPYL-3-METHYL-7-METHOXY
65 ★	2.66		306		C14H16O3	ETHYLACETOACETATE,2-METHYLBENZAL
6 ★	2.80		306		C14H16O3	ETHYLACETOACETATE,3-METHYLBENZAL
7 ★	2.63		306		C14H16O3	ETHYLACETOACETATE,4-METHYLBENZAL
8 ★	1.88		306		C14H16O4	ACETYLACETONE,2,4-DIMETHOXYBENZAL
9 ★	3.13		306		C14H16O4	DIETHYLMALONATE,BENZAL

	logP	pH	Ref.	Note	MF	Name / CAS / activity
10470 ★	2.39		306		C14H16O4	ETHYLACETOACETATE,2-METHOXYBENZAL
1 ★	2.34		306		C14H16O4	ETHYLACETOACETATE,3-METHOXYBENZAL
2 ★	2.46		306		C14H16O4	ETHYLACETOACETATE,4-METHOXYBENZAL
3 ★	2.50		2234		C14H16O6	ACETYLSALICYLIC ACID,PROPIONYLOXYMETHYL ESTER 118247-07-7
4	1.86		1479	459	C14H17Br1N4O2	2,4-DIAMINO-5-(2-BR-4-MEO-5-ETO-BENZYL)PYRIMIDINE
75	2.18		1479	459	C14H17Br1N4O2	2,4-DIAMINO-5-(2-BR-5-MEO-4-ETO-BENZYL)PYRIMIDINE
6	1.66		1479	459	C14H17Br1N4O3	2,4-DIAMINO-5-(2-BR-3,4,5-TRIMEO-BENZYL)PYRIMIDINE
7	-0.14	7.4	1204	1	C14H17Br2N3	N(2-ME-2BUTEN-1-YL)N(2,6-DIBRPH)AM)-2-IMIDAZOLINE
8	0.02	7.4	1204	1	C14H17Br2N3	N(3-ME-3BUTEN-1-YL)N(2,6-DIBRPH)AM)-2-IMIDAZOLINE
9	-0.06	7.4	1204	1	C14H17Br2N3	N(CYBUTYLME)N-(2,6-DIBRPH)AMINO)-2-IMIDAZOLINE
10480	-0.29	7.4	1204	1	C14H17Br2N3	N(CYPENTYL)-N-(2,6-DIBRPH)AMINO)-2-IMIDAZOLINE
1	-1.03	7.4	1204	1	C14H17Br2N3	N-CYPROYLME-N-(2,6-DIBRPH)AMINO)-2-4H-PYRIMIDINE
2	4.69		928		C14H17Cl1N1O4P1S2	DIALIFOR 10311-84-9 insecticide
3 ★	2.47		2237		C14H17Cl1N2O2	BENZOFURAN-3-AMINE,2,3-DIHYDRO-5-CHLORO-N-(1-PYRROLIDINYL)ACETYL
4 ★	1.55		2237		C14H17Cl1N2O3	BENZOFURAN-3-AMINE,2,3-DIHYDRO-5-CHLORO-N-(1-MORPHOLINO)ACETYL
85	3.30		2482		C14H17Cl2N3O1	HEXACONAZOLE
6	2.97		527		C14H17F3N2O1	1-CYCLOBUTYL-1-ET-3-(M-TRIFLUOROMETHYLPHENYL)-UREA
7	3.12	7.4	934	1	C14H17F3N2O1	N-(M-CF3-PHENYL)-3-N'-PIPERIDINOACETAMIDE
8	3.09	7.4	934	1	C14H17F3N2O1	N-(O-CF3-PHENYL)-3-N'-PIPERIDINOACETAMIDE
9	3.06	7.4	934	1	C14H17F3N2O1	N-(P-CF3-PHENYL)-3-N'-PIPERIDINOACETAMIDE
10490	1.79		1479	459	C14H17I1N4O3	2,4-DIAMINO-5-(2-IOD-3,4,5-TRIMEO-BENZYL)PYRIMIDINE
1	2.75		2633	459	C14H17N1	2-DELTA-OCTAHYDROPHENANTHREN-4A-AMINE nmda antagonist
2	0.04	7.4	2617	1	C14H17N1	N-METHYL-2D-HEXAHYDROFLUOREN-9A-AMINE
3	3.28		2617	2	"	N-METHYL-2D-HEXAHYDROFLUOREN-9A-AMINE
4	1.15	7.4	1990	1	C14H17N1O1	BUTYLAMINE,4-(1-NAPHTHOXY)
95	2.07	7.4	1990	1	C14H17N1O1	N,N-DIMETHYLETHYLAMINE,2-(1-NAPHTHOXY)
6	1.38	7.4	1443	1	C14H17N1O1	N,N-PENTAMETHYLENECINNAMAMIDE 5422-81-1
7 ★	2.74		572		"	N,N-PENTAMETHYLENECINNAMAMIDE
8	2.36	7.4	1163	1	C14H17N1O2	N(3,4-DIMEPH)-3,4-DIME-2-OH-5-OXO-2,5-H2-PYRROLON
9	0.67		2060		C14H17N1O2	SUCCININIMIDE,N-(P-T-BUTYLPHENYL)
10500 ★	2.77		1524		C14H17N1O3	1(3H)-ISOBENZOFURANONE-3-BUTYL-6-ACETYLAMINO
1 ★	1.95		2157		C14H17N1O3	PIPIRIDINE,1-(2-BENZOYLOXYACETYL)
2 ★	0.63		2157		C14H17N1O4	PIPERIDINE,1-(2-BENZOYLOXYACETYL)-4-HYDROXY
3 ★	1.56		2157		C14H17N1O5	2-(BENZOYLOXY)-(N-METHYL-N-ACETIC ACID,ETHYL ESTER)ACETAMIDE
4 ★	1.46		2234		C14H17N1O5	ACETYLSALICYLIC ACID,(2-(N-ACETYL,N-METHYL)AMINOETHYL ESTER 118247-08-8
5 ★	0.06		2234		C14H17N1O6	ACETYLSALICYLIC ACID,(N-HYDROXYETHYL)GLYCOLAMIDE ESTER 118247-05-5
6 ★	0.36		1912		C14H17N3	OCTAHYDROPYRAZINO(2',1':6,1)PYRIDO(3,4B)INDOLE
7	0.16	7.4	2331	1	C14H17N3	OCTAHYDROPYRIDO[3',2':4,5]PYRROLO[2,3-A]QUINOLIZINE 112114-05-3
8	1.30	7.4	416	1	C14H17N3O2	1-ME-3-PHENYL-5-DIMEAMINO-6-METHYLURACIL
9	1.19	7.4	416	1	C14H17N3O2	1-PHENYL-3-ME-5-DIMEAMINO-6-METHYLURACIL
10510	1.48	7.2	556		C14H17N3O2S1	5-I-BUTYL-2-BENZENESULFONAMIDOPYRIMIDINE
1	1.48	7.2	556		"	5-I-BUTYL-2-BENZENESULFONAMIDOPYRIMIDINE
2 ★	1.57		1136		C14H17N3O2S1	N1-(4-DIMETHYLAMINOPHENYL)SULFANILAMIDE 34392-61-5
3 ★	-2.21	7.0	594	314	C14H17N3O3	ALANYL-TRYPTOPHANE
4 ★	-1.98	7.0	2374		C14H17N3O3	TRYPTOPHANYLALANINE
15	1.41	7.2	556		C14H17N3O3S1	5-BUTOXY-2-BENZENESULFONAMIDOPYRIMIDINE
6 ★	0.16		2649		C14H17N3O4	BENZOQUINONE,2,5-BIS-AZIRIDNYL-3-METHYL-6-CARBAMYLOXYETHYL
7 ★	-2.41	7.0	594	314	C14H17N3O4	TRYPTOPHYL-SERINE
8	1.22		494		C14H17N3O4S1	3I-PR-2-I-PRNH-5(5-NO2-2-FURFURILIDINE)THIAZOL-4ONE
9	1.32		494		C14H17N3O4S1	3PR-2-PRIMINO-5-(5-NO2-2-FURFURILIDINE)THIAZOL-4ONE
10520	2.28		2279		C14H17N3O4S1	IMIDAZOLE,1-METHYL-4-NITRO-5-P-T-BUTYLPHENYLSULFONATO
1 ★	-0.20		401		C14H17N3O9	6-AZAURIDINETRIACETATE 2169-64-4 antipsoriatic
2 ✓	-1.35		2397	101	C14H17N4O4	IMIDAZOLIUM CHLORIDE,1-METHYL-2-HYDROXYIMINOMETHYL-3-(P-NITROBENZYLOXY)ETHYL
3	-1.51	7.4	2683	1	C14H17N5O1	TRYPTAMINE,5-(3-AMINOOXADIAZOLYLETHYL) 5ht-agonist
4	-1.04	7.4	1485	1	C14H17N5O3	2,4-DIAMINOPYRIMIDINE,5(3,5-DIMEO-4-CONH2)BENZYL
25 ★	-0.85	7.4	1485	2	"	2,4-DIAMINOPYRIMIDINE,5(3,5-DIMEO-4-CONH2)BENZYL
6	-2.17	6.9	2314	1	C14H17N5O3	PIPEMIDIC ACID 51940-44-4 antibacterial
7 ✓	-2.15	6.9	2314	2	"	PIPEMIDIC ACID
8	-1.52	7.2	2446	1	"	PIPEMIDIC ACID
9 ★	3.92		1481		C14H18Cl1N1O1	P-CHLOROCINNAMAMIDE,B-METHYL,N-S-BUTYL
10530 ★	2.57		565		C14H18Cl1N1O3	ETHYL-N-CHLOROACETYL-N-(2-ET-PHENYL)GLYCINATE
1 ★	3.77		2506		C14H18Cl1N2O3P1S1	PYRACLOFOS insecticide (cholinesterase inhib.)
2	-0.99	7.4	1204	1	C14H18Cl1N3	N-CYPROPYLME-N-(2-CL-6-MEPH)AMINO)-2-IMIDAZOLINE
3 ★	3.08		2506		C14H18Cl1N3O2	TRIADIMENOL 55219-65-3 fungicide, steroid demethylation inhib.
4 ✓	-0.94		2397	101	C14H18Cl1N4O3S1	IMIDAZOLIUM CHLORIDE,1-METHYL-2-HYDROXYIMINOMETHYL-3-(2-N-METHYL-P-CHLOROPHENYLSULFONAMIDO)ETHYL
35 ✓	1.01	7.6	2647	226	C14H18Cl2N2O3	SALICYLAMIDE,3,5-DICHLORO-6-METHOXY-N-(1-METHYLPIPERIDIN-4YL)
6	3.96		2662	459	C14H18Cl3N1O4	N-(3,4-DIETHOXYPHENYL)CARBAMATE,1,1,1-TRICHLOROPROP-2YL ESTER fungicide
7 ★	2.37		1439		C14H18N2O1	4H-PYRIDO(1,2-A)PYRIMIDIN-4-ONE,2-PR,3-ET,6-ME ✦
8	1.35		2089		C14H18N2O1	N-(3-DELTA-3,4-DIMETHYLPIPERIDIN-1-YL)BENZAMIDE 104642-68-4
9 ★	1.68	6.9	2175		C14H18N2O2	3-BUTYL-3-(4-PYRIDYL)PIPERIDINE-2,6-DIONE
10540 ★	1.76		2237		C14H18N2O2	BENZOFURAN-3-AMINE,2,3-DIHYDRO-N-(1-PYRROLIDINYL)ACETYL
1 ★	1.97		2649		C14H18N2O2	BENZOQUINONE,2,5-BIS-PYRROLIDIN-1YL
2	0.45		2089		C14H18N2O2	N-(3-DELTA-3-(1-HYDROXYETHYL)PIPERIDIN-1-YL)BENZAMIDE 104642-87-7
3	0.04		2089		C14H18N2O2	N-(3-DELTA-4-(1-HYDROXYETHYL)PIPERIDIN-1-YL)BENZAMIDE 104642-88-8
4 ★	0.81		2237		C14H18N2O3	BENZOFURAN-3-AMINE,2,3-DIHYDRO-N-(1-MORPHOLINO)ACETYL
45	3.30		871		C14H18N2O3	N-NITROSO-METHYLPHENIDATE
6 ★	-1.36	7.0	2374		C14H18N2O3	PROLINE,PHENYLALANYL 7669-65-0
7 ✓	-1.14	5.4	2017		"	PROLINE,PHENYLALANYL
8 ★	-2.07	7.0	2374		"	PROLINYLPHENYLALANINE
9	2.53	7.4	899	1	C14H18N2O3	REPOSAL 3625-25-0 ✦
10550	0.88		1205		C14H18N2O3S1	N,N-DIME-4(N-ME-N-2(HCC-PROPIONYL)BZSULFONAMIDE
1 ★	2.13		2142		C14H18N2O4	3,4-DIETHOXYPHENYLCARBAMATE,O-(2-CYANOETHYL)
2 ★	-2.03	7.0	594	314	C14H18N2O4	TYROSYL-PROLINE
3	-0.04	7.4	2542	1	C14H18N2O4S3	THIOPHENE-2-SULFONAMIDE,4-(I-PROPYLAMINOMETHYL)PHENYLSULFONYL
4	0.65		2495R1~314		C14H18N3O3P1	PHOSPHORIC-TRIAMIDE,N-M-ALLYLOXYBENZOYL-N',N'-DIETHYLENE
55	0.69	7.4	2495	1	C14H18N3O3P1	PHOSPHORIC-TRIAMIDE,N-P-ALLYLOXYBENZOYL-N',N'-DIETHYLENE
6 ✓	-2.39		2397	101	C14H18N3O4S1	IMIDAZOLIUM CHLORIDE,1-METHYL-2-HYDROXYIMINOMETHYL-3-(PHENYLSULFONYLETHOXYMETHYL 132566-69-9
7	2.00	7.4	1485	1	C14H18N4O1	2,4-DIAMINOPYRIMIDINE,5(3,5-DIME-4-MEO)BENZYL
8 ★	2.19	7.4	1485	2	"	2,4-DIAMINOPYRIMIDINE,5(3,5-DIME-4-MEO)BENZYL
9 ✓	2.28	13.0	579		"	2,4-DIAMINOPYRIMIDINE,5(3,5-DIME-4-MEO)BENZYL

	logP	pH	Ref.	Note	MF	Name / CAS / activity
10560 ★	2.82	7.4	2341	1	C14H18N4O1	3[1-(2-BENZOXAZOLYL)HYDROAZINO]PROPANENITRILE,5-T-BUTYL
1	-1.33	1.0	547	342	C14H18N4O2	2,4-DIAM-5(3-B-MEO-ETOBENZYL)PYRIMIDINE
2	1.20		1479	459	C14H18N4O2	2,4-DIAMINO-5-(3-ETO-4-MEO-BENZYL)PYRIMIDINE 73046-15-8
3	1.34		1479	459	C14H18N4O2	2,4-DIAMINO-5-(3-MEO-4-ETO-BENZYL)PYRIMIDINE
4	1.43	7.4	1485	1	C14H18N4O2	2,4-DIAMINOPYRIMIDINE,5(3,4-DIMEO-5-METHYL)BENZYL
65 ★	1.62	7.4	1485	2	"	2,4-DIAMINOPYRIMIDINE,5(3,4-DIMEO-5-METHYL)BENZYL
6	2.06	7.4	1485	1	C14H18N4O2	2,4-DIAMINOPYRIMIDINE,5-(3,5-DIMEO-4-ME)BENZYL
7 ★	2.14	12.0	1471			2,4-DIAMINOPYRIMIDINE,5-(3,5-DIMEO-4-ME)BENZYL
8 ★	2.25	7.4	1485	2	"	2,4-DIAMINOPYRIMIDINE,5-(3,5-DIMEO-4-ME)BENZYL
9 ★	3.43		2424		C14H18N4O2	BENZOTRIAZOLE,2-CYCLOPENTYL-5-(O-ETHYLCARBAMOYL)
10570	1.04	7.4	1485	1	C14H18N4O2	ORMETOPRIM 6981-18-6 antibacterial
1 ★	1.23	7.4	1485	2	"	ORMETOPRIM
2	1.38	7.4	1485	1	C14H18N4O2S1	METIOPRIM 68902-57-8 antibacterial
3 ★	1.54	7.4	1485	2	"	METIOPRIM
4 ★	1.52	13.0	579	224	C14H18N4O3	2,4-DIAMINOPYRIMIDINE,5-(2,3,4-TRIMETHOXYBENZYL)
75	0.94	7.4	1485	1	C14H18N4O3	2,4-DIAMINOPYRIMIDINE,5-(2,4,5-TRIMEO-)BENZYL
6 ★	1.17	7.4	1485	2	"	2,4-DIAMINOPYRIMIDINE,5-(2,4,5-TRIMEO-)BENZYL
7	0.23	7.4	1485	1	C14H18N4O3	2,4-DIAMINOPYRIMIDINE,5-(3,5-DIMEO-4-CH2OH)BENZYL
8 ★	0.42	7.4	1485	2	"	2,4-DIAMINOPYRIMIDINE,5-(3,5-DIMEO-4-CH2OH)BENZYL
9 ★	2.12		1641		C14H18N4O3	BENOMYL 17804-35-2 fungicide
10580	0.64	7.4	1485	1	C14H18N4O3	TRIMETHOPRIM 738-70-5 antibacterial
1	0.83	7.4	1485	2	"	TRIMETHOPRIM
2 ★	0.91		547			TRIMETHOPRIM
3 ★	0.11		2529		C14H18N4O4	DDI,5'-BUTYRATE ✚
4 ★	0.02		2529		C14H18N4O4	DDI,5'-ISOBUTYRATE
85 ★	2.40	4.0	360	2	C14H18N4O4S1	2,6-DIETHOXY-4-SULFANILAMIDOPYRIMIDINE
6	1.70	4.0	360	2	C14H18N4O4S1	2-ET-5,6-DIMEO-4-SULFANILAMIDOPYRIMIDINE
7	1.03	4.0	360	2	C14H18N4O4S1	5-I-PRO-6-MEO-4-SULFANILAMIDOPYRIMIDINE
8	1.52	4.0	360	2	C14H18N4O4S1	5-MEO-6-I-PRO-4-SULFANILAMIDOPYRIMIDINE
9	1.80	4.0	360	2	C14H18N4O4S1	5-MEO-6-PRO-4-SULFANILAMIDOPYRIMIDINE
10590 ✓	-1.89		2397	101	C14H18N5O5S1	IMIDAZOLIUM CHLORIDE,1-METHYL-2-HYDROXYIMINOMETHYL-3-(2-N-METHYL-O-NITROPHENYLSULFONAMIDO)ETHYL
1 ✓	-1.62		2397	101	C14H18N5O5S1	IMIDAZOLIUM CHLORIDE,1-METHYL-2-HYDROXYIMINOMETHYL-3-(2-N-METHYL-P-NITROPHENYLSULFONAMIDO)ETHYL
2 ★	0.70		2289		C14H18N6O4	TETRAZOLE,1-ETHOXYCARBONYLMETHYL-5-(4-DIMETHYLUREYL)PHENOXY 117121-37-6
3 ★	1.12		2289		C14H18N6O4	TETRAZOLE,2-ETHOXYCARBONYLMETHYL-5-(4-DIMETHYLUREYL)PHENOXY 117121-49-0
4	3.55		1524		C14H18O2	1(3H)-ISOBENZOFURANONE-3-HEXYL
95	2.01		1043		C14H18O2	2,3-BENZO-OCTAHYDRONAPHTHALENE-3,4-DIOH(3,4-DI-AX)
6	2.32		1043		"	2,3-BENZO-OCTAHYDRONAPHTHALENE-3,4-DIOH(3,4-DI-AX)
7	2.47		1043		"	2,3-BENZO-OCTAHYDRONAPHTHALENE-3,4-DIOH(3,4-DI-AX)
8	2.57		1043		"	2,3-BENZO-OCTAHYDRONAPHTHALENE-3,4-DIOH(3,4-DI-AX)
9	3.36		2404	459	C14H18O3	BENZOPYRAN,2,2-DIMETHYL-6-METHOXY-7-ETHOXY 65383-73-5
10600	3.79		501		C14H18O3	PHENOXYACETIC ACID,4-CYCLOHEXYL
1 ★	2.76		306		C14H18O4	BENZYLMALONIC ACID,DIETHYL ESTER
2 ★	2.83		1587		C14H18O4	DI-I-PROPYLPHTHALATE 605-45-8
3	3.27		1587		C14H18O4	DIPROPYLPHTHALATE 131-16-8
4	4.05		2431	459	"	DIPROPYLPHTHALATE
5 ★	1.67		2487		C14H18O4	OCT-7-ENOL-6-ONE,8-(3,4-DIHYDROXY)PHENYL
6	-1.02	7.4	2450	1	C14H18O4	TROLOX ✚
7	2.41	7.6	2007	1	C14H19Br1N2O2S1	BENZOYLTHIOFORMOHYDROXIMATE,4-BROMO-S-(2-DIETHYLAMINO)ETHYL 102941-47-9
8 ★	3.50		2142		C14H19Br2N1O4	3,4-DIETHOXYPHENYLCARBAMATE,O-2-(1,3-DIBROMOPROPYL)
9	0.16	7.4	1204	1	C14H19Br2N3	N(I-AMYL)-N-(2,6-BR2-PH)AMINO)-2-IMIDAZOLINE
10610	0.32	7.4	1204	1	C14H19Br2N3	N-AM-N(2,6-BR2-PH)AMINO)-2-IMIDAZOLINE
1 ★	2.88		2237		C14H19Cl1N2O2	BENZOFURAN-3-AMINE,2,3-DIHYDRO-5-CHLORO-N-DIETHYLAMINOACETYL
2	1.97	7.6	2007	1	C14H19Cl1N2O2S1	BENZOYLTHIOFORMOHYDROXIMATE,4-CHLORO-S-(2-DIETHYLAMINO)ETHYL 102941-49-1
3 ✓	-2.15		2397	101	C14H19Cl1N3O2S1	IMIDAZOLIUM CHLORIDE,1,2-DIMETHYL-3-(2-N-METHYL-P-CHLOROPHENYLSULFONAMIDO)ETHYL
4 ★	2.25		2289	459	C14H19Cl1N6O2	TETRAZOLE,1-T-BUTYL-5-(2-CHLORO-4-DIMETHYLUREYL)PHENOXY 117121-40-1
15	4.28		1050		C14H19Cl2N1O2	3-N,N-BIS(2-CL-PROP)AMINO-4-METHYL-BENZOIC ACID
6	0.93	7.9	536	314	C14H19Cl2N1O2	CHLORAMBUCIL,SODIUM SALT
7	1.47	7.4	594	1	C14H19Cl2N1O2	CHLORAMBUCIL 305-03-3 antineoplastic
8	1.70	7.4	401	1	"	CHLORAMBUCIL
9 ★	2.24		537		C14H19Cl2N1O3	CHLORAMPHENICOL,P-I-PROPYL ANALOG
10620 ★	-0.16		2518	122	C14H19F1N4O3	2'-DEOXY-3'-FLUORO-CYTIDINE,N4-(PYRROLIDIN-1YL)METHYLENE
1 ★	-0.68		2518	122	C14H19F1N4O4	2'-DEOXY-3'-FLUORO-CYTIDINE,N4-(MORPHOLIN-1YL)METHYLENE
2 ★	3.13	13.0	1623	224	C14H19N1	BENZOBICYCLO(2,2,1)HEPTENE,2-ENDO-PROPYLAMINO 86992-67-8
3 ★	3.30	13.0	1623	224	C14H19N1	BENZOBICYCLO(2,2,1)HEPTENE,2-EXO-PROPYLAMINO 86943-77-3
4	0.33	7.4	2617	1	C14H19N1	N-METHYLHEXAHYDROFLUOREN-4A-AMINE
25	0.43	7.4	2617	1	"	N-METHYLHEXAHYDROFLUOREN-4A-AMINE
6	0.54	7.4	2617	1	"	N-METHYLHEXAHYDROFLUOREN-4A-AMINE
7 ★	3.37		2617	2	"	N-METHYLHEXAHYDROFLUOREN-4A-AMINE
8	3.38		2617	2	"	N-METHYLHEXAHYDROFLUOREN-4A-AMINE
9	0.62	7.4	2617	1	C14H19N1	N-METHYLHEXAHYDROFLUOREN-9A-AMINE
10630	3.43		2617			N-METHYLHEXAHYDROFLUOREN-9A-AMINE
1	2.92		2633	459	C14H19N1	OCTAHYDROPHENANTHREN-4A-AMINE nmda antagonist
2	0.63	7.4	1443	1	C14H19N1O1	1-PHENYL-3-PIPERIDINOPROPAN-1-ONE 5703-17-3
3	0.86	7.4	2617	1	C14H19N1O1	2-HYDROXYMETHYLHEXAHYDROFLUOREN-4A-AMINE
4	0.93		2617	2	"	2-HYDROXYMETHYLHEXAHYDROFLUOREN-4A-AMINE
35 ★	2.88		1481		C14H19N1O1	A-METHYL-N-S-BUTYLCINNAMAMIDE
6	1.15	7.4	2535	1	C14H19N1O1	G-(2-ETHYLBENZOYL)PROPYL-ETHANOLAMINE (DEHYDRO-FUSED)
7	1.04	7.4	2535	1	C14H19N1O1	G-(PHENACETYL)PROPYL-(2-PROPANOL)AMINE (DEHYDRO-FUSED)
8	0.96	7.4	2633	1	C14H19N1O1	OCTAHYDRO-8-OXOPHENANTHREN-4A-AMINE,N-METHYL nmda antagonist
9	3.31		2633	459	"	OCTAHYDRO-8-OXOPHENANTHREN-4A-AMINE,N-METHYL
10640	3.65		1524		C14H19N1O2	1(3H)-ISOBENZOFURANONE-3-HEXYL-6-AMINO
1 ★	2.86		1481		C14H19N1O2	A-METHOXY-N-S-BUTYLCINNAMAMIDE
2	0.20	7.2	969	314	C14H19N1O2	METHYLPHENIDATE 113-45-1 stimulant (central)
3 ★	1.93	7.4	2204	1	C14H19N1O3	2H,5H-1-BENZOPYRANO[4,3,B]-1,4-OXAZIN-7-OL,3,4,4A,10B-TETRAHYDRO-4-PROPYL 116005-01-7 ✚
4 ★	1.97	7.4	2204	1	C14H19N1O3	2H,5H-1-BENZOPYRANO[4,3-B]-1,4-OXAZIN-9-OL,3,4,4A,10B-TETRAHYDRO4-PROPYL 112960-12-0
45 ★	2.76		2142		C14H19N1O4	3,4-DIETHOXYPHENYLCARBAMATE,O-ALLYL
6	3.03		1481		C14H19N1S1	A-METHYLTHIO-N-S-BUTYLCINNAMAMIDE
7	1.54	7.4	2633	1	C14H19N1S1	OCTAHYDRO-8-THIOPHENANTHREN-4A-AMINE,N-METHYL nmda antagonist
8	3.88		2633	459	"	OCTAHYDRO-8-THIOPHENANTHREN-4A-AMINE,N-METHYL
9 ★	2.68		559		C14H19N3O2	1-(5-CARBETHOXYPENTYL)BENZOTRIAZOLE

	logP	pH	Ref.	Note	MF	Name / CAS / activity
10650	3.17		559		C14H19N3O2	2-(5-CARBETHOXYPENTYL)BENZOTRIAZOLE
1 ★	3.32	5.6	2156	1	C14H19N3O2	BENZOTRIAZOLE,1-(2-HEXANOIC ACID,ETHYL ESTER)
2 ★	3.88	5.6	2165	1	C14H19N3O2	BENZOTRIAZOLE,2-(2-HEXANOIC ACID,ETHYL ESTER)
3 ★	2.50		1629		C14H19N3O3S2	N-(N"-ME-(4-ME-PHENYL)CARBAMYLTHIO)METHOMYL
4 ★	-0.79		2377		C14H19N3O4	AC-SER-PHE-N
55	2.50	7.4	1377	1	C14H19N3O6	2-(2,4-DINITROPHENYLAMINO)OCTANOIC ACID **31356-30-6**
6	1.35	7.4	1157	324	C14H19N3S1	METHAPYRILENE **91-80-5** *antihistaminic*
7	2.87		1157		"	METHAPYRILENE
8 ✓	-1.89		2397	101	C14H19N4O3S1	IMIDAZOLIUM CHLORIDE,1-METHYL-2-HYDROXYIMINOMETHYL-3-(3-PHENYLSULFONAMIDO)PROPYL **132567-02-3**
9 ✓	-1.72		2397	101	C14H19N4O3S1	IMIDAZOLIUM CHLORIDE,1-METHYL-2-HYDROXYIMINOMETHYL-3-(N-METHYL-PHENYLSULFONAMIDOETHYL)
10660 ★	2.45	7.4	1776	1	C14H19N5	N(2,5DIME-PYRROL-1YL)6-(4-N-PYRROLIDIN)-3-PYRIDAZINAM **75841-89-3** ✚
1	1.68		1479	459	C14H19N5O1	2,4-DIAMINO-5-(3-DIMEAMINO-4-MEO-BENZYL)PYRIMIDINE
2	1.87		1479	459	C14H19N5O1	2,4-DIAMINO-5-(4-DIMEAMINO-3-MEO-BENZYL)PYRIMIDINE
3 ★	1.94	7.4	1776	1	C14H19N5O1	MOPIDRALAZINE **75841-82-6** *antihypertensive*
4 ★	0.91	7.0	1793	1	C14H19N5O3S1	N(4-(6-PURINYLTHIO)VALERYL)GLYCINE,ETHYL ESTER **22181-94-8**
65 ★	-0.22		2070		C14H19N5O4	ALLOPURINOL,1-(N,N-DIETHYLSUCCINAMYL)OXYMETHYL
6 ★	3.30		1903		C14H19O6P1	CROTOXYPHOS **7700-17-6** *insecticide*
7	3.14	7.4	833	2	C14H20Cl1F1N2O2	3-CL-5-FL-2-MEO-BENZAMIDE,N-(2-DIETAMINOET) **61328-02-7**
8	3.16	7.4	1804	2	"	3-CL-5-FL-2-MEO-BENZAMIDE,N-(2-DIETAMINOET)
9 ★	3.18	7.4	1413	2	"	3-CL-5-FL-2-MEO-BENZAMIDE,N-(2-DIETAMINOET)
10670 ★	3.52		579		C14H20Cl1N1O2	ALACHLOR **15972-60-8** *herbicide*
1	2.97		2662	459	C14H20Cl1N1O4	N-(3,4-DIETHOXYPHENYL)CARBAMATE,CHLOROPROP-2YL ESTER *fungicide*
2 ★	1.05		547		C14H20Cl1N3O9	1-NO-1-CLET-3(TRIACETYLXYLOSYL)UREA **58845-59-3**
3 ★	3.96		2479	459	C14H20F1N3O2	PHENYLBIURET,2-FLUORO-5-T-BUTYL
4	2.59		2662	459	C14H20F1O4	N-(3,4-DIETHOXYPHENYL)CARBAMATE,FLUOROPROP-2YL ESTER *fungicide*
75	-0.59		2397	101	C14H20F3N4O4S1	IMIDAZOLIUM CHLORIDE ANALOG ✚
6	0.10		2358	200	(C14H20N1O5S2)2.Cd1	N-BENZYL-N-DITHIOCARBOXY-D-GLUCAMINE
7 ✓	0.41		2358	200	"	N-BENZYL-N-DITHIOCARBOXY-D-GLUCAMINE
8	2.95		1263		C14H20N2O1	3-PHENYL-1-CYCLOHEPTYLUREA **19095-79-5**
9	2.13		2032	459	C14H20N2O1	4-PYRIMIDONE,(5,6)-CYCLOHEXYL-(2,3)-CYCLOOCTYL
10680	2.32	7.4	934	1	C14H20N2O1	N-(M-TOLYL)-3-N'-PIPERIDINOACETAMIDE
1 ★	2.97	7.4	934	2	C14H20N2O1	N-(M-TOLYL)-3-N'-PIPERIDINOACETAMIDE
2	2.26	7.4	934	1	C14H20N2O1	N-(O-TOLYL)-3-N'-PIPERIDINOACETAMIDE
3	2.26	7.4	934	1	C14H20N2O1	N-(P-TOLYL)-3-N'-PIPERIDINOACETAMIDE **5429-42-5**
4	2.95	7.4	934	2	C14H20N2O1	N-(P-TOLYL)-3-N'-PIPERIDINOACETAMIDE
85 ★	2.96	7.4	1851	1	C14H20N2O1	N-(P-TOLYL)-3-N'-PIPERIDINOACETAMIDE
6	2.27		1147		C14H20N2O1	N-(PIPERIDINOACETYL)BENZYLAMINE
7	2.18		1725	459	C14H20N2O1	PYRIDO-PYRIMIDINE ANALOG (RINGS: 767)
8 ★	3.60		2094		C14H20N2O1	UREA,1-PHENYL-3-(1-METHYLCYCLOHEXYL) **1611-63-8**
9 ★	2.52	7.4	1851	1	C14H20N2O2	1-PIPERIDINEACETAMIDE,N-(4-METHOXYPHENYL) **58479-93-9**
10690 ★	2.20		2237		C14H20N2O2	BENZOFURAN-3-AMINE,2,3-DIHYDRO-N-(DIETHYLAMINOACETYL)
1	-0.86	7.0	477	1	C14H20N2O2	BUNITROLOL **34915-68-9** *anti-anginal*
2	-0.51	7.4	2465	41	"	BUNITROLOL
3	-0.40	7.4	2375	1	"	BUNITROLOL
4	-0.36	7.4	2501	1	"	BUNITROLOL
95 ★	1.91	10.2	2465	41	"	BUNITROLOL
6	1.93	7.4	934	1	C14H20N2O2	N-(M-METHOXYPHENYL)-3-N'-PIPERIDINOACETAMIDE
7 ★	2.71	7.4	934	2	C14H20N2O2	N-(M-METHOXYPHENYL)-3-N'-PIPERIDINOACETAMIDE
8	2.85	7.4	934	1	C14H20N2O2	N-(O-METHOXYPHENYL)-3-N'-PIPERIDINOACETAMIDE
9	1.11	7.3	1356	1	C14H20N2O2	N-PHENYLCARBAMIC ACID,O-2(N-PIPERIDINYL)ETHYL **55792-31-9**
10700	2.67	7.3	1356	2	"	N-PHENYLCARBAMIC ACID,O-2(N-PIPERIDINYL)ETHYL
1	-0.92	7.0	474	1	C14H20N2O2	PINDOLOL **13523-86-9** *vasodilator*
2	-0.70	7.0	1362	1	"	PINDOLOL
3	-0.70	7.0	2401	1	"	PINDOLOL
4	-0.40	7.4	1628	1	"	PINDOLOL
5	-0.39	7.4	2274	1	"	PINDOLOL
6	-0.39	7.4	2399	1	"	PINDOLOL
7	-0.33	7.4	1354	1	"	PINDOLOL
8	-0.33	7.4	2375	1	"	PINDOLOL
9	-0.30	7.4	2020	1	"	PINDOLOL
10710	-0.20	7.4	2598	1	"	PINDOLOL
1	-0.13	7.4	2399	1	"	PINDOLOL
2	-0.10	7.4	2465	41	"	PINDOLOL
3	-0.09	7.4	1362	1	"	PINDOLOL
4	-0.05	7.4	2178	1	"	PINDOLOL
15	0.08	7.4	2245	1	"	PINDOLOL
6	1.57	7.4	2245	2	"	PINDOLOL
7 ★	1.75		1984		"	PINDOLOL
8	-1.30	7.4	2274	1	C14H20N2O3	2H-INDOL-2-ONE,4-(3-I-PROPYLAMINO)-2-HYDROXYPROPOXY **40053-65-4**
9 ★	1.77		1439		C14H20N2O3	3-BUTOXYCARBONYL-PYRIDO(1,2A)PYRIMIDIN-4-ONE,4,6,7,8,9-H5,6-ME
10720	2.82	7.4	899	1	C14H20N2O3	HEPTABARBITAL,N-METHYL
1 ★	-1.68	7.0	594	314	C14H20N2O3	PHENYLALANYL-VALINE
2	0.49	7.6	2007	1	C14H20N2O3S1	BENZOYLTHIOFORMOHYDROXIMATE,4-METHOXY-S-(3-DIMETHYLAMINO)PROPYL **102941-31-1**
3 ★	-1.59	7.0	2374		C14H20N2O3S1	PHENYLALANYLMETHIONINE
4 ★	2.90	7.4	2004		C14H20N2O3S1	TOLCYCLAMIDE **664-95-9** *antidiabetic*
25 ★	-2.52	7.0	2374		C14H20N2O4	VALINYLTYROSINE
6 ★	-2.03	7.0	594	314	C14H20N2O4S1	METHIONYL-TYROSINE
7	1.16	7.4	2272	1	C14H20N2O4S1	MORPHOLINE-1-ACETAMIDE,N-METHYL-N-SULFONYL(P-TOLYL)
8	0.20		1205		C14H20N2O4S1	N,N-DIME-4(N-ME-N(PENTAN-1,4-DION)BENZENESULFONAMIDE
9	1.49		1018		C14H20N2O4S1	O(3-I-PR-PHEN)-O'-ME,N,N'-THIODICARBAMATE,N,N'-DIME
10730 ★	1.33		537	270	C14H20N2O5	TRIMETHYLAMPHENICOL
1	0.43		1205		C14H20N2O5S1	N,N-DIME-4(N-ME-N(2-MEO2C-PROPIONYL)BZSULFONAMIDE
2	0.26		1205		C14H20N2O5S1	N,N-DIME-4(N-ME-N(2CH3CO2-PROPIONYL)BENZSULFONAMIDE
3 ✓	-1.07		2397	101	C14H20N3O2	IMIDAZOLIUM CHLORIDE,1-METHYL-2-HYDROXYIMINOMETHYL-3-(2-ETHYNYLCYCLOHEXYLOXY)METHYL
4 ✓	-2.88		2397	101	C14H20N3O2S1	IMIDAZOLIUM CHLORIDE,1,2-DIMETHYL-3-(2-N-METHYL-PHENYLSULFONAMIDO)_ETHYL
35 ★	-0.47		2518	122	C14H20N4O3	2',3'-DIDEOXYCYTIDINE,N4-(PYRROLIDIN-1YL)METHYLENE ✚
6 ★	-0.97		2518	122	C14H20N4O4	2',3'-DIDEOXYCYTIDINE,N4-(MORPHOLIN-1YL)METHYLENE
7 ★	-1.48		562		C14H20N4O4	BENZOQUINONE,2,5-BIS-AZIRIDINYL-3,6-BIS-ETHANOLAMINO
8 ★	1.44	7.4	1776	1	C14H20N6	N(2,5DIME-PYRROL-1YL)6-(4-N-PIPERAZIN)-3-PYRIDAZINAM **75841-93-9**
9 ★	1.65		2289	459	C14H20N6O2	TETRAZOLE,1-T-BUTYL-5-(4-DIMETHYLUREYL)PHENOXY **117121-32-1**

	logP	pH	Ref.	Note	MF	Name / CAS / activity
10740 ★	2.08		2289		C14H20N6O2	TETRAZOLE,2-T-BUTYL-5-(4-DIMETHYLUREYL)PHENOXY 117121-44-5
1 ★	4.09		2430		C14H20O2	PHENYLCAPRYLIC ACID
2	2.24		458		C14H20O2Se1	P-T-BUTYLBENZYLSELENOPROPIONIC ACID
3 ★	3.37	3.5	2250	302	C14H20O3	BUTANOIC ACID,2-HYDROXY-2-(4-I-BUTYL)PHENYL
4 ★	3.29		2168		C14H20O3	ISOVALERIC ACID-2-ETHOXY,BENZYL ESTER
45 ★	4.83		508		C14H20O3	P-HYDROXYBENZOIC ACID,HEPTYL ESTER 1085-12-7
6 ★	4.89		2186		C14H20O4S2	DITHIOLANYLIDENEMALONIC ACID,DI-S-BUTYL ESTER
7 ★	0.91		2308		C14H20O5	BENZO-15-CROWN-5-ETHER 14098-44-3
8	-0.10		1294		C14H20O5S1	1-(2-PHENETHYLTHIO)-B-GALACTOPYRANOSIDE
9 ★	0.26		511		C14H20O6	GLUCOPYRANOSIDE,3,5-DIMETHYLPHENYL(BETA)
10750 ★	0.31		511		C14H20O6	GLUCOPYRANOSIDE,3-ETHYLPHENYL(BETA)
1 ★	2.63		594		C14H20O8	ETHENETETRACARBOXYLIC ACID,TETRAETHYL ESTER 6174-95-4
2 ★	3.20	7.4	1804	2	C14H21Br1N2O2	5-BR-2-METHOXYBENZAMIDE,N-(2-DIETAMINOETHYL) 71225-60-0
3	3.03	7.4	833	2	C14H21Cl1N2O2	5-CL-2-METHOXYBENZAMIDE,N-(2-DIETAMINOETHYL) 40256-75-5
4	3.10	7.4	1804	2	"	5-CL-2-METHOXYBENZAMIDE,N-(2-DIETAMINOETHYL)
55 ★	3.11	7.4	1413	2	"	5-CL-2-METHOXYBENZAMIDE,N-(2-DIETAMINOETHYL)
6	2.42	7.4	833	2	C14H21F1N2O2	5-FL-2-METHOXYBENZAMIDE,N-(2-DIETAMINOETHYL) 55236-14-1
7	2.60	7.4	1804	2	"	5-FL-2-METHOXYBENZAMIDE,N-(2-DIETAMINOETHYL)
8 ★	2.62	7.4	1413	2	"	5-FL-2-METHOXYBENZAMIDE,N-(2-DIETAMINOETHYL)
9 ★	0.22		2518	122	C14H21F1N4O3	2'-DEOXY-3'-FLUORO-CYTIDINE,N4-(DIETHYLAMINO)METHYLENE
10760	-0.11	1.0	547	342	C14H21N1O1	2-N-CYCLOHEXYLAMINO-1-PHENYLETHANOL
1	-0.11	1.0	547	342	"	2-N-CYCLOHEXYLAMINO-1-PHENYLETHANOL
2 ★	2.41		2410		C14H21N1O1	ACETANILIDE,2,6-DI-I-PROPYL
3	1.56	7.4	2535	1	C14H21N1O1	N,N,N-(G-BENZOYL)PROPYL-DIETHYL
4 ★	2.94		603		C14H21N1O1	N-HEPTYLBENZAMIDE
65 ★	3.79		536		C14H21N1O2	3,5-DI-I-PROPYL-N-METHYLPHENYLCARBAMATE
6	2.67		803		C14H21N1O2	3,5-DI-PROPYL-4-HYDROXYACETANILIDE
7 ★	-1.30		2018		C14H21N1O2	HEXANOIC ACID,6-(2-PHENYLETHYL)AMINO
8	1.40	7.4	2535	1	C14H21N1O2	N,N,N-(G-PHENACEYL)PROPYL-METHYL-ETHANOLAMINE
9 ★	2.20		594		C14H21N1O2	O-METHYLCARBAMATE,N-(2,6-DI-I-PROPYLPHENYL)
10770 ★	4.50		2630		C14H21N1O2	P-AMINOBENZOIC ACID,HEPTYL ESTER
1	2.43	7.4	2462	1	C14H21N1O2	P-PROPYLBENZOIC ACID,N,N-DIMETHYLAMINOETHYL ESTER
2 ★	3.63	7.4	2462	2	"	P-PROPYLBENZOIC ACID,N,N-DIMETHYLAMINOETHYL ESTER
3	3.37		866	838	C14H21N1O3	2-HEPTOXY-4-AMINOBENZOIC ACID
4 ★	4.80		547	20	C14H21N1O3	PHENOL,2,6-BIS-(S-BUTYL)-4-NITRO
75 ★	2.91		2142		C14H21N1O4	3,4-DIETHOXYPHENYLCARBAMATE,O-PROPYL
6 ★	2.82		2142		C14H21N1O4	DIETHOFENCARB ✚
7	-2.35	6.4	2017		C14H21N1O4	HEXANOIC ACID,6-(2-(2,4-DIHYDROXYPHENYL)ETHYL)AMINO
8 ★	2.31		2142		C14H21N1O5	3,4-DIETHOXYPHENYLCARBAMATE,O-2-(METHOXYETHYL)
9 ★	4.65		2506		C14H21N1S1	N,N-DIETHYL-S-BENZYLTHIOCARBAMATE herbicide
10780	-1.44	7.6	2007	1	C14H21N2O2S1.Cl1	BENZOYLTHIOFORMOHYDROXIMATE,4-METHYL-S-(2-TRIMETHYLAMMONIO)ETHYL CHLORIDE 102941-42-4
1	-1.96	7.6	2007	1	C14H21N2O3S1.Cl1	BENZOYLTHIOFORMOHYDROXIMATE,4-METHOXY-S-(2-TRIMETHYLAMMONIO)ETHYL CHLORIDE 102976-75-0
2	0.34	7.4	365	2	C14H21N3O1	N,N-DIME-PIPERAZINE-2-CARBOXANILIDE,2'-ME
3	1.04	7.4	365	2	C14H21N3O1	PIPERAZINE-2-CARBOXANILIDE,2'4'6'-TRIME
4	1.76		318		C14H21N3O2	1-(2-DIMEAMINOET)-3(M-MEOPHENYL)-2-IMIDAZOLIDINONE
85	1.95	7.4	1214	1	C14H21N3O2	3-HEXYL-1-(4-CARBOXYPHENYL)-3-METHYLTRIAZENE 74109-24-3 antineoplastic
6	1.97	3.3	2589	331	"	3-HEXYL-1-(4-CARBOXYPHENYL)-3-METHYLTRIAZENE
7	-1.17	7.4	2683	1	C14H21N3O2S1	SUMATRIPTAN 103628-46-2 5-ht1d agonist, antimigrane
8 ★	1.66		429		C14H21N3O3	KARBUTILATE 4849-32-5 herbicide
9 ★	2.24		2120		C14H21N3O3	OXAMNIQUINE 21738-42-1 antischistosomal
10790 ★	1.45	7.4	2004	1	C14H21N3O3S1	TOLAZAMIDE 1156-19-0 antidiabetic
1	-0.35		1205		C14H21N3O4S1	N,N-DIME-4(N-ME-N(2-ACETAMPROPIONYL)BENZSULFONAMIDE
2	1.64		2162	459	C14H21N3O4S1	SULMEPRIDE 57479-88-6 antidepressant
3 ★	-0.30	7.4	2350	1	C14H21N3O4S2	DEOXYSPARSOMYCIN,S-ETHYL ANALOG ✚
4	-1.19	7.0	1812	1	C14H21N3O5	5'-AMINO-3'O-BUTYRYL-5'-DEOXYTHYMIDINE
95	-0.83	7.0	1812	1	C14H21N3O5	5'-AMINO-3'O-ISOPROPANOYL-5'-DEOXYTHYMIDINE
6	0.10	7.0	801	1	C14H21N3O6	1-(5-PIVALYLARABINOFURANOSYL)CYTOSINE
7 ★	2.73	7.4	1776	1	C14H21N5	N(2,5-DIME-PYRROL-1YL)6-(DIETAMINO)-3-PYRIDAZINAMINE 75841-81-5
8 ★	2.16	7.4	1776	1	C14H21N5O1	N(2,5-DIME-PYRROL-1YL)6(4-N(ME)ETOME)3-PYRIDAZINAMINE 75842-04-5
9 ★	1.73	7.4	1776	1	C14H21N5O1	N(2,5-DIME-PYRROL-1YL)6(N(ME)(2-OHPR))-3PYRIDIAZINAM 89937-38-2
10800	-1.56	7.4	594	1	C14H21N5O1	S-TRIAZINE,4,6-DIAMINO-1,2-H2-2,2-DIMETHYL-1-(3'-PROPOXYPHENYL)
1 ★	1.06	7.4	1776	1	C14H21N5O2	N(2,5DIME-PYRROL-1YL)6-(N(ETOH)2)-3-PYRIDIAZINAMIN 75841-94-0
2 ★	1.27	8.0	2267	306	C14H21N5O3	ALLOPURINOL,1-N,N-DIPROPYLGLYCYLOXYMETHYL 98204-09-2
3	0.87		401		C14H21N5O3S1	BUTOCIN ✚
4	-1.12		1008	354	C14H21N5O5	NF-167 ✚
5	0.66	7.4	2375	1	C14H22Br1N3O2	BROMOPRIDE 4093-35-0 anti-emetic
6	2.82	7.4	833	2	"	BROMOPRIDE
7 ★	2.83	7.4	1413	2	"	BROMOPRIDE
8	0.57	7.0	474	1	C14H22Cl1N1O2	BUPRANOLOL 14556-46-8 anti-adrenergic (beta-receptor)
9	0.71	7.4	2375	1	"	BUPRANOLOL
10810	1.25	7.4	2245	1	"	BUPRANOLOL
1 ★	2.80		1802		"	BUPRANOLOL
2	3.45	7.4	2245	2	"	BUPRANOLOL
3	0.46	7.4	1599	1	C14H22Cl1N3O2	METOCLOPRAMIDE 364-62-5 anti-emetic
4	0.60	7.4	2375	1	"	METOCLOPRAMIDE
15 ★	2.62	7.4	1413	2	"	METOCLOPRAMIDE
6	2.64	7.4	833	2	"	METOCLOPRAMIDE
7	1.85	5.0	831	1	C14H22Cl1N5	N1-P-CHLOROPHENYL-N5-HEXYLBIGUANIDE
8 ★	2.28	7.4	1804	2	C14H22F1N3O2	3-FL-4-AM-6-MEO-BENZAMIDE,N-(2-DIETAMINOET)
9 ★	3.09	7.4	1413	2	C14H22I1N3O2	BENZAMID,N(DIETAMET),2-OME,4-AM,5-IODO
10820	1.29		1352		C14H22N2O1	4-AMINOHEXANO-2',6'-XYLIDIDE
1	1.41		1352		C14H22N2O1	4-AMINOHEXANO-2',6'-XYLIDIDE
2 ★	1.56		1622		C14H22N2O1	BENZAMIDE,N-(3-DIETHYLAMINOPROPYL) 66999-80-2
3	1.28	7.4	2320	1	C14H22N2O1	LIDOCAINE 137-58-6 anesthetic (topical)
4	1.38	7.4	1401	1	"	LIDOCAINE
25	1.63	7.4	2409	1	"	LIDOCAINE
6	1.65	7.4	1209	1	"	LIDOCAINE
7	1.84	7.4	1414	1	"	LIDOCAINE
8	2.04	7.4	2409	1	"	LIDOCAINE
9 ★	2.26	7.4	365	2	"	LIDOCAINE

	logP	pH	Ref.	Note	MF	Name / CAS / activity
10830 ★	3.15		29		C14H22N2O2	BENZOIC ACID,P-PR-AMINO,N,N-DIMETHYLAMINOETHYL ESTER
1	1.25	7.4	2540	1	C14H22N2O2	DIETHYLAMINOPROP-2YLCARBANILATE
2 ★	1.98		1622		C14H22N2O2	O-METHOXYBENZAMIDE,N-(2-DIETHYLAMINOETHYL) 65016-34-4
3	2.18	7.4	1804	2	"	O-METHOXYBENZAMIDE,N-(2-DIETHYLAMINOETHYL)
4 ★	0.22		2044		C14H22N2O3	ATENOLOL: META ISOMER
35	-2.52	7.0	1362	1	C14H22N2O3	ATENOLOL 29122-68-7 anti-adrenergic (beta-receptor)
6 ✓	-2.49	7.0	2401	1	"	ATENOLOL
7	-2.00	7.4	1382	1	"	ATENOLOL
8	-2.00	7.4	1746	1	"	ATENOLOL
9	-1.94	7.4	1155	1	"	ATENOLOL
10840	-1.92	7.4	1799	1	"	ATENOLOL
1	-1.82	7.4	1362	1	"	ATENOLOL
2	-1.78	7.4	1157	324	"	ATENOLOL
3	-1.74	7.4	1799	1	"	ATENOLOL
4	-1.70	7.4	2399	1	"	ATENOLOL
45	-1.64	7.4	2473	1	"	ATENOLOL
6	-1.61	7.4	2465	41	"	ATENOLOL
7	-1.60	7.4	2178	1	"	ATENOLOL
8	-1.42	7.4	2375	1	"	ATENOLOL
9	-1.40	7.4	2598	1	"	ATENOLOL
10850	-1.23	7.4	2399	1	"	ATENOLOL
1	-0.11	7.4	2245	1	"	ATENOLOL
2 ★	0.16	7.4	1700	2	"	ATENOLOL
3	0.23	7.4	1799	2	"	ATENOLOL
4	0.38	6.0	2284	1	C14H22N2O3	CARBISOCAINE,DES-HEPTYL ✚
55	1.25	7.4	2284		"	CARBISOCAINE,DES-HEPTYL
6 ★	0.74		2044		C14H22N2O3	PRACTOLOL: META ISOMER
7	-2.05	7.0	474	1	C14H22N2O3	PRACTOLOL 6673-35-4 anti-adrenergic (beta-receptor)
8	-1.66	7.5	1563	1	"	PRACTOLOL
9	-1.60	7.0	2401	1	"	PRACTOLOL
10860	-0.97	7.4	2473	1	"	PRACTOLOL
1 ★	0.79		163		"	PRACTOLOL
2	1.63		1205		C14H22N2O3S1	N,N-DIME-4(N-ME-N-PENTANOYL)BENZENESULFONAMIDE
3	0.79	7.4	2272	1	C14H22N2O3S1	P-TOLUENESULFONAMIDE,N-METHYL-N-((N,N-DIETHYLAMINOMETHYL)CARBONYL
4	2.60	5.9	1351		C14H22N2O3S1	PIPROZOLIN 17243-64-0 choleretic
65	2.65	7.4	1351	1	"	PIPROZOLIN
6	2.25	3.5	1267	1	C14H22N2O4	ETHYL-2-ME-2-CYHEXENYL-6-METHYLMALONURATE
7	-1.08	7.2	846	314	C14H22N2O4S1	6-(N-HEXANOYLAMINO)PENICILLANIC ACID
8 ★	2.37		1712		"	6-(N-HEXANOYLAMINO)PENICILLANIC ACID
9	1.80	3.5	1267	1	C14H22N2O5	METHOXYMETHYL-2-ME-2-CYHEXENYL-6-METHYLMALONURATE
10870 ✓	-1.42		2397	101	C14H22N3O1	IMIDAZOLIUM CHLORIDE,1-T-BUTYL-2-HYDROXYIMINOMETHYL-3-HEXYN-5YL
1 ✓	-0.62		2397	101	C14H22N3O2	IMIDAZOLIUM CHLORIDE,1-METHYL-2-HYDROXYIMINOMETHYL-3-(1,2-BIS-ACRYL)ETHOXYMETHYL 117983-18-3
2 ★	2.33		1967		C14H22N4	4-HEPTANONE,N-PHENYLGUANYLHYDRAZONE 97183-45-4
3 ★	3.69	6.6	2590	459	C14H22N4O1	1-(4-CARBAMOYLPHENYL)-3-METHYL-3-HEXYLTRIAZENE 89530-00-7 antineoplastic
4 ★	-0.06		2518	122	C14H22N4O3	2',3'-DIDEOXYCYTIDINE,N4-(DIETHYLAMINO)METHYLENE
75	1.84	7.4	2477	1	C14H22N4O3	A[(2-NITROIMIDAZOL-1-YL)METHYL]-9-AZABICYCLO[6.1.0]NONANE-9-ETHANOL 120277-96-5
6 ★	2.11	7.4	1413	2	C14H22N4O4	BENZAMIDE,N(DIETAMINOET),2-OME,4-AMINO,5-NITRO 4093-42-9
7	0.87	7.4	2491	1	C14H22N4O6S1	N(2(DIMETHYLAMINO)ETHYL),N-NE-4(1ME-1-NITRO-1-ET)-3-NITROBENZENESULFONAMIDE 126813-40-9
8 ★	1.42	7.4	2491	2	"	N(2(DIMETHYLAMINO)ETHYL),N-NE-4(1-ME-1-NITRO-1-ET)-3-NITROBENZENESULFONAMIDE
9	0.43		1045		C14H22N6O2	6(2,2-DIME-5-ME-1,3-DIOXAN-5YL)AM-2,4-DIAZIRID-TRIAZ. ✚
10880	-0.52		1045		C14H22N6O3	6(2,2-DIME-5-(HOME)1,3-DIOXAN-5YL)AM-2,4-DIAZIR-TRIAZ.
1 ★	4.36		594	401	C14H22O1	2,6-DI-S-BUTYLPHENOL 5510-99-6
2 ★	4.92		594		C14H22O1	PHENOL,2,6-DI-T-BUTYL 128-39-2
3 ★	5.09		1522		C14H22O2	(1R)-TR-CROTYL-CHRYSANTHEMATE ✚
4 ★	4.89		1522		C14H22O2	(1R)-TR-METHYLALLYL-CHRYSANTHEMATE
85	0.79		1937	459	C14H22O3	4-HYDROXY-2-PYRONE,3,5-DIBUTYL-6-METHYL 98393-94-3
6 ★	4.53		2011		C14H22O4S2	PROPANEDIOIC ACID,1,3-DITHIOLAN-2-YLIDINE,DI-I-BUTYL ESTER 50780-71-7
7 ★	4.01		2011		C14H22O4S2	PROPANEDIOIC ACID,1,3-DITHIOLAN-2-YLIDINE,DI-T-BUTYL ESTER 50780-72-8
8 ★	4.60		2011		C14H22O4S2	PROPANEDIOIC ACID,1,3-DITHIOLAN-2-YLIDINE,DIBUTYL ESTER 50780-69-3
9 ★	2.12		1589		C14H22O5S1	I-PENTYL-(3,4,5-TRIMETHOXY)PHENYL-SULFONE
10890	-0.18		2397	101	C14H23F3N5O6S1	IMIDAZOLIUM CHLORIDE ANALOG ✚
1	6.11		442	2	C14H23N1	P-NONYLPYRIDINE
2	0.72	1.0	547	342	C14H23N1O1	2-N,N-DIPROPYLAMINO-1-PHENYLETHANOL HYDROCHLORIDE
3 ★	3.79		2045		C14H23N1O1	MOXISYLYTE,DESACETYLOXY ✚
4	0.75	7.4	1382	1	C14H23N1O2	1(2,3-DIMEPHENOXY)-3-I-PROPYLAMINOPROPANOL-2
95	3.37	8.0	1011	2	C14H23N1O2	2,5-DIMETHOXY-4-PROPYLAMPHETAMINE
6	2.88	7.4	2685	1	C14H23N1O2	4-PYRIDONE,3-HYDROXY-2-METHYL-1-OCTYL iron chelator
7	0.28	7.4	2261	1	C14H23N1O2	DOPAMINE,N,N-DIPROPYL
8	0.25	7.0	477	1	C14H23N1O2	KO#707 ✚
9	0.73	7.4	1354	1	"	KO#707
10900 ★	2.93		2045		C14H23N1O2	MOXISYLYTE,DESACETYL
1	2.32	8.0	868	2	C14H23N1O3	3,5-DIMETHOXY-4-BUTOXYPHENETHYLAMINE
2	-0.44	7.4	1702	1	C14H23N1O3	T-BUTYLAMINE,N-(CH2CHOHOCH2-(2-MEO-PHENYL) 37708-25-1
3	-1.30	5.7	2241	1	C14H23N1O4	BRUFACAINE 36199-78-7 anesthetic (local)
4	0.88		1381		C14H23N2O1P1	N,N'BIS(ETHYLENE)-P-(1-ADAMANTYL)PHOSPHONICDIAMIDE
5	-0.14		301		C14H23N2O1.Cl1	N1-OCTYLNICOTINAMIDECHLORIDE
6 ✓	-0.70		1729	820	C14H23N2O1.I1	3-CARBAMOYLPYRIDINIUM IODIDE,N-OCTYL 35041-48-6
7	2.94	7.4	459	1	C14H23N3O2	1-CYCLOHEXYL-3-ME-5-DIMEAMINO-6-METHYLURACIL
8	3.02	7.4	459	1	C14H23N3O2	1-ME-3-CYCLOHEXYL-5-DIMEAMINO-6-METHYLURACIL
9 ★	1.40	7.4	1804	2	C14H23N3O2	N-(DIETAMINO-2-ETHYL)-2-MEO-4-NH2-BENZAMIDE 3761-48-6
10910	1.62	7.4	833	2	"	N-(DIETAMINO-2-ETHYL)-2-MEO-4-NH2-BENZAMIDE
1 ★	4.27		1101	421	C14H23O4P1	DIBUTYLPHENYLPHOSPHATE
2 ✓	-1.76		2397	101	C14H24F3N4O5S2	IMIDAZOLIUM CHLORIDE ANALOG ✚
3 ✓	-1.43		2397	101	"	IMIDAZOLIUM CHLORIDE ANALOG
4 ★	4.22		2506		C14H24N1O4P1S3	BENSULIDE 741-58-2 herbicide
15	-1.57		1635	820	C14H24N1.Cl1	BENZYLDIMETHYLPENTYLAMMONIUMCHLORIDE
6 ✓	-0.77		1729	820	C14H24N1.I1	2-METHYLPYRIDINIUM IODIDE,N-OCTYL 13515-66-7
7	1.47	11.0	533	322	C14H24N2	N,N-DIBUTYL-3-PYRIDYLMETHYLAMINE
8	3.08	7.4	899	1	C14H24N2O3	BARBITURIC ACID,5-OCTYL-5-ETHYL
9	1.45		1381		C14H24N3O1P1	P,P-BIS(1-AZIRIDINYL)N-ADAMANTYLPHOSPHINICAMIDE 53743-43-4

		logP	pH	Ref.	Note	MF	Name / CAS / activity
10920	✓	0.04		2397	101	C14H24N3O2	IMIDAZOLIUM CHLORIDE,1-METHYL-2-HYDROXYIMINOMETHYL-3-(1-CYCLOHEXYLETHOXY)METHYL **117941-62-5**
1	✓	-2.42		2397	101	C14H24N3O4S1	IMIDAZOLIUM CHLORIDE,1-ETHYL-2-HYDROXYIMINOMETHYL-3-(2-METHYLSULFONYLCYCLOHEXYLOXY)METHYL
2	✓	-0.83		2426		C14H24N4O2Pt1.(N1O3)2	PLATINUM,BIS(P-DIMETHYLAMINOPYRIDONIO)-DIAQUA-DINITRATE
3	✓	-0.92	3.0	594	314	C14H24N4O7	N-T-BOC-ALA-GLY-GLY-GLY
4		-0.25		964		C14H24N6O4	N2-(HO2CCH2CH2CO2CH2CH2)-PENTAMETHYLMELAMINE
25	★	4.36		1422	462	C14H25N1	1-AMINOADAMANTANE,3,5-DIETHYL
6	★	4.33	5.0	2425	1	C14H25N1O2	2-PYRROLIDONE,1-DECANOYL **33602-03-8**
7		0.80	7.0	2484	1	C14H25N2O1	2,5-DIHYDROPYRROLE,1-OXO-2,2,5,5-TETRAMETHYL-3-(PIPERIDIN-1-YL)METHYL
8		1.88	12.0	2484	1	"	2,5-DIHYDROPYRROLE,1-OXO-2,2,5,5-TETRAMETHYL-3-(PIPERIDIN-1-YL)METHYL
9		0.98	7.0	2484	1	C14H25N2O2	1,2,3,6-TETRAHYDROPYRIDINE,1-OXY-2,2,6,6-TETRAMETHYL-4-(MORPHOLIN-1YL)METHYL
10930		1.14	12.0	2484	1	"	1,2,3,6-TETRAHYDROPYRIDINE,1-OXY-2,2,6,6-TETRAMETHYL-4-(MORPHOLIN-1YL)METHYL
1	✓	-2.42		2397	101	C14H25N4O3	IMIDAZOLIUM CHLORIDE,1-PROPYL-2-HYDROXYIMINOMETHYL-3-(2-MORPHOLIN-1YLETHOXY)METHYL
2	✓	-0.63		2397	101	C14H25N4O4	IMIDAZOLIUM CHLORIDE,1-PROPYL-2-HYDROXYIMINOMETHYL-3-(1-ETHYL-2-METHYL-2-NITROPROPOXY)METHYL
3	✓	-2.60		2397	101	C14H25N4O6S1	IMIDAZOLIUM CHLORIDE,1-(4-METHYLSULFONYLBUTYL-2-HYDROXYIMINOMETHYL-3-(2-NITROPROPOXY)METHYL
4	✓	0.88		2397	101	C14H26N3O2	IMIDAZOLIUM CHLORIDE,1-METHYL-2-HYDROXYIMINOMETHYL-3-OCTYLOXYMETHYL **91900-14-0**
35	✓	0.08		2397	101	C14H26N3O2	IMIDAZOLIUM CHLORIDE,1-PROPYL-2-HYDROXYIMINOMETHYL-3-(1,2,2-TRIMETHYLPROPOXY)METHYL
6		0.88	7.6	2003	314	C14H26N3O2.Cl1	IMIDAZOLIUM CHLORIDE,1-OCTYLOXYMETHYL-2-HYDROXYIMINOMETHYL-3-METHYL **91900-14-0**
7	★	-0.60	7.2	2553		C14H26N4O4	AC-ALA-GLY-ALA-NTBU
8		-0.12	7.2	170	314	C14H26N6O2	DIAZINEDICARBOXYLICACIDBIS(N'-ETHYLPIPERAZIDE)
9	★	0.68		2377		C14H27N3O3	AC-LEU-ILE-N
10940	★	4.20		1629		C14H27N3O4S2	N-(N"-BUTYL-BUTOXYCARBAMYLTHIO)METHOMYL
1		4.08		723		C14H27N5O1	2,4-DIAMINO-6-DECYLAMINO-PYRIMIDINE-3-OXIDE
2	★	-0.43		2093		C14H27N5O3P1	TEPA,3-(2,2,6,6-TETRAMETHYL-1-OXYPIPERIDINYL)UREA **96662-64-5**
3	✓	2.79		567	820	C14H28N1.C6H2N3O7	DICYCLOHEXYLDIMETHYL AMMONIUM PICRATE
4		0.34	7.5	1230	1	C14H28N2Se1	BIS(B-PIPERIDINOETHYL)SELENIDE
45		0.93	8.0	1230	1	"	BIS(B-PIPERIDINOETHYL)SELENIDE
6		-0.82	7.0	2484	1	C14H28N3O1	2,5-DIHYDROPYRROLE,2,2,5,5-TETRAHYDRO-3-(DIMETHYLAMINOETHYL)N-METHYLAMINOMETHYL
7		0.91	12.0	2484	1	"	2,5-DIHYDROPYRROLE,2,2,5,5-TETRAHYDRO-3-(DIMETHYLAMINOETHYL)N-METHYLAMINOMETHYL
8		0.41		2083		C14H28N3O4P1S1	THIOTEPA,1,4,7,10-TETRAOXAAZACYCLOPENTADECANE ANALOG **101347-46-0**
9		0.12		2083		C14H28N3O5P1	TEPA,1,4,7,10-TETRAOXAAZACYCLOPENTADECANE ANALOG **101347-44-8**
10950	★	3.30		1629		C14H28N4O3S2	N-(N"DIME-N'-HEXYL)-URYLTHIO)METHOMYL
1	★	-0.80		2093		C14H28N5O2P1	TEPA,4-(2,2,6,6-TETRAMETHYLPIPERIDINYL)URYL **96662-66-7** ✚
2	★	-0.52		2093		C14H28N5O3P1	TEPA,4-(2,2,6,6-TETRAMETHYL-1-HYDROXYPIPERIDNYL)URYL **96662-65-6**
3		0.92		1084		C14H28N6O1	METHOXYTRIAZINOKII. ✚
4		2.13		1294		C14H28O5S1	1-OCTYLTHIO-B-GALACTOPYRANOSIDE
55		1.46		1294		C14H28O6	B-OCTYLGALACTOPYRANOSIDE
6		-0.47		2083		C14H29N2O3P1S1	DI(1,1 DIMETHYL)AZARIDINTHIOPHOSPHORAMIDATE, TRIOXYETHYLENE ESTER **101347-43-7**
7		0.22		2083		C14H29N2O4P1	DI(1,1 DIMETHYL)AZARIDINYL PHOSPHORAMIDATE, TRIOXYETHYLENE ESTER **101347-41-5**
8		7.20		1906		C14H30	TETRADECANE **629-59-4**
9		-1.30		524	820	C14H32N1.I1	TRIBUTYLETHYL AMMONIUM IODIDE
10960	✓	-0.35		2426	537	C14H34N2O2Pt1.(N1O3)2	PLATINUM,BIS-CYCLOHEPTYLAMMONIO-DIAQUA-DINITRATE ✚
1		6.60		1740	459	C14H42O5Si6	TETRADECAMETHYLHEXASILOXANE **107-52-8**
2		2.52	1.0	2471	459	C15H7N5O2	XANTHONE,6-(5-TETRAZOLYL)-2-CYANO
3		3.24		1973	459	C15H8F5N1O3	5-PENTAFLUOROBENZOYL-H2-PYRROLO-PYRROLE-1-CARBOXYLIC ACID **96327-60-5**
4		-4.60	7.4	2559	1	C15H8O9	CHROMONE-2,8-DICARBOXYLIC ACID,5-METHOXY ANALOG
65	★	2.88	7.4	1626	1	C15H9Cl1F2N2O1	1,4-BENZODIAZEPIN-2-ONE,5-(2,6-DIF-PH)-8-CL
6	★	2.68	7.4	1626	1	C15H9Cl1F2N2O1	1,4-BENZODIAZEPIN-2-ONE,5-(2,6-DIF-PH)-7-CL **28910-86-3**
7		4.26		2675	459	C15H9Cl1F3N1O5	ACIFLUORFEN-METHYL
8	★	3.63	7.4	1626	1	C15H9Cl2F1N2O1	1,4-BENZODIAZEPIN-2-ONE,5-(2-F-PH)-7,8-DICL
9	★	-0.62		2342		C15H9In1O9	TRIS(PYROMECONATO)INDIUM(III) **116699-24-2**
10970	★	4.26		594		C15H9N1	ANTHRACENE-9-CARBONITRILE **1210-12-4**
1	★	5.70		850		C15H9N1S1	2-ISOTHIOCYANOANTHRACENE
2	★	3.17	7.4	1626	1	C15H10Br1Cl1N2O1	1,4-BENZODIAZEPIN-2-ONE,5-(2-BR-PH)-7-CL **63574-83-4**
3	★	3.63		1730		C15H10Br1Cl1N2O1	7BR-1,2-H2-5(M-CLPH)1,4-BENZODIAZEPIN-2-ONE **65247-13-4**
4	★	3.30	7.4	874	1	C15H10Br1Cl1N2O1	7BR-1,2-H2-5(O-CLPH)1,4-BENZODIAZEPIN-2-ONE **51753-57-2**
75	★	3.78		1730		C15H10Br1Cl1N2O1	7BR-1,2-H2-5(P-CLPH)1,4-BENZODIAZEPIN-2-ONE **65247-14-5**
6		2.54	7.4	874	1	C15H10Br1Cl1N2O2	7-BR-5(O-CLPH)-3-OH-1,4-BENZODIAZEPIN-2-ONE
7	★	2.79		591		C15H10Br1Cl1N4S1	BROTIZOLAM **57801-81-7** *hypnotic, antagonist-paf*
8	★	2.98	7.4	594	1	C15H10Br1F1N2O1	1,4-BENZODIAZEPIN-2-ONE,1,3-DIHYDRO-7-BROMO-5-(2-FLUOROPHENYL) **2647-50-9**
9	★	3.88		1730		C15H10Br2N2O1	7BR-1,2-H2-5(M-BRPH)1,4-BENZODIAZEPIN-2-ONE **65247-11-2**
10980	★	3.30		1730		C15H10Br2N2O1	7BR-1,2-H2-5(O-BRPH)1,4-BENZODIAZEPIN-2-ONE **65247-10-1**
1	★	4.04		1730		C15H10Br2N2O1	7BR-1,2-H2-5(P-BRPH)1,4-BENZODIAZEPIN-2-ONE **65247-12-3**
2	★	2.90	7.4	1626	1	C15H10Cl1F1N2O1	1,4-BENZODIAZEPIN-2-ONE,5-(4-F-PH)-7-CL **1492-96-2**
3	★	2.70	7.4	1626	1	C15H10Cl1F1N2O1	7-CL-1,3-H2-5(2-FL-PH)1,4-BENZODIAZEPIN-2-ONE **2886-65-9**
4		1.81		1582	459	C15H10Cl1F1N2O2	7-CL-1,3-H2-3-OH-5(2-FLPH)-1,4-DIAZEPIN-2-ONE **17617-60-6**
85		2.41	7.4	1626	1	C15H10Cl1N3O3	CLONAZEPAM **1622-61-3** *anticonvulsant*
6	★	2.41	7.4	461	1	"	CLONAZEPAM
7	★	3.15	7.4	1626	1	C15H10Cl2N2O1	DELORAZEPAM **2894-67-9**
8		1.86	7.4	1688	1	C15H10Cl2N2O2	LORAZEPAM **846-49-1** *tranquilizer (minor)*
9		2.38	7.4	461	1	"	LORAZEPAM
10990	★	2.39	7.4	1626	1	"	LORAZEPAM
1	★	2.51		1017		"	LORAZEPAM
2	★	2.15	7.4	1626	1	C15H10F1N3O3	1,4-BENZODIAZEPIN-2-ONE,5-(2-F-PH)-7-NITRO **2558-30-7**
3		3.15	1.0	2471	459	C15H10N4O3	XANTHONE,6-(5-TETRAZOLYL)-2-METHOXY
4	★	1.61	7.4	1626	1	C15H10N4O5	1,4-BENZODIAZEPIN-2-ONE,5-(2-NO2-PH)-7-NITRO **4980-73-8**
95		3.06	7.2	1985	314	C15H10N6	PROPANEDINITRILE,4-PHENYLAZOPHENYLHYDRAZONE **66706-98-7**
6	★	4.25	3.0	1985		"	PROPANEDINITRILE,4-PHENYLAZOPHENYLHYDRAZONE
7	★	2.90		798		C15H10O2	2-PHENYL-1,3-INDANDIONE **83-12-5** *anticoagulant*
8	★	3.56		1941		C15H10O2	4H-1-BENZOPYRAN-4-ONE,2-PHENYL **525-82-6**
9	★	3.85	1.0	594		C15H10O2	ANTHRACENE-9-CARBOXYLIC ACID **723-62-6**
11000	★	3.20		1941		C15H10O3	4H-1-BENZOPYRAN-4-ONE,2-(4-HYDROXYPHENYL) **4143-63-9**
1	★	3.62		1941		C15H10O3	4H-1-BENZOPYRAN-4-ONE,2-PHENYL-7-HYDROXY **6665-86-7**
2		3.40		1572		C15H10O4S1	7-(METHYLTHIO)XANTHONE-2-CARBOXYLIC ACID **40363-76-6**
3		1.24	7.4	447	1	C15H10O5	APIGENIN **520-36-5**
4	★	1.74	2.0	447		"	APIGENIN
5		3.37	1.0	2471	459	C15H10O5	XANTHONE,6-CARBOXY-2-METHOXY
6		1.76		1572		C15H10O5S1	TIXANOX **40691-50-7** *anti-allergic*
7	★	1.54		599		C15H10O7	MORIN **480-16-0**
8		3.11	7.4	874	1	C15H11Br1N2O1	7-BR-1,2-H2-5-PH-H-1,4-BENZODIAZEPIN-2-ONE **2894-61-3**
9	★	3.11		1730		"	7-BR-1,2-H2-5-PH-3H-1,4-BENZODIAZEPIN-2-ONE

	logP	pH	Ref.	Note	MF	Name / **CAS** / *activity*
11010	1.53	7.4	874	1	C15H11Br1N2O2	7BR-1,2-H2-5-PH-3H-1,4-BENZODIAZEPIN-2-ONE,N-O
1	4.47		2661		C15H11Cl1F3N1O4	HALOXYFOP **69806-34-4**
2 ★	4.70	5.9	590		C15H11Cl1F3N1O4	OXYFLUORFEN **42874-03-3** *herbicide*
3	1.73	7.4	1688	1	C15H11Cl1N2O1	NORDAZEPAM **1088-11-5** *tranquilizer (minor)*
4	2.93	7.4	874	1	"	NORDAZEPAM
15 ★	2.93	7.4	1626	1	"	NORDAZEPAM
6	2.38		799		C15H11Cl1N2O1	QUINAZOLIN-2-ONE,1-ME-4-PHENYL-6-CHLORO
7	2.36		799		C15H11Cl1N2O1	QUINAZOLIN-2-ONE,1-ME-4-PHENYL-7-CHLORO
8	-0.62	2.8	2211	1	C15H11Cl1N2O1	QUINOLINE,7-CHLORO-4-(4-HYDROXYANILINO)
9	1.46	7.4	461	1	C15H11Cl1N2O2	DEMOXEPAM **963-39-3** *tranquilizer (minor)*
11020	1.49		1017		"	DEMOXEPAM
1	1.99	7.4	1688	1	C15H11Cl1N2O2	OXAZEPAM **604-75-1** *tranquilizer (minor)*
2	2.17	7.4	461	1	"	OXAZEPAM
3	2.24	7.4	874	1	"	OXAZEPAM
4 ★	2.24	7.4	1626	1	"	OXAZEPAM
25	1.23	2.8	2211	1	C15H11Cl1N2O2	QUINOLINE-1-OXIDE-7-CHLORO-4-(4-HYDROXYANILINO)
6	4.60		2040	459	C15H11Cl1O2S1	THIOPHENE,2-(PENT-1,3-DIYNYL)-5-(4-ACETYLOXY-3-CHLOROBUT-1-YNYL)
7 ★	2.29		1817		C15H11Cl2N1O1	3-PYRIDINEMETHANOL,A-2-PROPYNYL,A-2,4-DICLPHENYL **89242-84-2**
8 ★	2.42		1817		C15H11Cl2N1O1	3-PYRIDINEMETHANOL,A-2-PROPYNYL,A-3,4-DICLPHENYL **89242-85-3**
9	3.56		1973	459	C15H11Cl2N1O3	5-(2,4-DICHLOROBENZOYL)-H2-PYRROLO-PYRROLE-1-CARBOXYLIC ACID **96327-58-1**
11030 ★	2.38	7.4	1626	1	C15H11F1N2O1	1,4-BENZODIAZEPIN-2-ONE,5-(2-F-PH)
1 ★	2.32	7.4	1626	1	C15H11F1N2O1	1,4-BENZODIAZEPIN-2-ONE,5-PH-7-FLUORO **2648-00-2**
2	1.87		799		C15H11F1N2O1	QUINAZOLIN-2-ONE,1-ME-4-PHENYL-6-FLUORO
3	2.59		1937	459	C15H11F2N1O3	5-(2,4-DIFLUOROBENZOYL)-H2-PYRROLO-PYRROLE-1-CARBOXYLIC ACID **96327-57-0**
4	2.47		1973	459	C15H11F2N1O3	5-(2,6-DIFLUOROBENZOYL)-H2-PYRROLO-PYRROLE-1-CARBOXYLIC ACID **96327-56-9**
35	0.58	1.0	547	342	C15H11N1	2-PHENYLQUINOLINE HYDROCHLORIDE
6 ★	3.90		536		C15H11N1	2-PHENYLQUINOLINE
7	4.14	7.4	547	1	"	2-PHENYLQUINOLINE
8 ★	4.67		547		C15H11N1O1	OXAZOLE,2,5-DIPHENYL **92-71-7**
9	2.39	7.4	712	1	C15H11N1O2	A-METHYL-G-PYROPHTHALONE
11040	1.60	7.4	712	1	C15H11N1O2	N-METHYL-G-PYROPHTHALONE
1	2.57		163	459	C15H11N1O2S1	6-PHENYLSULFONYLQUINOLINE
2 ★	2.33		1519	459	C15H11N1O2S1	8-PHENYLSULFONYLQUINOLINE
3 ★	2.34		2209		C15H11N1O4	NAPHTHALENE-1,8-CARBODIIMIDE,N-ACETYLOXYMETHYL
4 ★	2.35		547		C15H11N1O4S1	N-(9-OXO-3-THIOXANTHENYL)ACETAMIDE,10,10-DIOXIDE ✚
45 ★	5.85		850		C15H11N1S1	4-ISOTHIOCYANOSTILBENE
6	2.12	7.4	461	1	C15H11N3O3	NITRAZEPAM **146-22-5** *anticonvulsant, hypnotic*
7	2.13	7.4	1626	1	"	NITRAZEPAM
8	2.16	7.4	874	1	"	NITRAZEPAM
9	2.21	7.4	2407	1	"	NITRAZEPAM
11050 ★	2.25		1017		"	NITRAZEPAM
1 ★	5.07		1190		C15H12	9-METHYLANTHRACENE **779-02-2**
2 ★	5.08		2158	459	C15H12	PHENANTHRENE,1-METHYL **610-48-0**
3 ★	5.15		2158	459	C15H12	PHENANTHRENE-2-METHYL
4 ★	5.10		2158	459	C15H12	PHENANTHRENE-3-METHYL **613-12-7**
55	3.60		1973	459	C15H12Br1N1O3	5-(4-BROMOBENZOYL)-H2-PYRROLO-PYRROLE-1-CARBOXYLIC ACID **96327-34-3**
6	2.71		1973	459	C15H12Cl1N1O3	5-(2-CHLOROBENZOYL)-H2-PYRROLO-PYRROLE-1-CARBOXYLIC ACID **66635-87-8**
7	3.48		1973	459	C15H12Cl1N1O3	5-(3-CHLOROBENZOYL)-H2-PYRROLO-PYRROLE-1-CARBOXYLIC ACID **66635-88-9**
8	3.49		1973	459	C15H12Cl1N1O3	5-(4-CHLOROBENZOYL)-H2-PYRROLO-PYRROLE-1-CARBOXYLIC ACID **66635-89-0**
9	3.45		1973	459	C15H12Cl1N1O3	5-(4-CHLOROBENZOYL)-H2-PYRROLO-PYRROLE-2-CARBOXYLIC ACID **96327-67-2**
11060 ★	3.21		547		C15H12Cl1N1O3	M-CHLOROPHENYLHIPPURATE
1	2.13		1017		C15H12Cl1N3O1	DESMETHYLCHLORDIAZEPOXIDE
2 ★	3.75		2198		C15H12Cl1N3O4	PHENYLUREA,4-CHLORO,N'-(2-METHYL-6-NITROBENZOYL)
3	3.40		2171		C15H12Cl1N5O4	DISPERSE DYE II
4 ★	4.88		2023		C15H12Cl2	CYCLOPROPANE,1,1-DIPHENYL-2,2-DICHLORO
65 ★	1.78		1817		C15H12F1O1	3-PYRIDINEMETHANOL,A-2-PROPYNYL,A-4-FLUOROPHENYL **89242-82-0**
6	2.90		1973	459	C15H12F1O3	5-(3-FLUOROBENZOYL)-H2-PYRROLO-PYRROLE-1-CARBOXYLIC ACID **96327-33-2**
7	2.82		1937	459	C15H12F1O3	5-(4-FLUOROBENZOYL)-H2-PYRROLO-PYRROLE-1-CARBOXYLIC ACID **66635-90-3**
8 ★	2.58		547		C15H12F1O3	M-FLUOROPHENYLHIPPURATE
9 ★	2.46		547		C15H12F1O3	P-FLUOROPHENYLHIPPURATE **29736-22-9**
11070 ★	3.84		1988		C15H12F3N1O1	ACETANILIDE,N-(3-TRIFLUORMETHYLPHENYL) **1939-21-5**
1	1.14	2.7	978	340	C15H12F3N1O2	N(3-CF3-PHENYL)3-ME-ANTHRANILIC ACID, SODIUM SALT
2	1.80	2.7	978	340	C15H12F3N1O2	N(3-CF3-PHENYL)5-ME-ANTHRANILIC ACID, SODIUM SALT
3	0.31	2.7	978	340	C15H12F3N1O3	N(3-CF3-PHENYL)3-MEO-ANTHRANILIC ACID, SODIUM SALT
4	2.98		1973	459	C15H12I1O3	5-(2-IODOBENZOYL)-H2-PYRROLO-PYRROLE-1-CARBOXYLIC ACID **96327-35-4**
75	0.93		1802		C15H12I3N1O4	LIOTHYRONINE **6893-02-3** *thyroid hormone*
6	-1.90		547	370	C15H12N1O6S1.Na1	P-SODIUMSULFONYLPHENYLHIPPURATE
7 ★	3.10		1519	459	C15H12N2	2-METHYL-4-PHENYLQUINAZOLINE
8 ★	2.18	7.4	1626	1	C15H12N2O1	1,2-H2-5-PH-3H-1,4-BENZODIAZEPIN-2-ONE **2898-08-0**
9	2.28	7.4	874	1	"	1,2-H2-5-PH-3H-1,4-BENZODIAZEPIN-2-ONE
11080 ★	3.50		1519	459	C15H12N2O1	2-METHOXY-4-PHENYLQUINAZOLINE
1	2.32	7.4	874	1	C15H12N2O1	4-PHENYL-6-METHYLQUINAZOLIN-2-ONE
2 ★	2.59		1439		C15H12N2O1	4H-PYRIDO(1,2-A)PYRIMIDIN-4-ONE,3-PHENYL,6-ME ✚
3 ★	2.75		1988		C15H12N2O1	ACETANILIDE,N-(3-CYANOPHENYL) **89246-40-2**
4 ★	2.82		1988		C15H12N2O1	ACETANILIDE,N-(4-CYANOPHENYL) **89246-38-8**
85 ★	2.45	7.4	2277	1	C15H12N2O1	CARBAMAZEPINE **298-46-4** *analgesic, anticonvulsant*
6 ★	1.79		799		C15H12N2O1	QUINAZOLIN-2-ONE,1-METHYL-4-PHENYL
7	1.45		675		C15H12N2O2	3,5-PYRAZOLIDINDIONE,1,2-DIPHENYL
8	1.96	7.0	1405	1	C15H12N2O2	HYDANTOIN,5,5-DIPHENYL **57-41-0** *anticonvusant*
9	2.46	7.0	2230	1	"	HYDANTOIN,5,5-DIPHENYL
11090 ★	2.47		505		"	HYDANTOIN,5,5-DIPHENYL
1	1.72		799		C15H12N2O2	QUINAZOLIN-2-ONE,1-ME-4-PHENYL-6-OH
2	1.67	7.4	447	1	C15H12N2O2S1	QUINOLINE-8-SULFONANILIDE
3 ★	1.75	2.0	447		"	QUINOLINE-8-SULFONANILIDE
4	2.63		2470	306	C15H12N2O2S1	TETRAHYDROQUINAZOLINE-4-OXO-3-THIO,3-(3-METHOXYPHENYL)
95 ★	3.08		595		C15H12N2O3	FLUORENE,2-ACETAMIDO-7-NITRO
6	1.67	7.0	1405	1	C15H12N2O3	HYDANTOIN,5-PHENYL-5-P-OH-PHENYL
7	1.72		1405			HYDANTOIN,5-PHENYL-5-P-OH-PHENYL
8	2.47		1973	459	C15H12N2O5	5-(3-NITROBENZOYL)-H2-PYRROLO-PYRROLE-1-CARBOXYLIC ACID **96327-45-6**
9	2.48		1937	459	C15H12N2O5	5-(4-NITROBENZOYL)-H2-PYRROLO-PYRROLE-1-CARBOXYLIC ACID **96327-46-7**

	logP	pH	Ref.	Note	MF	Name / CAS / activity
11100 ★	2.46		547		C15H12N2O5	M-NITROPHENYLHIPPURATE **2979-53-5**
1	1.67		555		C15H12N4O3	1-P-PYRIDINYLME-2-(5-NO2-2-FURFURILIDENE)IMIDAZOLE
2 ★	2.23		538		C15H12N4O5	5-(1-AZIRIDINYL)-2,4-DINITRO-N-PHENYLBENZAMIDE
3 ★	3.08		306		C15H12O1	BENZALACETOPHENONE **94-41-7**
4 ★	3.12		1524		C15H12O2	1(3H)-ISOBENZOFURANONE-3-BENZYL **7011-98-5**
5 ★	2.78		631		C15H12O2	9-CARBOXY-9,10-DIHYDROANTHRACENE
6	3.88		2040	459	C15H12O2	BENZENE,(7-ACETYLOXYHEPT-5-ENE-1,3-DIYNYL)
7 ★	2.88		2234		C15H12O4	ACETYLSALICYLIC ACID,PHENYL ESTER **134-55-4**
8 ★	3.51	3.4	1708		C15H13Br1O3	PHENYLACETIC ACID,3-BROMO-4-PHENYLMETHOXY **5884-48-0**
9 ★	4.35		2198		C15H13Cl1N2O3	PHENYLUREA,4-CHLORO,N'-(2-METHOXYBENZOYL)
11110 ★	2.60	3.5	2250	302	C15H13Cl1O3	PHENYLACETIC ACID,2-(4-CHLORO)BENZYLOXY
1 ★	3.43	3.5	1609	1	C15H13Cl1O3	PHENYLACETIC ACID,3-CHLORO-4-PHENYLMETHOXY **60736-83-6**
2	4.69		260		C15H13Cl1O3S1	2-OH-3-CARBOXY-5-ME-BENZTHIO-2'-CL-PHENYLETHER
3	3.20		2482		C15H13Cl2F4N3O3	M-14303 ✚
4 ★	4.85		1792		C15H13Cl2N1O1	2-PHENYLPROPANEAMIDE,N-(3,4-DICHLOROPHENYL)
15	1.69	7.4	2325	1	C15H13Cl2N1O2	PHENYLACETIC ACID,2-(2',6'-DICHLORO-3-METHYL)ANILINO **15307-71-8**
6 ★	4.93	7.4	2325	2	C15H13Cl2N1O2	PHENYLACETIC ACID,2-(2',6'-DICHLORO-3-METHYL)ANILINO
7	1.18	7.4	2325	1	C15H13Cl2N1O3	PHENYLACETIC ACID,2-(2',6'-DICHLORO-3'-METHOXY)ANILINO **127792-26-1**
8	4.63	7.4	2325	2	C15H13Cl2N1O3	PHENYLACETIC ACID,2-(2',6'-DICHLORO-3'-METHOXY)ANILINO
9	1.24	7.4	2325	1	C15H13Cl2N1O3	PHENYLACETIC ACID,2-(2',6'-DICHLORO-4-METHOXY)ANILINO **118409-80-6**
11120 ★	4.44	7.4	2325	2	C15H13Cl2N1O3	PHENYLACETIC ACID,2-(2',6'-DICHLORO-4-METHOXY)ANILINO
1	0.25	7.4	2325	1	C15H13Cl2N1O4	PHENYLACETIC ACID,2-(2',6'-DICHLORO-3'-HYDROXY-4'-METHOXY)ANILINO **106610-60-0**
2	0.51	7.4	2325	1	C15H13Cl2N1O4	PHENYLACETIC ACID,2-(2',6'-DICHLORO-3'-METHOXY-4'-HYDROXY)ANILINO **118423-38-4**
3	0.91	7.4	1687	1	C15H13F1O2	FLURBIPROFEN **5104-49-4** analgesic, anti-inflammatory, inhibitor bone loss (dental)
4 ★	4.16	7.4	1687	2	"	FLURBIPROFEN
25	4.96		331		C15H13F4N3O3	4,5-(-OCF2CF2O-)C6H3NHN=C(CN)CO-T-BU
6 ★	5.44		331		C15H13F6N3O1	3,5-DI-CF3-C6H3NHN=C(CN)CO-T-BU
7 ★	4.87		1453	459	C15H13N1	2-(3-TOLYL)INDOLIZINE
8 ★	4.96		1453	459	C15H13N1	2-(4-TOLYL)INDOLIZINE
9 ★	4.14		1453	459	C15H13N1O1	2-(3-METHOXYPHENYL)INDOLIZINE
11130 ★	4.20		1453	459	C15H13N1O1	2-(4-METHOXYPHENYL)INDOLIZINE
1 ★	1.67		1817		C15H13N1O1	3-PYRIDINEMETHANOL,A-2-PROPYNYL,A-PHENYL **89242-81-9**
2 ★	3.61		306		C15H13N1O1	CINNAMANILIDE
3	2.61		2161	468	C15H13N1O2	AZETIDINE-2-ONE,4-(4-BIPHENYLYLOXY)
4 ★	2.09		594		C15H13N1O2	BENZAMIDE,2-BENZOYL-N-METHYL **32557-55-4**
35	0.97	7.4	693	1	C15H13N1O2S1	METIAZINIC ACID **13993-65-2** anti-inflammatory, antirheumatic
6	1.71	1.4	693	2	"	METIAZINIC ACID
7	2.57		1973	459	C15H13N1O3	5-BENZOYL-H2-PYRROLO-PYRROLE-2-CARBOXYLIC ACID **96327-66-1**
8	2.72		1973	459	C15H13N1O3	KETOROLAC **74103-06-3** analgesic
9	1.21	7.0	426	1	C15H13N1O3	N-PHENYLANTHRANILIC ACID,3'-ACETYL
11140 ★	4.31	2.0	2236		C15H13N1O3	N-PHENYLANTHRANILIC ACID,3'-ACETYL
1 ★	2.31		547		C15H13N1O3	PHENYLHIPPURATE **2979-54-6**
2	2.94		1937	459	C15H13N1O4	5-(2-HYDROXYBENZOYL)-H2-PYRROLO-PYRROLE-1-CARBOXYLIC ACID **96327-41-2**
3	2.18		1973	459	C15H13N1O4	5-(3-HYDROXYBENZOYL)-H2-PYRROLO-PYRROLE-1-CARBOXYLIC ACID **96327-42-3**
4	2.17		1973	459	C15H13N1O4	5-(4-HYDROXYBENZOYL)-H2-PYRROLO-PYRROLE-1-CARBOXYLIC ACID **96327-43-4**
45 ★	1.99		501		C15H13N1O4	PHENOXYACETIC ACID,M-BENZAMIDO
6	2.58	1.0	547	342	C15H13N1O5	5AC-3(1-PHAM)ETHYLIDINE-4-OH-2H-PYRAN-2,6-DIONE
7 ★	3.58		1519	459	C15H13N3	2-METHYLAMINO-4-PHENYLQUINAZOLINE
8 ★	1.00	7.4	1626	1	C15H13N3O1	7-NH2-1,2-H2-5-PH-3H-1,4-BENZODIAZEPIN-2-ONE **4928-02-3**
9	1.50	7.4	874	1	"	7-NH2-1,2-H2-5-PH-3H-1,4-BENZODIAZEPIN-2-ONE
11150 ★	2.17		2602		C15H13N3O1S1	QUINAZOLINE-4-ONE,2,6-DIMETHYL-5-(PYRID-4YL)THIO thymidylate synthase inhibitor
1	-0.05	7.4	2640	1	C15H13N3O4S1	PIROXICAM **36322-90-4** anti-inflammatory
2	0.26	7.4	1368	1	"	PIROXICAM
3 ★	3.06	7.4	1368	2	"	PIROXICAM
4	1.15		1973	459	C15H13N4O3	5-(4-TRIAZOBENZOYL)-H2-PYRROLO-PYRROLE-1-CARBOXYLIC ACID **96327-48-9**
55	5.43		2601	300	C15H14Cl1N1O2	4-CHLOROANTHRANILIC ACID,N-(3,4-DIMETHYLPHENYL)
6	5.48		2601	300	C15H14Cl1N1O2	4-CHLOROANTHRANILIC ACID,N-(3,5-DIMETHYLPHENYL)
7	0.82	2.7	978	340	C15H14Cl1N1O2	N(2,3-DIME-PH)3-CL-ANTHRANILIC ACID, SODIUM SALT
8	1.87	2.7	978	340	C15H14Cl1N1O2	N(2,3-DIME-PH)4-CL-ANTHRANILIC ACID, SODIUM SALT
9	1.59	7.4	2325	1	C15H14Cl1N1O2	PHENYLACETIC ACID,2-(2'-CHLORO-3'-METHYL)ANILINO **127792-27-2**
11160	1.17	7.4	2325	1	C15H14Cl1N1O2	PHENYLACETIC ACID,2-(2'-CHLORO-6-METHYL)ANILINO **23189-28-8**
1	4.37	7.4	2325	2	C15H14Cl1N1O2	PHENYLACETIC ACID,2-(2'-CHLORO-6-METHYL)ANILINO
2	1.54	7.4	2325	1	C15H14Cl1N1O2	PHENYLACETIC ACID,2-(2'-METHYL-3'-CHLORO)ANILINO **37984-36-4**
3 ★	4.64	7.4	2325	2	C15H14Cl1N1O2	PHENYLACETIC ACID,2-(2'-METHYL-3'-CHLORO)ANILINO
4	1.40	7.5	1344	1	C15H14Cl1N1O6	3-DESETHYL-QUINCARBATE
65	0.76	7.5	1344	1	C15H14Cl1N1O6	9-HYDROXYMETHYLQUINCARBATE
6	4.85	7.4	901	2	C15H14Cl1N3O2	P(3-(P-CL-BENZYL)-3-ME-1-TRIAZENO)BENZOIC ACID
7 ★	-1.79	1.0	2196		C15H14Cl1N3O4S1	CEFACLOR **53994-73-3** antibacterial
8	-1.76		2464	41	"	CEFACLOR
9	1.31	6.5	321	1	C15H14Cl1N3O4S3	BENZTHIAZIDE **91-33-8** diuretic, antihypertensive
11170 ★	1.19	6.5	321	2	C15H14F3N3O4S2	3-BENZYLHYDROFLUMETHAZIDE **73-48-3** diuretic, antihypertensive
1 ★	4.29		547		C15H14N1O2P1S1	SURECIDE **13067-93-1** insecticide
2	2.39	7.4	2636	1	C15H14N2O2	2(2,4-DIMETHOXYPHENYL)-1H-PYRROLO[3,2-C]PYRIDINE
3 ★	3.50	7.4	2636	2	"	2(2,4-DIMETHOXYPHENYL)-1H-PYRROLO[3,2-C]PYRIDINE
4 ★	2.59	8.0	2005		C15H14N2O2	BENZIMIDAZOLE,2-(2,4-DIMETHOXYPHENYL)
75 ★	2.15	7.4	2636	1	C15H14N2O2S1	ISOMAZOLE,BENZIMIDAZOLE ANALOG
6 ★	3.12		1624		C15H14N2O3	3-PHENOXYBENZALDEHYDE,O-((MEAMINO)CARBONYL)OXIME
7	-0.02	7.4	1722	1	C15H14N2O3	CPA **3810-51-3** ✚
8 ★	1.30		547		C15H14N2O3	M-AMINOPHENYLHIPPURATE
9	1.01		547		C15H14N2O3	P-AMINOPHENYLHIPPURATE **30022-13-0**
11180	0.28	2.7	978	340	C15H14N2O4	N-(2,3-DIME-PHENYL)-3-NITROANTHRANILIC ACID,SODIUM SALT
1	1.14	2.7	978	340	C15H14N2O4	N-(2,3-DIME-PHENYL)-4-NITROANTHRANILICAC.,NA+
2 ★	0.84		547		C15H14N2O5S1	M-SULFONAMIDOPHENYLHIPPURATE
3 ★	3.82		538		C15H14N4	P-(3-BENZYL-3-METHYL-1-TRIAZENO)BENZONITRILE
4	0.11	7.5	652	1	C15H14N4O2S1	SULFAPHENAZOLE **526-08-9** antibacterial
85 ★	1.52		480	601	"	SULFAPHENAZOLE
6 ★	4.03		538		C15H14N4O4	P-(3-(P-NITROBENZYL)-3-ME-1-TRIAZENO)BENZOIC ACID ✚
7	-1.48	7.2	846	314	C15H14N4O8S1	6-(N-3,5-DINITROBENZOYLAMINO)PENICILLANIC ACID
8 ★	2.69		631		C15H14O2	A,A-DIPHENYLPROPIONIC ACID
9 ★	4.01		594	400	C15H14O2	PHENETHYL BENZOATE **94-47-3**

	logP	pH	Ref.	Note	MF	Name / CAS / activity
11190	0.83	7.4	693	1	C15H14O3	FENOPROFEN **31879-05-7** *anti-inflammatory, analgesic*
1 ★	2.00	3.5	2250	302	C15H14O3	PHENYLACETIC ACID,2-BENZYLOXY
2 ★	2.72	3.5	1609	1	C15H14O3	PHENYLACETIC ACID,P-PHENYLMETHOXY **6547-53-1**
3	0.73		1937	459	C15H14O4	4-HYDROXY-2-PYRONE,3,5-DIMETHYL-6-BENZOYLMETHYL **98393-96-5**
4 ★	0.36		2648		C15H14O6	CIANIDOL **154-23-4** *tmt of hepatic dis.*
95 ★	0.15		2648	"		CIANIDOL
6 ★	1.40		2506		C15H15Cl1F3N3O1	TRIFLUMIZOLE **68694-11-1** *fungicide, steroid demthylation inhib.*
7	2.92		1019		C15H15Cl1N2O2	N'-BENZYL-(4-CHLOROPHENOXY)ACETIC ACID HYDRAZIDE
8	3.20		2506		C15H15Cl1N2O2	TENORAN **1982-47-4** *herbicide*
9	2.04	7.4	455	1	C15H15Cl1N2S1	DIDESMETHYLCHLORPROMAZINE
11200	0.36	7.0	2418	2	C15H15Cl1N4O6S1	CHLORIMURON-ETHYL **94365-91-0**
1	2.51	5.0	2418	1	"	CHLORIMURON-ETHYL
2 ★	2.67	5.2	2566	302	C15H15Cl1N6S2	2-ACETYLPYRIDINE-THIOCARBONOHYDRAZONE,N-(2-CHLOROPHENYL)
3	4.25		2572	459	C15H15Cl2N1O3	O-PHENYLPIPERIDINOCARBAMATE,2,4-DICHLORO-5-(PROPYN-2-YL)OXY **87374-78-5**
4 ★	3.53		870		C15H15N1	2,6-DIMETHYLBENZOPHENONEIMINE **22627-02-7**
5	2.82		260		C15H15N1	DIBENZAZOCINE ✢
6 ★	3.31		870		C15H15N1	O,O'-DIMETHYLBENZOPHENONEIMINE
7 ★	3.71		870		C15H15N1	P,P'-DIMETHYLBENZOPHENONEIMINE
8 ★	3.14		1988		C15H15N1O1	ACETANILIDE,N-(3-METHYLPHENYL) **50916-16-0**
9 ★	3.05		1988		C15H15N1O1	ACETANILIDE,N-(4-METHYLPHENYL) **6876-65-9**
11210 ★	3.91		1523		C15H15N1O2	1,1-DIPHENYL-2-NITROPROPANE
1 ★	2.68		1987		C15H15N1O2	ACETANILIDE,N-(2-METHOXYPHENYL)
2 ★	2.86		1988		C15H15N1O2	ACETANILIDE,N-(3-METHOXYPHENYL) **50916-19-3**
3 ★	2.49		1988		C15H15N1O2	ACETANILIDE,N-(4-METHOXYPHENYL) **50916-21-7**
4 ★	2.55		2451		C15H15N1O2	BENZOQUINONE,2,3,5-TRIMETHYL-6-(PYRID-3YL)METHYL
15	0.82	2.7	978	340	C15H15N1O2	N-(2,3-DIME-PHENYL)ANTHRANILIC ACID,NA.SALT
6 ★	2.26		1838		C15H15N1O2	N-METHYLPHENOXYACETANILIDE
7	2.00	7.4	1838	1	C15H15N1O2	N-PHENYLANTHRANILIC ACID,2',3'-DIMETHYL **61-68-7** *anti-inflammatory, analgesic*
8	2.05	7.4	993	1	C15H15N1O2	N-PHENYLANTHRANILIC ACID,2',3'-DIMETHYL
9	2.37	7.0	426	1	C15H15N1O2	N-PHENYLANTHRANILIC ACID,2',3'-DIMETHYL
11220 ★	5.12	2.0	1838		C15H15N1O2	N-PHENYLANTHRANILIC ACID,2',3'-DIMETHYL
1	1.86		1158		C15H15N1O2	O-(P-METHYLBENZYL)BENZAMIDE
2	3.88		822		C15H15N1O3	4-AMINOSALICYLIC ACID,2,6-DIMETHYLPHENYL ESTER
3	-0.03	7.4	2282	1	C15H15N1O3	DOPAMINE,3-BENZOYL
4 ★	1.63		1158		C15H15N1O3	O-(P-METHOXYBENZYL)BENZAMIDE
25	-0.98	7.4	1838	1	C15H15N1O3	TOLMETIN **26171-23-3** *anti-inflammatory*
6 ★	2.79	2.0	1838		"	TOLMETIN
7 ★	2.95	7.4	2272	1	C15H15N1O3S1	P-TOLUENESULFONAMIDE,N-METHYL-N-BENZOYL
8 ★	2.32		1988		C15H15N1O4S1	ACETANILIDE,N-(3-SULFONYLOXYPHENYL)
9 ★	2.66	7.4	2636	1	C15H15N3O2	ISOMAZOLE,1-METHYL-2',4'-DIMETHOXY ANALOG
11230 ★	2.68	7.4	2636	1	C15H15N3O2	ISOMAZOLE,3-METHYL-2',4'-DIMETHOXY ANALOG
1	1.07	3.3	2590	331	C15H15N3O2	P-(3-BENZYL-3-ME-1-TRIAZENO)BENZOIC ACID **65587-38-4**
2	4.04		538		"	P-(3-BENZYL-3-ME-1-TRIAZENO)BENZOIC ACID
3 ★	3.46	6.0	2560	1	C15H15N3O2S1	THIABENDAZOLE,1-ISOBUTOXYCARBONYL
4	0.52		1571	702	C15H15N3O3	2'-HYDROXY-5'-PROPYL-4-NITROAZOBENZENE
35 ★	2.09		2105		C15H15N3O3	HYDRAZINECARBOXAMIDE,1-ACETYL-N-(4-PHENOXYPHENYL)
6 ★	1.49		2031	457	C15H15N3O3S1	IMIDAZO(1,2,A)PYRAZINE,2-(2-METHOXY-4-METHYLSULFOXYPHENYL)-4-METHOXY
7	0.87	7.4	2024	1	C15H15N3O5	1,2-AZIRIDINOMITOSINE,7-METHOXY **5047-66-5**
8 ★	0.40	8.0	2267	306	C15H15N5O3	ALLOPURINOL,1-PHENYLALANYLOXYMETHYL **98204-05-8**
9 ★	3.43		2198		C15H16	PROPANE,1,3-DIPHENYL **25167-94-6**
11240	-1.52	7.4	2018		C15H16Cl1N3O4S1	TIARAMIDE CARBOXYLIC ACID ✢
1 ★	-0.71	5.1	2018		"	TIARAMIDE CARBOXYLIC ACID
2	1.11	6.5	321	2	C15H16Cl1N3O4S3	HYDROBENTIZIDE **13957-38-5** *diuretic*
3 ★	0.21	2.1	2018		C15H16Cl1N3O5S1	TIARAMIDE-N-OXIDE,ACETIC ACID ANALOG ✢
4 ✓	0.47		2397	101	C15H16Cl2N3O4	IMIDAZOLIUM CHLORIDE,1-METHYL-2-HYDROXYIMINOMETHYL-3-PROPANOL,3,5-DICHLOROSALICYL ESTER
45 ★	4.60		2420		C15H16Cl3N3O2	PROCHLORAZ **67747-09-5** *fungicide*
6 ★	2.80		603		C15H16N2O1	1,1-DIPHENYL-3,3-DIMETHYLUREA
7 ★	2.31		1988		C15H16N2O1	ACETANILIDE,N-(3-METHYLAMINOPHENYL)
8 ★	2.16		1019		C15H16N2O2	N'-BENZYL-PHENOXYACETIC ACID HYDRAZIDE
9 ★	2.98		594		C15H16N2O3	CARBANILIDE,2,2'-DIMETHOXY
11250 ★	2.19		480	601	C15H16N2O3S1	XYLOYLSULFANILAMIDE ✢
1 ★	1.01	3.0	1638		C15H16N2O4S3	CEPHALOTHIN ANALOG(3-CH2-S-CH3) **26722-85-0**
2	1.71	7.4	606	1	C15H16N2O5	7-METHOXYMITOSENE ✢
3	1.85	2.0	300		C15H16N2O5	BARBITURIC ACID,5-ET-5-(CH2)-2-(1,4-BENZODIOXANE)
4	-0.88	7.2	846	314	C15H16N2O5S1	6-(N-PHENYLCARBAMINYLAMINO)PENICILLANIC ACID
55	1.10	7.4	606	1	C15H16N2O6	1-HYDROXY-7-METHOXYMITOSENE
6	1.46	7.4	455	1	C15H16N2S1	DIDESMETHYLPROMAZINE
7 ★	2.18		1967		C15H16N4	ACETOPHENONE,N-PHENYLGUANYLHYDRAZONE **97183-44-3**
8 ★	2.76		538		C15H16N4O1	P-(3-BENZYL-3-METHYL-1-TRIAZENO)BENZAMIDE **59708-24-6** *antineoplastic*
9	0.60	7.4	2683	1	C15H16N4O1	TRYPTAMINE,5-(3-CYCLOPROPYL-OXADIAZOLYL) *5ht-agonist*
11260 ★	2.93	7.4	2029	459	C15H16N4O2	XANTHINE,1,3-DIETHYL-8-PHENYL **75922-48-4**
1 ★	1.88		2031	457	C15H16N4O2S1	IMIDAZO(1,2,A)PYRAZINE,2-(2-METHOXY-4-METHYLSULFOXYPHENYL)-4-METHYLAMINO
2	-0.97	7.4	2029	1	C15H16N4O5S1	XANTHINE,1,3-DIETHYL-8-(4-SULFOPHENYL) **89073-47-2**
3	-0.64	7.0	801	1	C15H16N4O6	1-(5-NICOTINYLARABINOSYL)CYTOSINE
4	2.63	7.4	2028	459	C15H16N6	1,2,4-TRIAZOLE,3-AMINO-5-(2-P-METHYLBENZYLAMINO)PYRIDIN-4-YL **77314-73-9**
65	2.63	7.4	2028	459	C15H16N6	1,2,4-TRIAZOLE,3-AMINO-5-(2-PHENYLETHYLAMINO)PYRIDIN-4-YL **77314-69-3**
6 ★	3.32		893		C15H16O5	4,4'-I-PROPYLIDENE-DIPHENOL **80-05-7**
7	0.31		636		C15H16O5	VERNOLEPIN **18542-37-5**
8	-0.14		636		C15H16O5	VERNOMENIN **20107-26-0**
9	-0.64		636		C15H16O6	ELEPHANTOL ✢
11270 ★	4.23		579		C15H17Br1N4O1	2(5-BR-2-PYRIDYLAZO)-5-(DIETHYLAMINO)PHENOL **14337-53-2**
1 ✓	-0.07		2397	101	C15H17Cl1N3O4	IMIDAZOLIUM CHLORIDE,1-METHYL-2-HYDROXYIMINOMETHYL-3-PROPANOL,P-CHLOROSALICYL ESTER
2	-1.48	7.4	2018	1	C15H17Cl1N3O6S2.Na1	TIARAMIDE SULFATE,SODIUM SALT
3 ★	2.94		2506		C15H17Cl1N4	MYCLOBUTANIL **61019-78-1** *fungicide, steroid demethylation inhib.*
4	4.67		2572	459	C15H17Cl2N1O4	O-PHENYLPYRROLIDINOCARBAMATE,2,4-DICHLORO-5-I-PROPOXYCARBONYL **133636-96-1**
75 ★	3.41	7.0	2292	1	C15H17Cl2N3O2	DIOXOLANE,2-(2,4-DICHLOROPHENYL)-2-(1,2,4-TRIAZOL-1-YL)METHYL-4-ETHYL-5-METHYL
6 ★	3.10		2300	20	C15H17Cl2N3O2	ETACONAZOLE **60207-93-4** *pesticide*
7 ★	3.50		2420		C15H17Cl2N3O2	PROPICONAZOLE **60207-90-1** *fungicide (agric.)*
8 ★	2.10	7.0	2292	1	C15H17Cl2N3O3	1,3-DIOXANE,2-(2,4-DICHLOROPHENYL)-2-(1,2,4-TRIAZOL-1-YL)METHYL-4-METHYL-4-HYDROXYMETHYL
9	-2.15	7.2	2446	1	C15H17F1N4O3	ENOXACIN **74011-58-8** *antibacterial*

	logP	pH	Ref.	Note	MF	Name / CAS / activity
11280	-0.66	7.4	2530	1	"	ENOXACIN
1 ✓	-0.21		2397	101	C15H17F3N3O2	IMIDAZOLIUM CHLORIDE,1-METHYL-2-HYDROXYIMINOMETHYL-3-(2-O-TRIFLUOROMETHYLPHENYLETHOXY)METHYL
2 ★	3.35		1508		C15H17N1	BENZENEETHANEAMINE,A-BENZYL 4275-43-8
3	0.12	1.0	602	342	C15H17N1O1	2-DIPHENYLMETHOXYETHYLAMINE
4	1.80		2331	1	C15H17N1O1	2H-BENZO[B]FURO[2,3-A]QUINOLIZINE,1,3,4,6,7,12B-HEXAHYDRO 112114-09-7
85 ★	3.06		2467		C15H17N1O2S1	ANILINE,P-(2,4,6-TRIMETHYLPHENYL)SULFONYL
6 ★	2.34		2157		C15H17N1O3	2-(BENZYLOXY)-(N,N-DIALLYL)ACETAMIDE
7	2.10		656		C15H17N1O3	A,A-DIALLYLPHENYLACETAMIDE,3,4-DIOXYMETHYLENE
8	2.82		572		C15H17N1O3	N-(3,4-METHYLENEDIOXYCINNAMOYL)PIPERIDINE 23434-86-8
9 ★	2.72		2487		C15H17N1O3	NAPHTHALENE,2,3-DIHYDROXY-6-(N-BUTYL)CARBOXAMIDO
11290 ★	1.42		2157		C15H17N1O5	PYRROLIDINE,N-(2-BENZOYLOXYACETYL)-2-METHOXYCARBONYL
1 ★	1.03		2467		C15H17N1O5S1	ANILINE,P-(2,4,6-TRIMETHOXYPHENYL)SULFONYL
2 ★	0.30		2234		C15H17N1O6	ACETYLSALICYLIC ACID,(MORPHOLINEDYL)GLYCOLAMIDE ESTER 116482-80-5
3 ★	0.99		2234		C15H17N1O7	ACETYLSALICYLIC ACID,(N-(2-ETHOXYACETYL)GLYCOLAMIDE) ESTER 118247-03-3
4	2.09	7.4	2331	1	C15H17N1S1	2H-BENZO[B]THIENO[2,3-A]QUINOLIZINE,1,3,4,6,7,12B-HEXAHYDRO 29970-79-4
95	0.26	7.4	606	1	C15H17N3O6	1-HYDROXY-2-AMINO-7-METHOXYMITOSENE
6	0.61	7.0	2418	1	C15H17N3O6S1	BENSULFURON-METHYL 104466-83-3
7	2.19	5.0	2418	1	"	BENSULFURON-METHYL
8 ✓	-0.71		2397	101	C15H17N4O6	IMIDAZOLIUM CHLORIDE,1-METHYL-2-HYDROXYIMINOMETHYL-3-PROPANOL,5-NITROSALICYL ESTER
9	0.53	7.4	1485	1	C15H17N5O3	2,4-DIAMPYRIMIDINE,5-(3,5-DIMEO-4-OCH2CN)BENZYL
11300 ★	0.72	7.4	1485	2	"	2,4-DIAMPYRIMIDINE,5-(3,5-DIMEO-4-OCH2CN)BENZYL
1	-2.16		1525		C15H17N5O7S3	7B-(3-ME-THIADIAZ-5YL)THIOACETYLAMINO-CEPHAMYCIN
2 ★	-0.60		1980		C15H17N7O5S3	CEFMETAZOLE 56796-20-4 antibacterial
3	2.22		1069		C15H17O2P1S1	O-ETHYL-O-(P-TOLYL)-PHENYLPHOSPHONOTHIOATE 15453-18-6
4 ★	3.03		2421		C15H18Cl1N3O1	1-(4-CHLOROPHENYL)-2-(1,2,4-TRIAZOL-1YL)CYCLOHEPTANOL
5 ★	2.90		2482		C15H18Cl1N3O1	CYPROCONAZOLE fungicide,steroid demethylation inhib.
6	0.87	7.4	2018		C15H18Cl1N3O3S1	TIARAMIDE 32527-55-2 anti-asthmatic
7 ★	1.04	10.8	2018		"	TIARAMIDE
8 ★	0.05	7.4	2018		C15H18Cl1N3O4S1	TIARAMIDE-1-OXIDE
9 ★	-1.48	7.4	2018		C15H18Cl1N3O6S2	TIARAMIDE-O-SULPHATE
11310	4.80		2506		C15H18Cl2N2O3	OXADIAZON 19666-30-9 herbicide
1	0.75		536	800	C15H18Cl2N2O4	CYCLOPROPYLCARBOXAMINOAMPHENICOL
2	-1.87		534		C15H18N1.Br1	P-BIPHENYLTRIMETHYLAMMONIUMBROMIDE
3	-0.86	2.0	1039	210	C15H18N2	DESMETHYLPHENIRAMINE
4	1.83	7.4	2331	1	C15H18N2	INDOLO-QUINOLIZINE,HEXAHYDRO 4802-79-3
15	1.16	7.4	785	1	C15H18N2	N,N-DIPHENYLPROPYLENEDIAMINE
6 ★	2.58		1727	459	C15H18N2O3	5-PH,5-ETHYLBARBITURIC ACID,N1-I-PROPYL
7 ★	1.79		1727	459	C15H18N2O3	5-PH,5-ETHYLBARBITURIC ACID,O2-I-PROPYL
8 ★	2.22		1727	459	"	5-PH,5-ETHYLBARBITURIC ACID,O2-I-PROPYL
9 ★	3.14	0.7	962		C15H18N2O3	BARBITURIC ACID,1-ME,5-ET,5-(1-PHENYLETHYL)
11320	1.18		2089		C15H18N2O3	N-(3-DELTA-3-CARBOXYETHYLPIPERIDIN-1-YL)BENZAMIDE 104642-83-3
1	1.00		2089		C15H18N2O3	N-(3-DELTA-4-CARBOXYETHYLPIPERIDIN-1-YL)BENZAMIDE 104642-86-6
2	1.81	7.4	606	1	C15H18N2O5	1-ET-2,6-DIME-3-METHOXYAMIDO-5-MEO-INDOLOQUINONE
3	-1.29	7.2	846	314	C15H18N2O5S2	6-(N-P-TOLUENESULFONYLAMINO)PENICILLANIC ACID
4 ★	0.98	7.4	2542	1	C15H18N2O5S3	THIOPHENE-2-SULFONAMIDE,4-(4-MORPHOLIN-1-YL)METHYLPHENYLSULFONYL
25 ✓	-0.55		2397	101	C15H18N3O2	IMIDAZOLIUM CHLORIDE,1-METHYL-2-HYDROXYIMINOMETHYL-3-(3-PHENYLALLYL)OXYMETHYL 117983-39-8
6 ✓	-1.00		2397	101	C15H18N3O4	IMIDAZOLIUM CHLORIDE,1-METHYL-2-HYDROXYIMINOMETHYL-3-PROPANOL,SALICYLIC ESTER
7 ★	2.98		2506		C15H18N4	FERIMZONE fungicide
8	1.16		1713		C15H18N4O2	6-PH-2-(3-(N3-ME-UREIDO)PROPYL)-3(2H)-PYRIDAZINONE
9	1.21		1713		C15H18N4O2	6-PH-2-(4-UREIDO)BUTYL)-3(2H)-PYRIDAZINONE
11330	2.73		2424	459	C15H18N4O2	BENZOTRIAZOLE,2-CYLCOPENTYL-5-(N,N-DIACETYL)AMINO
1	0.91	7.4	1485	1	C15H18N4O3	2,4-DIAMINOPYRIMIDINE,5-(3,5-DIMEO-4-ACETYL)BENZYL
2 ★	1.10	7.4	1485	2	"	2,4-DIAMINOPYRIMIDINE,5-(3,5-DIMEO-4-ACETYL)BENZYL
3	0.76	7.4	1485	1	C15H18N4O4	2,4-DIAMINOPYRIMIDINE,5-(3,5-DIMEO-4-CO2ME)BENZYL
4 ★	0.95	7.4	1485	2	"	2,4-DIAMINOPYRIMIDINE,5-(3,5-DIMEO-4-CO2ME)BENZYL
35 ★	-2.72	7.0	2374		C15H18N4O4	TRYPTOPHANYLGLYCYLGLYCINE
6	-0.47	7.4	401	1	C15H18N4O5	MITOMYCIN-C 50-07-7 antineoplastic
7 ★	-0.40		1532		"	MITOMYCIN-C
8	-0.38	7.4	1689	1	"	MITOMYCIN-C
9	-0.38	7.4	606	1	"	MITOMYCIN-C
11340	1.03		636		C15H18O3	AMBROSIN 509-93-3
1	1.17		636		C15H18O3	AROMATICIN 5945-42-6
2 ★	3.22		306		C15H18O4	DIETHYLMALONATE,2-METHYLBENZAL
3 ★	3.55		306		C15H18O4	DIETHYLMALONATE,3-METHYLBENZAL
4 ★	2.71		306		C15H18O4	ETHYLACETOACETATE,3-ETHOXYBENZAL
45	0.87		636		C15H18O4	HELENALIN 6754-13-8
6	0.36		636		C15H18O4	MEXICANIN-1 5945-41-5
7	0.77		636		C15H18O4	PARTHENIN 508-59-8
8	1.37		2140	459	"	PARTHENIN
9 ★	3.29		306		C15H18O5	DIETHYLMALONATE,2-METHOXYBENZAL
11350 ★	3.18		306		C15H18O5	DIETHYLMALONATE,3-METHOXYBENZAL
1 ★	3.28		306		C15H18O5	DIETHYLMALONATE,4-METHOXYBENZAL
2 ★	2.21		306		C15H18O5	ETHYLACETOACETATE,2,3-DIMETHOXYBENZAL
3 ★	2.41		306		C15H18O5	ETHYLACETOACETATE,2,4-DIMETHOXYBENZAL
4 ★	2.02		306		C15H18O5	ETHYLACETOACETATE,3,4-DIMETHOXYBENZAL
55 ★	2.89		306		C15H18O6	DIETHYLMALONATE,3-METHOXY-4-HYDROXYBENZAL
6	2.31		1479	459	C15H19Br1N4O2	2,4-DIAMINO-5-(2-BR-4-MEO-5-PRO-BENZYL)PYRIMIDINE
7	2.84		1479	459	C15H19Br1N4O2	2,4-DIAMINO-5-(2-BR-5-MEO-4-PRO-BENZYL)PYRIMIDINE
8	0.32	7.4	1204	1	C15H19Br2N3	N(CYPENTYLME)N-(2,6-DIBRPH)AMINO)-2-IMIDAZOLINE
9	0.14	7.4	1204	1	C15H19Br2N3	N-CYHEXYL-N-(2,6-DIBRPH)AMINO)-2-IMIDAZOLINE
11360 ★	3.01		2237		C15H19Cl1N2O2	BENZOFURAN-3-AMINE,2,3-DIHYDRO-5-CHLORO-N-(1-PIPERIDINYL)ACETYL
1 ★	3.81		2506		C15H19Cl2N3O1	DICLOBUTRAZOL 75736-33-3 fungicide, steroid demethylation inhib.
2 ✓	-1.33		2397	101	C15H19F3N3O3S1	IMIDAZOLIUM CHLORIDE,1-METHYL-2-(A-HYDROXYBENZYL)-3-(2-N-METHYL-TRIFLUOROMETHYLSULFONAMIDO)ETHYL
3	1.89	7.4	2633	1	C15H19N1	2-DELTA-OCTAHYDROPHENANTHREN-4A-AMINE,9-METHYL nmda antagonist
4	3.40		2633	459	"	2-DELTA-OCTAHYDROPHENANTHREN-4A-AMINE,9-METHYL
65	0.76	7.4	2617	1	C15H19N1	9-METHYLENE-N-METHYLHEXAHYDROFLUOREN-4A-AMINE
6	3.79		2617	2	"	9-METHYLENE-N-METHYLHEXAHYDROFLUOREN-4A-AMINE
7	3.07		2633	459	C15H19N1	DELTA-2-OCTAHYDROPHENANTHREN-4A-AMINE,N-METHYL nmda antagonist
8 ★	2.09		1817		C15H19N1O1	3-PYRIDINEMETHANOL,A-2-PROPYNYL,A-CYCLOHEXYL 89242-79-5
9	1.85	7.4	1990	1	C15H19N1O1	N,N-DIMETHYLPROPYLAMINE,3-(1-NAPHTHOXY)

	logP	pH	Ref.	Note	MF	Name / **CAS** / *activity*
11370	0.78	7.4	2633	1	C15H19N1O1	OCTAHYDROPHENANTHREN-4A-AMINE,N-METHYL-9-OXO *nmda antagonist*
1	2.41		2633	459	"	OCTAHYDROPHENANTHREN-4A-AMINE,N-METHYL-9-OXO
2	1.41	7.4	1990	1	C15H19N1O1	PENTYLAMINE,5-(1-NAPHTHOXY)
3	1.17	7.4	2473	1	C15H19N1O1	PRONETALOL **54-80-8** *adrenergic blocker*
4 ★	3.00		1984		"	PRONETALOL
75 ★	2.81		2157		C15H19N1O3	2-(BENZOYLOXY)-(N-CYCLOHEXYL)ACETAMIDE
6 ★	2.30		2157		C15H19N1O3	2-(BENZOYLOXY)-(N-HEXAMETHYLENE)ACETAMIDE
7 ★	2.68		2142		C15H19N1O4	3,4-DIETHOXYPHENYLCARBAMATE,O-3-(3-METHYLBUTYNYL)
8	2.54		2662	459	C15H19N1O4	N-(3,4-DIETHOXYPHENYL)CARBAMATE,4-BUTYNYL ESTER *fungicide*
9 ★	1.16		2234		C15H19N1O5	ACETYLSALICYLIC ACID,N,N-DIETHYLGLYCOLAMIDE ESTER **116482-56-5**
11380	1.80	7.4	416	1	C15H19N3O2	1-ETHYL-3-PHENYL-5-DIMEAMINO-6-METHYLURACIL
1	2.47	7.4	459	1	"	1-ME-3-BENZYL-5-DIMEAMINO-6-METHYLURACIL
2	1.20	7.4	416	1	C15H19N3O2	1-ME-3-PHENYL-5-PROPYLAMINO-6-METHYLURACIL
3	2.29	7.4	416	1	C15H19N3O2	1-PHENYL-3-ETHYL-5-DIMETHYL-AMINO-6-METHYLURACIL
4	1.70	7.4	416	1	C15H19N3O2	1-PHENYL-3-ME-5-PROPYLAMINO-6-METHYLURACIL
85	-0.95	7.4	651	1	C15H19N3O2	4-(N,N-DIMETHYLAMINOACETYL)ANTIPYRENE
6	0.85	7.4	416	1	C15H19N3O3	1-(2-OHET)-3-PHENYL-5-DIMEAMINO-6-ME-URACIL
7	1.96	7.4	416	1	C15H19N3O3	1-(P-MEO-PHENYL)-3-ME-5-DIMEAMINO-6-METHYURACIL
8	1.12	7.4	416	1	C15H19N3O3	1-PHENYL-3-(2-OHET)-5-DIMEAMINO-6-ME-URACIL
9 ★	-0.15		2649		C15H19N3O5	CARBOQUONE **24279-91-2** *antineoplastic*
11390	0.00	7.2	1797	314	"	CARBOQUONE
1 ★	-0.80		1739		C15H19N3O8	O-TRIACETYL-ARA-C **6742-07-0**
2 ★	0.85	7.0	2292	1	C15H19N5O2	DIOXOLANE,2-(4-AMIDINOPHENYL)-2-(1,2,4-TRIAZOL-1-YL)METHYL-4-ETHYL
3 ★	4.38		1481		C15H20Cl1N1O1	P-CHLOROCINNAMAMIDE,B-ETHYL,N-S-BUTYL
4	2.75		565		C15H20Cl1N1O4	ET-N-CHLOROACETYL-N-(2-I-PROPOXYPHENYL)GLYCINATE
95 ★	3.20		2506		C15H20Cl1N3O1	PACLOBUTRAZOL **76738-62-0** *plant growth regulator*
6 ✓	-0.57		2397	101	C15H20Cl1N4O3S1	IMIDAZOLIUM CHLORIDE ANALOG ✚
7 ✓	-0.58		2397	101	C15H20Cl1N4O3S1	IMIDAZOLIUM CHLORIDE,1-ETHYL-2-HYDROXYIMINOMETHYL-3-(2-N-METHYL-P-CHLOROPHENYLSULFONAMIDO)ETHYL
8 ★	1.33	7.4	2647	226	C15H20Cl2N2O3	RACLOPRIDE **84225-95-6** *antipsychotic*
9	3.52		1167	459	C15H20F1O2	P-FLUOROCINNAMIC ACID,DIETHYLAMINOETHYL ESTER
11400 ★	2.26	6.8	2175	1	C15H20N2O2	3-PENTYL-3-(4-PYRIDYL)PIPERIDINE-2,6-DIONE
1 ★	1.07		1481		C15H20N2O2	A-ACETYLAMINO-N-S-BUTYL-CINNAMAMIDE
2 ★	2.24		2237		C15H20N2O2	BENZOFURAN-3-AMINE,2,3-DIHYDRO-5-METHYL-N-(1-PYRROLIDINYL)ACETYL
3 ★	2.30		2237		C15H20N2O2	BENZOFURAN-3-AMINE,2,3-DIHYDRO-N-(1-PIPERIDINYL)ACETYL
4 ★	3.30		1969		C15H20N2O2	DIETHYLCARBAMATE,2-(INDOLE-3YL)ETHYL ESTER **98154-95-1**
5	1.12		2089		C15H20N2O2	N-(3-DELTA-3-(3-HYDROXYPROPYL)PIPERIDIN-1-YL)BENZAMIDE **66611-58-3**
6	1.22		2089		C15H20N2O2	N-(3-DELTA-4-(3-HYDROXYPROPYL)PIPERIDIN-1-YL)BENZAMIDE **104642-70-8**
7 ★	1.38		2237		C15H20N2O3	BENZOFURAN-3-AMINE,2,3-DIHYDRO-5-METHYL-N-(1-MORPHOLINO)ACETYL
8 ★	1.92		2237		C15H20N2O3	BENZOFURAN-3-AMINE,2,3-DIHYDRO-6-METHOXY-N-(1-PYRROLIDINYL)ACETYL
9	3.15	7.4	899	1	C15H20N2O3	REPOSAL,N-METHYL
11410 ★	0.92		2237		C15H20N2O4	BENZOFURAN-3-AMINE-2,3-DIHYDRO-6-METHOXY-N-(1-MORPHOLINO)ACETYL
1	2.74	3.5	1267		C15H20N2O4	BENZYL-2,2-DIETHYLMALONURATE
2 ★	2.44	7.4	2004	1	C15H20N2O4S1	ACETOHEXAMIDE **968-81-0** *antidiabetic*
3 ★	2.70		1629		C15H20N2O4S1	CARBOFURAN,N-(N'ME,METHYLCARBAMIDOTHIO)
4	1.17	7.4	2542	1	C15H20N2O4S3	THIOPHENE-2-SULFONAMIDE,4-(2-I-BUTYLAMINOMETHYL)PHENYLSULFONYL
15	0.61	7.4	2542	1	C15H20N2O4S3	THIOPHENE-2-SULFONAMIDE,4-(3-I-BUTYLAMINOMETHYL)PHENYLSULFONYL
6	0.32	7.4	2542	1	C15H20N2O4S3	THIOPHENE-2-SULFONAMIDE,4-(4-BUTYLAMINOMETHYL)PHENYLSULFONYL
7	0.81	7.4	2542	1	C15H20N2O4S3	THIOPHENE-2-SULFONAMIDE,4-(4-BUTYLAMINOMETHYL)PHENYLSULFONYL
8	-0.14	7.4	2542	1	C15H20N2O4S3	THIOPHENE-2-SULFONAMIDE,4-(4-T-BUTYLAMINOMETHYL)PHENYLSULFONYL
9	0.48	7.4	2542	1	"	THIOPHENE-2-SULFONAMIDE,4-(4-T-BUTYLAMINOMETHYL)PHENYLSULFONYL
11420 ★	3.00		1629		C15H20N2O5S1	N(THIO-N,O-DIMETHYLCARBAMYL)CARBOFURAN
1	0.43	7.4	2542	1	C15H20N2O5S3	THIOPHENE-2-SULFONAMIDE,4-(3-HYDROXY-4-DIETHYLAMINOMETHYL)PHENYLSULFONYL
2	-0.09	7.4	2542	1	C15H20N2O5S3	THIOPHENE-2-SULFONAMIDE,4-(4-HYDROXY-3-DIETHYLAMINOMETHYL)PHENYLSULFONYL
3	2.74		2428	459	C15H20N2O6S1	N,N'-DIMETHYLTHIODICARBAMATE,O-3,4-METHYLENEDIOXYPHENYL-O'-BUTYL
4	0.14	2.0	1039	210	C15H20N2S1	METHAPHENILENE **493-78-7** *antihistaminic*
25 ✓	-0.45		2397	101	C15H20N3O2	IMIDAZOLIUM CHLORIDE,1-METHYL-2-HYDROXYIMINOMETHYL-3-(3-PHENYLPROPOXY)METHYL **91900-16-2**
6 ✓	-0.57		2397	101	C15H20N3O2	IMIDAZOLIUM CHLORIDE,1-PROPYL-2-HYDROXYIMINOMETHYL-3-BENZYLOXYMETHYL
7	0.85	7.6	2003	314	C15H20N3O2.Cl1	IMIDAZOLIUM CHLORIDE,1-PHENYLPROPOXYMETHYL-2-HYDROXYIMINOMETHYL-3-METHYL **91900-16-2**
8	-1.68		2397	101	C15H20N3O4S1	IMIDAZOLIUM CHLORIDE,1-METHYL-2-BENZYLOXYIMINOMETHYL-3-(3-PROPANESULFONIC ACID
9	2.79		1479	459	C15H20N4O1	2,4-DIAMINO-5-(3-I-PR-4-MEO-BENZYL)PYRIMIDINE **73264-20-7**
11430	3.14		1479	459	C15H20N4O1	2,4-DIAMINO-5-(3-MEO-4-IPR-BENZYL)PYRIMIDINE **73264-21-8**
1 ★	2.72	12.0	1471		C15H20N4O2	2,4-DIAM-5(3',5'-DIMEO-4-ET-BENZYL)PYRIMIDINE
2	1.72		1479	459	C15H20N4O2	2,4-DIAMINO-5-(3,4-DIETHOXY-BENZYL)PYRIMIDINE
3	2.52	13.0	547	224	C15H20N4O2	2,4-DIAMINO-5-(3,5-DIETHOXYBENZYL)PYRIMIDINE
4	2.02		1479	459	C15H20N4O2	2,4-DIAMINO-5-(3-MEO-4-PRO-BENZYL)PYRIMIDINE
35	1.88		1479	459	C15H20N4O2	2,4-DIAMINO-5-(3-PRO-4-MEO-BENZYL)PYRIMIDINE
6	1.19	7.4	1485	1	C15H20N4O2	2,4-DIAMINOPYRIMIDINE,5-(3-MEO-4-COHME2)BENZYL
7 ★	1.38	7.4	1485	2	"	2,4-DIAMINOPYRIMIDINE,5-(3-MEO-4-COHME2)BENZYL
8	-0.37	7.4	651	1	C15H20N4O2	4-(3-DIMEAMINO-ACETYLAMINO)ANTIPYRENE
9	1.83	7.4	2298	1	C15H20N4O2S1	1,2,4-OXADIAZOLE-3-(DIETHYLAMINOETHYL)THIOHYDROXIMATE,5-PHENYL **103499-12-3**
11440 ★	2.13	7.4	2298	2	"	1,2,4-OXADIAZOLE-3-(DIETHYLAMINOETHYL)THIOHYDROXIMATE,5-PHENYL
1	1.78	7.4	1769	212	C15H20N4O2S1	1,2,4-OXADIAZOLE-5-PHE-3-C(=NOH)SCH2CH2N(ET)2 **90507-29-2**
2 ★	2.88	7.4	2298	2	"	1,2,4-OXADIAZOLE-5-PHE-3-C(=NOH)SCH2CH2N(ET)2
3	1.60	7.4	1485	1	C15H20N4O2S1	2,4-DIAMINOPYRIMIDINE,5(3,5-DIMEO-4-ETS-)BENZYL
4 ★	1.79	7.4	1485	2	"	2,4-DIAMINOPYRIMIDINE,5(3,5-DIMEO-4-ETS-)BENZYL
45	1.13	7.6	1769	212	C15H20N4O2S1	5-PHENYL-1,3,4-OXADIAZOLE-3-C(=NOH)S2N(ET)2 **90507-27-0**
6	2.43		1769	2	"	5-PHENYL-1,3,4-OXADIAZOLE-3-C(=NOH)S2N(ET)2
7	1.15	7.4	1485	1	C15H20N4O3	2,4-DIAMINOPYRIMIDINE,5-(3,5-DIMEO-4-ETO-)BENZYL
8 ★	1.34	7.4	1485	2	"	2,4-DIAMINOPYRIMIDINE,5-(3,5-DIMEO-4-ETO-)BENZYL
9	1.00	7.4	1485	1	C15H20N4O3	2,4-DIAMINOPYRIMIDINE,5-(3,5-DIMEO-4-MEO-CH2-)BENZYL
11450	1.19	7.4	1485	2	"	2,4-DIAMINOPYRIMIDINE,5-(3,5-DIMEO-4-MEO-CH2-)BENZYL
1	-1.87	1.0	594		C15H20N4O3	2,4-DIAMINOPYRIMIDINE,5-(3-METHOXY-4-(2-METHOXY)ETHOXYBENZYL
2 ★	2.83	1.0	2196		C15H20N4O3	PYRAZOLE,3-METHYL-4-NITRO-5-(N-ADAMANT-1-YLCARBOXAMIDO)
3 ★	-1.14		2377		C15H20N4O4	AC-ASN-PHE-N
4 ★	0.60		2529		C15H20N4O4	DDI,5'-PIVALATE
55 ★	0.65		2529		C15H20N4O4	DDI,5'-VALERATE
6	1.90	4.0	360	2	C15H20N4O4S1	5-BUO-6-MEO-4-SULFANILAMIDOPYRIMIDINE
7 ★	1.82		2070		C15H20N4O5	ALLOPURINOL,1,5-DI-(BUTANOYLOXYMETHYL)
8 ★	1.60		2070		C15H20N4O5	ALLOPURINOL,2,5-DI-(BUTANOYLOXYMETHYL)
9	3.38		2140	459	C15H20O2	ALANTOLACTONE **546-43-0** *anthelmintic (nematodes)*

	logP	pH	Ref.	Note	MF	Name / CAS / activity
11460 ★	2.09		547		C15H20O2	COSTUNOLIDE 553-21-9
1	3.42		2140	459	C15H20O2	ISO-ALANTOLACTONE 470-17-7
2	1.23		636		C15H20O3	TAMAULIPINA 19888-11-0
3	1.40		636		C15H20O3	TAMAULIPINB 18045-83-5
4	0.66		636		C15H20O4	3-HYDROXYADAMSIN
65	1.62		2140	459	C15H20O4	BIPINNATIN 33649-13-7
6	0.83		636		C15H20O4	CORONOPILIN 2571-81-5
7	1.42		2140	459	"	CORONOPILIN
8	0.23		636		C15H20O4	DESACETYLCONFERTIFLORIN ✚
9 ★	1.13		2330		C15H20O4	HELENALIN,2,3-DIHYDRO ✚
11470	1.70		2140	459	C15H20O4	HYMENOLIN 20555-05-9
1 ★	0.53		2330		C15H20O4	PLENOLIN 34257-95-9
2	3.00	7.5	2651	28	C15H21Br1N2O2	BENZAMIDE,2-METHOXY-5-BROMO-N-(1-ETHYLPYRROLIDIN-2YL)METHYL
3	2.46	7.5	2651	28	C15H21Br1N2O3	BENZAMIDE,2,3-DIMETHOXY-5-BROMO-N-(1-METHYLPYRROLIDIN-2YL)METHYL
4	2.88	7.5	2651	28	C15H21Br1N2O3	BENZAMIDE,2-METHOXY-3-HYDROXY-5-BROMO-N(1-ETHYLPYRROLIDIN-2YL)METHYL
75	2.46		2216		C15H21Cl1N2O1	BUTYRAMIDE-2-(O-CHLOROPHENYL)-4-(N-PIPERIDINO)
6	2.76	7.3	1900	314	C15H21Cl1N2O2	1-PIPERIDINEETHANOL,A-METHYL,3-CLPHENYLCARBAMATE
7	2.86	7.5	2651	28	C15H21Cl1N2O2	BENZAMIDE,2-METHOXY-5-CHLORO-N-(1-ETHYLPYRROLIDIN-2YL)METHYL
8 ✓	-1.86		2397	101	C15H21Cl1N3O2S1	IMIDAZOLIUM CHLORIDE,1-METHYL-2-ETHYL-3-(2-N-METHYL-P-CHLOROPHENYLSULFONAMIDO)ETHYL
9 ★	0.36		2518	122	C15H21F1N4O3	2'-DEOXY-3'-FLUORO-CYTIDINE,N4-(PIPERIDIN-1YL)METHYLENE
11480	1.91	7.4	2651	1	C15H21I1N2O2	IODOPRIDE antagonist-dopamine-2 (antipsychotic)
1	3.23	7.5	2651	28	"	IODOPRIDE
2	2.69	7.5	2651	1	C15H21I1N2O3	BENZAMIDE,2,3-DIMETHOXY-5-IODO-N-(1-METHYLPYRROLIDIN-2YL)METHYL
3	3.44		2617	2	C15H21N1	1,N-DIMETHYLHEXAHYDROFLUOREN-4A-AMINE
4	3.80		2617	2	C15H21N1	4-METHYL-N-ETHYLHEXAHYDROFLUOREN-4A-AMINE
85	0.66	7.4	2617	1	C15H21N1	N-METHYLOCTAHYDROHOMOFLUOREN-4-AMINE
6	3.77		2617	2	"	N-METHYLOCTAHYDROHOMOFLUOREN-4-AMINE
7	1.61	7.4	2633	1	C15H21N1	OCTAHYDROPHENANTHREN-4A-AMINE,9-METHYL nmda antagonist
8	3.59		2633	459	"	OCTAHYDROPHENANTHREN-4A-AMINE,9-METHYL
9	3.25		2633	459	C15H21N1	OCTAHYDROPHENANTHREN-4A-AMINE,N-METHYL nmda antagonist
11490	0.17	7.4	2617	1	C15H21N1O1	2-HYDROXYMETHYL-N-METHYLHEXAHYDROFLUOREN-4A-AMINE
1	1.41		2617	2	"	2-HYDROXYMETHYL-N-METHYLHEXAHYDROFLUOREN-4A-AMINE
2	-1.54		534		C15H21N1O1	2-METHYL-5-ET-2'-OH-6,7-BENZOMORPHAN
3	-1.28		307	1	C15H21N1O1	2-METHYL-5-ET-2'-OH-6,7-BENZOMORPHAN
4	0.65	7.4	2617	1	C15H21N1O1	6-METHOXY-N-METHYLHEXAHYDROFLUOREN-4A-AMINE
95	3.54		2617	2	"	6-METHOXY-N-METHYLHEXAHYDROFLUOREN-4A-AMINE
6	0.21	7.4	2617	1	C15H21N1O1	7-METHOXY-N-METHYLHEXAHYDROFLUOREN-4A-AMINE
7	3.51		2617	2	"	7-METHOXY-N-METHYLHEXAHYDROFLUOREN-4A-AMINE
8	1.73	7.4	2535	1	C15H21N1O1	G-(2,4,6-TRIMETHYLBENZOYL)PROPYL-ETHANOLAMINE (DEHYDRO-FUSED)
9	0.47	7.4	2378	1	C15H21N1O1	METAZOCINE 3734-52-9 analgesic
11500	-0.14	7.4	2633	1	C15H21N1O1	OCTAHYDROPHENANTHREN-4A-AMINE,N-METHYL-3-HYDROXY nmda antagonist
1	2.37		2633	459	"	OCTAHYDROPHENANTHREN-4A-AMINE,N-METHYL-3-HYDROXY
2	0.73	7.4	2633	1	C15H21N1O1	OCTAHYDROPHENANTHREN-4A-AMINE,N-METHYL-9-HYDROXY nmda antagonist
3	1.84		2633	459	"	OCTAHYDROPHENANTHREN-4A-AMINE,N-METHYL-9-HYDROXY
4	0.31	7.4	2633R1~314		C15H21N1O1	OCTHYDROPHENANTHREN-4A-AMINE,N-METHYL-6-HYDROXY nmda antagonist
5	2.56		2633	459	"	OCTHYDROPHENANTHREN-4A-AMINE,N-METHYL-6-HYDROXY
6 ★	3.20		1481		C15H21N1O2	A-ETHOXY-N-S-BUTYL-CINNAMAMIDE
7	3.61		1167	459	C15H21N1O2	CINNAMIC ACID,DIETHYLAMINOETHYL ESTER 10369-88-7
8	2.22	7.4	2204	1	C15H21N1O2	HEXAHYDRO-1,4-NAPHTHOXAZINE,4-PROPYL-7-HYDROXY
9 ★	2.52	7.4	2204	2	"	HEXAHYDRO-1,4-NAPHTHOXAZINE,4-PROPYL-7-HYDROXY
11510	0.70	7.4	1354	1	C15H21N1O2	INDENOLOL 60607-68-3 beta-adrenergic blocker
1	0.40	7.4	2480	1	C15H21N1O2	KETOBEMIDONE 469-79-4 analgesic (narcotic)
2 ★	1.02	7.4	2480	2	"	KETOBEMIDONE
3	1.28	7.4	682	1	C15H21N1O2	MEPERIDINE 57-42-1 analgesic (narcotic)
4	1.30	7.1	554		"	MEPERIDINE
15	1.59	7.4	554	1	"	MEPERIDINE
6	1.68	7.5	554	1	"	MEPERIDINE
7	1.77	7.6	554	212	"	MEPERIDINE
8	1.86	7.7	554	1	"	MEPERIDINE
9 ★	2.45	7.7	554	2	"	MEPERIDINE
11520	2.09	7.4	2204	1	C15H21N1O2	NAXAGOLIDE 88058-88-2 antiparkinsonian, dopamine agonist
1 ★	2.44	7.4	2204	2	"	NAXAGOLIDE
2 ★	2.56		2157		C15H21N1O3	2-(BENZOYLOXY)-(N,N-DI-ISOPROPYL)ACETAMIDE
3 ★	2.65		2157		C15H21N1O3	2-(BENZOYLOXY)-(N,N-DIPROPYL)ACETAMIDE
4 ★	3.37		2487		C15H21N1O3	CINNAMAMIDE,3,4-DIHYDROXY,N-HEXYL
25	1.53	7.4	2062	1	C15H21N1O4	L-DOPA,CYCLOHEXYL ESTER
6 ★	1.79	7.4	2062	2	"	L-DOPA,CYCLOHEXYL ESTER
7 ★	1.65		1501		C15H21N1O4	METALAXYL 57837-19-1
8 ★	3.36		2487		C15H21N1O4	PROPENAMIDE,N-HEXYL,BETA-(3,4-DIHYDROXY)PHENOXY
9 ★	1.28		2157		C15H21N1O5	2-(BENZOYLOXY)-(N,N-DI-(METHOXYETHYL))ACETAMIDE
11530 ★	0.66		2157		C15H21N1O5	2-(BENZOYLOXY)-(N-DI-(2-PROPANOL))ACETAMIDE
1	0.28	7.4	2062	1	C15H21N1O5	L-DOPA,2-PYRANYLMETHYL ESTER
2 ★	0.47	7.4	2062	2	"	L-DOPA,2-PYRANYLMETHYL ESTER
3 ★	3.40		1481		C15H21N1S1	A-ETHYLTHIO-N-S-BUTYL-CINNAMAMIDE
4	-0.57	7.4	2095	1	C15H21N3O2	IMIDAZOLE,2-(4-(3-ISOPROPYLAMINO-2-HYDROXY)PROPOXY)PHENYL
35	1.58		505		C15H21N3O2	PHYSOSTIGMINE 57-47-6 cholinergic (ophthalmic)
6	2.10	7.3	1900	314	C15H21N3O2	1-PIPERIDINEETHANOL,A-METHYL,3-NITROPHENYLCARBAMATE
7	-1.76		1525		C15H21N3O7S2	7B-PROPYLTHIOACETYLAMINO-CEPHAMYCIN 57793-05-2
8 ✓	-1.46		2397	101	C15H21N4O3S1	IMIDAZOLIUM CHLORIDE,1-ETHYL-2-HYDROXYIMINOMETHYL-3-(N-METHYL-PHENYLSULFONAMIDO)ETHYL
9 ✓	-1.39		2397	101	C15H21N4O3S1	IMIDAZOLIUM CHLORIDE,1-METHYL-2-HYDROXYIMINOMETHYL-3-(2-N-ETHYL-PHENYLSULFONAMIDO)ETHYL
11540 ✓	-1.59		2397	101	C15H21N4O3S1	IMIDAZOLIUM CHLORIDE,1-METHYL-2-HYDROXYIMINOMETHYL-3-(3-N-METHYL-PHENYLSULFONAMIDOPROPYL
1 ✓	1.82	13.0	579		C15H21N5	2,4-DIAMINO-5-(3,5-DIETHYL-4-AMINOBENZYL)PYRIMIDINE
2 ✓	1.77	13.0	579		C15H21N5	2,4-DIAMINO-5-(3-METHYL-4-AMINO-5-ISOPROPYLBENZYL)PYRIMIDINE
3 ★	3.13	7.4	1776	1	C15H21N5	N(2,5-DIME-PYRROL-1YL)6(4-PIPERIDNYL)3-PYRIDIAZINAMINE 75841-90-6
4 ★	1.70	7.4	1776	1	C15H21N5O1	N(2,5-DIME-PYRROL-1YL)6(4(4-OH-PIPRIDIN)3PYRIDAZINAM 75841-98-4
45 ★	2.21	7.4	1776	1	C15H21N5O1	N(2,5DIME-PYRROL-1YL)N-ME-6(4-MORPHOLIN)3-PYRIDAZINAM 75841-99-5
6	1.50	7.4	1485	1	C15H21N5O2	2,4-DIAMINOPYRIMIDINE,5(3,5-DIETO-4-AMINO)BENZYL
7 ★	1.69	7.4	1485	2	"	2,4-DIAMINOPYRIMIDINE,5(3,5-DIETO-4-AMINO)BENZYL
8 ★	1.51	13.0	547	224	C15H21N5O2	ADITOPRIM 56066-63-8 antibacterial
9	-3.30		2397	101	C15H21N5O2	IMIDAZOLIUM CHLORIDE,1-METHYL-2-HYDROXYIMINOMETHYL-3-(4-P-CARBOXAMIDOPYRIDINIUM)BUTYL

	logP	pH	Ref.	Note	MF	Name / CAS / activity
11550 ★	1.51	7.0	1793	1	C15H21N5O3S1	N(4-(6-PURINYLTHIO)VALERYL)ALANINE,ETHYL ESTER 23374-51-8
1 ★	1.37	7.0	1793	1	C15H21N5O3S1	N(4-(6-PURINYLTHIO)VALERYL)GLYCINE,PROPYL ESTER 28610-09-5
2 ★	3.46		2468		C15H22Br1N1O1	BROMOBUTIDE
3 ★	3.13		1211		C15H22Cl1N1O2	METOLACHLOR 51218-45-2 herbicide
4 ★	2.82		2142		C15H22Cl1N1O5	3,4-DIETHOXYPHENYLCARBAMATE,O-2-(1-METHOXY-3-CHLOROPROPYL)
55	-0.42	7.6	2007	1	C15H22Cl1N2O2S1.Cl1	BENZOYLTHIOFORMOHYDROXIMATE,4-CHLORO-S-(2-DIETHYLMETHYLAMMONIO)ETHYL CHLORIDE 102941-35-5
6 ★	1.32		547		C15H22Cl1N3O9	N1(2-CLET)N1-NO-N3-(6-DEOXYTRIACETYLGLUCOSYL)UREA ✚
7	-1.17	7.6	2007	1	C15H22F1N2O2S1.Cl1	BENZOYLTHIOFORMOHYDROXIMATE,4-FLUORO-S-(2-DIETHYLMETHYLAMMONIO)ETHYL CHLORIDE 102941-40-2
8 ★	4.46		2479		C15H22F1N3O2	PHENYLBIURET,2-FLUORO-5-(3-PENTYL)
9	-1.42	7.4	2617	1	C15H22N2	2-AMINOMETHYL-N-METHYLHEXAHYDROFLUOREN-4A-AMINE
11560	1.84		2617	2	"	2-AMINOMETHYL-N-METHYLHEXAHYDROFLUOREN-4A-AMINE
1 ★	3.83		2612		C15H22N2O1	1,1-DIMETHYL-3-(4-CYCLOHEXYLPHENYL)UREA 88132-41-6
2 ★	2.74		1147		C15H22N2O1	1-(N-((N',N'-DIETHYLAMINO)ACETYL)AMINO)INDANE
3 ★	3.52	7.4	1851		C15H22N2O1	1-PIPERIDINEACETAMIDE,N-(4-ETHYLPHENYL) 65446-98-2
4	2.43		2032	459	C15H22N2O1	4-PYRIMIDONE,(5,6)-CYCLOHEPTYL-(2,3)-CYCLOOCTYL
65	2.41		2032	459	C15H22N2O1	4-PYRIMIDONE,(5,6)-CYCLOOCTYL-(2,3)-CYCLOHEPTYL
6	1.27	7.4	1209	1	C15H22N2O1	MEPIVACAINE 96-88-8 anesthetic (local)
7	1.32	7.4	2409	1	"	MEPIVACAINE
8	1.62	7.4	2409	1	"	MEPIVACAINE
9	1.75	7.4	1209	2	"	MEPIVACAINE
11570 ★	1.95		2409	2	"	MEPIVACAINE
1	1.78		2162	459	C15H22N2O1	PIQUINDONE 78541-97-6 antipsychotic
2 ★	4.08		2612		C15H22N2O2	1-(4-CYCLOHEXYLPHENYL)-3-METHOXY-3-METHYLUREA 88132-28-9
3	0.17	7.0	477	1	C15H22N2O2	1-(O-CYANOPHENOXY)-3-S-AM-AMINOPROPANOL-2
4	1.58	7.3	1900	314	C15H22N2O2	1-PIPERIDINEETHANOL,A-METHYL,PHENYLCARBAMATE 38473-70-0
75	0.28	7.5	2651	1	C15H22N2O2	BENZAMIDE,2-METHOXY-N-(1-ETHYLPYRROLIDIN-2-YL)METHYL
6	1.97	7.4	2651	28	"	BENZAMIDE,2-METHOXY-N-(1-ETHYLPYRROLIDIN-2-YL)METHYL
7 ★	2.69		2237		C15H22N2O2	BENZOFURAN-3-AMINE-5-METHYL-N-DIETHYLAMINOACETYL
8	-0.08	7.4	2274	1	C15H22N2O2	MEPINDOLOL 23694-81-7 adrenergic (beta receptor)
9	0.05	7.4	2501	1	"	MEPINDOLOL
11580 ★	2.30	13.0	2501		"	MEPINDOLOL
1	2.39	7.4	934	1	C15H22N2O2	N-(M-ETHOXYPHENYL)-3-N'-PIPERIDINOACETAMIDE
2 ★	3.17	7.4	934	2	C15H22N2O2	N-(M-ETHOXYPHENYL)-3-N'-PIPERIDINOACETAMIDE
3	3.04	7.4	934	1	C15H22N2O2	N-(O-ETHOXYPHENYL)-3-N'-PIPERIDINOACETAMIDE
4	2.14	7.4	934	1	C15H22N2O2	N-(P-ETHOXYPHENYL)-3-N'-PIPERIDINOACETAMIDE 58479-94-0
85 ★	2.82	7.4	934	2	C15H22N2O2	N-(P-ETHOXYPHENYL)-3-N'-PIPERIDINOACETAMIDE
6	2.83	7.4	1851	1	C15H22N2O2	N-(P-ETHOXYPHENYL)-3-N'-PIPERIDINOACETAMIDE
7	1.13	7.3	1356	314	C15H22N2O3	2-MEO-CARBANILATE-O-2(N-PIPERIDINYL)ETHYL 55792-05-7
8	2.69	7.3	1356	2	"	2-MEO-CARBANILATE-O-2(N-PIPERIDINYL)ETHYL
9	1.33	7.3	1356	314	C15H22N2O3	3-MEO-CARBANILATE-O-2(N-PIPERIDINYL)ETHYL 55792-06-8
11590	2.89	7.3	1356	2	"	3-MEO-CARBANILATE-O-2(N-PIPERIDINYL)ETHYL
1	0.97	7.3	1356	314	C15H22N2O3	4-MEO-CARBANILATE-O-2(N-PIPERIDINYL)ETHYL 55792-07-9
2	2.53	7.3	1356	2	"	4-MEO-CARBANILATE-O-2(N-PIPERIDINYL)ETHYL
3	-0.85	7.4	2274	1	C15H22N2O3	INDOLE-2-METHANOL-4-(3-I-PROPYLAMINO)PROPANOL-2 27748-17-0
4 ★	-1.15	7.0	2197		C15H22N2O3	LEUCINYLPHENYLALANINE 3063-05-6
95 ★	-1.17	7.0	2197		C15H22N2O3	PHENYLALANYLLEUCINE 3303-55-7
6	-0.60	7.4	2020	1	C15H22N2O3	PINDOLOL ANALOG: N-(1,1-DIMETHYL)ETHANOL 102573-75-1
7	1.43	7.6	2007	1	C15H22N2O3S1	BENZOYLTHIOFORMOHYDROXIMATE,4-METHOXY-S-(2-DIETHYLAMINO)ETHYL 102941-28-6
8 ★	-1.94	7.0	2374		C15H22N2O4	LEUCINYLTYROSINE
9 ★	-1.75	7.0	2374		C15H22N2O4	TYROSINYLLEUCINE
11600 ★	1.71		537	270	C15H22N2O5	DIETHYLAMPHENICOL ✚
1	0.77		1205		C15H22N2O5S1	N,N-DIME-4(N-ME-N(2-ETCO2-PROPIONYL)BZSULFONAMIDE
2	0.84		1205		C15H22N2O5S1	N,N-DIME-4(N-ME-N-2(ETO2C-PROPIONYL)BZSULFONAMIDE
3 ✓	-2.26		2397	101	C15H22N3O2S1	IMIDAZOLIUM CHLORIDE,1-METHYL-3-(3-N-METHYL-PHENYLSULFONAMIDO)PROPYL
4 ✓	-2.59		2397	101	C15H22N3O2S1	IMIDAZOLIUM CHLORIDE,1-METHYL-2-ETHYL-3-(2-N-METHYL-PHENYLSULFONAMIDO)ETHYL
5	2.64	7.2	1162	314	C15H22N4	2,4-DIAMINO-5-(1-ADAMANTYL)-6-METHYLPYRIMIDINE 31935-08-7
6 ★	2.64		1820			2,4-DIAMINO-5-(1-ADAMANTYL)-6-METHYLPYRIMIDINE
7 ★	1.94	7.4	1413	2	C15H22N4O2	BENZAMIDE,N(DIETAMINOET),2-OME,4-AMINO,5-CYANO 65016-46-8
8 ★	0.06		2518	122	C15H22N4O3	2',3'-DIDEOXYCYTIDINE,N4-(PIPERIDIN-1YL)METHYLENE
9	1.13	7.4	2477	1	C15H22N4O5	ETHYL 7[(2-HYDROXY-3-(2-NITROIMIDAZOL-1-YL)PROPYL)-7-AZABICYCLO[4.1.0]HEPTANE-3-CARBOXYLATE
11610	0.44		1404		C15H22N4O7	2(2-NO2-1-IMIDAZOLYL)ACETAMIDE,N,N(PROPIONYLET)
1	1.41	7.4	1485	1	C15H22N6	2,4-DIAMINOPYRIMIDINE,5-(3,5-BIS-DIMETHYLAMINO)BENZYL
2 ★	1.60	7.4	1485	2	"	2,4-DIAMINOPYRIMIDINE,5-(3,5-BIS-DIMETHYLAMINO)BENZYL
3 ★	1.94	7.4	1776	1	C15H22N6	N(2,5DIME-PYRROL-1YL)6(4(4-MEPIPERAZIN)-3-PYRIDAZINAM 75841-91-7
4	1.37	7.4	1485	1	C15H22N6O1	2,4-DIAMINOPYRIMIDINE,5-(3-MEAM-4-MEO-5-NME2)BENZYL
15 ★	1.56	7.4	1485	2	"	2,4-DIAMINOPYRIMIDINE,5-(3-MEAM-4-MEO-5-NME2)BENZYL
6 ★	3.30	3.4	1708		C15H22O4	PHENYLACETIC ACID,3-METHOXY-4-I-HEXYLOXY
7 ★	3.66		1335		C15H22O5	OCTYLGALLATE 1034-01-1
8	0.38		1294		C15H22O5S1	1-(3-PHENPROPYLTHIO)-B-GALACTOPYRANOSIDE
9	0.65		511		C15H22O6	GLUCOPYRANOSIDE,3-ISOPROPYLPHENYL(BETA)
11620	2.66		2162	459	C15H23Cl1N4O4S1	BENZAMIDE,4-CHLORO-2-METHOXY-5-(N-SULFONUREYL)-N-(1-ETHYLPYRROLID-2-YL)METHYL 68256-25-7
1	2.39	7.4	2204	1	C15H23N1O2	1-BENZOPYRAN,3-DIPROPYLAMINO-6-HYDROXY 116005-03-9 agonist-serotonin (5ht-1a)
2	3.17	7.4	2204	2	"	1-BENZOPYRAN,3-DIPROPYLAMINO-6-HYDROXY
3	2.44	7.4	2204	1	C15H23N1O2	1-BENZOPYRAN,3-DIPROPYLAMINO-8-HYDROXY
4	3.30	7.4	2204		"	1-BENZOPYRAN,3-DIPROPYLAMINO-8-HYDROXY
25	0.51	7.0	474	1	C15H23N1O2	ALPRENOLOL 13655-52-2 anti-adrenergic (beta-receptor)
6	0.76	7.5	1563	1	"	ALPRENOLOL
7	0.77	7.4	1628	1	"	ALPRENOLOL
8	0.85	7.5	1563	1	"	ALPRENOLOL
9	0.93	7.4	2375	1	"	ALPRENOLOL
11630	0.98	7.4	2399	1	"	ALPRENOLOL
1	1.00	7.4	1354	1	"	ALPRENOLOL
2	1.09	7.4	2399	1	"	ALPRENOLOL
3	1.16	7.4	2473	1	"	ALPRENOLOL
4	1.23	7.4	2598	1	"	ALPRENOLOL
35	1.34	7.4	2245	1	"	ALPRENOLOL
6 ★	3.10	13.0	1144	224	"	ALPRENOLOL
7	3.57	7.4	2245	2	"	ALPRENOLOL
8 ★	5.02		2630		C15H23N1O2	P-AMINOBENZOIC ACID,OCTYL ESTER
9	2.86	7.4	2462	1	C15H23N1O2	P-BUTYLBENZOIC ACID,N,N-DIMETHYLAMINOETHYL ESTER

	logP	pH	Ref.	Note	MF	Name / CAS / activity
11640 ★	4.06	7.4	2462	2	"	P-BUTYLBENZOIC ACID,N,N-DIMETHYLAMINOETHYL ESTER
1	0.84	7.4	2501	1	C15H23N1O2	PROCINOLOL **27325-36-6** *anti-adrenercic (beta-blocker)*
2 ★	3.07	13.0	2501		"	PROCINOLOL
3 ★	2.78	11.0	2195		C15H23N1O2	PROPANOL,3-(3-(N-PIPERIDINYLMETHYL)PHENOXY
4	3.77		866	838	C15H23N1O3	2-OCTOXY-4-AMINOBENZOIC ACID
45	-0.49	7.0	1982	1	C15H23N1O3	4-HYDROXYALPRENOLOL
6	-0.37	7.0	474	1	C15H23N1O3	OXPRENOLOL **6452-71-7** *vasodilator (coronary)*
7	-0.29	7.0	1362	1	"	OXPRENOLOL
8	0.09	7.4	2375	1	"	OXPRENOLOL
9	0.10	7.4	1628	1	"	OXPRENOLOL
11650	0.12	7.4	1354	1	"	OXPRENOLOL
1	0.13	7.4	2465	41	"	OXPRENOLOL
2	0.20	7.4	1702	1	"	OXPRENOLOL
3	0.20	7.4	2399	1	"	OXPRENOLOL
4	0.36	7.4	1362	1	"	OXPRENOLOL
55	0.37	7.4	2178	1	"	OXPRENOLOL
6	0.52	7.4	2399	1	"	OXPRENOLOL
7	0.67	7.4	2473	1	"	OXPRENOLOL
8	0.72	7.4	2245	1	"	OXPRENOLOL
9 ★	2.10	10.2	2465	41	"	OXPRENOLOL
11660	2.37	7.4	1700	2	"	OXPRENOLOL
1	2.83	7.4	2245	2	"	OXPRENOLOL
2	2.71	7.4	2462	1	C15H23N1O3	P-BUTOXYBENZOIC ACID,N,N-DIMETHYLAMINOETHYL ESTER
3 ★	3.91	7.4	2462	2	"	P-BUTOXYBENZOIC ACID,N,N-DIMETHYLAMINOETHYL ESTER
4 ★	3.46		2142		C15H23N1O4	3,4-DIETHOXYPHENYLCARBAMATE,O-BUTYL
65 ★	3.36		2142		C15H23N1O4	3,4-DIETHOXYPHENYLCARBAMATE,O-I-BUTYL
6 ★	0.55		505		C15H23N1O4	CYCLOHEXIMIDE **66-81-9** *antipsoriatic–fungicide*
7	3.31		2662	459	C15H23N1O4	N-(3,4-DIETHOXYPHENYL)CARBAMATE,S-BUTYL ESTER *fungicide*
8	3.31		2662	459	C15H23N1O4	N-(3,4-DIETHOXYPHENYL)CARBAMATE,T-BUTYL ESTER *fungicide*
9 ★	3.22		2413	459	C15H23N1O4	TETRAHYDROPYRAN-2,4-DIONE,3[1(-ETHOXYIMINO)BUTYL]-6,6-SPIRO-CYCLOPENTYL
11670 ★	3.28		2413	459	C15H23N1O4	TETRAHYDROPYRAN-2,4-DIONE,3[1-(ETHOXYIMINO)BUTYL]-6-METHYL-6-CYCLOPROPYL
1 ★	2.90		2413	459	C15H23N1O4S1	TETRAHYDROPYRAN-2,4-DIONE,3[1-(ETHOXYIMINO)BUTYL]-6,6-SPIRO-(4-THIOPYRAN) ✚
2	2.56		2662	459	C15H23N1O5	N-(3,4-DIETHOXYPHENYL)CARBAMATE,ETHOXYPROP-2YL ESTER *fungicide*
3 ★	1.29		1589		C15H23N1O6S1	(N-MORPHOLINYL)ET-(3,4,5-TRIMETHOXY)PHENYL-SULFONE
4 ★	4.60		2506		C15H23N1S1	ESPROCARB *herbicide* ✚
75	6.02		1361	459	C15H23N1S1	TENOCYCLIDINE **21500-98-1** *antagonist-nmda*
6	-0.21	7.4	821	1	C15H23N3	2-(2,6-DI-I-PROPYLPHENYLIMINO)IMIDAZOLIDINE
7	3.04	7.4	821	2	"	2-(2,6-DI-I-PROPYLPHENYLIMINO)IMIDAZOLIDINE
8	0.88	7.4	365	2	C15H23N3O1	N,N-DIET-PIPERAZINE-2-CARBOXANILIDE
9	1.02	7.4	365	2	C15H23N3O1	N,N-DIME-PIPERAZINE-2-CARBOXANILIDE,2'6'-DIME
11680	4.19		260		C15H23N3O2	2-(DIETAMINOME)-6-ME-7-NITROTETRAHYDROQUINOLINE
1	2.37	6.3	2589	459	C15H23N3O2	3-HEPTYL-1-(4-CARBOXYPHENYL)-3-METHYLTRIAZENE **74109-25-4** *antineoplastic*
2	2.58	7.4	1214	1	"	3-HEPTYL-1-(4-CARBOXYPHENYL)-3-METHYLTRIAZENE
3 ★	0.64	7.4	1804	2	C15H23N3O4S1	2-MEO-5-SO2NH2-BENZAMIDE,N(1-ETPYR-3YL)CH2NHCO ✚
4	1.31		872	459	C15H23N3O4S1	CYCLACILLIN **3485-14-1** *antibacterial*
85	-1.15	7.4	1599	1	C15H23N3O4S1	SULPIRIDE **15676-16-1** *antidepressant, d-2 antagonist*
6	-0.74	8.0	2375	1	"	SULPIRIDE
7	-0.50	6.0	1148	807	"	SULPIRIDE
8	0.52		2646	448	"	SULPIRIDE
9	2.14		2162	459	"	SULPIRIDE
11690	-0.48	7.0	1812	1	C15H23N3O5	5'-AMINO-3'O-T-BUTANOYL-5'-DEOXYTHYMIDINE
1	-0.57	7.0	1812	1	C15H23N3O5	5'-AMINO-3'O-VALEROYL-5'-DEOXYTHYMIDINE
2 ★	-0.67	7.4	2350	1	C15H23N3O5S2	SPARSOMYCIN,S-PROPYL ANALOG ✚
3	-0.46		594	314	C15H23N5	1,2-DIHYDROTRIAZINE,2,4-DIAMINO-2,2-DIMETHYL-1-(3-T-BUTYL)PHENYL
4	2.66	2.7	984	340	C15H23N5	4,6-DIAM-1,2-DIH-2,2-DIME-(1-P-BUPH)-S-TRIAZINE
95	-1.00	7.4	594	1	C15H23N5O1	4,6-DIAM-2,2-DIME-1(3-BUTOXYPH)-S-TRIAZINE
6 ★	2.03	13.0	547	224	"	4,6-DIAM-2,2-DIME-1(3-BUTOXYPH)-S-TRIAZINE
7 ★	1.30	7.4	1776	1	C15H23N5O2	N(2,5DIME-PYRROL-1YL)6(N(OHET)(2OHPR))3PYRIDIAZINAM **75841-95-1**
8	6.38		2253	459	C15H24	BENZENE,NONYL **1081-77-2**
9 ★	4.12		1693		C15H24FN1O4P1S1	ISOFENPHOS **25311-71-1** *insecticide*
11700	2.73		1352		C15H24N2O1	1-(DIETHYLAMINO)PROPIONYL-2',6'-XYLIDIDE
1	2.10	7.4	1414	1	C15H24N2O1	2(DIETAMINOET)-N-(2,6-DIMEPHENYL)PROPIONAMIDE
2	2.21		1352		C15H24N2O1	2-(DIETHYLAMINO)PROPIONYL-2',6'-XYLIDIDE
3	-0.70	4.8	1020		C15H24N2O1	FORMAMIDE,N(2,6-DIMEPH)-N-(2-DIETHYLAMINOETHYL)
4	3.11		1147		C15H24N2O1	N-(N',N'-DIPROPYLAMINO)ACETYL)BENZYLAMINE
5	1.64	7.3	1378	314	C15H24N2O1	TRIMECAINE **616-68-2** *anesthetic (local)*
6	1.69	7.4	2540	1	"	TRIMECAINE
7	2.41	7.3	1378	2	"	TRIMECAINE
8 ★	1.90	7.1	1591		C15H24N2O2	ADAMANTYLALANIN-AMIDE,N-ACETYL
9 ★	1.76		1622		C15H24N2O2	O-METHOXYBENZAMIDE,N-(3-DIETHYLAMINOPROPYL)
11710	-0.34	7.0	345		C15H24N2O2	TETRACAINE **94-24-6** *anesthetic (topical)*
1	-0.16	7.4	345		"	TETRACAINE
2	1.00	7.4	2217	1	"	TETRACAINE
3	2.34	7.4	2409	1	"	TETRACAINE
4	2.73	7.4	2409	1	"	TETRACAINE
15 ★	3.73		29		"	TETRACAINE
6	2.37	7.2	1150		C15H24N2O3	2-ME-4(O-PR)CARBANILATE-O-2(N,N-DIMETHYLAMINO)ET
7 ★	0.34		2044		C15H24N2O3	BENZYLAMINE,N-ACETYL-3-(2-HYDROXY-3-I-PROPYLAMINOPROPOXY)
8 ★	0.32		2044		C15H24N2O3	BENZYLAMINE,N-ACETYL-4-(2-HYDROXY-3-I-PROPYLAMINOPROPOXY
9	-0.22	5.7	2241	1	C15H24N2O3	CORNECAINE **3686-68-8** *anesthetic (local)*
11720	0.71		1205		C15H24N2O3S1	N,N-DIME-4(N-ME-N-HEX-4-ONYL)BENZENESULFONAMIDE
1	2.16		1205		C15H24N2O3S1	N,N-DIME-4(N-ME-N-HEXANOYL)BENZENESULFONAMIDE
2	2.93	7.4	1351	1	C15H24N2O3S1	N-PR(5(1-PIPERIDINO)THIAZOLIDON-2-YLIDENE)AC.AC.ET.
3 ★	0.35		2044		C15H24N2O4	PHENOXYACETAMIDE,N-METHYL-3-(2-HYDROXY-3-I-PROPYLAMINOPROPOXY
4 ★	0.56		2044		C15H24N2O4	PHENOXYACETAMIDE,N-METHYL-4-(2-HYDROXY-3-I-PROPYLAMINOPROPOXY)
25	-1.08	7.4	1599	1	C15H24N2O4S1	BENZAMIDE,N(DIETAMINOET),2-OME,5-SO2-ME **51012-32-9** *neuroleptic, anti-emetic*
6	0.90	7.4	1804	2	"	BENZAMIDE,N(DIETAMINOET),2-OME,5-SO2-ME
7 ★	0.90	7.4	1413	2	"	BENZAMIDE,N(DIETAMINOET),2-OME,5-SO2-ME
8	1.26		1205		C15H24N2O4S1	N,N-DIME-4(N-ME-N(2-PRO-PROPIONYL)BENZENESULFONAMID
9	1.14	7.4	2461	1	C15H24N2O5S3	THIENO[2,3-B]THIOPHENE-2-SULFONAMIDE,5-(N-METHOXYETHOXYETHYL-N-METHOXYETHYL)AMINOMETHYL

	logP	pH	Ref.	Note	MF	Name / **CAS** / *activity*
11730 ★	4.20	6.6	2590	459	C15H24N4O1	1-(4-CARBAMOYLPHENYL)-3-METHYL-3-HEPTYLTRIAZENE **89530-01-8** *antineoplastic*
1 ★	0.66	7.4	1804	2	C15H24N4O4S1	2-MEO-4-AM-5-SO2AM-BENZAMIDE,N(1-ETPYR-3YL)CH2NHCO ✚
2	2.02		2162	459	C15H24N4O4S1	BENZAMIDE,2-METHOXY-5-(N-SULFONUREYL)-N-(1-ETHYLPYRROLID-2-YL)METHYL **68256-26-8**
3	0.85		1045		C15H24N6O2	6(2,2-DIME-5-ET-1,3-DIOXAN-5YL)AM-2,4-DIAZIRID-TRIAZ.
4 ★	2.15	7.4	2481	1	C15H24N6O2S1	CIMETIDINE,1-BUTOXYCARBONYL
35 ★	1.60	7.4	2481	1	C15H24N6O2S1	CIMETIDINE,1-BUTYRYLOXYMETHYL ANALOG ✚
6	0.22		1045		C15H24N6O3	6(2-IPR-5(HOME)1,3-DIOXAN-5YL)AM-2,4-DIAZIRID.TRIAZIN
7	0.58		1045		"	6(2-IPR-5(HOME)1,3-DIOXAN-5YL)AM-2,4-DIAZIRID.TRIAZIN
8	4.17		1135	537	C15H24O1	BUTYLATED HYDROXYTOLUENE **128-37-0** *antioxidant*
9	5.76		2316	337	C15H24O1	PHENOL,2-NONYL
11740	5.61		2316	337	C15H24O1	PHENOL,3-NONYL
1	5.76		2316	337	C15H24O1	PHENOL,4-NONYL
2	4.86		2316	337	C15H24O2	1,2-DIHYDROXYBENZENE,3-NONYL
3	5.54		2316	337	C15H24O2	1,2-DIHYDROXYBENZENE,4-NONYL
4	4.10		2316	337	C15H24O2	1,3-DIHYDROXYBENZENE,4-NONYL
45	1.58		1396	459	C15H24O5	ARTEMISININE(ALPHA)
6	2.19		1396	459	"	ARTEMISININE(ALPHA)
7 ★	4.65		1564		C15H25Br1N2O3	PYRIMIDINE-2,4-DIONE,3-DECYLOXY-5-BROMO-6-ME **77317-79-4**
8 ✓	0.20		2397	101	C15H25F3N5O6S1	IMIDAZOLIUM CHLORIDE ANALOG ✚
9	0.56	7.0	477	1	C15H25N1O2	1-(3,4,5-TRIMEPHENOXY)-3-I-PRAMINOPROPANOL-2
11750	4.00	8.0	1011	2	C15H25N1O2	2,5-DIMETHOXY-4-BUTYLAMPHETAMINE
1	3.91	8.0	1011	2	C15H25N1O2	2,5-DIMETHOXY-4-T-BUTYLAMPHETAMINE
2	0.33	7.4	2261	1	C15H25N1O2	DOPAMINE,3-METHOXY-N,N-DIPROPYL
3	-0.21	7.4	1702	1	C15H25N1O2	T-BUTYLAMINE,N-(CH2CHOHC2H4-(2-MEO-PHENYL) **89789-92-4**
4	-0.82	7.0	1362	1	C15H25N1O3	METOPROLOL **37350-58-6** *anti-adrenergic (beta-receptor)*
55	-0.74	7.0	1982	1	"	METOPROLOL
6	-0.37	7.4	2375	1	"	METOPROLOL
7	-0.31	7.4	1354	1	"	METOPROLOL
8	-0.30	7.4	2399	1	"	METOPROLOL
9	-0.26	7.4	2465	41	"	METOPROLOL
11760	-0.25	7.4	1157	324	"	METOPROLOL
1	-0.01	7.4	1362	1	"	METOPROLOL
2	0.03	7.4	1746	1	"	METOPROLOL
3	0.04	7.4	1155	1	"	METOPROLOL
4	0.05	7.4	2178	1	"	METOPROLOL
65	0.07	7.4	2598	1	"	METOPROLOL
6	0.13	7.4	2473	1	"	METOPROLOL
7	0.16	7.4	2399	1	"	METOPROLOL
8	0.20	7.4	2245	1	"	METOPROLOL
9 ★	1.88	7.4	1700	2	"	METOPROLOL
11770	2.49	7.4	2245	2	"	METOPROLOL
1	-1.52	7.4	401	1	C15H25N1O6	INDICINE,N-OXIDE
2	0.46		301		C15H25N2O1.Cl1	N1-NONYLNICOTINAMIDECHLORIDE
3	-1.31	7.4	2224	1	C15H25N3O1	NORPACE,DES-PHENYL
4 ★	1.81		2093		C15H25N4O2P1	TEPA,1-ADAMANTYLAMINOCARBONYLAMIDE **65101-39-5**
75	-0.45		541		C15H25N5O6	N3-METHYL-5-(N-ME-N-BUTYLTRIAZINYL)URIDINE
6 ★	1.34		2218	20	C15H25N7S2	GUANIDINE,N,N'-BIS(5-METHYLIMIDAZOL-4YL)METHYLTHIOETHYL
7 ✓	-1.36		2397	101	C15H26F3N4O5S2	IMIDAZOLIUM CHLORIDE ANALOG ✚
8	-1.53		527	820	C15H26N1.Br1	BENZYLDIMETHYLHEXYLAMMONIUMBROMIDE
9	-0.72		527	404	C15H26N1.Br1	DECYLPYRIDINIUMBROMIDE
11780	-1.87		1635	820	C15H26N1.Cl1	BENZYLDIMETHYLHEXYLAMMONIUMCHLORIDE
1 ✓	-0.57		1729	820	C15H26N1.I1	2,4-DIMETHYLPYRIDINIUM IODIDE,N-OCTYL
2 ✓	-0.41		1729	820	C15H26N1.I1	3,4-DIMETHYLPYRIDINIUM IODIDE,N-OCTYL **53242-40-3**
3 ✓	-0.17		1729	820	C15H26N1.I1	3,5-DIMETHYLPYRIDINIUM IODIDE,N-OCTYL
4 ✓	0.11		1729	820	C15H26N1.I1	N-DECYLPYRIDINIUM IODIDE **7295-91-2**
85	3.30	7.4	899	1	C15H26N2O3	BARBITURIC ACID,5-NONYL-5-ETHYL
6 ★	1.73	7.4	1956	2	C15H26N2O3	PILOCARPIC ACID, BUTYL ESTER **92598-80-6**
7 ✓	0.22		2397	101	C15H26N3O2	IMIDAZOLIUM CHLORIDE,1-METHYL-2-HYDROXYIMINMETHYL-3-(1-CYCLOPENTYLBUTOXY)METHYL **117941-63-6**
8 ★	4.48	1.0	2196		C15H26N4O3	PYRAZOLE,3-METHYL-4-NITRO-5-(N-DECYLCARBOXAMIDO)
9	1.12	7.4	2244	1	C15H26N4O4S1	TIMOLOL,O-ACETYL
11790	2.54		508		C15H26O6	GLYCEROL,TRI-BUTYRATE
1	0.42	7.0	2484	1	C15H27N2O1	1,2,3,6-TETRAHYDROYRIDINE-1-OXY-2,2,6,6-TETRAMETHYL-4-(PIPERIDIN-1-YL)METHYL
2	1.58	12.0	2484	1	"	1,2,3,6-TETRAHYDROYRIDINE-1-OXY-2,2,6,6-TETRAMETHYL-4-(PIPERIDIN-1-YL)METHYL
3 ✓	-0.16		2397	101	C15H27N4O4	IMIDAZOLIUM CHLORIDE,1-BUTYL-2-HYDROXYIMINOMETHYL-3-(1-ETHYL-2-METHYL-2-NITROPROPOXY)METHYL
4 ✓	-1.73		2397	101	C15H27N4O6S1	IMIDAZOLIUM CHLORIDE ANALOG ✚
95 ★	5.64		401	335	C15H28Cl1N3O2	1-(2-CHLOROETHYL)-3-CYCLODODECYL-1-NITROSOUREA
6 ✓	-1.69		2397	101	C15H28N3O4	IMIDAZOLIUM CHLORIDE ANALOG ✚
7 ★	-0.51	7.2	2553		C15H28N4O4	AC-ALA-ALA-ALA-NTBU
8 ★	-0.67		2377		C15H28N4O4	AC-VAL-ALA-VAL-N
9 ★	-0.45		2377		C15H28N4O4	AC-VAL-ILE-GLY-N
11800 ★	-2.03	7.0	2374		C15H29N3O4	LEUCINYLALANYLLEUCINE
1 ★	-2.64	7.0	2374		C15H29N3O5	ISOLEUCINYLSERINYLISOLEUCINE
2 ★	-2.28	7.0	2374		C15H29N3O5	ISOLEUCINYLSERINYLLEUCINE
3 ★	-2.35	7.0	2374		C15H29N3O5	LEUCINYLSERINYLLEUCINE
4 ★	-1.99	7.0	2374		C15H29N3O5	SERINYLLEUCINYLISOLEUCINE
5 ★	-2.03	7.0	2374		C15H29N3O5	SERINYLLEUCINYLLEUCINE
6 ★	-1.97	7.0	2374		C15H29N3O5	THREONYLVALINYLLEUCINE
7	1.77	2.0	530		C15H31N1	N-DECYLPIPERIDINEHYDROCHLORIDE
8 ★	5.68		2506		C16H8Cl2F6N2O3	HEXAFLUMURON *insecticide (chitin synth.inhib.)*
9	4.27		2675	459	C16H9Cl1F3N1O7	BENZOFLUORFEN
11810 ★	5.16		2293	400	C16H10	FLUORANTHENE **206-44-0**
1 ★	4.88		536		C16H10	PYRENE **129-00-0**
2 ★	3.19	7.4	1626	1	C16H10Cl1F3N2O1	1,4-BENZODIAZEPIN-2-ONE,5-(2-CF3-PH)-7-CL **3864-49-1**
3	4.22		520		C16H10Cl1N1O2	1,4-NAPHTHOQUINONE,3-ANILINO-2-CHLORO
4	5.11		1369	459	C16H10F2O1	2,5-DI(P-FLUOROPHENYL)FURAN
15 ★	2.51	7.4	1626	1	C16H10F3N3O3	1,4-BENZODIAZEPIN-2-ONE,5-(2-CF3-PH)-7-NITRO
6 ★	3.55		306		C16H10N2	MALONONITRILE,A-PHENYLBENZAL
7 ★	3.72		579		C16H10N2O2	INDIGO **482-89-3**
8	4.40		520	20	C16H10O2	1,4-NAPHTHOQUINONE,2-PHENYL
9	2.14		46		C16H10O4S1	NAPHTHOQUINONE,2-PHENYLSULFONYL

	logP	pH	Ref.	Note	MF	Name / **CAS** / *activity*
11820 ★	3.04	7.4	1626	1	C16H11Cl3N2O1	1,4-BENZODIAZEPIN-2-ONE,1-ME-5-(2,6-DICL-PH)-7-CL
1 ★	3.18	7.4	594	1	C16H11F3N2O1	1,4-BENZODIAZEPIN-2-ONE,1,3-DIHYDRO-5-PHENYL-8-TRIFLUOROMETHYL **2730-04-3**
2 ★	2.47	7.4	1626	1	C16H11F3N2O1	1,4-BENZODIAZEPIN-2-ONE,5-(2-CF3-PH) **2730-05-4**
3 ★	3.19	7.4	1626	1	C16H11F3N2O1	1,4-BENZODIAZEPIN-2-ONE,5-(3-CF3-PH) **2285-16-7**
4 ★	3.34	7.4	1626	1	C16H11F3N2O1	1,4-BENZODIAZEPIN-2-ONE,5-(4-CF3-PH) **3894-63-1**
25 ★	3.10	7.4	1626	1	C16H11F3N2O1	1,4-BENZODIAZEPIN-2-ONE,5-PHENYL-7-CF3 **2285-16-7**
6	2.69		799		C16H11F3N2O1	QUINAZOLIN-2-ONE,1-ME-4-PHENYL-6-TRIFLUOROMETHYL
7 ★	4.20	7.4	594	1	C16H11N1	FLUORANTHENE,3-AMINO **2693-46-1**
8 ★	4.31	7.4	594	1	C16H11N1	PYRENE,1-AMINO **1606-67-3**
9 ★	2.84		520		C16H11N1O2	1,4-NAPHTHOQUINONE,2-ANILINO
11830	-0.99		520		C16H11N1O5S1	1,4-NAPHTHOQUINONE,2-ANILINO-3-SULFONATE, POTASSIUM SALT
1 ★	1.82	7.4	1626	1	C16H11N3O1	1,4-BENZODIAZEPIN-2-ONE,5-PH-7-CYANO **17562-53-7**
2 ✓	0.55		1412		C16H11N7O6	1,2,4-TRIAZOLE,3,5-DI(P-NITROBENZAMIDO)
3 ★	4.92		594	400	C16H12	1,2-DIHYDROPYRENE
4 ★	3.10	7.4	1626	1	C16H12Cl1F1N2O1	1,4-BENZODIAZEPIN-2-ONE,1-ME-5-(4-CL-PH)-7-F
35 ★	2.75	7.4	1626	1	C16H12Cl1F1N2O1	7CL-1,3-H2-1-ME-5(2-FLPH)1,4-BENZODIAZEPIN-2-ON **3900-31-0** *tranquilizer (minor)*
6	2.09		1582	459	C16H12Cl1F1N2O2	FLUTEMAZEPAM **52391-89-6**
7 ★	2.56	7.4	594	1	C16H12Cl1N3O3	1,4-BENZODIAZEPIN-2-ONE,1-METHYL-7-NITRO-5-(2-CHLOROPHENYL)
8 ★	2.72	7.4	1626	1	C16H12Cl1N3O3	MECLONAZEPAM **58662-84-3** *antischistosomal*
9 ★	3.12	7.4	1626	1	C16H12Cl2N2O1	7-CL-1,3-H2-1-ME-5(2-CL-PH)1,4-BENZODIAZEPIN-2-ONE **2894-68-0**
11840	2.50		1582	459	C16H12Cl2N2O2	LORMETAZEPAM **848-75-9** *sedative, hypnotic*
1	1.13		910		C16H12Cl2N4	1,2-BIS(5-CHLORO-2-BENZIMIDAZOLYL)ETHANE
2	1.21		910		C16H12Cl2N4O2	1,2-DIHYDROXY-1,2-BIS(5-CL-2-BENZIMIDAZOLYL)ETHANE
3 ★	3.14	7.4	1626	1	C16H12F1N2O1	1,4-BENZODIAZEPIN-2-ONE,1-ME-5-(2-F-PH)-7-IODO **34932-78-0**
4 ★	1.68	7.4	1688	1	C16H12F1N3O3	FLUNITRAZEPAM **1622-62-4** *hypnotic*
45 ★	2.06	7.4	1626	1	"	FLUNITRAZEPAM
6 ★	2.41	7.4	594	1	C16H12F2N2O1	1,4-BENZODIAZEPIN-2-ONE,1,3-DIHYDRO-1-METHYL-7-FLUORO-5-(2-FLUOROPHENYL) **2024-34-2**
7	3.69		1973	459	C16H12F3N1O3	5-(3-TRIFLUOROMETHYLBENZOYL)-H2-PYRROLO-PYRROLE-1-CARBOXYLIC ACID **96327-51-4**
8	3.71		1973	459	C16H12F3N1O3	5-(4-TRIFLUOROMETHYLBENZOYL)-H2-PYRROLO-PYRROLE-1-CARBOXYLIC ACID **96327-52-5**
9 ★	3.30		547		C16H12F3N1O3	M-TRIFLUOROMETHYLPHENYLHIPPURATE
11850	4.06		401		C16H12N2	5-METHYL-6H-PYRIDO(4,3-B)CARBAZOLE
1 ★	3.41		306		C16H12N2O1	BENZALCYANOACETANILIDE
2	3.27		897	459	C16H12N2O1	PYRAZINE-N-OXIDE,2,5-DIPHENYL **34046-78-1**
3	0.64	7.4	1722	1	C16H12N2O1	SUDANI **842-07-9**
4	1.70		2470	306	C16H12N2O2S1	TETRAHYDROQUINAZOLINE-4-OXO-2-THIO,3-(4-ACETYLPHENYL)
55	2.14		1973	459	C16H12N2O3	5-(3-CYANOBENZOYL)-H2-PYROLLO-PYRROLE-1-CARBOXYLIC ACID **96327-49-0**
6	2.21		1973	459	C16H12N2O3	5-(4-CYANOBENZOYL)-H2-PYRROLO-PYRROLE-1-CARBOXYLIC ACID **96327-50-3**
7 ★	2.19		1964		C16H12N2O3	BARBITURIC ACID,5,5-DIPHENYL **21914-07-8**
8 ★	2.11		547		C16H12N2O3	M-CYANOPHENYLHIPPURATE
9 ★	2.10		547		C16H12N2O3	P-CYANOPHENYLHIPPURATE
11860	-0.79	7.4	1722	1	C16H12N2O7S2	ORANGEG **1936-15-8**
1	4.06	1.0	2471	459	C16H12N4O2S1	XANTHONE,6-(5-TETRAZOLYL)-2-ETHYLTHIO
2	3.67	1.0	2471	459	C16H12N4O3	XANTHONE,6-(5-TETRAZOLYL)-2-ETHOXY
3	2.91	1.0	2471	459	C16H12N4O3	XANTHONE,6-(5-TETRAZOLYL)-2-METHOXYMETHYL
4	2.31	1.0	2471	459	C16H12N4O4	XANTHONE,6-(5-TETRAZOLYL)-2-(2-HYDROXYETHOXY)
65	1.29		910		C16H12N6O4	1,2-BIS(5-NITRO-2-BENZIMIDAZOLYL)ETHANE
6	1.45		910		C16H12N6O6	1,2-DIHYDROXY-1,2-BIS(5-NO2-2-2ENZIMIDAZOLYL)ETHANE
7 ★	3.16		306		C16H12O2	1,2-DIBENZOYLETHYLENE
8 ★	3.92		1941		C16H12O2	4H-1-BENZOPYRAN-4-ONE,2-PHENYL-3-METHYL **71972-66-2**
9 ★	4.37		2293	400	C16H12O2	ANTHRAQUINONE,2-ETHYL **84-51-5**
11870 ★	2.97		575		C16H12O3S1	TIOPINAC **61220-69-7** *anti-inflammatory, analgesic*
1	4.36	1.0	2471	459	C16H12O4S1	XANTHONE,6-CARBOXY-2-ETHYLTHIO
2	3.99	1.0	2471	459	C16H12O5	XANTHONE,6-CARBOXY-2-ETHOXY
3	3.12	1.0	2471	459	C16H12O5	XANTHONE,6-CARBOXY-2-METHOXYMETHYL
4	2.46	1.0	2471	459	C16H12O6	XANTHONE,6-CARBOXYL-2-(2-HYDROXYETHOXY)
75 ★	2.90		2393	345	C16H13Cl1F1O3	FLAMPROP **58667-63-3**
6	4.07		2661		C16H13Cl1F3N1O4	HALOXYFOP METHYL ESTER **69806-40-2**
7 ★	3.12	7.4	1626	1	C16H13Cl1N2O1	1,4-BENZODIAZEPIN-2-ONE,5-(O-TOLYL)-7-CL **5358-35-0**
8	3.25	7.4	874	1	C16H13Cl1N2O1	7-CL-5-PHENYL-3-METHYL-1,4-BENZODIAZEPIN-2-ONE **4699-82-5**
9 ★	3.33	7.4	1626	1	"	7-CL-5-PHENYL-3-METHYL-1,4-BENZODIAZEPIN-2-ONE
11880	2.49	7.4	1688	1	C16H13Cl1N2O1	DIAZEPAM **439-14-5** *sedative*
1	2.66	7.4	461	1	"	DIAZEPAM
2	2.80	7.4	1626	1	"	DIAZEPAM
3	2.82	7.4	919	1	"	DIAZEPAM
4	2.86	7.0	2230	1	"	DIAZEPAM
85	2.88	7.5	1510	1	"	DIAZEPAM
6 ★	2.99	7.4	2407	314	"	DIAZEPAM
7 ★	2.99	7.4	1626	1	C16H13Cl1N2O1S1	1,4-BENZODIAZEPIN-2-ONE,5-(2-METHIO-PH)-7-CL
8 ★	2.63	7.4	1626	1	C16H13Cl1N2O2	1,4-BENZODIAZEPIN-2-ONE,5-(2-MEO-PH)-7-CL **3023-44-7**
9 ★	3.14	7.4	1626	1	C16H13Cl1N2O2	1,4-BENZODIAZEPIN-2-ONE,5-(3-MEO-PH)-7-CL **5358-92-9**
11890	0.95	7.4	1688	1	C16H13Cl1N2O2	CLOBAZAM **22316-47-8** *tranquilizer (minor)*
1	1.79	7.4	1688	1	C16H13Cl1N2O2	TEMAZEPAM **846-50-4** *tranquilizer (minor)*
2 ★	2.19		1017		"	TEMAZEPAM
3	0.53	7.4	461	1	C16H13Cl1N2O4	CHLORAZEPATE,DIPOTASSIUM **57109-90-7**
4	-0.69		547	370	C16H13F1O2S1.Br1	N-(4-SO2F-BENZYL)-QUINOLINIUMBROMIDE
95 ★	2.30		2420		C16H13F2N3O1	FLUTRIAFOL **76674-21-0** *fungicide*
6	4.20		1104		C16H13N1	N-PHENYL-1-NAPHTHYLAMINE **90-30-2**
7 ★	3.51	7.4	595	314	C16H13N1O1	QUINOLINE,8-BENZYLOXY
8	3.01	7.4	712	1	C16H13N1O2	A,A'-DIMETHYL-G-PYROPHTHALONE
9	1.90	7.4	712	1	C16H13N1O2	A-METHYL-N-METHYL-G-PYROPHTHALONE
11900	2.12	7.4	712	1	C16H13N1O2	N-ETHYL-G-PYROPHTHALONE
1	2.96		759		C16H13N1O2S1	2-(4,4'-DIHYDROXYDIPHENYLMETHYL)THIAZOLE
2	1.83		2161	468	C16H13N1O3	AZETIDINE-2-ONE,4-(2-BENZOYL)PHENOXY
3	2.14		2161	468	C16H13N1O3	AZETIDINE-2-ONE,4-(BENZOYL)PHENOXY
4	2.55		594		"	AZETIDINE-2-ONE,4-(BENZOYL)PHENOXY
5	-0.70	7.3	2259	1	C16H13N1O3	BENZENEACETIC ACID,P-N-(1,3-DIHYDRO-1-OXO-ISOINDOLYL)
6	2.90	7.3	2259	2	"	BENZENEACETIC ACID,P-N-(1,3-DIHYDRO-1-OXO-ISOINDOLYL)
7	-1.01	7.3	2259	1	C16H13N1O4	PHENOXYACETIC ACID,P-N-(1,3-DIHYDRO-1-OXO-ISOINDOLYL)
8	3.58	7.3	2259	2	"	PHENOXYACETIC ACID,P-N-(1,3-DIHYDRO-1-OXO-ISOINDOLYL)
9 ★	2.76		547		C16H13N1O4S1	N-(7-ME-9-OXO-3-THIOXANTHENYL)ACETAMIDE-10,10-O2

	logP	pH	Ref.	Note	MF	Name / CAS / activity
11910	3.61		1973	459	C16H13N1O5	5-(3,4-DIOXYMETHYLBENZOYL)-H2-PYRROLO-PYRROLE-1-CARBOXYLIC ACID 96327-59-2
1	2.33	7.4	594	1	C16H13N3O3	1,4-BENZODIAZEPIN-2-ONE,1,3-DIHYDRO-9-METHYL-7-NITRO-5-PHENYL 4941-45-1
2 ★	2.44		594	400	"	1,4-BENZODIAZEPIN-2-ONE,1,3-DIHYDRO-9-METHYL-7-NITRO-5-PHENYL
3	1.55		555		C16H13N3O3	1-BENZYL-2-(5-NO2-2-FURFURILIDENE)IMIDAZOLE
4 ★	2.83	5.0	2674	302	C16H13N3O3	MEBENDAZOLE 31431-39-7 anthelmintic
15 ★	2.16	7.4	1626	1	C16H13N3O3	NIMETAZEPAM 2011-67-8 anticonvulsant, muscle relaxant
6	1.38	7.2	556		C16H13N3O3S1	5-PHENOXY-2-BENZENESULFONAMIDOPYRIMIDINE
7 ★	3.04		2602		C16H13N3O3S1	QUINAZOLINE-4-ONE,2,6-DIMETHYL-5-(P-NITROPHENYL)THIO thymidylate sythanse inhibitor
8 ✓	1.25		1412		C16H13N5O2	1,2,4-TRIAZOLE,3,5-DIBENZAMIDO
9 ★	5.85		2158	459	C16H14	ANTHRACENE,2-ETHYL 52251-71-5
11920 ★	5.69		2158	459	C16H14	ANTHRACENE,9,10-DIMETHYL 781-43-1
1	2.61		1973	459	C16H14Cl1N1O3	3-(3-CHLOROBENZOYL)-H4-PYRROLO-PYRIDINE-1-CARBOXYLIC ACID 96327-68-3
2	3.60		1973	459	C16H14Cl1N1O3	3-(4-CHLOROBENZOYL)-H4-PYRROLO-PYRIDINE-8-CARBOXYLIC ACID 88777-67-7
3	1.45	7.4	1688	1	C16H14Cl1N3O1	CHLORDIAZEPOXIDE 58-25-3 sedative, tranquilizer (minor)
4 ★	2.44		505		"	CHLORDIAZEPOXIDE
25	2.48	7.4	919	1	"	CHLORDIAZEPOXIDE
6	2.50	7.4	461	1	"	CHLORDIAZEPOXIDE
7	2.96		1159		C16H14Cl2N2O2	N-PH,N-(1-ACETANILIDO)DICHLOROACETAMIDE
8 ★	5.50		1555	459	C16H14Cl2O1	PROCLONOL 14088-71-2 anthelmintic, antifungal
9 ★	4.80		1903		C16H14Cl2O4	DICLOFOPMETHYL 51338-27-3 herbicide
11930 ★	1.30	7.4	1626	1	C16H14F1N3O1	1,4-BENZODIAZEPIN-2-ONE,1-ME-5-(2-F-PH)-7-AMINO
1	2.94		1973	459	C16H14F1O3	3-(4-FLUOROBENZOYL)-H4-PYRROLO-PYRIDINE-1-CARBOXYLIC ACID 88777-59-7
2	0.27	7.4	2635	1	C16H14F3N1O3S1	TOLRESTAT 82964-04-3 inhibitor-aldose reductase, antidiabetic
3	-0.82		547	370	C16H14N1.Br1	N-BENZYLQUINOLINIUMBROMIDE
4 ★	2.62	7.4	1626	1	C16H14N2O1	7-ME-1,2-H2-5-PH-3H-1,4-BENZODIAZEPIN-2-ONE 5571-63-1
35	2.71	7.4	874	1	"	7-ME-1,2-H2-5-PH-3H-1,4-BENZODIAZEPIN-2-ONE
6 ★	2.50	7.4	579	1	C16H14N2O1	METHAQUALONE 72-44-6 hypnotic, sedative
7 ★	2.92	7.4	1626	1	C16H14N2O1S1	1,4-BENZODIAZEPIN-2-ONE,5-PHENYL-7-METHIO 2891-12-5
8	2.93	7.4	594	1	"	1,4-BENZODIAZEPIN-2-ONE,5-PHENYL-7-METHIO
9	2.16		2470	306	C16H14N2O1S1	TETRAHYDROQUINAZOLINE-4-OXO-2-THIO,3-(2,3-DIMETHYLPHENYL)
11940	2.24	7.4	1626	1	C16H14N2O2	1,4-BENZODIAZEPIN-2-ONE,5-PH-7-METHOXY 5358-96-3
1	-0.67	7.4	675	1	C16H14N2O2	3,5-PYRAZOLIDINDIONE,1,2-DIPHENYL-4-METHYL
2	1.13	7.4	874	1	C16H14N2O2	7-ME-1,2-H2-5-PH-3H-1,4-BENZODIAZEPIN-2-ONE,N-O
3 ★	3.65		2166		C16H14N2O2	FURAZAN-2-OXIDE,3,4-DIBENZYL
4	1.91		799		C16H14N2O2	QUINAZOLIN-2-ONE,1-METHY-4-PHENYL-6-METHOXY
45 ★	0.80	7.4	594	1	C16H14N2O2S1	1,4-BENZODIAZEPIN-2-ONE,1,3-DIHYDRO-7-METHYLSULFINYL-5-PHENYL 5571-51-7
6 ★	3.23		2507		C16H14N2O2S1	MEFENACET 73250-68-7
7	-1.02	7.4	675	1	C16H14N2O3	3,5-PYRAZOLIDINDIONE,1-PH,2(4-OH-PH),4-METHYL
8	0.59		799		C16H14N2O3S1	QUINAZOLIN-2-ONE,1-ME-4-PHENYL-6-METHYLSULFONYL
9 ★	1.20		547		C16H14N2O4	M-CARBOXAMIDOPHENYLHIPPURATE
11950	1.14		910		C16H14N4O2	1,2-DIHYDROXY-1,2-BIS(2-BENZIMIDAZOLYL)ETHANE
1 ★	3.60		538		C16H14N4O2	P-(3-(P-CYANOBENZYL)-3-ME-1-TRIAZENO)BENZOIC ACID
2	0.52	4.0	360	2	C16H14N4O2S1	5-PHENYL-4-SULFANILAMIDOPYRIMIDINE
3	2.58		260		C16H14O1	DIBENZOCYCLOOCTANE-5-ONE
4	0.59	7.4	1838	1	C16H14O3	FENBUFEN 36330-85-5 anti-inflammatory
55	0.63	7.4	1687	1	"	FENBUFEN
6 ★	3.20	3.5	2179	302	"	FENBUFEN
7	3.53	7.4	1687	2	"	FENBUFEN
8	-0.25	7.4	1838	1	C16H14O3	KETOPROFEN 22071-15-4 anti-inflammatory
9	-0.01	7.4	693	1	"	KETOPROFEN
11960 ★	3.12	2.0	1838		"	KETOPROFEN
1	3.39		2021	459	C16H15Br1O6	NAPHTHALENE,1,4-DIACETYLOXY-2,3-DIMETHOXY-6-BROMO 91814-13-0
2	-0.05	7.4	1204	1	C16H15Br2N3	N-BENZYL-N-(2,6-DIBRPH)AMINO)-2-IMIDAZOLINE
3	4.05	7.4	461	1	C16H15Cl1N2	MEDAZEPAM 2898-12-6 tranquilizer (minor)
4 ★	4.41		1017		"	MEDAZEPAM
65 ★	3.49	7.4	2407	2	C16H15Cl1N2O1S1	CLOTIAZEPAM 33671-46-4 tranquilizer
6	2.32		1159		C16H15Cl1N2O2	N-PHENYL,N-(1-ACETANILIDO)CHLOROACETAMIDE
7	2.93		2572	459	C16H15Cl1N2O4	RH-O710 143121-09-9 ✚
8 ★	3.53	3.5	1609	1	C16H15Cl1O3	PHENYLACETIC ACID,3-CHLORO,4-(2-PHENYLETHOXY)
9 ★	2.45	3.5	2250	302	C16H15Cl1O4	PHENYLACETIC ACID,2-(3'-CHLORO-4'-METHOXY)BENZYLOXY
11970	3.26		2021	459	C16H15Cl1O6	LONAPALENE 91431-42-4 antipsoriatic
1	3.16		2021	459	C16H15Cl1O6	NAPHTHALENE,1,4-DIACETYLOXY-2,3-DIMETHOXY-5-CHLORO 91814-10-7
2 ★	4.80		2520		C16H15Cl2N1O2	CLOMEPROP ✚
3	3.57		434		C16H15Cl3	METHYLCHLOR
4	2.90		434		C16H15Cl3O1S1	METHOXY-METHIOCHLOR
75 ★	5.08		1487		C16H15Cl3O2	METHOXYCHLOR 72-43-5
6	3.85		434		C16H15Cl3S2	METHIOCHLOR
7	2.75		2021	459	C16H15F1O6	NAPHTHALENE,1,4-DIACETYLOXY-2,3-DIMETHOXY-6-FLUORO 91814-12-9
8 ★	3.70		2420		C16H15F2N3Si1	FLUSILAZOLE 85509-19-9 fungicide
9	1.31		542	800	C16H15F3N2O3	DI-DECHLORO-TRIFLUORO-4-(2-PYRIDYL)-AMPHENICOL
11980	2.46	7.2	2416	1	C16H15F3N2O4	BAY-K-8644 71145-03-4 activator-ca-channel ✚
1 ★	4.76		1523		C16H15F3O2	1,1,1-TRIFLUORO-2,2-DI(P-METHOXYPHENYL)ETHANE 384-97-4
2	2.10	7.4	2280	1	C16H15N1	DIZOCILPINE 77086-21-6 anticonvulsant, neuroprotective agent
3 ★	1.83		1817		C16H15N1O1	3-PYRIDINEMETHANOL,A-2-PROPYNYL,A-4-TOLYL 89242-87-5
4 ★	1.87		1817		C16H15N1O1	3-PYRIDINEMETHANOL,A-2-PROPYNYL,A-BENZYL 89242-78-4
85 ★	2.62		1988		C16H15N1O2	ACETANILIDE,N-(3-ACETYLPHENYL) 72116-69-9
6 ★	2.75		1988		C16H15N1O2	ACETANILIDE,N-(4-ACETYLPHENYL) 89246-39-9
7	0.95	7.4	564	1	C16H15N1O3	2,10,11-TRIHYDROXYNORAPORPHINE
8	2.90		1973	459	C16H15N1O3	5-(2-METHYLBENZOYL)-H2-PYRROLO-PYRROLE-1-CARBOXYLIC ACID 66635-84-5
9	3.23		1973	459	C16H15N1O3	5-(3-METHYLBENZOYL)-H2-PYRROLO-PYRROLE-1-CARBOXYLIC ACID 96327-32-1
11990	3.24		1973	459	C16H15N1O3	5-(4-METHYLBENZOYL)-H2-PYRROLO-PYRROLE-1-CARBOXYLIC ACID 66635-73-2
1	2.52		1973	459	C16H15N1O3	5-(PHENACETYL)-H2-PYRROLO-PYRROLE-1-CARBOXYLIC ACID 96327-65-0
2 ★	2.37		1988		C16H15N1O3	ACETANILIDE,N-(3-ACETOXYPHENYL)
3 ★	2.48		2487		C16H15N1O3	CINNAMAMIDE,3,4-DIHYDROXY-N-BENZYL
4 ★	2.72		547		C16H15N1O3	M-METHYLPHENYLHIPPURATE
95 ★	2.83		547		C16H15N1O3	P-METHYLPHENYLHIPPURATE
6	3.68		1973	459	C16H15N1O3S1	5-(4-METHYLTHIOBENZOYL)-H2-PYRROLO-PYRROLE-1-CARBOXYLIC ACID 76786-67-9
7	2.74		1973	459	C16H15N1O4	5-(3-METHOXYBENZOYL)-H2-PYRROLO-PYRROLE-1-CARBOXYLIC ACID 96327-36-5
8	2.81		1973	459	C16H15N1O4	ANIROLAC 66635-85-6 anti-inflammatory, analgesic
9 ★	2.28		547		C16H15N1O4	P-METHOXYPHENYLHIPPURATE 29736-21-8

	logP	pH	Ref.	Note	MF	Name / **CAS** / *activity*
12000	1.08		1973	459	C16H15N1O4S1	5-(4-METHYLSULFOXYBENZOYL)-H2-PYRROLO-PYRROLE-1-CARBOXYLIC ACID **76786-74-8**
1	1.41		1973	459	C16H15N1O5S1	5-(4-METHYLSULFONYLBENZOYL)-H2-PYRROLO-PYRROLE-1-CARBOXYLIC ACID **96327-44-5**
2 ★	3.22		1519	459	C16H15N3	4-PHENYL-1-PHTHALAZINAMINE,N,N-DIMETHYL **23099-85-6**
3 ★	1.46	7.4	1626	1	C16H15N3O1	1,4-BENZODIAZEPIN-2-ONE,1-ME-5-PH-7-AMINO
4 ★	2.45		2602		C16H15N3O1S1	QUINAZOLINE-4-ONE,2,6-DIMETHYL-5-(3-METHYLPYRID-4YL)THIO *thymidylate synthase inhibitor*
5	-0.19	7.4	2573	1	C16H15N3O9S1	CEPHALOTHIN,5-NITROFURYL ANALOG
6 ★	3.70		579		C16H16	PARACYCLOPHANE **1633-22-3**
7 ★	1.15	7.4	1978	1	C16H16Br1N3O6	1-ACETYL-6-DEMETHYL-6-BROMO-7-AMINO-N-METHYLMITOSENE **96000-48-5**
8 ★	3.53		2031		C16H16Cl1N3O2	1H-(4,5,C)PYRIDINE,2-(4-METHOXY-2-CHLOROPROPOXYPHENYL)
9 ★	4.10		2612		C16H16Cl2N2O2	1,1-DIMETHYL-3-(2,4-DICL-BENZYLOXY)PHENYLUREA **88132-42-7**
12010 ★	3.30		1712		C16H16Cl2N2O4S1	2,4-DICHLOROBENZYLPENICILLIN
1 ★	4.47		1523		C16H16Cl2O2	METHOXYCHLOR-DDD **7388-31-0**
2 ★	3.78		2479	459	C16H16F1N3O2	PHENYLBIURET,2-FLUORO-5-PHENYL
3	-1.04	7.4	2037	1	C16H16F1N5O2	ISOGUANINE,1-METHYL-7-(3-(P-FLUOROBENZOYL)PROPYL
4	0.44	7.4	2037	1	C16H16F1N5O2	ISOGUANINE,1-METHYL-9-(3-P-FLUORO)PHENYL)PROPYL
15 ★	2.59		2612		C16H16N2O2	1,1-DIMETHYL-3-(4-BENZOYLPHENYL)UREA **61706-06-7**
6 ★	3.67		580		C16H16N2O2	1-(2-FLUORENYL)-3-METHOXY-3-METHYLUREA
7	2.01		1159		C16H16N2O2	N-PHENYL,N-(1-ACETANILIDO)ACETAMIDE
8 ★	2.74	6.4	2018		C16H16N2O2	NAFAZATROM **59040-30-1** *platelet anti-aggregatory*
9 ★	2.86		2612		C16H16N2O3	1-(4-BENZOYLPHENYL)-3-METHOXY-3-METHYLUREA
12020	2.24		1159		C16H16N2O3	N-PH,N-(1-ACETANILIDO)CARBAMIC ACID,METHYL ESTER
1 ★	3.39		2506		C16H16N2O4	DESMEDIPHAM **13684-56-5** *herbicide*
2 ★	3.59		2506		C16H16N2O4	PHENMEDIPHAM **13684-63-4** *herbicide*
3	1.63	3.5	537		C16H16N2O5	N-BENZOYL-A-OH-ME-B-OH-P-NITROPHENETHYLAMINE
4 ★	0.46	3.5	508		C16H16N2O5S1	CEPHALOSPORANIC ACID,7-(D-MANDELAMIDO)-DESACETOXY
25	-2.20	7.4	2464	41	C16H16N2O6S2	CEPHALOTHIN **153-61-7** *antibacterial*
6 ★	-0.41	2.0	2464	41	"	CEPHALOTHIN
7	0.56	7.4	2573	1	"	CEPHALOTHIN
8 ★	0.72	7.4	1978	1	C16H16N2O7	1-ACETYL-6-DEMETHYL-7-METHOXYMITOSENE **96000-42-9**
9	0.09	7.4	2573	1	C16H16N2O7S1	CEPHALOTHIN,FURYL ANALOG
12030	-0.36	7.6	2003	314	C16H16N3O2.Cl1	IMIDAZOLIUM CHLORIDE,1-(1-NAPHTHOXY)METHYL-2-HYDROXYIMINOMETHYL-3-METHYL **91900-17-3**
1	0.30	7.4	2024	1	C16H16N4O4	1,2-AZIRIDINOMITOSINE,7-AZIRIDINO **103422-20-4**
2	-1.91	7.4	2464	41	C16H16N4O8S1	CEFUROXIME **55268-75-2** *antibacterial*
3 ★	-0.16		1980		"	CEFUROXIME
4 ★	0.00	3.0	1638		C16H16N6O4S3	CEPHALOTHIN ANALOG(3-(1-ME-TETRAZ-5YL)S-CH2) **32924-66-6**
35	0.20	7.4	2591	1	C16H16N6O5	AZT,5'-3-NICOTINYL **116333-43-8** *antiretroviral prodrug*
6	2.19		1593	459	C16H16N6O7S4	CP-1351 **69712-43-2** ✚
7	-0.70		1409		C16H16N8O6	PYRAZINE-1,4-H2-N,N-BIS(1-(2-ME-5-NO2-IMIDAZOLE)) **80348-51-2**
8 ★	2.78	5.5	2403	2	C16H16O3	BUTANOIC ACID,2-(4-BIPHENYLYL)-3-HYDROXY **93371-36-9**
9 ★	3.37	3.5	1609	1	C16H16O3	PHENYLACETIC ACID,3-METHYL,4-PHENYLMETHOXY
12040 ★	2.85	2.6	2018	2	C16H16O4	CYCLOHEXANO(6,7)CHROMONE-2-PROPIONIC ACID ✚
1 ★	2.59	3.5	1609	1	C16H16O4	PHENYLACETIC ACID,3-METHOXY,4-PHENYLMETHOXY
2 ★	2.96		2058		C16H16O5	PYRROL-3-ONE,2-(P-METHOXYBENZILIDINE)-4-CARBETHOXY-5-METHYL ✚
3	2.39		2021	459	C16H16O6	NAPHTHALENE,1,4-DIACETYLOXY-2,3-DIMETHOXY **61601-23-8**
4	2.76	7.4	461	1	C16H17Cl1N2O1	TETRAZEPAM **10379-14-3** *tranquilizer*
45 ★	3.20		1017		"	TETRAZEPAM
6 ★	2.30		1712		C16H17Cl1N2O4S1	A-CHLOROBENZYLPENICILLIN
7	2.30	7.4	455	1	C16H17Cl1N2S1	MONODESMETHYLCHLORPROMAZINE
8	2.30	7.4	678	1	"	MONODESMETHYLCHLORPROMAZINE
9 ★	1.28	7.0	2292	1	C16H17Cl1N6O2	DIOXOLANE,2-(2-CHLORO-4-(1,2,4-TRIAZOL-1-YL)PHENYL)-2-1,2,4-TRIAZOL-1-YL)METHYL-4-ETHYL
12050	4.20		1190		C16H17Cl1O1	2-S-BUTYL-4-CHLORODIPHENYLOXIDE
1 ★	-1.13		2681	226	C16H17F2N3O3	8-FLUORO-NORFLOXACIN
2 ★	1.08	7.0	2292	1	C16H17F3N4O3	DIOXOLANE,2-(4-TRIFLUOROMETHYLACETAMIDO)PHENYL-2-(1,2,4-TRIAZOL-1-YL)METHYL-4-ETHYL ✚
3 ★	3.47		1988		C16H17N1O1	ACETANILIDE,N-(3-ETHYLPHENYL)
4 ★	3.37		2077		C16H17N1O1	BENZYLIDINEANILINE-N-OXIDE,4-I-PROPYL
55 ★	3.23		1988		C16H17N1O2	ACETANILIDE,N-(3-ETHOXYPHENYL)
6	0.84	2.7	978	340	C16H17N1O2	N(2,3-DIME-PHENYL)3-ME-ANTHRANILIC ACID, SODIUM SALT
7	1.03	7.4	2325	1	C16H17N1O2	PHENYLACETIC ACID,2-(2',3'-DIMETHYL)ANILINO **64758-92-5**
8	1.21	7.4	2325	1	C16H17N1O2	PHENYLACETIC ACID,2-(2',6'-DIMETHYL)ANILINO **23189-27-7**
9	4.54	7.4	2325	2	C16H17N1O2	PHENYLACETIC ACID,2-(2',6'-DIMETHYL)ANILINO
12060	4.28		822		C16H17N1O3	4-AMINOSALICYLIC ACID,2-I-PROPYLPHENYL ESTER
1	4.12		822		C16H17N1O3	4-AMINOSALICYLIC ACID,4-I-PROPYLPHENYL ESTER
2	-0.58	7.2	2323	468	C16H17N1O3	BENZYLTYROSINE
3	-0.32		978		C16H17N1O3	N(2,3-DIME-PHENYL)3-MEO-ANTHRANILIC ACID, SODIUM SALT
4 ★	3.55		1681		C16H17N1O3	N,N-DIMETHYLCARBAMATE,M-PHENOXYBENZYL ESTER **84640-36-8**
65	2.07		520		C16H17N1O3S1	1,4-NAPHTHOQUINONE,2-ACETAMIDO,3-BUTYLTHIO
6 ★	3.57		1523		C16H17N1O4	1,1-DI(P-METHOXYPHENYL)-2-NITROETHANE
7	1.36	7.4	2062	1	C16H17N1O4	L-DOPA,BENZYL ESTER
8 ★	1.50	7.4	2062	2	"	L-DOPA,BENZYL ESTER
9 ★	2.06	7.4	2268	1	C16H17N1O4	NAPROXEN,GLYCOLAMIDE ESTER
12070	2.01		546		C16H17N1O4S1	N,N-DIMETHYL-P-TOLUENESULFONYLSALICYLAMIDE
1	2.81	7.4	2621	1	C16H17N1O5	MITOSENE(5,5,6),METHYL-ACETYLOXY ANALOG
2	0.89	7.4	2621	1	C16H17N1O6	MITOSENE, HYDROXY-ACETYLOXY ANALOG
3	0.66	7.4	2633	1	C16H17N1S1	OCTAHYDROPHENANTHREN-4A-AMINE-(2,3)-THIENYL ANALOG *nmda antagonist*
4	3.02		2633	459	"	OCTAHYDROPHENANTHREN-4A-AMINE-(2,3)-THIENYL ANALOG
75	4.75	7.4	901	2	C16H17N3O2	P-(3(P-ME-BENZYL)3-ME-1-TRIAZENO)BENZOIC ACID
6	4.43	7.4	901	2	C16H17N3O3	P-(3(P-MEO-BENZYL)-3-ME-1-TRIAZENO)BENZOIC ACID
7	0.99	1.3	548	1	C16H17N3O4	P-(3-(M-NITROPHENOXY)PROPOXY)BENZAMIDINE HCL
8	0.82		548	370	C16H17N3O4	P-(4-NITROPHENOXYPROPOXY)BENZAMIDINEHCL
9	-2.40	7.4	2464	41	C16H17N3O4S1	CEPHALEXIN **15686-71-2** *antibacterial*
12080	-1.80	6.5	2464	41	"	CEPHALEXIN
1	-1.64	4.5	2464	41	"	CEPHALEXIN
2	-1.10	7.4	2192	1	"	CEPHALEXIN
3	-0.85	7.2	917	314	"	CEPHALEXIN
4	0.65		872	459	"	CEPHALEXIN
85	0.83		548	370	C16H17N3O4.C6H6O3S1	P-(4-NO2-PHENOXYPROPOXY)BENZAMIDINE
6	1.28	7.4	2024	1	C16H17N3O5	1,2-(N-METHYLAZIRIDINO)MITOSINE,7-METHOXY **15973-07-6**
7 ★	0.67	7.0	2440	1	C16H17N3O5	2'-DEOXYCYTIDINE,N4-BENZOYL
8	-3.40	7.4	2464	41	C16H17N3O5S1	CEFADROXIL **50370-12-2** *antibacterial*
9 ★	-2.06	1.0	2196		"	CEFADROXIL

	logP	pH	Ref.	Note	MF	Name / CAS / activity
12090	0.50	7.0	801	1	C16H17N3O6	1-(5-BENZOYLARABINOFURANOSYL)CYTOSINE
1 ★	0.30	7.0	2440	1	C16H17N3O6	CYTIDINE,N4-BENZOYL
2	-1.89	7.4	2464	41	C16H17N3O7S2	CEFOXITIN **35607-66-0** *antibacterial*
3 ★	-0.02		1980		"	CEFOXITIN
4	0.53	7.4	2433	1	C16H17N5O3	ALLOPURINOL,N-(4-DIMETHYLAMINOMETHYL)BENZOYLOXYMETHYL ✚
95 ★	1.05		2433	2	"	ALLOPURINOL,N-(4-DIMETHYLAMINOMETHYL)BENZOYLOXYMETHYL
6	-1.70	7.4	2464	41	C16H17N5O7S2	CEFOTAXIME **63527-52-6** *antibacterial*
7 ★	-1.36	2.0	2464	41	"	CEFOTAXIME
8	1.25		1593	459	C16H17N7O9S5	CP-1363 **69712-50-1** ✚
9	-1.54		2530	1	C16H17N9O5S3	CEFMENOXIME **65085-01-0** *antibacterial*
12100	-2.00	7.2	2446	1	C16H18F1N3O3	NORFLOXACIN **70458-96-7** *antibacterial*
1 ★	-1.03		2681	226	"	NORFLOXACIN
2	-1.04	7.4	2037	1	C16H18F1N5O2	2H-PURIN-2-ONE,1,3-DIHYDRO-1-METHYL-6-AMINO-7-(P-FLUOROBENZOYL)PROPYL
3 ★	2.59		1988		C16H18N2O1	ACETANILIDE,N-(3-ETHYLAMINOPHENYL)
4 ★	2.83		2094		C16H18N2O1	UREA,1,1-DIMETHYL-3-(3-BENZYL)PHENYL **102587-48-4**
5 ★	3.15	7.4	2026	1	C16H18N2O1S1	BENZAMIDE,2-(P-TOLYLTHIO)ETHYLAMINO
6	-0.30	7.4	455	1	C16H18N2O1S1	MONODESMETHYLPROMAZINESULFOXIDE
7 ★	2.94	7.4	2026	1	C16H18N2O2	BENZAMIDE,2-(O-TOLYL)OXYETHYLAMINO
8	2.46		1019		C16H18N2O2	N'-(1-METHYLBENZYL)PHENOXYACETIC ACID HYDRAZIDE
9	2.32	7.6	2007	1	C16H18N2O2S1	2-NAPHTHOYLTHIOFORMOHYDROXIMATE,S-(2-DIMETHYLAMINO)ETHYL **102941-50-4**
12110	3.11		2612		C16H18N2O3	1-(3'-BENZYLOXYPHENYL)-3-MEO-3-METHYLUREA **74109-81-2**
1 ★	3.11		2612		C16H18N2O3	1-(4-BENZYLOXYPHENYL)-3-METHOXY-3-METHYLUREA **25998-87-2**
2	-1.82	7.2	846	314	C16H18N2O4S1	BENZYLPENICILLIN **61-33-6** *antibacterial*
3	-1.81	7.4	1945	1	"	BENZYLPENICILLIN
4	-1.70	7.0	1295	1	"	BENZYLPENICILLIN
15 ★	1.83		515		"	BENZYLPENICILLIN
6	1.11	7.4	2542	1	C16H18N2O4S2	THIOPHENE-2-SULFONAMIDE,4-(MORPHOLIN-1-YL-METHYLBENZOYL)
7	1.18	7.4	2542	1	C16H18N2O4S2	THIOPHENE-2-SULFONAMIDE,4-(MORPHOLIN-1YLMETHYLBENZOYL)
8	2.30	2.0	300		C16H18N2O5	BARBIT.ACID,N-ME-5-ET-5-(CH2)-2-(1,4-BENZODIOXANE)
9	-2.23	7.2	846	314	C16H18N2O5S1	6-(N-PHENYLACETYLAMINO)PENICILLANIC ACID-SO
12120 ★	1.40		515		C16H18N2O5S1	PENICILLIN,A-HYDROXYBENZYL
1	-1.54	7.2	846	314	C16H18N2O5S1	PHENOXYMETHYLPENICILLIN **87-08-1** *antibacterial*
2 ★	2.09		515		"	PHENOXYMETHYLPENICILLIN
3	-1.86	7.2	846	314	C16H18N2O6S1	PHENOXYMETHYLPENICILLIN-SULFOXIDE
4	-0.22		1712		C16H18N2O7S2	SULBENICILLIN **34779-28-7** *antibacterial*
25	0.59		872	459	"	SULBENICILLIN
6	1.58	7.4	455	1	C16H18N2S1	MONODESMETHYLPROMAZINE
7	1.58	7.4	678	1	"	MONODESMETHYLPROMAZINE
8	3.53	8.0	550	1	C16H18N2S1	PHENETAZINE **522-24-7** *antihistaminic*
9	4.20	7.4	550	2	"	PHENETAZINE
12130 ✓	-0.71		2397	101	C16H18N3O2	IMIDAZOLIUM CHLORIDE,1-BENZYL-2-HYDROXYIMINOMETHYL-3-BUTYN-3YLOXYMETHYL
1 ✓	-1.05		2397	101	C16H18N3O2	IMIDAZOLIUM CHLORIDE,1-METHYL-2-HYDROXYIMINOMETHYL-3-(1-PHENYLBUTYN-3YLOXY)METHYL **117983-43-4**
2 ★	0.87	7.4	534	1	C16H18N4O2	NIALAMIDE **51-12-5** *antidepressant*
3	2.15	7.4	2477	1	C16H18N4O3	TETRAHYDRO-A[(2-NITROIMIDAZOL-1-YL)METHYL]NAPHTH[1,2B]AZIRINE-1-ETHANOL **120277-98-7**
4	3.05	7.4	2028	459	C16H18N6	1,2,4-TRIAZOLE,3-AMINO-5-(2-PHENYLPROPYLAMINO)PYRIDIN-4-YL **98087-97-9**
35 ★	4.52		1523		C16H18O2	1,1-DI-(P-METHOXYPHENYL)ETHANE **10543-21-2**
6 ★	3.51	7.4	2268	1	C16H18O3	NAPROXEN,ETHYL ESTER
7	1.60	7.4	2559	1	C16H18O5	CHROMONE-2-CARBOXYLIC ACID,5-HYDROXY-6,8-DIPROPYL
8	-0.90	7.4	2559	1	C16H18O6	CHROMONE-2-CARBOXYLIC ACID,5-(2-HYDROXYPROPOXY)-8-PROPYL
9 ★	0.76		511		C16H18O6	GLUCOPYRANOSIDE,2-NAPHTHYL(BETA)
12140 ★	-0.78		2648	448	C16H18O8	4-METHYLUMBELLIFERYL-B-D-GALACTOPYRANOSIDE
1 ★	-0.66		2648	448	C16H18O8	4-METHYLUMBELLIFERYL-B-D-GLUCOPYRANOSIDE
2	0.24	2.0	1039	210	C16H19Br1N2	BROMPHENIRAMINE **132-21-8** *antihistaminic*
3	6.33		2404	459	C16H19Br1O2	(7,8-DIHYDROPYRANYL)BENZOPYRAN,2,2,2',2'-TETRAMETHYL-6-BROMO
4	2.71	7.0	1056	1	C16H19Br2N1O2	2,4-DIBROMOPROPRANOLOL
45	1.38	7.4	1209	1	C16H19Cl1N2	CHLORPHENIRAMINE **25523-97-1** *antihistaminic*
6 ★	3.39		1835		"	CHLORPHENIRAMINE
7	-0.17	2.0	1039	210	C16H19Cl1N2O1	CARBINOXAMINE **486-16-8** *antihistaminic*
8	1.10	7.4	591	1	"	CARBINOXAMINE
9	-0.28	2.0	1039	210	C16H19Cl1N2O1	ROTOXAMINE **5560-77-0**
12150	1.18	7.4	591	1	"	ROTOXAMINE
1	6.14		2404	459	C16H19Cl1O2	(7,8-DIHYDROPYRANYL)BENZOPYRAN,2,2,2',2'-TETRAMETHYL-6-CHLORO
2 ★	2.03	7.0	2292	1	C16H19Cl2N3O4	DIOXOLANE,2-(2,4-DICHLOROPHENYL)-2-(1,2,4-TRIAZOL-1-YL)METHYL-4-METHOXYETHOXYMETHYL
3	0.05	7.4	2681	1	C16H19F1N4O3	AMIFLOXACIN **86393-37-5** *antibacterial*
4 ★	0.23		2681	226	"	AMIFLOXACIN
55	0.53	7.2	1703	314	C16H19F2N3O8S2	CEPHALOSPORIN ANALOG
6	-1.26		1082		C16H19I3N2O8	FLW-9 ✚
7	0.74	7.4	785	1	C16H19N1	4,4-DIPHENYLBUTYLAMINE
8	1.87	7.5	1588	1	C16H19N1	PHENETHYLAMINE,N-METHYL-A-BENZYL
9	3.80	7.4	1588	2	"	PHENETHYLAMINE,N-METHYL-A-BENZYL
12160	0.65	1.0	870	342	C16H19N1O1	2-(P-METHYL-DIPHENYLMETHOXY)ETHYLAMINE
1	-0.01	1.0	870	342	C16H19N1O1	ETHANAMINE,N-METHYL-2-DIPHENYLMETHOXY
2	0.06	1.0	602	342	"	ETHANAMINE,N-METHYL-2-DIPHENYLMETHOXY
3 ★	1.77		591		C16H19N1O4	ROLETAMIDE **10078-46-3** *hypnotic*
4	1.36		634		C16H19N1O6	6-MEO-7-METHOXYETHOXY-4-OH-3-QUINOLINE-CO2ET
65 ★	1.67		2234		C16H19N1O7	ACETYLSALICYLIC ACID,N-(2-ETHOXYACETYL)GLYCOLAMIDE ESTER **116482-77-0**
6	1.12	7.4	2235	1	C16H19N1Si1	SILA-PIPERIDINE ✚
7	-0.92		2627	841	C16H19N2.Cl1	N-METHYL-4-(4'-DIMETHYLAMINOSTYRYL)-PYRIDINIUM CHLORIDE
8 ★	3.52		2150	459	C16H19N3	ADAMANTANE,1-(1-BENZOTRIAZOLYL)
9 ★	4.30		2150	459	C16H19N3	ADAMANTANE,1-(2-BENZOTRIAZOLYL)
12170	-0.03	5.3	1365		C16H19N3	FENAPANIL **61019-78-1**
1	1.44	7.4	416	1	C16H19N3O2	1-ALLYL-3-PHENYL-5-DIMEAMINO-6-METHYLURACIL
2	2.26	7.4	416	1	C16H19N3O2	1-ME-3-PHENYL-5-PYRROLIDINYL-6-METHYLURACIL
3	1.98	7.4	416	1	C16H19N3O2	1-PHENYL-3-ALLYL-5-DIMEAMINO-6-METHYLURACIL
4	2.68	7.4	459	1	C16H19N3O2	1-PHENYL-3-ME-5-PYRROLIDINO-6-METHYLURACIL
75	0.69	7.4	416	1	C16H19N3O3	1-ME-3-PHENYL-5-MORPHOLINO-6-METHYLURACIL
6 ★	-0.98		594	314	C16H19N3O3	TRPTOPHYL-PROLINE
7	-2.68	7.2	846	314	C16H19N3O4S1	AMPICILLIN **800-79-3** *antibacterial*
8	-2.60	7.4	2530	1	"	AMPICILLIN
9 ★	-1.13	1.0	2196		"	AMPICILLIN

	logP	pH	Ref.	Note	MF	Name / CAS / activity
12180	-0.64	7.2	917	314	"	AMPICILLIN
1	-2.10	7.4	2464	41	C16H19N3O4S1	CEPHRADINE **38821-53-3** *antibacterial*
2	-1.90	4.5	2464	41	"	CEPHRADINE
3	-1.75	6.5	2464	41	"	CEPHRADINE
4	-1.50	7.4	2373	1	"	CEPHRADINE
85	-1.15	7.4	2192	1	"	CEPHRADINE
6	1.09		872	459	"	CEPHRADINE
7 ★	2.60		593		C16H19N3O5	2-CYANO-3-(4-NITROANILINO)-2-PENTENOIC ACID,2-ETHOXYETHYL ESTER
8 ★	-1.99	1.0	2196		C16H19N3O5S1	AMOXICILLIN **61336-70-7** *antibacterial*
9	-1.52	7.4	1289	1	"	AMOXICILLIN
12190	-1.17	7.4	1945	1	"	AMOXICILLIN
1 ★	-2.27	1.0	2196		C16H19N3O5S1	CEFROXADINE **51762-05-1** *antibacterial*
2	0.26	7.4	606	1	C16H19N3O6	MITOMYCIN-A **4055-39-4**
3	0.11	7.4	606	1	C16H19N3O6	MITOMYCIN-B **4055-40-7**
4	0.64	6.0	695		C16H19N3S1	ISOTHIPENDYL **482-15-5** *antihistaminic*
95 ✓	-1.26		2397	101	C16H19N4O3S1	IMIDAZOLIUM CHLORIDE ANALOG ✚
6 ★	5.55		2150	459	C16H20	ADAMANTANE,1-PHENYL **30176-62-6**
7	2.00	7.0	1056	1	C16H20Cl1N1O2	4-CHLOROPROPRANOLOL
8 ✓	0.06		2397	101	C16H20F3N4O3S1	IMIDAZOLIUM CHLORIDE ANALOG **132566-92-8** ✚
9 ✓	0.33		2397	101	C16H20F3N4O3S1	IMIDAZOLIUM CHLORIDE,1-ETHYL-2-HYDROXYIMINOMETHYL-3-(N-BENZYL-TRIFLUOROMETHYLSULFONAMIDO)ETHYL
12200 ✓	-1.00		2397	101	C16H20F3N4O3S1	IMIDAZOLIUM CHLORIDE,1-METHYL-2-BENZYLOXYIMINOMETHYL-3-(2-(N-METHYL)TRIFLUOROMETHYLSULFONAMIDO)ETHYL
1 ★	-2.27		1682		C16H20I3N3O7	M-BENZAMIDE,N,N'BIS(CH(CH2OH)2),5-ACETAMIDO,I3 **76350-23-7**
2	-2.52		1506		C16H20N1.Cl1	DIBENZYL-DIMETHYLAMMONIUMCHLORIDE **100-94-7**
3	-1.23	7.4	921	22	C16H20N1.I1	DIBENZYLDIMETHYL AMMONIUM IODIDE **52111-74-7**
4	-0.80		1506		C16H20N1.C7H7O3S1	DIBENZYL-DIMETHYLAMMONIUMTOSYLATE **73566-67-3**
5	1.06	7.4	785	1	C16H20N2	N',N-DIPHENYL-N-METHYLPROPYLENEDIAMINE
6	-0.76	2.0	1039	210	C16H20N2	PHENIRAMINE **86-21-5** *antihistamine*
7	0.81	7.4	1074	324	C16H20N2O2	4-(P-DIMETHYLANILINOMETHYL)PYRIDOXOL
8 ★	3.00		2467		C16H20N2O2S1	ANILINE,P-(4-DIETHYLAMINOPHENYL)SULFONYL **51688-32-5**
9	0.74		2089		C16H20N2O3	N-(3-DELTA-3-(METHOXYCARBONYLETHYL)PIPERIDIN-1-YL)BENZAMIDE **104642-73-1**
12210	0.95		2089		C16H20N2O3	N-(3-DELTA-4-METHOXYCARBONYLETHYLPIPERIDIN-1-YL)BENZAMIDE **104642-74-2**
1	0.42	7.4	2542	1	C16H20N2O3S2	THIOPHENE-2-SULFONAMIDE,4-(3-I-BUTYLAMINOMETHYL)BENZOYL
2	0.57	7.4	2542	1	C16H20N2O3S2	THIOPHENE-2-SULFONAMIDE,4-(BUTYLAMINOMETHYLBENZOYL)
3	0.70	7.4	2542	1	C16H20N2O3S2	THIOPHENE-2-SULFONAMIDE,4-(I-BUTYLAMINOMETHYLBENZOYL)
4	2.17		1043		C16H20N2O4	2,3-BENZO-8H-NAPHTHALENE-3,4-DICARBAMYLOXY(3,4-DI-EQ)
15	2.19		1043		"	2,3-BENZO-8H-NAPHTHALENE-3,4-DICARBAMYLOXY(3,4-DI-EQ)
6	2.39		1043		"	2,3-BENZO-8H-NAPHTHALENE-3,4-DICARBAMYLOXY(3,4-DI-EQ)
7	2.76		1277		C16H20N2O4S2	PYRITINOL **1098-97-1** *neurotropic agent*
8	0.36	7.4	2542	1	C16H20N2O4S2	THIOPHENE-2-SULFONAMIDE,4-(4-HYDROXY-3-DIETHYLAMINOMETHYL)BENZOYL
9 ✓	-0.56		2397	101	C16H20N3O4	IMIDAZOLIUM CHLORIDE,1-METHYL-2-HYDROXYIMINOMETHYL-3-BUTANOL,SALICYLIC ESTER **132540-08-0**
12220 ✓	-0.74		2397	101	C16H20N3O5	IMIDAZOLIUM CHLORIDE,1-METHYL-2-HYDROXYIMINOMETHYL-3-PROPANOL,P-METHOXYSALICYL ESTER
1 ★	1.85	12.0	2408		C16H20N4O1	MINAPRINE,4-DESMETHYL ✚
2	2.07	13.0	547	224	C16H20N4O2	2,4-DIAM-5-(3,5-DIMEO-4-I-PROPENYLBENZYL)PYRIMIDINE
3	1.86	7.4	1485	1	C16H20N4O2	2,4-DIAMINOPYRIMIDINE,5(3,5-DIMEO-4-I-PRENYL)BENZYL
4 ★	2.05	7.4	1485	2	"	2,4-DIAMINOPYRIMIDINE,5(3,5-DIMEO-4-I-PRENYL)BENZYL
25	1.61		1713		C16H20N4O2	6-PH-2-(3-(N3-DIME-UREIDO)PROPYL)-3(2H)-PYRIDAZINONE
6	1.68		1713		C16H20N4O2	6-PH-2-(3-(N3-ET-UREIDO)PROPYL)-3(2H)-PYRIDAZINONE
7	1.23		1713		C16H20N4O2	6-PH-2-(4-(N3-ME-UREIDO)BUTYL)-3(2H)-PYRIDAZINONE
8	1.08		1713		C16H20N4O2	6-PH-2-(5-UREIDO-PENTYL)-3(2H)-PYRIDAZINONE
9	1.03	7.4	693	1	C16H20N4O2	AZAPROPAZONE **13539-59-8** *anti-inflammatory*
12230	0.45	7.4	2637	1	C16H20N4O3S1	TORASEMIDE **56211-40-6** *anti-inflammatory, diuretic*
1	1.06	7.4	1485	1	C16H20N4O4	2,4-DIAMINOPYRIMIDINE,5-(3,5-DIMEO-4-CO2ET)BENZYL
2 ★	1.25	7.4	1485	2	"	2,4-DIAMINOPYRIMIDINE,5-(3,5-DIMEO-4-CO2ET)BENZYL
3	0.36	7.4	1485	1	C16H20N4O5	2,4-DIAMPYRIMIDINE,5-(3,5-DIMEO-4-OCH2CO2ME)BENZYL
4 ★	0.55	7.4	1485	2	"	2,4-DIAMPYRIMIDINE,5-(3,5-DIMEO-4-OCH2CO2ME)BENZYL
35	0.03	7.4	606	1	C16H20N4O5	PORFIROMYCIN **801-52-5** *antibacterial, antineoplastic*
6	-0.50	7.4	1485	1	C16H20N4O5S1	2,4-DIAMPYRIMIDINE,5(3,5-DIMEO-4-COCH2SO2ME)BENZYL
7 ★	-0.31	7.4	1485	2	"	2,4-DIAMPYRIMIDINE,5(3,5-DIMEO-4-COCH2SO2ME)BENZYL
8 ★	-0.02	7.4	594	1	C16H20N4O6	DIAZIQUONE **57998-68-2** *antineoplastic*
9	0.11		1045		C16H20N6O4	6(2(2-FURYL)5(HOME)1,3-DIOXAN-5YL)AM-2,4-DIAZ.TRIAZIN
12240	0.27		1045		"	6(2(2-FURYL)5(HOME)1,3-DIOXAN-5YL)AM-2,4-DIAZ.TRIAZIN
1 ★	4.92		594		C16H20O2Si1	DIPHENYLDIETHOXYSILANE **2553-19-7**
2 ★	3.34		306		C16H20O4	ETHYLACETOACETATE,3-PROPOXYBENZAL
3 ★	3.39		306		C16H20O5	DIETHYLMALONATE,3-ETHOXYBENZAL
4	1.09		636		C16H20O5	GAILLARDIN **14682-46-3**
45	2.89		306		C16H20O6	DIETHYLMALONATE,3,4-DIMETHOXYBENZAL
6 ★	5.96		1693		C16H20O6P2S3	TEMEFOS **3383-96-8** *insecticide~ectoparasiticide(vet)*
7	0.27	7.4	1204	1	C16H21Br2N3	N(CYHEXYLME)-N-(2,6-DIBRPH)AMINO)-2-IMIDAZOLINE
8	4.38		509		C16H21F3N2	BENZIMIDAZOLE,5-OCTYL-2-(TRIFLUOROMETHYL)
9	2.92	7.3	1900	314	C16H21F3N2O2	1-PIPERIDINEETHANOL,A-METHYL,3-CF3-PHENYLCARBAMATE
12250	-1.07		307	1	C16H21N1	MORPHINANHBR
1	1.79	7.4	1990	1	C16H21N1O1	HEXYLAMINE,6-(1-NAPHTHOXY)
2	1.94	7.4	1990	1	C16H21N1O1	N,N-DIMETHYLBUTYLAMINE,4-(1-NAPHTHOXY)
3	0.73	7.0	1362	1	C16H21N1O2	PROPRANOLOL **525-66-6** *cardiac depressant (anti-arrhythmic), anti-adrenergic (beta-receptor)*
4	0.73	7.0	2230	1	"	PROPRANOLOL
55	0.73	7.0	474	1	"	PROPRANOLOL
6	0.93	7.4	1628	1	"	PROPRANOLOL
7	1.03	7.5	1563	1	"	PROPRANOLOL
8	1.07	7.4	2465	41	"	PROPRANOLOL
9	1.08	7.0	1056	1	"	PROPRANOLOL
12260	1.11	7.5	1563	1	"	PROPRANOLOL
1	1.13	7.4	2178	1	"	PROPRANOLOL
2	1.15	7.4	2288	1	"	PROPRANOLOL
3	1.17	7.4	2375	1	"	PROPRANOLOL
4	1.17	7.4	487	1	"	PROPRANOLOL
65	1.18	7.4	1411	1	"	PROPRANOLOL
6	1.18	7.4	2018	1	"	PROPRANOLOL
7	1.21	7.4	1354	1	"	PROPRANOLOL
8	1.22	7.4	2092	1	"	PROPRANOLOL
9	1.25	7.4	1216	1	"	PROPRANOLOL

	logP	pH	Ref.	Note	MF	Name / **CAS** / *activity*
12270	1.25	7.4	1758	1	"	PROPRANOLOL
1	1.26	7.2	2016		"	PROPRANOLOL
2	1.26	7.2	2416	1	"	PROPRANOLOL
3	1.28	7.4	2399	1	"	PROPRANOLOL
4	1.31	7.4	1362	1	"	PROPRANOLOL
75	1.33	7.4	1157	324	"	PROPRANOLOL
6	1.41	7.4	1155	1	"	PROPRANOLOL
7	1.42	7.4	1746	1	"	PROPRANOLOL
8	1.49	7.4	2245	1	"	PROPRANOLOL
9	1.51	7.4	2473	1	"	PROPRANOLOL
12280	1.54	7.4	2598	1	"	PROPRANOLOL
1 ★	2.98	10.2	2465	41	"	PROPRANOLOL
2	3.09	7.4	487	2	"	PROPRANOLOL
3	3.21	7.4	1700	2	"	PROPRANOLOL
4	3.54	7.4	2245	2	"	PROPRANOLOL
85	-0.45	2.0	551		C16H21N1O2.H1Cl1	PROPRANOLOLHYDROCHLORIDE
6	1.66	7.4	1216	1	C16H21N1O3	2'-HYDROXYPROPRANOLOL
7	0.57	7.4	1216	1	C16H21N1O3	3'-HYDROXYPROPRANOLOL
8	0.49	7.4	2018	1	C16H21N1O3	4'-HYDROXYPROPRANOLOL **10476-53-6**
9	0.53	7.4	1216	1	C16H21N1O3	4'-HYDROXYPROPRANOLOL
12290	0.54	7.4	1411	1	C16H21N1O3	4'-HYDROXYPROPRANOLOL
1	1.23	8.1	2018		C16H21N1O3	4'-HYDROXYPROPRANOLOL
2	0.56	7.4	1216	1	C16H21N1O3	5'-HYDROXYPROPRANOLOL
3	0.46	7.4	1216	1	C16H21N1O3	6'-HYDROXYPROPRANOLOL
4	0.62	7.4	1216	1	C16H21N1O3	7'-HYDROXYPROPRANOLOL
95	1.40	7.4	1216	1	C16H21N1O3	8'-HYDROXYPROPRANOLOL
6 ★	2.90		2157		C16H21N1O3	PIPERIDINE,1-(2-BENZOYLOXYACETYL)-2,6-DIMETHYL
7 ★	2.62		2157		C16H21N1O3	PIPERIDINE,1-(2-BENZOYLOXYACETYL)-2-ETHYL
8	-0.12	7.4	2501	1	C16H21N1O4	BEFUNOLOL **39552-01-7** *anti-adrenergic (beta-receptor)*
9	-0.08	7.4	1354	1	"	BEFUNOLOL
12300 ★	2.02	13.0	2501		"	BEFUNOLOL
1	3.07		2662	459	C16H21N1O4	N-(3,4-DIETHOXYPHENYL)CARBAMATE,1-PENTYN-3YL ESTER *fungicide*
2	2.86		2662	459	C16H21N1O4	N-(3,4-DIETHOXYPHENYL)CARBAMATE,1-PENTYN-4YL ESTER *fungicide*
3 ★	0.67		2234		C16H21N1O5	ACETYLSALICYLIC ACID,2-(N,N-DIETHYLPROPANAMIDYL) ESTER **118247-09-9**
4	1.12	7.4	2018	1	C16H21N1O5S1	PROPRANOLOL-O-SULPHATE
5 ★	1.16	2.8	2018		"	PROPRANOLOL-O-SULPHATE
6	-1.81	3.0	2018		C16H21N1O6S1	PROPRANOLOL-4-SULPHATE
7	-1.77	7.4	2018	1		PROPRANOLOL-4-SULPHATE
8 ★	1.87		1898		C16H21N1O7	(VANILLINOXIME-O-ETACETYL)ACETIC ACID,ETHYL ESTER ✚
9	-0.23	2.0	1039	210	C16H21N3	TRIPELENNAMINE **91-81-6** *antihistaminic*
12310	1.69	7.4	416	1	C16H21N3O2	1-ME-3-PHENYL-5-BUTYLAMINO-6-METHYLURACIL
1	2.11	7.4	416	1	C16H21N3O2	1-ME-3-PHENYL-5-DIETAMINO-6-METHYLURACIL
2	1.48	7.4	416	1	C16H21N3O2	1-ME-3-PHENYL-5-I-BUTYLAMINO-6-METHYLURACIL
3	2.19	7.4	416	1	C16H21N3O2	1-PHENYL-3-METHYL-5-DIETHYLAMINO-6-METHYLURACIL
4	-0.78	7.4	651	1	C16H21N3O2	4-(2-N,N-DIMETHYLAMINOPROPIONYL)ANTIPYRENE
15 ★	2.27		593		C16H21N3O3	2-CYANO-3-(4-AMINOANILINO)-2-PENTENOIC ACID,2-ETHOXYETHYL ESTER
6 ★	-1.50	7.0	594	314	C16H21N3O3S1	METHIONYL-TRYPTOPHANE
7	1.13	7.4	2542	1	C16H21N3O4S3	THIOPHENE-2-SULFONAMIDE,4-(2-N-METHYLPIPERAZIN-1YLMETHYL)PHENYLSULFONYL
8	0.38	7.4	2542	1	C16H21N3O4S3	THIOPHENE-2-SULFONAMIDE,4-(4-METHYLPIPERAZIN-1YLMETHYL)PHENYLSULFONYL
9 ✓	-1.49		2397	101	C16H21N4O3S1	IMIDAZOLIUM CHLORIDE,1-METHYL-2-HYDROXYIMINOMETHYL-3-(1-PHENYLSULFONYLPYRROLIDIN-2YL)METHYL
12320 ★	1.70	7.4	1776	1	C16H21N5	N(2,5-DIME-PYRROL-1YL)6(DIALLYLAMINO)3-PYRIDIAZINAMINE **75841-87-1**
1 ★	0.95	7.4	1804	2	C16H21N5O2	1H-BENZOTRIAZINE,6-MEO-5-(1-ALLYLPYRR-3YL)CH2NHCO ✚
2	0.58	7.4	2375	1	C16H21N5O2	ALIZAPRIDE **59338-93-1** *neuroleptic, anti-emetic*
3	3.61		2162	459	"	ALIZAPRIDE
4 ★	1.64	7.4	1776	1	C16H21N5O2	N(2,5-DIME-PYRROL-1YL)6(4-MORPHOL)3PYRIDAZ-ACETAMIDE **89937-39-3** ✚
25	-0.40	7.4	1485	1	C16H21N5O3	2,4-DIAMINOPYRIMIDINE,5-(3,5-DIMEO-4-CONME2)BENZYL
6 ★	-0.21	7.4	1485	2	"	2,4-DIAMINOPYRIMIDINE,5-(3,5-DIMEO-4-CONME2)BENZYL
7	-0.31	7.4	651	1	C16H21N5O3	4-(3-N,N-DIMETHYLACETYLURYL)ANTIPYRENE
8 ★	0.62	7.0	2292	1	C16H21N5O3	DIOXOLANE,2-(4-N'-METHYLUREIDO)PHENYL-2-(1,2,4-TRIAZOL-1-YL)METHYL-4-ETHYL ✚
9	4.28	7.5	2651	28	C16H22Br1Cl1N2O3	SALICYLAMIDE,3-BROMO-5-CHLORO-6-ETHOXY-N-(1-ETHYLPYRROLIDIN-2YL)METHYL
12330	3.18	7.5	2651	28	C16H22Br1F1N2O3	BENZAMIDE,2,3-DIMETHOXY-5-BROMO-N-(1-FLUOROETHYLPYRROLIDIN-2YL)METHYL
1	3.22	7.5	2651	1	"	BENZAMIDE,2,3-DIMETHOXY-5-BROMO-N-(1-FLUOROETHYLPYRROLIDIN-2YL)METHYL
2	4.46	7.5	2651	28	C16H22Br1I1N2O3	SALICYLAMIDE,3-BROMO-5-IODO-6-ETHOXY-N-(1-ETHYLPYRROLIDIN-2YL)METHYL
3	0.20	7.4	2095	1	C16H22Br1N3O2	(5-BROMOIMIDAZOLE)-SPACED-ADRENOCEPTOR ANTAGONIST ✚
4	4.23	7.5	2651	28	C16H22Br2N2O3	SALICYLAMIDE,3,5-DIBROMO-6-ETHOXY-N-(1-ETHYLPYRROLIDIN-2YL)METHYL
35	3.19	7.5	2651	1	C16H22Cl1I1N2O3	ICLOPRIDE
6	4.79	7.5	2651	28	"	ICLOPRIDE
7 ★	4.03		1481		C16H22Cl1N1O1	P-CHLOROCINNAMAMIDE,B-I-PROPYL,N-S-BUTYL
8 ★	4.87		1481		C16H22Cl1N1O1	P-CHLOROCINNAMAMIDE,B-PROPYL,N-S-BUTYL
9 ★	3.60		2506		C16H22Cl1N1O3	DIETHATYL-ETHYL ✚
12340 ★	3.70		2506		C16H22Cl1N3O1	TEBUCONAZOLE *fungicide, steroid demethylation inhib.* ✚
1	3.91		2162	459	C16H22Cl1N3O1	ZETIDOLINE **51940-78-4** *neuroleptic*
2 ✓	-1.30		2172		C16H22Cl1N4O4	HYDROXY-CCNU,N-METHYLNICOTINIIC ACID ESTER
3	2.23	7.4	1938	1	C16H22Cl2N2O1	1,2-DIAMINOCYCLOHEXANE,N-DIMETHYL,N'-METHYL-(3,4-DICHLOROBENZOYL) **82657-23-6**
4	2.76	7.5	2651	1	C16H22Cl2N2O3	SALICYLAMIDE,3,5-DICHLORO-6-ETHOXY-N-(1-ETHYLPYRROLIDIN-2YL)METHYL
45	4.18	7.5	2651	28	"	SALICYLAMIDE,3,5-DICHLORO-6-ETHOXY-N-(1-ETHYLPYRROLIDIN-2YL)METHYL
6	4.56	7.5	2651	28	C16H22Cl2N2O3	SALICYLAMIDE,3,5-DICHLORO-6-METHOXY-N-(1-PROPYLPYRROLIDIN-2YL)METHYL
7	5.86		1369	459	C16H22Cl2O3	2,4-D,OCTYL ESTER **1928-44-5**
8	3.44	7.5	2651	28	C16H22F2F1N2O3	BENZAMIDE,2,3-DIMETHOXY-5-IODO-N-(1-FLUOROETHYLPYRROLIDIN-2YL)METHYL
9 ★	4.58		2479		C16H22F1N3O2	PHENYLBIURET,2-FLUORO-5-CYCLOHEXYL
12350	-0.19	7.0	1056	1	C16H22N2O2	1-(3-AMINOPROPYL)AM-3-(1-NAPHTHYLOXY)-2-PROPANOL
1 ★	2.94	6.8	2175	1	C16H22N2O2	3-HEXYL-3-(4-PYRIDYL)PIPERIDINE-2,6-DIONE
2	-0.27	7.0	1056	1	C16H22N2O2	4-AMINOPROPRANOLOL
3 ★	2.78		2237		C16H22N2O2	BENZOFURAN-3-AMINE,2,3-DIHYDRO-5-METHYL-N-(1-PIPERIDINYL)ACETYL
4	0.91		2089		C16H22N2O2	N-(3-DELTA-3-(3-METHOXYPROPYL)PIPERIDIN-1-YL)BENZAMIDE **104642-79-7**
55	1.49		2089		C16H22N2O2	N-(3-DELTA-4-(3-METHOXYPROPYL)PIPERIDIN-1-YL)BENZAMIDE **104642-80-0**
6 ★	2.41		2237		C16H22N2O2	BENZOFURAN-3-AMINE,2,3-DIHYDRO-6-METHOXY-N-(1-PIPERIDINYL)ACETYL
7	-0.51	7.0	1056	1	C16H22N2O4S1	4-SULFONAMIDOPROPRANOLOL
8 ★	3.20		1629		C16H22N2O4S1	CARBOFURAN,N-(N'ME,ETHYLCARBAMIDOTHIO) ✚
9 ★	3.50		1629		C16H22N2O5S1	N-THIO-(N'-ME-O-ETHYL-CARBAMYL)CARBOFURAN ✚

	logP	pH	Ref.	Note	MF	Name / **CAS** / *activity*
12360	1.48	7.4	2542	1	C16H22N2O5S3	THIOPHENE-2-SULFONAMIDE,4-(3-HYDROXY-4-N-METHYL-N-I-BUTYLAMINOMETHYL)PHENYLSULFONYL
1	0.68	7.4	2542	1	C16H22N2O5S3	THIOPHENE-2-SULFONAMIDE,4-(4-HYDROXY-3-N-METHYL-N-I-BUTYLAMINOMETHYL)PHENYLSULFONYL
2 ✓	-0.13		2397	101	C16H22N3O2	IMIDAZOLIUM CHLORIDE,1-METHYL-2-HYDROXYIMINOMETHYL-3-(1-PHENYLBUTOXY)METHYL **117983-36-5**
3 ✓	-0.04		2397	101	C16H22N3O2	IMIDAZOLIUM CHLORIDE,1-METHYL-2-HYDROXYIMINOMETHYL-3-(4-PHENYLBUTOXY)METHYL **117983-37-6**
4 ★	3.40		2424	459	C16H22N4O1	BENZOTRIAZOLE,2-CYCLOPENTYL-5-TRIMETHYLACETAMIDO
65	3.18	7.4	1485	1	C16H22N4O2	2,4-DIAMINOPYRIMIDINE,5-(3,5-DIMEO-4-I-PROPYL)BENZYL
6 ★	3.37	7.4	1485	2	"	2,4-DIAMINOPYRIMIDINE,5-(3,5-DIMEO-4-I-PROPYL)BENZYL
7	0.35	7.4	651	1	C16H22N4O2	4-(2-N,N-DIMEAMIDOETHYLAMINO)ANTIPYRENE
8	-0.05	7.4	651	1	C16H22N4O2	AMINOPROPYLON **3690-04-8** *analgesic*
9 ★	4.28		2424		C16H22N4O2	BENZOTRIAZOLE,2-CYCLOPENTYL-5-(O-ISOBUTYLCARBAMOYL)
12370	1.28	7.4	2298	1	C16H22N4O2S1	1,2,4-OXADIAZOLE-5-(DIETHYLAMINOETHYL)THIOHYDROXAMATE,3-BENZYL **103499-11-2**
1 ★	2.48	7.4	2298	2	"	1,2,4-OXADIAZOLE-5-(DIETHYLAMINOETHYL)THIOHYDROXAMATE,3-BENZYL
2	0.49	4.0	360	2	C16H22N4O2S1	2-BU-5,6-DIME-4-SULFANILAMIDOPYRIMIDINE
3	0.69	7.4	651	1	C16H22N4O2.C1H4	4-(1-ME-2-N,N-DIMEAMIDOETHYLAMINO)ANTIPYRENE
4	1.30	7.4	1485	1	C16H22N4O3	2,4-DIAMINOPYRIMIDINE,5-(3,5-DIMEO-4-COHME2)BENZYL
75 ★	1.49	7.4	1485	2	"	2,4-DIAMINOPYRIMIDINE,5-(3,5-DIMEO-4-COHME2)BENZYL
6 ★	3.82	2.1	2018		C16H22N4O3	TOMELUKAST **88107-10-2** *anti-asmatic (leukotriene antagonist)*
7 ★	-0.71		2377		C16H22N4O4	AC-ALA-GLY-PHE-N
8 ★	-1.14		2377		C16H22N4O4	AC-GLN-PHE-N
9 ★	-1.03		2377		C16H22N4O4	AC-PHE-GLN-N
12380 ★	1.00		2529		C16H22N4O4	DDI,5'-(2-METHYLVALERATE)
1	0.38	7.4	1485	1	C16H22N4O4	TETROXOPRIM **53808-87-0** *antibacterial*
2 ★	0.56	7.4	1485	2	"	TETROXOPRIM
3	0.34	7.4	1485	1	C16H22N6O1	2,4-DIAMPYRIMIDINE,5-(3-NHME-4-ME-5-NMECOME)BENZYL
4 ★	0.53	7.4	1485	2	"	2,4-DIAMPYRIMIDINE,5-(3-NHME-4-ME-5-NMECOME)BENZYL
85 ★	2.18		2289		C16H22N6O2	TETRAZOLE,1-CYLOHEXYL-5-(4-DIMETHYLUREYL)PHENOXY **117121-33-2**
6 ★	2.67		2289		C16H22N6O2	TETRAZOLE,2-CYCLOHEXYL-5-(4-DIMETHYLUREYL)PHENOXY **117121-45-6**
7 ★	-2.46	7.4	2427	1	C16H22N6O4	PROTIRELIN **24305-27-9** *prothyrotropin*
8 ★	0.50	7.0	1793	1	C16H22N6O4S1	N(4-(6-PURINYLTHIO)VALERYL)GLYCYLGLYCINE,ET.ESTER
9	0.34	7.4	2591	1	C16H22N6O6	AZT,5'-N-MORPHOLINOACETYL **125762-97-2** *antiretroviral prodrug*
12390	3.94		1524		C16H22O2	1(3H)-ISOBENZOFURANONE-3-OCTYL
1 ★	4.11		1587		C16H22O4	DI-I-BUTYLPHTHALATE **84-69-5**
2 ★	4.72		547		C16H22O4	O-DIBUTYLPHTHALATE **84-74-2** *repallant,insect*
3 ★	5.53		547		C16H22O4	P-DIBUTYLPHTHALATE **1962-75-0**
4 ★	3.35	3.4	1708		C16H22O4	PHENYLACETIC ACID,3-METHOXY-4-CY-HEXYLMETHOXY
95 ★	0.63		579		C16H22O11	A-D-GLUCOSEPENTAACETATE **604-68-2**
6	0.70	7.5	2651	1	C16H23Br1N2O3	REMOXIPRIDE **80125-14-0** *antipsychotic*
7	2.13	7.5	2651	28	"	REMOXIPRIDE
8	4.11	7.5	2651	28	C16H23Br1N2O3	SALICYLAMIDE,3-BROMO-6-ETHOXY-N-(1-ETHYLPYRROLIDIN-2YL)METHYL
9	1.53	7.4	1938	1	C16H23Cl1N2O1	1,2-DIAMINOCYCLOHEXANE,N-DIMETHYL,N'-METHYL-(P-CHLOROBENZOYL) **67579-11-7**
12400	2.61	7.4	2651	28	C16H23Cl1N2O3	BENZAMIDE,2,3-DIMETHOXY-5-CHLORO-N-(1-ETHYLPYRROLIDIN-2YL)METHYL
1	3.58	7.5	2651	28	C16H23Cl1N2O3	SALICYLAMIDE,5-CHLORO-6-ETHOXY-N-(1-ETHYLPYRROLIDIN-2YL)METHYL
2 ★	2.53		2172		C16H23Cl1N4O4	HYDROXY-CCNU,1-METHYL-1,4-DIHYDRONICOTINIC ACID ESTER
3	2.43	7.5	2651	R28~459	C16H23I1N2O3	BENZAMIDE,2,6-DIMETHOXY-5-IODO-N-(1-ETHYLPYRROLIDIN-2YL)METHYL
4	1.43	7.4	2651	1	C16H23I1N2O3	EPIDEPRIDE
5	1.85	7.5	2651	1	"	EPIDEPRIDE
6	3.01	7.5	2651	28	"	EPIDEPRIDE
7	2.78	7.5	2651	1	C16H23I1N2O3	SALICYLAMIDE,3-IODO-6-ETHOXY-N-(1-ETHYLPYRROLIDIN-2YL)METHYL
8	2.84	7.5	2651	1	"	SALICYLAMIDE,3-IODO-6-ETHOXY-N-(1-ETHYLPYRROLIDIN-2YL)METHYL
9	4.45	7.5	2651	28	"	SALICYLAMIDE,3-IODO-6-ETHOXY-N-(1-ETHYLPYRROLIDIN-2YL)METHYL
12410	1.60	7.4	2633	R1~314	C16H23N1	OCTAHYDROPHENANTHREN-4A-AMINE,N,9-DIMETHYL *nmda antagonist*
1	3.96		2633	459		OCTAHYDROPHENANTHREN-4A-AMINE,N,9-DIMETHYL
2	1.15	7.4	2633	1	C16H23N1	OCTAHYDROPHENANTHREN-4A-AMINE,N-ETHYL *nmda antagonist*
3	3.41		2633	459		OCTAHYDROPHENANTHREN-4A-AMINE,N-ETHYL
4	-1.52		534		C16H23N1O1	2,5-DIMETHYL-9-ET-2'-OH-6,7-BENZOMORPHAN
15	-1.12		307	1	C16H23N1O1	2,5-DIMETHYL-9-ET-2'-OH-6,7-BENZOMORPHAN
6	-1.31		534		C16H23N1O1	2,9-DIMETHYL-5-ET-2'-OH-6,7-BENZOMORPHAN
7	-0.97		307	1	C16H23N1O1	2,9-DIMETHYL-5-ET-2'-OH-6,7-BENZOMORPHAN
8	1.51	7.4	2633	1	C16H23N1O1	OCTAHYDROPHENANTHREN-4A-AMINE,N-METHYL-6-METHOXY *nmda antagonist*
9	3.75		2633	459		OCTAHYDROPHENANTHREN-4A-AMINE,N-METHYL-6-METHOXY
12420	0.99	7.4	2648	1	C16H23N1O2	4-PIPERIDINOL,1,2,3-TRIMETHYL-4-PHENYL,ACETATE
1	1.29	7.4	2648	1	"	4-PIPERIDINOL,1,2,3-TRIMETHYL-4-PHENYL,ACETATE
2 ★	2.51	7.4	2648	2	"	4-PIPERIDINOL,1,2,3-TRIMETHYL-4-PHENYL,ACETATE
3 ★	2.52	7.4	2648	2	"	4-PIPERIDINOL,1,2,3-TRIMETHYL-4-PHENYL,ACETATE
4	1.73	7.4	2648	448	C16H23N1O2	ALPHAPRODINE **77-20-3** *analgesic (narcotic)*
25 ★	2.83	7.4	2648	2	"	ALPHAPRODINE
6	1.55	7.4	2648	448	"	BETAPRODINE **468-59-7**
7 ★	2.43	7.4	2648	2	"	BETAPRODINE
8	1.80	7.4	2399	1	C16H23N1O2	BUFURALOL **54340-62-4** *anti-adrenergic (beta-receptor)*
9	3.50	7.4	1700	1	"	BUFURALOL
12430	0.72	5.7	2241	1	C16H23N1O2	HEXYLCAINE **532-77-4** *anesthetic (local)*
1 ★	2.99		2157		C16H23N1O3	2-(BENZOYLOXY)-(N-METHYL,N-HEXYL)ACETAMIDE
2	3.55		1167	459	C16H23N1O3	P-METHOXYCINNAMIC ACID,DIETHYLAMINOETHYL ESTER
3	-0.19	7.4	2501	1	C16H23N1O3	PARGOLOL **47082-97-3** *anti-adrenergic (beta-blocker)*
4 ★	2.32	13.0	2501		"	PARGOLOL
35 ★	1.34		1524		C16H23N1O4	1(3H)-ISOBENZOFURANONE-3-BUTYL-6-N,N-DI(2-HYDROXYET)
6 ★	3.33		2142		C16H23N1O4	3,4-DIETHOXYPHENYLCARBAMATE,O-CYCLOPENTYL
7	3.42		2662	459	C16H23N1O4	N-(3,4-DIETHOXYPHENYL)CARBAMATE,1-PENTEN-3YL ESTER *fungicide*
8	3.37		2662	459	C16H23N1O4	N-(3,4-DIETHOXYPHENYL)CARBAMATE,1-PENTEN-4YL ESTER *fungicide*
9 ★	3.20		2413	459	C16H23N1O4	TETRAHYDROPYRAN-2,4-DIONE,3[1-(ETHOXYIMINO)BUTYL]-6,6-SPIRO-2-CYCLOHEXENYL ✚
12440	0.83	7.4	1938	1	C16H23N3O3	1,2-DIAMINOCYCLOHEXANE,N-DIMETHYL,N'-METHYL-(P-NITROBENZOYL) **98587-45-2**
1 ★	0.43		2377		C16H23N3O3	AC-PHE-VAL-N
2 ★	3.40		1629		C16H23N3O3S2	N-((N"-I-PR)-4-ME-PHENYLCARBAMYLTHIO)METHOMYL ✚
3 ★	3.40		1629		C16H23N3O3S2	N-(N"-BU)-PHENYLCARBAMYLTHIO)METHOMYL
4 ★	-0.20		2377		C16H23N3O4	AC-TYR-VAL-N
45 ★	-2.33	7.0	2374		C16H23N3O4	PHENYLALANYLVALINYLGLYCINE
6	0.80	7.0	801	1	C16H23N3O6	1-(5-CYCLOHEXYLFORMYLARABINOSYL)CYTOSINE
7	-1.38		1525		C16H23N3O7S2	7B-(1-METHYLPROPYL)THIOACETYLAMINO-CEPHAMYCIN **57793-08-5**
8	-1.57		1525		C16H23N3O7S2	7B-BUTYLTHIOACETYLAMINO-CEPHAMYCIN
9 ✓	-1.64	7.4	2397	101	C16H23N4O1	IMIDAZOLIUM CHLORIDE,1-METHYL-2-BENZYLOXYIMINOMETHYL-3-(2-DIMETHYLAMINO)ETHYL

	logP	pH	Ref.	Note	MF	Name / **CAS** / *activity*
12450 ✓	-1.25		2397	101	C16H23N4O3S1	IMIDAZOLIUM CHLORIDE,1-ETHYL-2-HYDROXYIMINOMETHYL-3-(3-N-METHYL-PHENYLSULFONAMIDO)PROPYL
1 ✓	-1.27		2397	101	C16H23N4O3S1	IMIDAZOLIUM CHLORIDE,1-METHYL-2-HYDROXYIMINOMETHYL-3-(3-(N-ETHYL)PHENYLSULFONAMIDO)PROPYL
2 ✓	-0.94		2397	101	C16H23N4O3S1	IMIDAZOLIUM CHLORIDE,1-METHYL-2-HYDROXYIMINOMETHYL-3-(N-PROPYL-PHENYLSULFONAMIDO)ETHYL
3 ✓	-1.03		2397	101	C16H23N4O3S1	IMIDAZOLIUM CHLORIDE,1-PROPYL-2-HYDROXYIMINOMETHYL-3-(N-METHYL-SULFONAMIDO)ETHYL
4	2.43		1479	459	C16H23N5O1	2,4-DIAMINO-5-(3-DIETAMINO-4-MEO-BENZYL)PYRIMIDINE
55 ★	2.71	7.4	1776	1	C16H23N5O1	N(2,5-DIET-PYRROL-1YL)6(N-MORPHOLIN)3-PYRIDIAZINAM **75841-80-4**
6 ★	2.59	7.4	1776	1	C16H23N5O1	N(2,5DIME-PYRROL-1YL)6(N(2,6DIMEMORPHOL)3PYRIDAZINAM **89937-40-6**
7 ★	1.39	7.0	1793	1	C16H23N5O3S1	N(4-(6-PURINYLTHIO)VALERYL)GLYCINE,BUTYL ESTER **23374-49-4**
8	0.67		547		C16H24Cl1N3O10	N1-NO-N1-CLET-N3-3(1-ME-TRIACETYLGLUCOSIDYL)UREA
9	-2.65	7.4	2617	1	C16H24N2	2-METHYLAMINOMETHYL-N-METHYLHEXAHYDROFLUOREN-4A-AMINE
12460	2.48		2617	2	"	2-METHYLAMINOMETHYL-N-METHYLHEXAHYDROFLUOREN-4A-AMINE
1	-1.14	7.4	2633	1	C16H24N2	OCTAHYDROPHENANTHREN-4A-AMINE,N-METHYL-2-AMINOMETHYL *nmda antagonist*
2	2.00		2633	459	"	OCTAHYDROPHENANTHREN-4A-AMINE,N-METHYL-2-AMINOMETHYL
3	0.40	7.4	1350	1	C16H24N2O1	XYLOMETAZOLINE **526-36-3** *adrenergic (vascoconstrictor)*
4	1.45		1407		C16H24N2O1	1,2,3,6-H4-PYRIDINE,N-(1-ADAMANTYL)-CONH-
65	-0.38	7.4	2320	1	C16H24N2O1	2,6-BIS(1-PYRROLIDINYLMETHYL)PHENOL
6	-0.46	7.4	1398	1	C16H24N2O1	OXYMETAZOLINE **1491-59-4** *adrenergic (vasoconstrictor)*
7	-0.17	7.4	1097	1	"	OXYMETAZOLINE
8	2.72		1725	459	"	PYRIDO-PYRIMIDINE ANALOG (RINGS:868)
9	0.35	7.4	1938	1	C16H24N2O2	1,2-DIAMINOCYCLOHEXANE,N-DIMETHYL-N'-METHYL-(P-HYDROXYBENZOYL) **98587-47-4**
12470	2.12	7.3	1900	314	C16H24N2O2	1-PIPERIDINEETHANOL,A-METHYL,3-METHYLPHENYLCARBAMATE
1	0.05	7.4	2274	1	C16H24N2O2	INDOLE,2-METHYL-4-(3-T-BUTYLAMINO-2-HYDROXY)PROPOXY **23869-98-9**
2	1.59	7.3	1356	314	C16H24N2O3	2-ETO-CARBANILATE-O-2(N-PIPERIDINYL)ETHYL **55792-08-0**
3	3.15	7.3	1356	2	"	2-ETO-CARBANILATE-O-2(N-PIPERIDINYL)ETHYL
4	1.77	7.3	1356	314	C16H24N2O3	3-ETO-CARBANILATE-O-2(N-PIPERIDINYL)ETHYL **55792-09-1**
75	2.89	7.3	1356	2	"	3-ETO-CARBANILATE-O-2(N-PIPERIDINYL)ETHYL
6	1.42	7.3	1356	314	C16H24N2O3	4-ETO-CARBANILATE-O-2(N-PIPERIDINYL)ETHYL **55792-10-4**
7	2.98	7.3	1356	2	"	4-ETO-CARBANILATE-O-2(N-PIPERIDINYL)ETHYL
8	1.88	7.5	2651	26	C16H24N2O3	BENZAMIDE,2,3-DIMETHOXY-N-(1-ETHYLPYRROLIDIN-2YL)METHYL
9	0.56	7.5	2651	1	C16H24N2O3	BENZAMIDE,2,5-DIMETHOXY-N-(1-ETHYLPYRROLIDIN-2YL)METHYL
12480	2.10	7.5	2651	28	"	BENZAMIDE,2,5-DIMETHOXY-N-(1-ETHYLPYRROLIDIN-2YL)METHYL
1	1.38	7.5	2651	28	C16H24N2O3	BENZAMIDE,2,6-DIMETHOXY-N-(1-ETHYLPYRROLIDIN-2YL)METHYL
2	-0.46	7.4	2178	1	C16H24N2O3	CARTEOLOL **51781-06-7** *anti-adrenergic (beta-receptor)*
3	0.70	7.3	1960	314	C16H24N2O3	PHENYLCARBANILIC ACID,2-METHOXYMETHYL,2-(1-PIPERIDYL)ETHYL ESTER ✚
4 ★	2.20	7.3	1960	2	"	PHENYLCARBANILIC ACID,2-METHOXYMETHYL,2-(1-PIPERIDYL)ETHYL ESTER
85	3.24	7.5	2651	28	C16H24N2O3	SALICYLAMIDE,6-METHOXY-N-(1-ETHYLPYRROLIDIN-2YL)METHYL
6	-1.21	7.4	2501	1	C16H24N2O4	DIACETOLOL **22568-64-5** *anti-adrenergic (beta-receptor)*
7	-1.10	7.4	1859	1	"	DIACETOLOL
8 ★	0.94	13.0	2501		"	DIACETOLOL
9	1.88		1018		C16H24N2O5S1	O(2-I-PRO-PHEN)-O'ET-N,N'-THIODICARBAMATE,N,N'-DIET
12490 ★	-0.37		562		C16H24N4O4	2,5DI(N-ME-OHETAMINO)3,6-DIAZIRIDINYLQUINONE
1	-1.97		562		C16H24N4O6	2,5DI(2,3-DIOHPRAMINO)3,6-DIAZIRIDINYLQUINONE
2	1.87	7.4	1485	1	C16H24N6	2,4-DIAMINOPYRIMIDINE,5(3,5-(NME2)2-4-ME)BENZYL
3 ★	2.06	7.4	1485	2	"	2,4-DIAMINOPYRIMIDINE,5(3,5-(NME2)2-4-ME)BENZYL
4	1.86	7.4	1485	1	C16H24N6O1	2,4-DIAMINOPYRIMIDINE,5(3,5-(NME2)2-4-MEO-)BENZYL
95 ★	2.05	7.4	1485	2	"	2,4-DIAMINOPYRIMIDINE,5(3,5-(NME2)2-4-MEO-)BENZYL
6 ★	4.05		2415	459	C16H24O2	CYCLOHEXANOL,3-(2-HYDROXY-4-T-BUTYL)PHENYL
7	1.74		1937	459	C16H24O4	4-HYDROXY-2-PYRONE,3,5-DIBUTYL-6-(2-PROP-2-ONYL) **98393-91-0**
8 ★	0.58		2308		C16H24O6	BENZO-18-CROWN-6-ETHER **14098-24-9**
9 ★	1.07		511		C16H24O6	GLUCOPYRANOSIDE,2-ISOPROPYL-5-MEPHENYL(BETA)
12500 ★	1.01		511		C16H24O6	GLUCOPYRANOSIDE,3-T-BUTYLPHENYL(BETA)
1 ★	1.18		511		C16H24O6	GLUCOPYRANOSIDE,4-T-BUTYL(BETA) **29074-02-2**
2	2.83		2216		C16H25Cl1N2O1	BUTYRAMIDE,2-(O-CHLOROPHENYL)-4-(DI-I-PROPYLAMINO)
3 ★	1.14		2518	122	C16H25F1N4O3	2'-DEOXY-3'-FLUORO-CYTIDINE,4N-DI-I-PROPYLAMINO)METHYLENE
4 ★	1.35		2518	122	C16H25F1N4O3	2'-DEOXY-3'-FLUORO-CYTIDINE,N4-(DIPROPYLAMINO)METHYLENE
5	1.26	7.4	1628	1	C16H25N1O1	BUTIDRINE **7433-10-5** *anti-adrenergic, antiarrhythmic*
6	1.53	7.4	2204	1	C16H25N1O1	TETRAHYDRONAPHTHALENE,2-DIPROPYLAMINO-5-HYDROXY *dopamine-agonist*
7	1.45	7.4	2204	1	C16H25N1O1	TETRAHYDRONAPHTHALENE,2-DIPROPYLAMINO-7-HYDROXY
8	1.63	7.4	2096	1	"	TETRAHYDRONAPHTHALENE,2-DIPROPYLAMINO-7-HYDROXY
9	3.18		803		C16H25N1O2	3,5-DI-T-BUTYL-4-HYDROXYACETANILIDE
12510 ★	3.23		547		C16H25N1O2	N,N-DIBUTYLPHENOXYACETAMIDE
1	0.77	7.4	2261	1	C16H25N1O2	TETRALIN,5,6-DIHYDROXY-2-DIPROPYLAMINO
2	0.67	7.4	2261	1	C16H25N1O2	TETRALIN,6,7-DIHYDROXY-2-DIPROPYLAMINO
3	0.66	7.4	2648	1	C16H25N1O2	TRAMADOL
4 ★	2.63	7.4	2648	2	"	TRAMADOL
15	0.71	7.4	2648	1	"	TRAMADOL **27203-92-5**
6 ★	2.51	7.4	2648	2	"	TRAMADOL
7 ★	3.17		2045		C16H25N1O3	MOXISYLYTE **54-32-0** *vasodilator (peripheral)*
8 ★	3.66		2413	459	C16H25N1O4	TETRAHYDROPYRAN-2,4-DIONE,3[1-(ETHOXYIMINO)BUTYL]-6,6-SPIRO-(3-METHYL)CYCLOPENTYL
9 ★	3.62		2413	459	C16H25N1O4	TETRAHYDROPYRAN-2,4-DIONE,3[1-(ETHOXYIMINO)BUTYL]-6,6-SPIRO-CYCLOHEXYL
12520 ★	3.40		2413	459	C16H25N1O4	TETRAHYDROPYRAN-2,4-DIONE,3[1-(ETHOXYIMINO)BUTYL]-6,METHYL-6-(2-BUTEN-2YL)
1 ★	1.60		1589		C16H25N1O5S1	(N-PIPERIDINYL)ET-(3,4,5-TRIMETHOXY)PHENYL-SULFONE
2	4.89		1361	459	C16H25N1S1	2(AX)METHOXY-THIOPHENCYCLIDINE(THPH=AX.) ✚
3	6.39		1361	459	C16H25N1S1	2(AX)METHOXY-THIOPHENCYCLIDINE(THPH=EQ.) ✚
4	7.29		1361	459	C16H25N1S1	2-(AX.)-METHYL-THIOPHENCYCLIDINE(THPH=AX.)
25	8.45		1361	459	C16H25N1S1	2-(AX.)-METHYL-THIOPHENCYCLIDINE(THPH=EQ.)
6 ★	4.40		2506		C16H25N1S1	TIOCARBAZIL **36756-79-3** *herbicide*
7	-1.60	7.6	2007	1	C16H25N2O3S1.Cl1	BENZOYLTHIOFORMOHYDROXIMATE,4-METHOXY-S-(2-DIETHYLMETHYLAMMONIO)ETHYL CHLORIDE **102976-76-1**
8	0.85	7.4	365	2	C16H25N3O1	N,N-DIET-PIPERAZINE-2-CARBOXANILIDE,2'ME
9	3.55	7.4	459	1	C16H25N3O2	1-CYCLOHEXYL-3-ME-5-PYRROLIDINO-6-METHYLURACIL
12530	2.77	7.4	1214	1	C16H25N3O2	3-OCTYL-1-(4-CARBOXYPHENYL)-3-METHYLTRIAZENE **66974-67-2** *antineoplastic*
1	2.96	6.3	2589	459	"	3-OCTYL-1-(4-CARBOXYPHENYL)-3-METHYLTRIAZENE
2	0.24	12.0	541		C16H25N3O2	SODIUM BENZOATE, O-(3-ME-3-OCTYLTRIAZINE)
3	2.39	7.4	416	1	C16H25N3O3	1-CYCLOHEXYL-3-ME-5-MORPHOLINO-6-METHYLURACIL
4	2.96		2162	459	C16H25N3O4S1	BENZAMIDE,4-AMINO-5-ETHYLSULFONYL-N-(1-METHYLPYRROLID-2-YL)METHYL
35	2.06		2162	459	C16H25N3O4S1	PROSULPRIDE **68556-59-2**
6 ★	0.82	7.4	2350	1	C16H25N3O4S2	DEOXYSPARSOMYCIN,S-BUTYL ANALOG ✚
7	-0.29	7.0	1812	1	C16H25N3O5	5'-AMINO-3'O-HEXOYL-5'-DEOXYTHYMIDINE
8 ★	0.14	7.4	2350	1	C16H25N3O5S2	SPARSOMYCIN,S-BUTYL ANALOG
9 ★	2.06	7.4	1776	1	C16H25N5O2	N(2,5-DIME-PYRROL-1YL)6(N(ETOME)2)3-PYRIDAZINAMIN **75842-02-3**

	logP	pH	Ref.	Note	MF	Name / CAS / activity
12540 ★	1.55	7.4	1776	1	C16H25N5O2	N(2,5DIME-PYRROL-1YL)6(N(PROH)2)-3-PYRIDIAZINAMIN **75841-97-3**
1	7.37		2253	459	C16H26	DECYLBENZENE **104-72-3**
2	5.75		401	335	C16H26Cl1N3O2	1-(2-CLET)-1-NO-3-(3,5,7-TRIME-1-ADAMANTYLUREA
3 ★	4.35	1.0	591	342	C16H26Cl1N5O1	OCTYLGUANIDINE,2-(3-P-CHLOROPHENYL)-1-UREYL
4	2.39	7.4	1414	1	C16H26N2O1	2(PR,ET-AMINOET)-N(2,6-DIMEPHENYL)PROPIONAMIDE
45	2.63		1352		C16H26N2O1	2-(DIETHYLAMINO)BUTYRYL-2',6'-XYLIDIDE
6	2.06		1352		C16H26N2O1	3-(DIETHYLAMINOBUTYRYL-2',6'-XYLIDIDE
7	-0.17	4.8	1020		C16H26N2O1	ACETAMIDE,N(2,6-DIMEPH),N-(2-DIETHYLAMINOETHYL)
8	0.73	7.4	253	1	C16H26N2O1	N-(3-PHENYL-3-AMIDOPROPYL)DI-I-PROPYLAMINE
9	-0.11	4.8	1020		C16H26N2O1	PROPIONAMIDE,N(2,6-DIMEPH),N-(3-DIMEAMINOPROPYL)
12550 ★	4.14		29		C16H26N2O2	BENZOIC ACID,P-AMYLAMINO,N,N-DIMEAMINOETHYL ESTER
1	2.96	7.2	1150		C16H26N2O3	2-ME-4(O-BU)CARBANILATE-O-2(N,N-DIMETHYLAMINO)ET
2	0.08	5.7	2241	1	C16H26N2O3	PROPARACAINE **499-67-2** anesthetic (topical, ophthalmic)
3	-1.03	7.4	2501	1	C16H26N2O4	CETAMOLOL **34919-98-7** anti-adrenergic (beta-receptor)
4 ★	1.36	13.0	2501	"		CETAMOLOL
55	0.11	7.2	846	314	C16H26N2O4S1	6-(N-OCTANOYLAMINO)PENICILLANIC ACID **525-97-3**
6	3.32		1261			6-(N-OCTANOYLAMINO)PENICILLANIC ACID
7 ★	4.70	6.6	2590	459	C16H26N4O1	1-(4-CARBAMOYLPHENYL)-3-METHYL-3-OCTYLTRIAZENE **66521-49-1** antineoplastic
8 ★	0.72		2518	122	C16H26N4O3	2',3'-DIDEOXYCYTIDINE,4N-DI-I-PROPYLAMINO)METHYLENE
9 ★	0.95		2518	122	C16H26N4O3	2',3'-DIDEOXYCYTIDINE,N4-(DIPROPYLAMINO)METHYLENE
12560	0.79	7.2	170	314	C16H26N6O2	DIAZINEDICARBOXYLICACIDBIS(N'-ALLYLPIPERAZIDE)
1	1.91	7.4	2310	2	C16H26N6O2	XANTHINE,1-METHYL-3-I-BUTYL-8-N-PIPERAZINYLETHYL
2 ★	1.81	7.4	2481	1	C16H26N6O2S1	CIMETIDINE,1-PIVALOYLMETHYL ANALOG ✚
3 ★	5.60		2011		C16H26O4S2	PROPANEDIOIC ACID,1,3-DITHIOLAN-2-YLIDINE,DIPENTYL ESTER **61782-00-1**
4	2.05		1396	459	C16H26O5	ARTEMISININE,METHYL ETHER (ALPHA) **71963-77-4** antimalarial ✚
65	2.86		1396	459	"	ARTEMISININE,METHYL ETHER (ALPHA)
6 ✦	0.76		2308		C16H26O6	1,2-BIS-(2-(2-METHOXYETHOXY)ETHOXY)BENZENE
7 ★	0.86		2308		C16H26O6	1,3-BIS-(2-(2-METHOXYETHOXY)ETHOXY)BENZENE
8 ★	0.76		2308		C16H26O6	1,4-BIS-(2-(2-METHOXYETHOXY)ETHOXY)BENZENE
9 ✓	0.38		2397	101	C16H27F3N5O6S1	IMIDAZOLIUM CHLORIDE ANALOG ✚
12570	0.46	1.0	547	342	C16H27N1	2-(N,N_DIBUTYLAMINO)-1-PHENYLETHANE HYDROCHLORIDE
1	0.86	1.0	547	342	C16H27N1O1	2-N,N-DIBUTYLAMINO-1-PHENYLETHANOL HYDROCHLORIDE
2 ★	3.16	13.0	547	224	C16H27N1O1	2-N,N-DIBUTYLAMINO-1-PHENYLETHANOL
3	4.43	8.0	1011	2	C16H27N1O2	2,5-DIMETHOXY-4-PENTYLAMPHETAMINE
4	1.17	7.4	2261	1	C16H27N1O2	DOPAMINE,N,N-DIBUTYL
75	1.19	7.0	477	1	C16H27N1O2	N-I-PR-3-(M-S-BUTYLPHENOXY)-2-PROPANOLAMINE
6 ★	2.05		1589		C16H27N1O5S1	(N-ME-BU)AMINOETHYL-(3,4,5-TRIMEO)PHENYL-SULFONE
7	1.12		301		C16H27N2O1.Cl1	N1-DECYLNICOTINAMIDECHLORIDE
8 ✓	-0.29		1729	820	C16H27N2O1.I1	3-CARBAMOYLPYRIDINIUM IODIDE,N-DECYL **35041-49-7**
9 ★	1.70	7.4	1804	2	C16H27N3O4S1	2-MEO-4-AM-5-SO2ET-BENZAMIDE,N-(2-DIETAMINOET)
12580	3.89		723		C16H27N5O1	2,4-DIAMINO-6-DICYCLOHEXYLAMINOPYRIMIDINE-3-OXIDE
1	-3.00		1506		C16H28N1.Cl1	BENZYL-TRIPROPYLAMMONIUMCHLORIDE **5197-87-5**
2	-1.26		1635	820	C16H28N1.Cl1	BENZYLDIMETHYLHEPTYLAMMONIUMCHLORIDE
3 ✓	0.19		1729	820	C16H28N1.I1	2-METHYLPYRIDINIUM IODIDE,N-DECYL **14402-23-4**
4	2.39	7.0	1970	1	C16H28N2O1S2.Tc1	2,9-DIMETHYL-2,9-DIMERCAPTO-5-(2-HYDROXYPHENYL)-4,7-DIAZADECANE,TECHNICIUM COMPLEX **96929-53-2**
85	2.41	7.0	1970	1	C16H28N2O1S3.Tc1	2,9-DIMETHYL-2,9-DIMERCAPTO-5-(2-METHOXYPHENYL)-4,7-DIAZADECANE,TECHNICIUM COMPLEX **96929-52-1**
6	1.69	7.4	1074	324	C16H28N2O2	4-(N,N-DIBUTYLAMINO)METHYLPYRIDOXOL
7	1.62	7.4	2244	1	C16H28N4O4S1	TIMOLOL,O-PROPIONYL
8 ★	-0.74	2.0	2553	1	C16H28N4O6	AC-ALA-ASP-ALA-NTBU
9 ★	5.05	5.0	2425	1	C16H28N4O2	2-PYRROLIDONE,1-DODECONOYL **66283-26-9**
12590 ✓	1.25		2397	101	C16H30N3O2	IMIDAZOLIUM CHLORIDE,1-HEXYL-2-HYDROXYIMINOMETHYL-3-(2,2-DIMETHYLPROPOXY)METHYL
1 ★	-0.14		2377		C16H30N4O4	AC-ALA-LEU-VAL-N
2 ★	-0.20		2377		C16H30N4O4	AC-ALA-VAL-ILE-N
3 ★	-0.21		2377		C16H30N4O4	AC-ILE-ALA-VAL-N
4	0.52	7.2	170	314	C16H30N6O2	DIAZINEDICARBOXYLICACIDBIS(N'-I-PR-PIPERAZIDE)
95	0.94	7.2	170	314	C16H30N6O2	DIAZINEDICARBOXYLICACIDBIS(N'-PROPYLPIPERAZIDE)
6	0.44		1966	459	C16H31Au1N3O6P1S1	GLUTATHIONYLTRIETHYLPHOSPHINE GOLD
7 ★	-2.10	7.0	2374		C16H31N3O4	LEUCINYLVALINYLVALINE
8 ★	-2.23	7.0	2374		C16H31N3O5	ISOLEUCINYLISOLEUCINYLTHREONINE
9 ★	-2.14	7.0	2374		C16H31N3O5	LEUCINYLISOLEUCINYLTHREONINE
12600 ★	-2.30	7.0	2374		C16H31N3O5	LEUCINYLTHREONYLISOLEUCINE
1 ★	-1.66	7.0	2374		C16H31N3O5	THREONYLLEUCINYLISOLEUCINE
2	0.75		2083		C16H32N3O5P1S1	THIOTEPA,1,3,7,10,13-PENTAAZACYCLOOCTADECANE ANALOG **101347-47-1**
3	-1.36		2083		C16H32N3O6P1	TEPA,1,4,7,10,13-PENTAAZACYCLOOCTADECANE ANALOG **101347-45-9**
4	2.70		541		C16H35S1.Br1	DIMETHYLTETRADECYLSULFONYLBROMIDE
5	0.72		1537	820	C16H36N1.C7H5O2	TETRABUTYL AMMONIUM BENZOATE
6 ✓	0.91		2288	"		TETRABUTYL AMMONIUM BENZOATE
7	0.39		2288		C16H36N1.C7H5O3	TETRABUTYL AMMONIUM SALICYLATE
8	1.43		1537	820	"	TETRABUTYL AMMONIUM SALICYLATE
9	1.03		1537	820	C16H36N1.C7H7O3S1	TETRABUTYL AMMONIUM P-TOLUENESULFONATE
12610	0.50	7.3	1230	314	C16H36N2Se1	BIS(B-(N,N-DIISOPROPYLAMINO)ETHYLSELENDIE
1	0.93	7.5	1230	1	"	BIS(B-(N,N-DIISOPROPYLAMINO)ETHYLSELENDIE
2	-4.02		549	60	C16H38N2.(Br1)2	DECAMETHONIUMBROMIDE **541-22-0** relaxant (skeletal muscle)
3	-2.70	7.4	921	22	"	DECAMETHONIUMBROMIDE
4	7.20		1740	459	C16H48O6Si7	HEXADECAMETHYLHEPTASILOXANE **541-01-5**
15 ★	3.36		2133		C17H11Cl1F4N2O1	2-OXOQUAZEPAM ✚
6 ★	4.03		2133		C17H11Cl1F4N2S1	QUAZEPAM **36735-22-5** sedative, hypnotic
7 ★	2.42	7.4	2437	1	C17H11N3O3S1	ZOPOLRESTAT,5-H-ANALOG ✚
8 ★	5.68		2158	459	C17H12	BENZO(A)FLUORENE
9 ★	5.77		2158	459	C17H12	BENZO(B)FLUORENE
12620	0.19	7.4	2635	1	C17H12Br1F1N2O3	PONALRESTAT **72702-95-5** aldose reductase inhibitor, antiretinopathic (diabetic)
1	2.68		2635	459	"	PONALRESTAT
2 ★	2.20	5.0	1824	1	C17H12Cl1F1N2O1	NUARIMOL **63284-71-9**
3	3.97		1582	459	C17H12Cl1F3N2O1	HALAZEPAM **23092-17-3** sedative
4	1.32	7.4	1687	1	C17H12Cl1N1O2S1	FENTIAZAC **18046-21-4** anti-inflammatory
25 ★	5.19	7.4	1687	2	"	FENTIAZAC
6 ★	3.60		2420		C17H12Cl2N2O1	FENARIMOL **60168-88-9**
7 ★	2.56	5.0	1824	1	C17H12Cl2N2O1	TRIARIMOL **26766-27-8**
8	1.63	7.4	1688	1	C17H12Cl2N4	TRIAZOLAM **28911-01-5** sedative, hypnotic, antagonist-paf
9 ★	2.42		2184	"		TRIAZOLAM

	logP	pH	Ref.	Note	MF	Name / CAS / activity
12630 ★	2.45	7.4	594	1	C17H12F3N3O3	1,4-BENZODIAZEPIN-2-ONE,1,3-DIHYDRO-1-METHYL-7-NITRO-5-(2-TRIFLUOROMETHYLPHENYL) **1959-37-1**
1	-3.50	7.4	2559	1	C17H12O8	PROBICROMIL **58805-38-2** *anti-allergic (prophylactic)*
2	3.60		2524	459	C17H13Br1F2N2O1	IMIDAZOLYLETHANOL,2',4'-DIFLUORO-3"-BROMO ANALOG
3	3.44		2524	459	C17H13Br1F2N2O1	IMIDAZOLYLETHANOL,2',4'-DIFLUORO-4"-BROMO ANALOG
4	2.76	7.4	874	1	C17H13Br1N2O3	7-BR-5-PH-3-ACETOXY-1,4-BENZODIAZEPIN-2-ONE
35	2.60	7.4	874	1	C17H13Cl1N2O3	7-CL-5-PH-3-ACETOXY-1,4-BENZODIAZEPIN-2-ONE
6	1.26	7.4	1688	1	C17H13Cl1N4	ALPRAZOLAM **28981-97-7** *sedative, tranquilizer*
7	1.22	7.4	1410	1	C17H13Cl1O3	ITANOXONE **58182-63-1** *antihyperlipoproteinemic*
8 ★	3.32	2.2	1410	1	"	ITANOXONE
9 ★	1.75	7.4	1626	1	C17H13F1N2O2	1,4-BENZODIAZEPIN-2-ONE,5-(2-F-PH)-7-ACETYL
12640	1.98	1.0	547	342	C17H13N1	2,6-DIPHENYLPYRIDINEHCL
1 ★	4.82		547		C17H13N1	2,6-DIPHENYLPYRIDINE
2	5.04	7.4	547	1	"	2,6-DIPHENYLPYRIDINE
3	3.08		2161	468	C17H13N1O2	AZETIDINE-2-ONE,4-PHENANTHREN-9-YLOXY
4	2.46	7.4	712	1	C17H13N1O2	N-ALLYL-G-PYROPHTHALONE
45	2.75	7.4	712	1	C17H13N1O2	N-PROPENYL-G-PYROPHTHALONE
6	3.33		1973	459	C17H13N1O3	5-(4-ETHYNYLBENZOYL)-H2-PYRROLO-PYRROLE-1-CARBOXYLIC ACID **96327-55-8**
7	2.87	7.4	712	1	C17H13N1O3	N-ACETOMETHYL-G-PYROPHTHALONE
8	2.04	7.4	1255	1	C17H13N1O4	7-BENZYLOXY-4-HYDROXYQUINOLINE-3-CARBOXYLIC ACID
9 ★	1.80	7.4	594	1	C17H13N3O1	1,4-BENZODIAZEPIN-2-ONE,1,3-DIHYDRO-1-METHYL3-CYANO-5-PHENYL **3489-59-6**
12650	3.87		2524	459	C17H14Br1Cl1N2O1	IMIDAZOLYLETHANOL,2'-CHLORO-4"-BROMO ANALOG
1	3.42		2524	459	C17H14Br1F1N2O1	IMIDAZOLYLETHANOL,4'-F,4"-BR ANALOG
2	3.30		2524	459	C17H14Br1F1N2O1	IMIDAZOLYLETHANOL,4'-FLUORO-4"-BROMO ANALOG
3	1.52		1582	459	C17H14Cl1F1N2O3	7CL-1,3H2-1-OHET-3OH-5(2-FLPH)1,4-DIAZEPIN-2-ON **40762-15-0** *sedative*
4 ★	3.40	7.4	1626	1	C17H14Cl2N2O1	1,4-BENZODIAZEPIN-2-ONE,1,3-DIME-5-(2-CL-PH)-7-CL
55 ★	3.23	7.4	1626	1	C17H14Cl2N2O2	1,4-BENZODIAZEPIN-2-ONE,1-MEOME-5-(2-CL-PH)-7-CL
6	2.16		1582	459	C17H14Cl2N2O3	7CL-1,3-H2-1OHET-3-OH-5(2-CLPH)1,4-DIAZEPIN-2-ONE **51230-35-4**
7	2.04	7.4	1242	1	C17H14Cl2N2O3	7CL-3(HOET)-5(O-CLPH)-1,4-BENZODIAZEPIN-2-ONE **40967-01-9**
8	1.88		910		C17H14Cl2N4	1,3-BIS(5-CHLORO-2-BENZIMIDAZOLYL)PROPANE
9	2.95		2524	459	C17H14F2N2O1	IMIDAZOLYLETHANOL,4',4"-DIFLUORO ANALOG
12660	3.78	3.5	2179	302	C17H14F2O3	BUTANOIC ACID-2-METHYL-4-OXO-4-(4-(2,4-DIFLUOROPHENYL)PHENYL)
1 ★	4.80		401		C17H14N2	ELLIPTICINE +
2	1.03	7.2	2446	1	C17H14N2O3	ROSOXACIN **40034-42-2** *antibacterial*
3	1.24		2470	306	C17H14N2O3S1	TETRAHYDROQUINAZOLINE-4-OXO,2-METHIO-3-PHENYL-5-METHOXYCARBONYL
4	4.28	1.0	2471	459	C17H14N4O2	XANTHONE,6-(5-TETRAZOLYL)-2-PROPYL
65	1.76		1582	459	C17H14N4O4	7-NO2-1,3-H2-1-(CONHCH3)-5-PH-1,4-BENZODIAZEPIN-2-ON **27016-91-7**
6	2.71	1.0	2471	459	C17H14N4O4	XANTHONE,6-(5-TETRAZOLYL)-2-(2-HYDROXYPROPOXY)
7 ★	0.97	7.4	2433	1	C17H14N6O3	ALLOPURINOL,3-(IMIDAZOL-1YL)METHYLBENZOYLOXYMETHYL
8	1.62		910		C17H14N6O4	1,3-BIS(5-NITRO-2-BENZIMIDAZOLYL)PROPANE
9	4.59	1.0	2471	459	C17H14O4	XANTHONE,6-CARBOXY-2-PROPYL
12670	2.87	1.0	2471	459	C17H14O6	XANTHONE,6-CARBOXYL-2-(2-HYDROXYPROPOXY)
1	3.79	7.4	874	1	C17H15Br1N2O1	7-BR-5-PHENYL-3-ETHYL-1,4-BENZODIAZEPIN-2-ONE
2	3.35		2524	459	C17H15Br1N2O1	IMIDAZOLYLETHANOL,4'-BROMO ANALOG
3 ★	3.21	7.4	1626	1	C17H15Cl1N2O1	1,4-BENZODIAZEPIN-2-ONE,1-ETHYL-5-PHENYL-7-CL **5571-65-5**
4 ★	3.03	7.4	1626	1	C17H15Cl1N2O1	7-CL-1,3-H2-1-ME-5(2-ME-PH)1,4-BENZODIAZEPIN-2-ONE **5358-34-9**
75	3.68	7.4	874	1	C17H15Cl1N2O1	7-CL-5-PHENYL-3-ETHYL-1,4-BENZODIAZEPIN-2-ONE
6 ★	3.32	7.4	1626	1	C17H15Cl1N2O1S1	1,4-BENZODIAZEPIN-2-ONE,1-METHIOME-5-PH-7-CL
7 ★	3.15	7.4	1626	1	C17H15Cl1N2O2	1,4-BENZODIAZEPIN-2-ONE,1-ME-5-(4-MEO-PH)-7-CL **72430-63-8**
8 ★	2.86	7.4	1626	1	C17H15Cl1N2O2	1,4-BENZODIAZEPIN-2-ONE,1-MEOME-5-PH-7-CL
9	1.99	7.4	1242	1	C17H15Cl1N2O3	7CL(H2)-3(B-HOET)-5-PH-1,4-BENZODIAZEPIN-2-ONE **40967-00-8**
12680	1.53		1582	459	C17H15Cl1N2O3	7CL-1,3-H2-1-OHET-3-OH-5PH-1,4-BENZODIAZEPIN-2-ON **51230-34-3**
1	0.82		1582	459	C17H15Cl1N2O3	7CL-1,3H2-1-OHET-5-PH-1,4-DIAZEPIN-2-ONE,N-OXIDE **56875-80-0**
2 ★	3.21		1898		C17H15Cl1O5	(4-HYDROXYETHYL)VANILLIN,2-CL-BENZYL ESTER
3 ★	3.67		1898		C17H15Cl1O5	(4-HYDROXYETHYL)VANILLIN,3-CL-BENZYL ESTER
4 ★	3.71		1898		C17H15Cl1O5	(4-HYDROXYETHYL)VANILLIN,4-CL-BENZYL ESTER
85 ★	2.92	7.4	1626	1	C17H15F1N2O1	1,4-BENZODIAZEPIN-2-ONE,5-(2-F-PH)-7-ETHYL **51307-86-9**
6	2.50	7.4	712	1	C17H15N1O2	A,A'-DIMETHYL-N-METHYL-G-PYROPHTHALONE
7	2.34	7.4	712	1	C17H15N1O2	A-METHYL-N-ETHYL-G-PYROPHTHALONE
8	2.44	7.4	712	1	C17H15N1O2	N-ISOPROPYL-G-PYROPHTHALONE
9	2.64	7.4	712	1	C17H15N1O2	N-PROPYL-G-PYROPHTHALONE
12690	3.34		759		C17H15N1O2S1	2-(4,4'-DIHYDROXYDIPHENYLMETHYL)-5-METHYLTHIAZOLE
1	3.10		759		C17H15N1O2S1	2-ME-4-(4,4'-DIHYDROXYDIPHENYLMETHYL)THIAZOLE
2	3.47		1973	459	C17H15N1O3	5-(4-ETHENYLBENZOYL)-H2-PYRROLO-PYRROLE-1-CARBOXYLIC ACID **96327-54-7**
3	-0.38	7.4	1838		C17H15N1O3	INDOPROFEN **31842-01-0** *analgesic, anti-inflammatory*
4	-0.21	7.3	2259	1	"	INDOPROFEN
95 ★	2.77	2.0	1838		"	INDOPROFEN
6	2.89	7.3	2259	2	"	INDOPROFEN
7	1.75	7.4	712	1	C17H15N1O3	N-(2-HYDOROXYPROPYL)-G-PYROPHTHALONE
8	2.11	7.4	712	1	C17H15N1O3	N-METHOXYETHYL-G-PYROPHTHALONE
9	2.62		2161	468	C17H15N1O3S1	AZETIDINE-2-ONE,4-(2-BENZYLOXYCARBONYL)PHENYLTHIO
12700	-0.74	7.3	2259	1	C17H15N1O4	2-PHENOXYPROPIONIC ACID,P-N-(1,3-DIHYDRO-1-OXO-ISOINDOLYL)
1	3.88	7.3	2259	2	"	2-PHENOXYPROPIONIC ACID,P-N-(1,3-DIHYDRO-1-OXO-ISOINDOLYL)
2	2.27		1973	459	C17H15N1O4	5-(4-ACETYLBENZOYL)-H2-PYRROLO-PYRROLE-1-CARBOXYLIC ACID **96327-53-6**
3	2.83		2161	468	C17H15N1O4	AZETIDINE-2-ONE,4-(3-BENZYLOXYCARBONYL)PHENYL
4	1.36	7.4	712	1	C17H15N1O4	N-(2,3-DIHDROXYPROPYL)-G-PYROPHTHALONE
5 ★	3.44		547		C17H15N1O4S1	N-(7-ET-9-OXO-3-THIOXANTHENYL)ACETAMIDE,10,10-O2
6 ★	2.15		2234		C17H15N1O5	BENORILATE **5003-48-5** *analgesic, anti-inflammatory*
7	2.51		2021	459	C17H15N1O6	NAPHTHALENE,1,4-DIACETYLOXY-2,3-DIMETHOXY-5-CYANO **91814-29-8**
8	2.28		2021	459	C17H15N1O6	NAPHTHALENE,1,4-DIACETYLOXY-2,3-DIMETHOXY-6-CYANO **91814-30-1**
9	1.37		555		C17H15N3O4	1-P-MEO-BENZYL-2-(5-NO2-2-FURFURILIDENE)IMIDAZOLE
12710 ★	2.05	7.4	1626	1	C17H15N3O4	MOTRAZEPAM **29442-58-8** *sedative*
1	0.55		1582	459	C17H15N3O5	7-NO2-1,3-H2-1-OHET-3-OH,-5-PH-1,4-DIAZEPIN-2-ONE **56875-94-6**
2 ★	0.99		1967		C17H15N5	QUININOLINE-2-CARBOXALDEHYDE,N-PHENYLGUANYLHYDRAZONE **74618-18-1**
3	1.73		1593	459	C17H15N7O7S4	CEPHALOSPORIN,CYANO ANALOG **69712-58-9** +
4	2.91	7.4	2621	1	C17H16Br1N1O7	MITOSENE(5,5,6),BROMO-DIACETYLOXY ANALOG
15 ★	2.97	2.5	2207	457	C17H16Cl1F1N2O2	PROGABIDE **62666-20-0** *anticonvulsant, relaxant (muscle)*
6 ★	3.40		2351	2	C17H16Cl1N1O4	MEGLITINIDE **54870-28-9**
7 ★	3.94		1898		C17H16Cl1N1O5	(4-HYDROXYETHYL)VANILLINOXIME,3-CL-BENZYL ESTER
8 ★	3.92		1898		C17H16Cl1N1O5	(4-HYDROXYETHYL)VANILLINOXIME,4-CL-BENZYL ESTER
9	3.90		2171		C17H16Cl1N5O3	AZOBENZENE,2-CHLORO-4-NITRO-4'-N,N-(HYDROXYETHYL-CYANOETHYL)AMINO

		logP	pH	Ref.	Note	MF	Name / **CAS** / *activity*
12720	★	1.31	7.4	1626	1	C17H16F1N3O2	1,4-BENZODIAZEPIN-2-ONE,1-MEOME-5-(2-F-PH)-7-AM
1	★	0.77		2440	1	C17H16F1N5O3	DIDEOXY-ARA-A,N6-BENZOYL-2'-FLUORO ✚
2		3.32		1973	459	C17H16F1O3	3-(4-FLUOROBENZOYL)-5H-PYRROLO(H4)AZEPINE-9-CARBOXYLIC ACID **96327-69-4**
3	★	3.70		2509		C17H16F3N1O2	FLUTOLANIL **66332-96-5**
4		2.17	7.4	2062	1	C17H16F3N1O4	L-DOPA,(3-TRIFLUOROMETHYL)BENZYL ESTER
25	★	2.29	7.4	2062	2	"	L-DOPA,(3-TRIFLUOROMETHYL)BENZYL ESTER
6	★	2.94	7.4	594	1	C17H16N2O1	1,4-BENZODIAZEPIN-2-ONE,1,3-DIHYDRO-7,9-DIMETHYL-5-PHENYL **5571-62-0**
7	★	2.95	7.4	1626	1	C17H16N2O1S1	1,4-BENZODIAZEPIN-2-ONE,1-ME-5-PH-7-METHIO **23193-98-8**
8		-0.11	7.4	675	1	C17H16N2O2	3,5-PYRAZOLIDINDIONE,1,2-DIPHENYL-4-ETHYL
9		1.65	7.4	847	1	C17H16N2O2	N-(3-AMINOPROPYL)-G-PYROPHTHALONE
12730	★	1.70		547		C17H16N2O4	M-ACETAMIDOPHENYLHIPPURATE
1		0.58	7.0	1405	1	C17H16N2O4S1	1-BENZENESULFONYL-5-ETHYL-5-PHENYLHYDANTOIN
2		2.44		1405		"	1-BENZENESULFONYL-5-ETHYL-5-PHENYLHYDANTOIN
3		-0.92	7.0	1526	1	C17H16N2O7S2	SULOSEMIDE **82666-62-4**
4	✓	0.52		2397	101	C17H16N3O3	IMIDAZOLIUM CHLORIDE,1-METHYL-2-HYDROXYIMINOMETHYL-3-(2-NAPHTHOYLOXY)METHYL
35		0.98	1.0	594		C17H16N4	2,4-DIAMINOPYRIMIDINE,5-(4-PHENYLBENZYL)
6		1.21	7.4	2024	1	C17H16N4O4	1,2-AZIRIDINOMITOSINE,7-(2-PROPYNYL)AMINO **103456-57-1**
7	★	0.66		2529		C17H16N4O4	DDI,5'-BENZOATE
8	★	0.67	3.0	1638		C17H16N4O4S4	CEPHALOTHIN ANALOG(3-(5-ME-DITHIAZ-2YL)SCH2) **26970-95-6**
9	★	3.40	3.5	2179	302	C17H16O3	METBUFEN **63472-04-8** *anti-inflammatory, analgetic*
12740	★	3.35	3.5	2179	302	C17H16O3	PENTANOIC ACID,5-OXO-5-(4-BIPHENYLYL)
1		3.02		1159		C17H17Cl1N2O2	N-PH,N-(1-ACETANILIDO)-1-CHLOROPROPIONAMIDE
2	★	2.08	7.4	1978	1	C17H17Cl1N2O7	1-ACETYL-6-DEMETHYL-6-CHLORO-7-METHOXY-N-METHYLMITOSENE **96000-44-1**
3	★	3.90	3.5	1609	1	C17H17Cl1O3	PHENYLACETIC ACID,3-CHLORO,4-(3-PHENYLPROPOXY)
4	★	2.69		2650	458	C17H17Cl1O3S1	DIETHYL KETONE,1-(4-CHLOROPHENYL)-2-BENZENESULFONYL
45	★	2.18		507		C17H17Cl1O6	GRISEOFULVIN **126-07-8** *antifungal*
6	★	2.86		537	800	C17H17Cl2N1O3	P-PHENYL-AMPHENICOL
7		4.00		2171		C17H17Cl2N5O4	AZOBENZENE,2-CHLORO-4-NITRO-1'-ACETYLAMINO-5'-CHLORO-4'-(2-PROPANOLAMINO)
8		3.96		434		C17H17Cl3O1	METHYL-ETHOXYCHLOR
9		1.73	7.4	2261	1	C17H17N1O2	APOMORPHINE **58-00-4** *emetic*
12750		1.76	7.4	564	1	"	APOMORPHINE
1		2.30	7.5	2254	2	"	APOMORPHINE
2		-0.28	7.3	2259	1	C17H17N1O3	2-PHENYLPROPIONIC ACID,P-N-(1,3,4,7-TETRAHYDRO-1-OXO-ISOINDOLYL)
3		2.58	7.3	2259	2	"	2-PHENYLPROPIONIC ACID,P-N-(1,3,4,7-TETRAHYDRO-1-OXO-ISOINDOLYL)
4	★	3.40		547		C17H17N1O3	3,5-DIMETHYLPHENYLHIPPURATE
55		3.35		1973	459	C17H17N1O3	3-(4-METHYLBENZOYL)-4H-PYRROLO-PYRIDINE-8-CARBOXYLIC ACID **88777-65-5**
6		3.17		1973	459	C17H17N1O3	3-BENZOYL-5H-PYRROLO(H4)AZEPINE-1-CARBOXYLIC ACID **88777-69-9**
7	★	2.72		2487		C17H17N1O3	CINNAMAMIDE,3,4-DIHYDROXY-N-PHENETHYL
8		3.54		1973	459	C17H17N1O3S1	3-(4-METHYLTHIOBENZOYL)-H4-PYRROLO-PYRIDINE-8-CARBOXYLIC ACID **88777-73-5**
9		-1.60	7.3	2259	1	C17H17N1O4	2-PHENYLPROPIONIC ACID,P-N-(3A,4,7,7A-TETRAHYDROPHTHALIMIDYL)
12760		2.27	7.3	2259	2	"	2-PHENYLPROPIONIC ACID,P-N-(3A,4,7,7A-TETRAHYDROPHTHALIMIDYL)
1		2.90		1973	459	C17H17N1O4	3-(4-METHOXYBENZOYL)-H4-PYRROLO-PYRIDINE-8-CARBOXYLIC ACID **88777-66-6**
2		3.31		1973	459	C17H17N1O4	5-(3-ETHOXYBENZOYL)-H2-PYRROLO-PYRROLE-1-CARBOXYLIC ACID **96327-37-6**
3		3.30		1973	459	C17H17N1O4	5-(4-ETHOXYBENZOYL)-H2-PYRROLO-PYRROLE-1-CARBOXYLIC ACID **66635-86-7**
4		1.41		1973	459	C17H17N1O4S1	3-(4-METHYLSULFOXYBENZOYL)-H4-PYRROLO-PYRIDINE-8-CARBOXYLIC ACID **88777-74-6**
65	★	2.94		1898		C17H17N1O5	(4-HYDROXYETHYL)VANILLINOXIME,BENZYL ESTER
6	★	2.69		547		C17H17N1O5	3,5-DIMETHOXYPHENYLHIPPURATE
7	★	2.30		2487		C17H17N1O5	CINNAMAMIDE,3,4-DIHYDROXY-N-(3,4-DIMETHOXY)PHENYL
8		1.53		1973	459	C17H17N1O5S1	3-(4-METHYLSULFONYLBENZOYL)-H4-PYRROLO-PYRIDINE-8-CARBOXYLIC ACID **88777-72-4**
9		1.15	7.4	2621	1	C17H17N1O7	MITOSENE(5,5,6),DI-ACETYLOXY ANALOG
12770	★	1.81	7.4	1626	1	C17H17N3O1	1,4-BENZODIAZEPIN-2-ONE,1-ET-5-PH-7-AMINO
1	★	2.43	7.4	1626	1	C17H17N3O1	1,4-BENZODIAZEPIN-2-ONE,5-PH-7-DIMEAMINO **30144-56-0**
2	★	1.92	7.4	1626	1	C17H17N3O1S1	1,4-BENZODIAZEPIN-2-ONE,1-METHIOME-5-PH-7-AMINO
3	★	1.32	7.4	1626	1	C17H17N3O2	1,4-BENZODIAZEPIN-2-ONE,1-MEOME-5-PH-7-AMINO
4	★	0.75	7.4	1626	1	C17H17N3O3	1,4-BENZODIAZEPIN-2-ONE,1-MEOME-3-OH-5-PH-7-AM
75		-1.59	7.0	2014		C17H17N3O3	SCEPTER *herbicide* ✚
6		0.64	4.0	2014		"	SCEPTER
7	★	2.40	5.4	2018		C17H17N3O3S1	PICOPRAZOLE **78090-11-6** *antisecretory*
8		-0.52	7.4	2542	1	C17H17N3O4S3	THIOPHENE-2-SULFONAMIDE,4-(2-PYRIDIN-2YLMETHYLAMINOMETHYL)PHENYLSULFONYL
9		1.06	7.4	2542	1	C17H17N3O4S3	THIOPHENE-2-SULFONAMIDE,4-(2-PYRIDYLMETHYLAMINOMETHYL)PHENYLSULFONYL
12780		-2.73	1.0	2196	1	C17H17N3O6S2	CEPHAPIRIN **21593-23-7** *antibacterial*
1		-1.93	2.0	2464	41	"	CEPHAPIRIN
2		-1.55	6.5	2464	41	"	CEPHAPIRIN
3	★	0.22	7.0	2440	1	C17H17N5O4	2'-DEOXYADENOSINE,N6-BENZOYL
4		3.82		2424	459	C17H17N5O4S1	BENZOTRIAZOLE,2-CYCLOPENTYL-5-(P-NITROBENZENESULFONAMIDO)
85	★	-0.10	7.0	2440	1	C17H17N5O5	ADENOSINE,N6-BENZOYL
6		-1.99		1525		C17H17N7O7S2	7B-(PURIN-6YL)THIOACETYLAMINO-CEPHAMYCIN **57792-96-8**
7		-1.72		1525		C17H17N7O7S2	7B-(PURIN-8YL)THIOACETYLAMINO-CEPHAMYCIN **57792-97-9**
8		-1.20	7.4	2373	1	C17H17N7O8S4	CEFOTETAN **69712-56-7** *antibacterial*
9		1.43		1593	459	"	CEFOTETAN
12790		3.21		2162	459	C17H18Cl1N1O1	SCH-23390 *antipsychotic, antagonist-d-1* ✚
1		2.12	7.4	2062	1	C17H18Cl1N1O4	L-DOPA,(P-CHLOROPHENYL)ETHYL ESTER
2	★	2.33	7.4	2062	2	"	L-DOPA,(P-CHLOROPHENYL)ETHYL ESTER
3		1.16	7.2	846	314	C17H18Cl2N2O5S1	CLOMETOCILLIN **1926-49-4** *antibacterial*
4	★	3.90		2479	459	C17H18F1N3O2	PHENYLBIURET,2-FLUORO-5-BENZYL
95		-1.70	7.2	2446	1	C17H18F1N3O3	CIPROFLOXACIN **85721-33-1** *antibacterial*
6		-1.11	7.4	2681	1	"	CIPROFLOXACIN
7	★	-1.08		2681	226	"	CIPROFLOXACIN
8		-0.70	7.2	2526	1	"	CIPROFLOXACIN
9		1.82	7.4	579	1	C17H18F3N1O1	FLUOXETINE **54910-89-3** *antidepressant, inhibitor-5ht-reuptake*
12800		-1.10	7.2	2446	1	C17H18F3N3O3	FLEROXACIN **79660-72-3** *antibacterial*
1		1.96		1407		C17H18N2O2	1,2,3,6-H4-PYRIDINE,N-(1-NAPHTHOXYME)CONH-
2	★	1.52		537	270	C17H18N2O5	PHENYLAMPHENICOL ✚
3		0.42	7.4	1754	320	C17H18N2O6	NIFEDIPINE **21829-25-4** *vasodilator (coronary)*
4		2.40	6.8	2337	1	"	NIFEDIPINE
5		2.86	7.2	2416		"	NIFEDIPINE
6		3.27	7.4	2018		"	NIFEDIPINE
7	★	1.13		936		C17H18N2O6S1	CARBENICILLIN **4697-36-3** *antibacterial*
8	★	1.19	7.4	1978	1	C17H18N2O7	1-ACETYL-6-DEMETHYL-7-METHOXY-N-METHYLMITOSENE **96000-43-0**
9	★	1.54	7.4	606	1	C17H18N2O7	1-ACETYL-7-METHOXYMITOSENE ✚

	logP	pH	Ref.	Note	MF	Name / CAS / activity
12810 ✓	-0.36		2397	101	C17H18N3O2	IMIDAZOLIUM CHLORIDE,1-METHYL-2-HYDROXYIMINOMETHYL-3-(1-NAPHTHYL)METHOXYMETHYL 91900-17-3
1	3.35		538		C17H18N4O3	P-(3-(P-ACETYLAM-BENZYL)-3-ME-1-TRIAZENO)BENZOIC ACID
2	2.88	8.4	2195	1	C17H18N4O3	PYRROLE,3-NITRO-2-(3-(5-DIMETHYLAMINOMETHYL)FURAN-2-YL)PHENYLAMINO
3 ★	3.10		579		C17H18N4O3S1	ENVIROXIME 72301-79-2 antiviral
4	0.71	7.4	2024	1	C17H18N4O4	1,2-(N-METHYLAZIRIDINO)MITOSINE,7-AZIRIDINO 103422-21-5
15	0.82	7.4	2024	1	C17H18N4O4	1,2-AZIRIDINOMITOSINE,7-(2-METHYLAZIRIDINO) 103422-22-6
6 ★	2.38		2463			BENZOTRIAZOLE,2-METHYL-5-(3,4,5-TRIMETHOXY)BENZAMIDO
7 ★	1.70		2289	459	C17H18N6O2	TETRAZOLE,1-BENZYL-5-(4-DIMETHYLUREYL)PHENOXY 117121-36-5
8 ★	2.25		2289	459	C17H18N6O2	TETRAZOLE,2-BENZYL-5-(4-DIMETHYLUREYL)PHENOXY 117121-48-9
9	2.60		1593	459	C17H18N6O7S5	CEPHALOSPORIN,METHYLTHIO ANALOG 69712-37-4
12820	2.00		1593	459	C17H18N6O8S4	CEPHALOSPORIN,METHOXY ANALOG 69712-35-2
1 ★	3.25	5.5	2403	2	C17H18O3	BUTANOIC ACID,2-(4-BIPHENYLYL)-3-HYDROXY-3-METHYL 93371-55-2
2 ★	1.82		2650	458	C17H18O3S1	DIETHYL KETONE,1-PHENYL-2-BENZENESULFONYL
3	2.94		1636	40	C17H18O4	2-HYDROXY-3-NAPHTHOXYPROPYLMETH ACRYLATE 62146-91-2
4	2.10	7.4	2559	1	C17H18O5	CHROMONE-2-CARBOXYLIC ACID,5-HYDROXY-6-PROPYL-7,8-TETRAMETHYLENE
25	1.60	7.4	2559	1	C17H18O5	PROXICROMIL 60400-92-2 anti-allergic
6	1.72	7.0	2144		"	PROXICROMIL
7 ★	4.40	1.5	2144			PROXICROMIL
8	2.85		2021	459	C17H18O6	NAPHTHALENE,1,4-DIACETYLOXY-2,3-DIMETHOXY-6-METHYL 91814-42-5
9	-0.20	7.4	2559	1	C17H18O6	PROXICROMIL-7-HYDROXY
12830	0.79		636		C17H18O6	VERNOLEPINACETATE
1	0.25		636		C17H18O6	VERNOMENINACETATE
2	2.43		2021	R459	C17H18O7	NAPHTHALENE,1,4-DIACETYLOXY-2,3,5-TRIMETHOXY 91814-24-3
3	2.39		2021	459	C17H18O7	NAPHTHALENE,1,4-DIACETYLOXY-2,3,6-TRIMETHOXY 91814-40-3
4	-0.66		232	820	C17H19Cl1N2O1S1	CHLORPROMAZINE-SULFOXIDE.HCL
35	2.12	7.4	455	1	C17H19Cl1N2O1S1	CHLORPROMAZINEN-OXIDE
6	-0.82	6.0	695		C17H19Cl1N2O1S1	CHLORPROMAZINESULFOXIDE
7	0.79	7.4	455	1	"	CHLORPROMAZINESULFOXIDE
8	1.23		232	820	C17H19Cl1N2S1	1-CHLORPROMAZINEHYDROCHLORIDE
9	1.79		232	820	C17H19Cl1N2S1	3-CHLORPROMAZINEHYDROCHLORIDE
12840	1.51		232	820	C17H19Cl1N2S1	CHLORPROMAZINEHYDROCHLORIDE 69-09-0
1	1.90	7.4	2217	1	C17H19Cl1N2S1	CHLORPROMAZINE 50-53-3 anti-emetic, antipsychotic
2	2.93	7.0	468	1	"	CHLORPROMAZINE
3	3.15	7.4	1552	1	"	CHLORPROMAZINE
4	3.22	7.4	1209	1	"	CHLORPROMAZINE
45	3.25	7.4	678	1	"	CHLORPROMAZINE
6	3.28	7.4	2407	1	"	CHLORPROMAZINE
7	3.50	7.4	880	1	"	CHLORPROMAZINE
8	5.16	7.4	268	2	"	CHLORPROMAZINE
9 ★	5.19		2407	2	"	CHLORPROMAZINE
12850	5.23	7.0	468	701	"	CHLORPROMAZINE
1	5.35	7.4	550	2	"	CHLORPROMAZINE
2	2.98		2162	459	C17H19F1N4S1	FLUMEZAPINE 61325-80-2 antipsychotic
3	-0.21	7.4	2681	1	C17H19F2N3O3	8-FLUORO-PEFLOXACIN
4 ★	-0.09		2681	226	"	8-FLUORO-PEFLOXACIN
55	-1.03	7.4	2681	1	C17H19F2N3O3	LOMEFLOXACIN 98079-51-7 anti-infective
6 ★	-0.80		2681	226		LOMEFLOXACIN
7	0.49	7.4	2235	1	C17H19N1	4,4-DIPHENYLPIPERIDINE
8	-0.04	1.0	602	342	C17H19N1O1	N-AZIRIDYLETHYL-DIPHENYLMETHYLETHER
9	0.30	7.4	2378	1	C17H19N1O2	DEOXYMORPHINE 51269-51-3
12860	-1.22		307	1	C17H19N1O2	DESOXYMORPHINE-C
1 ★	3.66		2506		C17H19N1O2	MEPRONIL fungicide
2	0.40	7.4	2378	1	C17H19N1O2	MORPHINE ANALOG 32295-31-1
3	4.21		822		C17H19N1O3	4-AMINOSALICYLIC ACID,2-T-BUTYLPHENYL ESTER
4	4.64		822		C17H19N1O3	4-AMINOSALICYLIC ACID,4-T-BUTYLPHENYL ESTER
65	-0.16	7.3	2259	1	C17H19N1O3	INDOPROFEN,4,5,6,7-TETRAHYDRO
6 ★	3.07	7.3	2259	2	"	INDOPROFEN,4,5,6,7-TETRAHYDRO
7	-0.40	7.4	2582	1	C17H19N1O3	MORPHINE 57-27-2 analgesic (narcotic)
8	-0.12	7.4	709	1	"	MORPHINE
9	-0.10	7.1	554		"	MORPHINE
12870	-0.06	7.4	2485	1	"	MORPHINE
1	0.11	7.4	2378	1	"	MORPHINE
2	0.15	7.4	554	1	"	MORPHINE
3	0.22	7.5	554	1	"	MORPHINE
4	0.30	7.6	554	212	"	MORPHINE
75	0.36	7.7	554	1	"	MORPHINE
6 ★	0.76		503		"	MORPHINE
7	0.79	7.7	554	2	"	MORPHINE
8	1.03	7.4	484	2	"	MORPHINE
9	-0.23	7.4	2378	1	C17H19N1O3	MORPHONE 81165-08-4
12880 ★	3.27		1681		C17H19N1O3	N,N-DIMETHYLCARBAMATE,P-BENZYLOXYBENZYL ESTER
1 ★	3.87		1523		C17H19N1O4	1,1-BIS(P-METHOXYPHENYL)-2-NITROPROPANE 34197-26-7
2	-1.64	7.3	2259	1	"	2-PHENYLPROPIONIC ACID,P-N-(HEXAHYDROPHTHALIMIDYL)
3	2.37	7.3	2259	2	"	2-PHENYLPROPIONIC ACID,P-N-(HEXAHYDROPHTHALIMIDYL)
4 ★	4.30		2506		C17H19N1O4	FENOXYCARB 72490-01-8
85 ★	2.61		547		C17H19N1O4	FURALAXYL 57646-30-7
6	1.57	7.4	2062	1	C17H19N1O4	L-DOPA,PHENYLETHYL ESTER
7 ★	1.76	7.4	2062	2	"	L-DOPA,PHENYLETHYL ESTER
8	-0.42	7.4	2378	1	C17H19N1O4	OXYMORPHONE 76-41-5 analgesic (narcotic)
9	-0.28	7.1	554		"	OXYMORPHONE
12890	0.00	7.4	554	1	"	OXYMORPHONE
1	0.08	7.5	554	1	"	OXYMORPHONE
2	0.16	7.6	554	212	"	OXYMORPHONE
3	0.23	7.7	554	1	"	OXYMORPHONE
4 ★	0.83	7.7	554	2	"	OXYMORPHONE
95	1.35	7.4	2062	1	C17H19N1O5	L-DOPA,2-(PHENOXY)ETHYL ESTER
6 ★	1.51	7.4	2062	2	"	L-DOPA,2-(PHENOXY)ETHYL ESTER
7	0.31	7.5	1563	1	C17H19N3O1	PHENTOLAMINE 50-60-2 adrenergic blocker, antihypertensive
8 ★	2.23	6.4	2018		C17H19N3O3S1	OMEPRAZOLE 73590-58-6 depressant (gastric acid secretory)
9	-0.59		548		C17H19N3O4	P-(4-NITROPHENOXYBUTOXY)BENZAMIDINEHCL

	logP	pH	Ref.	Note	MF	Name / CAS / activity
12900	1.25		548	370	"	P-(4-NITROPHENOXYBUTOXY)BENZAMIDINEHCL
1	-0.45	7.0	1812	1	C17H19N3O5	5'-AMINO-3'O-BENZOYL-5'-DEOXYTHYMIDINE
2	-1.08	7.4	546	1	C17H19N5	4,6-DIAM-1,2-DIH-2,2-DIME-1(P-DIPH)-S-TRIAZINE
3	-1.03	6.0	546		"	4,6-DIAM-1,2-DIH-2,2-DIME-1(P-DIPH)-S-TRIAZINE
4 ★	1.89	12.0	2408		C17H19N5O1	BAZINAPRINE 94011-82-2 inhibitor-mao-a
5 ★	1.63	7.4	1485	1	C17H19N5O2	2,4-DIAMINOPYRIMIDINE,5(3,5-DIMEO-4-(1-PYRRYL)BENZYL
6 ★	1.78	7.4	1485	2		2,4-DIAMINOPYRIMIDINE,5(3,5-DIMEO-4-(1-PYRRYL)BENZYL
7	1.77	7.4	594	1	C17H19N5O2	PIRITREXIM 72732-56-0 antiproliferative agent
8 ★	2.13	13.0	547	224		PIRITREXIM
9 ★	2.48		2424			BENZOTRIAZOLE,2-CYCLOPENTYL-5-(P-AMINOBENZENESULFONAMIDO)
12910	-1.10	7.4	2034	1	C17H19N7O3	TRIAMTERENE,PARA-CARBOXYBUTOXY
1 ★	4.02	7.4	2029	459	C17H20Cl1N5O2	XANTHINE,1,3-DIPROPYL-8-(2-AMINO-4-CHLOROPHENYL) 85872-51-1
2	-0.80	7.4	2681	1	C17H20F1N3O3	8-DESFLUORO-LOMEFLOXACIN
3 ★	-0.75		2681	226	"	8-DESFLUORO-LOMEFLOXACIN
4	0.12	7.2	2446	1	C17H20F1N3O3	PEFLOXACIN 70458-92-3 antibacterial
15	0.18	7.4	2681	1	"	PEFLOXACIN
6 ★	0.27		2681	226		PEFLOXACIN
7	-0.75	5.0	2375	1	C17H20F6N2O3	FLECAINIDE 54143-55-4 cardiac depressant (anti-arrhythmic)
8	1.14	7.4	2320	1	"	FLECAINIDE
9	3.64		2150	459	C17H20N2	ADAMANTANE,1-(1-BENZIMIDAZOLYL)
12920 ★	4.33		2150	459	C17H20N2	ADAMANTANE,1-(1-INDAZOLYL)
1	3.68		2150	459	C17H20N2	ADAMANTANE,1-(2-INDAZOLYL)
2 ★	3.87		594		C17H20N2O1	MICHLER'S KETONE
3 ★	3.21		2094		C17H20N2O1	UREA,1,1-DIMETHYL-3-(3-PHENYLETHYL)PHENYL 102587-49-5
4	1.86	7.4	455	1	C17H20N2O1S1	2-HYDROXYPROMAZINE
25 ★	1.90	7.4	455	1	C17H20N2O1S1	PROMAZINE-N-OXIDE
6	0.15	7.4	455	1	C17H20N2O1S1	PROMAZINESULFOXIDE
7 ★	3.39		2612		C17H20N2O2	1,1-DIMETHYL-3-(3-PHENYLETHOXYPHENYL)UREA 70859-35-7
8	2.87		1019		C17H20N2O2	N'-(1-METHYLBENZYL)-4-TOLYLOXYACETIC ACID HYDRAZIDE
9 ★	4.50		580		C17H20N2O3	1-METHYL-1-METHOXY-3-(P-PHENYLETHOXY)UREA
12930	2.39		1019		C17H20N2O3	N'-(1-MEBENZYL)-4-METHOXYPHENOXYACETIC ACID HYDRAZIDE
1 ★	1.44		2467		C17H20N2O3S1	ANILINE,P-(4-N,N-DIETHYLCARBOXAMIDOPHENYL)SULFONYL
2	1.24	7.4	1702	1	C17H20N2O4	SALICYLAMIDE,5(OC2H4NHCHOH-(3-MEO-PHENYL) 89789-87-7
3 ★	2.28		515		C17H20N2O5S1	PENICILLIN,A-PHENOXYETHYL 1752-26-7
4 ★	2.20		935		C17H20N2O5S1	PHENETHICILLIN 147-55-7 antibacterial
35	1.82	7.4	2621	1	C17H20N2O6	MITOSENE(5,5,6),METHYL-HYDROXY-N-ETHYLCARBAMOYL ANALOG
6	1.34	7.4	2621	1	C17H20N2O6	MITOSENE,ETHYLCARBAMOYL-HYDROXY ANALOG
7	-2.41	7.2	846	314	C17H20N2O6S1	PENICILLIN,A-(2,6-DIMETHOXYPHENYL) 61-32-5 antibacterial
8	-2.00	7.0	1295	1	"	PENICILLIN,A-(2,6-DIMETHOXYPHENYL)
9 ★	1.22		515		"	PENICILLIN,A-(2,6-DIMETHOXYPHENYL)
12940	0.91	7.0	232	820	C17H20N2S1	PROMAZINEHYDROCHLORIDE 53-60-1
1	2.15	7.0	468	1	C17H20N2S1	PROMAZINE 58-40-2 anticholinergic
2	2.48	7.4	1209	1	"	PROMAZINE
3	2.55	7.4	678	1	"	PROMAZINE
4	3.57	7.0	468	701	"	PROMAZINE
45 ★	4.55	7.4	268	2	"	PROMAZINE
6	2.59	7.0	919	1	C17H20N2S1	PROMETHAZINE 60-87-7 anti-emetic, antihistaminic
7	2.85	7.4	1209	1	"	PROMETHAZINE
8	2.88	7.4	678	1	"	PROMETHAZINE
9	3.10	7.4	2407	1	"	PROMETHAZINE
12950 ★	4.81		2407	2	"	PROMETHAZINE
1 ✓	-0.05	7.4	2397	101	C17H20N3O2	IMIDAZOLIUM CHLORIDE,1-METHYL-2-HYDROXYIMINO-3-(2-METHYL-4-PHENYLBUTYN-3-YL)OXYMETHYL
2 ✓	-0.05		2397	101	C17H20N3O2	IMIDAZOLIUM CHLORIDE,1-METHYL-2-HYDROXYIMINOMETHYL-3-(1,1-DIMETHYL-3-PHENYLPROPYN-2YLOXY)METHYL
3 ✓	0.31		2397	101	C17H20N3O2	IMIDAZOLIUM CHLORIDE,1-METHYL-2-HYDROXYIMINOMETHYL-3-(1-ETHYL-3-PHENYLPROPYN-2YLOXY)METHYL
4	2.08		1967		C17H20N4	1-PHENYLBUTANONE,N-PHENYLGUANYLHYDRAZONE 97183-51-2
55	-0.03	7.4	2029	459	C17H20N4O5S1	XANTHINE,1,3-DIPROPYL-8-(4-SULFOPHENYL) 89073-57-4
6	-1.46		1200		C17H20N4O6	RIBOFLAVIN 83-88-5 vitamin (enzyme co-factor)
7	3.08	7.4	2028	459	C17H20N6	1,2,4-TRIAZOLE,3-AMINO-5-(2-(3,4-DIMETHYLPHENYL)ETHYLAMINO)PYRIDIN-4-YL 77314-84-2
8 ★	1.82	13.0	547	224	C17H20N6O2	3(3'-ACETAMIDOPHENOXYMETHYL)PHENYLBIGUANIDE ✚
9	1.85	7.4	547	1	C17H20N6O2	3(3'-ACETAMIDOPHENOXYMETHYL)PHENYLBIGUANIDE
12960	0.86	7.4	2591	1	C17H20N6O5	AZT,5'-(1-METHYL-1,4-DIHYDRONICOTINYL) 116333-41-6 antiretroviral prodrug
1 ★	1.65		1589		C17H20O6S1	2-OH-2-PH-ETHYL-(3,4,5-TRIMETHOXY)PHENYL-SULFONE
2	2.16	7.0	708	1	C17H21Cl2N1O2	6(3,4-DICLBENZOYLOXY)2-DECAHYDRO-I-QUINOLINE-C
3	2.10	7.0	708	1	C17H21Cl2N1O2	6(3,4-DICLBENZOYLOXY)2-DECAHYDRO-I-QUINOLINE-T
4	0.68	7.4	785	1	C17H21N1	4,4-DIPHENYL-N-METHYLBUTYLAMINE
65	0.58	1.0	870	342	C17H21N1O1	2-(P-METHYL-DIPHENYLMETHOXY)-N-METHYLETHYLAMINE
6	0.20	1.0	602	342	C17H21N1O1	2-DIPHENYLMETHOXY-N-ETHYLETHYLAMINE
7	1.35	7.4	2378	1	C17H21N1O1	DIDEOXY-DIHYDROMORPHINE 55592-68-2
8	1.61	7.4	1157	324	C17H21N1O1	DIPHENHYDRAMINE 58-73-1 antihistaminic
9 ★	3.27		505		"	DIPHENHYDRAMINE
12970	3.33		505		C17H21N1O1.C6H8O7	PHENYLTOLOXAMINECITRATE
1 ★	3.36		2506		C17H21N1O2	NAPROPAMIDE 15299-99-7 herbicide
2	-0.22	7.3	2259	1	C17H21N1O3	2-PHENYLPROPIONIC ACID,P-N-(1-OXO-OCTAHYDRO-ISOINDOLYL)
3	3.15	7.3	2259	2	"	2-PHENYLPROPIONIC ACID,P-N-(1-OXO-OCTAHYDRO-ISOINDOLYL)
4	2.40	8.0	868	2	C17H21N1O3	3,5-DIMETHOXY-4-BENZYLOXYPHENTHYLAMINE
75	-1.05	7.4	709	1	C17H21N1O3	DIHYDRO-B-ISOMORPHINE 509-60-4
6	-0.51	7.4	2378	1	"	DIHYDRO-B-ISOMORPHINE
7	0.21	7.4	1056	1	C17H21N1O4	1(3-CARBOXYPROPYL)AM-3-(1-NAPHTHYLOXY)-2-PROPANOL
8	-0.83	7.3	2259	1	C17H21N1O4	2-PHENYLPROPIONIC ACID,P-N-(OCTAHYDROPHTHALIMIDYL)
9	2.95	7.3	2259	2	"	2-PHENYLPROPIONIC ACID,P-N-(OCTAHYDROPHTHALIMIDYL)
12980	1.05	7.4	484	1	C17H21N1O4	COCAINE 50-36-2 anesthetic (topical)
1 ★	2.30	7.4	484	2	"	COCAINE
2 ★	5.20		1521		C17H21N1O4	M-NITROBENZYL(1R)-TR-CHRYSANTHEMATE
3	1.24		505	20	C17H21N1O4	SCOPOLAMINE 51-34-3 anticholinergic (ophthalmic)
4 ★	3.36		2506		C17H21N1O4S4	BENSULTAP 17606-31-4 insecticide
85	1.70		634		C17H21N1O6	6-ETO-7-METHOXYETHOXY-4-OH-3-QUINOLINE-CO2ET
6	2.03	7.4	2235	1	C17H21N1Si1	SILA-MEDIPINE ✚
7	-0.55	7.4	722	1	C17H21N3O1	N-DIMETHYL-DESDI-I-PROPYL-DISOPYRAMIDE
8	2.76	7.4	416	1	C17H21N3O2	1-ME-3-PHENYL-5-PIPERIDYL-6-METHYLURACIL
9	3.21	7.4	459	1	C17H21N3O2	1-PHENYL-3-ME-5-PIPERIDINO-6-METHYLURACIL

	logP	pH	Ref.	Note	MF	Name / CAS / activity
12990	0.37	7.4	2542	1	C17H21N3O3S2	THIOPHENE-2-SULFONAMIDE,4-(4-METHYLPIPERAZIN-1YL)METHYLBENZOYL
1 ★	2.64		593		C17H21N3O5	2-CYANO-3-(4-NITROBENZYLAMINO)2-PENTENOIC ACID,2-ETHOXYETHYL ESTER
2 ★	0.95		1712		C17H21N3O6S2	(BENZYL-A-METHYLSULFONAMIDO)PENICILLIN ✚
3 ★	-0.16		1739		C17H21N3O9	O-TRIACETYL-N-ACETYL-ARA-C 15981-93-8
4 ★	0.63		1551		C17H21N3O10S1	1(2,3,4,6-O-AC-GLUCOTHIOPYRANOSYL)2-NITROIMIDAZOLE 83116-90-9
95 ★	0.52		1551		C17H21N3O11	1(2,3,4,6-O-ACETYL-GLUCOPYRANOSYL)2-NITROIMIDAZOLE 67774-11-2
6 ★	1.86	12.0	2408		C17H21N5O2	MINAPRINE,4-CARBOXAMIDO ANALOG ✚
7	-0.52		1407		C17H21N5O2	PYRIDINE,2,6-BIS-(1,2,3,6-H4-PYRIDINE,1-NHCO-)
8 ★	3.54	7.4	2029	459	C17H21N5O2	XANTHINE,1,3-DIPROPYL-8-(2-AMINOPHENYL) 96445-34-0
9	3.18	7.4	1938	1	C17H22Cl2N2O1	1,2-DIAMINOCYCLOHEXANE,N-METHYL-N-ALLYL,N'-(3,4-DICHLOROBENZOYL) 98717-08-9
13000	1.70	7.0	708	1	C17H22Cl2N2O1	6(3,4-DICLBENZAMIDO)2-METHYLDECAHYDRO-I-QUINOLINE,CIS
1	1.66	7.0	708	1	C17H22Cl2N2O1	6(3,4-DICLBENZAMIDO)2-METHYLDECAHYDRO-I-QUINOLINE,TRANS
2 ✓	0.35		2397	101	C17H22F3N4O3S1	IMIDAZOLIUM CHLORIDE ANALOG ✚
3 ★	-1.86		1682		C17H22I3N3O7	M-BENZAMIDE,N,N-BIS(2-OHET),5(N-2,3-DIOHPR)ACAM,I3 ✚
4 ★	-2.06		1682		C17H22I3N3O7	M-PHTHALAMIDE,N,N-BIS(2,3-DIHYDROXYPROPYL)-3-N-METHYLACETAMIDO-TRIIODO 76350-28-2
5 ★	-2.42	7.4	2476	1	C17H22I3N3O8	IOPAMIDOL 60166-93-0 diagnostic aid (radiopaque)
6 ★	-2.64	7.4	2476	1	C17H22I3N3O9	P569 ✚
7	2.67	7.4	785	1	C17H22N2	N',N'-DIPHENYL-N,N-DIMETHYLPROPYLENEDIAMINE
8	0.31	7.0	1056	1	C17H22N2O3	1(3-CARBOXAMIDOPR)AM-3-(1-NAPHTHYLOXY)-2-PROPANOL
9	1.49		2089		C17H22N2O3	N-(3-DELTA-3-ACETOXYPROPYLPIPERIDIN-1-YL)BENZAMIDE 104642-81-1
13010	1.52		2089		C17H22N2O3	N-(3-DELTA-4-(3-METHOXYCARBONYLPROPYL)PIPERIDIN-1-YL)BENZAMIDE 104642-82-2
1	1.03	7.4	2542	1	C17H22N2O3S2	THIOPHENE-2-SULFONAMIDE,4-(3-I-PENTYLAMINOMETHYL)BENZOYL
2	1.23	7.4	2542	1	C17H22N2O4S2	THIOPHENE-2-SULFONAMIDE,4-(4-HYDROXY-3-N-METHYL-N-I-BUTYLAMINO)METHYLBENZOYL
3	1.64		606		C17H22N2O6	1-ET-2,6-DIME-3-(N-HOETAMIDO)MEO-5-MEO-INDOLOQUINONE
4	0.50	2.0	1039	210	C17H22N2S1	THENALIDINE 86-12-4 antihistaminic
15 ✓	0.07		2397	101	C17H22N3O2	IMIDAZOLIUM CHLORIDE,1-METHYL-2-HYDROXYIMINOMETHYL-3-(TETRAHYDRONAPHTH-1-YL)METHOXYMETHYL
6 ✓	-0.18		2397	101	C17H22N3O4	IMIDAZOLIUM CHLORIDE,1-METHYL-2-HYDROXYIMINOMETHYL-3-PENTANOL,SALICYLIC ESTER
7 ★	2.03	12.0	2408		C17H22N4O1	MINAPRINE 25905-77-5 antidepressant
8	2.65		1479	459	C17H22N4O2	2,4-DIAMINO-5-(3-MEO-4-CY-PENTOXY-BENZYL)PYRIMIDINE
9	2.46	7.4	1485	1	C17H22N4O2	2,4-DIAMINOPYRIMIDINE,5(3-MEO-4-I-PRENYL-5-ETO)BENZYL
13020 ★	2.65	7.4	1485	2	"	2,4-DIAMINOPYRIMIDINE,5(3-MEO-4-I-PRENYL-5-ETO)BENZYL
1	1.58		1713		C17H22N4O2	6-PH-2-(4-(N3-DIME-UREIDO)BU)-3(2H)-PYRIDAZINONE
2	1.83		1713		C17H22N4O2	6-PH-2-(4-(N3-ET-UREIDO)BUTYL)-3(2H)-PYRIDAZINONE
3	1.26		1713		C17H22N4O2	6-PH-2-(5-N3-ME-UREIDO-PENTYL)-3(2H)-PYRIDAZINONE
4	1.82		1713		C17H22N4O2	6-PH-2-(6-UREIDO-HEXYL)-3(2H)-PYRIDAZINONE
25	1.85	7.4	1485	1	C17H22N4O3	2,4-DIAMINOPYRIMIDINE,5(3,5-ETO-4-ACETYL)BENZYL
6 ★	2.04	7.4	1485	2	"	2,4-DIAMINOPYRIMIDINE,5(3,5-ETO-4-ACETYL)BENZYL
7	1.53	7.4	1485	1	C17H22N4O4	2,4-DIAMINOPYRIMIDINE,5(3,5-DIMEO-4-CO2-I-PR)BENZYL
8 ★	1.72	7.4	1485	2	"	2,4-DIAMINOPYRIMIDINE,5(3,5-DIMEO-4-CO2-I-PR)BENZYL
9	0.35	7.4	1493	1	C17H22N4O5	7-(ETHYLAMINO)MITOSANE 4117-84-4
13030	-1.40	7.4	2034	1	C17H22N8O2	TRIAMTERENE,P-(3-DIMETHYLAMINO-2-HYDROXY)PROPOXY
1	6.20		2404	459	C17H22O2	(7,6-HYDROPYRANYL)BENZOPYRAN,2,2,8,2',2'-PENTAMETHYL
2 ★	5.49		1521		C17H22O2	BENZYL(1R)-T-CHRYSANTHEMATE 64312-78-3
3	4.01		2404	459	C17H22O3	(6,5-HYDROPYRANYL)BENZOPYRAN,2,2,2',2'-TETRAMETHYL-7-METHOXY
4	4.47		2404	459	C17H22O3	(7,6-BENZOPYRANYL)BENZOPYRAN,2,2,2',2'-TETRAMETHYL-8-METHOXY 85434-27-1
35	3.12		306		C17H22O4	ETHYLACETOACETATE,3-BUTOXYBENZAL
6 ★	1.47		2330		C17H22O4	HELENALINYL-3,3-DIMETHYL ACRYLATE ✚
7 ★	4.00		306		C17H22O5	DIETHYLMALONATE,3-PROPOXYBENZAL
8 ★	0.30		2330		C17H22O5	TENULIN 19202-92-7
9	2.99	7.5	2651	28	C17H23Br1N2O3	BENZAMIDE,2,3-DIMETHOXY-5-BROMO-N-(ALLYLPYRROLIDIN-2-YL)METHYL
13040	1.23	7.0	708	1	C17H23Cl1N2O1	6-(4-CL-BENZAMIDO)-2-MEDECAHYDRO-I-QUINOLINE-C
1	1.23	7.0	708	1	C17H23Cl1N2O1	6-(4-CL-BENZAMIDO)-2-METHYLDECAHYDRO-I-QUINOLINE,CIS
2	2.72	7.4	2651	1	C17H23I1N2O3	NALEPRIDE
3	3.19	7.5	2651	28	"	NALEPRIDE
4	2.90		1479	459	C17H23I1N4O3	2,4-DIAM-5-(2-I-3,4,5-TRIETHOXYBENZYL)PYRIMIDINE
45	-1.34		307	22	C17H23N1	N-METHYLMORPHINAN PHOSPHATE
6	-1.95		307	22	C17H23N1O1	LEVORPHANOL TARTRATE 77-07-6
7	0.84	7.1	554		C17H23N1O1	LEVORPHANOL 77-07-6 analgesic (narcotic)
8	1.03	7.4	2378	1	"	LEVORPHANOL
9	1.14	7.4	554	1	"	LEVORPHANOL
13050	1.23	7.5	554	1	"	LEVORPHANOL
1	1.33	7.6	554	212	"	LEVORPHANOL
2	1.43	7.7	554	1	"	LEVORPHANOL
3 ★	3.11	7.7	554	2	"	LEVORPHANOL
4	1.58	7.4	2378	1	C17H23N1O1	N-ALLYL-N-NORMETAZOCINE
55	1.73	7.4	2378	1	"	N-ALLYL-N-NORMETAZOCINE
6	2.08	7.0	1056	1	C17H23N1O2	1-BUTYLAMINO-3-(1-NAPHTHYLOXY)-2-PROPANOL
7	1.51	7.4	1411	1	C17H23N1O2S1	4-METHIOPROPRANOLOL
8	1.28	7.4	1056	1	C17H23N1O3	1-METHOXYPROPYLAMINO-3-(1-NAPHTHYLOXY)-2-PROPANOL
9	1.28	7.4	1411	1	C17H23N1O3	4-METHOXYPROPRANOLOL
13060	-0.44	7.4	1155	1	C17H23N1O3	ATROPINE 51-55-8
1	-0.05	7.4	1157	324	"	ATROPINE
2 ★	1.83		505		"	ATROPINE
3	0.43	7.4	2480	1	C17H23N1O3	KETOBEMIDONE,ACETYL ESTER
4 ★	1.05	7.4	2480	2	"	KETOBEMIDONE,ACETYL ESTER
65	0.57	7.4	2480	1	C17H23N1O4	KETOBEMIDONE,METHYL CARBONATE ✚
6 ★	0.75	7.4	2480	2	"	KETOBEMIDONE,METHYL CARBONATE
7 ★	3.12		2415	459	C17H23N1O4S1	D8-THC,3-METHYLSULFONAMIDO ANALOG
8	2.59		2415	459	C17H23N1O4S1	O-195 ✚
9 ★	2.03		2234		C17H23N1O5	ACETYLSALICYLIC ACID,(N,N-DI-I-PROPYL)GLYCOLAMIDE ESTER 116482-76-9
13070 ★	2.09		2234		C17H23N1O5	ACETYLSALICYLIC ACID,N,N-DIPROPYLGLYCOLAMIDE ESTER 116482-75-8
1 ★	2.39		1898		C17H23N1O7	(4-HOET)VANILLINOXIME-O-ETACETYL,PROPIONYL ESTER ✚
2	0.45	7.4	1938	1	C17H23N3O1	1,2-DIAMINOHEXANE,N-DIMETHYL,N'-METHYL-(P-CYANOBENZOYL) 67579-77-5
3	0.80	7.4	1758	1	C17H23N3O1	MEPYRAMINE 91-84-9 antihistaminic
4 ★	3.27		1835		"	MEPYRAMINE
75	2.20	7.4	416	1	C17H23N3O2	1-ETHYL-3-PHENYL-5-DIETHYLAMINO-6-METHYLURACIL
6	2.78	7.4	416	1	C17H23N3O2	1-PHENYL-3-ET-5-DIETHYLAMINO-6-METHYLURACIL
7	-0.24	7.4	651	1	C17H23N3O2	4-(2-N,N-DIMETHYLAMINOBUTYRYL)ANTIPYRENE
8 ★	-0.73	7.0	2374		C17H23N3O3	TRYPTOPHANYLLEUCINE
9 ✓	-1.01		2397	101	C17H23N4O3S1	IMIDAZOLIUM CHLORIDE,1-METHYL-2-HYDROXYIMINOMETHYL-3-(1-PHENYLSULFONYLPIPERIDIN-2-YL)METHYL

	logP	pH	Ref.	Note	MF	Name / CAS / activity
13080 ✓	-1.18		2397	101	C17H23N4O3S1	IMIDAZOLIUM CHLORIDE,1-METHYL-2-HYDROXYIMINOMETHYL-3-(1-PHENYLSULFONYLPIPERIDIN-3YL)METHYL
1	0.84	3.3	2590	331	C17H23N5O2S1	N-(PYRID-2YL)BENZENESULFONAMIDE,4-(3-METHYL-3-PENTYL) *antineoplastic*
2	3.47	6.6	2590	459	"	N-(PYRID-2YL)BENZENESULFONAMIDE,4-(3-METHYL-3-PENTYL)
3	0.12	7.4	651	1	C17H23N5O3	4-(2-N,N-DIMEAMINO-3-PROPIONYLURYL)ANTIPYRENE
4	2.48	7.5	2651	1	C17H24Br1F1N2O3	BENZAMIDE,2,3-DIMETHOXY-5-BROMO-N-{1-(3-FLUOROPROPYL)PYRROLIDIN-2YL)METHYL
85	2.60	7.4	2651	1	"	BENZAMIDE,2,3-DIMETHOXY-5-BROMO-N-{1-(3-FLUOROPROPYL)PYRROLIDIN-2YL)METHYL
6	2.86	7.5	2651	28	"	BENZAMIDE,2,3-DIMETHOXY-5-BROMO-N-{1-(3-FLUOROPROPYL)PYRROLIDIN-2YL)METHYL
7	1.04		547		C17H24Cl1N3O11	N1-NO-N1-CLET-N3-1(TETRAACETYLGLUCOSYL)UREA
8	0.52	6.9	1199		C17H24Cl1N3S1	DAZOLICINE *anti-arrhythmic*
9	1.01	7.4	1199	1	"	DAZOLICINE
13090	1.52	8.0	1199	1	"	DAZOLICINE
1	5.15	7.4	1199	2	"	DAZOLICINE
2 ★	5.15		1064		"	DAZOLICINE **61477-97-2**
3	4.63	7.5	2651	28	C17H24Cl2N2O3	SALICYLAMIDE,3,5-DICHLORO-N-(1-BUTYLPYRROLIDIN-2YL)METHYL
4	1.75	7.5	2651	1	C17H24F1N2O3	FIDA-1
95	2.95	7.5	2651	28	"	FIDA-1
6	2.69	7.5	2651	1	C17H24F1N2O3	FIDA-3
7	3.22	7.5	2651	28	"	FIDA-3
8	0.47	7.0	708	1	C17H24N2O1	6-BENZAMIDO-2-METHYLDECAHYDROISOQUINOLINE-C
9	0.38	7.0	708	1	C17H24N2O1	6-BENZAMIDO-2-METHYLDECAHYDROISOQUINOLINE-T
13100	2.30	5.7	2241	1	C17H24N2O1	QUINISOCAINE **86-80-6** *anesthetic (local)*
1 ★	3.44	6.9	2175	1	C17H24N2O2	3-HEPTYL-3-(4-PYRIDYL)PIPERIDINE-2,6-DIONE
2	1.24	7.3	873	314	C17H24N2O2	O-3(N-PYRROLIDINO)CYHEXYL-N-PHENYLCARBAMATE
3 ★	3.60		1629		C17H24N2O4S1	CARBOFURAN,N-(N'ME,I-PROPYLCARBAMIDOTHIO)
4 ★	3.80		1629		C17H24N2O4S1	CARBOFURAN,N-(N'ME,PROPYLCARBAMIDOTHIO)
5 ✓	0.73		2397	101	C17H24N3O2	IMIDAZOLIUM CHLORIDE,1-BENZYL-2-HYDROXYIMINOMETHYL-3-)2,2-DIMETHYLPROPOXY)METHYL
6	1.33	1.0	2297		C17H24N4	2,4-DIAMINO-5-(3,4,5-TRIETHYLBENZYL)PYRIMIDINE **36821-85-9**
7	1.55	1.0	547	342	"	2,4-DIAMINO-5-(3,4,5-TRIETHYLBENZYL)PYRIMIDINE
8 ✓	3.05	13.0	579		C17H24N4O1	2,4-DIAMINO-5-(3-METHYL-4-METHOXY-5-T-BUTYLBENZYL)PYRIMIDINE
9	2.59	7.4	1485		C17H24N4O2	2,4-DIAMINOPYRIMIDINE,5(2-ME-4-MEO-5-BUO-)BENZYL
13110 ★	2.78	7.4	1485	2	"	2,4-DIAMINOPYRIMIDINE,5(2-ME-4-MEO-5-BUO-)BENZYL
1	-0.80	7.4	651	1	C17H24N4O2	4-(1-METHYL-3-DIMEAMINOPROPIONYLAMINO)ANTIPYRENE
2	2.08		1479	459	C17H24N4O3	2,4-DIAMINO-5-(3,4,5-TRIETHOXY-BENZYL)PYRIMIDINE **39711-86-9**
3	1.92	7.4	1485	1	C17H24N4O3	2,4-DIAMINOPYRIMIDINE,5(3-ETO-4-COHME2-5-MEO)BENZYL
4 ★	2.11	7.4	1485	2	"	2,4-DIAMINOPYRIMIDINE,5(3-ETO-4-COHME2-5-MEO)BENZYL
15	-1.89	1.0	547	342	C17H24N4O4	2,4-DIAM-5(3,4-DI(B-MEO-ETO)BENZYL)PYRIMIDINE
6 ★	2.50		2070		C17H24N4O5	ALLOPURINOL,1,5-DI-(PIVALOYLOXYMETHYL)
7 ★	2.34		2070		C17H24N4O5	ALLOPURINOL,2,5-DI-(PIVALOYLOXYMETHYL)
8	-1.00	7.0	801	1	C17H24N4O6	1-(5-3'-QUINUCLIDINOYLARABINOSYL)CYTOSINE
9	0.86	7.4	1485		C17H24N6O2	2,4-DIAMINOPYRIMIDINE,5-(3NME2-4-MEO-5-NMECOME)BENZYL
13120 ★	1.02	7.4	1485	2	"	2,4-DIAMINOPYRIMIDINE,5-(3NME2-4-MEO-5-NMECOME)BENZYL
1	1.02	3.3	2590	331	C17H24N6O2S1	N-(4-METHYLPYRIMIDIN-2YL)BENZENESULFONAMIDE,4-(3-METHYL-3-PENTYLTRIAZENYL) *antineoplastic*
2	3.65	6.6	2590	459	"	N-(4-METHYLPYRIMIDIN-2YL)BENZENESULFONAMIDE,4-(3-METHYL-3-PENTYLTRIAZENYL)
3 ★	-1.18	13.0	547	224	C17H24N6O3	NSC#140020 ✚
4	1.01	7.4	2591	1	C17H24N6O7	AZT-5'-ISOLEUCINYL **125780-96-3** *antiretroviral prodrug*
25	6.77		2404	459	C17H24O2	BENZOPYRAN,2,2-DIMETHYL-6-T-BUTYL-7-ETHOXY **101185-74-4**
6 ✓	2.01		536		C17H24O4	M-DECANOYLOXY BENZOATE,SODIUM SALT
7 ✓	-0.28		2308		C17H24O8	BENZO-18-CROWN-6-ETHER,M-CARBOXY **60835-75-8**
8	3.11	7.5	2651	28	C17H25Br1N2O3	BENZAMIDE,2,3-DIMETHOXY-5-BROMO-N-(1-PROPYLPYRROLIDIN-2YL)METHYL
9	3.01	7.5	2651	28	C17H25Br1N2O3	BENZAMIDE,2-ETHOXY-3-METHOXY-5-BROMO-N-(1-ETHYLPYRROLIDIN-2YL)METHYL
13130	3.34	7.5	2651	28	C17H25Br1N2O3	BENZAMIDE,2-METHOXY-3-ETHOXY-5-BROMO-N-(1-ETHYLPYRROLIDIN-2YL)METHYL
1	2.01	7.5	2651	28	C17H25Br1N2O4	BENZAMIDE,2,3,6-TRIMETHOXY-5-BROMO-N-(1-ETHYLPYRROLIDIN-2YL)METHYL
2	3.65	7.5	2651	28	C17H25Br1N2O4	SALICYLAMIDE,3-BROMO-5-METHOXY-6-ETHOXY-N-(1-ETHYLPYRROLIDIN-2YL)METHYL
3	3.19	7.5	2651	28	C17H25Br1N2O4	SALICYLAMIDE,3-METHOXY-5-BROMO-6-ETHOXY-N-(1-ETHYLPYRROLIDIN-2YL)METHYL
4	3.39	7.5	2651	28	C17H25Cl1N2O4	SALICYLAMIDE,3-CHLORO-5-METHOXY-6-ETHOXY-N-(1-ETHYLPYRROLIDIN-2YL)METHYL
35	2.28	7.5	2651	28	C17H25F1N2O2	BENZAMIDE,2-METHOXY-5-FLUOROETHYL-N-(1-ETHYLPYRROLIDIN-2YL)METHYL
6	2.19	7.5	2651	1	C17H25I1N2O3	HOMEP-1
7	3.35	7.5	2651	28	"	HOMEP-1
8	2.17	7.5	2651	1	C17H25I1N2O3	HOMEP-2
9	3.31	7.5	2651	28	"	HOMEP-2
13140	2.14	7.5	2651	1	C17H25I1N2O3	HOMEP-3
1	2.32	7.5	2651	1	"	HOMEP-3
2	3.39	7.5	2651	28	"	HOMEP-3
3	2.53	7.5	2651	28	C17H25I1N2O4	BENZAMIDE,2,3,6-TRIMETHOXY-5-IODO-N-(1-ETHYLPYRROLIDIN-2YL)METHYL
4	2.52	7.5	2651	1	C17H25I1N2O4	IOXIPRIDE
45	4.06	7.5	2651	28	"	IOXIPRIDE
6	3.38	7.5	2651	28	C17H25I1N2O4	SALICYLAMIDE,3-METHOXY-5-IODO-6-ETHOXY-N-(1-ETHYLPYRROLIDIN-2YL)METHYL
7	-0.98		307	1	C17H25N1	2-METHYL-5,9-DIETHYL-6,7-BENZOMORPHAN-HCL
8	1.56	7.4	2617	1	C17H25N1	PHENCYCLIDINE **77-10-1** *anesthetic*
9	1.68	7.4	579	1	"	PHENCYCLIDINE
13150 ★	3.63		579		"	PHENCYCLIDINE
1 ★	4.07		1481		C17H25N1O1	A-BUTYL-N-S-BUTYL-CINNAMAMIDE
2	1.72	7.4	2633	1	C17H25N1O1	OCTAHYDROPHENANTHREN-4A-AMINE,N,8-DIMETHYL-6-METHOXY *nmda antagonist*
3	4.02		2633	459	"	OCTAHYDROPHENANTHREN-4A-AMINE,N,8-DIMETHYL-6-METHOXY
4	1.44	7.4	2648	1	C17H25N1O2	ALPHAPRODINE,1,2,3-TRIMETHYL ANALOG
55	1.62	7.4	2648	1	"	ALPHAPRODINE,1,2,3-TRIMETHYL ANALOG
6	1.68	7.4	2648	1	"	ALPHAPRODINE,1,2,3-TRIMETHYL ANALOG
7 ★	2.97	7.4	2648	2	"	ALPHAPRODINE,1,2,3-TRIMETHYL ANALOG
8 ★	3.00	7.4	2648	2	"	ALPHAPRODINE,1,2,3-TRIMETHYL ANALOG
9 ★	3.13	7.4	2648	2	"	ALPHAPRODINE,1,2,3-TRIMETHYL ANALOG
13160	1.85	7.4	682	1	C17H25N1O2	N-PROPYLNORMEPERIDINE
1	2.02	7.4	2648	1	C17H25N1O2	TRIMEPERIDINE
2 ★	3.13	7.4	2648	2	"	TRIMEPERIDINE
3	1.88	7.4	2648	1	"	TRIMEPERIDINE **64-39-1**
4 ★	2.93	7.4	2648	2	"	TRIMEPERIDINE
65 ★	3.80		2157		C17H25N1O3	2-(BENZOYLOXY)-(N,N-DI-ISOBUTYL)ACETAMIDE
6 ★	3.91		2157		C17H25N1O3	2-(BENZOYLOXY)-(N,N-DIBUTYL)ACETAMIDE
7 ★	2.40	7.4	1700	5	C17H25N1O3	LEVOBUNOLOL **47141-42-4** *anti-adrenergic (beta-receptor)*
8	2.88		1167	459	C17H25N1O4	3,4-DIMEO-CINNAMIC ACID,DIETHYLAMINOETHYL ESTER
9	0.43	7.4	2556	1	C17H25N1O4	IBOPAMINE **66195-31-1** *dopaminergic (peripheral), cardiotonic*

	logP	pH	Ref.	Note	MF	Name / CAS / activity
13170 ★	0.32		2377		C17H25N3O4	AC-TYR-LEU-N
1 ★	-2.19	7.0	2374		C17H25N3O4	PHENYLALANYLVALINYLALANINE
2 ★	3.43		2506		C17H25N3O4S2	ALANYCARB insecticide (cholinesterase inhib.) ✚
3	0.35	7.4	2375	1	C17H25N3O5S1	VERALIPRIDE 66644-81-3 tmt of menopause
4	-1.31		1525		C17H25N3O7S2	7B-PENTYLTHIOACETYLAMINO-CEPHAMYCIN 57793-06-3
75 ✓	-0.92		2397	101	C17H25N4O3S1	IMIDAZOLIUM CHLORIDE,1-ETHYL-2-HYDROXYIMINOMETHYL-3-(3-N-ETHYL-PHENYLSULFONAMIDO)PROPYL
6 ✓	-0.79		2397	101	C17H25N4O3S1	IMIDAZOLIUM CHLORIDE,1-METHYL-2-HYDROXYIMINOMETHYL-3-(3-N-PROPYL-PHENYLSULFONAMIDO)PROPYL
7 ✓	-0.83		2397	101	C17H25N4O3S1	IMIDAZOLIUM CHLORIDE,1-PROPYL-2-HYDROXYIMINOMETHYL-3-(3-N-METHYL-PHENYLSULFONAMIDO)PROPYL
8 ★	1.89	7.0	1793	1	C17H25N5O3S1	N(4-(6-PURINYLTHIO)VALERYL)VALINE,ETHYL ESTER 23401-43-6
9	-2.00	7.4	2591	1	C17H25N7O7	AZT,5'-LYSINYL 125780-84-9 antiretroviral prodrug
13180 ★	4.50		2300	20	C17H26Cl1N1O2	BUTACHLOR 23184-66-9 herbicide
1 ★	4.08		2506		C17H26Cl1N1O2	PRETILACHLOR herbicide ✚
2	2.62		2162	459	C17H26Cl1N3O4S1	BENZAMIDE,4-CHLORO-2-METHOXY-5-(N-METHYLMETHANESULFONAMIDO)-N-(1-ETHYLPYRROLID-2-YL)METHYL
3	2.06	7.4	2409	1	C17H26N2O1	ROPIVACAINE 84057-95-4 anesthetic (local)
4 ★	2.90		2409	2	"	ROPIVACAINE
85	0.75	7.4	1938	1	C17H26N2O2	1,2-DIAMINOCYCLOHEXANE-N-DIMETHYL,N'-METHYL-(3-METHOXYBENZOYL) 67579-34-4
6	1.64	7.5	2651	1	C17H26N2O2	BENZAMIDE,2-METHOXY-5-ETHYL-N-(1-ETHYLPYRROLIDIN-2YL)METHYL
7	2.99	7.5	2651	28	"	BENZAMIDE,2-METHOXY-5-ETHYL-N-(1-ETHYLPYRROLIDIN-2YL)METHYL
8 ★	2.15	11.3	2195		C17H26N2O2	N-PROPYLACETAMIDE,3-(3-(N-PIPERIDINYLMETHYL)PHENOXY)
9	1.22	7.3	1960	314	C17H26N2O3	2-ETHOXYMETHYLCARBANILIC ACID,2-(1-PIPERIDINYL)ETHYL ESTER
13190 ★	2.72	7.3	1960	2	"	2-ETHOXYMETHYLCARBANILIC ACID,2-(1-PIPERIDINYL)ETHYL ESTER
1	2.23	7.3	1356	314	C17H26N2O3	2-PROPOXY-CARBANILATE-O-2(N-PIPERIDINYL)ETHYL 55792-11-5
2	3.79	7.3	1356	2	"	2-PROPOXY-CARBANILATE-O-2(N-PIPERIDINYL)ETHYL
3	2.48	7.3	1356	314	C17H26N2O3	3-PROPOXY-CARBANILATE-O-2(N-PIPERIDINYL)ETHYL 52205-61-5
4	4.04	7.3	1356	2	"	3-PROPOXY-CARBANILATE-O-2(N-PIPERIDINYL)ETHYL
95	2.20	7.3	1356	314	C17H26N2O3	4-PROPOXY-CARBANILATE-O-2(N-PIPERIDINYL)ETHYL 55792-12-6
6	3.76	7.3	1356	2	"	4-PROPOXY-CARBANILATE-O-2(N-PIPERIDINYL)ETHYL
7	2.20	7.5	2651	28	C17H26N2O3	BENZAMIDE,2-ETHOXY-5-METHOXY-N-(1-ETHYLPYRROLIDIN-2YL)METHYL
8	2.21	7.2	1150		C17H26N2O4	2-ME-4(O-PR)CARBANILATE-O-2(N-MORPHOLINO)ETHYL
9	1.06	7.5	2651	28	C17H26N2O4	BENZAMIDE,2,3,5-TRIMETHOXY-N-(1-ETHYLPYRROLIDIN-2YL)METHYL
13200	2.82	7.5	2651	28	C17H26N2O4	SALICYLAMIDE,3-METHOXY-6-ETHOXY-N-(1-ETHYLPYRROLIDIN-2YL)METHYL
1	0.94	7.5	2651	1	C17H26N2O4	SALICYLAMIDE,5-METHOXY-6-ETHOXY-N-(1-ETHYLPYRROLIDIN-2YL)METHYL
2	3.00	7.5	2651	28	"	SALICYLAMIDE,5-METHOXY-6-ETHOXY-N-(1-ETHYLPYRROLIDIN-2YL)METHYL
3 ★	1.31	7.4	1804	2	C17H26N2O4S1	2-MEO-5-SO2ET-BENZAMIDE,N(1-ETPYR-3YL)CH2NHCO
4	-0.70	7.4	2375	1	C17H26N2O4S1	SULTOPRIDE 53583-79-2 antidepressant
5	-0.62	7.4	1599	1	"	SULTOPRIDE
6	2.58		2162	459	"	SULTOPRIDE
7	0.42	7.4	2375	1	C17H26N4O4S1	ALPIROPRIDE 81982-32-3 antimigrane
8	2.47		2162	459	"	ALPIROPRIDE
9 ★	4.32		2415	459	C17H26O2	CYCLOHEXANOL,3-(2-HYDROXY-4-DIMETHYLPROPYL)PHENYL
13210 ★	6.06		2579	220	C17H26O3	4-DECYLOXYBENZOIC ACID
1	2.08		1396		C17H26O6	ARTEMISININE,ACETYL ESTER(ALPHA)
2	4.00	7.4	2462	1	C17H27N1O2	P-HEXYLBENZOIC ACID,N,N-DIMETHYLAMINOETHYL ESTER
3 ★	5.20	7.4	2462	2	"	P-HEXYLBENZOIC ACID,N,N-DIMETHYLAMINOETHYL ESTER
4	1.15	7.4	2450	1	C17H27N1O2	TROLOX (DIMETHYLAMINOETHYL ANALOG)
15	1.84	5.7	2241	1	C17H27N1O3	PRAMOXINE 140-65-8 anesthetic (topical)
6	1.48	5.7	2241	1	C17H27N1O3	STADACAINE 2350-32-5 anesthetic (local)
7	0.45	7.4	1628	1	C17H27N1O4	METHYPRANOL 22664-55-7
8	-0.35	1.2	2501	1	C17H27N1O4	METIPRANOLOL 22664-55-7 antihypertensive, antiarrhythmic, antiglaucoma
9	-0.17	7.0	2401	1	"	METIPRANOLOL
13220	0.36	7.4	2375	1	"	METIPRANOLOL
1	0.43	7.4	2501	1	"	METIPRANOLOL
2	1.04		2359	1	"	METIPRANOLOL
3	2.28	13.0	2501		"	METIPRANOLOL
4	-2.10	7.0	1362	1	C17H27N1O4	NADOLOL 42200-33-9 anti-adrenergic (beta receptor)
25	-1.31	7.4	2399	1	"	NADOLOL
6	-1.30	7.4	2465	41	"	NADOLOL
7	-1.18	7.4	1362	1	"	NADOLOL
8	-1.05	7.4	2178	1	"	NADOLOL
9 ★	0.71		1984		"	NADOLOL
13230	0.93	7.4	1700	2	"	NADOLOL
1 ★	3.90		2413	459	C17H27N1O4	TETRAHYDROPYRAN-2,4-DIONE,3[1-(ETHOXYIMINO)BUTYL]-6,6-SPIRO-(2-METHYL)CYCLOHEXYL
2 ★	4.07		2413	459	C17H27N1O4	TETRAHYDROPYRAN-2,4-DIONE,3[1-(ETHOXYIMINO)BUTYL]-6,6-SPIRO-(4-METHYL)CYCLOHEXYL
3	1.27	7.4	365	2	C17H27N3O1	N,N-DIET-PIPERAZINE-2-CARBOXANILIDE,2'6'-DIME
4	3.68	7.4	459		C17H27N3O2	1-CYCLOHEXYL-3-ME-5-PIPERIDINO-6-METHYLURACIL
35	-0.50	7.4	2375	1	C17H27N3O4S1	AMISULPRIDE 71675-85-9 antipsychotic
6	2.91		2162	459	"	AMISULPRIDE
7	2.29		2162	459	C17H27N3O4S1	BENZAMIDE,2-METHOXY-5-(N-METHYL-METHANESULFONAMIDO)-N-(1-ETHYLPYRROLID-2-YL)METHYL 90763-18-1
8	0.27	7.0	1812	1	C17H27N3O5	5'-AMINO-3'O-HEPTANOYL-5'-DEOXYTHYMIDINE
9	0.41	7.4	2350	1	C17H27N3O5S1	SPARSOMYCIN ANALOG ✚
13240 ★	0.60	7.4	2350	1	C17H27N3O5S2	SPARSOMYCIN,S-PENTYL ANALOG ✚
1	1.60	7.0	801	1	C17H27N3O6	1-(5-OCTANOYLARABINOSYL)CYTOSINE
2	1.20	7.0	801	1	C17H27N3O6	1-(5-TRIETHYLACETYLARABINOFURANOSYL)CYTOSINE
3	0.13	6.0	581		C17H27N5	4,6-DIAM-1,2H-SYM-TRIAZINE,1-M-HEXYLPHENYL
4	-0.69	6.0	581		C17H27N5O1	4,6-DIAM-1,2H-2,2-DIME-SYM-TRIAZINE,1-M-HEXYLOXYPH
45	1.41		1966	459	C17H28Au1O9P1S1	TETRAACETYLGLUCOSE,2-TRIMETHYLPHOSPHIN-AUROTHIO
6	2.66		1352		C17H28N2O1	3-(DIETHYLAMINO)PENTANO-2',6'-XYLIDIDE
7	2.42		1352		C17H28N2O1	4-(DIETHYLAMINO)PENTANO-2',6'-XYLIDIDE
8	2.90	7.4	2409	1	C17H28N2O1	ETIDOCAINE 36637-18-0 anesthetic (local)
9	3.27	7.4	2409	1	"	ETIDOCAINE
13250 ★	3.69		2409	2	"	ETIDOCAINE
1	1.16		844	837	C17H28N2O2	ENDOMID 4582-18-7
2	3.18	7.2	1150		C17H28N2O3	2-ME-4(O-PR)CARBANILATE-O-2(N,N-DIETHYLAMINO)ET
3	3.41	7.2	1150		C17H28N2O3	2-ME-4-PENTYLOXYCARBANILATE-O-2-(N,N-DIMETHYLAMINO)ETHYL
4	0.81	5.7	2241	1	C17H28N2O3	3-BUTOXYPROCAINE 99-43-4 anesthetic (topical)
55	0.83	6.0	2284	1	C17H28N2O3	CARBISOCAINE,PROPOXY ANALOG
6	2.07	7.4	2284	1	"	CARBISOCAINE,PROPOXY ANALOG
7	0.99	7.4	2461	1	C17H28N2O6S3	THIENO[2,3-B]THIOPHENE-2-SULFONAMIDE,5-(N,N-BIS-METHOXYETHOXYETHYL)AMINOMETHYL 133445-78-0
8	-1.11		1901	820	C17H28N3O2.I1	ACETYLPROCAINAMIDEETHO IODIDE
9	1.74	7.4	2244	1	C17H28N4O4S1	TIMOLOL,O-CYCLOPROPANOYL

	logP	pH	Ref.	Note	MF	Name / CAS / activity
13260	4.07	7.4	2028	459	C17H28N6	1,2,4-TRIAZOLE,3-AMINO-5-(2-DECYLAMINO)PYRIDIN-4-YL **77314-60-4**
1 ★	3.10	7.4	2481	1	C17H28N6O2S1	3(H)-CIMETIDINE,3-HEXYLOXYCARBONYL ANALOG ✚
2	7.17		2316	337	C17H28O1	PHENOL,2-UNDECYL
3	6.98		2316	337	C17H28O1	PHENOL,3-UNDECYL
4	7.18		2316	337	C17H28O1	PHENOL,4-UNDECYL
65	6.30		2316	337	C17H28O2	1,2-DIHYDROXYBENZENE,3-UNDECYL
6	6.94		2316	337	C17H28O2	1,2-DIHYDROXYBENZENE,4-UNDECYL
7	5.43		2316	337	C17H28O2	1,3-DIHYDROXYBENZENE,4-UNDECYL
8	3.29		1396	459	C17H28O5	ARTEMISININEETHYLETHER(BETA) **75887-54-6** *antimalarial*
9	1.64	7.0	477	1	C17H29N1O2	PROPYLAMINE,3-(4-T-AMYLPHENOXY)-2-HYDROXY-N-ISOPROPYL
13270	-0.29		527	820	C17H30N1.Br1	BENZYLDIMETHYLOCTYLAMMONIUMBROMIDE
1	0.44		527	820	C17H30N1.Br1	DODECYLPYRIDINIUMBROMIDE
2	-0.26		1635	820	C17H30N1.Cl1	BENZYLDIMETHYLOCTYLAMMONIUMCHLORIDE
3 ✓	0.54		1729	820	C17H30N1.I1	2,4-DIMETHYLPYRIDINIUM IODIDE,N-DECYL
4 ✓	0.66		1729	820	C17H30N1.I1	3,4-DIMETHYLPYRIDINIUM IODIDE,N-DECYL
75 ✓	0.76		1729	820	C17H30N1.I1	3,5-DIMETHYLPYRIDINIUM IODIDE,N-DECYL
6 ✓	2.15		1729	820	C17H30N1.I1	N-DODECYLPYRIDINIUM IODIDE **3026-66-2**
7 ★	2.71	7.4	1956	2	C17H30N2O3	PILOCARPIC ACID,HEXYL ESTER **96914-11-3**
8 ★	-0.39	7.2	2553		C17H30N4O4	AC-ALA-PRO-ALA-NTBU
9	2.08	7.4	2244	1	C17H30N4O4S1	TIMOLOL,O-BUTYRYL
13280	2.19	7.4	2244	1	C17H30N4O4S1	TIMOLOL,O-ISOBUTYRYL
1 ★	-2.83		2377		C17H30N4O5	VAL-PRO-GLY-VAL
2 ★	-0.67	2.0	2553	1	C17H30N4O6	AC-ALA-GLU-ALA-NTBU
3	1.77	12.3	343		C17H31Br1N2O5S1	N-DEMETHYL-BRINDAMYCIN
4	1.69	12.3	343		C17H31Cl1N2O5S1	N-DEMETHYL-CLINDAMYCIN
85	1.28		213		C17H31N1O2	N-DODECANOYLCYCLOBUTANECARBOXAMIDE
6 ★	-1.58	7.0	2374		C17H31N3O4	LEUCINYLLEUCINYLPROLINE
7 ★	-1.56	7.0	2374		C17H31N3O4	LEUCINYLPROLINYLLEUCINE
8 ★	-1.64	7.0	2374		C17H31N3O4	PROLINYLLEUCINYLLEUCINE
9 ★	-0.40		2469		C17H32N2O6S1	LINCOMYCIN,4'-ETHYL ANALOG ✚
13290	0.04	12.3	343		C17H32N2O6S1	N-DEMETHYL-LINCOMYCIN
1 ✓	1.43		2397	101	C17H32N3O2	IMIDAZOLIUM CHLORIDE,1-HEXYL-2-HYDROXYIMINOMETHYL-3-1,2,2-TRIMETHYLPROPOXY)METHYL
2 ✓	2.02		2397	101	C17H32N3O2	IMIDAZOLIUM CHLORIDE,1-HEXYL-2-HYDROXYIMINOMETHYL-3-HEXYLOXYMETHYL
3	1.36		1084		C17H32N6O1	METHOXYTRIAZINOKI. ✚
4 ★	-1.57	7.0	2197		C17H33N3O4	LEUCINYLVALINYLLEUCINE **58337-01-2**
95 ★	6.20		1629		C17H33N3O4S2	N-(N"-METHYL-DECOXYCARBAMYLTHIO)METHOMYL
6 ★	0.51		2093		C17H34N5O3P1	DIMETHYLTEPA,3-(2,2,6,6-TETRAMETHYL-1-HYDROXYPYRROLIDINYL)URYL **103981-96-0**
7 ★	0.45		2093		C17H34N5O3P1	DIMETHYLTEPA,3-(2,2,6,6-TETRAMETHYL-1-OXYPYRROLIDINYL)URYL **103981-93-7**
8	6.13		2182	459	C17H37O3P1	METHYLPHOSPHONIC ACID,DIOCTYL ESTER **1832-68-4**
9	-0.22		524	820	C17H38N1.I1	TRIPENTYL-ETHYL-AMMONIUM IODIDE
13300	2.06		541		C17H38P1.Br1	TRIMETHYLTETRADECYLPHOSPHONIUMBROMIDE
1	0.05	7.4	2437	1	C18H11F2N3O3S1	ZOPOLRESTAT,5,7-DIFLUORO ANALOG ✚
2	0.09	7.4	2635	1	"	ZOPOLRESTAT,5,7-DIFLUORO ANALOG
3	1.72		2635	459	"	ZOPOLRESTAT,5,7-DIFLUORO ANALOG
4	3.01		2437		"	ZOPOLRESTAT,5,7-DIFLUORO ANALOG
5 ★	5.79		2158	459	C18H12	BENZ(A)ANTHRACENE **56-55-3**
6 ★	5.73		2158	459	C18H12	CHRYSENE **218-01-9**
7 ★	5.90		1190		C18H12	NAPHTHACENE **92-24-0**
8 ★	5.49		2158	459	C18H12	TRIPHENYLENE
9 ★	3.35		2437		C18H12Br1N3O3S1	ZOPOLRESTAT,5-BROMO ANALOG
13310 ★	-0.31		2342		C18H12Cl3In1O9	TRIS(CHLOROKOJATO)INDIUM(III) **116699-25-3**
1	1.30		329		C18H12Co1N2O2	COBALT OXINATE
2	2.70		329		C18H12Cu1N2O2	COPPEROXINATE
3	-0.30	7.4	2437	1	C18H12F1N3O3S1	ZOPOLRESTAT,5-FLUORO ANALOG
4	1.86		2635	459	"	ZOPOLRESTAT,5-FLUORO ANALOG
15	2.95		2437		"	ZOPOLRESTAT,5-FLUORO ANALOG
6	0.60		329		C18H12Mn1N2O2	MANGANOUS OXINATE
7 ★	4.31		1742		C18H12N2	2,2'-BIQUINOLINE **119-91-5**
8	1.90		329		C18H12N2Ni1O2	NICKEL OXINATE
9	-1.26	7.4	2076	1	C18H13Br1N4O2S1	4-QUINAZOLINONE,2-METHYL-3-(4-OXO-3-(4-BROMO)PHENYL-THIAZOLIDINYL)IMINO
13320	1.53	7.4	1688	1	C18H13Cl1F1N3	MIDAZOLAM **59467-70-8** *anesthetic (injection), hypnotic*
1 ★	3.65		2506		C18H13Cl1F3N1O7	FLUOROGLYCOFEN-ETHYL *herbicide (protoporphyrinogen oxidase inhib.)* ✚
2	-1.60	7.4	2076	1	C18H13Cl1N4O2S1	4-QUINAZOLINONE,2-METHYL-3-(4-OXO-3-(4-CHLORO)PHENYL-THIAZOLIDINYL)IMINO
3	-1.36	7.4	2076	1	C18H13F1N4O2S1	4-QUINAZOLINONE,2-METHYL-3-(4-OXO-3-(4-FLUORO)PHENYL-THIAZOLIDINYL)IMINO
4 ★	4.98		1742		C18H13N1	6-AMINOCHRYSENE **2642-98-0** *produces leukopenia*
25 ★	6.44		850		C18H13N1S1	PHENYL-1-NAPHTHYLMETHYLISOTHIOCYANATE
6 ★	6.03		1814	459	C18H14	4-TERPHENYL **26140-60-3**
7	2.96		401		C18H14N2O1	4(1H)-QUINAZOLINONE,2,3-DI-H-2(1-NAPHTHYL)
8 ★	1.73		520		C18H14N2O3	1,4-NAPHTHOQUINONE,2-ACETAMIDO-3-ANILINO
9	-1.06	7.4	2076	1	C18H14N4O2S1	4-QUINAZOLINONE,2-METHYL-3-(4-OXO-3-PHENYLTHIAZOLIDINYL)IMINO
13330	1.51	7.0	1220	1	C18H14O2	1-PHENYL-2-NAPHTHALENEACETIC ACID
1	1.23	7.0	1220	1	C18H14O2	2-PHENYL-1-NAPHTHALENEACETIC ACID
2	1.53	7.0	1220	1	C18H14O2	3-PHENYL-1-NAPHTHALENEACETIC ACID
3	1.26	7.0	1220	1	C18H14O2	3-PHENYL-1-NAPHTHALENEACETIC ACID
4	1.56	7.0	1220	1	C18H14O2	4-PHENYL-1-NAPHTHALENEACETIC ACID **4022-43-9**
35	1.45	7.0	1220	1	C18H14O2	4-PHENYL-2-NAPHTHALENEACETIC ACID
6	1.54	7.0	1220	1	C18H14O2	5-PHENYL-1-NAPHTHALENEACETIC ACID **4022-51-9**
7	1.43	7.0	1220	1	C18H14O2	5-PHENYL-2-NAPHTHALENEACETIC ACID
8	1.62	7.0	1220	1	C18H14O2	6-PHENYL-1-NAPHTHALENEACETIC ACID
9	1.56	7.0	1220	1	C18H14O2	6-PHENYL-2-NAPHTHALENEACETIC ACID
13340	1.52	7.0	1220	1	C18H14O2	7-PHENYL-1-NAPHTHALENEACETIC ACID
1	1.46	7.0	1220	1	C18H14O2	7-PHENYL-2-NAPHTHALENEACETIC ACID
2	1.38	7.0	1220	1	C18H14O2	8-PHENYL-2-NAPHTHALENEACETIC ACID
3 ★	2.50	7.4	1966	1	C18H15Au1Cl1P1	CHLOROTRIPHENYLPHOSPHINO GOLD
4	4.54		2675	R459	C18H15Cl1F3N1O6	RH-4638(R)
45	4.53		2675	459	C18H15Cl1F3N1O6	RH-4639(S)
6	4.08		1959		C18H15Cl1Sn1	CHLOROSTANNANE,TRIPHENYL **639-58-7**
7	0.41		547	370	C18H15F1O2S1.Br1	N-(SO2F-BENZYL)-4-PHENYLPYRIDINIUM BROMIDE
8 ★	-0.01		2342		C18H15In1O9	TRIS(2-METHYLPYROMECONATO)INDIUM(III) **116781-97-6**
9 ★	-0.92		2342		C18H15In1O12	TRIS(KOJATO)INDIUM(III) **116699-28-6**

	logP	pH	Ref.	Note	MF	Name / **CAS** / *activity*
13350 ★	5.74		562		C18H15N1	TRIPHENYLAMINE **603-34-9**
1	2.84		759		C18H15N1O2	2-(4,4'-DIHYDROXYDIPHENYLMETHYL)PYRIDINE
2	3.13		759		C18H15N1O2	3-(4,4'-DIHYDROXYDIPHENYLMETHYL)PYRIDINE
3	3.24		759		C18H15N1O2	4-(4,4'-DIHYDROXYDIPHENYLMETHYL)PYRIDINE
4	2.54	7.4	712	1	C18H15N1O2	A-METHYL-N-ALLYL-G-PYROPHTHALONE
55	2.60	7.4	712	1	C18H15N1O2	A-METHYL-N-PROPENYL-G-PYROPHTHALONE
6	2.56		2161	468	C18H15N1O3	AZETIDINE-2-ONE,4-(STYRYLOYL)PHENOXY
7	4.19		543		C18H15N1O3	B-(4,5-DIPHENYLOXAZOL-2YL)PROPIONIC ACID
8	3.63		1973	459	C18H15N1O3	5-(4(2-PROPYNYLOXY)BENZOYL)-H2-PYRROLO-PYRROLE-1-CARBOXYLIC ACID **96327-40-1**
9	-0.30	7.3	2259	1	C18H15N1O4	BENZENEBUTANOIC ACID-4-OXO,P-N-(1,3-DIHYDRO-1-OXO-ISOINDOLYL)
13360	1.92	7.3	2259	2	"	BENZENEBUTANOIC ACID-4-OXO,P-N-(1,3-DIHYDRO-1-OXO-ISOINDOLYL)
1 ★	2.47	5.0	2674	302	C18H15N3O5	MEBENDAZOLE,N-METHOXYCARBONYL
2 ★	3.19	7.4	2348	1	C18H15N5	ADENINE,9-BENZHYDRYL
3 ★	2.83		541		C18H15O1P1	TRIPHENYLPHOSPHINEOXIDE **791-28-6**
4 ★	4.59		547		C18H15O4P1	TRIPHENYLPHOSPHATE **115-86-6**
65 ★	5.69		594	401	C18H15P1	TRIPHENYLPHOSPHINE **603-35-0**
6	2.65		2019		C18H15Sn1	CHLOROSTANNANE,TRIPHENYL **639-58-7**
7 ★	1.51	7.4	1626	1	C18H16Cl1F1N2O3	PROFLAZEPAM **52829-30-8**
8	4.13		2478	459	C18H16Cl1N1O3	ISOXAZOLINYLDIOXEPIN DIASTEREOMER, CHLORINE ANALOG
9	2.92		2478		C18H16Cl1N1O3	ISOXAZOLINYLDIOXEPIN DIASTEREOMER,CHLORINE ANALOG
13370 ★	4.12		2506		C18H16Cl1N1O5	FENOXAPROP-ETHYL *herbicide (fatty acid synth.inhib.)* ✚
1	1.96	7.4	1242	1	C18H16Cl2N2O4	LORAZEPAM,3-GLYCERYL ANALOG **56057-75-1**
2 ★	1.66	3.0	1638		C18H16Cl2N2O6S2	CEPHALOTHIN ANALOG(7(2,4-DICLPH)S-ACETYLAMINO) **59521-86-7**
3	1.85		910		C18H16Cl2N4	1,4-BIS(5-CHLORO-2-BENZIMIDAZOLYL)BUTANE
4	5.32		2424	459	C18H16Cl2N4O1	BENZOTRIAZOLE,2-CYCLOPENTYL-5-(3,4-DICHLOROBENZAMIDO)
75 ★	1.81	7.4	1626	1	C18H16F1N2O3	1,4-BENZODIAZEPIN-2-ONE,1(2,3-DIOHPR)5(2-F-PH)7-I
6	2.35		2478		C18H16F1O3	ISOXAZOLINYLDIOXEPIN DIASTEREOMER, FLUORINE ANALOG ✚
7 ★	3.45		2478		C18H16F1O3	ISOXAZOLINYLDIOXEPINDIASTEREOMER,FLUORINE ANALOG ✚
8	0.04		547	370	C18H16N1.Br1	N-BENZYL-4-PHENYLPYRIDINIUMBROMIDE
9	4.69	1.0	2471	459	C18H16N4O3	XANTHONE,6-(5-TETRAZOLYL)-2-BUTOXY
13380	-0.83	7.4	1722	1	C18H16N4O10S3	AZOSULFAMIDE **133-60-8**
1	2.45		910		C18H16N6O4	1,4-BIS(5-NITRO-2-BENZIMIDAZOLYL)BUTANE
2 ★	4.45		1167	459	C18H16O2	CINNAMIC ACID,3-PHENYLPROP-2-ENYL ESTER **122-69-0**
3	1.23	7.4	2438	1	C18H16O3	PHENPROCOUMON **435-97-2** *anticoagulant*
4 ★	3.62		1144		"	PHENPROCOUMON
85	5.00	1.0	2471	459	C18H16O5	XANTHONE,6-CARBOXY-2-BUTOXY
6	2.88	7.0	564	1	C18H16O7	USNIC ACID **125-46-2**
7	2.62	1.0	2471	459	C18H16O7	XANTHONE,6-CARBOXY-2-HYDROXYETHOXYETHOXY
8	4.18	7.4	874	1	C18H17Br1N2O1	7-BR-5-PHENYL-3-I-PR-1,4-BENZODIAZEPIN-2-ONE
9	3.55		2524	459	C18H17Br1N2O1	IMIDAZOLYLETHANOL,4'-METHYL-4''-BROMO ANALOG
13390	4.02		2524	459	C18H17Br1N2O2	IMIDAZOLYLETHANOL,4'-METHOXY-4''-BROMO ANALOG
1	4.00	7.4	874	1	C18H17Cl1N2O1	7-CL-5-PHENYL-3-I-PR-1,4-BENZODIAZEPIN-2-ONE
2	4.14	7.4	874	1	C18H17Cl1N2O1	7-CL-5-PHENYL-3-PROPYL-1,4-BENZODIAZEPIN-2-ONE
3	1.68		1582	459	C18H17Cl1N2O3	7-CL-1,3-H2-1-OHPR-3OH-5-PH-1,4-DIAZEPIN-2-ONE **56875-85-5**
4	1.98	7.4	1242	1	C18H17Cl1N2O3	7CL-1-ME-3(B-HOET)5-PH-1,4-BENZODIAZEPIN-2-ONE **40966-99-2**
95	1.71	7.4	1242	1	C18H17Cl1N2O4	OXAZEPAM,3-GLYCERYL ANALOG **40967-05-3**
6	4.11		2424	459	C18H17Cl1N4O1	BENZOTRIAZOLE,2-CYCLOPENTYL-5-(P-CHLOROBENZOYL)
7	2.33	7.4	712	1	C18H17N1O2	A-METHYL-N-ISOPROPYL-G-PYROPHTHALONE
8	2.63	7.4	712	1	C18H17N1O2	A-METHYL-N-PROPYL-G-PYROPHTHALONE
9	3.05	7.4	712	1	C18H17N1O2	N-BUTYL-G-PYROPHTHALONE
13400	2.76	7.4	712	1	C18H17N1O2	N-ISOBUTYL-G-PYROPHTHALONE
1	2.54		1059		C18H17N1O2S1	5,6-H2-2-ME-N(2-BIPHEN)1,4-OXATHIIN-3-CARBOXAMIDE
2	2.48		2161	468	C18H17N1O3	AZETIDINE-2-ONE,4-(PHENETHYLCARBONYL)PHENOXY
3	2.34		2478		C18H17N1O3	ISOXAZOLINYLDIOXEPIN DIASTEREOMER
4	3.42		2478	459	"	ISOXAZOLINYLDIOXEPIN DIASTEREOMER
5	3.48		1973	459	C18H17N1O4	5-(4-ALLYLOXYBENZOYL)-H2-PYRROLO-PYRROLE-1-CARBOXYLIC ACID **96327-39-8**
6	-1.20	7.3	2259	1	C18H17N1O4	INDOPROFEN ANALOG ✚
7	2.31	7.3	2259	2	"	INDOPROFEN ANALOG
8 ★	3.51		547		C18H17N1O4S1	N-(7-PR-9-OXO-3-THIOXANTHENYL)ACETAMIDE-10,10-O2
9	-0.07		2080		C18H17N2O1.H2O4P1	9-HYDROXY-2-N-METHYLELLIPTICINE PHOSPHATE
13410	0.41		2080		"	9-HYDROXY-2-N-METHYLELLIPTICINE PHOSPHATE
1 ★	0.96	7.4	1626	1	C18H17N3O5	1,4-BENZODIAZEPIN-2-ONE,1-(2,3-DIOHPR)-5-PH-7-NO2 **61554-12-9**
2 ★	1.47		537		C18H17N3O5	A-CYANO-PHENYLAMPHENICOL
3 ★	3.75		2424		C18H17N5O3	BENZOTRIAZOLE,2-CYCLOPENTYL-5-(P-NITROBENZAMIDO)
4 ★	-0.52	3.0	1638		C18H17N5O6S3	CEPHALOTHIN ANALOG ✚
15	1.62	7.4	1004	1	C18H17N6O1S1	C6H5OP(S)(NHN=CH(2-PYRIDYL))2
6	1.28	7.4	1004	1	C18H17N6O3P1S1	C6H5OP(S)(NHN=CH(2-PYRIDYL-N-OXIDE))2
7	3.37		565		C18H18Cl1N1O3	ETHYL-N-CHLOROACETYL-N-(2-BIPHENYLYL)GLYCINATE
8 ★	2.70	3.0	1038		C18H18Cl1N1O4	CLANOBUTIN **30544-61-7** *chloretic*
9 ★	2.40		2506		C18H18Cl1N1O5	BENZOXIMATE **29104-30-1** *acaricide*
13420	2.62	6.0	1148		C18H18Cl1N1S1	CHLORPROTHIXENE-TR
1	4.60	6.0	1148	807	"	CHLORPROTHIXENE-TR
2	2.67	7.0	468	1	C18H18Cl1N1S1	CHLORPROTHIXENE **113-59-7** *antipsychotic*
3	3.37	7.0	468	701	"	CHLORPROTHIXENE
4 ★	5.18		1598	40	"	CHLORPROTHIXENE
25	2.67	7.0	468	1	"	CHLORPROTHIXENE
6	3.37	7.0	468	701	"	CHLORPROTHIXENE
7 ★	5.18		1598	40	"	CHLORPROTHIXENE
8 ★	2.72	7.4	594	1	C18H18Cl1N3O1	1,4-DIAZEPIN-2-ONE,1,3-DIHYDRO-1-METHYL-7-DIMETHYLAMINO-5-(2-CHLOROPHENYL) **30144-75-3**
9	0.09	7.4	675	1	C18H18N2O2	3,5-PYRAZOLIDINDIONE,1,2-DIPHENYL-4-PROPYL
13430	1.59	7.4	2676	1	C18H18N2O2	CYCLO-PHE-PHE
1	1.65	7.4	847	1	C18H18N2O2	N-(3-(N-MEAMINO)PROPYL)-G-PYROPHTHAOLONE
2	1.99	7.4	847	1	C18H18N2O2	N-(DIMETHYLAMINOETHYL)-G-PYROPHTHALONE
3 ★	0.20	3.0	1638		C18H18N2O6S1	CEFALORAM **859-07-4** *antibacterial*
4	2.66		2428	459	C18H18N2O6S1	N,N'-DIMETHYLTHIODICARBAMATE,O-3,4-METHYLENEDIOXPHENYL-O'-M-TOLYL
35 ★	-0.44	3.0	1638		C18H18N2O7S1	CEPHALOSPORANIC ACID,7(D-MANDELAMIDO) **27910-26-5**
6 ★	0.26	3.0	1638		C18H18N2O7S1	CEPHALOTHIN ANALOG(7-PHENOXYACETYLAMINO) **10390-44-0**
7	1.05		547	342	C18H18N2O8	5AC23(OH)2PROPANAMPHAM-ETHYLIDINE-4OH-PYRANDIONE
8	1.39		910		C18H18N4	1,2-BIS(5-METHYL-2-BENZIMIDAZOLYL)ETHANE
9	1.59		910		C18H18N4	1,4-BIS-(2-BENZIMIDAZOLYL)BUTANE

	logP	pH	Ref.	Note	MF	Name / CAS / activity
13440	0.53	1.0	547	342	C18H18N4O1	2,4-DIAMINO-5-(3-BENZYLOXYBENZYL)PYRIMIDINE
1	0.49	1.0	594		C18H18N4O1	2,4-DIAMINOPYRIMIDINE,5-(4-BENZYLOXYBENZYL)
2 ★	3.85		2424		C18H18N4O1	BENZOTRIAZOLE,2-CYCLOPENTYL-5-BENZAMIDO
3	1.14		910		C18H18N4O2	1,2-BIS(5-METHOXY-2-BENZIMIDAZOLYL)ETHANE
4	1.27		910		C18H18N4O2	1,2-DIHYDROXY-1,2-BIS(5-ME-2-BENZIMIDAZOLYL)ETHANE
45	0.76		910		C18H18N4O3	1,2-BIS(5-METHOXY-2-BENZIMIDAZOLYL)ETHANOL
6	1.05		910		C18H18N4O4	1,2-DIHYDROXY-1,2-BIS(5-MEO-2-BENZIMIDAZOLYL)ETHANE
7 ★	1.02		2529		C18H18N4O4	DDI,5-(2-METHYL BENZOATE)
8	-2.40	7.4	2464	41	C18H18N6O5S2	CEFAMANDOLE 34444-01-4 antibacterial
9	-1.90	6.5	2464	41	"	CEFAMANDOLE
13450	-1.47	4.5	2464	41	"	CEFAMANDOLE
1	0.25	1.0	2196		"	CEFAMANDOLE
2	0.50		1980		"	CEFAMANDOLE
3 ★	-2.24	1.0	2196		C18H18N6O5S2	CEFATRIZINE 51627-14-6 antibacterial
4	-4.30	7.4	2530	1	C18H18N8O7S3	CEFTRIAXONE 73384-59-5 antibacterial
55	-3.53	6.5	2464	41	"	CEFTRIAXONE
6	-1.61	1.0	2196		"	CEFTRIAXONE
7	-1.23	7.4	2464	41	"	CEFTRIAXONE
8	-0.60	7.4	2373	1	"	CEFTRIAXONE
9	3.82	3.5	2179	302	C18H18O3	PENTANOIC ACID-3-METHYL-5-OXO-5-(P-BIPHENYLYL)
13460	0.37	2.0	1039	210	C18H19Cl1N2	CYCLIRAMINE 47128-12-1 antihistaminic
1	1.86	6.0	1148		C18H19Cl1N4	CLOZAPINE 5786-21-0 sedative, antipsychotic
2	2.99	7.4	1209	1	"	CLOZAPINE
3	3.23		2162	459	"	CLOZAPINE
4	3.32	6.0	1148	807	"	CLOZAPINE
65 ★	3.43		2650	458	C18H19Cl1O3S1	DIETHYL KETONE,1-(4-CHLOROPHENYL)-2-(4-TOLUENESULFONYL)
6 ★	1.87	7.0	2292	1	C18H19Cl2N3O3	1,3-DIOXANE SPIRO FUNGICIDE ✚
7	0.33	7.4	594	1	C18H19Cl2N5O1	1,2-DIHYDRO-S-TRIAZINE,2,2-DIMETHYL-4,6-DIAMINO-1-(3(3,4-DICHLOROBENZYLOXY)PHENYL)
8	3.07		434		C18H19Cl3O2	ETHOXYCHLOR
9	1.90	7.4	2020	1	C18H19F1N2O2	PINDOLOL ANALOG:N-(P-FLUOROBENZYL) 102573-82-0
13470	2.00	7.4	2020	1	C18H19F1N2O2	PINDOLOL ANALOG: N-(O-FLUOROBENZYL) 102573-83-1
1	0.48	7.2	2526	1	C18H19F1N4O3	OXONAPHTHYRIDINECARBOXYLIC ACID,N-CYCLOPROPYL-7-(2,5-DIAZABICYCLOHEPTLY) ANALOG
2	1.78		232	820	C18H19F3N2S1	TRIFLUPROMAZINEHYDROCHLORIDE 1098-60-8
3	3.39	7.4	678	1	C18H19F3N2S1	TRIFLUPROMAZINE 146-54-3 antipsychotic
4 ★	5.19	7.4	268	2	"	TRIFLUPROMAZINE
75	0.57	5.0	1440	1	C18H19N1	BENZOCTAMINE 17243-39-9 sedative, relaxant (muscle)
6	1.18	7.4	785	1	C18H19N1	N-DESMETHYLPROTRIPTYLINE
7 ★	2.16		1817		C18H19N1O1	3-PYRIDINEMETHANOL,A-2-PROPYNYL,A-4-I-PROPYLPHENYL 89242-88-6
8	0.65	1.0	547	342	C18H19N1O1	9-(2-DIMETHYLAMINO-1-HYDROXYET)PHENANTHRENE
9	0.99	1.0	547	342	C18H19N1O1	9-(2-ETHYLAMINO-1-HYDROXYETHYL)PHENANTHRENE
13480	2.84	7.4	564	1	C18H19N1O2	APOCODEINE 641-36-1
1	3.92		1973	459	C18H19N1O3S1	3-(4-METHYLTHIOBENZOYL)-5H-PYRROLO(H4)AZEPINE-9-CARBOXYLIC ACID 96327-70-7
2	3.38		1973	459	C18H19N1O4	5-(4-I-PROPOXYBENZOYL)-H2-PYRROLO-PYRROLE-1-CARBOXYLIC ACID 96327-38-7
3 ★	1.99		2487		C18H19N1O5	CINNAMAMIDE,3,4-DIHYDROXY-N-(3,4-DIMETHOXY)BENZYL
4 ★	2.84		1898		C18H19N1O6	(4-HYDROXYETHYL)VANILLINOXIME,2-METHOXYBENZYL ESTER
85 ★	2.10		1898		C18H19N1O6	(4-HYDROXYETHYL)VANILLINOXIME,PHENOXYACETYL ESTER
6	2.12	7.4	2621	1	C18H19N1O7	MITOSENE(5,5,6),METHYL-DIACETYLOXY ANALOG
7	1.38	7.4	2621	1	C18H19N1O7	MITOSENE,DIACETYLOXY ANALOG
8	2.00		2021	459	C18H19N1O7	NAPHTHALENE,1,4-DIACETYLOXY-2,3-DIMETHOXY-5-ACETYLAMINO 91814-31-2
9 ★	2.63	7.4	1626	1	C18H19N3O1	1,4-BENZODIAZEPIN-2-ONE,1-ME-5-PH-7-DIMEAMINO 2891-09-0
13490	-1.05	7.4	2192	1	C18H19N3O6S1	CEPHALOGLYCIN 3577-01-3 antibacterial
1	-0.59	7.2	917	314	"	CEPHALOGLYCIN
2	-0.55	7.0	1526	1	C18H19N3O6S2	N-FURFURYL-4(N-MEANILINO)-5-SULFAMOYLORTHANILICAC.
3	2.88		1479	459	C18H19N5O1	2,4-DIAMINO-5-(3-ANILINO-4-MEO-BENZYL)PYRIMIDINE
4	3.42		1479	459	C18H19N5O1	2,4-DIAMINO-5-(4-ANILINO-3-MEO-BENZYL)PYRIMIDINE
95 ★	1.12	7.4	2433	1	C18H19N5O4	ALLOPURINOL,N-(3-(MORPHOLINOMETHYL)BENZOYLOXY)METHYL
6 ★	1.13	7.4	2433	1	C18H19N5O4	ALLOPURINOL,N-(4-(MORPHOLINOMETHYL)BENZOYLOXY)METHYL
7 ★	0.62	7.4	1689	1	C18H19N5O5S1	N7-(2-THIAZOLYL)-MITOMYCIN-C 78327-27-2
8	1.82		1593	459	C18H19N7O8S4	CEPHALOSPORIN,N-ME-CARBOXAMIDO ANALOG 69712-55-6
9 ★	3.52	6.0	2644	2	C18H20Cl1N1O2	FOMOCAINE ANALOG-26 (P-CL-PHENYL)
13500	2.93	7.4	2325	1	C18H20Cl1N1O2	PHENYLACETIC ACID,5-CHLORO-2-(2',3',5',6'-TETRAMETHYL)ANILINO 127792-28-3
1	0.65	7.5	1344	1	C18H20Cl1N1O8	3-DESETHYL-QUINCARBATE-3(CH2CHOHCH2OH)
2 ★	2.26		2292	1	C18H20Cl1N5O2	1,3-DIOXANE-2-(2-CHLORO-4-(IMIDAZOL-1-YL)PHENYL-2-(1,2,4-TRIAZOL-1-YL)METHYL-4,4-DIMETHYL
3	-0.48	7.2	2446	1	C18H20F1N3O4	OFLOXACIN 82419-36-1 antibacterial
4	-0.44	7.4	2681	1	"	OFLOXACIN
5 ★	-0.39		2681	226	"	OFLOXACIN
6	-0.36	7.4	2530	1	"	OFLOXACIN
7 ★	3.50		2439		C18H20N1O3P1S1	FOSTEDIL 75889-62-2 vasodilator, calcium channel blocker
8 ★	2.33		2517		C18H20N2O2	DIBENZOYLHYDRAZINE,N-T-BUTYL
9	2.09	7.0	2309	1	C18H20N2O2	N-ACETYL-2,2-DIBENZYLGLYCINAMIDE
13510	-1.29	7.4	547	1	C18H20N2O3	PHENYLALANYLPHENYLALANINE 2577-40-4
1 ★	-0.85	7.0	2197		"	PHENYLALANYLPHENYLALANINE
2 ★	-1.68	7.0	2374		C18H20N2O4	PHENYLALANYLTYROSINE
3 ★	-1.87	7.0	2374		C18H20N2O5	TYROSINE,TYROSYL 1050-28-8
4	0.97	7.4	1754	320	C18H20N2O6	NITRENDIPINE 39562-70-4 antihypertensive
15 ★	2.39	7.4	1978	1	C18H20N2O7	1-ACETYL-7-METHOXY-N-METHYLMITOSENE 96000-46-3
6 ★	5.23		550		C18H20N2S1	METHDILAZINE 1982-37-2 antipruritic
7	1.36	7.4	594	1	C18H20N4O2	NITRACRINE 4533-39-5 antineoplastic
8	3.91		2424	459	C18H20N4O2S1	BENZOTRIAZOLE,2-CYCLOPENTYL-5-(P-TOLUENESULFONAMIDO)
9	1.23	7.4	2024	1	C18H20N4O3	1,2-(N-METHYLAZIRIDINO)MITOSINE,7-(2-METHYLAZIRIDINO) 103422-23-7
13520 ★	0.08	7.4	1689	1	C18H20N6O5	N7-(3-PYRAZOLYL)-MITOMYCIN-C 84397-45-5
1	2.86		1593	459	C18H20N6O7S5	CEPHALOSPORIN,ETHYLTHIO ANALOG 69712-38-5
2 ★	5.07		505		C18H20O2	4,4'-STILBENEDIOL,A,A'-DIETHYL 56-53-1 estrogen
3 ★	3.75	5.5	2403	2	C18H20O3	HEXANOIC ACID,2-(4-BIPHENYLYL)-3-HYDROXY 93371-38-1
4 ★	3.84	5.5	2403	2	C18H20O3	PENTANOIC ACID,2-(4-BIPHENYLYL)3-HYDROXY-4-METHYL 93371-56-3
25	4.36		260		C18H20O3S1	2-OH-3-CARBOXY-5-ME-BENZTHIO-2'-I-PROPYLPHENYLETHER
6 ★	2.70		2650	458	C18H20O3S1	DIETHYL KETONE,1-(4-METHYLPHENYL)-2-BENZENESULFONYL
7 ★	2.45		2650	458	C18H20O3S1	DIETHYL KETONE,1-PHENYL-2-(4-TOLUENESULFONYL)
8	4.91		260		C18H20O4	2-OH-3-CARBOXY-5-ME-BENZYL-2'-I-PROPYLPHENYLETHER
9 ★	2.18		2650	458	C18H20O4S1	DIETHYL KETONE,1-(4-METHOXYPHENYL)-2-BENZENESULFONYL

	logP	pH	Ref.	Note	MF	Name / **CAS** / *activity*
13530 ★	2.21		2308		C18H20O5	2,3-DIBENZO-15-CROWN-5-ETHER **14262-60-3**
1	0.60	7.4	2559	1	C18H20O5	CHROMONE-2-CARBOXYLIC ACID,5-I-PENTOXY-8-ALLYL
2	3.14		2021	459	C18H20O6	NAPHTHALENE,1,4-DIACETYLOXY-2,3-DIMETHOXY-6,7-DIMETHYL **102632-22-4**
3	3.44		2021	459	C18H20O6	NAPHTHALENE,1,4-DIPROPIONYLOXY-2,3-DIMETHOXY **91814-56-1**
4	4.42	7.4	2090	459	C18H21Br1N2O3	5-(5-(2-BROMO-4-(4,5-DIHYDRO-2-OXAZOLYL)PHENOXY)PENTYL-3-METHYISOXAZOLE **105639-04-1**
35	3.45	7.4	591	1	C18H21Cl1N2	CHLORCYCLIZINE **82-93-9** *antihistaminic*
6	4.30	7.4	2090	459	C18H21Cl1N2O3	5-(5-(2-CHLORO-4-(4,5-DIHYDRO-2-OXAZOLYL)PHENOXY)PENTYL-3-METHYLISOXAZOLE **98033-68-2**
7	-0.41	7.2	2526	1	C18H21F1N4O3	OXONAPHTHYRIDINECARBOXYLIC ACID,7-(2,5-DIAZABICYCLOHEPTANE) ANALOG
8	1.38	7.4	2235	1	C18H21N1	PIPERIDINE,1-METHYL-4,4-DIPHENYL
9 ★	3.75	8.0	2644	2	C18H21N1O1	FOMOCAINE ANALOG-19 (PYRROLIDINE)
13540	0.04	1.0	602	342	C18H21N1O1	N-AZETIDYL-ETHYL-DIPHENYLMETHYLETHER
1	0.48	1.0	870	342	C18H21N1O1	N-AZIRIDYLETHYL-(P-ME-DIPHENYLMETHYL)ETHER
2	0.41	7.2	969	314	C18H21N1O1	PIPRADROL **467-60-7** *cns stimulant*
3	0.78	7.4	2493	1	C18H21N1O2	2-PROPANONE,1,1-DIPHENYL-1-METHOXY-3-DIMETHYLAMINO
4 ★	2.75	8.0	2644	2	C18H21N1O2	FOMOCAINE ANALOG-22 (ISOMORPHOLINE)
45 ★	4.41		2168		C18H21N1O2	ISOVALERIC ACID-2-PHENYLAMINO,BENZYL ESTER
6 ★	4.91		1521		C18H21N1O2	P-CYANOBENZYL(1R)-TR-CHRYSANTHEMATE
7	1.95	7.4	2325	1	C18H21N1O2	PHENYLACETIC ACID,2-(2',3',5',6'-TETRAMETHYL)ANILINO **127792-25-0**
8	5.15	7.4	2325	2	C18H21N1O2	PHENYLACETIC ACID,2-(2',3',5',6'-TETRAMETHYL)ANILINO
9	1.91	7.4	2325	1	C18H21N1O2	PHENYLACETIC ACID,2-(2',6'-DIETHYL)ANILINO **127792-29-4**
13550	5.11	7.4	2325	2	C18H21N1O2	PHENYLACETIC ACID,2-(2',6'-DIETHYL)ANILINO
1	0.10	7.1	554		C18H21N1O3	CODEINE **76-57-3** *antitussive, analgesic (narcotic)*
2	0.21	7.4	2680	324	"	CODEINE
3	0.25	7.4	2378	1	"	CODEINE
4	0.36	7.4	554	1	"	CODEINE
55	0.44	7.5	554	1	"	CODEINE
6	0.52	7.6	554	212	"	CODEINE
7	0.59	7.7	554	1	"	CODEINE
8	0.61	7.4	919	1	"	CODEINE
9	0.61	7.5	1510	1	"	CODEINE
13560 ★	1.14	7.7	554	2	"	CODEINE
1 ★	4.45		1523		C18H21N1O4	1,1-DI-(P-METHOXYPHENYL)-2-NITROBUTANE
2 ★	3.29		2142		C18H21N1O4	3,4-DIETHOXYPHENYLCARBAMATE,O-BENZYL
3	0.84	10.1	401	1	C18H21N1O4	CEPHALOTAXINE
4 ★	2.05	7.4	2268	1	C18H21N1O4	NAPROXEN,N,N-DIMETHYLGLYCOLAMIDE ESTER
65	1.50	7.4	2062	1	C18H21N1O5	L-DOPA,(P-METHOXYPHENYL)ETHYL ESTER
6 ★	1.69	7.4	2062	2	"	L-DOPA,(P-METHOXYPHENYL)ETHYL ESTER
7 ★	1.30	6.8	2337	1	C18H21N2O7P1	DHP-218 *antihypertensive* ✚
8	1.71	7.4	880	1	C18H21N3O1	DIBENZEPIN **4498-32-2** *antidepressant*
9	2.50	8.5	2381	2	"	DIBENZEPIN
13570	2.59	7.4	416	1	C18H21N3O2	1-PHENYL-3-ME-5-DIALLYLAMINO-6-METHYLURACIL
1 ★	1.13		1592		C18H21N3O2	PHENYLALANYLPHENYLALANAMIDE **15893-46-6**
2	3.49	7.4	2090	459	C18H21N3O5	5-(5-(2-NITRO-4-(4,5-DIHYDRO-2-OXAZOLYL)PHENOXY)PENTYL-3-METHYLISOXAZOLE **105639-10-9**
3 ★	0.90		1712		C18H21N3O5S1	(BENZYL-A-ACETYLAMINO)PENICILLIN
4	-1.33	7.4	594	1	C18H21N5O1	1,2-DIHYDROTRIAZINE,2,2-DIMETHYL-4,6-DIAMINO-1-(3'-PHENYLCARBINOL)PHENYL
75 ★	1.86	13.0	547	224	C18H21N5O1	4,6-DIAM-2,2-DIME1(3'-PHO-ME)PH-S-TRIAZINE
6	-0.15		508		C18H22Cl1N2S1.Cl1	CHLORPROMAZINE METHOCHLORIDE
7	0.78		1552	820	C18H22Cl1N2S1.I1	CHLORPROMAZINEMETH IODIDE
8 ★	-1.89		848		C18H22I3N3O8	METRIZAMIDE **31112-62-6** *diagnostic aid (radiopaque)*
9	-1.41	7.4	2476	1	"	METRIZAMIDE
13580	-0.03	2.0	1039	210	C18H22N2	CYCLIZINE **82-92-8** *antihistaminic, antiemetic*
1	1.31	7.4	785	1	C18H22N2	DESIPRAMINE **50-47-5** *antidepressant*
2	1.45	7.4	1209	1	"	DESIPRAMINE
3	1.48	7.4	268	1	"	DESIPRAMINE
4	1.70	7.0	468	1	"	DESIPRAMINE
85 ★	4.90	7.0	468	701	"	DESIPRAMINE
6 ★	3.49		592		"	BENZANILIDE,4-DIETHYLAMINOMETHYL
7	2.50	7.4	678	1	C18H22N2O1S1	METHOPROMAZINE **61-01-8** *neuroleptic*
8 ★	4.90	7.4	268	2	"	METHOPROMAZINE
9	1.11	7.0	2401	1	C18H22N2O2	CARAZOLOL **57775-29-8** *anti-adrenergic (beta-receptor)*
13590	1.32	7.4	1628	1	"	CARAZOLOL
1	1.50	7.4	2501	1	"	CARAZOLOL
2 ★	3.59	13.0	2501		"	CARAZOLOL
3	0.76	5.7	2241	1	C18H22N2O2	PHENACAINE **101-93-9** *anesthetic (topical, ophthalmic)*
4	-0.75	2.0	919		C18H22N2O2S1	OXOMEMAZINE **3689-50-7** *tranquilizer*
95	-0.70	4.3	904	1	"	OXOMEMAZINE
6	1.90	7.8	550		"	OXOMEMAZINE
7	2.43	8.0	919		"	OXOMEMAZINE
8	2.51		550		"	OXOMEMAZINE
9 ★	5.18		2519		C18H22N2O2S1	PYRIBUTICARB ✚
13600	3.50		508		C18H22N2O2S2	PHENOTHIAZINE,2-MESULFONYL-10-(3-DIMEAMINOPROPYL)
1 ★	3.81		2612		C18H22N2O3	1(4-(2'(4"MEPH)ETHOXY)PH)-3-MEO-3-METHYLUREA **68358-79-2**
2	3.83	7.4	2090	459	C18H22N2O3	5-(5-(4-(4,5-DIHYDRO-2-OXAZOLYL)PHENOXY)PENTYL)-3-METHYLISOXAZOLE **98034-30-1**
3	1.73		1205		C18H22N2O3S1	N,N-DIME-4(N-ME-N(2-PHENYLPROPIONYL)BZSULFONAMIDE
4 ★	2.76		515		C18H22N2O5S1	ISOPROPICILLIN **4780-24-9** *antibacterial*
5	-0.03	7.2	846	314	C18H22N2O5S1	PROPICILLIN **551-27-9**
6 ★	2.65		515		"	PROPICILLIN
7	1.82	7.4	1941	1	C18H22N2O6	BIS-(3-PHENOXYPROPAN-@-OL)AMINE,P'-NITRO
8 ★	2.62		2164		C18H22N2O6	BIS-(3-PHENOXYPROPAN-@-OL)AMINE,P'-NITRO
9	0.70	5.0	2375	1	C18H22N2S1	TRIMEPRAZINE **84-96-8** *antipruritic*
13610	0.93	4.3	882	1	"	TRIMEPRAZINE
1	1.96	6.0	1897	1	"	TRIMEPRAZINE
2	2.91	7.4	678	1	"	TRIMEPRAZINE
3	3.44	7.8	550		"	TRIMEPRAZINE
4	4.71		550		"	TRIMEPRAZINE
15 ✓	0.21		2397	101	C18H22N3O2	IMIDAZOLIUM CHLORIDE,1-METHYL-2-HYDROXYIMINO-3-(1-ETHYL-4-PHENYLBUTYN-3YL)OXYMETHYL
6 ✓	0.71		2397	101	C18H22N3O2	IMIDAZOLIUM CHLORIDE,1-METHYL-2-HYDROXYIMINOMETHYL-3-(1-I-PROPYL-3-PHENYLPROPYN-2YLOXY)METHYL
7	0.62		1045		C18H22N6O3	6(2-PH-5(HOME)1,3-DIOXAN-5YL)AM-2,4-DIAZIRID.TRIAZINE
8 ★	1.80	7.4	2206	306	C18H22N6O3	BENZIMIDAZOLE,1-(3-(N-MORPHOLINO)PROPYL)-2-(1-METHYL-5-NITRO-IMIDAZOL-2-YL)
9 ★	3.13		2460		C18H22O2	ESTRONE **53-16-7** *estrogen*

	logP	pH	Ref.	Note	MF	Name / CAS / activity
13620	6.67		2616	459	C18H22O3	FURAMETHRIN 23031-38-1
1	1.80	7.4	2559	1	C18H22O4	TERBUCROMIL 37456-21-6 anti-allergic
2	0.10	7.4	2559	1	C18H22O5	TERBUCROMIL-6-T-BUTANOL ANALOG
3 ★	4.05		579		C18H23Cl1N2O3	2-PENTENOIC ACID-2-CYANO-3-(P-CHLOROBENZYL)AMINO-4-METHYL,2-ETHOXYETHYL ESTER
4 ★	2.69	7.4	1956	2	C18H23Cl1N2O3	PILOCARPIC ACID,P-CHLOROBENZYL ESTER 92598-89-5
25	-0.74		848		C18H23I3O12	2,4,6-TRIIODORESORCYL-1,3-DIGLUCOSIDE
6	2.30	7.4	785	1	C18H23N1	4,4-DIPHENYL-N,N-DIMETHYLBUTYLAMINE
7	0.58	2.0	1039	210	C18H23N1	TOLPROPAMINE 5632-44-0 antihistamine, antipruritic
8	0.14	1.0	870	342	C18H23N1O1	2-(DIPHENYLMETHOXY)ETHYL-N-ME-N-ETHYLAMINE
9	0.72	1.0	870	342	C18H23N1O1	2-(P-ME-DIPHENYLMETHOXY)-N-ET-ETHYLAMINE
13630	0.62	1.0	870	342	C18H23N1O1	2-DIPHENYLMETHOXY-N-PROPYL-ETHYLAMINE
1	0.41	1.0	602	342	C18H23N1O1	4-METHYLDIPHENHYDRAMINE 19804-27-4 antihistaminic
2 ★	3.77		2132			ORPHENADRINE 83-98-7 relaxant (skeletal muscle), antihistaminic
3	0.96		307	1	C18H23N1O2	DIHYDRODESOXYCODEINE-D ✚
4 ★	3.04	8.0	2644	2	C18H23N1O2	FOMOCAINE ANALOG-23 (ETHYL-ETHANOLAMINE)
35 ★	2.33	8.0	2644	2	C18H23N1O3	FOMOCAINE ANALOG-18 (DIETHANOLAMINE)
6	0.62	4.0	2655	1	C18H23N1O3	PROPRANOLOL,ACETYL ESTER
7	1.73	7.4	1941	1	C18H23N1O4	BIS-(3-PHENOXYPROPAN-2-OL)AMINE
8 ★	2.52		2164		"	BIS-(3-PHENOXYPROPAN-2-OL)AMINE
9	1.86		634		C18H23N1O6	6-I-PRO-7-METHOXYETHOXY-4-OH-3-QUINOLINE-CO2ET
13640	1.70		634		C18H23N1O6	6-PRO-7-METHOXYETHOXY-4-OH-3-QUINOLINE-CO2ET
1	-0.66		1901	820	C18H23N2O1S1.I1	THIAZINAMIUMSULFOXIDE
2	1.60		1901	820	C18H23N2S1.I1	THIAZINAMIUM IODIDE
3	-0.66		2627	841	C18H23N2.Cl1	N-METHYL-4-(4'-DIETHYLAMINOSTYRYL)-PYRIDINIUM CHLORIDE
4	-0.84	7.4	722	1	C18H23N3O1	DESISOPROPYL-DISOPYRAMIDE
45	2.83	7.4	2090	459	C18H23N3O3	5-(5-(2-AMINO-4-(4,5-DIHYDRO-2-OXAZOLYL)PHENOXY)PENTYL-3-METHLISOXAZOLE 105639-05-2
6	3.65	7.4	1938	1	C18H24Cl2N2O1	1,2-DIAMINOCYCLOHEXANE,N-METHYL-N-ALLYL-N'-METHYL-(3,4-DICHLOROBENZOYL) 67579-46-8
7 ✓	0.78		2397	101	C18H24F3N4O3S1	IMIDAZOLIOUM CHLORIDE ANALOG ✚
8 ★	-1.31		848		C18H24I2O12	2,4-DIIODORESORCYL-1,3-DIGLUCOSIDE
9 ★	-0.25		848		C18H24I2O12	4,6-DIIODORESORCYL-1,3-DIGLUCOSIDE
13650 ★	-2.17		1682		C18H24I3N3O7	M-BENZAMIDE,N,N'BIS(2,3-OHPR)BIS-ME,5-ACETAMIDO,I3
1 ★	-2.05	7.4	2476	1	C18H24I3N3O8	IOPROMIDE 73334-07-3 diagnostic aid (radiopaque)
2 ★	-2.33		1682		C18H24I3N3O8	M-BENZAMIDE,N,N-BIS(CH(CH2OH)2),5-N-OHET-ACAM,I3
3 ★	-2.47		1682		C18H24I3N3O8	M-PHTHALAMIDE,N,N-BIS(2,3-DIHYDROXYPROPYL)-3-N-(HYDROXYETHYL)ACETAMIDO-TRIIODO 66108-93-8
4 ★	-2.98	7.4	2476	1	C18H24I3N3O9	IOVERSOL 87771-40-2 diagnostic aid (radiopaque)
55 ★	-2.55	7.4	2476	1	C18H24I3N3O9	P530 ✚
6 ★	1.97	7.4	1956	2	C18H24N2O3	PILOCARPIC ACID,BENZYL ESTER 92598-82-8
7	-4.80	7.4	2449	1	C18H24N2O5	ENALAPRILAT 76420-72-9 antihypertensive
8 ✓	-0.74		2018		"	ENALAPRILAT
9 ✓	0.38		2397	101	C18H24N3O2	IMIDAZOLIUM CHLORIDE-1-METHYL-2-HYDROXYIMINOMETHYL-3-(A-CYCLOPENTYLBENZYLOXY)METHYL
13660 ★	2.53	12.0	2408		C18H24N4O1	MINAPRINE,4-ETHYL ANALOG
1	2.89	7.4	1485	1	C18H24N4O2	2,4-DIAMINOPYRIMIDINE,5(3,5-DIETO-4(2-PROPENYL)BENZYL
2 ★	3.08	7.4	1485	2	"	2,4-DIAMINOPYRIMIDINE,5(3,5-DIETO-4(2-PROPENYL)BENZYL
3	1.72		1713		C18H24N4O2	6-PH-2-(3-(N3-DIET-UREIDO)PROPYL)-3(2H)-PYRIDAZINONE
4	1.91		1713		C18H24N4O2	6-PH-2-(4-(N3-I-PR-UREIDO)BUTYL)-3(2H)-PYRIDAZINONE
65	1.71		1713		C18H24N4O2	6-PH-2-(4-(N3-PR-UREIDO)BUTYL)-3(2H)-PYRIDAZINONE
6	1.74		1713		C18H24N4O2	6-PH-2-(5-N3-DIME-UREIDO-PENTYL)-3(2H)-PYRIDAZINONE
7	1.76		1713		C18H24N4O2	6-PH-2-(5-N3-ET-UREIDO-PENTYL)-3(2H)-PYRIDAZINONE
8	1.31		1713		C18H24N4O2	6-PH-2-(6-(N3-ME-UREIDO)HEXYL)-3(2H)-PYRIDAZINONE
9 ★	0.73		2377		C18H24N4O3	AC-TRP-VAL-N
13670 ★	2.09	7.4	1485	1	C18H24N4O4	2,4-DIAMINOPYRIMIDINE,5(3,5-DIETO-4-CO2ET)BENZYL
1 ★	2.28	7.4	1485	2	C18H24N4O4	2,4-DIAMINOPYRIMIDINE,5(3,5-DIETO-4-CO2ET)BENZYL
2 ★	0.52		562		C18H24N4O4	2,5DI(N-MORPHOLIN)3,6-DIAZIRIDINQUINONE
3	0.60	7.4	1485	1	C18H24N4O5S1	2,4-DIAMPYRIMIDINE,5(3,5-DIETO-4-COCH2SO2ME)BENZYL
4 ★	0.79	7.4	1485	2	"	2,4-DIAMPYRIMIDINE,5(3,5-DIETO-4-COCH2SO2ME)BENZYL
75	-1.75	7.4	2357	200	C18H24N5O8P1	BUCLADESINE 362-74-3 cardiotonic
6	-1.36	7.4	2357	200	C18H24N5O9P1	CYCLIC GMP,DIBUTYROYL
7 ★	-1.88	7.4	2442	1	C18H24N6O6	P-GLU-HIS(N-METHOXYCARBONYL)-PRO-NH2
8	6.89		2404	459	C18H24O2	(7,6-BENZOPYRANYL)BENZOPYRAN,2,2,2',2'-TETRAMETHYL-8-ETHYL
9	6.97		2404	459	C18H24O2	(7,8-DIHYDROPYRANYL)BENZOPYRAN,2,2,2',2'-TETRAMETHYL-6-ETHYL
13680	2.29	7.4	1076	320	C18H24O2	B-ESTRADIOL 57-91-0 estrogen
1 ★	4.01		579		"	B-ESTRADIOL
2 ★	2.45		2460		C18H24O3	ESTRIOL 50-27-1 estrogen
3	4.03		306		C18H24O5	DIETHYLMALONATE,3-BUTOXYBENZAL
4	-1.11		307		C18H25N1O1	3-HYDROXY-N-ETHYLMORPHINAN TARTRATE
85	-0.68		307	1	C18H25N1O1	3-METHOXY-N-METHYLMORPHINAN HYDROBROMIDE-(LEV)
6	-0.58		307	1	"	3-METHOXY-N-METHYLMORPHINAN HYDROBROMIDE-(LEV)
7	0.97	7.4	2378	1	C18H25N1O1	CYCLAZOCINE 63903-61-7 analgesic, narcotic antagonist
8	1.03	7.1	554		"	CYCLAZOCINE
9	1.33	7.4	554	1	"	CYCLAZOCINE
13690	1.43	7.5	554	1	"	CYCLAZOCINE
1	1.53	7.6	554	212	"	CYCLAZOCINE
2	1.63	7.7	554	1	"	CYCLAZOCINE
3	3.31	7.7	554	2	"	CYCLAZOCINE
4	2.97	7.4	1990	1	C18H25N1O1	N,N-DIMETHYLHEXYLAMINE,6-(1-NAPHTHOXY)
95	2.97	7.4	1990	1	C18H25N1O1	OCTYLAMINE,8-(1-NAPHTHOXY)
6	2.29	7.4	2648	1	C18H25N1O2	ALLYLPRODINE
7 ★	2.97	7.4	2648	2	"	ALLYLPRODINE
8	2.26	7.4	2648	1	"	ALLYLPRODINE 25384-17-2
9 ★	2.96	7.4	2648	2	"	ALLYLPRODINE
13700	1.91	7.4	1411	1	C18H25N1O2S1	4-ETHYLTHIOPROPRANOLOL
1	4.37		572		C18H25N1O3	N,N-DI-I-BUTYL-3,4-METHYLENEDIOXYCINNAMAMIDE 23795-32-6
2	1.11	7.4	2480	1	C18H25N1O4	KETOBEMIDONE,ETHYL CARBONATE
3 ★	1.29	7.4	2480	2	"	KETOBEMIDONE,ETHYL CARBONATE
4	-1.64		2629		C18H25N1O13	P-NITROPHENYLMALTOSIDE
5 ★	-1.39		564		"	P-NITROPHENYLMALTOSIDE
6	-0.57	7.4	651	1	C18H25N3O2	4-(2-N,N-DIETHYLAMINOPROPIONYL)ANTIPYRENE
7	0.34	7.4	651	1	C18H25N3O2	4-(2-N,N-DIMEAMINO-3-MEBUTYRYL)ANTIPYRENE
8	2.42		2162	459	C18H25N3O4S1	BENZAMIDE,2-METHOXY-5-(N-2-PROPYNYL)SULFONAMIDO-N-(N-ETHYLPYRROLID-2-YL)METHYL
9	-0.71	7.4	2375	1	C18H25N3O5	CAROCAINIDE 66203-00-7 antiarrhythmic

	logP	pH	Ref.	Note	MF	Name / **CAS** / *activity*
13710 ★	1.38	7.4	2375	2	"	CAROCAINIDE
1 ✓	-0.83		2397	101	C18H25N4O3S1	IMIDAZOLIUM CHLORIDE,1-METHYL-2-HYDROXYIMINOMETHYL-3-(1-PHENYLSULFONYLPIPERIDIN-2YL)ETHYL
2 ★	1.33	7.0	1793	1	C18H25N5O5S1	N(4-(6-PURINYLTHIO)VALERYL)ASPARTICAC.,DIET.ESTER
3 ★	0.21	7.0	1793	1	C18H25N7O5S1	N(4-(6-PURINYLTHIO)VALERYL)GLYGLYGLYCINE,ET.ESTER
4	3.01	7.5	2651	28	C18H26Br1F1N2O3	BENZAMIDE,2-(3-FLUOROPROPOXY)-3-METHOXY-5-BROMO-N-(1-ETHYLPYRROLIDIN-2YL)METHYL
15	1.23	7.4	1155	1	C18H26Cl1N3	CHLOROQUINE **54-05-7** *anti-amebic, antimalarial, antirheumatic*
6	1.44	7.4	1569	1	"	CHLOROQUINE
7 ★	4.63		1155		"	CHLOROQUINE
8	3.03		2162	459	C18H26Cl1N3O2	BENZAMIDE,4-AMINO-5-CHLORO-2-METHOXY-N-(OCTAHYDRO-6-METHYL-2H-QUINOLIZIN-2-YL) **99390-76-8**
9	-2.22	10.0	1940		C18H26N1O3.N1O3	METHYLATROPINE NITRATE **52-88-0** *anticholinergic*
13720	2.04	7.4	2096	1	C18H26N2	6,7,8,9-TETRAHYDRO-N,N-DIPROPYL-3H-BENZ(E)INDOL-8-AMINE
1	1.77	7.4	1938	1	C18H26N2O2	1,2-DIAMINOCYCLOHEXANE-N-METHYL-N-ALLYL-N'-METHYL-(P-HYDROXYBENZOYL) **98717-06-7**
2 ★	4.04	6.9	2175	1	C18H26N2O2	3-OCTYL-3-(4-PYRIDYL)PIPERIDINE-2,6-DIONE
3	0.44	7.0	708	1	C18H26N2O2	6-(4-MEO-BENZAMIDO)-2-MEDECAHYDRO-I-QUINOLINE-C
4	1.92	7.3	873	314	C18H26N2O2	O-3(N-PIPERIDINO)CYCLOHEXYL-N-PHENYLCARBAMATE
25 ★	4.60		1629		C18H26N2O5S1	N-THIO-(N'-BU-O-ME-CARBAMYL)CARBOFURAN
6 ★	4.70		1629		C18H26N2O5S1	N-THIO-(N'-ME-O-BUTYL-CARBAMYL)CARBOFURAN
7 ★	4.30		1629		C18H26N2O5S1	N-THIO-(N'-ME-O-T-BUTYL-CARBAMYL)CARBOFURAN
8 ✓	0.90		2397	101	C18H26N3O2	IMIDAZOLIUM CHLORIDE,1-BENZYL-2-HYDROXYIMINOMETHYL-3-(1,2,2-TRIMETHYLPROPOXY)METHYL
9 ✓	1.40		2397	101	C18H26N3O2	IMIDAZOLIUM CHLORIDE,1-BENZYL-2-HYDROXYIMINOMETHYL-3-HEXYLOXYMETHYL
13730 ✓	0.94		2397	101	C18H26N3O2	IMIDAZOLIUM CHLORIDE,1-HEXYL-2-HYDROXYIMINOMETHYL-3-BENZYLOXYMETHYL
1 ✓	3.47	13.0	579		C18H26N4O1	2,4-DIAMINO-5-(3,5-DIISOPROPYL-4-METHOXYBENZYL)PYRIMIDINE
2	-0.35	7.4	651	1	C18H26N4O2	4-(1-ME-3-DIMEAMINOBUTYRYLAMINO)ANTIPYRENE
3	2.30	7.4	1485	1	C18H26N4O3	2,4-DIAMINOPYRIMIDINE,5(3,5-DIETO-4-COHME2)BENZYL
4 ★	2.46	7.4	1485	2	"	2,4-DIAMINOPYRIMIDINE,5(3,5-DIETO-4-COHME2)BENZYL
35	2.27	7.4	2244	1	C18H26N4O4S2	TIMOLOL,O-(3-THIENYL)
6 ★	3.83		2506		C18H26O2	CINMETHYLIN **87818-31-3** *herbicide*
7 ★	2.62		1520		C18H26O2	TESTOSTERONE,19-NOR **434-22-0** *anabolic*
8	4.85		1587		C18H26O4	DIPENTYLPHTHALATE **131-18-0**
9	2.82		1937	459	C18H26O5	4-ACETYL-2-PYRONE,3,5-DIBUTYL-6-(2-PROP-2-ONYL) **98393-91-0**
13740	2.17		1937	459	C18H26O5	4-ACETYL-2-PYRONE,3,5-DIBUTYL-6-(PROPYLEN-1-OXIDE) **98393-89-6**
1 ★	-0.05		2308		C18H26O7	BENZO-18-CROWN-6-ETHER,M-ACETYL
2	3.52	7.5	2651	28	C18H27Br1N2O3	BENZAMIDE,2,3-DIETHOXY-5-BROMO-N-(1-ETHYLPYRROLIDIN-2YL)METHYL
3 ★	4.44		2237		C18H27Cl1N2O2	BENZOFURAN-3-AMINE,2,3-DIHYDRO-5-CHLORO-N-DIBUTYLAMINOACETYL
4	5.26	7.5	2651	28	C18H27Cl1N2O3	SALICYLAMIDE,3-ETHYL-5-CHLORO-6-ETHOXY-N-(1-ETHYLPYRROLIDIN-2YL)METHYL
45	2.86	7.5	2651	28	C18H27F1N2O2	BENZAMIDE,2-METHOXY-5-(3-FLUOROPROPYL)-N-(1-ETHYLPYRROLIDIN-2YL)METHYL
6	3.73	7.5	2651	1	C18H27I1N2O3	ITOPRIDE
7	5.53	7.5	2651	28	"	ITOPRIDE
8	6.75		1361		C18H27N1	2-(AX)-METHYLPHENCYCLIDINE (PH=AX)
9	8.57		1361		C18H27N1	2-METHYLPHENCYCLIDINE
13750	5.86		1361	459	C18H27N1	3-(EQ.)-METHYL-PHENCYCLIDINE(PHEN=AX.)
1	8.65		1361	459	C18H27N1	3-(EQ.)-METHYL-PHENCYCLIDINE(PHEN=EQ.)
2	5.66		1361	459	C18H27N1	4-(EQ.)-METHYL-PHENCYCLIDINE(PHEN=AX.)
3	7.33		1361	459	C18H27N1	4-(EQ.)-METHYL-PHENCYCLIDINE(PHEN=EQ.)
4	4.73		1361	459	C18H27N1O1	2-(AX.)-METHOXY-PHENCYCLIDINE(PHEN=AX.)
55	6.52		1361	459	C18H27N1O1	2-(AX.)-METHOXY-PHENCYCLIDINE(PHEN=EQ.)
6	2.13	7.4	682	1	C18H27N1O2	N-BUTYLNORMEPERIDINE
7	2.84		1167	459	C18H27N1O5	3,4,5-TRIMEO-CINNAMIC ACID,DIETHYLAMINOETHYLEST.
8	-0.65	7.4	2095	1	C18H27N3O2	(5-METHYLIMIDAZOLE)-SPACED ADRENOCEPTOR ANTAGONIST
9	0.95	7.4	2095	1	C18H27N3O2	(ISOPROPYLIMIDAZOLE)ADRENOCEPTOR ANTAGONIST
13760 ★	4.20		1629		C18H27N3O3S2	N-(N''-HEXYL)-PHENYLCARBAMYLTHIO)METHOMYL
1 ★	-2.04	7.0	2374		C18H27N3O5	ALANYLTYROSINYLISOLEUCINE
2 ✓	-0.63		2397	101	C18H27N4O3S1	IMIDAZOLIUM CHLORIDE,1-I-PROPYL-2-HYDROXYIMINOMETHYL-3-(3-N-ETHYL-PHENYLSULFONAMIDO)PROPYL
3 ✓	-0.52		2397	101	C18H27N4O3S1	IMIDAZOLIUM CHLORIDE,1-PROPYL-2-HYDROXYIMINOMETHYL-3-(3-N-ETHYL-PHENYLSULFONAMIDO)PROPYL
4	0.18	4.8	1020		C18H28N2O1	2-BUTENAMIDE,N(2,6-DIMEPH)N(2-DIETAMINOETHYL)
65	2.54	7.4	2409	1	C18H28N2O1	BUPIVACAINE **2180-92-9** *anesthetic (local)*
6	2.75	7.4	2409	1	"	BUPIVACAINE
7 ★	3.41		2409	2	"	BUPIVACAINE
8 ★	4.15		2237		C18H28N2O2	BENZOFURAN-3-AMINE,2,3-DIHYDRO-N-DIBUTYLAMINOACETYL
9 ★	2.76		2382		C18H28N2O3	1,1-DIMETHYL-3-PHENYLUREA,3'-METHYL-4-(5-(2,2-DIETHYL)-1,3-DIOXANYL)
13770	2.99	7.3	1900	314	C18H28N2O3	1-PIPERIDINEETHANOL,A-METHYL,3-OPR-PHENYLCARBAMATE
1	2.79	7.3	1356	314	C18H28N2O3	2-BUTOXYCARBANILATE-O-2(N-PIPERIDINYL)ETHYL **55792-13-7**
2	4.35	7.3	1356	2	"	2-BUTOXYCARBANILATE-O-2(N-PIPERIDINYL)ETHYL
3	3.32	7.2	1150		C18H28N2O3	2-ME-4(O-PR)CARBANILATE-Ö-2(N-PIPERIDINYL)ETHYL
4	1.84	7.3	1960	314	C18H28N2O3	2-PROPOXYMETHYLCARBANILIC ACID,2-(1-PIPERIDINYL)ETHYL ESTER
75 ★	3.34	7.3	1960	2	"	2-PROPOXYMETHYLCARBANILIC ACID,2-(1-PIPERIDINYL)ETHYL ESTER
6	3.04	7.3	1356	314	C18H28N2O3	3-BUTOXYCARBANILATE-O-2(N-PIPERIDINYL)ETHYL **55792-14-8**
7	4.60	7.3	1356	2	"	3-BUTOXYCARBANILATE-O-2(N-PIPERIDINYL)ETHYL
8	2.69	7.3	1356	314	C18H28N2O3	4-BUTOXYCARBANILATE-O-2(N-PIPERIDINYL)ETHYL **55792-15-9**
9	4.25	7.3	1356	2	"	4-BUTOXYCARBANILATE-O-2(N-PIPERIDINYL)ETHYL
13780	1.07	7.5	2651	1	C18H28N2O3	BENZAMIDE,2,3-DIMETHOXY-5-ETHYL-N-(1-ETHYLPYRROLIDIN-2YL)METHYL
1	2.69	7.5	2651	28	"	BENZAMIDE,2,3-DIMETHOXY-5-ETHYL-N-(1-ETHYLPYRROLIDIN-2YL)METHYL
2	2.20	7.5	2651	28	C18H28N2O3	BWNZAMIDE,2,6-DIMETHOXY-3-ETHYL-N-(1-ETHYLPYRROLIDIN-2YL)METHYL
3 ★	4.51		2612		C18H28N2O3	N1-MEO-N1-ME-N3-(P-2-CYHEXYL-1-ME-ETHOXYPHENYLUREA **88132-29-0**
4	4.66	7.5	2651	28	C18H28N2O3	SALICYLAMIDE,3-ETHYL-6-ETHOXY-N-(1-ETHYLPYRROLIDIN-2YL)METHYL
85	2.82	7.2	1150		C18H28N2O4	2-ME-4(O-BU)CARBANILATE-O-2(N-MORPHOLINO)ETHYL
6	-0.77	7.0	1362	1	C18H28N2O4	ACEBUTOLOL **37517-30-9** *anti-adrenergic (beta-receptor)*
7	-0.49	7.4	2375	1	"	ACEBUTOLOL
8	-0.28	7.4	1354	1	"	ACEBUTOLOL
9	-0.21	7.4	1859	1	"	ACEBUTOLOL
13790	-0.17	7.4	1362	1	"	ACEBUTOLOL
1	-0.02	7.4	2399	1	"	ACEBUTOLOL
2	0.00	7.4	2178	1	"	ACEBUTOLOL
3	0.09	7.4	2473	1	"	ACEBUTOLOL
4	0.48	7.4	2245	1	"	ACEBUTOLOL
95 ★	1.71	10.2	2465	41	"	ACEBUTOLOL
6	1.77	7.4	1700	2	"	ACEBUTOLOL
7	2.75	7.4	2245	2	"	ACEBUTOLOL
8 ★	3.96		2289		C18H28N6O2	TETRAZOLE,1-OCTYL-5-(4-DIMETHYLUREYL)PHENOXY **117121-34-3**
9 ★	4.40		2289	459	C18H28N6O2	TETRAZOLE,2-OCTYL-5-(4-DIMETHYLUREYL)PHENOXY **117121-46-7**

	logP	pH	Ref.	Note	MF	Name / CAS / activity
13800 ★	4.82		2415	459	C18H28O2	CYCLOHEXANOL,3-(2-HYDROXY-4-(1,1-DIMETHYL)BUTYL)PHENYL
1	1.92		1937	459	C18H28O5	4-ACETYL-2-PYRONE,3,5-DIBUTYL-6-(2-HYDROXYPROPYL) 98420-36-1
2 ★	2.26		2308		C18H28O5	BENZO-15-CROWN-5-ETHER,M-T-BUTYL
3	2.33		1396	459	C18H28O6	ARTEMISININE,PROPIONYL ESTER(ALPHA)
4 ★	0.98		2308		C18H28O6	BENZO-18-CROWN-6-ETHER,M-ETHYL 81117-71-7
5	2.37		1396	459	C18H28O7	ARTEMISININE,ETHYLCARBONATE(ALPHA)
6 ★	0.57		2308		C18H28O7	BENZO-21-CROWN-7-ETHER
7	2.86		2359	1	C18H29N1O2	EXAPROLOL 55837-19-9 anti-adrenergic (beta receptor)
8	1.59	7.0	968	1	C18H29N1O2	PENBUTOLOL 38363-40-5 anti-adrenergic (beta-receptor)
9	1.92	7.4	2375	1	"	PENBUTOLOL
13810	1.93	7.4	1861	1	"	PENBUTOLOL
1	1.97	7.4	2501	1	"	PENBUTOLOL
2	2.27	7.4	2178	1	"	PENBUTOLOL
3	3.83	7.4	1861	2	"	PENBUTOLOL
4 ★	4.15	7.4	1700	2	"	PENBUTOLOL
15	0.55	7.4	2501	1	C18H29N1O3	BETAXOLOL 63659-18-7 anti-anginal, antihypertensive
6	1.13	7.4	2473	1	"	BETAXOLOL
7 ★	2.81	13.0	2501		"	BETAXOLOL
8 ★	4.48		2413	459	C18H29N1O4	TETRAHYDROPYRAN-2,4-DIONE,3[1(ETHOXYIMINO)BUTYL)-6,6-SPIRO-(4-ETHYL)CYCLOHEXYL
9 ★	4.48		2413	459	C18H29N1O4	TETRAHYDROPYRAN-2,4-DIONE,3[1(ETHOXYIMINO)BUTYL)-6,6-SPIRO-3,5-DIMETHYL)CYCLOHEXYL
13820	1.68	7.4	2278	1	C18H29N1Si1	SILA-PROCYCLIDINE 104549-76-0
1	1.08	7.4	365	2	C18H29N3O1	N,N-DIET-PIPERAZINE-2-CARBOXANILIDE,2'4'6'-TRIME
2	1.08	7.4	365	2	C18H29N3O1	N,N-DIPR-PIPERAZINE-2-CARBOXANILIDE,2'-ME
3	2.57		2162	459	C18H29N3O4S1	BENZAMIDE,4-METHYL-2-METHOXY-5-(N-METHYLMETHANESULFONAMIDO)-N-(1-ETHYLPYRROLID-2-YL)
4	0.34	7.0	1812	1	C18H29N3O5	5'-AMINO-3'O-OCTANOYL-5'-DEOXYTHYMIDINE
25	1.02	7.4	2350	1	C18H29N3O5S1	SPARSOMYCIN ANALOG ✚
6	-1.17	7.4	2450	1	C18H30N1O2.C7H7O3S1	MDL-734O4 ✚
7	-0.60	7.4	2450	1	"	MDL-734O4
8	2.64		1352		C18H30N2O1	4-(DIETHYLAMINO)HEXANO-2',6'-XYLIDIDE
9	2.95		1352		C18H30N2O1	5-(DIETHYLAMINO)HEXANO-2',6'-XYLIDIDE
13830	0.53	5.7	2241	1	C18H30N2O2	BUTACAINE 149-16-6 anesthetic (local)
1	3.68	7.2	1150		C18H30N2O3	2-ME-4(O-BU)CARBANILATE-O-2(N,N-DIETHYLAMINO)ET
2	4.00	7.2	1150		C18H30N2O3	2-ME-4(O-HEX)CARBANILATE-O-2(N,N-DIMETHYLAMINO)ET
3	1.39	7.1	1150		C18H30N2O3	2-ME-4(O-PR)CARBANILATE-O-2(N,N-DIETAMINO)1-ME-ET
4	2.22	7.4	2244	1	C18H30N4O4S1	TIMOLOL,O-(1'-METHYLCYLOPROPANOYL)
35	2.26	7.4	2244	1	C18H30N4O4S1	TIMOLOL,O-(2'-METHYLCYCLOPROPANOYL)
6	2.36	7.4	2244	1	C18H30N4O4S1	TIMOLOL,O-CYCLOBUTANOYL
7	-0.48	7.0	2553	1	C18H30N6O4	AC-ALA-HIS-ALA-NTBU
8	7.91		2316	337	C18H30O1	PHENOL,4-DODECYL
9	3.49		1396	459	C18H30O5	ARTEMISININE,I-PROPYLETHER(BETA)
13840	3.65		1396	459	C18H30O5	ARTEMISININEPROPYLETHER(BETA)
1	1.20	1.0	547		C18H31N1O1	2-N,N-DIPENTYLAMINO-1-PHENYLETHANOL HYDROCHLORIDE
2 ★	2.86	13.0	547	224	C18H31N1O1	2-N,N-DIPENTYLAMINO-1-PHENYLETHANOL
3	-0.23	7.4	2501	1	C18H31N1O4	BISOPROLOL 66722-44-9 antihypertensive (beta-blocker)
4 ★	1.87	13.0	2501		"	BISOPROLOL
45	1.41	7.4	1094	1	C18H31N2O1.I1	2((HYDROXYIMINO)ME)-1-DODECYLPYRIDINIUM IODIDE
6 ✓	0.60		1729	820	C18H31N2O1.I1	3-CARBAMOYLPYRIDINIUM IODIDE,N-DODECYL 35096-55-0
7	0.15	7.4	2284	1	C18H31N2O3.I1	CARBISOCAINE,PROPOXY ANALOG,METH IODIDE
8 ★	2.21		2238		C18H32Cl1N4O3	NITROSOUREA,1-(2-CHLOROETHYL)-3-CYCLOHEXYL-3-(1-OXO-2,2,6,6-TETRAMETHYLPIPERIDIN-4-YL)
9	0.03		1635	820	C18H32N1.Cl1	BENZYLDIMETHYLNONYLAMMONIUMCHLORIDE
13850 ✓	1.32		1729	820	C18H32N1.I1	2-METHYLPYRIDINIUM IODIDE,N-DODECYL 14402-22-3
1	-1.19	7.4	1026	1	C18H32N2O4	1,1'(O-PHENYLENEDIOXY)BIS(3-I-PR-AMINO-2-PROPANOL
2	-1.20	7.4	487	1	C18H32N2O4	M-BIS(2-OH-3-I-PRAM-PR)PHENYLDIETHER
3	0.73	7.4	487	2	"	M-BIS(2-OH-3-I-PRAM-PR)PHENYLDIETHER
4	-1.19	7.4	487	1	C18H32N2O4	O-BIS(2-OH-3-I-PRAM-PR)PHENYLDIETHER
55	0.63	7.4	487	1	"	O-BIS(2-OH-3-I-PRAM-PR)PHENYLDIETHER
6	-1.19	7.4	487	1	C18H32N2O4	P-BIS(2-OH-3-I-PRAM-PR)PHENYLDIETHER
7	0.67	7.4	487	2	"	P-BIS(2-OH-3-I-PRAM-PR)PHENYLDIETHER
8 ✓	1.48		2397	101	C18H32N3O2	IMIDAZOLIUM CHLORIDE,1-METHYL-2-HYDROXYIMINOMETHYL-3-CYCLODODECYLOXYMETHYL 117941-59-0
9	2.68	7.4	2244	1	C18H32N4O4S1	TIMOLOL,O-PIVALOYL
13860	2.67	7.4	2244	1	C18H32N4O4S1	TIMOLOL,O-VALERYL
1 ★	-0.04		2377	821	C18H32N4O6	AC-ILE-LEU-ASP-N
2 ★	0.16		2377	821	C18H32N4O6	AC-LEU-ASP-LEU-N
3	2.16	12.3	343		C18H33Cl1N2O5S1	CLINDAMYCIN 18323-44-9 antibacterial
4	4.30		1959		C18H33Cl1Sn1	CHLOROSTANNANE,TRICYCLOHEXYL
65	1.38	8.0	1678	1	C18H34N1O4	3-OXYOXAZOLIDINE,2-BUTYL,2-NONANOIC ACID 84130-84-7
6 ★	0.20		2469		C18H34N2O6S1	LINCOMYCIN 154-21-2 antibacterial
7 ★	0.49		2377		C18H34N4O4	AC-VAL-VAL-ILE-N
8 ★	0.24		2377		C18H34N4O4	AC-LEU-THR-LEU-N
9 ★	-2.82		2377		C18H34N4O5	ILE-ALA-ALA-ILE
13870 ★	5.39		1178		C18H34O1Sn1	CYHEXATIN 13121-70-5 acaricide
1 ★	-1.11	7.0	2374		C18H35N3O4	LEUCINYLISOLEUCINYLISOLEUCINE
2 ★	-0.94	7.0	2197		C18H35N3O4	LEUCINYLLEUCINYLLEUCINE 10329-75-6
3	0.40		2093		C18H35N5O3P1	DIMETHYLTEPA,4-(2,2,6,6-TETRAMETHYL-1-OXYPIPERIDINYL)URYL 103981-92-6
4	0.55		2093		C18H36N5O2P1	DIMETHYLTEPA,4-(2,2,6,6-TETRAMETHYLPIPERIDINYL)URYL 103981-97-1
75 ★	0.45		2093		C18H36N5O3P1	DIMETHYLTEPA,3-(2,2,6,6-TETRAMETHYL-1-OXYPYRROLIDINYL)UREA 103981-93-7
6 ★	0.48		2093		C18H36N5O3P1	DIMETHYLTEPA,4-(2,2,6,6-TETRAMETHYL-1-HYDROXYPIPERIDINYL)URYL 103981-94-8
7 ✓	0.08	7.3	2085		C18H40N4.O2Tc1	N-OCTYLCYCLAM,TECHNICIUM OXIDE COMPLEX
8	0.73		1966	459	C18H45Au3P3S1.N1O3	TRIS(TRIETHYLPHOSPHINAURO) SULFONIUM NITRATE
9	7.70		1740		C18H54O7Si8	OCTADECAMETHYLOCTASILOXANE
13880 ★	5.78		2158	459	C19H12	1H-BENZO(CD)FLUORANTHENE 42126-84-1
1	0.55	7.4	2437	1	C19H12F3N3O3S1	ZOPOLRESTAT inhibitor,aldose reductase ✚
2	2.37		2635	459	"	ZOPOLRESTAT
3	3.43		2437		"	ZOPOLRESTAT
4	4.82		1814	459	C19H12O2	A-NAPHTHOFLAVONE 604-59-1
85	4.94		1814	459	C19H12O2	B-NAPHTHOFLAVONE 6051-87-2
6	1.83	7.4	2557	1	C19H12O6	DICOUMAROL 66-76-2 anticoagulant
7	1.90	7.4	2438	1	"	DICOUMAROL
8 ★	2.07		1135	537	"	DICOUMAROL
9 ★	3.16	7.4	579	1	C19H14F3N1O1	FLURIDONE 59756-60-4

	logP	pH	Ref.	Note	MF	Name / CAS / activity
13890	4.25	7.4	547	1	C19H14N2	9-ANILINOACRIDINE 3340-22-5
1 ★	3.02		2646	448	C19H14O5S1	PHENOLSULFONPHTHALEIN 143-74-8 diagnostic aid (renal function)
2	2.24	7.4	2459	1	C19H15Br1O4	WARFARIN,3'-BROMO
3	2.24	7.4	2459	1	C19H15Br1O4	WARFARIN,4'-BROMO
4	4.81		2675	459	C19H15Cl1F3N1O7	LACTOFEN
95	-0.76		1284		C19H15Cl1N1O4	INDOMETHACINANION
6	0.66		1284		C19H15Cl1N1O4.K1	INDOMETHACIN, K SALT
7	1.16		1284		C19H15Cl1N1O4.Na1	INDOMETHACIN, SODIUM SALT
8	1.50		1582	459	C19H15Cl1N2O5	BENZODIAZEPIN ANALOG#33 4700-56-5
9	2.02	7.4	2459	1	C19H15Cl1O4	WARFARIN,2'-CHLORO
13900	2.00	7.4	2459	1	C19H15Cl1O4	WARFARIN,3'-CHLORO
1	2.03	7.4	2459	1	C19H15Cl1O4	WARFARIN,4'-CHLORO rodenticide
2 ★	3.73		594	401	C19H15F2N1	TRITYLDIFLUOROAMINE ✚
3	-0.46		2389		C19H15F3N4O3	TOSUFLOXACIN TOSYLATE 115964-29-9 antibacterial
4 ★	2.14		1817		C19H15N1O1	3-PYRIDINEMETHANOL,A-2-PROPYNYL,A-2-NAPHTHYL 89242-89-7
5	3.97		1973	459	C19H15N1O3	5-(2-NAPHTHAZOYL)-H2-PYRROLO-PYRROLE-1-CARBOXYLIC ACID 96327-62-7
6	0.90	7.4	2438	1	C19H15N1O6	ACENOCOUMAROL 152-72-7 anticoagulant
7	0.98	7.4	2459	1	"	ACENOCOUMAROL
8	1.98	8.0	2114		"	ACENOCOUMAROL
9	3.24		1415		"	ACENOCOUMAROL
13910	0.92	7.4	2459	1	C19H15N1O6	WARFARIN,3'-NITRO
1	-0.76		1284		C19H16Cl1N1O4	INDOMETHACIN ANION
2	1.70		1284		C19H16Cl1N1O4	INDOMETHACIN, AMMONIUM SALT
3	0.66		1284		C19H16Cl1N1O4	INDOMETHACIN, POTASSIUM SALT
4	1.16		1284		C19H16Cl1N1O4	INDOMETHACIN,SODIUM SALT
15	1.18		1284		C19H16Cl1N1O4	INDOMETHACIN,TRIETHANOLAMINE SALT
6	-1.00	7.4	693	1	C19H16Cl1N1O4	INDOMETHACIN 53-86-1 anti-inflammatory
7	-0.52	7.4	1368	1	"	INDOMETHACIN
8	0.91	7.4	1838	1	"	INDOMETHACIN
9	1.00	7.4	993	1	"	INDOMETHACIN
13920 ★	4.27	2.0	1838		"	INDOMETHACIN
1	1.70		1284		C19H16Cl1N1O4.H3N1	INDOMETHACIN, AMMONIUM SALT
2	1.18		1284		C19H16Cl1N1O4.C6H15N1O3	INDOMETHACIN,TRIETHANOLAMINE SALT
3	3.00		2478	459	C19H16F3N1O3	ISOXAZOLINYLDIOXELPIN DIASTEREOMER,CF3-ANALOG
4 ★	4.31		2478		C19H16F3N1O3	ISOXAZOLINYLDIOXEPIN DIASTEREOMER,CF3-ANALOG
25 ★	3.28		603		C19H16N2O1	TRIPHENYLUREA
6	-1.12	7.4	2076	1	C19H16N4O2S1	4-QUINAZOLINONE,2-METHYL-3-(4-OXO-3-(4-METHYL)PHENYL-THIAZOLIDINYL)IMINO
7	0.48	6.0	2560	1	C19H16N4O2S1	THIABENDAZOLE,1-(P-AMINOMETHYLBENZOYL)METHYL
8	-1.23	7.4	2076	1	C19H16N4O3S1	4-QUINAZOLINONE,2-METHYL-3-(4-OXO-3-(4-METHOXY)PHENYL-THIAZOLIDNINYL)IMINO
9 ★	3.68		579		C19H16O1	TRIPHENYLMETHANOL 76-84-6
13930 ★	4.30		759		C19H16O2	4,4'-DIHYDROXYTRIPHENYLMETHANE
1 ★	4.67		1941		C19H16O2	4H-1-BENZOPYRAN-4-ONE,2-PHENYL-3-METHYL-8-ALLYL
2	1.04	7.4	2438	1	C19H16O4	WARFARIN 81-81-2 anticoagulant~rodenticide
3	1.20	7.4	2459	1	"	WARFARIN
4	1.46	7.0	2230	1	"	WARFARIN
35 ★	2.70		1694		"	WARFARIN
6 ★	2.61		935		C19H17Cl1F1N3O5S1	FLOXACILLIN 5250-39-5 antibacterial
7 ★	2.23		1260		C19H17Cl1F1N3O6S1	5-HYDROXYFLOXACILLIN ✚
8	3.31		1017		C19H17Cl1N2O1	PRAZEPAM 2955-38-6 sedative
9	3.73	7.4	461	1	"	PRAZEPAM
13940	2.45		1582	459	C19H17Cl1N2O4	7CL-1,3-H2-1-OHET-3-ACO-5-PH-1,4-DIAZEPIN-2-ONE 56875-84-4
1 ★	4.28		2214		C19H17Cl1N2O4	QUIZALOFOP-ETHYL herbicide
2 ★	3.23		2506		C19H17Cl1N4	4(4-CLPHENYL)-2-PH-2-(1H-1,2,4-TRIAZOL-1YLMETHYL)BUTYRONITRILE
3 ★	2.91		935		C19H17Cl2N3O5S1	DICLOXACILLIN 3116-76-5 antibacterial
4 ★	2.57		1260		C19H17Cl2N3O6S1	5-HYDROXYDICLOXACILLIN 52248-39-2
45 ★	3.65		2214		C19H17F1N2O4	2(4(2(6-FLUOROQUINOXALINYLOXY)PHENOXYPROPIONIC ACID,ETHYL ESTER
6 ★	4.95		2424		C19H17F3N4O1	BENZOTRIAZOLE,2-CYCLOPENTYL-5-(P-TRIFLUOROMETHYLBENZAMIDO)
7 ★	3.16		759		C19H17N1O2	4,4'-DIHYDROXY-2''-AMINOTRIPHENYLMETHANE
8 ★	2.22	3.0	2018	2	C19H17N1O7	NEDOCROMIL 69049-73-6 anti-allergic (prophylactic)
9 ★	-0.21		2646	448	C19H17N3	PARAROSANILINE 569-61-9 disinfectant
13950	-1.70	2.0	2464	41	C19H17N3O4S2	CEPHALORIDINE 50-59-9 antibacterial
1 ✓	-1.62	7.4	2464	41	"	CEPHALORIDINE
2	-1.52	7.4	2192	1	"	CEPHALORIDINE
3	1.85		872	459	"	CEPHALORIDINE
4 ★	2.93	5.0	2674	302	C19H17N3O5	MEBENDAZOLE,ETHOXYCARBONYL
55	4.51		1101	421	C19H17O4P1	CRESYLDIPHENYLPHOSPHATE
6	-1.82	7.4	2530	1	C19H18Cl1N3O5S1	CLOXACILLIN 61-72-3 antibacterial
7 ★	2.43		935		"	CLOXACILLIN
8 ★	2.00		1260		C19H18Cl1N3O6S1	5-HYDROXYCLOXACILLIN 55390-39-1
9	3.53		1582	459	C19H18Cl2N2O2	7CL-1,3H2-1-CLET-3-ETO-5PH-1,4-BENZODIAZEPIN-2-ON 56875-86-6
13960	1.54	7.4	675	1	C19H18Cl2N2O2	PHENYLBUTAZONE,P,P'-DICHLORO
1	1.98		910		C19H18Cl2N4	1,5-BIS(5-CHLORO-2-BENZIMIDAZOLYL)PENTANE
2 ★	5.85		2463		C19H18Cl2N4O1	BENZOTRIAZOLE,2-CYCLOHEXYL-5-(3,4-DICHLOROBENZAMIDO)
3	1.72	7.4	2542	1	C19H18N2O3S2	THIOPHENE-2-SULFONAMIDE,4-(P-BENZYLAMINOMETHYLBENZOYL)
4 ★	3.48		2214		C19H18N2O4	2(4(2-QUINOXALINYLOXY)PHENOXYPROPIONIC ACID, ETHYL ESTER
65	1.32	7.4	2683	1	C19H18N4O1	TRYPTAMINE,5-(3-BENZYLOXADIAZOLYL) 5ht-agonist
6	1.33	7.4	2024	1	C19H18N4O5	1,2-AZIRIDINOMITOSINE,7-(2-FURANYL)METHYLAMINO 103422-24-8
7	1.97		910		C19H18N6O4	1,5-BIS(5-NITRO-2-BENZIMIDAZOLYL)PENTANE
8	2.44		1582	459	C19H19Cl1N2O3	7CL-1,3H2-1-OHET-3-ETO-5PH-1,4-BENZODIAZEPIN-2-ON 56875-88-8
9	1.74	7.4	1242	1	C19H19Cl1N2O4	TEMAZEPAM,3-GLYCERYL ANALOG 40967-03-1
13970	0.73	2.0	1039	210	C19H19N1	PHENINDAMINE 82-88-2 antitussive, antihistaminic
1	2.82	7.4	712	1	C19H19N1O2	A-METHYL-N-ISOBUTYL-G-PYROPHTHALONE
2	3.27	7.4	712	1	C19H19N1O2	N-AMYL-G-PYROPHTHALONE
3	2.72		2478		C19H19N1O3	ISOXAZOLINYLDIOXEPIN DIASTEREOMER,METHYL ANALOG ✚
4	4.00		2478	459	"	ISOXAZOLINYLDIOXEPIN DIASTEREOMER,METHYL ANALOG
75	-1.24	7.2	846	314	C19H19N3O5S1	OXACILLIN 66-79-5 antibacterial
6	-1.15	7.0	1295	1	"	OXACILLIN
7 ★	2.38		936		"	OXACILLIN
8 ★	1.80		1260		C19H19N3O6S1	5-HYDROXYOXACILLIN 4914-62-9
9 ★	4.57		2463		C19H19N5O3	BENZOTRIAZOLE,2-CYCLOHEXYL-5-P-NITROBENZAMIDO

	logP	pH	Ref.	Note	MF	Name / **CAS** / *activity*
13980 ★	4.46		1481		C19H20Cl1N1O1	P-CHLOROCINNAMAMIDE,B-PHENYL,N-S-BUTYL
1	2.50	7.4	2020	1	C19H20Cl2N2O2	PINDOLOL ANALOG:N-(3,4-DICHLOROPHENETHYL) **102573-81-9**
2 ★	4.50		2506		C19H20F3N1O4	FLUAZIFOP-BUTYL *herbicide (fatty acid synth.inhib.)* ✚
3	0.89	2.0	1039	210	C19H20N2	MEBHYDROLINE **524-81-2** *antihistaminic*
4 ★	3.40	7.4	2277	1	C19H20N2O1	DENZIMOL **73931-96-1** *anticonvulsant*
85	0.97	7.4	847	1	C19H20N2O2	N-(3-(N-DIMEAMINO)PROPYL)-G-PYROPHTHALONE
6	0.68	7.4	693	1	C19H20N2O2	PHENYLBUTAZONE **50-33-9** *antirheumatic*
7	0.70	7.4	675	1	"	PHENYLBUTAZONE
8	0.80	7.4	1838	1	"	PHENYLBUTAZONE
9	1.25	7.0	919	1	"	PHENYLBUTAZONE
13990 ★	3.16	2.0	1838		"	PHENYLBUTAZONE
1	-0.06	7.4	675	1	C19H20N2O3	OXYPHENBUTAZONE **129-20-4** *anti-inflammatory, antirheumatic*
2	0.16	7.4	2205	1	"	OXYPHENBUTAZONE
3 ★	2.72	2.0	2205		"	OXYPHENBUTAZONE
4	2.57		1973	459	C19H20N2O4	5-(3-1-METHYLPROPIONYLAMINOBENZOYL)-H2-PYRROLO-PYRROLE-1-CARBOXYLIC ACID **96327-47-8**
95	0.10	7.4	675	1	C19H20N2O4	PHENYLBUTAZONE,P,P'-DIHYDROXY
6	1.75	7.4	2621	1	C19H20N2O6	MITOSENE(5,5,6),METHYL,AZIRIDINYL-DIACETYLOXY ANALOG
7 ✓	0.37		2397	101	C19H20N3O2	IMIDAZOLIUM CHLORIDE,1-BENZYL-2-HYDROXYIMINOMETHYL-3-BENZYLOXYMETHYL
8	1.56		910		C19H20N4	1,5-BIS(2-BENZIMIDAZOLYL)PENTANE
9 ★	4.45		2463		C19H20N4O1	BENZOTRIAZOLE,2-CYCLOHEXYL-5-BENZAMIDO
14000	1.75		910		C19H20N4O2	1,3-BIS-(5-MEO-2-BENZIMIDAZOLYL)PROPANE
1	2.47	7.4	1485	1	C19H20N4O2	2,4-DIAMINOPYRIMIDINE,5(3-MEO-4-PHCH2O-)BENZYL
2 ★	2.66	7.4	1485	2	"	2,4-DIAMINOPYRIMIDINE,5(3-MEO-4-PHCH2O-)BENZYL
3	2.25	7.4	1485	1	C19H20N4O2	2,4-DIAMINOPYRIMIDINE,5(4-MEO-3-PHCH2O-)BENZYL
4 ★	2.44	7.4	1485	2	"	2,4-DIAMINOPYRIMIDINE,5(4-MEO-3-PHCH2O-)BENZYL
5 ★	3.77		2424		C19H20N4O2	BENZOTRIAZOLE,2-CYCLOPENTYL-5-(P-METHOXYBENZAMIDO)
6 ★	4.91		1587		C19H20O4	BUTYLBENZYLPHTHALATE **85-68-7**
7	1.56		636		C19H20O6	VERNOLEPINMETH ACRYLATE
8 ★	0.69		636		C19H20O7	ELEPHANTOPIN ✚
9 ★	4.82		2513		C19H21Cl1N2O1	MONCEREN **66063-05-6** *pesticide*
14010	1.79	7.4	2092	1	C19H21Cl2N1O3	6,7-DICHLORO-1-(3,4,5-TRIMETHOXYBENZYL)-1,2,3,4-TETRAHYDROISOQUINOLINE **103959-65-5**
1	2.00	7.4	2020	1	C19H21F1N2O2	PINDOLOL ANALOG:N-(P-FLUOROPHENETHYL) **102573-80-8**
2	0.73	7.2	2526	1	C19H21F1N4O3	OXONAPHTHYRIDINECARBOXYLIC ACID,N-CYCLOPROOPYL-7-(3-METHYL-2,5-DIAZABICYCLOHEPTYL) ANALOG
3	1.00	7.2	2526	1	"	OXONAPHTHYRIDINECARBOXYLIC ACID,N-CYCLOPROOPYL-7-(3-METHYL-2,5-DIAZABICYCLOHEPTYL) ANALOG
4	0.71	7.2	2526	1	C19H21F1N4O3	OXONAPHTHYRIDINECARBOXYLIC ACID,N-CYCLOPROPYL-7-(6-METHYL-2,5-DIAZABICYCLOHEPTYL) ANALOG
15	0.71	7.2	2526	1	"	OXONAPHTHYRIDINECARBOXYLIC ACID,N-CYCLOPROPYL-7-(6-METHYL-2,5-DIAZABICYCLOHEPTYL) ANALOG
6	4.38	7.4	2090	459	C19H21F3N2O3	5(5(2-CHF2-4-(4,5-H2-2-OXAZOLYL)PHENOXY)PENTYL-3-METHYLISOXAZOLE **105639-06-3**
7	2.94	7.5	2545	1	C19H21N1	N-PHENETHYL-PTP
8	3.98	7.5	2545	2	"	N-PHENETHYL-PTP
9	1.31	5.0	1440	1	C19H21N1	NORTRIPTYLINE **72-69-5** *antidepressant*
14020	1.71	7.4	189	1	"	NORTRIPTYLINE
1	1.80	7.0	468	1	"	NORTRIPTYLINE
2	1.98	7.4	1209	1	"	NORTRIPTYLINE
3	1.18	5.0	1440	1	C19H21N1	PROTRIPTYLINE **438-60-8** *antidepressant*
4	1.18	7.4	189	1	"	PROTRIPTYLINE
25	1.20	7.4	785	1	"	PROTRIPTYLINE
6	1.40	7.0	468	1	"	PROTRIPTYLINE
7	1.43	1.0	547	342	C19H21N1O1	9-(2-PROPYLAMINO-1-HYDROXYETHYL)PHENANTHRENE
8 ★	4.05		1481		C19H21N1O1	A-PHENYL-N-S-BUTYL-CINNAMAMIDE
9	0.67	5.0	1440	1	C19H21N1O1	DOXEPIN **1668-19-5** *antidepressant*
14030	1.05	6.0	1058		"	DOXEPIN
1	2.22	7.4	1209	1	"	DOXEPIN
2	2.37	7.4	189	1	"	DOXEPIN
3	2.43	7.4	564	1	C19H21N1O3	N-(2-HYDROXYETHYL)APOCODEINE
4	1.22	7.4	2378	1	C19H21N1O3	NALORPHINE **62-67-9** *narcotic antagonist*
35	1.24	7.1	554		"	NALORPHINE
6	1.45	7.4	554	1	"	NALORPHINE
7	1.51	7.5	554	1	"	NALORPHINE
8	1.56	7.6	554	212	"	NALORPHINE
9	1.61	7.7	554	1	"	NALORPHINE
14040 ★	1.86	7.7	554	2	"	NALORPHINE
1	1.06	7.4	2378	1	C19H21N1O4	NALOXONE (5R,9R,13R,14S) **465-65-6** *antagonist (narcotic)*
2	1.29	7.1	554		"	NALOXONE (5R,9R,13R,14S)
3	1.53	7.4	554	1	"	NALOXONE (5R,9R,13R,14S)
4	1.60	7.5	554	1	"	NALOXONE (5R,9R,13R,14S)
45	1.66	7.6	554	212	"	NALOXONE (5R,9R,13R,14S)
6	1.72	7.7	554	1	"	NALOXONE (5R,9R,13R,14S)
7 ★	2.09	7.7	554	2	"	NALOXONE (5R,9R,13R,14S)
8 ★	2.07		2487		C19H21N1O5	CINNAMAMIDE,3,4-DIHYDROXY-N-(3,4-DIMETHOXY)PHENETHYL
9	0.26		564		C19H21N1O5	TRIMETHYLCOLCHICINIC ACID
14050	0.66	7.4	564	1	"	TRIMETHYLCOLCHICINIC ACID
1	0.74	7.0	2678	1	"	TRIMETHYLCOLCHICINIC ACID
2 ★	4.36		1481		C19H21N1S1	A-PHENYLTHIO-N-SEC-BUTYLCINNAMAMIDE
3	2.76	7.4	268	1	C19H21N1S1	DOSULEPINE **113-53-1** *antidepressant*
4 ★	4.49		680	462	"	DOSULEPINE
55	1.50	7.4	2020	1	C19H21N3O4	PINDOLOL ANALOG:N-(P-NITROPHENETHYL) **102573-78-4**
6	-0.64	7.4	880	1	C19H21N5O2	PIRENZEPINE **28797-61-7** *anti-ulcerative*
7 ★	0.10		880		"	PIRENZEPINE
8	0.90	12.0	1471		C19H21N5O3	2,4-DIAM-5(3',5'-DIMEO-4'-(P-AMPHO))PYRIMIDINE
9	2.11	8.5	2101	1	C19H21N5O3S1	OXMETIDINE **72830-39-8** *antagonist (histamine h2 receptors)*
14060	-0.30	7.4	1689	1	C19H21N5O5S1	N7-(5-METHYLISOTHIAZOL-3-YL)-MITOMYCIN-C **88854-64-2**
1 ★	1.16	7.4	1689	1	C19H21N5O5S1	N7-(5-METHYLTHIAZOL-2-YL)-MITOMYCIN-C **84397-36-4**
2	2.01		1593	459	C19H21N7O8S4	CEPHALOSPORIN,N,N-DIME-CARBOXAMIDO ANALOG **69712-47-6**
3	1.80	7.4	2217	1	C19H22Cl1N5O1	TRAZODONE **19794-93-5** *antidepressant*
4 ★	3.30		783		C19H22F1N3O1	AZAPERONE **1649-18-9** *antipsychotic*
65	4.59	7.4	2090	459	C19H22F2N2O3	5-(5-(2-DIFLUOROMETHYL-4-(4,5-DIHYDRO-2-OXAZOLYL)PHENOXY)PENTYL-3-METHYLISOXAZOLE **105639-15-4**
6	1.86	7.4	1940	1	C19H22N2	TRIPROLIDINE **486-12-4** *antihistaminic*
7 ★	3.92		260		"	TRIPROLIDINE
8 ★	3.64		1147		C19H22N2O1	9(N((N',N'-DIETHYLAMINO)ACETYL)AMINO)FLUORENE
9	1.73	7.4	2648	1	C19H22N2O1	CINCHONIDINE **485-71-2** *antimalarial*

	logP	pH	Ref.	Note	MF	Name / **CAS** / *activity*
14070 ★	2.68	7.4	2648	2	"	CINCHONIDINE
1	1.85	7.4	2648	1	C19H22N2O1	CINCHONINE **118-10-5** *antimalarial*
2 ★	2.82	7.4	2648	2	"	CINCHONINE
3	0.79	5.0	1440	1	C19H22N2O1	NOXIPTILINE **3362-45-6** *antidepressant*
4	2.34	7.4	678	1	C19H22N2O1S1	ACEPROMAZINE **61-00-7** *sedative (vet)*
75	1.70	7.4	2020	1	C19H22N2O2	PINDOLOL ANALOG:N-PHENETHYL **102573-79-5**
6 ★	1.37	7.4	2268	1	C19H22N2O5	NAPROXEN,(N-METHYL-N-AMINOCARBONYLMETHYL)GLYCOLAMIDE ESTER
7	2.92	7.4	2298	1	C19H22N4O2S1	1,2,4-OXADIAZOLE-5-(DIETHYLAMINOETHYL)THIOHYDROXAMATE,3-(1-NAPHTHYL) **103499-11-2**
8 ★	3.87	7.4	2298	2	"	1,2,4-OXADIAZOLE-5-(DIETHYLAMINOETHYL)THIOHYDROXAMATE,3-(1-NAPHTHYL)
9	1.29	7.4	2206	306	C19H22N4O4	TETRAHYDRONAPHTHALENE,1-OXO-6-(2-DIMETHYLAMINO)ETHOXY-2-(1-METHYL-5-NITRO-IMIDAZOL-2YL)METHYLENE
14080	-2.20	7.4	594	1	C19H22N6O2	1,2-DIHYDRO-S-TRIAZINE,2,2-DIMETHYL-4,6-DIAMINO-1-(3(4-CARBOXAMIDOBENZOYLOXY)PHENYL
1	0.97	7.4	2433	1	C19H22N6O3	ALLOPURINOL,N-(3-(4-METHYLPIPERIZINYLMETHYL)BENZOYL)OXYMETHYL
2 ★	1.49		2433	2	"	ALLOPURINOL,N-(3-(4-METHYLPIPERIZINYLMETHYL)BENZOYL)OXYMETHYL
3	0.99	7.4	2433	1	C19H22N6O3	ALLOPURINOL,N-(4-(4-METHYLPIPERAZINYLMETHYL)BENZOYLOXY)METHYL
4 ★	1.47	7.4	2433	2	"	ALLOPURINOL,N-(4-(4-METHYLPIPERAZINYLMETHYL)BENZOYLOXY)METHYL
85 ★	1.45	4.6	2043	2	C19H22N8O6S3	CEFTEZOLE,PIVALOYLOXYMETHYL ESTER
6 ★	3.30		2650	458	C19H22O3S1	DIETHYL KETONE,1-(4-METHYLPHENYL)-2-(4-TOLUENESULFONYL)
7 ★	2.47		2650	458	C19H22O4S1	DIETHYL KETONE,1-(4-METHOXYPHENYL)-2-(4-TOLUENESULFONYL)
8 ★	0.24		572		C19H22O6	GIBBERELLIC ACID **77-06-5**
9	3.44		2021	459	C19H22O7	NAPHTHALENE,1,4-DIPROPIONYLOXY-2,3,5-TRIMETHOXY **91814-25-4**
14090	3.28		2021	459	C19H22O7	NAPHTHALENE,1,4-DIPROPIONYLOXY-2,3,6-TRIMETHOXY **91814-27-6**
1	2.80	7.0	468	1	C19H23Cl1N2	CHLORIMIPRAMINE **303-49-1** *antidepressant*
2	3.24	7.4	1209	1	"	CHLORIMIPRAMINE
3	3.32	7.4	268	1	"	CHLORIMIPRAMINE
4	3.58	7.4	2407	1	"	CHLORIMIPRAMINE
95 ★	5.19		680	462	"	CHLORIMIPRAMINE
6	1.76	4.3	882	1	C19H23Cl1N2S1	CHLORPROETHAZINE **84-01-5** *muscle relaxant, tranquilizer*
7	-0.37	7.2	2526	1	C19H23F1N4O3	OXONAPHTHRIDINECARBOXYLIC ACID,7-(1-METHYL-2,5-DIAZABICYCLOHEPTYL) ANALOG
8	0.21	7.2	2526	1	C19H23F1N4O3	OXONAPHTHRIDINECARBOXYLIC ACID,7-(2-METHYL-2,5-DIAZABICYCLOHEPTYL) ANALOG
9	-0.12	7.2	2526	1	C19H23F1N4O3	OXONAPHTHRYIDINECARBOXYLIC ACID,7-(6-METHYL-2,5-DIAZABICYCLOHEPTYL) ANALOG
14100	0.10	7.2	2526	1	"	OXONAPHTHRYIDINECARBOXYLIC ACID,7-(6-METHYL-2,5-DIAZABICYCLOHEPTYL) ANALOG
1	-0.02	7.2	2526	1	C19H23F1N4O3	OXONAPHTHRYIDINECARBOXYLIC ACID,7-(3-METHYL-2,5-DIAZABICYCLOHEPTYL) ANALOG ✚
2	0.17	7.2	2526	1	"	OXONAPHTHRYIDINECARBOXYLIC ACID,7-(3-METHYL-2,5-DIAZABICYCLOHEPTYL) ANALOG
3	0.02	7.2	2526	1	C19H23F1N4O3	OXONAPHTHRYIDINECARBOXYLIC ACID,7-(4-METHYL-2,5-DIAZABICYCLOHEPTYL) ANALOG
4	0.23	2.0	1039	210	C19H23N1O1	DIPHENYLPYRALINE **147-20-6** *antihistaminic*
5	0.56	1.0	870	342	C19H23N1O1	N-AZETIDYLETHYL-(P-ME-DIPHENYL)METHYLETHER
6	0.25	1.0	602	342	C19H23N1O1	N-PYRROLIDYLETHYL-DIPHENYLMETHYLETHER
7	1.40	7.4	2493	1	C19H23N1O2	2-PROPANONE,1,1-DIPHENYL-1-ETHOXY-3-DIMETHYLAMINO
8	2.68		2378	1	C19H23N1O2	6,7-BENZOMORPHAN,N-(FURAN-2YL)METHYL-1,5,9-S **38047-74-4**
9	2.49		2378	1	C19H23N1O2	6,7-BENZOMORPHAN,N-(FURAN-2YL)METHYL-1,5,9-R
14110	0.87	7.4	2378	1	C19H23N1O2	DEOXY-DIHYDRONALORPHINE **69663-74-7**
1 ★	2.77	8.0	2644	2	C19H23N1O2	FOMOCAINE ANALOG-25 (4-HYDROXYPIPERIDINE)
2	2.00	7.4	2378	1	C19H23N1O2	FUR-3YLMETAZOCINE
3	2.25	7.4	2378	1	"	FUR-3YLMETAZOCINE **56498-83-0**
4	0.29	7.4	2680	324	C19H23N1O3	ETHYLMORPHINE **76-58-4** *antitussive, analgesic*
15	3.56		2662	459	C19H23N1O4	N-(3,4-DIETHOXYPHENYL)CARBAMATE,PHENETH-2YL ESTER *fungicide*
6 ★	2.01	7.4	2268	1	C19H23N1O5	NAPROXEN,(N-METHYL-N-HYDROXYETHYL)GLYCOLAMIDE ESTER
7 ★	5.50		2506		C19H23N3	AMITRAZ **33089-61-1** *scabicide, insecticide*
8	0.08	7.4	722	1	C19H23N3O2	MORPHOLINO-DES(N-DIIPR)DISOPYRAMIDE
9	0.30	7.4	2020	1	C19H23N3O2	PINDOLOL ANALOG:N-(P-AMINOPHENETHYL) **102573-76-2**
14120	1.18	7.4	2095	1	C19H23N3O2S1	(5-(2-THIENYL-IMIDAZOLE)-ADRENOCEPTOR ANTAGONIST
1	1.41	7.4	2095	1	C19H23N3O2S1	5-(THIEN-2YL)IMIDAZOLE,2-ALKOXYPHENYL ANALOG *adrenoceptor antagonist*
2 ★	0.95	1.0	2196		C19H23N3O4S1	HETACILLIN **3511-16-8** *antibacterial*
3 ✓	-0.68		2397	101	C19H23N4O3S1	IMIDAZOLIUM CHLORIDE, NAPHTHYL ANALOG
4 ✓	-0.13		2397	101	"	IMIDAZOLIUM CHLORIDE, NAPHTHYL ANALOG
25	-0.84		594	314	C19H23N5O2	1,2-DIHYDROTRIAZINE,2,6-DIAMINO-2,2-DIMETHYL-1-(3-PHENOXYETHOXY)PHENYL
6	2.74	7.4	1485	1	C19H23N5O2	2,4-DIAMINOPYRIMIDINE,5-(3,5-DIETO-4(1-PYRRYL)BENZYL **73090-70-7**
7 ★	2.89	7.4	1485	2	"	2,4-DIAMINOPYRIMIDINE,5-(3,5-DIETO-4(1-PYRRYL)BENZYL
8	0.84	7.4	594	1	C19H23N5O3	TRIMETREXATE **52128-35-5** *antineoplastic*
9	0.90	7.0	1200	1	"	TRIMETREXATE
14130 ★	2.55	13.0	1200	224	"	TRIMETREXATE
1 ★	1.76	6.2	1703		C19H23N5O7S2	CEFTIZOXIME,PIVALOYLOXYMETHYL ESTER
2 ★	2.42	4.6	2043	2	"	CEFTIZOXIME,PIVALOYLOXYMETHYL ESTER
3 ★	0.90	13.0	547	224	C19H23N7O2	4,6-DIAM-2,2-DIME-1(3'(3"URIDOPHO)MEPH)S-TRIAZINE
4	0.80	7.4	2034	1	C19H23N7O3	TRIAMTERENE,PARA-CARBOXYBUTOXY,ETHYL ESTER
35	0.04	4.0	2644	1	C19H24Cl1N1O1	FOMOCAINE ANALOG-30 (T-BUTYL-ME-AMINE)
6	1.78	6.0	2644	1	C19H24Cl1N1O1	FOMOCAINE ANALOG-30 (T-BUTYL-ME-AMINE)
7	-0.10		817		C19H24N1O3.Br1	N(2-BENZILOYLOXYET-N-TRIMETHYLAMMONIUMBROMIDE
8	0.80	2.0	1039	210	C19H24N2	BAMIPINE **4945-47-5** *antihistaminic*
9	0.64	2.0	1039	210	C19H24N2	HISTAPYRRODINE **493-80-1** *antihistaminic*
14140	2.16	7.0	2230	1	C19H24N2	IMIPRAMINE **50-49-7** *antidepressant*
1	2.30	7.0	468	1	"	IMIPRAMINE
2	2.40	7.4	880	1	"	IMIPRAMINE
3	2.49	7.4	1209	1	"	IMIPRAMINE
4	2.52	7.4	268	1	"	IMIPRAMINE
45	2.65	7.4	785	1	"	IMIPRAMINE
6 ★	4.80	7.0	468	701	"	IMIPRAMINE
7 ★	3.31		592		C19H24N2O1	BENZANILIDE,4-DIETHYLAMINOMETHYL,2'-METHYL
8 ★	3.05	8.0	2644	2	C19H24N2O1	FOMOCAINE ANALOG-21 (ME-PIPERIDINE)
9	3.39	7.8	550		C19H24N2O1S1	METHOTRIMEPRAZINE **60-99-1** *analgesic*
14150 ★	4.68		550		"	METHOTRIMEPRAZINE
1	4.40	7.4	2090	459	C19H24N2O3	5-(5-(2-METHYL-4-(4,5-DIHYDRO-2-OXAZOLYL)PHENOXY)PENTYL-3-METHYLISOXAZOLE **105639-02-9**
2	0.66	7.0	1362	1	C19H24N2O3	LABETALOL **36894-69-6** *anti-adrenergic (alpha & beta receptors)*
3	0.85	7.4	1055	1	"	LABETALOL
4	0.88	10.2	2465	41	"	LABETALOL
55	1.06	7.4	1362	1	"	LABETALOL
6	1.09	7.4	2465	41	"	LABETALOL
7	3.44	7.4	2090	459	C19H24N2O4	5-(5-(2-METHOXY-4-(4,5-DIHYDRO-2-OXAZOLYL)PHENOXY)PENTYL-3-METHYLISOXAZOLE **98033-66-0**
8 ★	3.57		547		C19H24N2O4	N1-MEO-N1-ME-N3-(4(4-PHENOXY)BUTOXY)PHENYLUREA
9	1.52	7.4	2399	1	C19H24N2O4	TOLAMOLOL

	logP	pH	Ref.	Note	MF	Name / **CAS** / *activity*
14160 ★	1.10		2531		C19H24N2O5	PHENOXYACETAMIDE ANALOG **139733-55-4**
1	1.14	7.4	1702	1	C19H24N2O6	SALICYLAMIDE,5-(OC2H4NHCHOHOCH2-(3-MEO-PHENYL)) **76813-08-6**
2	3.48	7.8	550		C19H24N2S1	ISOTHAZINE **522-00-9** *antiparkinsonian*
3	4.77		550		"	ISOTHAZINE
4	1.59	4.3	882	1	C19H24N2S2	METHIOMEPRAZINE **7009-43-0** *anti-emetic*
65	1.51		1711		C19H24N4O7	PROPYLOXYCARBONYL-MITOMYCIN-C
6 ★	1.14		636		C19H24O6	CHAMMISSONINDIACETATE ✚
7	1.12	1.0	870	342	C19H25N1O1	2-(DIPHENYLMETHOXY)ETHYL-N-BUTYLAMINE
8	0.26	1.0	870	342	C19H25N1O1	2-(DIPHENYLMETHOXY)TRIETHYLAMINE
9	0.66	1.0	870	342	C19H25N1O1	2-(P-ME-DIPHENYLMETHOXY)ETHYL-N-ME-N-ETHYLAMINE
14170	1.09	1.0	870	342	C19H25N1O1	2-(P-ME-PHENYL)-2-PHENYLMETHOXY-N-PR-ETHYLAMINE
1	0.45	1.0	602	342	C19H25N1O1	2-DIPHENYLMETHOXY-N-ME-N-PROPYL-ETHYLAMINE
2	-0.50		307	1	C19H25N1O1	3-HYDROXY-N-ALLYLMORPHINAN HYDROBROMIDE
3	2.07	7.1	554		C19H25N1O1	3-HYDROXY-N-ALLYLMORPHINAN **152-02-3** *antagonist (to narcotics)*
4	2.15	7.4	2378	1	"	3-HYDROXY-N-ALLYLMORPHINAN
75	2.35	7.4	554	1	"	3-HYDROXY-N-ALLYLMORPHINAN
6	2.44	7.5	554	1	"	3-HYDROXY-N-ALLYLMORPHINAN
7	2.53	7.6	554	212	"	3-HYDROXY-N-ALLYLMORPHINAN
8	2.62	7.7	554	1	"	3-HYDROXY-N-ALLYLMORPHINAN
9 ★	3.48	7.7	554	2	"	3-HYDROXY-N-ALLYLMORPHINAN
14180 ★	3.83		536		C19H25N1O1	PROPOXYPHENECARBINOL
1	1.06	4.0	2655	1	C19H25N1O3	PROPRANOLOL,PROPIONYL ESTER
2	2.38	7.4	1941	1	C19H25N1O4	METHYLAMINE,N,N-DI-(3-PHENOXY-2-HYDROXYPROPYL
3 ★	2.92	13.0	1941		"	METHYLAMINE,N,N-DI-(3-PHENOXY-2-HYDROXYPROPYL
4	1.43	5.7	2241	1	C19H25N1O4	PSICAINE-NEU **55608-72-5** *anesthetic (local)*
85 ★	4.73		2209		C19H25N1O4	TETRAMETHRIN **7696-12-0** *insecticide*
6 ★	2.18		1589		C19H25N1O5S1	(N-ME-BENZYL)AMINOET-(3,4,5-TRIMEO)PHENYL-SULFONE
7	1.72		634		C19H25N1O6	6-BUO-7-METHOXYETHOXY-4-OH-3-QUINOLINE-CO2ET
8	1.84		634		C19H25N1O6	6-I-BUO-7-METHOXYETHOXY-4-OH-3-QUINOLINE-CO2ET
9	2.07	7.4	2235	1	C19H25N1Si1	SILA-PRODIPINE ✚
14190	0.49	4.3	882	1	C19H25N3O2S2	DIMETIOTIAZIN **7456-24-8** *serotonin inhibitor*
1	2.83	7.8	550		C19H25N3S1	AMINOPROMAZINE **58-37-7** *anti-spasmodic*
2 ★	4.76		550		"	AMINOPROMAZINE
3 ★	-0.38		2377		C19H25N5O4	AC-TRP-ALA-ALA-N
4 ★	-3.05		1682		C19H26I3N3O9	IOHEXOL **66108-95-0** *diagnostic aid (radiopaque)*
95	-2.71	7.4	2476	1	"	IOHEXOL
6 ★	-2.80		1682		C19H26I3N3O9	M-BZAMID,N,N(CH(CH2OH)2)2,5-N(2,3-DIOHPR)ACAM,I3
7 ★	1.81	7.1	2155	2	C19H26N2O2	RAUWOLFINE
8 ★	2.23	7.4	1956	2	C19H26N2O3	PILOCARPIC ACID,ALPHA-METHYLBENZYL ESTER **92598-81-7**
9 ★	2.42	7.4	1956	2	C19H26N2O3	PILOCARPIC ACID,O-METHYLBENZYL ESTER **92598-92-0**
14200 ★	2.46	7.4	1956	2	C19H26N2O3	PILOCARPIC ACID,P-METHYLBENZYL ESTER **92598-93-1**
1 ★	2.31	7.4	1956	2	C19H26N2O3	PILOCARPIC ACID,PHENETHYL ESTER **92598-86-2**
2	3.55	7.5	2254	2	C19H26N2S1	PERGOLIDE **66104-22-1** *dopamine agonist, antiparkinsonian*
3	2.20		1713		C19H26N4O2	6-PH-2-(4-(N3-BU-UREIDO)BUTYL)-3(2H)-PYRIDAZINONE
4	1.89		1713		C19H26N4O2	6-PH-2-(4-(N3-DIET-UREIDO)BU)-3(2H)-PYRIDAZINONE
5	2.07		1713		C19H26N4O2	6-PH-2-(4-(N3-T-BU-UREIDO)BUTYL)-3(2H)-PYRIDAZINONE
6	1.45		1713		C19H26N4O2	6-PH-2-(6-(N3-DIME-UREIDO)HEXYL)-3(2H)-PYRIDAZINONE
7	1.81		1713		C19H26N4O2	6-PH-2-(6-(N3-ET-UREIDO)HEXYL)-3(2H)-PYRIDAZINONE
8	1.06		1713		C19H26N4O2	6-PH-2-(7-UREIDO-HEPTYL)-3(2H)-PYRIDAZINONE
9 ★	2.57	7.4	2029	459	C19H26N6O4S1	XANTHINE,1,3-DIETHYL-8-(4-(DIMETHYLAMINOETHYL)AMINOSULFONYL **89073-49-4**
14210 ★	-1.30	7.4	2442	1	C19H26N6O6	P-GLU-HIS(N-ETHOXYCARBONYL)-PRO-NH2
1 ★	2.75		542		C19H26O2	4-ANDROSTENE-3,17-DIONE **63-05-8**
2 ★	4.78		1521		C19H26O3	(1R)-TR-(S)-ALLETHRIN **584-79-2**
3	4.50		2616	459	C19H26O3	DIMETHRIN **70-38-2**
4	0.91	7.4	401	1	C19H26O7	ANGUIDINE **2270-40-8** *antineoplastic*
15	-0.83		307	1	C19H27N1O1	3-HYDROXY-N-PROPYLMORPHINANHBR
6	1.75	7.1	554		C19H27N1O1	PENTAZOCINE **359-83-1** *analgesic*
7	2.04	7.4	554	1	"	PENTAZOCINE
8	2.14	7.5	554	1	"	PENTAZOCINE
9	2.18	7.4	2407	1	"	PENTAZOCINE
14220	2.24	7.6	554	212	"	PENTAZOCINE
1	2.34	7.7	554	1	"	PENTAZOCINE
2 ★	3.31		2407	2	"	PENTAZOCINE
3	3.81	7.7	554	2	"	PENTAZOCINE
4	1.54	7.4	2480	1	C19H27N1O4	KETOBEMIDONE,I-PROPYL CARBONATE
25 ★	1.72	7.4	2480	2	"	KETOBEMIDONE,I-PROPYL CARBONATE
6	-0.02	7.4	651	1	C19H27N3O2	4-(-2-N,N-DIETHYLAMINOPROPIONYL)ANTIPYRENE
7 ★	4.64	6.9	2175	1	C19H28N2O2	3-NONYL-3-(4-PYRIDYL)PIPERIDINE-2,6-DIONE
8	0.30	7.0	708	1	C19H28N2O3	6-(3,4-DIMETHOXYBENZAMIDO)2-METHYLDECAHYDRO-I-QUINOLINE,CIS
9	0.27	7.0	708	1	C19H28N2O3	6-(3,4-DIMETHOXYBENZAMIDO)2-METHYLDECAHYDRO-I-QUINOLINE,TRANS
14230	3.20	7.5	2651	28	C19H28N2O3	BENZAMIDE,2,3-DIMETHOXY-5-(1D-PROPYL)-N-(1-ETHYLPYRROLIDIN-2YL)METHYL
1	2.95	7.5	2651	28	C19H28N2O3	BENZAMIDE,2,3-DIMETHOXY-5-ALLYL-N-(1-ETHYLPYRROLIDIN-2YL)METHYL
2	0.86	7.4	2274	1	C19H28N2O4	CARPINDOLOL **39731-05-0**
3 ★	2.88		1592	220	C19H28N2O5	T-BOC-PHE-VAL
4	4.67		2428	459	C19H28N2O6S1	N,N'-DIMETHYLTHIODICARBAMATE,O-3,4-METHYLENEDIOXYLPHENYL-O'-OCTYL
35	-0.24	7.4	651	1	C19H28N4O2	4-(1-ME-3-DIETHYLAMINOPROPIONYLAMINO)ANTIPYRENE
6 ★	0.60		2377		C19H28N4O4	AC-PHE-GLY-LEU-N
7 ★	0.34		2377		C19H28N4O4	AC-PHE-ILE-GLY-N
8 ★	0.06		2377		C19H28N4O4	AC-VAL-PHE-ALA-N
9 ★	3.60		2558		C19H28O2	ANDROSTANEDIONE ✚
14240 ★	3.23		2460		C19H28O2	PRASTERONE **53-43-0** *androgen*
1	2.99		1607	459	C19H28O2	TESTOSTERONE,19-NOR-17-A-METHYL **514-61-4** *androgen*
2	2.72	7.4	1076	320	C19H28O2	TESTOSTERONE **58-22-0** *androgen*
3 ★	3.32		300		"	TESTOSTERONE
4 ★	5.15		580		C19H29Cl1N2O3	2-CL-4(NHCOME2)BENZOICAC.,4(2,6-DIMEHEPTYL) ESTER
45 ★	5.15		2612		C19H29Cl1N2O3	UREA,1,1-DIMETHYL-3-(3-CHLORO-4-BENZOIC ACID),2,4-DIMETHYLPENT-3YL ESTER **88132-43-8**
6	1.38	7.5	2651	1	C19H29F1N2O3	FPMB
7	2.73	7.5	2651	28	"	FPMB
8	1.94	7.5	2651	1	C19H29F1N2O4	SALICYLAMIDE,3-(2-FLUOROETHYL)-5-METHOXY-6-ETHOXY-N-(1-ETHYLPYRROLIDIN-2YL)METHYL
9	3.75	7.5	2651	28	"	SALICYLAMIDE,3-(2-FLUOROETHYL)-5-METHOXY-6-ETHOXY-N-(1-ETHYLPYRROLIDIN-2YL)METHYL

	logP	pH	Ref.	Note	MF	Name / CAS / activity
14250	6.39		1361	459	C19H29N1	4,4-DIMETHYL-PHENCYCLICINE(PHEN=AX.)
1	1.94	7.4	2278	1	C19H29N1O1	PROCYCLIDINE **77-37-2** *antiparkinsonian, relaxant (skeletal muscle)*
2	2.53	7.4	2501	1	C19H29N1O2	BORNAPROLOL **66451-06-7** *anti-adrenergic (beta-receptor)*
3 ★	4.17	13.0	2501		"	BORNAPROLOL
4	1.56	7.4	2450	1	C19H29N1O3	TROLOX ACETATE (DIMETHYLAMINOETHYL ANALOG)
55	1.16	7.4	2556	1	C19H29N1O4	DIPIVALOYL EPININE
6	-1.89	7.0	1056	1	C19H29N2O2.Cl1	1(3-ME3-N+PROPYL)AM-3(1-NAPHTHYLOXY)2-PROPANOL
7 ★	-1.95	7.0	2374		C19H29N3O5	PHENYLALANYLISOLEUCINYLTHREONINE
8 ★	4.43		2237		C19H30N2O2	BENZOFURAN-3-AMINE-2,3-DIHYDRO-5-METHYL-N-DIBUTYLAMINOACETYL
9	2.44	7.3	1960	314	C19H30N2O3	2-BUTOXYMETHYLCARBANILIC ACID,2-(1-PIPERIDINYL)ETHYL ESTER
14260 ★	3.94	7.3	1960	2	"	2-BUTOXYMETHYLCARBANILIC ACID,2-(1-PIPERIDINYL)ETHYL ESTER
1	4.10	7.2	1150		C19H30N2O3	2-ME-4(O-BU)CARBANILATE-O-2(N-PIPERIDINYL)ETHYL
2	1.85	7.1	1150		C19H30N2O3	2-ME-4(O-PR)CARBANILATE-O-(1-ME-2(PIPERIDINYL)ET
3	2.90	7.3	2051	1	C19H30N2O3	2-PENTOXYCARBANILATE-O-2(N-PIPERIDINYL)ETHYL **55792-16-0**
4	4.46	7.3	1356	2	"	2-PENTOXYCARBANILATE-O-2(N-PIPERIDINYL)ETHYL
65	2.23	7.3	1960	314	C19H30N2O3	3-BUTOXYMETHYLCARBANILIC ACID,2-(1-PIPERIDINYL)ETHYL ESTER
6 ★	3.73	7.3	1960	2	"	3-BUTOXYMETHYLCARBANILIC ACID,2-(1-PIPERIDINYL)ETHYL ESTER
7	3.50	7.3	1356	314	C19H30N2O3	3-PENTOXYCARBANILATE-O-2(N-PIPERIDINYL)ETHYL **52205-62-6**
8	5.06	7.3	1356	2	"	3-PENTOXYCARBANILATE-O-2(N-PIPERIDINYL)ETHYL
9	3.31	7.3	1356	314	C19H30N2O3	4-PENTOXYCARBANILATE-O-2(N-PIPERIDINYL)ETHYL **55792-17-1**
14270	4.87	7.3	1356	2	"	4-PENTOXYCARBANILATE-O-2(N-PIPERIDINYL)ETHYL
1	3.28	7.5	2651	28	C19H30N2O3	BENZAMIDE,2,3-DIMETHOXY-5-PROPYL-N-(1-ETHYLPYRROLIDIN-2YL)METHYL
2 ★	4.24		2237		C19H30N2O3	BENZOFURAN-3-AMINE,2,3-DIHYDRO-6-METHOXY-N-DIBUTYLAMINOACETYL
3	2.12	7.0	1517	1	C19H30N2O3	CARBAMICAC,N-ET-N-(M-PROPH),2(N-PIPERIDINYL)ET.EST **52205-64-8**
4	2.12	7.1	1357		"	CARBAMICAC,N-ET-N-(M-PROPH),2(N-PIPERIDINYL)ET.EST
75	3.38	7.2	1150		C19H30N2O4	2ME-4(O-PEN)CARBANILATE-O-2(N-MORPHOLINO)ETHYL
6	0.52	7.5	2651	1	C19H30N2O4	BENZAMIDE,2,3,6-TRIMETHOXY-5-ETHYL-N-(1-ETHYLPYRROLIDIN-2YL)METHYL
7	2.07	7.5	2651	28	"	BENZAMIDE,2,3,6-TRIMETHOXY-5-ETHYL-N-(1-ETHYLPYRROLIDIN-2YL)METHYL
8	1.57	7.5	2651	28	C19H30N2O4	BENZAMIDE,2,3-DIMETHOXY-5-(3-HYDROXYPROPYL)-N-(1-ETHYLPYRROLIDIN-2YL)METHYL
9	2.14	7.5	2651		C19H30N2O4	SALICYLAMIDE,3-ETHYL-5-METHOXY-6-ETHOXY-N-(1-ETHYLPYRROLIDIN-2YL)METHYL
14280	4.26	7.5	2651	28	"	SALICYLAMIDE,3-ETHYL-5-METHOXY-6-ETHOXY-N-(1-ETHYLPYRROLIDIN-2YL)METHYL
1	3.33	7.5	2651	28	C19H30N2O4	SALICYLAMIDE,3-METHOXY-5-ETHYL-6-ETHOXY-N-(1-ETHYLPYRROLIDIN-2YL)METHYL
2 ★	3.55		2460		C19H30O2	5-ANDROSTENE-3,17-DIOL **521-17-5** *anabolic*
3 ★	3.69		2460		C19H30O2	ANDROSTERONE **53-41-8**
4 ★	5.43		2415	459	C19H30O2	CYCLOHEXANOL,3-(2-HYDROXY-4(1,1-DIMETHYL)PENTYL)PHENYL
85 ★	3.66		2460		C19H30O2	STANOLONE **521-18-6** *androgen*
6 ★	4.75		2506		C19H30O5	PIPERONYL BUTOXIDE *synergist (insecticide)*
7	2.99		1396	459	C19H30O6	ARTEMISININE,BUTYRYL ESTER(ALPHA)
8	2.78		1396	459	C19H30O7	ARTEMISININE,PROPYLCARBONATE(ALPHA)
9	3.11		1396	459	"	ARTEMISININE,PROPYLCARBONATE(ALPHA)
14290	2.60	7.4	1743	1	C19H31N1O1	BENCYCLANE **2179-37-5** *vasodilator, spasmolytic*
1	5.06		2662	459	C19H31N1O4	N-(3,4-DIETHOXYPHENYL)CARBAMATE,S-OCTYL ESTER *fungicide*
2	1.35	7.4	365	2	C19H31N3O1	N,N-DIPR-PIPERAZINE-2-CARBOXANILIDE,2'6'-DIME
3	1.00	4.8	1020		C19H32N2O1	2ET-BUTYRAMIDE,N(2,6-DIMEPH),N(3-DIMEAMINOPROPYL)
4	3.48		1352		C19H32N2O1	6-(DIETHYLAMINO)HEPTANO-2',6'-XYLIDIDE
95	2.18	7.1	1150		C19H32N2O3	2-ME-4(O-BU)CARBANILATE-O-2(N,N-DIETAMINO)1-ME-ET
6	4.26	7.2	1150		C19H32N2O3	2-ME-4(O-PENT)CARBANILATE-O-2(N,N-DIETAMINO)ETHYL
7	2.25	7.0	1517	1	C19H32N2O3	N-ME-N(M-PENTOXYPH)CARBAMATE,2-(DIETAMINO)ETHYL **59731-48-5**
8	2.25	7.1	1357		"	N-ME-N(M-PENTOXYPH)CARBAMATE,2-(DIETAMINO)ETHYL
9	2.75	7.4	2244	1	C19H32N4O4S1	TIMOLOL,O-CYCLOPENTANOYL
14300	8.61		2316	337	C19H32O1	PHENOL,2-TRIDECYL
1	8.84		2316	337	C19H32O1	PHENOL,3-TRIDECYL
2	8.20		2316	337	C19H32O1	PHENOL,4-TRIDECYL
3	7.75		2316	337	C19H32O2	1,2-DIHYDROXYBENZENE,3-TRIDECYL
4	8.42		2316	337	C19H32O2	1,2-DIHYDROXYBENZENE,4-TRIDECYL
5	6.84		2316	337	C19H32O2	1,3-DIHYDROXYBENZENE,4-TRIDECYL
6 ★	4.16		2460		C19H32O2	ANDROSTANEDIOL
7 ★	3.55		1589		C19H33N1O5S1	(N-DIBUTYL)AMINOET-(3,4,5-TRIMEO)PHENYL-SULFONE
8	1.98		2093		C19H33N4O2P1	DIMETHYLTEPA,1-ADAMANTYLAMINOCARBONYLAMIDE **103981-98-2**
9	0.60	6.0	1678		C19H34N1O6	3-OXYOXAZOLIDINE,2,2-BIS(HEPTANOIC ACID) **39078-03-0**
14310	-0.08		527	820	C19H34N1.Br1	BENZYLAMMONIUMBROMIDE,N-METHYL,N-DECYL
1	1.32		527	820	C19H34N1.Br1	TETRADECYLPYRIDINIUMBROMIDE
2	0.33		1635	820	C19H34N1.Cl1	DECYLDIMETHYLBENZYLAMMONIUMCHLORIDE
3	-1.60		1506	820	C19H34N1.Cl1	TRIBUTYLAMMONIUMCHLORIDE,N-BENZYL **23616-79-7**
4 ✓	1.41		1729	820	C19H34N1.I1	2,4-DIMETHYLPYRIDINIUM IODIDE,N-DODECYL
15 ✓	1.75		1729	820	C19H34N1.I1	3,4-DIMETHYLPYRIDINIUM IODIDE,N-DODECYL
6 ✓	1.91		1729	820	C19H34N1.I1	3,5-DIMETHYLPYRIDINIUM IODIDE,N-DODECYL
7 ✓	3.56		1729	820	C19H34N1.I1	N-TETRADECYLPYRIDINIUM IODIDE **27593-10-8**
8	-1.17	7.4	1026	1	C19H34N2O4	1,1'(3-ME-O-PHEN-DIOXY)BIS(3-I-PR-AM-2-PROPANOL)
9	-1.15	7.4	1026	1	C19H34N2O4	1,1'(4-ME-O-PHEN-DIOXY)BIS(3-I-PR-AM-2-PROPANOL)
14320	3.26	7.4	2244	1	C19H34N4O4S1	TIMOLOL,O-(2-ETHYLBUTYRYL)
1	3.09	7.4	2244	1	C19H34N4O4S1	TIMOLOL,O-(3,3-DIMETHYLBUTYRYL)
2	3.35	7.4	2244	1	C19H34N4O4S1	TIMOLOL,O-HEXANOYL
3 ★	-1.69		2377		C19H34N4O5	ILE-PRO-GLY-ILE
4	0.07		2377	821	C19H34N4O6	AC-LEU-ILE-GLU-N
25 ★	5.50		1903		C19H34O3	METHOPRENE **40596-69-8** *insect growth regulator*
6	2.74	12.3	343		C19H35Cl1N2O5S1	MIRINCAMYCIN **31101-25-4** *antibacterial, antimalarial*
7 ★	2.51	12.3	343		C19H35Cl1N2O5S1	N-ETHYL-CLINDAMYCIN
8	0.56		2469		C19H36N2O6S1	LINCOMYCIN,4'-BUTYL ANALOG ✚
9 ★	1.02	12.3	343		C19H36N2O6S1	N-ETHYL-LINCOMYCIN
14330	3.23		1714	820	C19H42N1.Cl1	CETYLTRIMETHYLAMMONIUMCHLORIDE **112-02-7**
1	1.36		2173		C19H42N1.I1	CETYLTRIMETHYL AMMONIUM IODIDE **7192-88-3**
2 ✓	0.53	7.3	2085		C19H42N4.O2Tc1	N-NONYLCYCLAM,TECHNICIUM OXIDE COMPLEX
3	0.65		1165	121	C20H2Cl4I4O5	ROSEBENGAL **632-69-9**
4	0.81		579		"	ROSEBENGAL
35	-1.46		1165	121	C20H6Br4Cl2O5	TETRABROMODICHLOROFLUORESCEIN, SODIUM SALT **6641-77-6**
6	-2.18		1293		C20H8Br2N2O9	DIBROMODINITROFLUORESCEIN **548-24-3**
7	-1.66		1165	121	"	DIBROMODINITROFLUORESCEIN
8	-1.46		1165	121	C20H8Br4O5	TETRABROMOFLUORESCEIN,SODIUM SALT **548-26-5**
9	-0.59		1165	121	C20H8I4O5	ERYTHROSINE-B **568-63-8**

	logP	pH	Ref.	Note	MF	Name / CAS / activity
14340	-0.15		1293		"	ERYTHROSINE-B
1	-0.81		1165	121	C20H10Br2O5	DIBROMOFLUORESCEIN,SODIUM SALT **4372-02-5**
2	0.11		1165	121	C20H10Cl2O5	DICHLOROFLUORESCEIN,SODIUM SALT **3474-67-7**
3	-0.21		1165	121	C20H10O5.(Na1)2	FLUORESCEIN,SODIUM SALT **518-47-8** *diagnostic aid (corneal indicator)*
4 ★	5.97		1696		C20H12	BENZO(A)PYRENE **50-32-8**
45 ★	5.82		2158	459	C20H12	PERYLENE **198-55-0**
6	-0.21	6.8	2263	1	C20H12O5	FLUORESCEIN **2321-07-5** *indicator, corneal trauma*
7 ★	3.46		579		C20H13Cl1F3N1O3	FLOXACRINE **53966-34-0** *antimalarial*
8 ★	6.40		1742		C20H13N1	13H-DIBENZO(A,I)CARBAZOLE **239-64-5**
9 ★	6.01		2158	459	C20H14	ANTHRACENE,9-PHENYL
14350	2.67	7.4	847	1	C20H14N2O2	N-(2-PYRIDINYLMETHYL)-G-PYROPHTHALONE
1	2.47	7.4	847	1	C20H14N2O2	N-(3-PYRIDINYLMETHYL)-G-PYROPHTHALONE
2	-0.67	7.4	1722	1	C20H14N2O7S2	BRILLIANTCRYSTALSCARLET
3	-0.75	7.4	1722	1	C20H14N2O10S3	AMARANTH **915-67-3**
4	2.41	7.4	772	1	C20H14O4	PHENOLPHTHALEIN **77-09-8** *laxative*
55	0.43	7.4	2557	1	C20H14O6	ETHYLIDENEBIS-4-HYDROXYCOUMARIN *anticoagulant*
6	0.64	7.4	2459	1	C20H15N1O4	WARFARIN,4'-CYANO
7 ★	7.02		850		C20H15N1S1	TRIPHENYLMETHYLISOTHIOCYANATE
8 ★	5.80		1486		C20H16	7,12-DIMETHYLBENZ(A)ANTHRACENE **57-97-6**
9 ★	3.69		2508		C20H16Cl2N2O3	PYRAZOXYFEN **71561-11-0**
14360	2.38	7.4	2573	1	C20H16Cl2N2O5S1	3-METHYLCEPHALOSPORIN,5-(2,6-DICHLOROPHENYL)FURYL ANALOG
1	3.72		562	400	C20H16N2O2	TEREPHTHALANILIDE **7154-31-6**
2	1.74	4.2	401		C20H16N2O4	CAMPTOTHECIN **7689-03-4** *antineoplastic*
3	1.85	7.4	594	1	"	CAMPTOTHECIN
4	2.61		759		C20H16O4	4,4'-DIHYDROXY-2"-CARBOXYTRIPHENYLMETHANE **81-90-3**
65	2.38	7.4	2573	1	C20H17Cl1N2O5S1	3-METHYLCEPHALOSPORIN,5-(P-CLOROPHENYL)FURYL ANALOG
6	-0.66	7.4	1838	1	C20H17F1O3S1	SULINDAC **38194-50-2** *anti-inflammatory*
7 ★	3.42	2.0	1838		"	SULINDAC
8 ★	4.37		2214		C20H17F3N2O4	2(4(2(6-TRIFLUOROMETHYLQUINOXALINYLOXY)PHENOXYPROPIONIC ACID,ETHYL ESTER
9	2.99	7.4	547	1	C20H17N3O2S1	9-ANILINOACRIDINE,1'-METHANSULFONAMIDO
14370 ★	4.74		1814	459	C20H18Cl1N1O6	OCHRATOXIN-A **303-47-9**
1	-1.28		1768	820	C20H18N1O4.Cl1	BERBERINECHLORIDE **633-65-8**
2	-2.12		777		C20H18N1O4.H1O1	BERBERINE **2086-83-1**
3	1.66	7.4	2573	1	C20H18N2O5S1	3-METHYLCEPHALOSPORIN,5-PHENYLFURYL ANALOG
4	1.68		1950		C20H18N2O6S2	BENZOTHIENYL-CEPHALOSPORIN
75	2.11	7.4	547	1	C20H18N4O2S1	4'(9-ANILINOACRIDINE)2-AM-1'-METHANESULFONAMIDO
6	0.85	7.4	547	1	C20H18N4O2S1	4'(9-ANILINOACRIDINE)3-AM-1'-METHANESULFONAMIDO
7	3.17	7.4	547	1	C20H18N4O2S1	4'(9-ANILINOACRIDINE)4-AM-1'-METHANESULFONAMIDO
8	3.73	1.0	2471	459	C20H18N4O5	XANTHONE,6-(5-TETRAZOLYL)-2-(2-BUTYROYLOXY)ETHOXY
9	3.27		759		C20H18O3	PHENOLPHTHALOL **81-92-5** *cathartic*
14380	1.92	7.4	2459	1	C20H18O4	WARFARIN,2'-METHYL
1	1.73	7.4	2459	1	C20H18O4	WARFARIN,3'-METHYL
2	1.71	7.4	2459	1	C20H18O4	WARFARIN,4'-METHYL
3	1.59	7.4	2459	1	C20H18O5	WARFARIN,2'-METHOXY
4	1.20	7.4	2459	1	C20H18O5	WARFARIN,3'-METHOXY
85	1.25	7.4	2459	1	C20H18O5	WARFARIN,4'-METHOXY
6	4.00	1.0	2471	459	C20H18O7	XANTHONE,6-CARBOXY-2-(2-BUTYRYLOXY)ETHOXY
7 ★	3.67		401		C20H19N1O3	ACRONYCINE *antineoplastic*
8	-0.19	7.5	638	1	C20H19N1O8	4-DEDIMETHYLAMINOTETRACYCLINE
·9	0.69	6.6	907	314	"	4-DEDIMETHYLAMINOTETRACYCLINE
14390	1.16	5.6	638	41	"	4-DEDIMETHYLAMINOTETRACYCLINE
1	1.05	7.4	2676	1	C20H19N3O3	CYCLO-TRP-TYR
2 ★	3.43	5.0	2674	302	C20H19N3O5	MEBENDAZOLE,N-PROPOXYCARBONYL
3	4.10		2171		C20H20Br1Cl1N4O6	AZOBENZENE,2-BROMO-6-CHLORO-4-NITRO-4'-N,N-DI(METHOXYCARBONYLETHYL)AMINO
4	2.62	7.4	2196	1	C20H20Cl1N1O3	DIMEFLINE: 4'-CHLORO ANALOG
95	-2.43	2.8	2211	1	C20H20Cl1N3O1	AMOPYROQUIN **550-81-2** *antimalarial*
6	2.57	7.4	1569	1	"	AMOPYROQUIN
7	-1.51	2.8	2211	1	C20H20Cl1N3O2	QUINOLINE-1-OXIDE-7-CHLORO-4-(4-HYDROXY-3-N-PYRROLIDINYLMETHYL)ANILINO
8 ★	4.90		1629		C20H20Cl2N2O4S1	CARBOFURAN,N-(N'ME,2,4-DICL-PH-CARBAMIDOTHIO)
9	1.38		910		C20H20Cl2N4	1,6-BIS(5-CHLORO-2-BENZIMIDAZOLYL)HEXANE
14400	-1.52	7.5	1476	1	C20H20Cl2N8O5	DICHLOROMETHOTREXATE **528-74-5** *antineoplastic*
1 ★	-0.10	2.2	401	1	"	DICHLOROMETHOTREXATE
2	1.97	7.4	2196	1	C20H20F1O3	DIMEFLINE: 4'-FLUORO ANALOG
3	1.26	7.4	675	1	C20H20N2O2	3,5-PYRAZOLIDINDIONE,1,2-DIPHENYL,4-CYCLOPENTYL
4	2.31	7.4	847	1	C20H20N2O2	N-(N-PYRROLIDINYLETHYL)-G-PYROPHTHALONE
5	1.25	7.4	693	1	C20H20N2O2	PRENAZONE **30748-29-9** *anti-inflammatory*
6	2.01	7.4	847	1	C20H20N2O3	N-(N-MORPHOLINOETHYL)-G-PYROPHTHALONE
7 ★	4.11		2214		C20H20N2O4	2(4(2-(6-METHYLQUINOXALINYLOXY)PHENOXYPROPIONIC ACID, ETHYL ESTER
8	1.73	7.4	2196	1	C20H20N2O5	DIMEFLINE: 4'-NITRO ANALOG ✚
9	0.42	7.4	2683	1	C20H20N4O1	TRYPTAMINE,5-(3-BENZYLOXADIAZOLYLMETHYL) *5ht-agonist*
14410	1.51		910		C20H20N6O4	1,6-BIS(5-NITRO-2-BENZIMIDAZOLYL)HEXANE
1	-3.45	7.4	2530	1	C20H20N6O9S1	MOXALACTAM **64952-97-2** *anti-infective*
2 ★	-0.58		1706		"	MOXALACTAM
3	3.65		1898		C20H20O6	(4-HYDROXYETHYL)VANILLIN,2-METHOXYSTYRYL ESTER
4	3.24		2125		C20H21Cl1F1O2	HALOPERIDOL,DESMETHYLENE
15 ★	4.10		1629		C20H21Cl1N2O4S1	CARBOFURAN,N-(N'ME,2-CL-PHENYLCARBAMIDOTHIO)
6 ★	4.00		1629		C20H21Cl1N2O4S1	CARBOFURAN,N-(N'ME,4-CL-PHENYLCARBAMIDOTHIO)
7	1.58	1.0	547	342	C20H21N1O1	9-(A-PIPERIDIN-4-HYDROXYMETHYL)PHENANTHRENE
8	0.70	7.4	2493	1	C20H21N1O2	2-PROPANONE,1,1-DIPHENYL-1-PROPYNYLOXY-3-DIMETHLAMINO
9	1.93	7.4	2196	1	C20H21N1O3	DIMEFLINE **1165-48-6** *stimulant (respiratory)*
14420 ★	3.65	13.0	2196		"	DIMEFLINE
1 ★	3.56		1898		C20H21N1O6	(4-HYDROXYETHYL)VANILLINOXIME,2-METHOXYSTYRYL ESTER
2 ★	3.00		2439		C20H21N2O3P1S1	FOSTEDIL,2-OXOPYRROLIDINO ANALOG ✚
3 ★	3.71	7.0	2292	1	C20H21N3O2	DIOXOLANE,2-P-BIPHENYL-4-ETHYL-2-(1,2,4-TRIAZOL-1-YL)METHYL
4 ★	-0.79	7.0	594	314	C20H21N3O3	PHENYLALANYL-TRYPTOPHANE
25 ★	-0.47	7.0	2374		C20H21N3O3	TRYPTOPHANYLPHENYLALANINE
6 ★	-1.13	7.0	2374		C20H21N3O4	TRYPTOPHANYLTYROSINE
7 ★	0.52	7.4	1689	1	C20H21N5O5	N7-(3-PYRIDYL)-MITOMYCIN-C **78142-97-9**
8 ★	2.85		2125		C20H22Cl1F1N2O2	R30072 ✚
9	3.55	7.4	1155	1	C20H22Cl1N1	1-CHLOROAMITRIPTYLINE

	logP	pH	Ref.	Note	MF	Name / CAS / activity
14430	5.55		1155		"	1-CHLOROAMITRIPTYLINE
1	-1.65	2.8	2211	1	C20H22Cl1N3O1	AMODIAQUINE **86-42-0** *antimalarial*
2	3.01	7.4	1569	1	"	AMODIAQUINE
3	-0.53	2.8	2211	1	C20H22Cl1N3O2	QUINOLINE-1-OXIDE-7-CHLORO-4-(4-HYDROXY-3-DIETHYLAMINOMETHYL)ANILINO
4 ★	5.70		2506		C20H22N2O1	FENAZAQUIN *acaricide*
35 ★	2.47		1481		C20H22N2O2	A-BENZOYLAMINO-N-S-BUTYL-CINNAMAMIDE
6	1.15	7.4	847	1	C20H22N2O2	B-ME-N-(3-(N-DIMEAMINO)PROPYL)-G-PYROPHTHALONE
7	2.31	7.4	847	1	C20H22N2O2	N-(DIETHYLAMINOETHYL)-G-PYROPHTHALONE
8	0.90	7.4	675	1	C20H22N2O3	3,5-PYRAZOLIDINDIONE,1(3-OH-PH)-2(4-TOLYL)-4-BU
9 ★	3.50		1629		C20H22N2O4S1	CARBOFURAN,N-(N'ME,PHENYLCARBAMIDOTHIO)
14440	-0.78	7.0	2682	1	C20H22N2O5	PHENYLPROPIONIC ACID,7-OXABICYCLOHEPTANEOXAZOLEAMIDE ANALOG *txa2-antagonist*
1	2.70		2428	459	C20H22N2O7S1	N,N'-DIMETHYLTHIODICARBAMATE,O-3,4-METHYLENEDIOXYPHENYL-O'-(2-I-PROPOXYPHENYL
2	1.96	7.4	579	1	C20H22N2S1	MEQUITAZINE **29216-28-2** *antihistaminic*
3	1.64		910		C20H22N4	1,4-BIS(5-METHYL-2-BENZIMIDAZOLYL)BUTANE
4	1.66		910		C20H22N4	1,6-BIS(2-BENZIMIDAZOLYL)HEXANE
45	1.42		910		C20H22N4O2	1,4-BIS(5-MEO-2-BENZIMIDAZOLYL)BUTANE
6	2.50	7.4	1485	1	C20H22N4O2	2,4-DIAMINOPYRIMIDINE,5(2-ME-4-MEO-5-PHCH2O-)BENZYL
7 ★	2.69	7.4	1485	2	"	2,4-DIAMINOPYRIMIDINE,5(2-ME-4-MEO-5-PHCH2O-)BENZYL
8	1.00	1.0	579	342	C20H22N4O2	2,4-DIAMONOPYRIMIDINE,5-(3-ETHOXY-4-BENZYLOXY)BENZYL
9	0.00		2397	101	C20H22N4O2	IMIDAZOLIUM CHLORIDE,1-METHYL-2-HYDROXYIMINOMETHYL-3-(3-M-BENZOYLPYRIDINIUM-1YL)PROPYL
14450	2.18	7.4	1485	1	C20H22N4O3	2,4-DIAMINOPYRIMIDINE,5(3,4-DIMEO-5-PHCH2O-)BENZYL
1 ★	2.37	7.4	1485	2	"	2,4-DIAMINOPYRIMIDINE,5(3,4-DIMEO-5-PHCH2O-)BENZYL
2	2.27	7.4	1485	1	C20H22N4O3	2,4-DIAMINOPYRIMIDINE,5(3,5-DIMEO-4-PHCH2O-)BENZYL
3 ★	2.46	7.4	1485	2	"	2,4-DIAMINOPYRIMIDINE,5(3,5-DIMEO-4-PHCH2O-)BENZYL
4 ★	0.89	7.2	1703	314	C20H22N4O10S1	CEFUROXIME AXETIL **64544-07-6** *antibacterial*
55 ★	-0.32	7.4	1689	1	C20H22N6O5	N7-(6-AMINO-PYRID-3-YL)-MITOMYCIN-C **88854-59-5**
6	0.52	7.4	2591	1	C20H22N6O7	AZT,5'-PHENYLANANYL **125780-80-5** *antiretroviral prodrug*
7	-0.03	7.4	2591	R1~314	C20H22N6O8	AZT,5'-TYROSINYL **125780-82-7** *antiretroviral prodrug*
8	-2.59	2.2	1200		C20H22N8O5	METHOTREXATE **59-05-2** *antineoplastic*
9	-2.52	7.5	1476	1	"	METHOTREXATE
14460 ✓	-1.85	2.2	401		"	METHOTREXATE
1 ★	1.05		636		C20H22O7	ELEPHANTIN
2	2.98		2652	459	C20H23Cl1F1N3O2	FLUOROCLEBOPRIDE
3	1.12		1531	820	C20H23N1	AMITRIPTYLINEHYDROCHLORIDE
4	1.87	7.4	2217	1	C20H23N1	AMITRIPTYLINE **50-48-6** *antidepressant*
65	2.30	7.0	468	1	"	AMITRIPTYLINE
6	2.50	7.4	1209	1	"	AMITRIPTYLINE
7	3.17	7.4	189	1	"	AMITRIPTYLINE
8	4.70	7.0	468	701	"	AMITRIPTYLINE
9	4.92	7.4	268	2	"	AMITRIPTYLINE
14470 ★	5.04		680	462	"	AMITRIPTYLINE
1	0.67	1.0	2099	342	C20H23N1	MAPROTILINE **10262-69-8** *antidepressant*
2	1.18	1.0	2099	342	"	MAPROTILINE
3	1.32	5.0	1440	1	"	MAPROTILINE
4	1.42	7.4	593	1	"	MAPROTILINE
75	1.44	7.4	2099	1	"	MAPROTILINE
6	2.85	9.0	2099		"	MAPROTILINE
7	2.41	7.4	785	1	C20H23N1	N-METHYLPROTRIPTYLINE
8	1.87	1.0	547	342	C20H23N1O1	9-(2-BUTYLAMINO-1-HYDROXYETHYL)PHENANTHRENE
9	0.94	1.0	547	342	C20H23N1O1	9-(2-DIETHYLAMINO-1-HYDROYET)PHENANTHRENE
14480 ★	3.99		1481		C20H23N1O1	A-BENZYL-N-SEC-BUTYLCINNAMAMIDE
1	3.09	7.4	2143	459	C20H23N1O2	2-(N-PYRROLIDINOMETHYL)BENZOIC ACID,2,6-DIMETHYLPHENYL ESTER
2	3.18	7.4	2143	459	C20H23N1O2	3-(N-PYRROLIDINOMETHYL)BENZOIC ACID,2,6-DIMETHYLPHENYL ESTER
3	3.20	7.4	2143	459	C20H23N1O2	4-(N-PYRROLIDINOMETHYL)BENZOIC ACID,2,4-DIMETHYLPHENYL ESTER
4	3.48	7.4	2143	1	C20H23N1O3	2-(MORPHOLINOMETHYL)BENZOIC ACID,2,6-DIMETHYLPHENYL ESTER
85	3.91	7.4	2143	1	C20H23N1O3	3-(N-MORPHOLINOMETHYL)BENZOIC ACID,2,6-DIMETHYLPHENYL ESTER
6	3.92	7.4	2143	1	C20H23N1O3	4-(N-MORPHOLINOMETHYL)BENZOIC ACID,2,6-DIMETHYLPHENYL ESTER
7 ★	3.40		2506		C20H23N1O3	BENALAXYL *fungicide* ✚
8	1.15	7.4	2274	1	C20H23N1O3	FLUORENONE,1-(3-T-BUTYLAMINO)-2-HYDROXYPROPOXY
9	0.66	7.4	2485	1	C20H23N1O4	MORPHINE,3-PROPIONYL
14490	0.52	7.4	2485	1	C20H23N1O4	MORPHINE,6-PROPIONYL
1	0.46	7.4	2378	1	C20H23N1O4	NALTREXONE **16590-41-3** *antagonist (narcotic)*
2	0.85	7.1	554		"	NALTREXONE
3	1.12	7.4	554	1	"	NALTREXONE
4	1.20	7.5	554	1	"	NALTREXONE
95	1.28	7.6	554	212	"	NALTREXONE
6	1.35	7.7	554	1	"	NALTREXONE
7 ★	1.10	7.4	564	1	C20H23N1O5	N-DESACETYLCOLCHICINE **49720-72-1** *antineoplastic*
8	2.70		1898		C20H23N1O8	(4-HYDROXYET)VANILLINOXIME,3,4,5-TRIMEOBENZYL ESTER
9	0.22	7.4	847	1	C20H23N2O2.C1H3O4S1	N-(3-TRIMETHYLAMMONIOPROPYL)-G-PYROPHTHALONE,METHANE SULFATE
14500 ★	4.16		2564		C20H23N3O2	BITERTANOL **55179-31-2**
1	4.18		2162	459	C20H23N3O2	OXIPEROMIDE **5322-53-2** *antipsychotic*
2 ★	0.54		2377		C20H23N3O2	AC-TYR-PHE-N
3 ★	0.92		2239	1	C20H23N3O4	EPANOLOL **86880-51-5** *anti-adrenergic (beta-receptor)*
4 ★	-1.33	7.0	2374		C20H23N3O4	GLYCYLPHENYLALINYLPHENYLALANINE
5 ★	-1.96	7.0	2374		C20H23N3O5	GLYCINYLPHENYLALANYLTYROSINE
6 ★	-1.86	7.0	2374		C20H23N3O5	TYROSINYLGLYCYLPHENYLALANINE
7 ★	1.40	7.4	1978	1	C20H23N3O6	1-ACETYL-7-(1-PYRROLIDYL)-N-METHYLMITOSENE **96000-47-4**
8 ✓	-0.50		2397	101	C20H23N4O3S1	IMIDAZOLIUM CHLORIDE ANALOG **132567-01-2** ✚
9 ★	1.13	7.0	1793	1	C20H23N5O3S1	N(4-(6-PURINYLTHIO)VALERYL)PHENYLGLYCINE,ET.ESTER
14510 ★	1.29	7.4	1689	1	C20H23N5O5S1	N7-(4,5-DIMETHYLTHIAZOL-2-YL)-MITOMYCIN-C **88854-60-8**
1	3.33		2162	459	C20H24Cl1N3O2	CLEBOPRIDE: ANILIDE ANALOG *antagonist-central da*
2	2.99	7.4	1599	1	C20H24Cl1N3O2	CLEBOPRIDE **55905-53-8** *anti-emetic, stimulant-peristaltic*
3	3.58		2162	459	"	CLEBOPRIDE
4	2.20	6.0	1897		C20H24Cl1N3S1	PROCHLORPERAZINE **58-38-8** *anti-emetic*
15	2.40	7.0	468	1	"	PROCHLORPERAZINE
6	3.55	7.0	468	701	"	PROCHLORPERAZINE
7	4.88		2162	459	"	PROCHLORPERAZINE
8	1.17	7.4	1599	1	C20H24F1N3O4S1	FLUBEPRIDE **56488-61-0** *neuroleptic*
9	2.46		2162	459	"	FLUBEPRIDE

	logP	pH	Ref.	Note	MF	Name / CAS / activity
14520	3.29	7.0	2644	1	C20H24F1O1S1	FOMOCAINE ANALOG-24 (ORTHO CH.4-F)
1 ★	3.69	7.0	2644	2	"	FOMOCAINE ANALOG-24 (ORTHO CH.4-F)
2	3.05	7.0	2644	1	C20H24F1O1S1	FOMOCAINE ANALOG-33 (P-FLUORO)
3 ★	3.51	7.0	2644	2	"	FOMOCAINE ANALOG-33 (P-FLUORO)
4 ★	2.70		1802		C20H24N2	DIMETINDEN 5636-83-9 antihistaminic
25	1.62	7.4	2143	1	C20H24N2O1	BENZANILIDE,2-(1-PYRROLIDINOMETHYL),2',6'-DIMETHYL
6	1.29	7.4	2143	1	C20H24N2O1	BENZANILIDE,3-(1-PYRROLIDINOMETHYL),2',6'-DIMETHYL
7 ★	3.10	13.0	592	224	C20H24N2O1	BENZANILIDE,3-(1-PYRROLIDINOMETHYL),2',6'-DIMETHYL
8	1.35	7.4	2143	1	C20H24N2O1	BENZANILIDE,4-(N-PYRROLILDINOMETHYL),2',6'-DIMETHYL
9 ★	3.14	13.0	2143		C20H24N2O1	BENZANILIDE,4-(N-PYRROLILDINOMETHYL),2',6'-DIMETHYL
14530	-0.36	7.4	2375	1	C20H24N2O1	INDECAINIDE 74517-78-5 cardiac depressant (anti-arrhythmic)
1	2.09	1.0	547	342	C20H24N2O1S1	MIRACILD 479-50-5 antischistosomal
2	2.52	7.4	592	1	C20H24N2O2	BENZANILIDE,2-(1-MORPHOLINOMETHYL),2',6'-DIMETHYL
3	2.08	7.4	2143	1	C20H24N2O2	BENZANILIDE,3-(1-MORPHOLINOMETHYL),2',6'-DIMETHYL
4 ★	2.14	13.0	592	224	C20H24N2O2	BENZANILIDE,3-(1-MORPHOLINOMETHYL),2',6'-DIMETHYL
35	1.76	7.4	2143	459	C20H24N2O2	BENZANILIDE,4-(N-MORPHOLINOMETHYL),2',6'-DIMETHYL
6 ★	2.00	13.0	2143		C20H24N2O2	BENZANILIDE,4-(N-MORPHOLINOMETHYL),2',6'-DIMETHYL
7	2.07	7.4	1209	1	C20H24N2O2	QUINIDINE 56-54-2
8	2.11	7.4	806	1	"	QUINIDINE
9	2.22	7.4	2648	1	"	QUINIDINE
14540 ★	2.88	7.4	2648	2	"	QUINIDINE
1	1.90	7.4	2648	1	C20H24N2O2	QUININE 130-95-0 antimalarial, anti-arrhythmic
2	2.14	7.0	919	1	"	QUININE
3 ★	2.64	7.4	2648	2	"	QUININE
4 ★	2.70		401		C20H24N2O2S1	HYCANTHONE 3105-97-3 antischistosomal
45	4.65	7.4	2090	459	C20H24N2O3	5(5(2-VINYL-4(4,5-H2-2-OXAZOLYL)PHENYOXY)PENTYL-3-METHYLISOXAZOLE 105663-64-7
6	3.28	7.4	2090	459	C20H24N2O4	5(5-(2-AC-4(4,5-H2-2-OXAZOLYL)PHENOXY)PENTYL-3-METHYLISOXAZOLE 105639-12-1
7	0.88	7.0	708	1	C20H24N2O6	6(3,4,5-TRIMEO-BENZAMIDO)2-ME-6H-I-QUINOLINE
8	1.58	7.2	2016		C20H24N2O6	NISOLDIPINE 63675-72-9 vasodilator (coronary)
9	-1.61	7.4	547	1	C20H24N6O2	4,6-DIAM-2,2-DIME-1(3'-3(ACETAMIDOPHO)MEPH)S-TRIAZIN
14550 ★	1.15	13.0	547	224	C20H24N6O2	4,6-DIAM-2,2-DIME-1(3'-3(ACETAMIDOPHO)MEPH)S-TRIAZIN
1 ★	1.34	4.6	2043	2	C20H24N8O6S3	CEFAZOLIN,PIVALOYLOXYMETHYL ESTER
2 ★	3.67		599		C20H24O2	ETHINYL ESTRADIOL 57-63-6 estrogen
3 ★	2.20		2308		C20H24O6	3,3-DIBENZO-18-CROWN-6-ETHER 14187-32-7
4	4.61		2021	459	C20H24O6	NAPHTHALENE,1,4-DI-BUTYRYLOXY-2,3-DIMETHOXY 91814-58-3
55	4.64		2021	459	C20H24O6	NAPHTHALENE,1,4-DIBUTYRYLOXY-2,3-DIMETHOXY 91814-57-2
6	1.69		636		C20H24O7	EUPATUNDIN 20071-53-8
7	1.48	7.2	2416	1	C20H25Cl1N2O5	AMLODIPINE 88150-42-9 antihypertensive (ca-antagonist)
8	1.30	7.2	2526	1	C20H25F1N4O3	OXONAPHTHYRIDINECARBOXYLIC ACID,7-(6-METHYL-2,5-DIAZABICYCLOHEPTYL) ANALOG
9	1.47	7.4	2235	1	C20H25N1	PRODIPINE 31314-38-2 antiparkinsonian
14560	3.37	7.4	2143	1	C20H25N1O1	2,6-DIMETHYLPHENYL 2-(N-PYRROLIDINOMETHYL)BENZYL ETHER
1	2.74	7.4	2143	1	C20H25N1O1	2,6-DIMETHYLPHENYL 3-(N-PYRROLIDINOMETHYL)BENZYL ETHER
2	2.70	7.4	2143	1	C20H25N1O1	2,6-DIMETHYLPHENYL 4-(N-PYRROLIDINOMETHYL)BENZYL ETHER
3	1.03	1.0	870	342	C20H25N1O1	2-(DIPHENYLMETHOXY)ETHYL-N-CYCLOPENTYLAMINE
4	-0.06	4.0	2644	210	C20H25N1O1	FOMOCAINE ANALOG-20 (AZEPINE)
65	2.87	8.0	2644	1	"	FOMOCAINE ANALOG-20 (AZEPINE)
6	0.77	1.0	602	342	C20H25N1O1	N-PYRROLIDYLETHYL-(P-ME-PH)PHENYLMETHYL-ETHER
7	0.48	1.0	602	342	C20H25N1O1	PERASTINE 4960-10-5
8	4.16	7.2	2143	459	C20H25N1O2	2,6-DIMETHYLPHENYL 2-(N-MORPHOLINOMETHYL)BENZYL ETHER
9	3.74	7.4	2143	1	C20H25N1O2	2,6-DIMETHYLPHENYL 3-(N-MORPHOLINOMETHYL)BENZYL ETHER
14570	3.76	7.4	2143	1	C20H25N1O2	2,6-DIMETHYLPHENYL 4-(N-MORPHOLINOMETHYL)BENZYL ETHER
1	2.98	7.4	2143	1	C20H25N1O2	2-(DIETHYLAMINOMETHYL)BENZOIC ACID,2,6-DIMETHYLPHENYL ESTER
2	1.73	7.4	2493	1	C20H25N1O2	2-PROPANONE,1,1-DIPHENYL-1-PROPOXY-3-DIMETHYLAMINO
3	2.99	7.4	2143	459	C20H25N1O2	3-(DIETHYLAMINOMETHYL)BENZOIC ACID,2,6-DIMETHYLPHENYL ESTER
4	3.29	7.4	2143	1	C20H25N1O2	4-(DIETHYLAMINOMETHYL)BENZOIC ACID,2,6-DIMETHYLPHENYL ESTER
75	2.05	5.7	2241	1	C20H25N1O2	FOMOCAINE 17692-39-6 anesthetic (local)
6	0.50	7.4	2680	324	C20H25N1O3	3-O-PROPYLMORPHINE 74886-10-5
7 ★	2.83	6.0	2644	2	C20H25N1O3	FOMOCAINE ANALOG-27 (P-METHOXYBENZYL)
8	1.16	4.0	2655	1	C20H25N1O3	PROPRANOLOL,CYCLOPROPANOYL ESTER
9 ★	3.32	7.4	2268	1	C20H25N1O4	NAPROXEN,N,N-DIETHYLGLYCOLAMIDE ESTER
14580	0.59	4.0	2655	1	C20H25N1O5	PROPRANOLOL,METHYLMALONYL ESTER
1 ★	1.70	7.4	2268	1	C20H25N1O6	NAPROXEN,(N,N-DI-HYDROXYETHYL)GLYCOLAMIDE ESTER
2	2.03		634		C20H25N1O7	6-ETACETOYLOXY-N-I-BUO-4-OH-3-QUINOLINE-CO2ET
3	3.14	7.0	2644	1	C20H25N1S1	FOMOCAINE ANALOG-31 (MORPH-H)
4 ★	3.62	7.0	2644	2	"	FOMOCAINE ANALOG-31 (MORPH-H)
85	3.32	7.0	2644	1	C20H25N1S1	FOMOCAINE ANALOG-32 (ORTHO CH.)
6 ★	3.80	7.0	2644	2	"	FOMOCAINE ANALOG-32 (ORTHO CH.)
7	1.66	4.0	2644	210	C20H25N1S2	FOMOCAINE ANALOG-38 (THIOMORPHOLINE)
8	2.09		538	1	C20H25N3O1	LYSERGIC ACID,DIETHYLAMIDE 50-37-3 halllucinogenic (5-ht2-agonist)
9 ★	2.95		538		"	LYSERGIC ACID,DIETHYLAMIDE
14590	1.78	7.4	2095	1	C20H25N3O3S1	(5-(2-THIENYL)IMIDAZOLE) ANDRENOCEPTOR ANTAGONIST
1	0.98	5.0	2375	1	C20H25N3S1	PERAZINE 84-97-9
2	1.13	4.3	882	1	"	PERAZINE
3	2.90	7.4	678	1	"	PERAZINE
4 ✓	-0.25		2397	101	C20H25N4O3S1	IMIDAZOLIUM CHLORIDE,NAPHTHYL ANALOG
95	1.73	7.4	594	1	C20H25N5O2	ACRIDINE,1-NITRO-4-N,N-DIMETHYLAMINO-9-(N,N-DMETHYLPROPANEDIAMINO)
6	0.78	7.4	2037	1	C20H26Cl1N7O1	ISOGUANINE,1-METHYL-7-(3-(1-(4-(P-CHLORO)BENZYL)PIPERIZINYL)PROPYL
7	1.16	7.4	2037	1	C20H26Cl1N7O1	ISOGUANINE,1-METHYL-9-(1-(3-(P-CHLORO)BENZYL)PIPERAZINYL)PROPYL
8 ★	2.31	6.8	1703	360	C20H26F2N6O7S3	CEPHALOSPORIN ANALOG +
9	-1.57		848		C20H26I3N1O13	2,4,6-I3-5-N-MECARBOXAMIDORESORCYL-1,3-DIGLUCOSIDE
14600	0.15		817		C20H26N1O3.Br1	N(2-BENZILOYLOXYET-N-ET-N,N-DIMETHYLAMMONIUMBR
1	1.58	5.0	1440	1	C20H26N2	DIMETACRIN 4757-55-5 antidepressant
2	1.33	5.0	1440	1	C20H26N2	TRIMIPRAMINE 739-71-9 antidepressant
3	1.82	7.4	2143	1	C20H26N2O1	BENZANILIDE,2-DIETHYLAMINOMETHYL,2',6'-DIMETHYL
4	1.30	7.4	2143	1	C20H26N2O1	BENZANILIDE,3-DIETHYLAMINOETHYL,2',6'-DIMETHYL
5 ★	3.09	13.0	2143		C20H26N2O1	BENZANILIDE,3-DIETHYLAMINOETHYL,2',6'-DIMETHYL
6	1.63	7.4	2143	1	C20H26N2O1	BENZANILIDE,4-DIETHYLAMINOMETHYL,2',6'-DIMETHYL
7 ★	3.49		592		C20H26N2O1	BENZANILIDE,4-DIETHYLAMINOMETHYL,2',6'-DIMETHYL
8	4.71	7.4	2090	459	C20H26N2O3	5(5-(2-ET-4(4,5-H2-2-OXAZOLYL)PHENOXY)PENTYL-3-ME-ISOXAZOLE 105639-07-4
9	4.79	7.4	2223	1	C20H26N2O3	DISOXARIL 87495-31-6 antiviral

		logP	pH	Ref.	Note	MF	Name / **CAS** / *activity*
14610		0.64	7.4	1941	1	C20H26N2O5	BIS-(3-PHENOXYPROPAN-2-OL)AMINE,P'-ACETYLAMINO
1	★	1.39		2164		C20H26N2O5	BIS-(3-PHENOXYPROPAN-2-OL)AMINE,P'-ACETYLAMINO
2		1.56	7.4	1702	1	C20H26N2O5	SALICYLAMIDE,5(OC2H4NHCHOHC2H4-(3-MEOPHENYL)) **89789-96-8**
3		-4.30	7.4	2449		C20H26N2O5S2	SPIRAPRILAT **83602-05-5** *antihypertensive (ace inhibitor)*
4	✓	0.27		2449	226	"	SPIRAPRILAT
15		0.50	7.4	722	1	C20H26N4O1	4-MEPIPERAZINO-DES(N-DIIPR)DISOPYAMIDE
6		0.89		1045		C20H26N6O5	6(2(3,4-DIMEOPH)5(HOME)1,3-DIOXAN-5YL)AM2,4DIAZ.TRIAZ
7		1.11		1045		"	6(2(3,4-DIMEOPH)5(HOME)1,3-DIOXAN-5YL)AM2,4DIAZ.TRIAZ
8		2.97		2286		C20H26O2	NORETHINDRONE **68-22-4** *progestin*
9		2.90	7.4	2378	1	C20H27F3N2O1	1,2-DIAMINOCYCLOHEXANE,N-METHYL-N-(CYCLOPROPYL)METHYL-N'-METHYL-(4-TRIFLUOROMETHYLBENZOYL)
14620		3.59	7.4	2143	1	C20H27N1O1	2,6-DIMETHYLPHENYL 2-(DIETHYLAMINOMETHYL)BENZYL ETHER
1		2.90	7.4	2143	1	C20H27N1O1	2,6-DIMETHYLPHENYL 3-(DIETHYLAMINOMETHYL)BENZYL ETHER
2		3.11	7.4	2143	1	C20H27N1O1	2,6-DIMETHYLPHENYL 4-(DIETHYLAMINOMETHYL)BENZYL ETHER
3		0.67	1.0	870	342	C20H27N1O1	2-(DIPHENYLMETHOXY)-N-ET,N-PROPYLAMINE
4		0.78	1.0	602	342	C20H27N1O1	P-ME-2-DIPHENYLMETHOXY-N,N-DIETHYLETHYLAMINE
25		0.97	1.0	602	342	C20H27N1O1	P-ME-2-DIPHENYLMETHOXY-N-ME-N-PROPYL-ETHYLAMINE
6	★	3.31	8.0	2644	2	C20H27N1O2S1	FOMOCAINE ANALOG-39
7		1.54	4.0	2655	1	C20H27N1O3	PROPRANOLOL,BUTYRYL ESTER
8		1.48	4.0	2655	1	C20H27N1O3	PROPRANOLOL,I-BUTYRYL ESTER
9	★	3.00	7.4	1700	2	C20H27N1O4	BEVANTOLOL **59170-23-9** *anti-anginal, antihypertensive, anti-arrhythmic*
14630		2.85	7.4	1941	1	C20H27N1O4	BIS-(3-O-TOLYLOXYPROPAN-2-OL)AMINE
1	★	3.67		2164		"	BIS-(3-O-TOLYLOXYPROPAN-2-OL)AMINE
2		2.18		634		C20H27N1O5	6,7-DI-I-BUTOXY-4-HYDROXY-3-QUINOLINE-CO2ET **5486-03-3** *coccidiostat (poultry)*
3		2.08	7.4	2235	1	C20H27N1Si1	SILA-BUDIPINE
4		-0.26		2627	841	C20H27N2.Cl1	N-METHYL-4-(4'-DIPROPYLAMINOSTYRYL)-PYRIDINIUM CHLORIDE
35		1.28		1901	820	C20H27N2.I1	N-METHYLIMIPRAMINIUM IODIDE
6		2.49	7.2	2195	1	C20H27N3O1	2-PYRIDINEAMINE,N-3-(3-N-PIPERIDINYLMETHYL)PHENOXYPROPYL *antimigrane*
7		4.29	7.2	2195	2	"	2-PYRIDINEAMINE,N-3-(3-N-PIPERIDINYLMETHYL)PHENOXYPROPYL
8		2.76	7.0	1955	1	C20H27N5O2	CILOSTAZOL **73963-72-1** *antithrombotic*
9		5.73		1101	421	C20H27O4P1	OCTICIZER **1241-94-7** *pharmaceutic aid (plasticizer)*
14640	★	-2.28		1682		C20H28I3N3O8	M-BZAMID,N,N(CH(CH2OH)2)2ME2,5-N-OHET-ACETAM,I3 **31122-82-4**
1		0.95		1901	820	C20H28N1.I1	EMEPRONIUM IODIDE
2		1.07	7.4	2224	1	C20H28N2	SC-26000 (NORPACE ANALOG) ✚
3		-2.00	7.4	2449	1	C20H28N2O5	ENALAPRIL **75847-73-3** *antihypertensive*
4		-0.52	2.4	2017		"	ENALAPRIL
45		-0.41	5.8	2017		"	ENALAPRIL
6	✓	-0.07	4.9	2017		"	ENALAPRIL
7		0.84		1713		C20H28N4O2	6-PH-2-(8-UREIDO-OCTYL)-3(2H)-PYRIDAZINONE
8		2.55	7.4	2244	1	C20H28N4O4S1	TIMOLOL,O-BENZOYL
9	★	-0.80	7.4	2427	1	C20H28N6O6	P-GLU-HIS(N-ISOPROPOXYCARBONYL)-PRO-NH2
14650		-0.92	7.4	2591		C20H28N6O9	AZT,5'-T-BOC-GLUTAMYL **125780-86-1** *antiretroviral prodrug*
1	★	-0.23	7.0	1793	1	C20H28N8O6S1	N(4-(6-PURINYLTHIO)VALERYL)GLYGLYGLYGLY,ET.ESTER
2		8.17		2404	459	C20H28O2	(7,8-DIHYDROPYRANYL)BENZOPYRAN,2,2,2',2'-TETRAMETHYL-6-T-BUTYL
3		2.80	7.0	2230	1	C20H28O2	ISOTRETINOIN **4759-48-2** *keratolytic*
4		6.00	7.0	2230	2	"	ISOTRETINOIN
55	★	6.26		2415	459	C20H28O2	O-178 ✚
6		4.47	7.4	579	1	C20H28O2	TRETINOIN **302-79-4**
7	★	6.30	1.0	579	400	"	TRETINOIN
8		2.38		401	459	C20H29F1O3	FLUOXYMESTERONE **76-43-7** *androgen*
9		-1.05		307	22	C20H29N1O5	3-HYDROXY-N-BUTYLMORPHINAN TARTRATE
14660		3.75	7.4	1990	1	C20H29N1O1	N,N-DIMETHYLOCTYLAMINE,8-(1-NAPHTHOXY)
1		2.11	7.4	2480	1	C20H29N1O3	KETOBEMIDONE,PIVALIC ESTER
2	★	2.59	7.4	2480	2	"	KETOBEMIDONE,PIVALIC ESTER
3		2.24	7.4	2480	1	C20H29N1O4	KETOBEMIDONE,BUTYL CARBONATE
4	★	2.42	7.4	2480	2	"	KETOBEMIDONE,BUTYL CARBONATE
65		2.20	7.4	2480	1	C20H29N1O4	KETOBEMIDONE,I-BUTYL CARBONATE
6	★	2.38	7.4	2480	2	"	KETOBEMIDONE,I-BUTYL CARBONATE
7		0.89	7.0	708	1	C20H29N1O5	6(3,4,5-TRIMEO-BENZOYLOXY)2-ME-10H-I-QUINOLINE-C
8		0.84	7.0	708	1	C20H29N1O5	6(3,4,5-TRIMEO-BENZOYLOXY)2-ME-10H-I-QUINOLINE-T
9		0.47		450		C20H29N1O5	C-5-(3,4,5-TRIMEO-BENZOYLOXY)-2-MEDECAHYDRO-I-QUINOL.
14670		0.58		450		C20H29N1O5	T-5-(3,4,5-TRIMEO-BENZOYLOXY)2-MEDECAHYDRO-I-QUINOL.
1		1.53	7.4	2217	1	C20H29N3O2	DIBUCAINE **85-79-0** *anesthetic (local)*
2		2.91	7.4	1209	1	"	DIBUCAINE
3	★	4.40		505		"	DIBUCAINE
4		2.18	7.0	532	1	C20H29N3O2.C2H4O2	DIBUCAINEACETATE
75		1.97	4.0	532		C20H29N3O2.C8H6O4	DIBUCAINEPHTHALATE
6		2.80		2162	459	C20H29N3O4S1	TINISULPRIDE **69387-87-7** *dopamine antagonist*
7		-1.13	1.0	1012	342	C20H29N5O3	URAPIDIL **34661-75-1** *antihypertensive*
8		1.60	10.1	1012	1	"	URAPIDIL
9		2.51	7.4	2244	1	C20H29N5O4S1	TIMOLOL,O-(2-AMINOBENZOYL)
14680		-0.43	7.4	2320	1	C20H30N2O3	1,6-BIS(1-PYRROLIDINYLMETHYL)PHENOL,4-ISOPROPOXYCARBONYL
1		-0.74	7.4	2320	1	C20H30N2O3	2,6-BIS(1-PYRROLIDINYLMETHYL)PHENOL-4-ACETIC ACID,ETHYL ESTER
2		2.91	7.3	873	314	C20H30N2O3	O-3(N-PYRROLIDINO)CYHEX-N(M-OC3H7-PH)CARBAMATE
3		2.56	7.3	873	314	C20H30N2O3	O-3(N-PYRROLIDINO)CYHEX-N(P-OC3H7-PH)CARBAMATE
4		-0.01	7.0	708	1	C20H30N2O4	6(3,4,5-TRIMETHOXYBENZAMIDO)2-METHYLDECAHYDRO-I-QUINOLINE,CIS-TRANS
85		0.06	7.0	708	1	C20H30N2O4	6(3,4,5-TRIMETHOXYBENZAMIDO)2-METHYLDECAHYDRO-I-QUINOLINE,CIS
6		0.11	7.0	708	1	C20H30N2O4	6(3,4,5-TRIMETHOXYBENZAMIDO)2-METHYLDECAHYDRO-I-QUINOLINE,TRANS
7		2.09	7.5	2651	28	C20H30N2O4	BENZAMIDE,2,3-DIMETHOXY-5-(3-HYDROXYPROPYL)-N-(1-ALLYLPYRROLIDIN-2YL)METHYL
8		-0.64		450		C20H30N2O4	C-5-(3,4,5-TRIMEO-BENZAMIDO)-2-ME-DECAHYDRO-I-QUINOL.
9		4.66	7.5	2651	28	C20H30N2O4	SALICYLAMIDE,3-ALLYL-5-METHOXY-6-ETHOXY-N-(1-ETHYLPYRROLIDIN-2YL)METHYL
14690		-0.30		450		C20H30N2O4	T-5-(3,4,5-TRIMEO-BENZAMIDO)-2-ME-DECAHYDRO-I-QUINOL.
1		2.54	1.0	579	342	C20H30N4O2	2,4-DIMANOPYRIMIDINE,5-(3-OCTYLOXY-4-METHOXY)BENZYL
2	★	0.52		2377		C20H30N4O5	AC-ILE-PHE-ALA-N
3	★	-0.04		2377		C20H30N4O5	AC-ALA-TYR-LEU-N
4	★	0.23		2377		C20H30N4O5	AC-LEU-SER-PHE-N
95	★	-1.78		2377		C20H30N4O5	ILE-ALA-GLY-PHE
6	★	-1.91		2377		C20H30N4O5	VAL-ALA-ALA-PHE
7	★	5.68		579	401	C20H30O1	RETINOL **68-26-8** *vitamin (antixerophthalmic)*
8		3.19		1607	459	C20H30O2	MIBOLERONE **3704-09-4** *anabolic, androgen*
9	★	3.36		542		C20H30O2	TESTOSTERONE,17-A-METHYL **58-18-4** *androgen*

	logP	pH	Ref.	Note	MF	Name / **CAS** / *activity*
14700	5.80		2015	459	C20H30O4	DIHEXYLPHTHALATE **84-75-3**
1	1.82	7.5	2651	1	C20H31F1N2O4	SALICYLAMIDE,3-(3-FLUOROPROPYL)-5-METHOXY-6-ETHOXY-N-(1-ETHYLPYRROLIDIN-2YL)METHYL
2	2.62	7.5	2651	1	"	SALICYLAMIDE,3-(3-FLUOROPROPYL)-5-METHOXY-6-ETHOXY-N-(1-ETHYLPYRROLIDIN-2YL)METHYL
3	4.27	7.5	2651	28	"	SALICYLAMIDE,3-(3-FLUOROPROPYL)-5-METHOXY-6-ETHOXY-N-(1-ETHYLPYRROLIDIN-2YL)METHYL
4	6.35		1361	459	C20H31N1	3,3,5-(EQ.)-TRIMETHYL-PHENCYCLIDINE(PHEN=AX.)
5	3.17	7.4	2407	1	C20H31N1O1	TRIHEXYPHENIDYL **144-11-6** *anticholinergic, antiparkinsonian*
6 ★	4.49		2407		"	TRIHEXYPHENIDYL
7	2.78	7.4	682	1	C20H31N1O2	N-HEXYLNORMEPERIDINE
8 ★	3.63		2045		C20H31N1O3	MOXISYLYTE ANALOG: 2-CYCLOHEXAN-2-0NYL ESTER ✚
9 ✓	5.00		2413	459	C20H31N1O4	TETRAHYDROPYRAN-2,4-DIONE,3[1-(ETHOXYIMINO)BUTYL]-SPIRO-1-PERHYDRONAPHTHYL
14710 ★	-1.03	7.0	2374		C20H31N3O4S1	METHIONYLLEUCINYLPHENYLALANINE
1 ★	-1.77	7.0	2374		C20H31N3O5	ISOLEUCINYLTYROSINYLVALINE
2 ★	-1.45	7.0	2374		C20H31N3O5	TYROSINYLLEUCINYLVALINE
3	0.70	7.4	2095	1	C20H31N5O3	(5-METHYLIMIDAZOLE)ADRENOCEPTOR ANTAGONIST ✚
4	-1.64	7.4	2450	1	C20H32N1O3.Br1	MDL-74270-BROMIDE ✚
15	-0.18	7.4	2450	1	C20H32N1O3.C7H7O3S1	MDL-74270 ✚
6	3.86	7.3	1900	314	C20H32N2O3	1-PIPERIDINEETHANOL,A-METHYL,3-OPENT-PHENYLCARBAMATE
7	3.73	7.3	1356	314	C20H32N2O3	2-HEXOCARBANILATE-O-2(N-PIPERIDINYL)ETHYL **55792-18-2**
8	5.29	7.3	1356	2	"	2-HEXOCARBANILATE-O-2(N-PIPERIDINYL)ETHYL
9	2.65	7.1	1150		C20H32N2O3	2-ME-4(O-BU)CARBANILATE-O-(1-ME-2(PIPERIDINYL)ET
14720	2.92	7.3	1960	314	C20H32N2O3	2-PENTOXYMETHYLCARBANILIC ACID,2-(1-PIPERIDINYL)ETHYL ESTER
1 ★	4.42	7.3	1960	2	"	2-PENTOXYMETHYLCARBANILIC ACID,2-(1-PIPERIDINYL)ETHYL ESTER
2	4.56	7.2	1150		C20H32N2O3	2ME-4(O-PEN)CARBANILATE-O-2(N-PIPERIDINYL)ETHYL
3	3.28	7.3	1149	714	C20H32N2O3	3(OPENT)CARBANILATE,O-(2-N,N-DIMEAMINO)CYCLOHEXYL **60093-64-3**
4	3.95	7.3	1356	314	C20H32N2O3	3-HEXOCARBANILATE-O-2(N-PIPERIDINYL)ETHYL **55792-19-3**
25	5.51	7.3	1356	2	"	3-HEXOCARBANILATE-O-2(N-PIPERIDINYL)ETHYL
6	2.70	7.3	1960	314	C20H32N2O3	3-PENTOXYMETHYLCARBANILIC ACID,2-(1-PIPERIDINYL)ETHYL ESTER
7 ★	4.20	7.3	1960	2	"	3-PENTOXYMETHYLCARBANILIC ACID,2-(1-PIPERIDINYL)ETHYL ESTER
8	3.85	7.3	1356	314	C20H32N2O3	4-HEXOCARBANILATE-O-2(N-PIPERIDINYL)ETHYL **55792-20-6**
9	5.41	7.3	1356	2	"	4-HEXOCARBANILATE-O-2(N-PIPERIDINYL)ETHYL
14730	2.73	7.0	1517	1	C20H32N2O3	CARBAMICAC,N-ET-N-(M-BUOPH),2(N-PIPERIDINYL)ET.EST.
1	2.73	7.1	1357		"	CARBAMICAC,N-ET-N-(M-BUOPH),2(N-PIPERIDINYL)ET.EST.
2	2.18	7.0	1517	1	C20H32N2O3	CARBAMICAC,N-ET-N-(M-PROPH),2(1-PIPERIDINYL)PR.EST
3	2.18	7.1	1357		"	CARBAMICAC,N-ET-N-(M-PROPH),2(1-PIPERIDINYL)PR.EST
4	2.20		2506		C20H32N2O3S1	CARBOSULFAN **55285-14-8** *insecticide (cholinesterase inhb.)*
35	3.70		1946		C20H32N2O4	2-PROPENOIC N-(1-METHYLDECYL)AMIDE,3-(5-NITRO-2-FURANYL)
6	3.90	7.2	1150		C20H32N2O4	2ME-4(O-HEX)CARBANILATE-O-2(N-MORPHOLINO)ETHYL
7	2.15	7.5	2651	28	C20H32N2O4	BENZAMIDE,2,3-DIMETHOXY-5-(3-HYDROXYPROPYL)-N-(PROPYLPYRROLIDIN-2YL)METHYL
8	2.02	7.5	2651	28	C20H32N2O4	BENZAMIDE,2-ETHOXY-3-METHOXY-5-(3-HYDROXYPROPYL)-N-(1-ETHYLPYRROLIDIN-2YL)METHYL
9	4.93	7.5	2651	28	C20H32N2O4	SALICYLAMIDE,3-PROPYL-5-METHOXY-6-ETHOXY-N-(1-ETHYLPYRROLIDIN-2YL)METHYL
14740	3.00	7.5	2651	28	C20H32N2O5	SALICYLAMIDE,3-(3-HYDROXYPROPYL)-5-METHOXY-6-ETHOXY-N-(1-ETHYPYRROLIDIN-2YL)METHYL
1 ★	5.97		2415	459	C20H32O2	CYCLOHEXANOL,3-(2-HYDROXY-4-(1,1-DIMETHYL)HEXYL)PHENYL
2	3.99		1607	459	C20H32O2	DROMOSTANOLONE *antineoplastic*
3 ★	2.82		2184		C20H32O5	PROSTAGLANDINE-2 **363-24-6** *oxytocic, abortifacient*
4 ★	1.71		2308		C20H32O6	BENZO-18-CROWN-6-ETHER,M-T-BUTYL **14098-26-1**
45 ★	2.73		511		C20H32O6	GLUCOPYRANOSIDE,3,5-DI(T-BUTYL)PHENYL(BETA)
6 ★	0.45		2308		C20H32O8	BENZO-24-CROWN-8-ETHER
7 ★	-0.57		1236		C20H32O8	DESACYLASEBOTOXINVII ✚
8	1.06	7.4	365	2	C20H33N3O1	N,N-DIBU-PIPERAZINE-2-CARBOXANILIDE,2'-ME
9	0.89		579		C20H33N5	4,6-DIAM-1,2H-2,2-DIME-SYM-TRIAZINE,1-M-NONYLPHENYL
14750	1.94	6.0	581		"	4,6-DIAM-1,2H-2,2-DIME-SYM-TRIAZINE,1-M-NONYLPHENYL
1	2.00	6.0	581		C20H33N5O1	4,6-DIAM-1,2H-2,2-DIME-SYM-TRIAZINE,1-M-NONYLOXYPH
2	4.85	7.2	1150		C20H34N2O3	2-ME-4(O-HEXL)CARBANILATE-O-2(N,N-DIETAMINO)ETHYL
3	2.80	7.1	1150		C20H34N2O3	2ME-4(O-PEN)CARBANILATE-O-2(N,N-DIETAMINO)1-ME-ET
4	3.34	7.0	1517	1	C20H34N2O3	N-ME-N(M-HEXOXYPH)CARBAMATE,2-(DIETAMINO)ETHYL **59731-49-6**
55	3.34	7.1	1357		"	N-ME-N(M-HEXOXYPH)CARBAMATE,2-(DIETAMINO)ETHYL
6 ★	6.58	6.3	2589	459	C20H34N4O1	1-(4-CARBAMOYLPHENYL)-3-METHYL-3-DODECYLTRIAZENE **89530-02-9** *antineoplastic*
7	3.30	7.4	2244	1	C20H34N4O4S1	TIMOLOL,O-CYCLOHEXANOYL
8	4.10		579	20	C20H34N4S2	2,2'-DITHIOBIS(4-T-BUTYL-1-I-PROPYL)IMIDAZOLE
9	7.60		2011		C20H34O4S2	PROPANEDIOIC ACID,1,3-DITHIOLAN-2-YLIDINE,DIHEPTYL ESTER **78432-88-9**
14760 ★	1.51		1236		C20H34O5	A-DIHYDROGRAYANOTOXINII ✚
1	4.07		1396	459	C20H34O5	ARTEMISININE,I-AMYLETHER(BETA)
2	2.00	6.0	929	314	C20H34O5	PROSTAGLANDIN E-1 **745-65-3** *vasodilator*
3 ★	2.72		2184		C20H34O5	PROSTAGLANDIN-F2-ALPHA **551-11-1** *oxytocic, abortifacient*
4 ★	0.27		1236		C20H34O6	GRAYANOTOXINIII ✚
65 ★	3.51		1396	459	C20H34O7	ARTEMISININE,MEO-ETO-ETHYLETHER(BETA)
6 ★	0.38		2308		C20H34O8	1,2-BIS(METHOXYETHOXYETHOXYETHOXY)BENZENE
7	2.03	1.0	547	342	C20H35N1O1	2-N,N-DIHEXYLAMINO-1-PHENYLETHANOL HYDROCHLORIDE
8 ✓	1.14		1729	820	C20H35N2O1.I1	3-CARBAMOYLPYRIDINIUM IODIDE,N-TETRADECYL **35129-56-7**
9	0.44		1635	820	C20H36N1.Cl1	UNDECYLDIMETHYLBENZYLAMMONIUMCHLORIDE
14770 ✓	2.27		1729	820	C20H36N1.I1	2-METHYLPYRIDINIUM IODIDE,N-TETRADECYL **74639-30-8**
1	3.18	12.3	343		C20H37Cl1N2O5S1	N-ETHYL-1-ETHYLTHIO-CLINDAMYCI
2 ★	3.21	12.3	343		C20H37Cl1N2O5S1	N-ETHYL-4'-BUTYL-CLINDAMYCIN
3	2.44	8.0	1678	1	C20H38N1O4	3-OXYOXAZOLIDINE,2-HEXYL,2-NONANOIC ACID **74156-68-6**
4 ★	0.93		2469		C20H38N2O6S1	LINCOMYCIN,4'-PENTYL ANALOG
75 ★	1.76	12.3	343		C20H38N2O6S1	N-ETHYL-4'-BUTYL-LINCOMYCIN
6 ★	1.91	12.3	343		C20H38N2O6S1	N1-ET,S-ET-LINCOMYCIN
7 ★	-1.74		2377		C20H38N4O5	ALA-VAL-LEU-LEU
8	-0.26		568	466	C20H44N1.Br1	TETRAPENTYLAMMONIUMBROMIDE **866-97-7**
9 ✓	1.26	7.3	2085		C20H44N4.O2Tc1	N-DECYLCYCLAM,TECHNICIUM OXIDE COMPLEX
14780	8.30		1740	459	C20H60O8Si9	EICOSOMETHYLNONASILOXANE
1	2.25	7.0	1405	1	C21H15Cl1N2O4S1	1-(P-CL-BENZENESULFONYL)5,5-DIPHENYLHYDANTOIN
2	3.48		1405		"	1-(P-CL-BENZENESULFONYL)5,5-DIPHENYLHYDANTOIN
3	2.99	7.4	712	1	C21H15N1O2	N-BENZYL-G-PYROPHTHALONE
4	3.12		579		C21H15N1O3	BORONAL **81-48-1**
85	0.62	7.0	353	1	C21H15N1O5	O,O'-DIBENZOYLRESORCYLAMIDE
6	1.57	7.0	1405	1	C21H15N3O6S1	1-(P-NO2-BENZENESULFONYL)5,5-DIPHENYLHYDANTOIN
7	4.04		1405		"	1-(P-NO2-BENZENESULFONYL)5,5-DIPHENYLHYDANTOIN
8 ★	6.42		1486		C21H16	3-METHYLCHOLANTHRENE **56-49-5**
9	-0.69	7.4	594	1	C21H16Br2O5S1	BROMCRESOL PURPLE **115-40-2**

	logP	pH	Ref.	Note	MF	Name / **CAS** / *activity*
14790	0.61	7.4	675	1	C21H16N2O2	3,5-PYRAZOLIDINDIONE,1,2,4-TRIPHENYL
1	1.58		2470	1	C21H16N2O3S1	TETRAHYDROQUINAZOLINE-4-OXO,2-METHIO-3-(2-NAPHTHYL)-5-METHOXYCARBONYL
2	1.93	7.0	1405	1	C21H16N2O4S1	1-BENZENESULFONYL-5,5-DIPHENYLHYDANTOIN
3	4.04		1405		"	1-BENZENESULFONYL-5,5-DIPHENYLHYDANTOIN
4	1.62	7.0	1405	1	C21H16N2O5S1	1-(P-OH-BENZENESULFONYL)5,5-DIPHENYLHYDANTOIN
95	3.58		1405		"	1-(P-OH-BENZENESULFONYL)5,5-DIPHENYLHYDANTOIN
6	2.02	7.4	2557	1	C21H16O6	PROPYLIDENEBIS-4-HYDROXYCOUMARIN
7	1.26	7.4	2438	1	C21H16O7	COUMETAROL **4366-18-1** *anticoagulant*
8	1.38	7.4	2557	1	"	COUMETAROL
9	1.67		2524	459	C21H17F1N4O1	1,2,4-TRIAZOLYLETHANOL,4'-FLUORO-4"-(4-PYRIDYL) ANALOG
14800 ★	3.31	5.5	2403	2	C21H17F1O3	PROPIONIC ACID,2-(4-BIPHENYLYL)-3-HYDROXY-3-(M-FLUOROPHENYL) **93371-51-8**
1 ★	3.77	5.5	2403	2	C21H17F1O3	PROPIONIC ACID,2-(4-BIPHENYLYL)-3-HYDROXY-3-(O-FLUOROPHENYL) **93371-50-7**
2	4.27		1973	459	C21H17N1O3	5-(4-PHENYLBENZOYL)-H2-PYRROLO-PYRROLE-1-CARBOXYLIC ACID **96327-61-6**
3	1.38	7.0	1405	1	C21H17N3O4S1	1-(P-NH2-BENZENESULFONYL)5,5-DIPHENYLHYDANTOIN
4	3.27		1405		"	1-(P-NH2-BENZENESULFONYL)5,5-DIPHENYLHYDANTOIN
5 ★	1.28	6.2	1703		C21H17N5O7S2	CEPHALASPORIN ANALOG ✦
6	0.09		2389		C21H18F3N3O3	TOSUFLOXIN ANALOG **114676-82-3**
7	0.03		2389		"	TOSUFLOXIN ANALOG **114676-83-4**
8	-0.98		2389		"	TOSUFLOXIN ANALOG **114676-51-6**
9	2.83		2428	459	C21H18N2O6S1	N,N'-DIMETHYLTHIODICARBAMATE,O-3,4-METHYLENEDIOXYPHENYL-O'-1-NAPHTHYL
14810	3.01		759		C21H18O3	2-CARBOXY-4'-HYDROXY-4"-METHYL-TRIPHENYLMETHANE
1	2.74		541		C21H18O9	NSC#109,351 ✦
2	2.74		1582	459	C21H19Cl1N2O5	7CL-1,3H2-1-ACOET-3-ACO-5-PH-1,4-DIAZEPIN-2-ONE **56875-82-2**
3	0.36		1582	459	C21H19Cl1N2O6	7CL-1,3H2-1-ETOCOETCO2H-5PH1,4-DIAZEPIN-2-ONE,N-O **56875-81-1**
4	0.81	7.4	2686	1	C21H19N2	N,N'-DIMETHYLISOCYANINE *noradrenaline transport inhibitor*
15	3.48	7.4	547	1	C21H19N3O2S1	4'(9-ANILINOACRIDINE)2-ME-1'-METHANESULFONAMIDO
6	3.21	7.4	547	1	C21H19N3O2S1	9-ANILINOACRIDINE,2'-ME-4'-METHANSULFONAMIDO
7	0.54	1.0	547	342	C21H19N3O3S1	AMSACRINE HYDROCHLORIDE
8	2.85		401		C21H19N3O3S1	AMSACRINE METHANESULFONATE
9	1.88	6.5	547	1	C21H19N3O3S1	AMSACRINE **51264-14-3** *antineoplastic*
14820	2.59	7.4	547	1	"	AMSACRINE
1	2.85	7.4	401	2	"	AMSACRINE
2	2.28		1593	459	C21H19N7O7S4	CEPHALOSPORINM-PYRIDINYL ANALOG **69712-51-2**
3	2.50		2171		C21H20Br1N7O6	AZOBENZENE,2,4-DINITRO-6-BROMO-1'ACETYLAMINO-5'-METHOXY-4'-N,N-(ALLYL-CYANOETHYL)AMINO
4	2.48	7.4	1242	1	C21H20Cl2N2O4	LORAZEPAM,3-(GLYCERYLACETONIDE) **56057-74-0**
25 ★	6.50		1679		C21H20Cl2O3	PERMETHRIN **52645-53-1** *ectoparasiticide~insecticide*
6 ★	2.61		2494	400	C21H20F3N1O5S1	COLCHICINE,3-HYDROXY-7-TRIFLUOROACETAMIDO-10-METHIO ANALOG **123643-51-6**
7 ★	1.82		2494	400	C21H20F3N1O6	COLCHICEINE,7-TRIFLUOROACETAMIDO ANALOG **71295-34-6**
8 ★	1.67		2494	400	C21H20F3N1O6	COLCHICINE,2-HYDROXY-7-TRIFLUOROACETAMIDO ANALOG **86436-46-6**
9	5.06		260		C21H20O3S1	1-(2-I-PROPYLPHENYLTHIOME)-3-CARBOXY-B-NAPHTHOL
14830	4.26		1135	537	C21H20O6	CURCUMIN **458-37-7** *inhibitor-leukotriene*
1	2.61		1582	459	C21H21Cl1N2O4	7CL-1,3H2-1ACOET-3-ETO-5PH1,4-BENZODIAZEPIN-2-ON **56875-87-7**
2	2.14	7.4	1242	1	C21H21Cl1N2O4	OXAZEPAM,3-(GLYCERYLACETONIDE) **41006-73-9**
3	-1.30	7.5	1831	1	C21H21Cl1N2O8	DEMECLOCYCLINE **127-33-3** *antibacterial*
4	-1.30	7.5	638	1	"	DEMECLOCYCLINE
35	-1.22	7.0	1134	1	"	DEMECLOCYCLINE
6	-0.60	5.6	638	41	"	DEMECLOCYCLINE
7	-0.60	6.6	907	314	"	DEMECLOCYCLINE
8	-0.50	5.5	1086	314	"	DEMECLOCYCLINE
9	3.11	7.4	1209	1	C21H21N1	CYPROHEPTADINE **129-03-3** *antihistaminic, antiprurtic*
14840	3.20	7.4	880	1	"	CYPROHEPTADINE
1 ★	4.69		1835		"	CYPROHEPTADINE
2	1.60	7.0	2678	1	C21H21N1O6	CORNIGERINE
3	0.72		547	370	C21H21N2O2.Br1	N-(4-NO2-BENZYL)-4-PHENYLPROPYLPYRIDINIUMBR
4	1.44	7.4	459	1	C21H21N3O3	1-PHENYL-3-PHENYL-5-MORPHOLINO-6-METHYLURACIL
45	0.97	7.3	1299	314	C21H21N3O4	2-NITROSTRYCHNINE **29854-52-2**
6 ★	1.61	7.4	1299	2	"	2-NITROSTRYCHNINE
7	-0.85	6.6	907	314	C21H21N3O9	9-NITRO-6-DEMETHYL-6-DEOXYTETRACYCLINE
8	-0.11	6.6	907	314	C21H21N3O9	NITROCYCLINE **5585-59-1** *antibacterial*
9 ★	2.17	7.0	2292	1	C21H21N5O5	DIOXOLANE,2-(4-P-NITROBENZAMIDO)PHENYL-2-(1,2,4-TRIAZOL-1-YL)METHYL-4-ETHYL
14850	5.31		1101	421	C21H21O4P1	I-PROPYLPHENYLDIPHENYLPHOSPHATE
1 ★	5.11		1101	421	C21H21O4P1	TRICRESYLPHOSPHATE
2	4.61		2162	459	C21H22Cl1F1N4O2	HALOPEMIDE **59831-65-1** *antipsychotic*
3	2.25	7.4	675	1	C21H22N2O2	3,5-PYRAZOLIDINDIONE,1,2-DIPHENYL,4-CYCLOHEXYL
4	-0.01		1943	1	C21H22N2O2	BW-825C ✦
55	3.02	7.4	847	1	C21H22N2O2	N-(N-PIPERIDINYLETHYL)-G-PYROPHTHALONE
6	0.68	7.3	1299	314	C21H22N2O2	STRYCHNINE **57-24-9** *cns stimulant (vet)*
7	1.68	7.4	1299	2	"	STRYCHNINE
8 ★	1.93		505		"	STRYCHNINE
9	0.62	7.3	1299	314	C21H22N2O3	PSEUDOSTRYCHNINE **465-62-3**
14860 ★	1.82	7.4	1299	2	"	PSEUDOSTRYCHNINE
1	-0.08	5.6	638	41	C21H22N2O7	SANCYCLINE *antibacterial*
2	-0.08	7.5	638	1	"	SANCYCLINE
3	0.00	6.6	907	314	"	SANCYCLINE
4	2.32		1713		C21H22N4O2	6-PH-2-(4-(N3-PHENYL-UREIDO)BU)-3(2H)-PYRIDAZINONE
65 ★	1.30	7.4	1689	1	C21H22N4O5	N7-PHENYL-MITOMYCIN-C **14896-01-6**
6	-1.80		1820		C21H22N6O5	METHASQUIN **18921-69-2**
7	0.75	6.8	1703	360	C21H22N6O7S2	CEPHALOSPORIN ANALOG ✦
8	2.35	7.4	461	1	C21H23Cl1F1N3O1	FLURAZEPAM **17617-23-1** *anticonvulsant, hypnotic, sedative, relaxant (muscle)*
9	1.89		1582	459	C21H23Cl1F1N3O2	7CL-1,3H2-1-DIETAMET-3OH-5FPH1,4-DIAZEPIN-2-ON **19011-82-6**
14870	1.72	7.4	1441	1	C21H23Cl1F1O2	HALOPERIDOL **52-86-8** *antipsychotic, antidiskinetic (for giles de la tourette's dis.)*
1	2.69	7.4	2407	1	"	HALOPERIDOL
2 ★	3.23		2407	2	"	HALOPERIDOL
3	2.98	7.4	1569	1	C21H23Cl1N4O1	PHENOL,4(7-CL-4-QUINYL)AM-2(4-ME-1-PIPERIDINYL)
4	1.76		1582	459	C21H23Cl2N3O2	7CL-1,3H2-1-DIETAMET-3OH-5(2CLPH)1,4-DIAZEPIN-2-O **52039-74-4**
75	0.85	7.4	2493	1	C21H23N1O2	2-PROPANONE,1,1-DIPHENYL-1-(2-BUTYNYL)OXY-3-DIMETHYLAMINO
6	2.41	7.4	2196	1	C21H23N1O3	DIMEFLINE: 4'METHYL ANALOG
7	1.02	7.0	2678	1	C21H23N1O5	3-DEMETHYLCOLCHICINE
8	0.66	7.5	1510	1	C21H23N1O5	HEROIN **561-27-3** *analgesic (narcotic)*
9 ★	1.43		2494	400	C21H23N1O5S1	COLCHICINE,3-HYDROXY-10-METHIO ANALOG **87424-25-7**

Exploring QSAR: Hydrophobic, Electronic, and Steric Constants

	logP	pH	Ref.	Note	MF	Name / CAS / activity
14880	1.70	7.0	2678	1	"	COLCHICINE,3-HYDROXY-10-METHIO ANALOG
1 ★	1.20		2494	400	C21H23N1O6	COLCHICEINE **477-27-0**
2 ★	0.48		2494	400	C21H23N1O6	COLCHICINE,2-HYDROXY ANALOG **7336-36-9**
3	2.83	7.8	550		C21H23N3O1S1	PERICIAZINE **2622-26-6** *antipsychotic*
4 ★	3.52		550		"	PERICIAZINE
85	2.40	7.4	459	1	C21H23N3O2	1-PHENYL-3-PHENYL-5-I-BUAMINO-6-METHYLURACIL
6	-0.47	7.3	1299	314	C21H23N3O2	2-AMINOSTRYCHNINE **3215-24-5**
7 ★	0.54	7.4	1299	2	"	2-AMINOSTRYCHNINE
8	-1.05	6.6	907	314	C21H23N3O7	7-AMINO-6-DEMETHYL-6-DEOXYTETRACYCLINE
9	-0.68	6.6	907	314	C21H23N3O7	AMICYCLINE **5874-95-3** *antibacterial*
14890	-1.01	1.0	2196	1	C21H23N3O7S1	LENAMPICILLIN **86273-18-9** *antibacterial*
1 ★	0.59	7.4	1689	1	C21H23N5O6	N7-(6-METHOXY-PYRID-3-YL)-MITOMYCIN-C **84397-27-3**
2	4.00		2652	459	C21H24Br1F1N2O3	N-(1-(P-FLUOROBENZYL)PIPERIDIN-4YL)BENZAMIDE,2,3-DIMETHOXY-5-BROMO
3	3.96		2162	459	C21H24Cl1F1N2O1	INDANE,6-CHLORO-1-(4-ETHANOLPIPERAZINE)-3-(4-FLUORO)PHENYL
4	-2.62	2.8	2211	1	C21H24Cl1N3O1	O-METHYLAMODIAQUINE
95	2.85	7.4	1569	1	"	O-METHYLAMODIAQUINE
6	3.62	7.8	550		C21H24Cl1N3O1S1	PIPAMAZINE **84-04-8** *anti-emetic*
7 ★	4.44		550		"	PIPAMAZINE
8	1.70		1582	459	C21H24Cl1N3O2	7CL-1,3-H2-1-DIETAMET-3-OH-5-PH-1,4-DIAZEPIN-2-ON **56875-95-7**
9	-1.90	2.8	2211	1	C21H24Cl1N3O2	QUINOLINE-1-OXIDE-7-CHLORO-4-(4-METHOXY-3-DIETHYLAMINOMETHYL)ANILINO
14900	2.42	7.4	547	1	C21H24Cl2F1N5O2S1	BAKER'SANTIFOLATE#2 **31368-48-6** *antineoplastic*
1	4.22	7.5	2651	1	C21H24F1N2O3	FIDA-2
2	4.81	7.5	2651	28	"	FIDA-2
3	3.30		2162	459	C21H24F2N2O1	INDANE,6-FLUORO-1-(4-ETHANOLPIPERAZINE)-3-(4-FLUORO)PHENYL
4	1.69		232	820	C21H24F3N3S1	TRIFLUOPERAZINEHYDROCHLORIDE **440-17-5**
5	3.90	7.0	468	1	C21H24F3N3S1	TRIFLUOPERAZINE **117-89-5** *antipsychotic, sedative*
6	4.01	7.4	2288	1	"	TRIFLUOPERAZINE
7 ★	5.03	7.0	468	701	"	TRIFLUOPERAZINE
8 ★	-1.59	7.4	579	1	C21H24Ga1N3O6	GALLIUM TRIS(1,2-DIMETHYLPYRID-4-ONE-3-OXY)
9	0.67	7.3	1299	314	C21H24N2O2	21,22,-DIHYDROSTRYCHNINE **15006-14-1**
14910 ★	1.62	7.4	1299	2	"	21,22,-DIHYDROSTRYCHNINE
1	1.93	7.4	675	1	C21H24N2O2	PHENYLBUTAZONE,P,P'-DIMETHYL
2	1.80	7.4	675	1	C21H24N2O3	3,5-PYRAZOLIDINDIONE,1-PH,2(4-OH-PH),4(3,3DIMEBU)
3	1.62	7.4	675	1	C21H24N2O3	3,5-PYRAZOLIDINDIONE,1-PH,2(4-OH-PH),4-HEXYL
4	1.84		799		C21H24N2O4	QUINAZOLIN-2-ONE,1-METHYL-4-PHENYL-6-TRIETHOXY
15 ★	0.30		2494	400	C21H24N2O5	COLCHICINE,10-AMINO ANALOG **3123-89-5**
6	-0.11	7.0	2682	1	C21H24N2O5	PHENYLPROPIONIC ACID,7-OXABICYCLOHEPTANEOXAZOLEAMIDE,N-METHYL ANALOG *txa2-antagonist*
7	3.87		2424	459	C21H24N4O4	BENZOTRIAZOLE,2-CYCLOPENTYL-5-(3,4,5-TRIMETHOXYBENZAMIDO)
8	4.02		2125		C21H25Cl1F1O1	R-2835 (BRANCHED CHAIN HALOPERIDOL) ✦
9	4.17		2125		C21H25Cl1F1O1S1	R-5986 (HALOPERIDOL-THIOETHER) ✦
14920	3.26		2125		C21H25Cl1F1O2	R2572 (HALOPERIDOL ANALOG) ✦
1 ★	1.70		2611	403	C21H25Cl1N2O3	CETIRIZINE **83881-51-0** *antihistaminic*
2	-2.46		1010		C21H25Cl1N6O2	BAKER'S ANTIFOL
3	-2.46	7.4	547	314	C21H25Cl1N6O2	BAKER'S ANTIFOL
4	-1.91	7.4	594	314	C21H25Cl1N6O2	BAKER'S ANTIFOL
25	-1.84	2.2	1200		C21H25Cl1N6O2	BAKER'S ANTIFOL
6	-1.06	7.4	401	314	C21H25Cl1N6O2	BAKER'S ANTIFOL
7	-0.57	7.0	1200	1	C21H25Cl1N6O2	BAKER'S ANTIFOL
8	2.95		2162	459	C21H25F1N2O1	INDANE,1-(4-ETHANOLPIPERAZINE)-3-(4-FLUORO)PHENYL
9	2.83		2652	459	C21H25F1N2O3	N-(1-(P-FLUOROBENZYL)PIPERIDIN-4YL)BENZAMIDE,2,3-DIMETHOXY
14930	-0.06	7.4	401	1	C21H25F1N6O3S1.C2H6O3S1	NSC#113,423 ✦
1	1.60	7.2	2416	1	C21H25I1N2O6	BAY-P-8857 **97856-31-0**
2	1.73	5.0	1440	1	C21H25N1	MELITRACEN **5118-29-6** *antidepressant*
3	-1.70	2.0	1443		C21H25N1O1	1,3-DIPHENYL-4-PIPERIDINOBUTAN-2-ONE
4	2.00	7.4	1443	1	"	1,3-DIPHENYL-4-PIPERIDINOBUTAN-2-ONE
35	3.50	7.4	2143	459	C21H25N1O2	2',6'-DIMETHYLPHENYL-4-PIPERIDINOMETHYL BENZOATE
6	3.83	7.4	592	1	C21H25N1O2	2',6'-DIMETHYLPHENYL-4-PIPERIDINOMETHYL BENZOATE
7	3.17	7.4	2143	1	C21H25N1O2	2-(N-PIPERIDINOMETHYL)BENZOIC ACID,2,6-DIMETHYLPHENYL ESTER
8	1.40	7.4	2493	1	C21H25N1O2	2-PROPANONE,1,1-DIPHENYL-1-ETHOXY-3-(N-PYRROLIDINYL)
9	3.47	7.4	2143	1	C21H25N1O2	3-(N-PIPERIDINOMETHYL)BENZOIC ACID,2,6-DIMETHYLPHENYL ESTER
14940	1.08	7.4	2493	1	C21H25N1O3	2-PROPANONE-1,1-DIPHENYL-1-ETHOXY-3-(N-MORPHOLINO)
1	2.77	7.4	2143	1	C21H25N1O3	CARBONIC ACID,2,6-DIMETHYLPHENYL,3-(N-PYRROLIDINOMETHYL)BENZYL ESTER
2	2.75	7.4	2143	1	C21H25N1O3	CARBONIC ACID,2,6-DIMETHYLPHENYL,4-(N-PYRROLIDINOMETHYL)BENZYL ESTER
3	3.47	7.4	2143	1	C21H25N1O4	CARBONIC ACID,2,6-DIMETHYLPHENYL,3-(N-MORPHOLINOMETHYL)BENZYL ESTER
4	3.34	7.4	2143	1	C21H25N1O4	CARBONIC ACID,2,6-DIMETHYLPHENYL,4-(N-MORPHOLINOMETHYL)BENZYL ESTER
45	1.22		157		C21H25N1O4	GLAUCINE-D ✦
6	1.15	7.4	2485	1	C21H25N1O4	MORPHINE,3-I-BUTANOYL
7	1.31	7.0	2678	1	C21H25N1O5	DEMECOLCINE **477-30-5** *antineoplastic*
8 ★	1.37	7.4	594	1	"	DEMECOLCINE
9	1.24	7.4	847	1	C21H25N2O2.C1H3O4S1	N-(DIET,METHYLAMMONIOETHYL)G-PYROPHTHALONE
14950	0.88	7.4	722	1	C21H25N3O1	N-DIALLYL-DESDI-I-PROPYL-DISOPYRAMIDE
1	-0.37	7.4	722	1	C21H25N3O1	SEARLE#SC-27829 ✦
2 ✓	-0.12		2397	101	C21H25N4O3S1	IMIDAZOLIUM CHLORIDE,1-METHYL-2-HYDROXYIMINOMETHYL-3-(3-N-BENZYL-PHENYLSULFONAMIDO)PROPYL
3 ★	2.58	7.4	2195	1	C21H25N5O2	ICOTIDINE **71351-79-6** *antagonist (h1 & h2 histamine receptors)*
4	0.78	7.4	2095	1	C21H25N5O3	(IMIDAZOLE)ADRENOCEPTOR ANTAGONIST
55	4.50		2171		C21H25N5O4	DISPERSE DYE X
6	0.10	4.0	2644	210	C21H26Br1N1O2	FOMOCAINE ANALOG-15 (4-BR)
7	-0.03	4.0	2644	210	C21H26Cl1N1O2	FOMOCAINE ANALOG-14 (4-CL)
8	2.96	7.0	468	1	C21H26Cl1N3O1S1	PERPHENAZINE **58-39-9** *antipsychotic*
9	3.24	7.0	550	1	"	PERPHENAZINE
14960	3.82	7.0	468	701	"	PERPHENAZINE
1 ★	4.20		550		"	PERPHENAZINE
2	3.51	7.4	1599	1	C21H26Cl1N3O2	N(1-BZYL-2ME-3PYRROLIDYL)5-CL-2-MEO-4-MEAMBNZAMID *antipsychotic*
3	3.98		2162	459	"	N(1-BZYL-2ME-3PYRROLIDYL)5-CL-2-MEO-4-MEAMBNZAMID
4 ★	3.30	8.0	2644	2	C21H26F1O2	FOMOCAINE ANALOG-13 (4-F)
65	0.38	4.0	2644	210	C21H26I1O2	FOMOCAINE ANALOG-16 (4-I)
6	2.36	7.4	2143	1	C21H26N2O1	BENZANILIDE,2-(1-PIPERIDINOMETHYL),2',6'-DIMETHYL
7	1.69	7.4	2143	1	C21H26N2O1	BENZANILIDE,3-(N-PIPERIDINOMETHYL),2',6'-DIMETHYL
8 ★	3.47	13.0	2143		C21H26N2O1	BENZANILIDE,3-(N-PIPERIDINOMETHYL),2',6'-DIMETHYL
9	1.94	7.4	2143	1	C21H26N2O1	BENZANILIDE,4-(1-PIPERIDINOMETHYL),2',6'-DIMETHYL

	logP	pH	Ref.	Note	MF	Name / **CAS** / *activity*
14970 ★	3.86	13.0	592	224	C21H26N2O1	BENZANILIDE,4-(1-PIPERIDINOMETHYL),2',6'-DIMETHYL
1	1.81	7.4	1209	1	C21H26N2O1S2	MESORIDAZINE **5588-33-0** *antipsychotic*
2	1.57	7.4	2143	1	C21H26N2O2	2,6-DIMETHYLPHENYLCARBAMATE,4-(N-PYRROLIDINOMETHYL)BENZYL ESTER
3	1.95	7.4	2143	459	C21H26N2O2	2,6-DIMETHYLPHENYLCARBAMIC ACID,2-(N-PYRROLIDINOMETHYL)BENZYL ESTER
4	1.73	7.4	2143	1	C21H26N2O2	2,6-DIMETHYLPHENYLCARBAMIC ACID,3-(N-PYRROLIDINOMETHYL)BENZYL ESTER
75	2.67	7.4	2143	1	C21H26N2O2	2-(N-PYRROLIDINOMETHYL)BENZYLCARBAMIC ACID,2,6-DIMETHYLPHENYL ESTER
6	1.84	7.4	2143	1	C21H26N2O2	3-(N-PYRROLIDINOMETHYL)BENZYLCARBAMIC ACID,2,6-DIMETHYLPHENYL ESTER
7	1.67	7.4	2143	1	C21H26N2O2	4-(N-PYRROLIDINOMETHYL)BENZYLCARBAMIC ACID,2,6-DIMETHYLPHENYL ESTER
8	-0.06	5.0	2375	1	C21H26N2O2S2	SULFORIDAZINE **14759-06-9** *neuroleptic*
9	2.80	7.4	2143	1	C21H26N2O3	2,6-DIMETHYLPHENYLCARBAMIC ACID,2-(N-MORPHOLINOMETHYL)BENZYL ESTER
14980	1.85	7.4	2143	1	C21H26N2O3	2,6-DIMETHYLPHENYLCARBAMIC ACID,3-(N-MORPHOLINOMETHYL)BENZYL ESTER
1	2.20	7.4	2143	1	C21H26N2O3	2,6-DIMETHYLPHENYLCARBAMIC ACID,4-(N-MORPHOLINOMETHYL)BENZYL ESTER
2	3.54	7.4	2143	1	C21H26N2O3	2-(N-MORPHOLINOMETHYL)BENZYLCARBAMIC ACID,2,6-DIMETHYLPHENYL ESTER
3	2.91	7.4	2143	1	C21H26N2O3	3-(N-MORPHOLINOMETHYL)BENZYLCARBAMIC ACID,2,6-DIMETHYLPHENYL ESTER
4	2.70	7.4	2143	1	C21H26N2O3	4-(N-MORPHOLINOMETHYL)BENZYLCARBAMIC ACID,2,6-DIMETHYLPHENYL ESTER
85	4.93	7.4	2090	459	C21H26N2O3	5-(5-(2-ALLYL-4-(4,5-DIHYDRO-2-OXAZOLYL)PHENOXY)PENTYL)-3-METHYLISOXAZOLE **105639-11-0**
6 ★	2.94	7.1	2155	2	C21H26N2O3	CORYNANTHINE **483-10-3**
7	0.09	7.5	1563	1	C21H26N2O3	HYDROXYBENZYLPINDOLOL **69925-27-5**
8 ★	2.54	7.1	2155	2	C21H26N2O3	YOHIMBINE,ALPHA-ISOMER
9 ★	2.73	7.1	2155	2	C21H26N2O3	YOHIMBINE **146-48-5**
14990	1.10	7.4	2020	1	C21H26N2O4	PINDOLOL ANALOG:N-(3,4-DIMETHOXYPHENETHYL) **102573-77-3**
1	2.86	7.2	2016		C21H26N2O7	NIMODIPINE **66085-59-4** *vasodilator* ✚
2 ★	5.90	7.4	550	2	C21H26N2S2	THIORIDAZINE **50-52-2** *antipsychotic, sedative*
3 ★	3.61	7.4	1776	1	C21H26N6O1	N(2,5DIME-PYRROL-1YL)6(N(4MEOHPIPERAZIN)3PYRIDAZINAM **75842-08-9**
4	-0.63	7.4	401	1	C21H26N6O1.C2H6O3S1	NSC#133,072 ✚
95 ★	1.36	4.6	2043	2	C21H26N8O6S3	CEFOTIAM,PIVALOYLOXYMETHYL ESTER ANALOG: METHYL TETRAZOLYL
6	6.23		2415	459	C21H26O2	CANNABINOL **521-35-7**
7 ★	1.46		401		C21H26O5	PREDNISONE **53-03-2** *glucocorticoid*
8 ★	1.37		1520		C21H27F1O5	CORTISONE-9-A-FLUORO
9 ★	1.51		1520		C21H27F1O5	PREDNISOLONE-9-A-FLUORO *anti-inflammatory*
15000 ★	1.16		569		C21H27F1O6	TRIAMCINOLONE **124-94-7** *glucocorticoid*
1	1.48	7.4	2235	1	C21H27N1	BUDIPINE **57982-78-2** *antiparkinsonian*
2	4.01	7.4	2143	1	C21H27N1O1	2,6-DIMETHYLPHENYL 2-(N-PIPERIDINOMETHYL)BENZYL ETHER
3	3.30	7.4	2143	1	C21H27N1O1	2,6-DIMETHYLPHENYL 3-(N-PIPERIDINOMETHYL)BENZYL ETHER
4	1.47	1.0	870	342	C21H27N1O1	2-(DIPHENYLMETHOXY)ETHYL-N-CYCLOHEXYLAMINE
5	1.59	1.0	870	342	C21H27N1O1	2-(P-ME-DIPHENYLMETHOXY)ETHYL-N-CYCLOPENTYLAMINE
6	3.49	7.4	2143	1	C21H27N1O1	4-PIPERIDINOMETHYLBENZYL,2',6'-DIMETHYLPHENYLETHER
7	1.77	7.1	554		C21H27N1O1	METHADONE **76-99-3** *analgesic (narcotic)*
8	2.07	7.4	554	1	"	METHADONE
9	2.16	7.5	554	1	"	METHADONE
15010	2.26	7.6	554	212	"	METHADONE
1	2.36	7.7	554	1	"	METHADONE
2 ★	3.93	7.7	554	2	"	METHADONE
3	0.85	1.0	602	342	C21H27N1O1	N-HEXAHYDROAZEPINYLETHYL-DIPHENYLMETHYLETHER
4	1.01	1.0	602	342	C21H27N1O1	N-PIPERIDYLETHYL-(P-ME-PH)PHENYLMETHYL-ETHER
15 ★	3.21	8.0	2644	2	C21H27N1O2	FOMOCAIANE ANALOG-1
6	1.00	7.4	2680	324	C21H27N1O3	3-O-BUTYLMORPHINE **74886-09-2**
7	2.69	7.4	2143	1	C21H27N1O3	CARBONIC ACID,2,6-DIMETHYLPHENYL,3-(DIETHYLAMINOMETHYL)BENZYL ESTER
8	2.73	7.4	2143	1	C21H27N1O3	CARBONIC ACID,2,6-DIMETHYLPHENYL,4-(DIETHYLAMINOMETHYL)BENZYL ESTER
9	0.01	5.0	2375	1	C21H27N1O3	PROPAFENONE **54063-53-5** *anti-arrhythmic*
15020	1.58	4.0	2655	1	C21H27N1O3	PROPRANOLOL,CYCLOBUTANOYL ESTER
1	0.72	4.0	2655	1	C21H27N1O5	PROPRANOLOL,ETHYLMALONYL ESTER
2	0.65	4.0	2655	1	C21H27N1O5	PROPRANOLOL,METHYLSUCCINYL ESTER
3 ★	3.67	8.0	2644	2	C21H27N1S1	FOMOCAINE ANALOG-35 (4-HYDROXYPIPERAZINE)
4 ★	3.85	8.0	2644	2	C21H27N1S1	FOMOCAINE ANALOG-36 (ORTHO CH.-HYDROX-PIPERAZINE)
25	0.56	7.4	722	1	C21H27N3O1	4-ME-PIPERIDINYL-DES-N-DI-I-PROPYL-DISOPYRAMIDE
6	0.47	7.4	2095	1	C21H27N3O2S1	(5-THIENYL-IMIDAZOLE)-SPACED-ADRENOCEPTOR ANTAGONIST
7	2.37	7.4	459	1	C21H27N3O3	1-CYCLOHEXYL-3-PHENYL-5-MORPHOLINO-6-METHYLURACIL
8	3.05	7.4	459	1	C21H27N3O3	1-PHENYL-3-CYCLOHEXYL-5-MORPHOLINO-6-METHYLURACIL
9	-0.45	7.4	2375	1	C21H27N3O3	NICAINOPROL **76252-06-7** *antiarrhythmic, antidiarrheal*
15030 ★	1.63	7.4	2375	2	"	NICAINOPROL
1	-0.70	2.0	1289		C21H27N3O6S1	SARMOXICILLIN **67337-44-4** *antibacterial*
2	-0.22	3.0	1289		"	SARMOXICILLIN
3	0.11	7.4	1289	1	"	SARMOXICILLIN
4	0.30	6.0	1289		"	SARMOXICILLIN
35	-0.26	1.0	2196	1	C21H27N3O7S1	BACAMPICILLIN **50972-17-3** *antibacterial*
6	2.06	8.2	2101	1	C21H27N5O2S1	LUPITIDINE **83903-06-4** *antagonist (to h2-histamine receptors)(vet)*
7	2.33	8.3	2218			LUPITIDINE
8 ★	1.91		2351	2	C21H27N5O4S1	GLIPIZIDE **29094-61-9** *antidiabetic*
9 ★	1.41	7.4	2242	1	C21H28F1O4	DEXAMETHASONE,17-BETA-CARBOXAMIDE ANALOG
15040	1.20	4.8	1020		C21H28N2O1	PHENACETAMIDE,N(2,6-DIMEPH)N(3-DIMEAMINOPROPYL)
1	1.85	7.4	2143	1	C21H28N2O2	2,6-DIMETHYLPHENYLCARBAMIC ACID,3-(DIETHYLAMINOMETHYL)BENZYL ESTER
2	1.70	7.4	2143	1	C21H28N2O2	2,6-DIMETHYLPHENYLCARBAMIC ACID,4-(DIETHYLAMINOMETHYL)BENZYL ESTER
3	2.61	7.4	2143	1	C21H28N2O2	2-(DIETHYLAMINOMETHYL)BENZYLCARBAMIC ACID,2,6-DIMETHYLPHENYL ESTER
4	1.91	7.4	2143	1	C21H28N2O2	3-(DIETHYLAMINOMETHYL)BENZYLCARBAMIC ACID,2,6-DIMETHYLPHENYL ESTER
45	1.84	7.4	2143	1	C21H28N2O2	4-(DIETHYLAMINOMETHYL)BENZYLCARBAMIC ACID,2,6-DIMETHYLPHENYL ESTER
6	4.95	7.4	2223	1	C21H28N2O3	DISOXARIL,4'-METHYL
7	3.16	7.4	2102	1	C21H28N2O4	PILOCARPIC ACID,O-ACETYL,P-METHYLBENZYL ESTER
8 ★	3.31		2102	2	"	PILOCARPIC ACID,O-ACETYL,P-METHYLBENZYL ESTER
9 ✓	0.69		2018	226	C21H28N2O5	RAMIPRILAT **87269-97-4** *antihypertensive*
15050	0.89	7.4	1599	1	C21H28N2O5	TIGAN **138-56-7** *anti-emetic*
1	2.57		2162	459	"	TIGAN
2 ★	7.00		2420		C21H28N2S2	BUTHIOBATE **51308-54-4** *fungicide*
3	1.99		1713		C21H28N4O2	6-PH-2-(4-(N3-CY-HEX-UREIDO)BU)-3(2H)-PYRIDAZINONE
4	2.45		1711		C21H28N4O7	PENTYLOXYCARBONYLMITOMYCINC
55	3.11		2286		C21H28O2	ETHISTERONE **434-03-7** *progestin*
6	7.67		2404	459	C21H28O3	5,6,7,8-BIS(DIHYDROPYRANYL)BENZOPYRAN,2,2,2',2',2'',2''-HEXAMETHYL
7	5.80		2404	459	C21H28O3	6,5,7,8-BIS(DIHYDROPYRANYL)BENZOPYRAN,2,2,2',2',2''',2''-HEXAMTHYL ✚
8	3.51		2415	1	C21H28O4	9-CARBOXY-D8-THC
9 ★	1.47		505		C21H28O5	4-PREGNENE,17-A,21-DIOL,3,11,20-TRIONE **53-06-5**

	logP	pH	Ref.	Note	MF	Name / CAS / activity
15060 ★	1.08	7.4	1076	320	C21H28O5	ALDOSTERONE **52-39-1** *mineralocorticoid*
1	1.55	7.0	2002	1	C21H28O5	PREDNISOLONE **50-24-8** *glucocorticoid*
2 ★	1.62		1146		"	PREDNISOLONE
3 ★	1.67		1520		C21H29F1O5	9-A-FLUOROHYDROCORTISONE **127-31-1**
4	1.73	7.0	2002	1	"	9-A-FLUOROHYDROCORTISONE
65 ★	5.10		1092		C21H29N1	2(4-T-BUTYLBENZHYDRYL)ETHYLAMINE,N,N-DIMETHYL
6	1.05	1.0	870	342	C21H29N1O1	2-(DIPHENYLMETHOXY)-N,N-DIPROPYLAMINE
7	1.19	1.0	870	342	C21H29N1O1	2-(P-ME-DIPHENYLMETHOXY)-N-ET,N-PROPYLAMINE
8	2.83	7.4	2407	1	C21H29N1O1	BIPERIDEN **514-65-8** *anticholinergic, antiparkinsonian*
9 ★	4.25		2407	2	"	BIPERIDEN
15070	1.95	4.0	2655	1	C21H29N1O3	PROPRANOLOL,I-VALERYL ESTER
1	1.79	4.0	2655	1	C21H29N1O3	PROPRANOLOL,PIVALOYL ESTER
2	1.99	4.0	2655	1	C21H29N1O3	PROPRANOLOL,VALERYL ESTER
3	2.17		634		C21H29N1O6	6-I-BUO-7-I-PROPOXYET-4-OH-3-QUINOLINE-CO2ET
4	-0.14	7.4	253	1	C21H29N3O1	B-PSEUDO-DIISOPYRAMIDE
75	-0.68	7.4	2375	1	C21H29N3O1	DISOPYRAMIDE **3737-09-5** *cardiac depressant (anti-arrhythmic)*
6	-0.18	7.4	806	1	"	DISOPYRAMIDE
7	2.71		2216		"	DISOPYRAMIDE
8	-0.32	7.4	253	1	C21H29N3O1	G-PSEUDO-DIISOPYRAMIDE
9	0.91	7.4	722	1	C21H29N3O1	N-DIPROPYL-DESDI-I-PROPYL-DISOPYRAMIDE
15080	3.27	7.4	459	1	C21H29N3O2	1-CYCLOHEXYL-3-PHENYL-5-I-BU-AMINO-6-METHYLURACIL
1	3.38	7.3	1378	314	C21H29N3O3	PENTACAIN **38198-35-5**
2	4.95	7.3	1378	2	"	PENTACAIN
3 ★	0.36		2377		C21H29N5O4	AC-TRP-ALA-VAL-N
4 ★	0.62		2377		"	AC-TRP-ILE-GLY-N
85	1.07	7.4	1441	1	C21H30F1N3O2	PIPAMPERONE **1893-33-0** *antipsychotic*
6 ★	2.02		1441	2	"	PIPAMPERONE
7 ★	5.34		1092		C21H30N2	N-PH-N-(4-T-BUTYLBENZYL)-N-(2-N,N-DIME-ETHYL)AMINE
8	0.30	7.4	2375	1	C21H30N4O1	QUINACAINOL **86024-64-8** *anti-arrhythmic*
9	-0.91	7.4	608	1	C21H30N4O1	DIISOPYRAMIDE,N1-NH2
15090	1.85		1713		C21H30N4O2	6-PH-2-(6-(N3-DIET-UREIDO)HEXYL)-3(2H)-PYRIDAZINONE
1	3.02	7.4	2244	1	C21H30N4O4S1	TIMOLOL,O-(2-METHYLBENZOYL)
2	3.11	7.4	2244	1	C21H30N4O4S1	TIMOLOL,O-(4-METHYLBENZOYL)
3	2.51	7.4	2244	1	C21H30N4O5S1	TIMOLOL,O-(2-METHOXYBENZOYL)
4	2.65	7.4	2244	1	C21H30N4O5S1	TIMOLOL,O-(4-METHOXYBENZOYL)
95 ★	0.61		2377	821	C21H30N4O6	AC-ASP-ILE-PHE-N
6 ★	0.40		2377	821	C21H30N4O6	AC-ASP-PHE-LEU-N
7 ★	0.39		2377	821	C21H30N4O6	AC-PHE-ASP-LEU-N
8 ★	3.32	7.4	2029	459	C21H30N6O4S1	XANTHINE,1,3-DIPROPYL-8-(4-(DIMETHYLAMINOETHYL)AMINOSULFONYL **89073-58-5**
9 ★	-0.47	7.4	2427	1	C21H30N6O6	P-GLU-HIS(N-BUTOXYCARBONYL)-PRO-NH2
15100 ★	-0.44	7.4	2427	1	C21H30N6O6	P-GLU-HIS(N-ISOBUTOXYCARBONYL)-PRO-NH2
1	5.79		2415	459	C21H30O2	CANNABIDIOL **13956-29-1**
2 ★	7.41		2415	459	C21H30O2	D8-TETRAHYDROCANNABINOL
3 ★	6.97		2415	459	C21H30O2	D9-TETRAHYDROCANNABINOL **1972-08-3** *anti-emetic*
4	3.26	7.4	1076	320	C21H30O2	PROGESTERONE **57-83-0** *progestin*
5 ★	3.87		300		"	PROGESTERONE
6 ★	5.33		2415	459	C21H30O3	11-HYDROXY-D9-THC
7 ★	2.70		2319		C21H30O3	17-BETA-ESTRADIOL,3-HYDROXYPROPOXY **21830-21-7**
8 ★	4.24		2415	459	C21H30O3	8-HYDROXY-D(9-11)THC ✚
9 ★	2.88		300		C21H30O3	DEOXYCORTICOSTERONE **64-85-7**
15110 ★	2.36		1146		C21H30O3	PROGESTERONE,11-A-HYDROXY **312-90-3**
1 ★	3.17		2460		C21H30O3	PROGESTERONE,17-A-HYDROXY **68-96-2**
2 ★	2.37		2319		C21H30O3.C3Cr1O3	17-BETA-ESTRADIOL,3-HYDROXYPROPOXY,CHROMIUM TRICARBONYL COMPLEX **93061-16-6**
3 ★	2.04		2319		"	17-BETA-ESTRADIOL,3-HYDROXYPROPOXY,CHROMIUM TRICARBONYL COMPLEX **93061-17-7**
4 ★	2.52		579		C21H30O4	CORTEXOLONE **152-58-9** *anti-inflammatory*
15	1.82	7.4	1076	320	C21H30O4	CORTICOSTERONE **50-22-6** *adrenocorticosteroid*
6 ★	1.94		1146		"	CORTICOSTERONE
7	1.20	7.4	1758	1	C21H30O5	HYDROCORTISONE **50-23-7** *glucocorticoid*
8	1.54	7.0	2002	1	"	HYDROCORTISONE
9	1.54	7.4	1076	320	"	HYDROCORTISONE
15120 ★	1.61		569		"	HYDROCORTISONE
1	-0.20		307	1	C21H31N1O1	3-HYDROXY-N-AMYLMORPHINAN HYDROCHLORIDE
2	4.32		627		C21H31N1O3	3(N,N-DIMEAMME-2-NORBORNANYL)4-BUOXY BENZOATE
3	4.35		627		"	3(N,N-DIMEAMME-2-NORBORNANYL)4-BUOXY BENZOATE
4	2.55	7.4	2480	1	C21H31N1O3	KETOBEMIDONE,(3,3-DIMETHYLBUTYRYL) ESTER
25 ★	3.17	7.4	2480	2	"	KETOBEMIDONE,(3,3-DIMETHYLBUTYRYL) ESTER
6	1.01	7.4	2274	1	C21H31N1O3	SPIRENDOLOL **65429-87-0** *adrenoreceptor blocker (b2)*
7	-3.40	7.4	2449	1	C21H31N3O5	LISINOPRIL **76547-98-3** *antihypertensive*
8 ✓	-2.86		2018	226	"	LISINOPRIL
9	3.04	7.4	2244	1	C21H31N5O4S1	TIMOLOL,O-(2-METHYLAMINOBENZOYL)
15130	0.17	7.4	2320	1	C21H32N2O3	2,6-BIS(1-PYRROLIDINYLMETHYL)PHENOL,4-ISOBUTYLOXYCARBONYL
1	-0.41	7.4	2320	1	C21H32N2O3	2,6-BIS(1-PYRROLIDINYLMETHYL)PHENOL-4-ACETIC ACID,I-PROPYL ESTER
2	-0.37	7.4	2320	1	C21H32N2O3	2,6-BIS(1-PYRROLIDINYLMETHYL)PHENOL-4-ACETIC ACID,PROPYL ESTER
3	0.05	7.4	2320	1	C21H32N2O3	2,6-BIS(1-PYRROLININYLMETHYL)PHENOL,4-S-BUTYLOXYMETHYL
4	3.19	7.3	1149	714	C21H32N2O3	3(OPENT)CARBANILATE,O-(2-N-PYRROLIDYL)CYCLOPENTYL **60093-71-2**
35	3.40	7.3	873	314	C21H32N2O3	O-3(N-PIPERIDINO)C-C6H10-N(P-OC3H7-C6H4)CARBAMATE
6	3.02	7.3	873	314	C21H32N2O3	O-3(N-PYRROLIDINO)CYHEX-N(M-OC4H9-PH)CARBAMATE
7	2.90	7.3	873	314	C21H32N2O3	O-3(N-PYRROLIDINO)CYHEX-N(P-OC4H9-PH)CARBAMATE
8 ★	1.01	7.2	2553		C21H32N4O4	AC-ALA-PHE-ALA-NTBU
9	2.11	7.4	1837	1	C21H32N6O3	ALFENTANIL **71195-58-9** *analgesic (narcotic)*
15140 ★	2.16	9.2	1837	1	"	ALFENTANIL
1 ★	4.22		2460		C21H32O2	PREGNENOLONE **145-13-1**
2 ★	2.00		2558		C21H32O4	4-PREGNENE-17A,20A,21-TRIOL-3-ONE
3	6.67		1361	459	C21H33N1	4-(EQ)-T-BUTYL-PHENCYCLIDINE(PHEN=AX.)
4	0.73		1361	459	C21H33N1	4-(EQ)-T-BUTYL-PHENCYCLIDINE(PHEN=EQ.)
45 ★	-1.34	7.0	2374		C21H33N3O5	LEUCINYLLEUCINYLTYROSINE
6	1.18	4.8	1020		C21H34N2O1	2CYHEX-ACETAMIDE,N(2,6-DIMEPH)N(6-DIMEAMPROPYL)
7	1.17	4.8	1020		C21H34N2O1	2CYHEX-ACETAMIDE,N(2,6-DIMEPH)N-(1-ME-2-DIMEAMET)
8	4.15	7.3	1900	314	C21H34N2O3	1-PIPERIDINEETHANOL,A-METHYL,3-OHEX-PHENYLCARBAMATE
9	4.06	7.3	1356	314	C21H34N2O3	2-HEPTOXYCARBANILATE-O-2(N-PIPERIDINYL)ETHYL **55792-21-7**

	logP	pH	Ref.	Note	MF	Name / **CAS** / *activity*
15150	5.62	7.3	1356	2	"	2-HEPTOXYCARBANILATE-O-2(N-PIPERIDINYL)ETHYL
1	3.39	7.3	1960	314	C21H34N2O3	2-HEXOXYMETHYLCARBANILIC ACID, 2-(1-PIPERIDINYL)ETHYL ESTER
2	5.18	7.2	1150		C21H34N2O3	2ME-4(O-HEX)CARBANILATE-O-2(N-PIPERIDINYL)ETHYL
3	3.27	7.1	1150		C21H34N2O3	2ME-4(O-PEN)CARBANILATE-O-(1-ME-2(PIPERIDINYL)ET
4	3.83	7.3	1149	714	C21H34N2O3	3(OPENT)CARBANILATE,O-(2-N-DIETAMINO)CYCLOPENTYL **60093-70-1**
55	3.75	7.3	1149	714	C21H34N2O3	3(OPENT)CARBANILATE,O-(2-N-PROPAMINO)CYCLOHEXYL **60093-67-6**
6	3.97	7.3	2051	1	C21H34N2O3	3-HEPTOXYCARBANILATE-O-2(N-PIPERIDINYL)ETHYL **55792-22-8**
7	5.53	7.3	1356	2	"	3-HEPTOXYCARBANILATE-O-2(N-PIPERIDINYL)ETHYL
8	3.47	7.3	1960	314	C21H34N2O3	3-HEXOXYMETHYLCARBANILIC ACID,2-(1-PIPERIDINYL)ETHYL ESTER
9 ★	4.97	7.3	1960	2	"	3-HEXOXYMETHYLCARBANILIC ACID,2-(1-PIPERIDINYL)ETHYL ESTER
15160	4.31	7.3	1356	314	C21H34N2O3	4-HEPTOXYCARBANILATE-O-2(N-PIPERIDINYL)ETHYL **55792-23-9**
1	5.87	7.3	1356	2	"	4-HEPTOXYCARBANILATE-O-2(N-PIPERIDINYL)ETHYL
2	2.42	7.1	1517	1	C21H34N2O3	CARBAMICAC,N-ET-N-(M-BUOPH),2(1-PIPERIDINYL)PR.EST.
3	2.42	7.1	1357		"	CARBAMICAC,N-ET-N-(M-BUOPH),2(1-PIPERIDINYL)PR.EST.
4	3.12	7.0	1517	1	C21H34N2O3	CARBAMICAC,N-ET-N-(M-PENOPH),2(N-PIPERIDINYL)ET.EST
65	3.12	7.1	1357		"	CARBAMICAC,N-ET-N-(M-PENOPH),2(N-PIPERIDINYL)ET.EST
6	2.75	7.0	1517	1	C21H34N2O3	CARBAMICAC,N-PR-N-(M-BUOPH),2(N-PIPERIDINYL)ET.EST
7	2.75	7.1	1357		"	CARBAMICAC,N-PR-N-(M-BUOPH),2(N-PIPERIDINYL)ET.EST
8	5.43		1607	459	C21H34O2	17-B-METHOXY-2-A-METHYL-5-A-ANDROSTAN-3-ONE
9 ★	6.68		2415	459	C21H34O2	CYCLOHEXANOL,3-(2-HYDROXY-4-(1,1-DIMETHYL)HEPTYL)PHENYL
15170 ★	5.30		2413	459	C21H35N1O4	TETRAHYDROPYRAN,2,4-DIONE,3[1(ETHOXYIMINO)BUTYL]-SPIRO-(4-T-BUTYL)CYCLOHEXYL
1	1.32	7.4	365	2	C21H35N3O1	N,N-DIBU-PIPERAZINE-2-CARBOXANILIDE,2'6'-DIME
2	-1.96		1118		C21H36N1O1.C2H5O3S1	TRIDIHEXETHYLETHYLSULFONATE
3	-1.26		1118		C21H36N1O1.C4H9O3S1	TRIDIHEXETHYLBUTYLSULFONATE
4	-1.62		1118		C21H36N1O1.C4H9O4S1	TRIDIHEXETHYLBUTYLSULFATE ✦
75	-1.09		1118		C21H36N1O1.C5H9O2	TRIDIHEXETHYLPENTANOATE
6	-0.85		1118		C21H36N1O1.C6H13O3S1	TRIDIHEXETHYLHEXYLSULFONATE
7	-0.51		1118		C21H36N1O1.C7H13O2	TRIDIHEXETHYLHEPTANOATE
8	0.30		1118		C21H36N1O1.C8H17O3S1	TRIDIHEXETHYLOCTYLSULFONATE
9	0.26		1118		C21H36N1O1.C8H17O4S1	TRIDIHEXETHYLOCTYLSULFATE
15180	0.11		1118		C21H36N1O1.C9H17O2	TRIDIHEXETHYLNONANOATE
1	0.99		1118		C21H36N1O1.C10H21O3S1	TRIDIHEXETHYLDECYLSULFONATE
2	0.95		1118		C21H36N1O1.C10H21O4S1	TRIDIHEXETHYLDECYLSULFATE
3	0.70		1118		C21H36N1O1.C11H21O2	TRIDIHEXETHYLUNDECANOATE
4	1.51		1118		C21H36N1O1.C12H25O3S1	TRIDIHEXETHYLDODECYLSULFONATE
85	3.48	7.1	1150		C21H36N2O3	2-ME-4(O-PR)CARBANILATE-O-2(N,N-DIBUAMINO)ETHYL
6	3.34	7.1	1150		C21H36N2O3	2ME-4(O-HEX)CARBANILATE-O-2(N,N-DIETAMINO)1-ME-ET
7	2.90	6.0	2284	1	C21H36N2O3	CARBISOCAINE **68931-03-3** *anesthetic (local)*
8	4.12	7.4	2284	1	"	CARBISOCAINE
9	10.11		2316	337	C21H36O1	PHENOL,3-PENTADECYL
15190	10.07		2316	337	C21H36O1	PHENOL,PENTADECYL
1	9.27		2316	337	C21H36O2	1,2-DIHYDROXYBENZENE,3-PENTADECYL
2	9.90		2316	337	C21H36O2	1,2-DIHYDROXYBENZENE,4-PENTADECYL
3	8.29		2316	337	C21H36O2	1,3-DIHYDROXYBENZENE,4-PENTADECYL
4 ★	1.56		2238		C21H38Cl1N5O4	NITROSOUREA,1-(2-CHLOROETHYL)-3,3-BIS(1-OXO-2,2,6,6-TETRAMETHYLPIPERIDIN-4-YL)
95	1.83		527	820	C21H38N1.Br1	HEXADECYLPYRIDINIUMBROMIDE
6	0.95		1635	820	C21H38N1.Cl1	BENZODODECINIUM CHLORIDE **10328-35-5**
7	1.71		527	820	C21H38N1.Cl1	HEXADECYLPYRIDINIUMCHLORIDE
8 ✓	2.47		1729	820	C21H38N1.I1	2,4-DIMETHYLPYRIDINIUM IODIDE,N-TETRADECYL
9 ✓	2.93		1729	820	C21H38N1.I1	3,4-DIMETHYLPYRIDINIUM IODIDE,N-TETRADECYL
15200 ✓	2.90		1729	820	C21H38N1.I1	3,5-DIMETHYLPYRIDINIUM IODIDE,N-TETRADECYL
1	-0.75	7.4	1026	1	C21H38N2O4	1,1'(3-I-PR-O-PHEN-DIOXY)BIS(3-I-PR-AM-2-PROPANOL
2	-0.83	7.4	1026	1	C21H38N2O4	1,1'(4-I-PR-O-PHEN-DIOXY)BIS(3-I-PR-AM-2-PROPANOL
3	4.66	7.4	2244	1	C21H38N4O4S1	TIMOLOL,O-OCTANOYL
4 ★	-1.91		2377		C21H38N4O5	VAL-PRO-VAL-LEU
5	2.00	7.4	1363	1	C21H45N3	HEXETIDINE **141-94-6** *antifungal*
6	2.05		1363			HEXETIDINE
7 ★	6.63		2158	459	C22H12	BENZO(GHI)PERYLENE **191-24-2**
8 ★	6.50		1486		C22H14	1,2,5,6-DIBENZANTHRACENE **53-70-3**
9 ★	6.17		579	401	C22H14	1,2:3,4-DIBENZANTHRACENE **215-58-7**
15210 ★	6.81		2158	459	C22H14	3,4-BENZNAPHTHACENE
1	3.38		520		C22H15N1O4S1	1,4-NAPHTHOQUINONE,2-ANILINO-3-PHENYLSULFONYL
2	2.11		2646	448	C22H16N4O1	SUDAN III **85-86-9**
3	1.32	7.4	2557	1	C22H16O8	ETHYL BISCOUMACETATE **548-00-5** *anticoagulant*
4	1.37	7.4	2438	1	"	ETHYL BISCOUMACETATE
15	3.65		2524	459	C22H17F2N3O1	IMIDAZOLYLETHANOL,2',4'-DIFLUORO-4"-(2-PYRIDYL) ANALOG
6	3.31		2524	459	C22H17F2N3O1	IMIDAZOLYLETHANOL,2',4'-DIFLUORO-4"-(3-PYRIDYL) ANALOG ✦
7	2.54		2524	459	C22H17F2N3O1	IMIDAZOLYLETHANOL,2',4'-DIFLUORO-4"-(4-PYRIDYL) ANALOG
8	3.30	7.4	712	1	C22H17N1O2	N(2-PHENETHYL)-G-PYROPHTHALONE
9	3.37		2524	459	C22H18Cl1N3O1	IMIDAZOLYLETHANOL,2'-CHLORO-4"-(4-PYRIDYL) ANALOG
15220	3.27		2524	459	C22H18Cl1N3O1	IMIDAZOLYLETHANOL,4'-CHLORO-4"-(4-PYRIDYL) ANALOG
1	2.78	7.4	2573	1	C22H18Cl2N2O7S1	CEPHALOTHIN,5-(2,6-DICHLOROPHENYL)FURYL ANALOG ✦
2	2.47		2524	459	C22H18F1N3O1	DUP-983 *anti-inflammatory(topical)* ✦
3	3.26		2524	459	C22H18F1N3O1	IMIDAZOLYLETHANOL,4'-FLUORO-2"-(3-PYRIDYL) ANALOG
4	2.51		2524	459	C22H18F1N3O1	IMIDAZOLYLETHANOL,4'-FLUORO-2"-(4-PYRIDYL) ANALOG *anti-inflammatory (topical)*
25	3.34		2524	459	C22H18F1N3O1	IMIDAZOLYLETHANOL,4'-FLUORO-3"-(2-PYRIDYL) ANALOG
6	3.22		2524	459	C22H18F1N3O1	IMIDAZOLYLETHANOL,4'-FLUORO-3"-(3-PYRIDYL) ANALOG
7	2.91		2524	459	C22H18F1N3O1	IMIDAZOLYLETHANOL,4'-FLUORO-3"-(4-PYRIDYL) ANALOG
8	3.58		2524	459	C22H18F1N3O1	IMIDAZOLYLETHANOL,4'-FLUORO-4"-(2-PYRIDYL) ANALOG
9	2.14		2524	459	C22H18F1N3O1	PYRAZOLYLETHANOL, 4'-FLUORO-4"-(4-PYRIDYL) ANALOG
15230 ★	4.77		579		C22H18N2	BIFONAZOLE **60628-96-8** *antifungal (topical)*
1 ★	3.81		2602		C22H18N2O3S2	QUINAZOLINE-4-ONE,2,6-DIMETHYL-5-(P-PHENYLSULFONYLPHENYL)THIO *thymidylate synthase inhibitor*
2	2.05	7.0	1405	1	C22H18N2O4S1	1-(O-ME-BENZENESULFONYL)-5,5-DIPHENYLHYDANTOIN
3	4.00		1405		"	1-(O-ME-BENZENESULFONYL)-5,5-DIPHENYLHYDANTOIN
4	2.31	7.4	2557	1	C22H18O6	BUTYLIDENEBIS-4-HYDROXYCOUMARIN
35 ★	6.20		1903		C22H19Br2N1O3	DECAMETHRIN **52918-63-5**
6	2.42	7.4	2573	1	C22H19Cl1N2O7S1	CEPHALOTHIN,5-(2-CHLOROPHENYL)FURYL ANALOG
7	2.76	7.4	2573	1	C22H19Cl1N2O7S1	CEPHALOTHIN,5-(P-CHLOROPHENYL)FURYL ANALOG
8 ★	6.05		1903		C22H19Cl2N1O3	CYPERMETHRIN,CIS **52315-07-8**
9 ★	6.05		1903		C22H19Cl2N1O3	CYPERMETHRIN,TRANS

	logP	pH	Ref.	Note	MF	Name / CAS / activity
15240	-0.14	7.2	2526	1	C22H19F3N4O3	OXONAPHTHYRIDINECARBOXYLIC ACID,N-(2,4-DIFLUOROPHENYL)-7-(6-METHYL-2,5-DIAZABICYCLOHEPTYL) ANALOG
1	1.50	7.4	2282	1	C22H19N1O4	DOPAMINE,DIBENZOYL
2	2.29		2524	459	C22H19N3O1	IMIDAZOLYLETHANOL,4'-(4-PYRIDYL) ANALOG
3	2.49		2524	459	C22H19N3O2	IMIDAZOLYLETHANOL,4'-HYDROXY-4"-(4-PYRIDYL) ANALOG
4 ★	4.69		2519		C22H20Cl2N2O3	BENZOFENAP ✚
45	2.06	7.4	2573	1	C22H20N2O7S1	CEPHALOTHIN,5-PHENYLFURYL ANALOG
6	-1.98	1.0	564	342	C22H20N6O2	TPA,3,3"-BISAMIDINO
7	-1.95	1.0	564	342	"	TPA,3,3"-BISAMIDINO
8	-1.88	7.4	564	1	"	TPA,3,3"-BISAMIDINO
9	-0.57	1.0	564	342	C22H20N6O2	TPA,4,4"-BISAMIDINO ✚
15250	-0.52	1.0	564	342	"	TPA,4,4"-BISAMIDINO
1 ★	3.59	5.5	2403	2	C22H20O4	PROPIONIC ACID,2-(4-BIPHENYLYL)-3-HYDROXY-3-(P-METHOXYPHENYL) 93371-46-1
2	3.08		1582	459	C22H21Cl1N2O5	7CL-1,3-H2-1-ACOPR-3-ACO-5-PH-1,4-DIAZEPIN-2-ONE 56875-83-3
3	1.05	7.4	401	1	C22H21N1O2S1	3-TRITYLTHIO-L-ALANINE 2799-07-7 antineoplastic
4	2.06	5.3	527	600	"	3-TRITYLTHIO-L-ALANINE
55	0.14	7.4	447	1	C22H21N1O7	CETOCYCLINE 29144-41-1 antibacterial
6	-1.21	1.0	564	342	C22H21N7O2	TPA,2-AMINO-4,4"-BISAMIDINO
7	2.21	7.4	1441	1	C22H22Cl1F4N1O2	SEPERIDOL 10457-91-7 antipsychotic
8	2.75	7.4	2196	1	C22H22Cl1N1O3	DIMEFLINE: 4'-CHLORO-N-PYRROLIDINYL ANALOG
9	3.94	7.4	2196	1	C22H22Cl1N1O4	DIMEFLINE: 4'-CHLORO-N-MORPHOLINO ANALOG
15260 ★	4.60		2506		C22H22Cl1N3O5	PROPAQUIZAFOP herbicide (fatty acid synth.inhib.)
1 ★	3.50		1781	459	C22H22F1N3O2	DROPERIDOL 548-73-2 antipsychotic
2	2.13	7.4	2196	1	C22H22F1O3	DIMEFLINE: 4'-FLUORO-N-PYRROLIDINYL ANALOG ✚
3	3.26	7.4	2196	1	C22H22F1O4	DIMEFLINE: 4'-FLUORO-N-MORPHOLINO ANALOG
4 ★	1.72		2494	400	C22H22F3N1O6	COLCHICINE,7-TRIFLUOROACETAMIDO ANALOG 26195-65-3
65 ★	2.08		2494	400	C22H22F3N1O6	ISOCOLCHICINE,7-TRIFLUOROACETAMIDO 71324-48-6
6	0.54		401		C22H22N1O4.Cl1	CORALYNE ✚
7	1.83	7.4	2196	1	C22H22N2O5	DIMEFLINE: 4'-NITRO-N-PYRROLIDINYL ANALOG
8	3.02	7.4	2196	1	C22H22N2O6	BENZOPYRAN-4-ONE,8-(N-MORPHOLINO)METHYL-7-METHOXY-3-METHYL-2-(4-NITROPHENYL)
9	0.78	6.6	907	314	C22H22N2O8	5A(6)ANHYDROTETRACYCLINE ✚
15270	-0.37	7.5	638	1	C22H22N2O8	METHACYCLINE 914-00-1 antibacterial
1	-0.35	7.5	911	1	"	METHACYCLINE
2	-0.04	5.6	638	41	"	METHACYCLINE
3	-0.45	7.4	547	1	C22H22N4O3	L-TRYPTOPHYL-L-TRYPTOPHAN 20696-60-0
4 ★	-0.27		2197		"	L-TRYPTOPHYL-L-TRYPTOPHAN
75	1.33		1532	459	C22H22N4O6	BENZOYLMITOMYCIN-C
6	1.89		1711		"	BENZOYLMITOMYCIN-C
7	-1.73	2.0	2464	41	C22H22N6O7S2	CEFTAZIDIME 78439-06-2 antibacterial
8 ✓	-1.60	7.4	2464	41	"	CEFTAZIDIME
9	-2.14	1.0	564	342	C22H22N8O2	TPA,4,4"-BISGUANIDINO ✚
15280 ★	2.01		564		C22H22O8	PODOPHYLLOTOXIN 518-28-5 antineoplastic
1 ★	5.20	7.4	593	1	C22H23Cl1N2O2	LORATIDINE 79794-75-5 antihistaminic
2	2.37	7.4	1242	1	C22H23Cl1N2O4	TEMAZEPAM,3-(GLYCERYLACETONIDE) 40967-02-0
3	-1.24	6.6	907	314	C22H23Cl1N2O8	5A(11)DEHYDRO-7-CHLOROTETRACYCLINE
4	-0.89	7.5	1831	1	C22H23Cl1N2O8	CHLORTETRACYCLINE 64-72-2 antibacterial, antiprotozoal
85	-0.89	7.5	638	1	"	CHLORTETRACYCLINE
6	-0.80	7.0	1134	1	"	CHLORTETRACYCLINE
7	-0.62		1135	537	"	CHLORTETRACYCLINE
8	-0.51	7.0	1295	1	"	CHLORTETRACYCLINE
9	-0.43	6.6	907	314	"	CHLORTETRACYCLINE
15290	-0.39	5.6	638	41	"	CHLORTETRACYCLINE
1	-0.36	5.5	1086	314	"	CHLORTETRACYCLINE
2 ★	2.84	7.4	1999	1	C22H23Cl1N4O4	INDANDIONE,2-(4-PYRIDYLIDENE,2,6-DIMETHYL,1-PROPYL-CNU)
3	1.44	7.4	1441	1	C22H23F2N1O2	LENPERONE 24678-13-5 antipsychotic
4	1.56	7.4	1441	1	C22H23F4N1O2	TRIFLUPERIDOL 749-13-3 antipsychotic
95 ★	5.70		1903		C22H23N1O3	FENPROPATHRIN 39515-41-8 ✚
6	3.25	7.4	2196	1	C22H23N1O4	DIMEFLINE: N-MORPHOLINO ANALOG
7	1.95		634		C22H23N1O6	6-MEOETO-7-BENZYLOXY-4-OH-3-QUINOLINE-CO2ET
8	0.35	7.8	967		C22H23N1O7	DAUNORUBICIN-MO-ME-A-AGLYCONE
9	3.24	7.4	459	1	C22H23N3O2	1-PHENYL-3-PHENYL-5-PIPERIDINO-6-METHYLURACIL
15300	5.12		1101	421	C22H23O4P1	T-BUTYLPHENYL-DIPHENYLPHOSPHATE
1	2.66	7.4	2196	1	C22H24Cl1N1O3	DIMEFLINE: 4'-CHLORO-N,N-DIETHYL ANALOG
2	3.14	7.4	1569	1	C22H24Cl1N3O1	O-ETHYLAMOPYROQUIN
3	4.05		2162	459	C22H24Cl1N5O2	DOMPERIDONE 57808-66-9 anti-emetic
4	1.69	7.5	1476		C22H24Cl2N8O5	DICHLOROMETHOTREXATE,DIMETHYL ESTER
5	3.91	7.4	591	1	C22H24F1N3O2	BENPERIDOL 2062-84-2 antipsychotic
6	4.17		2162	459	"	BENPERIDOL
7	1.99	7.4	2196	1	C22H24F1O3	DIMEFLINE: 4'-FLUORO-N,N-DIETHYL ANALOG
8	4.39		2162	459	C22H24F4N2O1	TEFLUDAZINE 80680-06-4 anti-serotonergic, anti-dopaminergic, 5-ht2/d2 antagonist
9	-0.53	7.4	921	22	C22H24N1.I1	TRIBENZYLMETHYL AMMONIUM IODIDE
15310	0.06		1506		C22H24N1.C7H7O3S1	TRIBENZYL-METHYLAMMONIUMTOSYLATE 73566-66-2
1	0.10	7.4	579	324	C22H24N2O2	ACRIVASTINE 87848-99-5 antihistaminic (non-sedative)
2	2.83	7.4	847	1	C22H24N2O2	B-ME-N-(N-PIPERIDINYLETHYL)-G-PYROPHTHALONE
3	1.77	7.4	847	1	C22H24N2O2	N-(3-N-PIPERIDINYLPROPYL)-G-PYROPHTHALONE
4	0.74	7.3	1299	314	C22H24N2O3	B-COLUBRINE 509-36-4
15 ★	1.85		1299	2	"	B-COLUBRINE
6	1.70	7.4	2196	1	C22H24N2O5	DIMEFLINE: 4'-NITRO-N,N-DIETHYL ANALOG
7	-0.22	7.5	1831	1	C22H24N2O8	DOXYCYCLINE 564-25-0 antibacterial
8	-0.22	7.5	638	1	"	DOXYCYCLINE
9	-0.22	7.5	911	1	"	DOXYCYCLINE
15320	-0.12	7.0	1134	1	"	DOXYCYCLINE
1	-0.06	7.4	2530	1	"	DOXYCYCLINE
2	-0.02	5.5	1086	314	"	DOXYCYCLINE
3	-0.02	5.6	638	41	"	DOXYCYCLINE
4	-1.47	6.9	505	314	C22H24N2O8	TETRACYCLINE 60-54-8 antibacterial, anti-amebic, antirickettsial
25	-1.44	7.5	1831	1	"	TETRACYCLINE
6	-1.44	7.5	638	1	"	TETRACYCLINE
7	-1.43	7.0	1134	1	"	TETRACYCLINE
8	-1.42	7.5	911	1	"	TETRACYCLINE
9	-1.39	7.0	940	1	"	TETRACYCLINE

	logP	pH	Ref.	Note	MF	Name / CAS / activity
15330	-1.37	7.4	447	1	"	TETRACYCLINE
1	-1.25	5.6	638	41	"	TETRACYCLINE
2	-1.15	7.0	1295	1	"	TETRACYCLINE
3	-1.13	5.5	1086	314	"	TETRACYCLINE
4	-1.05	5.3	940		"	TETRACYCLINE
35	-1.05	6.6	907	314	"	TETRACYCLINE
6	-1.60	7.0	1134	1	C22H24N2O9	OXYTETRACYCLINE **79-57-2** *antibacterial*
7	-1.60	7.5	1831	1	"	OXYTETRACYCLINE
8	-1.60	7.5	638	1	"	OXYTETRACYCLINE
9	-1.12	5.6	638	41	"	OXYTETRACYCLINE
15340	-0.92	6.6	907	314	"	OXYTETRACYCLINE
1	-0.89	5.5	1086	314	"	OXYTETRACYCLINE
2 ★	3.38	12.0	2408		C22H24N4O1	MINAPRINE,4-PHENYL ANALOG ✚
3	1.77		1713		C22H24N4O4S1	6-PH-2-(4-(N3-TOLSULFUREIDO)BU)-3(2H)-PYRIDAZINONE
4	1.59		1711		C22H24N4O5	BENZYLMITOMYCIN-C
45	2.31	7.4	2095	1	C22H25Br1Cl1N3O4	(5-BROMOIMIDAZOLY-3-CHLOROPHENYL)ADRENOCEPTOR ANTAGONIST
6 ★	3.61	8.4	2022	2	C22H25Br1N4O1	SK&F 93,944 ✚
7	3.63	1.0	547	342	C22H25Br2N1O1	2,6-BR2-9(2-DIPRAMINO-1-OH-ET)PHENANTHRENE
8	2.22	5.0	1440	1	C22H25Cl1N2O1S1	CLOPENTIXOL **982-24-1** *antipsychotic, neuroleptic*
9	2.30	7.0	468	1	"	CLOPENTIXOL
15350	3.00	7.0	468	701	"	CLOPENTIXOL
1	4.18		2162	459	"	CLOPENTIXOL
2	3.31	1.0	547	342	C22H25Cl2N1O1	2,6-CL2-9(2-DIPRAMINO-1-OH-ET)PHENANTHRENE
3	2.98	7.4	712	1	C22H25N1O2	N-ISOOCTYL-G-PYROPHTHALONE
4	1.14	7.4	1994	1	C22H25N1O2	PIPERIDINE-3-CARBOXYLIC ACID,1-(4,4-DIPHENYL3-BUTENYL) **85375-85-5** *inhibitor-gaba uptake*
55	1.99	7.4	2196	1	C22H25N1O3	DIMEFLINE: N,N-DIETHYL ANALOG
6	2.41	7.4	2480	1	C22H25N1O3	KETOBEMIDONE,BENZOYL ESTER
7 ★	3.03	7.4	2480	2	"	KETOBEMIDONE,BENZOYL ESTER
8	3.28	7.4	1941	1	C22H25N1O4	3-(1-NAPHTHOXYPROPAN-2-OL)AMINE,N-(3-PHENOXYPROPAN-2-OL)
9 ★	3.87		2164		"	3-(1-NAPHTHOXYPROPAN-2-OL)AMINE,N-(3-PHENOXYPROPAN-2-OL)
15360	1.60	7.4	2684	1	C22H25N1O4	PHENYLACETIC ACID ESTER,3-PYRROLIDINYL ANALOG *antiarrhythmic (vitro)*
1	1.34	7.4	2684	1	C22H25N1O4	PHENYLACETIC ACID ESTER,4-PYRROLIDINYL ANALOG *antiarrhythmic (vitro)*
2 ★	1.75	7.0	2678	1	C22H25N1O4S2	9-THIODEOXOTHIOCOLCHICINE
3	1.56	7.4	2684	1	C22H25N1O5	P-HYDROXYPHENYLACETIC ACID ESTER,3-PYRROLIDINYLMETHYL ANALOG *antiarrhythmic (vitro)*
4 ★	1.03		505		C22H25N1O6	COLCHICINE **64-86-8** *suppressant (gout)*
65	1.30	7.0	2678	1	"	COLCHICINE
6	1.30	7.4	594	1	"	COLCHICINE
7 ★	1.00		2494	400	C22H25N1O6	ISOCOLCHICINE **518-12-7**
8	0.33	7.0	2678	1	C22H25N1O7	COLCHICILINE
9 ★	3.27		1898		C22H25N1O8	(4-HOET)VANILLINOXIME-O-ETACETYL,2-MEOBENZYL ESTER
15370	1.18	7.4	847	1	C22H25N2O2.C1H3O4S1	N-(N-MEPIPERIDONYLETHYL)-G-PYROPHTHALONE
1	2.27	7.4	1209	459	C22H25N3O1	INDORAMIN **26844-12-2** *antihypertensive*
2	2.31	7.4	1209	1	"	INDORAMIN
3 ★	3.43	9.8	2039		C22H25N3O4	R-28935 **55806-43-4** ✚
4 ★	2.98	5.0	2375	2	C22H25N3O4S1	MORICIZINE **31883-05-3** *cardiac depressant (anti-arrhythmic)*
75	1.24	7.4	2095	1	C22H26Br1N3O4	(5-BROMOIMIDAZOLE)ADRENOCEPTOR ANTAGONIST
6 ★	2.61		1520		C22H26Cl1F1O4	CLOBETASONE **54063-32-0** *anti-inflammatory*
7	3.86		2162	459	C22H26Cl1N3O2	BENZAMIDE,4-AMINO-5-CHLORO-2-METHOXY-N-(8-BENZYL-TROPAN-4-YL) **105846-97-7**
8	1.61	7.4	2095	1	C22H26Cl1N3O4	(5-CHLOROIMIDAZOLE)ADRENOCEPTOR ANTAGONIST ✚
9	1.37	7.4	1441	1	C22H26F1O2	MOPERONE **1050-79-9** *tranquilizer*
15380	3.50	7.0	468	1	C22H26F3N3O1S1	FLUPHENAZINE **69-23-8** *antipsychotic*
1 ★	4.36	7.0	468	701	"	FLUPHENAZINE
2	1.17	6.0	581		C22H26F3N5O2	4,6-DIAM-2,2DIME-SYM-TRIAZINE,1(3(M-CF3-PHO-BUO)PH)
3	1.71	7.4	675	1	C22H26N2O3	3,5-PYRAZOLIDINDIONE,1-PH,2(4-OH-PH),4(4,4DIMEPE)
4	0.51	7.4	2095	1	C22H26N2O4S1	(THIAZOLE) ADRENOCEPTOR ANALOG
85	2.70	7.0	1694		C22H26N2O4S1	DILTIAZEM **42399-41-7** *vasodilator (coronary)*
6	1.60	7.4	2095	1	C22H26N2O5	(OXAZOLE) ADRENOCEPTOR ANTAGONIST
7 ★	2.43		1592		C22H26N2O6	BENZYLOXYCARBONYL-VALINYLTYROSINE **862-26-0**
8 ★	1.57	4.6	2043	2	C22H26N2O8S2	CEPHALOTHIN,PIVALOYLOXYMETHYL ESTER
9	1.76		910		C22H26N4	1,6-BIS(5-METHYL-2-BENZIMIDAZOLYL)HEXANE
15390 ★	4.45		2463		C22H26N4O4	BENZOTRIAZOLE,2-CYCLOHEXYL-5-(3,4,5-TRIMETHOXY)BENZAMIDO
1 ★	-1.51		2377		C22H26N4O5	PHE-GLY-GLY-PHE
2	0.54	7.5	1476	1	C22H26N8O5	METHOTREXATE,DIMETHYL ESTER
3 ★	6.14		2661		C22H26O3	BIORESMETHRIN **28434-01-7** *insecticide*
4	0.76		636		C22H26O8	EUPAROTINACETATE ✚
95 ★	1.19		636		C22H26O8	LIATRIN **34175-79-6**
6 ★	2.58		569		C22H27Cl1F2O5	HALOMETASONE **50629-82-8** *anti-inflammatory*
7	-0.06	5.0	2375	1	C22H27Cl1N2O1	LORCAINIDE **59729-31-6** *cardiac depressant (anti-arrhythmic)*
8	-0.77		559		C22H27Cl1N3S1.I1	3-H2-2(ET2-ME-N+ETS-)4(O-CLPH)BENZODIAZEPINI
9	-0.87		559		C22H27Cl1N3S1.I1	3-H2-2(ET2-ME-N+ETS-)4(P-CLPH)BENZODIAZEPINI
15400	1.42		636		C22H27Cl1O8	EUPACHLORINACETATE
1	3.57		2162	459	C22H27F1N2O1	INDANE,6-METHYL-1-(4-ETHANOLPIPERAZINE)-3-(4-FLUORO)PHENYL
2	3.80		2162	459	C22H27F1N2O1S1	INDANE,6-METHYLTHIO-1-(4-ETHANOLPIPERAZINE)-3-(4-FLUORO)PHENYL
3	3.04		2162	459	C22H27F1N2O2	INDANE,6-METHOXY-1-(4-ETHANOLPIPERAZINE)-3-(4-FLUORO)PHENYL
4	1.69	1.0	547	342	C22H27N1O1	9-(2-DIPROPYLAMINO-1-HYDROXYET)PHENANTHRENE
5	1.69	7.4	2493	1	C22H27N1O2	2-PROPANONE,1,1-DIPHENYL-1-ETHOXY-3-(N-PIPERIDYL)
6	3.14	7.4	2143	1	C22H27N1O3	4-PIPERIDINOMETHYLBENZYLCARBONATE,2',6'-DIMETHYLPHENYL
7	3.10	7.4	2143	1	C22H27N1O3	CARBONIC ACID,2,6-DIMETHYLPHENYL,3-(N-PIPERIDINOMETHYL)BENZYL ESTER
8	4.00	7.4	2621	1	C22H27N1O7	MITOSENE(5,5,6),METHYL-DIBUTYRYLOXY ANALOG
9	1.46	5.0	2195	1	C22H27N3O1S1	ZOLANTIDINE ✚
15410	2.88		2162	459	C22H27N3O4S1	BENZAMIDE,2-METHOXY-5-SULFONAMIDO-N-(8-BENZYLTROPAN-4-YL)
1	0.53	7.4	2274	1	C22H27N3O5	INDOLE,2-CARBOXAMIDE-4-(3-(3,4-DIMETHOXYPHENETHYLAMINO)-2-HYDROXYPROPOXY
2 ✓	0.11		2397	101	C22H27N4O3S1	IMIDAZOLIUM CHLORIDE,1-ETHYL-2-HYDROXYIMINOMETHYL-3-(3-N-BENSYL-PHENYLSULFONAMIDO)PROPYL
3 ★	2.62	4.6	2043	2	C22H27N5O9S2	CEFOTAXIME,PIVALOYLOXYMETHYL ESTER
4	0.96	7.4	2591	1	C22H27N7O5	AZT,5'-(4-PHENYLPIPERAZINE-1-ACETYL) **125762-96-1** *antiviral prodrug*
15 ★	1.74	6.6	1703	314	C22H27N9O7S3	CEFMENOXIME,PIVALOYLOXYMETHYL ESTER
6 ★	2.48		1520		C22H28Cl1F1O4	CLOBETASOL **25122-41-2** *anti-inflammatory*
7 ★	1.94		569		C22H28F2O5	6-A-FLUORO-DEXAMETHASONE **2135-17-3** *glucocorticoid*
8	-4.35		875	242	C22H28N1O3.Br1	BENZILONIUM BROMIDE **1050-48-2** *anticholinergic*
9	-0.82		875	242	C22H28N1O3.C6H13O3S1	BENZILONIUMHEXYLSULFONATE

	logP	pH	Ref.	Note	MF	Name / CAS / activity
15420	-0.55		1768	820	"	BENZILONIUMHEXYLSULFONATE
1	-0.03		1768	820	C22H28N1O3.C7H15O3S1	BENZILONIUMHEPTANESULFONATE
2	0.23		875		C22H28N1O3.C8H17O3S1	BENZILONIUMOCTYLSULFONATE
3	0.94		875	242	C22H28N1O3.C10H21O3S1	BENZILONIUMDECYLSULFONATE
4	1.85		875	242	C22H28N1O3.C12H25O3S1	BENZILONIUMDODECYLSULFONATE
25	1.59	4.8	1020		C22H28N2O1	CINNAMAMIDE,N(2,6-DIMEPH),N(3-DIMEAMINOPROPYL)
6	1.58	4.8	1020		C22H28N2O1	CINNAMAMIDE,N(2,6-DIMEPH)N-(1-ME-2-DIMEAMINOET)
7	2.86	7.4	547	1	C22H28N2O1	FENTANYL 437-38-7 analgesic (narcotic)
8 ★	2.98	7.4	1837	1	"	FENTANYL
9 ★	4.05		1837			FENTANYL
15430	2.73	7.4	2143	1	C22H28N2O2	2,6-DIMETHYLPHENYLCARBAMIC ACID,2-(N-PIPERIDINOMETHYL)BENZYL ESTER
1	2.14	7.4	2143	1	C22H28N2O2	2,6-DIMETHYLPHENYLCARBAMIC ACID,3-(N-PIPERIDINOMETHYL)BENZYL ESTER
2	3.12	7.4	2143	1	C22H28N2O2	2-(PIPERIDINOMETHYL)BENZYLCARBAMIC ACID,2,6-DIMETHYLPHENYL ESTER
3	2.33	7.4	2143	1	C22H28N2O2	3-(N-PIPERIDINOMETHYL)BENZYLCARBAMIC ACID,2,6-DIMETHYLPHENYL ESTER
4	2.25	7.4	2143	1	C22H28N2O2	4-PIPERIDINOMETHYLBENYL CARBAMATE,2',6'-DIMETHYLPHENYL
35	2.31	7.4	2143	1	C22H28N2O2	4-PIPERIDINOMETHYLBENZYLCARBAMATE,O-2',6'-DIMETHYLPHENYL
6	2.45	7.5	2195	1	C22H28N2O2	BENZAMIDE,N-3-(3-(N-PIPERIDINYLMETHYL)PHENOXY)
7	2.50		2609	468	C22H28N2O3	DIMETHYLETHER,BIS-(N-BENZYL-N-ETHYL-AMIDO)
8	0.07		547		C22H28N4O6	MITOXANTRONE 70476-82-3 antineoplastic
9	0.70	7.5	2624	1	"	MITOXANTRONE
15440	1.39		1937	459	C22H28O3	4-HYDROXY-2-PYRONE,3,5-DIBUTYL-6-(1-PHENYLPROP-1-ENYL) 98393-86-3
1 ★	2.68	7.4	1252	1	C22H28O3	PREGN-4,6-DIENE-21-CO2H-17-OH-3-OXO-G-LACTONE 976-71-6 aldosterone antagonist
2 ★	1.89	7.4	1252	1	C22H28O4	PREGN-4ENE-21-CO2H,9,11A-EPOXY-17-OH-3OXO-G-LACTONE 60687-48-1
3	1.96	7.4	1252	1	"	PREGN-4ENE-21-CO2H,9,11A-EPOXY-17-OH-3OXO-G-LACTONE
4	5.69		2021	459	C22H28O6	NAPHTHALENE,1,4-DI-T-PENTANOYLOXY-2,3-DIMETHOXY 91814-09-4
45	3.38		1396	459	C22H28O7	ARTEMISININE,PHENYLCARBONATE(ALPHA)
6	4.41		1396	459	"	ARTEMISININE,PHENYLCARBONATE(ALPHA)
7 ★	1.16		636		C22H28O7	EUPACUNIN 33854-15-8
8 ★	2.35		1520		C22H29F1O4	BETAMETHASONE-21-DESOXY 1879-77-2
9 ★	2.00		1758		C22H29F1O4	FLUOROMETHOLONE 426-13-1 glucocorticoid
15450 ★	2.40	7.4	1252	1	C22H29F1O4	PREGN-4ENE-21-CO2H,9A-F-11B,17A-OH-3OXO-G-LACTONE 1582-75-8
1 ★	1.94		1146		C22H29F1O5	BETAMETHASONE 378-44-9
2	1.72	7.0	2002	1	C22H29F1O5	DEXAMETHASONE 50-02-2 glucocorticoid
3 ★	1.83		1146		"	DEXAMETHASONE
4	1.94	1.0	870	342	C22H29N1O1	2-(P-ME-DIPHENYLMETHOXY)ETHYL-N-CYCLOHEXYLAMINE
55	1.39	1.0	602	342	C22H29N1O1	N-HEXAHYDROAZEPINYLETHYL-P-ME-DIPHENYLMEETHER
6 ★	3.50	8.0	2644	2	C22H29N1O2	FOMOCAINE ANALOG-2 (4-ME)
7	2.06	4.0	2644	1	C22H29N1O2	FOMOCAINE ANALOG-29 (P-T-BUTYL)
8	1.41	7.4	2217	1	C22H29N1O2	PROPOXYPHENE 469-62-5 analgesic
9	2.36	7.4	1209	1	"	PROPOXYPHENE
15460 ★	4.18	9.2	536	2	"	PROPOXYPHENE
1 ★	3.58	8.0	2644	2	C22H29N1O2S1	FOMOCAINE ANALOG-37 (ORTHO CH.2-MEO)
2	1.29	7.4	2680	324	C22H29N1O3	3-O-PENTYLMORPHINE 74886-08-1
3	1.88	4.0	2644	1	C22H29N1O3	FOMOCAINE ANALOG-28 (P-BUTOXY)
4 ★	2.82	8.0	2644	2	C22H29N1O3	FOMOCAINE ANALOG-4 (2-OME)
65 ★	3.35	8.0	2644	2	C22H29N1O3	FOMOCAINE ANALOG-5 (3-OME)
6 ★	3.10	8.0	2644	2	C22H29N1O3	FOMOCAINE ANALOG-6 (4-OME)
7	1.72	4.0	2655	1	C22H29N1O3	PROPRANOLOL,CYCLOPENTANOYL ESTER
8	0.87	4.0	2655	1	C22H29N1O5	PROPRANOLOL,METHYLGLUTARYL ESTER
9 ★	5.41		1598	40	C22H29N3S2	THIETHYLPERAZINE 1420-55-9 anti-emetic
15470 ★	0.03	13.0	547	224	C22H29N7O5	PUROMYCIN 53-79-2 antineoplastic, antiprotozoal (trypanosoma)
1	0.86	7.4	547	1	"	PUROMYCIN
2	0.08	5.0	831	1	C22H30Cl2N10	CHLORHEXIDINE 55-56-1 anti-infective (topical)
3 ★	1.54	7.4	2242	1	C22H30F1O4	DEXAMETHASONE,17-B-(N-METHYLCARBOXAMIDE) ANALOG ✚
4	1.44	7.4	2501	1	C22H30F1O4	FLUSOXOSOL 84057-96-5
75 ★	3.70	13.0	2501		"	FLUSOXOSOL
6	1.15		817		C22H30N1O3.Br1	N(2-BENZILOYLOXYET-N-BU-N,N-DIMETHYLAMMONIUMBR
7	-0.09	5.0	2375	1	C22H30N2	APRINDINE 37640-71-4 cardiac depressant (anti-arrhythmic)
8	0.57	7.4	608	1	C22H30N2O1	ACETONE,1-PH-1(2-PYRIDYL)-1-(N,N-DI-I-PR-AMINOETHYL)
9 ★	3.66		2216		C22H30N2O1	BUTYRAMIDE,2-(P-BIPHENYL)-4-DI-I-PROPYLAMINO
15480	3.24	7.4	1837	1	C22H30N2O2S1	SUFENTANIL 56030-54-7 analgesic
1 ★	3.95	10.8	1837	1	"	SUFENTANIL
2	5.32	7.4	2223	1	C22H30N2O3	DISOXARIL,4'-ETHYL ✚
3	3.63	7.4	2102	1	C22H30N2O4	PILOCARPIC ACID,O-BUTYROYL,BENZYL ESTER
4 ★	3.78		2102	2	"	PILOCARPIC ACID,O-BUTYROYL,BENZYL ESTER
85 ★	2.40		2531		C22H30N2O5	PHENOXYACETAMIDE ANALOG,N-I-PROPYL 139733-61-2
6 ★	2.40		2531		C22H30N2O5	PHENOXYACETAMIDE ANALOG,N-PROPYL 139733-59-8
7	-1.10	7.4	2449	1	C22H30N2O5S2	SPIRAPRIL 83647-97-6 antihypertensive (ace-inhibitor)
8 ✓	0.87		2449	226	"	SPIRAPRIL
9 ★	1.70		2531		C22H30N2O6	PHENOXYACETAMIDE ANALOG,N-(2-METHOXYETHYL) 139892-81-2
15490	0.50		2531		C22H30N2O6	PHENOXYACETAMIDE ANALOG,N-(3-HYDROXYPROPYL) 139733-77-0
1	2.03		1713		C22H30N4O2	6-PH-2-(5-N3-CYHEX-UREIDO-PENTYL)-3(2H)-PYRIDAZINONE
2	0.83	4.3	882	1	C22H30N4O2S2	THIOPROPERAZINE 316-81-4 neuroleptic, anti-emetic
3	1.21	7.4	2244	1	C22H30N4O6S1	TIMOLOL,O-(2-ACETYLOXY)BENZOYL
4 ★	2.75	7.4	401	1	C22H30O3	6-A-ME-PREGN-4-ENE-3,11,20-TRIONE 3642-85-1 antineoplastic
95 ★	2.98	7.4	1252	1	C22H30O3	DESACETYLTHIOSPIRONOLACTONE 976-70-5
6	4.20		1396	459	C22H30O5	ARTEMISININEBENZYLETHER(BETA)
7 ★	5.72		1092		C22H31N1	3(4-T-BUTYLBENZHYDRYLPROPYL)AMINE,N,N-DIMETHYL
8	1.53	1.0	602	342	C22H31N1O1	P-ME-2-DIPHENYLMETHOXY-N,N-DIPROPYLETHYLAMINE
9	1.22	4.0	2644	210	C22H31N1O3	FOMOCAINE ANALOG-24 (BIS-ETHOXYETHYLAMINE)
15500	3.38	7.4	1941	1	C22H31N1O4	2-METHYLPROPYLAMINE,N,N-DI-(3-PHENOXY-2-HYDROXYPROPYL)
1 ★	3.46	13.0	1941		"	2-METHYLPROPYLAMINE,N,N-DI-(3-PHENOXY-2-HYDROXYPROPYL)
2	3.37	7.4	1941	1	C22H31N1O4	T-BUTYLAMINE,N,N-DI-(3-PHENOXY-2-HYDROXYPROPYL)
3 ★	3.99	13.0	1941		"	T-BUTYLAMINE,N,N-DI-(3-PHENOXY-2-HYDROXYPROPYL)
4	0.62		2627	841	C22H31N2.Cl1	N-METHYL-4-(4'-DIBUTYLAMINOSTYRYL)-PYRIDINIUM CHLORIDE
5	-0.13	7.4	608	1	C22H31N3O1	SEARLE#7 ✚
6	-1.00	7.4	608	1	C22H31N3O2	DIISOPYRAMIDE,N1-CH2OH
7 ✓	0.55	4.9	2018		C22H31N3O5	CILAZAPRIL 88768-40-5 antihypertensive
8	-0.37	7.4	2449	1	C22H31N3O5S1	UTIBAPRIL 109683-61-6 ace inhibitor
9 ★	1.62	3.0	2018		"	UTIBAPRIL

	logP	pH	Ref.	Note	MF	Name / **CAS** / *activity*
15510	5.44		1101	421	C22H31O4P1	I-DECYLDIPHENYLPHOSPHATE
1	-1.52	7.5	2018		C22H32N2O2	DOPEXAMINE **86197-47-9** *cardiovascular agent*
2 ✓	1.88		2018		"	DOPEXAMINE
3 ★	3.67	7.4	1956	2	C22H32N2O3	PILOCARPIC ACID,P-T-BUTYL ESTER **92598-99-7**
4	-0.53	7.4	608	1	C22H32N4O1	DIISOPYRAMIDE,N1-(NH-METHYL)
15	1.32	7.0	594	314	C22H32N4O4	VALYL-TRYPTOPHYL-ISOLEUCINE
6 ★	0.43		2377	821	C22H32N4O6	AC-GLU-ILE-PHE-N
7 ★	0.32		2377	821	C22H32N4O6	AC-GLU-PHE-LEU-N
8 ★	0.48		2377	821	C22H32N4O6	AC-LEU-GLU-PHE-N
9 ★	-1.65		2377		C22H32N4O6	TYR-PRO-GLY-ILE
15520 ★	7.62		2415	459	C22H32O2	1-METHOXY-D(9-11)THC
1 ★	8.06		2415	459	C22H32O2	1-METHOXY-D8-THC
2	3.72		2631	538	C22H32O3	TESTOSTERONEPROPIONATE **57-85-2** *androgen*
3	0.02		307	1	C22H33N1O1	3-HYDROXY-N-HEXYLMORPHINAN HCL
4	4.39	7.4	1990	1	C22H33N1O1	N,N-DIMETHYLDECYLAMINE,10-(1-NAPHTHOXY)
25	2.08	5.7	2241	1	C22H33N1O3	CYCLOMETHYCAINE **139-62-8** *anesthetic (topical)*
6	3.20	7.4	2480	1	C22H33N1O4	KETOBEMIDONE,HEXYLCARBONATE
7 ★	3.38	7.4	2480	2	"	KETOBEMIDONE,HEXYLCARBONATE
8 ★	1.07	7.0	1793	1	C22H33N7O5S1	N(4-(6-PURINYLTHIO)VALERYL)GLYGLYLEUCINE,ET.ESTER
9	0.18	7.4	2320	1	C22H34N2O3	2,6-BIS(1-PYRROLIDINYLMETHYL)PHENOL-4-ACETIC ACID,BUTYL ESTER
15530	0.10	7.4	2320	1	C22H34N2O3	2,6-BIS(1-PYRROLIDINYLMETHYL)PHENOL-4-ACETIC ACID,S-BUTYL ESTER
1	-0.21	7.4	2320	1	C22H34N2O3	2,6-BIS(1-PYRROLIDINYLMETHYL)PHENOL-4-PROPIONIC ACID,I-PROPYL ESTER
2	-0.11	7.4	2320	1	C22H34N2O3	2,6-BIS(1-PYRROLIDINYLMETHYL)PHENOL-4-PROPIONIC ACID,PROPYL ESTER
3	0.10	7.4	2320	1	C22H34N2O3	2,6-BIS-(1-PYRROLIDINYLMETHYL)PHENOL,4-ACETIC ACID,I-BUTYL ESTER
4	4.20	7.3	1149	714	C22H34N2O3	3(OPENT)CARBANILATE,O-(2-N-PIPERIDINYL)CYCLOPENTL **60093-72-3**
35	3.69	7.3	1149	714	C22H34N2O3	3(OPENT)CARBANILATE,O-(2-N-PYRROLIDYL)CYCLOHEXYL **38198-35-5** *anesthetic (local)*
6	3.38	7.3	873	314	C22H34N2O3	O-3(N-PIPERIDINO)C-C6H10-N(M-OC4H9-C6H4)CARBAMATE
7	3.92	7.3	873	314	C22H34N2O3	O-3(N-PIPERIDINO)C-C6H10-N(P-OC4H9-PH)CARBAMATE
8	3.38	7.3	873	314	C22H34N2O3	O-3(N-PYRROLIDINO)CYHEX-N(M-OC5H11-PH)CARBAMATE
9	3.04	7.3	873	314	C22H34N2O3	O-3(N-PYRROLIDINO)CYHEX-N(P-OC5H11-PH)CARBAMATE
15540	-0.81	7.4	1026	1	C22H34N2O4	(2,3-NAPHTHALENEDIOXY)BIS(3-I-PR-AM-2-PROPANOL)
1	-0.76	7.4	487	1	C22H34N2O4	1,2-BIS(2-OH-3-I-PRAMPR)NAPHTHYLDIETHER
2	1.07	7.4	487	2	"	1,2-BIS(2-OH-3-I-PRAMPR)NAPHTHYLDIETHER
3	-0.94	7.4	487	1	C22H34N2O4	1,3-BIS(2-OH-3-I-PRAMPR)NAPHTHYLDIETHER
4	0.99	7.4	487	2	"	1,3-BIS(2-OH-3-I-PRAMPR)NAPHTHYLDIETHER
45	-1.02	7.4	487	1	C22H34N2O4	1,4-BIS(2-OH-3-I-PRAMPR)NAPHTHYLDIETHER ✚
6	0.93	7.4	487	2	"	1,4-BIS(2-OH-3-I-PRAMPR)NAPHTHYLDIETHER
7	-0.92	7.4	487	1	C22H34N2O4	1,5-BIS(2-OH-3-I-PRAMPR)NAPHTHYLDIETHE
8	0.98	7.4	487	2	"	1,5-BIS(2-OH-3-I-PRAMPR)NAPHTHYLDIETHE
9	-1.11	7.4	487	1	C22H34N2O4	1,6-BIS(2-OH-3-I-PRAMPR)NAPHTHYLDIETHER
15550	0.87	7.4	487	2	"	1,6-BIS(2-OH-3-I-PRAMPR)NAPHTHYLDIETHER
1	-0.49	7.4	487	1	C22H34N2O4	1,8-BIS(2-OH-3-I-PRAMPR)NAPHTHYLDIETHER
2	1.37	7.4	487	2	"	1,8-BIS(2-OH-3-I-PRAMPR)NAPHTHYLDIETHER
3	5.70	7.3	1149	714	C22H34N2O4	3(OPENT)CARBANILATE,O-(2-N-MORPHOLINYL)CYCLOHEXYL **60093-69-8**
4	-1.77		1383		C22H34N2.(I1)2	1,1'-DIHEXYL-4,4'-BIPYRIDYLIUM IODIDE
55	-0.77	12.5	562	1	C22H34N6O4	2,5BIS(1(4-HOET)PIPERIZYL)3,6-DIAZIRIDINYLQUINONE ✚
6	-0.73		562		"	2,5BIS(1(4-HOET)PIPERIZYL)3,6-DIAZIRIDINYLQUINONE
7 ★	0.51		2377	821	C22H35N5O4	AC-LYS-PHE-VAL-N
8	4.82	7.3	1356	314	C22H36N2O3	2-OCTYLOXYCARBANILATE-O-2(N-PIPERIDINYL)ETHYL **55792-24-0**
9	6.36	7.3	1356	2	"	2-OCTYLOXYCARBANILATE-O-2(N-PIPERIDINYL)ETHYL
15560	3.85	7.1	1150		C22H36N2O3	2ME-4(O-HEX)CARBANILATE-O-(1-ME-2(PIPERIDINYL)ET
1	3.55	7.3	1149	714	C22H36N2O3	3(OPENT)CARBANILATE,O-(2-N,N-DIETAMINO)CYCLOHEXYL **38198-31-1**
2	4.29	7.3	1149	714	C22H36N2O3	3(OPENT)CARBANILATE,O-(2-N-BUTYL)CYCLOHEXYL **60093-68-7**
3	4.26	7.3	1356	314	C22H36N2O3	3-OCTYLOXYCARBANILATE-O-2(N-PIPERIDINYL)ETHYL **55792-25-1**
4	5.82	7.3	1356	2	"	3-OCTYLOXYCARBANILATE-O-2(N-PIPERIDINYL)ETHYL
65	4.78	7.3	1356	314	C22H36N2O3	4-OCTYLOXYCARBANILATE-O-2(N-PIPERIDINYL)ETHYL **55948-94-2**
6	6.44	7.3	1356	2	"	4-OCTYLOXYCARBANILATE-O-2(N-PIPERIDINYL)ETHYL
7	3.42	7.0	1517	1	C22H36N2O3	CARBAMICAC,N-ET-N-(M-PENOPH),2(1-PIPERIDINYL)PR.EST ✚
8	3.42	7.1	1357		"	CARBAMICAC,N-ET-N-(M-PENOPH),2(1-PIPERIDINYL)PR.EST
9	3.54		1517	1	C22H36N2O3	CARBAMICAC,N-PR-N-(M-PENOPH),2(N-PIPERIDINYL)ET.ES
15570	3.54	7.1	1357		"	CARBAMICAC,N-PR-N-(M-PENOPH),2(N-PIPERIDINYL)ET.ES
1 ★	7.18		2415	459	C22H36O2	CYCLOHEXANOL,3-(2-HYDROXY-4-(1,1-DIMETHYL)OCTYL)PHENYL
2 ★	7.49		2415	459	C22H36O2	CYCLOHEXANOL,4-METHYL-3-(2-HYDROXY-4-(1,1-DIMETHYL)HEXYL)PHENYL
3	8.01		2415	459	"	CYCLOHEXANOL,4-METHYL-3-(2-HYDROXY-4-(1,1-DIMETHYL)HEXYL)PHENYL
4 ★	0.23		2308		C22H36O9	BENZO-27-CROWN-9-ETHER
75 ✓	3.34		1698	820	C22H37N2O3.Br1	2-HEPTOXYPHCARBAMICAC.,PIPERIDINOET.EST.,N-MEBR
6 ✓	3.35		1720	820	C22H37N2O3.Br1	N(O-HEPTOXYPH)CARBAM.AC,2(N-MEPIPERIDINIUM)ET.EST
7 ✓	4.00		1720	820	C22H37N2O3.I1	N(O-HEPTOXYPH)CARBAM.AC,2(N-MEPIPERIDINIUM)ET.ES.I
8 ✓	2.03		1698	820	C22H37N2O3.H2O4P1	2-HEPTOXYPHCARBAMICAC.,PIPERIDINOET.EST.,N-MEPHOS
9	4.11	7.1	1150		C22H38N2O3	2-ME-4(O-BU)CARBANILATE-O-2(N,N-DIBUAMINO)ETHYL
15580	-0.73	7.4	1026	1	C22H38N2O4	(TETRALINENE-2,3-DIOXY)BIS(3-I-PR-AM-2-PROPANOL)
1	2.18	1.0	547	342	C22H39N1O1	2-N,N-DIHEPTYLAMINO-1-PHENYLETHANOL HYDROCHLORIDE
2 ✓	1.44		1729	820	C22H39N2O1.I1	3-CARBAMOYLPYRIDINIUM IODIDE,N-HEXADECYL **36846-92-1**
3	0.82	7.4	2284	1	C22H39N2O3.I1	CARBISOCAINE METH IODIDE
4	0.08		1506		C22H40N1.Cl1	BENZYL-TRIPENTYLAMMONIUMCHLORIDE **74900-79-1**
85 ★	-1.41		2377		C22H42N4O5	ILE-ILE-VAL-VAL
6 ★	-1.23		2377		C22H42N4O5	VAL-LEU-VAL-LEU
7 ✓	1.88	7.3	2085		C22H48N4.O2Tc1	N-DODECYLCYCLAM,TECHNICIUM OXIDE COMPLEX
8	8.90		1740	459	C22H66O9Si10	DOCOSOMETHYLDECASILOXANE
9	1.90		2114		C23H16O3	DIPHENADIONE **82-66-6** *anticoagulant~rodenticide*
15590	-4.80	7.4	2559	1	C23H16O11	CROMOLYN **16110-51-3** *antiasthmatic*
1 ★	1.92	1.2	2017	2	"	CROMOLYN
2	3.11		2524	459	C23H18F3N3O1	IMIDAZOLYLETHANOL,4'-TRIFLUOROMETHYL-4"-(4-PYRIDYL) ANALOG
3 ★	3.62		579		C23H18N2O2	2-DIPHENYLACETYL-1,3-INDANDIONE-1-HYDRAZONE **5102-79-4**
4	6.02		2616	459	C23H19Cl1F3N1O3	CYHALOTHRIN **68085-85-8** *insecticide, acaricide*
95	4.23		2524	459	C23H19F1N2O1	IMIDAZOLYLETHANOL,4'-FLUORO-4"-PHENYL ANALOG
6	2.35		2524	459	C23H19F1N2O1	PYRROLYLETHANOL,4'-FLUORO-4"-(4-PYRIDYL) ANALOG
7	2.38		2524	459	C23H19N3O2	IMIDAZOLYLETHANOL,4'-CARBOXALDEHYE-4"-(4-PYRIDYL) ANALOG
8	1.08	7.4	675	1	C23H20N2O2S1	3,5-PYRAZOLIDINDIONE-1,2-DIPH-4-(2-S-PH-ETHYL)
9	0.63	7.4	675	1	C23H20N2O3S1	3,5-PYRAZOLIDINDIONE,1-PH,2-(4-OH-PH)4-ET-S-PH

	logP	pH	Ref.	Note	MF	Name / CAS / activity
15600	-0.08	7.4	675	1	C23H20N2O3S1	SULFINPYRAZONE 57-96-5 uricosuric
1 ★	2.30		508			SULFINPYRAZONE
2	-0.39	7.4	675	1	C23H20N2O4S1	3,5-PYRAZOLIDINDIONE,1-PH,2-(4-OH-PH)4-ET-SO-PH
3	2.98		1593	459	C23H20N6O8S4	CEPHALOSPORIN,PHENYLCARBONYL ANALOG 69712-36-3
4	2.69	7.4	2557		C23H20O6	PENTYLIDENEBIS-4-HYDROXYCOUMARIN
5	1.63	7.4	2556	1	C23H21N1O4	M-DIBENZOYLEPININE
6	3.13		2524	459	C23H21N3O1	IMIDAZOLYLETHANOL,4'-METHYL-4''-(4-PYRIDYL) ANALOG
7	2.40		2524	459	C23H21N3O2	IMIDAZOLYLETHANOL,4'-HYDROXYMETHYL-4''-(4-PYRIDYL) ANALOG
8 ★	3.55		2524	459	C23H21N3O2	IMIDAZOLYLETHANOL,4'-METHOXY-4''-(4-PYRIDYL) ANALOG
9 ★	6.00		2506	20	C23H22Cl1F3O2	BIFENTHRIN insecticide
15610 ★	2.96		935		C23H22N2O6S1	CARBENICILLIN,PHENYL 27025-49-6 antibacterial
1 ★	4.10		572		C23H22O6	ROTENONE 83-79-4 insecticide
2	3.84		2524	459	C23H23F1N2O1	1-PYRROLIDINYLETHANOL,4'-FLUORO-4''-(4-PYRIDYL) ANALOG
3	2.82		2524	459	C23H23F1N2O2	1-MORPHOLINOETHANOL,4'-FLUORO-4''-(4-PYRIDYL) ANALOG ✦
4	1.57	7.4	2493	1	C23H23N1O2	2-PROPANONE,1,1-DIPHENYL-1-PHENOXY-3-DIMETHYLAMINO
15	0.74		401		C23H23N1O5	NITIDINE,12,13-DIHYDRO,13-ETHOXY
6	0.86	7.4	2686	1	C23H23N2	N,N'-DIETHYLISOCYANINE noradrenaline transport inhibitor
7	0.92	7.4	2686	1	C23H23N2	N-METHYL-N'-I-PROPYLISOCYANINE noradrenaline transport inhibitor
8	3.46	7.4	2196	1	C23H24Cl1N1O3	DIMEFLINE: 4'-CHLORO-N-PIPERIDINYL
9	3.79	1.0	547	342	C23H24Cl2F3N1O1	2,3-DICL-7-CF3-9(DIPRAM-1-OH-ET)PHENANTHRENE
15620	2.71	7.4	2196	1	C23H24F1O3	DIMEFLINE: 4'-FLUORO-N-PIPERIDINYL ANALOG ✦
1	2.53	7.4	2196	1	C23H24N2O5	BENZOPYRAN-4-ONE,8-(N-PIPERIDINYL)METHYL-7-METHOXY-3-METHYL-2-(4-NITROPHENYL)
2	1.90	7.4	2095	1	C23H24N4O2S1	(5-THIENYL-IMIDAZOLE)ADRENOCEPTOR ANTAGONIST
3 ★	1.15		1532		C23H24N4O6	BENZYLCARBONYLMITOMYCINC
4 ★	1.26		1532	459	C23H24N4O7	BENZOYLOXYMETHYLMITOMYCINC
25 ★	1.34		1532		C23H24N4O7	BENZYLOXYCARBONYLMITOMYCINC
6	4.00		2171		C23H24N6O4	AZOBENZENE,2,6-DICYANO-4-NITRO-1'-METHYL-4'-N,N-(BUTYL-METHOXYCARBONYLETHYL)AMINO
7	4.31		2616	459	C23H24O4S1	KADETHRIN 58769-20-3
8	4.94		1582	459	C23H25Cl1N2O3	BENZODIAZEPIN ANALOG#34 20622-25-7
9	2.96	7.0	468	1	C23H25F3N2O1S1	FLUPENTIXOL 2709-56-0 antipsychotic
15630	3.83	7.0	468	701	"	FLUPENTIXOL
1 ★	4.51		1598	40	"	FLUPENTIXOL
2	4.25	6.0	1148	807	"	FLUPENTIXOL
3	4.25	6.0	1148	807	"	FLUPENTIXOL
4	4.81		2162	459	"	FLUPENTIXOL
35	4.91		2162	459		FLUPENTIXOL
6	2.50	7.4	2196	1	C23H25N1O3	DIMEFLINE: 4'-METHYL-N-PYRROLIDINYL ANALOG
7	2.63	7.4	2196	1	C23H25N1O3	DIMEFLINE: N-PIPERIDINYL ANALOG
8	3.71	7.4	2196	1	C23H25N1O4	DIMEFLINE: 4'-METHYL-N-MORPHOLINO ANALOG
9 ★	4.75		2413	459	C23H25N1O4	TETRAHYDROPYRAN-2,4-DIONE,3[1-(ETHOXYIMINO)BUTYL]-6,6-DIPHENYL
15640	0.62		530		C23H25N2.Cl1	MALACHITEGREEN 569-64-2 fungicide & parasiticide (fish)
1	0.62	7.0	1295	1	"	MALACHITEGREEN
2	1.12	7.4	722	1	C23H25N3O1	N-ME-N-BENZYL-DES(DI-I-PROPYL)DISOPYRAMIDE
3 ★	1.80		1912		C23H25N5O3	H8PYRAZINO(2',1':6,1)PYRID(3,4B)INDOL,2-NO2-BENZAMET
4 ★	3.03	6.0	1148	807	C23H26F1N3O2	SPIPERONE 749-02-0 antipsychotic
45	4.47		2162	459	C23H26F4N2O1S1	TEFLUTIXOL 55837-23-5 neuroleptic
6	0.26	7.3	1299	314	C23H26N2O4	BRUCINE 357-57-3 stimulant (central)
7 ★	0.98	7.4	1299	2	"	BRUCINE
8 ★	1.52		1912		C23H26N4O1	H8-PYRAZINO(2',1':6,1)PYRID(3,4B)INDOL,BENZAMIDOETHYL
9	3.20		1209	459	C23H26N4O1	INDORAMIN ANALOG
15650 ★	3.65	12.0	2408		C23H26N4O1	MINAPRINE,4-BENZYL ANALOG ✦
1 ★	-1.00	7.0	2374		C23H26N4O4	TRYPTOPHANYLPHENYLALANYLALANINE
2 ★	3.84		2125		C23H27Cl1F1O3	R-29808 ✦
3	3.45	7.4	712	1	C23H27N1O2	N-NONYL-G-PYROPHTHALONE
4	2.46	7.4	2196	1	C23H27N1O3	DIMEFLINE: 4'-METHYL-N,N-DIETHYL ANALOG
55 ★	4.25		2045		C23H27N1O4	MOXISYLYTE,DEACETYL ANALOG: N-(TETRAHYDRONAPHTHOQUIN0N-2YL)METHYL
6	1.66	7.4	2485	1	C23H27N1O5	DIPROPIONYL MORPHINE
7	1.30	7.0	2678	1	C23H27N1O7	N-DEACETYL-N-ETHOXYCARBONYLCOLCHICINE
8	0.52	7.4	847	1	C23H27N2O2.C1H3O4S1	N(3(N-MEPIPERIDONYL)PROPYL)G-PYROPHTHALONE
9	0.52	7.4	847	1	"	N(3(N-MEPIPERIDONYL)PROPYL)G-PYROPHTHALONE
15660	-0.09	6.6	907	314	C23H27N3O7	9-DIMEAMINO-6-DEMETHYL-6-DEOXYTETRACYCLINE
1	-0.04	5.6	638	41	"	9-DIMEAMINO-6-DEMETHYL-6-DEOXYTETRACYCLINE
2	0.04	7.0	1134	1	C23H27N3O7	MINOCYCLINE 10118-90-8 antibacterial
3	0.04	7.5	1831	1	"	MINOCYCLINE
4	0.05	5.6	638	41	"	MINOCYCLINE
65	0.17	6.6	907	314	"	MINOCYCLINE
6 ★	1.63	4.6	2043	2	C23H27N3O8S2	CEPHAPIRIN,PIVALOYLOXYMETHYL ESTER
7	-3.30	7.4	2530	1	C23H27N5O7S1	PIPERACILLIN 61477-96-1 antibacterial
8 ★	0.50		1980		"	PIPERACILLIN
9	1.98	7.4	2095	1	C23H28Br1N3O4	(4-BROMO-5-METHYLIMIDAZOLE)ADRENOCEPTOR ANTAGONIST
15670	1.60	7.4	2095	1	C23H28Br1N3O4	(5-METHYLIMIDAZOLYL-3-BROMOPHENYL)ADRENOCEPTOR ANTAGONIST
1	2.04	4.3	882	1	C23H28Cl1N3O2S1	THIOPROPAZATE 84-06-0 antipsychotic
2	1.88	7.4	2095	1	C23H28Cl1N3O4	(4-CHLORO-5-METHYLIMIDAZOLE)ADRENOCEPTOR ANTAGONIST
3	3.08		2351	2	C23H28Cl1N3O5S1	GLYBURIDE 10238-21-8 antidiabetic
4	0.88	4.3	882	1	C23H28F3N3O1S1	HOMOFENAZINE 3833-99-6 sedative
75	3.01		2059	459	C23H28N2O2	3-METHYLFENTANYL ANALOG: N-(BENZYOLMETHYL)
6	1.09	7.4	847	1	C23H28N2O2	N-(3-(DI-I-PROPYLAMINO)PROPYL)-G-PYROPHTHALONE
7	3.24		2162	459	C23H28N2O3	TROPAPRIDE 76352-13-1 dopamine antagonist
8	0.21	7.3	1299	314	C23H28N2O4	21,22-DIHYDROBRUCINE 28879-93-8
9 ★	0.84	7.4	1299	2	"	21,22-DIHYDROBRUCINE
15680	1.38	7.4	2684	1	C23H28N2O5	P-HYDROXYPHENYLACETIC ACID ESTER,3-(4-METHYLPIPERAZINYL)METHYL ANALOG antiarrhythmic (vitro)
1	0.05	7.0	2682	1	C23H28N2O5	PHENYLPROPIONIC ACID,7-OXABICYCLOHEPTANEOXAZOLEAMIDE,N-I-PROPYL ANALOG txa2-antagonist
2	0.40	7.0	2682	1	C23H28N2O5	PHENYLPROPIONIC ACID,7-OXABICYCLOHEPTANEOXAZOLEAMIDE,N-PROPYL ANALOG txa2-antagonist
3	-0.02	7.0	594	314	C23H28N2O6	BOC-PHE-TYR
4 ★	2.79	1.0	594		"	BOC-PHE-TYR
85	0.28	7.4	2095	1	C23H28N4O5	(5-CARBOXAMIDO-IMIDAZOLE)ADRENOCEPTOR ANTAGONIST ✦
6 ★	0.26		2377		C23H28N4O5	AC-ALA-TYR-PHE-N
7	-2.70	1.0	564	342	C23H28N8O2	NSC-70689, IMIDAZOLIN-PHTHALANILIDE ANALOG ✦
8 ★	2.43		1520		C23H28O6	PREDNISONE-17-ACETATE
9 ★	1.57	7.4	2242	1	C23H29F1N2O4	DEXAMETHSONE,17-B-(N-CYANOMETHYL)CARBOXAMIDE) ANALOG

	logP	pH	Ref.	Note	MF	Name / CAS / activity
15690 ★	1.67		1252		C23H29N1O3	PREGN-4-ENE-21-CO2H,7A-CN-17-OH-3-OXO-G-LACTONE
1	2.04	7.4	2485	1	C23H29N1O4	MORPHINE,3-HEXANOYL
2	0.71	5.0	1440	1	C23H29N3O1	OPIPRAMOL 315-72-0 antidepressant, antipsychotic
3	2.01	6.0	695		"	OPIPRAMOL
4	3.00	8.5	2381	2	"	OPIPRAMOL
95	0.87	7.4	722	1	C23H29N3O1	SEARLE#SC-13209 ✚
6	2.62	7.4	1209	1	C23H29N3O2S1	ACETOPHENAZINE 2751-68-0 antipsychotic
7 ★	3.78		1598	40	C23H29N3O2S2	THIOTHIXENE 5591-45-7 antipsychotic
8	0.88	7.4	2095	1	C23H29N3O4	(1-METHYLIMIDAZOLE)-ADRENOCEPTOR ANTAGONIST
9 ★	-0.76	7.0	2374		C23H29N3O4	PHENYLALANYLVALINYLPHENYLALANINE
15700	0.00	7.4	2095	1	C23H29N3O5	(5-HYDROXYMETHYLIMIDAZOLE)ADRENOCEPTOR ANTAGONIST
1	0.66	7.4	2095	1	C23H29N3O5	IMIDAZOLE-(METHOXY-SPACED)ADRENOCEPTOR ANTAGONIST
2 ★	-1.37	7.0	2374		C23H29N3O5	TYROSINYLVALINYLPHENYLALANINE
3 ★	-1.50	7.0	2374		C23H29N3O5	VALINYLPHENYLALANYLTYROSINE
4	1.91	7.4	1209	1	C23H30Cl1N3O1	MEPACRINE 83-89-6 anthelmintic, antimalarial
5	-1.44	7.4	2242	1	C23H30F1O6	DEXAMETHASONE,17-B-(N-ACETIC ACID)CARBOXAMIDE ANALOG ✚
6	-0.80		559	820	C23H30N3S2.I1	3-H2-2(ET2-ME-N+ETS)-4(P-MES-PH)BENSODIAZEPIN IODIDE ✚
7 ★	0.41		1551		C23H30N4O14	ME-2-DEO-2(2-NO2-IMIDAZOLYL)4,7,8,9-OAC-N-AC-NEURAMI 83107-50-0
8 ★	2.54	7.4	1252	1	C23H30O3	6,7-B-METHYLENE-DESACETYLTHIOSPIRONOLACTONE
9 ★	3.08	7.4	1252	1	"	6,7-B-METHYLENE-DESACETYLTHIOSPIRONOLACTONE
15710	-0.52	7.4	1252	1	C23H30O5	7A-CARBOXY-DESACETYLTHIOSPIRONOLACTONE 41020-69-3
1 ★	2.10		1146		C23H30O6	CORTISONE ACETATE 50-04-4 glucocorticoid
2 ★	2.45		1520		C23H30O6	CORTISONE-17-ACETATE
3 ★	2.40		1146		C23H30O6	PREDNISOLONEACETATE 52-21-1 glucocorticoid
4 ★	2.73	7.4	2242	1	C23H31Cl1F1O4	DEXAMETHASONE,17-B-(N-CHLOROETHYLCARBOXAMIDE) ANALOG
15	6.12		1092		C23H31N1	N-(4-T-BUTYLBENZHYDRYLETHYL)PYRROLIDINE
6	2.75	7.1	554		C23H31N1O2	ACETYLMETHADOL 509-74-0 analgesic (narcotic)
7	3.03	7.4	554	1	"	ACETYLMETHADOL
8	3.13	7.5	554	1	"	ACETYLMETHADOL
9	3.22	7.6	554	212	"	ACETYLMETHADOL
15720	3.31	7.7	554	1	"	ACETYLMETHADOL
1 ★	4.27	7.7	554	2	"	ACETYLMETHADOL
2 ★	3.75	8.0	2644	2	C23H31N1O2	FOMOCAINE ANALOG-3 (4-ET)
3 ★	4.65		260		C23H31N1O2	PROADIFEN 302-33-0 synergist (non-specific)
4 ★	3.42	8.0	2644	2	C23H31N1O3	FOMOCAINE ANALOG-10 (4-OET)
25 ★	3.10	8.0	2644	2	C23H31N1O3	FOMOCAINE ANALOG-8 (2-ETO)
6 ★	3.50	8.0	2644	2	C23H31N1O3	FOMOCAINE ANALOG-9 (3-OET)
7	2.05	4.0	2655	1	C23H31N1O3	PROPRANOLOL,CYCLOHEAXANOYL ESTER
8	1.70	6.0	581		C23H31N5O1	4,6-DIAM-2,2-DIME-SYM-TRIAZINE,1(3(P-PENT)PHOXYMEPH
9 ★	2.07	7.4	2242	1	C23H32F1O4	DEXAMETHASONE,17-B-(N-ETHYLCARBOXAMIDE) ANALOG
15730 ★	1.51	7.4	2242	1	C23H32F1O5	DEXAMETHASONE,17-B-(N-HYDROXYETHYL)CARBOXAMIDE) ANALOG
1	5.41		1092		C23H32N2	N-2(N'-PH-N'-4-T-BUBENZYL)ETHYLPYRROLIDINE
2	1.02	4.8	1020		C23H32N2O1	PHENACETAMIDE,N(2,6-DIMEPH)N(3-DIETAMINOPROPYL)
3 ★	6.00		2506		C23H32N2O1S1	DIAFENTHIURON insecticide
4	0.93	7.4	608	1	C23H32N2O2	ETACETATE,2-PH-2(2-PYRIDYL)2(N,N-DI-I-PR-ETHYL)
35	4.19	7.4	2102	1	C23H32N2O4	PILOCARPIC ACID,O-BUTYRYL,P-METHYLBENZYL ESTER
6 ★	4.33		2102	2	"	PILOCARPIC ACID,O-BUTYRYL,P-METHYLBENZYL ESTER
7	-0.75		608	1	C23H32N4O2	DIISOPYRAMIDE,N1-(NH-ACETYL)
8 ★	0.60	7.4	2427	1	C23H32N6O6	P-GLU-HIS(N-CYCLOHEXYLOXYCARBONYL)-PRO-NH2
9	2.52		1036	812	C23H32O3	14,15-ANHYDRODIGITOXIGENIN
15740	3.08		1146		C23H32O4	DEOXYCORTICOSTERONEACETATE 56-47-3 adrenocortical steroid (salt-regulating)
1 ★	2.30		1520		C23H32O6	HYDROCORTISONE-17-ACETATE
2 ★	2.19		1146		C23H32O6	HYDROCORTISONEACETATE 50-03-3 glucocorticoid
3	0.56	7.4	2285	1	C23H32O6	STROPHANTHIDIN 66-28-4
4 ★	0.64		2405		"	STROPHANTHIDIN
45	-0.63	7.4	2242	1	C23H33F1N2O4	DEXAMETHASONE-17-B-(N-AMINOETHYL-CARBOXAMIDE) ANALOG
6	3.12		2045		C23H33N1O5	MOXISYLYTE,DESACETYL ANALOG: N-METHYL,N-2(2,6-DIMETHOXYPHENOXY)ETHYL
7	0.94		1537	820	C23H33N2O1.C6H5O3S1	ISOPROPAMIDEP-TOLUENESULFONATE
8	0.41		1537	820	C23H33N2O1.C7H5O2	ISOPROPAMIDE BENZOATE
9	1.07		1537	820	C23H33N2O1.C7H5O3	ISOPROPAMIDE SALICYLATE
15750	0.03	7.4	608	1	C23H33N3O1	DIISOPYRAMIDE,N1-ETHYL
1	4.06	7.3	1378	314	C23H33N3O3	HEPTACAIN ✚
2	5.64	7.3	1378	2	"	HEPTACAIN
3 ★	1.25	7.2	2553		C23H33N5O4	AC-ALA-TRP-ALA-NTBU
4	5.53		1092		C23H34N2	N-PH-N-(4-T-BUBENZYL)-N-2(N,N-DIMEAMINOETHYL)AMINE
55	-0.77	7.4	608	1	C23H34N4O1	DIISOPYRAMIDE,N1-(N-DIMETHYL)
6 ★	0.71	7.4	2427	1	C23H34N6O6	P-GLU-HIS(N-HEXOXYCARBONYL)-PRO-NH2
7	2.43	7.4	2285	1	C23H34O4	DIGITOXIGENIN 143-62-4
8 ★	2.64		2405		"	DIGITOXIGENIN
9	1.10		1036	812	C23H34O5	DIGOXIGENIN 1672-46-4
15760	1.13	7.4	2285	1	"	DIGOXIGENIN
1 ★	1.60		1036	812	C23H34O5	GITOXIGENIN 545-26-6
2	1.64	7.4	2285	1	"	GITOXIGENIN
3 ★	3.85	7.0	2503	314	C23H34O5	MEVASTATIN 73573-88-3 potential antihypercholesteremic
4	4.86		2404	459	C23H34O6	(7,8-DIHYDROPYRANYL)BENZOPYRAN,2,2,2',2'-TETRAMETHYL-6-(METHOXY-TRIETHOXY)
65 ★	2.42	5.0	2503	314	C23H34O6	PRAVASTATIN LACTONE
6	2.45	7.0	2503	314	"	PRAVASTATIN LACTONE
7	-0.21		1036	812	C23H34O7S1	DIGITOXIGENIN-3-B-SULFATE
8 ✓	-0.14		1036	812	"	DIGITOXIGENIN-3-B-SULFATE
9 ★	-0.02		2405		C23H34O8	OUABAGENIN 508-52-1
15770	4.37		2045		C23H35N1O5S1	MOXISYLYTE ANALOG: 2-BORNANONEMETHYLSULFATE ESTER
1 ✓	3.68		1720	820	C23H36Br1N2O3.Br1	N(O-HEPTOXYPH)CARBAM.AC,2(N-BRALPIPERIDIUM)ET.EST
2 ★	3.03		599		C23H36N2O2	PROSCAR 98319-26-7 5-a-reductase inhib.(prostate reduction)
3	0.19	7.4	2320	1	C23H36N2O3	2,6-BIS(1-PYRROLIDINYLMETHYL)PHENOL-4-PROPIONIC ACID,I-BUTYL ESTER
4	0.30	7.4	2320	1	C23H36N2O3	2,6-BIS(1-PYRROLIDINYLMETHYL)PHENOL-4-PROPIONIC ACID,BUTYL ESTER
75	0.19	7.4	2320	1	C23H36N2O3	2,6-BIS(1-PYRROLIDINYLMETHYL)PHENOL-4-PROPIONIC ACID,S-BUTYL ESTER
6	4.59	7.3	1149	714	C23H36N2O3	3(OPENT)CARBANILATE, O-(2-N-PIPERIDINYL)CYCLOHEXYL 38198-41-3
7	2.49	7.3	2671	1	C23H36N2O3	N,N-DIETHYL-2-(2-METHOXYPHENYLCARBAMOYLOXY)BORNAN-3-YLMETHYLAMINE
8	4.03	7.3	873	314	C23H36N2O3	O-3(N-PIPERIDINO)C-C6H10-N(M-OC5H11-PH)CARBAMATE
9	4.15	7.3	873	314	C23H36N2O3	O-3(N-PIPERIDINO)C-C6H10-N(P-OC5H11-PH)CARBAMATE

	logP	pH	Ref.	Note	MF	Name / **CAS** / *activity*
15780	3.20	7.3	873	314	C23H36N2O3	O-3(N-PYRROLIDINO)CYHEX-N(M-OC6H13-PH)CARBAMATE
1	3.08	7.3	873	314	C23H36N2O3	O-3(N-PYRROLIDINO)CYHEX-N(P-OC6H13-PH)CARBAMATE
2	7.39		2428	459	C23H36N2O6S1	N,N'-DIMETHYLTHIODICARBAMATE,O-3,4-METHYLENEDIOXYPHENYL-O'-DODECYL
3	3.42		579		C23H36N2S1	2-AMINO-4-PHENYL-5-TETRADECYLTHIAZOLE
4 ★	-0.99		2377		C23H36N4O5	ILE-ILE-GLY-PHE
85 ★	-0.42		2377		C23H36N4O5	LEU-LEU-GLY-PHE
6 ★	-1.49		2377		C23H36N4O6	TYR-ILE-LEU-GLY
7	1.18	7.0	2503	1	C23H36O6	MEVASTATIN,5-HYDROXY ACID
8 ★	3.60	2.0	2503		"	MEVASTATIN,5-HYDROXY ACID
9	-0.23	7.0	2503	1	C23H36O7	PRAVASTATIN **81093-37-0** *antihyperlipoproteinemic*
15790	1.32	5.0	2503	1	"	PRAVASTATIN
1 ★	1.14		2377	821	C23H37N5O4	AC-LEU-LYS-PHE-N
2 ★	0.65		2377	821	C23H37N5O4	AC-LYS-ILE-PHE-N
3 ★	0.98		2377	821	C23H37N5O4	AC-LYS-PHE-LEU-N
4 ✓	-0.04		2377	821	C23H37N7O4	AC-ARG-ILE-PHE-N
95	-1.94		2377	1	C23H37N7O4	AC-ARG-PHE-LEU-N
6 ✓	-0.03		2377	821	"	AC-ARG-PHE-LEU-N
7 ✓	-0.05		2377	821	C23H37N7O4	AC-ILE-PHE-ARG-N
8 ✓	-0.06		2377	821	C23H37N7O4	AC-LEU-ARG-PHE-N
9 ★	-0.12		2377	821	C23H37N7O4	AC-LEU-PHE-ARG-N
15800	4.78		2216		C23H38Cl1N3O1	DISOBUTAMIDE **68284-69-5** *cardiac depressant (anti-arrhythmic)*
1	1.44	4.8	1020		C23H38N2O1	2CYHEX-ACETAMIDE,N(2,6-DIMEPH)(3-DIETAMINOPROPYL
2	4.21	7.3	2051	1	C23H38N2O3	2-NONYLOXYCARBANILATE-O-2(N-PIPERIDINYL)ETHYL **55792-26-2**
3	5.77	7.3	1356	2	"	2-NONYLOXYCARBANILATE-O-2(N-PIPERIDINYL)ETHYL
4	4.04	7.3	1356	314	C23H38N2O3	3-NONYLOXYCARBANILATE-O-2(N-PIPERIDINYL)ETHYL **55792-27-3**
5	5.60	7.3	1356	2	"	3-NONYLOXYCARBANILATE-O-2(N-PIPERIDINYL)ETHYL
6	4.40	7.3	1356	314	C23H38N2O3	4-NONYLOXYCARBANILATE-O-2(N-PIPERIDINYL)ETHYL **55792-28-4**
7	5.96	7.3	1356	2	"	4-NONYLOXYCARBANILATE-O-2(N-PIPERIDINYL)ETHYL
8	4.20	7.0	1517	1	C23H38N2O3	CARBAMICAC,N-ET-N-(M-HEXOPH),2(1-PIPERIDINYL)PR.EST
9	4.20	7.1	1357		"	CARBAMICAC,N-ET-N-(M-HEXOPH),2(1-PIPERIDINYL)PR.EST
15810	3.81	7.0	1517	1	C23H38N2O3	CARBAMICAC,N-PR-N-(M-PENOPH),2(1-PIPERIDINYL)PR.EST
1	3.81	7.1	1357		"	CARBAMICAC,N-PR-N-(M-PENOPH),2(1-PIPERIDINYL)PR.EST
2 ★	8.16		2415	459	C23H38O2	CYCLOHEXANOL,3-(2-HYDROXY-4(1,1-DIMETHYL)NONYL)PHENYL
3 ✓	3.56		1698	820	C23H39N2O3.Br1	2-HEPTOXYPHCARBAMICAC.,PIPERIDINOET.EST.,N-ETBR
4 ✓	3.56		1720	820	C23H39N2O3.Br1	N(O-HEPTOXYPH)CARBAM.AC,2(N-ETPIPERIDINIUM)ET.EST
15 ✓	2.12		1698	820	C23H39N2O3.H2O4P1	2-HEPTOXYPHCARBAMICAC.,PIPERIDINOET.EST.,N-ETPHOS
6	3.11	7.4	594	1	C23H39N5	1-(3-DODECYLPHENYL)-2,2-DIMETHYL-4,6-DIAMINO-S-TRIAZINE
7	3.38	7.4	594	314	C23H39N5O1	1,2-DIHYDROTRIAZINE,4,6-DIAMINO-2,2-DIMETHYL-1-(3-DODECYLOXY)PHENYL
8	4.62	7.1	1150		C23H40N2O3	2ME-4(O-PEN)CARBANILATE-O-2(N,N-DIBUAMINO)ETHYL
9	2.72		527		C23H42N1.Br1	OCTADECYLPYRIDINIUMBROMIDE
15820	1.81		1635	820	C23H42N1.Cl1	MIRISTALKONIUM CHLORIDE **139-08-2** *disinfectant*
1 ✓	3.90		1729	820	C23H42N1.I1	2,4-DIMETHYLPYRIDINIUM IODIDE,N-HEXADECYL
2 ✓	3.16		1729	820	C23H42N1.I1	3,4-DIMETHYLPYRIDINIUM IODIDE,N-HEXADECYL
3 ✓	3.35		1729	820	C23H42N1.I1	3,5-DIMETHYLPYRIDINIUM IODIDE,N-HEXADECYL
4 ★	-1.18		2377		C23H42N4O5	LEU-LEU-LEU-PRO
25 ★	-1.00		2377		C23H42N4O5	LEU-LEU-PRO-LEU
6 ★	-0.92		2377		C23H42N4O5	LEU-PRO-LEU-LEU
7 ★	-1.06		2377		C23H42N4O5	PRO-LEU-LEU-LEU
8 ★	-0.51		2377		C23H44N4O5	LEU-LEU-LEU-VAL
9 ★	-0.49		2377		C23H44N4O5S1	MET-ILE-LEU-ILE
15830	5.40		2129	401	C24H12	CORONENE **191-07-1**
1	6.70		2129	471	"	CORONENE
2	-3.00	7.4	2476	1	C24H21I6N5O8	IOXAGLIC ACID **59017-64-0** *diagnostic aid (radiopaque)*
3	2.41		2524	459	C24H21N3O2	IMIDAZOLYLETHANOL,4'-ACETYL-4"-(4-PYRIDYL) ANALOG
4	2.61		2524	459	C24H21N3O3	IMIDAZOLYLETHANOL,4'-ACETYLOXY-4"-(4-PYRIDYL) ANALOG
35	1.45	7.4	675	1	C24H22N2O2	3,5-PYRAZOLIDINDIONE,1,2-DIPHENYL-4-W-PHENPROPYL
6	1.67	7.4	675	1	C24H22N2O2S1	3,5-PYRAZOLIDINDIONE,1,2-DIPHENYL,4-PR-S-PHENYL
7	4.77		547		C24H23Cl2F6N1O1	2,4(CF3)2-6,7-CL2-9(2-DIPRAM-1-OHET)PHENANTHRENE
8	-0.08	1.0	2196	1	C24H23N3O6S1	TALAMPICILLIN **47747-56-8** *antibacterial*
9	3.29		2524	459	C24H24N4O1	IMIDAZOLYLETHANOL,4'-DIMETHYLAMINO-4"-(4-PYRIDYL) ANALOG
15840	-1.36	1.0	564	342	C24H24N6O2	TPA,3,3"-BIS(N'-MEAMIDINO) ✚
1	-1.36	1.0	564	342	C24H24N6O2	TPA,3,3"-BIS(N'-METHYLAMIDINO) ✚
2	-2.06	1.0	564	342	C24H24N6O2	TPA,4,4"-BIS(METHYLAMIDINO)
3	-1.42	1.0	564	342	C24H24N6O2	TPA,4,4"-BIS-ACETAMIDINO
4	4.09		2162	459	C24H25F4N1S1	PIFLUTIXOL **54341-02-5**
45	1.60	7.4	2493	1	C24H25N1O2	2-PROPANONE,1,1-DIPHENYL-1-BENZYLOXY-3-DIMETHYLAMINO
6	4.05	7.4	547	1	C24H25N3O3S1	9-ANILINOACRIDINE,2'-MEO-4'-BUTANSULFONAMIDO
7	0.72	6.0	581		C24H25N5O1	4,6-DIAM-2,2-DIME-SYM-TRIAZINE,1(3(M-DIPHENYL)-OMEPH)
8	-3.62		562	350	C24H26N8	TEREPHTHALAMIDINE,N,N-BIS(P-CH3N=)PHENYL **2053-23-8**
9	-3.55	1.0	562	342	"	TEREPHTHALAMIDINE,N,N-BIS(P-CH3N=)PHENYL
15850	-3.62		562	350	C24H26N8	TEREPHTHALAMIDINE *antineoplastic*
1	-3.55	1.0	562	342	"	TEREPHTHALAMIDINE
2	0.56		564		C24H26O12	USNIC ACID-GLYCOSIDE
3	3.13	7.4	2196	1	C24H27N1O3	DIMEFLINE: 4'-METHYL-N-PIPERIDINYL ANALOG
4	2.90	7.4	2102	1	C24H27N3O4	PILOCARPIC ACID,O-NICOTINOYL,BENZYL ESTER
55 ★	3.05		2102	2	"	PILOCARPIC ACID,O-NICOTINOYL,BENZYL ESTER
6 ★	0.99		2377		C24H27N5O4	AC-TRP-GLY-PHE-N
7	5.63		1101	421	C24H27O4P1	TRIXYLENEYLPHOSPHATE
8	3.43	7.4	1569	1	C24H28Cl1N5O2	AMOPYROQUIN ANALOG ✚
9	4.33				C24H28Cl2F2O5	2,21-DICHLOROFLUMETHASONE-17-ACETATE
15860 ★	-0.11	7.4	2442	1	C24H28N6O6	P-GLU-HIS(N-BENZYLOXYCARBONYL)-PRO-NH2
1 ★	0.85	4.6	2043	2	C24H28N6O7S2	CEFAMANDOLE,PIVALOYLOXYMETHYL ESTER
2	2.99		569		C24H29Cl1F2O6	2-CHLORO-FLUOCINOLONEACETONIDE
3 ★	3.26		1898		C24H29N1O10	(4-HOET)VANILLINOXIME-O-ETACETYL,3,4,5-TRIMEOBZ.EST
4	0.69	7.4	2095	1	C24H29N3O5	(5-ACETYLIMIDAZOLE)ADRENOCEPTOR ANTAGONIST
65	3.23	1.0	547	342	C24H30Br1N1O1	9-(2-DIBUTYLAMINO-1-OHET)10BR-PHENANTHRENE
6	4.20		2125		C24H30Cl1N1O2	R-2693 (DESFLUOROHALOPERIDOL, I-PROPYL) ✚
7 ★	2.00	7.4	2242		C24H30F1O4	DEXAMETHASONE,17-B-(N-PROPYNYLCARBOXAMIDE)ANALOGS
8	2.48		569		C24H30F2O6	FLUOCINOLONEACETONIDE **67-73-2** *glucocorticoid*
9	4.49		430		C24H30F3N3O2S1	TRIFLUOPERAZINE ANALOG: R(10)4-ETHOXYETHANOL **3093-23-0**

	logP	pH	Ref.	Note	MF	Name / CAS / activity
15870	3.00		1856	459	C24H30N2O1	1(2-TETRAHYDRONAPHTHYL)-4(N-PROPIONANILIDO)PIPERIDINE
1	0.08	1.0	1107	342	C24H30N2O3	N(1(B-PHET)PIPERIDIN-4YL)N-PR-ANTHRANIL.AC.ME.EST
2	-0.02	7.0	2682	1	C24H30N2O5	PHENYLPROPIONIC ACID,7-OXABICYCLOHEPTANEOXAZOLEAMIDE,N-T-BUTYL ANALOG txa2-antagonist
3 ★	0.98	7.5	1476	1	C24H30N8O5	METHOTREXATE,DIETHYL ESTER
4	3.83		1396	459	C24H30O6	ARTEMISININE,CINNAMYL ESTER(ALPHA)
75	1.50		1758		C24H31F1O5	FLUOROMETHOLONEACETATE 3801-06-7 anti-inflammatory
6 ★	2.77		1520		C24H31F1O6	BETAMETHASONE-21-ACETATE 987-24-6
7 ★	2.91		1146		C24H31F1O6	DEXAMETHASONE-21-ACETATE 1177-87-3 glucocorticoid
8	2.30	7.4	2306	1	C24H31F1O6	TRIAMCINOLONEACETONIDE 76-25-5 glucocorticoid
9 ★	2.53		569		"	TRIAMCINOLONEACETONIDE
15880	2.57	1.0	547	342	C24H31N1O1	9-(2-DIBUTYLAMINO-1-HYDROXYET)PHENANTHRENE
1	0.32	4.0	2644	210	C24H31N1O2	FOMOCAINE ANALOG-11 (2-ALLYL)
2	0.15	4.0	2644	210	C24H31N1O2	FOMOCAINE ANALOG-12 (3-ALLYL)
3	3.78	7.4	1199	2	C24H31N1O3	PIPOXIZINE 55837-21-3
4	0.96	7.4	722	1	C24H31N3O1	SEARLE#SC-13733 ✚
85	1.02	7.4	2095	1	C24H31N3O4	(4,5-DIMETHYLIMIDAZOLE)ADRENOCEPTOR ANTAGONIST
6	1.72		2095		"	(4,5-DIMETHYLIMIDAZOLE)ADRENOCEPTOR ANTAGONIST
7	0.30	7.4	2095	1	C24H31N3O4	(5-METHYLIMIDAZOLE)-SPACED-ADRENOCEPTOR ANTAGONIST
8	0.62	7.4	2095	1	C24H31N3O5	(5-METHOXYMETHYLIMIDAZOLE)-ADRENOCEPTOR ANTAGONIST
9	0.97	7.4	2095	1	C24H31N3O5	(5-METHYLIMIDAZOLYL-3-METHOXYPHENYL)ADRENOCEPTOR ANTAGONIST ✚
15890	0.63	7.4	2095	1	C24H31N3O5	IMIDAZOLE-(METHOXYMETHYL-SPACED)-ADRENOCEPTOR ANTAGONIST
1 ★	-1.38	7.0	2374		C24H31N3O6	TYROSINYLTYROSINYLLEUCINE
2 ★	1.69	7.4	2242	1	C24H32F1O6	DEXAMETHASONE,17-B-(N-METHOXYCARBOXYMETHYL-CARBOXAMIDE) ANALOG
3	1.83	4.8	1020		C24H32N2O1	CINNAMAMIDE,N(2,6-DIMEPH),N(3-DIETHYLAMINOPROPYL)
4	1.30	4.8	1020		C24H32N2O1	PHENACETAMIDE,N(2,6-DIMEPH),N(3N-PIPERIDYLPROPYL)
95	3.04		2059	459	C24H32N2O2	3-METHYLFENTANYL ANALOG: N-(1-METHYL-2-HYDROXY-2-PHENYL)ETHYL
6 ★	2.60		2531		C24H32N2O2	PHENOXYACETAMIDE ANALOG,N,N-PENTAMETHYLENE 139734-21-7
7	-0.13		559	820	C24H32N3S2.I1	3-H2-2(ET2-ME-N+ETS-)4(P-ETS-PH)BENZODIAZEPINI
8	2.19	7.4	1254	1	C24H32O4S1	SPIRONOLACTONE,7-A 52-01-7
9 ★	2.78	7.4	1252	1	"	SPIRONOLACTONE,7-A
15900 ★	2.26	7.4	1252	1	C24H32O4S1	SPIRONOLACTONE,7-BETA diuretic, aldosterone antagonist
1	1.20	7.4	1254	1	C24H32O5	PREGN-4ENE-7,21(CO2H)2,17-OH-3-OXOLACTONE,A-ME-EST. 41020-65-9
2 ★	1.94	7.4	1252	1	"	PREGN-4ENE-7,21(CO2H)2,17-OH-3-OXOLACTONE,A-ME-EST.
3 ★	2.18	7.4	1252	1	"	PREGN-4ENE-7,21(CO2H)2,17-OH-3-OXOLACTONE,A-ME-EST.
4 ★	2.11		2308		C24H32O8	4,4-DIBENZO-24-CROWN-8-ETHER 14174-09-5
5	6.43		1092		C24H33N1	N-(4-T-BUTYLBENZHYDRYLETHYL)PIPERIDINE
6	1.55	7.4	2680	324	C24H33N1O3	3-O-HEPTYLMORPHINE 74886-12-7
7 ★	2.60	8.0	2644	2	C24H33N1O5	FOMOCAINE ANALOG-7 (TRI-MEO)
8	0.69	7.4	722	1	C24H33N3O1	2,2,4,6-4MEPIPERIDINYL-DES(N-DIIPR)DISOPYRAMIDE
9 ★	1.57	4.6	2043	2	C24H33N9O6S3	CEFOTIAM,PIVALOYLOXYMETHYL ESTER
15910	-0.42		462	820	C24H33O5.C23H33N2O1	ISOPROPAMIDEDEHYDROCHOLATE
1	1.85		1966	459	C24H34Au1O9P1S1	TETRAACETYLGLUCOSE,2-DIETHYL,PHENYLPHOSPHINOAUROTHIO ✚
2 ★	2.43	7.4	2242	1	C24H34F1O4	DEXAMETHASONE,17-B-(N-ISOPROPYLCARBOXAMIDE) ANALOG
3 ★	2.59	7.4	2242	1	C24H34F1O4	DEXAMETHASONE,17-B-(N-PROPYLCARBOXAMIDE) ANALOG
4	-2.22		1082		C24H34I2O18	DTB
15	2.18		817		C24H34N1O3.Br1	N(2-BENZILOYLOXYET-N-HEXYL-N,N-DIMETHYLAMMONIUMBR
6	5.92		1092		C24H34N2	N-2(N'-PH-N'-(4-T-BUBENZYL)ETHYLPIPERIDINE
7	2.00	7.4	1754	320	C24H34N2O1	BEPRIDIL 74764-40-2 vasodilator
8	4.40	7.4	2102	1	C24H34N2O4	PILOCARPIC ACID,O-(2,2-DIMETHYLPROPIONYL),P-METHYLBENZYL ESTER
9 ★	4.55		2102	2	"	PILOCARPIC ACID,O-(2,2-DIMETHYLPROPIONYL),P-METHYLBENZYL ESTER
15920	4.60	7.4	2102	1	C24H34N2O4	PILOCARPIC ACID,O-HEXANOYL,BENZYL ESTER
1 ★	4.75		2102	2	"	PILOCARPIC ACID,O-HEXANOYL,BENZYL ESTER
2 ✓	1.97		2018	226	C24H34N2O5	INDOLAPRIL 80876-01-3 antihypertensive
3	4.00	7.4	608	1	C24H34N4O1	DIISOPYRAMIDE,N1-(N=CME2)
4 ★	2.70		1520		C24H34O6	HYDROCORTISONE-17-PROPIONATE
25 ★	2.80		1520		C24H34O6	HYDROCORTISONE-21-PROPIONATE
6	1.88		2627	841	C24H35N2.Cl1	N-METHYL-4-(4'-DIPENTYLAMINOSTYRYL)-PYRIDINIUM CHLORIDE
7	-1.00	7.4	487	1	C24H36N2O4	2,2'-BIS(2-OH-3-I-PRAMPR)BIPHENYLDIETHER
8	0.97	7.4	487	2	C24H36N2O4	2,2'-BIS(2-OH-3-I-PRAMPR)BIPHENYLDIETHER
9	0.12	7.4	608	1	C24H36N4O1	DIISOPYRAMIDE,N1-ME,N1-(DIMETHYL)
15930 ✓	0.62	7.3	2085		C24H36N4.O2Tc1	CYCLAM-DIMETHYL-DIPHENYL,TECHNICIUM OXIDE COMPLEX
1 ★	4.26	5.0	2503	314	C24H36O5	LOVASTATIN 75330-75-5 antihypercholesteremic, inhibitor (hmg-coa-reductase)
2	4.27	7.0	2503	314	"	LOVASTATIN
3	0.86	7.4	2320	1	C24H38N2O3	2,6-BIS(1-PYRROLIDINYLMETHYL)PHENOL,4-BUTYRIC ACID,I-BUTYL ESTER
4	1.11	7.4	2320	1	C24H38N2O3	2,6-BIS(1-PYRROLIDINYLMETHYL)PHENOL-4-BUTYRIC ACID,BUTYL ESTER
35	0.86	7.4	2320	1	C24H38N2O3	2,6-BIS(1-PYRROLIDINYLMETHYL)PHENOL-4-BUTYRIC ACID,S-BUTYL ESTER
6	2.90	7.4	2671	1	C24H38N2O3	N,N-DIETHYL-2-(2-ETHOXYPHENYLCARBAMOYLOXY)BORNAN-3YLMETHYLAMINE
7	4.00	7.3	873	314	C24H38N2O3	O-3(N-PIPERIDINO)C-C6H10-N(M-OC6H13-PH)CARBAMATE
8 ★	4.08	7.3	873	314	C24H38N2O3	O-3(N-PIPERIDINO)C-C6H10-N(P-OC6H13-PH)CARBAMATE
9 ★	-1.00		2377		C24H38N4O5	LEU-LEU-ALA-PHE
15940 ★	-1.32		2377		C24H38N4O6	VAL-PHE-LEU-THR
1	-2.20		2229		C24H38N8O14	DES-TRP-DSIP
2 ★	3.26		2319		C24H38O2Si1	17-BETA-ESTRADIOL,3-(T-BUTYLDIMETHYL)SILYLOXY 57441-02-8
3 ✓	2.60		2319		C24H38O2Si1.C3Cr1O2S1	17-BETA-ESTRADIOL,3-(T-BUTYLDIMETHYL)SILYLOXY,CHROMIUM DICARBONYLTHIOCARBONYL COMPLEX
4 ★	7.45		2293	400	C24H38O4	DI(2-ETHYLHEXYL)PHTHALATE 117-81-7
45	3.98		1190		C24H38O4	DI-2-ETHYLHEXYLPHTHALATE 117-81-7
6	5.55		579		C24H38O4	M-DIOCTYLPHTHALATE 117-84-0
7	5.22		579		C24H38O4	O-DIOCTYLPHTHALATE 117-84-0
8	8.06		1369	459	"	O-DIOCTYLPHTHALATE
9	1.70	7.0	2503	1	C24H38O6	LOVASTATIN,5-HYDROXY ACID
15950 ★	4.04	2.0	2503		"	LOVASTATIN,5-HYDROXY ACID
1	2.31		1236		C24H38O7	14,16-DI-O-ACETYL-A-DIHYDROGRAYANOTOXINII
2 ★	4.23		2216		C24H39Cl1N2O1	BUTYRAMIDE,2-(O-CHLOROPHENYL)-2-CYCLOPROPYLETHYL-4-DI-I-PROPYLAMINO
3	-0.20		462	820	C24H39O4.C23H33N2O1	ISOPROPAMIDEDEOXYCHOLATE
4	-0.54		462	820	C24H39O5.C23H33N2O1	ISOPROPAMIDECHOLATE
55	4.43	7.3	1356	314	C24H40N2O3	2-DECYLOXYCARBANILATE-O-2(N-PIPERIDINYL)ETHYL 55792-29-5
6	5.99	7.3	1356	2	"	2-DECYLOXYCARBANILATE-O-2(N-PIPERIDINYL)ETHYL
7	5.18	7.3	1149	714	C24H40N2O3	3(OPENT)CARBANILATE,O-(2-N,N-DIPRAMINO)CYCLOHEXYL 60093-65-4
8	4.18	7.3	1356	314	C24H40N2O3	3-DECYLOXYCARBANILATE-O-2(N-PIPERIDINYL)ETHYL 55792-30-8
9	5.74	7.3	1356	2	"	3-DECYLOXYCARBANILATE-O-2(N-PIPERIDINYL)ETHYL

	logP	pH	Ref.	Note	MF	Name / CAS / activity
15960	4.34	7.3	1356	314	C24H40N2O3	4-DECYLOXYCARBANILATE-O-2(N-PIPERIDINYL)ETHYL
1	5.90	7.3	1356	2	"	4-DECYLOXYCARBANILATE-O-2(N-PIPERIDINYL)ETHYL
2	4.14	7.0	1517	1	C24H40N2O3	CARBAMICAC,N-BU-N-(M-PENOPH),2(1-PIPERIDINYL)PR.EST
3	4.14	7.1	1357		"	CARBAMICAC,N-BU-N-(M-PENOPH),2(1-PIPERIDINYL)PR.EST
4 ★	6.11		2415	459	C24H40O3	CYCLOHEXANOL,4-HYDROXYPROPYL-3-(2-HYDROXY-4-(1,1-DIMETHYL)HEPTYL)PHENYL
65 ★	6.13		2415	459	"	CYCLOHEXANOL,4-HYDROXYPROPYL-3-(2-HYDROXY-4-(1,1-DIMETHYL)HEPTYL)PHENYL
6	1.52		1937	459	C24H40O4	4-HYDROXY-2-PYRONE,3,5-DIBUTYL-6-(1-BUTYL-1-FORMYLHEXYL) **98393-85-2**
7	2.52		1937	459	C24H40O4	4-HYDROXY-2-PYRONE,3,5-DIBUTYL-6-(2-BUTYLHEPTAN-2-ONYL) **68112-21-0**
8	2.20		2453	200	C24H40O4	CHENODIOL **474-25-9** *anticholelithogenic*
9	2.25		2453	200	"	CHENODIOL
15970	3.00		2453	225	"	CHENODIOL
1	3.28		2453	225	"	CHENODIOL
2	3.84	8.0	1102	1	"	CHENODIOL
3	4.15		1102	226	"	CHENODIOL
4	2.65		2453	200	C24H40O4	DEOXYCHOLIC ACID **83-44-3**
75	3.50		2453	225	"	DEOXYCHOLIC ACID
6	2.28		2453	200	C24H40O4	HYODEOXYCHOLIC ACID **83-49-8**
7	3.08		2453	225	"	HYODEOXYCHOLIC ACID
8	2.20		2453	200	C24H40O4	URSODEOXYCHOLIC ACID **128-13-2**
9	3.00		2453	225	"	URSODEOXYCHOLIC ACID
15980	1.10		2453	200	C24H40O5	CHOLIC ACID **81-25-4** *choleretic*
1	2.02		2453	225	"	CHOLIC ACID
2	1.84		2453	200	C24H40O5	HYOCHOLIC ACID
3	2.80		2453	225	"	HYOCHOLIC ACID
4	0.00		2453	200	C24H40O5	URSOCHOLIC ACID
85	0.92		2453	225	"	URSOCHOLIC ACID
6 ★	0.03		2308		C24H40O10	BENZO-30-CROWN-10-ETHER
7 ✓	3.87		1698	820	C24H41N2O3.Br1	2-HEPTOXYPHCARBAMICAC.,PIPERIDINOET.EST.,N-PRBR
8 ✓	3.89		1720	820	C24H41N2O3.Br1	N(O-HEPTOXYPH)CARBAM.AC,2(N-PRPIPERIDINIUM)ET.EST
9 ✓	2.36		1698	820	C24H41N2O3.H2O4P1	2-HEPTOXYPHCARBAMICAC.,PIPERIDINOET.EST.,N-PRPHOS
15990	3.35	7.4	594	314	C24H41N5O1	1,2-DIHYDROTRIAZINE,4,6-DIAMINO-2,2-DIMETHYL-1-(3-TRIDECYLOXY)PHENYL
1	2.53	1.0	547	342	C24H43N1O1	2-N,N-DIOCTYLAMINO-1-PHENYLETHANOL HYDROCHLORIDE
2 ✓	1.93		1729	820	C24H43N2O1.I1	3-CARBAMOYLPYRIDINIUM IODIDE,N-OCTADECYL **35096-56-1**
3	6.18		1323	820	C24H52N1.I1	TETRAHEXYL AMMONIUM IODIDE **2138-24-1**
4	4.24		1323	820	C24H52N1.C2H4N1O2	GLYCINE,TETRAHEXYL AMMONIUM SALT **36368-88-4**
95	4.26		1323	820	C24H52N1.C3H6N1O2	ALANINE,TETRAHEXYL AMMONIUM SALT **36368-90-8**
6	4.32		1323	820	C24H52N1.C3H6N1O3	SERINE,TETRAHEXYL AMMONIUM SALT **36368-89-5**
7	4.79		1323	820	C24H52N1.C5H10N1O2	VALINE,TETRAHEXYL AMMONIUM SALT **36368-91-9**
8	4.89		1323	820	C24H52N1.C5H10N1O2S1	METHIONINE,TETRAHEXYL AMMONIUM SALT **36368-92-0**
9	5.29		1323	820	C24H52N1.C6H12N1O2	LEUCINE,TETRAHEXYL AMMONIUM SALT **36368-94-2**
16000	4.98		1323	820	C24H52N1.C6H13N2O2	LYSINE,TETRAHEXYL AMMONIUM SALT **36368-93-1**
1	5.72		1323	820	C24H52N1.C9H10N1O2	PHENYLALANINE,TETRAHEXYL AMMONIUM SALT **36368-95-3**
2	5.77		1323	820	C24H52N1.C11H11N2O2	TRYPTOPHAN,TETRAHEXYL AMMONIUM SALT **36368-96-4**
3	3.84		1957		C24H54O1Sn2	BIS(TRIBUTYLTIN)OXIDE **56-35-9**
4	9.50		1740	459	C24H72O10Si11	TETRACOSOMETHYLUNDECASILOXANE
5	3.75		1582	459	C25H21Cl1N2O6	7CL-1,3H2-3(3,4,5-TRIMEOPHCOO)5PH-1,4DIAZEPIN-2-ON **18035-92-2**
6 ★	6.20		1679		C25H22Cl1N1O3	FENVALERATE **51630-58-1** *insecticide*
7	-0.38	7.4	401	1	C25H22N4O8	STREPTONIGRIN **3930-19-6** *antineoplastic*
8	-2.67	1.0	564	342	C25H23N7O2	NSC#63697 ✚
9	2.31		2506		C25H24F6N4	HYDRAMETHYLNON **67485-29-4** *insecticide*
16010	-0.96		555		C25H24N3O5.Br1	1-DI(P-MEOBENZYL)2(5-NO2-2-FURFURILIDENE)IMIDAZONE
1	-0.56		2389		C25H25F3N4O3	TOSUFLOXIN ANALOG **114691-38-2**
2	4.93	7.4	2102	1	C25H27Cl1N2O4	PILOCARPIC ACID,O-(3-CHLOROBENZOYL),BENZYL ESTER
3 ★	5.08		2102	2	"	PILOCARPIC ACID,O-(3-CHLOROBENZOYL),BENZYL ESTER
4	4.75	7.4	2102	1	C25H27Cl1N2O4	PILOCARPIC ACID,O-BENZOYL,P-CHLOROBENZYL ESTER
15 ★	4.90		2102	2	"	PILOCARPIC ACID,O-BENZOYL,P-CHLOROBENZYL ESTER
6	1.16	7.4	2686	1	C25H27N2	N,N'-DI-I-PROPYLISOCYANINE *noradrenaline transport inhibitor*
7	-2.12	7.4	2464	41	C25H27N9O8S2	CEFOPERAZONE **62893-19-0** *antibacterial*
8 ★	-0.74		1980		"	CEFOPERAZONE
9	-0.49	7.4	2373	1	"	CEFOPERAZONE
16020	4.22	7.4	2102	1	C25H28N2O4	PILOCARPIC ACID,O-BENZOYL,BENZYL ESTER
1 ★	4.37		2102	2	"	PILOCARPIC ACID,O-BENZOYL,BENZYL ESTER
2	3.88	7.4	2310	2	C25H28N6O2	XANTHINE,1,3-DIMETHYL-8-(4-DIPHENYLMETHYL)PIPERAZINYLMETHYL
3 ★	7.05		2506		C25H28O3	ETOFENPROX **80844-07-1** *insecticide*
4	1.59	7.0	2682	1	C25H29F3N2O5	PHENYLPROPIONIC ACID,7-OXABICYCLOHEPTANEOXAZOLEAMIDE,N-TRIFLUOROPENTYL ANALOG *txa2-antagonist*
25	0.72	6.0	2190	1	C25H29I2N1O3	AMIODARONE **1951-25-3** *cardiac depressant (anti-arrhythmic, ventricular)*
6	2.54	7.2	2416	1	"	AMIODARONE
7 ★	3.55		1520		C25H30Br1F1O5	BROBETASONE-17-PROPIONATE
8 ★	3.46		1520		C25H30Cl1F1O5	CLOBETASONE-17-PROPIONATE
9	3.95	7.4	1569		C25H30Cl1N5O2	PHENOL,4(7-CL-4-QUINLIN)AM)2-(4-DIETAMCARB)PIPERID
16030	-1.22	7.4	2335	314	C25H30N1O3.Cl1	TROSPIUM CHLORIDE **10405-02-4** *anticholinergic*
1	-1.05	7.4	2335	314	C25H30N1O3.C5H11O3S1	TROSPIUM PENTANESULFONATE
2	-0.33	7.4	2335	314	C25H30N1O3.C6H13O3S1	TROSPIUM HEXANESULFONATE
3	0.07	7.4	2335	314	C25H30N1O3.C6H13O4S1	TROSPIUM HEXYLSULFATE
4	0.05	7.4	2335	314	C25H30N1O3.C7H15O3S1	TROSPIUM HEPTANESULFONATE
35	0.44	7.4	2335	314	C25H30N1O3.C7H15O4S1	TROSPIUM HEPTYLSULFATE
6	0.47	7.4	2335	314	C25H30N1O3.C8H17O3S1	TROSPIUM OCTANESULFONATE
7	0.84	7.4	2335	314	C25H30N1O3.C8H17O4S1	TROSPIUM OCTYLSULFATE
8	1.19	7.4	2335	314	C25H30N1O3.C9H19O4S1	TROSPIUM NONYLSULFATE
9	1.52	7.4	2335	314	C25H30N1O3.C10H21O4S1	TROSPIUM DECYLSULFATE
16040	2.77	7.8	2335	314	C25H30N1O3.C12H25O4S1	TROSPIUM DODECYLSULFATE
1	0.51		2646	448	C25H30N3.Cl1	GENTIANVIOLET **548-62-9**
2	0.96		530	260	"	GENTIANVIOLET
3	1.16	7.0	1295	1	"	GENTIANVIOLET
4	-1.43		1592		C25H30N4O5	TRP-VAL-TYR
45 ★	0.65		2377	821	C25H30N4O6	AC-PHE-GLU-PHE-N
6	2.99		569		C25H31Cl1F2O6	2-CHLOROFLUMETHASONE-17-PROPIONATE
7 ★	3.30		1520		C25H31Cl1O5	CLOBETASONE-17-PROPIONATE-DES-F
8 ★	3.09		1520		C25H31F1O5	CLOBETASONE-17-PROPIONATE-DES-CL
9 ★	1.92		1146		C25H31F1O8	TRIAMCINOLONEDIACETATE **67-78-7** *glucocorticoid*

	logP	pH	Ref.	Note	MF	Name / **CAS** / *activity*
16050	3.35	6.0	1148		C25H31N1O1	BUTACLAMOL **51152-91-1** *antipsychotic*
1	4.35	6.0	1148		"	BUTACLAMOL
2	5.20		2162	459		BUTACLAMOL
3	2.60	7.4	2485	1	C25H31N1O5	MORPHINE,DI-I-BUTYRYOYL
4 ★	1.40	13.0	547	224	C25H31N1O7S1 ✚	NSC-248896 ✚
55	1.28	7.4	2095	1	C25H31N3O5	(4-ACETYL-5-METHYLIMIDAZOLE)ADRENOCEPTOR ANTAGONIST
6	1.41	7.4	2095	1	C25H31N3O6	(5-ETHOXYCARBONYL-IMIDAZOLE)ADRENOCEPTOR ANTAGONIST
7 ★	3.83		569		C25H32Cl1F1O5	CLOBETASOL PROPIONATE **25122-46-7** *anti-inflammatory*
8 ★	3.28		1520		C25H32Cl2O5	CLOBETASOL-17-PROPIONATE-9-A-CL
9	2.76		1856	459	C25H32N2O1	1(2(4H)NAPHTHYL)-4(N-PROPIONANILIDO)-3-ME-PIPERIDINE
16060	3.00		1856	459	"	1(2(4H)NAPHTHYL)-4(N-PROPIONANILIDO)-3-ME-PIPERIDINE
1	0.58	1.0	547	342	C25H32N2O1	2-PH-4-(2-DIBUTYLAMINO-1-HYDROXYET)QUINOLINE HYDROCHLORIDE
2	3.00	7.4	547	1	"	2-PH-4-(2-DIBUTYLAMINO-1-HYDROXYET)QUINOLINE HYDROCHLORIDE
3	3.16	7.4	1837	1	C25H32N2O3	LOFENTANIL **61380-40-3** *analgesic (narcotic)*
4 ★	4.22	9.5	1837		"	LOFENTANIL
65	0.51	1.0	1107	342	C25H32N2O3	N(1(B-PHET)PIPERIDIN-4YL)N-PR-ANTHRANIL.AC.ET.EST
6	0.51	7.0	2682	1	C25H32N2O5	PHENYLPROPIONIC ACID,7-OXABICYCLOHEPTANEOXAZOLEAMIDE,N-NEOPENTYL ANALOG *txa2-antagonist*
7	1.61	7.0	2682	1	C25H32N2O5	PHENYLPROPIONIC ACID,7-OXABICYCLOHEPTANEOXAZOLEAMIDE,N-PENTYL ANALOG *txa2-antagonist*
8	1.07	7.0	2682	1	C25H32N2O6	PHENYLPROPIONIC ACID,7-OXABICYCLOHEPTANEOXAZOLEAMIDE,N-HYDROXYPENTYL ANALOG *txa2-antagonist*
9 ★	-0.51		2377		C25H32N4O5	VAL-GLY-PHE-PHE
16070 ★	3.34		1520		C25H33Cl1O5	CLOBETASOL-17-PROPIONATE-DES-F
1 ★	3.09		1520		C25H33F1O5	BETAMETHASONE-21-DESOXY-17-PROPIONATE
2 ★	3.06		1520		C25H33F1O6	BETAMETHASONE-21-PROPIONATE
3	-1.37	2.0	534		C25H33N1O4	ETORPHINE HYDROCHLORIDE
4	1.86	7.4	709	1	C25H33N1O4	ETORPHINE *analgesic (narcotic)*
75	1.79	7.4	2095	1	C25H33N3O4	(5-ISOPROPYLIMIDAZOLE)ANDRENOCEPTOR ANTAGONIST
6	1.00	7.4	2095	1	C25H33N3O5	(3-ETHOXYMETHYLIMIDAZOLE)ADRENOCEPTOR ANTAGONIST
7	0.79	7.4	2095	1	C25H33N3O5	(5-METHOXYETHYLIMIDAZOLE)ANDRENOCEPTOR ANTAGONIST
8	0.84	7.4	2095	1	C25H33N3O5	(5-METHYL-4-METHOXYMETHYLIMIDAZOLE)ADRENOCEPTOR ANTAGONIST ✚
9	1.47	7.4	1209	1	C25H34N2O3	TILORONE **27591-97-5** *antiviral*
16080	-0.02		559	820	C25H34N3S2.I1	3-H2-2(ET2-ME-N+ETS-)4(P-I-PRSPH)BENZODIAZEPIN IODIDE ✚
1	0.20		559	820	C25H34N3S2.I1	3-H2-2(ET2-ME-N+ETS-)4(P-PRS-PH)BENZODIAZEPIN IODIDE
2	1.21		1937	459	C25H34O4	4-HYDROXY-2-PYRONE,3,5-DIBUTYL-6-(1-BENZOYLPENTYL) **98393-95-4**
3 ★	2.54	7.4	1252	1	C25H34O5	PREGN-4ENE-7,21(CO2H)2,17-OH-3-OXOLACTONE,A-ET-EST. **41020-77-3**
4	1.68	7.4	2680	324	C25H35N1O3	3-O-OCTYLMORPHINE **74886-13-8**
85	2.98	7.4	2656	1	C25H35N1O5S1	2,6-BIS(I-PROPYL)PHENYL-(2,6-BIS(I-PROPYL)PHENOXYSULFONYLCARBAMATE *hypocholesterolemic*
6 ★	3.10	7.4	2242	1	C25H36F1O4	DEXAMETHASONE,17-B-(N-BUTYLCARBOXAMIDE) ANALOG
7 ★	3.01	7.4	2242	1	C25H36F1O4	DEXAMETHASONE,17-B-(N-ISOBUTYLCARBOXAMIDE) ANALOG
8 ★	3.20	7.4	2242	1	C25H36F1O4	DEXAMETHASONE,17-B-(N-T-BUTYLCARBOXAMIDE) ANALOG
9	3.56		1711		C25H36N4O7	NONYLOXYCARBONYLMITOMYCINC
16090 ★	8.02		2415	459	C25H36O3	11-OXO-D9-THC-DIMETHYLHEXYL
1 ★	3.18		1520		C25H36O6	HYDROCORTISONE-17-BUTYRATE **13609-67-1** *glucocorticoid*
2	2.91		1520		C25H36O6	HYDROCORTISONE-21-BUTYRATE
3 ★	1.93		1036	812	C25H36O6	OLEANDRIGENIN
4	-0.72	7.4	2242	1	C25H37F1N2O4	DEXAMETHASONE,17-B-(N-(4-AMINOBUTYL)CARBOXAMIDE) ANALOG
95 ★	0.06		2377		C25H37N5O5	TRP-GLY-LEU-LEU
6	6.08		1092		C25H38N2	N-PH-N-(4-T-BUBENZYL)-N-2(N,N-DIPROPYLAMINOETHYL)AMINE
7 ★	1.82	7.4	2427	1	C25H38N6O6	P-GLU-HIS(N-2-ETHYLHEXYLOXYCARBONYL)-PRO-NH2
8 ★	1.88	7.4	2427	1	C25H38N6O6	P-GLU-HIS(N-OCTYLOXYCARBONYL)-PRO-NH2
9 ★	9.96		2415	459	C25H38O2	D8-THC-DIMETHYLHEPTYL
16100	8.30		2415	459	C25H38O2	DMHP **32904-22-6**
1 ★	7.44		2415	459	C25H38O3	11-HYDROXY-8D-THC-DIMETHYLHEPTYL
2 ★	7.05		2415	459	C25H38O3	11-HYDROXY-9D-THC-DIMETHYLHEXYL
3 ★	4.68	7.0	2503	314	C25H38O5	SIMVASTATIN **79902-63-9** *antihyperlipidemic*
4 ★	8.38		2415	459	C25H39N1O2	1-(4-AMINOBUTOXY)-D8-THC
5	3.37	7.3	2671	1	C25H40N2O3	N,N-DIETHYL-2-(2-PROPOXYPHENYLCARBAMOYLOXY)BORNAN-3YLMETHYLAMINE
6 ★	-0.63		2377		C25H40N4O5S1	VAL-MET-PHE-ILE
7	5.97		2415	459	C25H40O3	(+)ACD
8 ★	6.13		2415	459	C25H40O3	(-)ACD
9	2.06	7.0	2503	1	C25H40O6	SIMVASTATIN,5-HYDROXY ACID
16110 ★	4.47	2.0	2503		"	SIMVASTATIN,5-HYDROXY ACID
1 ★	8.42		2415	459	C25H42O2	CYCLOHEXANOL,3-(2-HYDROXY-4-(1,1-DIMETHYL)UNDECYL)PHENYL
2 ★	6.68		2415	459	C25H42O3	CYCLOHEXANOL,4-HYDROXYBUTYL-3-(2-HYDROXY-4-(1,1-DIMETHYLHEPTYL)PHENYL
3 ✓	4.26		1698	820	C25H43N2O3.Br1	2-HEPTOXYPHCARBAMICAC.,PIPERIDINOET.EST.,N-BUBR
4 ✓	4.26		1720	820	C25H43N2O3.Br1	N(O-HEPTOXYPH)CARBAM.AC,2(N-BUPIPERIDINIUM)ET.EST
15 ✓	4.96		1720	820	C25H44N2O3.I1	N(O-HEPTOXYPH)CARBAM.AC,2(N-BUTPIPERIDINIUM)ET.EST
6 ✓	2.69		1698	820	C25H44N2O3.H2O4P1	2-HEPTOXYPHCARBAMICAC.,PIPERIDINOET.EST.,N-BUPHOS
7	2.10	7.0	801	1	C25H43N3O6	1-(5-PALMITYLARABINOFURANOSYL)CYTOSINE ✚
8	2.14	1.3	401	1	"	1-(5-PALMITYLARABINOFURANOSYL)CYTOSINE
9	5.03		401		"	1-(5-PALMITYLARABINOFURANOSYL)CYTOSINE
16120	3.64		594	314	C25H43N5O1	1,2-DIHYDROTRIAZINE,4,6-DIAMINO-2,2-DIMETHYL-1-(3-TETRADECLYOXY)PHENYL
1	3.28		527	820	C25H46N1.Br1	BENZYLDIMETHYLHEXADECYLAMMONIUMBROMIDE **3529-04-2**
2	2.11		1635	820	C25H46N1.Cl1	CETALKONIUM CHLORIDE **122-18-9** *anti-infective (topical)*
3	1.95	7.4	2557	1	C26H18O6	BENZYLIDENEBIS-4-HYDROXYCOUMARIN
4	2.06		2524	459	C26H20F1N3O1	BENZIMIDAZOLYLETHANOL,4'-FLUORO-4"-(4-PYRIDYL) ANALOG ✚
25	5.50		2300		C26H22Cl1F3N2O3	FLUVALINATE **69409-94-5** *insecticide*
6	-2.16	1.0	564	342	C26H23Cl1N6O2	ITPA,2-CL-4,4"-BIS-IMIDAZOLIN-2-YL
7	-1.00	1.0	564	342	C26H23Cl1N6O2	ITPA,3-CL-4,4"-BIS-IMIDAZOLIN-2-YL
8	-1.49	1.0	564	342	C26H23F1N6O2	TPA,2-FL-4,4"-BIS-IMIDAZOLIN-2-YL
9 ★	6.20		1679		C26H23F2N1O4	PAYOFF **70124-77-5**
16130	-1.55	1.0	564	342	C26H23N7O4	TPA,2-NITRO-4,4"-BISIMIDAZOLIN-2-YL ✚
1	-1.76	1.0	564	342	C26H24N6O2	ISOTIC **5262-40-8** *antineoplastic*
2	-1.75	7.4	564	1	"	ISOTIC
3	-1.72		564	350	"	ISOTIC
4	-1.72	1.0	564	342	C26H24N6O2	NSC-63687, IMIDAZOLIN-PHTHALANILIDE ANALOG
35	-2.25	1.0	564	342	C26H24N6O2	TPA,4,4"-BIS-IMIDAZOLIN-2-YL, DIHYDROCHLORIDE **2545-16-6**
6	-1.44	1.0	564	342	C26H24N6O2	TPA,4,4"-BIS-IMIDAZOLIN-2YL
7	2.52	7.0	2682	1	C26H25Cl1N2O5	PHENYLPROPIONIC ACID,7-OXABICYCLOHEPTANEOXAZOLEAMIDE,N-P-CHLOROPHENYL ANALOG *txa2-antagonist*
8	-1.92	1.0	564	342	C26H25N7O2	ITPA,2-AMINO-4,4"-BIS-IMIDAZOLIN-2-YL
9	-3.41	1.0	564	342	C26H25N7O2	ITPA,5-AMINO-4,4"-BIS-IMIDAZOLIN-2-YL

	logP	pH	Ref.	Note	MF	Name / CAS / activity
16140	-2.08	1.0	564	342	C26H25N7O2	NSC-75990, IMIDAZOLIN-PHTHALANILIDE ANALOG
1	0.47	3.0	1126		C26H26I6N2O10	IODOXAMIC ACID **31127-82-9** *diagnostic aid (radiopaque)*
2	1.42	1.0	1126	342	"	IODOXAMIC ACID
3	1.19	7.0	2682	1	C26H26N2O5	PHENYLPROPIONIC ACID,7-OXOBICYCLOHEPTANEOXAZOLEAMIDE, N-PHENYL ANALOG *txa2-antagonist*
4 ★	3.77		935		C26H26N2O6S1	CARBENICILLIN,INDANYL **35531-88-5** *antibacterial*
45	3.90	7.4	1485	1	C26H26N4O3	2,4-DIAMPYRIMIDINE,5-(3-MEO-4,5(PHCH2O)2)BENZYL
6 ★	4.09	7.4	1485	2	"	2,4-DIAMPYRIMIDINE,5-(3-MEO-4,5(PHCH2O)2)BENZYL
7	3.70	7.4	1485	1	C26H26N4O3	2,4-DIAMPYRIMIDINE,5-(4-MEO-3,5(PHCH2O)2)BENZYL
8 ★	3.89	7.4	1485	2	"	2,4-DIAMPYRIMIDINE,5-(4-MEO-3,5(PHCH2O)2)BENZYL
9	-2.34	1.0	564	342	C26H26N8O2	NSC-57142, IMIDAZOLIN-PHTHALANILIDE ANALOG
16150	-2.41	1.0	564	342	C26H26N8O2	TPA,4,4"-BIS(2-IMIDAZOLIN-2-YL-AMINO)
1	-2.40	1.0	562	342	C26H26N8O2	TPA-4,4"-BIS(2-IMIDAZOLIN-2YL-AMINO)
2	3.21	1.0	1941		C26H27N1O4	BIS-(3-(1-NAPHTHYLOXY)2-HYDROXYPROPYL)AMINE
3	0.96	7.0	2304		C26H27N1O9	DOXORUBICIN,4-DEMETHOXY-4'-DEOXY
4	1.63	7.5	2624	1	C26H27N1O9	IDARUBICIN,7R,9R
55	0.68	7.0	2304		C26H27N1O9	IDARUBICIN **58957-92-9** *antineoplastic*
6	0.88	7.0	2372	324	"	IDARUBICIN
7	0.89	7.0	2304	459	"	IDARUBICIN
8	1.12	7.5	2624	1	"	IDARUBICIN
9	0.29	7.5	2624	1	C26H27N1O10	11-DEOXY-4-O-DEMETHYL-DOX
16160	0.96	7.0	2304	459	C26H27N1O10	CARUBICIN **50935-04-1** *antineoplastic*
1	0.74	7.0	2304		C26H27N1O10	MEDORUBICIN **64314-52-9** *antibiotic*
2	3.45		2524	459	C26H27N3O3	IMIDAZOLYLETHANOL,4'-DIMETHYLKETAL-4"-(4-PYRIDYL) ANALOG
3	2.34	7.4	2095	1	C26H28Cl1N3O4S1	(THIENYL-CHLOROPHENYL)ADRENOCEPTOR ANTAGONIST ✚
4 ★	4.34	7.0	2230	2	C26H28Cl2N4O4	KETOCONAZOLE **65277-42-1** *antifungal*
65	-2.06	1.0	564	342	C26H28N6O2	TPA,3,3"-BIS(N,N'-DIMETHYLAMIDINO) HYDROCHLORIDE
6	-1.70	1.0	564	342	C26H28N6O2	TPA,4,4"-BIS(N'-EHTYLAMIDINE),DIHYDROCHLORIDE
7	-2.05	1.0	564	342	C26H28N6O2	TPA,4,4"BIS-(N,N'-DIMETHYLAMIDINO)
8	-1.83	1.0	547	342	C26H28N6O2	TPA,4,4"BIS-(N,N'-DIMETHYLAMIDINO)
9	4.03	7.4	594	1	C26H29N1O1	TAMOXIFEN **10540-29-1** *anti-estrogen, antineoplastic*
16170	4.70	7.4	2102	1	C26H30N2O4	PILOCARPIC ACID,O-BENZOYL,P-METHYLBENZYL ESTER
1 ★	4.85		2102	2	"	PILOCARPIC ACID,O-BENZOYL,P-METHYLBENZYL ESTER
2	4.60		2102	1	C26H30N2O4	PILOCARPIC ACID,O-BENZOYL,PHENETHYL ESTER
3	3.85	7.4	2102	1	C26H30N2O4	PILOCARPIC ACID,O-PHENYLACETYL,BENZYL ESTER
4 ★	4.00		2102	2	"	PILOCARPIC ACID,O-PHENYLACETYL,BENZYL ESTER
75 ★	3.20		2531		C26H30N2O5	PHENOXYACETAMIDE ANALOG,N-BENZYL **139733-95-2**
6	-1.85	1.0	564	342	C26H30N6O2	NSC#72378 ✚
7	-1.85	1.0	564	342	"	NSC#72378
8	4.18		2310	2	C26H30N6O2	XANTHINE,1,3-DIMETHYL-8-(4-DIMETHYLPHENYL)PIPERAZINYLETHYL
9	-2.20	1.0	562	342	C26H30N8O2	TPA-4,4"BIS(3-ETHYLGUANIDINO)
16180 ★	3.76		1520		C26H32Cl1F1O5	CLOBETASONE BUTYRATE **25122-57-0** *anti-inflammatory*
1 ★	3.19		569		C26H32F2O7	FLUOCINONID **356-12-7** *glucocorticoid*
2	0.48	7.4	253	1	C26H32N2O1	N-(3-NAPHTHYL-3-PHENYL-3-AMIDOPROPYL)DI-I-PR-AMINE
3	0.89	7.0	2682	1	C26H32N2O5	PHENYLPROPIONIC ACID,7-OXABICYCLOHPTANEOXAZOLEAMIDE, N-CYCLOHEXYL ANALOG *txa2-antagonist*
4	2.29	7.4	564	1	C26H32N2O7	APOCODEINE ANALOG: A2-OH, R6-CO(CH2)2N(ETOH)2, R11-ACETYLOXY
85	0.23	7.8	967		C26H32N4O6	DAUNORUBICIN-DI-ME-D-AGLYCONE
6	1.90		2609	468	C26H32N4O6	18-TETRALACTAM-DIETHER,N1,N2-DIBENZYL
7	1.25	7.4	564	1	C26H33N1O9S1	10-S-DEACETYLCOLCHICINE,N-GLUCOPYRANOSYL
8 ★	3.63		1520		C26H34Cl1F1O5	CLOBETASOL-17-BUTYRATE
9	1.06	1.0	1107	342	C26H34N2O3	N(1(B-PHET)PIPERIDIN-4YL)N-PR-ANTHRANIL.AC.PR.EST
16190	1.23	7.0	2682	1	C26H34N2O5	PHENYLPROPIONIC ACID,7-OXABICYCLOHEPTANEOXAZOLEAMIDE, N-DIMETHYLBUTYL ANALOG *txa2-antagonist*
1	1.00	7.0	2682	1	C26H34N2O5	PHENYLPROPIONIC ACID,7-OXABICYCLOHEPTANEOXAZOLEAMIDE, NMETHYL, N-PENTYL ANALOG *txa2-antagonist*
2 ★	3.82		1520		C26H34O6	PREDNISONE-17-VALERATE
3	3.59		1135	537	C26H34O7	FUMAGILLIN **23110-15-8** *antibiotic (amebicide)*
4 ★	3.55		1520		C26H35F1O6	BETAMETHASONE-21-BUTYRATE
95	3.58	1.0	547	342	C26H35N1O1	9-(2-DIPENTYLAMINO-1-HYDROXYET)PHENANTHRENE
6	2.09	7.4	2095	1	C26H35N3O4	(T-BUTYLIMIDAZOLE)ADRENOCEPTOR ANTAGONIST
7	2.20	7.4	2095	1	"	(T-BUTYLIMIDAZOLE)ADRENOCEPTOR ANTAGONIST
8	-0.05		1686	272	C26H36F1O6	DEXAMETHASONE-21-G-AMINOBUTYRATE **89231-65-2**
9	2.30	4.8	1020		C26H36N2O2	CINNAMAMIDE,N(2,6-DIME-4-BUOPH)N(3-DIMEAMINOPROP)
16200	2.19	7.0	965	1	C26H36N2O3	D-557(KNOLL) ✚
1	2.02	7.0	965	1	C26H36N2O4	PR-23(KNOLL) ✚
2	4.00	7.5	2651	28	C26H36N2O6S1	BENZAMIDE,2,3-DIMETHOXY-5-(3-TOSYLOXYPROPYL)-N-(1-ETHYLPYRROLIDIN-2YL)METHYL
3	0.52		559	820	C26H36N3S2.I1	3-H2-2(ET2-ME-N+ETS-)2(P-BUS-PH)BENZODIAZEPINI
4	-3.20	7.4	921	22	C26H36N4O4.(Cl1)2	MORFAMQUATDICHLORIDE
5 ★	2.82	7.4	1252	1	C26H36O5	DICIRENONE **41020-79-5** *hypotensive, aldosterone antagonist*
6 ★	1.63		2308		C26H36O9	2,7-DIBENZO-27-CROWN-9-ETHER
7	1.86	7.4	2680	324	C26H37N1O3	3-O-NONYLMORPHINE **74886-14-9**
8 ★	3.74	7.4	2242	1	C26H38F1O4	DEXAMETHASONE,17-B-(N-ISOPENTYLCARBOXAMIDE) ANALOG
9 ★	3.64	7.4	2242	1	C26H38F1O4	DEXAMETHASONE,17-B-(N-PENTYLCARBOXAMIDE) ANALOG
16210	3.52		817		C26H38N1O3.Br1	N(2-BENZILOYLOXYET-N-OCTYL-N,N-DIMETHYLAMMONIUMBR
1 ★	3.79		1520		C26H38O6	HYDROCORTISONE-17-VALERATE **57524-89-7** *glucocorticoid*
2 ★	3.62		1520		C26H38O6	HYDROCORTISONE-21-VALERATE
3	0.67	7.4	2242	1	C26H39F1N2O4	DEXAMETHASONE,17-B-(N-PROPYLAMINOETHYL-CARBOXAMIDE) ANALOG
4	3.83	7.3	2671	1	C26H42N2O3	N,N-DIETHYL-2-(2-BUTOXYPHENYLCARBAMOYLOXY)BORNAN-3YLMETHYLAMINE
15	-0.72		1383		C26H42N2.(Br1)2	1,1'-DIOCTYL-4,4'-BIPYRIDYLIUMBR
6 ★	-0.25		2377		C26H42N4O5	LEU-LEU-VAL-PHE
7 ★	-1.09		2377		C26H42N4O6	ILE-TYR-ILE-VAL
8	3.53		1937	459	C26H42O5	4-ACETYL-2-PYRONE,3,5-DIBUTYL-6-(2-BUTYLHEPTAN-2-ONYL) **74583-84-9**
9	-0.36		462	820	C26H44N1O7S1.C23H33N2O1	ISOPROPAMIDETAUROCHOLATE
16220	5.51	7.3	1149	714	C26H44N2O3	3(OPENT)CARBANILATE,O-(2-N,N-DIBUAMINO)CYCLOHEXYL **60093-66-5**
1	4.92	7.1	1357		C26H44N2O3	CARBAMICAC,N-HX-N-(M-PENOPH),2(1-PIPERIDINYL)PR.EST
2	5.26	7.0	1517	1	"	CARBAMICAC,N-HX-N-(M-PENOPH),2(1-PIPERIDINYL)PR.EST
3 ★	-0.09		2308		C26H44O11	BENZO-33-CROWN-11-ETHER
4 ✓	4.83		1698	820	C26H45N2O3.Br1	2-HEPTOXYPHCARBAMICAC.,PIPERIDINOET.EST.,N-PNBR
25 ✓	4.83		1720	820	C26H45N2O3.Br1	N(O-HEPTOXYPH)CARBAM.AC,2(N-PNPIPERIDINIUM)ET.EST
6 ✓	3.65		1698	820	C26H45N2O4P1	2-HEPTOXYPHCARBAMICAC.,PIPERIDINOET.EST.,N-PNPHOS
7 ★	2.32		2238		C26H46Cl2N8O6	ETHANE,1,2-BIS(3-NITROSOUREA-1-(2-CHLOROETHYL)-3-(1-OXO-2,2,6,6-TETRAMETHYLPIPERIDN-4-YL)
8 ✓	0.49		1966		C26H49Au2O9P2S1.Cl1	TETRAACETYLGLUCOSE,2-BIS(TRIETHYLPHOSPHO-GOLD)SULFONIUM CHLORIDE
9 ✓	-0.25		1966	820	C26H49Au2O9P2S1.N1O3	TETRAACETYLGLUCOSE,2-BIS(TRIETHYLPHOSPHINAURO)SUFONIUM NITRATE

	logP	pH	Ref.	Note	MF	Name / **CAS** / *activity*
16230	4.30	5.2	2538	2	C27H22Cl2N4	CLOFAZIMINE **2030-63-9** *antibacterial (tuberculo- & lepro-static)*
1	4.54	5.2	2538	2	"	CLOFAZIMINE
2	-1.57	1.0	564	342	C27H26N6O2S1	TPA,2-SME-4,4"-BISIMIDAZOLIN-2-YL
3	1.39	7.4	594	1	C27H28Br2O5S1	BROMTHYMOLBLUE **76-59-5**
4	2.78		505		"	BROMTHYMOLBLUE
35	1.68	7.0	2304		C27H28F1O10	4'-FLUORODOXORUBICIN
6	1.49	7.0	2304		C27H28I1O10	DOXORUBICIN,4'-IODO
7	1.50	7.0	2372	324	C27H28I1O10	DOXORUBICIN,4'-IODO
8	1.53	7.0	2304	459	C27H28I1O10	DOXORUBICIN,4'-IODO
9	0.95	7.0	2682	1	C27H28N2O6	PHENYLPROPIONIC ACID,7-OXABICYCLOHEPTANEOXAZOLEAMIDE,N-P-METHOXYPHENYL ANALOG *txa2-antagonist*
16240	1.34	7.5	2624	1	C27H28O11	DAUNORUBICIN,4'-O-(2',6'-DIDEOXYARABINO)
1	1.75	7.0	2372	324	C27H28O12	EPIRUBICIN,3'-DEAMINO-3'-HYDROXY
2	2.04	7.5	2624	1	C27H29N1O9	DAUNORUBICIN,4-DEMETHOXY-4'-O-METHYL
3	0.00	7.0	2304		C27H29N1O10	4'-DEOXY-ADRIAMYCIN **63521-85-7** *antineoplastic*
4	0.00	7.0	2372	324	C27H29N1O10	4'-DEOXY-ADRIAMYCIN
45	0.01	7.0	2304	459	C27H29N1O10	4'-DEOXY-ADRIAMYCIN
6	0.80	7.5	2624	1	C27H29N1O10	4'-DEOXY-ADRIAMYCIN
7	0.18	7.0	2372	324	C27H29N1O10	4'-EPI-DAUNORUBICIN **20830-81-3** *antineoplastic*
8	0.19	7.0	2304		C27H29N1O10	4'-EPI-DAUNORUBICIN
9	0.37	7.0	2304	459	C27H29N1O10	4'-EPI-DAUNORUBICIN
16250	0.44	7.8	967		C27H29N1O10	4'-EPI-DAUNORUBICIN
1	0.66	7.4	1977	1	C27H29N1O10	4'-EPI-DAUNORUBICIN
2	0.66	7.4	564	1	C27H29N1O10	4'-EPI-DAUNORUBICIN
3	0.89	7.5	2624	1	C27H29N1O10	4'-EPI-DAUNORUBICIN
4	1.78		562	400	C27H29N1O10	4'-EPI-DAUNORUBICIN
55	1.83	7.4	401	2	C27H29N1O10	4'-EPI-DAUNORUBICIN
6	0.39	7.0	2304	459	C27H29N1O10	DOXORUBICIN,4-DEMETHOXY-4-O-METHYL
7	0.04	7.0	2372	324	C27H29N1O11	4'-EPI-ADRIAMYCIN **56420-45-2**
8	0.05	7.0	2304	459	C27H29N1O11	4'-EPI-ADRIAMYCIN
9	0.06	7.0	2304		C27H29N1O11	4'-EPI-ADRIAMYCIN
16260	0.81	7.5	2624	1	C27H29N1O11	4'-EPI-ADRIAMYCIN
1	-0.32	7.0	2304	459	C27H29N1O11	ADRIAMYCIN **23214-92-8** *antineoplastic*
2	-0.32	7.8	967		"	ADRIAMYCIN
3	-0.30	7.0	2372	324	"	ADRIAMYCIN
4	-0.28	7.0	2304		"	ADRIAMYCIN
65	0.07	7.4	1977	1	"	ADRIAMYCIN
6	0.09	7.5	2624	1	"	ADRIAMYCIN
7	0.10	7.4	564	1	"	ADRIAMYCIN
8	0.26	7.3	2225	1	"	ADRIAMYCIN
9	1.27	7.4	401	2	"	ADRIAMYCIN
16270 ★	-0.02	7.0	2374		C27H29N3O4	PHE-PHE-PHE **2578-81-6**
1	1.90		2559	1	C27H30O9	FPL-55712 *antagonist-leukotriene (anti-asthmatic)* ✚
2	0.51	5.0	2375	1	C27H31N1O3	ASOCAINOL **77400-65-8** *anti-arrhythmic*
3	0.69	7.5	2624	1	C27H31N1O10	DAUNORUBICINOL **28008-55-1**
4 ★	-0.34		2648	448	C27H31N2O2.Cl1	N-BENZYLCINCHONIDINE CHLORIDE
75 ★	-0.55		2648	448	C27H31N2O2.Cl1	N-BENZYLCINCHONINE CHLORIDE
6	1.58	7.4	2095	1	C27H31N3O4S1	(5-THIENYL-IMIDAZOLE)-SPACED-ADRENOCEPTOR ANTAGONIST ✚
7	2.76	7.4	2095	1	C27H31N3O4S1	(THIENYL-METHYLPHENYL)ADRENOCEPTOR ANTAGONIST
8 ★	-1.25		2377		C27H31N5O6	TYR-PRO-GLY-TRP
9	0.02		2286		C27H32N1O3.I1	NORETHINDRONE,NICOTINYL ESTER,METH IODIDE
16280	0.87	7.0	2682	1	C27H32N2O5	PHENYLPROPIONIC ACID,7-OXABICYCLOHEPTANEOXAZOLEAMIDE,N-HEPTYNYL ANALOG *txa2-antagonist*
1	3.73	7.4	2310	2	C27H32N6O2	XANTHINE,1,3-DIMETHYL-8-(4-DIPHENYLMETHYL)PIPERAZINYLPROPYL
2	4.44		2286		C27H33N1O3	NORETHINDRONE,N-METHYL-1,4-DIHYDRONICOTINYL ESTER
3	-1.89	7.4	1298	1	C27H33N2O7S2.Na1	ISOSULFANBLUE **68238-36-8** *diagnostic aid (lymphangiography)*
4	1.47	7.4	608	1	C27H33N3O1	DIISOPYRAMIDE,N1-PHENYL
85	4.73		569		C27H35Cl2F1O6	2,9-CL2-6A-FL-16A-METHYLPREDISOLONE-21-PIVALATE
6	5.74		2415	459	C27H35N1O4	DEXTRONANTRADOL
7	5.70		2415	459	C27H35N1O4	LEVONANTRADOL **71048-87-8** *analgesic*
8 ★	3.86		569		C27H36F2O6	FLUMETHASONE-21-PIVALATE **2002-29-1** *glucocorticoid*
9	2.88	7.0	2682	1	C27H36N2O5	PHENYLPROPIONIC ACID,7-OXABICYCLOHEPTANEOXAZOLEAMIDE,N-HEPTYL ANALOG *txa2-antagonist*
16290	0.21	7.4	2095	1	C27H36N4O5	(5-N-MORPHOLINOMETHYL-IMIDAZOLE)ADRENOCEPTOR ANTAGONIST
1 ★	3.60		1520		C27H37F1O6	BETAMETHASONE VALERATE **2152-44-5** *glucocorticoid*
2 ★	3.87		1520		C27H37F1O6	BETAMETHASONE-21-VALERATE
3 ★	3.54	7.4	2242	1	C27H38F1O4	DEXAMETHASONE,17-B-(N-CYCLOHEXYL-CARBOXAMIDE) ANALOG
4	-0.31	7.4	2242	1	C27H38F1O6	DEXAMETHASONE-17-B-(N-HEXANOIC ACID-CARBOXAMIDE) ANALOG
95	2.36	7.0	965		C27H38N2O2	D-559(KNOLL)-VERAPAMIL ANALOG ✚
6	1.83	7.4	1754	320	C27H38N2O4	VERAPAMIL **52-53-9** *vasodilator (coronary)*
7	2.15	7.0	965	1	"	VERAPAMIL
8 ★	3.79	9.0	2017		"	VERAPAMIL
9	4.29	7.5	2651	28	C27H38N2O6S1	BENZAMIDE,2-ETHOXY-3-METHOXY-5-(3-TOSYLOXYPROPYL)-N(-1-ETHYLPYRROLIDIN-2YL)METHYL
16300	0.89		559	820	C27H38N3S2.I1	3-H2-2(ET2-ME-N+ETS)-4(P-IPNSPH)BENZODIAZEPINI
1	3.30		2151		C27H38O6	CORTISONE-21-HEXANOATE
2 ★	0.11	7.0	2002	1	C27H38O10	PREDNISOLONE-21-GALACTOSIDE **92901-24-1**
3 ★	0.27	7.0	2002	1	C27H38O10	PREDNISOLONE-21-GLUCOSIDE
4	0.25	7.0	2002	1	C27H39F1O10	FLUDROCORTISONE-21-GALACTOSIDE **92901-26-3**
5	0.84	7.0	2002	1	C27H39F1O10	FLUDROCORTISONE-21-GLUCOSIDE
6 ★	4.54	7.4	2242	1	C27H40F1O4	DEXAMETHASONE,17-B-(N-HEXYLCARBOXAMIDE) ANALOG
7 ★	3.93	7.4	2242	1	C27H40F1O4	DEXAMETHASONE,17-B-(N-I-HEXYLCARBOXAMIDE) ANALOG
8 ★	2.65	7.4	2242	1	C27H40F1O6	DEXAMETHASONE,17-B-(N-DIETHOXYETHYL-CARBOXAMIDE) ANALOG
9	3.70		2151		C27H40O6	HYDROCORTISONE-21-CAPROATE
16310 ★	0.15	7.0	2002		C27H40O10	HYDROCORTISONE-21-GALACTOSIDE **92901-25-2**
1 ★	0.44	7.0	2002	1	C27H40O10	HYDROCORTISONE-21-GLUCOSIDE
2	-0.63	7.4	2242	1	C27H41F1N2O4	DEXAMETHASONE,17-B-(N-(6-AMINOHEXYL)CARBOXAMIDE) ANALOG
3	1.20		579		C27H42N1O2.Cl1	BENZETHONIUMCHLORIDE **121-54-0** *anti-infective (topical), preservative*
4	4.28	7.3	2671	1	C27H42N2O3	N,N-DIETHYL-2-(2-PENTOXYPHENYLCARBAMOYLOXY)BORNAN-3YLMETHYLAMINE
15	10.51		2428	459	C27H44N2O6S1	N,N'-DIMETHYLTHIODICARBAMATE,O-3,4-METHYLENEDIOXYPHENYL-O'-HEXADECYL
6 ★	0.24		2377		C27H44N4O5	LEU-LEU-LEU-PHE
7 ✓	5.38		1698	820	C27H47N2O3.Br1	2-HEPTOXYPHCARBAMICAC.,PIPERIDINOET.EST.,N-HXBR
8 ✓	5.38		1720	820	C27H47N2O3.Br1	N(O-HEPTOXYPH)CARBAM.AC,2(N-HXPIPERIDINIUM)ET.EST
9 ✓	4.18		1698	820	C27H47N2O3.H2O4P1	2-HEPTOXYPHCARBAMICAC.,PIPERIDINOET.EST.,N-HXPHOS

	logP	pH	Ref.	Note	MF	Name / CAS / activity
16320	3.23		1635	820	C27H50N1.Cl1	OCTADECYLDIMETHYLBENZYLAMMONIUMCHLORIDE
1	3.86		2524	459	C28H23N3O1	IMIDAZOLYLETHANOL,4'-PHENYL-4"-(4-PYRIDYL) ANALOG ✦
2	7.10	7.4	783		C28H28Cl1F2N3O1	CLOPIMOZIDE 53179-12-7 antipsychotic
3	-2.75		564	1	C28H28N6O2	NSC-55158, IMIDAZOLIN-PHTHALANILIDE ANALOG
4	-2.62	1.0	564	342	"	NSC-55158, IMIDAZOLIN-PHTHALANILIDE ANALOG
25	-2.56	1.0	562	342	C28H28N6O2	TPA,4,4"-BIS(1,4,5,6-H-2-PYRIMIDINYL) 73-53-0 antineoplastic
6	-2.33	1.0	564	342	"	TPA,4,4"-BIS(1,4,5,6-H-2-PYRIMIDINYL)
7	-2.34	1.0	564	342	C28H28N6O2	TPA,4,4"-BIS(1-ME)-2-IMIDAZOLIN-2-YL),DIHYDROCHLORIDE
8	-1.20	1.0	564	342	C28H28N6O2	TPA,4,4"-BIS(4-ME-2-IMIDAZOLIN-2-YL)
9	-1.76	1.0	564	342	C28H28N6O3	ITPA,4-OET-4,4"-BIS-IMIDAZOLIN-2-YL
16330	-2.23	1.0	564	342	C28H28N6O4	NSC#77885 ✦
1	1.55	7.0	2682	1	C28H29Cl1N2O5	PHENYLPROPIONIC ACID,7-OXABICYCLOHEPTANEOXAZOLEAMIDE,N-P-CHLOROPHENETHYL ANALOG
2 ★	6.30		783		C28H29F2N3O1	PIMOZIDE 2062-78-4 antipsychotic
3	-0.90	1.0	564	342	C28H29N7O2	NSC-67735, IMIDAZOLIN-PHTHALANILIDE ANALOG
4 ★	3.58		547		C28H30N2O4	NSC-309697 ✦
35	0.98	7.0	2682	1	C28H30N2O5	PHENYLPROPIONIC ACID,7-OXABICYCLOHEPTANEOXAZOLEAMIDE,N-PHENETHYL ANALOG txa2-antagonist
6	-1.81	1.0	564	342	C28H30N8O2	TPA,2,5-DIME-4,4"BIS(2-IMIDAZOLIN-2YL)AMINO
7	4.10	7.4	591	1	C28H31F1N4O1	ASTEMIZOLE 68844-77-9 anti-allergic, antihistaminic
8	0.94	7.5	2624	1	C28H31N1O11	DOXORUBICIN,4'O-METHYL
9 ★	3.60		2531		C28H32N2O5	PHENOXYACETAMIDE ANALOG (N IN TETRAHYDRO-I-QUINOLINE) 139734-30-8
16340	0.44		559	820	C28H32N3S2.I1	TIBEZONIUM IODIDE 54663-47-7 antibacterial
1 ✓	-0.14	0.3	1766		C28H32N4O2	CHROMOPYRAZOLE HYDRO IODIDE
2 ✓	-0.80	0.3	1766		C28H32N4O2	CHROMOPYRAZOLE HYDROBROMIDE
3 ✓	-1.38	0.3	1766		C28H32N4O2	CHROMOPYRAZOLE HYDROCHLORIDE
4	-1.54	0.3	1766		C28H32N4O2	CHROMOPYRAZOLE HYDROFLUORIDE
45 ✓	0.38	0.3	1766		C28H32N4O2	CHROMOPYRAZOLETHIOCYANATE
6	-1.19	1.0	564	342	C28H32N6O2	TPA,4,4"-BIS(N'-PROPYLAMIDINO),DIHYDROCHLORIDE
7 ✓	-1.59	1.0	564	342	C28H32N6O2	TPA,4,4"-BIS(N-ETHYL-N'-METHYLAMIDINO),DIHYDROCHLORIDE
8	-2.45	1.0	564	342	C28H32N6O2	TPA,4,4"-BIS(TRIMETHYLAMIDINO)
9 ★	3.44	7.4	2242	1	C28H33F1N2O6	DEXAMETHASONE,17-B-(N-(P-NITROBENZYL)CARBOXAMIDE) ANALOG
16350	0.94	4.0	2644	210	C28H33N1O2	FOMOCAINE ANALOG-17 (4-BENZYL)
1	1.53	7.4	2095	1	C28H33N3O4S1	(5-THIENYL-IMIDAZOLE)-SPACED-ADRENOCEPTOR ANTAGONIST ✦
2	0.36	7.5	2624	1	C28H33N3O9	9-(FORMYLHYDRAZONE-1-ETHYL)-DNR
3 ★	0.30	7.4	2374		C28H34N5O4	TRYPTOPHANYTRYPTOPHANYLLEUCINE
4 ★	3.28	7.4	2242	1	C28H34F1O4	DEXAMETHASONE,17-B-(N-BENZYLCARBOXAMIDE) ANALOG
55	-0.03		2286		C28H34N1O3.I1	ETHISTERONE,NICOTINYL ESTER,METH IODIDE
6	-0.51	7.8	967		C28H34N2O9	DAUNORUBICIN-MO-ME-A 61481-24-1
7	4.37	7.4	2244	1	C28H34N4O6S1	TIMOLOL, O-(2-BENZOYLOXYMETHYL)BENZOYL
8	3.94	7.4	2310	2	C28H34N6O2	XANTHINE, 1,3,7-TRIMETHYL-8-(4-DIPHENYLMETHYL)PIPERAZINYLPROPYL
9	5.23	7.4	2310	2	C28H34N6O2	XANTHINE, 1-METHYL-3-I-BUTYL-8-(4-DIPHENYLMETHYL)PIPERAZINYLMETHYL
16360	4.58	7.4	2310	2	C28H34N6O2	XANTHINE,3-I-BUTYL-8-(4-DIPHENYLMETHYL)PIPERAZINYLETHYL
1	-2.25	1.0	564	342	C28H34N8O2	TPA,4,4"BIS-N-ME-N'-DIME HYDRAZIDE
2	4.22		569		C28H35Cl1F2O7	2-CHLOROFLUMETHASONE-17,21-DIPROPIONATE
3	4.49		2286		C28H35N1O3	ETHISTERONE,N-METHYL-1,4-DIHYDRONICOTINYL ESTER
4	0.85	7.4	2684	1	C28H35N3O6	P-HYDROXYPHENYLACETIC ACID ESTER,PYRROLIDINAMIDO ANALOG
65	0.98	7.4	2684	1	C28H36N2O5	P-HYDROXYPHENYLACETIC ACID ESTER,3-(4-PIPERIDIN-1-YL-PIPERIDINYL)METHYL antiarrhythmic (vitro)
6	1.03	7.0	2682	1	C28H36N2O5	PHENYLPROPIONIC ACID,7-OXABICYCLOHEPTANEOXAZOLEAMIDE,N-CYCLOHEXYLETHYL ANALOG txa2-antagonist
7	0.50	7.0	2682	1	C28H36N2O7	7-OXABICYCLOHEPTANE-OXAZOLE,DIMETHYLHEXANOIC ACID ANALOG txa2-antagonist
8	2.68	6.5	1277	1	C28H36N2O8S2	PYRITHIOXINTETRAPROPIONATE
9 ★	1.66		547		C28H36O11	BRUCEANTIN 41451-75-6 antineoplastic
16370	4.07		569		C28H37F1O7	BETAMETHASONE-17,21-DIPROPIONATE 5593-20-4 glucocorticoid
1	5.73		2415	459	C28H37N1O4	DEXTRONANTRADOL,N-METHYL
2	5.72		2415	459	C28H37N1O4	LEVONANTRADOL,N-METHYL
3	0.65		401		C28H37N1O9	HARRINGTONINE ✦
4	3.41	7.0	2682	1	C28H38N2O5	PHENYLPROPIONIC ACID,7-OXABICYCLOHEPTANOXAZOLEAMIDE,N-DIMETHYLHEXYL ANALOG txa2-antagonist
75	0.82		559		C28H38N3S2.I1	3-H2-2(ET2-ME-N+ETS-)4(P-C6H11-S-PH)BENZODIAZEPINI
6	1.65	6.0	695		C28H38N4O1	CARPIPRAMINE 5942-95-0 psychotropic
7 ★	0.49	7.0	2002	1	C28H39F1O10	DEXAMETHASONE-21-GALACTOSIDE 92901-23-0
8 ★	0.59	7.0	2002	1	C28H39F1O10	DEXAMETHASONE-21-GLUCOSIDE
9	4.40	1.0	547	342	C28H39N1O1	9-(2-DIHEXYLAMINO-1-HYDROXYET)PHENANTHRENE
16380	2.12	7.0	965	1	C28H40N2O5	GALLOPAMIL 16662-47-8 calcium channel blocker
1 ★	1.82		2308		C28H40O10	2,8-DIBENZO-30-CROWN-10-ETHER
2 ★	1.84		2308		C28H40O10	5,5-DIBENZO-30-CROWN-10-ETHER 17455-25-3
3 ✓	3.95		1720	820	C28H41N3O3.Cl1	N(O-HEPTOXYPH)CARBAM.AC,2(N-BNZPIPERIDINIUM)ET.EST
4	0.62		2241	1	C28H41N3O3	OXETHAZAINE 126-27-2 anesthetic (topical)
85	4.57		817		C28H42N1O3.Br1	N(2-BENZILOYLOXYET-N-DECYL-N,N-DIMETHYLAMMONIUMBR
6 ★	0.23		2377		C28H43N5O5	TRP-LEU-LEU-VAL
7	1.07		1403		C28H44N4O8	SPECTINOMYCYLAMINE,4-(P-HEXAMIDO)PHENYLACYL
8	4.60	7.3	2671	1	C28H46N2O3	N,N-DIETHYL-2-(2-HEXOXYPHENYLCARBAMOYLOXY)BORNAN-3YLMETHYLAMINE
9 ✓	5.91		1698	820	C28H49N2O3.Br1	2-HEPTOXYPHENYLCARBAMIC ACID,PIPERIDINIUM-ET.ESTER,N-HEPTYL ✦
16390 ✓	5.68		1720	820	C28H49N2O3.I1	N(O-HEPTOXYPH)CARBAM.AC,2(N-HEXPIPERIDINIUM)ET.EST
1 ✓	4.78		1698	820	C28H49N2O3.H2O4P1	2-HEPTOXYPHCARBAMICAC.,PIPERIDINOET.EST.,N-HPTPHOS
2 ✓	3.40	7.3	2085		C28H60N4.O2Tc1	N-OCTADECYLCYCLAM,TECHNICIUM OXIDE COMPLEX
3 ★	1.19		1551		C29H23N3O9	1-(2,3,5-TRIBENZOYL-D-RIBOFURANOSYL)-2-NO2-IMIDAZOLE
4	-2.55	1.0	564	342	C29H27N5O1.(Br1)2	SN-6999 ✦
95	1.48	7.5	2624	1	C29H28F3N1O12	DOXORUBICIN,N-TRIFLUOROACETYL
6	-0.63		401		C29H28N6O2.(C2H5O4S1)2	NSC140026 ✦
7	-1.60		527	820	C29H29N7O1.(C7H7O3S1)2	NSC 114347
8	2.28	7.4	2025	1	C29H30N2O10	DAUNOMYCIN,N"-CYNAMOMETHYL 103450-88-0
9	1.20	7.4	2025	1	C29H30N2O11	ADRIAMYCIN,N"-CYANOMETHYL 103450-92-6
16400	5.86		2162	459	C29H31F2N3O1	FLUSPIRILENE 1841-19-6 antipsychotic
1	1.85	7.5	2624	1	C29H31N1O11	DAUNORUBICIN,N-ACETYL
2	0.43	7.4	2025	1	C29H31N1O12	DAUNOMYCIN,N"-CARBOXYMETHYL 103620-81-1
3	1.27	7.5	2624	1	C29H31N1O12	DOXORUBICIN,N-ACETYL
4	0.86	7.4	1210	1	C29H32Cl1N1O10	3"-DEAMIDO-8-DEME-8-CL-NOVOBIOCIN
5	1.65	7.4	2025	1	C29H32N2O11	DAUNOMYCIN,N"-AMINOCARBONYLMETHYL 103620-82-2
6	0.64	7.8	967		C29H32N2O11	DAUNORUBICIN-GLYCINE
7 ★	0.17		2377		C29H32N4O5	PHE-PHE-GLY-PHE
8 ★	0.60	7.4	547	1	C29H32O13	ETOPSIDE 33419-42-0 antineoplastic
9	1.37	7.5	2624	1	C29H33N1O10	DAUNORUBICIN,N,N-DIMETHYL

	logP	pH	Ref.	Note	MF	Name / CAS / activity
16410	1.56	7.5	2624	1	C29H33N1O11	DOXORUBICIN,N-ACETYL-13-DIHYDRO
1	4.16	7.4	2310	2	C29H34F2N6O2	XANTHINE,1-METHYL-3-I-BUTYL-8-(4-DI-P-FLUOROPHENYLMETHYL)PIPERAZINYLETHYL
2	0.29		559		C29H34N3S2.I1	3-H2-2(ET2-ME-N+ETS)-4(P-PHCH2SPH)BENZODIAZEPINI
3 ★	1.66		2330		C29H34O7	MICROLENIN ✚
4 ★	3.66	7.4	2242	1	C29H36F1O4	DEXAMETHASONE,17-B-(2-PHENETHYLCARBOXAMIDE) ANALOG
15 ★	2.89	7.4	2242	1	C29H36F1O5	DEXAMETHASONE,17-B-(N-(1-PHENYLETHANOL)CARBOXAMIDE) ANALOG
6 ★	3.27	7.4	2242	1	C29H36F1O5	DEXAMETHASONE,17-B-(N-(M-METHOXYBENZYL)CARBOXAMIDE) ANALOG
7	4.02	7.4	2310	2	C29H36N6O2	XANTHINE,1-METHYL-3-I-BUTYL-8-(4-DIPHENYLMETHYL)PIPERAZINYLETHYL
8	5.21	7.4	2310	2	C29H36N6O3	XANTHINE,3-I-BUTYL-1-METHYL-8-(2-(4-DIPHENYLMETHYL)PIPERAZINYL)ETHANOL
9 ★	3.37	7.4	594	1	C29H37N1O5	CYTOCHALASEN-B 14930-96-2
16420 ★	0.82		401		C29H39N1O9	HOMOHARRINGTONINE 26833-87-4 antineoplastic
1	0.53	7.0	2682	1	C29H39N3O5	PHENYLPROPIONIC ACID,7-OXABICYCLOHEPTANOXAZOLEAMIDE,N-PIPERIDINYLBUTYL ANALOG txa2-antagonist
2	3.24	10.1	401	1	C29H40N2O4	EMETINEHCL 316-42-7
3	2.36	7.0	965	1	C29H42N2O4	D-594(KNOLL)-VERAPAMIL ANALOG ✚
4	4.22		2151		C29H42O6	CORTISONE-21-OCTANOATE
25 ★	0.30		1036	812	C29H42O9	HELVETICOSIDE 630-64-8
6 ★	0.64	7.4	2285	1	C29H42O9	K-STROPHANTHIN-ALPHA
7	-0.48		1036	812	C29H42O10	CONVALLOTOXIN 508-75-8 cardiotonic
8	4.37		2151		C29H44O6	HYDROCORTISONE-21-OCTANOATE
9 ★	2.51	7.4	2285	1	C29H44O7	DIGITOXIGENIN-MONODIGITOXOSIDE
16430 ★	1.10	7.4	2285	1	C29H44O8	DIGOXIGENIN-MONODIGITOXOSIDE
1 ★	2.10		1036	812	C29H44O8	EVOMONOSIDE 508-93-0
2 ★	1.78	7.4	2285	1	C29H44O8	GITOXIGENIN-MONODIGITOXOSIDE
3	-0.64		1036	812	C29H44O9	DIGITOXIGENIN-3-GLUCORONIDE
4 ★	0.15		1036	812	C29H44O9	HELVETICOSOL 18695-02-8
35 ★	-0.62		1036	812	C29H44O10	CONVALLOTOXOL 3253-62-1
6 ★	-1.70	7.4	2285	1	C29H44O12	STROPHANTHIN-G 630-60-4 cardiotonic
7	0.48	7.4	2242	1	C29H45F1N2O4	DEXAMETHASONE,17-B-(N-(8-OCTYLAMINO)CARBOXAMIDE) ANALOG
8	-0.70	7.4	2285	1	C29H46O12	OUABAIN,DIHYDRO
9	4.90	7.3	2671	1	C29H48N2O3	N,N-DIETHYL-2-(2-HEPTOXYPHENYLCARBAMOYLOXY)BORNAN-3YLMETHYLAMINE
16440	2.13		2176		C29H50N2O6	1-LINOLENOYL-2,3-BIS(4-AMINOBUTYRYL)PROPANE-1,2,3-TRIOL 93383-17-6
1 ✓	6.51		1698	820	C29H51N2O3.Br1	2-HEPTOXYPHCARBAMICAC.,PIPERIDINOET.EST.,N-OCTBR
2 ✓	5.08		1698	820	C29H51N2O3.H2O4P1	2-HEPTOXYPHCARBAMICAC.,PIPERIDINOET.EST.,N-OCTPHOS
3 ✓	0.62		2176		C29H52N2O6.(C2H4O2)2	1-LINOLEOYL-2,3-BIS(4-AMINOBUTYRYL)PROPANE-1,2,3-TRIOL-DIACETATE 108920-65-6
4 ✓	0.78		2176		C29H54N2O6.(C2H4O2)2	1-OLEOYL-2,3-BIS(4-AMINOBUTYL)PROPANE-1,2,3-TRIOL-DIACETATE 108920-63-4
45 ✓	0.93		2176		C29H56N2O6.(C2H4O2)2	1-STEAROYL-2,3-BIS(4-AMINO)BUTYRYL)PROPANE-1,2,3-TRIOL-DIACETATE 108920-61-2
6	0.37	7.4	672	1	C30H22O12	XANTHOMEGNIN ✚
7	-1.81	7.0	564	342	C30H32N6O2	NSC72583 ✚
8	3.52	7.0	2682	1	C30H33Cl1N2O5	PHENYLPROPIONIC ACID,7-OXABICYCLOHEPTANEOXAZOLEAMIDE,N-P-CHLOROPHENYLBUTYL ANALOG
9	2.17	7.0	2682	1	C30H33F1N2O5	PHENYLPROPIONIC ACID,7-OXABICYCLOHEPTANEOXAZOLEAMIDE,N-P-FLUOROPHENYLBUTYL ANALOG
16450	1.80	7.0	2682	1	C30H34N2O5	PHENYLPROPIONIC ACID,7-OXABICYCLOHEPTANEOXAZOLEAMIDE,N-PHENYLBUTYL ANALOG txa2-antagonist
1	1.49	7.0	2682	1	C30H34N2O6	PHENYLPROPIONIC ACID,7-OXABICYCLOHEPTANEOXAZOLEAMIDE,N-P-HYDROXYPHENYLBUTYL ANALOG
2	0.78	7.4	1210	1	C30H35N1O10	3"-DEAMIDONOVOBIOCIN
3	4.97	7.4	2310	2	C30H36F2N6O2	XANTHINE,1,7-DIMETHYL-3-I-BUTYL-8-(4-DI-P-FLUOROPHENYLMETHYL)PIPERAZINYLETHYL
4	-0.08	1.0	564	342	C30H36N6O2	TPA,4,4"-BIS(N'-BUTYLAMIDINO) ✚
55	-1.34	1.0	564	342	C30H36N6O2	TPA,4,4"-BIS(N'-METHYL-N'-PROPYLAMIDINO), DIHYDROCHLORIDE
6	5.10	7.4	2310	2	C30H37N5O2	XANTHINE,1-METHYL-3-I-BUTYL-8-(4-DIPHENYLMETHYL)PIPERIDIN-1-YLETHYL
7	-0.10	7.8	1811		C30H38O2O10	DAUNORUBICIN-9-(1-(3-HYDROXYPRANYL)AMINOET)
8	5.62	7.4	2310	2	C30H38N6O2	XANTHINE,1,7-DIMETHYL-3-I-BUTYL-8-(4-DIPHENYLMETHYL)PIPERAZINYLETHYL
9	5.39	7.4	2310	2	C30H38N6O2	XANTHINE,1,7-DIMETHYL-3-PROPYL-8-(4-DIPHENYLMETHYL)PIPERAZINYLPROPYL
16460	4.29	7.4	2310	2	C30H38N6O2	XANTHINE,1-METHYL-3-I-BUTYL-8-(4-DIPHENYLMETHYL)PIPERAZINYLPROPYL
1	1.25	7.4	2591	1	C30H39N5O5	AZT,5'-RETINOYL 125780-97-4 antiretroviral prodrug
2	3.93	7.0	2682	1	C30H40N2O5	PHENYLPROPIONIC ACID,7-OXABICYCLOHEPTANEOXAZOLEAMIDE,N-CYCLOHEXYLBUTYL ANALOG txa2-antagonist
3	3.86	7.0	2682	1	C30H42N2O5	PHENYLPROPIONIC ACID,7-OXABICYCLOHEPTANEOXAZOLEAMIDE,N-DECYL ANALOG txa2-antagonist
4	1.33	7.5	1476	1	C30H42N8O5	METHOTREXATE,DIAMYL ESTER
65	2.48		1036	812	C30H42O8	PROSCILLARIDIN 466-06-8 cardiotonic
6	5.51	1.0	547	342	C30H43N1O1	9-(2-DIHEPTYLAMINO-1-HYDROXYET)PHENANTHRENE
7	0.64		1036		C30H44O9	CYMARIN 508-77-0
8	1.05		1036	812	C30H44O9	PERUVOSIDE 1182-87-2 cardiotonic
9 ★	1.45		2308		C30H44O11	4,7-DIBENZO-33-CROWN-11-ETHER
16470	4.77		2629		C30H46O4	GLYCYRRHETINIC ACID 471-53-4 anti-inflammatory (topical)
1	2.37		1036	812	C30H46O8	16-DESACETYLOLEANDRIN
2 ★	2.10		1036	812	C30H46O8	NERIIFOLIN 466-07-9 cardiotonic
3	0.59		1036	812	C30H46O8	STROSPESIDE 595-21-1
4 ★	0.56		1036	812	C30H46O9	CYMAROL 465-84-9
75	0.89	7.4	2242	1	C30H47F1N2O4	DEXAMETHASONE,17-B-(N-(9-NONYLAMINO)CARBOXAMIDE) ANALOG
6 ★	2.45		594		C30H53N1O11	BENZONATATE 104-31-4 antitussive
7 ✓	7.20		1698	820	C30H53N2O3.Br1	2-HEPTOXYPHCARBAMICAC.,PIPERIDINOET.EST.,N-NONBR
8 ✓	5.28		1698	820	C30H53N2O3.H2O4P1	2-HEPTOXYPHCARBAMICAC.,PIPERIDINOET.EST.,N-NONPHOS
9	4.39	5.2	2538	2	C31H31Cl2N5	CLOFAZIMINE ANALOG: 3-(3-DIETHYLAMINO)PROPYLIMINO
16480	4.69	5.2	2538	2	"	CLOFAZIMINE ANALOG: 3-(3-DIETHYLAMINO)PROPYLIMINO
1	0.03	7.4	2025	1	C31H32N2O12	DAUNOMYCIN,N"-CYANOETHYL,N"-ACETIC ACID 103620-84-4
2	2.64	7.4	2025	1	C31H34N2O11	DAUNOMYCIN,N"-2-(3-METHOXY)PROPIONITRILE 103667-18-1
3	0.92	7.4	2025	1	C31H34N2O12	ADRIAMYCIN,N"-2-(3-METHOXY)PROPIONITRILE 103620-79-7
4 ★	2.31	7.4	1977	1	C31H35N1O11	MORPHOLINO-DAUNORUBICIN 80844-67-3
85 ★	1.73	7.4	1977	1	C31H35N1O12	MORPHOLINO-ADRIAMYCIN 80790-68-7
6	1.66	7.0	2682	1	C31H36N2O5S1	PHENYLPROPIONIC ACID,7-OXABICYCLOHEPTANEOXAZOLEAMIDE,N-P-METHIOPHENYLBUTYL ANALOG
7	2.49	7.0	2682	1	C31H36N2O6	PHENYLPROPIONIC ACID,7-OXABICYCLOHEPTANEOXAZOLEAMIDE,N-P-METHOXYPHENYLBUTYL ANALOG
8	0.32	7.0	2682	1	C31H36N2O7S1	PHENYLPROPIONIC ACID,7-OXABICYCLOHEPTANEOXAZOLEAMIDE,N-P-METHYLSULFONYLPHENYLBUTYL ANALOG
9	0.23	7.4	564	1	C31H36N2O11	14-MORPHOLINODAUNOMYCIN
16490	1.26	7.8	967		C31H36N2O11	DAUNORUBICIN-N,N-DIMETHYLGLYCINE
1 ★	1.80	7.4	1977	1	C31H36N2O11	MORPHOLINO-DOXORUBICIN,12-IMINO 89196-05-4
2	0.72	7.4	1210	1	C31H36N2O11	NOVOBIOCIN 303-81-1 antibacterial
3	1.58	7.4	547	1	"	NOVOBIOCIN
4 ★	2.03	7.4	1977	1	C31H37N1O11	MORPHOLINO-DAUNORUBICIN,13-DIHYDRO 79951-58-9
95	4.81	7.4	2310	2	C31H38F2N6O2	XANTHINE,1,7-DIMETHYL-3-I-BUTYL-8-(4-DI-P-FLUOROPHENYLMETHYL)PIPERAZINYLPROPYL
6	2.59	7.0	965	1	C31H38N2O4	D-490(KNOLL)-VERAPAMIL ANALOG ✚
7 ★	1.43	7.4	1977	1	C31H38N2O11	MORPHOLINO-DOXORUBICIN,12-IMINO-13-DIHYDRO 89164-73-8
8	5.88	7.4	2310	2	C31H40F2N6O2	XANTHINE,1-METHYL-3-I-BUTYL-8-(DI-P-FLUOROPHENYLMETHYL)AMINOETHYLAMINOETHYL,N,N'-DIETHYL
9	0.63	7.8	967		C31H40N2O9	DAUNORUBICIN-BU-A 61481-26-3

	logP	pH	Ref.	Note	MF	Name / CAS / activity
16500	4.93	7.4	2310	2	C31H40N6O2	XANTHINE,1,7-DIMETHYL-3-I-BUTYL-8-(4-(DIPHENYLMETHYL)PIPERAZINYLPROPYL
1	5.02	7.4	2310	2	C31H40N6O2	XANTHINE,1-ETHYL-3-I-BUTYL-8-(4-(DIPHENYLMETHYL)PIPERAZINYLPROPYL
2	1.18	7.0	2682	1	C31H41N3O6	7-OXABICYCLOHEPTANE-OXAZOLE ANALOG txa2-antagonist
3	3.89	7.0	2682	1	C31H42N2O5	PHENYLPROPIONIC ACID,7-OXABICYCLOHEPTANEOXAZOLEAMIDE,N-METHYL,N-CYCLOHEXYLBUTYL ANALOG
4 ★	3.40	7.4	2242	1	C31H44F1O6	DEXAMETHASONE,17-(N-(2-PHENYLPROPIONIC ACID)CARBOXAMIDE) ANALOG
5	2.81		2454	700	C31H47N1O4	CHOLYLAMIDE-N-BENZYL
6	2.81		2454	700	"	CHOLYLAMIDE-N-BENZYL
7	3.54		2454	700	"	CHOLYLAMIDE-N-BENZYL
8	3.54		2454	700	"	CHOLYLAMIDE-N-BENZYL
9	4.95		2454	700	C31H48Cl1N1O1	LITHOCHOLYL-24-(M-CHLOROBENZYL)AMINE
16510	5.05		2454	700	C31H48Cl1N1O1	LITHOCHOLYL-24-(O-CHLOROBENZYL)AMINE
1	5.00		2454	700	C31H48Cl1N1O1	LITHOCHOLYL-24-(P-CHLOROBENZYL)AMINE
2	4.90		2454	700	C31H49N1O1	LITHOCHOLYL-24-BENZYLAMINE
3	4.20		2454	700	C31H49N1O2	CHENODEOXYCHOLYL-24-BENZYLAMINE
4	4.80		2454	700	C31H49N1O2	DEOXYCHOLYL-24-BENZYLAMINE
15	4.15		2454	700	C31H49N1O2	HYODEOXYCHOLYL-24-BENZYLAMINE
6	4.00		2454	700	C31H49N1O2	URSODEOXYCHOLYL-24-BENZYLAMINE
7	3.60		2454	700	C31H49N1O3	CHOLYL-24-BENZYLAMINE
8	3.60		2454	700	"	CHOLYL-24-BENZYLAMINE
9	3.82		2454	700	"	CHOLYL-24-BENZYLAMINE
16520	3.82		2454	700	"	CHOLYL-24-BENZYLAMINE
1	2.04		1686	272	C31H53N1O2	3-CHOLESTERYL-4-AMINOBUTYRATE 89210-68-4
2 ✓	7.56		1698	820	C31H55N2O3.Br1	2-HEPTOXYPHCARBAMICAC.,PIPERIDINOET.EST.,N-DECBR
3	5.15	7.4	1698	820	C31H55N2O3.H2O4P1	2-HEPTOXYPHCARBAMICAC.,PIPERIDINOET.EST.,N-DECPHOS
4	0.69	7.4	672	1	C32H26O12	O-METHYLXANTHOMEGNIN
25	2.16	7.4	564	1	C32H32N2O10S1	14-MERCAPTO(2")PYRIDINO-DAUNOMYCIN
6	1.80	7.0	564	1	C32H32O14	CHARTREUSIN ✦
7 ★	2.59	7.4	1977	1	C32H34N2O11	2-CYANOMORPHOLINO-DAUNORUBICIN 89164-74-9
8 ★	1.98	7.4	1977	1	C32H34N2O12	2-CYANOMORPHOLINODOXORUBICIN 89196-07-6
9	1.83	7.0	2304	459	C32H34N2O12	3'-DEAMINO-3'-(3-CYANO-4-MORPHOLINYL)DOXORUBICIN
16530 ★	1.97	7.4	1977	1	C32H35N3O11	2-CYANOMORPHOLINO-DOXORUBICIN,12-IMINO 89164-79-4
1 ★	1.56	7.4	1977	1	C32H36N2O12	2-CYANOMORPHOLINO-DOXORUBICIN,13-DIHYDRO 89164-77-2
2 ★	-0.32		2377		C32H38N4O6	PHE-VAL-TYR-PHE
3 ✓	0.15	0.3	1766		C32H40N4O2	CHROMOETHYLPYRAZOLE HYDRO IODIDE
4 ✓	-0.39	0.3	1766		C32H40N4O2	CHROMOETHYLPYRAZOLE HYDROBROMIDE
35 ✓	-0.82	0.3	1766		C32H40N4O2	CHROMOETHYLPYRAZOLE HYDROCHLORIDE
6 ✓	-0.96	0.3	1766		C32H40N4O2	CHROMOETHYLPYRAZOLE HYDROFLUORIDE
7 ✓	0.57	0.3	1766		C32H40N4O2	CHROMOETHYLPYRAZOLETHIOCYANATE
8	-0.60	1.0	564	342	C32H40N6O2	TPA,4,4"-BIS(N-BUTYL-N'-METHYLAMIDINO),DIHYDROCHLORIDE
9	-2.08	1.0	562	342	C32H40N6O4	TPA-4,4"BIS(N-(3-MEOPROPYL)N'-MEAMIDINO
16540	3.13	1.0	579	342	C32H41N1O2	TERFENADINE 50679-08-8 antihistaminic
1	3.22	6.1	579	1	"	TERFENADINE
2	4.74	7.4	2310	2	C32H42N6O2	XANTHINE,1-ETHYL-3-I-BUTYL-7-METHYL-8-(4-DIPHENYLMETHYL)PIPERAZINYLPROPYL
3	-1.00	7.8	1811		C32H43N3O9	DAUNORUBICIN-DI-ME-D 61481-25-2
4	-0.62	7.8	967		"	DAUNORUBICIN-DI-ME-D
45	5.26		1520		C32H45F1O7	BETAMETHASONE-17-21-DIVALERATE
6 ★	1.65		1036	812	C32H46O9	4'-DEHYDROOLEANDRIN ✦
7 ★	2.26		1036	812	C32H48O9	4'-EPIOLEANDRIN
8 ★	2.53		1036	812	C32H48O9	4'-EPIOLEANDRIN
9	1.78		1036	812	C32H49N1O8	4'-DESOXY-4'-AMINOOLEANDRIN
16550	1.83		1036	812	C32H49N1O8	4'-DESOXY-4'-AMINOOLEANDRIN
1 ✓	4.10	7.4	2574	1	C32H50N6O6	GLYCOL PEPTIDE,AZIDE ANALOG renin inhibitor(oral)
2 ★	1.24	7.4	547	1	C33H34O13S1	TENIPOSIDE 29767-20-2 antineoplastic
3	4.27	7.4	2310	2	C33H36N6O2	XANTHINE,1,7-DIMETHYL-3-PHENYL-8-(4-DIPHENYLMETHYL)PIPERAZINYLPROPYL
4 ★	1.33		2330		C33H36O10	BIS-HELENALINYLMALONATE ✦
55	2.04	7.4	2025	1	C33H39N1O12	DAUNOMYCIN,N"-(4-BUTYRIC ACID),ETHYL ESTER 103620-83-3
6	1.12	7.8	967		C33H39N3O12	DAUNORUBICIN-N,N-DIMETHYLGLYCYLGLYCINE
7 ★	0.76		2330		C33H44O10	BIS-TETRAHYDRO-HELENALINYLMALONATE
8 ★	-0.56	7.0	2002	1	C33H48O15	PREDNISOLONE-21-CELLOBIOSIDE 92901-31-0
9	0.76		967		C34H35N3O10	ZORUBICIN 54083-22-6 antineoplastic
16560	1.04	7.4	401	1	"	ZORUBICIN
1	1.22	7.4	564	1	"	ZORUBICIN
2	1.37	7.5	2624	1	"	ZORUBICIN
3	1.53	7.4	401	2	"	ZORUBICIN
4 ✓	1.38	0.3	1766		C34H36N4O2	N-BENZYL-CHROMOPYRAZOLE HYDRO IODIDE
65 ✓	1.00	0.3	1766		C34H36N4O2	N-BENZYL-CHROMOPYRAZOLE HYDROBROMIDE
6 ✓	0.69	0.3	1766		C34H36N4O2	N-BENZYL-CHROMOPYRAZOLE HYDROCHLORIDE
7 ✓	0.55	0.3	1766		C34H36N4O2	N-BENZYL-CHROMOPYRAZOLE HYDROFLUORIDE
8 ✓	1.69	0.3	1766		C34H36N4O2	N-BENZYL-CHROMOPYRAZOLETHIOCYANATE
9 ★	11.29		2415	459	C34H40O2	O-155 ✦
16570	2.20		2609	468	C34H41N5O6	18-TETRALACTAM-DIETHER-AMINE,N1,N2-DIBENZYL-N'-PHENYL
1 ★	1.52		2423		C34H42O13	BACCATIN III-13-LACTATE ✦
2	-1.64	1.0	564	342	C34H44N6O2	TPA,4,4"-BIS(TRIETHYLAMIDINO),DIHYDROCHLORIDE
3	0.98	1.0	564	342	C34H44N6O2	TPA,4,4"-N-HEXYLAMIDINO
4 ★	1.99		547		C34H46Cl1N3O10	MAYTANSINE 35846-53-8 antineoplastic
75 ★	1.94	7.4	2576	1	C34H50N6O7	OXAZOLIDINONE-PEPTIDE,HYDROXYETHYLAMINO-TERM, ANALOG renin inhibitor(oral)
6 ★	2.85	7.4	2576	1	C34H50N6O7	OXAZOLIDINONE-PEPTIDE,N-TERM-I-PROPYL ANALOG renin inhibitor(oral)
7 ★	2.95		1036	812	C34H50O10	ACETYLOLEANDRIN
8 ✓	3.89	7.4	2575	1	C34H52N4O6S2	SULFO-GLYCOL PEPTIDE,1-METHYLAZETIDIN-3YL ANALOG renin inhibitor(oral)
9	3.25		2112		C34H53Br1O8	X537 ANTIBIOTIC,6-BROMO ANALOG
16580	3.12		2112		C34H53Cl1O8	X537 ANTIBIOTIC, 6-CHLORO ANALOG
1	3.19		2112		C34H53I1O8	X537 ANTIBIOTIC, 6-IODO ANALOG
2	2.46		2112		C34H53N1O10	X537 ANTIBIOTIC,6-NITRO ANALOG ✦
3	2.85		2112		C34H54O8	LASALOCID 25999-31-9 coccidiostat (poultry)
4	0.28		2112		C34H55N1O8	X537 ANTIBIOTIC,6-AMIO ANALOG
85	2.27		1403		C34H56N4O8	SPECTINOMYCYLAMINE,4-(P-DODECAMIDO)PHENYLACYL
6	1.79		212		C34H59N3O15	3-AZIDO-3'-DE(DIMEAMINO)-4'-HYDROXYERYTHROMYCIN
7	-2.49		579		C35H27N5O9S2.(Na1)2	DIRECTYELLOW62 6409-90-1
8	1.78		1210	202	C35H37Cl1N2O11	CLOROBIOCIN
9	2.21	7.4	1210	1	"	CLOROBIOCIN

	logP	pH	Ref.	Note	MF	Name / **CAS** / *activity*
16590	1.16	7.4	1210	1	C35H38N2O11	8-DECHLORO-CLOROBIOCIN
1 ★	3.18		594		C35H38N6O5	TETRAPEPTIDE ANALOG ✚
2 ✓	2.79	7.4	1210	1	C35H39Cl1N2O11	DIHYDROCLOROBIOCIN
3 ★	1.37		2330		C35H40O10	BIS-HELENALINYL GLUTARATE
4	-2.03		2229		C35H48N10O15	DSIP
95 ✓	2.83	7.4	2574	1	C35H49N5O7S1	OXAZOLIDINONE PEPTIDE,THIAZOL-4YL ANALOG *renin inhibitor(oral)*
6 ✓	2.20	7.4	2574	1	C35H50N6O7	OXAZOLIDINONE PEPTIDE,IMIDAZOL-4YL ANALOG *renin inhibitor(oral)*
7 ✓	2.49	7.4	2574	1	C35H50N6O7	OXAZOLIDINONE PEPTIDE,PYRAZOL-3YL ANALOG *renin inhibitor(oral)*
8 ★	2.20	7.4	2576	1	C35H50N6O7	OXAZOLIDINONE-PEPTIDE,MORPHOLINE-TERM ANALOG *renin inhibitor(oral)*
9 ✓	2.11	7.4	2576	1	C35H51N7O8	OXAZOLIDINONE-PEPTIDE,N-TERM-MORPHOLIN-4YL ANALOG *renin inhibitor(oral)*
16600 ★	1.22	7.4	2576	1	C35H51N7O9	OXAZOLIDINONE-PEPTIDE,N-TERM-DI-OH-PYRROLIDINE ANALOG *renin inhibitor(oral)*
1 ✓	4.27	7.4	2574	1	C35H52N4O6S1	GLYCOL PEPTIDE,THIAZOL-4YL ANALOG *renin inhibitor(oral)*
2 ★	2.45	7.4	2576	1	C35H52N6O7	OXAZOLIDINONE-PEPTIDE,MEOETAMINO-TERM ANALOG *renin inhibitor(oral)*
3 ★	-0.96	7.4	2285	1	C35H52O12	K-STROPHANTHIN-BETA
4 ✓	3.79	7.4	2574	1	C35H53N5O6	GLYCOL PEPTIDE,IMIDAZOL-1YL ANALOG *renin inhibitor(oral)*
5 ✓	3.51	7.4	2574	1	C35H53N5O6	GLYCOL PEPTIDE,IMIDAZOL-4YL ANALOG *renin inhibitor(oral)*
6 ✓	4.01	7.4	2574	1	C35H53N5O6	GLYCOL PEPTIDE,PYRAZOL-3YL ANALOG *renin inhibitor(oral)*
7 ★	1.42	7.4	2576	1	C35H53N7O9	OXAZOLIDINONE-PEPTIDE,N-TERM-GLYCOLAMINE *renin inhibitor(oral)*
8 ★	2.76	7.4	2285	1	C35H54O10	DIGITOXIGENIN-BISDIGITOXOSIDE ✚
9 ★	1.24	7.4	2285	1	C35H54O11	DIGOXIGENIN-BISDIGITOXOSIDE
16610	-1.28		1036	812	C35H54O12	DIGITOXIGENIN-DIGITOXOSIDE-4'-GLUCURONIDE
1 ✓	4.57	7.4	2575	1	C35H55N5O6S2	SULFO-GLYCOL PEPTIDE,4-METHYLPIPERAZIN-1YL ANALOG *renin inhibitor(oral)*
2	-0.59	1.0	564	342	C36H32N6O2	TPA,4,4"-N'-METHYL-N-PHENYLAMIDINO,DIHYDROCHLORIDE
3 ★	1.63		2377		C36H38N4O5	PHE-PHE-PHE-PHE
4	2.89	7.4	1210	1	C36H39Cl1N2O11	HOMOCLOROBIOCIN ✚
15 ✓	3.59	7.4	2574	1	C36H50N4O7S1	OXAZOLIDINONE PEPTIDE,THIOPHEN-2YL ANALOG *renin inhibitor(oral)*
6	-1.77		2229		C36H50N10O15	D-ALA(3)-DSIP
7	-1.99		2229		C36H50N10O15	D-ALA(4)-DSIP
8	-1.47		2229		C36H51N11O14	D-ALA(4)-DSIP-NH2
9 ✓	2.16	7.4	2574	1	C36H52N6O7	OXAZOLIDINONE PEPTIDE,1-METHYLIMIDAZOL-4YL ANALOG *renin inhibitor(oral)*
16620 ★	2.49	7.4	2576	1	C36H54N6O8	OXAZOLIDINONE-PEPTIDE,MOMOETAMINO-TERM ANALOG *renin inhibitor(oral)*
1	2.16	7.5	1476		C36H54N8O5	METHOTREXATE,DIOCTYL ESTER
2	1.49		2112		C36H55Br1O9	X537 ANTIBIOTIC,6-BROMO,ACETYL ANALOG
3 ✓	4.12	7.4	2574	1	C36H55N5O5S1	METHYLPIPERIDINE-GLYCOL PEPTIDE,THIAZOL-4YL ANALOG *renin inhibitor(oral)*
4 ✓	3.71	7.4	2574	1	C36H55N5O6	GLYCOL PEPTIDE,1-METHYLIMIDAZOL-4YL ANALOG *renin inhibitor(oral)*
25 ★	2.30	7.4	2576	1	C36H55N7O9	OXAZOLIDINONE-PEPTIDE,N-TERM-MOMO(CH2)2N(ME) *renin inhibitor(oral)*
6 ✓	3.61	7.4	2575	1	C36H56N4O6S2	SULFO-GLYCOL PEPTIDE,1-METHYLPIPERIDIN-4YL ANALOG *renin inhibitor(oral)*
7 ✓	1.66	7.4	2574	1	C36H56N6O5	METHYLPIPERIDINE-GLYCOL PEPTIDE,IMIDAZOL-4YL ANALOG *renin inhibitor(oral)*
8 ✓	4.39	7.4	2574	1	C36H56N6O5	METHYLPIPERIDINE-GLYCOL PEPTIDE,PYRAZOL-1YL ANALOG *renin inhibitor(oral)*
9 ✓	4.02	7.4	2574	1	C36H56N6O5	METHYLPIPERIDINE-GLYCOL PEPTIDE,PYRAZOL-3YL ANALOG *renin inhibitor(oral)*
16630	0.95		2112		C36H56O9	X537 ANTIBIOTIC,O-ACETYL ANALOG
1 ★	3.32		2308		C36H56O10	5,5-DIBENZO-30-CROWN-10-ETHER,DI-(M-T-BUTYL)
2	1.04		2112		C36H57N1O9	X537 ANTIBIOTIC,6-ACETAMIDO ANALOG
3 ✓	3.11	7.4	2575	1	C36H57N5O5S1	N-TERMEXTENDED-GLYCOL PEPTIDE,DIMETHYLAMINO ANALOG *renin inhibitor(oral)*
4	4.80	6.5	2532	1	C36H60N2O7	ABBOTT NON-PEPTIDE RENIN INHIB.#1 ✚
35	4.80	7.4	2532	1	"	ABBOTT NON-PEPTIDE RENIN INHIB.#1
6	1.26		212		C36H65N1O13	ERYTHROMYCINC
7	2.21		212		C36H65N1O12	N-DESMETHYLERYTHROMYCIN
8	8.70		2609	468	C36H68N4O6	18-TETRALACTAM-DIETHER,N1,N2-DIDODECYL
9	2.06	7.4	1210	1	C37H39N1O11	3"-DEAMIDO-BENZOYLNOVOBIOCIN
16640	2.36	7.4	1210	1	C37H43Cl1N2O12	4'(2-HYDROXYETHOXY)DIHYDROCLOROBIOCIN
1	0.78	7.4	401	1	C37H47N1O12	RIFAMYCIN **6998-60-3** *antibacterial*
2	0.94	7.0	1295	1	"	RIFAMYCIN
3 ✓	4.01	7.4	2575	1	C37H54N6O5S1	N-TERMEXTENDED-GLYCOL PEPTIDE,IMIDAZOL-1YL ANALOG *renin inhibitor(oral)*
4 ✓	4.44	7.4	2574	1	C37H56N4O5S1	METHYLPIPERIDINE-GLYCOL PEPTIDE,THIOPHEN-2YL ANALOG *renin inhibitor(oral)*
45 ✓	2.83	7.4	2575	1	C37H59N5O5S1	N-TERMEXTENDED-GLYCOL PEPTIDE,DIMETHYLAMINOMETHYL ANALOG *renin inhibitor(oral)*
6	3.07	8.0	327	2	C37H67N1O12	BERYTHROMYCIN **527-75-3** *anti-amebic, antibacterial*
7	3.12		212		"	BERYTHROMYCIN
8	0.66	7.4	1945	1	C37H67N1O13	ERYTHROMYCIN **114-07-8** *antibacterial*
9	0.94	7.7	1169	1	"	ERYTHROMYCIN
16650	1.26	7.4	594	1	"	ERYTHROMYCIN
1 ★	2.54	8.0	327	2	"	ERYTHROMYCIN
2	1.44		212		C37H67N1O14	4'-HYDROXYERYTHROMYCIN
3	-2.68	1.0	564	342	C38H36N10O2	TPA,4,4"BIS(N'(P-N'MEAMIDINO)PHENYLAMIDINE)
4	-2.18	4.0	527		C38H42N2O6	D-TETRANDRINE **5956-77-4** *antineoplastic, analgesic, antipyretic*
55	2.75	7.4	401	1	"	D-TETRANDRINE
6	-2.11	7.4	921	22	C38H44N2O6.(I1)2	DIMETHYLTUBOCURARINEDIIODIDIE
7 ★	2.21	7.4	2576	1	C38H58N6O9	OXAZOLIDINONE-PEPTIDE,MEMOETAMINO-TERM ANALOG *renin ihibitor(oral)*
8 ✓	4.15	7.4	2575	1	C38H59N5O6S1	N-TERMEXTENDED-GLYCOL PEPTIDE,MORPHOLIN-4YL ANALOG *renin inhibitor(oral)*
9 ★	2.03	7.4	2576	1	C38H59N7O10	OXAZOLIDINONE-PEPTIDE,N-TERM-MEMO(CH2)2N(ME) *renin inhibitor(oral)*
16660	3.11		212	835	C38H65N1O14	ERYTHROMYCIN-9,11-CARBONATE-6,9-HEMIKETAL
1	3.05	8.0	327	2	C38H67N1O13	4"-FORMYLDEOXYERYTHROMYCIN
2	2.64	8.0	327	2	C38H67N1O14	4"-FORMYLERYTHROMYCIN-A
3	0.69	4.0	2610	1	C38H69N1O13	CLARITHROMYCIN **81103-11-9** *antibacterial*
4	0.86	6.0	2610	1	"	CLARITHROMYCIN
65	1.24	6.5	2610	1	"	CLARITHROMYCIN
6	1.68	8.0	2610	1	"	CLARITHROMYCIN
7	-2.14	1.0	564	342	C39H40N10O3	NSC-66764, IMIDAZOLIN-PHTHALANILIDE ANALOG
8 ✓	4.22	7.4	2575	1	C39H55N5O5S1	N-EXTENDED-GLYCOL PEPTIDE,PYRIDIN-2YL ANALOG *renin inhibitor(oral)*
9 ★	2.31	7.4	2576	1	C39H58N6O10	OXAZOLIDINONE-PEPTIDE,DI-MOMPYRROLIDINE-TERM ANALOG *renin inhibitor(oral)*
16670 ★	2.06	7.4	2576	1	C39H59N7O11	OXAZOLIDINONE-PEPTIDE,N-TERM-DI-MOMOPYRROLIDINE ANALOG *renin inhibitor(oral)*
1 ✓	3.58	7.4	2575	1	C39H62N6O5S1	N-TERMEXTENDED-GLYCOL PEPTIDE,4-METHYLPIPERAZIN-1YL ANALOG *renin inhibitor(oral)*
2	2.08		2112		C39H62O9	X537 ANTIBIOTIC,O-PENTANOYL ANALOG
3	2.75	6.5	2554		C39H64N4O8	3-AZAGLUTARAMIDE, ANALOG-A29
4	2.80	7.4	2554		"	3-AZAGLUTARAMIDE, ANALOG-A29
75	2.79	8.0	327	2	C39H67N1O14	4",11-DIFORMYL-DEOXYERYTHROMYCIN(26)
6 ★	3.50	7.4	2554		C39H67N5O6	3-AZAGLUTARAMIDE, ANALOG-A28
7	3.30	8.0	327	2	C39H69N1O13	11-O-ACETYLDEOXYERYTHROMYCIN
8	3.32		212		"	11-O-ACETYLDEOXYERYTHROMYCIN
9	3.23	8.0	327	2	C39H69N1O13	4"-ACETYL-DEOXYERYTHROMYCIN

	logP	pH	Ref.	Note	MF	Name / CAS / activity
16680	2.85	8.0	327	2	C39H69N1O14	4"-ACETYLERYTHROMYCIN-A
1 ✓	1.69	0.3	1766		C40H40N4O2	N,N'-DIBENZYL-CHROMOPYRAZOLE HYDRO IODIDE
2 ✓	1.28	0.3	1766		C40H40N4O2	N,N'-DIBENZYL-CHROMOPYRAZOLE HYDROBROMIDE
3 ✓	0.96	0.3	1766		C40H40N4O2	N,N'-DIBENZYL-CHROMOPYRAZOLE HYDROCHLORIDE
4 ✓	0.90	0.3	1766		C40H40N4O2	N,N'-DIBENZYL-CHROMOPYRAZOLE HYDROFLUORIDE
85 ✓	1.92	0.3	1766		C40H40N4O2	N,N'-DIBENZYL-CHROMOPYRAZOLETHIOCYANATE
6	0.38	1.0	564	342	C40H44N4O6	TPA,4,4"BIS(N'-ME-N-PHENETAMIDINO)
7	3.80		2609	468	C40H44N4O6	18-TETRALACTAM-DIETHER,N1,N2,N3,N4-TETRABENZYL
8 ★	2.90		2423		C40H46O13	BACCATIN III-13-PHENYLLACTATE
9	2.72		1858	459	C40H48N6O9	RA-V
16690 ★	0.52		2308		C40H64O16	8,8-DIBENZO-48-CROWN-16-ETHER
1 ★	1.86		2469		C40H65N1O14	NIDDAMYCIN
2	2.80	6.5	2554		C40H66N4O8	3-AZAGLUTARAMIDE, ANALOG-A40
3	3.30	7.4	2554		"	3-AZAGLUTARAMIDE, ANALOG-A40
4 ★	3.30	7.4	2554		C40H69N5O6	3-AZAGLUTARAMIDE, ANALOG-A39
95 ★	1.75		2423		C41H47N1O14	BACCATINIII-13-(N-BENZOYLISOSERINATE)
6 ★	1.92		2423		"	BACCATINIII-13-(N-BENZOYLISOSERINATE)
7	-1.27	6.5	400	342	C41H48N2O8	THALICARPINE 5373-42-2 antineoplastic
8	-1.14		430		"	THALICARPINE
9	3.32	7.4	401	2	"	THALICARPINE
16700	3.17		1858	459	C41H50N6O9	RA-VII
1	3.48		2112		C41H57Br1O9	X537 ANTIBIOTIC,6-(P-BROMOBENZOYL) ANALOG
2	0.86		2112		C41H59N1O8	X537 ANTIBIOTIC,6-BENZALIMINE ANALOG
3	-1.62		1757		C41H62O16	DIGOXIN-17-CARBOXYLATE
4 ★	2.83	7.4	2285	1	C41H64O13	DIGITOXIN 71-63-6 cardiotonic
5	1.85		1036	812	"	DIGITOXIN
6	1.26	7.4	2285	1	C41H64O14	DIGOXIN 20830-75-5 cardiotonic
7 ★	1.26		1036		"	DIGOXIN
8 ★	1.68	7.4	2285	1	C41H64O14	GITOXIN 4562-36-1
9 ★	2.90	7.4	2285	1	C41H66O13	DIGITOXIN,DIHYDRO
16710 ★	1.36		1036		C41H66O14	DIHYDRODIGOXIN
1	1.40	7.4	2285	1	"	DIHYDRODIGOXIN
2	4.14		2554		C41H68N8O6	3-AZAGLUTARAMIDE, ANALOG-A22 ✚
3	5.00	7.4	2554		"	3-AZAGLUTARAMIDE, ANALOG-A22
4 ★	4.11	7.4	2554		C41H68N8O7	3-AZAGLUTARAMIDE, ANALOG-A23
15 ★	4.20	7.4	2554		C41H69N9O6	3-AZAGLUTARAMIDE, ANALOG-A24
6 ★	3.80	7.4	2554		C41H71N5O6	3-AZAGLUTARAMIDE, ANALOG-A35
7	3.80		2609	468	C42H48N4O7	18-TETRALACTAM-TRIETHER,N1,N2,N3,N4-TETRABENZYL
8	3.19		1858	459	C42H50N6O10	RA-V-ACETATE
9	1.12		401		C42H56N4O12	DEMETHYL-RIFAMPICIN
16720 ★	2.04	7.4	2285	1	C42H64O15	GITALOXIN 3261-53-8 cardiotonic
1 ★	3.11	7.4	2285	1	C42H66O13	DIGITOXIN,BETA-METHYL
2 ★	1.75	7.4	2285	1	C42H66O14	DIGOXIN,ALPHA-METHYL
3 ★	1.80	7.4	2285	1	C42H66O14	DIGOXIN,BETA-METHYL
4	2.21		2112		C42H68O9	X537 ANTIBIOTIC,O-OCTANOYL ANALOG
25	1.85	7.4	1945	1	C42H69N1O15	JOSAMYCIN 56689-45-3 antibacterial
6	2.39		2318		"	JOSAMYCIN
7 ★	3.08		2318		C42H69N1O15	ROKITAMYCIN 74014-51-0 antibacterial
8 ★	3.50	7.4	2554		C42H73N5O7	3-AZAGLUTARAMIDE, ANALOG-A38
9 ★	5.66		594		C43H44N6O7	PEPTIDE,BENZYLOXYCARBONYL ANALOG
16730	1.46		401		C43H54N2O13	RIFAMYCIN6339
1	1.23	8.0	1169	1	C43H58N4O12	RIFAMPIN 13292-46-1 antibacterial
2	1.29		401		"	RIFAMPIN
3	1.32	7.5	1169	1	"	RIFAMPIN
4	1.76		564		C43H59N3O12	RIFAMYCIN,8-N'-ME-N-PIPERAZINYLMETHYL
35	3.88		1711		C43H62N4O7	CHOLESTERYLOXYCARBONYLMITOMYCINC
6 ★	1.87	7.4	2576	1	C43H66N6O12	OXAZOLIDINONE-PEPTIDE,DI-MEMOPYRROLIDIN-TERM ANALOG renin inhibitor(oral)
7 ★	2.16	7.4	2285	1	C43H66O15	16-ACETYLGITOXIN
8 ★	1.71		1036		C43H66O15	A-ACETYLDIGOXIN
9	1.74	7.4	2285	1	"	A-ACETYLDIGOXIN
16740 ★	1.97		1036		C43H66O15	B-ACETYLDIGOXIN cardiotonic
1 ★	2.29	7.4	2285	1	C43H66O15	DIGOXIN,12-ACETYL
2 ★	2.00	7.4	2285	1	C43H66O15	DIGOXIN,BETA-ACETYL
3 ★	1.67	7.4	2576	1	C43H67N7O13	OXAZOLIDINONE-PEPTIDE,N-TERM-DI-MEMOPYRROLIDINE ANALOG renin inhibitor(oral)
4 ★	2.14	7.4	2285	1	C43H68O14	DIGOXIN,ALPHA-BETA-DIMETHYL
45 ★	4.07		2554		C43H73N5O8	3-AZAGLUTARAMIDE, ANALOG-A-26
6	3.83		564		C44H61N3O12	RIFAMYCIN,8-N,N-DIPROPYLFORMIMIDOYL
7	0.51	7.0	1642	262	C44H63N3O9.Ga1	LINEAR-DIOCTYL-TRICATACHOLAMIDE,GALLIUMCOMPLEX
8 ★	2.87	7.4	2285	1	C44H68O15	DIGOXIN,12-ACETYL-BETA-METHYL
9	2.75		2112		C44H72O9	X537 ANTIBIOTIC, O-DECONOYL ANALOG
16750 ★	3.60	7.4	2554		C44H75N5O8	3-AZAGLUTARAMIDE, ANALOG-A37
1 ★	2.45		1036		C45H68O16	A,B-DIACETYLDIGOXIN
2 ★	2.59	7.4	2285	1	C45H68O16	DIGOXIN,ALPHA-BETA-DIACETYL
3	2.82		401		C46H56N4O10	VINCRISTINESULFATE 2068-78-2
4	2.57	7.4	827	2	C46H56N4O10	VINCRISTINE 57-22-7 antineoplastic
55	2.80		1010		"	VINCRISTINE
6	3.72		401		C46H58N4O9	VINCALEUKOBLASTINESULFATE 143-67-9
7	3.69	7.4	827	2	C46H58N4O9	VINCALEUKOBLASTINE 865-21-4 antineoplastic
8	3.40		827		C46H58N4O9	VINLEUROSINE
9	3.08	7.4	827	2	C46H58N4O9	VINROSIDINE 15228-71-4 antineoplastic
16760 ★	3.16		2285		C46H64O19	GITOFORMATE 7685-23-6 cardiac glycoside
1 ★	3.16	7.4	2285		C46H64O19	GITOXIN,PENTAFORMYL
2	4.83		564		C47H58N2O13	RIFAMYCIN,8-N-4-PHENYLPROPOXYFORMIMIDOYL
3 ★	-1.10		1036	812	C47H72O18	DIGITOXIN-16'-GLUCURONIDE
4 ★	-1.77		1036	812	C47H72O19	DIGOXIN-16'-GLUCURONIDE
65	5.20		2609	468	C48H52F1N5O6	18-TETRALACTAM-DIETHER-AMINE,N1,N2,N3,N4,N'-(P-FLUOROPHENYL)
6	5.20		2609	468	C48H53N5O6	18-TETRALACTAM-DIETHER-AMINE,N1,N2,N3,N4-TETRABENZYL,N'-PHENYL
7	4.02		827		C48H63N5O9	VINGLYCINATE antineoplastic
8	4.05		564		C48H67N3O12	RIFAMYCIN,8-N,N-(CH2)10-FORMIMIDOYL
9	3.47		564		C48H69N3O12	RIFAMYCIN,8-N,N-DIPENTYLFORMIMIDOYL

	logP	pH	Ref.	Note	MF	Name / **CAS** / *activity*
16770	2.21	7.0	1642	262	C48H71N3O9.Ga1	LINEAR-DIDECYL-TRICATACHOLAMIDE,GALLIUMCOMPLEX
1	1.23		1036	812	C49H76O19	LANATOSIDE-A **17575-20-1**
2 ★	0.07		1036	812	C49H76O20	LANATOSIDEC **17575-22-3** *cardiotonic*
3 ★	2.85		1036	812	C51H74O19	PENTAACETYLGITOXIN
4	-0.25	7.4	547	1	C52H76O24	MITHRAMYCIN **18378-89-7** *antineoplastic*
75	1.04	2.0	1631		"	MITHRAMYCIN
6	3.60		564		C54H81N3O12	RIFAMYCIN,8-N,N-DIOCTYLFORMIMIDOYL
7	2.92	7.4	2676	1	C62H111N11O12	CYCLOSPORIN-A *immunosuppressant*

F0   Preferred Value
F1   NOT ion-corrected
F1   Not ion-corrected
F2   Ion-corrected
F2   Ion-corrected
F3   Unbuffered but ion-corrected (?)
F4   Value is upper limit
F5   See ref. in above
F6   From log P/pH profile
F7   H-zero scale for lower pKas
F8   H'-R scale
F9   From chem.shift in dilute DMSO
F10   Highest confidence value
F11   Good confidence value
F12   Uncertain or Unreliable
F13   of monoprotonated solute
F14   Protonation on carbon
F15   Conc. dependent (extrap.0 conc.)
F16   Electrometric
F17   Optometric
F18   % keto tautomer unknown
F19   % enol tautomer unknown
F20   Approximate value
F21   Distribution ratio
F22   Apparent logP (= log D)
F23   % thiol unknown
F24   % thione unknown
F25   H-A scale
F26   Value probably low
F27   Value probably high
F28   Ion corr.,some pKas calc.
F40   Other values also given in reference
F41   Other pH(meas) values also given in reference
F42   Measurements at other conc. &/or temp. in reference
F50   As carbon acid
F58   Assay procedure: J. Agr. & Food Chem., 8, 460 (1960)
F60   Replaces earlier doubtful value
F100   Ratio of solubilities in separate, not-mutually-sat'd phases
F101   As chloride ion, but in 0.1M Phosphate Buffer, pH=7.4
F102   As iodide ion-pair in 0.1N KI
F103   By pKa Measurement
F121   Artificial sea water as aqueous phase
F122   Saline as aqueous phase
F140   Nitrogen instead of air (vs. water)
F160   In D2O & D2O sat'd octanol
F161   DOC8H17/D2O partitioning
F200   Ionic form
F202   [Na+] salt
F210   Protonated form
F211   Approx half of phenothiazine ring nitrogens protonated
F212   Where 2 N in side chain, some di-protonation probable
F220   Neutral form present
F221   0.0001% neutral form (approx.)
F222   0.01% neutral form (approx.)
F223   0.1% neutral form (approx.)
F224   Only neutral form present
F225   For neutral solute
F226   Zero net charge
F230   As dimer
F231   May be dimerized in organic phase
F240   5 mg/ml initial aqueous solute concentration
F241   0.001 M initial aqueous solute concentration
F242   Initial solute conc. = .002M
F243   0.002 M initial aqueous solute concentration
F244   0.005 M initial aqueous solute concentration
F250   1 uM metal ion concentration
F251   10 uM metal ion concentration
F252   100 uM metal ion concentration
F260   Commercial material, 96% pure
F261   Some lactone also present
F262   Addn cpd. also partitioning
F264   Impure sample, probable lipophilic impurity
F270   Any subst. is in acetyl alpha-C
F271   Attachment at 21- position probable but not positive
F272   Radiolabeled solute, likely to give low values
F297   Large conc. dependence, Av. value taken

F298   Slight conc. dependence, av. value taken
F299   Conc.dependent, extrap. to zero conc
F300   Not buffered, % ion unknown
F301   MOPS (3-morpholinopropanesulfonic acid)buffer
F302   Acetate buffer
F303   Aspartate Buffer
F304   Bicarbonate buffer
F305   Propionate buffer
F306   Borate buffer
F307   Glycine buffer
F308   Carbonate buffer
F309   Krebs Buffer
F310   Carbonate-bicarbonate buffer
F312   Citrate Buffer
F314   Phosphate buffer
F316   Phosphate or Ringer buffer
F318   Phosphate-citrate buffer
F320   Ringer's buffer
F321   Clark & Lubs Buffer
F322   Sorenson's buffer
F324   Tris buffer
F330   N-pentyl acetate
F331   0.37M Acetic acid
F332   Acetic acid (0.04)
F334   Acetonitrile (60%) / PH 9.9 Buffer (40%)
F335   Acetonitrile 7.41 buffer (30:70)
F336   1M BaCl2
F337   acetonitrile/buffer mobile phase
F338   1M [Cl-] ion
F340   0.001M CF3CO2H
F342   Using HCl
F343   5% HCl aqueous phase
F344   0.2N HCl OR H2SO4
F345   0.5M H2SO4
F346   0.5N HCl
F348   1.0N HCl
F350   0.1N HCl phase contains 5% methanol to aid initial solution, small effect on logP
F352   Hexadecylamine
F354   0.1N KCl
F356   1M KCl
F357   1M. NaCl
F358   2M KOH
F360   Dissolved in H2O, KOH added to iso-electric point
F362   Leucine (0.004)
F364   2.5% DMSO
F365   0.1M Na2CO3
F366   0.25M Na2CO3
F368   0.3M Na2CO3
F370   0.1N NaCl
F371   0.15 M NaCl
F372   1M NaCl
F374   1M NaCl & 0.1M Phosphate
F375   Aq.phase = 0.9% NaCl
F376   0.2M NaOH
F378   0.5M NaOH
F380   Octadecylamine
F382   p-Toluenesulfonic acid (0.06M)
F384   Cremophor or Tween-80 emulsifier (0.0001M)
F390   0.1 Na2S2O3 ionic strength
F391   0.2 ionic strength
F392   Ionic strength 1.0 M
F393   10mM [Na+] ionic strength
F394   100mM [Na+] ionic strength
F400   Both Phases Analyzed
F401   Slow-stir method (ref. 2129)
F402   P = ratio of counts(organic/water)
F403   By two-phase titration
F404   Analyzed in plastic containers
F405   Ratio of Ostwald sol.coef.
F406   Under inert gas
F407   Phases stirred 30 min.
F420   Aqueous Phase from partitioning re-extracted for analysis
F421   Aqueous Phase re-extracted with CH2Cl2 for analysis
F423   Successive re-extraction of lipid phase
F424   log P=1.51 + 1.98k'

F447 HPLC on (ODP),logP(oct)=0.83logK(ODP) -0.06
F448 by Centrifugal Partition Chromatography
F449 HPLC-octanol column
F450 Assay procedure: J. Agr. & Food Chem., 8, 460 (1960)
F451 AKUFVE analysis
F452 Atomic absorbtion analysis
F453 C-14 in methyl group analysis
F454 C.C.D. analysis
F455 Counter-current extraction analysis
F456 Electron capture gas chromotography analysis
F457 Filter probe analysis
F458 Generator column analysis
F459 From HPLC,regr.from log k
F460 HPLC analysis on both phases
F461 HPLC analysis on ODS, elut. 70% alc
F462 Brandstrom analysis (refs 29 & 680)
F463 Ninhydrin analysis
F464 Segmented flow analysis
F465 Concurrent partitioning analysis with internal reference compounds
F466 Coulometric titration analysis on both phases
F467 Calc. from TSA (Total Surface Area)
F468 From Rm or Rf values
F469 by CPC: Centrifugal Partition Chromatography
F470 Calc. from reverse-phase thin layer chromatography
F471 HPLC: MeOH/H2O extrap to 100% H2O, +dev.for H-accpt, -dev.H-donr.
F472 by GLC,+0.41 for correct log P of 2,4,5-Cl3 analog
F500 0 deg C
F501 At phase-transition temperature
F510 10 deg C
F515 15 deg.C
F520 20 deg. C
F524 24 deg.C
F525 25 deg.C
F530 30 deg.C
F534 34 deg. C
F536 36 deg.C
F537 37 deg.C
F538 38 deg.C
F540 40 Deg. C
F545 45 deg. C.
F555 55 deg. C
F560 60 deg C
F565 65 deg.C
F581 105 deg C
F598 polar phase = protein-free milk ultrafiltrate
F599 Measured over large temperature range
F600 At isoelectric point
F601 At pH where only neutral form present
F602 pH 2.0 units below pKa
F603 pH = PI
F604 pH = pKa calc. as neutral form
F605 pH = pKa calc. to pH = 7.4
F606 pH about 1.0
F700 Absolute values not reliable, but comparison within series valid
F701 Amphetamine corrections used pka = 9.60
F702 Anomalous set of azobenzenes
F703 Apparent initial value, hydrolysis products also partitioning
F704 Classification by regression equation appears anomalous
F705 For alkylpyridinium series, adsorbtion to glass gave values lower by 0.15
F706 Intra-mol. h-bonds indicated
F707 Lipid phase as surface layer, values range up to 2 units above this
F709 Octanol & ether p values more reliable
F710 Possible "salting out" which incr. logP
F711 Salting out
F712 Values 30% lower than without sulfoxide
F713 Values differ from those reported, eqs. 5 & 8 recalculated
F714 Values lower than calc - difference may be due to folded conform
F715 Values reported relative to benzene (benzene=2.30)
F716 Two Phases NOT Pre-saturated
F800 "Amphenicol" = de-nitro-chloramphenicol
F801 Because of the extra H-bonding capabilities of the boronic acids, the intercept in the regression equation appears to be increased from 1.46 to 2.15. Because electronegativity of ring substituents could affect this H-bonding, no logP octanol/water ..

F802 Both hydration & dimerization in lipid phase taken into account
F803 Calc. by converting (C-B)1/2 to (C-O)mon by: (C-O)mon=(C-B)1/2/(2KD)1/2. KD values for butyric to heptanoic acids est. from reference 667
F804 Calc. by pi or fragment constants differs by 0.8 or more
F805 Calc. from mol fraction partition coefficient, p-mf, by the expression P = P-MF 18(DO)/ MWO where DO = dens. org. solv. and MWO = mol.wt. org.sol. expression holds only for low conc. of solute
F806 Calc. logP-enol = 1.48, logP-keto = 0.04
F807 Calc. to ph = 8.0
F808 Calc. assuming only mono ion partitions at ph=12.5, Diff. between logP phenol and phenolate ion taken as = 3.70, mono-ion logP=3.84
F809 Calculated from extraction constant, e. logP = (pe + 2)
F810 Complete dissoc. in h2o assumed
F811 Corrected for ionization & dimerization by method of r-176
F812 Corresponding rm values (tlc) reported
F813 Cpds. with active hydrogens show unusually high logP-benzene
F814 Dimerization constants also reported
F815 Entered twice: once as enol, once as keto tautomer
F816 Est. pka = 4.9
F817 From pi values, sulfonilamide logP(chcl3)=-1.40,(oct)=-0.70
F818 From rm values in series well correlated with std. oct/h2o shake flask val
F819 From: log(p/ka) where pka assumed = 2.74 as with benzylpenicillin
F820 Ion pair logE (extraction constant) values reported. Complete ionization in aqueous phase is assumed, and log(E)=C3/C1C2 where C3=conc. ion pair in organic solvent and C1 & C2 = conc. of ions in aqueous phase
F821 Ionization suppressed
F822 Large logP increase with temp
F823 LogP at infinite dilution calc. by regression analysis, s=.03, r=.995
F824 LogP from pi values
F825 LogP phenol taken as 1.46
F826 LogP=log conc.(CHCl3) - 2log conc.(H2O)
F827 LogP(obs.) = 0.19 pH -1.9
F828 Low value probably due to non-equilibrium, diffusion-controlled conditions
F829 Min. charge, max. p
F830 P(unionized) calc. from P = P*(1-x) where x = deg. of dissoc. from pka. values differ widely from koizumi
F831 Pka measured in acetonitrile which accentuates base strength
F832 Reference 208 reports logP-benzene = 0.50
F833 Same logP for three xylene isomers
F834 Single nucleotides entered by mol formula & name. di-& poly-nucleotides are entered by mol formula of head group, followed by remainder. The term 'polynuc' precedes their common abbreviations: a=adenosine, g=guanine, c=cytosin u=uracil, py=pyrimidine. The wln of the head group is followed by the man-trap symbol for polymer_&7, followed by the number of units in the polymer
F835 Subject of u.s. patent 3,417,077 issued to Eli Lilly & Co
F836 The large difference between 3 & 4 isomers expl. in r-147
F837 This value appears "out of line" & was not used in regression equation
F838 Using eq.: logP = 0.95 rm - 1.11, n = 7, r = .994
F839 log Kd/Ka, where Ka is acid dissoc.const
F840 Ionized in both phases, Na+ or K+
F841 Ion-pair in octanol, free ions in H2O.

R1 Aboul-Seoud,A. & El-Hady,A., Recueil, (1962) 81, 958
R2 Adams,R., Rideal,E.K., Burnett,W.B., Jenkins,R.L., & Dreger,E.E., J. Am. Chem. Soc., (1926) 48, 1758
R3 Aiello,G., Biochem. Z., (1921) 124, 1921
R3 Dedek, W., Monatsber. Deut. Adad. Berlin, 4, 225 (1962), Chem. Abs., 59, 5194 (196
R4 Aksnes,G., Acta Chem. Scand., (1960) 14, 1447
R5 Albert,A., J. Chem. Soc., (1951) 1376
R6 Albert,A., Goldacre,R. & Heymann,E., J. Chem. Soc., (1943) 651
R7 Almquist,H.J., J. Phys. Chem., (1933) 37, 991
R8 Angadji,E. & Colleter,J.C., Bull. Soc. Pharm. Bordeaux, (1962) 101, 147, Chem. Abs., 58, 8851g
R9 Archibald,R.C., J. Am. Chem. Soc., (1932) 54, 3178
R10 Auerbach,Fr. & Zeglin,H., Z. Physik. Chem., (1922) 103, 200
R12 Babko,A.K. & Mikhel'son,P.B., Ukrain. Khim. Zhur., (1955) 21,

388, Chem. Abs., 50, 3847i

R14 Badgett,C.O., Ind. Eng. Chem., (1950) 42, 2530

R15 Baggesgaard,H., & Martins,I., Arch. Pharm., (1930) 269, 1

R16 Balt,S. & Van Dalen,E., Anal. Chim. Acta, (1962) 27, 188

R17 Balt,S. & Van Dalen,E., Anal. Chim. Acta., (1964) 30, 434

R18 Bankovskis,J., Cera,L. & Ievins,A., Zh. Anal. Khim., (1963) 18, 555

R19 Bankovskis,J., Zaruma,D., Ievins,A. & Labrence,I., Latvijas PSR Zinatnu Akad. Vestis, Kim. Ser. (1965) 4, 464, Chem. Abs., (1966) 64, 2809d

R20 Bashilova,V.M. & Figurovski,N.A., Aptechnoe Delo, (1959) 8, 20, Chem. Abs., (1956) 50, 6147f

R21 Bekturov,A., Zh. Obshch. Khim., (1939) 9, 419

R22 Bell,F.K., O' Neill, J.J. & Burgison, R.M., J. Pharm. Sci., (1963) 52, 637

R23 Bhattacharyya, P.K., J. Indian Chem. Soc., (1955) 32, 387, Chem. Abs., 50, 6147f

R24 Bodansky,M. & Meigs,A.V., J. Phys. Chem., (1932) 36, 814

R25 Bodansky,M., J. Biol. Chem., (1928) 79, 241

R26 Bodnya,V.A. & Alimarin,I.P., Vestn. Mosk. Univ., (1967) 22, 57, Chem. Abs., (1967) 67, 15400n

R27 Bowen,C.V., Ind. Eng. Chem., (1949) 41, 1295

R28 Brand,L., Mark,L.C., Snell,M.M., Vrindten,P. & Dayton,P.G., Anesthes., (1963) 24, 331

R29 Brandstrom,A., Acta Pharm. Suecica, (1964) 1, 159

R30 Brown,F.S. & Bury,C.R., J. Chem. Soc., (1923) 123, 2430

R31 Brunzell,A., J. Pharm. Pharmacol., (1956) 8, 329

R32 Buchi,J. & Perlia,X., Arzneim.-Forsch., (1960) 10, 930

R33 Buchi,J., Doulakas,J. & Perlia,X., Arzneim.-Forsch., (1969) 19, 578

R34 Burton,D.E., Clarke,K. & Gray,G.W., J. Chem. Soc., (1964) 1314

R35 Carpenter,F.H., McGregor,W.H., & Close,J.A., J. Am. Chem. Soc., (1959) 81, 849

R36 Carstensen,H., Acta Chem. Scand., (1955) 9, 1026

R37 Casy,A.F. & Wright,J., J. Pharm. Pharmac., (1966) 18, 677

R38 Chandler,E.E., J. Am. Chem. Soc., (1908) 30, 696

R39 Collander,R., Physiol. Plant., (1954) 7, 420

R40 Collander,R., Acta Chem. Scand., (1949) 3, 717

R41 Collander,R., Acta Chem. Scand., (1950) 4, 1085

R42 Collander,R., Acta Chem. Scand., (1951) 5, 774, (= Ref. 510)

R43 Courtemanche,P. & Merlin,J.C., Compt. Rend., (1965) 260, 3053

R44 Covello,M., Rend. Accad. Sci., (1932) 2, 73, Chem. Abs., (1934) 28, 1028

R45 Cymerman-Craig,J. & Warburton,W.K., Australian J. Chem., (1956) 9, 294

R46 Currie,D.J., Lough,C.E., Silver,R.F & Holmes,H.L., Can. J. Chem. 44, (1966) 1035

R47 Daniels,T.C. & Lyons,R.E., J. Phys. Chem., (1931) 35, 2049

R48 Davies,M., Jones,P., Patnaik,D. and Moelwyn-Hughes,E.A., J. Chem. Soc., (1951)1249

R49 Davson, H., J. Physiol., 110, 416 (1950)

R50 De Ligny,C.L., Kreutzer,H.J.H. and Visserman,G.F., Rec. Trav. Chim., (1966) 85, 5

R51 Dedek,J., Monatsber. Deut. Akad. Wiss. Berlin, (1962) 4, 225, Chem. Abs., 5194d

R52 Deniges,G., Bull. Soc. Pharm. Bordeaux, (1940) 78, 61, Chem. Abs., (1940) 34, 5724

R53 Dermer,O. & Dermer,V., J. Am. Chem. Soc., (1943) 65, 1653

R54 Dermer,O.C., Markham,W.G. and Trimble,H.M., J. Am. Chem. Soc., (1941) 63, 3524

R55 Deitzel,R. & Rosenbaum,E., Biochem. Z., (1927) 185, 275 & (1927) 189, 348

R56 Dietzel,R. & Schmitt,P., Z. Untersuch. Lebensm., (1932) 63, 369,R69 Chem. Abs., (1933) 27, 886

R57 Draber,W., Buchel,K.H., Dickore,K., Prog. In Photosyn. Res., (1969) 3, 1789

R58 Dyrssen,D. & Hay,L.D., Acta Chem. Scand., (1960) 14, 1091

R59 Dyrssen,D. & Petkovic,D.J., J. Inorg. Nucl. Chem., (1965) 27, 1381

R60 Dyrssen,D., Acta Chem. Scand., (1954) 8, 1394

R61 Dyrssen,D., Acta Chem. Scand., (1957) 11 1771

R62 Dyrssen,D., Ekberg,S. & Liem,D.H., Acta Chem. Scand., (1964) 18 135

R63 Emery,W.O. & Wright,C.D., J. Am. Chem. Soc., (1921) 43, 2323

R64 Endo,K., Bull. Chem. Soc. Japan, (1926) 1, 25

R65 England Jr.,A. & Cohn,E.J., J. Am. Chem. Soc., (1935) 57, 626

R66 Foster,A.G. & Siddiqi,I.R., J. Chem. Soc., (1961) 4906

R67 Fourneau,E. & Florence,G., Bull. Soc. Chim., (1928) 43, 1027

R68 Fresco,J. & Freiser,H., Anal. Chem., (1964) 36, 631

R69 Gaunder,R.G. & Hoffman,W.A., J. Sci. Lab., (1965) 46, 125, Chem. Abs., (1966) 64, 13729d

R70 Georgievics,G.V., Monatsh. Chem., (1915) 36, 391

R71 Gier,T.E. & Hougen,J.O., Ind. Eng. Chem., (1953) 45, 1362

R72 Golumbic,C. & Goldbach,G., J. Am. Chem. Soc., (1951) 73, 3966

R73 Golumbic,C. & Orchin,M., J. Am. Chem. Soc., (1950) 72, 4145

R74 Golumbic,C., Orchin,M. & Weller,S., J. Am. Chem. Soc., (1949) 71, 2624

R75 Gordon,K.F., Ind. Eng. Chem., (1953) 45, 1813

R76 Gordon,N.E. & Reid,E.E., J. Phys. Chem., (1922) 26, 773

R77 Green,R.W. & Alexander,P.W., Aust. J. Chem., (1965) 18, 329

R78 Green,R.W. & Le Mesurier,E.L., Aust. J. Chem., (1966) 19, 229

R79 Greene,R. & Black,A., J. Am. Chem. Soc., (1937) 59, 1820

R80 Greenfield,B.F. & Hardy,C.J., J. Inorg. Nucl. Chem., (1961) 21, 359

R81 Hadaway,A.B. & Barlow,F., Bull. Entom. Res., (1966) 56, 569

R82 Hagiwara,Z., Tech. Repts. Tohoku Univ. (1953) 18, 16, Chem. Abs., (1954) 48, 8118f

R83 Halmekoski,J. & Nissema,A., Suomen Kemi., (1962) 35b, 188

R84 Hammick,D.L. & Mason,S.F., J. Chem. Soc., (1950) 345

R85 Herz,A., Teschemacher,H., Hofstetter,A. and Kurz,H., Int. J. Neuropharm., (1965) 4, 207

R86 Herz,W. & Fischer,H., Chem. Ber., (1904) 37, 4746

R87 Herz,W. & Rathmann,W., Z. Electrochem., (1913) 19, 552

R88 Herz,W. & Stanner,E., Z. Physik. Chem., (1927) 128, 399

R89 Hjort,A.M., DeBeer,E.J., Buck,J.S. & Ide, W.S., J. Pharmacol. Exp. Ther., (1935) 55, 152

R90 Hok,B., Svensk Kem. Tidskr., (1953) 65, 182

R91 Hopkins,P.D. & Douglas,B.E., Inorg. Chem., (1964) 3, 357

R92 Irikura,T., Yakugaku Zasshi, (1962) 82, 356, Chem. Abs., (1963) 58, 3550

R93 Ishimori,T. & Fujino,T., Nippon Genshiryoku Gakkaishi, (1961) 3, 276, Chem. Abs., (1963) 58, 3948d

R94 Ivanov,B.I. & Makeikina,V.V., Tr. Vses. Nauchn.- (1964) 13, 171, Chem. Abs., (1965) 62, 15896b

R95 Jacobs,M.L. & Jenkins,G.L., J. Am. Pharm. Assoc., (1937) 26, 599

R96 Kakemi,K., Arita,T., Hori,R. & Konishi,R., Chem. Pharm. Bull. Japan, (1967) 15, 1705

R97 Karpfen,F.M. & Randles,J.E.B., Trans. Faraday Soc. (1953) 49, 823

R98 Kato,T., Tokai Denkyoku Giho, (1963) 23, 1, Chem. Abs. (1964) 60, 8701g

R99 Kemp,D.M., Anal. Chim. Acta, (1962) 27, 480

R100 Kemula,W. & Buchowski,H., J. Phys. Chem., (1959) 63, 155

R101 Kemula,W., Buchowski,H. & Teperek,J., Bull. Acad. Polon. Sci., (1964) 12, 347

R102 Kemula,W., Buchowski,H. and Lewandowski,R., Bull. Acad. Polon. Sci., (1964) 12, 267

R103 Kemula,W., Buchowski,H. & Teperek,J., Bull.Acad.Polon.Sci., (1964) 12, 343

R104 Kemula,W., Buchowski,H. & Pawlowski,W., Bull. Acad. Polon. Sci., (1964) 12, 491

R105 Keys,A. & Brugsch,J., J. Am. Chem. Soc., (1938) 60, 2135

R106 King,E.L. & Reas,W.H., J. Am. Chem. Soc., (1951) 73, 1804

R107 Klein,A., Roczniki Chem., (1925) 5, 101, Chem. Abs., (1926) 20, 1016

R108 Knudsen,L.F. & Grove,D.C., Ind. Eng. Chem, Anal. Ed., (1942) 14, 556

R109 Koizumi,T., Arita,T. & Kakemi,K., Chem. Pharm. Bull. Japan, (1964) 12, 413

R110 Kolossowsky,N., Bull. Soc. Chim., (1925) 37, 372

R111 Kolossowsky,N. & Megenine,I., Bull. Soc. Chim., (1932) 51, 1000

R112 Kolossowsky,N.A. & Kulikow,F.S., Z. Physik. Chem., (1934) A169, 459

R113 Kolossowsky,N., Kulikow,F. & Bekturow,A., Bull. Soc. Chim., (1935) 2, 460

R116 Komar,N.P. & Khukhryanskii,A.K., Zh. Neorgan. Khim., (1966) 11, 1148, Chem. Abs., (1966) 65, 6523d

R117 Komar,N.P. & Manzhelii,L.S., Chem. Abs., 61, 3736c (1964)

R118 Lindberg,J.J., Soc. Sci. Fennica, Commentationes Phys.-Math., (1958) 21, 1, Chem. Abs., (1960) 54, 14877g

R119 Lindenberg,A., Compt. Rend. Soc. Biol. Strassbourg, (1935) 118, 1086

R120 Lindenberg,A.B. & Massin, M.M., J. Chim. Phys., (1964) 61, 112

R121 Lindenberg,A., Soc. De Biol. De Strasbourg, (1935) 118, 1405
R122 Lough,C.E., Silver,R.F. & McClusky,F.K., Can. J. Chem., (1968) 46, 1943
R123 Lubieniecki,M., Chem. Abs., (1967) 63, 13927
R124 MacDonald,J.Y., J. Am. Chem. Soc., (1935) 57, 771
R125 Macy,R., J. Ind. Hyg. Toxicol., (1948) 30, 140
R126 Marden,J.W., J. Ind. Eng. Chem., (1914) 6, 315
R127 Marrian,G.F. & Sneddon,A., Biochem. J., (1960) 74, 430
R128 Marvel,C.S. & Richards,J.C., Anal. Chem., (1949) 21, 1480
R129 McClellan,B.E. & Freiser,H., Anal. Chem., (1964) 36, 2262
R130 Meeussen,E. & Huyskens,P., J. Chim. Phys., (1966) 63, 845
R131 Meyer,K.H. & Gottlieb-Billroth,H., Z. Physiol. Chem., (1921) 112, 55
R132 Meyer,K.H. & Hemmi,H., Biochem. Z., (1935) 277, 39
R134 Mindowicz,J. & Biallozor,S., Chem. Abs., (1964) 60, 3543
R135 Mindowicz,J. & Uruska,I., Chem Abs., (1964) 60, 4854
R136 Moore,T.S. & Winmill,T.F., J. Chem. Soc., (1912) 101, 1635
R137 Mottola,H.A. & Freiser,H., Talanta, (1966) 13, 55
R138 Munck,A., Scott,J.F. & Engle,L.L., Biochim. Biophys. Acta, (1957) 26, 397
R140 Oksne,S., Acta Chem. Scand., (1959) 13, 1814
R141 Pearson,D.E. & Levine,M., J. Org. Chem., (1952) 17, 1351
R142 Perschke,W. & Chufarov, Z. Anorg. Allgemchem., (1926) 151, 121
R143 Philbrick,F.A., J. Am. Chem. Soc., (1934) 56, 2581
R144 Plaut,G.W., Kuby,S.A. & Lardy,H.A., J. Biol. Chem., (1950) 184, 243
R145 Pressman,D., Brewer,L. & Lucas,H.J., J. Am. Chem. Soc., (1942) 64, 1117
R146 Pyatnitskii,I. & Kharchenko,R., Chem. Abs., (1964) 60, 7433
R147 Quintana,R.P., J. Pharm. Sci., (1965) 54, 462
R149 Rosenmund,K., & Karg,E., Chem. Ber., (1942) 75b, 1850
R150 Ruigh,W.L. & Erickson,A.E., Anesthesiology, (1941) 2, 546
R151 Rydberg,J., Svensk Kem. Tidskr., (1950) 62, 179
R152 Saha,N.C., Bhattacharjee,A., Basals,N. & Lahiri,A., J. Chem. Eng. Data, (1963) 8, 405
R153 Sandell,K.B., Monatsh. Chem., (1961) 92, 1066
R154 Sandell,K.B., Naturwissen., (1962) 49, 12
R155 Sandell,K.B., Naturwissen., (1962) 49, 348
R156 Sandell,K.B., Naturwissen., (1966) 53, 330
R157 Holmes,H., "Structure-Activity Relationships", II, Defense Res. Est., Ralston, Can
R158 Schanker,L.S., Johnson,J.M. & Jeffrey,J.J., Am. J. Physiol., (1964) 207, 503
R159 Schaumann,O., Arch. Expl. Path., Pharm., (1938) 190, 30
R160 Scholtan,W., Arzneim. Forsch., (1968) 18, 505
R161 Scribner,W., Treat,W.J., Weis,J.D. & Moshier,R.W., Anal. Chem., (1965) 37, 1136
R162 Korenman,I.M. & Chernorukova,Z.G., Zh. Prikl. Khim., (1975) 47, 2595
R163 Mirrlees,M.S. & Taylor,P.J., J. Med. Chem., (1976) 19, 615
R164 Davis,S.S. & Elson,G., J. Pharm. Pharmacol., (1974) 26(s), 90P
R165 Hala,J., Coll. Czech. Chem. Comm., (1974) 39, 3475
R166 Tanaka,M. & Kojima,I., J. Inorg. Nucl. Chem., (1967) 29, 1769
R167 Seidell,A., "Sol. Of Org. Cpds." Vol. 2, 3rd Ed., D. Van Nostrand (1941) P. 530
R168 Seidell,A., "Sol. Of Org. Cpds.", Vol. 2, 3rd Ed. Van Nostrand (1941) P. 674
R169 Buchi,J., Kestermann,H. & Perlia,X., Arzneim-Forsch., (1974) 24, 485
R170 Kosower,N.S., Kosower,E.M., Saltoun,G. & Levi,L., Biochem. Biophys. Res. Comm., (1975) 62, 98
R171 Kessel,D. & McElhinney,R.S., Mol. Pharmacol., (1975) 11, 298
R172 Houston,J.B., Upshall,D.G. & Bridges,J.W., J. Pharmacol. Exp. Ther., (1974) 189, 244
R173 Seidell,A., "Sol. Of Org. Cpds.", Vol. 2, 3rd Ed., D. Van Nostrand (1941) P. 813
R174 Sekera,A., Borovansky,A. & Vrba,C., Ann. Pharm. Franc., (1958) 16, 525
R175 Shindo,H., Okamoto,K. & Totsu,J.I., Chem. Pharm. Bull. Japan, (1967) 15, 295
R176 Smith,H.W. & White,T.A., J. Phys. Chem., (1929) 33, 1953
R177 Smith,H.W., J. Phys. Chem., (1921) 25, 204 & 605
R178 Smith,H.W., J. Phys. Chem., (1922) 26, 256
R179 Soloway,A.H., Whitman,B. & Messer,J.R., J. Pharmacol. Exp. Ther., (1960) 129, 310
R180 Sovostina,V., Astakhova,E. & Peshkova,V., Chem. Abs., (1963) 59, 1062
R181 Starnik,I., Ampelogova,N. & Kuznetsov,B., Radiokhimiya, (1964) 6, 519
R182 Stary,I. & Rudenko,N., Chem. Abs., (1959) 53, 5828
R183 Stevancevic,D. & Antonijevic,V., Chem. Abs., (1965) 63, 17215
R184 Stewart,D.C. & Crandall,H.W., J. Am. Chem. Soc., (1951) 73, 1377
R185 Synge,R.L.M., Biochem. J., (1939) 33, 1913
R186 Szabo,E. & Szabon,J., Acta Chim. Acad. Sci. Hung., (1966) 48, 299
R187 Szyszkowski,B., Z. Phys. Chem., (1927) 131, 175
R188 Tabern,D.L. & Shelberg,E.F., J. Am. Chem. Soc., (1933) 55, 328
R189 Elonen,E., Med. Biol., (1974) 52, 415
R190 Thies,H. & Ermer,E., Naturwissen., (1962) 49, 37
R191 Tokarev,B. & Sharkov,V., Chem. Abs., (1963) 58, 662
R192 Tsuzuki,Y., Bull. Chem. Soc. Japan, (1938) 13, 337
R193 Turyan,Y., Zaitsev,P. & Zaitseva,Z., Chem. Abs., (1963) 58, 70
R194 Umezawa,S., Suami,T., Maeda,K. & Nakada,S., J. Chem. Soc. Japan, (1949) 52, 30
R195 Unmack,A., Chem. Zentr., (1934) 2, 1862
R196 Vaubel,W., J. Prakt. Chem., (1903) 67, 473
R197 Vogt,H.J. & Geankoplis,C.J., Ind. Eng. Chem., (1953) 45, 2119
R198 Wakahayashi,T., Oki,S., Omori,T. & Suzuki,N., J. Inorg. Nucl. Chem., (1964) 26, 2255
R199 Weibull,B., Arkiv. Kemi, (1951) 3, 225
R200 Werkman,C.H., Ind. Eng. Chem. Anal. Ed., (1930) 2, 302
R201 Wilkinson,J.H., Biochem. J., (1953) 54, 485
R202 Wittenberger,W., Angew. Chem., (1949) 61, 412
R203 Walter,W. & Weidemann,H., Montsh. Chem., (1962) 93, 1235
R204 Wroth,B.B. & Reid,E.E., J. Am. Chem. Soc., (1916) 38, 2316
R206 Ziolkowski, Z., Respondek, J., & Olszowski, A., Chem. Abs., 60, 5743 (1964)
R207 Zolotov, Y. & Lambrev, V., Chem. Abs., 65, 9808 (1966)
R208 Zozulya,A.P., Mezentseva,N.N., Peshkova,V.M. & Yurev,Y.K., Zhur. Anal. Khim., (1959) 14, 15
R209 Zucal,R.H., Dean,J.A. & Handley,T.H., Anal. Chem., (1963) 35, 988
R210 Audrieth,L.F. & Gibbs,C.F., Inorg. Synthesis, (1939) 1, 77
R211 Hendrixson,W.S., Z. Anorg. Chem., (1897) 13, 73
R212 Abbott Laboratories, Private Communication
R213 Zirvi,K.A. & Jarboe,C.H., J. Med. Chem., (1969) 12, 923
R214 Rall,D.P., Stabenau,J.R. & Zubrod,C.G., J. Pharmacol. Exp. Ther., (1959) 125, 185
R215 Mark,L.C., Burns,J.J., Brand,L., Campomanes,I., Trousof,N., Papper,E.M. & Brodie,B.B., J. Pharmacol. Exp. Ther., (1958) 123, 70
R216 Mao,T.S.S. & Noval,J.J., Biochem. Pharmacol., (1966) 15, 501
R217 Fritz,J.S. & Hedrick,C.E., Anal. Chem., (1965) 37, 1015
R218 Thorne,C.B. & Peterson,W.H., J. Biol. Chem., (1948) 176, 413
R219 Elliott,D.F., Biochem. J., (1949) 45, 429
R220 Morello,V.S. & Beckmann,R.B., Ind & Eng. Chem., (1950) 42, 1078
R221 Hogben,C.A.M., Tocco,D.J., Brodie,B.B., Schanker,L.S., J. Pharmacol. Exp. Ther., (1959) 125, 275
R222 Bickel,M.H. & Weder,H.J., J. Pharm. Pharmacol., (1969) 21, 160
R223 Keston,A.S., Udenfriend,S. & Levy,M., J. Am. Chem. Soc., (1950) 72, 748
R224 Rohmann,C. & Eckert,T., Archiv. Der Pharm., (1958) 291 450
R225 Harrass,P., Arch. Int. Pharmacol Et Therap., (1903) 11, 431
R226 Eeckhout,A., Arch. Expl. Path. Pharmacol., (1907) 57, 338
R227 Brann, A., Dissertation, U. Of Wisc., 1914 (see Ref. 393)
R228 Bierick,R., Arch. Physiol., (1919) 174, 202
R229 Siebeck,R., Arch. Expl. Path. Pharmacol., (1922) 95, 93
R230 Hadaway,A.B., Barlow,F., Grose,J.E.H., Turner,C.R. & Flower,L.S., Bull. Wld. Hlth. Org., (1970) 42, 369
R231 Doerr,R.C. & Fiddler,W., J. Agr. Food Chem., (1970) 18, 937
R232 Murthy,K.S. & Zografi,G., J. Pharm. Sci., (1970) 59, 1281
R233 Butler,K., Howes,H.L., Lynch,J.E. & Pirie,D.K., J. Med. Chem., (1967) 10, 891
R234 Vold,R.D. & Washburn,E.R., J. Am. Chem. Soc., (1932) 54, 4217
R235 Davies,J.T., J. Phys. Chem., (1950) 54, 185
R236 Hutchinson,E., J. Phys. Chem., (1948) 52, 897
R237 Ting,H.P., Bertrand,G.L. & Sears,D.F., Biophys. J., (1966) 6, 813
R238 Banewicz,J.J., Reed,C.E. & Levitch,M.E., J. Am. Chem. Soc., (1957) 79, 2693
R239 Kang,S. & Green,J.P., Nature, (1969) 222, 794

R240 Gerlsma,S.Y., J. Biol. Chem., (1968) 243, 957

R241 Feltkamp,H., Arzneim. Forsch., (1965) 15, 238

R242 Mazel,P. & Henderson,J.F., Biochem. Pharmac., (1965) 14, 92

R243 McMahon,R.E., J. Med. & Pharm. Chem., (1961) 4, 67

R244 Lindenberg,B.A., J. Chim. Phys., (1951) 48, 350

R245 Fieser,L.F., Ettlinger,M.G. & Fawaz,G., J. Am. Chem. Soc., (1948) 70, 3228

R246 Dudley,K. & Miller,W., J. Med. Chem., (1970) 13, 535

R247 Courtier,M.A.J., Bull. Soc. Chim. Fr., (1948) 15, 528

R248 Craig,L.C., Anal. Chem., (1950) 22, 1346

R249 Fieser,L.F., J. Am. Chem. Soc., (1948) 70, 3237

R250 Crowdy,S.H., Grove,J.F. & McCloskey,P., Biochem., (1959) 72, 241

R251 Jaglan,P.S. & Gunther,F.A., Analyst, (1970) 95, 763

R252 Yamazaki,M., Kakeya,N., Morishita,T., Kamada,A., Aoki,M., Chem. Pharm. Bull. Japan, (1970) 18, 708

R253 Chien,Y.W., Lambert,H.J. & Lin,T.K., J. Pharm. Sci., (1975) 64, 961

R254 Das Gupta,V. & Cadwallader,D.E., J. Pharm. Sci., (1968) 57, 2140

R255 Currie,D.J. & Holmes,H.L., Can. J. Chem., (1970) 48, 1340

R256 Lindenberg,B.A., J. Chem. Phys., (1951) 48, 350

R257 Flynn,G.L., J. Pharm. Sci., (1971) 60, 345

R258 Glennon,R.A. & Gessner,P.K., J. Med. Chem., (1979) 22, 428

R259 Tollenaere,J., Janssen Pharmaceutica, Private Communication

R260 Tute,M., Pfizer Corp., Private Communication

R261 Hyman,E.S., Biophys. J., (1966) 6, 405

R262 Quintana,R.P. & Smithfield,W.R., J. Med. Chem., (1967) 10, 1178

R263 Englehardt,A. & Wick,H., Arzneim. Forsch., (1957) 7, 217

R264 Kakeya,N., Yata,N., Kamamda,A. & Aoki,M., Chem. Pharm. Bull. Japan, (1969) 17, 2558

R265 Faull,J.H., J. Am. Chem. Soc., (1934) 56, 522

R266 Felsing,W.A. & Buckley,S.E., J. Phys. Chem., (1933) 37, 779

R267 Freundlich,H. & Kruger,D., Z. Electrochem., (1930) 36, 305

R268 Glasser,H. & Krieglstein,J., Naunyn-Sch. Arch. Pharmak., (1970) 265, 321

R269 Gross,P. & Schwarz,K., Monatsh. Chem., (1930) 55, 287

R270 Grossfield,J. & Miermeister,A., Z. Anal. Chem., (1931) 85, 321

R271 Grossfield,J. & Miermeister,A., Z. Anal. Chem., (1932) 87, 241

R272 Ichikawa,Y., Yamano,T. & Fujishima,H., Biochim. Biophys. Acta, (1969) 171, 32

R273 Fujita,T., Kamoshita,K., Nishioka,T. & Nakajima,M., Agr. Biol. Chem., (1974) 38, 1521

R274 Baur,E.W., J. Pharmacol. Exp. Ther., (1971) 177, 219

R275 Needleman,P. & Hunter,F.E.Jr., Mol. Pharmacol., (1966) 2, 134

R276 Mayer,S., Maickel,R.P. & Brodie,B.B., J. Pharmacol. Exp. Ther., (1959) 127, 205

R277 Buchi,J., Perlia,X. & Strassle,A., Arzneim. Forsch., (1966) 16, 1657

R278 Garrett,E.R., Mielck,J.B., Seydel,J.K. & Kessler,H.J., J. Med. Chem., (1969) 12, 740

R279 Nogami,H., Hasegawa,J., Nakatsuka,S. & Noda,K., Chem. Pharm. Bull., (1969) 17, 228

R280 Leake,C.D. & Chen,M., Soc. Exp. Biol. Med., (1930) 28, 151

R281 Krantz,J.C. Jr., Carr,C.J. & Evans,W.E., Anesthesiology, (1944) 5, 291

R282 Cone,N.M., Forman,S.E. & Krantz,J.C. Jr., Proc. Soc. Exp. Biol. Med., (1941) 48, 461

R283 McCulloch,A.C. & Stock,B.H., Australian J. Pharm., (1966) 48, S14

R284 Mantica,L., Ciceri,R., Cassagne,J.P. & Mascitelli-Coriandoli,E., Arzneim. Forsch., (1970) 20, 109

R285 Kakemi,K., Sezaki,H., Suzuki,E. & Nakano,M., Chem. Pharm. Bull., (1969) 17, 242

R286 Ritter,P. & Jermann,M., Arzneim. Forsch., (1966) 16, 1647

R287 Herz,A., Holzhauser,H. & Teschemacher,H., Arch. Exp. Path. & Phar., (1966)

R288 Hultin,E., Acta Chem. Scand., (1961) 15, 879

R289 Huq,A.K.M. & Lodhi,S.A.K., J. Phys. Chem., (1966) 70, 1354

R290 Herrero,G., An. Soc. Espan. Fisica. Quin., (1931) 29, 616

R291 Ho,B.T., Fritchie,G.E., Kralik,P.M., Tansey,L.W., Walker,K.E. & McIsaac,W.M., J. Pharm. Sci., (1969) 58, 1423

R292 Teschemacher,H.J., Arch. Pharmacol. Exp. Path., (1966) 255, 85

R293 Kolassa,N., Pfleger,K. & Rummel,W., Euro. J. Pharmacol., (1970) 9, 265

R294 Kemula,W., Buchowski,H. & Pawlowski,W., Roczniki Chem., (1968) 42, 1951

R295 Jakowkin,A.A., Z. Physik. Chem., (1895) 18, 585

R296 Jakowkin, A. Z. Physik. Chem., 29, 613 (1899)

R297 Flurscheim,B., J. Chem. Soc., (1910) 97, 84

R298 Verebely,K., Kutt,H., Sohn,Y.J., Levitt,B. & Raines,A., Euro. J. Pharm., (1970) 10, 106

R299 McGowan, J., Atkinson, P. & Ruddle, L., J. Appl. Chem., 16, 99 (1966)

R300 Kutter,E., Unpublished Analysis 7/31/70

R301 Reynard,A.M., J. Pharmacol. Exp. Ther., (1968) 163, 461

R302 Hosein,E.A., Rambaut,P., Chabrol,J.G. & Orzeck,A., Arch. Biochem. Biophys., (1965) 111, 540

R303 Plakogiannis,F.M., Pharm. Acta Helv., (1971) 46, 236

R304 Buchi,J., Hetterich,K.H. & Perlia,X., Arzneim. Forsch., (1968) 18, 791

R305 Hussain,M.H. & Lien,E.J., J. Med. Chem., (1971) 14, 138

R306 Lough,C.E., Suffield Memorandum No. 17/ 71, Defense Res. Establishment

R307 Holmes,H.L., Suffield Technical Paper No. 373, Defense Res. Establishment

R308 Roth,S. & Seeman,P., Biochim. Biophys. Acta, (1972) 255, 207

R309 Machleidt,H., Roth,S. & Seeman,P., Biochim. Biophys. Acta, (1972) 255, 178

R310 Lough,C.E., Suffield Memorandum No. 9/ 70, DRES, Ralston, Alberta

R311 Lough,C.E., Suffield Memorandum No. 122/ 68, DRES

R312 Lough,C.E., Suffield Memorandum No. 28/69, DRES

R313 Von Wittenau,M.S. & Delahunt,C.S., J. Pharmacol. Exp. Ther., (1966) 152, 164

R314 Von Wittenau,M.S. & Yeary,R., J. Pharmacol. Exp. Ther., (1963) 140, 258

R315 Leach,B.E. & Teeters,C.M., J. Am. Chem. Soc., (1951) 73, 2794

R316 Leach,B.E., DeVries,W.H., Nelson,H.A., Jackson,W.G. & Evans,J.S., J. Am. Chem. Soc., (1951) 73, 2797

R317 Barfknecht,C.F., Smith,R.V., Nichols,D.E., Leseney,J.L., Long,J.P. & Engelbrecht,J.A., J. Pharm. Sci., (1971) 60, 799

R318 Lien,E.J., Hussain,M., Golden,M.P., J. Med. Chem., (1970) 13, 623

R319 Du Vigneaud,V., Gish,D.T., Katsoyannis,P.G. & Hess,G.P., J. Am. Chem. Soc., (1958) 80, 3355

R320 Katsoyannis,P.G. & Du Vigneaud,V., J. Biol. Chem., (1958) 233, 1352

R321 Novello,F.C. & Sprague,J.M., C.R. XXXVI Congr. Intern. Chem. Indust., (1967) 3, 1

R322 Levy,J.V., Europ. J. Pharmacol., (1968) 2, 250

R323 Laubender,W. & Hunn,L., Arzneim. Forsch., (1964) 14, 445

R324 Schultz,O.E., Jung,C. & Moller,K.E., Z. Naturforsch., (1970) 25b, 1024

R325 Delaney,A.D., Currie,D.J. & Holmes,H.L., Can. J. Chem., (1969) 47, 3273

R326 Holmes,H.L., DRES Ralston, Alberta, Private Communication

R327 Martin,Y.C., Jones,P.H., Perun,T.J., Grundy,W.E., Bell,S., Bower,R.R. & Shipkowitz,N.L., J. Med. Chem., (1972) 15, 355

R328 Seydel,J.K. & Wempe,E., Arzneim. Forsch., (1971) 21, 187

R329 Fujita,T., Nishioka,T. & Kano,M., Kyoto University, Unpublished Results

R330 Sanders,E. & Maren,T.H., Mol. Pharmacol., (1967) 3, 204

R331 Draber,W., Buchel,K.H. & Schafer,G., Z. Naturforsch., (1972) 27 b, 159

R332 Benthe,H.F., Private Communication Via E. Kutter

R333 Morgan,J.L.R. & Benson,H.K., Z. Anorg. Chem., (1907) 55, 356

R334 Rieder,J., Arzneim. Forsch., (1963) 13, 81

R335 Kreighbaum,W.E., Grunwald,F.A., Harrison,E.F., LaBudde,J.A. & Larsen,A.A., J. Med. Chem., (1970) 13, 247

R336 Neely,W.B., Allison,W.E., Crummett,W.B., Kauer,K. & Reifschneider,W., J. Agr. Food Chem., (1970) 18, 45 & Priv. Commun

R337 Garrett,E.R. & Chemburkar,P.B., J. Pharm. Sci., (1968) 57, 1401

R338 Nelson,E., J. Med. & Pharm. Chem., (1962) 5, 211

R339 Liberti,J.P. & Rogers,K.S., Biochim. Biophys. Acta, (1970) 222, 90

R340 Herzog,K.A. & Swarbrick,J., J. Pharm. Sci., (1971) 60, 1666

R341 Philip,J.C. & Clark,C.H.D., J. Chem. Soc., (1925) 127, 1274

R342 Hill,J.O., Worsley,I.G. & Hepler,L.G., J. Phys. Chem., (1968) 72, 3695

R343 Morozowich,W. & Metzler,C., Upjohn Co., Kalamazoo, Mich., Private Commumication

R344 Reese,D.R., Irwin,G.M., Dittert,L.W., Chong,C.W. &

Swintosky,J.V., J. Pharm. Sci., (1964) 53, 591

R345 Oguma,T., Nagai,T. & Nogami,H., Chem. Pharm. Bull., (1971) 19, 124

R346 Rich,S. & Horsfall,J.G., Phytopathology, (1952) 42, 457

R347 Riedel,R., Z. Physik. Chem., (1906) 56, 243 (landholt. B.)

R348 Rosano,H.L., Duby,P. & Schulman,J.H., J. Phys. Chem., (1961) 65, 1704

R349 Ross,R.G. & Ludwig,R.A., Can. J. Bot., (1957) 35, 65

R350 Kutter,E., Herz,A., Teschemacher,H.J. & Hess,R., J. Med. Chem., (1970) 13, 801

R351 Plakogiannis,F.M., Lien,E.J. & Biles,J.A., J. Med. Chem., (1971) 14, 430

R352 Hober,R. & Hober,J., J. Cell. & Comp. Physiol., (1937) 10, 401

R353 Leader,J.E. & Whitehouse,M.W., Biochem. Pharmacol., (1966) 15, 1379

R354 Neely,W.B., Dow Chemical Co., Midland Mich., Unpublished Results

R355 Karr,C. Jr., Estep,P.A. & Hirst,L.L. Jr., Anal. Chem., (1960) 32, 463

R356 Pressman,B. & Haynes,D.H., "The Molecular Basis of Membrane Function", Prentice-Hall, New Jersey, Tosteson,D.C., Ed. 1968

R357 Terada,H., Chem. Pharm. Bull., (1972) 20, 765

R358 Khan,A.H. & Ross,W.C.J., Chem. Biol. Interactions, (1969) 1, 27

R359 Stockdale,M. & Selwyn,M.J., Eur. J. Biochem., (1971) 21, 565

R360 Elofsson,R., Nilsson,S.O., Kluczykowska,B., Acta Pharm. Suecica, (1971) 8, 465

R361 Dayton,P.G., Weiss,M.M. & Perel,J.M., J. Med. Chem., (1966) 9, 941

R362 Perel,J.M., Snell,M.M., Chen,W., Dayton P.G., Biochem. Pharmacol., (1964) 13, 1305

R363 Terada,H. & Muraoka,S., Mol. Pharmacol., (1971) 8, 95

R364 Gary-Bobo,C.M., Dipolo,R. & Solomon,A.K., J. Gen. Physiol., (1969) 54, 369

R365 McKenzie,W.L. & Foye,W.O., J. Med. Chem., (1972) 15, 291

R366 Modin,R. & Tilly,A., Acta Pharm. Suecica, (1968) 5, 311

R367 Gustavii,K., Acta Pharm. Suecica, (1967) 4, 233

R368 Modin,R. & Schill,G., Acta Pharm. Suecica, (1967) 4, 301

R369 Borg,K.O. & Westerlund,D., Z. Anal. Chem., (1970) 252, 275

R370 Modin,R. & Schroder-Nielsen,M., Acta Pharm. Suecica, (1971) 8, 573

R371 Borg,K.O. & Schill,G., Acta Pharm. Suecica, (1968) 5, 323

R372 Gustavii, K. & Schill, G., Acta Pharm. Suecica, 3, 259 (1966)

R373 Persson, B. -A. & Eksborg, S. Acta Pharm. Suecica, 7, 353 (1970)

R374 Westerlund D. & Borg, K., Acta Pharm. Suecica, 7, 267 (1970)

R375 Westerlund, D., Borg, K. & Lagerstrom, P., Acta Pharm. Suecica, In Press

R376 Schill,G., Acta Pharm. Suecica, (1965) 2, 13

R377 Linden E. & Schill, G. Acta Pharm. Suecica, 4, 327 (1967)

R378 Eksborg,S. & Persson,B.A., Acta Pharm. Suecica, (1971) 8, 205

R379 Persson, B. -A., Acta Pharm. Suecica, 5, 335 (1968)

R380 Schill,G., Modin,R. & Persson,B.A., Acta Pharm. Suecica, (1965) 2, 119

R381 Persson, B. -A. & Schill, G., Acta Pharm. Suecica, 3, 281 (1966)

R382 Modin,R. & Back,S., Acta Pharm. Suecica, (1971) 8, 585

R383 Garel,J.P., Jordan,J.C. & Mandel,P., J. Chromatog., (1972) 67, 277

R384 Schweitzer,G.K. & Van Willis,W., Advan. Anal. Chem., (1965) 5, 169

R385 Buchi,J., Perlia,X. & Tinani,M., Arzneim. Forsch. (1971) 12, 2074

R386 Persson,B.A., Acta Pharm. Suecica, (1971) 8, 193

R387 Durley,R.C. & Pharis,R.P., Phytochemistry, (1972) 11, 317

R388 Munro,D.C., J. Chem. Soc., (1961) 5381

R389 Titus,E. & Fried,J., J. Biol. Chem., (1948) 174, 57

R390 Tschugajeff,L. & Lukaschuk,A.J., Z. Anorg. Allg. Chem., (1928) 172, 223

R391 Velluz,M.L., Compt. Rend., (1926) 182, 1178

R392 Wall,F.T., J. Am. Chem. Soc., (1942) 64, 472

R393 Walton,J.H. & Lewis,H.A., J. Am. Chem. Soc., (1916) 38, 633

R394 Weissenberger,G., Schuster,F. & Piatti,L., Z. Anorg. Allg. Chem., (1926) 151, 77

R395 Tilly, A., Acta Pharm. Suecica, 9, In Press

R396 Washburn, E., & Spencer, H., J. Am. Chem. Soc., 56, 361 (1934)

R397 Woodman,R.M. & Corbet,A.S., J. Chem. Soc., (1925), 2461

R398 Yost,D.M. & Stone,W.E., J. Am. Chem. Soc., (1933) 55, 1889

R399 Yost,D.M. & White,R.J., J. Am. Chem. Soc., (1928) 50, 81

R400 National Cancer Institute, Drug Development Branch (unpublished)

R401 Midwest Research Institute (under contract with NCI, unpublished)

R402 Southern Research Institute (under Contract With Nci, Unpubl.)

R403 Hess,R., Teschemacher,H.J. & Herz,A., N-S. Arch. Pharm. & Ex. Path., (1968) 261, 469

R404 Gruner,J., Krieglstein,J. & Rieger,H. N-S. Arch. Pharmacol., (1973) 277, 333

R405 Klein,R.A., Moore,M.J. & Smith,M.W., Biochim. Biophys. Acta, (1971) 233, 420

R406 Borg,K.O., Holgersson,H. & Lagerstrom,P.O., J. Pharm. Pharmacol., (1970) 22, 507

R407 Modin,R. & Johansson,M., Acta Pharm. Suecica, (1971) 8, 561

R408 Apsitis,A., Oskaja,V. & Kulikova,L.D., Chem. Abs., (1971) 74, 277

R409 Hartung,R. & Klinger,W., Environ. Sci. & Technol., (1970) 4, 407

R410 Kovalova,A.G., Chem. Abs., (1971) 74, 197

R411 Bartz,Q.R., J. Biol. Chem., (1948) 172, 445

R412 Reuning,R.H. & Levy,G., J. Pharm. Sci., (1968) 57, 1342

R413 Divatia,G.J. & Biles,J.A., J. Pharm. Sci., (1961) 50, 916

R414 Hull,R.L. & Biles,J.A., J. Pharm. Sci., (1964) 53 869

R415 Dearden,J.C. & Tomlinson,E., J. Pharm. Pharmac. (1971) 23, 73S

R416 Ogino,A., Tsuchiya,S., Kobayashi,N., Kitano,M., Ogino, A., Nippon Shinyaku, Ltd., Private Communication

R417 Gustavii,K., Brandstrom,A. & Allansson,S., Acta Chem. Scand., (1971) 25, 77

R418 Buchi,J., Oey,L.T. & Perlia,X., Arzneim. Forsch., (1972) 22, 1071

R419 Pitman,I.H., Uekama,K., Higuchi,T. & Hall,W.E., J. Am. Chem. Soc., (1972) 94, 8147

R420 Shindo,H., Sankyo Co., Private Communication

R421 Kiso,M., Fujita,T., Kurihara,N., Uchida,M. Tanaka,K. & Nakajima,M., Pest. Biochem. Physiol., (1978) 8, 33

R422 Schanker,L.S., Nafpliotis,P.A. & Johnson,J.M., J. Pharmacol. Exp. Ther., (1961) 133, 325

R423 Hemker,H.C., Biochim. Biophys. Acta, (1962) 63, 46

R424 Nasim,K., Meyer,M.C. & Autian,J., J. Pharm. Sci., (1972) 61, 1775

R425 Dillingham,E.O., Mast,R.W., Bass,G.E. & Autian,J., J. Pharm. Sci., (1973) 62, 22

R426 Terada,H., Muraoka,S. & Fujita,T., J. Med. Chem., (1974) 17, 330

R427 Jansson,I., Orrenius,S., Ernster,L., Schenkman,J.B., Arch. Biochem. Biophys, (1972) 151, 391

R428 Ward,T.M. & Holly,K., J. Coll. & Interface Sci., (1966) 22, 221

R429 Giacobbe, T. J., Dow Chemical Co., Private Communication

R430 Higbee,I., Drug Development Branch Natl. Cancer Inst., Private Communication

R431 Hoskin,F.C., Science, (1971) 172, 1243

R432 Weisbrodt,N.W., Kienzle,M. & Cooke,A.R., Proc. Soc. Exp. Biol. Med., (1973) 142, 450

R433 Nakagaki,M. & Nara,K., Yakugaku Zasshi, (1963) 83, 781

R434 Kapoor,I.P., Metcalf,R.L., Hirwe,A.S., Coats,J.R. & Khalsa,M.S., J. Agr. Food Chem., (1973) 21, 310

R435 Anderson,J.E., Hoffman,S.J. & Peters,C.R., J. Phys. Chem., (1972) 76, 4006

R436 Gomez,A., Mullens,J. & Huyskens,P., J. Phys. Chem., (1972) 76, 4011

R437 Hanssens,I., Mullens,J., Deneuter,C. & Huyskens,P., Bull. Soc. Chim. Fr., (1968) 3942

R438 Corner,E.D.S. & Sparrow,B.W., J. Mar. Biol. Assoc. U.K., (1957) 36, 459

R439 Dearden,J.C. & Weeks,G.R., J. Pharm. Sci., (1973) 62, 843

R440 Smith,R. & Tanford,C., Proc. Nat. Acad. Sci. USA, (1973) 70, 289

R441 Wiley,R.A., Faraj,B.A., Jantz,A. & Hava,M.M., J. Med. Chem., (1972) 15, 374

R442 Yeh,K.C. & Higuchi,W.I., J. Pharm. Sci., (1972) 61, 1648

R443 Javidan,S. & Mrtek,R.G., J. Pharm. Sci., (1973) 62, 424

R444 Rollo,I.M., Can. J. Physiol. Pharmacol., (1972) 50, 976

R445 Seeman,P., Chau-Wong,M. & Moyyen,S., Can. J. Physiol. Pharmacol., (1972) 50, 1193

R446 Badgett,C.O., Eisner,A. & Walens,H.A., J. Am. Chem. Soc., (1952) 74, 4096

R447 Lynn,K.R., Abbott Laboratories, Private Communication

R448 Skidmore,I.F. & Whitehouse,M.W., Biochem. Pharmacol., (1965) 14, 547

R449 Whitehouse,M.W. & Dean,P.D.G., Biochem. Pharmacol. (1965) 14, 557

R450 Mathison,I.W., Morgan,P.H., Tidwell,R.R. & Handorf,C.R., J. Pharm. Sci., (1972) 61, 637

R451 Bird,A.E. & Marshall,A.C., J. Chromatogr., (1971) 63, 313

R452 Okano,T., Maenosono,J., Kano,T. & Onoda,I., Gann, (1973) 64,

227

R453 Fastrez,J. & Fersht,A.R., Biochem., (1973) 12, 1067

R454 Wiethold,G., Hellenbrecht,D., Lemmer,B. & Palm,D., Biochem. Pharmacol., (1973) 22, 1437

R455 Krieglstein,J., Lier,F. & Michaelis,J., N-S Arch. Pharmacol., (1972) 272, 121

R456 Ruhland,W. & Heilmann,U., Planta, (1951) 39, 91

R457 Cull,G.A.G. & Scott,N.C., Br. J. Pharmacol., (1973) 47, 819

R458 Schwarz,K. & Porter,L.A., Anals. N. Y. Acad. Sci., (1972) 192, 200

R459 Inoue,S., Ogino,A., Kise,M., Kitano,M., Tsuchiya,S. & Fujita,T., Chem. Pharm. Bull. Japan, (1974) 22, 2064

R460 Fujita,T., Kyoto University, Unpublished Results

R461 Muller,W. & Wollert,U., N.-S. Arch. Pharmacol. (1973) 278, 301

R462 Gaginella,T.S., Bass,P., Perrin,J.H. & Vallner,J.J., J. Pharm. Sci., (1973) 62, 1121

R463 Sekine,T., Suzuki,Y. & Ihara,N., Bull. Chem. Soc. Japan, (1973) 46, 995

R464 Seeman,P., Chau-Wong,M. & Moyyen,S., Can. J. Physiol. Pharmacol., (1972) 50, 1181

R465 Sciarra,J.J., Patel,J.M. & Kapoor,A.L., J. Pharm. Sci., (1972) 61, 219

R466 Kramer,S.F. & Flynn,G.L., J. Pharm. Sci., (1972) 61, 1896

R467 Stehle,R.G. & Higuchi,W.I., J. Pharm. Sci., (1972) 61 1931

R468 Frisk-Holmberg,M. & van der Kleijn,E., Europ. J. Pharmacol., (1972) 18, 139

R469 Buchi,J., Perlia,X. & Studach,S.P., Arzneim. Forsch., (1967) 17, 1012

R470 Buchi,J., Bruhin,H.K. & Perlia,X., Arzneim. Forsch., (1971) 21, 1003

R471 Brodin,A. & Agren,A., Acta Pharm. Suecica, (1971) 8, 609

R472 Schmidt,H.L., Moller,M.R. & Weber,N., Biochem. Pharmacol., (1973) 22, 2989

R473 Busch,N., Moleyre,J., Bondivenne,R. & Labrid,C., Chim. Therap., (1973) 13, 7

R474 Hellenbrecht,D., Lemmer,B., Wiethold,G. & Grobecker,H., N-S. Arch. Pharmacol., (1973) 277, 211

R475 Schumacher,G.E. & Nagwekar,J.B., J. Pharm. Sci., (1974) 63, 245

R476 Coburn,R.A. & Glennon,R.A., J. Pharm. Sci., (1973) 62, 1785

R477 Hellenbrecht,D. & Muller,K.F., Experientia, (1973) 29, 1255

R478 Israili,Z.H., Perel,J.M., Cunningham,R.F., Dayton,P.G., Yu,T.F., Gutman,A.B., Long,K.R., Long,R.C. Jr. & Goldstein,J.H., J. Med. Chem., (1972) 15, 709

R479 Kitao,K., Kubo,K., Morishita,T., Yata,N. & Kamada,A., Chem. Pharm. Bull. Japan, (1973) 21, 2417

R480 Morishita,T., Yamazaki,M., Yata,N. & Kamada,A., Chem. Pharm. Bull. Japan, (1973) 21, 2309

R481 Cymerman-Craig,J. & Diamantis,A.A., J. Chem. Soc., (1953), 1619

R482 Chaudry,M.A.Q. & James,K.C., J. Med. Chem., (1974) 17, 157

R483 Arcamone,F., Franceshi,G., Minghetti,A., Penco,S., Redaelli,S., Di Marco,A., Casazza,A.M., Dasdia,T., Di Fronzo,G., Guiliani,F., Lenaz,L., Necco,A. & Soranzo,C., J. Med. Chem., (1974) 17, 335

R484 Carmichael,F.J. & Israel,Y., J. Pharmacol. Exp. Ther., (1973) 186, 253

R485 Maxwell,D.R., Gray,W.R. & Taylor,E.M., Brit. J. Pharmacol., 17, 310 (1961)

R486 Black,J.W., Durant,G.J., Emmett,J.C. & Ganellin,C.R., Nature, (1974) 248, 65

R487 Zaagsma,J. & Nauta,W.T., J. Med. Chem., (1974) 17, 507

R488 Weinbach, E. & Garbus, J., J. Biol. Chem., 240, 1811 (1965)

R489 Nazareth,R.I., Sokoloski,T.D., Witiak,D.T. & Hopper,A.T., J. Pharm. Sci., (1974) 63, 203

R490 Niazi,S. & Chiou,W.L., J. Pharm. Sci., (1974) 63, 532

R491 Vane,J.R., Brit. J. Pharm., (1959) 14, 87

R492 Scheuplein,R.J., Blank,I.H., Brauner,G.J. & MacFarlane,D.J., J. Investig. Dermat., (1969) 52, 63

R493 Schumacher,G.E. & Nagwekar,J.B., J. Pharm. Sci., (1974) 63, 240

R494 Akerblom,E.B., J. Med. Chem., (1974) 17, 609

R495 Chapman,J.D., Raleigh,J.A., Borsa,J., Webb,R.G. & Whitehouse,R., Int. J. Radiat. Biol., (1972) 21, 475

R496 Michailova,D., Natscheva,R. & Owtscharov,R., Pharmazie, (1973) 28, 208

R497 Brodin,A. & Nilsson,M., Acta Pharm. Suecica, (1973) 10, 187

R498 Takeda,J., Morikawa,H., Kinoshita,H. & Senda,M., Plant & Cell

Physiol., (1971) 12, 949

R499 Tukamoto,T., Ozeki,S., Hattori,F. & Ishida,T., Chem. Pharm. Bull. Japan, (1974) 22, 385

R500 Leo,A. & Hansch,C., Pomona College, Unpublished Results

R501 Fujita,T., Iwasa,J. & Hansch,C., J. Am. Chem. Soc., (1964) 86, 5175

R502 Iwasa,J., Fujita,T. & Hansch,C., J. Med. Chem., (1965) 8, 150

R503 Hansch,C. & Anderson,S.M., J. Org. Chem., (1967) 32, 2583

R504 Hansch,C., Steward,A.R., Anderson,S.M. & Bentley,D., J. Med. Chem., (1968) 11, 1

R505 Anderson,S. & Hansch,C., Pomona College, Unpublished Analysis

R506 Helmer,F. & Hansch,C., Pomona College, Unpublished Analysis

R507 Lien,E. & Hansch,C., Pomona College, Unpublished Analysis

R508 Church,C. & Hansch,C., Pomona College, Unpublished Analysis

R509 Buchel,K.H. & Draber,W., Prog. In Photosyn. Res., (1969) 3, 1777

R510 Collander,R., Acta Chem. Scand., 5, 774 (1951), (= Ref. 42)

R511 Poretz,R. & Goldstein,I., Arch. Biochem. & Biophys., (1968) 125, 1034

R512 Lee,V., M. S. Thesis, San Jose State College, Aug. 1967

R513 Iwasa,J. & Hansch,C., Pomona College, Unpublished Analysis

R514 Maloney,P. & Hansch,C., Pomona College, Unpublished Analysis

R515 Bird,A.E. & Marshall,A.C., Biochem. Pharm., (1967) 16, 2275

R516 Kurihara,N., Matazaemon,U., Fujita,T. & Nakajima,M., Pest. Biochem. Physiol., (1973) 2, 383

R517 Hansch,C. & Fujita,T., J. Am. Chem. Soc., (1964) 86, 1616

R518 Rogers,K.S. & Cammarata,A., Biochem. Biophys. Acta, (1969) 193, 22

R519 Rogers,K.S. & Cammarata,A., J. Med. Chem., (1969) 12, 692

R520 Currie,Lough, Silver & Holmes, Can. J. Chem., 44, 1035 (1966), (= Ref. 46)

R521 Liu,M. & Hansch,C., Pomona College, unpublished results

R522 Topliss,J.G. & Yudis,M.D., Schering Corp., Private Communication

R524 Eldefrawi,M.E. & O'Brien,R.D., J. Exp. Biol., (1967) 46, 1

R525 Coats,E. & Hansch,C., Pomona College, Unpublished Analysis

R526 Schaeffer,H.J., Johnson,R.N., Odin,E. & Hansch,C., J. Med. Chem., (1970) 13, 452

R527 Soderberg,D. & Hansch,C., Pomona College, Unpublished Anlaysis

R528 Kerely,R. & Hansch,C., Pomona College, Unpublished Analysis

R529 Gehring,P.J., Torkelson,T.R. & Oyen,F., Toxicol. Appl. Pharmacol., (1967) 11, 361

R530 Glave,W. & Hansch,C., Pomona College, Unpublished Analysis

R531 Tichy,M. & Bocek,K., Private Communication

R532 Dunn,W.J.III, & Hansch,C., Pomona College, Unpublished analysis

R533 Fujita,T., Soeda,Y., Yamamoto,I. & Nakajima,M., Unpublished Analysis

R534 Schaeffer,J. & Hansch,C., Pomona College, Unpublished Analysis

R535 Mackintosh,D. & Hansch,C., Pomona College, Unpublished Analysis

R536 Nikaitani,D. & Hansch,C., Pomona College, Unpublished Analysis

R537 Hansch, C. Gorin, M. Et. Al., J. Med. Chem., 16, 917 (1973)

R538 Dunn, W. J.,III, U. of Illinois Medical School, Private Communication

R539 Dunn & Bajanowski, U. Of Illinois Medical School, Private Communication

R540 Clayton, J. & Hansch, C., Pomona College, Unpublished Analysis

R541 Kim, K. & Hansch, C., Pomona College, Unpublished Analysis

R542 Briggs, M. & Hansch, C., Pomona College, Unpublished Analysis

R543 Dunn,W.J.III, J. Med. Chem., (1973) 16, 484

R544 Breen,M.P., Bojanowski,E.M., Cipolle,R.J., Dunn,W.J.III, Frank,E. & Gearien,J.E., J. Pharm. Sci., (1973) 62, 847

R545 Coats,E.A., U. Of Cincinnati, Private Communication

R546 Silipo,C. & Hansch,C., Pomona College, Unpublished Analysis

R547 Jow,P. & Hansch,C., Pomona College, Unpublished Analysis

R548 Yoshimoto,M. & Hansch,C., Pomona College, Unpublished Analysis

R549 Poindexter,T. & Hansch,C., Pomona College, Unpublished Analysis

R550 Azzaro,M., Lab. Of Physical Org. Chem., U. Of Nice, France, Private Comm

R551 Schaeffer,J., UCLA Dept. Pharmacology, Private Communication

R552 Vittoria,A., Silipo,C. & Hansch,C., Unpublished Results

R553 Kutter,E., Machleidt,H., Reuter,W., Sauter,R. & Wildfeuer,A.,

Arzneim. Forsch., (1972) 22, 1045

R554 Kaufman,J.J., Koski,W.S., Benson,D.W. & Semo,N., Drug & Alc. Depend., (1975-76) 1, 103

R555 Cory,M. & Henry,D., Stanford Research Inst., Private Communication

R556 Seydel,J.K., Ahrens,H. & Losert,W., J. Med. Chem., (1975) 18, 234

R557 Rockwell,S. & Hansch,C., Pomona College, Unpublished Results

R558 Fujita,T., Nishioka,T. & Nakajima,M., Kyoto Univ., Private Communication

R559 Sparatore,F., Grieco,C., Silipo,C. & Vittoria,A., Farmaco, (1979) 34, 11

R560 Krishnamurthy,T. & Wasik,S.P., J. Environ. Sci. Health, (1978) A13, 595

R561 Lien,E.J., Lien,L.L. & Tong,G.L., J. Med. Chem., (1971) 14, 846

R562 Yamakawa,M. & Hansch,C., Pomona College, Unpublished Results

R563 Hatheway,G.J., Hansch,C., Kim,K.H., Milstein,S.R., Schmidt,C.L., Smith,R.N. & Quinn,F.R., J. Med. Chem., (1978) 21, 563

R564 Panthananickal,A. & Hansch,C., Pomona College, Unpublished Results

R565 Fujinami,A., Satomi,T., Mine,A. & Fujita,T., Pesticide Biochem. Physiol., (1976) 6, 287

R566 Unger,S.H., Cook,J.R. & Hollenberg,J.S., J. Pharm. Sci., (1978) 67, 1364

R567 Takayama,C., Fujita,T. & Nakajima,M., Private Communication

R568 Fink,S. & Hansch,C., Pomona College, Unpublished Analysis

R569 Moser,P., Ciba-Geigy Co., Private Communication

R570 Pratesi,P., Villa,L., Ferri,V., DeMicheli,C., Grana,E., Grieco,C., Silipo,C. & Vittoria,A., Il Farmaco, (1979) 34, 580

R571 Good,P. & Hansch,C., Pomona College, Unpublished Results

R572 Chan,T. & Hansch,C., Pomona College, Unpublished Results

R573 Fauchere,J.L., Do,K.Q., Jow,P.Y.C. & Hansch,C., Experientia, (1980) 36, 1203

R574 Shinozaki,S. & Yoshimoto,M., Sankyo Co. Ltd, Tokyo, Private Communication

R575 Unger,S.H. & Feuerman,T.F., J. Chromatog., (1979) 176, 426

R576 Ogino,A., Matsumura,S. & Fujita,T., J. Med. Chem., (1980) 23, 437

R577 Li,R. & Hansch,C., Pomona College, Unpublished Results

R578 Guo,Z.-R. & Hansch,C., Pomona College, Unpublished Results

R579 Gould,G. & Hansch,C., Pomona College, Unpublished Results

R580 Strong,C. & Hansch,C., Pomona College, Unpublished Results

R581 Hathaway,B. & Hansch,C., Pomona College, Unpublished Results

R588 Takayama,C., Akamatsu, M. & Fujita, T., QSAR, 4, 149 (1985)

R589 Elison,C., Lien,E.J., Zinger,A.P., Hussain,M., Tong,G.L. & Golden,M., J. Pharm. Sci., (1971) 60, 1058

R590 Huang,R. & Hansch,C., Pomona College, Unpublished Results

R591 Bjorkroth,J.P. & Hansch,C., Pomona College, Unpublished Results

R592 Recanatini,M. & Hansch,C., Univ. of Bologna & Pomona College, Unpublished Results

R593 Compadre,R.L. & Hansch,C., Pomona College, Unpublished Results

R594 Debnath,G. & Hansch,C., Pomona College, Unpublished Results

R595 Debnath,A. & Hansch,C., Pomona College, Unpublished Data

R596 Low,C. & Hansch,C., Pomona College, Unpublished Data

R597 Solensky,R. & Hansch,C., Pomona College, Unpublished Results

R599 Vasanwala, R., & Hansch, C., Pomona College, Unpublished Results

R600 Hussain,M., Fukuto,T.R. & Reynolds,H.T., J. Agr. Food Chem., (1974) 22, 225

R601 Sellers,E.M. & Koch-Weser,J., Biochem. Pharmocol., (1974) 23, 553

R602 Nauta,W.T., Bultsma,T., Rekker,R.F. & Timmerman,H., Med. Chem. Spec. Contrib. -Milan, Pratesi, Ed., 1972, p.125

R603 Rekker,R., Gist-Brocades, Netherlands, Private Communication

R604 Mattox,V.R., Goodrich,J.E. & Vrieze,W.D., Steroids, (1971) 18, 147

R605 Asada,S., Fujita,R. & Shirakura,Y., Yakugaku Zasshi, (1974) 94, 80

R606 Remers,W.A. & Schepman,C.S., J. Med.. Chem., (1974) 17, 729

R607 Trepka,R.D., Harrington,J.K. & Belisle,J.W., J. Org. Chem., (1974) 39, 1094

R608 Lin,T.K., Chien,Y.W., Dean,R.R., Dutt,J.E., Sause,H.W., Yen,C.H. & Yonan,P.K., J. Med. Chem., (1974) 17, 751

R609 Aveyard,R. & Mitchell,R.W., Trans. Farad. Soc., (1970) 66, 37

R610 Aveyard,R. & Mitchell,R.W., Trans. Farad. Soc., (1969) 65, 2645

R611 Lin,Y., Hulbert,P.B., Bueding,E. & Robinson,C.H., J. Med. Chem., (1974) 17, 835

R612 Bacher,J.E. & Allen,F.W., J. Biol. Chem., (1951) 188, 59

R613 Vree,T.B., PhD Thesis, Katholieke U. Nijmegen, 1973, p.61

R614 Clarke,K., Cowen,R.A., Gray,G.W. & Osborne,E.H., J. Chem. Soc., (1963), 168

R615 Baldinger,L.H. & Nieuwland,J.A., J. Amer. Pharm. Assoc., (1933) 22, 711

R616 Peter,A., Arzneim. Forsch., (1974) 24, 874

R617 Beckett,A.H. & Dwuma-Badu,D., J. Pharm. Pharmacol., (1969) 21, 162S

R618 Beckett,A.H. & Moffat,A.C., J. Pharm. Pharmacol., (1969) 21, 144S

R619 Ho,Y., Zakrzewski,S. & Mead,L., Biochem., (1973) 12, 1003

R620 Baum,F., Arch. Exp. Path. Pharmakol., (1898) 42, 119

R621 Behrens,W.U., Z. Anal. Chem., (1926) 69, 97

R622 Bittenbender,W.A. & Degering,E.F., J. Amer. Pharm. Assoc., (1939) 28, 514

R623 Biles,J.A., Plakogiannis,F.M., Wong,B.J. & Biles,P.M., J. Pharm. Sci., (1966) 55, 909

R624 Bock,R. & Beilstein,G.M., Z. Anal. Chem., (1963) 192, 44

R625 Bock,R. & Hummel,C., Z. Anal. Chem., (1963) 198, 176

R626 Bock,R. & Jainz,J., Z. Anal. Chem., (1963) 198, 315

R627 Boots,M.R. & Boots,S.G., J. Pharm. Sci., (1969) 58, 553

R628 Brockmann,H., Bauer,K. & Borchers,I., Chem. Ber., (1951) 84, 700

R629 Brodie,B.B., Kurz,H. & Schanker,L.S., J. Pharm. Exp. Ther., (1960) 130, 20

R630 Brown,C.P. & Mathieson,A.R., J. Phys. Chem., (1954) 58, 1057

R631 Bowden,K. & Young,R.C., J. Med. Chem., (1970) 13, 225

R632 Buchi,J., Fischer,G., Mohs,M. & Perlia,X., Arzneim. Forsch., (1969) 19, 1183

R633 Bugarszky, S., Z. Physik. Chem., (1910) 71, 747

R634 Knobloch, E., Et. Al., Czech. Farmacie, 20, 337 (1971)

R635 Calvert,H.T., Z. Physik. Chem., (1901) 38, 513

R636 Kupchan,S.M., Eakin,M.A. & Thomas,A.M., J. Med. Chem., (1971) 14, 1147

R637 Gill,E.W., Jones,G. & Lawrence,D.K., Biochem. Pharmacol., (1973) 22, 175

R638 Colaizzi,J.L. & Klink,P.R., J. Pharm. Sci., (1969) 58, 1184

R639 Kemula,W., Buchowski,H. & Pawlowski,W., Rocz. Chem., (1969) 43, 1555

R640 Oettmeier,W. & Grewe,R., Z. Naturforsch., (1974) 29c, 545

R641 Takiura,K., Honda,S., Yamamoto,M., Takai,H., Kii,M. & Yuki,H., Chem. Pharm. Bull. Japan, (1974) 22, 1618

R642 Crook,E.H., Fordyce,D.B. & Trebbi,G.F., J. Coll. Sci., (1965) 20, 191

R643 Craig,L.C., Hogeboom,G.H., Carpenter,F.H. & DuVigneaud,V., J. Biol. Chem., (1947) 168, 665

R644 Crevar,G.E. & Goettsch,R.W., J. Pharm. Sci., (1966) 55, 446

R645 Cymerman-Craig,J., Rubbo,S.D. & Pierson,B.J., Brit. J. Exp. Pathol., (1954) 35, 478

R646 Cymerman-Craig,J., Rubbo,S.D. & Pierson,B.J., Brit. J. Exp. Pathol., (1955) 36, 254

R647 Cymerman-Craig,J., Rubbo,S.D., Loder,J.W. & Pierson,B.J., Brit. J. Exp. Pathol., (1955) 36, 261

R648 Davies,M. & Griffiths,D.M.L., J. Chem. Soc., (1955), 132

R649 Davies,M. & Griffiths,D.M.L., Z. Physik. Chem. N. F., (1954) 2, 353

R650 Buchowski,H. & Pawlowski,W., Chem. Analityczna, (1972) 17, 557

R651 Okada,J., Esaki,T. & Fujieda,K., Chem. Pharm. Bull., (1976) 24, 61

R652 Abel,G., Connors,T.A., Goddard,P., Hoellinger,H., Nguyen-Hoang-Nam, Pichat,L., Ross,W.C.J. & Wilman,D.E.V., Europ. J. Cancer, (1975) 11, 787

R653 Korenman,Y., Russ. J. Phys. Chem., (1972) 46, 42

R654 Dieckmann,W. & Hardt,A., Chem. Ber., (1919) 52, 1134

R655 Korenman,Y.I. & Koroleva,T.P., Zh. Prikl. Khim., (1975) 48, 1413

R656 Druckrey,E., Schwarz,H. & Leditschke,H., Chimie Therap., (1972) 3, 188

R657 Mohler, W. & Soder, A., Arzneim. Forsch., 21, 1159 (1971)

R658 Drucker, K., Z. Physik. Chem., 49, 563 (1904)

R659 Korenman,Y.I. & Makarova,T.V., Zh. Prikl. Khim., (1974) 47, 1624

R660 Czapkiewicz,J. & Czapkiewicz-Tutaj,B., Roczniki Chem., (1973)

47, 565

R661 Czapkiewicz-Tutaj,B. & Czapkiewicz,J., Roczniki Chem., (1975) 49, 1353

R662 Korenman,Y., Zh. Prikl. Khim., (1974) 47, 1816

R663 Eisenbrand,J. & Picher,H., Arch. Pharm., (1938) 276, 1

R664 Farmer,R.C. & Warth,F.J., J. Chem. Soc., (1904) 85, 1713

R665 Forbes,G.S. & Coolidge,A.S., J. Am. Chem. Soc., (1919) 41, 150

R666 Georgievics,G.V., Z. Physik. Chem., (1915) 90, 47

R667 Goodman,D.S., J. Am. Chem. Soc., (1958) 80, 3887

R668 Goshorn,R.H. & Degering,E.F., J. Amer. Pharm. Assoc., (1938) 27, 865

R669 Grieger,P.F. & Kraus,C.A., J. Am. Chem. Soc., (1949) 71, 1455

R670 Gross,G.C., Degering,E.F. & Tetrault,P.A., Proc. Indiana Acad. Sci., (1939) 42

R671 Greenwald,H.L., Kice,E.B., Kenly,M. & Kelly,J., Anal. Chem., (1961) 33, 465

R672 Kawai,K., Akita,T., Nishibe,S., Nozawa,Y., Ogihara,Y. & Ito,Y., J. Biochem., (1976) 79, 145

R673 Mrongovius,R.I., Eur. J. Med. Chem., (1975) 10, 474

R674 Golumbic,C. & Weller,S., Anal. Chem., (1950) 22, 1418

R675 Moser,P., Jakel,K., Krupp,P., Menasse,R. & Sallmann,A., Eur. J. Med. Chem., (1975) 10, 613

R676 Korenman,Y., Zh. Prikl. Khim., (1974) 47, 2079

R677 Oldendorf,W.H., Proc. Soc. Expl. Biol. Med., (1974) 147, 813

R678 Krieglstein,J., Meiler,W. & Staab,J., Biochem. Pharmacol., (1972) 21, 985

R679 Yeh,K.C. & Higuchi,W.I., J. Pharm. Sci., (1976) 65, 80

R680 Seiler,P., Eur. J. Med. Chem., (1974) 9, 663

R681 Johnson,J.R., Kilpatrick,P.J., Christian,S.D. & Affsprung,H.E., J. Phys. Chem., (1968) 72, 3223

R682 Larson,D.L. & Portoghese,P.S., J. Med. Chem., (1976) 19, 16

R683 Handschumacher,R.E. & Vane,J.R., J. Pharmac. Chemother., (1967) 29, 105

R684 Hantzsch,A. & Sebaldt,F., Z. Physik. Chem., (1899) 30, 258

R685 Hantzsch,A. & Vagt,A., Z. Physik. Chem., (1901) 38, 705

R686 Herz,W. & Fischer,H., Chem. Ber., (1905) 38, 1138

R687 Herz,W. & Lewy,M., Z. Fur Elektrochem., (1905) 46, 818

R688 Herz,W. & Schuftan,P., Z. Physik. Chem., (1922) 101, 269

R689 Herz,W. & Kurzer,A., Zeit. Fur Elektrochem., (1910) 16, 240 & 869

R690 Johnson,H.L., Tsakotellis,P. & Degraw,J., J. Pharm. Sci., (1970) 59, 278

R691 Jones,O.T.G. & Watson,W.A., Biochem. J., (1967) 102, 564

R692 Williams,C.H. & Lawson,J., Biochem. Pharmacol., (1975) 24, 1889

R693 Lombardino,J.G., Otterness,I.G. & Wiseman,E.H., Arzneim. Forsch., (1975) 25, 1629

R694 Sanvordeker,D.R., Chien,Y.W., Lin,T.K. & Lambert,H.J., J. Pharm. Sci., (1975) 64, 1797

R695 Nambu,N., Sakurai,S. & Nagai,T., Chem. Pharm. Bull. Japan, (1975) 23, 1404

R696 Kakemi,K., Sezaki,H., Muranishi,S. & Tsujimura,Y., Chem. Pharm. Bull., (1969) 17, 1650

R697 Kakemi,K., Sezaki,H., Okumura,K. & Ashida,S., Chem. Pharm. Bull., (1969) 17, 1332

R698 Katz,M. & Shaikh,Z.I., J. Pharm. Sci., (1965) 54, 591

R699 Kakemi,K., Arita,T., Kitazawa,S., Kawamura,M. & Takenaka,H., Chem. Pharm. Bull., (1967) 15, 1819

R700 Kakemi,K., Arita,T., Kitazawa,S. & Sagawa,Y., Chem. Pharm. Bull., (1967) 15, 1828

R701 Klobbie,E.A., Z. Physik. Chem., (1897) 24, 615

R702 Houston,J.B., Upshall,D.G. & Bridges,J.W., J. Pharmacol. Exp. Ther., (1975) 195, 67

R703 Sherrill,B.C. & Dietschy,J.M., J. Membrane Biol., (1975) 23, 367

R704 Gilbert,D., Goodford,P.J., Norrington,F.E., Weatherley,B.C. & Williams,S.G., Br. J. Pharmac., (1975) 55, 117

R705 Siekierski,S. & Olszer,R., J. Inorg. Nucl. Chem., (1963) 25, 1351

R706 Scheuplein,R.J., J. Invest. Dermat., (1965) 45, 334

R707 Knaffl-Lenz,E., Arch. Exptl. Path. Pharmacol., (1919) 84, 66

R708 Mathison,I.W. & Tidwell,R.R., J. Med. Chem., (1975) 18, 1227

R709 Herz,A. & Teschemacher,H.J., Adv. Drug Res., (1971) 6, 79

R710 Kolosovskii, N. & Andryushchenko, S., J. Gen. Chem., 481070 (1934)

R711 Gadalla,M., Saleh,A.M. & Motawi,M.M., Pharmazie, (1974) 29, 111

R712 Ploquin,J., Sparfel,L., LeBaut,G., Floc'h,R., Welin,L., Petit,J.Y. & Henry,N., Eur. J. Med. Chem., (1974) 9, 526

R713 Kolosovskii,N. & Ponomareva,N., J. Gen. Chem., (1934) 4, 1077

R714 Krieglstein,J. & Kuschinsky,G., N. S. Arch. Pharmak. Exp. Path., (1969) 262, 1

R715 Braddock,P.D., Connors,T.A., Jones,M., Khokhar,A.R., Melzack,D.H. & Tobe,M.L., Chem. Biol. Interactions, (1975) 11, 145

R716 Miko,M. & Chance,B., Biochim. Biophys. Acta, (1975) 396, 165

R717 Landholt-Bornstein, Zahlenwerte & Funktionen, Vol 2, Pt 2, 698

R718 Lawrence,J.H., Loomis,W.F., Tobias,C.A. & Turpin,F.H., J. Physiol., (1946) 105, 197

R719 Lukens,R.J. & Horsfall,J.G., Phytopathol., (1967) 57, 876

R720 Stoughton,R.B., Clendenning,W.E. & Kruse,D.,

R721 Feldman,S., DeFrancisco,M. & Cascella,P.J., J. Pharm. Sci., (1975) 64, 1713

R722 Chien,Y.W., Akers,M.J. & Yonan,P.K., J. Pharm. Sci., (1975) 64, 1632

R723 McCall,J.M., J. Med. Chem., (1975) 18, 549

R724 Mandel,H.G., Alpen,E.L., Winters,W.D. & Smith,P.K., J. Biol. Chem., (1951) 193, 63

R725 Seydel,J.K., Schaper,K.J., Wempe,E. & Cordes,H.P., J. Med. Chem., (1976) 19, 483

R726 Umeyama,H., Nagai,T. & Nogami,H., Chem. Pharm. Bull. Japan, (1971) 19, 1714

R727 Korenman,Y., Zh. Prikl. Khim., (1972) 45, 2031

R728 Martin,D., Birkhahn,H.J. & Niclas,H.J., J. Prakt. Chem., (1975) 317, 177

R729 Kuznetsova,E.M. & Gurarii,L.L., Russ. J. Phys. Chem., (1971) 45, 1761

R730 Melchakova,N., Et. Al., Chem. Abs., 57, 2910 (1962)

R731 Rich,D.H., Gesellchen,P.D., Tong,A., Cheung,A. & Buckner,C.K., J. Med. Chem., (1975) 18, 1004

R732 Meyer,H., N. S. Archiv. Exp. Path. Pharmakol., (1901) 46, 338

R733 Michaelis,A.F. & Higuchi,T., J. Pharm. Sci., (1969) 58, 201

R734 Mukerjee,P., J. Phys. Chem., (1965) 69, 2821

R735 Merki,F., Buchi,J. & Perlia,X., Arzneim. Forsch., (1975) 25, 1233

R736 Levy,J.V., Br. J. Pharmac., (1970) 38, 743

R737 Plakogiannis,F.M., Lien,E.J., Harris,C. & Biles,J.A., J. Pharm. Sci., (1970) 59, 197

R738 Nakano,M. & Patel,N.K., J. Pharm. Sci., (1970) 59, 77

R739 Odaira,I., Mem. Coll. Sci. (kyoto), (1916) 1, 324

R740 Panova,M. & Levin,V., Radiokhimiya, (1960) 2, 568, Chem. Abs., (1962) 57, 126

R741 Parker,V.H., Biochem. J., (1965) 97, 658

R742 Portoghese,P.S., Mikhail,A.A. & Kupferberg,H.J., J. Med. Chem., (1968) 11, 219

R743 Pinney,R.J. & Walters,V., J. Pharm. Pharmacol., (1969) 21, 415

R744 Pinnow,J., Z. Anal. Chem., (1915) 54, 321

R745 Pinnow,J., Z. Anal. Chem., (1911) 50, 162

R746 Renkin,E.M., Am. J. Physiol., (1952) 168, 538

R747 Richmond,D.V., Somers,E. & Zaracovitis,C., Nature, (1964) 204, 1329

R748 Robinson,R.A., J. Chem. Soc., (1952) 253

R749 Rothmund,V. & Drucker,K., Z. Physik. Chem., (1903) 46, 827

R750 Rothmund,V. & Wilsmore,N.T.M., Z. Phys. Chem., (1902) 40, 611

R751 Ross,E.J., J. Physiol., (1951) 112, 229

R752 Sandberg,F., Acta Physiol. Scand., (1951) 24, 7

R753 Sandell,K.B., Monatsh. Chem., (1958) 89, 36

R754 Sandell,K., Naturwissen., (1964) 51, 336

R755 Schilow,N. & Lepin,L., Z. Physik. Chem., (1922) 101, 353

R756 Belskii,V.E. & Motygullin,G.Z., Russ. J. Phys. Chem., (1969) 43, 1353

R757 Yaguzhinsky,L.S., Smirnova,E.G., Ratnikova,L.A., Kolesova,G.M. & Krasinskaya,I.P., Bioenergetics, (1973) 5, 163

R758 Katz,Y. & Diamond,J.M., J. Membr. Biol., (1974) 17, 101

R759 Schultz,O.E., Fedders,S., Holm,W.D. & Schulze,V., Arzneim. Forsch., (1974) 24, 1933

R760 Scholtan,W., Schlomann,K. & Rosenkranz,H., Arzneim.Forsch., (1968) 18, 767

R761 Schafer,G. & Bojanowski,D., Eur. J. Biochem., (1972) 27, 364

R762 Huyskens,P.L. & Tack,J.J., J. Phys. Chem., (1975) 79, 1654

R763 Korenman,Y. & Sotnikova,N.G., Zh. Prikl. Khim., (1975) 48, 195

R764 Seeman,P., Staiman,A. & Chau-Wong,M., J. Pharm. Exp. Ther., (1974) 190, 123

R765 Korenman,Y.I. & Udalova,V.Y., Zh. Fizich. Khim., (1974) 48, 1223

R766 Tichy,M., Coll. Czech. Chem. Comm., (1974) 39, 935

R767 Korenman,Y.I., Zh. Prikl. Khim., (1970) 43, 1410

R768 Korenman,I.M. & Dobromyslova,T.M., Zh. Prikl. Khim., (1975) 48, 2711

R769 Apsitis,A.A., Dorfman,K.Y. & Oshkaya,V.P., Zh. Obshch. Khim., (1975) 45, 2074

R770 Bracha,P. & O'Brien,R.D., J. Econ. Entomol., (1966) 59, 1255

R771 Weiner,I.M., Washington,J.A. & Mudge,G.H., Johns Hopkins Hospital Bull., (1960) 106, 333

R772 Illing,H.P.A. & Benford,D., Biochim. Biophys. Acta, (1976) 429, 768

R773 King,R.W. & Burgen,A.S.V., Proc. R. Soc. London, (1976) B193, 107

R774 Iguchi,A., Kagaku Sochi, (1975) 17, 80

R775 Millner,O.E. & Purcell,W.P., J. Pharm. Sci., (1976) 65, 910

R776 Taraszka,M.J., J. Pharm. Sci., (1970) 59, 873

R777 Pavelka,S. & Kovar,J., Coll. Czeck. Chem. Comm., (1975) 40, 753

R778 Wakabayashi,T., Bull. Chem. Soc. Japan, (1967) 40, 2836

R779 Omori,T., Wakahayashi,T., Oki,S. & Suzuki,N., J. Inorg. Nucl. Chem., (1964) 26, 2265

R780 Mottola,H.A. & Freiser,H., Talanta, (1967) 14, 864

R781 Gibbons,L.K., Koldenhoven,E.F., Nethery,A.A., Montgomery,R.E. & Purcell,W.P., J. Agric. Food Chem., (1976) 24, 203

R782 Dearden,J.C. & Wootton,J.C., J. Pharm. Pharmacol., (1975) 27, 30p

R783 Laduron,P., J. Pharm. Pharmacol., (1976) 28, 250

R784 Sumarokov, V. & Volodutskaya, Z., Chem. Abs., 54, 8225 (1960)

R785 Maxwell,R.A., Ferris,R.M., Burcsu,J., Woodward,E.C., Tang,D. & Williard,K., J. Pharmacol. Exp. Ther., (1974) 191, 418

R786 Kuo,K.H., Fukuto,T.R., Miller,T.A. & Bruner,L.J., Biophys. J., (1976) 16, 143

R787 Waring,M.J., Wakelin,L.P.G. & Lee,J.S., Biochim. Biophys. Acta, (1975) 407, 200

R788 Temple,D.L. & Comer,W.T., Ferguson,H.C. & Allen,L.E., J. Med. Chem., (1976) 19, 626

R789 Taubmann,A., Z. Physik. Chem., (1932) 161, 141

R790 Tinker,J.F. & Brown,G.B., J. Biol. Chem., (1948) 173, 585

R791 Traube,J., Archiv. Fur Physiologie, (1904) 105, 541

R792 Bachur,N.R., Steele,M., Meriwether,W.D. & Hildebrand,R.C., J. Med. Chem., (1976) 19, 651

R793 Vree,T.B., Muskens,A., Van Rossum,J.M., J. Pharm. Pharmacol., (1969) 21, 774

R794 Valette, G. & Etcheverry, J., Compt. Rend. Soc. Biol., 152, 315 (1958)

R795 Van Duyne,R., Taylor,S.A., Christian,S.D. & Affsprung,H.E., J. Phys. Chem., (1967) 71, 3427

R796 Voerman,S., Bull. Environ. Contam. & Toxicol., (1969) 4, 64

R797 Williams,G. & Soper,F.G., J. Chem. Soc., (1930), 2469

R798 Whitehouse,M.W. & Leader,J.E., Biochem. Pharmacol., (1967) 16, 537

R799 Wulfert,E., Bolla,P. & Mathieu,J., Bull. De Chimie Theraput., (1969) 4, 257

R800 Washitake,M., Anmo,T., Tanaka,I., Arita,T. & Nakano,M., J. Pharm. Sci., (1975) 64, 397

R801 Wechter,W.J., Johnson,M.A., Hall,C.M., Warner,D.T., Berger,A.E., Wenzel,A.H., Gish,D.T. & Neil,G.L., J. Med. Chem., (1975) 18, 339

R802 Albert,A. Et. Al., Brit. J. Exp. Pathol., 35, 75 (1954)

R803 Dearden,J.C. & O'Hara,J.H., Eur. J. Med. Chem., (1978) 13, 415

R804 Tanaka,T., Kobayashi,H., Okumura,K., Muranishi,S. & Sezaki,H., Chem. Pharm. Bull. Japan, (1974) 22, 1275

R805 Ozeki,S. & Tejima,K., Chem. Pharm. Bull. Japan, (1974) 22, 1297

R806 Chien,Y.W., Lambert,H.J. & Karim,A., J. Pharm. Sci., (1974) 63, 1877

R807 Takiura,K., Yamamoto,M., Miyaji,Y., Takai,H., Honda,S. & Yuki,H., Chem. Pharm. Bull. Japan, (1974) 22, 2451

R808 Takiura,K., Yuki,H., Okamoto,Y., Takai,H. & Honda,S., Chem. Pharm. Bull. Japan, (1974) 22, 2263

R809 Stewart,J.M., Ryan,J.W. & Brady,A.H., J. Med. Chem., (1974) 17, 537

R810 Drobnica,L., Chance,B. & Scarpa,A., Johnson Res. Foundation, Private Commun

R811 Plakogiannis,F.M. & Lien,E.J., Acta Pharm. Suecica, (1974) 11, 191

R812 Brodin,A., Acta Pharm. Suecica, (1974) 11, 141

R813 Falk,M., Furst,W., Hannig,E. & Bohm,R., Pharmazie, (1974) 29, 542

R814 Murphy,F.R., Krupa,V. & Marks,G.S., Biochem. Pharmacol.,

R815 Buchi,J., Koller,R.J. & Perlia,X., Arzneim. Forsch., (1975) 25, 14

R816 Bowen,D.B., James,K.C. & Roberts,M., J. Pharm. Pharmacol., (1970) 22, 518

R817 Lippold,B.C. & Schneider,G.F., Arzneim. Forsch., (1974) 24, 1952

R818 Franke,R., Zschiesche,W., Augsten,K., Guttner,J. & Hesse G., Res. J. Reticuloendothelial Soc., (1974) 16, 87

R819 Cho,A.K. & Miwa,G.T., Drug Metab. & Disposition, (1974) 2, 477

R820 Al-Gailany,K.A.S., Bridges,J.W. & Netter,K.J., Biochem. Pharmacol., (1975) 24, 867

R821 Rouot,B., Leclerc,G., Wermuth,C.G., Miesch,F. & Schwartz,J., J. Pharmacol., (1977) 8, 95

R822 Bechgaard,H. & Lund-Jensen,C., Eur. J. Med. Chem., (1975) 10, 103

R823 Boudier,H.S., de Boer,J., Smeets,G., Lien,E. & van Rossum,J., Life Sci., (1976) 17, 377

R824 Hauska,G., Febs Letters, (1972) 28, 217

R825 Holmes,H.L. & Lough,C.E., Suffield Tech. Note# 365, DRES, Ralston Canada (1976)

R826 Warner,V.D., Musto,J.D., Turesky,S.S. & Soloway,B., J. Pharm. Sci., (1975) 64, 1563

R827 Owellen,R.J., Donigian,D.W., Hartke,C.A. & Hains,F.O., Biochem. Pharmacol., (1977) 26, 1213

R828 Beckett,A.H., Gorrod,J.W. & Taylor,D.C., J. Pharm. Pharmacol., (1972) 24, 65P

R829 Beckett,A.H., Taylor,D.C. & Gorrod,J.W., J. Pharm. Pharmacol., (1975) 27, 588

R830 Loggia,R.D., Furlani,A., Savastano,F. & Scarcia,V., Experientia, (1976) 32, 636

R831 Warner,V.D., Lynch,D.M. & Ajemian,R.S., J. Pharm. Sci., (1976) 65, 1070

R832 Yang,C.F. & Sun,Y.P., Arch. Envir. Contam. Toxicol., (1977) 6, 325

R833 Van Damme,M., Hanocq,M., Topart,J. & Molle,L., Eur. J. Med. Chem., (1976) 11, 209

R834 Fowler,J.S., Gallagher,B.M., MacGregor,R.R. & Wolf,A.P., J. Pharm. Exp. Ther., (1976) 198, 133

R835 Miller,M., Howes,Jr.,H.L., Kasubick,R.V & English,A.R., J. Med. Chem., (1970) 13, 849

R837 Mullens,J., Hanssens,I. & Huyskens,P., Bull. Soc. Chim. Belges., (1971) 79, 539

R838 Burstein,S., Science, (1956) 124, 1030

R839 Abrams,D.S. & Prausnitz,J.M., J. Chem. Thermodynamics, (1975) 7, 61

R840 Bridges,J.W., Walker,S.R. & Williams,R.T., Biochem. J., (1969) 111, 173

R841 Nathansohn,G., Pasqualucci,C.R., Radaelli,P. & Schiatti,P., Steroids, (1968) 13, 365

R842 Engel,L.L., Physical Properties Of Steroid Hormones, MacMillan, (1963) Vol. 3, p.1

R843 Mrongovius,R.I., Gaff,G.A. & Rand,M.J., Clin. Exp. Pharmacol. Physiol., (1976) 3, 443

R844 Koch,H. & Bodmann,R., Arch. Pharm., (1976) 309, 812

R845 Azaz,E. & Donbrow,M., J. Coll. & Interface Sci., (1976) 57, 11

R846 Sobotka,P. & Safanda,J., J. Molecular Med., (1976) 1, 151

R847 Ploquin,J., Sparfel,L., LeBaut,G., Floc'h,R., Welin,L., Petit,J.Y. & Henry,N., Eur. J. Med. Chem., (1976) 11, 407

R848 Weitl,F.L., Sovak,M., Williams,T.M. & Lang,J.H., J. Med. Chem., (1976) 19, 1359

R849 Mirvish,S.S., Issenberg,P. & Sornson,H.C., J. Natl. Cancer Inst., (1976) 56, 1125

R850 Drobnica,L., Augustin,J. & Nemec,P., Proc. Qsar Conf., Prague, M.Tichy Ed., Birkhauser (1976)

R851 Hudson,R.F., Timmis,G.M. & Marshall,R.D., Biochem. Pharmacol., (1958) 1, 48

R852 Korenman,I.M. & Gorokhov,A.A., Zh. Obshch. Khim., (1969) 41, 948

R853 Norrington,F.E., Hyde,R.M., Williams,S.G. & Wootton,R., J. Med. Chem., (1975) 18, 604

R854 Dorsey,W.S. & Lucas,H.J., J. Am. Chem. Soc., (1956) 78, 1665

R855 Eberz,W.F., Welge,H.J., Yost,D.M. & Lucas,H.J., J. Am. Chem. Soc., (1937) 59, 45

R856 Winstein,S. & Lucas,H.J., J. Am. Chem. Soc., (1938) 60, 836

R857 Trueblood,K.N. & Lucas,H.J., J. Am. Chem. Soc., (1952) 74, 1338

R858 Lucas,H.J., Moore,R.S. & Pressman,D., J. Am. Chem. Soc., (1943) 65, 227

(1975) 24, 883

R859 Hepner,F.R., Trueblood,K.N. & Lucas,H.J., J. Am. Chem. Soc., (1952) 74, 1333

R860 Maren,T.H., Handbook Expl. Pharmacol., (1969) 24, 13

R861 Buchi,J., Perlia,X. & Portmann,R., Arzneim. Forsch., (1967) 17, 613

R862 Buchi,J., Cordes,G. & Perlia,X., Arzneim. Forsch., (1964) 14, 161

R863 Hulshoff,A. & Perrin,J.H., J. Chromatog., (1976) 129, 263

R864 Korenman,Y.I., Kotelyanskaya,E.B. & Nefedova,T.A., Zh. Prikl, Khim., (1976) 49, 1112

R865 Butler,T.C., Dudley,K.H., Johnson,D. & Roberts,S.B., J. Pharmacol. Exp. Ther., (1976) 199, 82

R866 Seydel,J.K. & Butte,W., J. Med. Chem., (1977) 20, 439

R867 Manabe,M., Koda,M. & Shirahama,K., Bull. Chem. Soc. Japan, (1975) 48, 3553

R868 Nichols,D.E. & Dyer,D.C., J. Med. Chem., (1977) 20, 299

R869 Garrett,E.R. & Chemburkar,P.B., J. Pharm. Sci., (1968) 57, 949

R870 Rekker, R., "The Hydrophobic Fragmental Constant", Elsevier (1977) 60, 146, 157

R871 Singer,G.M., Taylor,H.W. & Lijinsky,W., Chem. Biol. Interactions, (1977) 19, 133

R872 Yamana,T., Tsuji,A., Miyamoto,E. & Kubo,O., J. Pharm. Sci., (1977) 66, 747

R873 Pesak,M., Kopecky,F., Celechovsky,J., Benes,L. & Borovansky,A., Pharmazie, (1976) 31, 24

R874 Smulskii,S.P., Bogat-skii,A.A.V., Andronati,S.A., Vikhlyaev,Y.I. & Zhilina,Z.I., Dokl. Akad. Nauk. SSSR, (1977) 235, 369

R875 Lippold,B.C. & Schneider,G.F., Pharmazie, (1976) 31, 237

R876 Filippov,M.P., Zaitseva,L.F., Zaitseva,Z.V. & Chukur,A.P., Zh. Anal. Khim., (1965) 20, 132, 118ee

R877 Kojima,I., Yoshida,M. & Tanaka,M., J. Inorg. Nucl. Chem., (1970) 32, 987

R878 Schonenberger,H. & Bastug,T., Arzneim. Forsch., (1970) 20, 386

R879 Korenman,I.M., Russ. J. Phys. Chem., (1971) 45, 795ee

R880 Eberlein,W., Schmidt,G., Reuter,A. & Kutter,E., Arzneim. Forsch., (1977) 27, 356

R881 Tomita,A., Ebina,N. & Tamai,Y., J. Am. Chem. Soc., (1977) 99, 5725

R882 Thoma,K. & Arning,M., Arch. Pharm., (1976) 309, 945

R883 Julkunen,R.J.K., Med. Biology, (1977) 55, 41

R884 Nichol,C.A., Cavallito,J.C., Woolley,J.L. & Sigel,C.W., Cancer Treat. Rpts., (1977) 61, 559

R885 Hiles,R.A., Fd. Cosmet. Toxicol., (1977) 15, 205

R886 Korenman,Y.I., Zh. Prikl. Khim., (1973) 46, 380, 394ee

R887 Korenman,Y.I., Zh. Prikl. Khim., (1973) 46, 1305, 1391ee

R888 Korenman,I.M. & Gryaznova,M.I., Zh. Prikl. Khim., (1971) 44, 2352, 2415ee

R889 Korenman,Y.I., Zh. Prikl. Khim., (1973) 46, 2599, 2754ee

R890 Korenman,Y.I. & Pazynin,V.T., Zh. Anal. Khim., (1972) 27, 814, 723ee

R891 Korenman,I.M., Gurevich,N.Y. & Kulagina,T.G., Russ. J. Phys. Chem., (1972) 46, 1523ee

R892 Korenman,Y.I., Petrov,A.T. & Anokhina,N.A., Zh. Prikl. Khim., (1977) 50, 139, 132ee

R893 Korenman,Y.I. & Gorokhov,A.A., Zh. Prikl. Khim., (1973) 46, 2597, 2751ee

R894 Korenman,I.M. & Karyakina,L.N., Russ. J. Phys. Chem., (1973) 47, 850ee

R895 Korenman,I.M., Gurev,I.A. & Gureva,Z.M., Russ. J. Phys. Chem., (1973) 47, 682ee

R896 Korenman,I.M. & Gryaznova,M.I., Zh. Anal. Khim., (1974) 29, 964, 823ee

R897 Taylor,P., ICI Pharmaceuticals, Private Communication

R898 Miller,K. & Korten,K., Harvard Medical School, Private Communication

R899 Yih,T.D. & Van Rossum,J.M., J. Pharm. Exper. Therap., (1977) 203, 184, Biochem. Phar., (1977) 26, 2117

R900 Seidell,A., "Solubilities Of Organic Compounds", Van Nostrand, N.Y. 1941

R901 Dunn III,W.J. & Greenberg,M.J., J. Pharm. Sci., (1977) 66, 1416

R902 Seawright,J.A., Bowman,M.C. & Lofgren,C.S., J. Econ. Entom., (1973) 66, 613

R903 Howard,R.A., Sherwood,E., Erck,A., Kimball,A.P. & Bear,J.L., J. Med. Chem., (1977) 20, 943

R904 Thoma,K. & Arning,M., Arch. Pharm., (1976) 309, 872

R905 Hirose,K. & Tanaka,M., J. Inorg. Nucl. Chem., (1976) 38, 2285

R906 Seebald,H. & Forth,W., Arzneim. Forsch., (1977) 27, 624

R907 Miller,G.H., Smith,H.L., Rock,W.L. & Hedberg,S., J. Pharm. Sci., (1977) 66, 88

R908 Druckrey,H., Preussmann,R., Ivankovic,S. & Schmahl,D., Zeit. Krebsforsch., (1967) 69, 103

R909 Hille,B., J. Gen. Physiol., (1977) 69, 475

R910 Akihama,S., Meiji Yakka Daigaku Kenkyu Kiyo, (1975) 5, 5

R911 Cooke,D.T. & Gonda,I., J. Pharm. Pharmacol., (1977) 29, 190

R912 Oliver-Droz,P. & Fernandez,J., Helv. Chim. Acta, (1977) 60, 454

R913 Busev,A.I, Tereschchenko,A.P., Krotova,N.B., Nol'de,T.V., Byr'ko,V.M. & Puzanova,O.G., Zh. Anal. Khim., (1973) 28, 858, 765ee

R914 Sallee,V.L., J. Lipid Res., (1974) 15, 56

R915 Yalkowsky,S.H., Flynn,G.L. & Slunick,T.G., J. Pharm. Sci., (1972) 61, 852

R916 Al-Niaimi,N.S., Al-Karaghouli,R. & Aliwi,S.M., J. Chem. Eng. Data, (1973) 18, 182

R917 Purich,E.D., Colaizzi,J.L. & Poust,R.I., J. Pharm. Sci., (1973) 62, 545

R918 Flis,B.I. & Mukhin,N.N., Zh. Prikl. Khim., (1973) 46, 45, 42ee

R919 Lepetit,G., Pharmazie, (1977) 32, 289

R920 Otomo,M. & Kodama,K., Bull. Chem. Soc. Japan, (1973) 46, 2421

R921 Hirom,P.C., Hughes,R.D. & Millburn,P., Biochem. Soc. Trans.,(1974) 2, 327

R922 Kuznetsova,E.M., Medvedeva,L.V. & Soloveichik,S.A., Zh. Fizich. Khim., (1975) 49, 452, 261ee

R923 Iwachido,T., Bull. Chem. Soc. Japan, (1973) 46, 2761

R924 Ashton,B. & Taylor,P., ICI Pharmaceuticals, Private Communication

R925 Anderson,N., ICI Plant Protection, Private Communication

R926 Foye,W.O., "Principles of Medicinal Chemistry", Lea & Febiger, Philadelphia (1974) P. 146

R927 Raventos,J., Brit. J. Pharmacol., (1956) 11, 394

R928 Chiou,C.T., Freed,V.H., Schmedding,D.W. & Kohnert,R.L., Environ. Sci. & Techn., (1977) 11, 475

R929 Ho,N.F.H., Park,J.Y., Morozowich,W. & Higuchi,W.I., "Design Of Biopharm. Prop. Thru Prodrugs & Analogs" Roche, Ed. APA (1977) P. 136

R930 Bohringer,H. & Vogt,H., Arch. Pharm., (1977) 310, 235

R931 Suarez,C., Di Parsia,M.T. & Marquez,V.E., J. Heterocyclic Chem., (1978) 15, 1093

R932 Schroder-Nielson,M., Acta Pharm. Sucecica, (1976) 13, 133

R933 Tisdale,M.J., Elson,L.A. & Ross,W.C.J., Europ. J. Cancer, (1973) 9, 89

R934 Hagemann,R., Meyer,A., Barth,A. & Franke,R., Pharmazie, (1977) 32, 526

R935 Tsuji,A., Kubo,O., Miyamoto,E. & Yamana,T., J. Pharm. Sci., (1977) 66, 1675

R936 Ryrfeldt,A., J. Pharm. Pharmacol, (1971) 23, 463

R937 Inui,K-I., Tabara,K., Hori,R., Kaneda,A., Muranishi,S. & Sezaki,H., J. Pharm. Pharmacol., (1977) 29, 22

R938 Stillwell,W., BioSystems, (1976) 8, 111

R939 Bennett,L.J. & Miller,K.W., Harvard Medical School, Private Communication

R940 Terada,H. & Inagi,T., Chem. Pharm. Bull. Japan, (1975) 23, 1960

R941 Kiwan,A.M. & Kassim,A.Y., Talanta, (1975) 22, 931

R942 Won,K.W. & Prausnitz,J.M., J. Chem. Thermod., (1975) 7, 661

R943 Korenman,I.M., Gur'ev,I.A. & Denisova,G.S., Zh. Fizich. Khim., (1976) 50, 401, 235ee

R944 Tanifuji,Y., Eger,E.I. & Terrell,R.C., Anesth. & Analg. Curr. Res., (1977) 56, 387

R945 Mundy,R.L., Bowman,M.C., Farmer,J.H., Haley,T.J., Arch.Toxicol., 41, 111 (1978)

R946 Nakano,M., Uematsu,Y. & Arita,T., Chem. Pharm. Bull. Japan, (1977) 25, 1109

R947 Nakamura,J., Muranushi,N., Kimura,T., Muranishi,S. & Sezaki,H., Chem. Pharm. Bull. Japan, (1977) 25, 851

R948 Suzuki,E., Tsukigi,M., Muranishi,S., Sezaki,H. & Kakemi,K., J. Pharm. Pharmacol, (1972) 24, 138

R949 Malik,S.N., Canaham,D.H. & Gouda,M.W., J. Pharm. Sci., (1975) 64, 987

R950 Gabbay,E.J., Grier,D., Fingerle,R.E., Reimer,R., Levy,R., Pearce,S.W. & Wilson,W.D., Biochemistry, (1976) 15, 2062

R951 Lu,P-Y. & Metcalf,R.L., Envir. Health Perspect. (1975) 10, 269

R952 Radecki,A., Kaczmarek,B. & Grzybowski,J., J. Chem. Eng. Data, (1975) 20, 163

R953 Jun,H.W., Iturrian,W.B., Stewart,J.T. & Lee,B.H., J. Pharm. Sci.,

(1975) 64, 1843

R954 Kuznetsova,E.M., Zh. Fizich. Khim., (1975) 49, 408, 239ee

R955 Arro,J. & Molder,L., Zh. Fizich. Khim., (1975) 49, 1077, 635ee

R956 Apsitis,A., Bychkova,N.N. & Oskaja,V., Latv. Psr. Zinat. Akad. Ves. Kim. Ser., (1974) 5, 545,Chem. Abs., (1975) 82, 57324f

R957 Vlacil,F., Sayeh,B.M. & Koucky,J., Coll. Czech. Chem. Comm., (1975) 40, 1345

R958 Levina,K.S. & Zheleznyak,A.S., Zh. Fizich. Khim., (1975) 49, 849, 499ee

R959 Korenman,Y.I., Butyaeva,I.I. & Gel'fand,M.M., Zh. Prikl. Khim., (1974) 47, 473, 475ee

R960 Spivakov,B.Y., Stoyanov,E.S. & Zolotov,Y.A., Dokl. Akad. Nauk. SSSR, (1975) 220, 392, 87ee

R961 Pugach,L.M., Ogorodnikov,S.K. & Idlis,G.S., Zh. Prikl. Khim., (1974) 47, 1856, 1903ee

R962 Asada,S., Kashimoto,A. & Yamada,T., Yakugaku Zasshi, (1976) 96, 1169

R963 Saikawa,I., Yasuda,T., Taki,H., Tai,M., Watanabe,Y., Sakai,H., Takano,S., Yoshida,C. & Kasuya,K., Yakugaku Zasshi, (1977) 97, 987

R964 Cumber,A.J. & Ross,W.C.J., Chem. Biol. Interactions, (1977) 17, 349

R965 Mannhold,R., Steiner,R., Haas,W. & Kaufmann,R., N-S Arch. Pharmacol, (1978) 302, 217

R966 Strahl,N.R. & Lopez,S., J. Pharm. Sci., (1978) 67, 1041

R967 Gabbay,E.J., Int. J. Quantum Chem: Quantum Biol. Symp., (1976) 3, 217

R968 Hellenbrecht,D. & Gortner,L., Pol. J. Pharmacol. Pharm., (1976) 28, 625

R969 Granik,V.G., Altshuler,R.A., Persianova,I.V., Mashkovsky,M.D. & Sheinker,Y.N., Farmak. Toksikol., (1977) 40, 31

R970 Mahju,M.A. & Maickel,R.P., Biochem. Pharmacol., (1969) 18, 2701

R971 Sanvordeker,D.R., Pophristov,S. & Christensen,A., Drug Devel. Ind. Pharm., (1977) 3, 149

R972 Watanabe,J. & Kozaki,A., Chem. Pharm. Bull. Japan, (1978) 26, 665

R973 Makitra,R.G., Pirig,Y.N., Zeliznyj,A.M., de Aguar,M.A.D., Mikolajev,V.L. & Romanov,V.A., Org. Reactivity, (1977) 14, 421

R974 Orchin,M. & Golumbic,C., J. Am. Chem. Soc., (1949) 71, 4151

R975 Sato,A. & Nakajima,T., Jap. J. Ind. Health, (1977) 19, 194

R976 Timmermans,P., Brands,A. & van Zwieten,P.A., N-S. Arch. Pharmacol., (1977) 300, 217

R977 Zeller,W.J., Eisenbrand,G. & Fiebig,H.H., J. Natl. Cancer Inst., (1978) 60, 345

R978 Burkartsmaier,A. & Mutschler,E., Arch. Pharm. (1978) 311, 161

R979 Morimoto,K., Yamaha,T., Nakadate,M. & Suzuki,I., Gann, (1978) 69, 139

R980 Stary,J. & Kratzer,K., Radiochem. Radioanal. Letters, (1977) 28, 53

R981 Kozhukhov,A.N., Meiren,D.V. & Gilev,A.P., Bull. Eksper. Biol. Med., (1977) 83, 734, 872ee

R982 Nakano,M., Yamamoto,M. & Arita,T., Chem. Pharm. Bull. Japan, (1978) 26, 1505

R983 Sikk,P.F., Abduvakhabov,A.A. & Aaviksaar,A.A., Org. Reactivity, (1975) 12, 421

R984 Walsh,R.J.A., Wooldridge,K.R.H., Jackson,D. & Gilmour,J., Eur. J. Med. Chem., (1977) 12, 495

R985 Voss,G., Analyst, (1978) 103, 233

R986 Leonard,R.H., Peterson,W.H. & Johnson,M.J., Ind. Eng. Chem., (1948) 40, 57

R987 Whitehead,K.E. & Geankoplis,C.J., Ind. Eng. Chem., (1955) 47, 2114

R988 Moeller,T. & Pundsack,F.L., J. Am. Chem. Soc., (1953) 75, 2258

R989 Gibson,N.A. & Weatherburn,D.C., Anal. Chim. Acta, (1972) 58, 159

R990 Pagel,H.A. & McLafferty,F.W., Anal. Chem., (1948) 20, 272

R991 Schweitzer,G.K. & Morris,D.K., Anal. Chim. Acta, (1969) 45, 65

R992 Pearson,R.G. & Vogelsong,D.C., J. Am. Chem. Soc., (1958) 80, 1038

R993 Menasse,R., Hedwall,P.R., Kraetz,J., Pericin,C., Riesterer,L., Sallmann,A., Ziel,R. & Jaques,R., Scand. J. Rheumat. Suppl., (1978) 22, 5

R994 Gregory,M.D., Christian,S.D. & Affsprung,H.E., J. Phys. Chem., (1967) 71, 2283

R995 Korenman,Y.I., Nefedova,T.A. & Dyukova,R.I., Zh. Prikl. Khim.,

(1977) 50, 2736, 2605ee

R996 Slobodian,E. & Levy,M., J. Biol. Chem., (1953) 201, 371

R997 Korenman,I.M., Lunicheva,E.V., Pomerantseva,E.G. & Khairulina,F.D., Zh. Prikl. Khim., (1974) 47, 2365, 2432ee

R998 Blake,M. & Harris,L.E., J. Am. Pharm. Assoc., (1952) 41, 521

R999 Newton,D. & Kluza, R., Drug Intelligence & Clinical Pharmacy, 12, 546 (1978)

R1002 Korenman,Y.I., Nefedova,T.E. & Dyukova,R.I., Zh. Fiz. Khim., (1977) 51, 1242, 734ee

R1003 Brown,J.M., Yu,N.Y., Cory,M.J., Bicknell,R.B. & Taylor,D.L., Br. J. Cancer, (1978) 37, Suppl. III, 206

R1004 Cates,L.A., Good,D.J., Jones,G.S. & Lemke,T.L., J. Med. Chem., (1978) 21, 1146

R1006 Korenman,I.M. & Klyukvina,T.D., Zh. Prikl. Khim., (1972) 45, 2572, 2697ee

R1008 Chin,J.B., Sheinin,D.M.K. & Rauth,A.M., Mutation Research, (1978) 58, 1

R1010 Levin,V.A., J. Med. Chem., (1980) 23, 682

R1011 Nichols,D.E., Shulgin,A.T. & Dyer,D.C., Life Sciences, (1977) 21, 569

R1012 Klemm,K., Prusse,W. & Kruger,U., Arzneim. Forsch., (1977) 27, 1895

R1013 Wolfenden,R., Biochemistry, (1978) 17, 201

R1015 Bartholow,R.M., Eiden,L.E., Ruth,J.A., Grunewald,G.L., Siebert,J. & Rutledge,C.O., J. Pharm. Expl. Therap., (1977) 202, 532

R1016 Harrison,I.T., Kurz,W., Massey,I.J. & Unger,S.H., J. Med. Chem., (1978) 21, 588

R1017 Graf,E. & El-Menshawy,M., Pharm. Uns. Zeit, (1977) 6, 171

R1018 Fahmy,M.A.H., Mallipudi,N.M. & Fukuto,T.R., J. Agric. Food Chem., (1978) 26, 550

R1019 Fulcrand,P., Berge,G., Noel,A.M., Chevallet,P., Castel,J. & Orzalesi,H., Eur. J. Med. Chem., (1978) 13, 177

R1020 Courriere,P., Paubel,J.P., Niviere,P. & Foussard-Blanpin,O., Eur. J. Chem., (1978) 13, 121

R1022 Adams,G.E., Clarke,E.D., Flockhart,I.R., Jacobs,R.S., Sehmi,D.S., Stratford,I.J., Wardman,P. & Watts,M., Int. J. Radiat. Biol., (1979) 35, 133

R1026 Zaagsma,J. & Nauta,W.T., J. Med. Chem., (1977) 20, 527

R1027 Magee,P.S., Chevron Chem., Private Communication

R1035 Yoshida,M., Yakugaku Zasshi, (1973) 93, 1452

R1036 Cohnen,E., Flasch,H., Heinz,N. & Hempelmann,F.W., Arzneim. Forsch., (1978) 28, 2179

R1038 Klemm,K., Krastinat,W. & Kruger,U., Arzneim. Forsch., (1979) 29, 1

R1039 Testa,B. & Murset-Rossetti,L., Helv. Chim. Acta, (1978) 61, 2530

R1043 Nelson,W.L. & Sherwood,B.E., J. Med. Chem., (1974) 17, 904

R1045 Ivin,B.A., Filov,V.A., Kraiz,B.O., Belogorodskii,V.V., Malyugina,L.L., Pol'kina,R.I. & Kozlovskii,Y., Khim. Farmat. Zh., (1978) 12, 67, 481ee

R1049 Zerba,E. & Fukuto,T.R., J. Agric. Food Chem., (1978) 26, 1365

R1050 Niculescu-Duvaz,I., Feyns,V. & Dobre,V., Cancer Treat. Rpts., (1978) 62, 2045

R1051 Glowinski,I.B., Radtke,H.E. & Weber,W.W., Molec. Pharmacol., (1978) 14, 940

R1054 Korenman,Y.I., Lineva,G.S. & Gezikova,N.N., Zh. Prilk. Khim., (1978) 51, 900, 869ee

R1055 Bradshaw,J., Glaxo-Allenburys Research, Private Communication

R1056 Rauls,D.O. & Baker,J.K., J. Med. Chem., (1979) 22, 81

R1057 Kramer,C-R., Z. Chem., (1978) 18, 348

R1058 Gescher,A. & Po,A., J. Pharm. Pharmac., (1978) 30, 353

R1059 Mathre,D.E., J. Agr. Food Chem., (1971) 19, 872

R1060 Branson, D. Astm Stp 634, 44 (1977)

R1064 Van Damme,M., Hanocq,M., Auquier,H. & Molle,L., Anal. Letters, (1979) 12, 357

R1068 Sugiura,K., Ito,N., Matsumoto,N., Mihara,Y., Murata,K., Tsukakoshi,Y. & Goto,M., Chemosphere, (1978) 9, 731

R1069 Steurbaut,W., Dejonckheere,W. & Kips,R.H., J. Chromatog., (1978) 160, 37

R1070 Erkell,L.J. & Walum,E., Febs Letters, (1979) 104, 401

R1073 Harris,R.L.N. & Geissler,A.E., Aust. J. Plant Physiol., (1977) 4, 235

R1074 Frater-Schroder,M., Alder,S. & Zbinden,G., Proc. Eur. Soc. Toxicol., (1976) 17, 277

R1075 Illing,H.P.A., Biochem. Soc. Trans., (1978) 6, 1211

R1076 Pardridge,W.M. & Mietus,L.J., J. Clin. Invest., (1979) 64, 145

R1077 Armstrong,N.A., James,K.C. & Wong,C.K., J. Pharm. Pharmacol.,

(1979) 31, 627

R1080 Anderson,R.F. & Patel,K.B., Br. J. Cancer, (1979) 39, 705

R1082 Sovak,M., Ranganathan,R., Lang,J.H. & Lasser,E.C., Ann. Radiol., (1978) 21, 283

R1084 Zsuzsanna,H.K., Private Communication

R1086 Schumacher,G.E. & Linn,E.E., J. Pharm. Sci., (1978) 67, 1717

R1087 Rapoport,S., NIH Natl.Inst.on Aging, Private Communication

R1092 Tute,M.S. & Canas-Rodriguez,A., Quant. Struct. Act. Anal., Akad. Verl. (Berlin), (1978), 53

R1093 Lazare,A., Lazare,C., Bastide,J. & Couquelet,J., Ann. Pharm. Franc., (1974) 32, 677

R1094 Benschop,H., VandenBerg,G., VanHooidonk,C., DeJong,L., Kientz,C., Berends,F., Kepner,L., Meeter,E. & Visser,R., J. Med. Chem., (1979) 22, 1306

R1097 Kobinger,W. & Pichler,L., N.-S. Arch. Pharmacol, (1975) 291, 175

R1101 Saeger,V.W., Hicks,O., Kaley,R.G., Michael,P.R., Mieure,J.P. & Tucker,E.S., Envir. Sci. Techn., (1979) 13, 840

R1102 Klemm,K. & Kruger,U., Arzneim. Forsch., (1979) 29, 2

R1104 Ozeki,S. & Tejima,K., Chem. Pharm. Bull. Japan, (1979) 27, 638

R1105 Bowden,K. & Henry,M.P., Biol. Correl.-Hansch Approach, Adv. Chem., (1972) 114, 130

R1107 Burkartsmaier,A. & Mutschler,E., Arch. Pharm., (1978) 311, 843

R1108 Sikk,P., Aaviksaar,A. & Abduvakhabov,A., Eesti Nsv Teaduste Akad. Toimetised, Keemia Geol., (1977) 26, 242

R1113 Korenman,Y.I., Tishchenko,E.M. & Kombarova,L.A., Zh. Prikl. Khim., (1977) 50, 941, 899ee

R1118 Lippold,B.C. & Lettenbauer,W.A., Pharmazie, (1978) 33, 221

R1120 Blazek,L., Kalal,J. & Hrabak,F., Coll. Czech. Chem. Comm., (1972) 37, 4025

R1126 Felder,E., Pitre,D. & Grandi,M., Il Farmaco, (1973) 28, 925

R1134 Kellaway,I.W. & Marriott,C., Can. J. Pharm. Sci., (1978) 13, 90

R1135 Freese,E., Levin,B.C., Pearce,R., Sreevalsan,T., Kaufman,J.J., Koski,W.S. & Semo,N.M., Teratology, (1979) 20, 413

R1136 Klaus,J., J. Chem. Res., (1979), 4473

R1137 Kramer,C.R., Z. Chem., (1979) 19, 347

R1140 Kurihara,N., Yamakawa,K., Fujita,T. & Nakajima,M., J. Pesticide Sci., (1980) 5, 93

R1142 Lichtenwalner,M.R. & Speaker,T.J., J. Pharm. Sci., (1980) 69, 337

R1143 Konemann,H., Zelle,R. & Busser,F., J. Chromatog., (1979) 178, 559

R1144 Lullmann,H., Timmermans,P.B. & Ziegler,A., Eur. J. Pharmacol., (1979) 60, 277

R1145 Korenman,Y.I., Russ. J. Phys. Chem., (1978) 52, 252

R1146 Tomida,H., Yotsuyanagi,T. & Ikeda,K., Chem. Pharm. Bull., (1978) 26, 2832

R1147 Heymans,F., Therizien,L. & Godfroid,J.J., J. Med. Chem., (1980) 23, 184

R1148 Norman,J.A., Drummond,A.H. & Moser,P., Mol. Pharmacol., (1979) 16, 1089

R1149 Pesak,M., Benes,L., Borovansky,A. & Kopacova,L., Ceskoslovenska Farmacie, (1978) 27, 140

R1150 Pesak,M., Stankovicova,M., Lukas,A., Borovansky,A. & Kopacova,L., Ceskoslovenska Farmacie, (1977) 26, 104

R1153 Korenman,Y.I., Tishchenko,E.M. & Nadenenko,I.V., Zh. Prikl. Khim., (1979) 52, 885, 848ee

R1155 Lullmann,H. & Wehling,M., Biochem. Pharmacol., (1979) 28, 3409

R1156 Takayama,C. & Fujinami,A., Pest. Biochem. Physiol., (1979) 12, 163

R1157 Wang,P.-H. & Lien,E.J., J. Pharm. Sci., (1980) 69, 662

R1158 Thakur,A., Bombay College Of Pharm., Private Communication

R1159 Zsuzsanna,H.K. & Nora,K.K., Nehezvegyipari Kutato Intezet Kozlemenyei, (1979) 8, 139

R1161 Lien,E.J., Mayer,K., Wang,P.H. & Tong,G.L., Acta Pharm. Jugoslavica, (1979) 29, 181

R1162 Greco,W.R. & Hakala,M.T., J. Pharmacol. Expl. Therap., (1980) 212, 39

R1163 Brugnoni,G.P., Moser,P. & Trebst,A., Z. Naturforsch., (1979) 34, 1028 + Private Communication

R1165 Pooler,J.P. & Valenzeno,D.P., Photochem. Photobiol., (1979) 30, 491

R1167 Xu,X., Xu,H., Lu,Y. & Ghen,J., Acta Pharm. Sinica, (1979) 14, 246

R1168 Mack,F. & Bonisch,H., N.-S. Arch. Pharmacol., (1979) 310, 1

R1169 Stephen,K.W., McCrossan,J., Mackenzie,D., Macfarlane,C.B., & Speirs,C.F., Br. J. Clin. Pharmac., (1980) 9, 51

R1173 Mitchell,R. & Ganellin,C., S.K. & F. Lab. Ltd., England, Private Communication

R1174 Korenman,Y.I., Lineva,G.S. & Tishchenko,E.M., Zh. Fiz. Khim., (1979) 53, 2929, 1678ee

R1176 Korenman,Y.I., Polumestnaya,E.I. & Mezentseva,T.P., Zh. Prikl. Khim., (1979) 52, 2629, 2494ee

R1178 Kenaga,E.E., Envirn. Sci. Tech., (1980) 14, 553

R1179 Banerjee,S., Yalkowsky,S.H. & Valvani,S.C., Envirn. Sci. Tech., (1980) 14, 1227

R1182 Le Therizien,L., Heymans,F., Redeuilh,C., Godfroid,J.-J. & Busch,N., Eur. J. Med. Chem., (1980) 15, 311

R1184 Parker,G.R., Lemke,T.L. & Moore,E.C., J. Med. Chem., (1977) 20, 1221

R1185 Tomida,H., Yotsuyanagi,T. & Ikeda,K., Chem. Pharm. Bull. Japan, (1978) 26, 2824

R1188 Rao,P.S.C. & Davidson,J.M., " Envir. Impact Non-Pt Source Pollut.", Ann Arbor Publ. (1980)

R1189 Metcalf,R. & Lu,P., EPA Report , In Press

R1190 Kenaga,E.E. & Goring,C.A.I., Am. Soc. Test. Mat., 3rd Aq. Tox. Symp., New Orleans (1978)

R1198 Korenman,Y.I., Lineva,G.S. & Tishchenko,E.M., Russ. J. Phys. Chem., (1979) 53, 1678

R1199 Van Damme,M., Hanocq,M., Auquier,H. & Molle,L., Anal. Let., (1979) 12, 357

R1200 Nahum,A. & Horvath,C., J. Chromatog., (1980) 192, 315

R1204 Stahle,H., Daniel,H., Kobinger,W., Lillie,C. & Pichler,L., J. Med. Chem., (1980) 23, 1217

R1205 Van De Waterbeemd,J.T.M., Thesis Leyden Univ., (1980)

R1207 Yunger,L.M. & Cramer III,R.D., Mol. Pharmacol., (1981) 20, 602

R1209 Unger,S.H. & Chiang,G.H., J. Med. Chem., (1981) 24, 262

R1210 Coulson,C.J. & Smith,V.J., J. Pharm. Sci., (1980) 69, 799

R1211 Ellgehausen,H., Guth,J.A. & Esser,H.O., Ecotox. Environ. Safety, (1980) 4, 134

R1213 Dunn III,W.J. & Marrs,P., U. Of Illinois, Private Communication

R1214 Wilman,D.E.V. & Goddard,P.M., J. Med. Chem., (1980) 23, 1052

R1216 Oatis Jr.,J.E., Russell,M.P., Knapp,D.R. & Walle,T., J. Med. Chem., (1981) 24, 309

R1220 Kaltenbronn,J.S., J. Med. Chem., (1977) 20, 596

R1225 James,R., ICI Pharmaceuticals, Private Communication

R1230 Kung,H. & Blau,M., J. Med. Chem., (1980) 23, 1127

R1231 Koshimura,H. & Okubo,T., Anal. Chim. Acta, (1973) 67, 331

R1232 Pang, K.Y.Y., Braswell,L.M., Chang,L., Sommer,T.J. & Miller,K.W., Mol. Pharmacol., (1980) 18, 84

R1233 Fauchere,J.L. & Petermann,C., Helv. Chim. Acta., (1980) 63, 824

R1236 Masutani,T., Et. Al., J. Pharmacol. Expt. Therrap., (1981) 217, 812

R1237 Levin,V., Univ. Of Cal. San Francisco, Private Communication

R1242 Kovac,T. & Kolbah,D., J. Med. Chem., (1979) 22, 1093

R1252 Chien,Y.W., Hofmann,L.M., Lambert,H.J. & Tao,C., J. Pharm. Sci., (1976) 65, 1337

R1254 Ng,A.S., Kluza,R.B. & Newton,D.W., J. Pharm. Sci., (1980) 69, 30

R1255 Shah,K.J. & Coats,E.A., J. Med. Chem., (1977) 20, 1001

R1257 Pratesi,P., Villa,L., Ferri,V., DeMicheli,C., Grana,E., Santagostino Barbone,M.G., Silipo,C., Vittoria,A., Il Farmaco, (1981) 36, 749

R1258 Lentzen,H. & Philippu,A., Biochem. Pharmacol., (1981) 30, 1759

R1259 Lacko,L., Wittke,B. & Zimmer,G., Biochem. Pharmacol., (1981) 30, 1425

R1260 Thijssen,H.H.W., Eur. J. Med. Chem., (1981) 16, 449

R1261 Bird,A.E. & Nayler,J.H.C., Drug Design, Vol. II, E. Ariens Ed., Acad. Press, London, (1971), p.308

R1262 Thomas,E.L., Biochemistry, (1981) 20, 3273

R1263 Briggs,G.G., J. Agric. Food Chem., (1981) 29, 1050

R1265 DeVoe,H., Miller,M.M. & Wasik,S.P., J. Res. Natl. Bur. Stds., (1981) 86, 361

R1266 Bundgaard,H. & Larsen,C., Int. J. Pharmaceut., (1980) 7, 169

R1267 Bundgaard,H., Bagger Hansen,A. & Larsen,C., Int. J. Pharmaceut., (1979) 3, 341

R1268 Johansen,M. & Bundgaard,H., Arch. Pharm. Chemi, Sci. Ed., (1980) 8, 141

R1269 Bundgaard,H., Bagger Hansen,A. & Larsen,C., Arch. Pharm. Chemi, Sci. Ed., (1979) 7, 193

R1273 Suzuki,N., Akiba,K., Kanno,T. & Wakahayashi,T., J. Inorg. Nucl. Chem., (1968) 30, 2521

R1277 Kitao,K., Yata,N., Yamazaki,M., Iwane,J. & Kamada,A., Chem. Pharm. Bull., (1977) 25, 1350

R1283 Brown,D.S. & Flagg,E.W., J. Environ. Qual., (1981) 10, 382

R1284 Inagi,T., Muramatsu,T., Nagai,H. & Terada,H., Chem. Pharm. Bull., (1981) 29, 2330

R1289 Smyth,R.D., Pfeffer,M., VanHarken,D.R., Cohen,A. &

Hottendorf,G.H., Antimicrob. Agents Chemoth., (1981) 19, 1004

R1293 Levitan,H., Proc. Natl. Acad. Sci. USA, (1977) 74, 2914

R1294 De Bruyne,C.K. & Yde,M., Carbohyd. Res., (1977) 56, 153

R1295 Nikaido,H., Biochim. Biophys. Acta, (1976) 433, 118

R1298 Newton,D.W., Breen,P.J., Brown,D.E., Mackie Jr.,J.F. & Kluza,R.B., J. Pharm. Sci., (1981) 70, 122

R1299 Mackerer,C.R., Kochman,R.L., Shen,T.F. & Hershenson,F.M., J. Pharmacol. Expl. Therap., (1977) 201, 326

R1315 Paljk,S., Klofutar,C., Krasovec,F. & Suhac,M., Mikrochim. Acta, (1975) II, 485

R1322 Johnson,H.G. & Piret,E.L., Ind. Eng. Chem., (1948) 40, 743

R1323 James,H.J., Carmack,G.P. & Freiser,H., Anal. Chem., (1972) 44, 853

R1327 Kojima,S., Tanaka,R. & Hamada,C., Chem. Pharm. Bull., (1976) 24, 1555

R1335 Boyd,I. & Beveridge,E.G., Microbios, (1979) 24, 173

R1342 Defrise-Quertain,F., Chatelain,P., Ruysschaert,J.M. & Delmelle,M., Biochim. Biophys. Acta, (1980) 628, 57

R1343 Horstmann,H., Moller,E., Wehinger,E. & Meng,K., ACS Symposium Ser., (1978) 83, 125

R1344 Boschman,T.A.C., Van Dijk,J., Hartog,J. & Walop,J.N., ACS Symposium Ser., (1978) 83, 140

R1346 Ezumi,K. & Kubota,T., Chem. Pharm. Bull., (1980) 28, 85

R1347 Orzalesi,G., Mari,F., Bertol,E., Selleri,R. & Pisaturo,G., Arzneim. Forsch., (1980) 30, 1607

R1350 Timmermans,P., de Jonge,A., van Meel,J.C.A., Slothorst-Grisdijk,F.P., Lam,E. & van Zwieten,P.A., J. Med. Chem., (1981) 24, 502

R1351 Satzinger,G., Arzneim. Forsch., (1977) 27, 1742

R1352 Tenthorey,P.A., Block,A.J., Ronfeld,R.A., McMaster,P.D. & Byrnes,E.W., J. Med. Chem., (1981) 24, 798

R1353 Braumann,T. & Grimme,L.H., J. Chromatog., (1981) 206, 7

R1354 Sada,H. & Ban,T., Experienta, (1981) 37, 171

R1355 Barnish,I.T., Cross,P.E., Dickinson,R.P., Gadsby,B., Parry,M.J., Randall,M.J. & Sinclair,I.W., J. Med. Chem., (1980) 23, 117

R1356 Pesak,M., Kopecky,F., Cizmarik,J. & Borovansky,A., Pharmazie, (1980) 35, 150

R1357 Pesak,M., Zaloudkova,D., Borovansky,A. & Benes,L., Czeskol. Farm., (1980) 29, 32

R1358 Cossum,P.A. & Roberts,M.S., Eur. J. Clin. Pharmacol., (1981) 19, 181

R1361 Kamenka,J-M. & Geneste,P., Eur. J. Med. Chem., (1981) 16, 213

R1362 Woods,P.B. & Robinson,M.L., J. Pharm. Pharmacol., (1981) 33, 172

R1363 Satzinger,G., Herrmann,W. & Zimmermann,F., Arzneim. Forsch., (1975) 25, 1849

R1364 Kanazawa,J., Pestic. Sci., (1981) 12, 417

R1365 Martin,R.A. & Edgington,L.V., Pestic. Biochem. Physiol., (1981) 16, 87

R1368 Wiseman,E.H., Chang,Y-H. & Lombardino,J.G., Arzneim. Forsch., (1976) 26, 1300

R1369 McDuffie,B., Chemosphere, (1981) 10, 73

R1375 Bluth,R., Pharmazie, (1981) 36, 698

R1377 Zaslavsky,B.Y., Miheeva,L.M. & Rogozhin,S.V., J. Chromatog., (1981) 212, 13

R1378 Pesak,M., Stolc,S. & Benes,L., Ceskosl. Farm., (1980) 29, 272

R1381 Cates,L.A., Cramer,M.B. & Williams,L., J. Med. Chem., (1978) 21, 143

R1382 Harada,S., Ban,T., Fujita,T. & Koshiro,A., Arch. Int. Pharmacodyn., (1981) 252, 262

R1383 Ross,J.H. & Krieger,R.I., J. Agric. Food Chem., (1980) 28, 1026

R1389 Platford,R.F., Bull. Environm. Contam. Toxicol., (1979) 21, 68

R1394 Kramer,C-R. & Beck,L., Z. Chem., (1981) 21, 411

R1396 Wu,J. & Ji,R., Acta Pharmacol. Sin., (1982) 3, 55

R1397 Cornford,E.M., J. Membrane Biol., (1982) 64, 217

R1398 McCulloch,M.W., Medgett,I.C., Rand,M.J. & Story,D.F., Br. J. Pharmac., (1980) 69, 397

R1400 Butte,W., Fooken,C., Klussmann,R. & Schuller,D., J. Chromatog., (1981) 214, 59

R1401 Kowaluk,E.A., Roberts,M.S., Blackburn,H.D. & Polack,A.E., Am. J. Hosp. Pharm., 38, (1981) 1308

R1402 Tanii,H. & Hashimoto,K., Toxicol. Let., (1982) 11, 125

R1403 Werner,R.G., Lechner,U.L. & Goeth,H., Antimicrob. Agents Chemotherap., (1982) 21, 101

R1404 Brown,D.M, Parker,E. & Brown,J.M., Radiation Res., (1982) 90, 98

R1405 Fujioka,H. & Tan,T., J. Pharm. Dyn., (1981) 4, 759

R1407 Yeung, J., Corleto, L. & Knaus, E., J. Med. Chem., 25, 191 (1982)

R1409 Guerra,M.C., Barbaro,A.M., Forti,G.C., Foffani,M.T., Biagi,G.L., Borea,P.A. &Fini,A., J. Chromatog., (1981) 216, 93

R1410 Cousse,H., Mouzin,G., Ribet,J-P. & Vezin,J-C., J. Pharm. Sci., (1981) 70, 1245

R1411 Russell,M.P., Privitera,P.J., Walle,T., Peet,N.P., Halushka,P.V. & Gaffney,T.E., J. Pharmacol. Exper. Therap., (1981) 219, 685

R1412 Alhaider,A.A., Selassie,C.D., Chua,S-O. & Lien,E.J., J. Pharm. Sci., (1982) 71, 89

R1413 Verbiese-Genard,N., Hanocq,M., van Damme, M. & Molle,L., Internat. J. Pharmaceut., (1981) 9, 295

R1414 Cascella,P.J. & Feldman,S., J. Pharm. Sci., (1980) 69, 643

R1415 Van der Giesen,W.F. & Janssen,L.H.M., Int. J. Pharmaceut., (1982) 12, 231

R1420 Newton,D.W. & Murray,W.J., Anal. Chim. Acta, (1982) 135, 343

R1422 Henkel,J.G., Hane,J.T. & Gianutsos,G., J. Med. Chem., (1982) 25, 51

R1426 Thakur,A.B., Nimbkar,A.Y. & Tipnis,H.P., Indian J. Pharm. Sci., (1980) 42, 71

R1428 Workman,P. & Brown,J.M., Cancer Chemother. Pharmacol., (1981) 6, 39

R1439 Novak-Hanko,K., Szasz,G., Papp,O., Vamos,J. & Hermecz,I., Acta Pharm. Hungar., (1981) 51, 246

R1440 Thoma,K. & Albert,K., Pharm. Acta Helv., (1981) 56, 69

R1441 Aimoto,T., Masunari,T. & Murata,T., Yakugaku Zasshi, (1979) 99, 106

R1443 Kimura,K., Nagaoka,M., Agatsuma,H. & Ohgiya,S., Yakugaku Zasshi, (1979) 99, 235

R1453 Lins,C.L.K., Block,J.H., Doerge,R.F. & Barnes,G.J., J. Pharm. Sci., (1982) 71, 614

R1456 Korenman,Y.I., Sotnikova,N.G., Lineva,G.S. & Tuzhikova,N.I., Zh. Fiz. Khim., (1980) 54, 2072, 1176ee

R1463 Terada,H., Kitagawa,K., Yoshikawa,Y. & Kametani,F., Chem. Pharm. Bull., (1981) 29, 7

R1464 Takayama,C., Fujinami,A., Kirino,O. & Hisada,Y., Agric. Biol. Chem., (1982) 46, 2755

R1466 Pratesi,P., Villa,L., Ferri,V., De Micheli,C., Grana,E., Barbone,M.G.S., Grieco, C., Silipo,C. & Vittoria,A., Il Farmaco, (1980) 35, 621

R1467 Brown,J.M., Cancer Treat. Rpts., (1981) 65, 95

R1470 Sehgal,R.K. & Agrawal,K.C., J. Pharm. Sci., (1982) 71, 1203

R1471 Roth,B., Aig,E., Rauckman,B.S., Strelitz,J.Z., Phillips,A.P., Ferone,R., Bushby,S.R.M. & Sigel,C.W., J. Med. Chem., (1981) 24, 933

R1472 Wooldridge,K.R.H., May & Baker Ltd., Private Communication

R1473 Mayer,J.M., Testa,B., van de Waterbeemd,H. & Bornand-Crausaz,A., Eur. J. Med. Chem., (1982) 17, 461

R1474 Mayer,J.M., Testa,B., van de Waterbeemd,H. & Bornand-Crausaz,A., Eur. J. Med. Chem., (1982) 17, 453

R1476 Johns,D.G., Farquhar,D., Wolpert,M.K., Chabner,B.A. & Loo,T.L., Drug Metab. Disposition, (1973) 1, 580

R1477 Beezer,A.E., Hunter,W.H. & Storey,D.E., J.Pharm.Pharmacol., (1980) 32, 815

R1478 Bradshaw,J. & Latter,D.F., Glaxo Research, England, Private Communication

R1479 Calas,M., Barbier,A. & Giral,L., Eur. J. Med. Chem., (1982) 17, 497

R1480 Halmekoski,J. & Hannikainen,H., Suomen Kem., (1964) 37, 221

R1481 Li,R., Peking Medical College, Private Communication

R1485 Seiler,P., Bischoff,O. & Wagner,R., Arzneim. Forsch., (1982) 32, 711

R1486 Means,J.C., Wood,S.G., Hassett,J.J. & Banwart,W.L., Env. Sci. Techn., (1980) 14, 1524

R1487 Karickhoff,S.W., Brown,D.S. & Scott,T.A., Water Res., (1979) 13, 241 (1979)

R1493 Iyengar,B.S., Sami,S.M., Remers,W.A., Bradner,W.T. & Schurig,J.E., J. Med. Chem., (1983) 26, 16

R1497 Workman,P. & Twentyman,P.R., Br. J. Cancer, (1982) 46, 249

R1498 Mendel,J., Thust,R. & Schwarz,H., Arch. Geschwulstforsch., (1982) 52, 371

R1499 Stratford,I.J., Williamson,C., Hoe,S. & Adams,G.E., Radiation Res., (1981) 88, 502

R1500 Rasey,J.S., Krohn,K.A. & Freauff,S., Radiation Res., (1982) 91, 542

R1501 Ellgehausen,H., D'Hondt,C. & Fuerer,R., Pestic. Sci., (1981) 12,

219

R1505 Hammers,W.E., Meurs,G.J. & DeLigny,C.L., J. Chromatog., (1982) 247, 1

R1506 Bellando,M. & Trotta,A., Plant Sci. Let., (1980) 17, 467

R1508 Lindeke,B., Paulsen-Sorman,U., Hallstrom,G., Khuthier,A-H., Cho,A.K. & Kammerer,R.C., Drug Metab. Dispos., (1982) 10, 700

R1510 Oldendorf,W. & Cornford,E., V. A. Med. Center, L. A., Private Communication

R1517 Pesak,M., Zaloudkova,D., Borovansky,A. & Benes,L., Ceskosl. Farm., (1980) 29, 32

R1519 Lewis,S.J., Mirrlees,M.S. & Taylor,P.J., Quant. St-Act. Rel., (1983) 2, 100

R1520 Caron,J.C. & Shroot,B., J. Pharm. Sci. (1984) 73, 1703

R1521 Nakagawa,S., Okajima,N., Kitahaba,T., Nishimura,K., Fujita,T. & Nakajima,M., Pest. Biochem. Physiol., (1982) 17, 243

R1522 Nishimura,K., Okajima,N., Fujita,T. & Nakajima,M., Pest. Biochem. Physiol., (1982) 18, 341

R1523 Nishimura,K. & Fujita,T., J. Pest. Sci., (1983) 8, 69

R1524 Guo,Z., Beijing Univ., Private Communication

R1525 Yoshimoto,M., Ann. Rept. Sankyo Res. Lab., (1982) 34, 1

R1526 Sturm,K., Muschaweck,R. & Hropot,M., J. Med. Chem., (1983) 26, 1174

R1528 Newton,D.W., Ratanamaneichatara,S. & Murray,W.J., Int. J. Pharmaceut., (1982) 11, 209

R1529 Shaw,G. & Connell,D.W., Aust. J. Mar. Freshw. Res., (1982) 33, 1057

R1531 Irwin,W.J. & Po,A.L.W., Int. J. Pharmaceut., (1979) 4, 47

R1532 Sasaki,H., Mukai,E., Hashida,M., Kimura,T. & Sezaki,H., Int. J. Pharmaceut., (1983) 15, 49

R1533 Janjic,T.J., Milosavljevic,E.B. & Nanayakkara,W., Analysis, (1982) 10, 197

R1537 Shim,C.K., Nishigaki, R., Iga,T. & Hanano,M., Int. J. Pharmaceut., (1981) 8, 143

R1542 Kuchar,M., Rejholec,V., Brunova,B., Grimova,J., Matousova,O., Nemecek,O. & Cepelakova,H., Coll. Czech. Chem. Commun., (1982) 47, 2514

R1547 Timmermans,P., de Jonge,A., van Zwieten,P.A., de Boer,J.J.J. & Speckamp,N.W., J. Med. Chem., (1982) 25, 1122

R1549 Korenman,Y.I., Russ. J. Phys. Chem., (1982) 56, 748, 458ee

R1551 Sakaguchi,M., Webb,M.W. & Agrawal,K.C., J. Med. Chem., (1982) 25, 1339

R1552 Elferink,J.G.R., Biochem. Pharmacol., (1977) 26, 2411

R1553 Kurihara,N. & Fujita,T., Bull. Inst. Chem. Res., Kyoto Univ., (1983) Vol. 61, 89

R1554 Saarikoski,J. & Viluksela,M., Ecotox. Envir. Safety, (1982) 6, 501

R1555 Hermens,J. & Leeuwangh,P., Ecotox. Envir. Safety, (1982) 6, 302

R1556 Tong,L.K.J. & Glesmann,M.C., J. Am. Chem. Soc., (1957) 79, 4305

R1558 Wynne, Hj., Et. Al., Experientia, 38, 655 (1982)

R1559 Anderson,R.F., Patel,K.B. & Sehmi,D.S., Rad. Res., (1981) 85, 496

R1560 Tewari,Y.B., Miller,M.M., Wasik,S.P. & Martire,D.E., J. Chem. Eng. Data, (1982) 27, 451

R1563 Dax,E.M. & Partilla,J.S., Mol. Pharmacol., (1982) 22, 5

R1564 Brown,B.T., Phillips,J.N. & Rattigan,B.M., J. Agric. Food Chem., (1981) 29, 719

R1566 Dall'Acqua,F., Vedaldi,D., Bordin,F., Baccichetti,F., Carlassare,F, Tamaro,M., Rodigheiro,P., Pastorini,G., Guiotto,A., Recchia,G. & Cristofolini,M., J. Med. Chem., (1983) 26, 870

R1568 Banerjee,S., Et. Al., Env. Sci. Technol. (1980) 14, 1227 & Private Communication

R1569 Go,M.L., Ngiam,T.L., Becket,G. & Leung,S.L., S.E. Asian J. Trop. Med. Pub. Hlth., (1982) 13, 658

R1570 Wu,G-P., Bai,F-X., Leng,Z-K. & Gu,X-C., Acta Pharm. Sinica, (1982) 17, 821

R1571 Korenman,Y,I., Bolotov,V.M. & Proshunina,T.V., Russ. J. Phys. Chem., (1980) 54, 2714, 1556ee

R1572 Chowhan,Z.T. & Amaro,A.A., J. Pharm. Sci., (1976) 65, 1669

R1578 Chiou,C.T., Porter,P.E. & Schmedding,D.W., Env. Sci. Tech., (1983) 17, 227

R1582 Biagi,G.L., Barbaro,A.M., Guerra,M.C., Babbini,M., Gaiardi,M., Bartoletti,M. & Borea,P.A., J. Med. Chem., (1980) 23, 193

R1586 Bernabei,M.T., Forni,F., Bellei,S. & Cameroni,R., Atti. Soc. Mat. di Modena, (1980) 111, 63

R1587 Leyder,F. & Boulanger,P., Bull. Env. Contam. Tox., (1983) 30, 152

R1588 Duncan,J.D., Hallstrom,G., Paulsen-Sorman,U., Lindeke,B. &

Cho,A.K., Drug Metab. Disp., (1983) 11, 15

R1589 Foussard-Blanpin,O., Uchida-Ernouf,G., Moreau,R., Adam,Y. & Ray-Drouet,M., Ann. Pharm. Franc., (1978) 36, 581

R1590 Pliska,V., Schmidt,M. & Fauchere,J-L., J. Chromatog., (1981) 216, 79

R1591 Fauchere,J-L. & Pliska,V., Eur. J. Med. Chem., (1983) 18, 369

R1592 Moser,P., Ciba-Geigy, Private Communication

R1593 Matsuda,H., Furuya,T., Ichimura,T. & Kobayashi,Y., Rpt. Yamanouchi Res. Lab., (1980) 4, 107

R1598 Tollenaere,J.P., Moereels,H. & Koch,M.H.J., Eur. J. Med. Chem., (1977) 12, 199

R1599 Fleminger,S., Van De Waterbeemd,H., Rupniak,N.M.J., Reavill,C., Testa,B., Jenner,P. & Marsden,C.D., J. Pharm. Pharmacol., (1983) 35, 363

R1600 Watarai,H., Tanaka,M. & Suzuki,N., Anal. Chem., (1982) 54, 702

R1604 Korenman,Y.I. & Polumestnaya,E.I., Zh. Prikl. Khim., (1982) 55, 399, 364ee

R1607 Lobl,T.J., Tindall,D.J., Cunningham,G.R., Kemp,P.L. & Campbell,J.A., Ann. N.Y. Acad. Sci., (1982) 383, 477

R1609 Kuchar,M., Rejholec,V., Roubal,Z. & Matousova,O., Coll. Czech. Chem. Comm., (1983) 48, 1077

R1616 Giuliani,E., Lembo,S., Sasso,V., Sorrentino,L., Silipo,C. & Vittoria,A., Il Farmaco Ed. Sc., (1983) 38, 847

R1621 Miyashita,Y., Seki,T., Yotsui,Y., Yamazaki,K., Sano,M., Abe,H. & Sasaki,S., Bull. Chem. Soc. Jpn, (1982) 55, 1489

R1622 Anker,L., Testa,B., van de Waterbeemd,H. & Bornand-Crausaz,A., Theodorou,A., Jenner,P. & Marsden,C.D., Helv. Chim. Acta, (1983) 66, 542

R1623 Pleiss,M.A. & Grunewald,G.L., J. Med. Chem., (1983) 26, 1760

R1624 Briggs,G.G., Bromilow,R.H. & Evans,A.A., Pestic. Sci., (1982) 13, 495

R1625 Taylor,P.J., ICI, England, Private Communication

R1626 Seiler,P. & Zimmermann,I., Arzneim. Forsch., (1983) 33, 1519

R1627 Taylor,P.J., ICI, England, Private Communication

R1628 Facino,R.M. & Lanzani,R., Pharm. Res. Commun., (1979) 11, 433

R1629 Drabek,J. & Bachmann,F., IUPAC Pestic. Chem., Miyamoto,J. Ed., Pergamon (1983) p. 271

R1631 Egutkin,N.L., Maidanov,V.V., Nikitin,Y.E., Zhdanovich,Y.V. & Kuzovkov,A.D., Antibiotiki, (1983) 28, 246

R1634 Weinkam,R.J. & Lin,H-S., Adv. Pharmacol. Chemotherapy, (1982) 19, 1

R1635 Daoud,N.N., Dickinson,N.A. & Gilbert,P., Microbios, (1983) 37, 73

R1636 Fujisawa,S. & Masuhara,E., Sh. Riko. Zasshi (J. Dental Res.), (1981) 22, 277

R1638 Boyd,D.B., Lilly Res. Lab., Private Communication

R1639 Edelist,G., Morley,M.M. & Eger,E.I., Anesthes., (1964) 25, 223

R1641 Austin,D.J. & Briggs,G.G., Pestic. Sci., (1976) 7, 201

R1642 Moerlein,S.M., Dischino,D.D., Raymond,K.N., Weitl,F.L. & Welch,M.J., J. Label. Cpds. Radiopharm., (1982) 19, 1421

R1649 Briggs,G.G., Proc. 1981 Brit. Crop Protect. Conf., p. 701

R1661 Rapaport,R.A. & Eisenreich,S.J., Environ. Sci. Technol., (1984) 18, 163

R1662 Woodburn,K.B., Doucette,W.J. & Andren,A.W., Environ. Sci. Technol., (1984) 18, 457

R1663 Bruggeman, W. Et. Al., Chromatogr., 238, 335 (1982)

R1670 Golynko,Z.S., Skorovarov,D.I., Smirnov,V.F. & Skvortsov,N.V., Zh. Prikl. Khim., (1965) 38, 271, 273ee

R1678 Gaffney,B.J., Willingham,G.L. & Schepp,R.S., Biochem., (1983) 22, 881

R1679 Schimmel,S.C., Garnas,R.L., Patrick Jr.,J.M. & Moore,J.C., J. Agric. Food Chem., (1983) 31, 104

R1681 Yamagami,C., Sonoda,C., Takao,N., Tanaka,M., Yamada,J., Horisaka,K. & Fujita,T., Chem. Pharm. Bull., (1982) 30, 4175

R1682 Haavaldsen,J., Nordal,V. & Kelly,M., Acta Pharm. Suec., (1983) 20, 219

R1683 Canton,J.H. & Wegman,R.C.C., Water Res., (1983) 17, 743

R1685 Schoenwald,R.D., Eller,M.G., Dixson,J.A. & Barfknecht,C.F., J. Med. Chem., (1984) 27, 810

R1686 Shashoua,V.E., Jacob,J.N., Ridge,R., Campbell,A. & Baldessarini,R.J., J. Med. Chem., (1984) 27, 659

R1687 Chiarini,A., Tartarini,A. & Fini,A., Arch. Pharm., (1984) 317, 268

R1688 Greenblatt,D.J., Arendt,R.M., Abernethy,D.R., Giles,H.G., Sellers,E.M. & Shader,R.I., Br. J. Anaesth., (1983) 55, 985

R1689 Sami,S.M., Iyengar,B.S., Tarnow,S.E., Remers,W.A., Bradner,W.T., Schurig,J.E., J. Med. Chem., (1984) 27, 701

R1690 Banerjee,S., Howard,P.H., Rosenberg,A.M., Dombrowski,A.E.,

Sikka,H. & Tullis,D.L., Environ. Sci. Technol., (1984) 18, 416

R1693 Bowman,B.T. & Sans,W.W., J. Environ. Sci. Health, (1983) B18 (6), 667

R1694 Illum,L., Bundgaard,H. & Davis,S.S., Int. J. Pharmaceut., (1983) 17, 183

R1696 Mallon,B.J. & Harrison,F.L., Bull. Environ. Contam. Toxicol., (1984) 32, 316

R1697 Miller,M.M., Ghodbane,S., Wasik,S.P., Tewari,Y.B. & Martire,D.E., J. Chem. Eng. Data, (1984) 29, 184

R1698 Pesak,M., Slechtova,L., Cizmarik,J., Borovansky,A. & Svec,P., Chem. Zvesti, (1983) 37, 237

R1700 Schoenwald,R.D. & Huang,H-S., J. Pharm. Sci., (1983) 72, 1266

R1701 Akamatsu,M., Kyoto Univ., Private Communication

R1702 Fuhrer,W., Ostermayer,F., Zimmermann,M., Meier,M. & Muller,H., J. Med. Chem., (1984) 27, 831

R1703 Ezumi,K., Shionogi Research, Private Communication

R1704 Pratesi,P., Villa,L., Ferri,V., De Micheli,C., De Amici,M., Grana,E., Barbone,M.G.S., Silipo,C., Vittoria,A. & Cappello,B., Il Farmaco, (1984) 39, 171

R1705 Taylor,P.J., ICI, England, Private Communication

R1706 Kubota,T. & Ezumi,K., Chem. Pharm. Bull., (1980) 28, 3673

R1708 Kuchar,M., Rejholec,V., Kraus,E., Miller,V. & Rabek,V., J. Chromatog., (1983) 280, 279

R1709 Beltrame,P., Beltrame,P.L. & Carniti,P., Chemosph., (1984) 13, 3

R1710 Kaiser,K.L.E., Chemosph., (1983) 12, 1159

R1711 Sasaki,H., Takakura,Y., Hashida,M., Kimura,T. & Sezaki,H., J. Pharm. Dyn., (1984) 7, 120

R1712 Yoshimura,Y. & Kakeya,N., Int. J. Pharmaceut., (1983) 17, 47

R1713 Yamada,T., Tsukamoto,Y., Shimamura,H., Banno,S. & Sato,M., Eur. J. Med. Chem., (1983) 18, 209

R1714 Motomizu,S., Hamada,S. & Toei,K., Bunseki Kagaku, (1983) 32, 648

R1716 Hanamura,J., Kobayashi,K., Kano,K. & Kubota,T., Chem. Pharm. Bull., (1983) 31, 1357

R1717 Zon,G., Ludeman,S.M., Ozkan,G., Chandrasegaran,S., Hammer,C.F., Dickerson,R., Mizuta,K. & Egan,W., J. Pharm. Sci., (1983) 72, 687

R1718 Acree Jr.,W.E., Bacon,W.E. & Leo,A.J., Int. J. Pharmaceut., (1984) 20, 209

R1720 Cizmarik,J., Trupl,J. & Pesak,M., Pharmazie, (1983) 38, 789

R1722 Shargel,L., Banijamali,A.R. & Kuttab,S.H., J. Pharm. Sci., (1984) 73, 161

R1723 Broughton,B.J., Chaplen,P., Knowles,P., Lunt,E., Marshall,S.M., Pain,D.L. & Wooldridge,K.R.H., J. Med. Chem., (1975) 18, 1117

R1724 Papp,O., Jozan,M., Valko,K., Hankone-Novak,K., Hermecz,I. & Szasz,G., Acta Pharm. Hung., (1983) 53, 215

R1725 Szasz,G., Papp,O., Vamos,J. & Hanko-Novak,K., J. Chromatog., (1983) 269, 91

R1727 Menez,J-F., Bourn,M., Colombel,M-C., Benamou,C., Larousse,C. & Bardou,L., Eur. J. Med. Chem., (1983) 18, 521

R1728 Singer,G.M., Frederick Cancer Research, Private Communication

R1729 Egawa,H., Endo,M. & Ogura,H., J. Antibact. Antifung. Agents, (1983) 11, 563

R1730 Andronati,S.A., Chepelev,V.M., Yakubovskaya,L.N., Val'dman,A.V., Voronina,T.A., Rozhanets,V.V., Zhulin,V.V. & Korotkov,K.O., Bioorgan. Khim., (1983) 9, 1357, 747ee

R1731 De,A.U., Sengupta,C., Pal,D., Mandal,A. & Chatterjee,J., Indian J. Pharm. Sci., (1983) 45, 123

R1732 Yano,K., Katayama,H. & Takemoto,K., Cancer Res., (1984) 44, 1027

R1734 El Tayar,N., van de Waterbeemd,H., Gryllaki,M., Testa,B. & Trager,W.F., Int. J. Pharmaceut., (1984) 19, 271

R1735 Ehrenkaufer,R.L.E., Agranoff,B.W., Bieszki,J., Frey,K., Hays,S. & Jewett,D., J. Label. Cpds. Radiopharm., (1984) 21, 87

R1736 Geyer,H., Politzki,G. & Freitag,D., Chemosph., (1984) 13, 269

R1737 Kishida,K. & Otori,T., Jpn. J. Ophthalmol., (1980) 24, 251

R1738 Xie,T.M., Hulthe,B. & Folestad,S., Chemosph., (1984) 13, 445

R1739 Farghali,H., Novotny,L., Ryba,M., Berank,J. & Janku,I., Biochem. Pharmacol., (1984) 33, 655

R1740 Bruggeman,W.A., Weber-Fung,D., Opperhuizen,A., Van Der Steen,J., Wijbenga,A. & Hutzinger,O., Toxicol. Environ. Chem., (1984) 7, 287

R1742 Means,J., Et. Al., EPA-600/ 3-80-041, NTIS

R1743 Nakajima,T., Sunagawa,M. & Hirohashi,T., Chem. Pharm. Bull., (1984) 32, 401

R1746 Ros,F.E., Innemee,H.C. & Van Zwieten,P.A., Docum.

Ophthalmol., (1979) 48, 291

R1754 Pang,D.C. & Sperelakis,N., Biochem. Pharmacol., (1984) 33, 821

R1757 Hinderling,P.H., J. Pharm. Sci., (1984) 73, 1042

R1758 Grass,G.M. & Robinson,J.R., J. Pharm. Sci., (1984) 73, 1021

R1764 Mitchell,R.C. & Dunkley,D., J. Pharm. Pharmacol., (1984) 36, 331

R1766 Chelnokova, M. & Gertsen, T., Zh. Anal. Khim., 38, 2148, 1650ee (1983)

R1768 van de Waterbeemd,H., van Bakel,P. & Jansen,A., J. Pharm. Sci., (1981) 70, 1081

R1769 Kenley,R.A., Bedford,C.D., Dailey Jr.,O.D., Howd,R.A. & Miller,A., J. Med. Chem., (1984) 27, 1201

R1772 Eadsforth,C.V. & Moser,P., Chemosphere, (1983) 12, 1459

R1776 Bellasio,E., Campi,A., Di Mola,N. & Baldoli,E., J. Med. Chem., (1984) 27, 1077

R1780 Yamazaki,M., Suzuka,T., Ito,Y., Itoh,S., Kitamura,M., Ohashi,K., Takeda,Y., Kamada,A., Orita,Y. & Nakahama,H., Chem. Pharm. Bull., (1984) 32, 2380

R1781 Hafkenscheid,T.L. & Tomlinson,E., J. Chromatog., (1984) 292, 305

R1783 Hartman,G.D., Hartman,R.D., Schwering,J.E., Saari,W.S., Engelhardt,E.L., Jones,N.R., Wardman,P., Watts,M.E. & Woodcock,M., J. Med. Chem., (1984) 27, 1634

R1792 Hayashi,Y. & Fujita,T., Private Communication

R1793 Bachrata,M., Bezakova,Z., Pesak,M. & Blesova,M., Acta Facult. Pharmaceut., (1983) 37, 89

R1797 Tsukada,K., Ueda,S. & Okada,R., Chem. Pharm. Bull., (1984) 32, 1929

R1799 Taylor,P.J. & Cruickshank,J.M., J. Pharm. Pharmacol., (1984) 36, 118

R1802 Lippold,B.H. & Lichey,J.F., Acta Pharm. Technolog., (1984) 30, 140

R1804 Van Damme,M., Hanocq,M. & Molle,L., Pharm. Acta Helv., (1984) 59, 235

R1806 Grunewald,G.L., Pleiss,M.A., Gatchell,C.L., Pazhenchevsky,R. & Rafferty,M.F., J. Chromatog., (1984) 292, 319

R1808 Sarna,L.P., Hodge,P.E. & Webster,G.R.B., Chemosph., (1984) 13, 975

R1810 Scherrer,R.A., Riker/3M, Private Communication

R1811 Yen,S-F., Wilson,W.D., Pearce,S.W. & Gabbay,E.J., J. Pharm. Sci., (1984) 73, 1575

R1812 Lin,T-S., J. Pharm. Sci., (1984) 73, 1568

R1814 Garst,J.E., J. Pharm. Sci., (1984) 73, 1623

R1815 Keane,P.E., Simiand,J., Mendes,E., Santucci,V. & Morre,M., Neuropharmacol., (1983) 22, 875

R1816 Haberfield,P., Kivuls,J., Haddad,M. & Rizzo,T., J. Phys. Chem., (1984) 88, 1913

R1817 Arnoldi,A., Betto,E., Ceresa,L., Farina,G., Formigoni,A., Galli,R. & Scaglioni,L., Pestic. Sci., (1983) 14, 576

R1818 Audran,M., Chanal,J.L., Berge,G., Marignan,R. & Cousse,H., Int. J. Pharmaceut., (1984) 20, 129

R1819 Funasaki,N., Hada,S., Neya,S. & Machida,K., J. Phys. Chem., (1984) 88, 5786

R1820 Hamrell,M.R., Oncology, (1984) 41, 343

R1821 Serrentino,R., Citti,L., Gervasi,P.G. & Turchi,G., Ital. J. Biochem., (1983) 32, 111

R1822 Xie,T.M. & Dyrssen,D., Anal. Chim. Acta, (1984) 160, 21

R1824 Barak,E. & Dinoor,A. & Jacoby,B., Pestic. Sci., (1983) 14, 213

R1829 Toon,S. & Rowland,M., J. Pharmacol. Expl. Therap., (1983) 225, 752

R1831 Toon,S. & Rowland,M., J. Pharm. Pharmacol., (1979) 31, 43P

R1835 Clarke,F.H., J. Pharm. Sci., (1984) 73, 226

R1837 Mather,L.E., Clin. Pharmacokin., (1983) 8, 422

R1838 La Rotonda,M.I., Amato,G., Barbato,F., Silipo,C. & Vittoria,A., Quant. Struct. Act. Relat., (1983) 2, 168

R1851 Sunada,H., Furukawa,K., Ishino,R., Otsuka,A., Sugiura,M. & Hamada,Y., Chem. Pharm. Bull., (1984) 32, 4084

R1853 Ganellin,C.R. & Durant,G.J., "Burger's Med. Chem." IV, Part III, Wolff,M.E., Ed., p.487

R1856 Fifer,E.K., Davis,W.M. & Borne,R.F., Eur. J. Med. Chem., (1984) 19, 519

R1858 Itokawa,H., Takeya,K., Mori,N., Sonobe,T., Serisawa,N., Hamanaka,T. & Mihashi,S., Chem. Pharm. Bull., (1984) 32, 3216

R1859 Coombs,T.J., Coulson,C.J. & Smith,V.J., Br. J. Clin. Pharmac., (1980) 9, 395

R1861 Hajdu,P. & Damm,D., Arzneim. Forsch., (1979) 29, 602

R1873 Rodriguez,L., Zecchi,V. & Cini,M., Il Farmaco, (1979) 34, 371

R1874 Rodriguez,L., Chiarini,A. & Zecchi,V., Il Farmac., (1981) 36, 304

R1875 Rogers,J.A. & Wong,A., Int. J. Pharmaceut., (1980) 6, 339

R1881 Krasovec,F., Ostanek,M. & Klofutar,C., Anal. Chim. Acta, (1966) 36, 431

R1890 Sanderson,M.D. & Williams,D.R., J. Inorg. Nucl. Chem., (1977) 39, 641

R1894 Draber,W., Bayer AG, Private Communication (7/85)

R1895 DeMeere,A., "Drug Absorpt.Thru Atrif.& Biol. Membranes", Amsterdam, 1985

R1896 Tanii,H. & Hashimoto,K., Arch. Toxicol., (1984) 55, 47

R1897 Ahmed,A.M.S., Farah,F.H. & Kellaway,I.W., Pharm. Res., (1985), 119

R1898 Erhua,W., Weijia,J., Shuyue,W. & Zhenxiang,Y., Medic. Industry (China), (1984) 7, 13

R1900 Pesak,M., Kopecky,F. & Borovansky,A., "QSAR in Design of Bioactive Cpds.", J. Prous, (1984), p. 209

R1901 Neef,C. & Meijer,D.K.F., NS Arch. Pharmacol., (1984) 328, 111

R1903 Briggs,G.G., Rothamsted Expt. Sta., Harpenden, Herts., Private Communication

R1906 Coates,M., Connell,D.W. & Barron,D.M., Environ. Sci. Technol., (1985) 19, 628

R1908 Debord,J., Labadie,M., Bollinger,J-C. & Yvernault,T., Phosph. & Sulfur, (1985) 22, 121

R1909 Bronaugh,R.L. & Congdon,E.R., J. Invest. Dermat., (1984) 83, 124

R1912 Saxena,S., Dhaon,M.K., Ram,S., Saxena,M., Cjain,P., Patnaik,G.K. & Anand,N., Indian J. Chem., (1983) 22B, 1224

R1920 Scherrer,R.A., "Pestic. Synth. thru Rational Approaches", ACS#255, (1984), 225

R1924 Petermann,C. & Fauchere,J-L., Helv. Chim. Acta, (1983) 66, 1513

R1927 Villegas-Navarro,A. & Rodriguez,V.M., "QSAR in Design of Bioactive Cpds.", J.R. Prous, (1984), 229

R1930 Schmidt-Bleek,F., Haberland,W., Klein,A.W. & Caroli,S., Chemosph., (1982) 11, 383

R1935 Staroscik,R. & Blaskiewicz,T., Pharmazie, (1985) 40, 248

R1937 Spencer,R.W., Copp,L.J. & Pfister,J.R., J. Med. Chem., (1985) 28, 1828

R1938 Cheney,B.V., Szmuszkovicz,J., Lahti,R.A. & Zichi,D.A., J. Med. Chem., (1985) 28, 1853

R1940 Livingstone,D.J. & Hill,A.P., Wellcome Research, U.K., Private Communication

R1941 Recanatini,M., Univ. Bologna, Private Communication

R1942 Muller,N., Purdue Univ., Private Communication

R1943 Beauchamp,L.M., Wellcome Research, U.S. & U.K., Private Communication

R1945 Wildfeuer,A. & Lemme,J-D., Arzneim. Forsch., (1985) 35, 639

R1946 Balaz,S., Kuchar,A., Sturdik,E., Rosenberg,M., Stibranyi,L. & Ilavsky,D., Coll. Czech. Chem. Comm., (1985) 50, 1642

R1947 Augustin,J., Balaz,S., Hanes,J. & Sturdik,E., Chem. Papers (Slovak Acad. Sci.), (1987) 41, 401

R1948 Caccia,S., Fong,M.H. & Urso,R., J. Pharm. Pharmacol., (1985) 37, 567

R1949 Young,R.C., Durant,G.J., Emmett,J.C., Ganellin,C.R., Graham,M.J., Mitchell,R.C., Prain,H.D. & Roantree,M.L., J. Med. Chem., (1986) 29, 44

R1950 Nikaido,H., Rosenberg,E.Y. & Foulds,J., J. Bacteriol., (1983) 153, 232

R1952 Funasaki,N., Hada,S. & Neya,S., J. Phys. Chem., (1985) 89, 3046

R1953 Manners,C. & Payling,D., Fisons Pharmaceuticals, U.K., Private Communication

R1955 Shimizu,T., Osumi,T., Niimi,K. & Nakagawa,K., Arzneim. Forsch., (1985) 35, 1117

R1956 Bundgaard,H., Falch,E., Larsen,C. & Mikkelson,T.J., J. Pharm. Sci., (1986) 75, 36

R1957 Laughlin Jr.,R.B., Guard,H.E. & Coleman III,W.M., Environ. Sci. Technol., (1986) 20, 201

R1958 Maguire,R.J., Carey,J.H. & Hale,E.J., J. Agric. Food Chem., (1983) 31, 1060

R1959 Wulf,R.G. & Byington,K.H., Arch. Biochem. Biophys., (1975) 167, 176

R1960 Kopecky,F., Pesak,M., Macek,J. & Borovansky,A., Pharmazie, (1985) 40, 572

R1962 Dagorn,M., Huet,J. & Burgot,J.L., Ann. Pharmac. France, (1985) 43, 165

R1963 Huang,J-X., Bouvier,E.S.P., Stuart,J.D., Melander,W.R. & Horvath,C., J. Chromatog., (1985) 330, 181

R1964 Huguenard,J.R. & Wilson,W.A., J. Pharmacol. Exper. Therap.,

(1985) 234, 821

R1966 Hempel,J. & Sutton,B., Smith, Kline & French, Philadelphia, Private Communication

R1967 Jira,T., Fleischmann,B., Burghardt,G., Beyrich,T., Martin,E. & Kuhmstedt,H., Pharmazie, (1985) 40, 34

R1969 Tanaka,M., Horisaka,K., Yamagami,C., Takao,N. & Fujita,T., Chem. Pharm. Bull., (1985) 33, 2403

R1970 Kung,H.F., Yu,C.C., Billings,J., Molnar,M. & Blau,M., J. Med. Chem., (1985) 28, 1280

R1973 Muchowski,J.M., et.al., J. Med. Chem., (1985) 28, 1037

R1975 Gupta,R.P., Larroquette,C.A., Agrawal,K.C., Grodkowski,J. & Neta,P., J. Med. Chem., (1985) 28, 987

R1977 Acton,E.M., Tong,G.L., Mosher,C.W. & Wolgemuth,R.L., J. Med. Chem., (1984) 27, 638

R1978 Casner,M.L., Remers,W.A. & Bradner,W.T., J. Med. Chem., (1985) 28, 921

R1979 Novotny,L., Farghali,H., Ryba,M., Janku,I. & Beranek,J., Cancer Chemother. Pharmacol., (1984) 13, 195

R1980 Yoshimura,F. & Nikaido,H., Antimicr. Agents Chemotherap., (1985) 27, 84

R1982 Appelgren,C., Borg,K.O., Elofsson,R. & Johansson,K.A., Acta Pharm. Suecica, (1974) 11, 325

R1983 Broughton,A., Grant,A.O., Starmer,C.F., Klinger,J.K., Stambler,B.S. & Strauss,H.C., Circulation Res., (1984) 55, 513

R1984 Cruickshank,J.M., Am. Heart J., (1980) 100, 160

R1985 Sturdik,E., Balaz,S., Antalik,M. & Sulo,P., Coll. Czech. Chem. Comm., (1985) 50, 538

R1986 Eller,M.G., Schoenwald,R.D., Dixson,J.A., Segarra,T. & Barfknecht,C.F., J. Pharm. Sci., (1985) 74, 155

R1987 Yamagami,C., Takao,N., Tanaka,M., Horisaka,K., Asada,S., Fujita,T., Chem. Pharm. Bull., (1984) 32, 5003

R1988 Yamagami,C., Takami,H., Yamamoto,K., Miyoshi,K. & Takao,N., Chem. Pharm. Bull., (1984) 32, 4994

R1990 Unger,S.H., "QSAR in Design of Bioactive Compounds", J.R. Prous Publ., 1984, p.1

R1991 Dall'Acqua,F., Vedaldi,D., Baccichetti,F., Rodighiero,G. & Gennaro,A., "QSAR in Design of Bioactive Compounds", J. Prous Publ., 1984, p.87

R1992 Hoefke,W., Gaida,W. & Stahle,H., Arzneim. Forsch., (1985) 35, 424

R1994 Ali,F.E., Bondinell,W.E., Dandridge,P.A., Frazee,J.S., Garvey,E., Girard,G.R., Kaiser,C., Ku,T.W., Lafferty,J.J., Moonsammy,G.I., Oh,H.-J., Rush,J.A., Setler,P.E., Stringer,O.D., Venslavsky,J.W., Volpe,B.W., Yunger,L.M. & Zirkle,C.L., J. Med. Chem., (1985) 28, 653

R1996 Donetti,A., Trummlitz,G., Bietti,G., Cereda,E., Bazzano,C. & Wagner,H.-U., Arzneim. Forsch., (1985) 35, 306

R1997 Nishimura,K.-I., Nozaki,Y., Yoshimi,A., Nakamura,S., Kitagawa,M., Kakeya,N. & Kitao,K., Chem. Pharm. Bull, (1985) 33, 282

R1999 Letourneux,Y., Sparfel,L., Roussakis,C., Piessard,S. & Le Baut,G., Eur. J. Med. Chem., (1984) 19, 535

R2000 Chiou,C.T., Environ. Sci. Tech. (1985) 19, 57

R2001 Miller,M.M. & Wasik,S.P., Environ. Sci. Technol., (1985) 19, 522

R2002 Friend,D.R., Chang,G.W., J. Med. Chem., (1985) 28, 51

R2003 Bedford,C.D., Harris III,R.N., Howd,R.A., Miller,A., Nolen III,H.W. & Kenley,R.A., J. Med. Chem., (1984) 27, 1431

R2004 Goto,S., Yoshitomi,H. & Nakase,M., Chem. Pharm. Bull., (1978) 26, 472

R2005 Kutter,E., Boehringer-Ingelheim, Private Communication

R2007 Bedford,C.D., Miura,M., Bottaro,J.C., Howd,R.A. & Nolen III,H.W., J. Med. Chem., (1986) 29, 1689

R2008 Burkhard,L.P. & Kuehl,D.W., Chemosphere, (1986) 15, 163

R2009 Buur,A. & Bundgaard,H., Int. J. Pharmaceutics, (1985) 23, 209

R2010 Buur,A. & Bundgaard,H., J. Pharm. Sci., (1986) 75, 522

R2011 Uchida,M., J. Pestic. Sci.,(JPN) (1984) 9, 559

R2012 Chamberlain,K., Butcher,D.N. & White,J.C., Pestic. Sci., (1986) 17, 48

R2013 Govers,H., Ruepert,C., Stevens,T. & van Leeuwen,C.J., Chemosphere, (1986) 15, 383

R2014 Reider,M.L. & Shaner,D.L., Plant Physiol., (1984) 75, 50

R2015 Howard,P.H., Banerjee,S. & Robillard,K.H., Environ. Toxicol. Chem., (1985) 4, 653

R2016 Herbette,L.G., Chester,D.W. & Rhodes,D.G., Biophys. J., (1986) 49, 91

R2017 Manners,C.N. & Payling,D., Fisons, Private Communication

(1986)

R2018 Manners,C.N. & Payling,D., Fisons, Xenobiot. 19, 1387 (1989) & Private Communication (1986)

R2019 Wong,P.T.S., Chau,Y.K., Kramar,O. & Bengert,G.A., Can. J. Fish. Aquat. Sci., (1982) 39, 483 & Private Communication

R2020 Tejani-Butt,S.M. & Brunswick,D.J., J. Med. Chem., (1986) 29, 1524

R2021 Jones,G.H., Venuti,M.C., Young,J.M., Murthy,D.V.K., Loe,B.E., Simpson,R.A., Berks,A.H., Spires,D.A., Maloney,P.J., Kruseman,M., Rouhafza,S., Kappas,K.C., Beard,C.C., Unger,S.H. & Cheung,P.S., J. Med. Chem., (1986) 29, 1504

R2022 Mitchell,R.C., Smith Kline & French Research Limited, UK, Private Communicatio (1986)

R2023 Nishimura,K., Hirayama,K., Kobayashi,T., Fujita,T., Holan,G., Pestic. Biochem. Physiol., (1986) 25, 153

R2024 Iyengar,B.S., Remers,W.A. & Bradner,W.T., J. Med. Chem., (1986) 29, 1864

R2025 Acton,E.M., Tong,G.L., Taylor,D.L., Streeter,D.G., Filppi,J.A. & Wolgemuth,R.L., J. Med. Chem., (1986) 29, 2074

R2026 Lisciani,R., Lembo,S., Cozzolino,S., La Rotonda,M.I., Silipo,C. & Vittoria,A., Il Farmaco, (1986) 41, 89

R2027 Cantelli-Forti,G., Guerra,M.C., Barbaro,A.M., Hrelia,P., Biagi,G.L. & Borea,P.A., J. Med. Chem., (1986) 29, 555

R2028 Lipinski,C.A., LaMattina,J.L. & Hohnke,L.A., J. Med. Chem., (1985) 28, 1628

R2029 Hamilton,H.W., Ortwine,D.F., Worth,D.F., Badger,E.W., Bristol,J.A., Bruns,R.F., Haleen,S.J. & Steffen,R.P., J. Med. Chem., (1985) 28, 1071

R2030 Singer,G.M., Andrews,A.W. & Guo,S.-m., J. Med. Chem., (1986) 29, 40

R2031 Livingstone,D.J. & Hill,A.P., Wellcome Res. Labs., Private Communication (1986)

R2032 Shalaby,A., Budvari-Barany,Z.S., Szasz,G.Y. & Hermecz,I., Acta Pharmac. Hung., (1985) 55, 249

R2033 Pratesi,P., Grana,E., Santagostino Barbone,M.G., La Rotonda,M.I., Silipo,C. & Vittoria,A., Il Farmaco, (1986) 41, 335

R2034 Priewer,H., Kraft,H. & Mutschler,E., Arzneim. Forsch., (1985) 35, 1819

R2035 Knauf,H., Mutschler,E., Volger,K.-D. & Wais,U., Arzneim. Forsch., (1978) 28, 1417

R2036 Duffel,M.W., Ing,I.S., Segarra,T.M., Dixson,J.A., Barfknecht,C.F. & Schoenwald,R.D., J. Med. Chem., (1986) 29, 1488

R2037 Carney,C. & Graham,E., Arzneim. Forsch., (1985) 35, 228

R2038 Thessen,R., Rhone-Poulenc, Private Communication

R2039 Loonen,A.J.M., Soudijn,W., Van Rooy,H.H. & Van Vijngaarden,I., Eur. J. Pharmacol., (1977) 45, 281

R2040 McLachlan,D., Arnason,T. & Lam,J., Biochem. Systematics Ecol., (1986) 14, 17

R2043 Yoshimura,Y., Hamaguchi,N. & Yashiki,T., Intern. J. Pharmaceut., (1985) 23, 117

R2044 Coleman,A.J., Paterson,D.S. & Somerville,A.R., Biochem. Pharmacol., (1979) 28, 1011

R2045 Dallet,P., Dubost,J.-P., Colleter,J.-C., Audry,E. & Creuzet,M.-H., Eur. J. Med. Chem., (1985) 20, 551

R2046 Katayama,M., Yamamoto,M., Kobayashi,S., Oguchi,K. & Yasuhara,H., J. Phar. Pharmacol., (1986) 38, 382

R2047 Tanii,H., Tsuji,H. & Hashimoto,K., Toxicol. Letters, (1986) 30, 13

R2049 Cohen,G., UCLA, Private Communication

R2050 Munson,R., A.H. Robins, Private Communication

R2051 Kozlovsky,J., Cizmarik,J., Pesak,M., Inczinger,F. & Borovansky,A., Arzneim. Forsch., (1982) 32, 1032

R2058 Galdino,S.L., Pitta,I.R. & Luu-Duc,C., Il Farmaco, (1986) 41, 59

R2059 Zhu,Y-C., Wu,J-A. & Xiu-Rong,X., Acta Pharm. Sinica, (1985) 20, 267

R2060 Yang,D.J., Lahoda,E.P., Brown,P.I. & Rankin,G.O., Toxicol., (1985) 36, 23

R2061 Laznicek,M. & Kvetina,J., QSAR, (1988) 7, 234

R2062 Marrel,C., Boss,G., Van De Waterbeemd,H. & Testa,B., Eur. J. Med. Chem., (1985) 20, 459

R2069 Tanii,H. & Hashimoto,K., Arch. Toxicol., (1985) 57, 88

R2070 Bundgaard,H. & Falch,E., Intern. J. Pharmaceut., (1985) 24, 307

R2071 Halbach,A., Arch. Toxicol., (1985) 57, 139

R2072 Monzani,A., Gamberini,G., Braghiroli,D., Di Bella,M., Raffa,L. & Sandrini,M., Arch. Pharm. (Weinheim), (1985) 318, 299

R2073 Roberts,M.S., Cossum,P.A., Kowaluk,E.A. & Polack,A.E., Intern. J. Pharmaceut., (1983) 17, 145

R2076 Buyuktimkin,S., Arch. Pharm. (Weinheim), (1985) 318, 496

R2077 Kirchner,J.J., Acree Jr.,W.E., Leo,A. & Gelli,G., J. Pharm. Sci., (1985) 74, 1129

R2080 Ali-Osman,F., Rosenblum,M.L., Giannini,D.D. & Levin,V.A., Cancer Res., (1985) 45, 2988

R2083 Sosnovsky,G., Lukszo,J. & Rao,N.U.M., J. Med. Chem., (1986) 29, 1250

R2084 Medeiros,D.M., Cadwell,L.L. & Preston,R.L., Bull. Envir. Contam. Toxicol., (1980) 24, 97

R2085 Ketring,A.R., Troutner,D.E., Hoffman,T.J., Stanton,D.K., Volkert,W.A. & Holmes,R.A., Intern. J. Nucl. Med. Biol., (1984) 11, 113

R2088 Ponticello,G.S., Freedman,M.B., Habecker,C.N., Lyle,P.A., Schwam,H., Varga,S.L., Christy,M.E., Randall,W.C. & Baldwin,J.J., J. Med. Chem., (1987) 30, 591

R2089 Yeung,J.M. & Knaus,E.E., J. Med. Chem., (1987) 30, 104

R2090 Diana,G.D., Oglesby,R.C., Akullian,V., Carabateas,P.M., Cutcliffe,D., Mallamo,J.P., Otto,M.J., McKinley,M.A., Maliski,E.G. & Michalec,S.J., J. Med. Chem., (1987) 30, 383

R2092 Kaiser,C., Oh,H-J., Garcia-Slanga,B.J., Sulpizio,A.C., Hieble,J.P., Wawro,J.E. & Kruse,L.I., J. Med. Chem., (1986) 29, 2381

R2093 Sosnovsky,G., Rao,N.U.M. & Li,S.W., J. Med. Chem., (1986) 29, 2225

R2094 Mitsutake,K-i., Iwamura,H., Shimizu,R. & Fujita,T., J. Agric. Food Chem., (1986) 34, 725

R2095 Baldwin,J.J., Christy,M.E., Denny,G.H., Habecker,C.N., Freedman,M.B., Lyle,P.A., Ponticello,G.S., Varga,S.L., Gross,D.M. & Sweet,C.S., J. Med. Chem., (1986) 29, 1065

R2096 Asselin,A.A., Humber,L.G., Voith,K. & Metcalf,G., J. Med. Chem., (1986) 29, 648

R2097 Calvino,R., Fruttero,R., Garrone,A. & Gasco,A., Quant. Struct.-Act. Relat., (1988) 7, 26

R2098 Grunewald,G., University of Kansas, Private Communication

R2099 Stahl,P. & Moser,P., Ciba-Geigy, Private Communication

R2101 Lee,R.M. & McDowall,R.D., J. Clin. Hosp. Pharm., (1986) 11, 389

R2102 Bundgaard,H., Falch,E., Larsen,C., Mosher,G.L. & Mikkelson,T.J., J. Pharm. Sci., (1986) 75, 775

R2105 Stein,J., Z. Chem., (1986) 26, 258

R2107 Guy,R.H., Carlstrom,E.M., Bucks,D.A.W., Hinz,R.S. & Maibach,H.I., J. Pharm. Sci., (1986) 75, 968

R2111 Ritschel,W.A. & Hammer,G.V., Int. J. Clin. Pharmac., (1980) 18, 298

R2112 Westley,J.W., Oliveto,E.P., Berger,J., Evans Jr.,R.H., Glass,R., Stempel,A., Toome,V. & Williams,T., J. Med. Chem., (1973) 16, 397

R2114 Sengupta,C. & Mondal,A.K., Ind. J. Pharm. Sci., (1986) 48, 51

R2120 Kofitsekpo,W.M., Drugs Exptl. Clin. Res., (1980) 6, 421

R2123 Soman,G., Narayanan,J., Martin,B.L. & Graves,D.J., Biochem., (1986) 25, 4113

R2125 Sherman,M.A., Linthicum,D.S. & Bolger,M.B., Molec. Pharmacol., (1986) 29, 589

R2126 Funasaki,N., Hada,S. & Neya,S., J. Chromatog., (1986) 361, 33

R2128 Parkin,J.E., J. Chromatog., (1986) 351, 532

R2129 Brooke,D.N., Dobbs,A.J. & Williams,N., Ecotox. Environ. Safety, (1986) 11, 251

R2132 De Biasi,V., Lough,W.J. & Evans,M.B., J. Chromatog., (1986) 353, 279

R2133 Hilbert,J.M., Gural,R.P., Symchowicz,S. & Zampaglione,N., J. Clin. Pharmacol., (1984) 24, 457

R2134 Saarikoski,J., Lindstrom,R., Tyynela,M. & Viluksela,M., Ecotox. Environ. Safety, (1986) 11, 158

R2136 Scherrer,R.A. & Howard,S.M., J. Med. Chem. (1977) 20, 53

R2137 Buchel,K.H., Draber,W., Trebst,A. & Pistorius,E., Z. Naturforsch., (1966) 21b, 243

R2138 Schultz,T.W. & Moulton,B.A., Environ. Toxicol. Chem., (1985) 4, 353

R2139 Schultz,T.W. & Applehans,F.M., Ecotox. Environ. Safety, (1985) 10, 75

R2140 Arnason,J.T., Philogene,B.J.R., Duval,F., McLachlan,D., Picman,A.K., Towers,G.H.N. & Balza,F., J. Natl. Prod., (1985) 48, 581

R2141 Balaz,S., Ilavsky,D., Sturdik,E. & Kovac,J., Folia Microbiol., (1985) 30, 34

R2142 Takahashi,J., Kirino,O., Takayama,C. & Kamoshita,K., J. Chromatog., (1988) 436, 316

R2143 Recanatini,M., Valenti,P. & Da Re,P., QSAR, (1988) 7, 12

R2144 Lee,G., Swarbrick,J., Kiyohara,G. & Payling,D.W., Int. J. Pharmaceut., (1985) 23, 43

R2145 Bundgaard,H., Klixbull,U. & Falch,E., Int. J. Pharmaceut., (1986) 30, 111

R2146 Doucette,W.J. & Andren,A.W., Chemosphere, (1988) 17, 345

R2149 Hawker,D.W. & Connell,D.W., Environ. Sci. Technol., (1988) 22, 382

R2150 Claramunt,R.M., Sanz,D., Elguero,J., Alvarez-Builla,J. & Gago,F., Farmaco, (1987) 42, 915

R2151 Saket,M.M., James,K.C. & Kellaway,I.W., Int. J. Pharmaceut., (1985) 27, 287

R2153 Negro,A., Mendez,R. & Salto,F., J. Liq. Chromatog., (1987) 10, 2789

R2155 Chiang,C-H., Huang,H-S. & Schoenwald,R.D., J. Taiwan Pharm. Assoc., (1986) 38, 67

R2156 Sparatore,F., La Rotonda, M.I., Caliendo,G., Novellino,E., Silipo,C. & Vittoria,A., Il Farmaco-Ed.Sci., (1988) 43, 141

R2157 Nielsen,N.M. & Bundgaard,H., J. Pharm. Sci., (1988) 77, 285

R2158 Wang,L., Wang,X., Xu,O. & Tian,L., Acta Sci.Circumstantiae, (1986) 6, 491

R2159 McCrady,J.K., McFarlane,C. & Lindstrom,F.T., J. Expl. Bot., (1987) 38, 1875

R2161 Arnoldi,A., Grasso,S., Meinardi,G. & Merlini,L., Eur. J. Med. Chem., (1988) 23, 149

R2162 El Tayar,N., Kilpatrick,G.J., Van De Waterbeemd,H., Testa,B., Jenner,P. & Marsden,C.D., Eur. J. Med. Chem., (1988) 23, 173

R2163 Tsantili-Kakoulidou,A., El Tayar,N., Van De Waterbeemd,H. & Testa,B., J. Chromatog., (1987) 389, 33

R2164 Recanatini,M., Private Communication (1988)

R2165 Sparatore,F., La Rotonda,M.I., Caliendo,G., Novellino,E., Silipo,C. & Vittoria,A., Il Farmaco, Ed. Sc., (1988) 43, 29

R2166 Calvino,R., Gasco,A. & Leo, A., J.Chem.Soc.Perk.Trans., 2, 1644 (1992)

R2168 Yang,H-Z., Nishimura,M., Nishimura,K., Kuroda,S. & Fujita,T., Pestic. Biochem. Physiol., (1987) 29, 217

R2169 Opperhuizen,A., Toxicol. Environ. Chem., (1987) 15, 249

R2170 Kramarczyk,K., Studia Biophysica, (1987) 117, 49

R2171 Anliker,R. & Moser,P., Ecotox. Environ. Safety, (1987) 13, 43

R2172 Raghavan,K.S., Shek,E. & Bodor,N., Anti-Cancer Drug Design, (1987) 2, 25

R2173 Tsubaki,H., Nakajima,E., Shigehara,E., Komai,T. & Shindo,H., J. Pharmacobio-Dyn., (1986) 9, 737

R2175 Leung,C-S., Rowlands,M.G., Jarman,M., Foster,A.B., Griggs,L.J. & Wilman,D.E.V., J. Med. Chem., (1987) 30, 1550

R2176 Jacob,J.N., Hesse,G.W. & Shashoua,V.E., J. Med. Chem., (1987) 30, 1573

R2177 Dal Pozzo,A., Acquasaliente,M., Donzelli,G., DeMaria,P. & Nicoli,M.C., J. Med. Chem., (1987) 30, 1674

R2178 Nieder,M., Strosser,W. & Kappler,J., Arzneim. Forsch., (1987) 37, 549

R2179 Kuchar,M., Maturova,E., Brunova,B., Grimova,J., Tomkova,H. & Holubek,J., Coll. Czech. Chem. Commun., (1988) 53, 1862

R2180 De Amici,M., De Micheli,C., Pratesi,P., Grana,E., Barbone, M.G.S., Cappello,B., Silipo,C., Vittoria,A., Il Farmaco, Ed. Sc., (1987) 42, 409

R2182 Krikorian,S.E., Chorn,T.A. & King,J.W., Quant. Struct.-Act. Relat., (1987) 6, 65

R2184 Yalkowsky,S.H., Valvani,S.C. & Roseman,T.J., J. Pharm. Sci., (1983) 72, 866

R2185 De Kock,A.C. & Lord,D.A., Chemosphere, (1987) 16, 133

R2186 Murakami,N., Uchida,M., Sugimoto,T. & Nakatsugawa,T., Xenobiotica, (1987) 17, 241

R2188 Kurtz,A.P. & Durden Jr.,J.A., J. Agric. Food Chem., (1987) 35, 115

R2190 Bonati,M., Gaspari,F., D'Aranno,V., Benfenati,E., Neyroz,P., Galletti,F. & Tognoni,G., J. Pharm. Sci., (1984) 73, 829

R2191 Gaspari,F. & Bonati,M., J. Pharm. Pharmacol., (1987) 39, 252

R2192 Irwin,V.P., Quigley,J.M. & Timoney,R.F., Int. J. Pharmaceut., (1987) 34, 241

R2195 Young,R.C., Mitchell,R.C., Brown,T.H., Ganellin,C.R., Griffiths,R., Jones,M., Rana,K.K., Saunders,D., Smith,I.R., Sore,N.E. & Wilks,T.J., J. Med. Chem., (1988) 31, 656

R2196 Recanatini,M., Univ. of Bologna, Private Communication (1988)

R2197 Asao,M., Iwamura,H., Akamatsu,M. & Fujita,T., J. Med. Chem., (1987) 30, 1873

R2198 Sotomatsu,T., Nakagawa,Y. & Fujita,T., Pest. Biochem. Physiol., (1987) 27, 156

R2200 Grunbaum,Z., Freauff,S.J., Krohn,K.A., Wilbur,D.S., Magee,S. & Rasey,J.S., J. Nucl. Med., (1987) 28, 68

R2201 Martin-Villodre,A., Pla-Delfina,J.M., Moreno,J., Perez-Buendia,D., Miralles,J., Collado,E.F., Sanchez-Moyano,E. & del Pozo,A., J. Pharmacokin. Biopharmaceut., (1986) 14, 615

R2202 Wong,O. & McKeown,R.H., J. Pharm. Sci., (1988) 77, 926

R2203 Klein,W., Kordel,W., WeiB,M. & Poremski,H.J., Chemosphere, (1988) 17, 361

R2204 Dijkstra,D., Mulder,T.B.A., Rollema,H., Tepper,P.G., Van der Weide,J. & Horn,A.S., J. Med. Chem., (1988) 31, 2178

R2205 La Rotonda,M.I., Cappello,B., Grimaldi,M., Silipo,C. & Vittoria,A., Il Farmaco, (1988) 43, 439

R2206 De Ranter,C.J., Katholieke Univ. Leuven, Private Communication

R2207 Farraj,N.F., Davis,S.S., Parr,G.D. & Stevens,H.N.E., Int. J. Pharmaceut., (1988) 46, 231

R2209 Nishimura,K., Kitahaba,T., Ikemoto,Y. & Fujita,T., Pestic. Biochem. Physiol., (1988) 31, 155

R2211 Go,M.L. & Ngiam,T.L., Chem. Pharm. Bull., (1988) 36, 1393

R2212 Balaz,S., Sturdik,E., Durcova,E., Antalik,M. & Sulo,P., Biochim. Biophys. Acta, (1986) 851, 93

R2213 Marley,R.J., Miner,L.L., Wehner,J.M. & Collins,A.C., J. Pharmacol. Expl. Therap., (1986) 238, 1028

R2214 Makino,K., Sakata,G., Kawamura,Y. & Ikai,T., J. Pesticide Sci., (1986) 11, 469

R2215 Miyoshi,H., Maeda,H., Tokutake,N. & Fujita,T., Bull. Chem. Soc. Jpn., (1987) 60, 4357

R2216 Yeh,J.Z. & TenEick,R.E., Biophys. J., (1987) 51, 123

R2217 Lympany,P., Cassidy,S.L. & Henry,J.A., Mechan. Models Toxicol., Arch. Toxicol. Suppl., (1987) 11, 329

R2218 Shankley,N.P., Black,J.W., Ganellin,C.R. & Mitchell,R.C., Br. J. Pharmacol., (1988) 94, 264

R2219 Deneer,J.W., PhD. Thesis, Rijksuniversity, 4/25/88

R2220 Neumann,W., Laasch,H. & Urbach,W., Pestic. Biochem. Physiol., (1987) 27, 189

R2221 Bradshaw,J. & Taylor,P.J., Quant. Struct.-Act. Relat., (1989) 8, 279

R2222 Leahy,D.E., Taylor,P.J. & Wait,A.R., Quant. Struct.-Act. Relat., (1989) 8, 17

R2223 Maliski,E.G., Sterling-Winthrop, Private Communication (1987)

R2224 Chien,Y.,Communicated by K. Koehler, Searle Res. Devel., Unpublished

R2225 Dalmark,M. & Johansen,P., Mol. Pharmacol., (1982) 22, 158

R2226 Wolfenden,R. Andersson,L., Cullis,P.M. & Southgate,C.C.B., Biochem., (1981) 20, 849

R2228 Peips,M.M., Sikk,P.F. & Aaviksaar,A.A., Org. Reactivity, (1986) 23, 261

R2229 Banks,W.A., Kastin,A.J., Coy,D.H., Angulo,E., Brain Res. Bull., (1986) 17, 155

R2230 Walter,K. & Kurz,H., J. Pharm. Pharmacol., (1988) 40, 689

R2233 Anderson,W.K., Bhattacharjee,D. & Houston,D.M., J. Med. Chem., (1989) 32, 119

R2234 Nielsen,N.M. & Bundgaard,H., J. Med. Chem., (1989) 32, 727

R2235 Stasch,J-P., Rub,H., Schacht,U., Witteler,M., Neuser,D., Gerlach,M., Leven,M., Kuhn,W., Jutzi,P. & Przuntek,H., Arzneim. Forsch., (1988) 38, 1075

R2236 Terada,H., Kosuge,Y., Murayama,W., Nakaya,N., Nunogaki,Y. & Nunogaki,K.-I., J. Chromatog., (1987) 400, 343

R2237 Turan-Zitouni,G., Berge,G., Noel-Artis,A.M., Chevallet,P., Fulcrand,P. & Castel,J., Il Farmaco, Ed. Sci., (1987) 43, 643

R2238 Sosnovsky,G., Li,S.W. & Rao,N.U.M., Z. Naturforsch., (1987) 42c, 921

R2239 Bilski,A.J., Hadfield,S.E. & Wale,J.L., J. Cardiovasc. Pharmacol., (1988) 12, 227

R2241 Thoma,K. & Kasper,F.R., Pharm. Acta Helv., (1988) 63, 155

R2242 Maes,P., Formstecher,P., Lustenberger,P. & Dautrevaux,M., J. Chromatog., (1988) 445, 409

R2244 Bundgaard,H., Buur,A., Chang,S.-C. & Lee,V.H.L., Int. J. Pharmaceut., (1988) 46, 77

R2245 Betageri,G.V. & Rogers,J.A., Int. J. Pharmaceut., (1988) 46, 95

R2247 Gago,F., Alvarez-Builla,J. & Elguero,J., J. Chromatog., (1988) 449, 95

R2248 Miyake,K., Mizuno,N. & Terada,H., Chem. Pharm. Bull., (1986) 34, 4787

R2250 Kuchar,M., Kraus,E., Rejholec,V. & Miller,V., J. Chromatog., (1988) 449, 391

R2252 Balzarini,J., Cools,M. & De Clercq,E., Biochem. Biophys. Res.

Commun., (1989) 158, 413

R2253 Sherblom,P.M. & Eganhouse,R.P., J. Chromatog., (1988) 454, 37

R2254 Kilpatrick,G.J., El Tayar,N., Van de Waterbeemd,H., Jenner,P., Testa,B. & Marsden,C.D., Mol. Pharmacol., (1986) 30, 226

R2256 Wang,L., Wei,P., Tian,L., Xu,O., Zhang,Z., Environ. Chem. (China), (1987) 6, 51

R2259 Sunada,H., Sugimoto,M., Yonezawa,Y. & Otsuka,A., Yakuzaigaku, (1988) 48, 58

R2261 Feenstra,M.G.P., Homan,J.W., Everts,R., Rollema,H. & Horn,A.S., N.-S. Arch. Pharmacol., (1984) 326, 203

R2263 Van Bree,J.B., De Boer,A.G., Danhof,M., Ginsel,L.A. & Breimer,D.D., J. Pharmacol. Expt. Therap., (1988) 247, 1233

R2264 Winkler,D.A., CSIRO, private communication

R2265 Carotti,A., Raguseo,C., Campagna,F., Langridge,R. & Klein,T.E., Quant. Struct.-Act. Relat., (1989) 8, 1

R2266 Buur,A. & Bundgaard,H., Intern. J. Pharmaceut., (1987) 36, 41

R2267 Bundgaard,H. & Falch,E., Intern. J. Pharmaceut., (1985) 25, 27

R2268 Bundgaard,H. & Nielsen,N.M., Intern. J. Pharmaceut., (1988) 43, 101

R2269 Buur,A., Bundgaard,H. & Falch,E., Acta Pharm. Suec., (1986) 23, 205

R2270 Buur,A. & Bundgaard,H., Intern. J. Pharmaceut., (1984) 21, 349

R2271 Buur,A. & Bundgaard,H., Arch. Pharm. Chem., Sci. Ed., (1984) 12, 37

R2272 Larsen,J.D., Bundgaard,H. & Lee,V.H.L., Intern. J. Pharmaceut., (1988) 47, 103

R2273 Shibamoto,Y., Sakano,K., Kimura,R., Nishidai,T., Nishimoto,S-I., Ono,K., Kagiya,T. & Abe,M., Int. J. Radiation Oncology Bio. Phys., (1986) 12, 1063

R2274 Lemaire,M. & Tillement,J.P., Biochem. Pharmacol., (1982) 31, 359

R2277 Dal Pozzo,A., Donzelli,G., Rodriquez,L. & Tajana,A., Int. J. Pharmaceut., (1989) 50, 97

R2278 Tacke,R., Pikies,J., Linoh,H., Rohr-Aehle,R. & Gonne,S., Ann. Chem.(Liebigs), (1987), 51

R2279 Fitzpatrick,D.A., Heindel,N.D., Egolf,R.A. & Walton,H., Radiation Res., (1989) 117, 47

R2280 Wieland,D.M., Kilbourn,M.R., Yang,D.J., Laborde,E., Gildersleeve,D.L., Van Dort,M.E., Pirat,J-L., Ciliax,B.J. & Young,A.B., Appl. Radiat. Isot., (1988) 39, 1219

R2282 Tejani-Butt,S.M., Hauptmann,M., D'Mello,A., Frazer,A., Marcoccia,J.M. & Brunswick,D.J., N-S. Arch. Pharmacol., (1988) 338, 497

R2284 Horakova,L., Stolc,S. & Szocsova,H., Pharmazie, (1988) 43, 213

R2285 Dzimiri,N., Fricke,U. & Klaus,W., Br. J. Pharmac., (1987) 91, 31

R2286 Brewster,M.E., Estes,K.S. & Bodor,N., Pharmac. Res., (1986) 3, 278

R2288 Clarke,F.H. & Cahoon,N.M., J. Pharm. Sci., (1987) 76, 611

R2289 Camilleri,P., Kerr,M.W., Newton,T.W. & Bowyer,J.R., J. Agric. Food Chem., (1989) 37, 196

R2292 Ebert,E., Eckhardt,W., Jakel,K., Moser,P., Sozzi,D. & Vogel,C., Z. Naturforsch., (1989) 44c, 85

R2293 De Bruijn,J., Busser,F., Seinen,W. & Hermens,J., Environ. Tox. Chem., (1989) 8, 499

R2294 Kataoka,T., Kai,H., Ishizuka,I., Hatta,T. & Ogata,M., J. Pestic. Sci., (1987) 12, 445

R2297 Roth,B. & Aig,E., J. Med. Chem., (1987) 30, 1998

R2298 Bedford,C.D., Howd,R.A., Dailey,O.D., Miller,A., Nolen III,H.W., Kenley,R.A., Kern,J.R. & Winterle,J.S., J. Med. Chem., (1986) 29, 2174

R2300 Ohori,Y. & Ihashi,Y., Mitsubishi Chemical R&D Review, (1987) 1, 22

R2302 Young,R.C., Graham,M.J. & Roantree,M.L., "QSAR in Drug Design & Toxicology", Hadzi & Jerman-Blazic Eds., Elsevier, Amsterdam, 1987, p.91

R2303 Hermens,J. & De Bruijn,J., "QSAR in Drug Design & Toxicology", Hadzi & Jerman-Blazic Eds., Elsevier, Amsterdam, 1987, p.343

R2304 Facchetti,I. & Vigevani,A., "QSAR in Drug Design & Toxicology", Hadzi & Jerman-Blazic Eds., Elsevier, 1987, p.138

R2305 Deneer,J.W., Sinnige,T.L. & Hermens,J.L.M., "QSAR in Drug Design & Toxicology", Hadzi & Jerman-Blazic Eds., Elsevier, Amsterdam, 1987, p.352

R2306 Grass,G.M., Cooper,E.R. & Robinson,J.R., J. Pharm. Sci., (1988) 77, 24

R2307 Sauerberg,P., Fjalland,B., Larsen,J-J., Bach-Lauritsen,T., Falch,E. & Krogsgaard-Larsen,P., Eur. J. Pharmacol., (1986) 130, 125

R2308 Stolwijk,T.B., Vos,L.C., Sudholter,E.J.R. & Reinhoudt,D.N., Rec. Trav. Chim. Pays-Bas, (1989) 108, 103

R2309 Cotton,R., Hardy,P.M. & Langran-Goldsmith,A.E., Int. J. Peptide Protein Res., (1986) 28, 230

R2310 Walther,B., Carrupt,P-A., El Tayar,N. & Testa,B., Helv. Chim. Acta, (1989) 72, 507

R2312 Hanai,T. & Hubert,J., J. Chromatog., (1982) 239, 527

R2314 Izumi,T. & Kitagawa,T., Chem. Pharm. Bull., (1989) 37, 742

R2316 Itokawa,H., Totsuka,N., Nakahara,K., Maezuru,M., Takeya,K., Kondo,M., Inamatsu,M. & Morita,H., Chem. Pharm. Bull., (1989) 37, 1619

R2317 Schmiedel-Jakob,I., Breuninger,V. & Hatt,H., Chemical Senses, (1988) 13, 619

R2318 Muto,Y., Bandoh,K., Watanabe,K., Katoh,N. & Ueno,K., Antimicrob. Agents Chemother., (1989) 33, 242

R2319 Vessieres,A., Top,S., Ismail,A.A., Butler,I.S., Louer,M. & Jaouen,G., Biochemistry, (1988) 27, 6659

R2320 Stout,D.M., Black,L.A., Barcelon-Yang,C., Matier,W.L., Brown,B.S., Quon,C.Y. & Stampfli,H.F., J. Med. Chem., (1989) 32, 1910

R2323 Fauchere,J-L., Charton,M., Kier,L.B., Verloop,A. & Pliska,V., Int. J. Peptide Prot.Res., (1988) 32, 269

R2325 Moser,P., Sallmann,A. & Wiesenberg,I., J. Med. Chem., (1990) 33, 2358

R2326 Borea,P.A., Pietrogrande,M.C. & Biagi,G.L., Biochem. Pharmacol., (1988) 37, 3953

R2327 Garren,K.W. & Repta,A.J., J. Pharm. Sci., (1989) 78, 160

R2328 Moody,R.P., Carroll,J.M. & Kresta,A.M.E., Toxicol. Indust. Health, (1987) 3, 479

R2329 Schwarzenbach,R.P., Stierli,R., Folsom,B.R. & Zeyer,J., Environ. Sci. Technol., (1988) 22, 83

R2330 Page,J.D., Chaney,S.G., Hall,I.H., Lee,K.H. & Holbrook,D.J., Biochim. Biophys. Acta, (1987) 926, 186

R2331 Huff,J.R., Baldwin,J.J., deSolms,S.J., Guare Jr.,J.P., Hunt,C.A., Randall,W.C., Sanders,W.S., Smith,S.J., Vacca,J.P. & Zrada,M.M., J. Med. Chem., (1988) 31, 641

R2334 Degorre,F., Kiffer,D. & Terrier,F., J. Med. Chem., (1988) 31, 757

R2335 Langguth,P. & Mutschler,E., Arzneim. Forsch., (1987) 37, 1362

R2336 Laznicek,M., Kvetina,J., Mazak,J. & Krch,V., J. Pharm. Pharmacol., (1987) 39, 79

R2337 Morita,I., Kunimoto,K., Tsuda,M., Tada,S-I., Kise,M. & Kimura,K., Chem. Pharm. Bull., (1987) 35, 4144

R2340 Callahan,M.A., Slimak,M.W., Gabel,N.W., May,I.P., Fowler,C.F., Freed,J.R., Jennings,P., Durfee,R.L., Whitmore,F.C., Maestri,B., Mabey,W.R., Holt,B.R. & Gould,C., EPA-440/4-79-029a (1979)

R2341 Kim,K.H., Martin,Y.C., Norris,B. & Haviv,F., J. Pharm. Sci., (1989) 78, 494

R2342 Matsuba,C.A., Nelson,W.O., Rettig,S.J. & Orvig,C., Inorg. Chem., (1988) 27, 3935

R2343 Obraztsov,V.V., Shekhtman,D.G., Sklifas,A.N. & Makarov,K.N., Biokhim., (1988) 53, 613, 535ee

R2344 Frobe,Z., Drevenkar,V. & Stengl,B., Toxicol. Environ. Chem., (1989) 19, 69

R2347 Xu,Z-J., Love,M.L., Ma,L.Y.Y., Blum,M., Bronskill,P.M., Bernstein,J., Grey,A.A., Hofmann,T., Camerman,N. & Wong,J., J. Biol. Chem., (1989) 264, 4304

R2348 Lam,S.P., Barlow,D.J. & Gorrod,J.W., J. Pharm. Pharmacol., (1989) 41, 373

R2349 Robins,M.J., Wood,S.G., Dalley,N.K., Herdewijn,P., Balzarini,J. & De Clercq,E., J. Med. Chem., (1989) 32, 1763

R2350 van den Broek,L.A.G.M., Lazaro,E., Zylicz,Z., Fennis,P.J., Missler,F.A.N., Lelieveld,P., Garzotto,M., Wagener,D.J.T., Ballesta,J.P.G. & Ottenheijm,H.C.J., J. Med. Chem., (1989) 32, 2002

R2351 Panten,U., Burgfeld,J., Goerke,F., Rennicke,M., Schwanstecher,M., Wallasch,A., Zunkler,B.J. & Lenzen,S., Biochem. Pharmacol., (1989) 38, 1217

R2352 Cloux,J.L., Crommen,J., Delarge,J., Pirard,M.L. & Thunus,L., J. Pharm. Belg., (1988) 43, 141

R2353 Rekka,E. & Kourounakis,P., Eur. J. Med. Chem., (1989) 24, 179

R2354 Betageri,G. & Rogers,J.A., Pharm. Res., (1989) 6, 399

R2357 Nakatsu,K. & Diamond,J., Can. J. Physiol. Pharmacol., (1989) 67, 251

R2358 Gale,G.R., Atkins,L.M., Smith,A.B., Jones,S.G., Basinger,M.A. & Jones,M.M., Arch. Toxicol., (1988) 62, 428

R2359 Nosal,R., Drabikova,K., Pecivova,J. & Ondrias,K., Agents & Actions, (1989) 27, 36

R2360 Zeman,E.M., Baker,M.A., Lemmon,M.J., Pearson,C.I., Adams,J.A., Brown,J.M., Lee,W.W. & Tracy,M., Int. J. Radiat. Oncol. Biol. Phys., (1989) 16, 977

R2361 Shibamoto,Y., Nishimoto,S-I., Shimokawa,K., Hisanaga,Y., Zhou,L., Wang,J., Sasai,K., Takahashi,M., Abe,M., & Kagiya,T., Int. J. Rad. Oncol. Biol. Phys., (1989) 16, 1045

R2362 Berger,S.G., Waser,P.G. & Hofmann,A., Arzneim. Forsch., (1989) 39, 762

R2363 Wang,W., Wang,L., Tian,L., Zhang,Z. & Qiu,J., Chem. Abst., (1988) 108, 206, 2089h

R2366 Green,M.A., Klippenstein,D.L. & Tennison,J.R., J. Nucl. Med., (1988) 29, 1549

R2367 Yoshikawa,Y. & Terada,H., Chem. Pharm. Bull., (1988) 36, 2759

R2371 Burgot,G. & Burgot,J.J., Chem. Abst. (1987) 107, 16, 51380j

R2372 Arcamone,F., Cancer Res., (1985) 45, 5995

R2373 Gosland,M.P., Lum,B.L. & Sikic,B.I., Cancer Res., (1989) 49, 6901

R2374 Akamatsu,M., Yoshida,Y., Nakamura,H., Asao,M., Iwamura,H. & Fujita,T., Quant. Struct.-Act. Relat., (1989) 8, 195

R2375 Mannhold,R., Dross,K.P. & Rekker,R.F., Quant. Struct.-Act. Relat., (1990) 9, 21

R2377 Akamatsu,M. and Fujita,T., Private Communication

R2378 Cheney,B.V., J. Med. Chem., (1988) 31, 521

R2379 Guo,Z., Private Communication

R2380 Meulemans,A., Vicart,P., Mohler,J., Henzel,D. & Vulpillat,M., Chemother. (1986) 32, 486

R2381 Betschart,H.R., Jondorf,W.R. & Bickel,M.H., Xenobiot., (1988) 18, 113

R2382 Camilleri,P., Bowyer,J.R., Gilkerson,T., Odell,B. & Weaver,R.C., J. Agric. Food Chem., (1987) 35, 479

R2387 Suwinski,J, Salwinska,E., Watras,J. & Widel,M., Acta Polon Pharm., (1985) 42, 352

R2389 Rosen,T., Chu,D.T.W., Lico,I., Fernandes,P., Marsh, K., Shen,L., Cepa,V., and Pernet,A., J.Med.Chem., (1988) 31, 1598

R2390 Polasek,M., Kohoutkova,D. & Waisser,K., Anal. Chim. Acta, (1988) 212, 279

R2391 Abbott,F.S. & Acheampong,A.A., Neuropharmacol., (1988) 27, 287

R2392 Gombar,C.T., Demarinis,R.M., Wise,M. & Mico,B.A., Drug Metab. Dispos., (1988) 16, 367

R2393 Briggs,G.G., Rigitano,R.L.O. & Bromilow,R.H., Pestic. Sci., (1987) 19, 101

R2395 Sijm,D.T., Wever,H., de Vries,P.J. & Opperhuizen,A., Chemosph., (1989) 19, 263

R2397 Private Communication from Dept. of Medicinal Chemistry, Div. of Expt.Therapeutics, Walter Reed Army Institute of Research, Meas. by SRI

R2399 Hinderling,P.H., Schmidlin,O. & Seydel,J.K., J. Pharmacokin. Biopharmaceut., (1984) 12, 263

R2400 de Jonge,A., Timmermans,P. & van Zwieten,P., Quant. Struct.-Act. Relat., (1984) 3, 138

R2401 Hellenbrecht,D. & Enenkel,J., Arzneim. Forsch., (1984) 34, 980

R2402 Hurtado,I., Urbina,C., Suarez,C., Beyer,B. & Magro,A., Int. J. Immunopharmac., (1985) 7, 635

R2403 De Maria,P., Fini,A., Guarnieri,A. & Varoli,L., Archiv Pharmazie, (1984) 317, 877

R2404 Camps,F., Colomina,O., Messeguer,A. & Sanchez,F.J., J. Liq. Chromatog., (1986) 9, 23

R2405 Biagi,G.L., Barbaro,A.M., Guerra,M.C., Borea,P.A. & Recanatini,M., J. Chromatog., (1990) 504, 163

R2406 Tait,A., Gaberini,G., Giovannini,M.G. & Di Bella,M., Il Farmaco, (1989) 44, 1129

R2407 Yokogawa,K., Nakashima,E., Ishizaki,J., Maeda,H., Nagano,T. & Ichimura,F., Pharmaceut. Res., (1990) 7, 691

R2408 Arnaud-Neu,F., Schwing-Weill,M-J., Spiess,B. & Yahya,R., J. Chem. Soc. Perkin Trans. (1990) 2, 1191

R2409 Strichartz,G.R., Sanchez,V., Arthur,G.R., Chafetz,R. & Martin,D., Anesth. Analg., (1990) 71, 158

R2410 Nakagawa,Y., Izumi,K., Oikawa,N., Sotomatsu,T., Shigemura,M. & Fujita,T., Environ. Toxicol. Chem., 11, 901 1992

R2411 Shirasaka,T., Murakami,K., Ford Jr.,H., Kelley,J.A., Yoshioka,H., Kojima,E., Aoki,S., Broder,S. & Mitsuya,H., Proc. Natl. Acad. Sci. USA, (1990) 87, 9426

R2412 Yamagami,C., Takao,N. & Fujita,T., Quant. Struct.-Act. Relat.,

(1990) 9, 313

R2413 Winkler,D.A., Liepa,A.J., Anderson-McKay,J.E. & Hart,N.K., Pestic Sci., (1989) 27, 45

R2415 Thomas,B.F., Compton,D.R. & Martin,B.R., J. Pharmacol. Expt. Therap., (1990) 255, 624

R2416 Mason,R.P., Rhodes,D.G. & Herbette,L.G., J. Med. Chem., (1991) 34, 869

R2417 Chien,Y.W. & Mizuba,S.S., J. Med. Chem., (1978) 21, 374

R2418 Hay,J.V., Pestic. Sci., (1990) 29, 247

R2419 Andrea,T.A., Stranz,D.D., Yang,A., Kleier,D.A., Patel,K.M., Powell,J.E., Price,T.P. & Marynick,D.S., Pestic. Sci., (1990) 28, 49

R2420 Bateman,G.L., Nicholls,P.H. & Chamberlain,K., Pestic. Sci., (1990) 29, 109

R2421 Kataoka,T., Hayase,Y., Hatta,T., Hayashi,Y., Murabayashi,A., Makisumi,Y. & Fujita,T., Pest. Biochem. Physiol., (1989) 34, 228

R2423 Swindell,C.S., Krauss,N.E., Horwitz,S.B. & Ringel,I., J. Med. Chem., (1991) 34, 1176

R2424 Caliendo,G., Novellino,E., Sagliocco,G., Santagada,V., Silipo,C. & Vittoria,A., Eur. J. Med. Chem., (1990) 25, 343

R2425 Sasaki,H., Mori,Y., Nakamura,J. & Shibasaki,J., J. Med. Chem., (1991) 34, 628

R2426 Souchard,J-P., Ha,T.T.B., Cros,S. & Johnson,N.P., J. Med. Chem., (1991) 34, 863

R2427 Moss,J., Buur,A. & Bundgaard,H., Int. J. Pharmaceut., (1990) 66, 183

R2428 Wallace,G.C. & Zerba,E.N., Pestic. Sci., (1989) 26, 215

R2430 Garrigues,T.M., Perez-Varona,A.T., Climent,E.,Bermejo,M.V., Martin-Villodre,A. & Pla-Delfina,J.M., Int. J. Pharmaceut., (1990) 64, 127

R2431 Hayward,D.S., Kenley,R.A. & Jenke,D.R., Int. J. Pharmaceut., (1990) 59, 245

R2433 Bundgaard,H., Jensen,E., Falch,E. & Pedersen,S.B., Int. J. Pharmaceut., (1990) 64, 75

R2434 Rubino,J.T. & Thomas,E., Int. J. Pharmaceut., (1990) 65, 141

R2436 Sasaki,H., Takahashi,T., Mori,Y., Nakamura,J. & Shibasaki,J., Int. J. Pharmaceut., (1990) 60, 1

R2437 Mylari,B.L., Larson,E.R., Beyer,T.A., Zembrowski,W.J., Aldinger,C.E., Dee,M.F., Siegel,T.W. & Singleton,D.H., J. Med. Chem., (1991) 34, 108

R2438 Otagiri,M., Fokkens,J.G., Hardee,G.E. & Perrin,J.H., Pharm. Acta Helv., (1978) 53, 241

R2439 Yoshino,K., Goto,K., Yoshiizumi,K., Morita,T. & Tsukamoto,G., J. Med. Chem., (1990) 33, 2192

R2440 Barchi Jr.,J.J., Marquez,V.E., Driscoll,J.S., Ford Jr.,H., Mitsuya,H., Shirasaka,T., Aoki,S. & Kelley,J.A., J.Med.Chem., 34, 1647 (1991)

R2441 Kramer,C-R. & Henze,U., Z. Phys. Chem.(Leipzig), (1990) 271, 503

R2442 Bundgaard,H. & Moss,J., Pharmac. Res., (1990) 7, 885

R2444 Yamagami,C., Ogura,T. & Takao,N., J. Chromatog., (1990) 514, 123

R2446 Hirai,K., Aoyama,H., Irikura,T., Iyobe,S. & Mitsuhashi,S., Antimicrob. Agents Chemoth., (1986) 29, 535

R2447 Furst,W., Neubert,R., Jurkschat,T. & Lucke,L., Int. J. Pharmaceut., (1990) 61, 43

R2449 Bennion,C., Brown,R.C., Cook,A.R., Manners,C.N., Payling,D.W. & Robinson,D.H., J. Med. Chem., (1991) 34, 439

R2450 Grisar,J.M., Petty,M.A., Bolkenius,F.N., Dow,J., Wagner,J., Wagner,E.R., Haegele,K.D. & De Jong,W., J. Med. Chem., (1991) 34, 257

R2451 Ohkawa,S., Terao,S., Terashita,Z-i., Shibouta,Y. & Nishikawa,K., J. Med. Chem., (1991) 34, 267

R2452 Rich,P.R. & Harper,R., FEBS Let., (1990) 269, 139

R2453 Roda,A., Minutello,A., Angellotti,M.A. & Fini,A., J. Lipid Res., (1990) 31, 1433

R2454 Fini,A., Roda,A., Bellini,A.M., Mencini,E. & Guarneri,M., J. Pharm. Sci., (1990) 79, 603

R2456 Jafvert,C.T., Westall,J.C., Grieder,E. & Schwarzenbach,R.P., Environ. Sci. Technol., (1990) 24, 1795

R2457 Okmoto,H., Hashida,M. & Sezaki,H., J.Pharm.Sci., 80, 39 (1991)

R2458 Lipinski,C., Fiese,E. & Korst,R., QSAR, 10, 109 (1991)

R2459 Baars,L., Schepers,M., Hermans,J., Dahlmans,H. & Thijssen,H., J.Pharm.Pharmacol., 42, 861 (90)

R2460 Leszczynski,D. & Schafer,R., Lipids, 25, 711 (1990)

R2461 Prugh,J., Hartman,G., Mallorga,P., McKeever,B., Michelson,S., Murcko,M., Schwam,H., Smith,R., Sondey,J.,et.al., J.Med.Chem.,

34, 1805, (1991)

R2462 Amaral,A., Miyazaki,Y., Stachissini,A., Caprara,L. & Oliveira,A., "QSAR:Rat.Appr.Des.Bioact.Cpds.", Ed. Silipo, Elsevier,(1991) p.509

R2463 Caliendo,G., Novellino,E., Santagada,V., Silipo,C. & Vittoria,A., "QSAR:Rat.Appr.Des.Bioact.Cpds." Ed. Silipo & Vittoria, Elsevier(1991)p401

R2464 Barbato,F., Rotonda,M. & Morrica,P., "QSAR:Rat.Appr.Des.Bioact.Cpds.", Eds. Silipo & Vittoria, Elsevier, (1991) p.99

R2465 Barbato,F., Caliendo,G., Cappello,B. & Rotonda,M., "QSAR:RAT.Appr.Des.Bioact.Cpds., Eds. Silipo & Vittoria, Elsevier (1991) p.95

R2466 Avdeef,A., "QSAR:Rational Appr.Des.Bioact.Cpds." Eds. Silipo & Vittoria, Elsevier, (1991) p119

R2467 Altomare,C., Tsai,R-S., El Tayar,N., Testa,B., Carotti,A., Cellamare,S. & DeBenedetti,P., J.Pharm.Pharmacol., 43, 191 (1991)

R2468 Takayama,C.,Sumitomo Chemical C., private communication

R2469 Martin,Y. & Lynn,K., J.Med.Chem., 14, 1162 (1971)

R2470 Rulke,H., Martin,E., Kottke,K., Kuhmstedt,H., Pharmazie, 46, 137 (1991)

R2471 Hersey,A., Hyde,R., Livingstone,D. & Rahr,E., J.Pharm.Sci., 80, 333 (1991)

R2472 Porter,J., Morgan,J., Hoyes,K., Burke,L., Huehns,E. & Hider,R., Blood, 76, 2389 (1990)

R2473 Street,J. & Walsh,A., Eur.J.Pharmacol., 102, 315 (1984)

R2474 Adams,G., Stratford,I., Wallace,R., Wardman,P. & Watts,M., J.NCL.64, 555 (1980)

R2475 Hatanaka,T., Inuma,M., Sugibayashi,K. & Morimoto,Y., Chem.Pharm.Bull., 38, 3452 (1990)

R2476 Bonnemain,B., Meyer,D., Schaefer,M., Dugast-Zrihen,M., Legreneur,S. & Doucet,D., Invest.Radiol., 25, S104 (1990)

R2477 Suto,M., Stier,M., Werbel,L., Arundel-Suto,C., Leopold,W., Elliott,W., & Sebolt-Leopold,J., J.Med.Chem., 34, 2484 (1991)

R2478 Camilleri,P., Munro,D., Weaver,K., Williams,D. & Rzepa,H., J.Chem.Soc. Perkin Trans II, 1935 (1989)

R2479 Camilleri,P., Barker,M., Kerr,M., Whitehouse,M., Bowyer,J.& Lewis,R., J.Agric.Food Chem., 37, 1509 (1989)

R2480 Hansen,L., Christrup,L. & Bundgaard,H., Acta Pharm.Nord., 3, 77 (1991)

R2481 Buur,A. & Bundgaard,H., Acta Pharm.Nord., 3, 51 (1991)

R2482 Gozzo,F., Garavaglia,C. & Mirenna,L., Pest.Biochem.Physiol., 40, 68 (1991)

R2483 Fuchs,J., Mehlhorn,R. & Packer,L., J.Invest.Dermatol., 93, 633 (1989)

R2484 Fuchs,J., Nitschmann,W., Packer,L., Hankovszky,,O. & Hideg,K., Free Rad.Res.Comms., 10, 315 (1990)

R2485 Drustrup,J., Fullerton,A., Christrup,L. & Bundgaard,H., Int.J.Pharmaceut., 71, 105 (1991)

R2487 Naito,Y., Sugiura,M., Yamaura,Y., Fukaya,C., Yokoyama,K., Nakagawa,Y., Ikeda,T., Senda,M. & Fujita,T., Chem.Pharm. Bull., 39, 1736 (1991)

R2488 Morita,K., Nagare,T. & Hayashi,Y., Agric.Biol.Chem., 51, 1955 (1987)

R2489 Dunlop,R., Duncan,J. & Ayrey,G., Pestic.Sci., 11, 53 (1980)

R2490

Shepard,K.,Graham,S.,Hudcosky,R.,Michelson,S.,Scholz,T.,Sch wam,H.,Smith,A.,Sondey,J.,Strohmaier,K.,Smith,R.& Sugrue,M.,J.Med.Chem.34, 3098(91)

R2491 Saari,W., Schwering,J., Lyle,P., Engelhardt,E., Sartorelli,A.& Rockwell,S., J.Med.Chem., 34, 3132 (1991)

R2492 Desai,M., Thadeio,P., Lipinski,C., Liston,D., Spencer,R. & Williams,I., Bioorg.Med.Chem.Let., 1, 411, (1991)

R2493 Kourounakis,P. & Chilliard,N., Pharmazie, 38, 388 (1983)

R2494 Ringel,I., Jaffe,D., Alerhand,S., Boye,O., Muzaffar,A. & Brossi,A., J.Med.Chem., 34, 3334 (1991)

R2495 Kabankin,A., Boldeskul,I., Trinus,F., Protsenko,L. & Landau,M., Khim.-farm. Zh., 16, 961, 626ee (1982)

R2499 Yamagami,C., Takao,N. & Fujita,T., J.Pharm.Sci., 80, 772 (1991)

R2501 Recanatini,M., J.Pharm.Pharmacol., 44, 68 (1992)

R2503 Serajuddin,A., Ranadive,S. & Mahoney,E., J.Pharm.Sci., 80, 830 (1991)

R2504 Kruse,L., et. al., J.Med.Chem., 30, 486 (1987)

R2505 Binder,T. and Duffel,M., Molec.Pharmacol., 33, 477 (1988)

R2506 Fujita, T., EMIL Project, private communication

R2507 J.Pestic.Sci.,(Technical Inform.) 13, 391 (1988)

R2508 J.Pestic.Sci., (Technical Inform.) 13, 167 (1988)

R2509 Uchida,M., Nishizawa,R., Suzuki,T., J. Pestic.Sci., 7, 397 (1982)

R2510 Stevens,P., Baker,E. & Anderson,N., Pestic. Sci., 24, 31 (1988)

R2512 J.Pestic.Sci.,(Technical Inform.), 13, 395 (1988)

R2513 J.Pestic.Sci., (Technical Inform.) 13, 163 (1988)

R2515 J.Pestic.Sci. (Technical Inform.), 13, 629 (1988)

R2516 J.Pestic.Sci. (Technical Inform.), 12, 775 (1987)

R2517 Aller,H. & Ramsay,J., Brighton Crop Protection Conf., Pest & Diseases, 1988, V.1, 511

R2518 Kerr,S. & Kalman,T., J. Med. Chem. 35, 1996 (1992)

R2519 J.Pestic.Sci.,(Technical Infor.) 15, 125 & 503 & 641 (1990)

R2520 J.Pestic.Sci.(Technical Infor.) 16, 125 & 349 (1991)

R2521 Siddiqui,M., Driscoll,J., Marquez,V., Roth,J., Shirasaka,T., Mitsuya,H., Barchi,J. & Kelley,J., J.Med.Chem., 35, 2195 (1992)

R2522 Da,Yong-Zhong, Ito,K. & Fujiwara,H., J.Med.Chem., 35, 3382 (1992)

R2523 Yamagami,C. & Takao, N., Chem.Pharm.Bull., 40, 925 (1992)

R2524 Wright,S., Harris,R., Collins,R., Corbett,R., Green, A., Wadman, E. & Batt, D., J.Med.Chem., 35, 3148 (1992)

R2525 Hartman,G., Halczenko,W., Prugh,J., Smith,R., Sugrue,M., Mallorga,P., Michelson,S., Randall,W., Schwam,H. & Sondy,J., J.Med.Chem., 3027,(1992)

R2526 Remuzon,P., Bouzard,D., Guiol,C. & Jacquet,J-P., J.Med.Chem., 35, 2898 (1992)

R2527 Garrone,A., Marengo,E., Fornatto,E. & Gasco,A., QSAR, 11, 171 (1992)

R2529 Anderson,B., Galinsky,R., Baker,D., Chi,S., Hoesterey,B., Morgan,M., Murakami,K. & Mitsuya,H., J.Control.Release, 19, 219 (1991)

R2530 Meulemans,A., Vicart,P., Mohler,J. & Vulpillat,M., Chemotherapy, 34, 90 (1988)

R2531 Howe,R., Rao,B., Holloway,B. & Stribling,D., J.Med.Chem., 35, 1759 (1992)

R2532 Boyd,S., Fung,A., Baker,W., Et.Al., J.Med.Chem., 35, 1735 (1992)

R2535 Hadjipavlou-Litina,D., Rekka,E., Hadjipetrou-Kourounakis,L. & Kourounakis,P., Eur.J.Med.Chem., 27, 1 (1992)

R2536 Risby,T., Hsu,T-B., Sehnert,S. & Bhan,P., Environ.Sci.Technol., 24, 1680 (1990)

R2537 Hasegawa,T. et al., Int.J.Pharmaceut., 58, 129 (1990)

R2538 Quigley,J., Fahelelbom,K., Timoney,R. & Corrigan,O., Int.J.Pharmaceut., 58, 107 (1990)

R2539 Lamba,S., Buch,K. & Lewis,III, H., Pyton, 49, 67 (1989)

R2540 Stolc,S., Nemecek,V. & Szocsova,H., Eur.J.Pharmacol., 164, 249 (1989)

R2541 Leahy,D., DeMeere,A., Wait,A., Taylor,P., Tomenson,J. & Tomlinson,E., Int.J.Pharmaceut., 50, 117 (1989)

R2542 Hartman,G., Halczenko,W., Smith,R., Sugrue,M., Mallorga,P., Michelson,S., Randall,W., Schwam,H., & Soncey,J., J.Med.Chem., 35, 3822 (1992)

R2544 Nowick,J., Chemistry Dept., U.C. Irvine, private communication

R2545 Altomare,C., Carrupt,P-A., Gaillard,P., El Tayar,N., Testa, B., & Carotti,A., Chem.Res.Toxicol., 5, 366 (1992)

R2547 Sugita,T., Masuoka,M., Nishikawa,Y., Nishimoto,S., Zhou,L., Sasai,K. & Kagiya,T., Anti-Cancer Drug.Des., 7, 277 (1992)

R2550 Finizio,A., DiGuardo,A., Arnoldi,A., Vighi,M. & Fanelli, R., Chemosphere, 23, 801 (1991)

R2553 Kim,A. & Szoka,F.,Jr., Pharm.Res., 9, 504 (1991)

R2554 Boyd,S., Fung,A., Baker,W. et al., Poster, XIIth Intern.Symp.Med.Chem., Basel 9/13/92.

R2555 Xiang,T., Chen,X. & Anderson,B., Biophys.J., 63, 78 (1992)

R2556 Scriba,G. & Borchardt,R., J.Neurochem., 53, 610 (1989)

R2557 Rahman,M., Miyoshi,T., Sukimoto,K., Takadate,A. & Otagiri,M., J.Pharmacobio-Dyn., 15, 7 (1992)

R2558 Plesiat,P. & Nikaido,H., Molec.Microbiol., 6, 1323 (1992)

R2559 Smith,D., Brown,K. & Neale,M., Drug Metab.Rev., 16, 365 (1985)

R2560 Nielsen,L., Bundgaard,H. & Falch,E., Acta Pharm.Nord., 4, 43 (1992)

R2561 Nagasawa,H., Bando,M., Hori,H., Satoh,T., Tada,T., Onoyama,Y. & Inayama,S., Int.J.Radiat.Oncol.Biol.Phys., 22, 561 (1992)

R2562 Tanii,H., Saito,S. & Hashimoto,K., Arch.Toxicol., 66, 368 (1992)

R2563 Murayama,C., Suzuki,A., Sato,C. et al., Int.J.Rad.Oncol.Biol.Phys., 22, 557 (1992)

R2564 Schreiber,L. & Schonherr,J., Environ.Sci.Technol., 26, 153 (1992)

R2566 Blumenkopf,T., Harrington,J., Kolbe,C., Bankston,D., Morrison,R., Bigham,E., Styles,V. & spector,T., J.Med.Chem., 35, 2306 (1992)

R2567 Nishimoto,S-I.,et al, Int.J.Radiat.Oncol.Biol.Phys., 22, 601 (1992) & private communication

R2570 Vera,A., Montes,M., Usero,J. & Casado,J., J.Pharm.Sci., 81, 791 (1992)

R2571 Thumm,W., Bruggemann,R., Freitag,D. & Kettrup,A., Chemosph., 24, 1835 (1992)

R2572 Nandihali,U., Duke,M. & Duke,S., J.Agric.Food Chem., 40, 1993 (1992)

R2573 Skvareninova,K., Balaz,S., Sturdik,E., Veverka,M., Adamcova,J. & Olexova,J., Coll.Czech.Chem.Comm., 57, 1739 (1992)

R2574 Rosenberg,S., Spina,K., Woods,K., et al., J.Med.Chem., 36, 449 (1993)

R2575 Rosenberg,S., Spina,K., Condon,S. et al., J.Med.Chem., 36, 460 (1993)

R2576 Rosenberg,S., Woods,K., Sham,H. et al, J.Med.Chem., 33, 1962 (1990)

R2579 Avdeef,A., Comer,J. & Thomson,S., Anal.Chem., 65, 42 (1993)

R2581 Wong,C., Chin,Y. & Geshwend,P., Geochem.Cos Acta, 56, 3923 (1992)

R2582 Wishart,G. & Campbell,M., Biochem.J., 178, 443 (1979)

R2585 Brownlee,B., Carey,J., MacInnis,G. & Pellizzari,I., Environ.Toxicol.Chem., 11, 1153 (1992)

R2586 Korenman,Ya. & Danilov,V., Zh.Prikl.Khim., 63, 125, 111ee (1990)

R2589 Wilman,D., Cox,P., Goddard,P., Hart,L., Merai,K. & Newell,D., J.Med.Chem., 27, 870 (1984)

R2590 Wilman,D., Goddard,P. & Heales,B., J.Biopharm.Sci., 2, 101 (1991)

R2591 Aggarwal,S., Gogu,S., Rangan,S. & Agrawal,K., J.Med.Chem., 33, 1505 (1990)

R2592 Kruglik,A., Leshchev,S., Rakhmanko,E., Bubel,O. & Asratyan,G., Zh.Prikl.Khim., 64, 1721, 1576ee (1991)

R2596 Boikov,Y., Shibaev,V., Gaidamaka,V. & Vyunov,K., Zh.Prikl.Khim., 62, 177, 163ee (1989)

R2598 Tavakoli-Saberi,M. & Audus,K., Int.J.Pharmaceut.,56, 135 (1989)

R2600 Li,N-Y., Li,Y. & Gorrod,J., Med.Sci.Res., 20, 901 (92)

R2601 Gaidukevich,A., Svechnikova,E. & Kostina,T., Khim-farm.Zh., 26, 87, 96ee (1992)

R2602 Webber,S., Bleckman,T., et al., J.Med.Chem., 36, 733 (1993)

R2605 Rutherford,D. & Chiou, C., Environ.Sci.Technol., 26, 965 (1992)

R2606 Alvarez-Builla,J., Crespo,T., Lopez,R., Elguero,J., Toiron,C. & Yranzo,G., J.Pharm.Sci., 81, 577 (1992)

R2607 Varnavas,A., Nisi,C., Lassiani,L., Sava,G., Perissin,L., & Boccu,E., Arzneim.Forsch., 41, 1168 (1991)

R2609 Pigot,T., Duriez,M-C., Picard,C., Cazaux,L. & Tisnes,P., Tetrahedron, 48, 4359 (1992)

R2610 Nakagawa,Y., Itai,S., Yoshida,T. & Nagai,T., Chem.Pharm.Bull., 403, 725 (1992)

R2611 Lipinski,C., Pfizer Corp. Private Communication

R2612 Kakkis,E., Palmire,V.,Jr., Strong,C., Bertsch,W., Hansch,C., & Schirmer,U., J.Agric.Food Chem., 32, 133 (1984)

R2613 Yamagami,C., Takao,N. & Fujita,T., J.Pharm.Sci., 82, 155 (1993)

R2615 Gasco, A., Univ. of Turin, private communication

R2616 Juhua,H. & Xinfu,L., Chromatography (China), 10, 344 (1990)

R2617 Hays,S., Novak,P., Ortwine,D., Bigge,C.,et al., J.Med.Chem., 36, 654 (1993)

R2619 Yamagami,C. & Takao,N., Chemistry Express, 7, 385 (1992)

R2620 Suzuki,Y., Tsuchiya,M., Safadi,A., Kagan,V. & Packer,L., Free Rad.Biol.Med., 13, 517 (1992)

R2621 Maliepaard,M., de Mol,N., Janssen,L., van der Neut,W., Verboom,W.,& Reinhoudt,D., Anti-Cancer Drug. Des., 7, 415 (1992)

R2623 Jenke,D., J.Pharm.Sci., 82, 617 (1993)

R2624 Friche,E., Jensen,P., Roed,H., Skovsgaard,T. & Nissen, N., Biochem.Pharmacol., 39 1721 (1990).

R2625 Ilchmann,A., Wienke,G., Meyer,T. & Gmehling,J., Chem-Ing-Tech., 65, 72 (1993)

R2626 Ohsako,M., Matsumoto,Y. & Goto,S., Biol.Pharm.Bull., 16, 154 (1993)

R2627 Irion,G., Ochsenfeld,L., Naujok,A. & Zimmermann,H., Histochem., 99, 75 (1993)

R2629 Matsumoto,Y., Ohsako,M., Takadate,A. & Goto,S., J.Pharm.Sci., 82, 399 (1993)

R2630 Liang,W. & Lin, W., Acta Pharm.Sinica, 27, 684 (1992)

R2631 Al-Hindawi,M., James,K. & Nicholls,P., J.Pharm.Pharmacol., 33, 65p. (1981)

R2633 Bigge,C., Malone,T., Ortwine,D. et al., J.Med.Chem., 36, 1977

(1993)

R2635 Mylari,B. & Zembrowski,W., Pfizer, Groton Co., Private Communication

R2636 Barraclough,P., Livingstone,D., et al., Eur.J.Med.Chem., 25, 467 (1990)

R2637 Masereel,B., Lohrmann,E., Schynts,M., Pirotte,B., Greger,R. & Delarge,J., J.Pharm.Pharmacol., 44, 589 (1992)

R2638 Lipinski,C., J.Med.Chem., 26, 1 (1983)

R2639 Peeters,J. Private Communication, Janssen Research

R2640 Tsai,R-S., Carrupt,P-A., El Tayar,N., Giroud,Y., Andrade,P. & Testa,B., Helv.Chim.Acta, 76, 842 (1993)

R2641 Alcorn,C., Simpson,R., Leahy,D. & Peters,T., Biochem.Pharmacol., 45, 1775 (1993)

R2642 Yamagami,C. & Takao,N., Chem.Pharm.Bull., 40, 925 (1992)

R2644 Takacs-Novak,K., Oelschlager,H., Budvari-Barany,Z. & Szasz,G.,Pharmazie, 47, 587 (1992)

R2645 Sotomatsu,T., Shigemura,M., Murata,Y. & Fujita,T., J.Pharm.Sci., 82, 776 (1993)

R2646 El Tayar,N., Tsai,R-S., Vallat,P., Altomare,C. & Testa,B., J.Chromatog., 556, 181 (1991)

R2647 Tsai,R-S., Carrupt,P-A., Testa,B., Gaillard,P., El Tayar,N. & Hogberg,T., J.Med.Chem., 36, 196 (1993)

R2648 Tsai,R-S., Carrupt,P-A., Testa,B., El Tayar,N., Gruenewald,G. & Casy,A., J.Chem.Research (M),1901 (1993)

R2649 Morel,E., Univ. of Utrecht, Thesis, 1993

R2650 Wang,L., Li,H. & Wang,X., Huajing Huaxue, 12, 151 (1993)

R2651 Schmidt,D., Votaw,J., Kessler,R. & DePaulis,T., J.Pharm.Sci., 83, 305 (1994)

R2652 Mach,R., Luedtke,R., Unsworth,C.,et al., J.Med.Chem., 36, 3707 (1993

R2653 Kraak,J., Van Rooij,H. & Thus,J., J.Chromatog., 352, 455 (1986)

R2654 Tsai,R-S., Testa,B., El Tayar,N. & Carrupt,P-A., J.Chem.Soc.,Perkin Trans.2, 1798 (1991)

R2655 Shameem,M., Imai,T. & Otagiri,M., J.Pharm.Pharmacol., 45, 246 (1993)

R2656 Sliskovic,D., Krause,B., Picard,J. et al., J.Med.Chem., 37, 560 (1994)

R2659 Yamagami,C. & Takao,N., Chem.Pharm.Bull., 41, 694 (1993)

R2660 Bromilow,R., Rigitano,R., Briggs,G. & Chamberlain,K., Pestic.Sci., 19, 85 (1987)

R2661 Noble,A., J.Chromatog., 642, 3 (1993)

R2662 Takahashi,J., Kirino,O., Takayama,C., Nakamura,S., Noguchi,N., Kato,T. & Kamoshita,K., J.Pestic.Sci., 13, 587 (1988)

R2665 Patil,S., Nicholls,P., Chamberlain,K., Briggs,G. & Bromilow,R., Pestic.Sci., 22, 333 (1988)

R2667 Lichtner,F., Plant Physiol., 71, 307 (1983)

R2669 Kristl,A., Vesnaver,G., Mrhar,A. & Kozjek,F., Pharmazie, 48, 608 (1993)

R2671 Gregan,F. & Polasek,E., Cesk.Farm., 41, 303 (1992) & private communication

Hammett Sigmas

Empirical	Structural	π	σm	σp	MR	Es	additional σ values
As1O3	AsO3 (-2)						★S.PARA- -0.11 640 ★ES-CH 4.30 2
At1	At		★0.25 559	★0.18 559			F 0.27 559 R -0.07 559
B1Cl2	BCl2			★0.71 535			★S.INDUC 0.04 67 S.ZE.RS 0.28 217 ★ES-CH 3.40 2; S.INDUC 0.31 217 ★S.ZE.RS 0.30 86; ★S.PARA+ 0.86 217 S.ZE.RS 0.31 67
B1F2	BF2		★0.32 30	★0.48 30			★S.INDUC 0.16 30 ★S.ZE.RS 0.32 30 F 0.26 4; ★S.ZE.RS 0.24 217 ★ES-CH 2.60 2 R 0.22 4
Br1	Br	0.60 57; 0.75 220; 0.76 54; 0.86; 0.92 42	0.37 46; 0.39 10; 0.39 41; 0.39 121; 0.39 190; ★0.39 21; 1.00 128	0.12 41; 0.22 190; 0.22 254; 0.23 121; ★0.23 21; 0.27 46; 0.29 208; 0.30 10; 0.80 128	0.89	-1.34 35; ★-1.16 267	S.INDUC 0.42 99; ★S.INDUC 0.44 20; S.INDUC 0.44 119; S.INDUC 0.44 205; S.INDUC 0.45 142; S.INDUC 0.45 192; S.INDUC 0.47 6; S.INDUC 0.47 182; S.INDUC 0.50 17; S.INDUC 0.63 16; S.INDUC 0.64 12; S.PRIME 0.45 159; S.PRIME 0.45 586; S-INDQ 2.65 58; S.STAR 2.62 61; S.STAR 2.79 10; ★S.STAR 2.80 289; S.STAR 2.84 3; ★S.ZMTFT 0.32 95; S.ZMTFT 0.39 63; S.META+ 0.35 88; ★S.META+ 0.41 43; S.META+ 0.41 858; S.META- 0.24 114; ★S.META- 0.34 28; S.ZPTFT 0.26 95; ★S.ZPTFT 0.28 112; S.ZPTFT 0.30 63; S.PARA- 0.24 28; ★S.PARA- 0.25 181; S.PARA- 0.28 211; ★S.PARA+ 0.10 308; S.PARA+ 0.14 65; S.PARA+ 0.14 406; ★S.PARA+ 0.15 43; S.PARA+ 0.17 88; S.PARA+ 0.18 329; S.PARNO 0.30 65; S.ORTHO 0.44 432; S.ORTHO 0.58 158; S.ORTHO 0.70 53; S.ORTHO 0.71 104; S.ORTHO 0.88 335; S.ORTHO 1.10 186; S.ORTHO 1.35 78; S.ORTH+ 0.22 243; ★S.ZE.RS -0.25 6; S.ZE.RS -0.23 19; S.ZE.RS -0.16 12; S.ZE.RS -0.16 67; S.BA.RS -0.19 20; S.AN.RS -0.19 20; S.RES.+ -0.30 20; ★S.RES.+ -0.19 6; S.RES.+ -0.16 10; ES-CH 1.30 2; ES-V 0.65 7; L-STM 3.82 1; B1-STM 1.95 1; B5-STM 1.95 1; F 0.45 4; R -0.22 4; R+ -0.30 4; R- -0.20 4; S.NEOF 0.40 545; S.ZTWST -0.23 22; ★S.P.RAD 0.12 206; S.P.RAD 0.17 263; S.P.RAD 0.26 255; S-M(C+) 0.29 239; S-M(C+) 0.33 256; S-P(C+) -0.19 256; S-P(C+) -0.10 239; K-CHGTR 0.00 77; GRP.DPL -1.97 9; GRP.DPL -1.57 8; GR.ELCT 0.32 13; O-STER -0.82 90
Br1Hg1	HgBr						S.INDUC -0.05 16 ★S.STAR 0.64 61; ★S.INDUC 0.44 340 ★S.ZE.RS 0.00 340
Br1Mg1	MgBr						S.INDUC -0.75 16 GR.ELCT -0.64 13
Br2P1	PBr2						S.INDUC 0.32 16
Br3Ge1	GeBr3		★0.66 144	★0.73 144	3.63		★S.INDUC 0.59 144 ES-CH 5.20 2 B5-STM 4.17 1; S.STAR 2.26 61 ES-V 1.73 7 F 0.61 5; ★S.STAR 3.70 3 L-STM 4.67 1 R 0.12 5; S.ZE.RS 0.14 18 B1-STM 3.07 1
Br3Si1	SiBr3		★0.48 144	0.27 153; ★0.57 144	3.28		S.INDUC 0.39 73 S.ZE.RS 0.18 73 F 0.44 66; ★S.INDUC 0.39 144 ES-CH 5.10 2 R 0.13 5; S.STAR 1.39 216 ES-V 1.69 7 R+ -0.03 4; ★S.STAR 2.40 3 L-STM 4.55 1 S.M.RAD 0.41 218; ★S.PARA+ 0.41 136 B1-STM 3.03 1 S.P.RAD 0.61 218; S.ZE.RS 0.18 18 B5-STM 4.09 1 GR.ELCT -0.19 13
Br3Sn1	SnBr3						★S.STAR 2.06 61 L-STM 4.94 1 B5-STM 4.35 1; ★ES-CH 5.60 2 B1-STM 3.16 1
Br3Te1	TeBr3			★0.62 103			★S.INDUC 0.51 103 ★S.ZE.RS 0.11 103
Cl1	Cl	0.36 588; 0.39 57; 0.53 54; 0.62 220; 0.68 42; 0.71	0.30 41; 0.30 128; 0.35 36; 0.36 26; 0.37 10; 0.37 29; 0.37 46; 0.37 190; ★0.37 21; 0.38 121	0.06 41; 0.14 254; 0.21 29; 0.22 190; 0.23 36; 0.23 121; ★0.23 21; 0.24 46; 0.25 236; 0.26 208; 0.28 10; 0.40 128	0.60	-1.14 35; ★-0.97 267	S.INDUC 0.43 119; S.INDUC 0.44 99; S.INDUC 0.45 142; S.INDUC 0.46 20; S.INDUC 0.46 26; S.INDUC 0.46 205; S.INDUC 0.47 6; S.INDUC 0.47 192; ★S.INDUC 0.47 611; S.INDUC 0.48 182; S.INDUC 0.51 17; S.INDUC 0.52 12; S.INDUC 0.58 16; S.PRIME 0.45 159; S-INDQ 2.51 58; S.STAR 2.68 37; S.STAR 2.68 61; S.STAR 2.85 10; ★S.STAR 2.94 289; S.STAR 2.96 3; S.ZMTFT 0.31 95; S.ZMTFT 0.37 63; ★S.META+ 0.40 43; S.META+ 0.40 88; S.META- 0.36 28; S.ZPTFT 0.25 95; S.ZPTFT 0.28 63; ★S.ZPTFT 0.28 112; ★S.PARA- 0.19 28; S.PARA- 0.27 211; S.PARA- 0.30 26; S.PARA+ 0.09 308; S.PARA+ 0.10 406; ★S.PARA+ 0.11 43; S.PARA+ 0.11 65; S.PARA+ 0.14 88; S.PARA+ 0.16 329; S.PARNO 0.29 65; S.ORTHO 0.40 432; S.ORTHO 0.40 457; S.ORTHO 0.60 257; S.ORTHO 0.63 335; S.ORTHO 0.67 104; S.ORTHO 0.68 53; S.ORTHO 0.79 158; S.ORTHO 1.20 186; S.ORTHO 1.28 78; S.ORTH+ 0.18 243; ★S.ZE.RS -0.25 6; S.ZE.RS -0.22 22; S.ZE.RS -0.22 288; S.ZE.RS -0.22 387; S.ZE.RS -0.18 26; S.ZE.RS -0.18 67; S.ZE.RS -0.16 12; S.BA.RS -0.23 20; ★S.AN.RS -0.23 20; S.AN.RS -0.17 26; S.RES.+ -0.36 20; ★S.RES.+ -0.21 6; S.RES.+ -0.19 10; ES-CH 1.20 2; ES-V 0.55 7; L-STM 3.52 1; B1-STM 1.80 1; B5-STM 1.80 1; F 0.42 5; R -0.19 5; S.PHOSP 0.93 59; S.NEOF 0.18 386; S.NEOF 0.38 545; S.IND.P 2.48 39; S.RES.P -0.43 39; S.ZTWST -0.21 22; S.M.RAD -0.07 79; S.P.RAD 0.06 453; S.P.RAD 0.08 263; ★S.P.RAD 0.10 206; S.P.RAD 0.11 79; S.P.RAD 0.18 255; S-M(C+) 0.34 239; S-M(C+) 0.36 256; S-P(C+) -0.24 256; S-P(C+) -0.07 239; K-CHGTR -0.01 77; GRP.DPL -1.93 9; GRP.DPL -1.59 8; GR.ELCT 0.37 13; O-STER -0.54 90
Cl1D3N1	ND3+Cl-						★S.ZE.RS -0.19 47
Cl1Hg1	HgCl		★0.33 3	★0.35 3			★S.INDUC 0.17 18 S.STAR 0.83 61 S.ZE.RS 0.01 539; S.INDUC 0.17 173 S.STAR 1.90 3 F 0.33 5

Empirical	Structural	π	σ_m	σ_p	MR	E_s	additional σ values		
Cl1Hg1	HgCl						★ S.INDUC 0.31 *18* S.ZE.RS 0.01 *173* GR.ELCT -0.63 *13*		
Cl1Mg1	MgCl						S.INDUC -0.76 *16*		
Cl1O1S1	S=O(Cl)				1.39		★ S.INDUC 0.68 *31* ★ S.STAR 4.25 *3* ★ S.ZE.RS 0.14 *31*	★ ES-CH 3.40 *2* L-STM 3.89 *1* B1-STM 1.40 *1*	B5-STM 3.46 *1*
Cl1O2S1	SO2(Cl)		0.87 *121* 0.92 *3* ★ 1.20 *194*	0.98 *197* 1.00 *121* 1.04 *3* ★ 1.11 *194* 1.22 *38*	1.37		S.INDUC 0.81 *12* ★ S.INDUC 0.86 *67* S.STAR 5.00 *3* S.PARA- 1.31 *252* S.ZE.RS 0.11 *19* ★ S.ZE.RS 0.17 *67*	★ S.ZE.RS 0.24 *31* S.ZE.RS 0.28 *12* ES-CH 4.40 *2* L-STM 3.89 *1* B1-STM 2.03 *1* B5-STM 3.46 *1*	F 1.16 *4* R -0.05 *264* GRP.DPL -2.28 *9* GR.ELCT 0.25 *13*
Cl1O3	ClO3		★ 0.85 *3*				S.PARA- 1.03 *53*		
Cl1S1	SCl		★ 0.44 *3*	★ 0.48 *3*	1.38		★ S.INDUC 0.40 *31* ★ S.STAR 2.50 *3* ★ S.ZE.RS 0.08 *31* ★ ES-CH 2.40 *2*	L-STM 4.06 *1* B1-STM 1.70 *1* B5-STM 3.77 *1* F 0.42 *5*	R 0.06 *5* GRP.DPL -2.00 *9*
Cl2I1	ICl2		★ 1.10 *18*	★ 1.11 *18*			S.INDUC 1.08 *273* ★ S.INDUC 1.30 *729* S.STAR 6.80 *3*	★ S.ZE.RS 0.03 *273* S.ZE.RS 0.12 *19* ES-CH 4.10 *2*	F 1.03 *4* R 0.08 *4*
Cl2N1S1	N=SCl2						S.INDUC 0.17 *123*		
Cl2O1P1	P(O)Cl2		★ 0.78 *3* 0.80 *105*	0.43 *143* ★ 0.90 *3* 0.91 *111*	2.02		S.INDUC 0.51 *12* ★ S.INDUC 0.65 *69* S.STAR 4.10 *3* ★ S.PARA+ 0.38 *143* S.ZE.RS 0.12 *447*	★ S.ZE.RS 0.17 *69* S.ZE.RS 0.27 *12* ES-CH 4.60 *2* L-STM 4.14 *1* B1-STM 2.38 *1*	B5-STM 3.64 *1* F 0.90 *5* R+ 0.20 *5* R+ -0.32 *4* GR.ELCT -0.02 *13*
Cl2O2P1	OP(O)Cl2						S.INDUC 0.79 *12* S.ZE.RS -0.16 *12*		
Cl2P1	PCl2		0.53 *105* ★ 0.54 *41*	0.61 *143* ★ 0.61 *41* 0.69 *111*	2.14		S.INDUC 0.32 *16* ★ S.INDUC 0.46 *40* S.INDUC 0.49 *69* S.STAR 2.84 *3* S.PARA+ 0.61 *649* ★ S.PARA+ 0.62 *143*	S.ZE.RS 0.06 *86* S.ZE.RS 0.07 *482* ★ S.ZE.RS 0.11 *69* S.ZE.RS 0.16 *18* ES-CH 3.60 *2* L-STM 4.14 *1*	B1-STM 1.40 *1* B5-STM 3.64 *1* F 0.50 *5* R 0.11 *5* R+ 0.12 *4* GR.ELCT 0.01 *13*
Cl2P1S1	P(S)Cl2		★ 0.70 *3* 0.73 *105*	0.39 *143* ★ 0.80 *3* 0.83 *111*	2.83		S.INDUC 0.32 *16* ★ S.INDUC 0.43 *69* S.STAR 3.70 *3* ★ S.PARA+ 0.33 *143* S.ZE.RS 0.17 *69*	ES-CH 4.80 *2* L-STM 4.14 *1* B1-STM 2.70 *1* B5-STM 3.62 *1* F 0.63 *5*	R 0.17 *5* R+ -0.30 *4* GRP.DPL -3.00 *9* GR.ELCT -0.12 *13*
Cl3Ge1	GeCl3		★ 0.71 *144*	0.39 *60* 0.60 *153* ★ 0.79 *144*	2.58		★ S.INDUC 0.63 *144* S.INDUC 0.63 *242* S.STAR 2.32 *61* ★ S.STAR 3.90 *3* ★ S.PARA+ 0.57 *60* S.ZE.RS 0.16 *18*	ES-CH 4.90 *2* ES-V 1.53 *7* L-STM 4.47 *1* B1-STM 2.85 *1* B5-STM 3.87 *1* F 0.65 *5*	R 0.14 *5* R+ -0.08 *4* GRP.DPL -3.15 *8* GR.ELCT -0.04 *13*
Cl3N1P1	N=P(Cl)3						S.INDUC -0.18 *123*		
Cl3P1	+PCl3						S.ZE.RS 0.21 *447*		
Cl3Si1	SiCl3		★ 0.48 *144*	0.39 *60* 0.43 *153* ★ 0.56 *144*	2.38		S.INDUC 0.14 *599* S.INDUC 0.38 *73* ★ S.INDUC 0.39 *144* S.INDUC 0.40 *67* S.INDUC 0.45 *242* S.STAR 1.55 *61* ★ S.STAR 1.77 *216* S.STAR 2.40 *3* S.META+ 0.19 *136*	S.PARA+ 0.20 *136* S.PARA+ 0.46 *383* ★ S.PARA+ 0.57 *60* S.ZE.RS 0.09 *47* S.ZE.RS 0.11 *67* ★ S.ZE.RS 0.17 *18* S.ZE.RS 0.19 *73* ES-V 1.50 *7* L-STM 4.35 *1*	B1-STM 2.80 *1* B5-STM 3.79 *1* F 0.44 *66* R 0.12 *5* R+ 0.13 *4* S.P.RAD 0.24 *218* S.P.RAD 0.35 *218* GRP.DPL -2.40 *8* GR.ELCT -0.14 *13*
Cl3Sn1	SnCl3						★ S.INDUC 0.42 *198* ★ S.STAR 2.26 *61* S.ZE.RS -0.13 *365*	★ S.ZE.RS 0.07 *198* ★ ES-CH 5.30 *2* L-STM 4.74 *1*	B1-STM 2.94 *1* B5-STM 4.06 *1*
Cl3Te1	TeCl3			★ 0.66 *103*			★ S.INDUC 0.55 *103* ★ S.ZE.RS 0.11 *103*		
Cl4P1	PCl4						S.ZE.RS 0.07 *447*		
D1	D		★ 0.00 *799*				★ S.INDUC 0.00 *799*		
D2N1	ND2						★ S.ZE.RS -0.47 *19* ★ ES-CH 1.00 *2*		
F1	F	-0.17 *57* 0.14	0.00 *1004* 0.27 *26* 0.33 *46* 0.34 *10* 0.34 *190* ★ 0.34 *21* 0.36 *17*	0.05 *17* ★ 0.06 *21* 0.15 *46* 0.20 *128* 0.21 *10*	0.09	★ -0.55 *35* -0.46 *267*	S.INDUC 0.40 *657* S.INDUC 0.42 *119* S.INDUC 0.45 *142* S.INDUC 0.50 *20* S.INDUC 0.50 *26* S.INDUC 0.50 *205* ★ S.INDUC 0.52 *611* S.INDUC 0.54 *6* S.INDUC 0.62 *12* S.INDUC 0.84 *16* S.PRIME 0.43 *159* S-INDQ 2.57 *58* S.STAR 1.55 *494* S.STAR 3.08 *289* ★ S.STAR 3.19 *87* S.STAR 3.21 *3* S.ZMTFT 0.33 *95* S.ZMTFT 0.34 *63* S.META+ 0.35 *43* S.META+ 0.38 *88* S.META- 0.34 *28*	S.PARA 0.12 *26* S.PARA+ -0.09 *88* S.PARA+ -0.08 *65* S.PARA+ -0.08 *308* S.PARA+ -0.08 *406* ★ S.PARA+ -0.07 *43* S.PARNO 0.18 *65* S.ORTHO 0.25 *432* S.ORTHO 0.47 *104* S.ORTHO 0.54 *53* S.ORTHO 0.93 *78* S.ORTHO 1.20 *186* S.ORTH+ 0.06 *243* S.ORTH- 0.21 *1012* ★ S.ZE.RS -0.48 *6* S.ZE.RS -0.34 *19* S.ZE.RS -0.31 *26* S.ZE.RS -0.31 *67* S.ZE.RS -0.30 *12* S.BA.RS -0.45 *20* S.AN-RS -0.45 *20*	L-STM 2.65 *1* B1-STM 1.35 *1* B5-STM 1.35 *1* F 0.45 *5* R -0.39 *5* R+ -0.52 *4* R- -0.48 *4* S.PHOSP 0.56 *59* S.NEOF 0.34 *545* S.IND.P 2.71 *39* S.RES.P -1.17 *39* S.M.RAD -0.11 *79* S.M.RAD -0.09 *79* S.P.RAD -0.25 *263* S.P.RAD -0.07 *854* S.P.RAD 0.07 *453* S.P.RAD 0.12 *255* S-M(C+) 0.29 *239* S-M(C+) 0.35 *256* S-P(C+) -0.40 *256* S-P(C+) -0.24 *239*

Empirical	Structural	π	σm	σp	MR	Es	additional σ values
F1	F						S.ZPTFT 0.16 112 — S.AN-RS -0.38 26 — K-CHGTR -0.16 77 S.ZPTFT 0.17 95 — S.RES.+ -0.57 20 — GRP.DPL -1.90 9 S.ZPTFT 0.21 63 — ★S.RES.+ -0.37 6 — GRP.DPL -1.43 8 S.PARA- -0.06 28 — S.RES.+ -0.26 10 — GR.ELCT 1.10 13 ★S.PARA- -0.03 614 — ES-CH 0.80 2 S.PARA- 0.05 211 — ES-V 0.27 7
F1Hg1	HgF		★ 0.34 3	★ 0.33 3			★S.INDUC 0.18 18 — ★S.STAR 2.10 3 — F 0.35 5 S.INDUC 0.34 18 — ★S.ZE.RS -0.01 18 — R -0.02 5
F1O1S1	S=O(F)		★ 0.74 3	★ 0.83 3	0.89		★S.INDUC 0.66 31 — ES-CH 3.00 2 — B5-STM 2.70 1 ★S.STAR 4.12 3 — L-STM 3.33 1 — F 0.67 4 ★S.ZE.RS 0.17 31 — B1-STM 1.40 1 — R 0.16 4
F1O2S1	SO2(F)	0.05	0.78 10 ★ 0.80 696 0.98 18 0.98 30	0.91 10 ★ 0.91 696 1.08 18 1.08 30	0.86		★S.INDUC 0.75 89 — ★S.ZE.RS 0.23 80 — B5-STM 2.70 1 S.INDUC 0.82 80 — S.ZE.RS 0.30 12 — F 0.72 5 S.INDUC 0.87 12 — S.BA.RS 0.21 18 — R 0.19 5 S.INDUC 0.88 18 — S.BA.RS 0.23 80 — R- 0.82 4 S.INDUC 0.88 30 — S.AN-RS 0.49 80 — GRP.DPL -4.59 8 S.STAR 4.70 3 — S.RES.+ 0.00 10 — GRP.DPL -3.39 9 S.META- 0.99 297 — ES-CH 4.00 2 ★S.PARA- 1.54 297 — L-STM 3.33 1 S.ZE.RS 0.21 30 — B1-STM 2.01 1
F2I1	IF2		★ 0.85 18	★ 0.83 18			★S.INDUC 0.86 273 — ★S.ZE.RS -0.03 273 — F 0.82 4 ★S.STAR 5.40 3 — ★ES-CH 3.30 2 — R 0.01 4
F2O1P1	P=O(F)2		★ 0.81 41	★ 0.89 41	0.96		★ES-CH 3.80 2 — B1-STM 2.08 1 — F 0.74 5 L-STM 3.58 1 — B5-STM 2.88 1 — R 0.15 5
F2P1	PF2		0.26 41 0.48 18 ★ 0.49 30	0.59 18 ★ 0.59 30 0.61 41	1.10		S.INDUC 0.37 69 — S.ZE.RS 0.21 18 — B5-STM 2.79 1 ★S.INDUC 0.38 40 — ★S.ZE.RS 0.21 30 — F 0.44 5 S.STAR 2.37 3 — ES-CH 2.80 2 — R 0.15 5 S.PARA+ 0.70 649 — L-STM 3.55 1 — GR.ELCT 0.12 13 S.ZE.RS 0.16 69 — B1-STM 1.40 1
F3Ge1	GeF3		★ 0.85 144	★ 0.97 144	0.69		★S.INDUC 0.74 144 — ES-V 1.06 7 — F 0.76 5 S.STAR 4.60 3 — L-STM 3.88 1 — R 0.21 5 S.ZE.RS 0.23 18 — B1-STM 2.20 1 — GR.ELCT 0.08 13 ES-CH 3.70 2 — B5-STM 3.03 1
F3S1	SF3		0.70 30 ★ 0.70 18	0.80 30 ★ 0.80 18	0.83		★S.INDUC 0.60 31 — ES-CH 3.60 2 — F 0.63 4 S.STAR 3.75 3 — L-STM 3.30 1 — R 0.17 4 S.ZE.RS 0.20 18 — B1-STM 1.99 1 ★S.ZE.RS 0.20 31 — B5-STM 2.61 1
F3Si1	SiF3		0.54 144 ★ 0.54 41	0.51 153 0.66 144 ★ 0.69 41	0.76		S.INDUC 0.42 73 — ★S.ZE.RS 0.24 18 — R 0.22 5 ★S.INDUC 0.42 144 — S.ZE.RS 0.24 73 — S.M.RAD 0.25 218 S.INDUC 0.45 67 — S.ZE.RS 0.25 144 — S.P.RAD 0.30 218 S.STAR 2.62 3 — ES-CH 3.60 2 — GRP.DPL -2.72 8 S.PARA+ 0.01 136 — ES-V 1.01 7 — GR.ELCT -0.02 13 S.ZE.RS 0.22 67 — F 0.47 66
F4I1	IF4		★ 1.07 18	★ 1.15 18			★S.INDUC 1.00 273 — ★S.ZE.RS 0.15 273 — F 0.98 4 ★S.STAR 6.30 3 — ES-CH 4.90 2 — R 0.17 4
F4P1	PF4		0.63 30 ★ 0.63 18	0.80 30 ★ 0.80 18	1.26		★S.INDUC 0.45 40 — S.ZE.RS 0.35 30 — R 0.26 4 ★S.STAR 2.80 3 — ES-CH 4.40 2 — GRP.DPL -2.55 9 ★S.ZE.RS 0.35 18 — F 0.54 4
F5N3P3	pentafl-cyc-triphosphazyl						★S.PARA+ 0.74 537 — ★ES-CH 4.00 2
F5O1S1	OSF5			★ 0.44 214	1.17		ES-CH 6.20 2 — B1-STM 1.35 1 L-STM 4.78 1 — B5-STM 4.38 1
F5S1	SF5	1.23	0.59 17 0.60 10 ★ 0.61 622	0.67 17 ★ 0.68 622 0.69 10	0.99		S.INDUC 0.42 16 — S.PARA- 0.86 28 — ES-V 1.37 7 S.INDUC 0.53 31 — ★S.PARA- 0.86 743 — L-STM 4.65 1 ★S.INDUC 0.57 20 — S.ZE.RS 0.00 19 — B1-STM 2.47 1 S.INDUC 0.59 6 — S.ZE.RS 0.03 6 — B5-STM 2.92 1 S.INDUC 0.59 17 — S.ZE.RS 0.06 20 — F 0.56 5 S.STAR 3.56 3 — ★S.ZE.RS 0.08 31 — R 0.12 5 S.META- 0.61 744 — S.BA.RS 0.06 20 — R- 0.30 4 S.META- 0.62 28 — S.AN-RS 0.20 20 — GRP.DPL -3.44 8 S.META- 0.63 743 — S.RES.+ 0.06 20 — GR.ELCT 0.46 13 S.PARA- 0.70 744 — ES-CH 5.20 2
F6N2P1	NN+PF6-						★S.INDUC 1.53 24 — ★S.ZE.RS 0.69 24 — ★ES-CH 2.00 2
Ge1I3	Ge(I)3						ES-CH 6.40 2 — L-STM 4.94 1 — B5-STM 4.56 1 ES-V 1.96 7 — B1-STM 3.37 1
Hg1I1	HgI						S.INDUC 0.24 340 — ★S.STAR 0.45 61 — ★S.ZE.RS 0.04 340
I1	I	1.00 57 1.12	0.34 17 0.34 46 0.35 190 ★ 0.35 21 0.50 41	0.18 21 0.19 41 0.19 121 0.21 190 0.24 254 0.27 892 0.28 23 0.28 46 0.30 208	1.39	★ -1.62 35 -1.40 267	S.INDUC 0.39 20 — S.ZPTFT 0.30 63 — S.AN-RS -0.11 20 ★S.INDUC 0.39 32 — ★S.ZPTFT 0.31 112 — S.RES.+ -0.25 20 S.INDUC 0.39 205 — ★S.PARA- 0.27 28 — S.RES.+ -0.18 10 S.INDUC 0.40 6 — S.PARA+ 0.12 88 — ES-CH 1.70 2 S.INDUC 0.40 182 — S.PARA+ 0.13 65 — ES-V 0.78 7 S.INDUC 0.42 119 — S.PARA+ 0.13 406 — L-STM 4.23 1 S.INDUC 0.43 142 — ★S.PARA+ 0.14 43 — B1-STM 2.15 1 S.INDUC 0.44 17 — S.PARNO 0.31 65 — B5-STM 2.15 1 S.INDUC 0.54 16 — S.ORTHO 0.46 432 — F 0.42 5 S.INDUC 0.70 12 — S.ORTHO 0.63 53 — R -0.24 5 S-INDQ 2.34 58 — S.ORTHO 0.70 104 — R+ -0.28 4 ★S.STAR 2.22 61 — S.ORTHO 1.30 186 — R- -0.15 4 S.STAR 2.38 10 — S.ORTHO 1.34 78 — S.ZTWST -0.22 22 S.STAR 2.38 289 — S.ORTH+ 0.20 243 — S.P.RAD 0.12 206 S.STAR 2.46 3 — ★S.ORTH- 0.64 97 — S.P.RAD 0.16 263 S.ZMTFT 0.34 63 — S.ZE.RS -0.22 19 — S.P.RAD 0.31 255 S.ZMTFT 0.38 95 — ★S.ZE.RS -0.16 6 — K-CHGTR 0.01 77 S.META+ 0.36 43 — S.ZE.RS -0.16 20 — GRP.DPL -1.79 9

Empirical	Structural	π	σm	σp	MR	Es	additional σ values
I1	I						S.META- 0.20 114 · S.ZE.RS -0.14 12 · GRP.DPL -1.36 8 S.META- 0.33 28 · S.ZE.RS -0.12 142 · GR.ELCT 0.15 13 S.ZPTFT 0.27 95 · S.BA.RS -0.16 20
I1O1	IO	-3.74	★ 0.58 178	★ 0.62 178	3.91		ES-CH 2.70 2 · F 0.55 178 · R 0.07 178
I1O2	IO2	-3.46	0.63 748 ★ 0.68 17 0.70 23	0.69 748 0.76 23 ★ 0.78 17	6.35		★ S.INDUC 0.68 17 · L-STM 4.25 1 · F 0.61 5 S.STAR 3.71 3 · B1-STM 2.15 1 · R 0.17 5 ES-CH 3.70 2 · B5-STM 3.66 1 · GR.ELCT 0.49 13
I3Si1	Si(I)3						ES-CH 6.30 2 · ES-V 1.93 7
Li1	Li						S.INDUC -0.75 16 · ★ ES-CH 1.00 2 ★ S.ZE.RS 0.14 86 · GR.ELCT -0.85 13
Li1O1	O-Li						S.INDUC 0.13 16
N1O1	NO	-0.12	★ 0.62 747	0.12 23 ★ 0.91 747 1.46 904	0.52		S.INDUC 0.33 24 · ★ S.PARA 1.63 899 · B5-STM 2.44 1 ★ S.INDUC 0.34 89 · S.ZE.RS 0.07 19 · F 0.49 5 S.STAR 2.08 3 · S.ZE.RS 0.25 47 · R 0.42 5 S.PARA 0.15 609 · S.ZE.RS 0.31 24 · R- 1.14 4 S.PARA- 1.37 252 · ★ S.ZE.RS 0.32 161 · GRP.DPL -3.09 8 S.PARA- 1.46 247 · ES-CH 2.00 2 · GRP.DPL -2.30 9 S.PARA- 1.57 870 · L-STM 3.44 1 · GR.ELCT 0.84 13 S.PARA- 1.60 181 · B1-STM 1.70 1
N1O1S1	N=S=O			★ 0.44 997			S.ZE.RS -0.09 732
	NSO						★ S.ZE.RS -0.09 19 · ★ ES-CH 3.20 2
N1O2	NO2	-0.85 57 -0.30 42 -0.28 -0.17 42 -0.05 42 0.01 455 0.45 454 0.54 680	0.61 254 0.66 36 0.67 40 0.71 29 0.71 46 ★ 0.71 21 0.72 26 0.74 190 3.80 128	0.57 254 0.73 36 0.77 40 0.77 236 0.78 29 0.78 190 ★ 0.78 21 0.81 46 0.88 208 7.80 128	0.74	★ -2.52 682 -1.02 683	S.INDUC 0.56 161 · S.PARA- 1.24 181 · S.BA.RS 0.15 20 S.INDUC 0.57 24 · S.PARA- 1.24 907 · S.AN.RS 0.41 26 S.INDUC 0.63 12 · S.PARA- 1.26 104 · S.AN.RS 0.46 20 ★ S.INDUC 0.64 34 · S.PARA- 1.26 160 · S.RES.+ 0.00 10 S.INDUC 0.65 20 · ★ S.PARA- 1.27 23 · S.RES.+ 0.15 20 S.INDUC 0.65 26 · S.PARA- 1.27 247 · ES-CH 3.00 2 S.INDUC 0.65 119 · S.PARA- 1.27 258 · ES-V 0.35 341 S.INDUC 0.65 142 · S.PARA- 1.30 28 · ES-V 1.39 474 S.INDUC 0.65 205 · ★ S.PARA+ 0.79 43 · L-STM 3.44 1 S.INDUC 0.67 6 · S.PARA+ 0.86 308 · B1-STM 1.70 1 S.INDUC 0.76 32 · S.PARA+ 0.94 88 · B5-STM 2.44 1 S.INDUC 0.85 99 · S.PARNO 0.78 641 · F 0.65 5 S.PRIME 0.64 159 · S.ORTHO 0.10 254 · R 0.13 5 S.INDQ 3.52 58 · S.ORTHO 0.95 457 · R+ 0.14 4 S.STAR 4.25 3 · S.ORTHO 0.99 257 · R- 0.62 4 ★ S.STAR 4.66 10 · S.ORTHO 1.05 432 · S.NEOF 0.78 386 S.ZMTFT 0.68 95 · S.ORTHO 1.40 53 · S.P.RAD 0.27 263 S.ZMTFT 0.71 63 · S.ORTHO 1.72 104 · ★ S.P.RAD 0.41 206 S.META+ 0.67 43 · S.ORTHO 1.99 78 · S.P.RAD 0.73 453 S.META+ 0.70 88 · S.ORTHO 2.00 186 · S.P.RAD 0.76 255 S.META- 0.76 28 · S.ZE.RS 0.10 6 · S-CNTIG 1.27 139 S.ZPTFT 0.80 63 · ★ S.ZE.RS 0.15 20 · K-CHGTR 0.26 77 ★ S.ZPTFT 0.82 112 · S.ZE.RS 0.15 26 · GRP.DPL -4.13 981 S.ZPTFT 0.84 95 · S.ZE.RS 0.17 19 · GRP.DPL -3.59 832 S.PARA- 1.04 26 · S.ZE.RS 0.21 24 · GR.ELCT 0.75 13 S.PARA- 1.17 165 · S.ZE.RS 0.22 161 S.PARA- 1.18 211 · S.ZE.RS 0.27 12
N1O3	ONO2		★ 0.55 3	★ 0.70 3			★ S.INDUC 0.66 6 · ES-CH 4.00 2 · F 0.48 66 ★ S.STAR 3.77 10 · L-STM 4.46 1 · R 0.22 5 S.STAR 3.86 3 · B1-STM 1.35 1 · GRP.DPL -3.08 9 ★ S.RES.+ 0.00 10 · B5-STM 3.62 1
N2.B1F4	+(N=-N)(B1F4)-		1.11 36 ★ 1.65 29	1.32 36 ★ 1.79 29			★ S.ZPTFT 1.32 36 · ★ S.PARA- 2.99 29 · F 1.48 4 S.ZPTFT 2.20 29 · ★ S.ZE.RS 0.30 19 · R 0.31 4 S.PARA- 2.76 36 · ES-CH 2.00 2
N2	+N=-N		★ 1.76 608	1.83 631 ★ 1.91 608			★ S.INDUC 1.34 161 · ★ S.PARA- 3.43 897 · F 1.58 5 S.META- 1.80 252 · S.PARA+ 1.88 631 · R 0.33 5 S.PARA- 1.87 247 · S.ZE.RS 0.15 631 · R- 1.85 4 S.PARA- 2.63 252 · ★ S.ZE.RS 0.64 161 · GR.ELCT 1.71 13 S.PARA- 3.04 608 · ES-CH 2.00 2
N2O1	N=NO (-)			★ -0.22 72			S.ZE.RS -0.15 72
N2O2	NNO2-		★ 0.00 415	★ -0.43 415			ES-CH 4.00 2 · F 0.20 4 · R -0.63 4
N3	NNN	0.46	0.27 56 ★ 0.37 504	★ 0.08 504 0.15 56	1.02		S.INDUC 0.37 24 · ★ S.PARA- 0.11 505 · F 0.48 5 ★ S.INDUC 0.42 32 · S.ZE.RS -0.31 6 · R -0.40 5 S.INDUC 0.43 6 · S.ZE.RS -0.23 24 · R- -0.37 4 S.INDUC 0.46 805 · ES-CH 3.00 2 · GRP.DPL -2.17 9 S.STAR 2.62 3 · L-STM 4.62 1 · GRP.DPL -1.56 8 ★ S.META- 0.33 505 · B1-STM 1.50 1 S.PARA- 0.08 247 · B5-STM 4.18 1
	NNN	0.46	0.27 56 ★ 0.37 504	★ 0.08 504 0.15 56	1.02		S.INDUC 0.37 24 · ★ S.PARA- 0.08 247 · B1-STM 1.50 1 ★ S.INDUC 0.42 32 · ★ S.PARA- 0.11 505 · B5-STM 4.18 1 S.INDUC 0.43 6 · S.ZE.RS -0.31 6 · F 0.48 66 S.INDUC 0.46 805 · S.ZE.RS -0.23 24 · R -0.40 66 S.STAR 2.62 3 · ★ ES-CH 3.00 2 · GRP.DPL -2.17 9 ★ S.META- 0.33 505 · L-STM 4.62 1 · GRP.DPL -1.56 8
O1	oxide(anion)	-3.87	-0.87 279 -0.71 23 ★ -0.47 330	-1.86 637 -0.82 279 ★ -0.81 330 -0.52 23			★ S.INDUC -0.16 89 · S.PARA- -0.66 53 · S.ZE.RS -0.59 19 S.STAR -2.78 3 · S.PARA- -0.58 36 · S.RES.+ -3.10 71 S.ZMTFT -1.38 100 · S.PARA+ -4.27 100 · ES-CH 1.00 2 S.META- -1.15 100 · S.PARA+ -3.92 71 · F -0.35 66 S.META- -0.49 29 · ★ S.PARA- -2.30 65 · R -0.49 66 S.META- -0.22 36 · S.PARA+ -1.60 917 · ★ S.PHOSP 0.00 37 S.ZPTFT -2.67 100 · S.PARNO -0.50 65 · GR.ELCT 0.60 13 S.ZPTFT -0.75 306 · S.ORTHO -1.10 53

Empirical	Structural	π	σ_m	σ_p	MR	E_s	additional σ values					
O1	oxide(anion)						★S.PARA-	-0.82 *29*	★S.ZE.RS	-0.60 *134*		
O2S1	SOO-		★ -0.02 *702*	★ -0.05 *702*			★S.PARA-	0.08 *623*	L-STM	3.33 *1*	R	-0.08 *4*
							S.ORTHO	0.75 *980*	B1-STM	1.40 *1*	R-	0.05 *4*
							S.ZE.RS	0.00 *732*	B5-STM	2.70 *1*		
							ES-CH	3.20 *2*	F	0.03 *4*		
O3P1	phosphonate di-anion						S.INDUC	0.05 *825*	S.META-	0.07 *36*	ES-CH	4.20 *2*
							S.STAR	0.33 *825*	★S.PARA-	-0.26 *29*	F	0.02 *66*
							★S.META-	-0.40 *29*	S.PARA-	-0.16 *585*	R	-0.18 *66*
							★S.META-	-0.22 *138*	S.PARA-	-0.14 *138*		
							S.META-	-0.02 *585*	S.PARA-	0.15 *36*		
O3S1	sulfiteanion	-4.76	0.05 *21*	0.09 *21*			S.INDUC	0.15 *6*	S.PARA-	0.39 *307*	S.RES.+	0.23 *6*
			0.08 *29*	0.18 *29*			S.INDUC	0.23 *31*	S.PARA-	0.40 *28*	ES-CH	4.20 *2*
			0.15 *194*	0.30 *194*			★S.INDUC	0.25 *89*	S.PARA-	0.40 *181*	★ES-V	0.99 *7*
			★ 0.30 *976*	0.35 *10*			S.STAR	0.81 *3*	S.PARA-	0.46 *29*	L-STM	3.33 *1*
			0.31 *266*	★ 0.35 *279*			S.STAR	1.93 *87*	S.PARA-	0.48 *369*	B1-STM	2.03 *1*
			0.31 *764*	0.37 *764*			S.META-	0.23 *28*	S.PARA-	0.58 *765*	B5-STM	2.70 *1*
			0.41 *293*	0.38 *23*			★S.META-	0.28 *593*	★S.PARA-	0.58 *934*	F	0.29 *5*
			0.53 *36*	0.57 *36*			S.META-	0.31 *29*	S.PARA-	0.67 *36*	R	0.06 *5*
				1.40 *128*			S.META-	0.39 *765*	S.ORTHO	0.33 *279*	R-	0.29 *4*
							S.META-	0.55 *36*	S.ZE.RS	0.00 *19*	S-L	0.15 *7*
							S.ZPTFT	-0.12 *29*	S.ZE.RS	0.07 *31*	GR.ELCT	0.36 *13*
							S.ZPTFT	0.68 *36*	S.ZE.RS	0.33 *6*		
S1	sulfide anion		★ -0.36 *3*	★ -1.21 *637*			★S.STAR	-0.42 *87*	ES-CH	1.20 *2*	F	0.03 *4*
							★S.PARA+	-2.62 *71*	L-STM	3.47 *1*	R	-1.24 *4*
							★S.ZE.RS	-0.33 *19*	B1-STM	1.70 *1*	R+	-2.56 *4*
							S.RES.+	-2.10 *71*	B5-STM	1.70 *1*	GR.ELCT	0.03 *13*
Se1	Se-			★ -0.98 *637*			ES-CH	1.30 *2*				
H1	H	0.00 *130*	0.00 *26*	0.00 *128*	0.10	0.00 *62*	S.INDUC	-0.03 *99*	★S.PARA+	0.00 *43*	★S.PHOSP	0.00 *59*
		0.00	0.00 *46*	★ 0.00 *397*		★ 0.00 *50*	S.INDUC	0.00 *16*	S.ORTHO	-0.06 *186*	S.NEOF	0.00 *386*
			0.00 *128*			1.34 *311*	S.INDUC	0.00 *26*	S.ZE.RS	0.00 *26*	S.P.RAD	0.00 *79*
			★ 0.00 *397*				★S.INDUC	0.00 *89*	S.AN-RS	0.00 *26*	S.P.RAD	0.00 *206*
							★S.PRIME	0.00 *159*	ES-CH	0.00 *2*	S.P.RAD	0.00 *255*
							S-INDQ	0.00 *58*	ES-V	0.00 *7*	S-M(C+)	0.00 *666*
							★S.STAR	0.49 *289*	L-STM	2.06 *1*	S-P(C+)	0.00 *666*
							S.PARA-	-0.01 *211*	B1-STM	1.00 *1*	S-CNTIG	0.00 *139*
							★S.PARA-	0.00 *23*	B5-STM	1.00 *1*	K-CHGTR	0.00 *77*
							S.PARA-	0.00 *26*	F	0.00 *66*	GRP.DPL	0.03 *8*
							S.PARA+	-0.03 *203*	R	0.00 *66*	GR.ELCT	0.00 *13*
							S.PARA+	-0.03 *277*	R+	0.00 *4*	O-STER	0.00 *90*
							S.PARA+	-0.02 *88*	R-	0.00 *4*		
							S.PARA+	-0.02 *308*	S-L	0.00 *7*		
H1As1O3	AsO3H(-)		★ 0.00 *3*	★ -0.02 *23*			★S.INDUC	0.01 *32*	★S.PARA-	0.46 *640*	R	-0.06 *4*
							S.STAR	0.14 *3*	★S.BA.RS	-0.03 *34*	R-	0.42 *4*
							S.META-	0.13 *28*	ES-CH	4.30 *2*	★S-L	0.03 *7*
							S.PARA-	0.13 *28*	F	0.04 *4*		
H1Be1	BeH						GR.ELCT	-0.49 *13*				
H1Cl2Si1	SiHCl2						S.STAR	1.10 *393*				
H1N1	NH(-)						S.ZPTFT	-3.98 *100*	S.PARA+	-6.23 *100*	S.PARA+	-4.70 *989*
H1N2O2	NHNO2		★ 0.91 *415*	★ 0.57 *415*			ES-CH	4.30 *2*	B1-STM	1.35 *1*	F	0.99 *4*
							L-STM	4.50 *1*	B5-STM	3.66 *1*	R	-0.42 *4*
H1N3	3,4-N-NH-N-			★ 0.19 *616*			★S.PARA-	0.34 *908*	★ES-CH	2.00 *2*		
H1O1	OH	-1.12 *57*	0.01 *603*	-0.42 *254*	0.28	★ -0.55 *267*	S.INDUC	0.18 *99*	★S.PARA-	-0.37 *28*	S.ZE.RS	-0.42 *118*
		-0.67	0.02 *46*	-0.38 *190*			S.INDUC	0.23 *182*	S.PARA-	-0.31 *104*	S.ZE.RS	-0.40 *19*
		0.07 *54*	★ 0.12 *21*	★ -0.37 *21*			S.INDUC	0.24 *6*	S.PARA-	0.03 *53*	S.RES.+	-1.01 *71*
			0.13 *190*	-0.22 *46*			S.INDUC	0.25 *192*	S.PARA+	-1.60 *1009*	S.RES.+	-0.64 *6*
			0.20 *128*	0.70 *128*			S.INDUC	0.25 *210*	S.PARA+	-1.19 *71*	ES-CH	1.00 *2*
							★S.INDUC	0.29 *89*	S.PARA+	-1.06 *88*	ES-V	0.32 *7*
							S.INDUC	0.29 *118*	★S.PARA+	-0.92 *43*	L-STM	2.74 *1*
							S.INDUC	0.29 *119*	S.PARA+	-0.92 *243*	B1-STM	1.35 *1*
							S.INDUC	0.29 *205*	S.PARA+	-0.92 *467*	B5-STM	1.93 *1*
							S.INDUC	0.41 *574*	S.PARA+	-0.91 *65*	F	0.33 *5*
							S.INDUC	0.63 *16*	S.PARA+	-0.90 *322*	R	-0.70 *5*
							S.PRIME	0.22 *159*	S.PARNO	-0.12 *65*	R+	-1.25 *4*
							S.PRIME	0.28 *586*	★S.ORTHO	-0.20 *457*	R-	-0.70 *4*
							S-INDQ	1.74 *58*	S.ORTHO	-0.09 *104*	S-L	0.25 *7*
							S.STAR	1.31 *37*	S.ORTHO	0.04 *53*	S.PHOSP	-0.39 *59*
							S.STAR	1.34 *3*	S.ORTHO	0.60 *186*	S.IND.P	1.48 *39*
							★S.STAR	1.37 *10*	S.ORTHO	0.76 *158*	S.RES.P	-0.83 *39*
							S.STAR	1.55 *289*	S.ORTHO	1.22 *78*	S.P.RAD	0.17 *206*
							S.STAR	1.55 *494*	S.ORTH+	-0.80 *243*	K-CHGTR	0.40 *77*
							S.STAR	1.80 *183*	S.ORTH-	-0.41 *97*	GRP.DPL	-1.66 *9*
							S.META+	-0.04 *88*	S.ZE.RS	-0.62 *6*	GRP.DPL	-1.59 *8*
							S.META-	0.12 *28*	S.ZE.RS	-0.43 *12*	GR.ELCT	0.79 *13*
							S.ZPTFT	-0.16 *63*	★S.ZE.RS	-0.43 *134*		
H1O2P1	PO2H			★ 0.14 *435*			ES-CH	3.00 *2*	B1-STM	1.40 *1*		
							L-STM	4.22 *1*	B5-STM	2.88 *1*		
H1O2S1	S(O)OH		★ -0.04 *3*	★ -0.07 *3*			F	0.01 *5*	R	-0.08 *5*		
H1O3P1	PO3H (-)		-0.17 *29*	-0.09 *29*			★S.STAR	1.41 *3*	B5-STM	2.88 *1*		
			★ 0.20 *21*	0.17 *585*			S.ORTHO	0.40 *78*	F	0.19 *5*		
			0.25 *585*	★ 0.26 *21*			ES-CH	4.20 *2*	R	0.07 *5*		
			0.26 *10*	0.29 *36*			L-STM	4.22 *1*	GR.ELCT	0.03 *13*		
			0.27 *36*				B1-STM	2.12 *1*				
H1O3S1	SO3H	-2.68	★ 0.55 *194*		1.04		★S.INDUC	0.50 *16*	L-STM	3.97 *1*	B5-STM	2.70 *1*
			0.56 *943*				ES-CH	4.20 *2*	B1-STM	2.03 *1*		

Empirical	Structural	π	σm	σp	MR	Es	additional σ values					
H1O4P1	OPO3H-		★ 0.29 *3*	★ 0.00 *3*			F	0.41 *4*	R	-0.41 *4*		
H1S1	SH	0.28 *57*	0.23 *17*	0.01 *41*	0.92	★ -1.07 *267*	S.INDUC	0.18 *12*	S.PARA+	-0.03 *71*	★ES-V	0.60 *7*
		0.39	0.23 *41*	★ 0.15 *21*			S.INDUC	0.19 *89*	S.ORTHO	0.50 *78*	L-STM	3.47 *1*
			★ 0.25 *21*	0.16 *17*			S.INDUC	0.23 *118*	S.ZE.RS	-0.24 *126*	B1-STM	1.70 *1*
							S.INDUC	0.25 *102*	S.ZE.RS	-0.23 *102*	B5-STM	2.33 *1*
							★S.INDUC	0.26 *32*	S.ZE.RS	-0.20 *19*	F	0.30 *5*
							S.INDUC	0.26 *99*	S.ZE.RS	-0.19 *47*	R	-0.15 *5*
							S.INDUC	0.27 *182*	S.ZE.RS	-0.18 *12*	R+	-0.33 *4*
							S.INDUC	0.32 *126*	S.ZE.RS	-0.15 *118*	S-L	0.27 *14*
							S.STAR	1.52 *10*	★S.ZE.RS	-0.15 *134*	GRP.DPL	-1.51 *9*
							S.STAR	1.68 *3*	S.RES.+	-0.12 *71*	GRP.DPL	-1.33 *8*
							S.ZPTFT	0.07 *112*	ES-CH	1.20 *2*	GR.ELCT	0.17 *13*
H1Se1	Se(H)						★S.INDUC	0.18 *12*	★ES-V	0.70 *7*	B5-STM	2.56 *1*
							★S.ZE.RS	-0.18 *12*	L-STM	3.76 *1*		
							ES-CH	1.30 *2*	B1-STM	1.85 *1*		
H2As1O3	As(O)(OH)2	-2.13	★ 0.61 *10*				★S.PARA-	0.97 *640*	ES-CH	4.30 *2*	R-	0.93 *4*
							★S.RES.+	0.00 *10*	F	0.04 *4*		
H2B1	BH2						GR.ELCT	-0.17 *13*				
H2B1O2	B(OH)2	-0.55	★ -0.01 *17*	★ 0.12 *17*	1.10		★S.INDUC	-0.08 *17*	S.ZE.RS	0.13 *217*	R	0.15 *5*
			0.01 *23*	0.17 *643*			S.INDUC	-0.08 *89*	S.ZE.RS	0.23 *86*	R+	0.41 *4*
			0.03 *10*	0.45 *23*			S.INDUC	-0.05 *217*	S.RES.+	0.00 *10*	GRP.DPL	-1.16 *9*
							S.STAR	0.95 *3*	ES-CH	3.00 *2*	GR.ELCT	0.01 *13*
							S.PARA+	0.38 *217*	F	-0.03 *5*		
H2Cl1Si1	SiH2Cl			★ 0.26 *25*			★S.STAR	0.80 *393*				
				0.31 *238*								
H2N1	NH2	-1.42 *454*	-0.20 *41*	-0.94 *661*	0.54	★ -0.61 *267*	S.INDUC	0.01 *24*	S.PARA-	-0.24 *416*	S.BA.RS	-0.82 *20*
		-1.31 *680*	★ -0.16 *21*	-0.75 *251*			S.INDUC	0.01 *193*	S.PARA-	-0.17 *26*	S.AN.RS	-0.48 *20*
		-1.23	-0.15 *17*	-0.71 *41*			S.INDUC	0.01 *292*	★S.PARA-	-0.15 *290*	S.AN.RS	-0.24 *26*
		-1.19 *57*	-0.14 *10*	-0.66 *17*			S.INDUC	0.04 *102*	S.PARA+	-1.80 *811*	S.RES.+	-1.61 *20*
		-0.50 *54*	-0.09 *46*	★ -0.66 *21*			S.INDUC	0.09 *12*	S.PARA+	-1.74 *71*	S.RES.+	-1.43 *71*
		-0.42 *42*	-0.09 *251*	-0.62 *812*			S.INDUC	0.10 *26*	S.PARA+	-1.47 *65*	S.RES.+	-1.23 *10*
		-0.07 *220*	-0.07 *26*	-0.58 *93*			S.INDUC	0.11 *17*	S.PARA+	-1.33 *322*	S.RES.+	-1.10 *6*
			-0.07 *930*	-0.57 *3*			★S.INDUC	0.12 *20*	S.PARA+	-1.31 *467*	ES-CH	1.00 *2*
			0.00 *3*	-0.57 *316*			S.INDUC	0.14 *142*	★S.PARA+	-1.30 *43*	ES-V	0.35 *7*
			0.00 *316*	-0.52 *236*			S.INDUC	0.17 *6*	S.PARA+	-1.10 *811*	L-STM	2.78 *1*
			0.02 *330*	-0.43 *506*			S.INDUC	0.19 *119*	S.PARNO	-0.24 *65*	B1-STM	1.35 *1*
				-0.38 *10*			S-INDQ	1.06 *58*	S.ORTHO	-0.35 *457*	B5-STM	1.97 *1*
				-0.30 *46*			★S.STAR	0.62 *3*	S.ORTHO	-0.27 *158*	F	0.08 *5*
				-0.29 *330*			S.STAR	1.03 *10*	S.ORTHO	-0.17 *186*	R	-0.74 *5*
				-0.16 *208*			S.STAR	1.20 *170*	S.ORTHO	0.00 *104*	R+	-1.38 *4*
							S.META+	-0.16 *43*	★S.ORTHO	0.03 *53*	R-	-0.23 *4*
							S.META-	-0.18 *28*	S.ZE.RS	-0.80 *6*	S-L	0.17 *14*
							S.META-	-0.02 *416*	S.ZE.RS	-0.54 *12*	K-CHGTR	0.66 *77*
							S.ZPTFT	-0.37 *112*	S.ZE.RS	-0.50 *26*	GRP.DPL	-1.35 *9*
							S.ZPTFT	-0.36 *63*	★S.ZE.RS	-0.50 *67*	GRP.DPL	1.53 *8*
							S.ZPTFT	-0.34 *95*	S.ZE.RS	-0.48 *24*	GR.ELCT	0.47 *13*
							S.PARA-	-0.63 *28*	S.ZE.RS	-0.48 *292*	O-STER	-0.93 *90*
							S.PARA-	-0.29 *53*	S.ZE.RS	-0.44 *102*		
H2N1O1	NH(OH)	-1.34	★ -0.04 *23*	★ -0.34 *23*	0.72		★S.INDUC	0.12 *24*	ES-CH	2.00 *2*	F	0.11 *5*
							S.STAR	0.30 *3*	L-STM	3.87 *1*	R	-0.45 *5*
							S.ZE.RS	-0.39 *24*	B1-STM	1.35 *1*	GRP.DPL	-0.80 *9*
							★S.ZE.RS	-0.22 *19*	B5-STM	2.63 *1*		
	ONH2						ES-CH	2.00 *2*	★S-L	0.16 *14*		
H2N1O2S1	SO2(NH2)	-1.82	0.46 *21*	0.55 *10*	1.23		S.INDUC	0.44 *31*	S.PARA-	0.89 *28*	B5-STM	3.05 *1*
		-1.24 *54*	0.51 *10*	0.57 *21*			S.INDUC	0.44 *80*	★S.PARA-	0.94 *369*	F	0.49 *5*
			★ 0.53 *17*	0.58 *46*			★S.INDUC	0.46 *89*	S.ZE.RS	0.05 *80*	R	0.11 *5*
			0.55 *266*	★ 0.60 *17*			S.INDUC	0.53 *17*	S.BA.RS	0.05 *80*	R-	0.45 *4*
			0.58 *46*	0.62 *23*			S.STAR	2.61 *3*	S.RES.+	0.00 *10*	★S-L	0.48 *7*
				0.62 *266*			S.META-	0.58 *28*	ES-CH	4.20 *2*	K-CHGTR	1.24 *436*
							S.META-	0.65 *369*	L-STM	4.02 *1*	GRP.DPL	-4.60 *9*
							S.PARA-	0.80 *104*	B1-STM	2.04 *1*	GR.ELCT	0.25 *13*
H2N1O3S1	OSO2NH2						L-STM	4.66 *1*	B1-STM	1.35 *1*	B5-STM	3.94 *1*
H2O3P1	P(O)(OH)2	-1.59	★ 0.36 *3*	★ 0.42 *3*	1.26		★S.ZE.RS	0.08 *86*	B1-STM	2.12 *1*	R	0.08 *5*
							ES-CH	4.20 *2*	B5-STM	2.88 *1*		
							L-STM	4.22 *1*	F	0.34 *5*		
H2P1	Ph2		0.05 *105*	★ 0.05 *3*	1.22		S.INDUC	0.06 *16*	L-STM	3.13 *1*	R+	-0.03 *4*
			★ 0.06 *3*	0.24 *111*			★S.PARA+	0.06 *649*	B1-STM	1.40 *1*	S-L	0.18 *14*
							S.ZE.RS	-0.05 *47*	B5-STM	2.22 *1*	GRP.DPL	-1.11 *8*
							★S.ZE.RS	-0.05 *482*	F	0.09 *5*	GR.ELCT	-0.09 *13*
							ES-CH	1.20 *2*	R	-0.04 *5*		
H2P2	PPH2						★S.ZE.RS	-0.08 *47*	★ES-CH	2.40 *2*		
H3B1O3	B(OH)3-		★ -0.48 *643*	★ -0.44 *643*			★S.INDUC	-0.36 *89*	F	-0.42 *5*		
							ES-CH	4.00 *2*	R	-0.02 *5*		
H3Ge1	GeH3		★ 0.00 *3*	★ 0.01 *153*	0.76		S.INDUC	0.13 *242*	ES-V	0.72 *7*	B5-STM	2.45 *1*
				0.17 *238*			★S.STAR	0.70 *3*	L-STM	3.45 *1*	F	0.03 *4*
							ES-CH	1.30 *2*	B1-STM	1.73 *1*	R	-0.04 *4*
H3N1	+NH3	-4.19 *57*	0.63 *23*	0.49 *506*			S.INDUC	0.40 *126*	S.STAR	2.38 *87*	S.ZE.RS	-0.04 *24*
			0.64 *506*	0.53 *3*			S.INDUC	0.58 *24*	★S.STAR	3.61 *360*	ES-CH	1.00 *2*
			0.67 *3*	★ 0.60 *330*			S.INDUC	0.58 *89*	S.STAR	3.76 *3*	L-STM	2.78 *1*
			0.70 *10*	1.70 *831*			S.INDUC	0.59 *310*	S.META-	0.65 *28*	B1-STM	1.49 *1*
			★ 0.86 *330*				★S.INDUC	0.61 *32*	S.PARA-	0.45 *28*	B5-STM	1.97 *1*
			0.98 *316*				S.INDUC	0.72 *12*	S.PARA-	0.56 *104*	F	0.92 *5*
			1.13 *831*				S.INDUC	0.77 *99*	S.ORTHO	1.23 *104*	R	-0.32 *5*

Empirical	Structural	π	σm	σp	MR	Es	additional σ values
H3N1	+NH3						S.INDUC 0.79 24 — S.ORTHO 2.15 78 — GR.ELCT 0.93 13 S.INDUC 0.87 119 — S.ZE.RS -0.14 126 S.STAR 2.34 554 — S.ZE.RS -0.05 12
H3N1O1	ONH3+		★ 0.53 3				★S.INDUC 0.47 32 — ★ES-CH 2.00 2 — B1-STM 1.35 1 ★S.STAR 2.92 3 — L-STM 3.89 — B5-STM 2.96 1
H3N2	NHNH2	-0.88	★ -0.02 23	★ -0.55 23	0.84		S.INDUC 0.05 12 — S.ZE.RS -0.43 24 — F 0.22 5 S.INDUC 0.14 24 — ★S.ZE.RS -0.43 134 — R -0.77 5 ★S.INDUC 0.14 89 — ES-CH 2.00 2 — GRP.DPL -1.82 9 ★S.STAR 0.40 3 — L-STM 3.47 1 — GRP.DPL 1.80 8 S.ZE.RS -0.50 12 — B1-STM 1.35 1 — GR.ELCT 0.50 13 S.ZE.RS -0.49 19 — B5-STM 2.97 1
H3N2O2S1	NHSO2NH2	-2.40 57 -1.73			1.63		ES-CH 5.20 2 — B1-STM 1.35 1 L-STM 4.74 1 — B5-STM 3.72 1
	SO2NHNH2	-2.04			1.70		
H3O3Si1	Si(OH)3						★S.INDUC -0.22 25
H3Si1	SiH3		★ 0.05 18	0.07 153 ★ 0.10 18	0.83		S.INDUC 0.01 18 — S.ZE.RS 0.05 359 — B1-STM 1.69 1 ★S.INDUC 0.09 73 — ★S.ZE.RS 0.09 18 — B5-STM 2.37 1 S.INDUC 0.15 242 — S.ZE.RS 0.09 73 — F 0.06 4 S.INDUC 0.16 359 — ES-CH 1.20 2 — R 0.04 4 S.STAR 0.40 393 — ★ES-V 0.70 7 — R+ 0.08 4 ★S.PARA+ 0.14 383 — L-STM 3.33 1
H4N3O4S2	NHSO2nhso2NH2	-2.11			2.84		ES-CH 5.20 2 — B1-STM 1.35 1 L-STM 6.70 1 — B5-STM 4.99 1
H5Si2	SiH2sih3						S.STAR 1.10 393
C1B1Cl3N1	CN.BCl3		★ 0.95 41	★ 0.86 41			ES-CH 3.00 2 — F 0.93 264 — R -0.05 264
C1Br1Cl2	CCl2Br						S.INDUC -0.07 574 — S.ZE.RS 0.10 12
C1Br1O1	C=O(Br)						S.INDUC 0.43 12 — L-STM 4.03 1 — B5-STM 3.71 1 S.ZE.RS 0.33 12 — B1-STM 1.60 1
C1Br2Cl1	CClBr2						S.INDUC -0.16 574 — S.ZE.RS 0.10 12
C1Br3	CBr3	1.51 57	★ 0.28 40	★ 0.29 40	2.88	★ -3.67 50 -3.26 817	S.INDUC -0.03 127 — S.ZE.RS 0.02 110 — B1-STM 2.86 1 S.INDUC 0.25 110 — ★S.ZE.RS 0.03 64 — B5-STM 3.75 1 ★S.INDUC 0.26 40 — S.ZE.RS 0.12 127 — F 0.28 5 ★S.STAR 2.43 3 — ES-CH 4.90 2 — R 0.01 5 S.STAR 2.50 464 — ★ES-V 1.56 7 — GR.ELCT 0.20 13 S.ZE.RS 0.00 19 — L-STM 4.09 1
C1Br3O1	OCBr3						L-STM 5.74 1 — B1-STM 1.35 1 — B5-STM 4.69 1
C1Br3S1	SCBr3						L-STM 6.06 1 — B1-STM 1.70 1 — B5-STM 4.97 1
C1Cl1F2	CClF2		★ 0.42 3	★ 0.46 3			S.STAR 2.37 3 — B1-STM 1.99 1 — F 0.40 4 L-STM 3.89 1 — B5-STM 3.46 1 — R 0.06 4
C1Cl1F2O1	OCClF2						★S.INDUC 0.26 49 — ★S.ZE.RS -0.09 49 — ES-CH 4.80 2
C1Cl1N4	5-chloro-1-tetrazolyl	-0.65	★ 0.60 40	★ 0.61 40	2.32		★S.INDUC 0.58 40 — ES-CH 4.20 2 — R- 0.12 4 ★S.META- 0.72 165 — F 0.58 66 ★S.PARA- 0.70 165 — R 0.03 5
C1Cl1O1	C=O(Cl)		0.46 121 0.48 925 ★ 0.51 41 0.53 18	0.52 121 ★ 0.61 41 0.66 194 0.69 18	1.04		S.INDUC 0.33 16 — S.ZE.RS 0.06 102 — B5-STM 3.42 1 S.INDUC 0.36 102 — ★S.ZE.RS 0.21 19 — F 0.46 5 S.INDUC 0.38 18 — S.ZE.RS 0.31 18 — R 0.15 5 ★S.INDUC 0.44 34 — S.ZE.RS 0.32 12 — R+ 0.33 4 S.INDUC 0.49 12 — ★S.BA.RS 0.19 34 — R- 0.78 4 S.STAR 2.37 3 — ES-CH 3.20 2 — S.ZTWST 0.00 22 ★S.PARA- 1.24 252 — L-STM 3.83 1 — GRP.DPL -2.48 9 ★S.PARA+ 0.79 382 — B1-STM 1.60 1 — GR.ELCT 0.30 13
C1Cl1O2	OCOCl						★S.INDUC 0.36 49 — ★S.ZE.RS -0.11 49 — ES-CH 4.20 2
C1Cl2F1O1	OCCl2F						★S.INDUC 0.32 49 — ★S.ZE.RS -0.13 49 — ES-CH 5.20 2
C1Cl2N1	N=CCl2	0.41	★ 0.21 40	★ 0.13 40	1.83		★S.INDUC 0.29 40 — L-STM 5.65 1 — F 0.26 5 ★S.STAR 1.81 3 — B1-STM 1.70 1 — R -0.13 5 ES-CH 4.40 2 — B5-STM 4.54 1
C1Cl3	CCl3	1.31 57	0.32 40 0.40 6 ★ 0.40 17 0.41 10	0.33 40 0.42 197 0.44 6 ★ 0.46 17 0.47 10	2.01	★ -3.30 50 -2.87 35	S.INDUC -0.01 574 — ★S.STAR 2.65 52 — ES-CH 4.60 2 S.INDUC 0.13 127 — S.STAR 2.65 129 — ES-V 1.38 7 S.INDUC 0.29 110 — S.STAR 3.06 554 — L-STM 3.89 1 ★S.INDUC 0.31 40 — S.ZE.RS 0.00 19 — B1-STM 2.64 1 S.INDUC 0.31 185 — S.ZE.RS 0.02 110 — B5-STM 3.46 1 S.INDUC 0.36 6 — ★S.ZE.RS 0.03 64 — F 0.38 5 S.INDUC 0.40 17 — S.ZE.RS 0.03 185 — R 0.09 5 S.INDUC 0.43 16 — S.ZE.RS 0.07 47 — S-L 0.36 14 S.INDUC 0.43 34 — S.ZE.RS 0.08 6 — ★S.PHOSP 0.30 59 S.STAR 2.48 10 — S.ZE.RS 0.09 127 — GRP.DPL -2.03 8 S.STAR 2.55 580 — S.ZE.RS 0.10 12 — GRP.DPL -1.84 9 S.STAR 2.61 74 — S.BA.RS 0.02 34 — GR.ELCT 0.26 13
C1Cl3O1	OCCl3		★ 0.43 3	★ 0.35 3			S.INDUC 0.23 49 — ★S.ZE.RS -0.17 18 — L-STM 5.44 1 ★S.INDUC 0.51 18 — S.ZE.RS -0.17 102 — B1-STM 1.35 1 S.INDUC 0.51 234 — S.ZE.RS -0.17 234 — B5-STM 4.41 1 ★S.INDUC 0.55 102 — ★S.ZE.RS -0.08 49 — F 0.46 5 ★S.STAR 3.19 3 — ES-CH 5.60 2 — R -0.11 4
C1Cl3O2S1	SO2CCl3						L-STM 5.57 1 — B1-STM 2.03 1 — B5-STM 4.50 1
C1Cl3S1	SCCl3	1.65			2.83		ES-CH 5.80 2 — B1-STM 1.70 1 L-STM 5.76 1 — B5-STM 4.71 1
C1D3	Cd3						★S.INDUC 0.02 192 — ★S.STAR -0.01 800
C1F1O1	C=O(F)		★ 0.55 18	★ 0.70 18	0.54		S.INDUC 0.24 16 — S.ZE.RS 0.31 12 — B1-STM 1.60 1 ★S.INDUC 0.39 89 — ★S.ZE.RS 0.31 18 — B5-STM 2.57 1 S.INDUC 0.56 12 — ★ES-CH 2.80 2 — F 0.48 4

Empirical	Structural	π	σm	σp	MR	Es	additional σ values

C1F1O1 — C=O(F)
- additional σ values: ★S.STAR 2.44 _3_ | L-STM 3.53 _1_ | R 0.22 _4_

C1F2O2 — OCF2O* (3,4)
- σm ★0.36 _499_ | σp ★0.36 _499_ | MR 0.90
- additional σ values: ★ES-CH 0.00 _2_ | F 0.36 _5_ | R 0.00 _5_

C1F3 — CF3
- π: 0.10 _854_ ; 0.88 ; 1.07 _857_
- σm: 0.39 _132_ ; 0.41 _251_ ; 0.41 _716_ ; ★0.43 _21_ ; 0.44 _10_ ; 0.44 _156_ ; 0.46 _6_ ; 0.46 _17_ ; 0.46 _46_ ; 0.48 _26_ ; 0.48 _41_
- σp: 0.41 _89_ ; 0.49 _132_ ; 0.49 _156_ ; 0.51 _6_ ; 0.51 _10_ ; 0.53 _17_ ; 0.53 _716_ ; 0.54 _46_ ; ★0.54 _21_ ; 0.60 _41_
- MR: 0.50
- Es: ★-2.40 _50_ ; -1.90 _35_
- additional σ values:
  - S.INDUC 0.33 _16_ | S.ZPTFT 0.54 _95_ | ES-V 0.91 _7_
  - S.INDUC 0.34 _12_ | S.ZPTFT 0.67 _676_ | L-STM 3.30 _1_
  - S.INDUC 0.35 _127_ | S.PARA- 0.59 _26_ | B1-STM 1.99 _1_
  - S.INDUC 0.38 _132_ | S.PARA- 0.61 _160_ | B5-STM 2.61 _1_
  - S.INDUC 0.39 _110_ | S.PARA- 0.62 _558_ | F 0.38 _5_
  - S.INDUC 0.39 _156_ | S.PARA- 0.65 _28_ | R 0.16 _5_
  - ★S.INDUC 0.40 _6_ | ★S.PARA- 0.65 _477_ | R+ 0.23 _4_
  - S.INDUC 0.40 _119_ | S.PARA- 0.74 _717_ | R- 0.27 _4_
  - S.INDUC 0.42 _67_ | ★S.PARA+ 0.61 _43_ | S-L 0.40 _14_
  - S.INDUC 0.42 _142_ | S.PARNO 0.53 _324_ | ★S.PHOSP 0.70 _59_
  - S.INDUC 0.43 _26_ | ★S.ORTH- 0.81 _97_ | S.NEOF 0.46 _545_
  - S.INDUC 0.46 _17_ | S.ZE.RS 0.08 _20_ | S.NEOF 0.57 _386_
  - S.PRIME 0.38 _159_ | S.ZE.RS 0.08 _26_ | S.IND.P 2.21 _39_
  - S.STAR 2.00 _61_ | S.ZE.RS 0.10 _127_ | S.M.RAD -0.17 _79_
  - S.STAR 2.56 _10_ | ★S.ZE.RS 0.10 _224_ | S.P.RAD -0.09 _79_
  - S.STAR 2.61 _3_ | S.ZE.RS 0.11 _6_ | S-M(C+) 0.51 _239_
  - S.STAR 2.85 _900_ | S.ZE.RS 0.11 _19_ | S-M(C+) 0.56 _256_
  - S.STAR 3.00 _935_ | S.ZE.RS 0.15 _12_ | S-P(C+) 0.71 _239_
  - S.ZMTFT 0.48 _95_ | S.BA.RS 0.08 _20_ | S-P(C+) 0.79 _256_
  - S.META+ 0.52 _43_ | S.AN-RS 0.15 _26_ | K-CHGTR -0.09 _77_
  - S.META+ 0.53 _88_ | S.AN-RS 0.17 _20_ | GRP.DPL -2.61 _8_
  - S.META+ 0.57 _982_ | S.RES.+ 0.00 _10_ | GRP.DPL -1.94 _9_
  - S.META- 0.41 _717_ | S.RES.+ 0.08 _20_ | GR.ELCT 0.47 _13_
  - S.META- 0.43 _28_ | S.RES.+ 0.15 _6_
  - S.META- 0.49 _622_ | ES-CH 3.40 _1_

C1F3Hg1 — HgCF3
- σm ★0.29 _3_ | σp ★0.32 _3_
- additional σ values:
  - S.INDUC 0.06 _18_ | ★S.ZE.RS 0.01 _18_ | GR.ELCT -0.63 _13_
  - ★S.INDUC 0.27 _18_ | F 0.29 _4_
  - ★S.STAR 1.70 _3_ | R 0.03 _4_

C1F3Hg1O2S1 — HgOSOCF3
- additional σ values: ★S.INDUC 0.50 _340_ | ★S.ZE.RS 0.03 _340_

C1F3Hg1S1 — HgSCF3
- σm ★0.39 _3_ | σp ★0.42 _3_
- additional σ values:
  - S.INDUC 0.18 _18_ | ★S.STAR 2.30 _3_ | F 0.38 _5_
  - S.INDUC 0.37 _18_ | ★S.ZE.RS 0.02 _18_ | R 0.04 _5_

C1F3I1N1O2S1 — I=NSO2CF3
- σm ★1.30 _178_ | σp ★1.35 _178_
- additional σ values:
  - ★S.INDUC 1.24 _178_ | F 1.20 _4_
  - ★S.ZE.RS 0.11 _178_ | R 0.15 _4_

C1F3N2 — N=NCF3
- σm 0.43 _72_ ; ★0.56 _3_ | σp ★0.68 _3_ ; 0.74 _304_ ; 0.81 _72_
- additional σ values:
  - S.INDUC 0.45 _72_ | L-STM 5.45 _1_ | R 0.18 _4_
  - ★S.STAR 2.75 _3_ | B1-STM 1.70 _1_ | R+ 0.24 _4_
  - ★S.PARA+ 0.74 _72_ | B5-STM 3.48 _1_
  - S.ZE.RS 0.24 _72_ | F 0.50 _4_

C1F3O1 — OCF3
- π 1.04
- σm 0.35 _132_ ; 0.36 _370_ ; 0.37 _17_ ; ★0.38 _224_ | σp 0.32 _132_ ; 0.32 _370_ ; 0.35 _17_ ; ★0.35 _224_ | MR 0.79
- additional σ values:
  - ★S.INDUC 0.39 _224_ | S.PARA- 0.25 _28_ | ES-CH 4.40 _2_
  - S.INDUC 0.41 _17_ | S.PARA- 0.25 _160_ | L-STM 4.57 _1_
  - S.INDUC 0.43 _598_ | S.PARA- 0.26 _621_ | B1-STM 1.35 _1_
  - S.INDUC 0.50 _132_ | ★S.PARA- 0.27 _462_ | B5-STM 3.61 _1_
  - S.INDUC 0.53 _234_ | S.ZE.RS -0.25 _19_ | F 0.39 _5_
  - S.INDUC 0.55 _80_ | S.ZE.RS -0.21 _80_ | R -0.04 _5_
  - S.INDUC 0.57 _846_ | S.ZE.RS -0.18 _234_ | R- -0.12 _4_
  - S.META- 0.40 _621_ | ★S.ZE.RS -0.04 _224_ | GRP.DPL -2.36 _8_
  - S.META- 0.46 _28_ | S.BA.RS -0.31 _80_ | GR.ELCT 0.79 _13_
  - S.META- 0.47 _462_ | S.AN-RS -0.27 _80_

C1F3O1S1 — S=O(CF3)
- σm ★0.63 _132_ ; 0.66 _10_ ; 0.67 _6_ | σp 0.64 _6_ ; ★0.69 _132_ ; 0.72 _10_ | MR 1.31
- additional σ values:
  - S.INDUC 0.64 _20_ | ★S.PARA- 1.05 _160_ | ES-CH 5.60 _2_
  - ★S.INDUC 0.64 _132_ | S.ZE.RS -0.03 _6_ | L-STM 4.70 _1_
  - S.INDUC 0.67 _6_ | ★S.ZE.RS 0.08 _20_ | B1-STM 1.40 _1_
  - S.INDUC 0.68 _31_ | S.ZE.RS 0.13 _31_ | B5-STM 3.70 _1_
  - S.INDUC 0.85 _125_ | S.ZE.RS 0.34 _125_ | F 0.58 _5_
  - S.STAR 4.30 _3_ | ★S.BA.RS 0.08 _20_ | R 0.11 _5_
  - S.META- 0.76 _297_ | S.RES.+ 0.08 _20_ | R- 0.47 _4_

C1F3O1Se1 — Se=O(CF3)
- σm ★0.81 _3_ | σp ★0.83 _3_
- additional σ values:
  - ★S.INDUC 0.76 _856_ | ★S.ZE.RS 0.10 _18_ | F 0.76 _5_
  - ★S.STAR 4.75 _3_ | ES-CH 5.70 _2_ | R 0.07 _5_

C1F3O2S1 — SO2(CF3)
- π 0.55
- σm: 0.76 _132_ ; 0.79 _214_ ; 0.79 _370_ ; 0.80 _6_ ; 0.80 _17_ ; ★0.83 _176_ ; 0.86 _10_ ; 0.90 _294_ ; 0.92 _48_
- σp: 0.91 _17_ ; 0.92 _6_ ; 0.93 _214_ ; 0.95 _10_ ; 0.96 _176_ ; ★0.96 _132_ ; 1.03 _370_ ; 1.06 _294_ ; 1.08 _48_
- MR 1.29
- additional σ values:
  - S.INDUC 0.71 _6_ | S.META- 1.00 _28_ | S.ZE.RS 0.31 _294_
  - S.INDUC 0.73 _68_ | S.META- 1.00 _297_ | S.AN-RS 0.57 _80_
  - S.INDUC 0.74 _132_ | S.ZPTFT 0.93 _487_ | S.RES.+ 0.00 _10_
  - S.INDUC 0.75 _294_ | S.PARA- 1.36 _541_ | ES-CH 6.60 _2_
  - S.INDUC 0.77 _48_ | S.PARA- 1.63 _160_ | L-STM 4.70 _1_
  - ★S.INDUC 0.78 _31_ | S.PARA- 1.63 _164_ | B1-STM 2.03 _1_
  - S.INDUC 0.80 _17_ | S.PARA- 1.65 _865_ | B5-STM 3.70 _1_
  - S.INDUC 0.84 _80_ | S.PARA- 1.67 _28_ | F 0.74 _5_
  - S.INDUC 0.84 _598_ | S.ZE.RS 0.21 _6_ | R 0.22 _5_
  - S.STAR 4.32 _10_ | S.ZE.RS 0.24 _80_ | R- 0.89 _4_
  - S.STAR 4.50 _3_ | S.ZE.RS 0.31 _31_ | GR.ELCT 0.44 _13_
  - S.META- 0.92 _621_ | S.ZE.RS 0.31 _48_
  - S.META- 0.98 _164_ | S.ZE.RS 0.31 _68_

C1F3O2Se1 — SeO2CF3
- σm ★1.08 _3_ | σp ★1.21 _3_ | MR 1.54
- additional σ values:
  - ★S.INDUC 0.96 _18_ | ★ES-CH 6.70 _2_
  - S.INDUC 1.19 _856_ | F 0.97 _5_
  - ★S.ZE.RS 0.25 _18_ | R 0.24 _5_

C1F3O3S1 — OSO2CF3
- σm ★0.56 _522_ | σp ★0.53 _522_ | MR 1.45
- additional σ values:
  - S.INDUC 0.58 _522_ | S.BA.RS -0.05 _522_ | B5-STM 3.24 _1_
  - ★S.INDUC 0.84 _414_ | ★ES-CH 5.20 _2_ | F 0.56 _5_
  - ★S.STAR 4.37 _3_ | L-STM 5.23 _1_ | R -0.03 _5_
  - ★S.PARA- 0.49 _160_ | B1-STM 1.35 _1_ | R- -0.07 _4_

C1F3S1 — SCF3
- π 1.44
- σm: 0.30 _176_ ; 0.35 _370_ ; 0.36 _132_ ; 0.37 _17_ ; 0.38 _3_ ; 0.40 _224_ | σp: 0.38 _370_ ; 0.42 _17_ ; 0.42 _176_ ; 0.44 _6_ ; 0.46 _48_ ; 0.46 _132_ | MR 1.38
- additional σ values:
  - S.INDUC 0.31 _214_ | S.META- 0.45 _28_ | S.BA.RS 0.04 _20_
  - S.INDUC 0.35 _370_ | S.META- 0.46 _462_ | S.BA.RS 0.17 _18_
  - S.INDUC 0.40 _48_ | S.PARA- 0.57 _541_ | S.AN-RS 0.14 _20_
  - S.INDUC 0.40 _132_ | S.PARA- 0.61 _160_ | S.RES.+ 0.04 _20_
  - S.INDUC 0.42 _31_ | S.PARA- 0.62 _164_ | ES-CH 4.60 _2_
  - ★S.INDUC 0.42 _40_ | S.PARA- 0.63 _28_ | L-STM 4.89 _1_

Empirical	Structural	π	σm	σp	MR	Es	additional σ values
C1F3S1	SCF3		★ 0.40 214 0.42 10 0.43 48 0.46 6	0.50 3 0.50 224 ★ 0.50 214			S.INDUC 0.43 598 — ★S.PARA- 0.64 462 — B1-STM 1.70 1 S.INDUC 0.44 320 — S.ZE.RS -0.01 6 — B5-STM 3.94 1 S.INDUC 0.45 6 — S.ZE.RS 0.00 19 — F 0.36 5 S.INDUC 0.50 102 — S.ZE.RS 0.04 20 — R 0.14 5 S.STAR 2.70 10 — S.ZE.RS 0.06 48 — R- 0.28 4 S.STAR 2.75 3 — ★S.ZE.RS 0.06 224 — GRP.DPL -2.50 8 S.META- 0.44 164 — S.ZE.RS 0.06 320 — GR.ELCT 0.20 13
C1F3Se1	SeCF3		0.32 451 0.44 48 0.44 619 ★ 0.44 451	0.38 451 0.45 619 ★ 0.45 451 0.46 48	1.63		★S.INDUC 0.28 451 — S.ZE.RS 0.04 48 — B5-STM 4.09 1 S.INDUC 0.42 48 — S.ZE.RS 0.10 30 — F 0.43 5 S.INDUC 0.42 451 — ★S.BA.RS 0.10 18 — R 0.02 5 S.INDUC 0.42 619 — ES-CH 4.70 2 — R- 0.10 4 ★S.STAR 2.62 3 — L-STM 4.50 1 — GRP.DPL -2.48 8 ★S.PARA- 0.53 160 — B1-STM 1.85 1 — GR.ELCT 0.19 13
C1Hg1N1	HgCN		★ 0.28 3	★ 0.34 3			S.INDUC 0.07 18 — ★S.ZE.RS 0.02 18 — GR.ELCT -0.62 13 ★S.INDUC 0.23 18 — F 0.27 5 ★S.STAR 1.40 3 — R 0.08 5
C1Hg1N1S1	HgSCN						★S.INDUC 0.20 340 — ★S.ZE.RS 0.00 340
C1I3	CI3					★ -3.74 35	ES-CH 6.10 2 — L-STM 4.36 1 — B5-STM 4.15 1 ES-V 1.79 7 — B1-STM 3.16 1
C1N1.B1Br3	CN.BBr3		★ 0.61 41	★ 0.48 41			ES-CH 3.00 2 — F 0.64 5 — R -0.16 5
C1N1.B1F3	CN.BF3		★ 0.72 41	★ 0.66 41			ES-CH 3.00 2 — F 0.71 264 — R -0.05 264
C1N1	CN	-0.84 57 -0.67 54 -0.57 -0.24 42 -0.15 220	0.52 266 ★ 0.56 21 0.61 17 0.61 718 0.62 10 0.62 46 0.65 26 0.65 41 0.68 23	0.61 41 0.63 23 0.66 21 ★ 0.66 718 0.68 236 0.69 17 0.70 65 0.71 46 0.72 10 0.73 208	0.63	★ -0.51 267	S.INDUC 0.22 16 — S.ZPTFT 0.71 112 — S.ZE.RS 0.21 320 S.INDUC 0.48 320 — S.ZPTFT 0.72 95 — S.ZE.RS 0.21 334 S.INDUC 0.48 334 — S.PARA- 0.83 26 — S.BA.RS 0.13 20 S.INDUC 0.49 12 — S.PARA- 0.88 181 — S.AN.RS 0.23 26 ★S.INDUC 0.53 40 — S.PARA- 0.88 442 — S.AN.RS 0.33 20 S.INDUC 0.56 20 — S.PARA- 0.89 211 — S.RES.+ 0.00 10 S.INDUC 0.56 119 — S.PARA- 0.89 514 — S.RES.+ 0.13 20 S.INDUC 0.56 205 — S.PARA- 0.98 614 — ES-CH 2.00 2 S.INDUC 0.58 26 — S.PARA- 0.99 443 — ES-V 0.40 7 S.INDUC 0.58 182 — S.PARA- 0.99 723 — L-STM 4.23 1 S.INDUC 0.59 99 — ★S.PARA- 1.00 247 — B1-STM 1.60 1 S.INDUC 0.59 192 — S.PARA- 1.00 369 — B5-STM 1.60 1 S.INDUC 0.60 142 — S.PARA- 1.00 749 — F 0.51 5 S.INDUC 0.61 17 — S.PARA- 1.02 28 — R 0.15 5 S.INDUC 0.63 6 — S.PARA- 1.02 165 — R+ 0.15 4 S.PRIME 0.56 159 — S.PARA+ 0.63 406 — R- 0.49 4 S.PRIME 0.58 586 — ★S.PARA+ 0.66 43 — S.ZTWST 0.10 22 S-INDQ 3.04 58 — S.PARA+ 0.71 308 — S.M.RAD -0.26 79 S.STAR 3.30 3 — S.PARA+ 0.82 88 — S.P.RAD 0.24 206 ★S.STAR 3.64 289 — S.PARNO 0.67 324 — S.P.RAD 0.34 263 S.STAR 3.72 420 — S.ORTHO 1.06 78 — S.P.RAD 0.40 79 S.ZMTFT 0.62 95 — S.ORTHO 1.20 186 — S.P.RAD 0.71 255 S.ZMTFT 0.65 63 — S-CNTIG 1.32 53 — S-CNTIG 0.88 139 S.META+ 0.56 43 — ★S.ORTH- 1.18 97 — S-CNTIG 0.99 486 S.META+ 0.56 88 — S.ZE.RS 0.04 114 — K-CHGTR 0.23 77 S.META- 0.60 514 — S.ZE.RS 0.08 6 — GRP.DPL -4.08 8 S.META- 0.65 28 — S.ZE.RS 0.08 26 — GRP.DPL -3.63 9 S.META- 0.65 609 — ★S.ZE.RS 0.08 67 — GR.ELCT 0.61 13 S.META- 0.68 369 — S.ZE.RS 0.09 19 — O-STER -0.89 90 S.ZPTFT 0.68 63 — S.ZE.RS 0.18 12
	NC		★ 0.48 3	★ 0.49 3			S.INDUC 0.43 16 — ★S.ZE.RS 0.02 161 — B5-STM 1.60 1 S.INDUC 0.47 24 — ES-CH 2.00 2 — F 0.47 5 ★S.INDUC 0.47 161 — L-STM 4.23 1 — R 0.02 5 S.ZE.RS 0.02 24 — B1-STM 1.60 1 — ★S-L 0.63 14
C1N1O1	NCO		★ 0.27 40 0.30 512 0.43 405	★ 0.19 40 0.24 512 0.35 405 0.42 194	0.88		S.INDUC 0.35 24 — S.ZE.RS -0.40 19 — R -0.12 5 ★S.INDUC 0.36 40 — S.ZE.RS -0.17 24 — R+ -0.50 5 S.INDUC 0.36 161 — ★S.ZE.RS -0.17 161 — GRP.DPL -3.93 8 S.INDUC 0.38 16 — S.ZE.RS 0.40 114 — GRP.DPL -2.81 9 S.STAR 2.25 3 — ES-CH 3.00 2 — GR.ELCT 0.83 13 ★S.PARA+ -0.19 382 — F 0.31 5
	OCN		★ 0.67 3	★ 0.54 3	0.70		S.INDUC 0.79 540 — ★S.ZE.RS -0.26 31 — B5-STM 4.01 1 S.INDUC 0.80 31 — ES-CH 3.00 2 — F 0.69 5 ★S.STAR 5.00 3 — L-STM 3.87 1 — R -0.15 5 S.ZE.RS -0.31 540 — B1-STM 1.35 1
C1N1O2S1	SO2(CN)		★ 1.10 3	★ 1.26 3	1.32		★S.INDUC 0.94 31 — ES-CH 5.20 2 — B5-STM 4.12 1 ★S.STAR 5.90 3 — L-STM 3.99 1 — F 0.97 5 S.ZE.RS 0.32 31 — B1-STM 2.03 1 — R 0.29 5
C1N1S1	NCS	1.15	0.32 762 0.48 873 ★ 0.48 56 0.49 10	0.32 762 0.34 890 0.37 10 ★ 0.38 56	1.72		★S.INDUC 0.42 24 — S.ZE.RS -0.07 24 — F 0.51 5 ★S.STAR 2.62 3 — ★S.ZE.RS -0.06 134 — R -0.13 5 S.META- 0.34 763 — ES-CH 3.20 2 — R- -0.17 4 S.PARA- 0.34 763 — L-STM 4.29 1 — GRP.DPL -2.91 8 S.PARA- 0.34 874 — B1-STM 1.50 1 — GR.ELCT 0.80 13 S.ZE.RS -0.35 19 — B5-STM 4.24 1
	SCN	0.03 57 0.41	0.41 213 ★ 0.51 6 0.53 701	0.40 6 0.51 459 ★ 0.52 21 0.70 23	1.34		S.INDUC 0.42 540 — S.ZPTFT 0.70 676 — L-STM 4.08 1 S.INDUC 0.51 210 — ★S.PARA- 0.59 28 — B1-STM 1.70 1 S.INDUC 0.55 32 — S.PARA- 0.60 272 — B5-STM 4.45 1 S.INDUC 0.56 6 — S.PARA- 0.60 290 — F 0.49 5 S.INDUC 0.63 31 — S.ZE.RS -0.16 6 — R 0.03 5 ★S.INDUC 0.64 31 — S.ZE.RS -0.05 31 — R- 0.23 4 S.STAR 3.26 10 — S.ZE.RS 0.09 540 — GRP.DPL -3.89 9 ★S.STAR 3.43 3 — ES-CH 3.20 2 — GRP.DPL -3.01 8
C1N1Se1	SeCN		★ 0.61 555	★ 0.66 23	1.68		★S.INDUC 0.58 32 — B1-STM 1.85 1 — GRP.DPL -4.01 8 S.STAR 3.61 3 — B5-STM 4.68 1 — GRP.DPL -3.91 9 ★S.BA.RS 0.08 34 — F 0.57 66 — GR.ELCT 0.22 13

Empirical	Structural	π	σ_m	σ_p	MR	E_s	additional σ values
C1N1Se1	SeCN						ES-CH 3.30 2; L-STM 4.31 1; R 0.09 5; ★S-L 0.56 14
C1N3	N=NCN		★0.71 72	★1.03 72			S.INDUC 0.68 72; S.PARA+ 1.03 72; S.ZE.RS 0.32 72; F 0.56 4; R 0.47 4; R+ 0.47 4
C1N3O1	N(=O)=NCN		★0.78 179	★0.89 179			★S.INDUC 0.72 179; ★S.ZE.RS 0.06 179; F 0.70 4; R 0.19 4; ★S.RES.P 0.17 179
C1N3O6	C(NO2)3		★0.72 3	★0.82 291	2.27		S.STAR 3.00 986; S.STAR 4.54 291; ★S.STAR 4.60 3; S.STAR 5.19 157; ★ES-CH 9.99 2; L-STM 4.59 1; B1-STM 2.55 1; B5-STM 3.72 1; F 0.65 5; R 0.17 5
C1N3S1	5-(1,2,3,4-thiatriazolyl)		★0.30 398	★0.19 398			★S.INDUC 0.42 398; ★S.STAR 2.62 3; F 0.33 66; R -0.11 66
C1N7	5-azido-1-tetrazolyl		★0.54 40	★0.54 40	2.69		★S.INDUC 0.55 40; ★ES-CH 4.00 2; F 0.53 5; R 0.01 5
C1O2	CO2-	-4.67 57; -4.36	-0.20 279; -0.16 149; -0.12 10; -0.10 29; ★-0.10 21; -0.09 6; 0.01 316; 0.02 330; 0.09 506; 0.10 23; 0.10 266; 0.36 36	-0.20 10; -0.12 149; -0.12 279; -0.05 893; ★0.00 21; 0.04 6; 0.04 29; 0.11 330; 0.13 23; 0.20 208; 0.43 36	0.60		S.INDUC -0.35 89; ★S.INDUC -0.19 6; S.INDUC -0.18 149; S.PRIME -0.18 159; S.PRIME -0.12 149; S.INDQ 0.61 58; S.STAR -1.46 360; S.STAR -1.06 3; S.STAR 0.75 10; S.STAR 0.92 757; S.STAR 0.98 554; S.ZMTFT -0.19 100; S.META+ -0.10 100; S.META+ -0.03 43; S.META- -0.08 138; S.META- 0.02 29; S.META- 0.26 36; S.ZPTFT -0.16 100; S.ZPTFT -0.05 306; S.PARA- 0.15 138; S.PARA- 0.24 181; S.PARA- 0.28 29; ★S.PARA- 0.31 307; S.PARA- 0.37 894; S.PARA- 0.49 36; S.PARA+ -0.41 100; ★S.PARA+ -0.02 43; S.ORTHO -0.91 78; S.ORTHO -0.69 279; S.ORTHO -0.13 104; S.ORTHO -0.05 158; S.ZE.RS 0.00 149; S.ZE.RS 0.23 6; S.RES.+ 0.00 10; S.RES.+ 0.06 6; ES-CH 3.00 2; L-STM 3.53 1; B1-STM 1.60 1; B5-STM 2.66 1; F -0.10 5; R 0.10 5; R+ 0.08 4; R- 0.41 4; S-L -0.20 7; GR.ELCT 0.33 13
C1O2R1	C=O -O		0.36 17; ★0.37 21	0.43 17; ★0.45 21			S.INDUC 0.30 20; ★S.INDUC 0.34 32; S.INDUC 0.35 17; S.STAR 2.11 10; S.ZMTFT 0.36 95; S.META+ 0.37 43; S.ZPTFT 0.46 95; ★S.PARA+ 0.48 43; S.ZE.RS 0.14 20; S.BA.RS 0.14 20; S.AN-RS 0.34 20; S.RES.+ 0.14 20; ★ES-CH 4.00 2; F 0.33 66; R 0.15 66
C1O2S1	-OC(=O)S-(1,2)	0.05					ES-CH 3.20 2
C1H1Br1Cl1	CHBrCl						ES-CH 3.50 2; L-STM 4.09 1; B1-STM 1.89 1; B5-STM 3.75 1
C1H1Br1Cl1O1	OCHBrCl						L-STM 3.98 1; B1-STM 1.35 1; B5-STM 4.69 1
C1H1Br1F1	CHBrF						ES-CH 3.10 2; L-STM 4.09 1; B1-STM 1.73 1; B5-STM 3.75 1
C1H1Br1F1O1	OCHBrF						L-STM 3.98 1; B1-STM 1.35 1; B5-STM 4.69 1
C1H1Br1F1S1	SCHBrF						L-STM 4.30 1; B1-STM 1.70 1; B5-STM 4.97 1
C1H1Br2	CHBr2		★0.31 3	★0.32 3	1.68	★-3.10 50; -1.88 35	S.INDUC 0.05 12; S.INDUC 0.22 127; S.INDUC 0.26 185; S.INDUC 0.29 110; ★S.INDUC 0.30 64; S.STAR 1.95 10; S.STAR 1.96 3; S.ZE.RS 0.00 19; ★S.ZE.RS 0.02 64; S.ZE.RS 0.02 110; S.ZE.RS 0.03 185; S.ZE.RS 0.04 127; S.ZE.RS 0.06 12; ES-CH 3.60 2; ★ES-V 0.89 7; L-STM 4.09 1; B1-STM 1.92 1; B5-STM 3.75 1; F 0.31 4; R 0.01 4; GRP.DPL -1.90 9; GR.ELCT 0.18 13
C1H1Br2O1	OCHBr2						L-STM 3.98 1; B1-STM 1.35 1; B5-STM 4.69 1
C1H1Br2O2S1	SO2CHBr2						L-STM 4.11 1; B1-STM 2.03 1; B5-STM 4.77 1
C1H1Br2S1	SCHBr2						L-STM 4.30 1; B1-STM 1.70 1; B5-STM 4.97 1
C1H1Br3N1	NHCBr3						L-STM 5.78 1; B1-STM 1.35 1; B5-STM 4.72 1
C1H1Cl1F1	CHCIF						ES-CH 3.00 2; L-STM 3.89 1; B1-STM 1.73 1; B5-STM 3.46 1
C1H1Cl1F1O1	OCHCIF						L-STM 3.98 1; B1-STM 1.35 1; B5-STM 4.41 1
C1H1Cl1F1S1	SCHCIF						L-STM 4.30 1; B1-STM 1.70 1; B5-STM 4.71 1
C1H1Cl2	CHCl2	1.10 42	0.24 197; ★0.31 3	0.19 326; ★0.32 3; 0.41 197	1.53	★-2.78 50; -1.70 35	S.INDUC 0.29 110; S.INDUC 0.29 127; ★S.INDUC 0.30 64; S.INDUC 0.31 185; S.INDUC 0.32 16; S.STAR 1.93 74; S.STAR 1.94 52; S.STAR 1.94 129; S.STAR 1.94 494; S.STAR 1.94 554; S.STAR 1.94 580; ★S.STAR 2.01 10; S.STAR 2.10 595; S.ZE.RS 0.00 19; S.ZE.RS 0.02 47; ★S.ZE.RS 0.02 64; S.ZE.RS 0.02 110; S.ZE.RS 0.02 127; S.ZE.RS 0.02 185; ES-CH 3.40 2; ES-V 0.81 7; L-STM 3.89 1; B1-STM 1.88 1; B5-STM 3.46 1; F 0.31 4; R 0.01 4; ★S.PHOSP 0.27 59; S.IND.P 1.79 39; GRP.DPL -1.96 9; GR.ELCT 0.22 13
C1H1Cl2O1	OCHCl2		★0.38 3	★0.26 3			★S.INDUC 0.49 18; S.INDUC 0.49 234; ★S.STAR 3.06 3; ★S.ZE.RS -0.23 18; S.ZE.RS -0.23 234; ES-CH 4.40 2; L-STM 3.98 1; B1-STM 1.35 1; B5-STM 4.41 1; F 0.43 5; R -0.17 5
C1H1Cl2O2S1	SO2CHCl2						L-STM 4.11 1; B1-STM 2.03 1; B5-STM 4.50 1
C1H1Cl2S1	SCHCl2						L-STM 4.30 1; B1-STM 1.70 1; B5-STM 4.71 1
C1H1Cl3N1	NHCCl3						L-STM 5.48 1; B1-STM 1.35 1; B5-STM 4.44 1
C1H1D2O1	Cd2OH						★S.INDUC 0.10 192
C1H1F2	CHF2		★0.29 18; 0.32 3	★0.32 18; 0.35 3	0.52	-1.91 50; ★-1.44 35	S.INDUC 0.25 127; S.INDUC 0.26 16; S.INDUC 0.26 110; ★S.INDUC 0.29 64; S.STAR 2.05 52; S.ZE.RS 0.06 64; ★S.ZE.RS 0.06 110; S.ZE.RS 0.06 185; L-STM 3.30 1; B1-STM 1.71 1; B5-STM 2.61 1; F 0.29 4

Empirical	Structural	π	σ_m	σ_p	MR	E_s	additional σ values						
C1H1F2	CHF2						S.INDUC	0.29 185	S.ZE.RS	0.07 127	R	0.03 4	
							S.INDUC	0.32 32	ES-CH	2.60 2	S-L	0.32 14	
							S.STAR	2.01 10	ES-V	0.68 7	GR.ELCT	0.35 13	
C1H1F2O1	OCHF2	0.31	★ 0.31 318	★ 0.18 318	0.79		★S.INDUC	0.36 700	S.ZE.RS	-0.24 125	F	0.37 5	
							S.INDUC	0.45 234	★S.ZE.RS	-0.24 700	R	-0.19 5	
							S.INDUC	0.51 125	ES-CH	3.60 2	R-	-0.26 4	
							S.STAR	2.81 3	L-STM	3.98 1	GRP.DPL	-2.46 8	
							★S.PARA-	0.11 160	B1-STM	1.35 1			
							S.ZE.RS	-0.27 234	B5-STM	3.61 1			
C1H1F2O1S1	SOCHF2		★ 0.54 699	★ 0.58 699	1.33		★S.INDUC	0.65 125	ES-CH	4.80 2	F	0.51 5	
			0.70 3	0.76 3			S.STAR	4.10 3	L-STM	4.70 1	R	0.07 5	
							★S.PARA-	0.93 160	B1-STM	1.40 1	R-	0.42 4	
							★S.ZE.RS	0.11 125	B5-STM	3.70 1	GRP.DPL	-3.93 8	
C1H1F2O2S1	SO2(CHF2)		0.75 17	0.86 17	1.31		S.INDUC	0.65 30	S.ZE.RS	0.19 30	B5-STM	3.70 1	
			★ 0.75 318	★ 0.86 318			S.INDUC	0.73 125	S.ZE.RS	0.30 125	F	0.67 5	
							★S.INDUC	0.75 17	ES-CH	5.80 2	R	0.19 5	
							★S.STAR	3.69 3	L-STM	4.11 1	R-	0.77 4	
							★S.PARA-	1.44 160	B1-STM	2.03 1	GRP.DPL	-4.08 8	
C1H1F2S1	SCHF2		0.33 318	0.36 318	1.38		★S.INDUC	0.33 17	★S.ZE.RS	0.27 30	B5-STM	3.94 1	
			★ 0.33 17	★ 0.37 17			S.INDUC	0.36 125	S.BA.RS	0.07 18	F	0.32 5	
							S.STAR	2.06 3	ES-CH	3.80 2	R	0.05 5	
							★S.PARA-	0.52 160	L-STM	4.30 1	GRP.DPL	-2.48 8	
							S.ZE.RS	0.03 125	B1-STM	1.70 1			
C1H1F3N1	NHCF3						L-STM	4.61 1	B1-STM	1.35 1	B5-STM	3.65 1	
C1H1F3N1O1S1	S(=O)(=NH)CF3		★ 0.72 245	★ 0.84 245			★S.INDUC	0.60 245	F	0.64 4			
							★S.ZE.RS	0.24 245	R	0.20 4			
C1H1F3N1O2S1	NHSO2CF3		★ 0.44 3	★ 0.39 3	1.75		★S.INDUC	0.49 68	★ES-CH	5.20 2	B5-STM	4.00 1	
							★S.STAR	3.10 3	L-STM	5.26 1	F	0.45 5	
							★S.ZE.RS	-0.10 68	B1-STM	1.35 1	R	-0.06 5	
C1H1I2	CHI2		★ 0.26 3	★ 0.26 3	3.15	★ -2.05 35	S.INDUC	0.00 127	S.ZE.RS	0.06 127	B1-STM	1.95 1	
							★S.INDUC	0.26 110	★ES-CH	4.40 2	B5-STM	4.15 1	
							S.STAR	1.62 3	★ES-V	0.97 7	F	0.27 4	
							★S.ZE.RS	0.00 110	L-STM	4.36 1	R	-0.01 4	
C1H1N1O2	CHNO2 (-)						S.ZPTFT	-0.50 100	S.PARA+	-2.19 100			
C1H1N2	NHCN	-0.26	★ 0.21 398	★ 0.06 398	1.01		S.INDUC	0.37 24	ES-CH	3.00 2	B5-STM	4.05 1	
							★S.INDUC	0.37 398	L-STM	3.90 1	F	0.28 5	
							S.ZE.RS	-0.31 24	B1-STM	1.35 1	R	-0.22 5	
C1H1N4	1-tetrazolyl	-1.04	★ 0.52 40	★ 0.50 40	1.83		★S.INDUC	0.54 40	★S.PARA-	0.57 165	B5-STM	3.12 1	
			0.60 3	0.52 226			S.INDUC	0.65 673	ES-CH	3.00 2	F	0.52 5	
				0.57 3			S.STAR	3.40 3	L-STM	5.28 1	R	-0.02 5	
				0.71 275			★S.META-	0.60 165	B1-STM	1.71 1			
	2-tetrazolyl			★ 0.59 226			★S.INDUC	0.62 673					
				0.62 275			ES-CH	3.00 2					
	5-(1-H)tetrazolyl						F	0.65 4	R	-0.09 4			
	5-(1H)-tetrazole		0.46 15	0.31 15			S.INDUC	0.41 673	S.STAR	2.82 15			
			★ 0.64 797	0.44 15			S.INDUC	0.45 15					
			0.68 15	★ 0.56 797			S.INDUC	0.49 96					
C1H1N4O1	5-OH-1-tetrazolyl		★ 0.39 40	★ 0.33 40	1.98		★S.INDUC	0.43 40	F	0.41 5			
							ES-CH	4.00 2	R	-0.08 5			
C1H1N4S1	(1,2,3,4-thiatriazol-5-yl)amino		★ 0.30 458	★ 0.19 458			★S.INDUC	0.42 685	F	0.35 4			
							★S.STAR	2.62 685	R	-0.16 4			
	5-SH-1-tetrazolyl		★ 0.45 40	★ 0.45 40	2.61		★S.INDUC	0.45 40	F	0.44 5			
							ES-CH	4.20 2	R	-0.01 5			
C1H1O1	CHO	-0.65	0.17 197	0.22 23	0.69		S.INDUC	0.21 16	S.PARA-	1.02 28	S.AN-RS	0.46 26	
			★ 0.35 23	★ 0.42 484			S.INDUC	0.25 334	★S.PARA-	1.03 307	ES-CH	2.00 2	
			0.36 697	0.43 697			★S.INDUC	0.27 67	S.PARA-	1.04 181	L-STM	3.53 1	
			0.38 266	0.45 409			S.INDUC	0.31 102	S.PARA-	1.13 23	B1-STM	1.60 1	
			0.41 46	0.47 46			S.INDUC	0.32 26	★S.PARA+	0.47 978	B5-STM	2.36 1	
			0.43 18	0.52 18			S.INDUC	0.36 34	S.PARA+	0.73 382	F	0.33 5	
			0.43 26	0.52 82			S.INDUC	0.37 119	S.ORTHO	0.75 53	R	0.09 5	
			0.48 10	0.57 750			S.INDUC	0.43 12	S.ORTHO	1.10 186	R+	0.40 4	
				0.82 197			S.STAR	2.15 3	S.ORTH-	1.02 97	R-	0.70 4	
							S.META-	0.48 593	S.ZE.RS	0.06 114	S-L	0.37 7	
							S.ZPTFT	0.49 306	S.ZE.RS	0.07 102	S.ZTWST	0.00 22	
							S.ZPTFT	0.54 487	S.ZE.RS	0.15 6	S-CNTIG	0.98 139	
							S.ZPTFT	0.56 95	S.ZE.RS	0.22 26	K-CHGTR	0.32 77	
							S.ZPTFT	0.60 248	S.ZE.RS	0.22 67	GRP.DPL	-3.02 8	
							S.PARA-	0.79 26	★S.ZE.RS	0.24 19	GRP.DPL	-2.58 9	
							S.PARA-	0.90 211	S.ZE.RS	0.27 12	GR.ELCT	0.40 13	
							S.PARA-	0.94 247	S.ZE.RS	0.27 134	O-STER	-2.36 90	
							S.PARA-	0.98 443	S.ZE.RS	0.27 334			
							S.PARA-	0.99 104	S.BA.RS	0.07 34			
C1H1O1S1	SCHO						ES-CH	3.20 2	B1-STM	1.70 1			
							L-STM	4.70 1	B5-STM	3.52 1			
C1H1O2	CO2H	-0.32	0.25 149	0.27 23	0.69		S.INDUC	0.09 16	S.META-	0.56 593	ES-V	0.50 341
		-0.02 54	0.26 41	0.39 41			S.INDUC	0.17 149	S.PARA-	0.61 618	ES-V	1.45 474
			0.30 316	0.41 6			★S.INDUC	0.30 6	S.PARA-	0.73 23	L-STM	3.91 1
			0.33 251	0.41 65			S.INDUC	0.32 119	★S.PARA-	0.77 181	B1-STM	1.60 1
			0.35 17	0.41 149			S.INDUC	0.34 17	S.PARA-	0.78 28	B5-STM	2.66 1
			0.36 6	0.42 17			S.INDUC	0.40 182	★S.PARA+	0.42 43	F	0.34 5
			0.36 23	0.44 46			S.PRIME	0.28 159	S.PARNO	0.41 324	R	0.11 5
			0.36 46	★ 0.45 21			S.PRIME	0.29 149	S.ORTHO	0.51 158	R+	0.08 4

Empirical	Structural	π	σm	σp	MR	Es	additional σ values		
C1H1O2	CO2H		★ 0.37 21	0.61 254			S.STAR 1.70 87	S.ORTHO 0.95 78	R- 0.43 4
							S.STAR 1.72 757	S.ORTHO 1.20 186	S-L 0.30 14
							S.STAR 2.08 3	S.ZE.RS 0.11 6	S.ZTWST 0.00 22
							S.STAR 2.90 129	S.ZE.RS 0.12 149	GRP.DPL -1.65 9
							★S.STAR 2.94 289	S.ZE.RS 0.29 19	GRP.DPL -1.30 8
							S.META+ 0.32 43	S.ZE.RS 0.29 114	GR.ELCT 0.36 13
							S.META- 0.51 28	ES-CH 3.00 2	
	OCHO						S.INDUC 0.60 16	L-STM 3.93 1	B5-STM 3.67 1
							ES-CH 3.00 2	B1-STM 1.35 1	GR.ELCT 0.80 13
C1H1O3	OCO2H						ES-CH 4.00 2	B1-STM 1.60 1	
							L-STM 4.71 1	B5-STM 1.60 1	
C1H2Br1	CH2Br	0.79	★ 0.12 17 0.13 29 0.17 6	0.10 6 0.11 29 ★ 0.14 17	1.34	★ -1.51 50 -1.36 35	★S.INDUC 0.12 17	S.STAR 1.00 847	ES-V 0.64 7
							S.INDUC 0.14 110	S.STAR 1.02 155	L-STM 4.09 1
							S.INDUC 0.16 64	S.STAR 1.08 10	B1-STM 1.52 1
							S.INDUC 0.16 185	S.PARA+ 0.02 351	B5-STM 3.75 1
							S.INDUC 0.18 16	S.ZE.RS -0.11 49	F 0.14 5
							S.INDUC 0.18 192	★S.ZE.RS -0.10 6	R 0.00 5
							S.INDUC 0.19 182	S.ZE.RS -0.04 127	R- -0.12 4
							S.INDUC 0.20 6	S.ZE.RS -0.03 110	S-L 0.20 7
							S.INDUC 0.23 127	S.ZE.RS -0.03 233	★S.PHOSP 0.00 59
							S.INDUC 0.25 233	S.ZE.RS -0.02 22	S.IND.P 1.17 39
							S.INDUC 0.39 49	S.ZE.RS -0.02 64	S.ZTWST -0.10 22
							S-INDQ 1.07 58	S.ZE.RS -0.02 185	GRP.DPL -1.97 9
							★S.STAR 1.00 52	S.ZE.RS 0.00 19	GRP.DPL -1.87 8
							S.STAR 1.00 183	ES-CH 2.30 2	GR.ELCT 0.16 13
C1H2Br1Cl1N1	NHCHBrCl						L-STM 4.02 1	B1-STM 1.35 1	B5-STM 4.72 1
C1H2Br1F1N1	NHCHBrF						L-STM 4.02 1	B1-STM 1.35 1	B5-STM 4.72 1
C1H2Br1O1	OCH2Br						L-STM 5.74 1	B1-STM 1.35 1	B5-STM 3.27 1
C1H2Br1S1	SCH2Br						L-STM 6.06 1	B1-STM 1.70 1	B5-STM 3.89 1
C1H2Br2N1	NHCHBr2						L-STM 4.02 1	B1-STM 1.35 1	B5-STM 4.72 1
C1H2Cl1	CH2Cl	0.17	0.03 197 ★ 0.11 17 0.15 6	0.09 6 ★ 0.12 17 0.18 23 0.23 197	1.05	-1.48 50 ★ -1.30 35	S.INDUC 0.11 17	S.STAR 1.05 74	S.ZE.RS 0.04 114
							S.INDUC 0.12 12	S.STAR 1.05 129	ES-CH 2.20 2
							S.INDUC 0.14 89	S.STAR 1.05 155	ES-V 0.60 7
							S.INDUC 0.14 110	S.STAR 1.05 494	L-STM 3.89 1
							S.INDUC 0.15 185	S.META+ 0.14 43	B1-STM 1.52 1
							S.INDUC 0.16 64	★S.PARA+ -0.01 43	B5-STM 3.46 1
							S.INDUC 0.16 192	S.PARA+ 0.02 351	F 0.13 5
							★S.INDUC 0.17 6	S.ZE.RS -0.08 6	R -0.01 5
							S.INDUC 0.17 127	S.ZE.RS -0.04 22	R+ -0.14 4
							S.INDUC 0.17 182	S.ZE.RS -0.04 64	S-L 0.17 7
							S.INDUC 0.18 1001	S.ZE.RS -0.04 233	★S.PHOSP -0.05 59
							S.INDUC 0.19 133	S.ZE.RS -0.03 110	S.IND.P 1.12 39
							S.INDUC 0.26 233	★S.ZE.RS -0.03 134	S.ZTWST -0.10 22
							S-INDQ 1.02 58	S.ZE.RS -0.03 185	GRP.DPL -1.93 9
							S.STAR 0.94 3	S.ZE.RS -0.02 12	GRP.DPL -1.83 8
							★S.STAR 1.01 10	S.ZE.RS -0.02 127	GR.ELCT 0.18 13
							S.STAR 1.04 183	S.ZE.RS -0.01 133	
							S.STAR 1.05 52	S.ZE.RS 0.00 19	
C1H2Cl1D3N1	CH2N+D3Cl-						★S.ZE.RS 0.00 19	★ES-CH 2.00 2	
							★S.ZE.RS 0.00 47	★S.ZTWST -0.10 22	
C1H2Cl1F1N1	NHCH(Cl)F						L-STM 4.02 1	B1-STM 1.35 1	B5-STM 4.44 1
C1H2Cl1Hg1	CH2HgCl			★ -0.11 270			S.INDUC -0.02 173	S.ZE.RS -0.14 173	
							★S.INDUC -0.02 539	★S.ZE.RS -0.14 539	
C1H2Cl1O1	OCH2Cl		★ 0.25 3	★ 0.08 3			★S.INDUC 0.41 18	S.ZE.RS -0.33 234	B5-STM 3.13 1
							S.INDUC 0.41 234	ES-CH 3.20 2	F 0.33 5
							★S.STAR 2.56 3	L-STM 5.44 1	R -0.25 5
							★S.ZE.RS -0.33 18	B1-STM 1.35 1	GRP.DPL -1.90 9
C1H2Cl1O2S1	SO2CH2Cl						L-STM 5.57 1	B1-STM 2.03 1	B5-STM 3.25 1
C1H2Cl1S1	SCH2Cl				1.84		★S.INDUC 0.32 125	★S.ZE.RS -0.05 125	B1-STM 1.70 1
							S.INDUC 0.38 102	ES-CH 3.40 2	B5-STM 3.72 1
							S.ZE.RS -0.22 102	L-STM 5.76 1	
C1H2Cl2N1	NHCHCl2						L-STM 4.02 1	B1-STM 1.35 1	B5-STM 4.44 1
C1H2Cl2O1P1	CH2P(O)Cl2						★S.INDUC 0.20 12	★S.ZE.RS -0.02 12	
C1H2D1	CH2D						★S.STAR 0.00 800		
C1H2F1	CH2F		0.11 3 ★ 0.12 18	0.10 3 ★ 0.11 18	0.54	-1.48 50 ★ -1.32 35	★S.INDUC 0.12 64	S.STAR 1.22 10	ES-V 0.62 7
							S.INDUC 0.12 185	S.META+ 0.10 998	L-STM 3.30 1
							S.INDUC 0.13 16	★S.ZE.RS -0.02 64	B1-STM 1.52 1
							S.INDUC 0.13 110	S.ZE.RS -0.02 110	B5-STM 2.61 1
							S.INDUC 0.18 34	S.ZE.RS -0.01 127	F 0.15 4
							S.INDUC 0.20 127	S.ZE.RS -0.01 185	R -0.04 4
							★S.STAR 1.10 52	S.BA.RS -0.04 34	GR.ELCT 0.24 13
							S.STAR 1.10 155	ES-CH 1.80 2	
C1H2F1O1	OCH2F		★ 0.20 3	★ 0.02 3			★S.INDUC 0.37 18	S.ZE.RS -0.35 234	B5-STM 3.07 1
							S.INDUC 0.37 234	ES-CH 2.80 2	F 0.29 5
							★S.STAR 2.31 3	L-STM 4.57 1	R -0.27 5
							★S.ZE.RS -0.35 18	B1-STM 1.35 1	
C1H2F1O2S1	SO2CH2F				1.33		★S.INDUC 0.66 125	★S.ZE.RS 0.04 125	B1-STM 2.03 1
							★S.STAR 3.44 3	ES-CH 5.00 2	B5-STM 3.17 1
							★S.PARA- 1.17 160	L-STM 4.70 1	
C1H2F1S1	SCH2F		★ 0.23 3	★ 0.20 3	1.33		★S.INDUC 0.27 125	★ES-CH 3.00 2	B5-STM 3.41 1
							★S.STAR 1.69 3	L-STM 4.89 1	F 0.25 5
							★S.ZE.RS -0.07 125	B1-STM 1.70 1	R -0.05 5

Empirical	Structural	π	σ_m	σ_p	MR	E_s	additional σ values					
C1H2F2N1	NHCHF2						L-STM 4.02 *1*	B1-STM 1.35 *1*	B5-STM 3.65 *1*			
C1H2Hg1I1	CH2-Hg-I						★ S.PARA+ -0.56 *203*					
C1H2I1	CH2I	1.50 *57*	★ 0.10 *17*	0.08 *6*	1.86	★ -1.61 *50*	S.INDUC 0.10 *17*	S.STAR 0.88 *155*	B5-STM 4.15 *1*			
			0.15 *6*	★ 0.11 *17*		-1.42 *817*	S.INDUC 0.14 *110*	S.STAR 0.92 *10*	F 0.12 *5*			
							★ S.INDUC 0.17 *6*	S.STAR 1.00 *3*	R -0.01 *5*			
							S.INDUC 0.18 *16*	★ S.ZE.RS -0.09 *6*	S-L 0.17 *14*			
							S.INDUC 0.21 *127*	S.ZE.RS -0.06 *127*	★ S.PHOSP -0.10 *59*			
							S.INDUC 0.40 *182*	S.ZE.RS -0.04 *110*	S.IND.P 1.10 *39*			
							S.INDQ 1.05 *58*	ES-CH 2.70 *658*	GRP.DPL -1.79 *9*			
							★ S.STAR 0.85 *52*	ES-V 0.67 *7*	GRP.DPL -1.60 *8*			
							S.STAR 0.85 *74*	L-STM 4.36 *1*	GR.ELCT 0.14 *13*			
							S.STAR 0.87 *287*	B1-STM 1.52 *1*				
C1H2I1O1	OCH2I						L-STM 6.15 *1*	B1-STM 1.35 *1*	B5-STM 3.47 *1*			
C1H2I1S1	SCH2I						L-STM 6.47 *1*	B1-STM 1.70 *1*	B5-STM 4.11 *1*			
C1H2K1	CH2K						S.INDUC -0.08 *16*					
C1H2N1O1	C=O(NH2)	-1.71 *57*	0.27 *10*	0.31 *65*	0.98		S.INDUC 0.21 *89*	S.PARA- 0.63 *23*	B1-STM 1.50 *1*			
		-1.49	0.28 *266*	0.33 *867*			★ S.INDUC 0.28 *6*	S.ORTHO 0.45 *78*	B5-STM 3.07 *1*			
		-1.07 *54*	★ 0.28 *23*	★ 0.36 *213*			S.INDUC 0.28 *192*	S.ORTHO 0.72 *53*	F 0.26 *5*			
			0.35 *121*	0.38 *10*			S.INDUC 0.29 *444*	★ S.ZE.RS 0.01 *134*	R 0.10 *5*			
				0.43 *82*			S.INDUC 0.33 *119*	S.ZE.RS 0.08 *6*	R- 0.35 *4*			
				0.43 *121*			S.INDUC 0.38 *210*	S.ZE.RS 0.11 *444*	S-L 0.28 *14*			
							S.PRIME 0.30 *159*	S.ZE.RS 0.13 *47*	K-CHGTR 1.00 *436*			
							S.INDQ 1.82 *58*	S.BA.RS 0.09 *34*	GRP.DPL -3.73 *9*			
							★ S.STAR 1.66 *10*	S.RES.+ 0.00 *10*	GRP.DPL -3.42 *8*			
							S.STAR 1.68 *3*	ES-CH 3.00 *2*	GR.ELCT 0.30 *13*			
							★ S.PARA- 0.61 *181*	L-STM 4.06 *1*				
	CH=NOH -C						ES-CH 3.00 *2*	B1-STM 1.60 *1*	GRP.DPL 0.85 *8*			
							L-STM 3.94 *1*	B5-STM 4.07 *1*				
	CH=NOH -T	-1.22 *57*	★ 0.22 *721*	★ 0.10 *721*	1.03		★ S.INDUC 0.20 *6*	L-STM 5.05 *1*	F 0.28 *5*			
		-0.38	0.61 *706*	0.73 *706*			S.RES.+ -0.12 *6*	B1-STM 1.60 *1*	R -0.18 *5*			
							ES-CH 3.00 *2*	B5-STM 2.88 *1*	GRP.DPL -0.87 *8*			
	NHCHO	-0.98	0.19 *6*	-0.07 *6*	1.03		S.INDUC 0.29 *24*	S.ZPTFT 0.10 *633*	L-STM 4.22 *1*			
		-0.45 *54*	★ 0.19 *56*	★ 0.00 *56*			S.INDUC 0.29 *161*	★ S.PARNO 0.19 *65*	B1-STM 1.35 *1*			
			0.20 *10*	0.17 *10*			S.INDUC 0.29 *634*	S.ZE.RS -0.40 *6*	B5-STM 3.61 *1*			
							S.INDUC 0.29 *776*	S.ZE.RS -0.24 *633*	F 0.28 *5*			
							★ S.INDUC 0.33 *6*	★ S.ZE.RS -0.24 *777*	R -0.28 *5*			
							S.INDUC 0.34 *633*	S.ZE.RS -0.22 *634*	S-L 0.33 *14*			
							S.INDUC 0.34 *777*	S.ZE.RS -0.22 *776*	GRP.DPL -3.86 *9*			
							★ S.STAR 1.62 *3*	S.ZE.RS -0.20 *24*	GRP.DPL -3.35 *8*			
							S.STAR 1.98 *10*	S.ZE.RS -0.20 *161*				
							S.ZPTFT 0.07 *634*	ES-CH 3.00 *2*				
C1H2N1O1S1	SC=O(NH2)		★ 0.34 *3*		1.68		★ S.INDUC 0.33 *32*	★ ES-CH 4.20 *2*	B5-STM 3.95 *1*			
							S.STAR 1.70 *10*	L-STM 5.22 *1*	★ S-L 0.33 *14*			
							★ S.STAR 2.07 *3*	B1-STM 1.70 *1*				
	SCH=NOH -C						L-STM 6.40 *1*	B1-STM 1.70 *1*	B5-STM 3.91 *1*			
	SCH=NOH -T						L-STM 5.65 *1*	B1-STM 1.70 *1*	B5-STM 4.93 *1*			
C1H2N1O2	C=O(NHOH)	-1.87			1.12		ES-CH 4.00 *2*	B1-STM 1.50 *1*				
							L-STM 4.73 *1*	B5-STM 3.35 *1*				
	CH2NO2	-0.38			1.20	★ -2.71 *44*	S.INDUC 0.25 *16*	S.STAR 1.83 *10*	B5-STM 3.64 *116*			
							S.STAR 1.40 *155*	ES-CH 4.00 *2*	★ GRP.DPL -3.29 *9*			
							★ S.STAR 1.69 *87*	L-STM 3.70 *116*	GR.ELCT 0.21 *13*			
							S.STAR 1.73 *686*	B1-STM 1.52 *116*				
	CH2ONO						GRP.DPL -2.10 *8*					
	NHCO2H						L-STM 5.01 *1*	B1-STM 1.35 *1*	B5-STM 3.61 *1*			
	OCH=NOH -C						L-STM 6.06 *1*	B1-STM 1.35 *1*	B5-STM 3.36 *1*			
	OCH=NOH -T						L-STM 5.52 *1*	B1-STM 1.35 *1*	B5-STM 4.36 *1*			
	OCONH2	-1.05			1.13	★ 4.00 *816*	L-STM 4.82 *1*	B1-STM 1.35 *1*	B5-STM 3.63 *1*			
C1H2N1O2S1	SCH2NO2						L-STM 1.70 *116*	L-STM 5.68 *116*	B5-STM 3.60 *116*			
C1H2N1O3	CH2ONO2						★ S.STAR 1.34 *686*	L-STM 5.35 *1*	B5-STM 3.10 *1*			
							ES-CH 3.00 *2*	B1-STM 1.52 *1*	★ S-L 0.20 *14*			
	OCH2NO2						L-STM 5.36 *116*	B1-STM 1.35 *116*	B5-STM 3.07 *116*			
C1H2N1S1	Cs(NH2)	-0.64	★ 0.25 *853*	★ 0.30 *853*			S.INDUC 0.19 *444*	L-STM 4.10 *1*	R 0.06 *4*			
							★ S.INDUC 0.29 *444*	B1-STM 1.64 *1*	GR.ELCT 0.18 *13*			
							★ S.ZE.RS 0.11 *830*	B5-STM 3.18 *1*				
							ES-CH 3.20 *2*	F 0.24 *4*				
C1H2N2	3,4-N=CH-NH- *		★ -0.15 *411*	★ -0.15 *411*		★ 0.00 *816*	F -0.10 *4*	R -0.05 *4*				
C1H2N3O1	N=NCONH2			★ 0.79 *72*			S.ZE.RS 0.24 *72*					
C1H2N3O2	N(=O)=NC(=O)NH2		★ 0.59 *179*	★ 0.63 *179*			★ S.INDUC 0.58 *179*	F 0.56 *4*	★ S.RES.P 0.05 *179*			
							★ S.ZE.RS 0.01 *179*	R 0.07 *4*				
C1H2N4	5-tetrazole-cation			★ 1.02 *15*								
C1H2O1	CH2O (-)						★ S.STAR 0.27 *155*					
C1H2O2	OCH2O(3,4)	-0.05	-0.27 *439*	-0.27 *439*	0.90		S.ZPTFT -0.02 *95*	F -0.11 *5*	★ S.PHOSP -0.55 *85*			
			-0.18 *785*	-0.18 *785*			S.PARA+ -0.68 *483*	R -0.05 *5*				
			★ -0.16 *681*	★ -0.16 *681*			ES-CH 0.00 *2*	R+ -0.57 *4*				
C1H2O2S1	-CH2S(O2)-		★ 0.15 *17*	★ 0.17 *17*			★ S.INDUC 0.15 *17*	★ ES-CH 5.20 *2*				
C1H2O3S1	CH2SO3-					★ -2.49 *44*	★ S.STAR 0.68 *87*	L-STM 4.79 *1*	B5-STM 3.78 *1*			
							ES-CH 5.20 *2*	B1-STM 1.52 *1*	★ S-L 0.01 *7*			
C1H3	CH3	0.31 *130*	-0.37 *254*	-0.24 *254*	0.56	-1.24 *302*	S.INDUC -0.11 *12*	S.PARA- -0.06 *26*	S.AN-RS -0.09 *26*			
		0.46 *220*	-0.30 *128*	-0.17 *518*		★ -1.24 *50*	S.INDUC -0.09 *385*	S.PARA+ -0.33 *175*	S.RES.+ -0.25 *20*			

Empirical	Structural	π	σ$_m$	σ$_p$	MR	E$_s$		additional σ values				
C1H3	CH3	0.50 *57*	-0.10 *17*	★ -0.17 *21*		-1.12 *62*	S.INDUC -0.08 *17*	S.PARA+ -0.33 *329*	ES-CH 1.00 *2*			
		0.56	-0.10 *518*	-0.16 *190*		0.00 *311*	S.INDUC -0.07 *64*	S.PARA+ -0.32 *65*	ES-V 0.52 *7*			
		0.58 *42*	★ -0.07 *21*	-0.16 *271*			★S.INDUC -0.04 *20*	★S.PARA+ -0.31 *43*	ES-HYBO 0.00 *225*			
			-0.06 *46*	-0.14 *17*			S.INDUC -0.04 *205*	S.PARA+ -0.31 *88*	L-STM 2.87 *1*			
			-0.05 *190*	-0.14 *46*			S.INDUC -0.04 *261*	S.PARA+ -0.31 *203*	B1-STM 1.52 *1*			
			-0.02 *26*	-0.14 *236*			S.INDUC -0.03 *182*	S.PARA+ -0.31 *243*	B5-STM 2.04 *1*			
			-0.02 *41*	-0.10 *128*			S.INDUC -0.02 *99*	S.PARA+ -0.31 *308*	F 0.01 *5*			
				0.01 *41*			S.INDUC -0.01 *6*	S.PARA+ -0.31 *406*	R -0.18 *5*			
							S.INDUC 0.01 *192*	S.PARA+ -0.30 *145*	R+ -0.32 *4*			
							S.INDUC 0.02 *142*	S.PARNO -0.10 *65*	R- -0.18 *4*			
							S.INDUC 0.05 *16*	S.ORTHO -0.36 *254*	S-L -0.01 *14*			
							S.PRIME 0.00 *159*	S.ORTHO -0.15 *432*	S.PHOSP -0.96 *59*			
							S-INDQ 0.11 *58*	S.ORTHO -0.13 *53*	S.NEOF -0.22 *386*			
							S.STAR 0.00 *61*	S.ORTHO -0.13 *158*	S.NEOF -0.07 *386*			
							S.STAR 0.00 *289*	S.ORTHO -0.12 *186*	S.ZTWST -0.10 *22*			
							S.ZMTFT -0.07 *63*	S.ORTHO -0.10 *457*	S.M.RAD 0.02 *79*			
							S.ZMTFT -0.07 *95*	S.ORTHO 0.10 *104*	S.P.RAD -0.02 *263*			
							S.META+ -0.13 *973*	S.ORTHO 0.29 *78*	★S.P.RAD 0.03 *206*			
							S.META+ -0.09 *145*	S.ORTH+ -0.27 *154*	S.P.RAD 0.15 *79*			
							S.META+ -0.07 *43*	S.ORTH+ -0.25 *243*	S.P.RAD 0.38 *453*			
							S.META+ -0.07 *88*	★S.ORTH- -0.13 *97*	S.P.RAD 0.39 *255*			
							S.META+ -0.07 *243*	S.ZE.RS -0.16 *6*	S-M(C+) -0.15 *239*			
							S.META- -0.09 *28*	S.ZE.RS -0.15 *64*	S-M(C+) -0.13 *256*			
							S.META- -0.03 *426*	★S.ZE.RS -0.14 *12*	S-P(C+) -0.67 *256*			
							S.ZPTFT -0.13 *95*	S.ZE.RS -0.14 *385*	S-P(C+) -0.49 *239*			
							S.ZPTFT -0.13 *112*	S.ZE.RS -0.13 *26*	K-CHGTR 0.11 *77*			
							S.ZPTFT -0.12 *63*	S.ZE.RS -0.13 *67*	GRP.DPL 0.00 *832*			
							S.PARA- -0.22 *28*	S.ZE.RS -0.13 *134*	GRP.DPL 0.36 *8*			
							★S.PARA- -0.17 *271*	S.ZE.RS -0.10 *19*	GR.ELCT 0.14 *13*			
							S.PARA- -0.15 *181*	S.ZE.RS -0.10 *365*	O-STER -0.73 *90*			
							S.PARA- -0.15 *211*	S.BA.RS -0.11 *20*				
							S.PARA- -0.10 *426*	S.AN-RS -0.11 *20*				
C1H3B1Cl3O1	O+(CH3)BCl3						★S.INDUC 0.54 *31*	★S.ZE.RS -0.06 *31*	ES-CH 6.60 *2*			
C1H3B1Cl3S1	S+(CH3)BCl3						★S.INDUC 0.72 *31*	★S.ZE.RS 0.10 *31*	ES-CH 6.80 *2*			
C1H3Br1Cl1Si1	Si(Me)(Cl)Br						★S.INDUC 0.25 *25*	★S.INDUC 0.25 *73*	★S.ZE.RS 0.15 *73*			
C1H3Br1N1	NHCH2Br						L-STM 5.78 *1*	B1-STM 1.35 *1*	B5-STM 3.34 *1*			
C1H3Br2Si1	Si(CH3)Br2			★ 0.29 *3*			★S.PARA+ 0.29 *136*	★ES-CH 4.80 *2*				
C1H3Br2Te1	Te(CH3)Br2			★ 0.59 *103*			★S.INDUC 0.53 *103*	★S.ZE.RS 0.06 *103*				
C1H3Cd1	Cd(CH3)						GR.ELCT -0.59 *13*					
C1H3Cl1D2N1	+ND2(CH3)Cl-						★S.ZE.RS -0.15 *19*	★ES-CH 2.00 *2*				
C1H3Cl1N1	NHCH2Cl						L-STM 5.48 *1*	B1-STM 1.35 *1*	B5-STM 3.19 *1*			
C1H3Cl2N1O1P1	N(Me)P(=O)Cl2						S.INDUC 0.54 *12*	S.ZE.RS -0.09 *12*				
C1H3Cl2Si1	SiCl2(CH3)		★ 0.31 *3*	0.36 *60*	2.35		★S.INDUC 0.24 *73*	★S.PARA+ 0.08 *136*	F 0.29 *4*			
				★ 0.39 *3*			★S.STAR 1.26 *216*	S.PARA+ 0.51 *60*	R 0.10 *4*			
				0.41 *153*			S.STAR 1.50 *3*	★S.ZE.RS 0.15 *73*	R+ -0.21 *4*			
							★S.META+ 0.09 *136*	ES-CH 4.60 *2*	GR.ELCT -0.17 *13*			
C1H3Cl2Sn1	SnCl2(CH3)						S.INDUC 0.47 *242*	★S.STAR 1.23 *61*	★ES-CH 5.40 *2*			
C1H3Cl2Te1	Te(CH3)Cl2			★ 0.56 *103*			★S.INDUC 0.50 *103*	★S.ZE.RS 0.06 *103*				
C1H3D1N1	ND(CH3)						★S.ZE.RS -0.52 *19*	★ES-CH 2.00 *2*				
C1H3F1N1	NHCH2F						L-STM 4.61 *1*	B1-STM 1.35 *1*	B5-STM 3.11 *1*			
C1H3F2Si1	SiF2(CH3)		★ 0.29 *3*	★ 0.23 *3*	1.25		★S.PARA+ 0.23 *136*	F 0.32 *5*	R+ -0.09 *4*			
			0.36 *221*	0.40 *153*			ES-CH 3.80 *2*	R -0.09 *5*				
C1H3Hg1	Hg(CH3)	-0.27	★ 0.43 *41*	★ 0.10 *41*	1.94		F 0.55 *264*	R -0.45 *264*	GR.ELCT -0.65 *13*			
C1H3I1N1	NHCH2I						L-STM 6.19 *1*	B1-STM 1.35 *1*	B5-STM 3.54 *1*			
C1H3I2Te1	Te(CH3)I2			★ 0.62 *103*			★S.INDUC 0.58 *103*	★S.ZE.RS 0.04 *103*				
C1H3N1O3S1	NHCH2SO3(-)		★ -0.10 *786*	★ -0.57 *786*			ES-CH 3.00 *2*	F 0.12 *4*	R -0.69 *4*			
C1H3N2	3,4-NH+=CH-NH-			★ 1.13 *616*			★S.PARA- 0.54 *616*					
	C=NH(NH2)			★ 0.65 *249*			ES-CH 3.00 *2*	B1-STM 1.60 *1*				
							L-STM 3.94 *1*	B5-STM 3.07 *1*				
C1H3N2O1	C=O(NHNH2)	-1.90					ES-CH 4.00 *2*	B1-STM 1.50 *1*				
		-1.15 *42*					L-STM 4.90 *1*	B5-STM 3.09 *1*				
	N=NOCH3			★ 0.13 *72*			S.ZE.RS -0.11 *72*					
	NHC=O(NH2)	-1.30	★ -0.03 *56*	-0.24 *6*	1.37		S.INDUC 0.19 *24*	ES-CH 4.00 *2*	R -0.33 *5*			
			0.00 *10*	★ -0.24 *56*			★S.INDUC 0.23 *6*	L-STM 5.06 *1*	S-L 0.23 *7*			
			0.06 *6*				S.STAR 1.31 *3*	B1-STM 1.35 *1*	GR.ELCT 0.50 *13*			
							S.ZE.RS -0.47 *6*	B5-STM 3.61 *1*				
							S.ZE.RS -0.30 *24*	F 0.09 *5*				
	NHCH=NOH -C						L-STM 6.10 *1*	B1-STM 1.35 *1*	B5-STM 3.42 *1*			
	NHCH=NOH -T						L-STM 5.54 *1*	B1-STM 1.35 *1*	B5-STM 4.41 *1*			
C1H3N2O1S1	SC(=O)NHNH2						L-STM 5.60 *1*	B1-STM 1.70 *1*	B5-STM 4.97 *1*			
C1H3N2O2	N(CH3)NO2		★ 0.49 *415*	★ 0.61 *415*			ES-CH 5.00 *2*	R 0.18 *4*				
							F 0.43 *4*	★S-L 0.39 *14*				
	NHCH2NO2						L-STM 5.40 *116*	B1-STM 1.35 *116*	B5-STM 3.11 *116*			
	OC(=O)NHNH2						L-STM 5.43 *1*	B1-STM 1.35 *1*	B5-STM 4.43 *1*			
C1H3N2S1	NHC=S(NH2)	-1.40	★ 0.22 *458*	★ 0.16 *458*	2.22		★S.INDUC 0.29 *458*	L-STM 5.06 *300*	R -0.10 *4*			
							S.INDUC 0.29 *746*	B1-STM 1.35 *300*	GRP.DPL -0.16 *990*			
							★S.STAR 1.80 *3*	B5-STM 4.18 *300*	GR.ELCT 0.50 *834*			
							★S.ZE.RS -0.13 *746*	F 0.26 *4*				
C1H3N4O2	NHC(=NH)NHNO2						★S.INDUC 0.08 *117*	★S.ZE.RS -0.27 *117*	ES-CH 4.00 *2*			

Empirical	Structural	π	σm	σp	MR	Es	additional σ values		
C1H3O1	CH2OH	-1.03	★ 0.00 293	-0.05 750	0.72	★ -1.21 44	S.INDUC 0.05 192	★ S.PARA 0.08 307	F 0.03 5
		-0.62 54	0.01 3	-0.01 210		-1.09 35	S.INDUC 0.10 16	★ S.PARA+ -0.04 351	R -0.03 5
		-0.04 130	0.10 6	★ 0.00 293			★ S.INDUC 0.11 6	S.ORTHO 0.04 53	R+ -0.07 4
		0.60 128	0.60 128	0.01 3			S.PRIME 0.04 159	S.ORTH- -0.07 97	R- 0.05 4
				0.04 6			S-INDQ 0.66 58	★ S.ZE.RS -0.07 6	S-L 0.11 14
							S.STAR 0.51 376	S.ZE.RS -0.06 22	S.PHOSP -0.55 59
							S.STAR 0.53 10	S.ZE.RS 0.00 19	S.IND.P 0.61 39
							S.STAR 0.55 494	S.ZE.RS 0.06 552	S.ZTWST -0.10 22
							★ S.STAR 0.56 52	★ S.BA.RS -0.03 34	K-CHGTR 0.03 77
							S.STAR 0.56 580	ES-CH 2.00 2	GRP.DPL 1.73 8
							S.STAR 0.62 3	ES-V 0.53 7	GR.ELCT 0.22 13
							S.META- -0.03 138	L-STM 3.97 1	O-STER -0.67 90
							S.META- 0.08 307	B1-STM 1.52 1	
							S.PARA- -0.10 138	B5-STM 2.70 1	
	OCH3	-0.02	0.06 26	-1.00 128	0.79	★ -0.55 267	S.INDUC 0.24 142	S.PARA- -0.14 514	S.AN-RS -0.29 26
		0.22 54	0.08 10	-0.28 190			S.INDUC 0.25 26	S.PARA- -0.13 181	S.RES.+ -1.02 20
		0.35 42	0.10 46	★ -0.27 21			S.INDUC 0.25 611	S.PARA- -0.11 290	S.RES.+ -0.71 10
		0.71 220	0.11 190	-0.26 17			S.INDUC 0.26 12	S.PARA- -0.10 416	S.RES.+ -0.66 6
			0.12 17	-0.21 236			S.INDUC 0.26 99	S.PARA- -0.06 26	ES-CH 2.00 2
			★ 0.12 21	-0.15 10			S.INDUC 0.26 119	S.PARA+ -0.79 65	ES-V 0.36 7
			0.20 128	-0.12 46			★ S.INDUC 0.27 20	★ S.PARA+ -0.78 43	L-STM 3.98 1
				-0.12 208			S.INDUC 0.27 205	S.PARA+ -0.78 88	B1-STM 1.35 1
							S.INDUC 0.28 118	S.PARA+ -0.78 243	B5-STM 3.07 1
							S.INDUC 0.30 6	S.PARA+ -0.78 483	F 0.29 5
							S.INDUC 0.30 182	S.PARA+ -0.78 513	R -0.56 5
							S.INDUC 0.33 17	S.PARA+ -0.74 277	R+ -1.07 4
							S.INDUC 0.68 16	S.PARA+ -0.73 308	R- -0.55 4
							S.PRIME 0.29 159	S.PARA+ -0.60 145	S-L 0.30 14
							S-INDQ 1.86 58	S.PARNO -0.11 324	S.PHOSP -0.12 59
							S.STAR 1.67 183	S.ORTHO 0.00 53	S.NEOF -0.50 386
							S.STAR 1.73 37	S.ORTHO 0.00 104	S.IND.P 1.62 39
							★ S.STAR 1.77 10	S.ORTHO 0.04 335	S.RES.P -0.64 39
							S.STAR 1.81 268	S.ORTHO 0.12 78	S.M.RAD -0.01 79
							S.ZMTFT 0.04 95	S.ORTHO 0.34 158	S.P.RAD -0.12 263
							S.ZMTFT 0.08 63	S.ORTHO 0.60 186	★ S.P.RAD 0.11 206
							S.META+ 0.05 43	S.ORTH+ -0.67 154	S.P.RAD 0.18 79
							S.META+ 0.09 88	S.ORTH+ -0.67 243	S.P.RAD 0.31 453
							S.META- 0.09 416	★ S.ORTH- -0.37 97	S.P.RAD 0.42 255
							S.META- 0.10 28	S.ZE.RS -0.58 6	S-P(C+) -2.02 256
							S.META- 0.13 290	S.ZE.RS -0.45 20	S-P(C+) -1.29 239
							S.ZPTFT -0.16 95	S.ZE.RS -0.44 12	K-CHGTR 0.44 77
							★ S.ZPTFT -0.15 112	S.ZE.RS -0.43 19	GRP.DPL -1.30 8
							S.ZPTFT -0.15 337	S.ZE.RS -0.43 118	GRP.DPL -1.27 9
							S.ZPTFT -0.11 63	★ S.ZE.RS -0.43 134	GR.ELCT 0.82 13
							S.PARA- -0.31 28	S.ZE.RS -0.42 26	O-STER -1.28 90
							★ S.PARA- -0.26 614	S.BA.RS -0.61 20	
							S.PARA- -0.16 211	S.AN-RS -0.45 20	
C1H3O1S1	S(OCH3)		★ 0.21 3	★ 0.17 3	1.58		★ S.INDUC 0.25 31	★ ES-CH 3.20 2	B5-STM 3.52 1
							★ S.STAR 1.56 3	L-STM 5.08 1	F 0.24 5
							★ S.ZE.RS -0.08 31	B1-STM 1.70 1	R -0.07 5
	S=O(CH3)	-1.85 57	0.39 26	0.38 10	1.37		S.INDUC 0.36 26	★ S.PARA- 0.73 728	L-STM 4.11 1
		-1.58	0.49 425	0.45 425			S.INDUC 0.43 16	S.ORTHO 1.04 53	B1-STM 1.40 1
			0.51 612	0.48 612			S.INDUC 0.49 67	★ S.ORTH- 0.92 97	B5-STM 3.17 1
			0.52 17	0.49 17			★ S.INDUC 0.49 89	S.ZE.RS -0.01 67	F 0.52 5
			★ 0.52 21	★ 0.49 21			S.INDUC 0.52 125	★ S.ZE.RS 0.00 6	R -0.03 5
			0.55 23	0.51 612			S.INDUC 0.58 17	S.ZE.RS 0.07 26	R- 0.21 4
				0.57 23			S.STAR 2.88 3	S.ZE.RS 0.20 125	S.P.RAD 0.18 79
							S.META- 0.45 461	S.BA.RS 0.00 20	GRP.DPL -3.98 8
							S.META- 0.53 728	S.AN-RS 0.23 26	GRP.DPL -3.88 9
							S.ZPTFT 0.57 63	S.RES.+ -0.10 6	GR.ELCT 0.31 13
							S.PARA- 0.57 26	S.RES.+ 0.00 20	
							S.PARA- 0.62 372	ES-CH 3.20 2	
	SCH2OH						ES-CH 3.20 2	B1-STM 1.70 1	
							L-STM 4.98 1	B5-STM 3.85 1	
C1H3O2	CH(OH)2						★ S-INDQ 1.23 58	L-STM 3.97 1	★ S-L 0.22 14
							★ S.STAR 1.37 3	B1-STM 1.72 1	
							ES-CH 3.00 2	B5-STM 2.70 1	
	OCH2OH						ES-CH 3.00 2	B1-STM 1.35 1	
							L-STM 4.66 1	B5-STM 3.26 1	
C1H3O2S1	OS(O)CH3		★ 0.44 3	★ 0.45 3			L-STM 4.66 1	B5-STM 4.10 1	R 0.02 5
							B1-STM 1.35 1	F 0.43 5	
	S=O(OCH3)		★ 0.50 3	★ 0.54 3	1.59		★ S.INDUC 0.45 31	ES-CH 4.20 2	R 0.07 5
							★ S.STAR 2.84 3	L-STM 4.78 1	R- 0.65 4
							★ S.PARA- 0.89 623	B1-STM 1.40 1	S.P.RAD 0.16 79
							S.ZE.RS 0.07 47	B5-STM 3.09 1	GRP.DPL -2.83 9
							★ S.ZE.RS 0.09 31	F 0.47 5	
	SO2CH3	-1.63	0.56 266	0.68 266	1.35		S.INDUC 0.55 31	S.PARA- 0.88 26	S.ZE.RS 0.22 12
			★ 0.60 21	0.69 10			S.INDUC 0.55 68	S.PARA- 0.92 307	S.BA.RS 0.12 20
			0.61 713	0.70 6			S.INDUC 0.56 126	S.PARA- 0.98 711	S.AN-RS 0.30 26
			0.63 10	0.71 425			S.INDUC 0.57 12	S.PARA- 0.99 258	S.AN-RS 0.38 20
			0.64 17	0.72 709			S.INDUC 0.58 26	S.PARA- 1.05 477	S.RES.+ 0.00 10
			0.64 26	★ 0.72 21			★ S.INDUC 0.59 32	S.PARA- 1.08 28	S.RES.+ 0.12 20
			0.65 6	0.73 17			S.INDUC 0.60 67	S.PARA- 1.09 160	ES-CH 4.20 2
			0.65 23	0.73 23			S.INDUC 0.60 125	S.PARA- 1.10 536	L-STM 4.11 1
			0.65 709	0.73 46			S.INDUC 0.61 99	★ S.PARA- 1.13 710	B1-STM 2.03 1

Empirical	Structural	π	σm	σp	MR	Es	additional σ values
C1H3O2S1	SO2CH3		0.67 *425* 0.69 *46*	0.73 *713*			S.INDUC 0.64 *17*; S-INDQ 3.26 *58*; S.STAR 3.68 *3*; S.ZMTFT 0.70 *63*; S.META- 0.52 *461*; S.META- 0.67 *28*; S.META- 0.68 *369*; S.META- 0.69 *710*; S.META- 0.70 *711*; S.ZPTFT 0.67 *306*; ★S.ZPTFT 0.75 *63*; S.PARA- 0.78 *372*; S.PARA- 0.82 *618*; S.PARA- 1.14 *369*; S.PARA- 1.15 *888*; S.PARA- 1.15 *889*; S.PARNO 0.69 *324*; S.ZE.RS 0.07 *19*; S.ZE.RS 0.07 *114*; S.ZE.RS 0.08 *126*; ★S.ZE.RS 0.11 *6*; S.ZE.RS 0.12 *20*; S.ZE.RS 0.12 *26*; S.ZE.RS 0.16 *31*; S.ZE.RS 0.16 *68*; S.ZE.RS 0.18 *125*; B5-STM 3.17 *1*; F 0.53 *5*; R 0.19 *5*; R- 0.60 *4*; S-L 0.59 *14*; S.ZTWST 0.07 *22*; S.P.RAD 0.05 *79*; S.P.RAD 0.12 *659*; S-CNTIG 0.99 *139*; K-CHGTR 1.23 *436*; GRP.DPL -4.75 *8*; GRP.DPL -4.26 *9*; GR.ELCT 0.41 *13*
	SOOCH3						★S.PARA- 0.89 *623*; ★ES-CH 3.20 *2*
C1H3O2S2	SSO2CH3		★0.43 *3*	★0.54 *3*			F 0.38 *5*; R 0.16 *5*
C1H3O3S1	CH2SO3H						★S.META+ 0.64 *301*; ★S.PARA+ 0.59 *301*; ES-CH 5.20 *2*
	OSO2CH3	-0.88	★0.39 *56* 0.40 *10* 0.47 *6*	0.31 *6* 0.33 *414* 0.36 *10* ★0.36 *56*	1.70		★S.INDUC 0.55 *6*; S.INDUC 0.61 *414*; S.STAR 3.62 *3*; ★S.PARA+ 0.16 *478*; ★S.ZE.RS -0.26 *19*; S.ZE.RS -0.24 *6*; ES-CH 5.20 *2*; L-STM 4.66 *1*; B1-STM 1.35 *1*; B5-STM 4.10 *1*; F 0.40 *5*; R -0.04 *5*; R+ -0.24 *4*; GRP.DPL -4.16 *9*; GRP.DPL -3.77 *8*; GR.ELCT 0.78 *13*
	SO2(OCH3)			★0.90 *38*			S.ZE.RS 0.07 *114*; ★S.ZE.RS 0.09 *19*; ES-CH 5.20 *2*; L-STM 4.78 *1*; B1-STM 2.04 *1*; B5-STM 3.09 *1*; S.P.RAD 0.13 *79*; ★GRP.DPL -4.18 *9*
	SO2CH2OH						L-STM 4.79 *1*; B1-STM 2.03 *1*; B5-STM 3.38 *1*
C1H3S1	CH2SH	1.54 *130*	★0.08 *3*		1.39		★S.STAR 0.47 *155*; S.STAR 0.58 *87*; S.STAR 0.62 *3*; ★S.PARA+ -0.10 *351*; ★ES-CH 2.20 *2*; L-STM 4.47 *1*; B1-STM 1.52 *1*; B5-STM 3.41 *1*; ★S-L 0.12 *7*; GRP.DPL -1.52 *9*
	SCH3	0.61	0.12 *10* 0.14 *17* 0.14 *23* 0.14 *46* 0.14 *425* 0.14 *502* ★0.15 *21* 0.17 *6*	-0.08 *6* -0.05 *23* -0.03 *425* -0.01 *502* ★0.00 *21* 0.01 *17* 0.06 *46* 0.08 *10* 0.14 *208*	1.38	★-1.07 *267*	S.INDUC 0.07 *12*; S.INDUC 0.16 *125*; S.INDUC 0.17 *118*; S.INDUC 0.22 *17*; S.INDUC 0.23 *20*; S.INDUC 0.23 *142*; S.INDUC 0.24 *99*; ★S.INDUC 0.25 *32*; S.INDUC 0.27 *126*; S.INDUC 0.30 *6*; S-INDQ 1.66 *58*; S.STAR 1.47 *37*; S.STAR 1.56 *3*; S.STAR 1.66 *10*; S.ZMTFT 0.12 *63*; S.META+ 0.16 *43*; S.META- 0.17 *28*; S.META- 0.19 *461*; S.META- 0.23 *290*; S.META- 0.24 *114*; S.ZPTFT -0.04 *337*; ★S.ZPTFT -0.02 *112*; S.ZPTFT 0.02 *306*; S.ZPTFT 0.08 *63*; S.PARA- 0.00 *389*; S.PARA- 0.04 *28*; ★S.PARA- 0.06 *883*; S.PARA- 0.08 *104*; S.PARA- 0.16 *601*; S.PARA- 0.17 *372*; S.PARA- 0.21 *290*; S.PARA- 0.21 *307*; S.PARA+ -0.97 *389*; S.PARA+ -0.62 *65*; ★S.PARA+ -0.60 *43*; S.PARA+ -0.55 *145*; S.PARA+ -0.54 *775*; S.PARA+ -0.41 *88*; S.PARNO 0.09 *65*; S.ORTHO 0.28 *158*; S.ORTHO 0.30 *53*; S.ORTHO 0.52 *78*; S.ORTH+ -0.52 *154*; S.ORTH- 0.21 *97*; S.ZE.RS -0.38 *6*; S.ZE.RS -0.25 *19*; S.ZE.RS -0.23 *126*; S.ZE.RS -0.20 *20*; S.ZE.RS -0.19 *12*; S.ZE.RS -0.17 *125*; ★S.ZE.RS -0.17 *134*; S.ZE.RS -0.16 *118*; S.BA.RS -0.35 *34*; S.BA.RS -0.32 *20*; S.AN-RS 0.14 *20*; S.RES.+ -0.71 *10*; ES-CH 2.20 *2*; ES-V 0.64 *7*; L-STM 4.30 *1*; B1-STM 1.70 *1*; B5-STM 3.26 *1*; F 0.23 *5*; R -0.23 *5*; R+ -0.83 *4*; R- -0.17 *4*; S.PHOSP 0.15 *59*; S.IND.P 1.48 *39*; S.RES.P -0.23 *39*; S.ZTWST -0.05 *22*; ★S.P.RAD 0.24 *659*; S.P.RAD 0.63 *79*; K-CHGTR 0.40 *77*; GRP.DPL -1.45 *9*; GRP.DPL -1.34 *9*; GR.ELCT 0.16 *13*
C1H3S2	SSCH3		★0.22 *3*	★0.13 *3*			F 0.27 *5*; R -0.14 *5*
C1H3Se1	SeCH3	0.74	★0.10 *21*	★0.00 *21*	1.70		★S.STAR 0.95 *3*; ES-CH 2.30 *2*; L-STM 4.52 *1*; B1-STM 1.85 *1*; B5-STM 3.63 *1*; F 0.16 *5*; R -0.16 *5*; ★GRP.DPL -1.41 *9*; ★GRP.DPL -1.31 *8*; GR.ELCT 0.16 *13*
C1H3Te1	TeCH3			★0.04 *103*			S.INDUC 0.12 *103*; ★S.ZE.RS -0.08 *103*; GR.ELCT 0.01 *13*
C1H3Zn1	Zn(CH3)						GR.ELCT -0.47 *13*
C1H4Cl1N2	C=NH(NH2).HCl	-3.72					ES-CH 3.00 *2*
C1H4N1	CH2NH2	-1.04	★-0.03 *3*	★-0.11 *3*	0.91		★S.INDUC 0.00 *89*; S.INDUC 0.08 *16*; S.INDUC 0.08 *32*; S-INDQ 0.52 *58*; ★S.STAR 0.40 *155*; S.STAR 0.48 *929*; S.STAR 0.50 *3*; ★S.ZE.RS -0.15 *134*; ES-CH 2.00 *2*; L-STM 4.02 *1*; B1-STM 1.52 *1*; B5-STM 3.05 *1*; F 0.04 *5*; R -0.15 *5*; GRP.DPL -1.35 *9*; GR.ELCT 0.18 *13*
	NHCH3	-0.67 *57* -0.47	-0.30 *23* ★-0.21 *41* -0.19 *10*	-0.84 *21* ★-0.70 *41* -0.64 *236* -0.59 *23*	1.03		S.INDUC 0.02 *12*; ★S.INDUC 0.13 *6*; S.INDUC 0.40 *49*; S-INDQ 0.87 *58*; S.STAR -1.59 *796*; S.STAR -0.81 *3*; S.STAR 0.94 *10*; S.STAR 1.15 *170*; S.ZPTFT -0.46 *112*; S.ZPTFT -0.43 *337*; ★S.PARA+ -1.81 *71*; S.ZE.RS -0.60 *12*; S.ZE.RS -0.59 *49*; ★S.ZE.RS -0.52 *19*; S.RES.+ -1.48 *71*; ES-CH 2.00 *2*; ES-V 0.39 *7*; L-STM 3.53 *1*; B1-STM 1.35 *1*; B5-STM 3.08 *1*; F 0.03 *5*; R -0.73 *5*; R+ -1.78 *4*; S-L 0.13 *14*; K-CHGTR 0.73 *77*; GRP.DPL -1.01 *9*; GRP.DPL 1.69 *8*; GR.ELCT 0.50 *13*
C1H4N1O1	NHCH2OH						ES-CH 3.00 *2*; L-STM 4.70 *1*; B1-STM 1.35 *1*; B5-STM 3.32 *1*
	NHOCH3						★S.STAR 1.65 *10*
C1H4N1O1S1	NHS(=O)CH3						L-STM 4.70 *1*; B1-STM 1.35 *1*; B5-STM 4.13 *1*
C1H4N1O2S1	CH2SO2NH2						L-STM 4.18 *1*; B1-STM 1.52 *1*; B5-STM 4.01 *1*

Empirical	Structural	π	σm	σp	MR	Es	additional σ values
C1H4N1O2S1	NHSO2CH3	-1.18	★ 0.20 56 0.25 10	★ 0.03 56	1.82		S.INDUC 0.42 24 · ES-CH 5.20 2 · F 0.28 5 ★S.INDUC 0.42 68 · L-STM 4.70 1 · R -0.25 5 S.ZE.RS -0.21 24 · B1-STM 1.35 1 · K-CHGTR 1.30 436 ★S.ZE.RS -0.21 68 · B5-STM 4.13 1 · ★GRP.DPL -4.60 8
	SO2NHCH3						L-STM 4.83 1 · B1-STM 2.03 1 · B5-STM 3.11 1
C1H4N3	NHC=NH(NH2)			★ -0.02 249	1.54		★S.INDUC -0.01 117 · ★ES-CH 4.00 2 · B1-STM 1.35 1 ★S.ZE.RS -0.37 117 · L-STM 4.97 1 · B5-STM 3.96 1
C1H4N3O1	NHC(=O)NHNH2						L-STM 5.46 1 · B1-STM 1.35 1 · B5-STM 4.49 1
C1H4S1	+SHMe						S-P(C+) 1.54 844
C1H5N1	+NH2CH3		★ 0.96 23				★S.INDUC 0.60 32 · S.STAR 2.35 87 · ★S.STAR 3.74 3 S.INDUC 0.81 99 · S.STAR 3.70 360 · ES-CH 2.00 2
	CH2NH3+	-4.09	0.14 36 ★ 0.32 3 0.59 29	0.13 36 ★ 0.29 3 0.53 29		★ -3.54 44	S.INDUC 0.25 89 · S.ZPTFT 0.88 29 · B1-STM 1.52 1 S.INDUC 0.31 310 · ★S.ORTHO 0.41 53 · B5-STM 3.05 1 ★S.INDUC 0.36 32 · S.ZE.RS 0.00 134 · F 0.59 5 ★S.STAR 2.24 3 · ES-CH 2.00 2 · R -0.06 5 S.ZMTFT 0.99 29 · L-STM 4.02 1 · GR.ELCT 0.23 13
C1H5N3	NHC(NH2)=NH2+			★ 0.44 953			S.INDUC 0.34 117 · S.PARA- 0.38 955 · ES-CH 4.00 2 ★S.PARA- 0.32 954 · S.ZE.RS -0.14 117
C1H5Si1	CH2SiH3						S.PARA+ -0.27 285
	SiH2CH3						★S.INDUC 0.12 359 · S.STAR 0.20 393 · ★S.ZE.RS 0.04 359
C1H5Sn1	Sn(Me)H2						S.INDUC 0.09 242
C2Cl3	CCl=CCL2						S.STAR 2.25 3
C2Cl3O1	COCCl3						★S.STAR 3.74 129 · L-STM 5.52 1 · B5-STM 4.46 1 ES-CH 6.60 2 · B1-STM 1.88 1
C2F2N1O2	N(COF)2		★ 0.58 30	★ 0.57 30			★S.INDUC 0.58 30 · ★ES-CH 6.60 2 · B5-STM 3.52 1 ★S.STAR 3.60 3 · L-STM 4.45 1 · F 0.57 4 ★S.ZE.RS -0.01 30 · B1-STM 1.35 1 · R 0.00 4
C2F3	CF=CF2						S.STAR 1.94 3
C2F3Hg1O2	HgOC=O(CF3)		★ 0.50 3	★ 0.52 3			S.INDUC 0.22 18 · ★S.STAR 3.00 3 · F 0.48 5 ★S.INDUC 0.48 18 · ★S.ZE.RS 0.04 18 · R 0.04 5
C2F3O1	C=O(CF3)	0.02	★ 0.63 18	★ 0.80 18	1.12		★S.INDUC 0.45 32 · ★S.PARA+ 0.85 382 · B5-STM 3.67 1 S.INDUC 0.45 68 · S.ZE.RS 0.23 67 · F 0.54 5 S.INDUC 0.47 18 · S.ZE.RS 0.32 12 · R 0.26 5 S.INDUC 0.50 67 · ★S.ZE.RS 0.33 68 · R+ 0.31 4 S.INDUC 0.66 12 · ES-CH 5.40 2 · R- 0.55 4 S.STAR 3.70 129 · L-STM 4.65 1 · S-CNTIG 1.09 139 ★S.PARA- 1.09 258 · B1-STM 1.70 1 · GR.ELCT 0.44 13
C2F3O1S1	SC=O(CF3)		★ 0.48 3	★ 0.46 3	1.82		★S.INDUC 0.51 31 · L-STM 5.55 1 · F 0.48 5 ★S.STAR 3.19 3 · B1-STM 1.70 1 · R -0.02 5 ES-CH 4.20 2 · B5-STM 4.51 1
C2F3O2	trifluoroacetoxy		★ 0.56 3	★ 0.46 3	1.19		S.INDUC 0.65 31 · S.ZE.RS -0.19 31 · B1-STM 1.35 1 ★S.INDUC 0.67 707 · ★S.ZE.RS -0.19 134 · B5-STM 3.90 1 S.INDUC 0.73 12 · S.ZE.RS -0.13 12 · F 0.58 5 ★S.STAR 4.06 3 · ES-CH 4.00 2 · R -0.12 5 S.ZE.RS -0.23 19 · L-STM 5.17 1
C2F4N1O1	N(CF3)C=O(F)		★ 0.50 30 0.56 3	★ 0.50 30 0.56 3			★S.INDUC 0.50 30 · ★S.ZE.RS 0.00 30 · F 0.49 5 ★S.STAR 3.50 3 · ES-CH 7.20 2 · R 0.01 5
C2F4O1	CF2OC*F2 (3,4)		★ 0.81 499	★ 0.81 499	1.02		ES-CH 0.00 2 · F 0.77 5 · R 0.04 5
C2F5	CF2CF3	1.89 57	★ 0.47 156 0.50 3 0.52 18	★ 0.52 156 0.69 18	0.92		★S.INDUC 0.41 156 · S.ZE.RS 0.08 19 · B5-STM 3.64 1 S.STAR 2.56 3 · S.ZE.RS 0.11 30 · F 0.44 5 S.STAR 2.83 566 · ES-CH 6.00 2 · R 0.08 5 ★S.META- 0.52 165 · L-STM 4.11 1 · R- 0.25 4 ★S.PARA- 0.69 165 · B1-STM 1.99 1
C2F5O1	OCF2CF3		★ 0.48 3	★ 0.28 3	1.08		S.INDUC 0.52 30 · S.PARA- 0.27 541 · B1-STM 1.35 1 ★S.META- 0.43 541 · ★S.PARA- 0.28 477 · B5-STM 3.94 1 S.META- 0.48 28 · S.ZE.RS -0.25 30 · F 0.55 5 S.META- 0.48 297 · ES-CH 4.60 2 · R -0.27 66 S.PARA- 0.27 28 · L-STM 5.23 1 · R- -0.27 4
C2F5O2S1	SO2CF2CF3		0.83 838 ★ 0.92 48	0.99 838 ★ 1.08 48			★S.INDUC 0.76 48 · B1-STM 2.03 1 · R 0.27 4 ★S.ZE.RS 0.32 48 · B5-STM 4.07 1 L-STM 5.35 1 · F 0.81 4
C2F5S1	SCF2CF3		★ 0.44 48	★ 0.48 48			★S.INDUC 0.40 48 · F 0.42 4 ★S.ZE.RS 0.08 48 · R 0.06 4
C2F6N1	N(CF3)2		★ 0.40 224 0.47 3	0.53 3 ★ 0.53 224	1.43		S.INDUC 0.49 24 · S.ZE.RS 0.01 24 · L-STM 3.58 1 ★S.INDUC 0.49 40 · ★S.ZE.RS 0.01 224 · L-STM 4.01 1 S.STAR 3.10 3 · S.ZE.RS 0.13 906 · B1-STM 1.52 1 S.META- 0.47 165 · S.BA.RS 0.00 80 · F 0.35 264 ★S.PARA- 0.53 462 · S.AN.RS 0.00 80 · R 0.18 5 S.ZE.RS 0.00 80 · ES-CH 7.80 2 · R- 0.18 4
C2F6N1O2S2	S(CF3)=NSO2CF3		★ 1.18 245	★ 1.28 245			S.INDUC 1.08 245 · F 1.07 4 ★S.ZE.RS 0.20 245 · R 0.21 4
C2F6N1O3S2	S(=O)(CF3)=NSO2CF3		★ 1.23 245	★ 1.40 245			S.INDUC 1.06 245 · F 1.09 4 ★S.ZE.RS 0.34 245 · R 0.31 4
C2F6N1O4S2	N(SO2CF3)2		★ 0.61 751 0.75 3	0.80 3 ★ 0.83 751	3.03		S.INDUC 0.70 24 · S.ZE.RS 0.10 24 · F 0.50 4 ★S.INDUC 0.70 68 · ★S.ZE.RS 0.10 68 · R 0.33 4 ★S.STAR 4.40 3 · ES-CH 9.40 2
C2F6P1	P(CF3)2		★ 0.60 30	★ 0.69 30	1.99		★S.INDUC 0.50 40 · ES-CH 8.00 2 · B5-STM 3.86 1 ★S.STAR 3.12 3 · L-STM 4.96 1 · F 0.55 4 ★S.ZE.RS 0.19 30 · B1-STM 1.40 1 · R 0.14 4

Empirical	Structural	π	σm	σp	MR	Es	additional σ values
C2N1O1	C=O(CN)				1.18		★S.INDUC 0.66 *67*, ES-CH 4.00 *2*, B5-STM 4.08 *1*; ★S.STAR 3.40 *3*, L-STM 3.94 *1*, GR.ELCT 0.46 *13*; ★S.ZE.RS 0.26 *67*, B1-STM 1.60 *1*
C2N2P1	P(CN)2		★ 0.82 *3*	★ 0.90 *3*	2.08		★S.INDUC 0.58 *468*, L-STM 4.24 *1*, F 0.75 *5*; ★S.STAR 4.60 *3*, B1-STM 1.40 *1*, R 0.15 *5*; ES-CH 5.20 *2*, B5-STM 4.30 *1*, GR.ELCT 0.03 *13*
C2N3	N(CN)2						★S.INDUC 0.89 *24*, ES-CH 5.00 *2*, B1-STM 1.35 *1*; ★S.ZE.RS -0.22 *24*, L-STM 3.90 *1*, B5-STM 4.05 *1*
C2H1	CCH	0.40 0.48 *57*	★ 0.21 *562* 0.29 *6*	0.22 *641* ★ 0.23 *562* 0.25 *6* 0.28 *501*	0.95		S.INDUC -0.02 *12*, ★S.PARA- 0.53 *729*, B5-STM 1.60 *1*; S.INDUC 0.29 *6*, S.PARA+ 0.10 *382*, F 0.22 *5*; S.INDQ 1.64 *58*, ★S.PARA+ 0.18 *562*, R 0.01 *5*; S.STAR 0.91 *183*, S.ZE.RS -0.07 *19*, R+ -0.04 *4*; S.STAR 1.30 *464*, ★S.ZE.RS -0.04 *6*, R- 0.31 *4*; ★S.STAR 2.15 *10*, S.ZE.RS 0.02 *12*, S-L 0.29 *7*; S.STAR 2.18 *3*, ★S.BA.RS -0.09 *34*, GRP.DPL -0.78 *9*; S.STAR 2.30 *652*, ES-CH 2.00 *2*, GRP.DPL -0.77 *8*; S.META+ 0.33 *562*, ES-V 0.58 *7*, GR.ELCT 0.52 *13*; S.ZPTFT 0.16 *112*, L-STM 4.66 *1*; S.ZPTFT 0.23 *337*, B1-STM 1.60 *1*
C2H1Cl1F3O1	OCF2CHFCl		★ 0.35 *318*	★ 0.28 *318*	1.73		★S.INDUC 0.42 *30*, L-STM 5.82 *1*, F 0.38 *5*; ★S.BA.RS -0.12 *18*, B1-STM 1.35 *1*, R -0.10 *5*; ES-CH 4.60 *2*, B5-STM 4.78 *1*
C2H1Cl1F3O2S1	SO2CF2CHClF			★ 0.98 *18*			ES-CH 6.80 *2*, B1-STM 2.03 *1*; L-STM 5.94 *1*, B5-STM 4.91 *1*
C2H1Cl1N1O2	COC(Cl)=NOH						★S.PARA+ 0.80 *191*
C2H1Cl1N3	4-Cl-1-(1,2,3-triazolyl)						★S.INDUC 0.53 *15*, ★S.ZE.RS -0.10 *15*
	5-Cl-1-(1,2,3-triazolyl)						★S.INDUC 0.36 *15*, ★S.ZE.RS -0.04 *15*
C2H1Cl2	CH=CCl2		★ 0.11 *3*		2.09		★S.INDUC 0.18 *6*, L-STM 5.75 *1*, ★S-L 0.18 *14*; ★S.STAR 1.00 *3*, B1-STM 1.60 *1*; ES-CH 4.40 *2*, B5-STM 4.52 *1*
C2H1Cl2O1	COCHCl2						★S.STAR 3.19 *129*, L-STM 4.06 *1*, B5-STM 4.46 *1*; ES-CH 5.40 *2*, B1-STM 1.88 *1*
C2H1F2	CH=CF2						S.STAR 1.19 *3*
C2H1F3N1O1	NHC=O(CF3)	0.08	★ 0.30 *56* 0.36 *10*	★ 0.12 *56*	1.43		★S.INDUC 0.46 *68*, L-STM 5.62 *1*, F 0.38 *5*; ★S.ZE.RS -0.19 *68*, B1-STM 1.79 *1*, R -0.26 *5*; ES-CH 4.00 *2*, B5-STM 3.61 *1*, GR.ELCT 0.50 *13*
C2H1F3N1O2	CH(CF3)(NO2)						★S.STAR 2.67 *157*
C2H1F3N1O2S1	CH=NSO2CF3		★ 0.76 *178*	★ 1.00 *178*			★S.INDUC 0.53 *178*, F 0.63 *4*; ★S.ZE.RS 0.47 *178*, R 0.37 *4*
C2H1F4O1	OCF2CHF2		★ 0.34 *214*	★ 0.25 *214*	1.08		★S.INDUC 0.39 *214*, S.ZE.RS -0.14 *30*, B5-STM 3.94 *1*; S.META- 0.42 *28*, S.BA.RS -0.14 *18*, F 0.38 *5*; S.META- 0.43 *297*, ES-CH 4.60 *2*, R -0.13 *5*; S.PARA- 0.19 *28*, L-STM 5.23 *1*, R- -0.17 *4*; ★S.PARA- 0.21 *477*, B1-STM 1.35 *1*
C2H1F4O2S1	SO2CF2CHF2			★ 1.01 *18*			L-STM 5.35 *1*, B1-STM 2.03 *1*, B5-STM 4.07 *1*
C2H1F4S1	SCF2CHF2		★ 0.38 *214* 0.39 *17*	0.45 *17* ★ 0.47 *214*	1.84		S.INDUC 0.29 *214*, ★S.PARA- 0.61 *477*, B5-STM 4.55 *1*; ★S.INDUC 0.39 *17*, S.ZE.RS 0.20 *30*, F 0.35 *5*; S.META- 0.41 *28*, ★S.BA.RS 0.20 *18*, R 0.12 *5*; S.META- 0.42 *297*, L-STM 5.60 *1*, R- 0.26 *4*; S.PARA- 0.60 *28*, B1-STM 1.70 *1*
C2H1F5N1	NHCF2CF3						L-STM 5.26 *1*, B1-STM 1.35 *1*, B5-STM 4.00 *1*
C2H1N1	CHCN (-)						S.ZMTFT -0.81 *100*, S.ZPTFT -4.16 *100*; S.META+ -0.92 *100*, S.PARA+ -4.67 *100*
C2H1N1O4	CH(NO2)(COO-)						★S.STAR 1.83 *157*
C2H1N2O2	4-(1,3,4-oxadiazon-2-onyl						GRP.DPL -6.63 *8*
	N-sydnonyl						★S.PARA- 0.71 *919*, ★ES-CH 3.00 *2*
C2H1N8S2	1-(5,5-bis-tetrazolyl-SS)		★ 0.63 *40*	★ 0.64 *40*	4.92		★S.INDUC 0.62 *40*, F 0.60 *5*; ES-CH 4.20 *2*, R 0.04 *5*
C2H1O1	OC=-CH				0.99		S.INDUC 0.48 *12*, ★S.ZE.RS -0.30 *31*, B1-STM 1.35 *1*; ★S.INDUC 0.50 *31*, S.ZE.RS -0.26 *12*, B5-STM 4.37 *1*; S.INDUC 0.59 *102*, ES-CH 3.00 *2*; S.ZE.RS -0.34 *102*, L-STM 3.87 *1*
C2H1O2	CHC(=O)o(-2)						S.ZPTFT -3.85 *100*, S.PARA+ -7.17 *100*
C2H1O3	COCO2H	-1.41 *57*					★S.PARA+ 0.90 *191*
C2H1S1	SCCH		★ 0.26 *3*	★ 0.19 *3*	1.62		S.INDUC 0.32 *12*, ★S.ZE.RS -0.13 *31*, B5-STM 4.85 *1*; S.INDUC 0.32 *31*, S.ZE.RS -0.12 *12*, F 0.30 *5*; S.INDUC 0.53 *102*, ES-CH 3.20 *2*, R -0.11 *5*; ★S.STAR 2.00 *3*, L-STM 4.08 *1*, GRP.DPL -1.69 *9*; S.ZE.RS -0.19 *102*, B1-STM 1.70 *1*
C2H2Br1O1	OCH=CHBr -C						★S.INDUC 0.42 *31*, ★S.ZE.RS -0.31 *31*, ES-CH 3.00 *2*
	OCH=CHBr -T						L-STM 5.75 *1*, B1-STM 1.35 *1*, B5-STM 5.36 *1*
C2H2Cl1	C(Cl)=CH2						★S.INDUC 0.55 *6*, ★ES-CH 3.20 *2*, ★S-L 0.55 *14*
	CH=CHCl -C						★S.INDUC 0.18 *6*, ★S-L 0.18 *14*
	CH=CHCl -T				1.60		★S.INDUC 0.17 *6*, ES-CH 3.20 *2*, B5-STM 3.09 *1*; ★S.STAR 0.87 *3*, L-STM 5.75 *1*, ★S-L 0.17 *14*; ★S.STAR 0.90 *326*, B1-STM 1.60 *1*, ★GRP.DPL -1.64 *9*
C2H2Cl1O1	CH2COCl						★S.INDUC 0.04 *49*, ★S.ZE.RS 0.03 *49*, ES-CH 4.20 *2*
	COCH2Cl						★S.INDUC 0.35 *102*, ES-CH 4.20 *2*, B5-STM 3.21 *1*

Empirical	Structural	π	σm	σp	MR	Es	additional σ values		
C2H2Cl1O1	COCH2Cl						★ S.STAR 2.50 *129* ★ S.ZE.RS 0.11 *102*	L-STM 5.52 *1* B1-STM 1.60 *1*	GRP.DPL -2.27 *9*
C2H2Cl1S1	SCH=CHCl		★ 0.31 *168*	★ 0.24 *168*			★ ES-CH 3.20 *2*	F 0.34 *4*	R -0.10 *4*
C2H2Cl1Se1	SeCH=CHCl		★ 0.28 *168*	★ 0.26 *168*			ES-CH 3.30 *2*	F 0.30 *4*	R -0.04 *4*
C2H2Cl3	CH2CCl3		★ 0.06 *3*		2.48		S.INDUC 0.03 *500* ★ S.INDUC 0.14 *6* ★ S.STAR 0.75 *3*	ES-CH 5.60 *2* L-STM 5.57 *1* B1-STM 1.52 *1*	B5-STM 4.50 *1* S-L 0.14 *14* ★ GRP.DPL -1.84 *9*
C2H2F1	CF=CH2						S.STAR 1.56 *3*		
C2H2F1O1	C(=O)CH2F						L-STM 4.65 *1*	B1-STM 1.60 *1*	B5-STM 3.13 *1*
C2H2F3	CH2CF3		★ 0.12 *30* 0.16 *3*	★ 0.09 *30* 0.14 *3*	0.97		S.INDUC 0.03 *500* S.INDUC 0.14 *30* S.INDUC 0.15 *261* ★ S.INDUC 0.16 *6* S.STAR 0.87 *3* S.STAR 0.92 *52* S.META- 0.14 *297*	★ S.PARA- 0.14 *297* S.ZE.RS -0.05 *30* ★ S.ZE.RS -0.04 *80* ★ S.AN-RS -0.04 *80* ES-CH 4.40 *2* L-STM 4.70 *1* B1-STM 1.52 *1*	B5-STM 3.70 *1* F 0.15 *4* R -0.06 *4* R- -0.01 *4* S-L 0.16 *7* GR.ELCT 0.18 *13*
C2H2F3O1	OCH2CF3						★ S.INDUC 0.17 *49* S.INDUC 0.81 *16* S.ZE.RS -0.24 *49*	ES-CH 3.00 *2* L-STM 5.23 *1* B1-STM 1.35 *1*	B5-STM 3.94 *1* S.PHOSP -0.15 *449*
C2H2F3O1S1	CH2S=O(CF3)		★ 0.25 *288*	★ 0.24 *288*			★ S.INDUC 0.26 *288* S.STAR 1.47 *87* ★ S.STAR 1.62 *3* ★ S.ZE.RS -0.02 *18*	ES-CH 4.20 *2* L-STM 5.35 *1* B1-STM 1.52 *1* B5-STM 4.07 *1*	F 0.27 *4* R -0.03 *4*
C2H2F3O2S1	CH2SO2CF3		★ 0.29 *288*	★ 0.31 *288*	1.75		★ S.INDUC 0.28 *288* S.STAR 1.57 *87* ★ S.STAR 1.75 *3* S.ZE.RS 0.03 *18*	ES-CH 5.20 *2* L-STM 5.35 *1* B1-STM 1.52 *1* B5-STM 4.07 *1*	F 0.29 *5* R 0.02 *5* GR.ELCT 0.18 *13*
	SO2CH2CF3						L-STM 5.35 *1*	B1-STM 2.03 *1*	B5-STM 4.07 *1*
C2H2F3O2Si1	COOCH2SiF3						★ S.INDUC 0.61 *25*	★ S.ZE.RS 0.22 *25*	
C2H2F3S1	CH2SCF3		★ 0.12 *288*	★ 0.15 *288*	1.76		S.INDUC 0.15 *288* S.INDUC 0.19 *110* S.INDUC 0.21 *64* S.STAR 0.75 *3*	★ S.ZE.RS -0.07 *110* S.ZE.RS -0.06 *64* ES-CH 3.20 *2* L-STM 5.82 *1*	B1-STM 1.52 *1* B5-STM 4.10 *1* F 0.13 *5* R 0.02 *5*
	SCH2CF3						L-STM 5.60 *1*	B1-STM 1.70 *1*	B5-STM 4.55 *1*
C2H2N1	CH2CN	-0.57	★ 0.16 *17* 0.22 *29*	0.01 *23* 0.16 *29* ★ 0.18 *17*	1.01	★ -2.38 *44* -2.18 *726* -2.02 *35*	S.INDUC 0.16 *17* ★ S.INDUC 0.20 *6* S.INDUC 0.21 *16* S.INDUC 0.24 *89* S.INDUC 0.26 *64* S.INDUC 0.29 *12* S.INDUC 0.36 *233* S.STAR 1.25 *76* ★ S.STAR 1.30 *52* S.STAR 1.30 *74* S.STAR 1.30 *580* S.STAR 1.34 *10*	S.STAR 1.71 *420* ★ S.PARA- 0.11 *257* ★ S.PARA+ 0.16 *351* S.ZE.RS -0.10 *233* S.ZE.RS -0.09 *19* S.ZE.RS -0.08 *64* ★ S.ZE.RS -0.08 *134* S.ZE.RS -0.06 *12* S.ZE.RS 0.09 *114* ES-CH 3.00 *2* ES-V 0.89 *7* L-STM 3.99 *1*	B1-STM 1.52 *1* B5-STM 4.12 *1* F 0.17 *5* R 0.01 *5* R+ -0.01 *4* R- -0.06 *4* S-L 0.20 *7* K-CHGTR 0.39 *77* GRP.DPL -3.60 *8* GR.ELCT 0.20 *13*
C2H2N1O1	CH2NCO						★ S.STAR 0.81 *3*	★ S.ZE.RS -0.06 *47*	ES-CH 3.00 *2*
	OCH2CN						L-STM 6.15 *1*	B1-STM 1.35 *1*	B5-STM 3.07 *1*
C2H2N1O2	C(=O)NHCHO	-2.85 *57*							
	CH=CHNO2 -T	0.11	0.28 *180* ★ 0.32 *690*	★ 0.26 *690*	1.64		★ S.INDUC 0.24 *80* S.STAR 1.70 *326* ★ S.STAR 1.75 *3* ★ S.META- 0.37 *691* S.ZPTFT 0.50 *248* ★ S.PARA- 0.88 *691*	S.ZE.RS 0.13 *19* S.ZE.RS 0.13 *80* ES-CH 3.00 *2* L-STM 4.29 *1* B1-STM 1.60 *1* B5-STM 4.78 *1*	F 0.35 *5* R -0.09 *264* R- 0.52 *4* GRP.DPL -3.99 *9* GR.ELCT 0.35 *13*
	COCH=NOH	-0.73 *57*					★ S.PARA+ 0.71 *191*		
C2H2N1O2S1	SCH=CHNO2 -C						L-STM 7.14 *1*	B1-STM 1.70 *1*	B5-STM 4.23 *1*
	SCH=CHNO2 -T						L-STM 6.36 *1*	B1-STM 1.70 *1*	B5-STM 5.84 *1*
C2H2N1O3	OCH=CHNO2 -C						L-STM 6.82 *1*	B1-STM 1.35 *1*	B5-STM 3.65 *1*
	OCH=CHNO2 -T						L-STM 6.25 *1*	B1-STM 1.35 *1*	B5-STM 5.34 *1*
C2H2N1S1	CH2NCS						★ S.STAR 0.94 *3*	★ S.ZE.RS -0.07 *47*	ES-CH 3.00 *2*
	CH2SCN	-0.14	★ 0.12 *3*	★ 0.14 *210*	1.80		★ S.STAR 1.18 *87* ES-CH 3.20 *2* L-STM 6.63 *1*	B1-STM 1.52 *1* B5-STM 3.41 *1* F 0.14 *5*	R 0.00 *5*
	SCH2CN						L-STM 6.47 *1*	B1-STM 1.70 *1*	B5-STM 3.61 *1*
C2H2N1Se1	CH2SeCN						ES-CH 3.30 *2*	★ S-L 0.22 *14*	
C2H2N3	1-(1,2,3-triazolyl)			★ 0.40 *226* 0.48 *275*			★ S.INDUC 0.53 *91* ★ S.ZE.RS -0.10 *91*	ES-CH 3.00 *2* L-STM 5.44 *1*	B1-STM 1.71 *1* B5-STM 3.12 *1*
	1-(1,2,4-triazoylyl)			★ 0.37 *226* 0.44 *275*			★ S.INDUC 0.53 *91* ★ S.ZE.RS -0.12 *91*	ES-CH 3.00 *2* L-STM 5.44 *1*	B1-STM 1.71 *1* B5-STM 3.12 *1*
	2-(1,2,3-triazolyl)			★ 0.36 *226*			ES-CH 3.00 *2* L-STM 5.44 *1*	B1-STM 1.71 *1* B5-STM 2.82 *1*	
	4-(1,2,4-triazolyl)			★ 0.33 *226*			S.INDUC 0.66 *91* S.ZE.RS -0.10 *91*	ES-CH 3.00 *2* L-STM 5.28 *1*	B1-STM 1.71 *1* B5-STM 3.12 *1*
C2H2N3O6	CH2C(NO2)3				2.74		★ S.STAR 1.62 *291* ES-CH 5.00 *2*	L-STM 5.49 *1* B1-STM 1.52 *1*	B5-STM 5.00 *1*
C2H2N3S1	1(5-SH-1,2,3-triazolyl)						S.INDUC 0.54 *91*	S.ZE.RS -0.10 *91*	
C2H2O2	CH2CO2-		★ 0.07 *29*	★ -0.16 *29*			★ S.INDUC 0.01 *32*	S.PARA- 0.04 *36*	B1-STM 1.52 *1*

Empirical	Structural	π	σ_m	σ_p	MR	E_s	additional σ values
C2H2O2	CH2CO2-						★S.STAR -0.06 3 · ★S.PARA+ -0.53 100 · B5-STM 3.78 1 S.STAR 0.33 87 · ★S.BA.RS -0.23 34 · F 0.19 4 S.META- -0.13 29 · ES-CH 4.00 2 · R -0.35 4 S.ZPTFT -0.15 100 · L-STM 4.74 1
C2H2O3	OCH2COO-						S.ORTHO -0.27 78
C2H2S1	4-benzothienyl:						★S.META+ -0.11 412 · ★S.PARA+ -0.34 412 ★S.PARA+ -0.42 412 · ★S.PARA+ -0.25 412
C2H3	CH=CH2	0.82	0.01 134 ★0.06 6 0.08 180	-0.08 180 ★-0.04 6 -0.01 134	1.10	★-3.19 35	S.INDUC 0.01 89 · S.META+ 0.11 784 · L-STM 4.29 1 S.INDUC 0.02 26 · S.ZPTFT 0.00 112 · B1-STM 1.60 1 S.INDUC 0.06 823 · ★S.PARA+ -0.16 784 · B5-STM 3.09 1 S.INDUC 0.07 12 · S.PARA+ 0.10 382 · F 0.13 5 S.INDUC 0.10 16 · S.ZE.RS -0.15 6 · R -0.17 5 ★S.INDUC 0.11 6 · S.ZE.RS -0.05 19 · R+ -0.29 4 S-INDQ 0.60 58 · S.ZE.RS -0.04 12 · S-L 0.11 7 S.STAR 0.40 129 · ★S.ZE.RS -0.03 134 · ★S.PHOSP -0.68 59 S.STAR 0.40 464 · S.ZE.RS -0.03 193 · GRP.DPL -0.40 9 S.STAR 0.52 511 · S.BA.RS -0.11 34 · GRP.DPL 0.20 8 S.STAR 0.56 3 · ES-CH 2.00 2 · GR.ELCT 0.34 13 S.STAR 0.59 10 · ES-V 0.57 341 · O-STER -1.48 90 S.STAR 0.65 326 · ES-V 2.11 474
C2H3Br2	C(CH3)Br2					★-3.04 35	ES-CH 4.60 2 · L-STM 4.11 1 · B5-STM 3.75 1 ★ES-V 1.46 7 · B1-STM 2.81 1
	CHBrCH2Br				2.58		★S.STAR 1.38 223 · L-STM 5.02 1 · B5-STM 4.30 1 ★ES-CH 4.60 2 · B1-STM 1.92 1 · ★GRP.DPL -1.43 9
C2H3Cl1N1O1	NHC=O(CH2Cl)	-0.50	★0.17 56 0.20 6	-0.07 6 ★-0.03 56	1.98		S.INDUC 0.35 6 · L-STM 6.26 1 · R -0.30 5 ★S.STAR 2.06 · B1-STM 1.55 1 · ★S-L 0.36 7 S.ZE.RS -0.42 6 · B5-STM 4.26 1 ES-CH 4.00 2 · F 0.27 5
C2H3Cl1N1O2	CCl(CH3)(NO2)						★S.STAR 2.68 157
C2H3Cl2	CCl2CH3				2.10		★S.STAR 1.53 223 · L-STM 4.11 1 · B5-STM 3.46 1 ES-CH 4.40 2 · B1-STM 2.65 1 · ★GRP.DPL -2.33 9
	CH2CHCl2						★ES-CH 4.40 2
	CHClCH2Cl				2.10		★S.STAR 1.08 223 · L-STM 4.79 1 · B5-STM 4.05 1 ES-CH 4.40 2 · B1-STM 1.88 1 · ★GRP.DPL -1.79 9
C2H3Cl2Si1	SiCl2(CH=CH2)			★0.28 238			
C2H3F2	CH2CHF2						S.INDUC 0.09 261
C2H3F3N1	NHCH2CF3						★S.STAR 1.81 170 · L-STM 5.26 1 · B5-STM 4.00 1 ES-CH 3.00 2 · B1-STM 1.35 1
C2H3F3N1O2S1	N(CH3)SO2CF3		★0.46 3	★0.44 3	2.27		★S.INDUC 0.48 68 · ★ES-CH 6.20 2 · B5-STM 4.00 1 ★S.STAR 3.00 3 · L-STM 5.26 1 · F 0.46 4 ★S.ZE.RS -0.04 68 · B1-STM 1.54 1 · R -0.02 4
C2H3Hg1	Hg-CH=CH2						S.INDUC -0.17 16
C2H3Hg1O2	HgOC=O(CH3)	-1.42	★0.39 3	★0.40 3	2.23		S.INDUC -0.09 18 · S.STAR 2.40 3 · R 0.01 5 S.INDUC 0.16 18 · ★S.ZE.RS 0.00 18 · GR.ELCT -0.61 13 S.INDUC 0.22 353 · S.ZE.RS 0.01 353 ★S.INDUC 0.38 18 · F 0.39 5
C2H3N1O1	NC(=O)CH3 (-)						S.ZMTFT -0.93 100 · S.ZPTFT -0.96 100 S.META+ -0.75 100 · S.PARA+ -2.90 100
C2H3N2	NHCH2CN						L-STM 6.19 1 · B1-STM 1.35 1 · B5-STM 3.11 1
C2H3N2O1	N=NCOCH3			★0.89 72			S.ZE.RS 0.29 72
C2H3N2O2	C(=O)NHC(=O)NH2	-2.14 57			1.82		L-STM 5.85 1 · B1-STM 1.50 1 · B5-STM 4.45 1
	NHCH=CHNO2 -C						L-STM 6.86 1 · B1-STM 1.48 1 · B5-STM 3.71 1
	NHCH=CHNO2 -T						L-STM 6.27 1 · B1-STM 1.35 1 · B5-STM 5.40 1
C2H3N2O4	C(CH3)(NO2)2		★0.54 3	★0.61 291	2.17		★S.STAR 3.36 157 · B1-STM 2.55 1 · R 0.11 4 ES-CH 8.00 2 · B5-STM 3.72 1 L-STM 4.59 1 · F 0.50 4
C2H3N4	1-Me-5-tetrazolyl						S.INDUC 0.48 15 · S.STAR 2.99 15
	2-Me-5-tetrazolyl						S.INDUC 0.32 15 · S.STAR 1.99 15
	NH-4-(1,2,4-triazolyl)						L-STM 4.31 1 · B1-STM 1.50 1 · B5-STM 5.12 1
C2H3N4S1	4Me-tetrazole-5-S-						S.INDUC 0.53 495
C2H3O1	CH2CHO						★S.STAR 0.62 3 · L-STM 4.54 1 · ★GRP.DPL -2.23 9 ★S.ZE.RS -0.11 19 · B1-STM 1.52 1 ES-CH 3.00 2 · B5-STM 3.10 1
	COCH3	-0.62 57 -0.60 588 -0.55 -0.43 54 -0.17 42 0.18 220	0.28 18 0.31 23 0.31 266 0.33 26 0.35 502 0.36 17 0.36 46 0.38 10 ★0.38 21	0.38 18 0.43 502 0.44 17 0.47 46 0.48 409 0.49 10 ★0.50 21 0.52 23 0.55 208 0.56 236	1.12		S.INDUC 0.18 16 · S.PARA- 0.70 257 · S.RES.+ 0.00 10 S.INDUC 0.18 18 · S.PARA- 0.71 211 · S.RES.+ 0.06 6 S.INDUC 0.18 68 · S.PARA- 0.81 104 · S.RES.+ 0.16 20 S.INDUC 0.20 67 · S.PARA- 0.82 258 · ES-CH 3.00 2 S.INDUC 0.22 12 · S.PARA- 0.82 723 · ES-V 0.50 341 S.INDUC 0.22 26 · S.PARA- 0.84 181 · L-STM 4.06 1 S.INDUC 0.23 142 · ★S.PARA- 0.84 307 · B1-STM 1.60 1 S.INDUC 0.24 99 · S.PARA- 0.86 28 · B5-STM 3.13 1 S.INDUC 0.28 205 · S.PARA- 0.87 23 · F 0.33 5 ★S.INDUC 0.30 6 · S.PARA- 0.87 247 · R 0.17 5 S.INDUC 0.31 119 · S.PARA- 0.87 601 · R- 0.51 4 S.INDUC 0.34 17 · S.PARNO 0.50 324 · S-L 0.31 7 S-INDQ 1.69 58 · S.ORTHO 0.07 78 · S.ZTWST 0.00 22 ★S.STAR 1.65 10 · S.ZE.RS 0.15 114 · S.M.RAD 0.07 1003 S.STAR 1.65 52 · S.ZE.RS 0.16 20 · S.P.RAD 0.24 206 S.STAR 1.65 129 · S.ZE.RS 0.16 26 · S.P.RAD 0.53 263

Empirical	Structural	π	σ_m	σ_p	MR	E_s	additional σ values		
C2H3O1	COCH3						S.STAR 1.81 *3* / S.ZMTFT 0.35 *95* / S.ZMTFT 0.39 *63* / S.META- 0.32 *601* / S.META- 0.34 *28* / ★S.ZPTFT 0.49 *63* / S.ZPTFT 0.51 *95* / S.PARA- 0.66 *26*	S.ZE.RS 0.20 *6* / S.ZE.RS 0.20 *18* / ★S.ZE.RS 0.20 *68* / S.ZE.RS 0.22 *12* / S.ZE.RS 0.22 *19* / S.BA.RS 0.16 *20* / S.AN.RS 0.43 *26* / S.AN.RS 0.47 *20*	S.P.RAD 0.60 *79* / S-CNTIG 0.82 *139* / K-CHGTR 0.48 *77* / GRP.DPL -2.90 *8* / GRP.DPL -2.77 *9* / GR.ELCT 0.39 *13* / O-STER -2.24 *90*
	OCH=CH2		★ 0.21 *638*	★ -0.09 *638*	1.14		★S.INDUC 0.38 *31* / S.INDUC 0.40 *12* / S.ZE.RS -0.33 *12* / ★S.ZE.RS -0.33 *31*	S.ZE.RS -0.33 *260* / ES-CH 3.00 *2* / L-STM 4.98 *1* / B1-STM 1.35 *1*	B5-STM 3.65 *1* / F 0.34 *4* / R -0.43 *4*
	oxiranyl		★ 0.05 *3*	★ 0.03 *3*	1.09		S.INDUC 0.06 *339* / ★S.INDUC 0.07 *193* / S.ZE.RS -0.05 *339* / ★S.ZE.RS -0.04 *193*	ES-CH 3.00 *2* / L-STM 4.14 *1* / B1-STM 1.55 *1* / B5-STM 3.24 *1*	F 0.09 *5* / R -0.06 *5*
C2H3O1S1	S(O)CH=CH2						S.ZE.RS 0.07 *260*		
	SC=O(CH3)	0.10	0.37 *459* / 0.38 *10* / 0.39 *6* / 0.39 *17* / ★ 0.39 *21*	0.37 *6* / 0.42 *459* / ★ 0.44 *21* / 0.45 *17*	1.84		★S.INDUC 0.21 *31* / S.INDUC 0.32 *214* / S.INDUC 0.39 *6* / S.INDUC 0.39 *17* / S.STAR 2.29 *3* / ★S.PARA- 0.46 *290* / S.PARA- 0.47 *138*	S.ZE.RS -0.02 *6* / ★S.ZE.RS 0.01 *31* / S.ZE.RS 0.08 *19* / ES-CH 4.20 *2* / L-STM 5.11 *1* / B1-STM 1.70 *1* / B5-STM 4.01 *1*	F 0.37 *5* / R 0.07 *5* / R- 0.09 *4* / S.P.RAD 0.29 *79* / GR.ELCT 0.18 *13*
C2H3O2	C=O(OCH3)	-0.01 / 0.19 *42* / 0.50 *454*	0.26 *26* / 0.32 *23* / 0.32 *463* / 0.33 *316* / ★ 0.36 *251* / 0.37 *21* / 0.38 *6*	0.39 *463* / 0.43 *6* / 0.43 *93* / ★ 0.45 *251* / 0.50 *82*	1.29		S.INDUC 0.09 *16* / S.INDUC 0.13 *12* / S.INDUC 0.17 *26* / S.INDUC 0.21 *142* / S.INDUC 0.26 *119* / S.INDUC 0.30 *205* / S.INDUC 0.31 *99* / ★S.INDUC 0.32 *6* / S-INDQ 1.70 *58* / ★S.STAR 2.00 *52* / S.STAR 2.17 *87* / S.META+ 0.37 *43* / S.META- 0.32 *28* / S.PARA- 0.59 *26* / S.PARA- 0.64 *23* / S.PARA- 0.64 *463* / S.PARA- 0.71 *93* / S.PARA- 0.74 *442*	★S.PARA- 0.75 *104* / S.PARA- 0.78 *28* / S.PARA- 0.81 *443* / S.PARA+ -0.05 *913* / ★S.PARA+ 0.49 *43* / S.PARNO 0.39 *324* / S.ORTHO 0.51 *158* / S.ORTHO 0.63 *78* / S.ORTH- 0.88 *97* / S.ZE.RS 0.11 *6* / ★S.ZE.RS 0.16 *19* / S.ZE.RS 0.16 *26* / S.ZE.RS 0.16 *114* / S.ZE.RS 0.23 *12* / S.AN.RS 0.41 *26* / ES-CH 4.00 *2* / ES-V 0.50 *341* / ES-V 1.45 *474*	L-STM 4.73 *1* / B1-STM 1.64 *1* / B5-STM 3.36 *1* / F 0.34 *5* / R 0.11 *5* / R+ 0.15 *4* / R- 0.41 *4* / S-L 0.32 *14* / S.ZTWST 0.00 *22* / S.M.RAD -0.14 *79* / S.P.RAD 0.43 *79* / S-CNTIG 0.74 *139* / S-CNTIG 0.81 *486* / K-CHGTR 0.48 *77* / GRP.DPL -1.92 *232* / GRP.DPL -1.75 *9* / GR.ELCT 0.37 *13* / O-STER -1.04 *90*
	CH2CO2H	-0.77 *130* / -0.72 / -0.43 *54*		★ -0.07 *293*	1.19		S.INDUC 0.11 *34* / S.STAR 0.61 *87* / ★S.STAR 1.05 *52* / S.STAR 1.08 *3* / S.PARA- 0.05 *257*	★S.PARA+ -0.01 *351* / S.BA.RS -0.06 *34* / ES-CH 4.00 *2* / L-STM 4.74 *1* / B1-STM 1.52 *1*	B5-STM 3.78 *1* / GRP.DPL -1.86 *8* / GRP.DPL -1.68 *9* / GR.ELCT 0.17 *13*
	OC=O(CH3)	-0.64 / -0.35 *54* / -0.27 *57*	0.26 *56* / 0.26 *121* / 0.30 *6* / 0.37 *10* / ★ 0.39 *21*	0.15 *6* / 0.16 *56* / 0.16 *121* / ★ 0.31 *21* / 0.41 *10*	1.25		S.INDUC 0.29 *205* / S.INDUC 0.29 *444* / S.INDUC 0.32 *707* / ★S.INDUC 0.33 *89* / S.INDUC 0.33 *119* / S.INDUC 0.36 *444* / S.INDUC 0.37 *12* / S.INDUC 0.37 *192* / S.INDUC 0.38 *6* / S.INDUC 0.40 *99* / S.INDUC 0.42 *34* / S.INDUC 0.44 *17* / S-INDQ 2.12 *58* / S.STAR 1.29 *687* / S.STAR 2.45 *10*	S.STAR 2.56 *3* / S.META+ 0.23 *88* / S.META+ 0.26 *361* / S.META+ 0.27 *538* / S.META+ 0.34 *361* / ★S.PARA+ -0.19 *361* / S.PARA+ -0.13 *569* / S.PARA+ -0.12 *361* / S.PARA+ -0.06 *538* / S.PARA+ -0.01 *88* / S.ORTHO -0.37 *78* / ★S.ORTH- -0.01 *97* / S.ZE.RS -0.24 *19* / S.ZE.RS -0.23 *6* / ★S.ZE.RS -0.21 *134*	S.ZE.RS -0.19 *830* / S.ZE.RS -0.18 *12* / S.ZE.RS 0.27 *114* / S.BA.RS -0.12 *34* / ES-CH 4.00 *2* / L-STM 4.74 *1* / B1-STM 1.35 *1* / B5-STM 3.67 *1* / F 0.42 *5* / R -0.11 *5* / R+ -0.61 *4* / S.P.RAD -0.05 *79* / GRP.DPL -1.81 *9* / GRP.DPL -1.72 *8* / GR.ELCT 0.80 *13*
C2H3O2S1	SC(=O)OCH3						L-STM 6.04 *1*	B1-STM 1.70 *1*	B5-STM 4.83 *1*
	SO2CH=CH2						★S.INDUC 0.48 *1000* / ★S.STAR 0.17 *834* / ★S.ZE.RS 0.03 *956*	S.ZE.RS 0.21 *993* / ES-CH 5.20 *658* / L-STM 5.11 *1*	B1-STM 2.03 *1* / B5-STM 3.78 *1*
C2H3O3	OCH2CO2H	-0.87 / -0.79 / -0.42 *54*		★ -0.33 *194* / -0.18 *3*	1.40		ES-CH 3.00 *2* / L-STM 5.85 *1* / B1-STM 1.35 *1*	B5-STM 3.98 *1*	
	OCOOCH3						★S.INDUC 0.11 *49* / ★S.ZE.RS -0.16 *49*	★ES-CH 4.00 *2* / L-STM 5.86 *1*	B1-STM 1.35 *1* / B1-STM 4.28 *1*
C2H3S1	SCH=CH2		0.18 *3* / ★ 0.26 *760*	0.14 *3* / ★ 0.20 *168* / 0.25 *638*	1.77		★S.INDUC 0.21 *31* / S.INDUC 0.33 *102* / S.STAR 1.31 *3* / ★S.PARA- 0.23 *760* / S.ZE.RS -0.22 *102*	S.ZE.RS -0.11 *260* / ★S.ZE.RS -0.07 *31* / ES-CH 3.20 *2* / L-STM 5.33 *1* / B1-STM 1.70 *1*	B5-STM 4.23 *1* / F 0.29 *5* / R -0.09 *5* / GRP.DPL -1.38 *9*
	thiiranyl		★ 0.04 *3*	★ 0.01 *3*	1.70		★S.INDUC 0.07 *193* / ★S.ZE.RS -0.06 *193*	★ES-CH 3.20 *2* / F 0.08 *5*	R -0.07 *5*
C2H3S2	SC(=S)CH3						★S.STAR 3.00 *3*	★ES-CH 4.40 *2*	★S-L 0.45 *14*
C2H3Se1	SeCH=CH2		★ 0.26 *168*	★ 0.21 *168*			★S.ZE.RS -0.09 *260* / ★ES-CH 3.30 *2* / L-STM 4.47 *1*	B1-STM 1.85 *1* / B5-STM 5.00 *1* / F 0.29 *4*	R -0.08 *4*
C2H3Te1	TeCH=CH2						S.ZE.RS -0.08 *260*		
C2H4B1O2	-B(OCH2)2			★ 0.46 *535*					

Empirical	Structural	π	σ$_m$	σ$_p$	MR	E$_s$		additional σ values					
C2H4Br1	CH2CH2Br	0.96			1.80	-2.51 *44* ★ -2.24 *44*	S.INDUC 0.05 *6* ★ S.STAR 0.26 *74* S.STAR 0.36 *129* S.STAR 0.44 *223* ★ S.STAR 0.49 *287*	ES-CH 3.30 *2* L-STM 5.87 *1* B1-STM 1.52 *1* B5-STM 3.40 *1* S-L 0.05 *14*	★ S.PHOSP -0.80 *37* S.IND.P 0.53 *39* ★ GRP.DPL -1.97 *9*			
	CHBrCH3				1.80	-2.17 *44* ★ -1.93 *44*	S.INDUC 0.19 *6* S.STAR 0.90 *287* ★ S.STAR 1.00 *223* S.STAR 1.25 *3*	ES-CH 3.30 *2* L-STM 4.09 *1* B1-STM 1.91 *1* B5-STM 3.75 *1*	S-L 0.19 *14* GRP.DPL -2.08 *9*			
C2H4Br1O1	OCH2CH2Br				1.97		★ S.INDUC 0.35 *31* ★ S.ZE.RS -0.40 *31* ★ ES-CH 3.00 *2*	★ ES-V 0.58 *7* L-STM 6.02 *1* B1-STM 1.35 *1*	B5-STM 5.07 *1*			
	OCHBrCH3						L-STM 3.98 *1*	B1-STM 1.35 *1*	B5-STM 4.69 *1*			
C2H4Br1S1	SCH2CH2Br						L-STM 6.15 *1*	B1-STM 1.70 *1*	B5-STM 5.71 *1*			
	SCHBrCH3						L-STM 4.30 *1*	B1-STM 1.70 *1*	B5-STM 4.97 *1*			
C2H4Cl1	CH2CH2Cl	0.82			1.51	★ -2.14 *50*	★ S.INDUC -0.02 *500* S.INDUC 0.05 *34* S.INDUC 0.07 *6* ★ S.STAR 0.39 *52* S.STAR 0.39 *74*	S.BA.RS -0.10 *34* ES-CH 3.20 *2* ES-V 0.97 *7* L-STM 5.57 *1* B1-STM 1.52 *1*	B5-STM 3.25 *1* S-L 0.07 *14* GRP.DPL -1.93 *9*			
	CHClCH3				1.51	★ -1.74 *44*	S.INDUC 0.15 *6* S.STAR 0.94 *287* S.STAR 0.97 *74* ★ S.STAR 1.00 *3*	S.STAR 1.08 *223* ES-CH 3.20 *2* L-STM 3.89 *1* B1-STM 1.89 *1*	B5-STM 3.46 *1* S-L 0.15 *14* GRP.DPL -2.05 *9*			
C2H4Cl1O1	OCH2CH2Cl				1.68		★ ES-CH 3.00 *2* ★ ES-V 0.51 *7*	L-STM 5.82 *1* B1-STM 1.35 *1*	B5-STM 4.78 *1* ★ S.PHOSP 0.03 *37*			
	OCHClCH3						L-STM 3.98 *1*	B1-STM 1.35 *1*	B5-STM 4.41 *1*			
C2H4Cl1O2S1	SO2CH2CH2Cl						L-STM 5.94 *1*	B1-STM 2.03 *1*	B5-STM 4.91 *1*			
C2H4Cl1S1	SCH2CH2Cl						L-STM 5.97 *1*	B1-STM 1.70 *1*	B5-STM 5.41 *1*			
	SCHClCH3						L-STM 4.30 *1*	B1-STM 1.70 *1*	B5-STM 4.71 *1*			
C2H4F1	CH2CH2F	0.93					★ S.STAR 0.39 *4* L-STM 4.70 *1*	B1-STM 1.52 *1* B5-STM 3.17 *1*	F 0.06 *4*			
C2H4F1O1	OCH2CH2F	1.19					F 0.27 *4*					
C2H4F2N1	CH2CH2N(F)F						★ S.STAR 0.53 *590* ★ ES-CH 3.00 *2*	L-STM 5.26 *1* B1-STM 1.52 *1*	B5-STM 4.00 *1*			
C2H4I1	CH(I)CH3	2.32				★ -2.60 *44*	★ ES-CH 3.70 *2* L-STM 4.36 *1*	B1-STM 1.91 *1* B5-STM 4.15 *1*				
	CH2CH2I				2.32	★ -2.26 *44*	★ S.STAR 0.41 *223* ES-CH 3.70 *2* ★ ES-V 0.93 *7*	L-STM 6.28 *1* B1-STM 1.52 *1* B5-STM 3.60 *1*	★ GRP.DPL -1.79 *9*			
C2H4I1O1	OCH2CH2I						L-STM 6.30 *1*	B1-STM 1.35 *1*	B5-STM 5.47 *1*			
C2H4I1S1	SCH2CH2I						L-STM 6.41 *1*	B1-STM 1.70 *1*	B5-STM 6.11 *1*			
C2H4N1	1-aziridinyl		★ -0.07 *3*	★ -0.22 *3*	1.35		★ S.INDUC 0.07 *193* S.INDUC 0.08 *292* S.ZE.RS -0.38 *19* S.ZE.RS -0.30 *292*	★ S.ZE.RS -0.29 *193* ★ ES-CH 3.00 *2* L-STM 4.14 *1* B1-STM 1.35 *1*	B5-STM 3.24 *1* F 0.03 *5* R -0.25 *5*			
	2-aziridinyl		★ -0.06 *3*	★ -0.10 *3*	1.19		★ S.INDUC -0.02 *193* ★ S.ZE.RS -0.09 *193* ★ ES-CH 3.00 *2*	L-STM 4.14 *1* B1-STM 1.55 *1* B5-STM 3.24 *1*	F -0.01 *5* R -0.09 *5*			
	NHCH=CH2						★ ES-CH 3.00 *2* L-STM 5.02 *1*	B1-STM 1.35 *1* B5-STM 3.71 *1*				
C2H4N1O1	C(CH3)=NOH		★ 0.49 *293*				★ S.PARA- 0.61 *293* ★ S.PARA+ 0.16 *191*	★ ES-CH 4.00 *2* L-STM 5.05 *1*	B1-STM 1.79 *1* B5-STM 3.11 *1*			
	C=O(NHCH3)	-1.27	★ 0.35 *222*	★ 0.36 *222*	1.46		ES-CH 4.00 *2* L-STM 5.00 *1*	B1-STM 1.54 *1* B5-STM 3.16 *1*	F 0.35 *5* R -0.01 *5*			
	CH2C=O(NH2)	-1.68 -0.60 *130*	0.03 *6* ★ 0.06 *3*	-0.06 *6* ★ 0.07 *210*	1.44		★ S.INDUC 0.06 *6* S.STAR 0.31 *3* ★ S.STAR 0.65 *473* ★ S.ZE.RS -0.12 *6* ES-CH 4.00 *2*	L-STM 4.58 *1* B1-STM 1.52 *1* B5-STM 4.37 *1* F 0.08 *5* R -0.01 *5*	★ S-L 0.06 *7* GRP.DPL -3.75 *9* GR.ELCT 0.16 *13*			
	CH=N(CH3) -O						★ S.INDUC 0.20 *334*	★ S.ZE.RS 0.13 *334*	★ ES-CH 4.00 *2*			
C2H4N1O1.B1F3	CH=N(CH3)O.B1F3						★ S.INDUC 0.45 *334*	★ S.ZE.RS 0.44 *334*	★ ES-CH 4.00 *2*			
C2H4N1O1	CH=NOCH3 -C						L-STM 4.67 *1*	B1-STM 1.60 *1*	B5-STM 5.11 *1*			
	CH=NOCH3 -T						L-STM 6.33 *1*	B1-STM 1.60 *1*	B5-STM 3.97 *1*			
	CH=NOCH3	0.40	★ 0.37 *163*	★ 0.30 *163*	1.57		★ ES-CH 3.00 *2*	F 0.40 *5*	R 0.10 *5*			
	N-methyl-3-oxaziridnyl		★ 0.09 *3*	★ 0.12 *3*			★ S.INDUC 0.07 *193* ★ S.ZE.RS 0.05 *193*	★ ES-CH 4.00 *2* F 0.10 *5*	R 0.02 *5*			
	NHC=O(CH3)	-0.97 -0.22 *54* -0.15 *42* -0.04 *220*	-0.09 *194* 0.09 *399* 0.12 *56* 0.12 *121* 0.13 *10* 0.13 *40* 0.14 *46* 0.16 *6* 0.16 *222* 0.17 *197* ★ 0.21 *21*	-0.09 *56* -0.09 *121* -0.07 *6* -0.07 *222* -0.07 *399* -0.05 *197* 0.00 *46* ★ 0.00 *21* 0.02 *40* 0.08 *10*	1.49	★ 4.00 *2*	S.INDUC 0.19 *117* S.INDUC 0.20 *292* S.INDUC 0.24 *40* S.INDUC 0.24 *89* S.INDUC 0.24 *161* S.INDUC 0.26 *20* S.INDUC 0.28 *6* S.INDUC 0.28 *24* S.INDUC 0.28 *68* S.INDUC 0.28 *119* ★ S.INDUC 0.29 *17*	★ S.PARA- 0.00 *1016* S.PARA+ -0.69 *467* S.PARA+ -0.65 *967* S.PARA+ -0.64 *361* S.PARA+ -0.62 *361* S.PARA+ -0.60 *43* S.PARA+ -0.60 *277* ★ S.PARA+ -0.60 *322* S.PARA+ -0.59 *88* S.PARA+ -0.58 *65* S.PARA+ -0.58 *71*	★ S.ZE.RS -0.23 *68* S.ZE.RS -0.22 *161* S.BA.RS -0.36 *20* S.RES.+ -0.86 *20* S.RES.+ -0.72 *10* S.RES.+ -0.54 *71* S.RES.+ -0.47 *6* L-STM 5.09 *1* B1-STM 1.35 *1* B5-STM 3.61 *1* F 0.31 *5*			

Empirical	Structural	π	σ_m	σ_p	MR	E_s	additional σ values
C2H4N1O1	NHC=O(CH3)						S.INDUC 0.29 *99* · S.PARA+ -0.58 *513* · R -0.31 *5* · S.INDQ 1.65 *58* · S.PARNO 0.14 *65* · R+ -0.91 *4* · ★S.STAR 1.40 *3* · S.ORTH- -0.08 *97* · R- -0.77 *4* · S.STAR 1.62 *10* · S.ZE.RS -0.41 *19* · S-L 0.29 *7* · S.META+ 0.08 *361* · S.ZE.RS -0.35 *6* · K-CHGTR 1.06 *436* · S.META+ 0.13 *361* · S.ZE.RS -0.26 *117* · GRP.DPL -3.81 *9* · S.META- 0.12 *138* · S.ZE.RS -0.26 *292* · GRP.DPL -3.65 *8* · S.ZPTFT -0.09 *306* · S.ZE.RS -0.25 *20* · GR.ELCT 0.50 *13* · S.PARA- -0.46 *963* · S.ZE.RS -0.23 *24* · O-STER -1.93 *90*
C2H4N1O1S1	SCONHCH3						★ES-CH 4.20 *2* · B1-STM 1.70 *1* · L-STM 6.15 *1* · B5-STM 4.99 *1*
C2H4N1O2	CH(CH3)NO2						★S.STAR 1.63 *157*
	CH(NH3+)C(=O)o-						L-STM 4.54 *1* · B1-STM 1.81 *1* · B5-STM 3.64 *1*
	CH2CH2NO2	-0.05	★ 0.00 *3*		1.60		★S.STAR 0.50 *52* · L-STM 5.49 *116* · B5-STM 3.17 *116* · ★ES-CH 3.00 *2* · B1-STM 1.52 *116* · ★GRP.DPL -2.69 *9*
	NHCOCH2OH	-1.42					★ES-CH 4.00 *2*
	NHCOOMe		★ -0.02 *843*	★ -0.17 *843*			L-STM 5.84 *1* · B5-STM 3.99 *1* · R -0.24 *4* · B1-STM 1.45 *1* · F 0.07 *4*
	OCH2CONH2	-1.37			1.60		★ES-CH 3.00 *2*
	OCONHCH3	-0.53 *588* / -0.42 *54*			1.53		★ES-CH 4.00 *2* · B1-STM 1.35 *1* · L-STM 5.51 *1* · B5-STM 4.55 *1*
C2H4N1O2S1	SCH2CH2NO2						L-STM 5.81 *116* · B1-STM 1.70 *116* · B5-STM 5.47 *116*
C2H4N1O3	CH2CH2ONO2						S.STAR 0.49 *3*
	OCH2CH2NO2						L-STM 5.63 *116* · B1-STM 1.35 *116* · B5-STM 4.85 *116*
C2H4N1O3S1	SO2NHCOCH3			★ 1.00 *293*	2.28		★ES-CH 5.20 *2* · B1-STM 2.03 *1* · L-STM 5.67 *1* · B5-STM 4.75 *1*
C2H4N1S1	C=S(NHCH3)		★ 0.30 *222*	★ 0.34 *222*	2.23		★ES-CH 4.20 *2* · B1-STM 1.88 *1* · F 0.29 *5* · L-STM 5.00 *1* · B5-STM 3.18 *1* · R 0.05 *5*
	NHC=S(CH3)	-0.42	★ 0.24 *222*	★ 0.12 *222*	2.34		★ES-CH 4.20 *2* · B5-STM 4.38 *1* · GRP.DPL -4.28 *8* · L-STM 5.09 *1* · F 0.30 *5* · B1-STM 1.45 *1* · R -0.18 *5*
C2H4N1S2	NHC(=S)SCH3						★S.STAR 2.60 *3* · ★ES-CH 4.40 *2* · ★S-L 0.39 *14*
C2H4N3O1	CH=NNHCONH2	-0.89 *57* / -0.86			2.13		L-STM 7.16 *1* · B5-STM 4.66 *1* · B1-STM 1.60 *1*
C2H4N3O2	CONHNHCONH2	-2.63			2.19		
	NHCONHCONH2						★S.INDUC 0.38 *24* · ★ES-CH 4.00 *2* · B1-STM 1.45 *1* · ★S.ZE.RS -0.29 *24* · L-STM 7.44 *1* · B5-STM 4.46 *1*
C2H4N3S1	CH=NNHCSNH2	-0.27	★ 0.45 *163*	★ 0.40 *163*	2.96		★ES-CH 3.00 *2* · B1-STM 1.60 *1* · F 0.46 *5* · L-STM 7.16 *1* · B5-STM 5.41 *1* · R -0.06 *5*
C2H4O1	OCH2CH2* (4,3)						★S.PARA+ -0.98 *775* · ★ES-CH 0.00 *2* · S-P(C+) -2.40 *844*
C2H4O2	OCH2CH2O*		★ -0.12 *439*	★ -0.12 *439*			★ES-CH 2.00 *2* · R -0.04 *4* · F -0.08 *4* · ★S.PHOSP 0.11 *354*
C2H4O2Si1	SiH(Me)CO2H			★ 0.15 *238*			
C2H4O3S1	EtSO3-					★ -1.57 *44*	ES-CH 3.20 *2* · ★S-L -0.04 *7*
C2H4S2	CH2SSCH2*	0.98 *130*					
C2H5	CH2CH3	1.02	-0.50 *128* · -0.08 *46* · ★-0.07 *21* · -0.05 *6* · -0.04 *23*	-0.50 *128* · ★-0.15 *21* · -0.14 *418* · -0.13 *46* · -0.08 *208*	1.03	-1.62 *379* · -1.36 *302* · ★-1.31 *50* · -1.20 *62* · -0.15 *311*	S.INDUC -0.08 *12* · S.PARA+ -0.31 *65* · ES-HYBO -0.07 *225* · S.INDUC -0.05 *261* · ★S.PARA+ -0.30 *43* · L-STM 4.11 *1* · S.INDUC -0.03 *182* · S.PARA+ -0.29 *342* · B1-STM 1.52 *1* · S.INDUC -0.02 *64* · S.PARA+ -0.25 *277* · B5-STM 3.17 *1* · S.INDUC -0.02 *99* · S.PARNO -0.12 *65* · F 0.00 *5* · ★S.INDUC -0.01 *6* · S.ORTHO -0.17 *186* · R -0.15 *5* · S.INDUC 0.07 *16* · S.ORTHO -0.13 *158* · R+ -0.30 *4* · S.PRIME -0.01 *159* · S.ORTHO -0.09 *53* · R- -0.19 *4* · S.INDQ 0.03 *58* · S.ORTHO 0.05 *104* · S-L -0.01 *14* · ★S.STAR -0.10 *52* · S.ORTHO 0.41 *78* · S.PHOSP -1.10 *59* · S.STAR -0.10 *74* · ★S.ORTH- -0.15 *97* · S.ZTWST -0.10 *22* · S.STAR -0.10 *325* · S.ZE.RS -0.14 *64* · S.P.RAD 0.12 *79* · S.META+ -0.06 *326* · S.ZE.RS -0.13 *12* · K-CHGTR 0.13 *77* · S.META- -0.10 *138* · S.ZE.RS -0.13 *193* · GRP.DPL 0.00 *9* · S.ZPTFT -0.16 *112* · ★S.ZE.RS -0.10 *19* · GRP.DPL 0.39 *8* · S.ZPTFT -0.13 *63* · S.ZE.RS -0.10 *47* · GR.ELCT 0.15 *13* · ★S.PARA- -0.19 *138* · ES-CH 2.00 *2* · O-STER -1.08 *90* · S.PARA+ -0.32 *329* · ES-V 0.56 *7*
C2H5Br1N1	NHCH2CH2Br						L-STM 6.05 *1* · B1-STM 1.35 *1* · B5-STM 5.14 *1*
	NHCHBrCH3						L-STM 4.02 *1* · B1-STM 1.35 *1* · B5-STM 4.72 *1*
C2H5Br1N1S1	N=S(Me)(CH2Br)						S.INDUC -0.37 *123*
C2H5Cl1N1	NHCH2CH2Cl						L-STM 5.85 *1* · B1-STM 1.35 *1* · B5-STM 4.85 *1*
	NHCHClCH3						L-STM 4.02 *1* · B1-STM 1.35 *1* · B5-STM 4.44 *1*
C2H5Cl2Sn1	SnCl2(Et)						S.ZE.RS -0.07 *365* · S.ZE.RS 0.11 *564*
C2H5I1N1	NHCH2CH2I						L-STM 6.32 *1* · B1-STM 1.35 *1* · B5-STM 5.53 *1*
C2H5N2O1	CH2NHCONH2						★ES-CH 3.00 *2* · B1-STM 1.52 *1* · ★S-L 0.07 *7* · L-STM 5.58 *1* · B5-STM 4.65 *1*
	NHCONHCH3						★S.INDUC 0.14 *117* · ★ES-CH 4.00 *2* · B1-STM 1.35 *1* · ★S.ZE.RS -0.33 *117* · L-STM 5.89 *1* · B5-STM 3.68 *1*
C2H5N2O2	CH2N(NO2)(CH3)						★ES-CH 4.00 *2* · ★S-L 0.16 *14*
	NHCH2CH2NO2						L-STM 5.66 *116* · B1-STM 1.35 *116* · B5-STM 4.92 *116*
C2H5N2S1	CH2SC=NH(NH2)			★ 0.04 *249*			★ES-CH 3.20 *2* · B1-STM 1.56 *1* · L-STM 6.24 *1* · B5-STM 4.56 *1*

Empirical	Structural	π	σ_m	σ_p	MR	E_s	additional σ values
C2H5N2S1	NHC=S(NHCH3)			★ -0.10 249	2.70		★ES-CH 4.20 2, B1-STM 1.45 1; L-STM 5.89 1, B5-STM 4.38 1
C2H5N4O1	CH=NNHCONHNH2	-1.32	★ 0.22 163	★ 0.16 163	2.42		★ES-CH 3.00 2, B1-STM 1.60 1, F 0.26 5; L-STM 7.57 1, B5-STM 4.55 1, R -0.10 5
	NHCONHC(=NH)NH2	-1.48			2.35		
C2H5O1	CH(OH)(CH3)	0.26 130	0.00 197 ★ 0.08 3	★ -0.14 147 ★ -0.07 82	1.18	★ -1.15 44 -1.04 35	★S.INDUC 0.04 6, ES-CH 3.00 2, F 0.16 5; S.STAR 0.12 3, ES-V 0.50 7, R -0.23 5; S.STAR 0.46 74, L-STM 4.11 1, S-L 0.04 14; ★S.STAR 0.46 594, B1-STM 1.73 1, S.ZTWST -0.11 22; ★S.ZE.RS -0.08 22, B5-STM 3.17 1, GRP.DPL -1.69 9
	CH2CH2OH	-0.77		★ -0.06 147	1.18		★S.INDUC 0.00 34, ★ES-CH 3.00 2, B5-STM 3.38 1; ★S.STAR 0.21 606, ES-V 0.77 7, S-L 0.06 14; ★S.PARA- -0.15 28, L-STM 4.79 1, GRP.DPL -1.66 9; ★S.BA.RS -0.10 34, B1-STM 1.52 1, O-STER -0.86 90
	CH2OCH3	-0.78	-0.10 3 ★ 0.08 6	★ 0.01 6 0.03 210	1.21	★ -1.43 50	S.INDUC -0.01 64, S.STAR 0.68 10, B1-STM 1.52 1; S.INDUC 0.06 16, ★S.PARA+ -0.05 351, B5-STM 3.40 1; ★S.INDUC 0.11 6, S.ZE.RS -0.10 6, F 0.13 5; S.INDUC 0.11 233, S.ZE.RS -0.06 64, R -0.12 5; S.INDQ 0.68 58, S.ZE.RS -0.05 19, R+ -0.18 4; ★S.STAR 0.52 52, S.ZE.RS -0.05 233, S-L 0.11 14; S.STAR 0.52 87, S.ZE.RS 0.04 114, GRP.DPL -1.32 9; S.STAR 0.64 376, ES-CH 3.00 2, GR.ELCT 0.22 13; S.STAR 0.65 847, ★ES-V 0.63 7; S.STAR 0.66 155, L-STM 4.78 1
	OCH2CH3	0.03 57 0.38 0.51 455	-0.60 128 0.07 6 ★ 0.10 21 0.13 508	-1.50 128 -0.29 6 -0.25 23 -0.25 508 ★ -0.24 21	1.25		S.INDUC 0.23 12, S.ORTHO 0.02 104, R+ -1.07 4; ★S.INDUC 0.28 6, ★S.ORTH- -0.30 97, R- -0.54 4; S.INDUC 0.63 16, S.ZE.RS -0.57 6, S-L 0.29 7; S.STAR 1.64 37, S.ZE.RS -0.45 260, ★S.PHOSP -0.21 59; ★S.STAR 1.68 3, S.ZE.RS -0.44 12, S.IND.P 1.57 39; S.META- 0.12 28, ★S.ZE.RS -0.44 19, S.RES.P -0.62 39; S.ZPTFT -0.20 112, S.RES.+ -0.65 6, S.ZTWST -0.23 22; ★S.PARA- -0.28 28, ES-CH 3.00 2, K-CHGTR 0.39 77; ★S.PARA+ -0.81 483, ES-V 0.48 7, GRP.DPL -1.38 8; S.PARA+ -0.81 485, L-STM 4.80 1, GRP.DPL -1.27 9; S.PARA+ -0.72 979, B1-STM 1.35 1, GR.ELCT 0.82 13; S.PARNO -0.14 65, B5-STM 3.36 1, O-STER -1.36 90; S.ORTHO -0.08 53, F 0.26 5; S.ORTHO -0.01 78, R -0.50 5
C2H5O1S1	CH2S(O)CH3						★S.STAR 1.33 155
	OCH2SCH3						L-STM 6.24 1, B1-STM 1.35 1, B5-STM 4.67 1
	S(O)Et						S.ZE.RS 0.07 260, B1-STM 1.40 1; L-STM 4.92 1, B5-STM 3.49 1
	SCH(OH)CH3						L-STM 4.30 1, B1-STM 1.70 1, B5-STM 4.72 1
	SCH2CH2OH						L-STM 6.07 1, B1-STM 1.70 1, B5-STM 4.64 1
	SCH2OCH3						L-STM 6.09 1, B1-STM 1.70 1, B5-STM 4.87 1
C2H5O2	CH(OH)CH2OH				1.34	★ -2.05 44	ES-CH 4.00 2, B1-STM 1.73 1; L-STM 4.79 1, B5-STM 3.38 1
	OCH(OH)CH3						L-STM 3.98 1, B1-STM 1.35 1, B5-STM 4.35 1
	OCH2CH2OH	-0.97					★S.INDUC 0.27 49, ★ES-V 0.46 7, B5-STM 4.02 1; ★S.ZE.RS -0.55 49, L-STM 5.90 1; ★ES-CH 3.00 2, B1-STM 1.35 1
	OCH2och3						L-STM 5.91 1, B1-STM 1.35 1, B5-STM 4.32 1
C2H5O2S1	CH2SO2CH3		★ 0.18 3		1.81		★S.STAR 1.32 52, L-STM 4.92 1, ★GRP.DPL -4.40 9; S.STAR 1.38 155, B1-STM 1.52 1; ★ES-CH 5.20 2, B5-STM 3.78 1
	SO2Et		★ 0.66 3	★ 0.77 3	1.81		S.INDUC 0.50 31, S.ZE.RS 0.22 260, F 0.59 5; ★S.INDUC 0.60 32, ★S.BA.RS 0.13 34, R 0.18 5; S.INDUC 0.61 99, ★ES-CH 5.20 2, ★S-L 0.59 14; S.INDUC 0.63 89, L-STM 4.92 1, GRP.DPL -3.48 8; S.STAR 3.74 3, B1-STM 2.03 1; S.ZE.RS 0.18 31, B5-STM 3.49 1
C2H5O3S1	CH2CH2SO3H						★S.META+ 0.28 301, ★S.PARA+ -0.05 301, ★ES-CH 3.20 2
	OCH2SO2CH3						L-STM 6.05 1, B1-STM 1.35 1, B5-STM 4.42 1
	SO2CH2OCH3						L-STM 6.03 1, B1-STM 2.03 1, B5-STM 4.44 1
	SO2OCH2CH3	-0.90		★ 0.90 38			★ES-CH 5.20 2, ★GRP.DPL -4.99 8, GR.ELCT 0.30 13
C2H5S1	CH2SCH3				1.84	-1.68 44 ★ -1.58 50	S.INDUC 0.14 233, ★S.ZE.RS 0.10 552, B1-STM 1.52 1; ★S.STAR 0.42 155, ES-CH 3.20 2, B5-STM 3.53 1; S.STAR 0.59 10, ★ES-V 0.70 7, S-L 0.12 14; S.ZE.RS -0.08 233, L-STM 5.37 1, GRP.DPL 1.46 8
	S-Et	1.07	0.16 6 ★ 0.18 484 0.23 3	-0.04 6 ★ 0.03 21	1.84		S.INDUC 0.13 31, ★S.ZE.RS -0.16 260, R -0.23 5; ★S.INDUC 0.25 32, S.ZE.RS -0.11 31, ★S.PHOSP 0.09 59; S.INDUC 0.26 6, ES-CH 3.20 2, S.IND.P 1.48 39; ★S.STAR 1.44 37, ES-V 0.94 7, S.RES.P -0.23 39; S.STAR 1.56 3, L-STM 5.16 1, S.ZTWST -0.05 22; S.ZE.RS -0.30 6, B1-STM 1.70 1, GRP.DPL -4.08 8; S.ZE.RS -0.25 22, B5-STM 3.97 1, GR.ELCT 0.16 13; S.ZE.RS -0.19 19, F 0.26 5
C2H5S2	SCH2SCH3						L-STM 6.36 1, B1-STM 1.70 1, B5-STM 5.25 1
C2H5Se1	SeCH2CH3						S.ZE.RS -0.12 260, B1-STM 1.85 1; L-STM 5.37 1, B5-STM 5.19 1
C2H5Si1	SiH2(CH=CH2)			★ 0.21 238			

Empirical	Structural	π	σ_m	σ_p	MR	E_s	additional σ values					
C2H5Te1	TeCH2CH3						S.ZE.RS	-0.09 260				
C2H6Al1	Al(Me)2						GR.ELCT	-0.42 13				
C2H6As1	As(CH3)2						GR.ELCT	0.01 13				
C2H6B1O2	B(OMe)2						★S.INDUC -0.27 217	S.PARA+ 0.26 217				
							★S.STAR 0.62 358	★S.ZE.RS 0.12 217				
C2H6Br1Si1	Si(Br)(CH3)2			★ 0.10 3			★S.PARA+ 0.10 136	★ES-CH 4.50 2				
C2H6Br1Sn1	SnBr(CH3)2						★S.STAR -0.39 61	★ES-CH 5.00 2				
C2H6Cl1D1N1	+ND(CH3)2Cl-						★S.ZE.RS -0.14 19	★ES-CH 3.00 2				
C2H6Cl1N1P1	P(Cl)N(CH3)2		★ 0.38 41	★ 0.56 41	2.70		★ES-CH 5.40 2	B1-STM 1.49 1	F 0.31 5			
							L-STM 5.08 1	B5-STM 4.25 1	R 0.25 5			
C2H6Cl1N2S1	CH2C(NH2)2+Cl-		★ 0.13 121	★ 0.15 121			F 0.14 4	R 0.01 4				
	CH2SC(NH2)2+Cl-		★ 0.13 121	★ 0.15 121			★ES-CH 3.20 2	F 0.11 66	R 0.05 66			
C2H6Cl1Si1	Si(Cl)(CH3)2		★ 0.16 3	★ 0.21 3			S.INDUC 0.11 73	★S.PARA+ 0.02 136	F 0.16 5			
				0.30 153			S.STAR 0.57 216	★S.ZE.RS 0.10 73	R 0.05 5			
							★S.META+ 0.07 136	★ES-CH 4.40 2	R+ -0.14 4			
C2H6Cl1Sn1	SnCl(CH3)2						★S.STAR -0.32 61	★ES-CH 4.90 2				
C2H6F1Si1	Si(F)(CH3)2		★ 0.12 3	0.16 221	2.01		S.INDUC 0.08 73	★ES-CH 4.00 2	R+ 0.00 4			
			0.15 221	★ 0.17 3			★S.PARA+ 0.17 136	F 0.13 5				
				0.23 153			★S.ZE.RS 0.11 73	R 0.04 5				
C2H6Ga1	Ga(CH3)2						S.INDUC -0.35 16	GR.ELCT -0.27 13				
C2H6I1Sn1	SnI(CH3)2						★S.STAR -0.39 61	★ES-CH 5.40 2				
C2H6In1	In(CH3)2						GR.ELCT -0.39 13					
C2H6N1	CH2CH2NH2	-0.76 42		★ -0.06 147			S.PARA- 0.05 257	★ES-CH 3.00 2				
	CH2NHCH3	-0.26 42					★ES-CH 3.00 2	B1-STM 1.52 1				
							L-STM 4.83 1	B5-STM 3.42 1				
	N(CH3)2	-0.30 57	-0.21 17	-1.25 661	1.55		S.INDUC -0.05 12	S.PARA+ -1.95 88	ES-CH 3.00 2			
		0.18	-0.21 23	-0.83 437			★S.INDUC 0.06 20	S.PARA+ -1.74 483	ES-V 0.43 7			
		0.69 42	-0.16 6	★ -0.83 21			S.INDUC 0.06 26	★S.PARA+ -1.70 43	L-STM 3.53 1			
			★ -0.16 459	-0.82 17			S.INDUC 0.06 205	S.PARA+ -1.70 243	B1-STM 2.56 1			
			-0.15 977	-0.77 413			S.INDUC 0.10 24	S.PARA+ -1.70 308	B5-STM 1.35 1			
			-0.14 413	-0.72 459			S.INDUC 0.10 142	S.PARA+ -1.70 437	F 0.15 5			
			-0.12 10	-0.71 6			S.INDUC 0.10 292	S.PARA+ -1.67 65	R -0.98 5			
			-0.10 46	-0.70 475			S.INDUC 0.11 17	S.PARA+ -1.67 322	R+ -1.85 4			
			-0.09 26	-0.69 812			S.INDUC 0.17 6	S.PARA+ -1.50 467	R- -0.27 4			
				-0.63 236			S.INDUC 0.18 119	S.PARNO -0.24 65	S-L 0.17 14			
				-0.60 23			S.INDQ 1.05 58	S.ORTHO -0.36 53	★S.PHOSP -1.22 59			
				-0.44 10			S.STAR 0.32 3	S.ORTHO 0.30 186	S.IND.P 0.27 434			
				-0.32 46			S.STAR 0.62 37	★S.ORTH- -0.36 97	S.IND.P 0.80 39			
				-0.12 208			S.STAR 1.02 10	S.ZE.RS -0.88 6	S.RES.P -1.49 434			
							S.STAR 1.16 170	S.ZE.RS -0.62 12	S.RES.P -0.79 39			
							S.META- -0.04 416	S.ZE.RS -0.55 26	S.ZTWST -0.13 22			
							S.META- -0.02 138	S.ZE.RS -0.55 67	S.P.RAD 0.24 206			
							★S.META- 0.04 290	S.ZE.RS -0.54 24	S.P.RAD 0.61 255			
							S.ZPTFT -0.48 63	S.ZE.RS -0.54 292	K-CHGTR 0.90 77			
							★S.ZPTFT -0.48 112	★S.ZE.RS -0.53 19	GRP.DPL -1.26 9			
							S.ZPTFT -0.46 337	S.BA.RS -0.83 20	GRP.DPL 1.61 8			
							S.PARA- -0.32 138	S.AN.RS -0.34 20	GR.ELCT 0.48 13			
							S.PARA- -0.16 26	S.RES.+ -1.75 20	O-STER -2.32 90			
							★S.PARA- -0.12 290	S.RES.+ -1.45 10				
							S.PARA- -0.12 416	S.RES.+ -1.22 6				
	NH-Et	0.08	★ -0.24 23	★ -0.61 213	1.50		S.INDUC 0.03 12	★S.ZE.RS -0.52 19	B5-STM 3.42 1			
							S.STAR -0.62 3	ES-V 0.59 7	F -0.04 5			
							S.STAR 1.22 170	L-STM 4.83 1	R -0.57 5			
							S.ZE.RS -0.60 12	B1-STM 1.35 1	K-CHGTR 0.79 77			
C2H6N1O1	N(O)(CH3)2						★S.INDUC 0.54 69	★ES-CH 4.00 2				
							★S.ZE.RS 0.00 69	★S-L 0.58 14				
	NHCH(OH)CH3						L-STM 4.02 1	B1-STM 1.35 1	B5-STM 4.39 1			
	NHCH2CH2OH						★S.STAR 1.32 170	B1-STM 1.35 1				
							L-STM 5.93 1	B5-STM 4.08 1				
	NHCH2OCH3						L-STM 5.94 1	B1-STM 1.35 1	B5-STM 4.37 1			
	OCH2NHCH3						L-STM 5.95 1	B1-STM 1.35 1	B5-STM 4.35 1			
C2H6N1O1S1	S(O)N(CH3)2				2.21		★S.INDUC 0.30 31	★ES-CH 5.20 2	B5-STM 4.08 1			
							★S.STAR 1.87 3	L-STM 4.83 1				
							★S.ZE.RS 0.03 31	B1-STM 1.40 1				
C2H6N1O2S1	CH2NHSO2CH3			★ -0.01 147			★ES-CH 3.20 2					
	N(CH3)SO2CH3		★ 0.21 10	★ 0.24 3	2.34		★S.INDUC 0.34 68	L-STM 4.83 1	R 0.03 4			
			0.29 3				★S.STAR 2.10 3	B1-STM 1.35 1	★GRP.DPL -4.71 9			
							★S.ZE.RS -0.10 68	B5-STM 3.72 1				
							★ES-CH 6.20 2	F 0.21 4				
	NHCH2SO2CH3						L-STM 6.07 1	B1-STM 1.35 1	B5-STM 4.47 1			
	NHSO2C2H5						L-STM 6.07 1	B1-STM 1.35 1	B5-STM 4.47 1			
	SO2CH2NHCH3						L-STM 6.07 1	B1-STM 2.03 1	B5-STM 4.47 1			
	SO2N(CH3)2	-0.78	★ 0.51 3	0.63 304	2.19		★S.INDUC 0.42 31	★S.ZE.RS 0.12 31	R 0.21 5			
		0.03 42	0.53 578	★ 0.65 3			★S.STAR 2.62 3	ES-CH 6.20 2	R+ 0.42 4			
				0.70 578			S.META+ 0.68 578	L-STM 4.83 1	R- 0.55 4			
							S.PARA- 0.84 258	B1-STM 2.03 1	S-CNTIG 0.84 139			
							★S.PARA+ 0.99 886	B5-STM 4.08 1	K-CHGTR 1.24 77			
							★S.PARA+ 0.86 578	F 0.44 66	GRP.DPL -4.71 9			

Empirical	Structural	π	σm	σp	MR	Es	additional σ values
C2H6N1O4S2	N(SO2CH3)2	-1.51	★ 0.47 *3*	★ 0.49 *3*	3.12		S.INDUC 0.45 *24*; ★S.ZE.RS 0.04 *68*; B5-STM 3.72 *1*; ★S.INDUC 0.45 *68*; ★ES-CH 9.40 *2*; F 0.45 *4*; ★S.STAR 2.80 *3*; L-STM 4.83 *1*; R 0.04 *4*; S.ZE.RS 0.04 *24*; B1-STM 1.36 *1*
C2H6N1S1	N=S(CH3)2						S.INDUC -0.50 *123*
	NHCH2SCH3						L-STM 6.26 *1*; B1-STM 1.35 *1*; B5-STM 4.73 *1*
	S-N(CH3)2		★ 0.12 *3*	★ 0.09 *3*			★S.INDUC 0.15 *31*; ★S.ZE.RS -0.06 *31*; F 0.15 *5*; ★S.STAR 0.94 *3*; ★ES-CH 4.20 *2*; R -0.06 *5*
	SCH2NHCH3						L-STM 6.12 *1*; B1-STM 1.70 *1*; B5-STM 4.91 *1*
C2H6N3	3,4-N+(CH3)-N(CH3)-N- *		★ 1.17 *411*	★ 1.17 *411*			★ES-CH 0.00 *2*
	CH2NHC(=NH)NH2						★S.INDUC -0.05 *117*; ★S.ZE.RS -0.11 *117*; ★ES-CH 3.00 *2*
	N=NN(CH3)2	0.46	★ -0.05 *571*	★ -0.03 *571*	2.09		★S.INDUC 0.04 *571*; L-STM 5.68 *1*; F -0.01 *4*; ★S.PARA+ -0.46 *991*; B1-STM 1.77 *1*; R -0.01 *4*; ★S.ZE.RS -0.17 *571*; B5-STM 3.90 *1*; R+ -0.43 *4*
	NHC(=NCH3)NH2			★ -0.04 *249*			ES-CH 3.00 *2*; B1-STM 1.72 *1*; L-STM 5.94 *1*; B5-STM 4.01 *1*
C2H6N3O1	N(O)=NN(CH3)2	-0.33			2.33		★ES-CH 4.00 *2*
C2H6O1	O+(CH3)2						GR.ELCT 1.32 *13*
C2H6O1P1	P(O)(CH3)2		0.36 *107*; 0.42 *105*; ★0.43 *3*; 0.46 *141*; 0.51 *10*	0.47 *141*; 0.48 *107*; ★0.50 *3*; 0.62 *111*	1.99		S.INDUC 0.17 *69*; S.PARA- 0.62 *328*; B1-STM 2.39 *1*; ★S.INDUC 0.31 *45*; S.PARA- 0.72 *373*; B5-STM 3.32 *1*; S.INDUC 0.36 *141*; ★S.PARA- 0.74 *235*; F 0.40 *5*; S.INDUC 0.45 *107*; S.ZE.RS 0.03 *107*; R 0.10 *5*; S.INDUC 0.75 *654*; ★S.ZE.RS 0.10 *69*; R- 0.34 *4*; S.STAR 1.44 *10*; S.ZE.RS 0.11 *141*; S-L 0.30 *14*; S.STAR 2.81 *3*; S.RES.+ 0.00 *10*; GRP.DPL -4.39 *8*; ★S.META- 0.42 *328*; ★ES-CH 4.20 *2*; GRP.DPL -4.20 *9*; S.PARA- 0.60 *336*; L-STM 3.88 *1*
C2H6O2P1	P(OCH3)2				2.31		S.INDUC 0.09 *468*; L-STM 5.04 *1*; B5-STM 3.25 *1*; ★ES-CH 5.20 *2*; B1-STM 1.40 *1*
C2H6O3P1	P(O)(OCH3)2	-1.18	0.27 *105*; ★0.42 *207*; 0.43 *373*; 0.45 *6*	★0.53 *207*; 0.55 *3*; 0.57 *6*; 0.65 *111*	2.19		S.INDUC 0.36 *6*; L-STM 5.04 *1*; R 0.16 *5*; ★S.INDUC 0.36 *45*; B1-STM 2.42 *1*; S.ZE.RS 0.21 *6*; B5-STM 3.25 *1*; ES-CH 6.20 *2*; F 0.37 *5*
C2H6O3P1S1	OP(S)(OMe)2						S.INDUC 0.38 *395*; S.PARA+ -0.36 *395*; S.ZE.RS -0.25 *395*
C2H6O4P1	OP(O)(OMe)2	-0.57 *54*		★ 0.04 *143*	2.20		★S.PARA+ -0.21 *143*; ★ES-CH 5.20 *2*; B1-STM 1.35 *1*; S.PARA+ -0.21 *395*; L-STM 4.91 *1*; B5-STM 5.53 *1*
C2H6P1	P(CH3)2	0.44	0.03 *105*; ★0.03 *150*; 0.05 *3*; 0.10 *515*; 0.12 *10*; 0.15 *41*	0.03 *3*; ★0.06 *150*; 0.13 *515*; 0.18 *10*; 0.23 *111*; 0.31 *41*	2.12		S.INDUC 0.00 *150*; S.ZE.RS -0.08 *482*; F 0.05 *5*; S.INDUC 0.02 *69*; S.ZE.RS -0.02 *69*; R 0.01 *5*; ★S.INDUC 0.08 *468*; ★S.ZE.RS 0.06 *150*; R- 0.17 *4*; S.META- 0.09 *328*; ES-CH 3.20 *2*; GRP.DPL -1.31 *8*; S.PARA- 0.18 *336*; L-STM 3.88 *1*; GR.ELCT -0.05 *13*; ★S.PARA- 0.22 *328*; B1-STM 2.00 *1*; S.ZE.RS -0.08 *86*; B5-STM 3.32 *1*
C2H6P1S1	P(S)(CH3)2		0.43 *10*; ★0.43 *902*				S.INDUC 0.23 *69*; S.ZE.RS 0.10 *69*; B1-STM 2.76 *1*; ★S.META- 0.43 *328*; ★ES-CH 4.40 *2*; B5-STM 3.59 *1*; ★S.PARA- 0.62 *328*; L-STM 4.37 *1*
C2H6S1	+S(CH3)2		0.68 *36*; 1.00 *517*; ★1.00 *21*; 1.10 *510*; 1.14 *29*	0.30 *153*; 0.77 *36*; 0.90 *517*; ★0.90 *21*; 1.13 *510*; 1.18 *29*			S.INDUC 0.69 *126*; ★S.PARA- 0.83 *372*; ★ES-CH 3.20 *2*; ★S.INDUC 0.89 *31*; S.PARA- 1.02 *36*; L-STM 3.62 *1*; S.INDUC 0.90 *6*; S.PARA- 1.16 *426*; B1-STM 1.70 *1*; S-INDQ 4.50 *58*; S.PARA- 1.16 *613*; B5-STM 3.26 *1*; ★S.STAR 5.09 *3*; S.PARA- 1.25 *29*; F 0.98 *5*; S.META- 0.71 *461*; S.ZE.RS -0.09 *86*; R -0.08 *5*; S.META- 0.83 *36*; S.ZE.RS 0.08 *126*; R- -0.15 *4*; ★S.META- 1.00 *378*; S.ZE.RS 0.16 *333*; GR.ELCT 0.47 *13*; S.META- 1.00 *426*; ★S.ZE.RS 0.17 *31*; S.CH-P+ 1.02 *75*; S.META- 1.08 *29*; S.ZE.RS 0.24 *6*
C2H6S1.C7H7O3S1	S+(CH3)2TOS(-)		0.59 *36*; ★1.06 *29*	0.55 *36*; ★0.96 *29*			★ES-CH 3.20 *2*
C2H6Sb1	Sb(CH3)2						GR.ELCT -0.13 *13*
C2H7N1	(CH2)2NH3+		★ 0.23 *3*	0.01 *36*; ★0.17 *3*; 0.34 *29*		★ -3.06 *44*	S.INDUC 0.19 *310*; ES-CH 3.00 *2*; B5-STM 3.49 *1*; ★S.ORTHO 0.25 *158*; L-STM 4.92 *1*; F 0.27 *4*; ★S.ORTHO 0.28 *53*; B1-STM 1.52 *1*; R -0.10 *4*
	+NH(CH3)2		★ 0.84 *3*				S.INDUC 0.68 *310*; ★S.STAR 2.98 *87*; ★ES-CH 3.00 *2*; ★S.INDUC 0.70 *32*; S.STAR 3.14 *424*; L-STM 3.53 *1*; S.INDUC 0.90 *99*; S.STAR 4.17 *424*; B1-STM 1.86 *1*; S.INDUC 1.03 *119*; S.STAR 4.21 *360*; B5-STM 3.08 *1*; S.INDUC 1.25 *210*; ★S.STAR 4.36 *3*
	+NH2Et		★ 0.96 *23*				★S.INDUC 0.60 *32*; ★ES-CH 3.00 *2*; B1-STM 1.49 *1*; ★S.STAR 3.74 *3*; L-STM 4.83 *1*; B5-STM 3.42 *1*
	CH2NH2CH3+			0.20 *36*; ★0.60 *29*			★S.STAR 0.82 *87*; ★ES-CH 3.00 *2*; ★S.ORTHO 0.39 *158*
C2H7N1O1	N+OH(CH3)2			★ 0.80 *905*			★ES-CH 4.00 *2*
C2H7N1O3P1	OP(O)(OCH3)(NHCH3)	-0.61 *54*					L-STM 4.91 *1*; B1-STM 1.35 *1*; B5-STM 5.59 *1*
C2H7N2	NHCH2CH2NH2						★S.INDUC 0.20 *49*; ★S.ZE.RS -0.58 *49*; ★ES-CH 3.00 *2*
	NHCH2NHCH3						L-STM 5.98 *1*; B1-STM 1.35 *1*; B5-STM 4.41 *1*
	NHN(CH3)2						L-STM 4.73 *1*; B1-STM 1.35 *1*; B5-STM 4.01 *1*
C2H7N3	CH2NHC(=NH2+)NH2						★S.INDUC 0.08 *117*; ★S.ZE.RS -0.07 *117*; ★ES-CH 3.00 *2*
C2H7Si1	Si(CH3)2H		★ 0.01 *3*	★0.04 *3*; 0.05 *153*	2.03		S.INDUC -0.01 *73*; S.ZE.RS 0.06 *73*; F 0.03 *4*; ★S.INDUC 0.07 *359*; ★ES-CH 3.20 *2*; R 0.01 *4*

Empirical	Structural	π	σm	σp	MR	Es	additional σ values
C2H7Si1	Si(CH3)2H			0.07 60			S.STAR -0.60 393, ★S.PARA+ -0.04 60, ★S.ZE.RS 0.04 359; L-STM 4.09 1, B1-STM 2.11 1, B5-STM 3.47 1; R+ -0.07 4, GR.ELCT -0.23 13
	SiH2(CH2CH3)						S.STAR -1.50 393
C3Br3N2	3,4,5-tri-Br-1-pyrazyl						★S.ZE.RS -0.02 15
C3Cl2N3	2-(4,6-Cl2-S-triazinyl)						★S.PARA- 0.85 346, ★ES-CH 3.00 2
C3Cl5O1	CCl2COCCl3						★S.STAR 2.96 171, ★ES-CH 6.40 2
C3Cr1O3	Cr(CO)3						S.INDUC 0.27 840, S.ZE.RS 0.01 840; ★S.PARA+ 0.08 782, ★ES-CH 7.30 2
C3F3	CCCF3		★ 0.41 132	★ 0.51 132	1.41		★S.INDUC 0.35 132, ES-CH 3.00 2, F 0.37 5; S.STAR 1.94 3, L-STM 5.90 1, R 0.14 5; S.ZE.RS 0.17 30, B1-STM 1.99 1, GRP.DPL -3.38 8; S.BA.RS 0.17 18, B5-STM 2.61 1
C3F5	CF=CFCF3 -E		0.15 3; ★ 0.39 30	0.17 3; ★ 0.46 30			★S.INDUC 0.32 719, L-STM 5.53 1, R 0.10 4; ★S.PARA- 0.65 160, B1-STM 2.02 1, R- 0.29 4; ★S.ZE.RS 0.14 30, B5-STM 3.71 1; ★ES-CH 4.60 2, F 0.36 4
C3F6N1	N=C(CF3)2		★ 0.29 30	★ 0.23 30			★S.INDUC 0.35 30, ★S.ZE.RS -0.12 30, F 0.32 4; ★S.STAR 2.20 3, ★ES-CH 4.00 2, R -0.09 4
C3F7	(CF2)2CF3		★ 0.44 6	★ 0.48 6			★S.INDUC 0.39 6, ★ES-CH 6.20 2, ★S-L 0.39 14; S.STAR 2.83 566, F 0.42 4; ★S.ZE.RS 0.09 6, R 0.06 4
	C(F)(CF3)2		★ 0.37 156; 0.50 156; 0.55 10	0.52 156; ★ 0.53 156	1.34		★S.INDUC 0.48 156, ★S.ZE.RS 0.04 224, B1-STM 2.45 1; S.STAR 3.00 3, S.BA.RS 0.02 80, B5-STM 3.64 1; S.META- 0.52 165, S.AN-RS 0.12 80, F 0.31 5; ★S.PARA- 0.68 558, S.RES.+ 0.00 10, R 0.22 4; S.ZE.RS 0.02 19, ES-CH 8.60 2, R- 0.37 4; S.ZE.RS 0.02 80, L-STM 4.11 1, GRP.DPL -2.68 8
C3F7O2S1	SO2(CF2)2CF3		0.83 176; ★ 0.92 48	0.99 176; ★ 1.09 48			★S.INDUC 0.76 48, ★S.ZE.RS 0.33 48, R- 0.94 4; S.META- 0.98 164, F 0.81 4; ★S.PARA- 1.75 164, R 0.28 4
	SO2CF(CF3)2		★ 0.92 48; 0.98 176	1.01 176; ★ 1.10 48			★S.INDUC 0.77 48, ★S.ZE.RS 0.33 48, R- 0.96 4; S.META- 0.98 164, F 0.80 4; ★S.PARA- 1.76 164, R 0.30 4
C3F7S1	S(CF2)2CF3		0.30 176; ★ 0.45 48	0.47 176; ★ 0.48 48			★S.INDUC 0.41 48, ★S.ZE.RS 0.08 48, R- 0.22 4; S.META- 0.46 164, F 0.43 4; ★S.PARA- 0.65 164, R 0.05 4
	SCF(CF3)2		0.39 176; ★ 0.48 48	★ 0.51 48; 0.53 176			★S.INDUC 0.43 48, ★S.ZE.RS 0.09 48, R- 0.23 4; S.META- 0.49 164, F 0.46 4; ★S.PARA- 0.69 164, R 0.03 4
C3F7Te1	TeCF2CF2CF3		★ 0.46 561	★ 0.48 561			★S.INDUC 0.43 561, F 0.45 4; ★S.ZE.RS 0.06 561, R 0.03 4
C3H1Br2N2	3,4-diBr-1-pyrazyl						★S.ZE.RS -0.06 15
	3,5-diBr-1-pyrazyl						★S.ZE.RS -0.04 15
	4,5-diBr-1-pyrazyl						★S.ZE.RS -0.02 15
C3H1Cl2N2	3,5-diCl-1-pyrazyl						★S.INDUC 0.20 15, ★S.ZE.RS -0.03 15
C3H1Cl4O1	CCl2COCHCl2						★S.STAR 2.65 171, ★ES-CH 6.40 2
	CHClCOCCl3						★S.STAR 1.88 171, ★ES-CH 5.20 2
C3H1F6	CH(CF3)2						S.STAR 1.32 3, B1-STM 1.70 1; L-STM 5.54 1, B5-STM 4.74 1
C3H1F6O1	C(OH)(CF3)2	1.28	★ 0.29 156; 0.31 10; 0.35 3	0.30 3; ★ 0.30 156	1.52		★S.INDUC 0.28 156, ★S.ZE.RS 0.00 80, B1-STM 2.61 1; S.INDUC 0.35 80, S.ZE.RS 0.02 156, B5-STM 3.64 1; S.INDUC 0.38 30, S.ZE.RS 0.11 19, F 0.29 5; S.STAR 1.75 3, S.AN-RS 0.00 80, R 0.01 5; S.META- 0.35 30, S.RES.+ 0.00 10, R- 0.19 4; S.PARA- 0.38 30, ES-CH 8.80 2, GRP.DPL -1.71 8; ★S.PARA- 0.48 558, L-STM 4.11 1
	OCF2CHFCF3						★S.INDUC 0.50 551, ★S.ZE.RS -0.18 551, ★ES-CH 4.60 2
C3H1F6O2S1	SO2CF2CHFCF3			★ 1.03 18			★ES-CH 6.80 2, B1-STM 2.03 1; L-STM 6.76 1, B5-STM 5.05 1
C3H1F6S1	SCF2CHFCF3						★S.INDUC 0.42 551, ★S.ZE.RS 0.06 551, ★ES-CH 4.80 2
C3H1F6S2	CH(SCF3)2		★ 0.44 3	★ 0.44 3	3.05		★S.INDUC 0.44 64, S.ZE.RS -0.01 110, F 0.43 4; S.INDUC 0.44 110, ★S.ZE.RS 0.00 64, R 0.01 4; ★S.STAR 2.75 3, ★ES-CH 5.40 2
C3H1N2	CH(CN)2		★ 0.53 3	★ 0.52 3	1.43		★S.INDUC 0.55 64, L-STM 3.99 1, R 0.00 5; ★S.STAR 3.40 3, B1-STM 1.85 1, GR.ELCT 0.25 13; ★S.ZE.RS -0.03 64, B5-STM 4.12 1; ★ES-CH 5.00 2, F 0.52 5
C3H1N4	NHN=C(CN)2	0.54			2.22		
C3H1O1	C=O(CCH)						S.INDUC 0.33 12, S.ZE.RS 0.29 12
C3H2Br1N2	3-Br-1-pyrazyl						★S.ZE.RS -0.15 15
	5-Br-1-pyrazyl						★S.INDUC 0.15 15, ★S.ZE.RS -0.03 15
C3H2Cl1	C=-CCH2Cl						★S.STAR 1.02 183, L-STM 6.49 1, B5-STM 3.46 1; ★ES-CH 3.00 2, B1-STM 1.60 1
C3H2Cl1N2	3-Cl-1-pyrazyl						★S.INDUC 0.33 15, ★S.ZE.RS -0.16 15
	4-Cl-1-pyrazyl						★S.INDUC 0.28 15, ★S.ZE.RS -0.14 15
	5-Cl-1-pyrazyl						★S.INDUC 0.10 15, ★S.ZE.RS -0.02 15
C3H2Cl3	CH=CHCCl3 -E				3.01		★S.STAR 1.19 326, L-STM 6.12 1, B5-STM 4.55 1

Empirical	Structural	π	σ_m	σ_p	MR	E_s	additional σ values					
C3H2Cl3	CH=CHCCl3 -E						★ES-CH	3.00 2	B1-STM	1.96 1		
C3H2Cl3O1	CCl2COCH2Cl						★S.STAR	2.56 171	★ES-CH	6.40 2		
	CHClCOCHCl2						★S.STAR	1.66 171	★ES-CH	5.20 2		
C3H2Cl5	CCl2CH2CCl3						★S.STAR	2.16 171				
C3H2F3	C(CF3)=CH2						GRP.DPL	-2.25 8				
	CH=CHCF3 -C		★ 0.16 40	★ 0.17 40			★S.INDUC	0.14 40	S.ZE.RS	0.05 18	F	0.18 5
							S.INDUC	0.16 30	S.ZE.RS	0.14 30	R	0.01 5
							★S.ZPTFT	0.27 248	L-STM	4.29 1	R-	0.11 4
							★S.PARA-	0.29 165	B1-STM	1.60 1	GRP.DPL	-2.79 8
							S.PARA-	0.34 160	B5-STM	4.70 1		
	CH=CHCF3 -T		0.20 3	0.20 3	1.56		★S.INDUC	0.20 40	S.ZE.RS	0.09 18	B5-STM	3.71 1
			0.22 132	0.20 132			S.INDUC	0.22 132	S.BA.RS	-0.03 18	F	0.24 5
			★ 0.24 40	0.23 370			S.META-	0.20 165	ES-CH	3.00 2	R	0.03 5
				★ 0.27 40			★S.PARA-	0.34 165	L-STM	5.53 1	R-	0.10 4
							★S.ZE.RS	0.07 30	B1-STM	1.75 1		
C3H2F3O2S1	CH=CHSO2CF3		★ 0.31 3	★ 0.55 3	2.28		S.INDUC	0.31 18	S.ZE.RS	0.25 18	B5-STM	4.19 1
							★S.INDUC	0.34 719	ES-CH	3.20 2	F	0.22 5
							★S.STAR	1.94 3	L-STM	6.35 1	R	0.33 5
							★S.PARA-	0.83 160	B1-STM	1.77 1	R-	0.61 4
C3H2N1	CH=CHCN -T						L-STM	6.46 1	B1-STM	1.60 1	B5-STM	3.06 1
	CH=CHCN	-0.17	★ 0.24 180	★ 0.17 180	1.62		★S.ZPTFT	0.39 248	F	0.28 5	GRP.DPL	-4.12 820
							★ES-CH	3.00 2	R	-0.11 5	GRP.DPL	-3.54 8
C3H2N1O1S1	2-(5-H-4-oxothiazolyl)						★S.INDUC	0.46 96				
C3H2N1S1	2-thiazolyl						GRP.DPL	-1.21 8				
	4-thiazolyl						GRP.DPL	-1.33 8				
	5-thiazolyl						GRP.DPL	-1.89 8				
C3H2N1S1@	2-thiazolyl						★S.ORTH+	0.26 944	★S.ORTH-	1.64 124		
	3-isothiazolyl						★S.PARA+	0.65 739				
	4-isothiazolyl						★S.PARA+	-0.04 909				
	4-thiazolyl						★S.META+	-0.01 778				
	5-isothiazolyl						★S.PARA+	0.67 739	★ES-CH	3.20 2		
	5-thiazolyl						★S.META+	-0.18 778				
C3H2N3	2-S-triazinyl						S.INDUC	0.14 994	S.PARA-	0.70 996	★ES-CH	3.00 2
							S.INDUC	0.21 798	S.PARA-	0.78 995	GR.ELCT	0.43 13
							★S.PARA-	0.61 346	★S.ZE.RS	0.20 798		
	3-(1,2,4-triazinyl)		★ 0.35 15				S.INDUC	0.15 15	S.ZE.RS	0.12 15	S.RES.P	0.53 15
							S.PARA-	0.72 581	S.AN-RS	0.56 15		
	5-(1,2,4-triazinyl)		★ 0.48 15				S.INDUC	0.21 15	S.ZE.RS	0.13 15		
							S.PARA-	0.94 581	S.AN-RS	0.69 15		
	6-(1,2,4-triazinyl)		★ 0.39 15				S.INDUC	0.21 15	S.ZE.RS	0.07 15	S.AN-RS	0.50 15
							S.PARA-	0.72 581	S.ZE.RS	0.37 96		
C3H2O2	CH=CHC(=O)o- -E						★S.PARA-	0.24 920	L-STM	5.76 1	B5-STM	3.46 1
							★ES-CH	3.00 2	B1-STM	1.60 1		
C3H3	CCCH3		0.10 3	★ 0.03 6	1.41		S.INDUC	0.02 16	S.ZPTFT	0.05 112	F	0.29 4
			★ 0.21 6	0.09 501			★S.INDUC	0.30 6	★S.ZE.RS	-0.27 6	R	-0.26 4
				0.12 3			S.STAR	0.89 183	★ES-CH	3.00 2	S-L	0.30 14
							S.STAR	1.20 3	L-STM	5.47 1	GRP.DPL	-0.84 9
							S.STAR	1.30 488	B1-STM	1.60 1		
							★S.STAR	1.93 652	B5-STM	2.04 1		
	CH2CCH		★ 0.07 3		1.29		★S.INDUC	0.14 6	S.STAR	0.76 380	B1-STM	1.52 1
							S.STAR	0.56 262	S.STAR	0.81 3	B5-STM	4.49 1
							S.STAR	0.60 349	★ES-CH	3.00 2	★S-L	0.14 14
							★S.STAR	0.76 268	L-STM	3.99 1	GRP.DPL	-0.84 9
C3H3Br2	cy-Prop.2,2-Br2						★S.PARA+	-0.04 394				
C3H3Cl2	2,2-dichlorocyclopropyl						★S.PARA+	-0.02 352	★S.ZE.RS	-0.07 352		
							★S.PARNO	0.05 352	★ES-CH	5.40 2		
	CH2CH=CCl2		★ -0.06 3		2.41		★S.INDUC	0.05 6	L-STM	6.43 1	B5-STM	5.20 1
							★ES-CH	3.00 2	B1-STM	1.52 1	★S-L	0.05 7
	CH=CHCHCl2 -C				2.53		★S.STAR	0.88 326	★ES-CH	3.00 2		
	CH=CHCHCl2 -T						L-STM	6.12 1	B1-STM	1.65 1	B5-STM	4.20 1
	cy-Prop.2,2-Cl2						★S.PARA+	-0.04 394				
C3H3Cl2O1	CCl2COCH3						★S.STAR	2.33 171	★ES-CH	6.40 2		
	CH2COCHCl2						★S.STAR	0.68 171	★ES-CH	4.00 2		
	CHClCOCH2Cl						★S.STAR	1.52 171	★ES-CH	5.20 2		
C3H3Cl4	CCl2CH2CHCl2						★S.STAR	2.10 171	★ES-CH	5.40 2		
	CHClCH2CCl3						★S.STAR	1.33 171	★ES-CH	4.20 2		
C3H3F3N1O1	N(CH3)COCF3		★ 0.41 3	★ 0.39 3	1.95		★S.INDUC	0.43 68	L-STM	5.20 1	F	0.41 4
							★S.ZE.RS	-0.04 68	B1-STM	1.56 1	R	-0.02 4
							★ES-CH	5.00 2	B5-STM	3.96 1		
C3H3F4O1	OCH2CF2CHF2						S.PHOSP	-0.16 449				
C3H3N1	5-quinolinyl:			★ -0.06 417			★S.PARA-	0.45 232	★S.PARA+	0.15 664		
				0.24 417			★S.PARA-	0.47 926	★ES-CH	0.00 2		
				★ 0.24 530			★S.PARA+	-0.11 664				
				★ 0.37 417			★S.PARA+	0.07 664				
C3H3N2	1-imidazolyl			★ 0.24 226			★S.INDUC	0.51 91	★ES-CH	3.00 2	B5-STM	3.12 1
				0.45 275			S.INDUC	0.60 16	L-STM	1.71 1	★GRP.DPL	-3.14 8

Empirical	Structural	π	σm	σp	MR	Es	additional σ values					
C3H3N2	1-imidazolyl						★ S.ZE.RS -0.16 91	L-STM 5.44 1				
	1-pyrazyl			★ 0.19 226 / 0.23 275			★ S.INDUC 0.30 91 / ★ S.STAR 2.20 3 / ★ S.ZE.RS -0.06 91	★ ES-CH 3.00 2 / L-STM 5.44 1 / B1-STM 1.71 1	B5-STM 3.12 1 / ★ GRP.DPL -2.00 8			
	2-imidazolyl	-0.25					★ ES-CH 3.00 2 / L-STM 5.44 1	B1-STM 1.71 1 / B5-STM 3.12 1				
	3-pyrazyl						GRP.DPL -2.26 8					
	4-imidazolyl						S.INDUC 0.12 15					
C3H3N2O1	3-OH-1-pyrazyl						GRP.DPL -2.18 8					
	4-OH-1-pyrazyl						GRP.DPL -2.43 8					
	5-(3-Me-1,2,4-oxadiazolyl						★ S.INDUC 0.48 96					
	5-OH-pyrazyl						GRP.DPL -3.41 8					
C3H3N2O1S1	5Me-1,3,4-oxadiaz-2-S-						S.INDUC 0.51 495					
C3H3N2S1	1(2-thiol-imidazolyl)						S.INDUC 0.41 91	S.ZE.RS -0.07 91				
	4(2-aminothiazolyl)						★ S.PARA+ 0.06 191					
	5(3-Me-1,2,4-thiadiazolyl)						★ S.INDUC 0.40 96					
C3H3N2S2	5Me-1,3,4thiadiaz-2-S-						S.INDUC 0.47 495					
C3H3O1	1-cyclopropene-oxide						★ S.INDUC 0.04 339	★ S.ZE.RS -0.05 339	★ ES-CH 4.00 2			
	CH=CHCHO	-0.23	★ 0.24 180	★ 0.13 180	1.69		★ ES-CH 3.00 2 / L-STM 5.76 1 / B1-STM 1.60 1	B5-STM 3.46 1 / F 0.29 5 / R -0.16 5	★ GRP.DPL -2.71 8			
	COCH=CH2						★ S.INDUC 0.15 16 / ★ S.STAR 2.00 129	★ ES-CH 4.00 2 / L-STM 5.06 1	B1-STM 1.60 1 / B5-STM 3.74 1			
	OCH2CCH					★ -1.89 350	ES-CH 3.00 2 / L-STM 6.58 1	B1-STM 1.35 1 / B5-STM 3.07 1				
C3H3O2	CH=CHCO2H	0.00	★ 0.14 397	0.33 304 / 0.90 880	1.79		S.INDUC 0.26 34 / S.STAR 1.01 3 / ★ S.STAR 1.01 326 / ★ S.ZPTFT 0.30 248	★ S.PARA- 0.62 397 / S.BA.RS -0.05 34 / ES-CH 3.00 2 / L-STM 6.15 1	B1-STM 1.60 1 / B5-STM 3.46 1 / GRP.DPL -2.04 8			
	COCOCH3	-2.23 57					GRP.DPL -2.44 8					
C3H3O2S1	SCH=CHCO2H						L-STM 6.31 1	B1-STM 1.70 1	B5-STM 6.35 1			
C3H3O2Se1	SeCH=CHCO2H -C						GRP.DPL -1.69 8					
	SeCH=CHCO2H -T						GRP.DPL -2.27 8					
C3H3O3	OCH=CHCO2H						L-STM 6.21 1	B1-STM 1.35 1	B5-STM 5.85 1			
C3H3S1	SCH2CCH						★ ES-CH 3.20 2 / L-STM 6.89 1	B1-STM 1.70 1 / B5-STM 3.61 1				
C3H3Se1	Se-CCCH3						GRP.DPL -1.31 8					
C3H4Br1	C3H4-Br -C						★ S.PARA+ -0.20 394					
	C3H4-Br -T						★ S.PARA+ -0.28 394					
	CH2C(Br)=CH2						★ ES-CH 4.30 2 / L-STM 4.16 94	B1-STM 1.52 94 / B5-STM 4.62 94				
C3H4Br1N2O4	EtCBr(NO2)2				3.40		★ S.STAR 0.52 591 / ★ ES-CH 3.00 2	L-STM 6.64 1 / B1-STM 2.10 1	B5-STM 4.69 1			
C3H4Br1O1	OCH2C(Br)=CH2						L-STM 6.21 94	B1-STM 1.35 94	B5-STM 4.15 94			
C3H4Cl1	C3H4-Cl -C						★ S.PARA+ -0.23 394					
	C3H4-Cl -T						★ S.PARA+ -0.25 394					
C3H4Cl1Hg1	C(=CH2)CH2HgCl						★ S.INDUC 0.11 355	★ S.ZE.RS -0.04 355				
	CH=CHCH2HgCl						★ S.INDUC 0.02 355	★ S.ZE.RS -0.05 355				
C3H4Cl1N2O4	EtCCl(NO2)2				3.12		★ S.STAR 0.54 591 / ★ ES-CH 3.00 2	L-STM 6.64 1 / B1-STM 1.96 1	B5-STM 4.69 1			
C3H4Cl1O1	OC(CH3)=CHCl						S.IND.P 4.84 284	S.RES.P -3.44 284				
C3H4Cl3	CCl2CH2CH2Cl						★ S.STAR 2.02 171	★ ES-CH 5.40 2				
	CHClCH2CHCl2						★ S.STAR 1.26 171					
	EtCCl3	2.94					★ S.INDUC 0.07 6 / ★ S.STAR 0.25 3 / ★ S.STAR 3.00 2	L-STM 5.94 1 / B1-STM 1.52 1 / B5-STM 4.91 1	★ S-L 0.07 7			
C3H4Cl3O1	CH2CH(OH)CCl3						★ S.STAR 0.33 129 / ★ ES-CH 4.00 2	L-STM 5.94 1 / B1-STM 1.52 1	B5-STM 4.91 1			
C3H4F1N2O4	EtCF(NO2)2				2.61		★ S.STAR 0.56 591 / ★ ES-CH 3.00 2	L-STM 6.64 1 / B1-STM 1.52 1	B5-STM 4.69 1			
C3H4F3	EtCF3				1.43		★ S.STAR 0.32 52 / ★ ES-CH 3.00 2 / L-STM 5.35 1	B1-STM 1.52 1 / B5-STM 4.07 1 / S.PHOSP -0.40 449	★ GRP.DPL -1.94 9			
C3H4F3O1	OCH2CH2CF3						S.PHOSP -0.16 449					
C3H4N1	EtCN				1.48	★ -2.23 44 / -2.14 196	S.STAR 0.48 349 / S.STAR 0.49 74 / S.STAR 0.80 37 / S.STAR 0.80 420 / ★ S.STAR 0.87 3	ES-CH 3.00 2 / L-STM 6.28 1 / B1-STM 1.52 1 / B5-STM 3.17 1 / S-L 0.09 7	★ S.PHOSP -0.60 37 / S.IND.P 0.68 39 / GRP.DPL -3.51 9			
	NHCH2CCH						★ S.STAR 1.83 170 / ★ ES-CH 3.00 2	L-STM 6.62 1 / B1-STM 1.35 1	B5-STM 3.11 1			
C3H4N1O1	CH=CHC(=O)NH2						L-STM 6.29 1	B1-STM 1.60 1	B5-STM 4.16 1			
	OCH2CH2CN					★ -1.43 350	ES-CH 3.00 2 / L-STM 5.93 1	B1-STM 1.35 1 / B5-STM 5.44 1				

Empirical	Structural	π	σ$_m$	σ$_p$	MR	E$_s$	additional σ values						
C3H4N1O2	CH=C(Me)NO2						★S.PARA-	0.53 257					
	NHCH=CHCO2H -E						L-STM	6.23 1	B1-STM	1.35 1	B5-STM	5.90 1	
C3H4N1S1	SCH2CH2CN						L-STM	5.99 1	B1-STM	1.70 1	B5-STM	6.09 1	
C3H4N3	3-Me-1-(1,2,4-triazolyl)			★ 0.33 15									
	4-Me-1-(1,2,3-triazolyl)						★S.INDUC	0.37 15	★S.ZE.RS	-0.08 15			
	4-Me-2-(1,2,3-triazolyl)			★ 0.24 15									
	5-Me-1-(1,2,3-triazolyl)						★S.INDUC	0.34 15	★S.ZE.RS	0.01 15			
	5-Me-2-(1,2,3-triazolyl)						★S.INDUC	0.26 15	★S.ZE.RS	0.05 15			
	CH2-1-(1,2,4-triazolyl)						L-STM	4.41 94	B1-STM	1.52 94	B5-STM	5.42 94	
C3H4N3O6	EtC(NO2)3				3.20		★S.STAR 0.58 291 ★ES-CH 3.00 2		L-STM 6.64 1 B1-STM 2.00 1		B5-STM 4.69 1		
C3H4N3S1	1(5-MeS-1,2,3-triazolyl)						S.INDUC	0.48 91	S.ZE.RS	-0.43 91			
C3H4O2	CH2CH2CO2 (-)						★S.STAR 0.02 3 S.META- -0.20 29 ★S.META- -0.02 36		★S.PARA- -0.20 29 ★ES-CH 3.00 2 L-STM 5.58 1		B1-STM 1.52 1 B5-STM 4.11 1		
C3H5	C(CH3)=CH2		★ 0.09 6	★ 0.05 6 0.16 3	1.56		S.INDUC -0.03 173 ★S.INDUC 0.10 6 S.INDQ 0.60 58 ★S.STAR 0.48 511 S.PARA- 0.03 257 S.ZE.RS -0.06 173		★S.ZE.RS -0.05 6 ES-CH 3.00 2 ★ES-V 0.57 341 L-STM 4.29 1 B1-STM 1.73 1 B5-STM 3.11 1		F 0.13 851 R -0.08 4 S-L 0.10 14 GRP.DPL -0.34 9 GRP.DPL 0.77 8		
	CH2CH=CH2	1.10	★ -0.11 3	★ -0.14 3	1.45	★ -1.43 35	★S.STAR 0.23 516 S.INDUC -0.04 173 S.INDUC 0.02 6 S.INDUC 0.02 49 S.INDUC 0.07 16 S.STAR 0.00 3 S.STAR 0.02 511 ★S.STAR 0.12 155 S.STAR 0.15 606 S.STAR 0.17 770 S.STAR 0.21 349 S.STAR 0.23 37		S.STAR 0.26 779 S.META+ -0.02 460 ★S.PARA- -0.18 97 S.PARA+ -0.22 460 ★S.ORTH- -0.07 97 ★S.ZE.RS -0.14 6 S.ZE.RS -0.12 173 S.ZE.RS -0.09 49 ES-CH 3.00 2 ES-V 0.69 7		L-STM 5.11 1 B1-STM 1.52 1 B5-STM 3.78 1 F -0.06 5 R -0.08 4 R+ -0.16 4 R- -0.12 4 S-L 0.02 7 S.PHOSP -0.83 37 GRP.DPL -0.35 9		
	CH=CHCH3 -C						★ES-CH 3.00 2 L-STM 4.29 1		B1-STM 1.60 1 B5-STM 4.20 1				
	CH=CHCH3 -T	1.22	★ 0.02 6 0.05 3	★ -0.09 6 -0.04 501 0.02 3	1.56		S.INDUC -0.08 173 ★S.INDUC 0.07 6 S.STAR 0.17 511 S.STAR 0.26 488 ★S.STAR 0.36 52 S.ZE.RS -0.16 6		S.ZE.RS -0.07 173 ★S.BA.RS -0.13 34 ES-CH 3.00 2 L-STM 5.10 1 B1-STM 1.60 1 B5-STM 3.13 1		F 0.09 4 R -0.18 4 S-L 0.07 7 GRP.DPL -0.25 9 GR.ELCT 0.34 13		
	cyclopropyl	1.14	-0.10 152 ★ -0.07 774 -0.04 352	-0.30 271 -0.24 891 -0.22 352 ★ -0.21 774 -0.14 152	1.35	★ -2.21 35	S.INDUC -0.08 193 S.INDUC 0.01 6 S.INDUC 0.01 823 S.INDUC 0.09 16 S.STAR -0.15 124 S.STAR 0.04 488 S.STAR 0.11 464 S.META+ -0.04 152 ★S.PARA- -0.09 400 S.PARA+ -0.52 276 S.PARA+ -0.48 394 S.PARA+ -0.47 821		S.PARA+ -0.46 152 S.PARA+ -0.45 203 S.PARA+ -0.45 352 S.PARA+ -0.44 940 ★S.PARA+ -0.41 872 S.PARNO -0.10 352 S.ORTHO 0.07 78 S.ZE.RS -0.35 352 S.ZE.RS -0.19 6 S.ZE.RS -0.18 19 S.RES.+ -0.27 6 ES-CH 3.00 2		ES-V 1.06 7 L-STM 4.14 1 B1-STM 1.55 1 B5-STM 3.24 1 F 0.02 5 R -0.23 5 R+ -0.43 4 R- -0.11 4 S-L 0.01 7 GRP.DPL -0.14 9 GRP.DPL 0.51 8 GR.ELCT 0.15 13		
C3H5Cl2	CCl2CH2CH3						★S.STAR	1.91 171	★ES-CH	5.40 2			
	CH2CH2CHCl2						★S.STAR	0.25 171	★ES-CH	5.00 2			
C3H5Hg1O2	CH2HgOCOCH3						★S.INDUC	0.02 353	★S.ZE.RS	-0.16 353			
C3H5N2	NHCH2CH2CN						★S.STAR 1.69 170 ★ES-CH 3.00 2		L-STM 5.95 1 B1-STM 1.35 1		B5-STM 5.50 1		
C3H5N2O2	C=O(NHNHCOCH3)	-1.22 220					★ES-CH 4.00 2 L-STM 6.87 1		B1-STM 1.54 1 B5-STM 5.01 1				
	N=NC(O)OEt			★ 0.87 72			S.ZE.RS	0.30 72					
C3H5N2O3	N(=O)=NCOOEt			★ 0.68 179									
C3H5N2O4	C(Et)(NO2)2		★ 0.56 3	★ 0.64 291	3.66		★S.STAR 3.35 157 ★ES-CH 9.00 2 L-STM 4.92 1		B1-STM 2.55 1 B5-STM 3.72 1 F 0.51 5		R 0.13 5		
	CH2C(NO2)2(CH3)				3.66		★S.STAR 1.08 722 ★ES-CH 5.00 2		L-STM 5.49 1 B1-STM 1.52 1		B5-STM 4.19 1		
C3H5N3	3-Me-1-triazolinium						★S.INDUC	0.74 15	★S.ZE.RS	-0.06 15			
C3H5O1	(CH2)2CHO				1.49		★S.STAR 0.29 74 ★ES-CH 3.00 2		L-STM 5.58 1 B1-STM 1.52 1		B5-STM 3.81 1 ★GRP.DPL -2.23 9		
	C(OCH3)=CH2						S.INDUC	0.08 16					
	CH(OH)CH=CH2						★S.INDUC	0.07 173	★S.ZE.RS	-0.07 173			
	CH2COCH3	-0.69 -0.24 57			1.51	★ -1.99 44	★S.STAR 0.60 52 S.STAR 0.60 74 S.STAR 0.60 171 S.STAR 0.62 155		S.STAR 0.64 10 S.PARA+ 0.03 351 ES-CH 4.00 2 L-STM 4.54 1		B1-STM 1.52 1 B5-STM 4.39 1 ★GRP.DPL -2.80 9		
	CH=CHCH2OH				1.72		★S.INDUC 0.09 34 ★S.BA.RS -0.11 34		★ES-CH 3.00 2 L-STM 6.20 1		B1-STM 1.60 1 B5-STM 3.79 1		
	COEt	-0.21 57 0.06	★ 0.38 3	-0.12 194 ★ 0.48 409	1.58		★S.STAR 1.61 129 S.PARA+ 0.52 191		★S.BA.RS 0.09 34 ES-CH 4.00 2		F 0.34 4 R 0.14 4		

Empirical	Structural	π	σ_m	σ_p	MR	E_s	additional σ values		
C3H5O1	COEt	0.12 *42*					★S.ZE.RS 0.18 *6* S.ZE.RS 0.18 *114* S.ZE.RS 0.21 *22*	L-STM 4.87 *1* B1-STM 1.63 *1* B5-STM 3.45 *1*	S.ZTWST 0.00 *22* ★GRP.DPL -2.90 *8* GRP.DPL -2.79 *9*
	O- cyclopropyl						★ES-CH 4.00 *2* L-STM 5.00 *1*	B1-STM 1.35 *1* B5-STM 4.26 *1*	
	OC(CH3)=CH2						★ES-CH 4.00 *2* L-STM 4.98 *1*	B1-STM 1.35 *1* B5-STM 4.29 *1*	
	OCH2CH=CH2		★ 0.09 *168*	★ -0.25 *168*		★ -1.44 *350*	★S.ORTHO -0.35 *168* ES-CH 3.00 *2* L-STM 6.22 *1*	B1-STM 1.35 *1* B5-STM 4.42 *1* F 0.25 *4*	R -0.50 *4*
	OCH=CHCH3 -C						L-STM 5.93 *1*	B1-STM 1.35 *1*	B5-STM 3.65 *1*
	OCH=CHCH3 -T						★ES-CH 3.00 *2* L-STM 5.29 *1*	B1-STM 1.35 *1* B5-STM 4.77 *1*	
C3H5O1S1	2-thiacyclobutyl-O-						GRP.DPL -1.30 *8*		
	C(O)SCH2CH3						GRP.DPL -1.55 *8*		
	C(S)OCH2CH3						GRP.DPL -2.24 *8*		
	SC(=O)C2H5						L-STM 6.16 *1*	B1-STM 1.70 *1*	B5-STM 4.94 *1*
	SCH2C(=O)CH3						L-STM 6.43 *1*	B1-STM 1.70 *1*	B5-STM 4.32 *1*
C3H5O1S2	SC(=S)OCH2CH3						★S.STAR 2.75 *3* L-STM 5.04 *1*	★ES-CH 4.40 *2* B1-STM 1.72 *1*	★S-L 0.42 *14* B5-STM 3.09 *1*
C3H5O2	2-dioxolanyl								
	C=O(OEt)	0.22 *220* 0.51 0.74 *42* 1.01 *454*	0.20 *18* 0.31 *316* ★0.33 *603* 0.36 *6* ★0.37 *21* 0.40 *23*	0.30 *18* 0.40 *603* 0.41 *6* 0.41 *149* ★0.45 *21* 0.52 *23*	1.75		S.INDUC 0.11 *18* S.INDUC 0.20 *67* ★S.INDUC 0.21 *89* S.INDUC 0.29 *149* S.INDUC 0.30 *6* S.INDUC 0.30 *99* S.INDUC 0.32 *192* S.PRIME 0.28 *149* S.PRIME 0.29 *159* ★S.PRIME 0.30 *586* S-INDQ 1.70 *58* ★S.STAR 2.26 *3* S.META+ 0.37 *43* S.ZPTFT 0.52 *676*	S.PARA- 0.60 *211* S.PARA- 0.64 *181* S.PARA- 0.65 *618* S.PARA- 0.68 *625* S.PARA- 0.72 *104* ★S.PARA- 0.75 *28* ★S.PARA+ 0.48 *43* S.ZE.RS 0.11 *6* S.ZE.RS 0.13 *149* ★S.ZE.RS 0.16 *67* S.ZE.RS 0.17 *114* S.ZE.RS 0.18 *19* S.ZE.RS 0.19 *18* ES-CH 4.00 *2*	L-STM 5.95 *1* B1-STM 1.64 *1* B5-STM 4.41 *1* F 0.34 *5* R 0.11 *5* R+ 0.14 *4* R- 0.41 *4* S-L 0.30 *14* S.PHOSP 3.17 *648* S.ZTWST 0.00 *22* K-CHGTR 0.55 *77* GRP.DPL -1.85 *8* GRP.DPL -1.81 *9* O-STER -1.25 *90*
	CH(Me)CO2H						★S.PARA- 0.02 *257*		
	CH(OCH2)2						★ES-CH 3.00 *2* L-STM 4.78 *1*	B1-STM 1.90 *1* B5-STM 3.40 *1*	GRP.DPL 1.06 *8*
	CH2C=O(OCH3)	-0.30	★ 0.13 *3*		1.65		★S.INDUC 0.17 *32* ★S.STAR 1.06 *3* S.PARA- 0.07 *29* ★ES-CH 4.00 *2*	★ES-V 0.58 *7* L-STM 5.98 *1* B1-STM 1.52 *1* B5-STM 4.40 *1*	★S-L 0.19 *7* GRP.DPL -1.84 *9* GRP.DPL -1.81 *8*
	CH2OC=O(CH3)	-0.17	★ 0.04 *3* 0.13 *6*	★ 0.05 *210* 0.06 *6*	1.65		★S.INDUC 0.15 *6* S.INDUC 0.16 *192* S-INDQ 0.88 *58* ★S.STAR 0.70 *687* S.STAR 0.89 *155* S.STAR 0.91 *183* S.STAR 0.95 *10*	S.STAR 1.00 *3* S.ZE.RS -0.09 *6* ★S.BA.RS -0.05 *34* ES-CH 3.00 *2* L-STM 5.46 *1* B1-STM 1.52 *1* B5-STM 4.46 *1*	F 0.07 *851* R -0.02 *5* S-L 0.15 *14* GRP.DPL -1.84 *9* GRP.DPL -1.68 *8* GR.ELCT 0.22 *13*
	EtCO2H	-0.64 *130* -0.29	★ -0.03 *397*	★ -0.07 *397*	1.65	★ -2.21 *44*	★S.STAR 0.35 *3* ES-CH 3.00 *2* L-STM 5.97 *94*	B1-STM 1.52 *94* B5-STM 3.31 *94* F 0.02 *5*	R -0.09 *5*
	OCH2C(=O)CH3						L-STM 5.98 *1*	B1-STM 1.35 *1*	B5-STM 3.95 *1*
C3H5O2S1	SC(=O)OCH2CH3						L-STM 6.88 *1*	B1-STM 1.70 *1*	B5-STM 5.43 *1*
	SCH2C(=O)OCH3						L-STM 6.88 *1*	B1-STM 1.70 *1*	B5-STM 5.46 *1*
	SCH2CH2CO2H						L-STM 5.71 *1*	B1-STM 1.70 *1*	B5-STM 5.95 *1*
	SCH2OC(=O)CH3						L-STM 6.93 *1*	B1-STM 1.70 *1*	B5-STM 5.53 *1*
C3H5O3	OC(=O)OCH2CH3						L-STM 6.68 *1*	B1-STM 1.35 *1*	B5-STM 4.63 *1*
	OCH2C(=O)OCH3						L-STM 6.65 *1*	B1-STM 1.35 *1*	B5-STM 4.67 *1*
	OCH2CH2CO2H						L-STM 5.53 *94*	B1-STM 1.35 *94*	B5-STM 5.31 *94*
	OCH2OC(=O)CH3						L-STM 6.67 *1*	B1-STM 1.35 *1*	B5-STM 5.01 *1*
C3H5O4S1	SO2CH2OC(=O)CH3						L-STM 6.79 *1*	B1-STM 2.03 *1*	B5-STM 5.13 *1*
C3H5S1	S-cyclopropyl						★ES-CH 4.20 *2* L-STM 5.39 *1*	B1-STM 1.70 *1* B5-STM 4.67 *1*	
	SCH(CH3)=CH2						★ES-CH 4.20 *2* L-STM 5.33 *1*	B1-STM 1.70 *1* B5-STM 4.61 *1*	
	SCH2CH=CH2		★ 0.19 *168*	★ 0.12 *168*			★S.STAR 1.45 *3* ★S.ORTHO -0.01 *168* ★ES-CH 3.20 *2*	L-STM 6.42 *1* B1-STM 1.70 *1* B5-STM 5.02 *1*	F 0.23 *4* R -0.11 *4* ★S-L 0.27 *14*
	SCH=CHCH3 -C						L-STM 6.29 *1*	B1-STM 1.70 *1*	B5-STM 4.23 *1*
	SCH=CHCH3 -T						★ES-CH 3.20 *2* L-STM 5.40 *1*	B1-STM 1.70 *1* B5-STM 5.26 *1*	
C3H5S2	2-(1,3-dithiocyclopentane				2.88	★ -2.26 *115*	★S.STAR 1.02 *115* ES-CH 3.40 *2*	★ES-V 0.89 *7* L-STM 5.61 *1*	B1-STM 1.86 *1* B5-STM 3.41 *1*
C3H5S3	SC(=S)SCH2CH3						★S.STAR 3.00 *3*	★ES-CH 4.60 *2*	★S-L 0.46 *14*
C3H5Se1	SeCH2CH=CH2		★ 0.21 *168*	★ 0.12 *168*			★S.ORTHO 0.16 *168* ★ES-CH 3.30 *2*	F 0.26 *4* R -0.14 *4*	
C3H6	C+(CH3)2						GR.ELCT 0.58 *13*		
	CH2CH2CH2*	0.72 *130*	★ -0.26 *421*	★ -0.26 *421*	1.39		★S.STAR -0.24 *268*	ES-CH 2.00 *2*	GRP.DPL 0.55 *8*

Empirical	Structural	π	σm	σp	MR	Es	additional σ values
C3H6	CH2CH2CH2*	1.20					S.STAR 0.32 837 / ★S.PARA+ -0.41 27 / F -0.20 5 / R -0.06 5
C3H6Br1	C(CH3)2Br					★ -2.89 35	★ES-CH 4.30 2 / ★ES-V 1.39 7 / L-STM 4.11 1 / B1-STM 2.78 1 / B5-STM 3.75 1
	CH2CH2CH2Br						★S.INDUC 0.02 6 / ★ES-CH 3.00 2 / ★S-L 0.02 14
	CH2CHBrCH3				2.27		★S.STAR 0.44 223 / ★ES-CH 4.30 2
	CHBrEt				2.27		★S.STAR 1.02 223 / L-STM 4.92 1 / B5-STM 3.75 1 / ★ES-CH 4.30 2 / B1-STM 1.90 1
C3H6Cl1	C(Me)2Cl						★S.ZE.RS -0.06 22 / ★ES-CH 4.20 2 / ★S.ZTWST -0.10 22
	CH2CHClCH3				1.98		★S.STAR 0.35 223 / ★ES-CH 4.20 2
	CHClEt				1.98		★S.STAR 1.05 223 / L-STM 4.92 1 / B5-STM 3.49 1 / ★ES-CH 4.20 2 / B1-STM 1.88 1
	PrCl				1.98	★ -1.72 44	★S.STAR 0.14 223 / S.STAR 0.15 74 / ES-CH 3.00 2 / L-STM 5.94 1 / B1-STM 1.52 1 / B5-STM 4.91 1 / S-L 0.02 14 / ★S.PHOSP -0.72 37 / S.IND.P 0.46 39
C3H6Cl1O1	O(CH2)3Cl						★ES-CH 3.00 2 / ★ES-V 0.52 7 / L-STM 7.51 1 / B1-STM 1.35 1 / B5-STM 4.56 1
C3H6F1	(CH2)3F				1.48	★ -1.64 196	★S.STAR 0.14 4 / ES-CH 3.00 2 / L-STM 5.35 1 / B1-STM 1.52 1 / B5-STM 4.07 1 / F 0.02 4
C3H6F1O1	OCH2CH2CH2F				1.65		F 0.22 4
C3H6F3Sn1	Sn(CF3)(CH3)2						★S.STAR -0.07 61 / ★ES-CH 7.10 2
C3H6N1	1-azetidinyl						★S.ZE.RS -0.55 19 / ★ES-CH 3.00 2 / L-STM 4.77 1 / B1-STM 1.77 1 / B5-STM 3.82 1
	CH=N-Et						S.ORTHO -0.85 53
	NH+(CH2CH=CH2)					★ -1.44 215	
	NH-cyclopropyl			★ -0.81 475			★ES-CH 4.00 2 / B1-STM 1.35 1 / L-STM 5.05 1 / B5-STM 4.30 1
	NHC(CH3)=CH2						★ES-CH 4.00 2 / B1-STM 1.90 1 / L-STM 4.91 1 / B5-STM 1.90 1
	NHCH2CH=CH2		★ -0.17 794				L-STM 6.25 1 / B1-STM 1.35 1 / B5-STM 4.48 1
	NHCH=CHCH3 -C						★ES-CH 3.00 2 / B1-STM 1.35 1 / L-STM 5.97 1 / B5-STM 3.71 1
	NHCH=CHCH3 -T						L-STM 5.31 1 / B1-STM 1.35 1 / B5-STM 4.82 1
C3H6N1O1	C(=O)NHC2H5						L-STM 5.89 1 / B1-STM 1.63 1 / B5-STM 4.49 1
	C=NOH(Et)						★S.PARA+ 0.24 191
	CH2C(=O)NHCH3						L-STM 6.09 1 / B1-STM 1.52 1 / B5-STM 4.54 1
	CH2NHCOCH3		-0.04 3 / ★ 0.05 6	-0.05 6 / ★ -0.05 210	1.96		★S.INDUC 0.09 6 / ★S.STAR 0.43 3 / S.STAR 0.60 155 / S.ZE.RS -0.14 6 / ES-CH 3.00 2 / L-STM 5.67 1 / B1-STM 1.52 1 / B5-STM 4.75 1 / F 0.12 4 / R -0.17 4 / ★S-L 0.09 14 / GRP.DPL -3.55 9
	CH=NOCH2CH3						L-STM 7.24 1 / B1-STM 1.60 1 / B5-STM 4.12 1
	CON(CH3)2	-1.51					S.INDUC 0.25 119 / ★S.INDUC 0.28 6 / S.INDUC 0.28 192 / S.STAR 1.94 3 / ★S.PARA- 0.70 258 / ★ES-CH 5.00 2 / L-STM 4.77 1 / B1-STM 1.60 1 / B5-STM 4.04 1 / ★S-L 0.28 7 / S-CNTIG 0.70 139 / ★K-CHGTR 1.31 436 / GRP.DPL -3.81 9
	EtC=O(NH2)	-1.22 / -0.71 57 / -0.22 130	★ -0.06 3		1.96		★S.INDUC 0.03 32 / ★S.STAR 0.19 3 / ES-CH 3.00 2 / L-STM 6.11 1 / B1-STM 1.52 1 / B5-STM 3.53 1 / ★S-L 0.05 14 / ★GRP.DPL -3.78 9
	N(CH3)COCH3		0.28 10 / ★ 0.31 3	★ 0.26 3	1.96		S.INDUC 0.33 99 / ★S.INDUC 0.36 68 / ★S.STAR 2.25 3 / S.ZE.RS -0.41 19 / ★S.ZE.RS -0.10 68 / ES-CH 5.00 2 / L-STM 4.77 1 / B1-STM 1.35 1 / B5-STM 3.71 1 / F 0.34 5 / R -0.08 5 / GRP.DPL -3.86 9 / ★GRP.DPL -3.60 8
	NHC=O(Et)	-0.52	★ 0.23 3		1.96		★S.INDUC 0.25 32 / ★S.STAR 1.56 3 / ★ES-CH 4.00 2 / L-STM 6.45 1 / B1-STM 1.55 1 / B5-STM 3.97 1 / ★S-L 0.26 14 / GRP.DPL -3.55 9
	NHCH2C(=O)CH3						L-STM 6.02 1 / B1-STM 1.35 1 / B5-STM 3.99 1
	ON=C(CH3)2		★ 0.29 3		2.04		★S.INDUC 0.29 32 / ★S.STAR 1.81 3 / ★ES-CH 3.00 2 / L-STM 5.90 1 / B1-STM 1.35 1 / B5-STM 4.67 1 / ★S-L 0.30 14
C3H6N1O1S1	OC(=S)N(CH3)2						★S.STAR 3.12 3 / ★ES-CH 4.20 2 / ★S-L 0.47 14
	SCH2CH2C(=O)NH2						L-STM 5.70 94 / B1-STM 1.70 94 / B5-STM 6.09 94
	SCON(CH3)2						★S.STAR 2.06 3 / S.PARA+ 0.13 836 / ★ES-CH 4.20 2 / L-STM 5.87 1 / B1-STM 1.70 1 / B5-STM 4.62 1 / ★S-L 0.31 14
C3H6N1O2	C(NO2)(CH3)2		★ 0.18 3	★ 0.20 291	2.06		★S.STAR 1.10 464 / ★ES-CH 6.00 2 / L-STM 4.59 1 / B1-STM 2.58 1 / B5-STM 3.72 1 / F 0.19 5 / R 0.01 5
	CH(Et)(NO2)						★S.STAR 1.62 157
	CH2CH(NH3+)CO2-						L-STM 5.58 1 / B1-STM 1.52 1 / B5-STM 4.11 1
	CH2CH(NH3+)CO2-	-3.56 / -3.13 54					★ES-CH 4.00 2 / L-STM 5.58 1 / B1-STM 1.52 1 / B5-STM 4.22 1
	CH2NHCOCH2OH	-1.58					★ES-CH 3.00 2
	NHC=O(OEt)	0.17	0.03 512 / 0.07 3	-0.20 6 / ★ -0.15 56	2.12		★S.INDUC 0.28 6 / S.INDQ 1.56 58 / ES-CH 4.00 2 / L-STM 7.25 1 / R -0.38 5 / S-L 0.28 7

Empirical	Structural	π	σm	σp	MR	Es	additional σ values
C3H6N1O2	NHC=O(OEt)		0.08 10 0.11 6 ★ 0.11 399	0.04 512			S.STAR 1.62 10 · ★S.STAR 1.99 3 · ★S.ZE.RS -0.48 6 · B1-STM 1.35 1 · B5-STM 3.92 1 · F 0.23 5 · GRP.DPL -3.80 9
	NHCH2C(=O)OCH3						L-STM 6.69 1 · B1-STM 1.35 1 · B5-STM 4.75 1
	NHCH2CH2CO2H						L-STM 5.56 94 · B1-STM 1.35 94 · B5-STM 5.37 94
	NHCH2OC(=O)CH3						L-STM 6.70 1 · B1-STM 1.35 1 · B5-STM 5.06 1
	OC=O(NMe2)	-0.38 54					★S.STAR 2.87 3 · S.PARA+ 0.08 836 · ★ES-CH 4.00 2 · L-STM 5.41 1 · B1-STM 1.68 1 · B5-STM 3.98 1 · ★S-L 0.44 14 · ★GRP.DPL -3.80 9
	OCH2CH2C(=O)NH2						L-STM 5.58 94 · B1-STM 1.35 94 · B5-STM 5.45 94
C3H6N1O3S1	N(COCH3)SO2CH3	-1.84					★ES-CH 8.20 2
C3H6N1S1	C(=S)N(CH3)2						L-STM 1.88 1 · B1-STM 4.77 1 · B5-STM 4.04 1
	NHC(=S)CH2CH3						L-STM 5.99 1 · B1-STM 1.55 1 · B5-STM 4.38 1
C3H6N1S2	N(CH3)C(=S)SCH3						★S.STAR 2.90 3 · ★ES-CH 5.40 2 · ★S-L 0.44 14
	SC(=S)N(CH3)2						★S.STAR 2.31 3 · ★ES-CH 4.40 2 · ★S-L 0.37 7
C3H6O1	OCH2CH2CH2*						★ES-CH 2.00 2 · ★S.PHOSP -1.01 354
C3H6O2	OCH2CH(CH3)O*						★ES-CH 3.00 2 · ★S.PHOSP 1.21 479
	OCH2CH2CH2O*		★ 0.00 439	★ 0.00 439			★ES-CH 2.00 2 · R -0.03 4 · ★S.PHOSP -0.56 354 · F 0.03 4 · S.PHOSP -0.87 577
C3H7	CH(CH3)2	1.22 130 1.53	-0.08 46 -0.07 3 ★ -0.04 6	★ -0.15 21 -0.13 46 -0.10 271 -0.07 208	1.50	-2.32 379 -1.95 302 ★ -1.71 50 -1.60 62 -0.56 311	S.INDUC -0.08 12 · S.INDUC -0.06 261 · ★S.INDUC 0.01 6 · S.INDUC 0.10 16 · S.INDQ -0.08 58 · S.STAR -0.29 488 · ★S.STAR -0.19 52 · S.META- -0.09 28 · ★S.ZPTFT -0.18 112 · S.ZPTFT -0.16 63 · ★S.PARA- -0.16 138 · S.PARA- -0.09 271 · S.PARA+ -0.31 329 · S.PARA+ -0.29 65 · ★S.PARA+ -0.28 43 · S.PARNO -0.14 65 · S.ORTHO -0.23 53 · S.ORTHO -0.11 158 · S.ORTHO 0.03 104 · S.ORTHO 0.56 78 · S.ORTH- -0.15 97 · S.ZE.RS -0.16 12 · S.ZE.RS -0.13 12 · ★S.ZE.RS -0.12 19 · S.ZE.RS 0.12 114 · ES-CH 3.00 2 · ES-V 0.76 7 · ES-HYBO -0.66 225 · L-STM 4.11 1 · B1-STM 1.90 1 · B5-STM 3.17 1 · F 0.04 5 · R -0.19 5 · R+ -0.32 4 · R- -0.20 4 · S-L 0.01 14 · S.PHOSP -1.30 59 · S.P.RAD 0.03 206 · S.P.RAD 0.09 79 · S-M(C+) -0.14 666 · K-CHGTR 0.07 77 · GRP.DPL 0.08 9 · GRP.DPL 0.40 8 · GR.ELCT 0.15 13 · O-STER -1.44 90 · SAM(C+) -0.07 239
	Pr	1.45 42 1.55	★ -0.06 6 -0.05 3	-0.17 6 -0.17 82 -0.15 418 ★ -0.13 23	1.50	-1.91 379 -1.60 50 -1.56 302 ★ -1.43 62 -0.22 311	S.INDUC -0.12 12 · S.INDUC -0.05 261 · ★S.INDUC -0.01 6 · S.STAR -0.32 488 · ★S.STAR -0.12 52 · S.STAR -0.12 129 · ★S.PARA- -0.06 138 · ★S.PARA+ -0.29 76 · S.PARA+ -0.28 88 · S.ORTHO -0.25 53 · S.ORTHO -0.14 158 · S.ORTH+ -0.27 154 · ★S.ORTH- -0.16 97 · S.ZE.RS -0.16 6 · S.ZE.RS -0.12 12 · ★S.ZE.RS -0.11 19 · S.ZE.RS 0.12 114 · ES-CH 3.00 2 · ES-V 0.68 7 · ★ES-HYBO -0.56 225 · L-STM 4.92 1 · B1-STM 1.52 1 · B5-STM 3.49 1 · F 0.01 5 · R -0.14 5 · R+ -0.30 4 · R- -0.07 4 · S-L -0.01 14 · ★S.PHOSP -1.18 59 · S.ZTWST -0.10 22 · K-CHGTR 0.04 77 · ★GRP.DPL 0.00 9 · GR.ELCT 0.15 13
C3H7N2	3,4-N+(CH3)=CH-N(CH3)- *		★ 1.11 411	★ 1.11 411			★ES-CH 4.00 2 · F 1.05 4 · R 0.06 4
	CH=NN(CH3)2						★ES-CH 3.00 2 · B1-STM 1.83 1 · L-STM 5.91 1 · B5-STM 3.73 1
	N=CHN(CH3)2						★ES-CH 3.00 2 · B1-STM 1.70 1 · L-STM 5.80 1 · B5-STM 4.09 1
C3H7N2O1	(CH2)2NHCONH2						★ES-CH 3.00 2 · ★S-L 0.03 14
	NHC(=O)N(CH3)2	-1.15					L-STM 5.44 1 · B1-STM 1.64 1 · B5-STM 4.04 1
	NHC=O(NHC2H5)		★ 0.04 222	★ -0.26 222	2.32		★ES-CH 4.00 2 · B1-STM 1.45 1 · F 0.19 5 · L-STM 7.29 1 · B5-STM 3.98 1 · R -0.45 5
	NHCH2CH2C(=O)NH2						L-STM 5.60 94 · B1-STM 1.35 94 · B5-STM 5.51 94
	OCH=NN(CH3)2 -E						L-STM 6.33 1 · B1-STM 1.35 1 · B5-STM 5.60 1
C3H7N2O2	NHC(=O)N(CH3)(OCH3)	-0.84					
C3H7N2S1	CH2SC(=NCH3)NH2			★ -0.05 249			★ES-CH 3.20 2 · B1-STM 1.52 1 · L-STM 7.41 1 · B5-STM 4.56 1
	NHC(=NCH3)SCH3			★ -0.25 249	2.95		★ES-CH 3.00 2 · B1-STM 1.45 1 · L-STM 6.70 1 · B5-STM 4.26 1
	NHC(=S)N(CH3)2						★S.STAR 1.75 3 · L-STM 5.89 1 · B5-STM 4.38 1 · ★ES-CH 4.20 2 · B1-STM 1.45 1 · ★S-L 0.28 7
	NHC=S(NHC2H5)	-0.71	★ 0.30 222	★ 0.07 222	3.17		★ES-CH 4.20 2 · B1-STM 1.45 1 · F 0.40 5 · L-STM 7.22 1 · B5-STM 4.38 1 · R -0.33 5
	SCH=NN(CH3)2 -E						L-STM 6.39 1 · B1-STM 1.70 1 · B5-STM 6.04 1
C3H7N4O4	N(NO2)CH2CH2N(NO2)CH3						★ES-CH 6.00 2 · ★S-L 0.39 14
C3H7O1	C(OH)(CH3)2		★ 0.47 546	-0.12 660 ★ 0.60 546	1.64	★ -1.95 44	S.STAR 0.32 376 · ★S.STAR 0.35 74 · S.ZE.RS -0.08 22 · ES-CH 4.00 2 · L-STM 4.11 1 · B1-STM 2.40 1 · B5-STM 3.17 1 · F 0.41 615 · R 0.19 615 · S.ZTWST -0.12 22 · ★GRP.DPL -1.72 9
	CH(OCH3)CH3				1.67	★ -1.88 44	ES-CH 4.00 2 · B1-STM 1.90 1 · L-STM 4.78 1 · B5-STM 3.40 1
	CH(OH)C2H5			★ -0.16 147	1.64	★ -1.58 44	S.INDUC 0.01 173 · ★S.STAR 0.45 74 · ★S.ZE.RS -0.08 22 · S.ZE.RS -0.07 173 · ES-CH 3.00 2 · ★ES-V 0.71 7 · L-STM 4.92 1 · B1-STM 1.73 1 · B5-STM 3.49 1 · S.ZTWST -0.11 22

Empirical	Structural	π	σm	σp	MR	Es	additional σ values
C3H7O1	CH2CH2CH2OH				1.50		★S.STAR 0.08 4; F 0.00 4
	CH2CHOHCH3		★ -0.12 3	★ -0.17 147	1.64	★ -2.31 44	★S.INDUC 0.03 6; ES-CH 4.00 2; F -0.06 4; S.STAR -0.06 3; L-STM 4.92 1; R -0.11 4; ★S.STAR 0.16 594; B1-STM 1.52 1; S-L 0.03 14; ★S.BA.RS -0.13 34; B5-STM 3.78 1; GRP.DPL -1.77 9
	CH2OEt				1.67	-1.61 44; ★ -1.49 115	S.BA.RS 0.50 287; ES-CH 3.00 2; B5-STM 4.45 1; S.STAR 0.57 115; ★ES-V 0.61 7; GRP.DPL -1.27 9; ★S.STAR 0.58 3; L-STM 6.01 1; S.ZE.RS 0.06 552; B1-STM 1.52 1
	EtOCH3				1.67	★ -2.01 50	S.INDUC 0.00 261; L-STM 5.55 1; ★S-L 0.00 14; ★S.STAR 0.24 473; B1-STM 1.52 1; GRP.DPL -1.27 9; ★ES-V 0.89 7; B5-STM 4.49 1
	OCH(CH3)2	0.36 455; 0.92 54	0.05 3; ★ 0.10 21	★ -0.45 21; -0.29 23	1.71		S.INDUC 0.24 12; ★S.ZE.RS -0.43 19; R- -1.19 264; S.INDUC 0.26 32; ES-CH 4.00 2; S-L 0.28 7; ★S.STAR 1.51 37; ES-V 0.75 7; ★S.PHOSP -0.29 59; S.STAR 1.62 3; L-STM 4.80 1; S.IND.P 1.52 39; S.PARA+ -0.85 483; B1-STM 1.35 1; S.IND.P 2.04 284; ★S.PARA+ -0.85 485; B5-STM 4.10 1; S.RES.P -2.34 284; S.ORTHO -0.04 78; F 0.34 5; S.RES.P -0.52 39; S.ZE.RS -0.45 12; R -0.79 264; GRP.DPL -1.32 9
	OCH2CH2CH3	1.05	0.09 6; ★ 0.10 21	★ -0.27 23; ★ -0.25 21; -0.24 6	1.71		S.INDUC 0.23 12; S.ZE.RS -0.45 12; F 0.26 5; ★S.INDUC 0.28 6; S.ZE.RS -0.43 47; R -0.51 5; ★S.STAR 1.57 37; ES-CH 3.00 2; R+ -1.09 4; S.STAR 1.68 3; ES-V 0.56 7; ★S.PHOSP -0.32 59; ★S.PARA+ -0.83 485; L-STM 6.05 1; S.IND.P 1.57 39; S.ORTHO -0.04 78; B1-STM 1.35 1; S.RES.P -0.72 39; ★S.ZE.RS -0.52 6; B5-STM 4.42 1; GRP.DPL -1.32 9
C3H7O1S1	C(-)HS+(O)Me2 (ylide rad)						★ES-CH 5.20 2; ★S.IND.P -0.97 648; ★S.RES.P -0.17 648
	OCH2CH2SCH3						L-STM 7.31 1; B1-STM 1.35 1; B5-STM 4.83 1
	S(=O)C3H7						L-STM 6.17 1; B1-STM 1.40 1; B5-STM 4.54 1
C3H7O2	C(OOH)(CH3)2		★ 0.06 546	★ -0.14 546; -0.10 660			★ES-CH 5.00 2; R -0.31 4; F 0.17 4
	CH(OCH3)2				1.84		★S.INDUC -0.02 64; ★S.ZE.RS -0.03 64; ★S.ZE.RS 0.00 47; S.INDUC 0.08 16; ★S.ZE.RS 0.00 19; ★ES-CH 5.00 2
	OCH2CH2OCH3						★ES-CH 3.00 2; B1-STM 1.35 1; L-STM 6.71 1; B5-STM 4.71 1
C3H7O2S1	SO2CH(CH3)2				2.28		★S.INDUC 0.59 32; L-STM 4.11 1; ★S-L 0.57 14; ★S.STAR 3.68 3; B1-STM 2.03 1; GRP.DPL -4.50 9; ★ES-CH 6.20 2; B5-STM 4.45 1
	SO2Pr		★ 0.69 3		2.28		★S.INDUC 0.59 32; L-STM 6.17 1; ★S-L 0.57 14; ★S.STAR 3.68 3; B1-STM 2.03 1; ★ES-CH 5.20 2; B5-STM 4.54 1
C3H7O3S1	(CH2)3SO3H						★S.META+ 0.09 301; ★S.PARA+ -0.21 301; ★ES-CH 3.00 2
C3H7S1	CH2CH2SCH3	1.23 130					★ES-CH 3.20 2; L-STM 6.36 1; B5-STM 4.80 1; ★ES-V 0.78 7; B1-STM 1.52 1
	CH2SeT				2.41	★ -1.71 115	★S.STAR 0.56 115; ★ES-V 0.71 7; B1-STM 1.52 1; ES-CH 3.20 2; L-STM 6.65 1; B5-STM 4.60 1
	S-Pr		★ 0.22 3		2.41		★S.INDUC 0.24 32; L-STM 6.21 1; S.IND.P 1.43 39; ★S.STAR 1.38 37; B1-STM 1.70 1; S.RES.P -0.32 39; S.STAR 1.49 3; B5-STM 4.98 1; GRP.DPL -1.63 9; ES-CH 3.20 2; S-L 0.26 7; ES-V 1.07 7; ★S.PHOSP -0.06 59
	SCH(CH3)2		0.17 6; ★ 0.23 3	-0.01 6; ★ 0.07 21	2.41		★S.INDUC 0.26 6; ES-V 1.19 7; S-L 0.26 14; S.STAR 1.47 37; L-STM 5.16 1; ★S.PHOSP -0.06 59; ★S.STAR 1.56 3; B1-STM 1.70 1; S.IND.P 1.48 39; ★S.ZE.RS -0.27 6; B5-STM 4.41 1; S.RES.P -0.25 39; S.ZE.RS -0.16 47; F 0.30 4; GRP.DPL -1.61 9; ES-CH 4.20 2; R -0.23 4
C3H7S1Se1	SeCH2CH2SCH3						L-STM 7.61 1; B1-STM 1.85 1; B5-STM 5.88 1
C3H7S2	SCH2CH2SCH3						L-STM 7.36 1; B1-STM 1.70 1; B5-STM 5.70 1
C3H7Se1	SeC3H7						L-STM 6.45 1; B1-STM 1.85 1; B5-STM 5.19 1
C3H7Si1	SiH2(CH2CH=CH2)			★ 0.20 238			
C3H7Si2	CH(SiCH3)2						★S.ZE.RS -0.24 47; ★ES-CH 5.40 2
C3H8Br1S1	CH2S(CH3)2+ Br-			★ 0.12 249			★ES-CH 4.20 2
C3H8Br1Si1	Si(CH3)2CH2Br						★S.STAR 0.33 216
C3H8Cl1Si1	Si(CH3)2CH2Cl						★S.INDUC 0.06 73; ★S.STAR 0.23 216; ★S.ZE.RS 0.06 73
C3H8N1	CH(CH3)NHCH3				1.85		★S.STAR 0.16 74; L-STM 4.11 1; B5-STM 4.34 1; ★ES-CH 4.00 2; B1-STM 1.73 1
	CH2N(CH3)2	-0.16 42; -0.15	★ 0.00 3	★ 0.01 315	1.87		★S.INDUC 0.05 233; ★S.ZE.RS -0.06 233; F 0.03 5; S.STAR 0.22 155; ES-CH 4.00 2; R -0.02 5; ★S.STAR 0.24 74; L-STM 4.83 1; S.ZTWST -0.10 22; S.STAR 0.39 183; B1-STM 1.52 1; GR.ELCT 0.18 13; S.ZE.RS -0.10 22; B5-STM 4.08 1
	CH2NHCH2CH3	0.11 42					★ES-CH 3.00 2; B1-STM 1.52 1; L-STM 6.07 1; B5-STM 4.47 1
	N(CH3)Et						S.INDUC -0.01 12; ★ES-CH 4.00 2; B1-STM 1.35 1; ★S.STAR 1.08 170; ★ES-V 0.87 7; B5-STM 3.42 1; ★S.ZE.RS -0.65 12; L-STM 4.83 1
	NH(I-Pr)						★S.ZE.RS -0.53 19; ★ES-V 0.91 7; B1-STM 1.35 1

Empirical	Structural	π	σ_m	σ_p	MR	E_s	additional σ values
C3H8N1	NH(I-Pr)						★ES-CH 4.00 2 · L-STM 4.83 1 · B5-STM 4.13 1
	NH-propyl						★ES-CH 3.00 2 · L-STM 6.07 1 · B5-STM 4.47 1
							★ES-V 0.64 7 · B1-STM 1.35 1
C3H8N1O1	CH2N(O)(CH3)2						★ES-CH 5.00 2 · ★S-L 0.23 14
	NHOCH2CH2CH3		★ -0.03 794				
	OCH2N(CH3)2						L-STM 5.47 1 · B1-STM 1.35 1 · B5-STM 4.01 1
C3H8N1O2S1	CH2CH2NHSO2CH3			★ -0.11 147			★ES-CH 3.00 2
	SO2CH2N(CH3)2						L-STM 5.58 1 · B1-STM 2.03 1 · B5-STM 4.14 1
C3H8N1S1	SCH2N(CH3)2						L-STM 5.92 1 · B1-STM 1.70 1 · B5-STM 4.66 1
C3H8N3	NHC=NCH3(NHCH3)			★ -0.06 249	2.52		★ES-CH 3.00 2 · B1-STM 1.45 1
							L-STM 6.35 1 · B5-STM 4.26 1
	NHCH=NN(CH3)2 -E						L-STM 6.35 1 · B1-STM 1.35 1 · B5-STM 5.64 1
C3H8O1P1	CH2P(O)(CH3)2						★S.STAR 0.68 45 · L-STM 5.05 1 · B5-STM 4.76 1
							★ES-CH 5.20 2 · B1-STM 1.52 1
C3H8O2P1	P(O)(OC2H5)(CH3)						★S.INDUC 0.32 45 · L-STM 6.26 1 · B5-STM 4.63 1
							★ES-CH 5.20 2 · B1-STM 2.44 1 · ★S-L 0.31 14
C3H8O3P1	CH2P(O)(OCH3)2						★S.STAR 0.80 45 · L-STM 5.05 1 · B5-STM 5.81 1
							★ES-CH 5.20 2 · B1-STM 1.52 1
C3H8S1	CH2S+(CH3)2						L-STM 4.92 1 · B1-STM 1.52 1 · B5-STM 4.19 1
	S+(CH3)Et						★S.STAR 4.90 966 · ★S.STAR 4.90 974 · S.CH-P+ 1.06 75
C3H9As1	As+(CH3)3						S.CH-P+ 0.90 75
C3H9Ge1	CH2Ge(Et)(H)2						★S.PARA+ -1.05 51 · ★S.ZE.RS -0.24 51
	Ge(CH3)3		★ 0.00 3	-0.18 240	2.67		S.INDUC -0.10 133 · S.ZE.RS 0.01 126 · L-STM 4.21 1
				-0.08 338			S.INDUC 0.03 242 · S.ZE.RS 0.01 333 · B1-STM 2.81 1
				-0.06 153			S.INDUC 0.06 126 · S.ZE.RS 0.05 387 · B5-STM 3.56 1
				★ 0.00 21			S.STAR -1.32 446 · S.ZE.RS 0.07 133 · F 0.03 4
							S.ZE.RS -0.08 86 · ES-CH 4.30 2 · R -0.03 4
							S.ZE.RS -0.06 240 · ★ES-V 1.45 7 · GR.ELCT -0.10 13
C3H9Ge1O3	Ge(OCH3)3						S.INDUC 0.37 92 · S.ZE.RS 0.14 92
C3H9Ge1S3	Ge(SCH3)3						★S.PARA+ -0.27 51
C3H9N1	(CH2)3NH3+						S.INDUC 0.13 310
	+N(CH3)3	-5.96	0.54 36	0.55 36			S.INDUC 0.62 126 · S.META- 0.83 28 · S.PARNO 0.80 324
		-5.26 57	★ 0.88 21	★ 0.82 21			S.INDUC 0.71 310 · S.META- 0.83 613 · S.ORTHO 1.07 53
			0.90 23	0.86 23			S.INDUC 0.87 119 · S.META- 0.84 426 · S.ORTHO 1.95 279
			0.96 149	0.88 10			S.INDUC 0.93 24 · S.META- 0.85 715 · S.ZE.RS -0.15 19
			0.99 3	0.88 279			★S.INDUC 0.93 89 · S.META- 0.85 730 · ★S.ZE.RS -0.11 6
			0.99 766	0.88 517			S.INDUC 0.93 161 · S.META- 0.88 29 · S.ZE.RS -0.11 161
			1.02 714	0.88 714			S.INDUC 0.98 149 · S.META- 0.95 767 · S.ZE.RS -0.09 126
			1.03 29	0.90 141			S.INDUC 1.00 16 · S.META- 0.96 768 · S.ZE.RS -0.01 149
			1.03 510	0.92 149			S.INDUC 1.07 6 · ★S.ZPTFT 0.37 36 · S.RES.+ -0.31 6
			1.04 10	0.96 3			★S.PRIME 0.91 159 · S.ZPTFT 1.26 29 · ES-CH 4.00 2
			1.04 279	0.96 766			S.PRIME 0.93 149 · S.PARA- 0.54 36 · ES-V 1.22 7
			1.06 141	0.98 29			S-INDQ 4.15 58 · S.PARA- 0.64 28 · L-STM 4.02 1
			1.07 517	0.98 510			S.STAR 3.04 424 · ★S.PARA- 0.70 613 · B1-STM 2.57 1
				2.10 128			S.STAR 3.06 10 · S.PARA- 0.75 715 · B5-STM 3.11 1
							S.STAR 3.27 87 · S.PARA- 0.75 730 · F 0.86 5
							★S.STAR 4.16 424 · S.PARA- 0.76 426 · R -0.04 5
							S.STAR 4.38 360 · ★S.PARA- 0.77 29 · R+ -0.45 4
							S.STAR 4.55 3 · S.PARA- 0.77 181 · R- -0.09 4
							S.ZMTFT 0.47 36 · S.PARA- 0.85 768 · S-L 1.07 14
							S.ZMTFT 1.34 29 · S.PARA- 0.89 767 · GR.ELCT 0.89 13
							S.META+ 0.36 43 · ★S.PARA+ 0.41 43 · S.CH-P+ 0.55 75
							S.META- 0.61 36 · S.PARA+ 0.59 88 · S.CH-P+ 0.79 75
	CH2NH(CH3)2+		★ 0.40 3	★ 0.43 3			★S.STAR 1.04 87 · L-STM 4.02 1 · F 0.39 4
				0.44 315			★S.ORTHO 0.54 53 · B1-STM 1.52 1 · R 0.04 4
				0.57 210			ES-CH 4.00 2 · B5-STM 4.34 1
	N+H2CH2CH2CH3		★ 0.71 3				★S.INDUC 0.60 945 · L-STM 6.07 1 · B5-STM 4.47 1
							★S.STAR 3.74 3 · B1-STM 1.49 1
	Pr-NH3+						S.ORTHO 0.14 53
C3H9N1O3P1	N=P(OCH3)3			★ -0.51 345			★ES-CH 5.20 2
C3H9N2	NHCH2N(CH3)2						L-STM 5.49 1 · B1-STM 1.35 1 · B5-STM 4.08 1
C3H9O1Si1	OSi(CH3)3		★ 0.13 3	★ -0.27 3			★S.INDUC 0.28 12 · ★ES-CH 5.20 2 · F 0.31 5
							S.INDUC 0.36 16 · L-STM 5.27 1 · R -0.58 5
							★S.PARA+ -0.60 550 · B1-STM 1.35 1
							★S.ZE.RS -0.40 12 · B5-STM 4.66 1
	Si(CH3)2OCH3		★ 0.04 3	★ -0.02 3	2.59		★S.META+ 0.04 136 · ★ES-CH 5.20 2 · R -0.11 5
				0.12 153			★S.PARA+ -0.02 136 · F 0.09 5 · R+ -0.11 4
C3H9O2Si1	SiCH3(OCH3)2		★ 0.04 3	★ 0.10 3	2.68		★S.META+ 0.04 136 · ★ES-CH 6.20 2 · R 0.05 4
				0.18 153			★S.PARA+ 0.01 136 · F 0.05 4 · R+ -0.04 4
C3H9O3Si1	Si(OCH3)3		★ 0.09 3	★ 0.13 3	2.77		S.INDUC -0.07 92 · ★S.PARA+ 0.13 136 · ★ES-CH 7.20 2
				0.19 153			S.INDUC 0.00 73 · ★S.ZE.RS 0.08 140 · F 0.10 5
							S.INDUC 0.01 140 · S.ZE.RS 0.13 73 · R 0.03 5
							★S.STAR 0.02 140 · S.ZE.RS 0.13 92 · R+ 0.03 4
							S.META+ 0.09 136 · ★S.RES.+ 0.25 140
C3H9P1	P(CH3)3+		0.47 105	0.63 81			S.INDUC 0.43 81 · S.PARA- 1.14 373 · B1-STM 2.69 1
			0.50 3	★ 0.73 111			★S.INDUC 0.56 126 · ★S.ZE.RS 0.08 86 · B5-STM 3.33 1
			★ 0.74 373	0.80 3			★S.STAR 2.50 3 · S.ZE.RS 0.08 387 · F 0.71 4
			0.97 141	0.97 141			S.STAR 3.62 360 · S.ZE.RS 0.12 126 · R 0.02 4
							★S.META- 0.81 328 · S.ZE.RS 0.20 81 · R- 0.59 4

Empirical	Structural	π	σ$_m$	σ$_p$	MR	E$_s$	additional σ values
C3H9P1	P(CH3)3+						★S.PARA- 0.95 235 ; ES-CH 4.20 2 ; GR.ELCT 0.26 13 S.PARA- 1.02 328 ; L-STM 4.37 1 ; S.CH-P+ 0.94 75
C3H9Pb1	Pb(CH3)3						S.INDUC -0.12 133 ; ★S.ZE.RS -0.10 126 ; B1-STM 2.95 1 S.INDUC 0.08 242 ; S.ZE.RS 0.05 133 ; B5-STM 3.85 1 ★S.INDUC 0.12 126 ; L-STM 4.63 1 ; GR.ELCT -0.30 13
C3H9S1Si1	S-Si(CH3)3						★S.PARA- 0.14 389 ; ★ES-CH 5.40 2 ; B1-STM 1.70 1 ★S.PARA+ -0.44 389 ; L-STM 5.64 1 ; B5-STM 4.97 1
C3H9Si1	Si(CH3)3	2.59	-0.21 503 -0.12 23 -0.09 17 -0.08 246 -0.08 795 -0.05 10 -0.04 801 ★-0.04 21 -0.02 503 0.02 221 0.11 3	-0.11 240 -0.07 17 -0.07 23 -0.07 338 -0.07 602 -0.07 801 ★-0.07 21 -0.06 503 -0.05 153 -0.03 221 -0.02 503 -0.01 10 -0.01 795 0.00 3 0.02 246 0.07 60	2.50	★-2.91 35	S.INDUC -0.42 599 ; S.META- -0.04 321 ; S.ZE.RS 0.04 67 S.INDUC -0.24 12 ; S.META- -0.03 802 ; S.ZE.RS 0.04 240 S.INDUC -0.20 16 ; S.META- 0.00 761 ; S.ZE.RS 0.04 333 S.INDUC -0.14 99 ; S.PARA- 0.06 624 ; ★S.ZE.RS 0.05 12 ★S.INDUC -0.11 6 ; S.PARA- 0.06 780 ; S.ZE.RS 0.05 73 S.INDUC -0.11 26 ; S.PARA- 0.07 803 ; S.ZE.RS 0.05 126 S.INDUC -0.10 17 ; S.PARA- 0.07 884 ; S.ZE.RS 0.07 133 S.INDUC -0.10 20 ; S.PARA- 0.08 802 ; S.ZE.RS 0.09 114 S.INDUC -0.09 133 ; S.PARA- 0.08 885 ; S.ZE.RS 0.12 6 S.INDUC -0.04 73 ; S.PARA- 0.09 321 ; S.BA.RS 0.06 20 S.INDUC -0.04 142 ; S.PARA- 0.17 761 ; S.AN.RS 0.14 20 S.INDUC 0.02 359 ; S.PARA+ -0.22 299 ; S.RES.+ 0.06 20 S.INDUC 0.03 126 ; S.PARA+ -0.09 735 ; ES-CH 4.20 2 S.INDUC 0.03 242 ; S.PARA+ -0.05 136 ; ★ES-V 1.40 7 S.STAR -0.81 3 ; S.PARA+ -0.03 60 ; L-STM 4.09 1 S.STAR -0.74 87 ; S.PARA+ -0.03 602 ; B1-STM 2.76 1 S.STAR -0.15 216 ; S.PARA+ 0.00 550 ; B5-STM 3.48 1 S.STAR -0.14 446 ; ★S.PARA+ 0.02 43 ; F 0.01 4 S.STAR -0.14 568 ; S.PARA+ 0.03 383 ; R -0.08 5 S.STAR 0.28 377 ; S.PARA+ 0.04 175 ; R+ 0.01 4 S.META+ -0.16 735 ; S.ZE.RS -0.04 86 ; S-L -0.12 7 S.META+ 0.00 136 ; S.ZE.RS 0.02 365 ; S.ZTWST 0.03 22 S.META+ 0.01 43 ; S.ZE.RS 0.02 387 ; S.P.RAD 0.17 79 S.META- -0.08 803 ; S.ZE.RS 0.03 359 ; ★GR.ELCT -0.21 13
C3H9Sn1	Sn(CH3)3		★ 0.00 3 0.24 41	-0.26 240 -0.14 924 ★ 0.00 21 0.18 41	3.18		S.INDUC -0.11 133 ; S.ZE.RS -0.13 240 ; ES-CH 4.70 2 S.INDUC -0.04 142 ; S.ZE.RS -0.10 86 ; L-STM 4.48 1 S.INDUC 0.00 198 ; S.ZE.RS -0.02 126 ; B1-STM 2.90 1 S.INDUC 0.03 242 ; S.ZE.RS 0.00 114 ; B5-STM 3.75 1 S.INDUC 0.06 126 ; S.ZE.RS 0.00 142 ; F 0.03 5 S.INDQ -0.26 58 ; S.ZE.RS 0.01 198 ; R -0.03 5 ★S.STAR -1.29 61 ; S.ZE.RS 0.01 333 ; R+ -0.46 4 ★S.PARA+ -0.12 602 ; S.ZE.RS 0.05 133 ; GR.ELCT -0.22 13 S.ZE.RS -0.15 365 ; S.ZE.RS 0.07 387
C3H10Ge1O2P1	P(O)OHgEMe3						★S.ZE.RS 0.08 86 ; ★ES-CH 7.50 2
C3H10N1Si1	NHSi(Me)3						★S.INDUC 0.09 25 ; ★S.ZE.RS -0.78 25
C3H10O2P1Si1	P(O)OHSiMe3						★S.ZE.RS 0.09 86 ; ★ES-CH 7.40 2
C4F6I1O4	I(OCOCF3)2		★ 1.28 18	★ 1.34 18			★S.INDUC 1.22 273 ; ★S.ZE.RS 0.12 273 ; F 1.18 4 ★S.STAR 7.60 3 ; ★ES-CH 5.70 2 ; R 0.16 4
C4F7	cyc-C4F7		★ 0.48 30 0.49 3	★ 0.53 30			★S.INDUC 0.45 30 ; ★S.ZE.RS 0.08 30 ; F 0.45 5 ★S.STAR 2.81 3 ; ★ES-CH 7.00 2 ; R 0.08 5
C4F7O1	C=O(CF2CF2CF3)		★ 0.63 18	★ 0.79 18			★S.INDUC 0.46 18 ; ★ES-CH 5.60 2 ; R 0.24 4 ★S.ZE.RS 0.33 18 ; F 0.55 4
C4F9	C(CF3)3		0.35 520 ★ 0.55 48	0.52 520 ★ 0.55 48			S.INDUC 0.27 520 ; S.ZE.RS 0.26 520 ; F 0.53 852 S.INDUC 0.55 48 ; ★ES-CH 11.20 2 ; R 0.02 852 ★S.META- 0.39 731 ; L-STM 4.11 1 ; R- 0.42 4 ★S.PARA- 0.71 731 ; B1-STM 3.13 1 S.ZE.RS 0.00 48 ; B5-STM 3.64 1
	CF2CF2CF2CF3		★ 0.47 156 0.52 3	★ 0.52 156	1.76		★S.INDUC 0.39 156 ; ES-CH 6.20 2 ; F 0.44 5 S.STAR 2.44 3 ; L-STM 6.76 1 ; R 0.08 5 ★S.PARA- 0.73 558 ; B1-STM 1.99 1 ; R- 0.29 4 S.ZE.RS 0.11 18 ; B5-STM 5.05 1 ; GRP.DPL -2.86 8
C4F9O2S1	SO2C(CF3)3		★ 0.96 48 1.07 176	1.09 176 ★ 1.13 48			★S.INDUC 0.79 48 ; ★S.ZE.RS 0.34 48 ; R- 0.97 4 S.META- 1.02 164 ; F 0.84 4 ★S.PARA- 1.81 164 ; R 0.29 4
	SO2C4F9						S.INDUC 0.48 16
C4F9S1	SC(CF3)3		0.39 176 ★ 0.51 48	★ 0.58 48 0.83 176			★S.INDUC 0.44 48 ; ★S.ZE.RS 0.14 48 ; R- 0.32 4 S.META- 0.56 164 ; F 0.47 4 ★S.PARA- 0.79 164 ; R 0.11 4
C4F9S3	C(SCF3)3		★ 0.51 3	★ 0.53 3	4.40		★S.INDUC 0.49 64 ; S.ZE.RS 0.04 64 ; B5-STM 5.00 1 S.INDUC 0.49 320 ; ★ES-CH 7.60 2 ; F 0.49 5 ★S.STAR 3.06 3 ; L-STM 5.82 1 ; R 0.04 5 ★S.ZE.RS 0.03 320 ; B1-STM 3.32 1
C4F9Se1	SeC(CF3)3		★ 0.49 48	★ 0.54 48			★S.INDUC 0.44 48 ; F 0.46 4 ★S.ZE.RS 0.10 48 ; R 0.08 4
C4N3	C(CN)3		★ 0.97 40 0.98 3	★ 0.96 40 0.99 3	1.86		S.INDUC 0.94 67 ; ★S.ZE.RS 0.01 64 ; B5-STM 4.12 1 S.INDUC 0.98 40 ; S.ZE.RS 0.01 320 ; F 0.92 5 S.INDUC 1.00 320 ; ES-CH 7.00 2 ; R 0.04 5 S.STAR 6.10 3 ; L-STM 3.99 1 S.ZE.RS 0.00 67 ; B1-STM 2.87 1
C4H1	(C=-C)2H				1.81		★S.PARA- 0.72 860 ; L-STM 7.26 1 ; B5-STM 1.60 1 ★ES-CH 3.00 2 ; B1-STM 1.60 1
C4H1F6O1	A-OH-cyc-C4F6		★ 0.36 30 0.49 3	★ 0.37 30 0.53 3			★S.INDUC 0.35 30 ; ★S.ZE.RS 0.02 30 ; F 0.36 4 ★S.STAR 2.81 3 ; ★ES-CH 7.20 2 ; R 0.01 4
C4H1Fe1O5	OH/Fe(CO)4						S.INDUC 0.81 16
C4H1N2	CH=C(CN)2	0.05	0.45 294	0.70 40	1.97		★S.INDUC 0.43 40 ; S.ZE.RS 0.26 80 ; B5-STM 5.17 1

Empirical	Structural	π	σm	σp	MR	Es	additional σ values					
C4H1N2	CH=C(CN)2		0.55 3 ★ 0.66 163	★ 0.84 163			S.STAR 2.56 3 ★ S.META- 0.55 165 S.ZPTFT 0.80 248 ★ S.PARA- 1.20 165 ★ S.PARA+ 0.82 478 S.PARA+ 0.96 741	S.ZE.RS 0.29 294 S.AN-RS 0.68 80 S.AN-RS 0.75 294 ES-CH 4.00 2 L-STM 6.46 1 B1-STM 1.60 1	F 0.57 5 R 0.27 5 R+ 0.25 4 R- 0.63 4			
C4H1O4	CH=C(CO2)2(--)						★ S.PARA- 0.08 29	★ S.PARA- 0.46 36	★ ES-CH 4.00 2			
C4H2Br1N2	2-Br-4-pyrimidinyl						S.INDUC 0.34 108	S.ZE.RS 0.12 108				
	4-Br-2-pyrimidyl						★ S.INDUC 0.14 204	★ S.ZE.RS 0.13 204				
	6-Br-4-pyrimidinyl						S.INDUC 0.32 108	S.ZE.RS 0.11 108				
C4H2Br1O1	2-(5-bromofuryl)		★ 0.15 388	★ 0.00 295			★ ES-CH 3.00 2	F 0.23 4	R -0.23 4			
C4H2Br1S1@	2-(4-bromothienyl):						★ S.ORTHO 0.38 199					
	2-(5-bromothienyl):						S.STAR 1.29 331 ★ S.PARA+ -0.72 195	★ S.PARA+ -0.67 374 S.ORTHO 0.12 169	★ S.ORTHO 0.30 199			
C4H2Br1Se1@	2-(5-bromoselenienyl):						★ S.ORTHO 0.12 169					
C4H2Cl1N2	2-Cl-4-pyrimidinyl						S.INDUC 0.33 108	S.ZE.RS 0.12 108				
	2-Cl-pyrimidin-5yl						S.INDUC 0.32 113	S.ZE.RS -0.03 113				
	4-Cl-2-pyrimidyl						★ S.INDUC 0.15 204	★ S.ZE.RS 0.12 204				
	5-Cl-pyrimidin-2yl						S.INDUC 0.07 113	S.ZE.RS 0.11 113				
	6-Cl-4-pyrimidinyl						S.INDUC 0.30 108	S.ZE.RS 0.11 108				
C4H2Cl1S1@	2-(5-chlorothienyl)						★ S.STAR 1.26 331 S.ORTHO 0.13 169	★ S.ORTHO 0.30 199 S.ORTHO 0.63 675	★ ES-CH 3.20 2			
C4H2Cl1Se1@	2-(5-chloroselenienyl):						★ S.ORTHO 0.13 169					
C4H2F1N2	2-fluoro-4-pyrimidinyl						★ S.INDUC 0.32 15	★ S.ZE.RS 0.12 15				
	6-fluoro-4-pyrimidinyl						S.INDUC 0.28 15	S.ZE.RS 0.11 15				
C4H2F7	CH2CF2CF2CF3		★ 0.08 3		1.81		★ S.INDUC 0.14 32 ★ S.STAR 0.87 3 ★ ES-CH 4.60 2	L-STM 6.76 1 B1-STM 1.52 1 B5-STM 5.05 1	★ S-L 0.15 7			
C4H2F7O1	OCH2(CF2)2CF3						S.PHOSP -0.17 449					
C4H2I1O1	2-(5-iodofuryl)			★ -0.03 295			★ ES-CH 3.00 2					
C4H2I1S1@	2-(5-iodothienyl)						★ S.ORTHO 0.12 169					
C4H2I1Se1@	2-(5-iodoselenienyl):						★ S.ORTHO 0.11 169					
C4H2N1O2	D-3-pyrroline-2,5-dione		★ 0.33 212	★ 0.27 212			S.INDUC 0.38 212	★ S.ZE.RS -0.11 212	★ ES-CH 5.00 2			
C4H2N1O2S1	2-(5-nitrothienyl)						S.STAR 1.65 331	ES-CH 3.20 2				
C4H2N1O2S1@	2-(4-nitrothienyl):						★ S.ORTHO 0.66 199					
	2-(5-nitrothienyl):						S.ORTHO 0.29 169	★ S.ORTHO 0.81 199	S.ORTHO 0.94 675			
C4H2N1O2Se1@	2-(5-nitroselenienyl)						★ S.ORTHO 0.24 169					
C4H2N1O3	2-(5-nitrofuryl)			★ 0.20 295			★ ES-CH 3.00 2					
C4H3	CCCH=CH2						★ S.INDUC 0.35 6	★ ES-CH 3.00 2	★ S-L 0.35 14			
C4H3Br1	1-(5-Br-naphthyl)		★ 0.30 122 0.45 122	0.30 131			★ S.PARA+ -0.17 27 ★ ES-CH 1.00 2					
	1-(6-Br-naphthyl):		★ 0.28 55									
	1-(7-Br-naphthyl):			0.07 131								
	1-(8-Br-naphthyl):						S.PARA+ -0.20 27 ES-CH 2.30 2					
	2-(5-Br-naphthyl)			★ 0.27 55			★ ES-CH 0.00 2					
	2-(6-Br-naphthyl):		0.18 131	★ 0.26 55			★ ES-CH 0.00 2					
	2-(7-Br-naphthyl):		★ 0.11 55	★ 0.29 55			★ ES-CH 1.00 2					
	2-(8-Br-naphthyl)			★ 0.09 55			★ S.PARA+ -0.04 27					
C4H3Cl1	1-(5-Cl-naphthyl)		★ 0.30 122 0.44 55	0.29 131			★ S.PARA+ -0.17 27					
	1-(6-Cl-naphthyl)		0.17 131 ★ 0.25 55				★ S.PARA+ -0.18 27					
	1-(7-Cl-naphthyl)		★ 0.12 122 0.21 848				★ S.PARA+ -0.25 27 ★ ES-CH 1.00 2					
	1-(7-Cl-naphthyl)			★ 0.27 55			★ S.PARA+ -0.07 27					
	1-(8-Cl-naphthyl)						★ S.PARA+ -0.19 27	★ ES-CH 1.00 2				
	2-(5-Cl-naphthyl)						★ S.PARA+ -0.02 27	★ ES-CH 0.00 2				
	2-(6-Cl-naphthyl)			★ 0.24 55 0.30 939			S.PARA+ -0.16 27 ★ ES-CH 0.00 2					
	2-(8-Cl-naphthyl)			★ 0.08 635 0.09 55			★ S.PARA+ -0.06 27					
C4H3Cl1N1O2	N(4Cl-pyrroline-2,5-dione		★ 0.47 212	★ 0.46 212			★ ES-CH 5.00 2	F 0.47 4	R -0.01 4			
C4H3Cl1N3O1	4-MeO-6-Cl-2-triazinyl						★ S.INDUC 0.18 15	★ S.ZE.RS 0.22 15				
C4H3Cl2S1Si1	Si(2-thienyl)Cl2						★ S.STAR 1.38 216					
C4H3F1	1-(5-Fl-naphthyl)						★ S.PARA+ -0.19 27					
	1-(8-Fl-naphthyl)						★ S.PARA+ -0.20 27	★ ES-CH 1.00 2				
	2-(6-Fl-naphthyl)			★ 0.11 55			★ ES-CH 0.00 2					
	2-(7-Fl-naphthyl)			★ 0.22 55								
	2-(8-Fl-naphthyl)			★ 0.12 55			★ ES-CH 0.00 2					
C4H3I1	1-(8-I-naphthyl)						★ S.PARA+ -0.33 27	★ ES-CH 2.70 2				
	2-(6-I-naphthyl)			0.23 55			ES-CH 0.00 2					

Empirical	Structural	π	σm	σp	MR	Es	additional σ values
C4H3I1	2-(7-I-naphthyl)			0.27 55			
	2-(8-I-naphthyl)			0.07 55		ES-CH 0.00 2	
C4H3N1O2	1-(5-NO2-naphthyl)		★ 0.50 122, 0.55 848, 0.79 55	0.54 131		★ES-CH 1.00 2	
	1-(6-NO2-naphthyl)		★ 0.40 122, 0.41 131				
	1-(7-NO2-naphthyl)		0.39 122, ★ 0.55 55	0.36 131		★ES-CH 1.00 2	
	2-(5-NO2-naphthyl)			★ 0.36 122		★ES-CH 0.00 2	
	2-(6-NO2-naphthyl)			★ 0.54 122		★ES-CH 0.00 2	
	2-(7-NO2-naphthyl)			★ 0.36 122, 0.56 55			
	2-(8-NO2-naphthyl)			★ 0.40 122, 0.42 55			
C4H3N2	2-pyrimidinyl		★ 0.23 189	★ 0.53 189			S.INDUC 0.05 113; S.INDUC 0.06 189; S.STAR 1.67 3; S.ZE.RS 0.09 189; S.ZE.RS 0.10 113; S.AN-RS 0.45 189; L-STM 6.28 1; B1-STM 1.71 1; B5-STM 3.11 1; F 0.13 4; R 0.40 4
	3-pyridazinyl		★ 0.28 15	★ 0.48 15			★S.INDUC 0.18 15; ★S.INDUC 0.26 96; ★S.ZE.RS 0.30 15; ★S.AN-RS 0.04 15
	4-pyridazinyl		★ 0.36 15	★ 0.59 15			★S.INDUC 0.21 15; ★S.ZE.RS 0.02 15; ★S.AN-RS 0.35 15
	4-pyrimidinyl		★ 0.30 189	★ 0.63 189			S.INDUC 0.13 189; S.INDUC 0.20 108; S.ZE.RS 0.09 108; S.ZE.RS 0.09 189; S.AN-RS 0.51 189; L-STM 5.29 1; B1-STM 1.71 1; B5-STM 3.11 1; F 0.18 4; R 0.45 4
	5-pyrimidinyl		★ 0.28 189	★ 0.39 189			★S.INDUC 0.21 189; S.INDUC 0.30 113; ★S.ZE.RS -0.04 189; ★S.AN-RS 0.18 189; F 0.25 4; R 0.14 4
	C(CH3)(CN)2		★ 0.60 3	★ 0.57 3	1.90		★S.INDUC 0.63 64; ★S.STAR 3.94 3; ★S.ZE.RS -0.06 64; ★ES-CH 6.00 2; L-STM 4.11 1; B1-STM 2.81 1; B5-STM 4.12 1; F 0.59 4; R -0.02 4
	pyrazinyl						★S.INDUC 0.25 96
C4H3N2@	2-pyrimidinyl:						★S.ORTHO 1.50 410; ★S.ORTH- 1.98 231
	3-pyridazinyl:		★ 1.40 410				★S.ORTH- 1.66 231; F 0.21 4; R 0.27 4
	4-pyridazinyl:		★ 1.61 410				★S.META- 1.69 231
	4-pyrimidinyl:						★S.ORTHO 1.71 410; ★S.ORTH- 1.78 231
	5-pyrimidinyl:		1.30 410				S.META- 1.54 231
	pyrazinyl:		★ 1.40 410				★S.ORTH- 1.72 231
C4H3N2O1	3-(6-oxopyridazinyl)						★S.INDUC 0.27 96
	CH=C(CONH2)CN						★S.ZPTFT 0.57 248; ★ES-CH 4.00 2; L-STM 5.72 1; B1-STM 1.60 1; B5-STM 5.17 1
C4H3O1	2-furyl		0.05 228, ★ 0.06 524, 0.09 388	0.01 228, ★ 0.02 524	1.79, 1.81		★S.INDUC 0.04 32; S.INDUC 0.09 228; S.INDUC 0.15 16; S.STAR 0.25 3; ★S.STAR 1.08 331; ★S.META+ 0.10 524; S.META- 0.11 736; ★S.PARA- 0.21 736; ★S.PARA+ -0.39 524; S.ZE.RS -0.14 47; ★S.BA.RS 0.00 34; ★ES-CH 3.00 2; F 0.10 5; R -0.08 5; R+ -0.49 4; R- 0.11 4; S-L 0.17 14
	3-furyl					★ -3.45 35	★S.STAR 0.62 3; ★S.STAR 0.65 331; ★ES-CH 3.00 2
C4H3O1@	2-furyl:					★ -3.73 35	S.INDUC 0.60 667; S.ORTHO -0.30 430; S.ORTHO -0.19 548; S.ORTHO 0.10 408; S.ORTHO 0.27 570; S.ORTHO 0.57 364; S.ORTHO 0.61 627; S.ORTHO 0.63 533; S.ORTHO 1.04 471; S.ORTHO 1.08 259; S.ORTH+ -1.32 364; S.ORTH+ -1.03 528; S.ORTH+ -0.95 195; S.ORTH+ -0.93 509; S.ORTH+ -0.90 374; S.ORTH+ -0.89 529; S.ORTH+ -0.85 423; S.ORTH+ -0.75 404; S.ORTH+ -0.51 371; S.ORTH+ -0.51 430; S.ORTH+ -0.22 408; S.ORTH+ -0.13 480; S.ORTH- 0.10 148; S.ORTH- 0.18 124; S.ZE.RS 0.60 148
	3-furyl:		0.00 570, ★ 0.25 259, 0.27 533, 0.42 364				S.META+ -0.88 528; S.META+ -0.74 364; S.META+ -0.51 195; S.META+ -0.44 423; S.META+ -0.42 529
C4H3O3	CO2COCH=CH2						S.INDUC 0.24 16
C4H3O3S1	1-(6-SO3-naphthyl)-		★ 0.04 122				
	1-(7-SO3-naphthyl)-		★ 0.10 122				★ES-CH 1.00 2
	2-(6-SO3-naphthyl)			★ 0.12 122			
	2-(7-SO3-naphthyl)			★ 0.08 122			★ES-CH 0.00 2
	2-(8-SO3-naphthyl)		★ 0.10 122	★ 0.07 122, ★ 0.10 122			★ES-CH 0.00 2; ★ES-CH 1.00 2
C4H3O4	CH2CH(CO2)2 (--)						★S.PARA- -0.31 29; ★S.PARA- 0.06 36; ★ES-CH 4.00 2
C4H3S1	2-thienyl	1.61	0.08 228, ★ 0.09 497, 0.11 10	0.02 228, ★ 0.05 497	2.40		S.INDUC 0.12 228; ★S.INDUC 0.21 946; S.INDUC 0.24 16; S.STAR 0.93 331; S.STAR 1.31 3; S.META+ 0.16 523; ★S.META- 0.11 498; ★S.PARA- 0.19 498; ★S.PARA+ -0.43 523; S.PARA+ -0.33 471; S.ZE.RS -0.13 47; ES-CH 3.20 2; L-STM 6.53 1; B1-STM 1.64 1; B5-STM 3.37 1; F 0.13 5; R -0.08 5; R+ -0.56 4; R- 0.06 4; S-L 0.19 14; GRP.DPL -0.81 8
	3-thienyl	1.81 219	★ 0.03 497	★ -0.02 497	2.40		S.STAR 0.62 3; ★S.PARA- 0.13 498; R -0.10 5

Empirical	Structural	π	σm	σp	MR	Es	additional σ values
C4H3S1	3-thienyl		0.06 10				★S.STAR 0.65 331; S.META+ 0.08 523; ★S.META- 0.07 498 — ★S.PARA+ -0.38 523; ES-CH 3.20 2; F 0.08 5 — R+ -0.46 4; R- 0.05 4
C4H3S1@	2-thienyl:						S.INDUC 0.90 667; S.PARA- 0.58 530; S.ORTHO -0.21 430; S.ORTHO -0.07 548; S.ORTHO -0.06 408; S.ORTHO 0.02 169; S.ORTHO 0.36 627; S.ORTHO 0.37 533; S.ORTHO 0.37 570; S.ORTHO 0.50 364; S.ORTHO 0.63 1008; S.ORTHO 0.67 471 — ★S.ORTHO 0.72 259; ★S.ORTH+ -1.22 364; S.ORTH+ -1.06 528; S.ORTH+ -0.91 491; ★S.ORTH+ -0.85 195; S.ORTH+ -0.80 753; ★S.ORTH+ -0.79 374; S.ORTH+ -0.79 509; S.ORTH+ -0.79 529; S.ORTH+ -0.79 725; S.ORTH+ -0.76 423; S.ORTH+ -0.68 404 — S.ORTH+ -0.44 480; S.ORTH+ -0.38 371; S.ORTH+ -0.38 430; S.ORTH+ -0.15 408; S.ORTH+ 0.26 787; S.ORTH- 0.24 124; S.ORTH- 0.30 148; S.ORTH- 0.58 565; S.ZE.RS 0.40 148; O-STER -2.46 90
	3-thienyl:		-0.04 570; 0.07 627; 0.09 533; ★0.12 259				S.META+ -0.52 725; S.META+ -0.50 528; S.META+ -0.49 145; ★S.META+ -0.49 195; ★S.META+ -0.49 374 — S.META+ -0.47 753; S.META+ -0.44 423; S.META+ -0.38 529; S.META+ -0.24 404; S.META+ -0.10 480 — S.META+ -0.09 124; S.META+ -0.05 787; S.ZE.RS 0.20 148
C4H3Se1	2-selenienyl		★0.06 556; 0.09 228	0.01 228; ★0.04 556			S.INDUC 0.15 228; ★S.META- 0.16 557; ★S.PARA- 0.22 557 — S.ZE.RS -0.14 228; ★ES-CH 3.30 2; F 0.10 4 — R -0.06 4; R- 0.12 4
C4H3Se1@	2-selenienyl:						S.ORTHO -0.24 430; ★S.ORTHO -0.09 548; S.ORTHO 0.01 169; S.ORTHO 0.28 530 — S.ORTHO 0.55 364; ★S.ORTHO 0.60 471; S.ORTH+ -1.28 364; ★S.ORTH+ -0.88 509 — S.ORTH+ -0.43 371; S.ORTH+ -0.43 430
C4H3Te1	2-tellurienyl		★0.06 556	★0.03 556			★S.META- 0.10 557; ★S.PARA- 0.25 557
C4H3Te1@	2-tellurophen:						★S.PARA+ -0.92 509; ★S.ORTHO 0.23 471
C4H4	(CH)4 (2,3);2-naphthyl:	1.32	-0.19 804; 0.12 531; ★0.22 531		1.75		S.META+ -0.51 101; S.META+ -0.45 101; ★S.META+ -0.35 27; S.META+ -0.35 269 — ★S.META+ -0.35 583; S.META+ -0.14 456; ★S.ZPTFT 0.05 63; S.ORTHO 0.06 158 — S.ORTHO 0.24 104; S.ORTHO 0.28 53; S.ORTHO 0.50 78; ★ES-CH 1.00 2
	(CH)4(3,4)		0.01 804; 0.02 635; 0.03 531; 0.03 712; ★0.04 703; 0.17 421	0.02 635; 0.03 531; 0.03 712; ★0.04 703; 0.17 421	1.75		★S.ZPTFT 0.06 63; S.PARA- 0.05 514; S.PARA- 0.07 887; S.PARA- 0.11 307; ★S.PARA- 0.12 609; S.PARA- 0.14 28; S.PARA+ -0.51 101 — S.PARA+ -0.28 101; S.PARA+ -0.25 27; S.PARA+ -0.21 903; S.PARA+ -0.20 583; S.PARA+ -0.19 269; ★S.PARA+ -0.14 863; ★S.PARNO 0.07 65 — ES-CH 0.00 2; F 0.07 5; R -0.03 5; R+ -0.33 4; R- -0.07 4
C4H4Cl3	CH2CH=CHCCl3						S.STAR 0.19 3
C4H4N1	1-pyrryl	0.95	★0.47 163	-0.02 754; 0.10 226; 0.21 275; ★0.37 163	1.95		S.INDUC 0.35 91; S.ORTHO 0.56 754; S.ZE.RS -0.21 91; ES-CH 3.00 2 — L-STM 5.44 1; B1-STM 1.71 1; B5-STM 3.12 1; F 0.50 5 — R -0.13 5
	2-pyrryl						S.INDUC 0.17 15; S.STAR 0.46 15
C4H4N1@	2-pyrryl:						★S.ORTHO -0.24 259; S.ORTHO -0.15 758 — ★S.ORTH+ -2.10 371; ★S.ORTH+ -1.61 423 — S.ORTH+ -1.33 480; S.ORTH- -0.70 124
	3-pyrryl:		-0.75 758; ★-0.34 259				★S.META+ -1.20 423; ★S.META+ -0.54 480
C4H4N1O1	2-(4-Me-oxazolyl)						★S.INDUC 0.37 96
C4H4N1O2	1-pyrroline,2,5-dione		★0.34 212	★0.31 212			★S.INDUC 0.37 212; ★S.ZE.RS -0.05 212; L-STM 6.06 300 — B1-STM 1.71 300; B5-STM 3.79 300; F 0.36 4 — R -0.05 4
	pyrroline-2,5-dione		★0.34 212	★0.31 212			★S.ZE.RS -0.05 212; ★ES-CH 5.00 2; L-STM 6.06 1 — B1-STM 1.71 1; B5-STM 3.79 1; F 0.36 4 — R -0.09 4
C4H4N1S1	2-(4-Me-thiazolyl)						★S.INDUC 0.32 96
C4H4N3	2-NH2-pyrimidin-5yl						S.INDUC 0.21 113; S.ZE.RS -0.09 113
	5-NH2-pyrmidin-2yl						S.INDUC -0.04 113; S.ZE.RS 0.06 113
	6-(3-Me-1,2,4-triazinyl)						★S.INDUC 0.36 96
C4H4N3@	5(2-aminopyrimidine):						S.META+ 0.65 441
C4H4N3O1	4-MeO-2-triazinyl						★S.INDUC 0.15 15; ★S.ZE.RS 0.20 15
C4H4N3O2S1	S-(1,2,4-triazin-3yl-4-one-6-OH)						S.INDUC 0.52 495
C4H4O1	1-(5-OH-naphthyl)		★-0.10 55	-0.06 131			
	1-(6-OH-naphthyl)		★-0.12 55; -0.08 131				
	1-(7-OH-naphthyl)		★-0.15 55	-0.10 131			★ES-CH 1.00 2
	2-(5-OH-naphthyl)			★-0.04 55			★ES-CH 1.00 2
	2-(7-OH-naphthyl)			★-0.14 55			★ES-CH 0.00 2
	2-(8-OH-naphthyl)			★-0.22 55			★ES-CH 0.00 2
C4H4O2	CH=CHCH2CO2 (-)						★S.INDUC 0.02 6
C4H5	CCCH2CH3						★S.STAR 1.57 653; ★ES-CH 3.00 2
	CH2CCCH3						S.STAR 0.39 262; ★S.STAR 0.45 380; ★S.STAR 0.46 349 — ★ES-CH 3.00 2; L-STM 4.49 1; B1-STM 1.52 1 — B5-STM 5.05 1
	CH2CH=C=CH2						★S.STAR 0.23 349
	CH=CHCH=CH2 -E						★ES-CH 3.00 2; B1-STM 1.60 1; ★S-L 0.01 7

Empirical	Structural	π	σ$_m$	σ$_p$	MR	E$_s$	additional σ values		
C4H5	CH=CHCH=CH2 -E						L-STM 6.53 *1*	B5-STM 4.19 *1*	
	EtCCH				1.75		★ S.INDUC 0.05 *6* ★ S.STAR 0.28 *268*	★ ES-CH 3.00 *2* L-STM 6.71 *1*	B1-STM 1.52 *1* B5-STM 3.17 *1*
C4H5Cl1N2	2-Me-4-Cl-1-pyrazinium						★ S.INDUC 0.91 *15*	★ S.ZE.RS -0.05 *15*	
	2-Me-5-Cl-1-pyrazinium						★ S.INDUC 0.34 *15*	★ S.ZE.RS -0.06 *15*	
C4H5F3N1O2	CHOHCH2NHCOCF3	-0.39 *54*			2.51		★ ES-CH 4.00 *2* L-STM 7.67 *1*	B1-STM 1.73 *1* B5-STM 5.06 *1*	
C4H5N1	1-(5-NH2-naphthyl)		★ -0.20 *55*	-0.13 *131*					
	2-(5-NH2-naphthyl)			★ -0.06 *122*					
	2-(7-NH2-naphthyl)			★ -0.23 *55*					
	2-(8-NH2-naphthyl)			★ -0.20 *122* -0.01 *55*					
C4H5N2	2-(1-methylimidazolyl)						★ S.INDUC 0.26 *96*		
	2-(4-Me-imidazolyl)						★ S.INDUC 0.26 *96*		
	3-Me-1-pyrazyl						★ S.INDUC 0.26 *15*	★ S.ZE.RS -0.17 *15*	
	5-Me-1-pyrazyl						★ S.INDUC 0.01 *15*	★ S.ZE.RS -0.06 *15*	
	5-imidazolyl-Me	0.13 *130*							
C4H5N2@	2-(1-methylimidazolyl):						★ S.PARA+ -0.82 *620*		
	4-(1-methylimidazolyl):						★ S.PARA+ -1.01 *620*		
	5-(1-methylimidazolyl):						★ S.PARA+ -1.02 *620*		
C4H5N2O4	C(NO2)2(CH2CH=CH2)						★ S.STAR 3.54 *157*		
C4H5O1	1-cyclobutene-oxide						★ S.INDUC 0.03 *339*	★ S.ZE.RS -0.07 *339*	★ ES-CH 4.00 *2*
	C#CCH2OCH3						★ S.STAR 0.95 *183* ★ ES-CH 3.00 *2*	L-STM 7.38 *1* B1-STM 1.60 *1*	B5-STM 3.40 *1*
	CH=CHC(=O)CH3 -T						L-STM 6.29 *1*	B1-STM 1.60 *1*	B5-STM 4.16 *1*
	CH=CHCOCH3	-0.06	★ 0.21 *180*	★ -0.01 *180*	2.11		★ S.STAR 1.08 *460* ★ S.PARA+ 0.39 *191* ES-CH 3.00 *2*	F 0.31 *5* R -0.32 *5* R+ 0.08 *4*	GRP.DPL -2.89 *8*
	OCH2C#CCH3					★ -1.85 *350*	ES-CH 3.00 *2* L-STM 7.39 *1*	B1-STM 1.35 *1* B5-STM 3.07 *1*	
	OCH2CH=C=CH2					★ -1.70 *350*	ES-CH 3.00 *2*		
C4H5O1S1	SCH=CHC(=O)CH3 -E						L-STM 6.31 *1*	B1-STM 1.70 *1*	B5-STM 6.51 *1*
C4H5O2	CH=CHC=O(OCH3)				2.27		★ S.PARA- 0.50 *584* ★ S.ORTH- 0.62 *97*	★ ES-CH 3.00 *2* ★ GRP.DPL -2.13 *8*	
	CH=CHCH2CO2H						★ ES-CH 3.00 *2*	★ S-L 0.13 *14*	
	OCH=CHC(=O)CH3 -E						L-STM 6.21 *1*	B1-STM 1.35 *1*	B5-STM 6.02 *1*
C4H6Al1	Al(CH=CH2)2						S.INDUC -0.59 *16*		
C4H6As1	As(CH=CH2)2						S.INDUC 0.11 *16*		
C4H6B1	B(CH=CH2)2						S.INDUC -0.26 *16*		
C4H6F3	PrCF3				1.90		★ S.STAR 0.12 *52* ★ ES-CH 3.00 *2*	L-STM 6.76 *1* B1-STM 1.52 *1*	B5-STM 5.05 *1*
C4H6F4N1	NHCH(CH3)CF2CHF2						★ S.STAR 1.76 *170*	★ ES-CH 4.00 *2*	
C4H6Ga1	Ga(CH=CH2)2						S.INDUC -0.26 *16*		
C4H6I1O4	I(OCOCH3)2		★ 0.85 *18*	★ 0.88 *18*			★ S.INDUC 0.82 *273* ★ S.STAR 5.10 *3*	★ S.ZE.RS 0.06 *273* ★ ES-CH 5.70 *2*	F 0.80 *4* R 0.08 *4*
C4H6N1	(CH2)3CN				1.94	★ -1.74 *196*	★ S.STAR 0.30 *420* ES-CH 3.00 *2*	L-STM 6.04 *1* B1-STM 1.52 *1*	B5-STM 5.57 *1*
	C(CN)(CH3)2				1.91	-2.00 *999* ★ -2.00 *44*	ES-CH 5.00 *2* L-STM 4.11 *1*	B1-STM 2.77 *1* B5-STM 4.12 *1*	
C4H6N1O1	NHCH=CHC(=O)CH3 -E						L-STM 6.23 *1*	B1-STM 1.35 *1*	B5-STM 6.07 *1*
C4H6N1O2	N(COCH3)2		★ 0.35 *3*	★ 0.33 *3*	2.48		S.INDUC 0.37 *24* ★ S.INDUC 0.37 *68* ★ S.STAR 2.31 *3* S.ZE.RS -0.04 *24*	★ S.ZE.RS -0.04 *68* ★ ES-CH 7.00 *2* L-STM 4.45 *1* B1-STM 1.35 *1*	B5-STM 4.33 *1* F 0.36 *5* R -0.03 *5*
	NHCOCH2COCH3		★ -0.06 *194*		2.42		★ ES-CH 4.00 *2* L-STM 7.33 *1*	B1-STM 1.55 *1* B5-STM 4.68 *1*	
C4H6N2	2-Me-1-pyrazinium						★ S.INDUC 1.11 *15*	★ S.ZE.RS -0.08 *15*	
	N-Me-1-imidazinium						★ S.INDUC 0.93 *15*		
C4H6O2	CHC(=O)OEt(-)						S.ZPTFT -2.86 *100*	S.PARA+ -5.00 *100*	
C4H6P1	P(CH=CH2)2						S.INDUC 0.06 *16*		
C4H6Sb1	Sb(CH=CH2)2						S.INDUC 0.05 *16*		
C4H7	1-Me-cyclopropyl						S.PARA+ -0.53 *821*		
	CH2-cy-propyl				1.82		★ S.STAR 0.01 *516* ★ S.ZE.RS -0.12 *47*	★ ES-CH 4.00 *2* L-STM 5.14 *1*	B1-STM 1.52 *1* B5-STM 4.36 *1*
	CH2CH2CH=CH2				1.91		S.INDUC 0.02 *6* ★ S.PARA+ -0.26 *460* ES-CH 3.00 *2*	★ ES-V 0.75 *7* L-STM 6.35 *1* B1-STM 1.52 *1*	B5-STM 4.55 *1* ★ S-L 0.02 *14*
	CH2CH=CHCH3 -E		★ -0.04 *3*		1.91		★ S.INDUC 0.02 *6* S.STAR 0.00 *3* ★ S.STAR 0.13 *52* S.STAR 0.14 *349*	ES-CH 3.00 *2* L-STM 5.89 *1* B1-STM 1.52 *1* B5-STM 4.82 *1*	★ S-L 0.02 *14* GRP.DPL -0.34 *9*
	CH=C(CH3)2		★ -0.06 *3*		2.03		★ S.INDUC 0.05 *6* ★ S.STAR 0.19 *3* ★ ES-CH 4.00 *2*	L-STM 5.10 *1* B1-STM 1.60 *1* B5-STM 4.20 *1*	★ S-L 0.05 *14* GRP.DPL -0.34 *9*

Empirical	Structural	π	σ_m	σ_p	MR	E_s	additional σ values		
C4H7	CH=CHEt -E		★ -0.04 3		2.03		★S.INDUC 0.07 6 / ★S.STAR 0.31 3 / ★ES-CH 3.00 2	L-STM 6.35 1 / B1-STM 1.60 1 / B5-STM 4.19 1	★S-L 0.07 7 / GRP.DPL -0.34 9
	cyclobutyl		-0.13 152 / ★ -0.05 6	-0.15 271 / ★ -0.14 6 / -0.13 152	1.79	-1.30 50 / ★ -1.15 35	★S.STAR -0.15 124 / ★S.PARA- -0.07 400 / ★S.PARA+ -0.29 152 / S.ORTH+ -0.31 154 / S.ZE.RS -0.13 6	S.ZE.RS -0.12 19 / ES-CH 3.00 2 / ES-V 0.51 7 / L-STM 4.77 1 / B1-STM 1.77 1	B5-STM 3.82 1 / F 0.02 4 / R -0.16 4 / R+ -0.31 4 / R- -0.09 4
C4H7F4N2	EtC(NF2)2(CH3)						★S.STAR 0.37 590 / ★ES-CH 3.00 2	L-STM 6.17 1 / B1-STM 1.56 1	B5-STM 4.54 1
C4H7N2O2	CON(CH3)CONHCH3						★ES-CH 5.00 2 / L-STM 4.77 1	B1-STM 1.60 1 / B5-STM 5.82 1	
C4H7N2O2S1	SC(=O)N(CH3)C(=O)NHCH3						L-STM 6.15 1	B1-STM 1.70 1	B5-STM 6.54 1
C4H7N2O3	OC(=O)N(CH3)C(=O)NHCH3						L-STM 5.84 1	B1-STM 1.68 1	B5-STM 6.15 1
C4H7N2O4	C(Pr)(NO2)2						★S.STAR 3.34 157		
	EtC(CH3)(NO2)2				4.13		★S.STAR 0.35 291 / ★S.STAR 0.35 722	★ES-CH 3.00 2 / L-STM 6.64 1	B1-STM 1.52 1 / B5-STM 4.54 1
C4H7N2O5	C(NO2)2(CH2CH2CH2OH)						★S.STAR 3.53 157		
C4H7N3	3-4-diMe-1-triazolinium						★S.ZE.RS -0.06 15		
	3-5-diMe-1-triazolinium						★S.INDUC 0.81 15	★S.ZE.RS -0.03 15	
C4H7O1	CH2COCH2CH3						S.PARA+ -0.14 569		
	COC3H7					-1.64 35	★S.STAR 1.60 129 / ★S.ORTHO -0.40 78	★ES-CH 4.00 2 / L-STM 6.12 1	B1-STM 1.63 1 / B5-STM 4.50 1
	COCH(CH3)2		★ 0.38 3	★ 0.47 409			★S.ZE.RS 0.17 6 / ★S.ZE.RS 0.19 47 / ES-CH 5.00 2	L-STM 4.84 1 / B1-STM 1.99 1 / B5-STM 4.08 1	F 0.35 4 / R 0.12 4
	EtCOCH3				1.97	★ -2.13 44	★S.STAR 0.23 74 / ES-CH 3.00 2	L-STM 6.11 1 / B1-STM 1.52 1	B5-STM 4.04 1
	OCH2-cy-propyl						★ES-CH 3.00 2 / ★ES-V 0.48 7	L-STM 6.07 1 / B1-STM 1.35 1	B5-STM 4.57 1
	OCH2CH=CHCH3					★ -1.30 350	ES-CH 3.00 2		
C4H7O1S1	SC(=O)CH(CH3)2						L-STM 5.81 1	B1-STM 1.70 1	B5-STM 4.68 1
	SC(=O)Pr						L-STM 7.01 1	B1-STM 1.70 1	B5-STM 5.57 1
C4H7O2	(CH2)3CO2H				2.12	★ -1.65 44	ES-CH 3.00 2 / L-STM 5.65 1	B1-STM 1.52 1 / B5-STM 5.44 1	
	2,6-dioxocyclohexane						GRP.DPL -1.47 8		
	C(=O)OC3H7						S.ZE.RS 0.18 22 / ★ES-CH 4.00 2	L-STM 6.77 1 / B1-STM 1.64 1	B5-STM 4.83 1 / S.ZTWST 0.00 22
	C(O)OCH(CH3)2						★ES-CH 4.00 2 / L-STM 5.97 1	B1-STM 2.14 1 / B5-STM 3.43 1	★S-L 0.33 7 / ★GRP.DPL -1.82 8
	CH2C=O(OEt)				2.11		★S.INDUC 0.15 6 / S.STAR 0.82 901 / ★S.META+ -0.01 43 / S.META- 0.12 29 / ★S.PARA+ -0.16 43	S.PARA+ -0.06 569 / S.BA.RS -0.08 34 / ES-CH 4.00 2 / L-STM 4.54 1 / B1-STM 1.52 1	B5-STM 6.64 1 / S-L 0.16 7 / GRP.DPL -1.85 8
	CH2CH2CO2CH3						★S.STAR 0.26 3	★ES-CH 3.00 2	★S-L 0.07 14
	CH2CH2OC(=O)CH3						L-STM 6.79 1 / B1-STM 1.52 1	B5-STM 5.13 1 / GRP.DPL -1.86 8	
	CH2OC(O)CH3						GRP.DPL -1.80 8		
	OC(=O)CH(CH3)2						L-STM 5.50 1	B1-STM 1.35 1	B5-STM 4.03 1
C4H7O2S1	SC(=O)OPr						L-STM 7.96 1	B1-STM 1.70 1	B5-STM 6.42 1
	SCH2CO2CH2CH3						★ES-CH 3.20 2	★S-L 0.28 14	
C4H7O3	OC(=O)OPr						L-STM 7.94 1	B1-STM 1.35 1	B5-STM 5.66 1
	OCH2CO2CH2CH3			★ -0.18 304			L-STM 7.90 1	B1-STM 1.35 1	B5-STM 5.69 1
C4H7S1	SCH2CH2CH=CH2						★S.STAR 1.39 3	★ES-CH 3.20 2	★S-L 0.26 14
C4H7S2	2-(1,3-dithiocyclohexane				3.34	★ -2.93 115	★S.STAR 0.92 115	ES-CH 3.40 2	★ES-V 1.16 7
C4H8	(CH2)4 (2,3)						S.STAR 0.10 837 / ★S.ORTHO -0.13 53	★S.PHOSP -1.78 354 / ★S.PHOSP -1.65 941	GRP.DPL 0.73 8
	(CH2)4 (3,4)	1.39 587	★ -0.48 421	★ -0.48 421	1.86		★S.STAR -0.26 268 / ★S.PARA+ -0.41 27	ES-CH 0.00 2 / F -0.40 5	R -0.08 5
C4H8B1Cl2O2	B(OCH2CH2Cl)2						★S.STAR 0.75 358		
C4H8Cl2N1	N(CH2CH2Cl)2						★S.INDUC 0.04 49	★S.ZE.RS -0.19 49	★ES-CH 5.00 2
C4H8Cl2O3P1	PO(OCH2CH2Cl)2				4.08		★S.INDUC 0.52 34 / ★S.BA.RS 0.08 34	★ES-CH 6.20 2 / L-STM 7.72 1	B1-STM 2.52 1 / B5-STM 5.67 1
C4H8N1	1-pyrrolidinyl			★ -0.90 475			★S.ZE.RS -0.63 19 / ★S.ZE.RS -0.63 47	★ES-CH 3.00 2 / L-STM 4.90 1	B1-STM 1.90 1 / B5-STM 4.09 1
	2-pyrrolidinyl	-0.48 42					★ES-CH 3.00 2		
	CH=NCH(CH3)2						★ES-CH 3.00 2 / L-STM 6.00 1	B1-STM 1.70 1 / B5-STM 3.78 1	
C4H8N1O1	C(=O)NHPR						L-STM 6.76 1	B1-STM 1.63 1	B5-STM 5.20 1
	CH2C(=O)N(CH3)2						L-STM 6.02 1	B1-STM 1.52 1	B5-STM 3.78 1
	CH2CH2NHCOCH3			★ -0.06 147			★S.STAR 0.23 3	★ES-CH 3.00 2	
	CH=NOPr						L-STM 8.52 1	B1-STM 1.60 1	B5-STM 5.17 1
	Et-CONHCH3						S.STAR 0.25 3 / L-STM 6.98 1	B1-STM 1.52 1 / B5-STM 4.93 1	

Empirical	Structural	π	σ_m	σ_p	MR	E_s	additional σ values					
C4H8N1O1	N-morpholino			★ -0.50 475			★S.STAR 0.69 3	L-STM 5.29 1	B5-STM 3.42 1			
							★ES-CH 3.00 2	B1-STM 1.35 1				
	NHC=OCH(CH3)2	-0.18	★ 0.11 56	★ -0.10 56	2.42		★ES-CH 4.00 2	B1-STM 1.35 1	F 0.21 5			
							L-STM 5.53 1	B5-STM 4.09 1	R -0.31 5			
	OCH=NCH(CH3)2 -E						L-STM 6.39 1	B1-STM 1.35 1	B5-STM 5.58 1			
	Pr-CONH2		★ -0.08 3		2.37		★S.INDUC 0.02 32	L-STM 6.98 1	★S-L 0.04 7			
							★S.STAR 0.12 3	B1-STM 1.52 1				
							★ES-CH 3.00 2	B5-STM 5.23 1				
C4H8N1O2	CH(Pr)NO2						★S.STAR 1.61 157					
	NHCH2CH2OC(=O)CH3						L-STM 7.91 1	B1-STM 1.35 1	B5-STM 5.77 1			
	NHCH2CO2Et		★ -0.10 316	★ -0.68 3	2.69		★S.STAR 1.60 170	L-STM 7.91 1	B5-STM 5.77 1			
							★ES-CH 3.00 2	B1-STM 1.35 1				
	OCH2CON(CH3)2	-1.36			2.47		★ES-CH 3.00 2	B1-STM 1.35 1				
							L-STM 6.71 1	B5-STM 5.34 1				
C4H8N1O3	NHCH2C(=O)OCH2CH2OH						L-STM 8.59 1	B1-STM 1.35 1	B5-STM 6.11 1			
C4H8N1S1	SCH=NCH(CH3)2 -E						L-STM 6.44 1	B1-STM 1.70 1	B5-STM 6.14 1			
C4H8N3O2	NHC(=O)N(CH3)C(=O)NHCH3						L-STM 5.88 1	B1-STM 1.64 1	B5-STM 6.19 1			
C4H8O1	CH2CH2OCH2CH2*				2.04		★S.STAR 0.67 268					
	O(CH2)3CH2*						★S.PHOSP -1.44 354					
C4H8O2	C(Me)2CH2CO2 (-)						★S.PARA- -0.17 29	★S.PARA- -0.01 36	★ES-CH 5.00 2			
	OCH(CH3)CH(CH3)O*						★S.PHOSP 0.62 479					
	OCH(CH3)CH2CH2O*						S.PHOSP -0.75 577	S.PHOSP -0.75 674	S.PHOSP 1.06 479			
C4H9	Bu	2.13	★ -0.08 484	-0.19 418	1.96	-1.94 379	S.INDUC -0.06 1011	★S.PARA+ -0.29 76	R -0.15 5			
			-0.07 3	★ -0.16 23		-1.63 50	★S.INDUC -0.04 32	S.ZE.RS -0.12 656	R+ -0.28 4			
						-1.59 302	S.INDUC -0.01 6	S.ZE.RS -0.11 22	R- -0.11 4			
						-1.43 62	S.STAR -0.25 3	ES-CH 3.00 2	S-L -0.01 14			
						-0.23 311	★S.STAR -0.13 52	ES-V 0.68 7	S.PHOSP -1.22 59			
							S.STAR -0.13 74	L-STM 6.17 1	S.ZTWST -0.10 22			
							S.STAR -0.13 516	B1-STM 1.52 1	K-CHGTR 0.07 77			
							S.ZPTFT -0.15 337	B5-STM 4.54 1	GRP.DPL 0.08 9			
							S.PARA- -0.12 138	F -0.01 5				
	C(CH3)3	1.98	-0.20 518	-0.20 240	1.96	-3.70 379	S.INDUC -0.20 12	S.PARA- -0.15 211	S.ZE.RS 0.13 114			
			-0.12 23	-0.20 518		-3.11 302	★S.INDUC -0.07 32	★S.PARA- -0.13 181	S.RES.+ -0.13 6			
			★ -0.10 21	★ -0.20 21		★ -2.78 50	S.INDUC -0.07 77	S.PARA+ -0.37 543	ES-CH 4.00 2			
			-0.09 46	-0.19 338		-2.55 62	S.INDUC -0.07 261	S.PARA+ -0.31 145	ES-V 1.24 7			
				-0.15 46		-1.46 311	S.INDUC -0.01 6	S.PARA+ -0.30 329	ES-HYBO -4.22 225			
							S.INDUC 0.02 64	S.PARA+ -0.27 65	L-STM 4.11 1			
							S.INDUC 0.04 142	★S.PARA+ -0.26 43	B1-STM 2.60 1			
							S.INDUC 0.15 16	S.PARA+ -0.25 88	B5-STM 3.17 1			
							S-INDQ -0.18 58	S.PARNO -0.14 65	F -0.02 5			
							S.STAR -0.30 52	S.ORTHO -0.52 53	R -0.18 5			
							S.STAR -0.30 74	S.ORTHO -0.11 158	R+ -0.17 4			
							S.STAR -0.30 129	S.ORTHO 0.66 78	R- -0.04 4			
							S.STAR -0.30 325	S.ORTH+ -0.28 154	S-L -0.01 14			
							S.STAR -0.30 568	S.ORTH+ -0.23 822	S.PHOSP -1.55 59			
							S.META+ -0.19 543	S.ORTH- -0.08 97	S.ZTWST -0.13 22			
							S.META+ -0.06 43	S.ZE.RS -0.18 6	S.P.RAD 0.03 206			
							S.META- -0.09 28	S.ZE.RS -0.17 64	S.P.RAD 0.08 79			
							S.ZPTFT -0.17 63	S.ZE.RS -0.15 142	GRP.DPL 0.52 8			
							S.ZPTFT -0.16 112	S.ZE.RS -0.15 240	GR.ELCT 0.16 13			
							S.ZPTFT -0.14 95	S.ZE.RS -0.13 12	O-STER -3.94 90			
							S.PARA- -0.30 138	S.ZE.RS -0.13 19				
	CH(CH3)(Et)	1.80 130	★ -0.08 3	-0.19 418	1.96	-2.98 379	S.INDUC -0.10 12	S.ZPTFT -0.16 337	L-STM 4.92 1			
		2.04 455		-0.12 259		★ -2.37 50	S.INDUC -0.07 348	S.ORTH+ -0.28 154	B1-STM 1.90 1			
				★ -0.12 23		-2.24 302	S.INDUC -0.06 261	★S.ORTH- -0.18 97	B5-STM 3.49 1			
						-2.12 62	★S.INDUC -0.03 32	S.ZE.RS -0.12 12	F -0.02 4			
							S.INDUC -0.01 6	S.ZE.RS -0.12 19	R -0.10 4			
							★S.STAR -0.21 52	ES-CH 4.00 2	S-L -0.01 14			
							S.STAR -0.21 74	ES-V 1.02 7	S.PHOSP -1.36 59			
							S.STAR -0.19 3	ES-HYBO -1.15 225	GR.ELCT 0.15 13			
	CH2CH(CH3)2	1.70 130	★ -0.07 641	★ -0.12 23	1.96	-2.48 379	S.INDUC -0.06 348	★S.PARA- 0.01 138	B5-STM 4.45 1			
						★ -2.17 50	★S.INDUC -0.03 32	S.ZE.RS -0.12 47	F -0.01 4			
						-2.05 62	S.INDUC -0.01 6	ES-CH 4.00 2	R -0.11 4			
						-1.69 302	S.INDUC 0.10 16	ES-V 0.98 7	R- 0.00 4			
						-0.41 311	S.STAR -0.19 3	ES-HYBO -1.06 225	★S-L -0.01 14			
							S.STAR -0.19 568	L-STM 4.92 1	S.PHOSP -1.30 59			
							★S.STAR -0.13 52	B1-STM 1.52 1				
C4H9Br1N1O1S1	NHCOCH2S(CH3)2+Br-			★ -0.10 249			★ES-CH 4.00 2					
C4H9N1O1S1	NHC(=O)CH2S+(CH3)2						L-STM 7.39 1	B1-STM 1.55 1	B5-STM 5.11 1			
C4H9N2	1-piperazinyl						L-STM 6.17 1	B1-STM 1.91 1	B5-STM 3.49 1			
	N=N-T-butyl		★ 0.24 396	★ 0.28 396			★S.ZPTFT 0.30 396	★S.PARA+ 0.15 368				
	NHCH=NCH(CH3)2 -E						L-STM 6.40 1	B1-STM 1.35 1	B5-STM 5.64 1			
C4H9N2O1	(CH2)3C(=NOH)NH2						L-STM 7.69 1	B1-STM 1.52 1	B5-STM 5.93 1			
	C=O(NHNHCHMe2)	-0.28 42					★ES-CH 4.00 2	B1-STM 1.60 1				
							L-STM 6.68 1	B5-STM 5.35 1				
C4H9N2S1	CH2SC=NCH3(NHCH3)			★ -0.02 249	3.26		★ES-CH 3.20 2	B1-STM 1.35 1				
							L-STM 4.80 1	B5-STM 4.35 1				
C4H9N4O4	CH2N(NO2)EtN(NO2)CH3						★ES-CH 4.00 2	★S-L 0.14 14				
C4H9O1	(CH2)3OCH3				2.13	★ -1.66 44	★S.STAR 0.02 473	★ES-V 0.69 7	B1-STM 1.52 1			
							ES-CH 3.00 2	L-STM 6.83 1	B5-STM 4.85 1			

Empirical	Structural	π	σm	σp	MR	Es	additional σ values				
C4H9O1	C(CH3)2CH2OH						★ES-CH 5.00 2	B1-STM 3.16 1		
							L-STM 5.55 1	B5-STM 3.16 1		
	C(OCH3)(CH3)2				2.13	★ -2.67 44	ES-CH 5.00 2	B1-STM 2.54 1		
							L-STM 4.78 1	B5-STM 3.40 1		
	C(OH)(CH3)(C2H5)						★S.STAR 0.30 376	L-STM 4.92 1	B5-STM 3.49 1	
							★ES-CH 5.00 2	B1-STM 2.40 1		
	CH(OCH3)Et						★ES-CH 5.00 2	L-STM 4.92 1	B5-STM 4.29 1	
							★ES-V 1.22 7	B1-STM 1.90 1		
	CH(OEt)CH3				2.13	★ -1.93 44	ES-CH 4.00 2	B1-STM 1.90 1		
							L-STM 6.01 1	B5-STM 4.45 1		
	CH(OH)C3H7				2.11	-1.57 44	ES-CH 4.00 2	L-STM 6.17 1	B5-STM 4.54 1	
						★ -1.47 35	★ES-V 0.71 7	B1-STM 1.73 1		
	CH2C(OH)(CH3)2		★ -0.16 3	★ -0.17 147	2.11	★ -2.98 44	★S.INDUC -0.04 32	L-STM 4.92 1	F -0.11 4	
							★S.STAR -0.25 3	B1-STM 1.52 1	R -0.06 4	
							ES-CH 5.00 2	B5-STM 4.19 1	★S-L -0.03 7	
	CH2CHOHEt				2.11		★S.STAR 0.15 594	L-STM 6.17 1	B5-STM 4.54 1	
							★ES-CH 4.00 2	B1-STM 1.52 1		
	CH2OC3H7				2.13	★ -1.63 44	ES-CH 3.00 2	L-STM 6.82 1	B5-STM 4.87 1	
							★ES-V 0.65 7	B1-STM 1.52 1		
	CH2OCH(CH3)2						★ES-CH 3.00 2	L-STM 6.01 1	B5-STM 4.45 1	
							★ES-V 0.67 7	B1-STM 1.52 1		
	EtOEt				2.13	★ -2.21 44	★S.STAR 0.27 37	L-STM 6.85 1	★S.PHOSP -0.77 37	
							ES-CH 3.00 2	B1-STM 1.52 1	S.IND.P 0.50 39	
							★ES-V 0.89 7	B5-STM 4.81 1		
	O-Bu	1.03 57	★ -0.05 3	★ -0.32 21	2.17		★S.INDUC 0.28 6	ES-CH 3.00 2	★S.PHOSP -0.41 59	
		1.80 455	0.07 6	-0.30 6			S.INDUC 0.63 16	★ES-V 0.58 7	S.IND.P 1.57 39	
			★ 0.10 21				★S.STAR 1.55 37	L-STM 6.86 1	S.RES.P -0.81 39	
							S.STAR 1.68 3	B1-STM 1.35 1	GRP.DPL -1.26 9	
							S.PARA+ -0.81 485	B5-STM 4.79 1	GRP.DPL -1.19 8	
							S.ZE.RS -0.58 6	F 0.29 5		
							S.ZE.RS -0.43 47	R -0.61 5		
	O-T-butyl						S.INDUC 0.05 12	★ES-CH 5.00 2	B1-STM 1.35 1	
							★S.ZE.RS -0.34 47	★ES-V 1.22 7	B5-STM 4.35 1	
							S.ZE.RS -0.28 12	L-STM 4.80 1		
	OCH(CH3)Et		★ 0.25 3		2.17		★S.INDUC 0.26 32	★ES-V 0.86 7	B5-STM 4.42 1	
							S.STAR 1.62 3	L-STM 6.05 1	★S-L 0.29 7	
							ES-CH 4.00 2	B1-STM 1.35 1		
	OCH2CH(CH3)2				2.17		★S.STAR 1.52 37	L-STM 6.05 1	★S.PHOSP -0.30 85	
							★ES-CH 3.00 2	B1-STM 1.35 1	S.IND.P 1.43 39	
							★ES-V 0.62 7	B5-STM 4.42 1	S.RES.P -0.57 39	
C4H9O1S1	OCH2CH2SCH2CH3						L-STM 8.60 1	B1-STM 1.35 1	B5-STM 5.87 1	
	S(O)C(CH3)3						★S.ZE.RS 0.00 47	★ES-CH 6.20 2		
C4H9O1S2	SCH2CH2S(=O)Et						L-STM 8.12 1	B1-STM 1.70 1	B5-STM 6.59 1	
C4H9O2	CH2CH(OCH3)2						★S.STAR 0.46 473	L-STM 6.03 1	B5-STM 4.44 1	
							★ES-CH 4.00 2	B1-STM 1.52 1		
	CH2OCH2CH2OCH3						★ES-CH 3.00 2	★ES-V 0.57 7		
C4H9O2S1	OCH2CH2S(=O)Et						L-STM 8.11 1	B1-STM 1.35 1	B5-STM 5.81 1	
	SO2Bu						L-STM 6.97 1	B1-STM 2.03 1	B5-STM 4.94 1	
	SO2C(CH3)3						★S.ZE.RS 0.10 47	★ES-CH 7.20 2		
	SO2CH2CH(CH3)2						L-STM 6.17 1	B1-STM 2.03 1	B5-STM 4.54 1	
C4H9O2S2	SCH2CH2SO2Et						L-STM 8.12 1	B1-STM 1.70 1	B5-STM 6.59 1	
C4H9O2Si1	C(O)OSi(Me)3						S.INDUC 0.09 12	S.ZE.RS 0.23 12		
C4H9O3	C(OMe)3	0.14	★ -0.03 3	★ -0.04 3	2.48		S.INDUC -0.12 12	S.ZE.RS 0.04 12	B5-STM 4.29 1	
							★S.INDUC -0.12 92	S.ZE.RS 0.04 92	F 0.01 5	
							★S.INDUC -0.03 64	★ES-CH 7.00 2	R -0.05 5	
							★S.ZE.RS -0.01 64	L-STM 4.78 1		
							S.ZE.RS 0.00 19	B1-STM 2.56 1		
C4H9O3S1	(CH2)4SO3H						★S.PARA+ -0.30 301	★ES-CH 3.00 2		
	OCH2CH2SO2Et						L-STM 8.11 1	B1-STM 1.35 1	B5-STM 5.81 1	
C4H9S1	(CH2)2SC2H5						★S.STAR 0.28 278	★ES-V 0.79 7	B1-STM 1.52 1	
							★ES-CH 3.20 2	L-STM 7.21 1	B5-STM 5.42 1	
	CH(CH3)(SeT)				2.77	★ -2.77 115	★S.STAR 0.49 115	★ES-V 1.10 7	B1-STM 1.90 1	
							ES-CH 4.20 2	L-STM 6.65 1	B5-STM 4.60 1	
	S-Bu		★ 0.21 3		2.77		★S.INDUC 0.23 32	★ES-V 1.15 7	B5-STM 5.61 1	
							S.STAR 1.44 3	L-STM 7.06 1	★S-L 0.26 14	
							★ES-CH 3.20 2	B1-STM 1.70 1		
	SC(CH3)3						S.INDUC 0.20 16	★ES-V 1.60 7	B5-STM 4.72 1	
							★S.ZE.RS -0.07 47	L-STM 5.16 1		
							★ES-CH 5.20 2	B1-STM 1.70 1		
	SCH(CH3)Et						★S.STAR 1.32 3	L-STM 6.21 1	★S-L 0.25 14	
							★ES-CH 4.20 2	B1-STM 1.70 1		
							★ES-V 1.36 7	B5-STM 4.98 1		
	SCH2CH(CH3)2						★ES-CH 3.20 2	L-STM 6.42 1	B5-STM 4.98 1	
							★ES-V 1.15 7	B1-STM 1.70 1		
C4H9S1Se1	SeCH2CH2S-Et						L-STM 8.86 1	B1-STM 1.85 1	B5-STM 6.88 1	
C4H9S2	SCH2CH2SC2H5						L-STM 8.60 1	B1-STM 1.70 1	B5-STM 6.70 1	
C4H9Si1	SiCH=CH2(CH3)2		★ -0.04 3				★S.INDUC -0.03 73	★S.STAR 0.35 377	★ES-CH 5.20 2	
							★S.STAR -0.10 216	S.ZE.RS 0.05 73		

Empirical	Structural	π	σ_m	σ_p	MR	E_s	additional σ values
C4H9Si1	cy-Si(CH3)(CH2)3						★S.PARA- 0.18 624 ★S.PARA+ -0.20 299 ★ES-CH 4.20 2
C4H9Si3	C(SiCH3)3						★S.ZE.RS -0.24 47 ★ES-CH 7.20 2
C4H10As1	As(C2H5)2		★0.22 237	★0.00 237	3.29		S.INDUC 0.28 237 ★ES-CH 5.30 2 R- -0.24 4; ★S.PARA- 0.08 466 F 0.32 4; S.ZE.RS -0.28 237 R -0.32 4
C4H10As1O1	As(O)(Et)2		★0.57 237	★0.44 237			★S.INDUC 0.63 237 F 0.60 4; ★S.ZE.RS -0.34 237 R -0.16 4
C4H10As1S1	As(S)(Et)2		★0.52 237	★0.44 237			★S.INDUC 0.55 237 F 0.54 4; ★S.ZE.RS -0.23 237 R -0.10 4
C4H10B1N2	2(1,3-Me2-2-borimidazolyl)			★0.20 535			★ES-CH 5.00 2
C4H10B1O2	B(OEt)2						★S.STAR 0.50 358
C4H10Cl1Sn1	Sn(Et)2Cl						S.ZE.RS -0.01 365 S.ZE.RS 0.06 564
C4H10I1Sn1	Sn-I-(Et)2						S.ZE.RS -0.12 365 S.ZE.RS 0.05 564
C4H10N1	(CH2)4NH3+	-0.99 130					L-STM 6.89 1 B1-STM 1.52 1 B5-STM 4.87 1
	CH(CH3)N(CH3)2						★ES-CH 5.00 2 B1-STM 1.90 1; L-STM 4.11 1 B5-STM 4.34 1
	CH2CH2NHEt	-0.11 42					★ES-CH 3.00 2 B1-STM 1.52 1; L-STM 6.88 1 B5-STM 4.87 1
	CH2NHCH(CH3)2	0.14 42					★ES-CH 3.00 2 B1-STM 1.52 1; L-STM 6.07 1 B5-STM 4.47 1
	CH2NHCH2CH2CH3	0.25 42					★ES-CH 3.00 2 B1-STM 1.52 1; L-STM 6.88 1 B5-STM 4.87 1
	EtN(CH3)2	0.17 42		★-0.09 315	2.34		★S.STAR 0.13 473 L-STM 6.76 1 B5-STM 5.37 1; ★ES-CH 3.00 2 B1-STM 1.35 1
	N(Et)2	1.18	★-0.23 413 / -0.15 3	-1.42 661 / -0.90 413 / ★-0.72 236 / -0.70 194 / -0.53 3 / -0.51 335	2.48		S.INDUC 0.02 12 ★ES-V 1.37 7 S.IND.P -0.18 434; S.STAR 0.39 37 L-STM 4.83 1 S.IND.P 0.62 39; S.STAR 1.00 170 B1-STM 1.35 1 S.RES.P -1.36 434; S.ZPTFT -0.53 112 B5-STM 4.39 1 S.RES.P -0.60 39; ★S.PARA- -0.43 138 F 0.01 5 S.ZTWST -0.13 22; ★S.PARA+ -2.07 914 R -0.73 5 K-CHGTR 0.81 77; S.ORTHO -0.28 335 R+ -2.08 4 GR.ELCT 0.48 13; S.ZE.RS -0.57 19 R- -0.44 4; ES-CH 5.00 2 S.PHOSP -1.54 37
	NH-Bu	0.68 57 / 1.45	★-0.34 23	★-0.51 213	2.43		★S.STAR -1.08 3 L-STM 6.88 1 R -0.30 5; ★S.ZE.RS -0.54 19 B1-STM 1.35 1 ★GRP.DPL -1.27 9; ★ES-CH 3.00 2 B5-STM 4.87 1; ★ES-V 0.70 7 F -0.21 5
	NHC(CH3)3						★ES-CH 5.00 2 L-STM 4.83 1 B5-STM 4.39 1; ★ES-V 1.83 7 B1-STM 1.35 1
	NHCH(CH3)Et						★ES-CH 4.00 2 L-STM 6.07 1 B5-STM 4.47 1; ★ES-V 1.12 7 B1-STM 1.35 1
	NHCH2CH(CH3)2						★ES-CH 3.00 2 L-STM 6.07 1 B5-STM 4.47 1; ★ES-V 0.77 7 B1-STM 1.35 1
C4H10N1O1	OCH(CH3)N(CH3)2						L-STM 5.93 1 B1-STM 1.41 1 B5-STM 4.35 1
C4H10N1O2	N(CH2CH2OH)2						S.STAR 4.43 3
C4H10N1O2S1	CH2CH2N(CH3)SO2CH3			★-0.10 147			★ES-CH 3.00 2
	SO2N(Et)2						★K-CHGTR 1.31 77
C4H10N1S1	N=S(Et)2						S.INDUC -0.46 123
	SCH(CH3)N(CH3)2						L-STM 6.32 1 B1-STM 1.70 1 B5-STM 4.72 1
	SCH2CH2N(CH3)2						L-STM 7.02 1 B1-STM 1.70 1 B5-STM 5.86 1
C4H10N1Se1	SeCH2CH2N(CH3)2						L-STM 7.25 1 B1-STM 1.85 1 B5-STM 6.88 1
C4H10N2O3P1	N=NPO(OEt)2		★0.16 72	★0.74 72			S.INDUC 0.22 72 S.ZE.RS 0.27 72 R 0.79 4; S.PARA+ 0.67 72 F -0.05 4
C4H10N3	(CH2)3NHC(=NH)NH2	-1.01 130					L-STM 7.82 1 B1-STM 1.52 1 B5-STM 6.24 1
C4H10O1P1	P(O)(C2H5)2		★0.37 107 / 0.41 141	0.42 141 / ★0.47 107 / 0.53 911			★S.INDUC 0.28 45 ★S.PARA- -0.66 921 F 0.33 4; S.INDUC 0.32 141 S.ZE.RS 0.10 107 R 0.14 4; S.INDUC 0.37 141 S.ZE.RS 0.15 141 ★S-L 0.28 14; S.INDUC 0.75 654 ★ES-CH 6.20 2
C4H10O2P1	CH2P(O)(CH3)(OC2H5)						★S.STAR 0.70 45 L-STM 7.10 1 B5-STM 4.99 1; ★ES-CH 5.20 2 B1-STM 1.52 1
	P(O)(C2H5)(OC2H5)						★S.INDUC 0.30 45 L-STM 6.26 1 B5-STM 4.64 1; ★ES-CH 6.20 2 B1-STM 2.46 1 ★S-L 0.30 14
	P(OEt)2			★0.33 864	3.24		S.INDUC -0.17 69 S.ZE.RS 0.08 69 B1-STM 1.40 1; ★S.PARA+ 0.24 143 ★ES-CH 6.20 2 B5-STM 5.58 1; ★S.ZE.RS 0.06 482 L-STM 6.26 1
C4H10O2P1S1	P(S)(OC2H5)2						★S.INDUC -0.05 69 ★S.ZE.RS 0.18 69 ★ES-CH 6.40 2
C4H10O3P1	P=O(OEt)2		0.23 105 / 0.41 207 / 0.43 6 / 0.49 3 / 0.50 10 / ★0.55 693	0.52 207 / 0.55 143 / 0.56 6 / 0.57 10 / 0.60 111 / ★0.60 693	3.12		S.INDUC 0.02 16 S.PARA- 0.68 443 L-STM 6.26 1; S.INDUC 0.06 69 ★S.PARA- 0.73 695 B1-STM 2.52 1; S.INDUC 0.18 12 S.PARA- 0.83 373 B5-STM 5.58 1; S.INDUC 0.32 6 ★S.PARA- 0.84 694 F 0.52 5; ★S.INDUC 0.33 45 S.PARA- 0.85 28 R 0.08 5; S.INDUC 0.34 610 S.PARA+ 0.54 143 S-L 0.32 14; S.STAR 3.02 3 S.ZE.RS 0.16 69 S-CNTIG 0.58 139; S.META- 0.52 28 S.ZE.RS 0.20 12 S-CNTIG 0.68 486; S.META- 0.53 694 S.ZE.RS 0.24 6 GRP.DPL -3.04 8; S.META- 0.56 695 S.BA.RS 0.08 34 GRP.DPL -2.88 9; S.PARA- 0.58 442 ES-CH 6.20 2
C4H10O3P1S1	OP(S)(OEt)2						S.INDUC 0.38 395 S.PARA+ -0.36 395 S.ZE.RS -0.25 395

Empirical	Structural	π	σ_m	σ_p	MR	E_s	additional σ values					
C4H10O3P1S1	SP(=O)(OC2H5)2						L-STM	7.17 *1*	B1-STM	1.70 *1*	B5-STM	5.67 *1*
C4H10O4P1	OP(O)(OEt)2	-0.24 *54*			3.28		S.INDUC	0.49 *12*	S.ZE.RS	-0.24 *12*	B1-STM	1.35 *1*
							★ S.META+	0.32 *538*	★ ES-CH	5.20 *2*	B5-STM	5.68 *1*
							★ S.PARA+	-0.13 *538*	L-STM	6.99 *1*		
C4H10P1	P(Et)2		0.02 *150*	0.03 *969*	3.05		S.INDUC	-0.04 *150*	★ ES-CH	5.20 *2*	F	0.11 *5*
			★ 0.10 *150*	0.08 *150*			S.ZE.RS	-0.08 *47*	L-STM	5.17 *1*	R	0.02 *5*
			0.16 *41*	★ 0.13 *515*			★ S.ZE.RS	-0.08 *482*	B1-STM	1.40 *1*		
				0.35 *41*			S.ZE.RS	0.12 *150*	B5-STM	4.64 *1*		
C4H10P1S1	P(S)(Et)2		★ 0.39 *107*	★ 0.46 *107*			★ S.INDUC	0.31 *107*	F	0.36 *4*		
							★ S.ZE.RS	0.16 *107*	R	0.10 *4*		
C4H10S1	S+(C2H5)2						S.CH-P+	1.10 *75*				
	S+(I-Pr)(Me)						S.CH-P+	1.10 *75*				
	S+(Pr)(Me)						S.CH-P+	1.06 *75*				
C4H11Ge1	CH2Ge(CH3)3						★ S.INDUC	-0.10 *133*	★ S.PARA+	-0.61 *203*	★ S.ZE.RS	-0.15 *133*
							S.INDUC	-0.01 *172*	S.ZE.RS	-0.23 *51*	★ ES-CH	5.30 *2*
							S.PARA+	-1.01 *51*	S.ZE.RS	-0.18 *172*		
C4H11N1	(CH2)4NH3+	-0.99 *130*					S.INDUC	0.09 *310*				
	+CH2CH2NH(CH3)2		★ 0.24 *3*	★ 0.14 *3*			S.ORTHO	0.48 *53*	F	0.29 *4*	R	-0.15 *4*
	+NH(C2H5)2						★ S.STAR	3.13 *424*	★ S.STAR	4.14 *424*	★ ES-CH	5.00 *2*
	+NH2Bu		★ 0.71 *3*				★ S.INDUC	0.60 *32*	★ ES-CH	3.00 *2*	B1-STM	1.49 *1*
							★ S.STAR	3.74 *3*	L-STM	6.88 *1*	B5-STM	4.87 *1*
	+NH2C(CH3)3		★ 0.71 *3*									
	+NH2CH(CH3)(Et)					★ -2.34 *215*	★ S.INDUC	0.60 *32*	L-STM	6.07 *1*	B5-STM	4.47 *1*
							★ ES-CH	4.00 *2*	B1-STM	1.49 *1*		
	+NH2CH2CH(CH3)2						S.STAR	3.60 *3*				
	CH2N(CH3)3+		0.23 *36*	0.27 *36*		★ -4.13 *44*	★ S.INDUC	0.25 *34*	★ S.PARA-	0.57 *29*	B5-STM	4.08 *1*
			★ 0.40 *3*	★ 0.44 *315*			S.STAR	1.15 *87*	S.ORTHO	0.48 *53*	F	0.38 *4*
			0.68 *29*	0.67 *29*			★ S.STAR	1.90 *52*	S.ZE.RS	0.00 *387*	R	0.06 *4*
							S.STAR	1.90 *183*	S.ZE.RS	0.03 *22*	R-	0.19 *4*
							S.STAR	2.00 *155*	★ S.BA.RS	0.00 *34*	S-L	0.39 *14*
							S.META-	0.26 *36*	ES-CH	5.00 *2*	S.ZTWST	-0.10 *22*
							S.META-	0.50 *29*	L-STM	4.83 *1*		
							S.PARA-	0.35 *36*	B1-STM	1.52 *1*		
	CH2N+H2(I-Pr)						S.PHOSP	0.25 *367*				
	N+(CH3)2Et					★ -3.44 *215*						
C4H11N1O1	OCH2N+(CH3)3						L-STM	5.47 *1*	B1-STM	1.35 *1*	B5-STM	4.41 *1*
C4H11N1O3P1	NHP(O)(OC2H5)2						★ ES-CH	5.20 *2*	★ S-L	0.23 *14*		
C4H11N2	NHCH(CH3)N(CH3)2						L-STM	5.98 *1*	B1-STM	1.44 *1*	B5-STM	4.39 *1*
	NHCH2CH2N(CH3)2						L-STM	6.79 *1*	B1-STM	1.35 *1*	B5-STM	5.21 *1*
C4H11O1Si1	CH2OSi(CH3)3		★ -0.04 *3*	★ -0.05 *3*	3.12		★ ES-CH	3.20 *2*	R	-0.05 *4*		
				-0.04 *221*			F	0.00 *4*				
	CH2Si(Me)2OMe						S.PARA+	-0.45 *285*				
C4H11O2Si1	CH2SiMe(OMe)2						S.PARA+	-0.40 *285*				
C4H11O3Si1	CH2Si(OMe)3						S.PARA+	-0.24 *285*				
C4H11O4Si1	OCH2Si(OMe)3						S.INDUC	0.22 *12*	S.ZE.RS	-0.43 *12*		
C4H11Pb1	CH2Pb(CH3)3						★ S.INDUC	-0.12 *133*	S.PARA+	-0.92 *203*	★ S.ZE.RS	-0.19 *133*
							S.INDUC	-0.01 *172*	S.ZE.RS	-0.23 *172*		
C4H11S1Si1	SCH2Si(CH3)3						★ S.PARA-	-0.06 *389*	★ S.PARA+	-1.02 *389*		
C4H11Si1	CH2Si(C3H7)H2						★ S.STAR	-0.26 *87*	★ S.PARA+	-0.91 *60*	★ ES-CH	3.20 *2*
	CH2Si(CH3)3	2.00	-0.20 *246*	-0.29 *401*	2.96		S.INDUC	-0.19 *17*	S.PARA+	-0.65 *277*	S.RES.+	-0.30 *6*
			-0.19 *17*	-0.27 *3*			S.INDUC	-0.10 *133*	★ S.PARA+	-0.62 *276*	ES-CH	5.20 *2*
			-0.17 *3*	-0.26 *6*			S.INDUC	-0.08 *385*	S.PARA+	-0.55 *203*	ES-V	1.49 *7*
			★ -0.16 *21*	-0.26 *246*			★ S.INDUC	-0.07 *89*	S.PARA+	-0.54 *285*	L-STM	5.39 *1*
			-0.11 *6*	-0.22 *17*			S.INDUC	-0.03 *6*	S.PARA+	-0.54 *401*	B1-STM	1.52 *1*
				★ -0.21 *21*			S.INDUC	-0.01 *172*	S.PARA+	-0.50 *383*	B5-STM	4.75 *1*
				-0.16 *824*			S.STAR	-0.31 *3*	S.PARA+	-0.48 *175*	F	-0.09 *5*
							S.STAR	-0.27 *381*	S.PARA+	-0.22 *246*	R	-0.12 *5*
							S.STAR	-0.26 *52*	S.ORTH+	-0.45 *154*	R+	-0.62 *4*
							S.STAR	-0.18 *428*	S.ZE.RS	-0.23 *6*	R-	-0.22 *4*
							S.STAR	0.43 *841*	S.ZE.RS	-0.20 *134*	S-L	-0.04 *7*
							★ S.PARA-	-0.22 *321*	S.ZE.RS	-0.20 *387*	★ S.PHOSP	-1.60 *59*
							S.PARA+	-0.87 *60*	S.ZE.RS	-0.18 *172*	GRP.DPL	0.68 *8*
							S.PARA+	-0.76 *51*	S.ZE.RS	-0.17 *51*	GR.ELCT	0.13 *13*
							S.PARA+	-0.66 *299*	S.ZE.RS	-0.15 *133*		
	Si(Et)(Me)2						★ S.ZE.RS	0.02 *22*	★ ES-CH	5.20 *2*	★ S.ZTWST	0.02 *22*
C4H11Sn1	CH2Sn(CH3)3			★ -0.20 *824*			S.INDUC	-0.11 *133*	S.PARA+	-0.82 *203*	★ ES-CH	5.70 *2*
							S.INDUC	0.00 *172*	S.ZE.RS	-0.26 *185*	GR.ELCT	0.12 *13*
							★ S.PARA+	-0.92 *276*	S.ZE.RS	-0.23 *172*		
							★ S.PARA+	-0.90 *277*	S.ZE.RS	-0.19 *133*		
C4H12N2O1P1	P(O){N(CH3)2}2		0.30 *107*	0.40 *107*			★ S.INDUC	0.16 *45*	L-STM	6.21 *1*	S-L	0.16 *961*
			0.30 *141*	0.40 *141*			S.INDUC	0.18 *141*	B1-STM	1.35 *1*		
			★ 0.30 *107*	★ 0.40 *107*			S.INDUC	0.21 *141*	B5-STM	5.47 *1*		
							S.ZE.RS	0.19 *107*	F	0.27 *4*		
							S.ZE.RS	0.22 *141*	R	0.13 *4*		
							★ ES-CH	8.20 *2*	★ S-L	0.16 *14*		
C4H12N2O2P1	OP(=O)(N-Me)2						L-STM	6.21 *1*	B1-STM	1.35 *1*	B5-STM	5.47 *1*
C4H12N2P1	P(N(CH3)2)2		-0.03 *3*	-0.06 *3*			★ S.INDUC	0.00 *468*	B1-STM	1.99 *300*	R	0.08 *5*
			★ 0.18 *41*	★ 0.25 *41*			★ ES-CH	7.20 *2*	B5-STM	4.53 *300*	GR.ELCT	0.01 *13*

Empirical	Structural	π	σ_m	σ_p	MR	E_s	additional σ values				
C4H12N2P1	P(N(CH3)2)2						L-STM	5.08 _300_	F	0.17 _5_	
C5F9O1	CF2-(CFCF2CF2CF2O-)						S.STAR	2.74 _566_			
C5Mn1O5	Mn(CO)5						★S.INDUC	-0.01 _33_	★S.ZE.RS -0.24 _33_	★ES-CH 11.30 _2_	
C5N3	C(CN)=C(CN)2		★ 0.77 _3_ 0.83 _294_	★ 0.98 _40_			S.INDUC 0.61 _80_ ★S.INDUC 0.67 _40_ S.INDUC 0.67 _294_ S.STAR 4.20 _3_ ★S.META- 0.77 _165_ S.PARA- 1.68 _257_	★S.PARA- 1.70 _165_ S.ZE.RS 0.27 _80_ S.AN-RS 0.94 _80_ ES-CH 6.00 _2_ L-STM 6.46 _1_ B1-STM 1.61 _1_	B5-STM 5.17 _1_ F 0.65 _5_ R 0.33 _5_ R- 1.05 _4_ GRP.DPL -5.30 _8_		
C5H2N1O1	2-(5-cyanofuryl)		★ 0.25 _388_	★ 0.10 _295_			★ES-CH 3.00 _2_	F 0.32 _4_	R -0.22 _4_		
C5H2N3	2-CN-4-pyrimidinyl						S.INDUC 0.38 _108_	S.ZE.RS 0.14 _108_			
	2-CN-pyrimidin-5yl						S.INDUC 0.40 _113_	S.ZE.RS -0.01 _113_			
	4-CN-2-pyrimidyl						★S.INDUC 0.21 _204_	★S.ZE.RS 0.12 _204_			
	5-CN-pyrimidin-2yl						S.INDUC 0.16 _113_	S.ZE.RS 0.16 _113_			
	6-CN-4-pyrimidinyl						S.INDUC 0.41 _108_	S.ZE.RS 0.14 _108_			
C5H2O2S1@	2-(5-thienylcarboxylate):						★S.ORTHO 0.02 _199_				
C5H3	CCCCCH3						★S.INDUC 0.39 _6_	★ES-CH 3.00 _2_	★S-L 0.39 _14_		
C5H3F8O1	OCH2(CF2)3CHF2						★S.INDUC 0.12 _49_ ★S.ZE.RS -0.23 _49_	★ES-CH 3.00 _2_ S.PHOSP -0.15 _449_			
C5H3Fe1O3	CH=CH2/Fe(CO)3						S.INDUC 0.72 _16_				
C5H3N1	1-(5-CN-naphthyl)			0.46 _131_							
	1-(6-CN-naphthyl)		0.34 _131_								
	1-(7-CN-naphthyl)			0.31 _131_							
C5H3N2S1	2-imidazo(2,1B)thiazolyl						★S.PARA+ 0.10 _191_				
C5H3O1S1	2-thiophenyl-CO						GRP.DPL -3.45 _8_				
C5H3O2	2-(5-formylfuryl)		★ 0.22 _388_	★ -0.05 _295_			★S.STAR 0.25 _3_ ★ES-CH 3.00 _2_	F 0.34 _4_ R -0.39 _4_			
C5H3O2S1@	2-(5-carboxythienyl):						★S.ORTHO 0.59 _199_				
C5H3S2	2-thiophenyl-C(S)-						GRP.DPL -3.15 _8_				
C5H4As1@	2-arsabenzene		★ 0.40 _655_	★ 0.10 _655_			★S.ORTHO 0.30 _655_				
C5H4Br1N2@	3(2-amin-5-Br-pyridine):						S.META+ 0.56 _441_				
C5H4Cl1N2@	3(2-amin-5-Cl-pyridine):						S.META+ 0.54 _441_				
C5H4F8N1	NHCH3(CF2)3CHF2						★S.STAR 1.83 _170_	★ES-CH 5.60 _2_			
C5H4N1	2-pyridyl	0.50	0.17 _241_ ★ 0.33 _163_ 0.49 _298_	★ 0.17 _163_ 0.35 _241_ 0.69 _298_	2.30		S.INDUC 0.10 _228_ S.INDUC 0.11 _16_ S.INDUC 0.11 _241_ ★S.INDUC 0.40 _298_ ★S.PARA- 0.55 _258_ S.ZE.RS 0.01 _241_	★S.ZE.RS 0.22 _298_ ★ES-CH 3.00 _2_ L-STM 6.28 _1_ B1-STM 1.71 _1_ B5-STM 3.11 _1_ F 0.40 _5_	R -0.23 _5_ R- 0.15 _4_ S-CNTIG 0.55 _139_ GRP.DPL -1.94 _8_		
	3-pyridyl		★ 0.23 _241_	★ 0.25 _241_ 0.83 _298_			S.INDUC 0.19 _241_ S.INDUC 0.22 _16_ ★S.PARA- 0.58 _258_ S.ZE.RS -0.04 _241_	★ES-CH 3.00 _2_ F 0.24 _4_ R 0.01 _4_ R- 0.34 _4_	★S-CNTIG 0.73 _486_ ★S-CNTIG 0.81 _139_ ★GRP.DPL -2.28 _8_		
	4-pyridyl	0.46	★ 0.27 _241_ 0.49 _298_	★ 0.44 _241_ 0.67 _298_	2.30		S.INDUC 0.21 _241_ S.INDUC 0.24 _16_ S.INDUC 0.27 _298_ S.PARA- 0.73 _443_ ★S.PARA- 0.81 _442_ S.PARA- 1.07 _438_ ★S.PARA- 1.26 _493_	S.ZE.RS -0.04 _241_ S.ZE.RS 0.57 _298_ ★ES-CH 3.00 _2_ L-STM 5.92 _1_ B1-STM 1.71 _1_ B5-STM 3.11 _1_ F 0.21 _5_	R 0.23 _5_ R- 0.60 _4_ ★S-CNTIG 0.73 _139_ ★S-CNTIG 0.81 _139_ ★GRP.DPL -2.57 _8_		
C5H4N1@	2-pyridyl						S.ZPTFT 0.81 _645_ S.ZPTFT 0.84 _677_ S.ORTHO 0.37 _670_ S.ORTHO 0.56 _630_ ★S.ORTHO 0.75 _605_ S.ORTHO 0.75 _629_ S.ORTHO 0.81 _668_	S.ORTHO 0.88 _384_ S.ORTHO 1.00 _332_ S.ORTHO 1.10 _669_ ★S.ORTH+ 0.72 _507_ ★S.ORTH+ 0.75 _472_ S.ORTH- 0.58 _565_ S.ORTH- 0.62 _162_	S.ORTH- 0.62 _445_ S.ORTH- 0.85 _438_ S.ORTH- 1.00 _162_ S.ORTH- 1.00 _231_ S.ORTH- 1.00 _232_ S.ORTH- 1.50 _124_ O-STER -2.35 _90_		
	3-pyridyl		0.42 _642_ 0.55 _572_ 0.55 _670_ 0.60 _332_ 0.60 _630_ 0.62 _668_ 0.65 _605_ 0.65 _629_ ★ 0.73 _259_ 0.74 _384_ 1.30 _669_				S.ZMTFT 0.72 _645_ ★S.META+ 0.46 _507_ S.META+ 0.54 _472_ ★S.META- 0.40 _445_ S.META- 0.42 _162_ S.META- 0.53 _438_ S.META- 0.58 _231_ S.META- 0.59 _232_ S.META- 0.60 _162_ S.META- 0.72 _124_ ★S.META- 0.76 _565_	S.ZPTFT 0.63 _677_			
	4-pyridyl			0.80 _332_ ★ 0.83 _259_ 0.93 _668_ 0.94 _572_ 0.95 _670_ 0.96 _605_ 0.96 _629_ 1.11 _384_ 1.60 _669_			S.ZPTFT 0.90 _677_ S.ZPTFT 0.95 _645_ ★S.PARA- 0.83 _445_ S.PARA- 0.94 _162_ S.PARA- 1.16 _231_ S.PARA- 1.17 _162_ S.PARA- 1.17 _232_ ★S.PARA- 1.26 _565_ S.PARA- 1.86 _124_	S.PARA+ 1.16 _472_ ★S.PARA+ 1.17 _507_			
C5H4N1O1	2-(3,4-H-4oxopyrimidinyl						★S.INDUC 0.41 _96_				
	2-pyridine-oxy						GRP.DPL -1.96 _8_				

Empirical	Structural	π	σ_m	σ_p	MR	E_s	additional σ values					
C5H4N1O1	2-pyridyl-N-oxide						★ S.META-	0.23 469	★ S.PARA+	0.68 472		
							★ S.PARA-	0.27 469	★ ES-CH	4.00 2		
	2-pyrrol-CO						GRP.DPL	-1.64 8				
	4-pyridine-oxy						GRP.DPL	-2.46 8				
	4-pyridyl-N-oxide						★ S.PARA-	0.33 469	★ ES-CH	3.00 2	★ GRP.DPL	-4.52 8
C5H4N1O1@	2-pyridyl-N-oxide						★ S.ORTHO	-0.40 384	★ S.ORTH-	1.50 162	S.ORTH-	1.50 232
							★ S.ORTH-	1.44 162	S.ORTH-	1.50 218		
	3-pyridyl-N-oxide		0.70 332				S.META-	1.17 162				
			0.81 752				S.META-	1.18 232				
			1.18 534				S.META-	1.23 162				
			1.31 384				S.META-	2.25 628				
			★ 1.47 839				S.META-	2.30 532				
	4-pyridyl-N-oxide			0.02 752			★ S.PARA-	1.27 162				
				0.25 534			S.PARA-	1.53 232				
				0.40 332			S.PARA-	1.67 162				
				1.14 384			★ S.PARA+	0.45 472				
				★ 1.34 839								
C5H4N1O3	2(5-acetoxyisoxazolyl)						GRP.DPL	-2.95 8				
	CO-succinimido						★ GRP.DPL	-4.10 819				
C5H4S1@	+4-thiopyrylyl:						S.META+	3.20 931	★ S.PARA-	5.40 536	★ ES-CH	3.00 2
C5H5	CH2CCCH=CH2						★ S.STAR	0.66 380	L-STM	4.93 1	B5-STM	6.96 1
							★ ES-CH	3.00 2	B1-STM	1.52 1		
C5H5N1	1-pyridinium		★ 1.02 15	★ 0.96 15			★ S.INDUC	1.09 15	★ S.ZE.RS	-0.13 15		
	1-pyridinyl						S.INDUC	1.09 495				
	2-pyridinium+						★ S.PARA-	0.75 469	★ ES-CH	3.00 2		
	4-pyridinyl+						★ S.PARA-	0.65 469	★ ES-CH	3.00 2		
C5H5N1@	2-(pyridinium+):						★ S.ORTHO	0.88 384	S.ORTHO	3.11 992	S.ORTH-	2.49 232
							★ S.ORTHO	2.20 332	★ S.ORTHO	3.21 630		
	3-(pyridinium+):		1.90 332				★	2.18 259				
			2.10 532				★ S.META+	1.82 789				
			2.10 572				★ S.META-	2.02 927				
	4-(pyridinium+):			1.30 332			S.PARA-	2.32 232				
				2.30 532			★ S.PARA-	3.40 536				
				★ 2.42 259			S.PARA-	4.00 755				
				2.57 572			★ S.PARA-	4.06 628				
C5H5N1O1	4-pyridyl-N+-hydroxide						★ S.PARA-	3.40 628	★ S.PARA-	3.90 755		
							★ S.PARA-	3.90 536	★ ES-CH	3.00 2		
C5H5N1O1@	3-(pyridinium+)hydroxide:						★ S.META+	1.99 789	★ S.META+	2.30 532		
C5H5N2	2-(6-Me-pyrazinyl)						★ S.INDUC	0.24 96	★ S.INDUC	0.41 96		
	2-Me-4-pyrimidinyl						S.INDUC	0.16 108	S.ZE.RS	0.07 108		
	4-Me-2-pyrimidyl						★ S.INDUC	0.08 204	★ S.ZE.RS	0.08 204		
	6-Me-4-pyrimidinyl						S.INDUC	0.22 108	S.ZE.RS	0.07 108		
C5H5N2O1	2-MeO-4-pyrimidinyl						S.INDUC	0.25 108	S.ZE.RS	0.09 108		
	2-OMe-pyrimidin-5yl						S.INDUC	0.24 113	S.ZE.RS	-0.06 113		
	3-(1-Me-6-oxopyridazinyl)						★ S.INDUC	0.28 96				
	4-MeO-2-pyrimidyl						★ S.INDUC	0.08 204	★ S.ZE.RS	0.09 204		
	5-OMe-pyrimidin-2yl						S.INDUC	0.00 113	S.ZE.RS	0.07 113		
	6-OMe-4-pyrimidinyl						S.INDUC	0.20 108	S.ZE.RS	0.06 108		
C5H5N2O1S1	4(2-acetylam-thiazolyl)						★ S.PARA+	0.16 191				
C5H5O1	2-(5-methylfuryl)			★ -0.17 295			★ ES-CH	3.00 2				
	2-furylmethyl			★ -0.07 988		★ -1.46 35						
	3-cyclopentadienone						GRP.DPL	-4.10 8				
	3-furylmethyl					★ -1.86 35						
C5H5O1@	2-(5-methylfuryl)		★ 0.09 388		2.25		★ S.ORTH+	-1.34 195				
C5H5O1S1@	2-(5-MeO-thienyl):						★ S.ORTHO	-0.27 199				
C5H5O2	2-(5-hydroxymethylfuryl)			★ -0.12 295			★ ES-CH	3.00 2				
C5H5S1	+1-sulfino-benzene						GRP.DPL	-0.79 8				
	2-(3-methylthienyl)						★ GRP.DPL	-1.10 8	★ O-STER	-1.75 90		
	2-thienylmethyl		★ -0.04 3		2.87		★ S.INDUC	0.05	★ S.STAR	0.31 3	★ ES-CH	4.20 2
	4-Me-2-thiophenyl						GRP.DPL	-0.88 8				
C5H5S1@	2-(4-methylthienyl):						★ S.ORTHO	-0.07 199				
	2-(5-methylthienyl):				2.87		★ S.STAR	0.84 331	★ S.ORTHO	-0.03 169	★ S.ORTH+	-1.17 195
							★ S.ORTHO	-0.15 199	S.ORTHO	0.18 675	★ S.ORTH+	-1.05 374
C5H5Se1@	2-(5-methylselenieyl):						★ S.ORTHO	-0.03 169				
C5H6	1-(5-methylnaphthyl)		★ 0.02 55	0.01 131			★ S.PARA+	-0.39 27				
	1-(6-methylnaphthyl)		★ -0.07 55				★ S.PARA+	-0.36 27				
			-0.05 131									
	1-(7-methylnaphthyl)		★ -0.11 55	-0.07 131			★ S.PARA+	-0.40 27				
	1-(8-methylnaphthyl)						★ S.PARA+	-0.37 27				
	2-(5-methylnaphthyl)						★ S.PARA+	-0.27 145				
	2-(6-methylnaphthyl)			★ -0.08 55			★ S.PARA+	-0.39 27				
	2-(7-methylnaphthyl)			★ -0.07 55			★ S.PARA+	-0.27 27				
	2-(8-methylnaphthyl)			★ -0.10 55			★ S.PARA+	-0.30 27				

Empirical	Structural	π	σ_m	σ_p	MR	E_S	additional σ values
C5H6Cl1N4	4-diMeAm-6-Cl-2-triazinyl						★S.INDUC 0.13 15 ★S.ZE.RS 0.22 15
C5H6N1	1-(2-Me-pyrryl)						S.ZE.RS -0.43 15
C5H6N1@	2-(N-methylpyrryl):						★S.INDUC 0.30 667 ★S.PARA+ -1.90 195 ; ★S.PARA+ -2.07 371 ★S.ORTHO -1.44 548
C5H6N1O1	2-(4,5-diMe-oxazolyl)						★S.INDUC 0.35 96
C5H6N1O2	3(4-Me-5-MeO-isoxazyl)						GRP.DPL -2.82 8
	4(2,3-Me-5-CO-isoxazyl)						GRP.DPL -5.65 8
	4(3-Me-5-MeO-isoxazyl)						GRP.DPL -2.32 8
	5(1,4-Me-3-CO-isoxazyl)						GRP.DPL -5.70 8
C5H6N1S1	2-(4,5-diMe-thiazolyl)						★S.INDUC 0.31 96
C5H6N3	2(4,6-diMe-S-triazinyl)		★0.25 519	★0.39 519			★S.INDUC 0.15 519 ★ES-CH 3.00 2 R 0.18 850 ; ★S.ZE.RS 0.19 519 F 0.21 850
C5H6N3O2	2(4,6-di-OMe-S-triazinyl)						★S.PARA- 0.68 346 ★ES-CH 3.00 2
C5H6O1	1-(5-OCH3-naphthyl)		★-0.23 122	-0.01 131			★ES-CH 1.00 2
	1-(6-OCH3-naphthyl)		★-0.09 55 ; -0.06 131				
	1-(7-OCH3-naphthyl)		★-0.12 55	-0.08 131			★ES-CH 1.00 2
	2-(5-OCH3-naphthyl)			★-0.01 55			
	2-(6-OCH3-naphthyl)			-0.16 55 ; -0.16 122			
	2-(7-OCH3-naphthyl)			★-0.04 122 ; -0.01 55			★ES-CH 0.00 2
	2-(8-OCH3-naphthyl)			★-0.23 55			★ES-CH 0.00 2
C5H7	1-cyclopentenyl		★-0.06 152	★-0.05 152			★S.ZE.RS -0.05 22 B1-STM 1.91 1 R -0.02 4 ; ★ES-CH 3.00 2 B5-STM 3.08 1 ★S.ZTWST 0.00 22 ; L-STM 5.24 1 F -0.03 4
	C(CH3)2CCH						★S.STAR 0.57 253 ★ES-CH 5.00 2
	PRCCH				2.22		★S.INDUC 0.05 6 L-STM 6.04 1 ★S-L 0.05 14 ; ★S.STAR 0.08 268 B1-STM 1.52 1 ; ★ES-CH 3.00 2 B5-STM 5.94 1
C5H7Br3N1O3	CHOHCH(CH2OH)NHCOCBr3	0.32 54			5.51		★ES-CH 5.00 2 B1-STM 1.73 1 ; L-STM 8.54 1 B5-STM 6.00 1
C5H7Cl3N1O3	CHOHCH(CH2OH)NHCOCCl3	0.12 54			4.65		★ES-CH 5.00 2 B1-STM 1.73 1 ; L-STM 8.32 1 B5-STM 5.71 1
C5H7F3N1O3	CHOHCH(CH2OH)NHCOCF3	-0.78 54			3.13		★ES-CH 5.00 2 B1-STM 1.73 1 ; L-STM 7.67 1 B5-STM 5.09 1
C5H7N2	1-(3,5-dimethylpyrazolyl)						S.INDUC 0.28 91 S.ZE.RS -0.11 91
C5H7N2O1	CH2-1(3-Me-2-pyrazl-5-one						GRP.DPL -3.10 8
C5H7N2S1	1(3-Me-5-MeS-pyrazyl)						GRP.DPL -2.80 8
C5H7N4O1@	3-(1-Me2NCO)-triazolyl:		★0.89 642				★ES-CH 3.00 2
C5H7O1	1-cyclopentene-oxide						★S.INDUC 0.04 339 ★S.ZE.RS -0.09 339 ; ★S.STAR 0.28 3 ★ES-CH 4.00 2
	2-cyclopentene-oxide						S.STAR 0.19 3
	CH2CCCH2OCH3						★S.STAR 0.51 349 ★ES-CH 3.00 2 B5-STM 7.83 1 ; S.STAR 0.59 262 L-STM 4.49 1 ; ★S.STAR 0.68 380 B1-STM 1.52 1
C5H7O2	CH=CHC(=O)OEt -T						L-STM 8.19 1 B1-STM 1.60 1 B5-STM 5.48 1
	CH=CHCO2C2H5	0.86	★0.19 180	★0.03 180	2.72		★S.ZPTFT 0.31 248 F 0.27 5 GRP.DPL -1.73 8 ; S.ZE.RS 0.10 19 R -0.24 5 ; ES-CH 3.00 2 GRP.DPL -1.95 9
	OCH2CCCH2OCH3					★-1.54 350	ES-CH 3.00 2
C5H7O4	CH(O2CCH3)2						★S.PARA+ 0.22 741 ★ES-CH 5.00 2
C5H8Br2N1O3	CHOHCH(CH2OH)NHCOCHBr2	-0.47 54			4.74		★ES-CH 5.00 2 B1-STM 1.73 1 ; L-STM 8.54 1 B5-STM 6.00 1
C5H8Cl2N1O3	CHOHCH(CH2OH)NHCOCHCl2	-0.71 54			4.15		★ES-CH 5.00 2 B1-STM 1.73 1 ; L-STM 8.32 1 B5-STM 5.71 1
C5H8F2N1O3	CHOHCH(CH2OH)NHCOCHF2	-1.43 54			3.16		★ES-CH 5.00 2 B1-STM 1.73 1 ; L-STM 7.67 1 B5-STM 5.09 1
C5H8N1	(CH2)4CN						★S.STAR 0.06 420 ★ES-CH 3.00 2
	C=-CCH2N(CH3)2						★S.STAR 0.98 183 L-STM 7.43 1 B5-STM 4.08 1 ; ★ES-CH 3.00 2 B1-STM 1.60 1
C5H8N1O1	4,4-diMe-H2-oxazole						GRP.DPL -1.11 8
	N-valerolactamyl						GRP.DPL -3.96 8
C5H8N1O4	2(5-Me-5-NO2-1,3-dioxanyl						GRP.DPL -4.47 8
C5H9	C(CH3)2CH=CH2				22.49		S.PARA+ -0.21 859 L-STM 5.11 1 B5-STM 3.78 1 ; ES-CH 5.00 814 B1-STM 2.90 1
	CH(CH3)CH2CH=CH2						★ES-CH 4.00 2 L-STM 6.35 1 B5-STM 4.55 1 ; ★ES-V 1.04 7 B1-STM 1.91 1
	CH2CH2CH2CH=CH2					★-1.55 35	★ES-CH 3.00 2 L-STM 7.17 1 B5-STM 5.23 1 ; ★ES-V 0.75 7 B1-STM 1.52 1
	CH2CH=C(CH3)2				2.25		S.INDUC 0.00 6 ★ES-CH 3.00 2 B5-STM 4.82 1 ; ★S.STAR 0.00 779 L-STM 6.39 1 ★S-L 0.00 14 ; ★S.PARA+ -0.26 460 B1-STM 1.52 1
	CH=CHCH(CH3)2 -E						★ES-CH 3.00 2 B1-STM 1.89 1

Empirical	Structural	π	σm	σp	MR	Es	additional σ values
C5H9	CH=CHCH(CH3)2 -E						L-STM 6.35 1 B5-STM 3.33 1
	cyclopentyl	2.14 587	-0.15 152 ★ -0.05 6	★ -0.14 6 -0.13 152 -0.02 271	2.20	-1.75 50 ★ -1.53 35	S.STAR -0.20 52 ES-CH 3.00 2 R -0.16 5 ★ S.PARA- -0.18 400 ES-V 0.71 7 R+ -0.32 4 ★ S.PARA+ -0.30 152 L-STM 4.90 1 R- -0.20 4 S.ORTH+ -0.31 154 B1-STM 1.90 1 ★ S.PHOSP -1.25 37 S.ZE.RS -0.14 19 B5-STM 4.09 1 S.ZE.RS -0.13 6 F 0.02 5
C5H9Br1N1O3	CHOHCH(CH2OH)NHCOCH2Br	-1.19 54			3.97		★ ES-CH 5.00 2 B1-STM 1.73 1 L-STM 8.54 1 B5-STM 6.00 1
C5H9Cl1N1O3	CHOHCH(CH2OH)NHCOCH2Cl	-1.26 54			3.68		★ ES-CH 5.00 2 B1-STM 1.73 1 L-STM 8.32 1 B5-STM 5.71 1
C5H9F1N1O3	CHOHCH(CH2OH)NHCOCH2F	-1.70 54			3.18		★ ES-CH 5.00 2 B1-STM 1.73 1 L-STM 7.67 1 B5-STM 5.09 1
C5H9I1N1O3	CHOHCH(CH2OH)NHCOCH2I	-0.82 54			4.49		★ ES-CH 5.00 2 B1-STM 1.73 1 L-STM 8.83 1 B5-STM 6.38 1
C5H9N2O4	C(Bu)(NO2)2						★ S.STAR 3.34 157
	C(NO2)2(I-Bu)						★ S.STAR 3.33 157
	EtC(Et)(NO2)2				4.59		★ S.STAR 0.38 291 L-STM 6.97 1 B5-STM 4.94 1 ★ ES-CH 3.00 2 B1-STM 1.52 1
C5H9O1	1-OH-cyclopentyl				2.35	★ -1.58 44	★ S.STAR 0.36 376 ★ ES-CH 4.00 2
	C(=O)Bu						L-STM 6.92 1 B1-STM 1.63 1 B5-STM 4.90 1
	C(=O)CH2CH(CH3)2						L-STM 6.12 1 B1-STM 1.63 1 B5-STM 4.50 1
	CH2CH=CHCH2OCH3						★ S.STAR 0.17 349 ★ ES-CH 3.00 2
	CH2CH=CHOC2H5				2.41		★ S.STAR 0.34 74 ★ ES-CH 3.00 2
	CH2CH=CHOEt -T						L-STM 7.78 1 B1-STM 1.52 1 B5-STM 6.15 1
	COC(CH3)3		★ 0.27 3	★ 0.32 409 0.33 46			S.INDUC -0.01 16 ★ ES-CH 6.00 2 F 0.26 4 ★ S.STAR 1.45 129 L-STM 4.87 1 R 0.06 4 ★ S.ZE.RS 0.12 6 B1-STM 1.87 1 ★ S.ZE.RS 0.15 47 B5-STM 4.42 1
	OCH2-cy-butyl						★ ES-CH 3.00 2 ★ ES-V 0.52 7
	OCH=CHCH(CH3)2 -E						L-STM 6.55 1 B1-STM 1.35 1 B5-STM 5.51 1
	cyclopentoxy		★ 0.25 3		2.37		★ S.INDUC 0.26 32 ★ ES-V 0.77 7 B5-STM 5.21 1 S.STAR 1.62 3 L-STM 5.50 1 ★ S-L 0.28 7 ★ ES-CH 4.00 2 B1-STM 1.35 1 GRP.DPL -1.60 9
C5H9O2	(CH2)2CO2Et				2.58	★ -2.14 196	S.META- -0.02 36 L-STM 8.01 1 ★ S-L 0.08 14 S.PARA- -0.04 36 B1-STM 1.52 1 ES-CH 3.00 2 B5-STM 5.83 1
	(CH2)4CO2H				2.58	★ -1.56 44	ES-CH 3.00 2 B1-STM 1.52 1 L-STM 8.02 1 B5-STM 5.56 1
	C=O(OBu)	1.62 42			2.68		★ S.PARA- 0.67 625 L-STM 8.00 1 B5-STM 5.85 1 ★ ES-CH 4.00 2 B1-STM 1.64 1 ★ S-L 0.33 7
	OC(=O)Bu						L-STM 8.06 1 B1-STM 1.35 1 B5-STM 5.78 1
	OC(=O)CH2CH(CH3)2						L-STM 6.81 1 B1-STM 1.35 1 B5-STM 5.44 1
	OCH2CH=CHCH2OCH3					★ -1.33 350	ES-CH 3.00 2
C5H9O2S1	SCH(CH3)CO2CH2CH3						★ ES-CH 4.20 2 ★ S-L 0.25 14
C5H9O3	deoxypentofuranosyl	-0.50 42					★ ES-CH 3.00 2
C5H9O4	D-xylosidyl				2.85		★ S.PARA- 0.06 720 ★ ES-CH 4.00 2
	pentofuranosyl	-0.38 42					★ ES-CH 4.00 2
C5H9S1	S(CH2)3CH=CH2						★ S.STAR 1.36 3 ★ ES-CH 3.20 2 ★ S-L 0.26 14
	S-cyclopentyl						L-STM 5.81 1 B1-STM 1.70 1 B5-STM 5.63 1
	SCH=CHCH(CH3)2 -E						L-STM 6.63 1 B1-STM 1.70 1 B5-STM 6.09 1
C5H9S2	2-(1,3-dithiocycloheptane				3.81	★ -2.93 115	★ S.STAR 1.00 115 ES-CH 3.40 2
C5H10	(CH2)5*				2.42		★ S.STAR -0.18 268 S.STAR -0.14 314 ★ S.PHOSP -2.26 354
C5H10B1	B(CH2)5			★ 0.46 535			★ ES-CH 3.00 2
C5H10F2N1	EtC(NF2)(CH3)2						★ S.STAR 0.17 590 L-STM 6.17 1 B5-STM 4.54 1 ★ ES-CH 3.00 2 B1-STM 1.56 1
C5H10F3Si1	Si(Me)2CH2CH2CF3						★ S.INDUC 0.08 25 ★ S.ZE.RS 0.06 25
C5H10N1	1-piperidinyl	0.85		-0.57 475 -0.12 530 ★ -0.12 3			★ S.ZE.RS -0.47 19 B5-STM 3.49 1 L-STM 6.17 1 B1-STM 1.91 1
	2-(1-Me-pyrrolidinyl)	0.52 42					★ ES-CH 4.00 2
	2-piperidinyl	0.32 42					★ ES-CH 3.00 2
	CH=CHCH2CH2NHCH3 -C	0.48 42					★ ES-CH 3.00 2
	CH=CHCH2CH2NHCH3 -T						L-STM 8.31 1 B1-STM 1.60 1 B5-STM 5.54 1
	CH=N-Bu						S.ORTHO -0.95 53
	CH=NC(CH3)3						S.ORTHO -1.40 53
	Me-(1-pyrrolidinyl)	0.45 42					★ ES-CH 4.00 2 ★ GRP.DPL 1.80 8
	NH-cyclopentyl						★ ES-CH 4.00 2 B1-STM 1.35 1 L-STM 5.54 1 B5-STM 5.25 1
	NHCH=CHCH(CH3)2 -E						L-STM 6.57 1 B1-STM 1.35 1 B5-STM 5.57 1
C5H10N1O1	C=O(NEt2)	-0.32 42					S.INDUC 0.00 12 L-STM 6.02 1 ★ K-CHGTR 1.31 77 S.ZE.RS 0.04 12 B1-STM 1.64 1 ★ ES-CH 5.00 2 B5-STM 4.44 1
	CH=N(O)C(CH3)3	-0.87					

Empirical	Structural	π	σm	σp	MR	Es	additional σ values					
C5H10N1O1	CH=NOBu						L-STM	9.43 _1_	B1-STM	1.60 _1_	B5-STM	5.35 _1_
	Me-(1-morpholinyl)	-0.61 _42_					★ES-CH	4.00 _2_				
C5H10N1O2	NHCO2C4H9		★ 0.06 _405_	★ -0.05 _405_	3.05		★ES-CH	4.00 _2_	B1-STM	1.45 _1_	F	0.13 _5_
							L-STM	9.50 _1_	B5-STM	5.05 _1_	R	-0.18 _5_
C5H10N1O3	CHOHCH(CH2OH)NHCOCH3	-1.88 _54_			3.19		★ES-CH	5.00 _2_	B1-STM	1.73 _1_		
							L-STM	7.21 _1_	B5-STM	5.09 _1_		
C5H10O2	OCH(Et)CH2CH2O*						S.PHOSP	-0.67 _674_				
	OCH2C(CH3)2CH2O*						S.PHOSP	-0.70 _674_	S.PHOSP	-0.67 _577_	S.PHOSP	1.73 _479_
C5H11	Am		★ -0.08 _484_	★ -0.15 _484_	2.42	★ -1.64 _50_	S.INDUC	-0.06 _348_	S.ZE.RS	-0.12 _656_	F	-0.01 _5_
						-0.71 _403_	★S.INDUC	-0.03 _6_	S.ZE.RS	-0.11 _22_	R	-0.14 _5_
							S.STAR	-0.23 _3_	S.BA.RS	-0.03 _34_	R-	-0.18 _4_
							S.STAR	-0.16 _287_	ES-CH	3.00 _2_	S-L	-0.03 _7_
							S.STAR	-0.14 _544_	ES-V	0.68 _7_	★S.PHOSP	-1.21 _37_
							S.STAR	0.15 _770_	L-STM	6.97 _1_	S.ZTWST	-0.10 _22_
							S.ZPTFT	-0.16 _337_	B1-STM	1.52 _1_	GRP.DPL	0.10 _9_
							★S.PARA-	-0.19 _584_	B5-STM	4.94 _1_		
	C(Et)(CH3)2		★ -0.06 _6_	-0.21 _418_	2.42	-3.41 _771_	S.INDUC	-0.24 _12_	S.ZE.RS	-0.09 _656_	B5-STM	3.49 _1_
				-0.19 _23_		★ -3.40 _62_	S.INDUC	-0.08 _348_	ES-CH	5.00 _2_	F	0.03 _4_
				★ -0.18 _6_			S.STAR	-0.33 _348_	★ES-V	1.63 _7_	R	-0.21 _4_
							S.STAR	-0.31 _771_	★ES-HYBO	-4.72 _225_	★S.PHOSP	-1.54 _37_
							S.ZE.RS	-0.17 _6_	L-STM	4.92 _1_		
							S.ZE.RS	-0.13 _12_	B1-STM	2.60 _1_		
	CH(CH3)CH(CH3)2				2.42	★ -3.05 _391_	★ES-CH	5.00 _2_	★ES-HYBO	-1.65 _225_	B1-STM	1.90 _1_
							★ES-V	1.29 _7_	L-STM	4.92 _1_	B5-STM	4.19 _1_
	CH(CH3)CH2CH2CH3			★ -0.22 _418_		★ -2.14 _35_	★ES-CH	4.00 _2_	L-STM	6.17 _1_	B5-STM	4.54 _1_
							★ES-V	1.05 _7_	B1-STM	1.90 _1_		
	CH(Et)2				2.42	-3.83 _496_	S.INDUC	-0.16 _12_	ES-CH	5.00 _2_	B1-STM	2.13 _184_
						-3.22 _50_	★S.STAR	-0.23 _52_	★ES-V	1.51 _7_	B5-STM	4.01 _184_
						★ -3.12 _62_	S.ZE.RS	-0.12 _12_	L-STM	4.72 _184_		
	CH2C(CH3)3		-0.13 _3_	★ -0.17 _6_	2.42	-3.29 _607_	S.INDUC	-0.07 _133_	★S.PARA+	-0.31 _440_	B1-STM	1.52 _1_
			★ -0.05 _6_	-0.12 _3_		-2.98 _50_	S.INDUC	-0.07 _348_	S.PARNO	-0.10 _65_	B5-STM	4.18 _1_
						★ -2.75 _62_	S.INDUC	-0.05 _261_	S.ORTHO	-0.30 _53_	F	0.03 _4_
							S.INDUC	0.00 _6_	S.ZE.RS	-0.17 _6_	R	-0.14 _4_
							S.INDUC	0.00 _16_	S.ZE.RS	-0.06 _133_	R+	-0.22 _4_
							★S.STAR	-0.17 _52_	ES-CH	5.00 _2_	S-L	0.00 _7_
							S.STAR	-0.17 _568_	★ES-V	1.34 _7_	★S.PHOSP	-1.44 _59_
							S.STAR	-0.12 _3_	ES-HYBO	-1.84 _225_		
							S.STAR	1.42 _841_	L-STM	4.89 _1_		
	CH2CH(CH3)C2H5						★ES-CH	4.00 _2_	L-STM	6.17 _1_	B5-STM	4.54 _1_
							★ES-V	1.00 _7_	B1-STM	1.52 _1_		
	EtCH(CH3)2			★ -0.23 _23_	2.42	-1.90 _496_	★S.INDUC	-0.03 _34_	ES-V	0.68 _7_	S.PHOSP	-1.27 _59_
						-1.75 _196_	★S.STAR	-0.16 _287_	L-STM	6.17 _1_		
						-1.59 _50_	★S.BA.RS	-0.12 _34_	B1-STM	1.52 _1_		
						★ -1.44 _62_	ES-CH	3.00 _2_	B5-STM	4.54 _1_		
C5H11N1O1	OCH2CH2N*(I-Pr)						★S.PHOSP	-1.17 _354_				
	OCH=CHN+(CH3)3						L-STM	6.49 _1_	B1-STM	1.35 _1_	B5-STM	5.66 _1_
C5H11O1	(CH2)3OC2H5				2.60	★ -1.69 _44_	ES-CH	3.00 _2_	L-STM	8.06 _1_	B5-STM	5.87 _1_
							★ES-V	0.69 _7_	B1-STM	1.52 _1_		
	(CH2)4OCH3				2.60	★ -1.58 _44_	ES-CH	3.00 _2_	L-STM	8.09 _1_	B5-STM	5.85 _1_
							★ES-V	0.68 _7_	B1-STM	1.52 _1_		
	C(CH3)2CH2OCH3				2.55		L-STM	5.55 _1_	B1-STM	2.77 _1_	B5-STM	4.49 _1_
	CH(CH3)C(OH)(CH3)2				2.57	★ -4.35 _44_	ES-CH	6.00 _2_	B1-STM	1.90 _1_		
							L-STM	4.92 _1_	B5-STM	4.19 _1_		
	CH(OCH3)Pr					★ -2.52 _35_	★ES-CH	5.00 _2_	L-STM	6.17 _1_	B5-STM	4.54 _1_
							★ES-V	1.22 _7_	B1-STM	1.90 _1_		
	CH(OH)C(CH3)3				2.57	★ -3.45 _44_	ES-CH	6.00 _2_	B1-STM	1.73 _1_		
							L-STM	4.92 _1_	B5-STM	4.19 _1_		
	CH(OH)C4H9				2.57	★ -1.55 _44_	ES-CH	4.00 _2_	L-STM	6.97 _1_	B5-STM	4.94 _1_
							★ES-V	0.70 _7_	B1-STM	1.73 _1_		
	CH2CH2O-I-Pr						★ES-CH	3.00 _2_	L-STM	6.85 _1_	B5-STM	5.50 _1_
							★ES-V	0.87 _7_	B1-STM	1.52 _1_		
	CH2CH2OPr						★ES-CH	3.00 _2_	L-STM	8.10 _1_	B5-STM	5.83 _1_
							★ES-V	0.89 _7_	B1-STM	1.52 _1_		
	CH2OC4H9				2.60	★ -1.66 _44_	ES-CH	3.00 _2_	L-STM	8.04 _1_	B5-STM	5.89 _1_
							★ES-V	0.66 _7_	B1-STM	1.52 _1_		
	CH2OCH2CH(CH3)2				2.60	★ -1.71 _44_	ES-CH	3.00 _2_	L-STM	6.82 _1_	B5-STM	5.50 _1_
							★ES-V	0.62 _7_	B1-STM	1.52 _1_		
	O-Am		0.06 _6_	★ -0.34 _21_	2.63		S.STAR	0.95 _3_	L-STM	8.11 _1_	★S.PHOSP	-0.39 _59_
			★ 0.10 _21_	-0.32 _6_			★S.STAR	1.52 _37_	B1-STM	1.35 _1_	S.IND.P	1.43 _39_
							S.ZE.RS	-0.60 _6_	B5-STM	5.81 _1_	S.RES.P	-0.68 _39_
							ES-CH	3.00 _2_	F	0.29 _5_	GRP.DPL	-1.32 _9_
							★ES-V	0.58 _7_	R	-0.63 _5_		
	OC(CH3)2Et						★ES-CH	5.00 _2_	L-STM	6.05 _1_	B5-STM	4.42 _1_
							★ES-V	1.35 _7_	B1-STM	1.35 _1_		
	OCH(CH3)CH(CH3)2						★ES-CH	4.00 _2_	L-STM	6.05 _1_	B5-STM	4.42 _1_
							★ES-V	0.91 _7_	B1-STM	1.35 _1_		
	OCH(CH3)propyl						★ES-CH	4.00 _2_	L-STM	6.86 _1_	B5-STM	4.79 _1_
							★ES-V	0.90 _7_	B1-STM	1.35 _1_		
	OCH(Et)2						★ES-CH	4.00 _2_	L-STM	6.04 _184_	B5-STM	5.19 _184_

Empirical	Structural	π	σm	σp	MR	Es	additional σ values		
C5H11O1	OCH(Et)2						★ ES-V 1.00 7	B1-STM 1.35 184	
	OCH2C(CH3)3				2.63		★ S.STAR 1.51 37	L-STM 6.05 1	★ S.PHOSP -0.29 37
							★ ES-CH 3.00 2	B1-STM 1.35 1	S.IND.P 1.43 39
							★ ES-V 0.70 7	B5-STM 4.42 1	S.RES.P -0.56 39
	OCH2CH(CH3)Et						★ ES-CH 3.00 2	L-STM 6.86 1	B5-STM 4.79 1
							★ ES-V 0.62 7	B1-STM 1.35 1	
	OCH2CH2CH(CH3)2				2.63		★ S.STAR 1.52 37	L-STM 6.86 1	★ S.PHOSP -0.38 59
							ES-CH 3.00 2	B1-STM 1.35 1	S.IND.P 1.43 39
							ES-V 0.62 7	B5-STM 5.71 1	S.RES.P -0.65 39
C5H11O1S1	(CH2)3S(=O)Et						L-STM 6.17 1	B1-STM 1.52 1	B5-STM 7.00 1
	SC(CH3)2CH2OCH3						L-STM 6.98 1	B1-STM 1.70 1	B5-STM 5.09 1
C5H11O2	CH(CH3)OCH2CH2OCH3						★ ES-CH 4.00 2	L-STM 7.91 1	B5-STM 5.78 1
							★ ES-V 0.67 7	B1-STM 1.90 1	
	CH(OEt)2				2.77	★ -2.42 115	S.INDUC -0.17 49	ES-CH 5.00 2	★ S.ZTWST 0.00 22
							★ S.STAR 1.14 115	L-STM 6.01 1	GRP.DPL -1.27 9
							★ S.ZE.RS 0.00 22	B1-STM 1.90 1	GRP.DPL -1.23 8
							S.ZE.RS 0.02 49	B5-STM 5.40 1	
	CH2-O-o-C(CH3)3					★ -1.72 818			
	CH2O(CH2)3OCH3						★ ES-CH 3.00 2	L-STM 8.72 1	B5-STM 6.26 1
							★ ES-V 0.62 7	B1-STM 1.52 1	
	CH2OCH2CH2OEt						★ ES-CH 4.00 2	L-STM 8.72 1	B5-STM 6.22 1
							★ ES-V 0.56 7	B1-STM 1.52 1	
	O(CH2)4OCH3						★ ES-CH 3.00 2	L-STM 8.77 1	B5-STM 6.14 1
							★ ES-V 0.54 7	B1-STM 1.35 1	
	OC(CH3)2CH2OCH3						L-STM 6.71 1	B1-STM 1.35 1	B5-STM 4.35 1
C5H11O2S1	SO2-Am						L-STM 8.22 1	B1-STM 2.03 1	B5-STM 5.96 1
	SO2CH2CH2CH(CH3)2						L-STM 6.97 1	B1-STM 2.03 1	B5-STM 5.83 1
C5H11O3S1	(CH2)5SO3H						★ S.PARA+ -0.32 301	★ ES-CH 3.00 2	
C5H11S1	(CH2)3SC2H5						★ S.STAR 0.14 278	L-STM 8.70 1	B5-STM 6.02 1
							★ ES-CH 3.00 2	B1-STM 1.52 1	
	S(CH2)4CH3						S.STAR 1.35 3	B1-STM 1.70 1	GRP.DPL -1.63 9
							★ ES-CH 3.20 2	B5-STM 6.59 1	
							L-STM 8.12 1	S-L 0.26 14	
	SCH(Et)2						L-STM 6.28 184	B1-STM 1.70 184	B5-STM 5.39 184
	SCH2CH2CH(CH3)2						L-STM 7.06 1	B1-STM 1.70 1	B5-STM 6.29 1
C5H11S2	CH(SeT)2				4.00	★ -3.55 115	★ S.STAR 0.94 115	★ ES-V 1.39 7	B1-STM 1.90 1
							ES-CH 5.40 2	L-STM 6.65 1	B5-STM 5.81 1
C5H11Si1	SiCH2CH=CH2(CH3)2						★ S.STAR 0.32 377	★ ES-CH 3.20 2	
C5H12N1	(CH2)4NHCH3	0.26 42					★ ES-CH 3.00 2	B1-STM 1.52 1	
							L-STM 8.13 1	B5-STM 5.89 1	
	CH2N(C2H5)2	0.36 42			2.80		★ S.STAR 0.24 74	L-STM 6.07 1	B5-STM 4.47 1
							★ ES-CH 4.00 2	B1-STM 1.52 1	
	CH2NHBu	0.69 42					★ ES-CH 3.00 2	B1-STM 1.52 1	
							L-STM 8.13 1	B5-STM 5.89 1	
	NH(CH2)4CH3						★ ES-CH 3.00 2	L-STM 8.13 1	B5-STM 5.89 1
							★ ES-V 0.64 7	B1-STM 1.35 1	
	NHCH(Et)2						L-STM 6.07 184	B1-STM 1.35 184	B5-STM 5.21 184
	NHCH2CH2CH(CH3)2						★ ES-CH 3.00 2	L-STM 6.88 1	B5-STM 5.77 1
							★ ES-V 0.65 7	B1-STM 1.35 1	
	PrN(CH3)2	0.60		★ -0.13 315	2.80		★ ES-CH 3.00 2	B1-STM 1.52 1	
							L-STM 6.88 1	B5-STM 5.49 1	
C5H12N1O1	NHC(CH3)2CH2OCH3						L-STM 6.74 1	B1-STM 1.35 1	B5-STM 4.39 1
C5H12N1O2	CH2N(CH2CH2OH)2				3.11		★ S.STAR 0.34 74	L-STM 6.75 1	B5-STM 5.21 1
							★ ES-CH 4.00 2	B1-STM 1.52 1	
C5H12N3	NHC(N=H)NHC(CH3)3						★ S.INDUC -0.06 117	★ S.ZE.RS -0.38 117	★ ES-CH 4.00 2
C5H12O1P1	CH2P(O)(C2H5)2						★ S.STAR 0.62 45	L-STM 6.42 1	B5-STM 4.73 1
							★ ES-CH 5.20 2	B1-STM 1.52 1	★ S-L 0.14 14
C5H12O2P1	CH2P(O)(C2H5)(OC2H5)						★ S.STAR 0.67 45	L-STM 7.10 1	B5-STM 4.99 1
							★ ES-CH 5.20 2	B1-STM 1.52 1	
C5H12O3P1	CH2P=O(OEt)2		★ 0.12 6	★ 0.06 6	3.58		S.INDUC -0.01 12	S.ZE.RS -0.08 12	B5-STM 5.73 1
							S.INDUC 0.13 6	S.ZE.RS -0.07 6	F 0.17 4
							★ S.STAR 0.73 45	★ ES-CH 5.20 2	R -0.11 4
							★ S.STAR 0.78 74	L-STM 7.10 1	S-L 0.13 14
							★ S.STAR 0.78 435	B1-STM 1.52 1	S.PHOSP -0.47 367
C5H12S1	(CH2)2S+(CH3)C2H5						★ S.STAR 1.60 278	L-STM 6.97 1	B5-STM 4.94 1
							★ ES-CH 3.20 2	B1-STM 1.52 1	
	S+(I-Pr)(Et)						S.CH-P+ 1.14 75		
	S+(Pr)(Et)						S.CH-P+ 1.11 75		
C5H12S2	SCH2CH2S+(CH3)C2H5						L-STM 8.12 1	B1-STM 1.70 1	B5-STM 6.59 1
C5H13Ge1O2	Ge(Me)(OEt)2						S.STAR -0.24 446		
C5H13N1	(CH2)5NH3+						S.INDUC 0.08 310		
	EtN(CH3)3+		★ 0.16 3	-0.01 3		★ -3.23 44	S.META- 0.07 36	ES-CH 3.00 2	R -0.06 4
				0.01 36			S.META- 0.26 29	L-STM 5.58 1	R- -0.10 4
				★ 0.13 315			S.PARA- -0.08 36	B1-STM 1.52 1	★ S-L 0.18 14
				0.34 29			★ S.PARA- 0.09 29	B5-STM 4.53 1	
							★ S.ORTHO 0.11 53	F 0.19 4	

Empirical	Structural	π	σ_m	σ_p	MR	E_s		additional σ values					
C5H13N1	N+(CH3)2(I-Pr)					★ -4.14 215							
	N+(CH3)2Pr					★ -3.44 215							
	N+(Et)2Me		0.45 36			★ -4.24 215	ES-CH	6.00 2					
			★ 0.94 29										
	NH+(Me)(t-Bu)			0.47 36			★ES-CH	6.00 2					
				★ 0.89 29									
C5H13N1O1	OCH2CH2N+(CH3)3						L-STM	6.76 1	B1-STM	1.35 1	B5-STM	5.37 1	
C5H13N1O2P1	N=P(CH3)(OC2H5)2			★ -0.55 345			★ES-CH	5.20 2					
C5H13N1O3P1	N(CH3)P(O)(OC2H2)2						★S.IND.P	4.26 806	★S.RES.P	-4.72 806			
C5H13N1S1	SCH2CH2N+(CH3)3						L-STM	6.88 1	B1-STM	1.70 1	B5-STM	5.86 1	
C5H13N3	NHC(=NH2+)NHC(CH3)3						★S.INDUC	0.26 117	★S.ZE.RS	-0.12 117	★ES-CH	4.00 2	
C5H13O2Si1	Si(Me)(OEt)2						S.STAR	0.27 446					
C5H13Si1	CH2Si(H)(Et)2						★S.PARA+	-0.87 51	★S.ZE.RS	-0.23 51			
	Et-Si(CH3)3		★ -0.16 3	★ -0.17 246	3.43	★ -1.36 726	★S.INDUC	-0.04 32	★S.PARA-	-0.16 321	R	-0.06 4	
							S.STAR	-0.33 448	★S.ORTH+	-0.30 154	R-	-0.05 4	
							S.STAR	-0.25 3	ES-CH	3.20 2	S-L	-0.03 7	
							S.STAR	-0.19 428	F	-0.11 4			
C5H13Sn1	CH2CH2Sn(Me)3						★S.INDUC	-0.02 25					
C5H14N2O1P1	CH2P(=O){N(Me)2}2						★S.STAR	0.36 45					
C5H15O1Si2	OSi(CH3)2Si(CH3)3						★S.PARA+	-0.60 550	★ES-CH	5.40 2			
	Si(CH3)2OSi(CH3)3		★ 0.00 221	★ -0.01 221	4.36		★S.INDUC	-0.13 32	ES-CH	5.40 2	R	-0.05 5	
							S.STAR	-0.81 3	F	0.04 5	S-L	-0.12 7	
C5H15S1Si2	S-Si(CH3)2Si(CH3)3						★S.PARA-	0.17 972	★S.PARA+	-0.78 389	★ES-CH	5.60 2	
C5H15Si2	Si(CH3)2Si(CH3)3						S.PARA-	0.02 780	★S.PARA+	-0.23 550	★ES-CH	7.40 2	
							★S.PARA-	0.04 624	S.PARA+	-0.01 175			
							S.PARA+	-0.62 299	★S.ZE.RS	0.04 47			
C6Cl1F4O1	OC6F4-(4-Cl)						★S.INDUC	0.49 49	★S.ZE.RS	-0.27 49	★ES-CH	4.00 2	
C6Cl3F2O1	OC6Cl3-(3,5-F2)						★S.INDUC	0.37 49	★S.ZE.RS	-0.30 49	★ES-CH	4.00 2	
C6Cl5	C6Cl5		★ 0.24 3	★ 0.24 317	4.95		S.INDUC	0.25 18	ES-CH	5.40 2	B5-STM	4.48 1	
			★ 0.25 317				S.STAR	1.56 3	L-STM	7.74 1	F	0.27 5	
							S.ZE.RS	-0.01 18	B1-STM	1.81 1	R	-0.03 5	
C6F5	C6F5		-0.12 317	-0.03 317	2.40		S.INDUC	0.25 18	S.PARA-	0.66 647	L-STM	6.87 1	
			★ 0.26 317	★ 0.27 317			S.INDUC	0.31 6	★S.PARA+	0.23 650	B1-STM	1.71 1	
			0.34 317	0.41 317			S.STAR	2.00 61	S.PARA+	0.26 647	B5-STM	3.67 1	
							★S.STAR	4.07 37	S.ORTH+	0.26 650	F	0.27 5	
							S.META+	0.29 647	S.ZE.RS	-0.35 6	R	0.00 5	
							S.META+	0.29 650	S.ZE.RS	0.02 18	R+	-0.09 4	
							★S.PARA-	0.43 968	ES-CH	4.60 2	R-	0.11 4	
C6F5O1	OC6F5						★S.INDUC	-0.16 957	★S.ZE.RS	-0.18 49	★ES-CH	4.00 2	
C6F5S1	SC6F5						S.INDUC	0.53 12	S.ZE.RS	-0.07 12			
C6F9O6Si1	Si(OCOCF3)3						★S.INDUC	0.76 25	★S.ZE.RS	0.28 25			
C6F11	perfluoro-cyhex						S.STAR	2.25 566					
C6F14O1P1	P(O)(C3F7)2		★ 0.95 244	★ 1.10 244			★S.INDUC	0.79 244	F	0.84 4			
							★S.ZE.RS	0.31 244	R	0.26 4			
C6F14O2P1	OP(O)(C3F7)2		★ 0.66 244	★ 0.56 244			★S.INDUC	0.77 244	F	0.67 4			
							★S.ZE.RS	-0.21 244	R	-0.11 4			
C6H1F4	2,3,5,6-C6F4H						★ES-CH	4.60 2	★S-L	0.33 14			
C6H1F14N1O1P1	NHP(O)(C3H7)2		★ 0.28 244	★ 0.18 244			★S.INDUC	0.39 244	F	0.33 4			
							★S.ZE.RS	-0.21 244	R	-0.15 4			
C6H2Cl3S1	S-Ph,2,4,5-(Cl)3						L-STM	5.08 11	B1-STM	1.70 11	B5-STM	7.86 11	
C6H2Co1N5	CH2CO(CN)5 (-3)		★ -0.53 270	★ -0.68 270			★ES-CH	7.30 2	F	-0.39 5	R	-0.29 5	
C6H2Mn1O5	CH2Mn(CO)5		★ -0.14 270	★ -0.44 270			★ES-CH	7.30 2	F	-0.04 66	R	-0.40 66	
C6H2N3O6	C6H2(NO2)3		★ 0.26 17	★ 0.30 17	4.22		★S.INDUC	0.26 17	F	0.26 5			
			0.27 815	0.31 815			★S.STAR	1.62 3	R	0.04 5			
			0.43 793	0.41 793			ES-CH	5.00 2	★GRP.DPL	-1.19 9			
C6H2N3O6S1	SC6H2(NO2)3-2,4,6						★S.PARA+	0.30 76	★ES-CH	4.20 2			
C6H2N3O7	OC6H2(NO2)3-2,4,6						★S.PARA+	0.35 76	★ES-CH	4.00 2			
C6H3Br1Cl1N2	N=NC6H3,O-Cl,P-Br						★S.PARA-	0.81 83	★ES-CH	3.00 2			
C6H3Br1Cl1O1S1	SOC6H3-2-Cl-4-Br						★S.PARA-	0.80 106	★ES-CH	5.20 2			
	SOC6H3-2-Cl-5-Br						★S.PARA-	0.79 106	★ES-CH	5.20 2			
C6H3Br1Cl1O2S1	SO2C6H3-2-Cl-4-Br						★S.PARA-	1.00 106	★ES-CH	6.20 2			
	SO2C6H3-2-Cl-5-Br						★S.PARA-	1.01 106	★ES-CH	6.20 2			
C6H3Cl1I1N2	N=NC6H3,O-Cl,P-I						★S.PARA-	0.81 83	★ES-CH	3.00 2			
C6H3Cl1I1O1S1	SOC6H3-2-Cl-4-I						★S.PARA-	0.78 106	★ES-CH	5.20 2			
	SOC6H3-2-Cl-5-I						★S.PARA-	0.80 106	★ES-CH	5.20 2			
C6H3Cl1I1O2S1	SO2C6H3-2-Cl-4-I						★S.PARA-	1.01 106	★ES-CH	6.20 2			
	SO2C6H3-2-Cl-5-I						★S.PARA-	1.02 106	★ES-CH	6.20 2			
C6H3Cl1N1O3S1	SOC6H3-2-Cl-4-NO2						★S.PARA-	0.87 106	★ES-CH	5.20 2			
	SOC6H3-2-Cl-5-NO2						★S.PARA-	0.85 106	★ES-CH	5.20 2			
C6H3Cl1N1O4S1	SO2C6H3-2-Cl-4-NO2						★S.PARA-	1.11 106	★ES-CH	6.20 2			
	SO2C6H3-2-Cl-5-NO2						★S.PARA-	1.08 106	★ES-CH	6.20 2			
C6H3Cl1N3O2	N=NC6H3,2-Cl,5-NO2						★S.PARA-	0.94 83	★ES-CH	3.00 2			
	N=NC6H3,O-Cl,P-NO2						★S.PARA-	1.06 83	★ES-CH	3.00 2			

Empirical	Structural	π	σm	σp	MR	Es	additional σ values					
C6H3Cl2	Ph,2,4-(Cl)2						L-STM 7.74 *1*	B1-STM 1.80 *1*	B5-STM 4.48 *1*			
	Ph,3,4-(Cl)2						L-STM 7.74 *1*	B1-STM 1.80 *1*	B5-STM 4.48 *1*			
C6H3Cl2N2	N=N(C6H3)1,5-Cl2						★S.PARA- 0.83 *83*	★ES-CH 3.00 *2*				
	N=NC6H3,O,P-Cl2						★S.PARA- 0.79 *83*	★ES-CH 3.00 *2*				
C6H3Cl2O1	O-Ph-2,4-Cl2						S.STAR 3.17 *3*	GRP.DPL -2.77 *9*				
	OPh,2,6-(Cl)2						L-STM 4.51 *11*	B1-STM 1.35 *11*	B5-STM 5.89 *11*			
	OPh,3,4-(Cl)2						L-STM 5.20 *11*	L-STM 7.31 *11*	B1-STM 1.35 *11*			
C6H3Cl2O1S1	SOC6H3-2,4-Cl2						★S.PARA- 0.79 *106*	★ES-CH 5.20 *2*				
	SOC6H3-2,5-Cl2						★S.PARA- 0.79 *106*	★ES-CH 5.20 *2*				
C6H3Cl2O2S1	SO2C6H3-2,4-Cl2						★S.PARA- 1.00 *106*	★ES-CH 6.20 *2*				
	SO2C6H3-2,5-Cl2						★S.PARA- 1.02 *106*	★ES-CH 6.20 *2*				
C6H3Cl2S1	SPh,3,4-(Cl)2						L-STM 5.08 *11*	B1-STM 1.70 *11*	B5-STM 7.86 *1*			
C6H3N2O4	Ph-2,4-(NO2)2						S.STAR 1.88 *3*					
	Ph-3,5-(NO2)2						S.STAR 1.37 *3*					
C6H3N2O4S1	SC6H3(NO2)2-2,4						★S.PARA+ 0.10 *76*	★ES-CH 4.20 *2*				
C6H3N2O5	OC6H3(NO2)2-2,4						★S.PARA+ 0.19 *76*	★ES-CH 4.00 *2*				
C6H3N4O6	NHC6H2(NO2)3-2,4,6						★S.PARA+ 0.14 *76*	★ES-CH 4.00 *2*				
C6H3S2@	1-thieno(2,3-B)thiophen:						S.ORTH+ -1.03 *491*					
	1-thieno(3,2-B)thiophen:						S.ORTH+ -1.01 *491*					
	2-thieno(2,3-B)thiophen:						S.META+ -0.67 *491*					
	2-thieno(3,2-B)thiophen:						S.META+ -0.66 *491*					
C6H3Si1	Si(CCH)3						★S.INDUC -0.19 *25*	★S.ZE.RS 0.14 *25*				
C6H4	1-biphenylenyl:						★S.PARA+ -0.23 *27*	★ES-CH 2.00 *2*				
	2-biphenylenyl:				2.43		S.PARA+ -0.63 *862*	S.PARA+ -0.13 *269*				
							★S.PARA+ -0.47 *27*	★ES-CH 2.00 *2*				
C6H4Br1	C6H4Br -M	★ 0.09 *265*	★ 0.08 *265*	3.32			★S.STAR 0.86 *120*	R -0.04 *66*				
			0.09 *3*				ES-CH 3.00 *2*	★S.PHOSP -0.23 *724*				
			0.12 *82*				F 0.12 *5*					
	C6H4Br -O						S.INDUC 0.20 *16*	S.STAR 1.07 *3*				
	C6H4Br -P	★ 0.15 *6*	0.07 *82*	3.32			S.INDUC 0.15 *6*	S.ZE.RS -0.03 *6*	R -0.06 *4*			
			0.08 *265*				★S.STAR 0.74 *120*	S.RES.+ -0.14 *6*	R+ 0.38 *4*			
			★ 0.12 *6*				S.STAR 0.86 *3*	ES-CH 3.00 *2*	R- -0.17 *4*			
							S.STAR 0.90 *129*	L-STM 8.04 *1*	★S.PHOSP -0.25 *85*			
							★S.PARA- 0.03 *109*	B1-STM 1.95 *1*	GRP.DPL -1.91 *9*			
							★S.PARA+ -0.18 *145*	B5-STM 3.11 *1*				
							S.ZE.RS -0.14 *47*	F 0.18 *4*				
C6H4Br1N2	N=NC6H4-Br -M						★S.PARA+ 0.24 *368*					
	N=NC6H4-Br -P						★S.PARA+ 0.21 *368*	GRP.DPL -1.47 *8*				
C6H4Br1O1	O-Ph-2-Br						S.STAR 2.45 *3*	GRP.DPL -2.50 *9*	GRP.DPL -2.20 *8*			
	O-Ph-3-Br						S.STAR 2.48 *3*	GRP.DPL -2.05 *9*	GRP.DPL -1.78 *8*			
	OC6H4Br -P						★S.STAR 2.44 *3*	GRP.DPL -2.37 *9*				
							★S.PARA+ -0.28 *202*	★GRP.DPL -1.59 *8*				
C6H4Br1O1S1	S(O)Ph-3-Br						S.STAR 3.12 *3*					
	S(O)Ph-4-Br						S.STAR 3.14 *3*	GRP.DPL -3.26 *9*				
C6H4Br1O2S1	SO2-Ph-3-Br						S.STAR 3.35 *3*					
	SO2C6H4Br-P				4.11		★S.STAR 3.35 *3*	★S.PARA- 1.05 *272*	★ES-CH 6.20 *2*			
C6H4Br1O3S1	OSO2C6H4-Br -P						GRP.DPL -3.82 *8*					
C6H4Br1S1	S-Ph-3-Br						S.STAR 1.84 *3*	GRP.DPL -1.83 *9*				
	SC6H4Br -P						★S.STAR 1.83 *3*	L-STM 5.27 *11*	B5-STM 8.16 *11*			
							★S.PARA+ -0.15 *202*	B1-STM 1.70 *11*	★GRP.DPL -1.80 *9*			
C6H4Br1Se1	Se-Ph-2-Br						S.STAR 1.60 *3*					
	Se-Ph-4-Br						S.STAR 1.42 *3*					
C6H4Cl1	C6H4Cl -M	★ 0.15 *6*	0.09 *213*	3.04			S.INDUC 0.16 *6*	★S.PARA+ -0.15 *145*	F 0.19 *4*			
			★ 0.10 *6*				S.INDUC 0.17 *84*	S.ZE.RS -0.06 *6*	R -0.09 *4*			
							★S.STAR 0.85 *120*	S.ZE.RS -0.05 *84*	R+ -0.34 *4*			
							S.STAR 0.85 *342*	S.RES.+ -0.12 *6*	★S.PHOSP -0.22 *85*			
							S.STAR 0.98 *3*	ES-CH 3.00 *2*	GRP.DPL -1.82 *9*			
	C6H4Cl -O		★ 0.13 *213*	3.04			S.INDUC 0.20 *16*	★ES-CH 4.20 *2*				
							★S.STAR 1.05 *3*	★GRP.DPL -1.34 *9*				
	C6H4Cl -P	★ 0.15 *6*	0.05 *82*	3.04			S.INDUC 0.15 *6*	★S.PARA+ -0.19 *145*	B1-STM 1.80 *1*			
			0.08 *265*				S.INDUC 0.17 *84*	S.ZE.RS -0.14 *47*	B5-STM 3.11 *1*			
			★ 0.12 *6*				S.STAR 0.71 *61*	S.ZE.RS -0.06 *84*	F 0.18 *4*			
							★S.STAR 0.75 *120*	S.ZE.RS -0.03 *6*	R -0.06 *4*			
							S.STAR 0.75 *342*	S.RES.+ -0.15 *6*	R+ -0.37 *4*			
							S.STAR 0.87 *3*	ES-CH 3.00 *2*	★S.PHOSP -0.29 *85*			
							S.STAR 0.92 *325*	L-STM 7.74 *1*	GRP.DPL -1.90 *9*			
C6H4Cl1F1O1P1	P(O)Cl(C6H4-F) -M	★ 0.65 *105*		3.95			★ES-CH 6.40 *2*					
	P(O)Cl(C6H4-F) -P		★ 0.74 *111*	3.95			★ES-CH 6.40 *2*					
C6H4Cl1F1P1	PCl(C6H4-F) -M	★ 0.42 *105*		4.08			★ES-CH 5.40 *2*					
	PCl(C6H4-F) -P		★ 0.49 *111*	4.08			★ES-CH 5.40 *2*					
C6H4Cl1F1P1S1	P(S)Cl(C6H4-F) -M	★ 0.56 *105*		4.76			★ES-CH 6.60 *2*					
	P(S)Cl(C6H4-F) -P		★ 0.66 *111*	4.76			★ES-CH 6.60 *2*					
C6H4Cl1N2	N=NC6H4-M-Cl						★S.PARA- 0.78 *83*	★ES-CH 3.00 *2*				

Empirical	Structural	π	σ_m	σ_p	MR	E_s	additional σ values		
C6H4Cl1N2	N=NC6H4-P-Cl					★S.PARA- 0.74 83	★ES-CH 3.00 2	★GRP.DPL -1.56 8	
	N=NC6H4Cl-O					★S.PARA- 0.75 83	★ES-CH 3.00 2		
C6H4Cl1N2O2S1	N=S(Cl)(Ph-P-NO2)					S.INDUC 0.00 123			
C6H4Cl1O1	O-Ph-2-Cl					S.STAR 2.69 3			
	O-Ph-3-Cl					S.STAR 2.57 3	GRP.DPL -2.06 9		
	OC6H4Cl -P					★S.STAR 2.62 3 / ★S.PARA+ -0.26 202	L-STM 5.20 11 / B1-STM 1.35 11	B5-STM 7.31 11 / ★GRP.DPL -2.30 9	
C6H4Cl1O1S1	S(O)Ph-3-Cl					S.STAR 3.14 3			
	S(O)Ph-4-Cl					S.STAR 3.14 3	GRP.DPL -3.08 9		
	SOC6H4-2-Cl					★S.PARA- 0.83 106	★ES-CH 5.20 2		
C6H4Cl1O2S1	SO2-Ph-3-Cl					S.STAR 3.45 3			
	SO2C6H4-2-Cl					★S.PARA- 0.95 106	★ES-CH 6.20 2		
	SO2C6H4Cl -P			0.00 879	3.82	★S.STAR 3.49 3 / ★S.PARA- 0.96 272	★ES-CH 6.20 2 / L-STM 5.29 11	B1-STM 2.03 11 / B5-STM 7.44 11	
C6H4Cl1O3S1	SO2O-Ph,4-(Cl)					L-STM 9.65 11	B1-STM 2.03 11	B5-STM 3.52 11	
C6H4Cl1S1	S-Ph-2-Cl					S.STAR 2.12 3			
	S-Ph-3-Cl					S.STAR 2.02 3	GRP.DPL -1.83 9		
	S-Ph-4-Cl					S.STAR 1.97 3 / L-STM 5.08 11	B1-STM 1.70 11 / B5-STM 7.86 11	S.ZTWST -1.81 9	
C6H4Cl1Se1	Se-Ph-2-Cl					S.STAR 1.62 3			
	Se-Ph-3-Cl					S.STAR 1.51 3			
	Se-Ph-4-Cl					S.STAR 1.45 3			
C6H4Cl2N1	NH-Ph,3,5-(Cl)2					L-STM 4.68 11	B1-STM 1.50 11	B5-STM 6.71 11	
	NHC6H3Cl2 (O,P)			★ -0.37 147		★ES-CH 4.00 2			
C6H4Cl2N1S1	N=S(Cl)(Ph-P-Cl)					S.INDUC -0.08 123			
C6H4F1	C6H4F -P	★ 0.12 6	★ 0.06 6	2.52	★S.INDUC 0.13 6 / S.INDUC 0.16 84 / ★S.STAR 0.62 120 / S.STAR 0.81 3 / S.ZE.RS -0.07 6	S.ZE.RS -0.07 84 / ★ES-CH 3.00 2 / L-STM 6.87 1 / B1-STM 1.71 1 / B5-STM 3.11 1	F 0.17 4 / R -0.11 4 / ★S.PHOSP -0.49 526 / GRP.DPL -1.78 9	
	C6H4F-M	★ 0.15 6	★ 0.10 6	2.52	★S.INDUC 0.16 6 / ★S.STAR 0.82 120 / S.STAR 0.95 3	S.ZE.RS -0.06 6 / ★ES-CH 3.00 2 / F 0.19 4	R -0.09 4 / ★S.PHOSP -0.34 526 / GRP.DPL -1.78 9	
C6H4F1Hg1	HgC6H4F -P					★S.INDUC 0.08 353	S.ZE.RS 0.02 353		
C6H4F1N2	N=NC6H4F -C					★S.INDUC 0.37 24	★S.ZE.RS -0.04 24		
	N=NC6H4F -P -T					★S.INDUC 0.23 24 / ★S.PARA+ 0.18 368	★S.ZE.RS 0.08 24 / ★ES-CH 3.00 2		
C6H4F1N2O1	N(O)=NC6H4F -P					★S.INDUC 0.40 24	★S.ZE.RS 0.06 24	★ES-CH 4.00 2	
	N=N(O)C6H4F -P					★S.INDUC 0.28 24	★S.ZE.RS 0.08 24	★ES-CH 4.00 2	
C6H4F1O1	O-Ph-2-F					S.STAR 2.50 3			
	O-Ph-3-F					S.STAR 2.51 3			
	OC6H4F -P		★ -0.08 197	★ -0.10 197		★S.STAR 2.44 3 / ★ES-CH 4.00 2	F -0.03 4 / R -0.07 4		
C6H4F1O1S1	S(O)Ph-3-F					S.STAR 3.11 3			
	S(O)Ph-4-F					S.STAR 3.15 3			
C6H4F1O2S1	SO2-Ph-3-F					S.STAR 3.48 3			
	SO2C6H4F-P				3.31	★S.STAR 3.40 3	★S.PARA- 0.98 272	★ES-CH 6.20 2	
C6H4F1S1	S-Ph-3-F					S.STAR 1.87 3			
	S-Ph-4-F					S.STAR 1.77 3	GRP.DPL -1.64 9		
C6H4Fe1N1O6	O2CNHCH3/Fe(CO)4					S.INDUC 0.96 16			
C6H4I1	C6H4I -M	★ 0.13 6	★ 0.06 6		★S.INDUC 0.16 6 / ★S.STAR 0.90 3 / ★S.ZE.RS -0.10 6	★ES-CH 3.00 2 / L-STM 6.72 1 / B1-STM 1.84 1	B5-STM 5.15 1 / F 0.18 4 / R -0.12 4	
	C6H4I -O					S.STAR 1.08 3 / L-STM 6.28 1	B1-STM 1.84 1 / B5-STM 5.15 1		
	C6H4I -P	★ 0.14 6	★ 0.10 6		★S.INDUC 0.15 6 / ★S.STAR 0.87 3 / ★S.ZE.RS -0.05 6 / ★ES-CH 3.00 2	L-STM 8.45 1 / B1-STM 2.15 1 / B5-STM 3.11 1 / F 0.18 4	R -0.08 4 / GRP.DPL -1.76 9 / GRP.DPL -1.55 8	
C6H4I1O1	O-Ph-2-I					S.STAR 2.38 3	GRP.DPL -2.25 9	GRP.DPL -2.06 8	
	O-Ph-3-I					S.STAR 2.44 3	GRP.DPL -1.68 8		
	O-Ph-4-I					S.STAR 2.39 3	GRP.DPL -2.14 9	GRP.DPL 1.47 8	
C6H4I1O2S1	SO2C6H4I-P				4.61	★S.PARA- 1.05 272	★ES-CH 6.20 2		
C6H4I1S1	S-C6H4-I -O					GRP.DPL -2.38 8			
	S-C6H4-I -P					GRP.DPL -1.50 8			
	SC6H4-I -M					GRP.DPL -1.80 8			
C6H4N1O1	2-pyridoyl					★S.STAR 1.85 547	★ES-CH 5.00 2	★GRP.DPL -2.95 8	
	3-pyridoyl					★S.STAR 1.80 547	★ES-CH 5.00 2	★GRP.DPL -3.01 8	
	4-pyridoyl					GRP.DPL -3.06 8			
C6H4N1O2	C6H4(NO2) -M	★ 0.21 6	0.17 82 / 0.18 265 / ★ 0.20 6	3.17	★S.INDUC 0.19 34 / S.INDUC 0.20 6 / S.INDUC 0.26 84	S.ZE.RS -0.04 84 / S.ZE.RS 0.00 6 / ★S.BA.RS -0.01 34	F 0.23 4 / R -0.03 4 / S.PHOSP 0.10 85	

Empirical	Structural	π	σm	σp	MR	Es	additional σ values			
C6H4N1O2	C6H4(NO2)-M						★S.STAR 1.09 120 S.STAR 1.21 3	S.RES.+ -0.15 6 ES-CH 3.00 2	GRP.DPL -3.40 9	
	C6H4(NO2)-O			★ 0.17 213	3.17		★S.STAR 1.14 3	★ES-CH 4.00 2	★GRP.DPL -3.60 9	
	C6H4(NO2)-P	★ 0.25 6		0.23 82 0.23 265 ★ 0.26 6	3.17		★S.INDUC 0.23 6 S.INDUC 0.28 84 S.INDUC 0.33 119 ★S.STAR 1.14 120 S.STAR 1.26 3 S.META+ 0.18 632 ★S.PARA- 0.31 109 S.PARA+ 0.03 632	★S.PARA+ 0.04 305 S.ZE.RS -0.02 84 S.ZE.RS 0.03 6 S.RES.+ -0.18 6 ES-CH 3.00 2 L-STM 7.66 1 B1-STM 1.71 1 B5-STM 3.11 1	F 0.26 4 R 0.00 4 R+ -0.22 4 R- 0.05 4 S.PHOSP 0.13 85 GRP.DPL -4.43 9 O-STER -2.18 90	
C6H4N1O2S1	SC6H4NO2-M						★S.STAR 2.03 3 ★S.PARA+ -0.19 76	★ES-CH 4.20 2 ★GRP.DPL -4.04 8		
	SC6H4NO2-O						★S.STAR 2.47 3 ★S.PARA+ -0.14 76	★ES-CH 4.20 2 ★GRP.DPL -5.22 8	GRP.DPL -4.93 9	
	SC6H4NO2-P		★ 0.32 6	★ 0.24 6 0.25 375	4.11		S.INDUC 0.35 6 ★S.STAR 2.33 3 ★S.PARA- 0.56 109 ★S.PARA+ -0.17 76 S.ZE.RS -0.11 6	ES-CH 4.20 2 L-STM 4.92 11 B1-STM 1.70 94 B5-STM 7.86 11 F 0.36 4	R -0.12 4 R+ -0.53 4 R- 0.20 4 GRP.DPL -4.43 9	
C6H4N1O2Se1	Se-Ph-2-NO2						S.STAR 1.84 3			
	Se-Ph-3-NO2						S.STAR 1.65 3			
	SeC6H4NO2-P				4.39		★S.STAR 1.83 3 ★S.PARA- 0.41 109	★ES-CH 4.30 2 ★GRP.DPL -4.38 9		
C6H4N1O3	OC6H4NO2-M						★S.STAR 2.76 3 ★S.PARA+ -0.10 76	★ES-CH 4.00 2 ★GRP.DPL -4.04 8	GRP.DPL -4.00 9	
	OC6H4NO2-O						★S.STAR 2.78 3 ★S.PARA+ -0.04 76	★ES-CH 4.00 2 ★GRP.DPL -4.60 8	GRP.DPL -4.05 9	
	OC6H4NO2-P				3.45		★S.INDUC 0.47 6 ★S.STAR 2.91 3 ★S.PARA- 0.20 109	★S.PARA+ -0.08 76 ★ES-CH 4.00 2 L-STM 5.01 11	B1-STM 1.35 11 B5-STM 7.31 11 ★GRP.DPL -4.00 9	
C6H4N1O3S1	S(O)Ph-3-NO2						S.STAR 3.20 3			
	SOC6H4NO2-P		★ 0.58 6	★ 0.60 6 0.63 375	3.97		S.INDUC 0.54 6 ★S.STAR 3.24 3	S.ZE.RS 0.06 6 ES-CH 5.20 2	F 0.55 4 R 0.05 4	
C6H4N1O4S1	SO2C6H4NO2-M			★ 0.82 166	3.95		★S.INDUC 0.61 166 ★S.META- 0.74 167	★S.PARA- 1.04 167 ★ES-CH 6.20 2		
	SO2C6H4NO2-P			0.76 335 ★ 0.83 166	3.95		S.INDUC 0.61 6 ★S.INDUC 0.65 166 S.STAR 3.63 3 S.META- 0.75 167	S.PARA- 1.04 167 ★S.PARA- 1.06 600 ★S.PARA- 1.09 272 ES-CH 6.20 2	L-STM 5.11 11 B1-STM 2.03 11 B5-STM 7.44 11 GRP.DPL -2.80 9	
C6H4N1O5S1	OSO2-C6H4-NO2 -P						GRP.DPL -4.72 8			
	SO3-C6H4-NO2 -P						GRP.DPL -2.76 8			
C6H4N3	1-benzotriazolyl						S.INDUC 0.55 91	S.ZE.RS -0.10 91		
	2-benzotriazolyl		★ 0.49 3	★ 0.51 3	3.33		★S.META- 0.52 727 ★S.PARA- 0.57 727	★ES-CH 3.00 2 F 0.47 4	R 0.04 4 R- 0.10 4	
	2imidazo(1,2A)pyrimidinyl						★S.PARA+ 0.27 191			
C6H4N3O2	N=NC6H4-M-NO2						★S.PARA- 0.86 83	★ES-CH 3.00 2		
	N=NC6H4NO2-P				3.76		★S.PARA- 0.67 109 ★S.PARA- 0.96 83	★S.PARA+ 0.11 76 ★ES-CH 3.00 2		
C6H4N3O4	NHC6H3(NO2)2-2,4						★S.PARA+ -0.18 76	★ES-CH 4.00 2	★GRP.DPL -6.36 8	
C6H4Na1O2	CC-(CO2NA)cyclopropyl	-2.74					★ES-CH 3.00 2			
C6H4O1	2-dibenzofuryl:						★S.META+ -0.24 27 ★S.PARA+ -0.40 27 ★S.PARA+ -0.28 27	★S.PARA+ -0.25 27 ★ES-CH 0.00 2 ★ES-CH 1.00 2	★ES-CH 2.00 2	
C6H4O3S1	C6H4(SO3-)-M						★S.META+ 0.26 305	★S.PARA+ -0.20 305		
	C6H4(SO3-)-O						★S.META+ 0.10 305	★S.PARA+ 0.06 305		
	C6H4(SO3-)-P						★S.META+ 0.22 305	★S.PARA+ -0.17 305		
C6H4S1	3-dibenzothiophenyl:						★S.META+ -0.28 27 ★S.PARA+ -0.37 27 ★S.PARA+ -0.30 27	★S.PARA+ -0.29 27 ★ES-CH 0.00 2 ★ES-CH 1.20 2	★ES-CH 2.00 2	
C6H5	phenyl	1.80 42 1.96 2.15 57	0.01 10 0.04 46 0.06 17 ★ 0.06 21 0.09 6 0.10 527 0.22 23	-0.07 527 -0.02 265 ★ -0.01 21 0.01 6 0.01 23 0.03 17 0.04 10 0.05 46	2.54	-3.82 682 -3.43 35 -1.01 683	S.INDUC 0.08 17 S.INDUC 0.08 89 S.INDUC 0.10 20 S.INDUC 0.11 182 ★S.INDUC 0.12 6 S.INDUC 0.12 12 S.INDUC 0.13 99 S.INDUC 0.14 84 S.INDUC 0.17 119 S.INDUC 0.20 16 S.INDQ 0.85 58 S.STAR 0.60 52 S.STAR 0.60 183 S.STAR 0.60 325 S.STAR 0.60 342 S.STAR 0.62 61 S.STAR 0.66 87 S.STAR 0.75 3 S.ZMTFT 0.05 63	S.PARA- 0.47 258 S.PARA+ -0.35 476 S.PARA+ -0.34 912 S.PARA+ -0.25 362 S.PARA+ -0.24 145 S.PARA+ -0.22 329 S.PARA+ -0.21 65 S.PARA+ -0.21 476 ★S.PARA+ -0.18 43 S.PARA+ -0.18 513 S.PARA+ -0.12 322 S.PARA+ -0.07 88 S.PARNO 0.06 65 S.ORTHO 0.00 53 S.ORTHO 0.74 78 S.ORTH+ -0.35 476 S.ORTH+ -0.25 476 S.ORTH+ -0.22 362 S.ORTH+ -0.16 154	S.RES.+ -0.30 20 S.RES.+ -0.26 10 S.RES.+ -0.17 6 S.RES.+ -0.15 470 ES-CH 3.00 2 ES-V 0.57 341 ES-V 2.15 474 L-STM 6.28 1 B1-STM 1.71 1 B5-STM 3.11 1 F 0.12 5 R -0.13 5 R+ -0.30 4 R- -0.10 4 S-L 0.12 7 S.PHOSP -0.48 59 S.IND.P 0.80 39 S.RES.P -0.13 39 S.ZTWST -0.06 22	

Empirical	Structural	π	σm	σp	MR	Es	additional σ values
C6H5	phenyl						S.META+ 0.00 543 · S.ORTH+ -0.16 822 · S.P.RAD 0.12 263 S.META+ 0.11 43 · S.ORTH- 0.21 97 · ★S.P.RAD 0.39 453 S.META- 0.12 28 · S.ZE.RS -0.11 6 · S.P.RAD 0.40 659 S.ZPTFT -0.03 112 · S.ZE.RS -0.11 20 · S.P.RAD 0.42 255 S.ZPTFT 0.00 95 · S.ZE.RS -0.10 19 · S-CNTIG 0.47 139 S.ZPTFT 0.01 306 · S.ZE.RS -0.08 12 · K-CHGTR 0.45 77 S.ZPTFT 0.04 63 · S.ZE.RS -0.07 84 · GRP.DPL -0.38 9 ★S.PARA- 0.02 109 · S.ZE.RS 0.10 114 · GRP.DPL 1.00 8 S.PARA- 0.09 28 · S.BA.RS -0.11 20 · GR.ELCT 0.30 13 S.PARA- 0.10 211 · S.AN.RS 0.04 20 · O-STER -1.82 90
C6H5B1Na1	B(C6H5)NA						★S.ZE.RS -0.13 47 · ★ES-CH 4.00 2
C6H5Br2Ge1	GeBr2(C6H5)						★S.STAR 1.42 61 · ★ES-CH 6.90 2
C6H5Br2Te1	Te(C6H5)Br2			★ 0.60 103			★S.INDUC 0.54 103 · ★S.ZE.RS 0.06 103
C6H5Cd1	Cd(C6H5)						★S.ZE.RS 0.11 86 · ★ES-CH 4.70 2
C6H5Cl1N1O1S1	SOC6H3-2-Cl-4-NH2						★S.PARA- 0.66 106 · ★ES-CH 5.20 2
	SOC6H3-2-Cl-5-NH2						★S.PARA- 0.71 106 · ★ES-CH 5.20 2
C6H5Cl1N1O2S1	SO2C6H3-2-Cl-4-NH2						★S.PARA- 0.82 106 · ★ES-CH 6.20 2
	SO2C6H3-2-Cl-5-NH2						★S.PARA- 0.92 106 · ★ES-CH 6.20 2
C6H5Cl1N1S1	N=S(Cl)(Ph)						S.INDUC -0.08 123
C6H5Cl1N3	N=NC6H3,2-Cl,5-NH2						★S.PARA- 0.68 83 · ★ES-CH 3.00 2
	N=NC6H3,O-Cl,P-NH2						★S.PARA- 0.48 83 · ★ES-CH 3.00 2
C6H5Cl1P1	P(Cl)(C6H5)			★ 0.44 143	4.10		★S.INDUC 0.18 69 · ★ES-CH 5.40 2 · B5-STM 6.19 11 ★S.PARA+ 0.40 143 · L-STM 4.87 11 ★S.ZE.RS 0.05 69 · B1-STM 1.40 11
C6H5Cl2P1	+P(C6H5)Cl2						S.ZE.RS 0.22 447
C6H5Cl2Si1	SiCl2C6H5						★S.M.RAD 0.15 218 · ★S.P.RAD 0.11 218
C6H5Cl2Sn1	SnCl2(C6H5)						★S.STAR 1.29 61 · ★ES-CH 7.10 2
C6H5Cl2Te1	Te(C6H5)Cl2			★ 0.59 103			★S.INDUC 0.52 103 · S.ZE.RS 0.07 103
C6H5F1N1	NHC6H4F -P						★S.INDUC 0.27 24 · ★S.ZE.RS -0.36 24 · ★ES-CH 4.00 2
C6H5F1P1	Ph(C6H4-F) -M		★ 0.09 105		3.61		★ES-CH 4.20 2
	Ph(C6H4-F) -P			★ 0.25 111	3.61		S.INDUC 0.14 452 · S.ZE.RS 0.00 452 · ★ES-CH 4.20 2
C6H5F4N3P3	1-Ph-F4-cyc-triphosphazyl						★S.PARA+ 0.57 537 · ★ES-CH 6.20 2
	3-Ph-F4-cyc-triphosphazyl						★S.PARA+ 0.66 537 · ★ES-CH 4.00 2
C6H5Hg1	Hg(C6H5)						★S.ZE.RS 0.03 86
C6H5Hg1O1	O-Hg-C6H5						S.INDUC 0.22 118 · S.ZE.RS -0.50 118 · S.RES.+ -2.18 71 S.PARA+ -2.72 71 · S.ZE.RS -0.49 492
C6H5Hg1S1	S-Hg-Ph						S.INDUC 0.23 118 · S.ZE.RS -0.15 118 S.PARA+ -0.46 71 · S.RES.+ -0.45 71
C6H5I1	IC6H5+		★ 0.85 875				★S.PARA- 0.71 877 · ★S.ZE.RS 0.28 19 ★S.PARA- 1.06 876 · ★ES-CH 4.70 2
C6H5N2	CH=N-(3-pyridyl)						GRP.DPL -4.16 8 · GRP.DPL -2.98 8
	N=NC6H5	1.69	0.13 72 0.28 396 0.28 463 0.29 56 0.30 10 0.30 17 ★ 0.32 251	0.26 626 0.31 671 0.33 56 0.34 17 0.34 396 0.35 463 ★ 0.39 251 0.44 72 0.64 23	3.13		S.INDUC 0.19 24 · S.PARA- 0.65 869 · ES-CH 3.00 2 S.INDUC 0.19 72 · S.PARA- 0.66 247 · L-STM 8.43 1 S.INDUC 0.19 161 · S.PARA- 0.69 567 · B1-STM 1.70 1 ★S.INDUC 0.25 89 · S.PARA- 0.77 53 · B5-STM 4.31 1 S.INDUC 0.30 17 · S.PARA+ -0.19 76 · F 0.30 5 S.STAR 1.87 3 · S.PARA+ -0.15 626 · R 0.09 5 S.ZPTFT 0.36 95 · ★S.PARA+ 0.17 368 · R+ -0.49 4 ★S.ZPTFT 0.37 396 · S.PARA+ 0.26 72 · R- 0.15 4 S.PARA- 0.45 109 · S.PARA+ 0.32 88 · S.ZTWST 0.28 626 S.PARA- 0.57 104 · S.ZE.RS 0.06 19 · S.P.RAD 0.33 263 S.PARA- 0.62 28 · S.ZE.RS 0.08 24 · GR.ELCT 0.80 13 S.PARA- 0.64 83 · S.ZE.RS 0.08 72 ★S.PARA- 0.65 868 · S.ZE.RS 0.08 161
C6H5N2O1	3-pyridylamido	-0.40					★ES-CH 4.00 2
	N(O)=NC6H5				3.31		★S.PARA- 0.77 247 · ★ES-CH 4.00 2
	N=N(O)C6H5				3.31		★S.PARA- 0.60 247 · ★ES-CH 4.00 2 · ★GRP.DPL -1.73 8
	N=N-C6H4-OH -P						L-STM 9.11 1 · B5-STM 4.31 1 B1-STM 1.70 1 · GRP.DPL -1.66 8
	N=N-Ph-2-OH						S.ORTHO 0.82 53
	N=NOC6H5			0.31 72			S.ZE.RS -0.03 72
C6H5N2O2	NHC6H4NO2-M						★S.PARA+ -0.56 76 · ★ES-CH 4.00 2
	NHC6H4NO2-O						★S.PARA+ -0.52 76 · ★ES-CH 4.00 2
	NHC6H4NO2-P				3.64		★S.PARA- 0.05 109 · ★S.PARA+ -0.53 76 · ★ES-CH 4.00 2
C6H5N2O3S1	N(=O)=NSO2Ph		★ 0.69 179	★ 0.79 179			★S.INDUC 0.64 179 · F 0.62 4 · ★S.RES.P 0.15 179 ★S.ZE.RS 0.05 179 · R 0.17 4
C6H5O1	C6H4OH -M	1.46 219		★ 0.03 213	2.72		★ES-CH 3.00 2 · ★S.PHOSP -0.32 85
	C6H4OH -O	1.46 219		★ -0.09 213	2.72		★ES-CH 4.00 2 · ★GRP.DPL -1.63 8
	C6H4OH -P	1.44 57		-0.24 265 ★ -0.10 82	2.72		ES-CH 3.00 2 · ★GRP.DPL -1.34 8 ★S.PHOSP -0.65 85
	OC6H5	2.08	0.23 6 0.25 10 ★ 0.25 21 0.26 508	-0.32 21 -0.08 6 ★ -0.03 23 0.06 46 0.07 10 0.14 508	2.77		S.INDUC 0.29 34 · S.PARA+ -0.53 662 · L-STM 4.51 1 S.INDUC 0.34 99 · S.PARA+ -0.51 145 · B1-STM 1.35 1 S.INDUC 0.38 20 · ★S.PARA+ -0.50 43 · B5-STM 5.89 1 S.INDUC 0.38 205 · S.PARA+ -0.49 322 · F 0.37 5 ★S.INDUC 0.40 6 · S.PARA+ -0.48 485 · R -0.40 5 S.INDUC 0.40 182 · S.PARA+ -0.43 88 · R+ -0.87 4

Empirical	Structural	π	σ_m	σ_p	MR	E_s	additional σ values		
C6H5O1	OC6H5						S.INDUC 0.42 *12* S.PARA+ -0.37 *202* R- -0.47 *4*		
							S.INDUC 0.42 *17* S.PARNO 0.09 *65* ★S.PHOSP -0.06 *59*		
							S.INDUC 0.57 *16* S.ORTHO 0.67 *78* S.IND.P 2.11 *39*		
							S.STAR 2.24 *37* S.ORTH+ -0.46 *154* S.RES.P -0.97 *39*		
							S.STAR 2.42 *87* S.ZE.RS -0.48 *6* S.M.RAD -0.02 *79*		
							S.META+ 0.10 *145* S.ZE.RS -0.36 *19* S.P.RAD 0.13 *206*		
							S.ZPTFT -0.05 *112* S.ZE.RS -0.33 *12* S.P.RAD 0.18 *79*		
							S.ZPTFT 0.07 *63* S.ZE.RS -0.32 *67* GRP.DPL -1.38 *9*		
							★S.PARA -0.10 *109* S.BA.RS -0.58 *20* GRP.DPL -1.16 *8*		
							S.PARA+ -0.62 *829* S.RES.+ -0.87 *20* GR.ELCT 0.81 *13*		
							S.PARA+ -0.56 *329* S.RES.+ -0.68 *10*		
							S.PARA+ -0.56 *467* ES-CH 4.00 *2*		
C6H5O1S1	S=O(C6H5)	-0.07	★ 0.50 *6* 0.51 *3*	★ 0.44 *6* 0.46 *3* 0.47 *375*	3.34		★S.INDUC 0.51 *6* S.ZE.RS -0.07 *6* F 0.51 *5*		
							S.INDUC 0.51 *282* S.ZE.RS -0.01 *282* R -0.07 *5*		
							S.STAR 3.08 *10* S.ZE.RS 0.07 *19* R- 0.25 *4*		
							S.STAR 3.24 *3* ★S.BA.RS 0.05 *34* S-L 0.53 *14*		
							S.META- 0.52 *378* ES-CH 5.20 *2* S.P.RAD 0.26 *79*		
							S.PARA- 0.71 *378* L-STM 4.62 *11* S-CNTIG 0.76 *139*		
							S.PARA- 0.73 *372* B1-STM 1.40 *11* GRP.DPL -4.07 *8*		
							★S.PARA- 0.76 *442* B5-STM 6.02 *11* GR.ELCT 0.32 *13*		
C6H5O1S1@	2-(5-acetylthienyl):						★S.ORTHO 0.19 *169* ★S.ORTHO 0.55 *199* ★ES-CH 3.20 *2*		
C6H5O1Se1@	2-(5-acetylselenieyl):						★S.ORTHO 0.18 *169*		
C6H5O2	2-(5-acetylfuryl)		★ 0.24 *388*	★ 0.08 *295*			★ES-CH 3.00 *2* F 0.31 *4* R -0.23 *4*		
	2-(6-methylpyronyl)		★ 0.38 *227*	★ 0.43 *227*			★S.INDUC 0.33 *227* F 0.36 *4*		
							★ES-CH 3.00 *2* R 0.07 *4*		
	O-Ph-OH -O						S.STAR 2.60 *3* GRP.DPL -2.46 *9*		
C6H5O2S1	OS(O)C6H5		★ 0.23 *10*				★GRP.DPL -3.48 *9*		
	SO2C6H5	0.27	0.62 *17* ★ 0.62 *6*	★ 0.68 *6* 0.70 *166* 0.70 *375* 0.71 *17*	3.32		S.INDUC 0.43 *16* S.PARA- 0.90 *378* L-STM 5.86 *1*		
							S.INDUC 0.52 *282* S.PARA- 0.95 *272* B1-STM 2.03 *1*		
							★S.INDUC 0.56 *6* S.PARA- 0.95 *372* B5-STM 6.02 *1*		
							S.INDUC 0.56 *166* S.PARA- 1.00 *258* F 0.58 *5*		
							S.INDUC 0.62 *17* S.PARA- 1.01 *252* R 0.10 *5*		
							S.INDUC 0.62 *67* ★S.PARA- 1.21 *600* R- 0.65 *4*		
							S.INDUC 0.63 *99* S.ZE.RS 0.06 *19* S-L 0.58 *14*		
							S.STAR 3.25 *3* S.ZE.RS 0.09 *67* S.P.RAD 0.18 *79*		
							S.STAR 3.28 *87* S.ZE.RS 0.12 *6* S-CNTIG 1.00 *139*		
							★S.META- 0.62 *167* S.ZE.RS 0.14 *282* GRP.DPL -5.05 *8*		
							S.META- 0.62 *378* S.BA.RS 0.13 *34* GRP.DPL -4.75 *9*		
							S.PARA- 0.90 *167* ES-CH 6.20 *2* GR.ELCT 0.42 *13*		
C6H5O3S1	OSO2C6H5	0.93	★ 0.36 *56* 0.37 *10*	0.25 *335* ★ 0.33 *56* 0.34 *10*	3.67		S.STAR 3.62 *3* B5-STM 3.64 *1* GRP.DPL -4.72 *8*		
							ES-CH 5.20 *2* F 0.37 *5* GR.ELCT 0.78 *13*		
							L-STM 8.20 *1* R -0.04 *5*		
							B1-STM 1.35 *1* GRP.DPL -4.99 *9*		
	SO2C6H4OH-P			★ 0.64 *166*	3.47		★S.INDUC 0.53 *166* ★S.PARA- 0.86 *167*		
							★S.META- 0.56 *167* ★ES-CH 6.20 *2*		
	SO3C6H5			★ 0.51 *194*	3.67		S.INDUC 0.47 *16* ★S.PARA- 1.11 *916* ★ES-CH 5.20 *2*		
C6H5S1	C6H4SH -O						S.STAR 0.72 *3* GRP.DPL -1.16 *9*		
	CH=CH(2-thienyl)			★ 0.11 *304*					
	SC6H5	2.32	0.17 *3* ★ 0.23 *6* 0.30 *10*	0.02 *147* 0.07 *6* ★ 0.07 *375* 0.13 *3*	3.43		S.INDUC 0.20 *31* ★S.PARA+ -0.55 *76* B1-STM 1.70 *1*		
							S.INDUC 0.21 *282* S.PARA+ -0.54 *829* B5-STM 6.42 *1*		
							S.INDUC 0.25 *12* S.PARA+ -0.45 *145* F 0.30 *5*		
							S.INDUC 0.26 *99* S.PARA+ -0.25 *202* R -0.23 *5*		
							★S.INDUC 0.31 *6* S.ORTHO 0.72 *104* R+ -0.85 *4*		
							S.INDUC 0.36 *16* S.ORTH+ -0.41 *154* R- -0.12 *4*		
							S.INDUC 0.36 *102* S.ZE.RS -0.24 *6* S.PHOSP -1.01 *526*		
							S.STAR 1.82 *10* S.ZE.RS -0.21 *102* S.IND.P 1.71 *39*		
							S.STAR 1.87 *3* S.ZE.RS -0.19 *19* S.RES.P -0.52 *39*		
							S.META- 0.18 *378* S.ZE.RS -0.11 *12* S.ZTWST 0.15 *662*		
							★S.META- 0.20 *461* S.ZE.RS -0.08 *282* S.P.RAD 0.58 *79*		
							S.ZPTFT 0.01 *112* S.ZE.RS -0.07 *31* GRP.DPL -1.55 *8*		
							★S.PARA- 0.18 *109* S.BA.RS -0.22 *34* GRP.DPL -1.29 *9*		
							S.PARA- 0.29 *378* S.RES.+ -0.84 *470* GR.ELCT 0.17 *13*		
							S.PARA- 0.32 *372* ES-CH 4.20 *2*		
							S.PARA- 0.40 *833* L-STM 4.57 *1*		
C6H5S2	SS-C6H5						S.INDUC 0.23 *12* S.ZE.RS -0.09 *12* GRP.DPL -1.79 *8*		
C6H5Se1	SeC6H5			★ 0.13 *3*			S.INDUC 0.07 *12* S.PARA- 0.42 *833* ★ES-CH 4.30 *2*		
							S.STAR 1.38 *10* ★S.PARA+ -0.47 *492* ★S-L 0.26 *14*		
							S.STAR 2.30 *3* S.ZE.RS -0.19 *12* GR.ELCT 0.18 *13*		
							★S.PARA- 0.13 *109* S.ZE.RS -0.19 *86*		
C6H5Te1	TeC6H5			★ 0.34 *103*			★S.INDUC 0.38 *103* ★S.ZE.RS -0.04 *103*		
C6H5Zn1	Zn(C6H5)						★S.ZE.RS 0.11 *86* ★ES-CH 4.30 *2*		
C6H6B1O2	B(OCH2CCH)2						★S.STAR 0.70 *358*		
C6H6Br1Ge1	Ge(H)(Ph)Br			★ 0.14 *51*			★S.PARA+ 0.10 *51*		
C6H6Cl1Ge1	Ge(Ph)(H)(Cl)			★ 0.16 *51*			★S.PARA+ 0.13 *51*		
C6H6Hg1N1	NH-Hg-Ph						S.PARA+ -3.09 *71* S.RES.+ -2.47 *71*		
C6H6N1	2(pyridinyl)methyl						GRP.DPL -2.18 *8*		
	2(pyridinyl)methyl						GRP.DPL -1.89 *8*		
	2-(6-Me-pyridyl)						S.INDUC 0.17 *96*		
	4(pyridinyl)methyl						GRP.DPL -2.65 *8*		
	C6H4-NH2 -O						GRP.DPL 1.45 *8*		
	C6H4NH2 -M			★ -0.02 *82*	2.98		S.INDUC 0.07 *84* ★ES-CH 3.00 *2*		

Empirical	Structural	π	σ_m	σ_p	MR	E_s	additional σ values		
C6H6N1	C6H4NH2 -M						S.ZE.RS -0.07 84	★S.PHOSP -0.56 85	
	C6H4NH2 -P			-0.30 265	2.98		S.INDUC 0.07 84	ES-CH 3.00 2	★S.PHOSP -0.78 37
				★ -0.21 82			★S.PARA- -0.02 427	L-STM 7.00 1	★S.PHOSP -0.73 85
				-0.18 392			S.PARA- 0.05 187	B1-STM 1.71 1	
							S.ZE.RS -0.12 84	B5-STM 3.11 1	
	NHC6H5	1.37	-0.12 555	-0.59 147	3.00		★S.INDUC 0.02 555	S.ZE.RS -0.86 6	B5-STM 5.95 1
			★ -0.02 6	★ -0.56 6			S.INDUC 0.18 12	S.ZE.RS -0.50 19	F 0.22 5
				-0.45 236			S.INDUC 0.36 16	S.ZE.RS -0.42 12	R -0.78 5
				-0.40 555			S.ZPTFT -0.27 112	★S.ZE.RS -0.35 292	R+ -1.43 4
				-0.26 197			★S.PARA- -0.29 109	S.RES.+ -1.09 470	R- -0.32 4
							★S.PARA+ -1.40 43	ES-CH 4.00 2	GRP.DPL 1.11 8
							S.PARA+ -1.35 322	L-STM 4.53 1	
							S.ORTHO 0.20 78	B1-STM 1.35 1	
C6H6N1O1	2(pyridinoxide)CH2-						GRP.DPL -4.13 8		
	3(pyridineoxide)CH2-						GRP.DPL -4.61 8		
	4(pyridineoxide)CH2-						GRP.DPL -4.63 8		
	OC6H4NH2 -P			★ -0.36 392			★S.PARA- -0.17 427	★S.PARA- -0.06 187	★ES-CH 4.00 2
C6H6N1O2	CH=C(CO2C2H5)CN			★ 0.45 783	3.09		★S.ZPTFT 0.62 248	★ES-CH 4.00 2	B1-STM 1.60 1
							S.PARA- 1.01 257	L-STM 8.19 1	B5-STM 5.48 1
C6H6N1O2S1	C6H4SO2NH2-P				3.66		★ES-CH 3.00 2	★S.PHOSP 0.00 85	
	NHSO2C6H5	0.45	0.16 10	-0.03 6	3.79		★S.INDUC 0.33 6	ES-CH 5.20 2	R -0.23 5
			★ 0.16 56	-0.01 399			S.STAR 1.99 3	L-STM 8.24 11	R+ -1.22 4
			0.20 6	★ 0.01 56			★S.PARA+ -0.98 71	B1-STM 1.35 11	S-L 0.34 7
			0.20 399				S.ZE.RS -0.36 6	B5-STM 3.72 11	GRP.DPL -4.62 9
							S.RES.+ -0.85 71	F 0.24 5	GRP.DPL -4.58 8
	SO2C6H4NH2-M			★ 0.67 166	3.76		★S.INDUC 0.55 166	★S.PARA- 0.90 167	
							★S.META- 0.60 167	★ES-CH 6.20 2	
	SO2C6H4NH2-P			0.50 392	3.76		★S.INDUC 0.51 166	S.PARA- 0.80 167	★S.PARA- 1.33 600
				0.58 166			★S.META- 0.51 167	S.PARA- 0.83 272	ES-CH 6.20 2
				★ 0.58 375			S.PARA- 0.69 427	S.PARA- 0.92 187	
	SO2NHC6H5	0.45	★ 0.56 3	★ 0.65 304	3.78		L-STM 8.24 1	B5-STM 4.50 1	R 0.14 5
							B1-STM 2.03 1	F 0.51 5	
C6H6N1S1	S-C6H4-NH2 -O						GRP.DPL -1.87 8		
	SC6H4NH2 -P						★S.PARA- 0.09 427	★ES-CH 4.20 2	
							★S.PARA- 0.24 187	★GRP.DPL -2.44 8	
C6H6N1S2	SSC6H4NH2 -P						★S.PARA- 0.40 187	★ES-CH 3.40 2	
C6H6N2	NH+=NC6H5						★S.PARA- 2.04 252	★ES-CH 3.00 2	
C6H6N3	N=N-C6H4-NH2 -O						GRP.DPL 1.49 8		
	N=N-C6H4-NH2 -P						L-STM 9.15 1	B5-STM 4.31 1	
							B1-STM 1.70 1	GRP.DPL 2.50 8	
C6H7Ge1	Ge(C6H5)H2			★ 0.17 60			★S.PARA+ 0.15 60	★ES-CH 4.30 2	
C6H7N1	+NH2C6H5						★S.STAR 4.09 360	★S.STAR 4.37 3	
C6H7N1 @	2-(N-Me-pyridinium+):						S.ORTH- 1.75 445	★S.ORTH- 2.49 231	
							★S.ORTH- 2.49 162	★S.ORTH- 3.25 162	
	3-(N-Me-pyridinium+):						★S.META- 1.10 445	★S.META- 1.15 162	★S.META- 1.58 231
	4-(N-Me-pyridinium+):						S.PARA- 1.78 445	★S.PARA- 2.32 231	
							S.PARA- 2.16 162	★S.PARA- 2.59 162	
C6H7N1S1	S-Ph-4-NH3+						S.STAR 2.32 3		
C6H7N2	2-(4,6-diMe-pyridinyl)						★S.INDUC 0.21 96		
	2-(5,6-diMe-pyrazinyl)						★S.INDUC 0.23 96		
	C6H4NHNH2-M				3.28		★ES-CH 3.00 2	★S.PHOSP -0.47 85	
	C6H4NHNH2-P				3.28		★ES-CH 3.00 2	★S.PHOSP -0.79 85	
	NHNHC6H5						★S.ZE.RS -0.44 19	L-STM 8.15 11	B5-STM 3.49 11
							★ES-CH 3.00 2	B1-STM 1.35 11	
C6H7N2O2S1	NHC6H4SO2NH2 -P			★ -0.40 147			★ES-CH 4.00 2		
C6H7N3	N=N-C6H4-NH2 -M						GRP.DPL 1.71 8		
C6H7O1	2-(5-ethylfuryl)		★ 0.09 388	★ -0.13 295			★ES-CH 3.00 2	F 0.20 4	R -0.33 4
	CH2CCCH=CHOCH3 -C						★S.STAR 0.49 380	★ES-CH 3.00 2	
	CH2CCCH=CHOCH3 -T						L-STM 7.42 1	B1-STM 1.52 1	B5-STM 7.60 1
C6H7O2	CH2CCCH2CO2CH3						★S.STAR 0.75 380	L-STM 7.64 1	B5-STM 7.27 1
							★ES-CH 3.00 2	B1-STM 1.52 1	
C6H7S1 @	2-(5-ethylthienyl):						★S.ORTHO -0.16 199	★S.ORTHO -0.02 169	S.ORTHO -0.02 1014
C6H7Se1 @	2-(5-ethylselenieyl)						★S.ORTHO -0.02 169		
C6H7Si1	Si(C6H5)H2			★ 0.19 60			★S.PARA+ 0.19 60	★ES-CH 4.20 2	
C6H8Cl1O1	O-(2-clcyclohex-1-yl)						S.IND.P 4.50 284	S.RES.P -4.06 284	
C6H8N1	1-(2,5-diMe-pyrryl)		★ 0.49 163	★ 0.38 163	2.88		★ES-CH 5.00 2	F 0.52 5	R -0.14 5
C6H8N2 @	3(2-Am-5-Me-pyridinium)+						S.META+ 1.89 441		
C6H8N3	2-N(Me)2-pyrimidin-5yl						S.INDUC 0.19 113	S.ZE.RS -0.09 113	
	2-NMe2-4-pyrimidinyl						S.INDUC 0.18 108	S.ZE.RS 0.06 108	
	4-NMe2-2-pyrimidyl						★S.INDUC 0.02 204	★S.ZE.RS 0.06 204	
	6-NMe2-4-pyrimidinyl						S.INDUC 0.12 108	S.ZE.RS 0.04 108	
C6H8N3 @	5(6-Am-2,4-dimepyrimidn):						S.META+ 0.55 441		
C6H8N3O1 @	4-(1-(Me)2NCO-pyrazyl):			★ 0.32 642			★ES-CH 3.00 2		
C6H9	1-bicyclo(2.1.1)hexyl						★S.PARA+ -0.50 525	L-STM 5.45 1	B5-STM 3.19 1

Empirical	Structural	π	σ_m	σ_p	MR	E_s	additional σ values					
C6H9	1-bicyclo(2.1.1)hexyl						★ES-CH	4.00 2	B1-STM	2.73 1		
	1-bicyclohexyl(3.1.0)			★ 0.15 271	2.48		★S.PARA-	-0.14 400	★ES-CH	4.00 2	B1-STM	2.08 1
							★S.PARA+	-0.71 276	L-STM	5.69 1	B5-STM	3.90 1
	1-cyclohexenyl		★ -0.10 152	★ -0.08 152			★ES-CH	3.00 2	B5-STM	3.30 1	GR.ELCT	0.07 16
							L-STM	6.16 1	F	-0.07 4		
							B1-STM	2.23 1	R	-0.01 4		
	3-cyclohexenyl				2.96		★S.STAR	-0.12 740	★ES-CH	3.00 2		
	C(CH3)2CCCH3						★S.STAR	0.21 253	★ES-CH	5.00 2		
	CC(t-Bu)						★S.STAR	1.43 653				
	CCBu						★S.STAR	1.48 653				
	cyclopent-1-enylmethyl						★S.STAR	0.14 357				
	cyclopent-2-enylmethyl						★S.STAR	0.06 357				
	cyclopent-3-enylmethyl						★S.STAR	-0.03 357				
C6H9Ge1	Ge(CH=CH2)3						S.INDUC	-0.09 16				
C6H9N1	2-(6-NMe2-naphthyl)			★ -0.44 55								
	2-(7-NMe2-naphthyl)			★ -0.20 55			★ES-CH	0.00 2				
	2-(8-NMe2-naphthyl)			★ -0.07 55			★ES-CH	0.00 2				
C6H9N2O1	1(3,4,4-triMe-pyrazl-5-ON						GRP.DPL	-2.83 8				
	1(3-Me-5-EtO-pyrazolyl)						GRP.DPL	-2.65 8				
C6H9N2O3	CHOHCH(CH2OH)NHCOCH2CN	-2.07 54			3.46		★ES-CH	5.00 2	B1-STM	1.73 1		
							L-STM	8.54 1	B5-STM	6.32 1		
C6H9N3@	5(6-Am-2,4-dimepyridinium):						S.META+	1.95 441				
C6H9N4O1	2(4-NMe2,6-OMe-triazinyl)						★S.PARA-	0.59 346	★ES-CH	3.00 2		
C6H9O1	O-(cyclohex-1-yl)						S.IND.P	4.06 284	S.RES.P	-3.84 284		
C6H9O6Si1	Si(OCOCH3)3						★S.INDUC	0.07 25	★S.ZE.RS	0.20 25		
C6H9Si1	Si(CH=CH2)3						★S.INDUC	-0.19 25	S.INDUC	-0.18 16	★S.ZE.RS	0.03 25
C6H9Sn1	Sn(CH=CH2)3						S.INDUC	-0.12 16				
C6H10B1O2	B(OCH2CH=CH2)2						★S.STAR	0.68 358				
C6H10N1	(CH2)5CN						★S.STAR	-0.06 420	★ES-CH	3.00 2		
	CH2CCCH2N(Me)2						S.STAR	0.48 262				
C6H10N1O1	NHC(=O)-cyclopentyl						L-STM	7.29 1	B1-STM	1.55 1	B5-STM	4.71 1
C6H10N1O3	OCH2C=O(N-morpholinyl)	-1.39			3.41		★ES-CH	3.00 2				
C6H11	C(CH3)2CH=CHCH3 -E						L-STM	5.89 1	B1-STM	2.60 1	B5-STM	4.82 1
	cyclohexyl	2.76 454	-0.15 152	-0.22 271	2.67	-2.64 403	S.INDUC	-0.07 348	S.PARA-	-0.20 584	L-STM	6.17 1
		2.82 587	-0.14 3	★ -0.15 6		-2.34 302	S.INDUC	-0.06 500	★S.PARA-	-0.14 400	B1-STM	1.91 1
			★ -0.05 6	-0.13 152		-2.03 50	★S.INDUC	-0.02 32	S.PARA+	-0.38 543	B5-STM	3.49 1
						★ -1.81 35	S.INDUC	0.00 6	★S.PARA+	-0.29 152	F	0.03 5
						-1.02 311	S.INDUC	0.07 16	S.ORTH+	-0.30 154	R	-0.18 5
							S.STAR	-0.26 740	S.ORTH-	-0.17 97	R+	-0.32 4
							S.STAR	-0.20 348	S.ZE.RS	-0.15 6	R-	-0.17 4
							S.STAR	-0.18 3	S.ZE.RS	-0.13 19	S-L	0.00 14
							S.STAR	-0.15 52	S.ZE.RS	0.13 114	S.PHOSP	-1.19 59
							S.STAR	-0.10 576	ES-CH	3.00 2	GRP.DPL	0.00 9
							S.META+	-0.05 152	ES-V	0.87 7	GRP.DPL	0.62 8
	cyclopentylmethyl						★S.STAR	-0.11 357				
C6H11Cl1N1O3	CHOHCH(CH2OH)NHCOCH(Cl)Me	-0.87 54			4.14		★ES-CH	5.00 2	B1-STM	1.73 1		
							L-STM	8.50 1	B5-STM	5.37 1		
C6H11N2	CH=N-(N-piperidinyl)	0.91		★ 0.45 163			★ES-CH	3.00 2				
C6H11N2O4	C(pent)(NO2)2						★S.STAR	3.34 157				
C6H11N2S1	1(3,4,4-Me3-pyraz-5-thion						GRP.DPL	-3.16 8				
C6H11O1	4-hydroxycyclohexyl -C						GRP.DPL	-1.87 8				
	4-hydroxycyclohexyl -T						GRP.DPL	-1.56 8				
	C(=O)-C5H11						L-STM	8.17 1	B1-STM	1.63 1	B5-STM	5.91 1
	C(=O)CH2CH2CH(CH3)2						L-STM	6.92 1	B1-STM	1.63 1	B5-STM	5.79 1
	COC(CH3)2CH2CH3						★S.ZE.RS	0.15 47	★ES-CH	6.00 2		
	OCH2-cy-pentyl						★ES-CH	3.00 2	L-STM	6.84 1	B5-STM	5.21 1
							★ES-V	0.58 7	B1-STM	1.35 1		
	cyclohexyloxy		★ 0.29 3		2.85		S.INDUC	-0.07 49	L-STM	6.05 1	S.IND.P	2.07 284
							★S.INDUC	0.29 32	B1-STM	1.35 1	S.RES.P	-2.30 284
							S.STAR	1.81 3	B5-STM	5.71 1	S.RES.P	-0.71 39
							S.ZE.RS	-0.28 49	S-L	0.32 7	GRP.DPL	-1.68 9
							ES-CH	4.00 2	★S.PHOSP	-0.35 59	GRP.DPL	-1.55 8
							★ES-V	0.81 7	S.IND.P	1.66 39		
C6H11O2	(CH2)3CO2Et				3.04	★ -1.74 196	ES-CH	3.00 2	B1-STM	1.52 1	★S-L	0.04 14
							L-STM	8.85 1	B5-STM	6.23 1		
	C(O)o-pent						GRP.DPL	-1.99 8				
	Et-OC(O)CH2CH2CH3						GRP.DPL	1.85 8				
	OC(=O)-C5H11						L-STM	8.87 1	B1-STM	1.35 1	B5-STM	6.19 1
C6H11O2S1	SC(CH3)2CO2CH2CH3						★ES-CH	5.20 2	★S-L	0.23 14		
C6H11O5	?-glucosidyl						★S.PARA-	0.02 604				
	D-glucosidyl				3.47		★S.PARA-	0.04 720	★ES-CH	4.00 814		
C6H11O6	O-B-glucosyl	-2.84			3.65		★ES-CH	4.00 2				
C6H11S1	cyclohexylmercapto		★ 0.31 3		3.46		★S.INDUC	0.31 32	L-STM	6.21 1	★S-L	0.32 14
							★S.STAR	1.93 3	B1-STM	1.70 1		

Empirical	Structural	π	σm	σp	MR	Es	additional σ values				
C6H11S1	cyclohexylmercapto						★ES-CH 4.20 2	B5-STM 6.29 1			
C6H11Se1	cyclohexylseleno		★ 0.41 3		3.74		★S.INDUC 0.38 32	L-STM 6.45 1	★S-L 0.40 14		
							★S.STAR 2.37 3	B1-STM 1.85 1			
							★ES-CH 4.30 2	B5-STM 6.49 1			
C6H11Si1	CH2CCSi(Me)3						★S.STAR 0.74 433				
C6H12	hexamethylene						★S.STAR -0.35 433				
C6H12Ge1N1O3	Ge(OCH2CH2)3N						S.INDUC -0.23 92	S.ZE.RS 0.05 92			
C6H12N1	CH2CH2(1-pyrrolidyl)	0.69 42					★ES-CH 3.00 2				
	CH2piperidyl				3.06		★S.STAR 0.23 74	★ES-CH 4.00 2			
	CH=CHCH2CH2NMe2	1.03 42					★ES-CH 3.00 658	B1-STM 1.60 1			
							L-STM 8.31 1	B5-STM 5.54 1			
	CH=CHN(C2H5)2						★S.ZE.RS -0.31 47	★ES-CH 3.00 2			
	NH-cyclohexyl						★ES-CH 4.00 2	L-STM 6.07 1	B5-STM 5.77 1		
							★ES-V 0.92 7	B1-STM 1.35 1			
C6H12N1O1	(CH2)3C(=O)N(CH3)2						L-STM 8.07 1	B1-STM 1.52 1	B5-STM 5.86 1		
	CHOH(2-piperidinyl)	-0.85					★ES-CH 5.00 2				
C6H12N1O3Si1	Si(OCH2CH2)3N						S.INDUC -0.56 92	S.PARA+ -0.08 383	★S.RES.+ -0.09 140		
							★S.INDUC -0.40 140	S.ZE.RS 0.02 92	★ES-CH 4.20 2		
							S.STAR -0.89 140	★S.ZE.RS 0.02 140			
C6H12O2	OC(CH3)2C(CH3)2O*						★S.PHOSP 0.10 479				
C6H12O2P1S2	EtSSP(OCH(CH3)CH(CH3)O-)						★S.INDUC 0.22 137	★S.STAR 0.49 137	★ES-CH 3.20 2		
C6H13	(CH2)3CH(CH3)2				2.89	-1.71 196	ES-CH 3.00 2	B1-STM 1.52 1			
						-1.67 44	★ES-V 0.68 7	B5-STM 5.59 1			
						-1.43 35	L-STM 6.97 1	★S.PHOSP -1.21 37			
	C(CH3)(Et)2						L-STM 4.92 1	B1-STM 2.66 1	B5-STM 4.18 1		
	C(CH3)2CH(CH3)2				2.89	★ -4.66 62	ES-CH 6.00 2	L-STM 4.92 1	B5-STM 4.19 1		
							★ES-HYBO -5.21 225	B1-STM 2.60 1			
	C(CH3)2Pr						L-STM 6.17 1	B1-STM 2.60 1	B5-STM 4.54 1		
	CH(CH3)(I-butyl)						★ES-CH 5.00 2	L-STM 6.17 1	B5-STM 4.54 1		
							★ES-V 1.09 7	B1-STM 1.90 1			
	CH(CH3)Bu					★ -2.18 35	★ES-CH 4.00 2	L-STM 6.97 1	B5-STM 4.94 1		
							★ES-V 1.07 7	B1-STM 1.90 1			
	CH(CH3)C(CH3)3				2.89	★ -4.57 50	★S.STAR -0.29 52	★ES-HYBO -2.43 225	B5-STM 4.19 1		
						-4.33 62	ES-CH 6.00 2	L-STM 4.92 1			
							★ES-V 2.11 7	B1-STM 1.90 1			
	CH(Et)(I-propyl)					★ -4.35 62	ES-CH 6.00 2	L-STM 4.92 1	B5-STM 4.19 1		
							★ES-V 2.11 7	B1-STM 1.90 1			
	CH(Et)(Pr)					★ -3.12 35	L-STM 5.48 184	B1-STM 2.12 184	B5-STM 5.25 184		
	CH2C(CH3)2Et						★ES-CH 5.00 2	B1-STM 1.52 1			
							L-STM 6.17 1	B5-STM 4.54 1			
	EtC(CH3)3				2.89	-1.58 50	ES-CH 3.00 2	L-STM 6.17 1	B5-STM 4.54 1		
						★ -1.45 62	★ES-V 0.70 7	B1-STM 1.52 1			
	hexyl	★ -0.16 3			2.89	-1.68 196	★S.INDUC -0.04 32	S.ORTHO -0.35 53	B1-STM 1.52 1		
						★ -1.54 44	S.STAR -0.25 3	S.ZE.RS -0.11 22	B5-STM 5.96 1		
							★S.STAR -0.17 287	ES-CH 3.00 2	S-L -0.03 7		
							S.STAR -0.15 606	ES-V 0.73 7	★S.PHOSP -1.21 37		
							S.STAR 0.00 278	L-STM 8.22 1	S.ZTWST -0.10 22		
C6H13N1	CH2NH+(CH2)5						S.PHOSP 0.35 367				
	N-Me-piperidinium					★ -4.24 303					
C6H13N1O1	O(CH2)3N*(I-Pr)						★S.PHOSP -1.72 354				
C6H13N2O1	NHCH(I-Bu)C(=O)NH2						L-STM 6.02 1	B1-STM 1.63 1	B5-STM 5.46 1		
C6H13O1	(CH2)3OPr						★ES-CH 3.00 2	L-STM 8.87 1	B5-STM 6.33 1		
							★ES-V 0.70 7	B1-STM 1.52 1			
	(CH2)4OEt						★ES-CH 3.00 2	L-STM 8.90 1	B5-STM 6.27 1		
							★ES-V 0.67 7	B1-STM 1.52 1			
	CH(CH3)C(OH)(CH3)Et				3.04	★ -4.46 44	ES-CH 6.00 2	B1-STM 1.90 1			
							L-STM 6.17 1	B5-STM 4.54 1			
	CH(Et)C(OH)(CH3)2				3.04	★ -5.00 44	ES-CH 7.00 2	B1-STM 1.90 1			
							L-STM 4.92 1	B5-STM 4.45 1			
	CH(OCH3)Bu						★ES-CH 5.00 2	L-STM 6.97 1	B5-STM 4.94 1		
							★ES-V 1.20 7	B1-STM 1.90 1			
	CH(OH)C5H11				3.04	★ -1.58 44	ES-CH 4.00 2	L-STM 8.22 1	B5-STM 5.96 1		
							★ES-V 0.71 7	B1-STM 1.73 1			
	CH2CH2O-I-Bu						★ES-CH 3.00 2	★ES-V 0.89 7			
	CH2CH2OBu						★ES-CH 3.00 2	L-STM 8.92 1	B5-STM 6.25 1		
							★ES-V 0.89 7	B1-STM 1.52 1			
	CH2CH2OCH2CH(CH3)2						L-STM 8.10 1	B1-STM 1.52 1	B5-STM 5.83 1		
	O(CH2)5CH3				3.07		★ES-CH 3.00 2	L-STM 8.93 1	B5-STM 6.23 1		
							★ES-V 0.61 7	B1-STM 1.35 1	★S.PHOSP -0.32 37		
	OC(CH3)2propyl						★ES-CH 5.00 2	L-STM 6.86 1	B5-STM 4.79 1		
							★ES-V 1.39 7	B1-STM 1.35 1			
	OC(CH3)Et2						★ES-CH 4.00 2	L-STM 6.05 1	B5-STM 5.40 1		
							★ES-V 1.52 7	B1-STM 1.35 1			
	OCH(CH3)Bu						L-STM 8.11 1	B1-STM 1.35 1	B5-STM 5.81 1		
	OCH(CH3)C(CH3)3				3.07		★S.STAR 1.46 37	L-STM 6.05 1	★S.PHOSP -0.50 37		
							★ES-CH 4.00 2	B1-STM 1.35 1			

Empirical	Structural	π	σ_m	σ_p	MR	E_s	additional σ values					
C6H13O1	OCH(CH3)C(CH3)3						★ES-V	1.19 7	B5-STM	4.42 1		
	OCH(Et)I-propyl						★ES-CH	4.00 2	L-STM	6.05 1	B5-STM	5.40 1
							★ES-V	1.18 7	B1-STM	1.35 1		
	OCH(Et)propyl						★ES-CH	4.00 2	L-STM	6.86 1	B5-STM	5.40 1
							★ES-V	1.04 7	B1-STM	1.35 1		
	OCH2C(CH3)2ethyl						★ES-CH	3.00 2	L-STM	6.86 1	B5-STM	4.79 1
							★ES-V	0.78 7	B1-STM	1.35 1		
	OCH2CH(CH3)I-propyl						★ES-CH	3.00 2	L-STM	6.86 1	B5-STM	5.48 1
							★ES-V	0.64 7	B1-STM	1.35 1		
	OCH2CH(Et)2						★ES-CH	3.00 2	L-STM	6.86 1	B5-STM	4.79 1
							★ES-V	0.71 7	B1-STM	1.35 1		
	OCH2CH2-T-butyl						★ES-CH	3.00 2	L-STM	6.86 1	B5-STM	5.71 1
							★ES-V	0.53 7	B1-STM	1.35 1		
C6H13O2	C(Me)(OEt)2						S.STAR	0.62 446				
	CH2CH2-O-o-C(CH3)3					★ -2.05 818						
	CH2OCH2CH2CH2CH2OCH3						L-STM	9.95 1	B1-STM	1.52 1	B5-STM	7.24 1
	CH2OCH2CH2OPr						★ES-CH	3.00 2	L-STM	9.97 1	B5-STM	7.22 1
							★ES-V	0.56 7	B1-STM	1.52 1		
C6H13O2S1	SO2-hexyl						L-STM	9.03 1	B1-STM	2.03 1	B5-STM	6.39 1
	SO2CH2C(CH3)2Et						L-STM	6.97 1	B1-STM	2.03 1	B5-STM	4.94 1
	SO2CH2CH2C(CH3)3						L-STM	6.97 1	B1-STM	2.03 1	B5-STM	5.83 1
C6H13O3	CH2OCH2CH2OCH2CH2OCH3						★ES-CH	3.00 2	★ES-V	0.56 7		
C6H13O3S1	(CH2)6SO3H						★S.PARA+	-0.34 301	★ES-CH	3.00 2		
C6H13S1	(CH2)4SC2H5						★S.STAR	0.07 278	L-STM	9.26 1	B5-STM	6.87 1
							★ES-CH	3.00 2	B1-STM	1.52 1		
	S(CH2)5CH3						S.STAR	1.33 3	B1-STM	1.70 1	GRP.DPL	-1.56 9
							★ES-CH	3.20 2	B5-STM	7.25 1		
							L-STM	9.05 1	S-L	0.25 14		
	SCH2C(CH3)2Et						L-STM	7.06 1	B1-STM	1.70 1	B5-STM	5.61 1
	SCH2CH2C(CH3)3						L-STM	7.06 1	B1-STM	1.70 1	B5-STM	6.29 1
C6H13Si1	CH2Si(CH3)2CH2CH=CH2						★S.STAR	-0.15 381	★ES-CH	5.20 2		
C6H14B1O2	B(O-I-Pr)2						★S.STAR	0.53 358				
C6H14Cl1Ge1	Ge(H)(Cl)(N-hexyl)			★ 0.15 51			S.PARA+	-0.05 25	★S.PARA+	0.11 51		
C6H14N1	Bu-N(CH3)2	0.84 42	★ -0.08 3	★ -0.16 315	3.27		★ES-CH	3.00 2	B1-STM	1.52 1	F	-0.01 4
							L-STM	8.13 1	B5-STM	5.89 1	R	-0.15 4
	CH2CH2NEt2	0.58 42					★ES-CH	3.00 2	B1-STM	1.52 1		
							L-STM	7.11 1	B5-STM	4.86 1		
	N(C3H7)2		★ -0.26 413	★ -0.93 413			ES-CH	5.00 2	B1-STM	1.35 1	R	-0.99 5
							★ES-V	1.60 7	B5-STM	5.50 1	S.PHOSP	-1.35 526
							L-STM	6.07 1	F	0.06 5		
	N(I-propyl)2						★ES-CH	7.00 2	L-STM	4.83 1	B5-STM	4.39 1
							★ES-V	2.01 7	L-STM	2.10 1	★GRP.DPL	1.53 8
	NH(CH2)5CH3						★ES-CH	3.00 2	L-STM	8.93 1	B5-STM	6.33 1
							★ES-V	0.66 7	B1-STM	1.35 1		
	NHCH2C(CH3)2Et						L-STM	6.88 1	B1-STM	1.45 1	B5-STM	4.87 1
	NHCH2CH2C(CH3)3						L-STM	6.88 1	B1-STM	1.35 1	B5-STM	5.77 1
C6H14N1O1	OCH2CH2N(Et)2						L-STM	8.02 1	B1-STM	1.35 1	B5-STM	5.75 1
C6H14N1O2S1	SO2N(Pr)2						★K-CHGTR	1.33 77				
C6H14N1S1	SCH2CH2N(Et)2						L-STM	8.03 1	B1-STM	1.70 1	B5-STM	6.52 1
C6H14N1Se1	SeCH2CH2N(Et)2						L-STM	7.78 1	B1-STM	1.85 1	B5-STM	6.77 1
C6H14O1P1	CH2CH2P(O)(C2H5)2						★S.STAR	0.31 422	★ES-CH	3.20 2	★S-L	0.07 14
	P(O)(C3H7)2						S.INDUC	0.27 45	L-STM	6.42 1	B5-STM	5.74 1
							★ES-CH	6.20 2	B1-STM	2.52 1	★S-L	0.26 14
	P(O)(I-Pr)2		★ 0.37 107	★ 0.41 107			S.INDUC	0.32 107	F	0.36 4		
							★S.ZE.RS	0.09 107	R	0.05 4		
C6H14O1P1S2	CH2CH2SSPCH3(OPr)						★S.INDUC	0.16 137	★S.STAR	0.36 137	★ES-CH	3.20 2
C6H14O3P1	(CH2)2P(O)(OC2H5)2				4.05		★S.STAR	0.29 422	L-STM	8.31 1	★S-L	0.05 14
							★S.STAR	0.30 74	B1-STM	1.70 1	S.PHOSP	-0.76 367
							★ES-CH	3.20 2	B5-STM	6.05 1		
	CH(CH3)P(O)(OC2H5)2				4.05		★S.STAR	0.78 74	L-STM	7.10 1	B5-STM	5.73 1
							★ES-CH	6.20 2	B1-STM	1.90 1		
	P(O)(OPr)2		★ 0.38 207	★ 0.50 207	4.05		★ES-CH	6.20 2	B1-STM	2.52 1	F	0.33 5
							L-STM	7.07 1	B5-STM	6.90 1	R	0.17 5
C6H14O4P1	CH(CH3)OP(O)(OC2H5)2				4.14		★S.STAR	1.03 74	★ES-CH	4.20 2		
	CH2OCH2P(O)(OC2H5)2				4.21		★S.STAR	0.87 74	L-STM	8.97 1	B5-STM	6.59 1
							★ES-CH	3.00 2	B1-STM	1.52 1		
	OP(=O)(O-I-Pr)2						L-STM	6.99 1	B1-STM	1.35 1	B5-STM	6.71 1
	OP(=O)(OPr)2						L-STM	7.12 1	B1-STM	1.35 1	B5-STM	6.93 1
C6H14P1	P(I-Pr)2		★ 0.02 150	★ 0.06 150			S.INDUC	-0.03 150	F	0.04 4		
							S.ZE.RS	0.09 150	R	0.02 4		
C6H14P1S2	CH2CH2SSP(Et)2						★S.INDUC	0.15 137	★S.STAR	0.34 137	★ES-CH	3.20 2
C6H14S1	(CH2)3S+(CH3)C2H5						★S.STAR	0.91 278	L-STM	8.22 1	B5-STM	5.96 1
							★ES-CH	3.00 2	B1-STM	1.52 1		
	S+(I-Pr)2						S.CH-P+	1.18 75				

Empirical	Structural	π	σ_m	σ_p	MR	E_s	additional σ values					
C6H14S1	S+(Pr)2						S.CH-P+	1.12 75				
C6H14S2	SCH2CH2S+(Et)2						L-STM	8.12 1	B1-STM	1.70 1	B5-STM	6.59 1
C6H15As1	As+(Et)3						S.CH-P+	0.89 75				
C6H15Br1P1Pt1	Pt(Br)P(C2H5)3 -T						★ S.INDUC	-0.19 33	★ S.ZE.RS	-0.27 33		
C6H15Cl1P1Pt1	Pt(Cl)P(C2H5)3 -C						★ S.INDUC	-0.39 33	★ S.ZE.RS	-0.22 33		
	Pt(Cl)P(C2H5)3 -T						★ S.INDUC	-0.21 33	★ S.ZE.RS	-0.27 33		
C6H15Cl3P1Pt1Sn1	Pt(SnCl3)P(C2H5)3 -T						★ S.INDUC	0.12 33	★ S.ZE.RS	-0.24 33		
C6H15Ge1	Ge(Et)3		★ 0.00 3	-0.20 240 / ★ 0.00 21	4.06		★ S.STAR	-0.80 828	L-STM	5.51 1	R	-0.03 4
							S.ZE.RS	-0.07 240	B1-STM	3.16 1	★ S.ZTWST	0.05 22
							★ S.ZE.RS	0.05 22	B5-STM	4.58 1		
							★ ES-CH	7.30 2	F	0.03 4		
C6H15Ge1O3	Ge(OC2H5)3			★ 0.03 60			S.INDUC	0.24 242	★ S.PARA+	-0.12 60	★ ES-CH	7.30 2
							★ S.INDUC	0.29 92	S.ZE.RS	0.14 92		
C6H15I1P1Pt1	Pt(I)P(C2H5)3 -T						★ S.INDUC	-0.14 33	★ S.ZE.RS	-0.27 33		
C6H15N1	(CH2)2NH+(C2H5)2						★ S.STAR	0.72 936	★ ES-CH	3.00 2		
	N+(CH3)2(S-Bu)					★ -4.34 215						
	N+(CH3)2Bu					★ -3.44 215						
	N+(Et)3					★ -5.04 303						
	NH+(I-Pr)2					★ -5.14 215						
	Pr-N(CH3)3+	-4.15	★ 0.06 3	★ -0.01 3 / 0.02 315		★ -2.59 44	S.META-	0.00 36	★ S.ORTHO	0.02 53	B5-STM	5.49 1
							★ S.META-	0.16 29	ES-CH	3.00 2	F	0.12 1
							S.PARA-	-0.06 36	L-STM	6.88 1	R	-0.13 1
							★ S.PARA-	0.09 29	B1-STM	1.52 1	R-	0.03 4
C6H15N1O1	O(CH2)3N+(CH3)3						L-STM	7.53 1	B1-STM	1.35 1	B5-STM	5.75 1
	OCH(CH3)CH2N+(CH3)3						L-STM	6.76 1	B1-STM	1.35 1	B5-STM	5.37 1
C6H15N1O1P1	N=P(C2H5)2OC2H5			★ -0.59 345			★ ES-CH	5.20 2				
C6H15N1O2P1	N=P(C2H5)(OC2H5)2			★ -0.55 534			★ ES-CH	5.20 2				
C6H15N1O3P1	N=P(OC2H5)3			★ -0.52 345			★ ES-CH	5.20 2				
	N=P(OEt)3						S.INDUC	-0.40 123				
C6H15O1Sn1	O-Sn(Et)3						S.INDUC	0.08 118	S.ZE.RS	-0.54 492	S.ZE.RS	-0.48 118
C6H15O3Si1	Si(OC2H5)3		★ 0.02 3	★ 0.08 3	4.16		S.INDUC	-0.32 599	★ S.INDUC	-0.04 73	S.ZE.RS	0.12 73
							S.INDUC	-0.13 25	S.INDUC	0.00 242	S.ZE.RS	0.12 92
							S.INDUC	-0.13 92	S.STAR	-0.18 140	S.RES.+	0.21 140
							S.INDUC	-0.10 67	★ S.PARA+	0.17 383	ES-CH	7.20 2
							S.INDUC	-0.08 25	★ S.ZE.RS	0.08 47	F	0.03 5
							S.INDUC	-0.08 140	S.ZE.RS	0.08 140	R	0.05 5
							S.INDUC	-0.04 25	S.ZE.RS	0.10 67	R+	0.14 4
C6H15P1	P(Et)3+		★ 0.99 141	★ 0.98 141			S.INDUC	0.35 16	R	0.04 4		
							F	0.94 4	S.CH-P+	1.04 75		
C6H15Pb1	Pb(Et)3						L-STM	5.92 1	B5-STM	4.87 1		
							B1-STM	3.69 1	GR.ELCT	-0.30 13		
C6H15S1Sn1	S-Sn(Et)3						S.INDUC	0.23 118	S.ZE.RS	-0.17 118		
C6H15S3Si1	Si(S-Et)3						★ S.INDUC	0.06 25	★ S.ZE.RS	0.16 25		
C6H15Si1	PrSi(Me)3				3.89		S.STAR	-0.15 448	★ S.PARA-	-0.12 321	ES-CH	3.00 2
							S.STAR	-0.10 428	★ S.ORTH+	-0.28 154	★ S-L	-0.05 7
	Si(Et)3			-0.14 240 / ★ 0.00 21 / 0.03 246 / 0.11 60	3.89		★ S.INDUC	-0.24 92	★ S.ZE.RS	0.05 92	B5-STM	4.50 1
							★ S.STAR	-1.10 828	★ ES-CH	7.20 2	GR.ELCT	-0.21 13
							★ S.PARA+	0.04 60	L-STM	5.39 1		
							S.ZE.RS	0.04 240	B1-STM	3.12 1		
C6H15Sn1	Sn(Et)3		★ 0.00 3	★ 0.00 21	4.57		★ S.INDUC	0.02 198	S.ZE.RS	0.07 22	F	0.03 4
							S.ZE.RS	-0.15 365	ES-CH	7.70 2	R	-0.03 4
							S.ZE.RS	-0.13 86	L-STM	5.77 1	★ S.ZTWST	0.07 22
							★ S.ZE.RS	0.01 198	B1-STM	3.59 1	GR.ELCT	-0.21 13
							S.ZE.RS	0.01 333	B5-STM	4.77 1		
C6H17Si2	CH2Si(CH3)2Si(CH3)3						★ S.PARA+	-0.72 299	★ ES-CH	5.40 2		
C6H18B1O2Si2	B(OSi(CH3)3)2						★ S.STAR	0.71 358				
C6H18N1Si2	N(SiMe3)2						S.INDUC	-0.12 12	S.ZE.RS	-0.24 12		
C6H18N3O3Si1	Si(OCH2CH2NH2)3						★ S.INDUC	-0.10 25	★ S.ZE.RS	0.13 25		
C6H18N3Si1	Si(N(CH3)2)3		★ -0.04 3	★ -0.04 3 / 0.05 153			★ S.INDUC	-0.04 73	★ ES-CH	10.20 2	R	-0.04 4
							★ S.ZE.RS	0.00 73	F	0.00 4		
C6H18N4P1	N=P(N-Me2)3						S.INDUC	-0.56 123				
C7F5O1	COC6F5						★ S.INDUC	0.21 49	★ S.ZE.RS	0.22 49	★ ES-CH	5.00 2
C7F17N1O2P1S1	P(=NSO2CF3)(C3F7)2		★ 1.24 178	★ 1.37 178			★ S.INDUC	1.11 178	F	1.11 4		
							★ S.ZE.RS	0.26 178	R	0.26 4		
C7H1Br1F5	CH(Br)C6F5						★ S.INDUC	0.32 49	★ S.ZE.RS	-0.08 49	★ ES-CH	5.30 2
C7H2F5	CH2C6F5						★ S.INDUC	0.27 49	★ S.ZE.RS	-0.09 49	★ ES-CH	4.00 2
C7H2F5O1	CH(OH)C6F5						★ S.INDUC	0.14 49	★ S.ZE.RS	0.10 49	★ ES-CH	5.00 2
C7H3Cl1N3	N=NC6H3,2-Cl,5-CN						★ S.PARA-	0.96 83	★ ES-CH	3.00 2		
	N=NC6H3,O-Cl,P-CN						★ S.PARA-	0.97 83	★ ES-CH	3.00 2		
C7H3Cl1N3S1	N=NC6H3,O-Cl,P-SCN						★ S.PARA-	0.92 83	★ ES-CH	3.00 2		
C7H4Br1O1	COC6H4Br-P						★ S.STAR	2.39 129	★ ES-CH	5.00 2		
C7H4Cl1N2O1	N=NC6H3,2-Cl,5-CHO						★ S.PARA-	0.86 83	★ ES-CH	3.00 2		
C7H4Cl1N2O2	N=NC6H3,2-Cl,4-CO2H						★ S.PARA-	0.85 83	★ ES-CH	3.00 2		
	N=NC6H3,2-Cl,5-CO2H						★ S.PARA-	0.77 83	★ ES-CH	3.00 2		

Empirical	Structural	π	σ_m	σ_p	MR	E_s	additional σ values					
C7H4Cl1O1	COC6H4Cl -P			★ 0.40 194	3.53		★ES-CH	5.00 2				
C7H4Cl1O1S1	SC6H4(COCl) -P						★S.PARA+	0.11 202				
C7H4Cl1O2	OC(O)Ph-4-Cl						S.STAR	2.63 3	GRP.DPL	-1.94 9		
	OC6H4(COCl) -P						★S.PARA+	-0.12 202				
C7H4F1N2	N=C=NC6H4F -P						★S.INDUC	0.31 24	★S.ZE.RS	0.19 24	★ES-CH	3.00 2
C7H4F3	C6H4(CF3) -M		★ 0.12 82		2.94		★S.STAR	0.89 120	★ES-CH	3.00 2		
	C6H4CF3-P				2.94		★S.STAR	0.96 120	★ES-CH	3.00 2		
C7H4F3O1	OPh,4-(CF3)						L-STM	5.86 11	B1-STM	1.35 11	B5-STM	7.48 11
C7H4F3S1	S-Ph-3-CF3						S.STAR	2.02 3				
	SPh,4-(CF3)						L-STM	5.75 11	B1-STM	1.70 11	B5-STM	7.93 11
C7H4F4P1	P(CF3)C6H4F						★S.INDUC	0.41 18	★S.ZE.RS	0.06 18	★ES-CH	7.60 2
C7H4N1	C6H4CN -M				3.07		★S.INDUC	0.23 84	S.ZE.RS	-0.04 84		
							★S.STAR	0.98 120	★ES-CH	3.00 2		
	C6H5(CN) -P			★ 0.21 82	3.07		S.INDUC	0.36 84	S.ZE.RS	-0.02 84	★S.PHOSP	-0.04 37
							★S.STAR	1.05 120	★ES-CH	3.00 2		
C7H4N1O1	2-benzoxazolyl		★ 0.30 209	★ 0.33 209	3.27		★S.INDUC	0.28 209	S.ZE.RS	1.40 148	R-	0.38 4
							★S.PARA-	0.68 419	ES-CH	3.00 2	GRP.DPL	-1.22 8
							S.ORTH-	1.70 148	F	0.30 5		
							★S.ZE.RS	0.05 209	R	0.04 5		
	O-Ph-2-CN						S.STAR	2.67 3	GRP.DPL	-4.88 8		
	O-Ph-3-CN						S.STAR	2.59 3	GRP.DPL	-4.01 8		
	OPh,4-(CN)						S.STAR	2.73 3	B1-STM	1.35 11	GRP.DPL	-4.23 819
							L-STM	5.31 11	B5-STM	7.97 11		
C7H4N1O3	COC6H4NO2-P				3.67		★S.PARA-	0.84 109	★ES-CH	5.00 2		
C7H4N1O4	C(O)o-C6H4-NO2 -P						GRP.DPL	-4.43 8				
	OC(O)Ph-4-NO2						S.STAR	2.73 3	GRP.DPL	-3.48 9		
C7H4N1S1	2-benzthiazolyl	2.13	★ 0.27 209	★ 0.29 209	3.89		★S.INDUC	0.26 209	★S.ZE.RS	1.30 148	R-	0.38 4
							★S.PARA-	0.65 419	ES-CH	3.20 2	★GRP.DPL	-0.94 8
							★S.ORTH-	1.60 148	F	0.27 5		
							★S.ZE.RS	0.03 209	R	0.02 66		
	S-C6H4-CN -O						GRP.DPL	-5.04 8				
	S-Ph-4-CN						S.STAR	2.29 3	GRP.DPL	-4.14 8		
C7H4N1Se1 @	2-(benzoselenazolyl):						★S.ORTH-	1.80 148	★S.ZE.RS	1.20 148		
C7H4O2	C6H4CO2- -M						★ES-CH	3.00 2	★S.PHOSP	-0.42 85		
	C6H4CO2- -P						★ES-CH	3.00 2	★S.PHOSP	-0.42 85		
C7H5B1Cl3O1	C=O.BCl3(C6H5)			★ 1.21 450			★S.INDUC	0.62 450	★ES-CH	6.00 2		
C7H5B1F3O1	C=O.BF3(C6H5)			★ 0.99 450			★S.INDUC	0.49 450	★ES-CH	6.00 2		
C7H5Br1	7-(2-bromofluorenyl):						★S.PARA+	-0.40 27	★ES-CH	0.00 2		
C7H5Cl1	2-(7-Cl-fluorenyl):						★S.PARA+	-0.64 27	★ES-CH	0.00 2		
C7H5Cl2O1	OCH2Ph,-3,4-(Cl)2						L-STM	9.66 1	B1-STM	1.61 1	B5-STM	5.78 1
C7H5F3O1S1	S+(OCF3)C6H5						★S.INDUC	1.31 282	★S.ZE.RS	0.31 282	★ES-CH	6.20 2
C7H5Fe1O2	Fe(CO)2(C5H5 -PI)						★S.INDUC	-0.25 33	★S.ZE.RS	-0.29 33	★ES-CH	10.30 2
C7H5Fe1O3	1-butadienyliron(CO)3						★S.INDUC	0.10 742	★S.ZE.RS	-0.07 742	★ES-CH	6.30 2
C7H5N1O1	NC(=O)C6H5(-)						S.ZPTFT	-0.87 100	S.PARA+	-2.77 100		
C7H5N1O2	7-nitro-1-fluorenyl:						★S.PARA-	0.31 138				
	NHC6H4CO2- -P			★ -0.52 147			★ES-CH	4.00 2				
C7H5N2	1-(1,2-benzodiazolyl)						S.INDUC	0.36 91	S.ZE.RS	-0.15 91		
	1-benzamidazolyl			★ 0.38 226			★ES-CH	4.00 2				
				0.50 275								
	2-(1,2-benzimidazolyl)						S.INDUC	0.42 91	S.ZE.RS	-0.13 91		
	2-benzimidazolyl						★S.PARA-	0.48 419	★ES-CH	3.00 2		
	2-imidazo(1,2A)pyridinyl						★S.PARA+	0.22 191				
	NCNC6H5						★S.ZE.RS	-0.46 19	★ES-CH	3.00 2		
C7H5N2O1	N=NCOC6H5			★ 0.84 72			S.ZE.RS	0.28 72				
C7H5N2O2	N=NC(O)OC6H5			★ 0.94 72			S.ZE.RS	0.31 72				
	N=NC6H4-M-CO2H						★S.PARA-	0.73 83	★ES-CH	3.00 2		
	N=NC6H4-P-CO2H						★S.PARA-	0.79 83	★ES-CH	3.00 2		
C7H5O1	C=O(C6H5)	1.05	★ 0.34 737	0.40 450	3.03		S.INDUC	0.01 16	S.PARA-	0.86 970	L-STM	5.81 1
		1.23 220	0.36 6	0.41 6			S.INDUC	0.05 12	S.PARA-	0.88 252	B1-STM	1.60 1
		1.28 42	0.36 46	0.42 10			S.INDUC	0.14 99	S.PARA-	0.88 567	B1-STM	1.92 11
			0.39 10	★ 0.43 737			S.INDUC	0.19 68	S.PARA-	0.94 442	B5-STM	5.98 11
				0.46 46			★S.INDUC	0.20 450	S.PARA-	0.95 949	F	0.31 5
				0.46 409			S.INDUC	0.22 67	★S.PARA+	0.51 382	R	0.12 5
				0.53 82			S.INDUC	0.31 102	S.ORTHO	0.65 78	R+	0.20 4
							S.STAR	2.20 3	★S.ORTH-	0.93 97	R-	0.52 4
							S.STAR	2.20 129	S.ZE.RS	0.11 6	S-L	0.75 7
							S.ZPTFT	0.45 487	S.ZE.RS	0.13 67	S.ZTWST	0.00 22
							S.ZPTFT	0.47 306	S.ZE.RS	0.13 102	S.M.RAD	0.06 1002
							S.ZPTFT	0.50 95	S.ZE.RS	0.17 68	S.P.RAD	0.55 79
							S.ZPTFT	0.51 248	S.ZE.RS	0.19 19	S-CNTIG	0.81 486
							S.PARA-	0.59 109	S.ZE.RS	0.21 12	S-CNTIG	0.94 139
							S.PARA-	0.81 443	S.RES.+	0.00 10	GRP.DPL	-3.04 8
							S.PARA-	0.82 419	ES-CH	5.00 2	GRP.DPL	-2.90 9
							★S.PARA-	0.83 104	L-STM	4.57 11	GR.ELCT	0.27 13

Empirical	Structural	π	σ$_m$	σ$_p$	MR	E$_s$	additional σ values						
C7H5O1S1	C(=O)SC6H5			★ 0.48 *745*			★ES-CH 4.20 *2*	L-STM 8.62 *11*	B1-STM 1.96 *11*	B5-STM 3.58 *11*			
	OC(=S)C6H5			★ 0.19 *745*			★ES-CH 4.20 *2*						
	SC(=O)Ph						L-STM 8.45 *11*	B1-STM 1.70 *11*	B5-STM 5.21 *1*				
C7H5O2	2-hydroxybenzoyl						★S.STAR 1.84 *547*	★ES-CH 5.00 *2*					
	3,4-Me-dioxyphenyl				3.43		★ES-CH 3.00 *2*	★S-L 0.12 *14*	★S.PHOSP -0.55 *37*				
	C6H4(CO2H) -M			★ 0.10 *82*	3.13		★ES-CH 3.00 *2*	★S.PHOSP -0.18 *85*					
	C6H4(CO2H) -P			★ 0.13 *82*	3.13		★S.PARA+ 0.07 *305*	★ES-CH 3.00 *2*	★S.PHOSP -0.14 *85*				
	COOC6H5	1.46	★ 0.37 *121*	★ 0.44 *121*	3.02		S.INDUC 0.23 *12* / ★S.INDUC 0.42 *34* / S.ZE.RS 0.25 *12* / ★S.BA.RS 0.14 *34*	ES-CH 4.00 *2* / ★ES-V 0.50 *341* / L-STM 8.13 *11* / B1-STM 1.94 *11*	B5-STM 3.50 *11* / F 0.34 *5* / R 0.10 *5* / GRP.DPL -1.69 *9*				
	OC=O(C6H5)	1.46	★ 0.21 *56* / 0.33 *6* / 0.33 *827*	0.13 *6* / ★ 0.13 *56*	3.23		S.INDUC 0.32 *102* / S.INDUC 0.36 *846* / S.INDUC 0.43 *6* / S.INDUC 0.55 *12* / S.INDUC 0.60 *16* / ★S.STAR 2.40 *183* / S.STAR 2.58 *10* / S.META+ 0.23 *987*	★S.PARA+ -0.07 *569* / S.ZE.RS -0.30 *6* / S.ZE.RS -0.20 *12* / S.ZE.RS 0.09 *102* / ES-CH 4.00 *2* / L-STM 8.15 *11* / B1-STM 1.35 *11* / B1-STM 1.64 *1*	B5-STM 4.40 *11* / F 0.26 *5* / R -0.13 *5* / R+ -0.33 *4* / S.P.RAD 0.00 *79* / GRP.DPL -1.94 *9* / GRP.DPL -1.90 *8*				
C7H5O2S1	OC(=O)SC6H5						S.INDUC 0.51 *16*						
C7H5O3	O-Ph-2-CO2H						S.STAR 3.20 *3*						
	OC(O)-C6H4-OH -O						GRP.DPL -1.92 *8*						
	OC(O)Ph-4-OH						S.STAR 2.40 *3*						
C7H5O4S1	SO2C6H4CO2H-P			★ 0.62 *375*	4.01		★ES-CH 6.20 *2*						
C7H6	1-fluorenyl:						★S.META+ -0.15 *27*	★ES-CH 1.00 *2*					
	2-fluorenyl:				2.90		★S.ZPTFT 0.00 *63* / S.PARA- -0.01 *138*	★S.PARA+ -0.49 *456* / S.PARA+ -0.48 *27*	ES-CH 0.00 *2*				
	3-fluorenyl:						★S.PARA+ -0.24 *27*	★ES-CH 0.00 *2*					
	4-fluorenyl:						★S.PARA+ -0.20 *27*	★ES-CH 2.00 *2*					
C7H6Cl1	CH(Cl)C6H5						★ES-CH 5.20 *2* / ★ES-V 1.20 *7*	L-STM 4.62 *11* / B1-STM 1.88 *11*	B5-STM 6.02 *11*				
	CH2C6H4Cl-P				3.50		★S.STAR 0.28 *314* / ★ES-CH 4.00 *2*	L-STM 5.29 *11* / B1-STM 1.52 *11*	B5-STM 7.44 *11*				
C7H6Cl1N2	N=NC6H3,O-Cl,P-CH3						★S.PARA- 0.71 *83*	★ES-CH 3.00 *2*					
C7H6Cl1N2O1	N=NC6H3,O-Cl,P-OCH3						★S.PARA- 0.68 *83*	★ES-CH 3.00 *2*					
C7H6Cl1O1	CH2OC6H4Cl-P				3.68		★S.STAR 0.89 *314*	★ES-CH 3.00 *2*					
	OCH2Ph,4-(Cl)						L-STM 9.66 *1* / B1-STM 1.35 *11*	B1-STM 1.61 *1* / B5-STM 3.50 *11*	B5-STM 4.44 *1*				
C7H6Cl1S1	SCH2Ph,4-Cl						L-STM 9.96 *11*	B1-STM 1.70 *11*	B5-STM 4.12 *11*				
C7H6F1Hg1	CH2HgC6H4F -P						★S.INDUC -0.03 *353*	★S.ZE.RS -0.24 *353*					
C7H6F1N2O1	NHC(=O)NHC6H4F -P						★S.INDUC 0.22 *24*	★S.ZE.RS -0.29 *24*	★ES-CH 4.00 *2*				
C7H6N1	CH=CH-(3-pyridyl)						GRP.DPL -2.90 *8*						
	CH=CH-(4-pyridyl)						GRP.DPL -2.70 *8*						
	CH=NC6H5	-0.29	★ 0.35 *701*	★ 0.42 *671*	3.30		S.PARA- 0.41 *521* / ★S.PARA- 0.54 *567* / ES-CH 3.00 *2* / L-STM 8.50 *1*	B1-STM 1.70 *1* / B5-STM 4.07 *1* / F 0.33 *5* / R 0.09 *5*	R- 0.21 *4* / ★GRP.DPL -1.61 *8*				
	N=CHC6H5	-0.29	-0.14 *10* / ★ -0.08 *56*	★ -0.55 *56* / 0.00 *671*	3.30		S.PARA- 0.04 *252* / ★S.PARA- 0.22 *567* / S.PARA- 0.75 *521* / ES-CH 3.00 *2*	L-STM 8.40 *1* / B1-STM 1.70 *1* / B5-STM 4.65 *1* / F 0.14 *5*	R -0.69 *5* / R- 0.08 *4* / GRP.DPL 1.51 *8*				
C7H6N1O1	C=O(NHC6H5)	0.49	★ 0.23 *3* / 0.29 *809*	0.38 *809* / ★ 0.41 *82*	3.54		★S.INDUC 0.25 *32* / S.STAR 1.56 *10* / ★S.ORTH- 0.79 *97* / ES-CH 4.00 *2*	L-STM 8.24 *1* / B1-STM 1.63 *1* / B5-STM 4.85 *1* / F 0.17 *5*	R 0.24 *5* / ★S-L 0.26 *14* / GRP.DPL -3.62 *9*				
	CH=N-C6H4-OH -O						GRP.DPL -2.73 *8*						
	CH=N-C6H4-P-OH						GRP.DPL -1.94 *8*						
	CO(C6H4)NH2 -P			★ 0.34 *392*									
	N(Ph)CHO						GRP.DPL -3.44 *8*						
	NHCOC6H5	0.49 / 1.08 *42*	★ 0.02 *56* / 0.11 *6* / 0.12 *251* / 0.17 *399* / 0.21 *10* / 0.22 *23*	-0.19 *6* / ★ -0.19 *56* / -0.07 *251* / 0.08 *23*	3.46		★S.INDUC 0.28 *6* / S.INDUC 0.31 *68* / S.STAR 1.64 *10* / ★S.STAR 1.68 *3* / S.ZPTFT -0.14 *306* / ★S.PARA+ -0.60 *43* / S.PARA+ -0.58 *322* / S.PARA+ -0.58 *513*	S.ZE.RS -0.47 *6* / S.ZE.RS -0.22 *68* / S.RES.+ -0.34 *6* / ES-CH 4.00 *2* / L-STM 8.19 *11* / L-STM 8.40 *1* / L-STM 8.40 *1* / B1-STM 1.35 *11* / B1-STM 1.70 *1*	B5-STM 3.71 *11* / B5-STM 3.97 *1* / F 0.13 *5* / R -0.32 *5* / R+ -0.73 *4* / S-L 0.28 *14* / GR.ELCT 0.49 *13*				
	OCH=NHPh -E						L-STM 7.26 *1*	B1-STM 1.35 *1*	B5-STM 7.77 *1*				
C7H6N1O2	CH2C6H5NO2-P				3.63		★S.STAR 0.45 *3* / ★S.PARA- 0.09 *109*	★S.PARA+ -0.24 *76* / ★ES-CH 4.00 *2*					
	NHC6H4CO2H -P			★ -0.42 *147*			★ES-CH 4.00 *2*						
	NHCOOPh						S.STAR 1.46 *1007*						
	OC(=O)NHC6H5						S.INDUC 0.51 *16*						
C7H6N1O2S1	SCH2Ph,4-NO2						L-STM 9.88 *11*	B1-STM 1.70 *11*	B5-STM 4.11 *11*				
C7H6N1S1	SCH=NPh -E						L-STM 7.18 *1*	B1-STM 1.70 *1*	B5-STM 8.43 *1*				

Empirical	Structural	π	σ_m	σ_p	MR	E_s	additional σ values		
C7H6N2	6(N-Ph-benzopyrazyl):			★ 0.07 937			★ES-CH 0.00 2		
C7H7	C6H4(CH3) -M	2.69 219		★ 0.01 213	3.00		S.INDUC 0.09 84 ★S.STAR 0.48 120	★S.STAR 0.48 342 S.ZE.RS -0.07 84	★ES-CH 3.00 2 ★S.PHOSP -0.55 37
	C6H4(CH3) -O	2.69 219			3.00	★ -3.94 35	S.INDUC 0.12 16 ★S.STAR -0.03 213	★S.STAR 0.62 3 ★ES-CH 4.00 2	★GRP.DPL -0.54 9 O-STER -0.77 90
	C6H4(CH3) -P	2.69 219	★ 0.06 6	-0.07 82 -0.05 265 ★ -0.03 6	3.00		S.INDUC 0.10 6 S.INDUC 0.13 84 S.STAR 0.45 61 ★S.STAR 0.46 120 S.STAR 0.46 342 S.STAR 0.59 3 ★S.PARA+ -0.32 145	S.ZE.RS -0.14 47 S.ZE.RS -0.13 6 S.ZE.RS -0.08 84 S.RES.+ -0.20 6 ES-CH 3.00 2 L-STM 7.09 1 B1-STM 1.84 1	B5-STM 3.11 1 F 0.12 4 R -0.15 4 ★S.PHOSP -0.60 85 GRP.DPL -0.10 9
	CH2C6H5	1.79 130 2.01 2.69 219	★ -0.08 17 -0.05 46 -0.01 6	-0.11 82 -0.10 6 ★ -0.09 17 -0.06 46	3.00	-1.93 496 -1.69 115 -1.62 50 -1.61 230 ★ -1.51 35	★S.INDUC -0.08 17 S.INDUC 0.00 261 S.INDUC 0.03 6 S.INDUC 0.08 16 S.INDUC 0.11 151 S.STAR 0.20 759 S.STAR 0.22 52 S.STAR 0.22 74 S.STAR 0.22 250 S.STAR 0.23 115 S.STAR 0.25 10 S.STAR 0.26 516 S.META+ -0.07 145 S.PARA- -0.12 584 ★S.PARA- -0.09 109	★S.PARA+ -0.28 76 S.PARA+ -0.27 662 S.PARA+ -0.23 145 S.PARA+ -0.20 878 S.PARA+ -0.16 203 S.ORTHO 0.02 158 S.ORTH+ -0.19 154 S.ORTH- -0.08 97 S.ZE.RS -0.13 6 S.ZE.RS -0.12 19 S.ZE.RS 0.10 114 S.BA.RS -0.14 34 S.RES.+ -0.20 470 ES-CH 4.00 2 ES-V 0.70 7	L-STM 4.62 11 B1-STM 1.52 11 B5-STM 6.02 11 F -0.04 5 R -0.05 5 R+ -0.45 4 R- -0.26 4 S-L 0.03 7 S.PHOSP -0.69 59 S.IND.P 0.51 39 GRP.DPL -0.39 9 GRP.DPL 0.36 8 GR.ELCT 0.16 13 O-STER -1.16 90
C7H7Br1N1O1	4(2-Br-4-CN-cyclohexonyl)						GRP.DPL -4.21 8		
C7H7Cl1N1	N(CH3)C6H4Cl-M				4.03		★ES-CH 5.00 2	★S.PHOSP -0.83 274	
	N(CH3)C6H4Cl-P				4.03		★ES-CH 5.00 2	★S.PHOSP -0.87 274	
	NHCH2Ph,4-Cl						L-STM 9.70 11	B1-STM 1.35 11	B5-STM 3.53 11
C7H7Cl1N1S1	N=S(Cl)(P-tolyl)						S.INDUC -0.05 123		
C7H7F1N1	N(CH3)C6H4F-M				3.51		★ES-CH 5.00 2	★S.PHOSP -0.86 274	
C7H7F1O1P1	P(O)(CH3)(C6H4-F) -M		★ 0.40 105		3.94		★ES-CH 6.20 2		
	P(O)(CH3)(C6H4-F) -P			★ 0.56 111	3.94		★ES-CH 6.20 2		
	P(OCH3)(C6H4-F) -M		★ 0.33 105		4.17		★ES-CH 6.20 2		
	P(OCH3)C6H4F-P			★ 0.37 111	4.17		★ES-CH 6.20 2		
C7H7F1P1	PCH3(C6H4-F) -M		★ 0.20 105		4.06		★ES-CH 5.20 2		
	PCH3(C6H4F) -P			★ 0.28 111	4.06		★ES-CH 5.20 2		
C7H7F1P1S1	P(S)(CH3)(C6H4-F) -M		★ 0.47 105				★ES-CH 6.40 2		
	P(S)(CH3)(C6H4-F) -P			★ 0.54 111			★ES-CH 6.40 2		
C7H7Hg1	CH2HgC6H5						★S.PARA+ -1.20 276	S.PARA+ -0.97 203	
C7H7N2	CH=NNH-C6H5						GRP.DPL -2.03 8		
	N(Ph)CH=NH						GRP.DPL -2.20 8		
	N=NC6H4-M-CH3						★S.PARA- 0.64 83	★ES-CH 3.00 2	
	N=NPh,4-CH3						★S.PARA+ 0.13 368 L-STM 9.24 1	B1-STM 1.70 1 B5-STM 4.31 1	
	NHCH=NPh -E						L-STM 7.26 1	B1-STM 1.35 1	B5-STM 7.83 1
C7H7N2O1	CH2N(NO)C6H5	1.25					★ES-CH 4.00 2		
	N(NO)CH2C6H5	1.25					★ES-CH 5.00 2		
	N=N-C6H4-OCH3 -P						★S.PARA+ 0.03 368	GRP.DPL -1.54 8	
	N=NC6H3(5-CH3)(2-OH)		★ 0.27 56	★ 0.31 56	3.75		★ES-CH 3.00 2	F 0.26 5	R 0.05 5
	N=NPh,2-CH3-4-OH						L-STM 9.11 1	B1-STM 1.70 1	B5-STM 5.53 1
	N=NPh,3-CH3-4-OH						L-STM 9.11 1	B1-STM 1.70 1	B5-STM 4.31 1
	NHC(=O)NHC6H5						★S.INDUC 0.15 24 S.INDUC 0.20 117 ★S.ZE.RS -0.31 24	S.ZE.RS -0.31 117 ★ES-CH 4.00 2 L-STM 7.21 11	B1-STM 1.51 11 B5-STM 6.67 11
C7H7N2O2	2-CO2Et-pyrimidin-5yl						S.INDUC 0.36 113	S.ZE.RS -0.01 113	
	5-CO2Et-pyrimidin-2yl						S.INDUC 0.11 113	S.ZE.RS 0.14 113	
C7H7O1	C6H4(OCH3) -M			0.01 501 ★ 0.05 213	3.17		S.INDUC 0.11 84 S.STAR 0.62 325	★S.STAR 0.66 120 S.ZE.RS -0.06 84	ES-CH 3.00 2 ★S.PHOSP -0.45 37
	C6H4(OCH3) -O			★ 0.00 213	3.17		★ES-CH 4.00 2	★GRP.DPL -1.38 8	
	C6H4(OCH3) -P		★ 0.05 6	-0.09 82 -0.09 265 ★ -0.08 6	3.17		★S.INDUC 0.11 6 S.INDUC 0.12 84 ★S.STAR 0.36 120 S.STAR 0.36 342 S.STAR 0.42 325 S.STAR 0.60 3 S.ZE.RS -0.19 6	S.ZE.RS -0.15 47 S.ZE.RS -0.10 84 S.RES.+ -0.27 6 ★ES-CH 3.00 2 L-STM 7.71 1 B1-STM 1.80 1 B5-STM 3.11 1	F 0.13 4 R -0.21 4 ★S.PHOSP -0.57 85 GRP.DPL -1.23 9 O-STER -1.82 90
	CH2-Ph-2-OH						S.ORTHO 0.95 53		
	CH2-Ph-4-OH						S.STAR 0.93 3 S.ORTHO 0.73 53	L-STM 4.73 11 B1-STM 1.52 11	B5-STM 6.72 11
	CH2OC6H5	1.66 2.11 219	0.04 3 ★ 0.06 121	0.05 3 ★ 0.07 210	3.22	-1.86 115 -1.57 50 ★ -1.44 35	S.INDUC 0.12 6 ★S.STAR 0.85 52 S.STAR 0.87 3 S.STAR 0.92 10	ES-V 0.74 7 L-STM 8.19 1 B1-STM 1.52 11 B5-STM 3.53 11	S-L 0.12 14 ★S.PHOSP -0.20 37 S.IND.P 0.99 39 ★GRP.DPL -1.38 9

Empirical	Structural	π	σm	σp	MR	Es	additional σ values
C7H7O1	CH2OC6H5						S.STAR 0.95 115 F 0.08 5 ES-CH 3.00 2 R -0.01 5
	CHOHC6H5	0.54 0.87 220	★ 0.00 3	★ -0.03 82	3.15		S.INDUC 0.01 49 S.ZE.RS -0.06 49 F 0.05 5 ★S.INDUC 0.10 6 ES-CH 5.00 2 R -0.08 5 S.STAR 0.50 3 ★ES-V 0.69 7 S-L 0.10 7 ★S.STAR 0.76 52 L-STM 4.62 11 S.ZTWST -0.11 22 S.STAR 0.76 376 B1-STM 1.73 11 S.ZE.RS -0.08 22 B5-STM 6.02 11
	O-Ph-2-CH3						S.STAR 2.29 3 GRP.DPL -1.09 9
	O-Ph-3-CH3						S.STAR 2.33 3 GRP.DPL -1.25 9
	OC6H4CH3-P				3.17		★S.STAR 2.30 3 L-STM 5.29 11 ★S.PHOSP -0.12 898 ★S.PARA+ -0.41 202 B1-STM 1.35 11 GRP.DPL -1.23 9 ★ES-CH 4.00 2 B5-STM 6.98 11
	OCH2C6H5	1.66		-0.42 23 -0.41 3 ★ -0.23 922	3.22		S.STAR -0.62 796 L-STM 8.20 1 B5-STM 3.50 11 ★S.PARA+ -0.66 478 B1-STM 1.35 11 B5-STM 4.44 1 ★ES-CH 3.00 2 B1-STM 1.61 1
C7H7O1S1	S(O)Ph-3-CH3						S.STAR 2.99 3
	S(O)Ph-4-CH3						S.STAR 3.02 3
	S-Ph-2-OCH3						S.STAR 1.59 3
	S-Ph-3-OCH3						S.STAR 1.89 3 GRP.DPL -1.74 9
	SC6H4OCH3 -P						★S.INDUC 0.27 6 ★S.STAR 1.66 3 ★GRP.DPL -1.98 9
	SCH2OPh						L-STM 6.38 11 B1-STM 1.70 11 B5-STM 7.93 11
C7H7O1Se1	Se-Ph-2-OCH3						S.STAR 1.17 3
	Se-Ph-3-OCH3						S.STAR 1.38 3
	Se-Ph-4-OCH3						S.STAR 1.18 3
C7H7O2	O-Ph-2-OCH3						S.STAR 2.29 3 GRP.DPL -1.26 9
	O-Ph-3-OCH3						S.STAR 2.42 3 GRP.DPL -1.58 9
	OC6H4OCH3 -P						★S.INDUC 0.39 6 ★S.PARA+ -0.43 202 ★S.STAR 2.32 3 GRP.DPL -1.72 9
	OCH2O-C6H5						L-STM 6.47 11 B5-STM 7.21 11 B1-STM 1.35 11 GRP.DPL -1.16 8
C7H7O2S1	CH2SO2C6H5		★ 0.15 121 0.24 6	0.14 6 ★ 0.16 210	3.79		S.STAR 1.15 87 ES-CH 5.20 2 F 0.17 5 ★S.STAR 1.37 325 L-STM 8.33 11 R -0.01 5 S.ZE.RS -0.15 6 B1-STM 1.52 11 S.RES.+ 0.01 6 B5-STM 3.78 11
	S(O)C6H4OCH3 -P						★S.INDUC 0.50 6 ★S.STAR 3.00 3 ★GRP.DPL -4.24 9
	SO2-Ph-3-CH3						S.STAR 3.27 3
	SO2C6H4CH3-P			0.56 147 0.63 335 ★ 0.67 166	3.78		S.INDUC 0.55 166 S.PARA- 0.89 167 GRP.DPL -5.08 9 S.STAR 3.32 3 ★S.PARA- 0.93 272 ★S.META- 0.59 167 ES-CH 6.20 2
	SO2CH2C6H5						★S.PARA- 1.00 252 L-STM 8.33 11 B5-STM 3.58 11 ★ES-CH 5.20 2 B1-STM 2.03 11
C7H7O2S2	SC6H4(SO2CH3) -P						★S.PARA+ 0.16 202
C7H7O3S1	OC6H4(SO2CH3) -P						★S.STAR 2.85 3 L-STM 6.24 11 B5-STM 7.45 11 ★S.PARA+ -0.07 202 B1-STM 1.35 11
	OSO2C6H4(CH3) -P	1.49 589		★ 0.20 827 0.28 414	4.13		S.INDUC 0.54 414 ★S.PARA+ 0.16 478 ★S-L 0.58 7 ★S.PARA+ -0.66 478 ★ES-CH 5.20 2
	SO2-Ph-3-OCH3						S.STAR 3.24 3
	SO2C6H4OCH3-P			★ 0.65 166	4.00		★S.INDUC 0.54 166 ★S.META- 0.58 167 ES-CH 6.20 2 S.INDUC 0.71 6 S.PARA- 0.87 167 S.STAR 3.23 3 ★S.PARA- 0.89 272
	SO2CH2OPh						L-STM 6.56 11 B1-STM 2.03 11 B5-STM 7.35 11
	SO3-C6H4-CH3 -P						GRP.DPL -5.29 8
C7H7S1	C6H4SCH3 -P			★ 0.13 304			
	CH2SC6H5				3.79		★S.STAR 0.66 325 ★ES-V 0.82 7 B1-STM 1.52 11 ★ES-CH 3.20 2 L-STM 8.67 11 B5-STM 3.60 11
	S-Ph-2-CH3						S.STAR 1.90 3
	S-Ph-3-CH3						S.STAR 1.89 3 GRP.DPL -1.38 9
	SC6H4CH3 -P						★S.STAR 1.80 3 L-STM 5.34 11 B5-STM 8.19 11 ★S.PARA+ -0.27 202 B1-STM 1.70 11 ★GRP.DPL -1.49 9
	SCH2C6H5		★ 0.23 3		3.79		★S.INDUC 0.26 6 ES-CH 3.20 2 B5-STM 4.11 11 S.STAR 1.56 3 L-STM 8.50 11 ★S-L 0.26 14 ★S.RES.+ -0.36 6 B1-STM 1.70 11 GRP.DPL -1.46 9
C7H7S1Se1	Se-Ph-2-SCH3						S.STAR 1.27 3
	Se-Ph-4-SCH3						S.STAR 1.23 3
C7H7S2	S-Ph-2-SCH3						S.STAR 1.62 3
	S-Ph-3-SCH3						S.STAR 1.93 3
	S-Ph-4-SCH3						S.STAR 1.69 3 GRP.DPL -1.81 9
	SCH2S-C6H5						GRP.DPL -1.34 8
C7H7Se1	Se-Ph-2-CH3						S.STAR 1.33 3
	Se-Ph-3-CH3						S.STAR 1.30 3
	Se-Ph-4-CH3						S.STAR 1.23 3 GRP.DPL -1.46 9
C7H8Cl1N2S1	N=S(Ph-P-Cl)(NHCH3)						S.INDUC -0.35 123
C7H8N1	C6H4NHCH3-M				3.47		★ES-CH 3.00 2 ★S.PHOSP -0.65 85
	C6H4NHCH3-P				3.47		★ES-CH 3.00 2 ★S.PHOSP -0.81 85

Empirical	Structural	π	σ_m	σ_p	MR	E_s	additional σ values						
C7H8N1	CH2C6H4NH2 -P			★ -0.55 392			★ S.PARA-	-0.17 427	★ ES-CH	4.00 2			
							★ S.PARA-	0.00 187	★ GRP.DPL	1.84 8			
	CH2NHC6H5	1.00					★ ES-CH	3.00 2					
	N(CH3)C6H5				3.52		S.ZE.RS	-0.50 22	L-STM	4.53 11	★ S.PHOSP	-1.11 274	
							★ S.RES.+	-1.26 470	B1-STM	1.48 11	S.ZTWST	-0.13 22	
							★ ES-CH	5.00 2	B5-STM	5.95 11	★ GRP.DPL	1.24 8	
	NHC6H4CH3-P				3.47		★ ES-CH	4.00 2	★ S.PHOSP	-1.70 37			
	NHCH2C6H5	1.00					★ S.STAR	1.35 170	★ ES-V	0.62 7	B1-STM	1.35 11	
							★ ES-CH	3.00 2	L-STM	8.24 11	B5-STM	3.53 11	
C7H8N1O1	4(4-CN-cyclohexanonyl)						GRP.DPL	-3.63 8					
	NH-C6H4-OCH3 -M						GRP.DPL	-1.79 8					
	NHCH2OPh						L-STM	6.47 11	B1-STM	1.35 11	B5-STM	7.28 11	
C7H8N1O2	C(CH3)=C(CN)CO2C2H5			★ 0.37 783			★ ES-CH	5.00 2					
C7H8N1O2S1	N(Ph)SO2CH3						GRP.DPL	-4.41 8					
	NHSO2C6H4CH3 -P			★ 0.17 304									
C7H8N1O3S1	NHSO2-C6H4-OCH3 -P						GRP.DPL	-5.65 8					
	OCH2Ph,4-SO2NH2						L-STM	9.70 1	B1-STM	1.63 1	B5-STM	4.44 1	
	SO2NH-C6H4-OCH3 -P						GRP.DPL	-5.44 8					
C7H8N1S1	N=S(Me)(Ph)						S.INDUC	-0.45 123					
C7H8N3	N=N-C6H4-NHCH3 -P						GRP.DPL	2.91 8					
	NHC(=NH)NHC6H5						★ S.INDUC	0.08 117	★ S.ZE.RS	-0.36 117	★ ES-CH	4.00 2	
C7H8O1P1	P(O)(CH3)C6H5						★ S.INDUC	0.30 6	★ ES-CH	6.20 2			
							★ S.INDUC	0.30 45					
	P(OCH3)Ph			★ 0.32 3									
C7H8O2P1	P(O)(OCH3)C6H5						★ S.INDUC	0.32 45	★ S-L	0.32 14			
							★ ES-CH	7.20 2					
C7H8P1	P(Ph)CH3						GRP.DPL	-1.39 8					
C7H8S1	S+(Me)(Ph)						S.CH-P+	1.11 75					
C7H9	6-bicylo[3.2.0]-2-heptenyl						L-STM	6.34 1	B1-STM	1.82 1	B5-STM	3.79 1	
C7H9Mn1O7P1	Mn(CO)4P(OCH3)3						★ S.INDUC	-0.27 33	★ S.ZE.RS	-0.24 33	★ ES-CH	13.50 2	
C7H9N1	C6H4NH2CH3+-M						★ ES-CH	3.00 2	★ S.PHOSP	0.30 85			
	C6H4NH2CH3+-P						★ ES-CH	3.00 2	★ S.PHOSP	0.30 85			
	Ph-3-CH2NH3+						S.STAR	1.39 3					
	Ph-4-CH2NH3+						S.STAR	1.32 3					
C7H9N2O1	CO-5(1,3,4-Me3-pyrazolyl)						GRP.DPL	-2.80 8					
C7H9N3	NHC(=NH2+)NHC6H5						★ S.INDUC	0.30 117	★ S.ZE.RS	-0.15 117	★ ES-CH	4.00 2	
C7H9O1	CH=(cyclohexadiene-2-one						GRP.DPL	-3.23 8					
C7H9Si1	SiH(Me)(Ph)			★ 0.14 238									
C7H11	1-(2.2.1)bicycloheptyl-endo						★ S.PARA+	-0.36 440					
	1-(2.2.1)bicycloheptyl-exo						★ S.PARA+	-0.37 440					
	1-(2.2.1)bicycloheptyl						★ S.PARA+	-0.40 525	★ ES-CH	4.00 2			
	1-bicycloheptyl(4.1.0)			★ 0.01 271	2.94		★ S.PARA-	-0.07 400	★ ES-CH	4.00 2	B1-STM	2.08 1	
							★ S.PARA+	-0.60 525	L-STM	6.16 1	B5-STM	3.87 1	
	2-(4.1.0)bicycloheptyl						★ S.PARA+	-0.69 525	L-STM	6.45 1	B5-STM	3.87 1	
							★ ES-CH	3.00 2	B1-STM	2.20 1			
	4-cycloheptenyl						L-STM	6.03 1	B1-STM	1.87 1	B5-STM	4.26 1	
	C(CH3)2CCCH2CH3						★ S.STAR	0.25 253	★ ES-CH	5.00 2			
	CH2CC(t-Bu)						★ S.STAR	0.51 433					
	CH2CCC4H9						★ S.STAR	0.53 433					
	cyclohex-1-enylmethyl						★ S.STAR	0.08 357					
	cyclohex-2-enylmethyl						★ S.STAR	0.03 357					
	cyclohex-3-enylmethyl						★ S.STAR	-0.06 357					
C7H11N1	N+(CH3)2CH2CCCH=CH2					★ -3.04 303							
C7H11O4	CH(CO2Et)2						S.IND.P	3.17 790	S.RES.P	-3.46 790			
C7H12N1	CH=CH(N-piperidyl)						★ S.ZE.RS	-0.31 19	★ ES-CH	3.00 2			
C7H12N1O2	N(COCH3)COC(CH3)3						GRP.DPL	-3.59 8					
	OCH2C=O(N-piperidyl)	-0.32			3.70		★ ES-CH	3.00 2					
C7H12N5	2(4,6-di-NMe2-triazinyl)						★ S.PARA-	0.47 346	★ ES-CH	3.00 2			
C7H13	cycloheptyl				3.13	★ -2.34 50	★ S.STAR	-0.13 124	★ ES-V	1.00 7			
						-2.04 35	ES-CH	3.00 2					
	cyclohexyl-(1-CH3)				3.13	-4.19 403	★ S.STAR	-0.44 323					
						★ -3.27 323	ES-CH	4.00 2					
	cyclohexylmethyl		★ -0.17 3		3.13	★ -2.22 50	S.INDUC	-0.03 6	S.STAR	-0.06 576	B1-STM	1.52 1	
						-2.01 35	S.STAR	-0.31 3	ES-CH	4.00 2	B5-STM	5.42 1	
							S.STAR	-0.10 357	★ ES-V	0.97 7	★ S-L	-0.03 14	
							★ S.STAR	-0.06 52	L-STM	6.09 1			
C7H13Hg1	CH2Hg-cyhexyl						★ S.PARA+	-1.04 203					
C7H13N1	CH2CCCH2N(Me)3+						S.STAR	1.07 262					
	N+(CH3)2CH2CCC2H5					★ -3.04 201							
C7H13N1O2	CH2CH2N(COOEt)CH2CH2*				3.93		★ S.STAR	0.70 268					
C7H13O1	C(=O)CH2C(CH3)2Et						L-STM	6.92 1	B1-STM	2.11 1	B5-STM	4.90 1	

Empirical	Structural	π	σm	σp	MR	Es	additional σ values									
C7H13O1	C(=O)CH2CH2C(CH3)3						L-STM	6.92	1	B1-STM	1.76	1	B5-STM	5.97	1	
	C(=O)hexyl						L-STM	8.98	1	B1-STM	1.63	1	B5-STM	6.35	1	
	CH(Pr)CH2COCH3				3.36		★S.STAR	0.11	74	★ES-CH	5.00	2				
	COC(C2H5)2CH3						★S.ZE.RS	0.14	47	★ES-CH	6.00	2				
	cyclohexylmethoxy		★ 0.18	3		3.31		★S.INDUC	0.21	32	★ES-CH	3.00	2	★S-L	0.23	7
							S.STAR	1.31	3	★ES-V	0.65	7	GRP.DPL	-1.41	9
C7H13O2	(CH2)4CO2Et				3.51	★ -1.64 196	ES-CH	3.00	2	★S-L	0.02	14				
	2(tetrame-1,3-dioxolanyl)						GRP.DPL	-1.70	8							
	CH2C(O)o-pent						GRP.DPL	-2.13	8							
	OC(O)CH(Et)propyl						GRP.DPL	-2.06	8							
C7H14N1	CH2CH2-1-piperidinyl						L-STM	8.13	1	B1-STM	1.52	1	B5-STM	4.91	1
C7H14N1O1	Et-CONHC(CH3)3						S.STAR	0.25	3							
C7H14N1O2	N(I-Pr)CH2CO2Et						S.IND.P	0.90	434	S.RES.P	-2.08	434				
C7H14N1O3	CHOHCH(CH2OH)NHCOC3H7	-0.98 54			4.12		★ES-CH	5.00	2							
	CHOHCH(CH2OH)NHCOCH(CH3)2	-1.06 54			4.12		★ES-CH	5.00	2							
C7H14O2P1S2	(CH2)2SSP(OCH2C(CH3)2CH2O)						★S.INDUC	0.14	137	★S.STAR	0.30	137	★ES-CH	3.20	2
C7H15	C(CH3)2C(CH3)3				3.35	-6.52 62	ES-CH	7.00	2	B1-STM	2.60	1				
						-6.06 607	★ES-V	2.43	7	B5-STM	4.45	1				
						★ -5.14 50	L-STM	4.92	1							
	C(Et)3		★ -0.07 6	★ -0.20 6	3.35	★ -6.41 62	★S.ZE.RS	-0.19	6	L-STM	4.92	1	F	0.02	4
						-5.04 50	ES-CH	7.00	2	B1-STM	2.94	1	R	-0.22	4
							★ES-V	2.38	7	B5-STM	4.18	1			
	CH(CH3)CH2C(CH3)3				3.35	★ -3.09 50	ES-CH	4.00	2	L-STM	6.17	1	B5-STM	4.54	1
						-2.93 62	★ES-V	1.41	7	B1-STM	1.90	1			
	CH(Et)Bu					★ -3.15 35	★ES-CH	5.00	2	★ES-V	1.55	7				
	CH(I-Pr)2					★ -6.13 62	ES-CH	7.00	2	B1-STM	2.08	1				
							L-STM	4.92	1	B5-STM	4.19	1				
	CH(Pr)2				3.35	-3.35 50	ES-CH	5.00	2	L-STM	6.17	1	B5-STM	4.54	1
						★ -3.15 62	★ES-V	1.54	7	B1-STM	1.90	1			
	heptyl	-0.19 3	-0.16 6	3.35		S.INDUC	-0.06	32	★S.ZE.RS	-0.12	6	B5-STM	6.39	1
		★ -0.07 6					★S.INDUC	-0.04	6	★ES-CH	3.00	2	F	0.00	4
							S.STAR	-0.37	3	ES-V	0.73	7	R	-0.16	4
							S.STAR	-0.17	287	L-STM	9.03	1	S-L	-0.04	14
							S.STAR	-0.15	544	B1-STM	1.52	1			
C7H15N1	CH2CH2NH+(CH2)5						S.PHOSP	-0.37	367							
	CH2N+H2(cy-hexane)						S.PHOSP	0.23	367							
C7H15N1O1P1Pt1	Pt(OCN)P(C2H5)3 -T						★S.INDUC	-0.24	33	★S.ZE.RS	-0.26	33				
C7H15N1P1Pt1	Pt(CN)P(C2H5)3 -C						★S.INDUC	-0.32	33	★S.ZE.RS	-0.22	33				
	Pt(CN)P(C2H5)3 -T						★S.INDUC	-0.24	33	★S.ZE.RS	-0.24	33				
C7H15N1P1Pt1S1	Pt(SCN)P(C2H5)3 -T						★S.INDUC	-0.16	33	★S.ZE.RS	-0.26	33				
C7H15O1	(CH2)3 OBu						★ES-CH	3.00	2	★ES-V	0.71	7				
	C(OH)(I-Pr)2						S.INDUC	-0.51	12	S.ZE.RS	-0.06	12				
	CH(OH)C6H13						★S.ZE.RS	-0.08	22	★ES-CH	4.00	2	★S.ZTWST	-0.12	22
	O(CH2)6CH3						★ES-CH	3.00	2	B1-STM	1.35	1				
							L-STM	10.18	1	B5-STM	7.23	1				
	OC(CH3)2C(CH3)3						L-STM	6.03	1	B1-STM	1.35	1	B5-STM	4.47	1
	OCH(CH3)pentyl						★ES-CH	4.00	2	★ES-V	0.90	7				
	OCH2C(CH3)Et2						★ES-CH	3.00	2	★ES-V	0.82	7				
	OCH2CH(CH3)T-butyl						★ES-CH	3.00	2	★ES-V	0.66	7				
C7H15O1P1Pt1	Pt+(CO)P(C2H5)3 -T						★S.INDUC	0.20	33	★S.ZE.RS	-0.18	33				
C7H15O2	C(CH3)2OOC(CH3)3					★ -3.20 402	★S.STAR	0.52	402	ES-CH	5.00	2	★ES-V	1.49	7
	CH2OCH2CH2OBu						★ES-CH	3.00	2	★ES-V	0.55	7				
C7H15O2S1	SO2heptyl						L-STM	10.27	1	B1-STM	2.03	1	B5-STM	7.39	1
C7H15O3	C(OEt)3						S.INDUC	-0.19	92	S.ZE.RS	0.03	92				
C7H15S1	S(CH2)6CH3						★ES-CH	3.20	2	B1-STM	1.70	1				
							L-STM	10.02	1	B5-STM	8.21	1				
	SC(CH3)2C(CH3)3						L-STM	6.42	1	B1-STM	1.70	1	B5-STM	5.00	1
C7H16N1	CH2N(I-Pr)2	1.62 42					★ES-CH	4.00	2							
	CH2N(Pr)2	0.81 42					★ES-CH	4.00	2							
	NH(CH2)6CH3						★ES-CH	3.00	2	B1-STM	1.35	1				
							L-STM	10.18	1	B5-STM	7.32	1				
	NHC(CH3)2C(CH3)3						L-STM	6.07	1	B1-STM	1.36	1	B5-STM	4.53	1
C7H16N2O1	ON=CH(CH2)3N+(CH3)3						L-STM	9.75	1	B1-STM	1.35	1	B5-STM	6.50	1
C7H16O1P1	(CH2)3P(O)(C2H5)2						★S.STAR	0.17	422	★ES-CH	3.00	2	★S-L	0.04	14
	CH2P(O)(C3H7)2						★S.STAR	0.59	45	★ES-CH	5.20	2				
C7H16O1P1S2	CH2CH2SSPEt(OPr)						★S.INDUC	0.17	137	★S.STAR	0.37	137	★ES-CH	3.20	2
C7H16O3P1	(CH2)3P(O)(OC2H5)2				4.51		S.INDUC	-0.96	367	★S.STAR	0.13	422	★S-L	0.03	14
							★S.STAR	0.12	74	★ES-CH	3.00	2				
C7H16S1	(CH2)4S+(CH3)C2H5						★S.STAR	0.49	278	★ES-CH	3.00	2				
C7H17Ge1	CH2Ge(C2H5)3						★S.INDUC	-0.10	22	S.ZE.RS	-0.67	285	★ES-CH	5.30	2
							★S.PARA+	-1.10	60	★S.ZE.RS	-0.18	22				
C7H17Ge1O3	CH2Ge(OC2H5)3						★S.PARA+	-0.53	60	S.ZE.RS	-0.24	51	★ES-CH	5.30	2

Empirical	Structural	π	σ_m	σ_p	MR	E_s	additional σ values
C7H17N1	BuN(CH3)3+	-3.65 589		★ -0.04 315			★ES-CH 3.00 2
	CH2N+(Et)3			0.21 36 ★ 0.61 29			★ES-CH 5.00 2
	N+(CH3)2C5H11					★ -3.44 201	
C7H17Si1	BuSi(Me)3				4.36		★S.PARA- -0.09 321 ★ES-CH 3.00 2 ★S.ORTH+ -0.28 154 ★S-L -0.07 14
	CH2Si(Et)3			★ -0.24 246	4.36		★S.STAR -0.12 428 ★S.PARA+ -0.57 285 S.STAR 0.52 842 ★ES-CH 5.20 2
C7H18P1Pt1	Pt(CH3)P(C2H5)3 -T						★S.INDUC -0.47 33 ★S.ZE.RS -0.26 33
C7H19Si2	CH(SiMe3)2			★ -0.33 401	5.29		S.INDUC -0.03 385 ★S.PARA+ -0.62 401 S.ZE.RS -0.22 385 ★S.PARA+ -0.76 299 ★S.PARA+ -0.54 175 ★ES-CH 9.40 2
C7H21O2Si3	Si(Me)(OSiMe3)2		-0.07 3 -0.03 25 ★ -0.02 221	★ -0.01 221 0.01 25 0.02 3	6.23		★ES-CH 6.60 2 F 0.01 5 R -0.02 5
C7H21Si3	(SiMe2)2SiMe						S.PARA+ -0.01 175
	Si(CH3)(Si(CH3)3)2						★S.PARA- 0.04 299 S.PARA+ -0.07 175 ★ES-CH 10.60 2 ★S.PARA+ -0.81 299 ★S.ZE.RS 0.06 47
C8F5O2	OCCOC6F5						★S.INDUC 0.30 49 ★S.ZE.RS -0.17 49 ★ES-CH 3.00 2
C8H3S3@	dithieno-2-thiofen-A						S.ORTH+ -1.08 363
	dithieno-2-thiofen-B						S.ORTH+ -1.05 363
	dithieno-2-thiofen-C						S.ORTH+ -1.10 363
	dithieno-2-thiofen-D						S.ORTH+ -1.07 363
	dithieno-3-thiofen-A						S.META+ -0.72 363
	dithieno-3-thiofen-B						S.META+ -0.78 363
	dithieno-3-thiofen-C						S.META+ -0.75 363
	dithieno-3-thiofen-D						S.META+ -0.75 363
C8H4Br1	C=-CC6H4Br -M						★S.PARA- 0.44 356
	C=-CC6H4Br1 -P						★S.PARA- 0.42 356
C8H4Cl1	C=-CC6H4Cl -M						★S.PARA- 0.43 356
	C=-CC6H4Cl -P			★ 0.13 281			★S.PARA- 0.42 356 ★ES-CH 3.00 2
C8H4Cl1N2	4-Cl-2-quinazolinyl						S.INDUC 0.14 15 S.ZE.RS 0.12 15
C8H4D3N2@	2(1-Cd3)-1,3benzdiazolyl:						★S.ORTH+ 0.70 148 ★S.ORTH- 1.50 148 ★ES-CH 4.00 2
C8H4F5	CF2CF2C6H4F-P		★ 0.34 30	★ 0.39 30	3.37		★S.INDUC 0.29 30 ★ES-CH 6.20 2 R 0.07 4 ★S.ZE.RS 0.10 30 F 0.32 4
C8H4Mn1O3	cypentdienyl(CO)3-Mn						S.INDUC -0.17 286 S.ZE.RS -0.02 286
C8H4N1O2	C=-CC6H4NO2-P			★ 0.17 281	3.95		★S.PARA- 0.47 109 ★S.PARA- 0.49 356 ★ES-CH 3.00 2
	N-phthalimidyl			★ 0.19 304			
C8H4N3O6	CH=CHC6H2(NO2)3-2,4,6						★S.PARA+ 0.32 76 ★ES-CH 3.00 2
C8H4Re1O3	cypentdien(CO)3-Re						S.INDUC -0.22 286 S.ZE.RS -0.01 286
C8H5	CCC6H5	2.53 219 2.65	★ 0.14 17 0.16 429 0.26 6	0.12 6 0.12 281 ★ 0.16 17 0.19 429 0.55 672	3.32	★ -3.09 35	★S.INDUC 0.14 17 S.ZE.RS 0.07 19 R+ -0.18 4 S.INDUC 0.33 6 S.ZE.RS 0.15 552 R- 0.15 4 S.STAR 1.35 52 ES-CH 3.00 2 S-L 0.33 14 S.STAR 2.12 652 L-STM 8.88 1 ★S.PHOSP -0.28 37 ★S.PARA- 0.30 109 B1-STM 1.71 1 GRP.DPL 0.00 8 S.PARA- 0.39 356 B5-STM 3.11 1 GR.ELCT 0.53 13 ★S.PARA+ -0.03 429 F 0.15 5 S.ZE.RS -0.21 6 R 0.01 5
C8H5Cl1N1	5-Cl-2-indolyl						★S.INDUC 0.22 15 ★S.ZE.RS -0.06 15
C8H5F1	1-(9-Fl-phenanthryl)						★S.META+ -0.24 27 ★ES-CH 1.00 2
C8H5F6O2S1	S(OCF3)2C6H5						★S.INDUC 0.40 282 ★S.ZE.RS 0.09 282 ★ES-CH 8.20 2
C8H5N2	2-quinazolinyl						S.INDUC 0.06 15 S.ZE.RS 0.10 15 S.INDUC 0.23 96
	2-quinoxalinyl						★S.PARA+ 0.64 191
	N=C(CN)(C6H5)						★S.PARA- 0.35 958 ★ES-CH 4.00 2
C8H5N2O1	2(4-oxoquinoxalinyl)						★S.INDUC 0.40 96
	3(2-oxo-diH-quinoxalinyl)						★S.PARA+ 0.62 191
C8H5N2O2	5-nitro-2-indolyl						★S.INDUC 0.27 15 ★S.ZE.RS -0.04 15
C8H5N2O4	CH=CHC6H3(NO2)2-2,4						★S.PARA+ 0.18 76 ★ES-CH 3.00 2
C8H5O1	OC#CPh						L-STM 6.63 1 B1-STM 1.35 1 B5-STM 8.34 1
C8H5O1@	2-benzofuryl:						S.ORTH+ -0.65 404 S.ORTH- 0.80 148 ★S.ORTH+ -0.49 195 S.ZE.RS 0.70 148
	3-benzofuryl:						★S.ORTH+ -0.48 195
C8H5O2	COCOC6H5	1.25					S.INDUC 0.45 12 ★ES-CH 5.00 2 S.ZE.RS 0.29 12 ★GRP.DPL -3.71 8
C8H5O3	CO2COC6H5						GRP.DPL -3.30 8
C8H5O3W1	W(CO)3C5H5						★S.INDUC -0.17 431 ★S.ZE.RS -0.19 431
C8H5S1	SCCC6H5			★ 0.20 672			L-STM 6.51 1 B1-STM 1.70 1 B5-STM 8.94 1
C8H5S1@	2-benzothienyl:						S.ORTH+ -0.61 145 S.ORTH+ -0.49 412 S.ZE.RS 0.50 148 S.ORTH+ -0.53 404 ★S.ORTH+ -0.46 195 ★S.ORTH+ -0.51 374 S.ORTH- 0.60 148
	3-benzothienyl:						S.META+ -0.62 145 S.META+ -0.56 412 S.META+ -0.46 404 ★S.META+ -0.57 374 ★S.META+ -0.54 195 S.ZE.RS 0.50 148
C8H5Se1	Se-CCC6H5			★ 0.05 672			GRP.DPL 1.32 8

Empirical	Structural	π	σm	σp	MR	Es	additional σ values		
C8H6	1-anthryl:						★ S.META+ -0.45 542		
	1-phenanthryl:						★ S.META+ -0.55 101	★ S.META+ -0.43 101	★ S.META+ -0.34 27
	2-anthryl:	2.32			3.45		S.PARA- -0.39 269	★ S.PARA- -0.35 542	★ S.PARA- -0.30 456
	2-phenanthryl:	2.33			3.45		★ S.ZPTFT 0.13 63	S.PARA- -0.25 27	S.PARA- -0.20 583
							S.PARA- -0.33 101	S.PARA- -0.21 269	★ S.PARA- -0.12 456
	3-phenanthryl:						★ S.ZPTFT 0.11 63	S.PARA- -0.30 269	★ S.PARA- -0.20 456
							S.PARA+ -0.41 101	S.PARA- -0.29 27	
	4-phenanthryl						★ S.META+ -0.33 27	★ S.META+ -0.32 101	
C8H6Br1	CH=CHC6H4Br -M			★ -0.03 309			★ S.PARA- 0.20 390	★ S.PARA- 0.24 283	
	CH=CHC6H4Br -P			★ -0.04 309			★ S.PARA- 0.17 390	★ S.PARA- 0.22 283	★ GRP.DPL 1.85 8
C8H6Cl1	C(Cl)=CHC6H5 -C						GRP.DPL -1.68 8		
	C(Cl)=CHC6H5 -T						GRP.DPL -1.29 8		
	CH=CH-C6H4-Cl -O,C						GRP.DPL -1.56 8		
	CH=CH-C6H4-Cl -O,T						GRP.DPL -1.34 8		
	CH=CHC6H4Cl -M						★ S.PARA- 0.19 390	★ S.PARA- 0.24 283	★ GRP.DPL -1.66 820
	CH=CHC6H4Cl -P			-0.06 309			★ S.PARA- 0.17 390	★ ES-CH 3.00 2	
				★ -0.06 281			★ S.PARA- 0.22 283	★ GRP.DPL -1.73 8	
C8H6Cl3	CH(Ph)CCl3						GRP.DPL -1.82 8		
C8H6F1	CH=CH-C6H4-F -P						GRP.DPL -1.49 8		
C8H6I1	CH=CH-C6H4-I -P						GRP.DPL -1.80 8		
C8H6N1	1-indolyl						S.INDUC 0.33 573	S.STAR 0.28 826	
	2-indolizinyl						★ S.PARA+ 0.09 191		
	2-indolyl						★ S.PARA- 0.39 419	★ ES-CH 3.00 2	
	3-indolyl	2.14 219	★ -0.12 3		3.64		★ S.INDUC -0.01 32	★ ES-CH 4.00 2	
							★ S.STAR -0.06 3	★ S-L 0.00 7	
	C=-CC6H4NH2 -P						★ S.PARA- 0.34 356		
	CH2-Ph-4-CN						S.STAR 0.41 3		
C8H6N1@	3-indolyl:						★ S.PARA+ -2.25 371		
C8H6N1O1	CH2C6H4NCO-P		★ 0.04 405		3.78		★ ES-CH 4.00 2		
	CH=NC(O)Ph		★ 0.39 3	★ 0.51 3			F 0.34 5	R 0.17 5	
C8H6N1O2	CH=CHC6H4NO2 -M			★ -0.01 309					
	CH=CHC6H4NO2-P			0.01 309	4.05		S.PARA- 0.25 651	S.PARA- 0.41 283	★ ES-CH 3.00 2
				★ 0.02 281			★ S.PARA- 0.26 109	★ S.PARA+ -0.04 76	★ GRP.DPL -4.74 8
	CH=NOC(O)C6H5						GRP.DPL -3.32 8		
C8H7	C6H4-CH=CH2 -P						GRP.DPL 0.64 8		
	CH=CHC6H5	2.68	-0.05 769	-0.12 769	3.42		S.INDUC 0.02 34	S.PARA- 0.62 23	F 0.10 5
			★ 0.03 429	-0.11 281			★ S.STAR 0.41 52	★ S.PARA+ -1.00 429	R -0.17 5
			0.14 23	-0.08 82			S.PARA- 0.07 109	S.PARA+ -0.41 76	R+ -1.10 4
				-0.07 309			S.PARA- 0.10 971	S.BA.RS -0.12 34	R- 0.03 4
				★ -0.07 429			★ S.PARA- 0.13 651	S.RES.+ -0.30 6	S.PHOSP -0.58 37
							S.PARA- 0.17 283	ES-CH 3.00 2	GRP.DPL -0.77 9
	cy-octatetraenyl						S.INDUC 0.16 16		
C8H7Cl1N1O2S1	SOC6H3-2-Cl-4-NHAC						★ S.PARA- 0.74 106	★ ES-CH 5.20 2	
	SOC6H3-2-Cl-5-NHAC						★ S.PARA- 0.77 106	★ ES-CH 5.20 2	
C8H7Cl1N1O3S1	SO2C6H3-2-Cl-4-NHAC						★ S.PARA- 0.94 106	★ ES-CH 6.20 2	
	SO2C6H3-2-Cl-5-NHAC						★ S.PARA- 0.99 106	★ ES-CH 6.20 2	
C8H7Cl1N3O1	N=NC6H3,2-Cl,5-NHAC						★ S.PARA- 0.81 83	★ ES-CH 3.00 2	
	N=NC6H3,O-Cl,P-NHAC						★ S.PARA- 0.69 83	★ ES-CH 3.00 2	
C8H7F1N3O2	NHCONHCONHC6H4F -P						★ S.INDUC 0.28 24	★ S.ZE.RS -0.24 24	★ ES-CH 4.00 2
C8H7F3N1	N(CH3)C6H4CF3-P				3.92		★ ES-CH 5.00 2	★ S.PHOSP -0.76 274	
	NHCH(CF3)C6H5						★ S.STAR 1.97 170	★ ES-CH 4.00 2	
C8H7Fe1O2	CH2Fe(CO)2(PI-C5H5)		★ -0.26 270	★ -0.49 270			★ ES-CH 9.30 2	F -0.18 66	R -0.32 66
C8H7N2	2-N-methylbenzimidazolyl				3.77		★ S.PARA- 0.58 918	★ S.ZE.RS 0.04 209	★ ES-CH 4.00 2
	3-Me1-(1,2-benzodiazolyl)			★ 0.27 15					
	5-amino-2-indolyl						★ S.INDUC 0.15 15	★ S.ZE.RS -0.08 15	
	N(CH3)C6H4CN-P				4.05		★ ES-CH 5.00 2	★ S.PHOSP -0.49 274	
C8H7N2O1	CH=NNHCOC6H5	0.43	★ 0.39 163	★ 0.51 163	4.21		★ ES-CH 3.00 2	F 0.34 5	R 0.17 5
C8H7O1	C6H4(COCH3) -P			★ 0.14 82	3.49		★ S.PARA+ 0.18 305	★ ES-CH 3.00 2	★ GRP.DPL -3.11 8
	C=O(CH2Ph)	0.70 42					★ ES-CH 4.00 2		
	CH=CHC6H4OH -P						★ S.PARA- 0.09 390	★ S.PARA- 0.11 283	★ GRP.DPL -1.64 8
	CO-Ph-4-CH3						S.ORTHO 0.56 78		
C8H7O2	C6H4CO2CH3-P				3.64		S.INDUC 0.22 84	★ ES-CH 3.00 2	
							S.ZE.RS -0.04 84	★ S.PHOSP -0.24 37	
	C=O(OCH2C6H5)	1.84			3.72		★ S.PARA- 0.67 625	★ ES-CH 4.00 2	
							★ S.ORTH- 0.86 97	★ GRP.DPL -2.06 8	
	CH2COOPh						★ S.STAR 0.81 10		
	CH2O2CC6H5				3.72		★ S.INDUC 0.15 34	★ S.BA.RS -0.05 34	★ ES-CH 3.00 2
	O-Ph-4-COCH3						S.STAR 2.91 3	GRP.DPL -3.04 9	
C8H7O3	2-(OH)-4-(OCH3)benzoyl						★ S.STAR 1.77 547	★ ES-CH 5.00 2	
	CO2CH2-C6H4-OH -P						GRP.DPL -2.56 8		
	OC(O)Ph-4-OCH3						S.STAR 2.50 3		

Empirical	Structural	π	σ_m	σ_p	MR	E_s	additional σ values		
C8H7O3S1	SO2CH2COC6H5						★S.PARA- 1.02 252	★ES-CH 5.20 2	
C8H7S1Se1	SCH=CH-Se-C6H5						GRP.DPL -1.81 8		
C8H7Se1	SeCH=CHC6H5 -C						GRP.DPL -1.17 8		
	SeCH=CHC6H5 -T						GRP.DPL -1.06 8		
C8H8	1-(9,10-diH-phenanthryl):						★S.ORTH+ -0.23 915		
	1-dihydroanthryl:						★S.META- -0.30 27		
	2(9-methylfluorenyl):						★S.PARA+ -0.48 27		
	2-(9,10-diH-phenanthryl):						★S.PARA+ -0.39 27		
	2-dihydroanthryl:						★S.PARA+ -0.32 27		
	3-(9,10-diH-phenanthryl):						★S.PARA+ -0.26 773		
	4-(9,10-di-H-phenanthryl)						★S.META- -0.38 773		
	7(2-methylfluorenyl):						★S.PARA+ -0.55 27		
	9-anthryl:				3.79		★S.META- -1.25 101	★S.META+ -0.82 542	★S.ZPTFT 0.00 63
	9-phenanthryl:				3.79		★S.ZPTFT 0.11 63 S.PARA+ -0.55 101	★S.PARA+ -0.45 101 S.PARA+ -0.37 27	S.PARA+ -0.36 269 ★S.PARA+ -0.36 583
C8H8Cl1D2N2	NCD(CH3)=N+DC6H5Cl-						★S.ZE.RS -0.59 19	★ES-CH 4.00 2	
C8H8N1	CH=CHC6H4NH2 -P						★S.PARA- -0.18 427 ★S.PARA- 0.08 651	★S.PARA- 0.11 283 ★ES-CH 3.00 2	GRP.DPL 2.06 8
	N=CH-C6H4-CH3 -P						GRP.DPL -1.93 8		
C8H8N1O1	C(=O)N(CH3)Ph						L-STM 8.24 1	B1-STM 1.59 1	B5-STM 4.50 1
	C6H4NHCOCH3-P				3.93		★ES-CH 3.00 2	★S.PHOSP -0.49 85	
	CH2CONHC6H5		★ -0.11 3		3.93		★S.INDUC 0.00 32 ★S.STAR 0.00 3 ES-CH 4.00 2	L-STM 6.95 11 B1-STM 1.52 11 B5-STM 7.22 11	★S-L 0.02 7
	CH2N(CHO)(C6H5)	0.49					★ES-CH 4.00 2		
	N(C6H5)(COCH3)		★ 0.19 3		3.93		★S.INDUC 0.22 32 ★S.STAR 1.37 3	★ES-CH 7.00 2 ★S-L 0.24 7	GRP.DPL -3.63 9 GRP.DPL -3.61 8
	N(CH3)COC6H5		★ 0.26 10		3.93		★S.INDUC 0.28 68	★S.ZE.RS -0.12 68	★ES-CH 5.00 2
	N(CHO)(CH2C6H5)	0.49					★ES-CH 5.00 2		
	N=CH-C6H4-OCH3 -O						GRP.DPL -2.87 8		
	N=CHC6H4(OCH3) -P		★ -0.07 56	★ -0.54 56	3.93		★ES-CH 3.00 2	F 0.15 5	R -0.69 5
C8H8N1O1S1	SCH2C(=O)NHPh						L-STM 10.27 11	B1-STM 1.70 11	B5-STM 5.83 94
C8H8N1O2	EtC6H4-NO2-O						★S.PARA+ -0.28 76	★ES-CH 3.00 2	
	EtC6H4NO2-P				4.10		★S.PARA- -0.01 109	★S.PARA+ -0.28 76	★ES-CH 3.00 2
	NHCOC6H4(OCH3) -P		★ 0.09 56	★ -0.06 56	4.10		★ES-CH 4.00 2	F 0.17 5	R -0.23 5
	NHCOCH2OC6H5	0.60			4.16		★ES-CH 4.00 2		
	OC(O)Ph-4-NHCH3						S.STAR 2.00 3		
	OCH2CONHC6H5	0.60			4.16		★ES-CH 3.00 2 L-STM 10.18 11	B1-STM 1.35 11 B5-STM 4.90 11	
	OCH2Ph,4-CONH2						L-STM 10.21 1	B1-STM 1.61 1	B5-STM 4.44 1
C8H8N1O2S2	SCH=NSO2-Ph-4-CH3		★ 0.65 3	★ 0.70 3			F 0.61 5	R 0.09 5	
C8H8N3O1	N=N-C6H4-NHCOCH3 -M						GRP.DPL -3.71 8		
	N=N-C6H4-NHCOCH3 -O						GRP.DPL -3.47 8		
	N=N-C6H4-NHCOCH3 -P						GRP.DPL -3.72 8		
C8H9	C6H3(CH3)2 (2,6)						S.INDUC 0.03 16		
	C6H4-2-Et					★ -4.09 35			
	C6H4Et -M				3.47		★S.STAR 0.53 120	★ES-CH 3.00 2	
	C6H4Et -P		★ 0.07 6	★ -0.02 6	3.47		★S.INDUC 0.10 6 ★S.STAR 0.47 120 S.STAR 0.59 3	★S.ZE.RS -0.12 6 ★ES-CH 3.00 2 F 0.13 4	R -0.15 4
	CH(CH3)C6H5		★ -0.03 3		3.47	★ -2.43 50 -2.02 35	★S.INDUC 0.07 6 ★S.STAR 0.11 52 S.STAR 0.37 3	ES-CH 5.00 2 ★ES-V 0.99 7 S-L 0.07 14	GRP.DPL -0.40 9
	CH2C6H4CH3-M				3.47		★S.STAR 0.20 314	★ES-CH 4.00 2	
	CH2C6H4CH3-P				3.47		★S.STAR 0.17 314	★ES-CH 4.00 2	
	EtC6H5	2.66	★ -0.07 3	★ -0.12 3	3.46	-1.93 496 -1.79 196 -1.62 50 ★ -1.47 35	★S.INDUC 0.02 6 S.STAR -0.06 3 ★S.STAR 0.08 52 ★S.META+ -0.09 772 ★S.PARA- -0.12 109 ★S.PARA+ -0.28 76 S.PARA+ -0.26 772	S.ORTH+ -0.23 154 ES-CH 3.00 2 ES-V 0.70 7 L-STM 8.33 11 B1-STM 1.52 11 B5-STM 3.58 11 F -0.01 5	R -0.11 5 R+ -0.27 4 R- -0.11 4 S-L 0.02 14 ★S.PHOSP -1.06 37
C8H9Hg1	CH2HgCH2C6H5						★S.PARA+ -1.12 277	★S.PARA+ -0.95 285	
C8H9N1	9-Et-3-carbazoyl:				3.89		★S.PARA+ -1.14 882	★ES-CH 0.00 2	
	CH=CH(4-N-Me-pyridyl)						★S.PARA- 0.61 636	★S.ORTH+ 0.61 636	★ES-CH 3.00 2
	CH=N+(CH3)C6H5						★S.PARA- 1.36 521	★ES-CH 4.00 2	
	N+(CH3)=CHC6H5						★S.PARA- 1.71 521	★ES-CH 4.00 2	
C8H9N2	N=CCH3(NHC6H5)		★ 0.29 293	★ 0.08 293	4.23		★ES-CH 4.00 2	F 0.38 5	R -0.30 5
	N=NC6H3-2,6-(CH3)2			★ 0.31 396			★S.PARA+ 0.18 368		
C8H9N2O1	C8H9N2O1						L-STM 10.20 11	B1-STM 1.35 11	B5-STM 4.99 11
C8H9O1	(CH2)2OC6H5				3.64	★ -2.11 196	ES-CH 3.00 2		
	C(OH)(CH3)(C6H5)						★S.STAR 0.62 376	★ES-CH 6.00 2	

Empirical	Structural	π	σ_m	σ_p	MR	E_s	additional σ values					
C8H9O1	C6H4OC2H5-P				3.64		★ES-CH 3.00 2	★S.PHOSP -0.66 85				
	CH(OCH3)C6H5			★ -0.01 82	3.64		★ES-CH 6.00 2					
	CH2C6H4(OCH3) -P				3.64		★S.STAR 0.14 314	★ES-CH 4.00 2				
	CH2OC6H4(CH3) -M				3.64		★S.STAR 0.83 314	★ES-CH 3.00 2				
	CH2OC6H4(CH3) -P				3.64		★S.STAR 0.82 314	★ES-CH 3.00 2				
	OCH2CH2Ph						L-STM 6.58 11	B1-STM 1.35 11	B5-STM 7.33 11			
C8H9O1S1	CH2S(O)CH2C6H5						GRP.DPL -3.76 8					
C8H9O1Se1	Se-Ph-2-OEt						S.STAR 1.13 3					
	Se-Ph-4-OEt						S.STAR 1.18 3					
C8H9O2	CH2OC6H4(OCH3) -P				3.81		★S.STAR 0.80 314	★ES-CH 3.00 2				
	OCH2CH2OC6H5	1.68					★ES-CH 3.00 2					
	Ph,3,4-(OCH3)2						L-STM 7.71 1	B1-STM 1.80 1	B5-STM 5.21 1			
C8H9O2S1	CH2CH2SO2Ph						L-STM 6.67 11	B1-STM 1.52 11	B5-STM 7.47 11			
	CH2SO2CH2C6H5						GRP.DPL -4.25 8					
C8H9O3S1	CH2OSO2C6H4-PCH3						★S-INDQ 1.28 58	S.STAR 1.44 3	★S-L 0.22 7			
							S.STAR 1.31 962	★ES-CH 3.20 2	GRP.DPL -5.30 9			
C8H9S1	C6H4SC2H5-P				4.26		★ES-CH 3.00 2	★S.PHOSP -0.50 85				
	CH2CH2SPh						L-STM 7.93 11	B1-STM 1.52 11	B5-STM 7.88 11			
	CH2SCH2C6H5		★ -0.03 3		4.26		★S.INDUC 0.06 32	★S.STAR 0.68 74	★S-L 0.08 7			
							S.STAR 0.38 3	★ES-CH 3.20 2	GRP.DPL -1.34 8			
	CH2SPh,3-CH3						L-STM 8.67 1	B1-STM 1.77 1	B5-STM 5.65 1			
	SCH2CH2C6H5		★ 0.21 3		4.26		★S.INDUC 0.23 32	L-STM 6.47 11	★S-L 0.25 7			
							★S.STAR 1.44 3	B1-STM 1.70 11				
							★ES-CH 3.20 2	B5-STM 8.06 11				
C8H9S2	CH2SSCH2C6H5						GRP.DPL -1.87 8					
C8H10Cl1N2S1	N=S(Ph-P-Cl)(N-Me2)						S.INDUC -0.37 123					
C8H10F1Sn1	Sn(Me)2C6H4F -M						★S.ZE.RS -0.19 86	★S.ZE.RS 0.02 333	★ES-CH 6.70 2			
C8H10N1	C6H4N(CH3)2 -P			★ -0.56 493	3.99		★S.STAR 0.16 120	L-STM 7.75 1	★S.PHOSP -0.68 724			
							★S.STAR 0.16 342	B1-STM 1.79 1				
							★ES-CH 3.00 2	B5-STM 3.11 1				
	C6H4NHC2H5-M				3.99		★ES-CH 3.00 2	★S.PHOSP -0.62 85	★S.PHOSP -0.46 37			
	CH2N(CH3)C6H5	2.09					★ES-CH 4.00 2					
	N(C2H5)C6H5				3.99		★ES-CH 6.00 2	★S.PHOSP -1.30 274				
	N(CH3)C6H4CH3-M				3.99		★ES-CH 5.00 2	★S.PHOSP -1.18 274				
	N(CH3)C6H4CH3-P				3.99		★ES-CH 5.00 2	★S.PHOSP -1.22 274				
	N(CH3)CH2C6H5	2.09					★ES-CH 4.00 2					
	NHCH2CH2Ph						L-STM 6.58 11	B1-STM 1.35 11	B5-STM 7.40 11			
	NHPh,3,5-(CH3)2						L-STM 4.90 11	B1-STM 1.50 11	B5-STM 6.53 11			
C8H10N1O1	N(CH3)C6H4OCH3-P				4.17		★ES-CH 5.00 2	★S.PHOSP -1.32 274				
	NHCH2CH2OC6H5						★S.STAR 1.42 170	★ES-CH 3.00 2				
C8H10N1O2	OCH2CH2OPh,4-NH2						L-STM 10.84 1	B1-STM 1.35 1	B5-STM 5.79 1			
C8H10N1O2S2	S(CH3)NSO2C6H4CH3-P		★ 0.65 592	★ 0.70 592			★S.META- 0.80 592	F 0.61 5	★GRP.DPL -7.46 8			
							★S.PARA- 1.00 861	R 0.09 66				
							ES-CH 4.40 2	R- 0.39 4				
C8H10N3	N=N-C6H4-N(CH3)2 -P						GRP.DPL 2.82 8					
C8H10N3O2S1	N=S(Ph-P-NO2)(N-Me2)						S.INDUC -0.31 123					
C8H10O1P1	CH2P(O)(CH3)(C6H5)						★S.STAR 0.67 45	★ES-CH 5.20 2				
	P(O)(C2H5)(C6H5)						★S.INDUC 0.27 45	★ES-CH 7.20 2	★S-L 0.28 14			
	P(O)(CH3)C6H4CH3-P			★ 0.34 38			★ES-CH 6.20 2					
	P(OEt)(C6H5)			★ 0.32 143	4.66		★S.PARA+ 0.23 143	★ES-CH 6.20 2				
C8H10O2P1	CH2P(O)(OCH3)(C6H5)						★S.STAR 0.72 45	★ES-CH 5.20 2				
	P(O)(OC2H5)(C6H5)						★S.INDUC 0.31 45	★ES-CH 7.20 2	★S-L 0.30 14			
C8H11As1	As+(Me)2(Ph)						S.CH-P+ 0.94 75					
C8H11Ge1	CH2Ge(CH2Ph)(H)2						★S.PARA+ -0.95 51	★S.ZE.RS -0.23 51				
	Ge(CH3)2C6H5			★ 0.06 60			★S.PARA+ -0.06 60	★ES-CH 6.30 2				
C8H11Ge1O2	OGe(OPh)(Me)2						★S.PARA+ -0.86 51					
C8H11N1	+N(Ph)(CH3)2						S.CH-P+ 0.88 75					
	2,4,6-trimepyridinium		★ 0.62 366	★ 0.58 366			S.INDUC 0.67 366	F 0.61 4				
							S.ZE.RS -0.09 366	R -0.03 4				
	C6H4NH+(CH3)2-P						★ES-CH 3.00 2	★S.PHOSP 0.30 37				
	C6H4NH2C2H5+-M						★ES-CH 3.00 2	★S.PHOSP 0.30 85				
C8H11N2S1	N=S(Ph)(N-Me2)						S.INDUC -0.34 123					
C8H11N3	+NH=NC6H4N(CH3)2-P						★S.META- 0.49 733	★S.PARA- 0.54 733	★ES-CH 3.00 2			
C8H11O1Si1	Si(Me)(Ph)OMe						★S.INDUC -0.22 25	★S.ZE.RS 0.10 25				
C8H11O4	CH=C(COOC2H5)2	1.18 54			4.30		★S.ZPTFT 0.39 248	★ES-CH 4.00 2				
C8H11S1@	2-(5-T-butylthienyl)						★S.ORTHO -0.17 199					
C8H11Si1	Si-C6H5(CH3)2		★ 0.04 3	-0.03 338	4.46		★S.INDUC -0.14 32	S.STAR 0.70 10	F 0.06 4			
				★ 0.07 3			S.INDUC 0.02 73	S.PARA+ 0.06 60	R 0.01 4			
				0.12 60			★S.STAR -0.87 3	S.PARA+ 0.08 175	R+ 0.02 4			
							★S.STAR -0.02 216	S.ZE.RS 0.05 73	★S-L -0.13 7			

Empirical	Structural	π	σm	σp	MR	Es	additional σ values					
C8H11Si1	Si-C6H5(CH3)2						★ S.STAR	0.34 377	★ ES-CH	6.20 2	GR.ELCT	-0.20 13
	SiH(Et)(Ph)			★ 0.15 238								
	SiH(Me)CH2C6H5			★ 0.03 238								
C8H13	2-bicyclo(2.2.2)octyl					★ -2.36 391	L-STM	5.68 1	B1-STM	1.90 1	B5-STM	4.78 1
	bicyclo(2.2.2)oct-1-yl			★ -0.25 465	3.41		S.INDUC	-0.01 657	★ S.ZE.RS	-0.17 465	L-STM	6.17 1
							★ S.STAR	-0.18 124	S.ZE.RS	-0.17 657	B1-STM	2.77 1
							S.PARA+	-0.39 440	ES-CH	4.00 2	B5-STM	3.17 1
							★ S.PARA+	-0.27 465	ES-V	1.33 7		
C8H14N1	CH2CCCH2N(Et)2						S.STAR	0.48 262				
C8H15	CH2CH2-cy-hexyl					★ -1.46 35	★ S.INDUC	-0.04 6	L-STM	8.22 1	★ S-L	-0.04 14
							★ ES-CH	3.00 2	B1-STM	1.52 1		
							★ ES-V	0.70 7	B5-STM	4.94 1		
	cyclooctyl				3.60		★ S.STAR	-0.13 124	★ ES-CH	3.00 2		
C8H15Cr1O5P1	Cr(CO)2P(OC2H5)3						★ S.PARA+	-0.53 782	★ ES-CH	9.50 2		
C8H15N1	CH2CCCH2N(Me)2Et+						S.STAR	1.05 262				
C8H15N2O4	C(hept)(NO2)2						★ S.STAR	3.33 157				
C8H15O1	COC7H15						★ S.ZE.RS	0.22 22	L-STM	10.23 1	B5-STM	7.35 1
							★ ES-CH	4.00 2	B1-STM	1.63 1	★ S.ZTWST	0.00 22
C8H15Si1	CH2Si(CH3)(CH2CH=CH2)2						★ S.STAR	-0.03 381				
C8H16N1	(CH2)3-1-piperidinyl						L-STM	8.08 1	B1-STM	1.52 1	B5-STM	6.80 1
C8H16N1O3	CHOHCH(CH2OH)NHCOC(CH3)3	-0.52 54			4.58		★ ES-CH	5.00 2				
C8H16O2P1S2	EtSSP(OC(CH3)2C(CH3)2O-)						★ S.INDUC	0.19 137	★ S.STAR	0.43 137	★ ES-CH	3.20 2
C8H17	C(CH3)2CH2C(CH3)3				3.82	-3.60 62	ES-CH	5.00 2	L-STM	6.17 1	B5-STM	4.54 1
						★ -2.57 50	★ ES-V	1.74 7	B1-STM	2.60 1		
	C(Et)2I-Pr					★ -7.32 62	ES-CH	8.00 2				
	C(I-Pr)2CH3					★ -8.50 62	ES-CH	8.00 2				
	CH(t-Bu)(I-Pr)					★ -7.65 62	ES-CH	8.00 2				
	N-octyl				3.82	★ -1.57 50	★ S.STAR	-0.15 287	ES-CH	3.00 2	B5-STM	7.39 1
							S.STAR	-0.15 544	★ ES-V	0.68 7	★ S.PHOSP	-1.11 37
							S.STAR	0.00 278	L-STM	10.27 1		
							S.ZE.RS	-0.14 47	B1-STM	1.52 1		
C8H17N1	(CH2)3NH+(CH2)5						S.PHOSP	-0.62 367				
C8H17O1	C(CH3)2CH2OC(CH3)3				4.00	★ -2.81 402	★ S.STAR	-0.17 402	ES-CH	5.00 2	★ ES-V	1.30 7
	C(CH3)2OCH2C(CH3)3				4.00	★ -2.67 402	★ S.STAR	0.18 402	ES-CH	5.00 2	★ ES-V	1.23 7
	O(CH2)7CH3 -compact						★ ES-CH	3.00 2				
	O(CH2)7CH3						★ ES-V	0.61 7	B1-STM	1.35 1		
							L-STM	11.00 1	B5-STM	7.66 1		
	O-AmCH(CH3)2			★ -0.27 23	4.00		★ ES-CH	3.00 2				
	OCH(CH3)hexyl						★ ES-CH	4.00 2	★ ES-V	0.92 7		
	OCH2C(Et)3						★ ES-CH	3.00 2	★ ES-V	0.97 7		
	OCH2CH(Et)T-butyl						★ ES-CH	3.00 2	★ ES-V	0.96 7		
	OCH2CH(Et)butyl						★ ES-CH	3.00 2	★ ES-V	0.76 7		
	OCH2CH(I-Pr)2						★ ES-CH	3.00 2	★ ES-V	0.89 7		
C8H17O2S1	SO2-octyl						L-STM	11.08 1	B1-STM	2.03 1	B5-STM	7.84 1
C8H17O5Si1	COOCH2Si(OEt)3						★ S.INDUC	0.10 25	★ S.ZE.RS	0.22 25		
C8H17S1	(CH2)6SC2H5						★ S.STAR	0.02 278	★ ES-CH	3.00 2		
	S(CH2)7CH3 compact						★ ES-CH	3.20 2				
	S(CH2)7CH3						L-STM	10.87 1	B1-STM	1.70 1	B5-STM	8.89 1
C8H18B1	B(N-Bu)2						S.STAR	0.43 446				
C8H18B1O2	B(O-t-Bu)2						★ S.STAR	0.35 358				
C8H18N1	NH(CH2)7CH3 -compact						★ ES-CH	3.00 2				
	NH(CH2)7CH3						L-STM	10.99 1	B1-STM	1.35 1	B5-STM	7.78 1
C8H18N1S1	SCH2CH2N(I-Pr)2						L-STM	8.03 1	B1-STM	1.70 1	B5-STM	7.17 1
C8H18N1Se1	SeCH2CH2N(I-Pr)2						L-STM	8.05 1	B1-STM	1.85 1	B5-STM	7.39 1
C8H18O1P1	(CH2)4P(O)(C2H5)2		★ 0.31 107	★ 0.41 107			★ S.STAR	0.10 422	★ ES-CH	3.00 2		
	P(O)(t-Bu)2						F	0.28 4	R	0.13 4		
	P(O)Bu2		★ 0.35 207	★ 0.49 207	4.78		★ S.INDUC	0.24 610	S.ZE.RS	0.28 6	R	0.19 5
			0.37 6	0.53 6			S.INDUC	0.25 6	ES-CH	6.20 2	GRP.DPL	-4.31 8
							S.INDUC	0.25 45	F	0.30 5		
C8H18O2P1	P(O)(C4H9)(OC4H9)						★ S.INDUC	0.27 45	★ ES-CH	6.20 2	★ S-L	0.27 14
C8H18O2P1S2	CH2CH2SSP(O-Pr)2						★ S.INDUC	0.13 137	★ S.STAR	0.28 137	★ ES-CH	3.20 2
C8H18O3P1	(CH2)4P(O)(OC2H5)2						★ S.STAR	0.07 422	★ ES-CH	3.00 2		
	P=O(OBu)2		★ 0.41 6	0.54 207	4.97		S.INDUC	0.29 6	★ S.ZE.RS	0.28 6	R	0.22 4
				★ 0.57 6			S.INDUC	0.29 45	★ S.ZE.RS	6.20 2		
							★ S.STAR	1.77 3	F	0.35 4		
C8H18O4P1	OP(=O)(OBu)2						L-STM	7.76 1	B1-STM	1.35 1	B5-STM	9.21 1
C8H18P1	P(t-Bu)2		★ 0.01 150	★ 0.15 150			F	-0.01 4	R	0.16 4		
C8H18P1S2	CH2CH2SSP(I-Pr)2						★ S.INDUC	0.16 137	★ S.STAR	0.35 137	★ ES-CH	3.20 2
C8H18S1	S+(Bu)2						S.CH-P+	1.11 75				
C8H18S2	SCH2CH2S+(I-Pr)2						L-STM	8.12 1	B1-STM	1.70 1	B5-STM	7.28 1
C8H18S3	SCH2CH2S+(Et)SCH2CH2S(Et)						L-STM	11.06 1	B1-STM	1.70 1	B5-STM	9.38 1
C8H19Ge1O3	CH2CH2Ge(OC2H5)3						★ S.PARA+	-0.07 60	★ ES-CH	3.30 2		

Empirical	Structural	π	σ_m	σ_p	MR	E_s	additional σ values						
C8H19N1S1	SCH2CH2N+(Et)3						L-STM	8.03 _1_	B1-STM	1.70 _1_	B5-STM	6.52 _1_	
C8H19O2S1Si1	Si(O-t-Bu)2(SH)						★ S.STAR	1.50 _188_					
C8H19Si1	EtSi(Et)3			★ -0.15 _246_	4.82		S.STAR	-0.24 _448_	★ S.STAR	-0.16 _428_	★ ES-CH	3.20 _2_	
C8H20N2O1P1	P(O)[N(C2H5)2]2						★ S.INDUC	-0.02 _69_	★ S.ZE.RS	0.12 _69_	S-L	0.10 _14_	
							★ S.INDUC	0.10 _45_	★ ES-CH	8.20 _2_			
C8H20N2O2P1	N=P(OC2H5)2N(C2H5)2			★ -0.58 _534_			★ ES-CH	5.20 _2_					
	OP(=O)(N(Et)2)2						L-STM	7.02 _1_	B1-STM	1.35 _1_	B5-STM	6.78 _1_	
C8H20N2P1	P(NEt2)2						★ S.INDUC	-0.22 _69_	★ S.ZE.RS	-0.02 _69_	★ ES-CH	7.20 _2_	
C9H4Cr1O3	Ph-Cr(CO)3 (PI)	0.18 _791_	★ 0.14 _527_				S.INDUC	0.21 _951_	★ S.PARA+	-0.13 _632_	★ ES-CH	13.30 _2_	
			★ 0.29 _527_	0.15 _549_			S.ZPTFT	0.26 _549_	★ S.PARA+	-0.09 _549_			
				0.26 _791_			S.PARA	0.29 _952_	★ S.ZE.RS	-0.04 _549_			
C9H4N3	4-cyano-2-quinazolinyl						S.INDUC	0.19 _15_	S.ZE.RS	0.12 _15_			
C9H5N2	5-cyano-2-indolyl						★ S.INDUC	0.25 _15_	★ S.ZE.RS	-0.04 _15_			
C9H5O1S1	2-(benzo-4-thiopyronyl)		★ 0.34 _227_	★ 0.35 _227_			★ S.INDUC	0.33 _227_	★ ES-CH	3.00 _2_	R	0.01 _5_
							S.BA.RS	0.02 _15_	F	0.34 _5_			
	2-(benzothiopyronyl)		★ 0.48 _227_	★ 0.45 _227_			★ S.INDUC	0.36 _227_	★ ES-CH	3.20 _2_	R	-0.03 _5_
							S.ZE.RS	0.06 _15_	F	0.48 _5_			
C9H5O2	2-(benzo-1,4-pyronyl)		★ 0.41 _227_	★ 0.40 _227_			★ S.INDUC	0.40 _227_	★ ES-CH	3.00 _2_	R	-0.01 _5_
							★ S.PARA	0.70 _15_	F	0.41 _5_			
C9H6F6N1	C(CF3)2C6H4NH2-P			0.31 _392_			★ ES-CH	10.80 _2_					
				★ 0.32 _738_									
C9H6N1	2-isoquinolinyl						S.STAR	1.19 _15_					
	2-quinolinyl	1.77					★ S.INDUC	0.13 _228_	★ S.ORTH-	1.14 _438_	★ ES-CH	3.00 _2_	
	3-quinolinyl						S.STAR	1.22 _15_					
	6-quinolinyl						★ S.INDUC	0.17 _96_					
	CH=C(CN)C6H5						★ S.PARA	0.53 _950_	★ ES-CH	4.00 _2_			
C9H6N1@	2-quinolinyl:			★ 0.74 _417_			★ S.ORTH-	0.73 _663_	★ S.ORTH-	1.33 _232_			
	3-quinolinyl:			★ 0.52 _417_			★ S.META+	0.08 _663_	S.META-	0.70 _438_			
	4-quinolinyl:			★ 0.59 _417_			S.PARA-	1.20 _438_	★ S.PARA-	1.38 _232_	★ S.PARA+	0.75 _663_
C9H6N1O3	CH=CHCOC6H4(NO2)		★ 0.15 _180_	★ 0.05 _180_	4.57		★ ES-CH	3.00 _2_	F	0.21 _5_	R	-0.16 _5_
C9H6N3	6-(3-Ph-1,2,4-triazinyl)						★ S.INDUC	0.40 _96_					
C9H7	1(4H-cypen(DEF)phenanthr:						S.PARA+	-0.51 _563_					
	2(4H-cypen(DEF)phenanthr:						S.PARA+	-0.43 _563_					
	3(4H-cypen(DEF)phenanthr:						S.PARA+	-0.48 _563_					
	C=-CC6H4CH3 -P			★ 0.10 _281_			★ S.PARA-	0.38 _356_	★ ES-CH	3.00 _2_			
	CH2CCC6H5						S.STAR	0.63 _262_	★ S.STAR	0.66 _380_	★ ES-CH	3.00 _2_
C9H7Mo1O3	CH2Mo(CO)3(PI-C5H5)		★ -0.21 _270_	★ -0.45 _270_			★ ES-CH	10.70 _2_	F	-0.13 _66_	R	-0.33 _66_
C9H7N2	2-(4-Me-quinazolinyl)						★ S.INDUC	0.22 _96_					
	3-(1-Ph-pyrazole)						★ S.INDUC	0.21 _96_					
C9H7N2O1	4-MeO-2-quinazolinyl						S.INDUC	0.01 _15_	S.ZE.RS	0.11 _15_			
C9H7N2O2	3-benzyl-4-sydnonyl		★ 0.37 _646_				★ S.STAR	3.30 _646_	★ ES-CH	5.00 _2_			
C9H7O1	C=-CC6H4OCH3 -P			★ 0.09 _281_			★ ES-CH	3.00 _2_					
	CH=CHCOC6H5	0.95	★ 0.18 _180_	★ 0.05 _180_	4.03		★ ES-CH	3.00 _2_	F	0.25 _5_	R	-0.20 _5_
C9H7O1S1	CH=CHC(O)CH=CH-2-thienyl						GRP.DPL	-3.21 _8_					
	CH=CHCH=CHC(O)-2-thienyl						GRP.DPL	-3.50 _8_					
	COCH=CHCH=CH-(2-thienyl)						GRP.DPL	-3.25 _8_					
C9H7O2	CH=CHC(O)CH=CH-2-furyl						GRP.DPL	-3.29 _8_					
	CH=CHCH=CHC(O)-2-furyl						GRP.DPL	-3.27 _8_					
C9H7O3W1	CH2W(CO)3(PI-C5H5)			★ -0.44 _270_									
C9H7O4	OC(O)-C6H4-CO2CH3 -O						GRP.DPL	-2.54 _8_					
C9H8	8(4H-cypen(DEF)phenanthr:						S.PARA+	-0.44 _563_					
	9-(3-Me-phenanthryl):						S.PARA+	-0.51 _983_					
C9H8Cl1N2O2	N=NC6H3,2-Cl,5-CO2Et						★ S.PARA-	0.85 _83_	★ ES-CH	3.00 _2_			
	N=NC6H3,O-Cl,P-CO2Et						★ S.PARA-	0.94 _83_	★ ES-CH	3.00 _2_			
C9H8N1	1-(3-Me-indolyl)			0.25 _275_			★ ES-CH	4.00 _2_					
				★ 0.27 _226_									
	3-indolyl-Me	0.96 _130_											
	5-Me-2-indolyl						★ S.INDUC	0.17 _15_	★ S.ZE.RS	-0.08 _15_			
C9H8N1@	1-Me-2-indolyl:						★ S.PARA+	-1.17 _195_	★ S.ORTH-	0.50 _148_			
	3-(N-methylindolyl):						★ S.PARA+	-2.28 _371_	★ S.PARA+	-1.93 _195_			
C9H8N1O1	5-MeO-2-indolyl						★ S.INDUC	0.17 _15_	★ S.ZE.RS	-0.07 _15_			
C9H9	CH=CHC6H4CH3 -M			★ -0.08 _309_									
	CH=CHC6H4CH3 -P			★ -0.11 _281_			★ S.PARA-	0.12 _390_	★ ES-CH	3.00 _2_			
				-0.09 _309_			★ S.PARA-	0.15 _283_	★ GRP.DPL	0.59 _8_			
	homocubyl						★ S.PARA-	-0.75 _276_	★ ES-CH	5.00 _2_			
C9H9N2O2	N=NC6H4-P-CO2Et						★ S.PARA-	0.82 _83_	★ ES-CH	3.00 _2_			
C9H9O1	CH=CH-C6H4-OCH3 -O						GRP.DPL	-1.05 _8_					
	CH=CHC6H4OCH3 -P			★ -0.13 _281_			★ S.PARA-	0.11 _390_	★ ES-CH	3.00 _2_			
				-0.11 _309_			★ S.PARA-	0.15 _283_	★ GRP.DPL	-1.45 _8_			
C9H9O2	C6H4CO2Et -M						S.INDUC	0.18 _84_	S.ZE.RS	-0.06 _84_			

Empirical	Structural	π	σ_m	σ_p	MR	E_s	additional σ values					
C9H9O2	CH(C6H5)CO2CH3						★ES-CH	7.00 2	★S-L	0.11 14		
	CH2OC(O)CH2C6H5						GRP.DPL	-1.97 8				
C9H10	9,9-dimethylfluorenyl						★S.PARA+	-0.49 27	★ES-CH	0.00 2		
C9H10F3O1Si1	Si(Me)(Ph)OCH2CF3						★S.INDUC	-0.06 25	★S.ZE.RS	0.11 25		
C9H10N1O1	4-dimethylaminobenzoyl'				4.58		★S.PARA-	0.85 871	★ES-CH	5.00 2		
C9H10N1O2	CH(Et)C6H4-NO2 -M						★S.STAR	0.31 813				
	CH(Et)C6H4-NO2 -P						★S.STAR	0.38 813				
	N(CH3)COCH2OC6H5	0.12			4.63		★ES-CH	5.00 2				
	OCH2CON(CH3)(C6H5)	0.12			4.63		★ES-CH	3.00 2				
C9H10O2	OCH(Ph)CH2CH2O*						S.PHOSP	-0.67 577				
C9H11	C(CH3)2(C6H5)						★S.INDUC	0.05 6	★ES-CH	6.00 2	★S-L	0.05 14
	C6H2-2,4,6Me3						S.INDUC	0.02 835	★O-STER	-0.23 90		
	C6H4-2-Pr					★ -4.16 35						
	C6H4CH(CH3)2 -P		★ 0.08 6	★ 0.01 6			★S.INDUC	0.10 6	L-STM	7.84 1	R	-0.12 4
							★S.STAR	0.56 3	B1-STM	2.66 1	GRP.DPL	-0.15 9
							★S.ZE.RS	-0.09 6	B5-STM	3.14 1		
							★ES-CH	3.00 2	F	0.13 4		
	CH(CH3)CH2C6H5				3.93	-2.99 403	★S.STAR	0.03 323				
						★ -2.37 323	ES-CH	4.00 2				
						-1.56 35	★ES-V	0.98 7				
	CH(Et)C6H5				3.93	★ -2.74 50	★S.STAR	0.04 52	★ES-V	1.18 7		
						-2.46 35	ES-CH	6.00 2				
	Ph,4-Pr						L-STM	9.14 1	B1-STM	1.71 1	B5-STM	3.49 1
	PrC6H5	3.16 589	★ -0.12 3		3.93	-1.75 196	★S.INDUC	0.01 6	ES-CH	3.00 2	B5-STM	7.47 1
						-1.69 50	S.STAR	-0.06 3	ES-V	0.70 7	S-L	0.01 7
						★ -1.46 35	★S.STAR	0.02 52	L-STM	7.91 1		
							★S.PARA+	-0.28 76	B1-STM	1.52 1		
C9H11N2O6	uridin-5-yl						S.STAR	0.80 3				
C9H11O1	(CH2)3OC6H5				4.11	★ -1.75 196	ES-CH	3.00 2				
	O(CH2)3Ph						L-STM	10.27 11	B1-STM	1.35 11	B5-STM	4.70 11
C9H12Cl1O1Si1	Si(Me)(Ph)OCH2CH2Cl						★S.INDUC	-0.30 25	★S.ZE.RS	0.10 25		
C9H12N1	C(CH3)2C6H4NH2-P			★ -0.15 738			★S.PARA-	-0.02 187	★ES-CH	6.00 2		
C9H12O1P1	CH2P(O)(C2H5)(C6H5)						★S.STAR	0.63 45	★ES-CH	5.20 2		
	P(O)(C3H7)(C6H5)						★S.INDUC	0.27 45	★ES-CH	7.20 2	★S-L	0.27 14
	P(O)(Et)C6H4CH3-P			★ 0.23 38			★ES-CH	7.20 2				
C9H12O2P1	CH2P(O)(OC2H5)(C6H5)						★S.STAR	0.68 45	★ES-CH	5.20 2		
C9H13I1P1	P+(CH3)2(C6H4CH3-P)I-			★ 0.86 38			★S.INDUC	0.65 38	★S.ZE.RS	0.21 38	★ES-CH	6.20 2
C9H13N1	N+(CH3)2CH2Ph					★ -3.47 303						
	Ph-3-N(CH3)3+						S.STAR	1.65 3				
	Ph-4-N(CH3)3+						S.STAR	1.51 3				
C9H13N2S1	N=S(P-tolyl)(N-Me2)						S.INDUC	-0.40 123				
C9H13O1Si1	Si(Me)(Ph)OEt						★S.INDUC	-0.25 25	★S.ZE.RS	0.10 25		
C9H13Si1	CH2Si(CH2Ph)(H)(CH3)						★S.PARA+	-0.91 51	★S.ZE.RS	-0.21 51		
	CH2Si(CH3)2C6H5						★S.STAR	-0.11 381	★ES-CH	5.20 2		
	Si(CH3)2(CH2C6H5)						★S.STAR	0.06 216	★S.STAR	0.32 377	★ES-CH	5.20 2
C9H14N1	CH2CCCH2N(CH2)5						S.STAR	0.48 262				
C9H15	2,3-(Pr)2-cyclopropenyl			★ -0.43 928			★ES-CH	4.00 2				
C9H17	C(CH3)2cyclohexyl				4.06	-4.65 403	★S.STAR	-0.44 323	L-STM	6.17 1	B5-STM	5.83 1
						★ -3.73 323	ES-CH	6.00 2	B1-STM	2.60 1		
	CH2CH2CH2-cy-hexyl					★ -1.46 35	★ES-CH	3.00 2	L-STM	8.22 1	B5-STM	7.25 1
							★ES-V	0.71 7	B1-STM	1.52 1		
	cyclononyl				4.06		★S.STAR	-0.13 124	★ES-CH	3.00 2		
C9H17N2O4	C(oct)(NO2)2						★S.STAR	3.33 157				
C9H17O1	C(=O)octyl						L-STM	11.03 1	B1-STM	1.63 1	B5-STM	7.80 1
C9H17Si1	CH2CCSi(Et)3						★S.STAR	0.75 433				
C9H18N1O3	CHOHCH(CH2OH)NHCOCHEt2	-0.14 54			5.04		★ES-CH	5.00 2				
C9H19	C(Et)(I-Pr)2					★ -8.50 62	ES-CH	9.00 2				
	C(Et)2(t-Bu)					★ -8.33 62	ES-CH	9.00 2				
	C(Me)(I-Pr)(t-Bu)					★ -8.68 62	ES-CH	9.00 2				
	CH(I-Bu)2				4.29	-4.32 607	ES-CH	5.00 2	B1-STM	1.90 1		
						★ -3.71 50	★ES-V	1.70 7	B5-STM	5.47 1		
						-3.50 62	L-STM	6.17 1				
	CH(butyl)2					★ -3.20 62	ES-CH	5.00 2	L-STM	6.97 1	B5-STM	4.94 1
							★ES-V	1.56 7	B1-STM	1.90 1		
	CH(t-Bu)2					★ -8.09 62	S.INDUC	-0.03 16	L-STM	4.92 1	B5-STM	4.19 1
							ES-CH	9.00 2	B1-STM	2.08 1		
	CH2CH2CH(Et)T-butyl						★ES-CH	3.00 2	★ES-V	1.01 7		
	nonyl						L-STM	11.08 1	B1-STM	1.52 1	B5-STM	7.84 1
C9H19O1	OCH(I-butyl)2						★ES-CH	4.00 2	L-STM	6.86 1	B5-STM	6.58 1
							★ES-V	1.28 7	B1-STM	1.35 1		
C9H19Si1	CH2CH=CHSi(Et)3						★S.STAR	0.22 433				
C9H20Cl3N3P1	N=P(N-Et2)2(CCl3)						S.INDUC	-0.48 123				

Empirical	Structural	π	σ_m	σ_p	MR	E_s	additional σ values					
C9H20N1	CH2N(Bu)2	0.82 42					★ES-CH	4.00 2				
C9H20O1P1	CH2P(O)(C4H9)2						★S.STAR	0.56 45	★ES-CH	5.20 2		
C9H20O2P1	CH2P(O)(C4H9)(OC4H9)						★S.STAR	0.62 45	★ES-CH	5.20 2		
C9H20O3P1	CH2P=O(OBu)2				5.44		★S.STAR	0.62 435	★S.STAR	0.66 45	★ES-CH	5.20 2
C9H20S1	(CH2)6S+(CH3)C2H5						★S.STAR	0.28 278	★ES-CH	3.00 2		
C9H20Si1	CH2Si(t-Bu)3						★S.PARA+	-0.74 51	★S.ZE.RS	-0.22 51		
C9H21Ge1O3	Ge(O-I-Pr)3						S.INDUC	0.19 92	S.ZE.RS	0.13 92		
C9H21N1	N+(Pr)3					★ -5.34 215						
C9H21N1O3P1	N=P(OC3H7)3			★ -0.54 345			★ES-CH	5.20 2				
C9H21O3Si1	Si(O-I-Pr)3						★S.INDUC	-0.24 25	S.STAR	-0.36 140	S.ZE.RS	0.11 92
							S.INDUC	-0.24 92	S.STAR	1.50 188	★S.RES.+	0.16 140
							★S.INDUC	-0.16 25	★S.ZE.RS	0.08 25		
							★S.INDUC	-0.16 140	★S.ZE.RS	0.08 140		
	Si(OPr)3						★S.INDUC	-0.19 25	★S.ZE.RS	0.08 25	★S.ZE.RS	0.13 25
C9H21Si1	(CH2)3Si(Et)3						S.STAR	-0.24 448	★S.STAR	-0.08 428	★ES-CH	3.00 2
	Si(Pr)3						S.INDUC	-0.27 92	S.ZE.RS	0.05 92		
C9H21Sn1	Sn(I-Pr)3						L-STM	5.77 1	B1-STM	3.59 1	B5-STM	4.77 1
	Sn(Pr)3						L-STM	6.53 1	B1-STM	3.88 1	B5-STM	6.07 1
C9H22N2O1P1	CH2P(=O){N(Et)2}2						★S.STAR	0.22 45				
C9H27O3Si4	Si(OSiMe3)3		★ -0.09 221	★ -0.01 221	8.10		★ES-CH	7.80 2	R	0.07 4		
			★ -0.09 684	★ -0.01 684			F	-0.13 66	R	0.11 66		
			-0.06 25	0.02 25			F	-0.08 4				
C9H27Si4	Si(SiCH3)3						S.PARA+	-0.11 175	S.PARA+	-0.01 175		
C9H27Si5	cy-Si5(Me)9						S.PARA+	-0.07 175				
C10H5	CCCC-Ph						S.STAR	2.81 3				
C10H5Br1	8-(3-Br-fluoranthryl):						★S.PARA+	-0.36 27	★ES-CH	0.00 2		
C10H6	7-fluoranthryl:						★S.META+	-0.46 101	★S.META+	-0.27 27	★ES-CH	2.00 2
	8-fluoranthryl:						★S.META+	-0.49 933	S.META+	-0.41 27		
C10H6Br1	2-(4-Br-fluoranthryl):						★S.PARA+	0.13 27	★ES-CH	0.00 2		
	2-(8-Br-fluoranthryl):						★S.PARA+	-0.17 27	★ES-CH	0.00 2		
	2-(9-Br-fluoranthryl):						★S.PARA+	-0.16 27	★ES-CH	0.00 2		
	3-(4-Br-fluoranthryl):						★S.PARA+	-0.34 27	★ES-CH	2.30 2		
	3-(8-Br-fluoranthryl):						★S.PARA+	-0.41 27	★ES-CH	1.00 2		
	3-(9-Br-fluoranthryl):						★S.PARA+	-0.34 27	★ES-CH	1.00 2		
C10H6N3	4,5-Bnz-2-benzotriazolyl			★ 0.31 15								
C10H6N3@	2-benzotriazolyl:						★S.PARA-	0.49 419	★ES-CH	3.00 2		
C10H7	1-fluoranthryl:						★S.PARA+	-0.41 101	★S.PARA+	-0.28 27	★ES-CH	2.00 2
	1-naphthyl		★ 0.06 3		4.16	★ -1.64 692	★S.INDUC	0.14 6	★ES-CH	4.00 2	★S-L	0.14 7
							S.INDUC	0.21 16	L-STM	6.28 1	O-STER	-0.75 90
							★S.STAR	-0.02 692	B1-STM	1.71 1		
							S.STAR	0.75 3	B5-STM	5.50 1		
	1-pyrenyl:						★S.PARA+	-1.05 101	★S.PARA+	-0.67 27		
							★S.PARA+	-0.71 101	★ES-CH	1.00 2		
	2-fluoranthryl:						★S.PARA+	-0.21 27	★ES-CH	0.00 2		
	2-naphthyl		★ 0.06 3		4.16		★S.INDUC	0.13 6	★S.PARA-	0.13 493	B1-STM	1.71 1
							S.INDUC	0.14 16	★ES-CH	3.00 2	B5-STM	4.31 1
							★S.STAR	0.75 3	L-STM	8.35 1	★S-L	0.13 7
	2-pyrenyl:				4.29		S.PARA-	-0.22 27	★S.PARA+	-0.13 269	★ES-CH	0.00 2
	3-fluoranthryl:				4.29		★S.PARA+	-0.52 101	★S.PARA+	-0.45 27		
							★S.PARA+	-0.48 269	★ES-CH	1.00 2		
C10H7N2	4-(2-Ph-pyrimidinyl)						★S.INDUC	0.28 1013				
	4-Ph-2-pyrimidyl						★S.INDUC	0.13 204	★S.ZE.RS	0.09 204		
C10H7O2	C=-CCH2OOCC6H5						★S.STAR	1.03 183	★ES-CH	3.00 2		
C10H7S1@	2-(5-phenylthienyl):						★S.ORTHO	0.02 199				
C10H8	4-pyrenyl:				4.46		S.PARA+	-0.36 27	★S.PARA+	-0.34 269	★ES-CH	1.00 2
C10H8N1	NH-beta-naphthyl			★ -0.54 147			★ES-CH	4.00 2				
C10H8N3O2	2(4-OMe,6-OPh-triazinyl)						★S.PARA-	0.73 346	★ES-CH	3.00 2		
C10H9Fe1	ferricenium+		★ 0.29 756	★ 0.29 756			★S.INDUC	0.30 200	★S.STAR	1.85 200		
	ferrocenyl	2.46	★ -0.20 596	★ -0.42 596	4.82		S.INDUC	-0.22 781	S.META-	0.00 788	S.ZE.RS	-0.17 344
			-0.15 597	-0.18 597			S.INDUC	-0.12 959	S.PARA-	-0.05 960	S.ZE.RS	-0.11 286
			-0.12 344	-0.16 408			S.INDUC	-0.03 200	★S.PARA-	-0.03 698	S.ZE.RS	-0.11 845
			0.00 408	-0.10 200			S.INDUC	-0.03 286	S.PARA-	-0.03 788	S.ZE.RS	-0.05 200
				-0.05 344			S.INDUC	-0.01 344	S.PARA+	-1.30 596	S.RES.+	-0.50 200
							S.INDUC	-0.01 845	S.PARA+	-1.00 597	★ES-CH	8.30 2
							S.STAR	-0.33 200	★S.PARA+	-0.71 595	F	-0.15 66
							★S.META+	0.00 595	S.PARA+	-0.55 200	R	-0.04 66
							S.META-	0.00 698	S.ZE.RS	-0.29 781		
C10H9N2O2	3-benzyl-sydnonyl-4-CH2-						★S.STAR	1.20 646	★ES-CH	4.00 2		
C10H9O2	CH2OC(O)CH=CHC6H5						GRP.DPL	-2.27 8				
C10H9Os1	osmocenyl						S.INDUC	0.06 286	S.ZE.RS	-0.10 286		
C10H9Ru1	ruthenocenyl			★ -0.05 200			S.INDUC	-0.04 286	★S.PARA+	-0.44 200	S.RES.+	-0.48 200
							★S.INDUC	0.04 200	S.ZE.RS	-0.10 286	★ES-CH	8.70 2
							★S.STAR	0.26 200	S.ZE.RS	-0.09 200		

Empirical	Structural	π	σm	σp	MR	Es	additional σ values					
C10H10Fe1	ferrocenonium+		★ 0.05 *344*	★ 0.29 *344*			★S.INDUC	0.13 *344*	★S.ZE.RS	0.00 *344*		
C10H10N3	4-diMeAm-2-quinazolinyl						S.INDUC	-0.08 *15*	S.ZE.RS	0.09 *15*		
C10H11O1	CH=CH-C6H4-OEt -P						GRP.DPL	-1.66 *8*				
C10H11O1Si1	Si(Me)(Ph)OCH2CCH						★S.INDUC	-0.30 *25*	★S.ZE.RS	0.10 *25*		
C10H12	2,7-diMe-9-phenanthryl:						S.PARA+	-0.47 *489*				
	3,6-diMe-9-phenanthryl:						S.PARA+	-0.54 *489*				
	4,5-diMe-9-phenanthryl:						S.PARA+	-0.52 *489*				
C10H12N1	CH=CH-C6H5-N(CH3)2 -P						GRP.DPL	2.27 *8*				
C10H12N1O2	BuC6H4NO2-O						★S.PARA+	-0.28 *76*	★ES-CH	3.00 *2*		
	BuC6H4NO2-P						★S.PARA+	-0.28 *76*	★ES-CH	3.00 *2*		
	CH2C6H4NHCO2C2H5-P			★ 0.01 *507*	4.92		★ES-CH	4.00 *2*				
C10H12N1O3	CHOHCH(CH2OH)NHCOC6H5	-0.22 *54*			5.09		★ES-CH	5.00 *2*				
C10H13	Bu-C6H5				4.39	★ -1.45 *35*	★S.INDUC	0.00 *6*	★ES-CH	3.00 *2*	S-L	0.00 *7*
							★S.PARA+	-0.28 *76*	★ES-V	0.70 *7*		
	C6H4C(CH3)3 -P		★ 0.07 *6*	★ 0.01 *6*	4.39		★S.INDUC	0.09 *6*	★S.ZE.RS	-0.05 *6*	R	-0.11 *4*
							★S.STAR	0.43 *120*	★ES-CH	3.00 *2*	GRP.DPL	0.00 *9*
							S.STAR	0.52 *3*	F	0.12 *4*		
	CH2C(CH3)2C6H5				4.39	-3.25 *403*	★S.STAR	-0.04 *323*				
						★ -2.94 *323*	ES-CH	5.00 *2*				
C10H13O1	(CH2)4OC6H5				4.57	★ -1.64 *196*	ES-CH	3.00 *2*				
C10H13O1S1	SO2-Ph-4-C(CH3)3						S.STAR	2.97 *3*				
C10H13O2S1	SO2-Ph-4-C(CH3)3						S.STAR	3.23 *3*				
C10H13O3S1	(CH2)3OSO2Ph-P-CH3				5.02		★S.STAR	0.10 *4*	F	0.00 *4*		
C10H13S1	S-Ph-4-C(CH3)3						S.STAR	1.50 *3*				
C10H14F1N1P1	PN(Et)2(C6H4-F) -M		★ 0.22 *105*				★ES-CH	7.20 *2*				
	PN(Et)2C6H4F-P			★ 0.26 *111*			★ES-CH	7.20 *2*				
C10H14F1Sn1	Sn(Et)2C6H4F -M						★S.ZE.RS	-0.09 *86*	★S.ZE.RS	0.01 *333*	★ES-CH	8.70 *2*
C10H14N1	C6H4NHC4H9-M				4.86		★ES-CH	3.00 *2*	★S.PHOSP	-0.67 *85*		
	C6H4NHC4H9-P				4.86		★ES-CH	3.00 *2*	★S.PHOSP	-0.67 *37*		
C10H14O1P1	CH2P(O)(C3H7)(C6H5)						★S.STAR	0.62 *45*	★ES-CH	5.20 *2*		
	P(O)(C4H9)(C6H5)						★S.INDUC	0.27 *45*	★ES-CH	7.20 *2*	★S-L	0.27 *14*
	P(O)(I-C4H9)(C6H5)						★S.INDUC	0.27 *45*	★ES-CH	8.20 *2*	★S-L	0.26 *14*
C10H15	1-adamantyl	3.37	★ -0.12 *688*	★ -0.13 *688*			★S.INDUC	-0.09 *810*	S.PARA+	-0.25 *895*	B5-STM	3.49 *1*
							S.STAR	-0.26 *947*	★S.ZE.RS	-0.15 *47*	F	-0.07 *4*
							★S.STAR	-0.16 *881*	ES-V	1.33 *923*	R	-0.06 *4*
							★S.PARA-	-0.14 *964*	L-STM	6.17 *1*	R+	-0.31 *4*
							★S.PARA+	-0.38 *440*	B1-STM	3.16 *1*	R-	-0.07 *4*
	2-adamantyl	3.37 *57*	★ -0.12 *689*	-0.24 *465*	4.06	★ -2.76 *391*	S.INDUC	-0.09 *810*	S.PARA+	-0.38 *440*	ES-V	1.33 *7*
				★ -0.13 *689*			S.STAR	-0.26 *948*	S.PARA+	-0.25 *465*	F	-0.12 *66*
							★S.STAR	-0.16 *124*	S.ZE.RS	-0.15 *47*	R	-0.02 *66*
							S.PARA-	-0.14 *965*	★ES-CH	5.00 *2*		
C10H15Ge1	CH2Ge(H)(Et)(CH2Ph)						★S.PARA+	-0.91 *51*	★S.ZE.RS	-0.24 *51*		
C10H15Mn1O7P1	Mn(CO)4P(OC2H5)3						★S.INDUC	-0.32 *33*	★S.ZE.RS	-0.23 *33*	★ES-CH	13.50 *2*
C10H15N1P1	N=P(Et)2C6H5						★S.PARA-	-0.68 *910*	★S.PARA+	-1.55 *896*	★ES-CH	5.20 *2*
C10H15O1Si1	Si(Me)(Ph)O-I-Pr						★S.INDUC	-0.31 *25*	★S.ZE.RS	0.09 *25*		
	Si(Me)(Ph)OPr						★S.INDUC	-0.25 *25*	★S.ZE.RS	0.10 *25*		
C10H15S1	S-1-adamantyl						L-STM	6.16 *1*	B1-STM	1.70 *1*	B5-STM	6.63 *1*
C10H15Si1	CH2Si(CH3)2CH2C6H5						★S.STAR	-0.13 *381*	★ES-CH	5.20 *2*		
	CH2Si(H)(Et)(CH2Ph)						★S.PARA+	-0.84 *51*	★S.ZE.RS	-0.22 *51*		
C10H17	bornyl (exo)				4.33		★S.STAR	-0.17 *124*	★S.STAR	-0.15 *124*	★ES-CH	4.00 *2*
	decalinyl -C					★ -6.69 *391*						
	decalinyl -T					★ -8.13 *391*						
C10H17N1	CH2CCCH2N+(Me)(CH2)5						S.STAR	1.06 *262*				
C10H17O1	O-decahydnaphthyl						S.STAR	1.68 *3*				
C10H17Si2	(SiMe2)2C6H5						S.PARA+	0.01 *175*				
C10H17Sn1	CH2Sn(CH2CH=CH2)3						S.INDUC	0.06 *16*				
C10H19	4-cyclohexylbutyl		★ -0.19 *3*		4.53		★S.INDUC	-0.06 *32*	★S.STAR	-0.37 *3*	★ES-CH	3.00 *2*
	C6H11 -t-Bu -C						★S.STAR	-0.14 *124*				
	C6H11 -t-Bu -T				4.53		★S.STAR	-0.17 *124*	★ES-CH	4.00 *2*		
	cyclodecyl				4.53		★S.STAR	-0.13 *124*	★ES-CH	3.00 *2*		
	cyclohexyl(4-t-Bu) -C					★ -3.03 *391*						
	cycohexyl(4-t-Bu) -T					★ -1.85 *391*						
C10H19N1	CH2CCCH2N(Et)3+						S.STAR	1.07 *262*				
	N+(CH3)2CH2CH2CCC4H9					★ -3.84 *201*						
C10H19N1O1	N(t-Bu)C=C(t-Bu)O*						S.PHOSP	0.49 *1010*				
C10H19N2O4	C(nonyl)(NO2)2						★S.STAR	3.34 *157*				
C10H21	C(Et)(I-Pr)(t-Bu)					★ -7.74 *62*	ES-CH	10.00 *2*				
	C(I-Pr)3					★ -7.85 *62*	ES-CH	10.00 *2*				
	decyl						L-STM	12.33 *1*	B1-STM	1.52 *1*	B5-STM	8.83 *1*
C10H22O1P1	P(O)(C5H11)2						★S.INDUC	0.75 *654*				
C10H22O2P1S2	CH2CH2SSP(O-I-Bu)2						★S.INDUC	0.11 *137*	★S.STAR	0.24 *137*	★ES-CH	3.20 *2*

Empirical	Structural	π	σ_m	σ_p	MR	E_s	additional σ values
C10H23N1	(CH2)10NH3+						S.INDUC 0.04 310
C10H23O2S1Si1	Si(O-T-Am)2(SH)						★S.STAR 1.47 188
C10H23Si1	CH2Si(C4H9)2CH3						★S.STAR -0.13 644 S.STAR 0.39 842 ★ES-CH 5.20 2
	CH2Si(Pr)3						S.PARA+ -0.58 285
C10H25N3O1P1	N=P(OC2H5)(N(C2H5)2)2			★ -0.64 345			★ES-CH 5.20 2
C10H27Si3	C(SiMe3)3			★ -0.27 401	7.62		S.INDUC 0.13 385 S.PARA+ -0.51 175 B1-STM 4.30 1 ★S.PARA+ -0.79 299 S.ZE.RS -0.25 385 B5-STM 4.71 1 ★S.PARA+ -0.62 285 ES-CH 13.60 2 ★S.PARA+ -0.52 401 L-STM 5.20 1
C11H6Cl1	CH2CCCCC6H4Cl-P						★S.STAR 1.07 253 ★ES-CH 3.00 2
C11H7	CH2CCCCC6H5						★S.STAR 1.01 253 ★ES-CH 3.00 2
C11H7F2@	2(11-F2-1,6-CH2-annulene)						★S.PARA+ -0.41 639
	3(11-F2-1,6-CH2-annulene)						★S.PARA+ -0.25 639
C11H7N2	2-perimidinyl						★S.INDUC 0.24 15 ★S.ZE.RS 0.10 15
C11H8N1	2-(5-phenyl)pyridyl	2.69					★ES-CH 3.00 2
C11H8N2	2-perimidin-cation						★S.INDUC 0.31 15 ★S.BA.RS 0.25 15
C11H9	1-naphthylmethyl		★ -0.01 3		4.62		★S.INDUC 0.08 6 ★ES-CH 4.00 2 ★S.STAR 0.44 3 ★S-L 0.08 14
	2-(1,6-CH2-(10)annulene):						★S.PARA+ -0.80 639
C11H11N4O1	2(4-NMe2,6-OPh-triazinyl)						★S.PARA- 0.62 346 ★ES-CH 3.00 2
C11H13N1O1	CH2CH2N(COC6H5)CH2CH2*				5.17		★S.STAR 0.86 268
C11H14	4,5,7triMe-9-phenanthryl:						S.PARA+ -0.61 489
C11H14N1O1Si1	Si(Me)(Ph)O(CH2)3CN						★S.INDUC -0.18 25 ★S.ZE.RS 0.11 25
C11H14N1O3	CHOHCH(CH2OH)NHCOCH2C6H5	-0.33 54			5.55		★ES-CH 5.00 2
C11H14N3O1	5-BuO-6-Me-bnztriazolyl			★ 0.31 15			
C11H14N3O2	C=O(NHNHCH2CH2CONHCH2Ph)	0.22 42					★ES-CH 4.00 2
C11H15	C(C2H5)2C6H5						★ES-CH 8.00 2 B1-STM 3.10 1 L-STM 5.28 1 B5-STM 6.02 1
C11H16O1P1	CH2P(O)(C4H9)(C6H5)						★S.STAR 0.60 45 ★ES-CH 5.20 2
	CH2P(O)(I-C4H9)(C6H5)						★S.STAR 0.59 45 ★ES-CH 5.20 2
C11H17	CH2-adamantyl					-3.31 576	S.STAR -0.15 576
C11H17O1Si1	Si(Me)(Ph)O-I-Bu						★S.INDUC -0.25 25 ★S.ZE.RS 0.10 25
	Si(Me)(Ph)O-SeC-Bu						★S.INDUC -0.31 25 ★S.ZE.RS 0.09 25
	Si(Me)(Ph)OBu						★S.INDUC -0.25 25 ★S.ZE.RS 0.10 25
C11H19	(3-2,4,4-triMe-cyhexen)Et				4.95		★S.INDUC 0.00 34 ★S.BA.RS -0.15 34 ★ES-CH 3.00 2
	(CH2CH2CH=CMe)2CH3				3.98		★S.INDUC 0.01 34 ★S.BA.RS -0.14 34 ★ES-CH 3.00 2
C11H19N2	1(3,5-di-t-Bu-pyrazolyl)						S.INDUC 0.16 91 S.ZE.RS -0.03 91
C11H21	cycloundecyl				4.99		★S.STAR -0.13 124 ★ES-CH 3.00 2
C11H21N1	N+(CH3)2(CH2)3CCC4H9					★ -3.84 201	
	N+(Et)2CH2CCC4H9					★ -4.64 201	
	N+(Et)2CH2CH2CCC3H7					★ -5.44 201	
C11H23	(CH2)10CH3						★ES-CH 3.00 2 ★ES-V 0.68 7
	C(Me)(t-Bu)CH2C(CH3)3				4.75	★ -5.24 50	ES-CH 7.00 2
	CH(neopentyl)2				4.75	★ -4.42 50	ES-CH 5.00 2 -4.18 62 ★ES-V 2.03 7
C11H23O1	OC11H23						L-STM 14.32 1 B1-STM 1.35 1 B5-STM 10.07 1
C11H24N1P1Pt1	Pt+(CN)(t-Bu)P(C2H5)3 -T						★S.INDUC 0.00 33 ★S.ZE.RS -0.21 33
C11H25N1	N+(Et)2C7H15					★ -5.04 201	
C11H25Si1	(CH2)2Si(C4H9)2CH3						S.STAR -0.29 448 ★S.STAR -0.16 644 ★ES-CH 3.20 2
C11H33Si6	Si(Me)2-cy-Si5(Me)9						S.PARA+ 0.00 175
	cy-Si6(Me)11						S.PARA+ -0.05 175
C12H8	1-chrysenyl:						S.PARA+ -0.34 490
	1-tetrahelicene:						S.PARA+ -0.37 174
	1-triphenylenyl						★S.META- -0.46 101 ★S.META+ -0.32 542 ★S.META- -0.44 101 ★ES-CH 2.00 2
	2-chrysenyl:						S.PARA+ -0.26 490
	2-tetrahelicene:						S.PARA+ -0.35 174
	3-chrysenyl:						S.PARA+ -0.28 490
	3-tetrahelicene:						S.PARA+ -0.30 174
	4-chrysenyl:						S.PARA+ -0.33 490
	4-tetrahelicene:						S.PARA+ -0.38 174
C12H8As1	1-dibenzarsenyl		★ 0.19 313 0.21 177	★ 0.13 313 0.16 177			S.INDUC 0.23 177 ★S.ZE.RS -0.17 313 F 0.23 4 ★S.INDUC 0.26 313 S.ZE.RS -0.10 177 R -0.10 4
C12H8As1O1	1-dibenzoarsoxyl		★ 0.17 313 0.19 177	★ 0.09 313 0.14 177			S.INDUC 0.19 177 ★S.ZE.RS -0.19 313 F 0.22 4 ★S.INDUC 0.25 313 S.ZE.RS -0.08 177 R -0.13 4
C12H8Cl1N4	N=NC6H3,2-Cl,5-N=NC6H5						★S.PARA- 0.83 83 ★ES-CH 3.00 2
	N=NC6H3,O-Cl,P-N=NC6H5						★S.PARA- 0.73 83 ★ES-CH 3.00 2
C12H8Cl2O1P1	P(O)(C6H4Cl-P)2						★S.INDUC 0.30 45 ★ES-CH 8.20 2
C12H8F2N1	N(C6H4F -P)2						★S.INDUC 0.36 24 ★S.ZE.RS -0.29 24 ★ES-CH 7.00 2
C12H8F2O1P1	P(O)(C6H4-F)2 -M		★ 0.54 105		5.91		★ES-CH 8.20 2

Empirical	Structural	π	σ_m	σ_p	MR	E_s	additional σ values				
C12H8F2O1P1	P(O)(C6H4-F)2 -P			0.57 81 ★ 0.64 111	5.91		S.INDUC 0.45 81 S.ZE.RS 0.12 81	★ES-CH 8.20 2			
C12H8F2P1	P(C6H4-F)2 -M		★ 0.31 105		6.03		S.INDUC 0.23 452	S.ZE.RS 0.00 452	★ES-CH 7.20 2		
	P(C6H4F)2-P			0.25 81 ★ 0.34 111	6.03		S.INDUC 0.23 452 S.INDUC 0.26 81	S.ZE.RS -0.01 81 S.ZE.RS 0.00 452	★ES-CH 7.20 2		
C12H8N1	9-carbazolyl			★ 0.39 226 0.43 275			★S.INDUC 0.41 91 S.INDUC 0.48 16	★S.STAR 0.25 826 ★S.ZE.RS -0.12 91	★ES-CH 5.00 2		
C12H8N1O2	C6H4C6H4NO2-P				5.60		★S.PARA- 0.12 109	★ES-CH 3.00 2			
C12H8N1S1	1-phenothiazinyl						S.INDUC 0.22 573				
C12H8N2O5P1	P(O)(C6H4NO2-M)2						★S.INDUC 0.38 45	★ES-CH 8.20 2			
C12H9	C6H4(C6H5) -P	4.09 219		★ 0.02 82	4.97		★S.STAR 0.57 120 S.META+ -0.02 362 S.PARA- 0.05 109 ★S.PARA+ -0.28 362	S.ORTH+ -0.26 362 ★ES-CH 3.00 2 ★ES-V 0.60 7 L-STM 10.50 1	B1-STM 1.71 1 B5-STM 3.11 1		
	C6H4-2-Ph					★ -4.13 35	★S.PARA+ -0.22 362	★S.ORTH+ -0.18 362			
	C6H4-C6H5 -M						S.PARA+ -0.20 362				
	CH2CCCCC6H4CH3-P						★S.STAR 0.91 253	★ES-CH 3.00 2			
C12H9As1N1	5-10-di-H-phenarsazino				6.45		★S.PARA- 0.06 466 ★ES-CH 5.30 2	F 0.18 4 R -0.09 4			
	zinyl		★ 0.14 177	★ 0.09 177			★S.INDUC 0.15 177	★S.ZE.RS -0.07 177			
C12H9As1N1O1	phenarsazino-10-oxide				6.85		★S.PARA- -0.13 466	★ES-CH 6.30 2			
C12H9F1P1	P(C6H5)(C6H4-F) -M		★ 0.25 105		6.04		★ES-CH 7.20 2				
C12H9F1P1S1	P(S)C6H5(C6H4-F) -P			★ 0.40 111	6.73		★ES-CH 6.40 2				
	PSC6H5(C6H4-F) -M		★ 0.34 105		6.73		★ES-CH 6.40 2				
C12H9N2	1-Me-perimidinyl						★S.ZE.RS 0.04 15				
	C6H4N=NC6H5 -P				5.56		★S.PARA- 0.73 493	★S.PARA- 1.09 749	★ES-CH 3.00 2		
C12H9O1	C6H4OC6H5-M				5.25		★S.STAR 0.76 120	★ES-CH 3.00 2			
	C6H4OC6H5-P				5.25		★S.STAR 0.35 120	★ES-CH 3.00 2			
	CH2CCCCC6H4OCH3-P						★S.STAR 0.89 253	★ES-CH 3.00 2			
C12H10	12-1,2-benzanthracenyl						★S.META+ -1.40 101 ★S.META+ -1.11 101	★ES-CH 2.00 2 ★ES-CH 3.00 2			
	5-chrysenyl:						S.PARA+ -0.39 490				
C12H10Al1	Al(C6H5)2						★S.ZE.RS 0.11 86	★ES-CH 7.20 2			
C12H10As1	As(C6H5)2		★ 0.03 327 0.07 10	★ 0.09 327 0.13 10	6.30		★S.INDUC -0.01 327 S.INDUC 0.11 16 ★S.PARA- 0.29 466 S.ZE.RS -0.07 22	★S.ZE.RS -0.07 86 S.ZE.RS -0.01 98 ★ES-CH 7.30 2 F 0.04 5	R- 0.05 5 R- 0.25 4 S.ZTWST -0.05 22 GR.ELCT 0.03 13		
C12H10As1N1O1	10-hydroxyphenarsazino						★S.PARA- 0.36 466	★ES-CH 5.00 2			
C12H10As1O1	As(O)(C6H5)2		★ 0.54 10	0.60 327 ★ 0.64 10	6.70		★S.INDUC 0.43 327 S.RES.+ 0.00 10	★ES-CH 8.30 2 F 0.49 4	R 0.15 4		
C12H10B1	B(C6H5)2						★S.ZE.RS 0.22 86	★ES-CH 7.00 2			
C12H10B1Cl3P1	P(C6H5)2.BCl3		★ 0.67 855	★ 0.72 855			F 0.62 4	R 0.10 4			
	P(C6H5)2BCl3		★ 0.67 41	★ 0.72 41			★ES-CH 11.80 2	F 0.64 66	R 0.13 66		
C12H10Bi1	Bi(C6H5)2						★S.ZE.RS -0.12 98	★S.ZE.RS -0.10 86	★GR.ELCT -0.15 13		
C12H10Br1Ge1	Ge(Br)(C6H5)2			★ 0.35 25							
	GeBr(C6H5)2						★S.STAR 0.52 61	★ES-CH 8.60 2			
C12H10Cl1Ge1	Ge(C6H5)2Cl			★ 0.14 60			★S.PARA+ 0.10 60	★ES-CH 8.50 2			
C12H10Cl1P1	+P(C6H5)2Cl						S.ZE.RS 0.20 447				
C12H10Cl1Si1	Si(C6H5)2Cl			★ 0.23 60			★S.PARA+ 0.21 60 ★ES-CH 8.40 2	S.M.RAD 0.19 218 S.P.RAD 0.08 218			
C12H10Cl1Sn1	Sn(Cl)(Ph)2			★ 0.40 25							
	SnCl(C6H5)2						★S.STAR 0.45 61	★ES-CH 8.90 2			
C12H10Hg1N1O2S1	N(SO2Ph)-Hg-Ph						S.PARA+ -1.07 71	S.RES.+ -0.92 71			
C12H10I1	I(C6H5)2 (+)						★S.ZE.RS 0.28 47	★ES-CH 7.70 2			
C12H10N1	N(C6H5)2	3.50 219 3.61	-0.07 319 -0.03 10 ★ 0.00 579	-0.29 319 -0.28 3 -0.26 236 ★ -0.22 575	5.50		★S.INDUC 0.11 327 S.INDUC 0.16 12 S.INDUC 0.25 69 S.INDUC 0.36 16 S.ORTHO -0.89 38	S.ZE.RS -0.44 19 S.ZE.RS -0.33 12 S.ZE.RS -0.33 69 ES-CH 7.00 2 L-STM 5.77 1	B1-STM 1.35 1 B5-STM 5.95 1 F 0.12 5 R -0.34 5 GRP.DPL 0.70 8		
C12H10N1O1	N-acetyl-1-naphthylamino		★ 0.25 3		5.55		★S.INDUC 0.26 32 ★S.STAR 1.62 3	★ES-CH 7.00 2 ★S-L 0.28 7			
	N-acetyl-2-naphthylamino		★ 0.26 3		5.55		★S.INDUC 0.27 32 ★S.STAR 1.68 3	★ES-CH 7.00 2 ★S-L 0.30 7			
C12H10N1O2	OC6H4OC6H4NH2 -PP			★ -0.23 392			★S.PARA- 0.02 187	★ES-CH 4.00 2			
C12H10N1O2S1	N=S(O-Ph)2						S.INDUC -0.05 123				
	SO2N(Ph)2						S.INDUC 0.39 16				
C12H10N1S1	N=S(Ph)2						S.INDUC -0.32 123				
	S(Ph)=N-Ph						S.INDUC 0.36 16				
C12H10N2	1-Me-2-perimidin-cation						★S.INDUC 0.42 15	★S.BA.RS 0.15 15			
C12H10O1P1	P=O(C6H5)2	0.70	-0.09 38 0.38 207 ★ 0.38 319 0.40 6	0.05 38 0.42 81 0.44 679 0.50 560	5.93		S.INDUC 0.08 69 S.INDUC 0.11 16 S.INDUC 0.26 6 ★S.INDUC 0.26 45	★S.PARA- 0.68 407 S.PARA- 0.68 705 S.PARA- 0.84 373 S.PARA- 0.88 296	ES-CH 8.20 2 L-STM 5.40 1 B1-STM 2.68 1 B5-STM 6.19 1		

Empirical	Structural	π	σ_m	σ_p	MR	E_s	additional σ values						
C12H10O1P1	P=O(C6H5)2		0.42 704	0.53 143			S.INDUC	0.27 38	S.PARA-	0.99 146	F	0.32 5	
			0.43 10	0.53 207			★S.INDUC	0.27 327	★S.PARA+	0.52 143	R	0.21 5	
			0.43 105	★0.53 319			S.INDUC	0.30 81	S.ORTHO	0.41 38	R+	0.20 4	
			0.43 579	0.54 10			S.STAR	1.68 10	S.ZE.RS	0.07 22	R-	0.36 4	
			0.49 373	0.55 704			S.STAR	1.71 3	S.ZE.RS	0.12 81	S.ZTWST	0.07 22	
			0.50 41	0.58 575			S.META-	0.42 407	S.ZE.RS	0.14 69	S-CNTIG	0.58 139	
				0.60 6			S.META-	0.43 705	S.ZE.RS	0.15 98	GRP.DPL	-4.66 9	
				0.61 41			S.PARA-	0.58 258	S.ZE.RS	0.34 6	GRP.DPL	-4.49 8	
				0.62 111			S.PARA-	0.68 336	S.RES.+	0.00 10	GR.ELCT	0.11 13	
C12H10O3P1	P(O)(OC6H5)2						★S.INDUC	0.36 45	★S-L	0.36 14			
							★ES-CH	6.20 2	GR.ELCT	0.05 13			
C12H10O4P1	OP(O)(OC6H5)2	2.46					S.INDUC	0.71 16	★ES-CH	5.20 2			
C12H10P1	P(C6H5)2		0.11 150	-0.01 679	6.05		S.INDUC	0.04 69	S.PARA-	0.39 146	F	0.10 5	
			★0.11 319	0.16 81			S.INDUC	0.05 150	S.PARA-	0.43 235	R	0.09 5	
			0.15 10	0.19 150			★S.INDUC	0.05 327	★S.PARA+	0.70 143	R+	0.60 4	
			0.17 579	0.19 560			S.INDUC	0.17 81	S.ORTHO	0.20 38	R-	0.16 4	
			0.23 105	★0.19 319			★S.INDUC	0.17 468	S.ZE.RS	-0.06 86	S.ZTWST	0.02 22	
			0.39 41	0.22 10			★S.STAR	1.06 3	S.ZE.RS	-0.01 81	GRP.DPL	-1.52 8	
				0.24 575			S.META-	0.12 407	S.ZE.RS	0.00 69	GRP.DPL	-1.39 9	
				0.31 111			S.PARA-	0.26 296	S.ZE.RS	0.01 98	GR.ELCT	-0.02 13	
				0.42 41			S.PARA-	0.26 336	S.ZE.RS	0.14 150			
				0.68 143			★S.PARA-	0.26 407	ES-CH	7.20 2			
C12H10P1S1	P=S(C6H5)2		0.29 207	0.05 38	6.74		S.INDUC	0.07 69	S.ORTHO	0.22 38	GR.ELCT	0.04 13	
			★0.29 319	0.28 679			S.INDUC	0.11 16	S.ZE.RS	0.06 47			
			0.32 38	0.47 207			★S.INDUC	0.15 610	S.ZE.RS	0.13 69			
			0.33 10	★0.47 319			S.INDUC	0.25 38	S.ZE.RS	0.14 98			
			0.34 579	0.48 10			S.INDUC	0.27 45	S.ZE.RS	0.22 6			
			0.38 6	0.49 560			S.INDUC	0.28 6	S.RES.+	0.00 10			
			0.47 105	0.50 6			★S.META-	0.38 407	ES-CH	8.40 2			
			0.50 41	0.54 575			S.PARA-	0.63 407	F	0.23 5			
				0.58 111			★S.PARA-	0.73 296	R	0.24 5			
				0.60 41			S.PARA-	0.89 146	R-	0.50 4			
C12H10P1Se1	P(Se)(Ph)2						★S.ZE.RS	0.14 98	★ES-CH	8.50 2			
C12H10S1	S+(Ph)2						S.CH-P+	1.23 75					
C12H10Sb1	Sb(C6H5)2						★S.ZE.RS	-0.07 86	★ES-CH	7.70 2			
							★S.ZE.RS	-0.01 98	GR.ELCT	-0.10 13			
C12H11Ge1	Ge(H)(Ph)2			★0.16 25									
	Ge-H-(Ph)2			★0.11 51			★S.PARA+	0.04 51					
C12H11N1O1P1	P(O)(C6H5)(C6H4NH2 -P)						★S.PARA-	1.18 187	★ES-CH	8.20 2			
C12H11O1Si1	Si(OH)(C6H5)2						★S.PARA+	-0.04 25					
C12H11Si1	SiH(Ph)2			★0.20 238									
C12H11Sn1	Sn(H)(Ph)2			★0.20 25									
C12H12N1	9(2,3,4,5-H-carbazolyl)						★S.INDUC	0.28 15	★S.STAR	1.74 15			
C12H13N2O3	CHOHCH(CH2OH)NHCOCH(Ph)CN	-0.38 54			6.00		★ES-CH	5.00 2					
C12H15N1O1	N+(CH3)2CH2CCCH(OH)Ph					★-3.04 303							
C12H16	3,4,5,6Me-9-phenanthryl:						S.PARA+	-0.63 489					
C12H16Cl1Ge1O1	Ge(Ph)(Cl)(CH2)4COCH3			★0.14 51			★S.PARA+	0.10 51					
C12H16N1O2	CH2C6H4NHCO2C4H9-P			★-0.05 405	5.95		★ES-CH	4.00 2					
C12H17Hg1	CH2CHMeHgCHMeCH2C6H5						★S.PARA+	-0.25 277					
C12H17N1	N-Me-4-Ph-piperidinium					★-4.24 303							
C12H17O2	C(Me)2OOC(Me)2Ph			★-0.14 660									
C12H18F1N1P1	PN(Pr)2(C6H4-F) -M		★0.20 105	★0.24 111			★ES-CH	7.20 2	F	0.20 4	R	0.04 4
C12H19F1P1Pt1	Pt(C6H4F-M)P(C2H5)3 -C						★S.INDUC	-0.40 33	★S.ZE.RS	-0.26 33			
	Pt(C6H4F-M)P(C2H5)3 -T						★S.INDUC	-0.35 33	★S.ZE.RS	-0.26 33			
	Pt(C6H4F-P)P(C2H5)3 -C						★S.INDUC	-0.42 33	★S.ZE.RS	-0.27 33			
	Pt(C6H4F-P)P(C2H5)3 -T						★S.INDUC	-0.38 33	★S.ZE.RS	-0.25 33			
C12H19Si1	Si(C6H5)(Pr)2						★S.STAR	1.19 188					
C12H20P1Pt1	Pt(C6H5)P(C2H5)3 -C						★S.INDUC	-0.42 33	★S.ZE.RS	-0.27 33			
	Pt(C6H5)P(C2H5)3 -T						★S.INDUC	-0.40 33	★S.ZE.RS	-0.25 33			
C12H21O11	maltosyl	-3.27 54					★ES-CH	4.00 2					
C12H23	cyclododecyl				5.46		★S.STAR	-0.13 124	★ES-CH	3.00 2			
C12H23Si3	(SiMe2)3C6H5						S.PARA+	0.00 175					
C12H25	C12H25				5.68		★S.STAR	-0.16 544	B1-STM	1.52 1	★S.PHOSP	-1.24 37
							★ES-CH	3.00 2	B5-STM	10.27 1			
							L-STM	14.38 1	★S.PHOSP	-1.37 85			
C12H25O1	O(CH2)11CH3						★ES-CH	3.00 2	L-STM	15.13 1	B5-STM	10.52 1
							★ES-V	0.65 7	B1-STM	1.35 1			
C12H25O2S1	SO2-decyl						L-STM	15.19 1	B1-STM	2.03 1	B5-STM	10.75 1
C12H25S1	S-decyl						L-STM	14.69 1	B1-STM	1.70 1	B5-STM	12.17 1
C12H26N1	NH-decyl						L-STM	15.09 1	B1-STM	1.35 1	B5-STM	10.68 1
C12H27Ge1	Ge(C4H9)3						★S.ZE.RS	0.04 22	★ES-CH	7.30 2	★S.ZTWST	0.03 22
C12H27Ge1O3	Ge(O-t-Bu)3						S.INDUC	-0.01 92	S.ZE.RS	0.10 92			
C12H27N1	N+(Bu)3					★-5.74 215							
C12H27N1O3P1	N=P(OC4H9)3			★-0.55 345			★ES-CH	5.20 2					
C12H27N1P1	N=P(C4H9)4			★-0.69 345			★ES-CH	6.20 2					
C12H27O3Si1	Si(O-S-Bu)3						★S.STAR	1.44 188					

Empirical	Structural	π	σ_m	σ_p	MR	E_s	additional σ values					
C12H27O3Si1	Si(O-t-Bu)3						S.INDUC -0.43 _1015_	★ S.ZE.RS 0.08 _25_	S.ZE.RS 0.16 _25_			
	Si(O-t-Bu)3						S.INDUC -0.43 _92_	S.STAR 0.64 _188_	★ S.RES.+ 0.06 _140_			
							★ S.INDUC -0.30 _140_	★ S.ZE.RS 0.06 _140_				
							S.STAR -0.67 _140_	S.ZE.RS 0.08 _92_				
C12H27P1	P(C4H9)3+						★ S.STAR 2.37 _87_	★ ES-CH 7.20 _2_	S.CH-P+ 0.95 _312_			
							★ S.STAR 3.61 _3_	★ S-L 0.60 _14_	S.CH-P+ 1.04 _75_			
C12H27Si1	(CH2)3Si(C4H9)2CH3						S.STAR -0.15 _448_	★ S.STAR -0.07 _644_	★ ES-CH 3.00 _2_			
	Si(C4H9)3			★ 0.05 _60_			★ S.PARA+ -0.07 _60_	★ ES-CH 7.20 _2_				
	Si(I-Bu)3						S.INDUC -0.36 _92_	S.ZE.RS 0.05 _92_				
	Si(t-Bu)3			★ 0.05 _238_			★ S.ZE.RS 0.05 _25_					
C12H27Sn1	Sn(C4H9)3						★ S.INDUC 0.02 _198_	★ ES-CH 7.70 _2_	B1-STM 4.16 _1_			
							★ S.ZE.RS 0.00 _198_	L-STM 7.83 _1_	B5-STM 7.14 _1_			
	Sn(t-Bu)3						★ S.PARA+ -0.18 _25_					
C12H30Br1Ni1P2	Ni(PEt3)2Br						★ S.INDUC -0.20 _70_	★ S.ZE.RS -0.32 _70_	★ ES-CH 11.00 _2_			
C12H30Br1P2Pd1	Pd(PEt3)2Br						★ S.INDUC -0.11 _70_	★ S.ZE.RS -0.27 _70_	★ ES-CH 11.40 _2_			
C12H30Br1P2Pt1	Pt(PEt3)2Br						★ S.INDUC -0.24 _70_	★ S.ZE.RS -0.26 _70_				
C12H30Cl1Ni1P2	Ni(PEt3)2Cl						★ S.INDUC -0.23 _70_	★ S.ZE.RS -0.32 _70_	★ ES-CH 10.00 _2_			
C12H30Cl1P2Pd1	Pd(PEt3)2Cl						★ S.INDUC -0.14 _70_	★ S.ZE.RS -0.27 _70_	★ ES-CH 11.30 _2_			
C12H30Cl1P2Pt1	Pt(PEt3)2Cl						S.INDUC -0.27 _70_	★ S.ZE.RS -0.27 _553_				
							★ S.INDUC -0.21 _553_	S.ZE.RS -0.26 _70_				
C12H30Cl3P2Pt1	Pt(PEt3)2Cl3						★ S.INDUC -0.01 _553_	★ S.ZE.RS -0.18 _553_				
C12H30I1Ni1P2	Ni(PEt3)2I						★ S.INDUC -0.16 _70_	★ S.ZE.RS -0.32 _70_	★ ES-CH 11.40 _2_			
C12H30I1P2Pd1	Pd(PEt3)2I						★ S.INDUC -0.08 _70_	★ S.ZE.RS -0.27 _70_	★ ES-CH 11.80 _2_			
C12H30I1P2Pt1	Pt(PEt3)2I						★ S.INDUC -0.20 _70_	★ S.ZE.RS -0.26 _70_				
C12H30N4P1	N=P(N-Et2)3						S.INDUC -0.59 _123_					
C12H30P2Pt1	Pt+(P(C2H5)3)2 -T						★ S.INDUC 0.03 _33_	★ S.ZE.RS -0.20 _33_				
C13H7O2	2(5,6-Bnz-benzopyronyl)		★ 0.38 _15_	★ 0.38 _15_			★ S.INDUC 0.38 _15_	F 0.38 _4_				
							★ S.PARA- 0.59 _581_	R 0.00 _4_				
	2(7,8-Bnz-benzopyronyl)		★ 0.37 _15_	★ 0.38 _15_			★ S.INDUC 0.36 _15_	F 0.37 _4_	R 0.01 _4_			
C13H9	9-fluorenyl				5.16	★ -2.34 _230_	★ S.INDUC 0.22 _151_	★ S.STAR 0.50 _250_	★ ES-V 1.08 _7_			
						-2.27 _62_	S.STAR 0.30 _280_	ES-CH 5.00 _2_				
C13H9N2	2-N-phenylbenzimidazolyl		0.07 _3_	0.12 _3_	5.91		★ S.INDUC 0.14 _209_	★ ES-CH 7.00 _2_	R 0.04 _5_			
			★ 0.17 _209_	★ 0.21 _209_			★ S.ZE.RS 0.08 _209_	F 0.17 _5_				
C13H9O1	9-hydroxy-9-fluorenyl				5.32	★ -2.14 _151_	★ S.INDUC 0.47 _151_	ES-CH 6.00 _2_	★ ES-V 0.98 _7_			
	9-xanthene				5.46	★ -2.53 _151_	★ S.INDUC 0.31 _151_	ES-CH 5.00 _2_				
	Ph-COPh -P						★ S.PARA- 0.77 _493_					
C13H9O2	C6H4CO2C6H5-P				5.67		★ ES-CH 3.00 _2_	★ S.PHOSP -0.02 _37_				
C13H9S1	9-thioxanthene				6.08	★ -2.99 _151_	★ S.INDUC 0.25 _151_	ES-CH 5.00 _2_				
C13H10Cl1	C(CH3)2CCCCC6H4Cl-P						★ S.STAR 1.03 _253_	★ ES-CH 5.00 _2_				
C13H10Cl2O1P1	CH2P(O)(C6H4Cl-P)2						★ S.STAR 0.67 _45_	★ ES-CH 5.20 _2_				
C13H10F2P1	+P(C6H4F-para)2CH2-			★ 0.41 _81_			★ S.INDUC 0.29 _81_	★ S.ZE.RS 0.12 _81_	★ ES-CH 8.20 _2_			
C13H10N1	3-Me-9-carbazolyl						★ S.INDUC 0.23 _15_	★ S.STAR 1.45 _15_				
	3-Me-carbazolyl						S.INDUC 0.22 _573_					
C13H10N1O1	C6H4(CONHC6H5) -M			★ 0.04 _82_	5.97		★ ES-CH 3.00 _2_					
	C6H4(CONHC6H5) -P			★ -0.02 _82_	5.97		★ ES-CH 3.00 _2_					
	CON(C6H5)2			★ 0.35 _82_	5.85		★ ES-CH 5.00 _2_	B1-STM 1.66 _1_				
							L-STM 8.19 _1_	B5-STM 6.74 _1_				
C13H10N1O1S1	SC6H4(CONHC6H5) -P						★ S.PARA+ 0.06 _202_					
C13H10N1O2	OC6H4(CONHC6H5) -P						★ S.PARA+ -0.16 _202_					
C13H10N2O5P1	CH2P(O)(C6H4NO2-M)2						★ S.STAR 0.84 _45_	★ ES-CH 5.20 _2_				
C13H10N4S1	5-S-(2,3-diPh-tetrazolium						★ ES-CH 4.20 _2_	★ S-L 0.57 _14_				
C13H11	C(CH3)2CCCCC6H5						★ S.STAR 0.93 _253_	★ ES-CH 5.00 _2_				
	CH(C6H5)2		★ -0.03 _3_	★ -0.05 _147_	5.43	★ -3.00 _50_	S.INDUC 0.19 _151_	S.PARA+ -0.18 _938_	F 0.01 _4_			
				-0.04 _82_		-2.70 _151_	S.STAR 0.31 _280_	S.ZE.RS -0.11 _19_	R -0.06 _4_			
						-2.67 _230_	★ S.STAR 0.41 _52_	ES-CH 7.00 _2_	R+ -0.18 _4_			
						-2.62 _35_	S.STAR 0.41 _932_	★ ES-V 1.25 _7_	★ S.PHOSP -0.73 _59_			
							S.STAR 0.46 _10_	L-STM 5.15 _229_	GR.ELCT 0.18 _13_			
							S.META+ -0.04 _984_	B1-STM 2.01 _229_				
							★ S.PARA+ -0.19 _665_	B5-STM 6.02 _229_				
C13H11As1N1	1-dibnz-N-Me-arsazinyl		0.10 _313_	★ 0.07 _177_			★ S.INDUC 0.11 _177_	F 0.16 _4_				
			★ 0.12 _177_				★ S.ZE.RS -0.05 _177_	R -0.09 _4_				
C13H11F2P1	+PCH3(C6H4F-para)2			★ 0.87 _81_			★ S.INDUC 0.65 _81_	★ S.ZE.RS 0.22 _81_	★ ES-CH 8.20 _2_			
C13H11O1	OCH(Ph)2						S.INDUC 0.57 _16_	B1-STM 1.35 _229_				
							L-STM 8.20 _229_	B5-STM 6.82 _229_				
C13H11O2	CH(OC6H5)2				5.79		★ S.STAR 1.90 _115_	★ ES-CH 5.00 _2_				
C13H11S1	CH(C6H5)(SC6H5)				6.23		★ S.STAR 1.00 _115_	★ ES-CH 6.20 _2_				
	SCH(Ph)2						L-STM 8.50 _229_	B1-STM 1.70 _229_	B5-STM 6.96 _229_			
C13H11S2	CH(SC6H5)2				7.02	★ -4.53 _115_	★ S.STAR 1.54 _115_	ES-CH 5.40 _2_				
							★ S.STAR 1.56 _3_	★ GRP.DPL -1.72 _9_				
C13H12N1	NH(Ph)2						L-STM 8.24 _229_	B1-STM 1.35 _229_	B5-STM 6.83 _229_			
C13H12N2	1,3-diMe-2-perimidin-cation						★ S.INDUC 0.68 _15_	★ S.BA.RS 0.09 _15_				
C13H12O1P1	CH2P(O)(C6H5)2		★ 0.14 _6_	★ 0.01 _6_	6.39		S.INDUC 0.20 _6_	S.ZE.RS -0.19 _6_	R -0.20 _4_			

Empirical	Structural	π	σm	σp	MR	Es	additional σ values		
C13H12O1P1	CH2P(O)(C6H5)2						★ S.STAR 0.58 45 ★ S.STAR 0.60 435	★ ES-CH 5.20 2 F 0.21 4	★ S-L 0.20 14
	P(C6H5)(C6H4OCH3-P)		★ 0.18 146				★ S.PARA- 0.24 146	★ S.PARA- 0.86 296	★ ES-CH 7.20 2
	P(O)(C6H5)C6H4CH3-P		★ 0.13 38	★ 0.30 38			★ S.INDUC 0.27 38 ★ S.ORTHO 0.61 38	★ S.ZE.RS 0.03 38 ★ ES-CH 8.20 2	F 0.09 4 R 0.21 4
C13H12O1P1S1	P(S)(C6H5)(C6H4OCH3-P)		★ 0.65 146				★ S.PARA- 0.64 296	★ S.PARA- 0.89 146	★ ES-CH 8.40 2
C13H12O3P1	CH2P(O)(OC6H5)2						★ S.STAR 0.80 45	★ ES-CH 5.20 2	
C13H12P1	+P(C6H5)2CH2-			★ 0.54 81			★ S.INDUC 0.36 81	★ S.ZE.RS 0.18 81	★ ES-CH 8.20 2
	P(C6H5)(C6H4CH3-P)						★ S.ORTHO 0.23 38	★ ES-CH 7.20 2	
C13H12P1S1	CH2P(S)(C6H5)2						★ S.STAR 0.60 45	★ ES-CH 5.40 2	
	P(S)(C6H5)C6H4CH3-P		★ 0.09 38	★ 0.30 38			★ S.INDUC 0.25 38 ★ S.ORTHO 0.47 38	★ S.ZE.RS 0.05 38 ★ ES-CH 8.40 2	F 0.03 4 R 0.27 4
C13H13P1	+PCH3(C6H5)2		0.72 146 1.01 146 ★ 1.13 481	★ 0.83 81 1.01 975 ★ 1.18 481			★ S.INDUC 0.60 81 S.INDUC 0.71 481 S.PARA- 0.99 146 ★ S.PARA- 1.05 235 ★ S.PARA- 1.09 146	★ S.PARA- 1.28 866 S.ORTHO 0.01 481 S.ZE.RS 0.23 81 S.ZE.RS 0.47 481 ★ ES-CH 8.20 2	F 1.04 4 R 0.14 4 S.CH-P+ 1.12 75
C13H13Si1	Si(CH3)(C6H5)2		★ 0.10 3	0.01 338 ★ 0.13 3 0.16 238			★ S.INDUC 0.07 73 ★ S.STAR 0.19 216 ★ S.STAR 0.39 377 ★ S.STAR 1.22 188	★ S.PARA+ -0.04 25 ★ S.ZE.RS 0.06 73 ★ ES-CH 8.20 2 F 0.11 4	R 0.02 4 R+ -0.15 4
C13H14N1	9-(2,3,4,5-H4-6-Me-carbazolyl)						★ S.INDUC 0.22 15	S.INDUC 0.22 573	★ S.STAR 1.36 15
C13H15N1O2	1,3-diMe-6,7-diMeO-I-Quin		★ 0.65 15	★ 0.66 15					
C13H25N1	N+(Et)2(CH2)3CCC4H9					★ -5.24 201			
C13H25O1	C(=O)-decyl						L-STM 15.14 1	B1-STM 1.63 1	B5-STM 10.70 1
C13H27	(CH2)12CH3						★ ES-CH 3.00 2	★ ES-V 0.68 7	
C13H29N1	CH2N+(Bu)3			0.23 36 ★ 0.63 29			★ ES-CH 5.00 2		
C13H29P1	CH2P(Bu)3+						★ S.STAR 0.83 87	★ S.STAR 1.34 435	★ ES-CH 5.20 2
C13H29Si1	CH2Si(C4H9)3						★ S.PARA+ -0.85 60	★ ES-CH 5.20 2	
C13H30N1Ni1P2	NI(CN)(PEt3)2						★ S.INDUC -0.20 70	★ S.ZE.RS -0.29 70	★ ES-CH 11.70 2
C13H30N1P2Pd1	Pd(CN)(PEt3)2						★ S.INDUC -0.18 70	★ S.ZE.RS -0.24 70	★ ES-CH 12.10 2
C13H30N1P2Pt1	Pt(CN)(PEt3)2						★ S.INDUC -0.27 70	★ S.ZE.RS -0.23 70	
C13H33Ni1P2	NI(CH3)(PEt3)2						★ S.INDUC -0.50 70	★ S.ZE.RS -0.28 70	★ ES-CH 10.70 2
C13H33P2Pd1	Pd(CH3)(PEt3)2						★ S.INDUC -0.46 70	★ S.ZE.RS -0.24 70	★ ES-CH 11.10 2
C13H33P2Pt1	Pt(CH3)(PEt3)2						★ S.INDUC -0.51 70	★ S.ZE.RS -0.25 70	
C14H9	2-anthryl:						S.INDUC 0.36 835		
	3-perylenyl:						★ S.PARA+ -1.27 101	★ S.PARA+ -0.82 101	★ ES-CH 1.00 2
	6-(1,2-benzopyrenyl):						★ S.PARA+ -1.51 101	★ S.PARA+ -0.84 101	★ ES-CH 1.00 2
	9-anthryl						S.INDUC 0.06 16		
C14H10	1-tripticenyl:						★ S.META+ -0.19 27	★ ES-CH 1.00 2	
C14H10F2N3O2	N(CONHC6H4F -P)2						★ S.INDUC 0.39 24	★ S.ZE.RS -0.04 24	★ ES-CH 7.00 2
C14H11	9-anthracyl-9,10-dihydro				5.76	★ -2.53 230	★ S.INDUC 0.23 151 ★ S.STAR 0.23 280	★ S.STAR 0.40 250 ES-CH 5.00 2	
	9-methyl-9-fluorenyl				5.63	★ -2.97 230 -2.93 62	★ S.STAR 0.18 280 ★ S.STAR 0.32 250	ES-CH 6.00 2 ★ ES-V 1.41 7	
C14H11O1	C(=O)CH(Ph)2						L-STM 8.28 1	B1-STM 1.59 1	B5-STM 4.50 1
C14H11O2	CO2CH(C6H5)2		★ 0.36 251	★ 0.56 82	6.04		★ ES-CH 4.00 2	F 0.28 5	R 0.28 5
C14H12N1	3-Et-9-carbazolyl						★ S.INDUC 0.23 15	★ S.STAR 1.43 15	
	9-Et-2-carbazolyl						★ S.INDUC 0.02 15	★ S.STAR 0.15 15	
C14H12N3O2	NHCO(C6H4-M)CONHC6H4NH2						★ S.PARA- -0.05 187	★ ES-CH 4.00 2	
C14H13	C(CH3)(C6H5)2				5.90	★ -4.97 230 -4.85 35	★ S.STAR 0.27 250 ★ S.STAR 0.36 759	ES-CH 8.00 2 ★ ES-V 2.34 7	
	C(CH3)2CCCCC6H5CH3-P						★ S.STAR 0.87 253	★ ES-CH 5.00 2	
C14H13N1	CH=CH(4-N-benzyl-pyridyl)						★ S.ORTH- 0.70 636	★ ES-CH 3.00 2	
C14H13O1	C(CH3)2CCCCC6H5OCH3-P						★ S.STAR 0.86 253	★ ES-CH 5.00 2	
C14H14N1	C(CH3)(C6H5)(C6H4NH2-P)						★ S.PARA- 0.07 187	★ ES-CH 8.00 2	
C14H14O1P1	CH2CH2P(O)(C6H5)2						★ ES-CH 3.20 2	★ S-L 0.09 14	
	P(O)(C6H4CH3-P)2		★ 0.17 38	★ 0.30 38			S.INDUC 0.24 45 ★ S.INDUC 0.27 38 ★ S.ORTHO 0.61 38	★ S.ZE.RS 0.03 38 ★ ES-CH 8.20 2 F 0.14 4	R 0.16 4
C14H14O2P1	P(C6H4OCH3-P)2		★ 0.14 146				★ S.PARA- 0.19 146	★ S.PARA- 0.63 296	★ ES-CH 7.20 2
C14H14O2P1S1	P(S)(C6H4OCH3-P)2		★ 0.65 146				★ S.PARA- 0.50 296	★ S.PARA- 0.89 146	★ ES-CH 8.40 2
C14H14O3P1	P(O)(C6H4OCH3-P)2		★ 0.47 146				★ S.INDUC 0.23 45 ★ S.PARA- 0.60 296	★ S.PARA- 0.64 146 ★ ES-CH 8.20 2	
C14H14P1	P(C6H4CH3-P)2						★ S.ORTHO 0.25 38	★ ES-CH 7.20 2	
C14H14P1S1	P(S)(C6H4CH3-P)2		★ 0.20 38	★ 0.23 38			★ S.INDUC 0.25 38 ★ S.ORTHO 0.21 38	★ S.ZE.RS -0.02 38 ★ ES-CH 8.40 2	F 0.20 4 R 0.03 4
C14H14P1S2	CH2CH2SSP(C6H5)2						★ S.INDUC 0.11 137	★ S.STAR 0.25 137	★ ES-CH 3.20 2
C14H15Ge1O1	Ge(Ph)2(CH2OCH3)			★ 0.11 51			★ S.PARA+ 0.04 51		
C14H15I1P1	P+(CH3)(C6H5)(C6H4-P-CH3)(I-)		★ 1.09 38	★ 1.11 38			★ S.INDUC 0.71 38 ★ S.ORTHO 0.22 38	★ S.ZE.RS 0.40 38 ★ ES-CH 8.20 2	F 1.02 4 R 0.09 4
C14H15N1	NH+(CH2Ph)2					★ -1.94 215			

Empirical	Structural	π	σm	σp	MR	Es	additional σ values					
C14H15N1O1P1	P(O)(C6H5)(C6H4NMe2-P)						★ S.PARA- 0.80 235					
C14H15N1P1	P(C6H5)(C6H4NMe2-P)						★ S.PARA- 0.87 235					
C14H15N1P1S1	P(S)(C6H5)(C6H4NMe2-P)						★ S.PARA- 0.69 235					
C14H15O1Si1	Si(Me)(Ph)OCH2Ph						★ S.INDUC -0.24 25	★ S.ZE.RS 0.10 25				
C14H15Si1	CH2Si(CH3)(C6H5)2						★ S.STAR 0.07 381					
	Si(C6H5)2(Et)						★ S.STAR 1.22 188					
C14H19N1O1	N+(Et)2CH2CCCH(OH)Ph					★ -4.64 303						
C14H19O9	tetracetyl-D-galactosidyl				7.19		★ S.PARA- 0.12 604	★ S.PARA- 0.14 604	★ ES-CH 4.00 2			
C14H20P1Pt1	Pt(CCC6H5)P(C2H5)3 -T						★ S.INDUC -0.37 33	★ S.ZE.RS -0.24 33				
C14H22N1O1P1Pt1	Pt+(P-CH3OC6H4NC)P(C2H5)3						★ S.INDUC 0.02 33	★ S.ZE.RS -0.21 33				
C14H23P1	+P(Bu)2(C6H5)						S.CH-P+ 1.10 75					
C14H29O1	OC14H29						L-STM 17.20 1	B1-STM 1.35 1	B5-STM 11.95 1			
C15H10As1Cl3Co1O3Sn1	As(Ph)2CO(co)3SnCl3						★ S.ZE.RS 0.09 98	★ ES-CH 13.30 2				
C15H10Cl3Co1O3P1Sn1	P(Ph)2CO(co)3SnCl3						★ S.ZE.RS 0.13 98	★ ES-CH 13.20 2				
C15H10Cl3Co1O3Sb1Sn1	Sb(Ph)2CO(co)3SnCl3						★ S.ZE.RS 0.08 98	★ ES-CH 13.70 2				
C15H10N3O2	2-(4,6-OPh2-S-triazinyl)						★ S.PARA- 0.80 346	★ ES-CH 3.00 2				
C15H10Ni1O3P1	P(Ph)2Ni(CO)3						★ S.ZE.RS 0.06 98	★ ES-CH 11.50 2				
C15H11	2,3-(C6H5)2-cycropenyl			★ -0.52 734			★ ES-CH 5.00 2					
C15H13	2,3-diphenylcyclopropyl			★ -0.24 734			★ ES-CH 5.00 2					
	9-ethyl-9-fluorenyl				6.09	★ -3.21 230	★ S.STAR 0.15 280	ES-CH 7.00 2				
						-3.19 62	★ S.STAR 0.26 250	★ ES-V 1.53 7				
C15H15	C(Et)(C6H5)2				6.36	-5.67 35	★ S.STAR 0.21 250	ES-CH 9.00 2				
						★ -5.58 230	★ S.STAR 0.37 280	★ ES-V 2.75 7				
	CH(Ph-2-CH3)2						S.STAR 0.40 3					
C15H16O1P1	CH2P(O)(C6H4CH3-P)2						★ S.STAR 0.53 45	★ ES-CH 5.20 2				
C15H16O3P1	CH2P(O)(C6H4OCH3-P)2						★ S.STAR 0.52 45	★ ES-CH 5.20 2				
C15H17Ge1	CH2Ge(H)(CH2Ph)2						★ S.PARA+ -1.03 51	★ S.ZE.RS -0.24 51				
							★ S.PARA+ -0.95 60	★ ES-CH 4.30 2				
C15H17I1O2P1	P+(CH3)(C6H4OCH3-P).I(-)		★ 0.86 146				★ S.PARA- 1.09 296	★ S.PARA- 1.20 146	★ ES-CH 8.20 2			
C15H17I1P1	P+(CH3)(C6H4-P-CH3)2 (I-)		★ 1.13 38	★ 1.18 38			★ S.INDUC 0.71 38	★ S.ZE.RS 0.47 38	F 1.04 4			
							★ S.ORTHO 0.31 38	★ ES-CH 8.20 2	R 0.14 4			
C15H17Si1	CH2Si(CH2C6H5)2H						★ S.PARA+ -0.66 60	★ ES-CH 4.20 2				
	Si(C6H5)2(Pr)						★ S.STAR 1.19 188					
C15H18N1P1	P+(Me)(Ph)(C6H4NMe2-P)						★ S.PARA- 1.17 235					
C15H29N1	N+(Bu)2CH2CCC4H9					★ -4.94 201						
	N+(Bu)2CH2CH2CCC3H7					★ -5.74 201						
C15H30Mn1O9P2	Mn(CO)3(P(OC2H5)3)2						★ S.INDUC -0.60 33	★ S.ZE.RS -0.23 33	★ ES-CH 15.70 2			
C15H31	(CH2)14CH3						★ ES-CH 3.00 2	★ ES-V 0.68 7				
C15H33N1	N+(Bu)2C7H15					★ -5.64 201						
C15H33O3Si1	Si(O-S-Am)3						★ S.STAR 1.41 188					
	Si(O-S-I-Am)3						★ S.STAR 1.41 188					
C16H8N2O4P1	P(CCC6H4NO2-P)2						★ S.PARA- 0.56 336	★ ES-CH 5.20 2				
C16H8N2O5P1	P(O)(CCC6H4NO2-P)2						★ S.PARA- 0.97 336	★ ES-CH 6.20 2				
C16H10	1-pentahelicene:						S.PARA+ -0.40 174					
	2-pentahelicene:						S.PARA+ -0.39 174					
	3-pentahelicene:						S.PARA+ -0.34 985					
	4-pentahelicene:						S.PARA+ -0.44 174					
	7-1,2,5,6-dibenzanthrene:						★ S.META+ -1.10 101	★ ES-CH 3.00 2				
C16H10As1Br1Mn1O4	As(Ph)2Mn(CO)4Br						★ S.ZE.RS 0.07 98	★ ES-CH 13.90 2				
C16H10As1Fe1O4	As(Ph)2Fe(CO)4						★ S.ZE.RS 0.07 98	★ ES-CH 12.60 2				
C16H10Br1Mn1O4P1	P(Ph)2Mn(CO)4Br						★ S.ZE.RS 0.10 98	★ ES-CH 13.80 2				
C16H10Br1Mn1O4Sb1	Sb(Ph)2Mn(CO)4Br						★ S.ZE.RS 0.06 98	★ ES-CH 14.30 2				
C16H10Cl1Mn1O4P1	P(Ph)2Mn(CO)4Cl						★ S.ZE.RS 0.10 98	★ ES-CH 13.70 2				
C16H10Fe1O4P1	P(Ph)2Fe(CO)4						★ S.ZE.RS 0.10 98	★ ES-CH 12.50 2				
C16H10Fe1O4Sb1	Sb(Ph)2Fe(CO)4						★ S.ZE.RS 0.06 98	★ ES-CH 13.00 2				
C16H10O1P1	P(O)(CCC6H5)2						★ S.PARA- 0.86 336	★ ES-CH 6.20 2				
C16H10P1	P(CCC6H5)2						★ S.PARA- 0.44 336	★ ES-CH 5.20 2				
C16H11	6-anthanthrene:						★ S.PARA+ -0.87 101	★ ES-CH 2.00 2				
C16H12	5-Bz(A)NAP(1,2)anthracen:						S.PARA+ -0.46 174					
C16H15	9-I-propyl-9-fluorenyl				6.56	-4.58 62	★ S.STAR 0.08 280	ES-CH 8.00 2				
						★ -4.54 230	★ S.STAR 0.20 250	★ ES-V 2.21 7				
	C(C6H5)2(CH2CH=CH2)				6.78	★ -5.57 151	★ S.INDUC 0.17 151	ES-CH 5.00 2				
	CH2C(CH=CH2)(C6H5)2						★ ES-CH 5.00 2	★ ES-V 2.74 7				
C16H15O2	CH2C(C6H5)2CO2CH3						★ ES-CH 5.00 2	★ S-L 0.03 14				
C16H17Ge1	Ge(CH2CH=CHCH3)(Ph)2			★ 0.11 51			★ S.PARA- -0.10 25	★ S.PARA+ 0.04 51				
C16H19I1P1	P+(C2H5)(C6H4CH3-P)2			★ 1.25 38			★ S.INDUC 0.71 38	★ S.ZE.RS 0.54 38				
C16H19P1	+P(Bu)(C6H5)2						S.CH-P+ 1.16 75					
C16H19Si1	CH2Si(CH3)(CH2C6H5)2						★ S.STAR 0.02 381	S.ZE.RS -0.24 51				
							S.PARA+ -0.97 51	★ ES-CH 5.20 2				

Empirical	Structural	π	σm	σp	MR	Es	additional σ values					
C16H20N2O1P1	P(O)(C6H4NMe2-P)2						★ S.PARA- 0.65 235					
C16H20N2P1	P(C6H4NMe2-P)2						★ S.PARA- 0.68 235					
C16H20N2P1S1	P(S)(C6H4NMe2-P)2						★ S.PARA- 0.57 235					
C16H27Mn1O4P1	Mn(CO)4P(N-Bu)3						★ S.INDUC -0.29 33	★ S.ZE.RS -0.24 33	★ ES-CH 13.50 2			
C17H10As1Mo1O5	As(Ph)2Mo(CO)5						★ S.ZE.RS 0.04 98	★ ES-CH 14.00 2				
C17H10As1O5W1	As(Ph)2W(CO)5						★ S.ZE.RS 0.05 98					
C17H10Cr1O5P1	P(Ph)2Cr(CO)5						★ S.ZE.RS 0.07 98	★ ES-CH 13.50 2				
C17H10Mo1O5P1	P(Ph)2Mo(CO)5						★ S.ZE.RS 0.07 98	★ ES-CH 13.90 2				
C17H10Mo1O5Sb1	Sb(Ph)2Mo(CO)5						★ S.ZE.RS 0.03 98	★ ES-CH 14.40 2				
C17H10O5P1W1	P(Ph)2W(CO)5						★ S.ZE.RS 0.08 98					
C17H10O5Sb1W1	Sb(Ph)2W(CO)5						★ S.ZE.RS 0.04 98					
C17H13N1	4(2,6-diphenyl)pyridinyl						S.PARA- 0.72 678					
C17H17	9-T-butyl-9-fluorenyl	7.02				-5.44 62	★ S.STAR -0.66 280	ES-CH 9.00 2				
						★ -5.36 230	★ S.STAR 0.15 250	★ ES-V 2.63 7				
	9-butylfluorenyl	7.02				★ -3.33 151	★ S.INDUC 0.14 151	ES-CH 9.00 2				
C17H33N1	N+(Bu)2(CH2)3CCC4H9					★ -5.74 201						
C17H35	(CH2)16CH3						★ ES-CH 3.00 2	★ ES-V 0.68 7				
C18F13O2	OC6F4-C6F4-OC6F5						★ S.INDUC 0.44 49	★ S.ZE.RS -0.25 49	★ ES-CH 4.00 2			
C18F15Sn1	Sn(C6F5)3						★ S.STAR 1.48 61	★ ES-CH 10.70 2				
C18H5F10Sn1	Sn(C6H5)(C6F5)2						★ S.STAR 0.57 61	★ ES-CH 10.70 2				
C18H10	1-coronene:						S.PARA+ -0.51 476	S.PARA+ -0.44 942				
							S.PARA+ -0.44 174	ES-CH 1.00 2				
C18H10F5Sn1	Sn(C6F5)(C6H5)2						★ S.STAR 0.05 61	★ ES-CH 10.70 2				
C18H12Br3P1	P+(C6H4-Br-P)3						S.CH-P+ 1.19 312					
C18H12Cl3P1	P+(C6H4-Cl-P)3						S.CH-P+ 1.19 312	S.CH-P+ 1.31 75				
C18H12Cl3Si1	Si(C6H4Cl-P)3						★ S.STAR 1.74 188					
C18H12F3P1	P+(C6H4-F-M)3						S.CH-P+ 1.35 75					
C18H12F3Sn1	Sn(C6H4F-M)3						★ S.INDUC 0.26 198	★ S.ZE.RS 0.04 198				
							★ S.ZE.RS -0.12 564	★ ES-CH 10.70 2				
	Sn(C6H4F-P)3						★ S.INDUC 0.25 198	★ S.ZE.RS 0.03 198	★ ES-CH 10.70 2			
C18H14Br1N1P1	N=P(C6H5)2(C6H4Br-P)			★ -0.78 93			★ S.PARA+ -1.61 93	★ ES-CH 5.20 2				
C18H14Br1P1	P+(Ph)2(C6H4-Br-P)						S.CH-P+ 1.11 75					
C18H14Cl1N1P1	N=P(C6H5)2(C6H4Cl)			★ -0.77 93			★ S.PARA+ -1.61 93	★ ES-CH 5.20 2				
C18H14N1O3	(OC6H4)2C6H4NH2 -P,P,P						★ S.PARA- 0.00 187	★ ES-CH 4.00 2				
C18H15As1	+As(Ph)3						S.INDUC 0.36 16	★ ES-CH 10.30 2				
							★ S.ZE.RS 0.15 98	S.CH-P+ 0.99 75				
C18H15As1Au1	Au-As-(C6H5)3						★ S.STAR -3.22 61					
C18H15Au1O3P1	Au-P(OC6H5)3						★ S.STAR -2.78 61					
C18H15Au1P1	Au-P(C6H5)3						★ S.STAR -3.04 61					
C18H15Au1Sb1	Au-Sb(C6H5)3						★ S.STAR -5.40 61					
C18H15B1	+B(C6H5)3						★ S.ZE.RS -0.13 86	★ ES-CH 10.00 2				
C18H15F2N3P3	1,3,3(Ph)3-triphosphazyl						★ S.PARA+ 0.42 537	★ ES-CH 6.20 2				
C18H15Ge1	Ge(C6H5)3	★ 0.05 347		0.01 60			S.INDUC 0.00 16	★ ES-CH 10.30 2				
		0.15 347		0.08 347			★ S.STAR -0.45 61	F 0.07 4				
				★ 0.08 338			★ S.PARA+ -0.15 60	R 0.01 4				
				0.15 51			S.PARA+ 0.11 51	R+ -0.35 4				
				0.24 347			★ S.ZE.RS -0.08 47	GR.ELCT -0.07 13				
C18H15N1	1-pyridinium,2-Me,4,6-diphenyl	★ 0.65 849	★ 0.70 849									
	4(1-Me-2,6-diPh)pyridinyl						S.PARA- 0.70 678					
C18H15N1O3P1	N=P(OPh)3						S.INDUC -0.37 123					
C18H15N1P1	N=P(C6H5)3	★ -0.33 807	-0.86 437				S.INDUC -0.58 123	S.ZPTFT -0.50 487	★ ES-CH 5.20 2			
			-0.79 93				S.INDUC -0.20 24	★ S.PARA- -0.77 617	F -0.10 5			
			★ -0.77 807				S.INDUC -0.16 135	S.PARA+ -1.76 437	R -0.67 5			
			-0.75 617				S.INDUC -0.11 808	★ S.PARA+ -1.65 93	R+ -1.55 4			
							S.ZMTFT -0.33 487	S.ZE.RS -0.46 24	R- -0.67 4			
							★ S.META- -0.33 617	S.ZE.RS -0.45 135	S.RES.P -2.72 808			
C18H15O1Pb1	O-Pb(Ph)3						S.INDUC 0.06 118	S.ZE.RS -0.60 492	S.RES.+ -2.49 71			
							S.PARA+ -3.13 71	S.ZE.RS -0.46 118				
C18H15O1Sn1	O-Sn(Ph)3						S.INDUC 0.10 118	S.ZE.RS -0.45 118	S.RES.+ -1.89 71			
							S.PARA+ -2.34 71	S.ZE.RS -0.42 492				
C18H15O3Si1	Si(OPh)3						★ S.INDUC -0.05 25	★ S.ZE.RS 0.11 25				
C18H15P1	+P(C6H5)3						S.STAR 3.27 87	S.ZE.RS 0.14 447	S-L 0.75 14			
							S.STAR 4.46 3	★ S.ZE.RS 0.22 98	S.CH-P+ 1.08 312			
							S.STAR 4.48 360	★ S.ZE.RS 0.26 333	S.CH-P+ 1.22 75			
							★ S.ZE.RS 0.00 86	★ ES-CH 10.20 2				
C18H15Pb1	Pb(C6H5)3						★ S.STAR -0.39 61	★ S.ZE.RS 0.00 47				
C18H15Pb1S1	S-Pb(Ph)3						S.INDUC 0.00 118	S.ZE.RS -0.11 118				
							S.PARA+ -0.52 71	S.RES.+ -0.49 71				
C18H15S1Sn1	S-Sn(Ph)3						S.INDUC 0.10 118	S.ZE.RS -0.14 118				
							S.PARA+ -0.35 71	S.RES.+ -0.36 71				
C18H15Si1	Si(C6H5)3	-0.03 347	0.10 338		8.40		S.INDUC -0.18 16	★ S.PARA- 0.29 321	R+ 0.16 4			
		★ -0.03 246	0.10 347				★ S.INDUC 0.13 73	★ S.PARA+ 0.12 383	R- 0.33 4			
		0.12 347	★ 0.10 246				S.STAR -0.52 61	S.ZE.RS -0.04 47	S.P.RAD -0.03 218			

Empirical	Structural	π	σ_m	σ_p	MR	E_s	additional σ values						
C18H15Si1	Si(C6H5)3			0.27 347			★ S.STAR 0.44 377	★ S.ZE.RS 0.06 73	S.P.RAD 0.10 218				
							S.STAR 0.45 216	ES-CH 10.20 2	GR.ELCT -0.18 13				
							S.STAR 1.50 188	F -0.04 5					
							★ S.META- 0.10 321	R 0.14 615					
C18H15Sn1	Sn(C6H5)3				9.08		S.INDUC -0.12 16	S.STAR -0.45 61	S.ZE.RS 0.03 198				
							S.INDUC 0.18 173	★ S.ZE.RS -0.10 47	ES-CH 10.70 2				
							S.INDUC 0.20 198	S.ZE.RS 0.03 173					
C18H19N1O1	N+(CH3)2CH2CCC(OH)(Ph)					★ -3.04 303							
C18H21Ge1O1	Ge(Ph)2(CH2)4COCH3			★ 0.16 51			★ S.PARA+ 0.13 51						
C18H22B1	B(mesityl)2			0.42 1005									
				0.65 1006									
C18H23Ge1	Ge(N-hexyl)(Ph)2			★ 0.10 51			★ S.PARA+ 0.02 51						
C18H25N2P1	P+(Et)(C6H4NMe2-P)2						★ S.PARA- 1.05 235						
C18H30B1	B(9-bic(3.3.1)nonan)2-yl						S.INDUC 0.03 217	S.PARA+ 0.61 217	S.ZE.RS 0.22 217				
C18H31Ge1	Ge(C6H5)(N-hexyl)2			★ 0.09 51			★ S.PARA+ 0.00 51						
C18H33Au1P1	Au-P(cy-hexyl)3						★ S.STAR -3.45 61						
C18H33Ge1	Ge(CH=CHC4H9)3			★ 0.10 51			★ S.PARA+ 0.02 51						
C18H33P1	P+(C-hexyl)3						S.CH-P+ 1.07 75						
C18H33Sn1	Sn(cyclohexyl)3						L-STM 5.85 1	B1-STM 4.65 1	B5-STM 6.89 1				
C18H35Ni1P2	NI(C6H5)(PEt3)2						★ S.INDUC -0.42 70	★ S.ZE.RS -0.28 70	★ ES-CH 12.70 2				
C18H35P2Pd1	Pd(C6H5)(PEt3)2						★ S.INDUC -0.40 70	★ S.ZE.RS -0.24 70	★ ES-CH 13.10 2				
C18H35P2Pt1	Pt(C6H5)(PEt3)2						★ S.INDUC -0.44 70	★ S.ZE.RS -0.25 70					
C19H13	9-phenyl-9-fluorenyl				7.60	★ -4.32 230	★ S.STAR 0.33 280	ES-CH 8.00 2					
							★ S.STAR 0.59 250	★ ES-V 1.59 7					
C19H14F3N1P1	N=P(C6H5)2(C6H4CF3)			★ -0.77 93			★ S.INDUC -0.11 135	★ S.ZE.RS -0.44 135					
							★ S.PARA+ -1.60 93	★ ES-CH 5.20 2					
C19H14N2P1	N=P(C6H5)2(C6H4CN)			★ -0.75 93			★ S.PARA+ -1.58 93	★ ES-CH 5.20 2					
C19H15	C(C6H5)3		★ -0.01 708	0.02 338	7.87	-6.03 35	S.INDUC -0.21 12	S.META- -0.07 665	ES-V 2.92 7				
				0.02 708		★ -5.92 230	★ S.INDUC 0.25 151	★ S.PARA+ -0.21 665	F 0.01 5				
				0.08 82			S.STAR 0.40 3	S.ZE.RS -0.13 19	R 0.01 5				
							★ S.STAR 0.56 250	S.ZE.RS -0.04 12	R+ -0.22 4				
							S.STAR 0.65 280	ES-CH 10.00 2	★ S.PHOSP -1.02 37				
C19H15S1	SC(C6H5)3		★ 0.05 3		8.66		S.INDUC 0.11 32	★ ES-CH 5.20 2					
							★ S.STAR 0.69 3	★ S-L 0.12 7					
C19H16N1	C(C6H5)2(C6H4NH2-P)						★ S.PARA- 0.14 187	★ ES-CH 10.00 2					
C19H16N1O2P1	P(C6H5)2(CH2C6H4NO2)			★ 1.10 560			★ ES-CH 9.20 2						
C19H16P1	CH=P(C6H5)3			★ -1.58 343			★ S.PARA+ -3.10 343	★ ES-CH 5.20 2					
				-1.33 437			S.PARA+ -2.69 437						
C19H17Ge1	CH2Ge(C6H5)3						S.INDUC 0.03 172	★ S.PARA+ -0.60 276	S.ZE.RS -0.17 172				
							★ S.PARA+ -0.78 60	S.PARA+ -0.52 203	★ ES-CH 5.30 2				
C19H17N1P1	N=P(C6H5)2(C6H4-Me)			★ -0.79 93			★ S.INDUC -0.15 135	★ S.ZE.RS -0.46 135					
							★ S.PARA+ -1.65 93	★ ES-CH 5.20 2					
C19H17P1	CH2P(C6H5)3+						★ S.STAR 1.15 87	★ S.STAR 1.68 435	★ ES-CH 5.20 2				
C19H17Pb1	CH2Pb(C6H5)3						S.INDUC 0.06 172	★ S.PARA+ -1.00 276					
							★ S.PARA+ -1.08 277	S.ZE.RS -0.20 172					
C19H17Si1	CH2Si(C6H5)3						S.INDUC 0.01 172	★ S.PARA+ -0.40 276	S.ZE.RS -0.16 172				
							★ S.PARA+ -0.49 60	S.PARA+ -0.36 203	★ ES-CH 5.20 2				
C19H17Sn1	CH2Sn(C6H5)3						★ S.INDUC 0.00 355	S.PARA+ -0.69 203	★ ES-CH 5.70 2				
							S.INDUC 0.06 172	S.ZE.RS -0.21 172					
							★ S.PARA+ -0.75 276	★ S.ZE.RS -0.21 355					
C19H21N1O1	N+(Et)2CCC(OH)(Ph)2					★ -4.64 303							
C20H12	1-Bz(A)NAP(1,2)anthracen:						S.PARA+ -0.41 174						
	1-hexahelicene:						S.PARA+ -0.46 174						
	2-hexahelicene:						S.PARA+ -0.42 174						
	3-Bz(A)NAP(1,2)anthracen:						S.PARA+ -0.28 174						
	3-hexahelicene:						S.PARA+ -0.34 174						
	4-Bz(A)NAP(1,2)anthracen:						S.PARA+ -0.35 174						
	4-hexahelicene:						S.PARA+ -0.45 174						
C20H13F6N1P1	N=P(C6H5)(C6H4CF3)2			★ -0.75 93			★ S.INDUC -0.08 135	★ S.ZE.RS -0.43 135					
							★ S.PARA+ -1.56 93	★ ES-CH 5.20 2					
C20H15	9-benzyl-9-fluorenyl				8.06	★ -3.40 151	★ S.INDUC 0.23 151	ES-CH 7.00 2	★ ES-V 1.63 7				
C20H16N3O3	NHCOC6H4OC6H4CONHC6H4NH2						★ S.PARA- -0.03 187	★ ES-CH 4.00 2					
C20H17	CH2C(C6H5)3						★ S.PARA+ -0.21 60	★ ES-CH 5.00 2					
C20H17N1O2P1	N=P(C6H5)2(C6H4CO2Me)			★ -0.78 93			★ S.INDUC -0.13 135	★ S.ZE.RS -0.44 135					
							★ S.PARA+ -1.61 93	★ ES-CH 5.20 2					
C20H19I1P1	P+(C6H5)(C6H4P-CH3)2(I-)			★ 1.61 38			★ S.INDUC 0.76 38	★ S.ZE.RS 0.85 38	★ ES-CH 10.20 2				
C20H19O2P1	P+(Ph)(C6H4,P-OCH3)2						S.CH-P+ 1.01 312						
C20H19Pb1	CH2CH2Pb(C6H5)3						★ S.PARA+ -0.22 277						
C20H20N1P1	P+(Ph)2(C6H4,P-NMe2)						S.CH-P+ 1.11 75						
C20H20N2P1	N=P(C6H5)2(C6H4NMe2)			★ -0.82 93			★ S.INDUC -0.16 135	★ S.ZE.RS -0.47 135					
							★ S.PARA+ -1.72 93	★ ES-CH 5.20 2					
C21H12F9N1P1	N=P(C6H5CF3)3			★ -0.72 93			★ S.INDUC -0.01 135	★ S.ZE.RS -0.41 135					
							★ S.PARA+ -1.52 93	★ ES-CH 5.20 2					
C21H19Sn1	CH=CHCH2Sn(Ph)3						★ S.INDUC -0.01 355	★ S.ZE.RS -0.09 355					

Empirical	Structural	π	σm	σp	MR	Es	additional σ values
C21H21Au1P1	Au-P(C6H4CH3-P)3						★S.STAR -3.20 61
C21H21N1	N+(CH2Ph)3					★ -3.34 215	
C21H21N1P1	N=P(C6H4-Me)3			★ -0.81 93			★S.INDUC -0.16 135　★S.ZE.RS -0.47 135
							★S.PARA+ -1.70 93　★ES-CH 5.20 2
C21H21O3P1	+P(C6H4-OCH3-M)3						S.CH-P+ 1.09 312　S.CH-P+ 1.24 75
	+P(C6H4-OCH3-P)3						S.CH-P+ 0.98 312　S.CH-P+ 1.11 75
C21H21P1	P+(C6H4-CH3-M)3						S.CH-P+ 1.03 312　S.CH-P+ 1.17 75
	P+(C6H4-CH3-P)3						S.CH-P+ 1.03 312　S.CH-P+ 1.15 75
C21H21Si1	Si(C6H4CH3-P)3						★S.STAR 1.44 188
C21H41	(CH2)11CH=CH(CH2)7CH3 -C					★ES-V 0.67 7	
	(CH2)11CH=CH(CH2)7CH3 -T					★ES-CH 3.00 2　★ES-V 0.68 7	
C22H13F9P1	CH=P(C6H4CF3-P)3			★ -1.40 343		★S.PARA+ -2.77 343　★ES-CH 5.20 2	
C22H15As1Mn1O4	Mn(CO)4As(C6H5)3						★S.INDUC -0.20 33　★S.ZE.RS -0.24 33　★ES-CH 13.60 2
C22H15Mn1O4P1	Mn(CO)4P(C6H5)3						★S.INDUC -0.21 33　★S.ZE.RS -0.24 33　★ES-CH 13.50 2
C22H15Mn1O7P1	Mn(CO)4P(OC6H5)3						★S.INDUC -0.24 33　★S.ZE.RS -0.23 33　★ES-CH 13.50 2
C22H22P1	CH=P(C6H4CH3-P)3			★ -1.49 343		★S.PARA+ -2.95 343　★ES-CH 5.20 2	
C22H25N3P1	N=P(C6H5)(C6H4NMe2)2			★ -0.86 93			★S.INDUC -0.21 135　★S.ZE.RS -0.48 135
							★S.PARA+ -1.78 93　★ES-CH 5.20 2
C23H17N1	2,4,6-triPh-pyridinium	★ 0.34 366	★ 0.33 366				S.INDUC 0.36 366　F 0.35 4
							S.ZE.RS -0.03 366　R -0.02 4
	4(1,2,6-triPh)pyridinyl						S.PARA- 0.74 678
C23H20Ni1P1	NI-P(C6H5)3(C5H5 -PI)						★S.INDUC -0.44 33　★S.ZE.RS -0.29 33　★ES-CH 10.50 2
C24H18N1O4	(OC6H4)3C6H4NH2 -PPPP						★S.PARA- 0.05 187　★ES-CH 4.00 2
C24H20As1	As(C6H5)4						★S.ZE.RS -0.04 86　★ES-CH 13.30 2
C24H20O1Sb1	O-Sb(Ph)4						S.PARA+ -3.92 71　S.RES.+ -2.56 71
C24H20P1	P(C6H5)4						★S.ZE.RS -0.03 86　★ES-CH 13.20 2
C24H20S1Sb1	S-Sb(Ph)4						S.PARA+ -2.62 71　S.RES.+ -2.10 71
C24H20Sb1	Sb(C6H5)4						★S.ZE.RS -0.06 86　★ES-CH 13.70 2
C24H27O3Si1	Si(O-2,6-xylyl)3						★S.STAR 1.82 188
C24H30N4P1	N=P(C6H4NMe2)3			★ -0.87 93			★S.INDUC -0.24 135　★S.ZE.RS -0.50 135
							★S.PARA+ -1.80 93　★ES-CH 5.20 2
C24H30O3P2Pt1	Pt+P(OC6H5)3P(C2H5)3 -T						★S.INDUC 0.15 33　★S.ZE.RS -0.20 33
C24H47N1	N+(oct)2CH2CCC5H11					★ -4.94 201	
C24H51N1	N+(oct)3					★ -5.74 201	
C25H16F2P1	+P(Ph-F)2(9-fluoryliden			★ 0.71 81			★S.INDUC 0.56 81　★S.ZE.RS 0.15 81　★ES-CH 10.20 2
C25H17F2P1	+P(Ph-F)2(9-fluorenyl)			★ 0.87 81			★S.INDUC 0.65 81　★S.ZE.RS 0.22 81　★ES-CH 10.20 2
C25H20Mo1O2P1	Mo(CO)2(C5H5)P(C6H5)3						★S.INDUC -0.35 431　★S.ZE.RS -0.23 431　★ES-CH 14.90 2
C25H20O2P1W1	W(CO)2(C5H5)P(C6H5)3						★S.INDUC -0.44 431　★S.ZE.RS -0.20 431
C25H31N3P1	CH=P(C6H4NMe2-P)3			★ -1.57 343		★S.PARA+ -3.08 343　★ES-CH 5.20 2	
C28H45O2	cholesteryl			★ 0.36 792		★S.ZE.RS 0.11 792　★ES-CH 4.00 2	
C30H25Ge2	Ge(Ph)2(Ge-Ph3)						★S.PARA+ -0.26 25
C38H25Mn2O8P2	P(Ph)2MnMn(CO)8P(Ph)3						★S.ZE.RS 0.10 98
							S.INDUC 0.17 539　★S.ZE.RS 0.01 18　R 0.02 5

1 Verloop,A., Hoogenstraaten,W. and Tipker,J., "Drug Design", Ariens,E.,Ed., Academic Press, 1976, Vol. 7, Ch. 4, Revised in Private Communication. *Note:* Calculated from C.P.K. bond radii and angles

2 Austel,V., Kutter,E. and Kalbfleisch, Private Commun., Arzneim.-Forsch., 29, 585 (1979) *Note:* Steric constant calculated according to Arzneim.-Forsch.(1979) 29,585 *Note:* Apparent PKA of substituted acids (50% aq.-EtOH or water at 25 deg.)

3 Perrin,D.D., Dempsey,B. and Serjeant,E.P., "pKa Predictions for Organic Acids and Bases", Chapman and Hall, p.107 (1981) *Note:* Apparent PKA of substituted acids (50% aq.-EtOH or water at 25 deg.)

4 Hansch,C., Leo,A, Taft,R., Chem. Rev., 91, 165 (1991)

5 Hansch,C., Leo,A, Taft,R., Chem. Rev., 91, 165 (1991) *Note:* Modified 'F' and 'R' calculated

6 Charton,M., Prog. Phys. Org. Chem., 13, 119 (1981) *Note:* Miscellaneous

7 Charton,M., "Design of Biopharm. Prop. Thru Prodrugs and Analogs", Roche,E.,Ed., Am. Pharm. Assoc., 1977, Ch. 9, p. 269. *Note:* Vanderwal radii and secondary values

8 Lien,E.J., Guo,Z-r., Li,R-l. and Su,C-t., J. Pharm. Sci., 71, 641 (1982) *Note:* Majority from "Tables of Expl.Dipole Moments" vol.2, A. McClellan

9 Li,W-y., Guo,Z-r. and Lien,E.J., J. Pharm. Sci. (1983). *Note:* Aliphatic attached *Note:* From "Tables of Expl. Dipole Moments" Vol., Rahara Ent. ElCerito(1974)

10 Palm,V.A., Summary of Sci. and Technology, Ser. Gen. Quest. Org. Chem. Moscow, p.163 (1979) *Note:* Review

11 Verloop,A., Hoogenstraaten,W. and Tipker,J., "Drug Design", Ariens,E.,Ed., Academic Press, 1976, Vol. 7, Ch. 4, Revised in Private Communication. *Note:* Calculated from C.P.K. bond radii and angles *Note:* Torsion angle to ring = 90 deg.

12 Glukhikh,V.I. and Voronkov,M.G., Dokl. Akad. Nauk Sssr, 248, 142, 744EE (1979) *Note:* C-13 NMR

13 Inamoto,N. and Masuda,S., Chem. Let. Japan, 1007 (1982) *Note:* Gordy's electronegativities adjusted for atomic period

14 Charton,M., Prog. Phys. Org. Chem., in press. *Note:* Mean from several reaction systems

15 Mamaev,V.P., Shkurko,O.P. and Baram,S.G., Adv. Heterocyc. Chem., 42, 1 (1987)

16 Knorr,R., Tetrahed., 37, 929 (1981) *Note:* Inductive Sigma effect transmitted by Sigma bonds; by NMR coupling *Note:* H1-H1-NMR coupling constants olefins

17 Exner,O., Coll. Czech. Chem. Comm., 31, 65 (1966) *Note:* Apparent PKA of substituted acids (50% aq.-EtOH or water at 25 deg.)

18 Yagupol'skii,L.M., Il'chenko,A.Y. and Kondratenko,N.V., Russ. Chem. Rev., 43, 64, 32EE (1974) *Note:* Apparent PKA of substituted acids (50% aq.-EtOH or water at 25 deg.) *Note:* F-19 NMR shift (in CCl4)

19 Brownlee,R.T.C., Hutchinson,R.E.J., Katritzky,A.R., Tidwell,T.T. and Topsom,R.D., J. Am. Chem. Soc., 90, 1757 (1968) *Note:* Infrared intensities (sign not determined)

20 Dayal,S.K., Ehrenson,S. and Taft,R.W., J. Am. Chem. Soc., 94, 9113 (1972) *Note:* Thermodynamic PKA of substituted acids (benzoic acid, phenols or anilines unless otherwise specified, activities-not concentrations-used,50% aq.-EtOH or water at 25 deg.) *Note:* Rate of hydrolysis or solvolysis *Note:* F-19 NMR shift (in CCl4)

21 McDaniel,D.H. and Brown,H.C., J. Org. Chem., 23, 420 (1958) *Note:* Thermodynamic PKA of substituted acids (benzoic acid, phenols or anilines unless otherwise specified, activities-not concentrations-used,50% aq.-EtOH or water at 25 deg.)

22 Grindley,T.B., Johnson,K.F., Katritzky,A.R., Keogh,H.J., Thirkettle,C., Brownlee,T.C., Munday,J.A. and Topsom,R.D., J. Chem. Soc. Perkin Ii, 276 (1974) *Note:* IR *Note:* Substituted acetylenes

23 Jaffe,H.H., Chem. Rev., 53, 191 (1953) *Note:* Wide variety of methods, see review article for details

24 Weigert,F.J. and Sheppard,W.A., J. Org. Chem., 41, 4006 (1976) *Note:* F-19 NMR shift (in CCl4) *Note:* Assorted solvents

25 Egorochkin,A.N. and Razuvaev,G.A., Russ. Chem. Rev., 56, 846 (1987)

26 Fujio,M., McIver Jr., R.T. and Taft,R.W., J. Am. Chem. Soc., 103, 4017 (1981) *Note:* In gas phase *Note:* Gas phase acidities of phenols by ion cyclotron resonance

27 Baker,R., Eaborn,C. and Taylor,R., J. Chem. Soc. Perkin Ii, 97 (1972) *Note:* Multiple attachment; e.g. position may be ortho + meta *Note:* Rate of detritiation in CF3CO2H

28 Zeng,G-z., Acta Chim. Sinica, 32, 107 (1966) *Note:* From anilinium (Sigma -) or Sigma ortho *Note:* Apparent PKA of substituted acids (50% aq.-EtOH or water at 25 deg.)

29 Hoefnagel,A.J., Hoefnagel,M.A. and Wepster,B.M., J. Org. Chem., 43, 4720 (1978) *Note:* Acidic ester hydrolysis

30 Sheppard,W. and Sharts,C., "Organic Flourine Chem." W.A. Benjamin, N.Y., 1969, p. 348. *Note:* Apparent PKA of substituted acids (50% aq.-EtOH or water at 25 deg.) *Note:* F-19 NMR shift (in CCl4)

31 Sheppard,W.A. and Taft,R.W., J. Am. Chem. Soc., 94, 1919 (1972) *Note:* F-19 NMR shift (in CCl4) *Note:* Various solvents

32 Charton,M., J. Org. Chem., 29, 1222 (1964) *Note:* Apparent PKA of substituted acids (50% aq.-EtOH or water at 25 deg.) *Note:* Substituted acetic acids (various conditions, by regression)

33 Stewart,R.P. and Treichel,P.M., J. Am. Chem. Soc., 92, 2710 (1970) *Note:* I.R. and F-19 NMR

34 Charton,M., Prog. Phys. Org. Chem., 10, 81 (1973) *Note:* Apparent PKA of substituted acids (50% aq.-EtOH or water at 25 deg.) *Note:* Review

35 MacPhee,J.A., Panaye,A. and Dubois,J-e., Tetrahed., 34, 3553 (1978) *Note:* From acid catalytic esterification in MeOH, scaled to H = 0, see Tetrahed. Lett.(1978) 35,3293 *Note:* Acid catalyzed esterification rate in MeOH *Note:* See also: Tetrahed. Lett.(1978) 35,3293

36 Hoefnagel,A.J., Hoefnagel,M.A. and Wepster,B.M., J. Org. Chem., 43, 4720 (1978) *Note:* Bjerrum term for polar effect of charged substituent must be added, Sigma minus values from phenol system, aniline values given also in J. Org. Chem.(1978) 43,4720 *Note:* Acidic ester hydrolysis

37 Mastryukova,T.A. and Kabachnik,M.I., Russian Chem. Rev., 38, 795EE (1969) *Note:* PKA of phosphorous acids

38 Schiemenz,G.P., Angew. Chem. Internat. Edit., 7, 544 and 545 (1968) *Note:* NMR phosphorous substituted toluenes

39 Kasukhin,L.F. and Gololobov,Y.G., Org. React., 15, 463 (1978) *Note:* PKA substituted phosphoric acids

40 Sheppard,W.A., Transaction N.Y. Acad. Sci., Ser. Ii, 29, 700 (1967) *Note:* Apparent PKA of substituted acids (50% aq.-EtOH or water at 25 deg.) *Note:* F-19 NMR shift (in CCl4) *Note:* Benzene

41 Hogben,M.G. and Graham,W.A.G., J. Am. Chem. Soc., 91, 283 (1969) *Note:* F-19 NMR shift (in CCl4) *Note:* Pentafluorophenyl derivatives (coupling constant in benzene) *Note:* Derived new regression equation

42 N-aromatic ring--attachment not ortho

43 Brown,H.C. and Okamoto,Y., J. Am. Chem. Soc., 80, 4979 (1958) *Note:* Rate of hydrolysis or solvolysis *Note:* T-cumyl chlorides (90% acetone at 25 deg. and others)

44 Talvik,I. and Palm,V., Org. React. (USSR), 8, 445 (1971) *Note:* Ratio of acid vs. base catalysed hydrolysis of esters

45 Tsvetkov,E.N., Malevannaya,R.A., Petrovskaya,L.I. and Kabachnik,M.I., Zh. Obshch. Khim., 44, 1225, 1203EE (1974) *Note:* Apparent PKA of substituted acids (50% aq.-EtOH or water at 25 deg.)

46 Sjostrom,M. and Wold,S., Chem. Scripta, 9, 200 (1976) *Note:* Unified Sigma zero by statistical analysis

47 Katritzky,A.R. and Topsom,R.D., Chem. Rev., 77, 639 (1977) *Note:* I.R. intensities

48 Kondratenko,N.V., Popov,V.I., Kolomeitsev,A.A., Saenko,E.P., Prezhdo,V.V., Lutskii,A.E. and Yagupol'skii,L.M., Zh. Org. Khim., 16, 1215, 1049EE (1980) *Note:* F-19 NMR shift (in CCl4) *Note:* Other values in heptane

49 Pushkina,L.N., Stepanov,A.P., Zhukov,V.S. and Naumov,A.D., Organ. Magnet. Reson., 4, 607 (1972) *Note:* F-19 NMR *Note:* Reference contains more values

50 Newman, Ed., "Steric Effects in Organic Chemistry", Wiley and Sons, N.Y., 1956, p.598-604.

51 Sennikov,P.G., Skobeleva,S.E., Kuznetsov,V.A., Egorochkin,A.N., Riviere,P., Satge,J. and Richelme,S., J. Organomet. Chem., 201, 213 (1980) *Note:* I.R. of charge transfer complexes

52 Newman, Ed., "Steric Effects in Organic Chemistry", Wiley and Sons, N.Y., 1956, p.619.

53 Perrin,D.D., Dempsey,B. and Serjeant,E.P., "pKa Predictions for Organic Acids and Bases", Chapman and Hall, p.107 (1981) *Note:* From phenol (Sigma -) or Sigma ortho *Note:* Apparent PKA of substituted acids (50% aq.-EtOH or water at 25 deg.)

54 Nitrobenzene, meta or para attachment

55 Wells,P.R., Ehrenson,S. and Taft,R.W., Prog. Phys. Org. Chem., 6, 244 (1968) *Note:* Multiple attachment; e.g. position may be ortho + meta *Note:* Apparent PKA of substituted acids (50% aq.-EtOH or water at 25 deg.) *Note:* PKA's naphthoic acids

56 Exner,O. and Lakomy,J., Coll. Czech. Chem. Comm., 35, 1371 (1970) *Note:* Apparent PKA of substituted acids (50% aq.-EtOH or water at 25 deg.) *Note:* 80% Me-cellosolve and 50% EtOH

57 Aliphatic system (usually methyl or ethyl)

58 Grob,C.A., Schaub,B. and Schlageter,M.G., Helv. Chim. Acta, 63, 57 (1980) *Note:* Thermodynamic PKA of substituted acids (benzoic acid, phenols or anilines unless otherwise specified, activities-not concentrations-used,50% aq.-EtOH or water at 25 deg.) *Note:* Quinuclidines, also at .1 M KCl *Note:* Scaled by 0.166 to match Sigma induct.

59 Mastryukova,T.A. and Kabachnik,M.I., J. Org. Chem., 36, 1201 (1971) *Note:* PKA of phosphorus acids

60 Kuznetsov,V.A., Egorochkin,A.N., Skobeleva,S.E., Razuvaev,G.A., Pritula,N.A. and Zueva,G.Y., Zh. Obshch. Khim., 45, 2439, 2396EE (1975) *Note:* Absorption spectra of charge transfer complex

61 Ioganson,A.A., Antonova,A.B., Lokshin,B.V., Kolobova,N.E., Anisimov,K.N. and Nesmeyanov,A.N., Iz. Akad. Nauk. Ssr, Khim.,Z.Akad.Nauk.Ssr,Khim., 9, 1957, 1811EE (1969) *Note:* IR manganese carbonyls

62 MacPhee,J.A., Panaye,A. and Dubios,J-e., Tetrahed. Lett., 35, 3293 (1978) *Note:* From acid catalytic esterification in MeOH, scaled to H = 0, see Tetrahed. Lett.(1978) 35,3293 *Note:* Acid catalyzed esterification rate in MeOH

63 Yukawa,Y., Tsuno,Y. and Sawada,M., Bull. Chem. Soc. Japan, 45, 1198 (1972) *Note:* Rate of hydrolysis or solvolysis

64 Sheppard,W.A., Tetrahedron, 27, 945 (1971) *Note:* F-19 NMR shift (in CCl4)

65 Hoefnagel,A.J. and Wepster,B.M., J. Am. Chem. Soc., 95, 5357 (1973) *Note:* Thermodynamic PKA of substituted acids (benzoic acid, phenols or anilines unless otherwise specified, activities-not concentrations-used,50% aq.-EtOH or water at 25 deg.)

66 Hansch,C., Leo,A., Unger,S.H., Kim,K.H., Nikaitani,D. and Lien,E.J., J. Med. Chem., 16, 1207 (1973) *Note:* Modified 'F' and 'R' calculated

67 Bromilow,J., Brownlee,R.T.C., Lopez,V.O. and Taft,R.W., J. Org. Chem., 44, 4766 (1979) *Note:* C-13 NMR

68 Dronkina,M.I., Syrova,G.P., Gandel'sman,L.Z., Sheinker,Y.N. and Yagupol'skii,L.M., Zh. Org. Khim., 8, 9, 7EE (1972) *Note:* F-19 NMR shift (in CCl4) *Note:* In dichloroethane

69 Modro,T.A., Can. J. Chem., 55, 3681 (1977) *Note:* C-13 NMR

70 Parshall,G.W., J. Am. Chem. Soc., 96, 2360 (1974) *Note:* F-19 NMR

71 Epstein,L.M., Ashkinadze,L.D., Shubina,E.S., Kravtsov,D.N. and Kazitsyna,L.A., J. Organomet. Chem., 228, 53 (1982) *Note:* Solvent dependent; see reference for other values *Note:* I.R. in DMSO *Note:* Other values in CH2Cl2

72 Kazitsyna,L.A., Ustynyuk,Y.A., Gurman,V.S., Pergushov,V.I. and Gruzdneva,V.N., Dokl. Akad. Nauk. Sssr, 233, 866, 213EE (1977) *Note:* F-19 NMR

73 Lipowitz,J., J. Am. Chem. Soc., 94, 1582 (1972) *Note:* F-19 NMR

74 Bel'skii,V.E., Kudryavtseva,L.A. and Ivanov,B.E., Zh. Obshch. Khim., 42, 2427, 2421EE (1972) *Note:* P-31 NMR

75 Kabachnik,M.I. and Mastryukova,T.A., Zh. Obshch. Khim., 54, 2161, 1931EE (1984) *Note:* Acidities of phosphonium salts

76 Sharnin,G.P. and Falyakhov,I.F., Zh. Org. Khim., 9, 730, 751EE (1973) *Note:* Relative rates of nitration in acetic anhydride

77 Livingstone,D.J., Hyde,R.M. and Foster,R., Eur. J. Med. Chem., 14, 393 (1979) *Note:* Substituted benzenes complexed with trinitrobenzene in CCl4

78 Perrin,D.D., Dempsey,B. and Serjeant,E.P., "pKa Predictions for Organic Acids and Bases", Chapman and Hall, p.107 (1981) *Note:* For CO2H analogs *Note:* Apparent PKA of substituted acids (50% aq.-EtOH or water at 25 deg.)

79 Wayner,D.D.M. and Arnold,D.R., Can. J. Chem., 62, 1164 (1984) *Note:* Scaled by 10x from original reference values *Note:* ESR, values scaled by 10x from those in reference

80 Ehrenson,S., Brownlee,R.T.C. and Taft,R.W., Prog. Phys. Org. Chem., 10, 1 (1973) *Note:* Apparent PKA of substituted acids (50% aq.-EtOH or water at 25 deg.) *Note:* Review

81 Johnson,A.W. and Jones,H.L., J. Am. Chem. Soc., 90, 5232 (1968) *Note:* F-19 NMR in DMSO *Note:* Other values in TFA reported

82 Little,W.F., Reilley,C.N., Johnson,J.D. and Sanders,A.P., J. Am. Chem. Soc., 86, 1382 (1964) *Note:* Chronopotentiometric quarter wave shift in ferrocene derivatives *Note:* Acetonitrile at 25 deg.

83 Frankovskii,C.S. and Melamed,N.V., Zh. Org. Khim., 5, 108, 107EE (1969) *Note:* From phenol (Sigma -) or Sigma ortho *Note:* Apparent PKA of substituted acids (50% aq.-EtOH or water at 25 deg.)

84 Baram,S.G., Shkurko,O.P. and Mamaev,V.P., Iz. Akad. Nauk. Sssr S. Khim., 2, 294, 260EE (1983) *Note:* C-13 NMR

85 Kabachnik,M.I., Dokl. Akad. Nauk., 110, 393, 577EE (1956) *Note:* Rate of alkaline hydrolysis of (diethyl) substituted-phenyl phosphates *Note:* Ionization

86 Angelelli,J.M., Brownlee,R.T.C., Katritzky,A.R., Topsom,R.D. and Yakhontov,L., J. Am. Chem. Soc., 91, 4500 (1969) *Note:* From infra-red or other spectral measurements, *Note:* Infrared intensities (sign not determined)

87 Koppel,I.A., Karelson,M.M. and Palm,V.A., Organic React., 10, 497 (1973) *Note:* Apparent PKA of substituted acids (50% aq.-EtOH or water at 25 deg.)

88 Happer,D.A.R., Aust. J. Chem., 29, 2607 (1976) *Note:* C-13 NMR in DMSO *Note:* Other values in DCCl3 not entered

89 Taft,R.W., Price,E., Fox,I.R., Lewis,I.C., Andersen,K.K. and Davis,G.T., J. Am. Chem. Soc., 85, 709 (1963) *Note:* F-19 NMR shift (in CCl4) *Note:* M-substituted benzenes

90 Berg,U., Gallo,R., Klatte,G. and Metzger,J., J. Chem. Soc. Perkin Ii, 1350 (1980) *Note:* Methyliodide quaternization of pyridines

91 Elguero,J., Estopa,C. and Ilavsky,D., J. Chem. Res. (S), 364 (1981) *Note:* F-19 NMR

92 Glikhikh,V.I., Voronkov,M.G., Yarosh,O.G., Tandura,S.N., Alekseev,N.V., Khromova,N.Y. and Gar,T.K., Dokl. Akad. Nauk. Sssr, 258, 387, 402EE (1981) *Note:* C-13 NMR

93 Zhmurova,I.N., Yurchenko,R.I., Yurchenko,V.G. and Tukhar,A.A., Zh. Obshch. Khim., 43, 1036, 1028EE (1973) *Note:* Infra-red absorption: intensity or frequency (in CCl4) *Note:* Basicities

94 Verloop,A., Hoogenstraaten,W. and Tipker,J., "Drug Design", Ariens,E.,Ed., Academic Press, 1976, Vol. 7, Ch. 4, Revised in Private Communication. *Note:* Calculated from C.P.K. bond radii and angles *Note:* Torsion angle away from -CH2- is 90 deg.

95 Cohen,L.A. and Takahashi,S., J. Am. Chem. Soc., 95, 443 (1973) *Note:* Regression of ionization and kinetic data

96 Taylor,P.J. and Wait,A.R., J. Chem. Soc. Perkin Ii, 1765 (1986)

97 Tribble,M.T. and Traynham,J.C., J. Am. Chem. Soc., 91, 379 (1969) *Note:* OH- NMR shift in o-substituted phenols (in DMSO)

98 Wuyts,L.F., Van De Vondel,D.F. and Van Der Kelen,G.P., J. Organomet. Chem., 129, 163 (1977) *Note:* C-13 and P-31 NMR

99 Fischer,A., King,M.J. and Robinson,F.P., Can. J. Chem., 56, 3059 (1978) *Note:* Apparent PKA of substituted acids (50% aq.-EtOH or water at 25 deg.) *Note:* Alpha-substituted2-methylpyridinium ions

100 Binev,I.G., Kuzmanova,R.B., Kaneti,J. and Juchnovski,I.N., J. Chem. Sci. Perkin Ii, 1533 (1982) *Note:* I.R. frequency and intensity of nitriles

101 Stock,L.M. and Brown,H.C., Adv. Phys. Org. Chem., 1, 116 (1963) *Note:* Multiple attachment; e.g. position may be ortho + meta *Note:* Lower values from nitration rate *Note:* Higher from proton exchange

102 Kazakov,V.P. and Koreshkov,Y.D., Zh. Fiz. Khim., 53, 89, 47EE (1979) *Note:* Regression methods

103 Sadekov,I.D., Bushkov,A.Y., Yur'eva,V.S. and Minkin,V.I., Zh. Obshch. Khim., 47, 2541, 2321EE (1977) *Note:* F-19 NMR

104 Perrin,D.D., Dempsey,B. and Serjeant,E.P., "pKa Predictions for Organic Acids and Bases", Chapman and Hall, p.107 (1981) *Note:* From anilinium (Sigma -) or Sigma ortho *Note:* Apparent PKA of substituted acids (50% aq.-EtOH or water at 25 deg.)

105 Schindlbauer,H. and Prikoszovich,W., Chem. Ber., 102, 2914 (1969) *Note:* F-19 NMR shift (in CCl4) *Note:* Benzene

106 Frankovskii,C.S. and Katsnel'son,E.Z., Zh. Org. Khim., 4, 1631, 1567EE (1968) *Note:* From phenol (Sigma -) or Sigma ortho *Note:* Apparent PKA of substituted acids (50% aq.-EtOH or water at 25 deg.)

107 Tsvetkov,E.N., Malakhova,I.G. and Kabachnik,M.I., Zh. Obshch. Khim., 48, 1230, 1125EE (1978) *Note:* Apparent PKA of substituted acids (50% aq.-EtOH or water at 25 deg.)

108 Baram,S.G., Shkurko,O.P. and Mamaev,V.P., Publ. Siber. Sect. (Novosibirsk), 2, 111 (1983) *Note:* F-19 NMR

109 Litvinenko,L.M., Izv. Akad. Nauk. Khim., 10, 1737, 1653EE (1962) *Note:* From anilinium (Sigma -) or Sigma ortho *Note:* Reaction rates of substituted amines with nitrobenzoyl chloride and picryl chloride

110 Kondratenko,N.V., Matyushecheva,G.I. and Yagupol'skii,L.M., Zh. Org. Khim., 6, 1423, 1436EE (1970) *Note:* F-19 NMR shift (in CCl4)

111 Prikoszovich,W. and Schindlbauer,H., Chem. Ber., 102, 2922 (1969) *Note:* F-19 NMR shift (in CCl4) *Note:* Benzene

112 Berthelot,M., Laurence,C. and Wojtkowiak,B., J. Chim. Phys., 70, 1629 (1973) *Note:* Normalized

113 Baram,S.G., Shkurko,O.P. and Mamaev,V.P., Iz. Akad. Nauk. Sssr. S. Khim., 2, 299, 265EE (1983) *Note:* C-13 NMR

114 Katritzky,A.R., Pinzelli,R.F., Sinnott,M.V. and Topsom,R.D., J. Am. Chem. Soc., 92, 6861 (1970) *Note:* From infra-red or other spectral measurements, *Note:* IR intensities

115 Minamida,I., Ikeda,Y., Uneyama,K., Tagaki,W. and Oae,S., Tetrahedron, 24, 5293 (1968) *Note:* Apparent PKA of substituted acids (50% aq.-EtOH or water at 25 deg.) *Note:* Ratio of acid vs. base catalysed hydrolysis of esters

116 Verloop,A., Hoogenstraaten,W. and Tipker,J., "Drug Design", Ariens,E.,Ed., Academic Press, 1976, Vol. 7, Ch. 4, Revised in Private Communication. *Note:* Calculated from C.P.K. bond radii and angles *Note:* Torsion angle with nitro = 90 deg.

117 Heesing,A. and Schmaldt,W., Chem. Ber., 111, 320 (1978) *Note:* F-19 NMR

118 Nesmeyanov,A.N., Kravtsov,D.N., Kvasov,B.A.K., Rokhlina,E.M., Pachevskaya,V.M., Golovchenko,L.S. and Fedin,E., J. Organomet. Chem., 38, 307 (1972) *Note:* F-19 NMR

119 Adcock,W. and Abeywickrema,A.N., J. Org. Chem., 47, 2957 (1982) *Note:* F-19 and C-13 NMR, in CdCl3 *Note:* 6 other solvent values given

120 Nagai,Y., Matsumoto,H., Nakano,T. and Watanabe,H., Bull. Chem. Soc. Japan, 45, 2560 (1972) *Note:* NMR(Si-H), some calculated from: S*(XC6H4) = 0.725(X) + 0.58

121 Exner,O. and Bocek,K., Coll. Czech. Chem. Comm., 38, 50 (1973) *Note:* Infra-red absorption: intensity or frequency (in CCl4)

122 Wells,P.R. and Ward,E.R., Chem. and Ind., 528 as Rpt. in Ref. 187, p.222 (1958) *Note:* Multiple attachment; e.g. position may be ortho + meta *Note:* Mean from various reactions

123 Zabolotnaya,T.G., Tsymbal,I.F. and Boldeskul,I.E., Zh. Obshch. Khim., 53, 299, 259EE (1983) *Note:* Attached to S in PHSO2- *Note:* I.R. spectra of substituents attached to S of PHSO2

124 Zuman,P., "Substituent Effects in Organic Polarography", Plenum Press, N.Y., 1967, p.159 and 322.. *Note:* Polarography *Note:* Some are averages of widely deviating values

125 Syrova,G.P., Sedova,L.N., Gandel'sman,L.Z., Alekseeva,L.A. and Yagupol'skii,L.M., Zh. Org. Khim., 6, 2285, 2293EE (1970) *Note:* F-19 NMR shift (in CCl4) *Note:* In dichloroethane (-SCH2F in CCl4)

126 Adcock,W., Alste,J., Rizvi,S.Q.A. and Aurangzeb,M., J. Am. Chem. Soc., 98, 1701 (1976) *Note:* F-16 NMR of naphthalenes

127 Fong,C.W., Aust. J. Chem., 33, 1291 (1980) *Note:* C-13 NMR

128 Shim,S.C., Park,J.W. and Ham,H.S., Bull. Korean. Chem. Soc., 3, 13 (1982) *Note:* From thermal decomposition rate *Note:* PKA's of phenols in excited state

129 Kashik,T.V., Prokop'ev,B.V., Atavin,A.S., Mirskova,A.N., Al'pert,M.L., Pil'kevich,S.G. and Ponomareva,S.M., Zh. Org. Khim., 7, 1582, 1643EE (1971) *Note:* Also relative acid constant of CCl3CH(OH)NHCOR

130 Bzhezovskii,V.M., Kushnarev,D.F., Trofimov,B.A., Kalabin,G.A., Gusarova,N.K. and Efremova,G.G., Iz. Akad. Nauk. Sssr, S. Khim., 11, 2507, 2073EE (1981) *Note:* Side-chain in peptide *Note:* C-13 NMR

131 Dewar,M.J.S. and Grisdale,P.J., J. Am. Chem. Soc., 84, 3546 (1962) *Note:* Multiple attachment; e.g. position may be ortho + meta *Note:* PKA's 1-naphthoic acids in 50% EtOH

132 Bystrov,V.F., Yagupol'skii,L.M., Stepanyants,A.U. and Fialkov,Y.A., Doklady Akad. Nauk. Sssr, 153, 1321, 1019EE (1963) *Note:* Apparent PKA of substituted acids (50% aq.-EtOH or water at 25 deg.) *Note:* In 50% ethanol *Note:* F-19 NMR shift (in CCl4) *Note:* Heptane

133 Reynolds,W.F., Hamer,G.K. and Bassindale,A.R., J. Chem. Soc. Perkin Ii, 971 (1977) *Note:* H NMR and C-13 NMR

134 Taft,R.W., Price,E., Fox,I.R., Lewis,I.C., Andersen,K.K. and Davis,G.T., J. Am. Chem. Soc., 85, 3146 (1963) *Note:* F-19 NMR shift (in CCl4)

135 Zhmurova,N.N., Yurchenko,V.G., Yurchenko,R.I., Konovalov,E.V. and Egorov,Y.P., Zh. Obshch. Khim., 44, 2414, 2375EE (1974) *Note:* F-19 NMR

136 Mares,F., Plzak,Z., Hetflejs,J. and Chvalovsky,V., Coll. Czech. Chem. Comm., 36, 2957 (1971) *Note:* Bromination rate

137 Ovchinnikov,V., Kostakova,L., Chercasov,R. and Pudovik,N., Org. React., 15, 194 (1978) *Note:* Apparent PKA of substituted acids (50% aq.-EtOH or water at 25 deg.) *Note:* Dithiophosphorylated propionic acids

138 Zeng,G-z., Acta Chim. Sinica, 32, 107 (1966) *Note:* From phenol (Sigma -) or Sigma ortho *Note:* Apparent PKA of substituted acids (50% aq.-EtOH or water at 25 deg.)

139 Bradamante,S. and Pagani,G.A., J. Org. Chem., 45, 114 (1980) *Note:* Chemical shifts of O1H and N1H2 groups

140 Glukhikh,V.I., Tandura,S.N., Kuznetsova,G.A., Keiko,V.V., D'yakov,V.M. and Voronkov,M.G., Dokl. Akad. Nauk. Sssr, 239, 1129, 366EE (1978) *Note:* C-13 NMR

141 Tsvetkov,E.N., Malakhova,I.G. and Kabachnik,M.I., Zh. Obshch. Khim., 48, 1230, 1125EE (1978)

142 Reynolds,W.F., Gomes,A., Maron,A., MacIntyre,D.W., Tanin,A., Hamer,G.K. and Peat,I.R., Can. J. Chem., 61, 2376 (1983) *Note:* C-13 NMR

143 Retcofsky,H.L. and Griffin,C.E., Tetrahedron Lett., 18, 1975 (1966) *Note:* F-19 NMR shift (in CCl4) *Note:* Approximate values for NEAT, acetone and Cs2

144 Kondratenko,N.V., Syrova,G.P., Popov,V.I., Sheinker,Y.N. and Yagupol'skii,L.M., Zh. Obshch. Khim., 41, 2056, 2075EE (1971) *Note:* F-19 NMR shift (in CCl4) *Note:* Heptane, CCl4, and dichloroethane

145 Baker,R., Eaborn,C. and Taylor,R., J. Chem. Soc. Perkin Ii, 97 (1972) *Note:* Rate of detritiation in CF3CO2H

146 Schiemenz,G.P., Angew. Chem. Int. Ed., 5, 731 (1966) *Note:* Infra-red absorption: intensity or frequency (in CCl4) *Note:* C-O stretch in anisoles

147 Zuman,P., Coll. Czeck. Chem. Comm., 27, 2035 (1962) *Note:* Polarography

148 Zatsepina,N.N., Tupitsyn,I.F., Kaminskii,Y.L. and Kolodina,N.S., Organic. React., 6, 766, 327EE (1969) *Note:* Base-catalysed H-exchange

149 Wilcox Jr.,C.F. and McIntyre,J.S., J. Org. Chem., 30, 777 (1965) *Note:* PKA in 50% EtOH

150 Tsvetkov,E.N., Malakhova,I.G. and Kabachnik,M.I., Zh. Obshch. Khim., 51, 734, 598EE (1981) *Note:* Apparent PKA of substituted acids (50% aq.-EtOH or water at 25 deg.)

151 Bowden,K. and Young,R.C., Can. J. Chem., 47, 2775 (1969) *Note:* Rate of reaction with diphenyldiazomethane (EtOH, 30 deg.) *Note:* Ratio of acid vs. base catalysed hydrolysis of esters *Note:* MeOH at 60 deg.

152 Hahn,R.C., Corbin,T.F. and Shechter,H., J. Am. Chem. Soc., 90, 3404 (1968) *Note:* Rate of hydrolysis or solvolysis

153 Vo-Kim-Yen, Papouskova,Z., Schraml,J. and Chvalovsky,V., Coll. Czech. Chem. Comm., 38, 3167 (1973) *Note:* Dipole moment

154 Le Guen,M.M.J. and Taylor,R., J. Chem. Soc. Perkin Ii, 559 (1976) *Note:* Protodetritiation

155 Wells,P.R., Linear Free Energy Relationships, Acad. Press, 1968, p.38. *Note:* Rate of hydrolysis or solvolysis

156 Sheppard,W.A., J. Am. Chem. Soc., 87, 2410 (1965) *Note:* Apparent PKA of substituted acids (50% aq.-EtOH or water at 25 deg.) *Note:* F-19 NMR shift (in CCl4) *Note:* Various conditions

157 Slovetskii,V.I. and Fainzil'berg,A.A., Izvest. Akad. Nauk. Sssr, Ser. Khim., 5, 1048, 998EE (1968) *Note:* Ionization of nitro compounds

158 Perrin,D.D., Dempsey,B. and Serjeant,E.P., "pKa Predictions for Organic Acids and Bases", Chapman and Hall, p.107 (1981) *Note:* For pyridine, ortho position *Note:* Apparent PKA of substituted acids (50% aq.-EtOH or water at 25 deg.)

159 Baker,F.W., Parish,R.C. and Stock,L.M., J. Am. Chem. Soc., 89, 5677 (1967) *Note:* Thermodynamic PKA of substituted acids (benzoic acid, phenols or anilines unless otherwise specified, activities-not concentrations-used,50% aq.-EtOH or water at 25 deg.) *Note:* S.prime as defined in J. am. Chem. soc.(1953) 75,2167, but Rho in 50% W. EtOH = 1.65

160 Titov,E.V., Korzhenevskaya,N.G., Kapkan,L.M., Sedova,L.N., Gandel'sman,L.Z., Fialkov,Y.A. and Yagupol'skii,L.M., Zh. Org. Khim., 7, 2552, 2651EE (1971) *Note:* From anilinium (Sigma -) or Sigma ortho *Note:* F-19 NMR shift (in CCl4) *Note:* Of X-C6H4NH3+ in CCl4

161 Vaughan,L.G. and Sheppard,W.A., J. Am. Chem. Soc., 91, 6151 (1969) *Note:* F-19 NMR shift (in CCl4)

162 Jarvis,B.B. and Marien,B.A., J. Org. Chem., 42, 2676 (1977) *Note:* Halomethylpyridyl compounds with methoxide or triaryl phosphines

163 Hansch,C., Rockwell,S.D., Jow,P.Y.C., Leo,A. and Steller,E.E., J. Med. Chem., 20, 304 (1977) *Note:* Apparent PKA of substituted acids (50% aq.-EtOH or water at 25 deg.)

164 Kondratenko,N.V., Kolomeitsev,A.A., Popov,V.I., Il'chenko,A.Y., Korzhenevskaya,N.G., Pirgo,M.D., Titov,E.V. and Yagupol'skii,L.M., Zh. Obshch. Khim., 53, 2500, 2254EE (1983) Note: From anilinium (Sigma -) or Sigma ortho Note: Thermodynamic PKA of substituted acids (benzoic acid, phenols or anilines unless otherwise specified, activities-not concentrations-used,50% aq.-EtOH or water at 25 deg.) Note: F-19 NMR

165 Sheppard,W.A., Transaction N.Y. Acad. Sci., Ser. Ii, 29, 700 (1967) Note: From anilinium (Sigma -) or Sigma ortho Note: Apparent PKA of substituted acids (50% aq.-EtOH or water at 25 deg.) Note: F-19 NMR shift (in CCl4) Note: Benzene

166 Meyers,C.Y., Cremonini,B. and Maioli,L., J. Am. Chem. Soc., 86, 2944 (1964) Note: Apparent PKA of substituted acids (50% aq.-EtOH or water at 25 deg.)

167 Meyers,C.Y., Cremonini,B. and Maioli,L., J. Am. Chem. Soc., 86, 2944 (1964) Note: From phenol (Sigma -) or Sigma ortho Note: Apparent PKA of substituted acids (50% aq.-EtOH or water at 25 deg.)

168 Kargin,Y.M., Podkovyrina,T.A., Chmutova,G.A. and Mannafov,T.G., Zh. Obshch. Khim., 46, 2459, 2357EE (1976) Note: Polarography

169 Dell'erba,C., Spinelli,D., and Garbarino,G., Gazz. Chim. Ital., 100, 777 (1970) Note: Apparent PKA of substituted acids (50% aq.-EtOH or water at 25 deg.)

170 Condon,F.E., Reece,R.T., Shapiro,D.G., Thakkar,D.C. and Goldstein,T.B., J. Chem. Soc. Perkin Ii, 1112 (1974) Note: PKA of hydrazines

171 Geller,B., Skrunts,L. and Khaskin,I., Org. React., 13, 427 (1976) Note: NMR

172 Adcock,W., Cox,D.P. and Kitching,W., J. Organomet. Chem., 133, 393 (1977) Note: F-19 NMR of fluoronaphthalenes (in DMF)

173 Gubin,S.P., Rubezhov,A.Z. and Voronchikhina,L.I., J. Organomet. Chem., 149, 123 (1978) Note: F-19 NMR

174 Archer,W.J., Shafig,Y.E-d. and Taylor,R., J. Chem. Soc. Perkin Ii, 675 (1981) Note: Multiple attachment; e.g. position may be ortho + meta Note: Detritiation TFA-CHCl3 100 deg.

175 Brough,L.F. and West,R., J. Organomet. Chem., 229, 113 (1982) Note: C-13 NMR

176 Kondratenko,N.V., Kolomeitsev,A.A., Popov,V.I., Il'chenko,A.Y., Korzhenevskaya,N.G., Pirgo,M.D., Titov,E.V. and Yagupol'skii,L.M., Zh. Obshch. Khim., 53, 2500, 2254EE (1983) Note: Thermodynamic PKA of substituted acids (benzoic acid, phenols or anilines unless otherwise specified, activities-not concentrations-used,50% aq.-EtOH or water at 25 deg.) Note: F-19 NMR

177 Gavrilov,V.I. and Khusnutdinova,F.M., Zh. Obshch. Khim., 52, 602, 528EE (1982)

178 Yagupol'skii,L.M., Popov,V.I., Pavlenko,N.V., Maletina,I.I., Mironova,A.A., Gavrilova,R.Y. and Orda,V.V., Zh. Org. Khim., 22, 2169, 1947EE (1986)

179 Calvino,R., Fruttero,R., Garrone,A. and Gasco,A., Qsar, 7, 26 (1988)

180 Ellam,G.B. and Johnson,C.D., J. Org. Chem., 36, 2284 (1971) Note: Apparent PKA of substituted acids (50% aq.-EtOH or water at 25 deg.) Note: Substituted pyridines, regression with Sigma: Rho=6.01

181 Cohen,L.A. and Jones,W.M., J. Am. Chem. Soc., 85, 3397 and 3402 (1963) Note: From phenol (Sigma -) or Sigma ortho Note: Thermodynamic PKA of substituted acids (benzoic acid, phenols or anilines unless otherwise specified, activities-not concentrations-used,50% aq.-EtOH or water at 25 deg.) Note: Phenols

182 Bowden,K., Hardy,M. and Parkin,D.C., Can. J. Chem., 46, 2929 (1968) Note: Apparent PKA of substituted acids (50% aq.-EtOH or water at 25 deg.)

183 Gruglikova,R.I., Berestevich,B.K., Babaeva,L.G., Chernyshev,A.I., Yastrebov,V.V. and Unkovskii,B.V., Zh. Org. Khim., 10, 2479, 2499EE (1974) Note: PMR shift G-substituted propargyl alcohols and ethers

184 Verloop,A., Hoogenstraaten,W. and Tipker,J., "Drug Design", Ariens,E.,Ed., Academic Press, 1976, Vol. 7, Ch. 4, Revised in Private Communication. Note: Calculated from C.P.K. bond radii and angles Note: Torsion angle to ethyls (or propyl) = 120 deg.

185 McBee,E.T., Serfaty,I. and Hodgins,T., J. Am. Chem. Soc., 93, 5711 (1971) Note: F-19 NMR

186 Hess,R.E., Schaeffer Jr.,C.D. and Yoder,C.H., J. Org. Chem., 36, 2201 (1971) Note: C-13 coupling

187 Svetlichnyi,V.M., Kudryavtsev,V.V., Adrova,N.A. and Koton,M.M., Zh. Organ. Khim., 10, 1896, 1907EE (1974) Note: From anilinium (Sigma -) or Sigma ortho Note: Kinetics, reaction with PM and BZP

dianhydrides

188 Wojnowski,W. and Herman,A., Z. Anorg. Allg. Chem., 425, 91 (1976) Note: PKA silanthiols Note: Intercept from Krevoy(60)may shift all values upward

189 Shkuzko,O. and Mamayev,V., J. Org. Chem. (USSR), 15, 1737 (1979) Note: Apparent PKA of substituted acids (50% aq.-EtOH or water at 25 deg.)

190 Matsui,T., Ko,H.C. and Hepler,L.G., Can. J. Chem., 52, 2906 (1974) Note: Thermodynamic PKA of substituted acids (benzoic acid, phenols or anilines unless otherwise specified, activities-not concentrations-used,50% aq.-EtOH or water at 25 deg.)

191 Saldabol,N.O., Popelis,Y.Y. and Liepin'sh,E.E., Zh. Org. Khim., 16, 1494, 1285EE (1980) Note: Proton NMR

192 Palecek,J. and Hlavaty,J., Coll. Czech. Chem. Comm., 38, 1985 (1973)

193 Pews,R.G., J. Am. Chem. Soc., 89, 5605 (1967) Note: F-19 NMR shift (in CCl4)

194 Bray,P.J. and Barnes,R.G., J. Chem. Phys., 27, 551 (1957) Note: Quadrupole resonance

195 Hill, E. et.al., J. Am. Chem. Soc., 91, 7381 (1969) Note: Rate of hydrolysis or solvolysis Note: Hetero ring replaces phenyl

196 Lambert,F.L., J. Org. Chem., 31, 4184 (1966) Note: Ratio of acid vs. base catalysed hydrolysis of esters Note: Relates polarography half-wave to Sigma* and Es

197 Meyer,L.H. and Gutowsky,H.S., J. Phys. Chem., 57, 481 (1953) Note: F-19 NMR

198 Angelelli,J.M., Delmas,M.A., Maire,J.C. and Zahra,J.P., J. Organomet Chem., 128, 325 (1977) Note: F-19 NMR

199 Butler,A.R., J. Chem. Soc. (B), 867 (1970) Note: Apparent PKA of substituted acids (50% aq.-EtOH or water at 25 deg.)

200 Gubin,S.P. and Lubovich,A.A., J. Organomet. Chem., 22, 183 (1970) Note: Apparent PKA of substituted acids (50% aq.-EtOH or water at 25 deg.)

201 Andreev,V.P., Vuks,M., Kochetkova,E.V., Remizova,L.A. and Favorskaya,I.A., Zh. Organ. Khim., 15, 464, 410EE (1979) Note: From NMR measurments except after Es where = isosteric value for corresponding amines Note: Quaternization of tertiary amines with ETI, E-N referenced to H=0 as Taft Es Note: Iso-steric with carbon analog

202 Nemleva,S.A., Ivanova,V.M., Gitis,S.S., Vulakh,E.L. and Sidorov,G.V., Zh. Organ. Khim., 15, 2490, 2250EE (1979) Note: Reaction rate of substituted benzoyl chlorides with aniline in benzene

203 Davis,D.D., J. Organomet. Chem., 206, 21 (1981) Note: Charge-transfer frequencies

204 Baram,S.G., Shkurko,O.P. and Mamaev,V.P., Iz. Akad. Nauk Sssr, S. Khim., 8, 1781, 1263EE (1980) Note: F-19 NMR

205 Happer,D.A.R. and Wright,G.J., J. Chem. Soc. Perkin Ii, 694 (1979) Note: Derived from C-13 NMR Note: Other's data

206 Yamamoto,T. and Otsu,T., Chem. Ind. (London), 787 (1967) Note: From polymerization rate Note: No values in computer print-out Note: Radical

207 Tsvetkov,E.N., Lobanov,D.I., Izosenkova,L.A. and Kabachnik,M.I., Zh. Obshch. Khim., 39, 2177, 2126EE (1969) Note: Apparent PKA of substituted acids (50% aq.-EtOH or water at 25 deg.)

208 van Hooidonk,C. and Ginjaar,L., Rec. Trav. Chim., 86, 449 (1967) Note: Rate of alkaline hydrolysis of (diethyl) substituted-phenyl phosphates

209 Bystrov,V.F., Belaya,Z.N., Gruz,B.E., Syrova,G.P., Tolmachev,A.I., Shulezhko,L.M. and Yagupol'skii,L.M., Zh. Obshch. Khim., 38, 1001, 963EE (1968) Note: F-19 NMR shift (in CCl4) Note: Apparent PKA of substituted acids (50% aq.-EtOH or water at 25 deg.) Note: Rho = 1.5 in 50% EtOH, 1.72 in 80% Me-cellosolve

210 Exner,O. and Jonas,J., Coll. Czech. Chem. Comm., 27, 2296 (1962) Note: Apparent PKA of substituted acids (50% aq.-EtOH or water at 25 deg.) Note: Averaged--50% EtOH and 80% W. Me-cellosolve

211 Fischer,A., Leary,G.J., Topsom,R.D. and Vaughan,J., J. Chem. Soc. (B), 846 (1967) Note: From phenol (Sigma -) or Sigma ortho Note: Apparent PKA of substituted acids (50% aq.-EtOH or water at 25 deg.) Note: 4-substituted phenols and 2,6-di-Cl and 2,6-di-Me phenols

212 Romanenko,V.D., Kalibabchuk,N.N., Rositskii,A.A., Zalesskaya,V.G., Didkovskii,V.E. and Iksanova,S.V., Chem. Abstr., (1976) 85, 159203v, Chem. Heterocyc. Cpds., 906 (1976) Note: Apparent PKA of substituted acids (50% aq.-EtOH or water at 25 deg.)

213 Charton,M., J. Org. Chem., 28, 3121 (1963) *Note:* Calculated from linear relation between Sigma constant

214 Sheppard,W.A., J. Am. Chem. Soc., 85, 1314 (1963) *Note:* Apparent PKA of substituted acids (50% aq.-EtOH or water at 25 deg.)

215 Bogatkov,S.V., Popov,A.F. and Litvinenko,L.M., Reakt. Sposob. Organ. Soed., 6, 1011, 436EE (1969) *Note:* From NMR measurements except after Es where = isosteric value for corresponding amines *Note:* Quaternization of tertiary amines with ETI, E-N referenced to H=0 as Taft Es *Note:* Iso-steric with carbon analog

216 Pola,J. and Chvalovsky,V., Coll. Czech. Chem. Comm., 43, 746 (1978) *Note:* IR of phenolic OH with siloxy compounds

217 Ramsey,B.G. and Longmuir,K., J. Org. Chem., 45, 1322 (1980) *Note:* H-1 and C-13 MNR

218 Hradil,J. and Chvalovsky,V., Coll. Czech. Chem. Comm., 32, 171 (1967) *Note:* Chlorination and bromination of silyl substituted toluenes

219 Attached to hydrogen (i.e. = log P)

220 N-aromatic ring--attachment ortho

221 Plzak,Z., Mares,F., Hetflejs,J., Schraml,J., Papouskova,Z., Bazant,V., Rochow,E.G. and Chvalovsky,V., Coll. Czech. Chem. Comm., 36, 3115 (1971) *Note:* Rate of reaction with diphenyldiazomethane (EtOH, 30 deg.) *Note:* In dioxane, toluene and dime-formamide value reported

222 Nishiguchi,T. and Iwakura,Y., J. Org. Chem., 35, 1591 (1970) *Note:* Apparent PKA of substituted acids (50% aq.-EtOH or water at 25 deg.)

223 Moreau,C., Bull. Soc. Chim. Fr., 31 (1968) *Note:* Apparent PKA of substituted acids (50% aq.-EtOH or water at 25 deg.)

224 Fawcett,F.S. and Sheppard,W.A., J. Am. Chem. Soc., 87, 4341 (1965) *Note:* Apparent PKA of substituted acids (50% aq.-EtOH or water at 25 deg.) *Note:* F-19 NMR shift (in CCl4)

225 Fellous,R. and Luft,R., J. Am. Chem. Soc., 95, 5593 (1973) *Note:* Steric (Es), hydroboration of substituted ethylenes with disiamylborane

226 Bouchet,P., Coquelet,C. and Elguero,J., J. Chem. Soc. Perkin Ii, 449 (1974) *Note:* From NMR measurements except after Es where = isosteric value for corresponding amines *Note:* Infra-red (I) and NMR (N)

227 Tolmachev,A.I., Belaya,Z.N., Syrova,G.P., Shulezhko,L.M. and Sheinker,Y.N., Zh. Obshch. Khim., 43, 636, 633EE (1972) *Note:* Apparent PKA of substituted acids (50% aq.-EtOH or water at 25 deg.) *Note:* F-19 NMR shift (in CCl4)

228 Matyushecheva,G.I., Tolmachev,A.I., Shulezhko,A.A., Shulezhko,L.M. and Yagupol'skii,L.M., Zh. Obshch. Khim., 46, 162, 161EE (1976) *Note:* Apparent PKA of substituted acids (50% aq.-EtOH or water at 25 deg.) *Note:* F-19 NMR shift (in CCl4)

229 Verloop,A., Hoogenstraaten,W. and Tipker,J., "Drug Design", Ariens,E.,Ed., Academic Press, 1976, Vol. 7, Ch. 4, Revised in Private Communication. *Note:* Calculated from C.P.K. bond radii and angles *Note:* Torsion angle to 1st ring = 69 deg., to second = 130.25 deg.

230 Bowden,K., Chapman,N.B. and Shorter,J., J. Chem. Soc. London, 5239 (1963) *Note:* Rate of reaction with diphenyldiazomethane (EtOH, 30 deg.) *Note:* Averaged two conditions

231 Chan,T.L. and Miller,J., Aust. J. Chem., 20, 1595 (1967) *Note:* Exchange of Cl by nitrophenoxide

232 Zatsepina,N.N., Kirova,A.V. and Tupitsyn,I.F., Reakt. Sposob. Organ. Soed., 5, 70, 27EE (1968) *Note:* Basic deutero exchange

233 Kitching,W., Alberts,V., Adcock,W. and Cox, D.P., J. Org. Chem., 43, 4652 (1978) *Note:* C-13 NMR

234 Serfaty,I.W., Hodgins,T. and McBee,E.T., J. Org. Chem., 37, 2651 (1972) *Note:* F-19 NMR

235 Schiemenz,G.P., Angew. Chem. Int. Ed., 5, 129 (1966) *Note:* From anilinium (Sigma -) or Sigma ortho *Note:* PKA N,N-dimethylanilines

236 Herkstroeter,W.G., J. Am. Chem. Soc., 95, 8686 (1973) *Note:* Apparent PKA of substituted acids (50% aq.-EtOH or water at 25 deg.) *Note:* 50% ethanol

237 Gel'fond,A.S., Galyametdinov,Y.G., Ermilova,I.P. and Chernokal'skii,B.D., Zh. Obshch. Khim., 50, 559 447EE (1980) *Note:* Apparent PKA of substituted acids (50% aq.-EtOH or water at 25 deg.) *Note:* 2/1 ethanol/H2O

238 Zakomoldina,T.A., Sennikov,P.G., Kuznetsov,V.A., Egorochkin,A.N. and Reikhsfel'd,V.O., Zh. Obshch. Khim., 50, 898, 726EE (1980) *Note:* Charge transfer with tetracyanoethylene

239 Brown,H.C., Kelly,D.P. and Periasamy,M., J. Org. Chem., 46, 3170 (1981) *Note:* For alpha-di-alkyl carbocation, see J. Org. Chem.(1981) 46,3170 *Note:* C-13 NMR alpha-disubstituted carbocations

240 Razuvaev,G.A., Egorochkin,A.N., Skobeleva,S.E., Kuznetsov,V.A., Lopatin,M.A., Petrov,A.A., Zavgorodny,V.S. and Bogoradovsky,E.T., J. Organomet. Chem., 222, 55 (1981) *Note:* IR of acetylenides

241 Shkurko,O.P., Baram,S.G., Mamaev,V.P., Chem. Abs., 98, 197482m (1983) *Note:* C-13, H-1, and F-19 NMR

242 Adcock,W. and Aldous,G.L., J. Organomet. Chem., 202, 385 (1980) *Note:* C-13 and F-19 NMR

243 Koptyug,V.A., Salakhutdinov,N.F. and Detsina,A.N., Zh. Org. Khim., 20, 1143, 1039EE (1984) *Note:* Bromination of substituted benzenes

244 Yagupol'skii,L.M., Pavlenko,N.V., Ignat'ev,N.V., Matyushecheva,G.I. and Semenii,V.Y., Zh. Obshch. Khim., 54, 334, 297EE (1984)

245 Kondratenko,N.V., Popov,V.I., Radchenko,O.A., Ignat'ev,N.V. and Yagupol'skii,L.M., Zh. Org. Khim., 22, 1716, 1542EE (1986)

246 Chernyshev,E.A. and Tolstikova,N.G., Izv. Akad. Nauk. Khim., 3, 455, 419EE (1961) *Note:* Apparent PKA of substituted acids (50% aq.-EtOH or water at 25 deg.)

247 Miller,J. and Parker,A.J., Aust. J. Chem., 11, 302 (1958) *Note:* K for methoxide with 4-substituted 1-Cl-2-NO2-benzenes

248 Zhmurova,I.N., Kukhar,V.P., Tukhar,A.A. and Zolotareva,L.A., Zh. Obshch. Khim., 43, 82, 79EE (1973) *Note:* PKA anilines

249 Wang,C-c. and Shaw,E., Arch. Bioch. Biophys., 150, 259 (1972) *Note:* Rate of hydrolysis or solvolysis

250 Bowden,K., Chapman,N.B. and Shorter,J., J. Chem. Soc. London, 5239 (1963) *Note:* From diazomethane or phenyl azide reaction rate *Note:* Rate of reaction with diphenyldiazomethane (EtOH, 30 deg.) *Note:* Averaged two conditions

251 Little,W.F., Reilley,C.N., Johnson,J.D., Lynn,K.N. and Sanders,A.P., J. Am. Chem. Soc., 86, 1376 (1964) *Note:* Chronopotentiometric quarter wave shift in ferrocene derivatives *Note:* Acetonitrile at 25 deg.

252 Suhr,H., Ber. Bunsengesell., 68, 169 (1964) *Note:* F-19 and H-1 NMR *Note:* Other values in this reference

253 Koshkina,I.M., Remizova,L.A., Ermilova,E.V. and Favorskaya,I.A., Reakt. Sposob. Org. Soed., 7, 944, 427EE (1970) *Note:* PKA amines

254 Pressman,D. and Brown,D.H., J. Am. Chem. Soc., 65, 540 (1943) *Note:* PKA phenyl arsonic acids

255 Dincturk,S. and Jackson,R.A., J. Chem. Soc. Perkin Ii, 1121 and 1127 (1981) *Note:* From thermal decomposition rate *Note:* Thermal decomposition of dibenzyl mercurials

256 Brown,H.C., Periasamy,M. and Liu,K-t., J. Org. Chem., 46, 1646 (1981) *Note:* Carbocations in 'super' acids

257 Rackham,D.M. and Davies,G.L.O., Anal. Lett., 13, 447 (1980) *Note:* PKA P-substituted N-phenylpiperidines

258 Bradamante,S. and Pagani,G.A., J. Org. Chem., 45, 114 (1980) *Note:* C-13 and F-19 data by others

259 Barlin,G.B. and Perrin,D.D., Quart. Rev., 20, 75 (1966) *Note:* Apparent PKA of substituted acids (50% aq.-EtOH or water at 25 deg.)

260 Bzhezovskii,V.M., Kushnarev,D.F., Trofimov,B.A., Kalabin,G.A., Gusarova,N.K. and Efremova,G.G., Iz. Akad. Nauk. Sssr, S. Khim., 11, 2507, 2073EE (1981) *Note:* C-13 NMR

261 Taft,R.W., Taagepera,M., Abboud,J.L.M., Wolf,J.F., DeFrees,D.J., Hehre,W.J., Bartmess,J.E. and McIver Jr.,R.T., J. Am. Chem. Soc., 100, 7765 (1978) *Note:* In gas phase *Note:* Gas phase acidities *Note:* Pulsed ion cyclotron resonance equilibria

262 Kruglikova,R.I., Bogatkov,S.V., Zhestkova,L.N., Kundryutskova,L.A., Berestevich,B.K. and Unkovsky,B.V., Org. Reaktiv., 8, 1015 (1971) *Note:* Rate of hydrolysis or solvolysis

263 Fisher,T.H. and Meierhoefer,A.W., J. Org. Chem., 43, 224 (1978) *Note:* NBS bromination of 3-cyanotoluenes

264 Hansch,C., Leo,A, Taft,R., Chem. Rev., 91, 165 (1991) *Note:* Doubtful value

265 Berliner,E. and Liu,L.H., J. Am. Chem. Soc., 75, 2417 (1953) *Note:* Apparent PKA of substituted acids (50% aq.-EtOH or water at 25 deg.) *Note:* Rate of hydrolysis or solvolysis *Note:* Averaged (88.7% EtOH at 25 and 40 deg.)

266 Zollinger,H. and Wittwer,C., Helv. Chim. Acta, 39, 347 (1956) *Note:* Apparent PKA of substituted acids (50% aq.-EtOH or water at 25 deg.) *Note:* Ionic strength = 0.1

267 Kutter,E. and Hansch,C., J. Med. Chem., 12, 647 (1969) *Note:* Calculated from group radii

268 Hall Jr.,H.K., J. Am. Chem. Soc., 79, 5441 (1957) *Note:* Apparent PKA of substituted acids (50% aq.-EtOH or water at 25 deg.) *Note:* Primary, secondary and tertiary amines

269 Streitwieser Jr.,A., Hammond,H.A., Jagow,R.H., Williams,R.M., Jesaitis,R.G., Chang,C.J. and Wolf,R., J. Am. Chem. Soc., 92, 5141 (1970) *Note:* Multiple attachment; e.g. position may be ortho + meta *Note:* Rate of hydrolysis or solvolysis *Note:* Acetic acid at 40 deg., Rho = -5.71 for activating substituted

270 Johnson,M.D. and Winterton,N., J. Chem. Soc. (A), 507 (1970) *Note:* Apparent PKA of substituted acids (50% aq.-EtOH or water at 25 deg.) *Note:* Pyridinium PKA, Rho = 5.7

271 Shabarov,Y.S., Surikova,T.P. and Levina,R.Y., Zh. Org. Khim., 4, 1175, 1131EE (1968) *Note:* Apparent PKA of substituted acids (50% aq.-EtOH or water at 25 deg.)

272 Szmant,H.H. and Suld,G., J. Am. Chem. Soc., 78, 3400 (1956) *Note:* From phenol (Sigma -) or Sigma ortho *Note:* Apparent PKA of substituted acids (50% aq.-EtOH or water at 25 deg.)

273 Lyalin,V.V., Syrova,G.P., Orda,V.V., Alekseeva,L.A. and Yagupol'skii,L.M., Zh. Org. Khim., 6, 1420, 1433EE (1970) *Note:* F-19 NMR shift (in CCl4) *Note:* In dichloroethane, CCl4 and methylene chloride values also reported

274 Genkina,G.K., Korolev,B.A., Gilyarov,V.A. and Kabachnik,M.I., Zh. Obshch. Khim., 41, 80, 76EE (1971) *Note:* Basicity of imides of phosphorous acids

275 Bouchet,P., Coquelet,C. and Elguero,J., J. Chem. Soc. Perkin Ii, 449 (1974) *Note:* From infra-red or other spectral measurements, *Note:* Infra-red (I) and NMR (N)

276 Hartman,G.D. and Traylor,T.G., J. Am. Chem. Soc., 97, 6147 (1975) *Note:* Rate of hydrolysis or solvolysis

277 Hanstein,W., Berwin,H.J. and Traylor,T.G., J. Am. Chem. Soc., 92, 829 (1970) *Note:* Charge transfer frequencies

278 Yarv,Y.L., Pekhk,T.I., Aaviksaar,A.A., Lobanov,D.I. and Godovikov,N.N., Izv. Akad. Nauk Khim., 7, 1649, 1564EE (1976) *Note:* Rate of hydrolysis or solvolysis

279 Nummert,V.M., Org. Reactivity (USSR), 11, 621 (1975) *Note:* Hydrolysis rate phenyl tosylates

280 Bowden,K., Chapman,N.B. and Shorter,J., J. Chem. Soc., 3370 (1964) *Note:* Possible anomaly; check original article *Note:* Alkaline hydrolysis and diazomethane rates

281 Katritzky,A.R., Short,D.J. and Boulton,A.J., J. Chem. Soc., 1516 (1960) *Note:* Apparent PKA of substituted acids (50% aq.-EtOH or water at 25 deg.)

282 Kaplan,L.J. and Martin,J.C., J. Am. Chem. Soc., 95, 793 (1973) *Note:* F-19 NMR

283 Veschambre,H., Dauphin,G. and Kergomard,A., Bull. Soc. Chim. Fr., 2846 (1967) *Note:* From phenol (Sigma -) or Sigma ortho *Note:* Apparent PKA of substituted acids (50% aq.-EtOH or water at 25 deg.)

284 Kasukhin,L.F., Ponomarchuk,M.P., Kim,T.V., Ivanova,Z.M. and Gololobov,Y.G., Zh. Obshch. Khim., 48, 354, 318EE (1978) *Note:* From diazomethane or phenyl azide reaction rate *Note:* Reaction rate with azidobenzene

285 Eaborn,C., Hancock,A.R. and Stanczyk,W.A., J. Organomet. Chem., 218, 147 (1981) *Note:* C-13 NMR charge transfer frequency

286 Gubin,S.P., Koridze,A.A., Ogorodnikova,N.A., Bezrukova,A.A. and Kvasov,B.A., Iz. Akad. Nauk Sssr, S. Khim., 5, 1170, 929EE (1981) *Note:* F-19 NMR

287 Hoefelmeyer,A.B. and Hancock,C.K., J. Am. Chem. Soc., 77, 4746 (1955) *Note:* Rate of reaction with diphenyldiazomethane (EtOH, 30 deg.) *Note:* Toluene at 25 deg.

288 Orda,V.V., Yagupol'skii,L.M., Bystrov,V.F. and Stepanyants,A.U., Zh. Obshch. Khim., 35, 1628, 1631EE (1965) *Note:* Apparent PKA of substituted acids (50% aq.-EtOH or water at 25 deg.) *Note:* Alpha-substituted toluic and acetic acids *Note:* F-19 NMR shift (in CCl4) *Note:* Heptane

289 Newman,Ed., "Steric Effects in Organic Chemistry", Wiley and Sons, N.Y., 1956, p.591, p.615.

290 Bordwell,F.G. and Boutan,P.J., J. Am. Chem. Soc., 78, 854 (1956) *Note:* From phenol (Sigma -) or Sigma ortho *Note:* Apparent PKA of substituted acids (50% aq.-EtOH or water at 25 deg.)

291 Hine,J. and Bailey Jr.,W.C., J. Org. Chem., 26, 2098 (1961) *Note:* Rate of reaction with diphenyldiazomethane (EtOH, 30 deg.)

292 Yuzhakova,O.A., Bystrov,V.F. and Kostyanovskii,R.G., Izv. Akad. Nauk. Khim., 2, 240, 218EE (1966) *Note:* F-19 NMR shift (in CCl4)

293 Zuman,P., "Substituent Effects in Organic Polarography", Plenum Press, N.Y., 1967, p.76. *Note:* Polarography

294 Sheppard,W.A. and Henderson,R.M., J. Am. Chem. Soc., 89, 4446 (1967) *Note:* Thermodynamic PKA of substituted acids (benzoic acid, phenols or anilines unless otherwise specified, activities-not concentrations-used,50% aq.-EtOH or water at 25 deg.) *Note:* F-19 NMR shift (in CCl4)

295 Fisera,L., Sura,J., Kovac,J. and Lucky,M., Coll. Czech. Chem. Comm., 39, 1711 (1974) *Note:* Apparent PKA of substituted acids (50% aq.-EtOH or water at 25 deg.)

296 Schiemenz,G.P., Angew. Chem. Int. Ed., 5, 731 (1966) *Note:* From anilinium (Sigma -) or Sigma ortho *Note:* Infra-red absorption: intensity or frequency (in CCl4) *Note:* C-O stretch in anisoles

297 Sheppard,W. and Sharts,C., "Organic Flourine Chem." W.A. Benjamin, N.Y., 1969, p. 348. *Note:* From anilinium (Sigma -) or Sigma ortho *Note:* Apparent PKA of substituted acids (50% aq.-EtOH or water at 25 deg.) *Note:* F-19 NMR shift (in CCl4)

298 Pasternak,E.E. and Tomasik,P., Bull. Acad. Pol. Sci., 23, 57 (1975) *Note:* Polarographic reduction of azobenzenes in 60% EtOH

299 Traven,V.F., Korolev,B.A., Pyatkina,T.V. and Stepanov,B.I., Zh. Obshch. Khim., 45, 954 943EE (1975) *Note:* Basicity of organosilyl amines

300 Verloop,A., Hoogenstraaten,W. and Tipker,J., "Drug Design", Ariens,E.,Ed., Academic Press, 1976, Vol. 7, Ch. 4, Revised in Private Communication.

301 Cerfontain,H. and Schaasberg-Nienhuis,Z.R.H., J. Chem. Soc. Perkin Ii, 1780 (1976) *Note:* Rates of sulfonation

302 Jones,R.W.A. and Thomas,J.D.R., J. Chem. Soc. (B), 661 (1966) *Note:* Alkaline hydrolysis of esters in 70% acetone *Note:* Alkyl component

303 Bogatkov,S.V., Zaslavskii,V.G. and Litvinenko,L.M., Dokl. Akad. Nauk. Ssr, 210, 97, 351EE (1973) *Note:* From NMR measurments except after Es where = isosteric value for corresponding amines *Note:* Quaternization of tertiary amines with ETI, E-N referenced to H=0 as Taft Es *Note:* Iso-steric with carbon analog

304 Lifshits,E.B., Yagupol'skii,L.M., Naroditskaya,D.Y. and Kozlova,E.S., Reakt. Sposob. Org. Soedin., 6, 317, 133EE (1969) *Note:* PKA carbocyanine dyes

305 Kortekaas,T.A., Cerfontain,H. and Gall,J.M., J. Chem. Soc. Perkin Ii, 445 (1978) *Note:* Sulfonation of biphenylsulfonic acids

306 Gitis,S.S., Ivanov,A.V., Kaminski,A.J. and Kozina,Z.A., Org. Reactivity, 3, 142 (1966) *Note:* Alkaline hydrolysis esters of 2,4-dinitrophenol

307 Barlin,G.B. and Perrin,D.D., Quart. Rev., 20, 75 (1966) *Note:* From phenol (Sigma -) or Sigma ortho *Note:* Apparent PKA of substituted acids (50% aq.-EtOH or water at 25 deg.)

308 Robinson,C.N., Slater,C.D., Covington Iii,J.S., Chang,C.R., Dewey,L.S., Franceschini,J.M., Fritzsche,J.L., Hamilton,J.E., Irving Jr.,C.C., Morris,J.M., Norris,D.W., Rodman,L.E., Smith,V.I., Stablein,G.E. and Ward,F.C., J. Magnet. Reson., 41, 293 (1980) *Note:* From NMR measurments except after Es where = isosteric value for corresponding amines *Note:* C-13 NMR

309 Ananthakrishnanadar,P. and Rajasekaran,K., Ind. J. Chem., 19B, 324 (1980) *Note:* Rate of reaction of substituted pyridines with CH3I

310 Bijloo,G.J. and Rekker,R.F., Quant. Struct. Act. Rel., 2, 124 (1983) *Note:* Acetic acid PKA's from IUPAC

311 Cartledge,F.K., Organomet., 2, 425 (1983) *Note:* Es for silicon reactions only, see Organomet.(1983) 2,425 *Note:* Acid hydrolysis of Si-H compounds

312 Kabachnik,M.I. and Mastryukova,T.A., Zh. Obshch. Khim., 54, 2161, 1931EE (1984) *Note:* I Acidities of phosphonium salts

313 Gavrilov,V.I. and Khusnutdinova,F.M., Zh. Obshch. Khim., 57, 344, 295EE (1987)

314 Temple,R.D. and Leffler,J.E., Tet. Letters, Part 3, 1893 (1968) *Note:* Apparent PKA of substituted acids (50% aq.-EtOH or water at 25 deg.) *Note:* Aliphatic acids (RCH2-)

315 Smith,J.H. and Menger,F.M., J. Org. Chem., 34, 77 (1969) *Note:* Rate of hydrolysis or solvolysis *Note:* Basic hydrolysis methylbenzoates (MeOH at 25 deg.)

316 Serjeant,E.P., Aust. J. Chem., 22, 1189 (1969) *Note:* Apparent PKA of substituted acids (50% aq.-EtOH or water at 25 deg.)

317 Sheppard,W.A., J. Am. Chem. Soc., 92, 5419 (1970) *Note:* Apparent PKA of substituted acids (50% aq.-EtOH or water at 25 deg.) *Note:* F-19 NMR shift (in CCl4) *Note:* 75% MeOH at 39.8 deg., CCl3F

318 Van Poucke,R., Pollet,R. and De Cat,A., Bull. Soc. Chim. Belges, 75, 573 (1966) *Note:* Apparent PKA of substituted acids (50% aq.-

EtOH or water at 25 deg.) *Note:* 80% W. ethylene glycol monomethyl ether

319 Tsvetkov,E.N., Lobanov,D.I., Makhamatkhanov,M.M. and Kabachnik,M.I., Tetrahedron, 25, 5623 (1969) *Note:* Apparent PKA of substituted acids (50% aq.-EtOH or water at 25 deg.)

320 Yagupol'skii,L.M. and Kondratenko,N.V., Zh. Obshch. Khim., 39, 1755, 1719EE (1969) *Note:* F-19 NMR shift (in CCl4)

321 Chernyshev,E.A. and Tolstikova,N.G., Izv. Akad. Nauk. Khim., 3, 455, 419EE (1961) *Note:* From anilinium (Sigma -) or Sigma ortho *Note:* Apparent PKA of substituted acids (50% aq.-EtOH or water at 25 deg.)

322 Deno,N.C. and Evans,W.L., J. Am. Chem. Soc., 79, 5804 (1957) *Note:* Protonolysis of aryl silanes, average values

323 Roberts,D.D., J. Org. Chem., 29, 2714 (1964) *Note:* Ratio of acid vs. base catalysed hydrolysis of esters *Note:* Larger steric constant is corrected for hyperconjugation

324 Van Bekkum,H., Verkade,P.E. and Wepster,B.M., Rec. Trav. Chem., 78, 815 (1959) *Note:* Normalized on basis of 10 selected Sigma values--averaged

325 Chambers,A., and Stirling,C.J.M., J. Chem. Soc., 4558 (1965) *Note:* Infra-red absorption: intensity or frequency (in CCl4)

326 Hine,J. and Bailey Jr.,W.C., J. Am. Chem. Soc., 81, 2075 (1959) *Note:* Rate of reaction with diphenyldiazomethane (EtOH, 30 deg.) *Note:* T-3-substituted acrylic acids Sigma*=(log K +.663)/1.174

327 Tsvetkov,E.N., Lobanov,D.I., Kamai,G.K., Chadaeva,N.A. and Kabachnik,M.I., Zh. Obshch. Khim., 39, 2670, 2608EE (1969) *Note:* Apparent PKA of substituted acids (50% aq.-EtOH or water at 25 deg.)

328 Tsvetkov,E.N., Makhamatkhanov,M.M., Lobanov,D.I. and Kabachnik,M.I., Zh. Obshch. Khim., 42, 769, 761EE (1972) *Note:* From phenol (Sigma -) or Sigma ortho *Note:* Apparent PKA of substituted acids (50% aq.-EtOH or water at 25 deg.)

329 Packer,J., Vaughan,J. and Wilson,A.F., J. Org. Chem., 23, 1215 (1958) *Note:* Solvolysis diphenylcarbinyl chlorides

330 Hine,J., J. Am. Chem. Soc., 82, 4877 (1960) *Note:* Apparent PKA of substituted acids (50% aq.-EtOH or water at 25 deg.) *Note:* Rate of hydrolysis or solvolysis

331 Ten Thije,P.A. and Janssen,M.J., Recueil, 84, 1169 (1965) *Note:* Ratio of acid vs. base catalysed hydrolysis of esters

332 Katritzky,A.R. and Swinbourne,F.J., J. Chem. Soc., 6707 (1965) *Note:* Proton resonance NMR

333 Angelelli,J.M., Brownlee,R.T.C., Katritzky,A.R., Topsom,R.D. and Yakhontov,L., J. Am. Chem. Soc., 91, 4500 (1969) *Note:* From NMR measurements except after Es where = isosteric value for corresponding amines *Note:* Infrared intensities (sign not determined)

334 Minkin,V.I., Medyantseva,E.A., Andreeva,I.M. and Yur'eva,V.A., Zh. Org. Khim., 10, 2597, 2613EE (1974) *Note:* F-19 NMR shift (in CCl4)

335 Pelizzeti,E. and Verdi,C., J. Chem. Soc. Perkin Ii, 808 (1973) *Note:* Dissociation constant benzoylacetanilides

336 Chekunina,L.I., Bokanov,A.I. and Stepanov,B.I., Zh. Obshch. Khim., 42, 110, 105EE (1972) *Note:* From anilinium (Sigma -) or Sigma ortho *Note:* Apparent PKA of substituted acids (50% aq.-EtOH or water at 25 deg.) *Note:* Dimethylanilines

337 Laurence,C. and Wojtkowlak,B., Bull. Soc. Chim. Fr., 3833 (1971) *Note:* Infra-red absorption: intensity or frequency (in CCl4)

338 Steward,O.W., Dziedzic,J.E., Johnson,J.S. and Frohliger,J.O., J. Org. Chem., 36, 3480 (1971) *Note:* Apparent PKA of substituted acids (50% aq.-EtOH or water at 25 deg.)

339 Shabarov,Y.S., Blagodatskikh,S.A., Levina,M.I. and Fedotov,A.N., Zh. Org. Khim., 11, 1223, 1212EE (1975) *Note:* F-19 NMR shift (in CCl4)

340 Yagupol'skii,L.M., Popov,V.I., Kondratenko,N.V. and Konovalov,E.V., Zh. Org. Khim., 10, 277, 278EE (1974) *Note:* F-19 NMR shift (in CCl4)

341 Charton,M., "Design of Biopharm. Prop. Thru Prodrugs and Analogs", Roche,E.,Ed., Am. Pharm. Assoc., 1977, Ch. 9, p. 269. *Note:* Minimum dimension for steric effect *Note:* Vanderwal radii and secondary values

342 Nagai,Y., Ohtsuki,M-a., Nakano,T. and Watanabe,H., J. Organomet. Chem., 35, 81 (1972) *Note:* Si-H NMR shifts

343 Yurchenko,R.I., Voitsekhovskaya,O.M., Zhmurova,I.N. and Lysova,N.N., Zh. Obshch. Khim., 45, 1735, 1700EE (1975) *Note:* Spectra

344 Bitterwolf,T.E. and Ling,A.C., J. Organomet. Chem., 141, 355 (1977) *Note:* H-NMR

345 Genkina,G.K., Gilyarov,V.A. and Kabachnik,A.M.I., Dokl. Akad. Nauk Ssr, 188, 1288, 841EE (1969) *Note:* F-19 NMR P-F-phenylimidophosphates

346 Ohto,Y., Hashida,Y., Sekiguchi,S. and Matsui,K., Bull. Chem. Soc. Japan, 47, 1301 (1974) *Note:* From phenol (Sigma -) or Sigma ortho *Note:* Apparent PKA of substituted acids (50% aq.-EtOH or water at 25 deg.)

347 Benkeser,R.A., DeBoer,C.E., Robinson,R.E. and Sauve,D.M., J. Am. Chem. Soc., 78, 682 (1956) *Note:* Rate of reaction with diphenyldiazomethane (EtOH, 30 deg.) *Note:* Gives lower values, substituted dime-anilines gives higher values

348 Levitt,L.S. and Levitt,B.W., Tetrahedron, 29, 941 (1973) *Note:* Adiabatic photoionization

349 Babaeva,L.G., Bogatkov,S.V., Kruglikova,R.I. and Unkovskii,B.V., Zh. Org. Khim., 12, 1738, 1706EE (1976) *Note:* Alkaline hydrolysis and diazomethane rates

350 Babaeva,L.G., Bogatkov,S.V., Kruglikova,R.I. and Unkovskii,B.V., Zh. Org. Khim., 12, 1738, 1706EE (1976) *Note:* Possible anomaly; check original article *Note:* Alkaline hydrolysis and diazomethane rates

351 Cornish,A.J. and Eaborn,C., J. Chem. Soc. Perkin Ii, 874 (1975) *Note:* Rate of desilylation

352 Kusuyama,Y. and Ikeda,Y., Bull. Chem. Soc. Japan, 46, 204 (1973) *Note:* Apparent PKA of substituted acids (50% aq.-EtOH or water at 25 deg.) *Note:* Rate of hydrolysis or solvolysis

353 Kravtsov,D.N., Kvasov,B.A., Fedin,E.N., Faingor,B.A. and Golovchenko,L.S., Iz. Akad. Nauk. Ssr, Khim., 3, 536, 477EE (1969) *Note:* F-19 NMR

354 Matrosov,E.I., Kryuchkov,A.A., Nifant'ev,E.E. and Kabachnik,M.I., Izv. Nauk. Ssr Khim., 4, 791, 719EE (1977) *Note:* Apparent PKA of substituted acids (50% aq.-EtOH or water at 25 deg.)

355 Rubezhov,A.Z., Voronchikhina,L.I. and Gubin,S.P., Izv. Akad. Nauk Ssr, Ser. Khim., 328, 301EE (1979) *Note:* Proton NMR

356 Veschambre,H., Dauphin,G. and Kergomard,A., Bull. Soc. Chim. Fr., 134 (1967) *Note:* Apparent PKA of substituted acids (50% aq.-EtOH or water at 25 deg.)

357 Nikolic,G.S., Sokolov,M.T. and Muskatirovic,M.D., J. Chem. Soc. Perkin Ii, 23 (1979) *Note:* Hydrolysis of ethyl esters

358 Pola,J., Jakoubkova,M. and Chvalovsky,V., Coll. Czech. Chem. Comm., 44, 3688 (1979) *Note:* IR of phenolic H-bond complex in CCl4

359 Adcock,W., Aldous,G.L. and Kitching,W., Tetrahed. Lett., 36, 3387 (1978) *Note:* Apparent PKA of substituted acids (50% aq.-EtOH or water at 25 deg.) *Note:* From quinuclidines (not same scale as S-INDQ)

360 Palm,V.A., Summary of Sci. and Technology, Ser. Gen. Quest. Org. Chem. Moscow, p.163 (1979) *Note:* No electrostatic correction *Note:* Review

361 Taylor,R., J. Chem. Soc. Perkin Ii, 755 (1978) *Note:* Pyrolysis aryl-ethyl acetates

362 Shafig,Y. and Taylor,R., J. Chem. Soc. Perkin Ii, 1263 (1978) *Note:* Detritiation TFA 70 deg.

363 Archer,W.J. and Taylor,R., J. Chem. Soc. Perkin Ii, 301 (1982) *Note:* Detritiation TFA and TFA-acetic acid 70 deg.

364 Zatsepina,N.N., Kaminskii,Y.L. and Tupitsyn,I.F., Reakt. Spos. Org. Soed., 6, 778, 333EE (1969) *Note:* Statistics of 40 reaction series

365 Angelelli,J.M., Delmas,M.A., Chouteau,J., Guiliano,M. and Mille,G., J. Organomet. Chem., 188, 85 (1980) *Note:* From infra-red or other spectral measurements, *Note:* IR and NMR

366 Claramunt,R.M. and Elguero,J., Coll. Czech. Chem. Comm., 46, 584 (1981) *Note:* Rate of photochlorination of alkyl acetates *Note:* C-13, F-19 NMR

367 Bel'skii,V.E., Kurguzova,A.M., Kudryavtseva,L.A. and Ivanov,B.E., Iz. Akad. Nauk Sssr, S. Khim., 10, 2220, 1956EE (1982) *Note:* Kinetics alkaline hydrolysis

368 Byrne,C.J., Happer,D.A.R., Hartshorn,M.P. and Powell,H.K.J., J. Chem. Soc. Perkin Ii, 1649 (1987) *Note:* Solvolysis of 2-Cl-2-PH-propanes

369 Zollinger,H. and Wittwer,C., Helv. Chim. Acta, 39, 347 (1956) *Note:* From anilinium (Sigma -) or Sigma ortho *Note:* Apparent PKA of substituted acids (50% aq.-EtOH or water at 25 deg.) *Note:* Ionic strength = 0.1

370 Yagupol'skii,L.M. and Yagupol'skaya,L.N., Doklady Akad. Nauk. Sssr, 134, 1381, 1207EE (1960) *Note:* Apparent PKA of substituted acids (50% aq.-EtOH or water at 25 deg.)

371 Tsukerman,S.V., Kutulya,L.A. and Lavrushin,V.F., Chem. Abs., 72,

120872u (1970) *Note:* Basicity constant in H2SO4-HAC (hetero ring replaces phenyl)

372 Wehry,E.L., J. Am. Chem. Soc., 89, 41 (1967) *Note:* From phenol (Sigma -) or Sigma ortho *Note:* Apparent PKA of substituted acids (50% aq.-EtOH or water at 25 deg.) *Note:* Also values for photoexcited singlet and triplet

373 Bott,R.W., Dowden,B.F. and Eaborn,C., J. Chem. Soc., 4994 (1965) *Note:* Rate of hydrolysis or solvolysis *Note:* 39% W. MeOH at 50.3 deg.

374 Noyce,D.S. and Castenson,R.L., J. Am. Chem. Soc., 95, 1247 (1973) *Note:* Rate of hydrolysis or solvolysis *Note:* Trifluoroethanol at 75 deg.

375 Szmant,H.H. and Suld,G., J. Am. Chem. Soc., 78, 3400 (1956) *Note:* Apparent PKA of substituted acids (50% aq.-EtOH or water at 25 deg.)

376 Kruglikova,R.I. and Kalinina,G.R., Zh. Org. Khim., 7, 857, 871EE (1971) *Note:* Basicity alkynyl amino alcohols

377 Andrianov,K.A., Kopylov,V.M., Petrov,K.I., Syrtsova,Z.S., Ratomskaya,M.A. and Nuikin,A.Y., Zh. Obshch. Khim., 43, 1057, 1050EE (1973) *Note:* Infra-red absorption: intensity or frequency (in CCl4) *Note:* Basicity of organosilyl amines

378 Meyers,C.Y., Gazz. Chim. Ital., 93, 1206 (1963) *Note:* From phenol (Sigma -) or Sigma ortho *Note:* Apparent PKA of substituted acids (50% aq.-EtOH or water at 25 deg.)

379 Hancock,C.K., Meyers,E.A. and Yager,B.J., J. Am. Chem. Soc., 83, 4211 (1961) *Note:* Hancock's hyperconjugation correction for Es *Note:* LCAO-MO correction for hyperconjugation effects

380 Kruglikova,R.I., Vasil'ev,S.V., Kudryukova,L.A. and Kalinina,G.R., Zh. Obshch. Khim., 38, 1961, 1904EE (1968) *Note:* Apparent PKA of substituted acids (50% aq.-EtOH or water at 25 deg.) *Note:* Tertiary amines

381 Andrianov,K.A., Kopylov,V.M., Chernyshev,A.I., Andreeva,S.V. and Shragin,I.S., Zh. Obshch. Khim., 45, 351, 340EE (1975) *Note:* Basicity of organosilyl amines

382 Nelson,G.L., Levy,G.C. and Cargioli,J.D., J. Am. Chem. Soc., 94, 3089 (1972) *Note:* C-13 NMR

383 Rakita,P.E. and Worsham,L.S., J. Organomet. Chem., 137, 145 (1977) *Note:* C-13 NMR

384 Dimitrijevic,D.M., Tadic,Z.D., Misic-Vukovic,M.M. and Muskatirovic,M., J. Chem. Soc. Perkin Ii, 1051 (1974) *Note:* Reaction with diazodiphenylmethane

385 Bassindale,A.R., Eaborn,C. and Walton,D.R.M., J. Organomet. Chem., 21, 91 (1970) *Note:* F-19 NMR

386 Schadt Iii,F.L., Lancelot,C.J. and Schleyer,P., J. Am. Chem. Soc., 100, 228 (1978) *Note:* Rate of hydrolysis or solvolysis

387 Cutress,N.C., Katritzky,A.R., Eaborn,C., Walton,D.R.M. and Topsom,R.D., J. Organomet Chem., 43, 131 (1972) *Note:* IR

388 Fisera,L., Lesko,J., Kovac,J., Hrabovsky,J. and Sura,J., Coll. Czech. Chem. Comm., 42, 105 (1977) *Note:* PKA pyridines

389 German,M.I., Traven,V.F., Eismont,M.Y., Korolev,B.A. and Stepanov,B.I., Zh. Obshch. Khim., 47, 2392, 2187EE (1977) *Note:* Apparent PKA of substituted acids (50% aq.-EtOH or water at 25 deg.) *Note:* Absorption spectra

390 Veschambre,H., Dauphin,G. and Kergomard,A., Bull. Soc. Chim. Fr., 2846 (1967) *Note:* From anilinium (Sigma -) or Sigma ortho *Note:* Apparent PKA of substituted acids (50% aq.-EtOH or water at 25 deg.)

391 Panaye,A., MacPhee,J.A. and Dubois,J-e., Tetrahed. Lett., 21, 3485 (1980) *Note:* From acid catalytic esterification in MeOH, scaled to H = 0, see Tetrahed. Lett.(1978) 35,3293 *Note:* Acid catalyzed esterification rate in MeOH

392 Milyaev,Y.F., Khorishko,S.A. and Balyatinskaya,L.N., Zh. Obshch. Khim., 51, 192, 168EE (1981) *Note:* Apparent PKA of substituted acids (50% aq.-EtOH or water at 25 deg.) *Note:* Anilines

393 Egorochkin,A.N., Vyazankin,N.S., Khorshev,S.Y., Chernysheva,T.I. and Kuz'min,O.V., Iz. Akad. Nauk. Sssr, S. Khim., 5, 1194, 1139EE (1970) *Note:* IR of mono- and di-substituted silanes

394 Kulinkovich,O.F., Tishchenko,I.G., Reznikov,I.V. and Pap,A.A., Zh. Org. Khim., 17, 473, 396EE (1981) *Note:* Solvolysis of 2-Cl-2-PH-propanes

395 Shizong,Y., Zuoqin,W., Yasen,Y. and Yanwen,B., Kexue Tongbao (China), 29, 1490 (1984) *Note:* C-13 NMR

396 Byrne,C.J., Happer,D.A.R., Hartshorn,M.P. and Powell,H.K.J., J. Chem. Soc. Perkin Ii, 1649 (1987) *Note:* Apparent PKA of substituted acids (50% aq.-EtOH or water at 25 deg.)

397 Hammett,L.P., "Physical Organic Chemistry" McGraw-Hill, N.Y., 1940, p.188. *Note:* Thermodynamic PKA of substituted acids (benzoic acid, phenols or anilines unless otherwise specified, activities-not concentrations-used,50% aq.-EtOH or water at 25 deg.) *Note:* Apparent PKA of substituted acids (50% aq.-EtOH or water at 25 deg.)

398 Kauer,J.C. and Sheppard,W.A., J. Org. Chem., 32, 3580 (1967) *Note:* F-19 NMR shift (in CCl4) *Note:* In acetonitrile

399 Hall Jr.,H.K., J. Am. Chem. Soc., 78, 2570 (1956) *Note:* Apparent PKA of substituted acids (50% aq.-EtOH or water at 25 deg.)

400 Shabarov,Y.S., Surikova,T.P. and Levina,R.Y., Zh. Org. Khim., 4, 1175, 1131EE (1968) *Note:* From anilinium (Sigma -) or Sigma ortho *Note:* Apparent PKA of substituted acids (50% aq.-EtOH or water at 25 deg.)

401 Cook,M.A., Eaborn,C. and Walton,D.R.M., J. Organometal. Chem., 24, 293 (1970) *Note:* Apparent PKA of substituted acids (50% aq.-EtOH or water at 25 deg.)

402 Richardson,W.H., Smith,R.S., Snyder,G., Anderson,B. and Kranz,G.L., J. Org. Chem., 37, 3915 (1972) *Note:* Ratio of acid vs. base catalysed hydrolysis of esters *Note:* MeOH at 30 deg., 85% EtOH at 25 deg.

403 Roberts,D.D., J. Org. Chem., 29, 2714 (1964) *Note:* Hancock's hyperconjugation correction for Es *Note:* Ratio of acid vs. base catalysed hydrolysis of esters *Note:* Larger steric constant is corrected for hyperconjugation

404 Smith,G.G. and Kirby,J.A., J. Heterocycl. Chem., 8, 1101 (1971) *Note:* Rate of hydrolysis or solvolysis

405 Brock,F.H., J. Org. Chem., 24, 1802 (1959) *Note:* Reaction rate phenyl isocyanates with 1-butanol or ethanol

406 Illuminati,G., J. Am. Chem. Soc., 80, 4941 (1958) *Note:* Rate of bromination of substituted polymethylbenzenes

407 Tsvetkov,E.N., Makhamatkhanov,M.M., Lobanov,D.I. and Kabachnik,M.I., Zh. Obshch. Khim., 40, 2387, 2376EE (1970) *Note:* From phenol (Sigma -) or Sigma ortho *Note:* Apparent PKA of substituted acids (50% aq.-EtOH or water at 25 deg.)

408 Toma,S., Coll. Czech. Chem. Comm., 34, 2771 (1969) *Note:* Addition of Et-cyanobutyrate to chalcones

409 Bowden,K. and Shaw,M.J., J. Chem. Soc. (B), 161 (1971) *Note:* Apparent PKA of substituted acids (50% aq.-EtOH or water at 25 deg.)

410 Deady,L.W., Foskey,D.J. and Shanks,R.A., J. Chem. Soc. (B), 1962 (1971) *Note:* Rate of hydrolysis or solvolysis

411 Rudaya,L.I., Kvitko,I.Y. and Porai-Koshits,B.A., Zh. Org. Khim., 8, 13, 11EE (1972) *Note:* Multiple attachment; e.g. position may be ortho + meta *Note:* Thermodynamic PKA of substituted acids (benzoic acid, phenols or anilines unless otherwise specified, activities-not concentrations-used,50% aq.-EtOH or water at 25 deg.) *Note:* Infra-red absorption: intensity or frequency (in CCl4)

412 Noyce,D.S. and Forsyth,D.A., J. Org. Chem., 39, 2828 (1974) *Note:* Rate of hydrolysis or solvolysis

413 Price,C.C. and Belanger,W.J., J. Am. Chem. Soc., 76, 2682 (1954) *Note:* Rate of hydrolysis or solvolysis

414 Stang,P.J. and Anderson,A.G., J. Org. Chem., 41, 781 (1976) *Note:* Apparent PKA of substituted acids (50% aq.-EtOH or water at 25 deg.)

415 Thomas,A. and Tomasik,P., Bull. Acad. Polon. Sci., 23, 65 (1975) *Note:* Polarography

416 Chuchani,G. and Frohlich,A., J. Chem. Soc. (B), 1417 (1971) *Note:* From phenol (Sigma -) or Sigma ortho *Note:* S-2

417 Elderfield,R.C. and Siegel,M., J. Am. Chem. Soc., 73, 5622 (1951) *Note:* Apparent PKA of substituted acids (50% aq.-EtOH or water at 25 deg.)

418 Herbst Jr.,R.L. and Jacox,M.E., J. Am. Chem. Soc., 74, 3004 (1952) *Note:* Apparent PKA of substituted acids (50% aq.-EtOH or water at 25 deg.)

419 Durmis,J.,Karvas,M. and Manasek,Z., Coll. Czech. Chem. Comm., 38, 215 (1973) *Note:* From phenol (Sigma -) or Sigma ortho *Note:* Apparent PKA of substituted acids (50% aq.-EtOH or water at 25 deg.) *Note:* Phenols

420 Stevenson,G.W. and Williamson,D., J. Am. Chem. Soc., 80, 5943 (1958) *Note:* PKA cyanoamines

421 Jaffe,H.H., Chem. Rev., 53, 191 (1953) *Note:* Multiple attachment; e.g. position may be ortho + meta *Note:* Wide variety of methods, see review article for details

422 Tsvetkov,E.N., Malevannaya,R.A. and Kabachnik,M.I., Zh. Obshch. Khim., 45, 716, 706EE (1975) *Note:* PKA phosphinylacetic acids

423 Bruce,G.T., Cooksey,A.R. and Morgan,K.J., J. Chem. Soc. Perkin Ii,

551 (1975) *Note:* Rates of borohydride reduction

424 Nummert,V.M., Palm,V.A. and Uudam,M.K., Org. Reactivity (USSR), 12, 445 (1975) *Note:* PKA of quaternary substituted alcohols (note *Note:* Lower value includes electrostatic correct)

425 Price,C.C. and Hydock,J.J., J. Am. Chem. Soc., 74, 1943 (1952) *Note:* Rate of hydrolysis or solvolysis

426 Oae,S. and Price,C.C., J. Am. Chem. Soc., 80, 3425 (1958) *Note:* From phenol (Sigma -) or Sigma ortho *Note:* Apparent PKA of substituted acids (50% aq.-EtOH or water at 25 deg.)

427 Hansch,C., Regr. of Data in: Zh. Obshch. Khim. 43, 1556, 1540EE (1973) *Note:* From anilinium (Sigma -) or Sigma ortho *Note:* Apparent PKA of substituted acids (50% aq.-EtOH or water at 25 deg.)

428 Voronkov,M.G., Kashik,T.V., Deriglazova,E.S., Lukevits,E.Y., Pestunovich,A.E. and Sturkovich,R.Y., Zh. Obshch. Khim., 46, 1522, 1487EE (1976) *Note:* PKA heterocyclic amines

429 Kochi,J.K. and Hammond,G.S., J. Am. Chem. Soc., 75, 3452 (1953) *Note:* Apparent PKA of substituted acids (50% aq.-EtOH or water at 25 deg.) *Note:* Rate of hydrolysis or solvolysis *Note:* Substituted benzyl tosylates--averaged

430 Tsukerman,S.V., Kutulya,L.A., Lavrushin,V.F. and Tyau,N.M., Russ. J. Phys. Chem., 42, 1015 (1968) *Note:* Apparent PKA of substituted acids (50% aq.-EtOH or water at 25 deg.) *Note:* Chalcone analogs

431 Nesmeyanov,A.N., Makarova,L.G., Ustynyuk,N.A., Kvasov,B.A. and Bogatyreva,L.V., J. Organomet. Chem., 34, 185 (1972) *Note:* F-19 NMR

432 Smith,G.G. and Lum,K.K., Chem. Commun., 1208 (1968) *Note:* Miscellaneous

433 Voronkov,M.G., Kashik,T.V. and Deriglazova,E.S., Zh. Obshch. Khim., 49, 803, 697EE (1979) *Note:* Basicity of tertiary amines

434 Nesterova,L.I., Kasukhin,L.F., Ponomarchuk,M.P. and Gololobov,Y.G., Zh. Obshch. Khim., 48, 1061, 966EE (1978) *Note:* From diazomethane or phenyl azide reaction rate *Note:* Reaction rate with azidobenzene

435 Martin,D.J. and Griffin,C.E., J. Org. Chem., 30, 4034 (1965) *Note:* Apparent PKA of substituted acids (50% aq.-EtOH or water at 25 deg.) *Note:* Acetic acids Taft #3 equation for Sigma*

436 Livingstone,D.J., Hyde,R.M. and Foster,R., Eur. J. Med. Chem., 14, 393 (1979) *Note:* Calculated from other parameters *Note:* Substituted benzenes complexed with trinitrobenzene in CCl4

437 Zhmurova,I.N., Yurchenko,R.I., Tukhar,A.A., Yurchenko,V.G. and Voitsekhovskaya,O.M., Zh. Obshch. Khim., 49, 2401, 2119EE (1979) *Note:* Electronic spectra

438 Seconi,G., Eaborn,C. and Fischer,A., J. Organomet. Chem., 177, 129 (1979) *Note:* Basic cleavage of substituted trimethylsilanes

439 Perrin,D.D., Dempsey,B. and Serjeant,E.P., "pKa Predictions for Organic Acids and Bases", Chapman and Hall, p.107 (1981) *Note:* Multiple attachment; e.g. position may be ortho + meta *Note:* Apparent PKA of substituted acids (50% aq.-EtOH or water at 25 deg.)

440 Archer,W.J., Hossaini,M.A. and Taylor,R., J. Chem. Soc. Perkin Ii, 181 (1982) *Note:* Detritiation TFA 70 deg.

441 Katritzky,A.R., Kingsland,M. and Tee,O.S., J. Chem. Soc. (B), 1484 (1968) *Note:* Deuterium exchange in dilute acid

442 Bradamante,S. and Pagani,G.A., J. Org. Chem., 45, 114 (1980) *Note:* From anilinium (Sigma -) or Sigma ortho *Note:* C-13 and F-19 data by others

443 Bradamante,S. and Pagani,G.A., J. Org. Chem., 45, 114 (1980) *Note:* From phenol (Sigma -) or Sigma ortho *Note:* C-13 and F-19 data by others

444 Brownlee,R.T.C. and Sadek,M., Aust. J. Chem., 34, 1593 (1981) *Note:* Solvent dependent; see reference for other values *Note:* C-13 NMR

445 Bodeker,J., Bloedorn,W-d., Herz,S., Weber,F-g. and Radeglia,R., Z. Chem., 22, 262 (1982) *Note:* C-13 NMR

446 Pola,J. and Chvalovsky,V., Coll. Czech. Chem. Comm., 47, 613 (1982) *Note:* IR

447 Sergienko,L.M., Ratovskii,G.V., Dmitriev,V.I. and Timokhin,B.V., Zh. Obshch. Khim., 54, 64, 55EE (1984) *Note:* IR spectra

448 Voronkov,M.G., Kashik,T.V., Deriglazova,E.S., Kositsyna,E.I., Pestunovich,A.E. and Lukevits,E.Y., Zh. Obshch. Khim., 51, 375, 304EE (1981) *Note:* PKA of substituted cyclic amines

449 Matveeva,A.G., Kudryavtsev,I.Y., Grigor'eva,A.A., Matrosov,E.I., Kuznetsova,E.K., Zakharov,L.S. and Kabachnik,M.I., Izv. Akad. Nauk Sssr, 7, 1491, 1329EE (1982) *Note:* Apparent PKA of

substituted acids (50% aq.-EtOH or water at 25 deg.)

450 Pews,R.G., Tsuno,Y. and Taft,R.W., J. Am. Chem. Soc., 89, 2391 (1967) *Note:* F-19 NMR shift (in CCl4) *Note:* P-benzoyl derivatives (in CH2Cl2 at 25 deg.)

451 Boloshchuk,V.G., Yagupol'skii,L.M., Syrova,G.P. and Bystrov,V.F., Zh. Obshch. Khim., 3, 118, 105EE (1967) *Note:* Apparent PKA of substituted acids (50% aq.-EtOH or water at 25 deg.) *Note:* F-19 NMR shift (in CCl4) *Note:* In heptane

452 Louch,W.J. and Eaton,D.R., Inorg. Chim. Acta, 30, 215 (1978) *Note:* F-19 NMR

453 Dincturk,S., Jackson,R.A. and Townson,M., J. Chem. Soc. Chem. Comm., 172 (1979) *Note:* From thermal decomposition rate *Note:* Thermal decomposition substituted dibenzylmercury compounds alkaline solution

454 Phenol attachment para

455 Benzene with ester also attached

456 Brown,H.C. and Inukai,T., J. Am. Chem. Soc., 83, 4825 (1961) *Note:* Multiple attachment; e.g. position may be ortho + meta *Note:* Rate of hydrolysis or solvolysis *Note:* T-cumyl chlorides (90% acetone at 25 deg.)

457 Thompson,H.W. and Steel,G., Trans. Farady Soc., 52, 1451 (1956) *Note:* Infra-red absorption: intensity or frequency (in CCl4) *Note:* Substitueted benzonitriles (intensity)

458 Kauer,J.C. and Sheppard,W.A., J. Org. Chem., 32, 3580 (1967) *Note:* F-19 NMR shift (in CCl4)

459 Bordwell,F.G. and Boutan,P.J., J. Am. Chem. Soc., 78, 854 (1956) *Note:* Apparent PKA of substituted acids (50% aq.-EtOH or water at 25 deg.)

460 Jones,L.B. and Foster,J.P., J. Org. Chem., 35, 1777 (1970) *Note:* Rate of hydrolysis or solvolysis *Note:* 50% acetone at 30 deg.

461 Wehry,E.L., J. Am. Chem. Soc., 89, 41 (1967) *Note:* Apparent PKA of substituted acids (50% aq.-EtOH or water at 25 deg.) *Note:* Also values for photoexcited singlet and triplet

462 Fawcett,F.S. and Sheppard,W.A., J. Am. Chem. Soc., 87, 4341 (1965) *Note:* From anilinium (Sigma -) or Sigma ortho *Note:* Apparent PKA of substituted acids (50% aq.-EtOH or water at 25 deg.) *Note:* F-19 NMR shift (in CCl4)

463 Syz,M. and Zollinger,H., Helv. Chim. Acta, 48, 383 (1965) *Note:* Apparent PKA of substituted acids (50% aq.-EtOH or water at 25 deg.) *Note:* Rate of reaction with diphenyldiazomethane (EtOH, 30 deg.) *Note:* Rate of hydrolysis or solvolysis *Note:* Averaged

464 Brown,T.L., J. Am. Chem. Soc., 80, 6489 (1958) *Note:* Infra-red absorption: intensity or frequency (in CCl4) *Note:* Of O-H bond (intensity)

465 Broxton,T.J., Capper,G., Deady,L.W., Lenko,A. and Topsom,R.D., J. Chem. Soc., Perkin Ii, 1237 (1972) *Note:* Apparent PKA of substituted acids (50% aq.-EtOH or water at 25 deg.) *Note:* Rate of hydrolysis or solvolysis

466 Gel'fond,A.S., Gavrilov,V.I., Mironova,V.G. and Chernokal'skii,B.D., Zh. Obshch. Khim., 42, 2462, 2454EE (1972) *Note:* From anilinium (Sigma -) or Sigma ortho *Note:* Thermodynamic PKA of substituted acids (benzoic acid, phenols or anilines unless otherwise specified, activities-not concentrations-used,50% aq.-EtOH or water at 25 deg.) *Note:* Anilinium

467 Clementi,S. and Linda,P., J. Chem. Soc., Perkin Ii, 1887 (1973) *Note:* 'extended selectivity treatment' (EST) *Note:* 14 reactions

468 Rakshys,J.W., Taft,R.W. and Sheppard,W.A., J. Am. Chem. Soc., 90, 5236 (1968) *Note:* F-19 NMR shift (in CCl4)

469 Katritzky,A.R. and Simmons,P., J. Chem. Soc., 1511 (1960) *Note:* From anilinium (Sigma -) or Sigma ortho *Note:* Apparent PKA of substituted acids (50% aq.-EtOH or water at 25 deg.) *Note:* Anilines

470 Popova,R.S., Popov,A.F. and Litvinenko,L.M., Zh. Org. Khim., 6, 1049, 1053EE (1970) *Note:* Rate of amine acylation with P-NO2-benzenesulfonyl Br

471 Fringuelli,F., Marino,G. and Taticchi,A., J. Chem. Soc. Perkin Ii, 158 and 1738 (1972) *Note:* Apparent PKA of substituted acids (50% aq.-EtOH or water at 25 deg.)

472 Noyce,D.S., Virgilio,J.A. and Bartman,B., J. Org. Chem., 38, 2657 (1973) *Note:* Rate of hydrolysis or solvolysis *Note:* 80% EtOH

473 Hansch,C., Regr. From Data in J. Org. Chem., 35, 340 (1970) *Note:* Apparent PKA of substituted acids (50% aq.-EtOH or water at 25 deg.)

474 Charton,M., "Design of Biopharm. Prop. Thru Prodrugs and Analogs", Roche,E.,Ed., Am. Pharm. Assoc., 1977, Ch. 9, p. 269. *Note:* Maximum dimension for steric effect *Note:* Vanderwal radii and secondary values

475 Gum Jr.,W.F. and Joullie,M.M., J. Org. Chem., 32, 53 (1967) *Note:* Polarography

476 Stock,L.M. and Brown,H.C., Adv. Phys. Org. Chem., 1, 116 (1963) *Note:* Lower values from nitration rate *Note:* Higher from proton exchange

477 Sheppard,W.A., J. Am. Chem. Soc., 85, 1314 (1963) *Note:* From anilinium (Sigma -) or Sigma ortho *Note:* Apparent PKA of substituted acids (50% aq.-EtOH or water at 25 deg.)

478 Cornelis,A., Lambert,S. and Laszlo,P., J. Org. Chem., 42, 381 (1977) *Note:* NMR benzilidene malononitriles

479 Cherkasov,R.A., Ovchinnikov,V.V. and Pudovik,A.N., Zh. Obshch. Khim., 46, 957, 956EE (1976) *Note:* Apparent PKA of substituted acids (50% aq.-EtOH or water at 25 deg.)

480 Deady,L.W., Shanks,R.A. and Topsom,R., Tetrahedron Lett., 21, 1881 (1973) *Note:* IR spectra cyano substituted heterocycles

481 Schiemenz,G.P., Angew. Chem. Internat. Edit., 7, 544 and 545 (1968) *Note:* I.R. intensities

482 Kabachnik,M.I., Malakhov,I.G., Tsvetkov,E.N., Johnson,K.F., Katritzky,A.R., Sparrow,A.J. and Topsom,R.D., Aust. J. Chem., 28, 755 (1975) *Note:* From infra-red or other spectral measurements, *Note:* I.R. spectra

483 Brown,H.C., Rao,C.G. and Ravindranathan,M., J. Am. Chem. Soc., 100, 7946 (1978) *Note:* Rate of hydrolysis or solvolysis

484 Charton,M., J. Org. Chem., 30, 552 (1965) *Note:* Equation relating dipole moments of Tr-vinylene and acetylene

485 Seshadri,K.V. and Ganesan,R., Tetrahedron, 28, 3827 (1972) *Note:* Rate of bromination

486 Bradamante,S. and Pagani,G.A., J. Org. Chem., 45, 114 (1980) *Note:* Attached to oxygen (S-contig) *Note:* Chemical shifts of O1H and N1H2 groups

487 Kukhar,V.P., Petrashenko,A.A., Zhmurova,I.N., Tukhar,A.A. and Solodushenkov,S.N., Zh. Obshch. Khim., 40, 1696, 1682EE (1970) *Note:* Apparent PKA of substituted acids (50% aq.-EtOH or water at 25 deg.) *Note:* 95% ethanol

488 Kaminskii,Y.L., Reakt. Sposob. Organ. Soed., 6, 797, 344EE (1969) *Note:* Calculated from geometry and hybridization

489 Ansell,H.V. and Taylor,R., J. Org. Chem., 44, 4946 (1979) *Note:* Detritiation TFA 70 deg.

490 Archer,W.J., Taylor,R., Gore,P.H. and Kamounah,F.S., J. Chem. Soc. Perkin li, 1828 (1980) *Note:* Multiple attachment; e.g. position may be ortho + meta *Note:* Detritiation TFA-CHCl3 70 deg.

491 Archer,W.J. and Taylor,R., J. Chem. Soc. Perkin li, 295 (1982) *Note:* Detritiation TFA 70 deg.

492 Ashkinadze,L.D., Epshtein,L.M., Golovchenko,L.G., Pachevskaya,V.M. and Kravtsov,D.N., Iz. Akad. Nauk Sssr, S. Khim., 3, 552, 397EE (1981) *Note:* IR

493 Bodeker,J., Kockritz,P., Koppel,H., Radeglia,R., J. Prakt. Chem., 322, 735 (1980) *Note:* P-31 NMR

494 Neimysheva,A.A., Savchuk,V.I. and Knunyants,I.L., Zh. Obshch. Khim., 36, 500, 520EE (1966) *Note:* Rate of hydrolysis or solvolysis

495 Boyd,D.B., J. Med. Chem., 27, 63 (1984) *Note:* Apparent PKA of substituted acids (50% aq.-EtOH or water at 25 deg.)

496 Hancock,C.K. and Falls,C.P., J. Am. Chem. Soc., 83, 4214 (1961) *Note:* Hancock's hyperconjugation correction for Es *Note:* Rate of hydrolysis or solvolysis *Note:* 60% dioxane, correction for hyperconjugation

497 Fringuelli,F., Marino,G. and Taticchi,A., J. Chem. Soc. (B), 1595 (1970) *Note:* Apparent PKA of substituted acids (50% aq.-EtOH or water at 25 deg.) *Note:* Rate of hydrolysis or solvolysis

498 Fringuelli,F., Marino,G. and Taticchi,A., J. Chem. Soc. (B), 1595 (1970) *Note:* From phenol (Sigma -) or Sigma ortho *Note:* Apparent PKA of substituted acids (50% aq.-EtOH or water at 25 deg.) *Note:* Rate of hydrolysis or solvolysis

499 Yagupol'skii,L.M. and Yagupol'skaya,L.N., Doklady Akad. Nauk. Sssr, 134, 1381, 1207EE (1960) *Note:* Multiple attachment; e.g. position may be ortho + meta *Note:* Apparent PKA of substituted acids (50% aq.-EtOH or water at 25 deg.)

500 Levitt,L.S. and Levitt,B.W., Tetrahedron, 27, 3777 (1971) *Note:* Acid ionization constants of alcohols by conductivity

501 Charton,M. and Meislich,H., J. Am. Chem. Soc., 80, 5940 (1958) *Note:* Thermodynamic PKA of substituted acids (benzoic acid, phenols or anilines unless otherwise specified, activities-not concentrations-used,50% aq.-EtOH or water at 25 deg.) *Note:* Tr-3-substituted acrylic acids

502 Bordwell,F.G. and Cooper,G.D., J. Am. Chem. Soc., 74, 1058 (1952) *Note:* Apparent PKA of substituted acids (50% aq.-EtOH or water at 25 deg.)

503 Roberts,J.D., McElhill,E.A. and Armstrong,R., J. Am. Chem. Soc., 71, 2923 (1949) *Note:* Apparent PKA of substituted acids (50% aq.-EtOH or water at 25 deg.) *Note:* Rate of reaction with diphenyldiazomethane (EtOH, 30 deg.)

504 Smith,P.A.S., Hall,J.H. and Kan,R.O., J. Am. Chem. Soc., 84, 485 (1962) *Note:* Apparent PKA of substituted acids (50% aq.-EtOH or water at 25 deg.)

505 Smith,P.A.S., Hall,J.H. and Kan,R.O., J. Am. Chem. Soc., 84, 485 (1962) *Note:* From anilinium (Sigma -) or Sigma ortho *Note:* Apparent PKA of substituted acids (50% aq.-EtOH or water at 25 deg.)

506 Willi,A.V. and Meier,W., Helv. Chim. Acta, 39, 318 (1956) *Note:* Apparent PKA of substituted acids (50% aq.-EtOH or water at 25 deg.) *Note:* In water at 20 deg., ionic strength = 0.1

507 Broxton,T.J., Butt,G.L., Deady,L.W., Toh,S.H., Topsom,R.D., Fischer,A. and Morgan,M.W., Can. J. Chem., 51, 1620 (1973) *Note:* Rate of hydrolysis or solvolysis

508 Exner,O. and Svatek,E., Coll. Czech. Chem. Comm., 36, 534 (1971) *Note:* Infra-red absorption: intensity or frequency (in CCl4)

509 Fringuelli,F., Marino,G. and Taticchi,A., Gazz. Chim. Ital., 102, 534 (1972) *Note:* Rate of hydrolysis or solvolysis

510 Hojo,M., Utaka,M. and Yoshida,Z., Tetrahedron, 27, 4255 (1971) *Note:* Apparent PKA of substituted acids (50% aq.-EtOH or water at 25 deg.)

511 Evans,C.G. and Thomas,J.D.R., J.Chem.Soc. (B), 1502 (1971) *Note:* Rate of hydrolysis or solvolysis

512 Kaplan,M., J. Chem. Eng. Data, 6, 272 (1961) *Note:* Reaction rate phenyl isocyanates with 2-ethylhexanol

513 Brown,R.S. and Traylor,T.G., J. Am. Chem. Soc., 95, 8025 (1973) *Note:* Charge transfer spectra *Note:* 2-substituted cyclopropylbenzenes

514 Fickling,M.M., Fischer,A., Mann,B.R., Packer,J. and Vaughan,J., J. Am. Chem. Soc., 81, 4226 (1959) *Note:* From phenol (Sigma -) or Sigma ortho *Note:* Apparent PKA of substituted acids (50% aq.-EtOH or water at 25 deg.) *Note:* Also dimethylanilinium

515 Malakhova,I.G., Tsvetkov,E.N., Lobanov,D.I. and Kabachnik,M.I., Zh. Obshch. Khim., 41, 2807, 2837EE (1971) *Note:* Apparent PKA of substituted acids (50% aq.-EtOH or water at 25 deg.)

516 Rhodes,Y.E. and Vargas,L., J. Org. Chem., 38, 4077 (1973) *Note:* Apparent PKA of substituted acids (50% aq.-EtOH or water at 25 deg.)

517 Bordwell,F.G. and Boutan,P.J., J. Am. Chem. Soc., 78, 87 (1956) *Note:* Thermodynamic PKA of substituted acids (benzoic acid, phenols or anilines unless otherwise specified, activities-not concentrations-used,50% aq.-EtOH or water at 25 deg.)

518 Price,C.C. and Lincoln,D.C., J. Am. Chem. Soc., 73, 5836 (1951) *Note:* Rate of hydrolysis or solvolysis *Note:* Ethyl benzoates

519 Nyquist,H.L. and Wolfe,B., J. Org. Chem., 39, 2591 (1974) *Note:* Thermodynamic PKA of substituted acids (benzoic acid, phenols or anilines unless otherwise specified, activities-not concentrations-used,50% aq.-EtOH or water at 25 deg.) *Note:* F-19 NMR shift (in CCl4)

520 Yagupol'skii,L.M., Kondratenko,N.V., Delyagina,N.I., Dyatkin,B.L. and Knunyants,I.L., Zh. Org. Khim., 9, 649, 669EE (1973) *Note:* Apparent PKA of substituted acids (50% aq.-EtOH or water at 25 deg.) *Note:* F-19 NMR shift (in CCl4)

521 Minkin,V.I., Zhdanov,Y.A. and Medyantseva,E.A., Dokl. Akad. Nauk., 159, 1330, 1362EE (1964) *Note:* From phenol (Sigma -) or Sigma ortho *Note:* Apparent PKA of substituted acids (50% aq.-EtOH or water at 25 deg.)

522 Yagupol'skii,L.M. and Nazaretyan,V.P., Zh. Org. Khim., 7, 996, 1016EE (1971) *Note:* Apparent PKA of substituted acids (50% aq.-EtOH or water at 25 deg.)

523 Fringuelli,F., Marino,G. and Taticchi,A., J. Chem. Soc. (B), 2302 (1971) *Note:* Rate of hydrolysis or solvolysis

524 Fringuelli,F., Marino,G. and Taticchi,A., J. Chem. Soc. (B), 2304 (1971) *Note:* Apparent PKA of substituted acids (50% aq.-EtOH or water at 25 deg.) *Note:* Rate of hydrolysis or solvolysis

525 Clinton,N.A., Brown,R.S. and Traylor,T.G., J. Am. Chem. Soc., 92, 5228 (1970) *Note:* Rate of hydrolysis or solvolysis

526 Kabachnik,M.I., Phosphorus, 4, 247 (1974) *Note:* Rate of alkaline hydrolysis of (diethyl) substituted-phenyl phosphates *Note:* Ionization of substituted acids

527 Van Meurs,F., Baas,J.M.A. and Van Bekkum,H., J. Organomet. Chem., 129, 347 (1977) *Note:* IR

528 Maccarone,E., Musumarra,G. and Tomaselli,G.A., J. Chem. Soc. Perkin Ii, 906 (1976) *Note:* Reaction rates sulfonyl chlorides with aniline in MeOH

529 Taylor,R., J. Chem. Soc. (B), 1397 (1968) *Note:* Pyrolysis of arylethylacetates

530 Jaffe,H.H. and Jones,H.L., Adv. Heterocyc. Chem., 3, 209 (1964) *Note:* Apparent PKA of substituted acids (50% aq.-EtOH or water at 25 deg.) *Note:* Wolf-Kishner reduction

531 Oae,S. and Price,C.C., J. Am. Chem. Soc., 79, 2547 (1957) *Note:* Multiple attachment; e.g. position may be ortho + meta *Note:* Apparent PKA of substituted acids (50% aq.-EtOH or water at 25 deg.)

532 Jaffe,H.H., J. Am. Chem. Soc., 77, 4445 (1955) *Note:* Apparent PKA of substituted acids (50% aq.-EtOH or water at 25 deg.)

533 Ballistreri,A., Maccarone,E. and Musumarra,G., J. Chem. Soc. Perkin Ii, 984 (1977) *Note:* PKA sulfonamides

534 Shindo,H., Chem. Pharm. Bull. Japan, 6, 117 (1958) *Note:* IR

535 Simonnin,M.P. and Braun,J., Bull. Soc. Chim. Fr., 12, 4918 (1968) *Note:* H-NMR

536 Bellingham,P., Johnson,C.D. and Katritzky,A.R., J. Chem. Soc. (B), 866 (1968) *Note:* From phenol (Sigma -) or Sigma ortho *Note:* Acid catalysed H-exchange

537 Allen,C.W., J. Organomet. Chem., 125, 215 (1977) *Note:* C-13 NMR

538 Attia,S.Y., Berry,J.P., Koshy,K.M., Leung,Y-k., Lyznicki Jr.,E.P., Nowlan,V.J., Oyama,K. and Tidwell,T.T., J. Am. Chem. Soc., 99, 3401 (1977) *Note:* Rate of hydrolysis or solvolysis

539 Petrosyan,V.S. and Reutov,O.A., Dokl. Akad. Nauk Ssr, 180, 876, 514EE (1968) *Note:* F-19 NMR

540 Martin,D. and Brause,W.M., Chem. Ber., 102, 2508 (1969) *Note:* F-19 NMR and IR

541 Sheppard,W.A., J. Am. Chem. Soc., 85, 1314 (1963) *Note:* From phenol (Sigma -) or Sigma ortho *Note:* Apparent PKA of substituted acids (50% aq.-EtOH or water at 25 deg.)

542 Ansell,H.V., Hirschler,M.M. and Taylor,R., J. Chem. Soc. Perkin Ii, 353 (1977) *Note:* Multiple attachment; e.g. position may be ortho + meta *Note:* Protodetritiation

543 Glyde,E. and Taylor,R., J. Chem. Soc. Perkin Ii, 678 (1977) *Note:* Gas pyrolysis of acetates

544 Geiseler,G., Fruwert,J. and Gyalogh,F., Spectrochem. Acta, 22, 1165 (1966) *Note:* IR

545 Schadt Iii,F.L., Lancelot,C.J. and Schleyer,P., J. Am. Chem. Soc., 100, 228 (1978) *Note:* From thermal decomposition rate *Note:* Rate of hydrolysis or solvolysis

546 Ogata,Y. and Haba,M., J. Org. Chem., 38, 2779 (1973) *Note:* Oxidation with (O)

547 Butkiewicz,K., Electronanal. Chem., 39, 407 and 419 (1972) *Note:* Polarography

548 Tsukerman,S.V., Surov,Y.N. and Lavrushin,V.F., Zh. Obshch. Khim., 37, 364, 339EE (1967) *Note:* I.R. chalcone analogs

549 Gubin,S.P., Khandkarova,V.S. and Kreindlin,A.Z., J. Organomet. Chem., 64, 229 (1974) *Note:* Rate of hydrolysis or solvolysis

550 Sakurai,H., Deguchi,S., Yamagata,M., Morimoto,S-i., Kira,M. and Kumada,M., J. Organomet. Chem., 18, 285 (1969) *Note:* Proton NMR

551 Ishikawa,N., Nippon Kagaku Kaishi, 3, 520 (1974) *Note:* F-19 NMR shift (in CCl4)

552 Katritzky,A.R. and Topsom,R.D., Chem. Rev., 77, 639 (1977) *Note:* Sign undetermined *Note:* I.R. intensities

553 Coulson,D.R., J. Chem. Soc. Dalton, 2459 (1973) *Note:* F-19 NMR

554 Koppel,I., Karelson,M. and Palm,V., Org. React., 11, 101 (1974) *Note:* Apparent PKA of substituted acids (50% aq.-EtOH or water at 25 deg.)

555 Charton,M., J. Org. Chem., 30, 557 (1965) *Note:* Equation relating dipole moments of Tr-vinylene and acetylene

556 Fringuelli,F., Serena,B. and Taticchi,A., J. Chem. Soc. Perkin Ii, 971 (1980) *Note:* Apparent PKA of substituted acids (50% aq.-EtOH or water at 25 deg.) *Note:* NMR

557 Fringuelli,F., Serena,B. and Taticchi,A., J. Chem. Soc. Perkin Ii, 971 (1980) *Note:* From phenol (Sigma -) or Sigma ortho *Note:* Apparent PKA of substituted acids (50% aq.-EtOH or water at 25 deg.) *Note:* NMR

558 Sheppard,W.A., J. Am. Chem. Soc., 87, 2410 (1965) *Note:* From anilinium (Sigma -) or Sigma ortho *Note:* Apparent PKA of substituted acids (50% aq.-EtOH or water at 25 deg.) *Note:* F-19 NMR shift (in CCl4) *Note:* Various conditions

559 Visser,G.W.M., Diemer,E.L. and Kaspersen,F.M., J. Royal Neth. Chem. Soc., 99, 93 (1980) *Note:* Apparent PKA of substituted acids (50% aq.-EtOH or water at 25 deg.)

560 Schiemenz,G.P., Angew. Chem., 5, 595EE (1966) *Note:* Apparent PKA of substituted acids (50% aq.-EtOH or water at 25 deg.) *Note:* 67% aqueous MeOH

561 Kondratenko,N.V., Popov,V.I., Kolomeitsev,A.A., Saenko,E.P., Prezhdo,V.V., Lutskii,A.E. and Yagupol'skii,L.M., Zh. Org. Khim., 16, 1215, 1049EE (1980) *Note:* F-19 NMR shift (in CCl4)

562 Landgrebe,J.A. and Rynbrandt,R., J. Org. Chem., 31, 2585 (1966) *Note:* Apparent PKA of substituted acids (50% aq.-EtOH or water at 25 deg.) *Note:* Rate of hydrolysis or solvolysis *Note:* Benzyl chlorides

563 Archer,W.J. and Taylor,R., J. Chem. Soc. Perkin Ii, 1153 (1981) *Note:* Detritiation TFA 70 deg.

564 Angelelli,J.M., Delmas,M.A., Chouteau,J., Guiliano,M. and Mille,G., J. Organomet. Chem., 188, 85 (1980) *Note:* From NMR measurments except after Es where = isosteric value for corresponding amines *Note:* IR and NMR

565 Szmant,H.H. and Harmuth,C.M., J. Am. Chem. Soc., 86, 2909 (1964) *Note:* Rates of Wolff-Kishner reaction

566 Kashik,T.V., Rassolova,G.V., Ponomareva,S.M., Medvedeva,E.N., Yushmanova,T.I. and Lopyrev,V.A., Iz. Akad. Nauk. Sssr S. Khim., 10, 2230,1965EE (1982) *Note:* NH acidity of carboxylic acid hydrazides and CF3CONH2

567 Ryan,J.J. and Humffray,A., J. Chem. Soc. (B), 842 (1966) *Note:* From phenol (Sigma -) or Sigma ortho *Note:* Apparent PKA of substituted acids (50% aq.-EtOH or water at 25 deg.) *Note:* Various conditions

568 Bordeau,M., J. Organomet. Chem., 229, 203 (1982) *Note:* Optical anisotropy

569 Azzaro,M., Gal,J.F. and Geribaldi,S., J. Org. Chem., 47, 4981 (1982) *Note:* Bronsted and Lewis basicities of a,b-unsaturated ketones

570 Alberghina,G., Fisichella,S. and Musumarra,G., J. Chem. Soc. Perkin Ii, 1700 (1979) *Note:* Protonation of N,N-disubstituted carboxamides

571 Kazitsyna,L.A., Ustynyuk,Y.A., Gurman,V.S., Pergushov,V.I. and Gruzdneva,V.N., Dokl. Akad. Nauk. Sssr, 233, 866, 213EE (1977) *Note:* F-19 NMR shift (in CCl4)

572 Blanch,J.H., J. Chem. Soc. (B), 937 (1966) *Note:* Thermodynamic PKA of substituted acids (benzoic acid, phenols or anilines unless otherwise specified, activities-not concentrations-used,50% aq.-EtOH or water at 25 deg.) *Note:* Apparent PKA of substituted acids (50% aq.-EtOH or water at 25 deg.) *Note:* Rate of hydrolysis or solvolysis *Note:* Averaged

573 Filippova,T.A., Sukhoroslova,M.M., Lopatinskii,V.P., Filimonov,V.D., Chem. Abs., 98, 197508z. *Note:* PKA's acetic acids in 56% EtOH

574 Glukhikh,V.I. and Voronkov,M.G., Dokl. Akad. Nauk Sssr, 248, 142, 744EE (1979) *Note:* Possible anomaly; check original article *Note:* C-13 NMR

575 Tsvetkov,E., Lobanov,D.I. and Kabachnik,M.I., Chem. Abst., 64, 12523 (1966) *Note:* Apparent PKA of substituted acids (50% aq.-EtOH or water at 25 deg.)

576 Ivanova,L.P., Polis,Y.Y., Grava,I.Y., Raguel,B.P., Cherkasova,V.A., Khamad,K.D. and Kolobova,T.A., Zh. Organ. Khim., 17, 325, 272EE (1981) *Note:* Rate of oximation

577 Kryuchkov,A.A., Kozachenko,A.G., Matrosov,E.I. and Kabachnik,M.I., Iz. Akad. Nauk Sssr, 9, 1985, 1746EE (1978) *Note:* Solvent dependent; see reference for other values *Note:* Apparent PKA of substituted acids (50% aq.-EtOH or water at 25 deg.)

578 Hayashi,T. and Shibata,R., Bull. Chem. Soc. Jpn., 34, 1116 (1961) *Note:* IR shifts

579 Tsvetkov,E.N., Lobanov,D.I. and Kabachnik,M.I., Chem. Abst., 66, 28299r (1967) *Note:* Apparent PKA of substituted acids (50% aq.-EtOH or water at 25 deg.) *Note:* In 50% ethanol

580 Rotinov,A. and Chuchani,G., Int. J. Chem. Kinetics, 16, 1267 (1984) *Note:* Rate of gas-phase pyrolysis ethyl-alpha-substituted acetates

581 Mamaev,V.P., Shkurko,O.P. and Baram,S.G., Adv. Heterocyc. Chem., 42, 1 (1987) *Note:* From anilinium (Sigma -) or Sigma ortho

582 Hansch,C., Leo,A, Taft,R., Chem. Rev., 91, 165 (1991) *Note:* Calc.by CMR

583 van Leuwen,B.G. and Ouellette,R.J., J. Am. Chem. Soc., 90, 7056 (1968) *Note:* Multiple attachment; e.g. position may be ortho + meta *Note:* Rate of hydrolysis or solvolysis *Note:* Arylmethylmercuric.HClO4 in HAC

584 Tribble,M.T. and Traynham,J.C., J. Am. Chem. Soc., 91, 379 (1969) *Note:* From phenol (Sigma -) or Sigma ortho *Note:* OH- NMR shift in o-substituted phenols (in DMSO)

585 Jaffe,H.H., Freedman,L.D. and Doak,G.O., J. Am. Chem. Soc., 75, 2209 (1953) *Note:* Apparent PKA of substituted acids (50% aq.-EtOH or water at 25 deg.)

586 Roberts,J.D. and Moreland Jr.,W.T., J. Am. Chem. Soc., 75, 2167 (1953) *Note:* Apparent PKA of substituted acids (50% aq.-EtOH or water at 25 deg.) *Note:* 4-substituted bicyclooctane carboxylic acids

587 Phenoxyacetic acid system

588 Nitrobenzene, ortho attachment

589 Calculated from the methyl or ethyl homolog

590 Baum,K., J. Org. Chem., 35, 1203 (1970) *Note:* Apparent PKA of substituted acids (50% aq.-EtOH or water at 25 deg.)

591 Kaplan,L.A. and Pickard,H.B., J. Org. Chem., 35, 2044 (1970) *Note:* Rate of reaction with diphenyldiazomethane (EtOH, 30 deg.)

592 Kucsman,A., Ruff,F., Solyom,S. and Szirtes,T., Acta Chim. Acad. Sci. (Hung.), 57, 205 (1968) *Note:* Apparent PKA of substituted acids (50% aq.-EtOH or water at 25 deg.) *Note:* 48% V. EtOH at 25 deg.

593 Cohen,L.A. and Jones,W.M., J. Am. Chem. Soc., 85, 3397 and 3402 (1963) *Note:* Thermodynamic PKA of substituted acids (benzoic acid, phenols or anilines unless otherwise specified, activities-not concentrations-used,50% aq.-EtOH or water at 25 deg.) *Note:* Phenols

594 Perrault,G., Can. J. Chem., 45, 1063 (1967) *Note:* Apparent PKA of substituted acids (50% aq.-EtOH or water at 25 deg.) *Note:* Amines

595 Traylor,T.G. and Ware,J.C., J. Am. Chem. Soc., 89, 2304 (1967) *Note:* Rate of hydrolysis or solvolysis *Note:* 95% EtOH at 0 and 70 deg.

596 Arnett,E.M. and Bushick,R.D., J. Org. Chem., 27, 111 (1962) *Note:* Apparent PKA of substituted acids (50% aq.-EtOH or water at 25 deg.) *Note:* Extraction procedure in darkness

597 Nesmeyanov,A.N., Perevalova,E.G., Gubin,S.P., Grandberg,K.I. and Kozlovsky,A.G., Tetrahedron Lett., 22, 2381 (1966) *Note:* Apparent PKA of substituted acids (50% aq.-EtOH or water at 25 deg.) *Note:* 70% dioxane at 25 deg.

598 Yagupol'skii,L.M., Bystrov,V.F. and Utyanskaya,E.Z., Doklady Akad. Nauk Sssr, 135, 377, 1059EE (1960) *Note:* F-19 NMR shift (in CCl4) *Note:* NEAT

599 Knizek,J.,Horak,M. and Chvalovsky,V., Coll. Czech. Chem. Comm., 28, 3079 (1963) *Note:* Infra-red absorption: intensity or frequency (in CCl4)

600 Szmant,H.H. and Suld,G., J. Am. Chem. Soc., 78, 3400 (1956) *Note:* From anilinium (Sigma -) or Sigma ortho *Note:* Apparent PKA of substituted acids (50% aq.-EtOH or water at 25 deg.)

601 Bordwell,F.G. and Cooper,G.D., J. Am. Chem. Soc., 74, 1058 (1952) *Note:* From phenol (Sigma -) or Sigma ortho *Note:* Apparent PKA of substituted acids (50% aq.-EtOH or water at 25 deg.)

602 Buchman,O., Grosjean,M. and Nasielski,J., Helv. Chim. Acta, 47, 2037 (1964) *Note:* Apparent PKA of substituted acids (50% aq.-EtOH or water at 25 deg.)

603 Roberts,J.D. and Moreland Jr.,W.T., J. Am. Chem. Soc., 75, 2267 (1953) *Note:* Apparent PKA of substituted acids (50% aq.-EtOH or water at 25 deg.) *Note:* Rate of reaction with diphenyldiazomethane (EtOH, 30 deg.) *Note:* Rate of hydrolysis or solvolysis *Note:* Averaged

604 Zhdanov,Y.A. and Polenov,V.A., Zh. Obshch. Khim., 35, 589, 587EE (1965) *Note:* From anilinium (Sigma -) or Sigma ortho *Note:* Apparent PKA of substituted acids (50% aq.-EtOH or water at 25 deg.) *Note:* 70% EtOH

605 Campbell,A.D., et.al., Aust. J. Chem., 23, 203 (1970) *Note:* Rate of hydrolysis or solvolysis

606 Sestakova,J., Horak,V. and Zuman,P., Coll. Czech. Chem. Comm., 31, 3889 (1966) *Note:* Addition of primary amines to phenyl vinyl ketone, by polarography

607 Rosado-Lojo,O., Hancock,C.K. and Danti,A., J. Org. Chem., 31, 1899 (1966) *Note:* Hancock's hyperconjugation correction for Es *Note:* F-19 NMR shift (in CCl4)

608 Lewis,E.S. and Johnson,M.D., J. Am. Chem. Soc., 81, 2070 (1959) *Note:* Apparent PKA of substituted acids (50% aq.-EtOH or water at 25 deg.) *Note:* Water at 6 deg. also phenylacetic acid

609 Fickling,M.M., Fischer,A., Mann,B.R., Packer,J. and Vaughan,J., J. Am. Chem. Soc., 81, 4226 (1959) *Note:* From anilinium (Sigma -) or Sigma ortho *Note:* Apparent PKA of substituted acids (50% aq.-EtOH or water at 25 deg.) *Note:* Also dimethylanilinium

610 Tsvetkov,E.N., Malevannaya,R.A., Lobanov,D.I., Osipenko,N.G. and Kabachnik,M.I., Zh. Obshch. Khim., 39, 2429, 2369EE (1969) *Note:* Apparent PKA of substituted acids (50% aq.-EtOH or water at 25 deg.)

611 Taft Jr.,R.W. and Lewis,I.C., J. Am. Chem. Soc., 81 , 5343 and 5352 (1959) *Note:* F-19 NMR shift (in CCl4)

612 Bordwell,F.G. and Boutan,P.J., J. Am. Chem. Soc., 79, 717 (1957) *Note:* Apparent PKA of substituted acids (50% aq.-EtOH or water at 25 deg.)

613 Bordwell,F.G. and Boutan,P.J., J. Am. Chem. Soc., 78, 87 (1956) *Note:* From phenol (Sigma -) or Sigma ortho *Note:* Thermodynamic PKA of substituted acids (benzoic acid, phenols or anilines unless otherwise specified, activities-not concentrations-used,50% aq.-EtOH or water at 25 deg.)

614 Biggs,A.I. and Robinson,R.A., J. Chem. Soc., 388 (1961) *Note:* From anilinium (Sigma -) or Sigma ortho *Note:* Apparent PKA of substituted acids (50% aq.-EtOH or water at 25 deg.)

615 Hansch,C., Leo,A., Unger,S.H., Kim,K.H., Nikaitani,D. and Lien,E.J., J. Med. Chem., 16, 1207 (1973)

616 Rudaya,L.I., Kvitko,I.Y. and Porai-Koshits,B.A., Zh. Org. Khim., 8, 13, 11EE (1972) *Note:* Thermodynamic PKA of substituted acids (benzoic acid, phenols or anilines unless otherwise specified, activities-not concentrations-used,50% aq.-EtOH or water at 25 deg.) *Note:* Infra-red absorption: intensity or frequency (in CCl4)

617 Zhmurova,I.N., Kukhar,V.P., Tukhar,A.A. and Zolotareva,L.A., Zh. Obshch. Khim., 43, 82, 79EE (1973) *Note:* Infra-red absorption: intensity or frequency (in CCl4) *Note:* Basicities

618 Jarvis,B.B., Harper Jr.,R.L. and Tong,W.P., J. Org. Chem., 40, 3778 (1975) *Note:* From phenol (Sigma -) or Sigma ortho *Note:* Charge transfer frequencies

619 Boloshchuk,V.G., Yagupol'skii,L.M., Syrova,G.P. and Bystrov,V.F., Zh. Obshch. Khim., 3, 118, 105EE (1967) *Note:* F-19 NMR shift (in CCl4)

620 Noyce,D.S. and Stowe,G.T., J. Org. Chem., 38, 3762 (1973) *Note:* Rate of hydrolysis or solvolysis *Note:* 80% EtOH, imidazolyl-P-nitrobenzoates

621 Sheppard,W. and Sharts,C., "Organic Flourine Chem." W.A. Benjamin, N.Y., 1969, p. 348. *Note:* From phenol (Sigma -) or Sigma ortho *Note:* Apparent PKA of substituted acids (50% aq.-EtOH or water at 25 deg.) *Note:* F-19 NMR shift (in CCl4)

622 Sheppard,W.A., J. Am. Chem. Soc., 84, 3072 (1962) *Note:* Apparent PKA of substituted acids (50% aq.-EtOH or water at 25 deg.) *Note:* F-19 NMR shift (in CCl4)

623 Lindberg,B.J., Arkiv Kemi, 32, 317 (1970) *Note:* Polarography

624 Traven,V.F., Korolev,B.A., Pyatkina,T.V. and Stepanov,B.I., Zh. Obshch. Khim., 45, 954 943EE (1975) *Note:* From anilinium (Sigma -) or Sigma ortho *Note:* Basicity of organosilyl amines

625 Jaffe,H.H., Chem. Rev., 53, 191 (1953) *Note:* From phenol (Sigma -) or Sigma ortho *Note:* Wide variety of methods, see review article for details

626 Fisher,T.H. and Meierhoefer,A., Tetrahedron, 31, 2019 (1975) *Note:* Apparent PKA of substituted acids (50% aq.-EtOH or water at 25 deg.) *Note:* NBS bromination

627 Oae,S. and Price,C.C., J. Am. Chem. Soc., 79, 2547 (1957) *Note:* Apparent PKA of substituted acids (50% aq.-EtOH or water at 25 deg.)

628 Forsythe,P., Frampton,R., Johnson,C.D. and Katritzky,A.R., J. Chem. Soc. Perkin Ii, 671 (1972) *Note:* From anilinium (Sigma -) or Sigma ortho *Note:* PKA dimethyamino substituted heteroaromatics

629 Deady,L.W. and Shanks,R.A., Aust. J. Chem., 25, 2363 (1972) *Note:* Alkaline hydrolysis

630 Perrin,D.D., J. Chem. Soc., 5590 (1965) *Note:* Apparent PKA of substituted acids (50% aq.-EtOH or water at 25 deg.)

631 Ustynyuk,Y.A., Subbotin,O.A., Buchneva,L.M., Gruzdneva,V.N. and Kazitsyna,L.A., Dokl. Akad. Nauk Sssr, 227, 101, 175EE (1976) *Note:* C-13 NMR

632 Bly,R.S., Tse,K-k. and Bly,R.K., J. Organomet. Chem., 117, 35 (1976) *Note:* Rate of hydrolysis or solvolysis

633 Arnold,Z. and Krchnak,V., Tetrahedron Let., 5, 347 (1975) *Note:* Steric constant calculated according to Arzneim.-Forsch.(1979) 29,585 *Note:* F-19 NMR

634 Arnold,Z. and Krchnak,V., Tetrahedron Let., 5, 347 (1975) *Note:* Hancock's hyperconjugation correction for Es *Note:* F-19 NMR

635 Koshelev,Y.N., Kvitko,I.Y. and Efros,L.S., Zh. Org. Khim., 8, 1750, 1789EE (1972) *Note:* Multiple attachment; e.g. position may be ortho + meta *Note:* Apparent PKA of substituted acids (50% aq.-

EtOH or water at 25 deg.)

636 Gibson,H.W. and Bailey,F.C., J. Chem. Soc., Perkin Ii, 196 (1976) *Note:* From phenol (Sigma -) or Sigma ortho *Note:* Apparent PKA of substituted acids (50% aq.-EtOH or water at 25 deg.)

637 Kargin,Y.M., Kondranina,V.Z., Kataeva,L.M., Sorokina,L.A., Yanilkin,V.V. and Khysaenov,N.M., Zh. Obshch. Khim., 47, 672, 611EE (1977) *Note:* Electrochemical reduction

638 Kashik,T.V., Rassolova,G.V., Stepanova,Z.V. and Keiko,V.V., Zh. Org. Khim., 13, 1201, 1104EE (1977) *Note:* PKA aromatic amines

639 Taylor,R., J. Chem. Soc. Perkin Ii, 1287 (1975) *Note:* Rate of detritiation and desilylation

640 Nuallain,C.O. and Cinneide,S.O., J. Inorg. Nucl. Chem., 35, 2871 (1973) *Note:* PKA phenols and anilines

641 Charton,M., J. Chem. Soc., 1205 (1964) *Note:* No values in computer print-out *Note:* Polar effects in cyclopropane

642 Eaton,D.R., Benson,R.E., Bottomley,C.G. and Josey,A.D., J. Am. Chem. Soc., 94, 5996 (1972) *Note:* NMR shift with Ni chelates

643 Fischer,F.C. and Havinga,E., Rec. Trav. Chim., 93, 21 (1974) *Note:* Apparent PKA of substituted acids (50% aq.-EtOH or water at 25 deg.)

644 Voronkov,M.G., Kashik,T.V., Lukevits,E.Y., Deriglazova,E.S. and Pestunovich,A.E., Zh. Obshch. Khim., 45, 2200, 2162EE (1975) *Note:* Basicity of piperidines

645 Deady,L.W. and Shanks,R.A., Aust. J. Chem., 25, 431 (1972) *Note:* Hydrolysis pyridyl-methyl acetates

646 Borisov,E.V., Kholodov,L.E. and Yashunskii,V.G., Reakt. Sposob. Org. Soed., 7, 704, 314EE (1970) *Note:* PKA substituted acetic acids

647 Vlasov,V.M. and Yakobson,G.G., Russ. Chem. Rev., 43, 781 (1974) *Note:* Miscellaneous

648 Kasukhin,L.F., Ponomarchuk,M.P., Klepa,T.I., Lysenko,V.P., Repina,L.A. and Gololobov,Y.G., Dokl. Nauk. Ssr, 228, 1380, 579EE (1976) *Note:* Kinetic, with phenyl azide

649 Parr,W.J.E., J. Chem. Soc., Farad. Trans., 2, 603 (1978) *Note:* C-13 NMR

650 Taylor, J. Chem. Soc. Perkin Ii, 253 (1973) *Note:* Detritiation

651 Veschambre,H. and Kergomard,A., Mem. Pres. Soc. Chim., 336 (1965) *Note:* From anilinium (Sigma -) or Sigma ortho *Note:* Apparent PKA of substituted acids (50% aq.-EtOH or water at 25 deg.)

652 Feshin,V.P., Volkov,A.N., Nikitin,P.A., Kudyakova,R.N. and Voronkov,M.G., Dokl. Akad. Nauk. Sssr., 222, 650, 518EE (1975) *Note:* Cl-35 NQR spectra

653 Voronkov,M.G., Kashik,T.V. and Deriglazova,E.S., Zh. Obshch. Khim., 49, 1637, 1428EE (1979) *Note:* Miscellaneous

654 Kormachev,V.V., Vasil'eva,T.V., Solodova,K.V., Alekseeva,O.T. and Kukhtin,V.A., Zh. Obshch. Khim., 49, 600, 525EE (1979) *Note:* Reaction rate P-Cl-phenylphosphine oxides with Na-OMe

655 Ashe Iii,A.J. and Chan,W-t., J. Org. Chem., 45, 2016 (1980) *Note:* Thermodynamic PKA of substituted acids (benzoic acid, phenols or anilines unless otherwise specified, activities-not concentrations-used,50% aq.-EtOH or water at 25 deg.)

656 Broxton,T.J., Cameron,D.G. and Topsom,R.D., J. Chem. Soc. Perkin Ii, 256 (1974) *Note:* IR

657 Adcock,W. and Khor,T-c., J. Am. Chem. Soc., 100, 7799 (1978) *Note:* F-19 and C-13 NMR E = revised Taft Es *Note:* See also: Tetrahed. Lett.(1978) 35,3293

658 Austel,V., Kutter,E. and Kalbfleisch, Private Commun., Arzneim.-Forsch., 29, 585 (1979) *Note:* Steric constant calculated according to Arzneim.-Forsch.(1979) 29,585

659 Hansch,C. and Kerley,R., Chem. Ind., 294, Private Communication (1969) *Note:* Radical

660 Voronov,S.A., Puchin,V.A., Min'ko,S.S., Kiselev,E.M. and Dikii,M.A., Zh. Org. Khim., 16, 258, 228EE (1980) *Note:* Apparent PKA of substituted acids (50% aq.-EtOH or water at 25 deg.)

661 Ginzburg,O.F., Kvyat,E.I. and Idlis,G.S., Zh. Obshch. Khim., 32, 2633, 2593EE (1962) *Note:* Hydrolysis of dyes

662 Taylor,R., J. Chem. Soc. (B), 1450 (1971) *Note:* Pyrolysis aryl-ethyl acetates

663 Taylor,R., J. Chem. Soc. (B), 2382 (1971) *Note:* Pyrolysis aryl-ethyl acetates

664 Taylor,R., J. Chem. Soc. (B), 2382 (1971) *Note:* Multiple attachment; e.g. position may be ortho + meta *Note:* Pyrolysis aryl-ethyl acetates

665 Ansell,H.V. and Taylor,R., J. Chem. Soc. Perkin Ii, 751 (1978) *Note:* Detritiation TFA 70 deg.

666 Brown,H.C., Kelly,D.P. and Periasamy, Proc. Natl. Acad. Sci. Usa, 77, 6956 (1980) *Note:* C-13 shifts of carbocations

667 Tupitsyn,I.F., Zatsepina,N.N., Kolodina,N.S. and Kaminskii,Y.L., Reakt. Sposob. Org. Soed., 6, 458, 192EE (1969) *Note:* I.R. in CCl4 *Note:* Apparent PKA of substituted acids (50% aq.-EtOH or water at 25 deg.)

668 Falkner,P.R. and Harrison,D., J. Chem. Soc. B, 1171 (1960) *Note:* Alkaline hydrolysis 70% EtOH

669 Simonetta and Favini, Gazzetta, 84, 566 (1954) *Note:* Alkaline hydrolysis pyridyl methyl acetates 20 deg.

670 Favini, Rend. Ist. Lombardi Sci., 91, 162 (1957) *Note:* Alkaline hydrolysis pyridine-carboxamides at 64 deg.

671 Ryan,J.J. and Humffray,A., J. Chem. Soc. (B), 842 (1966) *Note:* Apparent PKA of substituted acids (50% aq.-EtOH or water at 25 deg.) *Note:* Various conditions

672 Kargin,Y.M., Latypova,V.Z., Yakovleva,O.G., Vafina,A.A., Ustyugov,A.N. and Il'yasov,A.V., Zh. Obshch. Khim., 52, 655, 572EE (1982) *Note:* Polarography

673 Poplavskii,V.S., Ostrovskii,V.A., Koldobskii,G.I., Kulikova,E.A., J. Heterocyc. Chem. (Russ.), 2, 264 (1982) *Note:* Apparent PKA of substituted acids (50% aq.-EtOH or water at 25 deg.)

674 Matrosov,E.I., Nifant'ev,E.E., Kryuchkov,A.A. and Kabachnik,M.I., Iz. Akad. Nauk Sssr, 3, 530, 512EE (1976) *Note:* Apparent PKA of substituted acids (50% aq.-EtOH or water at 25 deg.)

675 Price,C.C., Mertz,E.C. and Wilson,J., J. Am. Chem. Soc., 76, 5153 (1954) *Note:* Apparent PKA of substituted acids (50% aq.-EtOH or water at 25 deg.) *Note:* Rate of hydrolysis or solvolysis

676 Lozinskii,M.O., Kukota,S.N., Bratolyubova,A.G., Pel'kis,P.S. and Vasil'eva,Z.A., Zh. Obshch. Khim., 42, 2270, 2267EE (1972) *Note:* PKA's phenylhydrazones

677 Misic-Vukovic,M. and Jovanovic,B., Bull. Soc. Chim. Beograd, 49, 83 (1984) *Note:* Rate of reaction with diphenyldiazomethane (EtOH, 30 deg.)

678 Azvezdina,E., et al., Chem. Heterocyc. Cpds. (Russ), 8, 1025 (1976) *Note:* Apparent PKA of substituted acids (50% aq.-EtOH or water at 25 deg.)

679 Baldwin,R.A., Cheng,M.T. and Homer,G.D., J. Org. Chem., 32, 2176 (1967) *Note:* Apparent PKA of substituted acids (50% aq.-EtOH or water at 25 deg.) *Note:* In 60% tetrhydrofurane

680 Phenol attachment meta

681 Hammett,L.P., "Physical Organic Chemistry" McGraw-Hill, N.Y., 1940, p.188. *Note:* Multiple attachment; e.g. position may be ortho + meta *Note:* Thermodynamic PKA of substituted acids (benzoic acid, phenols or anilines unless otherwise specified, activities-not concentrations-used,50% aq.-EtOH or water at 25 deg.) *Note:* Apparent PKA of substituted acids (50% aq.-EtOH or water at 25 deg.)

682 Kutter,E. and Hansch,C., J. Med. Chem., 12, 647 (1969) *Note:* Maximum dimension for steric effect *Note:* Calculated from group radii

683 Kutter,E. and Hansch,C., J. Med. Chem., 12, 647 (1969) *Note:* Minimum dimension for steric effect *Note:* Calculated from group radii

684 Plzak,Z., Mares,F., Hetflejs,J., Schraml,J., Papouskova,Z., Bazant,V., Rochow,E.G. and Chvalovsky,V., Coll. Czech. Chem. Comm., 36, 3115 (1971) *Note:* Rate of reaction with diphenyldiazomethane (EtOH, 30 deg.)

685 Kauer,J.C. and Sheppard,W.A., J. Org. Chem., 32, 3580 (1967)

686 Slovetskii,V.I. and Fainzil'berg,A.A., Izv. Akad. Nauk. Khim., 8, 1488, 1435EE (1966) *Note:* Apparent PKA of substituted acids (50% aq.-EtOH or water at 25 deg.)

687 Soumillion,J. and Bruylants,A., Bull. Soc. Chim. Belges, 78, 169 (1969) *Note:* Rate of photochlorination of alkyl acetates

688 Alper,H., Keung,E.C.H. and Partis,R.A., J. Org. Chem., 36, 1352 (1971) *Note:* F-19 NMR and IR

689 Alper,H., Keung,E.C.H. and Partis,R.A., J. Org. Chem., 36, 1352 (1971) *Note:* IR and NMR

690 Stewart,R. and Walker,L.G., Can. J. Chem., 35, 1561 (1957) *Note:* Apparent PKA of substituted acids (50% aq.-EtOH or water at 25 deg.)

691 Stewart,R. and Walker,L.G., Can. J. Chem., 35, 1561 (1957) *Note:* From phenol (Sigma -) or Sigma ortho *Note:* Apparent PKA of substituted acids (50% aq.-EtOH or water at 25 deg.)

692 Packer,J., Vaughan,J. and Wong,E., J. Org. Chem., 23, 1373 (1958) *Note:* Ratio of acid vs. base catalysed hydrolysis of esters *Note:* Averaged (56% W. acetone, possible resonance component in Es

values)

693 Freedman,L.D. and Jaffe,H.H., J. Am. Chem. Soc., 77, 920 (1955) *Note:* Apparent PKA of substituted acids (50% aq.-EtOH or water at 25 deg.) *Note:* Water at 25 deg.

694 Freedman,L.D. and Jaffe,H.H., J. Am. Chem. Soc., 77, 920 (1955) *Note:* From anilinium (Sigma -) or Sigma ortho *Note:* Apparent PKA of substituted acids (50% aq.-EtOH or water at 25 deg.) *Note:* Water at 25 deg.

695 Freedman,L.D. and Jaffe,H.H., J. Am. Chem. Soc., 77, 920 (1955) *Note:* From phenol (Sigma -) or Sigma ortho *Note:* Apparent PKA of substituted acids (50% aq.-EtOH or water at 25 deg.) *Note:* Water at 25 deg.

696 Kalfus,K., Vecera,M. and Exner,O., Coll. Czech. Chem. Comm., 35, 1195 (1970) *Note:* Apparent PKA of substituted acids (50% aq.-EtOH or water at 25 deg.) *Note:* Averaged--80% Me-cellosolve and 50% EtOH

697 Humffray,A.A., Ryan,J.J., Warren,J.P. and Yung,Y.H., Chem. Comm., 23, 610 (1965) *Note:* Apparent PKA of substituted acids (50% aq.-EtOH or water at 25 deg.)

698 Nesmeyanov,A.N., Perevalova,E.G., Gubin,S.P., Grandberg,K.I. and Kozlovsky,A.G., Tetrahedron Lett., 22, 2381 (1966) *Note:* From anilinium (Sigma -) or Sigma ortho *Note:* Apparent PKA of substituted acids (50% aq.-EtOH or water at 25 deg.) *Note:* 70% dioxane at 25 deg.

699 Sedova,L.N., Gandel'sman,L.Z., Alekseeva,L.A. and Yagupol'skii,L.M., Zh. Obshch. Khim., 39, 2057, 2011EE (1969) *Note:* Apparent PKA of substituted acids (50% aq.-EtOH or water at 25 deg.)

700 Yagupol'skii,L.M., Sedova,L.N. and Alekseeva,L.A., Zh. Obshch. Khim., 39, 206, 190EE (1969) *Note:* F-19 NMR shift (in CCl4)

701 Weiss,C., Engewald,W. and Muller,H., Tetrahedron, 22, 825 (1966) *Note:* Deuterium exchange (MeOD at 100 deg.)

702 Lindberg,B., Acta Chem. Scand., 24, 2852 (1970)

703 McDaniel,D.H. and Brown,H.C., J. Org. Chem., 23, 420 (1958) *Note:* Multiple attachment; e.g. position may be ortho + meta *Note:* Thermodynamic PKA of substituted acids (benzoic acid, phenols or anilines unless otherwise specified, activities-not concentrations-used,50% aq.-EtOH or water at 25 deg.)

704 Monagle,J.J., Mengenhauser,J.V. and Jones Jr.,D.A., J. Org. Chem., 32, 2477 (1967) *Note:* Apparent PKA of substituted acids (50% aq.-EtOH or water at 25 deg.)

705 Monagle,J.J., Mengenhauser,J.V. and Jones Jr.,D.A., J. Org. Chem., 32, 2477 (1967) *Note:* From phenol (Sigma -) or Sigma ortho *Note:* Apparent PKA of substituted acids (50% aq.-EtOH or water at 25 deg.)

706 Zuman,P., "Substituent Effects in Organic Polarography", Plenum Press, N.Y., 1967, p.76. *Note:* Possible anomaly; check original article *Note:* Polarography

707 Peterson,P.E. and Allen,G., J. Org. Chem., 27, 2290 (1961) *Note:* F-19 NMR shift (in CCl4)

708 Benkeser,R.A. and Gosnell,R.B., J. Org. Chem., 22, 327 (1957) *Note:* Rate of reaction with diphenyldiazomethane (EtOH, 30 deg.)

709 Bordwell,F.G. and Andersen,H.M., J. Am. Chem. Soc., 75, 6019 (1953) *Note:* Apparent PKA of substituted acids (50% aq.-EtOH or water at 25 deg.)

710 Bordwell,F.G. and Andersen,H.M., J. Am. Chem. Soc., 75, 6019 (1953) *Note:* From anilinium (Sigma -) or Sigma ortho *Note:* Apparent PKA of substituted acids (50% aq.-EtOH or water at 25 deg.)

711 Bordwell,F.G. and Andersen,H.M., J. Am. Chem. Soc., 75, 6019 (1953) *Note:* From phenol (Sigma -) or Sigma ortho *Note:* Apparent PKA of substituted acids (50% aq.-EtOH or water at 25 deg.)

712 Fischer,A., Packer,J., Vaughan,J., Wilson,A.F. and Wong,E., J. Org. Chem., 24, 155 (1959) *Note:* Multiple attachment; e.g. position may be ortho + meta *Note:* Apparent PKA of substituted acids (50% aq.-EtOH or water at 25 deg.) *Note:* Ratio of acid vs. base catalysed hydrolysis of esters

713 Kloosterziel,H. and Backer,H.J., Rec. Trav. Chem., 71, 295 (1952) *Note:* Apparent PKA of substituted acids (50% aq.-EtOH or water at 25 deg.)

714 Roberts,J.D., Clement,R.A. and Drysdale,J.J., J. Am. Chem. Soc., 73, 2181 (1951) *Note:* Apparent PKA of substituted acids (50% aq.-EtOH or water at 25 deg.)

715 Roberts,J.D., Clement,R.A. and Drysdale,J.J., J. Am. Chem. Soc., 73, 2181 (1951) *Note:* From anilinium (Sigma -) or Sigma ortho *Note:* Apparent PKA of substituted acids (50% aq.-EtOH or water at 25

716 Roberts,J.D., Webb,R.L. and McElhill,E.A., J. Am. Chem. Soc., 72, 408 (1950) *Note:* Apparent PKA of substituted acids (50% aq.-EtOH or water at 25 deg.)

717 Roberts,J.D., Webb,R.L. and McElhill,E.A., J. Am. Chem. Soc., 72, 408 (1950) *Note:* From anilinium (Sigma -) or Sigma ortho *Note:* Apparent PKA of substituted acids (50% aq.-EtOH or water at 25 deg.)

718 Roberts,J.D. and McElhill,E.A., J. Am. Chem. Soc., 72, 628 (1950) *Note:* Apparent PKA of substituted acids (50% aq.-EtOH or water at 25 deg.)

719 Yagupol'skii,L.M., Bystrov,V.F., Stepanyants,A.U. and Failkov,Y.A., Zh. Obshch, Khim., 34, 3682, 3731EE (1964) *Note:* F-19 NMR shift (in CCl4)

720 Zhdanov,Y.A. and Polenov,V.A., Zh. Obshch. Khim., 35, 589, 587EE (1965) *Note:* Apparent PKA of substituted acids (50% aq.-EtOH or water at 25 deg.) *Note:* 70% EtOH

721 Cecchi,P., La Ricerca Sci., 28, 2526 (1958) *Note:* Apparent PKA of substituted acids (50% aq.-EtOH or water at 25 deg.) *Note:* Pyridines

722 Sitzmann,M.E., Adolph,H.G. and Kamlet,M.J., J. Am. Chem. Soc., 90, 2815 (1968) *Note:* Apparent PKA of substituted acids (50% aq.-EtOH or water at 25 deg.)

723 Hoefnagel,A.J. and Wepster,B.M., J. Am. Chem. Soc., 95, 5357 (1973) *Note:* From anilinium (Sigma -) or Sigma ortho *Note:* Thermodynamic PKA of substituted acids (benzoic acid, phenols or anilines unless otherwise specified, activities-not concentrations-used,50% aq.-EtOH or water at 25 deg.)

724 Gel'fond,A.S., Gavrilov,V.I., Mironova,V.G. and Chernokal'skii,B.D., Zh. Obshch. Khim., 42, 2462, 2454EE (1972) *Note:* Thermodynamic PKA of substituted acids (benzoic acid, phenols or anilines unless otherwise specified, activities-not concentrations-used,50% aq.-EtOH or water at 25 deg.) *Note:* Anilinium

725 Clementi,S., Linda,P. and Marino,G., J. Chem. Soc. (B), 1153 (1970) *Note:* Isomer distribution in halogenation

726 Pavelich,W.A. and Taft Jr.,R.W., J. Am. Chem. Soc., 79, 4935 (1957) *Note:* Rate of hydrolysis or solvolysis

727 Cepciansky,I. and Majer,J., Coll. Czech. Chem. Comm., 34, 72 (1969) *Note:* From phenol (Sigma -) or Sigma ortho *Note:* Apparent PKA of substituted acids (50% aq.-EtOH or water at 25 deg.)

728 Bordwell,F.G. and Boutan,P.J., J. Am. Chem. Soc., 79, 717 (1957) *Note:* From phenol (Sigma -) or Sigma ortho *Note:* Apparent PKA of substituted acids (50% aq.-EtOH or water at 25 deg.)

729 Eaborn,C., Thompson,A.R. and Walton,D.R.M., J. Chem. Soc. (B), 859 (1969) *Note:* Cleavage of C6H4SN(CH3)2

730 Bordwell,F.G. and Boutan,P.J., J. Am. Chem. Soc., 78, 87 (1956) *Note:* From anilinium (Sigma -) or Sigma ortho *Note:* Thermodynamic PKA of substituted acids (benzoic acid, phenols or anilines unless otherwise specified, activities-not concentrations-used,50% aq.-EtOH or water at 25 deg.)

731 Yagupol'skii,L.M., Kondratenko,N.V., Delyagina,N.I., Dyatkin,B.L. and Knunyants,I.L., Zh. Org. Khim., 9, 649, 669EE (1973) *Note:* From anilinium (Sigma -) or Sigma ortho *Note:* Apparent PKA of substituted acids (50% aq.-EtOH or water at 25 deg.) *Note:* F-19 NMR shift (in CCl4)

732 Brownlee,R.T.C., Hutchinson,R.E.J., Katritzky,A.R., Tidwell,T.T. and Topsom,R.D., J. Am. Chem. Soc., 90, 1757 (1968)

733 Korolev,B.A., Titova,S.P. and Ufimtsev,V.N., Zh. Org. Khim., 7, 1191, 1228EE (1971) *Note:* From anilinium (Sigma -) or Sigma ortho *Note:* Thermodynamic PKA of substituted acids (benzoic acid, phenols or anilines unless otherwise specified, activities-not concentrations-used,50% aq.-EtOH or water at 25 deg.)

734 Komendantov,M.I., Fomina,T.B., Kuzina,N.A. and Domnin,I.N., Zh. Org. Khim., 10, 215, 219EE (1974) *Note:* PKA trans acrylic acids

735 Glyde,E. and Taylor,R., J. Chem. Soc. Perkin Ii, 1632 (1973) *Note:* Rate of pyrolysis of 1-arylethyl acetates

736 Fringuelli,F., Marino,G. and Taticchi,A., J. Chem. Soc. (B), 2304 (1971) *Note:* From phenol (Sigma -) or Sigma ortho *Note:* Apparent PKA of substituted acids (50% aq.-EtOH or water at 25 deg.) *Note:* Rate of hydrolysis or solvolysis

737 White,W.N., Schlitt,R. and Gwynn,D., J. Org. Chem., 26, 3613 (1961) *Note:* Apparent PKA of substituted acids (50% aq.-EtOH or water at 25 deg.) *Note:* 75% MeOH at 39.8 deg.

738 Mirkind,L.A., Livshits,B.R., Kornienko,A.G. and Fioshin,M.Y., Zh. Obshch. Khim., 40, 1909, 1893EE (1970) *Note:* Polarography 1/2 wave

739 Noyce,D.S. and Sandel,B.B., J. Org. Chem., 40, 3381 (1975) *Note:* Rate of hydrolysis or solvolysis *Note:* 80% EtOH

740 Wilcox Jr.,C.F. and Chibber,S.S., J. Org. Chem., 27, 2210 (1962) *Note:* Ratio of acid vs. base catalysed hydrolysis of esters *Note:* 70% V. acetone at 25 deg.

741 Posner,T.B. and Hall,C.D., J. Chem. Soc. Perkin Ii, 729 (1976) *Note:* C-13 NMR

742 Nametkin,N.S., Nekhaev,A.I., Gubin,S.P. and Tyurin,V.D., Izv. Akad. Nauk Khim., 7, 1636, 1552EE (1976) *Note:* F-19 NMR shift (in CCl4)

743 Sheppard,W.A., J. Am. Chem. Soc., 84, 3072 (1962) *Note:* From anilinium (Sigma -) or Sigma ortho *Note:* Apparent PKA of substituted acids (50% aq.-EtOH or water at 25 deg.) *Note:* F-19 NMR shift (in CCl4)

744 Sheppard,W.A., J. Am. Chem. Soc., 84, 3072 (1962) *Note:* From phenol (Sigma -) or Sigma ortho *Note:* Apparent PKA of substituted acids (50% aq.-EtOH or water at 25 deg.) *Note:* F-19 NMR shift (in CCl4)

745 Hansch,C., Regr. From Data in J. Org. Chem., 35, 944 (1970) *Note:* H-abstraction by N-bromsuccinimide

746 Weigert,F.J. and Sheppard,W.A., J. Org. Chem., 41, 4006 (1976) *Note:* F-19 NMR shift (in CCl4)

747 Belyaev,E.Y., Tovbis,M.S. and Suboch,G.A., Zh. Org. Khim., 12, 1826, 1790EE (1976) *Note:* Apparent PKA of substituted acids (50% aq.-EtOH or water at 25 deg.)

748 Bothner-By,A.A. and Medalia,A.I., J. Am. Chem. Soc., 74, 4402 (1952) *Note:* Apparent PKA of substituted acids (50% aq.-EtOH or water at 25 deg.) *Note:* 50 deg.

749 Jaffe,H.H., Chem. Rev., 53, 191 (1953) *Note:* From anilinium (Sigma -) or Sigma ortho *Note:* Wide variety of methods, see review article for details

750 Girault,J-p. and Dana,G., J. Chem. Soc. Perkin Ii, 993 (1977) *Note:* C-13 NMR

751 Yagupol'skii,L.M., Gandel'sman,L.Z. and Nazaretyan,V.P., Zh. Org. Khim., 10, 889, 901EE (1974) *Note:* Apparent PKA of substituted acids (50% aq.-EtOH or water at 25 deg.)

752 Taylor,R., J. Chem. Soc. Perkin Ii, 277 (1975) *Note:* Rate of ester hydrolysis

753 Noyce,D.S., Lipinski,C.A. and Loudon,G.M., J. Org. Chem., 35, 1718 (1970) *Note:* Solvolysis of nitrobenzoates

754 Katritzky,A.R., Lewis,J., Musumarra,G. and Ogretir,C., Chim. Acta Turc., 4, 71 (1976) *Note:* PKA substituted pyridines

755 Jaffe,H.H., J. Am. Chem. Soc., 77, 4445 (1955) *Note:* From anilinium (Sigma -) or Sigma ortho *Note:* Apparent PKA of substituted acids (50% aq.-EtOH or water at 25 deg.)

756 Little,W.F., Berry,R.A. and Kannan,P., J. Am. Chem. Soc., 84, 2525 (1962) *Note:* PKA ferrocenylazobenzenes

757 Koppel,I., Karelson,M. and Palm,V., Org. Reactivity (USSR), 9, 101 (1974) *Note:* Apparent PKA of substituted acids (50% aq.-EtOH or water at 25 deg.)

758 Khan,M.K.A. and Morgan,K.J., Tetrahedron, 21, 2197 (1965) *Note:* Apparent PKA of substituted acids (50% aq.-EtOH or water at 25 deg.)

759 Bowden,K., Chapman,N.B. and Shorter,J., J. Chem. Soc., 3370 (1964) *Note:* Alkaline hydrolysis and diazomethane rates

760 Voronkov,M.G., Kashik,T.V., Ponomareva,S.M. and Abramova,N.D., Dokl. Akad. Nauk Ssr, 222, 350, 311EE (1975) *Note:* Apparent PKA of substituted acids (50% aq.-EtOH or water at 25 deg.)

761 Mares,F. and Streitwieser Jr.,A., J. Am. Chem. Soc., 89, 3770 (1967) *Note:* H-2 and H-3 exchange rates on subst toluene

762 Kristian,P., Antos,K., Vlachova,D. and Zahradnik,R., Coll. Czech. Chem. Comm., 28, 1651 (1963) *Note:* Apparent PKA of substituted acids (50% aq.-EtOH or water at 25 deg.)

763 Kristian,P., Antos,K., Vlachova,D. and Zahradnik,R., Coll. Czech. Chem. Comm., 28, 1651 (1963) *Note:* From anilinium (Sigma -) or Sigma ortho *Note:* Apparent PKA of substituted acids (50% aq.-EtOH or water at 25 deg.)

764 Zollinger,H., Buchler,W. and Wittwer,C., Helv. Chim. Acta, 36, 1711 (1953) *Note:* Apparent PKA of substituted acids (50% aq.-EtOH or water at 25 deg.)

765 Zollinger,H., Buchler,W. and Wittwer,C., Helv. Chim. Acta, 36, 1711 (1953) *Note:* From phenol (Sigma -) or Sigma ortho *Note:* Apparent PKA of substituted acids (50% aq.-EtOH or water at 25 deg.)

766 Willi,A.V., Z. Physik. Chem. N.F., 26, 42 (1960) *Note:* Thermodynamic PKA of substituted acids (benzoic acid, phenols or anilines unless otherwise specified, activities-not concentrations-used,50% aq.-EtOH or water at 25 deg.) *Note:* Apparent PKA of substituted acids (50% aq.-EtOH or water at 25 deg.)

767 Willi,A.V., Z. Physik. Chem. N.F., 26, 42 (1960) *Note:* From anilinium (Sigma -) or Sigma ortho *Note:* Thermodynamic PKA of substituted acids (benzoic acid, phenols or anilines unless otherwise specified, activities-not concentrations-used,50% aq.-EtOH or water at 25 deg.) *Note:* Apparent PKA of substituted acids (50% aq.-EtOH or water at 25 deg.)

768 Willi,A.V., Z. Physik. Chem. N.F., 26, 42 (1960) *Note:* From phenol (Sigma -) or Sigma ortho *Note:* Thermodynamic PKA of substituted acids (benzoic acid, phenols or anilines unless otherwise specified, activities-not concentrations-used,50% aq.-EtOH or water at 25 deg.) *Note:* Apparent PKA of substituted acids (50% aq.-EtOH or water at 25 deg.)

769 Cauzzo,G., Galiazzo,G., Mazzucato,U. and Mongiat,N., Tetrahedron, 22, 589 (1966) *Note:* Apparent PKA of substituted acids (50% aq.-EtOH or water at 25 deg.)

770 Kan,R.O., J. Am. Chem. Soc., 86, 5180 (1964) *Note:* P-NMR

771 Chapman,N.B., Lee,J.R. and Shorter,J., J. Chem. Soc. (B), 769 (1969) *Note:* Solvent effect on diazodiphenylmethane-phenylacetic acid reaction rate

772 Ansell,H.V. and Taylor,R., J. Chem. Soc. Perkin Ii, 866 (1977) *Note:* F-16 NMR of naphthalenes

773 Ansell,H.V. and Taylor,R., J. Chem. Soc. Perkin Ii, 866 (1977) *Note:* Multiple attachment; e.g. position may be ortho + meta *Note:* F-16 NMR of naphthalenes

774 Smejkal,J., Jonas,J. and Farkas,J., Coll. Czech. Chem. Comm., 29, 2950 (1964) *Note:* Thermodynamic PKA of substituted acids (benzoic acid, phenols or anilines unless otherwise specified, activities-not concentrations-used,50% aq.-EtOH or water at 25 deg.)

775 Brown,H.C., Rao,C.G. and Ravindranathan,M., J. Am. Chem. Soc., 99, 7663 (1977) *Note:* Rate of hydrolysis or solvolysis

776 Arnold,Z., Krchnak,A. and Trska,P., Tetrahedron Lett., 5, 347 (1975) *Note:* From polymerization rate *Note:* F-19 NMR

777 Arnold,Z., Krchnak,A. and Trska,P., Tetrahedron Lett., 5, 347 (1975) *Note:* From acid catalytic esterification in MeOH, scaled to H = 0, see Tetrahed. Lett.(1978) 35,3293 *Note:* F-19 NMR

778 Noyce,D.S. and Fike,S.A., J. Org. Chem., 38, 3316 (1973) *Note:* Multiple attachment; e.g. position may be ortho + meta *Note:* Rate of hydrolysis or solvolysis

779 Funabiki,T. and Tarama,K., Bull. Chem. Soc. Japan, 45, 2945 (1972) *Note:* I.R. of cyano derivatves

780 Sakurai,H., Deguchi,S., Yamagata,M., Morimoto,S-i., Kira,M. and Kumada,M., J. Organomet. Chem., 18, 285 (1969) *Note:* From phenol (Sigma -) or Sigma ortho *Note:* Proton NMR

781 Koridze,A.A., Gubin,S.P., Lubovich,A.A., Kvasov,B.A. and Ogorodnikova,N.A., J. Organomet. Chem., 32, 273 (1971) *Note:* F-19 NMR

782 Jaouen,G. and Simonneaux,G., J. Organomet. Chem., 61, C39 (1973) *Note:* IR spectra

783 Hirata,Y. and Nagai,W., Nagoya Kogyodaigaku Gakuho, 23, 159 (1971) *Note:* NMR

784 Peters,E.N., Polymer Lett., 13, 479 (1975) *Note:* Rate of hydrolysis or solvolysis

785 Lurik,B.B., Marinova,R.I. and Volkov,Y.P., Zh. Obshch. Khim., 45, 2287, 2246EE (1975) *Note:* Multiple attachment; e.g. position may be ortho + meta *Note:* Polarography

786 Kurono,Y., Ikeda,K. and Uekama,K., Chem. Pharm. Bull. Japan, 23, 340 (1975) *Note:* Apparent PKA of substituted acids (50% aq.-EtOH or water at 25 deg.)

787 Baciocchi,E., Mancini,V. and Perucci,P., J. Chem. Soc. Perkin Ii, 821 (1975) *Note:* Bromide and tosylate elimination by ethoxide

788 Perevalova,E.G., Grendberg,K.I., Zharikova,N.A., Gubin,S.P. and Nesmeyanov,A.N., Iz. Akad. Nauk. Khim., 5, 832, 796EE (1966) *Note:* From anilinium (Sigma -) or Sigma ortho *Note:* Apparent PKA of substituted acids (50% aq.-EtOH or water at 25 deg.)

789 Bean,G.P. and Katritzky,A.R., J. Chem. Soc. (B), 864 (1968) *Note:* Apparent PKA of substituted acids (50% aq.-EtOH or water at 25 deg.)

790 Kasukhin,L.F., Ponomarchuk,M.P., Klepa,T.I., Lysenko,V.P., Repina,L.A. and Gololobov,Y.G., Dokl. Nauk. Ssr, 228, 1380, 579EE (1976) *Note:* From diazomethane or phenyl azide reaction rate *Note:* Kinetic, with phenyl azide

791 Van Meurs,F., Hoefnagel,A.J., Wepster,B.M. and Van Bekkum,H.,

J. Organomet. Chem., 142, 299 (1977) *Note:* Apparent PKA of substituted acids (50% aq.-EtOH or water at 25 deg.)

792 Subbarao,S.N. and Bray,P.J., J. Chem. Phys., 67, 3947 (1977) *Note:* Nuclear quadrapole resonance

793 Ruskie,H.E. and Kaplan,L.A., J. Org. Chem., 30, 319 (1965) *Note:* Rate of reaction with diphenyldiazomethane (EtOH, 30 deg.) *Note:* Picryl substituted benzoic acids (EtOH at 30 deg.)

794 Bykhovskaya,T.N. and Vlasov,O.N., Reakt. Sposob. Organ. Soed., 4, 510, 210EE (1967) *Note:* Hydrolysis of Cl-S-triazines

795 Roberts,J.D. and Regan,C.M., J. Am. Chem. Soc., 75, 4102 (1953) *Note:* Apparent PKA of substituted acids (50% aq.-EtOH or water at 25 deg.) *Note:* Rate of reaction with diphenyldiazomethane (EtOH, 30 deg.) *Note:* Rate of hydrolysis or solvolysis *Note:* Averaged

796 Jaffe,H.H., J. Org. Chem., 23, 1790 (1958) *Note:* PKA pyridine-1-oxides

797 Kaczmarek,J., Smagowski,H. and Grzonka,Z., J. Chem. Soc., Perkin Ii, 1670 (1979) *Note:* Thermodynamic PKA of substituted acids (benzoic acid, phenols or anilines unless otherwise specified, activities-not concentrations-used,50% aq.-EtOH or water at 25 deg.)

798 Shkuzko,O.P., Khmeleva,E.P., Baram,S.G., Shakizov,M.M. and Mamayev,V.P., Chem. Heterocyc. Comp. (Latvia), 7, 996 (1978) *Note:* Miscellaneous

799 Streitwieser Jr.,A. and Humphrey Jr.,J.S., J. Am. Chem. Soc., 89, 3767 (1967) *Note:* Deuterium exchange in deuterated toluenes

800 Streitwieser Jr.,A. and Klein,H.S., J. Am. Chem. Soc., 85, 2759 (1963) *Note:* Conductivity vs. proton analogs

801 Benkesser,R.A. and Krysiak,H.R., J. Am. Chem. Soc., 75, 2421 (1953) *Note:* Thermodynamic PKA of substituted acids (benzoic acid, phenols or anilines unless otherwise specified, activities-not concentrations-used,50% aq.-EtOH or water at 25 deg.) *Note:* Apparent PKA of substituted acids (50% aq.-EtOH or water at 25 deg.)

802 Benkesser,R.A. and Krysiak,H.R., J. Am. Chem. Soc., 75, 2421 (1953) *Note:* From anilinium (Sigma -) or Sigma ortho *Note:* Thermodynamic PKA of substituted acids (benzoic acid, phenols or anilines unless otherwise specified, activities-not concentrations-used,50% aq.-EtOH or water at 25 deg.) *Note:* Apparent PKA of substituted acids (50% aq.-EtOH or water at 25 deg.)

803 Benkesser,R.A. and Krysiak,H.R., J. Am. Chem. Soc., 75, 2421 (1953) *Note:* From phenol (Sigma -) or Sigma ortho *Note:* Thermodynamic PKA of substituted acids (benzoic acid, phenols or anilines unless otherwise specified, activities-not concentrations-used,50% aq.-EtOH or water at 25 deg.) *Note:* Apparent PKA of substituted acids (50% aq.-EtOH or water at 25 deg.)

804 Pressman,D. and Brown,D.H., J. Am. Chem. Soc., 65, 540 (1943) *Note:* Multiple attachment; e.g. position may be ortho + meta *Note:* PKA phenyl arsonic acids

805 Wagner,P.J. and Scheve,B.J., J. Am. Chem. Soc., 101, 378 (1979) *Note:* Rate of photochemical reaction

806 Klepa,T.I., Kasukhin,L.F., Ponomarchuk,M.P. and Gololobov,Y.G., Zh. Obshch. Khim., 48, 1065, 970EE (1978) *Note:* Reaction rate with azidobenzene

807 Kukar,V.P., Zhmurova,I.N. and Yurchenko,R.I., Zh. Obshch. Khim., 42, 279, 268EE (1972) *Note:* Apparent PKA of substituted acids (50% aq.-EtOH or water at 25 deg.) *Note:* In DMF

808 Zhmurova,I.N., Martynyuk,A.P., Shtepanek,A.S., Zasorina,V.A. and Kukhar,V.P., Zh. Obshch. Khim., 44, 79, 76EE (1974) *Note:* PKA N-phenylphosphine imides

809 Gal'pern,G.M., Kreshkov,A.P., Ser'yanova,S.E., Mokrova,L.G. and Shchel'tsyn,V.K., Zh. Obshch. Khim., 47, 2104, 1920EE (1977) *Note:* Apparent PKA of substituted acids (50% aq.-EtOH or water at 25 deg.) *Note:* In glacial acetic acid or methanol

810 Jorgensen,F. and Snyder,J.P., J. Org. Chem., 45, 1015 (1980) *Note:* Photoelectron spectroscopy

811 Coburn,W. et.al., J. Org. Chem., 30, 1110 (1965) *Note:* Proton NMR of 2,6-disubstituted purines *Note:* Shows maximum variance in Sigma+ for NH2

812 van de Graaf,B., Hoefnagel,A.J. and Wepster,B.M., J. Org. Chem., 46, 653 (1981) *Note:* Thermodynamic PKA of substituted acids (benzoic acid, phenols or anilines unless otherwise specified, activities-not concentrations-used,50% aq.-EtOH or water at 25 deg.)

813 McKeown,R.H., J. Chem. Soc. Perkin Ii, 515 (1980) *Note:* Apparent PKA of substituted acids (50% aq.-EtOH or water at 25 deg.)

814 Austel,V., Kutter,E. and Kalbfleisch, Private Commun., Arzneim.-Forsch., 29, 585 (1979)

815 Glover,D.J., J. Org. Chem., 31, 1660 (1966) *Note:* Apparent PKA of substituted acids (50% aq.-EtOH or water at 25 deg.)

816 Sennikov,P.G., Skobeleva,S.E., Kuznetsov,V.A., Egorochkin,A.N., Riviere,P., Satge,J. and Richelme,S., J. Organomet. Chem., 201, 213 (1980) *Note:* Steric constant calculated according to Arzneim.-Forsch.(1979) 29,585 *Note:* I.R. of charge transfer complexes

817 Rackham,D.M. and Davies,G.L.O., Anal. Lett., 13, 447 (1980) *Note:* From acid catalytic esterification in MeOH, scaled to H = 0, see Tetrahed. Lett.(1978) 35,3293 *Note:* PKA P-substituted N-phenylpiperidines

818 Rakhimov,A.I. and Baklanov,A.V., Zh. Org. Khim., 16, 551, 476EE (1980) *Note:* Rate of hydrolysis or solvolysis

819 Lien,E.J., Guo,Z-r., Li,R-I. and Su,C-t., J. Pharm. Sci., 71, 641 (1982)

820 Lien,E.J., Guo,Z-r., Li,R-I. and Su,C-t., J. Pharm. Sci., 71, 641 (1982) *Note:* Steric constant calculated according to Arzneim.-Forsch.(1979) 29,585 *Note:* Majority from "Tables of Expl.Dipole Moments" vol.2, A. McClellan

821 Fischer,P. and Taylor,R., J. Chem. Soc. Perkin Ii, 781 (1980) *Note:* Detritiation TFA 70 deg.

822 Ansell,H.V., Le Guen,J. and Taylor,R., Tetrahed. Let., 1, 13 (1973) *Note:* Detritiation in TFA

823 Ree,B.R. and Martin,J.C., J. Am. Chem. Soc., 92, 1660 (1970) *Note:* Solvolysis of tosylates

824 Colonna,F.P., Danieli,R., Distefano,G. and Ricci,A., J. Chem. Soc. Perkin Ii, 306 (1976) *Note:* U.V.photoelectron spectroscopy

825 Shames,S.L. and Byers,L.D., J. Am. Chem. Soc., 103, 6177 (1981) *Note:* Alkaline hydrolysis phosphono-substituted carboxylic acids

826 Filimonov,V.D., Sukhoroslova,M.M., Novikov,V.T., Vidyagina,T.V., Chem. Abst., 96, 103520f (1982) *Note:* Apparent PKA of substituted acids (50% aq.-EtOH or water at 25 deg.)

827 Traynham,J.G. and Knesel,G.A., J. Org. Chem., 31, 3350 (1966) *Note:* F-19 NMR shift (in CCl4) *Note:* Substituted phenols (in DMSO)

828 Egorochkin,A.N., Khorshev,S.Y., Vyazankin,N.S. and Gladyshev,E.N., Iz. Akad. Nauk. Sssr, S. Khim., 8, 1863, 1733EE (1969) *Note:* Spectral shifts

829 McClelland,R.A. and Leung,M., J. Org. Chem., 45, 187 (1980) *Note:* Hydolyzed vinyl ethers and sulfides

830 Brownlee,R.T.C. and Sadek,M., Aust. J. Chem., 34, 1593 (1981) *Note:* C-13 NMR

831 Barlin,G.B. and Perrin,D.D., Quart. Rev., 20, 75 (1966) *Note:* Possible anomaly; check original article *Note:* Apparent PKA of substituted acids (50% aq.-EtOH or water at 25 deg.)

832 Li,W-y., Guo,Z-r. and Lien,E.J., J. Pharm. Sci. (1983). *Note:* From "Tables of Expl. Dipole Moments" Vol., Rahara Ent. ElCerito(1974)

833 Strom,E.T. and Norton,J.R., J. Am. Chem. Soc., 92, 2327 (1970) *Note:* ESR

834 Inamoto,N. and Masuda,S., Chem. Let. Japan, 1007 (1982)

835 Collins,N.C. and Glass,W.K., Spectrochim. Acta, 30A, 1335 (1974) *Note:* Inductive Sigma effect transmitted by Sigma bonds; by NMR coupling *Note:* H-1 NMR

836 Perjessy,A., Jones,R.G., McClair,S.L. and Wilkins,J.M., J. Org. Chem., 48, 1266 (1983) *Note:* IR

837 Ogata,T., Oikawa,K., Fujisawa,T., Motoyama,S., Izumi,T., Kasahara,A. and Tanaka,N., Bull. Chem. Soc. Jpn., 54, 3723 (1981) *Note:* Polarography and voltammetry

838 Kondratenko,N.V., Kolomeitsev,A.A., Popov,V.I., Il'chenko,A.Y., Korzhenevskaya,N.G., Pirgo,M.D., Titov,E.V. and Yagupol'skii,L.M., Zh. Obshch. Khim., 53, 2500, 2254EE (1983) *Note:* Apparent PKA of substituted acids (50% aq.-EtOH or water at 25 deg.)

839 Jaffe,H.H., J. Am. Chem. Soc., 76, 3527 (1954) *Note:* Apparent PKA of substituted acids (50% aq.-EtOH or water at 25 deg.) *Note:* Substituted pyridine oxides--hetero ring replaces phenyl

840 Bitterwolf,T.E., Polyhedron, 2, 675 (1983) *Note:* H-1, F-19, and P-31 NMR

841 Voronkov,M.G., Kashik,T.V., Deriglazova,E.S., Kositsyna,E.I., Pestunovich,A.E. and Lukevits,E.Y., Zh. Obshch. Khim., 51, 375, 304EE (1981) *Note:* PKA of substituted cyclic amines *Note:* Limited applicability

842 Voronkov,M.G., Kashik,T.V., Deriglazova,E.S., Kositsyna,E.I., Pestunovich,A.E. and Lukevits,E.Y., Zh. Obshch. Khim., 51, 375, 304EE (1981) *Note:* I PKA of substituted cyclic amines

843 Trub,E.P., Boitsov,E.N. and Tsigin,B.M., Zh. Organ. Khim., 19, 87, 78EE (1983) *Note:* H-1 NMR

844 Giansiracusa,J.J., Jenkins,M.J. and Kelly,D.P., Aust. J. Chem., 35,

443 (1982) *Note:* C-13 shift of cations

845 Koridze,A.A., Gubin,S.P. and Ogorodnikova,N.A., J. Organomet. Chem., 74, C37 (1974) *Note:* F-19 NMR

846 de la Mare,P. and Wilson,R.D., J. Chem. Soc. Perkin Ii, 2048 (1977) *Note:* Bromination rate in acetic acid

847 Odyakov,V.F., Org. Reaktiv., 3, 36 (1966) *Note:* Apparent PKA of substituted acids (50% aq.-EtOH or water at 25 deg.)

848 Price,C.C., Mertz,E.C. and Wilson,J., J. Am. Chem. Soc., 76, 5153 (1954) *Note:* Multiple attachment; e.g. position may be ortho + meta *Note:* Apparent PKA of substituted acids (50% aq.-EtOH or water at 25 deg.) *Note:* Rate of hydrolysis or solvolysis

849 Mamaev,V.P., Shkurko,O.P. and Baram,S.G., Adv. Heterocyc. Chem., 42, 1 (1987) *Note:* Apparent PKA of substituted acids (50% aq.-EtOH or water at 25 deg.)

850 Hansch,C., Leo,A, Taft,R., Chem. Rev., 91, 165 (1991) *Note:* Thermodynamic PKA of substituted acids (benzoic acid, phenols or anilines unless otherwise specified, activities-not concentrations-used,50% aq.-EtOH or water at 25 deg.) *Note:* F-19 NMR shift (in CCl4)

851 Hansch,C., Leo,A, Taft,R., Chem. Rev., 91, 165 (1991) *Note:* Apparent PKA of substituted acids (50% aq.-EtOH or water at 25 deg.)

852 Hansch,C., Leo,A, Taft,R., Chem. Rev., 91, 165 (1991) *Note:* Apparent PKA of substituted acids (50% aq.-EtOH or water at 25 deg.) *Note:* F-19 NMR shift (in CCl4)

853 Waisser,K., Machacek,M., Lebvoua,J., Hrbata,J. and Drsata,J., Coll. Czech. Chem., Commun., 53, 2957 (1988) *Note:* H-1 NMR

854 Cross, B.,et al., J.Agric.Food Chem., 31, 260 (1983)

855 Hogben,M.G. and Graham,W.A.G., J. Am. Chem. Soc., 91, 283 (1969)

856 Yagupol'skii,L.M., Syrova,G.P., Voloshchuk,V.G. and Bystrov,V.F., Zh. Obshch. Khim., 38, 2591, 2508EE (1968) *Note:* F-19 NMR shift (in CCl4) *Note:* Of ARXCF3 vs reference PHF

857 Benzene with carbonyl attatched meta

858 Leffler and Grunwald, "Rates and Equilibria of Organic Reactions", Wiley and Sons, N.Y., 1963, p.204. *Note:* No values in computer print-out

859 Jones,L.B. and Foster,J.P., J. Org. Chem., 35, 1777 (1970)

860 Eaborn,C., Thompson,A.R. and Walton,D.R.M., J. Chem. Soc. (B), 357 (1970) *Note:* Rate of hydrolysis or solvolysis *Note:* 71% MeOH at 50.1 deg.

861 Kucsman,A., Ruff,F., Solyom,S. and Szirtes,T., Acta Chim. Acad. Sci. (Hung.), 57, 205 (1968) *Note:* From phenol (Sigma -) or Sigma ortho *Note:* Apparent PKA of substituted acids (50% aq.-EtOH or water at 25 deg.) *Note:* 48% V. EtOH at 25 deg.

862 Blatchly,J. and Taylor,R., J. Chem. Soc. (B), 1402 (1968) *Note:* Multiple attachment; e.g. position may be ortho + meta *Note:* Ester pyrolysis (352 deg.)

863 Brown,H.C. and Okamoto,Y., J. Am. Chem. Soc., 80, 4979 (1958) *Note:* Multiple attachment; e.g. position may be ortho + meta *Note:* Rate of hydrolysis or solvolysis *Note:* T-cumyl chlorides (90% acetone at 25 deg. and others)

864 Retcofsky,H.L. and Griffin,C.E., Tetrahedron Lett., 18, 1975 (1966)

865 Shein,S.M., Ignatov,V.A., Kozorez,L.A. and Chervatyuk,L.F., Zh. Obshch. Khim., 37, 114, 101EE (1967) *Note:* From anilinium (Sigma -) or Sigma ortho *Note:* K for aromatic chloride and sodium methoxide

866 Tsvetkov,E.N., Lobanov,D.I., Makhamatkhanov,M.M. and Kabachnik,M.I., Tetrahedron, 25, 5623 (1969) *Note:* From phenol (Sigma -) or Sigma ortho *Note:* Apparent PKA of substituted acids (50% aq.-EtOH or water at 25 deg.)

867 Weinstein,J., Bluhm,A.L. and Sousa,J.A., J. Org. Chem., 31, 1983 (1966) *Note:* Rate of fading (photo-ind.color) 2-(2-nitro-4-substituted benzyl)pyridines

868 Syz,M. and Zollinger,H., Helv. Chim. Acta, 48, 383 (1965) *Note:* From anilinium (Sigma -) or Sigma ortho *Note:* Apparent PKA of substituted acids (50% aq.-EtOH or water at 25 deg.) *Note:* Rate of reaction with diphenyldiazomethane (EtOH, 30 deg.) *Note:* Rate of hydrolysis or solvolysis *Note:* Averaged

869 Syz,M. and Zollinger,H., Helv. Chim. Acta, 48, 383 (1965) *Note:* From phenol (Sigma -) or Sigma ortho *Note:* Apparent PKA of substituted acids (50% aq.-EtOH or water at 25 deg.) *Note:* Rate of reaction with diphenyldiazomethane (EtOH, 30 deg.) *Note:* Rate of hydrolysis or solvolysis *Note:* Averaged

870 Rae,I.D. and Dyall,L.K., Aust. J. Chem., 19, 835 (1966) *Note:* F-19 NMR shift (in CCl4) *Note:* Benzene and CHCl3 at 25 deg.

871 Rae,I.D. and Dyall,L.K., Aust. J. Chem., 19, 835 (1966) *Note:* From anilinium (Sigma -) or Sigma ortho *Note:* F-19 NMR shift (in CCl4) *Note:* Benzene and CHCl3 at 25 deg.

872 Jones,L.B. and Jones,V.K., Tetrahedron Lett., 14, 1493 (1966) *Note:* Rate of hydrolysis or solvolysis *Note:* 90% V. acetone at 25 deg.

873 Antos,K., Marton,A. and Kristian,P., Chem. Abs., 59, 14742a (1963) *Note:* Infra-red absorption: intensity or frequency (in CCl4)

874 Antos,K., Marton,A. and Kristian,P., Chem. Abs., 59, 14742a (1963) *Note:* From phenol (Sigma -) or Sigma ortho *Note:* Infra-red absorption: intensity or frequency (in CCl4)

875 Beringer,F.M. and Lillien,I., J. Am. Chem. Soc., 82, 5141 (1960) *Note:* Apparent PKA of substituted acids (50% aq.-EtOH or water at 25 deg.) *Note:* 94% acetonitrile and 30% EtOH

876 Beringer,F.M. and Lillien,I., J. Am. Chem. Soc., 82, 5141 (1960) *Note:* From anilinium (Sigma -) or Sigma ortho *Note:* Apparent PKA of substituted acids (50% aq.-EtOH or water at 25 deg.) *Note:* 94% acetonitrile and 30% EtOH

877 Beringer,F.M. and Lillien,I., J. Am. Chem. Soc., 82, 5141 (1960) *Note:* From phenol (Sigma -) or Sigma ortho *Note:* Apparent PKA of substituted acids (50% aq.-EtOH or water at 25 deg.) *Note:* 94% acetonitrile and 30% EtOH

878 Hassan,M., Mudawi,B. and Salama,A., J. Chem. Soc. (B), 928 (1970) *Note:* Rate of hydrolysis or solvolysis *Note:* 90% acetone

879 Bray,P.J. and Barnes,R.G., J. Chem. Phys., 27, 551 (1957) *Note:* Possible anomaly; check original article *Note:* Quadrupole resonance

880 Zuman,P., "Substituent Effects in Organic Polarography", Plenum Press, N.Y., 1967, p.76. *Note:* Multiple attachment; e.g. position may be ortho + meta *Note:* Polarography

881 Zuman,P., "Substituent Effects in Organic Polarography", Plenum Press, N.Y., 1967, p.159 and 322. *Note:* Polarography

882 Hill,E. et.al., J. Am. Chem. Soc., 91, 7381 (1969) *Note:* Multiple attachment; e.g. position may be ortho + meta *Note:* Rate of hydrolysis or solvolysis *Note:* Hetero ring replaces phenyl

883 Bordwell,F.G. and Cooper,G.D., J. Am. Chem. Soc., 74, 1058 (1952) *Note:* From anilinium (Sigma -) or Sigma ortho *Note:* Apparent PKA of substituted acids (50% aq.-EtOH or water at 25 deg.)

884 Buchman,O., Grosjean,M. and Nasielski,J., Helv. Chim. Acta, 47, 2037 (1964) *Note:* From anilinium (Sigma -) or Sigma ortho *Note:* Apparent PKA of substituted acids (50% aq.-EtOH or water at 25 deg.)

885 Buchman,O., Grosjean,M. and Nasielski,J., Helv. Chim. Acta, 47, 2037 (1964) *Note:* From phenol (Sigma -) or Sigma ortho *Note:* Apparent PKA of substituted acids (50% aq.-EtOH or water at 25 deg.)

886 Eleil,E.L. and Nelson,K.W., J. Org. Chem., 20, 1657 (1955) *Note:* Rate of hydrolysis or solvolysis *Note:* MeOH-meona at 25 deg.

887 Fischer,A., Packer,J., Vaughan,J., Wilson,A.F. and Wong,E., J. Org. Chem., 24, 155 (1959) *Note:* From anilinium (Sigma -) or Sigma ortho *Note:* Apparent PKA of substituted acids (50% aq.-EtOH or water at 25 deg.) *Note:* Ratio of acid vs. base catalysed hydrolysis of esters

888 Kloosterziel,H. and Backer,H.J., Rec. Trav. Chem., 71, 295 (1952) *Note:* From anilinium (Sigma -) or Sigma ortho *Note:* Apparent PKA of substituted acids (50% aq.-EtOH or water at 25 deg.)

889 Kloosterziel,H. and Backer,H.J., Rec. Trav. Chem., 71, 295 (1952) *Note:* From phenol (Sigma -) or Sigma ortho *Note:* Apparent PKA of substituted acids (50% aq.-EtOH or water at 25 deg.)

890 Pisarcik,M., Chem. Abst., 59, 12321d (1963) *Note:* Raman spectra

891 Levina,R.Y., Gembitskii,P.A., Guseva,L.P. and Agasyan,P.K., Zh. Obshch. Khim., 34, 146, 144EE (1964) *Note:* Apparent PKA of substituted acids (50% aq.-EtOH or water at 25 deg.)

892 Robinson,R.A. and Ang,K.P., J. Chem. Soc., 2314 (1959) *Note:* Apparent PKA of substituted acids (50% aq.-EtOH or water at 25 deg.)

893 Willi,A.V. and Stocker,J.F., Helv. Chim. Acta, 38, 1279 (1955) *Note:* Apparent PKA of substituted acids (50% aq.-EtOH or water at 25 deg.)

894 Willi,A.V. and Stocker,J.F., Helv. Chim. Acta, 38, 1279 (1955) *Note:* From phenol (Sigma -) or Sigma ortho *Note:* Apparent PKA of substituted acids (50% aq.-EtOH or water at 25 deg.)

895 Broxton,T.J., Capper,G., Deady,L.W., Lenko,A. and Topsom,R.D., J. Chem. Soc., Perkin Ii, 1237 (1972) *Note:* Apparent PKA of substituted acids (50% aq.-EtOH or water at 25 deg.)

896 Zhmurova,I.N., Yurchenko,R.I., Yurchenko,V.G., Tukhar,A.A. and Kirsanov,A.V., Zh. Obshch. Khim., 42, 779, 770EE (1972) *Note:* Absorption spectra

897 Lewis,E.S. and Johnson,M.D., J. Am. Chem. Soc., 81, 2070 (1959) *Note:* From anilinium (Sigma -) or Sigma ortho *Note:* Apparent PKA of substituted acids (50% aq.-EtOH or water at 25 deg.) *Note:* Water at 6 deg. also phenylacetic acid

898 Mastryukova,T.A., Melent'eva,T.A., Shipov,A.E. and Kabachnik,M.I., Zh. Obshch. Khim., 29, 2178, 2145EE (1959) *Note:* Rate of alkaline hydrolysis of (diethyl) substituted-phenyl phosphates

899 Fickling,M.M., Fischer,A., Mann,B.R., Packer,J. and Vaughan,J., J. Am. Chem. Soc., 81, 4226 (1959) *Note:* Apparent PKA of substituted acids (50% aq.-EtOH or water at 25 deg.) *Note:* Also dimethylanilinium

900 Holtz,D., Prog. Phys. Org. Chem., 8, 1 (1971) *Note:* Least squares fit and trend of F and Cl values

901 Tsvetkov,E.N., Makhamatkhanov,M.M., Lobanov,D.I. and Kabachnik,M.I., Zh. Obshch. Khim., 40, 2387, 2376EE (1970) *Note:* Apparent PKA of substituted acids (50% aq.-EtOH or water at 25 deg.)

902 Tsvetkov,E.N. and Kabachnik,M.I., Russ. Chem. Rev., 40, 97EE (1971) *Note:* Apparent PKA of substituted acids (50% aq.-EtOH or water at 25 deg.)

903 Packer,J., Vaughan,J. and Wilson,A.F., J. Org. Chem., 23, 1215 (1958) *Note:* Multiple attachment; e.g. position may be ortho + meta *Note:* Solvolysis diphenylcarbinyl chlorides

904 Habraken,C.L., Recueil, 87, 1241 (1968) *Note:* PKA substituted pyrazoles

905 Ellerhorst,R.H. and Jaffe,H.H., J. Org. Chem., 33, 4115 (1968) *Note:* Apparent PKA of substituted acids (50% aq.-EtOH or water at 25 deg.)

906 Brownlee,R.T.C., Hutchinson,R.E.J., Katritzky,A.R., Tidwell,T.T. and Topsom,R.D., J. Am. Chem. Soc., 90, 1757 (1968) *Note:* Sign undetermined *Note:* Infrared intensities (sign not determined)

907 Biggs,A.I. and Robinson,R.A., J. Chem. Soc., 388 (1961) *Note:* From phenol (Sigma -) or Sigma ortho *Note:* Apparent PKA of substituted acids (50% aq.-EtOH or water at 25 deg.)

908 Rudaya,L.I., Kvitko,I.Y. and Porai-Koshits,B.A., Zh. Org. Khim., 8, 13, 11EE (1972) *Note:* From anilinium (Sigma -) or Sigma ortho *Note:* Thermodynamic PKA of substituted acids (benzoic acid, phenols or anilines unless otherwise specified, activities-not concentrations-used,50% aq.-EtOH or water at 25 deg.) *Note:* Infrared absorption: intensity or frequency (in CCl4)

909 Yoshioka,M., Hamamoto,K. and Kubota,T., Bull. Chem. Soc. Japan, 35, 1723 (1962) *Note:* Apparent PKA of substituted acids (50% aq.-EtOH or water at 25 deg.)

910 Edel'man,T.G. and Stepanov,B.I., Zh. Obshch. Khim., 39, 712, 678EE (1969) *Note:* PKA in nitromethane

911 Stepanov,B.I., Bokanov,A.I. and Korolev,B.A., Zh. Obshch. Khim., 36, 762, 778EE (1966) *Note:* Thermodynamic PKA of substituted acids (benzoic acid, phenols or anilines unless otherwise specified, activities-not concentrations-used,50% aq.-EtOH or water at 25 deg.)

912 Noyce,D.S. and Fike,S.A., J. Org. Chem., 38, 2433 (1973) *Note:* Rate of hydrolysis or solvolysis *Note:* 80% EtOH, from 2-thiazolyl system

913 Posner,T.B. and Hall,C.D., J. Chem. Soc. Perkin Ii, 729 (1976) *Note:* Possible anomaly; check original article *Note:* C-13 NMR

914 Hansch,C., Regr. of Data in: Zh. Org. Khim., 11, 1991EE (1975) *Note:* PKA and tautomeric equilibrium constant in 30% dioxane

915 Le Guen,M.M.J. and Taylor,R., J. Chem. Soc. Perkin Ii, 559 (1976) *Note:* Multiple attachment; e.g. position may be ortho + meta *Note:* Protodetritiation

916 Hansch,C., Regr. of Data in J. Heterocyc. Chem., 4, 438 (1967)

917 Clementi,S. and Katritzky,A.R., J. Chem. Soc. Perkin Ii, 1077 (1973) *Note:* Acid catalyzed hydrogen exchange

918 Durmis,J.,Karvas,M. and Manasek,Z., Coll. Czech. Chem. Comm., 38, 215 (1973) *Note:* From anilinium (Sigma -) or Sigma ortho *Note:* Apparent PKA of substituted acids (50% aq.-EtOH or water at 25 deg.) *Note:* Phenols

919 Tin-Lok,C., Miller,J. and Stansfield,F., J. Chem., 1213 (1964) *Note:* K *Note:* Methanethiolysis

920 Miller,J., Austr. J. Chem., 9, 61 (1956) *Note:* Rate of hydrolysis or solvolysis

921 Bokanov,A.I., Korolev,B.A. and Stepanov,B.I., Zh. Obshch. Khim., 39, 373, 350EE (1969) *Note:* From phenol (Sigma -) or Sigma ortho *Note:* Apparent PKA of substituted acids (50% aq.-EtOH or water at 25 deg.) *Note:* Phenols

922 Hansch,C., Regr. From Data in J. Chem. Soc., 4375 (1957) *Note:* PKA X-pyridine-N-oxides

923 Charton,M., "Design of Biopharm. Prop. Thru Prodrugs and Analogs", Roche,E.,Ed., Am. Pharm. Assoc., 1977, Ch. 9, p. 269.

924 Schaeffer Jr.,C.D. and Zuckerman,J.J., J. Organomet. Chem., 78, 373 (1974) *Note:* Proton NMR

925 Vulakh,E.L., Freidlin,E.G. and Gitis,S.S., Zh. Org. Khim., 11, 1481, 1463EE (1975) *Note:* Reaction rate of thionyl chloride with substituted benzoic acids

926 Jaffe,H.H. and Jones,H.L., Adv. Heterocyc. Chem., 3, 209 (1964) *Note:* From phenol (Sigma -) or Sigma ortho *Note:* Apparent PKA of substituted acids (50% aq.-EtOH or water at 25 deg.) *Note:* Wolf-Kishner reduction

927 Forsythe,P., Frampton,R., Johnson,C.D. and Katritzky,A.R., J. Chem. Soc. Perkin Ii, 671 (1972) *Note:* Thermodynamic PKA of substituted acids (benzoic acid, phenols or anilines unless otherwise specified, activities-not concentrations-used,50% aq.-EtOH or water at 25 deg.) *Note:* PKA dimethyamino substituted heteroaromatics

928 Komendantov,M.I., Fomina,T.B. and Oglobin,K.A., Chem. Abstr., 85, 108195m (1976) *Note:* PKA substituted acrylic acids

929 Afanas'ev,I.B., Zh. Organich. Khim., 11, 1145, 1135EE (1975) *Note:* Reference also reports radical constant from ESR

930 Bryson,A., J. Am. Chem. Soc., 82, 4858 (1960) *Note:* Apparent PKA of substituted acids (50% aq.-EtOH or water at 25 deg.)

931 Bellingham,P., Johnson,C.D. and Katritzky,A.R., J. Chem. Soc. (B), 866 (1968) *Note:* Acid catalysed H-exchange

932 Bowden,K., Chapman,N.B. and Shorter,J., J. Chem. Soc., 3370 (1964) *Note:* From diazomethane or phenyl azide reaction rate *Note:* Alkaline hydrolysis and diazomethane rates

933 Stock,L.M. and Brown,H.C., Adv. Phys. Org. Chem., 1, 116 (1963) *Note:* Multiple attachment; e.g. position may be ortho + meta *Note:* Rate of detritiation in CF3CO2H

934 Zollinger,H., Buchler,W. and Wittwer,C., Helv. Chim. Acta, 36, 1711 (1953) *Note:* From anilinium (Sigma -) or Sigma ortho *Note:* Apparent PKA of substituted acids (50% aq.-EtOH or water at 25 deg.)

935 Hansch,C., Regr. of Data in: Izv. Akad. Nauk, 1504 (1962) *Note:* Apparent PKA of substituted acids (50% aq.-EtOH or water at 25 deg.)

936 Ostrovskii,V.A., Enin,A.S. and Koldobskii,G.I., Zh. Org. Khim., 9, 802, 827EE (1973) *Note:* Reaction rate alkyl-phenyl ketones with HN3

937 Koshelev,Y.N., Kvitko,I.Y. and Efros,L.S., Zh. Org. Khim., 8, 1750, 1789EE (1972) *Note:* Apparent PKA of substituted acids (50% aq.-EtOH or water at 25 deg.)

938 Heck,R. and Winstein,S., J. Am. Chem. Soc., 79, 3432 (1957) *Note:* Rate of hydrolysis or solvolysis

939 Wells,P.R. and Ward,E.R., Chem. and Ind., 528 as Rpt. in Ref. 187, p.222 (1958) *Note:* Mean from various reactions

940 Jones,L.B. and Eng,S.S., Tetrahedron Let., 12, 1431 (1968) *Note:* Hydration of substituted styrene

941 Ovchinnikov,V.V., Galkin,V.I., Cherkasov,R.A. and Pudovik,A.N., Zh. Obshch. Khim., 47, 290, 267EE (1977) *Note:* Apparent PKA of substituted acids (50% aq.-EtOH or water at 25 deg.)

942 Ansell,H.V., Hirschler,M.M. and Taylor,R., J. Chem. Soc. Perkin Ii, 353 (1977) *Note:* Protodetritiation

943 Hopkinson,A.C. and Wyatt,P.A.H., J. Chem. Soc. (B), 530 (1970) *Note:* Absorption spectra

944 Noyce,D.S. and Fike,S.A., J. Org. Chem., 38, 3316 (1973) *Note:* Rate of hydrolysis or solvolysis

945 Charton,M., J. Org. Chem., 29, 1222 (1964) *Note:* Apparent PKA of substituted acids (50% aq.-EtOH or water at 25 deg.)

946 Charton,M., J. Org. Chem., 29, 1222 (1964) *Note:* From phenol (Sigma -) or Sigma ortho *Note:* Apparent PKA of substituted acids (50% aq.-EtOH or water at 25 deg.) *Note:* Substituted acetic acids (various conditions, by regression)

947 Voronkov,M.G., Feshin,V.P. and Polis,J., Teor. Eksp. Khim., (1971) 7, 555, Chem. Abst. 76, 33456a (1972)

948 Voronkov,M.G., Feshin,V.P. and Polis,J., Teor. Eksp. Khim., (1971) 7, 555, Chem. Abst. 76, 33456a (1972) *Note:* Nuclear quadrapole resonance

949 Dyall,L.K., Aust. J. Chem., 17, 419 (1964) *Note:* From anilinium (Sigma -) or Sigma ortho *Note:* Proton NMR

950 Jerumanis,S. and Bruylants,A., Bull. Soc. Chim. Belg., 72, 69 (1963) *Note:* From anilinium (Sigma -) or Sigma ortho *Note:* Apparent PKA of substituted acids (50% aq.-EtOH or water at 25 deg.)

951 Kreindlin,A.Z., Khandkarova,V.S. and Gubin,S.P., J. Organomet. Chem., 92, 197 (1975) *Note:* Apparent PKA of substituted acids (50% aq.-EtOH or water at 25 deg.)

952 Kreindlin,A.Z., Khandkarova,V.S. and Gubin,S.P., J. Organomet. Chem., 92, 197 (1975) *Note:* From phenol (Sigma -) or Sigma ortho *Note:* Apparent PKA of substituted acids (50% aq.-EtOH or water at 25 deg.)

953 Koike,H., Nippon Kagaku Zasshi, (1962) 83, 917, Chem. Abst., 57, 13301e (1963) *Note:* Apparent PKA of substituted acids (50% aq.-EtOH or water at 25 deg.)

954 Koike,H., Nippon Kagaku Zasshi, (1962) 83, 917, Chem. Abst., 57, 13301e (1963) *Note:* From anilinium (Sigma -) or Sigma ortho *Note:* Apparent PKA of substituted acids (50% aq.-EtOH or water at 25 deg.)

955 Koike,H., Nippon Kagaku Zasshi, (1962) 83, 917, Chem. Abst., 57, 13301e (1963) *Note:* From phenol (Sigma -) or Sigma ortho *Note:* Apparent PKA of substituted acids (50% aq.-EtOH or water at 25 deg.)

956 Katritzky,A.R. and Topsom,R.D., Chem. Rev., 77, 639 (1977)

957 Pushkina,L.N., Stepanov,A.P., Zhukov,V.S. and Naumov,A.D., Organ. Magnet. Reson., 4, 607 (1972) *Note:* Possible anomaly; check original article *Note:* F-19 NMR *Note:* Reference contains more values

958 Feytmans-De Medicis,M.E. and Bruylants (Louvain),A., Bull. Soc. Chim. Belg., 72, 603 (1963) *Note:* Apparent PKA of substituted acids (50% aq.-EtOH or water at 25 deg.)

959 Perevalova,E.G., Grendberg,K.I., Zharikova,N.A., Gubin,S.P. and Nesmeyanov,A.N., Iz. Akad. Nauk. Khim., 5, 832, 796EE (1966) *Note:* Apparent PKA of substituted acids (50% aq.-EtOH or water at 25 deg.)

960 Perevalova,E.G., Grendberg,K.I., Zharikova,N.A., Gubin,S.P. and Nesmeyanov,A.N., Iz. Akad. Nauk. Khim., 5, 832, 796EE (1966) *Note:* From phenol (Sigma -) or Sigma ortho *Note:* Apparent PKA of substituted acids (50% aq.-EtOH or water at 25 deg.)

961 Charton,M., Prog. Phys. Org. Chem., in press.

962 Hine,J. and Ramsay,O.B., J. Am. Chem. Soc., 84, 973 (1962) *Note:* Base catalyzed deuterium exchange

963 Zeng,G-z., Acta Chim. Sinica, 32, 107 (1966) *Note:* 2-substituted cyclopropylbenzenes *Note:* Apparent PKA of substituted acids (50% aq.-EtOH or water at 25 deg.) *Note:* Doubtful value

964 Korolev,B.A., Khardin,A.P., Radchenko,S.S., Novakov,I.A. and Orlinson,B.S., Zh. Organ. Khim., 14, 1632, 1524EE (1978) *Note:* Apparent PKA of substituted acids (50% aq.-EtOH or water at 25 deg.)

965 Korolev,B.A., Khardin,A.P., Radchenko,S.S., Novakov,I.A. and Orlinson,B.S., Zh. Organ. Khim., 14, 1632, 1524EE (1978) *Note:* From anilinium (Sigma -) or Sigma ortho *Note:* Apparent PKA of substituted acids (50% aq.-EtOH or water at 25 deg.)

966 Jarv,J.L., Aaviksaar,A.A., Godovikov,N.N. and Morozova,N.A., Organic Reactiv., 9, 813 (1972) *Note:* Alkaline hydrolysis of phosphate esters

967 Csizmadia,V.M., Koshy,K.M., Lau,K.C.M., McClelland,R.A., Nowlan,V.J. and Tidwell,T.T., J. Am. Chem. Soc., 101, 974 (1979) *Note:* Acid hydrolysis of N-vinylacetamides

968 Rodionov,P., Kollegov,V.F. and Lel'kin,K.P., Acad. Sci. Sssr., Kinetics and Catal., 20, 1327 (1979) *Note:* F-19 NMR

969 Stepanov,B.I., Bokanov,A.I. and Korolev,B.A., Zh. Obshch. Khim., 37, 2139, 2029EE (1967) *Note:* PKA of phosphines

970 Ryan,J.J. and Humffray,A.A., J. Chem. Soc. (B), 1300 (1967) *Note:* From anilinium (Sigma -) or Sigma ortho *Note:* Rate of reaction with picryl chloride

971 Ryan,J.J. and Humffray,A.A., J. Chem. Soc. (B), 1300 (1967) *Note:* From phenol (Sigma -) or Sigma ortho *Note:* Rate of reaction with picryl chloride

972 Traven,V.F., German,M.I., Eismont,M.Y., Kostyuchenko,E.E. and Stepanov,B.I., Zh. Obshch. Khim., 48, 2232, 2027EE (1978) *Note:* From anilinium (Sigma -) or Sigma ortho *Note:* Apparent PKA of substituted acids (50% aq.-EtOH or water at 25 deg.)

973 Amin,H.B. and Taylor,R., J. Chem. Soc. Perkin Ii, 228 (1979) *Note:* Pyrolysis of 1-aryl-1-methylethyl acetates

974 Jarv,J.L., Aaviksaar,A.A., Godovikov,N.N. and Morozova,N.A., Org. React., 9, 813 (1972) *Note:* Rate of hydrolysis or solvolysis

975 Schiemenz,G.P., Angew. Chem., 5, 595EE (1966) *Note:* Apparent

976 Palm,V.A., Summary of Sci. and Technology, Ser. Gen. Quest. Org. Chem. Moscow, p.163 (1979) *Note:* Thermodynamic PKA of substituted acids (benzoic acid, phenols or anilines unless otherwise specified, activities-not concentrations-used,50% aq.-EtOH or water at 25 deg.)

977 Howard,J.C. and Lewis,J.P., J. Org. Chem., 31, 2005 (1966) *Note:* Apparent PKA of substituted acids (50% aq.-EtOH or water at 25 deg.)

978 Goering,H.L. and Jacobson,R.R., J. Am. Chem. Soc., 80, 3277 (1958) *Note:* Rate of O-Claisen rearrangement

979 Tice,B.B.P., Lee,I. and Kendall,F.H., J. Am. Chem. Soc., 85, 329 (1963) *Note:* Deuterium exchange in dialkyl anilines

980 Perrin,D.D., Dempsey,B. and Serjeant,E.P., "pKa Predictions for Organic Acids and Bases", Chapman and Hall, p.107 (1981)

981 Lien,E.J., Guo,Z-r., Li,R-l. and Su,C-t., J. Pharm. Sci., 71, 641 (1982) *Note:* Aliphatic attached *Note:* Majority from "Tables of Expl.Dipole Moments" vol.2, A. McClellan

982 Taylor,R., J. Chem. Soc. (B), 622 (1971) *Note:* Pyrolysis aryl-ethyl acetates

983 Ansell,H.V., Sheppard,P.J., Simpson,C.F., Stroud,M.A. and Taylor,R., J. Chem. Soc. Perkin Ii, 381 (1979) *Note:* Multiple attachment; e.g. position may be ortho + meta *Note:* Detritiation TFA 70 deg.

984 Ansell,H.V. and Taylor,R., J. Chem. Soc. Perkin Ii, 751 (1978) *Note:* Multiple attachment; e.g. position may be ortho + meta *Note:* Detritiation TFA 70 deg.

985 Archer,W.J., Shafig,Y.E-d. and Taylor,R., J. Chem. Soc. Perkin Ii, 675 (1981) *Note:* 50% acetone at 30 deg. *Note:* Bromination of substituted benzenes

986 Kwasny,M., Syczewski,M., Chem. Abstr., 95, 61164q (1981) *Note:* IR and NMR

987 Brittain,J.M., de la Mare,P.B.D. and Smith,J.M., J. Chem. Soc. Perkin Ii, 1629 (1981) *Note:* Chlorination in acetic acid 25 deg.

988 Hine,J. and Linden,S-m., J. Org. Chem., 46, 1635 (1981) *Note:* T-butoxide catalysis of isomerization

989 Epstein,L.M., Ashkinadze,L.D., Shubina,E.S., Kravtsov,D.N. and Kazitsyna,L.A., J. Organomet. Chem., 228, 53 (1982) *Note:* I.R. in DMSO *Note:* Other values in CH2Cl2

990 Li,W-y., Guo,Z-r. and Lien,E.J., J. Pharm. Sci. (1983).

991 Kazitsyna,L.A., Ustynyuk,Y.A., Gurman,V.S., Pergushov,V.I. and Gruzdneva,V.N., Dokl. Akad. Nauk. Sssr, 233, 866, 213EE (1977)

992 Blanch,J.H., J. Chem. Soc. (B), 937 (1966)

993 Bzhezovskii,V.M., Kushnarev,D.F., Trofimov,B.A., Kalabin,G.A., Gusarova,N.K. and Efremova,G.G., Iz. Akad. Nauk. Sssr, S. Khim., 11, 2507, 2073EE (1981)

994 Shkurko,O.P., Gogin,L.L., Baram,S.G. and Mamaev,V.P., Izv. Sib. Otd. Akad. Nauk., 95 (1980) *Note:* F-19 NMR

995 Shkurko,O.P., Gogin,L.L., Baram,S.G. and Mamaev,V.P., Izv. Sib. Otd. Akad. Nauk., 95 (1980) *Note:* From anilinium (Sigma -) or Sigma ortho *Note:* F-19 NMR

996 Shkurko,O.P., Gogin,L.L., Baram,S.G. and Mamaev,V.P., Izv. Sib. Otd. Akad. Nauk., 95 (1980) *Note:* From phenol (Sigma -) or Sigma ortho *Note:* F-19 NMR

997 Collins,N.C. and Glass,W.K., Spectrochim. Acta, 30A, 1335 (1974) *Note:* H-1 NMR

998 Gassman,P.G. and Guggenheim,T.L., J. Org. Chem., 47, 4002 (1982) *Note:* Solvolysis of T-benzyl chloride in 90% aqueous acetone

999 Biechler,S.S. and Taft Jr.,R.W., J. Am. Chem. Soc., 79, 4927 (1957) *Note:* Rate of aqueous alkaline hydrolysis

1000 Knorr,R., Tetrahed., 37, 929 (1981)

1001 Knorr,R., Tetrahed., 37, 929 (1981) *Note:* From anilinium (Sigma -) or Sigma ortho *Note:* H1-H1-NMR coupling constants olefins

1002 Wayner,D.D.M. and Arnold,D.R., Can. J. Chem., 62, 1164 (1984) *Note:* ESR, values scaled by 10x from those in reference

1003 Wayner,D.D.M. and Arnold,D.R., Can. J. Chem., 62, 1164 (1984) *Note:* From infra-red or other spectral measurements, *Note:* ESR, values scaled by 10x from those in reference

1004 Shim,S.C., Park,J.W. and Ham,H.S., Bull. Korean. Chem. Soc., 3, 13 (1982) *Note:* PKA's of phenols in excited state

1005 Glogowski,M.E. and Williams,J.L.R., J. Organomet. Chem., 218, 137 (1981) *Note:* PKA and spectra

1006 Glogowski,M.E. and Williams,J.L.R., J. Organomet. Chem., 218, 137 (1981) *Note:* From infra-red or other spectral measurements, *Note:* PKA and spectra

1007 Shames,S.L. and Byers,L.D., J. Am. Chem. Soc., 103, 6170 (1981) *Note:* Acyl transfer rate to thiolate, OH, and oxy dianions

1008 Alberghina,G., Amato,M.E., Fisichella,S. and Occhipinti,S., J. Chem. Soc. Perkin Ii, 1721 (1980) *Note:* U.V. spectra of protonation

1009 de la Mare,P. and Newman,P.A.., Tetrahed. Let., 23, 1305 (1982) *Note:* Rate of hydrolysis or solvolysis

1010 Kalyagin,G.A., Boldeskul,I.E., Egorov,Y.P. and Balitskii,Y.V., Chem. Abs., 94, 120688t (1981) *Note:* IR shifts

1011 Levitt,L.S. and Levitt,B.W., Chem. Indust., 990 (1970) *Note:* Ionization energies

1012 Rodionov,P.P., Org. React., 16, 366 (1979) *Note:* Rate of methoxide replacement of F

1013 Taylor,P.J. and Wait,A.R., J. Chem. Soc. Perkin Ii, 1765 (1986) *Note:* PKA's of subst. guanidinium ions

1014 Mamaev,V.P., Shkurko,O.P. and Baram,S.G., Adv. Heterocyc. Chem., 42, 1 (1987) *Note:* Ionization of substituted acids

1015 Egorochkin,A.N. and Razuvaev,G.A., Russ. Chem. Rev., 56, 846 (1987) *Note:* Possible anomaly; check original article

1016 Hansch,C., estimated value

Depiction of Solute Structures

957   logP = 1.24

1072   logP = -1.28

1190   logP = -1.31

1262   logP = -2.50

1361   logP = 0.96

1362   logP = -1.10

1447   logP = -0.21

1656   logP* = 0.04

1670   logP* = -0.22

1920   logP = -3.90

1939   logP* = 0.02

1987   logP* = 1.85

2100   logP* = 1.76

2388   logP* = 0.98

2395   logP* = 0.30

2464   logP* = 1.78

2468   logP* = 1.34

2510   logP* = -0.77

2598   logP* = 1.80

2600   logP* = 0.74

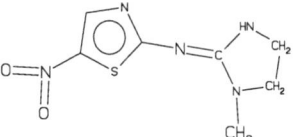

2605   logP* = -0.22

2642   logP = -3.10

2707   logP* = -0.94

2708   logP* = 2.31

2712   logP* = -0.17

2717   logP* = 1.33

2728   logP* = -0.24

2763   logP* = 0.48

2817   logP = -3.03

2819   logP = -2.70

2878   logP* = 2.42

2898   logP* = 0.74

I⁻

Br⁻

2900   logP* = 0.55

3023   logP* = 0.57

3508   logP* = -0.52

3518   logP* = -0.14

3580   logP* = 1.28

3581   logP* = 0.43

3625   logP* = 1.51

3639   logP* = 0.66

3640   logP* = 0.41

3642   logP* = -0.48

3779   logP* = 0.40

3800   logP = 2.15

3859   logP = -2.41

3871   logP* = 0.19

3885   logP* = 1.48

3986   logP = -2.10

Br⁻

3987   logP = -2.10

4030   logP = -1.71

4123   logP* = 1.06

4125   logP* = 1.71

I⁻

4216   logP = 2.61

4217   logP* = 2.72

4222   logP* = 1.16

4223   logP* = 0.86

4224   logP* = 1.44   4244   logP* = 1.90   4245   logP* = 2.89   4362   logP* = 1.00

4405   logP* = 3.41   4462   logP* = 1.27   4531   logP* = 1.35   4549   logP* = 1.41

4581   logP* = 2.75   4625   logP = -2.85   4637   logP* = 0.99   4641   logP* = 0.21

4723   logP* = 2.30   4747   logP* = -0.95   4782   logP* = 1.70   4859   logP* = -0.38

4895   logP* = 2.75   4896   logP* = 0.00   4898   logP* = 3.63   4915   logP* = 0.12

4937   logP = 0.26   4938   logP* = 0.56   4956   logP* = -1.73   4978   logP* = 1.88

4979   logP* = -0.10   5004   logP* = 0.47   5018   logP* = 1.89   5024   logP* = 0.76

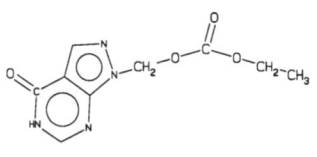
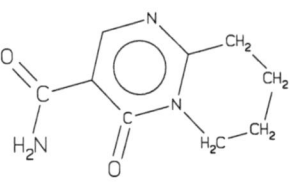
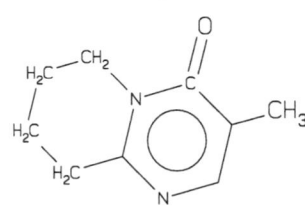

5025   logP* = 1.10

5137   logP* = 0.16

5189   logP* = -0.15

5277   logP = 0.74

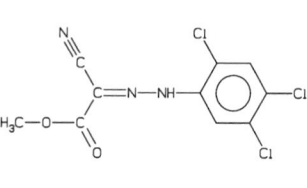

5285   logP* = 0.13

5294   logP* = 1.68

5301   logP* = 1.60

5423   logP* = 5.15

5442   logP* = 0.93

5480   logP* = 1.34

5506   logP* = 2.60

5539   logP = 1.66

5545   logP* = 1.38

5619   logP* = 1.59

5667   logP* = 1.44

5678   logP = 2.12

5734   logP* = 1.87

5774   logP = -0.34

5786   logP* = 2.42

5929   logP = -1.60

5942   logP* = 0.22

5969   logP* = -0.25

6207   logP = 1.05

6302   logP* = -0.66

6333   logP = 0.60

6380   logP = 0.03

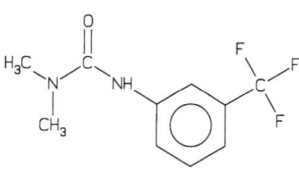

6392   logP = 1.28

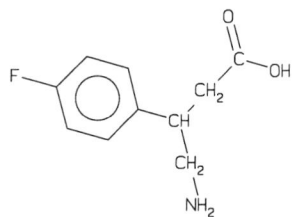

6422   logP* = -1.42

6626   logP* = 1.70

6636   logP* = -0.40

6637   logP* = -0.74

6647   logP = -2.35

I⁻

6662   logP* = 2.68

6663   logP* = 1.02

6665   logP = -0.41

6666   logP* = 0.71

6679   logP = 1.46

6680   logP = 1.62

6681   logP = 1.74

6752   logP* = 0.69

6754   logP = -1.90

6755   logP = -1.33

6768   logP* = 4.30

6769   logP = 2.08

6773   logP* = 4.42

6790   logP* = 3.20

6791   logP = 2.02

6867   logP* = 2.60

6904   logP = 1.19

6909   logP* = 0.25

6913   logP* = 1.46

6914   logP* = 1.75

6922   logP* = -1.40

6991   logP = 2.62

7006   logP* = 1.01

7022   logP* = 2.37

7035   logP* = 1.59

7112   logP* = 2.50

7219   logP = 0.95

7231   logP = 0.30

7232   logP* = 1.25

7235   logP = 1.28

7236   logP = 0.71

7251   logP* = 0.16

7335   logP* = 0.50

7375   logP = -1.46

7464   logP* = 1.04

7585   logP* = 2.90

7639   logP* = -0.87

7756   logP* = 2.00

8109   logP = 2.22

8110   logP* = 2.79

8224   logP = 2.30

8225   logP* = 2.48

8283   logP* = 1.97

8330   logP* = 2.80

8340   logP = 1.63

8342   logP = 1.61

8372   logP* = 1.98

8388   logP* = 2.56

8429   logP* = 0.64

8494   logP* = 0.30

8516   logP* = -0.90

8564   logP* = 2.71

8569   logP = 1.04

8596   logP = 0.05

8643   logP = 1.08

8648   logP = 1.19

8651   logP = 1.24

8653   logP = 1.17

8685   logP* = 0.80

8690   logP* = -0.08

8706   logP* = 2.33

8708   logP* = 3.93

8799   logP* = 2.26

8808   logP = 0.60

8829   logP = 0.45

8841   logP = 1.43

8958   logP = 1.78

8959   logP = 2.29

8971   logP = -1.52

9035   logP = 1.39

9061   logP = -1.67

9218   logP = 3.02

9232   logP = 3.10

9311   logP* = 2.23

9359   logP* = 1.17

9364   logP = 1.19

9369   logP = 0.06

9370   logP* = 1.35

9397   logP = 1.98

9418   logP* = 2.28

9419   logP* = 2.71

9497   logP = -2.40

9498   logP = -0.93

9524   logP* = 4.45

9531   logP* = 2.22

9578   logP = -0.07

9602   logP* = 4.26

9623   logP* = 3.18

9641   logP = -0.48

9708   logP = 1.60

9719   logP = 1.73

9770   logP* = 3.81

9797   logP* = 3.28

9800   logP* = 1.64

9833   logP* = -0.70

9864   logP* = 1.88

9931   logP* = 2.35

9961   logP = 0.01

9962   logP = -1.89

9963   logP = -1.17

9964   logP = -1.88

9989   logP = -1.99

9990   logP = -1.64

9998   logP = -2.54

10006   logP = -2.50

10007   logP = -2.47

10035   logP* = 1.57

10052   logP = 2.17

10089　logP = 2.03　　10151　logP = 3.42　　10183　logP* = 4.29　　10260　logP* = 2.90

10308　logP* = 2.57　　10311　logP* = 1.11　　10327　logP = -0.57　　10431　logP = 6.34

10537　logP* = 2.37　　10549　logP = 2.53　　10583　logP* = 0.11　　10606　logP = -1.02

10643　logP* = 1.93　　10660　logP* = 2.45　　10675　logP = -0.59　　10735　logP* = -0.47

10776　logP* = 2.82　　10793　logP* = -0.30　　10803　logP = 0.87　　10804　logP = -1.12

10854　logP = 0.38　　10855　logP = 1.25　　10879　logP = 0.43　　10883　logP* = 5.09

10890　logP = -0.18　　10893　logP* = 3.79　　10898　logP = 0.25　　10899　logP = 0.73

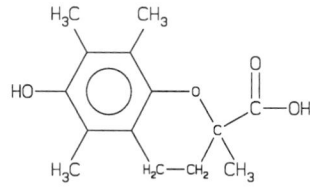
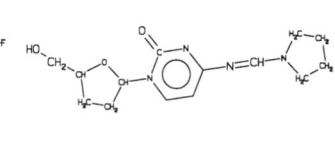
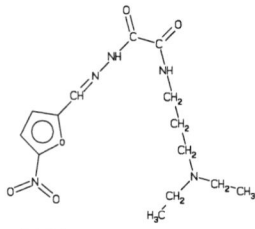

10912   logP = -1.76

10913   logP = -1.43

10951   logP* = -0.80

10953   logP = 0.92

10960   logP = -0.35

11044   logP* = 2.35

11082   logP* = 2.59

11113   logP = 3.20

11177   logP = -0.02

11186   logP* = 4.03

11205   logP = 2.82

11240   logP = -1.52

11241   logP* = -0.71

11243   logP* = 0.21

11250   logP* = 2.19

11252   logP = 1.71

11269   logP = -0.64

11396   logP = -0.57

11468   logP = 0.23

11469   logP* = 1.13

11556   logP* = 1.32

11600   logP* = 1.71

11671   logP* = 2.90

11674   logP* = 4.60

11683   logP* = 0.64

11692   logP* = -0.67

11731   logP* = 0.66

11735   logP* = 1.60

11748   logP = 0.20

11777   logP = -1.36

11794   logP = -1.73

11796   logP = -1.69

11967   logP = 2.93

11972   logP* = 4.80

11980   logP = 2.46

12036   logP = 2.19

12040   logP* = 2.85

12042   logP* = 2.96

12052   logP* = 1.08

12094   logP = 0.53

12095   logP* = 1.05

12098   logP = 1.25

12156   logP = -1.26

12166   logP = 1.12

12195   logP = -1.26

12198   logP = 0.06

12221   logP* = 1.85

12308   logP* = 1.87

12321   logP* = 0.95   12324   logP* = 1.64   12328   logP* = 0.62   12333   logP = 0.20

12339   logP* = 3.60   12340   logP* = 3.70   12358   logP* = 3.20   12359   logP* = 3.50

12439   logP* = 3.20   12442   logP* = 3.40   12483   logP = 0.70   12484   logP* = 2.20

12522   logP = 4.89   12523   logP = 6.39   12536   logP* = 0.82   12562   logP* = 1.81

12564   logP = 2.05   12565   logP = 2.86   12569   logP = 0.38   12615   logP* = 3.36

12617   logP* = 2.42

12661   logP* = 4.80

12713   logP = 1.73

12721   logP* = 0.77

12775   logP = -1.59

12776   logP = 0.64

12790   logP = 3.21

12802   logP* = 1.52

12809   logP* = 1.54

12958   logP* = 1.82

12959   logP = 1.85

12986   logP = 2.03

12992   logP* = 0.95

12996   logP* = 1.86

13002   logP = 0.35

13003   logP* = -1.86

13006   logP* = -2.64

13036   logP* = 1.47

13065   logP = 0.57

13066   logP* = 0.75

13068 logP = 2.59

13071 logP* = 2.39

13123 logP* = -1.18

13172 logP* = 3.43

13181 logP* = 4.08

13239 logP = 0.41

13240 logP* = 0.60

13261 logP* = 3.10

13289 logP* = -0.40

13293 logP = 1.36

13301 logP = 0.05

13302 logP = 0.09

13303 logP = 1.72

13304 logP = 3.01

13321 logP* = 3.65

13370 logP* = 4.12

13376 logP = 2.35

13377 logP* = 3.45

13406 logP = -1.20

13407 logP = 2.31

13412   logP* = 1.47   13414   logP* = -0.52   13466   logP* = 1.87   13567   logP* = 1.30

13599   logP* = 5.18   13633   logP = 0.96   13647   logP = 0.78   13655   logP* = -2.55

13825   logP = 1.02   13826   logP = -1.17   13827   logP = -0.60   13881   logP = 0.55

13882   logP = 2.37   13883   logP = 3.43   13902   logP* = 3.73   13937   logP* = 2.23

13973   logP = 2.72   13974   logP = 4.00   13982   logP* = 4.50   14008   logP* = 0.69

14101   logP = -0.02   14102   logP = 0.17   14166   logP* = 1.14   14189   logP = 2.07

14239   logP* = 3.60   14328   logP* = 0.56   14408   logP = 1.73   14422   logP* = 3.00

14428   logP* = 2.85   14487   logP* = 3.40   14508   logP = -0.50   14598   logP* = 2.31

14642   logP = 1.07   14655   logP* = 6.26   14708   logP* = 3.63   14713   logP = 0.70

14714   logP = -1.64   14715   logP = -0.18   14747   logP* = -0.57   14760   logP* = 1.51

14764   logP* = 0.27

14805   logP* = 1.28

14811   logP = 2.74

14854   logP = -0.01

14867   logP = 0.75

14918   logP = 4.02

14919   logP = 4.17

14920   logP = 3.26

14930   logP = -0.06

14945   logP = 1.22

14951   logP = -0.37

14991   logP = 2.86

14994   logP = -0.63

15057   logP = 5.80

15108   logP* = 4.24

15174   logP = -1.62

15216   logP = 3.31

15221   logP = 2.78

15222   logP = 2.47

15244   logP* = 4.69

15249   logP = -0.57   15250   logP = -0.52   15262   logP = 2.13   15266   logP = 0.54

15269   logP = 0.78   15279   logP = -2.14   15295   logP* = 5.70   15342   logP* = 3.38

15346   logP* = 3.61   15373   logP* = 3.43   15378   logP = 1.61   15394   logP = 0.76

15409   logP = 1.46   15473   logP* = 1.54   15482   logP = 5.32   15505   logP = -0.13

15545   logP = -1.02   15546   logP = 0.93   15555   logP = -0.77   15556   logP = -0.73

15567   logP = 3.42

15568   logP = 3.42

15613   logP = 2.82

15620   logP = 2.71

15650   logP* = 3.65

15652   logP* = 3.84

15685   logP = 0.28

15687   logP = -2.70

15695   logP = 0.87

15705   logP = -1.44

15706   logP = -0.80

15751   logP = 4.06

15752   logP = 5.64

15840   logP = -1.36

15841   logP = -1.36

15858   logP = 3.43

15866   logP = 4.20

15884   logP = 0.96

15889   logP = 0.97

15911   logP = 1.85

16008   logP = -2.67   16054   logP* = 1.40   16078   logP = 0.84   16080   logP = -0.02

16117   logP = 2.10   16118   logP = 2.14   16119   logP = 5.03   16124   logP = 2.06

16130   logP = -1.55   16163   logP = 2.34   16176   logP = -1.85   16177   logP = -1.85

16200   logP = 2.19   16201   logP = 2.02   16271   logP = 1.90   16276   logP = 1.58

16295   logP = 2.36   16321   logP = 3.86   16330   logP = -2.23   16334   logP* = 3.58

16351   logP = 1.53

16373   logP = 0.65

16389   logP = 5.91

16394   logP = -2.55

16396   logP = -0.63

16413   logP* = 1.66

16423   logP = 2.36

16446   logP = 0.37

16447   logP = -1.81

16454   logP = -0.08

16496   logP = 2.59

16526   logP = 1.80

16546   logP* = 1.65   16554   logP* = 1.33   16569   logP* = 11.29

16571   logP* = 1.52   16582   logP = 2.46   16591   logP* = 3.18

16608   logP* = 2.76   16614   logP = 2.89   16634   logP = 4.80

16635   logP = 4.80   16712   logP = 4.14   16713   logP = 5.00